T0310421

PLASMIDS

BIOLOGY AND IMPACT IN BIOTECHNOLOGY AND DISCOVERY

PLASMIDS

BIOLOGY AND IMPACT IN BIOTECHNOLOGY AND DISCOVERY

Edited by | **Marcelo E. Tolmasky**
Center for Applied Biotechnology Studies
Department of Biological Science
College of Natural Sciences and Mathematics
California State University, Fullerton

AND

Juan C. Alonso
Centro Nacional de Biotecnología, CSIC
Departamento de Biotecnología Microbiana,
Madrid, Spain

ASM Press, Washington, DC

Library of Congress Cataloging-in-Publication Data

Plasmids : biology and impact in biotechnology and discovery / edited by Marcelo Tolmasky, Department of Biological Sciences, California State University, Fullerton, and Juan Carlos Alonso, Centro Nacional de Biotecnología, Cantoblanco, Departamento de Biotecnología Microbiana, Madrid, Spain.
pages cm
Includes bibliographical references and index.
ISBN 978-1-55581-897-5 (hardcover : alk. paper) 1. Plasmids. I. Tolmasky, Marcelo, editor. II. Alonso, Juan Carlos (Biotechnologist), editor.
QR76.6.P563 2015
572.8'69—dc23

2015004787

eISBN: 978-1-55581-898-2 doi:10.1128/9781555818982

10 9 8 7 6 5 4 3 2 1

Printed in the United States of America

Address editorial correspondence to ASM Press, 1752 N St., N.W., Washington, DC 20036-2904, USA

Send orders to ASM Press, P.O. Box 605, Herndon, VA 20172, USA
Phone: 800-546-2416; 703-661-1593
Fax: 703-661-1501
E-mail: books@asmusa.org
Online: http://estore.asm.org

Cover images: on the front and back covers, the gold-colored ring structures in the background are atomic force microscopy (AFM) images showing one linear molecule and three covalently closed circular ones at different degrees of supercoiling (image provided by Sonia Trigueros); the spine has an electron microscopy (EM) image of a replicating plasmid (image produced by Jorge Crosa).

Cover design: Rings Leighton Design Group, Washington, DC

Dedicated to the memory of Jorge Crosa

Contents

Contributors

VICKI ADAMS
Department of Microbiology, Australian Research Council Centre of Excellence in Structural and Functional Microbial Genomics, Monash University, Clayton, Victoria 3800, Australia

JUAN C. ALONSO
Departamento de Biotecnología Microbiana, Centro Nacional de Biotecnología, CSIC, 28049 Madrid, Spain

KARSTEN ARENDS
Robert Koch-Institute, Nordufer 20, 13353 Berlin, Germany

SILVIA AYORA
Departamento de Biotecnología Microbiana, Centro Nacional de Biotecnología, CNB-CSIC, 28049 Madrid, Spain

CHANG-HO BAEK
Life Technologies, Carlsbad, CA 92008

FERNANDO BAQUERO
Ramón y Cajal University Hospital, IRYCIS, 28034 Madrid, Spain

JAMIE C. BAXTER
Department of Molecular Genetics, University of Toronto, Toronto, Ontario M5S 1A8, Canada

ILARIA BENEDETTI
Departamento de Biotecnología Microbiana, Centro Nacional de Biotecnología, CNB-CSIC, 3, Darwin Street, 28049 Madrid, Spain

LORENA BORDANABA-RUISECO
Centro de Investigaciones Biológicas, CSIC, Ramiro de Maeztu 9, 28040 Madrid, Spain

SABINE BRANTL
AG Bakteriengenetik, Philosophenweg 12, Friedrich-Schiller-Universität Jena, D-07743 Jena, Germany

ALICIA BRAVO
Centro de Investigaciones Biológicas, CSIC, 28040 Madrid, Spain

KATARZYNA BURY
Department of Molecular and Cellular Biology, Intercollegiate Faculty of Biotechnology, University of Gdansk and Medical University of Gdansk, Gdansk, Poland

MANEL CAMPS
Department of Microbiology and Environmental Toxicology, University of California, Santa Cruz, 1156 High Street, Santa Cruz, CA 95064

WAI TING CHAN
Centro de Investigaciones Biológicas, CSIC, 28040 Madrid, Spain

KENG-MING CHANG
Section of Molecular Genetics and Microbiology, Institute for Cellular and Molecular Biology, The University of Texas at Austin, Austin, Texas 78712

JONATHAN CHESNUT
Life Technologies, Carlsbad, CA 92008

PETER J. CHRISTIE
Department of Microbiology and Molecular Genetics, University of Texas Medical School at Houston, Houston, TX 77005

KEVIN CLANCY
Life Technologies, Carlsbad, CA 92008

MIQUEL COLL
Institute for Research in Biomedicine (IRB-Barcelona), and Institut de Biologia Molecular de Barcelona, CSIC, Baldiri Reixac 10-12, 08028 Barcelona, Spain

LAURA C.C. COOK
Department of Medicinal Chemistry, University of Illinois, Chicago, IL 60607

TERESA M. COQUE
Department of Microbiology, Ramón y Cajal University Hospital, IRYCIS, 28034 Madrid, Spain

FRANÇOIS CORNET
CNRS, Laboratoire de Microbiologie et Génétique Moléculaires, F-31062 Toulouse, France

ESTELLE CROZAT
UPS, Laboratoire de Microbiologie et Génétique Moléculaires, Université de Toulouse, F-31062 Toulouse, France

Fernando De La Cruz
Instituto de Biomedicina y Biotecnología de Cantabria (IBBTEC), Universidad de Cantabria, CSIC, 39011 Santander, Spain

Gloria del Solar
Centro de Investigaciones Biológicas, CSIC, Ramiro de Maeztu 9, 28040 Madrid, Spain

Eduardo Díaz
Department of Environmental Biology, Centro de Investigaciones Biológicas (CSIC), 28040 Madrid, Spain

Ramón Díaz-Orejas
Centro de Investigaciones Biológicas, CSIC, 28040 Madrid, Spain

Manuela Di Lorenzo
Department of Microbial Ecology, Netherlands Institute of Ecology (NIOO-KNAW), 6708 PB Wageningen, The Netherlands

Gary M. Dunny
Department of Microbiology, University of Minnesota, Minneapolis, MN 55455

Manuel Espinosa
Consejo Superior de Investigaciones Científicas, Centro de Investigaciones Biológicas, CSIC, Ramiro de Maeztu 9, 28040 Madrid, Spain

Hsiu-Fang Fan
Department of Life Sciences and Institute of Genome Sciences, National Yang-Ming University, Taipei 112, Taiwan

Cris Fernández-López
Centro de Investigaciones Biológicas, CSIC, 28040 Madrid, Spain

Andrea T. Feßler
Institute of Farm Animal Genetics, Friedrich-Loeffler-Institut (FLI), Neustadt-Mariensee, Germany

Patrick Forterre
Institut Pasteur, 75015 Paris, France

Florian Fournes
UPS, Laboratoire de Microbiologie et Génétique Moléculaires, Université de Toulouse, F-31062 Toulouse, France

Barbara E. Funnell
Department of Molecular Genetics, University of Toronto, Toronto, Ontario M5S 1A8, Canada

José L. García
Department of Environmental Biology, Centro de Investigaciones Biológicas, CSIC, 28040 Madrid, Spain

M. Pilar Garcillán-Barcia
Instituto de Biomedicina y Biotecnología de Cantabria (IBBTEC), Universidad de Cantabria, CSIC, 39011 Santander, Spain

DANIELLE A. GARSIN
Department of Microbiology and Molecular Genetics, The University of Texas Health Science Center at Houston, Houston, Texas

MICHAEL S. GILMORE
Department of Ophthalmology, Massachusetts Eye and Ear Infirmary, Boston, MA 02114

NIKOLAUS GOESSWEINER-MOHR
Institute of Molecular Biosciences, University of Graz, 8010 Graz, Austria

JAY E. GORDON
Department of Microbiology and Molecular Genetics, University of Texas Medical School at Houston, Houston, TX 77005

ELISABETH GROHMANN
Faculty of Biology, Microbiology, Albert-Ludwigs-University Freiburg, 79104 Freiburg, Germany

BERNARD HALLET
Institut des Sciences de la Vie, UC Louvain, 4/5 L7.07.06 Place Croix du Sud, B-1348 Louvain-la-Neuve, Belgium

ANA MARÍA HERNÁNDEZ-ARRIAGA
Centro de Investigaciones Biológicas, Consejo Superior de Investigaciones Científicas, 28040 Madrid, Spain

ANGELES HUESO
Departamento de Biotecnología Microbiana, Centro Nacional de Biotecnología, CNB-CSIC, 3, Darwin Street, 28049 Madrid, Spain

N. PATRICK HIGGINS
Department of Biochemistry and Molecular Genetics, University of Alabama at Birmingham, Birmingham, AL 35294

DAVID C. HOOPER
Massachusetts General Hospital, 55 Fruit Street, Boston, MA 02114

GEORGE A. JACOBY
Lahey Hospital and Medical Center, 41 Mall Road, Burlington, MA 01805

MAKKUNI JAYARAM
Department of Life Sciences and Institute of Genome Sciences, National Yang-Ming University, Taipei 112, Taiwan

SVEN JECHALKE
Julius Kühn-Institut, Federal Research Centre for Cultivated Plants (JKI), Institute for Epidemiology and Pathogen Diagnostics, Messeweg 11-12, 38104 Braunschweig, Germany

AASHIQ H. KACHROO
Section of Molecular Genetics and Microbiology, Institute for Cellular and Molecular Biology, The University of Texas at Austin, Austin, Texas 78712

KRISTINA KADLEC
Institute of Farm Animal Genetics, Friedrich-Loeffler-Institut (FLI), Neustadt-Mariensee, Germany

CLARENCE I. KADO
Plant Pathology, University of California, Davis, Davis, CA 95616

FEDERICO KATZEN
Life Technologies, Carlsbad, CA 92008

WALTER KELLER
Institute of Molecular Biosciences, University of Graz, 8010 Graz, Austria

EKATERINA KINNEAR
Mucosal Infection and Immunity Group, Section of Virology, Imperial College London, St Mary's Campus, London W2 1PG,United Kingdom

IGOR KONIECZNY
Department of Molecular and Cellular Biology, Intercollegiate Faculty of Biotechnology, University of Gdansk and Medical University of Gdansk, Gdansk, Poland

MART KRUPOVIC
Institut Pasteur, 75015 Paris, France

ANTONIO LAGARES
Departamento de Ciencias Biológicas, Facultad de Ciencias Exactas, IBBM, Instituto de Biotecnología y Biología Molecular, CONICET, Universidad Nacional de La Plata, (1900) La Plata, Argentina

VAL FERNÁNDEZ LANZA
Centro de Investigación en Red en Epidemiología y Salud Pública (CIBER-ESP), Melchor Fernández Almagro, 3-5, 28029 Madrid, Spain

JIHONG LI
Department of Microbiology and Molecular Genetics, University of Pittsburgh School of Medicine, 3550 Terrace Street, Pittsburgh, PA 15261

JOSHUA LILLY
Department of Microbiology and Environmental Toxicology, University of California, Santa Cruz, 1156 High Street, Santa Cruz, CA 95064

DAVID L. LIN
Department of Biological Science, College of Natural Sciences and Mathematics, Center for Applied Biotechnology Studies, California State University, Fullerton, 800 N. State College Blvd., Fullerton, CA 92831

MICHAEL LISS
Life Technologies, Carlsbad, CA 92008

YEN-TING LIU
Section of Molecular Genetics and Microbiology, Institute for Cellular and Molecular Biology, The University of Texas at Austin, Austin, Texas 78712

VÍCTOR DE LORENZO
Departamento de Biotecnología Microbiana, Centro Nacional de Biotecnología, CSIC, 3, Darwin Street, 28049 Madrid, Spain

FABIÁN LORENZO-DÍAZ
Instituto Universitario de Enfermedades Tropicales y Salud Pública de Canarias, Universidad de La Laguna, 38071 Laguna, Spain

CHIEN-HUI MA
Section of Molecular Genetics and Microbiology, Institute for Cellular and Molecular Biology, The University of Texas at Austin, Austin, Texas 78712

CRISTINA MACHÓN
Institute for Research in Biomedicine (IRB-Barcelona), and Institut de Biologia Molecular de Barcelona, CSIC, Baldiri Reixac 10-12, 08028 Barcelona, Spain

ALFONSO H. MAGADAN
Département de Biochimie, Microbiologie et Bio-Informatique, Groupe de Recherche en Écologie Buccale, Faculté des Sciences et de Génie, et de Médecine Dentaire, Félix d'Hérelle Reference Center for Bacterial Viruses, Université Laval, Quebec City, Quebec G1V 0A6, Canada

JOSÉ LUÍS MARTÍNEZ
Centro Nacional de Biotecnología, CNB, and Unidad de Resistencia a Antibióticos y Virulencia Bacteriana (HRYC-CSIC), Madrid, Spain

ESTEBAN MARTÍNEZ-GARCÍA
Departamento de Biotecnología Microbiana, Centro Nacional de Biotecnología, CNB-CSIC, 3, Darwin Street, 28049 Madrid, Spain

BRUCE A. MCCLANE
Department of Microbiology and Molecular Genetics, University of Pittsburgh School of Medicine, 3550 Terrace Street, Pittsburgh, PA 15261

SYLVAIN MOINEAU
Groupe de Recherche en Écologie Buccale, Faculté de Médecine Dentaire, Université Laval, Quebec City, Quebec G1V 0A6, Canada

LÁZARO MOLINA
CIDERTA, Laboratorio de Investigación y Control Agroalimentario (LICAH), Parque Huelva Empresarial, 21007 Huelva, Spain

GABRIEL A. MONTEIRO
Department of Bioengineering, Centre for Biological and Chemical Engineering, IBB, Institute for Biotechnology and Bioengineering, Instituto Superior Técnico, Universidade de Lisboa, 1049-001 Lisboa, Portugal

ROBERT J. MOORE
Department of Microbiology, Australian Research Council Centre of Excellence in Structural and Functional Microbial Genomics, Monash University, Clayton, Victoria 3800, Australia

MARIANO PISTORIO
Departamento de Ciencias Biológicas, Facultad de Ciencias Exactas, IBBM, Instituto de Biotecnología y Biología Molecular, CONICET, Universidad Nacional de La Plata, (1900) La Plata, Argentina

DUARTE MIGUEL F. PRAZERES
Department of Bioengineering, Instituto Superior Técnico, Centre for Biological and Chemical Engineering, IBB, Institute for Biotechnology and Bioengineering, Universidade de Lisboa, 1049-001 Lisboa, Portugal

MARIA S. RAMIREZ
Department of Biological Science, Center for Applied Biotechnology Studies, College of Natural Sciences and Mathematics, California State University, Fullerton, 800 N. State College Blvd., Fullerton, CA 92831

JUAN LUIS RAMOS
Environmental Protection Department, Profesor Albareda, Estación Experimental del Zaidin, CSIC, 18008 Granada, Spain

NIKOLAI V. RAVIN
Center of Bioengineering, Russian Academy of Sciences, Prosp. 60-let Oktiabria, Bldg. 7-1, Moscow 117312, Russia

KASIE RAYMANN
Institut Pasteur, 75015 Paris, France

JULIAN I. ROOD
Department of Microbiology, Australian Research Council Centre of Excellence in Structural and Functional Microbial Genomics, Monash University, Clayton, Victoria 3800, Australia

PHILIPPE ROUSSEAU
UPS, Laboratoire de Microbiologie et Génétique Moléculaires, Université de Toulouse, F-31062 Toulouse, France

PAUL A. ROWLEY
Section of Molecular Genetics and Microbiology, Institute for Cellular and Molecular Biology, The University of Texas at Austin, Austin, Texas 78712

JOSÉ A. RUIZ-MASÓ
Centro de Investigaciones Biológicas, CSIC, Ramiro de Maeztu 9, 28040 Madrid, Spain

SOFÍA RUIZ-CRUZ
Centro de Investigaciones Biológicas, CSIC, 28040 Madrid, Spain

JULIE E. SAMSON
Département de Biochimie, Microbiologie et Bio-Informatique, Groupe de Recherche en Écologie Buccale, Faculté des Sciences et de Génie, et de Médecine Dentaire, Félix d'Hérelle Reference Center for Bacterial Viruses, Université Laval, Quebec City, Quebec G1V 0A6, Canada

JUAN SANJUÁN
Departamento de Microbiología del Suelo y Sistemas Simbióticos. Estación Experimental del Zaidín, CSIC, Granada, Spain

SAUMITRA SAU
Section of Molecular Genetics and Microbiology, Institute for Cellular and Molecular Biology, The University of Texas at Austin, Austin, Texas 78712

STEFAN SCHWARZ
Institute of Farm Animal Genetics, Friedrich-Loeffler-Institut (FLI), Neustadt-Mariensee, Germany

ANA SEGURA
Environmental Protection Department, Profesor Albareda, Estación Experimental del Zaidin, CSIC, 1, 18008 Granada, Spain

JIANZHONG SHEN
Beijing Key Laboratory of Detection Technology for Animal-Derived Food Safety, College of Veterinary Medicine, China Agricultural University, Beijing 100193, P. R. China

KORNELIA SMALLA
Julius Kühn-Institut, Federal Research Centre for Cultivated Plants (JKI), Institute for Epidemiology and Pathogen Diagnostics, Messeweg 11-12, 38104 Braunschweig, Germany

NORA E. SOBERÓN
Departamento de Biotecnología Microbiana, Centro Nacional de Biotecnología, CNB-CSIC, 28049 Madrid, Spain

VIRTU SOLANO-COLLADO
Centro de Investigaciones Biológicas, CSIC, 28040 Madrid, Spain

NICOLAS SOLER
DynAMic, Université de Lorraine, UMR1128, INRA, Vandoeuvre-lès-Nancy, France

MICHIEL STORK
Process Development, Institute for Translational Vaccinology, 3720 AL Bilthoven, The Netherlands

JACOB STRAHILEVITZ
Hadassah-Hebrew University, Jerusalem 91120, Israel

ANA P. TEDIM
Department of Microbiology, Ramón y Cajal University Hospital, IRYCIS, 28034 Madrid, Spain

MARCELO E. TOLMASKY
Center for Applied Biotechnology Studies, Department of Biological Science, College of Natural Sciences and Mathematics, California State University, Fullerton, 800 N. State College Blvd., Fullerton, CA 92831

EVA M. TOPP
Department of Biological Sciences, University of Idaho, 875 Perimeter, MS 3051, Moscow, Idaho 83844-3051

MARÍA DE TORO
Instituto de Biomedicina y Biotecnología de Cantabria (IBBTEC), Universidad de Cantabria—CSIC, 39011 Santander, Spain

GERMAN M. TRAGLIA
Institute of Microbiology and Medical Parasitology, National Scientific and Technical Research Council (CONICET), University of Buenos Aires, Buenos Aires, Argentina

TUNG TRAN
Department of Biological Science, Center for Applied Biotechnology Studies, College of Natural Sciences and Mathematics, California State University, Fullerton, 800 N. State College Blvd., Fullerton, CA 92831

JOHN S. TREGONING
Mucosal Infection and Immunity Group, Section of Virology, Imperial College London, St Mary's Campus, London, W2 1PG, United Kingdom

FRANCISCO A. UZAL
California Animal Health and Food Safety Laboratory, San Bernardino Branch, School of Veterinary Medicine, University of California, Davis, San Bernardino, CA

DARIA VAN TYNE
Department of Ophthalmology, Massachusetts Eye and Ear Infirmary, and Department of Microbiology and Immunobiology, Harvard Medical School, Boston, MA 02114

ANDREA VOLANTE
Departamento de Biotecnología Microbiana, Centro Nacional de Biotecnología, CNB-CSIC, 28049 Madrid, Spain

ALEXANDER V. VOLOGODSKII
Department of Chemistry, New York University, New York, NY 10003

YANG WANG
Beijing Key Laboratory of Detection Technology for Animal-Derived Food Safety, College of Veterinary Medicine, China Agricultural University, Beijing 100193, P. R. China

ALEKSANDRA WAWRZYCKA
Department of Molecular and Cellular Biology, Intercollegiate Faculty of Biotechnology, University of Gdansk and Medical University of Gdansk, Gdansk, Poland

KATARZYNA WEGRZYN
Department of Molecular and Cellular Biology, Intercollegiate Faculty of Biotechnology, University of Gdansk and Medical University of Gdansk, Gdansk Poland

SARAH WENDLANDT
Institute of Farm Animal Genetics, Friedrich-Loeffler-Institut (FLI), Neustadt-Mariensee, Germany

JESSICA A. WISNIEWSKI
Australian Research Council, Centre of Excellence in Structural and Functional Microbial Genomics, Department of Microbiology, Monash University, Clayton, Victoria 3800, Australia

CONG-MING WU
Institute of Farm Animal Genetics, Friedrich-Loeffler-Institut (FLI), Neustadt-Mariensee, Germany

Preface

One of the biggest dreams of medicine from the 1940s, the complete defeat of infectious diseases caused by bacteria, was treated to a rough awakening with the rise and dissemination of antibiotic resistance, toxins, and pathogenicity functions. In the early 1960s it was found that this dissemination was usually associated with the acquisition of genes that were located in extrachromosomal elements analogous to those that Joshua Lederberg had called "plasmids" in 1952. The importance of the discovery led to intense research on plasmid biology, which in turn resulted in innumerable benefits to the development of science. The list of discoveries in the fields of cell and molecular biology is far too long to detail in this Preface. In addition, a monumental contribution of research on plasmids was instrumental in the development of molecular cloning and the biotechnology revolution that ensued. Their role in virulence and antibiotic resistance, together with the generalization of "omics" disciplines, has recently ignited a new wave of interest in plasmids. As models for understanding innumerable biological mechanisms of living cells, as tools for creating the most diverse therapies, and as invaluable helpers to understand the dissemination of microbial populations, plasmids continue to be at the center of research.

Marcelo E. Tolmasky
Juan C. Alonso

Introduction

I

Plasmids—Biology and Impact in Biotechnology and Discovery
Edited by Marcelo E. Tolmasky and Juan C. Alonso

doi:10.1128/microbiolspec.PLAS-0019-2013

Clarence I. Kado[1]

Historical Events That Spawned the Field of Plasmid Biology

1

INTRODUCTION

Extrachromosomal genetic elements, now widely known as plasmids, were recognized over 60 years ago. Historically, extrachromosomal genetic elements that transferred antibiotic resistance to recipient pathogenic bacteria were called R factors, and those that were conjugative were called T factors (1). Bacteria, particularly *Shigella* strains harboring R and T factors, were found in 1951 in Japan, then in Taiwan and Israel in 1960 (2), and in the United States and Europe in 1963 to 1968 (3). The F factor (for fertility) was the genetic element, also called the "sex factor," that was required for bacterial conjugation (4–8). The sex factor determined the ability of *Escherichia coli* strain K12 to conjugate and transfer genes to recipients.

All of these extrachromosomal elements .that propagated either autonomously in the cytoplasm or as an integral part of the host chromosome were called episomes (9). To avoid unnecessary confusion in the usage of a number of terms related to extrachromosomal elements such as plasmagenes, conjugons, pangenes, plastogenes, choncriogenes, cytogenes, proviruses, etc., Lederberg (10; Fig. 1) coined the term *plasmid* to represent any extrachromosomal genetic entity. This term has been widely accepted and used with the understanding that these genetic elements are not organelles, individual genes, parasites (viruses), or symbionts (11). Henceforth, *plasmid(s)* became the conventional term used today.

Based on the established fact that plasmids can reside in *E. coli* and *Shigella* spp., a number of workers began searching for plasmids in other enteric bacteria as well as in pseudomonads and Gram-positive bacteria. By 1977, over 650 plasmids were listed and classified into 29 incompatibility groups (12). Recently, through DNA sequence comparisons of 527 plasmids, there has appeared to be a great deal of interchange of genes between plasmids due to horizontal gene transfer events (13). Incompatibility is determined when two plasmids introduced into a single cell can both replicate and be maintained stably. If the plasmids coexist (replicate and be maintained stably), they are considered compatible. If the plasmids cannot coexist stably, their replication systems are incompatible (14). The incompatible plasmids cannot share a common replication system. Thus, a plasmid classification system was developed that allowed researchers to make logical comparisons of their work on similar plasmids. The classification also provided a system that helped prevent instituting a different name or number for identical plasmids worked on by separate laboratories.

From these early studies, several basic areas of research on plasmids evolved. Researchers focused on

[1]Plant Pathology, University of California Davis, Davis, CA 95616.

Figure 1 Joshua Lederberg.
doi:10.1128/microbiolspec.PLAS-0019-2013.f1

(i) analyzing the physical structure and locating genes on plasmids; (ii) identifying the replication system and the mechanism of replication of plasmids, including how they partition; (iii) determining the conjugative machinery and the mechanism and regulation of plasmid transfer; (iv) dissecting the genetic traits conferred by plasmids, such as metabolic TOL plasmids, bacteriocin-producing Col plasmids, tumor-inducing Ti and virulence plasmids, heavy metal resistance pMOL plasmids, radiation resistant plasmids, etc; (v) restructuring plasmids for utilitarian use, e. g., gene vector development, reporter systems, genetic engineering of mammals and plants; and (vi) surveying the epidemiology and horizontal gene transfer events and reconstructing the evolution of plasmids.

BIRTH OF THE FIELD OF PLASMID BIOLOGY

The term *plasmid biology* was conceived in 1990 at the Fallen Leaf Lake Conference on Promiscuous Plasmids in Lake Tahoe, California. International conferences on plasmid biology were henceforth launched, being held in different countries including Germany, Canada, Spain, the United States, Austria, Mexico, the Czech Republic, Greece, Poland, and Argentina. An example of the proceedings of one of these conferences was published in 2007 (15). An Asian venue is yet to be selected. The International Society for Plasmid Biology was established in 2004 and remains an active internationally recognized professional society (www.ISPB.org).

EARLY STRUCTURAL STUDIES AND GENETIC MAPPING OF PLASMIDS

Knowledge gained from a novel method of separating closed circular DNA from linear DNA in HeLa cells using dye-buoyant CsCl density gradient centrifugation (16) made it possible to examine plasmid DNA derived from bacteria. Earlier studies used analytical centrifugation and density gradient centrifugation on an *E. coli* "episomal element" (F-*lac*) that was conjugatively transferred to *Serratia marcescens*. The 8% difference in guanine plus cytosine content between the episome of *E. coli* (50% GC) vs. *S. marcescens* DNA (58% GC) was sufficient to neatly separate the episome from chromosomal DNA and established the fact that the episome was indeed made of DNA (17). Further physical evidence led to the suggestion that bacteriophage φX174 DNA was circular (18). This was confirmed by electron microscopy by Kleinschmidt et al. (19). Kleinschmidt carefully prepared and used the Langmuir trough technique and examined over 1,000 electron-micrographs to obtain a perfect photograph (A. K. Kleinschmidt, personal communication, 1964). Moreover, phage PM2 DNA was observed by electron microscopy to be a closed circular double-stranded molecule (20). These findings prompted researchers to examine by electron microscopy bacterial extrachromosomal elements of their particular interest and confirmed that plasmids are indeed circular DNA molecules (although linear plasmids also exist).

RECOGNITION OF PLASMID REPLICATION AND PARTITIONING SYSTEMS

Replication of plasmids requires DNA synthesis proteins encoded by chromosomal genes of the hosting bacterial cell. Between one and eight proteins can be involved, depending on the plasmid (Table 1). DNA replication of plasmids is initiated by the binding of the initiator protein to specific binding sites at the replicative origin. Initiator binding promotes the localized unwinding of a discrete region from the DNA origin.

Table 1 Plasmid initiator proteins

Initiator	Replication mode	Plasmid	Molecular mass	References
RepA	Theta type	R1, R100	33 kDa	74, 75
RepA1	Theta type	EntP307	40 kDa	76
RepA	Theta type	pSC101	37.5 kDa	77, 78
RepC	Theta type	RSF1010	31 kDa	79
RepE	Theta type	F	29 kDa	80
TrfA	Theta type	RK2	33 kDa	81
π (*pir*)	Theta type	R6K	35 kDa	82
RepA	Rolling circle	pA1	5.6 kDa	83
RepB	Rolling circle	pLS1	24.2 kDa	84
RepC	Rolling circle	pT181	38 kDa	85
RepD	Rolling circle	pC221	38 kDa	86

A helicase is then directed to the exposed single-stranded DNA region followed by a prepriming complex to initiate DNA synthesis (21). Initiation of DNA replication by the initiator binding to the origin sequence(s) is a critical function in plasmid survival as an extracellular genetic element.

As part of the plasmid replication process, specific plasmid concentrations (copy number) occur as the host bacterial cell initiates cell division. Partitioning and stable segregation of the plasmid are initiated. Partition systems are categorically classified based on ATPase proteins (22). Type I is characterized by Walker box ATPases, while a subset, type Ia, occurs when the nucleotide-binding P-loop is preceded by an N-terminal regulatory domain, and in type Ib this is not the case. The mechanisms that contribute to the stable segregation of plasmids F, P1, R1, NR1, pSC101, and ColE1 have been reviewed (23). The locus responsible for partitioning of pSC101 was designated "*par*" (24). The *par* locus is able to rescue unstable pSC101-derived replicons in the *cis*, but not the *trans*, configuration. It is independent of copy number control, does not specify plasmid incompatibility, and is not associated directly with plasmid replication functions. From phylogenetic analysis of *par* loci from plasmids and bacterial chromosomes, two trans-acting proteins form a nucleoprotein complex at a *cis*-acting centromere-like site (22). One these proteins, identified as an ATPase, functions to tether plasmids and chromosomal origin regions to specific poles of the dividing cells. Therefore, the mitotic stability of plasmids depends on a centromere, a centromere-binding protein, and an ATPase. In the case of plasmid F, two genes, *sopA* and *sopB*, and a centromeric target site, *sopC*, function to ensure that both daughter cells receive a daughter plasmid during cell division. The products of *sopA* and *sopB* stabilize the plasmid bearing a centromere-like sequence in *sopC* (25). SopA hydrolyzes ATP by binding DNA (26). The centromere-like region contains a 43-bp sequence that is repeated 12 times in the same orientation (27), and each element contains a 7-bp inverted repeat targeted by SopB (28). Like *sopA*, *sopB*, and *sopC* of plasmid F, plasmid P1 has counterpart partition genes (*parA*, *parB*) and a target site (*pars*) (29).

Some plasmids such as ColE1 are partitioned randomly at cell division, and their inheritance is proportional to the number of plasmids present in the cell (30). High-copy-number plasmids usually do not require an active *par* system for stable maintenance because random distribution ensures plasmid segregation to the two daughter cells at the time of cell division, while larger, low-copy plasmids such as F, R100, and P1 possess genes that encode inhibitors of host cell growth. In the case of plasmid F, the *ccdA* and *ccdB* (for coupled cell division) genes encode an 8.7-kDa and an 11.7-kDa protein, respectively, the latter of which inhibits cell growth (31). This inhibitor functions in cells that have lost their plasmid due to errors in replication or cell division. The action of the inhibitor is prevented by the CcdA protein, which loses stability in the absence of the plasmid and therefore no longer functions to inhibit the action of the CcdB protein. Plasmid biologists have referred to this interesting mechanism of controlling plasmid copy number as a "killing" function that specifically kills cells lacking a plasmid (or postsegregational killing).

LANDMARKS LEADING TO PLASMID-MEDIATED CONJUGATIVE TRANSFER

The historical experiments on plating together two different triple auxotrophic mutants leading to prototrophic bacterial colonies that propagated indefinitely on minimal medium was the classical laboratory event that led Lederberg and Tatum (6, 7) to conclude that there was sex in bacteria (32). Examination of single cell isolates of these prototrophic strains showed that they were indeed heterozygotes. Hayes (4) showed the heterothallic nature of conjugation whereby recombination is mediated by the one-way transfer of genetic material from donor to recipient bacteria. Self-transmissible plasmids such as F, R1, R100, and R6K encode the capacity to promote conjugation. They all possess related transfer (*tra*) genes. Plasmid F (called sex factor)-mediated conjugation has received the most attention. *E. coli* harboring this sex factor produce a filamentous organelle called the F pilus (Fig. 2) that was needed for conjugation between sex factor-bearing

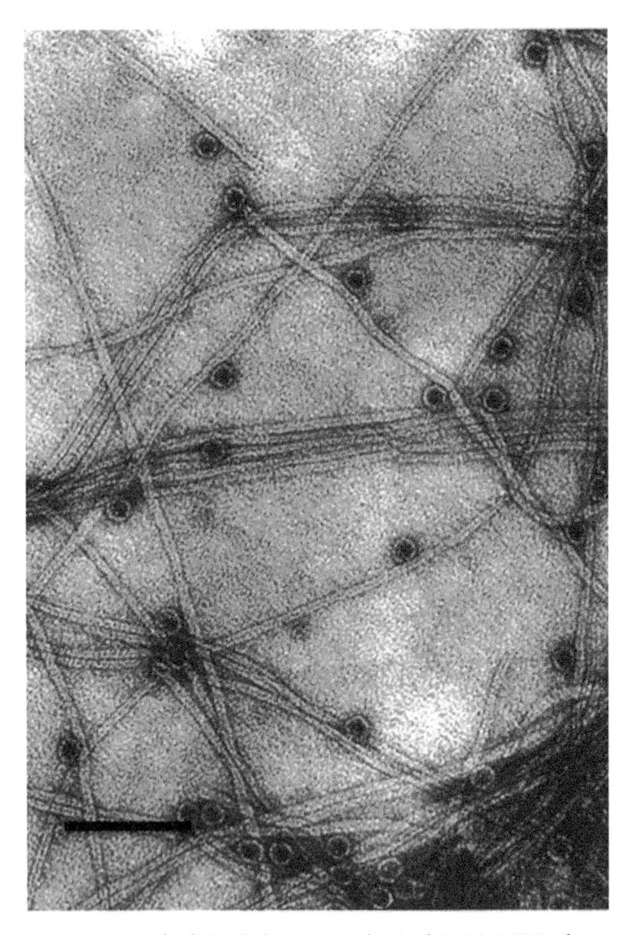

Figure 2 Purified F pili bearing spherical RNA MS2 phages. Electron micrograph courtesy of Professor Manabu Inuzuka, Fukui Medical University, Fukui, Japan. Bar = 2000 Å. doi:10.1128/microbiolspec.PLAS-0019-2013.f2

donors (known as F^+ donors) and F^- recipients. Historically, the F pilus (or "sex pilus," coined by Harden and Meynell [33] and reviewed by Tomoeda et al. [34]) was suggested by Brinton (35) to serve as a conduit through which DNA passes. Somewhat similar to bacteriophage (T phage) tail retraction, the F pilus was proposed to retract and bring together conjugating cells into wall-to-wall contact (36, 37). Although the F pilus is needed for initial contact between F^+ and F^- cells, it is not necessary for DNA transfer after the contacts have stabilized (38). The formation of mating pairs involves a complex apparatus bridging the donor cell envelope that assembles the conjugative pilus. The pilus interacts with the recipient cell and apparently retracts by depolymerization into the donor cell, culminating in intimate wall-to-wall contact during mating-pair stabilization (38, 39).

This type of intimate contact, termed the conjugational junction, between stabilized mating pairs was examined by electron microscopy of thin sections of the junction (40). No specific substructure such as a plasma bridge was observed. Interestingly, the F pilus of *E. coli* was claimed to support stable DNA transfer in the absence of wall-to-wall contact between cells (41). In earlier work using micromanipulation, Ou and Anderson (42) showed DNA transfer in the absence of direct cell-to-cell contact. More recently, the F pilus was observed in real-time visualization to mediate DNA transfer at considerable cell-to-cell distances (43). Most (96%) of the transferred DNA integrated by recombination in the distal recipient cells.

Genetic and sequence analyses have provided further insights to the mechanism of plasmid DNA transfer. With conjugative plasmids, the genes required for mating pair formation and DNA transfer are located in one or two clusters identified as the transfer (*tra*) regions (44). The proteins involved in the unidirectional transfer of single-stranded DNA from donor to recipient are encoded by the *tra* operon of the F plasmid. These proteins form the relaxosome, which processes plasmid DNA at the origin of transfer (*oriT*). Sequence similarities were recognized between pilin-encoding genes of F-like plasmids (45). Studies of the promiscuous DNA transfer system encoded by the Ti (for tumor-inducing) plasmid of *Agrobacterium tumefaciens* revealed that the *virB* operon encodes a sex pilus involved in T-DNA transfer to plants (46). Moreover, the *virB* operon of the Ti plasmid exhibits close homologies to genes that are known to encode the pilin subunits and pilin assembly proteins of other conjugative plasmids such as F, R388, RP4, and even the *ptl* operon of *Bordetella pertussis* (46, 47). The components of these plasmid transfer apparatuses became classified as members of the type IV secretion family (48). The F plasmid transfer apparatus has homologs to VirB proteins encoded by the *virB* operon of the type IV secretion system (49). In fact, the VirB2 propilin protein is similar to the TraA propilin of F and is processed into their respective pilin subunit of a size (50–53) similar to the T-pilus (51, 54). Posttranslational processing also occurs with VirB1, a pilin-associated protein (55). Interestingly, the type IV DNA-protein transfer system of the Ti plasmid is highly promiscuous by promoting transfer between the domain *Bacteria* to members of the domain *Eukarya* (56).

Based on the intensive and excellent studies on plasmid DNA transfer systems of narrow and broad-host-range conjugative plasmids by a large number of excellent researchers past and present (reviewed in 57, 58), it appears that the transmission or transfer of plasmids is essential to their survival (see below).

FUNCTIONAL ATTRIBUTES REQUIRED FOR PLASMID PERSISTENCE AND SURVIVAL

Conjugative transfer of plasmids reflects an indispensable trait required for their ensured survival as selfish DNA molecules (56, 59). Traits such as conferring antibiotic resistance were first recognized as being plasmid-borne in *Shigella* and *Salmonella* spp. as described in the introduction above. Antibiotic resistance conferred by plasmid genes provided survival value to pathogens that would otherwise be killed by the antibiotic(s). This in turn offered survival and maintenance of the plasmid itself in the antibiotic-resistant pathogenic bacterial host. Likewise, metabolic/catabolic plasmids confer on host bacteria the ability to survive in harsh environments such as in sediments from industrial waste and from mining exudates of silver, copper, cadmium, tellurite, etc. Unusual environments such as sites containing an abundance of substrates such as aromatic hydrocarbons, toluene, xylene, pesticides, herbicides, and organic waste products all provided specialized niches for bacteria that live under the auspices of specialized enzymes that degrade or modify one or more of these compounds. These bacteria harbor plasmids that confer on their host cell the ability to metabolize, degrade, or modify substances that otherwise would be toxic or lethal to the host bacterial cell. The catabolic TOL plasmid pWWO, first described by Williams and Murray (60), is one of the best studied for its catabolic enzymes and genetic structure (61).

The selfishness of plasmids is exemplified by plasmids encoding bacteriocins that kill susceptible bacterial cells not harboring the same or like plasmids. The lethal action of these antibacterial proteins occurs through puncturing plasma membranes, degrading nucleic acids, or cleaving peptidoglycans. Examples of bacteriocins are colicin encoded by plasmid ColE1 (62), cloacin encoded by plasmid CloDF13 (63), and nisin F encoded by plasmid pF10 (64).

Of medical and veterinary relevance are plasmids that confer virulence traits on their bacterial hosts. Various pathogenic *E. coli* strains harbor plasmids that confer interesting virulence traits (65). Loss of the virulence-conferring plasmid results in the loss of its pathogenic trait unless the pathogenicity island transposes into the chromosome of the bacterial host. Another member of the *Enterobacteriaceae* are *Shigella* spp. All invasive *Shigella flexneri* strains, regardless of serotype, harbor a large virulence plasmid, pWR110 (66). Mutagenesis or curing of the plasmid results in the loss of pathogenicity. Plasmid-conferred virulence is not restricted to Gram-negative bacteria. Indeed, the pathogenicity of *Staphylococcus aureus* is highly dependent on its resident plasmid (67). The genes conferring the pathogenic trait and antibiotic resistance are highly conserved, and their spread among *S. aureus* strains is restrained. A number of plant pathogens also harbor virulence plasmids, a number of which encode secretion machinery for injection into their host plants (reviewed in Kado [68]).

RECONSTRUCTION OF PLASMIDS FOR BIOTECHNOLOGY AND BIOMEDICAL APPLICATIONS

The development of recombinant DNA techniques (69) has led to a multitude of possibilities of designing plasmid vector systems useful in fundamental research and industrial, agricultural, and medical applications. Early vector systems were based on ColE1 derivatives that were primarily restricted to *E. coli* owing to their replication machinery. The introduction of broad-host-range plasmids such as RK2 and RSF1010 made it possible to introduce recombinant DNA technologies into bacteria other than members of the *Enterobacteriaceae*. In recent times, a number of plasmid shuttle vector systems have become commercially available, too numerous to list in this paper. Plasmids constructed as vectors for various purposes are reviewed elsewhere (70). Some examples of useful vector systems are listed in Table 2. Vectors designed for pharmaceutical and genetic engineering of mammalian and plant cells have been recently reviewed (71–73).

CONCLUSION AND FUTURE OF PLASMID BIOLOGY

Plasmids have provided the basic foundation for recombinant DNA technologies. Significant insights are being gained from genome sequencing and reconstruction by computer modeling of prospective enzymes (proteins) encoded by sequenced plasmid genes. The commercially available kits for plasmid isolation, DNA amplification, sequencing, and a large number of purified enzymes have made earlier laborious procedures part of history. However, at the same time, there is the loss of insightful knowledge due to the absence of on-hand experiences for isolating nucleic acids and proteins and seeing exactly what they do in reconstruction experiments.

In-depth studies of how plasmids are maintained and dispersed, and how they acquire or lose encoded traits, and of why they persist in the natural and even in man-made environments all are important questions that remain in the field of plasmid biology. Plasmid biologists

Table 2 Examples of plasmid vector systems and their uses

Vector	Application	References
pBR322	General cloning, provided basis for ColE1 cloning vector derivatives	87
pUC	Multiple cloning sites, open reading frame DNA as *lacZ* fusions controlled by *lac* regulatory elements	88–90
pHG175	Multiple cloning sites, promoter probe for tetracycline resistance	91
pKUN9	A pUC9 derivative modified whereby both strands of a cloned DNA fragment can be obtained in a single-stranded form for expeditious sequencing	92
pUCD2335	Mini-T DNA vector bearing a high-copy vir region for genetic engineering of plants	93
pBIN19	Binary vector system for genetic engineering of plants	72
pUCD607	Luciferase reporter of real-time infection by bacteria in higher cells	94
pUCD800	Vector for positive selection of transposons and insertion elements via sucrose sensitivity conferred by the *sacB* gene that encodes levan sucraseLethal to enteric bacteria	95
pUCD2715	*Vibrio* luciferase vector for genetic engineering of plants to make them glow in the dark	96
pWS233	*sacRB* bearing vector bearing gentamicin and tetracycline resistance genes and Mob functions of RP4	97
pUCD4121	Vector that generates unmarked deletions in bacterial chromosomes; bears a *sacB* lethality and neomycin resistance gene	98
pGKA10CAT	A Bluescript pKS(+) derivative for functional analysis of enhancer domains of a transcriptional regulatory region	99
pXL1635	Derived from pRK290, contains RP4 *par* fragment and deleted *oriT* of RK2; for industrial use	100
pJQ200 & pJQ210	Suicide vectors bearing *sacB*, *ori* of pACYC184, and *oriT* and *mob* of RP4	101
pUCD5140	Light sensitivity-producing vector derived from pUCD2335 containing a *rbcS3A* promoter-*gus* fusion and CaMV35S promoter driving a phytochrome A gene of *Avena sativa*	102
pJAZZ	Linear vector for *E. coli* cloning, contains phage N15 *ori*, minimizes formation of nonrecombinants	103
pHP45Ω	A pBR322 derivative for insertional mutagenesis, bearing Ω, and streptomycin/spectinomycin resistance genes flanked by inverted repeats with transcription/translation termination signals and synthetic polylinkers	104

who, "outside of the box" (e.g., replication, partitioning, conjugation) have far-sighted visions of the future prospects of the field of plasmid biology will be the key contributors to the science.

Acknowledgment. *Conflicts of interest: I disclose no conflicts.*

Citation. **Kado CI.** 2014. Historical events that spawned the field of plasmid biology. Microbiol Spectrum 2(5):PLAS-0019-2013.

References

1. **Mitsuhashi S, Kameda M, Harada K, Suzuki M.** 1969. Formation of recombinants between non-transmissible drug-resistance determinants and transfer factors. *J Bacteriol* **97**:1520–1521.
2. **Nakaya R, Nakamura A, Murata Y.** 1960. Resistance transfer agents in *Shigella. Biochem Biophys Res Commun* **3**:654–659.
3. **Mitsuhashi S.** 1977. Epidemiology of R factors, p 25–43. *In* Mitsuhashi S (ed), *R Factor, Drug Resistance Plasmid.* University Park Press, Baltimore, MD.
4. **Hayes W.** 1952. Recombination in *Bact. coli* K12: unidirectional transfer of genetic material. *Nature (London)* **169**:118–119.
5. **Hayes W.** 1953. Observations on a transmissible agent determining sexual differentiation in *Bact. coli. J Gen Microbiol* **8**:72–88.
6. **Lederberg J, Tatum EL.** 1946. Novel genotypes in mixed cultures of biochemical mutants of bacteria. *Cold Spring Harbor Symp Quant Biol* **11**:113–114.
7. **Lederberg J, Tatum EL.** 1946. Gene recombination in *Escherichia coli. Nature (London)* **158**:558.
8. **Lederberg J, Cavalli LL, Lederberg EM.** 1952. Sex compatibility in *Escherichia coli. Genetics* **37**:720–730.
9. **Jacob F, Wollman EL.** 1958. Les épisomes, elements génétiques ajoutés. *C R Hebd. Seances Acad Sci* **247**: 154–156.
10. **Lederberg J.** 1952. Cell genetics and hereditary symbiosis. *Physiol Rev* **32**:403–430.
11. **Lederberg J.** 1998. Plasmid (1952–1997). *Plasmid* **39**: 1–9.
12. **Bukhari AI, Shapiro JA, Adhya SL (ed).** 1977. *DNA Insertion Elements, Plasmids, and Episomes.* Cold Spring Harbor Laboratory, Cold Spring Harbor, NY.
13. **Zhou Y, Call DR, Broschat SL.** 2012. Genetic relationships among 527 Gram-negative bacterial plasmids. *Plasmid* **68**:133–141.
14. **Novick RP.** 1987. Plasmid incompatibility. *Microbiol Rev* **51**:381–395.
15. **Kado CI, Helinski DR.** 2007. Proceedings of the international symposium on plasmid biology. *Plasmid* **57**: 182–243.
16. **Radloff R, Bauer W, Vinograd J.** 1967. A dye-buoyant-density method for the detection and isolation of closed

circular duplex DNA: the closed circular DNA in HeLa cells. *Proc Natl Acad Sci USA* **57:**1514–1521.

17. **Marmur J, Rownd R, Falkow S, Baron LS, Schildkraut C, Doty P.** 1961. The nature of intergeneric episomal infection. *Proc Natl Acad Sci USA* **47:**972–979.
18. **Fiers W, Sinsheimer RL.** 1962. The structure of the DNA of bacteriophage X174 III. Ultracentrifugal evidence for a ring structure. *J Mol Biol* **5:**424–434.
19. **Kleinschmidt AK, Burton A, Sinsheimer RL.** 1963. Electron microscopy of the replicative form of the DNA of the bacteriophage phi-X174. *Science* **142:**961.
20. **Espejo RT, Canelo ES, Sinsheimer RL.** 1969. DNA of bacteriophage PM2: a closed circular double-stranded molecule. *Proc Natl Acad Sci USA* **63:**1164–1168.
21. **Bramhill D, Kornberg A.** 1988. Duplex opening by dnaA protein at novel sequences in initiation of replication at the origin of the *E. coli* chromosome. *Cell* **52:** 743–755.
22. **Gerdes K, Møller-Jensen J, Bugge Jensen R.** 2000. Plasmid and chromosome partitioning: surprises from phylogeny. *Mol Microbiol* **37:**455–466.
23. **Nordström K, Austin SJ.** 1989. Mechanisms that contribute to the stable segregation of plasmids. *Annu Rev Genet* **23:**37–69.
24. **Meacock PA, Cohen SN.** 1980. Partitioning of bacterial plasmids during cell division: a *cis*-acting locus that accomplishes stable plasmid inheritance. *Cell* **20:**529–542.
25. **Ogura T, Hiraga S.** 1983. Partition mechanism of F plasmid: two plasmid gene-encoded products and a *cis*-acting region are involved in partition. *Cell* **32:** 351–360.
26. **Ah-Seng Y, Lopez F, Pasta F, Lane D, Bouet J-Y.** 2009. Dual role of DNA in regulating ATP hydrolysis by the SopA partition protein. *J Biol Chem* **284:**30067–30075.
27. **Mori H, Kondo A, Ohshima A, Ogura T, Hiraga S.** 1986. Structure and function of F plasmid genes essential for partitioning. *J Mol Biol* **192:**1–15.
28. **Hayakawa Y, Murotsu T, Matsubara K.** 1985. Mini-F protein that binds to a unique region for partition of mini-F plasmid DNA. *J Bacteriol* **163:**349–354.
29. **Abeles AL, Snyder KM, Chattoraj DK.** 1984. P1 plasmid replication: replicon structure. *J Mol Biol* **173:**307–324.
30. **Summers DK, Sherratt DJ.** 1984. Multimerization of high copy number plasmids causes instability: ColE1 encodes a determinant essential for plasmid monomerization and stability. *Cell* **36:**1097–1103.
31. **Jaffé A, Ogura T, Hiraga S.** 1985. Effects of the ccd function of the F plasmid on bacterial growth. *J Bacteriol* **163:**841–849.
32. **Lederberg J, Tatum EL.** 1953. Sex in bacteria: genetic studies, 1945–1952. *Science* **118:**169–175.
33. **Harden V, Meynell E.** 1972. Inhibition of gene transfer by antiserum and identification of serotypes of sex pili. *J Bacteriol* **109:**1067–1074.
34. **Tomoeda M, Inuzuka M, Date T.** 1975. Bacterial sex pili. *Prog Biophys Mol Biol* **30:**23–56.
35. **Brinton CC Jr.** 1965. The structure, function, synthesis and genetic control of bacterial pili and a molecular model for DNA and RNA transport in Gram negative bacteria. *Trans NY Acad Sci* **27:**1003–1054.
36. **Curtiss R.** 1969. Bacterial conjugation. *Annu Rev Microbiol* **23:**69–136.
37. **Marvin DA, Hohn B.** 1969. Filamentous bacterial viruses. *Bacteriol Rev* **33:**172–209.
38. **Achtman M, Morelli G, Schwuchow S.** 1978. Cell-cell interactions in conjugating *Escherichia coli*: role of F pili and fate of mating aggregates. *J Bacteriol* **135:** 1053–1061.
39. **Achtman M, Kennedy N, Skurray R.** 1977. Cell-cell interactions in conjugating *Escherichia coli*: role of *traT* protein in surface exclusion. *Proc Natl Acad Sci USA* **74:**5104–5108.
40. **Durrenberger MB, Villiger W, Bachi T.** 1991. Conjugational junctions: morphology of specific contacts in conjugating *Escherichia coli* bacteria. *J Struct Biol* **107:** 146–156.
41. **Harrington LC, Rogerson AC.** 1990. The F pilus of *Escherichia coli* appears to support stable DNA transfer in the absence of wall-to-wall contact between cells. *J Bacteriol* **172:**7263–7264.
42. **Ou JT, Anderson TF.** 1970. Role of pili in bacterial conjugation. *J Bacteriol* **102:**648–654.
43. **Babic A, Lindner AB, Vulic M, Stewart EJ, Radman M.** 2008. Direct visualization of horizontal gene transfer. *Science* **319:**1533–1536.
44. **Frost LS, Ippen-Ihler K, Skurray RA.** 1994. Analysis of the sequence and gene products of the transfer region of the F sex factor. *Microbiol Rev* **58:**162–210.
45. **Frost LS, Finlay BB, Opgenorth A, Paranchych W, Lee JS.** 1985. Characterization and sequence analysis of pilin from F-like plasmids. *J Bacteriol* **164:**1238–1247.
46. **Kado CI.** 1994. Promiscuous DNA transfer system of *Agrobacterium tumefaciens:* role of the *virB* operon in sex pilus assembly and synthesis. *Mol Microbiol* **12:** 17–22.
47. **Shirasu K, Kado CI.** 1993. The *virB* operon of the *Agrobacterium tumefaciens* virulence regulon has sequence similarities to B, C and D open reading frames downstream of the pertussis toxin-operon and to the DNA transfer-operons of broad-host-range conjugative plasmids. *Nucleic Acids Res* **21:**353–354.
48. **Cascales E, Christie PJ.** 2003. The versatile bacterial type IV secretion systems. *Nat Rev Microbiol* **1:**137–149.
49. **Lawley TD, Klimke WA, Gubbin MJ, Frost LS.** 2003. F factor conjugation is a true type IV secretion system. *FEMS Microbiol Lett* **224:**1–15.
50. **Jones AL, Lai EM, Shirasu K, Kado CI.** 1996. VirB2 is a processed pilin-like protein encoded by the *Agrobacterium* Ti plasmid. *J Bacteriol* **178:**5706–5711.
51. **Lai EM, Kado CI.** 1998. Processed VirB2 is the major subunit of the promiscuous pilus of *Agrobacterium tumefaciens*. *J Bacteriol* **180:**2711–2717.
52. **Lai EM, Eisenbrandt R, Kalkum M, Lanka E, Kado CI.** 2002. Biogenesis of T pili in *Agrobacterium tumefaciens*

requires precise VirB2 propilin cleavage and cyclization. *J Bacteriol* **184:**327–330.

53. **Shirasu K, Kado CI.** 1993. Membrane location of the Ti plasmid VirB proteins involved in the biosynthesis of a pilin-like conjugative structure on *Agrobacterium tumefaciens*. *FEMS Microbiol Lett* **111:**287–294.
54. **Lai EM, Kado CI.** 2000. The T-pilus of *Agrobacterium tumefaciens*. *Trends Microbiol* **8:**361–369.
55. **Zupan JR, Ward D, Zambryski P.** 1998. Assembly of the VirB transport complex for DNA transfer from *Agrobacterium tumefaciens* to plant cells. *Curr Opin Microbiol* **1:**649–655.
56. **Kado CI.** 2009. Horizontal gene transfer: sustaining pathogenicity and optimizing host-pathogen interactions. *Mol Plant Pathol* **10:**143–150.
57. **Novick RP.** 1969. Extrachromosomal inheritance in bacteria. *Bacteriol Rev* **33:**210–235.
58. **Phillips G, Funnel B.** 2004. *Plasmid Biology*. ASM Press, Washington, DC.
59. **Kado CI.** 1998. Origin and evolution of plasmids. *Antonie van Leeuwenhoek* **73:**117–126.
60. **Williams PA, Murray K.** 1974. Metabolism of benzoate and the methylbenzoates by *Pseudomonas putida* (arvilla) mt-2: evidence for the existence of a TOL plasmid. *J Bacteriol* **120:**416–423.
61. **Burlage RS, Hooper SW, Sayler GS.** 1989. The TOL (pWW0) catabolic plasmid. *Appl Environ Microbiol* **55:** 1323–1328.
62. **Staudenbauer WL.** 1978. Structure and replication of the colicin E1 plasmid. *Curr Top Microbiol Immunol* **83:**93–156.
63. **Van Tiel-Menkvled GJ, Rezee A, De Graaf FK.** 1979. Production and excretion of cloacin DF13 by *Escherichia coli* harboring plasmid CloDF13. *J Bacteriol* **140:** 415–423.
64. **De Kwaadsteniet M, ten Doeschate K, Dicks LMT.** 2007. Characterization of the structural gene encoding Nisin F, a new lantibiotic produced by a *Lactococcus lactis* subsp. *lactis* isolate from freshwater catfish (*Claria gariepinus*). *Appl Environ Microbiol* **74:**547–549.
65. **Johnson TJ, Nolan LK.** 2009. Pathogenomics of the virulence plasmids of *Escherichia coli*. *Microbiol Molec Biol Rev* **73:**750–774.
66. **Sansonetti PJ, Kopecko DJ, Formal SB.** 1982. Involvement of a plasmid in the invasive ability of *Shigella flexneri*. *Infect Immun* **35:**852–860.
67. **McCarthy AJ, Lindsay JA.** 2012. The distribution of plasmids that carry virulence and resistance genes in *Staphylococcus aureus* is lineage associated. *BMC Microbiol* **12:**104. doi:10.1186/1471-2180-12-104.
68. **Kado CI.** 2010. *Plant Bacteriology*. APS Press, St. Paul, MN.
69. **Lobban P, Kaiser AD.** 1973. Enzymatic end-to-end joining of DNA molecules. *J Mol Biol* **79:**453–471.
70. **Rodriguez RL, Denhardt DT (ed).** 1988. *Vectors: A Survey of Molecular Cloning Vectors and Their Use*. Butterworths, London.
71. **Kaufman RJ.** 2000. Overview of vector design for mammalian gene expression. *Mol Biotechnol* **16:**151–160.
72. **Lee LY, Gelvin SB.** 2008. T-DNA binary vectors and systems. *Plant Physiol* **146:**325–332.
73. **Tolmachov OE.** 2011. Building mosaics of therapeutic plasmid gene vectors. *Curr Gene Ther* **11:**466–478.
74. **Masai H, Kaziro Y, Arai K.** 1983. Definition of *oriR*, the minimum DNA segment essential for initiation of R1 plasmid replication *in vitro*. *Proc Natl Acad Sci USA* **80:**6814–6818.
75. **Rosen J, Ryder T, Inokuchi H, Ohtsubo H, Ohtsubo E.** 1980. Genes and sites involved in replication and incompatibility of an R100 plasmid derivative based on nucleotide sequence analysis. *Mol Gen Genet* **179:** 527–537.
76. **Song H, Phillips SE, Parsons MR, Maas R.** 1996. Crystallization and preliminary crystallographic analysis of RepA1, a replication control protein of the RepFIC replicon of enterotoxin plasmid EntP307. *Proteins* **25:** 137–138.
77. **Churchward G, Linder P, Caro L.** 1983. The nucleotide sequence of replication and maintenance functions encoded by plasmid pSC101. *Nucleic Acids Res* **11:** 5645–5659.
78. **Vocke C, Bastia D.** 1983. DNA-protein interaction at the origin of DNA replication of the plasmid pSC101. *Cell* **35:**495–502.
79. **Scherzinger E, Haring V, Lurz R, Otto S.** 1991. Plasmid RSF1010 DNA replication *in vitro* promoted by purified RSF1010 RepA, RepB and RepC proteins. *Nucleic Acids Res* **19:**1203–1211.
80. **Komori H, Matsunaga F, Higuchi Y, Ishiai M, Wada C, Miki K.** 1999. Crystal structure of a prokaryotic replication initiator protein bound to DNA at 2.6 Å resolution. *EMBO J* **18:**4597–4607.
81. **Kongsuwan K, Josh P, Picault MJ, Wijffels G, Dalrymple B.** 2006. The plasmid RK2 replication initiator protein (TrfA) binds to the sliding clamp beta subunit of DNA derived from the amino-terminal portion of 33-kilodalton TrfA. *J Bacteriol* **188:**5501–5509.
82. **Germino J, Bastia D.** 1983. Interaction of the plasmid R6K-encoded replication initiator protein with its binding sites on DNA. *Cell* **34:**125–134.
83. **Vuicic M, Topisirovic L.** 1993. Molecular analysis of the rolling-circle replicating plasmid pA1 of *Lactobacillus plantarum* A112. *Appl Environ Microbiol* **59:**274–280.
84. **De la Campa AG, del Solar GH, Espinosa M.** 1990. Initiation of replication of plasmid pLS1: the initiator protein RepB acts on two distant DNA regions. *J Mol Biol* **213:**247–262.
85. **Koepsel RR, Murray RW, Rosenblum WD, Khan SA.** 1985. Purification of pT181-encoded RepC protein required for the initiation of plasmid replication. *J Biol Chem* **260:**8571–8577.
86. **Thomas CD, Baison DF, Shaw WV.** 1990. *In vitro* studies of the initiation of staphyloccal plasmid replication. Specificity of RepD for its origin (oriD) and characterization of the Rep-ori tyrosyl ester intermediate. *J Biol Chem* **265:**5519–5530.

87. **Balbás P, Soberón X, Merino E, Zurita M, Lomeli H, Valle F, Flores N, Bolivar F.** 1986. Plasmid vector pBR322 and its special-purpose derivatives: a review. *Gene* 50:3–40.
88. **Messing J.** 1983. New M13 vectors for cloning, p 20–78. *In* Wu R, Grossman L, Moldave K (ed), *Methods in Enzymology*, Academic Press, Orlando, FL.
89. **Norrander J, Kempe T, Messing J.** 1983. Construction of improved M13 vectors using oligo-deoxynucleotide-directed mutagenesis. *Gene* **16:**101–106.
90. **Vieira J, Messing J.** 1982. The pUC plasmids, an M13mp7-derived system for insertion mutagenesis and sequencing with synthetic universal primers. *Gene* **19:** 259–268.
91. **Stewart GSAB, Lubinsky-Mink S, Jackson CG, Cassel A, Kuhn J.** 1986. pHG165: a pBR322 copy number derivative of pUC8 for cloning and expression. *Plasmid* **15:**172–181.
92. **Peeters BPH, Schoenmakers JGG, Konings RNH.** 1986. Plasmid pKUN9, a versatile vector for the selective packaging of both DNA strands into single-stranded DNA-containing phage-like particles. *Gene* **41:**39–46.
93. **Zyprian E, Kado CI.** 1990. *Agrobacterium*-mediated plant transformation by novel mini-T vectors in conjunction with a high-copy *vir* region helper plasmid. *Plant Mol Biol* **15:**245–256.
94. **Shaw JJ, Kado CI.** 1986. Development of a *Vibrio* bioluminescence gene-set to monitor phytopathogenic bacteria during the ongoing disease process in a non-disruptive manner. *Nat Biotechnol* **4:**560–564.
95. **Gay P, LeCoq D, Steinmetz M, Berkelman T, Kado CI.** 1985. Positive selection procedure for entrapment of insertion sequence elements in Gram-negative bacteria. *J Bacteriol* **164:**918–921.
96. **Okumura K, Chlumsky L, Baldwin TO, Kado CI.** 1992. Enhanced stable expression of a *Vibrio* luciferase under the control of the Ω-3 translational enhancer in transgenic plants. *World J Microbiol Biotechnol* **8:**638–644.
97. **Selbitschka W, Niemann S, Pühler A.** 1993. Construction of gene replacement vectors for Gram$^-$ bacteria using a genetically modified *sacRB* gene as a positive selection marker. *Appl Microbiol Biotechnol* **38:**615–618.
98. **Kamoun S, Tola E, Kamdar H, Kado CI.** 1992. Rapid generation of directed and unmarked deletions in *Xanthomonas*. *Mol Microbiol* **6:**809–816.
99. **Kumar G.** 1992. Two *cat* expression vectors for cloning and generation of 3′- and 5′-deletion mutants. *Gene* **110:**101–103.
100. **Crouzet J, Lévy-Schil S, Cauchois L, Cameron B.** 1992. Construction of a broad-host-range non-mobilizable stable vector carrying RP4 par-region. *Gene* **110:**105–108.
101. **Quandt J, Hynes MF.** 1993. Versatile suicide vectors which allow direct selection for gene replacement in Gram-negative bacteria. *Gene* **127:**15–21.
102. **Kurata H, Furusaki S, Kado CI.** 1998. Light-enhanced target gene expression in tobacco BY-2 by the combination of overexpressed phytochrome and *rbcS3A* promoter. *Biotechnol Lett* **20:**463–468.
103. **Godiska R, Dhodda V, Gilbert V, Ravin N, Mead D.** 2007. Proceedings of the International Symposium on Plasmid Biology. *Plasmid* **57:**182–243.
104. **Prentki P, Krisch HM.** 1984. *In vitro* insertional mutagenesis with a selectable DNA fragment. *Gene* **29:** 303–313.

Plasmid Replication Systems and Their Control

II

Plasmids—Biology and Impact in Biotechnology and Discovery
Edited by Marcelo E. Tolmasky and Juan C. Alonso

doi:10.1128/microbiolspec.PLAS-0026-2014

Igor Konieczny,[1] Katarzyna Bury,[1] Aleksandra Wawrzycka,[1]
and Katarzyna Wegrzyn[1]

Iteron Plasmids

2

INTRODUCTION

Iteron plasmids are extrachromosomal genetic elements that can be found in all Gram-negative bacteria. Despite the fact that these plasmids bring antibiotic resistance to host bacterium, they can also bring other features, for example, genes for degradation of specific compounds or toxin production. Iteron plasmids possess characteristic directed repeats located within the origin of replication initiation that are called iterons. These plasmids became model systems for investigation of the molecular mechanisms for DNA replication initiation and for the analysis of mechanisms of control of plasmid copy number in bacterial cells. This research has provided our basic understanding of plasmid biology and the relationship between plasmid DNA and host cells. The control mechanisms utilized by iteron plasmids are based on the nucleoprotein complexes formed by the plasmid-encoded replication initiation protein (Rep). The Rep proteins interact with iterons, which initiates the process of plasmid DNA synthesis, but Rep proteins are also able to form complexes with iterons, which inhibits the replication initiation process. This inhibition is called "handcuffing." Also, Rep protein can interact with inverted repeated sequences, causing transcriptional auto-repression. Finally, various chaperone protein systems and proteases affect the Rep activity and, therefore, overall plasmid DNA metabolism.

STRUCTURE OF THE ORIGIN OF REPLICATION INITIATION

The origin region is one of the most important sequences within plasmid DNA; it ensures plasmid autonomous replication, independent of replication of the bacterial chromosome. As in other replicons, plasmid origins consist of characteristic motifs recognized by replication initiation proteins. In iteron-containing plasmids (Fig. 1), iterons that are directly repeated sequences play a crucial role during DNA replication initiation and are critical for plasmid copy number control (see also text below). They are quite short sequences, whose lengths vary from 17 bp in plasmid RK2 (1), 19 bp in plasmids F (2) and P1 (3), to 22 bp in R6K (4), pPS10 (5), and plasmids from the IncQ incompatibility group (6). Sometimes, such as in plasmid pXV2 from the IncW incompatibility group, direct repeats within the origin can vary in length. In pXV2 there are two 18-bp and two 19-bp repeats (7). The iteron number and spacing between iterons also can differ among iteron-containing plasmids. From the best-characterized plasmids the smallest iterons were identified in plasmid pSC101 (8), in which there are three iterons. In plasmids pPS10 (5) and F (2) there are four iterons; in RK2, 5 (1); and up to seven have been identified in R6K (4). In plasmids from the IncQ group there are three or four identical direct repeats, but sometimes

[1]Department of Molecular and Cellular Biology, Intercollegiate Faculty of Biotechnology of University of Gdansk and Medical University of Gdansk, Gdansk, Poland.

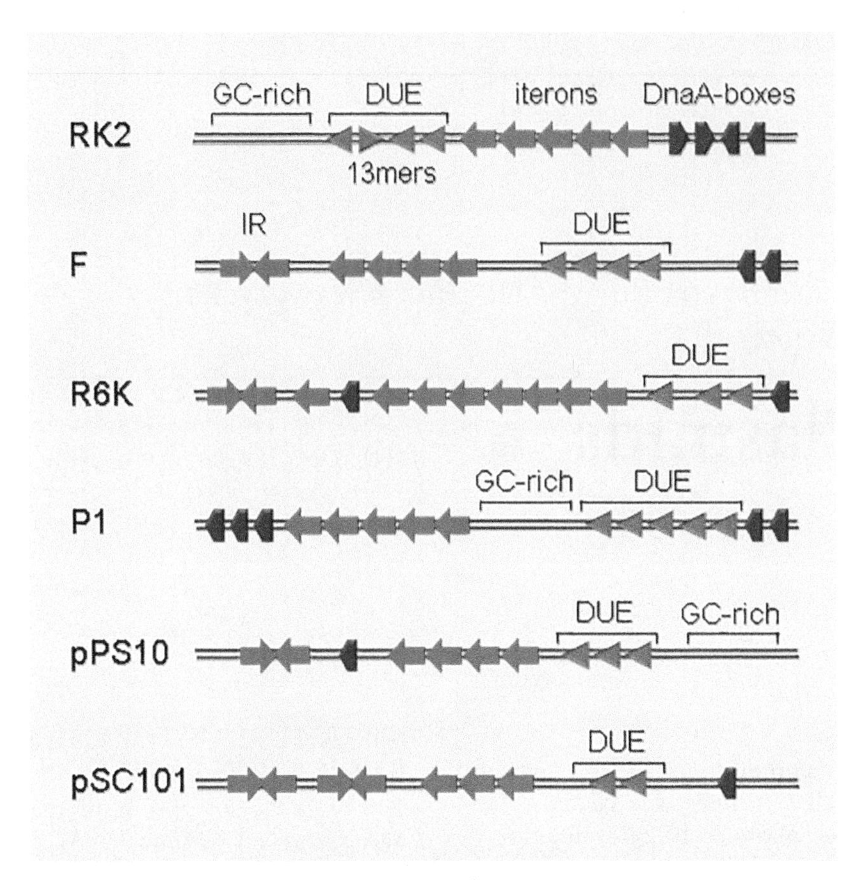

Figure 1 Scheme of the iteron-containing plasmid origin structure. The direct repeats—iterons—and inverted repeats (IR) are depicted as red arrows. The DUE region of each origin is marked, and repeated sequences within the region are depicted as green triangles. DnaA-box sequences are marked in blue. The region rich in guanidine and cytidine residues (GC-rich) is marked within the origins, if identified. The origins are not drawn to scale. doi:10.1128/microbiolspec.PLAS-0026-2014.f1

the functional origin contains more iterons that are partly deleted, contain point mutations, or are incorrectly spaced (6). Plasmid R478 from the IncHI2 incompatibility group even contains several iterons that differ in length (eight 18-bp and nine 76-bp iterons) and are separated by a sequence of *rep* genes (9). Iterons are recognized by a plasmid-encoded Rep protein, and they are bound by a Rep monomeric form (10–13) in a cooperative manner (14, 15). Mutations within an iteron sequence can abolish the binding of Rep protein and, in consequence, plasmid replication. This was shown for plasmid R6K, in which changes in a sequence of iterons made impossible the binding of the π protein *in vitro* and replication of plasmids with mutated origins *in vivo* (16). Similarly, mutations within an iteron sequence in the origin of plasmid P1 reduced or completely prevented origin activity (17). Negative effects on plasmid replication are also exerted by changes in spacers between iterons. The importance of sequences adjacent to iterons was shown for plasmids P1 (17), RK2 (14), and pSC101 (18). Also, disturbances in the position of iterons in relation to other motifs present in the origin region, especially changes in proper helical phasing, have a negative influence on plasmid replication activity (19).

The binding of the plasmid initiator to double-stranded DNA (dsDNA) containing iterons results in local destabilization of the DNA duplex. Plasmid Rep protein is very often accompanied in its action by host initiator DnaA protein. DnaA protein binds a specific motif called DnaA-box, also localized within the plasmid origin. DnaA-boxes are 9-bp-long sequences with consensus sequences that are varied depending on the host bacteria (20). DnaA-boxes can be localized upstream from iterons (e.g., plasmids RK2 and pPS10), downstream from the region rich in adenine and thymine residues (AT-rich), where local destabilization of the duplex occurs (e.g., plasmids F and pSC101), or in both these positions (e.g., plasmids R6K and P1) (21). In exceptional situations, such as in plasmid pXV2, DnaA-box overlaps with the first iteron (7). In some plasmid origins there is just one DnaA-box (e.g., plasmids pSC101, pPS10, and pXV2), and in others there are two (e.g., oriγ of plasmid R6K and plasmid F), four (e.g., plasmid RK2), or even five (e.g., plasmid P1) such motifs. Their length and sequence usually correspond to the consensus sequence of DnaA-boxes present in the origin of the *Escherichia coli* chromosome (*oriC*). If there are some deviations from consensus, they usually do not exceed point mutations. Examples include DnaA-boxes from plasmids P1 and RK2, which contain one or two mismatches. The position of DnaA-boxes is as important as their sequence. Insertions of more or less than a helical turn between DnaA-boxes and iterons within the plasmid RK2 origin resulted in inactivation of the origin's replication activity (19). The binding of DnaA protein to DnaA-boxes in the origin of broad-host-range iteron-containing plasmids can vary in different host bacteria. For instance, in the plasmid RK2 origin, DnaA-boxes 3 and 4 should be present when replication takes place in *E. coli* and *Pseudomonas putida* cells. However, they can be missed during plasmid replication in *Pseudomonas aeruginosa* (22). For *E. coli* chromosome *oriC* it was shown that beside DnaA-boxes, DnaA protein bound with ATP can interact with ATP-DnaA-boxes localized within AT-rich repeats (23). However, in iteron-containing plasmid origin regions, motifs for ATP-DnaA binding, similar to those observed in *oriC*, have not been identified to date.

The third motif, in addition to iterons and DnaA-boxes, that can be distinguished within the iteron-containing plasmid origin is the AT-rich region. This is the sequence, usually located near iterons, where local destabilization of the double-stranded helix occurs during the process of replication initiation. This region is therefore considered a DNA unwinding element (DUE) where single-stranded DNA (ssDNA) is created. Although the thermodynamic stability of the AT-rich region can differ in different origins, usually it has much lower free energy (ΔG) than the overall profile of adjacent sequences (21). In the AT-rich region, it is possible to discern short repeated sequences, usually oriented directly. The exception can be the origin of plasmid RK2, where one of the repeated sequences is inverted in relation to the other ones (24). Repeated sequences within the AT-rich region are located tandemly one after the other (e.g., origin of plasmids RK2 [24] and pSC101 [25]), or they are separated with spacers of different length (e.g., 7-, 1-, and 6-bp spacers between AT-rich repeats of plasmid F [25] and 29- and 9-bp spacers in oriγ of plasmid R6K [26]). The length of those repeated sequences can preserve 13 nucleotides (13-mers), as are present in the AT-rich region of *E. coli oriC* (e.g., plasmid RK2 [1] and pSC101 [25]). But more often they are shorter, such as in plasmids R6K (10 nucleotides [26, 27]), F (8 nucleotides [25]), and P1 (7 nucleotides [28, 29]). Also, the number of repeats can be different, and there can be two repeats in the AT-rich region of plasmid pSC101, four in plasmids RK2 and F, and up to five in plasmid P1 (21).

Although the consensus sequences for AT-rich repeats in different origins are difficult to identify, the consensus can be established for particular plasmid origins. The presence of all repeats within the AT-rich region, as well as their sequence, is very important for the proper replication activity of the origin. Even point mutations within these sequences can completely abolish plasmid replication (17, 30, 31). Also, substitution of one AT-rich repeat in a plasmid origin into a repeat from a bacterial chromosome origin results in a lack of replication activity *in vitro* and a decrease of activity *in vivo* (30). Although the presence and sequence of AT-rich repeats is critical for plasmid replication, the exact role of these motifs is still ambiguous.

The presence of binding sites for replication initiation proteins, iterons, and DnaA-boxes, as well as the region where duplex opening occurs, is very important for the replication initiation process. However, these motifs are not the only ones that can be distinguished within the origin of iteron-containing plasmids. In some plasmid origins the binding site for integration host factor (IHF) can be identified. Such a situation occurs, for instance, in the plasmid P1 origin, where the IHF binding site is located downstream from the cluster of three DnaA-boxes (32). The binding of IHF protein results in the bending of the DNA molecule; however, not only the bend but also its proper phasing for the downstream DNA is required for the activity of the origin (32). Insertions of less than a helical turn between IHF binding sites and DnaA-boxes in the P1 origin had a negative effect on origin activity. The IHF binding sites were also present in plasmids pSC101 (33) and R6K (34). This motif was identified as well in the plasmid RK2 origin, but the IHF deficiency in *E. coli* seemed not to alter plasmid replication efficiency or plasmid copy number control (35).

Other motifs that can be identified within some plasmid origins but are not directly involved during the replication process are sites, GATC motifs recognized by Dam methylotransferase. They are usually overlapped AT-rich repeated sequences (e.g., in plasmids P1 [36] and pSC101 [31]) or are located adjacent to these repeats (e.g., in plasmid P1 [36]). The methylated GATC sequence becomes hemimethylated during replication and in this form is recognized by the SeqA protein (37, 38), which sequestrates newly synthesized DNA (39). SeqA negatively regulates DNA replication by blocking the GATC sites and preventing replication proteins from binding. Apart from the GATC motif, a region rich in guanidine and cytidine residues (GC-rich) can be identified in some plasmids' origins (e.g., plasmids RK2, P1, pPS10, and IncQ). Its exact role is unknown, and in plasmid RK2 it can be deleted without any effects on origin activity (19). In plasmid P1, in which a GT-rich sequence plays the role of a spacer between iterons and AT-rich repeats, the sequence of this region can vary considerably, but its length must be preserved (36).

In a few plasmids identification of motifs other than those described here was reported. For example, in plasmids F and R1 the binding site for the IciA protein was detected (25). The IciA protein, which binds the site located in the AT-rich region of plasmid origins, probably, like in *E. coli oriC* (40), inhibits the unwinding process at the AT-rich region. In the origin of plasmid R6K, binding sites for other regulatory proteins, Fis (factor for inversion stimulation) were found (41). It was shown that plasmid replication depends on the Fis protein when the gene for the copy-up mutant of the π protein and the penicillin resistance gene were present on plasmid DNA (41).

It could be concluded that for the proper activity of the origin of iteron-containing plasmids, not only the presence and the sequence of essential motifs, such iterons, DnaA-boxes, and AT-rich repeats, is important. The appropriate location of these motifs in relation to each other also has a great impact on replication activity. In particular, changes in proper helical phasing have a negative influence on plasmid replication.

Rep PROTEIN STRUCTURE

Although many plasmid Rep proteins have been identified, the crystallographic data are limited to a few replicons. This is due to a high instability of the Rep proteins, so understanding the initiators' role in the structural context is a challenge. Plasmid replication initiators such as the RepA initiator of plasmid pPS10, RepE of plasmid F, and the π protein of plasmid R6K are best characterized in terms of structure. The RepA initiator of pPS10 was the first Rep protein whose structure was predicted to consist of two winged helix (WH) domains (42). These findings have been confirmed by the crystal structure of the monomer of a homologous RepE initiator of plasmid F, bound to iteron DNA (43). The other crystal structure of a plasmid Rep protein was determined for the monomeric form of the π initiator protein of plasmid R6K as a complex with a single copy of its cognate DNA-binding site (iteron) (44). The crystal structures of both RepE and π proteins are depicted in Fig. 2. Although the crystal structures of RepE and π proteins shed new light on the Rep monomers' interaction with DNA, the molecular nature of Rep activation remained unknown until the crystal structure of the dimeric N-terminal domain of the plasmid pPS10 initiator (dRepA) was resolved (45). Nonetheless, the crystallographic data obtained for plasmid Rep proteins are limited to the WH domain description. Rep proteins are composed of two WH domains—N-terminal WH1 and C-terminal WH2—that are responsible for interaction with DNA (42) (Fig. 2). The WH2 domain contains a putative helix-turn-helix motif, which is the main determinant of Rep binding to both the iteron sequences and the inverted repeats (partially homologous to the iteron sequence), which was shown for the RepE initiation protein of the mini-F plasmid and RepA of plasmid pPS10 (46, 47). A formation of nucleoprotein complex by Rep protein results in the bending of the DNA molecule. Iteron interaction with the WH1 and WH2 domains of the Rep monomer, or interaction of inverted repeats with both WH2 domains of Rep dimer, induce DNA bending (42, 48). In Rep monomers, the WH2 domain binds to the 3′-half of the iteron, while the WH1 domain changes structure and contacts the 5′-iteron end, through both the phosphodiester backbone and the minor grove (42).

In contrast to initiation proteins of replicons F, R6K, and pPS10, the crystal structures of the TrfA protein of RK2 as well as P1 RepA have not been determined. The structure prediction using fold-recognition homology modeling was carried out in both cases. The N-terminal part of TrfA does not show a unique three-dimensional structure with the absence of stabilizing factors; it seems to be disordered in solution as opposed to the C-terminal part of the protein, which is expected as two copies of WH domains. Helices of both WH structures interact with major grooves of the DNA phosphate backbone (49). A series of mutations located within the WH1WH2 domains have been found to affect the TrfA-DNA interaction (50, 51). The structure predicted for P1 RepA, similar to TrfA, contains WH domains. By means of fold-recognition programs, it was shown that despite the lack of sequence similarity, RepA shares structural homology with plasmid F RepE. The model predicted that RepA binds one half of the binding site through interactions with the N-terminal

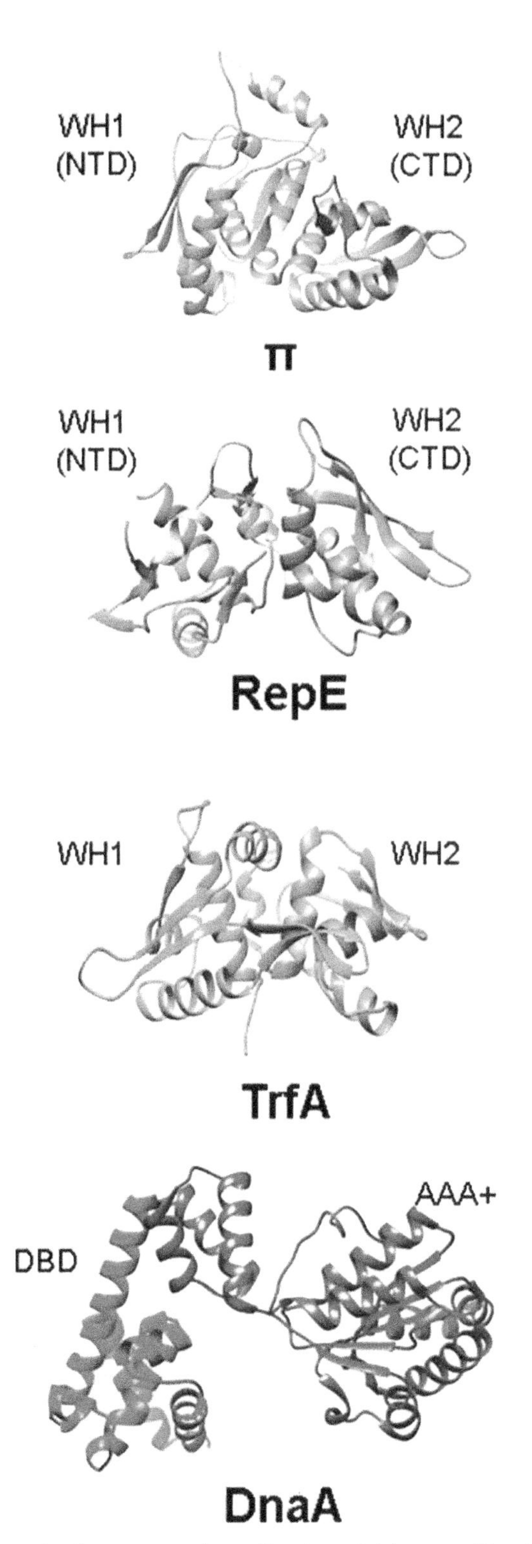

Figure 2 Structure of replication initiators. DnaA of *A. aeolicus*, RepE54 from *E. coli* mini-F plasmid, π from R6K, and the C-terminal part of the TrfA protein (190-382 aa) of plasmid RK2 are depicted. Structure of the DnaA, RepE54, and π are derived from crystallographic data (PDB entry 1L8Q, 1REP, and 2NRA, respectively). The TrfA model was developed based on homology modeling. The AAA+ domain is colored in blue, the DNA binding domain (DBD) is shown in red, and Winged-Helix domains (WH1 and WH2) are colored in yellow and green, respectively. References and detailed information for crystallographic data of the DnaA, RepE54, π, and TrfA model are given in the text.
doi:10.1128/microbiolspec.PLAS-0026-2014.f2

DNA binding domain (WH1) and the second half through interactions with the C-terminal domain (WH2) (52). Interestingly, the residues involved in Rep-DNA interactions located outside the WH domains have been determined with the use of RK2 initiator TrfA mutants (50, 51). These results assume the existence of an additional DNA binding motif, apart from WH1WH2 domains.

Like in plasmid-encoded Rep proteins, WH domains responsible for DNA binding were found in *Archaea* and *Eukaryota* initiators. However, the AAA+ domain (ATPases associated with various cellular activities) commonly present in *Archaea* and *Eukaryota* initiation proteins was not found in plasmid Reps (53) (Fig. 2). Thus, with regard to the DNA binding mechanism, the plasmid Rep proteins are similar to eukaryotic replication initiators. The results of biochemical and spectroscopic experiments revealed functional similarities between pPS10 RepA and archaeal/eukaryal initiators (53). The crystal structure determined for the archaeal initiator Cdc6 confirmed these findings (54). Interestingly, it was reported that similar to the mammalian proteins PrP and α-synuclein, the WH1 domain of the pPS10 RepA can assemble into amyloid fibers upon binding to DNA *in vitro* and in *E. coli* cells (55–58). It opens a direct means to untangle the general pathway (s) for protein amyloidosis in a host with reduced genome and proteome (59).

Plasmid Rep proteins exist in cells mostly as dimers (12, 60). The dissociation of dimers by the action of chaperones or interaction with iteron-containing DNA (see also text below) results in conformational changes in the Rep structure (61). A compact arrangement of the two WH domains, competent for binding to the inversely repeated sequences, becomes a more elongated form, which is suited for iteron binding (42). These conformational changes consist of a significant increase of the overall β-sheet at the expense of the α-helical one (61). The situation is different for the Rep dimers that interact with inversely repeated sequences. Binding of Rep dimers to the inverted repeats does not result in dissociation to monomeric forms or change in the dimers' conformation (61). Although only the monomeric form of Rep proteins is replication-active, dimers of Rep can bind to an inversely repeated sequence localized close to the promoter region of the *rep* gene, which results in transcription auto-repression (see text below). This was shown for the RepA initiator of the pSC101 (62) F RepE initiation protein (43) and the π initiator of the plasmid R6K (63). In the dimeric form of Rep, the WH2 domain binds to inverted repeats via the major groove, whereas the WH1 domain acts as the dimerization interface (61). Dimerization of pPS10 RepA is determined by interactions between β-sheets of the monomers that are originated due to a conformational change in the protein that involves a leucine zipper (LZ)-like motif (42). The LZ-like motif, present in several eukaryotic regulatory proteins (64), has also been found in the WH1 domain of RepA of pSC101(65), RepE of F (43), and π of R6K (66). The dimerization interface is also localized in the WH1 of the model predicted for plasmid RK2 TrfA replication initiation protein (Fig. 2). Similar to the proposal for pPS10 RepA (42), this interface is located on an extended antiparallel β-sheet forming two hairpins (49).

Besides the indirect effect of the LZ motif in Rep protein dimerization, the LZ-like motif was characterized as responsible for Rep interaction with host replication factors. The mutations, described either in pPS10 or in the *E. coli* chromosome, have revealed evidence of a WH1-mediated interaction between RepA and the chromosomal initiator DnaA (67). Nonetheless, protein-protein interaction of Reps are not restricted to the LZ-like region. The best evidence for this statement is a TrfA initiator of plasmid RK2 existing in two replicationally active forms of different molecular mass. The smaller, 33-kDa protein, TrfA-33, is the result of an independent in-frame translational start in the open reading frame used for the larger, 44-kDa protein, TrfA-44 (68–70). The mutation at the N-terminal end of the *trfA* gene (resulting in the availability of the TrfA-33 version only) changes the host range of plasmid RK2, but the binding of DNA remains unaffected. These results demonstrate that the N-terminal end of TrfA is involved in interaction with host replication factors (71). With the use of the evolution experiment, IncP1 plasmids were shown to specialize to a novel host due to the single mutations reported at the N-terminal region of replication initiation protein TrfA (72, 73). In *P. aeruginosa* the TrfA-44 residues between 20 and 30 are responsible for DnaB recruiting (71), and in *E. coli* TrfA-33 interacts *in vitro* with DnaB helicase (74). It also acts with the *E. coli* Hda regulator, which inactivates DnaA and this way prevents overinitiation of RK2 (75). In addition, the specific motif characteristic of proteins interacting with the β clamp of *E. coli* DNA polymerase III was reported in TrfA and TrfA/RepA orthologues from plasmids related to RK2 and pMLb (76), but the relevance of this interaction needs to be elucidated.

The replication of iteron-containing plasmids requires the plasmid-encoded replication initiator, but the host-encoded initiation protein is also involved. The chromosomal initiator, *E. coli* DnaA, is composed of four functional domains (77–79). Crystallographic data obtained for the DnaA conserved core domains III/IV of the thermophilic bacterium *Aquifex aeolicus* revealed that, in contrast to plasmid initiators, this protein is composed of the AAA+ and DBD (DNA binding domain) domains (79) (Fig 2). These domains are involved in DnaA oligomerization and DNA binding/remodeling functions, which are the critical aspects of origin processing. It is crucial for the interaction with ssDNA DUE at chromosomal replication origins and formation of filament structure (80–82). Since plasmid Rep does not possess an AAA+ domain is responsible for nucleotide binding, it could be considered that WH domains, responsible for the binding of iterons within the dsDNA origin, can also bind ssDNA arising after dsDNA melting.

MECHANISM OF ITERON PLASMID DNA REPLICATION INITIATION

Origin Recognition

Models presenting steps of DNA replication initiation of iteron-containing plasmid and bacterial chromosomes are presented in Fig. 3. The first step of replication initiation at the plasmid origin is the formation of an initial complex facilitated by the specific interaction of Rep proteins with iterons. It has been demonstrated that replication initiation of iteron plasmids usually requires cooperative interaction of Rep monomers with iterons. pPS10 RepA as well as RK2 TrfA initiators cooperatively bind iterons at the plasmid replication origins (14, 15). Although the pPS10 RepA dimers and monomers both interact with iterons, only monomers initiate DNA replication. It is noteworthy that the existence of an early transient complex between a dimeric pPS10 RepA and an iteron half has been reported, and based on this, a model for iteron-induced dimeric pPS10 RepA dissociation and conformational activation has been proposed (61). Also, the TrfA protein functionally interacts with plasmid RK2 iterons as a monomer (12). Similar to pPS10 and RK2, the origin of the narrow host range plasmid P1 is recognized by the

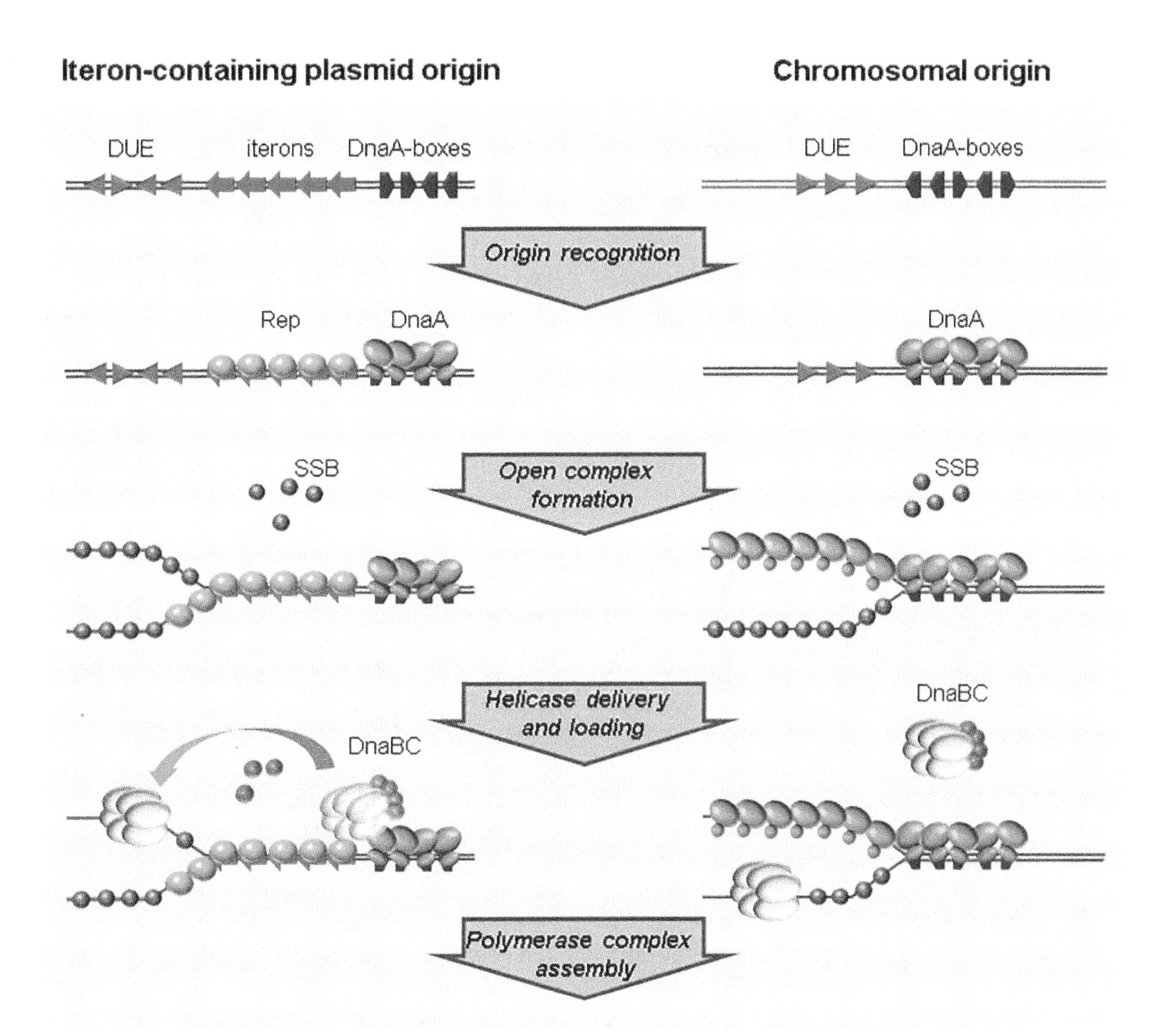

Figure 3 Model of replication initiation: comparison of the processes occurring on the iteron-containing plasmid origin with the replication initiation of bacterial chromosomes. The iteron-containing plasmid origin is recognized by the plasmid-encoded initiator (Rep), which binds cooperatively to the iterons. The interaction of Rep with iterons results in the formation of an open complex and destabilization of the DNA unwinding element (DUE), which creates ssDNA. In RK2, pPS10, F, R6K, P1, and pSC101 the formation of the open complex requires cooperation of the plasmid Rep and host DnaA proteins, while at the chromosomal origin the DnaA protein is sufficient for this process. During the chromosomal origin opening DnaA forms filament on the ssDNA. Helicase delivery and loading requires interaction with the replication initiators; in addition, in *E. coli* the DnaB helicase delivery at the chromosomal *oriC,* as well as at the plasmid RK2 *oriV,* requires the DnaC accessory protein. During the RK2 replication initiation in *E. coli* the host-encoded DnaBC helicase complex is delivered to the DnaA-box sequence through interaction with DnaA, and subsequently the plasmid initiator TrfA translocates the helicase to the opened plasmid origin. The interactions between *E. coli* DnaB and the R6K π protein, F RepE, and pSC101 RepA have also been established as essential for helicase complex formation at the plasmids' origins. The helicase unwinds the DNA double helix, and after a short RNA fragment is synthesized by a primase, a polymerase complex is assembled. Single-stranded DNA binding protein (SSB) is required for replication initiation of both chromosomal and iteron-containing plasmid DNA. The HU/IHF proteins' contribution in DNA replication initiation was omitted in the scheme. For a detailed description see the text.
doi:10.1128/microbiolspec.PLAS-0026-2014.f3

monomer of the P1 initiation protein RepA (83, 84). The interaction between the Rep protein and iterons has also been shown for RepE of plasmid F (85), RepA of plasmid pSC101 (86), and the π initiation protein of *E. coli* plasmid R6K (87). The narrow host range plasmid R6K contains three origins of replication, α, β, and γ, but only two elements, the *ori* and *pir* gene product π proteins, are required for a minimal replicon. The binding of seven iterons by the π initiator has been demonstrated as required for proper *ori* activity (88–91). The π initiator efficiently binds to *ori* iterons but not to the *ori* or to the *ori* iterons (92).

Origin Opening

It was determined that Rep plasmid interaction with iterons generates a localized strand destabilization of DUE, leading to an open complex formation at the origin of plasmid replication. Although the involvement of the plasmid initiator is essential, the host-encoded DnaA and histone-like proteins are also required for plasmid origin opening. It was demonstrated for pPS10 that mutations within the DnaA-box sequence affect the replication *in vivo* (5). DnaA is mainly needed for the enhancement or stabilization of the Rep plasmid-induced open complex formation and histone-like protein (HU and/or IHF) interaction with the DNA-enhanced DNA-bending process. It was determined with KMnO4 assay that TrfA interaction with iterons generates a localized strand destabilization, and *E. coli* DnaA protein enhanced the TrfA-induced open complex (24). It was shown that this reaction occurs only in the presence of the *E. coli* HU protein (24). Similar to RK2 initiator TrfA, the binding of the RepE initiator of plasmid F to iterons induces a localized opening in the origin region, with the assistance of HU (93). The addition of DnaA increases the opening of the F plasmid origin (93) and is also required for the pSC101 origin (94) and R6K *ori* (88, 89). The open complex formation by pSC101 RepA monomers in cooperation with host DnaA also requires the presence of the IHF protein (33, 95). The open complex at the R6K *ori* is formed as a result of cooperative π monomers binding to the iterons and host DnaA interaction with its cognate binding sites (15). KMnO4 footprinting has shown that, in contrast to the RK2 initiator TrfA and F RepE, the P1 RepA alone is not sufficient for *oriR* opening, but in the presence of DnaA, the addition of RepA increased the KMnO4 reactivity of the origin (96). The replication initiation of plasmid RK2 might occur in a DnaA-dependent or DnaA-independent way, depending on the host bacterium. In *E. coli* RK2 efficiently replicates and is maintained in the presence of TrfA and a host DnaA protein, while in *Pseudomonas* the longer form (44 kDa) of the replication initiator is required and DnaA is indispensable (97, 98). In *Caulobacter crescentus* both DnaA-dependent and DnaA-independent models of RK2 plasmid replication initiation are possible (99). Interestingly, the structure of DnaA protein itself might influence the host range of plasmids. Narrow-host-range plasmid pPS10 usually replicates only in the phytopathogen *Pseudomonas savastanoi* cells, due to the ability to bind DnaA-box in the pPS10 origin only by DnaA protein from this bacterium. It has been demonstrated that both the mutation in the LZ motif of pPS10 RepA and mutations in the sequence of *E. coli* DnaA promote the efficient establishment of plasmid pPS10 in the *E. coli* host (67, 100). These results suggest that mutations in plasmid and bacterial initiators that result in expanding the host range of the plasmid probably favor efficient and functional interactions between those proteins. Although the chromosomal initiator, DnaA protein, alone is insufficient for the efficient formation of an open complex at the origin of plasmids F, RK2, pSC101, and R6K (13, 24, 90, 93, 101), it has been shown to be both sufficient and indispensable in opening the AT-rich region at the origin of the bacterial chromosome (see Fig. 3). DnaA interaction with DnaA-box sequences localized within the origin of chromosomal replication (*oriC*) results in destabilization of the DUE, leading to open complex formation. The histone-like proteins HU and IHF stimulate the assembly of the open complex at *oriC* (102–104). This nucleoprotein structure formation requires ATP due to *E. coli* DnaA ATP-dependent conformational changes that promote the formation of the DnaA filament on ssDNA of DUE that is essential for the opening of the replication origin (81, 82, 105). The formation of an open complex at the plasmid origin, in contrast to *E. coli* chromosomal replication, is an ATP-independent process (24, 90, 93, 96, 106, 107), but the presence of ATP or its nonhydrolyzable analogue (ATPγS) promotes the extension of the open region (24). It is not known if plasmid Rep proteins can interact with the ssDNA and form filament structures to promote origin opening, like the DnaA replication initiator does.

Helicase Delivery and Loading

The origin opening generates ssDNA, which is a key element for replication complex assembly at the replication origin. The first step in the assembly of the replication complex is delivering helicase at the replication origin and loading it on ssDNA. While plasmids belonging to the IncP incompatibility group extensively use the replication proteins from the host cell for their own DNA synthesis, they utilize different host-specific mechanisms for helicase delivery and loading (71, 108, 109). Both *in vivo* (97, 98) and *in vitro* (108, 109) analysis with the use of purified proteins from *E. coli* and *Pseudomonas* sp. revealed different host-dependent requirements for RK2 replication initiation. In *E. coli* the DnaB helicase complex with DnaC is initially recruited by DnaA protein interaction (110). The DnaA bound at DnaA-boxes located at the plasmid origin recruits host helicase (111). Then, as a result of translocation into the AT-rich region of the plasmid origin and interaction with the 33-kDa version of the plasmid replication initiator, the helicase is activated for the unwinding of the plasmid dsDNA template. The mechanism of helicase recruitment and loading during the RK2 plasmid replication in *P. aeruginosa* is DnaA-independent and relies on the 44-kDa TrfA protein, while in *P. putida* cells two variants of TrfA protein can be utilized (108, 109). The helicase complex formation during RK2 replication in *C. crescentus* cells might proceed through two different modes: DnaA-independent employing TrfA-44 and DnaA-dependent relying on the shorter version of the replication initiator (99). *In vitro* activity of *C. crescentus* DnaB helicase on the RK2 DNA template was observed in the presence of TrfA-44, and *C. crescentus* DnaA was not required for this process. *In vivo* the mini-RK2 plasmid encoding only TrfA-33 was as stably maintained as those encoding TrfA-44 or both. In contrast, TrfA-33 in cooperation with *C. crescentus* DnaA *in vitro* was unable to activate *C. crescentus* DnaB. The homologue of the *E. coli* DnaC protein needed for proper helicase loading into the open complex might be required for *C. crescentus* DnaB helicase activation. To date, no data about this kind of protein either in *Pseudomonas* or in *Caulobacter* cells have been reported, and its identification requires further investigation (99).

Rep-Helicase Interaction

Similar to the RK2 plasmid initiator TrfA, the interactions between other iteron-containing plasmid Rep proteins and host-encoded helicases have also been reported. *E. coli* DnaB interacts with plasmid replication initiators as was shown for the R6K π protein (112) plasmid F RepE (113) and pSC101 RepA (114). These interactions have been established as essential for helicase complex formation at the mentioned plasmid origins. A DnaB mutant, which does not interact with pSC101 RepA, was unable to activate the replication initiation at the pSC101 origin. Nonetheless, this mutant was able to support *E. coli* chromosomal replication (114). The R6K π protein and pSC101 RepA have also been shown to form complexes with *E. coli* DnaA (90, 101). Similar to R6K and pSC101, the helicase complex formation at the origins of pPS10 and P1 replicons, in addition to the plasmid-encoded initiator, depends on host DnaA protein and requires other host-encoded factors such as DnaC and HU/IHF (67, 115, 116).

The lack of ability for stable complex formation between the plasmid Rep protein and a host helicase might be one of the reasons for plasmid host range restrictions as was shown for *E. coli* plasmid F. The helicase complex at the F origin composed of the replication proteins from the nonnative hosts (*P. aeruginosa* and *P. putida*) might be formed in the presence of F initiator RepE. However, the interactions between RepE and DnaB of *P. aeruginosa* and *P. putida* were unstable, contrary to RepE interaction with *E. coli* DnaB helicase (113).

Polymerase Complex Assembly

Synthesis of iteron-containing plasmid DNA depends on the initial activity of a plasmid replication initiator and utilization of host replication machinery. Because plasmids do not encode their own polymerases, the host bacterium polymerase is utilized for the plasmid DNA replication. The mechanism of the events leading to the formation of the polymerase complex at the plasmid origin of replication still needs to be elucidated. Even though the DNA replication of plasmids RK2 (111), R6K (117), and F (118) has been reconstituted *in vitro* with purified proteins, and specific requirements for this reaction have been identified, the molecular mechanism for the assembly of the polymerase complex at plasmid origins is still not known. The *in vitro* analysis showed that in addition to the plasmid Rep protein, the *E. coli* proteins DnaA, HU, DnaB helicase, DnaC, SSB, DnaG primase, DNA gyrase, and Pol III holoenzyme are required for plasmid DNA synthesis. Interestingly, the specific motif (QL[S/D]LF) determining interaction with the β clamp subunit of Pol III has been identified in plasmid Rep proteins (119), though the relevance of the interaction between the β clamp and Rep proteins has not been determined. The loading of the β clamp is a composite reaction involving clamp opening and then positioning around the DNA with the use of the γ-complex (reviewed in reference 120). β clamp interaction with primed DNA is the first of subsequent events leading to polymerase complex assembly at the chromosomal origin of replication (121). Although the direct involvement of a replication initiation protein in the process of polymerase recruitment has not been reported to date, the plasmid Rep protein interaction with specific Pol III holoenzyme subunits might determine the mechanism for an efficient recruitment of host-encoded replication machinery to the plasmid origin.

CONTROL MECHANISMS OF REPLICATION IN ITERON-CONTAINING PLASMIDS

The iteron-containing plasmid replicons have evolved a number of strategies to ensure their hereditary stability and maintenance at the specific copy number. These plasmids occur in a low-copy number per bacterial cell, so their maintenance requires tight regulation of replication. The main elements involved in the regulation of these plasmid replications are iterons.

Control by Handcuffing

"Handcuffing" is a mechanism of replication inhibition observed in iteron-containing plasmids. The handcuff structure formation is based on the ability of the initiator protein to couple two *ori* regions located on separate plasmid molecules. The *ori* coupling occurs via binding of the Rep protein to iterons. This pairing of iterons is believed to cause steric hindrance to their function that prevents a new round of replication initiation (Fig. 4) (122) by inhibiting origin melting (123). It is considered that handcuffing is a major mechanism that controls the plasmid copy number.

There are three alternative models of Rep-mediated handcuffing. The first one assumes that the handcuff structures are created by the action of Rep dimers, which can bridge two DNA particles. This model was proposed for the replication protein of plasmid R6K (124, 125). Here, the major role of π dimers in the creation of R6K handcuff complexes was detected by electron microscopy (124) and ligation enhancement assays (66, 126, 127). Both of these techniques enable detection of handcuff structures in reaction to the dimeric form of the π protein. In the ligation assay, the monomeric variant of Rep was less efficient in forming ligated products (125). In contrast, the mutant of the π initiator, which binds iterons exclusively as a dimer (13), handcuffed DNA more efficiently than the wild type of the π protein. To summarize, the π dimers have a greater affinity to participate in handcuff structure creation than π monomers. The indirect evidence supporting this model is the fact of handcuffing being counteracted by molecular chaperones (DnaK-J/GrpE triad), which mediate the dissociation of dimers to monomers (123, 128).

The handcuff structure creation in the second model assumes the participation of Rep monomers in the creation of such structures by direct interactions between two arrays of Rep monomers bound to iterons in two plasmid molecules (56). This model is based on the fact that monomers of Rep initiators have a higher affinity for the iteron repeats than the dimeric forms (42, 124). Moreover, it has been reported for plasmid pPS10 that the dimeric Rep mutant is unable to create handcuff structures (56), and iterons of this plasmid play an active role in displacing the equilibrium between Rep dimers and monomers (61).

The third model of handcuff structure is a combination of the other two models. In this model, two monomers bound to the iterons of two separate plasmid molecules, are bridged by the dimer of the Rep protein. Such a model was proposed for handcuffing of plasmids RK2 (129) and F (123). The evidence for this model was obtained in a purified *in vitro* replication system (123). The handcuffing was found to be most proficient only when monomeric and dimeric forms of Rep protein were present simultaneously. Models involving participation of Rep protein dimers are also supported by the fact that handcuffing-defective mutants (Rep monomers of RK2 and R6K) were found to have abnormally high copy numbers (130). Therefore, it can be concluded that the handcuffing has a substantial role in iteron-mediated plasmid copy number control.

If the role of the handcuff is to block the origin and inhibit the replication, then there must be a mechanism that acts in an opposite way and "uncuffs" the coupled origin structures, which enables the reinitiation of plasmid replication. However, the mechanism of handcuff reversal is still unclear. There are results suggesting the participation of the chaperones in handcuff structure disruption (128), showing that the efficiency of handcuffing decreases in the presence of chaperones. Those results indicate that an increasing ratio of monomers over dimers is predominantly responsible for handcuffing reversal. It has also been discovered that the efficiency of handcuff structure creation increases with increasing Rep-bound iteron concentration and decreases when the reaction mixture is diluted. However, the dilution did not decrease Rep binding to the iterons (128).

Control by Auto-Repression

A high concentration of Rep protein initiator may result in more frequent, uncontrolled initiation replication events. To prevent this, the control mechanism that limits the amount of Rep initiator in the cell has to exist. Transcriptional auto-repression is a well-known mechanism for maintaining levels of gene

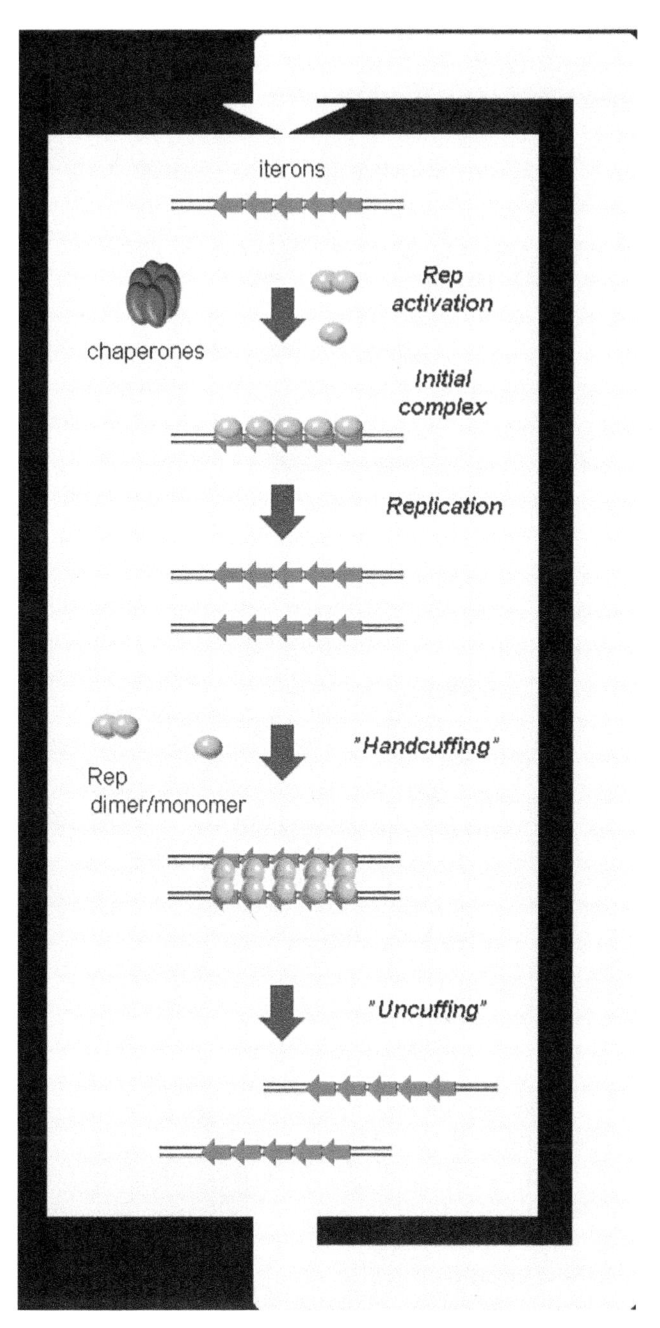

Figure 4 Regulation of iteron-containing plasmid replication initiation by the iterons. Rep protein activation occurs by the action of chaperones that convert the Rep dimer to the active monomeric form. Monomers bind to the iteron sequences and perform the initial complex that leads to replication of DNA. Rep protein may also act as a negative regulator of DNA replication by creating "handcuff" structures. Rep proteins couple origins of two separate plasmid particles in a process termed "handcuffing." In the literature suggestions of chaperone proteins' participation in the "uncuffing" process can be found, but the mechanism of the handcuff structures' reversal is still unclear. For details see the text.
doi:10.1128/microbiolspec.PLAS-0026-2014.f4

product within narrow limits (131, 132). In many plasmid systems (F, R6K, pPS10, and pSC101), autorepression is mediated by binding of the Rep dimer to inverted repeats located adjacent to the origin region (Fig. 5) (11, 46, 65, 133). A sequence of inverted repeats overlaps with the *rep* gene promoter. This kind of regulation mechanism inhibits transcription initiation starting from the *rep* gene promoter, and this effect is promoter-specific (63). The affinity of the Rep dimers is higher for inverted than direct repeats, so the Reps must have specific, dimeric conformation for binding to these sites (133). Symmetrical motifs in the Rep dimer recognize the symmetry of inverted repeats (134). The mechanism of auto-regulation appears to be one of steric hindrance. When the promoter site is occupied by Rep protein, the RNA polymerase cannot displace it from the binding site. However, it has been shown that the initiator proteins can displace RNA polymerase from the promoter, and the addition of the RNA polymerase before the Rep protein does not prevent binding of Rep protein to its binding site (63, 135, 136). This inhibition of RNA polymerase binding resembles typical repressor-polymerase competition and, in this model, the Rep dimer acts as a repressor. An explanation for this auto-regulation mechanism is a higher affinity of the initiator protein for DNA sequence than that of RNA polymerase for the same sequence (136).

Activation and Proteolysis of Rep

As mentioned above, Rep proteins exist in monomer-dimer equilibrium, but only the monomeric form of the proteins can bind specifically to the iterons (12). Saturation of iterons in the replication origin by Rep monomers allows replication initiation. To create such a complex, conformational activation of Rep proteins is required. Dissociation of the Rep dimers into monomers simultaneously changes the conformation of the proteins and makes them competent for the iteron binding. The dissociation may be spontaneous and could occur just by dilution to low/sub-micromolar concentration. This phenomenon has been found for P1 and pSC101 plasmids (62, 83, 137). However, those monomeric forms of Rep proteins require the chaperones for refolding into the active form and for DNA binding (62, 137). The conversion of a dimer to an active monomer can also be mediated by dissociation induced by interaction with iteron-containing DNA. It has been shown that micromolar amounts of DNA, which contain a single iteron, actively induce *in vitro* the dissociation dimers into both monomers and conformational changes (61, 138).

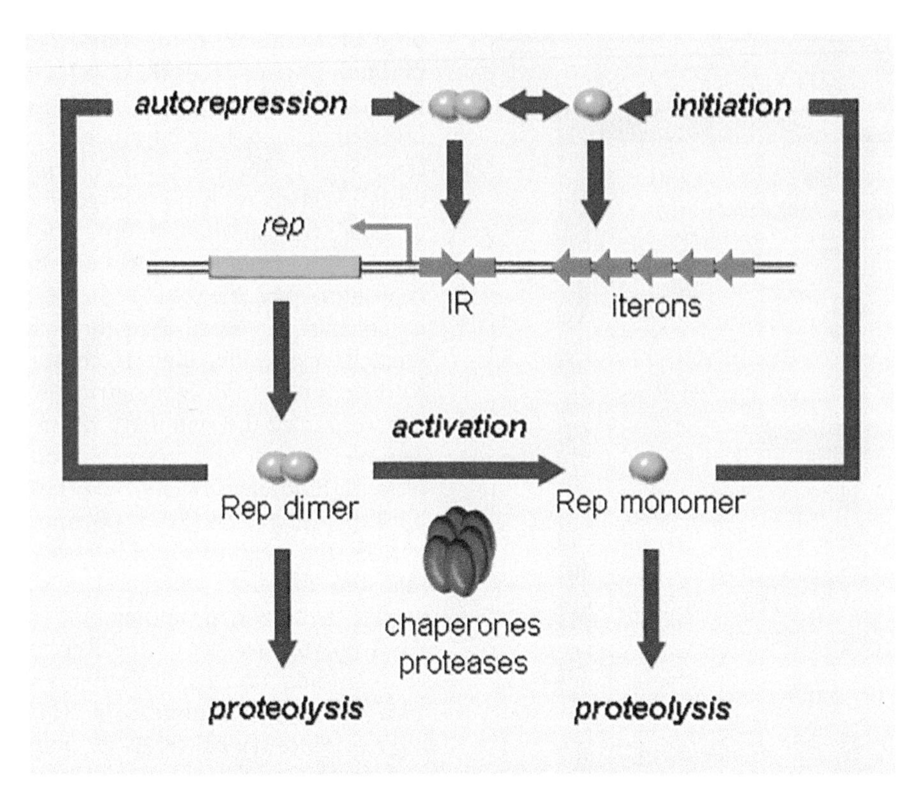

Figure 5 Regulation of iteron-containing plasmid replication initiation by the auto-repression mechanism. Binding of Rep dimers to inverted repeats inhibits the initiation of transcription starting from the *rep* gene promoter. This phenomenon is called auto-repression. An active, monomeric form of Rep protein arises as a result of the action of chaperones. It binds to the iteron sequences that lead to the initiation of DNA replication. Proteases are another factor that may influence the replication process. They limit the amount of both dimer and monomer forms of the Rep protein. For details see the text.
doi:10.1128/microbiolspec.PLAS-0026-2014.f5

The monomeric form of Rep may also arise by the action of molecular chaperons, which actively convert dimers to monomers (139–142). *In vitro* techniques demonstrated that both the ClpX chaperone (139) and the ClpB/DnaK/DnaJ/GrpE system (140) activate the plasmid RK2 replication initiation protein TrfA by converting inactive dimers to an active monomer form. It has been also shown that DnaK/DnaJ/GrpE heat shock proteins are required for the activation of Rep initiators of F, R6K, and P1 plasmids (10, 60, 83, 138, 142, 143). Monomerization of the P1 plasmid initiator may also occur by the action of the ClpA protein, which alone functions as a molecular chaperone (144, 145).

The proteases are other factors affecting iteron-containing plasmid metabolism. They may influence the replication process by proteolysis of the replication initiator. In *E. coli*, four cytosolic proteases have been identified to date: ClpXP, ClpAP, ClpYQ, and Lon (146). The proteases limit the half-life of Rep initiator proteins, which is important for replication initiation. It has been shown that initiator proteins of bacteriophages lambda and Mu and of plasmid RK2 are proteolyzed by *E. coli* ClpXP protease (49, 147, 148) and that ClpAP protease degrades the Rep initiator of plasmid P1 (144). Additionally, it has been described for the TrfA initiator of plasmid RK2 that DNA is a factor that stimulates TrfA proteolysis by ClpAP and Lon proteases (149). Moreover, the Lon protease degrades the TrfA protein only in the nucleoprotein complex, while ClpAP-dependent degradation of TrfA is substantially stimulated in the presence of iteron-containing plasmid DNA (149). This specific stimulation of proteolysis could be important in terms of understanding nucleoprotein complex stability. It may also have an effect on the iteron-containing plasmid copy number, by interaction with the nucleoprotein complex handcuff structure or the other complexes of Rep protein with iteron-containing plasmid DNA.

CONCLUSIONS

All the described mechanisms that affect plasmid metabolism are intended to control the plasmid replication frequency and thereby to control the plasmid copy number. The iteron-containing plasmids, as described above, predominantly use the limitation of Rep protein concentration to control initiation of replication. The limited amount of initiator is achieved by the autorepression mechanism. This kind of replication regulation was initially proposed to be the sole mechanism of replication control, but subsequent experiments showed the marginal effect of surplus initiator. This proved that this mechanism is insufficient (150–152). In iteron-containing plasmids, origin inactivation by handcuffing is an essential mechanism for effective replication regulation. It assumes that the iteron concentration, rather than the level of Rep expression, determines the rate of replication. Another critical parameter that influences the replication initiation is the dimer/monomer ratio of the Rep initiator. The efficient control of plasmid replication initiation requires a combination of all the above-mentioned regulatory mechanisms. Furthermore, it has been reported that all these mechanisms need to work in concert and no single mechanism alone is able to regulate plasmid replication effectively (130). Therefore, it seems to be clear why there are multiple modes of control and that all these modes appear to be cooperative rather than mutually exclusive, which explains why they have been conserved.

Acknowledgments. *This work was supported by the Polish National Science Center (grant number 2012/04/A/NZ1/00048]) and the Foundation for Polish Science (TEAM/2009-3/5). Conflicts of interest: We disclose no conflicts.*

Citation. **Konieczny I, Bury K, Wawrzycka A, Wegrzyn K.** 2014. Iteron plasmids. Microbiol Spectrum 2(6):PLAS-0026-2014.

References

1. **Stalker DM, Thomas CM, Helinski DR.** 1981. Nucleotide sequence of the region of the origin of replication of the broad host range plasmid RK2. *Mol Gen Genet* **181:**8–12.
2. **Murotsu T, Matsubara K, Sugisaki H, Takanami M.** 1981. Nine unique repeating sequences in a region essential for replication and incompatibility of the mini-F plasmid. *Gene* **15:**257–271.
3. **Abeles AL, Snyder KM, Chattoraj DK.** 1984. P1 plasmid replication: replicon structure. *J Mol Biol* **173:** 307–324.
4. **Filutowicz M, McEachern MJ, Mukhopadhyay P, Greener A, Yang SL, Helinski DR.** 1987. DNA and protein interactions in the regulation of plasmid replication. *J Cell Sci Suppl* **7:**15–31.
5. **Nieto C, Giraldo R, Fernandez-Tresguerres E, Diaz R.** 1992. Genetic and functional analysis of the basic replicon of pPS10, a plasmid specific for *Pseudomonas* isolated from *Pseudomonas syringae* patovar *savastanoi*. *J Mol Biol* **223:**415–426.
6. **Loftie-Eaton W, Rawlings DE.** 2012. Diversity, biology and evolution of IncQ-family plasmids. *Plasmid* **67:** 15–34.
7. **Wu LT, Tseng YH.** 2000. Characterization of the IncW cryptic plasmid pXV2 from *Xanthomonas campestris* pv. *vesicatoria*. *Plasmid* **44:**163–172.
8. **Churchward G, Linder P, Caro L.** 1984. Replication functions encoded by the plasmid pSC101. *Adv Exp Med Biol* **179:**209–214.
9. **Page DT, Whelan KF, Colleran E.** 2001. Characterization of two autoreplicative regions of the IncHI2 plasmid R478: RepHI2A and RepHI1A((R478)). *Microbiology* **147:**1591–1598.
10. **Wickner S, Hoskins J, McKenney K.** 1991. Monomerization of RepA dimers by heat shock proteins activates binding to DNA replication origin. *Proc Natl Acad Sci USA* **88:**7903–7907.
11. **Ishiai M, Wada C, Kawasaki Y, Yura T.** 1994. Replication initiator protein RepE of mini-F plasmid: functional differentiation between monomers (initiator) and dimers (autogenous repressor). *Proc Natl Acad Sci USA* **91:**3839–3843.
12. **Toukdarian AE, Helinski DR, Perri S.** 1996. The plasmid RK2 initiation protein binds to the origin of replication as a monomer. *J Biol Chem* **271**(12):7072–7078.
13. **Kruger R, Konieczny I, Filutowicz M.** 2001. Monomer/dimer ratios of replication protein modulate the DNA strand-opening in a replication origin. *J Mol Biol* **306:** 945–955.
14. **Perri S, Helinski DR.** 1993. DNA sequence requirements for interaction of the RK2 replication initiation protein with plasmid origin repeats. *J Biol Chem* **268:** 3662–3669.
15. **Bowers LM, Kruger R, Filutowicz M.** 2007. Mechanism of origin activation by monomers of R6K-encoded pi protein. *J Mol Biol* **368:**928–938.
16. **McEachern MJ, Filutowicz M, Helinski DR.** 1985. Mutations in direct repeat sequences and in a conserved sequence adjacent to the repeats result in a defective replication origin in plasmid R6K. *Proc Natl Acad Sci USA* **82:**1480–1484.
17. **Brendler TG, Abeles AL, Reaves LD, Austin SJ.** 1997. The iteron bases and spacers of the P1 replication origin contain information that specifies the formation of a complex structure involved in initiation. *Mol Microbiol* **23:**559–567.
18. **Ohkubo S, Yamaguchi K.** 1997. A suppressor of mutations in the region adjacent to iterons of pSC101 ori. *J Bacteriol* **179:**2089–2091.
19. **Doran KS, Konieczny I, Helinski DR.** 1998. Replication origin of the broad host range plasmid RK2. Positioning of various motifs is critical for initiation of replication. *J Biol Chem* **273:**8447–8453.

20. **Messer W.** 2002. The bacterial replication initiator DnaA. DnaA and oriC, the bacterial mode to initiate DNA replication. *FEMS Microbiol Rev* **26:**355–374.
21. **Rajewska M, Wegrzyn K, Konieczny I.** 2012. AT-rich region and repeated sequences: the essential elements of replication origins of bacterial replicons. *FEMS Microbiol Rev* **36:**408–434.
22. **Doran KS, Helinski DR, Konieczny I.** 1999. Host-dependent requirement for specific DnaA boxes for plasmid RK2 replication. *Mol Microbiol* **33:**490–498.
23. **Speck C, Messer W.** 2001. Mechanism of origin unwinding: sequential binding of DnaA to double- and single-stranded DNA. *EMBO J* **20:**1469–1476.
24. **Konieczny I, Doran KS, Helinski DR, Blasina A.** 1997. Role of TrfA and DnaA proteins in origin opening during initiation of DNA replication of the broad host range plasmid RK2. *J Biol Chem* **272:**20173–20178.
25. **Wei T, Bernander R.** 1996. Interaction of the IciA protein with AT-rich regions in plasmid replication origins. *Nucleic Acids Res* **24:**1865–1872.
26. **Bramhill D, Kornberg A.** 1988. A model for initiation at origins of DNA replication. *Cell* **54:**915–918.
27. **Stalker DM, Kolter R, Helinski DR.** 1979. Nucleotide sequence of the region of an origin of replication of the antibiotic resistance plasmid R6K. *Proc Natl Acad Sci USA* **76:**1150–1154.
28. **Chattoraj DK, Snyder KM, Abeles AL.** 1985. P1 plasmid replication: multiple functions of RepA protein at the origin. *Proc Natl Acad Sci USA* **82:**2588–2592.
29. **Abeles AL, Reaves LD, Austin SJ.** 1990. A single DnaA box is sufficient for initiation from the P1 plasmid origin. *J Bacteriol* **172:**4386–4391.
30. **Kowalczyk L, Rajewska M, Konieczny I.** 2005. Positioning and the specific sequence of each 13-mer motif are critical for activity of the plasmid RK2 replication origin. *Mol Microbiol* **57:**1439–1449.
31. **Rajewska M, Kowalczyk L, Konopa G, Konieczny I.** 2008. Specific mutations within the AT-rich region of a plasmid replication origin affect either origin opening or helicase loading. *Proc Natl Acad Sci USA* **105:**11134–11139.
32. **Fekete RA, Venkova-Canova T, Park K, Chattoraj DK.** 2006. IHF-dependent activation of P1 plasmid origin by dnaA. *Mol Microbiol* **62:**1739–1751.
33. **Stenzel TT, Patel P, Bastia D.** 1987. The integration host factor of *Escherichia coli* binds to bent DNA at the origin of replication of the plasmid pSC101. *Cell* **49:** 709–717.
34. **Filutowicz M, Appelt K.** 1988. The integration host factor of *Escherichia coli* binds to multiple sites at plasmid R6K gamma origin and is essential for replication. *Nucleic Acids Res* **16:**3829–3843.
35. **Shah DS, Cross MA, Porter D, Thomas CM.** 1995. Dissection of the core and auxiliary sequences in the vegetative replication origin of promiscuous plasmid RK2. *J Mol Biol* **254:**608–622.
36. **Brendler T, Abeles A, Austin S.** 1991. Critical sequences in the core of the P1 plasmid replication origin. *J Bacteriol* **173:**3935–3942.
37. **Brendler T, Abeles A, Austin S.** 1995. A protein that binds to the P1 origin core and the oriC 13mer region in a methylation-specific fashion is the product of the host seqA gene. *EMBO J* **14:**4083–4089.
38. **Slater S, Wold S, Lu M, Boye E, Skarstad K, Kleckner N.** 1995. *E. coli* SeqA protein binds oriC in two different methyl-modulated reactions appropriate to its roles in DNA replication initiation and origin sequestration. *Cell* **82:**927–936.
39. **Lu M, Campbell JL, Boye E, Kleckner N.** 1994. SeqA: a negative modulator of replication initiation in *E. coli*. *Cell* **77:**413–426.
40. **Hwang DS, Kornberg A.** 1990. A novel protein binds a key origin sequence to block replication of an *E. coli* minichromosome. *Cell* **63:**325–331.
41. **Wu F, Wu J, Ehley J, Filutowicz M.** 1996. Preponderance of Fis-binding sites in the R6K gamma origin and the curious effect of the penicillin resistance marker on replication of this origin in the absence of Fis. *J Bacteriol* **178:**4965–4974.
42. **Giraldo R, Andreu JM, Diaz-Orejas R.** 1998. Protein domains and conformational changes in the activation of RepA, a DNA replication initiator. *EMBO J* **17:** 4511–4526.
43. **Komori H, Matsunaga F, Higuchi Y, Ishiai M, Wada C, Miki K.** 1999. Crystal structure of a prokaryotic replication initiator protein bound to DNA at 2.6 A resolution. *EMBO J* **18:**4597–4607.
44. **Swan MK, Bastia D, Davies C.** 2006. Crystal structure of pi initiator protein-iteron complex of plasmid R6K: implications for initiation of plasmid DNA replication. *Proc Natl Acad Sci USA* **103:**18481–18486.
45. **Giraldo R, Fernandez-Tornero C, Evans PR, Diaz-Orejas R, Romero A.** 2003. A conformational switch between transcriptional repression and replication initiation in the RepA dimerization domain. *Nat Struct Biol* **10:** 565–571.
46. **Garcia de Viedma D, Serrano-Lopez A, Diaz-Orejas R.** 1995. Specific binding of the replication protein of plasmid pPS10 to direct and inverted repeats is mediated by an HTH motif. *Nucleic Acids Res* **23:**5048–5054.
47. **Matsunaga F, Kawasaki Y, Ishiai M, Nishikawa K, Yura T, Wada C.** 1995. DNA-binding domain of the RepE initiator protein of mini-F plasmid: involvement of the carboxyl-terminal region. *J Bacteriol* **177:**1994–2001.
48. **Diaz-Lopez T, Davila-Fajardo C, Blaesing F, Lillo MP, Giraldo R.** 2006. Early events in the binding of the pPS10 replication protein RepA to single iteron and operator DNA sequences. *J Mol Biol* **364:**909–920.
49. **Pierechod M, Nowak A, Saari A, Purta E, Bujnicki JM, Konieczny I.** 2009. Conformation of a plasmid replication initiator protein affects its proteolysis by ClpXP system. *Protein Sci* **18:**637–649.
50. **Lin J, Helinski DR.** 1992. Analysis of mutations in trfA, the replication initiation gene of the broad-host-range plasmid RK2. *J Bacteriol* **174:**4110–4119.
51. **Cereghino JL, Helinski DR, Toukdarian AE.** 1994. Isolation and characterization of DNA-binding mutants

of a plasmid replication initiation protein utilizing an *in vivo* binding assay. *Plasmid* **31:**89–99.

52. **Sharma S, Sathyanarayana BK, Bird JG, Hoskins JR, Lee B, Wickner S.** 2004. Plasmid P1 RepA is homologous to the F plasmid RepE class of initiators. *J Biol Chem* **279:**6027–6034.
53. **Giraldo R.** 2003. Common domains in the initiators of DNA replication in Bacteria, Archaea and Eukarya: combined structural, functional and phylogenetic perspectives. *FEMS Microbiol Rev* **26:**533–554.
54. **Liu J, Smith CL, DeRyckere D, DeAngelis K, Martin GS, Berger JM.** 2000. Structure and function of Cdc6/Cdc18: implications for origin recognition and checkpoint control. *Mol Cell* **6:**637–648.
55. **Giraldo R.** 2007. Defined DNA sequences promote the assembly of a bacterial protein into distinct amyloid nanostructures. *Proc Natl Acad Sci USA* **104:**17388–17393.
56. **Gasset-Rosa F, Mate MJ, Davila-Fajardo C, Bravo J, Giraldo R.** 2008. Binding of sulphonated indigo derivatives to RepA-WH1 inhibits DNA-induced protein amyloidogenesis. *Nucleic Acids Res* **36:**2249–2256.
57. **Fernandez-Tresguerres ME, de la Espina SM, Gasset-Rosa F, Giraldo R.** 2010. A DNA-promoted amyloid proteinopathy in *Escherichia coli*. *Mol Microbiol* **77:**1456–1469.
58. **Giraldo R, Moreno-Diaz de la Espina S, Fernandez-Tresguerres ME, Gasset-Rosa F.** 2011. RepA-WH1 prionoid: a synthetic amyloid proteinopathy in a minimalist host. *Prion* **5:**60–64.
59. **Giraldo R.** Amyloid assemblies: protein legos at a crossroads in bottom-up synthetic biology. *Chembiochem* **11:**2347–2357.
60. **Kawasaki Y, Wada C, Yura T.** 1990. Roles of *Escherichia coli* heat shock proteins DnaK, DnaJ and GrpE in mini-F plasmid replication. *Mol Gen Genet* **220:**277–282.
61. **Diaz-Lopez T, Lages-Gonzalo M, Serrano-Lopez A, Alfonso C, Rivas G, Diaz-Orejas R, Giraldo R.** 2003. Structural changes in RepA, a plasmid replication initiator, upon binding to origin DNA. *J Biol Chem* **278:**18606–18616.
62. **Ingmer H, Fong EL, Cohen SN.** 1995. Monomer-dimer equilibrium of the pSC101 RepA protein. *J Mol Biol* **250:**309–314.
63. **Filutowicz M, Davis G, Greener A, Helinski DR.** 1985. Autorepressor properties of the pi-initiation protein encoded by plasmid R6K. *Nucleic Acids Res* **13:**103–114.
64. **Landschulz WH, Johnson PF, McKnight SL.** 1988. The leucine zipper: a hypothetical structure common to a new class of DNA binding proteins. *Science* **240:**1759–1764.
65. **Ingmer H, Cohen SN.** 1993. Excess intracellular concentration of the pSC101 RepA protein interferes with both plasmid DNA replication and partitioning. *J Bacteriol* **175:**7834–7841.
66. **Miron A, Mukherjee S, Bastia D.** 1992. Activation of distant replication origins *in vivo* by DNA looping as revealed by a novel mutant form of an initiator protein defective in cooperativity at a distance. *EMBO J* **11:**1205–1216.
67. **Fernandez-Tresguerres ME, Martin M, Garcia de Viedma D, Giraldo R, Diaz-Orejas R.** 1995. Host growth temperature and a conservative amino acid substitution in the replication protein of pPS10 influence plasmid host range. *J Bacteriol* **177:**4377–4384.
68. **Shingler V, Thomas CM.** 1984. Analysis of the trfA region of broad host-range plasmid RK2 by transposon mutagenesis and identification of polypeptide products. *J Mol Biol* **175:**229–249.
69. **Kornacki JA, West AH, Firshein W.** 1984. Proteins encoded by the trans-acting replication and maintenance regions of broad host range plasmid RK2. *Plasmid* **11:**48–57.
70. **Toukdarian AE, Helinski DR, Perri S.** 1996. The plasmid RK2 initiation protein binds to the origin of replication as a monomer. *J Biol Chem* **271:**7072–7078.
71. **Zhong Z, Helinski D, Toukdarian A.** 2003. A specific region in the N terminus of a replication initiation protein of plasmid RK2 is required for recruitment of *Pseudomonas aeruginosa* DnaB helicase to the plasmid origin. *J Biol Chem* **278:**45305–45310.
72. **Hughes JM, Lohman BK, Deckert GE, Nichols EP, Settles M, Abdo Z, Top EM.** 2012. The role of clonal interference in the evolutionary dynamics of plasmid-host adaptation. *MBio* **3**(4):e00077-e00012.
73. **Sota M, Yano H, Hughes JM, Daughdrill GW, Abdo Z, Forney LJ, Top EM.** 2010. Shifts in the host range of a promiscuous plasmid through parallel evolution of its replication initiation protein. *ISME J* **4:**1568–1580.
74. **Pacek M, Konopa G, Konieczny I.** 2001. DnaA box sequences as the site for helicase delivery during plasmid RK2 replication initiation in *Escherichia coli*. *J Biol Chem* **276:**23639–23644.
75. **Kim PD, Banack T, Lerman DM, Tracy JC, Camara JE, Crooke E, Oliver D, Firshein W.** 2003. Identification of a novel membrane-associated gene product that suppresses toxicity of a TrfA peptide from plasmid RK2 and its relationship to the DnaA host initiation protein. *J Bacteriol* **185:**1817–1824.
76. **Kongsuwan K, Josh P, Picault MJ, Wijffels G, Dalrymple B.** 2006. The plasmid RK2 replication initiator protein (TrfA) binds to the sliding clamp beta subunit of DNA polymerase III: implication for the toxicity of a peptide derived from the amino-terminal portion of 33-kilodalton TrfA. *J Bacteriol* **188:**5501–5509.
77. **Sutton MD, Kaguni JM.** 1997. The *Escherichia coli* dnaA gene: four functional domains. *J Mol Biol* **274:**546–561.
78. **Messer W, Blaesing F, Majka J, Nardmann J, Schaper S, Schmidt A, Seitz H, Speck C, Tungler D, Wegrzyn G, Weigel C, Welzeck M, Zakrzewska-Czerwinska J.** 1999. Functional domains of DnaA proteins. *Biochimie* **81:**819–825.
79. **Erzberger JP, Pirruccello MM, Berger JM.** 2002. The structure of bacterial DnaA: implications for general mechanisms underlying DNA replication initiation. *EMBO J* **21:**4763–4773.

80. **Erzberger JP, Mott ML, Berger JM.** 2006. Structural basis for ATP-dependent DnaA assembly and replication-origin remodeling. *Nat Struct Mol Biol* **13:**676–683.
81. **Duderstadt KE, Chuang K, Berger JM.** 2011. DNA stretching by bacterial initiators promotes replication origin opening. *Nature* **478:**209–213.
82. **Ozaki S, Kawakami H, Nakamura K, Fujikawa N, Kagawa W, Park SY, Yokoyama S, Kurumizaka H, Katayama T.** 2008. A common mechanism for the ATP-DnaA-dependent formation of open complexes at the replication origin. *J Biol Chem* **283:**8351–8362.
83. **DasGupta S, Mukhopadhyay G, Papp PP, Lewis MS, Chattoraj DK.** 1993. Activation of DNA binding by the monomeric form of the P1 replication initiator RepA by heat shock proteins DnaJ and DnaK. *J Mol Biol* **232:** 23–34.
84. **Wickner S, Hoskins J, McKenney K.** 1991. Function of DnaJ and DnaK as chaperones in origin-specific DNA binding by RepA. *Nature* **350:**165–167.
85. **Kawasaki Y, Wada C, Yura T.** 1992. Binding of RepE initiator protein to mini-F DNA origin (ori2). Enhancing effects of repE mutations and DnaJ heat shock protein. *J Biol Chem* **267:**11520–11524.
86. **Manen D, Upegui-Gonzalez LC, Caro L.** 1992. Monomers and dimers of the RepA protein in plasmid pSC101 replication: domains in RepA. *Proc Natl Acad Sci USA* **89:**8923–8927.
87. **Germino J, Bastia D.** 1983. Interaction of the plasmid R6K-encoded replication initiator protein with its binding sites on DNA. *Cell* **34:**125–134.
88. **Kelley WL, Bastia D.** 1992. Activation *in vivo* of the minimal replication origin beta of plasmid R6K requires a small target sequence essential for DNA looping. *New Biol* **4:**569–580.
89. **Kelley WL, Patel I, Bastia D.** 1992. Structural and functional analysis of a replication enhancer: separation of the enhancer activity from origin function by mutational dissection of the replication origin gamma of plasmid R6K. *Proc Natl Acad Sci USA* **89:**5078–5082.
90. **Lu YB, Datta HJ, Bastia D.** 1998. Mechanistic studies of initiator-initiator interaction and replication initiation. *EMBO J* **17:**5192–5200.
91. **Kruger R, Filutowicz M.** 2000. Dimers of pi protein bind the A+T-rich region of the R6K gamma origin near the leading-strand synthesis start sites: regulatory implications. *J Bacteriol* **182:**2461–2467.
92. **Mukherjee S, Erickson H, Bastia D.** 1988. Enhancer-origin interaction in plasmid R6K involves a DNA loop mediated by initiator protein. *Cell* **52:**375–383.
93. **Kawasaki Y, Matsunaga F, Kano Y, Yura T, Wada C.** 1996. The localized melting of mini-F origin by the combined action of the mini-F initiator protein (RepE) and HU and DnaA of *Escherichia coli*. *Mol Gen Genet* **253:**42–49.
94. **Hasunuma K, Sekiguchi M.** 1977. Replication of plasmid pSC101 in *Escherichia coli* K12: requirement for dnaA function. *Mol Gen Genet* **154:**225–230.
95. **Gamas P, Burger AC, Churchward G, Caro L, Galas D, Chandler M.** 1986. Replication of pSC101: effects of mutations in the *E. coli* DNA binding protein IHF. *Mol Gen Genet* **204:**85–89.
96. **Mukhopadhyay G, Carr KM, Kaguni JM, Chattoraj DK.** 1993. Open-complex formation by the host initiator, DnaA, at the origin of P1 plasmid replication. *EMBO J* **12:**4547–4554.
97. **Durland RH, Helinski DR.** 1987. The sequence encoding the 43-kilodalton trfA protein is required for efficient replication or maintenance of minimal RK2 replicons in *Pseudomonas aeruginosa*. *Plasmid* **18:**164–169.
98. **Fang FC, Helinski DR.** 1991. Broad-host-range properties of plasmid RK2: importance of overlapping genes encoding the plasmid replication initiation protein TrfA. *J Bacteriol* **173:**5861–5868.
99. **Wegrzyn K, Witosinska M, Schweiger P, Bury K, Jenal U, Konieczny I.** 2013. RK2 plasmid dynamics in *Caulobacter crescentus* cells: two modes of DNA replication initiation. *Microbiology* **159:**1010–1022.
100. **Maestro B, Sanz JM, Faelen M, Couturier M, Diaz-Orejas R, Fernandez-Tresguerres E.** 2002. Modulation of pPS10 host range by DnaA. *Mol Microbiol* **46:** 223–234.
101. **Sharma R, Kachroo A, Bastia D.** 2001. Mechanistic aspects of DnaA-RepA interaction as revealed by yeast forward and reverse two-hybrid analysis. *EMBO J* **20:** 4577–4587.
102. **Skarstad K, Baker TA, Kornberg A.** 1990. Strand separation required for initiation of replication at the chromosomal origin of *E. coli* is facilitated by a distant RNA-DNA hybrid. *EMBO J* **9:**2341–2348.
103. **Ryan VT, Grimwade JE, Nievera CJ, Leonard AC.** 2002. IHF and HU stimulate assembly of pre-replication complexes at *Escherichia coli* oriC by two different mechanisms. *Mol Microbiol* **46:**113–124.
104. **Hwang DS, Kornberg A.** 1992. Opening of the replication origin of *Escherichia coli* by DnaA protein with protein HU or IHF. *J Biol Chem* **267:**23083–23086.
105. **Duderstadt KE, Mott ML, Crisona NJ, Chuang K, Yang H, Berger JM.** 2010. Origin remodeling and opening in bacteria rely on distinct assembly states of the DnaA initiator. *J Biol Chem* **285:**28229–28239.
106. **Schnos M, Zahn K, Inman RB, Blattner FR.** 1988. Initiation protein induced helix destabilization at the lambda origin: a prepriming step in DNA replication. *Cell* **52:**385–395.
107. **Park K, Mukhopadhyay S, Chattoraj DK.** 1998. Requirements for and regulation of origin opening of plasmid P1. *J Biol Chem* **273:**24906–24911.
108. **Jiang Y, Pacek M, Helinski DR, Konieczny I, Toukdarian A.** 2003. A multifunctional plasmid-encoded replication initiation protein both recruits and positions an active helicase at the replication origin. *Proc Natl Acad Sci USA* **100:**8692–8697.
109. **Caspi R, Pacek M, Consiglieri G, Helinski DR, Toukdarian A, Konieczny I.** 2001. A broad host range replicon with different requirements for replication initiation in three bacterial species. *EMBO J* **20:**3262–3271.

110. **Davey MJ, Fang L, McInerney P, Georgescu RE, O'Donnell M.** 2002. The DnaC helicase loader is a dual ATP/ADP switch protein. *EMBO J* **21:**3148–3159.

111. **Konieczny I, Helinski DR.** 1997. Helicase delivery and activation by DnaA and TrfA proteins during the initiation of replication of the broad host range plasmid RK2. *J Biol Chem* **272:**33312–33318.

112. **Ratnakar PV, Mohanty BK, Lobert M, Bastia D.** 1996. The replication initiator protein pi of the plasmid R6K specifically interacts with the host-encoded helicase DnaB. *Proc Natl Acad Sci USA* **93:**5522–5526.

113. **Zhong Z, Helinski D, Toukdarian A.** 2005. Plasmid host-range: restrictions to F replication in *Pseudomonas*. *Plasmid* **54:**48–56.

114. **Datta HJ, Khatri GS, Bastia D.** 1999. Mechanism of recruitment of DnaB helicase to the replication origin of the plasmid pSC101. *Proc Natl Acad Sci USA* **96:** 73–78.

115. **Giraldo R, Diaz R.** 1992. Differential binding of wild-type and a mutant RepA protein to oriR sequence suggests a model for the initiation of plasmid R1 replication. *J Mol Biol* **228:**787–802.

116. **Wickner SH, Chattoraj DK.** 1987. Replication of mini-P1 plasmid DNA *in vitro* requires two initiation proteins, encoded by the repA gene of phage P1 and the dnaA gene of *Escherichia coli*. *Proc Natl Acad Sci USA* **84:**3668–3672.

117. **Abhyankar MM, Zzaman S, Bastia D.** 2003. Reconstitution of R6K DNA replication *in vitro* using 22 purified proteins. *J Biol Chem* **278:**45476–45484.

118. **Zzaman S, Abhyankar MM, Bastia D.** 2004. Reconstitution of F factor DNA replication *in vitro* with purified proteins. *J Biol Chem* **279:**17404–17410.

119. **Dalrymple BP, Kongsuwan K, Wijffels G.** 2007. Identification of putative DnaN-binding motifs in plasmid replication initiation proteins. *Plasmid* **57:**82–88.

120. **Bloom LB.** 2009. Loading clamps for DNA replication and repair. *DNA Repair (Amst)* **8:**570–578.

121. **Simonetta KR, Kazmirski SL, Goedken ER, Cantor AJ, Kelch BA, McNally R, Seyedin SN, Makino DL, O'Donnell M, Kuriyan J.** 2009. The mechanism of ATP-dependent primer-template recognition by a clamp loader complex. *Cell* **137:**659–671.

122. **Park K, Han E, Paulsson J, Chattoraj DK.** 2001. Origin pairing ('handcuffing') as a mode of negative control of P1 plasmid copy number. *EMBO J* **20:**7323–7332.

123. **Zzaman S, Bastia D.** 2005. Oligomeric initiator protein-mediated DNA looping negatively regulates plasmid replication *in vitro* by preventing origin melting. *Mol Cell* **20:**833–843.

124. **Urh M, Wu J, Forest K, Inman RB, Filutowicz M.** 1998. Assemblies of replication initiator protein on symmetric and asymmetric DNA sequences depend on multiple protein oligomerization surfaces. *J Mol Biol* **283:**619–631.

125. **Kunnimalaiyaan S, Inman RB, Rakowski SA, Filutowicz M.** 2005. Role of pi dimers in coupling ("handcuffing") of plasmid R6K's gamma ori iterons. *J Bacteriol* **187:** 3779–3785.

126. **McEachern MJ, Bott MA, Tooker PA, Helinski DR.** 1989. Negative control of plasmid R6K replication: possible role of intermolecular coupling of replication origins. *Proc Natl Acad Sci USA* **86:**7942–7946.

127. **Miron A, Patel I, Bastia D.** 1994. Multiple pathways of copy control of gamma replicon of R6K: mechanisms both dependent on and independent of cooperativity of interaction of tau protein with DNA affect the copy number. *Proc Natl Acad Sci USA* **91:**6438–6442.

128. **Das N, Chattoraj DK.** 2004. Origin pairing ('handcuffing') and unpairing in the control of P1 plasmid replication. *Mol Microbiol* **54:**836–849.

129. **Toukdarian AE, Helinski DR.** 1998. TrfA dimers play a role in copy-number control of RK2 replication. *Gene* **223:**205–211.

130. **Paulsson J, Chattoraj DK.** 2006. Origin inactivation in bacterial DNA replication control. *Mol Microbiol* **61:** 9–15.

131. **Becskei A, Serrano L.** 2000. Engineering stability in gene networks by autoregulation. *Nature* **405:**590–593.

132. **Simpson ML, Cox CD, Sayler GS.** 2003. Frequency domain analysis of noise in autoregulated gene circuits. *Proc Natl Acad Sci USA* **100:**4551–4556.

133. **York D, Filutowicz M.** 1993. Autoregulation-deficient mutant of the plasmid R6K-encoded pi protein distinguishes between palindromic and nonpalindromic binding sites. *J Biol Chem* **268:**21854–21861.

134. **Pabo CO, Sauer RT.** 1992. Transcription factors: structural families and principles of DNA recognition. *Annu Rev Biochem* **61:**1053–1095.

135. **Vocke C, Bastia D.** 1985. The replication initiator protein of plasmid pSC101 is a transcriptional repressor of its own cistron. *Proc Natl Acad Sci USA* **82:**2252–2256.

136. **Kelley W, Bastia D.** 1985. Replication initiator protein of plasmid R6K autoregulates its own synthesis at the transcriptional step. *Proc Natl Acad Sci USA* **82:** 2574–2578.

137. **Chattoraj DK, Ghirlando R, Park K, Dibbens JA, Lewis MS.** 1996. Dissociation kinetics of RepA dimers: implications for mechanisms of activation of DNA binding by chaperones. *Genes Cells* **1:**189–199.

138. **Nakamura A, Wada C, Miki K.** 2007. Structural basis for regulation of bifunctional roles in replication initiator protein. *Proc Natl Acad Sci USA* **104:**18484–18489.

139. **Konieczny I, Helinski DR.** 1997. The replication initiation protein of the broad-host-range plasmid RK2 is activated by the ClpX chaperone. *Proc Natl Acad Sci USA* **94:**14378–14382.

140. **Konieczny I, Liberek K.** 2002. Cooperative action of *Escherichia coli* ClpB protein and DnaK chaperone in the activation of a replication initiation protein. *J Biol Chem* **277:**18483–18488.

141. **Kruklitis R, Welty DJ, Nakai H.** 1996. ClpX protein of *Escherichia coli* activates bacteriophage Mu transposase in the strand transfer complex for initiation of Mu DNA synthesis. *EMBO J* **15:**935–944.

142. **Zzaman S, Reddy JM, Bastia D.** 2004. The DnaK-DnaJ-GrpE chaperone system activates inert wild type pi initiator protein of R6K into a form active in replication initiation. *J Biol Chem* **279:**50886–50894.

143. **Sozhamannan S, Chattoraj DK.** 1993. Heat shock proteins DnaJ, DnaK, and GrpE stimulate P1 plasmid replication by promoting initiator binding to the origin. *J Bacteriol* **175:**3546–3555.

144. **Wickner S, Gottesman S, Skowyra D, Hoskins J, McKenney K, Maurizi MR.** 1994. A molecular chaperone, ClpA, functions like DnaK and DnaJ. *Proc Natl Acad Sci USA* **91:**12218–12222.

145. **Pak M, Wickner S.** 1997. Mechanism of protein remodeling by ClpA chaperone. *Proc Natl Acad Sci USA* **94:** 4901–4906.

146. **Dougan DA, Mogk A, Bukau B.** 2002. Protein folding and degradation in bacteria: to degrade or not to degrade? That is the question. *Cell Mol Life Sci* **59:**1607–1616.

147. **Levchenko I, Luo L, Baker TA.** 1995. Disassembly of the Mu transposase tetramer by the ClpX chaperone. *Genes Dev* **9:**2399–2408.

148. **Wojtkowiak D, Georgopoulos C, Zylicz M.** 1993. Isolation and characterization of ClpX, a new ATP-dependent specificity component of the Clp protease of *Escherichia coli*. *J Biol Chem* **268:**22609–22617.

149. **Kubik S, Wegrzyn K, Pierechod M, Konieczny I.** 2012. Opposing effects of DNA on proteolysis of a replication initiator. *Nucleic Acids Res* **40:**1148–1159.

150. **Tsutsui H, Fujiyama A, Murotsu T, Matsubara K.** 1983. Role of nine repeating sequences of the mini-F genome for expression of F-specific incompatibility phenotype and copy number control. *J Bacteriol* **155:**337–344.

151. **Pal SK, Chattoraj DK.** 1988. P1 plasmid replication: initiator sequestration is inadequate to explain control by initiator-binding sites. *J Bacteriol* **170:**3554–3560.

152. **Durland RH, Helinski DR.** 1990. Replication of the broad-host-range plasmid RK2: direct measurement of intracellular concentrations of the essential TrfA replication proteins and their effect on plasmid copy number. *J Bacteriol* **172:**3849–3858.

Plasmids—Biology and Impact in Biotechnology and Discovery
Edited by Marcelo E. Tolmasky and Juan C. Alonso

doi:10.1128/microbiolspec.PLAS-0029-2014

Joshua Lilly[1]
Manel Camps[1]

3 Mechanisms of Theta Plasmid Replication

INTRODUCTION

Plasmids have been used as convenient models for the study of molecular mechanisms of replication and DNA repair due to their small size, dispensability to the host, and easy manipulation. In addition, plasmids are key facilitators for the evolution and dissemination of drug resistance and for the evolution of complex interactions with animal or plant hosts. Understanding plasmid replication and maintenance therefore has significant practical implications for the clinic and for bioremediation.

Circular plasmids use a variety of replication strategies depending on the mechanism of initiation of DNA replication and depending on whether leading- and lagging-strand synthesis are coupled or uncoupled. This article focuses on replication of circular plasmids whose lagging strand is synthesized discontinuously, a mechanism known as theta replication because replication intermediates have the shape of the Greek letter θ (theta). Our discussion will focus on replication initiation, which informs different biological properties of plasmids (size, host range, plasmid copy number, etc.), and on how initiation is regulated in these plasmids. To highlight unique aspects of theta plasmid replication, this mode of replication will also be compared with another mode of circular plasmid replication, strand-displacement.

REPLICATION INITIATION

General Structure of Plasmid Origins of Replication

Replication initiation depends on a section of sequence known as the plasmid origin of replication (*ori*). *Basic replicon* refers to the minimal sequence that supports replication, preserving the regulatory circuitry. *Minimal replicon* refers to the minimal portion of sequence supporting plasmid replication even though replication may not be properly regulated, as seen in alterations in plasmid copy number or in the compatibility properties of the plasmids. Finally, there is an even narrower definition of *ori*, which refers to the portion of sequence that is targeted by replication initiation factors *in trans* to initiate replication. In this article we will use the term *origin of replication*, or *ori*, to refer to the *cis-ori*, and *replicon* to refer to basic or minimal replicons.

Rep proteins are plasmid-encoded initiators of replication, although some theta plasmids rely exclusively on host initiation factors for replication. Rep recognition

[1]Department of Microbiology and Environmental Toxicology, University of California Santa Cruz, Santa Cruz, CA 95064.

sites typically consist of direct repeats or *iterons*, whose specific sequence and spacing are important for initiator recognition. Spacing is critically relevant so that the distance matches the helical periodicity of the DNA double helix, allowing recognition of specific DNA sequences (1). Iterons are intrinsically bent, and iteron curvature is enhanced by Rep binding.

Rep proteins are essential and rate-limiting for plasmid replication initiation. Controlled expression of two Rep proteins (π of R6K and RepA of ColE2) can produce a wide range of plasmid copy numbers per cell (between 1 and 250 copies), providing a convenient system for gene dosage optimization of recombinant proteins (2).

Plasmid replicons have a modular structure. Replicons often have motifs that are recognized by plasmid-encoded Reps, A+T-rich areas, G+C-rich areas, methylation sites, and binding sites for host initiation and/or remodeling factors. *Rep* loci, when present, are typically upstream of the plasmid *ori*, immediately adjacent or in close proximity to it.

Replication Initiation: Duplex Melting and Replisome Assembly

Depending on the replicon, duplex melting can be either dependent on transcription or mediated by plasmid-encoded *trans*-acting proteins (Reps). Rep binding of *ori* iterons generally leads to the formation of a nucleoprotein complex that opens up the DNA duplex at the A+T-rich segment.

Opening of the DNA duplex is necessary for replisome assembly, which in theta-type plasmids can be DnaA-dependent or PriA-dependent. DnaA-dependent assembly closely resembles replication initiation at *oriC*, the site initiating chromosomal replication. By contrast, PriA-dependent assembly parallels replication restart following replication fork arrest, which depends on D-loop formation, with the extra DNA strand supplied by homologous recombination (3–5).

In theta-type plasmids, Rep-mediated duplex melting leads to loading of DnaB on the replication fork, often with DnaA assistance. In plasmids that instead rely on transcription for duplex melting, the transcript itself can be processed and becomes the primer for extension. Continuous extension of this primer initiates leading-strand synthesis, facilitating the formation of a displacement loop, or D-loop, as the nascent single-stranded DNA (ssDNA) strand separates the two strands of the DNA duplex and hybridizes with one of them. In this case, PriA (initiator of primosome assembly) can be recruited to the forked structure of the D-loop; alternatively, PriA can be recruited to a hairpin structure that forms when the double-stranded DNA opens (6). PriA promotes both the unwinding of the lagging-strand arm and assembly of two additional proteins (PriB and DnaT) to load DnaB onto the lagging strand template. Thus, in this case loading of DnaB is independent of DnaA.

After loading of DnaB, both DnaA-dependent and -independent modes of replication converge. In both cases, replisome assembly involves the following additional players: SSB (single-stranded binding protein), DnaB (helicase), DnaC (loading factor), the DnaG (primase), and the DNA polymerase III (Pol III) holoenzyme. SSB is recruited to exposed areas of ssDNA, stabilizing them. DnaB is loaded onto the replication fork in the form of a complex with DnaC and recruits DnaG (the primase), which distributively synthesizes RNA primers for lagging-strand synthesis (7). Replisome assembly is completed by loading of the Pol III holoenzyme (8). This holoenzyme contains a core (with α, a catalytic, and ε, a $3'\rightarrow5'$, catalytic subunit), a β_2 processivity factor, and a DnaX complex ATPase that loads β_2 onto DNA and recruits the Pol III core to the newly loaded β_2 (9). DnaB helicase activity is stimulated through its interaction with Pol III and modulated through its interaction with DnaG, facilitating the coordination of leading-strand synthesis with that of lagging-strand synthesis during slow primer synthesis on the lagging strand (10).

Unlike Gram-negative bacteria, which have a single replicative polymerase (Pol III), Gram-positive bacteria have two replicative polymerases: PolC and DnaE. PolC is a processive polymerase responsible for leading-strand synthesis, while DnaE extends DnaG-synthesized primers before handoff to PolC at the lagging strand (11, 12).

In theta plasmids, lagging-strand synthesis is discontinuous and coordinated with leading-strand synthesis. The replicase extends a free 3′-OH of an RNA primer, which can be generated by DnaG primase (in Gram-negative bacteria), by the concerted action of DnaE and DnaG primase (in Gram-positive bacteria), or by alternative plasmid-encoded primases. Discontinuous lagging-strand synthesis involves repeated priming and elongation of Okazaki fragments and is comparable in plasmids and chromosomes, although Okazaki fragments were found to be smaller in a ColE1-like plasmid, approximately one-third the length of Okazaki fragments in the chromosome (13).

DNA polymerase I (Pol I) contributes to plasmid replication in several ways. In ColE1 and ColE1-like plasmids, Pol I can extend a primer to initiate leading-strand synthesis and open the DNA duplex; this process can expose a hairpin structure in the lagging strand, known as a single-strand initiation (*ssi*) site or primosome assembly (*pas*) site, and/or generate a D-loop. Both hairpins and forked structures recruit PriA, which is the first step in the replisome initiation complex. Following replisome assembly, Pol I plays a critical role in discontinuous lagging-strand synthesis, removing RNA primers through its 5′→3′ exonuclease activity and filling in the remaining gap through its polymerase activity (14). In addition, two lines of evidence suggest that Pol I can functionally replace Pol III in *Escherichia coli*: (i) Pol I is essential for *polC* (Poll III-minus) strain viability, showing that both polymerases are functionally redundant (15). (ii) Mutations generated through error-prone Pol I replication of a ColE1-like plasmid *in vivo* strongly suggest that Pol I replicates both plasmid strands with similar frequency beyond the point where the switch to Pol III is expected, again suggesting that Pol I can be redundant with the Pol III replisome (16).

THETA PLASMID REPLICATION

Three modes of replication can be distinguished for circular plasmid replication: theta, strand-displacement, and rolling circle. This review focuses on theta. This mode of replication is similar to chromosomal replication in that the leading and lagging strands are replicated coordinately, with discontinuous lagging-strand synthesis. No DNA breaks are required for this mode of replication. Coordinated replication of both strands leads to the formation of bubbles in the early stages of replication, seen as the Greek letter θ under electron microscopy. Four classes of theta-type plasmids can be distinguished based on their mode of replication initiation, although the last two categories show hybrid features of the first two and will be discussed together (see theta replication section in Table 1).

Class A Theta Replication

Class A theta plasmids include R1, RK2, R6K, pSC101, pPS10, F, and P. All these plasmids depend on Rep proteins for replication initiation: RepA for R1, pSC101, pPS10, and P1; Trf1 for RK1; and π for R6K. Note that the name of these Reps is incidental, so sharing a name is not an indication of related structure or mode of action. Rep proteins bind direct repeats (iterons) in the plasmid *ori*. In class A, these iterons are rarely identical, although they frequently conform to a consensus motif. In plasmid P1, RepA monomers contact each iteron through two consecutive turns of the helix, leading to in-phase bending of the DNA, which wraps around RepA (17). Similarly, in R6K plasmids, π binding of its cognate iterons bends the DNA and generates a wrapped nucleoprotein structure (18).

There are two prominent exceptions to the presence of multiple iterons in class A theta plasmid *oris*: (i) Plasmid R1, which features two partial palindromic sequences instead of iterons; however, similar to other plasmids of this class, R1 palindromic sequences are recognized by RepA. (ii) The R6K plasmid, which has three *oris*, only one of which has multiple iterons: γ (with seven iterons), a second origin (α) with a single

TABLE 1 Comparison of the three basic modes of plasmid replication initiation in circular plasmids

	Leading-strand synthesis		Lagging-strand synthesis		
Type of replication	Plasmid initiation factors	Host factors	Coupling with leading strand	Plasmid factors	Host factors
Theta class A	Rep (duplex melting)	DnaA-replisome	Yes	No	Replisome
Theta class B	None	RNAP Pol I RNase H PriA-replisome	Yes	No	Replisome
Theta class C	Rep (duplex melting, primase)	Replisome	Yes	No	Replisome
Theta class D	Rep (duplex melting, RNA processing?)	RNAP PriA-replisome	Yes	No	Replisome
Strand-displacement	Rep A (helicase) Rep B (primase) Rep C (initiator)	Replisome (recruited by RepA)	No (simultaneous)	Rep A (helicase) Rep B (primase) Rep C (initiator)	None

iteron, and a third origin (β) with only half an iteron. It appears that the γ *ori* is an establishment origin, allowing replication initiation immediately following mobilization, when levels of π protein are low, whereas α and β *oris* would be maintenance origins in cells inheriting the plasmid by vertical transmission (19). In any case, γ ori acts as an enhancer, favoring the long-range activation of α and β oris by transfer of π. Thus, α and β *oris* are still dependent on the multiple iterons present in *ori* γ.

Rep binding of a cognate sequence in the plasmid *ori* mediates the earliest step in replication initiation: duplex DNA melting. A Rep-DnaA interaction is frequently involved, although the importance of this interaction varies between individual *oris*. In plasmid pSC101, RepA serves to stabilize DnaA binding to distant *dnaA* boxes, leading to strand melting (20). Plasmid P1's *ori* has two sets of tandem *dnaA* boxes at each end; DnaA binding loops up the DNA, leading to preferential loading of DnaB to one of the strands (21). By contrast, RK2's TrfA was shown to mediate open complex formation and DnaB helicase loading in the absence of *dnaA* boxes, although the presence of DnaA protein was still required (22).

As mentioned above, the double strand melts in response to iteron binding by Rep protein. Melting occurs at an AT-rich region. Similar to chromosomal *oriC*, AT-rich segments of sequence frequently have sites for host factors playing an architectural role such as histone-like protein, integration host factor, and factor for inversion stimulation. These host factors help with DNA melting and with the structural organization of the initiation complex (1, 23, 24).

Class B Theta Replication

Class B theta plasmids include ColE1 and ColE1-like plasmids, which are frequently used for recombinant gene expression. Unlike class A, class B plasmids rely exclusively on host factors for both double-strand melting and primer synthesis. The DNA duplex is opened in this case by transcription of a long (~600 bp) preprimer called RNA II, which is transcribed from a constitutive promoter P2. Constitutive expression from this promoter is enhanced by a 9-bp motif 5′-AAGATCTTC, which is located immediately upstream of the -35 box (25). The 3′ end of the preprimer RNA forms a stable hybrid with the 5′ end of the lagging-strand DNA template of *ori*. This stable RNA-DNA hybridization (R-loop formation) is facilitated by the pairing of a stretch of G-rich sequence on the transcript with a C-rich stretch on the lagging-strand DNA template and by a hairpin structure located between the G- and C-rich stretches (26). Following R-loop formation, the RNA preprimer is processed by RNase H (which recognizes the AAAAA motif in RNAII), producing a free 3′-OH end. Extension of this RNA primer by Pol I initiates leading-strand synthesis. The point where the RNA primer is extended (known as RNA/DNA switch) is considered the replication start point (reviewed in references 27–29).

As mentioned above, the nascent leading strand separates the two strands of the DNA duplex and can hybridize with the leading-strand template, forming a D-loop. PriA is recruited to the forked structure of the D-loop; alternatively, PriA can be recruited to hairpin structures forming on the lagging-strand template when the duplex opens. Indeed, *priA* strains do not support ColE1 plasmid replication, and hypomorphic mutations in *priA priB* result in a reduced ColE1 plasmid copy number (30–32).

When the Pol III holoenzyme is loaded (27, 28) this polymerase continues leading-strand synthesis and initiates lagging-strand synthesis. Pol III replication of the lagging strand toward the RNA II sequence is arrested 17 bp upstream of the DNA/RNA switch, at a site known at *terH*, ensuring unidirectional replication (33). Lagging-strand replication by Pol III appears to end a few hundred nucleotides upstream of the *terH* site (33), leaving a gap that is filled by Pol I (16).

The only step that is essential in this process of replication initiation is R-loop formation; deficits in RNase H and/or Pol I do not prevent initiation, although they have a substantial impact on the efficiency of replication initiation. In the absence of RNase H, unprocessed transcripts can still be extended with some frequency, and in the absence of Pol I, the Pol III replisome can still be loaded on an R-loop formed by the transcript and lagging-strand template (28).

R-loop formation can happen as a result of local supercoiling in the trail of the advancing RNA polymerase during transcription and is highly deleterious because R-loops block transcription and the elongation step during translation (34). Therefore, cells have mechanisms to suppress unscheduled R-loop formation. The most important ones are relaxation of the DNA template by type I topoisomerase activity, RNA degradation by RNase H, RecG dissociation of R-loops by branch migration, factor-dependent transcriptional termination, and coupling transcription to translation (reviewed in reference 35). Accordingly, titration of R-loop-suppressing factors through uncoupling transcription from translation (by starvation, temperature shift, or chloramphenicol treatment) results in increased ColE1 plasmid copy number (36), whereas

RecG overexpression dramatically suppresses replication initiation (37). However, loss of topoisomerase I and RNase H activity do not increase plasmid copy number despite inducing increased R-loop formation because these activities are also required for plasmid replication initiation (particularly RNase H).

Hybrid Classes of Theta Replication (Classes C and D)

Classes C and D have specialized priming mechanisms combined with elements of class A and class B replication. Like class A plasmids, class C and D plasmids have Rep proteins, located immediately upstream of *ori*. Like class B plasmids, however, both initiate leading-strand synthesis by Pol I extension of a free 3′-OH. Class C and D plasmids both have termination signals in the 3′ direction of lagging-strand synthesis, making replication of these plasmids unidirectional.

Class C and D theta plasmid replication is based on the evolution of more efficient ways to prime replication initiation. The evolution of plasmid-specific primases exploits the specificity provided by Rep interaction with *ori* to minimize the size of the *cis-ori* sequence. Such specificity is not possible when multiple primers are needed, as in the case of lagging-strand synthesis in the chromosome. Also, the evolution of specialized priming mechanisms broadens the host range of these plasmids by reducing dependence on host factors (38).

Class C includes ColE2 and ColE3 plasmids. The *ori*s for these two plasmids are the smallest described so far (32 bp for ColE2 and 33 bp for ColE3); these two *ori*s differ only at two positions, one of which determines plasmid specificity (39). ColE2 and ColE3 *oris* have two iterons and show two discrete functional subregions: one specializing in stable binding of the Rep protein (region I) and the other specializing in initiation of DNA replication (region III), with an area of overlap in between (region II) (40). Unlike class A initiator Rep proteins, the Rep protein in class C plasmids has primase activity, synthesizing a unique primer RNA (ppApGpA) that is extended by Pol I at a fixed site in the origin region (41). Class C replication is unidirectional, as the 3′ end of the lagging-strand DNA fragment was mapped to a specific site at the end of the *ori* region. The Rep protein may stay bound to the *ori* after initiation of replication, blocking progression of the replisome synthesizing the lagging strand (42).

Class D includes large, low-copy streptococcal plasmids that replicate in a broad range of Gram-positive bacteria. Examples include pAMβ1 from *Enterococcus faecalis*, pIP501 from *Streptococcus agalactiae*, and pSM19035 from *Streptococcus pyogenes*. In these plasmids, replication shares some features with class B theta replication, specifically a requirement for transcription across the *ori* sequence, Pol I extension and PriA-dependent replisome assembly (43). In this case, the transcript is generated from a promoter controlling expression of rep, which is immediately upstream of the *ori* (43). The replication process has been studied in detail for pAMβ1, although the Rep proteins (RepE for pAMβ1, RepF for pIP501, and RepS for pSM19035) are 97% identical for all three plasmids, and the three plasmids share a replisome structure, suggesting that they share mechanisms for replication initiation and termination. Replication depends on transcription through the origin. Rep binds specifically and rapidly to a unique site immediately upstream of the replication initiation site. This binding denatures an AT-rich sequence immediately downstream of the binding site to form an open complex (44). Compared to class A, this open complex is atypical on several counts: (i) the cognate sequence does not have multiple iterons, (ii) binding does not induce strong bending of the origin, and (iii) melting does not require additional host factors. In addition to opening of the double strand, RepE appears to have an active role in primer processing, as melting increases RepE binding and RepE can cleave transcripts from the repE operon in close proximity to the RNA/DNA switch (45).

Class D replisome assembly is PriA-dependent. A primosome assembly signal can be found 150 nucleotides (nt) downstream from the *ori* on the lagging-strand template. There is a site for replication arrest induced by Topb, a plasmid-encoded topoisomerase related to topo III, 190 nt downstream for the *ori* (46). A second replication arrest site can be found 230 nt downstream from the plasmid *ori*; in this case arrest is caused by collision with a site-specific resolvase, Resb, which is a plasmid-borne gene responsible for plasmid segregation stability (47). The presence of two independent checkpoints for Pol I progression in pAMβ1 is intriguing; this may be a mechanism that ensures Pol I availability for chromosomal replication and/or that facilitates recruitment of PriA, as PriA is known to be recruited to sites of replication fork arrest. In any case the two replication blocks appear to be largely redundant, as Topb is dispensable for pAMBβ1 replication (46).

COMPARISON OF THE THETA AND STRAND-DISPLACEMENT MODES OF PLASMID REPLICATION

Plasmids that replicate using the strand-displacement mode of replication include *E. coli* incompatibility group Q (IncQ) plasmids of γ-proteobacteria such as RSF1010. Strand-displacement replication depends on a specialized primase: RepB. In this case, the function of replication initiator function is provided by a different Rep (RepC). Similar to initiator Rep proteins in class A theta plasmids, Rep C binds cognate iteron sequences, bending the DNA and melting duplex DNA at an adjacent A+T-rich region. An additional plasmid-encoded protein (a helicase, RepA) helps melt the DNA, recruit Pol III, and support continuous replication of one strand. This single-stranded replication produces a daughter ssDNA strand, which separates the two strands of the DNA duplex and allows hybridization with one of them, creating a D-loop (hence the name of this mode of replication).

A model for strand-displacement replication is presented in Fig. 1. After RepC-induced melting of the duplex, RepA monomers assemble around the exposed ssDNA and catalyze bidirectional unwinding of the DNA. This exposes the two different *ssi* sites, which are adjacent and are both palindromic, resulting in inverted repeats on the two DNA strands. When these two sites are exposed in single-stranded configuration, base-pair complementarity favors the formation of two hairpins, one for each strand, (Fig. 1, panel II) (48). Hairpin formation is assisted by a slowdown in RepA progression at a G+C-rich region (reviewed in reference 49). The base of each hairpin contains the start point for DNA synthesis, which is recognized by Rep B, and primer synthesis ensues (50, 51). The Pol III holoenzyme extends off of the synthesized primer (Fig. 1, panels III to V). Initiation can occur at either site independently and is continuous. As replication progresses, facilitated by the RepA helicase, a theta-type intermediate forms (Fig. 1, panels III and IV). Ligation of the two daughter strands produces two double-stranded circles (Fig. 1, panel VI).

Unlike theta-type replication, strand-displacement replication initiation is independent of host factors. This autonomous replication initiation gives these plasmids a very broad range of operation (52). As mentioned above, strand-displacement replication initiation has some similarities to class C theta plasmid replication (with a specialized, plasmid-encoded primase) and similarities to class A theta plasmid replication (with a Rep initiator involved in melting the duplex), but strand displacement presents three major differences relative to theta plasmid replication: (i) no involvement of DnaBC, as RepA is loaded on *ssi* sites exposed in the ssDNA configuration, recruiting the replicase; (ii) priming is carried out by RepB, functionally replacing the host primase DnaG; and (iii) Pol III replicates each strand continuously, initiating at two single-stranded motifs located on opposite strands (*ssiA* and *ssiB*). Note that continuous replication includes the lagging strand, which in this case does not involve synthesis of Okazaki primers (53).

REGULATION OF REPLICATION INITIATION

The frequency of replication initiation is regulated by negative feedback loop mechanisms. These regulatory mechanisms allow for rapid expansion when plasmids colonize a new permissive cell (establishment phase) and later tune the frequency of replication so that, on average, there is one replicative event per plasmid copy number per cell cycle (steady state phase), minimizing fluctuations in copy number (54).

Types of Feedback Regulatory Mechanisms

Plasmid copy number regulation needs mechanisms to monitor the plasmid copy number through a "sensor" and mechanisms to modulate replication initiation in response to feedback through an "effector" (55). The sensor mechanism depends on molecules whose concentration in the cytoplasm is proportional to plasmid copy number. In theta plasmids, inhibition of replication occurs at the initiation step and depends on three types of mechanisms: (i) antisense RNAs that hybridize to a complementary region of an essential RNA (countertranscribed RNAs, or ctRNAs) – dual mechanisms involving ctRNA and an additional protein repressor also occur –; (ii) Rep binding of iterons located in the Rep promoter, suppressing transcription; and (iii) steric hindrance between plasmids by interaction between Rep initiator proteins bound to different plasmids, which "handcuffs" them. Note that in all three cases sensor and effector functions are performed by the same molecule.

Countertranscribed RNA Inhibition

These feedback mechanisms share the following elements: two promoters in opposite orientations, one directing the synthesis of an RNA essential for replication and the other directing the synthesis of an inhibitor ctRNA. The ctRNA is complementary to a region near the 5′ end of the essential RNA, is typically strongly expressed, and has a short half-life, whereas its target RNA is expressed at constitutive but low levels. Examples of targets include maturation of a primer required for replication initiation (ColE1 plasmids), inhibition of repA translation (R1), and premature termination of translation of a rep mRNA (class D

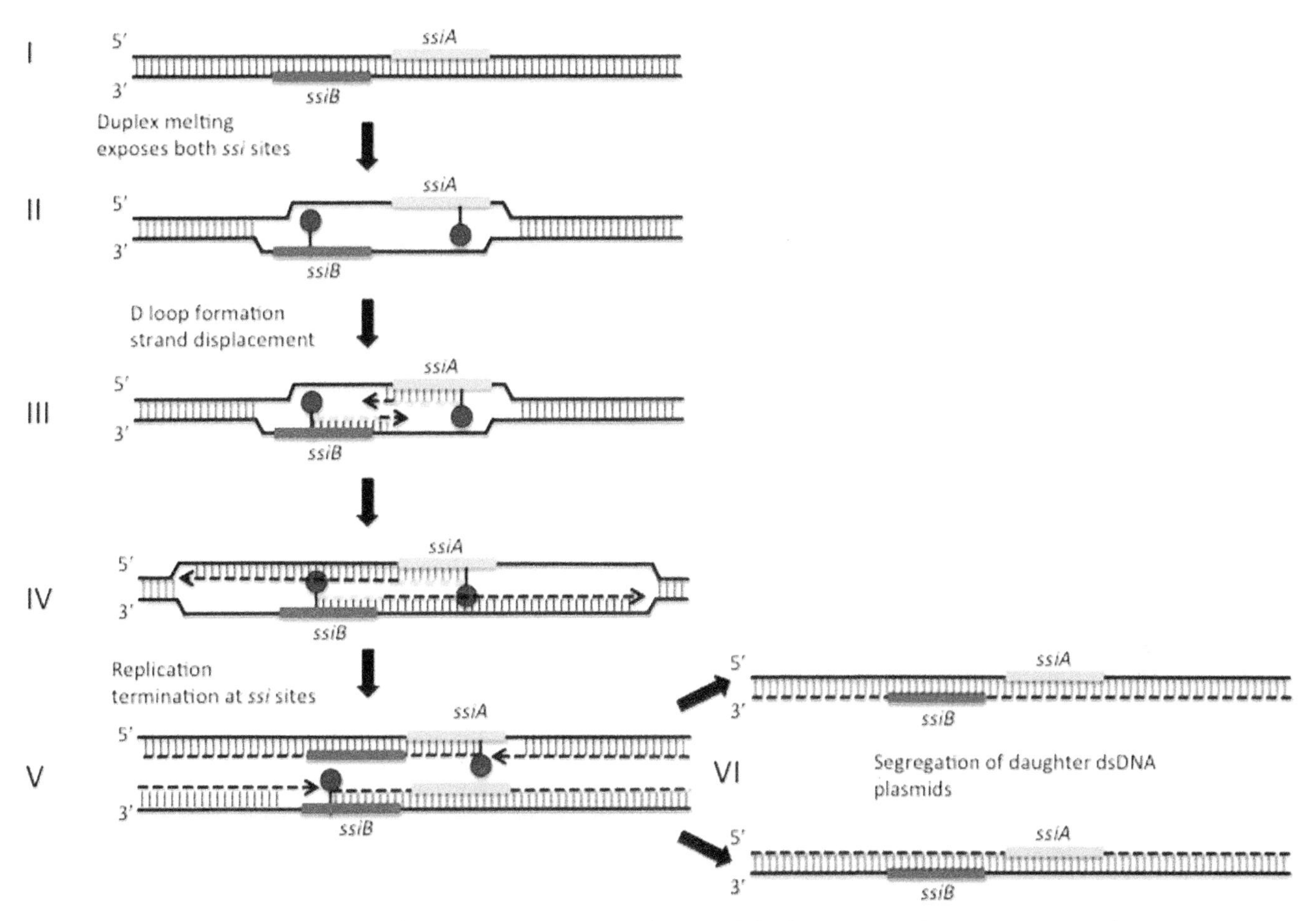

Figure 1 Model of plasmid replication by the strand-displacement mechanism. (I) Parental DNA duplex (solid black lines) depicting the two single-stranded replication initiation sites, *ssiA* (light gray box) and *ssiB* (dark gray box). Vertical lines show hybridization between DNA strands. (II) The DNA duplex is melted through binding of RepC (possibly in concert with the RepA helicase), allowing the two *ssi* sites to form hairpins (ball and stick). (III) The base of the hairpin is recognized by RepB′, which initiates the synthesis of an RNA primer (light gray dashed line). Extension of the free 3′-OH of the primer by Pol III (assisted by the RepA helicase) is shown as dashed black arrows. Two D-loops are formed, one for each direction of synthesis, as parental strands are displaced and dissociate from each other, leaving ssDNA intermediates. This is shown as areas where one of the strands has no hydrogen bonding. (IV) Synthesis continues in both directions, extending the area of D-loop formation. (V) Elongation is completed and termination of replication occurs on both strands at the *ssi* sites in which replication began. At this point, the *ssi* sites on the newly synthesized daughter strands are restored. (VI) Segregation: the two daughter strands are ligated, resulting in two DNA duplexes, each containing a parental strand (solid black line) and daughter strand (dashed black line). doi:10.1128/microbiolspec.PLAS-0029-2014.f1

plasmids). Antisense RNA regulation of plasmid replication has been extensively reviewed elsewhere (55–57).

RNAI (ColE1, ColE2) and CopA (R1) ctRNA molecules are highly structured. Given that the target preprimer and ctRNA sequence are complementary, higher-order structures for both RNAs are mirror images of each other. The first contact between sense and antisense RNAs occurs by pairing between complementary sequences at the loop portion of stem-loops, a rate-limiting step known as the "kissing complex formation" (58). Point mutations at the loop portion of stem-loops are frequently tolerated, as mutations in the template DNA introduce complementary changes in sense and ctRNA at the same time, preserving base-pairing. These mutations modulate the affinity of sense RNA–ctRNA interaction, with A-U pairs generally decreasing affinity relative to G-C pairs (for ColE1 plasmids reviewed in reference 27).

Several ctRNAs (ColE1 and ColE2 RNAI and R1 copA) have a short half-life due to the presence of an RNase E cleavage site, which consists of the U-rich sequence and a hairpin structure at the 3′ end. Conditional expression of a hyperactive variant of RNase E has been used for controlled overproduction of ColE1 plasmid DNA (59). RNase E cleavage produces monophosphorylated decay intermediates lacking short portions of the 5′ end. In the case of ColE1 and ColE2, these pRNAI cleavage intermediates are polyadenylated by PAPI, facilitating exonucleotidic digestion by PNPase (60, 61). Deletion of *pcnB*, the gene encoding PAPI, leads to increased cytoplasmic levels of pRNAI cleavage intermediates and to a 5- to 10-fold (ColE1) and 2-fold (ColE2) decrease in plasmid copy number (61, 62). RNase III has also been reported to degrade ColE1 RNAI upstream of RNase E (63). In ColE2, the differential stability between RNAI and its target rep-mRNA is partially due to differential exonuclease recruitment by RNase E (64).

Single Mechanisms Involving ctRNA Inhibition

In ColE1, the ctRNA (RNAI) is transcribed from P1, a promoter located 108 bp downstream from the sense promoter P2. Both preprimer and ctRNA form three stem-loops (SL1-3); the loop portion consists of six to seven unpaired residues. These residues are critical, as their pairing with their complementary counterparts initiates hybridization. Next, the 5′ end of RNAI (antitail) nucleates the hybridization between the two RNAs to form a duplex.

Hybridization between the preprimer and ctRNA leads to conformational changes in the preprimer, blocking R-loop formation further downstream, a phenomenon known as "action at a distance" (reviewed in references 27 and 65). This conformational change is mediated by the interaction of a sequence domain (β) in the preprimer with another sequence domain further downstream (γ), making the preprimer incompetent for R-loop formation. In addition to being short-lived, ColE1 RNA I has a short window of action, because as soon as RNAII is transcribed past position 200 downstream of the RNA/DNA switch, hybridization of the β domain with another sequence domain (α) forms a new loop (SL4), which makes RNAII refractory to RNAI inhibition.

SL1 to SL3 bear a structural resemblance to the cloverleaf structure of tRNAs and even have homology to the anticodon loops of 11 tRNAs (66). Competitive hybridization between tRNA and RNAI or RNAII appears to interfere with RNAI/RNAII hybrid formation (66). In addition, uncharged $tRNA^{ala}$ cleaves RNAI both *in vitro* and *in vivo* (67), and there is evidence suggesting that the 3′-CAA terminus of uncharged tRNAs hybridizes stably with RNAI (68). This functional cross-talk between RNAI and tRNAs may contribute to plasmid copy number deregulation associated with amino-acid starvation in *relA* strains used for recombinant gene expression; one of the key factors is the limiting yield of large-scale recombinant expression (69). Cross-talk between ctRNA and tRNAs may also explain the conservation of the 5′-UUGGCG-3′ sequence at the loop region of many of the antisense RNAs and their targets involved in regulation of replication, suggesting that this sequence is under common and strong selective pressure (70).

In ColE2 plasmids the ctRNA is also known as RNAI and has a complex secondary structure. In this case, RNAI is complementary to the 5′ end of rep mRNA containing an untranslated sequence. Given that the 5′ end portion of RNAI does not cover the initiation codon of Rep or its immediate vicinity, inhibition in this case appears to be caused by structural disruption of secondary or tertiary structures required for translation (70).

Dual Mechanisms Involving ctRNA

These mechanisms are plasmid copy number regulatory systems that include two elements: a ctRNA and a transcriptional repressor protein. In these systems, Rep expression is controlled by a strong, repressor-regulated promoter so that there is a high rate of Rep transcription when the repressor does not operate. The two best-studied examples are the R1 plasmid, where the ctRNA is CopA and the repressor is CopB, and pIP501, where the ctRNA is RNAIII and the repressor is CopR. These dual mechanisms may represent an advantage during the establishment phase, particularly for mobilizable plasmids such as class D plasmids.

In R1, repA can be transcribed from an upstream promoter P1 or from an alternative promoter further downstream, P2. Expression of repA is translationally coupled to that of tap, a small leader peptide. CopA inhibits repA expression by inhibiting translation of tap. The second element is a transcriptional repressor of P2, CopB. CopB expression is under the control of P1 but not P2. When levels of CopB are high, tap+repA are transcribed as polycystron copB-tap-RepA RNA from P1 (as P2 is silenced by CopB), but when they are low, the P2 promoter becomes derepressed and tap+repA can also be expressed from that alternative promoter, leading to a transient increase in tap+repA expression (71).

Class D plasmids have a cop-ctRNA-rep modular structure. In this case the two regulatory elements are RNAIII and a Cop protein. RNAIII is transcribed in the opposite orientation relative to its target DNA (5′ end of rep) from promoter pIII, whereas pI and pII control CopR and Rep expression, respectively, in the sense orientation. In pIP501 plasmids, RNAIII hybridization to its complementary sequence induces folding of RNA into a transcriptional terminator structure that prevents transcription of repR. This mechanism only operates on nascent (<260-nt-long) RNAs, as longer rep transcripts form an alternative secondary structure that is refractory to repR-induced transcriptional attenuation (72). CopR (whose levels reflect plasmid copy number in the cell) inhibits the sense promoter pII. A decreased plasmid copy number leads to pII derepression, resulting in increased RepR expression. In addition, induction of pII (*repR*) transcription results in a substantial decrease in pIII transcription because pIII is supercoiling-sensitive. In pAMBβ1, CopF (the equivalent of CopR), in addition to suppressing RepF transcription, decreases primer formation since CopF transcription generates the primer for replication initiation (see class D in the "Hybrid Classes of Theta Replication" section above).

Transcriptional Regulation by Rep Binding

In some class A theta plasmids, a different mechanism of regulation involves inhibiting Rep transcription by Rep itself. In these plasmids, iterons are located in the promoter of the Rep operon, outside the plasmid *ori*. Rep binding to these cognate sequences inhibits Rep expression and thus acts as an autoregulatory mechanism.

Rep binding of two alternative binding sites (Rep promoter and plasmid *ori*) involves changes in the conformation and oligomerization status of the Rep protein. These changes have been studied in detail in the RepA protein of pPS10 (73). This protein has two winged-helix domains (WH1 and WH2). When Rep A is in dimeric form, it acts as a transcriptional repressor, with the WH1 domain functioning as a dimerization interface. Low concentrations of RepA favor dissociation of Rep dimers into monomers, which are the only form that is active as an initiator. Monomerization involves conversion of the dimerization domain into a second origin-binding sequence and remodeling of the WH1 sequence to bind the opposite iteron end (73). In some cases, monomerization can be assisted by chaperones or by the allosteric effect of binding iterons at the *ori* (74–77).

Steric Hindrance

A different feedback mechanism, known as steric hindrance or handcuffing, was initially proposed for P1 and R6K plasmids (78, 79) but could operate in more iteron-containing plasmids. According to this model, as the number of plasmids in the cell increases, Rep molecules bound to iterons of one origin begin to interact with similar complexes generated in other origins. This pairing (known as handcuffing) produces plasmid pairs linked through Rep-Rep interactions, causing a steric hindrance to both origins that interferes with origin melting (80). Rep molecules are paired through zipping-up DNA-bound RepA monomers (78). A difference between this model and the autoregulation model is that the rate of replication depends on iteron concentration, not Rep expression level. Both mechanisms of autoregulation could be working together for initiators that are limiting (81).

CONCLUDING REMARKS

Plasmids contribute to the adaptation of bacterial hosts to an ever-changing environment through mobilization and amplification of selected genes. Different circular plasmids show differences in duplex melting, leading-strand priming, and lagging-strand synthesis. Learning more about the diversity of the replication mechanisms present in plasmids can help us understand the mechanisms that cells have available to replicate and repair their DNA. Organellar replication and restoration of replication after replication fork arrest are two examples of processes that occur in cells that are mechanistically closely related to plasmid replication. Also, learning more about these mechanisms will improve our understanding of plasmid biology, as mechanisms of replication limit plasmid size, host range, and mobilization capacity. Finally, maintaining a stable plasmid copy number is critical for the host, as loss of the plasmid entails losing the adaptive functions carried in the plasmid sequence, and runaway plasmid replication is lethal. Thus, mechanisms of plasmid replication regulation represent potential targets for antimicrobial intervention.

Acknowledgments. *We thank Saleem Khan and Douglas Rawlings for critical reading of the manuscript. This work was partially supported by a National Cancer Institute award (K08CA116429), by a National Institute of Environmental Health Sciences grant (R01ES019625) award, and by an NSF Advances in Biological Informatics Innovation grant (ABI1262435). Conflicts of interest: We disclose no conflicts.*

Citation. Lilly J, Camps M. 2015. Mechanisms of theta plasmid replication. Microbiol Spectrum 3(1):PLAS-0029-2014.

References

1. **Giraldo R, Fernandez-Tresguerres ME.** 2004. Twenty years of the pPS10 replicon: insights on the molecular mechanism for the activation of DNA replication in iteron-containing bacterial plasmids. *Plasmid* **52:** 69–83.

2. **Kittleson JT, Cheung S, Anderson JC.** 2011. Rapid optimization of gene dosage in *E. coli* using DIAL strains. *J Biol Eng* **5**:10.

3. **Gregg AV, McGlynn P, Jaktaji RP, Lloyd RG.** 2002. Direct rescue of stalled DNA replication forks via the combined action of PriA and RecG helicase activities. *Mol Cell* **9**:241–251.

4. **Kogoma T.** 1997. Stable DNA replication: interplay between DNA replication, homologous recombination, and transcription. *Microbiol Mol Biol Rev* **61**:212–238.

5. **Sandler SJ, Marians KJ.** 2000. Role of PriA in replication fork reactivation in *Escherichia coli*. *J Bacteriol* **182**: 9–13.

6. **Masai H, Arai K.** 1996. DnaA- and PriA-dependent primosomes: two distinct replication complexes for replication of *Escherichia coli* chromosome. *Front Biosci* **1**: d48–d58.

7. **Wu CA, Zechner EL, Marians KJ.** 1992. Coordinated leading- and lagging-strand synthesis at the *Escherichia coli* DNA replication fork. I. Multiple effectors act to modulate Okazaki fragment size. *J Biol Chem* **267**: 4030–4044.

8. **Johnson A, O'Donnell M.** 2005. Cellular DNA replicases: components and dynamics at the replication fork. *Annu Rev Biochem* **74**:283–315.

9. **McHenry CS.** 2011. DNA replicases from a bacterial perspective. *Annu Rev Biochem* **80**:403–436.

10. **Tanner NA, Hamdan SM, Jergic S, Loscha KV, Schaeffer PM, Dixon NE, van Oijen AM.** 2008. Single-molecule studies of fork dynamics in *Escherichia coli* DNA replication. *Nat Struct Mol Biol* **15**:998.

11. **Lenhart JS, Schroeder JW, Walsh BW, Simmons LA.** 2012. DNA repair and genome maintenance in *Bacillus subtilis*. *Microbiol Mol Biol Rev* **76**:530–564.

12. **Rannou O, Le Chatelier E, Larson MA, Nouri H, Dalmais B, Laughton C, Janniere L, Soultanas P.** 2013. Functional interplay of DnaE polymerase, DnaG primase and DnaC helicase within a ternary complex, and primase to polymerase hand-off during lagging strand DNA replication in *Bacillus subtilis*. *Nucleic Acids Res* **41**:5303–5320.

13. **Allen JM, Simcha DM, Ericson NG, Alexander DL, Marquette JT, Van Biber BP, Troll CJ, Karchin R, Bielas JH, Loeb LA, Camps M.** 2011. Roles of DNA polymerase I in leading and lagging-strand replication defined by a high-resolution mutation footprint of ColE1 plasmid replication. *Nucleic Acids Res* **39**:7020–7033.

14. **Patel PH, Suzuki M, Adman E, Shinkai A, Loeb LA.** 2001. Prokaryotic DNA polymerase I: evolution, structure, and "base flipping" mechanism for nucleotide selection. *J Mol Biol* **308**:823–837.

15. **Maki H, Bryan SK, Horiuchi T, Moses RE.** 1989. Suppression of dnaE nonsense mutations by pcbA1. *J Bacteriol* **171**:3139–3143.

16. **Troll C, Yoder J, Alexander D, Hernandez J, Loh Y, Camps M.** 2014. The mutagenic footprint of low-fidelity Pol I ColE1 plasmid replication in *E. coli* reveals an extensive interplay between Pol I and Pol III. *Curr Genet* **60**:123–134.

17. **Mukhopadhyay G, Chattoraj DK.** 1993. Conformation of the origin of P1 plasmid replication. Initiator protein induced wrapping and intrinsic unstacking. *J Mol Biol* **231**:19–28.

18. **Urh M, Wu J, Wu J, Forest K, Inman RB, Filutowicz M.** 1998. Assemblies of replication initiator protein on symmetric and asymmetric DNA sequences depend on multiple protein oligomerization surfaces. *J Mol Biol* **283**: 619–631.

19. **Filutowicz M, Dellis S, Levchenko I, Urh M, Wu F, York D.** 1994. Regulation of replication of an iteron-containing DNA molecule. *Prog Nucleic Acid Res Mol Biol* **48**:239–273.

20. **Stenzel TT, MacAllister T, Bastia D.** 1991. Cooperativity at a distance promoted by the combined action of two replication initiator proteins and a DNA bending protein at the replication origin of pSC101. *Genes Dev* **5**: 1453–1463.

21. **Park K, Chattoraj DK.** 2001. DnaA boxes in the P1 plasmid origin: the effect of their position on the directionality of replication and plasmid copy number. *J Mol Biol* **310**:69–81.

22. **Doran KS, Helinski DR, Konieczny I.** 1999. Host-dependent requirement for specific DnaA boxes for plasmid RK2 replication. *Mol Microbiol* **33**:490–498.

23. **del Solar G, Giraldo R, Ruiz-Echevarria MJ, Espinosa M, Diaz-Orejas R.** 1998. Replication and control of circular bacterial plasmids. *Microbiol Mol Biol Rev* **62**:434–464.

24. **Rakowski SA, Filutowicz M.** 2013. Plasmid R6K replication control. *Plasmid* **69**:231–242.

25. **Wu YC, Liu ST.** 2010. A sequence that affects the copy number and stability of pSW200 and ColE1. *J Bacteriol* **192**:3654–3660.

26. **Masukata H, Tomizawa J.** 1986. Control of primer formation for ColE1 plasmid replication: conformational change of the primer transcript. *Cell* **44**:125–136.

27. **Camps M.** 2010. Modulation of ColE1-like plasmid replication for recombinant gene expression. *Recent Pat DNA Gene Seq* **4**:58–73.

28. **Cesareni G, Helmer-Citterich M, Castagnoli L.** 1991. Control of ColE1 plasmid replication by antisense RNA. *Trends Genet* **7**:230–235.

29. **Wang Z, Yuan Z, Hengge UR.** 2004. Processing of plasmid DNA with ColE1-like replication origin. *Plasmid* **51**: 149–161.

30. **Lee EH, Kornberg A.** 1991. Replication deficiencies in priA mutants of *Escherichia coli* lacking the primosomal replication n' protein. *Proc Natl Acad Sci USA* **88**: 3029–3032.

31. **Sandler SJ, Samra HS, Clark AJ.** 1996. Differential suppression of priA2::kan phenotypes in *Escherichia coli* K-12 by mutations in priA, lexA, and dnaC. *Genetics* **143**:5–13.

32. **Jaktaji RP, Lloyd RG.** 2003. PriA supports two distinct pathways for replication restart in UV-irradiated *Escherichia coli* cells. *Mol Microbiol* **47**:1091–1100.

33. **Nakasu S, Tomizawa J.** 1992. Structure of the ColE1 DNA molecule before segregation to daughter molecules. *Proc Natl Acad Sci USA* **89**:10139–10143.

34. **Drolet M.** 2006. Growth inhibition mediated by excess negative supercoiling: the interplay between transcription elongation, R-loop formation and DNA topology. *Mol Microbiol* **59:**723–730.

35. **Drolet M, Broccoli S, Rallu F, Hraiky C, Fortin C, Masse E, Baaklini I.** 2003. The problem of hypernegative supercoiling and R-loop formation in transcription. *Front Biosci* **8:**d210–d221.

36. **Gowrishankar J, Harinarayanan R.** 2004. Why is transcription coupled to translation in bacteria? *Mol Microbiol* **54:**598–603.

37. **Fukuoh A, Iwasaki H, Ishioka K, Shinagawa H.** 1997. ATP-dependent resolution of R-loops at the ColE1 replication origin by *Escherichia coli* RecG protein, a Holliday junction-specific helicase. *EMBO J* **16:**203–209.

38. **del Solar G, Alonso JC, Espinosa M, Diaz-Orejas R.** 1996. Broad-host-range plasmid replication: an open question. *Mol Microbiol* **21:**661–666.

39. **Yasueda H, Horii T, Itoh T.** 1989. Structural and functional organization of ColE2 and ColE3 replicons. *Mol Gen Genet* **215:**209–216.

40. **Aoki K, Shinohara M, Itoh T.** 2007. Distinct functions of the two specificity determinants in replication initiation of plasmids ColE2-P9 and ColE3-CA38. *J Bacteriol* **189:** 2392–2400.

41. **Takechi S, Matsui H, Itoh T.** 1995. Primer RNA synthesis by plasmid-specified Rep protein for initiation of ColE2 DNA replication. *EMBO J* **14:**5141–5147.

42. **Takechi S, Itoh T.** 1995. Initiation of unidirectional ColE2 DNA replication by a unique priming mechanism. *Nucleic Acids Res* **23:**4196–4201.

43. **Bruand C, Ehrlich SD.** 1998. Transcription-driven DNA replication of plasmid pAMbeta1 in *Bacillus subtilis*. *Mol Microbiol* **30:**135–145.

44. **Le Chatelier E, Janniere L, Ehrlich SD, Canceill D.** 2001. The RepE initiator is a double-stranded and single-stranded DNA-binding protein that forms an atypical open complex at the onset of replication of plasmid pAMbeta 1 from Gram-positive bacteria. *J Biol Chem* **276:**10234–10246.

45. **Brantl S.** 2014. Antisense-RNA mediated control of plasmid replication: pIP501 revisited. *Plasmid*. [Epub ahead of print.] doi:10.1016/j.plasmid.2014.07.004.

46. **Bidnenko V, Ehrlich SD, Janniere L.** 1998. *In vivo* relations between pAMbeta1-encoded type I topoisomerase and plasmid replication. *Mol Microbiol* **28:**1005–1016.

47. **Janniere L, Bidnenko V, McGovern S, Ehrlich SD, Petit MA.** 1997. Replication terminus for DNA polymerase I during initiation of pAM beta 1 replication: role of the plasmid-encoded resolution system. *Mol Microbiol* **23:** 525–535.

48. **Sakai H, Komano T.** 1996. DNA replication of IncQ broad-host-range plasmids in Gram-negative bacteria. *Biosci Biotechnol Biochem* **60:**377–382.

49. **Loftie-Eaton W, Rawlings DE.** 2012. Diversity, biology and evolution of IncQ-family plasmids. *Plasmid* **67:**15–34.

50. **Honda Y, Akioka T, Takebe S, Tanaka K, Miao D, Higashi A, Nakamura T, Taguchi Y, Sakai H, Komano T.** 1993. Mutational analysis of the specific priming signal essential for DNA replication of the broad host-range plasmid RSF1010. *FEBS Lett* **324:**67–70.

51. **Miao DM, Honda Y, Tanaka K, Higashi A, Nakamura T, Taguchi Y, Sakai H, Komano T, Bagdasarian M.** 1993. A base-paired hairpin structure essential for the functional priming signal for DNA replication of the broad host range plasmid RSF1010. *Nucleic Acids Res* **21:**4900–4903.

52. **Rawlings DE, Tietze E.** 2001. Comparative biology of IncQ and IncQ-like plasmids. *Microbiol Mol Biol Rev* **65:**481–496.

53. **Tanaka K, Kino K, Taguchi Y, Miao DM, Honda Y, Sakai H, Komano T, Bagdasarian M.** 1994. Functional difference between the two oppositely oriented priming signals essential for the initiation of the broad host-range plasmid RSF1010 DNA replication. *Nucleic Acids Res* **22:**767–772.

54. **Nordstrom K, Wagner EG.** 1994. Kinetic aspects of control of plasmid replication by antisense RNA. *Trends Biochem Sci* **19:**294–300.

55. **del Solar G, Espinosa M.** 2000. Plasmid copy number control: an ever-growing story. *Mol Microbiol* **37:**492–500.

56. **Wagner EG, Altuvia S, Romby P.** 2002. Antisense RNAs in bacteria and their genetic elements. *Adv Genet* **46:** 361–398.

57. **Brantl S.** 2014. Plasmid replication control by antisense RNAs. *In* Tolmasky ME, Alonso JC (ed), *Plasmids: Biology and Impact in Biotechnology and Discovery*. ASM Press, Washington, DC. In press.

58. **Hjalt TA, Wagner EG.** 1995. Bulged-out nucleotides in an antisense RNA are required for rapid target RNA binding *in vitro* and inhibition *in vivo*. *Nucleic Acids Res* **23:**580–587.

59. **Go H, Lee K.** 2011. A genetic system for RNase E variant-controlled overproduction of ColE1-type plasmid DNA. *J Biotechnol* **152:**171–175.

60. **Nishio SY, Itoh T.** 2008. Replication initiator protein mRNA of ColE2 plasmid and its antisense regulator RNA are under the control of different degradation pathways. *Plasmid* **59:**102–110.

61. **Xu FF, Gaggero C, Cohen SN.** 2002. Polyadenylation can regulate ColE1 type plasmid copy number independently of any effect on RNAI decay by decreasing the interaction of antisense RNAI with its RNAII target. *Plasmid* **48:**49–58.

62. **Nishio SY, Itoh T.** 2008. The effects of RNA degradation enzymes on antisense RNAI controlling ColE2 plasmid copy number. *Plasmid* **60:**174–180.

63. **Binnie U, Wong K, McAteer S, Masters M.** 1999. Absence of RNASE III alters the pathway by which RNAI, the antisense inhibitor of ColE1 replication, decays. *Microbiology* **145**(Pt 11):3089–3100.

64. **Nishio SY, Itoh T.** 2009. Arginine-rich RNA binding domain and protein scaffold domain of RNase E are important for degradation of RNAI but not for that of the Rep mRNA of the ColE2 plasmid. *Plasmid* **62:**83–87.

65. **Polisky B.** 1988. ColE1 replication control circuitry: sense from antisense. *Cell* **55:**929–932.

66. **Yavachev L, Ivanov I.** 1988. What does the homology between *E. coli* tRNAs and RNAs controlling ColE1 plasmid replication mean? *J Theor Biol* **131:**235–241.

67. **Wang Z, Yuan Z, Xiang L, Shao J, Wegrzyn G.** 2006. tRNA-dependent cleavage of the ColE1 plasmid-encoded RNA I. *Microbiology* **152:**3467–3476.

68. **Wang Z, Le G, Shi Y, Wegrzyn G, Wrobel B.** 2002. A model for regulation of ColE1-like plasmid replication by uncharged tRNAs in amino acid-starved *Escherichia coli* cells. *Plasmid* **47:**69–78.

69. **Grabherr R, Nilsson E, Striedner G, Bayer K.** 2002. Stabilizing plasmid copy number to improve recombinant protein production. *Biotechnol Bioeng* **77:**142–147.

70. **Hiraga S, Sugiyama T, Itoh T.** 1994. Comparative analysis of the replicon regions of eleven ColE2-related plasmids. *J Bacteriol* **176:**7233–7243.

71. **Malmgren C, Engdahl HM, Romby P, Wagner EG.** 1996. An antisense/target RNA duplex or a strong intramolecular RNA structure 5′ of a translation initiation signal blocks ribosome binding: the case of plasmid R1. *RNA* **2:**1022–1032.

72. **Brantl S, Birch-Hirschfeld E, Behnke D.** 1993. RepR protein expression on plasmid pIP501 is controlled by an antisense RNA-mediated transcription attenuation mechanism. *J Bacteriol* **175:**4052–4061.

73. **Giraldo R, Fernandez-Tornero C, Evans PR, Diaz-Orejas R, Romero A.** 2003. A conformational switch between transcriptional repression and replication initiation in the RepA dimerization domain. *Nat Struct Biol* **10:**565–571.

74. **DasGupta S, Mukhopadhyay G, Papp PP, Lewis MS, Chattoraj DK.** 1993. Activation of DNA binding by the monomeric form of the P1 replication initiator RepA by heat shock proteins DnaJ and DnaK. *J Mol Biol* **232:**23–34.

75. **Diaz-Lopez T, Lages-Gonzalo M, Serrano-Lopez A, Alfonso C, Rivas G, Diaz-Orejas R, Giraldo R.** 2003. Structural changes in RepA, a plasmid replication initiator, upon binding to origin DNA. *J Biol Chem* **278:**18606–18616.

76. **Kawasaki Y, Wada C, Yura T.** 1990. Roles of *Escherichia coli* heat shock proteins DnaK, DnaJ and GrpE in mini-F plasmid replication. *Mol Gen Genet* **220:**277–282.

77. **Wickner S, Skowyra D, Hoskins J, McKenney K.** 1992. DnaJ, DnaK, and GrpE heat shock proteins are required in oriP1 DNA replication solely at the RepA monomerization step. *Proc Natl Acad Sci USA* **89:**10345–10349.

78. **Gasset-Rosa F, Diaz-Lopez T, Lurz R, Prieto A, Fernandez-Tresguerres ME, Giraldo R.** 2008. Negative regulation of pPS10 plasmid replication: origin pairing by zipping-up DNA-bound RepA monomers. *Mol Microbiol* **68:**560–572.

79. **Park K, Han E, Paulsson J, Chattoraj DK.** 2001. Origin pairing ('handcuffing') as a mode of negative control of P1 plasmid copy number. *EMBO J* **20:**7323–7332.

80. **Zzaman S, Bastia D.** 2005. Oligomeric initiator protein-mediated DNA looping negatively regulates plasmid replication *in vitro* by preventing origin melting. *Mol Cell* **20:**833–843.

81. **Das N, Valjavec-Gratian M, Basuray AN, Fekete RA, Papp PP, Paulsson J, Chattoraj DK.** 2005. Multiple homeostatic mechanisms in the control of P1 plasmid replication. *Proc Natl Acad Sci USA* **102:**2856–2861.

Plasmids—Biology and Impact in Biotechnology and Discovery
Edited by Marcelo E. Tolmasky and Juan C. Alonso

doi:10.1128/microbiolspec.PLAS-0035-2014

José A. Ruiz-Masó,[1] Cristina Machón,[2,3] Lorena Bordanaba-Ruiseco,[1]
Manuel Espinosa,[1] Miquel Coll,[2,3] and Gloria del Solar[1]

4

Plasmid Rolling-Circle Replication

GENERAL ASPECTS OF PLASMID ROLLING-CIRCLE REPLICATION

The main features that characterize rolling-circle replication (RCR) (see Fig. 1A) derive from its singular initiation mechanism, which relies on the sequence-specific cleavage, at the nick site of the double-strand origin (*dso*), of one of the parental DNA strands by an initiator Rep protein. This cleavage generates a 3′-OH end that allows the host DNA polymerases to initiate the leading strand replication. Therefore, the RCR initiation circumvents the synthesis of a primer RNA that is required in all other modes of replication of circular double-stranded DNA (dsDNA). Elongation of the leading strand takes place as the parental double helix is unwound by a host DNA helicase and the cleaved nontemplate strand is covered with the single-stranded DNA binding protein. Since the nascent DNA is covalently attached to the parental DNA, termination of a round of leading-strand replication implies a new cleavage event at the reconstituted nick site. This reaction is assumed to be catalyzed by the same Rep molecule that carried out the initiation cleavage and remained bound to the 5′ end of the parental strand while traveling along with the replication fork. A trans-esterification then occurs that joins this 5′ end to the 3′ end generated in the termination cleavage, releasing the displaced parental strand as a circular single-stranded DNA (ssDNA). This replicative intermediate serves as the template for the synthesis of the lagging strand, which depends solely on host-encoded enzymes and is initiated from a highly structured region of the ssDNA, termed the single-strand origin (*sso*).

Thus, the entire process of asymmetric RCR yields, in two separate steps (this is what asymmetric refers to), two circular dsDNAs containing either the newly synthesized leading or lagging strand and the complementary parental template strand. The DNA ligase and gyrase of the host cell next convert the new daughter DNA molecules in supercoiled forms indistinguishable from the rest of the plasmid pool. Generation of the ssDNA replicative intermediates is the hallmark of RCR, and detection of intracellular strand-specific plasmid ssDNA provides valuable clues about whether a given plasmid replicates by the rolling-circle mechanism (1, 2).

The basic catalytic mechanism operating in initiation and termination of RCR, i.e., the cleavage and rejoining

[1]Centro de Investigaciones Biológicas (CSIC), Ramiro de Maeztu 9, 28040 Madrid, Spain; [2]Institute for Research in Biomedicine (IRB-Barcelona), Baldiri Reixac 10-12, 08028 Barcelona, Spain; [3]Institut de Biologia Molecular de Barcelona (CSIC), Baldiri Reixac 10-12, 08028 Barcelona, Spain.

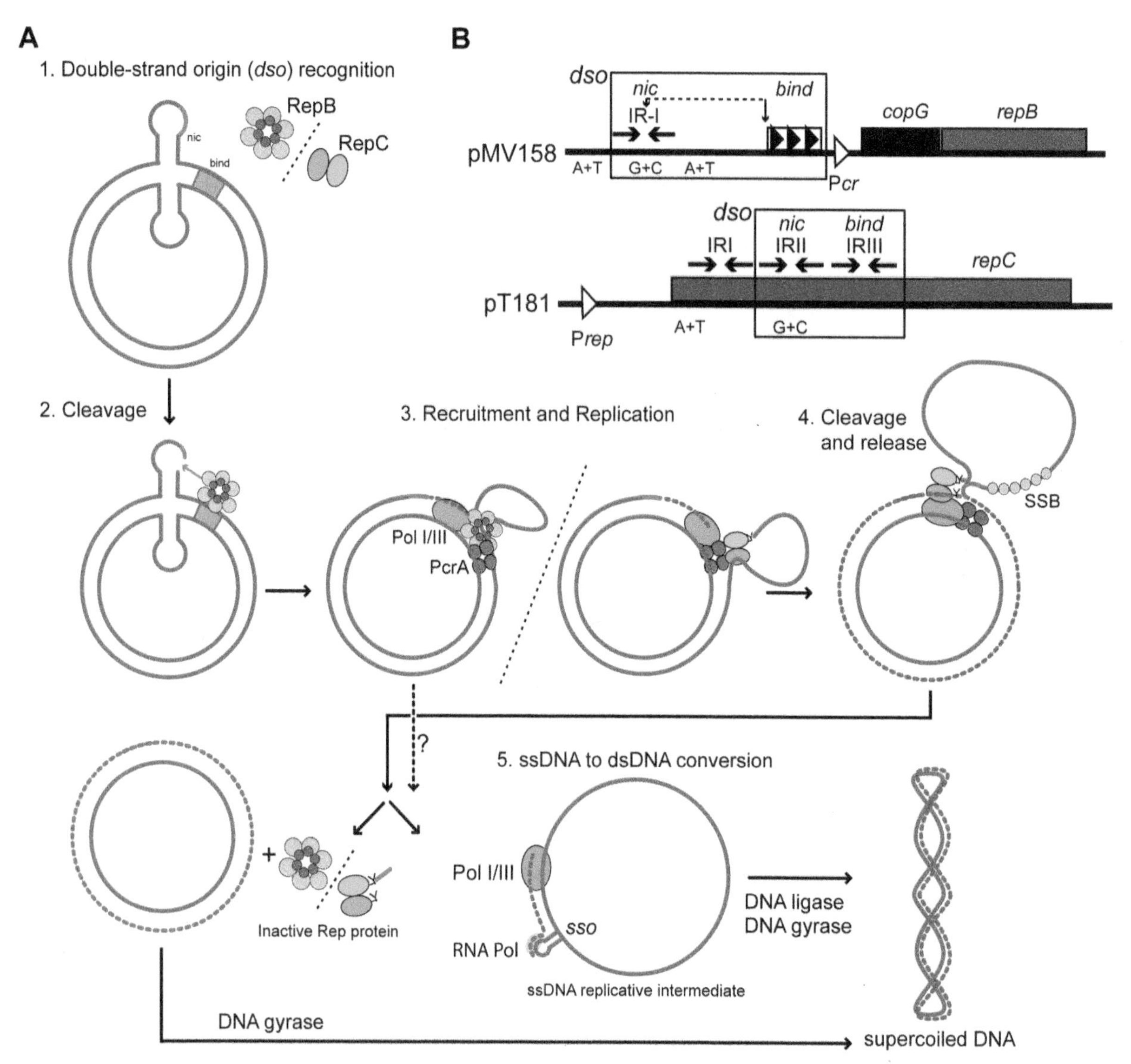

Figure 1 (A) A model for plasmid RCR based on pMV158 and pT181 replicons. Detailed information about the RCR process is given in the text. In the pMV158 replication model, a possible mechanism is shown in which, upon assembly and cleavage at the nick site, the hexameric ring of RepB encircles one of the plasmid strands within the central channel. As discussed in the text, the strand enclosure may confer high processivity to the replisome complex. The RepB-mediated mechanism that, at the termination step, yields the dsDNA replication product and the ssDNA intermediate, as well as the mechanism of RepB inactivation, remain undisclosed (dotted arrow with ? symbol). (B) Scheme of the *dso*s and of the adjacent regions of the pMV158 and pT181 RCR plasmids. The symbols used are as follows: direct repeats in the replication region are indicated by solid boxed arrows; the inverted arrows represent the two arms of the inverted repeat elements; promoters are indicated by open arrowheads. The AT- and GC-rich sequences (A+T and G+C, respectively) are also indicated. The dotted line above the pMV158 map indicates that the direct repeats of the *bind* locus are separated by 84 bp from the nick site. SSB, single-stranded DNA binding protein. doi:10.1128/microbiolspec.PLAS-0035-2014.f1

of ssDNA using an active-site Tyr that forms a transient 5′-phosphotyrosine bond with the cleaved DNA, is involved in a range of processes that take place in mobile genetic elements in all three domains of life. The enzymes that exhibit this catalytic mechanism are mainly included in the widespread His-bulky hydrophobic residue-His (HUH) endonuclease superfamily and have key roles in the replication of plasmids, bacteriophages, and plant and animal viruses; in plasmid conjugative transfer; and in transposition (3). RCR was discovered in ssDNA coliphage ΦX174 some 45 years ago (4–6). The pioneer characterization of gene A protein made the initiator of ΦX174 RCR the first member of the HUH endonuclease superfamily (7–10).

Plasmid RCR was first evidenced for the *Staphylococcus aureus* plasmid pT181 based on the characterization of the origin-specific nicking-closing activity of the purified pT181-encoded RepC protein (11). Shortly afterward, several other small plasmids from staphylococci, bacilli, streptococci, and streptomyces were also found to replicate by the RCR mechanism (12–14), which led to the assumption that most, if not all, small multicopy plasmids in Gram-positive bacteria use RCR. However, this premise proved inaccurate, as some small plasmids isolated from Gram-positive organisms were later reported to replicate by the theta mode (1). Moreover, although RCR plasmids are particularly abundant in Gram-positive bacteria, they have also been identified in various Gram-negative organisms, in archaea, and in mitochondria of the higher plant *Chenopodium album* (1, 2, 15).

Natural RCR plasmids range in size from as low as the 846 bp of the *Thermotoga* plasmids pRQ7, pMC24, and pRKU1 (16–18) to the almost 30 kb of pCG4 from *Corynebacterium glutamicum* (19). The nearly identical plasmids pRQ7, pMC24, and pRKU1 are the smallest found so far and consist of only the basic replicon, i.e., the backbone regions involved in replication and copy-number control. The basic replicon of RCR plasmids should include an essential module containing the *dso* and the genes that encode the initiator Rep protein and the replication control element(s), as well as at least one host-recognized *sso*, which, although not strictly essential, provides efficient synthesis of the lagging strand and hence is present in all natural RCR plasmids (Fig. 2). Homology in the essential module of the basic replicon has been the criterion used to classify RCR plasmids into replicon families (see below).

Apart from the basic replicon, some larger RCR plasmids contain additional backbone genes and elements that contribute to their maintenance or help them transfer between host cells (Fig. 2). Of special relevance, because of its frequent presence in RCR plasmids, is the MOB module, which is involved in the conjugative mobilization of the plasmid and consists of the transfer origin (*oriT*) and the *mob* gene(s) that encode the relaxase protein and, in some cases, auxiliary proteins (20). The apparent lack of active partition systems in RCR plasmids is consistent with the medium copy number (10 to 30 per chromosome equivalent) that they exhibit in their natural hosts. This feature ensures the stable inheritance of RCR plasmids by only random segregation to the daughter cells, providing that the replication control system efficiently corrects fluctuations of the plasmid copy number in single cells and that the plasmid molecules are maintained as individual copies. In this sense, the presence of homologs to components of toxin-antitoxin (TA) systems in some RCR plasmids is intriguing (21). It is noteworthy that whereas the TA systems were first proposed to play a role in plasmid stability through postsegregational killing of plasmid-free cells, the more recent competition hypothesis postulates that acquisition of these modules allows plasmids to exclude competing TA-free plasmids (22–24).

Some RCR plasmids also carry accessory genes that encode functions that can benefit the host cell under special conditions, thus reflecting the adaptation of the bacteria to their environment (Fig. 2). Antibiotic resistance determinants are among the most frequent traits encoded by RCR plasmids isolated from a variety of bacteria (25). Other accessory genes have been found to be relatively abundant in RCR plasmids from a given host. This is the case of small heat shock protein (*shsp*) genes carried by *Streptococcus thermophilus* plasmids belonging to the pC194 replicon family (26, 27). The presence of *shsp*-containing plasmids has been reported to increase cell survival at the high temperatures reached during different stages of fermentation in the dairy industry (27). Another striking example is the presence, in some *Bacillus thuringiensis* plasmids, of open reading frames encoding collagen-like proteins that are thought to play a role in aggregation formation or in adherence to other cells or substrates (28).

RCR plasmids are considered to contain promiscuous replicons, as many of them have been shown to replicate in species, genera, or even phyla other than those from which they were isolated (25). The simplicity of the RCR initiation, with only the plasmid-encoded Rep protein participating in recognition of the origin and priming of the leading strand synthesis, may underlie the usual promiscuity of these plasmids. The broadness of the host range of RCR plasmids would depend on the balanced expression of their essential

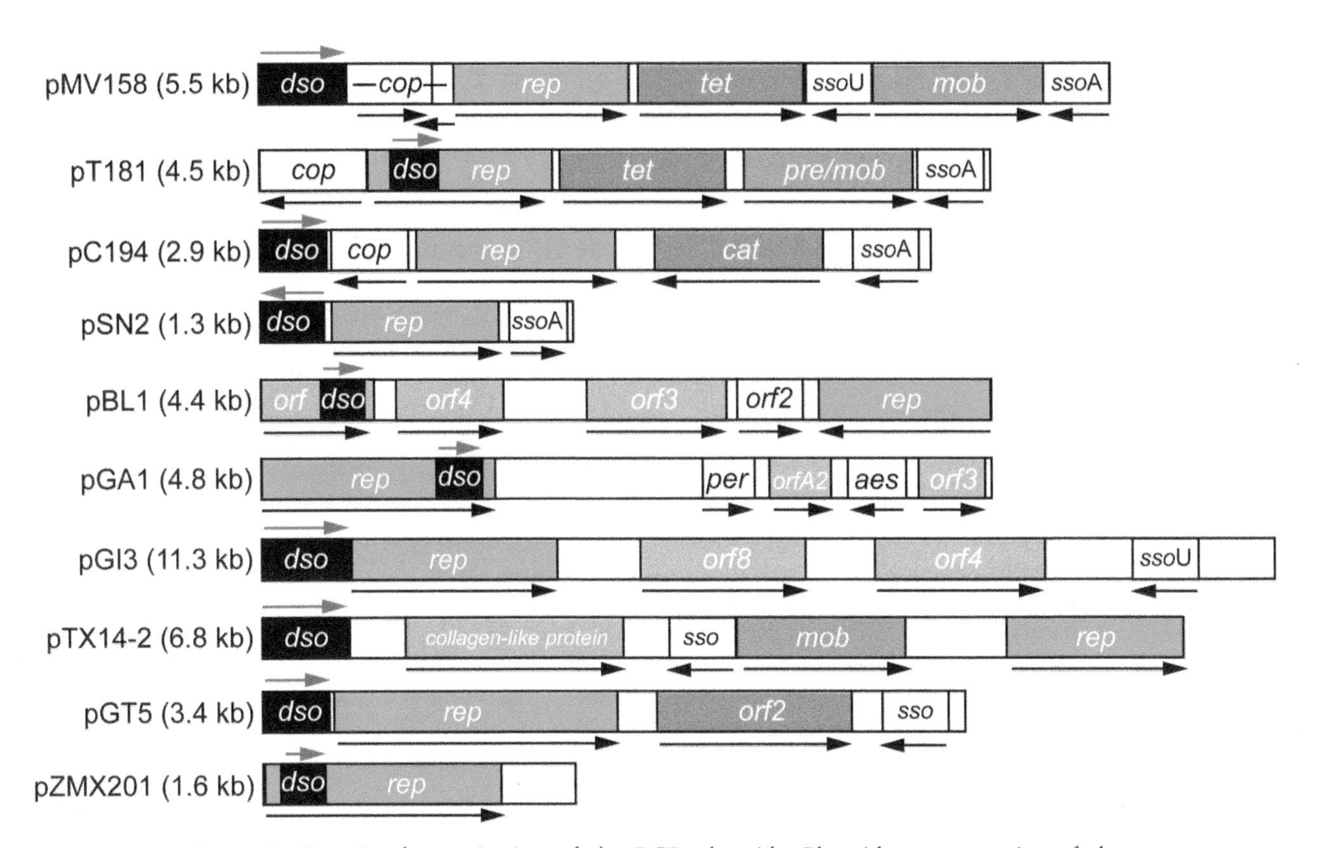

Figure 2 Functional organization of the RCR plasmids. Plasmids representative of the different families are shown. The arrows point to the direction of transcription (black) or the direction of replication (red) from the *dso* (leading strand) and *sso* (lagging strand). Inside the boxes, *rep* is the replication gene; *cop* represents the copy number control gene(s); *dso* is the double-strand origin of replication; *sso* is the single-strand origin of replication; *cat* and *tet* are chloramphenicol- and tetracycline-resistant genes, respectively; *pre/mob* represents the conjugative mobilization gene; *orf* indicates an open reading frame with unknown homology. The positions of the copy number control genes *per* and *aes* of pGA1, and of the *collagen-like protein* gene of pTX14-2 are also indicated.
doi:10.1128/microbiolspec.PLAS-0035-2014.f2

genes involved in initiation and control of replication as well as on the formation of a functional Rep-host helicase complex that can extensively unwind the plasmid DNA in a variety of bacteria (29–31). The broad host range of RCR plasmids is best exemplified by the pMV158-family prototype, which was initially isolated from *Streptococcus agalactiae* and subsequently transferred to a variety of *Firmicutes* (several *Streptococcus* and *Bacillus* species, *Listeria, S. aureus, Lactococcus lactis, Enterococcus faecalis, Clostridium*), *Actinobacteria* (*C. glutamicum, Brevibacterium*), and the γ-proteobacterium *Escherichia coli*. Moreover, the fact that members of each replicon family have been isolated from a variety of bacteria suggests the promiscuity of the ancestors from which these plasmids derive. In turn, plasmid adaptation to a new host can lead to the narrowing of the host range of the adapted plasmid. This seems to be the case for two *Mycoplasma mycoides* plasmids of the pMV158-replicon family, namely pADB201 and pKMK1, whose *rep* genes contain at least one UGA codon, which encodes tryptophan in this bacterium but is a stop codon in other bacteria, so that the host range of these plasmids is restricted to *Mycoplasma* species (32).

Due to their general smallness, high copy number, and promiscuity, RCR plasmids appear to be well suited for the construction of vectors for gene cloning and expression, provided a functional *sso* is present to minimize the generation of the recombinogenic ssDNA intermediates, which can lead to structural and segregational plasmid instability (33–37). Nevertheless, it has been reported that cloning of heterologous DNA in RCR plasmid vectors can result in the generation of linear high-molecular-weight (HMW) plasmid multimers in relative amounts that correlate positively with the size of the DNA insert (38, 39). The formation of HMW by RCR plasmids has also been implicated in both structural (40) and segregational (41) instability. The generation of HMW plasmid DNA was at first related to a replication defect, as plasmids lacking *sso* were prone to accumulate HMW DNA (42). Accumulation of HMW plasmid DNA was enhanced in the absence of the ExoV enzyme (RecBCD in Gram-negative or AddAB in Gram-positive bacteria) (41). Despite the potential instability problems, vectors based on RCR plasmids have been developed and successfully used in pneumococci, enterococci, lactococi, and corynebacteria (43–45), for which genetic and biotechnological tools are scarce and hence welcome. It is worth mentioning that most of the nonintegrative plasmid vectors available in *Streptococcus pneumoniae* are based on pMV158 and that inducible expression vector pLS1ROM and recombinant pLS1ROM-GFP (containing the *gfp* gene, encoding the *Aequorea victoria* green fluorescent protein, cloned under control of the maltose-inducible P_M promoter) have proved to be structurally and segregationally stable in pneumococcus, even under induction conditions (45). Similarly, most of the autonomously replicating vectors for the industrial microorganism *C. glutamicum* are based on plasmids pBL1, pCG1, and pGA1 from *C. glutamicum* or on the broad-host-range plasmid pNG2 from *Corynebacterium diphtheriae*, all of them replicating by the rolling circle mode (46). These RCR plasmid vectors were found to be stably maintained in *C. glutamicum* cells grown under nonselective conditions (47).

An aspect of recognized relevance when pursuing the biotechnological use of plasmid vectors is the metabolic cost that carriage of these extrachromosomal elements imposes on the host, since a significant burden can lead to the overgrowth of the culture by plasmid-free cells even though plasmid inheritance is quite stable. Little information is available on the burden caused by RCR plasmids, as this subject has only been analyzed for the pMV158 replicon. Small (4.4 kb), medium-copy-number (~20 copies per chromosome equivalent) pMV158 derivatives that are stably inherited in pneumococcus and harbor an *sso* element efficiently recognized in this host slightly burden the *S. pneumoniae* cells, causing a 7 to 8% increase in the bacterial doubling time (48). Nevertheless, fitness impairment of pneumococcal cells harboring pMV158 derivatives has not been found to negatively affect the segregational stability of pLS1ROM and pLS1ROM-GFP (45).

This chapter aims to provide an updated review of the major findings in the study of the RCR plasmids and to highlight the pending questions and challenges for the detailed understanding of this kind of plasmid replication. Most of these issues have been dealt with in previous reviews on this subject (1, 25, 32, 49, 50).

Apart from the above-referenced asymmetric RCR, which is initiated by the Rep-mediated cleavage of one parental strand, a different, recombination-dependent replication mechanism that also leads to σ-shaped circular intermediates consisting of a circular DNA attached to a growing linear DNA has been reported to play an essential role during the replication cycle of many dsDNA viruses. Single origin-dependent replication of bacterial genomes and of many dsDNA viruses with circular genomes proceeds by the θ (circle to circle) mechanism. The trade-off between different DNA transactions could lead to the stall or collapse of the replication machinery, so that origin-independent remodeling and assembly of a new replisome at the stalled fork is required to restart the replication process. In dsDNA viruses (e.g., bacteriophage lambda, SPP1, etc.), replication restart becomes dependent on recombination proteins with a switch from the origin-mediated θ type to a σ type recombination-dependent replication. The replication shift from θ to σ generates the concatemeric viral DNA substrate needed to produce mature viral particles. This RCR-like σ mode has been reviewed by Lo Piano et al. (51) and will not be addressed here.

THE DOUBLE-STRAND ORIGIN

Replication of the leading strand of RCR plasmids initiates and proceeds in a unidirectional manner from their *dso*, a plasmid DNA region highly specific for its cognate initiator protein that contains the sequences involved in the initiation and termination of the leading strand. The *dso*, along with the *rep* gene and the control elements, is part of an essential module that harbors the functions for plasmid replication. Based on the homologies found in this essential module, up to 17 RCR plasmid families have been defined. Only three of these plasmid families have been studied in depth, their prototypes being the staphylococcal plasmids pT181/pC221 (2 and references therein; 52) and pC194/pUB110 (53) and the streptococcal plasmid pMV158 (1). The following plasmid families have also been studied although less thoroughly: the staphylococcal plasmid pSN2 family (54), the pBL1 and pCG1 plasmid families from *C. glutamicum* (55 and references therein), the pSTK-1 and pTX14-2 plasmid families from *B. thuringiensis* (28 and references therein), and the pGRB1 (56) and pGT5 (57) plasmid families from archaea.

The *dsos* of RCR plasmids can be found located upstream of the *rep* gene (pC194, pMV158, and pSN2 families), embedded within the 5′ portion (pT181 family) or the 3′ portion (pCG1 family) of the sequence coding their respective Rep proteins, or even downstream from the *rep* gene stop codon (pTX14-2 family). The *dso* can be physically and functionally divided into two regions, namely *bind*, which contains the specific binding sequence for the initiator protein, and *nic*, where Rep specifically cleaves the DNA at the nick site. The two loci can be either adjacent to each other (pT181 and pC194 families) or separated by a spacer region of up to 100 bp (pMV158 family) (Fig. 1B). The *dsos* of plasmids of the same family are characterized by a high degree of conservation in the *nic* region and by the presence of a less well-conserved *bind* region. In fact, Rep proteins encoded by different plasmids of the same family can perform *in vitro* the nicking-closing reaction on the *dsos* of all the plasmids belonging to the same family, but there is little or no cross-interaction with the *bind* region, which is indicative of the replicon-specificity of the *bind* locus. Interestingly, the pT181-encoded RepC initiator has been shown to drive *in vitro* replication of plasmid pC221, although this was greatly reduced if a competing pT181-*dso* was present (58). In spite of such *in vitro* recognition and extensive homologies of the Rep proteins and the *dsos* of pT181 and pC221, there is no cross-reactivity between the Rep proteins and the *dsos* of these plasmids *in vivo*, unless the Rep proteins are overproduced (59).

In the case of the pMV158 family, the DNA sequence of the *bind* locus was reported to consist of two or three direct repeats (DRs), whose lengths ranged from 5 to 21 bp (35), separated from the nick sequence by an intervening sequence of variable length (Fig. 1B). The *dso* of pJB01, a member of the pE194 subfamily, contains as the Rep-binding site three 7-bp nontandem DRs located 77 bp downstream from the nick site (60). Interestingly, the existence of distant DRs has not been elucidated in some plasmids of this subfamily (unpublished observation). The role of the different regions of the pMV158-*dso* in the interaction with the plasmid-encoded RepB initiator protein has been addressed in a systematic study (35, 61–64). RepB binds with high affinity to the *bind* locus, which is made up of three 11-bp tandem DRs located 84 bp downstream from the nick site. These repeats do not constitute an incompatibility determinant toward pMV158 and seem to be essential for plasmid *in vivo* replication but not for *in vitro* relaxation of supercoiled DNA mediated by RepB. A second RepB binding site is located in a region around the nick site, within the *nic* locus. Characterization of the relative affinity of RepB for the *bind* and *nic* loci revealed that the three DRs of the *bind* locus constitute the primary binding site, whereas the weaker binding of RepB to the *nic* locus could be involved in recognition of the nick site during initiation of replication (64). In plasmids of the pT181 and pC194 families, the DNA sequences of the *bind* (IRIII) and *nic* (IRII) loci are located in contiguous inverted repeats (IR) (Fig. 1B). In pT181, both the spacing and the phasing of IRII to IRIII are crucial for origin functionality (65). In addition, the proximal arm and the central part of the IRIII are important for sequence-specific recognition (65). A similar picture is found in plasmids of the pC194 family.

A typical feature of the *nic* regions is the presence of secondary structures such as hairpins and cruciform. The Rep nick sequence is generally located on an unpaired region within these hairpins, as exemplified by IRII of pT181 and IR-I of pMV158, which accounts for the requirement of plasmid DNA supercoiling to render the cleavage sequence a suitable ssDNA substrate for replication (66–68). The presence of secondary structures is likely to be involved in efficient recruitment and utilization of the initiator protein. Additionally, binding of the initiator protein to the *nic* locus could promote the melting of the substrate nick sequence. This seems to be the case in pMV158, where the extrusion frequency of the cruciform involving IR-I is very low at the growth temperature of the plasmid host (37°C) (69). *In vitro* footprinting experiments performed with supercoiled pMV158 DNA showed that binding of RepB to the *nic* locus promotes the extrusion of the IR-I cruciform, which in turn indicates that initiation of replication would take place only when specific binding of RepB occurs (64). Genetic analysis of the pC194 *dso* pointed to the existence of a hairpin located downstream of the nick site (70) that was shown to be important for replication of the plasmid (71). In contrast, RepU, the initiator protein of pUB110, does not require the presence of hairpins for efficient recognition of the *oriU*. Hairpin II, located downstream from the nick site, seems to be dispensable for initiation of replication of pUB110, although its absence provokes the accumulation of multimers, which is indicative of the involvement of this structure in termination of replication (72).

Out of the three plasmid family prototypes that have been studied in more detail (pT181, pC194, and pMV158), available information regarding the characteristics of the *dso* is limited to a few plasmids of different families. In the *dso* sequences of pJV1, pIJ101, and pSN22, three plasmids belonging to the same subfamily inside the pC194 family, three conserved regions were identified: the conserved region I of about 100 bp, located upstream of the nick sequence, which is essential for replication, the nick sequence in region II, and region III, which overlaps with the *rep* start codon. A co-integration experiment between pJV1 and a pIJ101-derived vector allowed the identification of the nick site within the sequence 5′-CTAGGTA-3′ of pJV1, located 159 bp upstream of the start codon of the corresponding *rep* gene (73). In the case of pIJ101 and pSN22, the putative nick sequence identified in region II was 5′-CTT GGGA-3′, which is not identical to that of the pC194 group (5′-CTTGATA-3′) (74, 75).

Plasmid pGA1 isolated from *C. glutamicum* is the best-studied plasmid of the pCG1 family. The location of the *dso* was analyzed using the runoff DNA synthesis assay. The site- and strand-specific breakage of pGA1 dsDNA occurred within the nucleotide sequence 5′-CTGG/AT-3′ (where / indicates the nick site) in the distal part of the pGA1 *rep* gene, which is an atypical position among RCR plasmids (76).

The group of RCR plasmids isolated from hyperthermophilic archaea is constituted of pGT5, pRT1, and pTN1. Plasmid pGT5 is the first plasmid to be isolated from a hyperthermophilic organism, *Pyrococcus abyssi*, and presents similarities to plasmids from the pC194 family. A sequence of 11 nucleotides (nt) identical to that in the *dso* of pC194 and related plasmids is located at the 5′ region of the *rep* gene. The presence of pGT5 ssDNA (corresponding to the putative plus strand) in cell extracts of *P. abyssi* strongly suggests that pGT5 replicates via an RCR mechanism (57). Plasmid pTN1, isolated from *Thermococcus*, shows an identical nick sequence to that of pGT5 (5′-TTATCTTGATA-3′) and is also located in the 5′ region of the *rep74* gene (77). Similarly, the *dso* of pRT1, isolated from *Pyrococcus* sp. strain JT1, exhibited significant identity to the *dso*s of both pGT5 and pC194, further suggesting that the replication mode of this plasmid is via the RCR mechanism, which was confirmed with the detection of ssDNA replication intermediates (78). The position of the *dso* nick site in the plasmid pZMX201, considered the prototype of a family of plasmids isolated from halophilic archaea, was precisely determined in the sequence 5′-TCTC/GGC-3′ (where / denotes the nick site), which is conserved among the members of the family. Although the heptameric sequence is usually located in the stem region of an imperfect hairpin structure that, in turn, could serve as a target for recognition by the Rep protein, the nick site lies in an unpaired position or near an unpaired nucleotide (56). In addition, the use of a hybrid plasmid system revealed the role of the nucleotides of the conserved nick sequence in the RCR initiation and termination process.

THE REPLICATION INITIATOR REP PROTEINS

As mentioned above, Rep proteins involved in plasmid RCR initiation are mainly included in a vast superfamily of HUH endonucleases that catalyze cleavage and ligation of ssDNA by using particular recognition and reaction mechanisms. Besides the Rep class, which also includes Reps from ssDNA coliphages and animal and plant viruses, proteins involved in conjugative plasmid transfer (Mob class or relaxases) and in DNA transposition (transposases) also belong to this superfamily, and all of them exhibit a familial relationship based on several conserved protein motifs (79). The two most relevant motifs are the metal-binding HUH motif, composed of two His residues separated by a bulky hydrophobic residue, and the catalytic motif containing either one or two Tyr residues separated by several amino acids (Fig. 3). Characterization of the biochemical activities of several plasmid RCR initiator proteins has contributed to a better understanding of the molecular events during the initiation and termination of RCR. Initiation of plasmid RCR requires Rep-mediated nicking within the unpaired nick sequence of the *nic* locus in supercoiled DNA. Rep endonucleases exhibit DNA strand-transfer enzymatic activity and catalyze cleavage and rejoining of ssDNA using an active-site Tyr residue to make a transient 5′-phosphotyrosine bond with the DNA substrate and a free 3′-OH at the cleavage site. Moreover, the resultant 3′-OH not only serves to prime replication but also can act as the nucleophile for strand transfer to resolve the phosphotyrosine intermediate in the termination step of RCR.

The divalent metal ion required for the activity of the HUH enzymes probably coordinates one of the oxygen atoms of the scissile DNA phosphate, polarizing it and facilitating the nucleophilic attack of the hydroxyl group of the catalytic Tyr (80, 81). Curiously, the HUH motif for metal binding is not present in plasmids of the pT181 family or in ssDNA filamentous phages, although in both cases the presence of a divalent metal ion is required for the enzymatic activity. In the case of the pT181 family, it has been shown that the reactive Tyr188 of RepD, the initiator protein of pC221, cleaves the phosphodiester bond 5′-ApT-3′ and remains covalently attached to the 5′-P end generated by the cleavage reaction. The importance of the tyrosyl hydroxyl group was confirmed by substitution of Tyr188 by Phe, since this protein variant retains the sequence-specific

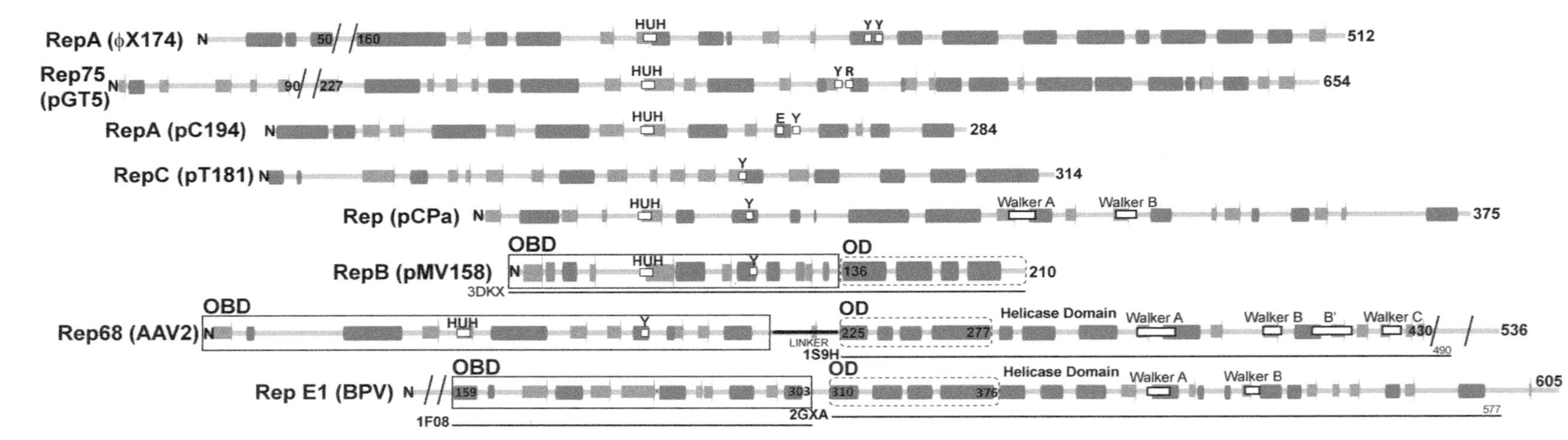

Figure 3 Domain structure of the Rep proteins from RCR plasmids. Predicted and observed secondary structures of the replication proteins of different RCR plasmids and of the Rep proteins from the adeno associated virus (AAV) and bovine papillomavirus (BPV). The amino-terminal end (N) and the number of amino acids are indicated for each of the proteins analyzed. The predicted or observed α-helices and β-strands are represented as red and green bars, respectively. The 3_{10}-helices are represented as blue bars. Conserved amino acid residues of the active site involved in metal binding (HUH) and in the endonucleolytic activity are indicated in the protein maps. The conserved Walker A, B, and C motifs are indicated in the proteins with a helicase domain. The limits of the origin binding domain (OBD) and of the oligomerization domain (OD) are indicated in the protein maps of RepB, Rep68, and E1. The additional line below the sequence of RepB, Rep68, and E1 shows the secondary structure present in the crystal structure of the protein (PDB entry code is given in the figure). Plasmidic Rep proteins were aligned by the metal binding HUH motif. However, viral Reps were aligned with RepB by the all-helical OD domain due to the structural similarity found in this region. doi:10.1128/microbiolspec.PLAS-0035-2014.f3

DNA-binding activities of wild-type RepD but is unable to attach covalently to the replication origin or participate in the nicking-closing reaction *in vitro* (52). Similarly, the initiator protein RepA of pC194 forms a 5′-phosphotyrosyl DNA link at the initiation step, and mutations in the catalytic Tyr214 drastically reduced its catalytic activity without affecting RepA binding affinity. RepB of pMV158 does not seem to generate a stable covalent tyrosylphosphodiester bond with its DNA target (62). However, there is experimental evidence showing that RepB, like the filamentous phage gpII, forms a transient covalent complex with the 5′-P end of the cleaved DNA (63, 82). A singular case among Rep proteins of RCR plasmids is represented by Rep75 of pGT5, a protein that exhibits a highly thermophilic nicking-closing activity *in vitro* combined with an unusual site-specific nucleotidyl terminal transferase (NTT) activity which has not been described for proteins of this type (83). Substitution of the catalytic Tyr448 by Phe caused a severe reduction of the nicking-closing activity *in vitro* and prevented the formation of the 5′-phosphotyrosyl DNA link without affecting the dsDNA binding activity of Rep75 (84). A second critical residue for the activity of Rep75 is Arg451, as the protein variant Rep75-Arg451Leu exhibits a reduced closing activity and has completely lost the NTT activity (84).

A modular structure based on the presence of at least two domains involved in origin nicking and specific recognition of the *dso* sequence has been assumed to be a general feature of the initiator proteins of RCR plasmids (85–88). This assumption could be valid for plasmids of the pT181 family, as the sequence-specific DNA binding and DNA relaxation activities of RepC, the initiator of pT181, are mutationally separable and lie on distant protein regions (86). Taking advantage of this property, the identification of the role of individual monomers of pT181-RepC in RCR was addressed by generating heterodimers of the initiator containing a combination of wild-type, DNA binding, and nicking mutants (89). The results demonstrated that a single monomer of RepC is sufficient for origin-specific binding and nicking. In addition, the monomer involved in sequence-specific binding to the *dso* must also nick the DNA to initiate replication (89).

In plasmids of the pMV158 family, a similar assumption was proposed on the basis of the higher degree of amino acid identity found at the N-terminal region than at the C terminus of their Rep proteins, which suggests that their conserved N-terminal moiety would be involved in endonuclease activities, whereas the C termini would be involved in specific *dso* recognition (32). However, resolution of the three-dimensional structure of RepB, the initiator protein of pMV158, by X-ray crystallography and by image reconstruction methods showed a different functional domain organization from that initially proposed. RepB is the first and to date the only published example of an atomic structure of a Rep protein from RCR plasmids or bacteriophages (3, 90). Purified full-length native RepB behaves as a hexamer in solution, as observed in analytical ultracentrifugation assays (91), and is crystallized in the same oligomeric state, forming a toroidal homohexameric ring (90). Each RepB protomer comprises an N-terminal origin binding domain (OBD), which retains the DNA binding capabilities as well as the nuclease and strand-transfer activities of RepB, and a C-terminal oligomerization domain (OD) that forms a cylinder with a 6-fold symmetry in the hexamer (Fig. 4A). Separate expression and purification of the RepB OD and OBD domains demonstrated that the enzymatic and dsDNA binding activities and the oligomerization potential can be uncoupled and confirmed the essentiality of OD for hexamerization (D. R. Boer, J. A. Ruiz-Masó, M. Rueda, M. Pethoukov, D. I. Svergun, M. Espinosa, M. Orozco, G. del Solar, and M. Coll, submitted for publication).

Resolution of the three-dimensional structure of the catalytic N-terminal domain of RepB confirmed the involvement of the conserved motifs in the enzymatic activity of the protein. Based on the protein sequence alignment of Reps of the pMV158 family, a set of five conserved motifs were designated for this particular family of proteins (32). Motifs I, III, and IV correspond to the conserved motifs 1, 2, and 3 of Ilyina and Koonin (79), whereas motifs II and V were new. A divalent metal ion, Mn^{2+}, required for the catalytic activity of RepB, is located at the active site coordinated by four amino acid ligands and a single-solvent molecule. Residues His39 and Asp42, included in motif II and with no function assigned initially, together with the His55 and His57 residues of motif III (HUH), provide the four amino acid ligands necessary for metal ion coordination. The conserved catalytic Tyr99 (motif IV) and the conserved Tyr115 (motif V), which interacts with the metal ligand Asp42 of motif II, complete the list of the conserved residues located in the active center. Residues in motif I are outside the active center and seem to play a structural role by forming part of the strand β1.

In spite of the presence of the *bind* locus dsDNA sequence in the cocrystals of pMV158-RepB, it was not possible to extract structural information on the DNA or on the protein elements interacting with it. Nevertheless, the presence of a single and large electroposi-

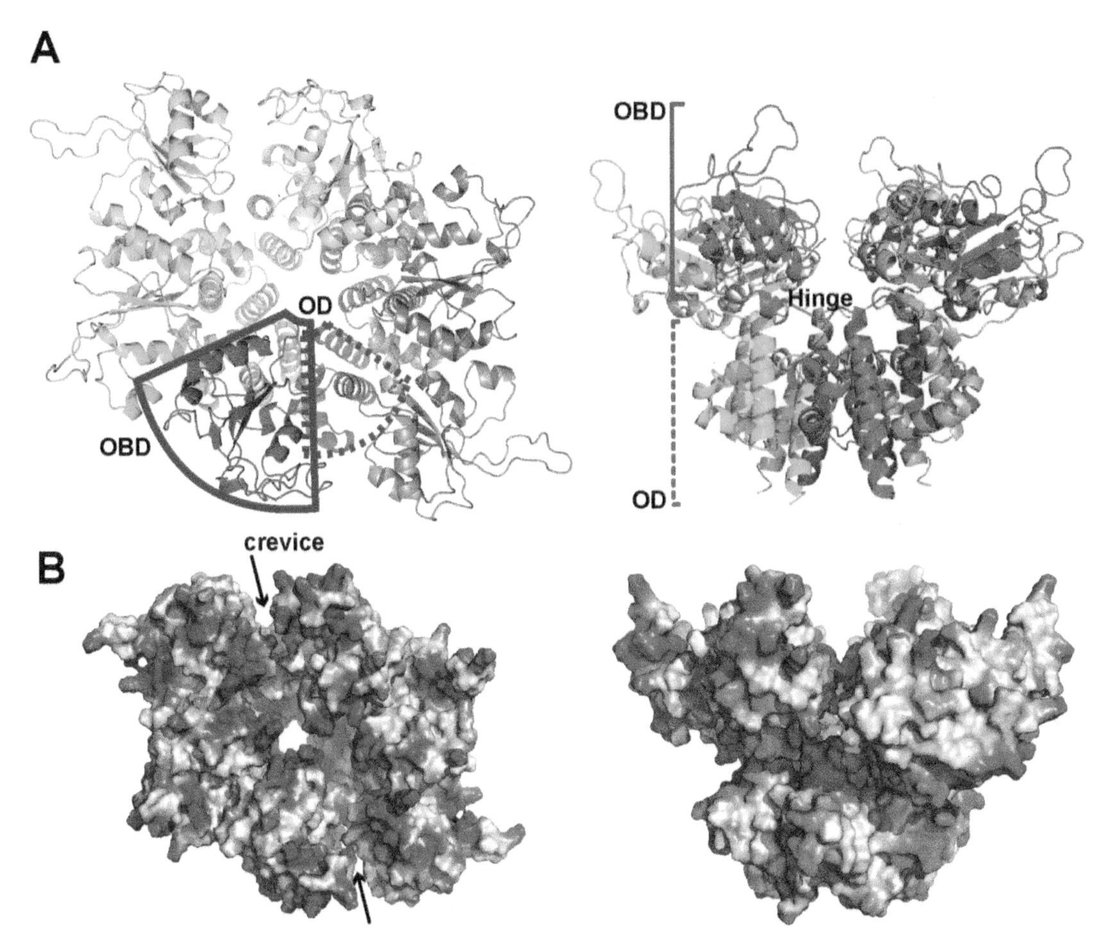

Figure 4 Cartoon representation of the structure of RepB obtained by X-ray crystallography. (A) Top (left) and side (right) views of the RepB hexamer. The locations of the OBD (continuous line) and of the OD (dotted line) are also indicated in the two views. The position of the hinge connecting both domains is indicated in the side view. (B) Top (left) and side (right) views of the electrostatic potential on the solvent-accessible surface of the RepB hexamer structure. The location of the crevice is indicated.
doi:10.1128/microbiolspec.PLAS-0035-2014.f4

tive region that covers the outer surface and crevice at the N-terminal of the RepB hexamer was consistent with binding to only one *bind* locus-containing DNA molecule (Fig. 4B). In addition, by using electron microscopy reconstruction methods, it was possible to observe a density that occludes the N-terminal region crevice in RepB hexameric particles exposed to the *bind* locus (90). These observations point to the location of the DNA-interacting surface in the N-terminal outer region and crevice of the hexamer. In fact, site-directed mutagenesis to some positively charged side chains of helix α2 resulted in OBD protein variants with a reduced dsDNA binding affinity but with an intact nicking-closing activity (90).

Two structures of the pMV158-RepB hexamer were obtained from crystals belonging to a trigonal and a tetragonal form, respectively. In both structures, a short hinge region connects the two protein domains. However, the OBDs do not follow the 6-fold symmetry of the ODs and the positions of the OBDs with respect to the ODs change significantly when the two crystal forms are compared. In fact, analysis of the two structures revealed that the N-terminal OBD domains are found in nine distinct orientations relative to the OD. The conformation plasticity of RepB has been explored by combining techniques such as X-ray crystallography, small-angle X-ray scattering, sedimentation experiments, and molecular simulations (D. R. Boer, J. A. Ruiz-Masó, M. Rueda, M. Pethoukov, D. I. Svergun, M. Espinosa, M. Orozco, G. del Solar, and M. Coll, submitted for publication). These studies revealed that the position and freedom of movement of the OBDs is mainly determined by the OBD-OD hinge region, since contacts between adjacent OBDs do not seem to play an important role in fixing their positions. A consequence of the loosely coupled domain arrangement observed in the RepB hexamer is the high level of conformational freedom of the OBDs, which is probably important for formation of a functional replisome.

The hexameric state of purified RepB is unique among plasmid replication initiators characterized so far. Most of these proteins are purified as monomers, as in the case of the Rep proteins of the pC194 family (53), or dimers, such as Reps of the pT181 family (52, 54, 92, 93). Interestingly, some of these proteins tend to form multimers upon binding to the *ori* DNA, suggesting that oligomerization could be involved in some of the biochemical activities of the initiators (93, 94).

The existence of tight interactions between the pMV158-RepB all-helical ODs enables the formation of a toroidal ring with near 6-fold rotational symmetry and an inner surface narrowing down from a maximum diameter of ~20 Å to a minimum diameter of ~13 Å (90). The search for fold similarities in the ODs from different atomic structures, in combination with a comparative analysis of Rep sequences, confirmed the existence of a RepB-like all-helical OD domain responsible for oligomerization in viral Rep proteins and replication initiators from plasmids of the pMV158 family (D. R. Boer, J. A. Ruiz-Masó, M. Rueda, M. Pethoukov, D. I. Svergun, M. Espinosa, M. Orozco, G. del Solar, and M. Coll, submitted for publication). The combination of hexameric OD ring and hinge-connected OBDs has been proposed as a general feature of hexameric replication initiators of the pMV158 family, although this configuration can also be found among hexameric initiators of animal and plant viruses (Fig. 3) (D. R. Boer, J. A. Ruiz-Masó, M. Rueda, M. Pethoukov, D. I. Svergun, M. Espinosa, M. Orozco, G. del Solar, and M. Coll, submitted for publication). In fact, the domain organization of RepB resembles that of the viral replication initiators, which suggests an evolutionary link between the two protein families. Structural similarities between the α-helical hexamerization domain of RepB and the equivalent domain of the papillomavirus E1 helicase have been reported (90). The crystal structure of E1 comprises the OD and helicase domains and was obtained in complex with a short ssDNA oligonucleotide in the central channel (Fig. 3) (95). Therefore, the resemblance of the E1 and RepB hexameric rings favors a mechanistic model in which the RepB ring might close around a DNA region that could have melted on assembly of the protein and/or cleavage at the nick site, thus encircling one of the plasmid strands within the central channel. Subsequent recruitment of a host helicase (perhaps PcrA) would allow further unwinding of the DNA and the concomitant progression of the hexamer along the plasmid. Enclosure of one plasmid strand may confer high processivity to the RepB/helicase/DNA polymerase replisome complex, thereby allowing replication of pMV158 in a broad range of bacterial hosts.

Ilyina and Koonin (79) hypothesized that geminiviruses would descend from bacterial replicons according to the limited sequence similarity of the three conserved motifs of the RCR Reps of geminiviruses and certain plasmids of Gram-positive bacteria. However, more recent phylogenetic analyses of various RCR Reps suggest that Rep proteins of geminiviruses share a most recent common ancestor with Reps encoded on plasmids of phytoplasmas (96). Several plasmids from phytoplasma have been sequenced and characterized to some extent. The Rep protein of pOYW, a plasmid isolated from onion yellow phytoplasma, is characterized by a chimerical nature containing an RCR plasmid Rep-like domain in the N-terminal region, which displays four out of the five conserved motifs characteristic of the pMV158-Rep family of proteins, and a virus-like helicase domain in the C-terminal region, which also includes Walker A and B nucleotide-binding motifs and shows great similarity to Reps from eukaryotic small DNA viruses or some RNA viruses (97). Similarly, the Rep protein of plasmid pCPa, isolated from *Candidatus* Phytoplasma australiense, also shows four out of the five protein motifs characteristic of the pMV158-Rep family of proteins in its N-terminal region and Walker A and B motifs typically found in geminiviruses in its C-terminal region (Fig. 3) (98).

CONTROL OF RCR

Mechanisms for Controlling Synthesis of the Rep Protein

Control of plasmid RCR is exerted via regulation of the synthesis of the replication initiator. Other general control mechanisms such as the origin inactivation by handcuffing, which involves coupling between plasmid molecules through Rep proteins bound to multiple initiator binding sites (iterons), has not been reported in RCR plasmids. Availability of the Rep protein determines the frequency of the leading-strand initiation, which is rate-limiting for the plasmid RCR process, and therefore the *rep* gene encoding the initiator is subjected to strict control. Transcriptional and translational inhibition mechanisms of *rep* expression are not specific to RCR plasmids and have been reviewed (1, 99, 100). Two classes of replication control systems have been described in RCR plasmids: those that use only antisense RNAs and those involving an antisense RNA in combination with a transcriptional repressor protein. An example of the first class is found in the replication control of pT181 by an antisense RNA-mediated transcriptional attenuation mechanism (101), which is also expected to control replication in most of the plasmids of the family. In this system, the antisense RNA transcribed from the *cop* region targets the *rep*-mRNA encoding the initiator protein and blocks its expression. The main consequence of the antisense-target interaction is the premature termination of the *rep*-mRNA transcription due to the formation of a ρ-independent transcriptional terminator (attenuator) located just 5′ to the *repC* start codon. Release of the antisense RNA from the mRNA target permits refolding and, therefore, silencing of the transcriptional terminator (101).

A second class of copy-number control system reported in RCR plasmids involves two trans-acting plasmid elements, namely a transcriptional repressor and an antisense RNA, that are involved in controlling the synthesis of the initiator at the transcriptional and translational levels, respectively. This dual mechanism was first discovered in pMV158 and seems to be widespread among plasmids from Gram-positive bacteria, including the RCR plasmids of the pMV158 family (99, 102). In pMV158, the transcriptional repressor CopG binds to its own promoter and represses the transcription of the *copG-repB* operon. The mechanism of transcriptional repression mediated by CopG has been elucidated by analyzing the interactions among the RNA polymerase (RNAP), CopG, and the promoter (103). CopG is able to bind specifically and cooperatively to its operator, which overlaps with the regulated promoter, in such a way that it not only prevents the binding of the RNAP, but also efficiently displaces the polymerase bound to the promoter. The second control element, the *rnaII* gene encoding RNAII, overlaps the intergenic region of the pMV158 *copG-repB* operon and is transcribed in the opposite direction to it. Therefore, the entire sequence of antisense RNAII is complementary to a *copG-repB* mRNA target region that contains the translation initiation signals for the essential *repB* gene (104). Small antisense RNAII (48-nt long) consists of a single-stranded 5′ tail followed by a hairpin and a 3′ poly(U) tail, the latter two elements constituting a very efficient intrinsic transcription terminator (105, 106). Additionally, RNAII is able to inhibit plasmid replication *in trans*, it has a short (1 to 2 min) half-life (P. Acebo et al., unpublished results), and it determines a strong incompatibility against pMV158, thus matching the features required for an efficient plasmid replication control element (102, 107).

In the antisense RNA-mediated systems of RCR plasmid replication control where the entire process has been dissected, the formation of the RNA/RNA duplex seems to be initiated by a "kissing" step that involves reversible base-pairing between complementary hairpin loops (108). However, wild-type antisense RNAII supplied *in trans* retains its inhibitory capacity on derivatives of pMV158 that encode a mutant *copG-repB* mRNA lacking the hairpin complementary to that in RNAII. These findings suggested that formation of a kissing complex was not strictly required for the RNA pairing mechanism of the pMV158 control system (107) and led to the proposal that *copG-repB* mRNA/RNAII binding initiates *via* a loop-linear pairing scheme (109). The existence of an antisense RNA that controls replication of pJB01 (a member of the pMV158 family) has also been proved, and the involvement of different regions of this RNA in translation inhibition of the *rep* gene was studied by mutational analysis. Copy number inspection of the various mutant plasmids suggested that the entire secondary structure of the antisense RNA was important for interaction with the target mRNA (110).

In addition to this dual control of *rep* expression that senses and corrects fluctuations in plasmid copy number, proper availability of Rep also relies on the adequate functionality of the transcription and translation initiation regulatory signals. In pMV158, translation of *repB* was proposed to be initiated from what was termed an atypical ribosome binding site (ARBS) located in the small intergenic region of the *copG-repB* operon encoding the antisense RNAII. The initiation signals that regulate *repB* translation were identified and characterized in reference 111. Translation of *repB* relies on its own initiation signals, which rules out a possible mechanism of translational coupling to the upstream *copG* gene. Only changes in the sequences involving the ARBS proximal box, thus named because of its proximity to the *repB* start codon, and the region 3′-adjacent to the ARBS proximal box result in a significant reduction of *in vitro* synthesis of *repB*. The results of this study demonstrate the importance of the region immediately upstream of the *repB* start codon in the efficiency of translation of *repB* and call into question the functionality of the postulated ARBS. Moreover, the conclusions of this study could be applicable to the majority of the plasmids of the pMV158 family due to the high degree of identity found at the *rep* translation initiation regions. The possible involvement of those features of the translation initiation signals of the *rep* genes in an additional mechanism to regulate the level of the Rep protein remains to be determined.

The existence of a singular element that positively influences the plasmid copy number and stability has been reported in plasmid pGA1 from *C. glutamicum*. On one hand, the copy number of pGA1 is negatively regulated by an antisense RNA at the translational level (112). Additionally, the IR1 sequence, located in the leader region of the *rep* mRNA, negatively influences the *rep* gene expression. On the other hand, the *per* gene, which encodes a positive effector of replication, positively affects the copy number and segregation stability of pGA1, though it was found to be dispensable for replication (47). In fact, deletion of the *per* gene results in a significant decrease of the pGA1 copy number in *C. glutamicum*. Furthermore, the related plasmid pSR1 encodes a similar gene product that can act *in trans* on pGA1 derivatives (47). Per protein has no effect on the expression from the *rep* promoter, and it has been hypothesized that it counteracts the inhibitory effect of the antisense RNA on the *rep* gene expression by interacting with it (112). Interestingly, pGA1 still codes for another accessory effector, the small *aes* gene, which was shown to increase the segregational stability of pGA1 derivatives in the presence of *per* (113). The genes *per* and *aes* are located nearby and transcribed convergently. The authors propose that a cooperative regulation of expression of *per* and *aes* genes could be on the basis of the control of the pGA1 plasmid copy number and of its stable maintenance in *C. glutamicum*.

In plasmid pUB110, expression of the replication initiator gene (*repU*) is controlled by two small and unstable antisense RNAs, transcribed from a major incompatibility region, that interfere with *rep*-mRNA translation by targeting the *repU* translation initiation signals (114). Interestingly, an additional control of plasmid copy number seems to exist in pUB110. In this plasmid, inactivation of a RepU molecule after a replication event has been proposed to occur by incorporation of a short oligonucleotide, in a way analogous to that described for pT181-RepC (see below). The inactive RepU-RepU* hetero-oligomer complex could form a large nucleoprotein structure at the *dso* region that interferes with transcription from the *repU* promoter (94). Therefore, even in the absence of a Cop-like transcriptional repressor, the amount of active RepU protein available for replication initiation could be subjected to a dual control at the translational and transcriptional level.

Mechanisms That Restrict the Use of Rep Molecules to a Single Replication Event

The replication and copy number of RCR plasmids are tightly regulated, and the mechanisms ensuring that the initiator proteins are unable to catalyze multiple rounds of DNA synthesis are critical to achieve such efficient control. In pT181, the RepC dimer is inactivated after the completion of a round of replication by the attachment of an approximately 10-nt oligonucleotide, representing sequences immediately 3′ to the initiation nick site, to the active tyrosine residue of one of its subunits, thus generating a heterodimer termed RepC/C*. Generation of RepC/C* occurs because once the replication fork reaches the reconstituted origin, the nascent leading strand is extended by ~10 additional nucleotides beyond the Rep nick site. Two site-specific transesterification reactions take place then: the first leads to the release of the circular ssDNA intermediate, and the second yields the RepC/C* heterodimer containing the short 3′ leading-strand extension as well as a dsDNA plasmid molecule that has the sealed new leading strand paired to the parental template strand (115, 116). Thus, the second transesterification allows completion of the termination process by avoiding recycling, i.e., continuous synthesis of the leading strand driven by the same initiator molecule (117). Analysis of the different interaction patterns of RepC/C and RepC/C* with the pT181-*dso* DNA by *in vitro* footprinting and binding-bending assays revealed that, although RepC/C* retains the ability to bind to DNA, it is unable to promote cruciform extrusion to expose the nick sequence in ssDNA form (118). This may explain why the formation of a RepC/RepC* heterodimer inactivates the protein in addition to uncoupling termination of leading-strand replication and initiation of a new replicative round.

A different mechanism preventing recycling upon termination of one replication round has been proposed for plasmid pC194. The pC194-encoded RepA protein has two active-site catalytic residues, namely Tyr214 and Glu210, which catalyze the DNA cleavage at the initiation and termination steps, respectively. In this plasmid, the termination reaction catalyzed by the glutamate residue is hydrolysis that does not generate a covalent complex, thus preventing continuous synthesis of the leading strand (53). By contrast, in gene A protein of phage ΦX174 two catalytic Tyr of a single protomer would perform alternative nicking and nicking-closing reactions following a so called "flip-flop" mechanism that would allow recycling of the initiator protein (119). These two alternative pathways at the termination step of leading-strand synthesis reflect the different lifestyles of phages and plasmids, with replication control featuring the latter. It is worth noting that the lack of formation of a Rep-DNA adduct in the cleavage reaction of the termination stage of the pC194 leading-strand replication implies that the yet-unknown mechanism of inactivation of the used RepA molecules would differ from that reported for the Rep proteins of the pT181/pC221-family plasmids. Remarkably, although the catalytic residues Glu and Tyr of RepA are conserved in the RepU initiator of pUB110, a replication-dependent modification of RepU implying the loss of its catalytic activity has been reported, and the modification has been suggested to consist, as in the case of RepC*, in the covalent attachment of a single-stranded oligonucleotide (94). In pMV158, an inactivation mechanism similar to that described for pT181-RepC does not seem to be applicable, since RepB has not been shown to form a stable covalent complex with the 5′-P end of the cleaved DNA (62). However, generation of the circular ssDNA intermediate during termination of the plasmid leading-strand synthesis requires that the 5′-P end to be ligated is covalently attached to RepB (62). Since this is the 5′-P end generated in the initiation cleavage, the feasibility of preservation of the RepB-DNA bond through the entire leading-strand replication process under certain circumstances is suggested. Therefore, a mechanism analogous to the flip-flop scheme, which involves alternating nicking and nicking-closing by two catalytic residues of a single protomer, is plausible for RepB. As an alternative, the second catalytic residue could be provided by the active site of an adjacent monomer of the RepB hexamer, which would imply that substrates are transferred between OBDs during the termination reactions. In either case, and despite the biochemical and structural data available about RepB, the identification of the second catalytic residue remains undisclosed.

The unusual NTT activity displayed by the pGT5-encoded Rep75 initiator appears to be directly related to the mechanism that prevents plasmid overreplication. The three reactions catalyzed by Rep75, namely nicking, closing, and NTT, can be uncoupled *in vitro*, although they share part of their mechanisms. Interestingly, replication initiation activities mediated by Rep75 are inhibited at the concentrations of ATP or dATP that promote the NTT activity of the protein. According to the proposed model, an adenine residue could be transferred to the 3′-OH generated after specific cleavage by the NTT of another Rep molecule, thus re-creating a putative nick site. After cleavage at this site, Rep75 becomes attached to an adenine residue through its active Tyr and, therefore, is inactivated. Hence, the role of the NTT activity might be to reduce the intracellular level of active Rep molecules. The authors propose that the levels of pGT5 replication could be regulated by an equilibrium between active and inactive Rep proteins, itself determined by the intracellular ATP and dATP pools and the availability of free 3′-OH and host replication proteins (84).

THE SINGLE-STRAND ORIGIN

Synthesis of the lagging strand of RCR plasmids initiates from the so-called single-strand origin (*sso*), a noncoding region containing long and imperfect inverted repeats that form complex secondary structures as the parental leading strand is displaced and becomes an ssDNA replicative intermediate (12, 120). In general, the *sso*s are located a short distance upstream of the *dso*s and, hence, these elements are among the latest plasmid regions to become single-stranded during replication of the leading strand. This relative position might play a role in preventing run-off synthesis of the lagging strand before closing of the displaced parental strand upon termination of the leading-strand synthesis. The *sso* is recognized by host factors (usually RNAP) that make a small RNA for priming the lagging-strand synthesis. Functionality of the *sso* is orientation-dependent, which points to a crucial role of unpaired sequences within the secondary structure of these elements.

Five main types of *sso* (*ssoU*, *ssoA*, *ssoT*, *ssoW*, and *ssoL*) have been reported that differ from each other in structure, in sequence motifs highly conserved among the members of the same group, and in the host range in which they are functional.

The *ssoA*-type origins consist of a single and long (~150-nt) hairpin structure containing internal and bulge loops, in addition to the terminal loop. These origins display a high degree of sequence heterogeneity, with only two well-conserved regions: the previously termed recombination site B (RS_B), which is present in the lower stem of the hairpin of all the *ssoA*s, and a 6-nt consensus sequence (CS-6) that is located in the terminal loop of the hairpin of most but not all of these origins (12, 120, 121). The RS_B has been shown to be involved in binding of RNAP to the *ssoA* (122). Mutations in the RS_B abolish almost completely the *ssoA*-dependent synthesis of the lagging strand both *in vitro* and *in vivo*, giving rise to the accumulation of very high numbers of plasmid ssDNA intermediates in the host cell (121, 123). The CS-6-containing terminal loop seems to function as a terminator of the primer RNA synthesis, and changes in this region lead to a moderate increase in the intracellular number of ssDNA plasmid forms without impairing *ssoA* binding by RNAP (121–123). Sequences similar to CS-6 have been identified in the terminal loops of hairpins located in DNA regions that have been either proposed or shown to act as *sso*s in plasmids from *Actinobacteria* (124–128). In addition to the RS_B and CS-6 conserved motifs, sequences resembling the consensus -35 and -10 promoter regions have been identified in various *ssoA*s (129). A given *ssoA* only functions efficiently in its natural host or in a few closely related species (12, 120, 129).

The *ssoW* was identified in lactococcal plasmid pWVO1 (130, 131). This origin is located in a 250-bp DNA fragment that contains two inverted repeats, IR I and IR II. IR I shows homology to the *ssoA*-type origins since it harbors the CS-6 sequence in the terminal loop as well as a sequence similar to the RS_B at the lower stem. Also, the upper stem of IR I shares remarkable sequence similarity with the ΦX174 minus-strand origin, which is recognized by the primosome for priming synthesis of the complementary strand (132). Full *ssoW* activity requires both IR I and IR II, and conversion of ssDNA into dsDNA from the entire element is only partially inhibited by rifampicin. IR II has no activity on its own, whereas IR I has a partial, RNAP-independent activity for complementary strand synthesis. Thus, priming of the lagging strand from *ssoW* seems to occur through two different pathways: one is catalyzed by RNAP and requires the entire origin, and the other, which was suggested to involve a primosomal complex, only requires IR I. Efficient functionality of the pWVO1-*ssoW* seems to be confined to lactococci (131).

The *ssoT*-type origins have been found in plasmids isolated from *Bacillus* spp. (133–135). In fact, the majority of the RCR plasmids from *Bacillus subtilis* harbor an *ssoT* (136). The minimal *ssoT*, as defined for the *B. subtilis* plasmid pBAA1, spans 120 to 190 bp, encompassing three imperfect palindromes that would give rise to hairpins I, II, and III on ssDNA. Results from mutation analysis suggest that the structure of hairpin III and both the structure and sequence of hairpin I are required for full activity of the *ssoT* (135). Comparison of the *ssoT* of pBAA1 with the homologous region of the *B. thuringiensis* pGI2 plasmid showed the existence of three conserved sequence motifs located in the loop of hairpin I (motifs 1 and 2) and in the intervening DNA between hairpins II and III (motif 3). Motif 1 has been shown to play an important role in the activity of the *ssoT*, while the role of motif 3 is controversial and that of motif 2 has not been proved. Initiation of lagging-strand synthesis from the *ssoT* is RNAP dependent, as can be inferred from the intracellular accumulation of ssDNA plasmid forms upon addition of rifampicin (135). It is also worth noting that the pBAA1-*ssoT* functions in both *B. subtilis* and *S. aureus*, hence showing a broader host range than the *ssoA*- and *ssoW*-type origins (135). Based on sequence similarity, the *ssoT*s of a number of *B. subtilis* plasmids have been classified into two groups, *palT1* and *palT2*, each including almost identical *ssoT* origins. The *palT1* group includes the *ssoT*s of plasmids pTA1015, pTA 1020, pTA1060, pLS11, and pBAA1, whereas the *palT2* group includes those of pTA1030, pTA1040, and pTA1050 (136). The DNA region involved in initiation of the lagging-strand synthesis in *B. thuringiensis* plasmid pTX14-3 was found to be homologous to the pBAA1 *ssoT* (137). Actually, sequences highly similar to motifs 1, 2, and 3 are present in the pTX14-3 origin. Curiously enough, the activity of the *sso* of pTX14-3 is at least partially resistant to rifampicin, which suggests the existence of a lagging-strand priming mechanism independent of host RNAP (138).

The *ssoU* is the most promiscuous *sso* origin characterized so far, as it seems to be fully functional in many, if not all, *Firmicutes* (121, 139–142). The *ssoU* origin was first identified in staphylococcal plasmid pUB110 (142), within a ~250-bp DNA fragment with the potential to form several hairpin structures containing symmetric and asymmetric internal loops in addition to the terminal one (139). Sequences nearly identical to the pUB110-*ssoU* were subsequently found in the streptococcal pMV158 and *Bacillus* pTB913 plasmids, and the involvement of these elements in lagging-strand synthesis was also proved (143). The high level of sequence identity among the *ssoU*s of plasmids isolated from different bacterial genera is consistent with this kind of *sso* being efficiently recognized in a broad range of hosts. The promiscuous activity of the *ssoU* has been suggested to be accounted for by the proven ability of this origin to bind efficiently to RNAP from different bacteria (144). No sequences with significant similarity to the canonical -35 and -10 promoter regions have been found in the *ssoU*. *In vitro*, RNAP binds to the left one of the two large hairpins of the *ssoU* origin, whereas the transcription initiation of the primer RNA and the transition from RNA to DNA synthesis occur, respectively, at the 3′ end of the right large hairpin and at the 3′ arm of its stem (144).

In spite of the sequence heterogeneity between the four types of *sso*s described above, they all exhibit an

RS_B-like motif partially unpaired in the bottom part of the stem of a hairpin structure. Remarkably, the RS_B-like motif of the *ssoU* origin overlaps the region of the hairpin contacted by RNAP (144).

Apart from the *sso* elements mentioned above, a number of RCR *Lactobacillus* plasmids carry lagging-strand origins showing high levels of similarity among each other but not with the *sso*s found in plasmids from other bacteria genera. This kind of origin (*ssoL*) spans ~100 bp, being considerably smaller than those of the other groups (145).

Unlike the *dso*, the *sso* is not essential for plasmid replication, provided that an alternative pathway exists for priming lagging-strand synthesis. In general, alternative plasmid-borne signals and/or bacterial mechanisms seem to exist that can partially overcome the lack of the genuine *sso* (12, 121, 139). Especially efficient appears to be the predominant *sso*- and RNAP-independent priming system revealed in *Streptomyces lividans* and proposed to result from the stabilization of RCR plasmids lacking *sso* in this bacterium (146). In a few cases, however, the indispensability of the genuine *sso* origin has been reported, as is the case with the *Nocardia* plasmid pYS1 (127).

Although, as stated above, the *sso*s are usually not strictly required for replication, removal of a DNA region encompassing them not only leads to an increase of the ssDNA replicative intermediates, but frequently results in a reduction of the copy number (measured as dsDNA forms) and in segregational instability of the plasmids (12, 147, 148). The stability function linked to the *sso*s does not seem, however, to rely on the efficient conversion of ssDNA to dsDNA, since stable inheritance is observed for plasmids whose *sso* lacks activity in a given host, in spite of the inefficient conversion of ssDNA replicative intermediates into dsDNA plasmid forms. Removal of the genuine *sso* region usually leads to unstable plasmid inheritance even in hosts where the lagging-strand origin is not functional at all (148). Thus, although efficient conversion of ssDNA to dsDNA does not guarantee plasmid stability (149), the presence of an *sso* element appears to contribute to the efficient replication and accurate inheritance of the plasmid molecules and, therefore, to plasmid fitness. Remarkably, all natural RCR plasmids contain at least one *sso* that functions efficiently in their natural host, and plasmids isolated from different bacteria and belonging to the same replicon family (i.e., sharing homology at their *dso*s and *rep* genes) show no conservation at their *sso*s. These observations suggest that acquisition of an active *sso* upon plasmid entrance in a new host may improve not only plasmid fitness itself, but also fitness of the plasmid-containing bacteria, hence enabling them to overgrow cells that contain plasmids without a functional *sso*. In this sense, it is noteworthy that intracellular accumulation of ssDNA, which can arise from the presence of a plasmid lacking a functional *sso*, has been reported to induce bacterial stress responses such as the *E. coli* SOS system (150, 151). This could result, in turn, in a decreased growth rate of the cells and hence impaired bacterial fitness.

Streptococcal plasmid pMV158 is unique, among many other things that have appeared throughout this chapter, in that it contains two lagging-strand origins, namely streptococcal-specific *ssoA* and promiscuous *ssoU*. It has been shown that whereas the pMV158-*ssoU* participates in the plasmid mobilization between different bacterial species, the pMV158-*ssoA* would be involved mainly in intraspecific transfer (140). The presence of both *sso*s might reflect the evolutionary story and lifestyle of this mobilizable, highly promiscuous plasmid.

HOST PROTEINS INVOLVED IN RCR

Participation of DNA and RNA Polymerases

In addition to the critical role frequently played by bacterial RNAP in recognition of the *sso* and priming of the lagging-strand synthesis (see above), several pieces of experimental evidence point to the direct participation of DNA polymerases (Pol) I and III in plasmid RCR.

Analyses of the involvement of Pol I in leading- and lagging-strand synthesis during RCR were performed on the pMV158 model system, by employing plasmid derivatives that carried the *ssoA* and lacked the *ssoU*. These studies took advantage of the previous characterization of Pol I of *S. pneumoniae* and the construction of pneumococcal *polA* mutants lacking the polymerase activity of this enzyme (152, 153). Mutant strains depleted in the 5′- 3′ exonuclease function of Pol I could not be obtained since this activity was found to be essential for cell viability of *S. pneumoniae* (152). Therefore, only the involvement of the polymerase function of Pol I in plasmid RCR could be tested.

Participation of the polymerase activity of Pol I in the initiation of the lagging-strand synthesis was inferred from the increased fraction of ssDNA plasmid forms accumulated within the cells of the *polA* mutant strains compared to the wild-type strain (154). The same conclusion was drawn from the analysis of the *in vitro* replication of *ssoA*-containing ssDNA in cell-free extracts prepared from wild-type or *polA*-deficient pneumococcal strains. Wild-type levels of *in vitro* replication of plasmid ssDNA in extracts from the *polA*-depleted strain were obtained only upon complementation with the entire pneumococcal Pol I but not when a protein variant lacking the 5′-3′ exonuclease activity was added, indicating that both the polymerizing and the exonuclease domains are required for efficient lagging-strand synthesis (122).

A role of Pol I during the termination of the lagging-strand synthesis was suggested from the comparative analysis of single-strand discontinuities detected in the DNA of the *ssoA*-containing pMV158 derivative extracted from the wild-type and *polA* mutant strains of *S. pneumoniae* (154). In the polymerase-deficient mutants, a discontinuity in the vicinity of the *ssoA* origin was observed that could arise from plasmid molecules in which the conversion of ssDNA to dsDNA was not completed as a consequence of the defective replacement of the primer of the lagging strand.

An additional DNA discontinuity only detected in the pMV158 derivative replicating in the pneumococcal *polA* mutants mapped at around the nick site of the plasmid *dso*, thus pointing to the participation of the polymerase activity of Pol I in an early step of the synthesis of the leading strand (154).

Finally, a critical role of the host Pol III replicase (the one containing the PolC polymerase) in plasmid RCR was inferred early from the inhibition of the *in vivo* replication of the pT181 DNA in *S. aureus* by hydroxyphenylhydrazinouracil, an antimicrobial agent that acts specifically on the Gram-positive bacterial PolC polymerase (155).

Role of Superfamily Group 1 Helicases and Their Interaction with the Plasmid-Encoded Rep Initiator

As previously described, plasmid RCR starts with the recognition of a *dso* by the replication initiator protein, which binds, cleaves, and remains covalently bound to the 5′-end of the nicked DNA strand. After these initial steps, several cellular-encoded proteins are recruited to continue the replication of the plasmid. The versatility of the interaction between these proteins and the replication initiator protein determines whether this plasmid is successfully replicated and maintained in a broad range of hosts or whether it is lost. One of the required proteins for the replication of the leading strand is a helicase, which is capable of unwinding the dsDNA ahead of the replication fork. The helicases responsible for RCR are PcrA for Gram-positive bacteria and UvrD in Gram-negative bacteria, both of which belong to the superfamily 1 group of helicases (156, 157). These proteins share 42% sequence identity and are structurally similar (158–161), both possessing the seven conserved helicase motifs. In both cases, there is some controversy regarding the active oligomeric state of the protein: some authors have suggested that PcrA and UvrD are monomeric helicases (162, 163), while others have shown that they are active only as a dimer (164, 165).

The involvement of PcrA in RCR was first reported by Iordanescu and Basheer (166), who found that plasmid pT181 was unable to replicate in a strain of *S. aureus* carrying a mutation (*pcrA3*) within the *pcrA* gene. Subsequently, a mutation in the gene encoding the replication initiator protein RepC (Asp57Tyr) was identified, which suppressed the *pcrA3* mutation and restored pT181 replication in the *pcrA3* mutant (167). From these data it was proposed that the *pcrA3* mutation may impede replication by disrupting the PcrA3-RepC wild-type interaction, thereby preventing separation of the dsDNA, while the interaction is restored by the RepC-Asp57Tyr mutant, hence rescuing plasmid replication. However, it has since been shown by pull-down experiments that the PcrA3 mutant protein was able to interact with both RepC wild type and the Asp57Tyr mutant (168). Rather, the authors showed that the *pcrA3* mutation leads to a threonine to isoleucine change in residue 61, which is located in the conserved motif Ia of the superfamily 1 helicases and results in weak ATPase activity, which prevents the unwinding of the pT181 DNA in the presence of both RepC wild type or Asp57Tyr mutant *in vitro* (168). The authors postulated that the viability of the *pcrA3* mutant and the capability of the double mutant *pcrA3*-*rep*CAsp57Tyr to replicate the pT181 plasmid *in vivo* may be explained by additional cellular factors required for the replication to take place in this strain or by the role of an alternative cellular helicase which may replace PcrA. However, these hypotheses have yet to be confirmed.

PcrA from *S. aureus* is able to hydrolyze ATP, dATP, dGTP, dCTP, and TTP, and its NTPase activity is increased in the presence of either ssDNA or RepC covalently bound to the *oriC* of pT181 (169). In addition, the unwinding of supercoiled plasmid pT181 is only achieved when RepC is covalently attached to the origin of replication and in the presence of ATP in the reaction (169). PcrA is only capable of unwinding a dsDNA when there is a 3′ or a 5′ single-stranded tail exposed. Hence, PcrA from *S. aureus* has a bipolar 3′-5′ and 5′-3′ helicase activity (169, 170).

Two hypotheses have been postulated for the direction of unwinding achieved by PcrA. As mentioned above, PcrA from *S. aureus* may have a dual helicase activity, whereas PcrA from *Bacillus stearothermophilus* has been shown to only unwind dsDNA in a 3′-5′ helicase direction (158). This disparity may be explained by several factors, such as the use of different constructs/protocols for protein purification, the absence *in vitro* of additional cofactors that would favor the unwinding in a specific direction, or the differences in sequence between PcrA from the two bacteria (59% sequence identity), among others.

PcrA from *B. stearothermophilus* is able to bind to a nicked dsDNA containing the *oriD* sequence from the pC221 plasmid, then engage the 3′-OH end and translocate in a 3′-5′ direction along the nicked strand, thus unwinding the DNA. However, in the presence of RepD, the helicase is loaded on the opposing strand (the continuous strand), translocating in the same direction (171). The directionality of PcrA translocation was also confirmed to be 3′-5′, using atomic force microscopy of a linearized plasmid containing the *oriD* sequence at different positions relative to the DNA ends. The *oriD* was nicked by RepD, and the unwinding of the DNA was followed by the appearance of condensed ssDNA (171). Later, the kinetic parameters for PcrA helicase activity were determined in bulk and single-molecule experiments, using lineal and supercoiled DNAs containing the *oriD* sequence, in the presence of RepD and saturating concentrations of ATP (172). Under these experimental conditions, the unwinding speed of PcrA was 30 bp s^{-1}, while the translocation rate on ssDNA was 99 bases s^{-1}. The unwinding rate is dependent on the amount of ATP in the reaction and on the presence of RepD: in the absence of the replication initiator protein the number of unwinding events was reduced by more than 10-fold (162). Moreover, the affinity of PcrA for partial duplex DNA increased by one order of magnitude when RepD was already bound to the DNA, from a K_d of 22 nM to 170 nM (172).

The recruitment of the helicase onto the replication initiation site has also been shown by footprinting experiments using exonuclease (Exo) III (173). RepD binding to *oriD* creates an area of protection to ExoIII digestion which extends beyond the *oriD* region: ~74 to 80 bp upstream of ICR I for the continuous strand and ~46 to 50 bp downstream of ICR III for the nicked strand (173), although further resistance points can be found within the *oriD*, after longer digestion times. When PcrA was incorporated into the reaction, the helicase was recruited upstream of ICR I, which served to stabilize the complex. The RepD-PcrA complex covered the region spanning from 80 bp upstream of ICR I on the continuous strand to the limit of ICR III in the nicked strand, and this complex was not displaced by ExoIII. However, a drastic change in the ternary complex was observed when a nonhydrolyzable nucleotide (ADPNP) was included in the reaction. Under these circumstances, the resistance to ExoIII digestion was located only on ICR II, indicating that there were important conformational changes at the *oriD* once the helicase had begun to unwind the plasmid. Furthermore, these three stages of protein loading on the DNA have been studied by atomic force microscopy, where RepD appears as a globular particle which bends the DNA fragment around 90° in 39% of the *oriD* fragments analyzed. However, when both proteins were bound to the DNA, the proportion of bent DNA increased up to 60%, while ADPNP decreased this percentage to 41% (173).

PcrA helicase has also been identified and characterized in other Gram-positive microorganisms such as *S. pneumoniae*, *Bacillus anthracis*, or *Bacillus cereus*. In all these strains, the role of PcrA in RCR has been studied using the broad host range pT181 plasmid as a reference (31, 174). A clear interaction of the helicase with the replication initiation protein was successfully observed by pull-down for all three PcrAs. However, they displayed different unwinding activity on the plasmid DNA in the presence of RepC: while PcrA from *B. anthracis* and *B. cereus* fully unwound the DNA, PcrA from *S. pneumoniae* failed to do so and produced only partial unwinding. This may indicate that the interaction between RepC from *S. aureus* and PcrA from *S. pneumoniae* is not sufficiently stable and, hence, the helicase is unable to continue DNA unwinding. In fact, the authors reported that it was not possible to maintain the pT181 plasmid in the *S. pneumoniae* strain (31).

In Gram-negative bacteria, a role of UvrD in plasmid RCR has been postulated, although its role has been much less studied than that of PcrA in Gram-positive bacteria. It has been shown that deletion of *uvrD* in *E. coli* results in the accumulation of nicked pC194 plasmid DNA in the cell and a lack of ssDNA intermediates of plasmid replication (175). Nevertheless, a direct interaction between the replication initiator protein and UvrD or its effect on the helicase activity has yet to be studied.

In summary, the effect of the replication initiator protein on PcrA ATPase/helicase activity has been thoroughly studied at the biochemical and molecular biology level. However, there are still some unanswered questions regarding how the replication machinery is loaded onto the replication initiation origin. For example, it has been reported that PcrA interacts with RNAP through its disordered but highly conserved C-terminal region (176), but it is not known which domains of PcrA and the replication initiator protein are implicated in the interaction between both proteins. In addition, the stability of the helicase-Rep complex is not known, nor is it known whether there are cycles of loading/unloading of the helicase during the entire round of plasmid replication. Furthermore, unlike pT181, plasmids belonging to the pC194 or pMV158 families are able to replicate in *pcrA3* bacterial mutants (177; Ruiz-Masó et al., unpublished results). This raises the question of whether the helicase activity of PcrA is responsible for the replication of these plasmids or if another helicase can cope with this task. Finally, despite a great deal of work on the helicase activity of UvrD alone (160, 163, 164, 178, 179), as previously indicated, little is known about its role in RCR and in the putative interaction with the replication initiator protein.

CONCLUDING REMARKS

In this article we have tried to compile several aspects of the biology of the small RCR plasmids that constitute, *per se*, an extremely interesting family of replicons. Indeed, they link the primitive forms of self-replicating molecules harboring just the information needed for their replication in the host to more sophisticated beings that are interconnected to the viral world. Although much is known, as we have reflected here, about the replicative mechanism and control of RCR plasmids, still a full horizon expands before we can truly assert that we do know these molecules. We envisage several approaches to the understanding of the biology of these molecules: (i) mechanistic studies; (ii) solution of three-dimensional structures of Rep proteins; (iii) characterization of regulatory nucleoprotein complexes; (iv) solution of ternary complexes such as DNA-Rep-PcrA or DNA-Rep-DNA polymerase; (v) involvement of host-encoded factors, other than the obvious roles of DNA- and RNA-polymerase, DNA helicases, and ssDNA binding proteins (with an assumed but not demonstrated participation in plasmid replication); and (vi) comprehension of the mechanisms of adaptation of plasmids to a new host. We could speculate that we have just started uncovering the biology of RCR plasmids. Their role in bacterial adaptation to a changing world manipulated by humans and their contribution to the fitness of bacteria to their niches and to the biodiversity in the microbial world is totally unknown. Let's not allow the funding agencies to say that there is no new relevant information in the plasmid world: it is fully untrue.

Acknowledgments. *We are very grateful to Luis Blanco, Juan Alonso, and the members of the REDEEX consortium for fruitful discussions and help.Funding by the Spanish Ministry of Economy and Competitiveness (grants CSD2008/00013-INTERMODS to M.E.; BFU2011-22588 to M.C.; Ramón and Cajal subprogramme RYC-2011-09071 to C.M.; BFU2010-19597 to G.S.; Complementary Action BFU2008-00179-E-REDEEX to G.S.) are acknowledged. Conflicts of interest: We disclose no conflicts.*

Citation. Ruiz-Masó JA, Machón C, Bordanaba-Ruiseco L, Espinosa M, Coll M, del Solar G. 2015. Plasmid rolling-circle replication. Microbiol Spectrum 3(1):PLAS-0035-2014.

References

1. **del Solar G, Giraldo R, Ruiz-Echevarría MJ, Espinosa M, Díaz-Orejas R.** 1998. Replication and control of circular bacterial plasmids. *Microbiol Mol Biol Rev* **62:** 434–464.
2. **Khan SA.** 2005. Plasmid rolling-circle replication: highlights of two decades of research. *Plasmid* **53:**126–136.
3. **Chandler M, de la Cruz F, Dyda F, Hickman AB, Moncalian G, Ton-Hoang B.** 2013. Breaking and joining single-stranded DNA: the HUH endonuclease superfamily. *Nat Rev Microbiol* **11:**525–538.
4. **Dressler D.** 1970. The rolling circle for phiX174 DNA replication. II. Synthesis of single-stranded circles. *Proc Natl Acad Sci USA* **67:**1934–1942.
5. **Dressler D, Wolfson J.** 1970. The rolling circle for phiX174 DNA replication. 3. Synthesis of supercoiled duplex rings. *Proc Natl Acad Sci USA* **67:**456–463.
6. **Gilbert W, Dressler D.** 1968. DNA replication: the rolling circle model. *Cold Spring Harb Symp Quant Biol* **33:**473–484.
7. **Eisenberg S, Kornberg A.** 1979. Purification and characterization of phiX174 gene A protein. A multifunctional enzyme of duplex DNA replication. *J Biol Chem* **254:** 5328–5332.
8. **Henry TJ, Knippers R.** 1974. Isolation and function of the gene A initiator of bacteriophage phiX174, a highly specific DNA endonuclease. *Proc Natl Acad Sci USA* **71:**1549–1553.
9. **Ikeda JE, Yudelevich A, Hurwitz J.** 1976. Isolation and characterization of the protein coded by *gene A* of bacteriophage phiX174 DNA. *Proc Natl Acad Sci USA* **73:** 2669–2673.
10. **van Mansfeld AD, van Teeffelen HA, Baas PD, Veeneman GH, van Boom JH, Jansz HS.** 1984. The bond in the bacteriophage phiX174 gene A protein–DNA complex is a tyrosyl-5′-phosphate ester. *FEBS Lett* **173:**351–356.
11. **Koepsel RR, Murray RW, Rosenblum WD, Khan SA.** 1985. The replication initiator protein of plasmid pT181 has sequence-specific endonuclease and topoisomerase-like activities. *Proc Natl Acad Sci USA* **82:**6845–6849.
12. **del Solar GH, Puyet A, Espinosa M.** 1987. Initiation signals for the conversion of single stranded to double stranded DNA forms in the streptococcal plasmid pLS1. *Nucleic Acids Res* **15:**5561–5580.

13. **Deng ZX, Kieser T, Hopwood DA.** 1988. "Strong incompatibility" between derivatives of the *Streptomyces* multi-copy plasmid pIJ101. *Mol Gen Genet* **214:**286–294.
14. **te Riele H, Michel B, Ehrlich SD.** 1986. Single-stranded plasmid DNA in *Bacillus subtilis* and *Staphylococcus aureus*. *Proc Natl Acad Sci USA* **83:**2541–2545.
15. **Backert S, Meissner K, Borner T.** 1997. Unique features of the mitochondrial rolling circle-plasmid mp1 from the higher plant *Chenopodium album* (L.). *Nucleic Acids Res* **25:**582–589.
16. **Yu JS, Noll KM.** 1997. Plasmid pRQ7 from the hyperthermophilic bacterium *Thermotoga* species strain RQ7 replicates by the rolling-circle mechanism. *J Bacteriol* **179:**7161–7164.
17. **Akimkina T, Ivanov P, Kostrov S, Sokolova T, Bonch-Osmolovskaya E, Firman K, Dutta CF, McClellan JA.** 1999. A highly conserved plasmid from the extreme thermophile *Thermotoga maritima* MC24 is a member of a family of plasmids distributed worldwide. *Plasmid* **42:**236–240.
18. **Nesbo CL, Dlutek M, Doolittle WF.** 2006. Recombination in *Thermotoga*: implications for species concepts and biogeography. *Genetics* **172:**759–769.
19. **Nesvera J, Hochmannova J, Patek M.** 1998. An integron of class 1 is present on the plasmid pCG4 from Gram--positive bacterium *Corynebacterium glutamicum*. *FEMS Microbiol Lett* **169:**391–395.
20. **Smith MC, Thomas CD.** 2004. An accessory protein is required for relaxosome formation by small staphylococcal plasmids. *J Bacteriol* **186:**3363–3373.
21. **Blanco M, Kadlec K, Gutierrez Martin CB, de la Fuente AJ, Schwarz S, Navas J.** 2007. Nucleotide sequence and transfer properties of two novel types of *Actinobacillus pleuropneumoniae* plasmids carrying the tetracycline resistance gene tet(H). *J Antimicrob Chemother* **60:**864–867.
22. **Cooper TF, Heinemann JA.** 2000. Postsegregational killing does not increase plasmid stability but acts to mediate the exclusion of competing plasmids. *Proc Natl Acad Sci USA* **97:**12643–12648.
23. **Cooper TF, Paixao T, Heinemann JA.** 2010. Within-host competition selects for plasmid-encoded toxin-antitoxin systems. *Proc Biol Sci* **277:**3149–3155.
24. **Hernández-Arriaga AM, Chan WT, Espinosa M, Díaz-Orejas R.** 2014. Conditional activation of toxin-antitoxin systems: postsegregational killing and beyond. *Microbiol Spec* **2.** doi:10.1128/microbiolspec.PLAS-0009-2013.
25. **del Solar G, Fernández-López C, Ruiz-Masó JA, Lorenzo-Díaz F, Espinosa M.** 2014. Rolling circle replicating plasmids. *In* Bell E (ed), *Molecular Life Sciences*. Springer Science+Business Media, New York. doi:10.1007/978-1-4614-6436-5_567-2.
26. **Petrova P, Miteva V, Ruiz-Masó JA, del Solar G.** 2003. Structural and functional analysis of pt38, a 2.9 kb plasmid of *Streptococcus thermophilus* yogurt strain. *Plasmid* **50:**176–189.
27. **Petrova PM, Gouliamova DE.** 2006. Rapid screening of plasmid-encoded small hsp-genes in *Streptococcus thermophilus*. *Curr Microbiol* **53:**422–427.
28. **Andrup L, Jensen GB, Wilcks A, Smidt L, Hoflack L, Mahillon J.** 2003. The patchwork nature of rolling-circle plasmids: comparison of six plasmids from two distinct *Bacillus thuringiensis* serotypes. *Plasmid* **49:**205–232.
29. **del Solar G, Alonso JC, Espinosa M, Díaz-Orejas R.** 1996. Broad-host-range plasmid replication: an open question. *Mol Microbiol* **21:**661–666.
30. **Iordanescu S.** 1995. Plasmid pT181 replication is decreased at high levels of RepC per plasmid copy. *Mol Microbiol* **16:**477–484.
31. **Ruiz-Masó JA, Anand SP, Espinosa M, Khan SA, del Solar G.** 2006. Genetic and biochemical characterization of the *Streptococcus pneumoniae* PcrA helicase and its role in plasmid rolling circle replication. *J Bacteriol* **188:**7416–7425.
32. **del Solar G, Moscoso M, Espinosa M.** 1993. Rolling circle-replicating plasmids from Gram-positive and -negative bacteria: a wall falls. *Mol Micobiol* **8:**789–796.
33. **Biswas I, Jha JK, Fromm N.** 2008. Shuttle expression plasmids for genetic studies in *Streptococcus mutans*. *Microbiology* **154:**2275–2282.
34. **Bron S, Meijer W, Holsappel S, Haima P.** 1991. Plasmid instability and molecular cloning in *Bacillus subtilis*. *Res Microbiol* **142:**875–883.
35. **del Solar G, Moscoso M, Espinosa M.** 1993. *In vivo* definition of the functional origin of replication (ori(+)) of the promiscuous plasmid pLS1. *Mol Gen Genet* **237:** 65–72.
36. **O'Sullivan TF, Fitzgerald GF.** 1999. Electrotransformation of industrial strains of *Streptococcus thermophilus*. *J Appl Microbiol* **86:**275–283.
37. **Turgeon N, Laflamme C, Ho J, Duchaine C.** 2006. Elaboration of an electroporation protocol for *Bacillus cereus* ATCC 14579. *J Microbiol Methods* **67:**543–548.
38. **Gruss A, Ehrlich SD.** 1988. Insertion of foreign DNA into plasmids from Gram-positive bacteria induces formation of high-molecular-weight plasmid multimers. *J Bacteriol* **170:**1183–1190.
39. **Kiewiet R, Kok J, Seegers JF, Venema G, Bron S.** 1993. The mode of replication is a major factor in segregational plasmid instability in *Lactococcus lactis*. *Appl Environ Microbiol* **59:**358–364.
40. **Leonhardt H, Alonso JC.** 1991. Parameters affecting plasmid stability in *Bacillus subtilis*. *Gene* **103:**107–111.
41. **Viret JF, Alonso JC.** 1987. Generation of linear multigenome-length plasmid molecules in *Bacillus subtilis*. *Nucleic Acids Res* **15:**6349–6367.
42. **Alonso JC, Trautner TA.** 1985. Generation of deletions through a cis-acting mutation in plasmid pC194. *Mol Gen Genet* **198:**432–436.
43. **Fernández-López C, Bravo A, Ruiz-Cruz S, Solano-Collado V, Garsin DA, Lorenzo-Díaz F, Espinosa M.** 2014. Mobilizable rolling-circle replicating plasmids from Gram-positive bacteria: a low-cost conjugative transfer. *Microbiol Spec* **2.** doi:10.1128/microbiolspec.PLAS-0008-2013.
44. **Ruiz-Cruz S, Solano-Collado V, Espinosa M, Bravo A.** 2010. Novel plasmid-based genetic tools for the study of promoters and terminators in *Streptococcus pneumoniae* and *Enterococcus faecalis*. *J Microbiol Methods* **83:**156–163.

45. **Ruiz-Masó JA, López-Aguilar C, Nieto C, Sanz M, Burón P, Espinosa M, del Solar G.** 2012. Construction of a plasmid vector based on the pMV158 replicon for cloning and inducible gene expression in *Streptococcus pneumoniae*. *Plasmid* **67:**53–59.
46. **Patek M, Nesvera J.** 2013. Promoters and plasmid vectors of *Corynebacterium glutamicum*. *In* Yukawa H, Inui M (ed), *Corynebacterium glutamicum*. Microbiology Monographs 23. Springer Verlag, Berlin.
47. **Nesvera J, Patek M, Hochmannova J, Abrhamova Z, Becvarova V, Jelinkova M, Vohradsky J.** 1997. Plasmid pGA1 from *Corynebacterium glutamicum* codes for a gene product that positively influences plasmid copy number. *J Bacteriol* **179:**1525–1532.
48. **Hernández-Arriaga AM, Espinosa M, del Solar G.** 2012. Fitness of the pMV158 replicon in *Streptococcus pneumoniae*. *Plasmid* **67:**162–166.
49. **Espinosa M, del Solar G, Rojo F, Alonso JC.** 1995. Plasmid rolling circle replication and its control. *FEMS Microbiol Lett* **130:**111–120.
50. **Khan SA.** 1997. Rolling-circle replication of bacterial plasmids. *Microbiol Mol Biol Rev* **61:**442–455.
51. **Lo Piano A, Martínez-Jiménez MI, Zecchi L, Ayora S.** 2011. Recombination-dependent concatemeric viral DNA replication. *Virus Res* **160:**1–14.
52. **Thomas CD, Balson DF, Shaw WV.** 1990. *In vitro* studies of the initiation of staphylococcal plasmid replication. Specificity of RepD for its origin (oriD) and characterization of the Rep-ori tyrosyl ester intermediate. *J Biol Chem* **265:**5519–5530.
53. **Noirot-Gros MF, Bidnenko V, Ehrlich SD.** 1994. Active site of the replication protein of the rolling circle plasmid pC194. *EMBO J* **13:**4412–4420.
54. **Novick RP.** 1989. Staphylococcal plasmids and their replication. *Annu Rev Microbiol* **43:**537–565.
55. **Tauch A, Puhler A, Kalinowski J, Thierbach G.** 2003. Plasmids in *Corynebacterium glutamicum* and their molecular classification by comparative genomics. *J Biotechnol* **104:**27–40.
56. **Zhou L, Zhou M, Sun C, Han J, Lu Q, Zhou J, Xiang H.** 2008. Precise determination, cross-recognition, and functional analysis of the double-strand origins of the rolling-circle replication plasmids in haloarchaea. *J Bacteriol* **190:**5710–5719.
57. **Erauso G, Marsin S, Benbouzid-Rollet N, Baucher MF, Barbeyron T, Zivanovic Y, Prieur D, Forterre P.** 1996. Sequence of plasmid pGT5 from the archaeon *Pyrococcus abyssi*: evidence for rolling-circle replication in a hyperthermophile. *J Bacteriol* **178:**3232–3237.
58. **Zock JM, Birch P, Khan SA.** 1990. Specificity of RepC protein in plasmid pT181 DNA replication. *J Biol Chem* **265:**3484–3488.
59. **Iordanescu S.** 1989. Specificity of the interactions between the Rep proteins and the origins of replication of *Staphylococcus aureus* plasmids pT181 and pC221. *Mol Gen Genet* **217:**481–487.
60. **Kim SW, Jeong EJ, Kang HS, Tak JI, Bang WY, Heo JB, Jeong JY, Yoon GM, Kang HY, Bahk JD.** 2006. Role of RepB in the replication of plasmid pJB01 isolated from *Enterococcus faecium* JC1. *Plasmid* **55:**99–113.
61. **de la Campa AG, del Solar GH, Espinosa M.** 1990. Initiation of replication of plasmid pLS1. The initiator protein RepB acts on two distant DNA regions. *J Mol Biol* **213:**247–262.
62. **Moscoso M, del Solar G, Espinosa M.** 1995. Specific nicking-closing activity of the initiator of replication protein RepB of plasmid pMV158 on supercoiled or single-stranded DNA. *J Biol Chem* **270:**3772–3779.
63. **Moscoso M, Eritja R, Espinosa M.** 1997. Initiation of replication of plasmid pMV158: mechanisms of DNA strand-transfer reactions mediated by the initiator RepB protein. *J Mol Biol* **268:**840.
64. **Ruiz-Masó JA, Lurz R, Espinosa M, del Solar G.** 2007. Interactions between the RepB initiator protein of plasmid pMV158 and two distant DNA regions within the origin of replication. *Nucleic Acids Res* **35:**1230–1244.
65. **Wang PZ, Projan SJ, Henriquez V, Novick RP.** 1993. Origin recognition specificity in pT181 plasmids is determined by a functionally asymmetric palindromic DNA element. *EMBO J* **12:**45–52.
66. **del Solar G, Díaz R, Espinosa M.** 1987. Replication of the streptococcal plasmid pMV158 and derivatives in cell-free extracts of *Escherichia coli*. *Mol Gen Genet* **206:**428–435.
67. **Moscoso M, del Solar G, Espinosa M.** 1995. *In vitro* recognition of the replication origin of pLS1 and of plasmids of the pLS1 family by the RepB initiator protein. *J Bacteriol* **177:**7041–7049.
68. **Noirot P, Bargonetti J, Novick RP.** 1990. Initiation of rolling-circle replication in pT181 plasmid: initiator protein enhances cruciform extrusion at the origin. *Proc Natl Acad Sci USA* **87:**8560–8564.
69. **Puyet A, del Solar G, Espinosa M.** 1988. Identification of the origin and direction of replication of the broad-host-range plasmid pLS1. *Nucl Acids Res* **16:**115–133.
70. **Michel B, Ehrlich SD.** 1986. Illegitimate recombination occurs between the replication origin of the plasmid pC194 and a progressing replication fork. *EMBO J* **5:**3691–3696.
71. **Gros MF, te Riele H, Ehrlich SD.** 1987. Rolling circle replication of single-stranded DNA plasmid pC194. *EMBO J* **6:**3863–3869.
72. **Alonso JC, Leonhardt H, Stiege CA.** 1988. Functional analysis of the leading strand replication origin of plasmid pUB110 in *Bacillus subtilis*. *Nucleic Acids Res* **16:**9127–9145.
73. **Servin-Gonzalez L.** 1993. Relationship between the replication functions of *Streptomyces plasmids* pJV1 and pIJ101. *Plasmid* **30:**131–140.
74. **Kataoka M, Kiyose YM, Michisuji Y, Horiguchi T, Seki T, Yoshida T.** 1994. Complete nucleotide sequence of the *Streptomyces nigrifaciens* plasmid, pSN22: genetic organization and correlation with genetic properties. *Plasmid* **32:**55–69.
75. **Suzuki I, Seki T, Yoshida T.** 1997. Nucleotide sequence of a nicking site of the *Streptomyces* plasmid pSN22

replicating by the rolling circle mechanism. *FEMS Microbiol Lett* **150:**283–288.

76. **Abrhamova Z, Patek M, Nesvera J.** 2002. Atypical location of double-strand origin of replication (nic site) on the plasmid pGA1 from *Corynebacterium glutamicum*. *Folia Microbiol (Praha)* **47:**307–310.
77. **Soler N, Justome A, Quevillon-Cheruel S, Lorieux F, Le Cam E, Marguet E, Forterre P.** 2007. The rolling-circle plasmid pTN1 from the hyperthermophilic archaeon *Thermococcus nautilus*. *Mol Microbiol* **66:**357–370.
78. **Ward DE, Revet IM, Nandakumar R, Tuttle JH, de Vos WM, van der Oost J, DiRuggiero J.** 2002. Characterization of plasmid pRT1 from *Pyrococcus* sp. strain JT1. *J Bacteriol* **184:**2561–2566.
79. **Ilyina TV, Koonin EV.** 1992. Conserved sequence motifs in the initiator proteins for rolling circle DNA replication encoded by diverse replicons from eubacteria, eucaryotes and archaebacteria. *Nucleic Acids Res* **20:** 3279–3285.
80. **Boer R, Russi S, Guasch A, Lucas M, Blanco AG, Pérez-Luque R, Coll M, de la Cruz F.** 2006. Unveiling the molecular mechanism of a conjugative relaxase: the structure of TrwC complexed with a 27-mer DNA comprising the recognition hairpin and the cleavage site. *J Mol Biol* **358:**857.
81. **Hickman AB, Ronning DR, Kotin RM, Dyda F.** 2002. Structural unity among viral origin binding proteins: crystal structure of the nuclease domain of adeno-associated virus Rep. *Mol Cell* **10:**327.
82. **Asano S, Higashitani A, Horiuchi K.** 1999. Filamentous phage replication initiator protein gpII forms a covalent complex with the 5′ end of the nick it introduced. *Nucl Acids Res* **27:**1882–1889.
83. **Marsin S, Forterre P.** 1998. A rolling circle replication initiator protein with a nucleotidyl-transferase activity encoded by the plasmid pGT5 from the hyperthermophilic archaeon *Pyrococcus abyssi*. *Mol Microbiol* **27:** 1183–1192.
84. **Marsin S, Forterre P.** 1999. The active site of the rolling circle replication protein Rep75 is involved in site-specific nuclease, ligase and nucleotidyl transferase activities. *Mol Microbiol* **33:**537–545.
85. **Dempsey LA, Birch P, Khan SA.** 1992. Six amino acids determine the sequence-specific DNA binding and replication specificity of the initiator proteins of the pT181 family. *J Biol Chem* **267:**24538–24543.
86. **Dempsey LA, Birch P, Khan SA.** 1992. Uncoupling of the DNA topoisomerase and replication activities of an initiator protein. *Proc Natl Acad Sci USA* **89:** 3083–3087.
87. **Thomas CD, Nikiforov TT, Connolly BA, Shaw WV.** 1995. Determination of sequence specificity between a plasmid replication initiator protein and the origin of replication. *J Mol Biol* **254:**381–391.
88. **Wang PZ, Projan SJ, Henriquez V, Novick RP.** 1992. Specificity of origin recognition by replication initiator protein in plasmids of the pT181 family is determined by a six amino acid residue element. *J Mol Biol* **223:**145–158.
89. **Chang T-L, Kramer MG, Ansari RA, Khan SA.** 2000. Role of individual monomers of a dimeric initiator protein in the initiation and termination of plasmid rolling circle replication. *J Biol Chem* **275:**13529–13534.
90. **Boer R, Ruiz-Masó JA, Gomez-Blanco JR, Blanco AG, Vives-Llàcer M, Chacón P, Usón I, Gomis-Rüth FX, Espinosa M, Llorca O, del Solar G, Coll M.** 2009. Plasmid replication initiator RepB forms a hexamer reminiscent of ring helicases and has mobile nuclease domains. *EMBO J* **28:**1666–1678.
91. **Ruiz-Masó JA, López-Zumel C, Menéndez M, Espinosa M, del Solar G.** 2004. Structural features of the initiator of replication protein RepB encoded by the promiscuous plasmid pMV158. *Biochim Biophys Acta* **1696:** 113–119.
92. **Rasooly A, Projan SJ, Novick RP.** 1994. Plasmids of the pT181 family show replication-specific initiator protein modification. *J Bacteriol* **176:**2450–2453.
93. **Zhao AC, Ansari RA, Schmidt MC, Khan SA.** 1998. An oligonucleotide inhibits oligomerization of a rolling circle initiator protein at the pT181 origin of replication. *J Biol Chem* **273:**16082–16089.
94. **Muller AK, Rojo F, Alonso JC.** 1995. The level of the pUB110 replication initiator protein is autoregulated, which provides an additional control for plasmid copy number. *Nucleic Acids Res* **23:**1894–1900.
95. **Enemark EJ, Joshua-Tor L.** 2006. Mechanism of DNA translocation in a replicative hexameric helicase. *Nature* **442:**270.
96. **Krupovic M, Ravantti JJ, Bamford DH.** 2009. Geminiviruses: a tale of a plasmid becoming a virus. *BMC Evol Biol* **9:**112.
97. **Oshima K, Kakizawa S, Nishigawa H, Kuboyama T, Miyata S, Ugaki M, Namba S.** 2001. A plasmid of phytoplasma encodes a unique replication protein having both plasmid- and virus-like domains: clue to viral ancestry or result of virus/plasmid recombination? *Virology* **285:**270–277.
98. **Tran-Nguyen LTT, Gibb KS.** 2006. Extrachromosomal DNA isolated from tomato big bud and *Candidatus Phytoplasma australiense* phytoplasma strains. *Plasmid* **56:**153–166.
99. **Brantl S.** 2014. Plasmid replication control by antisense RNAs. *Microbiol Spec* **2.** doi:10.1128/microbiolspec. PLAS-0001-2013.
100. **del Solar G, Espinosa M.** 2000. Plasmid copy number control: an ever-growing story. *Mol Microbiol* **37:**492–500.
101. **Novick RP, Iordanescu S, Projan SJ, Kornblum J, Edelman I.** 1989. pT181 plasmid replication is regulated by a countertranscript-driven transcriptional attenuator. *Cell* **59:**395–404.
102. **del Solar G, Acebo P, Espinosa M.** 1995. Replication control of plasmid pLS1: efficient regulation of plasmid copy number is exerted by the combined action of two plasmid components, CopG and RNA II. *Mol Microbiol* **18:**913–924.
103. **Hernández-Arriaga AM, Rubio-Lepe TS, Espinosa M, del Solar G.** 2009. Repressor CopG prevents access of

RNA polymerase to promoter and actively dissociates open complexes. *Nucleic Acids Res* **37:**4799–4811.

104. **del Solar G, Acebo P, Espinosa M.** 1997. Replication control of plasmid pLS1: the antisense RNA II and the compact rnaII region are involved in translational regulation of the initiator RepB synthesis. *Mol Microbiol* **23:**95–108.

105. **del Solar G, Espinosa M.** 2001. *In vitro* analysis of the terminator T(II) of the inhibitor antisense *rna II* gene from plasmid pMV158. *Plasmid* **45:**75–87.

106. **López-Aguilar C, del Solar G.** 2013. Probing the sequence and structure of *in vitro* synthesized antisense and target RNAs from the replication control system of plasmid pMV158. *Plasmid* **70:**94–103.

107. **del Solar G, Espinosa M.** 1992. The copy number of plasmid pLS1 is regulated by two trans-acting plasmid products: the antisense RNA II and the repressor protein, RepA. *Mol Microbiol* **6:**83–94.

108. **Brantl S, Wagner EG.** 2000. Antisense RNA-mediated transcriptional attenuation: an *in vitro* study of plasmid pT181. *Mol Microbiol* **35:**1469–1482.

109. **Franch T, Petersen M, Wagner EGH, Jacobsen JP, Gerdes K.** 1999. Antisense RNA regulation in prokaryotes: rapid RNA/RNA interaction facilitated by a general U-turn loop structure. *J Mol Biol* **294:**1115–1125.

110. **Kim SW, Jeong IS, Jeong EJ, Tak JI, Lee JH, Eo SK, Kang HY, Bahk JD.** 2008. The terminal and internal hairpin loops of the ctRNA of plasmid pJB01 play critical roles in regulating copy number. *Mol Cells* **26:**26–33.

111. **López-Aguilar C, Ruiz-Masó JA, Rubio-Lepe TS, Sanz M, del Solar G.** 2013. Translation initiation of the replication initiator *repB* gene of promiscuous plasmid pMV158 is led by an extended non-SD sequence. *Plasmid* **70:**69–77.

112. **Venkova-Canova T, Patek M, Nesvera J.** 2003. Control of *rep* gene expression in plasmid pGA1 from *Corynebacterium glutamicum*. *J Bacteriol* **185:**2402–2409.

113. **Venkova T, Patek M, Nesvera J.** 2001. Identification of a novel gene involved in stable maintenance of plasmid pGA1 from *Corynebacterium glutamicum*. *Plasmid* **46:** 153–162.

114. **Maciag IE, Viret JF, Alonso JC.** 1988. Replication and incompatibility properties of plasmid pUB110 in *Bacillus subtilis*. *Mol Gen Genet* **212:**232–240.

115. **Rasooly A, Novick RP.** 1993. Replication-specific inactivation of the pT181 plasmid initiator protein. *Science* **262:**1048–1050.

116. **Rasooly A, Wang P, Novick R.** 1994. Replication-specific conversion of the *Staphylococcus aureus* pT181 initiator protein from an active homodimer to an inactive heterodimer. *EMBO J* **13:**5245–5251.

117. **Novick RP.** 1998. Contrasting lifestyles of rolling-circle phages and plasmids. *Trends Biochem Sci* **23:**434–438.

118. **Jin R, Zhou X, Novick RP.** 1996. The inactive pT181 initiator heterodimer, RepC/C, binds but fails to induce melting of the plasmid replication origin. *J Biol Chem* **271:**31086–31091.

119. **Noirot-Gros MF, Ehrlich SD.** 1996. Change of a catalytic reaction carried out by a DNA replication protein. *Science* **274:**777–780.

120. **Gruss AD, Ross HF, Novick RP.** 1987. Functional analysis of a palindromic sequence required for normal replication of several staphylococcal plasmids. *Proc Natl Acad Sci USA* **84:**2165–2169.

121. **Kramer MG, del Solar G, Espinosa M.** 1995. Lagging-strand origins of the promiscuous plasmid pMV158: physical and functional characterization. *Microbiology* **141:**655–662.

122. **Kramer MG, Khan SA, Espinosa M.** 1997. Plasmid rolling circle replication: identification of the RNA polymerase-directed primer RNA and requirement for DNA polymerase I for lagging strand synthesis. *EMBO J* **16:**5784–5795.

123. **Kramer MG, Khan SA, Espinosa M.** 1998. Lagging-strand replication from the *ssoA* origin of plasmid pMV158 in *Streptococcus pneumoniae*: *in vivo* and *in vitro* influences of mutations in two conserved *ssoA* regions. *J Bacteriol* **180:**83–89.

124. **Farrar MD, Howson KM, Emmott JE, Bojar RA, Holland KT.** 2007. Characterisation of cryptic plasmid pPG01 from *Propionibacterium granulosum*, the first plasmid to be isolated from a member of the cutaneous propionibacteria. *Plasmid* **58:**68–75.

125. **Fernández-Gonzalez C, Cadenas RF, Noirot-Gros MF, Martin JF, Gil JA.** 1994. Characterization of a region of plasmid pBL1 of *Brevibacterium lactofermentum* involved in replication via the rolling circle model. *J Bacteriol* **176:**3154–3161.

126. **Nakashima N, Tamura T.** 2004. Isolation and characterization of a rolling-circle-type plasmid from *Rhodococcus erythropolis* and application of the plasmid to multiple-recombinant-protein expression. *Appl Environ Microbiol* **70:**5557–5568.

127. **Shibayama Y, Dabbs ER, Yazawa K, Mikami Y.** 2011. Functional analysis of a small cryptic plasmid pYS1 from *Nocardia*. *Plasmid* **66:**26–37.

128. **Zaman S, Radnedge L, Richards H, Ward JM.** 1993. Analysis of the site for second-strand initiation during replication of the *Streptomyces* plasmid pIJ101. *J Gen Microbiol* **139:**669–676.

129. **Kramer MG, Espinosa M, Misra TK, Khan SA.** 1998. Lagging strand replication of rolling-circle plasmids: specific recognition of the *ssoA*-type origins in different Gram-positive bacteria. *Proc Natl Acad Sci USA* **95:** 10505–10510.

130. **Leenhouts KJ, Tolner B, Bron S, Kok J, Venema G, Seegers JFML.** 1991. Nucleotide sequence and characterization of the broad-host-range lactococcal plasmid pWVO1. *Plasmid* **26:**55–66.

131. **Seegers JF, Zhao AC, Meijer WJ, Khan SA, Venema G, Bron S.** 1995. Structural and functional analysis of the single-strand origin of replication from the lactococcal plasmid pWV01. *Mol Gen Genet* **249:**43–50.

132. **Baas PD, Jansz HS.** 1988. Single-stranded DNA phage origins. *Curr Top Microbiol Immunol* **136:**31–70.

133. **Chang S, Chang SY, Gray O.** 1987. Structural and genetic analyses of a par locus that regulates plasmid partition in *Bacillus subtilis*. *J Bacteriol* **169:**3952–3962.

134. **Mahillon J, Seurinck J.** 1988. Complete nucleotide sequenceof pGI2, a *Bacillus thuringiensis* plasmid containing Tn4430. *Nucleic Acids Res* **16:**11827–11828.

135. **Seery L, Devine KM.** 1993. Analysis of features contributing to activity of the single-stranded origin of *Bacillus plasmid* pBAA1. *J Bacteriol* **175:**1988–1994.

136. **Meijer WJJ, de Boer AJ, van Tongeren S, Venema G, Bron S.** 1995. Characterization of the replication region of the *Bacillus subtilis* plasmid pLS20: a novel type of replicon. *Nucleic Acids Res* **23:**3214–3223.

137. **Madsen SM, Andrup L, Boe L.** 1993. Fine mapping and DNA sequence of replication functions of *Bacillus thuringiensis* plasmid pTX14-3. *Plasmid* **30:**119–130.

138. **Boe L, Nielsen TT, Madsen SM, Andrup L, Bolander G.** 1991. Cloning and characterization of two plasmids from *Bacillus thuringiensis* in *Bacillus subtilis*. *Plasmid* **25:**190–197.

139. **Boe L, Gros MF, te Riele H, Ehrlich SD, Gruss A.** 1989. Replication origins of single-stranded-DNA plasmid pUB110. *J Bacteriol* **171:**3366–3372.

140. **Lorenzo-Díaz F, Espinosa M.** 2009. Lagging-strand DNA replication origins are required for conjugal transfer of the promiscuous plasmid pMV158. *J Bacteriol* **191:** 720–727.

141. **Meijer WJ, van der Lelie D, Venema G, Bron S.** 1995. Effects of the generation of single-stranded DNA on the maintenance of plasmid pMV158 and derivatives in *Lactococcus lactis*. *Plasmid* **33:**91–99.

142. **Viret JF, Alonso JC.** 1988. A DNA sequence outside the pUB110 minimal replicon is required for normal replication in *Bacillus subtilis*. *Nucleic Acids Res* **16:** 4389–4406.

143. **van der Lelie D, Bron S, Venema G, Oskam L.** 1989. Similarity of minus origins of replication and flanking open reading frames of plasmids pUB110, pTB913 and pMV158. *Nucleic Acids Res* **17:**7283–7294.

144. **Kramer MG, Espinosa M, Misra TK, Khan SA.** 1999. Characterization of a single-strand origin, *ssoU*, required for broad host range replication of rolling-circle plasmids. *Mol Microbiol* **33:**466–475.

145. **Leer RJ, Luijk N, Posno M, Pouwels PH.** 1992. Structural and functional analysis of two cryptic plasmids from *Lactobacillus pentosus* MD353 and *Lactobacillus plantarum* ATCC 8014. *Mol Gen Genet* **234:**265–274.

146. **Suzuki I, Kataoka M, Yoshida T, Seki T.** 2004. Lagging strand replication of rolling-circle plasmids in *Streptomyces lividans*: an RNA polymerase-independent primer synthesis. *Arch Microbiol* **181:**305–313.

147. **Bron S, Luxen E.** 1985. Segregational instability of pUB110-derived recombinant plasmids in *Bacillus subtilis*. *Plasmid* **14:**235–244.

148. **del Solar G, Kramer G, Ballester S, Espinosa M.** 1993. Replication of the promiscuous plasmid pLS1: a region encompassing the minus origin of replication is associated with stable plasmid inheritance. *Mol Gen Genet* **241:**97–105.

149. **Hernández-Arriaga AM, Espinosa M, del Solar G.** 2000. A functional lagging strand origin does not stabilize plasmid pMV158 inheritance in *Escherichia coli*. *Plasmid* **43:**49–58.

150. **Gigliani F, Ciotta C, Del Grosso MF, Battaglia PA.** 1993. pR plasmid replication provides evidence that single-stranded DNA induces the SOS system *in vivo*. *Mol Gen Genet* **238:**333–338.

151. **Higashitani N, Higashitani A, Horiuchi K.** 1995. SOS induction in *Escherichia coli* by single-stranded DNA of mutant filamentous phage: monitoring by cleavage of LexA repressor. *J Bacteriol* **177:**3610–3612.

152. **Díaz A, Lacks SA, López P.** 1992. The 5′ to 3′ exonuclease activity of DNA polymerase I is essential for *Streptococcus pneumoniae*. *Mol Microbiol* **6:**3009–3019.

153. **López P, Martínez S, Díaz A, Espinosa M, Lacks SA.** 1989. Characterization of the *polA* gene of *Streptoco ccus pneumoniae* and comparison of the DNA polymerase I it encodes to homologous enzymes from *Escherichia coli* and phage T7. *J Biol Chem* **264:**4255–4263.

154. **Díaz A, Lacks SA, López P.** 1994. Multiple roles for DNA polymerase I in establishment and replication of the promiscuous plasmid pLS1. *Mol Micobiol* **14:**773–783.

155. **Majumder S, Novick RP.** 1988. Intermediates in plasmid pT181 DNA replication. *Nucleic Acids Res* **16:** 2897–2912.

156. **Gorbalenya AE, Koonin EV.** 1993. Helicases: amino acid sequence comparisons and structure-function relationships. *Curr Opin Struct Biol* **3:**419–429.

157. **Singleton MR, Dillingham MS, Wigley DB.** 2007. Structure and mechanism of helicases and nucleic acid translocases. *Annu Rev Biochem* **76:**23–50.

158. **Bird LE, Brannigan JA, Subramanya HS, Wigley DB.** 1998. Characterisation of *Bacillus stearothermophilus* PcrA helicase: evidence against an active rolling mechanism. *Nucleic Acids Res* **26:**2686–2693.

159. **Jia H, Korolev S, Niedziela-Majka A, Maluf NK, Gauss GH, Myong S, Ha T, Waksman G, Lohman TM.** 2011. Rotations of the 2B sub-domain of *E. coli* UvrD helicase/translocase coupled to nucleotide and DNA binding. *J Mol Biol* **411:**633–648.

160. **Lee JY, Yang W.** 2006. UvrD helicase unwinds DNA one base pair at a time by a two-part power stroke. *Cell* **127:**1349–1360.

161. **Velankar SS, Soultanas P, Dillingham MS, Subramanya HS, Wigley DB.** 1999. Crystal structures of complexes of PcrA DNA helicase with a DNA substrate indicate an inchworm mechanism. *Cell* **97:**75–84.

162. **Chisty LT, Toseland CP, Fili N, Mashanov GI, Dillingham MS, Molloy JE, Webb MR.** 2013. Monomeric PcrA helicase processively unwinds plasmid lengths of DNA in the presence of the initiator protein RepD. *Nucleic Acids Res* **41:**5010–5023.

163. **Fischer CJ, Maluf NK, Lohman TM.** 2004. Mechanism of ATP-dependent translocation of *E. coli* UvrD monomers along single-stranded DNA. *J Mol Biol* **344:**1287–1309.

164. **Ali JA, Maluf NK, Lohman TM.** 1999. An oligomeric form of *E. coli* UvrD is required for optimal helicase activity. *J Mol Biol* **293:**815–834.

165. **Yang Y, Dou SX, Ren H, Wang PY, Zhang XD, Qian M, Pan BY, Xi XG.** 2008. Evidence for a functional dimeric form of the PcrA helicase in DNA unwinding. *Nucleic Acids Res* **36:**1976–1989.

166. **Iordanescu S, Basheer R.** 1991. The *Staphylococcus aureus* mutation *pcrA3* leads to the accumulation of pT181 replication initiation complexes. *J Mol Biol* **221:** 1183–1189.

167. **Iordanescu S.** 1993. Plasmid pT181-linked suppressors of the *Staphylococcus aureus pcrA3* chromosomal mutation. *J Bacteriol* **175:**3916–3917.

168. **Anand SP, Chattopadhyay A, Khan SA.** 2005. The PcrA3 mutant binds DNA and interacts with the RepC initiator protein of plasmid pT181 but is defective in its DNA helicase and unwinding activities. *Plasmid* **54:**104–113.

169. **Chang TL, Naqvi A, Anand SP, Kramer MG, Munshi R, Khan SA.** 2002. Biochemical characterization of the *Staphylococcus aureus* PcrA helicase and its role in plasmid rolling circle replication. *J Biol Chem* **277:** 45880–45886.

170. **Anand SP, Khan SA.** 2004. Structure-specific DNA binding and bipolar helicase activities of PcrA. *Nucl Acids Res* **32:**3190–3197.

171. **Zhang W, Dillingham MS, Thomas CD, Allen S, Roberts CJ, Soultanas P.** 2007. Directional loading and stimulation of PcrA helicase by the replication initiator protein RepD. *J Mol Biol* **371:**336–348.

172. **Slatter AF, Thomas CD, Webb MR.** 2009. PcrA helicase tightly couples ATP hydrolysis to unwinding double-stranded DNA, modulated by the initiator protein for plasmid replication, RepD. *Biochemistry* **48:**6326–6334.

173. **Machón C, Lynch GP, Thomson NH, Scott DJ, Thomas CD, Soultanas P.** 2010. RepD-mediated recruitment of PcrA helicase at the *Staphylococcus aureus* pC221 plasmid replication origin, oriD. *Nucleic Acids Res* **38:**1874–1888.

174. **Anand SP, Mitra P, Naqvi A, Khan SA.** 2004. *Bacillus anthracis* and *Bacillus cereus* PcrA helicases can support DNA unwinding and *in vitro* rolling-circle replication of plasmid pT181 of *Staphylococcus aureus*. *J Bacteriol* **186:**2195–2199.

175. **Bruand C, Ehrlich SD.** 2000. UvrD-dependent replication of rolling-circle plasmids in *Escherichia coli*. *Mol Microbiol* **35:**204–210.

176. **Gwynn EJ, Smith AJ, Guy CP, Savery NJ, McGlynn P, Dillingham MS.** 2013. The conserved C-terminus of the PcrA/UvrD helicase interacts directly with RNA polymerase. *PLoS One* **8:**e78141. doi:10.1371/journal.pone.0078141.

177. **Petit MA, Dervyn E, Rose M, Entian KD, McGovern S, Ehrlich SD, Bruand C.** 1998. PcrA is an essential DNA helicase of *Bacillus subtilis* fulfilling functions both in repair and rolling-circle replication. *Mol Microbiol* **29:** 261–273.

178. **Maluf NK, Fischer CJ, Lohman TM.** 2003. A dimer of *Escherichia coli* UvrD is the active form of the helicase *in vitro*. *J Mol Biol* **325:**913–935.

179. **Tomko EJ, Fischer CJ, Lohman TM.** 2012. Single-stranded DNA translocation of *E. coli* UvrD monomer is tightly coupled to ATP hydrolysis. *J Mol Biol* **418:** 32–46.

Plasmids—Biology and Impact in Biotechnology and Discovery
Edited by Marcelo E. Tolmasky and Juan C. Alonso

doi:10.1128/microbiolspec.PLAS-0032-2014

Nikolai V. Ravin[1]

5 Replication and Maintenance of Linear Phage-Plasmid N15

THE FAMILY OF LINEAR N15-LIKE PHAGE-PLASMIDS

All cells with linear chromosomes must employ special mechanisms to replicate the extreme termini of their chromosomes, since DNA polymerases alone are unable to perform this function (1). Most eukaryotes have open-ended DNA and employ special "telomerase" enzymes for this purpose, but there are other solutions that ensure complete replication of linear DNA: protein priming, recombination, and covalently closed terminal hairpins (reviewed in reference 2). Prokaryotes usually posses circular plasmids and chromosomes, but examples of linear replicons are known. Bacteriophage N15 belongs to the small group of organisms known to replicate as linear DNA with covalently closed telomeres. Besides N15 and related phage-plasmids, only a few examples of such replicons from bacteria are known, including the linear plasmids and chromosomes common in the spirochete genus *Borrelia* (3–5) and one of the two chromosomes of *Agrobacterium tumefaciens* (6, 7). In this review I will summarize the most relevant work on N15 and related phages, with a special emphasis on the mechanism of replication, generation of hairpin telomeres, control of lysogeny, and plasmid prophage maintenance.

Phage N15 was isolated by Victor Ravin in 1964 (8). N15 belongs to the lambdoid phage family on the basis of cross-hybridization of their DNA (9) and is similar to phage λ with respect to the morphology of phage particles, latent period, burst size, and frequency of lysogenization (10). An unusual feature of phage N15 is that its prophage replicates extrachromosomally (9–11). Therefore, N15 could be considered a phage-plasmid, like phages P1 and P4 (12). Initially, it was supposed that the N15 prophage is a circular plasmid, like P1, but later Rybchin and colleagues showed (13, 14) that the N15 prophage is a linear plasmid with covalently closed ends (telomeres). It was the first example of linear DNA with covalently closed ends in prokaryotes. The mature N15 phage DNA has 12-bp, single-stranded cohesive ends, named *cosL* and *cosR*. The gene order in plasmid DNA is a circular permutation of that in the virion phage DNA, and the telomere-forming site *telRL* in phage DNA is a 56-bp inverted repeat (IR). The above data suggest that after infection of an *Escherichia coli* cell, the phage DNA becomes circularized via its cohesive termini. Then a special phage-encoded enzyme, protelomerase (prokaryotic telomerase), cuts the *telRL* sequence and joins the phosphodiester bonds, making covalently closed ends, *telL* and *telR* (12) (Fig. 1).

During the past two decades several other phages able to lysogenize their hosts as linear plasmids with covalently closed telomeres were discovered. All of them

[1]Center of Bioengineering, Russian Academy of Sciences, Prosp. 60-let Oktiabria, Bldg. 7-1, Moscow 117312, Russia.

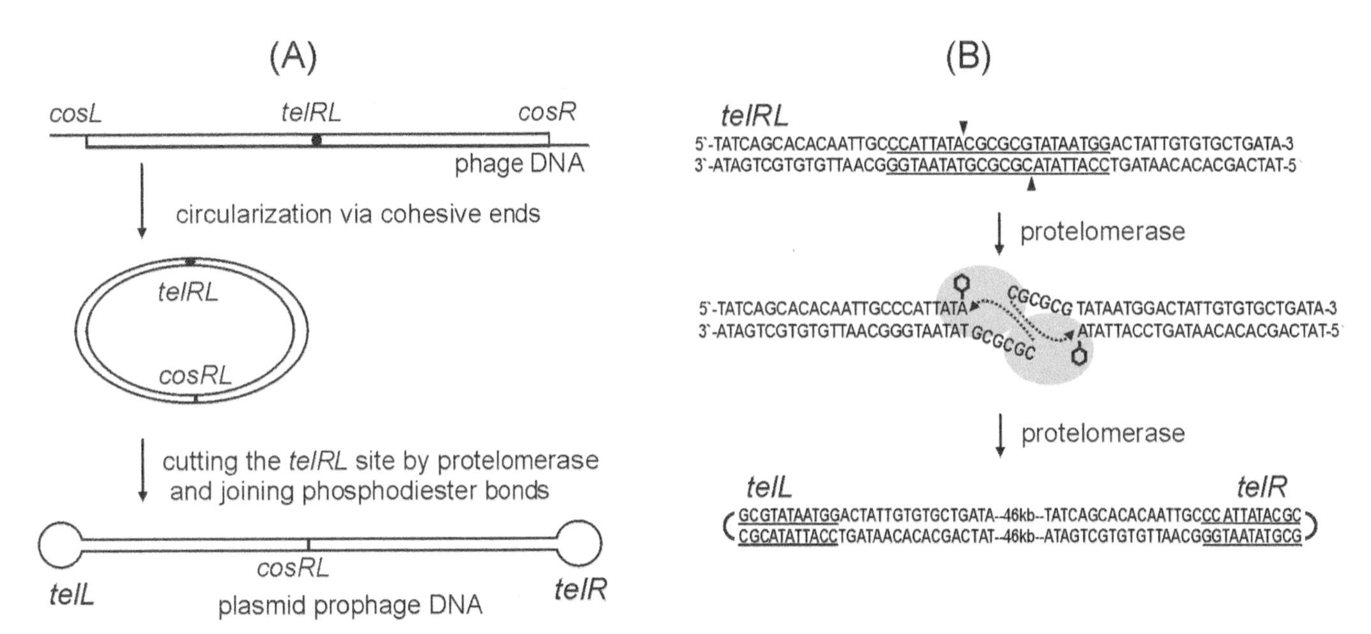

Figure 1 (A) Mechanism of conversion of phage DNA into linear plasmid. *cosL*, *cosR*, single stranded cohesive ends; *cosRL*, *cos* site after annealing and ligation of cohesive ends; *telRL*, uncut target site of protelomerase; *telL* and *telR*, left and right hairpin ends of the prophage created by protelomerase. (B) Hairpin formation reaction by the protelomerase. The positions of the cleavage sites are marked by a filled triangle. Catalytically active tyrosine is shown by a hexagon; the direction of refolding of single-stranded ends is shown by a dotted arrow. The protelomerase is shown by a gray oval.
doi:10.1128/microbiolspec.PLAS-0032-2014.f1

were identified in γ-proteobacteria, and they fall into two groups: the siphoviruses ϕKO2 of *Klebsiella oxytoca* (15) and PY54 of *Yersinia enterocolitica* (16), and the myoviruses of marine bacteria: phages VP882, Vp58.5, and vB_VpaM_MAR of *Vibrio parahaemolyticus* (17–19) and phage ΦHAP-1 of *Halomonas aquamarina* (20). All of them are related to N15 and carry a similar genetic determinants responsible for prophage replication and generation of covalently closed telomeres. Another marine phage, VHML of *Vibrio harveyi* (21), is closely related to VP58.5 (18) and contains an N15-like protelomerase. VHML was initially described as integrative (21), but later it was suggested that its prophage may be a linear plasmid as well (22).

ORGANIZATION OF THE GENOMES

About half of the 46,363-bp long genome of phage N15 (23) is similar to the bacteriophage λ genome sequence (Fig. 2). This is the right arm of the prophage genome (Fig. 2) that contains mostly the genes encoding the proteins required for virion head and tail assembly. The division between the left and right arms of the prophage genome is determined by the site (*cosRL*) formed upon joining the ends of mature phage DNA at infection. From N15 genes *1* through *21* there is a one to one correlation with the phage λ genes *A* through *J*. All of these genes are transcribed in a rightward direction. There is up to 90% amino acid sequence identity between the N15 and the λ head gene products. Some regions are more closely related to other lambdoid phages, and starting from gene *17* (the λ tail assembly gene *M* analog) to gene *25*, except for gene *24*, N15 better matches phages HK97 and HK022. The above observation that N15 encodes λ-like capsid proteins correlates well with the observation that N15 virion morphology is similar to that of lambdoid phages (10).

The N15 gene *24* is the homolog and functional analog of the *cor* gene of phage ϕ80 (24) and is responsible for the inability of N15 lysogenes to adsorb bacteriophages N15, T1, and ϕ80 (25). To the right of the head and tail gene cluster is gene *26*, a homolog of the *E. coli umuD* gene, which encodes a subunit of an error-prone DNA polymerase UmuC/UmuD that is involved in the repair of DNA damage during the SOS response. The promoter of this gene is overlapped by a potential LexA binding site, suggesting derepression of N15 *umuD* upon DNA damage. The next two genes in the

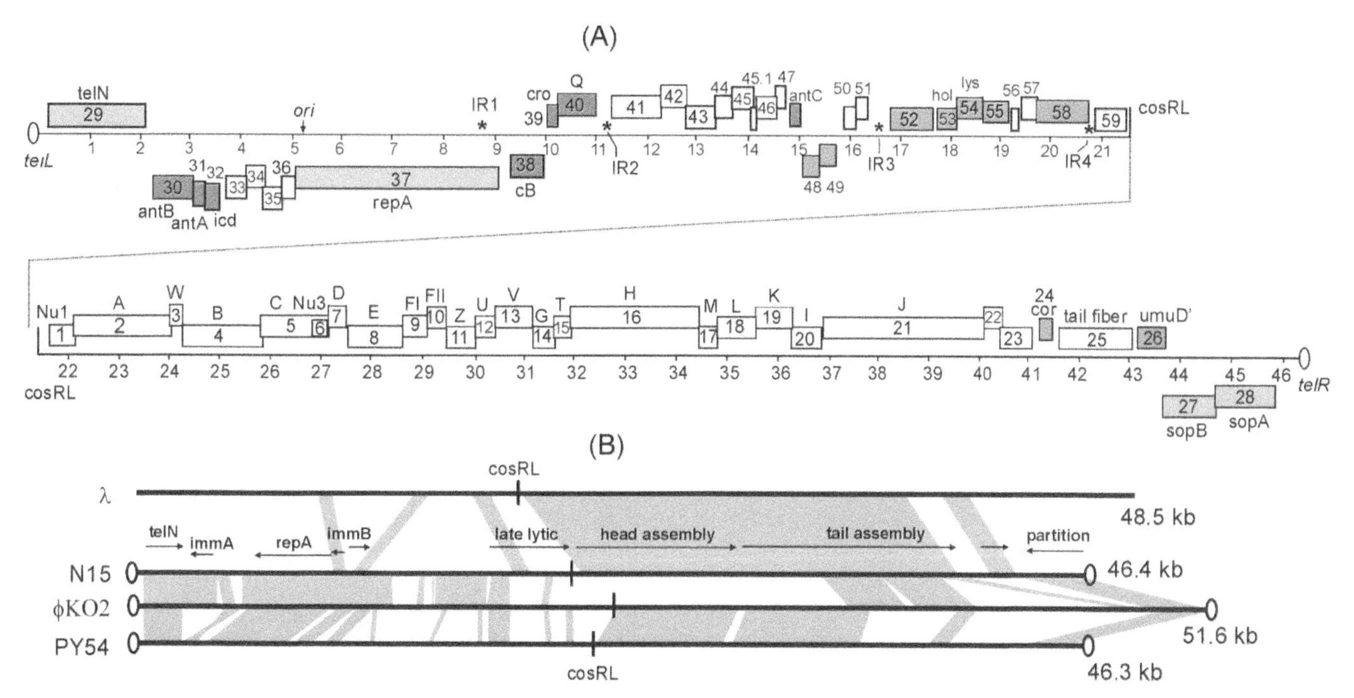

Figure 2 (A) Map of N15 plasmid prophage. The N15 linear prophage is shown with a scale in kilobase pairs. Rectangles immediately above and below the scale represent predicted genes that are transcribed rightward and leftward, respectively; their colors indicate functional assignments in the following way: genes encoding terminase and virion proteins (yellow), genes responsible for plasmid maintenance (green), genes responsible for the control of lysogeny (red), other genes with predicted functions (gray), and genes with unknown functions (white). The N15 gene names are given within or near the rectangles, and alternate descriptive names are indicated above or below. Asterisks (*) mark the position of the centromere sites involved in plasmid partition. *Ori*, replication initiation site. (B) Mosaic relationship between the whole genomes of phages λ, N15, ϕKO2, and PY54. Shaded areas between the genomes indicate the main regions of homology. The end of each phage's circularly permuted virion genome (cosRL) is marked by a black vertical line. doi:10.1128/microbiolspec.PLAS-0032-2014.f2

right half of the prophage are homologs of the *sopA* and *sopB* genes of the F plasmid and determine the segregation stability of the N15 prophage (see below).

Unlike the right-arm genes, only about one third of the prophage left-arm genes have homologs in lambdoid phages. Among them are the gene *38*, *39*, and *40* homologs to *cB*, *cro*, and *Q*, respectively; they are responsible for the control of lysogeny (see below). Genes *53*, *54*, *55*, and *55.1* are supposed to encode lysis function and also have homologs in the lambdoid phage family (23). Two operons located in the left arm are characteristics for N15-like viruses and reflect their linear plasmid lifestyle: the protelomerase gene (gene *29*) located at the left telomere (*telL*) and the counter-oriented replication operon (genes *33–37*) comprising gene *37* (*repA*), which encodes a multifunctional replication protein.

The nucleotide sequences of the genomes of N15-like phages ϕKO2 and pY54 have been determined (26, 27). Overall, the structures of the three genomes are similar and mosaically related (Fig. 2). The sequences of the virion protein genes of PY54 and ϕKO2 are very similar, and both are different from the set of N15 structural proteins. On the other hand, the regions of the ϕKO2 genome outside the late operon are largely similar to those of N15 rather than PY54 (Fig. 2). The prophage PY54 left arm contains N15-like protelomerase, replication, and lysogeny control regions; other N15-related open reading frames are not clustered but are scattered over the whole PY54 left arm, and there are a number of unrelated open reading frames lying in between. The similar overall genome organizations of phages N15, PY54, ϕKO2, and λ and relationships of their genes to each other and/or to genes of other lambdoid phages suggest that it is legitimate to include the former three within the lambdoid phage group, but in a subgroup that has a different strategy of lysogeny (26).

Analysis of genome sequences of N15-related marine viruses VHML, ΦHAP-1, VP882, Vp58.5, and vB_VpaM_MAR revealed that the overall genomic organization of the functional modules was similar across these phages and N15, although the sequence similarity is limited. The capsid proteins are similar to those from lambda-like siphoviruses, while tail proteins are similar to those from P2-like temperate myoviruses (20). Homologs of N15 protelomerase and replication protein RepA are present in these phages. As in phage N15, protelomerase and a counteroriented *repA*-like gene are located between the structural gene cluster and λ-type lysogeny control region with the *cI*-like repressor gene.

CONTROL OF LYSOGENY AND PLASMID REPLICATION

The plasmid nature of the N15 prophage requires controlled expression of not only the repressor function but also the genes responsible for prophage replication and maintenance. Analysis of N15 transcription patterns showed that about a half of the N15 genes are transcribed in the lysogen (23). This situation differs from that of phage λ and suggests the possibility of more complex regulatory mechanisms.

Three distinct loci are involved in the control of lysogeny (Fig. 3). The primary immunity region, *immB*, is structurally and functionally similar to the lambdoid phages' immunity regions (28). This locus, located between the divergent early left and early right operons, contains three genes (Fig. 3). Gene *38* (*cB*) encodes the protein homologous to the λ CI repressor. Clear plaque mutants, mapping at *immB*, were found in the *cB* gene, supporting its role as a primary repressor. Products of genes *39* and *40* are homologues to λ Cro repressor and Q transcription antiterminator. The *cB* gene is flanked by a complex array of divergent operator-promoter sites. The two operators to the left of the *cB* overlap the predicted promoter of the N15 *repA* gene, suggesting that binding of CB at these operators represses and regulates transcription of *repA* and thus plasmid replication (28). This assumption is supported by an observation that the N15-based miniplasmids lacking the *cB* gene have a higher copy number than similar plasmids with an intact *cB* (29). The three operators to the right of *cB* overlap the predicted promoter of the *cB* itself and promoters of the rightward operon containing *cro* and *Q*. It has been proposed (28) that CB, by binding to these operators, represses both its own transcription and transcription of *cro* and *Q* genes.

In addition to CB, two other factors could regulate the expression of *repA*. The leader region of *repA* contains a ρ-independent transcription terminator, suggesting the involvement of premature termination of transcription in the regulation of expression of *repA*. Also, in the leader region there is a putative counteroriented promoter *Pinc* that could initiate transcription of short RNA antisense to

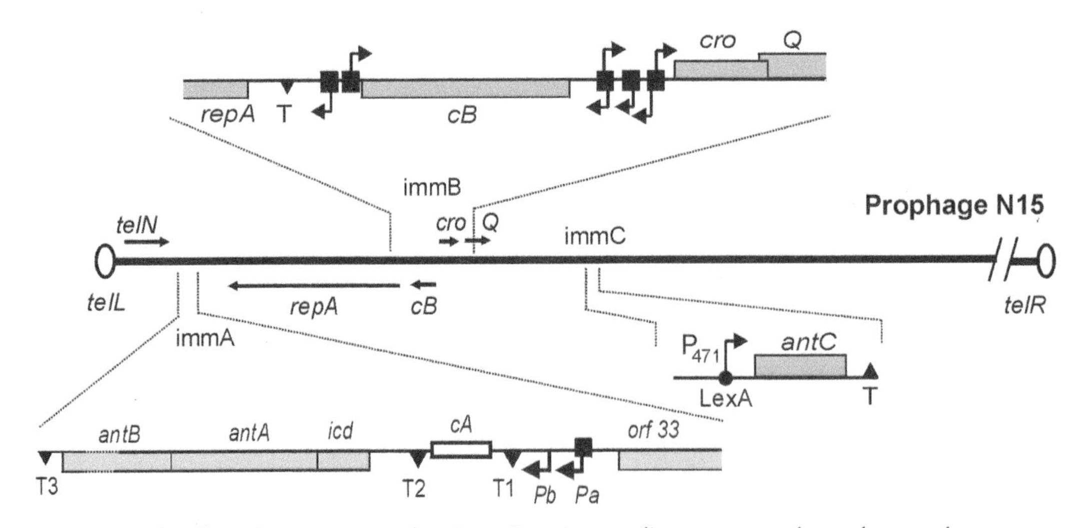

Figure 3 Three lysogeny control regions. Protein-encoding genes are shown by gray boxes; *cA*, which encodes an RNA, by an open box across the main line. Bent arrows indicate promoters (P). The positions of transcription terminators (T) are shown by solid triangles. CB binding sites (O) are shown by black rectangles; the LexA binding site at P471, by a black circle. Genes and sites shown above (below) the main line apply to transcription from left to right (right to left). doi:10.1128/microbiolspec.PLAS-0032-2014.f3

the leader sequence of *repA*, and this RNA may modulate transcription of *repA* (28). Modulation of replication gene expression by antisense RNA is a common strategy employed by other plasmids and, particularly, by bacteriophage P1, where it regulates transcription of the lytic replicon (30). Analysis of transcription patterns of the N15 prophage in the lysogenic culture by RNA-Seq revealed a more than 10-fold difference of RNA level upstream and downstream of the predicted terminator in the leader region of *repA* (N. Ravin, unpublished).

The secondary immunity region, *immA* (Fig. 3), is located between the protelomerase gene and replication operon (31). Three genes at the *immA* form an operon and encode an inhibitor of cell division (Icd), an antirepressor protein, (AntA) and a protein that may play an ancillary role in antirepression (AntB). Constitutive expression of *antA* prevents the establishment and maintenance of lysogeny. AntA counteracts the repression of promoters controlled by the primary N15 repressor, CB, resulting in activation of replication of N15-based miniplasmids and expression of late gene clusters. This *ant* operon may be transcribed from two promoters: the upstream promoter *Pa* could be repressed by the CB repressor, whereas the weaker downstream promoter *Pb* is constitutive. Full repression of the antirepressor operon is achieved by premature transcription termination elicited by a small RNA (CA RNA) produced by processing of the leader transcript of the operon—a mechanism similar to the one used in the lysogeny control regions of phages P1 and P4 (32, 33). The CA RNA thus acts as a secondary repressor, and clear plaque mutants mapped at *immA* were found within the *cA* sequence. The antirepressor functions encoded at the *immA* seem to be involved in the lysis-lysogeny decision of N15 early after infection. Analysis of the transcription patterns of the *immA* locus showed that the structural genes (*icd*, *antA*, and *antB*) of the *ant* operon can only be expressed soon after infection from the two promoters, before the CA RNA is produced. In the lysogen, the CB repressor turns off promoter *Pa*, while the second promoter, *Pb*, allows production of the immunity factor, CA RNA.

The third region involved in the control of lysogeny (34) contains a gene that encodes an antirepressor protein, AntC (Fig. 3). Like the case of *antA*, expression of this gene (*antC*) from a plasmid is sufficient to prevent lysogenization by an infecting phage and to induce lytic development in N15 lysogens. *antC* seems to be involved in the switch from lysogeny to lytic development in the process of prophage induction. Phage N15 mutants in *antC* can infect *E. coli* cells, as well as establish stable lysogens, but are deficient in prophage induction. *antC* expression is controlled by one of the major components of the SOS system, the LexA protein, whose binding site overlaps the *antC* promoter. Exposure of the host cell to DNA-damaging agents that challenge cell survival results in RecA-dependent autocleavage of LexA, derepression of the *antC* promoter, synthesis of this antirepressor protein, and finally, activation of lytic development. Thus, a cellular repressor whose activity is regulated by DNA damage controls N15 prophage induction (34).

In most lambdoid phages prophage induction upon DNA damage is achieved by inactivation of the main CI repressor by RecA-dependent autoproteolysis. However, the amino acid sequence of phage N15 CB does not contain the site of RecA-stimulated autocleavage, an Ala-Gly or Cys-Gly dipeptide sequence located within a "linker" region that joins the N-terminal DNA-binding domain and C-terminal dimerization/autocleavage domain. Thus, the Lex/Ant regulatory system of N15 seems to be an alternative to the λ model of the cleavable repressor (34).

Analysis of the nucleotide sequences of linear phage-plasmids ϕKO2 and pY54 revealed conservation of the regulatory modules described above. Genomes of both phages contains an N15-like primary immunity region comprising homologs of N15 genes *cB*, *cro*, and *Q*, located between their *repA* genes and early right operons (26, 27). Similar control regions are present in the genomes of marine phages VHML, ΦHAP-1, VP882, Vp58.5, and vB_VpaM_MAR (17–21), suggesting that this regulatory circuit is universally conserved in linear phage-plasmids.

Phages ϕKO2 and pY54 encode AntC-like antirepressors, and these genes may be controlled by LexA. In phage ϕKO2 the LexA binding site directly overlaps the antirepressor gene (gene *47*) promoter, while in phage PY54 the LexA operator controls the expression of a two-gene operon comprising gene *56*, which encodes a DinI-like protein and the antirepressor gene, *57* (27). The *immA* immunity region is present in the phage ϕKO2 genome in the same region, as in N15. However, in PY54 the corresponding operon is missing and the homologs of N15 genes *icd* and *antA* are separated and located in distinct areas of the phage PY54 genome. Perhaps this immunity region and the corresponding regulatory circuits are not closely related to the linear prophage lifestyle and could be dispensable.

PLASMID PROPHAGE PARTITIONING SYSTEM

The N15 prophage is maintained at three to five copies per bacterial chromosome and is very stable; its rate of spontaneous loss is less than 10^{-4} per generation (35). Like other low-copy-number plasmids, to be stably maintained, N15 should encode special stabilization machinery. Two principal mechanisms ensuring stable inheritance of bacterial plasmids are known: active partition of plasmid copies to daughter cells prior to division and addiction systems responsible for postsegregational killing of plasmid-free cells (36, 37).

Dziewit et al (38) suggested that a phage N15 operon consisting of genes *49* and *48* constitute a toxin-antitoxin module. After being inserted into a heterologous low-copy-number plasmid, this operon was able to stabilize its maintenance in *E. coli* (38). However, it seems unlikely that the addiction system is a primary factor of the stability of N15 since prophage-free cells are easily accumulated at the nonpermissive temperature in the N15 lysogens carrying "early" temperature-sensitive mutations (11). Likewise, linear N15 prophage carrying insertion mutations in a partitioning locus (39) and N15-based miniplasmid pG54 lacking this region (29) are unstable. Moreover, the rates of spontaneous loss of two linear N15-based miniplasmids that differ in the presence or absence of the gene *49* and *48* operons are similar (N. Ravin, unpublished data), indicating the limited influence of this locus on N15 maintenance. The similar loci are present in phages ϕKO2 and pY54, suggesting that their presence could provide some other benefits for the phage and/or lysogen.

The primary mechanism ensuring stable inheritance of the N15 prophage is the partitioning system (40, 41). The region near the right end of the N15 prophage is similar to the *sop* locus, which governs partition of F plasmid copies to daughter cells. PY54 and ϕKO2 also contain the N15-like *sop* loci close to the right ends of their prophages. Partition loci such as F *sop* and P1 *par* usually consist of a two-gene operon and an adjacent *cis*-acting site (36). The first gene (gene *28* = *sopA*) encodes a protein that binds to the promoter of the partition operon to repress transcription (the operon is thus negatively autoregulated) and also acts as a Walker-box ATPase directly in the partition process itself, while the product of the second gene (gene *27* = *sopB*) binds to the *cis* acting centromere site (C) to form a partition complex and acts as a corepressor of operon expression (70). The *sop* locus of N15 in fact determines the stability of the prophage since Sop proteins can stabilize the partition-defective N15 derivatives (40). The N15 and F partition functions appear to be partly interchangeable: N15 SopA and SopB can partly stabilize partition-defective mini F and repress the F *sop* promoter and vice versa (40, 42). However, the N15 partition system, although a functional analogue of the F *sop* system, differs from it in several important respects.

At first, while transcription of the F *sop* operon is driven from one negatively autoregulated promoter, N15 *sop* is transcribed from two major promoters (43). The first promoter, proximal to *sopA*, is similar to the F sop promoter and could be repressed by Sop proteins. The second, stronger promoter is located close to the right end of the prophage. It is insensitive to regulation by Sop proteins but is tightly repressed by N15 protelomerase. Therefore, there is a regulatory link between the processes of prophage partitioning and generation of covalently closed telomeres.

The most remarkable feature of the N15 partitioning system is that the centromere site of N15 is not composed of a cluster of short inverted repeats (IR) adjacent to the *sop* operon, as in the case of F, P1, and most other circular plasmids, but is represented by four IRs located in different regions of the N15 genome (40). Each of these sites binds SopB and acts as a centromere (41). Likewise, genomes of PY54 and ϕKO2 also contain multiple centromere IR sites in different regions of their genomes—10 and 4 sites, respectively.

Consistent with the expected importance of the centromere dispersion for linear phage-plasmids, it was found that a single IR site is sufficient to completely stabilize a circular, but not a linear, N15-based plasmid (44). The stability of linear N15 derivatives varying in centromere-site (IR) position, number, and spacing increased in proportion to the number of IR sites and the distance between IR sites (44). The plausible explanation for this phenomenon is that centromere dispersion could enable condensation of linear DNA through interaction of partition complexes formed at different centromeres and thus facilitate movement of segregating plasmid molecules. However, visualization of two IR sites on the same molecule revealed that their colocalization did not depend on SopB but resulted from spontaneous folding of the linear DNA (44). Therefore, it was suggested that the beneficial effects of IR number and spacing on partition stem not from condensation but from provision of more numerous and better arranged substrates for SopA action (44), but the molecular mechanisms of this phenomenon remain to be investigated.

N15 centromere sites are located in the regions of the N15 genome that are supposed to be essential for replication and control of gene expression. One site, IR1, is located within the coding sequence of the replication gene *repA*; the second, IR2, is located downstream of gene *Q*; the other two, IR3 and IR4, are located close to the late promoters of genes 52 and 59, respectively. This arrangement suggests that the N15 partition functions may be involved in the regulation of gene expression and replication.

It was found that following binding to IR4, the N15 Sop proteins could repress transcription from the P59 promoter (45). Such an effect that could prevent undesired expression of late genes in the N15 lysogen was not unexpected, since F Sop proteins are known to mediate the silencing of genes linked to the *sopC* centromere (46).

The first centromere, IR1, located about 400 bp downstream of the start codon of the *repA* gene, seems to play another role. Surprisingly, elevated expression of N15 Sop proteins results in a significant increase of the copy number of N15-based circular and linear plasmids carrying replication and control regions including the *repA* and *cB* genes (B.D. Dorokhov and N.V. Ravin, unpublished). The mechanisms of Sop-mediated regulation of N15 prophage replication remain to be determined.

The centromere site IR3 is located immediately upstream of the major late promoter, P52, followed by the transcription terminator located before the first gene of the late operon. As in phage lambda, expression of the late operon is controlled by Q-dependent antitermination (47). However, in N15, the Q gene is located 6 kb upstream of the major late promoter, in contrast to phage λ, where the Q gene is adjacent to this promoter. The importance of the location of IR2 and IR3 sites was evidenced by an observation that following prophage induction, N15 mutated in IR2 or IR3 showed a pronounced delay in lysis relative to that for wild-type N15. It was suggested that a pairing between Sop molecules bound to IR2 and IR3 on the same N15 molecule (*in cis*) would effectively deliver Q to its target site *qut* at P52, thus enhancing its effective concentration and improving the efficiency of transcription antitermination in the course of lytic development (47). Phages ϕKO2 and PY54 both carry centromeres near their presumed Q genes and late promoters, suggesting similar regulatory mechanisms of their Sop operons. The three prophages provide the raw material for addressing such intriguing questions as which function of the *sop* system came first and was coopted to the other: partition or regulation of gene expression?

GENERATION OF COVALENTLY CLOSED TELOMERES

The N15 protelomerase was first hypothesized by Rybchin (12) as an enzyme responsible for the formation of a linear hairpin prophage molecule from the circularized phage DNA upon infection (Fig. 1). In this model, the N15 protelomerase is a functional analogue of integrases of lambdoid phages.

The protelomerase gene, *telN*, was identified upon sequencing of the N15 genome. Its predicted product, TelN, has limited sequence homology with the tyrosine recombinases and type IB topoisomerases (12, 23). Similar telomere resolvases were identified in all linear phage-plasmids and in bacteria with linear chromosomes: ResT (gene BBB03) in *Borrelia burgdorferi* (48, 49) and TelA (gene AGR_C_4584) in *A. tumefaciens* (50).

The cleavage-joining activity of TelN and protelomerase TelK of phage ϕKO2 was analyzed *in vitro* (51–53), and a structure of the TelK in complex with the target site has been solved (54). Protelomerases and tyrosine recombinases have similar catalytic mechanisms for DNA cleavage and ligation, generating a 3′-phosphotyrosine DNA intermediate that enables the covalent rejoining of cleaved DNA strands without the use of a high-energy cofactor. Two protelomerase molecules bind as a dimer to a double-stranded DNA and generate a pair of transient staggered cleavages 6 bp apart from the axis of symmetry of the palindromic target site (53), as shown in Fig. 1. Protelomerase molecules form a pair of protein-linked DNA intermediates at each 3′ end of the cleaved openings, leaving a 5′-OH. Then the partners of the two initial openings are exchanged, and the transient breaks are resealed to generate hairpin ends (55).

Analysis of N15 protelomerase activity *in vivo* revealed that this enzyme is required for replication of linear (but not circular) N15-based plasmids (56). Protelomerase was found to be an end-resolving enzyme responsible for processing of replicative intermediates. Ravin et al. (56) constructed the N15 mutant carrying a deletion in the protelomerase gene and cloned the *telN* gene in the expression vector under the control of a regulatable promoter. The mutant may be maintained as a linear plasmid if the *telN* is expressed *in trans*; removal of protelomerase activity results in accumulation of unprocessed replicative intermediates that were found to be circular head-to-head dimer molecules (56). The *telN* gene and its target site, the *telRL* site, constitute an independent functional unit acting on other replicons independently of other phage genes; cloning of this module in circular mini-F and mini-P1 plasmids resulted in their linearization and further maintenance as linear plasmids with hairpin telomeres (56). Insertion of *telRL* in the *E. coli* chromosome results in its maintenance in a linear form if protelomerase is expressed (57). Functional independence of the protelomerase unit was used to develop a system to assemble BACs up to 100 kb as linear plasmids capped with N15 telomeres (58).

The telomere-forming site *telRL* is flanked by three IRs (tosL1/tosR1, tosL2/tosR2, and tosL3/tosR3) with the consensus sequence taacAgaACA (Fig. 1). In the prophage DNA the *tosL/R* sequences are located at the ends of the prophage. Protelomerase probably binds to *tosL/R* sites since it represses promoters overlapped by these sequences: the P2 promoter of the *sop* operon (43) and the promoter of the *telN* gene itself (59). Presumably, binding of TelN to telomere-proximal regions protects the hairpin telomeres of the N15 prophage from the host SbcCD nuclease (59), which is active against single-stranded DNA.

REPLICATION MECHANISM

The construction of various miniplasmids consisting of different fragments of N15 DNA and an antibiotic resistance gene (29, 60) made it possible to identify the minimal set of genes able to drive replication of the N15 prophage. The minimal circular self-replicating plasmid contained only gene *37* (*repA*), which is thus necessary and sufficient to drive replication (61). The shortest linear plasmid requires the presence of *repA* and a protelomerase module (the *telN* gene and the *telRL* site). RepA is a multifunctional protein combining primase, helicase, and origin-binding activities (23, 62), thus resembling the phage P4 α replication protein (63). The replication initiation site (*ori*) is located within *repA* toward its 3′ end; replication initiated at this site proceeds bidirectionally (61). Likewise, the *repA* gene of phage PY54 contains the *ori* site and functions as a circular minimal replicon in *E. coli* (64). Since the ϕKO2 and N15 RepA proteins are 92% identical and their putative replication origin sequences have 62 of 64 identical base pairs (26), it is likely that replication of ϕKO2 follows the same mechanisms.

All together, the data on protelomerase activity and replication of N15-based miniplasmids suggest the following model of N15 plasmid replication (Fig. 4) (61). Replication is initiated from the *ori* site located within *repA* and proceeds bidirectionally following the θ mode. After duplication of *telL*, protelomerase cuts this palindromic site, creating two hairpin ends, and thus a Y-shaped molecule is formed. Upon replication of the right telomere and subsequent cutting of the duplicated *telR* site by TelN, two linear molecules are produced. Alternatively, complete replication of the prophage, resulting in the formation of a head-to-head circular dimer molecule, may precede resolution of the telomeres (Fig. 4). The intermediates of replication predicted by the former model were detected by electron microscopy (61).

The above model also explains the mechanism of phage N15 replication in the course of lytic growth. In view of the strong similarity of N15 packaging and virion genes to those of phage λ, it is very likely that late

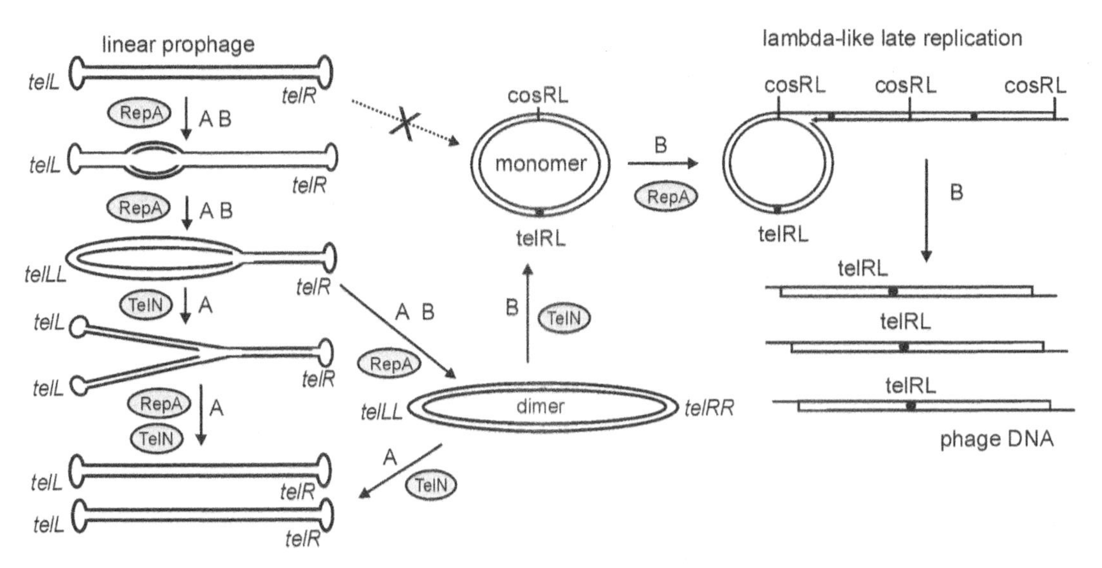

Figure 4 Model of the N15 plasmid prophage and lytic replication. A indicates replication of the N15 plasmid prophage. B shows lytic replication initiated in the lysogen. Note that the circular head-to-head dimer is supposed to be processed by protelomerase into two circular monomers (only one is shown). The known or suggested participation of the N15 proteins TelN and RepA at the individual steps is shown.
doi:10.1128/microbiolspec.PLAS-0032-2014.f4

steps of N15 lytic replication and encapsidation follow the λ model. The lytic growth could occur upon infection but could also be initiated in the lysogenic culture. Mardanov and Ravin (65) analyzed the structures of phage N15 DNA in the course of lytic development upon infection and found that phage DNA, circularized upon joining of the cohesive ends, is not used as a template for λ-like lytic replication but became converted into a linear plasmid. Replication of N15 DNA then follows the plasmid mode with the increase of the copy number; only at late steps did the circular unit-length molecules that could start λ-like late replication appear (Fig. 4, pathway B). These circular monomers became produced as a result of TelN-mediated resolution of a circular dimer into two circular molecules (instead of two linear molecules in the course of prophage replication) rather than being generated directly from a linear plasmid in a reverse "telomere fusion" reaction (65), as has been observed with the *Borrelia* ResT (66). The switch to circular plasmid formation in the course of lytic replication of N15 may result from depletion of protelomerase or modification of the protelomerase and/or its target site by some unknown phage-encoded factor at a late step of lytic growth (65).

Replication of other linear phage-plasmids seems to follow the N15 model since their genomes contain both key elements that determine the replication pathways: the N15-like protelomerase and *repA* genes. However, replication of these plasmid prophages has not yet been studied *in vivo*.

BIOTECHNOLOGICAL APPLICATIONS

One particular property of N15 prompted its biotechnological exploitation: linearity of the plasmid prophage, presumably resulting in the absence of supercoiling. Supercoiling in circular plasmids can generate cruciforms and other secondary structures that are substrates for deletion or rearrangements, mediated by activity of cellular nucleases, particularly in regions that contain numerous tandem or inverted repeats. N15-based linear miniplasmids have been used as cloning vectors (29), which appeared to be particularly suitable for cloning DNA sequences with repeated sequences (60). The next-generation N15-based linear cloning system, the "pJAZZ" series of transcription-free linear cloning vectors, developed by Lucigen Corp., can stably maintain templates that are difficult or impossible to clone in circular vectors, including AT-rich inserts of up to 30 kb and short tandem repeats of up to 2 kb (67). The pJAZZ vector shows decreased size bias in cloning, allowing more uniform representation of larger fragments in libraries. These vectors were successfully employed for cloning of a number of "difficult" templates (for example, reference 68).

N15 was used to develop an *E. coli* host/vector expression system, allowing a combination of two principles of regulation of protein synthesis: the use of an inducible promoter and regulation of the copy number of the vector (69). The pN15E vectors are low-copy-number circular N15-based miniplasmids, comprising the *repA* and *cB* genes, as well as inducible promoters to drive expression of the target gene. Regulation of the pN15E copy number is achieved through expression of antirepressor AntA, whose gene was integrated into the host strain chromosome under control of the arabinose-inducible promoter. The low copy number of these vectors ensures a very low basal level of expression, which allows the cloning of genes encoding toxic products, while simultaneous induction of the inducible promoter and elevation of the copy number of the vector allow high-level expression of the target protein.

ORIGIN OF N15-LIKE PHAGE-PLASMIDS

The N15-like group of viruses is unique among bacteriophages in its genetic organization. The similar overall genome organization and the presence of genes associated with the linear plasmid lifestyle (protelomerase, *repA*-like replicase, and *cB*-like repressor) in N15, ϕKO2, PY54, ΦHAP-1, VP882, Vp58.5, vB_VpaM_MAR, and VHML suggest that these phages may belong to a group diverged from a common ancestor. Phylogenetic analysis of both TelN and RepA proteins revealed three well-separated groups: N15, ϕKO2, and PY54; ΦHAP-1 and VP882; and Vp58.5, vB_VpaM_MAR, and VHML (Fig. 5). Notably, the RepA and TelN trees have similar topology, suggesting coevolution of these genes rather than their independent acquisition or exchange. *RepA*-like genes were found in a number of circular plasmids, while the origin of TelN-like protelomerases is less clear; such enzymes may originate from the phage integrases or type IB topoisomerases. Insertion of such a gene into a *repA*-carrying circular plasmid would result in a linear plasmid with hairpin ends. The common ancestor of N15-like phages must have arisen either through the accumulation of a *telN-repA* genetic module by a lambdoid progenitor or from a linear plasmid acquiring a lambdoid set of "virion" genes. Subsequent evolution occurred via the exchange of groups of functional genes between different phages and plasmids. Therefore, N15 and its relatives provide a very interesting model system

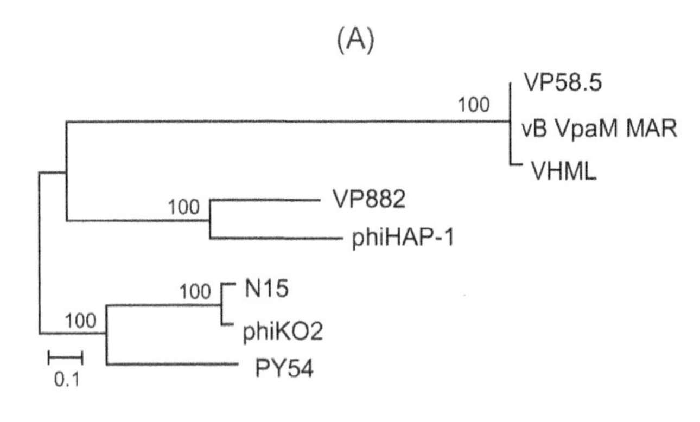

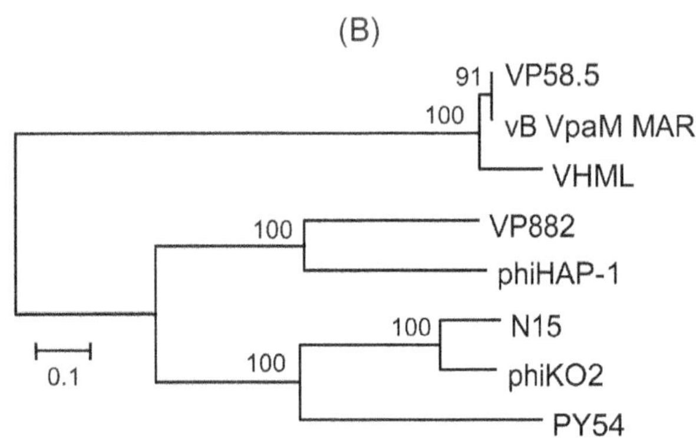

Figure 5 The maximum likelihood trees were calculated from the multiple sequence alignments of (A) RepA or (B) TelN proteins. The numbers above the nodes indicate bootstrap support values. The trees are drawn to scale, with branch lengths measured by the number of substitutions per site. doi:10.1128/microbiolspec.PLAS-0032-2014.f5

for the study of phage and plasmid evolution and interactions between phages, plasmids, and bacterial hosts.

Acknowledgments. *This work was supported by the Program for Molecular and Cell Biology of the Russian Academy of Sciences and by the Russian Foundation for Basic Research (grant 10-04-01204). Conflicts of interest: I disclose no conflicts.*

Citation. Ravin NV. 2015. Replication and maintenance of linear phage-plasmid N15. Microbiol Spectrum **3**(1):PLAS-0032-2014.

References

1. **Watson J.** 1972. Origin of concatemeric T7 DNA. *Nat New Biol* **239:**197–201.
2. **Casjens S.** 1999. Evolution of the linear DNA replicons of the *Borrelia* spirochetes. *Curr Opin Microbiol* **2:** 529–534.
3. **Barbour AG, Garon CF.** 1987. Linear plasmids of the *Borrelia burgdorferi* have covalently closed ends. *Science* **237:**409–411.
4. **Casjens S, Murphy M, DeLange M, Sampson L, van Vugt R, Huang WM.** 1997. Telomeres of the linear chromosomes of Lyme disease spirochaetes: nucleotide sequence and possible exchange with linear plasmid telomeres. *Mol Microbiol* **26:**581–596.
5. **Tourand Y, Deneke J, Moriarty TJ, Chaconas G.** 2009. Characterization and *in vitro* reaction properties of 19 unique hairpin telomeres from the linear plasmids of the Lyme disease spirochete. *J Biol Chem* **284:**7264–7272.
6. **Allardet-Servent AS, Michaux-Charachon E, Jumas-Bilak L, Karayan, Ramuz M.** 1993. Presence of one linear and one circular chromosome in the *Agrobacterium tumefaciens* C58 genome. *J Bacteriol* **175:**7869–7874.
7. **Goodner B, Hinkle G, Gattung S, Miller N, Blanchard M, Qurollo B, Goldman B, Cao Y, Askenazi M, Halling C, Mullin L, Houmiel K, Gordon J, Vaudin M, Iartchouk O, Epp A, Liu F, Wollam C, Allinger M, Doughty D, Scott C, Lappas C, Markelz B, Flanagan C, Crowell C, Gurson J, Lomo C, Sear C, Strub G, Cielo C, Slater S.** 2001. Genome sequence of the plant pathogen and biotechnology agent *Agrobacterium tumefaciens* C58. *Science* **294:**2323–2328.
8. **Golub EI, Ravin VK.** 1967. New system of phage mediated conversion. *Dokl Acad Nauk USSR* **174:**465–467.
9. **Ravin VK, Shulga MG.** 1970. The evidence of extrachromosomal location of prophage N15. *Virology* **40:**800–807.
10. **Ravin VK.** 1971. *Lysogeny*, p.106. Nauka Press, Moscow.
11. **Ravin VK.** 1968. The functioning of the genes of temperate bacteriophage in lysogenic cells. *Genetika* **4:**119–124.
12. **Rybchin VN, Svarchevsky AN.** 1999. The plasmid prophage N15, a linear DNA with covalently closed ends. *Mol Microbiol* **33:**895–903.
13. **Svarchevsky AN, Rybchin VN.** 1984. Physical mapping of plasmid N15 DNA. *Mol Gen Mikrobiol Virusol* **10:** 16–22.
14. **Malinin AY, Vostrov AA, Rybchin VN, Svarchevsky AN.** 1992. Structure of the linear plasmid N15 ends. *Mol Gen Mikrobiol Virusol* **5–6:**19–22.
15. **Stoppel RD, Meyer M, Schlegel HG.** 1995. The nickel resistance determinant cloned from the enterobacterium *Klebsiella oxytoca*: conjugational transfer, expression, regulation and DNA homologies to various nickel-resistant bacteria. *Biometals* **8:**70–79.
16. **Hertwig S, Klein I, Lurz R, Lanka E, Appel B.** 2003. PY54, a linear plasmid prophage of *Yersinia enterocolitica* with covalently closed ends. *Mol Microbiol* **48:** 989–1003.
17. **Lan SF, Huang CH, Chang CH, Liao WC, Lin IH, Jian WN, Wu YG, Chen SY, Wong HC.** 2009. Characterization of a new plasmid-like prophage in a pandemic *Vibrio parahaemolyticus* O3:K6 strain. *Appl Environ Mirobiol* **75:**2659–2667.
18. **Zabala B, Hammerl JA, Espejo RT, Hertwig S.** 2009. The linear plasmid prophage Vp58.5 of *Vibrio parahaemolyticus* is closely related to the integrating phage VHML and constitutes a new incompatibility group of telomere phages. *J Virol* **83:**9313–9320.
19. **Villa A, Kropinski AM, Abbasifar R, Griffiths MW.** 2012. Complete genome sequence of *Vibrio parahaemolyticus* bacteriophage vB_VpaM_MAR. *J Virol* **86:** 13138–13139.

20. **Mobberley JM, Authement RN, Segall AM, Paul JH.** 2008. The temperate marine phage PhiHAP-1 of *Halomonas aquamarina* possesses a linear plasmid-like prophage genome. *J Virol* **82:**6618–6630.
21. **Oakey HJ, Cullen BR, Owens L.** 2002. The complete nucleotide sequence of the *Vibrio harveyi* bacteriophage VHML. *J Appl Microbiol* **93:**1089–1098.
22. **Lima-Mendez G, Van Helden J, Toussaint A, Leplae R.** 2008. Reticulate representation of evolutionary and functional relationships between phage genomes. *Mol Biol Evol* **25:**762–777.
23. **Ravin V, Ravin N, Casjens S, Ford M, Hatfull G, Hendrix R.** 2000. Genomic sequence and analysis of the atypical bacteriophage N15. *J Mol Biol* **299:**53–73.
24. **Vostrov A, Vostrukhina O, Svarchevsky A, Rybchin V.** 1996. Proteins responsible for lysogenic conversion caused by coliphages N15 and phi80 are highly homologous. *J Bacteriol* **178:**1484–1486.
25. **Ravin V, Golub E.** 1967. A study of phage conversion in *Escherichia coli*. I. The aquisition of resistance to bacteriophage T1 as a result of lysogenization. *Genetika* **4:** 113–121.
26. **Casjens SR, Gilcrease EB, Huang WM, Bunny KL, Pedulla ML, Ford ME, Hourtz JM, Hatfull GF, Hendrix RW.** 2004. The pKO2 linear plasmid prophage of *Klebsiella oxytoca*. *J Bacteriol* **186:**1818–1832.
27. **Hertwig S, Klein I, Schmidt V, Beck S, Hammerl JA, Appel B.** 2003. Sequence analysis of the genome of the temperate *Yersinia enterocolitica* phage PY54. *J Mol Biol* **331:**605–622.
28. **Lobocka M, Svarchevsky AN, Rybchin VN, Yarmolinsky M.** 1996. Characterization of the primary immunity region of the *Escherichia coli* linear plasmid prophage N15. *J Bacteriol* **178:**2902–2910.
29. **Ravin NV, Ravin VK.** 1994. An ultrahigh-copy plasmid based on the mini-replicone of the temperate phage N15. *Mol Gen Mikrobiol Virusol* **1:**37–39.
30. **Heinrich J, Riedel HD, Ruckert B, Lurz R, Schuster H.** 1995. The lytic replicon of bacteriophage P1 is controlled by an antisense RNA. *Nucleic Acids Res* **23:**1468–1474.
31. **Ravin NV, Svarchevsky AN, Deho G.** 1999. The antiimunity system of phage-plasmid N15: identification of the antirepressor gene and its control by a small processed RNA. *Mol Microbiol* **34:**980–994.
32. **Citron M, Schuster H.** 1992. The c4 repressor of bacteriophage P1 is a processed 77 base antisense RNA. *Nucleic Acids Res* **20:**3085–3090.
33. **Deho G, Zangrossi S, Sabbattini P, Sironi G, Ghisotti D.** 1992. Bacteriophage P4 immunity controlled by small RNAs via transcription termination. *Mol Microbiol* **6:** 3415–3425.
34. **Mardanov AV, Ravin NV.** 2007. Antirepressor needed for induction of linear plasmid-prophage N15 belongs to the SOS regulon. *J Bacteriol* **189:**6333–6338.
35. **Svarchevsky AN, Rybchin VN.** 1984. Characteristics of plasmid properties of bacteriophage N15. *Mol Gen Mikrobiol Virusol* **10:**34–39.
36. **Hiraga S.** 1992. Chromosome and plasmid partition in *Escherichia coli*. *Annu Rev Biochem* **61:**283–306.
37. **Gerdes K, Jacobsen JS, Franch T.** 1997. Plasmid stabilization by post-segregational killing. *Genet Eng* **19:**49–61.
38. **Dziewit L, Jazurek M, Drewniak L, Baj J, Bartosik D.** 2007. The SXT conjugative element and linear prophage N15 encode toxin-antitoxin-stabilizing systems homologous to the tad-ata module of the *Paracoccus aminophilus* plasmid pAMI2. *J Bacteriol* **189:**1983–1997.
39. **Sankova TP, Svarchevsky AN, Rybchin VN.** 1992. Isolation, characterization and mapping of N15 plasmid insertion mutants. *Genetika* **28:**66–76.
40. **Ravin N, Lane D.** 1999. Partition of the linear plasmid, N15: functional interactions with the sop locus of the F plasmid. *J Bacteriol* **181:**6898–6906.
41. **Grigoriev PS, Lobocka MB.** 2001. Determinants of segregational stability of the linear plasmid-prophage N15 of *Escherichia coli*. *Mol Microbiol* **42:**355–368.
42. **Ravin NV, Rech J, Lane D.** 2003. Mapping of functional domains in F plasmid partition proteins reveals a bipartite SopB-recognition domain in SopA. *J Mol Biol* **329:** 875–889.
43. **Dorokhov BD, Lane D, Ravin NV.** 2003. Partition operon expression in the linear plasmid prophage N15 is controlled by both Sop proteins and protelomerase. *Mol Microbiol* **50:**713–721.
44. **Dorokhov B, Ravin N, Lane D.** 2010. On the role of centromere dispersion in stability of linear bacterial plasmids. *Plasmid* **64:**51–59.
45. **Mardanov AV, Lane D, Ravin NV.** 2010. Sop proteins can cause transcriptional silencing of genes located close to the centromere sites of linear plasmid N15. *Mol Biol (Mosk)* **44**(2):294–300(InRussian).
46. **Lynch AS, Wang JC.** 1995. SopB protein-mediated silencing of genes linked to the sopC locus of *Escherichia coli* F plasmid. *Proc Natl Acad Sci USA* **92:**1896–1900.
47. **Ravin NV, Rech J, Lane D.** 2008. Extended function of plasmid partition genes: Sop system of linear phage-plasmid N15 facilitates late gene expression. *J Bacteriol* **190:**3538–3545.
48. **Kobryn K, Chaconas G.** 2002. ResT, a telomere resolvase encoded by the Lyme disease spirochete. *Mol Cell* **9:** 195–201.
49. **Chaconas G, Kobryn K.** 2010. Structure, function, and evolution of linear replicons in *Borrelia*. *Annu Rev Microbiol* **64:**185–202.
50. **Huang WM, DaGloria J, Fox H, Ruan Q, Tillou J, Shi K, Aihara H, Aron J, Casjens S.** 2012. Linear chromosome generating system of *Agrobacterium tumefaciens* C58: protelomerase generates and protects hairpin ends. *J Biol Chem* **287:**25551–25563.
51. **Deneke J, Ziegelin G, Lurz R, Lanka E.** 2000. The protelomerase of temperate *Escherichia coli* phage N15 has cleaving-joining activity. *Proc Natl Acad Sci USA* **97:** 7721–7726.
52. **Deneke J, Ziegelin G, Lurz R, Lanka E.** 2002. Phage N15 telomere resolution: target requirements for recognition and processing by the protelomerase. *J Biol Chem* **277:** 10410–10419.
53. **Huang WM, Joss L, Hsieh T, Casjens S.** 2004. Protelomerase uses a topoisomerase IB/Y-recombinase

type mechanism to generate DNA hairpin ends. *J Mol Biol* **337**:77–92.

54. **Aihara H, Huang WM, Ellenberger T.** 2007. An interlocked dimer of the protelomerase TelK distorts DNA structure for the formation of hairpin telomeres. *Mol Cell* **27**:901–913.

55. **Shi K, Huang WM, Aihara H.** 2013. An enzyme-catalyzed multistep DNA refolding mechanism in hairpin telomere formation. *PLoS Biol* **11**:e1001472. doi:10.1317/journal.pbio.1001472.

56. **Ravin NV, Strakhova TS, Kuprianov VV.** 2001. The protelomerase of the phage-plasmid N15 is responsible for its maintenance in linear form. *J Mol Biol* **312**:899–906.

57. **Cui T, Moro-oka N, Ohsumi K, Kodama K, Ohshima T, Ogasawara N, Mori H, Wanner B, Niki H, Horiuchi T.** 2007. *Escherichia coli* with a linear genome. *EMBO Rep.* **8**:181–187.

58. **Ooi YS, Warburton PE, Ravin NV, Narayanan K.** 2008. Recombineering linear DNA that replicate stably in *E. coli. Plasmid* **59**:63–71.

59. **Dorokhov BD, Strakhova TS, Ravin NV.** 2004. Expression regulation of the protelomerase gene of the bacteriophage N15. *Mol Gen Mikrobiol Virusol* **2**:28–32.

60. **Ravin NV, Ravin VK.** 1999. Use of a linear multicopy vector based on the mini-replicon of temperate coliphage N15 for cloning DNA with abnormal secondary structures. *Nucleic Acids Res* **27**:e13.

61. **Ravin NV, Kuprianov VV, Gilcrease EB, Casjens SR.** 2003. Bidirectional replication from an internal ori site of the linear N15 plasmid prophage. *Nucleic Acids Res* **31**:6552–6560.

62. **Mardanov AV, Ravin NV.** 2006. Functional characterization of the *repA* replication gene of linear plasmid prophage N15. *Res Microbiol* **157**:176–183.

63. **Ziegelin G, Scherzinger E, Lurz R, Lanka E.** 1993. Phage P4 alpha protein is multifunctional with origin recognition, helicase and primase activities. *EMBO J* **12**: 3703–3708.

64. **Ziegelin G, Tegtmeyer N, Lurz R, Hertwig S, Hammerl J, Appel B, Lanka E.** 2005. The *repA* gene of the linear *Yersinia enterocolitica* prophage PY54 functions as a circular minimal replicon in *Escherichia coli. J Bacteriol* **187**:3445–3454.

65. **Mardanov AV, Ravin NV.** 2009. Conversion of linear DNA with hairpin telomeres into a circular molecule in the course of phage N15 lytic replication. *J Mol Biol* **391**:261–268.

66. **Kobryn K, Chaconas G.** 2005. Fusion of hairpin telomeres by the *B. burgdorferi* telomere resolvase ResT: implications for shaping a genome in flux. *Mol Cell* **17**: 783–791.

67. **Godiska R, Mead D, Dhodda V, Wu C, Hochstein R, Karsi A, Usdin K, Entezam A, Ravin N.** 2010. Linear plasmid vector for cloning of repetitive or unstable sequences in *Escherichia coli. Nucleic Acids Res* **38**(6):e88.

68. **Pfander C, Anar B, Schwach F, Otto TD, Brochet M, Volkmann K, Quail MA, Pain A, Rosen B, Skarnes W, Rayner JC, Billker O.** 2011 A scalable pipeline for highly effective genetic modification of a malaria parasite. *Nat Methods* **8**:1078–1082.

69. **Mardanov AV, Strakhova TS, Smagin VA, Ravin NV.** 2007. Tightly regulated, high-level expression from controlled copy number vectors based on the replicon of temperate phage N15. *Gene* **395**:15–21.

70. **Baxter JC, Funnell BE.** 2015. Plasmid partition mechanisms, p 135–172. *In* Tomalsky M, Alonso JC (ed), *Plasmids: Biology and Impact in Biotechnology and Discovery*. ASM Press, Washington, DC.

Plasmids—Biology and Impact in Biotechnology and Discovery
Edited by Marcelo E. Tolmasky and Juan C. Alonso

doi:10.1128/microbiolspec.PLAS-0001-2013

Sabine Brantl[1]

Plasmid Replication Control by Antisense RNAs

6

INTRODUCTION

Small regulatory RNAs (sRNAs) from bacterial chromosomes came into focus in 2001, when two groups independently discovered many small RNAs from intergenic regions of the *Escherichia coli* genome by a combination of computational and experimental approaches (1, 2). By now, >140 sRNAs have been found in *E. coli* and hundreds in other prokaryotic species, and it is estimated that an average bacterial genome encodes ≈200 to 300 riboregulators (3, 4). They can be classified into *cis*- and *trans*-encoded base-pairing sRNAs, sRNAs acting via protein binding, and sensory RNA modules like RNA thermometers and riboswitches.

sRNAs from accessory DNA elements, such as plasmids, phages, and transposons, have been known since 1981. Their role in the regulation of plasmid replication, maintenance, and conjugation, in fine tuning of the decision between phage lysis and lysogeny, or in transposition, has been investigated in great depth for more than 30 years. Plasmid-encoded sRNAs are *bona fide* antisense RNAs, i.e., they are *cis* encoded and act by a base-pairing mechanism. Usually, they are 50 to 150 nucleotides (nt) long, diffusible, highly structured (one to four stem-loops) molecules that act via sequence complementarity on target RNAs called sense RNAs. The sense RNAs are mostly mRNAs encoding proteins of important/essential function. In this article, antisense-RNA-mediated regulation of plasmid replication will be reviewed in detail.

Plasmids are selfish genetic elements that normally constitute a burden for the bacterial host cell. Therefore, the host cell tries to eliminate the "intruder." To prevent this elimination, plasmids have evolved copy number control systems and maintenance functions. Special systems control unavoidable copy number fluctuations and prevent great decreases or increases of copy numbers: Low copy numbers can lead to plasmid loss (5, 6), whereas high copy numbers may lead to "runaway replication" (7) that kills the host cell. Principally, two control modes can be distinguished: iteron-mediated control and antisense-RNA-mediated control (188).

Antisense-RNA control in plasmid replication works through a negative control circuit. Antisense RNAs are constitutively synthesized and metabolically unstable (one exception is pIP501; see below). Therefore, any change in plasmid concentration will be reflected in the corresponding concentration changes of the regulating antisense RNA. These concentration changes are "sensed," leading to altered replication frequencies. Increased plasmid copy numbers result in correspondingly increasing antisense-RNA concentrations, which,

[1]Friedrich-Schiller-Universität Jena, AG Bakteriengenetik, Philosophenweg 12, Jena D-07743, Germany.

in turn, cause increased inhibition of a function essential for replication (replication initiator protein or replication primer). On the other hand, decreased plasmid copy numbers entail decreasing concentrations of the inhibiting antisense RNA, thereby, increasing the replication frequency. Thus, the antisense RNA is both the measuring device and the regulator, and regulation occurs in all cases by inhibition.

This inhibition is achieved by a variety of mechanisms that will be discussed below in detail in the following order. Antisense-RNA-mediated transcriptional attenuation is a mechanism that affects the fate of the nascent target RNA during transcription. So far, it has been detected only in plasmids from Gram-positive bacteria (pT181 family and *inc18* family). By contrast, ColE1, a plasmid that does not need a plasmid-encoded replication initiator protein, uses the inhibition of primer formation. Another mechanism, antisense-RNA-mediated inhibition of pseudoknot formation that is required for efficient Rep translation, was found in the IncIα/IncB family of plasmids. Three different strategies can be applied to inhibit the translation of an essential replication initiator protein (Rep): The most trivial case is direct blockage of the *rep*-ribosome binding site (RBS) (pMV158 family). Alternatively, ribosome binding to the RBS of a leader peptide reading frame whose translation is required for efficient Rep translation can be prevented (plasmid R1). Furthermore, in translation attenuation, a *rep* mRNA conformation is induced that sequesters the RBS (plasmid pSK41). In some recently discovered cases (e.g., *repABC* plasmids) it is not yet clear if *rep* expression is controlled by transcriptional or translational attenuation. In ColE2, translational inhibition is most likely, and the Rep protein is, surprisingly, a primase that generates an unusual primer.

In some plasmids, the antisense RNA(s) act alone (pT181 family, IncB/IncIα family, ColE2). Here, the rate of synthesis per plasmid copy of the essential *rep* RNA needed for replication is constant (constitutive *rep* promoter) but rather low compared with that of the antisense RNA. In other plasmids, antisense RNAs are accompanied by regulatory proteins, which are either transcriptional repressors, Cop proteins (R1 and relatives, *inc18* and pMV158 families), or RNA-binding proteins (ColE1). These proteins can either play an auxiliary role, as in R1 or ColE1, or can be necessary for proper regulation (*inc18* family, pMV158 family). In the latter case, expression of the *rep* gene is directed by a strong and Cop-regulated promoter, so that, in the absence of Cop, the *rep* transcription rate is high. Interestingly, this is mainly found in mobilizable plasmids with a broad host range. A high potential to transcribe the essential *rep* gene is an advantage for these plasmids during the establishment stage, because they would replicate at high rates, thereby reducing the frequency of appearance of plasmid-free cells from newly colonized bacteria. Plasmids with auxiliary proteins involved in their replicational control may also share the same advantage during establishment (8).

For antisense-RNA-controlled plasmids that replicate by the theta mechanism, the results on origin characterization and replication mechanism are briefly summarized. For those that replicate by the rolling-circle mechanism see reference 189.

ANTISENSE-RNA-MEDIATED TRANSCRIPTIONAL ATTENUATION: THE *inc18* AND THE pT181 FAMILIES

Regulation of plasmid replication by antisense-RNA-mediated transcriptional attenuation has, so far, only been found in Gram-positive bacteria. In 1989, it was described by the Novick group for plasmid pT181 (9), the best characterized representative of a family of five related staphylococcal plasmids (10), which replicate via the rolling-circle mechanism (189). Subsequently, transcription attenuation was found for two plasmids belonging to the *inc18* family of broad-host-range streptococcal plasmids (11) that replicate unidirectionally via the theta mechanism (12, 13, 14), pIP501 (15), and pAMβ1 (16). Later, it was proposed for *Lactobacillus pentosus* plasmid p353-2 (17) and *Lactococcus lactis* plasmid pNE324 (18).

All these plasmids encode essential replication initiator proteins (Rep proteins) that bind to their respective replication origins (12, 19). The amount of the Rep proteins is rate limiting for replication (20, 21, 22). Consequently, the *rep* mRNA is the target for copy number control. During transcription, it can adopt two mutually exclusive structures depending on the presence or absence of the antisense RNA. (The corresponding antisense RNAs [≈85 nt RNAI and ≈150 nt RNAII for pT181 {23}; ≈140 nt RNAIII for pIP501 {24}; and pAMβ1 {16}] and their promoters have been detected and characterized previously). If the nascent *rep* mRNA encounters an antisense-RNA molecule, binding leads to formation of a rho-independent transcriptional terminator (attenuator) by base pairing between complementary sequences a and b upstream of the *rep*-RBS and, consequently, premature termination of *rep* mRNA transcription (Fig. 1A). Thus, only a short *rep* mRNA (in the case of pIP501, 260 nt) is synthesized that does not contain the information for the Rep protein. However, if the nascent *rep* mRNA escapes the antisense RNA during transcription, it may refold by alternative base pairing between sequences A and a (see Fig. 1A). In this case, a is no longer available for base pairing with b, and the transcriptional terminator cannot be formed. Transcription proceeds to a full-length *rep* mRNA (pIP501, 1.9 kb) that can be used for synthesis of the Rep protein. Thus, the antisense RNA affects gene expression by aborting a transcript for an essential protein.

Binding of the antisense RNA must occur within a short time frame in order to be effective. This time window, during which the *rep* mRNA is long enough to contain the target sequence for the antisense RNA, but short enough to not yet have reached the attenuator, has been experimentally estimated to be 10 to 20 s (25, 26). With the use of this estimate, the inhibition rate constant of the pIP501 antisense RNA (RNAIII) was calculated to be 1×10^6 to 2×10^6 M^{-1} s^{-1} which is ten times higher than the pairing rate constant of the sense/antisense RNA pair (1×10^5 to 2×10^5 M^{-1} s^{-1}), indicating that full duplex formation is not required for inhibition. Apparently, steps preceding formation of a complete duplex between sense and antisense RNA are sufficient for an efficient action of the antisense RNA (25, 27). In contrast to pIP501, inhibition and the pairing rate constants were in the same range in pT181 (26). pT181 has two antisense RNAs (RNAI, 85 nt, and RNAII, 144 nt long [9]). RNAII seems to be a read-through product of RNAI that contains two stem-loops of the identical sequence at its 5′ end, which are sufficient for its inhibitory function (26). One of two additional 3′ stem-loops is the transcription terminator. Pairing and inhibition rate constants of RNAII and RNAI were identical suggesting that both molecules fulfill the same function (26). For pIP501, intracellular concentrations and half-lives of sense and antisense RNAs were determined by quantitative Northern blotting (28), which led to two surprising results: (i) the antisense RNA (RNAIII) is unusually stable (half-life of ≈30 min), and (ii) the concentration of RNAIII is identical for *copR*$^+$ and *copR*$^-$ plasmids. These plasmids that contain or lack the second copy number control component CopR (29), respectively, replicate with 5 to 10 or 50 to 100 copies/cell. That is, in the absence of CopR, the ratio RNAIII/plasmid is unexpectedly low. The unusually long half-life of RNAIII is expected to cause problems upon downward fluctuations of copy numbers: the high concentration of the inhibitor would lead to further decreases of the replication frequency and eventually plasmid loss. However, plasmid pIP501 is stably maintained, indicating that a second control mechanism is operating.

This mechanism is provided by the dual function of transcriptional repressor CopR (30, 31). Almost identical transcriptional repressors with analogous functions have been found in the related *inc18* family plasmids pAMβ1 (CopF [32, 33]) and pSM19035 (CopS [34]). CopR is a small protein (10.6 kDa) that binds exclusively as a dimer at two consecutive sites in the major groove immediately upstream of the −35 box of the (sense) *repR* promoter pII, thereby inducing a slight bend in the operator DNA (20 to 25°) (35, 36, 37). Binding of CopR leads to 10- to 20-fold decreased *repR* transcription and, consequently, of the pIP501 copy number (29, 30). CopR does not autoregulate its own promoter (in contrast to CopG of pLS1; see below) and does not completely repress the *rep* promoter (30). A SELEX experiment showed that evolution of the *copR* operator was directed at maximal binding affinity (38). Recently, we demonstrated that CopR acts by inhibiting RNA polymerase binding (39).

The equilibrium dissociation constants for the CopR-DNA complex and the CopR dimers have been determined to be 0.4 nM and 1.45 μM, respectively. The intracellular CopR concentration in logarithmically grown *Bacillus subtilis* cells is 20 to 30 μM (36), suggesting that the protein also *in vivo* binds exclusively as a dimer. A three-dimensional model of the N-terminal 63 amino acids (aa) of CopR was constructed, and amino acids involved in DNA binding and dimerization were localized: R29 and R34 within the HTH motif are involved in specific recognition of the operator DNA (40). Eight amino acids are involved in forming the dimeric interface, and two of them required for correct folding of the monomer (41, 42). The structured acidic C terminus of CopR is important for protein stability and contains a stretch of alternating hydrophilic and hydrophobic amino acids that form a β-strand (43, 44). Its deletion does not impair the *in vivo* function of CopR, but decreases CopR half-life from 42 min to 4 to 5 min (43).

The second function of CopR is to prevent convergent transcription between sense and antisense RNA (31). In the absence of CopR, transcription from the strong sense promoter pII through the supercoiling sensitive antisense promoter pIII decreases transcription initiation at pIII. In the presence of CopR, repression of RNAII transcription leads to a decrease in convergent transcription, thereby indirectly increasing transcription initiation at pIII. Consequently, a ≈3- to 4-fold higher concentration of RNAIII is found in the presence of CopR (see above), and the ratio RNAIII/plasmid is higher than in its absence (high-copy-number *copR*$^-$ plasmid). The scenario for copy number control is as

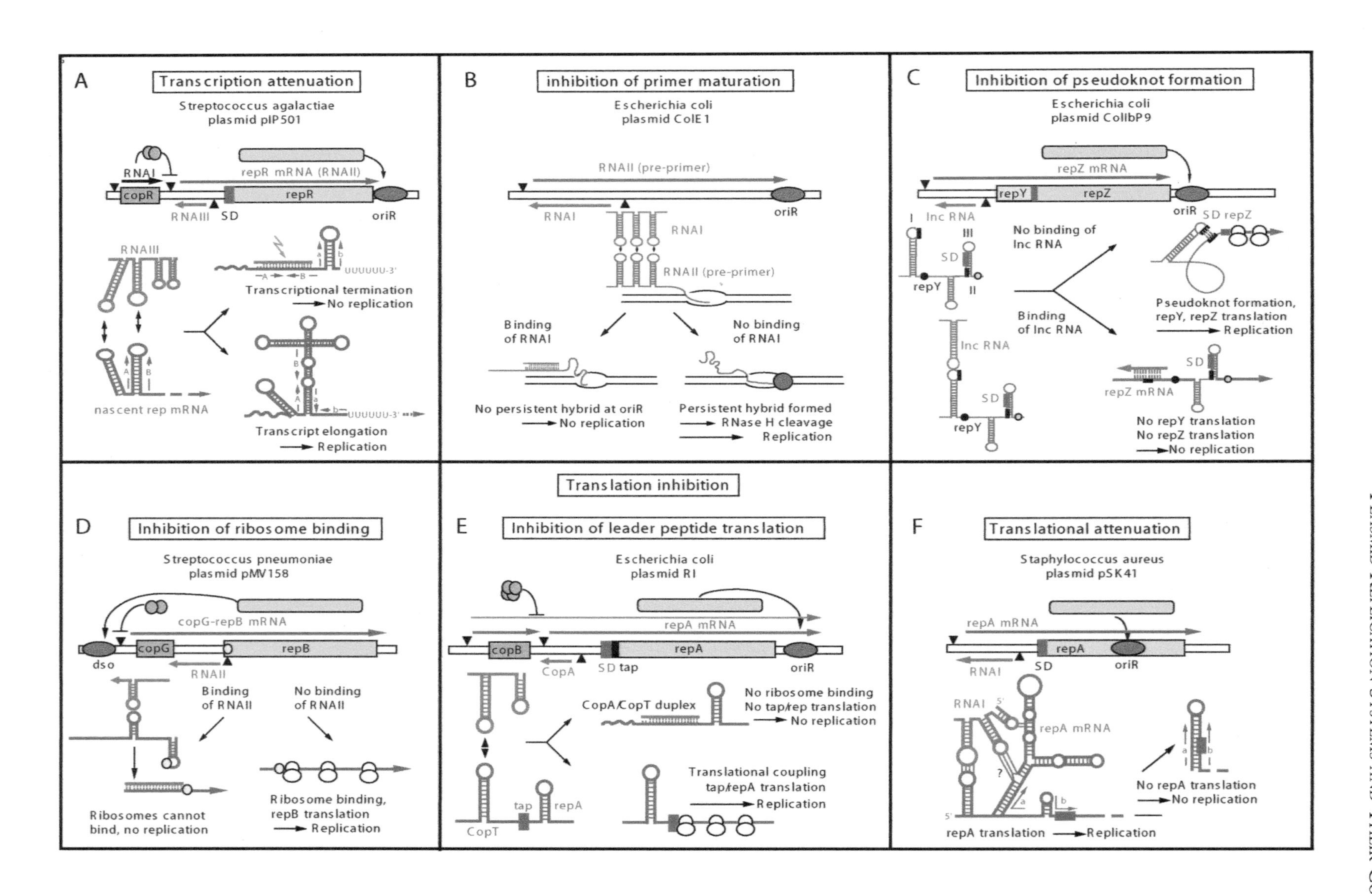
A
Transcription attenuation
Streptococcus agalactiae
plasmid pIP501
RNAI
copR
repR mRNA (RNAII)
repR
RNAIII
SD
oriR
RNAIII
UUUUUU-3'
Transcriptional termination
→ No replication
nascent rep mRNA
UUUUUU-3'
Transcript elongation
→ Replication
B
inhibition of primer maturation
Escherichia coli
plasmid ColE1
RNAII (pre-primer)
RNAI
oriR
RNAI
RNAII (pre-primer)
Binding
of RNAI
No binding
of RNAI
No persistent hybrid at oriR
→ No replication
Persistent hybrid formed
→ RNase H cleavage
→ Replication
C
Inhibition of pseudoknot formation
Escherichia coli
plasmid CollbP9
repZ mRNA
repY
repZ
oriR
SD repZ
Inc RNA
I
III
II
SD
repY
No binding of
Inc RNA
Binding
of Inc RNA
Pseudoknot formation,
repY, repZ translation
→ Replication
Inc RNA
SD
repY
SD
repZ mRNA
No repY translation
No repZ translation
→ No replication
Translation inhibition
D
Inhibition of ribosome binding
Streptococcus pneumoniae
plasmid pMV158
copG-repB mRNA
copG
repB
dso
RNAII
Binding
of RNAII
No binding
of RNAII
Ribosomes cannot
bind, no replication
Ribosome binding,
repB translation
→ Replication
E
Inhibition of leader peptide translation
Escherichia coli
plasmid R1
repA mRNA
copB
repA
CopA
SD tap
oriR
CopA/CopT duplex
No ribosome binding
No tap/rep translation
→ No replication
CopT
tap
repA
Translational coupling
tap/repA translation
→ Replication
F
Translational attenuation
Staphylococcus aureus
plasmid pSK41
repA mRNA
repA
RNAI
SD
oriR
RNAI
repA mRNA
No repA translation
→ No replication
repA translation → Replication

follows: If copy numbers increase, RNAIII is able to cope with the situation. Its constitutive synthesis, directly correlated with the plasmid concentration, leads to higher concentrations of the inhibitor. Thus, more *repR* mRNA is terminated prematurely, and the replication frequency decreases accordingly. When copy number decreases, however, the high stability of RNAIII presents a severe problem: "too much inhibition" would cause plasmid loss. Now, the CopR protein comes into the play as the second regulator: lower plasmid copy numbers entail a decrease in the intracellular CopR concentration, resulting in less repression of *repR* transcription, and, thus, an increase in the replication frequency. At the same time, convergent transcription from the *repR* promoter pII and antisense promoter pIII leads to reduced initiation at the antisense-promoter, causing decreased transcriptional attenuation. Hence, more full-length *repR* mRNA can be transcribed and the replication frequency increases. The now higher amounts of *repR* mRNA titrate the remaining long-lived RNAIII, further lowering the concentration of this inhibitor (31). In summary, pIP501, and also its relatives pAMβ1 and pSM19035 that show the same modular concept of *cop*-antisense-RNA-*rep* (29, 30), have evolved a very efficient mechanism to correct copy number fluctuations. The concerted action of two control components, an unusually long-lived antisense RNA and a Cop protein with a dual function, prevents plasmid loss at cell division.

For pT181 and related plasmids, no plasmid-encoded transcriptional repressor is needed to ensure an efficient regulation of plasmid replication, as the antisense RNAs are fairly unstable (half-life ≈3 to 5 min; R. P. Novick, personal communication). The number of molecules of the essential, unstable replication initiator protein RepC was determined to be 300/cell under wild-type conditions (9).

In the case of pIP501, *in vivo* and *in vitro* analysis of RNAIII mutants (28) showed that the 5′-terminal stem-loops L1 and L2 are not required for inhibition. By contrast, stem-loop L3 is the so-called recognition loop, where primary interactions (kissing) between RNAIII and RNAII occur. Nucleotide exchanges in this loop created new incompatibility groups. Another class of copy number mutations in the single-stranded region between L2 and L3 exhibited a shortened RNA III half-life. A subsequent analysis revealed that a simultaneous interaction between two complementary loop pairs, L3 and L4 of RNAIII and the corresponding loop pair of RNAII, is required for inhibition (45). Sequence and

Figure 1 Mechanisms of antisense-RNA-mediated plasmid copy number control. Antisense RNAs are drawn in red; sense RNAs are drawn in blue. ORFs encoding essential replication initiator proteins are shown as orange boxes; ORFs encoding transcriptional repressor proteins are shown as brown boxes. SD sequences for *rep* ORFs are blue rectangles. Promoters are symbolized by black triangles and replication origins by dark grey ovals. Arrows indicate positive interaction; black bars indicate repression. Ribosomes are in light yellow. (**A**) Transcriptional attenuation: plasmid pIP501. (**Upper part**) Working model on regulation of pIP501 replication. The minimal replicon with the *copR* and *repR* genes is shown, separated by the 329-nt-long leader region. CopR represses transcription from the *repR* promoter pII and, at the same time, indirectly increases transcription initiation from the antisense promoter pIII. The antisense RNA causes premature termination of *repR* (sense) RNA transcription at the attenuator (*att*). (**Lower part**) Mechanism of transcriptional attenuation. For details see text. Complementary sequence elements are designated A, B, a, and b. Green arrow, RNase III. (**B**) Inhibition of primer maturation: plasmid ColE1. (**Upper part**) Schematic representation of the minimal replicon. (**Lower part**) Mechanism of inhibition of primer maturation. Violet circle, RNA polymerase. For details, see text. (**C**) Inhibition of pseudoknot formation: plasmid ColIb-P9. (**Upper part**) The minimal replicon with the leader peptide *repY* (dark grey) and *repZ* genes is shown. White: leader region of *repZ* mRNA. (**Lower part**) Genes for *repY* and *repZ* are translationally coupled. On the mRNA, the *repY* SD sequence is exposed, whereas structure III sequesters both the *repZ* SD sequence (black rectangle) and the 5′-rCGCC-3′ sequence (thick black line) and, thereby, *repZ* translation. Inc is the region complementary to the antisense RNA; black circle, *repY* start codon; grey circle, *repY* stop codon. Unfolding of structure II by the ribosome stalling at the *repY* stop codon results in formation of a pseudoknot by base paring between the 5′-rGGCCG-3′ and 5′-CGCC-3′ (thick black lines) sequences distantly separated, and allows the ribosome to access the *repZ* RBS. Binding of Inc RNA to the loop of structure I of *repZ* RNA directly inhibits formation of the pseudoknot and the subsequent IncRNA-*repZ*-mRNA duplex formation inhibits *repY* translation. (**D**) Translation inhibition by inhibition of ribosome binding. (**Upper part**) Working model on regulation of plasmid pMV158 replication. (**Lower part**) The antisense RNA binds directly upstream of the extended non-SD sequence (light blue circle) 5′ of the *repB* start codon and prevents binding of the 30S ribosomal subunit. The CopG protein represses transcription from the *copG-repB*-promoter and from the *repB* promoter. (**E**) Inhibition of leader peptide translation. (**Upper part**) Working model on regulation of plasmid R1 replication. (**Lower part**) Translation of the leader peptide (black box) *tap* is required for efficient *repA* translation. The CopB protein represses transcription from the *repA*, but not from the *copB* promoter. (**F**) Translational attenuation. (**Upper part**) Working model on regulation of plasmid pSK41 replication. (**Lower part**) The antisense RNA interacts via three loops with the nascent *repA* mRNA resulting in a stem-loop structure that sequesters the ribosome binding site. In the absence of RNAI, the *repA* mRNA refolds into an alternative structure that exposes the ribosome binding site, allowing *repA* translation. doi:10.1128/microbiolspec.PLAS-0001-2013.f1.

length of the spacer connecting L3 and L4 are not important. Complex formation progresses into the lower stems of both loop pairs, but the complex is not a full duplex (45). Interestingly, the 5′ loop of the *repR*-mRNA (RNAII) contains a 5′ YUNR motif. In all antisense-RNA regulated plasmid systems either antisense or sense RNA carries such a loop sequence (46, 47), which yields a sharp bend that facilitates the interaction between sense and antisense RNA loops. Experiments demonstrated that the U-turn in RNAII is indeed formed and necessary for efficient interaction with RNAIII (48).

In contrast to plasmids from Gram-negative bacteria, deletion of either control component in pIP501, RNAIII or CopR, fails to give "runaway" replication but causes a significant increase in copy number (19, 29). The same holds true for pT181 (23). We hypothesize that in Gram-positive bacteria, a limiting host factor (e.g., a chromosomally encoded enzyme needed for plasmid replication) is responsible for this behavior (28).

Interestingly, antisense-RNA mediated transcriptional attenuation has not been found as a replication control mechanism for plasmids of Gram-negative bacteria. We showed that this mechanism principally functions in *E. coli*, ruling out significant differences in the transcriptional machineries, albeit at a much lower efficiency than in *B. subtilis* or *Staphylococcus aureus* (49). This might be due to some nucleolytic or processing activity present in *E. coli* but lacking in Gram-positive hosts, which could affect the concentrations of the interacting RNAs and the distribution of their inactive and active processing products. We suggest that transcriptional attenuation, which leads to a much broader copy number distribution than a control mechanism based on inhibition of translation (as, e.g., in R1 [see below]), can only be tolerated by plasmids that do not tend to "runaway replication."

In plasmid pSM19035, the ω-protein presents a link between copy number control and better-than-random segregation (50). It acts as transcriptional repressor at its own, the *copS* and the δ-promoter. Omega dimers repress *copS* transcription ≈8-fold by binding to 7-bp arrays with the sequence 5′-NATCACN in the major DNA groove by using an antiparallel β-sheet structure (51, 52). The crystal structure of the ribbon-helix-helix protein ω protein was solved (53). Although the ω open reading frame (ORF) is present in pIP501, it is not preceded by a promoter, and, therefore, not transcribed (S. Brantl, unpublished data). Apparently, pIP501 does not seem to need ω for proper copy number control.

Previously, it has been shown that pAMβ1 replication proceeds unidirectionally via the theta mechanism (12). Replication is initiated by DNA polymerase I (54) and depends on a transcription step through the origin that is proposed to generate the primer for DNA replication. Since termination of *repE* transcription within the origin is not very efficient, cleavage by either RepE itself, or by an RNA polymerase-associated RNase would be feasible (55). Interestingly, RepE has an RNase activity that can cleave free RNA molecules of *repE* mRNA polarity in close proximity to the initiation site of DNA synthesis (C. Canceill and E. Le Chatelier, personal communication). This argues for a direct involvement of RepE in primer synthesis from *repE* mRNA.

Two independent mechanisms involving (i) a protein/DNA complex that acts as a roadblock and (ii) a plasmid-encoded type I topoisomerase that produces topological constraints that impede fork progression, mediate the arrest of Pol I (56, 57). A primosome assembly site, located in the arrested D loop on the lagging-strand template, or the forked structure of the D loop, are sites for assembly of a PriA-dependent (restart) primosome formed of proteins PriA, DnaB, DnaD, and DnaI (55, 58, 59). This assembly is thought to recruit a replication fork in several steps including (i) loading of the DnaC helicase and DnaG primase on the lagging-strand template, (ii) initiation of lagging-strand synthesis, and (iii) the Pol I to PolC switching at the tip of the arrested leading strand (56, 57). The replication fork requires at least DNA polymerases PolC and DnaE, the processivity (clamp) factor DnaN, and the clamp-loading complex containing DnaX (13, 60). Surprisingly, DnaE polymerizes the plasmid lagging strand, and PolC the leading strand (60). A primosome assembly site (*ssi*) has been located on the lagging-strand template, ≈150 nt downstream of the origin. Lagging-strand synthesis is inefficient when any of the proteins involved in *ssiA* activity is mutated suggesting that normal plasmid replication requires primosome assembly. However, plasmid replication can occur efficiently in the absence of *ssiA* indicating that the primosome can also assemble elsewhere on the plasmid (61). The analysis of pAMβ1 RepE (62) revealed that it is a double-strand (ds) and single-strand (ss) DNA-binding protein. It is monomeric in solution and binds specifically, rapidly, and tightly to the origin at a unique binding site immediately upstream of the initiation site, thereby inducing a weak bend of 31°. RepE binding to the ds origin leads to denaturation of the AT-rich sequence immediately downstream of the binding site to form an atypical open complex. A model for successive steps of pAMβ1 replication initiation has been proposed (62).

Since RepR of pIP501 and RepS of pSM19035 are 97% identical to RepE, most likely, the same general mode of origin recognition and replication initiation/termination is used by these plasmids. Because of the combination of Rep dependence, DNA PolI dependence, and the lack of requirement of certain features in the origin like DnaA boxes, iterons, or AT-rich sequences, pAMβ1 and its relatives have been classified as a fourth class of theta-replicating plasmids (54).

INHIBITION OF PRIMER FORMATION: THE ColE1 REPLICON

ColE1 is a representative of many closely related high-copy number plasmids that replicate in *E. coli*. In contrast to all other plasmids, it does not require a plasmid-encoded replication initiator protein. The only essential plasmid-encoded component is an RNA primer, RNAII, which is the target for copy number control. First, a 550-nt-long preprimer is synthesized by host-RNA polymerase. During synthesis, this pre-primer undergoes specific conformational changes, which are required for its activity (63, 64, 65, 66). The active conformation of this RNA forms a persistent hybrid, which involves two regions of contact between RNAII and the DNA in the origin region (67). The RNA of the RNA-DNA hybrid is cleaved by host RNase H and converted to a mature primer for replication (64, 67, 68, 69) that delivers the free 3′-OH end required by DNA polymerase I, which extends it starting leading strand synthesis. Later, DNA polymerase I is replaced by DNA polymerase III holoenzyme (70). During the initial elongation of the leading strand by PolI, creating a D-loop structure of increasing size, a specific DNA sequence at the original lagging strand becomes single stranded. This so-called primosome assembly site (*pas*) is the functional equivalent of the single-strand origins of rolling-circle type plasmids. Once single-stranded, *pas* recruits the primosome complex to initiate lagging-strand synthesis. The DnaA protein, and the DnaA recognition site at the ColE1 origin of replication seem to be important for ColE1 replication. The effect of DnaA is enhanced when the *pas* site is defective suggesting that DnaA plays a role similar to that of the proteins i, n, n′, and n″ in directing primosome assembly (71).

Replication control is mediated by a 108-nt-long antisense-RNA, RNAI, which is transcribed constitutively from the complementary strand in the preprimer region (Fig. 1B) (72, 73). RNAI consists of three stem-loops and an unstructured 5′ tail (74). Binding of RNAI to RNAII prevents the refolding of the nascent pre-primer; the structure formed upon RNAI binding is incompatible with the formation of a persistent RNA-DNA hybrid within the origin region, and, consequently, primer maturation is prevented. As in pIP501, a time window exists, during which inhibition can occur. RNAI can bind to pre-primers of all lengths; however, only when it interacts with a target of 100 to 150 nt in length primer in formation is blocked (75). Binding at later stages does not result in inhibition. Mutations that affect copy numbers and result in new incompatibility groups, have been mapped to the loops of RNAI as the most important determinants for binding rate and specificity (76, 77).

Tomizawa has analyzed the stepwise conversion of initial RNAI/RNAII binding intermediates to progressively more stable structures (78, 79, 80, 81, 82). Binding follows a two-step pathway. It initiates between one or two loop pairs (out of three). A reversible unstable kissing complex, C_χ, whose structure is not known but which is likely to involve a single pair of stem-loops, initiates binding. By kinetic inhibition and RNase protection studies, the rate constant of formation of the more stable complex C* was determined at 6×10^6 M^{-1} s^{-1}. Subsequently, a kissing complex possibly involving all three RNAI loops is formed with $\approx 3 \times 10^6$ M^{-1} s^{-1}. Finally, stable complex formation (complex C_s) occurs at a rate constant of 10^6 M^{-1} s^{-1} (78, 79, 81, 82). Thus, the high rates characteristic of early steps are almost maintained throughout the binding pathway. RNAI lacking its 5′ tail is arrested at this stage, but inhibits primer formation *in vitro* and *in vivo*, which implies that a full RNA duplex is not required for control (e.g., 82, 83, 84). The duplex is formed very slowly, concomitant with stepwise loss of loop-loop contacts and unfolding of the stem-regions (85). It was concluded that all seven loop-bases are base paired to each other, creating a coaxial stack of the two stems bent at the loop-loop helix (78). Nuclear magnetic resonance studies, although with a loop-sequence inversion, suggested the same properties (85, 86). In summary, the ColE1 family represents a case where the antisense RNA does not affect the expression of a protein-coding gene, but the activity of a target RNA by induction of a nonfunctional conformation.

For degradation of RNAI, RNase E, PcnB, and PNPase play the decisive roles (87, 88). The initial event is cleavage by RNase E. PcnB adds a poly(A) tail to the 3′ end of RNAI, which greatly facilitates the ability of the exoribonucleases RNase II or PNPase to degrade RNAI. Deletion of PcnB had been observed earlier to yield ≈10-fold copy number down effects (reviewed in reference 88).

A second plasmid-encoded control component is the small Rop (repressor of primer) or Rom (RNA one modulator) protein (63 aa) encoded downstream of the origin. Homodimeric Rom promotes the interaction between RNAI and RNAII, i.e., the conversion of the unstable RNAI-RNAII complex to a stable complex, thereby increasing inhibition of replication (84, 89, 90). However, Rom deletion has only a minor effect on copy number control (ca. 2- to 3-fold increase in copy number in slowly growing cells, but no effect on copy number in fast-growing cells [91]). Extra *rom* copies do not cause incompatibillity showing that it is not a primary inhibitor of ColE1 replication, but only an auxiliary factor that exerts its maximum effect at wild-type concentration (8). The crystal structure of Rom, determined at 1.7-Å resolution, reveals a bundle of four tightly packed α-helices held together by hydrophobic interactions (92). Rom mutants with decreased activity were all clustered at the extremities of the α-helix bundle, with the exception of F14 (93). Amino acids involved in RNA recognition form a narrow stripe down one face of the bundle and are symmetrically arranged, with recognition centered around the two F14 residues that interact with the loop region of the hairpin pair, with additional interactions between eight polar residues and the RNA backbone (94). Rom recognizes the RNA in a structure- rather than sequence-dependent fashion.

An analysis of the high-copy number ColE1 derivatives pUC18/pUC19 used as cloning vectors in many laboratories has shown that they contain a single point mutation in RNAII, which can be phenotypically suppressed by Rom. This mutation seems to alter the secondary structure of RNAII and produces a temperature-dependent alteration of RNAII conformation. Rom may either promote normal folding of mutated RNAII or enable the interaction of suboptimally folded RNAII with RNAI (95).

The intracellular concentrations of RNAI, RNAII, and Rom, have been determined with 1 μM, 7 nM, and 1 μM, respectively, and the authors suggest that plasmid copy number is little affected by the rate of RNAII synthesis but is strongly dependent on that of RNA I (96). Two mathematical models of ColE1 copy number control have been published. In the first model, the plasmid copy number is greatly influenced by changes in the rate constant for formation of the initial unstable RNAI-RNAII complex, but is only slightly influenced by changes in the dissociation rate of this complex. The presence or absence of Rom does not seem to quantitatively alter the copy number control mechanism (97). The second, not experimentally tested, model (98) made three theoretical proposals to account for an important role of Rom: First, Rom concentration would be proportional to the copy number, so that the response in replication frequency to variations in the copy number would be sharper than in the presence of RNAI alone. This hypothesis requires that Rom is rapidly degraded, which has not been investigated. Second, Rom would cause the probability of plasmid replication to approach zero at high RNAI concentrations, because, in the absence of Rom, the intrinsic rate of RNAI/RNAII duplex formation would be too slow to ensure total inhibition of replication. Third, Rom could act as back-up system when the copy number is greatly reduced: under normal conditions, the replication frequency would not depend on small deviations in Rom concentration but, if this concentration decreases greatly, inhibition of primer formation would decrease, thus leading to an increased replication frequency. The presence of *rom*$^-$-ColE1 derivatives reduced bacterial growth in carbon source-poor medium, whereas *rom*$^+$ derivatives did not show such effects on cell growth (91). From these observations and from the fact that amplification of ColE1 derivatives in slowly growing cells is higher with *rom* mutant plasmids, a key role for Rom has been suggested: to prevent ColE1-type plasmids from representing a metabolic burden to their hosts in natural habitats where cells grow much slower than under laboratory conditions.

INHIBITION OF PSEUDOKNOT FORMATION: THE IncI/IncB CASE

The IncB, IncIα, IncZ, IncK, and IncL/M plasmids of Gram-negative bacteria form a family of low-copy number plasmids, which is similar to the IncFII family but uses another mechanism of antisense-RNA-mediated inhibition (99, 100). The two best-characterized examples are ColIb-P9 (Incα) and pMU720 (IncB). Here, an antisense RNA (only regulator; no transcriptional repressor needed) inhibits formation of a long-distance RNA pseudoknot that is required for efficient translation of the essential, rate-limiting replication initiator protein Rep (101, 102, 103). Additionally, a leader peptide ORF, *repY* in ColIb-P9, must be translated to permit synthesis of RepZ (104, 105). Figure 1C illustrates the regulatory circuit. Two stem-loop structures in the *repZ* mRNA that have been mapped *in vitro* (106) and are located upstream (structure I) of the *repY* RBS and in the middle (structure III) of the *repY* gene are necessary for replication control. Structure III occludes both the *repZ* RBS and a short sequence complementary to a region in the loop of structure I. Appropriate termination of *repY* translation unfolds structure III, which in turn allows the formation of a short helix between the target loop and the disrupted stem, located ≈100 nt apart, thus inducing the formation of the activator pseudoknot by intramolecular pairing of loop I with its complementary sequence. Pseudoknot formation facilitates ribosome binding to the *repZ* RBS. The pseudoknot could be mapped *in vitro* using mutations that disrupt structure III (106).

The indispensable copy number regulator is a ≈70-nt antisense-RNA (RNAI or Inc-RNA) encoded upstream of the *repY*-ORF (as CopA in R1) (107, 108). RNAI has a dual function. On the one hand, it blocks directly *repY* translation by sequestering the *repY* RBS, and, on the other hand, it prevents activation of *repZ* translation, since the site of RNAI/*repZ* mRNA interaction involves the nucleotides in structure I required for pseudoknot formation (109, 110). In this way, RNAI can repress *repZ* translation at the level of a transient interaction with its target before a complete duplex is formed, similar to the R1 and the pIP501 cases (see reference 27). In the case of the IncIα plasmids, a hexanucleotide, which includes the structure I sequence involved in the initial interaction presumably supports a U turn (see above). The early stages in pseudoknot formation and in the binding of the RNAI are similar. However, RNA I represses *repY* translation much less efficiently than *repZ* translation. Repression of *repZ* and *repY* expression are accomplished at different stages during the pairing between RNAI and *rep* mRNA (111). This differential repression allows RNAI to keep the total level of *repZ* expression constant thereby ensuring a constant copy number value.

Although a similar structure and regulatory mechanism as in IncIα plasmid ColIb-P9 exists in the IncB plasmid pMU720 (112, 113), some differences were found in the IncL/M group, represented by plasmid pMU604 (114). In contrast to ColIb-P9, the positioning of proximal pseudoknot bases involved in the expression of the essential *repA* gene is different, which may result in differences in their presentation thus affecting the process of pseudoknot formation. The requirement for pseudoknot formation in pMU604 could be obviated by mutations that improved the sequence of the *repA* RBS. The authors demonstrated that, although the pseudoknot was essential for expression of the *repA* gene, its presence interfered with translation of *repB* encoding the leader peptide. The spacing between the distal pseudoknot sequence and the *repA* RBS was shown to be suboptimal for maximal expression of *repA*. Since more optimal spacing increased the IncL/M plasmid copy number and pMU604 is a derivative of the large conjugative plasmid pMU407.1, suboptimal spacing may have evolved to ensure a lower copy number to reduce the metabolic burden on the host (114).

Antisense/sense RNA binding of the IncIα and IncB group plasmids occurs in a two-step pathway, very similar to that of CopA/CopT of plasmid R1 described below (e.g., references 111 and 115). The antisense RNAs carry only one major stem-loop with a hexanucleotide loop, a destabilized upper stem, and a 5′ tail. The loop sequence is identical in all IncB-related plasmids, and is also identical to R1. However, no incompatibility was observed between these plasmids because of sequence differences in the upper stem regions. The initial interaction between Inc RNA and its target occurs via loop-loop contacts. Subsequently, helix progression unfolds the upper stems, resulting in a four-helix junction structure. The proposed secondary structure resembles that of the CopA/CopT complex of plasmid R1, but differs in the position of the junction (111). Recent data suggest that antisense/sense RNA complexes of ColIb-P9, R1, and many other distantly related plasmids may share the same overall topology including the position of the junction (116). Apparently, efficient inhibitory antisense RNAs initiate interactions by loop-loop contacts, but the subsequent steps of helix progression to stable inhibitory complexes depend on topological constraints to keep topological stress at a minimum. In summary, the Incα/IncB-plasmid family uses antisense RNAs for translational inhibition, but with an unusual twist: inhibition works through prevention of an activator pseudoknot structure (Fig. 1C).

For IncB plasmid pMU720, sequence requirements for the origin and a novel essential *CIS* element have been studied. The replication origin was shown to contain a 5′-A/TANCNGCAAA/T-3′ motif which is also present on pMU407.1, but not on IncZ plasmids. This motif is repeated four and two times in the origins of IncB and IncL/M plasmids, respectively, and might represent the binding site of RepA. A DnaA box was shown to be inessential. However, its deletion reduced the copy number of the IncB replicon 3-fold. *CIS*, a 166-bp sequence separating the *repA* gene from the origin, contains two domains, a *repA* proximal domain with strong transcription termination activity, and a *repA* distal domain acting as a spacer to position sequences within *ori* on the correct face of the DNA helix. A model for RepA loading on the origin, which involves an initial interaction between nascent RepA and the RNA polymerase transcribing the *repA* mRNA was discussed (117). In a two-plasmid situation (*repA*$^{+}$*ori*$^{-}$ plasmid and *repA*$^{-}$*ori*$^{+}$ plasmid in the same cell), *CIS*—when present on a *repA*$^{+}$*ori*$^{-}$ plasmid—inhibited replication of the *ori*⁺ plasmid by interacting with the C-terminal 20 to 37 aa or RepA. In contrast, it had no effect when present on the *ori*⁺ plasmid. Initiation of replication from the *ori* in *trans* was independent of transcription into *CIS* (118).

TRANSLATIONAL INHIBITION

Inhibition of Translation of the *rep* mRNA

Plasmids of the pMV158 family (pMV158 from *Streptococcus pneumoniae*, pE194 from *S. aureus*, pADB201 from *Mycoplasma mycoides*, and pLB4 from *Lactobacillus plantarum*) apparently use the same copy number control mechanism (119, 120, 121). The replication of these rolling-circle-type plasmids is controlled by two components, an antisense RNA and a transcriptional repressor. Both elements control the synthesis of the essential replication initiator protein. The antisense RNA inhibits translation of the *rep* mRNA.

The best-characterized plasmid of this family is the promiscuous streptococcal plasmid pMV158 and its derivative pLS1. Here, a 50-nt-long antisense RNA, RNAII, is complementary to the intergenic region between the *copG* and the *repB* ORF encoding the regulator CopG and the essential 24.5-kDa replication initiation protein RepB, respectively (Fig. 1D). The *cop-rep* target region could be reduced to 21 nt and is complementary to the 5′-unstructured region of RNAII. RNAII is the main incompatibility determinant of the plasmid (119). Recently, the secondary structures of RNAII (one stem-loop) and two shortened (60 and 80 nt) *repB* mRNA (two stem-loops) species have been determined, and binding rate constants have been calculated (122). These were with 1.6×10^5 to 3.8×10^5 M^{-1} s^{-1}, in the same order of magnitude as those of other sense/antisense RNA pairs. Interestingly, the initiation of *repB* translation involves an extended non-Shine-Dalgarno (SD) sequence, i.e., not a typical SD sequence (123). Although the complementarity between RNAII and its target neither includes the *repB* start codon nor the extended non-SD, previous experiments had shown that RNAII inhibits *repB* translation (124). Possibly, RNAII binding upstream of the *repA* AUG codon interferes with binding of the 30S ribosomal subunit. Alternatively, a conformational change might be induced that prevents translation initiation, as has been found in *trans*-encoded sRNAs (reviewed in references 3 and 4). However, this can only be elucidated by structure probing of the RNAII/*cop-rep*-mRNA complex.

The second control component of pLS1 is the transcriptional repressor CopG (formerly RepA). By binding to its own promoter, the homodimeric protein CopG (45 aa, 5.1 kDa) represses its own synthesis and that of RepB (125). CopG is not essential, since its deletion affects neither plasmid replication nor maintenance. An increase in CopG dosage does not result in incompatibility toward pLS1 or in any significant reduction of its copy number. Hence, CopG is not able to efficiently correct major fluctuations in plasmid copy number, probably because of its autoregulatory role (119). Instead, the entire regulatory unit including RNAII and CopG proved to be the strongest incompatibility determinant. RNAII-defective plasmids were still regulated by CopG indicating that their copy number was not limited by a host function. Plasmids with *copG* mutations or deletions also replicated in a regulated way, and attempts to construct an *rnaII/copG* double mutant were unsuccessful (120). Some CopG features resemble those of CopB from plasmid R1 (see below). However, CopB, in contrast to CopG does not regulate its own synthesis, and the CopB-repressed promoter is totally silent, being activated only when the copy number drops dramatically. In contrast, the *copG-repB* promoter P_{cr} from pLS1 is never totally blocked and seems to be the only promoter involved in *repB* expression. CopG binds at two successive major grooves on one face of the DNA to a 13-bp element with a 2-fold rotational symmetry. Within this imperfect repeat element lies the −35 box of P_{cr} (125). The crystal structure of CopG has been solved both alone and in complex with a 19-bp oligodeoxyribonucleotide containing its DNA target (126). The CopG dimer has a ribbon-helix-helix structure resembling that of the P22 Arc repressor. One CopG tetramer binds at one face of a 19-bp oligonucleotide containing the pseudosymmetric element, with two β-ribbons inserted into the major groove. Thereby, the DNA is bent by 60° through compression of both major and minor grooves. In contrast to other repressors, CopG uses its HTH region for oligomerization instead of DNA recognition. Lately, the repression mechanism of CopG has been elucidated: CopG prevents access of the RNA polymerase to the *repB* promoter and actively dissociates open complexes (127).

For one more plasmid of the pMV158 family, pE194, it was also shown that replication control involves the concerted action of a Cop protein that acts as repressor at its own, but not the *repF* promoter, and a ≈65-nt-long countertranscript RNA with a predicted single stem-loop (128). The *cop* operator comprises a 28-bp inverted repeat. Increasing the proposed 6-nt RNA loop of the antisense RNA to 14 nt or decreasing it to 4 nt resulted in elevated copy numbers (128). The 6.1-kDa Cop protein of pE194 has been purified (129).

A mechanism that involves RNA-RNA interactions in a manner that interferes with translation was also suggested for pC194 and pUB110, two other staphylococcal RCR-type plasmids (130).

Recently, ≈72-nt antisense RNAs with two predicted stem-loop structures were found to be the major replication control elements in plasmids pCGR2 and pCG1 from *Corynebacterium glutamicum*. Inactivation of the antisense promoters resulted in significantly higher amounts of *repA* mRNA and 7-fold elevated copy numbers (131). Based on the complementarity of their immediate 5′ regions to the *rep* RBS and the lack of rho-independent terminator structures in the *rep* leader regions, a translation inhibition mechanism is conceivable. Surprisingly, the deletion of a second locus upstream of *repA*, *parB*, yielded a drastic 87-fold increased copy number in a pCGR2 derivative (132).

Inhibition of Leader Peptide Translation: R1 and Related Plasmids

The best-studied example for this type of replication control is the IncFII plasmid R1 that replicates in *E. coli* and closely related bacteria. However, plasmids of the IncFc, the FIII, and some other incompatibility groups exhibit the same genetic organization and use the same control mechanism. The basic replicon contains the ORFs *copB*, *tap*, and *repA* encoding the small transcriptional repressor CopB, a 24-aa leader peptide TAP, and the essential initiator protein RepA, respectively. The latter is rate limiting for replication. The replication origin *oriR* is located downstream of the *repA* gene (Fig. 1E) and was characterized previously (133): The minimal *oriR* is 188 bp long and separated from *repA* by a ≈170-bp-long sequence denoted CIS, which is required for efficient replication of a *repA-oriR* plasmid *in vivo*. CIS contains a rho-dependent transcriptional terminator, which terminates *repA* mRNA at position 1299 and is required for *cis* action of RepA *in vitro*. A DnaA box (TTATCCACA) is found at position 1427 to 1435, consistent with a DnaA requirement for replication. An AT-rich (87% AT) sequence, essential for *oriR* function, is located between nt 1513 and 1586. Three TCNTTTAAA repeats, separated by 23-bp intervals, and a putative integration host factor site are present in this region (however, integration host factor is dispensable). Replication depends on DNA gyrase and other host functions, such as DnaB, DnaC, DnaG, SSB, and DNA polymerase III, but does not require RNA polymerase. About 40 to 50 molecules of *cis*-active RepA and one or two DnaA monomers per template are necessary for initiation from *oriR*. It has been speculated that formation of a nucleoprotein structure, involving part of the *oriR* sequence, RepA and DnaA, is essential for the initiation of R1 replication. A consensus recognition sequence of RepA ("*repA* box") has been identified (reviewed in reference 133). Although RepA alone can promote opening of the helix and assembly of the replisome complex at *oriR*, the DnaA box and DnaA help optimize the initiation frequency (134). A model for the initiation of plasmid R1 replication was proposed in 1992 (135). *oriR* contains two sites with different affinities for RepA separated by 8 helical turns. The *oriR* sequence becomes bent upon RepA binding. Protein-protein interactions between RepA bound to both distal sites could be responsible for *oriR* looping. Two so-called Ter sites, which bind the *E. coli* Tus protein, have been located near *oriR* (136). Inactivation of the *tus* gene caused a great decrease in stability of maintenance of an R1 miniderivative. The downstream Ter site appears to stabilize the plasmid by preventing multimerization and affects a shift from theta to rolling-circle replication (136).

Within the gene segment encoding the leader region of the *repA* mRNA, a ≈90-nt-long antisense RNA, CopA, is transcribed from the complementary strand (Fig. 1E). Regulation occurs on two levels. The main control element is CopA, an RNA that contains two stem-loops and is unstable (1- to 2-min half-life [137]). Its target, CopT, is part of the *repA*-mRNA leader region (138). Binding of CopA to CopT sterically blocks initiation of translation of the Tap leader peptide (139) and also results in RNase III-dependent cleavage of both RNAs. Cleavage, however, has only minor effects on control (140). *Tap* translation is required for *repA* translation (translational coupling) since a stable RNA secondary structure blocks the *repA* RBS (141, 142, 143). Consequently, the CopA antisense RNA inhibits *repA* translation via inhibition of translation of the Tap leader peptide. The alternative hypothesis that an activator RNA pseudoknot, similar to that in IncIα/IncB plasmids (see above), may be needed for efficient *repA* translation was discarded (142). Copy number mutants map to the loop of the major stem-loop L1 (144, 145), and many of these mutations result in new incompatibility groups. Therefore, this stem-loop plays a central role in the rate-limiting step in binding and the efficiency of control (145, 146, 147, 148), as well as the specificity of target recognition. A thorough analysis of CopA showed that a loop size of 5 to 7 nt was optimal for efficient interaction with CopT, and bulges present in the upper stem of CopA were required for rapid binding of CopT *in vitro* and inhibition *in vivo* (149, 150). The binding pathway between CopA and CopT has been elucidated in detail. Binding starts with the interaction of two single loops of CopA and CopT. The low stability of the upper stem regions facilitates progression of this loop-loop interaction. Next, a partial duplex is formed that contains a four-helix junction (151, 152). This intermediate is converted into a stable inhibitory complex, which carries a fifth intermolecular helix (153). This structure is only slowly converted into a complete duplex (for a figure see reference 154). A full duplex is clearly not required for control (153; reviewed in reference 27). The interaction between two highly structured antisense and sense RNAs, initiating by defined loop-loop contacts as shown for plasmid R1, is a recurrent one and valid for most cases of plasmid replication control.

The degradation pathway of CopA has been studied in detail. As in ColE1, RNase E performs the initial cleavage—here, between stem-loops I and II—and the longer stem-loop is degraded directly by exoribonucleases RNase II or PNPase, or subject to polyadenylation by PcnB (also known as PAPI, a poly(A)polymerase of *E. coli*), which facilitates subsequent degradation by exoribonucleases (137, 155). In contrast to the ColE1 case (see above), the *pcnB* mutation has a smaller effect on CopA stability and leads only to a 2- to 3-fold copy number increase of plasmid R1.

The second copy number control element of plasmid R1 is the transcriptional repressor CopB. The deletion of *copB* results in an 8-fold copy number increase (156). The *copB* gene is cotranscribed with *tap* and *repA* from promoter pI. CopB is a small (11 kDa), basic protein (157) that binds as a tetramer to a DNA region of dyad symmetry overlapping the *repA* promoter. The binding site was narrowed down to 20 to 25 bp including an inverted repeat sequence, which overlaps the −35 box of the *repA* promoter. At steady state, the CopB-repressed *repA* promoter pII is almost entirely silent; binding occurs at an equilibrium dissociation constant of 0.1 nM. Approximately 1,000 molecules of CopB are present per cell. Under these conditions, *repA* is expressed almost exclusively from the *copB* promoter pI. Like CopR of pIP501, but in contrast to CopG, CopB does not autoregulate its own synthesis. The cloned *copB* gene does not exert incompatibility against wild-type R1, indicating that it is only an auxiliary control component. Although CopB prevents convergent transcription from *repA* and *copA* promoters (158), this has less serious consequences than in the case of pIP501, where the unusually long-lived antisense RNA needs a second control element upon downward fluctuations of copy number (see above). The main biological role of CopB appears to be a rescue device at dangerously low copy numbers and/or after conjugal transfer of R1 (159). During normal steady-state conditions, the unstable antisense RNA CopA is sufficient to correct copy number deviations.

Translational Attenuation: pSK41 and pSK1 Families

The first known example is *S. aureus* plasmid pSK41 (160, 161), the prototype of the pSK41 family of conjugative multiresistance staphylococcal plasmids. It is suggested that the pSK1-like multiresistant nonconjugative staphylococcal plasmids that encode similar Rep proteins and regulatory elements and use theta-type replication in a variety of staphylococcal species employ the same control mechanism (160, 162).

The proposed mechanism resembles that of a translational riboswitch: the *repA* mRNA can adopt two conformations, depending on the presence or absence of the antisense RNA (RNAI). In the presence of RNAI (83 nt), which is transcribed in the 240-nt *repA* leader region and is not complementary to the SD sequence, the interaction between three RNAI loops and three regions in *repA* mRNA results in a conformation that sequesters the SD in a thermodynamically stable stem-loop structure preventing translation (Fig. 1F). In the absence of RNAI, refolding occurs, making the SD sequence accessible to ribosomes (161). So far, no experimentally probed structures of antisense and sense RNA or their complex are available. A hypothetical transcription attenuation mechanism was excluded by mutation of a putative rho-independent terminator (inverted repeat IR-IV) overlapping the *repA* RBS: replacing the 3′-U stretch by UCACU or alteration of paired stem structures did not result in elevated *repA* expression. A slight additional (2- to 3-fold) effect of RNAI on *repA* transcription observed in reporter gene fusions could so far not be explained, as transcriptional interference could be excluded by providing RNAI in *trans*. One possible explanation is an influence of RNAI on *repA* mRNA degradation (161), but this could be a consequence of translational inhibition, as found in many *trans*-encoded sRNA systems (3, 4). Plasmid pSK1 also encodes an 80-nt antisense RNA within the 199-nt leader region upstream of the *rep* RBS. When absent, the copy number increased ≈8-fold (162), whereas for pSK41, a ≈35-fold increase was observed. The predicted RNA structures also suggest a translation attenuation mechanism for pSK1. Despite differences in sequence and predicted RNA structures, the conjugative and the nonconjugative multiresistance staphylococcal plasmids seem to utilize the same replication control mechanism.

The *repC* and *repABC* Plasmids: Transcriptional or Translational Attenuation?

Large plasmids in α-proteobacteria encode genes required for plant or animal pathogenesis or symbiosis. Most of these replicons encode *ABC* genes that are always in the same order. RepA and RepB are required for active segregation, whereas RepC is the replication initiator protein (reviewed in reference 163), also in the smaller *repC* plasmids. In several of these plasmids, e.g., *Rhizobium etli* p42d, *Sinorhizobium* megaplasmids SymA/SymB and *Agrobacterium tumefaciens* Ti plasmid and in the *repC* plasmids (164), short antisense RNAs are encoded upstream of *repC*. The 59-nt antisense RNA (ctRNA) of plasmid p42d was reported in 2004 (165) and has recently has been analyzed in detail (166). Secondary structures of antisense and target RNAs as well as their complex were determined and binding constants calculated. The ctRNA forms one stem-loop, and its single-stranded 5′ tail was found to be sufficient for inhibition. In the absence of the ctRNA that is encoded 110 bp upstream of the *rep* ORF, the *rep* RBS is single stranded and, thus, accessible to ribosomes. In the ctRNA/*rep* RNA complex, refolding results in a stem-loop structure (S element) that sequesters the RBS in a double-stranded region. The authors suggest a transcription attenuation mechanism. However, the S element lacks a 3′-U stretch typical for rho-independent transcription terminators, and no prematurely terminated *repA* mRNA has been detected *in vivo*. Therefore, a translation attenuation mechanism cannot be excluded (166).

In 2005, antisense RNA genes *incA* and *repE* were found in pSymA/pSymB (167) and Ti plasmid (168), respectively. For Ti, predicted secondary structures of the 206-bp *repB-repC* intergenic region in the presence and absence of antisense RNA RepE (54 nt) and a comparison of the amount of *repC* leader and *repC* ORF-RNA indicated a transcription attenuation mechanism (168). The authors hypothesized that only nonterminated transcripts are subject to additional translation control (169). However, as in p42d, folding of the *repC* RNA could also support a general translation attenuation mechanism.

Interestingly, in 2012 it was reported for *Salmonella* that a translational riboswitch controls Rho-dependent transcription termination (170). Because neither in Ti nor in p42d, a typical Rho-independent terminator was found, but different amounts *repC* mRNA leader were detected in the presence and absence of the antisense RNA, a similar overlap of translational attenuation and Rho-dependent transcription termination could be also conceivable for the *repABC* plasmid family.

THE ColE2 CASE

The initiator protein Rep (35 kDa) of the colicin E2 (ColE2) plasmid (10 to 15 copies/host chromosome) is the only plasmid-specified *trans*-acting factor required for initiation of plasmid replication (171). It was shown that Rep is a ColE2-specific plasmid-encoded primase with unique properties (171, 172). Host DNA polymerase I specifically uses the primer RNA 5′-ppApGpA generated by Rep to start DNA synthesis (171). Replication is unidirectional. Leading-strand synthesis initiates at a unique site in the origin, and lagging-strand synthesis terminates at another unique site in the origin (173). Expression of Rep is negatively controlled posttranscriptionally by a 115-nt-long antisense RNA, RNAI, which is complementary to the 5′-nontranslated region of the *rep* mRNA. The hybridized 5′ terminus of RNAI is located 16 nt upstream of the *rep* start codon and, therefore, does not cover the AUG and only part of the putative SD sequence (174). RNAI consists of two stem-loops. The 3′-terminal long stem-loop makes the initial contact with the corresponding loop in the *rep* mRNA, since many copy number mutants (173) have been mapped there. The binding rates constant between RNAI and *rep* mRNA is 3.1×10^6 M^{-1} s^{-1}, i.e., only slightly higher than in other antisense RNA regulatory systems (174). For efficient Rep expression, two regions of the *rep* mRNA are necessary: (i) a sequence 17 nt to 70 nt upstream of the *rep* start codon, and (ii) a sequence within the coding region (175). Furthermore, the *rep* gene lacks an efficient SD sequence and depends on a stem-loop and a purin-rich sequence in the leader region for efficient translation (176), which is reminiscent of pMV158. This is the same region to which the 5′ part of RNAI binds. Binding of RNAI to the *rep* mRNA might block certain sequence elements involved in interaction with the ribosome or other host factor(s) and/or it might block formation of a certain secondary or tertiary structure in the region of the *rep* mRNA required for efficient initiation of translation (174). Some years ago, authors suggested that pseudoknot formation might be involved in the translation of the Rep protein and/or in inhibition of Rep translation by RNAI (T. Itoh, personal communication).

By chimera analysis, specificity determinants in the interaction of the Rep proteins with the origin were found in the plasmids ColE2-P9 and the related ColE3-CA38 (177): Two regions, the C terminus of Rep (A and B) and two sites in the origins (a and b), were important for the determination of specificity. When each A/a and B/b pairs were from the same plasmids, replication was efficient. If only A/a was from the same plasmid, replication was inefficient. A seems to be a linker connecting the two domains of the Rep protein involved in DNA binding, and region B is part of the DNA-binding domain.

The main players for degradation of ColE2 RNAI are endoribonuclease E cleaving multiple positions within the 5′ part of RNAI and the 3′-5′-exoribonuclease PNPase (178, 179).

COMPARISON BETWEEN REGULATORY sRNAs INVOLVED IN PLASMID REPLICATION CONTROL AND THOSE ACTING ON CHROMOSOMALLY ENCODED TARGETS

In general, studies on sense/antisense RNA systems in plasmids have added a lot to our understanding of regulatory systems as such, and, in particular, of control mechanisms in bacteria. Combinations of *in vivo* and *in vitro* assays that have proven useful for the characterization of plasmid-encoded riboregulators have also been used successfully for the investigation of chromosome-encoded regulatory sRNAs. Many control mechanisms discovered for plasmid-encoded antisense RNAs have been later found for sRNAs from the chromosome. Among them, translation inhibition is the most frequently used mechanism employed in both cases (reviewed in references 3 and 4).

However, a comparison of plasmid-encoded antisense RNAs and chromosome-encoded sRNAs reveals five major differences: First, plasmid-encoded sRNAs are constitutively expressed, whereas chromosome-encoded riboregulators are frequently only expressed under certain environmental conditions. Second, antisense RNAs that control plasmid replication act solely as inhibitors of their targets, whereas chromosome-encoded sRNAs can either inhibit or activate gene expression. Third, although all antisense RNAs controlling plasmid replication are encoded in *cis*, they act in *trans*, in line with their function as incompatibility components. By contrast, sRNAs on bacterial chromosomes are encoded in *cis* or in *trans*, and one *cis*-encoded sRNA has been found to date that even acts only in *cis*, namely by transcriptional interference (180). Fourth, no antisense RNA-regulating plasmid replication impacts exclusively target RNA stability, whereas chromosome-encoded sRNAs primarily affect translation or degradation/processing of their target RNA(s), and, in most cases, both processes. Fifth, the abundant RNA chaperone Hfq plays an important role for either stability or function of many chromosomally *trans*-encoded sRNAs, but is not required for the action of the *cis*-encoded sRNAs regulating plasmid replication, because they have a long stretch of complementarity with their target RNAs.

Because of their small size, it is rather unlikely that, after >30 years of intensive research, completely novel regulatory principles will be found in sense/antisense systems located on plasmids. In contrast, in the large bacterial chromosomes that encode on average 200 to 300 sRNAs, new mechanisms of action and new classes of regulators can still be expected. It is conceivable that some RNAs might act in *cis* on one target and in *trans* on one or more other targets, thereby using different regulatory mechanisms on different targets. First indications have been reported recently in *S. aureus* (181) and in the archeon *Methanosarcina mazei* (182). Additionally, "dual function" sRNAs, i.e., sRNAs that act on some targets as base-pairing antisense RNAs and, on others, as peptide-encoding mRNAs, were identified (e.g., *B. subtilis* SR1 [183], reviewed in reference 184), and this so far small group will definitely expand in the next future. New unprecedented functions for peptides encoded by these sRNAs can be envisaged that will add a new layer to the interplay between peptides and RNA.

With regard to their biological functions, plasmid-encoded antisense RNAs control only three distinct processes: replication, conjugation, and segregational stability; and the genes for antisense RNAs and their targets are located on complementary strands of the same molecule. In one case, *E. coli* plasmid R1, three different antisense RNAs regulate the processes mentioned above: CopA (replication), FinP (conjugation), and Sok (maintenance) (154). On the contrary, chromosome-encoded sRNAs have usually several targets (reviewed in reference 3) and are involved in complex regulatory networks (185) that affect all aspects of life: metabolism, stress response, cell-wall composition, pathogenesis, etc. After ≈13 years of extensive research on chromosome-encoded antisense/target systems, it is evident that sRNAs are abundant and versatile regulators, and that we currently only see the tip of the iceberg. For instance, it can be expected that novel classes of very short or very long sRNAs, similar to si/miRNAs or lncRNAs in eukaryotes, with up-to-now unimaginable functions might come to our knowledge. First examples of long *cis*-encoded bacterial sRNAs were reported in the regulation of the *Clostridium acetobutylicum ubiG* operon (180), the control of YabE autolysin synthesis in *B. subtilis* (186), or of virulence in *Salmonella enterica* serovar Typhimurium (187), and many others with hitherto unknown functions were found in recent sequencing projects.

In summary, plasmids have been and still are outstanding systems to study all aspects of bacterial gene regulation, and numerous lessons have been learned from them for the investigation of chromosomal gene regulation.

Acknowledgment. *Conflicts of interest: I declare no conflicts.*

Citation. **Brantl S.** 2014. Plasmid replication control by antisense RNAs. Microbiol Spectrum 2(4):PLAS-0001-2013.

References

1. **Argaman L, Herschberg R, Vogel J, Bejerano G, Wagner EGH, Margalit H, Altuvia S.** 2001. Novel small RNA-encoding genes in the intergenic regions of *Escherichia coli*. *Curr Biol* **11:**941–950.
2. **Wassarman KM, Repoila F, Rosenow C, Storz G, Gottesman S.** 2001. Identification of novel small RNAs using comparative genomics and microarrays. *Genes Dev* **15:**1637–1651.
3. **Brantl S.** 2009. Bacterial chromosome-encoded regulatory RNAs. *Future Microbiol* **4:**85–103.
4. **Brantl S.** 2012. Acting antisense—plasmid- and chromosome-encoded small regulatory RNAs (sRNAs) from Gram-positive bacteria. *Future Microbiol* **7:**853–871.
5. **Nordström K, Molin S, Light J.** 1984. Control of replication of bacterial plasmids: genetics, molecular biology, and physiology of the plasmid R1 system. *Plasmid* **12:** 71–90.
6. **Wagner EG, Altuvia S, Romby P.** 2002. Antisense RNAs in bacteria and their genetic elements. *Adv Genet* **46:** 361–398.
7. **Uhlin BE, Nordström K.** 1978. A runaway-replication mutant of plasmid R1drd-19: temperature-dependent loss of copy number control. *Mol Gen Genet* **165:**167–179.
8. **Summers D.** 1996. *The Biology of Plasmids.* Blackwell Science, Oxford, United Kingdom.
9. **Novick RP, Iordanescu S, Projan SJ, Kornblum J, Edelman I.** 1989. pT181 plasmid replication is regulated by a countertranscript-driven transcriptional attenuator. *Cell* **59:**395–404.
10. **Projan S, Novick RP.** 1988. Comparative analysis of five related staphylococcal plasmids. *Plasmid* **19:**203–221.
11. **Brantl S, Behnke D, Alonso JC.** 1990. Molecular analysis of the replication region of the conjugative *Streptococcus agalactiae* plasmid pIP501 in *Bacillus subtilis.* Comparison with plasmids pAMβ1 and pSM19035. *Nucleic Acids Res* **18:**4783–4790.
12. **Bruand C, Ehrlich SD, Jannière L.** 1991. Unidirectional theta replication of the structurally stable *Enterococcus faecalis* plasmid pAMβ1. *EMBO J* **10:** 2171–2177.
13. **Ceglowski P, Lurz R, Alonso JC.** 1993. Functional analysis of pSM19035 derived replicons in *Bacillus subtilis.* *FEMS Microbiol Lett* **109:**145–150.
14. **Le Chatelier E, Ehrlich SD, Jannière L.** 1993. Biochemical and genetic analysis of the unidirectional theta replication of the *S. agalactiae* plasmid pIP501. *Plasmid* **29:**50–56.
15. **Brantl S, Birch-Hirschfeld E, Behnke D.** 1993. RepR protein expression on plasmid pIP501 is controlled by an antisense RNA-mediated transcription attenuation mechanism. *J Bacteriol* **175:**4052–4061.
16. **Le Chatelier E, Ehrlich SD, Jannière L.** 1996. Countertranscript-driven attenuation system of the pAMβ1 *repE* gene. *Mol Microbiol* **20:**1099–1112.
17. **Pouwels PH, van Luijk N, Leer RJ, Posno M.** 1994. Control of replication of the *Lactobacillus pentosus* plasmid p353-2: evidence for a mechanism involving transcriptional attenuation of the gene coding for the replication protein. *Mol Gen Genet* **242:**614–622.
18. **Duan K, Liu CQ, Supple S, Dunn NW.** 1998. Involvement of antisense RNA in replication control of the lactococcal plasmid pND324. *FEMS Microbiol Lett* **164:**419–426.
19. **Brantl S, Behnke D.** 1992. Characterization of the minimal origin required for replication of the streptococcal plasmid pIP501 in *Bacillus subtilis. Mol Microbiol* **6:** 3501–3510.
20. **Brantl S, Behnke D.** 1992. The amount of the RepR protein determines the copy number of plasmid pIP501 in *B. subtilis. J Bacteriol* **174:**5475–5478.
21. **Manch-Citron JN,Gennaro ML, Majumder S, Novick RP.** 1986. RepC is rate limiting for pT181 plasmid replication. *Plasmid* **16:**108–115.
22. **Bruand C, Ehrlich SD.** 1998. Transcription-driven DNA replication of plasmid pAMβ1 in *Bacillus subtilis. Mol Microbiol* **30:**135–145.
23. **Kumar CC, Novick RP.** 1985. Plasmid pT181 replication is regulated by two countertranscripts. *Proc Natl Acad Sci USA* **82:**638–642.
24. **Brantl S, Nuez B, Behnke D.** 1992. *In vitro* and *in vivo* analysis of transcription within the replication region of plasmid pIP501. *Mol Gen Genet* **234:**105–112.
25. **Brantl S, Wagner EG.** 1994. Antisense-RNA–mediated transcriptional attenuation occurs faster than stable antisense/target RNA pairing: an *in vitro* study of plasmid pIP501. *EMBO J* **13:**3599–3607.
26. **Brantl S, Wagner EG.** 2000. Antisense RNA-mediated transcriptional attenuation: an *in vitro* study of plasmid pT181. *Mol Microbiol* **35:**1469–1482.
27. **Wagner EG, Brantl S.** 1998. Kissing and RNA stability in antisense control of plasmid replication. *Trends Biochem Sci* **23:**451–454.
28. **Brantl S, Wagner EG.** 1996. An unusually long-lived antisense RNA in plasmid copy number control: *in vivo* RNAs encoded by the streptococcal plasmid pIP501. *J Mol Biol* **255:**275–288.
29. **Brantl S, Behnke D.** 1992. Copy number control of the streptococcal plasmid pIP501 occurs at three levels. *Nucleic Acids Res* **20:**395–400.
30. **Brantl S.** 1994. The *copR* gene product of plasmid pIP501 acts as a transcriptional repressor at the essential *repR* promoter. *Mol Microbiol* **14:**473–483.

31. **Brantl S, Wagner EG.** 1997. Dual function of the *copR* gene product of plasmid pIP501. *J Bacteriol* **179:**7016–7024.
32. **Le Chatelier E, Ehrlich SD, Jannière L.** 1994. The pAMβ1 CopF repressor regulates plasmid copy number by controlling transcription of the *repE* gene. *Mol Microbiol* **14:**463–471.
33. **Swinfield TJ, Oultram JD, Thompson DE, Brehm JK, Minton NP.** 1990. Physical characterisation of the replication region of the plasmid pAMβ1. *Gene* **87:**79–90.
34. **Ceglowski P, Alonso JC.** 1994. Gene organization of the *Streptococcus pyogenes* plasmid pDB101: sequence analysis of the *orf eta-copS* region. *Gene* **145:**33–39.
35. **Steinmetzer K, Brantl S.** 1997. Plasmid pIP501 encoded transcriptional repressor CopR binds asymmetrically at two consecutive major grooves of the DNA. *J Mol Biol* **269:**684–693.
36. **Steinmetzer K, Behlke J, Brantl S.** 1998. Plasmid pIP501 encoded transcriptional repressor CopR binds to its target DNA as a dimer. *J. Mol. Biol.* **283:**595–603.
37. **Steinmetzer K, Behlke J, Brantl S, Lorenz M.** 2002. CopR binds and bends its target DNA: A footprinting and fluorescence resonance energy transfer study. *Nucleic Acids Res* **30:**2052–2060.
38. **Freede P, Brantl S.** 2004. Transcriptional repressor CopR: use of SELEX to study the *copR* operator indicates that evolution was directed at maximal binding affinity. *J Bacteriol* **186:**6254–6264.
39. **Licht A, Freede P, Brantl S.** 2011. Transcriptional repressor CopR acts by inhibiting RNA polymerase binding. *Microbiology* **157:**1000–1008.
40. **Steinmetzer K, Hillisch A, Behlke J, Brantl S.** 2000. Transcriptional repressor CopR: Structure model based localization of the DNA binding motif. *Proteins* **38:** 393–406.
41. **Steinmetzer K, Hillisch A, Behlke J, Brantl S.** 2000. Transcriptional repressor CopR: amino acids involved in forming the dimeric interface. *Proteins* **39:**408–416.
42. **Steinmetzer K, Kuhn K, Behlke J, Golbik R, Brantl S.** 2002. Plasmid pIP501 encoded transcriptional repressor CopR: Single amino acids involved in dimerization are also important for folding of the monomer. *Plasmid* **47:** 201–209.
43. **Kuhn K, Steinmetzer K, Brantl S.** 2000. Transcriptional repressor CopR: the structured acidic C terminus is important for protein stability. *J Mol Biol* **300:**1021–1031.
44. **Kuhn K, Steinmetzer K, Brantl S.** 2001. Transcriptional repressor CopR: dissection of stabilizing motifs within the C terminus. *Microbiology* **14:**3387–3392.
45. **Heidrich N, Brantl S.** 2007. Antisense-RNA mediated transcriptional attenuation: the simultaneous interaction between two complementary loop pairs is required for efficient inhibition by the antisense RNA. *Microbiology* **153:**420–427.
46. **Franch T, Petersen M, Wagner EG, Jacobsen JP, Gerdes K.** 1999. Antisense RNA regulation in prokaryotes: Rapid RNA/RNA interaction facilitated by a general U-turn loop structure. *J Mol Biol* **294:**1115–1125.
47. **Franch T, Gerdes K.** 2000. U-turns and regulatory RNAs. *Curr Opin Microbiol* **3:**159–164.
48. **Heidrich N, Brantl S.** 2003. Antisense-RNA mediated transcriptional attenuation: importance of a U-turn loop structure in the target RNA of plasmid pIP501 for efficient inhibition by the antisense RNA. *J Mol Biol* **333:** 917–929.
49. **Brantl S, Wagner EG.** 2002. An antisense RNA-mediated transcription attenuation mechanism functions in *Escherichia coli*. *J Bacteriol* **184:**2740–2747.
50. **de la Hoz AB, Ayora S, Sitkiewicz I, Fernández S, Pankiewicz R, Alonso JC, Ceglowski P.** 2000. Plasmid copy-number control and better-than random segregation genes of pSM19035 share a common regulator. *Proc Natl Acad Sci USA* **97:**728–733.
51. **de laHoz AB, Pratto F, Misselwitz R, Speck C, Weihofen W, Welfle K, Saenger W, Welfle H, Alonso JC.** 2004. Recognition of DNA by omega protein from the broad-host range *Streptococcus pyogenes* plasmid pSM19035: analysis of binding to operator DNA with one to four heptad repeats. *Nucleic Acids Res* **32:**3136–3147.
52. **Welfle K, Pratto F, Misselwitz R, Behlke J, Alonso JC, Welfle H.** 2005. Role of the N-terminal region and of β-sheet residue Thr29 on the activity of the omega2 global regulator form the broad-host range *Streptococcus pyogenes* plasmid pSM19035. *Biol Chem* **386:** 881–894.
53. **Murayama K, Orth P, del la Hoz AB, Alonso JC, Saenger W.** 2001. Crystal Structure of ω transcriptional repressor encoded by *Streptococcus pyogenes* plasmid pSM19035 at 1.5 Å resolution. *J Mol Biol* **314:**789–796.
54. **Bruand C, Le Chatelier E, Ehrlich SD, Jannière L.** 1993. A fourth class of theta replicating plasmids. The pAMβ1 family from gram-positive bacteria. *Proc Natl Acad Sci USA* **90:**11668–11672.
55. **Bruand C, Farache M, McGovern S, Ehrlich SD, Polard P.** 2001. DnaB, DnaD and DnaI proteins are components of the *Bacillus subtilis* replication restart primosome. *Mol Microbiol* **42:**245–255.
56. **Bidnenko V, Ehrlich SD, Jannière L.** 1998. *In vivo* relations between pAMβ1-encoded type I topoisomerase and plasmid replication. *Mol Microbiol* **28:**1005–1016.
57. **Jannière L, Bidnenko V, McGovern S, Ehrlich SD, Petit MA.** 1997. Replication terminus for DNA polymerase I during initiation of pAMβ1 replication: role of the plasmid encoded resolution system. *Mol Microbiol* **23:**525–535.
58. **Marsin S, McGovern S, Ehrlich SD, Bruand C, Polard P.** 2001. Early steps of *Bacillus subtilis* primosome assembly. *J Biol Chem* **276:**45818–45825.
59. **Polard P, Marsin S, McGovern S, Velten M, Wigley DB, Ehrlich SD, Bruand C.** 2002. Restart of DNA replication in Gram-positive bacteria: functional characterisation of the *Bacillus subtilis* PriA initiator. *Nucleic Acids Res* **30:** 1593–1605.
60. **Dervyn E, Suski C, Daniel R, Chapuis J, Errington J, Jannière L, Ehrlich SD.** 2001. Two essential DNA polymerases at the bacterial replication fork. *Science* **294:** 1716–1719.

61. **Bruand C, Ehrlich SD, Jannière L.** 1995. Primosome assembly site in *Bacillus subtilis*. *EMBO J* **14:**2642–2650.
62. **Le Chatelier E, Jannière L, Ehrlich SD, Canceill C.** 2001. The RepE initiator is a double-stranded and single-stranded DNA-binding protein that forms an atypical open complex at the onset of replication of plasmid pAMβ1 from Gram-positive bacteria. *J Biol Chem* **276:** 10234–10246.
63. **Masukata H, Tomizawa J.** 1984. Effects of point mutations on formation and structure of the RNA primer for ColE1 DNA replication. *Cell* **36:**513–522.
64. **Masukata H, Tomizawa J.** 1986. Control of primer formation for ColE1 plasmid replication: conformational change of the primer transcript. *Cell* **44:**125–136.
65. **Polisky B, Tamm J,Fitzwater T.** 1985. Construction of ColE1 RNA I mutants and analysis of their function *in vivo*. *Basic Life Sci* **30:**321–333.
66. **Polisky B, Zhang XY, Fitzwater T.** 1990. Mutations affecting primer RNA interaction with the replication repressor RNA I in plasmid ColE1: potential RNA folding pathway mutants. *EMBO J* **9:**295–304.
67. **Masukata H, Tomizawa J.** 1990. A mechanism of formation of a persistent hybrid between elongating RNA and template DNA. *Cell* **62:**331–338.
68. **Itoh T, Tomizawa J.** 1980. Formation of an RNA primer for initiation of replication of ColE1 DNA by ribonuclease H. *Proc Natl Acad Sci USA* **77:**2450–2454.
69. **Itoh T, Tomizawa J.** 1982. Purification of ribonuclease H as a factor required for initiation of *in vitro* ColE1 DNA replication. *Nucleic Acids Res* **10:**5949–5965.
70. **Itoh T, Tomizawa J.** 1978. Initiation of replication of plasmid ColE1 DNA by RNA polymerase, ribonuclease H and DNA polymerase I. *Cold Spring Harbor Symp Quant Biol* **43:**409–417.
71. **Ma D, Campbell JL.** 1988. The effect of *dnaA* protein and n' sites on the replication of plasmid ColE1. *J Biol Chem* **263:**15008–15015.
72. **Lacatena RM, Cesareni G.** 1981. Base pairing of RNA I with its complementary sequence in the primer precursor inhibits ColE1 replication. *Nature* **294:**623–626.
73. **Tomizawa J, Itoh T.** 1981. Inhibition of ColE1 RNA primer formation by a plasmid-specified small RNA. *Proc Natl Acad Sci USA* **78:**1421–1425.
74. **Tamm J, Polisky B.** 1983. Structural analysis of RNA molecules involved in plasmid copy number control. *Nucleic Acids Res* **11:**6381–6397.
75. **Tomizawa J.** 1986. Control of ColE1 plasmid replication: binding of RNA I to RNA II and inhibition of primer formation. *Cell* **47:**89–97.
76. **Lacatena RM, Cesareni G.** 1983. Interaction between RNA I and the primer precursor in the regulation of ColE1 replication. *J Mol Biol* **170:**635–650.
77. **Muesing M, Tamm J, Shepard HM, Polisky B.** 1981. A single base-pair alteration is responsible for the DNA overproduction phenotype of a plasmid-copy-number mutant. *Cell* **24:**235–242.
78. **Eguchi Y, Itoh T, Tomizawa J.** 1991. Antisense RNA. *Annu Rev Biochem* **60:**631–652.
79. **Eguchi Y, Tomizawa J.** 1990. Complex formed by complementary RNA stem-loops and its stabilization by a protein: function of ColE1 Rom protein. *Cell* **60:**199–209.
80. **Tomizawa J.** 1985. Control of ColE1 plasmid replication: initial interaction of RNA I and the primer transcript is reversible. *Cell* **40:**527–535.
81. **Tomizawa J.** 1990. Control of ColE1 plasmid replication. Interaction of Rom protein with an unstable complex formed by RNA I and RNA II. *J Mol Biol* **212:** 695–708.
82. **Tomizawa J.** 1990. Control of ColE1 plasmid replication. Intermediates in the binding of RNA I and RNA II. *J Mol Biol* **212:**683–694.
83. **Tomizawa J.** 1984. Control of ColE1 plasmid replication: the process of binding of RNAI to the primer transcript. *Cell* **38:**861–870.
84. **Tomizawa J, Som T.** 1984. Control of ColE1 plasmid replication. Enhancement of binding of RNA I to primer transcript by the Rom protein. *Cell* **38:**871–878.
85. **Lee AJ, Crothers DM.** 1998. The solution structure of an RNA loop-loop complex: the ColE1 inverted loop sequence. *Structure* **6:**993–1005.
86. **Marino JP, Gregorian RSJ, Csankovszki G, Crothers DM.** 1995. Bent helix formation between RNA hairpins with complementary loops. *Science* **268:**1448–1454.
87. **Lin-Chao S, Cohen SN.** 1991. The rate of processing and degradation of antisense RNA I regulates the replication of ColE1-type plasmids *in vivo*. *Cell* **65:**1233–1242.
88. **He L, Söderbom F, Wagner EG, Binnie U, Binns N, Masters M.** 1993. PcnB is required for the rapid degradation of RNAI, the antisense RNA that controls the copy number of ColE1-related plasmids. *Mol Microbiol* **9:**1131–1142.
89. **Cesareni G, Muesing MA, Polisky B.** 1982. Control of ColE1 DNA replication. The *rop* gene product negatively affects transcription from the replication primer promoter. *Proc Natl Acad Sci USA* **79:**6313–6317.
90. **Som T, Tomizawa J.** 1983. Regulatory regions of ColE1 that are involved in determination of plasmid copy number. *Proc Natl Acad Sci USA* **80:**3232–3236.
91. **Atlung T, Christensen BB, Hansen FG.** 1999. Role of the Rom protein in copy number control of plasmid pBR322 at different growth rates in *Escherichia coli* K-12. *Plasmid* **41:**110–119.
92. **Banner DW, Kokkinidis M, Tsernoglou D.** 1987. Structure of the ColE1 Rop protein at 1.7 Å resolution. *J Mol Biol* **5:**657–675.
93. **Castagnoli L, Scarpa M, Kokkinidis M, Banner DW, Tsernoglou D, Cesareni G.** 1989. Genetic and structural analysis of the ColE1 Rop (Rom) protein. *EMBO J* **8:** 621–629.
94. **Predki PF, Nayak LM, Gottlieb MBC, Regan L.** 1995. Dissecting RNA-protein interactions: RNA-RNA recognition by Rop. *Cell* **80:**41–50.
95. **Lin-Chao S, Chen WT, Wong TT.** 1992. High copy number of the pUC plasmid results from a Rom/Rop-suppressible point mutation in RNAII. *Mol Microbiol* **6:** 3385–3393.

96. **Brenner M, Tomizawa J.** 1991. Quantitation of ColE1-encoded replication elements. *Proc Natl Acad Sci USA* **88:**405–409.
97. **Brendel V, Perelson AS.** 1993. Quantitative model of ColE1 plasmid copy number control. *J Mol Biol* **229:** 860–872.
98. **Paulson J, Nordström K, Ehrenberg M.** 1998. Requirements for rapid plasmid ColE1 copy number adjustments: a mathematical model of inhibition modes and RNA turnover rates. *Plasmid* **39:**215–234.
99. **Hama C, Takizawa T, Moriwaki H, Urasaki Y, Mizobuchi K.** 1990. Organization of the replication control region of plasmid ColIb-P9. *J Bacteriol* **172:** 1983–1991.
100. **Praszkier J, Wei T, Siemering K, Pittard AJ.** 1991. Comparative analysis of the replication regions of IncB, IncK and IncZ plasmids. *J Bacteriol* **173:**2393–2397.
101. **Asano K, Moriwaki H, Mizobuchi K.** 1991. An induced mRNA secondary structure enhances *repZ* translation in plasmid ColIb-P9. *J Biol Chem* **266:**24549–24556.
102. **Asano K, Kato A, Moriwaki H, Hama C, Shiba K, Mizobuchi K.** 1991. Positive and negative regulations of plasmid ColIb-P9-*repZ* gene expression at the translational level. *J Biol Chem* **266:**3774–3781.
103. **Wilson IW, Praszkier J, Pittard AJ.** 1993. Mutations affecting pseudoknot control of the replication of B group plasmids. *J Bacteriol* **175:**6476–6483.
104. **Hama C, Takizawa T, Moriwaki H, Mizobuchi K.** 1990. Role of leader peptide synthesis in *repZ* gene expression of the ColIb-P9 plasmid. *J Biol Chem* **265:** 10666–10673.
105. **Praszkier J, Wilson IW, Pittard AJ.** 1992. Mutations affecting translation coupling between the *rep* genes of an IncB miniplasmid. *J Bacteriol* **174:**2376–2383.
106. **Asano K, Mizobuchi K.** 1998. An RNA pseudoknot as the molecular switch for translation of the *repZ* gene encoding the replication initiator of IncIα plasmid ColIb-P9. *J Biol Chem* **273:**11815–11825.
107. **Praszkier J, Bird P, Nikoletti S, Pittard AJ.** 1989. Role of countertranscript RNA in the copy number control system of an IncB miniplasmid. *J Bacteriol* **171:**5056–5064.
108. **Shiba K, Mizobuchi K.** 1990. Posttranscriptional control of plasmid ColIb-P9 *repZ* gene expression by a small RNA. *J Bacteriol* **172:**1992–1997.
109. **Asano K, Mizobuchi K.** 1998. Copy number control of IncIα plasmid ColIb-P9 by competition between pseudoknot formation and antisense RNA binding at a specific RNA site. *EMBO J* **17:**5201–5213.
110. **Wilson IW, Siemering KR, Praszkier J, Pittard AJ.** 1997. Importance of structural differences between complementary RNA molecules to control of replication of an IncB plasmid. *J Bacteriol* **179:**742–753.
111. **Asano K, Mizobuchi K.** 2000. Structural analysis of late intermediate complex formed between plasmid ColIb-P9 Inc RNA and its target RNA. How does a single antisense RNA repress translation of two genes at different rates? *J Biol Chem* **275:**1269–1274.
112. **Siemering KR, Praszkier J, Pittard AJ.** 1993. Interaction between the antisense and target RNAs involved in the regulation of IncB plasmid replication. *J Bacteriol* **175:** 2895–2906.
113. **Wilson IW, Praszkier J, Pittard AJ.** 1994. Molecular analysis of RNAI control of *repB* translation in IncB plasmids. *J Bacteriol* **176:**6497–6508.
114. **Athanasopoulos V, Praszkier J, Pittard AJ.** 1999. Analysis of elements involved in pseudoknot-dependent expression and regulation of the *repA* gene of an IncL/M plasmid. *J Bacteriol* **181:**1811–1819.
115. **Siemering KR, Praszkier J, Pittard AJ.** 1994. Mechanism of binding of the antisense and target RNAs involved in the regulation of IncB plasmid replication. *J Bacteriol* **176:**2677–2688.
116. **Kolb FA, Westhof E, Ehresmann B, Ehresmann C, Wagner EG, Romby P.** 2001. Four-way junctions in antisense RNA-mRNA complexes involved in plasmid replication control: a common theme? *J Mol Biol* **309:** 605–614.
117. **Praszkier J, Pittard AJ.** 1999. Role of *CIS* in replication of an IncB plasmid. *J Bacteriol* **181:**2765–2772.
118. **Praszkier J, Murthy S, Pittard AJ.** 2000. Effect of *CIS* on activity *in trans* of the replication initiator protein of an IncB plasmid. *J Bacteriol* **182:**3972–3980.
119. **del Solar G, Espinosa M.** 1992. The copy number of plasmid pLS1 is regulated by two trans-acting plasmid products: the antisense RNA II and the repressor protein, RepA. *Mol Microbiol* **6:**83–94.
120. **del Solar G, Acebo P, Espinosa M.** 1995. Replication control of plasmid pLS1: efficient regulation of plasmid copy number is exerted by the combined action of two plasmid components, CopG and RNAII. *Mol Microbiol* **18:**913–924.
121. **del Solar G, Espinosa M.** 2000. Plasmid copy number control: an ever-growing story. *Mol Microbiol* **37:** 492–500.
122. **López-Aguilar C, del Solar G.** 2013. Probing the sequence and structure of *in vitro* synthesized antisense and target RNAs from the replication control system of plasmid pMV158. *Plasmid* **70:**94–103.
123. **López-Aguilar C, Ruiz-Masó JA, Rubio-Lepe TS, Sanz M, del Solar G.** 2013. Translation initiation of the replication initiator *repB* gene of promiscuous plasmid pMV158 is led by an extended non-SD sequence. *Plasmid* **70:**69–77.
124. **del Solar G, Acebo P, Espinosa M.** 1997. Replication control of plasmid pLS1: the antisense RNA II and the compact *rnaII* region are involved in translational regulation of the initiator RepB synthesis. *Mol Microbiol* **23:**95–108.
125. **del Solar G, Perez-Martin J, Espinosa M.** 1990. Plasmid-encoded RepA protein regulates transcription from *repAB* promoter by binding to a DNA sequence containing a 13 base pair symmetric element. *J Biol Chem* **265:**12569–12575.
126. **Gomis-Rüth FX, Sola M, Acebo P, Parraga A, Guasch A, Eritja R, Gonzalez A, Espinosa M, del Solar G, Coll M.** 1998. The structure of plasmid encoded transcriptional repressor CopG unliganded and bound to its operator. *EMBO J* **17:**7404–7415.

127. **Hernández-Arriaga AM, Rubio-Lepe TS, Espinosa M, del Solar G.** 2009. Repressor CopG prevents access of RNA polymerase to promoter and actively dissociates open complexes. *Nucleic Acids Res* **37:**4799–4811.

128. **Kwak JH, Weisblum B.** 1994. Regulation of plasmid pE194 replication: control of *cop-repF* operon transcription by Cop and of *repF* translation by countertranscript RNA. *J Bacteriol* **176:**5044–5051.

129. **Kwak JH, Kim J, Kim M-Y, Choi E-C.** 1998. Purification and characterization of Cop, a protein involved in the copy number control of plasmid pE194. *Arch Pharm Res* **3:**291–297.

130. **Alonso JC, Tailor RM.** 1987. Initiation of plasmid pC194 replication and its control in *Bacillus subtilis. Mol Gen Genet* **210:**476–484.

131. **Okibe N, Suzuki N, Inui M, Yukawa H.** 2010. Antisense-RNA-mediated plasmid copy number control in pCG1-family plasmids, pCGR2 and pCG1, in *Corynebacterium glutamicum. Microbiology* **156:**3609–3623.

132. **Okibe N, Suzuki N, Inui M, Yukawa H.** 2013. pCGR2 copy number depends on the *par* locus that forms a ParC-ParB-DNA partition complex in *Corynebacterium glutamicum. J Appl Microbiol* **115:**495–508.

133. **Masai H, Arai K.** 1988. R1 plasmid replication *in vitro*. RepA and dnaA-dependent initiation at *oriR*, p 113–121 *In* Moses RE, Summers KC (ed), *DNA Replication and Mutagenesis*. ASM Press, Washington, DC.

134. **Ortega-Jiménez R, Giraldo-Suárez ME, Fernández-Tresguerres ME, Berzal-Herranz A, Díaz-Orejas R.** 1992. DnaA dependent replication of plasmid R1 occurs in the presence of point mutations that disrupt the *dnaA* box of *oriR*. *Nucleic Acids Res* **20:**2547–2551.

135. **Giraldo R, Diaz R.** 1992. Differential binding of wild-type and a mutant RepA protein to *oriR* sequence suggests a model for the initiation of plasmid R1 replication. *J Mol Biol* **228:**787–802.

136. **Krabbe M, Zabielski J, Bernander R, Nordström K.** 1997. Inactivation of the replication-termination system affects the replication mode and causes unstable maintenance of plasmid R1. *Mol Microbiol* **24:**723–735.

137. **Söderbom F, Binnie U, Masters M, Wagner EG.** 1997. Regulation of plasmid R1 replication: PcnB and RNase E expedite the decay of the antisense RNA, CopA. *Mol Microbiol* **26:**493–504.

138. **Light J, Molin S.** 1982. The sites of action of the two copy number control functions of plasmid R1. *Mol Gen Genet* **187:**486–493.

139. **Malmgren C, Engdahl HM, Romby P, Wagner EG.** 1996. An antisense/target RNA duplex or a strong intramolecular RNA structure 5′ of a translation initiation signal blocks ribosome binding: the case of plasmid R1. *RNA* **2:**1022–1032.

140. **Blomberg P, Wagner EG, Nordström K.** 1990. Replication control of plasmid R1: the duplex between the antisense RNA, CopA, and its target, CopT, is processed specifically *in vivo* and *in vitro* by RNase III. *EMBO J* **9:**2331–2340.

141. **Blomberg P, Nordström K, Wagner EG.** 1992. Replication control of plasmid R1: RepA synthesis is regulated by CopA RNA through inhibition of leader peptide translation. *EMBO J* **11:**2675–2683.

142. **Blomberg P, Engdahl HM, Malmgren C, Romby P, Wagner EG.** 1994. Replication control of plasmid R1: disruption of an inhibitory RNA structure that sequesters the *repA* ribosome-binding site permits tap-independent RepA synthesis. *Mol Microbiol* **12:**49–60.

143. **Wu RP, Wang X, Womble DD, Rownd RH.** 1992. Expression of the *repA1* gene of IncFII plasmid NR1 is translationally coupled to expression of an overlapping leader peptide. *J Bacteriol* **174:**7620–7628.

144. **Brady G, Frey J, Danbara H, Timmis KN.** 1983. Replication control mutations of plasmid R6-5 and their effects on interactions of the RNA-I control element with its target. *J Bacteriol* **154:**429–436.

145. **Giskov M, Molin S.** 1984. Copy mutants of plasmid R1: effects of base pair substitutions in the *copA* gene on the replication control system. *Mol Gen Genet* **194:** 286–292.

146. **Persson C, Wagner EG, Nordström K.** 1988. Control of replication of plasmid R1: kinetics of *in vitro* interaction between the antisense RNA, CopA, and its target, CopT. *EMBO J* **7:**3279–3288.

147. **Persson C, Wagner EG, Nordström K.** 1990. Control of replication of plasmid R1: structures and sequences of the antisense RNA, CopA, required for its binding to the target RNA, CopT. *EMBO J* **9:**3767–3775.

148. **Persson C, Wagner EG, Nordström K.** 1990. Control of replication of plasmid R1: formation of an initial transient complex is rate-limiting for antisense RNA-target RNA pairing. *EMBO J* **9:**3777–3785.

149. **Hjalt TA, Wagner EG.** 1992. The effect of loop size in antisense and target RNAs on the efficiency of antisense RNA control. *Nucleic Acids Res* **20:**6723–6732.

150. **Hjalt TA, Wagner EG.** 1995. Bulged-out nucleotides in an antisense RNA are required for rapid target RNA binding *in vitro* and inhibition *in vivo*. *Nucleic Acids Res* **23:**580–587.

151. **Kolb FA, Engdahl HM, Slagter-Jäger JG, Ehresmann B, Ehresmann C, Westhof E, Wagner EG, Romby P.** 2000. Progression of a loop-loop complex to a four-way junction is crucial for the activity of a regulatory antisense RNA. *EMBO J* **19:**5905–5915.

152. **Kolb FA, Malmgren C, Westhof E, Ehresmann C, Ehresmann B, Wagner EG, Romby P.** 2000. An unusual structure formed by antisense-target RNA binding involves an extended kissing complex with a four-way junction and a side-by-side helical alignment. *RNA* **6:**311–324.

153. **Malmgren C, Wagner EG, Ehresmann C, Ehresmann B, Romby P.** 1997. Antisense RNA control of plasmid R1 replication. The dominant product of the antisense RNA-mRNA binding is not a full RNA duplex. *J Biol Chem* **272:**12508–12512.

154. **Brantl S.** 2007. Regulatory mechanisms employed by cis-encoded antisense RNAs. *Curr Opin Microbiol* **10:** 102–109.

155. **Söderbom F, Wagner EG.** 1998. Degradation pathway of CopA, the antisense RNA that controls replication of plasmid R1. *Microbiology* **144:**1907–1917.

156. **Riise E, Stougaard P, Bindslev B, Nordström K, Molin S.** 1982. Molecular cloning and functional characterization of a copy number control gene (*copB*) of plasmid R1. *J Bacteriol* **151:**1136–1145.

157. **Riise E, Molin S.** 1986. Purification and characterization of the CopB replication control protein, and precise mapping of its target site in the R1 plasmid. *Plasmid* **15:**163–171.

158. **Stougaard P, Light J, Molin S.** 1982. Convergent transcription interferes with expression of the copy number control gene, *copA*, from plasmid R1. *EMBO J* **1:** 323–328.

159. **Light J, Riise E, Molin S.** 1985. Transcription and its regulation in the basic replicon region of plasmid R1. *Mol Gen Genet* **198:**503–508.

160. **Kwong SM, Skurray RA,Firth N.** 2004. *Staphylococcus aureus* multiresistance plasmid pSK41: analysis of the replication region, initiator protein binding and antisense RNA regulation. *Mol Microbiol* **51:**497–509.

161. **Kwong SM, Skurray RA,Firth N.** 2006. Replication control of staphylococcal multiresistance plasmid pSK41: an antisense RNA mediates dual-level regulation of Rep expression. *J Bacteriol* **188:**4404–4412.

162. **Kwong SM,Lim R, LeBard RJ, Skurray RA, Firth N.** 2008. Analysis of the pSK1 replicon, a prototype from the staphylococcal multiresistance plasmid family. *Microbiology* **154:**3084–3094.

163. **Cevallos MA, Cervantes-Rivera R, Gutiérrez-Ríos RM.** 2008. The *repABC* plasmid family. *Plasmid* **60:** 19–37.

164. **Izquierdo J, Venkova-Canova T, Ramírez-Romero MA, Téllez-Sosa J, Hernández-Lucas I, Sanjuan J, Cevallos MA.** 2005. An antisense RNA plays a central role in the replication control of a *repC* plasmid. *Plasmid* **54:** 259–277.

165. **Venkova-Canova T, Soberón NE, Ramírez-Romero MA, Cevallos MA.** 2004. Two discrete elements are required for the replication of a *repABC* plasmid: an antisense RNA and a stem-loop structure. *Mol Microbiol* **54:**1431–1444.

166. **Cervantes-Rivera R, Romero-López C, Berzal-Herranz A, Cevallos MA.** 2010. Analysis of the mechanism of action of the antisense RNA that controls the replication of the *repABC* plasmid p42d. *J Bacteriol* **192:** 3268–3278.

167. **MacLellan SR, Smallbone LA, Sibley CD, Finan TM.** 2005. The expression of a novel antisense gene mediates incompatibility within the large *repABC* family of α-proteobacterial plasmids. *Mol Microbiol* **55:**611–623.

168. **Chai Y, Winans SC.** 2005. A small antisense RNA downregulates expression of an essential replicase protein of an *Agrobacterium tumefaciens* Ti plasmid. *Mol Microbiol* **56:**1574–1585.

169. **Pinto UM, Pappas KM, Winans SC.** 2012. The ABCs of plasmid replication and segregation. *Nat Rev Microbiol* **10:**755–765.

170. **Hollands K, Proshkin S, Sklyarova S, Epshtein V, Mironov A, Nudler E, Groisman EA.** 2012. Riboswitch control of Rho-dependent transcription termination. *Proc Natl Acad Sci USA* **109:**5376–5381.

171. **Takechi S, Matsui H, Itoh T.** 1995. Primer RNA synthesis by plasmid-specified Rep protein for initiation of ColE2 DNA replication. *EMBO J* **14:**5141–5147.

172. **Takechi S, Itoh T.** 1995. Initiation of unidirectional ColE2 DNA replication by a unique priming mechanism. *Nucleic Acids Res* **23:**4196–4201.

173. **Takechi S, Yasueda H, Itoh T.** 1994. Control of ColE2 plasmid replication: regulation of Rep expression by a plasmid-coded antisense RNA. *Mol Gen Genet* **244:** 49–56.

174. **Sugiyama T, Itoh T.** 1993. Control of ColE2 DNA replication: *in vitro* binding of the antisense RNA to the *rep* mRNA. *Nucleic Acids Res* **21:**5972–5977.

175. **Yasueda H, Takechi S, Sugiyama T, Itoh T.** 1994. Control of ColE2 plasmid replication: negative regulation of the expression of the plasmid-specified initiator protein, Rep, at a posttranscriptional step. *Mol Gen Genet* **244:** 41–48.

176. **Nagase T, Nishio S-Y, Itoh T.** 2007. Importance of the leader region of mRNA for translation initiation of ColE2 Rep protein. *Plasmid* **58:**249–260.

177. **Shinora M, Itoh T.** 1996. Specificity determinants in interaction of the initiator (Rep) proteins with the origins in the plasmids ColE2-P9 and ColE3-CA38 identified by chimera analysis. *J Mol Biol* **257:**290–300.

178. **Nishio SY, Itoh T.** 2008. Replication initiator protein mRNA of ColE2 plasmid and its antisense regulator RNA are under the control of different degradation pathways. *Plasmid* **59:**102–110.

179. **Nishio SY, Itoh T.** 2008. The effects of RNA degradation enzymes on antisense RNAI controlling ColE2 plasmid copy number. *Plasmid* **60:**174–180.

180. **André G, Even S, Putzer H, Burguière P, Croux C, Danchin A, Martin-Verstraete I, Soutourina O.** 2008. S-box and T-box riboswithces and antisense RNA control a sulphur metabolic operon of *Clostridium acetobutylicum*. *Nucleic Acids Res* **36:**5955–5969.

181. **Sayed N, Jousselin A, Felden B.** 2011. A cis-antisense RNA acts in trans in *Staphylococcus aureus* to control translation of a human cytolytic peptide. *Nat Struct Mol Biol* **19:**105–112.

182. **Jäger D, Pernitzsch SR, Richter AS, Backofen R, Sharma CM, Schmitz RA.** 2012. An archeaeal sRNA targeting cis- and trans-encoded mRNAs via two distinct domains. *Nucleic Acids Res* **40:**10964–10979.

183. **Gimpel M, Heidrich N, Mäder U, Krügel H, Brantl S.** 2010. A dual-function sRNA from *B. subtilis*: SR1 acts as a peptide encoding mRNA on the *gapA* operon. *Mol Microbiol* **76:**990–1009.

184. **Vanderpool CK, Balasubramanian D, Lloyd CR.** 2011. Dual-function RNA regulators in bacteria. *Biochimie* **93:**1943–1949.

185. **Beisel CL, Storz G.** 2010. Base pairing small RNAs and their roles in global regulatory networks. *FEMS Microbiol Rev* **34:**866–882.

186. **Eiamphungporn W, Helmann JD.** 2009. Extracytoplasmic function sigma factors regulate expression of

the *Bacillus subtilis yabE* gene via a cis-acting antisense RNA. *J Bacteriol* **191**:1101–1105.

187. **Lee EJ, Groisman EA.** 2010. An antisense RNA that governs the expression kinetics of a multifunctional virulence gene. *Mol Microbiol* **76**:1020–1033.

188. **Konieczny I, Bury K, Wawrzycka A, Wegrzyn K.** Iteron plasmids, p 15–32. *In* Tolmasky ME, Alonso JC (ed), *Plasmids: Biology and Impact in Biotechnology and Discovery*. ASM Press, Washington, DC.

189. **Ruiz-Masó JA, Machón C, Bordanaba L, Espinosa M, Coll M, del Solar G.** 2014. Plasmid rolling-circle replication, p 45–70. *In* Tolmasky ME, Alonso JC (ed), *Plasmids: Biology and Impact in Biotechnology and Discovery*. ASM Press, Washington, DC.

Plasmids—Biology and Impact in Biotechnology and Discovery
Edited by Marcelo E. Tolmasky and Juan C. Alonso

doi:10.1128/microbiolspec.PLAS-0036-2014

N. Patrick Higgins[1]
Alexander V. Vologodskii[2]

7 Topological Behavior of Plasmid DNA

DNA topology is a critical factor in essentially all *in vivo* chromosomal processes, including DNA replication, RNA transcription, homologous recombination, site-specific recombination, DNA repair, and integration of the abundant and mechanistically distinct forms of transposable elements. Plasmids can be invaluable tools to define the dynamic mechanisms of proteins that shape DNA, organize chromosome structure, and channel chromosome movement inside living cells. The advantages of plasmids include their ease of isolation and the ability to quantitatively measure DNA knots, DNA catenation, hemi-catenation between two DNA molecules, and positive or negative supercoils in purified DNA populations. Under ideal conditions, *in vitro* and *in vivo* results can be compared to define the complex mechanism of enzymes that move along and change DNA chemistry in living cells. Many techniques that can be easily done with plasmids are not feasible for the massive chromosome that carries most of the genetic information in *Escherichia coli* or *Salmonella typhimurium*. Whereas a large fraction of contemporary chromosomal "philosophy" is based on extrapolation of results from small plasmids such as pBR322 to the 4.6-Mb bacterial chromosome, the comparison is not always valid. One aim of this article is to explain how results derived from small plasmids can be misleading for understanding and interpreting the DNA structure of the large bacterial chromosome.

TOPOLOGY OF CIRCULAR DNA

Three levels of discrimination are needed to specify the topological state of double-stranded circular DNA. For the first and second levels one considers DNA as a simple curve that coincides with the DNA axis. The fist level describes the topology of an isolated closed curve that corresponds to an unknotted circle or to a knot of a particular type. If we have many DNA molecules, part of them can form topological links with others, and the second level specifies types of these links. An infinite number of different types of knots and links exist. Some simple examples are shown in Fig. 1. The third level of the description specifies links formed by complementary strands of the double helix. This component of DNA topology will be a major subject in this review.

[1]Department of Biochemistry and Molecular Genetics, University of Alabama at Birmingham, Birmingham, AL 35294; [2]Department of Chemistry, New York University, New York, NY 10003.

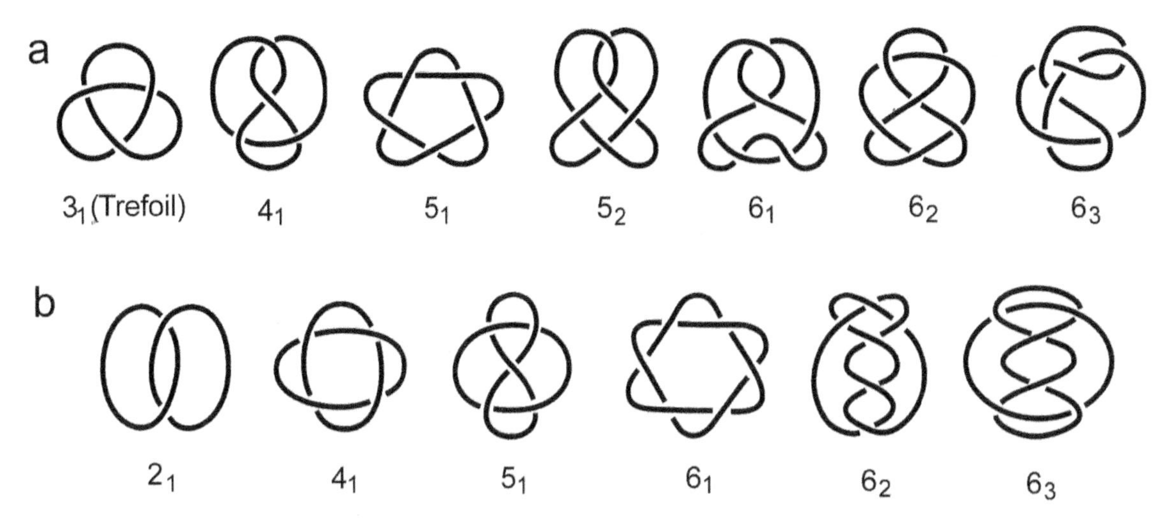

Figure 1 The simplest knots (a) and catenanes (b). DNA molecules are capable of adopting these and many more complex topological states.
doi:10.1128/microbiolspec.PLAS-0036-2014.f1

Supercoiling, Linking Number, and Linking Number Difference

In the initial studies of plasmid DNA structure, two predominant types of circular DNA molecules were isolated from cells. These types were designated as form I and form II. The more compact form I turned into form II when a single-stranded break was introduced into one chain of the double helix. Studies by Vinograd (1) connected the compactness of form I to negative supercoiling. Form I is also called the closed circular form since each strand that makes up the DNA molecule is closed on itself. A diagrammatic view of a model of closed circular DNA is presented in Fig. 2.

The strands of the double helix in closed circular DNA are linked. The quantitative description of this characteristic is called the linking number (*Lk*), which may be determined in the following way. One of the strands defines the edge of an imaginary surface (any such surface gives the same result). *Lk* is the algebraic (i.e., sign-dependent) number of intersections between the other strand and this spanning surface. By convention, the *Lk* of a closed circular DNA formed by a right-handed double helix is positive. *Lk* depends only on the topological state of the strands and hence is maintained through all conformational changes that occur in the absence of strand breakage.

Quantitatively, the linking number is close to N/γ, where N is the number of base pairs in the molecule and γ is the number of base pairs per double-helix turn in linear DNA under given conditions. However, these values are not equal, and the difference between *Lk* and N/γ (which is also denoted as Lk_O) defines most of the properties of closed circular DNA. A parameter that specifies this difference is called the linking number difference, ΔLk, and is defined as

$$\Delta Lk = Lk - N/\gamma \tag{1}$$

There are two inferences to be made from the above definition.

1. The value of ΔLk is not invariant; it depends on solution conditions that determine γ. Even though γ itself changes only slightly according to variable ambient conditions of temperature and ionic strength, such changes can substantially alter ΔLk because the right-hand part of equation 1 is a difference between two large quantities.
2. The value of *LK* is by definition an integer, whereas N/γ is not an integer. Hence ΔLk is not an integer either. However, the values of ΔLk for a closed circular DNA with a particular sequence can differ by an integer only. This follows from the fact that, whatever the prescribed conditions, all changes in ΔLk can only be due to changes in *Lk*, since the value of N/γ is the same for all molecules. (Of course, any change of *Lk* would involve a temporary violation of the integrity of a double-helix strand.) Molecules with the same chemical structure that differ only with respect to *Lk* are defined as topoisomers.

It is convenient to use the value of superhelical density, σ, which is ΔLk normalized for Lk_O:

$$\sigma = \Delta Lk / Lk_o = \gamma \cdot \Delta Lk / N \tag{2}$$

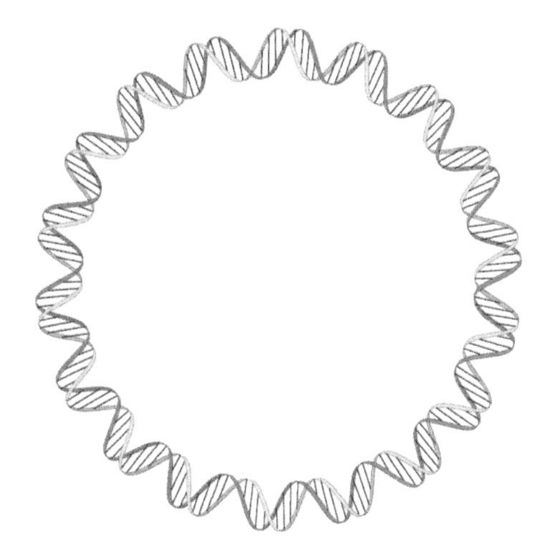

Figure 2 Diagram of closed circular DNA. The linking number, *Lk*, of the complementary strands is 18.
doi:10.1128/microbiolspec.PLAS-0036-2014.f2

Whenever $\Delta Lk \neq 0$, closed circular DNA is said to be supercoiled. Clearly, the entire double helix is stressed in a supercoiled condition. This stress can either lead to a change in the actual number of base pairs per helix turn in closed circular DNA or cause regular spatial deformation of the helix axis. The axis of the double helix then forms a helix of a higher order (Fig. 3). It is this deformation of the helix axis in closed circular DNA that gave rise to the term "superhelicity" or "supercoiling" (1). Circular DNA extracted from cells turns out to be always (or nearly always) negatively supercoiled and has a σ between –0.03 and –0.09, but typically is near the middle of this range (2).

Twist and Writhe

Supercoiling can be created in two ways: by deforming the molecular axis or by altering the twist of the double helix. This can be demonstrated in a simple experiment involving a rubber hose or flexible tubing. Take a length of hose and connect the ends by inserting a short rod into each end to join the ends of the hose; this is equivalent to a relaxed form of DNA. If one turns the hose several times around the rod axis, the hose will change into a helical band. By drawing longitudinal stripes on the hose prior to the experiment, one can see that reciprocal twisting of the ends causes torsional deformation of the hose. There is a quantitative relationship between the twist around the DNA axis and its deformation and *Lk* of the complementary strands of the double helix. The first mathematical treatment of the problem was presented by Calugareanu (3), who described the geometrical and topological properties of a closed ribbon. White (4) proved the theorem in its current form. Later, Fuller showed that the theorem applies to the analysis of circular DNA (5).

According to the theorem, the *Lk* of the edges of the ribbon is the sum of two values. One is the twist of the ribbon, *Tw*, and the second was a new concept, writhe *Wr*. Thus,

$$Lk = Tw + Wr \quad (3)$$

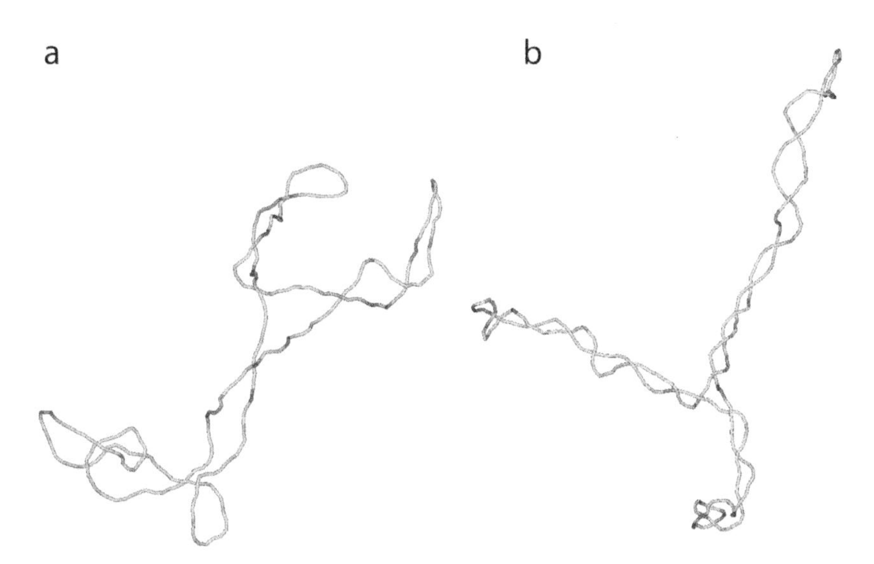

Figure 3 Typical simulated conformations of supercoiled DNA 4.4 kb in length. The conformations correspond to a DNA superhelix density of (a) –0.030 and (b) –0.060.
doi:10.1128/microbiolspec.PLAS-0036-2014.f3

Tw is a measure of the number of times one of the edges of the ribbon spins about its axis. The *Tw* of the entire ribbon is the sum of the *Tw* of its parts. The value of *Wr* is defined by the spatial course of the ribbon axis; i.e., it is a characteristic of a single closed curve, unlike *Lk* and *Tw*, which are properties of a closed ribbon. Thus, *Lk* can be represented as a sum of two values that characterize the available degrees of freedom: the twist around the ribbon axis and the deformation of this axis. To apply the theorem to circular DNA, the two strands of the double helix are considered as edges of a ribbon.

Wr describes a curve's net right-handed or left-handed asymmetry, i.e., its chirality, and is equal to zero for a planar curve (5). Unlike *Lk*, which can only be an integer, a curve's *Wr* can have any value. This value changes continuously with the curve's deformation that does not involve the intersection of segments. A curve's *Wr* does not change with a change in the curve's scale and depends solely on its shape. However, when a curve is deformed so that one of its parts passes through another, the writhe changes by 2 (or –2 for the opposite direction of the pass) (5). This property helped reveal the reaction mechanism of DNA gyrase (6).

Measuring Conformations of Supercoiled DNA

The critical result of the theory is that *Lk* can be structurally distributed in two ways: as a torsional deformation of the double helix and as a deformation of the DNA axis. Equation 3 states that any change in the twist of the double helix results in deformation of the helix axis, giving rise to a specific writhe. As the first to make a theoretical analysis of the shape of supercoiled DNA, Fuller (5) concluded that an interwound superhelix was favored over a simple solenoid superhelix from an energetic point of view. All available experimental and theoretical data indicate that supercoiled DNA adopts interwound conformations (7).

Electron microscopy (EM) is the most straightforward way to study conformations of supercoiled DNA. This method has been used extensively since the discovery of DNA supercoiling (1). The compact interwound supercoiled DNA form has been confirmed in numerous EM studies (8–10). It became clear, however, that labile DNA conformations can change during sample preparation for EM (7). A serious problem for interpreting EM results is the unspecified ionic conditions on the grid (11). Independent solution studies were required to confirm conclusions about these flexible objects (9, 12, 13).

Experimental solution methods, such as hydrodynamic and optical measurements, do not give direct, model-independent information about the three-dimensional structure of supercoiled DNA. These methods do, however, measure structure-dependent features of supercoiled DNA in a well-defined solution. Solution methods combined with computer simulation of supercoiled molecules were very productive in supercoiling studies (14, 15). The strategy involves calculating measurable properties of supercoiled DNA using a model of the double helix and comparing simulated results with experimental data. Simulated and actual conformations were in excellent agreement. Thus, computations can predict properties of the supercoiled molecules that are hard to measure directly. Fig. 3 shows simulated conformations of supercoiled molecules for two different values of σ: –0.03, which is close to the physiological level of unrestrained supercoiling (see below), and –0.06, which is close to the shape of plasmid DNAs stripped of all bound proteins. For DNA molecules with more than 2,500 base pairs, both computational and experimental data indicate that in "physiological" ionic conditions about three fourths of ΔLk is realized as bending deformation (*Wr*) and one fourth is realized as torsional deformation (*Tw*) (7).

Electrophoretic Separation of DNA Molecules with Different Topology

DNA molecules of a few thousand base pairs in length adopt many conformations in solution. However, exchange between conformations occurs in milliseconds, and during gel electrophoresis one observes the average mobility of any molecule. These average values depend on the topology of circular DNA, and thus a mixture of molecules with different *Lk* can be separated by gel electrophoresis. Such separations proved to be very powerful in studying DNA properties. Keller (16) was the first to separate topoisomers with different values of ΔLk in closed circular DNA. Since the values of ΔLk in any mixture of DNA topoisomers can differ only by an integer, under appropriate experimental conditions, molecules with different ΔLk form separate bands in the electrophoretic pattern (Fig. 4). If a DNA sample contains all possible topoisomers with ΔLk from 0 to some limiting value, and they are all well resolved with respect to mobility, one can find the value of ΔLk corresponding to each band simply by band counting. The band that corresponds to $\Delta Lk \approx 0$ (ΔLk is not an integer) can be identified through a comparison with the band for the nicked circular form. One should bear in mind the fact that topoisomer mobility is determined

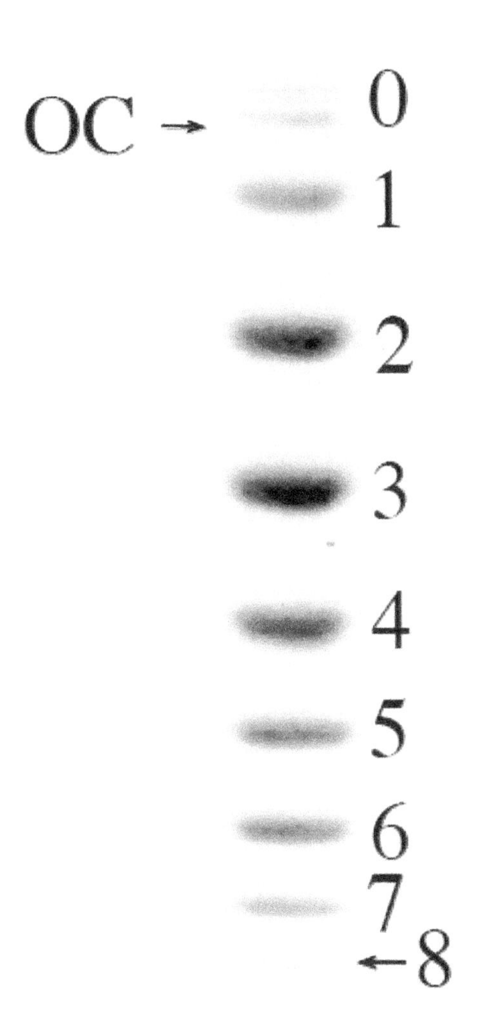

Figure 4 Electrophoretic separation of topoisomers of pUC19 DNA. The mixture of topoisomers covering the range of ΔLk from 0 to –8 was electrophoresed from a single well in 1% agarose that was run from top to bottom. The topoisomer with $\Delta Lk = 0$ has the lowest mobility: it moves slightly slower than the open circular form (OC). The value of ($-\Delta Lk$) for each topoisomer is shown.
doi:10.1128/microbiolspec.PLAS-0036-2014.f4

by the absolute value of ΔLk only, so the presence of topoisomers with both negative and positive ΔLk can make interpreting the electrophoresis profile difficult. Also, the mobility of topoisomers approaches a limiting value when $|\Delta Lk|$ increases, so a special method is required to separate topoisomers beyond this limit of resolution (16).

An elegant way to overcome the shortcoming of one-dimensional analysis is two-dimensional gel electrophoresis (17). A mixture of DNA topoisomers is loaded into a well at the top-left corner of a slab gel and electrophoresed along the left side of the gel. The bands corresponding to topoisomers with large ΔLk values merge into one spot. Then the gel is transferred to a buffer containing the intercalating ligand chloroquine and electrophoresed in the second horizontal direction (Fig. 5). This effectively shifts the value of ΔLk by a significant positive increment. As a result, the topoisomers with opposite values of ΔLk are well separated by the electrophoresis in the second direction. Also, the topoisomers with large negative ΔLk, which had identical mobility in the first direction, move with lower but different speed in the second direction. The number of topoisomers that can be resolved almost doubles in two-dimensional electrophoresis. See references 18 and 19 for applications of the electrophoretic separations of topoisomers.

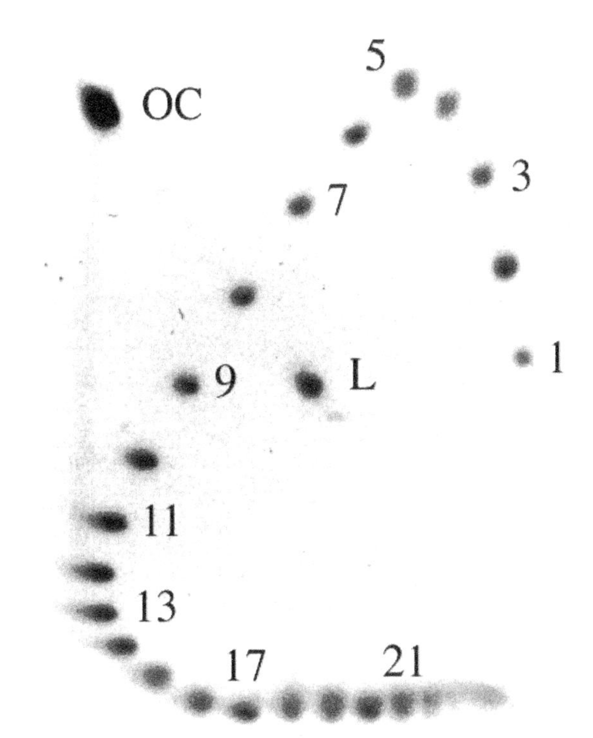

Figure 5 Separation of pUC19 DNA topoisomers by two-dimensional gel electrophoresis. Topoisomers 1 to 4 are positively supercoiled; the rest have negative supercoiling. After electrophoresis in the first direction, from top to bottom, the gel was saturated with ligand intercalating into the double helix. Upon electrophoresis from left to right in the second direction, the 2nd and 13th topoisomers turned out to migrate near the relaxed position in the second dimension. The spot in the top-left corner corresponds to the open circular form (OC); the spot near the middle of the gel corresponds to linear DNA (L). doi:10.1128/microbiolspec.PLAS-0036-2014.f5

Knots and links of different types formed by circular DNA molecules can also be separated by electrophoresis (Fig. 6). The method requires special calibration, because it is impossible to say in advance what position a particular topological structure must occupy relative to the unknotted circular DNA form. A large body of experimental results on the mobility of various topological structures has been accumulated (20, 21). To separate knotted and linked molecules by gel electrophoresis, the DNAs must have single-strand breaks, because otherwise mobility will also depend on the linking number of the complementary strands.

ENZYMES OF DNA TOPOLOGY

In enteric bacteria, four distinct topoisomerases are able to change the linking status of plasmid DNA molecules. When a DNA molecule is isolated from the cell, it is frozen in a particular topological state. However, plasmid populations exist in dynamic equilibrium inside the cell, and they can rapidly change topological structure (22). The known enzymes (DNA topoisomerases) that alter linking number include Topo I, DNA gyrase, Topo III, and Topo IV. DNA gyrase and Topo IV are related enzymes that break both strands of DNA simultaneously and are classified as type II enzymes. Topo I and Topo III are type I enzymes that break one strand at a time.

Topo I (also called ω protein) removes negative supercoils from covalently closed DNA and is an essential enzyme in *E. coli* (23, 24). Topo I conserves the energy of the DNA phosphodiester bond during the strand passage in a covalent phospho-tyrosine linkage (25). The conserved tyrosine residue and the phosphotyrosine intermediate are also found in Topo III and both type II topoisomerases, with the essential tyrosine being on the GyrA subunit of gyrase and the ParC subunit of Topo IV. This mechanism of breaking and rejoining DNA strands is also found in many site-specific recombinases, which use either tyrosine or serine as a high-energy phospho-protein link to DNA. Many enzymes including Hin, Gin, Tn3 Res, and XerC/D function as site-specific topoisomerases when a plasmid contains their cognate DNA recombination sequences (26–29).

DNA gyrase (Topo II) is a tetrameric protein made up of two GyrA and two GyrB subunits (30, 31). Gyrase is unique among all known topoisomerases for its ability to utilize the energy of ATP binding and hydrolysis to introduce negative supercoils into relaxed covalently closed DNA. The ATP binding domain is contained within the GyrB protein, and the drug novobiocin acts as a competitive inhibitor of ATP binding. Most of the DNA-binding site is formed by the GyrA subunit, and the potent antibiotics nalidixic acid and related fluoroquinolones such as ciprofloxacin and norfloxacin stabilize the covalent enzyme-DNA intermediate (30, 32). Such stabilized complexes lead to DNA breakage by either denaturation or by interactions with DNA replication machinery (33).

Topo III, the second type I enzyme discovered in *E. coli*, is not essential for cell viability, although it has important roles in normal DNA replication (34). Topo III can decatenate plasmids in the act of replication, and this enzyme requires a single-stranded region for it to separate replicating molecules. Topo III mutants accumulate small chromosomal deletions at positions where short repeats occur along the DNA sequence (35). This enzyme is conserved in eukaryotes, and a similar deletion phenotype is observed in *Saccharomyces cerevisiae*

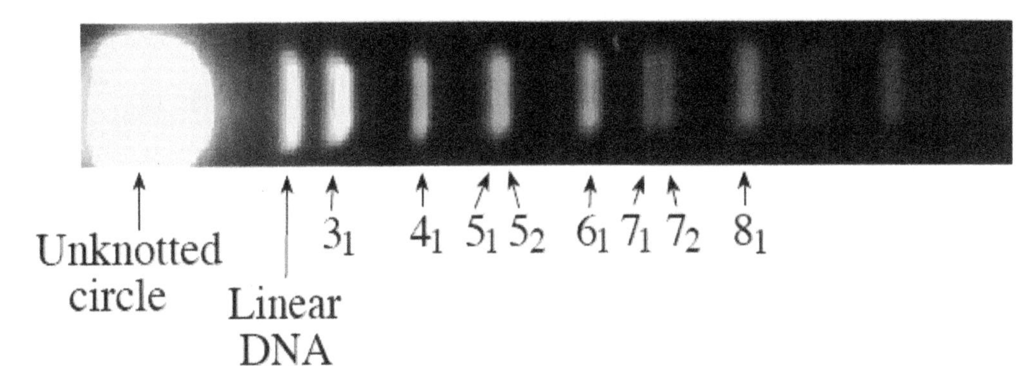

Figure 6 High-resolution gel electrophoresis of knotted forms of plasmid DNA that was run from left to right. Knot types are described in Fig. 1 (see reference 21). Reproduced from the *Journal of Molecular Biology* with permission from Elsevier.
doi:10.1128/microbiolspec.PLAS-0036-2014.f6

(36, 37). The discontinuously synthesized strand may provide the single-stranded substrate that is required for Topo III to function as a decatenase, and the enzyme can also disentangle hemicatenanes (see below).

Topo IV was the third topoisomerase found to be essential for cell viability in *E. coli* and *S. typhimurium* (38, 39). Topo IV is closely related to gyrase in subunit structure and catalytic mechanism. Topo IV controls the segregation of bacterial plasmids and the bacterial chromosome at the end of replication by completely unlinking and unknotting the replication products. Topo IV is a heteromeric tetramer made up of two ParE proteins, which have an ATP binding site, and two ParC subunits, which fashion most of the DNA binding site and include the essential tyrosine. Like gyrase, Topo IV is inhibited by both novobiocin and by fluoroquinolone antibiotics. Topo IV requires ATP to catalyze all strand-passing reactions, and it removes positive supercoils at a much higher rate than negative ones (40). Mutant strains have been constructed that allow the selective inhibition of either gyrase or Topo IV (41, 42). Topo IV provides the primary unknotting activity of the cell, as well as being the major decatenase, which explains its essential nature.

DYNAMIC TOPOLOGICAL EQUILIBRIUM

In vivo, the average value of σ has been determined for a relatively small number of large and small plasmids using a gel electrophoretic technique. Whereas results vary with growth conditions and genetic background (see Table 1) (43, 44), σ in actively growing wild type (WT) cells falls within a relatively narrow range of values from −0.05 to −0.07. *In vitro*, DNA gyrase can supercoil a small circular plasmid up to a much higher level of σ of −0.10. What controls gyrase-driven supercoiling inside cells?

The homeostatic supercoiling model of Menzel and Gellert (45) was developed to explain how bacteria maintain a relatively constant supercoiling level over a broad range of physiological conditions. This model was inspired by observing that transcriptional regulation of Topo I and gyrase responds to negative supercoiling in reciprocal ways. Expression of *gyrB* and *gyrA* genes increases when chromosomal DNA loses negative supercoiling. One example is a culture of bacterial cells treated with novobiocin, a compound that inhibits the binding of ATP to the GyrB subunit. Such cells increase expression of both gyrase subunits. Conversely, expression of the *topA* gene, which encodes Topo I, increases under conditions of elevated supercoiling. The homeostatic model posits that this expression pattern, combined with the opposing catalytic activities of the two enzymes, leads to an equilibrium where gyrase-induced negative supercoiling is balanced by Topo I-driven relaxation of negative supercoils. Important to the model is the fact that Topo I only removes supercoils from molecules with high levels of negative supercoiling, due to its requirement for an unpaired region of DNA, which is promoted by negative supercoiling.

Table 1 Constrained and unconstrained supercoiling in *E. coli* K12-derived strains[a]

E. coli strain	Relevant mutations	$-\sigma$ (Total)	$-\sigma_D$	$-\sigma_C$	% $-\sigma_D$
HB 101	WT	0.065	0.024	0.041	37
JTT1	WT	0.056	0.024	0.032	43
RS2	*top10*	0.072	0.030	0.042	42
SD-7	*top10 gyrB*	0.052	0.021	0.031	39

[a]Data from Jaworski et al. (86).

The Sternglanz lab generated an isogenic set of three *E. coli* strains that illustrate a basic pattern (23). Strain JTT1, with a full complement of WT topoisomerases, produces plasmids with an average σ of −0.056 (Table 1). Strain RS2 carries the *top10* allele, which makes a defective TopA protein and has increased plasmid supercoiling of σ equal to −0.072. Strain SD-7 is a double mutant with the *top10* mutation plus a *gyrB266* mutation, and its plasmids have an average σ of −0.052. The discovery of Topo III and Topo IV revealed that topological equilibrium is actually more complex than the Menzel and Gellert model predicted. Tests of all four enzymes indicate that Topo III does not normally contribute to the *in vivo* topology of plasmid DNA (46, 47), but Topo IV does (46, 48). However, overexpression of Topo III can suppress mutations in Topo I (49), so Topo III levels may determine how significantly it contributes to topological balance. Topo IV can remove negative supercoils from plasmid DNA *in vivo*, and topological balance inside living cells involves at least DNA gyrase, Topo I, and Topo IV (46–48). Both gyrase and Topo IV require ATP for their reactions, and they are influenced by the cellular ATP/ADP ratios (50, 51).

Whereas this homeostatic control model has been shown to work for many species of bacteria, experiments designed to modulate protein levels *in vivo* have shown that the average supercoiling level is not very sensitive to changes in the abundance of Topo I or gyrase. For example, increasing or decreasing *E. coli* Topo I or gyrase by 10% changes the average supercoil density by only 1.5% (52), which is below the detection limit of many techniques. Since there are 500 molecules of gyrase per cell, changing the protein concentration by 30 to 40% takes time, and when cells are shifted to new growth conditions, supercoiling is altered significantly within a minute or two. Why does supercoil density change rapidly?

In 1987, Liu and Wang (53) proposed a model for RNA transcription in which the DNA rotates around its axis during transcription rather than polymerases rotating around the flexible DNA template. Their model anticipated that this would create twin domains in which (–) supercoiling of DNA occurs upstream of the promoter, and (+) supercoiling is generated downstream from the transcription terminator. The twin domain effect was confirmed in *E. coli* using plasmids (54) and was confirmed in the bacterial chromosome of WT strains of *Salmonella* by measuring the supercoil density upstream and downstream of the highly transcribed *rrnG* ribosomal operon (55). In addition to causing a supercoil differential on opposing sides of highly transcribed genes, RNA polymerase creates a barrier to supercoil diffusion in the chromosome across the transcribed track (55–57). Mutant studies revealed that WT Topo I and gyrase both turn over processively at rates of 5 supercoils/sec at 30°, which is tuned to a rate of RNA polymerase elongation (50 nucleotides/sec at 30°) for *E. coli* and *Salmonella* (48). Moreover, growth rates, transcription rates, and average supercoil density levels vary significantly among different species of bacteria (48, 58–60).

SUPERCOIL-SENSITIVE DNA STRUCTURE

Negative supercoiling is associated with significant free energy that is stored in DNA. Negative supercoiling influences DNA structure, and like a spring, the free energy of supercoiling increases as a square of superhelix density (7, 61). Several unusual DNA conformations can be stabilized with negative supercoiling; well-characterized examples include left-handed Z-DNA, cruciforms, intramolecular triplexes or H-form DNA, and intermolecular triplexes or R-loops (Fig. 7). Each of these four alternative DNA conformations has a specific sequence requirement. Z-DNA is perhaps the best-characterized alternative DNA structure. Sequences adopting the left-handed conformation usually involve simple dinucleotide repeats of either dC-dG, or dT-dG. The equilibrium between right- and left-handed conformations depends on two things: (i) the level of negative superhelicity and (ii) the dCG or dTG repeat length. The longer the repeat length, the lower the supercoiling energy necessary to stabilize the left-handed conformation in plasmid DNA. A general quantitative description of the Z form formation in DNA has been obtained (62).

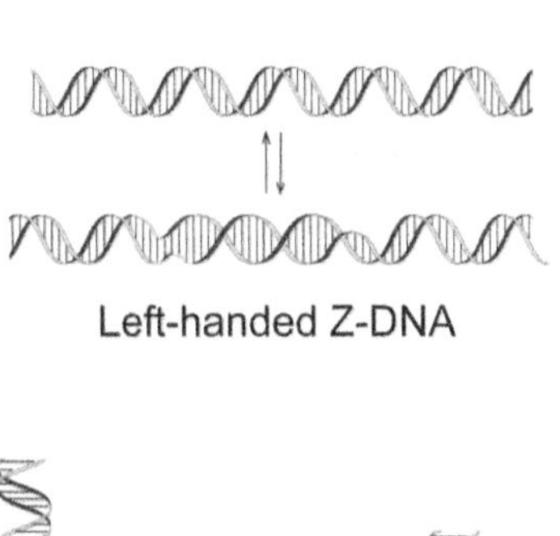

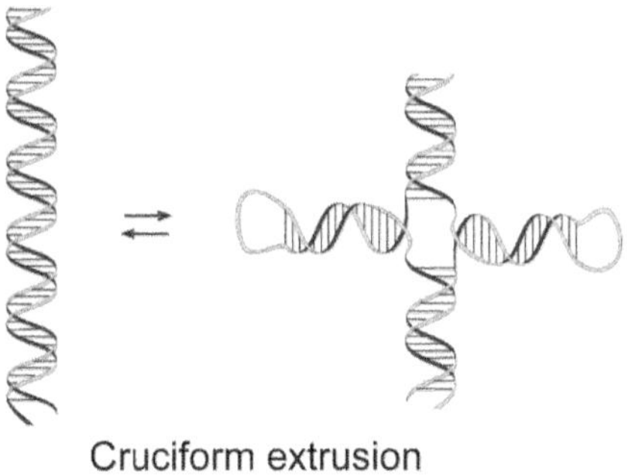

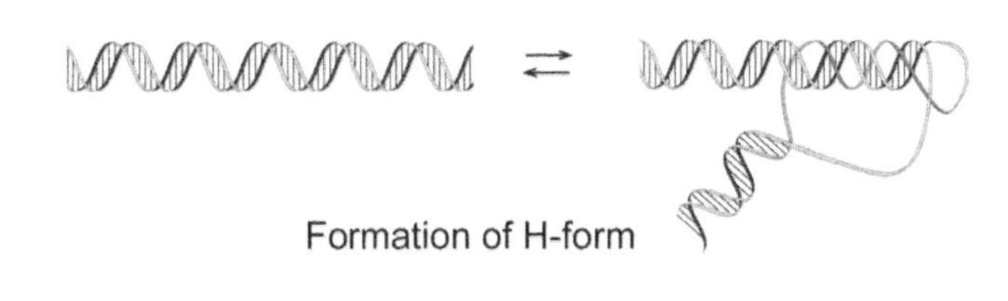

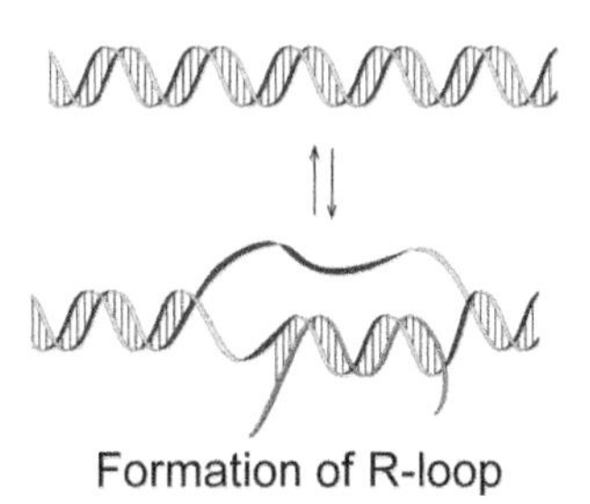

Figure 7 Alternative DNA structures that are stabilized by negative supercoiling.
doi:10.1128/microbiolspec.PLAS-0036-2014.f7

An example of a two-dimensional gel that illustrates Z-DNA formation is shown in Fig. 8. Panel A illustrates a series of topoisomers of plasmid pRW756 carrying a tract of repeating $(dC\text{-}dG)_{16}$ (63). At topoisomer number 15 (Fig. 8A) a break appears in the pattern. The break reflects the critical energy required for adopting a left-handed conformation of the 32-bp dC-dG insert. At this point, the plasmid mobility in the second dimension shifts backward because negative superhelicity is released from plasmids that change the conformation of the insert to the left-handed form. The control plasmid lacking the 32-bp dC-dG insert behaves as a continuous series of

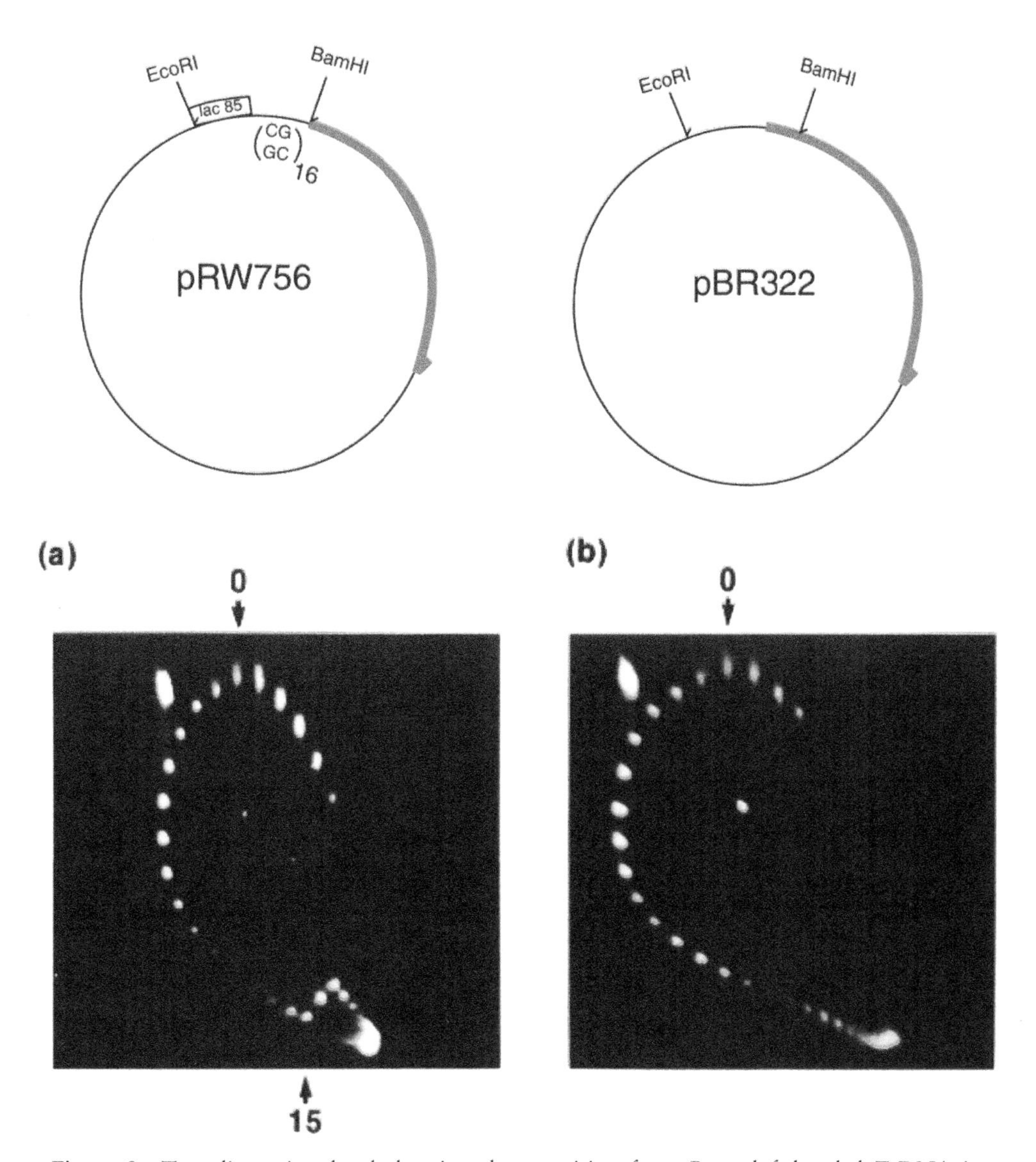

Figure 8 Two-dimensional gel showing the transition from B- to left-handed Z-DNA in plasmid DNA. This research was originally published in Kang DS, Wells RD. 1985. B-Z DNA junctions contain few, if any, nonpaired bases at physiological superhelical densities. *J Biol Chem* **260**:7783–7790. © the American Society for Biochemistry and Molecular Biology. doi:10.1128/microbiolspec.PLAS-0036-2014.f8

spots with increasing negative supercoiling (Fig. 8B). Plasmids with different repeat lengths have been engineered to monitor supercoiling torsional strength *in vivo* (64, 65).

Cruciforms can form in supercoiled DNA at positions where perfect or near perfect inverted repeat sequences occur. Two general mechanisms promote cruciform extrusion. First, supercoiled plasmids can adopt the cruciform conformation when sufficient supercoiling energy is present to destabilize the Watson-Crick structure at the tip of the hairpin (66). The simplest sequence that forms a cruciform structure is a long run of the dinucleotide repeat dA-dT. The sequence $(dA\text{-}dT)_{34}$ forms cruciforms with an unpaired adenine and thymine at the loop center in *E. coli* (67, 68). The second mechanism is coupled to DNA replication. As DNA becomes unwound at the replication fork, there is a potential for single-strand annealing. When a cruciform appears, it is a substrate for enzymes that stimulate Holiday branch migration, such as RuvAB, and it is also a substrate for Holiday-resolving enzymes: SbcC, SbcD, RuvB, and RuvC. These enzymes make long palindromes unstable in *E. coli* (69–72).

Intramolecular triplex DNA (H-DNA) may form at sequences containing long stretches of polypurine-polypyrimidine (73). In the H-form, half of either the purine- or pyrimidine-rich strand becomes unpaired, and its complement becomes triple stranded by forming Hoogsteen base pairs with purines in the major groove of the Watson-Crick base-paired segment (see Fig. 7). H-form DNA can be detected *in vivo*, but only under unusual circumstances (74). Intermolecular complexes can be formed with either RNA (75), which occasionally happens naturally (see below), during Clustered Regularly Interspaced Short Palindromic Repeats (CRISPR) reactions that cleave DNA (76), or with single-stranded oligonucleotides, which is useful for modifying gene expression patterns *in vivo* (77).

Another interesting case involves changes in supercoil density that modulates the efficiency of a plasmid recombination mechanism. Trigueros et al. (78) discovered that the multi-drug resistance plasmid pJHCMW1 contains a novel Xer recombination site (*mwr*) that converts dimeric or multimeric plasmids to the monomeric form. Other plasmids are known to contain a Xer site called *cer*, which works efficiently over a broad range of supercoil density. However, the *mwr* recombination efficiency was significantly lower when cells were grown under high-osmolarity conditions compared to low-osmolarity medium; the difference reflects a novel property of the *mwr* site that causes it to work better at high levels of DNA supercoiling that are achieved in low-osmolarity conditions. Many site-specific recombination reactions have been characterized, and some rely on negative supercoiling to achieve a functional synapse of directly repeated sites before strand exchanges. The Tn3/γδ resolution system is an example of a system with stringent requirements for supercoiling that spans a recombination efficiency of two orders of magnitude (79). This system has been developed so that one can pair a number of different mutant resolvases with a battery of supercoil sensors to efficiently and easily measure bacterial chromosome supercoil levels at many different sites *in vivo* (48, 55, 80, 81).

CONSTRAINED AND UNCONSTRAINED PLASMID TOPOLOGY *IN VIVO*

Circular plasmids such as the SV40 virus in human cells and pBR322 in bacterial cells have equivalent supercoil densities after the DNA is purified, but the basis of supercoiling is different in eukaryotes and prokaryotes. SV40 supercoils (82) are created by nucleosomes that bind DNA and wind almost two left-handed turns of DNA around every histone octomer (83). In *E. coli*, supercoiling is introduced by gyrase, a unique bacterial enzyme that catalytically introduces unrestrained supercoils in an ATP-dependent reaction.

A variety of studies show that in *E. coli* only half of plasmid and chromosomal supercoiling is unconstrained and diffusible *in vivo*, suggesting the possible existence of histone-like proteins in bacteria. In one of the first experiments, Pettijohn and Pfenninger (84) created single-strand breaks in the F plasmid using γ-irradiation and allowed the breaks to heal while DNA gyrase was inhibited with novobiocin to block supercoiling. They measured the F plasmid supercoiling levels with two different methods and showed that about half the supercoils were retained in the absence of gyrase activity. Because single-strand breaks allow DNA to relax to the energy minimum, these studies strongly suggested that half of bacterial supercoiling is constrained *in vivo*. This finding by the Pettijohn lab has been confirmed numerous times, and the interpretation was made clearer with a clever site-specific recombination experiment by Bliska and Cozzarelli (85). Using a plasmid substrate containing the *attB* and *attP* recombination sites of phage λ, Bliska characterized intramolecular recombination catalyzed by the Int protein. If supercoiling exists in a freely diffusible interwound conformation, recombination between inverted sites should trap these supercoils as catenane links between the recombinant circles (Fig. 9). By calibrating the reaction *in vitro* and then carrying assays *in vivo*, they demonstrated that approximately half of the plasmid topology existed as interwound supercoils inside living cells (Fig. 9).

Jaworski et al. (86) refined the estimate of constrained supercoiling by using plasmids with different segment lengths able to adopt the left-handed conformation (Table 1). Z-DNA conformation *in vivo* can be detected using several assays, including chemical reactivity of B-Z junction bases to osmium tetroxide (OsO_4) (87), the left-handed inhibition of *EcoRI*-dependent methylation (88), and by changes in the linking number of plasmids after DNA switches to a left-handed conformation (89). Independent estimates confirmed these results by analysis of increased A-T sensitivity to chemical modification (67) and by cruciform formation (68).

PROTEINS THAT CONSTRAIN TOPOLOGY IN *E. COLI*

What constrains half of a cell's topological structure inside a living cell? A number of abundant chromosome-bound proteins can constrain DNA topology (Table 2). In the section below, we provide snapshots of the most

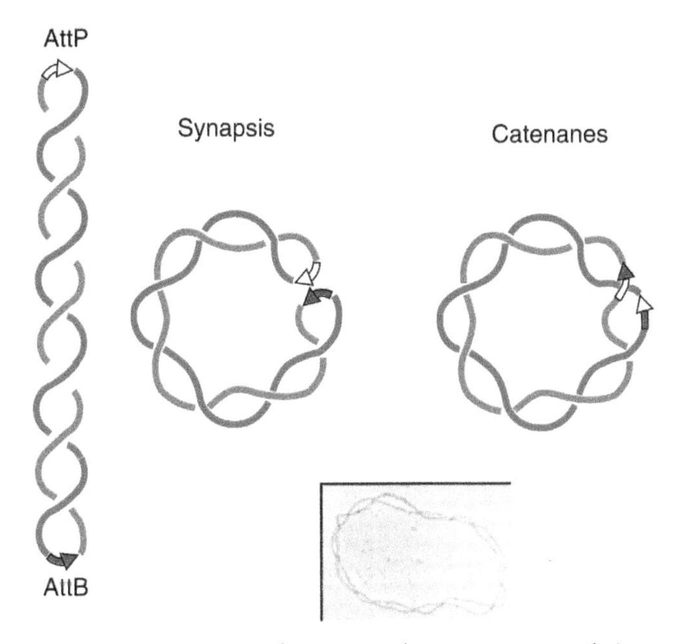

Figure 9 Conversion of interwound negative supercoils into catenanes linked by site-specific recombination. EM reprinted from Spergler SJ, Stasiak A, Cozzarelli NR. 1985. The stereo-structure of knots and catenanes produced by phage lambda integrative recombination: implications for mechanism and DNA structure. *Cell* **42**:325–334 with permission from Elsevier. doi:10.1128/microbiolspec.PLAS-0036-2014.f9

well-understood chromosome-associated proteins and a "back of the envelope" accounting of their potential contribution to constrained structure *in vivo* in *E. coli*. Assumptions about individual proteins are given in each section. Assumptions about the bacterium are as follows: for the model, we used the supercoiling values reported for strain JTT1 in Table 1. *E. coli* has a 4,639-kbp genome equivalent (GE) (90). Assuming 10.5-bp helical turns in B-form DNA, one GE will have an *Lk* value of 440,000 under physiological conditions. Extrapolating plasmid measures of supercoiling to the large *E. coli* chromosome, JTT1 (σ = 0.059) would have a *Lk* value of 24,500 per GE. For cells growing exponentially in rich medium, the average cell contains partially replicated copies of DNA amounting to 3 GE (91, 92). The total cell *Lk* would be 73,800, with unrestrained *Lk* equal to 31,000 and restrained ΔLk equal to 42,100. There are two problems with this accounting. First, nobody has measured all of the critical parameters of a specific protein's abundance and supercoil densities in the same *E. coli* strain under a uniform set of growth conditions. Second, attempts to experimentally confirm the calculation using mutants are frustrated by the facts that transcription contributes to supercoiling (see below) and that the group of proteins listed in Table 2 are all involved in regulating transcription in complex and interwoven ways (93). Elimination of one protein can be compensated for by changes in the expression patterns of other proteins, which makes it difficult to prove that any single protein accounts for a specific fraction of chromosome behavior in a WT cell. Nonetheless, making an educated estimate of this number from a growing body of data focused on specific DNA-protein complexes is a useful exercise (93, 94).

RNA polymerase is a pentameric protein made up of the proteins (α_2 $\beta\beta'\omega$) with a molecular mass of about 400 kDa. It is structurally conserved from bacteria to humans. About 3,000 molecules of RNA polymerase are present in *E. coli* cells growing exponentially in rich medium, and two thirds of these enzymes are engaged in transcription. One consequence of transcription is that each RNA polymerase molecule influences DNA writhe and unwinds a short segment of the DNA template, which results in a ΔLk of +1.7 supercoils per RNA polymerase (95). On a 5-kb plasmid, 3 transcribing polymerases constrain 5 negative supercoils, irrespective of any torsional effects, and induction of transcription can add 6 or more polymerases, producing a shift of 10 polymerase-constrained supercoils (22).

HU protein is a hetero-dimer composed of HupA and HupB protomers (94). Both subunits are related to each other and to the subunits of IHF (described below). All three dimers (A_2, AB, and B_2) bind double-stranded DNA, single-stranded DNA, and RNA (96). The double-stranded DNA binding site for HU-DNA complexes spans 36 bp. DNA-co-crystals and modeling experiments suggest that, like IHF, the HU protein bends DNA by >150° (93, 94). *In vivo* cross-linking experiments show that the predominant form of HU is AB, but A_2 and B_2 homo-dimers can be detected in the early exponential phase (97, 98). In exponential phase cells growing in Luria broth, HU is estimated to be

Table 2 Nucleoid-associated proteins[a]

Protein	Mol./cell in exponential phase[b]	% σ Total	% σ_c	Mol./cell in stationary phase
RNAP	3,000	5	8	
HU	30,000	16	30	15,000
IHF	6,000	0–3	0–6	27,500
FIS	30,000	9	15	1,000
H-NS	10,000	1.6	3	7,500
STPA	12,500	2	4	5,000
DPS	500	?	?	10,000

[a]Constrained supercoiling estimates for the most abundant chromosome-associated protein in *E. coli*.
[b]Mol., molecules.

present at about 30,000 molecules per cell (99). Of all the known small nucleoid-associated proteins, HU is clearly responsible for the largest effect on plasmid supercoiling (100, 101). HU constrains supercoils *in vitro* and, like nucleosomes, it can produce supercoiled plasmid DNA when incubated with either supercoiled or relaxed circular plasmid DNA plus an enzyme like the calf thymus Topo I enzyme, which removes either positive or negative supercoils (see above). At the optimum supercoiling ratio of protein and DNA, 2.5 AB HU dimers constrain one supercoil (102). Assuming this number reflects the conditions in a living cell, HU would account for 12,000 supercoils, or about 16% of total ΔLk and 40% of constrained topology. HU also promotes the circular ligation of DNA fragments shorter than the persistence length (103).

Strains carrying *hupAB* deletions show a slow growth phenotype (101) and are sensitive to γ- and UV-radiation (104). Analysis of plasmid DNA supercoiling in *hupAB* mutants of both *E. coli* and *S. typhimurium* agrees with the above calculation; the mutant exhibits a 15% loss of plasmid supercoiling and a broadening of the topoisomer distribution (101). Although it has never been measured using a torsion-sensitive assay (see below), the simplest explanation is that *hupAB* mutants lose primarily constrained supercoil structure.

Integration host factor (IHF) is a heterodimer encoded by two genes that are independently regulated: *ihfA* (*himA*) and *ihfB* (*hip*). IHF is closely related to HU in sequence and structure (93). IHF has a DNA binding site of 36 bp, and one can think of this protein as sequence-specific and less abundant HU. The consensus for IHF binding is a 13-bp sequence, WATCAA NNNNTTR, and IHF recognizes this site by interactions in the minor groove of DNA (105). When added alone to plasmid DNA, IHF does not show significant supercoiling activity in the Topo I assays. However, IHF dramatically bends DNA by 150° at its high-affinity sites and makes a structure that is optimally located at the end of a supercoiled loop (106–109). Bending by IHF is implicated in many genetic systems, as mutant strains show expression changes in about 100 *E. coli* genes (110). In the presence of HU, IHF can bind DNA that does not contain a consensus site, and single-molecule studies demonstrate that IHF has a large compaction effect in single-molecule experiments (111). Although the ability of IHF to induce supercoiling has not been reported, in Table 1, based on IHF abundance, we indicate a constrained supercoiling effect of between 0 and 6% of constrained supercoiling, with the upper limit set for IHF nonspecific supercoiling that would approach HU (1 supercoil per 2.5 dimers).

FIS (factor for inversion stimulation) is a homo-dimer encoded by the *fis* gene. FIS binding sites are complex; reports range from a footprint of 21 to 27 bp, with a consensus sequence of GNtYAaWWWtTRaNC (93). FIS is most abundant in cells growing exponentially in rich medium, with estimates ranging from 30,000 to 60,000 copies/cell (99, 112). FIS is capable of supercoiling DNA weakly in the Topo I supercoiling assay of relaxed plasmid DNA *in vitro*. FIS has several well-characterized roles in enhancing loop-dependent reactions that include phase inversion in *S. typhimurium* pilus type, G-inversion of the tail fiber of bacteriophage Mu, and upstream promotor (UP)-element stimulation of transcription of growth rate–regulated genes such as *rrn* P1 (113). Bending angles for specific FIS-DNA complexes are reported from 45 to 90°. Assuming FIS forms a DNA interaction that averages a 75° turn, or half the supercoiling potential of HU and IHF, 30,000 copies of FIS could account for 15% of constrained *in vivo* supercoiling. Single-molecule studies suggest that clusters of FIS might organize DNA segments into supercoiling domain barriers (114), although this has not been observed *in vivo*.

Histone-like nucleoid structuring (H-NS) protein is a homo-dimeric protein that, like HU, is able to bind double-stranded and single-stranded DNA as well as RNA (93). A high percentage of the protein exists as homodimers (115, 116), although higher oligomeric species are also found in solution (117–119). In log phase, estimates of H-NS concentrations vary from 10 to 20,000 molecules/cell (99). On double-stranded DNA, an H-NS homodimer is estimated to bind approximately 10 bp (120). Thus, unlike HU and IHF, which require 2 to 3 dimers to supercoil DNA, a much higher cooperative interaction is most likely necessary for H-NS to supercoil the DNA. H-NS binds preferentially to A/T-rich sequences and has the ability to supercoil DNA in the Topo I assay (121–123). H-NS is thought to regulate about 100 *E. coli* genes either directly or indirectly (123–125). It also is an element that enforces silencing of multiple operons including the *bgl* operon of *E. coli* (126), the *proU* operon of *Salmonella* and *E. coli* (120, 127), and bacteriophage Mu (128). Many of the genes regulated by H-NS are related to starvation and stress responses (129).

H-NS can spread along DNA sequences from a single high-affinity site (120, 130–133). In addition to forming homo-multimers, H-NS forms heteromeric complexes with a related protein, StpA (see below) (123, 134). Although H-NS has been proposed to be a regulator of DNA supercoiling, the collective analysis of plasmid supercoiling in *hns* strains shows no clear trend. Some plasmid data show increased supercoiling, some show decreased supercoiling, and some show no change (128, 135–140). This may be a case where plasmids mislead the interpretation of chromosome supercoiling (see below). Atomic force microscopy images show that unlike HU, IHF, and Fis, which make solenoidal structures, H-NS directly stabilizes an interwound form of DNA (141), and its ability to stabilize supercoils may depend on specific DNA sequences. A recent crystallographic structure shows that two-domain interactive bridging accounts for H-NS-dependent negative supercoiling (142). Molecular dynamics simulations indicate how nonspecific bridging interactions would drive molecular aggregation (143). Super-resolution fluorescent imaging of H-NS shows that H-NS forms novel clusters near the nucleoid center, which is different from all other nucleoid proteins (144). There is also evidence that ionic conditions may modulate the stability of H-NS structures (145–147). The sole study of chromosomal supercoiling using psoralen cross-linking concluded that *hns* mutants have slightly increased unconstrained supercoiling (148). However, this effect might be caused by the increased transcription around the genome that is observed in H-NS deletions. Assuming that 10,000 copies of H-NS can coat 20,000 bp of interwound DNA at a σ of –0.06, it could account for 3% of *in vivo* constrained supercoiling.

StpA (*stpA* 25,000 molecules/cell) was discovered as an *E. coli* gene that was involved in the splicing of a bacteriophage protein. Sequence analysis showed it to be closely related to H-NS (134), and subsequent studies demonstrated that these proteins share extensive structural similarity and form heterodimers. The significance of the homo- and hetero-dimeric species remains to be demonstrated, but cross talk seems likely. Nonetheless, several *hns* phenotypes are not altered in *stpA* strains, but others are. Like H-NS, the double-stranded DNA binding site is 10 bp, and like H-NS, homomeric ensembles of StpA spread along DNA as filaments that block access to the DNA (149). Slightly more abundant than H-NS, StpA could account for 4% of constrained supercoiling.

VARIATION OF CONSTRAINED AND UNCONSTRAINED SUPERCOILING *IN VIVO*

The ability to distinguish between constrained and unconstrained supercoiling is often necessary to fully explain topological changes that can be measured in plasmid DNA. Torsional strain (diffusible supercoiling) can be estimated using psoralen cross-linking in both plasmids and the bacterial chromosomes (84, 148, 150). Three cases illustrate the usefulness of understanding both constrained and unconstrained topology: (i) the importance of σ_D levels in biological control of closely related strains and species of bacteria, (ii) sequence-dependent topology of RNA-DNA triplexes, and (ii) *topA*'s critical role in controlling RNA-loop initiation.

Case 1

Although plasmid supercoil measurements have been performed using several *E. coli* strains and related bacterial species, experiments that discriminate between the fractional distribution of free interwound supercoils and constrained supercoils are rare. Z-DNA and cruciforms provide one method to perform such analyses. Plasmids are available in which different lengths of dG-dC are cloned at the same position in a pBR322-derived backbone, and the energy needed to adopt the left-handed form has been calibrated *in vitro* (89). By transforming bacterial strains with a panel of such plasmids, one can measure both the torsional strain, reported by the repeat length needed to adopt left-handed conformation, and the linkage state of the parental plasmid. Jaworski et al. (86) showed that both constrained and unconstrained supercoiling vary independently (Table 1). For example, torsional strain decreased in the presence of a mutation in *gyrB* and increased in a *topA* strain (Table 2). However, different *E. coli* strains (HB101 and JTT1) with a WT complement of topoisomerases and nucleoid binding proteins (see above) produced a significantly different balance of these two supercoiling states (37 and 43% $-\sigma_D$, respectively). *E. coli* generates more unconstrained supercoil tension than the closely related Gram-negative organisms. *S. typhimurium*, *Klebsiella*, *Enterobacter*, and *Morganella* all showed a significantly lower fraction of unconstrained supercoiling than *E. coli* (58, 86), which explains the different phenotypes for the *topA* genes in *E. coli* compared to *S. typhimurium* (23, 151–154). In addition, the phenotype of many nucleoid-associated proteins are different between *Salmonella* and *E. coli* (58). For example, deletion of the bacterial condensin (MukB) is tolerated in *E. coli* but is lethal under identical conditions in *Salmonella*. Also, deletions of SeqA are sick in *E. coli* but healthy in *Salmonella*, and specific mutations in GyrA and GyrB have significantly different effects in the two species. The phenotypes can be explained by the different supercoiling densities of the two species: *Salmonella* is 15% lower than *E. coli* (σ = 0.059 and –0.069, respectively) (58).

Case 2

DNA topology is sometimes influenced by unexpected factors. One example is the formation of an R-loop in plasmid DNA. An R-loop unwinds the Watson Crick helix and creates $+\sigma_C$. The subsequent release of RNA by alkali lysis leads to a compensating increase in (–) ΔLk. Early on, the R-loop effect was misinterpreted to be diffusible hyper-σ_D. R-loop formation occurs during transcription as the RNA polymerase opens a single-strand bubble in the Watson Crick double helix as it moves along DNA. Normally, polymerase rewinds Watson Crick strands back together, and the RNA transcript and rewound DNA exit through separate channels, but the displaced DNA strand is accessible (155). Under circumstances where the complementary strands do not rewind, RNA can remain hydrogen-bonded to the template in an R-loop. The earliest example of an RNA-plasmid heteroduplex involved the initiation of DNA replication in ColE-1 and its derivatives that include pBR322 and the pUC plasmids. Two RNA molecules regulate R-loop formation. RNAI is a 600-bp transcript that primes DNA synthesis, and RNAII is a regulatory RNA that controls the efficiency of the initiation process by binding to RNAI (156). Interactions between the 5′ end of the RNAI and the transcribing RNA polymerase cause the 3′ end to form an R-loop that initiates DNA replication. To form the RNA loop, a sequence at the 5′ end of RNAI interacts with the displaced DNA strand as RNA polymerase transcribes the plasmid *ori* region. When the 5′ end of RNAI and the displaced DNA strand interact, a stable RNA heteroduplex is produced at the origin (157). RnaseH and DNA PolI process the RNA-DNA complex to prime DNA replication.

When stable RNA-DNA structures arise, they can generate confusing data. One example is the chicken IgA immunoglobulin switch region. The switch region encodes the repeating 140-bp sequence $(AGGAG)_{28}$ that plays a functional role in switching immunoglobulin isotypes (158). When a plasmid containing this insert is transcribed by either *E. coli* or T7 polymerase (*in vitro* or *in vivo*) in the same direction as occurs in a chicken cell, the product is a remarkably stable 140-bp RNA-DNA hybrid that withstands melting temperatures up to 98° (159). The defined structure of the RNA-DNA complex remains unproven, but the most likely conformation is the typical R-loop (Fig. 10A), with a less likely possibility being the intermolecular triplex (Fig. 10B). If relaxed DNA is used as the template for transcription *in vitro*, the plasmid migrates near the position of a +12 topoisomer. Once this is formed *in vivo*, gyrase adds 12 additional (–) supercoils, and most popular kits for isolating plasmids exploit alkaline lysis, which removes the R-loop RNA, adding 12 more (–) supercoils to yield a hyper-negative plasmid. Unless the researchers know that there was a stable R-loop RNA formed in the plasmid, they can wrongly conclude that the plasmid existed under high negative supercoiling tension *in vivo* (see reference 160).

Another R-loop-inducing system is found for DNA triplet repeat sequences that expand and contract in the human genome and are often associated with neurologic diseases such as fragile X syndrome, Friedrich's ataxia, and amyotrophic lateral sclerosis (161). Long stretches of the triplet repeat sequence of CTG.CAG can trigger R-loop formation during active transcription in *E. coli* (162). The suggested mechanism is shown in Fig. 11. During transcription the CTG.CAG repeats can form hairpin loops on the nontemplate strand of DNA. This secondary structure impedes the rewinding of the nontemplate and transcribed strand, initiating an R-loop that can extend throughout the total length of the repeat sequence (163).

Case 3

Hyper-negative topology in plasmids isolated from *topA* mutants was initially considered to be an example of very high diffusible supercoiling. Pruss and Drlica discovered that plasmids like pBR322, which encode a membrane-bound tetracycline export pump, exhibited a hyper-negatively supercoiled phenotype in strains that lack a fully functional *topA* gene (164). Subsequent work demonstrated that this effect was due, in part, to association of the plasmid with secretion machinery that led to membrane insertion of the Tet protein while it was being transcribed. The hyper-supercoiling phenotype was not restricted to Tet and could be demonstrated for several proteins that were either cotranscriptionally inserted into the membrane or cotranscriptionally exported to the periplasm or outer membrane (165). Essential for the assay was the presence of a *topA* mutation, which encodes the protein that normally removes negative supercoils. The explanation was that torsional effects on plasmid DNA are mediated by membrane anchoring in the absence of full relaxing power of Top I. During transcription, gyrase replenished negative supercoils downstream of the transcription terminator, creating a super-abundance of negative supercoils (54, 166).

An important factor in resolving the true mechanism came from a plasmid with a Z-DNA-forming segment. From the known transition point of this plasmid, the researchers expected the plasmid to exist in the Z-DNA

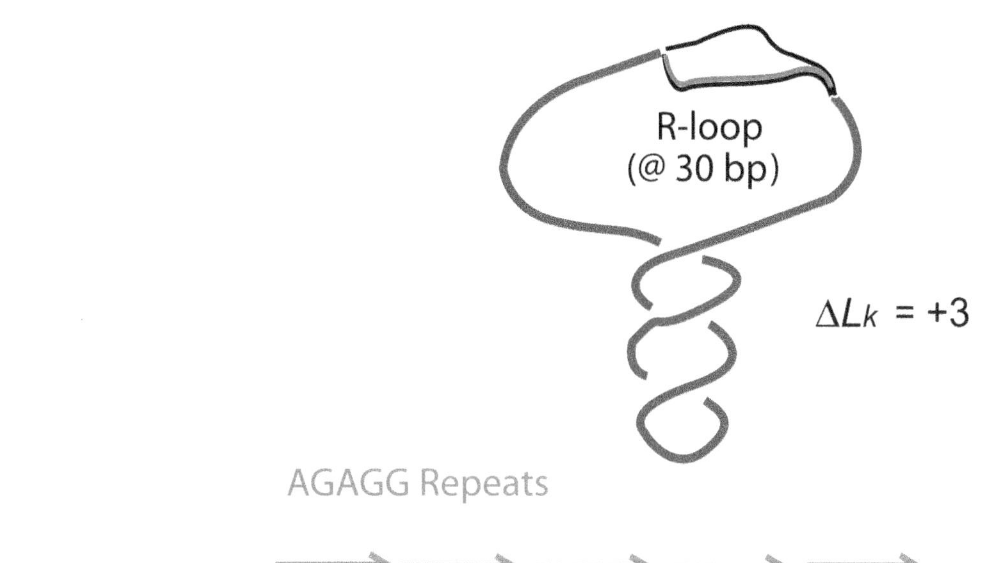

A.

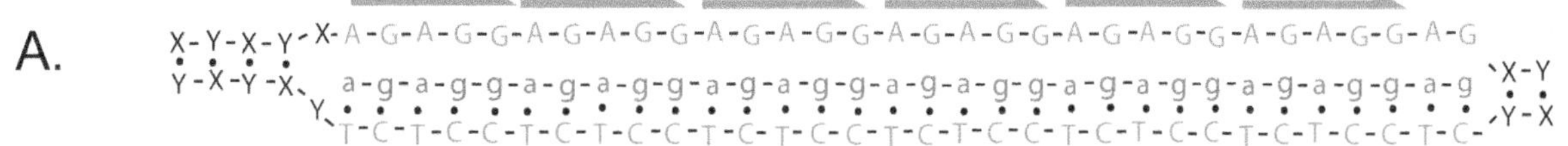

B.

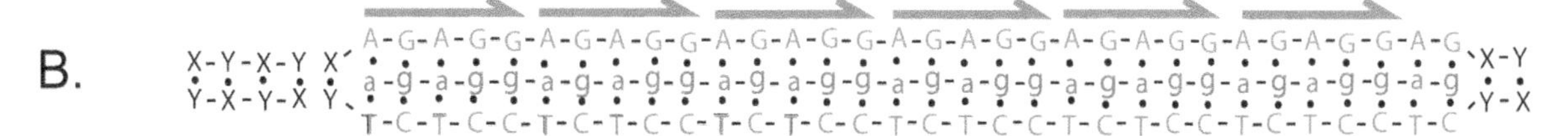

Figure 10 Alternative RNA-DNA structures that contribute to constrained supercoiling in a plasmid containing a fragment of the chicken IgA immunoglobulin switch region during transcription. The R-loop structure shown in (A) results in a displaced strand of DNA that constrains a ΔLk of about +1 for every 10 bp of RNA/DNA hybrid. (B) shows the structure of an intermolecular triplex in which Hoogsteen base pairing occurs in the major groove of the DNA strand (see reference 159). doi:10.1128/microbiolspec.PLAS-0036-2014.f10

in vivo. However, there was no sign of left-handed DNA, which suggested that the additional supercoils were constrained *in vivo* (160). Drolet later proved that hyper-negative supercoiling under these conditions was caused by plasmid R-loops (167, 168). Consistent with this theme is the observation that overexpression of RnaseH ameliorates the hyper-supercoiled phenotype of a *topA* mutant in *E. coli* (168).

Another protein that is important in controlling RNA-DNA interactions during transcription is the Rho protein, which exerts its effect by terminating transcription complexes that are not followed by translating ribosomes (169). The fact that *topA* is critical for preventing R-loop formation in plasmids suggests that this is an essential role for the protein, and the free unconstrained supercoil level of *E. coli* may require more TopA activity than other bacteria with lower unconstrained superhelix densities (see above). R-loop formation is most easily observed in plasmids, but this phenomenon occurs in large chromosomes as well.

DNA REPLICATION: REVERSIBLE FORKS, CATENANES, HEMICATENANES, AND KNOTS

Plasmids have been used in the dissection of many complex steps in DNA replication. The mechanism that initiates a round of replication was worked out first for the unidirectional ColE-1 origin (169), and plasmids containing the origin of phage λ and *E. coli oriC* sequences served as model substrates for establishing bidirectional replication *in vitro* (170, 171). When the genetic elements of the ter/Tus system were discovered, plasmids were used to study the mechanism of inducing a sequence-directed replication pause *in vitro* and *in vivo* (172). Plasmids continue to be useful in dissecting

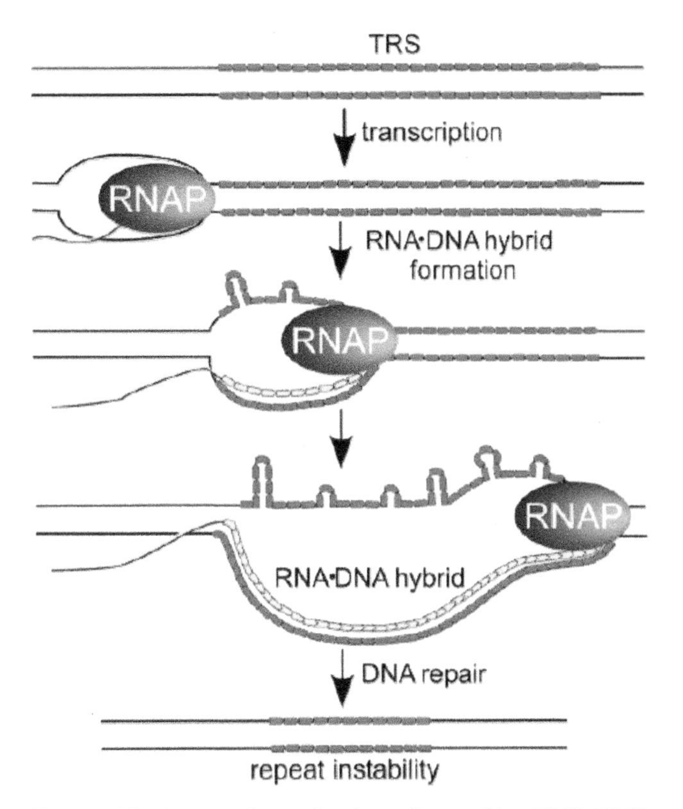

Figure 11 Proposed mechanism for stable RNA-DNA hybrids that can stimulate repeat instability. Transcription of DNA regions containing CG-rich trinucleotide repeats (red) favors formation of stable RNA-DNA hybrids. The displaced nontemplate DNA strand can adopt non-B DNA structures, such as CTG or CAG hairpins. The unpaired regions of the nontemplate strand are reactive to bisulfite modification. Reprinted from reference 162 with permission from the National Academy of Sciences.
doi:10.1128/microbiolspec.PLAS-0036-2014.f11

important stages of DNA replication. Three examples illustrate the power of plasmids for studies of DNA repair and topological reactions that occur in front of, or behind, a replisome (Fig. 12).

Fork Reversal

As DNA elongation proceeds from an initiation site, the machinery that carries out synthesis works in a semi-discontinuous pattern. The *E. coli* DnaB helicase unwinds the double helix, and on one side of the fork, DNA replication extends processively in the 5′ to 3′ direction. The opposite side of the fork is synthesized in a discontinuous mode that requires multiple reinitiation events and short replication tracks called Okazaki fragments. However, these two patterns are coordinated within the replisome, and if an impediment to DNA synthesis is encountered on either template strand, DNA replication machinery slows down or comes to a halt. DNA damage caused by chemical nucleotide modification is one example of a blocking lesion that can occur in DNA, and even under ideal conditions, chemical base damage is encountered in nearly every round of replication (173). Some DNA damage can be bypassed, but most damage must be repaired. Repairing DNA damage requires replication forks to sense damage and respond to different types of damage in a sophisticated way. About 40 years ago a model was proposed to explain how damage on the template for continuous synthesis could be bypassed after fork reversal (174) (Fig. 13). Branch migration at the fork can pair the two nascent strands, which forms a structure that can prime and carry out repair synthesis that will breach the block caused by a damaged nucleotide. Reversing the branch migration product, which can be carried out by a number of enzymes (175), allows

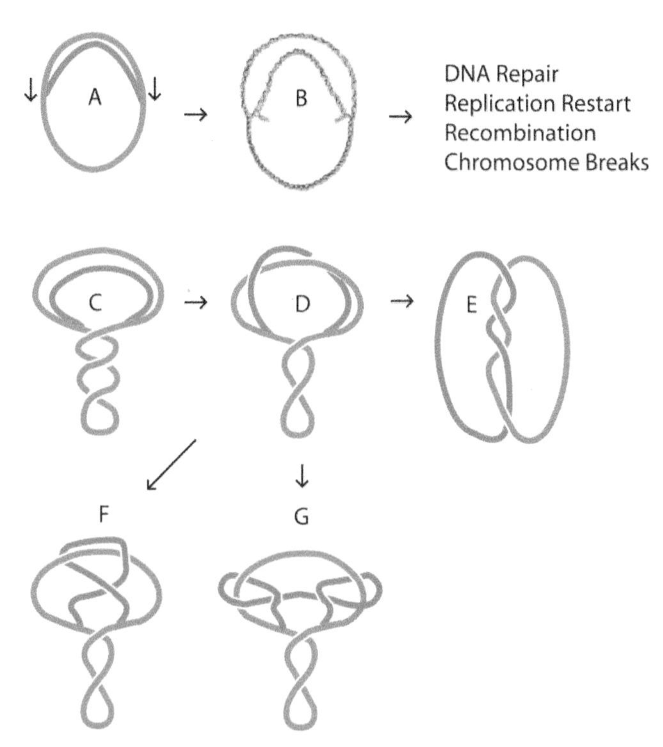

Figure 12 Replication intermediates identified in plasmid replication systems. (A) Replication initiated at a unique position leads to dual forms that move toward the terminus of replication. (B) Introduction of positive supercoils leads to replication fork reversal and formation of a four-way junction. (C) Negative supercoiling, which is generated by gyrase ahead of the fork, can be converted into precatenanes (D), which become catenanes (E) upon completion of DNA synthesis. (F, G) Topoisomerase activity in the replicated region can lead to complex knots.
doi:10.1128/microbiolspec.PLAS-0036-2014.f12

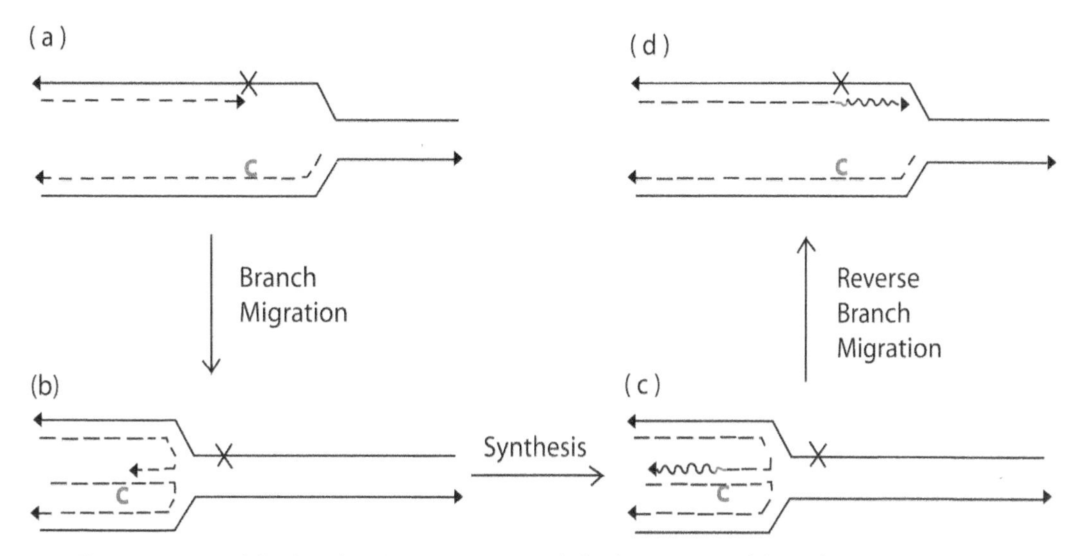

Figure 13 Model of replication repair. Strand displacement and branch migration create an alternative replication template allowing replication to bypass a lesion (X). Reproduced from reference 174 with permission from Elsevier.
doi:10.1128/microbiolspec.PLAS-0036-2014.f13

fork movement to restart and bypasses damage with high-fidelity polymerases (175). The original damage can then be repaired by excision repair enzymes post-synthesis.

Although the model was conceived in eukaryotic cells, evidence supports the role of fork reversal *in vitro* and *in vivo* in bacteria as well. Using a bidirectional replication system in a specialized plasmid that carries *oriC* and two *ter* sites to pause replication forks at specific points in a plasmid, Postow et al. demonstrated that fork reversal could be easily induced by incubating paused replication intermediates with ethidium bromide (176). The explanation for this result is that positive supercoiling induced by the intercalation of ethidium caused fork branch migration that generated a four-way junction. These structures were called chicken feet (177). Positive supercoiling might be generated ahead of a replication fork, but the question of whether reversed forks exist *in vivo* remained unanswered. Experiments by Courcelle et al. (178) provided physical evidence for reversed forks in *E. coli*. PBR322-containing cells were treated with controlled doses of DNA damage, which was followed by plasmid extraction and analysis by two-dimensional electrophoresis. Structures at the predicted position of reversed forks appeared as a cone above the normal two-dimensional plasmid profile (Fig. 14). The appearance and disappearance of these reverse forks closely followed a genetic repair response, which is strong evidence that fork reversal and repair do occur *in vivo*.

Catenanes and Hemicatenanes

Plasmids are powerful tools for studying DNA segregation. Replication intermediates are substrates for topoisomerases and repair enzymes, but what happens immediately after replication remains unclear (179, 180). The newly replicated chromosome segment (behind the forks) cannot be supercoiled because the nascent strands contain nicks that provide swivel points. Negative supercoiling introduced by DNA gyrase can exist in the unreplicated portion of the molecule (Fig. 12C) because the parental template strands provide a topologically closed system (181, 182). If the forks are not constrained by proteins or cotranslational attachment to the cytoplasmic membrane, negative supercoils may diffuse across the forks and exist as a mix of supercoils and links between the replicated daughter segments of the chromosomes (Fig. 12D).

Peter et al. analyzed the structure of replication intermediates accumulated by Tus-arrested replicating plasmids *in vitro* and *in vivo* (182). In the absence of a fork-blocking Tus/ter complex, replication produced catenated dimers (Fig. 15). When Tus/ter complexes were present, the daughter DNA segments were wound around each other, and linking was roughly evenly distributed between left-handed negative supercoils in the unreplicated segment of the molecule and left-handed links between daughter segments (Fig. 12C–D). These links can become catenane links between fully replicated daughter molecules at the completion of DNA synthesis (Fig. 12D–E); they are often called precatenanes. High-

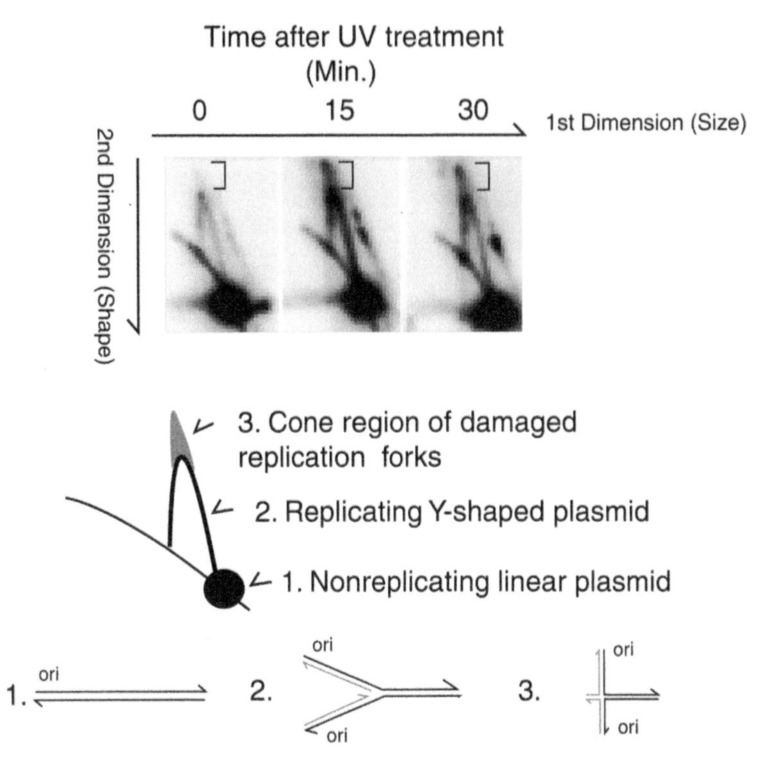

Figure 14 Replication fork reversal *in vivo* (see Fig. 12). Reprinted from reference 178 with permission from the American Association for the Advancement of Science.
doi:10.1128/microbiolspec.PLAS-0036-2014.f14

resolution gels indicated that 80% of neighboring plasmid topoisomers differed by two nodes rather than one (Fig. 15, center). This suggests that a type II topoisomerase(s) works behind forks since type I enzymes would generate distributions separated equally by steps of one (182) (Fig. 15). How (or whether) this type of strand movement occurs within a replicating bacterial chromosome remains an important question to be addressed.

Hemicatenanes represent the condition where two DNA duplexes are joined through a single-stranded

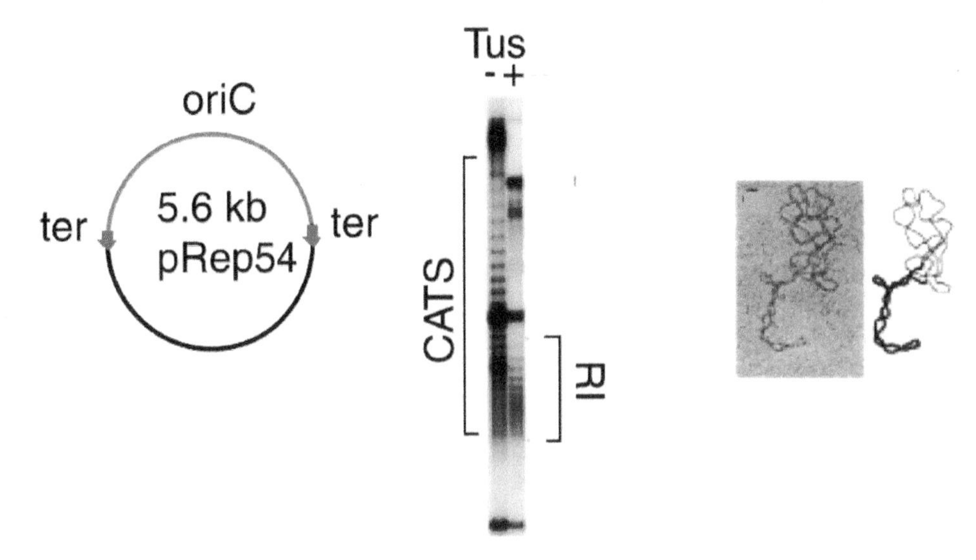

Figure 15 Resolution of catenane (CATS) and precatenane links (RI) in plasmid DNA (see Fig. 12). Reprinted from reference 182 with permission from Elsevier.
doi:10.1128/microbiolspec.PLAS-0036-2014.f15

interlock rather than the catenane discussed above that joins two DNA circles with linkages involving both strands of DNA. Hemicatenanes can occur and be stable between two linear DNA molecules, while catenanes separate when one circle becomes linearized. Historically, hemicatenanes have been observed but have been hard to study because they were hard to make. Recently, purification of the NeqTop3 enzyme from the hyperthermophilic archaeum, *Nanoarchaeum equitans*, has solved the problem (183). This thermophile encodes a type IA topoisomerase that works at 80°. At the high temperature and at high enzyme levels, NeqTop3 generates complex hemicatenane networks by binding and linking DNA at A-T rich regions that become unpaired by transient breathing. The enzyme removes negative supercoils from supercoiled DNA with the same enzyme mechanism as *E. coli* Top I and Top III, but at low enzyme concentrations NeqTop3 will dissolve the hemicatenated DNA that it makes at high enzyme levels. Three examples of situations where hemicatenanes form and are important *in vivo* (Fig. 16) include a replication fork blocked on the discontinuously synthesized strand (left), the point at which converging replication forks meet during bidirectional chromosome replication, and the final disposition of a double-Holiday junction in which branch points move together to result in a single hemicatenane linkage.

Knotted Bubbles

In the region behind the fork, topoisomerases can act to either remove precatenane links (see above), which would be equivalent to removing supercoils, or topoisomerases can introduce simple and complex knots in the precatenane portions of the plasmids (Fig. 12F, G). Recent work provides evidence for topoisomerase-mediated knots in the replicated precatenane portion of molecules *in vivo*. As in the studies by Peter mentioned above, Olavarrieta et al. exploited plasmids designed to

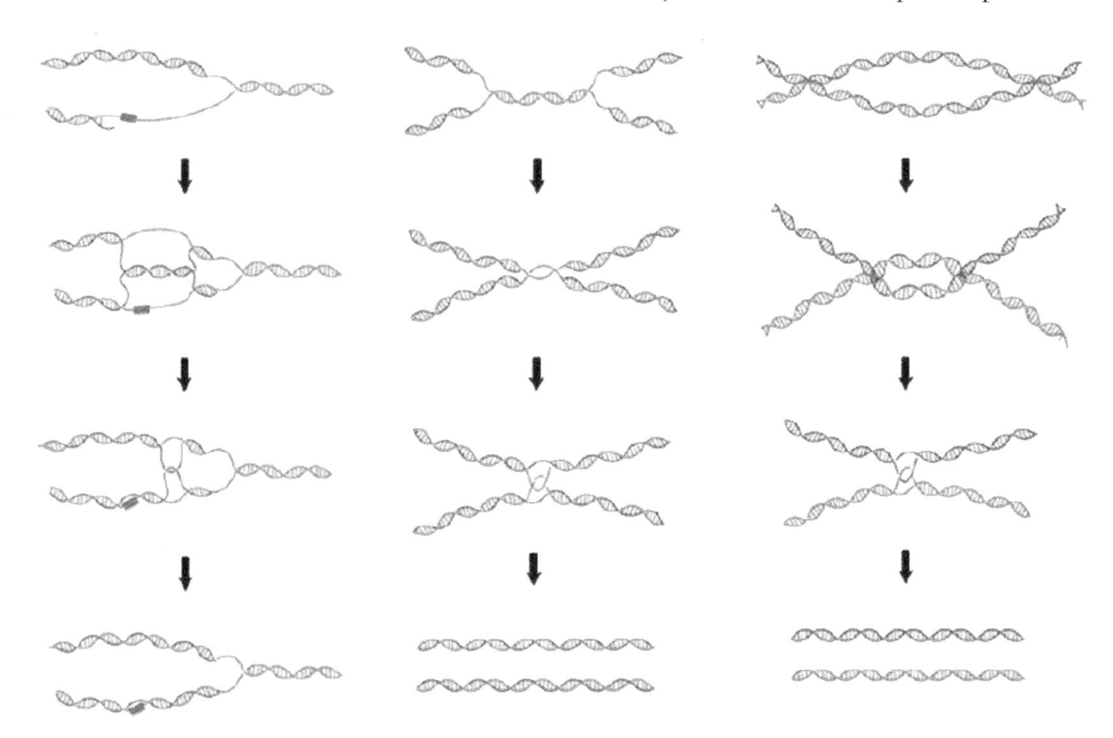

Figure 16 Schematic models for generating hemicatenanes during DNA replication. Three pathways to yield hemicatenane structures are shown. (Left) Lagging strand synthesis encounters a damage site, and the pairing of the lagging strand with the complementary leading strand can produce a pseudo-double Holliday structure. Dissolution of the pseudo-double Holliday structure leads to hemicatenanes and allows replication to bypass the damage site. (Center) Convergence of two replication forks at the final stage of replication can lead to either a single-strand catenane or hemicatenane conjoining two replicated duplexes. Both single-strand catenanes and hemicatenanes can be resolved by a type IA topoisomerase, allowing the segregation of the daughter chromosomes. (Right). Convergent branch migration of a double Holliday junction can generate a hemicatenane. Reproduced from reference 183 with permission from the National Academy of Sciences.
doi:10.1128/microbiolspec.PLAS-0036-2014.f16

arrest replication forks, with different segments of the plasmid represented by a replicated sector (184). A high-resolution two-dimensional gel analysis resolved mixtures of three plasmids: pTerE25, pTerE52, and pTerE81 (Fig. 17). Knotted bubbles provided a strong argument that a type II topoisomerase (Topo IV) must have been at work behind forks either during replication or while the forks were stalled at the Tus/ter complexes. It is not known whether the blocked replication forks, which are necessary for accumulating these populations, create intermediates that do not reflect structures occurring during unimpeded fork movement. Nonetheless, the experiments demonstrate the impressive ability of two-dimensional gels to analyze a wide range of biochemical functions in DNA metabolism.

Plasmid Data Can Obscure Conditions in a Bacterial Chromosome

Because supercoiling density in the chromosome is difficult to measure, investigators often rely on plasmids to gauge chromosomal supercoil levels and to interpret the effects of mutant enzymes or structural protein mutations on DNA structure. However, plasmids can mislead as well as inform. In addition to the R-loop issues discussed above, there are scenarios where one can make erroneous conclusions from plasmid data. One example is the practice of reporting chromosomal supercoiling density from the plasmid population. If the distribution is normal, that is a good sign, and the mean chromosomal situation may mirror the plasmid average. But when a broad distribution of plasmid topoisomers is seen, the common practice is to report the value of the center of plasmid distribution as the chromosomal supercoil density mean. The problem is that a broad or biphasic supercoil distribution is often a sign of a heterogeneous cell population (i.e., some fraction in a dramatically different physiological state than the normal WT). Sick or dying cells can have relaxed plasmids because of a low pool of ATP or a low ATP/ADP ratio, while healthy cells generate normal densities. Heterogeneous ensembles often mean that there are essentially no chromosomes in the population with a supercoil density indicated by the center of the plasmid distribution.

A second example comes from experiments in a novel gyrase mutant (*gyrB652*^{TS}). This mutation has a temperature-sensitive (TS) phenotype, but the GyrB652 enzyme does not lose activity at the restrictive temperature of 42°C. The *gyrB652* mutation produces an enzyme with a low catalytic turnover number relative to WT at all temperatures from 30°C to 42°C (185). GyrB652 strains stop growing at high temperatures because the supercoiling rate cannot keep up with the pace of (+) supercoiling generated by RNA transcription, which obeys the Q_{10} rule. The rule states that chemical reaction rates double for every increase of 10°C (see reference 48). In cultures of *Salmonella* growing at 30°C, a GyrB652 culture loses 95% of the normal σ_D, even at the permissive temperature of 30°C (σ_D = 0.005). Band-counting analysis of pUC19 DNA isolated from this mutant grown at either 30° or 42°C showed a σ_D of 0.030. The explanation is that, given time, GyrB655 gyrase will supercoil DNA up near the WT limit because pUC19 has no strong promoters, and negative supercoils generated by transcription cancel out by diffusing around the circle (54). We completely missed the large impact of this gyrase mutation on chromosomal supercoiling by assuming that pUC19

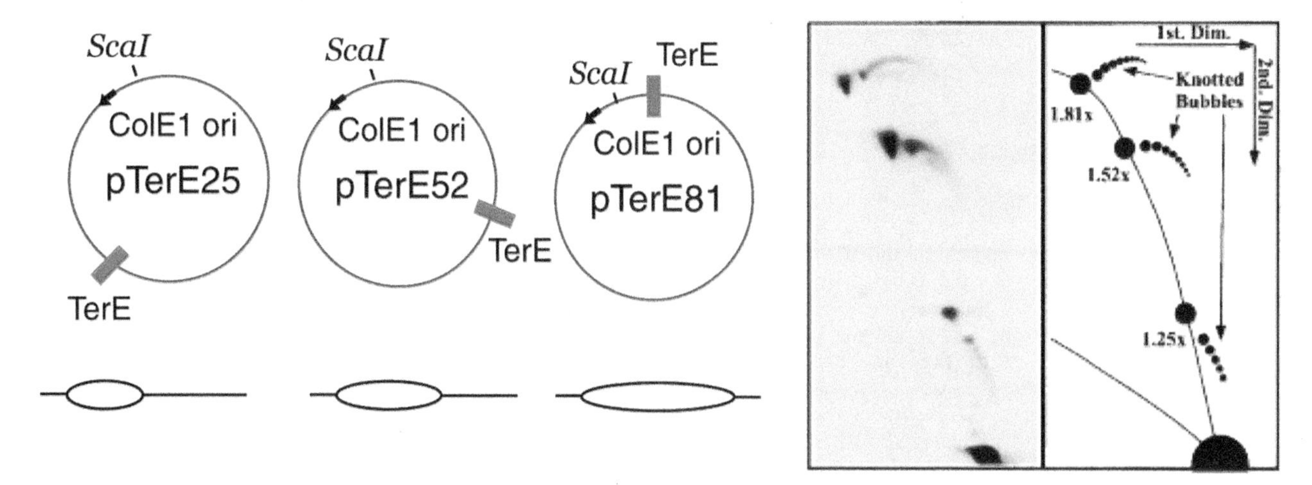

Figure 17 Knotting of replication bubbles *in vivo*. Reprinted from reference 184 with permission from Wiley. doi:10.1128/microbiolspec.PLAS-0036-2014.f17

was a good model for the chromosome. Even when plasmids have a strong promoter, the most significant topological effect of transcription is the constrained 1.7 supercoil/RNAP on the DNA (186).

CONCLUSION

The postgenomic scientific world is witnessing a dramatic shift from a strong interest in basic science to correlation science, with a theory that computer-assisted analysis of complex and sometimes questionable data is the future of modern medicine. Molecular structure/function studies of the biochemical processes that underpin major diseases, including cancer, diabetes, and emerging genetic, viral, and microbial infections, are disappearing from funding portfolios at the National Institutes of Health and the National Science Foundation. In the past year principal investigators dropped out of the NIH support portfolio, which is a record bad year. The teaching of basic science is also being cut, as medical schools join a stampede to cash in on "translational science," which is advertised to be the proven method for speeding discoveries from basic science to the bedside. This article's historical perspective is intended to illustrate the amazing nuance of DNA structure and function and to provide examples of powerful basic methods that have been developed to understand enzyme mechanics and measure the quantitative topological properties of DNA. As such, it is dedicated to those who chose to work on the difficult basic science front lines.

Acknowledgments. *Work in the laboratory of N.P.H. has been supported by NIH grant GM33143 from the U.S. National Institutes of Health. Work in the laboratory of A.V.V. has also been supported by the National Institutes of Health. Conflicts of interest: We declare no conflicts.*

Citation. Higgins NP, Vologodskii AV. 2015. Topological behavior of plasmid DNA. Microbiol Spectrum 3(2):PLAS-0036-2014.

References

1. **Vinograd J, Lebowitz J, Radloff R, Watson R, Laipis P.** 1965. The twisted circular form of polyoma viral DNA. *Proc Natl Acad Sci USA* **53:**1104–1111.
2. **Bauer WR, Crick FHC, White JH.** 1980. Supercoiled DNA. *Sci Am* **243:**100–113.
3. **Calugareanu G.** 1961. Sur las classes d'isotopie des noeuds tridimensionnels et leurs invariants. *Czech Math J* **11:**588–625.
4. **White JH.** 1969. Self-linking and the Gauss integral in higher dimensions. *Am J Math* **91:**693–728.
5. **Fuller FB.** 1971. The writhing number of a space curve. *Proc Natl Acad Sci USA* **68:**815–819.
6. **Brown PO, Cozzarelli NR.** 1979. A sign inversion mechanism for enzymatic supercoiling of DNA. *Science* **206:**1081–1083.
7. **Vologodskii AV, Cozzarelli NR.** 1994. Conformational and thermodynamic properties of supercoiled DNA. *Annu Rev Biophys Biomol Struct* **23:**609–643.
8. **Laundon CH, Griffith JD.** 1988. Curved helix segments can uniquely orient the topology of supertwisted DNA. *Cell* **52:**545–549.
9. **Adrian M, Wahli W, Stasiak AZ, Stasiak A, Dubochet J.** 1990. Direct visualization of supercoiled DNA molecules in solution. *EMBO J* **9:**4551–4554.
10. **Boles TC, White JH, Cozzarelli NR.** 1990. Structure of plectonemically supercoiled DNA. *J Mol Biol* **213:** 931–951.
11. **Vologodskii A, Cozzarelli NR.** 1994. Supercoiling, knotting, looping and other large-scale conformational properties of DNA. *Curr Opin Struct Biol* **4:**372–375.
12. **Bednar J, Furrer P, Stasiak A, Dubochet J, Egelman EH, Bates AD.** 1994. The twist, writhe and overall shape of supercoiled DNA change during counterion-induced transition from a loosely to a tightly interwound superhelix. Possible implications for DNA structure *in vivo*. *J Mol Biol* **235:**825–847.
13. **Lyubchenko YL, Shlyakhtenko LS.** 1997. Visualization of supercoiled DNA with atomic force microscopy *in situ*. *Proc Natl Acad Sci USA* **94:**496–501.
14. **Rybenkov VV, Vologodskii AV, Cozzarelli NR.** 1997. The effect of ionic conditions on the conformations of supercoiled DNA. I. Sedimentation analysis. *J Mol Biol* **267:**299–311.
15. **Vologodskii AV, Levene SD, Klenin KV, Frank-Kamenetskii M, Cozzarelli NR.** 1992. Conformational and thermodynamic properties of supercoiled DNA. *J Mol Biol* **227:**1224–1243.
16. **Keller W.** 1975. Determination of the number of superhelical turns in simian virus 40 DNA by gel electrophoresis. *Proc Natl Acad Sci USA* **72:**4876–4880.
17. **Lee C-H, Mizusawa H, Kakefuda T.** 1981. Unwinding of double-stranded DNA helix by dehydration. *Proc Natl Acad Sci USA* **78:**2838–2842.
18. **Wang JC.** 1986. Circular DNA, p 225–260. *In* Semlyen JA (ed), *Cyclic Polymers*. Elsevier, Essex, England.
19. **Vologodskii A.** 1998. Circular DNA. *Mol Biol* **35:**1–30.
20. **Wasserman SA, Cozzarelli NR.** 1986. Biochemical topology: application to DNA recombination and replication. *Science* **232:**951–960.
21. **Vologodskii AV,Crisona NJ, Laurie B, Pieranski P, Katritch V, Dubochet J, Stasiak A.** 1998. Sedimentation and electrophoretic migration of DNA knots and catenanes. *J Mol Biol* **278:**1–3.
22. **Cook DN, Ma D, Pon NG, Hearst JE.** 1992. Dynamics of DNA supercoiling by transcription in *Escherichia coli*. *Proc Natl Acad Sci USA* **89:**10603–10607.
23. **DiNardo S, Voelkel KA, Sternglanz R, Reynolds AE, Wright A.** 1982. *Escherichia coli* DNA topoisomerase I mutants have compensatory mutations in DNA gyrase genes. *Cell* **31:**43–51.

24. **Wang JC.** 1996. DNA topoisomerases. *Annu Rev Biochem* **65:**635–692.

25. **Stewart L, Redinbo MR, Qiu X, Hol WGJ, Champoux JJ.** 1998. A model for the mechanism of human topoisomerase I. *Science* **279:**1534–1541.

26. **Krasnow MA, Cozzarelli NR.** 1983. Site-specific relaxation and recombination by the Tn3 resolvase: recognition of the DNA path between oriented res sites. *Cell* **32:**1313–1324.

27. **Johnson RC, Bruist MF.** 1989. Intermediates in hin-mediated DNA inversion: a role for Fis and the recombinational enhancer in the strand exchange reaction. *EMBO J* **8:**1581–1590.

28. **Kanaar R, Klippel A, Shekhtman E, Dungan JM, Kahmann R, Cozzarelli NR.** 1990. Processive recombination by the phage Mu Gin system: implications for the mechanisms of DNA strand exchange, DNA site alignment, and enhancer action. *Cell* **62:**353–366.

29. **Perals K, Capiaux H, Vincourt J-B, Louarn J-M, Sherratt DJ, Cornet F.** 2001. Interplay between recombination, cell division and chromosome structure during chromosome dimer resolution in *Escherichia coli. Mol. Microbiol.* **39:**904–913.

30. **Higgins NP, Peebles CL, Sugino A, Cozzarelli NR.** 1978. Purification of the subunits of *Escherichia coli* DNA gyrase and reconstitution of enzymatic activity. *Proc Natl Acad Sci USA* **75:**1773–1777.

31. **Cozzarelli NR.** 1980. DNA gyrase and the supercoiling of DNA. *Science* **207:**953–960.

32. **Kampranis SC, Maxwell A.** 1998. The DNA gyrase-quinolone complex. ATP hydrolysis and the mechanism of DNA cleavage. *J Biol Chem* **273:**22615–22626.

33. **Hiasa H, Yousef DO, Marians KJ.** 1996. DNA strand cleavage is required for replication fork arrest by a frozen topoisomerase-quinolone-DNA ternary complex. *J Biol Chem* **271:**26424–26429.

34. **Hiasa H, Marians KJ.** 1994. Topoisomerase III, but not topoisomerase I, can support nascent chain elongation during theta-type DNA replication. *J Biol Chem* **269:** 32655–32659.

35. **Schofield M, Agbunag R, Miller J.** 1992. DNA inversions between short inverted repeats in *Escherichia coli. Genetics* **132:**295–302.

36. **Gangloff S, McDonald JP, Bendixen C, Arthur L, Rothstein R.** 1994. The yeast type I topoisomerase Top3 interacts with Sgs1, a DNA helicase homolog: a potential eukaryotic reverse gyrase. *Mol Cell Biol* **14:**8391–8398.

37. **Kim RA, Caron PR, Wang JC.** 1995. Effects of yeast DNA topoisomerase III on telomere structure. *Proc Natl Acad Sci USA* **92:**2667–2671.

38. **Kato J, Nishimura Y, Imamura R, Niki H, Hiraga S, Suzuki H.** 1990. New topoisomerase essential for chromosome segregation in *E. coli. Cell* **63:**393–404.

39. **Luttinger AL, Springer AL, Schmid MB.** 1991. A cluster of genes that affects nucleoid segregation in *Salmonella typhimurium. New Biol* **3:**687–697.

40. **Crisona NJ, Strick TR, Bensimon D, Croquette V, Cozzarelli NR.** 2000. Preferential relaxation of positively supercoiled DNA by *Escherichia coli* topoisomerase IV in single-molecule and ensemble measurements. *Genes Dev* **14:**2881–2892.

41. **Khodursky AB, Zechiedrich EL, Cozzarelli NR.** 1995. Topoisomerase IV is a target of quinolones in *Escherichia coli. Proc Natl Acad Sci USA* **92:**11801–11805.

42. **Hardy CD, Cozzarelli NR.** 2003. Alteration of *Escherichia coli* topoisomerase IV to novobiocin resistance. *Antimicrob Agents Chemother* **47:**941–947.

43. **Yamamoto N, Droffner ML.** 1985. Mechanisms determining aerobic or anaerobic growth in the facultative anaerobe *Salmonella typhimurium. Proc Natl Acad Sci USA* **82:**2077–2081.

44. **Dorman CJ, Barr GC, Bhriain NN, Higgins CF.** 1988. DNA supercoiling and the anaerobic growth phase regulation of tonB gene expression. *J Bacteriol* **170:**2816–2826.

45. **Menzel R, Gellert M.** 1983. Regulation of the genes for *E. coli* DNA gyrase: homeostatic control of DNA supercoiling. *Cell* **34:**105–113.

46. **Zechiedrich EL, Khodursky AB, Bachellier S, Schneider R, Chen D, Lilley DM, Cozzarelli NR.** 2000. Roles of topoisomerases in maintaining steady-state DNA supercoiling in *Escherichia coli. J Biol Chem* **275:**8103–8113.

47. **Khodursky AB, Peter BJ, Schmid MB, DeRisi J, Botstein D, Brown PO, Cozzarelli NR.** 2000. Analysis of topoisomerase function in bacterial replication fork movement: use of DNA microarrays. *Proc Natl Acad Sci USA* **97:**9419–9424.

48. **Rovinskiy N, Agbleke AA, Chesnokova O, Pang Z, Higgins NP.** 2012. Rates of gyrase supercoiling and transcription elongation control supercoil density in a bacterial chromosome. *PLoS Genet* **8:**e1002845. doi: 10.1371/journal.pgen.1002845.

49. **Broccoli S, Phoenix P, Drolet M.** 2000. Isolation of the *topB* gene encoding DNA topoisomerase III as a multicopy suppressor of *topA* null mutations in *Escherichia coli. Mol Microbiol* **35:**58–68.

50. **Hsieh LS, Burger RM, Drlica K.** 1991. Bacterial DNA supercoiling and ATP/ADP. Changes associated with a transition to anaerobic growth. *J Mol Biol* **219:**443–450.

51. **Hatfield GW, Benham CJ.** 2002. DNA topology-mediated control of global gene expression in *Escherichia coli. Annu Rev Genet* **36:**175–203.

52. **Snoep JL, van der Weijden CC, Andersen HW, Westerhoff HV, Jensen PR.** 2002. DNA supercoiling in *Escherichia coli* is under tight and subtle homeostatic control, involving gene-expression and metabolic regulation of both topoisomerase I and DNA gyrase. *Eur J Biochem* **269:**1662–1669.

53. **Liu LF, Wang JC.** 1987. Supercoiling of the DNA template during transcription. *Proc Natl Acad Sci USA* **84:** 7024–7027.

54. **Wu H-Y, Shyy S, Wang JC, Liu LF.** 1988. Transcription generates positively and negatively supercoiled domains in the template. *Cell* **53:**433–440.

55. **Booker BM, Deng S, Higgins NP.** 2010. DNA topology of highly transcribed operons in *Salmonella enterica* serovar Typhimurium. *Mol Microbiol* **78:**1348–1364.

56. **Deng S, Stein RA, Higgins NP.** 2004. Transcription-induced barriers to supercoil diffusion in the *Salmonella typhimurium* chromosome. *Proc Natl Acad Sci USA* **101:**3398–3403.
57. **Deng S, Stein RA, Higgins NP.** 2005. Organization of supercoil domains and their reorganization by transcription. *Mol Microbiol* **57:**1511–1521.
58. **Champion K, Higgins NP.** 2007. Growth rate toxicity phenotypes and homeostatic supercoil control differentiate *Escherichia coli* from *Salmonella enterica* serovar Typhimurium. *J Bacteriol* **189:**5839–5849.
59. **Tretter EM, Berger JM.** 2012. Mechanisms for defining supercoiling set point of DNA gyrase orthologs. I. A nonconserved acidic C-terminal tail modulates *Escherichia coli* gyrase activity. *J Biol Chem* **287:**18636–18644.
60. **Higgins NP.** 2014. RNA polymerase: chromosome domain boundary maker and regulator of supercoil density. *Curr Opion Microbiol* **22:**138–143.
61. **Depew RE, Wang JC.** 1975. Conformational fluctuations of DNA helix. *Proc Natl Acad Sci USA* **72:**4275–4279.
62. **Mirkin SM, Lyamichev VI, Kumarev VP, Kobzev VF, Nosikov VV, Vologodskii AV.** 1987. The energetics of the B-Z transition in DNA. *J Biomol Struct Dyn* **5:**79–88.
63. **Kang DS, Wells RD.** 1985. B-Z DNA junctions contain few, if any, nonpaired bases at physiological superhelical densities. *J Biol Chem* **260:**7783–7790.
64. **Klysik J, Stirdivant SM, Wells RD.** 1982. Left-handed DNA. Cloning, characterization, and instability of inserts containing different lengths of (dC-dG) in *Escherichia coli*. *J Biol Chem* **257:**10152–10158.
65. **Peck LJ, Nordheim A, Rich A, Wang JC.** 1982. Flipping of cloned d(pCpG)n.d(pCpG)n DNA sequences from right- to left-handed helical structure by salt, Co(III), or negative supercoiling. *Proc Natl Acad Sci USA* **79:** 4560–4564.
66. **Murchie AIH, Lilley DMJ.** 1987. The mechanism of cruciform formation in supercoiled DNA: initial opening of central basepairs in salt-dependent extrusion. *Nucleic Acids Res* **15:**9641–9654.
67. **McClellan JA, Boublikova P, Palecek E, Lilley DM.** 1990. Superhelical torsion in cellular DNA responds directly to environmental and genetic factors. *Proc Natl Acad Sci USA* **87:**8373–8377.
68. **Dayn A, Malkhosyan S, Duzhy D, Lyamichev V, Panchenko Y, Mirkin S.** 1991. Formation of $(dA\text{-}dT)_n$ cruciforms in *Escherichia coli* cells under different environmental conditions. *J Bacteriol* **173:**2658–2664.
69. **Chalker AF, Leach DR, Lloyd RG.** 1988. *Escherichia coli sbcC* mutants permit stable propagation of DNA replicons containing a long palindrome. *Gene* **71:**201–205.
70. **Connelly JC, Leach DR.** 1996. The *sbcC* and *sbcD* genes of *Escherichia coli* encode a nuclease involved in palindrome inviability and genetic recombination. *Genes Cells* **1:**285–291.
71. **Leach D, Lindsey J, Okely E.** 1987. Genome interactions which influence DNA palindrome mediated instability and inviability in *Escherichia coli*. *J Cell Sci* **7:**33–40.
72. **Leach DR, Okely EA, Pinder DJ.** 1997. Repair by recombination of DNA containing a palindromic sequence. *Mol Microbiol* **26:**597–606.
73. **Lyamichev VI, Mirkin SM, Frank-Kamenetskii MD.** 1986. Structures of homopurine-homopyrimidine tract in superhelical DNA. *J Biomol Struct Dyn* **3:**667–669.
74. **Mirkin SM, Frank-Kamenetskii MD.** 1994. H-DNA and related structures. *Annu Rev Biophys Biomol Struct* **23:**541–576.
75. **Frank-Kamenetskii MD, Mirkin SM.** 1995. Triplex DNA structures. *Annu Rev Biochem* **64:**65–96.
76. **Jinek M, Chylinski K, Fonfara I, Hauer M, Doudna JA, Charpentier E.** 2012. A programmable dual-RNA-guided DNA endonuclease in adaptive bacterial immunity. *Science* **337:**816–821.
77. **Vasquez KM, Wilson JH.** 1998. Triplex-directed modification of genes and gene activity. *Trends Biochem Sci* **23:**4–9.
78. **Trigueros S, Tran T, Sorto N, Newmark J, Colloms SD, Sherratt DJ, Tolmasky ME.** 2009. mwr Xer site-specific recombination is hypersensitive to DNA supercoiling. *Nucleic Acids Res* **37:**3580–3587.
79. **Benjamin KR, Abola AP, Kanaar R, Cozzarelli NR.** 1996. Contributions of supercoiling to Tn*3* resolvase and phage Mu Gin site-specific recombination. *J Mol Biol* **256:**50–65.
80. **Higgins NP, Yang X, Fu Q, Roth JR.** 1996. Surveying a supercoil domain by using the gd resolution system in *Salmonella typhimurium*. *J Bacteriol* **178:**2825–2835.
81. **Stein R, Deng S, Higgins NP.** 2005. Measuring chromosome dynamics on different timescales using resolvases with varying half-lives. *Mol Microbiol* **56:**1049–1061.
82. **Vinograd J, Lebowitz J.** 1966. Physical and topological properties of circular DNA. *J Gen Phys* **49:**103–125.
83. **Luger K, Mader AW, Richmond RK.** 1997. Crystal structure of the nucleosome core particle at 2.8 Å resolution. *Nature* **389:**251–260.
84. **Pettijohn DE, Pfenninger O.** 1980. Supercoils in prokaryotic DNA restrained *in vivo*. *Proc Natl Acad Sci USA* **77:**1331–1335.
85. **Bliska JB, Cozzarelli NR.** 1987. Use of site-specific recombination as a probe of DNA structure and metabolism *in vivo*. *J Mol Biol* **194:**205–218.
86. **Jaworski A, Higgins NP, Wells RD, Zacharias W.** 1991. Topoisomerase mutants and physiological conditions control supercoiling and Z-DNA formation *in vivo*. *J Biol Chem* **266:**2576–2581.
87. **Rajagopalan M, Rahmouni AR, Wells RD.** 1990. Flanking AT-rich tracts cause a structural distortion in Z-DNA in plasmids. *J Biol Chem* **265:**17294–17299.
88. **Jaworski A, Hsieh W-T, Blaho JA, Larson JE, Wells RD.** 1987. Left handed DNA *in vivo*. *Science* **238:** 773–777.
89. **Zacharias W, Jaworski A, Larson JE, Wells RD.** 1988. The B- to Z-DNA equilibrium *in vivo* is perturbed by biological processes. *Proc Natl Acad Sci USA* **85:** 7069–7073.

90. **Blattner FR, Plunkett G, Bloch CA, Perna NT, Burland V, Riley M, Collado-Vides J, Glasner JD, Rode CK, Mayhew GF, Gregor J, Davis NW, Kirkpatrick HA, Goeden MA, Rose DJ, Mau B, Shao Y.** 1997. The complete genome sequence of *Escherichia coli* K-12. *Science* **277:**1453–1474.

91. **Skarstad K, Steen HB, Boye E.** 1983. Cell cycle parameters of slowly growing *Escherichia coli* B/r studied by flow cytometry. *J Bacteriol* **154:**656–662.

92. **Skarstad K, Steen HB, Boye E.** 1985. *Escherichia coli* DNA distributions measured by flow cytometry and compared with theoretical computer simulations. *J Bacteriol* **163:**661–668.

93. **Johnson RC, Johnson LM, Schmidt JW, Gardner JF.** 2005. The major nucleoid proteins in the structure and function of the *E. coli* chromosome, p 65–132. *In* Higgins NP (ed), *The Bacterial Chromosome.* ASM Press, Washington, DC.

94. **Dillon SC, Dorman CJ.** 2010. Bacterial nucleoid-associated proteins, nucleoid structure and gene expression. *Nat Rev Microbiol* **8:**185–195.

95. **Gamper HB, Hearst JE.** 1982. A topological model for transcription based on unwinding angle analysis of *E. coli* RNA polymerase binary, initiation and ternary complexes. *Cell* **29:**81–90.

96. **Drlica K, Rouviere-Yaniv J.** 1987. Histonelike proteins of bacteria. *Microbiol Rev* **51:**301–319.

97. **Claret L, Rouviere-Yaniv J.** 1996. Regulation of HU alpha and HU beta by CRP and Fis in *Excherichia coli. J Mol Biol* **263:**126–139.

98. **Claret L, Rouviere-Yaniv J.** 1997. Variation in HU composition during growth of *Escherichia coli*: the heterodimer is required for long term survival. *J Mol Biol* **273:** 93–104.

99. **Ali Azam TA, Iwata A, Nishimura A, Ueda S, Ishihama A.** 1999. Growth phase-dependent variation in protein composition of the *Escherichia coli* nucleoid. *J Bacteriol* **181:**6361–6370.

100. **Huisman O, Faelen M, Girard D, Jaffe A, Toussaint A, Rouviere-Yaniv J.** 1989. Multiple defects in *Escherichia coli* mutants lacking HU protein. *J Bacteriol* **171:** 3704–3712.

101. **Hillyard DR, Edlund M, Hughes KT, Marsh M, Higgins NP.** 1990. Subunit-specific phenotypes of *Salmonella typhimurium* HU mutants. *J Bacteriol* **172:** 5402–5407.

102. **Broyles SS, Pettijohn DE.** 1986. Interaction of the *Escherichia coli* HU protein with DNA: evidence for the formation of nucleosome-like structures with altered DNA helical pitch. *J Mol Biol* **187:**47–60.

103. **Hodges-Garcia Y, Hagerman PJ, Pettijohn DE.** 1989. DNA ring closure mediated by protein HU. *J Biol Chem* **264:**14621–14623.

104. **Boubrik F, Rouviere-Yaniv J.** 1995. Increased sensitivity to gamma irradiation in bacterial lacking protein HU. *Proc Natl Acad Sci USA* **92:**3958–3962.

105. **Goodrich JA, Schwartz ML, McClure WR.** 1990. Searching for and predicting the activity of sites for DNA binding proteins: compilation and analysis of the binding sites for *Escherichia coli* integration host factor (IHF). *Nucleic Acids Res* **17:**4993–5000.

106. **Higgins NP, Collier DA, Kilpatrick MW, Krause HM.** 1989. Supercoiling and integration host factor change the DNA conformation and alter the flow of convergent transcription in phage Mu. *J Biol Chem* **264:**3035–3042.

107. **Thompson RJ, Mosig G.** 1988. Integration host factor (IHF) represses a *Chlamydomonas* chloroplast promoter in *E. coli. Nucleic Acids Res* **16:**3313–3326.

108. **Rice PA, Yang S, Mizuuchi K, Nash HA.** 1996. Crystal structure of an IHF-DNA complex: a protein-induced DNA U-turn. *Cell* **87:**1295–1306.

109. **Hillisch A, Lorenz M, Diekmann S.** 2001. Recent advances in FRET: distance determination in protein-DNA complexes. *Curr Opin Struct Biol* **11:**201–207.

110. **Arfin SM, Long AD, Ito ET, Tolleri L, Riehle MM, Paegle ES, Hatfield GW.** 2000. Global gene expression profiling in *Escherichia coli* K12. The effects of integration host factor. *J Biol Chem* **275:**29672–29684.

111. **Ali BM, Amit R, Braslavsky I, Oppenheim BA, Gileadi O, Stavans J.** 2001. Compaction of single DNA molecules induced by binding of integration host factor (IHF). *Proc Natl Acad Sci USA* **98:**10658–10663.

112. **Ball CA, Osuna R, Ferguson KC, Johnson RC.** 1992. Dramatic changes in Fis levels upon nutrient upshift in *Escherichia coli. J Bacteriol* **174:**8043–8056.

113. **Hirvonen CA, Ross W, Wozniak CE, Marasco E, Anthony JR, Aiyar SE, Newburn VH, Gourse RL.** 2001. Contributions of UP elements and the transcription factor FIS to expression from the seven rrn P1 promoters in *Escherichia coli. J Bacteriol* **183:**6305–6314.

114. **Skoko D, Yoo D, Bai H, Schnurr B, Yan J, McLeod SM, Marko JF, Johnson RC.** 2006. Mechanism of chromosome compaction and looping by the *Escherichia coli* nucleoid protein Fis. *J Mol Biol* **364:**777–798.

115. **Falconi M, Gualtieri MT, La Teana A, Losso MA, Pon CL.** 1988. Proteins from the prokaryotic nucleoid: primary and quaternary structure of the 15kD *Escherichia coli* DNA binding protein H-NS. *Mol Microbiol* **2:** 323–329.

116. **Williams RM, Rimsky S, Buc H.** 1996. Probing the structure, function, and interactions of *Escheerichia coli* H-NS and StpA proteins by using dominant negative derivatives. *J Bacteriol* **178:**4335–4343.

117. **Atlung T, Ingmer H.** 1997. H-NS: a modulator of environmentally regulated gene expression. *Mol Microbiol* **24:**7–17.

118. **Spurio R, Falconi M, Brandi A, Pon CL, Gualerzi CO.** 1997. The oligomeric structure of nucleoid protein H-NS is necessary for recognition of intrinsically curved DNA and for DNA bending. *EMBO J* **16:**1795–1805.

119. **Williams RM, Rimsky S.** 1997. Molecular aspects of the *E. coli* nucleoid protein, H-NS: a central controller of gene regulatory networks. *FEMS Microbiol Lett* **156:** 175–185.

120. **Lucht JM, Dersch P, Kempf B, Bremer E.** 1994. Interactions of the nucleotide-associated DNA-binding protein H-NS with the regulatory region of the osmotically

controled *proU* operon of *Escherichia coli*. *J Biol Chem* **269:**6578–6586.

121. **Spassky A, Rimsky S, Garreau H, Buc H.** 1984. H1a, an *E. coli* DNA-binding protein which accumulates in stationary phase, strongly compacts DNA in vitro. *Nucleic Acids Res* **12:**5321–5340.
122. **Tupper AE, Owen-Hughes TA, Ussery DW, Santos DS, Ferguson DJP, Sidebotham JM, Hinton JCD, Higgins CF.** 1994. The chromatin-associated protein H-NS alters DNA topology *in vitro*. *EMBO J* **13:**258–268.
123. **Zhang A, Rimsky S, Reaban ME, Buc H, Belfort M.** 1996. *Escherichia coli* protein analogs StpA and H-NS: regulatory loops, similar and disparate effects on nucleic acid dynamics. *EMBO J* **15:**1340–1349.
124. **Bertin P, Lejeune P, Laurent-Winter C, Danchin A.** 1990. Mutations in *bglY*, the structural gene for the DNA-binding protein H1, affect expression of several *Escherichia coli* genes. *Biochimie* **72:**889–891.
125. **Hommais F, Krin E, Laurent-Winter C, Soutourina O, Malpertuy A, Le Caer JP, Danchin A, Bertin P.** 2001. Large-scale monitoring of pleiotropic regulation of gene expression by the prokaryotic nucleoid-associated protein, H-NS. *Mol Microbiol* **40:**20–36.
126. **May G, Dersch P, Haardt M, Middendorf A, Bremer E.** 1990. The osmZ (bglY) gene encodes the DNA-binding protein H-NS, a component of the *Escherichia coli* K12 nucleoid. *Mol Gen Genet* **224:**81–90.
127. **Rajkumari K, Kusano S, Ishihama A, Mizuno T, Gowrishankar J.** 1996. Effects of H-NS and potassium glutamate on s^s- and s^{70}- directed transcription *in vitro* from osmotically regulated P1 and P2 promoters of *proU* in *Escherichia coli*. *J Bacteriol* **178:**4176–4181.
128. **Falconi M, McGovern V, Gualerzi C, Hillyard D, Higgins NP.** 1991. Mutations altering chromosomal protein H-NS induce mini-Mu transposition. *New Biologist* **3:**615–625.
129. **Hengge-Aronis R.** 1999. Interplay of global regulators and cell physiology in the general stress response of *Escherichia coli*. *Curr Opin Microbiol* **2:**148–152.
130. **Afflerbach H, Schroder O, Wagner R.** 1999. Conformational changes of the upstream DNA mediated by H-NS and FIS regulate *E. coli* RrnB P1 promoter activity. *J Mol Biol* **286:**339–353.
131. **Falconi M, Higgins NP, Spurio R, Pon CL, Gualerzi CO.** 1993. Expression of the gene encoding the major bacterial nucleoid protein H-NS is subject to transcriptional auto-repression. *Mol Microbiol* **10:**273–282.
132. **Rimsky S, Zuber F, Buckle M, Buc H.** 2001. A molecular mechanism for the repression of transcription by the H-NS protein. *Mol Microbiol* **42:**1311–1323.
133. **Schnetz K.** 1995. Silencing of *Escherichia coli bgl* promoter by flanking sequence elements. *EMBO J* **14:** 2545–2550.
134. **Zhang A, Belfort M.** 1992. Nucleotide sequence of a newly-identified *Escherichia coli* gene, *stpA*, encoding an H-NS-like protein. *Nucleic Acids Res* **20:**6735.
135. **Dorman CJ, Bhriain NN, Higgins CF.** 1990. DNA supercoiling and environmental regulation of virulence gene expression in *Shigella flexneri*. *Nature* **344:**789–792.
136. **Higgins CF, Dorman CJ, Stirling DA, Waddell L, Booth IR, May G, Bremer E.** 1988. A physiological role for DNA supercoiling in the osmotic regulation of gene expression in *S. typhimurium* and *E. coli*. *Cell* **52:** 569–584.
137. **Hulton CSJ, Seirafi A, Hinton JCD, Sidebotham JM, Waddell L, Pavitt GD, Owen-Hughes T, Spassky A, Buc H, Higgins CF.** 1990. Histone-like protein H1 (H-NS), DNA supercoiling, and gene expression in bacteria. *Cell* **63:**631–642.
138. **McGovern V, Higgins NP, Chiz S, Jaworski A.** 1994. H-NS over-expression induces an artificial stationary phase by silencing global transcription. *Biochimie* **76:** 1030–1040.
139. **Nieto JM, Mourino M, Balsalobre C, Madrid C, Prenafeta A, Munoa FJ, Juarez A.** 1997. Construction of a double *hha hns* mutant of *Escherichia coli*: effect on DNA supercoiling and alpha-haemolysin production. *FEMS Microbiol Lett* **155:**39–44.
140. **Ordnorff PE, Kawula TH.** 1991. Rapid site-specific DNA inversion in *Escherichia coli* mutants lacking the histonelike protein H-NS. *J Bacteriol* **173:**4116–4123.
141. **Dame RT, Wyman C, Goosen N.** 2000. H-NS mediated compaction of DNA visualised by atomic force microscopy. *Nucleic Acids Res* **28:**3504–3510.
142. **Arold ST, Leonard PG, Parkinson GN, Ladbury JE.** 2010. H-NS forms a superhelical protein scaffold for DNA condensation. *Proc Natl Acad Sci USA* **107:** 15728–15732.
143. **Brackley CA, Taylor S, Papantonis A, Cook PR, Marenduzzo D.** 2013. Nonspecific bridging-induced attraction drives clustering of DNA-binding proteins and genome organization. *Proc Natl Acad Sci USA* **110:** E3605–E3611.
144. **Wang W, Li GW, Chen C, Xie XS, Zhuang X.** 2011. Chromosome organization by a nucleoid-associated protein in live bacteria. *Science* **333:**1445–1449.
145. **Liu Y, Chen H, Kenney LJ, Yan J.** 2010. A divalent switch drives H-NS/DNA-binding conformations between stiffening and bridging modes. *Genes Dev* **24:** 339–344.
146. **Leonard PG, Parkinson GN, Gor J, Perkins SJ, Ladbury JE.** 2010. The absence of inorganic salt is required for the crystallization of the complete oligomerization domain of *Salmonella typhimurium* histone-like nucleoid-structuring protein. *Acta Crystallogr Sect F Struct Biol Cryst Commun* **66:**421–425.
147. **Lim CJ, Lee SY, Kenney LJ, Yan J.** 2012. Nucleoprotein filament formation is the structural basis for bacterial protein H-NS gene silencing. *Sci Rep* **2:**509.
148. **Mojica FJM, Higgins CF.** 1997. *In vivo* supercoiling of plasmid and chromosomal DNA in an *Escherichia coli hns* mutant. *J Bacteriol* **179:**3528–3533.
149. **Lim CJ, Whang YR, Kenney LJ, Yan J.** 2012. Gene silencing H-NS paralogue StpA forms a rigid protein filament along DNA that blocks DNA accessibility. *Nucleic Acids Res* **40:**3316–3328.
150. **Thompson RJ, Davies JP, Lin G, Mosig G.** 1990. Modulation of transcription by altered torsional stress,

upstream silencers, and DNA-binding proteins, p 227–240. *In* Drlica K, Riley M (ed), *The Bacterial Chromosome*. American Society for Microbiology, Washington, DC.

151. **Dubnau E, Margolin P.** 1972. Suppression of promoter mutations by the pleiotropic *supX* mutations. *Mol Gen Genet* **117:**91–112.
152. **Pruss GJ, Manes SH, Drlica K.** 1982. *Escherichia coli* DNA topoisomerase I mutants: increased supercoiling is corrected by mutations near gyrase genes. *Cell* **31:** 35–42.
153. **Margolin P, Zumstein L, Sternglanz R, Wang JC.** 1985. The *Escherichia coli supX* locus is *topA*, the structural gene for DNA topoisomerase I. *Proc Natl Acad Sci USA* **82:**5437–5441.
154. **Staczek P, Higgins NP.** 1998. DNA gyrase and topoisomerase IV modulate chromosome domain size *in vivo*. *Mol Micro* **29:**1435–1448.
155. **Tahirov TH, Temiakov D, Anikin M, Patlan V, McAllister WT, Vassylyev DG, Yokoyama S.** 2002. Structure of a T7 RNA polymerase elongation complex at 2.9 A resolution. *Nature* **420:**43–50.
156. **Itoh T, Tomizawa J.** 1980. Formation of an RNA primer for initiation of replication of ColE1 DNA by ribonuclease H. *Proc Natl Acad Sci USA* **77:**2450–2454.
157. **Masukata H, Tomizawa J.** 1990. A mechanism of formation of a persistent hybrid between elongating RNA and template DNA. *Cell* **62:**331–338.
158. **Reaban ME, Griffin JA.** 1990. Induction of RNA-stabilized DNA conformers by transcription of an immunoglobulin switch region. *Nature* **348:**342–344.
159. **Reaban ME, Lebowitz J, Griffin JA.** 1994. Transcription induces the formation of a stable RNA.DNA hybrid in the immunoglobulin alpha switch region. *J Biol Chem* **269:**21850–21857.
160. **Albert AC, Spirito F, Figueroa-Bossi N, Bossi L, Rahmouni AR.** 1996. Hyper-negative template DNA supercoiling during transcription of the tetracycline-resistance gene in topA mutants is largely constrained *in vivo*. *Nucleic Acids Res* **24:**3093–3099.
161. **Wojciechowska M, Bacolla A, Larson JE, Wells RD.** 2004. The myotonic dystrophy type 1 triplet repeat sequence induces gross deletions and inversions. *J Biol Chem* **280:**280.
162. **Lin Y, Dent SY, Wilson JH, Wells RD, Napierala M.** 2010. R loops stimulate genetic instability of CTG.CAG repeats. *Proc Natl Acad Sci USA* **107:**692–697.
163. **Iyer RI, Pluciennik A, Napierala M, Wells RD.** 2015. DNA triplet repeat expansion and mismatch repair. *Annu Rev Biochem*. [Epub ahead of print.] doi: 10.1146/annurev-biochem-060614-034010.
164. **Pruss G, Drlica K.** 1986. Topoisomerase I mutants: the gene on pBR322 that encodes resistance to tetracycline affects plasmid DNA supercoiling. *Proc Natl Acad Sci USA* **83:**8952–8956.
165. **Lynch AS, Wang JC.** 1993. Anchoring of DNA to the bacterial cytoplasmic membrane through cotranscriptional synthesis of polypeptides encoding membrane proteins or proteins for export: a mechanism of plasmid hypernegative supercoiling in mutants deficient in DNA topoisomerase I. *J Bacteriol* **175:**1645–1655.
166. **Wu H-Y, Liu LF.** 1991. DNA looping alters local DNA conformation during transcription. *J Mol Biol* **219:** 615–622.
167. **Masse E, Drolet M.** 1999. R-loop-dependent hypernegative supercoiling in *Escherichia coli topA* mutants preferentially occurs at low temperatures and correlates with growth inhibition. *J Mol Biol* **294:**321–332.
168. **Masse E, Drolet M.** 1999. *Escherichia coli* DNA topoisomerase I inhibits R-loop formation by relaxing transcription-induced negative supercoiling. *J Biol Chem* **274:**16659–16664.
169. **Li T, Panchenko YA, Drolet M, Liu LF.** 1997. Incompatibility of the *Escherichia coli rho* mutants with plasmids is mediated by plasmid-specific transcription. *J Bacteriol* **179:**5789–5794.
170. **Alfano C, McMacken R.** 1989. Ordered assembly of nucleoprotein structures at the bacteriophage l replication origin during the initiation of DNA replication. *J Biol Chem* **264:**10699–10708.
171. **Kaguni JM, Kornberg A.** 1984. Replication initiated at the origin (oriC) of the *E. coli* chromosome reconstituted with purified enzymes. *Cell* **38:**183–190.
172. **Hill TM, Tecklenburg ML, Pelletier AJ, Kuempel PL.** 1989. *tus*, the trans-acting gene required for termination of DNA replication in *Escherichia coli*, encodes a DNA-binding protein. *Proc Natl Acad Sci USA* **86:**1593–1597.
173. **Cox MM.** 2001. Historical overview: searching for replication help in all of the rec places. *Proc Natl Acad Sci USA* **98:**8173–8180.
174. **Higgins NP, Kato KH, Strauss BS.** 1976. A model for replication repair in mammalian cells. *J Mol Biol* **101:** 417–425.
175. **Michel B, Grompone G, Flores MJ, Bidnenko V.** 2004. Multiple pathways process stalled replication forks. *Proc Natl Acad Sci USA* **101:**12783–12788.
176. **Postow L, Ullsperger C, Keller RW, Bustamante C, Vologodskii AV, Cozzarelli NR.** 2001. Positive torsional strain causes the formation of a four-way junction at replication forks. *J Biol Chem* **276:**2790–2796.
177. **Postow L, Crisona NJ, Peter BJ, Hardy CD, Cozzarelli NR.** 2001. Topological challenges to DNA replication: conformations at the fork. *Proc Natl Acad Sci USA* **98:** 8219–8226.
178. **Courcelle J, Donaldson JR, Chow K-H, Courcelle CT.** 2003. DNA damage-induced replication fork regression and processing in *Escherichia coli*. *Science* **299:**1064–1067.
179. **Joshi MC, Bourniquel A, Fisher J, Ho BT, Magnan D, Kleckner N, Bates D.** 2011. *Escherichia coli* sister chromosome separation includes an abrupt global transition with concomitant release of late-splitting intersister snaps. *Proc Natl Acad Sci USA* **108:**2765–2770.
180. **Sherratt D.** 2013. Plasmid partition: sisters drifting apart. *EMBO J* **32:**1208–1210.
181. **Podtelezhnikov AA, Cozzarelli NR, Vologodskii AV.** 1999. Equilibrium distributions of topological states in circular DNA: interplay of supercoiling and knotting. *Proc Natl Acad Sci USA* **96:**12974–12979.

182. **Peter BJ, Ullsperger C, Hiasa H, Marians KJ, Cozzarelli NR.** 1998. The structure of supercoiled intermediates in DNA replication. *Cell* **94:**819–827.

183. **Lee SH, Siaw GE, Willcox S, Griffith JD, Hsieh TS.** 2013. Synthesis and dissolution of hemicatenanes by type IA DNA topoisomerases. *Proc Natl Acad Sci USA* **110:**E3587–E3594.

184. **Olavarrieta L, Martínez-Robles ML, Hernández P, Krimer DB, Schvartzman JB.** 2002. Knotting dynamics during DNA replication. *Mol Microbiol* **46:**699–707.

185. **Pang Z, Chen R, Manna D, Higgins NP.** 2005. A gyrase mutant with low activity disrupts supercoiling at the replication terminus. *J Bacteriol* **187:**7773–7783.

186. **Spirito F, Figueroa-Bossi N, Bossi L.** 1994. The relative contributions of transcription and translation to plasmid DNA supercoiling in *Salmonella typhimurium*. *Mol Microbiol* **11:**111–122.

Plasmid Maintenance, Transfer and Barriers

III

Plasmids—Biology and Impact in Biotechnology and Discovery
Edited by Marcelo E. Tolmasky and Juan C. Alonso

doi:10.1128/microbiolspec.PLAS-0023-2014

Jamie C. Baxter[1]
Barbara E. Funnell[1]

Plasmid Partition Mechanisms

8

PARTITION SYSTEMS IN BACTERIA

The stable maintenance of low-copy-number plasmids in bacteria is actively driven by partition mechanisms that are responsible for the positioning of plasmids inside the cell. Partition systems are ubiquitous in the microbial world and are encoded by most bacterial chromosomes as well as plasmids. Partition is generally the most important determinant of the stability of low-copy-number plasmids, which are common in bacteria. In contrast, high-copy-number plasmids typically do not encode partition systems because random segregation is sufficient for stability. There has been significant progress in the last several years in our understanding of partition mechanisms. Two general areas that have developed are (i) the structural biology of partition proteins and their interactions with DNA and (ii) the action of the partition ATPases that drive the process. In addition, systems that use tubulin-like GTPases to partition plasmids have recently been identified. In this chapter, we concentrate on these recent developments since the publication of the first edition of *Plasmid Biology* in 2004. We will briefly introduce the biology of plasmid partition systems to date; we refer the reader to the earlier chapter on plasmid partition for a comprehensive review of their biology (1).

Partition is a dynamic process; plasmids are moved and positioned inside the cell so that cell division separates at least one copy into each daughter cell. Although there is significant diversity among the types of plasmid systems, their genetic organization and components are remarkably conserved. Plasmid partition systems typically consist of two proteins and one or more partition, or *par*, sites, which are the DNA sites that direct the action of the segregation machinery. The *par* sites are considered prokaryotic centromeres because they are required in *cis* for plasmid stability and because they are the assembly sites of the segregation machinery. One partition protein is a site-specific DNA binding protein that recognizes the *par* site(s), and is often referred to as the centromere-binding protein, or CBP. The second protein is an ATPase or GTPase, which uses the energy of nucleotide binding and hydrolysis to move plasmid DNA inside the cell. The genes are typically arranged in an operon, and the proteins regulate their own expression. The location of the *par* site varies; it can be directly downstream of the *par* genes, upstream and close to the promoter for the operon, or in multiple locations on the plasmid.

Par systems can be divided into different classes, based on the properties of the NTPase. Type I systems encode Walker ATPases, and the ATP binding site contains a specific variant of the canonical Walker A motif (2–4). ATPases of type II systems contain actin-like ATP binding sites, and the proteins structurally and

[1]Department of Molecular Genetics, University of Toronto, Toronto, Ontario M5S 1A8, Canada.

enzymatically resemble eukaryotic actin (2, 5). Type III systems encode GTPases that resemble tubulin (6–8). Recent structural information for members from all of these classes has validated the classification based on primary sequence data. Therefore, it is likely that since both the sequence and structure of each class of NTPase are conserved, the properties of the NTPase also reflect similar mechanisms of action.

The CBPs fall into two general classes. The first are helix-turn-helix (HTH) DNA binding proteins, and the second are ribbon-helix-helix dimers (RHH_2). In addition to site-specific DNA binding activities, they assemble into higher-order oligomers at and around the *par* site. The interaction of this partition complex with the NTPase is essential for the dynamics and patterning/positioning activities of the NTPase.

To regulate *par* gene expression, one Par protein directly represses transcription and the other stimulates the repressor activity of its partner (reviewed in references 1, 2, 4). In most partition systems, the CBP is the direct repressor, which acts on operator sites that resemble or act also as centromere sites in the operon. In some type I systems, however, the ATPase is the repressor. Type I ATPases can be divided into two subgroups, Ia and Ib, based on domain analysis. Both contain the core partition activities, whereas type Ia ParAs contain an additional site-specific DNA binding domain, which is responsible for the transcriptional repressor activity.

Many bacterial chromosomes encode plasmid-like partition systems, which have been shown to contribute to bacterial chromosome segregation (reviewed in references 4, 9, 10). These are, so far, invariably type Ib Walker ATPases with an HTH CBP. Chromosomal systems have also been shown to support plasmid partition when cloned into a low-copy-plasmid vector (11). The chromosomal proteins also interact with host replication and other factors, such as DnaA initiator proteins and SMC (structural maintenance of chromosome) condensins, which add another level of complexity to chromosomal segregation in the bacterial cell cycle (12–14). The plasmid systems can be considered minimalist cassettes, which provide excellent context to study the basic biological mechanism of action of these proteins.

In this review we concentrate on recent developments in our understanding of these two broad steps: partition complex assembly at the *par* site and the mechanisms of the NTPases that move plasmid cargo inside the cell. There are several plasmid systems that have provided significant structural and mechanistic information in the past several years. The paradigms for type Ia partition are *Escherichia coli* plasmids P1 and F, for type Ib partition they are *Salmonella enterica* plasmid TP228 and *Streptococcus pyogenes* plasmid pSM19035, and for type II partition they are *E. coli* plasmid R1 and *Staphylococcus aureus* pSK41. In addition, *E. coli* plasmid pB171 encodes two partition systems, *par1* (type II) and *par2* (type Ib) (15). Both contribute to plasmid stability, and studies of each have added to our understanding of plasmid partition in bacteria. We concentrate on these systems to discuss CBPs and partition mechanisms, but not exclusively, because results from many systems have contributed to our current understanding (Table 1 summarizes specific nomenclature for the main systems discussed in this chapter). Finally, we describe the new class of partition systems that encode tubulin-like GTPases, primarily from studies of the virulence plasmids pXO1 of *Bacillus anthracis* and pBtoxis of *Bacillus thuringiensis*.

PARTITION COMPLEX RECOGNITION AND ASSEMBLY

The first step in plasmid partition is site-specific DNA binding of the CBP to the *par* site. There is considerable divergence in sequence among CBPs and *par* sites in plasmid partition systems. The sites can vary in sequence, organization, number, and arrangement on plasmid chromosomes. The sequence diversity presumably evolved to avoid competition among different, otherwise compatible, plasmids. The *par* site typically overlaps with the promoter of the *par* operon when the CBP is the repressor, which is the case for most type Ib, type II, and type III systems. In contrast, the *par* sites of type Ia systems are distinct from the operators. They can be downstream of the *par* genes, for example, in plasmids P1, P7, and F. The RepABC family of plasmids from *Agrobacterium* and *Rhizobium* genera (*repA* and *repB* are the partition genes) show both variable position and number of *par* sites (18). The N15 prophage/plasmid in *E. coli* has four *par* sites, and plasmid stability does depend on the number of sites (19, 20). It is unclear why some plasmids utilize more than one *par* site, but the answer presumably reflects the architecture of the partition complex.

Although DNA binding specificity differs among CBPs, there are several common themes in partition complex assembly following sequence recognition. First, binding of one CBP is not sufficient for partition, and all partition complexes contain multiple CBPs bound to their respective *par* sites. This large structure therefore presents many potential binding sites for the cognate NTPase. The modulation of many protein-protein interactions is likely important for the dynamics of partition,

Table 1 Plasmid partition system nomenclature

Plasmid(s)	Bacterial host	ATPase/GTPase	CBP[a]	*par* site	Reference[b]
Type I ATPase Systems					
P1, P7	*E. coli*	ParA	ParB	*parS*	27
F	*E. coli*	SopA	SopB	*sopC*	26
RK2, RP4	*E. coli/Pseudomonas aeruginosa*	IncC	KorB	O_B	29, 36
TP228	*S. enterica*	ParF	ParG	*parH*	57
pSM19035	*S. pyogenes*	δ	ω	*parS*	58
pB171 (*par2)*	*E. coli*	ParA	ParB	*parC*	15
Type II ATPase Systems					
R1	*E. coli*	ParM	ParR	*parC*	15, 22, 61
pB171 (*par1*)	*E. coli*				
pSK41	*S. aureus*				
pLS20	*B. subtilis*	Alp7A	Alp7R	*alp7C*	16, 65
pLS32	*B. subtilis*	AlfA	AlfB	*parN*	17, 75
Type III GTPase Systems					
pXO1	*B. anthracis*	TubZ	TubR	*tubC*	8, 125
	B. cereus				
pBtoxis	*B. thuringiensis*				

[a]CBP, Centromere-binding protein
[b]For simplicity, selected studies that define the nomenclature of the partition loci listed here and that are mentioned in this chapter are cited. References 1, 2, and 4 also provide a more extensive review of the original literature.

so that plasmids are never completely released from their interactions with the NTPase during movement. Second, the interaction of the CBP/plasmid partition complexes stimulates the NTPase activity of the partner, and this modulation of the ATP binding and hydrolysis cycles is necessary for plasmid dynamics. Third, the CBPs are thought to pair (or group) plasmids together. Pairing has been shown to occur in several systems (21–23), although the role of pairing is not understood.

HTH CBPs

CBPs with HTH DNA binding motifs are found in type Ia plasmid partition systems and in all known bacterial systems. Again, the sequence conservation among these CBPs is not high, but there is a conservation of domain organization (24–28). Dimerization is mediated by a C-terminal dimerization domain, adjacent to the central HTH DNA binding domain. The N-terminal region of the protein interacts with the ATPase and mediates oligomerization of the CBP at and around the *par* site. The HTH CBPs have been observed to spread away from their *par* sites, but the exact molecular nature of this activity is not clear.

The structures of the dimerization and DNA binding domains have been determined for three plasmid HTH ParBs: P1 ParB, F SopB, and RP4 KorB (Table 2) (29–33). All three DNA binding domain structures were solved in complex with their specific DNA binding sites (Fig. 1A–C). For KorB and SopB, the dimerization and DNA binding domains were solved separately. None of the N-termini of these CBPs were amenable to crystallization, consistent with data indicating that they are flexible and not stably folded in solution (24, 26, 34, 35).

The *par* sites of HTH CBPs contain inverted repeat recognition elements, as is typical of this type of DNA binding motif. RP4 KorB (and that from the related plasmid RK2) is a global transcriptional regulator as well as a CBP, and it binds to 12 related inverted repeat operator sequences called O_B in different locations in the plasmid (reviewed in reference 4). One, O_B3, is thought to act as the *par* site (36). The F *sopC* partition site consists of 12 copies of a 43-bp repeat, only one of which is essential for partition. A 16-bp inverted repeat within the 43-bp region is the specific binding site for SopB (33, 37), and the remaining sequence is postulated to be required for proper spacing of the inverted repeats so that they are arranged on the same face of the DNA helix (37). Plasmids such as P1 and P7 in *E. coli* are members of a group of plasmids with bipartite partition sites (reviewed in reference 1). These ParBs recognize two distinct sequence motifs; one is an inverted repeat (the A-box) and the second is a hexamer sequence (the B-box). These motifs are asymmetrically arranged around a binding site for *E. coli* IHF protein. IHF binding strongly stimulates ParB binding to *parS in vitro* and ParB activity in partition *in vivo*.

Table 2 Structures of plasmid partition proteins

	Plasmid protein[a]	Cofactor(s)	PDB identifier	Reference
Type I ATPase Systems				
ATPase	P1 ParA	None	3EZ7	97
		ADP	3EZ2, 3EZ6	
	P7 ParA	None	3EZ9, 3EZF	97
	pSM19035 δ	ATPγS	2OZE	58
	TP228 ParF	ADP	4DZZ, 4E03	98
		AMP-PCP	4E07, 4E09	
CBP	RP4 KorB dimer domain	none	1IGQ, 1IGU	29, 30
	RP4 DNA binding domain	17-mer O_B	1R71	
	P1 ParB	25-mer *parS*	1ZX4	31, 32
		16-mer *parS*	2NTZ	
	F SopB dimer domain	None	3KZ5	33
	F SopB DNA binding domain	18-mer *sopC*	3MKW, 3MKY, 3MKZ	
	pSM19035 ω	None	1IRQ	52, 54
		18-mer *PcopS*	2BNW, 2BNZ	
		17- and 18-mer *PcopS*	2CAX	
	TP228 ParG	None	1P94	53
	pCXC100 ParB	None	3NO7	55
Type II ATPase Systems				
ATPase	R1 ParM	None	1MWK, 3IKU[b], 3IKY[b]	63, 69, 70, 74
		ADP	1MWM, 2ZHC	
		AMP-PNP	4A61, 4A6J	
		GDP	2ZGY	
		GMP-PNP	2ZGZ	
		AMP-PNP, $ParR^{101-117}$	4A62	
	pSK41 ParM	None	3JS6	77
CBP	pB171 ParR	None	2JD3	60
	pSK41 ParR	20-mer *parC*	2Q2K	61
Type III GTPase Systems				
GTPase	pBtoxis TubZ	None	3M8K	125, 127
		GDP	2XKB 2XKA,	
		GTPγS	3M89	
	pXO1 TubZ	None	4EI8	128
		GDP	4EI7	
		GTPγS	4EI9	
CBP	pBtoxis TubR	None	3M8E, 3M8F, 3M9A	125, 129
		24-mer *tubC*	4ASO	
		26-mer *tubC*	4ASS	
	pBM400 TubR	None	4ASN	129

[a]Bacterial species for most plasmids are listed in Table 1, with the exception of pCXC100 (from *Leifsonia xyli*) and pBM400 (from *Bacillus megaterium*).
[b]Models of ParM filaments based on cryo-EM data (70).

The HTH structures of KorB, SopB, and ParB are all-helical domains that are highly similar to each other (Fig. 1A–C). As predicted from canonical HTH/DNA interactions, the HTH regions of KorB, SopB, and ParB bind to the inverted repeat sequences in DNA (P1 ParB HTH with box-A) such that the "recognition helix" sits in the major groove. Interestingly, the specificity determinants for DNA binding differ among these three proteins. P1 ParB makes its specific contacts with the inverted repeat via residues in this helix (α3) (31, 32). F SopB makes specific contacts via residues in this helix (α3) as well as with a residue, R219, outside of the HTH (33). Genetic and biochemical evidence confirms that R219 is required for sequence-specific binding by SopB (38). KorB makes only non-specific DNA contacts via the HTH motif (α3-α4), and specificity is determined by residues in helices α6 (T211) and α8 (R240) (30). All proteins make several contacts with the phosphate backbones of their respective sites.

The presence of another DNA recognition element in *parS*, the B-box, adds a unique and extra level of complexity to the P1 ParB family of proteins. The secondary DNA binding domain in ParB is an integral part of the C-terminal dimerization domain and requires dimerization to form (31). The dimer domain is a six-stranded β-sheet coiled coil; two loops between β1 and β2 and between β2 and β3 sit in the major groove and make sequence-specific contacts with the B-box sequence (Fig. 1A) (31, 32). Interestingly, the structure of the F SopB dimer domain is very similar to that of ParB except for these extended loops, which explains the lack of DNA binding by this region of SopB (33). Therefore, while the structure is conserved, the DNA binding function is not. In contrast, the dimer domain

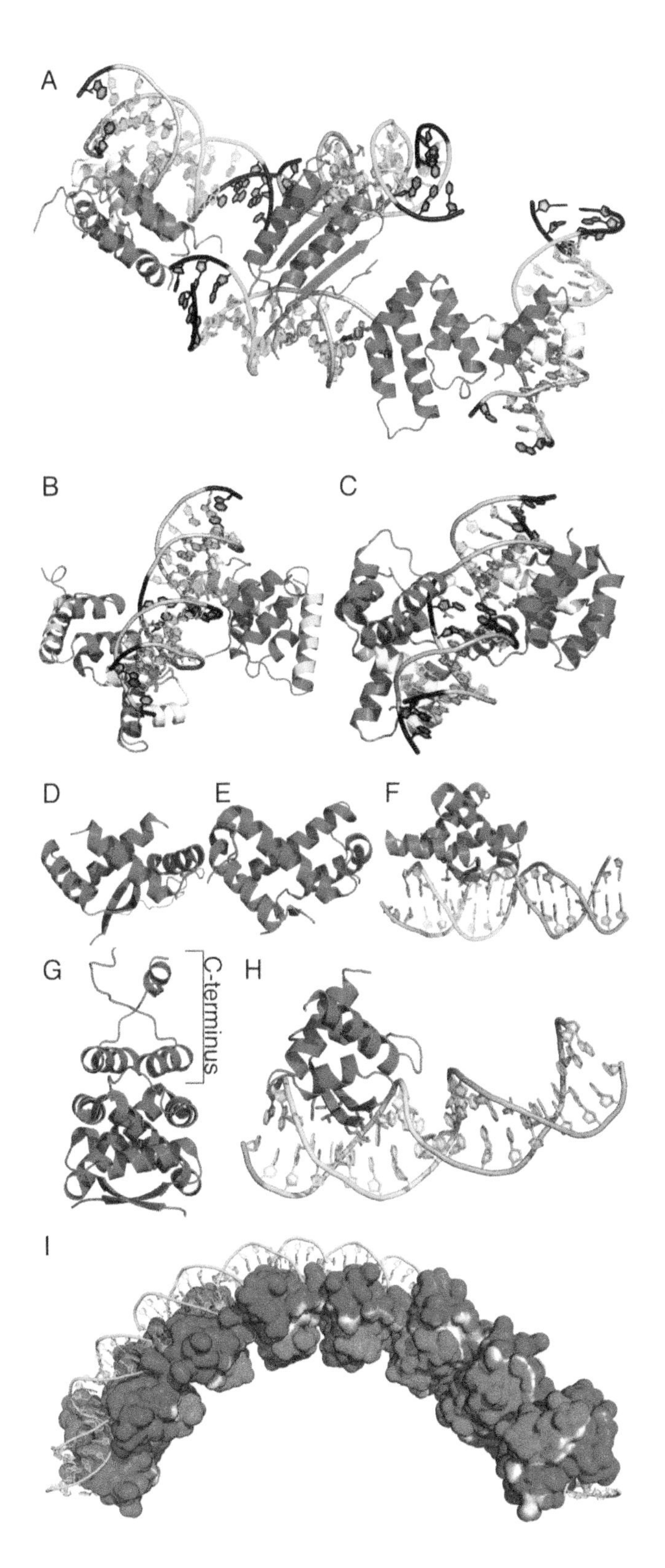

Figure 1 Structures of DNA binding domains of plasmid CBPs. (A–C) The DNA binding domains of HTH CBPs: (A) P1 ParB$^{142-333}$, (B) F SopB$^{157-271}$, and (C) RP4 KorB$^{139-152}$ are shown bound to their respective centromere sequences (gray). The HTH motifs are in yellow. Residues that make base-specific contacts with the centromere DNA are shown as sticks (red) and can be found both within the HTH (ParB and SopB) and in the adjacent four-helix bundle (KorB). The dimerization domain of ParB is its second DNA binding domain (to the B-box DNA motif, green DNA). The ParB structure bridges across four DNA molecules (31). Because the dimer domains are absent in the structures of SopB and KorB, two monomers are presented interacting with the one inverted repeat sequence. Additional monomer-monomer contacts (cyan) in the SopB structure suggest it may also bridge across DNA molecules (not shown) (33). (D–H) The DNA binding regions of RHH_2 CBPs. Type Ib CBPs of (D) TP228 ParG$^{1-76}$, (E) pCXC100 ParB$^{69-128}$, and (F) pSM19035 ω^{23-71} and type II ParRs of (G) pB171$^{6-96}$ and (H) pSK41$^{4-48}$ illustrate the simple β-strand dimerization and DNA binding [represented as red sticks in (F) and (H)] interface stabilized by α-helical interactions. Type-Ib CBPs contain a flexible N-terminal tail (not shown) that interacts with their cognate ATPase. Type II CBPs mediate their cognate ATPase contacts through the C-termini, present in the pB171 ParR structure (G). The C-terminal region of ParRs also promotes higher order assembly on DNA, yielding a super-helical structure, as illustrated by the crystal packing of the pSK41 ParR structure in (I). The electrostatic representation of the super-helical filament shows the electropositive surface (blue) interacting with the DNA and the electronegative surface (red) on the inside of the helical filament. The latter interface interacts with the predominantly electropositive surface of its cognate ATPase ParM and serves as a cap for stabilizing ParM filamentsin the cell. All structural images were generated using PyMOL v1.6.0.0 software (Schrödinger, LLC 2010). PDB information is listed in Table 2.
doi:10.1128/microbiolspec.PLAS-0023-2014.f1

of KorB is a five-stranded antiparallel all β-sheet structure that strongly resembles the SH3 (Src-homology 3) fold and not that of the ParB/SopB dimerization domains (29). In this case, function (dimerization) is conserved but structure is not.

Because the P1 ParB crystal structure contains both the dimer domain and the HTH domain, it provides unique information on the overall conformation of the protein. In ParB, a short flexible linker connects the HTH α-helical domain with the β-sheet dimer domain. Even with this flexibility, the HTH domains point away from each other, so it is not possible that they contact the same inverted repeat. Indeed, in the structure, a single dimer of ParB bridges across *parS* sites such that the HTH of one monomer binds to half of the inverted repeat on one DNA molecule while the HTH of the other monomer binds to the other half of the inverted repeat on a different DNA molecule (Fig. 1A). Interestingly, this cross-site arrangement is supported by the SopB/*sopC* crystal structure even though it did not contain the SopB dimerization domain (33). There were some dimer contacts within the DNA binding domains (Fig. 1B), so in this arrangement, the corresponding DNA binding domains bridged different DNA molecules.

Similarly, one dimer of P1 ParB is able to bridge across four sites due to the extra DNA binding sites (to box-B motifs) in its dimer domains (32). This arrangement gives ParB remarkable flexibility in binding to its various motifs in *parS* so that multiple ParB dimers can cooperate to bind to a single *parS* site or across paired sites (Fig. 1A) (31, 39). It is attractive to propose that this flexibility and the ability to pair across sites are conserved among HTH ParBs. Conformational data for KorB suggests similar flexible or disordered linkers (35). Indeed, this flexibility would explain the inability to obtain crystal information for both the HTH and dimer domains in the same structure except in the case of P1, in which the presence of DNA, and binding to both DNA boxes, would constrain this flexibility.

N-termini of HTH CBPs

In general, the N-termini of HTH CBPs contain the regions required for an interaction with the partition ATPase and for oligomerization with itself (24, 26, 27, 40). Interaction with the ATPase is mediated by residues close to the N-terminus, and where tested, a critical arginine is involved in stimulation of ParA ATPase activity. The latter property is shared with type Ib RHH_2 CBPs (see below). There are exceptions; for example, sequences in the center of RK2 KorB have also been implicated in interactions with IncC, its cognate ATPase (Table 1) (28). There are no structures of the N-termini of plasmid HTH CBPs, likely because these regions are flexible and somewhat disordered (24, 26, 35). The crystal structure of the N-terminus of an archaeal ParB, *Thermus thermophilus* Spo0J, is a dimer with an extended dimerization interface (41). The relevance of this structure for plasmid CBPs is unclear because this Spo0J fragment does not dimerize in biochemical assays and lacks the strong dimerization domain at its C-terminus (41). Nevertheless, it is attractive to consider that this structure represents the higher-order oligomerization interaction, which is weak in solution by itself but would normally occur in the context of the full-length protein after binding to its cognate *par* site (42).

After the HTH CBP recognizes and binds to its *par* site, more CBP molecules are recruited to form higher-order complexes containing many protein molecules. *In vivo*, these can be seen as bright foci of fluorescently labeled CBP that colocalize with plasmids (43–47). F SopB and P1 ParB have been shown to spread away from their binding sites (48, 49), and this is thought to be a general property of this class of CBPs. This activity likely represents the loading of multiple ParBs onto the plasmid DNA so that a highly concentrated focus of multiple proteins is available for interactions with ParA during the partition reaction. The minimal stoichiometry of ParB to plasmid necessary for partition is unknown, except that extensive spreading (>400 bp away from the site) is not required *in vivo* (50). How CBPs interact with each other and with DNA around the *par* site is an important and as yet unanswered question in partition complex assembly.

RHH Dimer CBPs

The RHH_2 structure is common in many CBPs (Fig. 1D–H). This DNA binding motif was originally characterized in the Arc/MetJ family of transcriptional repressors (51). Indeed, these CBPs are, in those systems that have been tested, also transcriptional repressors. As opposed to the α helix contacts with DNA by HTH proteins, RHH_2 proteins bind to DNA via their antiparallel β-sheets, which insert into the major groove of and make sequence-specific contacts with the DNA (51). The two α helices of each monomer interact with each other to form a tight dimer. In general, the RHH_2 CBPs are small proteins but otherwise diverge significantly in their primary sequence. They are common among type Ib and type II partition systems. The primary difference in domain organization among the CBPs of the latter two classes is in the position of the region that interacts with the ATPase. In type Ib CBPs, the RHH_2 domain is C-terminal, and these proteins interact with the ATPases via a small and flexible N-terminal region. The type II CBPs interact with DNA via an N-terminal RHH_2 domain and with the cognate ATPase via the C-terminus.

Type I CBPs

Structures of three RHH_2 CBPs from type I partition systems have been determined (Fig. 1D-F). They are TP228 ParG, pSM19035 ω (omega), and pCXC100 ParB (52–55). The first, TP228 ParG, is a 76-amino acid polypeptide consisting of a C-terminal (residues 33 to 76) DNA binding domain and an N-terminal mobile tail (residues 1 to 32). The structure of the C-terminal domain was determined by nuclear magnetic resonance (53) and contains the minimal DNA binding domain and dimerization determinant. The N-terminal extension of ParG is necessary for interactions with the ParF ATPase (56) and also modulates the binding of the C-terminal RHH_2 domain to DNA to improve specific over nonspecific DNA binding (57).

The RHH_2 structure of pSM19035 ω bound to its DNA site was solved by X-ray crystallography, providing a direct picture of the assembly of partition complexes (Fig. 1F) (54). ω is a global regulator of transcription as well as a CBP. It is a 71-amino acid protein that binds to series of 7-bp repeats, and affinity depends on the number and arrangement of these heptads. pSM19035 contains three *parS* sites, each with multiple heptad repeats. The crystal structure of $\omega^{20\text{-}71}$ in complex with two heptad repeats, both in direct and inverted orientations, confirmed base-specific contacts with the β-sheets in the DNA major groove. The crystal structure has led to a model in which ω_2 dimers bind to successive repeats by wrapping around the *parS* DNA as a left-handed helix, without bending or otherwise distorting the DNA (23, 54, 58). The stoichiometry is one ω_2 per heptad, and these partition complexes do not spread beyond the *par* site, in contrast to the HTH class of CBPs. As with ParG, the unstructured N-terminal tail of ω_2 is necessary for its interaction with the ATPase δ. Another feature that ω_2 shares with other CBPs is the capacity to pair plasmids at *parS* sites, which requires δ and ATP (23, 58).

The crystal structure of the DNA binding domain of pCXC100 ParB (residues 69 to 128 of 139 total) shows high structural similarity to those of ParG and ω_2 (Fig. 1E; Table 2) (55). Its N-terminus is also likely a flexible tail, as it was highly sensitive to proteolytic digestion.

Type II CBPs

Type II CBPs are typically named ParR, because the R1 plasmid ParM/ParR/*parC* system has been the paradigm for functional studies of this class of partition mechanism (Table 1). ParR binds to the partition site, *parC*, which in R1 consists of 10 11-bp direct repeats upstream of the *par* genes and overlaps with the promoter (59). ParR can also pair plasmids at *parC* sites, and pairing is stimulated by the ParM ATPase (22). Unfortunately, R1 ParR was not amenable to crystallization, but the crystal structures of two other ParRs have been solved: pB171 $ParR^{6\text{-}95}$ by itself, and pSK41 $ParR^{1\text{-}53}$ in complex with DNA (Fig. 1G, H) (60, 61). The pSK41 *par* site consists of two sets of five 10-bp repeats, and two repeats (20 bp) constitute the minimal DNA binding site for ParR (61). pSK41 ParR crystallized as a dimer of RHH_2 dimers bound to this 20-bp sequence. The exciting observation for both ParR structures was their super-structure crystal packing, which provides a detailed picture of the higher-order ParR/*parC* partition complex and how it may interact with the ParM ATPase (Fig. 1I) (60, 61). The proteins assembled into a superhelical structure containing 12 ParR dimers (or 6 pairs of dimers) per 360° turn. This superstructure is consistent with electron microscopy pictures of R1 ParR bound to *parC* (60). The DNA wraps on the outside of the helix on a positively charged outer surface of the protein helix. The C-termini of the ParR molecules point inward, toward the center cavity, and are negatively charged. Because ParM interacts with the C-termini of these ParR molecules, the resulting models propose that ParM monomers interact with the concave surface of this helical structure as they insert themselves into the growing actin-like ParM filament during partition.

PARTITION DYNAMICS PROMOTED BY PARTITION NTPases

ParM and Actin-Like Proteins

A major step toward our understanding of plasmid partition dynamics was first elucidated in the type II partition systems. Several elegant biochemical, cell biology, and structural studies have established that the ParM class of ATPases (named for the R1 plasmid ParM) work as "cytomotive" filaments that resemble the behavior of eukaryotic cytoskeletal elements (Fig. 2) (reviewed in reference 62).

ParM is structurally and biochemically similar to eukaryotic actin, although the overall sequence similarity is low (63). It shares an ATPase motif originally described in a variety of proteins, including actin, Hsp70, and hexokinase (5). This group includes MreB, a bacterial protein required for proper cell shape (64). Both ParM and MreB possess the actin-like ATPase

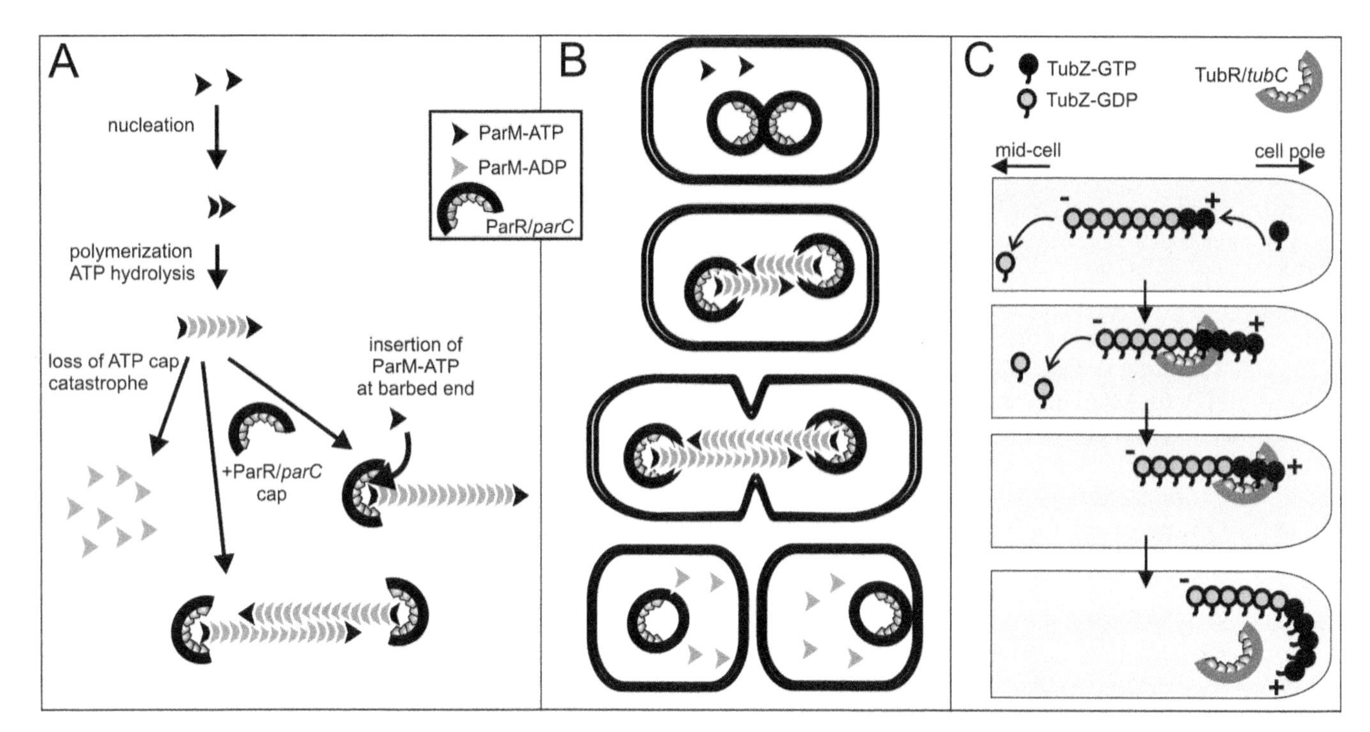

Figure 2 Filamentation mechanisms of plasmid partition. The paradigm for type II partition is the ParMRC system of plasmid R1, illustrated here for (A) properties of filament growth and catastrophe and (B) plasmid partition in the bacterial cell. A critical concentration of ParM nucleates polymerization into filaments. ParM within the filaments hydrolyzes ATP to ADP, but as long as the filaments are capped with ParM-ATP or with ParR/*parC* complexes, the filament is stable. Loss of the cap results in rapid depolymerization, or catastrophe. For ParM, each filament is a double-helical bundle of ParM. Antiparallel arrangement of these bundles results in bidirectional plasmid movement during partition. (C) Treadmilling by TubZ in type III partition systems. TubZ forms dynamic filaments, which grow at the plus (+) end by addition of TubZ-GTP and shrink at the minus (-) end by dissociation of TubZ-GDP. TubR associates with the C-terminal tail of TubZ. TubR/plasmid complexes move toward the plus direction as they are handed from one TubZ to the next in the polymer. Ni et al. have proposed that plasmids dissociate from the TubZ filament when it bends after contact with the curved cell pole (125).
doi:10.1128/microbiolspec.PLAS-0023-2014.f2

fold and subdomains and formed protofilaments reminiscent of F-actin (the polymerized form), and these observations led to the proposal that they represented prokaryotic cytoskeletal elements. A sequence search by Derman et al. has identified over 40 actin-like families of proteins in prokaryotes (65), which were generically called ALPs (actin-like proteins). Evidence is accumulating that these are related families of filament-forming, or cytoskeletal, proteins. They include bacterial, plasmid, and phage-encoded proteins.

The ALPs from several plasmids (R1 and pB171 in *E. coli*, pSK41 in *S. aureus*, and pLS32 and pLS20 in *Bacillus subtilis*) have been examined in detail in a variety of *in vivo* and *in vitro* studies. The results support an insertional polymerization, filament-growth mechanism for plasmid transport inside the cell, although there are several structural and mechanistic differences among the families.

The ParMs from R1 and pSK41 are closely related to each other and are the best understood of the type II partition ATPases (Fig. 3). ParM monomers polymerize in the presence of ATP or GTP to form left-handed two-start helices that are dynamically unstable (63, 66–70). The molecules in the filament hydrolyze ATP to ADP but are stable in the filament form as long as they are protected at the end by a cap of molecules in the ATP-bound conformation (Fig. 2A). Loss of this cap results in rapid depolymerization, or catastrophe. The superhelical ParR/*parC* structure (Fig. 1I) is thought to be the cap, such that ParM monomers interact with C-termini of ParR on the concave surface of ParR/*parC*. ParM polymerizes at this interface, effectively

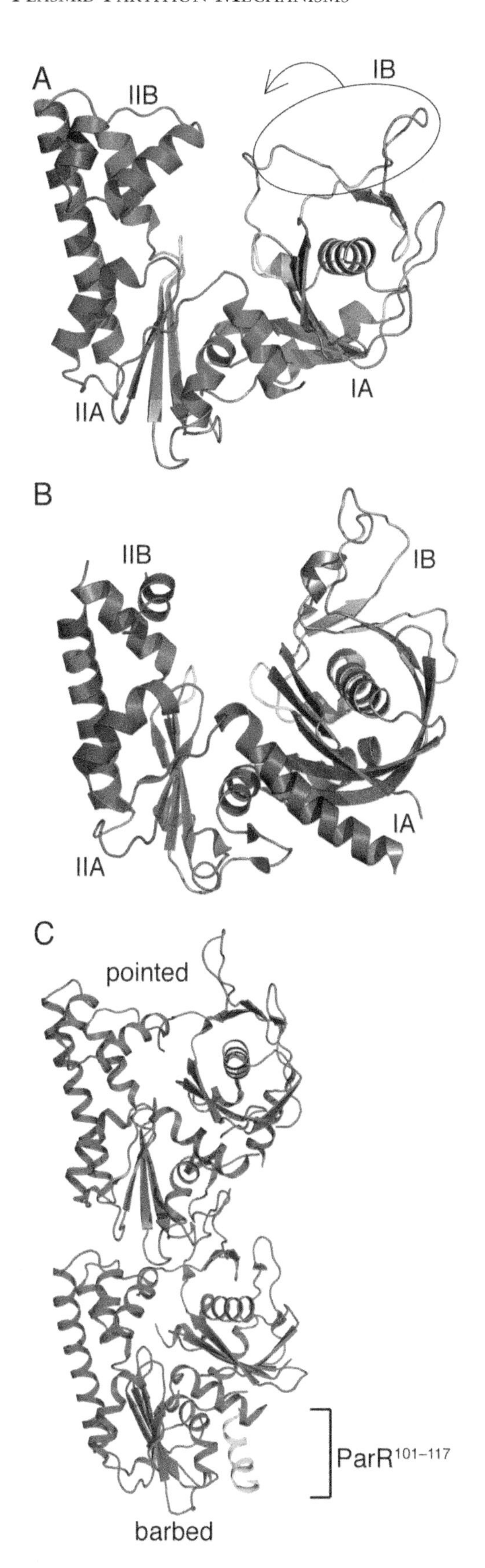

pushing the plasmid. Filaments with one end bound to ParR/*parC* are proposed to search for a second plasmid in cycles of polymerization and catastrophe (67). When the filament captures a second plasmid, and thus has ParB/*parC* bound at both ends, plasmid separation and movement occurs as the plasmids are pushed to opposite poles of the cell (Fig. 2A, B).

Both *in vivo* and *in vitro* experiments support the patterns and the dynamic behavior of ParM and ALPs. Using the R1 system, simultaneous labeling of plasmids and ParMs *in vivo* revealed filaments of ParM with plasmids attached at the tips, both by immunofluorescence (71) and in live cells (72). The filaments grew between plasmids and then disassembled, showing that polymerization pushes plasmids apart. ParM mutants unable to hydrolyze ATP formed long, stable filaments. This behavior was recapitulated *in vitro* by TIRF microscopy, which showed filament growth that was coupled to *parC*-coated beads (67). The filaments went through cycles of growth and shrinkage that depended on ATP binding and hydrolysis. ParM filaments also assemble as bundles, which are visible *in vivo* by fluorescent microscopy and cryo-electron microscopy (72, 73).

A central question is the arrangement of ParM filaments in the cell during partition. The individual filaments are polar; that is, they have a pointed end and a barbed end reflecting head to tail polymer arrangement

Figure 3 ParM structures. The ParM structures from (A) R1 and (B) pSK41 are shown in the absence of nucleotide, the open state of these proteins. ATP binds in the central cleft, and the loops important for binding ATP are highlighted in green. The rest of the structure is colored according to the structural conservation (blue, most conserved; red, least conserved; gray, little to no identifiable conservation) through alignment with the actin structure (PDB 1yag, not shown). Most of the conservation with actin is in domains IA, IIA, and IIB. Domain IB of R1 ParM closes over the ATP binding pocket when nucleotide is present [direction of motion illustrated in (A)]. This cap serves to close the structure, forming the pointed end (domains IB and IIB) of the protein. During filament growth, the pointed end interacts with the barbed end (domains IA and IIA), as shown with the closed forms of ParM in (C). Binding of a $ParR^{101-117}$ peptide induces a closer interaction between the pointed domains, which is proposed to either stabilize the filament to slow down ATP hydrolysis or prevent filament disassembly even in the presence of ATP hydrolysis. ParR binding is confined to the end of the filament because the interaction between the pointed and barbed interfaces of each monomer occludes the ParR binding site (74). All structural images were generated using PyMOL v1.6.0.0 software (Schrödinger, LLC 2010). PDB information is listed in Table 2.
doi:10.1128/microbiolspec.PLAS-0023-2014.f3

(Fig. 3C). How can these filaments grow bidirectionally, and how does ParR/*parC* interact with different ends? A recent study that examined ParM growth with TIRF microscopy as well as a cocrystal structure of ParM in complex with a peptide from ParR indicated that ParR/*parC* interacts with the barbed end, resulting in unidirectional growth as new ParM monomers are added to this end (74). In addition, ParM bundles contain antiparallel sets of ParM double-helical filaments (Fig. 2A, B). In this way, one plasmid can be attached at each end of the bundle, or spindle. This scenario further proposes that the pointed end of ParM filaments is protected from catastrophe by pairing with another ParR/*parC* bound filament.

Other ALPs

Studies of ParMs/ALPs of several other plasmids support the hypothesis that they promote plasmid movement by filamentation (65, 75–78). Interestingly, however, the details of structure and mechanism differ among these proteins. Based on primary sequence, ParM of pSK41, AlfA of pLS32, and Alp7A of pLS20 fall into different classes of prokaryotic ALPs than that of R1 ParM (and those of each other) (65). The proteins also show different architectures and dynamics of their filaments. For example, although the crystal structure of pSK41 ParM confirms its actin-like domains, the protein assembles into one-strand helices, and filaments are not dynamically unstable (77). AlfA forms filaments *in vivo* and *in vitro*, which are double-stranded but with a different architecture than that of R1 ParM and are not dynamically unstable (75, 76, 78). Alp7A forms filaments *in vivo* that colocalize with plasmids, grow and push plasmids apart, are dynamically unstable, but are also capable of treadmilling (65). These and other molecular differences in nucleotide specificity, stability, and the seed size for nucleation, for example, have been noted and indicate that the mechanisms by which plasmids use cytomotive filaments will also differ.

ParAs

By far the most common type of partition system that has been identified in bacterial plasmids uses Walker ATPases as the positioning protein. For simplicity, we will use the most common nomenclature and call these ATPases ParA, and their cognate CBPs ParB, when discussing general members of this class. ParAs contain a deviant version of the Walker A motif (KGGxxK[T/S]) (79), which is the P loop, part of the ATP binding site that interacts with the phosphates of ATP and ADP. The second lysine in the motif is conserved in all Walker ATPases, whereas the first lysine is often called the signature lysine for the ParA class of ATPases. ParA ATPases are also related in sequence, structure, and activity, to bacterial MinD proteins, which are involved in proper positioning of the cell division septum. ParA and MinD are members of a larger class of related ATPases, which include proteins involved in the positioning of other macromolecular structures in bacteria (reviewed in references 80, 81). Studies of MinD support the idea that the ParA/MinD (or "ParA-like") family of proteins shares common mechanisms of action.

In vivo, fluorescent ParA fusions display dynamic patterns that coincide with the bacterial nucleoid. They have been reported to form broad helical structures, patches, or foci that oscillate over nucleoids, for example (44, 46, 47, 82–85). The oscillatory behavior is also similar to that of *E. coli* MinD protein, except that the latter movement is on the membrane (86). Oscillation of ParAs requires the presence of cognate ParBs and *par* sites.

ParA Interactions with ATP, with DNA, and with itself

ATP binding and hydrolysis are involved in all properties and interactions of ParAs. Much of our recent understanding of the biochemistry of plasmid ParAs comes from studies with P1 ParA, F SopA, TP228 ParF, pB171 ParA, and pSM19035 δ protein (23, 26, 40, 56, 58, 87–96). Crystal structures of P1 ParA (97), pSM19035 δ (58), and TP228 ParF (98) (Table 2, Fig. 4) have also contributed insight into the action of this class of ParAs.

ParA ATPase activity is strongly stimulated by the cognate CBP and by DNA (40, 56, 58, 100–102). Often this stimulation is highest when the CBP is bound to the *par* site (40, 58, 101). An arginine residue close to the N-terminus of the CBP is necessary for ATPase stimulation, which has led to the proposal that an "arginine-finger" interaction is responsible (40, 56). In this model, the arginine is inserted into the ATPase active site of ParA to alter catalysis. However, a recent cocrystal structure of MinD with MinE (its stimulatory partner) peptide has challenged this model because the comparable region of MinE sits in a cleft on MinD and not within the ATPase active site (103). Nevertheless, the requirement for a critical arginine in the N-terminus of the CBP is emerging as a common feature among these ParAs.

All ParAs form dimers. For most, dimerization is influenced by adenine nucleotides, although the effects of ATP and ADP differ among different members of Walker partition ATPases. TP228 ParF as well as bacterial Soj and MinD are monomers with ADP and dimers with ATP (98, 104, 105). In contrast, pSM19035 δ, another type Ib ParA, forms dimers with or without ATP or ADP (58). Dimerization of type Ia ParAs is stimulated by both ATP and ADP but can occur at high protein concentration without nucleotide (92, 97, 106). Biochemical experiments with P1 ParA show that it exists in a monomer-dimer equilibrium that is shifted toward dimer in the presence of ATP or ADP (106).

Another important property of ParA ATPases, and one that distinguishes them from ALPs, is an ATP-dependent non-specific DNA (nsDNA) binding activity (58, 87, 92, 93). Again, this behavior is shared with *B. subtilis* Soj (107). The nsDNA binding activity maps to residues in the C-terminus of the ATPase, mutation of which destroys nsDNA binding *in vitro* and partition activity *in vivo* (92, 96, 107). In P1 ParA, however, ATP binding is necessary but not sufficient for the nsDNA binding activity. Following ATP binding, which is fast, ParA-ATP then undergoes a slow conformational change to a form that is able to bind DNA nonspecifically, called ParA-ATP* (87). A requirement for ATP to bind DNA nonspecifically is an unusual property of DNA binding proteins and suggests that the ability to control DNA binding is an essential feature of the partition mechanism. The slow conformational change from ParA-ATP to ParA-ATP* is thought to be a key timing step in the mechanism of ParA, and ParA-like, protein action (see below).

A feature of ParAs that is shared with ALPs is a propensity to polymerize (58, 88, 93–95, 97, 102, 108), but the role of this polymerization is not understood. Mutations in TP228 ParF that eliminate polymerization *in vitro* are defective for partition *in vivo*, supporting the idea that the activity is necessary for ParA mechanism (98). *In vitro*, ParAs polymerize, but the conditions and requirements for polymerization differ, and it has not been possible to generalize polymerization dynamics as it has for ParM (for a comparison, see reference 81). Does polymerization represent formation of structural filaments or the cooperative association of ParA dimers when ParA binds to nsDNA and/or to ParB-DNA complexes? *In vivo* no filaments of plasmid ParAs that resemble the dynamic behavior of ALPs have been reported. *In vitro* at physiological concentrations (≤ 5 μM), neither direct visualization by TIRF microscopy nor size exclusion chromatography of P1 ParA showed significant polymerization (87, 91). The simplest explanation is that cooperativity of ParA, perhaps as the formation of short polymers, is necessary for higher-order complex formation with ParB and DNA, but the resolution to this debate awaits further understanding of the role of this cooperativity in ParA action.

Higher-Order Complexes with ParBs

How does ParA interact with ParB on DNA? Biochemical experiments have demonstrated that ParA interacts with ParB bound to DNA, particularly ParB bound to the *par* site. ParA-ParB-DNA complexes have been detected and examined in electrophoretic mobility shift, light scattering, sucrose-density gradient, and electron microscopy assays (23, 58, 88, 96, 109). Complexes form with ATP and not with ADP. They are larger/more stable with ATPγS, indicating that ATP binding is necessary for assembly and ATP hydrolysis promotes disassembly. This conclusion is further strengthened by the behavior of ParA mutants that can bind but not hydrolyze ATP (89, 96). In addition, while these higher-order complexes form with both specific and nsDNA, they are more stable when DNA containing the *par* site is present (88, 96). Taken together, these observations suggest that these complexes represent the interaction of plasmids via the ParB/*parS* partition complexes, with ParA bound to the bacterial nucleoid (88, 96).

Structural Biology of ParAs

Crystal structures for four plasmid ParAs have been solved: P1 ParA (apoParA and ParA-ADP), P7 ParA (apoParA), pSM19035 δ (δ-ATPγS), and TP228 ParF (ParF-ADP and ParF-AMPPCP) (Table 2; Fig. 4) (58, 97, 98). In addition, structures of an archaeal ParA, *T. thermophilus* Soj, and bacterial MinD proteins have been determined (103–105, 110, 111). The results illustrate structural features that are common to the ParA-like class of ATPases.

As predicted from the conservation of the Walker A motifs, the structure of the ATPase core is similar among these proteins (Fig. 4G-I). The second lysine in Walker A (conserved in all ParA ATPases) contacts the phosphates of ADP/ATP, and the conserved aspartate residues in Walker A′ (also called switch 1) and Walker B (switch 2) coordinate the Mg^{++} ions. The nucleotide binding site sits at the dimer interface, and several structures are consistent with a "sandwich dimer" in that the nucleotide is bound by residues from both monomers. The signature lysine of ParA-like ATPases crosses the dimer interface to interact with the γ-phosphate of ATP that is bound by the other monomer for ParF, Soj, and MinD. In P1 ParA, this lysine is positioned at the dimer interface but is unoccupied because only ADP is bound. The exception is δ-ATPγS; however, the signature lysine is on a flexible loop, and in this structure it is not positioned to interact with ATP (Fig. 4B) (58).

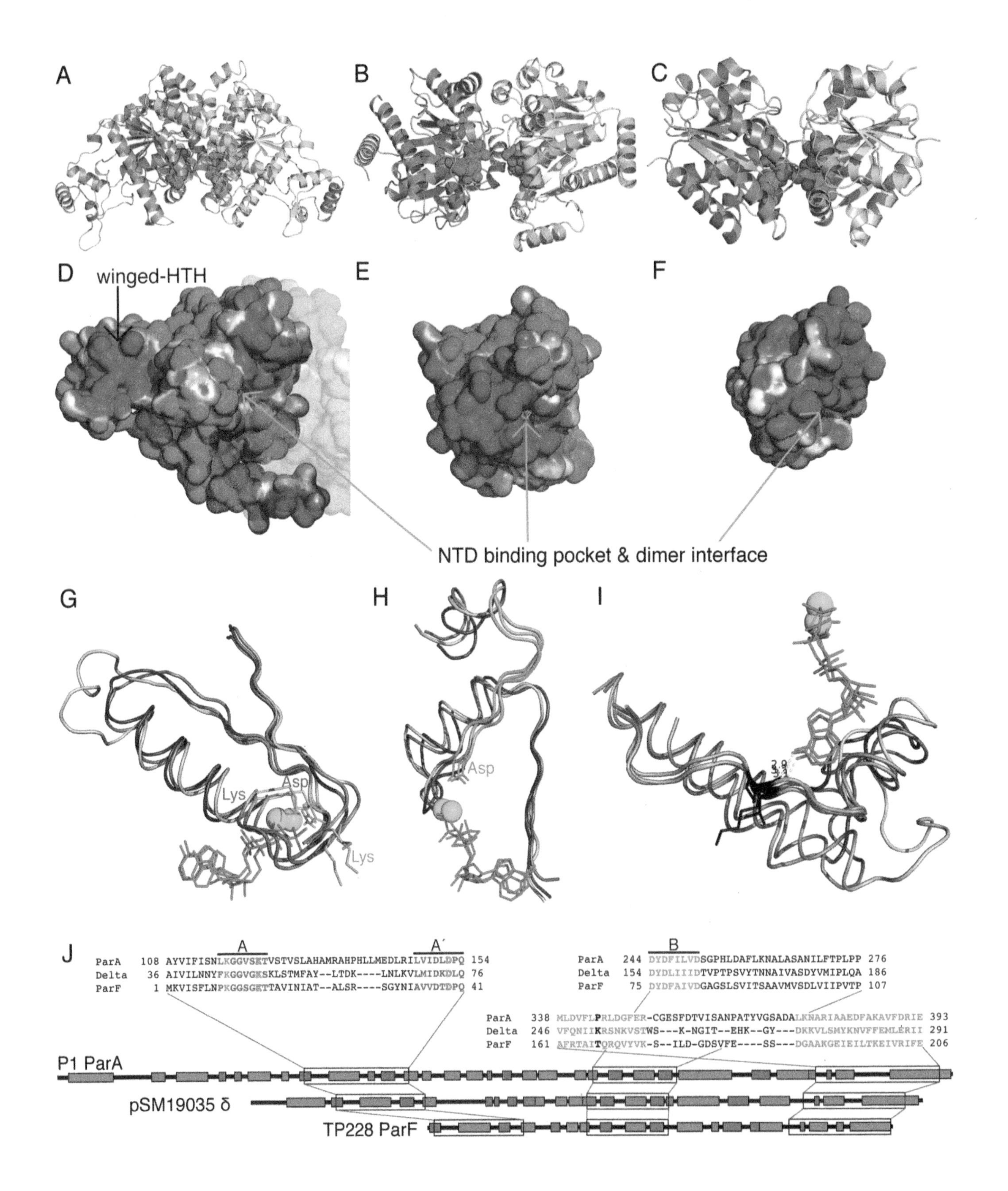
A
B
C
D
winged-HTH
E
F
NTD binding pocket & dimer interface
G
Lys
Asp
Lys
H
Asp
I
J
A
A´
ParA 108 AYVIFISNLKGGVSKTVSTVSLAHAMRAHPHLLMEDLRILVIDLDPQ 154
Delta 36 AIVILNNYFKGGVGKSKLSTMFAY--LTDK----LNLKVLMIDKDLQ 76
ParF 1 MKVISFLNPKGGSGKTTAVINIAT--ALSR----SGYNIAVVDTDPQ 41
B
ParA 244 DYDFILVDSGPHLDAFLKNALASANILFTPLPP 276
Delta 154 DYDLIIIDTVPTPSVYTNNAIVASDYVMIPLQA 186
ParF 75 DYDFAIVDGAGSLSVITSAAVMVSDLVIIPVTP 107
ParA 338 MLDVFLPRLDGFER-CGESFDTVISANPATYVGSADALKNARIAAEDFAKAVFDRIE 393
Delta 246 VFQNIIKRSNKVSTWS---K-NGIT--EHK--GY---DKKVLSMYKNVFFEMLERII 291
ParF 161 AFRTAITQRQVYVK-S--ILD-GDSVFE----SS---DGAAKGEIEILTKEIVRIFE 206
P1 ParA
pSM19035 δ
TP228 ParF

Although no DNA/ParA structures have been solved, the nsDNA binding region of ParA is in the C-terminus, based on mutagenesis studies (92, 96, 107). The surface charge distribution of the existing ParA structures is consistent with this prediction because areas of positive charge are localized on the surface of the C-termini (Fig. 4D–F).

For ParF, Soj, and MinD, the ADP forms are monomers, and the ATP forms are dimers (98, 104, 105). P1 and P7 ParAs, both type Ia ATPases, are dimers with and without nucleotides at the concentrations used for crystallography. However, the dimerization interfaces differ in the apo (P1 and P7) and ADP-bound (P1) forms (97). The structures indicate a flexible dimer interface in the apo form, allowing several different conformations of dimers. ADP binding appears to lock ParA into one dimer conformation. In addition, ADP promotes the folding of a winged HTH motif in the N-terminus of the protein that is responsible for site-specific DNA binding to the *par* operator sequence for the repressor activity of ParA.

The ParF-AMPPCP structure also provides intriguing data concerning the polymerization activity of ParAs (98). Within the crystal, linear ParF polymers resulted from dimer-dimer contacts that were rotated 90° relative to each other, rather than the head-to-tail polymerization seen with ALP/ParM proteins. Mutation of key residues in the dimer-dimer interface of ParF eliminated polymerization of ParF *in vitro*, measured by light scattering, and partition activity *in vivo*. It will be very interesting to see how this arrangement of ParF is involved in partition activity and whether it is shared by other ParAs as more crystal structures of these proteins become available.

The identification of additional ParA-like molecules in other uncharacterized plasmid systems has also begun to contribute to structural information. A sequence study of four large (≥40 kilobases) plasmids in cyanobacterium *Synechocystis* sp. PCC 6803 identified a *parA* homologue common to each plasmid (112), of which one was subsequently crystallized and deposited to the Protein Data Bank (*sll6036*, PDB ID 3cwq). This structure and its sequence show high similarity to those of TP228 ParF, suggesting a type Ib partition mechanism.

Type Ia ParAs and Transcriptional Regulation

The repressor activity of type Ia ParAs is due to a site-specific DNA binding domain that recognizes and binds an operator sequence in the promoter region of the *par* operons (100, 113–116). *In vivo*, however, ParA repressor activity is weak unless stimulated by its cognate ParB protein. *In vitro*, site-specific DNA binding requires ATP or ADP, which distinguishes this activity from nsDNA binding that strictly requires ATP. In the P1 system, evidence from ParAs with mutations in the ATP binding site indicate that ParA-ATP and ParA-ADP, but not ParA-ATP*, are the forms competent for transcriptional repression (87, 89, 114). Mutations that prevent the ParA-ATP* transition act as super-repressors; that is, they repress much more strongly than wild-type ParA represses, and their repression is insensitive to ParB. This observation also suggests an explanation for the corepressor activity of ParB; that is, ParB converts ParA-ATP* to ParA-ADP and/or

Figure 4 Structures of ParAs. (A–C) Structures of the nucleotide-bound forms of (A) P1 ParA (ADP), (B) pSM19035 δ (ATPγS), and (C) TP228 ParF (AMPPCP). Nucleotides are shown in red and magnesium, when present, in green. The structures are presented with their dimerization interfaces perpendicular to the viewing plane. (D–F) Surface charge electrostatics for monomer views of the above proteins are presented as electropositive (blue) or electronegative (red) and were generated with the Adaptive Poisson-Boltzmann Solver software (99). Structures were superimposed with the PyMOL software package, with similar orientations of each monomer [(D) ParA, (E) δ, and (F) ParF)] to expose the nucleotide binding pocket and dimerization surface. DNA binding regions are characteristically electropositive, such as the winged HTH in ParA (D). (G–I) Alignments of the nucleotide binding pockets of ParA, δ, and ParF. The structure of the protein backbone is shown, with the side-chains of critical residues represented as sticks. Nucleotides are in red and magnesium ions are in green. The phosphate-binding regions consist of three conserved motifs: (G) Walker A and A′, in purple and Walker B in orange in (H). Also shown in (G) is the signature lysine (blue) of ParA-like Walker A motifs. (I) The region highlighted in tan is structurally conserved among ParA proteins and provides multiple specific contacts with the adenine moiety. (J) The sequences of the nucleotide binding regions above are aligned with each other and with the overall secondary structure for ParA, δ, and ParF (α-helices in red, β-strands in blue). The Walker A, A′, and B motifs are indicated above the sequence alignment. All structural images were generated using PyMOL v1.6.0.0 software (Schrödinger, LLC 2010). PDB information is listed in Table 2. doi:10.1128/microbiolspec.PLAS-0023-2014.f4

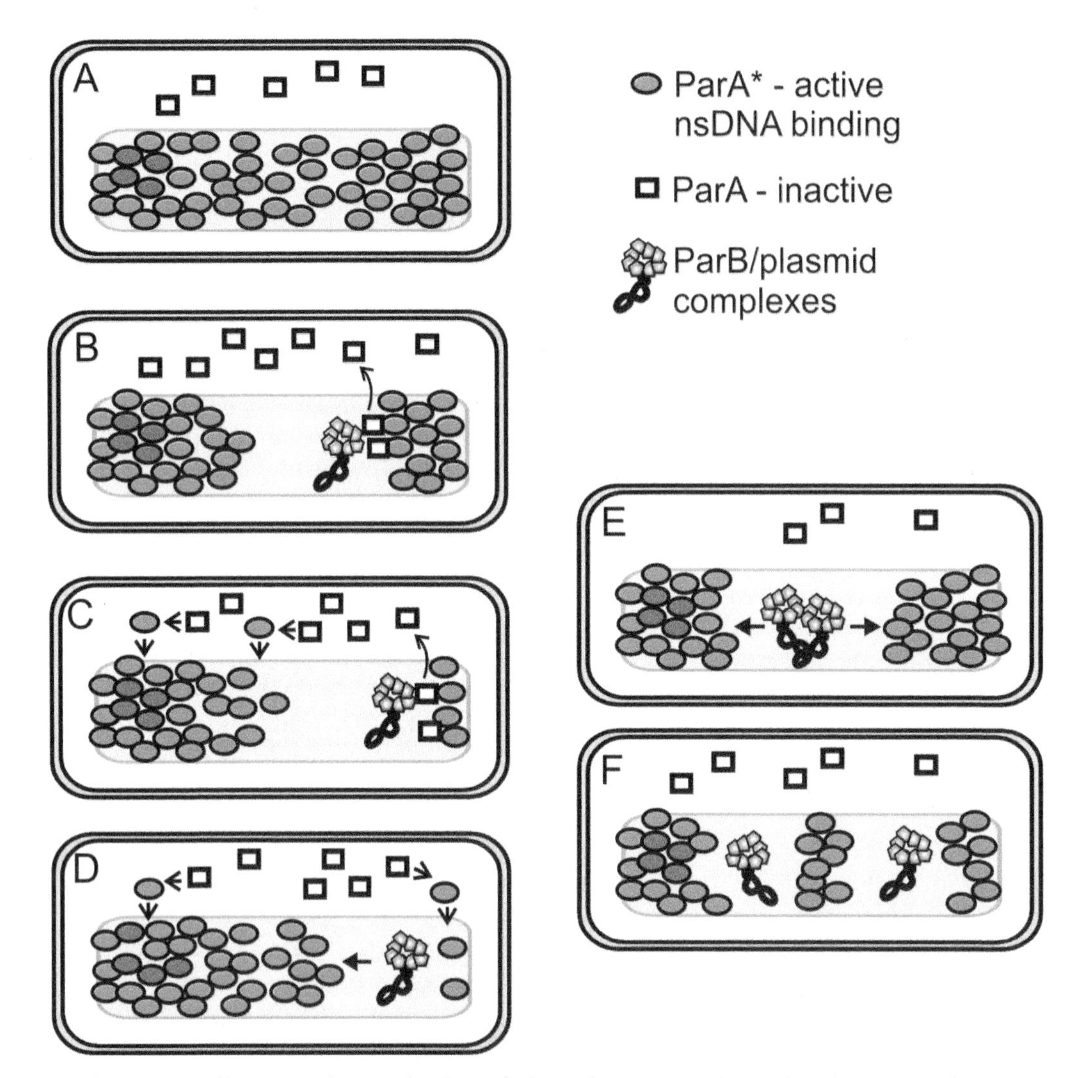

Figure 5 Diffusion-ratchet mechanism of plasmid partition. (A) ParA exists in two forms, one active to bind the nucleoid (ParA*, gray ovals) and one not, which is diffusible in the cytoplasm (ParA, white squares). The conversion between these two forms is slow and depends on the ATP binding cycle of ParA. In P1 ParA, the slow step is a specific conformational change after it binds ATP (ParA-ATP to ParA-ATP* [87], not shown). (B) ParB/plasmid complexes interact with ParA on the nucleoid, and this interaction stimulates the conversion of active ParA back to the diffusible form. Because the conversion to the active form is slow, the inactive form diffuses away from its original location, leaving a void of ParA on the nucleoid. (C–D) This movement continues as ParA rebinds ATP and is then converted back to the DNA binding active form. (E–F) When two plasmid complexes are present, they move toward the nearest, high concentration of ParA on the nucleoid, away from each other. doi:10.1128/microbiolspec.PLAS-0023-2014.f5

ParA-ATP. An intriguing question concerning the type Ia class of ParAs is why repressor activity is integrated into the ATP cycle of these proteins.

Models for ParA Action

What is the mechanistic force that drives plasmid segregation? Two general models have been proposed, which arise from (i) similarities among the ParA/MinD class of proteins and (ii) polymerization properties of ParA and ParM-like proteins.

The first model, based on reaction-diffusion-type mechanisms and the mechanism of MinD action, is supported by growing biochemical, genetic, and cell biology evidence. Reaction-diffusion mechanisms, originally proposed by Alan Turing (117), depend on the interactions between two components to set up biological patterns. In this model for partition, interactions of ParB with ParA set up dynamic patterns of ParAs on the bacterial nucleoid. In essence, ParB both creates and follows a wave of ParA on the nucleoid. This model, termed diffusion-ratchet, proposes that plasmids, via the ParB/*parS* complex, ratchet along the surface of the bacterial nucleoid following an oscillating wave of ParA (Fig. 5) (87).

The first critical parameter is the ATP-dependent nsDNA binding activity of ParAs. Because the bacterial nucleoid is the major nsDNA in the cell, it is proposed as the surface along which the plasmids move, in an ATP-dependent fashion. The second key parameter is the slow interconversion of the DNA binding and non-binding forms of ParA, which essentially creates two pools of ParA in the cell: one that is bound to the nucleoid and one that is free to diffuse in the cytoplasm. Many ParB molecules bind at and around *parS* on the plasmid, which creates a high local concentration of ParB. ParB on the plasmid binds to ParA on the nucleoid (a positive interaction) and then acts to remove ParA from the nucleoid (a negative interaction) via ATP hydrolysis. Because of the slow timing step, ParA cannot immediately rebind the nucleoid and is free to diffuse throughout the cell, losing its positional memory. The latter interaction creates a void of ParA on the nucleoid, which means that ParB must move toward adjacent ParA on the nucleoid for the next interaction. Repeated cycles of these interactions result in plasmid ratcheting along the surface of the nucleoid. In effect, ParB is following this gradient of ParA that it creates. When two plasmid/ParB complexes are present, they will move away from each other toward the nearest high concentration of ParA because the ParA between them has been depleted.

As in the above situation, a slow timing step is crucial for this type of patterning mechanism to yield two populations of ParA, bound and freely diffusing. This step has been identified biochemically for P1 ParA (the slow ParA-ATP to ParA-ATP* conformational transition), but not yet for other ParAs. In principle, it could be any step in the ATP cycle of the protein, so this property may vary among different ParAs. Indeed, this timing step is a critical feature for the dynamic mechanism of MinD movement and has been proposed to be at the step of nucleotide exchange in this system (118). Another possibility may be a slow dimerization step for those ParAs that dimerize only with ATP.

The oscillatory patterns of ParAs over the bacterial nucleoid seen *in vivo* are consistent with this model (46, 47, 85). The dynamic interactions of ParB/plasmid complexes with ParA-coated DNA (as a mimetic of the nucleoid) have also recently been recapitulated *in vitro* using TIRF microscopy (90, 91) and support the ATP-dependent steps proposed for this model of ParA-mediated plasmid partition.

The alternative model for ParA action proposes that filaments of ParA either push or pull plasmids and was suggested based on the polymerization properties of ParAs (85, 95). The energetics of movement would be provided by insertion or removal of ParA molecules at the end of a filament to which the ParB/plasmid complexes are attached, conceptually similar to ParM action. The model was subsequently modified to suggest that ParA filaments polymerized on DNA, to explain the involvement of nsDNA in ParA action, and that filament disassembly (retraction) pulled plasmids inside the cell (85). However, the *in vitro* and *in vivo* behavior of fluorescent plasmid ParA molecules does not detect discrete filaments (46, 47, 85, 87, 90, 91, 102). The debate about these two models continues (for example, see references 119, 120), and its resolution will likely depend on further *in vitro* reconstruction of the partition reaction as well as a better understanding of the role of polymerization in ParA action.

ParA-Like ATPases in Other Processes

ParA-like ATPases participate in the transport of a variety of subcellular structures and organelles, as well as DNA, in bacteria (reviewed in references 80, 81, 119). These ParA-like proteins, including MinD, use ATP-dependent patterning on the surface of either the nucleoid (ParA) or the membrane (MinD) to position cellular cargo. One example of protein transport is the localization of chemotaxis machinery, which also uses nsDNA binding by the ATPase to interact with the bacterial nucleoid (121, 122).

TYPE III (TUBULIN-LIKE) PARTITION SYSTEMS

TubZ

Relatively new to the realm of plasmid partition systems are members of the tubulin-like superfamily of bacterial proteins. The first of these proteins identified was RepX, a protein necessary for replication and maintenance of the virulence plasmid pXO1 of *B. anthracis* (6). RepX bore sequence homology to FtsZ, a GTPase and tubulin-like protein necessary for cell division in bacteria. A similar gene and function was characterized from the endotoxin-encoding plasmid pBtoxis of *B. thuringienses* subsp. *israelensis* (7). The protein, renamed TubZ (for tubulin), from both *B. anthracis* (Ba-TubZ) and *B. thuringiensis* (Bt-TubZ), was shown to form filaments *in vivo* and to be necessary for plasmid maintenance (8, 123). In pBtoxis, the genetic arrangement of *tubZ* also supported that it belonged to a partition system. An additional open reading frame, now called *tubR*, and a set of repeat sequences as a putative partition site, *tubC*, were upstream of *tubZ*. TubR is a small HTH protein that binds to *tubC* and interacts with TubZ (124, 125). The wrinkle in this system is its apparent requirement during plasmid DNA replication as well as segregation, which has yet to be explained. Nevertheless, the properties of TubRZ/*tubC* strongly suggest that it is a segregation system, and its coordination with replication of these plasmids is an intriguing unanswered question.

The biochemistry, cell biology, and structural biology of TubZ and TubR support a filamentation mechanism for partition of these plasmids (Fig. 2C). TubZ is a GTPase that polymerizes into filaments in the presence of GTP (8, 123, 124, 126). A critical concentration of TubZ is necessary to initiate filamentation, and mutations that prevent GTP hydrolysis lead to longer filaments and plasmid instability *in vivo*. Several crystal structures of TubZ (Table 2) have been reported and illustrate that these proteins do share the FtsZ and tubulin folds predicted from their sequences (125, 127, 128). The initial Bt-TubZ structure confirmed the expected nucleotide-binding Rossman fold seen in both FtsZ and tubulin structures (127). Electron microscopy of TubZ filaments along with the crystal structures of TubZ bound to GDP and GTPγS indicate that TubZ forms a parallel, double helical filament (126–128). TubZ in the filaments is almost all in the GDP form, suggesting that a GTP cap must be present to prevent depolymerization (126). This pattern is reminiscent of ParM filaments. However, there is one important distinction about filamentation dynamics. *In vivo*, Bt TubZ-GFP forms filaments that grow from one (the plus) end and disassemble from the other (minus) end, a behavior known as treadmilling (8) (Fig. 2C). The latter property has led to a cable-car model in which plasmids, via TubR, move along the growing TubZ filament (125) (Fig. 2C). Interestingly, *in vivo* when TubZ filament growth reaches a cell pole, it curves and continues along the inside of the cell. One proposal is that this bending stress acts as a switch for the plasmid to "hop off" the cable (125).

TubR/*tubC*

The *tubC* partition site in pBtoxis contains four imperfect 12-bp repeats, to which TubR binds (124, 125). A closer inspection based on TubR binding further refined the region to seven repeats in two sets, to which TubR binds cooperatively (129). This arrangement, two sets of direct repeats, is also reminiscent of the type II *par* site, R1 *parC*.

CBPs in type III partition systems take structural features from both type I and type II systems. Crystallization of TubR from both *Bacillus megaterium* pBM400 and *B. thuringiensis* pBtoxis revealed a compact, dimerized CBP that oligomerizes on centromeric DNA similarly to ParR of the type II partition systems (125, 129). Unlike the compact type Ib and type II CBPs, however, TubR possesses a winged-HTH structure instead of the RHH_2 motif. Interestingly, DNA recognition is different than that by canonical HTH proteins. The recognition helix of the HTH makes up a majority of the dimerization interface, and the N-termini of these paired helices project into a single major groove of *tubC*. A basic patch of residues on the wing and short helix of the HTH (Arg^{74}, Lys^{43}, and Lys^{79} in Bt-TubR) further contribute to DNA binding by interacting with the phosphate backbone; mutation of these residues abrogates DNA binding activity (125, 129). The positioning of dimers on DNA is such that the wings of one dimer are inserted into successive minor grooves, but the wings of adjacent dimers are tightly paired in the same minor groove (129). In this way, TubR forms an oligomerized helical filament with the DNA bound around the surface of the protein in a fashion also reminiscent of type II ParR (Fig. 2C).

WHAT NEXT?

Our understanding of plasmid partition has advanced considerably, yet many important questions still remain. At the forefront of these is the mechanism of type I ParAs. What is the role of ParA polymerization, for example? How does the ATP binding and hydrolysis cycle of these proteins dictate the steps in the patterning reactions? What is the role of ATP hydrolysis by ParA, and ParB stimulation thereof, in the partition mechanism? A recent study of the F Sop system concluded that SopB stimulation of SopA ATPase activity modulated, but was not required for, the oscillatory patterning of SopA in the cell (130). In addition, evidence suggested that this stimulation was involved in separation of plasmid pairs or complexes during the cell cycle. Indeed, how plasmids count to make sure the required number of plasmids are partitioned is unknown. How do ParA-ParB and ParB-DNA interactions mediate pairing and unpairing? What is the biochemical basis for longitudinal and bidirectional movement? Another important question concerns the role of the nucleoid and its shape on mechanism as well as on the cell biology patterns *in vivo*, since several studies have concluded that the nucleoid has a helical superstructure (131–133).

There are variations within any one class of partition system that we have discussed here, for example, the different architectures and dynamics of plasmid ALPs. Therefore, although general models have been developed, we anticipate there will also be significant variation in the molecular interactions involved. This variation is an exciting aspect of plasmid partition research. We expect that as more protein (and protein-DNA) structures are determined, as *in vitro* reconstitution of partition dynamics is achieved, and as additional genetic analyses identify novel phenotypes, our understanding of plasmid partition will progress rapidly.

Acknowledgments. *This work was supported by Canadian Institutes of Health Research grant 37997 (to BEF). Conflicts of interest: We disclose no conflicts.*

Citation. Baxter JC, Funnell BE. 2014. Plasmid partition mechanisms. Microbiol Spectrum **2**(6):PLAS-0023-2014.

References

1. **Funnell BE, Slavcev RA.** 2004. Partition systems of bacterial plasmids, p 81–103. *In* Funnell BE, Phillips GJ (ed), *Plasmid Biology*. ASM Press, Washington, DC.
2. **Gerdes K, Moller-Jensen J, Jensen RB.** 2000. Plasmid and chromosome partitioning: surprises from phylogeny. *Mol Microbiol* **37:**455–466.
3. **Koonin EV.** 1993. A superfamily of ATPases with diverse functions containing either classical or deviant ATP-binding motif. *J Mol Biol* **229:**1165–1174.
4. **Bignell C, Thomas CM.** 2001. The bacterial ParA-ParB partitioning proteins. *J Biotechnol* **91:**1–34.
5. **Bork P, Sander C, Valencia A.** 1992. An ATPase domain common to prokaryotic cell cycle proteins, sugar kinases, actin, and hsp70 heat shock proteins. *Proc Natl Acad Sci USA* **89:**7290–7294.
6. **Tinsley E, Khan SA.** 2006. A novel FtsZ-like protein is involved in replication of the anthrax toxin-encoding pXO1 plasmid in *Bacillus anthracis*. *J Bacteriol* **188:** 2829–2835.
7. **Tang M, Bideshi DK, Park H-W, Federici BA.** 2006. Minireplicon from pBtoxis of *Bacillus thuringiensis* subsp. *israelensis*. *Appl Environ Microbiol* **72:**6948–6954.
8. **Larsen RA, Cusumano C, Fujioka A, Lim-Fong G, Patterson P, Pogliano J.** 2007. Treadmilling of a prokaryotic tubulin-like protein, TubZ, required for plasmid stability in *Bacillus thuringiensis*. *Genes Dev* **21:**1340–1352.
9. **Reyes-Lamothe R, Nicolas E, Sherratt DJ.** 2012. Chromosome replication and segregation in bacteria. *Annu Rev Genet* **46:**121–143.
10. **Wang XD, Llopis PM, Rudner DZ.** 2013. Organization and segregation of bacterial chromosomes. *Nat Rev Genet* **14:**191–203.
11. **Yamaichi Y, Niki H.** 2000. Active segregation by the *Bacillus subtilis* partitioning system in *Escherichia coli*. *Proc Natl Acad Sci USA* **97:**14656–14661.
12. **Murray H, Errington J.** 2008. Dynamic control of the DNA replication initiation protein DnaA by Soj/ParA. *Cell* **135:**74–84.
13. **Gruber S, Errington J.** 2009. Recruitment of condensin to replication origin regions by ParB/SpoOJ promotes chromosome segregation in *B. subtilis*. *Cell* **137:** 685–696.
14. **Sullivan NL, Marquis KA, Rudner DZ.** 2009. Recruitment of SMC by ParB-*parS* organizes the origin region and promotes efficient chromosome segregation. *Cell* **137:**697–707.
15. **Ebersbach G, Gerdes K.** 2001. The double *par* locus of virulence factor pB171: DNA segregation is correlated with oscillation of ParA. *Proc Natl Acad Sci USA* **98:** 15078–15083.
16. **Derman AI, Nonejuie P, Michel BC, Truong BD, Fujioka A, Erb ML, Pogliano J.** 2012. Alp7R regulates expression of the actin-like protein Alp7A in *Bacillus subtilis*. *J Bacteriol* **194:**2715–2724.
17. **Tanaka T.** 2010. Functional analysis of the stability determinant AlfB of pBET131, a miniplasmid derivative of *Bacillus subtilis (natto)* plasmid pLS32. *J Bacteriol* **192:**1221–1230.
18. **Cevallos MA, Cervantes-Rivera R, Gutiérrez-Ríos RM.** 2008. The *repABC* plasmid family. *Plasmid* **60:**19–37.
19. **Grigoriev PS, Lobocka MB.** 2001. Determinants of segregational stability of the linear plasmid-prophage N15 of *Escherichia coli*. *Mol Microbiol* **42:**355–368.
20. **Dorokhov B, Ravin N, Lane D.** 2010. On the role of centromere dispersion in stability of linear bacterial plasmids. *Plasmid* **64:**51–59.
21. **Edgar R, Chattoraj DK, Yarmolinsky M.** 2001. Pairing of P1 plasmid partition sites by ParB. *Mol Microbiol* **42:** 1363–1370.
22. **Jensen RB, Lurz R, Gerdes K.** 1998. Mechanism of DNA segregation in prokaryotes: replicon pairing by *parC* of plasmid R1. *Proc Natl Acad Sci USA* **95:** 8550–8555.
23. **Pratto F, Suzuki Y, Takeyasu K, Alonso JC.** 2009. Single-molecule analysis of protein-DNA complexes formed during partition of newly replicated plasmid molecules in *Streptococcus pyogenes*. *J Biol Chem* **284:** 30298–30306.
24. **Surtees JA, Funnell BE.** 1999. P1 ParB domain structure includes two independent multimerization domains. *J Bacteriol* **181:**5898–5908.
25. **Surtees JA, Funnell BE.** 2001. The DNA binding domains of P1 ParB and the architecture of the P1 plasmid partition complex. *J Biol Chem* **276:**12385–12394.
26. **Ravin NV, Rech J, Lane D.** 2003. Mapping of functional domains in F plasmid partition proteins reveals a bipartite SopB-recognition domain in SopA. *J Mol Biol* **329:** 875–889.
27. **Radnedge L, Youngren B, Davis M, Austin S.** 1998. Probing the structure of complex macromolecular interactions

by homolog specificity scanning: the P1 and P7 plasmid partition systems. *EMBO J* **17:**6076–6085.
28. **Lukaszewicz M, Kostelidou K, Bartosik AA, Cooke GD, Thomas CM, JaguraBurdzy G.** 2002. Functional dissection of the ParB homologue (KorB) from IncP-1 plasmid RK2. *Nucleic Acids Res* **30:**1046–1055.
29. **Delbruck H, Ziegelin G, Lanka E, Heinemann U.** 2002. An Src homology 3-like domain is responsible for dimerization of the repressor protein KorB encoded by the promiscuous IncP plasmid RP4. *J Biol Chem* **277:** 4191–4198.
30. **Khare D, Ziegelin G, Lanka E, Heinemann U.** 2004. Sequence-specific DNA binding determined by contacts outside the helix-turn-helix motif of the ParB homolog KorB. *Nat Struct Mol Biol* **11:**656–663.
31. **Schumacher MA, Funnell BE.** 2005. ParB-DNA structures reveal DNA-binding mechanism of partition complex formation. *Nature* **438:**516–519.
32. **Schumacher MA, Mansoor A, Funnell BE.** 2007. Structure of a four-way bridged ParB-DNA complex provides insight into P1 segrosome assembly. *J Biol Chem* **282:** 10456–10464.
33. **Schumacher MA, Piro KM, Xu WJ.** 2010. Insight into F plasmid DNA segregation revealed by structures of SopB and SopB-DNA complexes. *Nucleic Acids Res* **38:** 4514–4526.
34. **Hanai R, Liu RP, Benedetti P, Caron PR, Lynch AS, Wang JC.** 1996. Molecular dissection of a protein SopB essential for *Escherichia coli* F plasmid partition. *J Biol Chem* **271:**17469–17475.
35. **Rajasekar K, Muntaha ST, Tame JRH, Kommareddy S, Morris G, Wharton CW, Thomas CM, White SA, Hyde EI, Scott DJ.** 2010. Order and disorder in the domain organization of the plasmid partition protein KorB. *J Biol Chem* **285:**15440–15449.
36. **Williams DR, Macartney DP, Thomas CM.** 1998. The partitioning activity of the RK2 central control region requires only *incC*, *korB* and KorB-binding site $O(_B)3$ but other KorB-binding sites form destabilizing complexes in the absence of $O(_B)3$. *Microbiology* **144:** 3369–3378.
37. **Pillet F, Sanchez A, Lane D, Leberre VA, Bouet JY.** 2011. Centromere binding specificity in assembly of the F plasmid partition complex. *Nucleic Acids Res* **39:** 7477–7486.
38. **Sanchez A, Rech J, Gasc C, Bouet JY.** 2013. Insight into centromere-binding properties of ParB proteins: a secondary binding motif is essential for bacterial genome maintenance. *Nucleic Acids Res* **41:**3094–3103.
39. **Vecchiarelli AG, Schumacher MA, Funnell BE.** 2007. P1 partition complex assembly involves several modes of protein-DNA recognition. *J Biol Chem* **282:**10944–10952.
40. **Ah-Seng Y, Lopez F, Pasta F, Lane D, Bouet J-Y.** 2009. Dual role of DNA in regulating ATP hydrolysis by the SopA partition protein. *J Biol Chem* **284:** 30067–30075.
41. **Leonard TA, Butler PJG, Lowe J.** 2004. Structural analysis of the chromosome segregation protein Spo0J from *Thermus thermophilus*. *Mol Microbiol* **53:**419–432.
42. **Murray H, Ferreira H, Errington J.** 2006. The bacterial chromosome segregation protein Spo0J spreads along DNA from *parS* nucleation sites. *Mol Microbiol* **61:** 1352–1361.
43. **Erdmann N, Petroff T, Funnell BE.** 1999. Intracellular localization of P1 ParB protein depends on ParA and *parS*. *Proc Natl Acad Sci USA* **96:**14905–14910.
44. **Adachi S, Hori K, Hiraga S.** 2006. Subcellular positioning of F plasmid mediated by dynamic localization of SopA and SopB. *J Mol Biol* **356:**850–863.
45. **Sengupta M, Nielsen HJ, Youngren B, Austin S.** 2010. P1 plasmid segregation: accurate re-distribution by dynamic plasmid pairing and separation. *J Bacteriol* **192:** 1175–1183.
46. **Hatano T, Yamaichi Y, Niki H.** 2007. Oscillating focus of SopA associated with filamentous structure guides partitioning of F plasmid. *Mol Microbiol* **64:** 1198–1213.
47. **Hatano T, Niki H.** 2010. Partitioning of P1 plasmids by gradual distribution of the ATPase ParA. *Mol Microbiol* **78:**1182–1198.
48. **Kim S-K, Wang JC.** 1999. Gene silencing via protein-mediated subcellular localization of DNA. *Proc Natl Acad Sci USA* **96:**8557–8561.
49. **Rodionov O, Lobocka M, Yarmolinsky M.** 1999. Silencing of genes flanking the P1 plasmid centromere. *Science* **283:**546–549.
50. **Rodionov O, Yarmolinsky M.** 2004. Plasmid partitioning and the spreading of P1 partition protein ParB. *Mol Microbiol* **52:**1215–1223.
51. **Schreiter ER, Drennan CL.** 2007. Ribbon-helix-helix transcription factors: variations on a theme. *Nat Rev Microbiol* **5:**710–720.
52. **Murayama K, Orth P, de la Hoz AB, Alonso JC, Saenger W.** 2001. Crystal structure of ω transcriptional repressor encoded by *Streptococcus pyogenes* plasmid pSM19035 at 1.5 Å resolution. *J Mol Biol* **314:** 789–796.
53. **Golovanov AP, Barillà D, Golovanova M, Hayes F, Lian L-Y.** 2003. ParG, a protein required for active partition of bacterial plasmids, has a dimeric ribbon-helix-helix structure. *Mol Microbiol* **50:**1141–1153.
54. **Weihofen WA, Cicek A, Pratto F, Alonso JC, Saenger W.** 2006. Structures of ω repressors bound to direct and inverted DNA repeats explain modulation of transcription. *Nucleic Acids Res* **34:**1450–1458.
55. **Huang L, Yin P, Zhu X, Zhang Y, Ye KQ.** 2011. Crystal structure and centromere binding of the plasmid segregation protein ParB from pCXC100. *Nucleic Acids Res* **39:**2954–2968.
56. **Barillà D, Carmelo E, Hayes F.** 2007. The tail of the ParG DNA segregation protein remodels ParF polymers and enhances ATP hydrolysis via an arginine finger-like motif. *Proc Natl Acad Sci USA* **104:** 1811–1816.
57. **Wu MY, Zampini M, Bussiek M, Hoischen C, Diekmann S, Hayes F.** 2011. Segrosome assembly at the pliable *parH* centromere. *Nucleic Acids Res* **39:** 5082–5097.

58. **Pratto F, Cicek A, Weihofen WA, Lurz R, Saenger W, Alonso JC.** 2008. *Streptococcus pyogenes* pSM19035 requires dynamic assembly of ATP-bound ParA and ParB on *parS* DNA during plasmid segregation. *Nucleic Acids Res* **36:**3676–3689.
59. **Breuner A, Jensen RB, Dam M, Pedersen S, Gerdes K.** 1996. The centromere-like *parC* locus of plasmid R1. *Mol Microbiol* **20:**581–592.
60. **Moller-Jensen J, Ringgaard S, Mercogliano CP, Gerdes K, Lowe J.** 2007. Structural analysis of the ParR/*parC* plasmid partition complex. *EMBO J* **26:**4413–4422.
61. **Schumacher MA, Glover TC, Brzoska AJ, Jensen SO, Dunham TD, Skurray RA, Firth N.** 2007. Segrosome structure revealed by a complex of ParR with centromere DNA. *Nature* **450:**1268–1271.
62. **Salje J, Gayathri P, Lowe J.** 2010. The ParMRC system: molecular mechanisms of plasmid segregation by actin-like filaments. *Nat Rev Microbiol* **8:**683–692.
63. **van den Ent F, Moller-Jensen J, Amos LA, Gerdes K, Lowe J.** 2002. F-actin-like filaments formed by plasmid segregation protein ParM. *EMBO J* **21:**6935–6943.
64. **van den Ent F, Amos LA, Lowe J.** 2001. Prokaryotic origin of the actin cytoskeleton. *Nature* **413:**39–44.
65. **Derman AI, Becker EC, Truong BD, Fujioka A, Tucey TM, Erb ML, Patterson PC, Pogliano J.** 2009. Phylogenetic analysis identifies many uncharacterized actin-like proteins (Alps) in bacteria: regulated polymerization, dynamic instability and treadmilling in Alp7A. *Mol Microbiol* **73:**534–552.
66. **Garner EC, Campbell CS, Mullins RD.** 2004. Dynamic instability in a DNA-segregating prokaryotic actin homolog. *Science* **306:**1021–1025.
67. **Garner EC, Campbell CS, Weibel DB, Mullins RD.** 2007. Reconstitution of DNA segregation driven by assembly of a prokaryotic actin homolog. *Science* **315:** 1270–1274.
68. **Rivera CR, Kollman JM, Polka JK, Agard DA, Mullins RD.** 2011. Architecture and assembly of a divergent member of the ParM family of bacterial actin-like proteins. *J Biol Chem* **286:**14282–14290.
69. **Popp D, Narita A, Oda T, Fujisawa T, Matsuo H, Nitanai Y, Iwasa M, Maeda K, Onishi H, Maeda Y.** 2008. Molecular structure of the ParM polymer and the mechanism leading to its nucleotide-driven dynamic instability. *EMBO J* **27:**570–579.
70. **Galkin VE, Orlova A, Rivera C, Mullins RD, Egelman EH.** 2009. Structural polymorphism of the ParM filament and dynamic instability. *Structure* **17:**1253–1264.
71. **Moller-Jensen J, Borch J, Dam M, Jensen RB, Roepstorff P, Gerdes K.** 2003. Bacterial mitosis: ParM of plasmid R1 moves plasmid DNA by an actin-like insertional polymerization mechanism. *Mol Cell* **12:** 1477–1487.
72. **Campbell CS, Mullins RD.** 2007. *In vivo* visualization of type II plasmid segregation: bacterial actin filaments pushing plasmids. *J Cell Biol* **179:**1059–1066.
73. **Salje J, Zuber B, Lowe J.** 2009. Electron cryomicroscopy of *E. coli* reveals filament bundles involved in plasmid DNA segregation. *Science* **323:**509–512.
74. **Gayathri P, Fujii T, Moller-Jensen J, van den Ent F, Namba K, Lowe J.** 2012. A bipolar spindle of antiparallel ParM filaments drives bacterial plasmid segregation. *Science* **338:**1334–1337.
75. **Becker E, Herrera NC, Gunderson FQ, Derman AI, Dance AL, Sims J, Larsen RA, Pogliano J.** 2006. DNA segregation by the bacterial actin AlfA during *Bacillus subtilis* growth and development. *EMBO J* **25:** 5919–5931.
76. **Popp D, Narita A, Ghoshdastider U, Maeda K, Maéda Y, Oda T, Fujisawa T, Onishi H, Ito K, Robinson RC.** 2010. Polymeric structures and dynamic properties of the bacterial actin AlfA. *J Mol Biol* **397:** 1031–1041.
77. **Popp D, Xu WJ, Narita A, Brzoska AJ, Skurray RA, Firth N, Goshdastider U, Maeda Y, Robinson RC, Schumacher MA.** 2010. Structure and filament dynamics of the pSK41 actin-like ParM protein. *J Biol Chem* **285:**10130–10140.
78. **Polka JK, Kollman JM, Agard DA, Mullins RD.** 2009. The structure and assembly dynamics of plasmid actin AlfA imply a novel mechanism of DNA segregation. *J Bacteriol* **191:**6219–6230.
79. **Motallebi-Veshareh M, Rouch DA, Thomas CM.** 1990. A family of ATPases involved in active partitioning of diverse bacterial plasmids. *Mol Microbiol* **4:** 1455–1463.
80. **Lutkenhaus J.** 2012. The ParA/MinD family puts things in their place. *Trends Microbiol* **20:**411–418.
81. **Vecchiarelli AG, Mizuuchi K, Funnell BE.** 2012. Surfing biological surfaces: exploiting the nucleoid for partition and transport in bacteria. *Mol Microbiol* **86:** 513–523.
82. **Marston AL, Errington J.** 1999. Dynamic movement of the ParA-like Soj protein of *B. subtilis* and its dual role in nucleoid organization and developmental regulation. *Mol Cell* **4:**673–682.
83. **Lim GE, Derman AI, Pogliano J.** 2005. Bacterial DNA segregation by dynamic SopA polymers. *Proc Natl Acad Sci USA* **102:**17658–17663.
84. **Ebersbach G, Gerdes K.** 2004. Bacterial mitosis: partitioning protein ParA oscillates in spiral-shaped structures and positions plasmids at mid-cell. *Mol Microbiol* **52:**385–398.
85. **Ringgaard S, van Zon J, Howard M, Gerdes K.** 2009. Movement and equipositioning of plasmids by ParA filament disassembly. *Proc Natl Acad Sci USA* **106:** 19369–19374.
86. **Raskin DM, de Boer PAJ.** 1999. Rapid pole-to-pole oscillation of a protein required for directing division to the middle of *Escherichia coli*. *Proc Natl Acad Sci USA* **96:**4971–4976.
87. **Vecchiarelli AG, Han YW, Tan X, Mizuuchi M, Ghirlando R, Biertümpfel C, Funnell BE, Mizuuchi K.** 2010. ATP control of dynamic P1 ParA–DNA interactions: a key role for the nucleoid in plasmid partition. *Mol Microbiol* **78:**78–91.
88. **Havey JC, Vecchiarelli AG, Funnell BE.** 2012. ATP-regulated interactions between P1 ParA, ParB and

non-specific DNA that are stabilized by the plasmid partition site, *parS*. *Nucleic Acids Res* **40:**801–812.

89. **Vecchiarelli AG, Havey JC, Ing LL, Wong EOY, Waples WG, Funnell BE.** 2013. Dissection of the ATPase active site of P1 ParA reveals multiple active forms essential for plasmid partition. *J Biol Chem* **288:** 17823–17831.
90. **Vecchiarelli AG, Hwang LC, Mizuuchi K.** 2013. Cell-free study of F plasmid partition provides evidence for cargo transport by a diffusion-ratchet mechanism. *Proc Natl Acad Sci USA* **110:**E1390–E1397.
91. **Hwang LC, Vecchiarelli AG, Han Y-W, Mizuuchi M, Harada Y, Funnell BE, Mizuuchi K.** 2013. ParA-mediated plasmid partition driven by protein pattern self-organization. *EMBO J* **32:**1238–1249.
92. **Castaing J-P, Bouet J-Y, Lane D.** 2008. F plasmid partition depends on interaction of SopA with non-specific DNA. *Mol Microbiol* **70:**1000–1011.
93. **Bouet J-Y, Ah-Seng Y, Benmeradi N, Lane D.** 2007. Polymerization of SopA partition ATPase: regulation by DNA binding and SopB. *Mol Microbiol* **63:**468–481.
94. **Machon C, Fothergill TJG, Barillà D, Hayes F.** 2007. Promiscuous stimulation of ParF protein polymerization by heterogeneous centromere binding factors. *J Mol Biol* **374:**1–8.
95. **Barillà D, Rosenberg MF, Nobbmann U, Hayes F.** 2005. Bacterial DNA segregation dynamics mediated by the polymerizing protein ParF. *EMBO J* **24:**1453–1464.
96. **Soberón NE, Lioy VS, Pratto F, Volante A, Alonso JC.** 2011. Molecular anatomy of the *Streptococcus pyogenes* pSM19035 partition and segrosome complexes. *Nucleic Acids Res* **39:**2624–2637.
97. **Dunham TD, Xu W, Funnell BE, Schumacher MA.** 2009. Structural basis for ADP-mediated transcriptional regulation by P1 and P7 ParA. *EMBO J* **28:**1792–1802.
98. **Schumacher MA, Ye QZ, Barge MT, Zampini M, Barilla D, Hayes F.** 2012. Structural mechanism of ATP-induced polymerization of the partition factor ParF. *J Biol Chem* **287:**26146–26154.
99. **Baker N, Sept D, Joseph S, Holst M, McCammon J.** 2001. Electrostatics of nanosystems: application to microtubules and the ribosome. *Proc Natl Acad Sci USA* **98:**10037–10041.
100. **Davis MA, Martin KA, Austin SJ.** 1992. Biochemical activities of the ParA partition protein of the P1 plasmid. *Mol Microbiol* **6:**1141–1147.
101. **Watanabe E, Wachi M, Yamasaki M, Nagai K.** 1992. ATPase activity of SopA, a protein essential for active partitioning of F-plasmid. *Mol Gen Genet* **234:** 346–352.
102. **Ebersbach G, Ringgaard S, Moller-Jensen J, Wang Q, Sherratt DJ, Gerdes K.** 2006. Regular cellular distribution of plasmids by oscillating and filament-forming ParA ATPase of plasmid pB171. *Mol Microbiol* **61:** 1428–1442.
103. **Park K-T, Wu W, Battaile Kevin P, Lovell S, Holyoak T, Lutkenhaus J.** 2011. The Min oscillator uses MinD-dependent conformational changes in MinE to spatially regulate cytokinesis. *Cell* **146:**396–407.
104. **Leonard TA, Butler PJ, Lowe J.** 2005. Bacterial chromosome segregation: structure and DNA binding of the Soj dimer: a conserved biological switch. *EMBO J* **24:** 270–282.
105. **Wu W, Park K-T, Holyoak T, Lutkenhaus J.** 2011. Determination of the structure of the MinD–ATP complex reveals the orientation of MinD on the membrane and the relative location of the binding sites for MinE and MinC. *Mol Microbiol* **79:**1515–1528.
106. **Davey MJ, Funnell BE.** 1997. Modulation of the P1 plasmid partition protein ParA by ATP, ADP and P1 ParB. *J Biol Chem* **272:**15286–15292.
107. **Hester CM, Lutkenhaus J.** 2007. Soj (ParA) DNA binding is mediated by conserved arginines and is essential for plasmid segregation. *Proc Natl Acad Sci USA* **104:** 20326–20331.
108. **Batt SM, Bingle LEH, Dafforn TR, Thomas CM.** 2009. Bacterial genome partitioning: N-terminal domain of IncC protein encoded by broad-host-range plasmid RK2 modulates oligomerisation and DNA binding. *J Mol Biol* **385:**1361–1374.
109. **Bouet J-Y, Funnell BE.** 1999. P1 ParA interacts with the P1 partition complex at *parS* and an ATP- ADP switch controls ParA activities. *EMBO J* **18:**1415–1424.
110. **Sakai N, Yao M, Itou H, Watanabe N, Yumoto F, Tanokura M, Tanaka I.** 2001. The three-dimensional structure of septum site-determining protein MinD from *Pyrococcus horikoshii* OT3 in complex with Mg-ADP. *Structure* **9:**817–826.
111. **Hayashi I, Oyama T, Morikawa K.** 2001. Structural and functional studies of MinD ATPase: implications for the molecular recognition of the bacterial cell division apparatus. *EMBO J* **20:**1819–1828.
112. **Kaneko T, Nakamura Y, Sasamoto S, Watanabe A, Kohara M, Matsumoto M, Shimpo S, Yamada M, Tabata S.** 2003. Structural analysis of four large plasmids harboring in a unicellular cyanobacterium, *Synechocystis* sp PCC 6803. *DNA Res* **10:**221–228.
113. **Davey MJ, Funnell BE.** 1994. The P1 plasmid partition protein ParA. A role for ATP in site-specific DNA binding. *J Biol Chem* **269:**29908–29913.
114. **Fung E, Bouet J-Y, Funnell BE.** 2001. Probing the ATP-binding site of P1 ParA: partition and repression have different requirements for ATP binding and hydrolysis. *EMBO J* **20:**4901–4911.
115. **Mori H, Mori Y, Ichinose C, Niki H, Ogura T, Kato A, Hiraga S.** 1989. Purification and characterization of SopA and SopB proteins essential for F plasmid partitioning. *J Biol Chem* **264:**15535–15541.
116. **Libante V, Thion L, Lane D.** 2001. Role of the ATP-binding site of SopA protein in partition of the F plasmid. *J Mol Biol* **314:**387–399.
117. **Turing AM.** 1952. The chemical basis of morphogenesis. *Philos Trans R Soc London B Biol Sci* **237:** 37–72.
118. **Huang KC, Meir Y, Wingreen NS.** 2003. Dynamic structures in *Escherichia coli*: spontaneous formation of MinE rings and MinD polar zones. *Proc Natl Acad Sci USA* **100:**12724–12728.

119. **Szardenings F, Guymer D, Gerdes K.** 2011. ParA ATPases can move and position DNA and subcellular structures. *Curr Opin Microbiol* **14:**712–718.

120. **Sherratt D.** 2013. Plasmid partition: sisters drifting apart. *EMBO J* **32:**1208–1210.

121. **Ringgaard S, Schirner K, Davis BM, Waldor MK.** 2011. A family of ParA-like ATPases promotes cell pole maturation by facilitating polar localization of chemotaxis proteins. *Genes Dev* **25:**1544–1555.

122. **Roberts MAJ, Wadhams GH, Hadfield KA, Tickner S, Armitage JP.** 2012. ParA-like protein uses nonspecific chromosomal DNA binding to partition protein complexes. *Proc Natl Acad Sci USA* **109:** 6698–6703.

123. **Akhtar P, Anand SP, Watkins SC, Khan SA.** 2009. The tubulin-like RepX protein encoded by the pXO1 plasmid forms polymers *in vivo* in *Bacillus anthracis*. *J Bacteriol* **191:**2493–2500.

124. **Tang M, Bideshi DK, Park H-W, Federici BA.** 2007. Iteron-binding ORF157 and FtsZ-like ORF156 proteins encoded by pBtoxis play a role in its replication in *Bacillus thuringiensis* subsp. *israelensis*. *J Bacteriol* **189:** 8053–8058.

125. **Ni LS, Xu WJ, Kumaraswami M, Schumacher MA.** 2010. Plasmid protein TubR uses a distinct mode of HTH-DNA binding and recruits the prokaryotic tubulin homolog TubZ to effect DNA partition. *Proc Natl Acad Sci USA* **107:**11763–11768.

126. **Chen YD, Erickson HP.** 2008. *In vitro* assembly studies of FtsZ/tubulin-like proteins (TubZ) from *Bacillus* plasmids: evidence for a capping mechanism. *J Biol Chem* **283:**8102–8109.

127. **Aylett CHS, Wang Q, Michie KA, Amos LA, Löwe J.** 2010. Filament structure of bacterial tubulin homologue TubZ. *Proc Natl Acad Sci USA* **107:**19766–19771.

128. **Hoshino S, Hayashi I.** 2012. Filament formation of the FtsZ/tubulin-like protein TubZ from the *Bacillus cereus* pXO1 plasmid. *J Biol Chem* **287:**32103–32112.

129. **Aylett CHS, Lowe J.** 2012. Superstructure of the centromeric complex of TubZRC plasmid partitioning systems. *Proc Natl Acad Sci USA* **109:**16522–16527.

130. **Ah-Seng Y, Rech J, Lane D, Bouet JY.** 2013. Defining the role of ATP hydrolysis in mitotic segregation of bacterial plasmids. *PLoS Genet* **9:**e1003956. doi: 10.1371/journal.pgen.1003956.

131. **Berlatzky IA, Rouvinski A, Ben-Yehuda S.** 2008. Spatial organization of a replicating bacterial chromosome. *Proc Natl Acad Sci USA* **105:**14136–14140.

132. **Hadizadeh Yazdi N, Guet CC, Johnson RC, Marko JF.** 2012. Variation of the folding and dynamics of the *Escherichia coli* chromosome with growth conditions. *Mol Microbiol* **86:**1318–1333.

133. **Fisher JK, Bourniquel A, Witz G, Weiner B, Prentiss M, Kleckner N.** 2013. Four-dimensional imaging of *E. coli* nucleoid organization and dynamics in living cells. *Cell* **153:**882–895.

Plasmids—Biology and Impact in Biotechnology and Discovery
Edited by Marcelo E. Tolmasky and Juan C. Alonso

doi:10.1128/microbiolspec.PLAS-0025-2014

Estelle Crozat,[1] Florian Fournes,[1] François Cornet,[2] Bernard Hallet,[3] and Philippe Rousseau[1]

9 Resolution of Multimeric Forms of Circular Plasmids and Chromosomes

One of the serious disadvantages of circular plasmids and chromosomes is their high sensitivity to rearrangements caused by homologous recombination. Odd numbers of recombinational exchanges occurring during or after replication of a circular replicon result in the formation of a dimeric molecule in which the two copies of the replicon are fused in a head-to-tail configuration (Fig. 1). If they are not converted back to monomers, the dimers of replicons may fail to correctly segregate at the time of cell division.

Resolution of multimeric forms of circular plasmids and chromosomes is mediated by site-specific recombination, an efficient and tightly controlled DNA breakage and joining reaction that occurs at specific DNA sequences (Fig. 1A). Site-specific recombinases, the enzymes that catalyze this type of reaction, fall into two families of proteins: the serine and tyrosine recombinase families (1).

This chapter, which is an update of a previous version (2), provides an overview of the variety of site-specific resolution systems found on circular plasmids and chromosomes. After an introduction about the formation and the incidence of replicon multimers, we present the two families of resolution systems, based on serine or tyrosine recombinases. The most recent advances in understanding of the molecular mechanisms that control the recombination reactions catalyzed by these systems will be discussed. These illustrate how different molecular actors have evolved to achieve the same function but also how some are devoted to plasmid dispersion, whereas some are devoted to chromosome or to mobile genetic element dispersion.

FORMATION AND INCIDENCE OF PLASMID MULTIMERS

Multimerization of circular DNA molecules is largely due to homologous recombination. Studies of high-copy-number plasmids, such as pBR322, revealed that multimerization primarily occurs by RecF-dependent recombination (3). RecBCD activity on double-strand ends would lead to rapid degradation and inefficient

[1]Université de Toulouse, UPS, Laboratoire de Microbiologie et Génétique Moléculaires, F-31062 Toulouse, France; [2]CNRS, Laboratoire de Microbiologie et Génétique Moléculaires, F-31062 Toulouse, France; [3]Institut des Sciences de la Vie, UCLouvain, 4/5 L7.07.06 Place Croix du Sud, B-1348 Louvain-la-Neuve, Belgium.

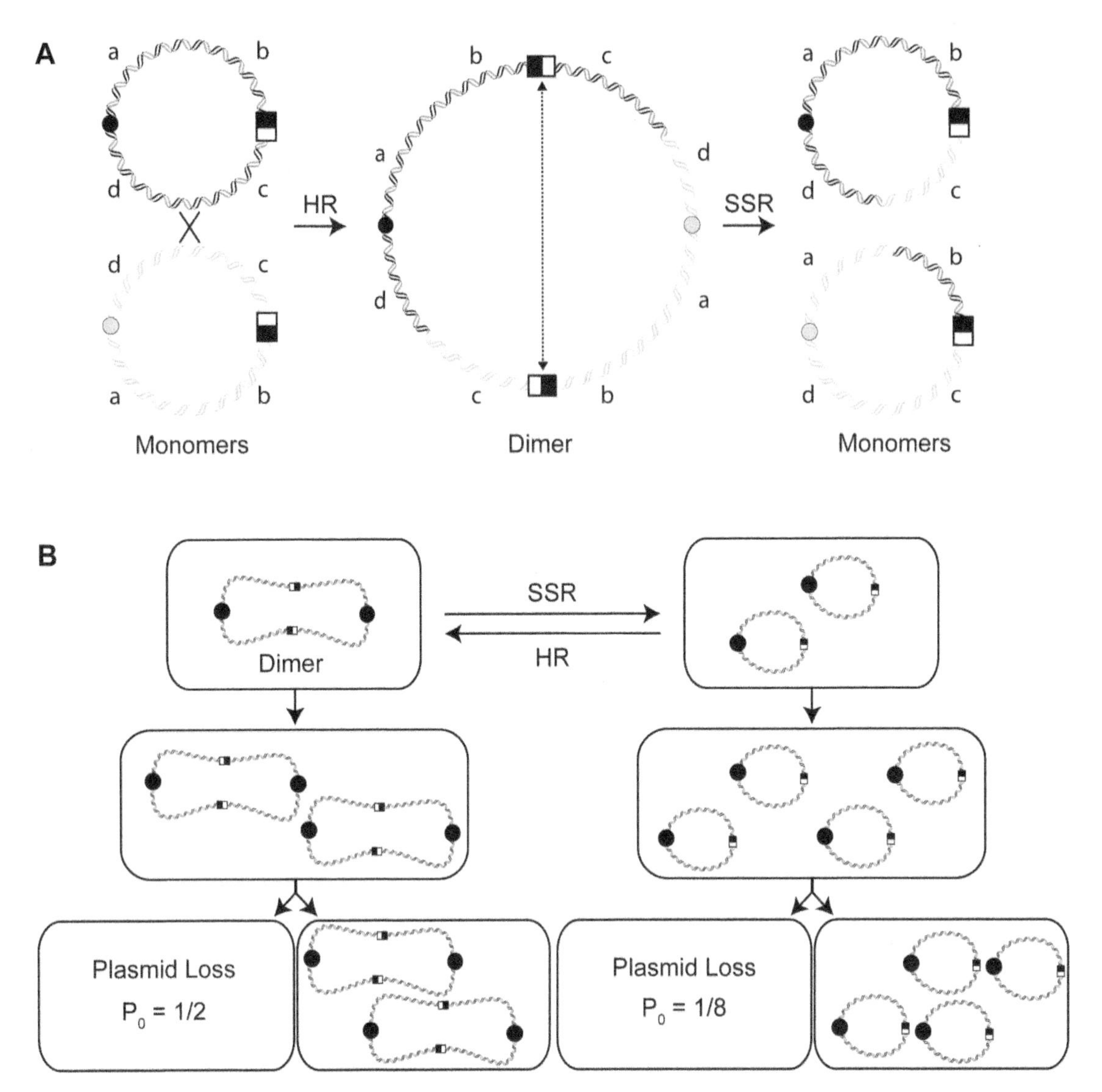

Figure 1 Formation of plasmid dimers and their stability in the cell. (**A**) Dimers of plasmids or chromosomes are formed by homologous recombination (HR) during replication and are later resolved by site-specific recombination (SSR). Origins are shown by a circle, and site-specific recombination sites by a black and white square. (**B**) The accumulation of plasmid dimers in the cell leads to an increase in plasmid loss compared to monomeric plasmids. doi:10.1128/microbiolspec.PLAS-0025-2014.f1

recombination of small DNA molecules. To our knowledge, similar studies have not been performed on large low-copy-number plasmids, although these plasmids harbor multimer resolution systems whose inactivation causes a defect in plasmid stability.

The net result of homologous recombination between sister plasmids is the production of multimers (Fig. 1A). However, the formation of multimers may have different effects depending on the mode of replication and partition of the replicon. Unresolved chromosome dimers are certainly lethal in most cases and are broken in a desperate effort of resolution due to the so-called guillotine effect. Plasmid multimers affect plasmid stability by lowering the number of independently segregating units at the time of cell division. They may also interfere with the control of replication and with the activity of the plasmid segregation system. High-copy-number plasmids segregate randomly so that the probability of generating plasmid-free cells is inversely proportional to their copy number at division time (4, 5) (Fig. 1B). Replication of these plasmids is controlled by origin counting, which means that a dimer counts as two plasmids for replication but only as a single unit for segregation. Thus, the formation of multimers lowers the number of freely segregating DNA molecules per origin, thereby raising the frequency of plasmid loss (4, 6, 7) (Fig. 1B). In addition, replication of high-copy-number plasmids follows a random copy choice mode so that a dimer replicates twice as frequently as a monomer. This replicative advantage of multimers causes their rapid accumulation in the progeny of the cells in which they appeared. This phenomenon, called the dimer catastrophe, is responsible for the largest part of the segregation defect due to plasmid multimerization since it leads to the formation of a subpopulation of cells that mostly contain multimers (4, 6, 7).

Although they are likely to have adverse effects on partition systems, multimers of actively partitioned replicons such as F or P1 have complex and poorly understood effects since the stability of these plasmids does not solely depend on the number of independently segregating units at cell division (for review see reference 8).

DNA SITE-SPECIFIC RECOMBINATION: THE GOOD RESOLUTION OF PLASMIDS AND CHROMOSOMES

Conservative DNA site-specific recombination is a carefully orchestrated reaction during which four DNA strands are broken, exchanged, and resealed to equivalent positions of separate sequences (see below). This is mediated by relatively simple molecular machines in which specialized enzymes, termed site-specific recombinases, catalyze the essential DNA breakage and joining reactions. The recombination reaction can lead to the integration, excision, or inversion of a DNA fragment, depending on the relative positioning of the recombination sites. However, recombination between directly repeated sites on a circular DNA molecule leads to deletion of the intervening DNA segment.

Besides its role in the stable inheritance of circular plasmids and chromosomes, site-specific recombination is exploited in a range of programmed DNA rearrangements in bacteria including integration and excision of temperate bacteriophages into and out of the genome of their host, the movement of different classes of mobile genetic elements (e.g., transposons, integrons, insertion sequences, and integrating conjugative elements [ICE]), the variable expression of virulence genes in pathogens (by means of simple or combinatorial DNA inversion switches), and the control of developmentally regulated genes (9–13).

Recombinases that mediate these different rearrangements, including dimer resolution, are often termed resolvases, integrases, transposases, or DNA invertases to designate the type of reaction they catalyze. These enzymes fall into two major families of unrelated proteins using different mechanisms to cleave and rejoin DNA molecules. These two groups of enzymes are now commonly referred to as the serine recombinase family and the tyrosine recombinase family according to the conserved residue that provides the primary nucleophile in the DNA cleavage reaction (1). Recombinases of both families mediate recombination at the level of short (~30 base pairs) DNA segments termed the core or crossover site onto which two recombinase molecules bind, usually by recognizing specific sequences with dyad symmetry. The recombinase recognition motifs are separated by a central region at the borders of which the DNA strands are cut and exchanged by the protein (see below and Figs. 2 and 3). With a few exceptions that will be outlined below, this minimal core site is usually insufficient to mediate recombination. The recombination sites of most characterized systems have a more complex organization, with additional binding sites for accessory proteins. These accessory sequences and proteins are used to control the recombination reaction, allowing recombination systems to achieve their biological function without generating undesirable and potentially deleterious DNA rearrangements. For plasmid and chromosome resolution systems this control is important to convert multimers to monomers and not vice versa.

Plasmid Resolution Systems of the Serine Recombinase Family

Recombinases belonging to the serine recombinase family are present on a number of plasmids from both Gram-negative and Gram-positive bacteria, but only a few of them have been shown to contribute to plasmid maintenance by converting multimers into monomers. Here, we report several characterized plasmid resolution systems of the serine recombinase family, most of them being clearly derived, during recent evolution, from the γδ and Tn3 resolvases from replicative transposons (for review see references 14 and 15).

Genetic Organization of Serine Recombinase-Catalyzed Resolution Systems

RK2/RP4 resolution systems: Tn3 and γδ recombinases

An example of an integrated plasmid resolution system is provided by the ParA/res system of the IncP-1α plasmid RK2 (identical to RP4) (15, 17). Plasmids of this family have a relatively low copy number (between five

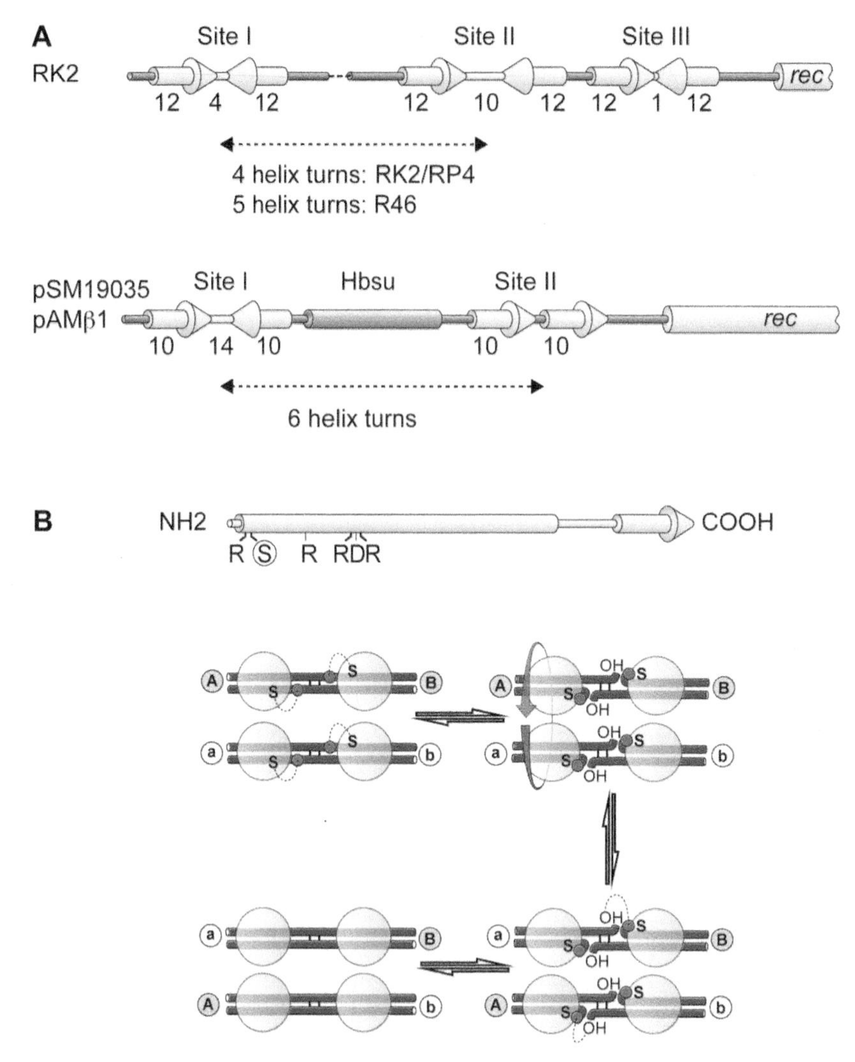

Figure 2 Site-specific recombination by serine recombinases. (**A**) Structure of the recombination sites of Tn3 and γδ recombinases (top) and the β recombinase and Sin family (bottom). Sequences are shown with arrows, and their length is (bp) indicated nearby. The genes coding for the recombinases are located downstream (rec). Hbsu (red) depicts a binding site for the accessory protein. (**B**) The primary structure of a typical serine recombinase (top), showing the catalytic serine in a circle; other amino acids implicated in catalysis are also indicated (bottom). The scheme depicts the four principal steps of recombination by serine recombinase: four recombinases bind on their sites and form the synapse; the four strands are cut and bound covalently to the catalytic serines (red dot), which is followed by a 180° rotation of two of the subunits and rejoining of the DNA strands.
doi:10.1128/microbiolspec.PLAS-0025-2014.f2

and eight copies per chromosome) and can be stably propagated in a wide range of Gram-negative bacteria (17). Important determinants of this stability are encoded in the 3.2-kb *par* locus of the plasmid. This locus is comprised of two divergently transcribed operons (*parCBA* and *parDE*), separated by a short intergenic region of about 180 bp. The *parDE* operon encodes a post-segregational killing system (toxin-antitoxin system) homologous to those found in other plasmids and in the chromosome of various bacterial species (19, 20). In addition to the serine recombinase ParA, the *parCBA* operon encodes two other proteins that have no apparent role in site-specific recombination. This region also contains the recombination site at which the ParA recombinase acts to resolve plasmid multimers (15, 20, 21). This site, designated RK2 *res*, has an

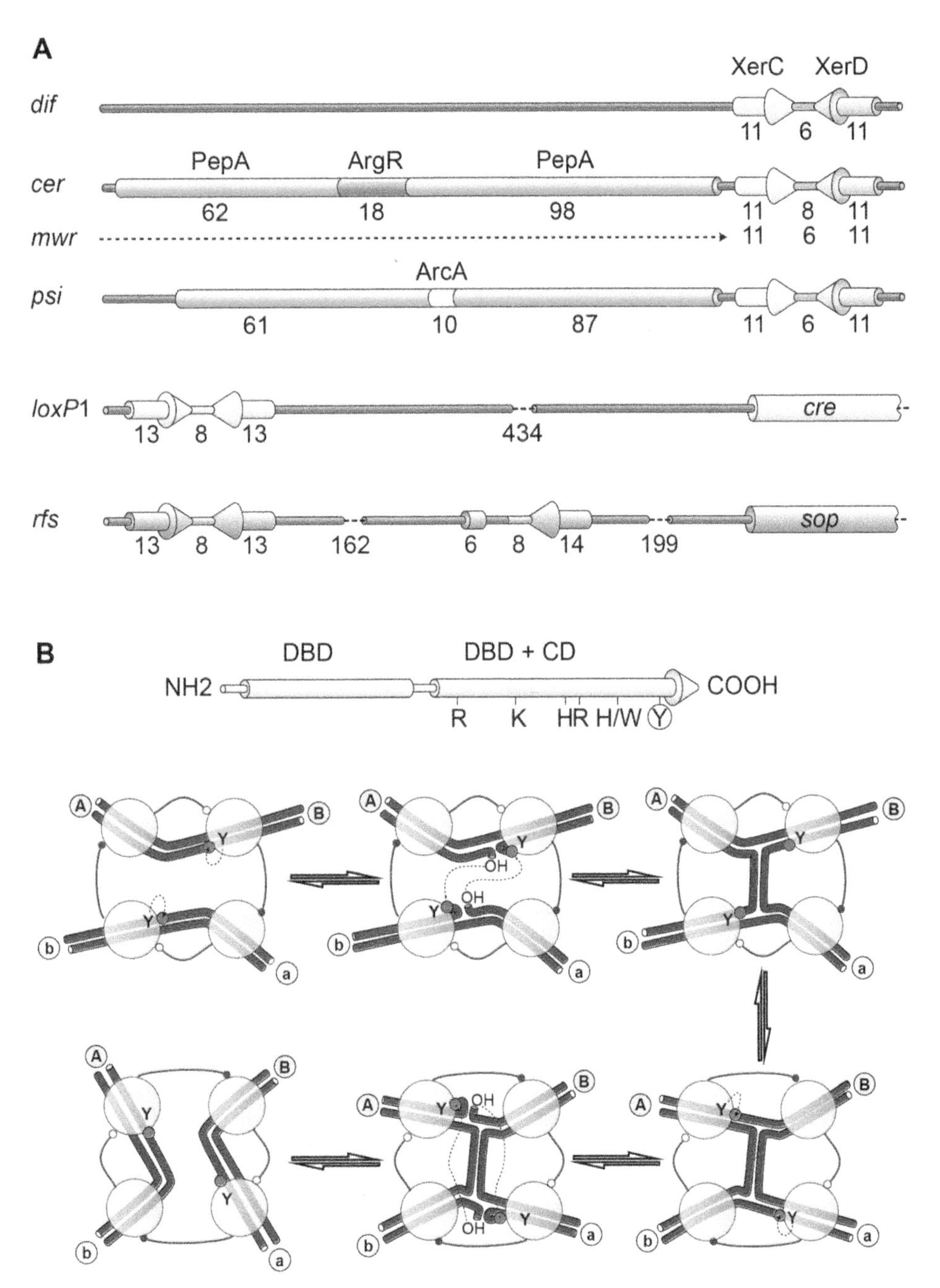

Figure 3 Site-specific recombination by tyrosine recombinases. (**A**) Structure of principal recombination sites from tyrosine recombinase family. Symbols are the same as in Fig. 2. (**B**) The primary structure of a typical tyrosine recombinase (top), showing the catalytic tyrosine in a circle; other amino-acids implicated in catalysis are also indicated. Two distinct protein domains are shown: the DNA binding domain (DBD) and another DNA binding domain containing the catalytic domain (DBD + CD) (bottom). The scheme depicts the principal steps of recombination by tyrosine recombinase: four recombinases bind on their sites and form the synapse; cleavage occurs first on only two DNA strands, catalyzed by one active pair of recombinases in the teramer (XerC or XerD in the case of *dif*, *cer*, *mwr*, and *psi*). A transient Holliday junction is formed after a first strand exchange and is isomerized to allow the cleavage of the two other strands by the other pair of recombinases. The second strand exchange takes place, and the complex dissociates.
doi:10.1128/microbiolspec.PLAS-0025-2014.f3

organization similar to that of many Tn3-family transposon *res* sites, with three inverted repeats being bound by the recombinase (17) (Fig. 2A).

The ParA/*res* recombination system significantly contributes to the segregational stability of the RK2 plasmid, but recombination was insufficient to account for this contribution. Resolution of dimers may thus not be the only function of the ParA/*res* system (21). In addition, its replacement by another multimer resolution system (i.e., the *Escherichia coli* Xer/*cer* system or the bacteriophage P1 Cre/*loxP* system; see below) only partially complements the defects in plasmid stability (22). Together, these observations suggest that the plasmid resolution function of the Par/*res* system forms a part of a more complex stabilization process involving additional host- and/or plasmid-encoded factors.

pSM19035 and pAMβ1 resolution system: β recombinase and Sin

Large theta-replicating plasmids of Gram-positive bacteria encode different subfamilies of serine recombinases, which together form the β recombinase family. This family is a group of relatively well-conserved proteins that are present on broad-host-range plasmids. The best-studied members of this family are the β protein of the *Streptococcus pyogenes* plasmid pSM19035 and the pAMβ1 Resβ recombinase from *Enterococcus faecalis* (for review see reference 23). These proteins are ~40% identical to Sin, the prototype of a second group of highly homologous recombinases found on large staphylococcal plasmids (2, 24, 25). Sin and the β recombinases have been characterized at a genetic and biochemical level. The recombination target site of these recombinases exhibits a similar two-subsite structure, with the protein recognizing inverted repeats at the crossover site I and direct repeats at the accessory site II (26, 27) (Fig. 2A). *In vitro* studies have shown that recombination by Sin and the β recombinases also requires the nucleoid-associated protein Hbsu or any equivalent nonspecific DNA-bending protein, such as HU from *E. coli* or the HMG proteins from eukaryotes (27, 28). This protein plays an architectural role in the formation of the synaptic complex by inducing substantial DNA bends in the recombining partners (see below) (27).

Mechanism of Serine Recombinase-Catalyzed Resolution: Concerted Double-Strand Breaks and Rotational Exchange of the Recombination Half-Sites

Serine recombinases are characterized by the presence of a relatively well-conserved catalytic domain of about 120 amino acid residues. This domain contains the catalytic serine and several other conserved residues clustered into specific motifs (Fig. 2B). Some of these residues are part of the active site pocket, whereas others are involved in protein-protein interactions between separate recombinase molecules in the recombination complex (29–32).

Although functional and structural diversity among serine recombinases is likely to reflect important variations in the molecular organization of the recombination complex, all the family members are thought to mediate recombination by using the same strand exchange mechanism (27, 30) (Fig. 2B). Recombination by serine recombinase is a concerted process in which all four DNA strands of the two recombination sites are cut, exchanged, and then rejoined in the recombinant configuration (for review see reference 1). The reaction starts when the active serine of each of the four recombinase molecules in the complex attacks the phosphodiester bond adjacent to its binding site. Cleavage of each duplex is staggered by two base pairs, generating protruding 3′-OH overhangs and recessed 5′ ends to which the recombinase is attached through a covalent phosphoseryl bond. Catalysis of these reactions is highly coordinated, and the complex in which the four cleaved half-sites are solely held together by protein-DNA and protein-protein interactions is an obligate intermediate in the recombination pathway (33–35). Topological changes occurring during recombination mediated by serine recombinase are consistent with the DNA strands being exchanged by a simple 180° right-handed rotation of one pair of cleaved ends relative to the other. In this mechanism, the extended two bases that are exchanged between the two duplexes must be complementary to allow base pairing between rejoining ends. Thus, the two base-pair overlap regions define the polarity of the recombination site (Fig. 2B).

Crystallographic studies on the resolvase of γδ, together with the biochemical characterization of specific recombinase mutants, have provided important insights on the molecular organization of serine recombinases and the protein-protein and protein-DNA interactions within the recombination complex supporting a detailed molecular model for the 180° rotation (36–39). The arrangement of the active site residues in the resolvase dimer is consistent with the previous finding that the recombinase cleaves the DNA in *cis*, by acting at the nearest position from its binding site (33). In this model, the strand exchange is coupled to a rotational rearrangement of the DNA-linked recombinase subunits within the tetramer (39). This mechanism requires that the dimer interface that holds the cleaved half-sites in the complex is transiently disrupted during the dissociation/reassociation process. To make this possible, a flat and hydrophobic interface is created within the tetramer, which permits the rotation of the synapsed subunits relative to each other (29, 39). This class of mechanism is supported by data suggesting that the DNA strands are at the outside of the synaptic complex and are thus too far apart to be exchanged without involving substantial motions of the proteins (30, 37). However, based on homology between serine recombinases and topoisomerases, an alternative model, without subunit rotation, has been recently proposed and is still under debate (40).

Plasmid Resolution Systems of the Tyrosine Recombinase Family

Genetic Organization of Tyrosine Recombinase-Catalyzed Resolution Systems

Site-specific resolution systems that function with a tyrosine recombinase exhibit variable levels of complexity. Large conjugative plasmids and replicative prophages generally encode their own recombinase adjacent to the recombination site, allowing them to be transferred with a fully functional resolution system among different bacteria. In contrast, small plasmids, such as those of the ColE1 family, utilize the chromosome dimer resolution system of their host. The recombination site of these different systems may be limited to a simple crossover sequence, as in the bacteriophage P1 Cre/*loxP* system, or may be more complex, containing additional DNA binding sites for regulatory proteins (Fig. 3A).

Xer: a universal and multipurpose recombination system for the resolution of replicon dimers

Xer recombination provides a good example illustrating how relatively simply DNA site-specific recombination mechanisms can be adapted to accomplish different biological functions. The Xer recombinase is unusual in that it functions as a heterotetramer comprising two proteins of the tyrosine recombinase family, XerC and XerD (41, 42). Homologues of these two proteins are found in the genome of virtually all bacteria and archaea harboring circular chromosomes, consistent with Xer recombination having an important and conserved function in chromosome segregation (43–46). The *xerC* and *xerD* genes are generally found in separate regions of the genome where they are sometimes associated with other genes involved in DNA repair and recombination (47). They were first demonstrated to increase the stability of naturally occurring multicopy plasmids, such as ColE1, by converting multimers into monomers (5, 48). This function was shown to require the presence of a specific site termed *cer* in ColE1. Related recombination sites were subsequently identified in a number of *E. coli* plasmids (49), as well as in pSC101 (*psi*) from *Salmonella typhimurium* (50) and pJHCMW1 (*mwr*) from *Klebsiella pneumoniae* (51, 52). It was later found that the primary function of the Xer system is to resolve dimers of the chromosome by acting at the *dif* site, located in the replication terminus region of the *E. coli* chromosome (see below) (53–55).

The different target sites for Xer recombination share a conserved ~30-bp core sequence containing 11-bp XerC and XerD recognition motifs separated by a central region of 6 to 8 bp (53, 56). In addition to XerC and XerD, recombination at plasmid resolution sites depends on additional host-encoded proteins and on the presence of ~160 to 180 bp of accessory sequences adjacent to the core (Fig. 3A). Interactions between the accessory sequences and proteins are important to control the outcome of recombination, ensuring that it will convert plasmid multimers to monomers and not the converse (see below and Fig. 4) (for a recent review see reference 57). A special section is devoted to the XerCD-*dif* chromosome dimer resolution (see below and Fig. 5).

Individual resolution systems: Cre/*loxP* of P1 and ResD/*rfs* of F

The Cre/*loxP* recombination system of *E. coli* bacteriophage P1 was discovered for its role in the circularization of the infecting phage genome by performing a recombination reaction between the terminally redundant P1 DNA ends (58–60). The recombination system was subsequently shown to contribute to the proper segregation of the prophage (61). In P1 derivatives lacking either the *loxP* site or a functional *cre* gene, products were lost 20 to 40 times more frequently than in the wild-type lysogenic form of the phage at each generation. This segregational defect disappeared when plasmids were propagated in a $RecA^-$ strain, indicating that it resulted from the formation of dimers and higher multimeric forms by homologous recombination. This finding provided the first demonstration that site-specific recombination can be used to counteract damaging effects caused by homologous recombination on circular replicons (61).

The Cre recombinase is distantly related to the integrase Int of bacteriophage λ, the archetype of the tyrosine recombinase family. This is consistent with

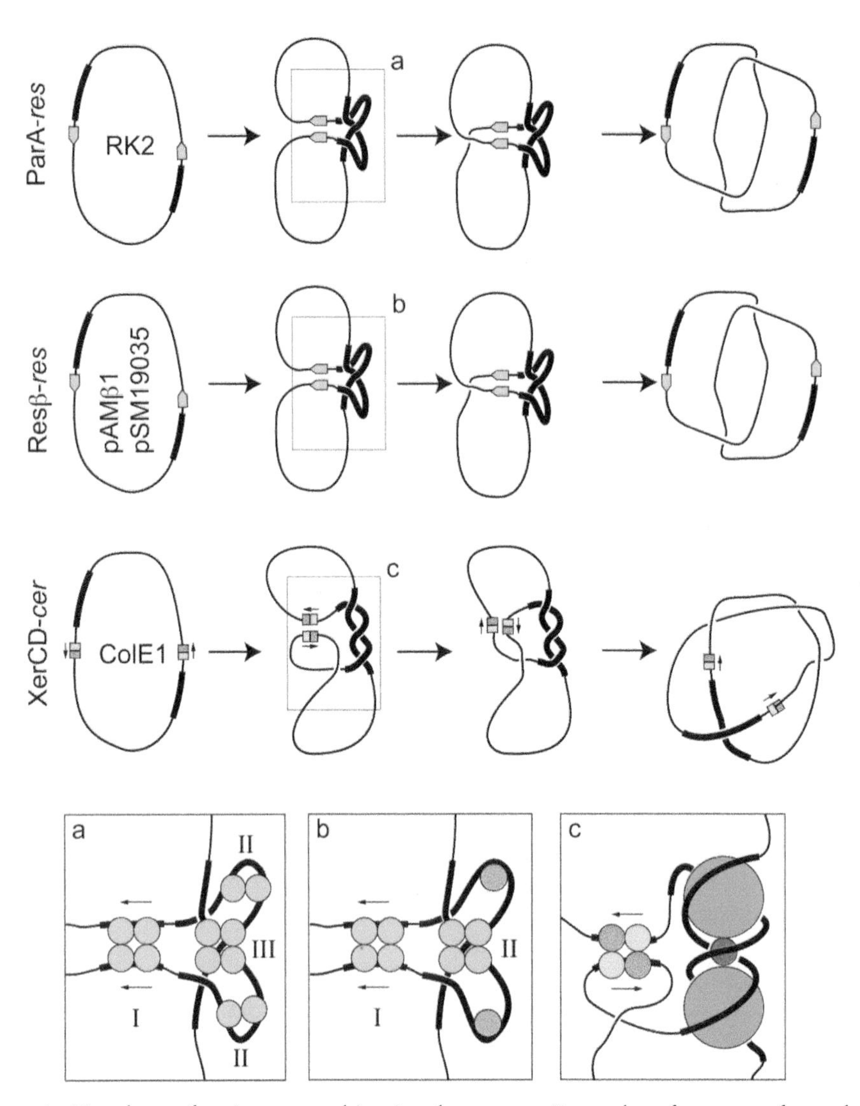

Figure 4 Topology of various recombinational synapses. Examples of synapses formed during recombination by two main families of serine recombinase (**a**, **b**) and one of tyrosine recombinase (**c**) are shown. Main recombination sites are colored (blue and green/purple), and the accessory sequences are shown as a thick line. The precise configuration of the synapses are drawn below, with (**a**, **b**) serine recombinase in blue, Hbsu in orange, (**c**) XerC and Xer in green and purple, PepA in dark green, and ArgR in red.
doi:10.1128/microbiolspec.PLAS-0025-2014.f4

biochemical data showing that both proteins use the same mechanism to cut and rejoin DNA molecules (for reviews see references 1, 62). The *loxP* site, which is located 434 bp upstream of the *cre* gene, has the minimal structure of a recombination core site, with two inversely oriented 13-bp recombinase-binding motifs flanking an asymmetrical sequence of 8 bp (Fig. 3A). Both *in vivo* and *in vitro* studies have demonstrated that this 34-bp *loxP* sequence is a sufficient substrate for Cre (63, 64). In addition, in contrast to most recombinases, Cre shows no strong preference for a particular arrangement of the recombination sites (direct repeat sites lead to deletion of the intervening DNA, whereas inverted repeat sites lead to its inversion) and has little requirement regarding the topology of the DNA, at least *in vitro* (65, 66). The reaction works equally well on supercoiled, relaxed, or linear DNA molecules, generating all possible intra- and intermolecular recombination products. This apparent lack of selectivity is consistent with the fact that the Cre/*loxP* system is required to mediate functionally distinct DNA rearrangements at separate stages of P1 development.

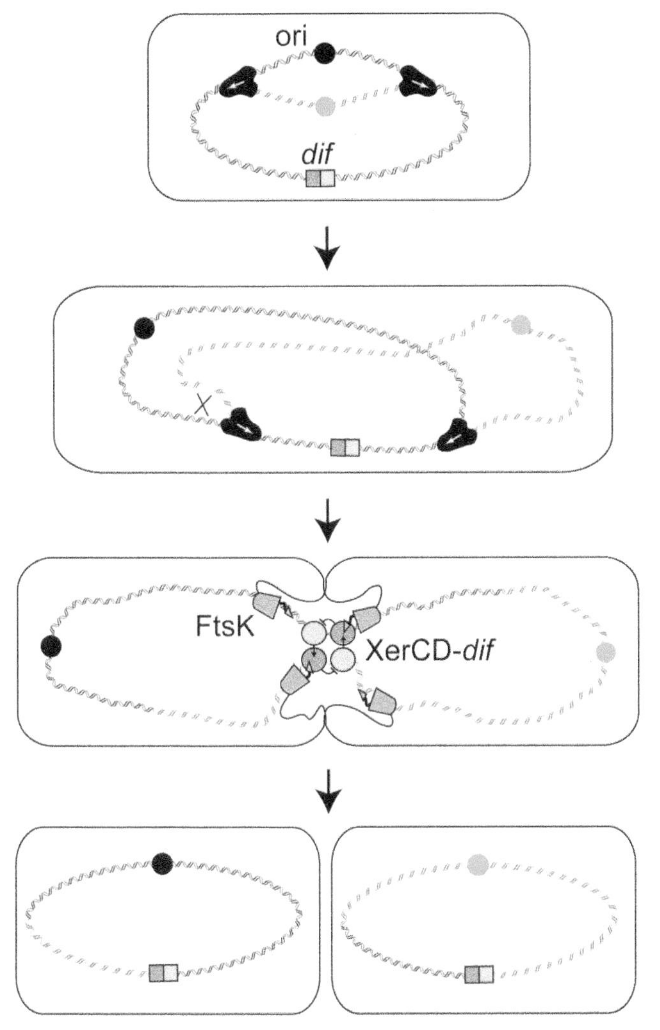

Figure 5 Resolution of chromosome dimers. During chromosome replication, single- or double-strand breaks appear that necessitate homologous recombination to be repaired. This leads to the formation of dimers of chromosomes, which are resolved by XerC (green) and XerD (purple) after the activation of XerD by FtsK (orange), localized at the septum. Cell division can then take place and maintain two integrate copies of the chromosome. Origins of replication are represented as green and black dots, replication forks as black complexes with white arrows indicating replication direction, and the *dif* site as a green and purple square.
doi:10.1128/microbiolspec.PLAS-0025-2014.f5

The relative simplicity of this system has also greatly facilitated biochemical and structural studies aiming at elucidating the molecular recombination mechanism of tyrosine recombinases (for recent reviews see references 1, 62).

The ResD/*rfs* resolution system of the *E. coli* F factor is much less well described. It is encoded within the RepFIA replication region of the plasmid. The recombinase gene, *resD*, is cotranscribed with ccdB and ccdA, the toxin and antitoxin of the plasmid post-segregational killing system (19). The RepFIA region also contains the *sopAB* operon that encodes the centromere-like partitioning system of F (67). The target site for ResD is the *rfs* locus located upstream of the *ccd* operon, in a DNA segment that overlaps with the *oriV1* replication origin of F. The fully functional resolution site of F is ill defined, but it seems to have a more complex organization than that described for Cre (Fig. 3A). Deletion analysis showed that recombination requires additional DNA sequences extending up to 200 bp away from the *rfs* crossover site. ResD was found to protect two discrete regions in this sequence, although binding to the core inverted repeats was undetectable (68, 69).

Multimer resolution systems, which are closely related to the F plasmid ResD/*rfsF* system, have been shown to contribute to the stable inheritance of other conjugative plasmids, but very little is known about them (2).

Mechanism of Tyrosine Recombinase-Catalyzed Resolution

Tyrosine recombinases differ widely in amino acid sequence. However, the ternary structure of the catalytic domain is remarkably conserved among different members of the family (70–75). The catalytic signature of the family is composed of six residues, RKHRH/WY, which are clustered in specific regions of the C-terminal catalytic domain of the proteins (Fig. 3B) (1, 76).

The crystal structures of protein-DNA complexes reported for Cre and Flp, together with biochemical studies of different recombination systems, have given conclusive information on how the recombination complex is organized and how it works (Fig. 3B) (1). The assembly of the synapse is crucial for the regulation of the reaction (see below and references 65, 66) within this complex; the two recombination sites are aligned in an antiparallel conformation, and the DNA is bent in a roughly square planar configuration. It has been proposed that the DNA is bent in such a way that only one pair of recombinases is ready to cut DNA. Indeed, DNA bending "hides" two of the four scissile phosphate groups, one in each DNA duplex, allowing nucleophilic attack by catalytic tyrosine residues on the two other scissile phosphates (Fig. 3B; red circles). This creates a 3′-phosphotyrosyl covalent link between the DNA and the protein and liberates a 5′-OH end at the cleaved strand. DNA strands are then exchanged by melting three or four nucleotides from the parental duplexes before re-annealing them to the complementary bases in the partner (Fig. 3B). Watson-Crick base-pairing between the cleaved and uncleaved DNA strands tests the homology between the two recombination sites and helps to orient the invading 5′-OH ends for ligation (77). This ligation is the result of the nucleophilic attack on the 3′ phosphotyrosyl group by the partner 5′-OH end after strand exchange and forms a first Holliday junction intermediate (HJ1). HJ1 isomerizes into a second Holliday junction intermediate, HJ2, within which the second pair of recombinases is activated (66, 78–80). The resolution of HJ2 into the recombination product is performed via a second cycle of cleavage, strand exchange, and ligation reactions (Fig. 3B) (for a more detailed description see reference 1).

The choice of the first active pair of recombinases is crucial for the control of the reaction. It is thought that this choice is made during the assembly of the synapse (66, 81). Within the Cre-*loxP* synapse, the four recombinase molecules forming the tetramer are held together by a cyclic network of protein-protein interactions involving both the C and N-terminal domains of the protein. This cyclical donor-acceptor interaction between Cre monomers connects the four recombinase active sites in the tetramer, providing a means of communication between them (71, 72). Deviation from the perfect 4-fold symmetry in the Cre-DNA complexes provides a simple model for the isomerization mechanism that leads to sequential activation (and reciprocal inactivation) of pairs of recombinase subunits during a complete recombination reaction. In this mechanism, which is also true for other tyrosine recombinases such as λint, small readjustments of the angles formed by the arms of the Holliday junction, coupled with little conformational changes in the C-terminal donor-acceptor interaction, suffice to explain the reciprocal activation/inactivation between pairs of recombinases (65, 66). Accordingly, in the case of Cre, this isomerization step recently appeared to be very rapid (79, 80, 82).

For XerCD-*dif* recombination, the synapse contains two units of XerC and two units of XerD, and the two pairs, XerC or XerD, are active sequentially (81, 83–85). In this case the model for the isomerization mechanism is supported by experiments showing that alteration of the recombinase C-terminus affects catalysis by the partner recombinase (84, 86).

A remarkably similar organization of the recombination complex is observed for Flp, although the N-terminal domains have totally different structures (87). A fundamental difference between Cre and Flp lies in the nature of the allosteric interactions that regulate the recombinase activity. In the Flp-DNA structures, the α-helix carrying the tyrosine nucleophile of one monomer is donated to the active site of an adjacent monomer of the complex. This active site sharing mechanism of Flp and the allosteric donor-acceptor interaction proposed for Cre and other cis-acting recombinases represent alternative, but functionally equivalent, strategies for coordinating catalysis in the recombination complex.

SYNAPTOSOME AND TOPOLOGY

Convergent Mechanisms To Impose Resolution Selectivity on the Recombination Reaction

As mentioned above, for most site-specific recombination systems, the minimal core site at which the cognate recombinase acts to catalyze the strand exchange reaction is usually insufficient to carry out normal recombination at a normal frequency. The target sites of these systems contain additional accessory sequences to which further recombinase molecules or other auxiliary proteins bind, thereby forming part of the functional recombination complex. Formation of this complex may be needed to facilitate the pairing of the core sites and/or to activate the recombinase catalytic activity within the tetramer. The specific architecture of the recombination complex provides a means of controlling the reaction, ensuring that recombination will only occur at the correct time or between appropriately positioned recombination sites. It is crucial for the biological function of site-specific resolution systems to selectively carry out recombination between directly repeated copies of the recombination site in order to prevent the formation of multimers and to avoid other undesirable DNA rearrangements. In these systems, discrimination between different arrangements of the recombination sites is achieved by imposing a specific local DNA geometry within the synaptic complex. Requirements for this complex act as a topological filter because it cannot readily assemble if the two recombination sites are in a wrong configuration or on separate DNA molecules. The recombination products of these systems are characterized by a single topology that is dictated both by the topology of the initial synapse and by the strand exchange mechanism used by the recombinases (for a comprehensive discussion on the mechanisms of topological selectivity, see reference 57).

A Common Synapse Architecture for Resolvases of the Serine Recombinase Family

The mechanism of topological selectivity was first established for the resolution systems of replicative transposons belonging to the Tn3 family (88). The resolvase is the only protein required to resolve cointegrates *in vitro*. Recombination only takes place if two complete copies of the transposon resolution *res* sites are present in an appropriate head-to-tail orientation on the same supercoiled DNA molecule. The major products of the reaction are two molecules interlinked twice (four-node catenane). This observation, combined with a number of biochemical and topological data from different laboratories, led to the conclusion that resolvase binding to the three subsites of *res* results in the formation of a specific synaptic complex, termed the synaptosome, in which the two recombination sites are interwrapped in a structure trapping three negative supercoils from the initial DNA substrate (Fig. 4). Assembly of this complex is a stepwise process that initiates with the antiparallel pairing of the accessory sites II and III of the two *res* sites. DNA wrapping around the resolvase accessory subunits acts as a checkpoint (i.e., topological filter) that dictates whether the reaction can proceed further by correctly positioning the subsite I-bound resolvase subunits for the strand exchange reaction (36, 88).

In the most recent model for the Tn3 synaptosome, the recombination sites are wrapped around a pair of interlocked protein filaments comprised of six resolvase dimers, each dimer being bound to a different *res* subsite (36) (Fig. 4). The site III-bound dimer forms the central unit of each filament, making equivalent contacts with its two neighbors. An important implication of this model is that the crossover sites lie at the outside of the protein core formed by the recombinase catalytic domains, which is consistent with the view that there must be substantial structural rearrangements within the complex to carry out strand exchange.

The topology of the recombination reaction catalyzed by other resolvases of the serine recombinase family, such as the ParA protein of RP4/RK2 (17), the Sin recombinase of *S. aureus* (31, 89) and the β recombinase of pSM19035 (90) (Fig. 4), was found to be same as that reported for the cointegrate resolution system of Tn3 and γδ, generating four-node catenane products. Consistently, a similar overall architecture of the synaptic complex has been proposed for the plasmid resolvases Sin and β (31, 35, 37, 90). This arrangement thus represents a common structural unit in the synaptic complex of different resolution systems. Hbsu plays an important role in the formation of the complex by stabilizing the DNA bends between the two regions that are bound by the recombinase (Fig. 4). Note that in the case of the β recombinase, alternative configurations of the synapse were proposed to account for its ability to mediate DNA inversion under certain circumstances (90).

Topological Selectivity in Xer Recombination of the Tyrosine Recombinase Family

To satisfy its dual role in chromosome and plasmid segregation, the Xer system has evolved separate mechanisms to control recombination at the chromosomal site *dif*, or at plasmid resolution sites such as *cer* and *psi* (for review see reference 13). In the case of *cer* and *psi*, binding of the accessory proteins (i.e., PepA and ArgR, and PepA and ArcA, respectively) to the recombination site accessory sequences promotes the assembly of a topologically defined synaptic complex in which, as in the resolvase synaptosome, three negative DNA supercoils are trapped (Fig. 4). A fourth node is introduced to align the recombination core sites in an antiparallel configuration. Consequently, recombination between directly repeated *psi* sites produces a four-noded catenane (Fig. 4). Recombination at *cer* follows the same pathway but stops after the first strand exchange, generating a catenated Holliday junction-containing molecule that is resolved to products by a Xer-independent mechanism *in vivo* (for review see reference 57).

Based on genetic, structural, and topological data, a model of the *cer* synaptic complex was proposed (Fig. 4) (91–93). In this model, one ArgR hexamer is sandwiched between two PepA hexamers, and the recombination sites wrap around the accessory proteins as a right-handed superhelix. The ArgR hexamer bridges the two recombination sites by binding to the ArgR box from either partner, whereas PepA specifies the topology of the complex by directing the DNA across three large grooves running from the lower face to the upper face of the hexamer. Recombination at *psi* is thought to occur within a similar synaptic structure, in which an oligomer of ArcA would take the place of ArgR (57). Formation of the synaptic complex at *psi* was found to position the recombinases in a specific configuration so as to activate the XerC protomers for the first strand exchange reaction (94).

A similar topological filter mechanism was more recently found to control the recombination reaction catalyzed by the tyrosine recombinase TnpI from the Tn3-family transposon Tn*4430*, providing another example of convergent strategies to mediate selectivity in plasmid multimers and transposition cointegrate resolution systems (95, 96). However, in the case of the Tn*4430* resolution system, additional subunits of the TnpI protein itself provide the architectural regulatory elements of the recombination complex without requiring any other host- or transposon-encoded proteins. In addition, recombination by TnpI produces two-noded catenane products instead of four-noded catenanes as observed for Xer, indicating that the topological organization of the recombination complexes differ between the two systems (95, 96).

RESOLUTION OF CHROMOSOME DIMERS

E. coli Model

The best-studied case of multimer formation is dimerization of circular chromosomes, and most of the work has been done in *E. coli*. Ten to fifteen percent of the chromosomes require Xer recombination for correct segregation (97, 98). This requirement is suppressed in a RecA-deficient strain, consistent with the view that the vast majority of chromosome dimers form by homologous recombination (97). It is generally admitted that most recombination events are a consequence of recombinational repair of stalled, broken, or collapsed replication forks. Collapse or processing of stalled forks creates DNA ends that need recombination to be resealed with the circular molecule, allowing replication to restart (99–101). In the case of an odd number of these events, this leads to the formation of a dimer of chromosomes. It is believed that unresolved dimers are trapped by the cell septum at division, leading to chromosome breakage by an unknown mechanism (100).

To avoid random cutting of its chromosome, *E. coli* uses the tyrosine recombinases XerC and XerD, but with a different control than for plasmid dimer resolution. The recombination site, called *dif*, possesses the same characteristics as a plasmid core recombination site (Figs. 3A and 5), with oppositely oriented XerC and XerD recognition elements separated by a 6-bp central region (Fig. 3A). However, the mechanism that controls recombination at the chromosome site *dif* and the plasmid resolution sites differs significantly. There is no need for accessory sequences around *dif*. Instead, the reaction is under the control of the DNA-translocase FtsK. This ATP-dependent translocase is anchored at the septum through its N-terminal transmembrane domain, which interacts with the early proteins of the divisome. This restricts FtsK action to a specific cellular location, the septum, and a short period of time following replication (97, 98, 102). Septal localization of FtsK is also important to recruit late divisome proteins that are required for septum completion and closure (103). Analogues of FtsK are found in almost all bacteria sequenced so far (104). The main ones studied are SpoIIIE and SftA from *Bacillus* (105, 106), but their role in site-specific recombination is still unclear (107). Recombination activation involves direct and specific interactions between the very C-terminal domain of FtsK, called the γ-domain, and XerD bound to *dif* (Fig. 5) (108–110). When activated, XerD cuts and exchanges the first pair of strands, and the resulting Holliday junction intermediate is then resolved by XerC.

Finding two *dif* sites within a dimer of chromosomes might not be an easy task for FtsK. In addition to interacting with XerD, the γ-domain of FtsK also recognizes 8-bp oriented sequences on DNA, called KOPS (for FtsK oriented polar sequences), which point toward the *dif* site (105, 111). FtsK recognizes these sequences in an oriented manner, and therefore translocates toward the direction of *dif* (112, 113). During its high-speed translocation (up to 17 kb/s), FtsK removes any protein bound to DNA (114, 115). It only stops when it reaches a *dif*-XerCD complex, but it is still unclear whether FtsK contributes to synapse assembly by recruiting another XerCD-*dif* complex or whether the synapse is already formed at this point (81, 85, 114).

It has been long questioned whether the translocase activity of FtsK could play a more general role in chromosome segregation, since the presence of the C-terminal motor does not appear to be essential for proper cell division (103). Recent results have shown that FtsK does indeed participate in chromosome segregation, by ordering segregation of the terminal domain (116), independently of the monomeric or dimeric state of the replicated chromosomes. Therefore, FtsK ensures that the last genetic locus to remain at the septum is *dif*, since the KOPS converge at this site.

Another remaining question concerns the ability of the FtsK/XerCD system to resolve dimers by generating only free products (not catenated). Experiments done on plasmids containing two *dif* sites show that the γ-domain of FtsK (free or linked to XerC or XerD) is sufficient to activate recombination. However, such reactions lead to products with complex topologies (catenated or knotted). These topologically complex products never arise when the reaction is carried out with a full-length FtsK motor (110). FtsK thus either prevents the formation of interlinked products before activating recombination and/or resolves them by further rounds of recombination. This latter hypothesis has been demonstrated *in vitro* and *in vivo* (117, 118).

Multichromosomal Bacteria

The only experimental work on multichromosomal bacteria has been done with *Vibrio cholerae*, which has two chromosomes (119) but only one XerC, XerD, and FtsK. This work revealed that chromosome dimer resolution in *V. cholerae* followed the same mechanism as in *E. coli*. The key point is the difference in the sequences of the two *dif* sites carried by the two chromosomes. The XerD binding motifs of both sites are conserved from other bacterial species, as are most of the XerC ones. The main difference lies in the central region: the central region of *dif1* is highly conserved with respect to the *dif* site of other bacteria, whereas the central region of *dif2* is much more divergent. This is also the case for most of the multichromosomal bacteria. A plausible explanation for this observation is selective pressure to avoid recombinational fusion between two, nonhomologous, chromosomes (43, 44). Despite this, chromosome fusions appear to happen sporadically within populations of *V. cholerae* (120), suggesting that this mechanism is not totally error-proof.

Alternative, Single-Recombinase Xer Systems in Bacteria

Despite the broad conservation of the XerCD system in bacteria, some bacteria have acquired another system that functions with a single recombinase. Examples include the firmicutes *Lactococci* and *Streptococci* and some of the ε-proteobacteria such as *Helicobacter* spp. and *Campylobacter* spp. The XerS/dif_{SL} and XerH/dif_{H} systems acting in the two groups of bacteria (121, 122) function on the same principle as XerCD/*dif*. However, protomers of the same recombinase bind the two arms of the *dif* sites, and the control imposed by FtsK seems to be looser than that described for canonical systems involving two different recombinases (122–124).

Resolution of Dimers in Archaea

Despite their eukaryotic-type machineries for replication and DNA repair, archaea also possess a Xer/dif recombination system to maintain their circular chromosomes. A survey covering about 20 archaeal genomes suggested that archaea possess only one Xer protein, later named XerA (45, 125). The XerA proteins from *Pyrococcus abyssi* and *Sulfolobus solfataricus* have been characterized biochemically (45, 46). Both display recombination activity on their respective *dif* sites without the help of any accessory protein, which is unusual for Xer-type proteins but is consistent with the lack of a FtsK homologue in archaea. It is therefore unclear yet how archaea couple the end of replication with cell division.

DIMER RESOLUTION AND HORIZONTAL GENE TRANSFER

Beyond their demonstrated role in the maintenance of genetic information, site-specific resolution systems have been found to be involved in horizontal gene transfer. As mentioned earlier, Tn3-family transposable elements use a replicative transposition mechanism, which generates a cointegrate structure made of the fusion of the donor and target DNA molecules. Resolution of this cointegrate by intramolecular-site-specific recombination between the two copies of the element regenerates the initial donor molecule and releases a copy of the target into which the transposon is inserted. For most Tn*3*-family transposons the recombination reaction is catalyzed by a serine recombinase, while in the case of Tn*4430* and a few other elements, resolution of the cointegrates is mediated by a tyrosine recombinase (2, 96, 126).

More recently, integrative mobile elements exploiting Xer recombination systems (13) have been described to hijack the chromosomal XerCD-*dif* dimer resolution system. First discovered for their involvement in toxic conversion in *V. cholerae* (127), it appears that integration/excision of these mobile genetic elements depends on XerC and/or XerD (128–130).

Acknowledgment. *Conflicts of interest: We disclose no conflicts.*

Citation. Crozat E, Fournes F, Cornet F, Hallet B, Rousseau P. 2014. Resolution of multimeric forms of circular plasmids and chromosomes. Microbiol Spectrum **2**(5):PLAS-0025-2014.

References

1. **Grindley NDF, Whiteson KL, Rice PA.** 2006. Mechanisms of site-specific recombination. *Annu Rev Genet* **75**:567–605.
2. **Hallet B, Vanhooff V, Cornet F.** 2004. DNA site-specific resolution systems, p 145–178. *In* Funnel BE, Phillips GJ (ed), *Plasmid Biology*. ASM Press, Washington, DC.
3. **James AA, Morrison PT, Kolodner R.** 1982. Genetic recombination of bacterial plasmid DNA. Analysis of the effect of recombination-deficient mutations on plasmid recombination. *J Mol Biol* **160**:411–430.
4. **Summers DK, Beton CW, Withers HL.** 1993. Multicopy plasmid instability: the dimer catastrophe hypothesis. *Mol Microbiol* **8**:1031–1038.
5. **Summers DK, Sherratt DJ.** 1984. Multimerization of high copy number plasmids causes instability: ColE1 encodes a determinant essential for plasmid monomerization and stability. *Cell* **36**:1097–1103.
6. **Hodgman TC, Griffiths H, Summers DK.** 1998. Nucleoprotein architecture and ColE1 dimer resolution: a hypothesis. *Mol Microbiol* **29**:545–558.
7. **Field CM, Summers DK.** 2011. Multicopy plasmid stability: revisiting the dimer catastrophe. *J Theor Biol* **291**:119–127.
8. **Salje J.** 2010. Plasmid segregation: how to survive as an extra piece of DNA. *Crit Rev Biochem Mol Biol* **45**: 296–317.

9. **Hallet B.** 2001. Playing Dr Jekyll and Mr Hyde: combined mechanisms of phase variation in bacteria. *Curr Opin Microbiol* **4:**570–581.
10. **Cambray G, Guerout A-M, Mazel D.** 2010. Integrons. *Annu Rev Genet* **44:**141–166.
11. **Wisniewski-Dyé F, Vial L.** 2008. Phase and antigenic variation mediated by genome modifications. *Antonie Van Leeuwenhoek* **94:**493–515.
12. **Wozniak RAF, Waldor MK.** 2010. Integrative and conjugative elements: mosaic mobile genetic elements enabling dynamic lateral gene flow. *Nat Rev Microbiol* **8:**552–563.
13. **Das B, Martínez E, Midonet C, Barre F-X.** 2013. Integrative mobile elements exploiting Xer recombination. *Trends Microbiol* **21:**23–30.
14. **Cortez D, Quevillon-Cheruel S, Gribaldo S, Desnoues N, Sezonov G, Forterre P, Serre M-C.** 2010. Evidence for a Xer/dif system for chromosome resolution in archaea. *PLoS Genet* **6:**e1001166.
15. **Curcio MJ, Derbyshire KM.** 2003. The outs and ins of transposition: from mu to kangaroo. *Nat Rev Mol Cell Biol* **4:**865–877.
16. **Deleted in proof.**
17. **Eberl L, Kristensen CS, Givskov M, Grohmann E, Gerlitz M, Schwab H.** 1994. Analysis of the multimer resolution system encoded by the parCBA operon of broad-host-range plasmid RP4. *Mol Microbiol* **12:** 131–141.
18. **Deleted in proof.**
19. **Gerdes K.** 2000. Toxin-antitoxin modules may regulate synthesis of macromolecules during nutritional stress. *J Bacteriol* **182:**561–572.
20. **Roberts RC, Ström AR, Helinski DR.** 1994. The parDE operon of the broad-host-range plasmid RK2 specifies growth inhibition associated with plasmid loss. *J Mol Biol* **237:**35–51.
21. **Sobecky PA, Easter CL, Bear PD, Helinski DR.** 1996. Characterization of the stable maintenance properties of the par region of broad-host-range plasmid RK2. *J Bacteriol* **178:**2086–2093.
22. **Easter CL, Sobecky PA, Helinski DR.** 1997. Contribution of different segments of the par region to stable maintenance of the broad-host-range plasmid RK2. *J Bacteriol* **179:**6472–6479.
23. **Lioy VS, Pratto F, la Hoz de AB, Ayora S, Alonso JC.** 2010. Plasmid pSM19035, a model to study stable maintenance in *Firmicutes*. *Plasmid* **64:**1–17.
24. **Paulsen IT, Gillespie MT, Littlejohn TG, Hanvivatvong O, Rowland SJ, Dyke KG, Skurray RA.** 1994. Characterisation of sin, a potential recombinase-encoding gene from *Staphylococcus aureus*. *Gene* **141:**109–114.
25. **Smith MCM, Thorpe HM.** 2002. Diversity in the serine recombinases. *Mol Microbiol* **44:**299–307.
26. **Rojo F, Alonso JC.** 1995. The beta recombinase of plasmid pSM19035 binds to two adjacent sites, making different contacts at each of them. *Nucleic Acids Res* **23:** 3181–3188.
27. **Mouw KW, Steiner AM, Ghirlando R, Li N-S, Rowland S-J, Boocock MR, Stark WM, Piccirilli JA, Rice PA.** 2010. Sin resolvase catalytic activity and oligomerization state are tightly coupled. *J Mol Biol* **404:**16–33.
28. **Stemmer C, Fernández S, López G, Alonso JC, Grasser KD.** 2002. Plant chromosomal HMGB proteins efficiently promote the bacterial site-specific beta-mediated recombination *in vitro* and *in vivo*. *Biochemistry* **41:** 7763–7770.
29. **Burke ME, Arnold PH, He J, Wenwieser SVCT, Rowland S-J, Boocock MR, Stark WM.** 2004. Activating mutations of Tn3 resolvase marking interfaces important in recombination catalysis and its regulation. *Mol Microbiol* **51:**937–948.
30. **Sarkis GJ, Murley LL, Leschziner AE, Boocock MR, Stark WM, Grindley ND.** 2001. A model for the gamma delta resolvase synaptic complex. *Mol Cell* **8:** 623–631.
31. **Rowland S-J, Stark WM, Boocock MR.** 2002. Sin recombinase from *Staphylococcus aureus*: synaptic complex architecture and transposon targeting. *Mol Microbiol* **44:**607–619.
32. **Yang W, Steitz TA.** 1995. Crystal structure of the site-specific recombinase gamma delta resolvase complexed with a 34 bp cleavage site. *Cell* **82:**193–207.
33. **Boocock MR, Zhu X, Grindley ND.** 1995. Catalytic residues of gamma delta resolvase act in cis. *EMBO J* **14:**5129–5140.
34. **Merickel SK, Haykinson MJ, Johnson RC.** 1998. Communication between Hin recombinase and Fis regulatory subunits during coordinate activation of Hin-catalyzed site-specific DNA inversion. *Genes Dev* **12:** 2803–2816.
35. **Mouw KW, Rowland S-J, Gajjar MM, Boocock MR, Stark WM, Rice PA.** 2008. Architecture of a serine recombinase-DNA regulatory complex. *Mol Cell* **30:** 145–155.
36. **Rice PA, Mouw KW, Montaño SP, Boocock MR, Rowland S-J, Stark WM.** 2010. Orchestrating serine resolvases. *Biochem Soc Trans* **38:**384–387.
37. **Keenholtz RA, Rowland S-J, Boocock MR, Stark WM, Rice PA.** 2011. Structural basis for catalytic activation of a serine recombinase. *Structure* **19:**799–809.
38. **Dhar G, Heiss JK, Johnson RC.** 2009. Mechanical constraints on Hin subunit rotation imposed by the Fis/enhancer system and DNA supercoiling during site-specific recombination. *Mol Cell* **34:**746–759.
39. **Li W, Kamtekar S, Xiong Y, Sarkis GJ, Grindley NDF, Steitz TA.** 2005. Structure of a synaptic gammadelta resolvase tetramer covalently linked to two cleaved DNAs. *Science* **309:**1210–1215.
40. **Yang W.** 2010. Topoisomerases and site-specific recombinases: similarities in structure and mechanism. *Crit Rev Biochem Mol Biol* **45:**520–534.
41. **Colloms SD, Sykora P, Szatmari G, Sherratt DJ.** 1990. Recombination at ColE1 cer requires the *Escherichia coli* xerC gene product, a member of the lambda integrase family of site-specific recombinases. *J Bacteriol* **172:** 6973–6980.
42. **Blakely G, May G, McCulloch R, Arciszewska LK, Burke M, Lovett ST, Sherratt DJ.** 1993. Two related

recombinases are required for site-specific recombination at dif and cer in *E. coli* K12. *Cell* **75:**351–361.

43. **Kono N, Arakawa K, Tomita M.** 2011. Comprehensive prediction of chromosome dimer resolution sites in bacterial genomes. *BMC Genomics* **12:**19.
44. **Carnoy C, Roten C-A.** 2009. The dif/Xer recombination systems in proteobacteria. *PLoS One* **4:**e6531.
45. **Cortez D, Quevillon-Cheruel S, Gribaldo S, Desnoues N, Sezonov G, Forterre P, Serre M-C.** 2010. Evidence for a Xer/dif system for chromosome resolution in archaea. *PLoS Genet* **6:**e1001166.
46. **Duggin IG, Dubarry N, Bell SD.** 2011. Replication termination and chromosome dimer resolution in the archaeon *Sulfolobus solfataricus*. *EMBO J* **30:**145–153.
47. **Recchia GD, Sherratt DJ.** 1999. Conservation of xer site-specific recombination genes in bacteria. *Mol Microbiol* **34:**1146–1148.
48. **Hakkaart MJ, van den Elzen PJ, Veltkamp E, Nijkamp HJ.** 1984. Maintenance of multicopy plasmid Clo DF13 in *E. coli* cells: evidence for site-specific recombination at parB. *Cell* **36:**203–209.
49. **Hiraga S, Sugiyama T, Itoh T.** 1994. Comparative analysis of the replicon regions of eleven ColE2-related plasmids. *J Bacteriol* **176:**7233–7243.
50. **Cornet F, Mortier I, Patte J, Louarn JM.** 1994. Plasmid pSC101 harbors a recombination site, psi, which is able to resolve plasmid multimers and to substitute for the analogous chromosomal *Escherichia coli* site dif. *J Bacteriol* **176:**3188–3195.
51. **Tolmasky ME, Colloms S, Blakely G, Sherratt DJ.** 2000. Stability by multimer resolution of pJHCMW1 is due to the Tn1331 resolvase and not to the *Escherichia coli* Xer system. *Microbiology* **146**(Pt 3):581–589.
52. **Pham H, Dery KJ, Sherratt DJ, Tolmasky ME.** 2002. Osmoregulation of dimer resolution at the plasmid pJHCMW1 mwr locus by *Escherichia coli* XerCD recombination. *J Bacteriol* **184:**1607–1616.
53. **Blakely G, Colloms S, May G, Burke M, Sherratt D.** 1991. *Escherichia coli* XerC recombinase is required for chromosomal segregation at cell division. *New Biol* **3:** 789–798.
54. **Clerget M.** 1991. Site-specific recombination promoted by a short DNA segment of plasmid R1 and by a homologous segment in the terminus region of the *Escherichia coli* chromosome. *New Biol* **3:**780–788.
55. **Kuempel PL, Henson JM, Dircks L, Tecklenburg M, Lim DF.** 1991. dif, a recA-independent recombination site in the terminus region of the chromosome of *Escherichia coli*. *New Biol* **3:**799–811.
56. **Hayes F, Sherratt DJ.** 1997. Recombinase binding specificity at the chromosome dimer resolution site dif of *Escherichia coli*. *J Mol Biol* **266:**525–537.
57. **Colloms SD.** 2013. The topology of plasmid-monomerizing Xer site-specific recombination. *Biochem Soc Trans* **41:**589–594.
58. **Sternberg N, Hamilton D, Austin S, Yarmolinsky M, Hoess R.** 1981. Site-specific recombination and its role in the life cycle of bacteriophage P1. *Cold Spring Harbor Symp Quant Biol* **45**(Pt 1):297–309.
59. **Segev N, Cohen G.** 1981. Control of circularization of bacteriophage P1 DNA in *Escherichia coli*. *Virology* **114:**333–342.
60. **Hochman L, Segev N, Sternberg N, Cohen G.** 1983. Site-specific recombinational circularization of bacteriophage P1 DNA. *Virology* **131:**11–17.
61. **Abremski K, Hoess R, Sternberg N.** 1983. Studies on the properties of P1 site-specific recombination: evidence for topologically unlinked products following recombination. *Cell* **32:**1301–1311.
62. **Ghosh K, Van Duyne GD.** 2002. Cre-loxP biochemistry. *Methods* **28:**374–383.
63. **Paul S, Summers D.** 2004. ArgR and PepA, accessory proteins for XerCD-mediated resolution of ColE1 dimers, are also required for stable maintenance of the P1 prophage. *Plasmid* **52:**63–68.
64. **MacDonald AI, Lu Y, Kilbride EA, Akopian A, Colloms SD.** 2008. PepA and ArgR do not regulate Cre recombination at the bacteriophage P1 loxP site. *Plasmid* **59:** 119–126.
65. **Ghosh K, Lau C-K, Gupta K, Van Duyne GD.** 2005. Preferential synapsis of loxP sites drives ordered strand exchange in Cre-loxP site-specific recombination. *Nat Chem Biol* **1:**275–282.
66. **Ghosh K, Guo F, Van Duyne GD.** 2007. Synapsis of loxP sites by Cre recombinase. *J Biol Chem* **282:** 24004–24016.
67. **Bouet J-Y, Bouvier M, Lane D.** 2006. Concerted action of plasmid maintenance functions: partition complexes create a requirement for dimer resolution. *Mol Microbiol* **62:**1447–1459.
68. **Disqué-Kochem C, Eichenlaub R.** 1993. Purification and DNA binding of the D protein, a putative resolvase of the F-factor of *Escherichia coli*. *Mol Gen Genet* **237:** 206–214.
69. **Lane D, de Feyter R, Kennedy M, Phua SH, Semon D.** 1986. D protein of miniF plasmid acts as a repressor of transcription and as a site-specific resolvase. *Nucleic Acids Res* **14:**9713–9728.
70. **Chen Y, Narendra U, Iype LE, Cox MM, Rice PA.** 2000. Crystal structure of a Flp recombinase-Holliday junction complex: assembly of an active oligomer by helix swapping. *Mol Cell* **6:**885–897.
71. **Guo F, Gopaul DN, Van Duyne GD.** 1997. Structure of Cre recombinase complexed with DNA in a site-specific recombination synapse. *Nature* **389:**40–46.
72. **Guo F, Gopaul DN, Van Duyne GD.** 1999. Asymmetric DNA bending in the Cre-loxP site-specific recombination synapse. *Proc Natl Acad Sci USA* **96:**7143–7148.
73. **Gopaul DN, Guo F, Van GD.** 1998. Structure of the Holliday junction intermediate in Cre-loxP site-specific recombination. *EMBO J* **17:**4175–4187.
74. **Hickman AB, Waninger S, Scocca JJ, Dyda F.** 1997. Molecular organization in site-specific recombination: the catalytic domain of bacteriophage HP1 integrase at 2.7 A resolution. *Cell* **89:**227–237.
75. **Subramanya HS, Arciszewska LK, Baker RA, Bird LE, Sherratt DJ, Wigley DB.** 1997. Crystal structure of the site-specific recombinase, XerD. *EMBO J* **16:**5178–5187.

76. **Gibb B, Gupta K, Ghosh K, Sharp R, Chen J, Van Duyne GD.** 2010. Requirements for catalysis in the Cre recombinase active site. *Nucleic Acids Res* **38:** 5817–5832.

77. **Rajeev L, Malanowska K, Gardner JF.** 2009. Challenging a paradigm: the role of DNA homology in tyrosine recombinase reactions. *Microbiol Mol Biol Rev* **73:** 300–309.

78. **Lee L, Chu LCH, Sadowski PD.** 2003. Cre induces an asymmetric DNA bend in its target loxP site. *J Biol Chem* **278:**23118–23129.

79. **Fan H-F.** 2012. Real-time single-molecule tethered particle motion experiments reveal the kinetics and mechanisms of Cre-mediated site-specific recombination. *Nucleic Acids Res* **40:**6208–6222.

80. **Pinkney JNM, Zawadzki P, Mazuryk J, Arciszewska LK, Sherratt DJ, Kapanidis AN.** 2012. Capturing reaction paths and intermediates in Cre-loxP recombination using single-molecule fluorescence. *Proc Natl Acad Sci USA* **109:**20871–20876.

81. **Zawadzki P, May PFJ, Baker RA, Pinkney JNM, Kapanidis AN, Sherratt DJ, Arciszewska LK.** 2013. Conformational transitions during FtsK translocase activation of individual XerCD-dif recombination complexes. *Proc Natl Acad Sci USA* **110:**17302–17307.

82. **Fan H-F, Ma C-H, Jayaram M.** 2013. Real-time single-molecule tethered particle motion analysis reveals mechanistic similarities and contrasts of Flp site-specific recombinase with Cre and λ Int. *Nucleic Acids Res* **41:** 7031–7047.

83. **McCulloch R, Coggins LW, Colloms SD, Sherratt DJ.** 1994. Xer-mediated site-specific recombination at cer generates Holliday junctions *in vivo*. *EMBO J* **13:** 1844–1855.

84. **Hallet B, Arciszewska LK, Sherratt DJ.** 1999. Reciprocal control of catalysis by the tyrosine recombinases XerC and XerD: an enzymatic switch in site-specific recombination. *Mol Cell* **4:**949–959.

85. **Diagne CT, Salhi M, Crozat E, Salomé L, Cornet F, Rousseau P, Tardin C.** 2014. TPM analyses reveal that FtsK contributes both to the assembly and the activation of the XerCD-dif recombination synapse. *Nucleic Acids Res* **42:**1721–1732.

86. **Arciszewska LK, Baker RA, Hallet B, Sherratt DJ.** 2000. Coordinated control of XerC and XerD catalytic activities during Holliday junction resolution. *J Mol Biol* **299:**391–403.

87. **Chen Y, Rice PA.** 2003. New insight into site-specific recombination from Flp recombinase-DNA structures. *Annu Rev Biophys Biomol Struct* **32:**135–159.

88. **Olorunniji FJ, Stark WM.** 2010. Catalysis of site-specific recombination by Tn3 resolvase. *Biochem Soc Trans* **38:**417–421.

89. **Rowland S-J, Boocock MR, Stark WM.** 2006. DNA bending in the Sin recombination synapse: functional replacement of HU by IHF. *Mol Microbiol* **59:** 1730–1743.

90. **Canosa I, López G, Rojo F, Boocock MR, Alonso JC.** 2003. Synapsis and strand exchange in the resolution and DNA inversion reactions catalysed by the beta recombinase. *Nucleic Acids Res* **31:**1038–1044.

91. **Reijns M, Lu Y, Leach S, Colloms SD.** 2005. Mutagenesis of PepA suggests a new model for the Xer/cer synaptic complex. *Mol Microbiol* **57:**927–941.

92. **Vazquez M, Colloms SD, Sumners DW.** 2005. Tangle analysis of Xer recombination reveals only three solutions, all consistent with a single three-dimensional topological pathway. *J Mol Biol* **346:**493–504.

93. **Sträter N, Sherratt DJ, Colloms SD.** 1999. X-ray structure of aminopeptidase A from *Escherichia coli* and a model for the nucleoprotein complex in Xer site-specific recombination. *EMBO J* **18:**4513–4522.

94. **Bregu M, Sherratt DJ, Colloms SD.** 2002. Accessory factors determine the order of strand exchange in Xer recombination at psi. *EMBO J* **21:**3888–3897.

95. **Bregu M, Sherratt DJ, Colloms SD.** 2002. Accessory factors determine the order of strand exchange in Xer recombination at psi. *EMBO J* **21:**3888–3897.

96. **Vanhooff V, Normand C, Galloy C, Segall AM, Hallet B.** 2010. Control of directionality in the DNA strand-exchange reaction catalysed by the tyrosine recombinase TnpI. *Nucleic Acids Res* **38:**2044–2056.

97. **Pérals K, Cornet F, Merlet Y, Delon I, Louarn JM.** 2000. Functional polarization of the *Escherichia coli* chromosome terminus: the dif site acts in chromosome dimer resolution only when located between long stretches of opposite polarity. *Mol Microbiol* **36:**33–43.

98. **Steiner WW, Kuempel PL.** 1998. Cell division is required for resolution of dimer chromosomes at the dif locus of *Escherichia coli*. *Mol Microbiol* **27:**257–268.

99. **Barre FX, Søballe B, Michel B, Aroyo M, Robertson M, Sherratt D.** 2001. Circles: the replication-recombination-chromosome segregation connection. *Proc Natl Acad Sci USA* **98:**8189–8195.

100. **Cox MM, Goodman MF, Kreuzer KN, Sherratt DJ, Sandler SJ, Marians KJ.** 2000. The importance of repairing stalled replication forks. *Nature* **404:**37–41.

101. **Lesterlin C, Barre F-X, Cornet F.** 2004. Genetic recombination and the cell cycle: what we have learned from chromosome dimers. *Mol Microbiol* **54:**1151–1160.

102. **Bigot S, Sivanathan V, Possoz C, Barre F-X, Cornet F.** 2007. FtsK, a literate chromosome segregation machine. *Mol Microbiol* **64:**1434–1441.

103. **Dubarry N, Possoz C, Barre F-X.** 2010. Multiple regions along the *Escherichia coli* FtsK protein are implicated in cell division. *Mol Microbiol* **78:**1088–1100.

104. **Iyer LM, Makarova KS, Koonin EV, Aravind L.** 2004. Comparative genomics of the FtsK-HerA superfamily of pumping ATPases: implications for the origins of chromosome segregation, cell division and viral capsid packaging. *Nucleic Acids Res* **32:**5260–5279.

105. **Ptacin JL, Nöllmann M, Becker EC, Cozzarelli NR, Pogliano K, Bustamante C.** 2008. Sequence-directed DNA export guides chromosome translocation during sporulation in *Bacillus subtilis*. *Nat Struct Mol Biol* **15:** 485–493.

106. **Marquis KA, Burton BM, Nöllmann M, Ptacin JL, Bustamante C, Ben-Yehuda S, Rudner DZ.** 2008.

SpoIIIE strips proteins off the DNA during chromosome translocation. *Genes Dev* **22:**1786–1795.

107. **Kaimer C, Schenk K, Graumann PL.** 2011. Two DNA translocases synergistically affect chromosome dimer resolution in *Bacillus subtilis. J Bacteriol* **193:** 1334–1340.

108. **Aussel L, Barre F-X, Aroyo M, Stasiak A, Stasiak AZ, Sherratt D.** 2002. FtsK is a DNA motor protein that activates chromosome dimer resolution by switching the catalytic state of the XerC and XerD recombinases. *Cell* **108:**195–205.

109. **Yates J, Aroyo M, Sherratt DJ, Barre F-X.** 2003. Species specificity in the activation of Xer recombination at dif by FtsK. *Mol Microbiol* **49:**241–249.

110. **Grainge I, Lesterlin C, Sherratt DJ.** 2011. Activation of XerCD-dif recombination by the FtsK DNA translocase. *Nucleic Acids Res* **39:**5140–5148.

111. **Grainge I, Lesterlin C, Sherratt DJ.** 2011. Activation of XerCD-dif recombination by the FtsK DNA translocase. *Nucleic Acids Res* **39:**5140–5148.

112. **Grainge I, Lesterlin C, Sherratt DJ.** 2011. Activation of XerCD-dif recombination by the FtsK DNA translocase. *Nucleic Acids Res* **39:**5140–5148.

113. **Lee JY, Finkelstein IJ, Crozat E, Sherratt DJ, Greene EC.** 2012. Single-molecule imaging of DNA curtains reveals mechanisms of KOPS sequence targeting by the DNA translocase FtsK. *Proc Natl Acad Sci USA* **109:** 6531–6536.

114. **Graham JE, Sherratt DJ, Szczelkun MD.** 2010. Sequence-specific assembly of FtsK hexamers establishes directional translocation on DNA. *Proc Natl Acad Sci USA* **107:**20263–20268.

115. **Crozat E, Grainge I.** 2010. FtsK DNA translocase: the fast motor that knows where it's going. *Chembiochem* **11:**2232–2243.

116. **Stouf M, Meile J-C, Cornet F.** 2013. FtsK actively segregates sister chromosomes in *Escherichia coli. Proc Natl Acad Sci USA* **110:**11157–11162.

117. **Grainge I, Bregu M, Vazquez M, Sivanathan V, Ip SCY, Sherratt DJ.** 2007. Unlinking chromosome catenanes *in vivo* by site-specific recombination. *EMBO J* **26:** 4228–4238.

118. **Shimokawa K, Ishihara K, Grainge I, Sherratt DJ, Vazquez M.** 2013. FtsK-dependent XerCD-dif recombination unlinks replication catenanes in a stepwise manner. *Proc Natl Acad Sci USA* **110:**20906–20911.

119. **Val M-E, Kennedy SP, Karoui El M, Bonné L, Chevalier F, Barre F-X.** 2008. FtsK-dependent dimer resolution on multiple chromosomes in the pathogen *Vibrio cholerae. PLoS Genet* **4:**e1000201.

120. **Val M-E, Kennedy SP, Soler-Bistué AJ, Barbe V, Bouchier C, Ducos-Galand M, Skovgaard O, Mazel D.** 2014. Fuse or die: how to survive the loss of Dam in *Vibrio cholerae. Mol Microbiol* **91:**665–678.

121. **Le Bourgeois P, Bugarel M, Campo N, Daveran-Mingot M-L, Labonté J, Lanfranchi D, Lautier T, Pages C, Ritzenthaler P.** 2007. The unconventional Xer recombination machinery of *Streptococci/Lactococci. PLoS Genet* **3:**e117.

122. **Debowski AW, Gauntlett JC, Li H, Liao T, Sehnal M, Nilsson H-O, Marshall BJ, Benghezal M.** 2012. Xer-cise in *Helicobacter pylori*: one-step transformation for the construction of markerless gene deletions. *Helicobacter* **17:**435–443.

123. **Nolivos S, Pages C, Rousseau P, Le Bourgeois P, Cornet F.** 2010. Are two better than one? Analysis of an FtsK/Xer recombination system that uses a single recombinase. *Nucleic Acids Res* **38:**6477–6489.

124. **Nolivos S, Touzain F, Pages C, Coddeville M, Rousseau P, El Karoui M, Le Bourgeois P, Cornet F.** 2012. Co-evolution of segregation guide DNA motifs and the FtsK translocase in bacteria: identification of the atypical *Lactococcus lactis* KOPS motif. *Nucleic Acids Res* **40:**5535–5545.

125. **Serre M-C, El Arnaout T, Brooks MA, Durand D, Lisboa J, Lazar N, Raynal B, van Tilbeurgh H, Quevillon-Cheruel S.** 2013. The carboxy-terminal αN helix of the archaeal XerA tyrosine recombinase is a molecular switch to control site-specific recombination. *PLoS One* **8:**e63010.

126. **Yano H, Genka H, Ohtsubo Y, Nagata Y, Top EM, Tsuda M.** 2013. Cointegrate-resolution of toluene-catabolic transposon Tn4651: determination of crossover site and the segment required for full resolution activity. *Plasmid* **69:**24–35.

127. **Val M-E, Bouvier M, Campos J, Sherratt D, Cornet F, Mazel D, Barre F-X.** 2005. The single-stranded genome of phage CTX is the form used for integration into the genome of *Vibrio cholerae. Mol Cell* **19:**559–566.

128. **Das B, Bischerour J, Barre F-X.** 2011. VGJphi integration and excision mechanisms contribute to the genetic diversity of *Vibrio cholerae* epidemic strains. *Proc Natl Acad Sci USA* **108:**2516–2521.

129. **Bischerour J, Spangenberg C, Barre F-X.** 2012. Holliday junction affinity of the base excision repair factor Endo III contributes to cholera toxin phage integration. *EMBO J* **31:**3757–3767.

130. **Domínguez NM, Hackett KT, Dillard JP.** 2011. XerCD-mediated site-specific recombination leads to loss of the 57-kilobase gonococcal genetic island. *J Bacteriol* **193:**377–388.

Plasmids—Biology and Impact in Biotechnology and Discovery
Edited by Marcelo E. Tolmasky and Juan C. Alonso

doi:10.1128/microbiolspec.PLAS-0009-2013

Ana María Hernández-Arriaga,[1] Wai Ting Chan,[1] Manuel Espinosa,[1] and Ramón Díaz-Orejas[1]

10

Conditional Activation of Toxin-Antitoxin Systems: Postsegregational Killing and Beyond

INTRODUCTION

Toxin-antitoxin (TA) genes are small genetic modules coding for a toxin and an antitoxin. Toxins inhibit cell proliferation or viability, and antitoxins neutralize this inhibition. The toxin (always a protein) is the stable component and the antitoxin (a protein or a regulatory RNA) is less stable, and this differential stability plays an important role in the conditional activation of TAs. The term *p*ost*s*egregational *k*illing (PSK) was introduced to define the toxin-dependent elimination of plasmid-free cells that occurs as a consequence of the loss of TA-containing plasmids at cell division. Conditional activation of the toxins in these cells requires a differential decay of the antitoxins compared with the toxins. This differential stability is due to the action of proteases or RNases on the antitoxin half-life. In plasmid-containing cells, the toxin is kept under control because the levels of the antitoxin are replenished by *de novo* synthesis. In plasmid-free cells, the toxins are activated as the consequence of the faster decay of the antitoxins, and this leads to the elimination of these cells from the population (PSK) and to an increase of the percentage of plasmid-containing cells (1, 2) (Fig. 1). Since the discovery of TA systems as auxiliary maintenance modules in plasmids (3), they have been found in phages and chromosomes of Bacteria and Archaea, often in multiple copies (4). Conditional activation of TA pairs has also been detected in chromosomal systems in response to particular signals. Furthermore, some of the chromosomal TA systems have the potential to stabilize plasmids via PSK, implying that the differential stability of toxins and antitoxins plays a role in their activation. Conditional activation of TA systems has consequences beyond plasmid stabilization, such as in plasmid competition, phage-abortive infection, stress response, stabilization of particular genomic regions, biofilm formation, and bacterial persistence (5). Most recently, TA systems have been found tightly associated with other defense systems that can be found in Archaea and in Bacteria and that include the so-called CRISPR-Cas immunity system (6).

[1]Centro de Investigaciones Biológicas, Consejo Superior de Investigaciones Científicas, 28040 Madrid, Spain.

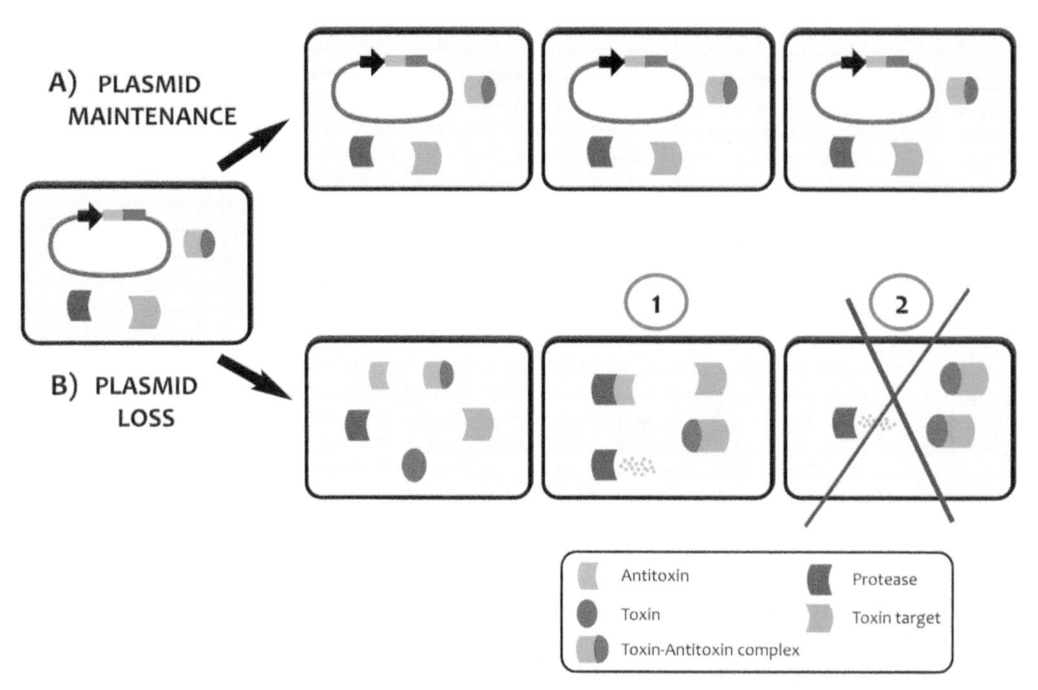

Figure 1 TAs determine plasmid maintenance by PSK. A bacterial population contains cells with a plasmid that encodes a TA system. (**A**) In plasmid-containing cells, both the antitoxin and the toxin will be continuously expressed. The inhibitory activity of the toxin will keep neutralized its cognate antitoxin. (**B**) In plasmid-free cells, a specific depletion of the antitoxin levels by cellular RNases or proteases activates the more stable toxin. This activation induces cell death or arrests the growth of plasmid-free cells (PSK) and increases the number of plasmid-containing cells in the growing population (plasmid maintenance phenotype). doi:10.1128/microbiolspec.PLAS-0009-2013.f1

In TA systems so far described, the antitoxins are either small RNAs or labile proteins; thus, RNases or proteases can lead to degradation of the antitoxin component and, therefore, to the activation of the cognate toxin (7, 8). To date, TAs are classified in five main types depending on the nature and activities of their antitoxins. In type I and type III systems, the antitoxins are small RNAs that inhibit translation of the toxin (type I) or neutralize its activity (type III). In type II TAs, the antitoxins are proteins that neutralize their cognate toxins by direct interactions (9). A basic representation of these three TA types is shown in Fig. 2. Two additional TA-type systems, named IV and V, have been described. In these systems, the antitoxins are also proteins but they do not interact directly with the toxin. Instead, in type IV systems (10), the antitoxin interacts with its target, protecting it from the toxin activity, and, in type V systems (11), the antitoxins cleave and inactivate the mRNA of the toxin. Toxins can act on different targets, but in all cases they inhibit cell growth or affect cell viability (9). Expression of TAs is regulated at the transcriptional and/or posttranscriptional levels. Depending on the similitude of their amino acid sequences or on their structural homologies, toxins and antitoxins are grouped into different families and superfamilies (12).

So far, proper TA pairs have not been found in eukaryotic cells. However, it has been proposed that toxin/immunity systems found in lineal plasmids of unicellular fungi could mediate plasmid stabilization via the PSK mechanism (13). These systems code for toxins that, as microcines, are secreted to the growth medium and can inhibit the proliferation of competitors. Cells losing these plasmids and therefore the immunity determinants are susceptible to the action of these toxins and, as a consequence, can be eliminated from the cell population by them. This potential extends the signature of PSK beyond the bacterial and archaeal kingdoms. Even more, the bacterial toxin-antitoxin systems can be engineered to mimic in eukaryotic cell lines the selective elimination of particular cell lines as the result of the differential expression of the toxin and the antitoxin (14, 15).

Owing to experimental and bioinformatic searches, the number of TAs is continuously increasing. In parallel, the interest in the basic and applied implications

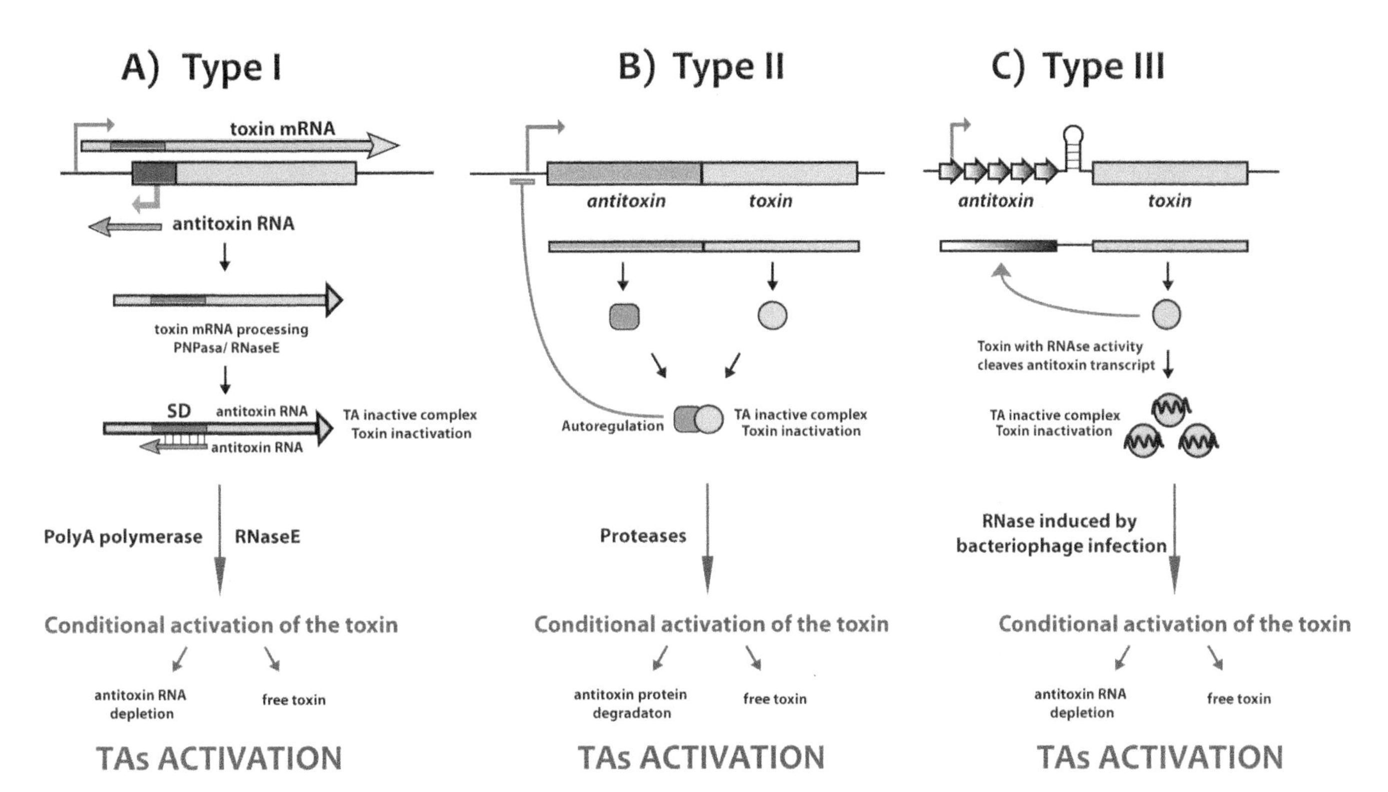

Figure 2 TA regulation and activation. TA systems are operons that codify a toxin (T) and an antitoxin (A). They share common features: (i) Expression of the operon is regulated at the transcriptional or posttranscriptional levels; (ii) the antitoxin binds and neutralizes the toxic activity of the toxin; and (iii) the antitoxin is unstable and the toxin is stable. The decay of the more unstable antitoxin leads to toxin activation. **A**, **B**, and **C** show the basic features of the regulation and activation of type I, II, and III TAs. (**A**) Type I TAs: the antitoxin is a small antisense RNA, and the toxin is a protein; processing of the toxin mRNA and cleavage of RNA-RNA hybrids regulate the activity of these systems. (**B**) Type II TAs: Both toxin and antitoxin are proteins; proteases targeting specifically the antitoxin regulate activation of the toxin. (**C**) Type III TAs: the antitoxin is an RNA that inactivates the toxin. Toxin activation can occur in response to bacteriophage infection leading to the elimination of these cells and thus preventing the spread of the infection.
doi:10.1128/microbiolspec.PLAS-0009-2013.f2

of these systems has developed to the point of establishing TAs as a new and fertile field of microbiology. A recent compilation of the contribution of the different TAs to the field can be found in reference 16.

Focus

Because of the large expansion of the number of putative TA systems and the broad nature and mechanism of action of toxins, we will concentrate on some well-defined TA systems present in plasmids or in chromosomes rather than lavishly covering the PSK subject. In the plasmid context, we will concentrate in several systems found in *Escherichia coli*: *hok-sok* (type I) and *kis-kid* (type II) systems of plasmid R1, *ccd* (type II) of plasmid F, *parDE* (type II) of plasmid RK2, and *phd-doc* (type II) of bacteriophage P1; we will also review the conditional activation of the *toxIN* system found in pECA1039 of the Gram-negative phytopathogen *Pectobacterium atrosepticum*, the prototype of type III TA systems. At the chromosomal level, we will focus on three type II TA systems: *parDE* of *Vibrio cholerae* and *yefM-yoeB* and *pezAT* of *Streptococcus pneumoniae*. We will also discuss the plasmid-chromosome cross-talk between TAs and the implications of PSK in these interactions. As a summary, the systems analyzed include the main three type TA systems found in plasmids, in bacteriophages, or in the bacterial chromosome and toxins that reach different targets; on the whole, they illustrate PSK modulated by RNA or protein decay. In addition to the reviews mentioned above, other excellent reviews covering different aspects of the TA field can be found in references 17, 18, 19, 20, 21, 22, 23, and 24.

PSK MEDIATED BY PLASMID-ENCODED TAS

PSK Mediated by a Fertility Factor: *ccd* TA of Plasmid F

The coupling cell division (*ccd*) system of the fertility factor F1 was the first TA plasmid maintenance system described (3). This system, which moderately stabilizes the plasmid (10-fold), codes for an antitoxin protein, CcdA, and a toxin protein, CcdB, that inhibits cell growth and/or cell viability. *ccd*-mediated stabilization was initially thought to be due to the inhibition of cell division occurring when the copy number of the plasmid was one per cell; this prevented the subsequent accumulation of plasmid-free cells (3). However, it was later found that these cells could indeed divide, giving rise to two cell types: plasmid-containing cells that continue dividing and plasmid-free cells in which its division was arrested after a few residual divisions (2). The result was an increase in the percentage of plasmid-containing cells in the population, i.e., a plasmid maintenance phenotype. These observations formulated the first definition of PSK: increased plasmid maintenance associated with the inhibition of proliferation of plasmid-free cells due to the conditional activation of the toxin in these cells. It was later found that this activation was due to the cleavage and inactivation of the antitoxin by the Lon protease (see below).

CcdB, the toxin of the system, targets and inhibits the dimeric subunit A of the DNA gyrase at a stage in which the DNA strands are cleaved and unsealed (25, 26); it stabilizes this cleavage complex and inhibits DNA gyrase activity (27). Broken DNA can be released from this complex probably owing to its interactions with transcription and/or replication machineries. As a result of this DNA damage, the SOS regulon is induced. SOS induction is not required for CcdB toxicity, but it increases genetic variation associated with DNA repair, mutagenesis, chromosomal rearrangements, and transposition events, thus speeding up genetic variation and adaptation (18).

CcdA antitoxin interacts with CcdB and neutralizes its toxic effect (28, 29). In addition, CcdA can remove the toxin from a CcdB-gyrase complex, thus "rejuvenating" the enzyme (30, 31). Adding to this control of toxin activity, CcdA represses transcription of the *ccd* operon (29), and this regulation is more efficient in the presence of the CcdB toxin. Interestingly, the efficient repressor is a $CcdA_2$-$CcdB_2$ heterotetramer. This repressor is formed in excess of the antitoxin and is disrupted when the toxin is in excess, a condition in which the main form observed is a $CcdA_2$-$CcdB_4$ heterohexamer (32). This implies that the relative dosage of toxin and antitoxin regulates in a reversible way the expression of the system. The molecular interactions regulating this switch involved two sites of TA interactions, one of high affinity and the other of low affinity (33). Transcriptional regulation and neutralization of the active toxin by the antitoxin are the first level at which PSK is regulated. An additional point of control is antitoxin stability. This stability is modulated by the action of Lon protease that degrades preferentially CcdA (34, 35). The preferential decay of CcdA is at the base of the activation of CcdB in plasmid-free cells and of the elimination of these cells (PSK). CcdB protects the CcdA antitoxin from the action of Lon and, in this way, also influences the efficiency of PSK. Presumably, CcdA is available for Lon-mediated degradation when it dissembles from the CcdA-CcdB complex. Alternatively, CcdA antitoxin is cleaved far less efficiently by this protease when it is in complex with the CcdB toxin.

PSK Mediated by a TA Present in a Multiresistance Conjugative Plasmid: *hok-sok* of Plasmid R1

parB or *hok-sok* locus of the low-copy-number plasmid R1 was the first type I TA maintenance module described (1) and the prototype of these important TA systems. *hok-sok* increased by 2 orders of magnitude the stability of plasmid R1 without changing its copy number. It stabilized the plasmid by promoting the elimination of plasmid-free segregants and contributed, jointly with *ccd*, to establishing the PSK concept. This locus contains two genes *hok* and *sok* encoding, respectively, a killing protein (Hok, *ho*st *k*illing) and an unstable RNA antitoxin (Sok, *s*uppressor *o*f *k*illing). Genes *sok* and *hok* are constitutively expressed; a strong promoter directs Sok-mRNA synthesis, while *hok*-mRNA synthesis is directed from a weak promoter. This ensures an increased dosage of the Sok over *hok*-mRNA. Sok, a small and unstable antisense RNA, inhibits the expression of Hok at the posttranscriptional level. The structure of the Hok toxin and the way in which it acts is not accurately known, but indirect evidence has suggested that it is a protein that disrupts the membrane potential. Cells killed by this toxin have a particular fragile aspect (ghost cells). The way in which this killing potential is maintained under control in plasmid-containing cells and activated in plasmid-free cells has been elegantly elucidated (36): full-length *hok*-mRNA is inactive but, owing to three processing steps and subsequent refolding, the mRNA precursor becomes an active mRNA. In plasmid-containing cells, this mRNA is silenced by interactions with antisense

Sok-RNA, and this RNA-RNA complex is removed because of RNase III cleavage. In plasmid-free cells, decay of the unstable antisense Sok-RNA leaves the active form of *hok*-mRNA available for translation; and the accumulation of the Hok toxins eliminate these cells (PSK). The *hok-sok* model of PSK remains a paradigm in the field, but, as could be expected, different mechanisms to prevent irreversible inactivation of toxin mRNA by RNA-RNA interactions do occur (37, 38).

A Second PSK System in an Antibiotic Resistance Factor: The *kis-kid* TA of Plasmid R1

In addition to *hok-sok*, plasmid R1 contains a TA close to its basic replicon: *kis-kid* or *parD* (39). This system, which was also found in plasmid R100 (40), encodes a toxin, Kid (*Ki*lling *d*eterminant), and an antitoxin, Kis (*Ki*lling *s*uppressor). Both of them are proteins. The wild-type *kis-kid* system shows low activity as a maintenance system. It remained undetected until its discovery, by serendipity, following the isolation of a mutation in *kis*, the antitoxin gene, that increased by over 3 orders of magnitude the stability of the plasmid (39). This mutation inhibited cell growth in rich medium, an effect that was enhanced at high temperature. Both phenotypes, maintenance and cell growth inhibition, were abolished by mutations in *kid*, the gene adjacent to *kis*, pointing to Kid as the toxin and Kis as the antitoxin of a plasmid stabilization system (41). The great increase in plasmid stabilization observed in the original mutant was not correlated with a parallel increase in plasmid copy number; this pointed to PSK mediated by the Kid toxin as the explanation of this increased maintenance. Following, the identification of this mutated system, it was also found that the wild-type system had a poor but detectable plasmid stabilization activity, similar or slightly lower than the stabilization determined by the *ccd* system (39, 42). Kid, the toxin of this system, is a site-specific RNase that cleaves mRNA and inhibits protein synthesis, and, therefore, cell growth is thwarted (43, 44). Kis antitoxin interacts with the toxin and prevents the binding and cleavage of the RNA substrate and the subsequent growth inhibition (45, 46). Expression of the Kis antitoxin up to 30 min after expression of the Kid toxin can rescue the growth of cells arrested by this toxin (our unpublished results), implying an initial bacteriostatic and reversible effect of the toxin. PSK activation determined by this system is regulated at the transcriptional level by the coordinated action of the toxin and the antitoxin (47, 48). Regulation is also modulated, at a posttranscriptional level by coupling translation of the toxin to the synthesis of the antitoxin and by the differential processing of the polycistronic *kis-kid*-mRNA (49); this posttranscriptional regulation can determine an increased dosage of the antitoxin compared with the toxin. The activation of the *kis-kid* system in response to inefficient replication (see "Coupling Plasmid Replication and TA Maintenance Modules" below) adds a new complexity to the regulation of this system. Another level of regulation is modulated by ClpAP protease which targets the Kis antitoxin but not the Kid toxin (50); this proteolysis also modulates the relative dosage of toxin and antitoxin and is at the base of the activation of the toxin. The antitoxin is protected from the protease action by interaction with the Kid toxin. Furthermore, this protection is more efficient when the toxin is in excess (50). The relative dosage of the toxin and antitoxin proteins modulates both transcription of this system and the antitoxin stability, thus influencing PSK in a complex way. Mutations that inactivate the Kid toxin or a deletion of the *clpP* gene that increased the stability of the Kis antitoxin decreased plasmid stability which is consistent with the role of Kid activity and of Kis proteolysis in PSK (50).

PSK by a TA Present in a Broad-Host-Range Plasmid: The *parDE* System of Plasmid RK2

parDE is the TA module of plasmid RK2/RP4, a broad-host-range replicon of Gram-negative bacteria (51, 52). ParD is the antitoxin of this system and ParE is the toxin. The *parDE* system increases 100-fold the maintenance of the RK2 replicon. PSK modulated by *parDE* is effective in *E. coli* host and also in other Gram-negative bacteria, thus mimicking the broad-host-range character of the replicon. As CcdA, ParE interacts with the DNA gyrase and inhibits this enzyme at a stage in which DNA strands are cleaved (cleavable complex) (53). It has been proposed that broken DNA can be released from this complex owing to collisions with the replication or transcription machineries; as a consequence, the SOS regulon is induced. Owing to its antitopoisomerase activity, ParE inhibits DNA replication *in vitro*, induces cell filaments, and affects cell growth and viability. Toxin expression and activity are regulated at different levels. In contrast to *ccd* and *kis-kid* in which efficient repression requires the coordinated action of the toxin and antitoxin, repression of the ParDE operon is basically modulated by the ParD antitoxin, being negligible the contribution of the ParE toxin (54, 55, 56, 57). This suggests that in this system conditional cooperativity is not involved in transcriptional regulation of the PSK potential. In addition, ParD interacts with ParE and neutralizes the activity of this toxin (56, 58). Furthermore, ParD-ParE interactions can rescue the ParE toxin when in complex with DNA gyrase, thus rejuvenating a ParE-poisoned DNA gyrase (53). This is an additional safety mechanism that contributes to neutralize the effects of possible uncontrolled release of ParE in plasmid-containing cells. In *parDE*, the differential stability of toxin and antitoxin required for PSK is due to the action of the Lon protease on the ParD antitoxin (52). In plasmid-containing cells, the levels of the ParD required to neutralize ParE are restored by *de novo* synthesis. In plasmid-free cells, the continuous decay of the antitoxin leads to activation of ParE and to plasmid stabilization by PSK. The fact that both ParE and CcdB toxins act on the same target but that the plasmid maintenance mediated by *parDE* is 10-fold higher than *ccd*, suggests that this difference could be due to a different initial load of TA proteins in plasmid-free cells or to a different decay of the antitoxin in these cells.

PSK by a TA Present in a Lysogenic Bacteriophage: The *phd-doc* System of Plasmid P1

The bicistronic *phd-doc* TA system was first discovered in P1 bacteriophage as a maintenance system acting by a PSK mechanism during lysogeny (59); *phd* codes an antitoxin, and *doc* codes a toxin. Interactions between PhD antitoxin and Doc toxin neutralize the inhibitory potential of this toxin (PhD, *P*revent *h*ost *d*eath; Doc, *D*eath *o*n *c*uring). This system makes the cell addicted to the antitoxin and, by extension, to the system and to the plasmid. Later on, homologs of this system were found on the chromosomes of many bacteria (60). *doc* encodes a toxin, Doc, that prevents translation elongation by interacting with the ribosome, inhibiting cell growth/viability (61). This activity is at the base of PSK modulated by this system. Interestingly, PSK mediated by *phd-doc* requires, in *E. coli*, the chromosomal *mazEF* TA, thus underlining complexities at the base of the PSK phenotype and the interaction between plasmidic and chromosomal TA systems (62). As in other systems, *phd-doc* is regulated at the transcriptional level by the coordinated interaction of the toxin and the antitoxin, and the antitoxin pilots the specific interactions at the promoter-operator region (63, 64). The translation initiation signals of *doc* overlap with *phd*, suggesting that synthesis of the toxin is coupled to the synthesis of the antitoxin. The regulatory complex is formed when the antitoxin is in excess and, as in *ccd*, the *phd-doc* system is de-repressed in excess of the Doc toxin. This protein-dosage-dependent regulation (conditional cooperativity) is due to a low- to high-affinity switch in the interaction between Phd and Doc (65). Thus, conditional cooperativity plays an important role in activation of the system and therefore in PSK. The ATP-serine protease ClpXP targets and degrades selectively the Phd antitoxin (66). The relevance of this activity in plasmid maintenance and in PSK is indicated by the fact that *rexB* gene, which product prevents the degradation of Phd by ClpXP, inhibits host death on curing of P1 (63, 67).

toxIN: PSK That Protects from Bacteriophage Infection

The *toxIN* locus is encoded on plasmid pECA1039 of the Gram-negative phytopathogen *P. atrosepticum*. This system was discovered because of its ability to confer resistance to bacteriophage, i.e., abortive infection (Abi) system (68, 69, 70). *toxIN* is the prototype of the type III TA systems. *toxN* encodes a protein (ToxN) which is a sequence-specific (AA/AU) endoribonuclease that cleaves the RNA substrate in the absence of ribosomes (71, 72). *toxI* encodes a small and noncoding RNA (ToxI) that interacts with ToxN and neutralizes the RNase activity of this protein (68, 71, 72). ToxI-RNA precursor contains a tandem array of direct repeats that are probably processed by the ToxN RNase to generate 36-nucleotide pseudoknot repeats that bind ToxN. The RNA-Protein interaction, a heterohexameric assembly of three pseudoknots and three ToxN dimers, inactivates the RNase potential of ToxN (71). Because of the differential stabilities of ToxN and ToxI components (68), antitoxin levels must be maintained by the continued expression of the *toxIN* operon. Transcription of the system occurs from a constitutive promoter (68) and is regulated by transcriptional termination at a short inverted repeat that precedes *toxN*. This signal determines the relative levels of the ToxI-RNA precursor and the *toxN* mRNA. ToxI RNA is degraded by an unknown mechanism shortly after infection; this releases ToxN RNase that arrests growth of the infected cell, thus preventing replication and propagation of the phage in the bacterial population (68, 69). It remains to be tested if, in addition to its function as an Abi system, ToxIN can also stabilize plasmids via PSK.

Lessons from the Comparative Analysis of PSK of Some of the Above TAs

In an attempt to obtain a general view on the stability mediated by PSK, the stabilization mediated by each of four early plasmidic TA systems described above was analyzed comparatively: the *hok-sok*, *kis-kid*, and *ccd* systems of narrow-host-range plasmids R1 and F, and the *parDE* system of plasmid RK2 (42). Based on the extensive knowledge of the R1 plasmid replication control (73), an elegant approach was developed to determine the individual contributions of the above TA systems to the maintenance of this replicon: the four selected systems were independently cloned in a transcription-free region of a TA-free mini-R1 vector and introduced into an *E. coli* host containing a compatible plasmid-expressing *copA* gene under the control of an inducible promoter. *copA* codes a small antisense RNA and inhibits, at the posttranscriptional level, the synthesis of RepA, the replication initiation protein of plasmid R1. The CopA RNA arrests very efficiently the replication of this plasmid. As a consequence, following *copA* transcription, the plasmid recombinants are eliminated. In this way, the effects of each of the four TAs on the maintenance of the same replicon and on the growth of plasmid-free cells could be followed in a comparative way. Consistently, with the maintenance effects of the individual systems, *hok-sok* or *parDE* stabilized the plasmid vector 100-fold while stabilization mediated by *ccd* and *kis-kid* was close to 10-fold, i.e., an order of magnitude lower. Further analysis of the cell growth, number of cells (particles), and viable cells suggested that stabilization mediated by the four systems occurred at the postsegregation level. Maintenance mediated by *ccd*, *hok-sok*, or *parDE* systems was associated to a decrease in the number of viable cells compared with the total number of cells following induction of *copA* and the loss of the plasmid. This suggested that the systems eliminate plasmid-free cells. For the *kis-kid* system, the effect on plasmid-free cells was a gradual inhibition of cell proliferation and cell division. This was indicated by (i) the parallel increase of cell counts and viable cells occurring during the first generations after *copA* induction and (ii) by an equivalent decrease of these parameters and by an increase in cell size occurring at later times. Both situations, killing or cell growth arrest of plasmid-free cells, led an increase of the percentage of plasmid-containing cells in the bacterial population, i.e., to a plasmid maintenance phenotype.

The analysis underlines the basic hallmark of plasmid stabilization mediated by PSK: (i) it is a specific feature of TA maintenance systems present in plasmids, (ii) it is associated with the loss of the plasmids at cell division and with the activation of the toxin due to the differential decay of the antitoxin in these cells, and (iii) it is the consequence of the nonproliferation of plasmid-free cells that either remain viable or die and of the proliferation of plasmid-containing cells.

Coupling Plasmid Replication and TA Maintenance Modules

A singularity of the *kis-kid* system is that the toxin of the system is activated in plasmid-containing cells when plasmid replication is inefficient; this coupling seems to be modulated by a decrease in the antitoxin levels and, surprisingly, can contribute to rescue plasmid replication (74, 75). This rescuing is due to the cleavage by Kid RNase of the *copB-repA* polycistronic mRNA at a preferred sequence located in the intergenic *copB-repA* region; the cleavage reduces the CopB repressor levels and increases transcription of the *repA* gene from an internal promoter, thus enhancing plasmid replication (76). Because this coupling is triggered by a decrease in plasmid gene dosage, it could play a role in situations in which this dosage is reduced, such as in plasmid propagation in nonfavorable hosts (75). Note that *kis-kid*-mediated replication rescue does not exclude stabilization via PSK and that the combined effects of PSK and replication rescue are not sufficient to eliminate completely plasmid-free segregants when the efficiency of replication is reduced. An interesting case of the coupling of a plasmid TA system to the plasmid replication functions, and also to the plasmid-partitioning system, occurs in the system of plasmid pSM19035 (see reference 126). The coupling of the basic maintenance modules of this plasmid results in a complete stabilization of this genetic element (77).

On the Success of PSK: Plasmid Stabilization or Plasmid Competition?

TA systems are widely present in plasmids and phages, mobile genetic elements, and chromosomes. It was first proposed that the selection of TAs in plasmids could be due to their contribution to their vertical stable inheritance (stability hypothesis). This hypothesis predicts that the number of plasmid-containing cells in a population should increase in the presence of TAs. However, an evaluation of this prediction made in independent cultures with or without TAs encoded in plasmids indicated that the total number of plasmid-containing cells remained similar in both types of cultures. An alternative hypothesis, the competition hypothesis, proposed that these modules were selected in plasmids because they allow them to exclude TA-free plasmids (78, 79). This last hypothesis implied that TA systems have been selected during horizontal plasmid propagation rather than associated with the vertical propagation. The competition hypothesis was strongly supported by the fact that a plasmid containing the *parDE* TA system excluded an isogenic plasmid devoid of this maintenance module (Fig. 3) (78). This same observation has been reported independently for the TA system formed by a restriction-modification pair (80, 81). The authors concluded that PSK does not increase plasmid stability but acts to mediate the exclusion of competing plasmids; they further proposed that the competition hypothesis could also apply to the evolution of virulence determinants or antibiotic resistance.

Some Evolutionary Considerations on TA Interactions

PSK associated with TA interactions could play a role in the evolution of TA systems. Some of these systems present in bacterial chromosomes could have been delivered by plasmids and thus acquired by horizontal gene transfer. A signature of this common origin has been detected in the *ccd* system present in the chromosome of *Erwinia chrysanthemi*. Because of the close similitude to the *ccd* system of plasmid F, the antitoxin of the chromosomal system protects the cell against the PSK associated with the loss of this plasmid (82). These authors suggested that this antiaddiction context could select plasmid toxin variants that no longer recognize the chromosomal antitoxin. The plasmid-encoded antitoxin could then evolve to achieve effective neutralization of this toxin variant. Eventually, this should originate independent TA systems. In fact, an intermediate situation in which a plasmidic CcdA antitoxin can neutralize the CcdB toxin of a *ccd* chromosomal system, but in which the chromosomal CcdA antitoxin does not neutralize the plasmidic CcdB toxin, has been reported. In this case, the host is not protected from PSK associated with the loss of the plasmid (see "Plasmid-Chromosome Cross-talk: One- or Two-Way Communications?" below).

An interesting study of the possible regions targeted by evolutionary process to avoid antiaddiction has been done in the *kis-kid* system of plasmid R1 and in the homologous *chpA* TA found in the *E. coli* chromosome. Functional studies indicated a residual activity of the ChpAI antitoxin on the Kid toxin that could only be detected at high gene dosage of the chromosomal system (83). Subsequent NMR analysis of Kis-Kid and ChpAI-Kid TA complexes clearly identified the regions involved in TA interactions, and native mass spectrometry defined the stoichiometry of these complexes (46). This functional and structural analysis indicated two different ways in which toxin neutralization occurs, one of them being efficient (self-TA interactions: heterohexamers in excess of the toxin) and the other being inefficient (crossed TA interactions: ChpAI-Kid heterotetramers). In self-Kis-Kid interactions, the C-terminal region of the antitoxin invaded the interprotomeric region of the dimeric toxin and disrupted the connection between the two protomers, inactivating one of the RNase cleavage sites and disrupting regions required for the binding of the RNA substrate. In the heterologous ChpAI-Kid interactions, the two C-terminal regions of

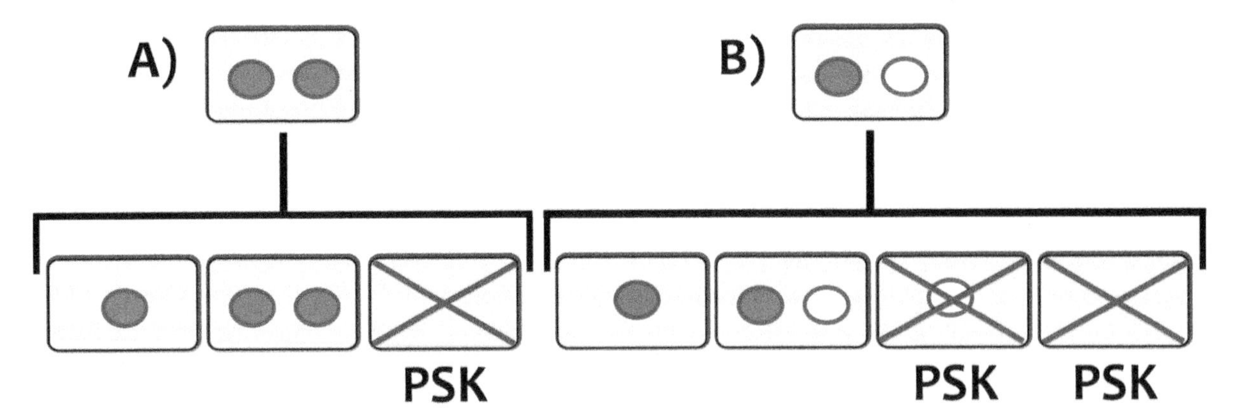

Figure 3 PSK in plasmid maintenance and in plasmid competition. The random distribution, at cell division, of two plasmid copies with or without a TA system, are shown, in (**A**) and (**B**), respectively. Filled circle, plasmid containing TA; open circle, plasmid without TA. Toxin is activated in cells that lose the TA plasmid, and this results in cell death or inhibition of cell proliferation (PSK). Elimination of plasmid-free cells (crossed cell) increases the proportion of plasmid-containing cells in the culture (maintenance phenotype). In **B**, only one of the two plasmids contains a TA system. Proliferation of cells containing a TA-free plasmid requires the presence of the TA plasmid. This gives a reproductive advantage to cells containing the TA plasmid (competition). doi:10.1128/microbiolspec.PLAS-0009-2013.f3

the dimeric antitoxin contact in a lateral way the two active sites of the dimeric toxin, thus preventing the access of the RNA to these sites. The interactions of the toxin and the amino-terminal region of the antitoxin that are crucial to orient the interaction of the C-terminal region of the antitoxin with the toxin in the Kis-Kid complex are lost in the ChpAI-Kid complex. This suggests that subtle changes in these interactions allowed efficient self-TA interactions and inefficient cross-TA interactions between homologous systems.

A signature of the evolution of TA systems from a common origin is found in members of the *kis-kid* and *ccd* systems of plasmid R1 and F, respectively. CcdB and Kid, the toxins of these two systems, have clearly different activities (RNase and antitopoisomerase, respectively) and share poor similitude at the amino acid sequencing level. Surprisingly, their structures are similar, indicating that they share a common structural module (84). The fact that critical regions for activity in CcdB and Kid toxins are located in different regions of this common module opens the possibility that the two toxins evolved from an ancestor protein containing both activities. This common ancestor remains to be identified. A comparative analysis of the toxins and antitoxins of these systems and their interactions suggested that the toxins and antitoxins of the *ccd* and *kis-kid* systems derive from a common ancestor (85). Furthermore, the analysis indicated that the activities of their toxins are clearly separated: there is no residual antitopoisomerase activity in Kid nor a residual RNase activity in CcdB. In addition, the analysis showed that the Kis antitoxin could neutralize the CcdB toxin, albeit very inefficiently. Furthermore, a distorted trait of CcdA-Kid interactions resulting in enhanced RNase activity of the Kid toxin was demonstrated. The complexes involved in self- and cross-interactions and their stoichiometry were clearly identified and defined by native mass spectrometry. This analysis also revealed distorted interactions between heterologous toxin-antitoxin pairs. On the whole, the analysis suggested that the two systems evolved from a common ancestor in a process in which the toxins reached different targets and the antitoxin coevolved with their toxins to neutralize them.

A variant of this situation occurs between ParE, the toxin of the *parDE* system of plasmid RK2, and RelE, the toxin of the *relBE* system of *E. coli*. ParE inactivates DNA gyrase as CcdB does, but its structure is similar to RelE (86) rather than to CcdB. The structures of the RelE and Kid toxins are also different, pointing to the evolution of *kis-kid*/*ccd* and *parDE*/*relBE* superfamilies from different TA ancestors. Note that Kid and RelE are both RNases, but that they cleave RNA in different contexts: RelE cleaves mRNA in a ribosomal (translation) context (87), and Kid is an RNase that cleaves RNA independently of ribosomes (43, 88).

PSK VIA CHROMOSOMALLY ENCODED TAs

Are Mobile Genetic Elements Stabilized by Chromosomally Encoded TA Systems?

The finding that copies of genes encoding toxin-antitoxin pairs were found in many of the genomic islands integrated within the chromosome of many bacteria showed that these mobile genetic elements (MGEs) could make use of the ways already "discovered" by plasmids as to not be eliminated from the cell population; thus, the islands would have benefited by the use of TA systems (89, 90). An immediate question derived from this idea is whether these TAs are involved in the maintenance of the islands through PSK: excision from the chromosome could represent a risk for the island if the host cell divides before it is integrated back to the chromosome, transferred to another recipient cell, or is replicated, increasing its copy number and thus enhancing the probability of stable inheritance. Would these possibilities (depicted as a simplified model in Fig. 4) condition the success of the mobile islands in spreading between its bacterial hosts? Would the presence of TAs on those islands, which are transferred by conjugation (without replication), be needed? Or would the TAs be operative as PSK systems for those that are lost without the chance of integrating back? An elegant, albeit partial, answer to these questions was provided by the work of Wozniak and Waldor on the *V. cholerae* 100-kb integrative and conjugative element (ICE) termed SXT (91). The authors showed that expression of the *mosAT* operon (a TA encoded within SXT) was increased when the mobile element was excised from the chromosome, thus leading to its stabilization by, most likely, PSK.

Superintegrons, which contain plenty of recombination sequences, surprisingly are found to be remarkably stable on the bacterial chromosome. Bioinformatics analyses revealed that genes encoding TAs were found on large superintegrons but not on the small ones. The stabilization of superintegron and a 165-kb dispensable chromosome region of *E. coli* by TAs were demonstrated, indicating that chromosomal TAs could limit the extensive genetic loss (92, 93). However, deletion of the entire chromosomal TA operon was feasible (94), which thus gives a hint that the presence of TA systems alone might not directly lead to stabilization of the bacterial genome. Perhaps the TAs expression levels or other yet to be discovered factors could influence the stable inheritance of some of these elements. An interesting finding showed that chromosomal TAs might exhibit PSK as plasmid TAs: *V. cholerae* contains two chromosomes and, strikingly, all the 13 annotated TAs were found on chromosome II within a 126-kb superintegron (4). Among them are three functional ParDE homologs, where *parDE1* (which is identical in both coding and promoter sequence with *parDE3*) and *parDE2* are regulated by their cognate ParD antitoxins, respectively, being ParE toxins only neutralized by their cognate ParD antitoxins. Expression of ParE toxin homologs, which inactivate DNA gyrase, led to DNA damage and SOS response in *V. cholerae*, presumably as a consequence of blocking cell division (95). More interestingly, the three ParE homologs were shown to degrade chromosome I in *V. cholerae* cells that had lost chromosome II, which indicates that chromosome missegregation stimulates ParE-mediated killing of aneuploid daughter cells (95), in the way similar to PSK by plasmid-encoded TAs. Nonetheless, although chromosome II of *V. cholerae* contains several essential genes that qualify it as a *bona fide* chromosome (96), this chromosome replicates like the plasmid/phage P1 (97). It is thus tempting to speculate that chromosome II could be evolved from an MGE.

In the case of the Gram-positive pathogenic bacterium *S. pneumoniae*, bioinformatics studies of the sequenced strains showed that it contained one or two mobile islands in many of them. One of these islands, a ~27-kb ICE termed Pneumococcal Pathogenicity Island 1 (PPI-1), was found to contain the *pezAT* TA (98, 99). Not all the *S. pneumoniae* strains contain *pezAT*: only 67% of 26 strains that had been analyzed encode *pezT* (98), which also coincide with our study on bioinformatics search of PezAT homologs found in 48 of the 48 annotated strains (65% harbor *pezAT*) (22). Strikingly, some of the strains carry two copies of *pezAT* on different ICEs: in strain ATCC700669, one copy of *pezAT* was found on PPI-1, and another copy of *pezAT* was discovered on an 81-kb Tn5253-like ICE (W. T. Chan and M. Espinosa, unpublished data). The absence of *pezAT* in some of the capsulated strains hinted that *pezAT* is not essential for virulence. However, the disruption of *pezT* resulted in strains with impaired virulence in mouse models of infection, showing its influence in virulence, despite the fact that *pezT* mutant did not show different growth patterns in laboratory broth, serum, or blood (98). Not all the reported ICEs from low-G+C bacteria harbor a cognate TA, as would be expected if the theory that the very existence of the ICE might be endangered by excision and lack of reintegration into the host in the case of lack of replication (see above and Fig. 4). We could speculate that replication of a TA-free ICE would generate more than one copy of it, which would increase the chances of the ICE to integrate back when there is no TA-mediated PSK. Although not mentioned by the authors, this could be, perhaps, the case of the ICEA genetic element from *Mycoplasma agalactiae*, a member of the *Mollicutes* class of bacteria related to lactococci and streptococci. We performed a search for TAs associated with this island and could not find any hint of a possible TA within the ICEA (100).

Chromosomally Encoded TAs: Is There Something Else?

As stated above, TAs were discovered on plasmids where they seem to function as ways to achieve successful plasmid maintenance via PSK (1, 21, 39, 41). In addition, more subtle processes, such as coupling plasmid replication and maintenance, were later shown for the R1-encoded *kis-kid* (75). However, when TAs are located on the chromosome their roles can, at least, be qualified as debatable, since different functions have been ascribed to them from stress response (101), to persistence and antibiotic tolerance (77, 102), and from programmed cell death and altruistic response (24) to ICE maintenance (91). However, it might be that chromosomally encoded TAs should be considered to have a different function depending on their genomic context or to provide any selective advantage to its host (22).

As for the pneumococcal YefM-YoeB TA, we have data showing that it is involved in biofilm formation (Moreno-Córdoba I, Chan WT, Moscoso M, Garcia E, Nieto C, Espinosa M, unpublished data), a role for TAs that has been shown for *E. coli* (103). The presence of a BOX element (a predicted mobile element present in more than 100 copies within the *S. pneumoniae* genome) placed upstream of the *yefM-yoeB* pneumococcal operon, had provided to the system an extra promoter that was independent of regulation by the YefM-YoeB protein complex (104). Location of BOX elements adjacent to genes related to competence and virulence has allowed us to speculate that at least some of the BOX elements might be involved in coordinating the expression and modulation of competence and virulence genes (22). Such a modulation could depend on the copy number of some of the BOX subelements and its orientation, as suggested by reference 105. It is worth noting that, just as with the TAs, competence and virulence can be also viewed as global responses of *S. pneumoniae* to stress.

PLASMID-CHROMOSOME CROSS-TALK: ONE- OR TWO-WAY COMMUNICATIONS?

To try to define a clear borderline between extra- and intrachromosomal elements can lead to confusion, since it is not so clear which is which. For instance, many rhizobia have a heavy genetic load of megaplasmids (considered as minichromosomes by some). Although undisputedly genetic loads to the bacteria, rhizobia are slow growers; some of the megaplasmids cannot be removed without loss of viability (106). Thus, the essentiality of the DNA cannot be considered as the basis to distinguish between extra- and intrachromosomal information: some megaplasmids cannot be considered as "dispensable" (107). A clearer example is, perhaps, provided by *V. cholerae* (see above), which has its genome divided into two chromosomes, chromosome I replicating like the *E. coli oriC*, whereas chromosome II replicates like the plasmid/phage P1 (97). Furthermore, bacterial genes including all contained within the bacteria-shared extrachromosomal gene pool, the so-called MGEs, are naturally dynamic entities, so that plasmid-plasmid and plasmid-chromosome interactions lead to exchanges of the gene cassettes; and these exchanges increase the overall bacteria biodiversity. Plasmids replicating by the rolling circle mechanism generate single-stranded DNA intermediates, a kind of molecule that is highly recombinogenic. Thus, these kinds of plasmids can cointegrate and excise by RecA-independent (illegitimate) recombination mechanisms using as little as 6- to 14 bp sequence homology (108, 109). Other MGEs, such as ICEs, integrative and mobilizable elements (IMEs), or pathogenicity islands (PIs), can promote transfer of chromosomal material to recipient cells, and many of these islands carry their own TAs.

In conclusion, dialogs between bacterial genomes and their mobilome do occur and can be facilitated by the horizontal transmission of MGEs and, as a result, novel adaptive functions will appear (110). Such new functions pertain to adaptation of bacteria to novel niches or to newly acquired external elements (be they independent like plasmids or integrated into the chromosome as "islands"). This would lead, in turn, to increase the modularity of the bacterial genomes so that the so-called variable genome is the one that would contribute to bacterial biodiversity and adaptation to changing environments (111). Under these circumstances, mutual dialogs based on control of plasmid genes by chromosome-encoded regulators, together with that of chromosomal functions regulated by plasmids, may certainly take place. Examples of these dialogs can be found (112), even though most of them have been uncovered in bacteria grown in laboratory conditions. Nevertheless, an impressive example was the finding that up to 4% of the genes encoded by the *E. coli* chromosome were influenced by the presence of plasmid F; and, furthermore, significant differences of the presence of F on different strains were found (113). These findings provided an even more exciting example that the plasmid-chromosome cross-talks not only take place, but also that they can vary between strains, perhaps as a result of differences in fitness of the strain-plasmid pair. Moreover, bioinformatics analyses performed with the cyanobacterium *Synechococcus* sp. PCC 7002 and six of its plasmids showed that there were a number of cross-talks taking place between the different plasmids as well as between plasmids and the chromosome of the bacterium (114).

For TAs, cross-talks have been reported to take place among chromosomally encoded operons (115). Measurements of the transcription levels of the *E. coli* of the *relBE* operon as a response to several ectopically expressed toxins (MazF, MqsR, YafQ, HicA, and HipA) showed that these toxins were able to activate the transcription of the former operon (115). Furthermore, ectopic expression of the VapC toxins from *Salmonella enterica* serovar Typhimurium LT2 chromosome and from *Shigella flexneri* 2a plasmid pMYSH6000 into *E. coli* led to inhibition of translation and, unexpectedly, to Lon-dependent *trans*-activation of the *E. coli* YoeB toxin, indicative of strong cross-interactions between different TAs from different bacteria and from different genomic origins as well (116). An even more exciting example was provided by the finding that the type II (proteic) TA operon *mqsR-mqsA* was able to control type V TA *ghoT-ghoS* by a complex mechanism involving recognition of the toxin MqsR of the unprotected mRNAs mostly at 5′-GCU sites (117). Under steady state (that is, unstressful conditions), MqsA antitoxin generates a stable and harmless complex with its cognate toxin MqsR. Similarly, and under the same growth conditions, GhoS antitoxin would prevent expression of *ghoT* by cleaving its transcript. However, when the cells are placed under stress, the intracellular Lon protease would degrade MqsA; this, in turn, would release free toxin MqsR, and further induce *mqsR* expression (118). The toxin MqsR will degrade all mRNAs not protected at 5′-GCU sites, like the 5′ end of *ghoST* mRNA. As a consequence, *ghoST* mRNA would be preferentially degraded in the *ghoS* coding region, while the *ghoT* coding region would remain uncleaved. This, in turn, would lead to increased levels of GhoT toxin and, ultimately, to disruption of the cell membrane. Then, this complex interaction between

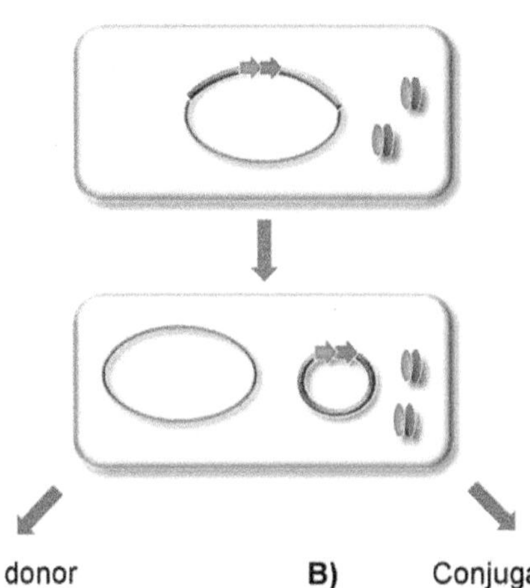

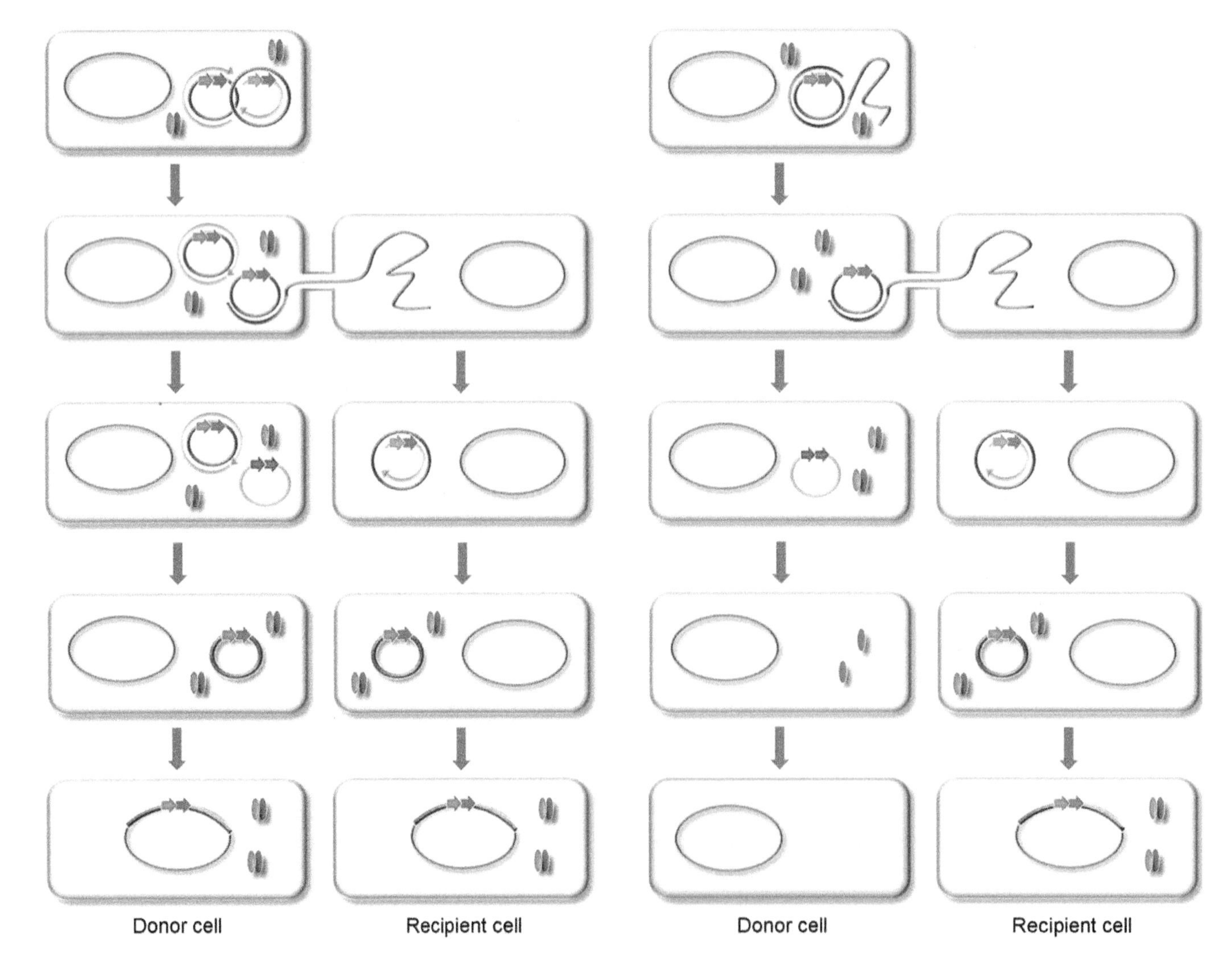
A) Conjugative transfer of ICE from a donor cell to a recipient cell after ICE replication
B) Conjugative transfer of ICE from a donor cell to a recipient cell without ICE replication
Donor cell
Recipient cell
Donor cell
Recipient cell

two very different TAs would lead to an increase in bacteria persistence (117). The above findings, in conjunction with those from Gerdes' group on the reduction of persisters as a consequence of the progressive deletion of 10 ribonuclease-encoding TAs (whereas deletion of individual operons had no detectable effects [119]), added another level of intricacy to the already complex scenario of the TA roles in the bacterial lifestyle. We conclude that it is important to take into account the possible cross-talks between the various TAs that exist within any given bacterium. In addition, scarce information, if any, is available on the possibility of cross-talks between plasmid-encoded and chromosomally encoded TAs.

In another context, bioinformatics approaches indicated that homologs of TAs are present within the genome of a bacterium (4); furthermore, the abundance of TAs of the same family in a single cell and whether they can be cross-talking to each other was somehow an intriguing possibility. A comprehensive functional analysis study reported that 20 VapBC homologs (of 47 putative VapBC homologs) of *Mycobacterium tuberculosis* were functional (120). In addition, a subset of four VapB antitoxin homologs were found to be able only to antagonize their own VapC toxicity but not the rest of the three VapC toxin homologs, indicating that cross-talk (in terms of counteracting individual toxicity) among the coexisting VapBC homologs in *M. tuberculosis* is unlikely (120).

In the case of the *ccd* TA, the chromosomally encoded ccd_{O157} and the ccd_F from F plasmid are coexisting in *E. coli* O157:H7 isolates in nature. Cross-interaction experiments showed that the chromosomal $\mathrm{CcdB_{O157}}$ toxin could be neutralized by the plasmidic $\mathrm{CcdA_F}$ antitoxin, but the plasmidic $\mathrm{CcdB_F}$ toxin was not able to be counteracted by the chromosomal $\mathrm{CcdA_{O157}}$ antitoxin (121). These findings would allow the coexistence or maintenance of plasmid F to be feasible despite the presence of the ccd_{O157} in the chromosome, because losing plasmid F would lead to PSK mediated by the plasmidic $\mathrm{Ccd_F}$ (121). On the other hand, the chromosomally encoded *Erwinia chrysanthemi* ccd_{Ech} could instead hinder the PSK exerted by the plasmidic Ccd_F (82). Because chromosomal $\mathrm{CcdA_{Ech}}$ antitoxin was able to inhibit the toxicity of plasmidic $\mathrm{CcdB_F}$ toxin, the chromosomal ccd_{Ech} was thus capable of protecting the cells from the invasion of MGEs that carry TA (ccd_F in this case) that mediate PSK, and, therefore, this chromosomal TA was termed antiaddiction module (82).

FURTHER THOUGHTS

It appears that TA homologs, or the same family of TAs that are encoded on plasmid, can also be found on chromosome, i.e., *ccd* (2, 121), albeit they may be more distantly related to each other (82). The function of plasmid-encoded TAs is usually associated with PSK and the chromosomally encoded TAs are suggested to play more complex roles in cell processes than PSK. These roles include general stress response, biofilm formation, persisters, stabilization of mobile elements in chromosome, development processes, virulence, or even essential for niches and survival (see above).

Do Chromosomally Encoded TAs Act Also as PSK?

The two chromosomally encoded *ccd* (ccd_{O157} and ccd_{Ech}) were shown to be unable to stabilize plasmid

Figure 4 Proposed model for a possible role of TA in stabilizing ICE by PSK during conjugative transfer. Under normal conditions, the TA genes (antitoxin gene is depicted as blue arrow; toxin gene is represented by red arrow) on the ICE (purple fragment) are expressed at a basal level within the chromosome. Toxin and antitoxin proteins (red and blue ovals, respectively) form tight complexes that are inert to the cell. During conjugative transfer, the ICE is excised from the chromosome and forms a circular mobilome. (**A**) The ICE replicates (one copy or more), and one copy of the ICE is transferred to the recipient cell through rolling circle (single-stranded DNA is transferred to the recipient cell and its complimentary DNA strand will be degraded gradually in the donor cell); another copy of the ICE remains in the donor cell. In the donor cell, the ICE is integrated back into the chromosome, while the transferred single-stranded DNA in the recipient cell will replicate to form an intact ICE, followed by integration into the chromosome. (**B**) The ICE is transferred to the recipient cell without replication in the donor cells. Since the donor cell has lost the TA-containing ICE, the remaining TA complexes will be triggered. The antitoxin proteins that are more susceptible to the degradation of the host proteases are degraded and not replenished owing to the loss of the TA-containing ICE, thus releasing the toxin activity that poisons the donor cell. On the other hand, the recipient cell, which has newly acquired a TA-containing ICE, will thus incorporate the ICE into the chromosome. This recipient cell is subject to the same fate as the donor cell if the ICE is lost. doi:10.1128/microbiolspec.PLAS-0009-2013.f4

(82, 122), and thus the chromosomal TAs were believed to play roles other than PSK. Conversely, the chromosomally encoded *relBE* locus of the *E. coli* was able to stabilize plasmid R1, although plasmid stabilization was not observed in a Lon-deficient strain, indicating that activation of RelBE-mediated PSK required the host Lon protease (123). In addition, as mentioned above, the ParE toxins within the superintegron of chromosome II of *V. cholerae* led to the PSK of cells that was missing chromosome II (124). Together with the examples of TAs stabilizing MGEs (91, 93), allows us to envisage a scenario in which a selfish invasion of a MGE to a bacterial cell should be counteracted by the invaded organism by means of a defensive behavior (6) or by an adaptive response, in which adaptation could lead to the acquisition of new selective advantages that could result in out-competing cognate and competitors, as well.

Do Plasmid-Encoded TAs Respond to Stress Like Chromosomally Encoded TAs?

An elegant study has shown an additional role of plasmid-encoded ccd_F as a transmissible persistence factor, besides PSK (125). A CcdB mutant, which is not toxic but still able to bind to the CcdA antitoxin, was found to be able to compete with the wild-type CcdB toxin to bind to its cognate CcdA antitoxin, and also de-repress the *ccd* promoter. With low expression (induced by 0.001% of arabinose) of CcdB mutant, cell growth inhibition was evident, and under this condition, the cells were exposed to lethal dosage of different antibiotics. As a consequence, the induced cells showed more surviving colonies in comparison with the uninduced ones, signifying that ccd_F increased tolerance and promote persistence (125).

Acknowledgments. *We gratefully acknowledge financial support from the Spanish MICINN (projects BFU 2008-01566 and CSD2008-00013). Ana María Hernández-Arriaga and Wai Ting Chan contributed equally to this chapter. Conflicts of interest: We declare no conflicts.*

Citation. **Hernández-Arriaga AM, Chan WT, Espinosa M, Díaz-Orejas R.** 2014. Conditional activation of toxin-antitoxin systems: postsegregational killing and beyond. Microbiol Spectrum **2**(5):PLAS-0009-2013.

References

1. **Gerdes K, Rasmussen PB, Molin S.** 1986. Unique type of plasmid maintenance function: postsegregational killing of plasmid-free cells. *Proc Natl Acad Sci USA* **83:** 3116–3120.
2. **Jaffe A, Ogura T, Hiraga S.** 1985. Effects of the *ccd* function of the F plasmid on bacterial growth. *J Bacteriol* **163:**841–849.
3. **Ogura T, Hiraga S.** 1983. Mini-F plasmid genes that couple host cell division to plasmid proliferation. *Proc Natl Acad Sci USA* **80:**4784–4788.
4. **Pandey DP, Gerdes K.** 2005. Toxin-antitoxin loci are highly abundant in free-living but lost from host-associated prokaryotes. *Nucleic Acids Res* **33:**966–976.
5. **Van Melderen L, Saavedra De Bast M.** 2009. Bacterial toxin-antitoxin systems: more than selfish entities? *PLoS Genet* **5:**1–6.
6. **Makarova KS, Wolf YI, Koonin EV.** 2013. Comparative genomics of defense systems in archaea and bacteria. *Nucleic Acids Res* **41:**4360–4377.
7. **Moller-Jensen J, Franch T, Gerdes K.** 2001. Temporal translational control by a metastable RNA structure. *J Biol Chem* **276:**35707–35713.
8. **Brzozowska I, Zielenkiewicz U.** 2013. Regulation of toxin-antitoxin systems by proteolysis. *Plasmid* **70:** 33–41.
9. **Hayes F, Van Melderen L.** 2011. Toxins-antitoxins: diversity, evolution and function. *Crit Rev Biochem Mol Biol* **46:**386–408.
10. **Masuda H, Tan Q, Awano N, Wu KP, Inouye M.** 2012. YeeU enhances the bundling of cytoskeletal polymers of MreB and FtsZ, antagonizing the CbtA (YeeV) toxicity in *Escherichia coli*. *Mol Microbiol* **84:** 979–989.
11. **Wang X, Lord DM, Cheng HY, Osbourne DO, Hong SH, Sanchez-Torres V, Quiroga C, Zheng K, Herrmann T, Peti W, Benedik MJ, Page R, Wood TK.** 2012. A new type V toxin-antitoxin system where mRNA for toxin GhoT is cleaved by antitoxin GhoS. *Nat Chem Biol* **8:**855–861.
12. **Leplae R, Geeraerts D, Hallez R, Guglielmini J, Dreze P, Van Melderen L.** 2011. Diversity of bacterial type II toxin-antitoxin systems: a comprehensive search and functional analysis of novel families. *Nucleic Acids Res* **39:**5513–5525.
13. **Satwika D, Klassen R, Meinhardt F.** 2012. Anticodon nuclease encoding virus-like elements in yeast. *Appl Microbiol Biotechnol* **96:**345–356.
14. **de la Cueva-Mendez G, Mills AD, Clay-Farrace L, Diaz-Orejas R, Laskey RA.** 2003. Regulatable killing of eukaryotic cells by the prokaryotic proteins Kid and Kis. *EMBO J* **22:**246–251.
15. **Slanchev K, Stebler J, de la Cueva-Mendez G, Raz E.** 2005. Development without germ cells: the role of the germ line in zebrafish sex differentiation. *Proc Natl Acad Sci USA* **102:**4074–4079.
16. **Gerdes K.** 2012. *In* Gerdes K (ed), *Prokaryotic Toxin-Antitoxins*, 1st ed. Springer, Heidelberg, Germany.
17. **Gerdes K, Christensen SK, Lobner-Olesen A.** 2005. Prokaryotic toxin-antitoxin stress response loci. *Nat Rev* **3:** 371–382.
18. **Couturier M, Bahassi el M, Van Melderen L.** 1998. Bacterial death by DNA gyrase poisoning. *Trends Microbiol* **6:**269–275.
19. **Buts L, Lah J, Dao-Thi MH, Wyns L, Loris R.** 2005. Toxin-antitoxin modules as bacterial metabolic stress managers. *Trends Biochem Sci* **30:**672–679.

20. **Condon C.** 2006. Shutdown decay of mRNA. *Mol Microbiol* **61:**573–583.

21. **Diago-Navarro E, Hernandez-Arriaga AM, Lopez-Villarejo J, Munoz-Gomez AJ, Kamphuis MB, Boelens R, Lemonnier M, Diaz-Orejas R.** 2010. *parD* toxin-antitoxin system of plasmid R1-basic contributions, biotechnological applications and relationships with closely-related toxin-antitoxin systems. *FEBS J* **277:** 3097–3117.

22. **Chan WT, Moreno-Córdoba I, Yeo CC, Espinosa M.** 2012. Toxin-antitoxin genes of the gram-positive pathogen *Streptococcus pneumoniae*: so few and yet so many. *Microbiol Mol Biol Rev* **76:**773–791.

23. **Yamaguchi Y, Park JH, Inouye M.** 2011. Toxin-antitoxin systems in bacteria and archaea. *Annu Rev Genet* **45:**61–79.

24. **Engelberg-Kulka H, Amitai S, Kolodkin-Gal I, Hazan R.** 2006. Bacterial programmed cell death and multicellular behavior in bacteria. *PLoS Genet* **2:**e135. doi: 10.1371/journal.pgen.0020135.

25. **Bernard P, Couturier M.** 1992. Cell killing by the F plasmid CcdB protein involves poisoning of DNA-topoisomerase II complexes. *J Mol Biol* **226:**735–745.

26. **Miki T, Park JA, Nagao K, Murayama N, Horiuchi T.** 1992. Control of segregation of chromosomal DNA by sex factor F in *Escherichia coli*. Mutants of DNA gyrase subunit A suppress letD (*ccdB*) product growth inhibition. *J Mol Biol* **225:**39–52.

27. **Bernard P, Kezdy KE, Van Melderen L, Steyaert J, Wyns L, Pato ML, Higgins PN, Couturier M.** 1993. The F plasmid CcdB protein induces efficient ATP-dependent DNA cleavage by gyrase. *J Mol Biol* **234:**534–541.

28. **Bernard P, Couturier M.** 1991. The 41 carboxy-terminal residues of the miniF plasmid CcdA protein are sufficient to antagonize the killer activity of the CcdB protein. *Mol Gen Genet* **226:**297–304.

29. **Madl T, Van Melderen L, Mine N, Respondek M, Oberer M, Keller W, Khatai L, Zangger K.** 2006. Structural basis for nucleic acid and toxin recognition of the bacterial antitoxin CcdA. *J Mol Biol* **364:**170–185.

30. **Maki S, Takiguchi S, Horiuchi T, Sekimizu K, Miki T.** 1996. Partner switching mechanisms in inactivation and rejuvenation of *Escherichia coli* DNA gyrase by F plasmid proteins LetD (CcdB) and LetA (CcdA). *J Mol Biol* **256:**473–482.

31. **Bahassi EM, O'Dea MH, Allali N, Messens J, Gellert M, Couturier M.** 1999. Interactions of CcdB with DNA gyrase. Inactivation of Gyra, poisoning of the gyrase-DNA complex, and the antidote action of CcdA. *J Biol Chem* **274:**10936–10944.

32. **Dao-Thi MH, Charlier D, Loris R, Maes D, Messens J, Wyns L, Backmann J.** 2002. Intricate interactions within the ccd plasmid addiction system. *J Biol Chem* **277:** 3733–3742.

33. **De Jonge N, Garcia-Pino A, Buts L, Haesaerts S, Charlier D, Zangger K, Wyns L, De Greve H, Loris R.** 2009. Rejuvenation of CcdB-poisoned gyrase by an intrinsically disordered protein domain. *Mol Cell* **35:** 154–163.

34. **Van Melderen L, Bernard P, Couturier M.** 1994. Lon-dependent proteolysis of CcdA is the key control for activation of CcdB in plasmid-free segregant bacteria. *Mol Microbiol* **11:**1151–1157.

35. **Van Melderen L, Thi MH, Lecchi P, Gottesman S, Couturier M, Maurizi MR.** 1996. ATP-dependent degradation of CcdA by Lon protease. Effects of secondary structure and heterologous subunit interactions. *J Biol Chem* **271:**27730–27738.

36. **Gerdes K, Gultyaev AP, Franch T, Pedersen K, Mikkelsen ND.** 1997. Antisense RNA-regulated programmed cell death. *Annu Rev Genet* **31:**1–31.

37. **Gerdes K, Wagner EG.** 2007. RNA antitoxins. *Currt Opin Microbiol* **10:**117–124.

38. **Weaver K.** 2013. TypeI toxin-antitoxin loci: *hok/sok* and *fst.*9-26. *In* Gerdes K (ed), *Prokariotic Toxin-Antitoxins*, 21st ed. Springer, Heidelberg, Germany.

39. **Bravo A, de Torrontegui G, Diaz R.** 1987. Identification of components of a new stability system of plasmid R1, ParD, that is close to the origin of replication of this plasmid. *Mol Gen Genet* **210:**101–110.

40. **Tsuchimoto S, Ohtsubo H, Ohtsubo E.** 1988. Two genes, *pemK* and *pemI*, responsible for stable maintenance of resistance plasmid R100. *J Bacteriol* **170:** 1461–1466.

41. **Bravo A, Ortega S, de Torrontegui G, Diaz R.** 1988. Killing of *Escherichia coli* cells modulated by components of the stability system ParD of plasmid R1. *Mol Gen Genet* **215:**146–151.

42. **Jensen RB, Grohmann E, Schwab H, Diaz-Orejas R, Gerdes K.** 1995. Comparison of ccd of F, parDE of RP4, and parD of R1 using a novel conditional replication control system of plasmid R1. *Mol Microbiol* **17:**211–220.

43. **Munoz-Gomez AJ, Lemonnier M, Santos-Sierra S, Berzal-Herranz A, Diaz-Orejas R.** 2005. RNase/anti-RNase activities of the bacterial *parD* toxin-antitoxin system. *J Bacteriol* **187:**3151–3157.

44. **Zhang J, Zhang Y, Zhu L, Suzuki M, Inouye M.** 2004. Interference of mRNA function by sequence-specific endoribonuclease PemK. *J Biol Chem* **279:** 20678–20684.

45. **Santos-Sierra S, Pardo-Abarrio C, Giraldo R, Diaz-Orejas R.** 2002. Genetic identification of two functional regions in the antitoxin of the *parD* killer system of plasmid R1. *FEMS Microbiol Lett* **206:** 115–119.

46. **Kamphuis MB, Monti MC, van den Heuvel RH, Santos-Sierra S, Folkers GE, Lemonnier M, Diaz-Orejas R, Heck AJ, Boelens R.** 2007. Interactions between the toxin Kid of the bacterial *parD* system and the antitoxins Kis and MazE. *Proteins* **67:**219–231.

47. **Ruiz-Echevarria MJ, Berzal-Herranz A, Gerdes K, Diaz-Orejas R.** 1991. The *kis* and *kid* genes of the *parD* maintenance system of plasmid R1 form an operon that is autoregulated at the level of transcription by the co-ordinated action of the Kis and Kid proteins. *Mol Microbiol* **5:**2685–2693.

48. **Monti MC, Hernandez-Arriaga AM, Kamphuis MB, Lopez-Villarejo J, Heck AJ, Boelens R, Diaz-Orejas R,**

van den Heuvel RH. 2007. Interactions of Kid-Kis toxin-antitoxin complexes with the *parD* operator-promoter region of plasmid R1 are piloted by the Kis antitoxin and tuned by the stoichiometry of Kid-Kis oligomers. *Nucleic Acids Res* **35:**1737–1749.

49. **Ruiz-Echevarria MJ, de la Cueva G, Diaz-Orejas R.** 1995. Translational coupling and limited degradation of a polycistronic messenger modulate differential gene expression in the parD stability system of plasmid R1. *Mol Gen Genet* **248:**599–609.
50. **Diago-Navarro E, Hernandez-Arriaga AM, Kubik S, Konieczny I, Diaz-Orejas R.** 2013. Cleavage of the antitoxin of the *parD* toxin-antitoxin system is determined by the ClpAP protease and is modulated by the relative ratio of the toxin and the antitoxin. *Plasmid* **70:**78–85.
51. **Roberts RC, Helinski DR.** 1992. Definition of a minimal plasmid stabilization system from the broad-host-range plasmid RK2. *J Bacteriol* **174:**8119–8132.
52. **Roberts RC, Strom AR, Helinski DR.** 1994. The *parDE* operon of the broad-host-range plasmid RK2 specifies growth inhibition associated with plasmid loss. *J Mol Biol* **237:**35–51.
53. **Jiang Y, Pogliano J, Helinski DR, Konieczny I.** 2002. ParE toxin encoded by the broad-host-range plasmid RK2 is an inhibitor of *Escherichia coli* gyrase. *Mol Microbiol* **44:**971–979.
54. **Davis TL, Helinski DR, Roberts RC.** 1992. Transcription and autoregulation of the stabilizing functions of broad-host-range plasmid RK2 in *Escherichia coli*, *Agrobacterium tumefaciens* and *Pseudomonas aeruginosa*. *Mol Microbiol* **6:**1981–1994.
55. **Eberl L, Givskov M, Schwab H.** 1992. The divergent promoters mediating transcription of the par locus of plasmid RP4 are subject to autoregulation. *Mol Microbiol* **6:**1969–1979.
56. **Roberts RC, Spangler C, Helinski DR.** 1993. Characteristics and significance of DNA binding activity of plasmid stabilization protein ParD from the broad host-range plasmid RK2. *J Biol Chem* **268:**27109–27117.
57. **Johnson EP, Strom AR, Helinski DR.** 1996. Plasmid RK2 toxin protein ParE: purification and interaction with the ParD antitoxin protein. *J Bacteriol* **178:**1420–1429.
58. **Oberer M, Zangger K, Gruber K, Keller W.** 2007. The solution structure of ParD, the antidote of the ParDE toxin antitoxin module, provides the structural basis for DNA and toxin binding. *Protein Sci* **16:**1676–1688.
59. **Lehnherr H, Maguin E, Jafri S, Yarmolinsky MB.** 1993. Plasmid addiction genes of bacteriophage P1: *doc*, which causes cell death on curing of prophage, and *phd*, which prevents host death when prophage is retained. *J Mol Biol* **233:**414–428.
60. **Makarova KS, Wolf YI, Koonin EV.** 2009. Comprehensive comparative-genomic analysis of type 2 toxin-antitoxin systems and related mobile stress response systems in prokaryotes. *Biol Direct* **4:**19. doi: 10.1186/1745-6150-4-19.
61. **Liu M, Zhang Y, Inouye M, Woychik NA.** 2008. Bacterial addiction module toxin Doc inhibits translation elongation through its association with the 30S ribosomal subunit. *Proc Natl Acad Sci USA* **105:**5885–5890.
62. **Hazan R, Sat B, Reches M, Engelberg-Kulka H.** 2001. Postsegregational killing mediated by the P1 phage "addiction module" *phd-doc* requires the *Escherichia coli* programmed cell death system mazEF. *J Bacteriol* **183:** 2046–2050.
63. **Magnuson R, Lehnherr H, Mukhopadhyay G, Yarmolinsky MB.** 1996. Autoregulation of the plasmid addiction operon of bacteriophage P1. *J Biol Chem* **271:** 18705–18710.
64. **Magnuson R, Yarmolinsky MB.** 1998. Corepression of the P1 addiction operon by Phd and Doc. *J Bacteriol* **180:**6342–6351.
65. **Garcia-Pino A, Balasubramanian S, Wyns L, Gazit E, De Greve H, Magnuson RD, Charlier D, van Nuland NA, Loris R.** 2010. Allostery and intrinsic disorder mediate transcription regulation by conditional cooperativity. *Cell* **142:**101–111.
66. **Lehnherr H, Yarmolinsky MB.** 1995. Addiction protein Phd of plasmid prophage P1 is a substrate of the ClpXP serine protease of *Escherichia coli*. *Proc Natl Acad Sci USA* **92:**3274–3277.
67. **Engelberg-Kulka H, Reches M, Narasimhan S, Schoulaker-Schwarz R, Klemes Y, Aizenman E, Glaser G.** 1998. *rexB* of bacteriophage lambda is an anti-cell death gene. *Proc Natl Acad Sci USA* **95:**15481–15486.
68. **Fineran PC, Blower TR, Foulds IJ, Humphreys DP, Lilley KS, Salmond GP.** 2009. The phage abortive infection system, ToxIN, functions as a protein-RNA toxin-antitoxin pair. *Proc Natl Acad Sci USA* **106:**894–899.
69. **Blower T, Fineran P, Johnson M, Toth I, Humphreys D, Salmond G.** 2009. Mutagenesis and functional characterization of the RNA and protein components of the toxIN abortive infection and toxin-antitoxin locus of Erwinia. *J Bacteriol* **191:**6029–6039.
70. **Chopin MC, Chopin A, Bidnenko E.** 2005. Phage abortive infection in lactococci: variations on a theme. *Curr Opin Microbiol* **8:**473–479.
71. **Blower TR, Pei XY, Short FL, Fineran PC, Humphreys DP, Luisi BF, Salmond GP.** 2011. A processed noncoding RNA regulates an altruistic bacterial antiviral system. *Nat Struct Mol Biol* **18:**185–190.
72. **Short FL, Pei XY, Blower TR, Ong SL, Fineran PC, Luisi BF, Salmond GP.** 2012. Selectivity and self-assembly in the control of a bacterial toxin by an antitoxic noncoding RNA pseudoknot. *Proc Natl Acad Sci USA* **110:**E241–E249.
73. **Nordstrom K.** 2006. Plasmid R1-replication and its control. *Plasmid* **55:**1–26.
74. **Ruiz-Echevarria MJ, de la Torre MA, Diaz-Orejas R.** 1995. A mutation that decreases the efficiency of plasmid R1 replication leads to the activation of *parD*, a killer stability system of the plasmid. *FEMS Microbiol Lett* **130:**129–135.
75. **Lopez-Villarejo J, Diago-Navarro E, Hernandez-Arriaga AM, Diaz-Orejas R.** 2012. Kis antitoxin couples plasmid R1 replication and *parD* (*kis*, *kid*) maintenance modules. *Plasmid* **67:**118–127.

76. **Pimentel B, Madine MA, de la Cueva-Mendez G.** 2005. Kid cleaves specific mRNAs at UUACU sites to rescue the copy number of plasmid R1. *EMBO J* **24:** 3459–3469.
77. **Lewis K.** 2010. Persister cells. *Annu Rev Microbiol* **64:** 357–372.
78. **Cooper TF, Heinemann JA.** 2000. Postsegregational killing does not increase plasmid stability but acts to mediate the exclusion of competing plasmids. *Proc Natl Acad Sci USA* **97:**12643–12648.
79. **Cooper TF, Paixão T, Heinemann JA.** 2010. Within-host competition selects for plasmid-encoded toxin-antitoxin systems. *Proc Biol Sci* **22:**3149–3155.
80. **Naito T, Kusano K, Kobayashi I.** 1995. Selfish behavior of restriction-modification systems. *Science (New York, NY)* **267:**897–899.
81. **Naito Y, Naito T, Kobayashi I.** 1998. Selfish restriction modification genes: resistance of a resident R/M plasmid to displacement by an incompatible plasmid mediated by host killing. *Biol Chem* **379:**429–436.
82. **Saavedra De Bast M, Mine N, Van Melderen L.** 2008. Chromosomal toxin-antitoxin systems may act as antiaddiction modules. *J Bacteriol* **190:**4603–4609.
83. **Santos-Sierra S, Giraldo R, Diaz-Orejas R.** 1997. Functional interactions between homologous conditional killer systems of plasmid and chromosomal origin. *FEMS Microbiol Lett* **152:**51–56.
84. **Hargreaves D, Santos-Sierra S, Giraldo R, Sabariegos-Jareno R, de la Cueva-Mendez G, Boelens R, Diaz-Orejas R, Rafferty JB.** 2002. Structural and functional analysis of the kid toxin protein from *E. coli* plasmid R1. *Structure* **10:**1425–1433.
85. **Smith AB, Lopez-Villarejo J, Diago-Navarro E, Mitchenall LA, Barendregt A, Heck AJ, Lemonnier M, Maxwell A, Diaz-Orejas R.** 2012. A common origin for the bacterial toxin-antitoxin systems *parD* and *ccd*, suggested by analyses of toxin/target and toxin/antitoxin interactions. *PloS One* **7:**e46499. doi: 10.1371/journal.pone.0046499.
86. **Francuski D, Saenger W.** 2009. Crystal structure of the antitoxin-toxin protein complex RelB-RelE from *Methanococcus jannaschii*. *J Mol Biol* **393:**898–908.
87. **Pedersen K, Zavialov AV, Pavlov MY, Elf J, Gerdes K, Ehrenberg M.** 2003. The bacterial toxin RelE displays codon-specific cleavage of mRNAs in the ribosomal A site. *Cell* **112:**131–140.
88. **Zhang Y, Inouye M.** 2009. The inhibitory mechanism of protein synthesis by YoeB, an *Escherichia coli* toxin. *J Biol Chem* **284:**6627–6638.
89. **Makarova KS, Wolf YI, Snir S, Koonin EV.** 2011. Defense islands in bacterial and archaeal genomes and prediction of novel defense systems. *J Bacteriol* **193:**6039–6056.
90. **Pandey DP, Gerdes K.** 2005. Toxin-antitoxin loci are highly abundant in free-living but lost from host-associated prokaryotes. *Nucleic Acids Res* **33:**966–976.
91. **Wozniak RA, Waldor MK.** 2009. A toxin–antitoxin system promotes the maintenance of an integrative conjugative element. *PLoS Genet* **5:**e1000439. doi: 10.1371/journal.pgen.1000439.
92. **Rowe-Magnus DA, Guerout AM, Biskri L, Bouige P, Mazel D.** 2003. Comparative analysis of superintegrons: engineering extensive genetic diversity in the Vibrionaceae. *Genome Res* **13:**428–442.
93. **Szekeres S, Dauti M, Wilde C, Mazel D, Rowe-Magnus DA.** 2007. Chromosomal toxin-antitoxin loci can diminish large-scale genome reductions in the absence of selection. *Mol. Microbiol* **63:**1588–1605.
94. **Christensen KS, Maenhauf-Michel G, Mine N, Gothesman S, Gerdes K, Van Melderen L.** 2004. Overproduction of the Lon protease triggers inhibition of translation in *Escherichia coli*: involvement of the *yefM-yoeB* toxin-antitoxin system. *Mol Microbiol* **51:**1705–1717.
95. **Yuan J, Yamaichi Y, Waldor MK.** 2011. The three vibrio cholerae chromosome II-encoded ParE toxins degrade chromosome I following loss of chromosome II. *J Bacteriol* **193:**611–619.
96. **Egan ES, Fogel MA, Waldor MK.** 2005. Divided genomes: negotiating the cell cycle in prokaryotes with multiple chromosomes. *Mol Microbiol* **56:**1129–1138.
97. **Srivastava P, Chattoraj DK.** 2007. Selective chromosome amplification in *Vibrio cholerae*. *Mol Microbiol* **66:**1016–1028.
98. **Brown JS, Gilliland SM, Spratt BG, Holden DW.** 2004. A locus contained within a variable region of pneumococcal pathogenicity island 1 contributes to virulence in mice. *Infect Immun* **72:**1587–1593.
99. **Khoo SK, Loll B, Chan WT, Shoeman RL, Ngoo L, Yeo CC, Meinhart A.** 2007. Molecular and structural characterization of the PezAT chromosomal toxin-antitoxin system of the human pathogen *Streptococcus pneumoniae*. *J Biol Chem* **282:**19606–19618.
100. **Dordet-Frisoni E, Marenda MS, Sagne E, Nouvel LX, Blanchard A, Tardy F, Sirand-Pugnet P, Baranowski E, Citti C.** 2013. ICEA of *Mycoplasma agalactiae*: a new family of self-transmissible integrative element that confers conjugative properties to the recipient strain. *Mol Microbiol* **89:**1226–1239.
101. **Christensen SK, Pedersen K, Hansen FG, Gerdes K.** 2003. Toxin-antitoxin loci as stress-response-elements: ChpAK/MazF and ChpBK cleave translated RNAs and are counteracted by tmRNA. *J Mol Biol* **332:**809–819.
102. **Moyed HS, Bertrand KP.** 1983. *hipA*, a newly recognized gene of *Escherichia coli* K-12 that affects frequency of persistence after inhibition of murein synthesis. *J Bacteriol* **155:**768–775.
103. **Kim Y, Wang X, Ma Q, Zhang XS, Wood TK.** 2009. Toxin-antitoxin systems in *Escherichia coli* influence biofilm formation through YjgK (TabA) and fimbriae. *J Bacteriol* **191:**1258–1267.
104. **Chan WT, Nieto C, Harikrishna JA, Khoo SK, Yasmin Othman R, Espinosa M, Yeo CC.** 2011. Genetic regulation of the *yefM-yoeB*$_{Spn}$ toxin-antitoxin locus of *Streptococcus pneumoniae*. *J Bacteriol* **193:**4612–4625.
105. **Knutsen E, Johnsborg O, Quentin Y, Claverys JP, Havarstein LS.** 2006. BOX elements modulate gene expression in *Streptococcus pneumoniae*: impact on the fine-tuning of competence development. *J Bacteriol* **188:**8307–8312.

106. **Petersen J, Frank O, Göker M, Pradella S.** 2013. Extrachromosomal, extraordinary and essential: the plasmids of the *Roseobacter* clade. *Appl Microbiol Biotechnol* **97:**2805–2815.

107. **González V, Santamaría RI, Bustos P, Hernández-González I, Medrano-Soto A, Moreno-Hagelsieb G, Janga SC, Ramírez MA, Jiménez-Jacinto V, Collado-Vides J, Dávila G.** 2006. The partitioned *Rhizobium etli* genome: Genetic and metabolic redundancy in seven interacting replicons. *Proc Natl Acad Sci USA* **103:** 3834–3839.

108. **Dempsey LA, Dubnau D.** 1989. Identification of plasmid and *Bacillus subtilis* chromosomal recombination sites used for pE194 integration. *J Bacteriol* **171:**2856–2865.

109. **Hahn J, Dubnau D.** 1985. Analysis of plasmid deletional instability in *Bacillus subtilis. J Bacteriol* **162:** 1014–1023.

110. **Baquero F.** 2004. From pieces to patterns: evolutionary engineering in bacterial pathogens. *Nat Rev Microbiol* **2:**510–518.

111. **Baquero F.** 2009. Environmental stress and evolvability in microbial systems. *Clin Microbiol Infect* **15:**5–10.

112. **Camacho EM, Serna A, Madrid C, Marqués S, Fernández R, de la Cruz F, Juárez A, Casadesús J.** 2005. Regulation of *finP* transcription by DNA adenine methylation in the virulence plasmid of *Salmonella enterica. J Bacteriol* **187:**5691–5699.

113. **Harr B, Schlötterer C.** 2006. Gene expression analysis indicates extensive genotype-specific crosstalk between the conjugative F-plasmid and the *E. coli* chromosome. *BMC Microbiol* **6:**80. doi:10.1186/1471-2180-6-80.

114. **Maida I, Fondi M, Papaleo MC, Perrin E, Fani R.** 2011. The gene flow between plasmids and chromosomes: insights from bioinformatics analyses. *Open Appl Informatics J* **5:**62–76.

115. **Kasari V, Mets T, Tenson T, Kaldalu N.** 2013. Transcriptional cross-activation between toxin-antitoxin systems of *Escherichia coli. BMC Microbiol* **13:**45. doi: 10.1186/1471-2180-13-45.

116. **Winther KS, Gerdes K.** 2009. Ectopic production of VapCs from *Enterobacteria* inhibits translation and trans-activates YoeB mRNA interferase. *Mol Microbiol* **72:**918–930.

117. **Wang X, Lord DM, Hong SH, Peti W, Benedik MJ, Page R, Wood TK.** 2013. Type II toxin/antitoxin MqsR/MqsA controls type V toxin/antitoxin GhoT/GhoS. *Environ Microbiol* **15:**1734–1744.

118. **Wang X, Kim Y, Hong SH, Ma Q, Brown BL, Pu M, Tarone AM, Benedik M, Peti W, Page R, Wood TK.** 2011. Antitoxin MqsA helps mediate the bacterial general stress response. *Nat Chem Biol* **7:**359–366.

119. **Maisonneuve E, Shakespeare LJ, Jørgensen MG, Gerdes K.** 2011. Bacterial persistence by RNA endonucleases. *Proc Natl Acad Sci USA* **108:**13206–13211.

120. **Ramage HR, Connolly LE, Cox JS.** 2009. Comprehensive functional analysis of *Mycobacterium tuberculosis* toxin-antitoxin systems: implications for pathogenesis, stress responses, and evolution. *PLoS Genet* **5:** e1000767. doi: 10.1371/journal.pgen.1000767.

121. **Wilbaux M, Mine N, Guerout AM, Mazel D, Van Melderen L.** 2007. Functional interactions between coexisting toxin-antitoxin systems of the ccd family in *Escherichia coli* O157:H7. *J Bacteriol* **189:**2712–2719.

122. **Wilbaux M, Mine N, Guerout A-M, Mazel D, Van Melderen L.** 2007. Functional interactions between coexisting toxin-antitoxin systems of the *ccd* family in *Escherichia coli* O157:H7. *J Bacteriol* **189:**2712–2719.

123. **Gronlund H, Gerdes K.** 1999. Toxin-antitoxin systems homologous with *relBE* of *Escherichia coli* plasmid P307 are ubiquitous in prokaryotes. *J Mol Biol* **285:** 1401–1415.

124. **Yuan J, Sterckx Y, Mitchenall LA, Maxwell A, Loris R, Waldor MK.** 2010. *Vibrio cholerae* ParE2 poisons DNA gyrase via a mechanism distinct from other gyrase inhibitors. *J Biol Chem* **285:**40397–40408.

125. **Tripathi A, Dewan PC, Barua B, Varadarajan R.** 2012. Additional role for the *ccd* operon of F-plasmid as a transmissible persistence factor. *Proc Natl Acad Sci USA* **109:**12497–12502.

126. **Volante A, Soberón NE, Ayora S, Alonso JC.** The interplay between different stability systems contributes to faithful segregation: *Streptococcus pyogenes* pSM19035 as a model, p 193–207. 2015. *In* Tolmasky ME, Alonso JC, *Plasmids: Biology and Impact in Biotechnology and Discovery*. ASM Press, Washington, DC.

Plasmids—Biology and Impact in Biotechnology and Discovery
Edited by Marcelo E. Tolmasky and Juan C. Alonso

doi:10.1128/microbiolspec.PLAS-0007-2013

Andrea Volante,[1] Nora E. Soberón,[1] Silvia Ayora,[1] and Juan C. Alonso[1]

11

The Interplay between Different Stability Systems Contributes to Faithful Segregation: *Streptococcus pyogenes* pSM19035 as a Model

INTRODUCTION

The long-term maintenance of low-copy plasmids cannot rely on random distribution to ensure their proper propagation to daughter cells at cell division. It relies largely on plasmid-encoded functions that promote their replication and ensure their faithful vertical transmission to daughter cells. Low-copy plasmids make use of several different genetic loci, which usually map outside of the minimal replicon, to ensure almost absolute plasmid stability (1, 2). A subset of plasmid-encoded functions ensures one or two rounds of replication per cell cycle and the distribution of the replicated copies to nascent daughter cells. Others halt proliferation of cells that do not receive a plasmid copy or regulate the interplay between the segregation (*seg*) loci among them and with the replication locus. To study this interplay pSM19035, a *Streptococcus pyogenes* low-copy number plasmid (2 ± 1 copies/cell) widely distributed in bacteria of the *Firmicutes* phylum, was used as a model system (Fig. 1) (3). This plasmid as well as *Streptococcus agalactiae* pIP501 (4), *Enterococcus faecalis* pAMβ1, pW9-2, and pRE25 (5, 6, 7), *Enterococcus faecium* pIP186, and pVEF series (pVEF1, pVEF2, and pVEF3) (8, 9) belong to the *inc18* incompatibility group (Fig. 2) (10, 11). With few exceptions, the nucleotide sequences of the minimal backbone region required for plasmid replication and stable segregation are highly conserved (>92% identity). The *inc18* plasmids, which are ubiquitous among erythromycin/lincomycin-, vancomycin-, or methicillin-resistant bacteria, can be divided in three large groups: (i) non-self-transmissible plasmids, with duplicated and inverted replication (IR) and stable maintenance regions, separated by long nonrepeated (NR) segments, represented by pSM19035 (Fig. 1A), pSM22095 and pSM10419 (3, 12); (ii) self-transmissible or non-self-transmissible, with directly repeated (DR) regions separated by long NR segments, represented by pRE25 self-transmissible or pIP816 and the plasmids

[1]Departamento de Biotecnología Microbiana, Centro Nacional de Biotecnología, CNB-CSIC, 28049 Madrid, Spain.

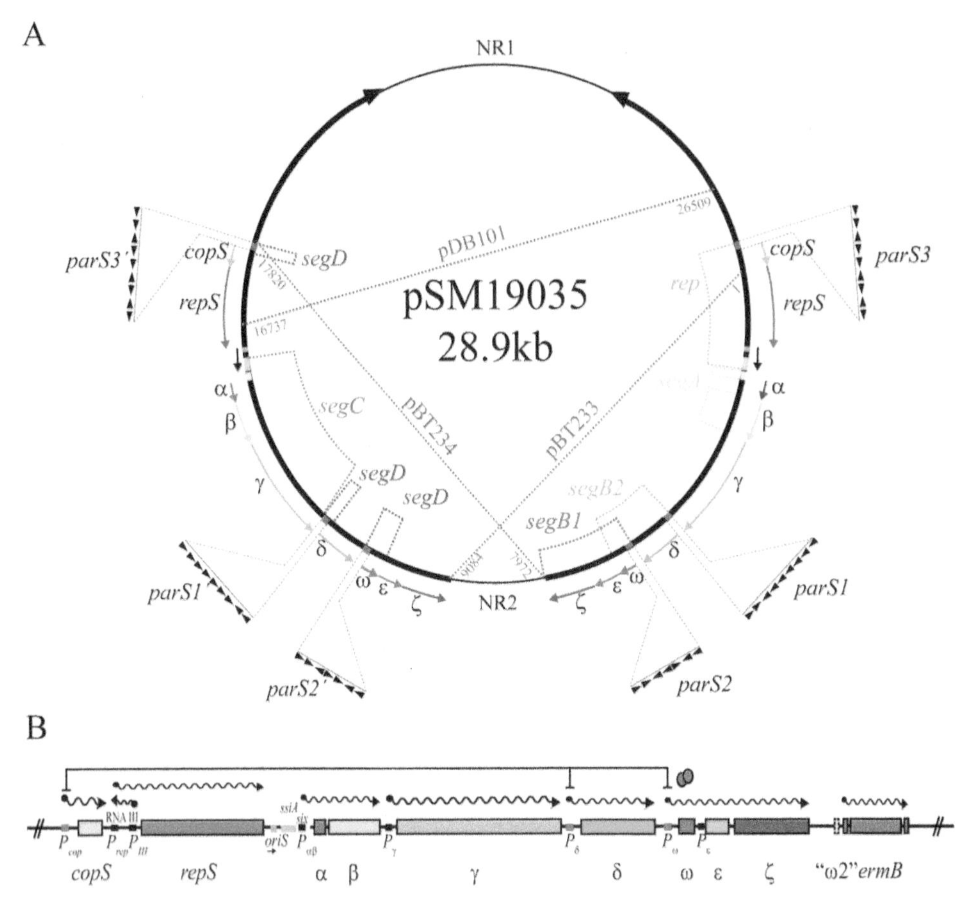

Figure 1 Genome organization of plasmid pSM19035 and its derivatives. (**A**) pSM19035 or pDB101 duplicated sequences, which comprise ~80 and 68% of the molecule, respectively, are indicated by thick arrows, and the unique NR1 and NR2 regions are indicated by thin lines. The region deleted in pDB101 is contained between residues 16737 and 26509 of pSM19035. pBT233 and pBT234 contain residues 1 to 9084 and 7972 to 17820 of pSM19035, respectively (represented by dashed lines in the map). Number 1 was chosen arbitrarily to be the first nucleotide of the ninth codon of *copS*. The loci involved in replication (*rep*) and segregation (*seg*) are indicated. The upstream region of the promoters of the *copS*, δ and ω genes, which constitute the six *cis*-acting centromere-like *parS* sites (red boxes), are blown up. A *parS* site consists of a variable number of contiguous 7-bp heptad repeats (iterons) symbolized by ▶ (in direct) or ◀ (invert orientation), and the number of repeats and their relative orientations are indicated. The colored outer thin arrows indicate the organization of the genes. For the sake of simplicity, the *rep* and *seg* loci are indicated only once although they are repeated twice. (**B**) The promoters (*P*), the mRNAs, and the genes are shown and denoted as boxes, wavy lanes, and rectangles, respectively. The leading-strand replication origin (*oriS*, orange), the lagging-strand replication origin (*ssiA*, light blue), the *six* sites (yellow), and the direction of replication (black arrows) are denoted. Protein ω_2-mediated transcriptional repression is indicated (red ovals). The plasmid region involved in replication is marked as *rep* (involving CopS, RepS, and *oriS* and *ssiA*). The regions involved in stable segregation are five: *segA* (β_2 and *six* site), *segB1* (ε_2 and ζ), *segB2* (δ_2, ω_2, and six *parS* sites), *segC* (α, β_2, γ, and *ssiA* and *six* sites) and *segD* (ω_2 and P_{cop}, P_δ, and P_ω sites [denoted as red boxes]). doi:10.1128/microbiolspec.PLAS-0007-2013.f1.

of the pVEF series (pVEF1, pVEF2, or pVEF3) non-self-transmissible (Fig. 2) (6, 9, 13); and (iii) self-transmissible, that contain neither IR nor DR segments, represented by pAMβ1 and pIP501 (Fig. 2) (4, 5, 14). These plasmids, which replicate "unidirectionally" via the θ or circle-to-circle replication mode (15, 16, 17, 18, 19), encode a sophisticated control to warrant that a sufficient number of copies populates daughter cells.

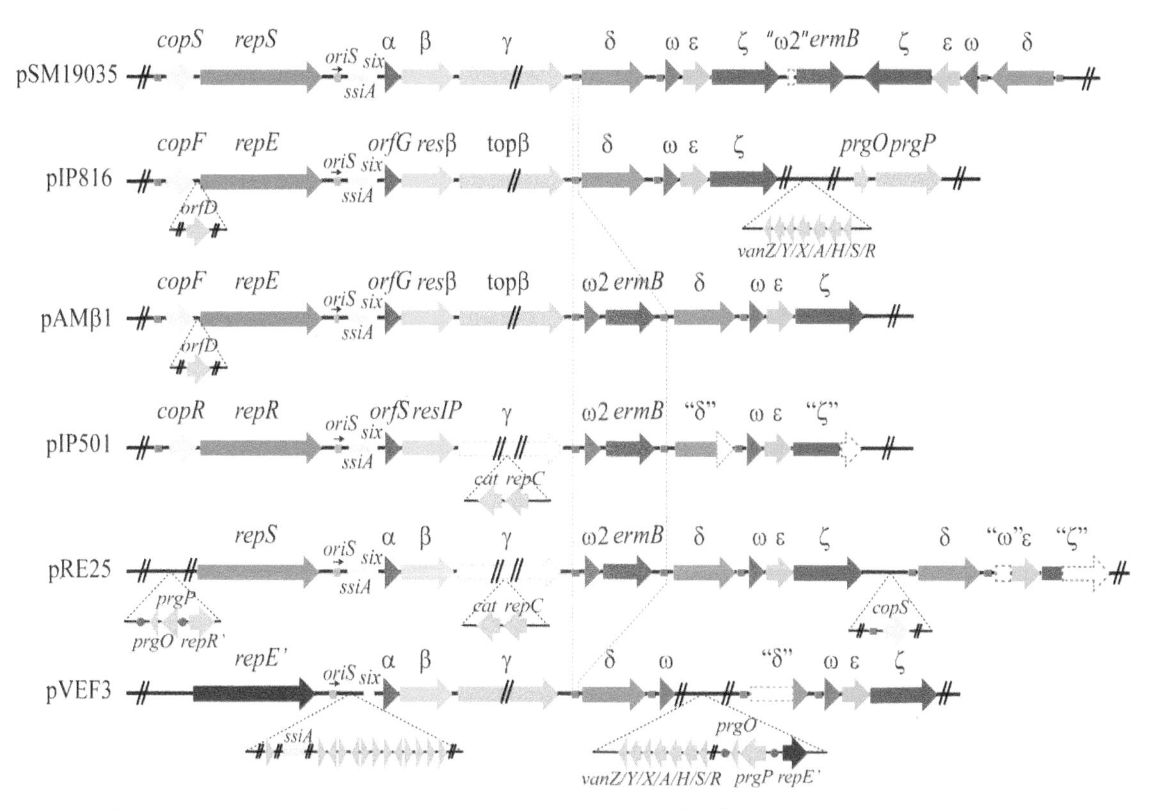

Figure 2 Genetic organization in plasmids of the *inc18* family. The genomic organization at the *rep* and *seg* loci of the relevant pSM19035, pIP816, pAMβ1, pIP501, pRE25, and pVEF3 (as representative of the pVEF series) plasmids is shown schematically. The conserved color code indicates that the gene products are highly conserved (>89% identity) within the family. The quotation marks surrounding a gene denote that this gene contains deletions and/or point mutations. Similar *parS* sites are linked by vertical broken lines. A double bar indicates that the corresponding gene/region is out of scale. In pRE25, the putative centromere sites of the second *par* system (*segE*) are indicated as red circles. doi:10.1128/microbiolspec.PLAS-0007-2013.f2.

A minimal replicon is defined as the smallest segment of a plasmid necessary and sufficient for autonomous replication with wild-type copy number. The smallest autonomous replicating segment of *inc18* plasmids does not qualify as a minimal replicon. This segment contains a bipartite replication origin (*oriS* and *ssiA*), the structural gene (*rep*) that codes for a replication initiation protein (e.g., RepS in pSM19035), and genetic information for control of *rep* gene expression (CopS, a countertranscribed antisense RNA [RNAIII] and a *cis*-acting element) (Fig. 1B) (10, 15, 17, 18, 20, 21, 106). In plasmids containing long IRs, functions other than the *rep* locus are essential for plasmid replication (Fig. 1A) (22). In contrast, in plasmids lacking IR regions, this minimal replicon requires these functions to keep the copy number of the parental plasmid. It is likely that functions mapping outside of the minimal replicon are necessary to overcome any potential roadblock, which might reduce the overall replication rate, or remove any remaining catenates (22). In the present report, the plasmid replication machinery will be discussed in the context of the interplay between stabilization and replication functions.

pSM19035 has evolved different stability systems that minimize plasmid loss at cell division, coordinate replication, segregation, and cell growth making plasmid maintenance extremely stable. Sequence analysis has revealed six different loci (*rep*, *segA*, *segB1*, *segB2*, *segC*, and *segD*) in the IR regions that are required for the structural and segregational stability of a derivative of pSM19035 (i.e., pDB101) in *Bacillus subtilis* (Fig. 1A) (21, 23). pSM19035-derivatives lacking IR regions (e.g., pBT233 and pBT234; Fig. 1A) replicate autonomously and are $>10^6$-fold more stable than expected for random segregation (21), suggesting the presence of discrete regions outside the minimal replicon that may contribute to the stabilization of the plasmid (*seg* loci). As we will describe in more detail later in this review, there are three systems that directly contribute to stable maintenance of low-copy-number plasmids (defined by functions encoded by the *segA*, *segB1*, and *segB2* loci) and an avenue involved in facilitating the processing of replication intermediates (*segC*) (22). There is also a system that contributes to coordinate copy number control and stable inheritance (*segD*). The interplay between the different *seg* loci facilitates the separation of sister plasmids and, in coordination with the replication machinery, allows *bona fide* replication and stable inheritance. Most of the studies described here have been performed using *B. subtilis* as a host, with derivatives of pSM19035 (pDB101, pBT233, etc.) (Fig. 1A).

THE *segA* LOCUS

An oligomeric plasmid molecule, generated by recombination-dependent crossover or by replication of oligomers, has to be resolved to monomers by conservative site-specific reciprocal recombination between two short directly oriented cognate DNA sequences (24, 25, 107). A *segA*-encoded function, which catalyzes the resolution of plasmid dimers (or higher-order oligomers) into monomers, maximizes the distribution of plasmid copies to nascent daughter cells and stabilizes the plasmids ~5-fold (21). The *segA* locus encodes α and β genes, and two inversely oriented *six* sites (Fig. 1A). Protein β, which exists as a dimer (β_2) (26), regulates the expression of the locus (27). The *orf*α, which has different names in other members of the *inc18* family (*orfG*, *orfS*, *etc.*; see Fig. 2), shows relatively low levels of identity (<45%) at the nucleotide level, when compared with the β gene or with the upstream untranslated region (which contains the lagging strand replication origin [*ssiA*] and the *six* site) that share >91% identity among the different members of *inc18* (22). The β_2 resolvoinvertase binds to a 90-bp *six* site, which can be divided in two adjacent subsites, one of them named synaptic subsite I where recombination takes place, and an accessory subsite II (24, 26, 28, 29, 30, 31). Two β_2 resolvoinvertases bind to each subsite; β_2 bound to IR sequences at subsite I enhances its binding to subsite II that contains only a half binding site (one arm of the IR region present in subsite I) (24, 26, 29, 30, 32, 33). Resolvoinvertase β_2 requires a sequence-independent DNA-bending protein, such as Hbsu histone-like protein, to stabilize a synaptic complex with a relative geometry of the *six* sites and for catalyzing both resolution and inversion activities at its cognate sites (29, 32, 34, 35). While the eukaryotic HMGB1 histone-like protein can efficiently replace Hbsu as β_2-accessory protein (36, 37, 38, 39, 40, 41, 42), this is not the case for numerous other bacterial DNA-bending proteins. For example, *Escherichia coli* encodes for at least six histone-like proteins: HU, IHF, Fis, H-NS, Hha, and Lrp. HU efficiently stimulated β_2-mediated recombination, the effect of IHF was partial, whereas the stimulatory effect of the chromatin-associated proteins Fis, H-NS, Hha, and *B. subtilis* LrpC was undetectable (35, 43).

Hbsu mainly binds between both subsites and stabilizes a synaptic complex with a relative topology of the *six* sites. When β_2 binds the subsite I it limits Hbsu binding. As a consequence, β_2 mediates resolution between two directly oriented *six* sites on a supercoiled dimeric DNA substrate (32; see below). DNA supercoiling helps to overcome the energetic barrier; resolution does not occur in relaxed DNA (32, 34, 35, 44).

A *segA* locus, contributes to genetic isolation by impairing horizontal gene transfer via plasmid transformation or plasmid transduction (45, 46, 47). Plasmid DNA, released from bacterial populations that harbored plasmids consisting of monomers-only due to *segA* activity, cannot transform natural competent cells (*B. subtilis*) or do so with very low efficiency (*Streptococcus pneumoniae*) (48, 49), whereas plasmid DNA, released from bacterial populations that harbor even a low proportion of plasmids dimers (~5%), can efficiently transform naturally competent cells (45). In *Firmicutes*, bacteriophage-mediated transduction is one of the main forms of plasmid horizontal gene transfer (45, 46, 47). A plasmid-transducing particle, which consists of an oligomeric head-to-tail molecule, can be generated either via viral-directed plasmid replication or by multimeric plasmid replication, as observed in the absence of end-processing functions (e.g., in the *addAB* [counterpart *recBCD*] context) (50, 51, 52). If the *segA* locus catalyzes the resolution of the plasmid oligomers on the donor strain, the concatemer should not have the required length for efficient packaging into a viral prohead of a generalizing transducing phage, and indirectly reduces plasmid transduction. However, following introduction of the transducing plasmid DNA, and after circularization, the *segA* locus positively contributes to resolve the concatemeric plasmid molecule (53, 54).

THE *segC* LOCUS

The *segC* locus comprises three *trans*-acting products (*orf*α, the resolvoinvertase β_2 [described as part of the *segA* locus], and the type I topoisomerase γ), and two inversely oriented *cis*-acting regions: *ssiA* and *six*

(Fig. 1A). The mode of action of the *segC* locus is briefly described in this section. For a more detailed description, the reader is referred to a recent review (22).

Plasmids pSM19035 and pDB101 have two perfect long IR segments, which encompass ~80 and 68% of the molecule, respectively, separated by two NR sequences (NR1 and NR2) of different lengths (Fig. 1A). Both plasmids lack genuine dedicated replication termination (*ter*) regions (18, 55). The essential role of the *segC* locus in these plasmids with IR regions is illustrated by mutations in the β gene, or a deletion of the *six* site, which abolish the intramolecular recombination event that causes the flipping of the NR2 region, leads to the accumulation of knotted replication intermediates, and causes gross genome rearrangements (27, 30, 55, 56). Replication of pSM19035 or pDB101 may proceed "bidirectionally" from each of the inversely oriented replication origins (a ~50-bp *oriS* region where leading strand synthesis begins, and the ~150-bp lagging strand *ssiA* site) located downstream of the *rep* gene and immediately upstream of the *segC* locus (15, 16, 17, 18). Upon firing unidirectional replication from one of the origins, β_2-mediated inversion between the two inversely oriented *six* sites flips the orientation of this replication fork with respect to the still unfired second origin (see reference 22). Firing of the second origin should lead to unidirectional replication with both forks traveling in the same direction around a circular monomer template (see reference 22). After one full round of replication, the forks undergo β_2-mediated inversion, and replication is terminated at the site where the two forks meet (see reference 22). The flip back should lead to two equimolar monomeric forms that differ from each other in the orientation of the two unique halves (the NR sequences; Fig. 1A) as detected in plasmid pDB101 purified from *B. subtilis* cells (18, 55).

The *segC* locus in plasmids lacking IR regions is partially dispensable and works as a helper of the *rep* locus (Fig. 2) (15, 16, 55, 57). In these plasmids, the *segC* locus modulates the synchronization of leading- and lagging-strand DNA synthesis, controls the termination of replication, and stabilizes by ~25-fold plasmid segregation by reducing a fitness cost (55). Indeed, deletion of the *ssiA* site, which lies ~50 bp downstream from *oriS* (15, 58), leads to the accumulation of linear high-molecular-weight DNA and high plasmid structural and segregational instability (20).

THE *segB* LOCI

Low-copy plasmids rely on segregation systems to ensure that each daughter cell harbors newly replicated plasmids or to induce the proliferation stop of those cells that lost all plasmid copies or did not receive at least a plasmid copy at cell division (1, 2, 59, 60, 108, 109). Accurate distribution of a newly replicated genome to daughter cells at cell division is a precise process, but subject to occasional error or inactivation by spontaneous mutations (expected to be 0.5×10^{-8} to 5×10^{-8}). pDB101 is lost at a frequency $<1 \times 10^{-7}$ per generation, suggesting that it has highly efficient segregation functions to warrant stable (better-than-random) segregation at cell division (21). The *segB* region, whose expression is under the control of the ω_2-transcriptional regulator, is divided in two discrete loci, *segB1* and *segB2* (Fig. 1A and 1B). The *segB1* locus encodes the ω_2 repressor, the dimeric ε (ε_2) antitoxin and the monomeric ζ toxin (21, 61, 62). When pBT233 or pBT234 variants were analyzed, the toxin-antitoxin [TA] system (Fig. 1A) stabilizes the plasmid >10,000-fold by inhibiting the proliferation of plasmid-free cells (21, 61). The *segB2* locus comprises two *trans*-acting proteins (δ_2-ATPase, and ω_2 acting as a centromeric-binding protein [CBP]) and six *parS* sites (21, 63, 64). The *segB2* locus is responsible for the partition of plasmid copies to daughter cells at cell division (Fig. 1), increasing by ~50-fold the stability of pBT233 miniderivatives (65). Some members of the *inc18* family encode a second ParAB-like system (*segE* locus) (Fig. 2). The analysis of this *segE* locus, the mechanisms that underlie the segregation of plasmids with two genuine partition systems (*segB2* and *segE*), and the interplay between them remain to be characterized.

The *segB1* Locus

Stabilization of low-copy-number plasmids in the bacterial population can be enhanced by the action of TA systems that contribute to plasmid stability through the postsegregational proliferation halt of the cell that has lost the plasmid. Five different types of TA systems have been described to date (reviewed in references 59, 66, 67, 68, and 109).

The TA system of *inc18* plasmids is genetically linked to antibiotic resistance genes. Indeed, the spread of the resistance to vancomycin, methicillin, gentamycin, erythromycin, linezolid, glycopeptide resistance, and multiresistant *cfr* among enterococci and staphylococci (7, 9, 13, 69, 70, 71) could be mainly attributed to their coexistence with the plasmid-borne ζ-ε TA module (22). Unlike the majority of TA systems that are autogenously controlled, the expression of the ζ-ε TA module is regulated by the ω_2-transcriptional regulator, which regulates its own synthesis and also controls plasmid copy number (see "The *segD* Locus"

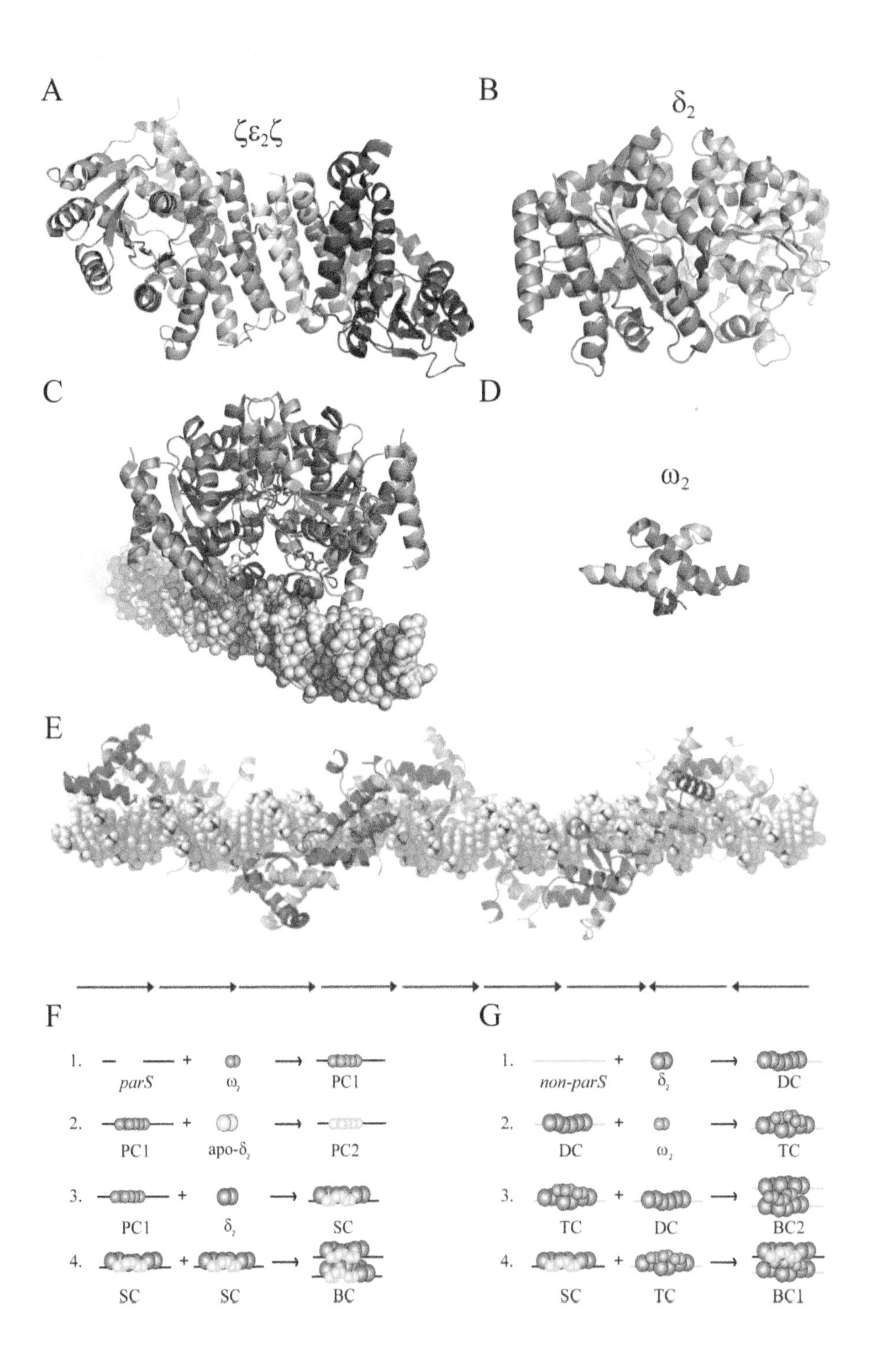
A
ζε₂ζ
B
δ₂
C
D
ω₂
E
F
1. parS ω₂ PC1
2. PC1 apo-δ₂ PC2
3. PC1 δ₂ SC
4. SC SC BC
G
1. non-parS δ₂ DC
2. DC ω₂ TC
3. TC DC BC2
4. SC TC BC1

below). Conditions that facilitate antitoxin ε_2 degradation, by the LonA and/or ClpXP protease, permit ζ toxin to act freely to block cell proliferation (61, 62, 72). Therefore, the potential threat that is posed by the accumulation of multiple resistance genes on plasmids with a TA module is of great concern.

The *Firmicutes* $\zeta\varepsilon_2\zeta$ TA module consists of two monomeric long-living 287-amino-acid ζ toxins (half-life > 60 min) separated by a short-lived 90-amino-acid-long ε_2 antitoxin (half-life < 20 min) (Fig. 3A) (72, 73, 74). The superfamily of ζ toxins, which are bacteriostatic by nature (61, 75), are among the most abundant in bacteria of the *Firmicutes* phylum (76). Massive ζ-toxin overexpression leads to cell death (77, 78).

The TA system is the ultimate stabilization function of pSM19035 (21, 23) and many other plasmids of the *inc18* family. This TA system was also found in the chromosome of *S. pneumoniae* (known as PezAT [79]) as well as in the chromosome of many bacteria of the *Enterococcus* genus. A minimal replicon carrying the *segB1* locus will be stably maintained by the contribution of a long-living plasmid-encoded toxin that inhibits the proliferation of plasmid-free daughter cells. However, this maintenance is not absolute. The rate of cured cells should be between the subpopulation of noninheritable toxin-resistant cells (1×10^{-5} to 5×10^{-5} toxin tolerants) and toxin-inactive mutants (0.5×10^{-8} to 5×10^{-8} spontaneous mutants) (61, 75).

Toxin ζ inhibits the first step of peptidoglycan biosynthesis by phosphorylating the 3′-OH group (3P) of the amino sugar moiety of uridine diphosphate-*N*-acetylglucosamine (UNAG), leading to the accumulation of a fraction of unreactive UNAG-3P *in vitro* (78). Within the first 15 min of toxin expression, physiological levels of free wild-type (wt) ζ or short-living ζ variant (ζY83C) alter the expression of about 2% of total genes (reduces the expression of essential genes involved in cell membrane synthesis and increases the expression of RelA), increases synthesis of (p)ppGpp with subsequent decrease of the GTP pool, and reversible induces dormancy (61, 62). Within the 30 to 90 min of toxin expression, ζY83C decreases macromolecule synthesis (DNA replication, RNA transcription, and protein translation), the intracellular pool of both ATP and GTP, and inhibits cell wall biosynthesis followed by the death of a small fraction (20 to 30%) of the cell population (22, 75). Expression of ε_2 antitoxin then reverses ζ-induced dormancy (75).

The *segB2* Locus

The plasmid or bacterial chromosomal partition machinery is functionally equivalent to a simplified mitotic apparatus and consists of a motor ATPase, a "centromeric" binding protein, and a *cis*-acting site considered to be analogue to the eukaryotic centromere (reviewed in references 2, 22, 80, 81, and 108). Most of the low-copy-number plasmids encode a partition system of the ParA type (reviewed in references 2, 80, 82, and 108). Plasmids of the *inc18* family can be classified into three discrete groups on the basis of their partition system: (i) those that have traces of a *segB2* partition system in their genome, as pIP501 (6, 9, 13, 23, 65, 83); (ii) those that encode a single ParAB system (*segB2*) as pSM19035, pAMβ1, and pVEF4 plasmids; and (iii) those that encode two different ParAB (*segB2* and *segE*) loci as pRE25, pIP816, and pVEF2. In

Figure 3 Protein structures. (A) Structure of the $\zeta\varepsilon_2\zeta$ inactive heterotetramer. The atomic coordinates of the inactive $\zeta\varepsilon_2\zeta$ heterotetramer (74) were obtained from the 1GVN Protein Data Bank (PDB) entry. (B) Structure of the $(\delta\text{·ATP}\gamma\text{S·Mg}^{2+})_2$ complex. The atomic coordinates of δ_2 were derived from the 2OZE PDB entry (65). (C) Model of the three-dimensional structure of δ_2 complexed with DNA. (D) The three-dimensional structure of the $\omega_2\Delta19$ protein. The atomic coordinates of $\omega_2\Delta19$ were derived from the 1IRQ PDB entry (88). (E) Three-dimensional structure model of $\omega_2\Delta19$ bound to *parS*2 was derived from the 2BNW, 2BNZ, and 2CAX PDB entries (89). *parS*2 DNA is space filling, $\omega_2\Delta19$ is orange/red ribbons, and the repeats are indicated below. The modeled structures were prepared with DeepView/Swiss-Pdb-Viewer 3.7. (F) Complexes formed by ω_2 and δ_2 upon binding to *parS* DNA. Protein ω_2 bound to *parS* DNA led to the formation of a short-living partition complex 1 (PC1) (condition 1); and ω_2 bound to *parS* DNA in the presence of δ_2 (out of scale), in the apo form, led to the formation of a long-living partition complex 2 (PC2) (2); δ_2 bound to PC1 led to segrosome complex (SC) formation (3); and the interaction of two SCs led to bridging complex (BC) formation (4). (G) Complexes formed by ω_2 and δ_2 upon binding to non-*parS* DNA. Protein δ_2 bound to DNA leading to dynamic complex (DC) formation (1); ω_2 binding to DC led to a transient complex (TC) (2); the interaction of one TC and one DC led to pseudo-bridging complex 2 (BC2) formation (3); and the interaction of one SC and one TC led to bridging complex 1 (BC1) formation (4).
doi:10.1128/microbiolspec.PLAS-0007-2013.f3.

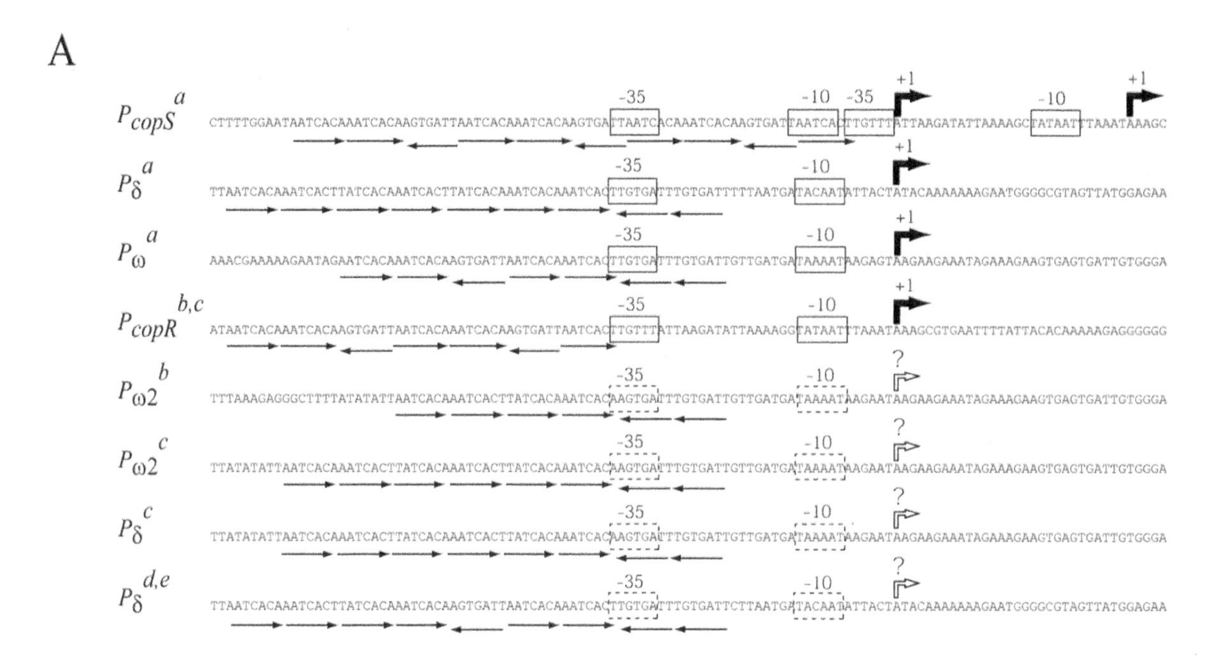

A
PcopS a
Pδ a
Pω a
PcopR b,c
Pω2 b
Pω2 c
Pδ c
Pδ d,e
-35
-10
+1
?

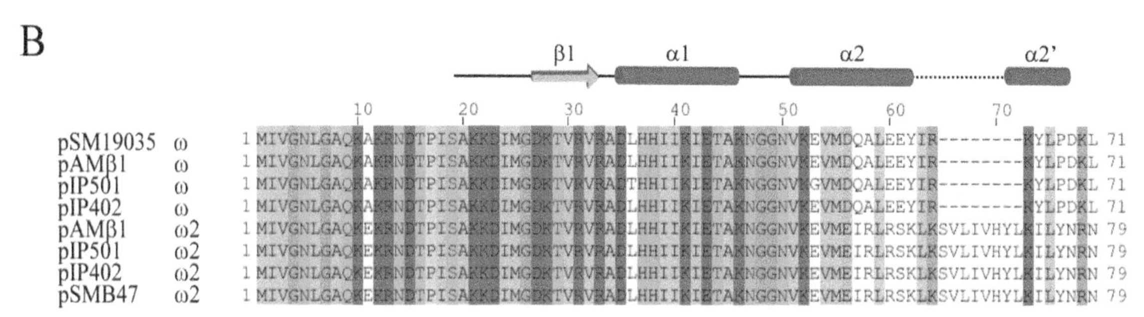

B
β1
α1
α2
α2'
pSM19035 ω
pAMβ1 ω
pIP501 ω
pIP402 ω
pAMβ1 ω2
pIP501 ω2
pIP402 ω2
pSMB47 ω2

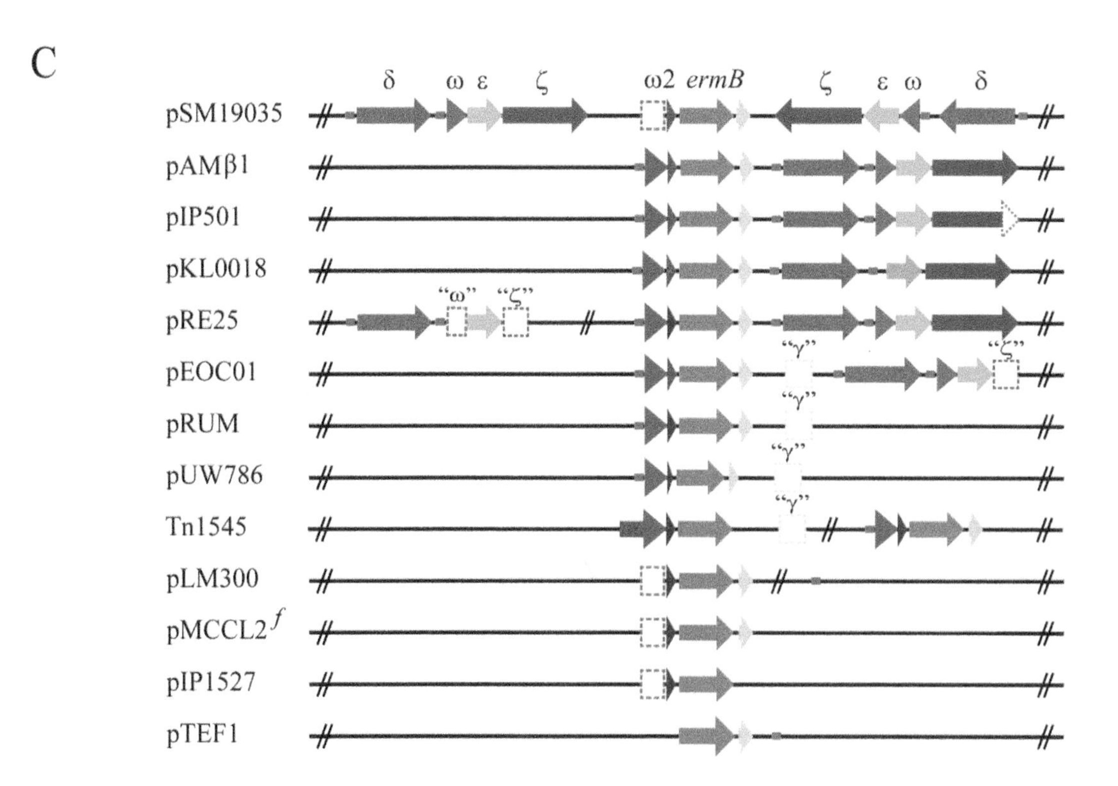

C
δ
ω ε
ζ
ω2 ermB
ζ
ε ω
δ
pSM19035
pAMβ1
pIP501
pKL0018
pRE25
"ω"
"ζ"
pEOC01
"γ"
"ζ"
pRUM
pUW786
Tn1545
pLM300
pMCCL2 f
pIP1527
pTEF1

pSM19035, the *segB2* locus includes two *trans*-acting proteins (the δ_2-ATPase, and the ω_2-CBP) and a variable number of *cis*-acting *parS* sites located outside of the minimal replicon (Fig. 1B and 3B through 3E). Unlike other known partition systems, in this case, the two genes are separately transcribed; however, they are coregulated by the action of the CBP ω_2-transcriptional regulator (63). Unlike the large type *Ia* ATPases (P1-ParA and F-SopA), which have an N-terminal extension needed for autoregulated expression, the small type *Ib* ATPases, as δ_2, do not function as transcriptional regulators of the *par* locus. The δ_2-ATPase contains an N-terminal deviant Walker A box domain, a C-terminal nonspecific DNA binding domain, and a poorly defined domain involved in the interaction with its regulator (ω_2) (Fig. 3B) (65, 84). δ_2, upon binding to ATP, binds DNA, and ATP hydrolysis is not required for this reaction (65, 84). Indeed, δ_2D60A·ATP·Mg^{2+}, which binds but fails to hydrolyze ATP, binds DNA with high efficiency, whereas δ_2K36A, a δ_2 variant, which neither binds nor hydrolyzes ATP, fails to bind DNA (84, 85).

The monomer of ω is a 71-residue polypeptide with an unstructured N-terminal domain (residues 1 to 19) and a ribbon-helix-helix (RHH) fold (residues 20 to 71) (Fig. 3D) (86, 87, 88, 89). Protein ω_2 or its variant $\omega_2\Delta$N19, which lacks the first 19 residues, transiently binds, with high affinity and cooperativity (apparent dissociation constant [K_{Dapp}] 5 ± 1 nM) to its cognate sites (forming partition complex 1, PC1) (Fig. 3E and 3F) (63, 64, 84, 89, 90). However, the interaction of the unfolded N-terminal end of ω_2 with δ_2, even in the apo form, significantly increases the binding affinity of ω_2 for *parS* DNA (K_{Dapp} 0.7 nM) leading to the long-living partition complex 2 (PC2) (Fig. 3F, condition 2) (84). In pSM19035, the ω_2-target sites are six *parS* sites: two *parS1*/*parS1´* sites, located in the promoter (*P*) region of δ gene (P_δ), two *parS2*/*parS2´* in P_ω, and two *parS3*/*parS3´* in P_{cop}, featuring, respectively, 9, 7, and 10 contiguous heptads (in direct → or inverse ← orientation) with the sequence 5´-WATCACW-3´ (where W is an A or a T) (Fig. 1A and 4A) (64). In other members of the family, the number and relative orientation of the highly conserved repeats varies (Fig. 4A). The minimal protein binding site consists of two contiguous heptads in direct (→→) or inverted (→←) orientations, to which two ω_2 molecules bind, and the complex is stabilized by cooperative interdimer contacts (64, 89). However, the binding site to two contiguous heptads in (←→) orientations was poor (64, 89). Protein ω_2 ($\omega2_2$, Fig. 4B) belongs to the family of proteins with a RHH_2 fold (87, 88, 91). The half DNA binding determinant is located at residues 28 to 32, which constitutes a β sheet, and is followed by two α helices (α1 [residues 34–46] and α2 [residues 51–66]) involved in the interaction with the phosphate backbone, with itself and with another dimer (88, 90, 92, 93). In ω_2 the two β strands are antiparallel and contact its cognate sequence in the major groove of DNA with high specificity and affinity, and the two α1 regions make nonspecific contacts with the DNA phosphate backbone (88, 89, 90, 93). The structure of ω_2 bound to a mini *parS* DNA (→→ or →←) was solved (Fig. 3E) (89). This structure shows that ω_2 binds symmetrically to this mini *parS* DNA with a 0.3 Å deviation with respect to the central C-G pair of each repetition (89). Residues Thr29 and Arg31 from the antiparallel β strands are essential for specific binding (Fig. 3C) (89, 90, 93).

The structure of pSM19035-δ_2, in the nucleotide-bound form, is U-shaped with each arm of the U

Figure 4 Conserved organization of the *segD* locus. (**A**) The P_{copS}, P_δ, and P_ω regions of plasmid pSM19035[a], P_{copR} or P_{copF} of pIP501[b] or pAMβ1[c], $P_{\omega 2}$ of pIP501[b] or pAMβ1[c], P_δ of pAMβ1[c], pRE25[d] and pVEF3[e] are indicated. The variable number of contiguous 7-bp heptad repeats (iterons) and their relative orientations (→ or ←) are shown. The P_ω region is highly conserved among plasmids of the *inc18* family. (**B**) Sequence alignment of the transcriptional repressors ω_2 and their highly relative $\omega2_2$. The alignment was done by using Clustal W2 and visualized with default coloring of the different residues by using Jalview v 14.0. (**C**) Conserved genetic organization of the *ermB* locus. The open reading frames and their relative orientations are represented by arrows. The *erm* gene coding for the leader peptide and *ermB* are denoted in dark gray (arrowhead and arrow, respectively), and the conserved downstream gene in light gray. The genes of the *segB1*, *segB2*, and *segC* loci are indicated. The ω_2 or $\omega2_2$ proteins are denoted by red and pink arrowheads, respectively, and the *P*s sensitive to ω_2 or $\omega2_2$ repression are denoted as red boxes. Traces of a given gene are denoted by broken-line squares with the name of the gene indicated (e.g., pSM19035 containing traces of $\omega2_2$). Fusions (e.g., pLM300 where ω2 was fused to an upstream gene) or deletions (e.g., pIP501 contains a deletion in the ζ gene) are also indicated. Similar organization was observed in pAM77 and pLUL631.
doi:10.1128/microbiolspec.PLAS-0007-2013.f4.

representing one monomer, and the ATP facing the cleft of the U (65, 84). The surface charge of δ_2 is negative near the bottom of the U, and positive at the tips of the arms of the U (65, 84). The positively charged residues are necessary to form small blobs on DNA (Fig. 3G) rather than filaments (65, 84). Upon binding to ATP, δ_2 undergoes conformational changes that make it proficient for DNA binding (Fig. 3G) (84). Like type *Ia* P1-ParA (94), δ_2 upon binding to DNA, in a sequence-independent manner, forms a transient δ_2-chromosomal DNA complex. In this binary complex, 10 to 20 δ_2 molecules are clustered on the DNA (65, 84, 85). Protein δ_2, upon interaction with ω_2, bound to *parS* (a cluster of up to 10 ω_2 and ~15 δ_2 molecules), as part of two discrete segregosome (*parS*·ω_2·δ_2) complexes (SC), lead to plasmid pairing or bridging complex formation (*parS*·ω_2·δ_2·ω_2·*parS*) (BC) (Fig. 3F, conditions 3 and 4).

Several studies revealed that pSM19035 partition is a multistep process similar to the diffusion ratchet model proposed by the Mizuuchi and Funnell groups for P1-ParAB- and F-SopAB–mediated partition (82, 94, 95, 96, 108). In the pSM19035, P1 and F partition systems the motive force for plasmid positioning does not directly rely on the polymerization of the motor protein ParA, but, instead, it is directed by a dynamic gradient of ParA in the cell. The uneven distribution of ParA molecules in the cell depends on the affinity of ParA for nonspecific host nucleoid DNA when bound to ATP, and the ability of the CBP ParB bound to the plasmid DNA to release ParA from the nucleoid by stimulating its ATPase activity.

THE *segD* LOCUS

In *B. subtilis*, the ratio between pBD101 and the *oriC* (origin of chromosomal replication) is approximately 2:1. The smallest possible increase in plasmid copy number, i.e., from 2 to 4, corresponds to a dramatic 100% change in concentration. Thus, the feedback control of low concentrations is notoriously difficult and easily leads to random oscillations. Furthermore, slight variations in the feedback control scheme result in markedly different efficiencies of noise suppression. The *segD* locus, which comprises the ω_2 transcriptional repressor and its cognate sites upstream the promoter regions of the *cop* (P_{copS}), δ (P_δ), and ω (P_ω) genes (also termed *parS1-parS3* and *parS1′-parS3′* sites, Fig. 1A), provides a sophisticated degree of interplay and a regulated coordination of plasmid copy number fluctuation and stable inheritance (Fig. 4A). The coupling between these functions stabilizes the plasmid ~10-fold (63) and contributes to minimize the metabolic burden that the plasmid might impose to the host cell. How widespread is this interplay strategy in nature? In plasmid RK2/RP4 of *Gammaproteobacteria* class, ParA-like (IncC), ParB-like (KorB), and the KorA transcriptional repressor are encoded in a central control operon (110). KorB in concert with KorA and TrfA coordinates plasmid replication, transfer, and stable maintenance functions without the need of increased repressor concentrations (97). It is likely that there is also an indirect physiological interplay of all active partition and TA systems, resulting in a near-ideal symbiosis when the systems combine, suggesting that such interplay contributes to almost absolute plasmid stability (98). This interplay is not restricted to plasmids. The host-encoded partition system coordinates replication initiation and chromosomal segregation (99, 100, 101).

Some plasmids of the *inc18* family encode two ω-like repressors: ω_2 (a 71-residue-long peptide) and $\omega2_2$ (a 79-residue-long peptide). Protein $\omega2_2$, which also belongs to the family of proteins with a RHH_2 fold (Fig. 3C), shares with ω_2 the first unstructured 25 residues, the DNA binding specificity determinants located in the β sheet (at residues 28 to 32), and protein-protein and DNA backbone interacting domains from the α1 helix (residues 34 to 46) (Fig. 4B). Both proteins repress the levels of CopS, and indirectly increase the supply of the *rep* mRNA (63; A. Volante, unpublished results). Once ω_2 or $\omega2_2$ bind the upstream P_{copS} and $P_{copS'}$ regions, they inhibit *copS* gene expression and indirectly correct any downward copy number fluctuation (63, 102, 103, 106). Indeed, repression of CopS synthesis by ω_2 or $\omega2_2$ correlates with an increase in plasmid copy number and indirectly ensures stable plasmid maintenance. It is likely that the interplay of CopS and ω_2 is part of a negative-feedback control system of the minimal replicon of *inc18*.

Protein ω_2 also couples the expression of genes of the *segB1* and *segB2* loci by repressing the utilization of *P* and *P* (Fig. 4A). After plasmid establishment or upon downward fluctuations in plasmid copy number, the low levels of ω_2 result in increased ParAB and TA and decreased CopS·RepS ratios with increased copy number. At high-copy-number conditions, ω_2 concentrations increase, which results in a rapid reduction in the levels of δ_2 and ω_2 (ParAB system) and of ε_2 and ζ (TA system) proteins required for active partitioning and control of cell proliferation, respectively (61, 63, 65, 72, 74, 85). Any stochastic decrease of the $\omega\varepsilon\zeta$ mRNA decreases the concentration of ω_2 and of the short half-living ε_2 antitoxin with relative increase of the long-living ζ toxin. Then, free ζ toxin halts cell proliferation (61, 72, 75). To overcome these disadvantages,

ω_2 promotes the synthesis of the ωεζ operon to exit ζ-mediated dormancy (61, 63).

Plasmids of the *inc18* family with IR encode two copies of ω gene (e.g., pSM19035, pRE25, pVEF3 with 100% identity among them), and plasmids that encode *ermB* also code for ω2 gene (e.g., pIP501, pAMβ1, pRE25) (Fig. 2). The ω2 gene is invariantly located upstream of the *ermB* gene (Fig. 2 and 4C). A truncated ω2 gene was detected upstream of the *ermB* gene also in pSM19035 (Fig. 2 and 4C; C. E. César, personal communication). The genetic linkage between $\omega_2$2 and the erythromycin resistance cassette remains to be unraveled, but suggests that both genes should be present in a common ancestor. The role of the *segD* locus in the spread of the *ermB* cassette is poorly understood.

The *segD* locus may participate also in the formation of a plasmid cointegrate. It has been proposed that ω_2 bound to *parS* DNA might constitute a barrier to DNA replication (22). This natural impediment, albeit with low efficiency, might lead to one-ended double-strand break, so that the PC1 complex may constitute a recombination hotspot upon plasmids pairing (see Fig. 2) (22). Then, homologous recombination functions in concert with ω_2-mediated site-specific BC formation might lead to the formation of cointegrates. Indeed, it was reported that a *Lactococcus lactis* recipient strain received a transconjugate, in a cointegrate form, and such transfer was blocked if RecA or the *segD* locus was absent in the donor strain (104, 105), suggesting that the PC1 complex, in concert with RecA, works as a pseudo *mob* region. Plasmid pairing is enhanced >20-fold in the presence of δ_2 (BC formation, Fig. 3F) (84, 85), but the contribution of δ_2 in this mobilization avenue remains to be characterized.

CONCLUDING REMARKS

Low-copy plasmids commonly contain several distinct loci to enhance their maintenance. In plasmids of the *inc18* family, there is an active interplay of the *rep* locus with the *segB1*, *segB2*, and *segD* loci to warrant plasmid segregational stability, and with the *segA* and *segC* loci to warrant plasmid structural stability without disadvantage to their host. The assessment of the significance of a particular locus in plasmid stabilization revealed that the *segB1* locus stabilizes the plasmid ~10,000-fold by impairing growth of plasmid-free segregants, and *segB2* stabilizes the plasmid ~50-fold by active partition (21, 61, 63, 65). The *segA* locus, which optimizes plasmid random segregation, and the *segA* and *segC* loci in concert, which correct errors in the shift from DNA polymerase I to the DNA replicase (PolC and DnaE) to replicate the DNA, stabilize the plasmid ~25-fold (17, 21, 24). The *segC* locus, which maps outside the minimal replicon, is essential for replication of plasmids with IR regions. The *segD* locus, which is responsible for reducing plasmid copy number fluctuations to less than two copies and for regulating expression of the *segB1* and *segB2* loci, stabilizes the plasmid ~10-fold (63).

Acknowledgments. *We thank Dolors Balsa, Detlef Behnke, Alexander Boitsov, Martin Boocock, Sabine Brantl, Piotr Ceglowski, Klauss Grasser, Carlos Martinez, Anton Meinhart, Fernando Rojo, Wolfram Saenger, Bernardo Schvartzman, Kunio Takeyasu, and Heinz Welfle for enjoyable cooperation and/or fruitful discussions over the years and Carolina E. César for the communication of unpublished results. Our sincere gratitude is extended to a number of graduate students (Ana B. de la Hoz, Inés Canosa, Gema López, Virginia S. Lioy, Florencia Pratto, and Mariangela Tabone) and postdoctoral fellows (Ana Camacho, María T. Martín) who do not coauthor this review. A.V. thanks the Comunidad de Madrid for its fellowship (CPI/0266/2008). The work reviewed here was supported by the Ministerio de Economía y Competividad Ciencia e Innovación through grants BFU2012-39879-C02-01 (to J.C.A.) and BFU2012-39879-C02-02 (to S.A.).Conflicts of interest: We disclose no conflicts.*

Citation. Volante A, Soberón NE, Ayora S, Alonso JC. 2014. The interplay between different stability systems contributes to faithful segregation: *Streptococcus pyogenes* pSM19035 as a model. Microbiol Spectrum **2**(4):PLAS-0007-2013.

References

1. **Nordstrom K, Austin SJ.** 1989. Mechanisms that contribute to the stable segregation of plasmids. *Annu Rev Genet* **23**:37–69.
2. **Ebersbach G, Gerdes K.** 2005. Plasmid segregation mechanisms. *Annu Rev Genet* **39**:453–479.
3. **Malke H.** 1974. Genetics of resistance to macrolide antibiotics and lincomycin in natural isolates of *Streptococcus pyogenes*. *Mol Gen Genet* **135**:349–367.
4. **Horodniceanu T, Bouanchaud DH, Bieth G, Chabbert YA.** 1976. R plasmids in *Streptococcus agalactiae* (group B). *Antimicrob Agents Chemother* **10**:795–801.
5. **Dunny GM, Clewell DB.** 1975. Transmissible toxin (hemolysin) plasmid in *Streptococcus faecalis* and its mobilization of a noninfectious drug resistance plasmid. *J Bacteriol* **124**:784–790.
6. **Schwarz FV, Perreten V, Teuber M.** 2001. Sequence of the 50-kb conjugative multiresistance plasmid pRE25 from *Enterococcus faecalis* RE25. *Plasmid* **46**:170–187.
7. **Liu Y, Wang Y, Schwarz S, Li Y, Shen Z, Zhang Q, Wu C, Shen J.** 2013. Transferable multiresistance plasmids carrying cfr in *Enterococcus* spp. from swine and farm environment. *Antimicrob Agents Chemother* **57**:42–48.
8. **Sletvold H, Johnsen PJ, Simonsen GS, Aasnaes B, Sundsfjord A, Nielsen KM.** 2007. Comparative DNA

analysis of two vanA plasmids from *Enterococcus faecium* strains isolated from poultry and a poultry farmer in Norway. *Antimicrob Agents Chemother* **51:** 736–739.

9. **Sletvold H, Johnsen PJ, Wikmark OG, Simonsen GS, Sundsfjord A, Nielsen KM.** 2010. Tn1546 is part of a larger plasmid-encoded genetic unit horizontally disseminated among clonal *Enterococcus faecium* lineages. *J Antimicrob Chemother* **65:**1894–1906.
10. **Brantl S, Behnke D, Alonso JC.** 1990. Molecular analysis of the replication region of the conjugative *Streptococcus agalactiae* plasmid pIP501 in *Bacillus subtilis.* Comparison with plasmids pAM beta 1 and pSM19035. *Nucleic Acids Res* **18:**4783–4790.
11. **Brantl S, Nowak A, Behnke D, Alonso JC.** 1989. Revision of the nucleotide sequence of the *Streptococcus pyogenes* plasmid pSM19035 *repS* gene. *Nucleic Acids Res* **17:**10110. doi:10.1093/nar/17.23.10110.
12. **Boitsov AS, Golubkov VI, Iontova IM, Zaitsev EN, Malke H, Totolian AA.** 1979. Inverted repeats on plasmids determining resistance to MLS antibiotics in group A streptococci. *FEMS Microbiol Lett* **6:**11–14.
13. **Sletvold H, Johnsen PJ, Hamre I, Simonsen GS, Sundsfjord A, Nielsen KM.** 2008. Complete sequence of *Enterococcus faecium* pVEF3 and the detection of an omega-epsilon-zeta toxin-antitoxin module and an ABC transporter. *Plasmid* **60:**75–85.
14. **Berryman DI, Rood JI.** 1995. The closely related ermB-ermAM genes from *Clostridium perfringens, Enterococcus faecalis* (pAMβ1), and *Streptococcus agalactiae* (pIP501) are flanked by variants of a directly repeated sequence. *Antimicrob Agents Chemother* **39:** 1830–1834.
15. **Bruand C, Ehrlich SD, Janniere L.** 1991. Unidirectional theta replication of the structurally stable *Enterococcus faecalis* plasmid pAMβ1. *EMBO J* **10:**2171–2177.
16. **Bruand C, Ehrlich SD, Janniere L.** 1995. Primosome assembly site in *Bacillus subtilis. EMBO J* **14:**2642–2650.
17. **Bruand C, Le Chatelier E, Ehrlich SD, Janniere L.** 1993. A fourth class of theta-replicating plasmids: the pAMβ1 family from gram-positive bacteria. *Proc Natl Acad Sci USA* **90:**11668–11672.
18. **Ceglowski P, Lurz R, Alonso JC.** 1993. Functional analysis of pSM19035-derived replicons in *Bacillus subtilis. FEMS Microbiol Lett* **109:**145–150.
19. **Le Chatelier E, Ehrlich SD, Janniere L.** 1993. Biochemical and genetic analysis of the unidirectional theta replication of the *S. agalactiae* plasmid pIP501. *Plasmid* **29:** 50–56.
20. **Alonso JC, Espinosa M.** 1993. *Plasmids from Gram-Positive Bacteria,* 2nd ed. IRL Press, Oxford, United Kingdom.
21. **Ceglowski P, Boitsov A, Karamyan N, Chai S, Alonso JC.** 1993. Characterization of the effectors required for stable inheritance of *Streptococcus pyogenes* pSM19035-derived plasmids in *Bacillus subtilis. Mol Gen Genet* **241:**579–585.
22. **Lioy VS, Pratto F, de la Hoz AB, Ayora S, Alonso JC.** 2010. Plasmid pSM19035, a model to study stable maintenance in *Firmicutes. Plasmid* **64:**1–17.
23. **Ceglowski P, Boitsov A, Chai S, Alonso JC.** 1993. Analysis of the stabilization system of pSM19035-derived plasmid pBT233 in *Bacillus subtilis. Gene* **136:**1–12.
24. **Canosa I, Rojo F, Alonso JC.** 1996. Site-specific recombination by the β protein from the streptococcal plasmid pSM19035: minimal recombination sequences and crossing over site. *Nucleic Acids Res* **24:**2712–2717.
25. **Summers DK, Sherratt DJ.** 1984. Multimerization of high copy number plasmids causes instability: ColE1 encodes a determinant essential for plasmid monomerization and stability. *Cell* **36:**1097–1103.
26. **Bhardwaj A, Welfle K, Misselwitz R, Ayora S, Alonso JC, Welfle H.** 2006. Conformation and stability of the *Streptococcus pyogenes* pSM19035-encoded site-specific β recombinase, and identification of a folding intermediate. *Biol Chem* **387:**525–533.
27. **Rojo F, Alonso JC.** 1994. The beta recombinase from the streptococcal plasmid pSM19035 represses its own transcription by holding the RNA polymerase at the promoter region. *Nucleic Acids Res* **22:**1855–1860.
28. **Rojo F, Weise F, Alonso JC.** 1993. Purification of the β product encoded by the *Streptococcus pyogenes* plasmid pSM19035. A putative DNA recombinase required to resolve plasmid oligomers. *FEBS Lett* **328:**169–173.
29. **Rojo F, Alonso JC.** 1995. The β recombinase of plasmid pSM19035 binds to two adjacent sites, making different contacts at each of them. *Nucleic Acids Res* **23:** 3181–3188.
30. **Canosa I, Ayora S, Rojo F, Alonso JC.** 1997. Mutational analysis of a site-specific recombinase: characterization of the catalytic and dimerization domains of the β recombinase of pSM19035. *Mol Gen Genet* **255:** 467–476.
31. **Orth P, Jekow P, Alonso JC, Hinrichs W.** 1999. Proteolytic cleavage of Gram-positive β recombinase is required for crystallization. *Protein Eng* **12:**371–373.
32. **Canosa I, Lurz R, Rojo F, Alonso JC.** 1998. β Recombinase catalyzes inversion and resolution between two inversely oriented six sites on a supercoiled DNA substrate and only inversion on relaxed or linear substrates. *J Biol Chem* **273:**13886–13891.
33. **Canosa I, Lopez G, Rojo F, Boocock MR, Alonso JC.** 2003. Synapsis and strand exchange in the resolution and DNA inversion reactions catalysed by the β recombinase. *Nucleic Acids Res* **31:**1038–1044.
34. **Alonso JC, Gutierrez C, Rojo F.** 1995. The role of chromatin-associated protein Hbsu in β-mediated DNA recombination is to facilitate the joining of distant recombination sites. *Mol Microbiol* **18:**471–478.
35. **Alonso JC, Weise F, Rojo F.** 1995. The *Bacillus subtilis* histone-like protein Hbsu is required for DNA resolution and DNA inversion mediated by the β recombinase of plasmid pSM19035. *J Biol Chem* **270:**2938–2945.
36. **Grasser KD, Ritt C, Krieg M, Fernandez S, Alonso JC, Grimm R.** 1997. The recombinant product of the *Chryptomonas* plastid gene *hlpA* is an architectural HU-like protein that promotes the assembly of complex nucleoprotein structures. *Eur J Biochem* **249:**70–76.
37. **Ritt C, Grimm R, Fernandez S, Alonso JC, Grasser KD.** 1998. Four differently chromatin-associated maize

HMG domain proteins modulate DNA structure and act as architectural elements in nucleoprotein complexes. *Plant J* **14:**623–631.

38. **Ritt C, Grimm R, Fernandez S, Alonso JC, Grasser KD.** 1998. Basic and acidic regions flanking the HMG domain of maize HMGa modulate the interactions with DNA and the self-association of the protein. *Biochemistry* **37:**2673–2681.
39. **Stemmer C, Fernandez S, Lopez G, Alonso JC, Grasser KD.** 2002. Plant chromosomal HMGB proteins efficiently promote the bacterial site-specific β-mediated recombination *in vitro* and *in vivo*. *Biochemistry* **41:**7763–7770.
40. **Diaz V, Rojo F, Martinez AC, Alonso JC, Bernad A.** 1999. The prokaryotic β-recombinase catalyzes site-specific recombination in mammalian cells. *J Biol Chem* **274:**6634–6640.
41. **Diaz V, Servert P, Prieto I, Gonzalez MA, Martinez AC, Alonso JC, Bernad A.** 2001. New insights into host factor requirements for prokaryotic β-recombinase-mediated reactions in mammalian cells. *J Biol Chem* **276:** 16257–16264.
42. **Servert P, Diaz V, Lucas D, de la Cueva T, Rodriguez M, Garcia-Castro J, Alonso J, Martinez AC, Gonzalez M, Bernad A.** 2008. *In vivo* site-specific recombination using the β-rec/six system. *Biotechniques* **45:**69–78.
43. **Tapias A, Lopez G, Ayora S.** 2000. *Bacillus subtilis* LrpC is a sequence-independent DNA-binding and DNA-bending protein which bridges DNA. *Nucleic Acids Res* **28:**552–559.
44. **Rojo F, Alonso JC.** 1994. A novel site-specific recombinase encoded by the *Streptococcus pyogenes* plasmid pSM19035. *J Mol Biol* **238:**159–172.
45. **Kidane D, Ayora S, Sweasy J, Graumann PL, Alonso JC.** 2012. The cell pole: the site of cross talk between the DNA uptake and genetic recombination machinery. *Crit Rev Biochem Mol Biol* **47:**531–555.
46. **Lo Piano A, Martinez-Jimenez MI, Zecchi L, Ayora S.** 2011. Recombination-dependent concatemeric viral DNA replication. *Virus Res* **160:**1–14.
47. **Viret JF, Bravo A, Alonso JC.** 1991. Recombination-dependent concatemeric plasmid replication. *Microbiol Rev* **55:**675–683.
48. **Canosi U, Morelli G, Trautner TA.** 1978. The relationship between molecular structure and transformation efficiency of some *S. aureus* plasmids isolated from *B. subtilis*. *Mol Gen Genet* **166:**259–267.
49. **de Vos WM, Venema G, Canosi U, Trautner TA.** 1981. Plasmid transformation in *Bacillus subtilis*: fate of plasmid DNA. *Mol Gen Genet* **181:**424–433.
50. **Bravo A, Alonso JC.** 1990. The generation of concatemeric plasmid DNA in *Bacillus subtilis* as a consequence of bacteriophage SPP1 infection. *Nucleic Acids Res* **18:**4651–4657.
51. **Leonhardt H, Lurz R, Alonso JC.** 1991. Physical and biochemical characterization of recombination-dependent synthesis of linear plasmid multimers in *Bacillus subtilis*. *Nucleic Acids Res* **19:**497–503.
52. **Viret JF, Alonso JC.** 1987. Generation of linear multigenome-length plasmid molecules in *Bacillus subtilis*. *Nucleic Acids Res* **15:**6349–6367.
53. **Alonso JC, Luder G, Trautner TA.** 1986. Requirements for the formation of plasmid-transducing particles of *Bacillus subtilis* bacteriophage SPP1. *EMBO J* **5:**3723–3728.
54. **Novick RP, Edelman I, Lofdahl S.** 1986. Small *Staphylococcus aureus* plasmids are transduced as linear multimers that are formed and resolved by replicative processes. *J Mol Biol* **192:**209–220.
55. **Ceglowski P, Alonso JC.** 1994. Gene organization of the *Streptococcus pyogenes* plasmid pDB101: sequence analysis of the orf η-*copS* region. *Gene* **145:**33–39.
56. **Viguera E, Hernandez P, Krimer DB, Boistov AS, Lurz R, Alonso JC, Schvartzman JB.** 1996. The ColE1 unidirectional origin acts as a polar replication fork pausing site. *J Biol Chem* **271:**22414–22421.
57. **Bidnenko V, Ehrlich SD, Janniere L.** 1998. In vivo relations between pAMβ1-encoded type I topoisomerase and plasmid replication. *Mol Microbiol* **28:** 1005–1016.
58. **Brantl S, Behnke D.** 1992. Characterization of the minimal origin required for replication of the streptococcal plasmid pIP501 in *Bacillus subtilis*. *Mol Microbiol* **6:** 3501–3510.
59. **Van Melderen L, Saavedra De Bast M.** 2009. Bacterial toxin-antitoxin systems: more than selfish entities? *PLoS Genet* **5:**e1000437. doi:10.1371/journal.pgen.1000437.
60. **Funnell BE.** 2005. Partition-mediated plasmid pairing. *Plasmid* **53:**119–125.
61. **Lioy VS, Martin MT, Camacho AG, Lurz R, Antelmann H, Hecker M, Hitchin E, Ridge Y, Wells JM, Alonso JC.** 2006. pSM19035-encoded ζ toxin induces stasis followed by death in a subpopulation of cells. *Microbiology* **152:**2365–2379.
62. **Lioy VS, Rey O, Balsa D, Pellicer T, Alonso JC.** 2010. A toxin-antitoxin module as a target for antimicrobial development. *Plasmid* **63:**31–39.
63. **de la Hoz AB, Ayora S, Sitkiewicz I, Fernandez S, Pankiewicz R, Alonso JC, Ceglowski P.** 2000. Plasmid copy-number control and better-than-random segregation genes of pSM19035 share a common regulator. *Proc Natl Acad Sci USA* **97:**728–733.
64. **de la Hoz AB, Pratto F, Misselwitz R, Speck C, Weihofen W, Welfle K, Saenger W, Welfle H, Alonso JC.** 2004. Recognition of DNA by ω protein from the broad-host range *Streptococcus pyogenes* plasmid pSM19035: analysis of binding to operator DNA with one to four heptad repeats. *Nucleic Acids Res* **32:** 3136–3147.
65. **Pratto F, Cicek A, Weihofen WA, Lurz R, Saenger W, Alonso JC.** 2008. *Streptococcus pyogenes* pSM19035 requires dynamic assembly of ATP-bound ParA and ParB on *parS* DNA during plasmid segregation. *Nucleic Acids Res* **36:**3676–3689.
66. **Engelberg-Kulka H, Amitai S, Kolodkin-Gal I, Hazan R.** 2006. Bacterial programmed cell death and multicellular behavior in bacteria. *PLoS Genet* **2:** e135. doi: 10.1371/journal.pgen.0020135.
67. **Gerdes K, Christensen SK, Lobner-Olesen A.** 2005. Prokaryotic toxin-antitoxin stress response loci. *Nat Rev Microbiol* **3:**371–382.

68. **Mruk I, Kobayashi I.** 2013. To be or not to be: regulation of restriction-modification systems and other toxin-antitoxin systems. *Nucleic Acids Res* **42:**70–86.
69. **Moritz EM, Hergenrother PJ.** 2007. Toxin-antitoxin systems are ubiquitous and plasmid-encoded in vancomycin-resistant enterococci. *Proc Natl Acad Sci USA* **104:** 311–316.
70. **Zhu W, Clark NC, McDougal LK, Hageman J, McDonald LC, Patel JB.** 2008. Vancomycin-resistant *Staphylococcus aureus* isolates associated with Inc18-like vanA plasmids in Michigan. *Antimicrob Agents Chemother* **52:**452–457.
71. **Rosvoll TC, Lindstad BL, Lunde TM, Hegstad K, Aasnaes B, Hammerum AM, Lester CH, Simonsen GS, Sundsfjord A, Pedersen T.** 2012. Increased high-level gentamicin resistance in invasive *Enterococcus faecium* is associated with aac(6′)Ie-aph(2′′)Ia-encoding transferable megaplasmids hosted by major hospital-adapted lineages. *FEMS Immunol Med Microbiol* **66:**166–176.
72. **Camacho AG, Misselwitz R, Behlke J, Ayora S, Welfle K, Meinhart A, Lara B, Saenger W, Welfle H, Alonso JC.** 2002. In vitro and *in vivo* stability of the $\varepsilon_2\zeta_2$ protein complex of the broad host-range *Streptococcus pyogenes* pSM19035 addiction system. *Biol Chem* **383:** 1701–1713.
73. **Meinhart A, Alings C, Strater N, Camacho AG, Alonso JC, Saenger W.** 2001. Crystallization and preliminary X-ray diffraction studies of the εζ addiction system encoded by *Streptococcus pyogenes* plasmid pSM19035. *Acta Crystallogr D Biol Crystallogr* **57:**745–747.
74. **Meinhart A, Alonso JC, Strater N, Saenger W.** 2003. Crystal structure of the plasmid maintenance system ε/ζ: functional mechanism of toxin ζ and inactivation by $\varepsilon_2\zeta_2$ complex formation. *Proc Natl Acad Sci USA* **100:** 1661–1666.
75. **Lioy VS, Machon C, Tabone M, Gonzalez-Pastor JE, Daugelavicius R, Ayora S, Alonso JC.** 2012. The ζ toxin induces a set of protective responses and dormancy. *PLoS One* **7:**e30282. doi:10.1371/journal.pone.0030282.
76. **Leplae R, Geeraerts D, Hallez R, Guglielmini J, Dreze P, Van Melderen L.** 2011. Diversity of bacterial type II toxin-antitoxin systems: a comprehensive search and functional analysis of novel families. *Nucleic Acids Res* **39:**5513–5525.
77. **Zielenkiewicz U, Ceglowski P.** 2005. The toxin-antitoxin system of the streptococcal plasmid pSM19035. *J Bacteriol* **187:**6094–6105.
78. **Mutschler H, Meinhart A.** 2011. epsilon/zeta systems: their role in resistance, virulence, and their potential for antibiotic development. *J Mol Med* **89:**1183–1194.
79. **Khoo SK, Loll B, Chan WT, Shoeman RL, Ngoo L, Yeo CC, Meinhart A.** 2007. Molecular and structural characterization of the PezAT chromosomal toxin-antitoxin system of the human pathogen *Streptococcus pneumoniae*. *J Biol Chem* **282:**19606–19618.
80. **Szardenings F, Guymer D, Gerdes K.** 2011. ParA ATPases can move and position DNA and subcellular structures. *Curr Opin Microbiol* **14:**712–718.
81. **Thanbichler M, Shapiro L.** 2008. Getting organized - how bacterial cells move proteins and DNA. *Nat Rev Microbiol* **6:**28–40.
82. **Vecchiarelli AG, Mizuuchi K, Funnell BE.** 2012. Surfing biological surfaces: exploiting the nucleoid for partition and transport in bacteria. *Mol Microbiol* **86:**513–523.
83. **Zuñiga M, Pardo I, Ferrer S.** 2003. Conjugative plasmid pIP501 undergoes specific deletions after transfer from *Lactococcus lactis* to *Oenococcus oeni*. *Arch Microbiol* **180:**367–373.
84. **Soberon NE, Lioy VS, Pratto F, Volante A, Alonso JC.** 2011. Molecular anatomy of the *Streptococcus pyogenes* pSM19035 partition and segrosome complexes. *Nucleic Acids Res* **39:**2624–2637.
85. **Pratto F, Suzuki Y, Takeyasu K, Alonso JC.** 2009. Single-molecule analysis of protein-DNA complexes formed during partition of newly replicated plasmid molecules in *Streptococcus pyogenes*. *J Biol Chem* **284:** 30298–30306.
86. **Misselwitz R, de la Hoz AB, Ayora S, Welfle K, Behlke J, Murayama K, Saenger W, Alonso JC, Welfle H.** 2001. Stability and DNA-binding properties of the ω regulator protein from the broad-host range *Streptococcus pyogenes* plasmid pSM19035. *FEBS Lett* **505:**436–440.
87. **Murayama K, de la Hoz AB, Alings C, Lopez G, Orth P, Alonso JC, Saenger W.** 1999. Crystallization and preliminary X-ray diffraction studies of *Streptococcus pyogenes* plasmid pSM19035-encoded ω transcriptional repressor. *Acta Crystallogr D Biol Crystallogr* **55:** 2041–2042.
88. **Murayama K, Orth P, de la Hoz AB, Alonso JC, Saenger W.** 2001. Crystal structure of ω transcriptional repressor encoded by *Streptococcus pyogenes* plasmid pSM19035 at 1.5 A resolution. *J Mol Biol* **314:**789–796.
89. **Weihofen WA, Cicek A, Pratto F, Alonso JC, Saenger W.** 2006. Structures of ω repressors bound to direct and inverted DNA repeats explain modulation of transcription. *Nucleic Acids Res* **34:**1450–1458.
90. **Welfle K, Pratto F, Misselwitz R, Behlke J, Alonso JC, Welfle H.** 2005. Role of the N-terminal region and of beta-sheet residue Thr29 on the activity of the ω_2 global regulator from the broad-host range *Streptococcus pyogenes* plasmid pSM19035. *Biol Chem* **386:**881–894.
91. **Schreiter ER, Drennan CL.** 2007. Ribbon-helix-helix transcription factors: variations on a theme. *Nat Rev Microbiol* **5:**710–720.
92. **Dostál L, Misselwitz R, Laettig S, Alonso JC, Welfle H.** 2003. Raman spectroscopy of regulatory protein ω from *Streptococcus pyogenes* plasmid pSM19035 and complexes with operator DNA. *Spectroscopy* **17:**435–445.
93. **Dostál L, Pratto F, Alonso JC, Welfle H.** 2007. Binding of regulatory protein ω from *Streptococcus pyogenes* plasmid pSM19035 to direct and inverted 7-base pair repeats of operator DNA. *J Raman Spectrosc* **38:**166–175.
94. **Vecchiarelli AG, Han YW, Tan X, Mizuuchi M, Ghirlando R, Biertumpfel C, Funnell BE, Mizuuchi K.** 2010. ATP control of dynamic P1 ParA-DNA interactions: a key role for the nucleoid in plasmid partition. *Mol Microbiol* **78:**78–91.
95. **Hwang LC, Vecchiarelli AG, Han YW, Mizuuchi M, Harada Y, Funnell BE, Mizuuchi K.** 2013. ParA-mediated plasmid partition driven by protein pattern self-organization. *EMBO J* **32:**1238–1249.

96. **Vecchiarelli AG, Hwang LC, Mizuuchi K.** 2013. Cell-free study of F plasmid partition provides evidence for cargo transport by a diffusion-ratchet mechanism. *Proc Natl Acad Sci USA* **110:**E1390–E1397.
97. **Thomas CM.** 2006. Transcription regulatory circuits in bacterial plasmids. *Biochem Soc Trans* **34:**1072–1074.
98. **Brendler T, Reaves L, Austin S.** 2004. Interplay between plasmid partition and postsegregational killing systems. *J Bacteriol* **186:**2504–2507.
99. **Gruber S, Errington J.** 2009. Recruitment of condensin to replication origin regions by ParB/SpoOJ promotes chromosome segregation in *B. subtilis*. *Cell* **137:**685–696.
100. **Sullivan NL, Marquis KA, Rudner DZ.** 2009. Recruitment of SMC by ParB-*parS* organizes the origin region and promotes efficient chromosome segregation. *Cell* **137:**697–707.
101. **Scholefield G, Errington J, Murray H.** 2012. Soj/ParA stalls DNA replication by inhibiting helix formation of the initiator protein DnaA. *EMBO J* **31:**1542–1555.
102. **Bingle LE, Thomas CM.** 2001. Regulatory circuits for plasmid survival. *Curr Opin Microbiol* **4:**194–200.
103. **del Solar G, Alonso JC, Espinosa M, Diaz-Orejas R.** 1996. Broad-host-range plasmid replication: an open question. *Mol Microbiol* **21:**661–666.
104. **Langella P, Le Loir Y, Ehrlich SD, Gruss A.** 1993. Efficient plasmid mobilization by pIP501 in *Lactococcus lactis* subsp. lactis. *J Bacteriol* **175:**5806–5813.
105. **Pratto F.** 2007. *Análisis del sistema de partición activa del plasmido pSM19035 de* Streptococcus pyogenes. *Doctoral dissertation*, Universidad Autómoma de Madrid, Madrid, Spain.
106. **Brantl S.** Plasmid replication control by antisense RNAs, p 83–103. 2015. *In* Tolmasky ME, Alonso JC (ed), *Plasmids: Biology and Impact in Biotechnology and Discovery*. ASM Press, Washington, DC.
107. **Crozat E, Fournes F, Cornet F, Hallet B, Rousseau P.** Resolution of multimeric forms of circular plasmids and chromosomes, p 157–174. 2015. *In* Tolmasky ME, Alonso JC (ed), *Plasmids: Biology and Impact in Biotechnology and Discovery*. ASM Press, Washington, DC.
108. **Baxter JC, Funnell BE.** Plasmid partition mechanisms, p 135–172. 2015. *In* Tolmasky ME, Alonso JC (ed), *Plasmids: Biology and Impact in Biotechnology and Discovery*. ASM Press, Washington, DC.
109. **Hernández-Arriaga AM, Chan WT, Espinosa M, Díaz-Orejas R.** Conditional activation of toxin-antitoxin (TA) systems: postsegregational killing and beyond, p 175–192. 2015. *In* Tolmasky ME, Alonso JC (ed), *Plasmids: Biology and Impact in Biotechnology and Discovery*. ASM Press, Washington, DC.
110. **Konieczny I, Bury K, Wawrzycka A, Wegrzyn K.** Iteron plasmids, p 15–32. 2015. *In* Tolmasky ME, Alonso JC (ed), *Plasmids: Biology and Impact in Biotechnology and Discovery*. ASM Press, Washington, DC.

Plasmids—Biology and Impact in Biotechnology and Discovery
Edited by Marcelo E. Tolmasky and Juan C. Alonso

doi:10.1128/microbiolspec.PLAS-0034-2014

Julie E. Samson[1]
Alfonso H. Magadan[1]
Sylvain Moineau[1]

12 The CRISPR-Cas Immune System and Genetic Transfers: Reaching an Equilibrium

INTRODUCTION

In 1987 Ishino et al. (1) sequenced the *Escherichia coli* alkaline phosphatase isozyme conversion gene (*iap*). Downstream of *iap*, they observed an array of short repeats (29 nucleotides) separated by nonrepetitive short sequences (spacers) (2). The terms "CRISPR," for clustered regularly interspaced short palindromic repeats, and "Cas," for CRISPR-associated genes, were first coined by Jansen et al. (3) in 2002 to describe the genetic structure of these loci. The increasing availability of genomic sequences in databases allowed Mojica et al. (4) to identify CRISPR as a specific family of repeats. Now we know that CRISPR-Cas systems are found in approximately 90% of archaeal and 40% of eubacterial sequenced genomes (5–7). In 2005, three groups independently reported similarities between spacer sequences and foreign mobile genetic elements (MGEs) such as phages and plasmids (8–10). These observations led to several hypotheses including that CRISPR-Cas systems may play a role in immunity and protect archaeal and bacterial cells from invasion by foreign DNA.

The immune mechanism of the CRISPR-Cas systems was experimentally demonstrated for the first time by Barrangou et al. in 2007 (11). These authors showed that the lactic acid bacterium *Streptococcus thermophilus* could acquire resistance against a bacteriophage by integrating a genome fragment (the protospacer) from this infectious bacterial virus into its CRISPR locus (spacer). Later studies discovered the expanded role of CRISPRs in preventing horizontal gene transfer (HGT) through conjugation and transformation (12–14). The CRISPR spacer sequences of archaeal species frequently match those of their own resident viruses or plasmids, suggesting a regulatory rather than inhibitory role for CRISPR-Cas. Indeed, it has been demonstrated that CRISPR-Cas systems may also play a role in transcriptional regulation (15, 16), DNA repair (17), pathogenesis (16, 18), modulation of biofilm production (19, 20), and sporulation (21).

CRISPR-Cas Organization

CRISPR loci are genomic DNA clusters that consist of a series of short repeat sequences (typically 24 to

[1]Département de Biochimie, Microbiologie et Bio-Informatique, Faculté des Sciences et de Génie, Groupe de Recherche en Écologie Buccale, Félix d'Hérelle Reference Center for Bacterial Viruses, Faculté de Médecine Dentaire, Université Laval, Québec City, Québec, G1V 0A6, Canada.

37 bp) separated by spacer sequences of similar length (Fig. 1) (5). Within a given locus, the length of the repeat and spacer sequences is typically conserved. Spacer sequences correspond mostly to fragments derived from viral genomes or MGEs. They appear to serve as a "genetic memory" of previous nucleic acid invasions and provide the specific CRISPR immunity (8–10). An adenine- and thymine-rich leader region containing 20 to 500 bp, including a transcriptional promoter, is present upstream of the CRISPR locus (3). Moreover, the *cas* genes, essential for the CRISPR-Cas machinery, are often encoded in the vicinity of the CRISPR locus (upstream or downstream) (3). Based on the presence of specific signature *cas* genes, CRISPR-Cas systems are divided into three main types (I, II, and III). These types are further divided into several subtypes (I-A, I-B, and so on), each of which expresses a different protein complex responsible for the CRISPR-Cas immunity mechanism (22).

Steps of the CRISPR-Cas Mechanism

CRISPR-Cas systems function in three general steps: (i) adaptation or immunization (involving the acquisition of spacers), (ii) biogenesis and maturation of CRISPR RNA (crRNA encoded by the repeat-spacer region), and (iii) interference (cleavage of invading nucleic acids) (23). These steps are summarized below (Fig. 1).

Spacers in the CRISPR locus are acquired from the DNA of invading plasmids or viruses in a process known as adaptation. New spacers are usually added at the 5′-leader region of the CRISPR locus (11, 24, 25) and come from defective or fragmented molecules (26). For type I and type II CRISPR-Cas systems, a conserved sequence motif in the vicinity of the protospacer, known as the protospacer-adjacent motif (PAM), is needed for spacer acquisition and interference (23, 25, 27). The incorporation of spacers in the CRISPR locus implies the action of two Cas proteins, Cas1 and Cas2, and sometimes the help of accessory elements specific for some CRISPR-Cas systems (28, 29). During this step, a repeat is also duplicated to conserve the genetic organization of the repeat-spacer region. Despite the "adaptive" mode of CRISPR-Cas systems, the molecular mechanism of this process still remains enigmatic.

The CRISPR locus is then transcribed as a long primary precursor CRISPR RNA (pre-crRNA) transcript, which is processed within the repeat sequences to produce a collection of short crRNAs (29). Further trimming can be done at the 3′ end or 5′ end to complete the maturation of crRNAs (29). This is a process known as crRNA biogenesis and maturation. Recently, an unusual crRNA maturation pathway was discovered in *Neisseria meningitidis*, in which crRNAs are transcribed from promoters embedded within each repeat, thus initiating the crRNA transcription independently within each repeat (13).

In conjunction with a set of Cas proteins, these crRNAs form the core of CRISPR-Cas ribonucleoprotein complexes. These complexes act as a guided-surveillance system and provide immunity against ensuing foreign nucleic acid invasion of DNA or RNA molecules complementary to each specific crRNA. On recognition of a matching target sequence, the plasmid or viral DNA is cleaved in a sequence-specific manner (30, 31) known as the interference step. The combination of recognition of an effective PAM (types I and II) and the base-pairing of a "seed" crRNA spacer region of 6 to 8 nucleotides identical to the viral or plasmid targeted protospacer are necessary to induce nucleic acid cleavage (5, 29). However, for some systems, the PAM sequence necessary for the acquisition step can diverge from the PAM recognized for the interference step, although they usually overlap considerably (32). Type I and type II CRISPR-Cas systems cleave DNA, while type III systems can cleave DNA or RNA (33). To discriminate between "self" (CRISPR) versus "non-self" (foreign) targets, the crRNA-Cas complexes are base-paired with sequences outside the protospacer (PAM or non-PAM) to avoid autoimmunity (34). In this way, CRISPR-Cas systems recognize and target invading nucleic acid molecules, avoiding the destruction of their own bacterial genomes.

CRISPR TARGETS MOBILE GENETIC ELEMENTS

Rapid evolution of bacteria is not just the consequence of rearrangements in the genome. HGT is also an important mechanism for bacterial evolution (35). Large proportions of bacterial genomes contain genes that come from the horizontal exchange of genetic material. This transfer of genetic material can occur by acquiring DNA directly from the environment (transformation) or by incorporating heterologous DNA using MGEs such as plasmids (conjugation) and bacteriophages (transduction) (Fig. 2). Generally, the outcome of horizontal gene transfer depends on the nature of the incoming DNA and on the genetic background of the host (36–38). Favorable DNA helping to cope with a specific environmental condition may no longer provide a selective advantage to the same bacteria that grow under other conditions and even can reduce host fitness (36, 37, 39, 40). To control the entry of new genetic material, bacteria possess systems that either

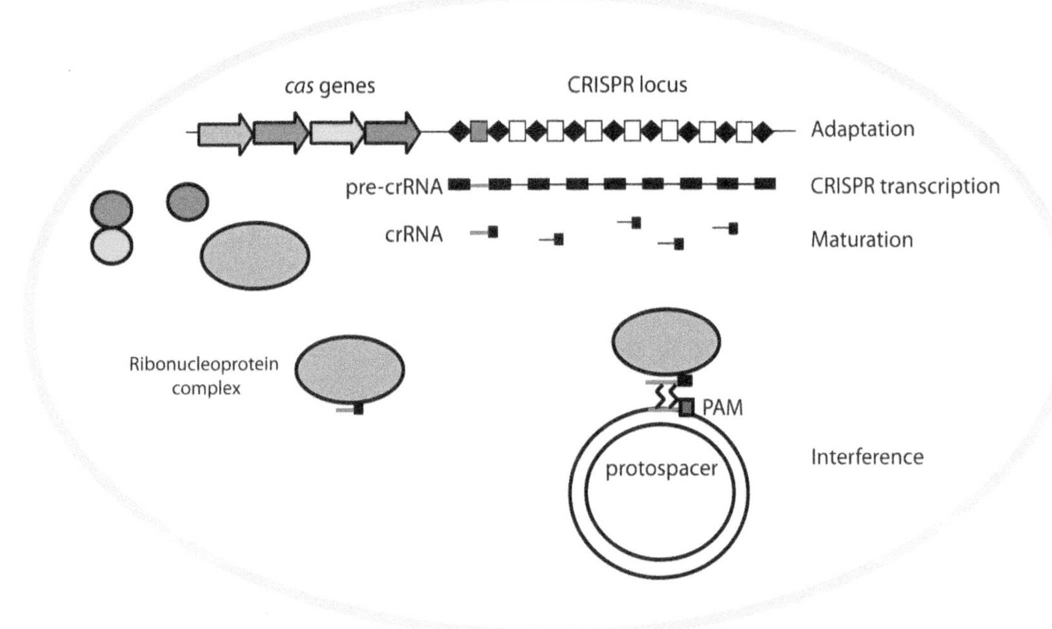

Figure 1 Genetic organization of a type II CRISPR-Cas system and its general steps of action. The CRISPR locus is composed of repeats (black diamonds) interspaced with spacers (red and white rectangles) of similar length. In the vicinity of the CRISPR array, *cas* genes (colored arrows) are coding for proteins necessary for the immunity process. During the adaptation step, a repeat and, most importantly, a new spacer (red rectangle) is acquired in the CRISPR locus, usually at the 5′ region. Transcription of the CRISPR locus leads to pre-crRNAs that are processed, leading to short crRNAs. These crRNAs are assembled with Cas protein(s) in ribonucleoprotein complexes that act as surveillance guides looking for matching invading sequences. If the crRNA sequence matches a protospacer found on the foreign and invading nucleic acid molecule and if a PAM (gray box) is present next to the protospacer (for type I and II systems), this leads to the cleavage of the invading molecule (interference step). doi:10.1128/microbiolspec.PLAS-0034-2014.f1

block the access to foreign DNA or remove MGE from the cell. Restriction-modification and CRISPR-Cas systems protect bacteria against foreign DNA by cleaving specific DNA sequences (41). Both systems are compatible in the same cells and efficiently protect bacterial genomes against phages and plasmids (42).

Incompatibility of Plasmids and CRISPR-Cas Systems

CRISPR-Cas systems can cleave foreign DNA and were expected to cause plasmid loss or block plasmid acquisition (Fig. 2). This phenomenon was demonstrated experimentally *in vivo* using a CRISPR-Cas system (type II-A) from *S. thermophilus*. When the CRISPR-Cas system acquired spacers from the artificially transformed plasmid, this led to plasmid loss (30). Moreover, some acquired spacers match an antibiotic resistance gene (present on the plasmid), which could interfere with the dissemination of other plasmids carrying the same antibiotic resistance marker. This antagonism between CRISPR-Cas interference and plasmid maintenance could explain why *S. thermophilus* naturally contains few plasmids as well as other bacteria (30).

E. coli produces a global transcriptional repressor, H-NS, that represses transcription of the CRISPR-Cas type I-E systems encoded in its genome (43, 44). A prolonged incubation time (around 1 to 2 weeks) of *E. coli* K12 Δ*hns* under nonselective conditions resulted in the loss of a high-copy plasmid concomitant with acquisition of new plasmid-targeting spacers in the CRISPR loci (25). The transformation efficiency (by heat shock) of the cured strain that has acquired new spacers targeting the plasmid dropped significantly (25). Others have used these PIMs (plasmid interfering bacterial mutants) to artificially transform them with an unrelated plasmid (45). Some spacers in the CRISPR loci match partly with this latter plasmid, but no spacer

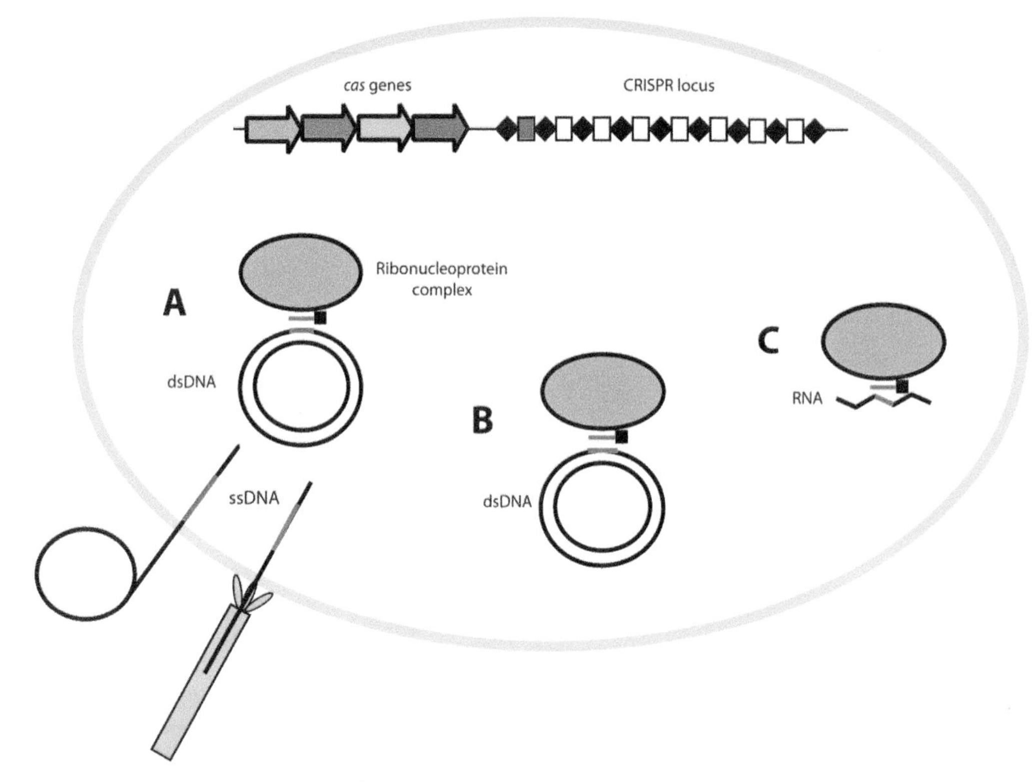

Figure 2 Probable action of CRISPR-Cas systems against invading plasmids. (A) Plasmids entering a bacterial cell via natural transformation (free DNA), conjugation (pilus not represented), and transduction (phagemids). Of note, to date, no CRISPR-Cas system has been identified that cleaves ssDNA molecules *in vivo*. However, after their entry in the bacteria, ssDNA are usually transformed into double-stranded DNA molecules (dsDNA) and maintained as plasmids or integrated within the chromosome. If the new dsDNA molecule contains a protospacer matching a crRNA sequence, it will be cleaved by the CRISPR-Cas machinery in a sequence-specific manner. (B) During artificial transformation of the bacterium with heat treatments or electroporation, dsDNA directly enters the cells. Thus, the CRISPR-Cas ribonucleoprotein complexes can directly target these dsDNA molecules to eliminate them. (C) Some CRISPR-Cas systems (type III) cleave RNA molecules. After transcription of plasmid genes, these molecules are silenced by the CRISPR-Cas system, and after a few rounds of bacterial replication, these plasmids may be lost.
doi:10.1128/microbiolspec.PLAS-0034-2014.f2

was 100% identical to the plasmid sequence. When the CRISPR possesses a plasmid-targeting spacer containing a sequence mismatch, it "primes" the strain to acquire new spacers from the invading plasmid, causing plasmid loss after two days of incubation (45). As demonstrated previously, CRISPR loci can acquire new spacers "naïvely" from the invading DNA, but at low frequency. An intact spacer sequence results in interference, while mutation in the protospacer can abolish the CRISPR interfering activity. However, for some systems, when the CRISPR already contains a spacer that matches the invading DNA with a protospacer mismatch in its seed sequence and/or mutation in the PAM, the CRISPR-Cas machinery is used to accelerate the spacer acquisition of the targeted foreign DNA element in a phenomenon called priming (45–47). In *E. coli*, the new spacers acquired by priming come from the same strand as the first spacer acquired (25, 45–47), while in other organisms, this strand bias has not been observed (48, 49).

In a similar way, when the CRISPR-Cas I-F systems of *Pectobacterium atrosepticum* contain spacers targeting an engineered plasmid (with the corresponding protospacer and a functional PAM), they cause plasmid loss after a few days of culturing (49). Moreover, the presence of this spacer primes the systems to incorporate new spacers that match the plasmid into the CRISPR loci, while no detectable spacer acquisition occurs with the control plasmid (without the protospacer sequence). These newly acquired spacers provide bacterial immunity against the transformation and the conjugation of similar plasmids. Surprisingly, one strain has added as many as nine new plasmid-targeting spacers in the two functional CRISPR loci of *P. atrosepticum* (CRISPR1 and CRISPR2). Interestingly, the more spacers a strain possesses against a plasmid, the more resistant the strain is to transformation and conjugation (49), probably the result of higher surveillance done by the CRISPR-Cas complexes, the cut at multiple sites within the plasmid DNA, and the difficulty of mutating several protospacers for escapees.

The *Staphylococcus aureus* strain USA300 contains three plasmids including the plasmid pUSA02, which confers resistance to the antibiotic tetracycline. A phagemid construction with Cas9, tracrRNA, and a repeat-spacer unit of *Streptococcus pyogenes* was designed to target the plasmid pUSA02 and was delivered into the cells by transduction using the phage pBD121 (50). Plasmid loss was achieved in 99.99% of cells, resulting in the restoration of sensitivity to tetracycline. Moreover, cells that have been pretreated with the engineered CRISPR system targeting pUSA02 could not acquire the plasmid by transduction, compared to the control without CRISPR immunization (50).

An antagonistic correlation between the presence of CRISPR-Cas systems and the multidrug-resistant phenotype of *Enterococcus* strains has been observed by Palmer and Gilmore (51). In these opportunistic pathogenic enterococci, the principal method of genome adaptation is *de novo* horizontal uptake of genes. The acquisition of new MGEs such as plasmids and transposons is frequently associated with the transfer of antibiotic resistance genes. These authors observed that many antibiotic-resistant enterococcal strains lack CRISPR-Cas systems. Conversely, bacteria with CRISPR possess spacers in their loci that match MGE sequences, suggesting that CRISPR-Cas systems regulate the flux of these elements (51). Thus, the extensive use of antibiotics appears to have enriched for enterococcal strains with lower defenses (no functional CRISPR-Cas systems) against MGE. However, this correlation between the presence of CRISPR-Cas activity and the antibiotic resistance phenotype or the presence of plasmids does not seem to apply for *E. coli* strains from clinical or animal samples (52). Some nonpathogenic *E. coli* carry additional CRISPR arrays (CRISPR3 and CRISPR4) with a significant abundance of spacers targeting plasmids compared to CRISPR1 and CRISPR2, which contain more phage-related spacers (53). Experiments demonstrating the link between the presence of plasmids and CRISPR activity should be done to confirm these bio-informatics results.

Bacteriophages escaping CRISPR-Cas systems usually have mutations in the targeted protospacer or PAM via point mutations or deletion (27). In contrast, when a CRISPR locus contains a spacer targeting a plasmid, especially if the targeted gene induces a strong selective advantage to the bacteria such as virulence factors or antibiotic resistance genes, it often results in the inactivation of the CRISPR-Cas system. This inactivation is caused by major modifications in the CRISPR-Cas locus such as the deletion or significant modification of the spacer, inactivation of a gene coding for an essential Cas protein, or deletion of the complete CRISPR-Cas locus (54–58). Therefore, bacteria possess CRISPR-Cas systems that protect the cells from invading DNA, but they must control CRISPR activity to permit incorporation of novel genes in their genomes.

CRISPR-Cas Systems Interfere with Transformation

Bacterial HGT involving plasmid exchange happens through one of two processes: transformation or conjugation. Transformation is the acquisition of free DNA molecules from the surrounding environment. Natural transformation occurs when bacterial cells are in a physiological state of competence. This time-limited state usually happens in response to specific environmental conditions or stresses such as altered growth conditions, limited nutrient access, cell density (monitored by quorum sensing), or starvation (35). Artificial transformation of weakened bacteria using electroporation or thermal shock is a commonly used tool in laboratories around the world to introduce DNA fragments into bacteria.

Evolutionary study of competent versus noncompetent strains of *Aggregatibacter actinomycetemcomitans* revealed that competent strains contain more CRISPR-Cas systems than noncompetent ones (59). Moreover, the remnant CRISPR-Cas systems found in the noncompetent strains often contain mutations or deletions in *cas* genes. The genetic diversity between both types of strains varies: competent strains have larger genomes with frequent rearrangements, while noncompetent strains have stable genomes with more extrachromosomal elements such as prophages and plasmids. This peculiarity is reflected in the spacer's content of the CRISPR loci. Spacers from competent strains more often target plasmids or phages sequences, while spacers from noncompetent cells more frequently match *A. actinomycetemcomitans* genomes (59).

To determine if the CRISPR-Cas system inhibits the acquisition of DNA by natural transformation in *Streptococcus pneumoniae*, *in vitro* and *in vivo* (in mice) assays were performed (54). No functional CRISPR-Cas system has been identified so far in *S. pneumoniae* strains for which the genome is available. Thus, the authors engineered the noncapsulated pneumococcal strain R6 to integrate the type II CRISPR system of *Streptococcus pyogenes* into its chromosome. The CRISPR locus was also engineered to add specific spacer sequences derived from prophage DNA or from the capsule biosynthesis gene (54). When a DNA molecule targeted by a CRISPR spacer was introduced into the bacteria by transformation *in vitro*, using the natural competency of *S. pneumoniae*, no transformants were obtained, indicating that the CRISPR-Cas system blocks the transformation or that the bacteria die because the transferred DNA molecule is destroyed and the antibiotic used as a selection marker is not detoxified. Then, the same authors tried to transfer a CRISPR-Cas locus containing a spacer targeting the chromosome of the recipient cell into bacteria without a CRISPR-Cas system (54). Similarly, no antibiotic-resistant transformant was obtained with a strain carrying a functional CRISPR-Cas system and containing a chromosomal-targeting spacer, reinforcing the idea that an active type II CRISPR-Cas system and its target cannot coexist in the same cell. Finally, the authors co-infected mice with noncapsulated and capsulated strains, each containing or missing the CRISPR-Cas system (with a spacer targeting the capsule biosynthesis gene) to see if the CRISPR system could prevent capsular gene transformation *in vivo* (54). Noncapsulated strains of *S. pneumoniae* cannot cause infections, while capsulated strains can avoid the immune system and cause an infection. Noncapsulated strains can naturally acquire a functional capsule gene (*cap3*) to produce a capsule and begin the infection. As expected, the strain without the CRISPR-Cas system acquired the *cap3* gene by HGT, produced a capsule, and caused an infection in mice. Conversely, the CRISPR-containing cells could not acquire the *cap3* gene *in vivo* and could not cause infection.

Using the native CRISPR-Cas system of *N. meningitidis* 8013 and the natural competency of this bacterium, Zhang and coworkers obtained similar results (13). They cloned different protospacers targeted by the endogenous CRISPR-Cas system of this strain into different plasmids or engineered genomic DNA (gDNA). They then transformed *N. meningitidis* using its natural competence and tested the transformation efficiency using these molecules. When the plasmid or gDNA contained a protospacer and a good PAM sequence, the transformation efficiency was lower (no transformants) than the controls without a protospacer or those containing a potential protospacer with a defective PAM (13). Thus, the CRISPR-Cas system of *N. meningitidis* also interferes with the natural transformation of this bacterium. Moreover, the presence of CRISPR-Cas systems could explain the rare presence of plasmids in this bacterium.

CRISPR-Cas systems can also inhibit artificial transformation done by electroporation or heat shock (Fig. 2). Transformation efficiency assays are often used to verify CRISPR-Cas activity during the adaptation step (acquisition of new spacers from the transformed plasmid) or interference (capacity to cut the plasmid DNA or eliminate the plasmid). Moreover, these plasmid-invader tests allowed the determination of possible protospacer mutations required to escape CRISPR-Cas systems or the identification of PAM sequences required for CRISPR adaptation/interference steps. These kinds of experiments have been used successfully with many different organisms such as *E. coli* (25, 47, 60–63), *Haloarcula hispanica* (48), *Haloferax volcanii* (56, 57, 64), *P. atrosepticum* (49), *Sulfolobus* spp. (58, 65), *Staphylococcus epidermidis* (14), *Streptococcus agalactiae* (66), *S. pyogenes* (67), *S. thermophilus* (30), and *Thermococcus kodakarensis* (68). Transformation interference is an important consideration when trying to introduce foreign DNA into industrially relevant cultures. In addition, plasmid stability can be compromised during long-term incubation or when using nonselective media.

The Role of CRISPR-Cas Systems in Conjugation Interference

Conjugation is the process by which a conjugative plasmid is transferred from a donor to a recipient cell. Conjugative plasmids can be grouped into at least six MOB families (genes required for mobilization) based mainly on the comparison of their relaxase genes (69). Since CRISPR-Cas systems defend bacteria against foreign DNA, it is not totally surprising that these systems can interfere with the conjugation process (Fig. 2).

Westra and coworkers screened public databases to analyze the spacer content of CRISPR loci related to conjugative plasmids (12). They found that targeted protospacers are not randomly distributed on conjugative plasmids: there is a MOB family-dependent bias. Indeed, MOB_P family plasmids are usually targeted within the lagging regions while the protospacers of the MOB_F family are mostly located in the leading region (the first plasmid section entering the cell). Using synthetic constructions, the authors demonstrated that the *E. coli* K12 type I-E CRISPR-Cas system interferes with the conjugation of the plasmid F (MOB_F family) (12). Surprisingly, the level of protection achieved by the CRISPR-Cas system is independent of the protospacer position (leading or lagging) and independent of the DNA strand (parental or newly synthesized strand). Thus, the bias distributions of the protospacers probably depend on the acquisition step of a new spacer in the CRISPR locus or on the first regions that go double-stranded as this CRISPR-Cas system target dsDNA.

Experimental studies demonstrating the conjugation-interfering role of CRISPR-Cas systems are not abundant. CRISPR-Cas systems can acquire spacers targeting conjugative plasmids, as shown when the archaeon *Sulfolobus solfataricus* was challenged with the SMV1 virus (70) and when *S. solfataricus* culture was subjected to freeze-thaw stress (71). The *Listeria monocytogenes* RliB-CRISPR array can interfere with plasmid conjugation using type I CRISPR-Cas machinery and a PNPase (polynucleotide phosphorylase) (72). Moreover, CRISPR-Cas systems can mediate resistance to conjugative plasmids in staphylococci (14, 73) and *P. atrosepticum* (49).

EFFECTS OF CRISPR-Cas SYSTEMS ON PLASMID REPLICATION AND PROTEIN PRODUCTION

Plasmids are useful genetic tools for analyzing the activities of the CRISPR-Cas immunity or to overexpress Cas proteins for characterization. They can also be used to better understand the role of specific Cas proteins. For example, the Cas3 protein (nuclease-helicase) from the type I-E CRISPR-Cas system of *E. coli* can increase the copy number of ColEI-replicon plasmids but has no effect on pRSF1010- or p15A-replicon plasmid yields (74). The Cas3 plasmid copy number-increasing activity is RNaseHI dependent and promotes the formation of plasmid concatemers or multimers (74).

In another study, Maier and coworkers tested the influence of plasmid copy number and the origin of replication on CRISPR-Cas type I-B interference in *H. volcanii* (64). Interestingly, they found that plasmid copy number has no effect on the plasmid transformation rate, but the origin of replication (ori) used has an effect. Plasmids with the pHV1 ori were degraded by the CRISPR-Cas system (few transformants), while plasmids with the pHV2 ori were successfully transformed into cells (64). This observation is analogous to the results obtained with the nonrandom MOB-family conjugative plasmid spacer acquisition in CRISPRs (12). Thus, spacer acquisition from plasmids is potentially dependent on their mode of entry into the cells or their mode of replication. Further experiments are needed to understand the mechanism.

Perez-Rodriguez et al. accidentally activated an *E. coli* CRISPR-Cas system while trying to overexpress the twin-arginine translocation (Tat) protein in cells with a deletion in the *dnaK* chaperone gene (75). This CRISPR activation silences the plasmid bearing the *tat* gene and is dependent on the two-component, extracytoplasmic stress system, BaeSR. Indeed, BaeSR activates transcription of the CRISPR-Cas operon by binding to the promoter upstream of *casA* (75). Henceforth, when trying to overexpress a protein, especially if the protein can induce cell envelope stress, it will be important to test for CRISPR activity if the bacteria inhibit protein production. Otherwise, the resident CRISPR-Cas systems can be activated and cure the constructed plasmid.

CONCLUSIONS

Bacteria must adapt to survive in ever-changing environments, including when they face phages or other antimicrobial agents. They mutate or acquire new advantageous genes that helps them to cope with these harmful conditions. The CRISPR-Cas adaptive immune system of bacteria and archaea can clearly interfere with the transfer of foreign DNA into the cells, including plasmids. This inhibition can also interfere with the maintenance of existing plasmids as observed initially in *S. thermophilus* (30). Future studies are still needed to better understand the spacer acquisition process, including its preference for specific DNA sequences or for the uptake of particular plasmid replicon sequences. Another area of research interest is to understand why only half of the bacteria for which a complete genomic sequence is available do not possess a CRISPR-Cas system. For example, clinical strains appear to contain fewer CRISPR-Cas systems (51, 76). Finally, it is largely unknown how so many bacterial strains are able to carry and maintain plasmids in the presence of seemingly functional CRISPR-cas systems.

Acknowledgments. *We thank Barb Conway and Alexander Hynes for editorial assistance. S.M. acknowledges funding from NSERC of Canada (Discovery program) and holds a Tier 1 Canada Research Chair in Bacteriophages. Conflicts of interest: We declare no conflicts.*

Citation. Samson JE, Magadan AH, Moineau S. 2015. The CRISPR-Cas immune system and genetic transfers: reaching an equilibrium. Microbiol Spectrum 3(1):PLAS-0034-2014.

References

1. **Ishino Y, Shinagawa H, Makino K, Amemura M, Nakata A.** 1987. Nucleotide sequence of the *iap* gene, responsible

for alkaline phosphatase isozyme conversion in *Escherichia coli*, and identification of the gene product. *J Bacteriol* **169**:5429–5433.

2. **Nakata A, Amemura M, Makino K.** 1989. Unusual nucleotide arrangement with repeated sequences in the *Escherichia coli* K-12 chromosome. *J Bacteriol* **171**: 3553–3556.
3. **Jansen R, Embden JD, Gaastra W, Schouls LM.** 2002. Identification of genes that are associated with DNA repeats in prokaryotes. *Mol Microbiol* **43**:1565–1575.
4. **Mojica FJ, Diez-Villasenor C, Soria E, Juez G.** 2000. Biological significance of a family of regularly spaced repeats in the genomes of Archaea, Bacteria and mitochondria. *Mol Microbiol* **36**:244–246.
5. **Deveau H, Garneau JE, Moineau S.** 2010. CRISPR/Cas system and its role in phage-bacteria interactions. *Annu Rev Microbiol* **64**:475–493.
6. **Terns MP, Terns RM.** 2011. CRISPR-based adaptive immune systems. *Curr Opin Microbiol* **14**:321–327.
7. **Wiedenheft B, Sternberg SH, Doudna JA.** 2012. RNA-guided genetic silencing systems in bacteria and archaea. *Nature* **482**:331–338.
8. **Bolotin A, Quinquis B, Sorokin A, Ehrlich SD.** 2005. Clustered regularly interspaced short palindrome repeats (CRISPRs) have spacers of extrachromosomal origin. *Microbiology* **151**:2551–2561.
9. **Mojica FJ, Diez-Villasenor C, Garcia-Martinez J, Soria E.** 2005. Intervening sequences of regularly spaced prokaryotic repeats derive from foreign genetic elements. *J Mol Evol* **60**:174–182.
10. **Pourcel C, Salvignol G, Vergnaud G.** 2005. CRISPR elements in *Yersinia pestis* acquire new repeats by preferential uptake of bacteriophage DNA, and provide additional tools for evolutionary studies. *Microbiology* **151**: 653–663.
11. **Barrangou R, Fremaux C, Deveau H, Richards M, Boyaval P, Moineau S, Romero DA, Horvath P.** 2007. CRISPR provides acquired resistance against viruses in prokaryotes. *Science* **315**:1709–1712.
12. **Westra ER, Staals RH, Gort G, Hogh S, Neumann S, de la Cruz F, Fineran PC, Brouns SJ.** 2013. CRISPR-Cas systems preferentially target the leading regions of MOBF conjugative plasmids. *RNA Biol* **10**:749–761.
13. **Zhang Y, Heidrich N, Ampattu BJ, Gunderson CW, Seifert HS, Schoen C, Vogel J, Sontheimer EJ.** 2013. Processing-independent CRISPR RNAs limit natural transformation in *Neisseria meningitidis*. *Mol Cell* **50**: 488–503.
14. **Marraffini LA, Sontheimer EJ.** 2008. CRISPR interference limits horizontal gene transfer in staphylococci by targeting DNA. *Science* **322**:1843–1845.
15. **Sampson TR, Weiss DS.** 2014. CRISPR-Cas systems: new players in gene regulation and bacterial physiology. *Front Cell Infect Microbiol* **4**:37.
16. **Sampson TR, Saroj SD, Llewellyn AC, Tzeng YL, Weiss DS.** 2013. A CRISPR/Cas system mediates bacterial innate immune evasion and virulence. *Nature* **497**:254–257.
17. **Babu M, Beloglazova N, Flick R, Graham C, Skarina T, Nocek B, Gagarinova A, Pogoutse O, Brown G, Binkowski A, Phanse S, Joachimiak A, Koonin EV, Savchenko A, Emili A, Greenblatt J, Edwards AM, Yakunin AF.** 2011. A dual function of the CRISPR-Cas system in bacterial antivirus immunity and DNA repair. *Mol Microbiol* **79**:484–502.
18. **Louwen R, Horst-Kreft D, de Boer AG, van der Graaf L, de Knegt G, Hamersma M, Heikema AP, Timms AR, Jacobs BC, Wagenaar JA, Endtz HP, van der Oost J, Wells JM, Nieuwenhuis EE, van Vliet AH, Willemsen PT, van Baarlen P, van Belkum A.** 2013. A novel link between *Campylobacter jejuni* bacteriophage defence, virulence and Guillain-Barre syndrome. *Eur J Clin Microbiol Infect Dis* **32**:207–226.
19. **Zegans ME, Wagner JC, Cady KC, Murphy DM, Hammond JH, O'Toole GA.** 2009. Interaction between bacteriophage DMS3 and host CRISPR region inhibits group behaviors of *Pseudomonas aeruginosa*. *J Bacteriol* **191**:210–219.
20. **Cady KC, O'Toole GA.** 2011. Non-identity-mediated CRISPR-bacteriophage interaction mediated via the Csy and Cas3 proteins. *J Bacteriol* **193**:3433–3445.
21. **Viswanathan P, Murphy K, Julien B, Garza AG, Kroos L.** 2007. Regulation of dev, an operon that includes genes essential for *Myxococcus xanthus* development and CRISPR-associated genes and repeats. *J Bacteriol* **189**: 3738–3750.
22. **Makarova KS, Haft DH, Barrangou R, Brouns SJ, Charpentier E, Horvath P, Moineau S, Mojica FJ, Wolf YI, Yakunin AF, van der Oost J, Koonin EV.** 2011. Evolution and classification of the CRISPR-Cas systems. *Nat Rev Microbiol* **9**:467–477.
23. **Marraffini LA, Sontheimer EJ.** 2010. CRISPR interference: RNA-directed adaptive immunity in bacteria and archaea. *Nat Rev Genet* **11**:181–190.
24. **Tyson GW, Banfield JF.** 2008. Rapidly evolving CRISPRs implicated in acquired resistance of microorganisms to viruses. *Environ Microbiol* **10**:200–207.
25. **Swarts DC, Mosterd C, van Passel MW, Brouns SJ.** 2012. CRISPR interference directs strand specific spacer acquisition. *PLoS One* **7**:e35888. doi:10.1371/journal.pone.0035888.
26. **Hynes AP, Villion M, Moineau S.** 2014. Adaptation in bacterial CRISPR-Cas immunity can be driven by defective phages. *Nat Commun* **5**:4399.
27. **Deveau H, Barrangou R, Garneau JE, Labonte J, Fremaux C, Boyaval P, Romero DA, Horvath P, Moineau S.** 2008. Phage response to CRISPR-encoded resistance in *Streptococcus thermophilus*. *J Bacteriol* **190**: 1390–1400.
28. **Yosef I, Goren MG, Qimron U.** 2012. Proteins and DNA elements essential for the CRISPR adaptation process in *Escherichia coli*. *Nucleic Acids Res* **40**:5569–5576.
29. **van der Oost J, Westra ER, Jackson RN, Wiedenheft B.** 2014. Unravelling the structural and mechanistic basis of CRISPR-Cas systems. *Nat Rev Microbiol* **12**:479–492.
30. **Garneau JE, Dupuis ME, Villion M, Romero DA, Barrangou R, Boyaval P, Fremaux C, Horvath P, Magadan AH, Moineau S.** 2010. The CRISPR/Cas bacterial immune system cleaves bacteriophage and plasmid DNA. *Nature* **468**:67–71.

31. **Magadan AH, Dupuis ME, Villion M, Moineau S.** 2012. Cleavage of phage DNA by the *Streptococcus thermophilus* CRISPR3-Cas system. *PLoS One* **7:**e40913. doi: 10.1371/journal.pone.0040913.
32. **Shah SA, Erdmann S, Mojica FJ, Garrett RA.** 2013. Protospacer recognition motifs: mixed identities and functional diversity. *RNA Biol* **10:**891–899.
33. **Barrangou R, Marraffini LA.** 2014. CRISPR-Cas systems: prokaryotes upgrade to adaptive immunity. *Mol Cell* **54:**234–244.
34. **Marraffini LA, Sontheimer EJ.** 2010. Self versus non-self discrimination during CRISPR RNA-directed immunity. *Nature* **463:**568–571.
35. **Thomas CM, Nielsen KM.** 2005. Mechanisms of, and barriers to, horizontal gene transfer between bacteria. *Nat Rev Microbiol* **3:**711–721.
36. **Humphrey B, Thomson NR, Thomas CM, Brooks K, Sanders M, Delsol AA, Roe JM, Bennett PM, Enne VI.** 2012. Fitness of *Escherichia coli* strains carrying expressed and partially silent IncN and IncP1 plasmids. *BMC Microbiol* **12:**53.
37. **Smith MA, Bidochka MJ.** 1998. Bacterial fitness and plasmid loss: the importance of culture conditions and plasmid size. *Can J Microbiol* **44:**351–355.
38. **De Gelder L, Ponciano JM, Joyce P, Top EM.** 2007. Stability of a promiscuous plasmid in different hosts: no guarantee for a long-term relationship. *Microbiology* **153:**452–463.
39. **Dahlberg C, Chao L.** 2003. Amelioration of the cost of conjugative plasmid carriage in *Eschericha coli* K12. *Genetics* **165:**1641–1649.
40. **Paulander W, Maisnier-Patin S, Andersson DI.** 2009. The fitness cost of streptomycin resistance depends on *rpsL* mutation, carbon source and RpoS (sigmaS). *Genetics* **183:**539–546.
41. **Johnston C, Martin B, Polard P, Claverys JP.** 2013. Postreplication targeting of transformants by bacterial immune systems? *Trends Microbiol* **21:**516–521.
42. **Dupuis ME, Villion M, Magadan AH, Moineau S.** 2013. CRISPR-Cas and restriction-modification systems are compatible and increase phage resistance. *Nat Commun* **4:**2087.
43. **Westra ER, Pul U, Heidrich N, Jore MM, Lundgren M, Stratmann T, Wurm R, Raine A, Mescher M, van Heereveld L, Mastop M, Wagner EG, Schnetz K, van der Oost J, Wagner R, Brouns SJ.** 2010. H-NS-mediated repression of CRISPR-based immunity in *Escherichia coli* K12 can be relieved by the transcription activator LeuO. *Mol Microbiol* **77:**1380–1393.
44. **Pougach K, Semenova E, Bogdanova E, Datsenko KA, Djordjevic M, Wanner BL, Severinov K.** 2010. Transcription, processing and function of CRISPR cassettes in *Escherichia coli. Mol Microbiol* **77:**1367–1379.
45. **Fineran PC, Gerritzen MJ, Suarez-Diez M, Kunne T, Boekhorst J, van Hijum SA, Staals RH, Brouns SJ.** 2014. Degenerate target sites mediate rapid primed CRISPR adaptation. *Proc Natl Acad Sci USA* **111:**E1629–1638.
46. **Datsenko KA, Pougach K, Tikhonov A, Wanner BL, Severinov K, Semenova E.** 2012. Molecular memory of prior infections activates the CRISPR/Cas adaptive bacterial immunity system. *Nat Commun* **3:**945.
47. **Savitskaya E, Semenova E, Dedkov V, Metlitskaya A, Severinov K.** 2013. High-throughput analysis of type I-E CRISPR/Cas spacer acquisition in *E. coli. RNA Biol* **10:** 716–725.
48. **Li M, Wang R, Zhao D, Xiang H.** 2014. Adaptation of the *Haloarcula hispanica* CRISPR-Cas system to a purified virus strictly requires a priming process. *Nucleic Acids Res* **42:**2483–2492.
49. **Richter C, Dy RL, McKenzie RE, Watson BN, Taylor C, Chang JT, McNeil MB, Staals RH, Fineran PC.** 2014. Priming in the type I-F CRISPR-Cas system triggers strand-independent spacer acquisition, bi-directionally from the primed protospacer. *Nucleic Acids Res* **42:** 8516–8526.
50. **Bikard D, Euler CW, Jiang W, Nussenzweig PM, Goldberg GW, Duportet X, Fischetti VA, Marraffini LA.** 2014. Exploiting CRISPR-Cas nucleases to produce sequence-specific antimicrobials. *Nat Biotechnol* **32:** 1146–1150.
51. **Palmer KL, Gilmore MS.** 2010. Multidrug-resistant enterococci lack CRISPR-*cas*. *mBio* **1:**e00227-10. doi: 10.1128/mBio.00227-10.
52. **Touchon M, Charpentier S, Pognard D, Picard B, Arlet G, Rocha EP, Denamur E, Branger C.** 2012. Antibiotic resistance plasmids spread among natural isolates of *Escherichia coli* in spite of CRISPR elements. *Microbiology* **158:**2997–3004.
53. **Touchon M, Rocha EP.** 2010. The small, slow and specialized CRISPR and anti-CRISPR of *Escherichia* and *Salmonella*. *PLoS One* **5:**e11126. doi:10.1371/journal.pone.0011126.
54. **Bikard D, Hatoum-Aslan A, Mucida D, Marraffini LA.** 2012. CRISPR interference can prevent natural transformation and virulence acquisition during *in vivo* bacterial infection. *Cell Host Microbe* **12:**177–186.
55. **Jiang W, Maniv I, Arain F, Wang Y, Levin BR, Marraffini LA.** 2013. Dealing with the evolutionary downside of CRISPR immunity: bacteria and beneficial plasmids. *PLoS Genet* **9:**e1003844. doi:10.1371/journal.pgen.1003844.
56. **Maier LK, Stoll B, Brendel J, Fischer S, Pfeiffer F, Dyall-Smith M, Marchfelder A.** 2013. The ring of confidence: a haloarchaeal CRISPR/Cas system. *Biochem Soc Trans* **41:**374–378.
57. **Fischer S, Maier LK, Stoll B, Brendel J, Fischer E, Pfeiffer F, Dyall-Smith M, Marchfelder A.** 2012. An archaeal immune system can detect multiple protospacer adjacent motifs (PAMs) to target invader DNA. *J Biol Chem* **287:** 33351–33363.
58. **Gudbergsdottir S, Deng L, Chen Z, Jensen JV, Jensen LR, She Q, Garrett RA.** 2011. Dynamic properties of the *Sulfolobus* CRISPR/Cas and CRISPR/Cmr systems when challenged with vector-borne viral and plasmid genes and protospacers. *Mol Microbiol* **79:**35–49.
59. **Jorth P, Whiteley M.** 2012. An evolutionary link between natural transformation and CRISPR adaptive immunity. *mBio* **3:**e00309-12. doi:10.1128/mBio.00309-12.

60. **Sapranauskas R, Gasiunas G, Fremaux C, Barrangou R, Horvath P, Siksnys V.** 2011. The *Streptococcus thermophilus* CRISPR/Cas system provides immunity in *Escherichia coli*. *Nucleic Acids Res* **39:**9275–9282.

61. **Semenova E, Jore MM, Datsenko KA, Semenova A, Westra ER, Wanner B, van der Oost J, Brouns SJ, Severinov K.** 2011. Interference by clustered regularly interspaced short palindromic repeat (CRISPR) RNA is governed by a seed sequence. *Proc Natl Acad Sci USA* **108:**10098–10103.

62. **Almendros C, Guzman NM, Diez-Villasenor C, Garcia-Martinez J, Mojica FJ.** 2012. Target motifs affecting natural immunity by a constitutive CRISPR-Cas system in *Escherichia coli*. *PLoS One* **7:**e50797. doi: 10.1371/journal.pone.0050797.

63. **Shmakov S, Savitskaya E, Semenova E, Logacheva MD, Datsenko KA, Severinov K.** 2014. Pervasive generation of oppositely oriented spacers during CRISPR adaptation. *Nucleic Acids Res* **42:**5907–5916.

64. **Maier LK, Lange SJ, Stoll B, Haas KA, Fischer S, Fischer E, Duchardt-Ferner E, Wohnert J, Backofen R, Marchfelder A.** 2013. Essential requirements for the detection and degradation of invaders by the *Haloferax volcanii* CRISPR/Cas system I-B. *RNA Biol* **10:**865–874.

65. **Deng L, Garrett RA, Shah SA, Peng X, She Q.** 2013. A novel interference mechanism by a type IIIB CRISPR-Cmr module in *Sulfolobus*. *Mol Microbiol* **87:** 1088–1099.

66. **Lopez-Sanchez MJ, Sauvage E, Da Cunha V, Clermont D, Ratsima Hariniaina E, Gonzalez-Zorn B, Poyart C, Rosinski-Chupin I, Glaser P.** 2012. The highly dynamic CRISPR1 system of *Streptococcus agalactiae* controls the diversity of its mobilome. *Mol Microbiol* **85:**1057–1071.

67. **Deltcheva E, Chylinski K, Sharma CM, Gonzales K, Chao Y, Pirzada ZA, Eckert MR, Vogel J, Charpentier E.** 2011. CRISPR RNA maturation by trans-encoded small RNA and host factor RNase III. *Nature* **471:**602–607.

68. **Elmore JR, Yokooji Y, Sato T, Olson S, Glover CV, Graveley BR, Atomi H, Terns RM, Terns MP.** 2013. Programmable plasmid interference by the CRISPR-Cas system in *Thermococcus kodakarensis*. *RNA Biol* **10:** 828–840.

69. **Garcillan-Barcia MP, Francia MV, de la Cruz F.** 2009. The diversity of conjugative relaxases and its application in plasmid classification. *FEMS Microbiol Rev* **33:** 657–687.

70. **Erdmann S, Garrett RA.** 2012. Selective and hyperactive uptake of foreign DNA by adaptive immune systems of an archaeon via two distinct mechanisms. *Mol Microbiol* **85:**1044–1056.

71. **Erdmann S, Shah SA, Garrett RA.** 2013. SMV1 virus-induced CRISPR spacer acquisition from the conjugative plasmid pMGB1 in *Sulfolobus solfataricus* P2. *Biochem Soc Trans* **41:**1449–1458.

72. **Sesto N, Touchon M, Andrade JM, Kondo J, Rocha EP, Arraiano CM, Archambaud C, Westhof E, Romby P, Cossart P.** 2014. A PNPase dependent CRISPR system in *Listeria*. *PLoS Genet* **10:**e1004065. doi:10.1371/journal.pgen.1004065.

73. **Hatoum-Aslan A, Maniv I, Samai P, Marraffini LA.** 2014. Genetic characterization of antiplasmid immunity through a type III-A CRISPR-Cas system. *J Bacteriol* **196:**310–317.

74. **Ivancic-Bace I, Radovcic M, Bockor L, Howard JL, Bolt EL.** 2013. Cas3 stimulates runaway replication of a ColE1 plasmid in *Escherichia coli* and antagonises RNaseHI. *RNA Biol* **10:**770–778.

75. **Perez-Rodriguez R, Haitjema C, Huang Q, Nam KH, Bernardis S, Ke A, Delisa MP.** 2011. Envelope stress is a trigger of CRISPR RNA-mediated DNA silencing in *Escherichia coli*. *Mol Microbiol* **79:**584–599.

76. **Hatoum-Aslan A, Marraffini LA.** 2014. Impact of CRISPR immunity on the emergence and virulence of bacterial pathogens. *Curr Opin Microbiol* **17:**82–90.

Plasmids—Biology and Impact in Biotechnology and Discovery
Edited by Marcelo E. Tolmasky and Juan C. Alonso

doi:10.1128/microbiolspec.PLAS-0031-2014

María de Toro[1]
M. Pilar Garcillán-Barcia[1]
Fernando de la Cruz[1]

13

Plasmid Diversity and Adaptation Analyzed by Massive Sequencing of *Escherichia coli* Plasmids

INTRODUCTION

There are two types of evolutionary events that impinge on bacterial genomes: microevolution (mutations and homologous recombination), which produces point mutations, and macroevolution, which results in the acquisition, loss, or rearrangement of large DNA segments (1–3). Macroevolutionary events ultimately arise from the import of adaptive traits by means of mobile genetic elements (MGEs). MGEs are basically plasmids, phages, and their integrated counterparts (integrative and conjugative elements [ICEs] and prophages). Genomic islands are fossil remnants of MGEs, since they generally lost mobility by rearrangement after the initial insertion event. Nevertheless, they can sometimes still be mobilized by the recombinogenic activity of IncF plasmids (4), among other possibilities. In summary, the dynamics of bacterial genomes can be interpreted as resulting from the microevolution of a core genome and the macroevolution of the accessory genome, i.e., the horizontal gene transfer events that give rise to the concept of a pangenome (5).

As often occurs in microbiology, a paradigmatic example of pangenome evolution can be found in the best-studied bacterial model, *Escherichia coli*, and more specifically, in the genomic analysis of pathogenic *E. coli* (6, 7). A recent phylogenomic analysis of *E. coli* ST131 isolates (a well-defined extraintestinal pathogenic *E. coli* lineage [8]) from different geographical areas emphasizes the interplay of micro- and macroevolutionary events in the fate of *E. coli* ST131 diversification (9). As shown in that report, more than 100 regions, totaling 0.94 Mb of the genome (representing roughly one fifth of the entire genome), were involved in recombination events within the ST131 lineage. Large recombinant regions were associated with prophages and genomic islands related to fimbrial adhesins and bacterial motility. In addition, the work by Lanza et al. underscores the importance of plasmid flux in

[1]Instituto de Biomedicina y Biotecnología de Cantabria (IBBTEC), Universidad de Cantabria, (CSIC), 39011 Santander, Spain.

ST131 lineage evolution (10). It corresponds to the analysis of the first collection listed in Table 1. The 10 genomes analyzed contained 39 plasmids, corresponding to 15 different plasmid groups. The most conspicuous was the MOB_{F12}/IncF plasmid group, represented by one (or two) plasmids for each genome. These plasmids carry multiple virulence and resistance determinants as well as several insertion sequence elements. As shown before (11), transposons and insertion sequence elements contained within IncF plasmids drive recombination events that result in plasmid rearrangements and may affect genome evolution. Not only MOB_{F12}/IncF plasmids, but also MOB_{H12}/IncA-C and Rep3/phage-like plasmids are present in *E. coli* ST131 genomes. These plasmids are involved in *E. coli* chromosome evolution by facilitating mobilization *in trans* of genetic islands or integrating new genetic material (4, 6).

Plasmids are the spearhead of bacterial evolution. If we judge by a contemporary macroevolutionary event, i.e., the appearance and dissemination of multiple antibiotic resistance, plasmids are the first MGE platforms that come into play (2). Plasmids disseminate not only antibiotic resistance, but also other adaptive genes (virulence, metabolic adaptations, competition weapons, etc.). Examples can be found in the processes of acquisition by the ST131 lineage of *E. coli* of the extended-spectrum beta-lactamase (ESBL) $bla_{CTX\text{-}M\text{-}15}$ gene (12, 13), fluoroquinolone resistance (14, 15), and carbapenemase resistance (16), among other antibiotic resistance genes. Furthermore, it has recently been shown that ESBL transmission from animals to human pathogens occurs by the transmission of IncI1-like conjugative plasmids carrying ESBL (17). Is the finding of these specific plasmids just a coincidence, or do they contain specific genetic peculiarities that make them more efficient transmission vectors? This is an important question that we will explore in this review.

Finally, research on the evolution of conjugative elements suggests a possible trend with important clinical consequences: adaptive traits carried initially by plasmids tend to shift from autonomous mobile elements to more stably integrated elements, such as ICEs and genomic islands (2). Thus, there is a temporal window of opportunity in which we can better fight antibiotic resistance. Where to act depends on our knowledge of which are the most important mobile elements and what are the main routes and scenarios where antibiotic resistance genes land in human pathogens. Again, this review will address which are the most relevant plasmids in the dissemination of antibiotic resistance. It will concentrate on proteobacterial systems and, more specifically, on *E. coli*, since our knowledge base is more advanced with this model organism.

DIVERSITY AND CLASSIFICATION OF PROTEOBACTERIAL TRANSFER SYSTEMS

The first need for a rigorous analysis of the importance of different conjugation systems in bacterial adaptation and evolution is a comprehensive classification system, which helps us to order the myriad of plasmids identified in natural systems.

Plasmids organize their genomes in three main architectures, according to their conjugal transmissibility: conjugative, mobilizable, and nonmobile elements. Conjugative plasmids are relatively large (>25 kb), since they carry a complete conjugation system. Genetic determinants for conjugative ability are complex, as shown by the example of the prototype plasmid R388, represented in Fig. 1. As can be seen in the figure, several operons and more than 20 genes are involved in this function and its regulation. The plasmid R388 conjugation module occupies almost two thirds of its 33-kb DNA, which underscores its relevance in the genetic organization of the plasmid. Besides the conjugation

Table 1 Genomes sequenced to establish the *E. coli* plasmid collection

Origin	Source [number]	Number of genomes	Number of plasmids	Reference
ST131 isolates from USA and Spain	Human clinical [9], human commensal [1]	10	48	10
ESBL-positive isolates from The Netherlands	Human commensal [17], chicken [11], pig [4]	32	147	17
Carbapenem resistant from Boston hospitals	Human clinical [19]	19	70	Bioproject PRJNA202876 and this work
Total	Human clinical [28], human commensal [18], chicken [11], pig [4]	61	255	

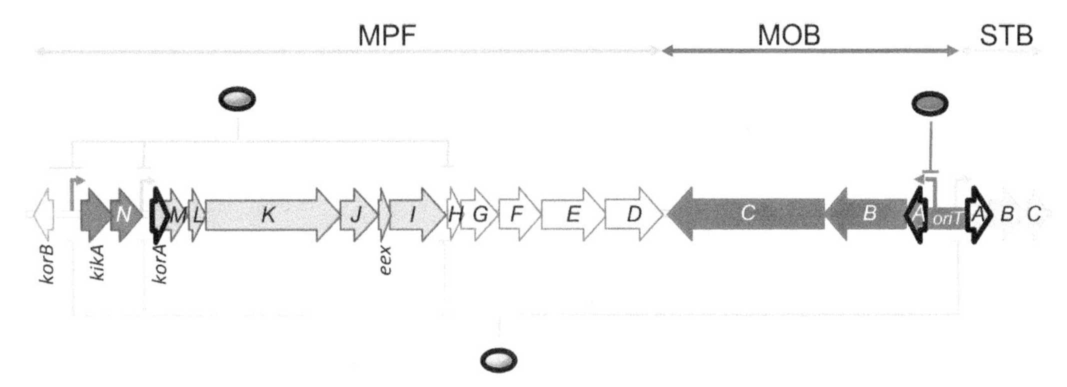

Figure 1 Genetic map of the conjugation genes of plasmid R388. Three modules related to conjugative transfer are depicted: STB, MOB, and MPF. Genes encoded in the 18-kb fragment comprised between *korB* and *stbC* (GenBank Acc. No. BR000038) are represented by arrows. Genes contained in the same operon are depicted using the same color pattern, as well as the corresponding promoters, which are represented by small arrows. Genes encoding transcription factors are outlined in black. The encoded repressors follow the same color pattern as their corresponding genes and are represented as ovals. The activity of the repressors on the promoters is indicated by lines.
doi:10.1128/microbiolspec.PLAS-0031-2014.f1

module, conjugative plasmids contain various stability systems and other propagation-related genes, such as antirestriction systems, avoidance of SOS response, single-strand DNA binding proteins, etc. This genetic load implies a minimum size of 25 kb for conjugative plasmids. Mobilizable plasmids, on the other hand, are relatively small (<20 kb), since they do not need to carry a complete transfer system. Their copy number is relatively higher, and thus they can also manage without sophisticated stability determinants. As a result of this dichotomy, plasmid size distribution in populations follows a bimodal distribution (18). This distribution is not casual. It is reproduced in our collection of *E. coli* plasmids (see Table 1), as shown in Fig. 2. Besides conjugation-transmissible plasmids, roughly 50% of the plasmids listed in the NCBI database seem to be nontransmissible by conjugation (18). In our collection, non-MOB plasmids represent only 27% of the total. Non-MOB plasmids also display a bimodal distribution, with peaks at 4 and 40 kb. It was suggested that those plasmids could spread by transformation (the small ones) or transduction (the large ones) (18).

Plasmid conjugative systems are composed of three main genetic modules, as shown in Fig. 1 for the prototype plasmid R388: STB, MOB, and MPF. The STB module is involved in plasmid segregation. Apparently, conjugation and segregation imply conflicting demands. Thus, the function of this operon seems to be to allow efficient partition while enabling an optimal rate of conjugation. StbA, a DNA binding protein, is essential for plasmid stability, while StbB is required for localization of plasmid copies for conjugation (19, 20). The MOB operon contains the origin of transfer (*oriT*) and the proteins required to process this DNA site for conjugative replication. The mechanisms of *oriT* processing have been recently reviewed (21). A plasmid classification system was developed based on the plasmid MOB systems, which is shared by conjugative and mobilizable plasmids (22–24). It uses relaxases as plasmid evolutionary clocks. As a result, six MOB families were described in proteobacteria, which correspond to six relaxase families. The crystal structure of both the relaxase (25) and type 4-coupling protein (26) of plasmid R388 have been solved. All relaxase families, except MOB_C, are related in structure to R388 relaxase TrwC and belong to the endonuclease HUH family (27). The MOB_C relaxase family is related to the *Bam*HI family of restriction endonucleases (28).

Besides MOB, all conjugative (but not mobilizable) plasmids contain an MPF system, composed of more than 10 genes, organized in one or more operons (Fig. 1). The three-dimensional structure of the MPF_T system of plasmid R388 has been recently elucidated by electron microscopy (29). Several recent reviews focus on the structure and dynamics of MPF systems (30–33), so they need not be reviewed in this work. Smillie et al. extended the MOB classification concept for the classification of MPF mobility systems (18). Four main MPF systems were described in proteobacteria (called MPF types F, T, I, and G). A scheme summarizing the types of MOB and MPF families and their most common combinations is shown in Fig. 3. Plasmid

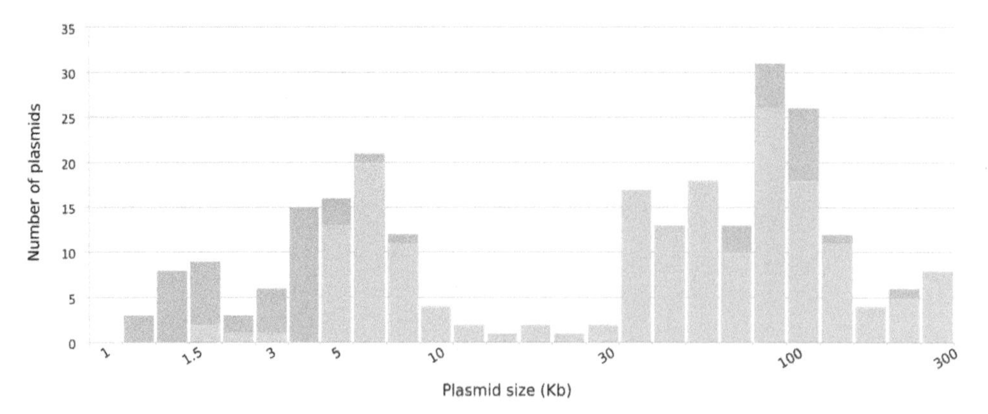

Figure 2 Distribution of plasmid sizes in the 255-*E. coli* plasmid collection. The figure represents the size distribution of plasmids in the collection described in Table 1. The histogram shows the total number of plasmids corresponding to each size class (in logarithmic scale). Plasmid distribution shows a trimodal abundance curve (with calculated median sizes of 1.6, 5.1, and 87 kb). Plasmids in which a relaxase gene was detected are shown in green (N = 187), and those in which it was not, in gray (N = 68).
doi:10.1128/microbiolspec.PLAS-0031-2014.f2

R388 corresponds to MOB_{F11} and MPF_T. Not all combinations between MOB and MPF types seem possible. As shown in Fig. 3, all MOB_F plasmids of γ-proteobacteria are combined with either MPF_F or MPF_T systems, while MOB_P plasmids usually associate with either MPF_I or MPF_T. The reasons for these preferences are not known but seem to be precise. Perhaps they are related to the details of the conjugation mechanism, or perhaps they are just constrained by evolutionary causes.

Genome database scanning explored the abundance of conjugative systems in bacteria and produced surprising results (24): ICEs were more abundant conjugative elements than plasmids, while mobilizable plasmids outnumbered conjugative ones. Although the MPF classes are evenly distributed in ICEs and plasmids, both types of self-transmissible elements contain members of all classes and can be found intermingled at larger evolutionary distances. The Guglielmini study provided a unitary view of conjugation, in which plasmid conjugative systems have often converted to ICEs and vice versa. Further phylogenetic analyses detailed the evolutionary pathway of dsDNA and ssDNA conjugation, systematically identified the protein families involved, and associated them with each MPF class. In addition, the homology between components of the different classes and the patterns of acquisition and loss of the protein families were determined (31, 34). These studies provided the basis for the classification of all kinds of conjugative systems and a guide for the experimental analysis of poorly characterized ones. A publicly available database was developed for conjugative system analysis. It is called ConjDB and can be accessed from http://conjdb.web.pasteur.fr/conjdb/_design/conjdb/index.html.

OTHER GENETIC DETERMINANTS AND PHENOTYPES ASSOCIATED WITH PLASMID PROPAGATION

Conjugation is a complex phenomenon, as shown not only by the complexity of the genetic determinants involved but also by the abundance of phenotypes related to conjugation. Those related phenotypes, which differ between different MOB and MPF types, might be responsible for the adaptation of specific conjugative systems to specific hosts of environmental conditions. This section will review some of these determinants and related phenotypes. Among the relevant concepts linked to conjugation are incompatibility, host range, fertility inhibition, and entry exclusion. The genetic determinants responsible for each of these properties will affect the infectivity rate of a plasmid and, thus, its fitness in specific hosts and environmental conditions. Incompatibility and host range are properties of a plasmid replication system, while fertility inhibition and entry exclusion are properties of the transfer system proper.

Incompatibility is the inability of two related plasmids to be stably inherited in the same bacterial cell (35, 36). The host range of a plasmid is limited by the effectiveness of its replication system more than by the conjugation system itself. A plasmid host range is modulated by additional stability determinants. All plasmids, and especially large conjugative plasmids that

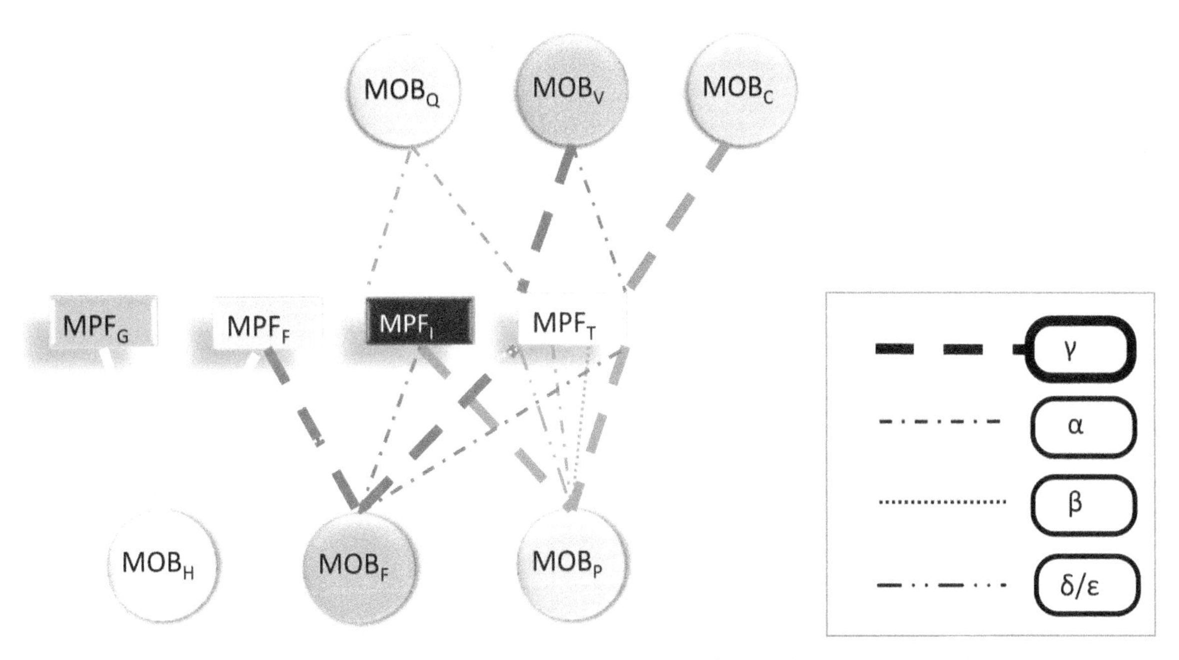

Figure 3 Most frequent combinations between MOB and MPF families in *Proteobacteria*. The six MOB families and four MPF families with representation in proteobacterial plasmids are shown by circles and rectangles, respectively. Existing combinations between MOB and MPF families are depicted by lines following the color code of the corresponding MOB family. The line pattern (boxed at the right of the figure) corresponds to the proteobacterial class in which the MOB-MPF element is hosted. Data were taken from Table S1 in reference 24. doi:10.1128/microbiolspec.PLAS-0031-2014.f3

exhibit low copy numbers, contain one or more stability determinants. Plasmid replication and stability has been a hot research topic for years. Comprehensive reviews are available (35, 37–40), so these concepts need not be further discussed here.

Transfer systems can be significantly more promiscuous than replication systems. Broad-host-range IncP1 plasmids illustrate this concept. They replicate in proteobacterial classes α, β, and γ (41, 42), but their conjugative capability goes further, being able to transfer genes to other bacterial phyla such as *Cyanobacteria* (43), *Actinobacteria* (44), and even yeast (45). This fact is exploited for transposon-mediated mutagenesis (46) and to introduce foreign genes in selected bacteria (see, for example, reference 47). Not all systems are equally proficient at delivering DNA to distantly related hosts. When three MPF types prevalent in proteobacteria (MPF_T, MPF_F F_T, and MPF_I) were tested for their ability to conjugate to the cyanobacterium *Synechococcus elongatus* PCC 7942 (47), plasmids belonging to the MPF_T class could introduce DNA, whereas those of the MPF_F and MPF_I classes failed. These and other data suggest that MPF_T is more promiscuous than the other MPF types.

Fertility inhibition is a term used to describe two different phenomena: (i) self-regulation of conjugative transfer and (ii) inhibition of the conjugative functions of a cohabiting plasmid as a result of the interaction between two transfer systems. Related to the first type of fertility inhibition is the concept of conjugative overshoot. When a repressed plasmid conjugates to a recipient cell, fresh transconjugants are better donors than the original plasmids. This behavior was first observed for both colicinogenic and antibiotic resistance factors and was called high-frequency transfer (48, 49). Concerning the second type of fertility inhibition, several unrelated mechanisms appear to be involved. The most conspicuous, in historical terms, were those affecting transfer of plasmid F. In the early 1960s, Watanabe's group reported that some R factors were able to suppress the transfer functions of the sex factor F (50), which led to the classification of antibiotic resistance plasmids as fi^+ or fi^- (51). Meynell and Datta pointed out that fertility inhibition was caused by a close relationship between F and R factors (52, 53). The isolation of mutants with increased donor ability led to the conclusion that a cytoplasmic repressor controlled transfer (54). Genetic approaches revealed that the transfer operon of F-like plasmids was positively regulated by TraJ and negatively regulated by the FinOP system (55–59). FinP is an antisense RNA that binds *traJ* mRNA helped by the chaperone FinO, impairing its translation (for a recent review, see reference 60). Finally, some plasmids were able to suppress transfer of apparently unrelated systems. For instance, IncP1 plasmids inhibit IncW transfer (61, 62) and are inhibited by IncN (63) or IncF plasmids (64). In turn, IncW plasmids inhibit *Agrobacterium tumefaciens* oncogenicity by preventing T-DNA transfer (65–67). In summary, the transfer systems of different plasmids appear to interact with each other in various ways, probably affecting the outcome of what plasmid survives in a given genetic exchange community (68).

From an evolutionary point of view, the process of entry exclusion is essential for conjugative systems (69). All conjugative plasmids contain one or more exclusion systems that impede the acquisition of other, similar plasmids by a bacterium already containing the given plasmid. Conjugation is also affected by environmental signals, such as subinhibitory concentrations of antibiotics (70), or by changes in the ecology of the environment, e.g., gut conditions of high osmolarity and microanaerobiosis such as those present in the ileum (71) or inflammation produced by pathogens, which accelerates horizontal gene transfer (72).

PLASMID RECONSTRUCTION FROM WHOLE-GENOME SEQUENCING (WGS) DATASETS

In the past two decades, epidemiological surveys focused on plasmid replication functions to identify and classify these MGEs (73, 74). More recently, a protocol based on MOB classification, called degenerate primer MOB typing (DPMT), was implemented to detect and assort transmissible plasmids and integrative elements from γ-proteobacteria (75). It resulted in a powerful technique that uncovers not only backbones related to previously classified elements, but also distant new members sharing a common evolutionary ancestor (76, 77).

Many reports describe specific plasmids that are prevalent in one or another collection. But there is not a global concept of plasmid diversity, mainly because plasmids are not directly visualized in WGS datasets, but are only inferred by searching for replication (PCR-based replicon typing schemes [73]) or relaxase genes (DPMT [75]). The advent of WGS, which allows massive analysis of bacterial isolates, is revolutionizing plasmid science. Among WGS strategies, Illumina is by far the most used, because of its simplicity of sample preparation and because it is the cheapest. Roche 454 sequencing is probably second. Although 454 provides longer reads, it is more expensive, sample preparation is more cumbersome, and its accuracy in mononucleotide runs is low. PacBio is promising, since it produces significantly longer reads and therefore a small number of final contigs (78, 79). The result of an Illumina bacterial genome sequencing project is a list of 100 to 400 contigs, with some additional data containing possible scaffolding connections among contigs.

Reconstructing plasmid genomes from Illumina data is therefore difficult. What is needed is a method to sort contigs belonging to each plasmid and separate them from chromosomal and other plasmid contigs. A method for plasmid reconstruction called PLACNET (plasmid constellation networks) (10) has recently been reported. As mentioned above, previous attempts at plasmid analysis from WGS data did not try to reconstruct plasmids. Even if researchers worked with plasmid-enriched DNA, they managed contigs only by matching them with reference sequences, without trying to reconstruct plasmids, as exemplified by reference 80. On the other hand, PLACNET allows comprehensive identification, visualization, and analysis of plasmids by creating a network of contig interactions. Three types of data are essential for PLACNET: assembly information (including scaffold links and coverage), comparison to reference sequences, and plasmid-diagnostic sequence features (relaxases and replication initiator proteins). The resulting network is pruned by expert analysis, to eliminate confounding data, and implemented in a Cytoscape-based (81) graphic representation. The result of a typical PLACNET plasmid reconstruction is shown in Fig. 4. The *E. coli* strain BIDMC43a corresponds to a carbapenem-resistant isolate from a Boston hospital (Bioproject PRJNA219249; Table 1). Its genome was sequenced with Illumina HiSeq 2000, giving 150 assembled contigs. Application of PLACNET resulted in the identification of five plasmids. The largest one, pBIDMC43a_1 (277 kb, 12 contigs), belongs to the MOB_{H11}/IncHI1 group and carries carbapenem bla_{KPC-2} and beta-lactam bla_{SHV-12} resistance genes. The second largest, pBIDMC43a_2 (34 kb, one contig) belongs to an atypical branch of the MOB_{P3}/IncX family. Additionally, there were three small (1.1, 1.2, and 1.3 kb) cryptic plasmids that encoded only for the RepA HTH-like replication protein. The copy numbers of these plasmids were inferred from sequence coverage to be 1, 2, 13, 20, and 22 chromosomal equivalents. PLACNET was recently applied to the analysis of three *E. coli* strain collections, shown in Table 1. The results obtained with the first two collections were published elsewhere (10, 17).

In summary, PLACNET analysis allowed the demonstration of two important concepts in plasmid

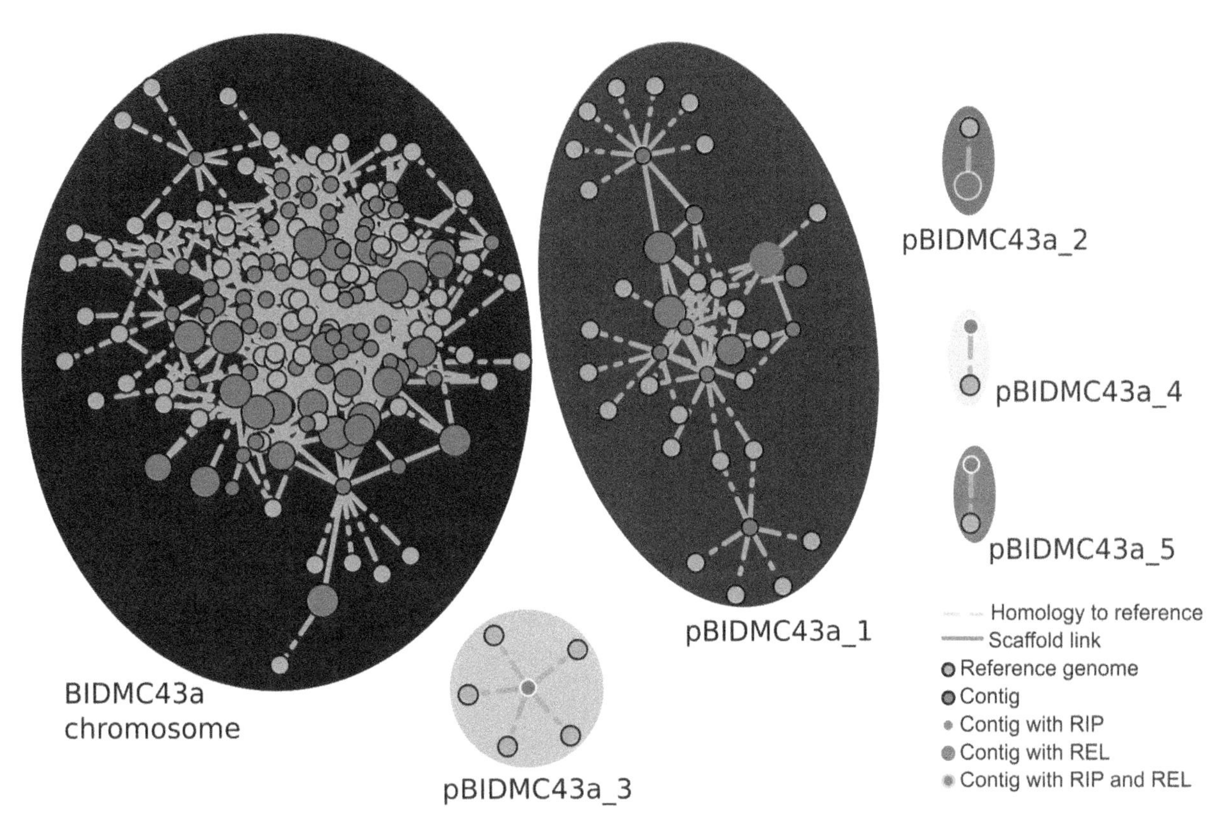

Figure 4 PLACNET reconstruction of the *E. coli* BIDMC43a genome. The dataset was taken from Bioproject PRJNA219249 and processed by PLACNET as detailed in reference 10. Contigs are represented by blue nodes, while gray nodes represent reference genomes. The sizes of contig nodes are proportional to the contig length, while those of reference nodes are fixed. Colored node outlines represent contigs containing plasmid-specific protein genes (yellow, RIP proteins; red, relaxases; green, both proteins). Solid edges represent scaffold interactions, while dotted edges represent homology to references. doi:10.1128/microbiolspec.PLAS-0031-2014.f4

evolution. Analysis of the ST131 collection (10) demonstrated that, within a given ST type, in this case ST131, there is rapid plasmid flux compared to accumulation of point mutations. Therefore, strains with almost identical core genome cgMLST (<100 SNP/Mb) contain different plasmids. In fact, nothing can be said about the plasmid load of a specific ST131 strain by just looking at the cgMLST. Plasmids thus provide very rapid means of adaptive evolution and become essential to understand the rapid adaptations of this *E. coli* sublineage. The ESBL+ Dutch collection (Table 1) was used to compare ESBL+ animal and human strains belonging to the same ST type, in order to investigate if identical strains had colonized both hosts (17). Surprisingly, the reason for ESBL transmission was not strain transfer but plasmid transfer. This was demonstrated because the relevant strains showed somewhat different cgMLST but identical plasmids (of the IncI and IncK types). Thus, plasmids were transferred from animals to humans, carrying with them the ESBL+ genes, contrary to previous assumptions (82, 83).

A third collection (Table 1) was analyzed as the previous two, and the analysis of its plasmidome is aggregated into this review to increase the total number of plasmid sequences available for comparative analysis. This collection was built on 19 carbapenem-resistant clinical isolates from Bioproject PRJNA202876, entitled “Carbapenem Resistance Initiative, Broad Institute” (broadinstitute.org). The 19 genomes were assembled and their plasmids reconstructed by PLACNET (10). The genome shown in Fig. 4 is one of these genomes. In total, 70 plasmids were found, distributed among 10 MOB and 12 Rep/Inc groups.

Reconstructed plasmids can be analyzed by using hierarchical clustering dendrograms (Fig. 5A and B). This type of dendrogram is useful for comparison. It compares the total gene content of a plasmid, allowing a visualization of relatedness between individual plasmids. Thus, plasmids branch together if they contain similar proteins. The closer they are, the more proteins they share. For each plasmid, panel A shows the size, REP, and MOB type. As can be observed, the dendrogram shows how plasmids self-arrange in branches that coincide with backbone MOB/Inc groups. Plasmids carrying beta-lactam resistance genes (present in 14 of the 19 carbapenem resistance strains) are highlighted by using different background colors. As shown in Fig. 5 (panel A), 10 MOB_{F11}/IncN plasmids harbored carbapenem resistance bla_{KPC-2} (pink). Each of them was coresident with a MOB_{H11}/IncHI2 plasmid encoding for beta-lactam resistance SHV-12 enzyme (orange). Seven MOB_{F12}/IncF plasmids were also found among these isolates: four of them carried a bla_{KPC-3} gene; and a fifth was $bla_{CTX-M-15}$-positive. Figure 5B shows the hierarchical clustering dendrogram of the beta-lactam and carbapenem resistance plasmids together with their closest reference plasmids. As can be observed, the plasmids found in this study cluster together within a branch of related plasmids, indicating that they correspond to specific subbranches of a given plasmid family. This implies, in turn, that specific plasmid configurations were selected during the acquisition of their specific antibiotic resistance. It can be speculated that these singular plasmids contain the genetic adaptation determinants that resulted in such selection.

The above three examples, taken together, indicate that plasmids are responsible for antibiotic resistance dissemination and possible routes by which pathogenic strains become antibiotic resistant. The results become obvious with the use of WGS and PLACNET analysis. They can rigorously answer questions that have lingered in microbiological research for years. PLACNET's utility resides in its allowing closer inspection of wide sets of bacterial plasmidomes. We are convinced that many more analyses, such as those described above, will help settle important questions in plasmid biology as well as in bacterial genome adaptation and evolution. The next section addresses one of these pressing questions in plasmid biology.

MOST PREVALENT PLASMID GROUPS IN *E. COLI*

Analysis of conjugative plasmids in the Murray collection (enterobacterial isolates collected from 1917 to 1957) revealed that plasmids of the preantibiotic era belong to the same Inc groups as plasmids in the present circulating pool after antimicrobial drug usage (84). This finding points at a scenario in which new cargo genes are preferentially inserted into preexisting plasmid backbones rather than previously rare plasmids being selected by antibiotic resistance gene acquisition. WGS technologies currently picture a similar setting, in which largely successful practically identical plasmid backbones are identified in unrelated bacterial strains isolated in different locations, usually carrying multiple antibiotic resistance determinants (10, 17, 85–87).

PLACNET analysis of just three *E. coli* strain collections (Table 1) allowed the reconstruction of 255 *E. coli* plasmids, thus outnumbering those collected up to now in NCBI DNA databases. The availability of this large number of plasmids makes possible some inferences about global plasmid statistics. The results are summarized in Fig. 6 and Fig. 7. As can be seen, analysis of the 255-plasmid collection, although necessarily biased, updates the most prevalent MOB and MPF subfamilies (some old, some new) in antibiotic resistance *E. coli* isolates. They are, in descending order of prevalence, MOB_{F12}/IncF (17%), MOB_{P12}/IncI (15%), MOB_{P3}/IncX (6%), and MOB_{F11}/IncN (6%) among the conjugative plasmids and MOB_{P5}/ColE (10%) and MOB_{Q4}/repQ4 (5%) among the mobilizable plasmids. As reported before, IncF and IncI plasmids occur frequently in *E. coli*. Why these plasmids and not others? As reported previously, at least 17 different plasmid families show significant occurrences in *E. coli* (75). We do not know what makes IncF and IncI plasmids the most prevalent in *E. coli*. We will highlight some characteristics of prevailing plasmid backbones and try to find some insights.

Plasmids belonging to the IncF, IncI1, and IncX groups are usually repressed for conjugation. Prototype plasmids R1, R64, and R6K exhibit up to a 10^4-fold increase in conjugation frequency in their derepressed variants R1*drd-19*, R64*drd11*, and R6K*drd1* (54, 88–90). Cojugative overshoot seems to play a role in multilevel selection, since it shifts the burden of gene expression on to the recipient cells (91). In fact, a derepressed plasmid invades a mixed population quickly but is ultimately displaced by the repressed variant, the cost of which is significantly lower to the host (92). It seems that IncF plasmids find a marginal advantage in the use of repressed systems with strong overshoots.

Not only that, but IncF plasmids have developed cross talk mechanisms with the host for the control of expression of their conjugation genes. In fact, IncF plasmids show a narrow host range, their replication being restricted to enterobacteria. The elevated number of host-encoded transcriptional regulators that control IncF plasmid transfer promoters reflects this long-term relationship. For example, the nucleoid structuring protein

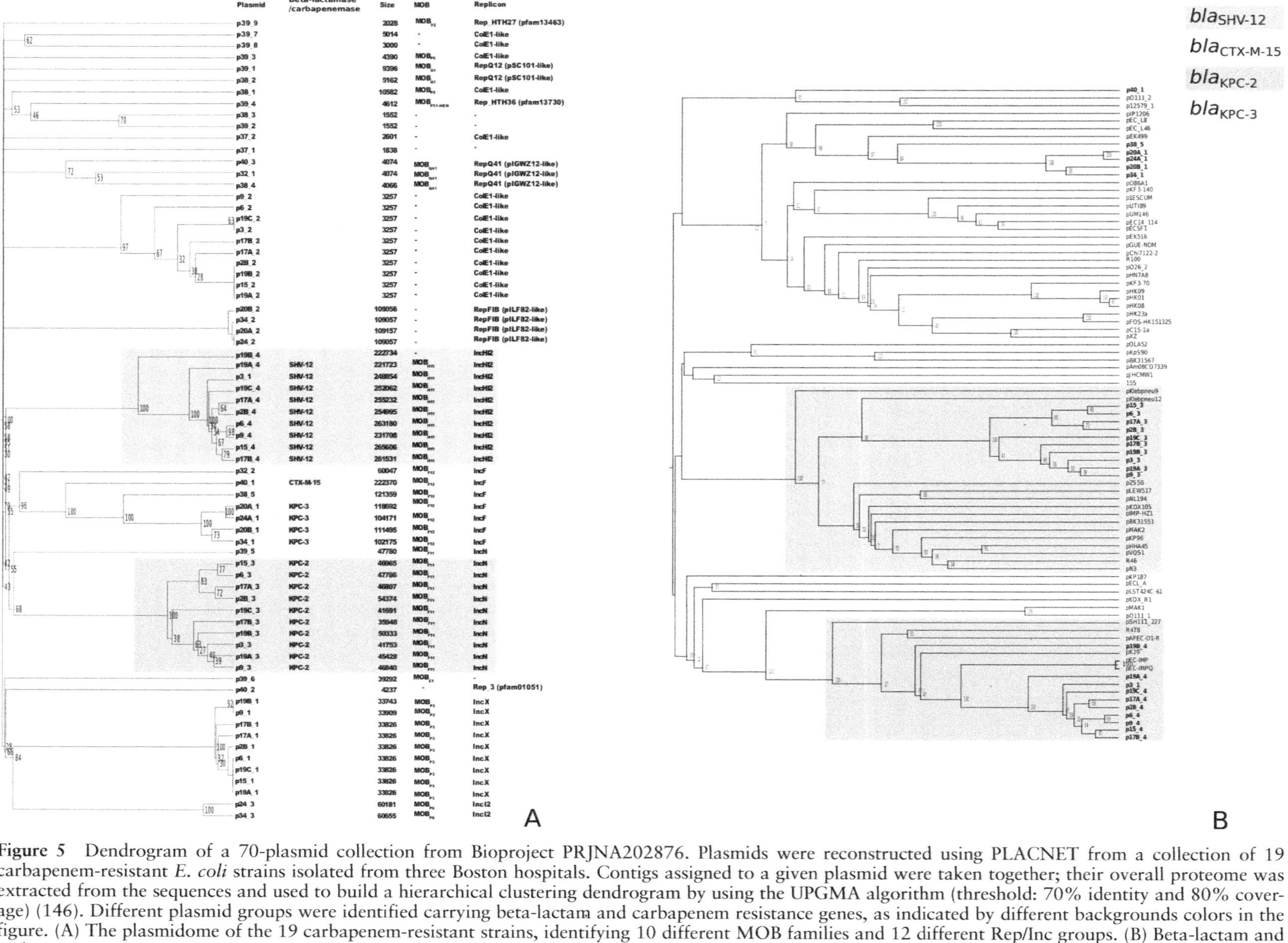

Figure 5 Dendrogram of a 70-plasmid collection from Bioproject PRJNA202876. Plasmids were reconstructed using PLACNET from a collection of 19 carbapenem-resistant *E. coli* strains isolated from three Boston hospitals. Contigs assigned to a given plasmid were taken together; their overall proteome was extracted from the sequences and used to build a hierarchical clustering dendrogram by using the UPGMA algorithm (threshold: 70% identity and 80% coverage) (146). Different plasmid groups were identified carrying beta-lactam and carbapenem resistance genes, as indicated by different backgrounds colors in the figure. (A) The plasmidome of the 19 carbapenem-resistant strains, identifying 10 different MOB families and 12 different Rep/Inc groups. (B) Beta-lactam and carbapenem resistance plasmids compared to their closely related reference plasmids. doi:10.1128/microbiolspec.PLAS-0031-2014.f5

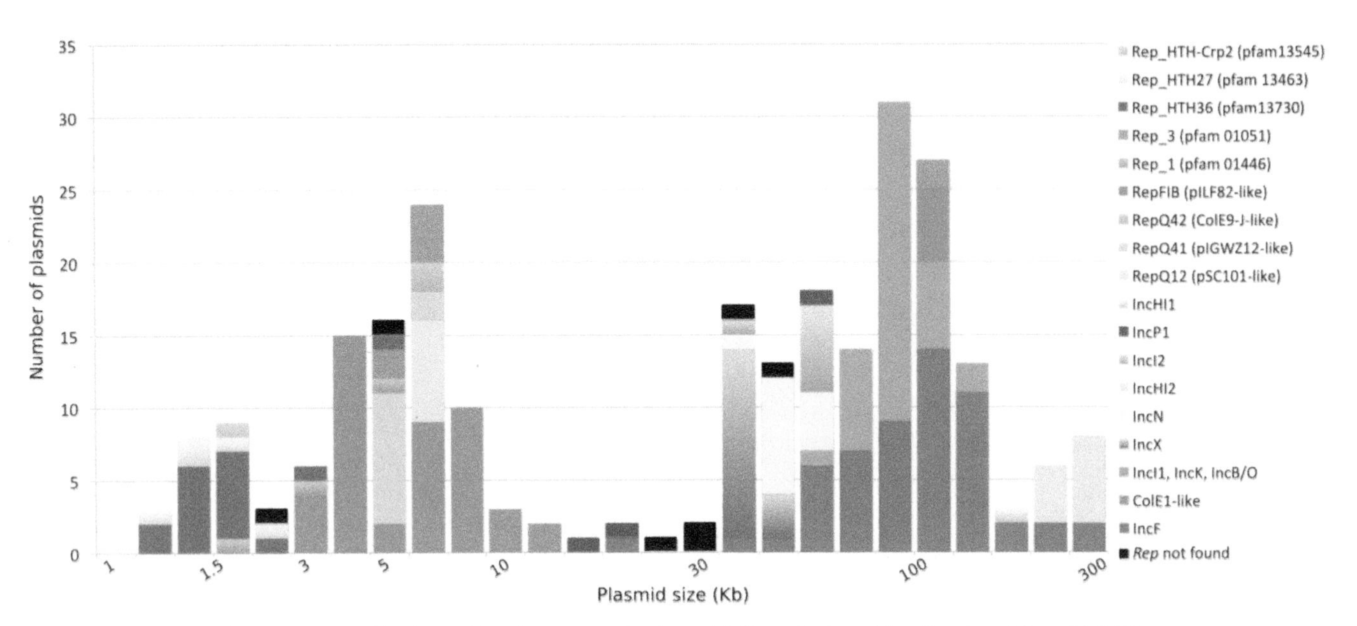

Figure 6 Distribution of Inc/Rep types in the 255–*E. coli* plasmid collection. Plasmid REP types were identified as reported in references 73 and 147. doi:10.1128/microbiolspec.PLAS-0031-2014.f6

H-NS negatively affects the P_J, P_Y, and P_M promoters (93). Cyclic Amp receptor protein and leucine-responsive regulatory protein are activators of P_J, and ArcA is an activator of the P_Y promoter (94–99). Dam methylation acts as a conjugation repressor by activating FinP RNA synthesis (100). Posttranslational control exerted by host proteins also modulates conjugation. TraJ proteolysis promotes a significant decrease in F-like plasmid transfer. In response to extracytoplasmic stimuli, such as misfolding of cell envelope components, the ClpX regulon activates and drives HslVU-mediated proteolysis of TraJ (101). For an overview of F plasmid *tra* operon regulatory factors, see reference 102. In summary, IncF plasmid transfer functions are densely regulated by the host, resulting in better adaptation, at the cost of a narrower host range. Finally, IncF plasmids carry an ample repertoire of transposons and insertion sequences, which facilitate their recombination with the chromosome, easily becoming ICEs. The evolutionary implications of chromosomal insertions have been reported (103, 104).

A particular plasmid imposes different burdens depending on the host. For a given IncP1 plasmid, its fitness differs not only when hosted in different host species (105, 106), but also among strains of the same species (107, 108). Thus, the fact that plasmid RP1 had less fitness impact on *E. coli* strains isolated from pigs than on those of human origin (108) correlates with the low prevalence of IncP1α plasmids in human isolates of *E. coli* resistant to ESBL and carbapenems (85, 109). On the other hand, practically identical IncN backbones containing different cargoes show marked fitness differences (108).

Less epidemiological attention is normally paid to small mobilizable plasmids. However, they are also carriers of antibiotic resistance genes (110–113), as well as competition weapons, such as bacteriocins (113–116) and restriction-modification enzymes (117–120). They rely on conjugative plasmids as helpers for their conjugative transmission. Not all possible combinations of conjugative and mobilizable plasmids render productive mobilization (121), biasing the presence of mobilizable plasmids in specific hosts. Two groups of mobilizable plasmids were found most frequently in PLACNET analysis of the 255-plasmid collection: MOB_{P5} and MOB_{Q4}. The first mainly includes ColE1-like plasmids (for a review of ColE1-like plasmid replication, see reference 122), but association of MOB_{P5} relaxases with other types of replication different from the RNAII-RNAI system of ColE1 also exists (23). The antisense RNA mechanism that regulates ColE1 replication is highly dynamic, adjusting plasmid metabolic burden to the physiological state of the host (123–127). As stated above, MOB_{P5} plasmids encode relevant phenotypic traits. In fact, plasmid ColE1 was named after the bacteriocin it encodes: colicin E1. MOB_{P5} plasmids are efficiently mobilized by the two most prevalent conjugative plasmid groups in our PLACNET analysis: MOB_{F12}/IncF and MOB_{P12}/IncI plasmids (121). The

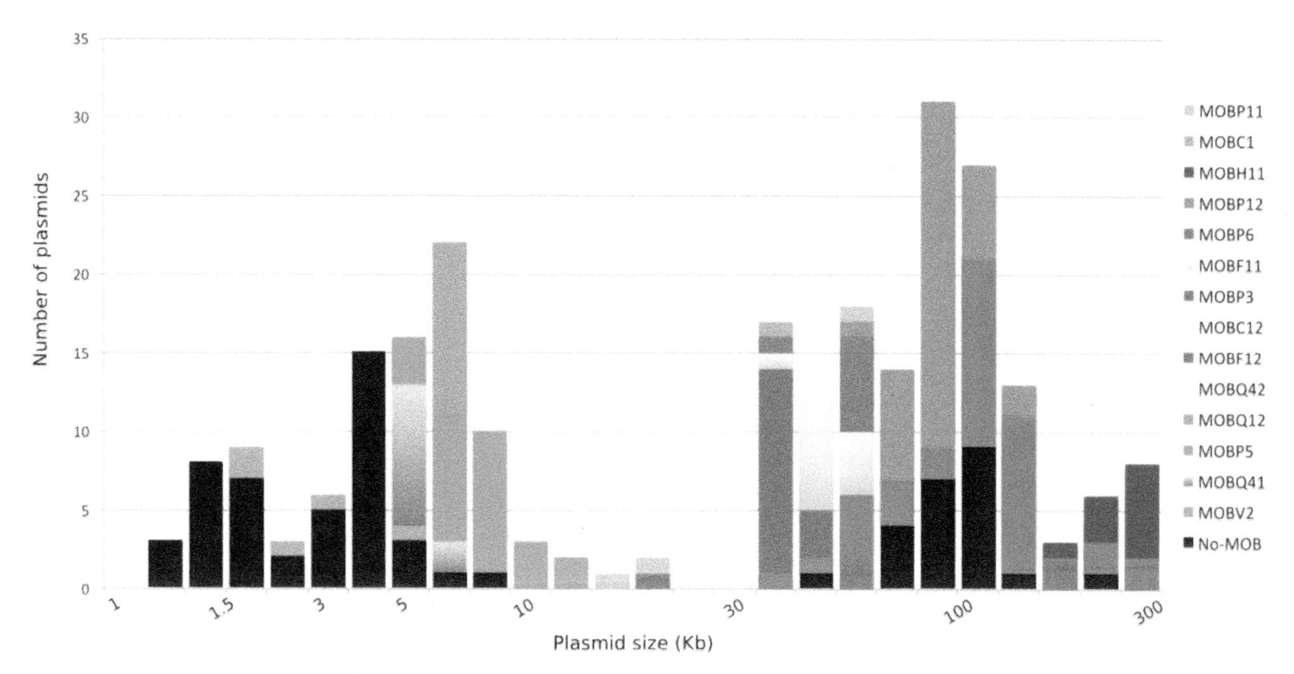

Figure 7 Distribution of MOB types in the 255–*E. coli* plasmid collection. Plasmid MOB types were identified as reported in references 23 and 75.
doi:10.1128/microbiolspec.PLAS-0031-2014.f7

second (MOB_{Q4}) is a new emerging group of small mobilizable plasmids that has been unnoticed by PCR-based replicon typing. A pair of degenerate primers (*Qu*) to detect this group was included in the DPMT plasmid identification method, which revealed the frequent presence of MOB_{Q4} plasmids in clinical isolates of enterobacteria (75, 77). Two replicon types were associated with the MOB_{Q4} module. This differential association matches the split of the MOB_{Q4} clade into two subgroups: Q41 and Q42. Q41 plasmids, exemplified by pIGWZ12, encode a Rep_3 replication initiation protein. They are cryptic and highly stable (128). Q42 are highly stable, colicin-encoding plasmids that replicate by a ColE2-like mechanism (129, 130).

IS IT POSSIBLE TO LIMIT PLASMID DISSEMINATION?

As stated by the World Health Organization (WHO) in "Antimicrobial Resistance Global Report on Surveillance 2014" (http://www.who.int/topics/drug_resistance/en/), "a post-antibiotic era—in which common infections and minor injuries can kill—far from being an apocalyptic fantasy, is instead a very real possibility for the 21st Century" (131). Bacteria are becoming resistant to most antimicrobials, while the development of new antibiotics has drastically slowed (131, 132). Innovations to achieve rapid diagnostic tests and development of alternatives to antibiotics are urgently demanded by institutions worldwide (http://ec.europa.eu/health/antimicrobial_resistance/docs/antimicrobial_cons01_report_en.pdf; http://apps.who.int/iris/bitstream/10665/112642/1/9789241564748_eng.pdf?ua=1; http://www.longitudeprize.org/challenge/antibiotics).

As mentioned in the introduction, there is a window of opportunity to combat the dissemination of antibiotic resistance. We should act now while the most "damaging" antibiotic resistance genes are still carried by plasmids, probably not yet well adapted to the human pathogens in which they were isolated and where they cause health problems. As reviewed in reference 2, adaptive genes such as those that cause antibiotic resistance usually enter human pathogens by horizontal gene transfer, carried by plasmid platforms. Adaptive genes carried on plasmids, like the plasmids themselves, produce a small but significant burden to their host cells. As a result, research suggests that, after decades or maybe less time, these adaptive genes tend to be progressively incorporated into the main bacterial chromosome, where they are more resilient to potential genome-cleaning operations. Thus, it will be ideal to fight the dissemination of antibiotic resistance genes while they reside in plasmids and are thus vulnerable.

Plasmids cannot be completely eliminated from bacterial populations at large, nor can bacterial human pathogens. Contrary to common thought, drastic reduction of antibiotic use was revealed as an ineffective strategy for washing out antibiotic resistance-encoding plasmids (133–135). There will always be bacteria and plasmids thriving around us. Besides, recipient bacteria cannot avoid receiving plasmids by conjugation (136). Nevertheless, what we can aim to alter are the conditions that result in a given ecological equilibrium. This is the basis for a number of so-called eco-evo strategies that have been proposed to fight antibiotic resistance (137). Among the suggested new eco-evo approaches, we might act to minimize the opportunities for plasmid infectivity in human pathogens by acting against the dissemination of transmissible plasmids.

Several decades ago, bacteriophages such as M13 were found to inhibit conjugation (138, 139) and promote the loss of F-like plasmids (140, 141). Only recently was the mechanism of such conjugation inhibition elucidated. It involved the occlusion of the conjugative pilus by the phage particles, primarily mediated by phage coat protein Gp3 (142). The relaxase protein has been a target to block plasmid transfer by expressing intracellular antibodies against it in recipient cells (143). Some inhibitors of the type IV secretion system have been found by high-throughput assays (144, 145). Some of them were natural bioactive compounds able to inhibit plasmid conjugation (144) that can be used for testing in relevant environments.

In summary, technological and conceptual advances allow us to develop new strategies to curb the increase of antibiotic resistance. We need to do it, because failure will result in resistance stabilization and, consequently, in progressive morbidity and mortality caused by antibiotic-resistant infections.

Acknowledgments. *Work at the FdlC laboratory was financed by the Spanish Ministries of Economy and Competitivity (BFU2011-26608) and Health (REIPI, RD12/0015/0019), as well as by the European Seventh Framework Program (289326/FP7-KBBE-2011-5, 612146/FP7-ICT-2013-10, and 282004/FP7-HEALTH-2011-2.3.1-2). Conflicts of interest: We disclose no conflict.*

Citation. De Toro M, Garcillán-Barcia MP, de la Cruz F. 2014. Plasmid diversity and adaptation analyzed by massive sequencing of *Escherichia coli* plasmids. Microbiol Spectrum 2(6):PLAS-0031-2014.

References

1. **Ochman H, Lawrence JG, Groisman EA.** 2000. Lateral gene transfer and the nature of bacterial innovation. *Nature* **405:**299–304.
2. **de la Cruz F, Davies J.** 2000. Horizontal gene transfer and the origin of species: lessons from bacteria. *Trends Microbiol* **8:**128–133.
3. **Jain R, Rivera MC, Moore JE, Lake JA.** 2003. Horizontal gene transfer accelerates genome innovation and evolution. *Mol Biol Evol* **20:**1598–1602.
4. **Schubert S, Darlu P, Clermont O, Wieser A, Magistro G, Hoffmann C, Weinert K, Tenaillon O, Matic I, Denamur E.** 2009. Role of intraspecies recombination in the spread of pathogenicity islands within the *Escherichia coli* species. *PLoS Pathog* **5:**e1000257. doi: 10.1371/journal.ppat.1000257.
5. **Lapierre P, Gogarten JP.** 2009. Estimating the size of the bacterial pan-genome. *Trends Genet* **25:**107–110.
6. **Rasko DA, Rosovitz MJ, Myers GS, Mongodin EF, Fricke WF, Gajer P, Crabtree J, Sebaihia M, Thomson NR, Chaudhuri R, Henderson IR, Sperandio V, Ravel J.** 2008. The pangenome structure of *Escherichia coli*: comparative genomic analysis of *E. coli* commensal and pathogenic isolates. *J Bacteriol* **190:**6881–6893.
7. **Polz MF, Alm EJ, Hanage WP.** 2013. Horizontal gene transfer and the evolution of bacterial and archaeal population structure. *Trends Genet* **29:**170–175.
8. **Nicolas-Chanoine MH, Bertrand X, Madec JY.** 2014. *Escherichia coli* ST131, an intriguing clonal group. *Clin Microbiol Rev* **27:**543–574.
9. **Petty NK, Ben Zakour NL, Stanton-Cook M, Skippington E, Totsika M, Forde BM, Phan MD, Gomes Moriel D, Peters KM, Davies M, Rogers BA, Dougan G, Rodriguez-Bano J, Pascual A, Pitout JD, Upton M, Paterson DL, Walsh TR, Schembri MA, Beatson SA.** 2014. Global dissemination of a multidrug resistant *Escherichia coli* clone. *Proc Natl Acad Sci USA* **111:**5694–5699.
10. **Lanza VF, de Toro M, Garcillán-Barcia MP, Mora A, Blanco J, Coque TM, de la Cruz F.** 2014. Plasmid flux in *Escherichia coli* ST131 sublineages, analyzed by plasmid constellation network (PLACNET), a new method for plasmid reconstruction from whole genome sequences. *PLoS Genet* [Epub ahead of print.] doi:10:1371/journal.pgen.1004766.
11. **Partridge SR, Zong Z, Iredell JR.** 2011. Recombination in IS26 and Tn2 in the evolution of multiresistance regions carrying blaCTX-M-15 on conjugative IncF plasmids from *Escherichia coli. Antimicrob Agents Chemother* **55:**4971–4978.
12. **Nicolas-Chanoine MH, Blanco J, Leflon-Guibout V, Demarty R, Alonso MP, Canica MM, Park YJ, Lavigne JP, Pitout J, Johnson JR.** 2008. Intercontinental emergence of *Escherichia coli* clone O25:H4-ST131 producing CTX-M-15. *J Antimicrob Chemother* **61:**273–281.
13. **Coque TM, Novais A, Carattoli A, Poirel L, Pitout J, Peixe L, Baquero F, Canton R, Nordmann P.** 2008. Dissemination of clonally related *Escherichia coli* strains expressing extended-spectrum beta-lactamase CTX-M-15. *Emerg Infect Dis* **14:**195–200.
14. **Rogers BA, Sidjabat HE, Paterson DL.** 2011. *Escherichia coli* O25b-ST131: a pandemic, multiresistant, community-associated strain. *J Antimicrob Chemother* **66:**1–14.
15. **Johnson JR, Johnston B, Clabots C, Kuskowski MA, Pendyala S, Debroy C, Nowicki B, Rice J.** 2010.

Escherichia coli sequence type ST131 as an emerging fluoroquinolone-resistant uropathogen among renal transplant recipients. *Antimicrob Agents Chemother* **54:** 546–550.

16. **Morris D, McGarry E, Cotter M, Passet V, Lynch M, Ludden C, Hannan MM, Brisse S, Cormican M.** 2012. Detection of OXA-48 carbapenemase in the pandemic clone *Escherichia coli* O25b:H4-ST131 in the course of investigation of an outbreak of OXA-48-producing *Klebsiella pneumoniae*. *Antimicrob Agents Chemother* **56:**4030–4031.
17. **de Been M, Lanza VF, de Toro M, Sharringa J, Dohmen W, Du Y, Hu J, Lei Y, Li N, Heederik DJJ, Fluit AC, Bonten MJM, Willems RJL, de la Cruz F, van Schaik W.** 2014. Dissemination of cephalosporin resistance genes between *Escherichia coli* strains from farm animals and humans by specific plasmid lineages. *PLoS Genet* [Epub ahead of print.] doi:10:1371/journal.pgen.1004776.
18. **Smillie C, Garcillan-Barcia MP, Francia MV, Rocha EP, de la Cruz F.** 2010. Mobility of plasmids. *Microbiol Mol Biol Rev* **74:**434–452.
19. **Guynet C, Cuevas A, Moncalian G, de la Cruz F.** 2011. The stb operon balances the requirements for vegetative stability and conjugative transfer of plasmid R388. *PLoS Genet* **7:**e1002073. doi:10.1371/journal.pgen.1002073.
20. **Guynet C, de la Cruz F.** 2011. Plasmid segregation without partition. *Mob Genet Elements* **1:**236–241.
21. **de la Cruz F, Frost LS, Meyer RJ, Zechner EL.** 2010. Conjugative DNA metabolism in Gram-negative bacteria. *FEMS Microbiol Rev* **34:**18–40.
22. **Francia MV, Varsaki A, Garcillan-Barcia MP, Latorre A, Drainas C, de la Cruz F.** 2004. A classification scheme for mobilization regions of bacterial plasmids. *FEMS Microbiol Rev* **28:**79–100.
23. **Garcillan-Barcia MP, Francia MV, de la Cruz F.** 2009. The diversity of conjugative relaxases and its application in plasmid classification. *FEMS Microbiol Rev* **33:** 657–687.
24. **Guglielmini J, Quintais L, Garcillan-Barcia MP, de la Cruz F, Rocha EP.** 2011. The repertoire of ICE in prokaryotes underscores the unity, diversity, and ubiquity of conjugation. *PLoS Genet* **7:**e1002222. doi:10.1371/journal.pgen.1002222.
25. **Guasch A, Lucas M, Moncalian G, Cabezas M, Perez-Luque R, Gomis-Ruth FX, de la Cruz F, Coll M.** 2003. Recognition and processing of the origin of transfer DNA by conjugative relaxase TrwC. *Nat Struct Biol* **10:**1002–1010.
26. **Gomis-Ruth FX, Moncalian G, Perez-Luque R, Gonzalez A, Cabezon E, de la Cruz F, Coll M.** 2001. The bacterial conjugation protein TrwB resembles ring helicases and F1- ATPase. *Nature* **409:**637–641.
27. **Chandler M, de la Cruz F, Dyda F, Hickman AB, Moncalian G, Ton-Hoang B.** 2013. Breaking and joining single-stranded DNA: the HUH endonuclease superfamily. *Nat Rev Microbiol* **11:**525–538.
28. **Francia MV, Clewell DB, de la Cruz F, Moncalian G.** 2013. Catalytic domain of plasmid pAD1 relaxase TraX defines a group of relaxases related to restriction endonucleases. *Proc Natl Acad Sci USA* **110:**13606–13611.
29. **Low HH, Gubellini F, Rivera-Calzada A, Braun N, Connery S, Dujeancourt A, Lu F, Redzej A, Fronzes R, Orlova EV, Waksman G.** 2014. Structure of a type IV secretion system. *Nature* **508:**550–553.
30. **Cabezon E, Ripoll-Rozada J, Pena A, de la Cruz F, Arechaga I.** 2014. Towards an integrated model of bacterial conjugation. *FEMS Microbiol Rev.* [Epub ahead of print.] doi:10.1111/1574-6976.12085.
31. **Guglielmini J, Neron B, Abby SS, Garcillan-Barcia MP, de la Cruz F, Rocha EP.** 2014. Key components of the eight classes of type IV secretion systems involved in bacterial conjugation or protein secretion. *Nucleic Acids Res* **42:**5715–5727.
32. **Trokter M, Felisberto-Rodrigues C, Christie PJ, Waksman G.** 2014. Recent advances in the structural and molecular biology of type IV secretion systems. *Curr Opin Struct Biol* **27C:**16–23.
33. **Christie PJ, Whitaker N, Gonzalez-Rivera C.** 2014. Mechanism and structure of the bacterial type IV secretion systems. *Biochim Biophys Acta* **1843:**1578–1591.
34. **Guglielmini J, de la Cruz F, Rocha EP.** 2013. Evolution of conjugation and type IV secretion systems. *Mol Biol Evol* **30:**315–331.
35. **Bouet JY, Nordstrom K, Lane D.** 2007. Plasmid partition and incompatibility: the focus shifts. *Mol Microbiol* **65:**1405–1414.
36. **Novick RP.** 1987. Plasmid incompatibility. *Microbiol Rev* **51:**381–395.
37. **Schumacher MA.** 2012. Bacterial plasmid partition machinery: a minimalist approach to survival. *Curr Opin Struct Biol* **22:**72–79.
38. **Diago-Navarro E, Hernandez-Arriaga AM, Lopez-Villarejo J, Munoz-Gomez AJ, Kamphuis MB, Boelens R, Lemonnier M, Diaz-Orejas R.** 2010. parD toxin-antitoxin system of plasmid R1: basic contributions, biotechnological applications and relationships with closely-related toxin-antitoxin systems. *FEBS J* **277:** 3097–3117.
39. **del Solar G, Giraldo R, Ruiz-Echevarria MJ, Espinosa M, Diaz-Orejas R.** 1998. Replication and control of circular bacterial plasmids. *Microbiol Mol Biol Rev* **62:** 434–464.
40. **del Solar G, Espinosa M.** 2000. Plasmid copy number control: an ever-growing story. *Mol Microbiol* **37:**492–500.
41. **Kolatka K, Kubik S, Rajewska M, Konieczny I.** 2010. Replication and partitioning of the broad-host-range plasmid RK2. *Plasmid* **64:**119–134.
42. **Yano H, Deckert GE, Rogers LM, Top EM.** 2012. Roles of long and short replication initiation proteins in the fate of IncP-1 plasmids. *J Bacteriol* **194:**1533–1543.
43. **Wolk CP, Vonshak A, Kehoe P, Elhai J.** 1984. Construction of shuttle vectors capable of conjugative transfer from *Escherichia coli* to nitrogen-fixing filamentous cyanobacteria. *Proc Natl Acad Sci USA* **81:**1561–1565.
44. **Dominguez W, O'Sullivan DJ.** 2013. Developing an efficient and reproducible conjugation-based gene

transfer system for bifidobacteria. *Microbiology* **159:** 328–338.

45. **Bates S, Cashmore AM, Wilkins BM.** 1998. IncP plasmids are unusually effective in mediating conjugation of *Escherichia coli* and *Saccharomyces cerevisiae*: involvement of the tra2 mating system. *J Bacteriol* **180:** 6538–6543.
46. **Martinez-Garcia E, Calles B, Arevalo-Rodriguez M, de Lorenzo V.** 2011. pBAM1: an all-synthetic genetic tool for analysis and construction of complex bacterial phenotypes. *BMC Microbiol* **11:**38.
47. **Encinas D, Garcillan-Barcia MP, Santos-Merino M, Delaye L, Moya A, de la Cruz F.** 2014. Plasmid conjugation from proteobacteria as evidence for the origin of xenologous genes in cyanobacteria. *J Bacteriol* **196:** 1551–1559.
48. **Watanabe T.** 1963. Episome-mediated transfer of drug resistance in *Enterobacteriaceae*. VI. High-frequency resistance transfer system in *Escherichia coli*. *J Bacteriol* **85:**788–794.
49. **Stocker BAD, Smith SM, Ozeki H.** 1963. High infectivity of *Salmonella typhimurium* newly infected by the colI factor. *J Gen Microbiol* **30:**201–221.
50. **Watanabe T, Fukasawa T.** 1962. Episome-mediated transfer of drug resistance in *Enterobacteriaceae*. IV. Interactions between resistance transfer factor and F-factor in *Escherichia coli* K-12. *J Bacteriol* **83:**727–735.
51. **Watanabe T, Nishida H, Ogata C, Arai T, Sato S.** 1964. Episome-mediated transfer of drug resistance in *Enterobacteriaceae*. VII. Two types of naturally occurring R factors. *J Bacteriol* **88:**716–726.
52. **Meynell E, Datta N.** 1965. Functional homology of the sex-factor and resistance transfer factors. *Nature* **207:** 884–885.
53. **Meynell E, Meynell GG, Datta N.** 1968. Phylogenetic relationships of drug-resistance factors and other transmissible bacterial plasmids. *Bacteriol Rev* **32:**55–83.
54. **Meynell E, Datta N.** 1967. Mutant drug resistant factors of high transmissibility. *Nature* **214:**885–887.
55. **Finnegan D, Willetts N.** 1973. The site of action of the F transfer inhibitor. *Mol Gen Genet* **127:**307–316.
56. **Finnegan DJ, Willetts NS.** 1971. Two classes of Flac mutants insensitive to transfer inhibition by an F-like R factor. *Mol Gen Genet* **111:**256–264.
57. **Timmis KN, Andres I, Achtman M.** 1978. Fertility repression of F-like conjugative plasmids: physical mapping of the R6–5 finO and finP cistrons and identification of the finO protein. *Proc Natl Acad Sci USA* **75:** 5836–5840.
58. **Willetts N.** 1977. The transcriptional control of fertility in F-like plasmids. *J Mol Biol* **112:**141–148.
59. **Willetts NS.** 1974. The kinetics of inhibition of Flac transfer by R100 in *E. coli*. *Mol Gen Genet* **129:**123–130.
60. **Glover JNM, Chaulk SG, Edwards RA, Arthur D, Lu J, Frost LS.** 2014. The FinO family of bacterial RNA chaperones. *Plasmid* [Epub ahead of print.] doi: 10.1016/j.plasmid.2014.07.003.
61. **Fong ST, Stanisich VA.** 1989. Location and characterization of two functions on RP1 that inhibit the fertility of the IncW plasmid R388. *J Gen Microbiol* **135:** 499–502.
62. **Goncharoff P, Saadi S, Chang CH, Saltman LH, Figurski DH.** 1991. Structural, molecular, and genetic analysis of the kilA operon of broad-host-range plasmid RK2. *J Bacteriol* **173:**3463–3477.
63. **Winans SC, Walker GC.** 1985. Fertility inhibition of RP1 by IncN plasmid pKM101. *J Bacteriol* **161:**425–427.
64. **Santini JM, Stanisich VA.** 1998. Both the fipA gene of pKM101 and the pifC gene of F inhibit conjugal transfer of RP1 by an effect on traG. *J Bacteriol* **180:** 4093–4101.
65. **Cascales E, Atmakuri K, Liu Z, Binns AN, Christie PJ.** 2005. *Agrobacterium tumefaciens* oncogenic suppressors inhibit T-DNA and VirE2 protein substrate binding to the VirD4 coupling protein. *Mol Microbiol* **58:** 565–579.
66. **Chen CY, Kado CI.** 1994. Inhibition of *Agrobacterium tumefaciens* oncogenicity by the osa gene of pSa. *J Bacteriol* **176:**5697–5703.
67. **Farrand S, Kado CI, Ireland CR.** 1981. Suppression of tumorigenicity by the IncW R plasmid pSa in *Agrobacterium tumefaciens*. *Mol Gen Genet* **181:**44–51.
68. **Skippington E, Ragan MA.** 2011. Lateral genetic transfer and the construction of genetic exchange communities. *FEMS Microbiol Rev* **35:**707–735.
69. **Garcillan-Barcia MP, de la Cruz F.** 2008. Why is entry exclusion an essential feature of conjugative plasmids? *Plasmid* **60:**1–18.
70. **Schuurmans JM, van Hijum SA, Piet JR, Handel N, Smelt J, Brul S, ter Kuile BH.** 2014. Effect of growth rate and selection pressure on rates of transfer of an antibiotic resistance plasmid between *E. coli* strains. *Plasmid* **72:**1–8.
71. **Garcia-Quintanilla M, Ramos-Morales F, Casadesus J.** 2008. Conjugal transfer of the *Salmonella enterica* virulence plasmid in the mouse intestine. *J Bacteriol* **190:** 1922–1927.
72. **Stecher B, Denzler R, Maier L, Bernet F, Sanders MJ, Pickard DJ, Barthel M, Westendorf AM, Krogfelt KA, Walker AW, Ackermann M, Dobrindt U, Thomson NR, Hardt WD.** 2012. Gut inflammation can boost horizontal gene transfer between pathogenic and commensal *Enterobacteriaceae*. *Proc Natl Acad Sci USA* **109:** 1269–1274.
73. **Carattoli A, Bertini A, Villa L, Falbo V, Hopkins KL, Threlfall EJ.** 2005. Identification of plasmids by PCR-based replicon typing. *J Microbiol Methods* **63:**219–228.
74. **Gotz A, Pukall R, Smit E, Tietze E, Prager R, Tschape H, van Elsas JD, Smalla K.** 1996. Detection and characterization of broad-host-range plasmids in environmental bacteria by PCR. *Appl Environ Microbiol* **62:** 2621–2628.
75. **Alvarado A, Garcillan-Barcia MP, de la Cruz F.** 2012. A degenerate primer MOB typing (DPMT) method to

classify gamma-proteobacterial plasmids in clinical and environmental settings. *PLoS One* **7**:e40438. doi: 10.1371/journal.pone.0040438.

76. **Garcillan-Barcia MP, de la Cruz F.** 2013. Ordering the bestiary of genetic elements transmissible by conjugation. *Mob Genet Elements* **3**:e24263. doi:10.4161/mge.24263.

77. **Garcillan-Barcia MP, Ruiz del Castillo B, Alvarado A, de la Cruz F, Martinez-Martinez L.** 2014. Degenerate primer MOB typing of multiresistant clinical isolates of *E. coli* uncovers new plasmid backbones. *Plasmid* [Epub ahead of print.] doi:10:1016/j.plasmid.2014.11.003.

78. **Miyamoto M, Motooka D, Gotoh K, Imai T, Yoshitake K, Goto N, Iida T, Yasunaga T, Horii T, Arakawa K, Kasahara M, Nakamura S.** 2014. Performance comparison of second- and third-generation sequencers using a bacterial genome with two chromosomes. *BMC Genomics* **15**:699.

79. **Liu L, Li Y, Li S, Hu N, He Y, Pong R, Lin D, Lu L, Law M.** 2012. Comparison of next-generation sequencing systems. *J Biomed Biotechnol* **2012**:251364.

80. **Brolund A, Franzen O, Melefors O, Tegmark-Wisell K, Sandegren L.** 2013. Plasmidome-analysis of ESBL-producing *Escherichia coli* using conventional typing and high-throughput sequencing. *PLoS One* **8**:e65793. doi:10.1371/journal.pone.0065793.

81. **Smoot ME, Ono K, Ruscheinski J, Wang PL, Ideker T.** 2011. Cytoscape 2.8: new features for data integration and network visualization. *Bioinformatics* **27**:431–432.

82. **Leverstein-van Hall MA, Dierikx CM, Cohen Stuart J, Voets GM, van den Munckhof MP, van Essen-Zandbergen A, Platteel T, Fluit AC, van de Sande-Bruinsma N, Scharinga J, Bonten MJ, Mevius DJ.** 2011. Dutch patients, retail chicken meat and poultry share the same ESBL genes, plasmids and strains. *Clin Microbiol Infect* **17**:873–880.

83. **Kluytmans JA, Overdevest IT, Willemsen I, Kluytmans-van den Bergh MF, van der Zwaluw K, Heck M, Rijnsburger M, Vandenbroucke-Grauls CM, Savelkoul PH, Johnston BD, Gordon D, Johnson JR.** 2013. Extended-spectrum beta-lactamase-producing *Escherichia coli* from retail chicken meat and humans: comparison of strains, plasmids, resistance genes, and virulence factors. *Clin Infect Dis* **56**:478–487.

84. **Datta N, Hughes VM.** 1983. Plasmids of the same Inc groups in *Enterobacteria* before and after the medical use of antibiotics. *Nature* **306**:616–617.

85. **Carattoli A.** 2013. Plasmids and the spread of resistance. *Int J Med Microbiol* **303**:298–304.

86. **Carattoli A, Zankari E, Garcia-Fernandez A, Voldby Larsen M, Lund O, Villa L, Moller Aarestrup F, Hasman H.** 2014. *In silico* detection and typing of plasmids using PlasmidFinder and plasmid multilocus sequence typing. *Antimicrob Agents Chemother* **58**: 3895–3903.

87. **Hoffmann M, Zhao S, Pettengill J, Luo Y, Monday SR, Abbott J, Ayers SL, Cinar HN, Muruvanda T, Li C, Allard MW, Whichard J, Meng J, Brown EW, McDermott PF.** 2014. Comparative genomic analysis and virulence differences in closely related *Salmonella enterica* serotype Heidelberg isolates from humans, retail meats, and animals. *Genome Biol Evol* **6**:1046–1068.

88. **Avila P, Nunez B, de la Cruz F.** 1996. Plasmid R6K contains two functional oriTs which can assemble simultaneously in relaxosomes *in vivo*. *J Mol Biol* **261**: 135–143.

89. **Egawa R, Hirota Y.** 1962. Inhibition of fertility by multiple drug-resistance in *Escherichia coli* K-12. *Jpn J Genet* **37**:66–69.

90. **Edwards S, Meynell GG.** 1968. General method for isolating de-repressed bacterial sex factors. *Nature* **219**: 869–870.

91. **Fernandez-Lopez R, Del Campo I, Revilla C, Cuevas A, de la Cruz F.** 2014. Negative feedback and transcriptional overshooting in a regulatory network for horizontal gene transfer. *PLoS Genet* **10**:e1004171. doi: 10.1371/journal.pgen.1004171.

92. **Haft RJ, Mittler JE, Traxler B.** 2009. Competition favours reduced cost of plasmids to host bacteria. *ISME J* **3**:761–769.

93. **Will WR, Frost LS.** 2006. Characterization of the opposing roles of H-NS and TraJ in transcriptional regulation of the F-plasmid tra operon. *J Bacteriol* **188**: 507–514.

94. **Silverman PM, Rother S, Gaudin H.** 1991. Arc and Sfr functions of the *Escherichia coli* K-12 arcA gene product are genetically and physiologically separable. *J Bacteriol* **173**:5648–5652.

95. **Strohmaier H, Noiges R, Kotschan S, Sawers G, Hogenauer G, Zechner EL, Koraimann G.** 1998. Signal transduction and bacterial conjugation: characterization of the role of ArcA in regulating conjugative transfer of the resistance plasmid R1. *J Mol Biol* **277**: 309–316.

96. **Camacho EM, Casadesus J.** 2002. Conjugal transfer of the virulence plasmid of *Salmonella enterica* is regulated by the leucine-responsive regulatory protein and DNA adenine methylation. *Mol Microbiol* **44**:1589–1598.

97. **Camacho EM, Casadesus J.** 2005. Regulation of traJ transcription in the *Salmonella* virulence plasmid by strand-specific DNA adenine hemimethylation. *Mol Microbiol* **57**:1700–1718.

98. **Starcic M, Zgur-Bertok D, Jordi BJ, Wosten MM, Gaastra W, van Putten JP.** 2003. The cyclic AMP-cyclic AMP receptor protein complex regulates activity of the traJ promoter of the *Escherichia coli* conjugative plasmid pRK100. *J Bacteriol* **185**:1616–1623.

99. **Starcic-Erjavec M, van Putten JP, Gaastra W, Jordi BJ, Grabnar M, Zgur-Bertok D.** 2003. H-NS and Lrp serve as positive modulators of traJ expression from the *Escherichia coli* plasmid pRK100. *Mol Genet Genomics* **270**:94–102.

100. **Camacho EM, Serna A, Madrid C, Marques S, Fernandez R, de la Cruz F, Juarez A, Casadesus J.** 2005. Regulation of finP transcription by DNA adenine methylation in the virulence plasmid of *Salmonella enterica*. *J Bacteriol* **187**:5691–5699.

101. **Lau-Wong IC, Locke T, Ellison MJ, Raivio TL, Frost LS.** 2008. Activation of the Cpx regulon destabilizes the F plasmid transfer activator, TraJ, via the HslVU protease in *Escherichia coli*. *Mol Microbiol* **67:**516–527.

102. **Wong JJ, Lu J, Glover JN.** 2012. Relaxosome function and conjugation regulation in F-like plasmids: a structural biology perspective. *Mol Microbiol* **85:**602–617.

103. **Toleman MA, Walsh TR.** 2011. Combinatorial events of insertion sequences and ICE in Gram-negative bacteria. *FEMS Microbiol Rev* **35:**912–935.

104. **Parks AR, Peters JE.** 2009. Tn7 elements: engendering diversity from chromosomes to episomes. *Plasmid* **61:**1–14.

105. **De Gelder L, Ponciano JM, Joyce P, Top EM.** 2007. Stability of a promiscuous plasmid in different hosts: no guarantee for a long-term relationship. *Microbiology* **153:**452–463.

106. **Heuer H, Fox RE, Top EM.** 2007. Frequent conjugative transfer accelerates adaptation of a broad-host-range plasmid to an unfavorable *Pseudomonas putida* host. *FEMS Microbiol Ecol* **59:**738–748.

107. **Dahlberg C, Chao L.** 2003. Amelioration of the cost of conjugative plasmid carriage in *Eschericha coli* K12. *Genetics* **165:**1641–1649.

108. **Humphrey B, Thomson NR, Thomas CM, Brooks K, Sanders M, Delsol AA, Roe JM, Bennett PM, Enne VI.** 2012. Fitness of *Escherichia coli* strains carrying expressed and partially silent IncN and IncP1 plasmids. *BMC Microbiol* **12:**53.

109. **Carattoli A.** 2009. Resistance plasmid families in *Enterobacteriaceae*. *Antimicrob Agents Chemother* **53:** 2227–2238.

110. **de Toro M, Rojo-Bezares B, Vinue L, Undabeitia E, Torres C, Saenz Y.** 2010. *In vivo* selection of aac (6′)-Ib-cr and mutations in the gyrA gene in a clinical qnrS1-positive *Salmonella enterica* serovar Typhimurium DT104B strain recovered after fluoroquinolone treatment. *J Antimicrob Chemother* **65:**1945–1949.

111. **Garcia-Fernandez A, Fortini D, Veldman K, Mevius D, Carattoli A.** 2009. Characterization of plasmids harbouring qnrS1, qnrB2 and qnrB19 genes in *Salmonella*. *J Antimicrob Chemother* **63:**274–281.

112. **San Millan A, Escudero JA, Gutierrez B, Hidalgo L, Garcia N, Llagostera M, Dominguez L, Gonzalez-Zorn B.** 2009. Multiresistance in *Pasteurella multocida* is mediated by coexistence of small plasmids. *Antimicrob Agents Chemother* **53:**3399–3404.

113. **Lorenzo-Diaz F, Fernandez-Lopez C, Garcillan-Barcia MP, Espinosa M.** 2014. Bringing them together: plasmid pMV158 rolling circle replication and conjugation under an evolutionary perspective. *Plasmid* **74:**15–31.

114. **Rijavec M, Budic M, Mrak P, Muller-Premru M, Podlesek Z, Zgur-Bertok D.** 2007. Prevalence of ColE1-like plasmids and colicin K production among uropathogenic *Escherichia coli* strains and quantification of inhibitory activity of colicin K. *Appl Environ Microbiol* **73:**1029–1032.

115. **Tan Y, Riley MA.** 1997. Nucleotide polymorphism in colicin E2 gene clusters: evidence for nonneutral evolution. *Mol Biol Evol* **14:**666–673.

116. **Watson RJ, Vernet T, Visentin LP.** 1985. Relationships of the Col plasmids E2, E3, E4, E5, E6, and E7: restriction mapping and colicin gene fusions. *Plasmid* **13:** 205–210.

117. **Gregorova D, Pravcova M, Karpiskova R, Rychlik I.** 2002. Plasmid pC present in *Salmonella enterica* serovar Enteritidis PT14b strains encodes a restriction modification system. *FEMS Microbiol Lett* **214:**195–198.

118. **Zakharova MV, Beletskaya IV, Denjmukhametov MM, Yurkova TV, Semenova LM, Shlyapnikov MG, Solonin AS.** 2002. Characterization of pECL18 and pKPN2: a proposed pathway for the evolution of two plasmids that carry identical genes for a type II restriction-modification system. *Mol Genet Genomics* **267:**171–178.

119. **Mruk I, Sektas M, Kaczorowski T.** 2001. Characterization of pEC156, a ColE1-type plasmid from *Escherichia coli* E1585-68 that carries genes of the EcoVIII restriction-modification system. *Plasmid* **46:**128–139.

120. **Miller CA, Cohen SN.** 1978. Phenotypically cryptic EcoRI endonuclease activity specified by the ColE1 plasmid. *Proc Natl Acad Sci USA* **75:**1265–1269.

121. **Cabezon E, Sastre JI, de la Cruz F.** 1997. Genetic evidence of a coupling role for the TraG protein family in bacterial conjugation. *Mol Gen Genet* **254:**400–406.

122. **Camps M.** 2010. Modulation of ColE1-like plasmid replication for recombinant gene expression. *Recent Pat DNA Gene Seq* **4:**58–73.

123. **Fukuoh A, Iwasaki H, Ishioka K, Shinagawa H.** 1997. ATP-dependent resolution of R-loops at the ColE1 replication origin by *Escherichia coli* RecG protein, a Holliday junction-specific helicase. *EMBO J* **16:** 203–209.

124. **Xu FF, Gaggero C, Cohen SN.** 2002. Polyadenylation can regulate ColE1 type plasmid copy number independently of any effect on RNAI decay by decreasing the interaction of antisense RNAI with its RNAII target. *Plasmid* **48:**49–58.

125. **Dasgupta S, Masukata H, Tomizawa J.** 1987. Multiple mechanisms for initiation of ColE1 DNA replication: DNA synthesis in the presence and absence of ribonuclease H. *Cell* **51:**1113–1122.

126. **Jung YH, Lee Y.** 1995. RNases in ColE1 DNA metabolism. *Mol Biol Rep* **22:**195–200.

127. **Colloms SD, McCulloch R, Grant K, Neilson L, Sherratt DJ.** 1996. Xer-mediated site-specific recombination *in vitro*. *EMBO J* **15:**1172–1181.

128. **Zaleski P, Wolinowska R, Strzezek K, Lakomy A, Plucienniczak A.** 2006. The complete sequence and segregational stability analysis of a new cryptic plasmid pIGWZ12 from a clinical strain of *Escherichia coli*. *Plasmid* **56:**228–232.

129. **Takechi S, Matsui H, Itoh T.** 1995. Primer RNA synthesis by plasmid-specified Rep protein for initiation of ColE2 DNA replication. *EMBO J* **14:**5141–5147.

130. **Takechi S, Yasueda H, Itoh T.** 1994. Control of ColE2 plasmid replication: regulation of Rep expression by a plasmid-coded antisense RNA. *Mol Gen Genet* **244:** 49–56.

131. **Butler MS, Cooper MA.** 2011. Antibiotics in the clinical pipeline in 2011. *J Antibiotics* **64:**413–425.

132. **Cooper MA, Shlaes D.** 2011. Fix the antibiotics pipeline. *Nature* **472:**32.

133. **Brolund A, Sundqvist M, Kahlmeter G, Grape M.** 2010. Molecular characterisation of trimethoprim resistance in *Escherichia coli* and *Klebsiella pneumoniae* during a two year intervention on trimethoprim use. *PLoS One* **5:**e9233. doi:10.1371/journal.pone.0009233.

134. **Sundqvist M, Geli P, Andersson DI, Sjolund-Karlsson M, Runehagen A, Cars H, Abelson-Storby K, Cars O, Kahlmeter G.** 2009. Little evidence for reversibility of trimethoprim resistance after a drastic reduction in trimethoprim use. *J Antimicrob Chemother* **65:**350–360.

135. **Yates CM, Shaw DJ, Roe AJ, Woolhouse ME, Amyes SG.** 2006. Enhancement of bacterial competitive fitness by apramycin resistance plasmids from non-pathogenic *Escherichia coli. Biol Lett* **2:**463–465.

136. **Perez-Mendoza D, de la Cruz F.** 2009. *Escherichia coli* genes affecting recipient ability in plasmid conjugation: are there any? *BMC Genomics* **10:**71.

137. **Baquero F, Coque TM, de la Cruz F.** 2011. Ecology and evolution as targets: the need for novel eco-evo drugs and strategies to fight antibiotic resistance. *Antimicrob Agents Chemother* **55:**3649–3660.

138. **Novotny C, Knight WS, Brinton CC Jr.** 1968. Inhibition of bacterial conjugation by ribonucleic acid and deoxyribonucleic acid male-specific bacteriophages. *J Bacteriol* **95:**314–326.

139. **Ou JT.** 1973. Inhibition of formation of *Escherichia coli* mating pairs by f1 and MS2 bacteriophages as determined with a Coulter counter. *J Bacteriol* **114:**1108–1115.

140. **Palchoudhury SR, Iyer VN.** 1969. Loss of an episomal fertility factor following the multiplication of coliphage M13. *Mol Gen Genet* **105:**131–139.

141. **Cullum J, Collins JF, Broda P.** 1978. Factors affecting the kinetics of progeny formation with F′lac in *Escherichia coli* K12. *Plasmid* **1:**536–544.

142. **Lin A, Jimenez J, Derr J, Vera P, Manapat ML, Esvelt KM, Villanueva L, Liu DR, Chen IA.** 2011. Inhibition of bacterial conjugation by phage M13 and its protein g3p: quantitative analysis and model. *PLoS One* **6:**e19991.

143. **Garcillan-Barcia MP, Jurado P, Gonzalez-Perez B, Moncalian G, Fernandez LA, de la Cruz F.** 2007. Conjugative transfer can be inhibited by blocking relaxase activity within recipient cells with intrabodies. *Mol Microbiol* **63:**404–416.

144. **Fernandez-Lopez R, Machon C, Longshaw CM, Martin S, Molin S, Zechner EL, Espinosa M, Lanka E, de la Cruz F.** 2005. Unsaturated fatty acids are inhibitors of bacterial conjugation. *Microbiology* **151:**3517–3526.

145. **Smith MA, Coincon M, Paschos A, Jolicoeur B, Lavallee P, Sygusch J, Baron C.** 2012. Identification of the binding site of *Brucella* VirB8 interaction inhibitors. *Chem Biol* **19:**1041–1048.

146. **Zhou Y, Call DR, Broschat SL.** 2013. Using protein clusters from whole proteomes to construct and augment a dendrogram. *Adv Bioinformatics* **2013:**191586.

147. **Villa L, Garcia-Fernandez A, Fortini D, Carattoli A.** 2010. Replicon sequence typing of IncF plasmids carrying virulence and resistance determinants. *J Antimicrob Chemother* **65:**2518–2529.

Plasmids—Biology and Impact in Biotechnology and Discovery
Edited by Marcelo E. Tolmasky and Juan C. Alonso

doi:10.1128/microbiolspec.PLAS-0004-2013

Nikolaus Goessweiner-Mohr,[1] Karsten Arends,[2] Walter Keller,[1] and Elisabeth Grohmann[3,4]

14 Conjugation in Gram-Positive Bacteria

INTRODUCTION

Conjugative transfer is an important driver in evolution, enabling bacteria to acquire new traits (1, 2, 3, 4, 5). During conjugative transfer, DNA translocation across the cell envelopes of two cells forming a mating pair is mediated by two types of mobile genetic elements: conjugative plasmids and integrating conjugative elements (ICEs) (1, 5, 6, 7, 8). Most conjugative plasmids apply a sophisticated multiprotein secretion apparatus, the so-called type IV secretion system (T4SS) to transfer DNA to a recipient cell (9, 10, 11, 12, 13). Conjugative T4SSs of Gram-positive (G+) bacteria exhibit considerable similarities to their Gram-negative (G–) counterparts; the first steps processing the plasmid DNA to be transferred with the relaxase, covalently attached to its 5′ end, are virtually identical (11, 14, 15, 16). However, the actual DNA translocation process including the passage of the cell envelope of the donor and the recipient cell appears to differ considerably between G+ and G– bacteria. This might be due to the differences in the structure of the cell envelope: cytoplasmic membrane followed by a thick multilayered peptidoglycan (PG) in G+ bacteria versus a two-membrane configuration with periplasmic space and thin PG layer between the two membranes in G– bacteria. Therefore, it is not surprising that homologs of VirB7, VirB9, and VirB10 proteins identified as actual G– T4SS channel components (17, 18, 19, 20, 21, 22, 23) have not been detected so far in G+ T4SSs.

Another type of conjugative DNA translocation machinery appears to have evolved in the order of *Actinomycetales*, it appears to be unique to conjugative plasmids and ICEs of multicellular bacteria belonging to this group of high G + C G+ bacteria (24). Their conjugative system resembles the machinery that promotes the segregation of chromosomal DNA during bacterial cell division and sporulation and requires a single FtsK-homologous protein to transfer double-stranded DNA to the recipient cell (24, 25, 26). Key factors of both translocation systems have been identified and the first models for conjugative plasmid transfer in G+ bacteria via both distinct modes have been presented (17, 25, 27, 28, 29).

The purpose of this article is to summarize the current state of knowledge of conjugative plasmid transfer in G+ bacteria explaining the distinct concepts as far as understood to date on the basis of three prominent model systems, the broad-host-range plasmids of

[1]Institute of Molecular Biosciences, University of Graz, 8010 Graz, Austria; [2]Robert Koch-Institute, Nordufer 20, 13353 Berlin, Germany; [3]Faculty of Biology, Microbiology, Albert-Ludwigs-University Freiburg, 79104 Freiburg, Germany; [4]Division of Infectious Diseases, University Medical Centre Freiburg, 79106 Freiburg, Germany.

the Inc18 family, the *Enterococcus* sex pheromone-responsive plasmids, and the *Streptomyces* plasmids. In addition, we present a new classification system based on a secondary structural homology prediction that can detect new T4SS components and assign the detected proteins to protein families postulating their function in the T4S process. Finally, we discuss the importance of conjugative plasmid transfer in different environments and end with emerging tools to quantify conjugative plasmid transfer *in situ*.

TWO DIFFERENT CONJUGATIVE TRANSFER MECHANISMS

In G+ bacteria two distinct conjugative mechanisms have evolved, most likely dependent on the tendency to live as "unicellular bacteria" versus "multicellular bacteria," the latter reminiscent of multicellular eukaryotic organisms, as is the case for streptomycetes (11). The great majority of G+ bacteria seem to conjugate via passage of single-stranded DNA through T4SSs. Nevertheless, the G+ T4SSs appear to be more simply organized than the better characterized G– counterparts. Multicellular G+ bacteria like *Streptomyces* seem to use a completely different mechanism that is reminiscent of the machinery involved in bacterial cell division or spore formation. Moreover, once a *Streptomyces* cell has acquired a plasmid molecule it is easily transferred to the cells in the vicinity via a process called spreading (11, 30, 31).

SINGLE-STRANDED DNA TRANSFER IN GRAM-POSITIVE BACTERIA

The G+ T4S mechanism and its peculiarities will be presented on the basis of the broad-host-range conjugative plasmid pIP501 belonging to the incompatibility group Inc18 (32) and the *Enterococcus* sex pheromone-responsive plasmid pCF10 (33, 34).

Conjugative Transfer of Broad-Host-Range Plasmids

The conjugative broad-host-range model system in G+ bacteria is the promiscuous plasmid pIP501 studied in our groups. Similarities and peculiarities with respect to the most extensively studied G– systems, as exemplified by the prototype *Agrobacterium tumefaciens* T-DNA transfer system, are highlighted. All major protein families encountered in G– T4SSs have also been detected in the G+ bacteria (17, 27, 35). T4SSs promoting conjugative DNA transfer encode member(s) of an additional protein family—not present in T4SSs dedicated to effector transport—the so-called DNA-processing enzymes or relaxases. Relaxases are required for the preparation of the single-stranded plasmid DNA molecule that is translocated via the mating pair formation (Mpf) complex constituted by the other T4SS proteins to the recipient cell. The reaction catalyzed by the relaxase is mechanistically identical in G+ and G– bacteria (recently reviewed by Zechner et al. [16]). For plasmid pIP501, originally isolated from *Streptococcus agalactiae* (36), for which self-transfer to virtually all G+ bacteria and additionally to *Escherichia coli* was demonstrated (37), the TraA relaxase has been characterized in some biochemical detail (38, 39, 40). In addition to its function as a site-specific DNA-nicking enzyme, it has been shown to negatively regulate the expression of all T4SS proteins encoded in the *tra* operon thereby autoregulating its own expression (Fig. 1). Interestingly, in contrast to most other conjugative T4SSs (9, 16, 41, 42, 43) no auxiliary factors required for TraA-mediated $oriT_{pIP501}$ cleavage have been identified so far (39).

The other protein families also represented in G– T4SSs are as follows: (i) the motor protein family, (ii) the PG hydrolase family, (iii) the T4SS channel/putative core component family, and (iv) the surface factor/adhesin family. Representatives of these protein families from plasmid pIP501 will be introduced in the following section.

Motor Protein Family

The pIP501 transfer (*tra*) region codes for two putative ATPases (TraE, TraJ) containing the NTP-binding motifs typically found in proteins with ATPase activity. The TraE protein is composed of 653 amino acids (aa), it shows similarities to VirB4 (COG3451) at position 121 to 570 (GenBank AAA99470, e-value $1.56e^{-13}$) and Walker A and Walker B motifs are predicted at positions 251 to 258 and 510 to 516, respectively. TraJ consists of 551 aa; similarities to the VirD4 coupling protein (COG3505) are predicted between positions 34 and 442 (GenBank 44390, e-value $3.54e^{-51}$), as well as Walker A and Walker B motifs at positions 39 to 46 and 291 to 295, respectively. For both proteins, ATP binding and hydrolysis activity have been shown (TraJ; E.-K. Celik, A. Guridi et al., unpublished data; TraE; E.-K. Celik, M.-Y. Abajy et al., unpublished data).

Yeast-two-hybrid and pull-down experiments showed that TraE strongly interacts with itself (27), confirming the formation of oligomers that were predicted based on its similarity to T4SS ATPases of G– and G+ origin (23, 44, 45, 46, 47). Moreover, TraE was shown to bind to

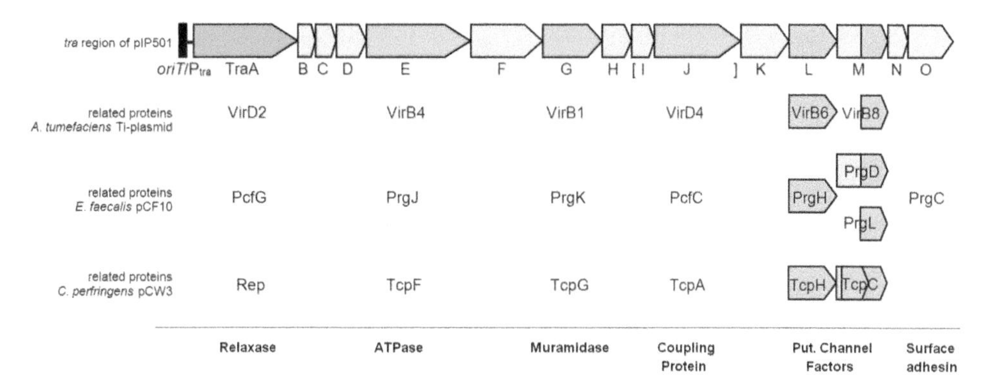

Figure 1 Genetic organization of the pIP501 *tra* operon. Proteins with sequence similarities with the corresponding *A. tumefaciens* Ti-plasmid VirB/D4, *E. faecalis* pCF10, and *C. perfringens* pCW3 T4SS proteins are in blue; the potential two-protein-coupling protein (consisting of $TraI_{pIP501}$ and $TraJ_{pIP501}$) is indicated with brackets; relations based on structure (TraM C-terminal domain, VirB8-like [54]) and domain prediction (TraL, VirB6-like) based similarities are in yellow; the gene encoding the putative relaxase is in green. The respective protein families are indicated. P_{tra}, *tra* operon promoter. The genes of the pIP501 *tra* region are drawn to scale. Put., putative. doi:10.1128/microbiolspec.PLAS-0004-2013.f1.

the PG hydrolase TraG and to TraN, a cytosolic double-stranded DNA-binding protein (27). Recent data suggest two further candidates for a close interaction with TraE: TraH and TraJ (N. Goessweiner-Mohr et al., unpublished data). These interactions point to a crucial role of TraE in the T4S process in addition to its putative function as motor protein fueling the DNA translocation process and/or the assembly of the T4S machinery. We postulate that it could be directly involved in the build-up of the transfer channel and/or the transport of the substrate.

For TraJ, we postulate that it might act as a T4SS coupling protein in the pIP501 plasmid transfer process, likely aided by TraI, which would position TraJ to the cell membrane through protein-protein interaction (N. Goessweiner-Mohr et al., unpublished data). If this assumption holds true, TraI and TraJ will represent a novel two-partner T4SS coupling protein as postulated by Alvarez-Martinez and Christie (35). Surprisingly, no self-interaction was detected for TraJ (27). It is likely that a stable oligomerization, as reported for related T4SS coupling proteins (29, 48, 49, 50, 51), requires the presence of TraI. TraJ was found to interact with the relaxase TraA (27), further confirming its putative role as a coupling protein, which links the relaxosome protein-DNA complex to the actual transfer channel. Furthermore, TraJ showed binding affinity for the double-domain, membrane-spanning protein TraF and for the PG hydrolase TraG (27); recently, binding to the membrane protein TraH and the cytosolic DNA-binding protein TraN have been suggested (N. Goessweiner-Mohr et al., unpublished data).

The Peptidoglycan Hydrolase Family

The pIP501 T4SS codes for TraG, a 369-aa protein with PG cleaving activity. TraG has been shown to be indispensable for pIP501 transfer in *Enterococcus faecalis* (52). It is the first PG-hydrolyzing protein characterized so far for which absolute requirement in the T4SS process has been demonstrated. For TraG, a modular architecture is anticipated: at the amino terminus of the protein (positions 20 to 36, GenBank: CAD44387.1; HMMTOP [53]) a transmembrane helix (TMH) is predicted, postulated to be required for its proper location in the cell envelope, followed by a specific lytic transglycosylase (SLT) domain at position 77 to 157 (LT_GEWL domain, acc.number: cd00254, e-value $1.32e^{-7}$). At the carboxy terminus positions 249 to 359 (pfam05257, e-value $5.04e^{-31}$), a cysteine, histidine-dependent amidohydrolases/peptidases (CHAP) domain is anticipated. Furthermore, TraG contains *N*-acetyl-D-glucosamine binding sites within the SLT domain at positions 87, 99, 122, and 144, likely required for a proper linkage of the protein to its PG substrate. TraG-mediated cleavage has been shown for PG isolated from both *E. faecalis* and *E. coli* thereby underpinning the broad host range of pIP501 transfer (52). Moreover, the TraG domains were expressed separately, and both domains, SLT as well as CHAP, degraded PG isolated from *E. faecalis*. We postulate that the SLT_{TraG} domain acts as a lytic transglycosylase because it was efficiently inhibited by the lytic transglycosylase inhibitors, bulgecin A and hexa-*N*-acetylchitohexaose, respectively (52). Furthermore, an exchange of the conserved glutamate residue in the putative catalytic center of the SLT domain to glycine (E87G) nearly completely abolished PG cleavage activity, which is consistent with this part of TraG being a lytic transglycosylase (52).

In yeast-two-hybrid, pull-down assays and Thermofluor-based protein-protein interaction studies, TraG showed self-interaction (it is postulated to be a dimer) as well as interactions with six other pIP501 T4SS proteins, namely with TraB, TraE, TraH, TraI, TraJ, and TraN ([27]; N. Goessweiner-Mohr et al., unpublished data). We postulate that TraG locally punches holes in the PG meshwork. Through its interactions with the VirB4 homolog, TraE, the VirB6-like protein, TraL (17, 35), and the putative two-component coupling protein TraI/TraJ, TraG might recruit them to their proper location in the T4S complex, thereby possibly positioning the T4S complex at discrete foci on the cell surface (K. Arends and E. Grohmann, unpublished data). In the working model of pIP501 DNA transfer (Fig. 2), the putative key role of TraG in the DNA transfer process,

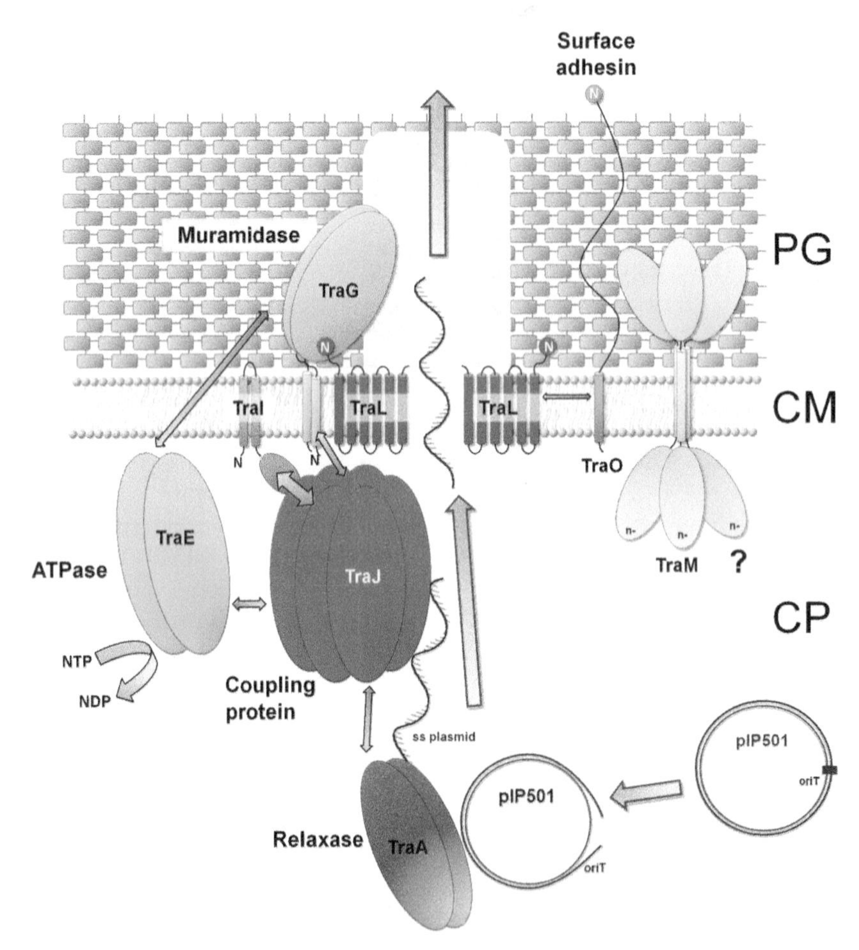

Figure 2 Model of the pIP501 DNA transfer pathway. First, $oriT_{pIP501}$ is bound by the relaxase TraA. After being nicked, the single-stranded plasmid is recruited to the putative transfer channel (modified from reference 17) via the putative two-protein coupling protein TraJ. Decreased shading of PG symbolizes TraG-mediated local opening of PG. The localization and orientation of the T4SS proteins is based on *in silico* predictions and localization studies (52, 54). The N terminus of the T4SS proteins is marked (N). Arrows indicate protein-protein interactions determined by yeast-two-hybrid studies and validated by pull-down assays (27), as well as interactions found by using the Thermofluor method (Goessweiner-Mohr et al., unpublished data). The thickness of the arrows marks the strength of the detected interactions. The putative function of key members of the pIP501 *tra* operon in the DNA secretion process is indicated. PG, peptidoglycan; CM, cytoplasmic membrane; CP, cytoplasm. doi:10.1128/microbiolspec.PLAS-0004-2013.f2.

its interactions with other T4SS components, and the putative way of the pIP501 DNA through the T4S complex are depicted.

The T4SS Channel/Putative Core Complex Component Family

The pIP501 T4SS comprises two putative channel components, the VirB8-like protein TraM (54) and the putative VirB6 homolog TraL (17, 35).

Bhatty and colleagues postulated a key role for the G+ VirB8 homologs as putative cell wall-spanning components in the T4S complex (17). For the pIP501 system, TraM was predicted to act as a VirB8 homolog (54). For TraM, a possible scaffolding role in the T4S complex has been suggested based upon its structural similarities to the G– VirB8 proteins (55, 56) and to TcpC from *Clostridium perfringens* (57). However, its largely different domain composition points toward a divergent role in the T4S process (54). Nevertheless, we have no experimental evidence for the key role of TraM in the secretion process, because no interactions with other pIP501 T4SS proteins have been detected so far ([27]; N. Goessweiner-Mohr et al., unpublished data). This could be explained, in part, by the fact that all interaction studies have been conducted with the soluble variant of TraM; the C-terminal domain TraMΔ, as the full-length protein, could not be expressed in suitable amounts so far. It is likely that the interaction domain of TraM is located in the amino-terminal region which contains a transmembrane domain (TMD); studies in this direction are currently being performed in our laboratories.

In the G+ T4SS model postulated by the Christie group (17), the VirB6 homolog is part of the cytoplasmic translocon/ATPase complex that is presented based upon findings from G– T4SSs. For $\text{TraL}_{\text{pIP501}}$, only interactions with the cytoplasmic protein TraN and the putative surface adhesin, TraO, have been demonstrated so far (27). Nevertheless, because VirB6-like proteins are highly conserved among G– and G+ T4SSs alike, they are assumed to play an important role in the scaffolding or even the makeup of the inner membrane secretion channel, as suggested for G– transfer systems (58, 59, 60, 61, 62). This is an especially appealing argument for G+ T4SSs, because no sequence-based or structural homologs were detected for the actual G– channel proteins, VirB7, 9, and 10 (17, 35, 63).

Surface Factor/Adhesin Family

Based on our current knowledge, the only pIP501 T4SS protein that could act as a surface adhesin is the TraO protein (29.9 kDa, GenBank CAD44395). TraO is a 282-aa protein containing a signal peptide with a putative cleavage site at positions 24 to 25 (SignalP 4.1). A Pro-rich domain is found at the amino terminus of the protein with a highly repetitive region [$(\text{Asp-Pro-Val})_7$-$(\text{Glu-Pro-Thr})_{37}$] at positions 54 to 184. An LPxTG cell wall anchor motif (64) was found at the carboxy-terminal part of the protein (aa 252 to 256) followed by mainly hydrophobic and positively charged amino acids and a putative transmembrane helix (TMH) at positions 261 to 278 (HMMTOP). The C-terminal part of TraO is therefore most likely anchored to the cell wall via a covalent bond between Thr of the LPxTG motif and the PG mediated by the transpeptidase sortase A, and the N terminus is consequently exposed to the surface of the enterococcal cell wall (65, 66).

TraO is related to the putative cell wall-anchored surface protein PrgC, encoded by the sex pheromone-responsive plasmid pCF10 (35). In agreement with TraO, PrgC also contains highly repetitive sequence motifs of a three-residue periodicity comprising Pro-uncharged-Glu/Asp residues (17, 35). Such repeat regions are also present in other G+ surface adhesins, such as the IgA-binding proteins of *S. agalactiae*, and *Staphylococcus aureus* and streptococcal fibronectin-binding proteins (17, 67). These characteristics make a role of $\text{TraO}_{\text{pIP501}}$ and $\text{PrgC}_{\text{pCF10}}$ in establishing the contact with the recipient cell likely (17). In preliminary data from the Grohmann laboratory, deletion of the TraO LPxTG cell wall anchor motif appeared to abolish pIP501 transfer in *E. faecalis* (K. Arends and E. Grohmann, unpublished data).

Conjugative Transfer of *Enterococcus* Sex Pheromone Plasmids

In the sex pheromone-responsive DNA transfer systems, the contact between donor and recipient cells is mediated via a sophisticated, tightly controlled system based upon small peptides, the so-called sex pheromones, which are secreted by potential plasmid recipients. pAD1 and pCF10 are sex pheromone-responsive plasmids present in many *E. faecalis* strains (33, 34, 68, 69). Genetic and physiological studies on pAD1 transfer have been performed by the Clewell group (68, 70). Recently, interesting features of the catalytic domain of the pAD1 relaxase, TraX, have been published by Francia and coworkers (71). For pCF10, conjugative plasmid transfer has been studied in some mechanistic detail by the Christie and Dunny groups. Upon sensing of a specific peptide pheromone, *E. faecalis* efficiently transfers plasmid pCF10 through a T4SS to recipient cells (72). The 67.6-kb plasmid has so far been found only in the enterococci, but it was shown to mobilize *oriT*-containing plasmids to *Lactococcus lactis* and *S. agalactiae* (72, 73). The tetracycline resistance plasmid pCF10 codes for pheromone sensing and response functions, conjugation factors, and a surface adhesin termed aggregation substance that is important both for conjugation and virulence (72). Putative key proteins of the pCF10 T4SS are presented in comparison to the pIP501 system.

Motor Protein Family

Similarly to pIP501, pCF10 encodes two proteins with putative ATPase activity, PcfC, a membrane-bound putative ATPase related to the coupling proteins of G– T4SSs (72), and PrgJ, a VirB4-like ATPase (46). A PcfC Walker A NTP binding site mutant fractionated with the *E. faecalis* membrane and formed foci, whereas PcfC devoid of its N-terminal putative TMD distributed uniformly throughout the cytoplasm. PcfC wild-type and mutant proteins were shown to bind only the processed form of the pCF10 plasmid *in vivo* (72). Reminiscent of observations in the *A. tumefaciens* T4SS, formation of the pCF10-PcfC formaldehyde cross-link required the relaxase PcfG and its accessory factor, PcfF, but not the Mpf channel components or ATP hydrolysis by PcfC (17, 72). Chen and colleagues presented a model in which the PcfC coupling protein initiates DNA transfer through the pCF10 T4S channel in an NTP-dependent mode (72).

PrgJ is a member of the VirB4 family of ATPases that is found in virtually all T4SSs. Purified PrgJ dimers were demonstrated to bind and hydrolyze ATP (46). A PrgJ NTP-binding site mutation slightly diminished ATP binding but abolished ATP hydrolysis *in vitro* and blocked pCF10 transfer *in vivo*. PrgJ wild type and the mutant protein interacted with the coupling protein PcfC, the relaxase PcfG, and the accessory factor PcfF (46). Moreover, PrgJ and its mutant protein bound single-stranded and double-stranded DNA substrates without sequence specificity *in vitro*, and *in vivo* by a mechanism dependent on an intact pCF10 *oriT* sequence and cosynthesis of PcfC, PcfF, and PcfG. Li and colleagues presented a model in which the PcfC coupling protein coordinates with the PrgJ ATPase to drive early steps of pCF10 transfer. In this model, PrgJ catalyzes DNA substrate transfer to the membrane translocase, thereby pointing to a novel function of VirB4-type ATPases in mediating early steps of T4S (46).

The Peptidoglycan Hydrolase Family

The hydrolases associated with G+ T4SSs differ from prototypical VirB1 from *A. tumefaciens*, which is secreted to the periplasm and contains a single hydrolase domain. The G+ PG hydrolases instead contain amino-proximal TMDs and thus are likely anchored in the membrane (17, 35).

$PrgK_{pCF10}$ is a large, 871-aa protein with an N-proximal TMD and three predicted hydrolase domains of the LytM, gp13/SLT, and CHAP (NlpC/P60) families (17). Recent studies at the Christie laboratory have demonstrated that the latter two domains are catalytically active and that one, but not both of them, is required for assembly of a functional T4SS. Deletions or catalytic site mutations of both gp13/SLT and CHAP domains completely abolished pCF10 transfer (17). As demonstrated for the pIP501 PG hydrolase TraG (52), PrgK is absolutely required for $T4SS_{pCF10}$ function (132).

Bhatty and colleagues presented a working model for T4S in G+ bacteria in which the G+ VirB1 homologs form a crucial part of the T4SS membrane complex. In the model, the amino-terminal TMD of the PG hydrolase forms part of the membrane translocation complex, and the carboxy-terminal hydrolase domain and the VirB8-like domains (PrgL in the case of pCF10) extend across the cell wall and form a channel or fiber through or along which the secretion substrates pass (17).

The T4SS Channel/Putative Core Complex Component Family

Only two of the conserved components of the G+ T4SSs are predicted to localize to the exterior face of the cytoplasmic membrane, the VirB1-like PG hydrolase and the carboxy-terminal domain of the VirB8-like subunit, PrgL in the case of pCF10 (17). Together with the polytopic VirB6-like protein, PrgH, they could form the major components of the T4SS channel spanning the G+ cell envelope (17). Interestingly, pCF10 encodes a second VirB8-like protein, which resembles in terms of its domain composition $TraM_{pIP501}$ (54). The presence of a second VirB8-like protein in the pCF10 T4SS might indicate distinct roles for these two proteins, reflected in their diverse structural composition. Alternatively, one of the proteins might be a remnant of an ancestral T4SS, a prototypic conjugation machinery from which the respective G– and G+ T4SSs might have branched off earlier in evolution.

Surface Factor/Adhesin Family

The G– T4SSs elaborate conjugative pili, but to date no G+ system has been demonstrated to generate similar structures. Instead, the G+ systems enable target cell attachment, at least in part, through the generation of surface factors or adhesins (17). The best-characterized adhesin is the aggregation substance, PrgB encoded by plasmid pCF10 (74, 75). PrgB is a large approximately 137-kDa protein with homologs found predominantly in related pheromone-inducible conjugation systems of enterococci. PrgB contains an amino-terminal signal sequence, a carboxy-terminal LPxTG cell wall anchor motif, cell adhesion RGD motifs, and a glucan-binding or aggregation domain resembling those of the *Streptococcus* glucan-binding protein C and surface protein antigen (Spa)-family proteins (17, 76, 77). A *prgB* mutant was able to transfer pCF10 at significant frequencies in solid-surface matings, indicating that other surface factors or the secretion channel itself also can promote target cell contacts (17). The pCF10 *tra* region codes for two other cell wall-anchored surface proteins, PrgA and PrgC (35). PrgA was demonstrated to be involved in surface exclusion, but it might also participate in the formation of mating junctions (78). PrgC is related to TraO of pIP501. Both proteins are predicted to mediate specific contacts with recipient cells (17, 27, 35) and possibly also with eukaryotic cells (17).

Secondary Structure Homology-Based Classification Scheme For T4SS Proteins

Although proteins may only show limited sequence identity, and thus would not be identified as being related, a comparison of their respective secondary structure content and domain composition can reveal relations that would not have been detected otherwise. To broaden our understanding of the structural relations between T4S components of G+ and G− origin, we performed an extended search for putative PG hydrolases, VirB6-like and VirB8-like proteins in a broad spectrum of conjugative plasmids, transposons, ICEs, and genomic islands (GIs). Based on their predicted secondary structure content and domain composition, we categorized the homologous proteins into distinct classes. In the following sections, the results are given in detail.

The VirB1 Protein Family

The putative PG hydrolases were categorized into four distinct classes (Fig. 3). The prediction-based comparisons are presented in Table 1 and in detail in Fig. 4.

All analyzed putative lytic transglycosylases found in putative T4SSs of G+ origin (12 plasmids and 10 T4SSs located on the chromosome, including three ICEs and five transposons) contain an SLT and a CHAP(-like) domain. These proteins can be further divided into three distinct classes. Class ALPHA proteins (e.g., TraG of pIP501) contain a very short (15 to 20 aa) N-terminal sequence, followed by a TMH, a short linker, and an SLT domain. The C-terminal CHAP domain is connected to the core part of the protein via another linker region. Class BETA proteins (e.g., TcpG of

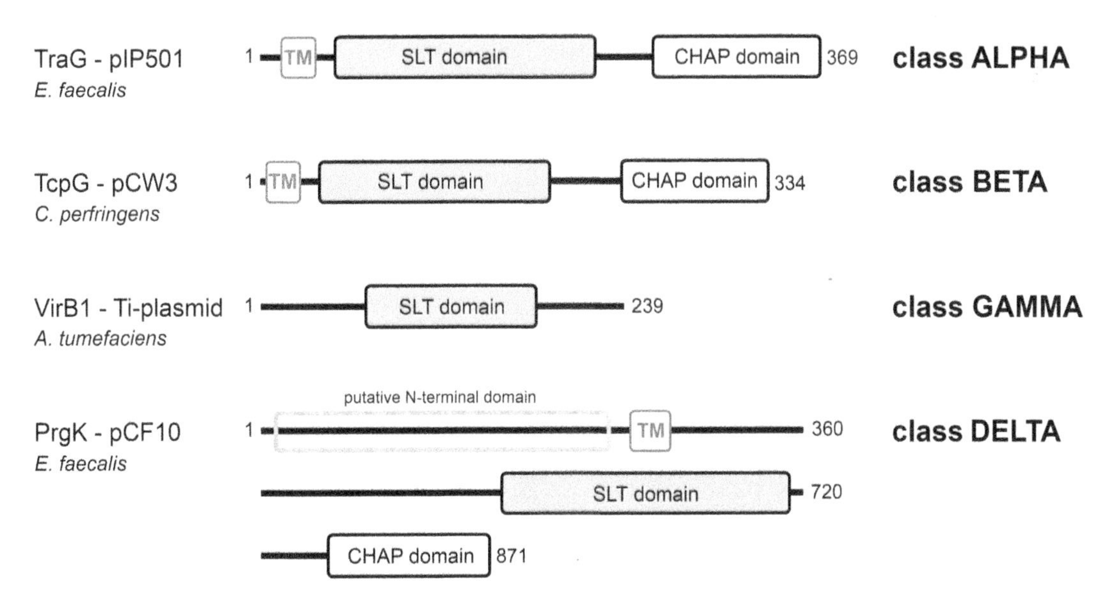

Figure 3 Comparison of the domain arrangement and classification of putative lytic transglycosylases from G+ and G− putative T4SSs. TraG (pIP501), TcpG (pCW3), VirB1 (Ti-plasmid), and PrgK (pCF10) were chosen as representatives of their respective classes. The potential SLT and CHAP(-like) domains were assigned according to secondary structure predictions with PSIPred (131). TMHs were annotated with HMMTOP (53). doi:10.1128/microbiolspec.PLAS-0004-2013.f3.

Table 1 Classification of TraG-like and other lytic transglycosylases

Gram reaction	Species	Location	Protein[c]	Classification
+	*Enterococcus faecalis*	pRE25	Orf30	α
+	*E. faecalis*	pIP501	TraG	α
+	*Staphylococcus aureus*	pSK41	TraG	α
+	*S. aureus*	pGO1	TrsG	α
+	*Bacillus thuringiensis*	pAW63	006	α
+	*B. thuringiensis*	pBT9727	0007	α
+	*Bacillus anthracis*	pXO2	B0007	α
+	*S. aureus*	ICE*6013*	Orf5	α
+	*Clostridium perfringens*	pCW3	TcpG	β
+	*C. perfringens*	pCPF5603	69	β
+	*C. perfringens*	pCPF4969	25	β
+	*C. perfringens*	pJIR26	TcpG	β
+	*C. perfringens*	CW459[a]	Orf14	β
+	*E. faecalis*	strain EF62[a]	0531	β
+	*E. faecalis*	Tn*916*	Orf14	β
+	*Streptococcus parauberis*	Tn*916*-like	O14	β
+	*Streptococcus pneumoniae*	Tn*916*-like	Orf14	β
+	*S. pneumoniae*	Tn*5253*	Orf32	β
+	*S. pneumoniae*	Tn*5251*	Orf14	β
+	*Bacillus subtilis*	ICE*Bs1*	YddH	β
–	*Agrobacterium tumefaciens*	Ti-plasmid	VirB1	γ
–	*Escherichia coli*	pKM101	TraL	γ
–	*E. coli*	R1	Orf19	γ
–	*E. coli*	R100	Orf169	γ
–	*E. coli*	F-plasmid	Orf169	γ
–	*E. coli*	R6K	Pilx1	γ
–	*E. coli*	pOLA52	Pilx1	γ
–	*E. coli*	R388	TrwN	γ
–	*E. coli*	pR721	TraB	γ
–	*Brucella suis*	VirB operon; chromosome[a]	VirB1	γ
–	*Brucella abortus*	Chromosome[a]	VirB1	γ
–	*Haemophilus influenzae*	pF3031	Bpl11	γ
–	*Salmonella typhi*	R27	BfpH	γ
–	*Pseudomonas aeruginosa*	pADP-1	TrbN	γ
–	*Enterobacter aerogenes*	pR751	TrbN	γ
–	*P. aeruginosa*	RP4	TrbN	γ
–	*Salmonella enterica*	ICE*Se3*	0069	γ
–	*Helicobacter pylori*	CagPAI[b]	Cag-gamma	γ
+	*E. faecalis*	pCF10	PrgK	δ
+	*Streptococcus suis*	ICE*Ssu32457*	Orf42	δ

[a]Chromosome, not a further characterized location on the genome.
[b]PAI, within the pathogenicity island.
[c]The proteins are classified based on their predicted secondary structure composition.

pCW3 from *C. perfringens*) share this composition, but lack the N-terminal sequence in front of the TMH. Thus, these proteins are smaller than the class ALPHA proteins. Only two proteins were found for class DELTA: PrgK from *E. faecalis* conjugative plasmid pCF10 and Orf42 from *Streptococcus suis* ICE*Ssu32457*. These two proteins show an exceptional length (871 and 933 aa, respectively), both contain a TMH, an SLT, and a CHAP-like domain but in addition they contain a large N-terminal domain of unknown function, facing toward the cytoplasm.

All analyzed putative lytic transglycosylases found in putative T4SSs of G– origin (15 plasmids and three T4SSs located on the chromosome, including one pathogenicity island [PAI] and one ICE) contain only an SLT domain and most of them were predicted to lack a TMH. Because these proteins lack a CHAP-like domain, they are significantly shorter than class ALPHA and class BETA proteins of the G+ T4SSs and were all classified into class GAMMA.

Besides an universally shared SLT domain, enzymes from G+ bacteria comprise an additional catalytic domain at the C terminus (CHAP) and a TMH at the N terminus, whereas all T4S-related G– lytic transglycosylases lack these two additional characteristics (class GAMMA). We propose that the significant difference between lytic transglycosylases of G– and G+ T4SSs is due to the disparate composition of the cell envelopes. G– cell walls comprise two membranes, separated by a thin and to a certain extent cross-linked PG layer. In contrast, G+ bacteria possess a single membrane, coated by a much thicker, highly cross-linked PG layer, which itself may be covered by a protein or glycoprotein outer layer (79). In the case of G– bacteria, the activity of the SLT domain might be sufficient to open up the thin PG layer between the two membranes (80, 81). Possibly because of the highly cross-linked PG in G+ bacteria, the PG-degrading enzymes possess an additional catalytic domain at the C terminus to facilitate local opening of the PG meshwork.

The VirB6 Protein Family

An extensive comparison of VirB6-like polytopic membrane proteins of G– and G+ origin showed significant variability of their secondary structure motifs (N. Goessweiner-Mohr et al., unpublished data). VirB6-like proteins originating from G+ T4SSs can be grouped according to their number of amino acids. A first set of proteins shows a length of 265 to 335 aa, a second group of candidates comprises 720 to 1120 aa, respectively (N. Goessweiner-Mohr et al., unpublished data). The *E. faecalis*-derived proteins $PrgH_{pCF10}$ and $TraL_{pIP501}$ belong to the group of smaller proteins, which resemble the classical *A. tumefaciens* VirB6 (35) composition in size and number of predicted transmembrane (TM) motifs, whereas $TcpH_{pCW3}$ belongs to the group of larger VirB6-likes (Fig. 5). Similar to the G+ VirB6-like proteins, the respective proteins of G– origin can also be graded. Again, this classification depends on the overall number of amino acids and the observed secondary structure motifs (N. Goessweiner-Mohr et al., unpublished data). The classical VirB6 proteins show a limited variability and possess about 295 to 350 aa (e.g., VirB6 of the VirB/D operon, *A. tumefaciens*; and TraD of pKM101, *E. coli*). Another group of potential VirB6-like proteins consists of proteins of about 570 to 940 aa, respectively (e.g., TraG of plasmid R1, *E. coli*). A last group of putative VirB6-likes (e.g., TrbP of pBP136, *Bordetella pertussis*) comprises much shorter candidates (about 240 aa).

The VirB6-like proteins contain up to eight TM motifs with diverse domains as linkers between distinct TMHs. Members of shorter VirB6 protein classes only show short linker regions between the individual TMH motifs, whereas large domains are found among other VirB6-likes. The exact function of these domains in the course of the conjugative transfer is still a matter of debate. Most likely, they enable VirB6-like proteins to establish protein-protein interactions or to form stable oligomers, and these interactions might be necessary for the buildup of an inner-membrane transfer channel and the subsequent DNA transport (35, 82, 83). In the case of the large putative VirB6-likes of G– T4SSs (e.g., plasmids R1/R100-1/F of *E. coli*), it has been argued that the C-terminal part of the respective protein could serve as an entry exclusion factor (84, 85, 86, 87).

The VirB8 Protein Family

Through a secondary structure prediction-based search we were able to identify numerous previously undetected proteins of G– and G+ T4SSs alike, which share the VirB8 (NTF2-like) fold (54). Furthermore, a comparison of their buildup revealed striking differences in the composition of the VirB8-like proteins. Based on our observation, we were able to propose a new classification of VirB8-like T4S proteins. Classic VirB8-like proteins found in G–bacteria, as well as in five T4SSs of G+ origin (e.g., pCF10 from *E. faecalis*), contain only one NTF2-like domain (class ALPHA). Proteins with two successive NTF2-like domains, such as TcpC from *C. perfringens*, belong to class BETA. Class GAMMA proteins (e.g., TraM of pIP501 from *E. faecalis*) consist of a single, C-terminal NTF2-like domain plus a large cytoplasmic N-terminal domain of yet unknown fold. We further propose that the significant variation between VirB8-like proteins of G– and G+ bacteria is due to the distinct composition of the cell envelopes. Another explanation for the existence of the three protein classes may be the functional adaptation of the respective proteins to the requirements of the diverging conjugation systems.

DOUBLE-STRANDED DNA TRANSFER IN MULTICELLULAR GRAM-POSITIVE BACTERIA

Conjugative Transfer in *Streptomyces*

The antibiotic producers of the genus *Streptomyces* represent a huge natural reservoir of antibiotic resistance genes for the spread of resistance within the soil community. *Streptomyces* plasmids encode a unique

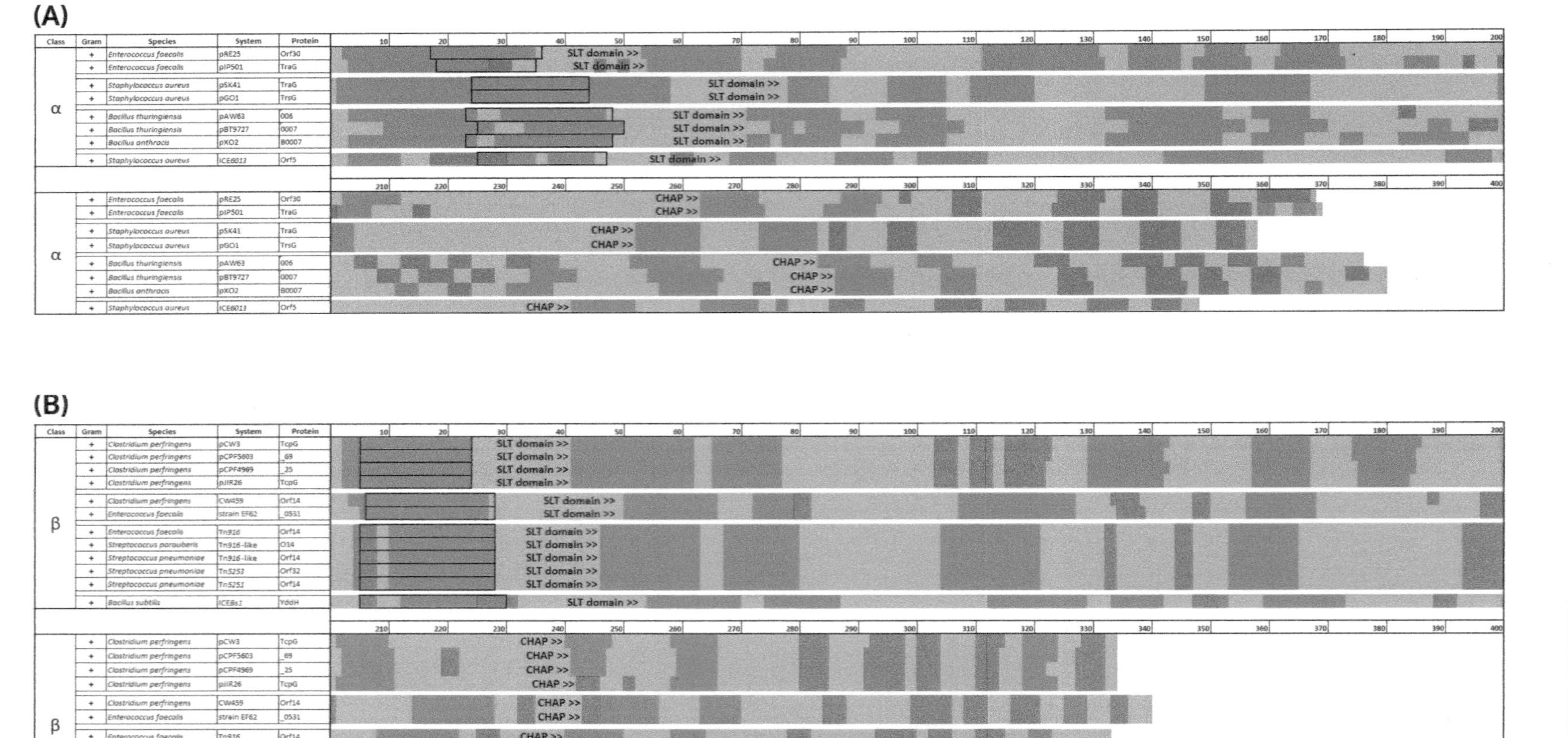

Figure 4 Secondary structure-based classification of lytic transglycosylases. A, class α, B, class β, C, class γ, D, class δ lytic transglycosylases. Secondary structure (PSIpred) and TM motif (HMMTOP) prediction for G– and G+ lytic transglycosylases from conjugative plasmids, transposons, ICEs, and GIs; alpha helices (blue), beta strands (red), and TM motifs (boxes) are highlighted; the putative N-terminal ends of the SLT and CHAP(-like) domains are indicated. doi:10.1128/microbiolspec.PLAS-0004-2013.f4.

Figure 4 *continues on next page*

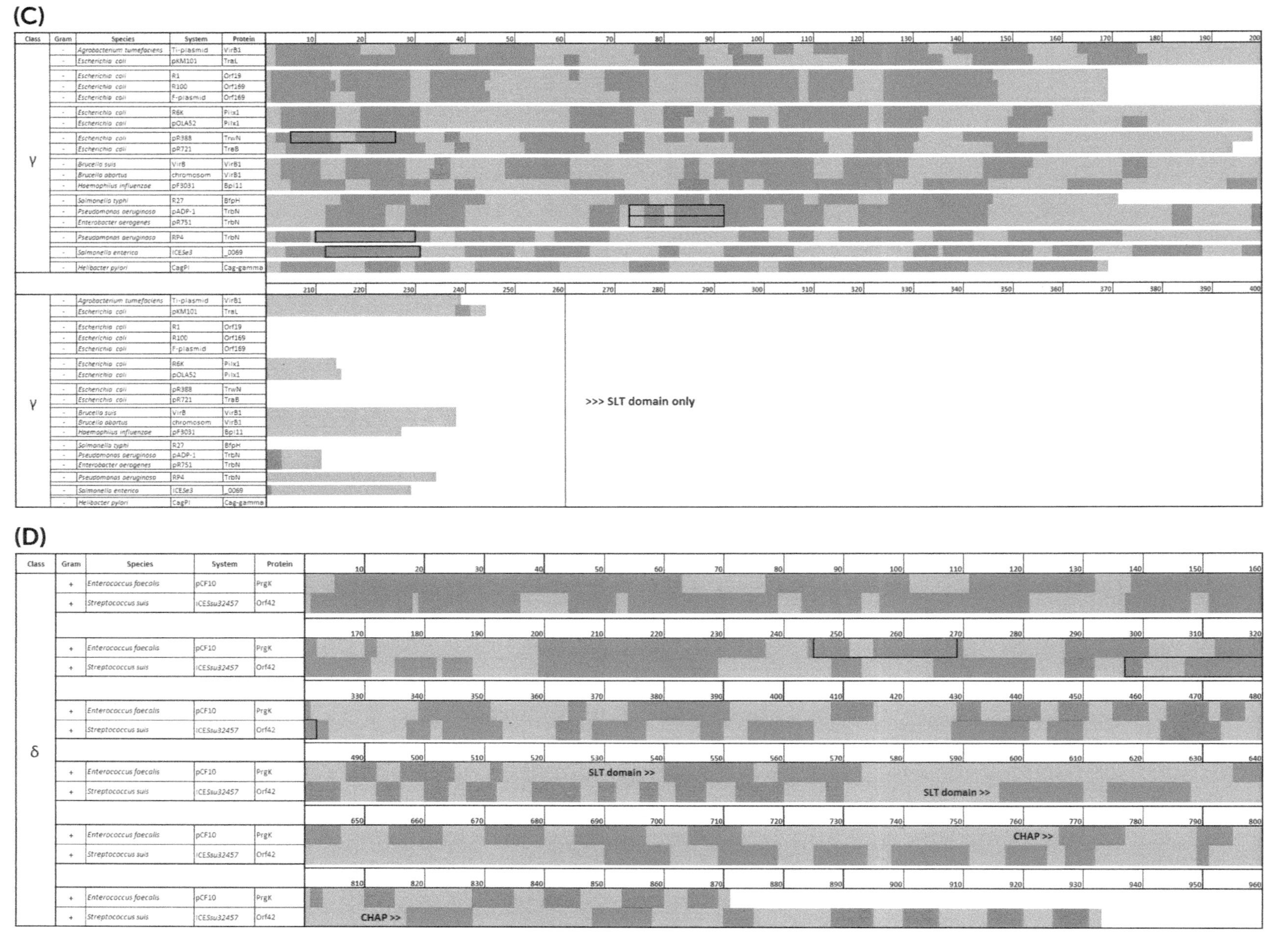

Figure 4 *continued*

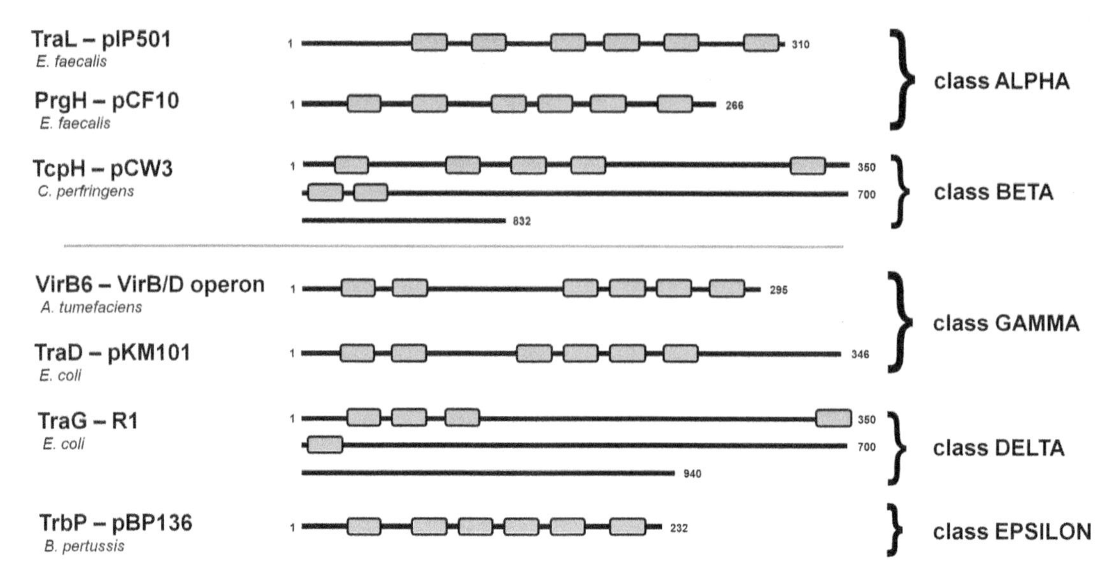

Figure 5 Comparison of the domain arrangement and classification of VirB6-like proteins from G+ and G– putative T4SSs. TraL (pIP501), PrgH (pCF10), TcpH (pCW3), VirB6 (Ti-plasmid), TraD (pKM101), TraG (R1 plasmid), and TrbP (pBP136) were selected as representatives of their respective classes. Predicted TMHs (HMMTOP [53]) are represented as gray boxes. doi:10.1128/microbiolspec.PLAS-0004-2013.f5.

conjugative DNA transfer system clearly distinct from the classical conjugative transfer systems involving a single-stranded DNA molecule and a T4SS. In *Streptomyces* matings, only a single plasmid-encoded protein, TraB, is required to translocate a double-stranded circular DNA molecule into the recipient (88). Recently a potentially distinct mode for the transfer of double-stranded linear plasmid DNA in *Streptomyces* has been postulated (89).

The Classical *Streptomyces* Mode: Circular Double-Stranded DNA Transfer

Streptomycetes feature a unique conjugative DNA transfer system relying on double-stranded DNA translocation that is performed by a single protein, TraB, related to FtsK, mediating chromosome segregation during bacterial cell division (25). The first experimental evidence for the distinct transfer mechanism of *Streptomyces* plasmids came from the work of Possoz et al. (90) demonstrating that conjugation of the *Streptomyces ambofaciens* plasmid pSAM2 was sensitive to the presence of the *Sal*I restriction/modification system in the recipient indicating that the transferred incoming DNA must be double stranded (90). Probably owing to the toxic effects of the transfer determinants, plasmid transfer in *Streptomyces* becomes apparent by the formation of so-called pock structures with a diameter of 1 to 3 mm. Pocks are generated when donor spores germinate on a lawn of a plasmid-free recipient. The pocks are temporally retarded growth inhibition zones and indicate the area where the recipient mycelium has acquired a plasmid by conjugative transfer (88). Formation of pock structures has been interpreted as the result of intramycelial plasmid spreading via the septal cross-walls of the recipient mycelium (11, 88, 91).

In intramycelial plasmid spreading additional plasmid-encoded factors, termed Spd proteins, appear to be involved (88). The characteristics of the TraB and Spd proteins are presented in the next paragraphs.

TraB Is Sufficient for Circular Double-Stranded DNA Transfer

The TraB protein family—up to now, only members in *Streptomyces* plasmids have been detected—was shown to be a highly diverse family of proteins with only very limited sequence similarity (88). However, secondary structure predictions showed identical domain architecture for all TraB homologs resembling that of FtsK: an amino-terminal membrane association domain that is followed by a translocase/ATPase domain with Walker A and B motifs and a carboxy-terminal winged helix-turn-helix fold (26). ATPase activity and membrane association have been demonstrated for TraB proteins of various plasmids (88, 92, 93, 94). Inactivation of the ATP binding site of TraB from the *Streptomyces nigrifaciens* plasmid pSN22 confirmed that the ATPase activity is essential for conjugative transfer (92).

TraB of the *Streptomyces venezuelae* plasmid pSVH1 was expressed and purified in *Streptomyces lividans* (94). Chemical cross-linking revealed higher oligomeric structures that were also observed when the membrane association domain of TraB was deleted. Ring-shaped TraB particles were detected by electron microscopy, the images revealed symmetric hexamers of approximately 12 nm in diameter containing a central pore (88). This structure fits with a predicted TraB-DNA-translocase structure obtained by homology modeling with the *P. aeruginosa* FtsK translocase domain crystal structure as a template (26). Both structures contained a central pore of 3.0 and 3.1 nm, respectively, which is sufficiently wide to accommodate a double-stranded DNA molecule (88).

To accomplish DNA transfer from the donor to the recipient cell, the plasmid DNA has to pass the cell envelopes of two *Streptomyces* cells, meaning the cytoplasmic membrane and PG layers of both partners (88). PG binding by TraB was demonstrated by Vogelmann et al. (26). For the membrane passage, a pore structure has been postulated for TraB (88). Studies using planar lipid bilayers demonstrated that TraB inserted into the membrane at various voltages and formed pores with an opening time of about 47 to 81 ms (positive voltage applied) and 105 to 200 ms, respectively, when a negative voltage was applied (26).

Only a short DNA region, denominated *clt* locus, consisting of 8-bp repeats, is required *in cis* on the plasmid for TraB-mediated DNA transfer (88). Electrophoretic mobility shift assays (EMSAs) showed specific binding of TraB to a plasmid region at the 3′ end of *traB*, which represents the *clt* region of pSVH1 (94). The clt_{pSVH1} region contains nine imperfectly conserved copies of the GACCCGGA sequence. In other *Streptomyces* plasmids also, specific 8-bp repeats have been detected in the (predicted) *clt* regions (26, 88, 95). With the exception of two plasmids, the *clt* locus localizes in all *Streptomyces* plasmids to the 3′ end of *traB*, forming a 2.5-kb *tra* module consisting of *traB* and *clt* next to it (88). EMSAs applying covalently closed circular (ccc) DNA as a substrate revealed that TraB binds noncovalently to plasmid DNA and that the plasmid DNA molecule was not processed by TraB binding (94).

Intramycelial DNA Spread Requires Plasmid-Encoded Spd Proteins

Whereas plasmid transfer from a donor into the recipient requires only TraB, plasmid spreading involves five to seven plasmid-encoded Spd proteins in addition to TraB. This likely reflects the challenge to cross the *Streptomyces* septal cross-walls. The Spd proteins show no significant similarity to any functionally characterized protein (88). Inactivation of a single *spd* gene diminishes the size of the pock structures (88, 96, 97, 98, 99). In addition, the genetic organization of the *spd* genes with overlapping stop and start codons and detection of protein-protein interactions indicated that the Spd proteins form a multiprotein complex together with TraB (31, 88). How this putative multiprotein complex promotes intramycelial plasmid spreading remains to be elucidated.

A POTENTIAL NOVEL LINEAR DNA TRANSFER MECHANISM IN *STREPTOMYCES*

The ability of the *Streptomyces* plasmid pIJ101 transfer apparatus to promote conjugative transfer of double-stranded circular versus double-stranded linear versions of the same plasmid was compared by Wang and Pettis (89). While the pIJ101 *tra* locus comprising the translocase gene *traB* and *clt* site readily transferred the circular form of the plasmid, the linear version was transferred orders of magnitude less efficiently and all plasmids isolated from the transconjugants were circular, regardless of their original configuration in the donor (89). Moreover, relatively rare circularization of linear plasmids was observed in the donor cells, which coincides with the notion that circularization is a prerequisite for transfer mediated by $\mathrm{TraB}_{\mathrm{pIJ101}}$. Interestingly, the linear version of the same replicon transferred efficiently from donors harboring the conjugative linear plasmid SLP2 (89).

Only a little information is available about conjugative transfer of *Streptomyces* linear plasmids. Some linear plasmids encode a SpoIIIE/FtsK homolog, which, in the case of SLP2, allows efficient conjugative transfer not only of linear SLP2, but also of its circularized derivatives (100). For the transfer of linear plasmids in their natural configuration, an end-first model has been postulated where transfer is initiated from a terminal protein-capped end (101). It is unknown whether linear plasmids also contain *clt* loci. However, additional functions potentially unique to the transfer of linear plasmids have been identified (100, 102, 103). These functions include *ttrA*, a putative helicase gene present near the end of the SLP2 genome (103, 104). It was demonstrated that the *ttrA* gene of SLP2 was important for SLP2 transfer (103).

Wang and Pettis suggested that functions that were sufficient for efficient transfer of circular DNA in

Streptomyces were insufficient for effective transfer of linear DNA, strengthening the notion that the conjugative transfer mechanisms of linear versus circular DNA in *Streptomyces* are inherently different (89).

CONJUGATIVE PLASMIDS AND BIOFILMS

Biofilms are the predominant mode of life for bacteria in nature. Bacteria living in biofilms have been shown to be better protected from harmful impacts from their environment than their planktonic counterparts (105, 106) and adapt more readily to environmental changes via specialized communication systems, denominated quorum sensing (107, 108). The formation of a stable mature biofilm is the product of social interactions that have evolved through a series of adaptations (109). In past decades, a growing number of studies showed that biology of conjugative plasmids and biofilm community structure and functions are intertwined through many complex interactions, ranging from the genetic level to the community level (109). There is growing evidence that conjugative plasmids can promote the formation of biofilms or at least increase or accelerate their formation through genetic traits encoded on their genomes (109, 110, 111).

Conjugative plasmid transfer in biofilms was experimentally shown for the first time by Hausner and Wuertz in 1999 (112). In 2001, Ghigo claimed that conjugative plasmids themselves can encode traits that induce biofilm formation of planktonic bacteria (113). Numerous experimental data provided evidence that conjugative transfer occurs at higher frequencies between members of biofilm communities than between bacteria in a planktonic state (109, 112, 114).

Conjugation between Gram-Positive Bacteria in Biofilms

Cook and colleagues demonstrated that growth in biofilms can alter the induction of conjugative transfer by a sex pheromone in *E. faecalis* harboring plasmid pCF10. Variations in pCF10 copy number and a bimodal response to the induction of conjugative transfer in populations of pCF10-harboring donor cells were observed in biofilms. Cook and coworkers argued that the *Enterococcus* pheromone system may have evolved such that donors in biofilms are only induced to transfer when they are in extremely close proximity to potential recipients in the biofilm (115).

There is growing evidence for conjugative plasmid transfer among G+ bacteria in natural biofilms. Sedgley and coworkers provided evidence that the conjugative erythromycin-resistance plasmid pAM81 could transfer efficiently between *Streptococcus gordonii* and *E. faecalis* in biofilms in the root canals of human teeth (116).

In general, higher plasmid transfer frequencies are observed in biofilms. Nevertheless, there exist spatial constraints within biofilms that may hinder the spread of plasmids in an already-established biofilm (117). Krol and coworkers showed how the transfer of an IncP-1 plasmid has spatial and nutritional constraints and occurred predominantly in the aerobic zone in an *E. coli* biofilm (118). Madsen and colleagues speculated that a prerequisite for successful transfer of certain plasmids in a biofilm community might be that the plasmid is present in the initial phases of biofilm formation. This can be accomplished if the biofilm priming traits are encoded by the plasmid itself (109).

MONITORING OF CONJUGATIVE PLASMID TRANSFER

A variety of monitoring techniques, either PCR based or fluorescence microscopy based, have been designed in the past two decades to quantitatively follow conjugative plasmid transfer under different conditions and in distinct environments.

In complex microbial communities with a high background of antibiotic resistance genes, the detection of conjugative transfer of resistance genes is challenging. One option to overcome the problem is labeling the antibiotic resistance gene. This approach was performed by Haug et al. (119). The conjugative multiresistance plasmid pRE25, originally isolated from *E. faecalis,* whose transfer region is virtually identical to that of pIP501 (120), was tagged with a 34-bp random sequence marker spliced by *tet*(M). The plasmid construct, denominated pRE25*, was transformed into *E. faecalis* CG110/*gfp*, containing a *gfp* gene as a chromosomal marker. pRE25* was shown to be fully functional compared with its parental pRE25 and could be transferred to *Listeria monocytogenes* and *Listeria innocua* at frequencies of 6×10^{-6} to 8×10^{-8} transconjugants per donor. Different markers on the chromosome and the plasmid enabled independent quantification of donor and plasmid via quantitative PCR. Haug and coworkers concluded that *E. faecalis* CG110/*gfp*/pRE25* is a potent tool for the study of conjugative resistance transfer in complex environments such as biofilms or food matrices (119).

Over the past decade, advances in reporter gene technology have provided new insights into the extent and spatial frequency of conjugative transfer *in vitro* and in natural environments (114, 121). This methodology involves the integration of genes encoding reporter proteins such as GFP in the conjugative plasmid of interest. In this way, the fate of plasmids in a bacterial community can be monitored *in situ* nondestructively. By this approach, conjugative spread of different plasmids was monitored in a variety of environments including agar-surface grown colonies (122, 123, 124), biofilm model systems (112, 118, 122, 125), freshwater microcosms (126), or plant leaves (127). Nancharaiah and coworkers were the first to use a dual-labeling technique involving GFP and the red fluorescent protein (DsRed) for *in situ* monitoring of conjugation (128). However, most monitoring tools have been developed for G– bacteria and consequently most studies have been performed among G– bacteria.

Recently, fluorescence monitoring tools for conjugative transfer among G+ bacteria have been developed. Arends and coworkers used a dual-fluorescence approach comprising two differently labeled plasmids, a mobilizable GFP-labeled plasmid based on the *tra* region of broad-host-range plasmid pIP501 and an RFP-labeled nonmobilizable plasmid ([129]; K. Arends, K. Schiwon, and E. Grohmann, unpublished data). With the use of this approach, conjugative transfer among distinct G+ bacteria and from G+ *E. faecalis* to *E. coli* could be visualized. A similar approach was applied for monitoring conjugative plasmid transfer among different G+ bacteria by the use of a GFP-labeled mobilizable plasmid, a CFPopt-labeled nonmobilizable plasmid, and a YFP-labeled conjugative plasmid (P. Modrie and J. Mahillon, unpublished results).

CONCLUSIONS AND PERSPECTIVES

Due to the steady increase of antibiotic-resistant bacteria, which is not restricted to hospitals and health care centers, the elucidation of the mechanistic details of conjugative resistance transfer is an issue of primordial importance. Through the application of new structural biology, cell biology, and microscopic tools, enormous progress has been made toward an understanding of the assembly and functioning of the G– T4SSs required for conjugative DNA spread. Considerable progress has been also made in the characterization of DNA transfer systems originating from G+ bacteria, particularly on the plasmid-encoded T4SSs of *Enterococcus* and *Clostridium* origin, both have evolved into prototype systems and serve as reference for studies in G+ pathogens such as the streptococci in which a multitude of strains with so-called minimized T4SSs have been detected recently (17, 130).

In recent biochemical and structural studies, significant differences between the G+ and G– transfer machineries concerning the putative membrane channel became evident. Most of them might be explained by the distinct structure and composition of the G+ and G– cell envelopes. Clear evidence for this argument has been provided by the lack of the G– core complex components, VirB7, VirB9, and VirB10 in the G+ systems analyzed so far. In this respect, to advance our understanding of the DNA transport through the G+ cell envelope, the identification and structure solution of the G+ T4SS core complex are urgently required.

In past years, a revival of interest in plasmid transfer and its relevance in nature, particularly in complex microbial communities such as biofilms, was noticed. Based on the advances in reporter gene technology and high-resolution microscopic tools for spatial and temporal analysis of microbial communities, it became possible to quantify plasmid transfer in complex bacterial communities which enabled the demonstration of high-frequency plasmid transfer, e.g., in biofilms, in freshwater microcosms, and in dental plaque of human teeth.

In summary, we expect that in the near future a global picture of the mechanism and the hot spots of conjugative transfer in the community and in the environment will emerge that will aid in designing countermeasures to fight antibiotic resistance spread to maintain the efficiency of antimicrobial drugs.

Acknowledgments. *We regret that not all valuable contributions of colleagues in the field could have been included due to space limitation. We thank Guenther Koraimann for his valuable comments on the manuscript. Conflicts of interest: We disclose no conflicts.*

Citation. **Goessweiner-Mohr N, Arends K, Keller W, Grohmann E.** 2014. Conjugation in Gram-positive bacteria. Microbiol Spectrum **2**(4):PLAS-0004-2013.

References

1. **Babic A, Berkmen MB, Lee CA, Grossman AD.** 2011. Efficient gene transfer in bacterial cell chains. *mBio* **2**: e00027-11. doi:10.1128/mBio.00027-11.
2. **Frost LS, Leplae R, Summers AO, Toussaint A.** 2005. Mobile genetic elements: the agents of open source evolution. *Nat Rev Microbiol* **3**:722–732.
3. **Ochman H, Lawrence JG, Groisman EA.** 2000. Lateral gene transfer and the nature of bacterial innovation. *Nature* **405**:299–304.
4. **Thomas CM, Nielsen KM.** 2005. Mechanisms of, and barriers to, horizontal gene transfer between bacteria. *Nat Rev Microbiol* **3**:711–721.
5. **Wozniak RAF, Waldor MK.** 2010. Integrative and conjugative elements: mosaic mobile genetic elements enabling dynamic lateral gene flow. *Nat Rev Microbiol* **8**: 552–563.

6. **Burrus V, Pavlovic G, Decaris B, Guédon G.** 2002. Conjugative transposons: the tip of the iceberg. *Mol Microbiol* **46:**601–610.
7. **Burrus V, Waldor MK.** 2004. Shaping bacterial genomes with integrative and conjugative elements. *Res Microbiol* **155:**376–386.
8. **Roberts AP, Mullany P.** 2009. A modular master on the move: the Tn*916* family of mobile genetic elements. *Trends Microbiol* **17:**251–258.
9. **de la Cruz F, Frost LS, Meyer RJ, Zechner EL.** 2010. Conjugative DNA metabolism in Gram-negative bacteria. *FEMS Microbiol Rev* **34:**18–40.
10. **Fernández-López R, Garcillán-Barcia MP, Revilla C, Lázaro M, Vielva L, de la Cruz F.** 2006. Dynamics of the IncW genetic backbone imply general trends in conjugative plasmid evolution. *FEMS Microbiol Rev* **30:**942–966.
11. **Grohmann E, Muth G, Espinosa M.** 2003. Conjugative plasmid transfer in gram-positive bacteria. *Microbiol Mol Biol Rev* **67:**277–301.
12. **Schröder G, Lanka E.** 2005. The mating pair formation system of conjugative plasmids—a versatile secretion machinery for transfer of proteins and DNA. *Plasmid* **54:**1–25.
13. **Smillie C, Garcillan-Barcia MP, Francia MV, Rocha EPC, de la Cruz F.** 2010. Mobility of Plasmids. *Microbiol Mol Biol Rev* **74:**434–452.
14. **Grohmann E, Guzmán L, Espinosa M.** 1999. Mobilisation of the streptococcal plasmid pMV158: interactions of MobM protein with its cognate *oriT* DNA region. *Mol Gen Genet* **261:**707–715.
15. **Guzman LM, Espinosa M.** 1997. The mobilization protein, MobM, of the streptococcal plasmid pMV158 specifically cleaves supercoiled DNA at the plasmid *oriT*. *J Mol Biol* **266:**688–702.
16. **Zechner EL, Lang S, Schildbach JF.** 2012. Assembly and mechanisms of bacterial type IV secretion machines. *Philos Trans R Soc Lond B Biol Sci* **367:**1073–1087.
17. **Bhatty M, Laverde Gomez JA, Christie PJ.** 2013. The expanding bacterial type IV secretion lexicon. *Res Microbiol* **164:**620–639.
18. **Chandran V, Fronzes R, Duquerroy S, Cronin N, Navaza J, Waksman G.** 2009. Structure of the outer membrane complex of a type IV secretion system. *Nature* **462:**1011–1015.
19. **Fronzes R, Schafer E, Wang L, Saibil HR, Orlova EV, Waksman G.** 2009. Structure of a type IV secretion system core complex. *Science* **323:**266–268.
20. **Rivera-Calzada A, Fronzes R, Savva CG, Chandran V, Lian PW, Laeremans T, Pardon E, Steyaert J, Remaut H, Waksman G, Orlova EV.** 2013. Structure of a bacterial type IV secretion core complex at subnanometer resolution. *EMBO J* **32:**1195–1204.
21. **Vincent CD, Friedman JR, Jeong KC, Buford EC, Miller JL, Vogel JP.** 2006. Identification of the core transmembrane complex of the *Legionella* Dot/Icm type IV secretion system. *Mol Microbiol* **62:**1278–1291.
22. **Waksman G, Fronzes R.** 2010. Molecular architecture of bacterial type IV secretion systems. *Trends Biochem Sci* **35:**691–698.
23. **Wallden K, Williams R, Yan J, Lian PW, Wang L, Thalassinos K, Orlova EV, Waksman G.** 2012. Structure of the VirB4 ATPase, alone and bound to the core complex of a type IV secretion system. *Proc Natl Acad Sci USA* **109:**11348–11353.
24. **Bordeleau E, Ghinet MG, Burrus V.** 2012. Diversity of integrating conjugative elements in actinobacteria: coexistence of two mechanistically different DNA-translocation systems. *Mob Genet Elements* **2:**119–124.
25. **Sepulveda E, Vogelmann J, Muth G.** 2011. A septal chromosome segregator protein evolved into a conjugative DNA-translocator protein. *Mob Genet Elements* **1:** 225–229.
26. **Vogelmann J, Ammelburg M, Finger C, Guezguez J, Linke D, Flötenmeyer M, Stierhof Y-D, Wohlleben W, Muth G.** 2011. Conjugal plasmid transfer in *Streptomyces* resembles bacterial chromosome segregation by FtsK/SpoIIIE. *EMBO J* **30:**2246–2254.
27. **Abajy MY, Kopec J, Schiwon K, Burzynski M, Doring M, Bohn C, Grohmann E.** 2007. A type IV-secretion-like system is required for conjugative DNA transport of broad-host-range plasmid pIP501 in gram-positive bacteria. *J Bacteriol* **189:**2487–2496.
28. **Li J, Adams V, Bannam TL, Miyamoto K, Garcia JP, Uzal FA, Rood JI, McClane BA.** 2013. Toxin plasmids of *Clostridium perfringens*. *Microbiol Mol Biol Rev* **77:** 208–233.
29. **Steen JA, Bannam TL, Teng WL, Devenish RJ, Rood JI.** 2009. The putative coupling protein TcpA interacts with other pCW3-encoded proteins to form an essential part of the conjugation complex. *J Bacteriol* **191:**2926–2933.
30. **Brolle D-F, Pape H, Hopwood DA, Kieser T.** 1993. Analysis of the transfer region of the *Streptomyces* plasmid SCP2*. *Mol Microbiol* **10:**157–170.
31. **Tiffert Y, Gotz B, Reuther J, Wohlleben W, Muth G.** 2007. Conjugative DNA transfer in *Streptomyces*: SpdB2 involved in the intramycelial spreading of plasmid pSVH1 is an oligomeric integral membrane protein that binds to dsDNA. *Microbiology* **153:**2976–2983.
32. **Brantl S, Behnke D, Alonso JC.** 1990. Molecular analysis of the replication region of the conjugative *Streptococcus agalactiae* plasmid pIP501 in *Bacillus subtilis*. Comparison with plasmids pAMβ1 and pSM 19035. *Nucleic Acids Res* **18:**4783–4790.
33. **Dunny GM.** 2007. The peptide pheromone-inducible conjugation system of *Enterococcus faecalis* plasmid pCF10: cell-cell signalling, gene transfer, complexity and evolution. *Philos Trans R Soc Lond B Biol Sci* **362:** 1185–1193.
34. **Kozlowicz BK, Shi K, Gu Z-Y, Ohlendorf DH, Earhart CA, Dunny GM.** 2006. Molecular basis for control of conjugation by bacterial pheromone and inhibitor peptides. *Mol Microbiol* **62:**958–969.
35. **Alvarez-Martinez CE, Christie PJ.** 2009. Biological diversity of prokaryotic type IV secretion systems. *Microbiol Mol Biol Rev* **73:**775–808.
36. **Jacob A, Hobbs S.** 1974. Conjugal transfer of plasmid-borne multiple antibiotic resistance in *Streptococcus faecalis* var. *zymogenes*. *J Bacteriol* **117:**360–372.

37. **Kurenbach B, Bohn C, Prabhu J, Abudukerim M, Szewzyk U, Grohmann E.** 2003. Intergeneric transfer of the *Enterococcus faecalis* plasmid pIP501 to *Escherichia coli* and *Streptomyces lividans* and sequence analysis of its *tra* region. *Plasmid* **50:**86–93.

38. **Kopec J, Bergmann A, Fritz G, Grohmann E, Keller W.** 2005. TraA and its N-terminal relaxase domain of the Gram-positive plasmid pIP501 show specific *oriT* binding and behave as dimers in solution. *Biochem J* **387:** 401–409.

39. **Kurenbach B, Grothe D, Farias ME, Szewzyk U, Grohmann E.** 2002. The *tra* region of the conjugative plasmid pIP501 is organized in an operon with the first gene encoding the relaxase. *J Bacteriol* **184:**1801–1805.

40. **Kurenbach B, Kopéc J, Mägdefrau M, Andreas K, Keller W, Bohn C, Abajy MY, Grohmann E.** 2006. The TraA relaxase autoregulates the putative type IV secretion-like system encoded by the broad-host-range *Streptococcus agalactiae* plasmid pIP501. *Microbiology* **152:**637–645.

41. **Chen Y, Staddon JH, Dunny GM.** 2007. Specificity determinants of conjugative DNA processing in the *Enterococcus faecalis* plasmid pCF10 and the *Lactococcus lactis* plasmid pRS01. *Mol Microbiol* **63:**1549–1564.

42. **Parker C, Meyer RJ.** 2007. The R1162 relaxase/primase contains two, type IV transport signals that require the small plasmid protein MobB. *Mol Microbiol* **66:** 252–261.

43. **Ragonese H, Haisch D, Villareal E, Choi J-H, Matson SW.** 2007. The F plasmid-encoded TraM protein stimulates relaxosome-mediated cleavage at *oriT* through an interaction with TraI. *Mol Microbiol* **63:**1173–1184.

44. **Dang TA, Zhou XR, Graf B, Christie PJ.** 1999. Dimerization of the *Agrobacterium tumefaciens* VirB4 ATPase and the effect of ATP-binding cassette mutations on the assembly and function of the T-DNA transporter. *Mol Microbiol* **32:**1239–1253.

45. **Durand E, Waksman G, Receveur-Brechot V.** 2011. Structural insights into the membrane-extracted dimeric form of the ATPase TraB from the *Escherichia coli* pKM101 conjugation system. *BMC Struct Biol* **11:**4. doi:10.1186/1472-6807-11-4.

46. **Li F, Alvarez-Martinez C, Chen Y, Choi K-J, Yeo H-J, Christie PJ.** 2012. *Enterococcus faecalis* PrgJ, a VirB4-like ATPase, mediates pCF10 conjugative transfer through substrate binding. *J Bacteriol* **194:**4041–4051.

47. **Pena A, Matilla I, Martin-Benito J, Valpuesta JM, Carrascosa JL, de La Cruz F, Cabezon E, Arechaga I.** 2012. The hexameric structure of a conjugative VirB4 protein ATPase provides new insights for a functional and phylogenetic relationship with DNA translocases. *J Biol Chem* **287:**39925–39932.

48. **Gomis-Rüth FX, Moncalían G, de la Cruz F, Coll M.** 2001. Conjugative plasmid protein TrwB, an integral membrane type IV secretion system coupling protein. *J Biol Chem* **277:**7556–7566.

49. **Gomis-Rüth FX, Solà M, de la Cruz F, Coll M.** 2004. Coupling factors in macromolecular type-IV secretion machineries. *Curr Pharm Des* **10:**1551–1565.

50. **Haft RJF, Gachelet EG, Nguyen T, Toussaint L, Chivian D, Traxler B.** 2007. In vivo oligomerization of the F conjugative coupling protein TraD. *J Bacteriol* **189:**6626–6634.

51. **Vecino AJ, de La Arada I, Segura RL, Goñi FM, de la Cruz F, Arrondo JL, Alkorta I.** 2011. Membrane insertion stabilizes the structure of TrwB, the R388 conjugative plasmid coupling protein. *Biochim Biophys Acta Biomembr* **1808:**1032–1039.

52. **Arends K, Celik E-K, Probst I, Goessweiner-Mohr N, Fercher C, Grumet L, Soellue C, Abajy MY, Sakinc T, Broszat M, Schiwon K, Koraimann G, Keller W, Grohmann E.** 2013. TraG encoded by the pIP501 type IV secretion system is a two domain peptidoglycan degrading enzyme essential for conjugative transfer. *J Bacteriol* **195:**4436–4444.

53. **Tusnady GE, Simon I.** 2001. The HMMTOP transmembrane topology prediction server. *Bioinformatics* **17:** 849–850.

54. **Goessweiner-Mohr N, Grumet L, Arends K, Pavkov-Keller T, Gruber CC, Gruber K, Birner-Gruenberger R, Kropec-Huebner A, Huebner J, Grohmann E, Keller W.** 2013. The 2.5 A structure of the *Enterococcus* conjugation protein TraM resembles VirB8 type IV secretion proteins. *J Biol Chem* **288:**2018–2028.

55. **Bailey S, Ward D, Middleton R, Grossmann JG, Zambryski PC.** 2006. *Agrobacterium tumefaciens* VirB8 structure reveals potential protein-protein interaction sites. *Proc Natl Acad Sci USA* **103:**2582–2587.

56. **Terradot L, Bayliss R, Oomen C, Leonard GA, Baron C, Waksman G.** 2005. Structures of two core subunits of the bacterial type IV secretion system, VirB8 from *Brucella suis* and ComB10 from *Helicobacter pylori*. *Proc Natl Acad Sci USA* **102:**4596–4601.

57. **Porter CJ, Bantwal R, Bannam TL, Rosado CJ, Pearce MC, Adams V, Lyras D, Whisstock JC, Rood JI.** 2012. The conjugation protein TcpC from *Clostridium perfringens* is structurally related to the type IV secretion system protein VirB8 from Gram-negative bacteria. *Mol Microbiol* **83:**275–288.

58. **Jakubowski SJ, Cascales E, Krishnamoorthy V, Christie PJ.** 2005. *Agrobacterium tumefaciens* VirB9, an outer-membrane-associated component of a type IV secretion system, regulates substrate selection and T-pilus biogenesis. *J Bacteriol* **187:**3486–3495.

59. **Judd PK, Kumar RB, Das A.** 2005. Spatial location and requirements for the assembly of the *Agrobacterium tumefaciens* type IV secretion apparatus. *Proc Natl Acad Sci USA* **102:**11498–11503.

60. **Judd PK, Mahli D, Das A.** 2005. Molecular characterization of the *Agrobacterium tumefaciens* DNA transfer protein VirB6. *Microbiology* **151:**3483–3492.

61. **Judd PK, Kumar RB, Das A.** 2005. The type IV secretion apparatus protein VirB6 of *Agrobacterium tumefaciens* localizes to a cell pole. *Mol Microbiol* **55:** 115–124.

62. **Villamil Giraldo AM, Sivanesan D, Carle A, Paschos A, Smith MA, Plesa M, Coulton J, Baron C.** 2012. Type IV secretion system core component VirB8 from *Brucella*

binds to the globular domain of VirB5 and to a periplasmic domain of VirB6. *Biochemistry* **51**:3881–3890.

63. **Wallden K, Rivera-Calzada A, Waksman G.** 2010. Microreview: type IV secretion systems: versatility and diversity in function. *Cell Microbiol* **12**:1203–1212.
64. **Navarre WW, Schneewind O.** 1999. Surface proteins of gram-positive bacteria and mechanisms of their targeting to the cell wall envelope. *Microbiol Mol Biol Rev* **63**:74–229.
65. **Dawson P, Clancy KW, Melvin JA, McCafferty DG.** 2010. Sortase transpeptidases: insights into mechanism, substrate specificity, and inhibition. *Biopolymers* **94**: 385–396.
66. **Hendrickx AP, Willems RJ, Bonten MJ, van Schaik W.** 2009. LPxTG surface proteins of enterococci. *Trends Microbiol* **17**:423–430.
67. **Krishnan V, Narayana SV.** 2011. Crystallography of gram-positive bacterial adhesins. *Adv Exp Med Biol* **715**:175–195.
68. **Clewell DB.** 2007. Properties of *Enterococcus faecalis* plasmid pAD1, a member of a widely disseminated family of pheromone-responding, conjugative, virulence elements encoding cytolysin. *Plasmid* **58**:205–227.
69. **Francia MV, Varsaki A, Garcillán-Barcia MP, Latorre A, Drainas C, de La Cruz F.** 2004. A classification scheme for mobilization regions of bacterial plasmids. *FEMS Microbiol Rev* **28**:79–100.
70. **Francia MV, Clewell DB.** 2002. Transfer origins in the conjugative *Enterococcus faecalis* plasmids pAD1 and pAM373: identification of the pAD1 *nic* site, a specific relaxase and a possible TraG-like protein. *Mol Microbiol* **45**:375–395.
71. **Francia MV, Clewell DB, de La Cruz F, Moncalian G.** 2013. Catalytic domain of plasmid pAD1 relaxase TraX defines a group of relaxases related to restriction endonucleases. *Proc Natl Acad Sci USA* **110**:13606–13611.
72. **Chen Y, Zhang X, Manias D, Yeo H-J, Dunny GM, Christie PJ.** 2008. *Enterococcus faecalis* PcfC, a spatially localized substrate receptor for type IV secretion of the pCF10 transfer intermediate. *J Bacteriol* **190**: 3632–3645.
73. **Staddon JH, Bryan EM, Manias DA, Chen Y, Dunny GM.** 2006. Genetic characterization of the conjugative DNA processing system of enterococcal plasmid pCF10. *Plasmid* **56**:102–111.
74. **Hirt H, Manias DA, Bryan EM, Klein JR, Marklund JK, Staddon JH, Paustian ML, Kapur V, Dunny GM.** 2005. Characterization of the pheromone response of the *Enterococcus faecalis* conjugative plasmid pCF10: complete sequence and comparative analysis of the transcriptional and phenotypic responses of pCF10-containing cells to pheromone induction. *J Bacteriol* **187**:1044–1054.
75. **Olmsted SB, Kao S-M, van Putte LJ, Gallo JC, Dunny GM.** 1991. Role of the pheromone-inducible surface protein AsclO in mating aggregate formation and conjugal transfer of the *Enterococcus faecalis* plasmid pCF10. *J Bacteriol* **173**:7665–7672.
76. **Waters CM, Dunny GM.** 2001. Analysis of functional domains of the *Enterococcus faecalis* pheromone-induced surface protein aggregation substance. *J Bacteriol* **183**: 5659–5667.
77. **Waters CM, Wells CL, Dunny GM.** 2003. The aggregation domain of aggregation substance, not the RGD motifs, is critical for efficient internalization by HT-29 enterocytes. *Infect Immun* **71**:5682–5689.
78. **Olmsted SB, Erlandsen SL, Dunny GM, Wells CL.** 1993. High-resolution visualization by field emission scanning electron microscopy of *Enterococcus faecalis* surface proteins encoded by the pheromone-inducible conjugative plasmid pCF10. *J Bacteriol* **175**:6229–6237.
79. **Vollmer W, Seligman SJ.** 2010. Architecture of peptidoglycan: more data and more models. *Trends Microbiol* **18**:59–66.
80. **Hoppner C.** 2005. The putative lytic transglycosylase VirB1 from *Brucella suis* interacts with the type IV secretion system core components VirB8, VirB9 and VirB11. *Microbiology* **151**:3469–3482.
81. **Hoppner C, Liu Z, Domke N, Binns AN, Baron C.** 2004. VirB1 orthologs from *Brucella suis* and pKM101 complement defects of the lytic transglycosylase required for efficient type IV secretion from *Agrobacterium tumefaciens*. *J Bacteriol* **186**:1415–1422.
82. **Cascales E, Christie PJ.** 2004. Definition of a bacterial type IV secretion pathway for a DNA substrate. *Science* **304**:1170–1173.
83. **Jakubowski SJ, Krishnamoorthy V, Cascales E, Christie PJ.** 2004. *Agrobacterium tumefaciens* VirB6 domains direct the ordered export of a DNA substrate through a type IV secretion system. *J Mol Biol* **341**:961–977.
84. **Anthony KG, Klimke WA, Manchak J, Frost LS.** 1999. Comparison of proteins involved in pilus synthesis and mating pair stabilization from the related plasmids F and R100-1: insights into the mechanism of conjugation. *J Bacteriol* **181**:5149–5159.
85. **Audette GF, Manchak J, Beatty P, Klimke WA, Frost LS.** 2007. Entry exclusion in F-like plasmids requires intact TraG in the donor that recognizes its cognate TraS in the recipient. *Microbiology* **153**:442–451.
86. **Firth N, Skurray R.** 1992. Characterization of the F plasmid bifunctional conjugation gene, *traG*. *Mol Gen Genet* **232**:145–153.
87. **Manning PA, Morelli G, Achtman M.** 1981. TraG protein of the F sex factor of *Escherichia coli* K-12 and its role in conjugation. *Proc Natl Acad Sci USA* **78**: 7487–7491.
88. **Thoma L, Muth G.** 2012. Conjugative DNA transfer in *Streptomyces*. *FEMS Microbiol Lett* **337**:81–88.
89. **Wang J, Pettis GS.** 2010. The *tra* locus of streptomycete plasmid pIJ101 mediates efficient transfer of a circular but not a linear version of the same replicon. *Microbiology* **156**:2723–2733.
90. **Possoz C, Ribard C, Gagnat J, Pernodet J-L, Guérineau M.** 2001. The integrative element pSAM2 from *Streptomyces*: kinetics and mode of conjugal transfer. *Mol Microbiol* **42**:159–166.

91. **Hopwood DA, Kieser T.** 1993. Conjugative plasmids of *Streptomyces*, p 293–311. *In* Clewell DB (ed), *Bacterial Conjugation*. Plenum Press, New York.

92. **Kosono S, Kataoka M, Seki T, Yoshida T.** 1996. The TraB protein, which mediates the intermycelial transfer of the *Streptomyces* plasmid pSN22, has functional NTP-binding motifs and is localized to the cytoplasmic membrane. *Mol Microbiol* **19:**397–405.

93. **Pettis GS, Cohen SN.** 1996. Plasmid transfer and expression of the transfer (*tra*) gene product of plasmid pIJ101 are temporally regulated during the *Streptomyces lividans* life cycle. *Mol Microbiol* **19:**1127–1135.

94. **Reuther J, Gekeler C, Tiffert Y, Wohlleben W, Muth G.** 2006. Unique conjugation mechanism in mycelial *Streptomycetes*: a DNA-binding ATPase translocates unprocessed plasmid DNA at the hyphal tip. *Mol Microbiol* **61:**436–446.

95. **Franco B, Gonzalez-Ceron G, Servin-Gonzalez L.** 2003. Direct repeat sequences are essential for function of the *cis*-acting locus of transfer (*clt*) of *Streptomyces phaeochromogenes* plasmid pJV1. *Plasmid* **50:**242–247.

96. **Kataoka M, Kiyose YM, Michisuji Y, Horiguchi T, Seki T, Yoshida T.** 1994. Complete nucleotide sequence of the *Streptomyces nigrifaciens* plasmid, pSN22: genetic organization and correlation with genetic properties. *Plasmid* **32:**55–69.

97. **Kieser T, Hopwood DA, Wright HM, Thompson C.** 1982. pIJ101, a multi-copy broad host-range *Streptomyces* plasmid: functional analysis and development of DNA cloning vectors. *Mol Gen Genet* **185:**223–228.

98. **Reuther J, Wohlleben W, Muth G.** 2006. Modular architecture of the conjugative plasmid pSVH1 from *Streptomyces venezuelae*. *Plasmid* **55:**201–209.

99. **Servin-Gonzalez L, Sampieri A, Cabello J, Galvan L, Juarez V, Castro C.** 1995. Sequence and functional analysis of the *Streptomyces phaeochromogenes* plasmid pJV1 reveals a modular organization of *Streptomyces* plasmids that replicate by rolling circle. *Microbiology* **141:**2499–2510.

100. **Xu M-X, Zhu Y-M, Shen M-J, Jiang W-H, Zhao G-P, Qin Z-J.** 2006. Characterization of the essential gene components for conjugal transfer of *Streptomyces lividans* linear plasmid SLP2. *Prog Biochem Biophys* **33:**986–993.

101. **Chen CW.** 1996. Complications and implications of linear bacterial chromosomes. *Trends Genet* **12:**192–196.

102. **Bentley SD, Brown S, Murphy LD, Harris DE, Quail MA, Parkhill J, Barrell BG, McCormick JR, Santamaria RI, Losick R, Yamasaki M, Kinashi H, Chen CW, Chandra G, Jakimowicz D, Kieser HM, Kieser T, Chater KF.** 2004. SCP1, a 356,023 bp linear plasmid adapted to the ecology and developmental biology of its host, *Streptomyces coelicolor* A3(2). *Mol Microbiol* **51:** 1615–1628.

103. **Huang C-H, Chen C-Y, Tsai H-H, Chen C, Lin Y-S, Chen CW.** 2003. Linear plasmid SLP2 of *Streptomyces lividans* is a composite replicon. *Mol Microbiol* **47:** 1563–1576.

104. **Bey S-J, Tsou M-F, Huang C-H, Yang C-C, Chen CW.** 2000. The homologous terminal sequence of the *Streptomyces lividans* chromosome and SLP2 plasmid. *Microbiology* **146:**911–922.

105. **Costerton JW, Stewart PS, Greenberg EP.** 1999. Bacterial biofilms: a common cause of persistent infections. *Science* **284:**1318–1322.

106. **Hogan D, Kolter R.** 2002. Why are bacteria refractory to antimicrobials? *Curr Opin Microbiol* **5:**472–477.

107. **Parsek MR, Greenberg E.** 2005. Sociomicrobiology: the connections between quorum sensing and biofilms. *Trends Microbiol* **13:**27–33.

108. **Schuster M, Sexton DJ, Diggle SP, Greenberg EP.** 2012. Acyl-homoserine lactone quorum sensing: from evolution to application. *Annu Rev Microbiol* **67:**43–63.

109. **Madsen JS, Burmølle M, Hansen LH, Sørensen SJ.** 2012. The interconnection between biofilm formation and horizontal gene transfer. *FEMS Immunol Med Microbiol* **65:**183–195.

110. **D'Alvise PW, Sjøholm OR, Yankelevich T, Jin Y, Wuertz S, Smets BF.** 2010. TOL plasmid carriage enhances biofilm formation and increases extracellular DNA content in *Pseudomonas putida* KT2440. *FEMS Microbiol Lett* **312:**84–92.

111. **May T, Okabe S.** 2008. *Escherichia coli* harboring a natural IncF conjugative F plasmid develops complex mature biofilms by stimulating synthesis of colanic acid and curli. *J Bacteriol* **190:**7479–7490.

112. **Hausner M, Wuertz S.** 1999. High rates of conjugation in bacterial biofilms as determined by quantitative *in situ* analysis. *Appl Environ Microbiol* **65:**3710–3713.

113. **Ghigo J-M.** 2001. Natural conjugative plasmids induce bacterial biofilm development. *Nature* **412:**442–445.

114. **Sørensen SJ, Bailey M, Hansen LH, Kroer N, Wuertz S.** 2005. Studying plasmid horizontal transfer *in situ*: a critical review. *Nat Rev Microbiol* **3:**700–710.

115. **Cook L, Chatterjee A, Barnes A, Yarwood J, Hu W-S, Dunny G.** 2011. Biofilm growth alters regulation of conjugation by a bacterial pheromone. *Mol Microbiol* **81:**1499–1510.

116. **Sedgley CM, Lee EH, Martin MJ, Flannagan SE.** 2008. Antibiotic resistance gene transfer between *Streptococcus gordonii* and *Enterococcus faecalis* in root canals of teeth ex vivo. *J Endod* **34:**570–574.

117. **Merkey BV, Lardon LA, Seoane JM, Kreft J-U, Smets BF.** 2011. Growth dependence of conjugation explains limited plasmid invasion in biofilms: an individual-based modelling study. *Environ Microbiol* **13:**2435–2452.

118. **Krol JE, Nguyen HD, Rogers LM, Beyenal H, Krone SM, Top EM.** 2011. Increased transfer of a multidrug resistance plasmid in *Escherichia coli* biofilms at the air-liquid interface. *Appl Environ Microbiol* **77:**5079–5088.

119. **Haug MC, Tanner SA, Lacroix C, Meile L, Stevens MJ.** 2010. Construction and characterization of *Enterococcus faecalis* CG110/gfp/pRE25*, a tool for monitoring horizontal gene transfer in complex microbial ecosystems. *FEMS Microbiol Lett* **313:**111–119.

120. **Teuber M, Schwarz F, Perreten V.** 2003. Molecular structure and evolution of the conjugative multiresistance plasmid pRE25 of *Enterococcus faecalis* isolated from a raw-fermented sausage. *Int J Food Microbiol* **88:** 325–329.

121. **Reisner A, Wolinski H, Zechner EL.** 2012. In situ monitoring of IncF plasmid transfer on semi-solid agar surfaces reveals a limited invasion of plasmids in recipient colonies. *Plasmid* **67:**155–161.

122. **Christensen B, Sternberg C, Andersen J, Eberl L, Moller S, Givskov M, Molin S.** 1998. Establishment of new genetic traits in a microbial biofilm community. *Appl Environ Microbiol* **64:**2247–2255.

123. **Fox RE, Zhong X, Krone SM, Top EM.** 2008. Spatial structure and nutrients promote invasion of IncP-1 plasmids in bacterial populations. *ISME J* **2:**1024–1039.

124. **Krone SM, Lu R, Fox R, Suzuki H, Top EM.** 2007. Modelling the spatial dynamics of plasmid transfer and persistence. *Microbiology* **153:**2803–2816.

125. **Seoane J, Yankelevich T, Dechesne A, Merkey B, Sternberg C, Smets BF.** 2011. An individual-based approach to explain plasmid invasion in bacterial populations. *FEMS Microbiol Ecol* **75:**17–27.

126. **Dahlberg C, Bergström M, Hermansson M.** 1998. In situ detection of high levels of horizontal plasmid transfer in marine bacterial communities. *Appl Environ Microbiol* **64:**2670–2675.

127. **Normander B, Christensen BB, Molin S, Kroer N.** 1998. Effect of bacterial distribution and activity on conjugal gene transfer on the phylloplane of the bush bean (*Phaseolus vulgaris*). *Appl Environ Microbiol* **64:** 1902–1909.

128. **Nancharaiah YV, Wattiau P, Wuertz S, Bathe S, Mohan SV, Wilderer PA, Hausner M.** 2003. Dual labeling of *Pseudomonas putida* with fluorescent proteins for in situ monitoring of conjugal transfer of the TOL Plasmid. *Appl Environ Microbiol* **69:**4846–4852.

129. **Arends K, Schiwon K, Sakinc T, Hubner J, Grohmann E.** 2012. Green fluorescent protein-labeled monitoring tool to quantify conjugative plasmid transfer between gram-positive and gram-negative bacteria. *Appl Environ Microbiol* **78:**895–899.

130. **Zhang W, Rong C, Chen C, Gao GF, Schlievert PM.** 2012. Type-IVC secretion system: a novel subclass of type IV secretion system (T4SS) common existing in gram-positive genus *Streptococcus*. *PLoS One* **7:** e46390. doi:10.1371/journal.pone.0046390.

131. **Jones DT.** 1999. Protein secondary structure prediction based on position-specific scoring matrices. *J Mol Biol* **292:**195–202.

132. **Laverde Gomez JA, Bhatty M, Christie PJ.** 2014. Prgk, a multidomain peptidoglycan hydrolase, is essential for conjugative transfer of the pheromone-responsive plasmid pCF10. *J Bacteriol* **196:**527–539.

Plasmids—Biology and Impact in Biotechnology and Discovery
Edited by Marcelo E. Tolmasky and Juan C. Alonso

doi:10.1128/microbiolspec.PLAS-0008-2013

Cris Fernández-López,[1] Alicia Bravo,[1] Sofía Ruiz-Cruz,[1] Virtu Solano-Collado,[1] Danielle A. Garsin,[2] Fabián Lorenzo-Díaz,[3] and Manuel Espinosa[1]

15 Mobilizable Rolling-Circle Replicating Plasmids from Gram-Positive Bacteria: A Low-Cost Conjugative Transfer

INTRODUCTION

Bacteria are everywhere simply because they can colonize and adapt to different ecological niches in a very short-term period. One important molecular mechanism underlying the abilities of bacteria to colonize new niches is the acquisition of novel traits by conjugative DNA transfer. Under these circumstances, the so-called variable genome (as opposed to the core genome), which encodes an array of accessory functions (such as antibiotic resistance, specific degradation pathways, symbiosis, and virulence, to name a few), is freely exchanged among bacteria (1). These newly acquired DNA pieces are represented by intra- or extrachromosomal elements, which may or may not have self-replication and/or auto-transferable capacities. However, all of them participate in the fitness of the bacteria to colonize and to adapt to new niches; thus, they contribute to create new evolutionary patterns (2). Mobile genetic elements (MGEs) constitute a reservoir of DNA that is shared among bacterial species (3), and being so, they contribute to the virulence and to the colonization of different niches by their bacterial hosts. Among MGEs, bacterial plasmids play a key role in horizontal gene transfer and thus are important in the coevolution and fitness of the bacterial-plasmid pair. Bacterial conjugation (described in depth elsewhere in this book) involves the unidirectional transfer of plasmid DNA from a donor to a recipient cell through physical contact (4). In the donor cell, the prerequisite for transfer is the assembly of the plasmid-encoded relaxase and other plasmid- or host-encoded proteins on a specific *cis*-acting DNA plasmid region, the origin of transfer (*oriT*). This protein-DNA complex, the relaxosome, initiates the DNA transfer to the recipient cell by the relaxase-mediated cleavage of a phosphodiester bond at a specific dinucleotide (the nick site) within the *oriT* (5). It has been proposed that the relaxosome is already preformed on supercoiled DNA even

[1]Centro de Investigaciones Biológicas, CSIC, Madrid, Spain 28040; [2]Department of Microbiology and Molecular Genetics, The University of Texas Health Science Center at Houston, Houston, Texas; [3]Instituto Universitario de Enfermedades Tropicales y Salud Pública de Canarias, Universidad de La Laguna, Laguna, Spain 38071.

before the transfer signals reach the donor-recipient cell pair (6). However, this hypothesis poses a yet unsolved question when the plasmid replicates by the rolling circle (RC) mechanism: in these RC-replicating (RCR) plasmids, initiation of replication and initiation of conjugative transfer are exerted by two different plasmid-encoded initiation proteins—Rep in the case of replication and Mob for transfer (7). Each of these proteins recognizes a different origin on the plasmid DNA (Rep recognizes the origin of double-stranded replication, *dso*, whereas Mob recognizes the *oriT*); both proteins require that their DNA substrate is supercoiled (7–9). If the relaxosome was already preformed, the plasmid DNA would be relaxed and the Rep-initiator could not recognize its cognate *dso*. As a result, the plasmid could not replicate. On the other hand, if the Rep-initiator nicked the plasmid DNA to initiate the replication, the relaxosome could not be formed, and thus transfer would be hindered.

The initiation of transfer involves a Mob relaxase-mediated DNA cleavage at the *oriT*; the result of the cleavage is one covalent amino acyl-DNA adduct that would be actively pumped into the recipient cell by the plasmid-encoded coupling protein (CP) and the transferosome, a type IV secretion system (T4SS) (10–13). Thus, like almost all the nucleotidyl transferase cation-dependent enzymes, relaxases leave a free 3′-OH end, while the protein remains stably bound to the 5′-phosphate product. Once in the recipient, the relaxase-DNA intermediate restores the original circular plasmid molecule after termination of transfer by a reversion of the strand transfer reaction. Such reactions resemble the RCR termination mechanism elegantly solved in Novick's lab (reviewed in reference 14), although no indication of one modified relaxase with a short oligonucleotide covalently bound to it (as the by-product of the termination reaction) has been shown for conjugation (see references 8, 14 for a detailed explanation of the termination mechanism). The last stage of transfer, the conversion of the single-stranded (ss) DNA intermediates into double-stranded (ds) plasmid molecules would take place in the recipient cell by conjugative lagging strand replication through transcription of a small primer RNA (15–18).

Since the *oriT* is the only *cis*-acting element required for DNA transfer (6), every *oriT*-containing plasmid should be transferred by conjugation if the protein resources (relaxase, CP, and T4SS) are available *in trans* from any compatible self-transferable element, a phenomenon defined as *mobilization*. This interaction is therefore an important factor that can determine the frequency with which a given mobilizable element and its associated genes are involved in horizontal gene transfer (19). In general, self-transferable elements are transferred at higher frequencies than their mobilizable partner elements (20–21). A wide variety of mobilizable elements harboring the *oriT* and the relaxase- and CP-codifying genes have been found (reviewed in references 20, 21). Further, many small plasmids contain a single gene cassette (*oriT* and relaxase-encoding gene) that allows them to be mobilized with the aid of the machinery provided by helper (auxiliary) plasmids. This is the case of many small RCR-plasmids from Gram-positive bacteria that can be mobilized by their Mob relaxase when they coreside with an auxiliary self-transferable element (22–24). Whether the relationship between mobilizable and conjugative elements is considered parasitic or altruistic is arguable; it seems reasonable to propose mobilization as a strategy to travel around the microbial world at low cost.

NATURE AND DIVERSITY OF MOBILIZABLE ELEMENTS

Many MGEs share the ability to be transferred by conjugation between bacteria when they coreside with a compatible auto-transferable element in the donor cell. Furthermore, due to their modular structure and their dynamic genetic nature, any MGE can be considered a platform where new events of bidirectional mobilization/integration (in and out of the MGE) of other gene cassettes can occur, making it difficult to determine its original genomic location. Since the discovery of the first mobile element in 1953 (25), the diversity of the entire mobilome that one could expect in nature has been found to be very rich (21, 26). Several aspects can be considered to study the diversity of mobilizable elements, leaving aside the bacteriophages. Depending on the location of the mobile elements, they can be classified into extra- (plasmids), or intrachromosomal elements. In the former case, they constitute the so-called plasmidome (27), whereas within the MGEs that have an intrachromosomal location two categories can be distinguished: (i) integrative and mobilizable elements (28) and (ii) mobilizable conjugative transposons, which usually show nonspecific integration/excision (29). The adaptive functions provided by the MGE are important because a large number of MGEs encode one or several genes involved in resistance to antibiotics, heavy metals, or metabolic pathways (21, 30).

Concerning their replication autonomy, plasmids keep self-replication machinery, whereas integrative and mobilizable elements and mobilizable conjugative transposons retain the ability to integrate and excise from the host chromosome and, consequently, they do not necessarily have to be self-replicative. The extra-chromosomal MGEs can be generally classified considering their replication strategies, namely RC, theta, and strand displacement, although some different replication strategies can be found in linear plasmids (8, 31). Analysis of the genetic structure of the plasmid transfer-cassette shows that some plasmids contain only the *oriT* (plasmid pCI411), the *oriT* and genes encoding relaxosome components (plasmids pMV158 and pC221), or the *oriT* and the genes encoding the relaxase and the CP (plasmid CloDF13) (32–34). An important aspect to consider in plasmids is their host range, because replication, stability, and transfer mechanisms contribute to the promiscuity of a plasmid as shown for pMV158 (7) and for IncQ plasmids (35).

THE MOBILIZABLE RCR-ELEMENTS: SMALL PLASMIDS COULD DO IT

The finding that genetic elements could be transferred by conjugation between Gram-positive bacteria took a long while after the discovery of conjugation in *Escherichia coli* (36). In fact, the first reports on conjugation in Gram-positive bacteria were overcautious about naming "conjugation" to the transfer of genetic elements encoding antibiotic-resistance determinants (the current integrative and conjugative elements [ICEs]), so they were referred to as "DNase-resistant transfer in filter mating assays" (37). Once it was accepted that the observed genetic transfer was due to genuine conjugation, it was soon apparent that small plasmids, needing helper plasmids of the Inc18 family (such as pIP501 or pAMβ1), could be also transferred by conjugation between *Streptococcus pneumoniae* strains (38). It was unclear which the mechanism was, since the generally believed process for transfer of these small plasmids was their cointegration with the helper, a process termed conduction ("piggyback"). Three relevant findings at the time were that (i) all RCR-plasmids studied generated large amounts of ssDNA intermediates, which were shown to be highly recombinogenic; (ii) some RCR-plasmids such as the staphylococcal plasmids pE194 and pT181 and, later on, pUB110 and pMV158 shared conserved sequences termed recombination site A (RS_A) (22, 39); and (iii) plasmid-encoded proteins, termed Pre, could be responsible for the plasmid cointegration. These findings led to the proposal that the interaction of the RS_A site and the Pre protein could play a role not only in plasmid maintenance but also in the distribution of small antibiotic-resistance plasmids among Gram-positive bacteria (22, 39). These Pre proteins were claimed to contain positively charged amino acids, which were considered to be probably involved in the binding of the Pre protein to the RS_A site, as it is still found in some databases (http://www.uniprot.org/uniprot/P13015; http://www.ebi.ac.uk/interpro/entry/IPR001668). In spite of this, it was also clear that at least the Pre protein of the streptococcal plasmid pMV158 was needed for its transfer (24), so the term *Pre/Mob* to name these proteins is still retained. However, demonstration that (i) the pMV158-encoded MobM protein was able to cleave supercoiled plasmid DNA and that (ii) MobM behaved like a bona fide relaxase led to the conclusion that interplasmidic recombination was a by-product of the primary mobilization function (34, 40).

The only Mob proteins of RCR-plasmids that have been characterized in some detail so far are the pMV158-MobM (34, 41–42) and the mobilization proteins from plasmid pC221 (43–45). However, there are nearly 30 Mob proteins that show a high degree of similarity to MobM, including Mob proteins from well-characterized staphylococcal plasmids such as pUB110 and pE194 (Table 1). Curiously enough, the staphylococcal plasmid pC194, which is closely related to pUB110, does not appear to carry a clear mobilization cassette (see below). Cross-recognition of heterologous *oriT*s by Mob proteins could play a role in the plasmid cassettes shuffling and spreading between bacterial species, and, in fact, we have recently demonstrated the *in vitro* cross-recognition of *oriT*s of three RCR-plasmids by the pMV158-encoded MobM relaxase (41).

Historically, mobilizable plasmids have been studied in greater detail than the intrachromosomal MGEs. In 2004, Francia and colleagues presented the first comprehensive classification of plasmids based on the *oriT*-relaxase module and the homology of the relaxases identified in small mobilizable plasmids (26). Later on, this classification was updated and extended by analyzing the available sequences of conjugative plasmids on the GenBank database; six MOB families were defined: MOB_F, MOB_H, MOB_Q, MOB_C, MOB_P, and MOB_V (20–21). Some remarkable conclusions from these studies were that (i) 39% of the prokaryotic plasmids could be conjugative (although it has been experimentally demonstrated only for some of them), 61.5% of which may be mobilizable; (ii) the majority of mobilizable plasmids are small in size (mean peak of 5 kb), but some could be larger than 1 Mb; (iii) small mobilizable plasmids have a high copy number, making them more promiscuous than the conjugative elements; (iv) relaxase MOB typing can be useful to classify conjugative

Table 1 RCR-plasmids from Gram-positive bacteria belonging to the MOB_V family[a]

Plasmid[b]	Accession number	Size (bp)[c]	Bacterial source	Mob (aa)[c]	Reference
pMV158	NC_010096	5,540	*Streptococcus agalactiae*	494	47
pGB2001	NC_015973	4,967	*S. agalactiae*	500	137
pGB2002	NC_015971	6,825	*S. agalactiae*	500	137
p5580	NC_019370	4,950	*Streptococcus dysgalactiae*	500	138
pSBO2	AB021465	3,582	*Streptococcus equinus*	527	139
pSMA198	NC_016750	12,728	*Streptococcus macedonicus*	410	140
pRW5	NC_010423	4,968	*Streptococcus pyogenes*	500	83
pGA2000	NC_019252	4,967	*S. pyogenes*	500	137
pSSU1	NC_002140	4,975	*Streptococcus suis*	499	141
pSMQ172	NC_004958	4,230	*Streptococcus thermophilus*	499	57
pER13	NC_002776	4,139	*S. thermophilus*	499	81
pE194	NC_005908	3,728	*Staphylococcus aureus*	403	142
pCPS49	NC_019142	5,292	*S. aureus*	409	143
pLA106	NC_004985	2,862	*Lactobacillus acidophilus*	411	144
pCD034-2	NC_016034	2,707	*Lactobacillus buchneri*	363	145
pRCEID2.9	NC_017466	2,952	*Lactobacillus casei*	436	146
pWCZ	NC_019669	3,078	*Lactobacillus paracasei*	436	Unpublished
pTXW	NC_013952	3,178	*L. paracasei*	432	147
pMRI 5.2	NC_019900	5,206	*Lactobacillus plantarum*	408	148
pLFE1	NC_012628	4,031	*L. plantarum*	83	149
pLB4	M33531	3,547	*L. plantarum*	361	150
pA1	NC_010098	2,820	*L. plantarum*	103	55
pPB1	NC_006399	2,899	*L. plantarum*	355	58
pLS55	NC_010375	5,031	*Lactobacillus sakei*	439	151
pYSI8	NC_010936	4,973	*L. sakei*	403	152
pGL2	NC_016981	4,572	*Lactococcus garvieae*	504	153
pBM02	NC_004930	3,854	*Lactococcus lactis*	304	154
pMBLR00	NC_019353	3,370	*Leuconostoc mesenteroides*	361	155
pMA67	NC_010875	5,030	*Paenibacillus larvae*	435	156

[a]Data were collected by BLASTP, using as query the pMV158-initiator of replication RepB protein. We extracted 76 plasmids, and out of these, the 29 plasmids showed in this table harbor an open reading frame (ORF) that could encode a relaxase.
[b]In plasmid pSBO2 the ORF-Mob is annotated as a protein of 144 residues (corresponding to coordinates 3149 to 3582). Extending the analyzed DNA sequence from coordinates 1 to 1150, a larger ORF-Mob (527 residues) could be identified. Plasmids pGA2000, pGB2001, pGB2002, and p5580 share the same replicon and mobilization cassettes (137–138).
[c]Plasmids and Mob protein sizes are indicated in number of base pairs (bp) and amino acids (aa), respectively.

plasmids (46); and (v) MOB families are distributed differentially among bacterial phyla: for instance, MOB_V is overrepresented in *Firmicutes* but rare in *Proteobacteria* and totally absent in *Actinobacteria.* Whereas MOB_F and MOB_H are almost absent, MOB_V is the most frequent module in mobilizable plasmids, followed by MOB_C, MOB_P, and MOB_Q. The following sections summarize the representative systems belonging to mobilizable RCR-plasmid families.

THE MOB_V FAMILY

After years of research on mobilization of pMV158 and on its MobM-relaxase (24, 34), MobM is today the prototype of the MOB_V family of relaxases (clade MOB_{V1}) (20). Plasmid pMV158 (5540 bp) was originally isolated from *Streptococcus agalactiae*, strain MV158; historically, it was one of the first useful cloning tools since it specifies resistance to tetracycline (*tetL*-type determinant), which is selectable in many Gram-positive and Gram-negative bacteria (47–50). The promiscuity of plasmid pMV158 is very high, and so far it has been established in more than 20 bacterial species. In fact, it is naturally mobilizable between Gram-positive bacteria by functions supplied by auxiliary plasmids such as pIP501 (51) or even between Gram-positive and Gram-negative bacteria helped by plasmids RP4 or R388 (40). In addition, its replication by the RC mode (52), and the presence of four origins, namely *dso* for leading strand replication, *ssoA* and *ssoU* for lagging strand synthesis, and $oriT_{pMV158}$ for mobilization, has made it an ideal molecule to study macromolecular interactions between the RepB initiator of replication and the MobM initiator of transfer with their cognate origins, *dso* and $oriT_{pMV158}$, respectively (reviewed in references 7, 8, 53).

During the writing of this review, we performed an in-depth search on the entire database of sequenced plasmids and bacterial genomes and found that the family of plasmids sharing the organization of the pMV158 replicon (54) has increased to 76 members. The search was based on homologies with the RepB-initiator of replication and with its cognate dso_{pMV158}. Out of the 76 replicons, 29 harbor a MOB cassette that includes a putative relaxase gene or an experimentally verified relaxase (Table 1). There are a few plasmids, such as pFX2, pCI411, and pXY3, that contain just an *oriT* but lack any recognizable relaxase gene. Another interesting exception was the case of plasmid pA1 from *Lactobacillus plantarum*, which appears to harbor a relaxase pseudogene that could codify a truncated Mob protein of 103 residues. However, no promoter could be identified, and synthesis of the truncated protein was not detected by *in vitro* assays (55). Plasmid pLFE1 also seems to encode a putative truncated Mob protein (83 amino acids), which would belong to the MOB_V family and harbors an *oriT* sequence similar to that of pMV158. Whether these pseudogenes exist as such or are due to sequence errors remains to be determined. In the rest of the 29 plasmids, the organization of the mobilization region is similar: the *oriT* is placed upstream of the Mob-encoding gene. Moreover, such plasmids harbor one or two lagging-strand origins, *ssos*; in the latter case, the two *ssos* flank the MOB cassette (Figs. 1A and 2). Out of the 29 RCR-plasmids that harbor a MOB cassette, pE194 and pMV158 are the only ones for which the nick site at the *oriT* has been determined *in vivo* (56). Both nick sites coincided with the one previously mapped for pMV158 *in vitro*, namely 5′-TAGTGTG↓TTA-3′, "↓" being the dinucleotide cleaved by the Mob proteins (Fig. 1B) (34). Putative nick sites for other replicons of the MOB_V family, such as the *Streptococcus thermophilus* plasmid pSMQ172 (57) or the *L. plantarum* plasmid pPB1 (58)

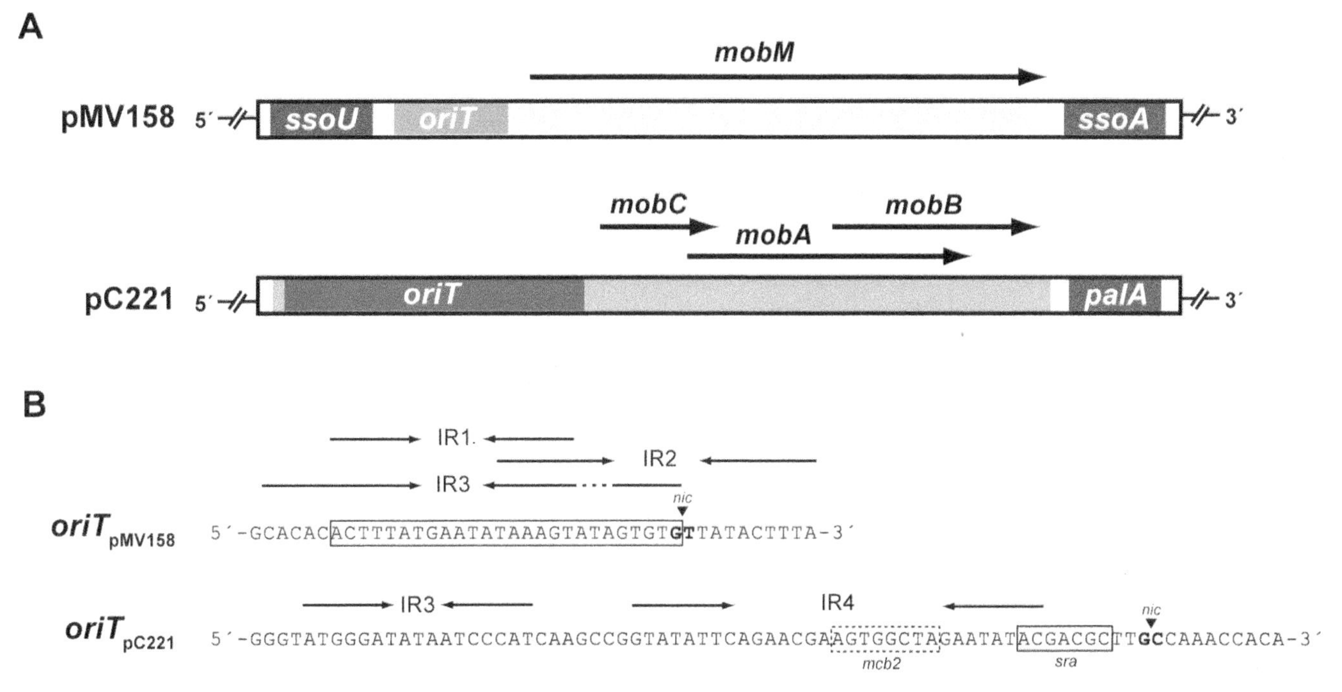

Figure 1 (**A**) The MOB modules of plasmids pMV158 and pC221. The MOB cassette of pMV158, which contains a single gene encoding the MobM relaxase, is flanked by two *ssos* of the types U and A, respectively (see Fig. 2). In the case of pC221, there is a single *ssoA* (also termed *palA*). Its MOB cassette contains three genes encoding the MobC, MobA, and MobB proteins. (**B**) Relevant features of the *oriTs* of plasmids pMV158 and pC221. The nucleotide sequences of the $oriT_{pMV158}$ (coordinates 3564 to 3605) and $oriT_{pC221}$ (coordinates 3084 to 3160) are shown. The inverted repeats (IRs, arrows) and the nick site (*nic*, arrowhead) are indicated. Demonstrated minimal binding site for MobM (lined box), and MobA recognition site (*sra*) and one of the MobC binding sites (*mcb2*) are also specified.
doi:10.1128/microbiolspec.PLAS-0008-2013.f1

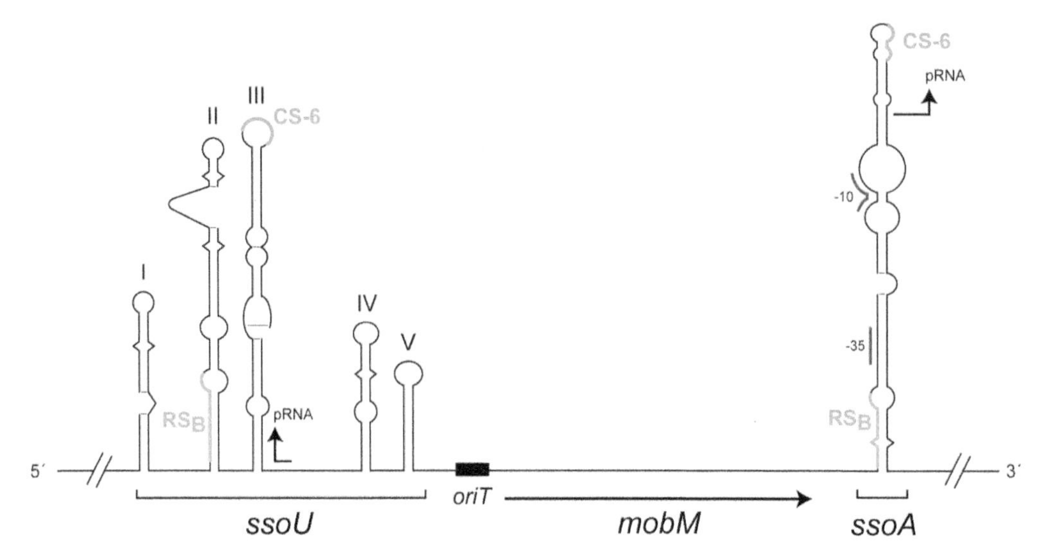

Figure 2 Genetic organization of the MOB cassette in pMV158. Plasmid-*oriT* and the *mobM* gene are flanked by two lagging-strand origins of replication (*ssoA* and *ssoU*). Both *sso*s have the possibility of generating long hairpin-loop structures exhibited as "ssDNA promoters" (15, 18) where the RNAP-binding site (RS_B) is recognized by the RNAP to synthesize a short RNA primer (pRNA). A consensus sequence (CS-6) located in the loop of the hairpin acts as the termination point for the pRNA synthesis. The pRNA is then used by DNA polymerase I for limited extension synthesis, followed by replication of the lagging strand by DNA Pol III. doi:10.1128/microbiolspec.PLAS-0008-2013.f2

have been proposed on the basis of the pMV158 nick site. However, whereas the proposed nick site of pSMQ172 (AGTGNG↓TT) coincides with that of pMV158 (AGTGTG↓TT), the nick site of pPB1 (AGTGG↓GTT) was proposed to be located just one nucleotide upstream in the plasmid sequence. We believe that these nick sites should be experimentally determined or revised.

THE MOB$_P$ FAMILY

The closely related plasmids pC221 and pC223 (4.6 kb) were originally isolated from *Staphylococcus aureus* and confer chloramphenicol resistance (59–60). They belong to the MOB$_P$ family (clade MOB$_{P7}$) (20) and can be mobilized by a coresident self-transmissible plasmid such as pGO1 (61). The transfer module of pC221/pC223 contains the *oriT* and three overlapping genes required for the plasmid DNA processing and mobilization. The MobC and MobA reading frames overlap 16 nucleotides, whereas MobA and MobB overlap 266 nucleotides (43, 45). These genes encode the MobA relaxase and the MobB and MobC accessory proteins (Fig. 1A). Nicking of supercoiled plasmid DNA by MobA was shown to be dependent on the presence of MobC and divalent metal ions (Mg^{2+}) both *in vitro* and *in vivo* (43, 45). Whereas the MobC proteins of pC221 and pC223 are interchangeable, the MobA proteins display specificity toward their cognate DNAs (43). The *oriT*s of pC221 and pC223 span a 77-bp sequence (Fig. 1B) that includes, in addition to the nick site, the MobA binding site (*sra*), which only differs in 4 bp between both plasmids (5′-A**CGA**C**G**C-3′ and 5′-**CCAGT**GC-3′ in pC221 and pC223, respectively; differences in boldface letters). Remarkably, the substrate specificity of MobA could be effectively exchanged by swapping such a small region between the two plasmid *oriT*s (44). Little is known about the function of the MobB accessory protein, but more information is available for MobC. Protein MobC recognizes three different sites that share a consensus sequence: two sites within the $oriT_{pC221}$ (*mcb1* and *mcb2*, at 106 and 15 bp upstream of the nick site, respectively) and another site downstream of the $oriT_{pC221}$ and within the *mobC* gene (*mcb3*). All the MobC binding sites are required for an efficient plasmid mobilization, but only the one closest to the nick site is necessary for DNA nicking, which supports the idea that MobC works at more than one level, such as (i) assisting MobA in its nicking activity, probably by altering the substrate structure (i.e., nick site melting) and (ii) positioning MobA on its recognition sequence and mediating in a high-order protein-complex structure to yield a mobilizable substrate (44).

OTHER MOBILIZABLE RCR-PLASMIDS

Detailed experimental information on the features of other RCR-plasmids that can be mobilized is scarce, except for early reports on (i) transfer of staphylococcal plasmids (62–63), (ii) those referring to the transfer of pUB110 from *Bacillus* (64), and (iii) the transfer of the streptococcal plasmid pVA380-1 (65; reviewed in reference 9). Joint transfer of the staphylococcal plasmids pC194 and pUB110 by Tn*916* from *B. subtilis* to *B. thuringiensis* has been reported (66). However, it appeared that pC194 was either not stable or nonreplicative in the recipient host, which raises the question of whether pC194 was conducted by the transposable element. Recently, the mobilization of pC194 (and two other plasmids lacking mobilization cassettes) by ICE*Bs1* was also reported, achieving higher transfer frequencies than with Tn*916*. These findings led the authors to propose that the initiator of replication of the plasmid could also act as a conjugative relaxase (67). Whether conduction was operating in these transfers remains to be evaluated in depth.

PLASMID pMV158

As stated above, the streptococcal plasmid pMV158 is the representative of the RepB-family of replicons and also of the MOB_V family of relaxases. Due to its promiscuity (7) and to the numerous genetic tools based on it and developed for use in low G+C *Firmicutes* (see below), we will describe here some of the main features of the pMV158 MOB-cassette (Fig. 1A).

The $oriT_{pMV158}$ is located within a 41-bp region upstream of the *mobM* gene. The DNA sequence spanning this region exhibits a high A+T content and includes three inverted repeats (IR1, IR2, and IR3) that partially overlap (68). Such organization would permit the generation of three mutually exclusive hairpin-loop structures in which the position of the dinucleotide cleaved by MobM (G/T at coordinates 3595-3596 in the pMV158 sequence; accession number NC_010096) would be different. There is a 6-bp direct repeat (5′-ACTTTA-3′), which is placed in the left arm of IR1/IR3 and in the right arm of IR2 (Fig. 1B). These features confer the $oriT_{pMV158}$ a particularly complex configuration since most plasmids studied harbor only one or two IR sequences at their *oriT*s (69). How would MobM recognize the different IRs? This may depend on the contextual DNA topology, since, for example, extrusion of the hairpin generated by IR3 would exclude extrusion of the hairpin generated by IR2 (68). In addition, the affinity of MobM for binding to supercoiled DNA or ssDNA is much higher than for binding to linear dsDNA. Further, MobM binds to the ssDNA-encompassing IR1/3 with higher affinity ($K_d \sim 60$ nM) than to ssDNA-encompassing IR2 ($K_d >$ 320 nM). In fact, we have shown that on ssDNA substrates the minimal $oriT_{pMV158}$ sequence for an efficient MobM-binding spans 26 nucleotides that are located just upstream of the nick site (68). This minimal origin includes IR1 and eight adjacent nucleotides (Fig. 1B). Thus, differences in MobM binding could be due to the relative position of the nick site, which would be located eight nucleotides downstream of the IR1 hairpin, within the loop of the IR2 hairpin structure, or just at the 3′-end of the stem of the IR3 hairpin (68).

The MobM protein (494 residues) is a dimer of identical subunits with two main domains: the N-terminal moiety harbors the DNA binding and cleaving activities (the relaxase domain), whereas the C-terminal moiety is involved in dimerization (probably through a leucine zipper motif) and, most likely, in the interaction with other proteins of the conjugative machinery (CP and association to membrane components) through coiled-coil regions (33, 41). The MobM N-terminal region has been studied in detail, and three conserved motifs have been defined: (i) motif I (HxxR) of yet unknown function, (ii) motif II (NYEL), which contains the proposed catalytic tyrosine, and (iii) motif III (HxDExxPHuH), also known as the 3H motif of the HUH proteins, involved in the coordination of a divalent metal (70). The catalytic residue of MobM was predicted to be Tyr44 (20, 33). However, in the case of the Mob relaxase of plasmid pBBR1 (a broad-host-range theta replicating plasmid from *Bordetella bronchiseptica*), which belongs to the MOB_V family (clade MOB_{V2}), it was shown that none of its Tyr residues was involved in transfer (71). This finding suggested that perhaps the MOB_V relaxases might present a new mechanism of conjugative DNA transfer mediated by a non-Tyr catalytic residue or by a water-mediated nucleophilic attack. This is a subject that awaits the resolution of the structure of any member of the MOB_V family of relaxases.

Interestingly, MobM was able to relax supercoiled DNA of other RCR-plasmids from the MOB_V family (41). However, the efficiency of cleavage depended upon the homologies of the *oriT*s: the efficiency of MobM-mediated cleavage was similar for pMV158 and pUB110 (50% of nicking), whereas it was lower for plasmid pDL287 and pE194 (40% and 30% of nicking, respectively) (41). Previous experiments performed with the pMV158-RepB initiator of replication protein showed that the degree of superhelicity of the DNA substrate was essential for successful cleavage (72).

This held true also for the RepB activity on other RCR-plasmids with origins of replication similar to the pMV158-*dso* (72). We postulate that, in the case of MobM, the activity of the protein would also depend on the degree of superhelicity of the DNA substrate. Biochemical studies have been performed with the full length MobM protein and with two truncated derivatives that conserved the first 199 (MobMN199) or the first 243 (MobMN243) residues (41, 73). The results showed that the specificity of MobM for the DNA substrate also depends on the protein length. The three proteins relaxed supercoiled DNA with the same efficiency. However, compared to MobMN199, the 44 additional residues present in MobMN243, which are predicted to form an extra α-helical region, were found to be essential for efficient ssDNA cleavage (73). Biophysical studies showed that the thermal stability of MobMN199 increased in the presence of Mn^{2+} and/or DNA and that the protein retained its enzymatic activity after thermal denaturation and renaturation (68). These findings are consistent with a model in which the relaxase-DNA complex needs the protein to be unfolded to enter the recipient cell through the T4SS (74–75).

We have shown that MobM is able to negatively regulate the transcription of its own gene (76). This does not seem to be the case of several other plasmids, which require an accessory protein for transcriptional regulation of the relaxase synthesis (77). We have found that the two main protein players interacting at the $oriT_{\text{pMV158}}$ region are the host RNA polymerase (RNAP) and the plasmid-encoded MobM. However, depending on the host in which the plasmid replicates, transcription of the *mobM* gene is controlled by different promoters (Fig. 3). Two promoters placed nearly in tandem, termed P_{mob1} and P_{mob2}, are used by the host RNAP: the latter is preferentially used in *E. coli* (76), whereas the former is used in *Lactococcus lactis* (51). Whereas promoter P_{mob1} was mapped to reside within $oriT_{\text{pMV158}}$, promoter P_{mob2} was found to be adjacent to it; yet both promoters were subjected to self-regulation by MobM in *E. coli* (76). Given the relative positions of RNAP and MobM proteins in the region encompassing the two promoters (Fig. 3), we could speculate that control of the *mobM* gene expression would be exerted by the relaxase hindering the accessibility of the RNAP to the promoters, which, in turn, could also depend on the efficiency of the host RNAP to recognize one or the other promoter. Which of these two promoters is used in *S. pneumoniae* is a subject under investigation. Control of gene expression exerted by only a protein seems to be inefficient when a fast response to intracellular changes is required. The control role is better performed by a short-lived antisense RNA than by a relatively long-lived protein. In the case of RCR-plasmids, which are apparently devoid of any partition system, control of the synthesis of the essential Rep initiator protein is exerted at a posttranscriptional level by one or two antisense RNAs (78). However, there are instances, such as replication of pMV158 or of pIP501, in which synthesis of the initiator is controlled by the joint contribution of an antisense RNA and a transcriptional regulator protein (79–80). We have no indication of the existence of an antisense RNA to control expression of the *mobM* gene, and we do not yet know the half-lives of the

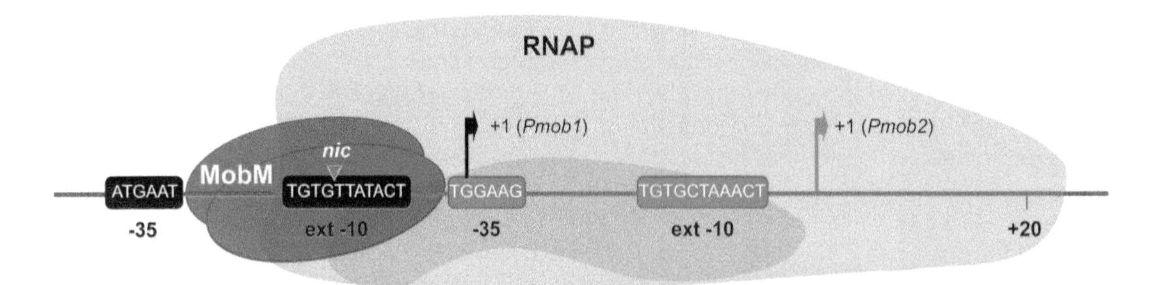

Figure 3 Autoregulation of *mobM* gene expression. The region of pMV158 that includes the *Pmob1* and *Pmob2* promoters of the *mobM* gene is shown. In *E. coli*, *mobM* transcription is mostly initiated from promoter *Pmob2* (blue) (76). Promoter *Pmob1* (black), which is located within the *oriT* sequence, is used in *L. lactis* (51). The main sequence elements of both promoters and the nick site (*nic*) are indicated. Transcription start sites are also indicated with arrows. The regions recognized by MobM and *E. coli* RNA polymerase (RNAP) were determined by DNase I footprinting assays using linear double-stranded DNAs (42, 76). MobM is able to repress *in vivo* the transcription initiated from both promoters, P_{mob1} and P_{mob2} (76). doi:10.1128/microbiolspec.PLAS-0008-2013.f3

MobM-DNA and of the RNAP-P_{mob1} or -P_{mob2} complexes. We also lack information on the competition of the two proteins for binding to the *oriT*$_{pMV158}$ promoter region; thus, it is presently difficult to evaluate the efficiency of MobM as a transcriptional self-regulator. Several streptococcal plasmids such as pER13 (81), pSMQ172 (57), pVA380-1 (82), and pRW35 (83) maintain a similar structure at their *oriT* regions, with two putative tandem promoters for expression of the relaxase gene. Thus, we propose that an autoregulation mechanism, similar to that of pMV158, would also be involved in the synthesis of their respective relaxases.

FACTORS CONTRIBUTING TO THE PROMISCUITY OF RCR-PLASMIDS

The RCR-plasmids are constituted of a small number of gene modules and, in general, show unusual promiscuity (84). Their capacity to colonize new hosts depends on their ability to (i) express their Rep initiator protein and trigger overreplication until they surpass their average copy number (around 20 to 30 plasmid copies per bacterial genome equivalent at the steady state), (ii) control efficiently their replication once they are established in the new host and the average number of copies is reached, (iii) have an efficient mechanism to convert ssDNA intermediates into dsDNA molecules, and (iv) use the host machinery for their own replication and/or transfer. Thus, the philosophy of "travel light, not constituting a genetic load, and making a clever use of the host resources" applies to RCR-plasmids (84). Their replication needs the Rep initiator that is provided by the plasmid, while the rest of the machinery is provided by the host: RNAP, DNA polymerases (I and III), ssDNA-binding protein, DNA gyrase, and a DNA helicase, most likely PcrA (85–87). These host proteins are synthesized in sufficient quantities that the plasmid would not be a heavy genetic load to its host (88). Conversion of ssDNA intermediates into dsDNA forms is initiated by the host RNAP and would require the same proteins used for leading strand synthesis (15, 89).

It was soon demonstrated that small RCR-plasmids were able to replicate in several Gram-positive bacterial hosts (49, 90), and also in Gram-negative bacteria such as *E. coli* (50, 91–92). Some of the RCR-plasmids, such as pC194, could even replicate in yeasts (93). However, some of these plasmids exhibited either structural (deletions) or segregational (loss of plasmid in the absence of selective pressure) instability. This was, most likely, associated with their replication functions (94–98), especially the functionality of the plasmid *sso* (99–100). The lagging strand origins of the RCR-plasmids are located at specific noncoding regions that can fold into imperfect hairpin structures on the ssDNA plasmid intermediates (Fig. 2). They operate in an orientation-dependent manner, indicative of the need of unpaired sequences within the hairpins that would be recognition sites of the host-encoded machinery (99, 101–102). Sequence homology analyses led to the description of various types of *sso* (63): *ssoA* (in pT181 and pC194), *ssoU* (in pUB110), *ssoT* (in pTA1060 and pBAA1), and *ssoW* (in pWV01 and closely related replicons) (97). The streptococcal plasmid pMV158 harbors both *ssoA* and *ssoU* (22, 99). Deletions affecting the *ssoA* functionality lead to plasmid copy number decrease, segregational instability, and accumulation of large amounts of intracellular ssDNA plasmid intermediates (22, 99, 102). Analysis of the DNA sequence and structure of the *ssoA* of several plasmids showed the presence of conserved unpaired regions (63) and homologies between *ssoA* and *ssoU* sequences (16, 103). One of the conserved regions encompasses part of the so-called plasmid recombination site B (Pre_RS_B), which is highly conserved in nearly all the reported *ssoAs*. Later on, it was demonstrated that the RS_B site was the binding site of the host RNAP for a class of promoters located on ssDNA (15, 18). There is another unpaired region, which was termed CS-6 (consensus sequence 6: 5′-TAGCGt/a-3′), which is a transcriptional terminator for the synthesis of a small primer RNA by the host RNAP (15; see Fig. 2). Lagging strand synthesis is continued by DNA polymerase I, followed by DNA polymerase III and DNA gyrase to complete the plasmid replication (8, 89). An indication of a good adaptation of RCR-plasmids to their hosts is provided by the ratio of ss/ds DNA, which should be very low because accumulation of intracellular ssDNA can be harmful for the host bacteria (16). The MOB cassette of pMV158 (*oriT*$_{pMV158}$ and *mobM* gene) is flanked by the *ssoU* and *ssoA* origins (Fig. 2). Transfer of pMV158 between *S. pneumoniae* and *Enterococcus faecalis* allowed us to study which pMV158-*sso* is used in homologous and heterologous hosts (Table 2). Determination of the frequencies of transfer, plasmid copy number in the two hosts, and the efficiency of replication measured as the ratio of ss/ds DNA showed that (i) either *sso* supported transfer between strains of *S. pneumoniae* with the same efficiency as the parental pMV158, (ii) only the *ssoU* efficiently supported transfer when it was tested between *S. pneumoniae* and *E. faecalis*, and (iii) a functional *ssoU* is a critical plasmid region in the colonization of hosts by pMV158 (16).

Table 2 Indicators of pMV158 promiscuity: relative efficiencies of conjugative transfer from donor *S. pneumoniae* to recipients *S. pneumoniae* and *E. faecalis*

	S. pneumoniae			*E. faecalis*		
pMV158	Transfer[a]	Copy number[b]	ss/dsDNA[c]	Transfer[a]	Copy number[b]	ss/dsDNA[c]
Wild type	1	30	1	10	20	1
Δ*ssoA*	1	30	1	10	20	1
Δ*ssoU*	1	30	1	0.01	5	15
Δ*ssoA*/*ssoU*	0.01	10	50	0.01	3	30

[a]Relative transfer in comparison to the pMV158 wild-type plasmid.
[b]Number of plasmid molecules per chromosomal equivalent.
[c]Ratio of ssDNA/dsDNA, indicative of efficiency of conversion.

INHIBITORS OF CONJUGATIVE TRANSFER

Antibiotic resistance genes and virulence genes are commonly associated with MGEs that can be transferred among bacteria sharing the same niche. Many human activities related to hospital use and to the industrial waste of antibiotics constitute a major driving force in the selection of MGEs carrying multiple antibiotic-resistance genes (104). Thus, strategies for inhibiting conjugation may be crucial to preserve the effectiveness of antibiotics and to avoid the spread of resistance traits. It must be considered that many mobilizable plasmids not only are promiscuous but also harbor one or more resistance genes. Considering the molecular mechanism of conjugative transfer, several components of the relaxosome/CP/T4SS machinery could be looked at as potential targets for the development of inhibitory drugs (conjugation inhibitors, or COINS) (105). For instance, it has been shown that unsaturated fatty acids are effective inhibitors of plasmid R388 conjugation (105). The underlying mechanism involved in this inhibition is presently unknown, but it might be related to either direct effects on the transfer machinery or to modifications of plasmatic membrane fluidity. Remarkably, whereas R388 plasmid conjugation was inhibited by unsaturated fatty acids, R388-mediated mobilization of plasmid CloDF13 was not affected, suggesting a possible effect of unsaturated fatty acids on the assembly of the R388 relaxosome. A second example of possible COINS has been provided: conjugation of plasmid F was shown to be inhibited by filamentous bacteriophages (such as M13), a family of ssDNA phages that attach to the tip of the conjugative pilus (106). The proposed model contemplates that the phage coat protein gp3 occludes or reduces the assembly of the T4SS components. Therefore, these ssDNA phage coat proteins could be considered a natural source of proteins that can compete with the conjugative T4SS, thus acting like bona fide COINS. Using plasmid F again, a third example of COINS was provided by showing that conjugative DNA relaxase was inhibited by bisphosphonates (107). However, this approach was directed against plasmids, such as F, that encode multityrosine relaxases and that are represented by the MOB_F family (20). This may not be the general feature for all plasmid-encoded relaxases, so bisphosphonates may be considered limited-value COINS. A fourth approach was designed to inhibit the activity of the TrwC_R388 relaxase, and it was done by blocking its relaxation activity once the relaxase-DNA complex was within the recipient cell. This was achieved by employment of single-chain antibodies (108). All the above approaches may not be of wide usefulness, so it would be interesting to search for COINS that target general transfer processes such as relaxosome assembly in the donor or lagging strand synthesis in the recipient bacteria.

APPLICATIONS OF THE pMV158 REPLICON

Due to their small size, it was soon envisaged that many of the RCR-plasmids could be used to construct useful cloning and expression vectors for Gram-positive bacteria, especially because they were scarce or nonexistent compared to those developed for Gram-negative bacteria (48). It was also apparent that generation of ssDNA intermediates, as a consequence of the RCR mechanism, could result in plasmid structural instability (94, 98, 109–111). Thus, deletions of cloned DNA happened with a certain frequency, especially when the recombinant plasmids were transferred among hosts of different species (112). Nevertheless, cloning of DNA fragments up to 10 kb in size was shown to be feasible, and the recombinant plasmids were structurally stable in the homologous host (48). Furthermore, employment of mobilizable RCR-plasmids to assess the horizontal gene transfer as well as their use as possible tools to test novel COINS has been important in the development of novel strategies to deal with the spread of antibiotic-resistance traits. To date, many different genetic tools based on RCR-plasmids have been constructed. However, small sized plasmids replicating by the theta mechanism are frequently the preferred vectors for employment in the dairy industry (113). We review below the applications of some of the plasmid vectors (mobilizable or not) that have been recently constructed using the pMV158 replicon.

Construction of Green Fluorescent Bacteria by Using the pMV158 Replicon

The plasmid pMV158 has been employed to construct useful genetic tools for Gram-positive bacteria in both versions, mobilizable and nonmobilizable vectors. In the latter case, a derivative of pMV158 lacking the *mobM* gene (plasmid pLS1) was used to construct vectors that harbor a variant of the *gfp* gene, which encodes a green fluorescent protein (GFP) that carries the F64L and S65T mutations (114). The F64L mutation increases GFP solubility, while the S65T mutation increases GFP fluorescence and causes a red shift in the excitation spectrum. In addition, this variant of the *gfp* gene contains translation initiation signals that are optimal for its expression in prokaryotes (114). In the pLS1GFP vector, the *gfp* gene is under the control of a pneumococcal promoter (P_M) that is inducible by maltose. The MalR transcriptional repressor controls the activity of the P_M promoter in *S. pneumoniae* (115). A second version, termed pLS1RGFP, has the *malR* gene placed *in cis* to provide an increased gene dosage of the repressor (116). Pneumococcal cells harboring either of these plasmids would be unable to synthesize GFP by growing them in a sucrose-containing medium (repression conditions). However, when the culture medium is supplemented with maltose, the MalR repressor will be inactivated and the strong P_M promoter will direct synthesis of GFP (see Fig. 4).

The mobilizable pMV158GFP vector is a pMV158 derivative that also carries the *gfp* gene fused to the maltose-inducible promoter P_M (117). Thus, in pneumococcal cells carrying pMV158GFP and the chromosomal *malR* gene, expression of *gfp* depends on the carbon source added to the culture medium (repression or activation in the presence of sucrose or maltose, respectively). Nevertheless, in other Gram-positive bacteria (such as *E. faecalis*), the *gfp* gene is expressed constitutively (Fig. 4). The plasmid pMV158GFP has been mobilized not only between strains of *S. pneumoniae* but also from *S. pneumoniae* to *L. lactis* or *E. faecalis* (117) and to *S. aureus* (118). Plasmid pMV158GFP was shown to be a powerful tool for the construction of green fluorescent bacteria, which have been used for many different aims, such as (to name a few) (i) investigation of the phagocytosis and bactericidal activity of granulocytes against live *S. pneumoniae* (119), (ii) analysis of the intracellular localization of pneumococcal cells within murine microglial cells (120), (iii) detection of *L. lactis* during cheese production (121), (iv) visualization of the binding of *Enterococcus durans* to human Caco-2 cells (122), (v) quantitative analysis of the intra- and interspecies conjugal transfer of pMV158 (123), and (vi) quantification of adhesion of GFP producing *S. aureus* (118). More interestingly, pMV158GFP has been used in the development of *Caenorhabditis elegans* as an infection model for *E. faecalis*. *E. faecalis* harboring the plasmid pMV158GFP allowed for visualization of the infection in gut of the transparent animal (Fig. 5) and can be used to quantify colonization under different conditions (D.A.G., not shown). Because the P_M promoter drives constitutive expression in *E. faecalis* so well, it was used in the construction of an *E. faecalis* chromosomal integration vector to drive *gfp* expression in stable integrants. The expression of *gfp* in such a strain (SD234) was very even and stable (124).

Construction of a Promoter-Probe Vector Based on the pMV158 Replicon

Promoter-probe vectors have been shown to be very useful for studies of the regulation of gene expression. Although many algorithms have been developed for the prediction of promoters in bacterial genomes (125–126), definitive identification of promoter sequences requires the use of several experimental approaches, both *in vivo* and *in vitro* (127). Among these approaches, the use of promoter-probe plasmid vectors facilitates the identification of sequence elements that are essential for promoter activity. Moreover, such vectors make it possible to investigate promoter activity in a variety of genetic backgrounds and under diverse environmental stimuli. Compared to *E. coli*, promoter-probe vectors are still scarce in many Gram-positive bacteria. Ruiz-Cruz et al. (128) designed the pAST plasmid vector (5456 bp) (Fig. 6) that derives from a nonmobilizable pMV158 replicon (plasmid pLS1). The vector pAST contains a highly efficient transcriptional terminator signal (the tandem terminators *T1* and *T2* of the *E. coli rrnB* operon) just upstream of a multiple cloning site (MCS). The MCS is followed by a promoter-less *gfp* reporter gene (114). The pAST vector was shown to be suitable to assess the activity of homologous and heterologous promoters in *S. pneumoniae* and

A *S. pneumoniae*

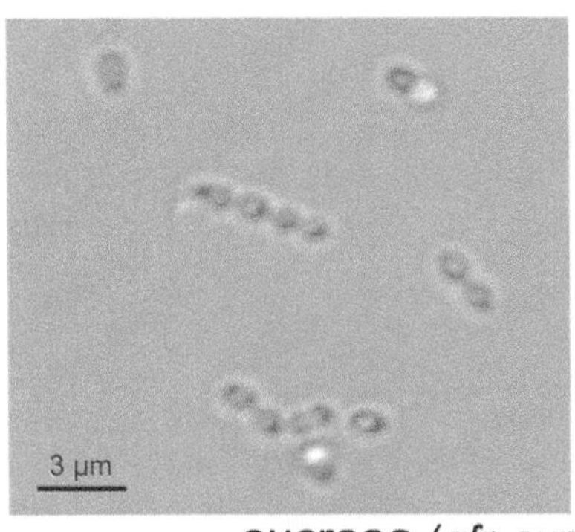

sucrose (*gfp* expression repressed)

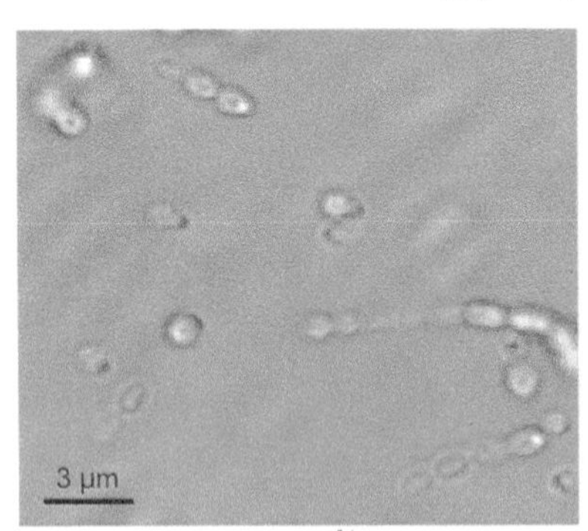

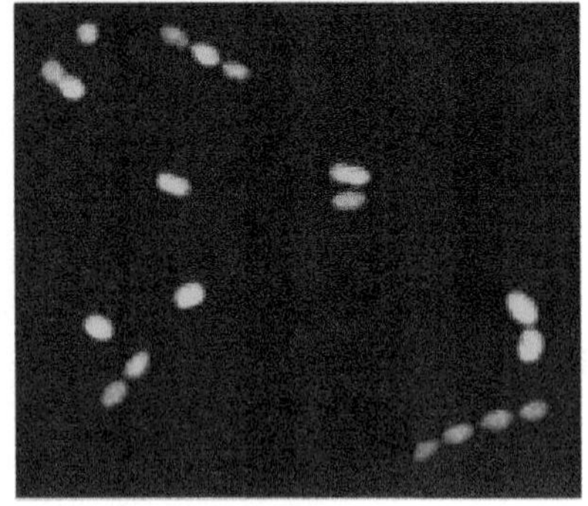

maltose (*gfp* expression induced)

B *E. faecalis*

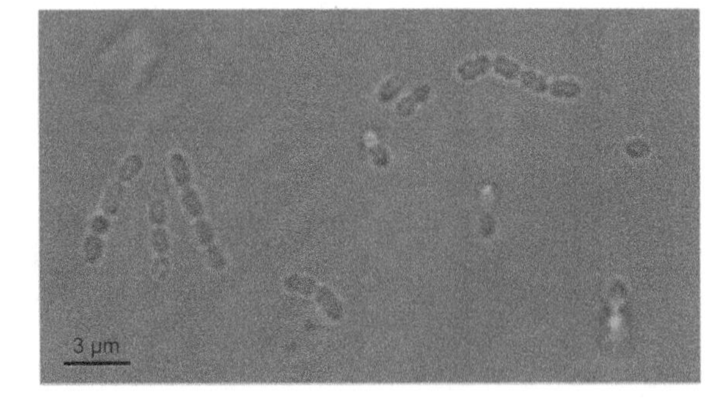

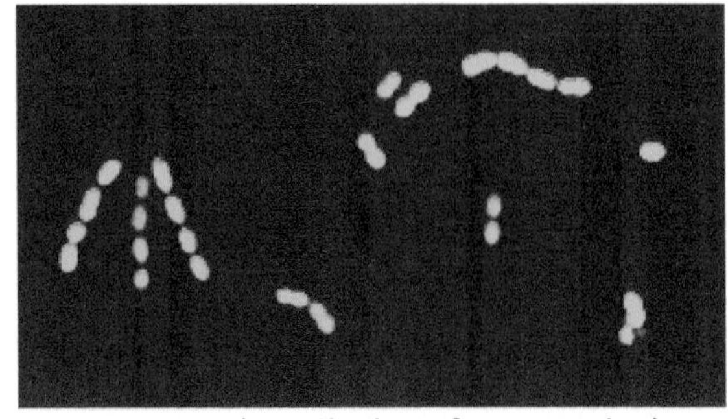

sucrose (constitutive *gfp* expression)

E. faecalis. It is very likely that it can be used in numerous Gram-positive bacteria because pMV158 replicates autonomously in streptococci, enterococci, staphylococci, bacilli, and lactococci. In addition, pAST is a valuable tool for the design of novel expression vectors (see below).

Expression Vectors Based on the pMV158 Replicon

The vector pLS1ROM (6805 bp) is a nonmobilizable pMV158 derivative that was designed for regulated gene expression in *S. pneumoniae* (129). It carries the *malR* gene under the control of a constitutive promoter and the MalR-repressed promoter P_M, which is followed by an MCS. The P_M promoter is activated when bacteria are grown in maltose-containing media. Under such induction conditions, pLS1ROM was shown to be structurally and segregationally stable. The functionality of this system was tested using the *gfp* reporter gene.

The plasmids pAST-*Pung* and pAST-*PfcsK* are a new category of plasmid pAST-derivatives (128) that can be also used as expression vectors in *S. pneumoniae* and that contain useful pneumococcal promoters. The plasmid pAST-*Pung* (5615 bp) was constructed by cloning the promoter region (*Pung*) of the pneumococcal *ung* gene (uracil-DNA glycosylase) into the *Bam*HI site of the pAST promoter-probe vector (Fig. 6). The promoter *Pung* was shown to increase 10-fold the expression of the *gfp* gene in pneumococcal cells (fluorescence assays). The plasmid pAST-*Pung* has an MCS between the *Pung* promoter and the *gfp* gene. Thus, promoterless genes inserted into the MCS can be expressed in pneumococcus. For the construction of the plasmid pAST-*PfcsK* (5573 bp), the *PfcsK* promoter of the fucose operon (130) was inserted into the *Xba*I site of the pAST vector (Fig. 6). Expression of *gfp* from the *PfcsK* promoter increased about 5-fold (as measured by fluorescence assays) when bacteria were grown in the presence of fucose (128). Since the putative fucose regulator gene *fcsR* is widely conserved in *S. pneumoniae* strains (131), it can be expected that the plasmid pAST-*PfcsK* will be valuable as an inducible-expression vector in pneumococcus, and the promoter-less gene of interest can be inserted into the MCS positioned between the *PfcsK* promoter and the *gfp* gene.

HIGH-THROUGHPUT ASSESSMENT OF THE MOBILITY OF RCR-PLASMIDS

Conjugation implies physical contact between the pair of donor/recipient cells, which can be established by flexible retractile pili (exemplified by F), by means of rigid pili (R388 and RP4) or, like in some Gram-positive bacteria, by surface proteins of the sortase-type (132) or peptidoglycan degrading enzymes (133). Thus, depending on the specific conjugative system involved, cell mating can take place in liquid media or on solid/semisolid surfaces. Several high-throughput formats have been developed to study plasmid transfer in liquid (134), semisolid (105), or solid surfaces (123) using fluorescent as well as bioluminescent detection technologies. Such approaches permit testing plasmid transfer in single multiwell plates under the same or different conditions, reducing experimental variations and hands-on time. For instance, mobilization of the plasmid pMV158GFP between different Gram-positive bacteria has been analyzed by using microtiter plates coupled to a filtration device with sterile 0.22-μm filters (Fig. 7) to quantify (i) intraspecies plasmid transfer, where the R61 pneumococcal donor strain harboring pIP501 was mated with another pneumococcal recipient strain (carrying different selectable markers), and (ii) interspecies plasmid transfer, using *S. pneumoniae* as donors and *E. faecalis* as recipients (123). Detection of transconjugants was performed by measurement of the fluorescence provided by pMV158GFP (*gfp* expression) and by plating suitable dilutions on selectable-marker-containing plates followed by colony-forming-unit counting. The frequency of transfer was high enough to be quantified: on average, 1 out of 10,000 recipient cells received pMV158GFP. A further advantage of the employment of this format is the use of microaerophilic bacteria (such as *S. pneumoniae*) as donors and aerobic bacteria (i.e., *E. faecalis*) as recipients. This improves counterselection of donors when testing the number of transconjugants and makes the assays much easier. Therefore, this large-scale filter mating assay could be employed to analyze the parameters involved in interspecies RCR-plasmid transfers as well as to test lead compounds that could act as COINS.

Figure 4 Expression of gene *gfp* in cells harboring plasmid pMV158GFP. Phase-contrast and fluorescence microscopy of (**A**)*S. pneumoniae* cells expressing *gfp* after growth in sucrose (maltose promoter P_M repressed in pneumococci; top panel) or in maltose (P_M induced; bottom panel) and (**B**)*E. faecalis* cells expressing constitutive *gfp* after growth in sucrose. See text for details. doi:10.1128/microbiolspec.PLAS-0008-2013.f4

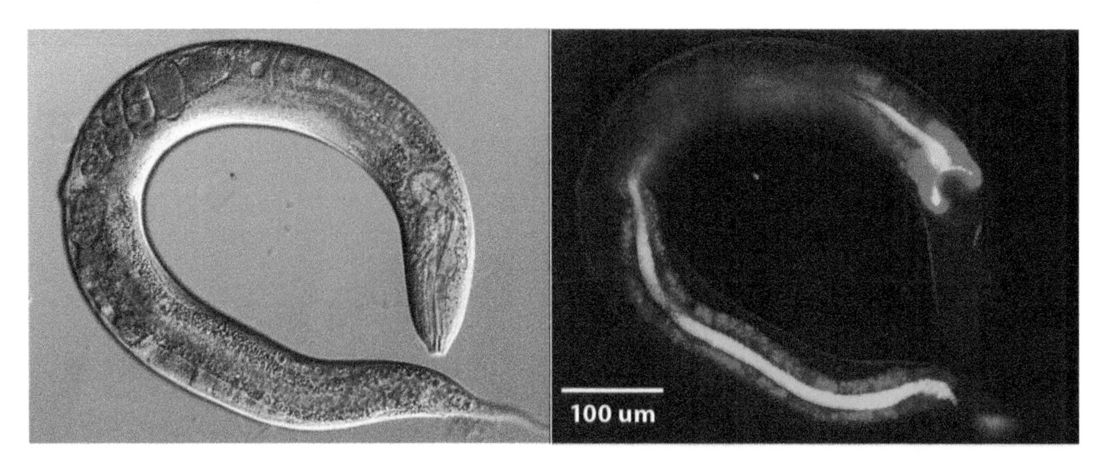

Figure 5 *Enterococcus faecalis* OG1RF expressing green fluorescent protein (GFP) colonizes the intestine of *C. elegans*. L4 larvae were exposed for 3 hours to *E. faecalis* OG1RF carrying plasmid pMV158GFP. Subsequently, worms were washed with M9 medium and anesthetized with 0.25 mM levamisol before imaging. DIC and fluorescence microscopy were used to visualize the worms. GFP expressing bacteria and auto-fluorescence generated by lipofuscin granules in the body of the worm were observed using FITC and DAPI filters, respectively. doi:10.1128/microbiolspec.PLAS-0008-2013.f5

FINAL CONSIDERATIONS

Within the present "ome" era, we are confronted with the still scarce knowledge and attention that it is given to the mobilome as a whole because the spread of the antibiotic resistance is due to the presence of mobile elements among bacteria. The discovery that many Gram-positive bacteria encode relatively simple T4SSs (4, 13, 135), which are accompanied by surface proteins involved in the DNA traffic during conjugation or genetic competence for transformation (136), raised the concern about the massive transfer of resistance genes to many pathogens. Since these "minimal" T4SSs are

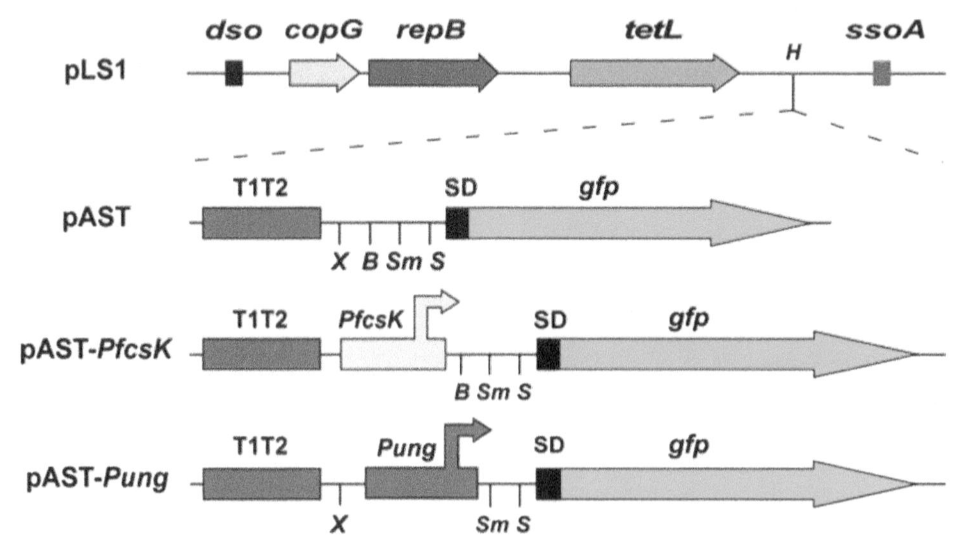

Figure 6 Genetic maps of plasmid pLS1 (a nonmobilizable pMV158 replicon) and its derivatives pAST, pAST-*PfcsK*, and pAST-*Pung* (128). Only relevant features are indicated. *copG* and *repB* genes are involved in plasmid DNA replication. The location of the replication origins *dso* (double-strand origin) and *ssoA* (single-strand origin) is indicated. The *tetL* gene confers resistance to tetracycline. T1T2, tandem terminators *T1* and *T2* of the *E. coli rrnB* rRNA operon; SD, translation initiation signals optimized for the expression of the *gfp* gene in prokaryotes (114). The positions of the *PfcsK* and *Pung* promoters are indicated. *H*, *Hind*III; X, *Xba*I; B, *Bam*HI; *Sm*, *Sma*I; *S*, *Sac*I.
doi:10.1128/microbiolspec.PLAS-0008-2013.f6

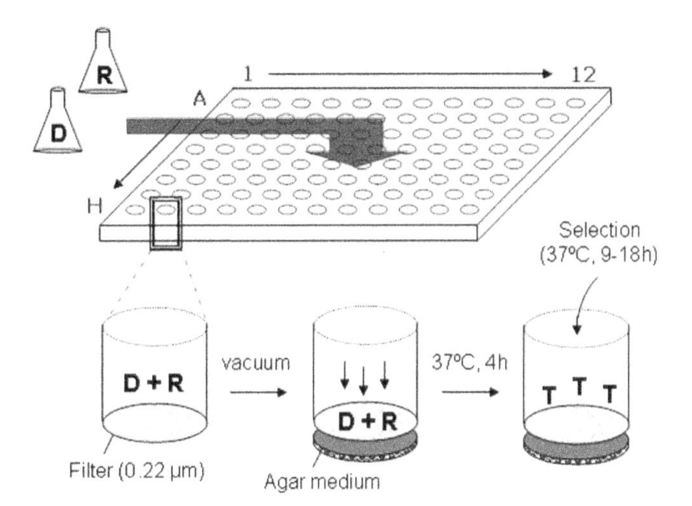

Figure 7 Strategy for high-throughput plasmid transfer detection. Plasmid-containing donor (D) and recipient (R) cells are mixed (it can be done at different ratios), filtered in a 96-filter-well plate (0.22 m), and placed on a layer of solid conjugation medium (containing 5 g/ml DNase I). After 4 h at 37°C, selection for transconjugants (T) is applied by adding antibiotic-containing medium and application of gentle vacuum. Thus, for each well the transfer frequencies can be assessed by a number of methods, such as plating on selective medium, fluorescent confocal microscopy (*gfp*-expressing plasmids), LacZ measurements, or quantitative PCR. (Modified from Lorenzo-Díaz F, Espinosa M, 2009. Large-scale filter mating assay for intra- and inter-specific conjugal transfer of the promiscuous plasmid pMV158 in Gram-positive bacteria. *Plasmid* **61:**65–70 [123]. Copyright 2009, with permission from Elsevier.)
doi:10.1128/microbiolspec.PLAS-0008-2013.f7

generally associated with MGEs (such as Tn*916*, ICEs, pathogenicity islands, and plasmids), the risks of explosive outbreaks of Gram-positive-mediated bacterial infectious diseases are evident. Due to their small size and their promiscuity, we cannot but be astonished that only two RCR-plasmids (pC221 and pMV158) have merited study of their transfer properties. One should not take lightly the assumption that *all* mobilizable RCR-plasmids will be transferred by the *same* mechanism or will have only one type of relaxase or CP, so that "once known one, we know them all" is a poor scientific philosophy.

Acknowledgments. *Research was financed by the Spanish Ministry of Economy and Competitiveness (grants CSD-2008-00013-INTERMODS to M.E. and BFU2009-11868 to A.B.), by the Spanish National Research Council (grant CSIC-PIE-201320E028 to A.B.), by the National Institute of Allergy and Infectious Disease of the National Institutes of Health under award number R01AI076406 and R56AI 093699 to D.A.G., and by the 7th Framework Programme (IMBRAIN project FP7-REGPOT-2012-CT2012-31637-IMBRAIN [Capacities]) to F.L.D. Conflicts of interest: We disclose no conflicts.*

Citation. Fernández-López C, Bravo A, Ruiz-Cruz S, Solano-Collado V, Garsin DA, Lorenzo-Díaz F, Espinosa M. 2014. Mobilizable rolling-circle replicating plasmids from gram-positive bacteria: a low-cost conjugative transfer. Microbiol Spectrum 2(5):PLAS-0008-2013.

References

1. **Osborn MA, Böltner D.** 2002. When phage, plasmids, and transposons collide: genomic islands, and conjugative- and mobilizable-transposons as a mosaic continuum. *Plasmid* **48:**202–212.
2. **Baquero F.** 2004. From pieces to patterns: evolutionary engineering in bacterial pathogens. *Nat Rev Microbiol* **2:**510–518.
3. **Thomas CM.** 2000. *The Horizontal Gene Pool.* Harwood Academic Publishers, Amsterdam.
4. **Zechner EL, Lang S, Schildbach JF.** 2012. Assembly and mechanisms of bacterial type IV secretion machines. *Philos Trans R Soc B Biol Sci* **367:**1073–1087.
5. **Pansegrau W, Lanka E.** 1996. Enzymology of DNA strand transfer by conjugative mechanisms. *Prog Nucl Acid Res Mol Biol* **54:**197–251.
6. **Lanka E, Wilkins BM.** 1995. DNA processing reactions in bacterial conjugation. *Annu Rev Biochem* **64:**141–169.
7. **Espinosa M.** 2013. Plasmids as models to study macromolecular interactions: the pMV158 paradigm. *Res Microbiol* **164:**199–204.
8. **del Solar G, Giraldo R, Ruiz-Echevarría MJ, Espinosa M, Díaz-Orejas R.** 1998. Replication and control of circular bacterial plasmids. *Microbiol Mol Biol Rev* **62:** 434–464.
9. **Grohmann E, Muth G, Espinosa M.** 2003. Conjugative plasmid transfer in Gram-positive bacteria. *Microbiol Mol Biol Rev* **67:**277–301.
10. **de la Cruz F, Frost LS, Meyer RJ, Zechner EL.** 2010. Conjugative DNA metabolism in Gram-negative bacteria. *FEMS Microbiol Rev* **34:**18–40.
11. **Llosa M, Gomis-Rüth FX, Coll M, de la Cruz F.** 2002. Bacterial conjugation: a two-step mechanism for DNA transport. *Mol Microbiol* **45:**1–8.
12. **Christie PJ, Atmakuri K, Krishnamoorthy V, Jakubowski S, Cascales E.** 2005. Biogenesis, architecture, and function of bacterial type IV secretion systems. *Annu Rev Microbiol* **59:**451–485.
13. **Bhatty M, Laverde Gomez JA, Christie PJ.** 2013. The expanding bacterial type IV secretion lexicon. *Res Microbiol* **164:**620–639.
14. **Novick RP.** 1998. Contrasting lifestyles of rolling-circle phages and plasmids. *Trends Biochem Sci* **23:**434–438.
15. **Kramer MG, Khan SA, Espinosa M.** 1997. Plasmid rolling circle replication: identification of the RNA polymerase-directed primer RNA and requirement of DNA polymerase I for lagging strand initiation. *EMBO J* **16:**5784–5795.
16. **Lorenzo-Díaz F, Espinosa M.** 2009. Lagging strand DNA replication origins are required for conjugal transfer of the promiscuous plasmid pMV158. *J Bacteriol* **191:**720–727.

17. **Parker C, Meyer R.** 2005. Mechanisms of strand replacement synthesis for plasmid DNA transferred by conjugation. *J Bacteriol* **187:**3400–3406.
18. **Masai H, Arai K.** 1997. *Frpo*: a novel single-stranded DNA promoter for transcription and for primer RNA synthesis of DNA replication. *Cell* **89:**897–907.
19. **Thomas CM, Nielsen KM.** 2005. Mechanisms of, and barriers to, horizontal gene transfer between bacteria. *Nat Rev Microbiol* **3:**711–721.
20. **Garcillán-Barcia MP, Francia MV, de la Cruz F.** 2009. The diversity of conjugative relaxases and its application in plasmid classification. *FEMS Microbiol Rev* **33:** 657–687.
21. **Smillie C, Garcillan-Barcia MP, Francia MV, Rocha EPC, de la Cruz F.** 2010. Mobility of plasmids. *Microbiol Mol Biol Rev* **74:**434–452.
22. **van der Lelie D, Bron S, Venema G, Oskam L.** 1989. Similarity of minus origins of replication and flanking open reading frames of plasmids pUB110, pTB913 and pMV158. *Nucleic Acids Res* **17:**7283–7294.
23. **Abajy MY, Kopeć J, Schiwon K, Burzynski M, Döring M, Bohn C, Grohmann E.** 2007. A type IV-secretion-like system is required for conjugative DNA transport of broad-host-range plasmid pIP501 in Gram-positive bacteria. *J Bacteriol* **189:**2487–2496.
24. **Priebe SD, Lacks SA.** 1989. Region of the streptococcal plasmid pMV158 required for conjugative mobilization. *J Bacteriol* **171:**4778–4784.
25. **Lederberg J, Tatum E.** 1953. Sex in bacteria; genetic studies, 1945–1952. *Science* **118:**169–175.
26. **Francia MV, Varsaki A, Garcillan-Barcia MP, Latorre A, Drainas C, de la Cruz F.** 2004. A classification scheme for mobilization regions of bacterial plasmids. *FEMS Microbiol Rev* **28:**79–100.
27. **Walker A.** 2012. Welcome to the plasmidome. *Nat Rev Microbiol* **10:**379.
28. **Waldor MK.** 2010. Mobilizable genomic islands: going mobile with *oriT* mimicry. *Mol Microbiol* **78:**537–540.
29. **Burrus V, Waldor MK.** 2004. Shaping bacterial genomes with integrative and conjugative elements. *Res Microbiol* **155:**376–386.
30. **Garcillán-Barcia MP, Alvarado A, de la Cruz F.** 2011. Identification of bacterial plasmids based on mobility and plasmid population biology. *FEMS Microbiol Rev* **35:**936–956.
31. **Espinosa M, Cohen S, Couturier M, del Solar G, Diaz-Orejas R, Giraldo R, Jannière L, Miller C, Osborn M, Thomas CM.** 2000. Plasmid replication and copy number control, p 1–47. *In* Thomas CM (ed), *The Horizontal Gene Pool.* Harwood Academic, Amsterdam.
32. **Cabezon E, Sastre JI, de la Cruz F.** 1997. Genetic evidence of a coupling role for the TraG protein family in bacterial conjugation. *Mol Gen Genet* **254:**400–406.
33. **de Antonio C, Farias ME, de Lacoba MG, Espinosa M.** 2004. Features of the plasmid pMV158-encoded MobM, a protein involved in its mobilization. *J Mol Biol* **335:** 733–743.
34. **Guzmán L, Espinosa M.** 1997. The mobilization protein, MobM, of the streptococcal plasmid pMV158 specifically cleaves supercoiled DNA at the plasmid *oriT*. *J Mol Biol* **266:**688–702.
35. **Meyer R.** 2009. Replication and conjugative mobilization of broad host-range IncQ plasmids. *Plasmid* **62:** 57–70.
36. **Lederberg J, Tatum E.** 1946. Gene recombination in *E. coli*. *Nature* **158:**558.
37. **Shoemaker NB, Smith MD, Guild WR.** 1980. DNase-resistant transfer of chromosomal *cat* and *tet* insertions by filter mating in Pneumococcus. *Plasmid* **3:**80–87.
38. **Smith MD, Shoemaker NB, Burdett V, Guild WR.** 1980. Transfer of plasmids by conjugation in *Streptococcus pneumoniae*. *Plasmid* **3:**70–79.
39. **Gennaro ML, Kornblum J, Novick RP.** 1987. A site-specific recombination function in *Staphylococcus aureus* plasmids. *J Bacteriol* **169:**2601–2610.
40. **Farías ME, Espinosa M.** 2000. Conjugal transfer of plasmid pMV158: uncoupling of the pMV158 origin of transfer from the mobilization gene *mobM*, and modulation of pMV158 transfer in *Escherichia coli* mediated by IncP plasmids. *Microbiology* **146:**2259–2265.
41. **Fernández-López C, Lorenzo-Díaz F, Pérez-Luque R, Rodríguez-González L, Boer R, Lurz R, Bravo A, Coll M, Espinosa M.** 2013. Nicking activity of the pMV158 MobM relaxase on cognate and heterologous origins of transfer. *Plasmid* 70(1):120–130.
42. **Grohmann E, Guzmán LM, Espinosa M.** 1999. Mobilisation of the streptococcal plasmid pMV158: interactions of MobM protein with its cognate *oriT* DNA region. *Mol Gen Genet* **261:**707–715.
43. **Caryl JA, Smith MCA, Thomas CD.** 2004. Reconstitution of a staphylococcal plasmid-protein relaxation complex *in vitro*. *J Bacteriol* **186:**3374–3383.
44. **Caryl JA, Thomas CD.** 2006. Investigating the basis of substrate recognition in the pC221 relaxosome. *Mol Microbiol* **60:**1302–1318.
45. **Smith MCA, Thomas CD.** 2004. An accessory protein is required for relaxosome formation by small staphylococcal plasmids. *J Bacteriol* **186:**3363–3373.
46. **Alvarado A, Garcillán-Barcia MP, de la Cruz F.** 2102. A degenerate primer MOB typing (DPMT) method to classify Gamma-proteobacterial plasmids in clinical and environmental settings. *PLoS One* **7:**e40438
47. **Burdett V.** 1980. Identification of tetracycline-resistant R-plasmids in *Streptococcus agalactiae* (group B). *Antimicrob Agents Chemother* **18:**753–760.
48. **Stassi DL, Lopez P, Espinosa M, Lacks SA.** 1981. Cloning of chromosomal genes in *Streptococcus pneumoniae*. *Proc Natl Acad Sci USA* **78:**7028–7032.
49. **Espinosa M, López P, Pérez-Ureña MT, Lacks SA.** 1982. Interspecific plasmid transfer between *Streptococcus pneumoniae* and *Bacillus subtilis*. *Mol Gen Genet* **188:**195–201.
50. **Lacks SA, López P, Greenberg B, Espinosa M.** 1986. Identification and analysis of genes for tetracycline resistance and replication functions in the broad-host-range plasmid pLS1. *J Mol Biol* **192:**753–765.

51. **Farias ME, Grohmann E, Espinosa M.** 1999. Expression of the *mobM* gene of the streptococcal plasmid pMV158 in *Lactococcus lactis* subsp. *lactis*. *FEMS Microbiol Lett* **176:**403–410.
52. **Puyet A, del Solar G, Espinosa M.** 1988. Identification of the origin and direction of replication of the broad-host-range plasmid pLS1. *Nucleic Acids Res* **16:** 115–133.
53. **del Solar G, Hernández-Arriaga AM, Gomis-Rüth FX, Coll M, Espinosa M.** 2002. A genetically economical family of plasmid-encoded transcriptional repressors in control of plasmid copy number. *J Bacteriol* **184:** 4943–4951.
54. **del Solar G, Moscoso M, Espinosa M.** 1993. Rolling circle-replicating plasmids from Gram-positive and Gram-negative bacteria: a wall falls. *Mol Microbiol* **8:** 789–796.
55. **Vujcic M, Topisirovic L.** 1993. Molecular analysis of the rolling-circle replicating plasmid pA1 of *Lactobacillus plantarum* A112. *Appl Environ Microbiol* **59:** 274–280.
56. **Grohmann E, Zechner EL, Espinosa M.** 1997. Determination of specific DNA strand discontinuities with nucleotide resolution in exponentially growing bacteria harbouring rolling circle-replicating plasmids. *FEMS Microbiol Lett* **152:**363–369.
57. **Turgeon N, Moineau S.** 2001. Isolation and characterization of a *Streptococcus thermophilus* plasmid closely related to the pMV158 family. *Plasmid* **45:**171–183.
58. **de las Rivas B, Marcobal A, Muñoz R.** 2004. Complete nucleotide sequence and structural organization of pPB1, a small *Lactobacillus plantarum* cryptic plasmid that originated by modular exchange. *Plasmid* **52:** 203–211.
59. **Iordanescu S, Surdeanu M.** 1978. Interactions between small plasmids in *Staphylococcus aureus*. *Arch Roum Pathol Exp Microbiol* **37:**155–160.
60. **Projan SJ, Novick RP.** 1988. Comparative analysis of five related staphylococcal plasmids. *Plasmid* **19:** 203–221.
61. **Projan SJ, Archer GL.** 1989. Mobilization of the *Staphylococcus aureus* plasmid pC221 by the conjugative plasmid pGO1 involves three pC221 loci. *J Bacteriol* **171:**1841–1845.
62. **Novick RP.** 1987. Plasmid incompatibility. *Microbiol Rev* **51:**381–395.
63. **Novick RP.** 1989. Staphylococcal plasmids and their replication. *Annu Rev Microbiol* **43:**537–565.
64. **Oskam L, Hillenga DJ, Venema G, Bron S.** 1991. The large *Bacillus* plasmid pTB19 contains two integrated rolling-circle plasmids carrying mobilization functions. *Plasmid* **26:**30–39.
65. **Le Blanc DJ, Chen Y-Y, Lee LN.** 1993. Identification and characterization of a mobilization gene in the streptococcal plasmid pVA380–1. *Plasmid* **30:**296–302.
66. **Naglich JG, Andrews REJ.** 1988. *Tn916*-dependent conjugal transfer of pC194 and pUB110 from *Bacillus subtilis* into *Bacillus thuringiensis* subsp. *israelensis*. *Plasmid* **20:**113–126.
67. **Lee CA, Thomas J, Grossman AD.** 2012. The *Bacillus subtilis* conjugative transposon ICEBs1 mobilizes plasmids lacking dedicated mobilization functions. *J Bacteriol* **194:**3165–3172.
68. **Lorenzo-Díaz F, Dostál L, Coll M, Schildbach JF, Menendez M, Espinosa M.** 2011. The MobM-relaxase domain of plasmid pMV158: thermal stability and activity upon Mn^{2+}- and DNA specific-binding. *Nucl Acids Res* **39:**4315–4329.
69. **González-Pérez B, Carballeira JD, Moncalián G, de la Cruz F.** 2009. Changing the recognition site of a conjugative relaxase by rational design. *Biotechnol J* **4:**554–557.
70. **Chandler M, de la Cruz F, Dyda F, Hickman AB, Moncalian G, Ton-Hoang B.** 2013. Breaking and joining single-stranded DNA: the HUH endonuclease superfamily. *Nat Rev Microbiol* **11:**625–538.
71. **Szpirer CY, Faelen M, Couturier M.** 2001. Mobilization function of the pBHR1 plasmid, a derivative of the broad-host-range plasmid pBBR1. *J Bacteriol* **183:** 2101–2110.
72. **Moscoso M, del Solar G, Espinosa M.** 1995. *In vitro* recognition of the replication origin of pLS1 and of plasmids of the pLS1 family by the RepB initiator protein. *J Bacteriol* **177:**7041–7049.
73. **Fernández-López C, Pluta R, Pérez-Luque R, Rodríguez-González L, Espinosa M, Coll M, Lorenzo-Díaz F, Boer R.** 2013. Functional properties and structural requirements of the plasmid pMV158-encoded MobM relaxase domain. *J Bacteriol* **195:**3000–3008.
74. **Draper O, Cesar CE, Machon C, de la Cruz F, Llosa M.** 2005. Site-specific recombinase and integrase activities of a conjugative relaxase in recipient cells. *Proc Natl Acad Sci USA* **102:**16385–16390.
75. **Gonzalez-Prieto C, Agundez L, Linden RM, Llosa M.** 2013. HUH site-specific recombinases for targeted modification of the human genome. *Trends Biotech* **31:** 305–312.
76. **Lorenzo-Díaz F, Solano-Collado V, Lurz R, Bravo A, Espinosa M.** 2012. Autoregulation of the synthesis of the MobM relaxase encoded by the promiscuous plasmid pMV158. *J Bacteriol* **194:**1789–1799.
77. **Varsaki A, Moncalián G, Garcillán-Barcia MdP, Drainas C, de la Cruz F.** 2009. Analysis of ColE1 MbeC unveils an extended ribbon-helix-helix family of nicking accessory proteins. *J Bacteriol* **191:**1446–1455.
78. **Kumar C, Novick RP.** 1985. Plasmid pT181 replication is regulated by two countertranscripts. *Proc Natl Acad Sci USA* **82:**638–642.
79. **del Solar G, Espinosa M.** 1992. The copy number of plasmid pLS1 is regulated by two trans-acting plasmid products: the antisense RNA II and the repressor protein, RepA. *Mol Microbiol* **6:**83–94.
80. **Brantl S, Wagner EGH.** 1996. An unusually long-lived antisense RNA in plasmid copy number control: *in vivo* RNAs encoded by the streptococcal plasmid pIP501. *J Mol Biol* **255:**275–288.
81. **Somkuti GA, Steinberg DH.** 2007. Molecular organization of plasmid pER13 in *Streptococcus thermophilus*. *Biotechnol Lett* **29:**1991–1999.

82. **LeBlanc DJ, Chen YYM, Lee LN.** 1993. Identification and characterization of a mobilization gene in the streptococcal plasmid pVA380-1. *Plasmid* **30:**296–302.
83. **Woodbury RL, Klammer KA, Xiong Y, Bailiff T, Glennen A, Bartkus JM, Lynfield R, Van Beneden C, Beall BW, for the Active Bacterial Core Surveillance Team.** 2008. Plasmid-borne erm(T) from invasive, macrolide-resistant *Streptococcus pyogenes* strains. *Antimicrob Agents Chemother* **52:**1140–1143.
84. **del Solar G, Alonso JC, Espinosa M, Díaz-Orejas R.** 1996. Broad host range plasmid replication: an open question. *Mol Microbiol* **21:**661–666.
85. **Anand SP, Khan SA.** 2004. Structure-specific DNA binding and bipolar helicase activities of PcrA. *Nucl Acids Res* **32:**3190–3197.
86. **Anand SP, Mitra P, Naqvi A, Khan SA.** 2004. *Bacillus anthracis* and *Bacillus cereus* PcrA helicases can support DNA unwinding and *in vitro* rolling-circle replication of plasmid pT181 of *Staphylococcus aureus*. *J Bacteriol* **186:**2195–2199.
87. **Ruiz-Masó JA, Anand SP, Espinosa M, Khan SA, del Solar G.** 2006. Genetic and biochemical characterization of the *Streptococcus pneumoniae* PcrA helicase and its role in plasmid rolling circle replication. *J Bacteriol* **188:**7416–7425.
88. **Hernández-Arriaga AM, Espinosa M, Del Solar G.** 2012. Fitness of the pMV158 replicon in *Streptococcus pneumoniae*. *Plasmid* **67:**162–166.
89. **Khan SA.** 2005. Plasmid rolling-circle replication: highlights of two decades of research. *Plasmid* **53:**126–136.
90. **Gruss AD, Ehrlich SD.** 1989. The family of highly interrelated single-stranded deoxyribonucleic avid plasmids. *Microbiol Rev* **53:**231–241.
91. **del Solar G, Díaz R, Espinosa M.** 1987. Replication of the streptococcal plasmid pMV158 and derivatives in cell-free extracts of *Escherichia coli*. *Mol Gen Genet* **206:**428–435.
92. **Goze A, Ehrlich SD.** 1980. Replication of plasmids from *Staphylococcus aureus* in *Escherichia coli*. *Proc Natl Acad Sci USA* **77:**7333–7337.
93. **Goursot R, Goze A, Niaudet B, Ehrlich SD.** 1982. Plasmids from *Staphylococcus aureus* replicate in yeast *Saccharomyces cerevisiae*. *Nature* **298:**488–490.
94. **Bron S, Luxen E, Swart P.** 1988. Instability of recombinant pUB110 plasmids in *Bacillus subtilis*: plasmid-encoded stability function and effects of DNA inserts. *Plasmid* **19:**231–241.
95. **Meijer WJJ, van der Lelie D, Venema G, Bron S.** 1995. Effects of the generation of single-stranded DNA on the maintenance of plasmid pMV158 and derivatives in *Lactococcus lactis*. *Plasmid* **33:**91–99.
96. **Meijer WJJ, van der Lelie D, Venema G, Bron S.** 1995. Effects of the generation of single-stranded DNA on the maintenance of plasmid pMV158 and derivatives in different *Bacillus subtilis* strains. *Plasmid* **33:**79–89.
97. **Seegers JFML, Zhao AC, Meijer WJJ, Khan SA, Venema G, Bron S.** 1995. Structural and functional analysis of the single-strand origin of replication from the lactococcal plasmid pWV01. *Mol Gen Genet* **249:**43–50.
98. **Ballester S, Lopez P, Espinosa M, Alonso JC, Lacks SA.** 1989. Plasmid structural instability associated with pC194 replication functions. *J Bacteriol* **171:**2271–2277.
99. **del Solar G, Puyet A, Espinosa M.** 1987. Initiation signals for the conversion of single stranded to double stranded DNA forms in the streptococcal plasmid pLS1. *Nucleic Acids Res* **15:**5561–5580.
100. **Hernández-Arriaga AM, Espinosa M, del Solar G.** 2000. A functional lagging strand origin does not stabilize plasmid pMV158 inheritance in *Escherichia coli*. *Plasmid* **43:**49–58.
101. **del Solar G, Kramer G, Ballester S, Espinosa M.** 1993. Replication of the promiscuous plasmid pLS1: a region encompassing the minus origin of replication is associated with stable plasmid inheritance. *Mol Gen Genet* **241:**97–105.
102. **Gruss AD, Ross HF, Novick RP.** 1987. Functional analysis of a palindromic sequence required for normal replication of several staphylococcal plasmids. *Proc Natl Acad Sci USA* **84:**2165–2169.
103. **Kramer MG, Espinosa M, Misra TK, Khan SA.** 1999. Characterization of a single-strand origin, *ssoU*, required for broad host range replication of rolling-circle plasmids. *Mol Microbiol* **33:**466–475.
104. **Wellington EM, Boxall AB, Cross P, Feil EJ, Gaze WH, Hawkey PM, Johnson-Rollings AS, Jones DL, Lee NM, Otten W, Thomas CM, Williams AP.** 2013. The role of the natural environment in the emergence of antibiotic resistance in Gram-negative bacteria. *Lancet Infect Dis* **13:**155–165.
105. **Fernandez-Lopez R, Machon C, Longshaw CM, Martin S, Molin S, Zechner EL, Espinosa M, Lanka E, de la Cruz F.** 2005. Unsaturated fatty acids are inhibitors of bacterial conjugation. *Microbiology* **151:**3517–3526.
106. **Lin A, Jimenez J, Derr J, Vera P, Manapat ML, Esvelt KM, Villanueva L, Liu DR, Chen IA.** 2011. Inhibition of bacterial conjugation by phage M13 and its protein g3p: quantitative analysis and model. *PLoS One* **6:** e19991
107. **Lujan SA, Guogas LM, Ragonese H, Matson SW, Redinbo MR.** 2007. Disrupting antibiotic resistance propagation by inhibiting the conjugative DNA relaxase. *Proc Natl Acad Sci USA* **104:**12282–12287.
108. **Garcillán-Barcia MP, Jurado P, González-Pérez B, Moncalián G, Fernández LA, de la Cruz F.** 2007. Conjugative transfer can be inhibited by blocking relaxase activity within recipient cells with intrabodies. *Mol Microbiol* **63:**404–416.
109. **Bron S, Bosma P, van Belkum MJ, Luxen E.** 1987. Stability function in the *Bacillus subtilis* plasmid pTA1060. *Plasmid* **18:**8–15.
110. **Janniere L, Bruand C, Ehrlich SD.** 1990. Structurally stable *Bacillus subtilis* cloning vectors. *Gene* **87:**53–61.
111. **Jannière L, Gruss A, Ehrlich SD.** 1993. Plasmids, p 625–644. *In* Sonenshein AL, Hoch JA, Losick R (ed), *Bacillus subtilis and Other Gram-positive Bacteria: Biochemistry, Physiology and Molecular Genetics*. American Society for Microbiology, Washington, DC.

112. **López P, Espinosa M, Greenberg B, Lacks SA.** 1984. Generation of deletions in pneumococcal *mal* genes cloned in *Bacillus subtilis. Proc Natl Acad Sci USA* **81:** 5189–5193.

113. **Leenhouts KJ, Tolner B, Bron S, Kok J, Venema G, Seegers JFML.** 1991. Nucleotide sequence and characterization of the broad-host-range lactococcal plasmid pWV01. *Plasmid* **26:**55–66.

114. **Miller WG, Lindow SE.** 1997. An improved GFP cloning cassette designed for prokaryotic transcriptional fusions. *Gene* **191:**149–153.

115. **Puyet A, Ibañez AM, Espinosa M.** 1993. Characterization of the *Streptococcus pneumoniae* maltosaccharide regulator MalR, a member of the LacI-GalR family of repressors displaying distinctive genetic features. *J Biol Chem* **268:**25402–25408.

116. **Nieto C, Fernández de Palencia P, López P, Espinosa M.** 2000. Construction of a tightly regulated plasmid vector for *Streptococcus pneumoniae*: controlled expression of the green fluorescent protein. *Plasmid* **43:** 205–213.

117. **Nieto C, Espinosa M.** 2003. Construction of the mobilizable plasmid pMV158GFP, a derivative of pMV158 that carries the gene encoding the green fluorescent protein. *Plasmid* **49:**281–285.

118. **Li J, Busscher HJ, van der Mei HC, Sjollema J.** 2012. Surface enhanced bacterial fluorescence and enumeration of bacterial adhesion. *Biofouling* **29:**11–19.

119. **Letiembre M, Echchannaoui H, Bachmann P, Ferracin F, Nieto C, Espinosa M, Landmann R.** 2005. Toll-like receptor 2 deficiency delays pneumococcal phagocytosis and impairs oxidative killing by granulocytes. *Infect Immun* **73:**8397–8401.

120. **Ribes S, Ebert S, Regen T, Agarwal A, Tauber SC, Czesnik D, Spreer A, Bunkowski S, Eiffert H, Hanisch U-K, Hammerschmidt S, Nau R.** 2010. Toll-like receptor stimulation enhances phagocytosis and intracellular killing of nonencapsulated and encapsulated *Streptococcus pneumoniae* by murine microglia. *Infect Immun* **78:** 865–871.

121. **Fernández de Palencia P, de la Plaza M, Mohedano ML, Martínez-Cuesta MC, Requena T, López P, Peláez C.** 2004. Enhancement of 2-methylbutanal formation in cheese by using a fluorescently tagged Lacticin 3147 producing *Lactococcus lactis* strain. *Int J Food Microbiol* **93:**335–347.

122. **Fernández de Palencia P, Fernández M, Mohedano ML, Ladero V, Quevedo C, Alvarez MA, López P.** 2011. Role of tyramine synthesis by food-borne *Enterococcus durans* in adaptation to the gastrointestinal tract environment. *Appl Environ Microbiol* **77:**699–702.

123. **Lorenzo-Díaz F, Espinosa M.** 2009. Large-scale filter mating assay for intra- and inter-specific conjugal transfer of the promiscuous plasmid pMV158 in Gram-positive bacteria. *Plasmid* **61:**65–70.

124. **DebRoy S, van der Hoeven R, Singh KV, Gao P, Harvey BR, Murray BE, Garsin DA.** 2012. Development of a genomic site for gene integration and expression in *Enterococcus faecalis. J Microbiol Methods* **90:**1–8.

125. **Askary A, Masoudi-Nejad A, Sharafi R, Mizbani A, Parizi SN, Purmasjedi M.** 2009. N4: a precise and highly sensitive promoter predictor using neural network fed by nearest neighbors. *Genes Genet Syst* **84:**425–430.

126. **Jacques PE, Rodrigue S, Gaudreau L, Goulet J, Brzezinski R.** 2006. Detection of prokaryotic promoters from the genomic distribution of hexanucleotide pairs. *BMC Bioinformatics* **7:**423–436.

127. **Ross W, Gourse RL.** 2009. Analysis of RNA polymerase-promoter complex formation. *Methods* **47:**13–24.

128. **Ruiz-Cruz S, Solano-Collado V, Espinosa M, Bravo A.** 2010. Novel plasmid-based genetic tools for the study of promoters and terminators in *Streptococcus pneumoniae* and *Enterococcus faecalis. J Microb Methods* **83:** 156–163.

129. **Ruiz-Masó JA, López-Aguilar C, Nieto C, Sanz M, Burón P, Espinosa M, del Solar G.** 2012. Construction of a plasmid vector based on the pMV158 replicon for cloning and inducible gene expression in *Streptococcus pneumoniae. Plasmid* **67:**53–59.

130. **Chan PF, O'Dwyer KM, Palmer LM, Ambrad JD, Ingraham KA, So C, Lonetto MA, Biswas S, Rosenberg M, Holmes DJ, Zalacain M.** 2003. Characterization of a novel fucose-regulated promoter (PfcsK) suitable for gene essentiality and antibacterial mode-of-action studies in *Streptococcus pneumoniae. J Bacteriol* **185:**2051–2058.

131. **Weng L, Biswas I, Morrison DA.** 2009. A self-deleting Cre-*lox-ermAM* cassette, Cheshire, for marker-less gene deletion in *Streptococcus pneumoniae. J Microbiol Methods* **79:**353–357.

132. **Ton-That H, Schneewind O.** 2004. Assembly of pili in Gram-positive bacteria. *Trends Microbiol* **12:**228–234.

133. **Arends K, Celik E-K, Probst I, Goessweiner-Mohr N, Fercher C, Grumet L, Soellue C, Abajy MY, Sakinc T, Broszat M, Schiwon K, Koraimann G, Keller W, Grohmann E.** 2013. TraG encoded by the pIP501 type IV secretion system is a two-domain peptidoglycan-degrading enzyme essential for conjugative transfer. *J Bacteriol* **195:**4436–4444.

134. **Johnsen AR, Kroer N.** 2007. Effects of stress and other environmental factors on horizontal plasmid transfer assessed by direct quantification of discrete transfer events. *FEMS Microbiol Ecol* **59:**718–728.

135. **Zhang W, Rong C, Chen C, Gao GF, Schlievert PM.** 2012. Type-IVC secretion system: a novel subclass of type IV secretion system (T4SS) common existing in Gram-positive genus *Streptococcus. PLoS One* **7:**e46390.

136. **Melville S, Craig L.** 2013. Type IV pili in Gram-positive bacteria. *Microbiol Mol Biol Rev* **77:**323–341.

137. **DiPersio LP, DiPersio JR, Beach JA, Loudon AM, Fuchs AM.** 2011. Identification and characterization of plasmid-borne erm(T) macrolide resistance in group B and group A *Streptococcus. Diagn Microbiol Infect Dis* **71:**217–223.

138. **Palmieri C, Magi G, Creti R, Baldassarri L, Imperi M, Gherardi G, Facinelli B.** 2013. Interspecies mobilization of an *erm*(T)-carrying plasmid of *Streptococcus dysgalactiae* subsp. *equisimilis* by a coresident ICE of the ICESa2603 family. *J Antimicrob Chemother* **68:**23–26.

139. **Nakamura M, Ogata K, Nagamine T, Tajima K, Matsui H, Benno Y.** 2000. Characterization of the cryptic plasmid pSBO2 isolated from *Streptococcus bovis* JB1 and construction of a new shuttle vector. *Curr Microbiol* **41:** 27–32.
140. **Papadimitriou K, Ferreira S, Papandreou NC, Mavrogonatou E, Supply P, Pot B, Tsakalidou E.** 2012. Complete genome sequence of the dairy isolate *Streptococcus macedonicus* ACA-DC 198. *J Bacteriol* **194:** 1838–1839.
141. **Takamatsu D, Osaki M, Sekizaki T.** 2000. Sequence analysis of a small cryptic plasmid isolated from *Streptococcus suis* serotype 2. *Curr Microbiol* **40:**61–66.
142. **Horinouchi S, Weisblum B.** 1982. Nucleotide sequence and functional map of pE194, a plasmid that specifies inducible resistance to macrolide, lincosamine, and streptogramin type B antibiotics. *J Bacteriol* **150:**804–814.
143. **Kadlec K, Pomba CF, Couto N, Schwarz S.** 2010. Small plasmids carrying *vga*(A) or *vga*(C) genes mediate resistance to lincosamides, pleuromutilins and streptogramin A antibiotics in methicillin-resistant *Staphylococcus aureus* ST398 from swine. *J Antimicrob Chemother* **65:** 2692–2693.
144. **Sano K, Otani M, Okada Y, Kawamura R, Umesaki M, Ohi Y, Umezawa C, Kanatani K.** 1997. Identification of the replication region of the *Lactobacillus acidophilus* plasmid pLA106. *FEMS Microbiol Lett* **148:**223–226.
145. **Heinl S, Spath K, Egger E, Grabherr R.** 2011. Sequence analysis and characterization of two cryptic plasmids derived from *Lactobacillus buchneri* CD034. *Plasmid* **66:**159–168.
146. **Panya M, Lulitanond V, Tangphatsornruang S, Namwat W, Wannasutta R, Suebwongsa N, Mayo B.** 2012. Sequencing and analysis of three plasmids from *Lactobacillus casei* TISTR1341 and development of plasmid-derived *Escherichia coli-L. casei* shuttle vectors. *Appl Microbiol Biotechnol* **93:**261–272.
147. **Zhang H, Hao Y, Zhang D, Luo Y.** 2011. Characterization of the cryptic plasmid pTXW from *Lactobacillus paracasei* TXW. *Plasmid* **65:**1–7.
148. **Cho GS, Huch M, Mathara JM, van Belkum MJ, Franz CM.** 2013. Characterization of pMRI 5.2, a rolling-circle-type plasmid from *Lactobacillus plantarum* BFE 5092 which harbours two different replication initiation genes. *Plasmid* **69:**160–171.
149. **Feld L, Bielak E, Hammer K, Wilcks A.** 2009. Characterization of a small erythromycin resistance plasmid pLFE1 from the food-isolate *Lactobacillus plantarum* M345. *Plasmid* **61:**159–170.
150. **Bates EEM, Gilbert HJ.** 1989. Characterization of a cryptic plasmid from *Lactobacillus plantarum. Gene* **85:**253–258.
151. **Ammor MS, Gueimonde M, Danielsen M, Zagorec M, van Hoek AHAM, de los Reyes-Gavilan CG, Mayo B, Margolles A.** 2008. Two different tetracycline resistance mechanisms, plasmid-carried tet(L) and chromosomally located transposon-associated tet(M), coexist in *Lactobacillus sakei* Rits 9. *Appl Environ Microbiol* **74:**1394–1401.
152. **Zhai Z, Hao Y, Yin S, Luan C, Zhang L, Zhao L, Chen D, Wang O, Luo Y.** 2009. Characterization of a novel rolling-circle replication plasmid pYSI8 from *Lactobacillus sakei* YSI8. *Plasmid* **62:**30–34.
153. **Aguado-Urda M, Gibello A, Blanco MM, Lopez-Campos GH, Cutuli MT, Fernandez-Garayzabal JF.** 2012. Characterization of plasmids in a human clinical strain of *Lactococcus garvieae. PLoS One* **7:** e40119.
154. **Sanchez C, Mayo B.** 2003. Sequence and analysis of pBM02, a novel RCR cryptic plasmid from *Lactococcus lactis* subsp *cremoris* P8-2-47. *Plasmid* **49:**118–129.
155. **Chae HS, Lee JM, Lee JH, Lee PC.** 2013. Isolation and characterization of a cryptic plasmid, pMBLR00, from *Leuconostoc mesenteroides* subsp. *mesenteroides* KCTC 3733. *J Microbiol Biotechnol* **23:**837–842.
156. **Murray KD, Aronstein KA.** 2006. Oxytetracycline-resistance in the honey bee pathogen *Paenibacillus larvae* is encoded on novel plasmid pMA67. *J Apicult Res* **45:**207–214.

Specific Plasmid Systems

IV

Plasmids—Biology and Impact in Biotechnology and Discovery
Edited by Marcelo E. Tolmasky and Juan C. Alonso

doi:10.1128/microbiolspec.PLAS-0005-2013

Antonio Lagares[1]
Juan Sanjuán[2]
Mariano Pistorio[1]

16

The Plasmid Mobilome of the Model Plant-Symbiont *Sinorhizobium meliloti*: Coming up with New Questions and Answers

EXTRACHROMOSOMAL REPLICONS IN *S. MELILOTI*: PLASMID CONTENT AND DIVERSITY

Bacteria grouped within the *Rhizobiaceae*, *Phyllobacteriaceae*, and *Bradyrhizobiaceae* families, collectively known as rhizobia, inhabit the soil under free-living conditions and are associated in symbiosis with the root of legumes as nitrogen-fixing organisms. Rhizobia do not form a single taxonomic cluster. Instead, they are distributed within distantly related lineages amongst the alpha- and beta-subdivisions of the proteobacteria. The superb nitrogen-fixing capacity of the *rhizobia*-legume symbioses argues for the use of rhizobia for the introduction of nitrogen into agricultural soils, as a means of avoiding the massive use of chemical fertilizers (1, 2).

A common feature of the rhizobial genome is that, in addition to the chromosome, plasmids are usually present that carry genetic material encoding widely diverse functions (3, 4). The plasmids carrying genes involved in the symbiotic process have been named symbiotic plasmids or *pSyms* (5, 6, 7). In addition to these symbiotic elements, rhizobia may carry other plasmids—referred to as nonsymbiotic plasmids, *non-pSyms* or *the functionally cryptic* or *accessory* compartment—that are not indispensable for symbiosis or have simply not yet been assigned a specific function. That any of these plasmids may be lost, recovered, or also change their copy number or the contents of their genetic information is consistent with the plasmid's role as transient vectors of genomic structural change(s) within a given bacterial population (8).

[1]IBBM, Instituto de Biotecnología y Biología Molecular, CONICET - Departamento de Ciencias Biológicas, Facultad de Ciencias Exactas, Universidad Nacional de La Plata, (1900) La Plata, Argentina; [2]Departamento de Microbiología del Suelo y Sistemas Simbióticos. Estación Experimental del Zaidín, Consejo Superior de Investigaciones Científicas (CSIC), Granada, Spain.

In many rhizobia, plasmid DNA could constitute a substantially high percentage of the bacterial DNA, as in *Sinorhizobium meliloti* where the contribution of (mega)plasmid DNA is ca. 45% of the total. *S. meliloti* is a model species for studying plant-bacteria interactions and, in particular, rhizobia-legume symbiosis and symbiotic nitrogen fixation. In the type strain *S. meliloti* 1021 the genome consists of one chromosome of 3.65 Mbp and two symbiotic megaplasmids, pSymA and pSymB, that are 1.36 and 1.68 Mbp, respectively. In many other strains non-pSym (i.e., cryptic) plasmids of different sizes have also frequently been reported (9, 10, 11, 12, 13, 14) (also, see further on).

SYMBIOTIC (MEGA) PLASMIDS (pSyms)

As mentioned above, pSyms are those plasmids that bear essential genes for the establishment of a fully functional symbiotic nodule in the legume host. The complete genetic sequencing of the model strain *S. meliloti* 1021 more than 10 years ago enabled a knowledge of the gene content of both these pSym plasmids (15, 16). The availability of such information furthered progress in the development of new tools for the functional study of symbiosis, which, in turn, shed additional light on the structure and modular conformation of both megaplasmids. This novel information deepened our understanding of how a symbiosis such as this model association could have evolved (15, 16). pSymA carries genes required for nodulation (*nod*), for the nitrogenase (*nif*), and for nitrogen fixation (*fix* [15]). In addition, pSymA carries genes likely involved in nitrogen and carbon metabolism, transport, and stress and resistance responses that give *S. meliloti* an advantage in its specialized niches (15). The determination of the *S. meliloti* 1021 complete genomic sequence indicated that pSymA could have been acquired by horizontal gene transfer (HGT) because of the megaplasmid's lower GC content (60.4%) and its different codon usage compared with the other replicons (i.e., the bacterial chromosome and pSymB).

pSymB was found to bear genes encoding solute-uptake systems along with the enzymes involved in both polysaccharide biosynthesis and catabolic activities. pSymB also exhibited many features of a typical chromosome such as the presence of essential housekeeping genes including those encoding the arginine tRNA, *minCDE* cell division, and the asparagine-synthetic pathway. In contrast, no essential gene could be mapped on pSymA.

Giuntini et al. (17) applied comparative genomic hybridization on an oligonucleotide microarray in order to estimate genetic variations in four natural strains of *S. meliloti* compared with the type strain 1021. For the analysis, they used two strains obtained from Italian agricultural soils and two isolated from a desert soil in the Aral Sea region. The assay showed that the largest amount of genetic polymorphism—i.e., the genes that were found to be more variable in the comparison between the type strain 1021 and the field isolates—was present in pSymA (17). Gene diversity was not randomly distributed among the three replicons (pSymA, pSymB, and chromosome) in the four natural isolates. Consistent with this, pSymA showed a higher density of insertion sequence elements and phage sequences compared with pSymB (15). In general, transposable elements tend to accumulate in regions where they do not disrupt essential genes. All these data suggest that pSymA is a plasmid-like replicon, whereas pSymB has several chromosomal characteristics. This conclusion was also supported by the observation that pSymA could be cured in *S. meliloti* 2011, whereas pSymB could not (18). The pSymA-cured derivative strain showed no difference in its growth behavior compared with the wild-type strain in both complex and defined media, but otherwise was unable to use a number of substrates as a sole source of carbon on defined media (18). Slater et al. (19) proposed a common mechanism of secondary chromosome formation in *Rhizobiaceae* and other bacteria. A prerequisite for this evolution was proposed to be the intracellular presence of a second replicon capable of stably and efficiently replicating large DNA molecules. Current items of evidence taken together support the notion of a near chromosome status for pSymB (i.e., one of a so-called *chromid* [20, 21]). The genes present in pSymB were recently observed to be more widespread in different taxa than those encoded by the other *S. meliloti* replicons (20). Within such a context, pSymB was proposed to have a role in intraspecies differentiation through positive gene selection, whereas pSymA was suggested to contribute to the emergence of new functions (20). Another recent investigation compared the full genomic sequence of three strains of *S. meliloti*: 1021, AK83, and BL225C (22). The pSymB showed a high resistance to genomic rearrangements, demonstrating an almost perfect synteny. Only strain AK83 presented few rearrangements in certain regions of its pSymB. Within pSymA, however, a very low degree of synteny was observed, thus indicating a much greater rate of rearrangements. When the *S. meliloti* SM11 genome (23) was included in the analysis, plasmid pSmeSM11c (the equivalent to the pSymA of strain 1021) presented the most diverse information in comparison with the pSymA homologs from strains AK83 and BL225C (22). Another interesting feature of pSmeSM11c was that large gene regions proved to be closely related to regions in different plasmids from strain *Sinorhizobium medicae* WSM419. In addition, pSmeSM11c carries other novel gene regions, such as additional plasmid-survival genes; the *acdS* gene involved in modulating the level of the phytohormone ethylene; and a set of genes with predicted functions associated with degradative capabilities, stress response, and amino-acid metabolism and associated pathways (23).

The higher structural variation in pSymA is also consistent with its potential self-transmissibility via conjugation (24), compared with the pSymB, whose mobilization depends on the type IV secretion system present in the pSymA megaplasmid (25). Later on in this review, we will discuss the regulatory aspects associated with the conjugative mobilization of both pSyms.

NON-pSym (*CRYPTIC* OR *ACCESSORY*) PLASMID REPLICONS IN *S. MELILOTI*

The size of the non-pSym plasmids is highly variable, ranging from ca. 7 kb (26) to up to 260 kb (22). Certain regions of these plasmids may be involved in the emergence of genetic rearrangements (27, 28) owing to the presence of insertion sequences, the generation of deletions, or plasmid cointegrations. The information available, mainly stemming from sequence data, indicates that the non-pSyms may encode phenotypic features providing the rhizobia with adaptive advantages. Non-pSym plasmids are thought to encode functions that most likely impact the environmental fitness of the rhizobia. For this reason, the observation that the phenotypic changes resulting from gene knockouts in these plasmids proved to be difficult to study under laboratory conditions was certainly logical. Stiens et al. (11) isolated the plasmid pSmeSM11a from a dominant indigenous *S. meliloti* subpopulation within the context of a long-term field-release experiment involving genetically modified *S. meliloti* strains in Germany. An analysis of the plasmid sequence revealed that approximately two-thirds of pSmeSM11a was occupied by accessory genetic modules that could be providing adaptive advantages or broadening the host bacterium's responsive spectrum. Therefore, the authors postulated that the presence of pSmeSM11a might have been responsible for the dominance of strain SM11. The finding that non-pSym plasmids may encode information for the metabolism of soil compounds or root exudates suggested that these plasmids could play a significant role in the life cycle of the rhizobia and generated speculation on future possibilities for using the plasmid information for enhancing rhizobial capabilities to colonize rhizospheric environments. In addition, reports have indicated that the presence of non-pSym plasmids may be related to nonessential symbiotic capacities. For example, Bromfield et al. (29) found that the presence of the cryptic plasmid pTA2 stimulated competitiveness for nodulation, and genes directly involved in nodulation efficiency have been described on the plasmid pRmeGR4b of *S. meliloti* GR4 (30, 31, 32). By contrast, Velázquez et al. (33) reported that the development of fully effective root nodules in alfalfa by *S. meliloti* strain SAF22 was attenuated by the presence of the cryptic plasmid pRmSAF22c, which was interfering with the normal nodular development necessary to sustain a fully effective nitrogen-fixing symbiosis. Thus, an understanding of how these rhizobia become adapted to changing environments (soil and plant host) will clearly require a more solid characterization of the non-pSym mobilome, because those plasmids appear to be a main cellular *trans* source of new functions.

THE DIVERSITY OF NON-pSym PLASMIDS IN *S. MELILOTI*

Over the years, different investigations were aimed at examining the genomic diversity of bacteria associated with the genus *Medicago*, paying special attention to the presence of cryptic plasmids in the associated rhizobial strains. The most convenient technique for examining large plasmids—such as those borne by *S. meliloti* —was developed by Eckhardt (34) and was based on the lysis of the bacterial cells within an agarose gel (lysis *in situ*). Since under those conditions of minimal mechanical stress (i.e., the shear forces applied) little physical damage to the plasmids occurs, they are released intact in the lysate before migration in the gel. Several modifications of the original technique have been described (7, 35, 36) and have been used accordingly to screen for the presence of large plasmids in rhizobia.

Within the context of the genomic analysis of a collection of *S. meliloti* isolates from Germany by IS fingerprinting, Kosier et al. (12) demonstrated the existence of a high diversity of plasmid profiles in Eckhardt gels. In that study, 78 *S. meliloti* field strains were first grouped into 18 different plasmid profiles with 96% of the isolates containing cryptic plasmids. The most common plasmid profile exhibited a ca. 250 kb band that was present in 18 strains (23%). In another study aimed at investigating the symbiotic characteristics of *S. meliloti* populations present in soils with different agricultural histories, Velázquez et al. (37) isolated rhizobia from three different sites in Salamanca, Spain. In each soil, 7 plasmid profiles were found with 13 different plasmid profiles being observed overall. None of these earlier studies, however, characterized whether or not the plasmid profiles obtained were representative of the existing plasmid diversity within the populations investigated. Pistorio et al. (13) gathered a collection of *S. meliloti* isolates that, while remaining manageable and practical from an experimental point of view, contained a good representation of the plasmid diversity present in the soils sampled. Those authors used *S. meliloti* isolates recovered from 25 different soils of central Argentina to investigate whether any correlation existed between the overall genomic constitution of the rhizobia and the content of cryptic plasmids present in those same bacteria (see the following section). The analysis by Eckhard-like gels of the plasmid content of 64 isolates resulted in 22 different plasmid profiles with at least 38 discrete plasmid bands. In order to evaluate to what extent the local diversity of plasmid profiles was represented in the collection of those isolates, the cumulative Shannon-Weaver index (38) was calculated, and the resulting values for a given number of strains (n >49) reached a *plateau*. The asymptotic behavior of the cumulative Shannon-Weaver index, for the region of higher n values, indicated that the different plasmid profiles from the *S. meliloti* collection constituted a satisfactory representation of the existing plasmid diversity (13).

REPLICATION SYSTEMS PRESENT IN THE *S. MELILOTI* pSym AND NON-pSym PLASMIDS

The best studied replication genes in soil bacteria belong to the *repABC* system. More specifically, within the *Rhizobiaceae* family these *repABC*-type replicons predominate in the plasmids whose replication regions have been investigated (39, 40). The system is named according to the characteristic genetic arrangement of *repA-repB-repC* that, as such, conforms an operon (41). The RepA and RepB have predicted amino acid sequences similar to those of proteins related to plasmid partition; the *repA* genes encode ATPases of the ParA family, while the *repB* genes encode for the *parS*-binding proteins ParB. The *parS* stretch is a centromere-like sequence and can be located at different positions within the *repABC* operon. One general model suggests that the ParA (RepA) proteins polymerize into filaments, and ParB (RepB) binds both the *parS* region and ParA, thus acting as an adaptor between the plasmids and the filaments that are responsible for the segregation process (42). An alternative mechanism suggests a diffusion-ratchet model, where ParB associates with the plasmid chases and redistributes the ParA gradient on the nucleoid, which in turn mobilizes the plasmid (43). The RepC is essential for plasmid replication as a replication-initiator protein (39, 41). The large intergenic sequence between *repB* and *repC* contains a gene encoding a small antisense RNA (ctRNA). This RNA is a *trans*-incompatibility factor, modulating the RepC levels and thereby the resulting plasmid copy number (44, 45, 46, 47). Current genomic data indicate that most *S. meliloti* plasmids described so far bear *repABC* replication cassettes. Other plasmids bear a *repC* replication cassette without partition components and are evolutionarily related to the *repABC* family, because they bear a RepC replication-initiator protein (48, 49). In this family, an antisense RNA also plays a central role as a negative regulator of the expression of *repC*, and is a *trans*-incompatibility determinant as well (49).

Other plasmid-replication systems have been found in *S. meliloti*. One includes only one member, the 7.2-kb plasmid pRm1132f isolated from strain *S. meliloti* 1132. This replication system belongs to group III of the rolling-circle-replication plasmids (26). Finally, Watson and Heys (50) isolated a plasmid replication region from strain *S. meliloti* MB19 that contained genes similar to those associated with several broad-host-range plasmids. The Rep protein in this instance was related to the one present in the broad-host-range plasmid pVS1 from *Pseudomonas aeruginosa*.

CONJUGATIVE PROPERTIES OF THE NON-pSym PLASMID MOBILOME IN *S. MELILOTI*: THE FREQUENCY AT WHICH NON-pSym PLASMIDS ARE FOUND TO BE TRANSMISSIBLE—EVIDENCE IN SUPPORT OF A STRIKINGLY ACTIVE MOBILOME

HGT, homologous recombination, and gene conversion appear to be central mechanisms contributing to the shaping of genomic structures in rhizobia (13, 24, 25, 51, 52, 53, 54, 55, 56, 57, 58, 59). As mentioned above, *S. meliloti* isolates frequently bear a varying and significant amount of non-pSym-plasmid DNA (12), which in several instances has been reported to be either mobilizable or self-transmissible via conjugation (9, 14). In direct relevance to these considerations, a publication by Bailly et al. (60) aimed at investigating the diversity of bacteria associated with plants of the genus *Medicago* provided strong evidence in support of an active HGT involved in modeling the genomes of the symbionts *S. meliloti* and *S. medicae*. All examples of this type made the transferable character of the less studied non-pSym plasmids in *S. meliloti* an issue of central concern for investigation because of their possible role as vehicles of adaptation, evolution, and diversification. Thus, using a collection of *S. meliloti* isolates recovered from humic soils of Argentina, Pistorio et al. (13) studied the percentage of rhizobia that carried transmissible plasmids, the presence of helper functions, and the incidence of phenomena restricting plasmid entry. In that work, the quantitative estimation of diversity was numerically analyzed by means of the Shannon-Weaver (61) and the Simpson (62) indices through the use of different operational taxonomic units (OTUs) that were indicative of: (i) the kind of plasmid profiles present, (ii) the type of PCR fingerprints (i.e., genomic variants), and (iii) possible variants with respect to the combinations of plasmid profiles and genomic fingerprints within the collection of isolates. Interestingly, no clear evidence of association (linkage) was obtained between particular genomic backgrounds and their accompanying plasmids. This result thus argued for the existence of an active horizontal exchange of plasmids, the occurrence of active recombinational events, or both. In order to investigate how active the plasmid transfer within the collected diversity was, a systematic search for transmissible plasmids was performed by screening representative *S. meliloti* isolates from each of the 22 different plasmid profiles (i.e., the plasmid OTUs) identified. Transmissible plasmids could be found in 14% of the isolates representing the different plasmid OTUs, indicating the presence of either self-transmissible or mobilizable plasmids (13). Of the isolates representing one plasmid OTU each, 29% were demonstrated to be capable of mobilizing the model plasmid pSmeLPU88b (9) to a third bacterial strain, thus providing strong evidence for the ubiquitousness of compatible helper functions within the *S. meliloti* isolates (13). Remarkably, such a conjugative system could also be found in a strain from a distant geographic origin as well, as demonstrated by the ability of the European *S. meliloti* GR4 to mobilize plasmid pSmeLPU88b (9). Strain GR4 bears the self-transmissible plasmid pSmeGR4a, and the mobilizable plasmid pSmeGR4b (14, 63). Figure 1 presents a summary of the different events observed when plasmid pSmeLPU88b was transferred to an *S. meliloti* final recipient through the use of representative isolates from the different plasmid-diversity groups as potential donors. Certain isolates (14%), although refractory to receiving the plasmid pSmeLPU88b, accepted other broad-host-range plasmids. Such behavior could be the consequence of surface exclusion and/or either plasmid-replication or recipient-strain DNA-restriction system incompatibility. Irrespective of these observations, the results taken together argued for a strikingly high proportion of the *S. meliloti* genotypes (36%) bearing either transmissible plasmids or helper functions (see Fig. 1). The available evidence indicates that the non-pSym compartment in *S. meliloti* is a highly active plasmid mobilome.

Contrasting with the information on the conjugative character of several *S. meliloti* plasmids, little research has been done to characterize transfer frequencies of pSym and non-pSym plasmids in soil samples and in the field. Studies for the *S. meliloti* cryptic plasmid pSmeLPU88b (64), and those reported by Kinkle and Schmidt for the symbiotic plasmid (pSym) pJB5I from a pea symbiont *Rhizobium leguminosarum* bv. *viciae* (65), indicate that rhizobial plasmids with detectable mobilization frequencies in the laboratory are transferred in nonsterile rhizospheric soil at rates between 10^4 and 10^6 events/recipient bacteria, irrespective of their symbiotic or nonsymbiotic character. The evaluation of plasmid transfer in nature has long been considered a key issue to understand the relevance of the horizontal gene flow in the adaptive responses of the bacteria. Precise estimations of plasmid mobilization frequencies with analysis of the transient host bacteria (including nonrhizobia) will be required, both in soil and the rhizosphere, in support of any future attempt to develop predictive models of conjugative gene dispersion within the natural rhizobial communities.

GENETICS OF CONJUGATIVE MODULES IN *S. MELILOTI*: pSym- AND NON-pSym-PLASMIDS

Considerable efforts have been made to understand the genetics of the rhizobial conjugative systems (for a comprehensive summary, see the review by Ding and Hynes [66]). The available genomic information shows that the conjugative regions in rhizobia can be divided into four Dtr (*D*NA *t*ransfer and *r*eplication) types (I to IV) and three Mpf (*M*ating *p*air *f*ormation) types (I, II, and IV) according to their gene structure plus their associated regulatory mechanisms (66, 67, 68). While the conjugative activities of the plasmids bearing type I Dtr/Mpf proved to be regulated by *quorum sensing* (e.g., plasmids pRetCFN42a from *Rhizobium etli* CFN42, pNGR234a from *Sinorhizobium fredii* NGR234, and pTiC58 from *Agrobacterium tumefaciens* C58), plasmids with type II Dtr/Mpf were found to have an RctA-

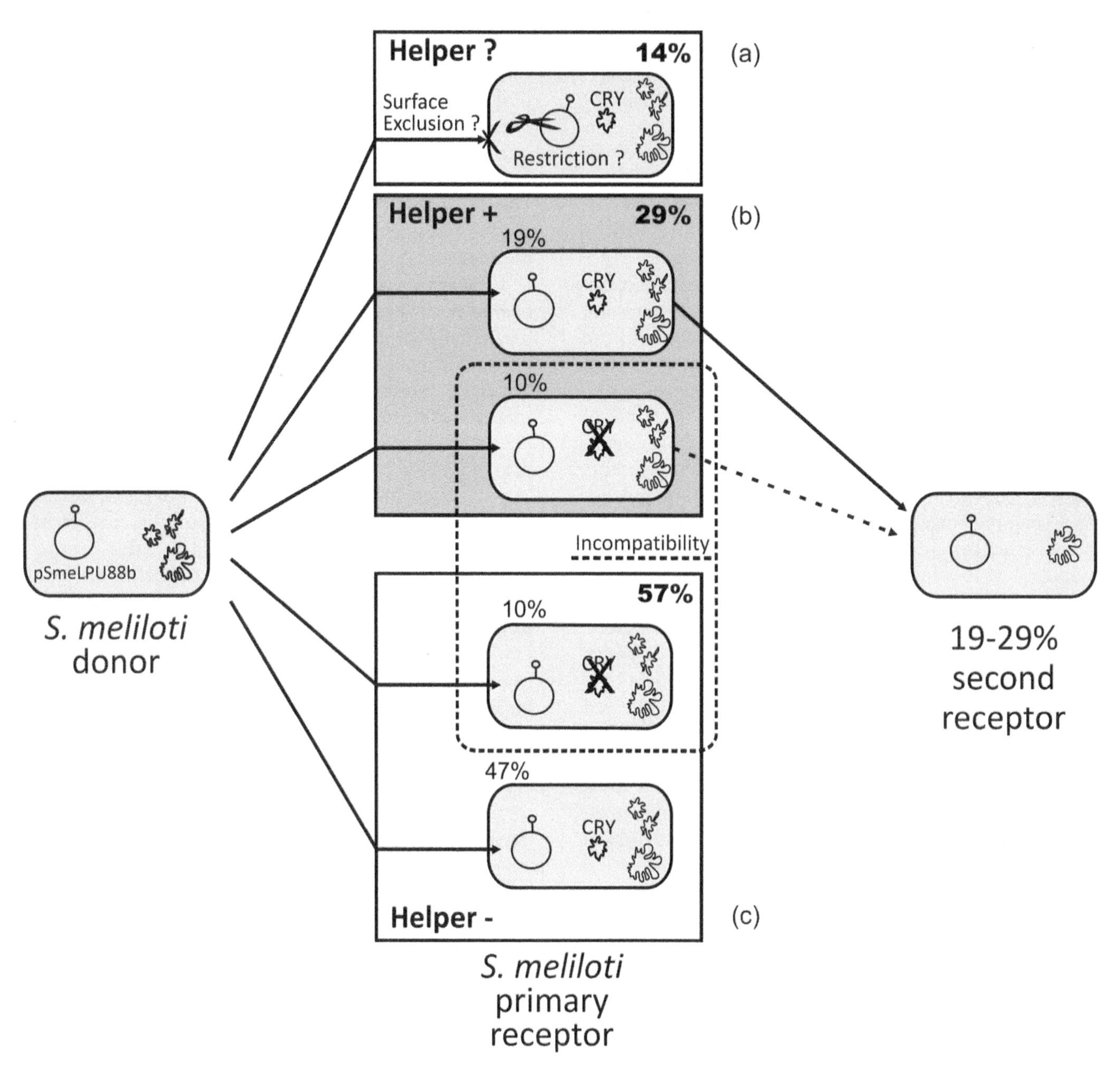

Figure 1 Scheme of the different possible destinies of the model plasmid pSmeLPU88b::Tn5 when transferred to *S. meliloti* from different plasmid-diversity groups (plasmid OTU types; see reference 13). The percentage assigned to each event was calculated upon consideration of an abundance of plasmid OTU types in which all are equally represented. Box A represents those OTUs that proved unable to host plasmid pSmeLPU88b::Tn5 because of either surface-exclusion phenomena or plasmid restriction within the recipient cell. Box B represents OTUs bearing helper functions able to mobilize the model plasmid. Box C represents isolates that, under our experimental conditions, were not able to mobilize plasmid pSmeLPU88b::Tn5. Some isolates belonging to the helper (+) class (Box B) or the helper (–) class (Box C) bear cryptic plasmids that display a replication incompatibility with plasmid pSmeLPU88b::Tn5 (dotted box). CRY, cryptic plasmid. The percent values on the right denote the maximal expected transfer frequency to a third bacterial strain (Reprinted from *FEMS Microbiol Ecol* [13] with permission of the publisher).
doi:10.1128/microbiolspec.PLAS-0005-2013.f1

repressed conjugative regulation (e.g., the megaplasmid pSymA in strain *S. meliloti* 1021, and plasmids pRet42d and pAtC58). Finally, a strong association has been observed between each of the Dtr types (I through IV) and a specific kind of (cognate) relaxase, a correspondence that provides a strong indication of the coevolution of the genes within these conjugative elements in rhizobia. Figure 2 shows the phylogeny of representative rhizobial relaxases and how they correlate with each of the Dtr types.

In the specific case of *S. meliloti*, plasmids may bear a Dtr/Mpf belonging to any of the four types described in rhizobia (Fig. 2). The megaplasmids pSymA and pSymB from the model strain *S. meliloti* 1021 bear a type II Dtr/Mpf (we will describe below how conjugation is regulated in these symbiotic megaplasmids). The pSymA homologs in strains *S. meliloti* SM11 and AK83 both carry, in addition to the type II, a type IV Dtr/Mpf first described for a non-pSym plasmid (67) and recently reclassified as type IVB (68). With respect to the non-pSym *S. meliloti* plasmids, most functional analyses have concentrated on the model plasmid systems pSmeLPU88a/b and pSmeGR4a/b, both of which were also found to carry type IVB Dtrs/Mpfs (Fig. 2) (67, 69). The remarkable ubiquitousness of the Mpf functions able to mobilize pSmeLPU88b (13, 67) points to the *S. meliloti* type IVB Dtr as being a key component for gene exchange within the species. Interestingly, this type IVB Dtr has also been found in plasmids from other rhizospheric and soil bacteria that include *Agrobacterium* (e.g., pATS4a), *Ochrobactrum* (e.g., pOANT01), and *Chelativorans* (e.g., pBNC-01 [67]). In all these organisms, the type IVB Dtr presented a highly conserved synteny with a gene sequence (*parA*-like–*oriT-mobC-mobZ*) (67). The *oriT* in plasmid pSmeLPU88b could be functionally mapped on a 278-bp fragment slightly overlapping with the *mobC* locus (67).

As occurs with the Dtrs I, II, and III, the Dtrs IVB also encode cognate relaxases whose phylogenies reveal a clearly distinct branch (MOB_{P0}), close to the MOB_{P3} and MOB_{P4} enzymes and separated from all other known MOB_Q-like rhizobial relaxases (67). Sequence analysis of the type IVB relaxase from plasmid pSmeLPU88b (designed MobZ) revealed the presence of three conserved motifs (I, II, and III) as in other relaxases (70). Within Motif I (amino acids 109 to 129) of the relaxase, as expected, a tyrosine (Y115) was present consistent with the requirement of this residue for the catalytic activity of the enzyme. Upon a profile-profile analysis and fold recognition, the residue Y115 aligned—within a predicted α-helix—with the experimentally confirmed catalytic residue Y25 of the relaxase MobA from plasmid R1162 (67).

The bacteria carrying plasmids with relaxases of the MOB_{P0} group thus far reported inhabit soils, plants, or aquatic environments; with the sole exception of *Ochrobactrum anthropi* species (although also present in soil) has been found mainly in humans (71). The close phylogenetic relationship among the relaxases of the MOB_{P0}, MOB_{P3}, and MOB_{P4} clades suggests that their host plasmids might constitute frequent HGT vehicles among bacteria with overlapping environments (e.g., among soil- and/or plant-associated bacteria). At the moment, no evidence has been found for conjugative-transfer regulation in plasmids bearing the type IVB Dtrs/Mpfs. Recent evidence indicated that the conjugative transfer of plasmid pRleVF39b carrying a type IVA Dtr/Mpf is under repression control by a transcriptional-regulator protein (TrbR) whose sequence includes a helix-turn-helix xenobiotic-response element (66). Nevertheless, the existing genetic differences between types IVA and IVB Dtrs/Mpfs—plus the absence of a *trbR* ortholog in the fully sequenced plasmid pSmeGR4a—both suggest that this regulation is likely not to be present in the *S. meliloti* plasmids bearing type IVB Dtrs/Mpfs.

Current genomic data on the *S. meliloti* extrachromosomal replicons revealed an increasing sequence variation from pSymB homologs → pSymA homologs → non-pSym plasmids. Interestingly, the increasing genetic variation in each of these replicons parallels their likewise increasing conjugative-mobilization rates: from a strict conjugative control in the pSym plasmids to the lack of any evident conjugative regulation in some of the non-pSym *S. meliloti* plasmids (i.e., plasmids bearing type IVB Dtrs/Mpfs). Such a gradient of replicon plasticity strongly suggests that novel genes might likely reach rhizobia through the non-pSym mobilome, with only some of those genes being incorporated (after mid- and long-term selection) into structurally more stable replicons. An adaptive strategy of this type recalls the mosaic character of the rhizobial plasmid mobilome. A comparison of the *nod*-gene locations among the plasmids (see black dots in Fig. 2) with the phylogenetic topology of the rhizobial relaxases reveals that the symbiotic genes may be found on plasmids bearing any of the four Dtr types, thus indicating that pSym plasmids arise as independent chimeras not specifically associated with any particular mobilization system. A deeper sequencing of the *S. meliloti* plasmid mobilome will be necessary to better understand how gene exchange among replicons (and their further selection) operates over an evolutionary time scale.

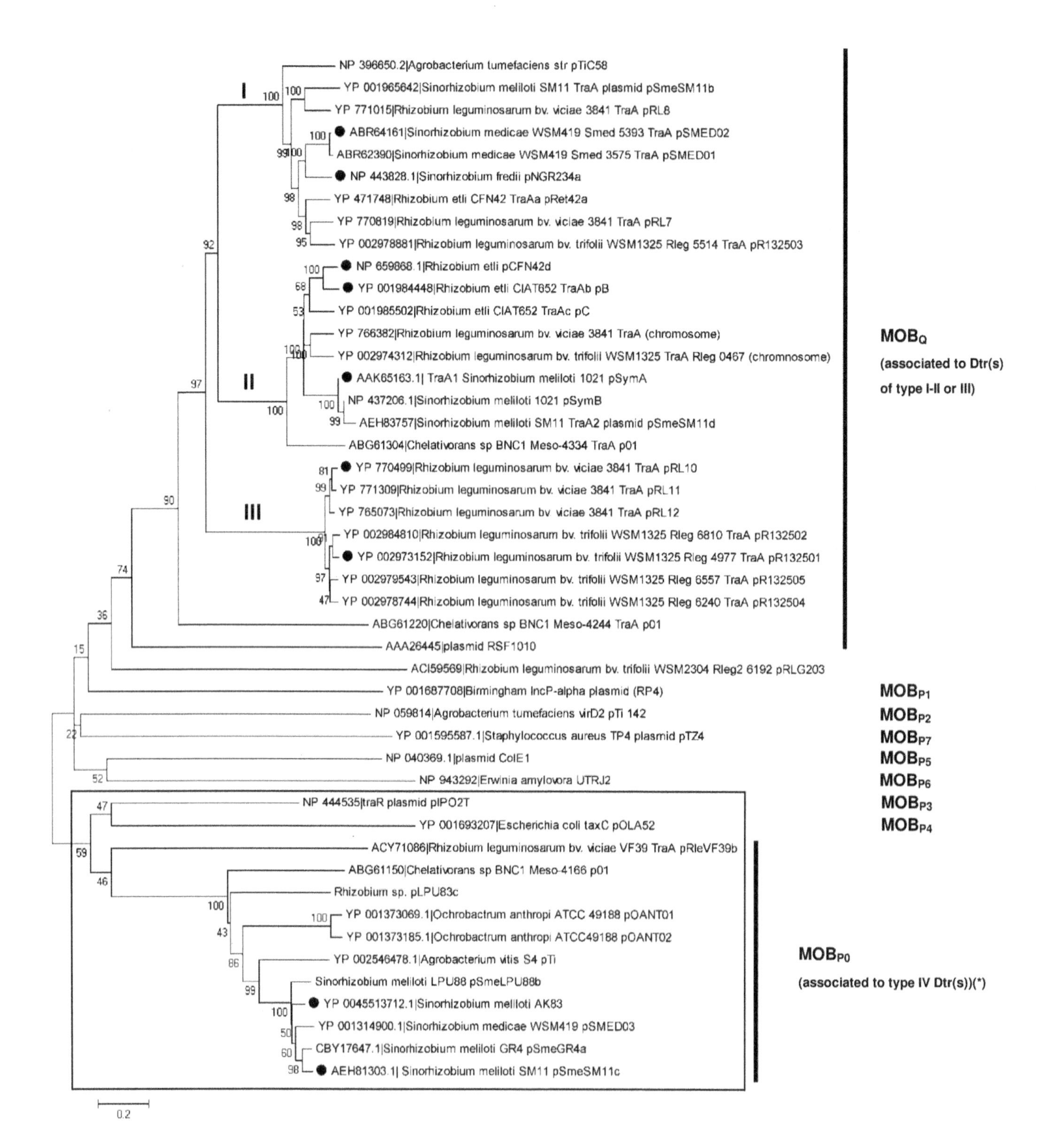

NP 396650.2|Agrobacterium tumefaciens str pTiC58
YP 001965642|Sinorhizobium meliloti SM11 TraA plasmid pSmeSM11b
YP 771015|Rhizobium leguminosarum bv. viciae 3841 TraA pRL8
ABR64161|Sinorhizobium medicae WSM419 Smed 5393 TraA pSMED02
ABR62390|Sinorhizobium medicae WSM419 Smed 3575 TraA pSMED01
NP 443828.1|Sinorhizobium fredii pNGR234a
YP 471748|Rhizobium etli CFN42 TraAa pRet42a
YP 770819|Rhizobium leguminosarum bv. viciae 3841 TraA pRL7
YP 002978881|Rhizobium leguminosarum bv. trifolii WSM1325 Rleg 5514 TraA pR132503
NP 659868.1|Rhizobium etli pCFN42d
YP 001984448|Rhizobium etli CIAT652 TraAb pB
YP 001985502|Rhizobium etli CIAT652 TraAc pC
YP 766382|Rhizobium leguminosarum bv. viciae 3841 TraA (chromosome)
YP 002974312|Rhizobium leguminosarum bv. trifolii WSM1325 TraA Rleg 0467 (chromnosome)
AAK65163.1| TraA1 Sinorhizobium meliloti 1021 pSymA
NP 437206.1|Sinorhizobium meliloti 1021 pSymB
AEH83757|Sinorhizobium meliloti SM11 TraA2 plasmid pSmeSM11d
ABG61304|Chelativorans sp BNC1 Meso-4334 TraA p01
YP 770499|Rhizobium leguminosarum bv. viciae 3841 TraA pRL10
YP 771309|Rhizobium leguminosarum bv. viciae 3841 TraA pRL11
YP 765073|Rhizobium leguminosarum bv. viciae 3841 TraA pRL12
YP 002984810|Rhizobium leguminosarum bv. trifolii WSM1325 Rleg 6810 TraA pR132502
YP 002973152|Rhizobium leguminosarum bv. trifolii WSM1325 Rleg 4977 TraA pR132501
YP 002979543|Rhizobium leguminosarum bv. trifolii WSM1325 Rleg 6557 TraA pR132505
YP 002978744|Rhizobium leguminosarum bv. trifolii WSM1325 Rleg 6240 TraA pR132504
ABG61220|Chelativorans sp BNC1 Meso-4244 TraA p01
AAA26445|plasmid RSF1010
ACI59569|Rhizobium leguminosarum bv. trifolii WSM2304 Rleg2 6192 pRLG203
YP 001687708|Birmingham IncP-alpha plasmid (RP4)
NP 059814|Agrobacterium tumefaciens virD2 pTi 142
YP 001595587.1|Staphylococcus aureus TP4 plasmid pTZ4
NP 040369.1|plasmid ColE1
NP 943292|Erwinia amylovora UTRJ2
NP 444535|traR plasmid pIPO2T
YP 001693207|Escherichia coli taxC pOLA52
ACY71086|Rhizobium leguminosarum bv. viciae VF39 TraA pRleVF39b
ABG61150|Chelativorans sp BNC1 Meso-4166 p01
Rhizobium sp. pLPU83c
YP 001373069.1|Ochrobactrum anthropi ATCC 49188 pOANT01
YP 001373185.1|Ochrobactrum anthropi ATCC49188 pOANT02
YP 002546478.1|Agrobacterium vitis S4 pTi
Sinorhizobium meliloti LPU88 pSmeLPU88b
YP 0045513712.1|Sinorhizobium meliloti AK83
YP 001314900.1|Sinorhizobium medicae WSM419 pSMED03
CBY17647.1|Sinorhizobium meliloti GR4 pSmeGR4a
AEH81303.1| Sinorhizobium meliloti SM11 pSmeSM11c
I
II
III
MOBQ
(associated to Dtr(s)
of type I-II or III)
MOBP1
MOBP2
MOBP7
MOBP5
MOBP6
MOBP3
MOBP4
MOBP0
(associated to type IV Dtr(s))(*)
0.2

THE NEED FOR ADDITIONAL GENETIC DETERMINANTS OTHER THAN THE Dtr AND Mpf REGIONS AS REVEALED BY THE STRAIN-DEPENDENT REQUIREMENT FOR MOBILIZATION OF PLASMID pSmeLPU88b IN *S. MELILOTI*

The mobilization of pSmeLPU88b by plasmid pSmeLPU88a has recently been shown to require the newly identified gene *rptA* (for *Rhizobium-p*lasmid *t*ransfer) in strain *S. meliloti* 2011, but not in strain LPU88 (72). A mutant carrying a disrupted copy of *rptA* in plasmid pSmeLPU88a was still able to facilitate the mobilization of plasmid pSmeLPU88b from strain *S. meliloti* LPU88, but not from strain 2011. Strikingly, no extra copies of the *rptA* gene were found by hybridization in strain LPU88. These observations raise new questions as to (i) what the alternative active components are that complement the deficiencies in *rptA* in strain LPU88 (but not in strain 2011), and (ii) whether or not such components operate in association with the Dtr or the Mpf conjugative regions. Interestingly, syntenic regions containing *rptA* homologs could also be identified in plasmids from the strains *S. meliloti* SM11, AK83, and GR4; *S. medicae* WSM419; *S. fredii* HH103; and *Agrobacterium vitis* S4; thus suggesting a more extensive requirement of *rptA* for plasmid mobilization (72). Analysis of plasmids in these strains suggests that the presence of *rptA* is associated to the type IVB Dtr-bearing plasmids.

REGULATION OF THE CONJUGATIVE TRANSFER OF MEGAPLASMIDS pSymA AND pSymB IN THE MODEL STRAIN *S. MELILOTI* 1021

As mentioned in the previous sections, studies on conjugative transfer in rhizobia have so far identified at least two different regulatory transfer systems: (i) those associated with plasmids containing type I Dtr/Mpf and exhibiting transfer regulation through a *quorum sensing* that responds positively to population density via *N*-acylhomoserine lactones (73, 74, 75, 76), and (ii) systems present in plasmids bearing type II Dtr/Mpf characterized by a so-called RctA-repressed conjugative transfer that occurs only under conditions disturbing the activity of the transcriptional repressor RctA. The *rctA* gene is usually associated with a *virB*-type T4SS (type IV secretion system); on the basis of this feature, several symbiotic plasmids of *R. etli*, *S. fredii*, and *S. meliloti* would belong to this category, although experimental data have been obtained so far only for the 372-kb symbiotic plasmid of *R. etli* CFN42, a symbiont of *Phaseolus vulgaris* (the common bean), and the 1,354-Kb pSymA of *S. meliloti* 1021 (24). Thus far, no clues have been discerned concerning the conditions governing the conjugative transfer of plasmids either with type III Dtrs—which group includes mainly *Rhizobium leguminosarum* plasmids that do not contain an Mpf system and therefore are not self-transmissible (66)—or with type IVB Dtrs/Mpfs, except for the cryptic plasmid pRmeGR4a of *S. meliloti* GR4, where conjugation efficiency was shown to be modulated by the nitrogen source (63). Care should be taken, however, when extrapolating phylogenetic data to mechanistic hypotheses. For instance, the plasmid pSmeSM11c—the pSymA homolog from *S. meliloti* SM11—was initially found to bear a type IVB Dtr/Mpf based on the phylogeny of the putative relaxase gene SM11_pC0230. However, this plasmid was also found to contain another putative relaxase (SM11_pC1037) with an associated coupling protein (SM11_pC1041) and a typical *rctA* gene (SM11_pC0937) linked to a *virB*-like Mpf operon. Thus, plasmid pSmeSM11c likely carries two different conjugative systems.

Figure 2 Phylogenetic (neighbor-joining) tree showing the relationships between different relaxases of the $MOB_{P/Q}$ cluster as inferred from their complete protein sequences. The bootstrap-consensus tree inferred from 1,000 replicates is taken to represent the evolutionary history of the proteins (corresponding to the indicated taxa) analyzed (80). The percentage of replicate trees in which the associated taxa clustered together in the bootstrap test (1,000 replicates) is shown next to the branches. The tree is drawn to scale, with the branch lengths in proportion to the evolutionary distances used to construct the phylogenetic tree. Those distances were computed by means of the Poisson-correction method (81) with the units being the number of amino acid substitutions per site. The protein sequences used were obtained from GenBank under the accession numbers indicated in parentheses before the name of each plasmid replicon. The box over the tree indicates the clades that include the close MOB_{P0}, MOB_{P3}, and MOB_{P4} families of relaxases. *See text for description of type IV Dtrs. The black dots denote those plasmid replicons that bear nodulation (*nod*) genes (Reprinted from *Plasmid* [67] with permission of the publisher).
doi:10.1128/microbiolspec.PLAS-0005-2013.f2

Among the pSyms, only the conjugative ability of pSymA (type II Dtr/Mpf) of *S. meliloti* 1021 has been established; whereas, as stated above, the pSymB behaves like a typical mobilizable element that depends on Mpf functions encoded in pSymA (25). Conjugation is regulated through the interplay of the pSymA transcriptional repressor RctA and the antirepressor gene *rctB*, similar to the control that operates in the *R. etli* CFN42 pSym (24). Under conditions nonoptimal for conjugation, RctA maintains the Dtr and Mpf genes silenced. When conditions are favorable for plasmid conjugation, a transcriptional activation of *rctB* occurs and RctA becomes inhibited, thereby allowing expression of the conjugative genes. The mechanism of action of *rctB* is unknown and its predicted product shows no significant homology to proteins in existing databases. In strain *S. meliloti* 1021, activation of *rctB* has recently been found to require the prior inhibition of another transcriptional repressor, RctR, that belongs to the GntR family of one-component regulatory systems (59). The DNA-binding activity of this family of regulators is usually modulated through interaction with a chemical ligand (77). According to available data, RctR does not directly repress *rctB* transcription. Instead, RctR represses the transcription of two adjacent gene operons, Sma0950/53, encoding an ABC transporter, and Sma0956/61, which encodes for several enzymes plus the transcriptional activator *rctC* (Sma0961; Fig. 3). The RctC gene product, in turn, appears to activate transcription of the antirepressor *rctB* (59). It has

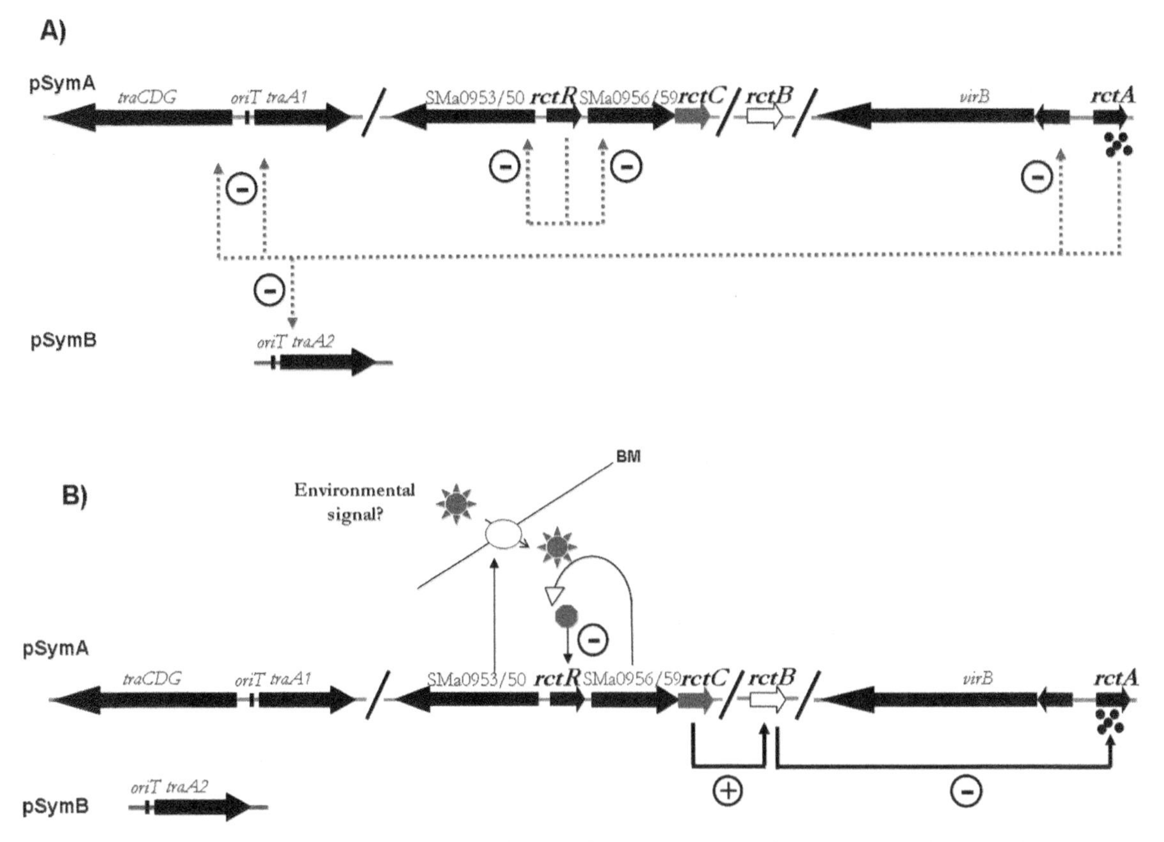

Figure 3 Proposed model for the regulation of conjugative transfer of the *S. meliloti* symbiotic plasmids. (A) Under nonfavorable conditions (i.e., standard laboratory media), RctR silences the Sma0953/50- and SMa0956-58-59-*rctC* operons, so that *rctB* remains inactive and RctA represses conjugative transfer in both pSyms. (B) Conjugative transfer is achieved after an unknown compound—putatively imported by the ABC transporter encoded by SMa0953-50 and modified by the SMa0956-58-59 gene products—binds RctR, thereby relieving transcription repression of *rctC*. The resulting *rctC* transcription ends up with the activation of *rctB* to counteract RctA, the conjugative-transfer operons become expressed, and plasmid conjugation is able to proceed. (+) activation; (–) repression or inhibition; BM, bacterial membrane (Reprinted from *Environ Microbiol* [59] with permission of the publisher). doi:10.1128/microbiolspec.PLAS-0005-2013.f3

been suggested that an as-yet-unknown compound would be imported by the *rctR*-adjacent ABC transporter, and that that signal compound or a metabolic derivative would bind the repressor RctR. That binding of the compound to RctR would thus relieve repression of the adjacent operons Sma0950/53 and Sma0956/61, including the activator gene *rctC*. Once produced, the RctC protein activates transcription at the antirepressor locus *rctB* and the RctA repressor becomes inhibited, thus allowing plasmid conjugation to proceed. Whether or not phosphorylation of the response-regulator protein RctC is needed for conjugative transfer or whether the active form of RctC binds directly or not to the promoter region of *rctB* have not yet been investigated. The nature and origin of the signal compound(s) derepressing conjugative functions are still unknown. The RctR-RctC system is also present in other RctA plasmids—i.e., the pSym of strain *S. meliloti* GR4, the symbiotic plasmids of different *S. fredii* strains (e.g., NGR234, HH103, and USDA257), in addition to the nontumorigenic plasmid pAtC58 of *A. tumefaciens*. This system, however, is not conserved in other RctA plasmids such as the pSym of *R. etli* CFN42 or several others of *S. meliloti*. Therefore, in these other plasmids, *rctB* activation must proceed through a different regulatory cascade.

Although no experimental evidence has been found that *quorum-sensing* mechanisms might be controlling the conjugation of RctA-containing plasmids, some indications exist of a putative relationship between the *rctA/B* system and population density. First, all the pSymA conjugative-transfer genes of *S. meliloti* strain 8530 (an $ExpR^+$ derivative of 1021) appear overexpressed in a *mucR* mutant (78). MucR has accordingly arisen as a global regulator involved in the repression at low population densities of a number of energy-consuming processes, at least some of which are also activated by *quorum sensing* through the ExpR/SinI system (78). In addition, the pSyms of several *S. meliloti* strains (e.g., SM11, AK83, BL225C, and Rm41)—those also likely harboring RctA-type plasmids but not containing the above described RctR-RctC pathway to activate *rctB*—instead possess *luxI*-like and/or *luxR*-like genes adjacent to a putative *rctB* (J. Sanjuán, unpublished observations). The conservation of putative *quorum-sensing*-like genes in the vicinity of the *rctB* locus suggests a relationship between *rctB* activation and population density that certainly deserves to be investigated. What seems clear is that the primary signal(s) that drive the conjugation of the RctA-containing plasmids may be different for several of these. This qualification together with a possible connection with population-density cues suggests a complexity of conditions that likely need to be satisfied for activating the conjugative transfer of these plasmids.

DIVING INTO THE NON-pSym PLASMID COMPARTMENT AT AN OMIC SCALE

In a recent communication, plasmids of up to 200 to 300 kb from 18 *S. meliloti* strains were separated from the chromosomal DNA and ca. 1.5 Mb of nonredundant sequence could be collected by means of high-throughput technologies (López JL, Giusti MA, Martini MC, et al., presented at the International Plasmid Biology Conference, Santander, Spain, September 12–16, 2012). The work constituted the first deep-sequencing approach aimed at the massive characterization of the non-pSym information in *S. meliloti*. The sequence analysis of plasmid replication and mobilization elements evidenced an intricate phylogenetic distribution when compared with homologous elements from other rhizobia, thus pointing to the existence of an interspecific horizontal plasmid exchange. Interestingly, the mobilome sequence revealed a significant diversity of relaxases, in addition to the MOB_{P0} subfamily described previously. As to the kind of open reading frames present in the non-pSym mobilome, the proportion of COG classes showed clear differences compared with those from the chromosome, pSymA, and pSymB (A. Lagares and M. Pistorio, unpublished results); indicating clear functional differentiations among all these four genetic compartments. The experimental approach used proved to be feasible, with deep penetration down to the scale of single-gene analysis. The thousands of genes collected other than those associated with the basic plasmid biology (replication/maintenance/transfer) represent, certainly, the most valuable portion of the new information available.

CONCLUDING REMARKS

The split configuration of the bacterial genome qualitatively into chromosome(s) along with plasmids as intercellular mobile genetic vectors is a cogent physical indication of an efficient and extensive ancestral mechanism of communal evolution and adaptation in prokaryotes. The separation of the essential housekeeping functions resident in the chromosomes from those associated with adaptive responses encoded in plasmid enables bacteria to access and manage, at the population level, an enormous and diverse amount of (horizontally) shareable and accessible information (i.e., the mobilome) that could not otherwise be handled by single bacterial cells, with the final result being that the community as a whole evolves (79). Rhizobia, in general, and *S. meliloti*, in particular, are soil bacteria whose plasmid replicons are known to provide the genetic basis both for speciation events (i.e., HGT of symbiotic traits) and for environmental adaptation. The available information demonstrates that the *S. meliloti* plasmid compartment includes replicons with a wide range of sizes, diverse gene contents, and different degrees of conservation. Among the different strains, diverse kinds of conjugative genes along with their associated regulatory mechanisms have been described, with clear evidences of coevolution. A solid understanding of the conjugative mechanisms that mediate both transient and long-term genomic changes in *S. meliloti* will be necessary to unravel fundamental questions such as how these rhizobia were born as plant symbionts, how their symbioses and specificities evolved, how active the exchange of genetic information with other rhizobia is, and — related to this last question — how ubiquitous the *S. meliloti* conjugative systems are in other plant-associated sympatric rhizospheric-soil bacteria. Upcoming research is expected to focus with much more intensity on the analysis of the (still mostly hidden) genes and functions encoded in the *S. meliloti* non-pSym-plasmid pool. Information of that sort is expected to further our understanding of the diversity of processes that contribute to the symbiotic and free-living lifestyles of these rhizobia.

Acknowledgments. *A.L. and M.P. are funded by CONICET (Consejo Nacional de Investigaciones Científicas y Técnicas—Argentina), MinCyT (Ministerio de Ciencia Tecnolología e Innovación Productiva—MinCyT-Argentina; PICT 2005-31937, PICT 2008-0736, PICT 2012-0102; PICT 2012-0518; PICT 2012-1719), and the European Commission (FP7, Metaexplore, KBBE-222625). J.S. is funded by CSIC (Consejo Superior de Investigaciones Científicas, Spain), Consejería de Economía, Innovación, Ciencia y Empleo de la Junta de Andalucía, Spain (grant AGR-258), and Plan Andaluz de Investigación, Spain.Conflicts of interest: We declare that we have no conflicts.*

Citation. Lagares A, Sanjuán J, Pistorio M. 2014. The plasmid mobilome of the model plant-symbiont *Sinorhizobium meliloti*: coming up with new questions and answers. Microbiol Spectrum **2**(5):PLAS-0005-2013.

References

1. **Peoples MB, Craswell ET.** 1992. Biological nitrogen fixation: investments, expectations and actual contributions to agriculture. *Plant Soil* **141:**13–39.
2. **Peoples MB, Herridge DF.** 1990. Nitrogen fixation by legumes in tropical and sub-tropical agriculture. *Adv Agron* **44:**155–224.
3. **García de los Santos A, Brom S, Romero D.** 1996. *Rhizobium* plasmids in bacteria-legume interactions. *World J Microbiol Biotechnol* **12:**119–125.
4. **Mercado-Blanco J, Toro N.** 1996. Plasmids in *Rhizobia*: the role of nonsymbiotic plasmids. *Mol Plant Microbe Interact* **9:**535–545.
5. **Banfalvi Z, Kondorosi E, Kondorosi A.** 1985. *Rhizobium meliloti* carries two megaplasmids. *Plasmid* **13:**129–138.
6. **Hynes MF, Simon R, Müller P, Niehaus K, Labes M, Pühler A.** 1986. The two megaplasmids of *Rhizobium meliloti* are involved in the effective nodulation of alfalfa. *Mol Gen Genet* **202:**356–362.
7. **Banfalvi Z, Sakanyan V, Koncz C, Kiss A, Dusha I, Kondorosi A.** 1981. Location of nodulation and nitrogen fixation genes on a high molecular weight plasmid of *Rhizobium meliloti*. *Mol Gen Genet* **184:**318–325.
8. **Smets BF, Barkay T.** 2005. Horizontal gene transfer: perspectives at a crossroads of scientific disciplines. *Nat Rev Microbiol* **3:**675–678.
9. **Pistorio M, Del Papa MF, Balague LJ, Lagares A.** 2003. Identification of a transmissible plasmid from an Argentine *Sinorhizobium meliloti* strain which can be mobilised by conjugative helper functions of the European strain *S. meliloti* GR4. *FEMS Microbiol Lett* **225:** 15–21.
10. **Roumiantseva ML, Andronov EE, Sharypova LA, Dammann-Kalinowski T, Keller M, Young JP, Simarov BV.** 2002. Diversity of *Sinorhizobium meliloti* from the Central Asian alfalfa gene center. *Appl Environ Microbiol* **68:**4694–4697.
11. **Stiens M, Schneiker S, Keller M, Kuhn S, Pühler A, Schlüter A.** 2006. Sequence analysis of the 144-kilobase accessory plasmid pSmeSM11a, isolated from a dominant *Sinorhizobium meliloti* strain identified during a long-term field release experiment. *Appl Environ Microbiol* **72:**3662–3672.
12. **Kosier B, Pühler A, Simon R.** 1993. Monitoring the diversity of *Rhizobium meliloti* field and microcosm isolates with a novel rapid genotyping method using insertion elements. *Mol Ecol* **2:**35–36.
13. **Pistorio M, Giusti MA, Del Papa MF, Draghi WO, Lozano MJ, Tejerizo GT, Lagares A.** 2008. Conjugal properties of the *Sinorhizobium meliloti* plasmid mobilome. *FEMS Microbiol Ecol* **65:**372–382.
14. **Mercado-Blanco J, Olivares J.** 1993. Stability and transmissibility of the cryptic plasmids of *Rhizobium meliloti* GR4. Their possible use in the construction of cloning vectors of rhizobia. *Arch Microbiol* **160:**477–485.
15. **Barnett MJ, Fisher RF, Jones T, Komp C, Abola AP, Barloy-Hubler F, Bowser L, Capela D, Galibert F, Gouzy J, Gurjal M, Hong A, Huizar L, Hyman RW, Kahn D, Kahn ML, Kalman S, Keating DH, Palm C, Peck MC, Surzycki R, Wells DH, Yeh KC, Davis RW, Federspiel NA, Long SR.** 2001. Nucleotide sequence and predicted functions of the entire *Sinorhizobium meliloti* pSymA megaplasmid. *Proc Natl Acad Sci USA* **98:**9883–9888.
16. **Finan TM, Weidner S, Wong K, Buhrmester J, Chain P, Vorholter FJ, Hernandez-Lucas I, Becker A, Cowie A, Gouzy J, Golding B, Puhler A.** 2001. The complete

sequence of the 1,683-kb pSymB megaplasmid from the N_2-fixing endosymbiont *Sinorhizobium meliloti*. *Proc Natl Acad Sci USA* **98:**9889–9894.

17. **Giuntini E, Mengoni A, De Filippo C, Cavalieri D, Aubin-Horth N, Landry CR, Becker A, Bazzicalupo M.** 2005. Large-scale genetic variation of the symbiosis-required megaplasmid pSymA revealed by comparative genomic analysis of *Sinorhizobium meliloti* natural strains. *BMC Genomics* **6:**158. doi:10.1186/1471-2164-6-158.
18. **Oresnik IJ, Liu SL, Yost CK, Hynes MF.** 2000. Megaplasmid pRme2011a of *Sinorhizobium meliloti* is not required for viability. *J Bacteriol* **182:**3582–3586.
19. **Slater SC, Goldman BS, Goodner B, Setubal JC, Farrand SK, Nester EW, Burr TJ, Banta L, Dickerman AW, Paulsen I, Otten L, Suen G, Welch R, Almeida NF, Arnold F, Burton OT, Du Z, Ewing A, Godsy E, Heisel S, Houmiel KL, Jhaveri J, Lu J, Miller NM, Norton S, Chen Q, Phoolcharoen W, Ohlin V, Ondrusek D, Pride N, Stricklin SL, Sun J, Wheeler C, Wilson L, Zhu H, Wood DW.** 2009. Genome sequences of three agrobacterium biovars help elucidate the evolution of multi-chromosome genomes in bacteria. *J Bacteriol* **191:**2501–2511.
20. **Galardini M, Pini F, Bazzicalupo M, Biondi EG, Mengoni A.** 2013. Replicon-dependent bacterial genome evolution: the case of *Sinorhizobium meliloti*. *Genome Biol Evol* **5:**542–558.
21. **Harrison PW, Lower RP, Kim NK, Young JP.** 2010. Introducing the bacterial 'chromid': not a chromosome, not a plasmid. *Trends Microbiol* **18:**141–148.
22. **Galardini M, Mengoni A, Brilli M, Pini F, Fioravanti A, Lucas S, Lapidus A, Cheng JF, Goodwin L, Pitluck S, Land M, Hauser L, Woyke T, Mikhailova N, Ivanova N, Daligault H, Bruce D, Detter C, Tapia R, Han C, Teshima H, Mocali S, Bazzicalupo M, Biondi EG.** 2011. Exploring the symbiotic pangenome of the nitrogen-fixing bacterium *Sinorhizobium meliloti*. *BMC Genomics* **12:**235. doi:10.1186/1471-2164-12-235.
23. **Schneiker-Bekel S, Wibberg D, Bekel T, Blom J, Linke B, Neuweger H, Stiens M, Vorholter FJ, Weidner S, Goesmann A, Puhler A, Schluter A.** 2011. The complete genome sequence of the dominant *Sinorhizobium meliloti* field isolate SM11 extends the *S. meliloti* pan-genome. *J Biotechnol* **155:**20–33.
24. **Perez-Mendoza D, Sepulveda E, Pando V, Munoz S, Nogales J, Olivares J, Soto MJ, Herrera-Cervera JA, Romero D, Brom S, Sanjuan J.** 2005. Identification of the *rctA* gene, which is required for repression of conjugative transfer of rhizobial symbiotic megaplasmids. *J Bacteriol* **187:**7341–7350.
25. **Blanca-Ordóñez H, Oliva-García JJ, Pérez-Mendoza D, Soto MJ, Olivares J, Sanjuán J, Nogales J.** 2010. pSymA-dependent mobilization of the *Sinorhizobium meliloti* pSymB megaplasmid. *J Bacteriol* **192:**6309–6312.
26. **Barran LR, Ritchot N, Bromfield ES.** 2001. *Sinorhizobium meliloti* plasmid pRm1132f replicates by a rolling-circle mechanism. *J Bacteriol* **183:**2704–2708.
27. **Hahn M, Hennecke H.** 1987. Mapping of a *Bradyrhizobium japonicum* DNA region carrying genes for symbiosis and an asymmetric accumulation of reiterated sequences. *Appl Environ Microbiol* **53:**2247–2252.
28. **Kaluza K, Hahn M, Hennecke H.** 1985. Repeated sequences similar to insertion elements clustered around the *nif* region of the *Rhizobium japonicum* genome. *J Bacteriol* **162:**535–542.
29. **Bromfield ES, Lewis DM, Barran LR.** 1985. Cryptic plasmid and rifampin resistance in *Rhizobium meliloti* influencing nodulation competitiveness. *J Bacteriol* **164:** 410–413.
30. **García-Rodriguez FM, Toro N.** 2000. *Sinorhizobium meliloti nfe* (nodulation formation efficiency) genes exhibit temporal and spatial expression patterns similar to those of genes involved in symbiotic nitrogen fixation. *Mol Plant Microbe Interact* **13:**583–591.
31. **Soto MJ, Zorzano A, Garcia-Rodriguez FM, Mercado-Blanco J, Lopez-Lara IM, Olivares J, Toro N.** 1994. Identification of a novel *Rhizobium meliloti* nodulation efficiency nfe gene homolog of *Agrobacterium* ornithine cyclodeaminase. *Mol Plant Microbe Interact* **7:**703–707.
32. **Soto MJ, Zorzano A, Mercado-Blanco J, Lepek V, Olivares J, Toro N.** 1993. Nucleotide sequence and characterization of *Rhizobium meliloti* nodulation competitiveness genes *nfe*. *J Mol Biol* **229:**570–576.
33. **Velázquez E, Mateos PF, Pedrero P, Dazzo FB, Martinez-Molina E.** 1995. Attenuation of symbiotic effectiveness by *Rhizobium meliloti* SAF22 related to the presence of a cryptic plasmid. *Appl Environ Microbiol* **61:**2033–2036.
34. **Eckhardt T.** 1978. A rapid method for the identification of plasmid desoxyribonucleic acid in bacteria. *Plasmid* **1:** 584–588.
35. **Wheatcroft R, McRae DG, Miller RW.** 1990. Changes in the *Rhizobium meliloti* genome and the ability to detect supercoiled plasmids during bacteroid development. *Mol Plant Microbe Interact* **3:**9–17.
36. **Plazinski J, Cen YH, Rolfe BG.** 1985. General method for the identification of plasmid species in fast-growing soil microorganisms. *Appl Environ Microbiol* **49:** 1001–1003.
37. **Velázquez E, Mateos PF, Velasco N, Santos F, Burgos PA, Villadas P, Toro N, Martínez-Molina E.** 1999. Symbiotic characteristics and selection of autochthonous strains of *Sinorhizobium meliloti* populations in different soils. *Soil Biol Biochem* **31:**1039–1047.
38. **Coutinho HLC, Oliveira VM, Lovato A, Maia AHN, Manfio GP.** 1999. Evaluation of the diversity of rhizobia in Brazilian agricultural soils cultivated with soybeans. *Appl Soil Ecol* **13:**159–167.
39. **Cevallos MA, Cervantes-Rivera R, Gutierrez-Rios RM.** 2008. The *repABC* plasmid family. *Plasmid* **60:**19–37.
40. **Pinto UM, Pappas KM, Winans SC.** 2012. The ABCs of plasmid replication and segregation. *Nat Rev Microbiol* **10:**755–765.
41. **Ramírez-Romero MA, Soberón N, Pérez-Oseguera A, Téllez-Sosa J, Cevallos MA.** 2000. Structural elements required for replication and incompatibility of the *Rhizobium etli* symbiotic plasmid. *J Bacteriol* **182:**3117–3124.

42. **Salje J.** 2010. Plasmid segregation: how to survive as an extra piece of DNA. *Crit Rev Biochem Mol Biol* **45:** 296–317.
43. **Hwang LC, Vecchiarelli AG, Han YW, Mizuuchi M, Harada Y, Funnell BE, Mizuuchi K.** 2013. ParA-mediated plasmid partition driven by protein pattern self-organization. *EMBO J* **32:**1238–1249.
44. **Cevallos MA, Porta H, Izquierdo J, Tun-Garrido C, Garcia-de-los-Santos A, Davila G, Brom S.** 2002. *Rhizobium etli* CFN42 contains at least three plasmids of the *repABC* family: a structural and evolutionary analysis. *Plasmid* **48:**104–116.
45. **Venkova-Canova T, Soberón NE, Ramírez-Romero MA, Cevallos MA.** 2004. Two discrete elements are required for the replication of a *repABC* plasmid: an antisense RNA and a stem-loop structure. *Mol Microbiol* **54:**1431–1444.
46. **Chai Y, Winans SC.** 2005. A small antisense RNA downregulates expression of an essential replicase protein of an *Agrobacterium tumefaciens* Ti plasmid. *Mol Microbiol* **56:**1574–1585.
47. **MacLellan SR, Smallbone LA, Sibley CD, Finan TM.** 2005. The expression of a novel antisense gene mediates incompatibility within the large *repABC* family of alpha-proteobacterial plasmids. *Mol Microbiol* **55:**611–623.
48. **Mercado-Blanco J, Olivares J.** 1994. The large nonsymbiotic plasmid pRmeGR4a of *Rhizobium meliloti* GR4 encodes a protein involved in replication that has homology with the RepC protein of *Agrobacterium* plasmids. *Plasmid* **32:**75–79.
49. **Izquierdo J, Venkova-Canova T, Ramirez-Romero MA, Tellez-Sosa J, Hernandez-Lucas I, Sanjuan J, Cevallos MA.** 2005. An antisense RNA plays a central role in the replication control of a *repC* plasmid. *Plasmid* **54:** 259–277.
50. **Watson RJ, Heys R.** 2006. Replication regions of *Sinorhizobium meliloti* plasmids. *Plasmid* **55:**87–98.
51. **Castellanos M, Romero D.** 2009. The extent of migration of the Holliday junction is a crucial factor for gene conversion in *Rhizobium etli*. *J Bacteriol* **191:**4987–4995.
52. **Hernandez-Salmeron JE, Santoyo G.** 2011. Phylogenetic analysis reveals gene conversions in multigene families of rhizobia. *Genet Mol Res* **12:**1383–1392.
53. **Tian CF, Young JP, Wang ET, Tamimi SM, Chen WX.** 2010. Population mixing of *Rhizobium leguminosarum* bv. viciae nodulating Vicia faba: the role of recombination and lateral gene transfer. *FEMS Microbiol Ecol* **73:** 563–576.
54. **Torres Tejerizo G, Del Papa MF, Giusti MA, Draghi W, Lozano M, Lagares A, Pistorio M.** 2010. Characterization of extrachromosomal replicons present in the extended host range *Rhizobium* sp. LPU83. *Plasmid* **64:** 177–185.
55. **van Berkum P, Terefework Z, Paulin L, Suomalainen S, Lindstrom K, Eardly BD.** 2003. Discordant phylogenies within the *rrn* loci of rhizobia. *J Bacteriol* **185:**2988–2998.
56. **Cervantes L, Bustos P, Girard L, Santamaría RI, Dávila G, Vinuesa P, Romero D, Brom S.** 2011. The conjugative plasmid of a bean-nodulating *Sinorhizobium fredii* strain is assembled from sequences of two *Rhizobium* plasmids and the chromosome of a *Sinorhizobium* strain. *BMC Microbiol* **11:**149. doi:10.1186/1471-2180-11-149.
57. **Mazur A, Majewska B, Stasiak G, Wielbo J, Skorupska A.** 2011. *repABC*-based replication systems of *Rhizobium leguminosarum* bv. trifolii TA1 plasmids: incompatibility and evolutionary analyses. *Plasmid* **66:**53–66.
58. **Suominen L, Roos C, Lortet G, Paulin L, Lindstrom K.** 2001. Identification and structure of the *Rhizobium galegae* common nodulation genes: evidence for horizontal gene transfer. *Mol Biol Evol* **18:**907–916.
59. **Nogales J, Blanca-Ordonez H, Olivares J, Sanjuan J.** 2013. Conjugal transfer of the *Sinorhizobium meliloti* 1021 symbiotic plasmid is governed through the concerted action of one- and two-component signal transduction regulators. *Environ Microbiol* **15:**811–821.
60. **Bailly X, Olivieri I, Brunel B, Cleyet-Marel JC, Béna G.** 2007. Horizontal gene transfer and homologous recombination drive the evolution of the nitrogen-fixing symbionts of *Medicago* species. *J Bacteriol* **189:**5223–5236.
61. **Shannon CE, Weaver W.** 1949. Evolution and measurement of species diversity. *Taxon* **21:**213–251.
62. **Simpson EH.** 1949. Measurement of diversity. *Nature* **163:**688.
63. **Herrera-Cervera JA, Olivares J, Sanjuan J.** 1996. Ammonia inhibition of plasmid pRmeGR4a conjugal transfer between *Rhizobium meliloti* strains. *Appl Environ Microbiol* **62:**1145–1150.
64. **Giusti MA, Lozano MJ, Torres Tejerizo GA, Martini MC, Salas ME, López JL, Draghi WO, Del Papa MF, Pistorio M, Lagares A.** 2013. Conjugal transfer of a *Sinorhizobium meliloti* cryptic plasmid evaluated during a field release and in soil microcosms. *Eur J Soil Biol* **55:** 9–12.
65. **Kinkle BK, Schmidt EL.** 1991. Transfer of the pea symbiotic plasmid pJB5JI in nonsterile soil. *Appl Environ Microbiol* **57:**3264–3269.
66. **Ding H, Hynes MF.** 2009. Plasmid transfer systems in the rhizobia. *Can J Microbiol* **55:**917–927.
67. **Giusti MA, Pistorio M, Lozano MJ, Torres Tejerizo GA, Salas ME, Martini MC, López JL, Draghi WO, Del Papa MF, Pérez-Mendoza D, Sanjuán J, Lagares A.** 2012. Genetic and functional characterization of a yet-unclassified rhizobial Dtr (DNA-transfer-and-replication) region from the ubiquitous plasmid conjugal system present in *Sinorhizobium meliloti*, *Sinorhizobium medicae*, and in other Gram-negative bacteria. *Plasmid* **67:**199–210.
68. **Ding H, Yip CB, Hynes MF.** 2013. Genetic characterization of a novel rhizobial plasmid conjugation system in *Rhizobium leguminosarum* bv. viciae strain VF39SM. *J Bacteriol* **195:**328–339.
69. **Herrera-Cervera JA, Sanjuan-Pinilla JM, Olivares J, Sanjuan J.** 1998. Cloning and identification of conjugative transfer origins in the *Rhizobium meliloti* genome. *J Bacteriol* **180:**4583–4590.
70. **Garcillan-Barcia MP, Francia MV, de la Cruz F.** 2009. The diversity of conjugative relaxases and its application

in plasmid classification. *FEMS Microbiol Rev* **33:** 657–687.

71. **Holmes B, Poppoff M, Kiredjian M, Kersters K.** 1988. *Ochrobactrum anthropi* gen. nov., sp. nov. from human clinical specimens and previously known as group Vd. *Int J Syst Evol Bacteriol* **38:**406–416.
72. **Pistorio M, Torres Tejerizo GA, Del Papa MF, de Los Angeles Giusti M, Lozano M, Lagares A.** 2013. *rptA*, a novel gene from *Ensifer* (*Sinorhizobium*) meliloti involved in conjugal transfer. *FEMS Microbiol Lett* **345:** 22–30.
73. **He X, Chang W, Pierce DL, Seib LO, Wagner J, Fuqua C.** 2003. Quorum sensing in *Rhizobium* sp. strain NGR234 regulates conjugal transfer (*tra*) gene expression and influences growth rate. *J Bacteriol* **185:**809–822.
74. **Danino VE, Wilkinson A, Edwards A, Downie JA.** 2003. Recipient-induced transfer of the symbiotic plasmid pRL1JI in *Rhizobium leguminosarum* bv. viciae is regulated by a quorum-sensing relay. *Mol Microbiol* **50:**511–525.
75. **Wilkinson A, Danino V, Wisniewski-Dye F, Lithgow JK, Downie JA.** 2002. N-acyl-homoserine lactone inhibition of rhizobial growth is mediated by two quorum-sensing genes that regulate plasmid transfer. *J Bacteriol* **184:** 4510–4519.
76. **Tun-Garrido C, Bustos P, Gonzalez V, Brom S.** 2003. Conjugative transfer of p42a from *Rhizobium etli* CFN42, which is required for mobilization of the symbiotic plasmid, is regulated by quorum sensing. *J Bacteriol* **185:**1681–1692.
77. **Hoskisson PA, Rigali S.** 2009. Chapter 1: variation in form and function: the helix-turn-helix regulators of the GntR superfamily. *Adv Appl Microbiol* **69:**1–22.
78. **Mueller K, Gonzalez JE.** 2011. Complex regulation of symbiotic functions is coordinated by MucR and quorum sensing in *Sinorhizobium meliloti*. *J Bacteriol* **193:** 485–496.
79. **Woese CR.** 2002. On the evolution of cells. *Proc Natl Acad Sci USA* **99:**8742–8747.
80. **Felsenstein J.** 1985. Confidence limits on phylogenies: an approach using the bootstrap. *Evolution* **39:**783–791.
81. **Zuckerkandl E, Pauling L.** 1965. Evolutionary divergence and convergence in proteins, p 97–166. *In* Bryson V, Vogel HJ (ed), *Evolving Genes and Proteins*. Academic Press, New York, NY.

Plasmids—Biology and Impact in Biotechnology and Discovery
Edited by Marcelo E. Tolmasky and Juan C. Alonso

doi:10.1128/microbiolspec.PLAS-0010-2013

Jay E. Gordon[1]
Peter J. Christie[1]

The *Agrobacterium* Ti Plasmids 17

NOMENCLATURE AND TYPES OF Ti PLASMIDS

Agrobacterium species that are pathogenic on plants, including *Agrobacterium tumefaciens*, *A. vitis*, *A. rubi*, and *A. rhizogenes*, all carry megaplasmids. By contrast, nonpathogenic strains either lack these plasmids entirely or carry mutant forms of plasmids. A strict requirement of the Ti plasmid for virulence was established through mutational analyses and by a demonstration that the introduction of Ti plasmids into *Rhizobium* or *Phyllobacterium* spp. converts these nonpathogenic species into tumor-inducing pathogens (2, 3). Ti plasmids induce a disease called crown gall, which is typified by the formation of undifferentiated plant tumors at the plant crown (the subterranean-to-aerial transition zone). The related root-inducing or Ri megaplasmids carried by *A. rhizogenes* instead induce hairy root disease, which is typified by the formation of entangled masses of roots at the infection site (4).

As discussed in more detail below, Ti plasmid-carrying strains of agrobacteria induce not only plant tumor formation but also the production of various amino acid and sugar phosphate derivatives termed opines. The transformed plant cell secretes opines, which can then be taken up and catabolized for use as a food source by the infecting bacterium. The Ti plasmid carries the genes for opine synthesis by plant cells as well as the corresponding catabolism genes. Ti plasmids traditionally have been classified by opine type, and here we will retain this classification scheme with a focus mainly on the two best-characterized Ti plasmids designated as the octopine and nopaline types. Several octopine-type (pTiA6, B6, Ach5, 15955, R10) and nopaline-type (pTiC58, pTi37) Ti plasmids have been extensively characterized, and many have been sequenced at this time. All Ti plasmids code for functions associated with (i) plasmid replication and maintenance, (ii) conjugative transfer, (iii) virulence, (iv) opine utilization, and (v) sensory perception of exogenous signals released by the plant host and neighboring agrobacterial cells at the site of infection (see reference 5). Genes encoding each of these functions are generally clustered on the Ti plasmid, with the exception of two spatially distinct regions, the virulence or *vir* region and the transfer-DNA or T-DNA required for infection of plants, and the *tra* and *trb* regions required for conjugative plasmid transfer (Fig. 1). The following sections summarize our current understanding of the Ti plasmid-encoded functions.

Ti PLASMID MAINTENANCE

repABC

The Ti plasmids belong to the *repABC* family of replicons, whose members are widely distributed among

[1]Department of Microbiology and Molecular Genetics, University of Texas Medical School at Houston, Houston, TX 77005.

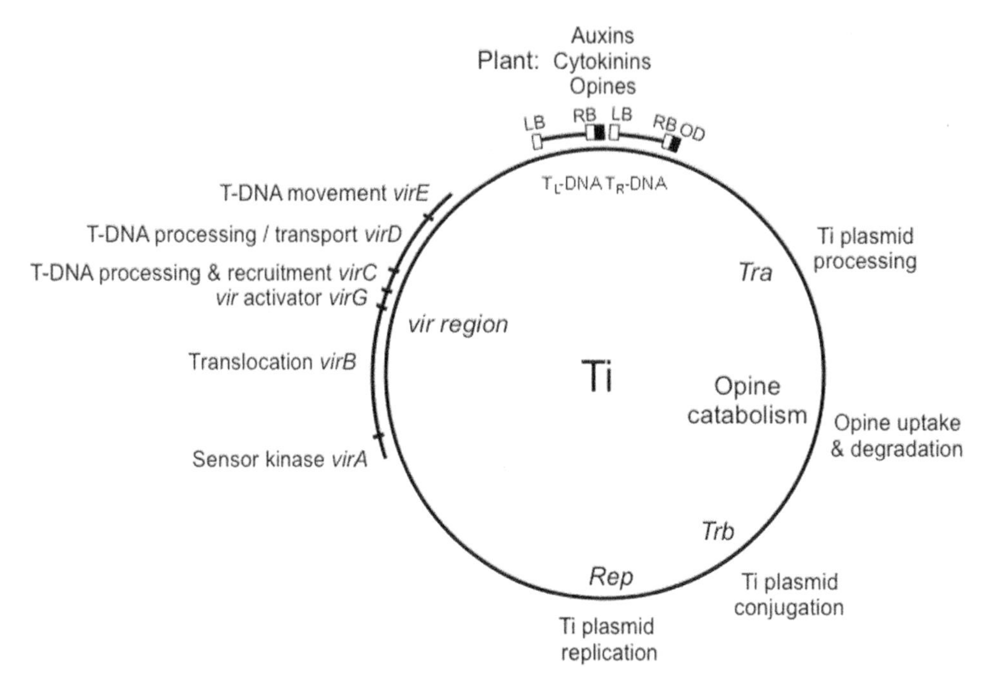

Figure 1 Schematic of octopine-type Ti plasmid pTiA6 showing locations of genes coding for plasmid maintenance (*rep*), infection of plant cells (*vir* region, T-DNA), cell survival in the tumor environment (opine catabolism), and conjugative transfer of the Ti plasmid to recipient agrobacteria (*tra* and *trb*). The various contributions of the *vir* gene products to T-DNA transfer are listed. T_L-DNA and T_R-DNA are delimited by *oriT*-like border sequences (black boxes; RB, right border; LB, left border); OD, *overdrive* sequence (white boxes) enhances VirD2 relaxase nicking at the T-DNA border sequences. When delivered to plant cells and integrated into the plant nuclear genome, T-DNAs code for biosynthesis of auxins and cytokinins, resulting in the proliferation of plant tissues, and production of opines that serve as nutrients for the infecting bacterium. (Adapted from reference 117 [Christie PJ, *Agrobacterium* and Plant Cell Transformation, *In*: *Encyclopedia of Microbiology*, 2009], copyright 2009, with permission from Elsevier.)
doi:10.1128/microbiolspec.PLAS-0010-2013.f1

many species of *Alphaproteobacteria* (1). This replicon family is composed mainly of extrachromosomal plasmids and some secondary chromosomes. The *repABC* cassette was identified nearly 25 years ago as essential for replication and partitioning of an octopine-type Ti plasmid (6). The cassette is composed of three genes (*repA*, *repB*, *repC*), a *cis*-acting partitioning site (*parS*), and an origin of replication *oriV* (Fig. 2). *repA* and *repB* code for a partitioning system, and *repC* encodes the replication initiator protein (1, 7). The RepA/RepB partitioning system closely resembles the ParA/ParB systems harbored by many plasmids, phages, and chromosomes in diverse bacterial species (8, 9). These proteins function together with *parS* sequence to ensure faithful plasmid partitioning during cell division (119). The RepC initiator protein is uniquely associated with *repABC* cassettes in *Alphaproteobacteria* and is unrelated in sequence to other well-characterized Rep proteins (1). RepC binds a putative *oriV* sequence *in vitro* (10), and cloned *repC* genes support autonomous replication of associated replicons (10, 11). The *in vivo* findings strongly indicate that RepC functions as the initiator protein and further suggest that the *oriV* replication origin is located within the *repC* gene.

Plasmid Partitioning

The RepA/RepB partitioning system resembles that of the well-characterized ParA/ParB systems and likely has a similar mechanism of action (Fig. 2) (1, 119). In brief, these partitioning systems consist of three components: a weak NTPase (ParA), a DNA-binding protein (ParB), and the *cis*-acting *parS* sequence to which ParB binds. RepA is a member of the ParA/MinD/Soj superfamily of ATPases, and, as shown for many ParA homologs, the RepA proteins of pTiR10 and other Ti plasmids negatively autoregulate *rep* gene expression by binding an upstream promoter (12, 13). In the Par systems, ADP-bound ParA is active in autorepression via a

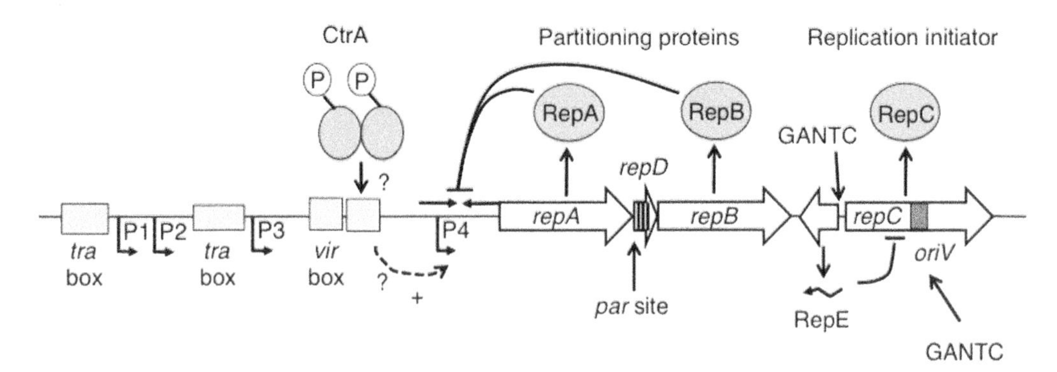

Figure 2 Regulation of the *repABC* operon of octopine-type Ti plasmids. Transcription of the *repABC* operon is inhibited by autorepression mediated by RepA-RepB complexes at the operator region downstream of P4 (a region of dyad symmetry is denoted by inverted arrows) and at the partitioning (*parS*) site located between *repA* and *repB*. Expression of *repC* is inhibited transcriptionally and posttranscriptionally by the countertranscribed RNA RepE. Tumor-inducing (Ti) plasmids are maintained as single copies in the absence of external signals. Additional regulation of Ti plasmid replication during the cell cycle may be provided by phosphorylated CtrA and by CcrM methylation at GANTC motifs within *repA* and upstream of *repE*. Sensory perception of two exogenous signals results in elevated transcription of the *repABC* cassette and increased plasmid copy number. Plant-released phenolic compounds are detected by the VirA-VirG two-component system; Phospho-VirG binds a *vir* box to activate transcription from promoter P4. TraR–3-oxo-octanoylhomoserine lactone complexes bind *tra* boxes, activating the *repABC* operon through promoters P1, P2, P3, and P4. (Adapted from reference 1 with permission from Macmillan Publishers Ltd. [Pinto UM, Pappas KM, Winans SC, The ABCs of plasmid replication and segregation. *Nature Rev Microbiol* **10**:755–765, 2012, doi:10.1038/nrmicro2882], copyright 2012.)
doi:10.1128/microbiolspec.PLAS-0010-2013.f2

DNA-binding activity of its N-terminal domain. In contrast, the ATP-bound form of ParA is dimeric and fulfills another function through cooperative but nonspecific binding to DNA. ParA-ATP binds DNA dynamically, oscillating over the nucleoid. This oscillatory behavior is thought to influence the positioning of associated DNA in the predivisional cell or DNA translocation during cell segregation (14, 15, 16). This dynamic activity and the capacity of ParA proteins to form filaments or spiral structures are reminiscent of eukaryotic cytoskeletal or motor proteins and suggest that the partitioning system resembles a form of bacterial mitosis (17). ParA also binds ParB. ParB specifically binds the centromere-like *parS* sequence via a helix-turn-helix motif located in its C terminus. Thus, ParA foci or filamentous structures are postulated to bind cognate ParB-*parS* partitioning complexes to provide a pulling or ratcheting force necessary for spatial localization of the associated plasmids in predivisional cells.

Although most characterized *parAB* loci are autoregulated, they usually function separately from the replication and copy number control system of the hosting replicon. In contrast, the *repABC* genes associated with the alphaproteobacterial megaplasmids, e.g., symbiotic plasmid p42d and octopine-type Ti plasmids, are cotranscribed and constitute a single operon (Fig. 2) (1). This genetic context places expression of the partitioning genes and the replication initiator protein under the same regulatory control. Therefore, RepA-ADP-mediated negative autoregulation is partly responsible for the low copy number of these plasmids. Additionally, in the *repABC* cassette of Ti plasmids, a small gene located between *repB* and *repC* encodes a small nontranslocated countertranscribed RNA that downregulates expression of *repC* (18). This short transcript maintains the Ti plasmid at a low copy number and also functions to inhibit replication of coresident replicons bearing the same *repABC* cassette in a mechanism termed incompatibility (1). The *repABC* cassettes have two other noteworthy features (Fig. 2). First, the DNA sequence GANTC is overrepresented in the putative replication origin and in the promoter of the countertranscribed *repE* RNA. These sequences are potential substrates for the DNA methylase CcrM, which in *Caulobacter crescentus* contributes to cell cycle timing (1). Second, a binding site for another *C. crescentus* cell cycle factor, the two-component response regulator CtrA, was identified upstream of the *repABC* promoter region of plasmid pTiR10 (1). These findings warrant further studies defining the contribution of

DNA methylation, CrtA, and possibly other factors in temporal control of Ti plasmid replication during the cell cycle.

Plasmid Replication

As mentioned above, the RepC-type proteins have been found only in *Alphaproteobacteria*, and the presumptive RepC-binding target, *oriV*, resides within *repC*. In contrast to other plasmid replication systems in which the initiator protein binds directly repeated DNA sequences called iterons to target the replication machinery to the origin of replication, *repABC* origins lack such repeat sequences. The *repC* genes do, however, contain AT-rich sequences of ~150 nucleotides near the middle of their sequences, which is commonly encountered with other replication origins (Fig. 2) (1). The RepC protein from pTiR10 was purified and shown to bind a region of imperfect dyad symmetry within the AT-rich segment, lending support to the proposal that this sequence conforms to the origin of replication (10). RepC has two domains, an N-terminal domain (NTD) that exhibits DNA-binding activity and a C-terminal domain (CTD) whose function is currently unknown. The NTD has structural similarity to members of the DnaD family of replication proteins found in low GC-content Gram-positive bacteria as well as to members of the MarR family of transcriptional factors (1). While the molecular details are unknown, it is reasonable to predict that RepC-*oriV* binding serves to recruit other replication proteins to *oriV* to build the replisome.

Another interesting feature of the RepC proteins encoded by plasmids pTiR10 and p42D is that they appear to function only in *cis* (10, 11). That is, they act only on the origin of replication embedded within the *repC* gene itself and not on origins located within *repABC* cassettes of coresident plasmids. Consistent with this activity, overproduction of RepC results in an increase in the copy number of the plasmid encoding the protein but has no effect on the copy numbers of coresident *repABC* replicons. Some alphaproteobacterial species can have as many as six *repABC* family replicons (1). In such cells, the *cis*-acting function of RepC proteins could serve to ensure the fidelity of replication and copy number control of cognate replicons without interfering effects on heterologous replicons.

Ti PLASMID-ENCODED TYPE IV SECRETION SYSTEMS

Ti plasmids carry genes for elaboration of two DNA conjugation systems, one (Tra/Trb) responsible for conjugative transfer of the Ti plasmid and the second (VirB/VirD4) dedicated to the delivery of a segment of the Ti plasmid called the T-DNA as well as several effector proteins to plant cells during the infection process (Fig. 1) (5). Studies of the Ti plasmid transfer system have focused mainly on defining the functions of regulatory factors in controlling *tra/trb* gene expression (for example, see reference 19), whereas investigations of the T-DNA transfer system have explored the biogenesis, mechanism of action, and architecture of the conjugation channel and associated T pilus (see references 20, 21). All bacterial conjugation systems are now grouped together with a set of ancestrally related translocation systems in pathogenic bacteria that are dedicated to the delivery of effector proteins into eukaryotic cells during the course of infection. Collectively, these translocation systems are called the type IV secretion systems (T4SSs) (22). The VirB/VirD4 system has emerged as a paradigm for this T4SS superfamily (20), and, here, we will use the nomenclature associated with this system when discussing the functions of the individual subunits. The following sections will summarize general features of T4SSs, as well as available mechanistic and structural information about the Tra/Trb and VirB/VirD4 systems. The overall mechanism of type IV secretion can be viewed as three biochemically distinct but spatially and temporally coupled reactions: (i) substrate processing as a translocation-competent transfer intermediate, (ii) substrate docking with a protein termed the type IV coupling protein (T4CP), and (iii) substrate transfer through the envelope-spanning translocation channel (Fig. 3).

DNA and Effector Protein Substrate Processing

The *tra* and *vir* genes, respectively, code for proteins responsible for processing the Ti plasmid and T-DNA. The overall conjugative DNA-processing reaction is as follows. The relaxase is the principal enzyme required for DNA processing. This enzyme binds a cognate origin of transfer (*oriT*) sequence and nicks the DNA strand (T strand) destined for transfer. Upon nicking, the relaxase remains covalently bound to the 5′ end of the T strand. Additional auxiliary or accessory factors, termed DNA transfer and replication (Dtr) proteins, also bind at the *oriT* sequence to form the relaxosome. The Dtr factors enhance relaxase binding and cleavage at *oriT*, and they can also participate in docking of the DNA substrate with the substrate receptor for the cognate T4SS channel. Upon nicking, the T strand is unwound from the template strand, the Dtr factors are dissociated from the relaxase-T-strand particle, and the

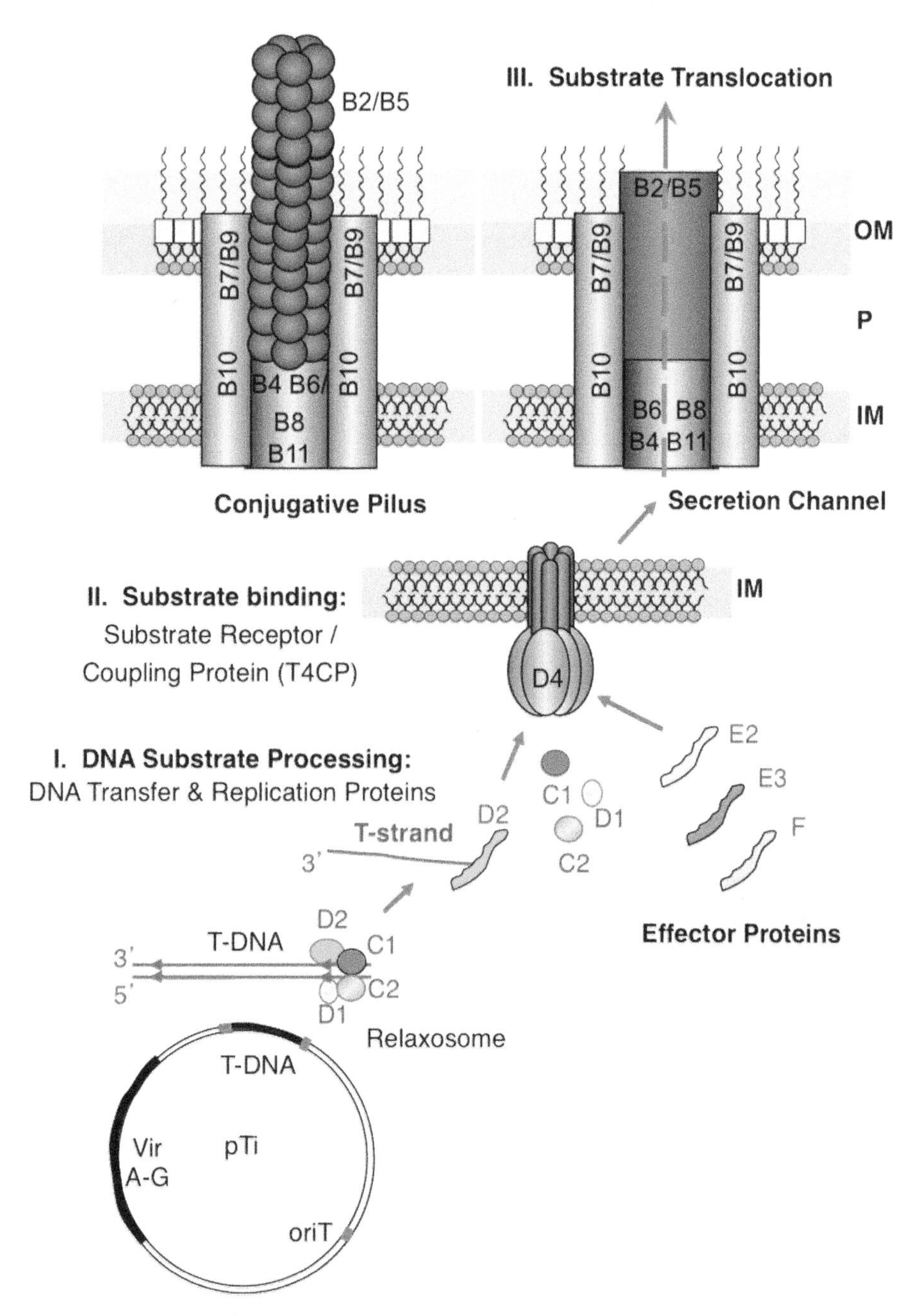

Figure 3 Schematic showing steps of type IV secretion, as presented for the Ti-encoded VirB/VirD4 transfer system. Step I: the DNA transfer and replication (Dtr) proteins bind the *oriT*-like right border repeat sequence (Ti plasmid, red squares flanking T-DNA) to form the relaxosome. VirD2 relaxase nicks the T strand, which is then unwound from the template strand of the pTi plasmid. Step II: ParA-like VirC1 and VirD2, and probably other factors, mediate binding of the VirD2-T-strand transfer intermediate with the VirD4 substrate receptor or type IV coupling protein (T4CP). Step III: The transfer intermediate is translocated across the cell envelope through a secretion channel composed of the VirD4 T4CP and the VirB mating pair formation (Mpf) proteins. Effector proteins, e.g., VirE2, VirE3, VirF, also dock with VirD4 and then are delivered independently of the T-DNA through the secretion channel. Independently of VirD4, the VirB proteins also assemble into a conjugative pilus, which is used to establish contact with a susceptible target cell. IM, inner membrane; P, periplasm; OM, outer membrane. doi:10.1128/microbiolspec.PLAS-0010-2013.f3

transfer intermediate is delivered to and through the transfer channel (23, 24).

Recent phylogenetic studies resulted in classification of the relaxases into 8 different mobilization (MOB) groups (25). The Ti-encoded TraA relaxase is closely related to the RSF1010 MobA relaxase and therefore grouped in the MOBQ family. Accordingly, the *oriT*-binding target of TraA is closely related to that of RSF1010 (5). By contrast, VirD2 is grouped in the MOBP family and its binding targets, the *oriT*-like border sequences that flank the T-DNA, resemble the *oriT* sequences of IncP plasmids (26). Although TraA and VirD2 cleave their DNA substrates by similar catalytic mechanisms, differences in their primary sequences and with the cognate Dtr factors likely confer specificity of the Ti plasmid as a substrate for the Tra/Trb T4SS and T-DNA as a substrate for the VirB/VirD4 system.

VirD2 binding and nicking at T-DNA border sequences is enhanced by the accessory factors VirD1, VirC1, and VirC2 (Fig. 3). VirD1 is important for VirD2 nicking on the supercoiled, double-stranded plasmid. VirC1 and VirC2 bind a sequence termed *overdrive* located immediately adjacent to the right border repeat sequences of octopine-type Ti plasmids (27, 28). This binding reaction stimulates T-DNA processing and results in accumulation of many copies of free VirD2-T-strand transfer intermediates in a cell (29). Interestingly, VirC1 is a member of the ParA family of ATPases that, as discussed above, mediate partitioning of chromosomes and plasmids during cell division (5). VirC1 was shown to localize at *A. tumefaciens* cell poles, to recruit the VirD2-T-strand complex to the cell poles, and to interact with VirD4, the substrate receptor for the VirB/D4 T4SS (29). Taken together, these findings prompted a model that *A. tumefaciens* adapted an ancestral Par-like function for the novel purposes of (i) stimulating a conjugative DNA-processing reaction and (ii) promoting DNA substrate docking with a cognate T4SS receptor. Both activities potentially mediate transfer of many copies of T-DNA to susceptible plant host cells, presumably for enhanced probability of infection (29).

How the T-DNA and Ti plasmid substrates engage with their respective T4SSs is presently not known at a molecular level, but some general features of this interaction have been defined. As can be surmised from the above, the protein components associated with the translocated DNA carry the recognition signals for the substrate-T4SS docking reaction. The VirC accessory factors facilitate contact between the VirD2 relaxase and the VirD4 receptor (29), but the VirD2 relaxase also possesses a translocation sequence that contributes to this interaction. VirD2 was shown to carry a translocation sequence at its C terminus by use of the Cre recombinase reporter assay for translocation (CRAfT) (30, 31). In this assay, full-length or fragments of protein substrates, e.g., VirD2, are fused at their N termini to Cre recombinase and translocation is monitored to a reporter bacterial or plant cell carrying a *lox* cassette whereby Cre-mediated excision confers a reporter activity, e.g., antibiotic resistance. Studies of the VirD2 translocation signal showed that a cluster of positively charged Arg residues was important for Cre transfer, leading to a proposal that C-terminally mediated ionic interactions are important for docking with the VirD4 receptor (31).

The VirB/VirD4 T4SS also translocates several effector proteins, including VirE2, VirE3, and VirF to target cells (Fig. 3) (30, 32). These proteins do not interact with the VirD2-T-strand intermediate in agrobacteria, but instead they are independently translocated through the T4SS into the plant cell. Functions of these translocated proteins are discussed briefly below. These effectors also carry positively charged C-terminal domains that are required for substrate-VirD4 engagement (33). Interestingly, the VirB/VirD4 T4SS is also capable of translocating plasmid RSF1010, a non-self-transmissible plasmid of the MOBQ family, to target cells (34, 35). In contrast to VirD2, the MobA relaxase carries two internal motifs designated as translocation signals 1 and 2 (TS1 and TS2), each of which can mediate transfer of the relaxase-T-strand complex through the T4SS (36). RSF1010 transfer also requires the accessory factor MobB, which is thought to function analogously to VirC1 and VirC2 in promoting DNA substrate-receptor docking. Thus, at least two distinct types of translocation signals, a positively charged C-terminal motif or internal motifs of unspecified sequence composition, can mediate transfer of DNA and protein substrates through the VirB/VirD4 T4SS.

The T4CP Receptor

All conjugation systems and nearly all T4SS effector translocation systems have a substrate receptor that is ancestrally related to the Ti plasmid-encoded VirD4 subunit (37). Receptor activities of these proteins have been demonstrated genetically (38, 39), through demonstration of relaxase-receptor interactions *in vitro* (40, 41), and with a ChIP-based, UV-cross-linking assay (42). By use of the latter assay, designated as *tr*ansfer DNA *i*mmuno*p*recipitation (TrIP), cross-linkable interactions were identified between the translocating T-DNA substrate and components of the *A. tumefaciens* VirB/VirD4 T4SS. Confirming its role as the T-DNA receptor,

VirD4 was shown to form a cross-linkable contact with the T-DNA substrate even in a strain lacking the VirB channel subunits (42). VirD4-like receptors are also termed type IV coupling proteins (T4CPs) because they functionally couple the DNA-processing and transfer reactions (43). T4CPs contain Walker A and B nucleotide-binding motifs, which are essential for nucleotide binding and hydrolysis, and mutations in these motifs abolish translocation indicating that one or more stages of transfer are energized by NTP hydrolysis. T4CPs are tethered to the inner membrane by an N-terminal membrane anchor sequence (37). An X-ray structure of the soluble, ~50-kDa cytoplasmic domain of the TrwB T4CP encoded by the conjugative plasmid R388 revealed a globular hexameric assembly in which each subunit is composed of two distinct domains, a nucleotide-binding domain (NBD) and a 7-helix motif called the all-α-domain (AAD) that faces the cytoplasm (44, 45). The six TrwB protomers assemble to form a globular ring that is ~110 Å in diameter and 90 Å in height, with a ~20-Å-wide channel in the center that constricts to 8 Å at the cytoplasmic pole. The N terminus of the TrwB hexamer spans the inner membrane. TrwB undergoes conformational changes in the central channel upon substrate binding and hydrolysis (46), suggesting that T4CPs might act as motor proteins during secretion, but precisely how the T4CP interacts with substrates and energizes substrate transfer is not yet defined.

The T4SS Channel

The TraG T4CP interacts with a translocation channel composed of the Trb proteins to mediate conjugative transfer of the Ti plasmid to agrobacterial recipient cells. Similarly, VirD4 interacts with the VirB channel to coordinate T-DNA and protein substrate transfer to plant cells. Both channels are assembled from at least 11 subunits whose stoichiometries for the most part are unknown (Fig. 4). The Ti plasmid-encoded Trb proteins were assigned the same names as their closest homologs in the database, the Trb proteins encoded by plasmid RP4 (47). Both sets of *trb* genes are arranged collinearly as a single operon, strongly indicating that the two systems share a common ancestry. The VirB proteins are most highly related to the Tra proteins encoded by plasmid pKM101, and the two gene sets are also arranged collinearly (48, 49). The VirB system lacks homologs for two Trb proteins, TrbJ and TrbK, the latter of which plays a role in entry exclusion. Conversely, the Trb system lacks a VirB8 homolog, which contributes to nucleation of machine assembly as well as substrate transfer. Other differences exist between the two Ti plasmid-encoded T4SSs. For example, polytopic TrbL whose counterpart is VirB6 in the VirB/VirD4

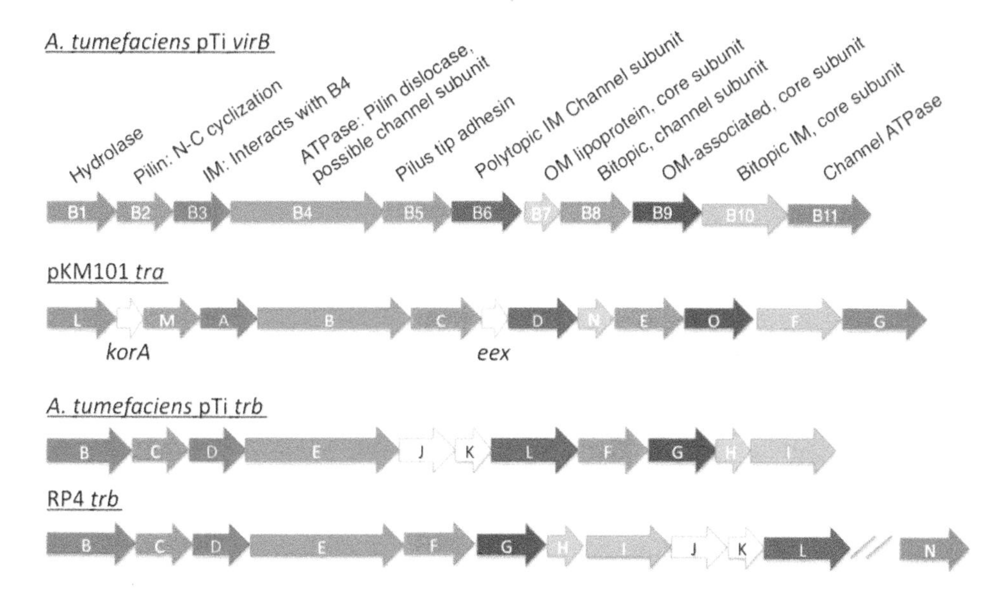

Figure 4 Genetic organization of the *A. tumefaciens* Ti plasmid-encoded *virB* and *trb* operons. The *virB* genes and some of the known functions of the encoded products are presented at the top. This T4SS is closely related in operon organization and subunit composition to a T4SS encoded by the *E. coli* conjugative plasmid pKM101. The Trb system is closely related in operon organization and subunit composition to a T4SS encoded by the *E. coli* conjugative plasmid RP4. Genes encoding protein homologs are identically color-coded. doi:10.1128/microbiolspec.PLAS-0010-2013.f4

system, contains a large hydrophilic C-terminal extension that is absent in VirB6. Such domains are found only in a subset of VirB6-like subunits, and there is some evidence for their extension through the channel and even across the envelopes of target cells (37). It should be noted, however, that despite important differences in subunit composition that likely impact overall machine architecture, recent phylogenetic studies have shown that the two systems in fact share a common ancestry (25, 50, 51). The T4SSs have been grouped into 8 clades, the largest of which is designated as MPF_T. The Ti plasmid-encoded VirB/VirD4 T4SS is the prototypical member of the MPF_T clade, but both the RP4- and Ti plasmid-encoded Trb systems also are members of this subgroup (25).

In light of their common ancestry, it is therefore not surprising that both of the Ti plasmid-encoded T4SSs are capable of translocating a common substrate, the non-self-transmissible IncQ plasmid RSF1010, to other agrobacterial cells. Moreover, in receptor-swapping experiments, it was shown that VirD4 is capable of interfacing with the Trb channel to mediate RSF1010 transfer (39). These observations underscore the commonality of function of the VirB and Trb T4SSs, and suggest that both systems use common mechanisms to mediate substrate transfer. Most of the structure-function studies of Gram-negative T4SSs to date have been performed with the VirB/VirD4 system and related systems encoded by the pKM101 and R388 plasmids. In the next section, we will briefly summarize a current view of how these machines are architecturally arranged and how they deliver their cargoes across the cell envelope.

The T4SSs of Gram-negative bacteria are composed of four sets of subunits distinguishable by function and subcellular localization at the cell envelope:

1. Energetic components: Two ATPases in addition to the T4CP are located at the cytoplasmic entrance to the secretion channel. In the VirB/VirD4 system, they are termed VirB4 and VirB11; and, in the Tra/TrB system, they are TrbE and TrbB, respectively (Fig. 3 and 4) (20, 47). Phylogenetic analyses established that the VirB4 ATPase family is ancestrally related to the VirD4 superfamily (25), leading to a proposal that these subunits also function as homohexamers. VirB4-like TrwK was found to assemble as a homohexamer (52), but other VirB4 homologs exist in solution as monomers and dimers (53, 54), raising the possibility that the oligomeric state of these ATPases might vary depending on membrane association, ATP binding or hydrolysis, or interactions with other T4S machine subunits. VirB11 is a member of a family of ATPases termed the "traffic ATPases" (21). These ATPases are associated with Gram-negative bacterial type II, type III, type IV, and type VI secretion systems (20, 55). VirB11-like ATPases are peripheral inner membrane proteins, but soluble forms also exist, suggesting that these ATPases might exist in a dynamic equilibrium with the membrane. Electron microscopy (EM) visualization of VirB11 homologs showed hexameric rings of ~100 to 120 Å in diameter (56, 57). The N- and C-terminal halves of the 6 protomers each form rings, giving rise to a double-stacked structure wherein the nucleotide-binding site is at the interface between the two domains. The EM studies have shown that the VirB11 hexamers undergo dynamic structural changes upon ATP binding and hydrolysis, although the functional importance of these transitions is not defined at this time (56, 57). The T4SS ATPases coordinate early steps of substrate transfer. As mentioned above, the *A. tumefaciens* VirD4 T4CP engages with the T-DNA substrate, as shown by TrIP (42). Further studies showed that VirD4 then delivers the DNA substrate to the VirB11 ATPase by a mechanism that does not require ATP hydrolysis by either subunit (42, 58). In the early studies, no evidence was obtained for interaction of the DNA substrate with the *A. tumefaciens* VirB4 ATPase (42), but the role of VirB4 subunits in substrate transfer might need to be revisited in light of more recent findings that VirB4 homologs associated with other T4SSs bind DNA *in vitro* (52, 53) and *in vivo* (53).
2. The inner membrane channel translocase: At this time, there is no structural information available for the inner membrane portion of the T4SS channel. In the VirB/VirD4 system, three integral membrane proteins, including VirB3, VirB6, and VirB8, are postulated to form this translocase on the basis of experimentally derived inner membrane topology models, protein-protein interaction data, and results of TrIP studies (Fig. 3 and 4) (20, 21, 42). VirB3 interacts with VirB4 (59) and might play a role in coordinating a biologically important interaction between this ATPase and other components of the translocase. VirB6 is unique among the VirB subunits in spanning the inner membrane 5 to 6 times, a topology common among subunits of inner membrane translocases and transporters (60). The

TrIP studies placed VirB6 and VirB8 at an intermediate point in the postulated T-DNA translocation pathway, dispensable for DNA transfer to VirD4 and VirB11, but necessary for transfer to VirB2 and VirB9 (42). VirB6 and VirB8 functionally interact, as evidenced by the finding that null mutations in *virB6* or *virB8*, respectively, block substrate transfer to VirB8 or VirB6. However, certain mutations in VirB6 were shown to permit T-DNA contacts with VirB6 but block contacts with VirB8 (60), giving rise to a proposal that the T-DNA substrate forms close contacts sequentially with VirB6 and then VirB8. Walker A mutations in each of the energetic components—VirD4, VirB4, and VirB11—block substrate transfer from VirB11 to VirB6 and VirB8, suggesting that ATP energy utilization is required for driving the DNA substrate through the inner membrane portion of the channel (42, 58). Additionally, other subunits including VirB7, VirB9, and VirB10 are required for this transfer step (58). These subunits form a stabilizing "core" complex that is envisaged to house and form critical contacts with the inner membrane translocase as summarized below.

3. The envelope-spanning core complex: In *A. tumefaciens*, the outer membrane lipoprotein VirB7, outer membrane-associated VirB9, and bitopic VirB10 stabilize each other as well as other VirB channel subunits (20), and evidence has been presented for assembly of these proteins as a ring-shaped complex now called the core complex (Fig. 3 and 4) (61). An analogous core complex composed of the corresponding VirB homologs encoded by the conjugative plasmid pKM101 was structurally resolved by cryoelectron microscopy (CryoEM) and a portion was further solved by X-ray crystallography (62, 63). The pKM101 core complex, composed of 14 copies each of VirB7-like TraN, VirB9-like TraO, and VirB10-like TraF, is a 1.05-MDa structure of 185 Å in width and height (62). It is composed of two layers (I and O layers) that form a double-walled ringlike structure. The I layer is composed of the N-terminal domains of TraO and TraF and forms a 55-Å-diamete ring at the inner membrane. The O layer is composed of TraN and the C-terminal domains of TraO and TraF and forms a main body and narrower cap with a central hole of 10 Å that is presumed to span the outer membrane. A crystal structure of the entire O layer further showed that TraF/VirB10 forms the outer membrane channel (63). Specifically, 14 copies of an α-helical domain termed "the antennae projection" or "AP" is thought to form the channel. In the assembled core complex, therefore, TraF/VirB10 subunits are predicted to span the entire cell envelope such that 14 N-terminal transmembrane helices form the ~55-Å inner membrane ring, a proline-rich region and β-barrel domain span the periplasm, and the AP forms the outer membrane pore (63, 64). A current model depicts this core complex as a structural scaffold for the translocation channel, wherein the ATPases are positioned at the base of the channel and VirB3, VirB6, and VirB8 are within the inner membrane ring. Other channel subunits, including the pilin subunit and a portion of VirB9, are postulated to form the distal portion of the channel within the core's central chamber (65). Finally, a VirB2 pilus structure or the TraF/VirB10 AP forms the channel through which substrates pass across the outer membrane (21).
4. The conjugative pilus: T4SSs elaborate conjugative pili in addition to the translocation channel (Fig. 3). Although these two organelles likely assemble as a single supramolecular structure, we depict them as physically distinct to convey the idea that they fulfill distinct functions. For example, the extended pilus initiates contact with target cells, but it is completely dispensable for intercellular translocation (20). This was demonstrated genetically through the isolation of "uncoupling" mutations. Such mutations block pilus biogenesis without affecting substrate transfer or, conversely, block substrate transfer without affecting pilus biogenesis. The isolation of such mutations strongly indicates that pili extending from the cell surface function mainly or exclusively to initiate the donor-target cell contact. The conjugative pili are distinguished by width, length, and flexibility. For example, F pili encoded by the *Escherichia coli* F plasmid are ~9 nm and flexible, and range in length up to 1 μm. By contrast, the VirB/VirD4 and Tra/Trb pili are both classified as P type. These pili are thicker (9 to 11 nm), more rigid, and shorter than F pili although length measurements are complicated by the fact that isolated pili are typically broken (66, 67). Plasmid RP4-encoded pili and probably those of the closely related Ti plasmid-encoded Tra/Trb system are thick and straight, whereas those elaborated by the *A. tumefaciens* VirB/VirD4 system are more flexuous (68, 69). P pili are abundantly present in the extracellular milieu, often as bundles,

and are rarely found associated with cells. Whether this is an artifact accompanying the preparation of cells for electron microscopy owing to their fragility or is a normally occurring process is not known (69, 70).

Pili are composed predominantly of a single pilin subunit. Both VirB2 and RP4-encoded TrbC are small (~7-kDa) proteins with hydrophilic N and C termini and two hydrophobic stretches of ~20 to 22 residues separated by a small central hydrophilic loop (71). These pilins are synthesized as pro-proteins with unusually long (~30 to 50 residues) leader peptides that are cleaved upon insertion into the inner membrane (71, 72). They are then processed further to yield a membrane pool of mature pilin subunits. Interestingly, VirB2 and TrbC from plasmid RP4 and probably also Ti plasmid-encoded TrbC are processed in part by a head-to-tail cyclization reaction whereby the N- and C-terminal residues of the pro-protein are covalently joined (73). Cyclization stabilizes the P-type pilins in the membrane and appears to be essential for pilus assembly.

Mature pilin monomers are thought to assemble as a pool in the inner membrane for use upon receipt of an unknown signal in building the conjugative pilus. Evidence has been presented for a role by the *A. tumefaciens* VirB4 ATPase in dislocation of VirB2 monomers from the inner membrane, and a pilus assembly pathway involving VirB4, VirB5, and VirB8 has been postulated (74, 75). VirB5 likely plays a critical role in pilus polymerization, as deduced from evidence that the conjugative pilus assembles from its base (76) and that VirB5 subunits are located at the tip of the polymerized pilus (77). The X-ray structure of VirB5-like TraC from pKM101 revealed a 3-helix bundle flanked by a smaller globular part (78). Further mutational analyses identified residues important for DNA transfer and binding of pilus-specific bacteriophages (78). Precisely how VirB5 interacts with VirB2 in the assembled pilus is unknown, but the present findings point to a role for the TraC/VirB5-like subunits in pilus nucleation and establishment of donor-recipient cell contacts.

T-DNA TRANSFER TO THE PLANT CELL

In nature, the Ti plasmid-encoded Tra/TrB and VirB/VirD4 transfer systems are repressed in the absence of plant-derived inducing signals. However, when *A. tumefaciens* cells encounter wounded plant tissue, sensory perception of plant-derived molecules serves to activate the expression of the Ti plasmid-encoded *vir* regulon. An important outcome of *vir* gene induction is the translocation of T-DNA and protein substrates to susceptible plant cells and the resulting formation of crown gall tumors.

Role of Cotransported Proteins in T-DNA Transfer and Plasmid Conjugation

As mentioned above, *A. tumefaciens* cells use the VirB/VirD4 T4SS to deliver the VirD2-T-strand and effector proteins to the plant target cell. One of these effectors is VirE2, a single-stranded DNA-binding protein (SSB). Upon translocation of VirE2, the SSB binds cooperatively along the length of the T strand to generate a VirD2-T-strand-VirE2 particle termed the T complex (79). VirD2 and VirE2 in turn contribute in various ways to successful translocation of the T complex to the plant nucleus. Importantly, both proteins carry nuclear localization sequences (NLSs) that help guide the T complex to the nucleus through specific interactions with plant proteins. For example, both VirD2 and VirE2 were shown to interact with one or more members of the importin-α family, which function as adaptor molecules by interacting with NLS motifs in cargo proteins and with the nuclear shuttle protein importin-β to promote nuclear uptake (80, 81). VirD2 is also a phosphoprotein, and several additional plant proteins that interact with VirD2 may play roles in phosphorylation/dephosphorylation (82). For example, VirD2 interacts with and is phosphorylated by the cyclin-dependent kinase-activating kinase CAK2Ms (83). CAK2Ms also phosphorylate RNA polymerase II large subunits, which in turn recruits a TATA-box binding protein important for transcription initiation. The binding of VirD2 to a TATA-box binding protein led to a suggestion that VirD2 phosphorylation may play an additional role in targeting T complexes to chromatin. VirD2 also interacts with plant cyclophilins, which contribute to the maintenance of protein folding, although the biological role of this interaction is not clear at this time (83, 84). Finally, intriguingly, VirE2 has been shown to form gated channels in black lipid membranes, leading to a proposal that VirE2 might promote entrance of the T complex across the plant cytoplasmic membrane through a channel-forming activity (85). While it is not immediately obvious how a protein could function dually as a channel for the T complex and an SSB that coats the length of the T strand, the findings are intriguing and warrant further study.

Once in the nucleus, T-DNA integrates into the plant nuclear genome by nonhomologous or "illegitimate" recombination. The T-DNA invades at nick gaps in the plant genome possibly generated as a consequence of active DNA replication. The invading ends of the single-stranded T-DNA are proposed to anneal via short regions of homology to the unnicked strand of the plant DNA. Once the ends of T-DNA are ligated to the target ends of plant DNA, the second strand of the T-DNA is replicated and annealed to the opposite strand of the plant DNA. VirD2 and VirE2 have been implicated in contributing to T-DNA integration but precisely how remains unclear (see reference 79).

T-DNA Genes Expressed in Plant Cells

Octopine-type Ti plasmids carry two T-DNA fragments designated T_L-DNA and T_R-DNA of 13 and 7.8 kb in length, respectively (see Fig. 1) (5). Once integrated into the plant nuclear genome, these T-DNAs encode 13 proteins with two main functions. One set of enzymes promotes the synthesis of two plant growth regulators, auxins and cytokinin zeatin. Production of these plant hormones results in a stimulation of cell division and a loss of cell growth control, ultimately leading to the formation of characteristic crown gall tumors. The second group of enzymes promotes the synthesis of novel amino acid and sugar derivatives called opines. Octopine-type T-DNA's code for synthesis of octopine which is a reductive condensation product of pyruvate with arginine. Octopine synthases also promote condensation of pyruvate with other amino acids to produce lysopine, histopine, or octopinic acid. Octopine-type T-DNAs also code for synthesis of other opines, including mannopine, mannopinic acid, agropine and agropinic acid, whereas T-DNAs carried by other Ti plasmids code for nopalines, which are derived from a-ketoglutarate and arginine as well as other classes of opines (5).

OPINE CATABOLISM AND THE "OPINE CONCEPT"

Plants cannot metabolize opines and instead release them into the extracellular milieu. By contrast, the Ti plasmid carries opine catabolism genes that are responsible for the active transport of opines across the agrobacterial envelope and their degradation in the cytoplasm. Over 40 genes coding for at least 6 ATP-binding cassette-type permeases and 12 opine catabolic enzymes are responsible for opine uptake and degradation (5). The capacity of infecting agrobacteria to degrade opines for use as carbon and energy is central to the evolution of these bacteria as phytopathogens. The "opine concept" was developed to rationalize the finding that *A. tumefaciens* evolved as a pathogen by acquiring the ability to transfer DNA to plant cells. According to this concept, *A. tumefaciens* adapted an ancestral DNA conjugation system for interkingdom DNA transport specifically to incite synthesis of opines by the plant host. The cotransfer of oncogenes ensures that transformed plant cells proliferate, resulting in enhanced opine synthesis. The environment of the tumor thus is a rich chemical environment favorable for growth and propagation of the infecting *A. tumefaciens* (86, 87). It is noteworthy that a given *A. tumefaciens* strain catabolizes only those opines that it incites plant cells to synthesize. This is thought to ensure a selective advantage of the infecting bacterium over other *A. tumefaciens* strains that are present in the vicinity of the tumor.

SIGNALING NETWORKS CONTROLLING Ti PLASMID FUNCTIONS

Most of the Ti plasmid-encoded functions summarized above are activated only in response to sensory perception of extracellular signals. These signals originate from wounded and transformed plant tissue, as well as other *A. tumefaciens* cells at the infection site. Upon encountering wounded plant tissues, *A. tumefaciens* integrates these signals into a regulatory cascade that culminates in the following sequence of events: (i) T-DNA transfer and oncogenesis, (ii) production and utilization of opine nutrients, (iii) elevated Ti plasmid copy number, and (iv) dissemination of the Ti plasmid. This regulatory network has been extensively characterized and many of its features are known in molecular detail, as summarized below.

Plant Wound Signals and VirA/VirG-Mediated Sensory Perception

A. tumefaciens induces its virulence regulon upon sensory perception of various plant-derived signals. These signals are present at the plant wound site and include specific classes of phenolic compounds, monosaccharides, low phosphorus levels, and acidic pH (88). Sensory perception is achieved through the Ti plasmid-encoded VirA/VirG two-component regulatory system (89, 90). VirA is a histidine sensor kinase that autophosphorylates at a conserved histidine residue and then transfers the phosphate group to a conserved aspartate residue on VirG. Phosphorylated VirG in turn activates transcription of the operons comprising the *vir* regulon that are responsible for T-DNA processing and translocation to plant cells. VirA senses the plant-derived signals, the most important of which are phenolic compounds that carry an *ortho*-methoxy group (91). The type of substitution at the *para* position distinguishes strong inducers such as acetosyringone from weaker inducers such as ferulic acid and acetovanillone (89). A variety of monosaccharides, including glucose, galactose, arabinose, and the acidic sugars D-galacturonic acid and D-glucuronic acid, strongly enhance *vir* gene induction. The presence of these compounds is a general feature of most plant wounds and likely contributes to the extremely broad host range of *A. tumefaciens*. VirA functions as a homodimer with an N-terminal transmembrane domain and large C-terminal cytoplasmic domain. Recent studies suggest that a cytoplasmic linker domain of VirA interacts directly with phenolics and imparts specificity for phenolics (92). The periplasmic sugar-binding protein ChvE binds the monosaccharide sugars and the ChvE-sugar complexes interact with the periplasmic domain of VirA, thereby inducing a conformational change that increases VirA's sensitivity to phenolic inducer molecules (90, 93, 94). The periplasmic domain of VirA also senses acidic pH, which is required for maximal induction of the *vir* genes, but the underlying mechanism is unknown (95). Phospho-VirG activates transcription of the *vir* genes by interacting with a *cis*-acting regulatory sequence (TNCAATTGAAAPy) called the *vir* box located upstream of each of the *vir* promoters (88).

Opine Signals and Regulation of Opine Catabolism

Upon delivery and incorporation of T-DNA into the plant nuclear genome, expression of the T-DNA genes results in synthesis and release of opines into the

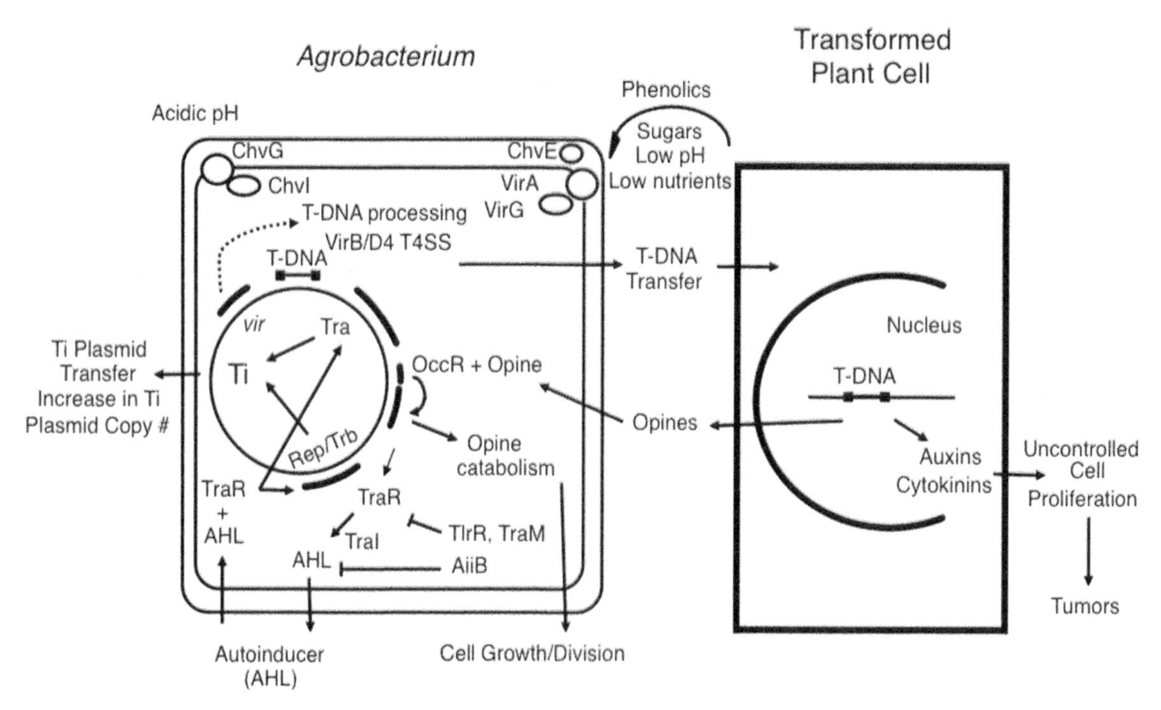

Figure 5 A schematic of chemical signaling events between *Agrobacterium* cells and transformed plant cells. Signals released from wounded plant cells initiate the infection process through the VirA/VirG/ChvE sensory response system, resulting in activation of the Ti plasmid-encoded *vir* genes. The Vir proteins mediate T-DNA processing, assembly of the VirB/VirD4 T4SS, and T-DNA translocation to susceptible plant cells. VirA/VirG also induce expression of the Ti plasmid *rep* genes resulting in elevated Ti plasmid copy number. Opines released from transformed plant cells activate opine catabolism functions for growth of infecting bacteria. Opines also activate synthesis of TraR which in turn induces production of the TraI homoserine lactone (AHL) synthase. TraR and AHL at a critical concentration activate the Ti plasmid replication and conjugation functions resulting in elevated Ti plasmid copy number and dissemination to neighboring agrobacterial cells. TlrR and TraM negatively regulate TraR activity, and AiiB negatively controls AHL levels. Adapted from reference 118. doi:10.1128/microbiolspec.PLAS-0010-2013.f5

milieu. The opines are taken up by agrobacterial cells for use not only as nutrients, but also as activators of gene expression (Fig. 5). Opines interact with opine-responsive transcriptional regulators that control the Ti plasmid-encoded opine catabolic functions (96). Opines derived from amino acids activate gene expression through binding a LysR-type transcriptional activator. For example, octopine binds LysR-like OccR (96). OccR positively regulates expression of the *occ* genes involved in octopine uptake and catabolism by inducing a bend in the DNA at the OccR-binding site. Octopine binding alters both the affinity of OccR for its target site and the angle of the DNA bend, establishing that octopine modulates OccR regulatory activity inside the bacterium (97, 98). Opines derived from sugars, e.g., agrocinopines and mannopine, function through binding a LacI-type repressor (99). These repressors limit catabolic gene function in the absence of opines but are inactivated in their presence allowing transcription and opine utilization.

N-Acylhomoserine Lactone Signals and Regulation of Ti Plasmid Conjugative Transfer

Besides regulating expression of opine catabolism genes, opines serve another important regulatory function. A regulatory cascade has been shown to activate Ti plasmid transfer under conditions of high cell density (100, 101, 102). This cascade initiates when *A. tumefaciens* cells import opines released from plant cells (Fig. 5). For example, binding of the octopine-OccR to the promoter located upstream of the *occ* operon induces expression of the octopine catabolism genes and a gene located at the 3′ end of the operon. This gene codes for the TraR transcriptional regulator, a protein related to LuxR shown over 20 years ago to regulate synthesis of quorum signals (QS) known as *N*-acylhomoserine lactones (AHLs) (5, 103). At low cell densities, quorum signals are present at a low concentration, whereas at high cell densities they accumulate in the surrounding environment and then passively diffuse back into the bacterial cell to activate transcription of a defined set of genes. In the case of *A. tumefaciens*, TraR responds to *N*-3-(oxooctonoyl)-L-homoserine lactone (3-oxo-C8-HSL), which is produced by the Ti plasmid-encoded TraI AHL synthase. This AHL acts in conjunction with TraR to activate transcription of the Ti plasmid-encoded *tra* genes as well as *traI*. Synthesis of TraR under conditions of high cell density thus creates a positive-feedback loop whereby the TraR-AHL complex induces synthesis of TraI, which in turn synthesizes more AHL (104). Ultimately, the regulatory cascade involving opine-mediated expression of *traR* and TraR-AHL-mediated expression of Ti plasmid transfer genes at high cell densities, results in enhanced Ti plasmid conjugative transfer to neighboring agrobacterial cells in the environment of the plant tumor. This signaling network is subject to additional levels of feedback regulation, for example, through synthesis of a TraR antiactivator (TraM) (105), a quorum-quenching AHL lactonase (AiiB) (106), and a truncated form of TraR (TrlR) that dimerizes with and poisons TraR activity (107). It is generally considered that *A. tumefaciens* evolved this complex regulatory system to maximize the number of Ti plasmid-carrying bacterial cells, and hence the potential for interkingdom transmission of opine-encoding T-DNA, in the vicinity of susceptible host tissue (100, 101, 102).

Plant Phenol, Opine, and AHL Regulatory Control of Ti Plasmid Replication

Most Ti plasmids exist in single copy within a cell as a result of tight regulatory control of the *repABC* gene cluster. For plasmid pTiR10, this tight control is achieved by autorepression at one of the four upstream promoters (the P4 promoter) by RepA and RepB, and by transcriptional and posttranscriptional inhibition of *repC* expression by the countertranscript RNA RepE (see Fig. 2) (1). However, expression of the *repABC* operon, and therefore the Ti plasmid copy number, is also subject to regulation by two diffusible chemical signals (Fig. 5). First, as discussed above, the VirA/VirG two-component system senses specific classes of plant phenolic compounds to activate *vir* gene expression through binding of phospho-VirG to upstream *vir* boxes. Phospho-VirG also has been shown to bind a *vir* box centered 71-bp upstream of the promoter P4 controlling *repABC* expression. Sensing of plant phenolics by the VirA/VirG system thus also activates expression of the *repABC* operon, resulting in a 3- to 4-fold increase in Ti plasmid copy number (108). Second, *repABC* operon is subject to regulatory control by the opine-activated QS system (109). In response to opine activation of TraR synthesis and sensory perception of AHL, TraR induces expression not only of the conjugation genes, but also of the *repABC* operon, resulting in an 8-fold increase in Ti plasmid copy number (109). Sensory perception of these two exogenous signals, plant phenolics and bacterial AHLs, thus results in enhanced *repABC* gene expression and an appreciable increase in Ti plasmid copy number at the site of infection. The resulting increase in Ti gene dosage correlates with enhanced Ti plasmid dissemination and virulence potential by invading agrobacterial cells (100).

FITNESS COST TO MAINTENANCE OF Ti PLASMIDS

From the above discussion, it is evident that Ti plasmids code for a myriad of biological functions in response to sensory perception of a complex array of signals of plant and bacterial origin. A recent study assessed the fitness cost accompanying the metabolic load associated with carriage of Ti plasmids (110). As might be expected, the findings indicated that under conditions of nutrient abundance, the low-copy Ti plasmids exert only a modest cost to the agrobacterial host. Under such nutrient replete conditions, Ti plasmidless strains do not significantly outcompete Ti plasmid-carrying strains in batch culture, suggesting that carriage of the Ti plasmid does not impose a large metabolic burden. The complex regulatory network described above operates to minimize expression of the Ti plasmid genes under such conditions. Indeed, the only Ti plasmid locus required under these conditions is the *repABC* cassette, which is tightly controlled for maintenance of the Ti plasmid at single copy.

Surprisingly, however, there is a significant fitness cost associated with the Ti plasmid under conditions of nutrient limitation, even without induction of the virulence genes (110). Under these conditions, the expression profile of Ti plasmid genes is not expected to vary appreciably from growth in a rich environment, establishing that even the maintenance of this large plasmid at low copy and its vertical transmission during cell division imposes an appreciable metabolic burden. Nutrient limitation is likely a common stress for agrobacteria in nature, but despite the carriage cost of Ti plasmids, Ti-plasmidless strains are uncommon. In part, this can be attributed to the coupled expression of replication and partitioning genes, the latter ensuring vertical transmission of the plasmid during cell division. Additionally, a Ti plasmid toxin-antitoxin system also was recently described. Cells that lose the plasmid during cell division are killed owing to antitoxin instability and toxin stability (111).

Under conditions of nutrient limitation but favorable for *vir* gene expression, a cascade of events culminates in elaboration of the VirB/VirD4 T4SS, processing and transfer of T-DNA and effector proteins, expression of opine catabolism genes, upregulation of the *repABC* cassette, synthesis of the Ti plasmid Trb/Tra T4SS, and processing and translocation of the Ti plasmid. This burst of biological activity has the effect of imposing a major metabolic burden, as demonstrated through competition assays between plasmid-carrying and plasmifree strains under *vir*-inducing conditions (110). Tight regulation of Ti plasmid genes under these conditions minimizes this metabolic load, yet the fitness cost accompanying *vir* gene expression begs the question of whether strains of agrobacteria called "cheaters" arise in nature. Such strains, for example, would carry variant forms of the Ti plasmid lacking the virulence genes while retaining the plasmid maintenance functions and, importantly, opine catabolism genes. Indeed, recent studies have shown that opine-catabolizing, avirulent strains of agrobacterial "cheaters" are widespread in nature, as are opine-catabolizing nonagrobacterial soil microbes (110, 112). Thus, virulent strains of agrobacteria face competition at the infection site with a spectrum of microbial "cheaters" or opportunists. Perhaps horizontal transmission of the Ti plasmid, activated by the cascade of plant, opine, and QS signals at the infection site, has evolved not just for dissemination of the Ti plasmid among coresident agrobacteria, but also as a means of forcing cooperation by agrobacterial cheaters (110). Clearly, there is a strong selective pressure in the framework of an intricate signaling network that serves to couple pathogenesis with Ti plasmid maintenance and dissemination at the infection site.

BIOTECHNOLOGICAL APPLICATIONS

One of the most interesting features of Ti plasmids is their capacity to mediate conjugative transfer of T-DNA across kingdom boundaries. This property is of central importance for *A. tumefaciens* pathogenesis, but the discovery of transkingdom sex also spawned the multibillion dollar industry of plant genetic engineering. It is now recognized that *A. tumefaciens* is capable of delivering any DNA flanked by T-DNA borders to an extremely broad range of plant hosts. Susceptible plant species include a wide range of gymnosperms and dicotyledonous species, as well as many monocotyledonous species of agricultural importance. Early problems encountered in transformation of monocots such as rice, corn, and wheat were overcome with the use of actively dividing cells such as immature embryos, preinduction of *A. tumefaciens* with phenolic inducers prior to infection, and screens for optimal plant genotype, type and age of plant tissue, bacterial strains, and T-DNA border-containing vectors. For rice and corn and many other species initially found to be recalcitrant to *A. tumefaciens* infection, most of these parameters have been optimized such that "agrotransformation" is now a routine technique (113).

For plant biologists, *A. tumefaciens*-mediated T-DNA transfer also has broader, more basic applications. For example, it is now possible to isolate novel plant genes by T-DNA tagging. Several variations to this methodology exist depending on the desired goals. For example, because insertions are generally randomly distributed throughout the plant genome, T-DNA is widely used as a mutagen for isolating plant genes with novel phenotypes. If the mutagenic T-DNA carries a bacterial origin of replication, the mutated gene of interest can easily be recovered in bacteria by suitable molecular techniques. Furthermore, if the T-DNA is engineered to carry a selectable or scorable gene near one of its ends, insertion downstream of a plant promoter will permit characterization of promoter activity. Conversely, if the T-DNA is engineered to carry an outward reading promoter, insertion can result in a modulation of gene expression with potentially interesting phenotypic consequences. Although random T-DNA insertion is a boon to investigators interested in characterizing plant genes, it is an undesired event for plant genetic engineering. In addition to the potential result that T-DNA will insert into an essential gene, insertion often is accompanied by rearrangements of flanking sequences, thus further increasing the chances that the insertion will have undesired consequences. Ideally, T-DNA could be delivered to a restricted number of sites in the plant genome. Progress toward this goal has involved the use of the bacteriophage P1 Cre/*lox* system for site-specific integration in the plant genome (114). The Cre site-specific recombinase catalyzes strand exchange between two *lox* sites, which for P1 results in circularization of the P1 genome upon infection of bacterial cells. For directed T-DNA insertion, both the plant and the T-DNA are engineered to carry *lox* sequences and the plant is also engineered to express the Cre protein. Upon entry of T-DNA into the plant cell, Cre was shown to catalyze the site-specific integration of T-DNA at the plant *lox* site. The frequency of directed insertion events is low in comparison with random insertion events, but additional manipulation of this system should enhance its general applicability.

Also of importance, the host range of *A. tumefaciens* now extends beyond the plant kingdom to include budding and fission yeast, many species of filamentous fungi, and even human cells (115, 116). The transformation of filamentous fungi with *A. tumefaciens* was an exciting advancement. *A. tumefaciens* was shown to efficiently deliver DNA to fungal protoplasts and fungal conidia and hyphal tissue. This DNA transfer system is especially valuable for species that are recalcitrant to transformation by other methods. Its overall simplicity and high efficiency make this gene delivery system an extremely useful tool for the genetic manipulation and characterization of fungi, and essentially all the current methodologies developed for basic studies of plants can now be applied to fungi. The discovery that *A. tumefaciens* can transform HeLa cells in laboratory culture expands the host range of this bacterium even further (116), enticing one to ask whether T-DNA transfer might form the basis of a viable gene delivery system for humans.

Finally, although the primary substrate of interest for the infecting bacterium and plant biotechnology is the T-DNA, Ti plasmid-encoded T4SSs as well as many other bacterial T4SSs also translocate protein substrates (22, 37). These translocation systems thus might be adaptable for the targeted delivery of therapeutic proteins to eukaryotic cell types, including human cells, of interest. The potential for protein therapy becomes more tangible as we learn more about type IV effector translocation signals, substrate-T4SS docking mechanisms, and T4SS-eukaryotic cell interactions.

Acknowledgments. *We thank members of the Christie laboratory for helpful discussions.This work was supported by National Institutes of Health grant GM48746 to P.J.C. Conflicts of interest: We declare no conflicts.*

Citation. Gordon JE, Christie PJ. 2014. The *Agrobacterium* Ti plasmids. Microbiol Spectrum 2(6):PLAS-0010-2013.

References

1. **Pinto UM, Pappas KM, Winans SC.** 2012. The ABCs of plasmid replication and segregation. *Nat Rev Microbiol* **10:**755–765.
2. **Teyssier-Cuvelle S, Oger P, Mougel C, Groud K, Farrand SK, Nesme X.** 2004. A highly selectable and highly transferable Ti plasmid to study conjugal host range and Ti plasmid dissemination in complex ecosystems. *Microb Ecol* **48:**10–18.
3. **Broothaerts W, Mitchell HJ, Weir B, Kaines S, Smith LM, Yang W, Mayer JE, Roa-Rodriguez C, Jefferson RA.** 2005. Gene transfer to plants by diverse species of bacteria. *Nature* **433:**629–633.
4. **Binns AN, Castantino P.** 1998. The *Agrobacterium* oncogenes. p 251–266. *In* Spaink HP, Kondorosi A, Hooykaas PJ (ed), *The* Rhizobiaceae: *Molecular Biology of Model Plant-Associated Bacteria.* Kluwer Academic Publishers, Dordrecht, The Netherlands.
5. **Zhu J, Oger PM, Schrammeijer B, Hooykaas PJ, Farrand SK, Winans SC.** 2000. The bases of crown gall tumorigenesis. *J Bacteriol* **182:**3885–3895.
6. **Tabata S, Hooykaas PJ, Oka A.** 1989. Sequence determination and characterization of the replicator region in the tumor-inducing plasmid pTiB6S3. *J Bacteriol* **171:** 1665–1672.
7. **Cevallos MA, Cervantes-Rivera R, Gutierrez-Rios RM.** 2008. The *repABC* plasmid family. *Plasmid* **60:** 19–37.

8. **Ghosh SK, Hajra S, Paek A, Jayaram M.** 2006. Mechanisms for chromosome and plasmid segregation. *Annu Rev Biochem* **75:**211–241.
9. **Gerdes K, Howard M, Szardenings F.** 2010. Pushing and pulling in prokaryotic DNA segregation. *Cell* **141:** 927–942.
10. **Pinto UM, Flores-Mireles AL, Costa ED, Winans SC.** 2011. RepC protein of the octopine-type Ti plasmid binds to the probable origin of replication within *repC* and functions only in *cis*. *Mol Microbiol* **81:**1593–1606.
11. **Cervantes-Rivera R, Pedraza-Lopez F, Perez-Segura G, Cevallos MA.** 2011. The replication origin of a *repABC* plasmid. *BMC Microbiol* **11:**158. doi:10.1186/1471-2180-11-158.
12. **Ramirez-Romero MA, Tellez-Sosa J, Barrios H, Perez-Oseguera A, Rosas V, Cevallos MA.** 2001. RepA negatively autoregulates the transcription of the *repABC* operon of the *Rhizobium etli* symbiotic plasmid basic replicon. *Mol Microbiol* **42:**195–204.
13. **Pappas KM, Winans SC.** 2003. The RepA and RepB autorepressors and TraR play opposing roles in the regulation of a Ti plasmid *repABC* operon. *Mol Microbiol* **49:**441–455.
14. **Havey JC, Vecchiarelli AG, Funnell BE.** 2012. ATP-regulated interactions between P1 ParA, ParB and non-specific DNA that are stabilized by the plasmid partition site, *parS*. *Nucleic Acids Res* **40:**801–812.
15. **Ringgaard S, Schirner K, Davis BM, Waldor MK.** 2011. A family of ParA-like ATPases promotes cell pole maturation by facilitating polar localization of chemotaxis proteins. *Genes Dev* **25:**1544–1555.
16. **Vecchiarelli AG, Han YW, Tan X, Mizuuchi M, Ghirlando R, Biertumpfel C, Funnell BE, Mizuuchi K.** 2010. ATP control of dynamic P1 ParA-DNA interactions: a key role for the nucleoid in plasmid partition. *Mol Microbiol* **78:**78–91.
17. **Ringgaard S, van Zon J, Howard M, Gerdes K.** 2009. Movement and equipositioning of plasmids by ParA filament disassembly. *Proc Natl Acad Sci USA* **106:** 19369–19374.
18. **Chai Y, Winans SC.** 2005. A small antisense RNA downregulates expression of an essential replicase protein of an *Agrobacterium tumefaciens* Ti plasmid. *Mol Microbiol* **56:**1574–1585.
19. **Su S, Khan SR, Farrand SK.** 2008. Induction and loss of Ti plasmid conjugative competence in response to the acyl-homoserine lactone quorum-sensing signal. *J Bacteriol* **190:**4398–4407.
20. **Christie PJ, Atmakuri K, Krishnamoorthy V, Jakubowski S, Cascales E.** 2005. Biogenesis, architecture, and function of bacterial type IV secretion systems. *Annu Rev Microbiol* **59:**451–485.
21. **Fronzes R, Christie PJ, Waksman G.** 2009. The structural biology of type IV secretion systems. *Nat Rev Microbiol* **7:**703–714.
22. **Cascales E, Christie PJ.** 2003. The versatile bacterial type IV secretion systems. *Nat Rev Microbiol* **1:**137–150.
23. **de la Cruz F, Frost LS, Meyer RJ, Zechner EL.** 2010. Conjugative DNA metabolism in Gram-negative bacteria. *FEMS Microbiol Rev* **34:**18–40.
24. **Zechner EL, Lang S, Schildbach JF.** 2012. Assembly and mechanisms of bacterial type IV secretion machines. *Philos Trans R Soc Lond B Biol Sci* **367:**1073–1087.
25. **Guglielmini J, de la Cruz F, Rocha EP.** 2012. Evolution of conjugation and type IV secretion systems. *Mol Biol Evol* **30:**315–331.
26. **Waters VL, Hirata KH, Pansegrau W, Lanka E, Guiney DG.** 1991. Sequence identity in the nick regions of IncP plasmid transfer origins and T-DNA borders of *Agrobacterium* Ti plasmids. *Proc Natl Acad Sci USA* **88:** 1456–1460.
27. **Toro N, Datta A, Yanofsky M, Nester E.** 1988. Role of the *overdrive* sequence in T-DNA border cleavage in *Agrobacterium*. *Proc Natl Acad Sci USA* **85:**8558–8562.
28. **Toro N, Datta A, Carmi OA, Young C, Prusti RK, Nester EW.** 1989. The *Agrobacterium tumefaciens virC1* gene product binds to *overdrive*, a T-DNA transfer enhancer. *J Bacteriol* **171:**6845–6849.
29. **Atmakuri K, Cascales E, Burton OT, Banta LM, Christie PJ.** 2007. *Agrobacterium* ParA/MinD-like VirC1 spatially coordinates early conjugative DNA transfer reactions. *EMBO J* **26:**2540–2551.
30. **Vergunst AC, Schrammeijer B, den Dulk-Ras A, de Vlaam CM, Regensburg-Tuink TJ, Hooykaas PJ.** 2000. VirB/D4-dependent protein translocation from *Agrobacterium* into plant cells. *Science* **290:**979–982.
31. **van Kregten M, Lindhout BI, Hooykaas PJ, van der Zaal BJ.** 2009. *Agrobacterium*-mediated T-DNA transfer and integration by minimal VirD2 consisting of the relaxase domain and a type IV secretion system translocation signal. *Mol Plant Microbe Interact* **22:** 1356–1365.
32. **Schrammeijer B, Dulk-Ras Ad A, Vergunst AC, Jurado Jacome E, Hooykaas PJ.** 2003. Analysis of Vir protein translocation from *Agrobacterium tumefaciens* using *Saccharomyces cerevisiae* as a model: evidence for transport of a novel effector protein VirE3. *Nucleic Acids Res* **31:**860–868.
33. **Vergunst AC, van Lier MC, den Dulk-Ras A, Grosse Stuve TA, Ouwehand A, Hooykaas PJ.** 2005. Positive charge is an important feature of the C-terminal transport signal of the VirB/D4-translocated proteins of *Agrobacterium*. *Proc Natl Acad Sci USA* **102:**832–837.
34. **Buchanan-Wollaston V, Passiatore JE, Cannon F.** 1987. The *mob* and *oriT* mobilization functions of a bacterial plasmid promote its transfer to plants. *Nature* **328:** 172–175.
35. **Fullner KJ.** 1998. Role of *Agrobacterium virB* genes in transfer of T complexes and RSF1010. *J Bacteriol* **180:** 430–434.
36. **Parker C, Meyer RJ.** 2007. The R1162 relaxase/primase contains two, type IV transport signals that require the small plasmid protein MobB. *Mol Microbiol* **66:** 252–261.

37. **Alvarez-Martinez CE, Christie PJ.** 2009. Biological diversity of prokaryotic type IV secretion systems. *Microbiol Mol Biol Rev* **73:**775–808.

38. **Cabezon E, Sastre JI, de la Cruz F.** 1997. Genetic evidence of a coupling role for the TraG protein family in bacterial conjugation. *Mol Gen Genet* **254:**400–406.

39. **Hamilton CM, Lee H, Li PL, Cook DM, Piper KR, von Bodman SB, Lanka E, Ream W, Farrand SK.** 2000. TraG from RP4 and TraG and VirD4 from Ti plasmids confer relaxosome specificity to the conjugal transfer system of pTiC58. *J Bacteriol* **182:**1541–1548.

40. **Szpirer CY, Faelen M, Couturier M.** 2000. Interaction between the RP4 coupling protein TraG and the pBHR1 mobilization protein Mob. *Mol Microbiol* **37:**1283–1292.

41. **Chen Y, Zhang X, Manias D, Yeo HJ, Dunny GM, Christie PJ.** 2008. *Enterococcus faecalis* PcfC, a spatially localized substrate receptor for type IV secretion of the pCF10 transfer intermediate. *J Bacteriol* **190:** 3632–3645.

42. **Cascales E, Christie PJ.** 2004. Definition of a bacterial type IV secretion pathway for a DNA substrate. *Science* **304:**1170–1173.

43. **Gomis-Ruth FX, Sola M, de la Cruz F, Coll M.** 2004. Coupling factors in macromolecular type-IV secretion machineries. *Curr Pharm Des* **10:**1551–1565.

44. **Gomis-Ruth FX, Moncalian G, Perez-Luque R, Gonzalez A, Cabezon E, de la Cruz F, Coll M.** 2001. The bacterial conjugation protein TrwB resembles ring helicases and F1-ATPase. *Nature* **409:**637–641.

45. **Hormaeche I, Alkorta I, Moro F, Valpuesta JM, Goni FM, De La Cruz F.** 2002. Purification and properties of TrwB, a hexameric, ATP-binding integral membrane protein essential for R388 plasmid conjugation. *J Biol Chem* **277:**46456–46462.

46. **Gomis-Ruth FX, Moncalian G, de la Cruz F, Coll M.** 2002. Conjugative plasmid protein TrwB, an integral membrane type IV secretion system coupling protein. Detailed structural features and mapping of the active site cleft. *J Biol Chem* **277:**7556–7566.

47. **Alt-Morbe J, Stryker JL, Fuqua C, Li PL, Farrand SK, Winans SC.** 1996. The conjugal transfer system of *Agrobacterium tumefaciens* octopine-type Ti plasmids is closely related to the transfer system of an IncP plasmid and distantly related to Ti plasmid *vir* genes. *J Bacteriol* **178:**4248–4257.

48. **Christie PJ, Vogel JP.** 2000. Bacterial type IV secretion: conjugation systems adapted to deliver effector molecules to host cells. *Trends Microbiol* **8:**354–360.

49. **Winans SC, Walker GC.** 1985. Conjugal transfer system of the IncN plasmid pKM101. *J Bacteriol* **161:**402–410.

50. **Smillie C, Garcillan-Barcia MP, Francia MV, Rocha EPC, de la Cruz F.** 2010. Mobility of Plasmids. *Microbiol Mol Biol Rev* **74:**434–452.

51. **Ding H, Hynes MF.** 2009. Plasmid transfer systems in the rhizobia. *Can J Microbiol* **55:**917–927.

52. **Pena A, Matilla I, Martin-Benito J, Valpuesta JM, Carrascosa JL, de la Cruz F, Cabezon E, Arechaga I.** 2012. The hexameric structure of a conjugative VirB4 protein ATPase provides new insights for a functional and phylogenetic relationship with DNA translocases. *J Biol Chem* **287:**39925–39932.

53. **Li F, Alvarez-Martinez C, Chen Y, Choi KJ, Yeo HJ, Christie PJ.** 2012. *Enterococcus faecalis* PrgJ, a VirB4-like ATPase, mediates pCF10 conjugative transfer through substrate binding. *J Bacteriol* **194:**4041–4051.

54. **Wallden K, Williams R, Yan J, Lian PW, Wang L, Thalassinos K, Orlova EV, Waksman G.** 2012. Structure of the VirB4 ATPase, alone and bound to the core complex of a type IV secretion system. *Proc Natl Acad Sci USA* **109:**11348–11353.

55. **Savvides SN.** 2007. Secretion superfamily ATPases swing big. *Structure* **15:**255–257.

56. **Yeo HJ, Savvides SN, Herr AB, Lanka E, Waksman G.** 2000. Crystal structure of the hexameric traffic ATPase of the *Helicobacter pylori* type IV secretion system. *Mol Cell* **6:**1461–1472.

57. **Savvides SN, Yeo HJ, Beck MR, Blaesing F, Lurz R, Lanka E, Buhrdorf R, Fischer W, Haas R, Waksman G.** 2003. VirB11 ATPases are dynamic hexameric assemblies: new insights into bacterial type IV secretion. *EMBO J* **22:**1969–1980.

58. **Atmakuri K, Cascales E, Christie PJ.** 2004. Energetic components VirD4, VirB11 and VirB4 mediate early DNA transfer reactions required for bacterial type IV secretion. *Mol Microbiol* **54:**1199–1211.

59. **Mossey P, Hudacek A, Das A.** 2010. *Agrobacterium tumefaciens* type IV secretion protein VirB3 is an inner membrane protein and requires VirB4, VirB7, and VirB8 for stabilization. *J Bacteriol* **192:**2830–2838.

60. **Jakubowski SJ, Krishnamoorthy V, Cascales E, Christie PJ.** 2004. *Agrobacterium tumefaciens* VirB6 domains direct the ordered export of a DNA substrate through a type IV secretion system. *J Mol Biol* **341:**961–977.

61. **Sarkar MK, Husnain SI, Jakubowski SJ, Christie PJ.** 2013. Isolation of bacterial type IV machine subassemblies. *Methods Mol Biol* **966:**187–204.

62. **Fronzes R, Schafer E, Wang L, Saibil HR, Orlova EV, Waksman G.** 2009. Structure of a type IV secretion system core complex. *Science* **323:**266–268.

63. **Chandran V, Fronzes R, Duquerroy S, Cronin N, Navaza J, Waksman G.** 2009. Structure of the outer membrane complex of a type IV secretion system. *Nature* **462:**1011–1015.

64. **Jakubowski SJ, Kerr JE, Garza I, Krishnamoorthy V, Bayliss R, Waksman G, Christie PJ.** 2009. *Agrobacterium* VirB10 domain requirements for type IV secretion and T pilus biogenesis. *Mol Microbiol* **71:**779–794.

65. **Christie PJ.** 2009. Structural biology: translocation chamber's secrets. *Nature* **462:**992–994.

66. **Paranchych W, Frost LS.** 1988. The physiology and biochemistry of pili. *Adv Microb Physiol* **29:**53–114.

67. **Bradley DE.** 1980. Morphological and serological relationships of conjugative pili. *Plasmid* **4:**155–169.

68. **Schröder G, Lanka E.** 2005. The mating pair formation system of conjugative plasmids—a versatile secretion machinery for transfer of proteins and DNA. *Plasmid* **54:**1–25.

69. **Sagulenko E, Sagulenko V, Chen J, Christie PJ.** 2001. Role of *Agrobacterium* VirB11 ATPase in T-pilus assembly and substrate selection. *J Bacteriol* **183:**5813–5825.

70. **Worobec EA, Frost LS, Pieroni P, Armstrong GD, Hodges RS, Parker JM, Finlay BB, Paranchych W.** 1986. Location of the antigenic determinants of conjugative F-like pili. *J Bacteriol* **167:**660–665.

71. **Kalkum M, Eisenbrandt R, Lurz R, Lanka E.** 2002. Tying rings for sex. *Trends Microbiol* **10:**382–387.

72. **Silverman PM, Clarke MB.** 2010. New insights into F-pilus structure, dynamics, and function. *Integr Biol (Camb)* **2:**25–31.

73. **Eisenbrandt R, Kalkum M, Lai EM, Lurz R, Kado CI, Lanka E.** 1999. Conjugative pili of IncP plasmids, and the Ti plasmid T pilus are composed of cyclic subunits. *J Biol Chem* **274:**22548–22555.

74. **Kerr JE, Christie PJ.** 2010. Evidence for VirB4-mediated dislocation of membrane-integrated VirB2 pilin during biogenesis of the *Agrobacterium* VirB/VirD4 type IV secretion system. *J Bacteriol* **192:**4923–4934.

75. **Yuan Q, Carle A, Gao C, Sivanesan D, Aly KA, Hoppner C, Krall L, Domke N, Baron C.** 2005. Identification of the VirB4-VirB8-VirB5-VirB2 pilus assembly sequence of type IV secretion systems. *J Biol Chem* **280:** 26349–26359.

76. **Clarke M, Maddera L, Harris RL, Silverman PM.** 2008. F-pili dynamics by live-cell imaging. *Proc Natl Acad Sci USA* **105:**17978–17981.

77. **Aly KA, Baron C.** 2007. The VirB5 protein localizes to the T-pilus tips in *Agrobacterium tumefaciens*. *Microbiology* **153:**3766–3775.

78. **Yeo H-J, Yuan Q, Beck MR, Baron C, Waksman G.** 2003. Structural and functional characterization of the VirB5 protein from the type IV secretion system encoded by the conjugative plasmid pKM101. *Proc Natl Acad Sci USA* **100:**15947–15952.

79. **Gelvin SB.** 2012. Traversing the cell: *Agrobacterium* T-DNA's journey to the host genome. *Front Plant Sci* **3:** 52. doi:10.3389/fpls.2012.00052.

80. **Ballas N, Citovsky V.** 1997. Nuclear localization signal binding protein from *Arabidopsis* mediates nuclear import of *Agrobacterium* VirD2 protein. *Proc Natl Acad Sci USA* **94:**10723–10728.

81. **Bhattacharjee S, Lee LY, Oltmanns H, Cao H, Veena, Cuperus J, Gelvin SB.** 2008. IMPα-4, an *Arabidopsis* importin alpha isoform, is preferentially involved in *Agrobacterium*-mediated plant transformation. *Plant Cell* **20:**2661–2680.

82. **Tao Y, Rao PK, Bhattacharjee S, Gelvin SB.** 2004. Expression of plant protein phosphatase 2C interferes with nuclear import of the *Agrobacterium* T-complex protein VirD2. *Proc Natl Acad Sci USA* **101:**5164–5169.

83. **Bako L, Umeda M, Tiburcio AF, Schell J, Koncz C.** 2003. The VirD2 pilot protein of *Agrobacterium*-transferred DNA interacts with the TATA box-binding protein and a nuclear protein kinase in plants. *Proc Natl Acad Sci USA* **100:**10108–10113.

84. **Deng W, Chen L, Wood DW, Metcalfe T, Liang X, Gordon MP, Comai L, Nester EW.** 1998. *Agrobacterium* VirD2 protein interacts with plant host cyclophilins. *Proc Natl Acad Sci USA* **95:**7040–7045.

85. **Dumas F, Duckely M, Pelczar P, van Gelder P, Hohn B.** 2001. An *Agrobacterium* VirE2 channel for T-DNA transport into plant cells. *Proc Natl Acad Sci USA* **98:** 485–490.

86. **Guyon P, Chilton MD, Petit A, Tempe J.** 1980. Agropine in "null-type" crown gall tumors: evidence for generality of the opine concept. *Proc Natl Acad Sci USA* **77:**2693–2697.

87. **Palanichelvam K, Veluthambi K.** 1996. Octopine- and nopaline-inducible proteins in *Agrobacterium tumefaciens* are also induced by arginine. *Curr Microbiol* **33:** 156–162.

88. **Winans SC.** 1990. Transcriptional induction of an *Agrobacterium* regulatory gene at tandem promoters by plant-released phenolic compounds, phosphate starvation, and acidic growth media. *J Bacteriol* **172:** 2433–2438.

89. **Winans SC.** 1991. An *Agrobacterium* two-component regulatory system for the detection of chemicals released from plant wounds. *Mol Microbiol* **5:**2345–2350.

90. **Cangelosi GA, Ankenbauer RG, Nester EW.** 1990. Sugars induce the *Agrobacterium* virulence genes through a periplasmic binding protein and a transmembrane signal protein. *Proc Natl Acad Sci USA* **87:** 6708–6712.

91. **Melchers LS, Regensburg-Tuink AJ, Schilperoort RA, Hooykaas PJ.** 1989. Specificity of signal molecules in the activation of *Agrobacterium* virulence gene expression. *Mol Microbiol* **3:**969–977.

92. **Gao R, Lynn DG.** 2007. Integration of rotation and piston motions in coiled-coil signal transduction. *J Bacteriol* **189:**6048–6056.

93. **Shimoda N, Toyoda-Yamamoto A, Aoki S, Machida Y.** 1993. Genetic evidence for an interaction between the VirA sensor protein and the ChvE sugar-binding protein of *Agrobacterium*. *J Biol Chem* **268:**26552–26558.

94. **Peng WT, Lee YW, Nester EW.** 1998. The phenolic recognition profiles of the *Agrobacterium tumefaciens* VirA protein are broadened by a high level of the sugar binding protein ChvE. *J Bacteriol* **180:**5632–5638.

95. **Gao R, Lynn DG.** 2005. Environmental pH sensing: resolving the VirA/VirG two-component system inputs for *Agrobacterium* pathogenesis. *J Bacteriol* **187:**2182–2189.

96. **Habeeb LF, Wang L, Winans SC.** 1991. Transcription of the octopine catabolism operon of the *Agrobacterium* tumor-inducing plasmid pTiA6 is activated by a LysR-type regulatory protein. *Mol Plant Microbe Interact* **4:** 379–385.

97. **Wang L, Helmann JD, Winans SC.** 1992. The *A. tumefaciens* transcriptional activator OccR causes a bend at a target promoter, which is partially relaxed by a plant tumor metabolite. *Cell* **69:**659–667.

98. **Wang L, Winans SC.** 1995. High angle and ligand-induced low angle DNA bends incited by OccR lie in the same plane with OccR bound to the interior angle. *J Mol Biol* **253:**32–38.

99. **Beck von Bodman S, Hayman GT, Farrand SK.** 1992. Opine catabolism and conjugal transfer of the nopaline Ti plasmid pTiC58 are coordinately regulated by a single repressor. *Proc Natl Acad Sci USA* **89:**643–647.
100. **White CE, Winans SC.** 2007. Cell-cell communication in the plant pathogen *Agrobacterium* tumefaciens. *Philos Trans R Soc Lond B Biol Sci* **362:**1135–1148.
101. **Pappas KM.** 2008. Cell-cell signaling and the *Agrobacterium tumefaciens* Ti plasmid copy number fluctuations. *Plasmid* **60:**89–107.
102. **Venturi V, Fuqua C.** 2013. Chemical signaling between plants and plant-pathogenic bacteria. *Annu Rev Phytopathol* **51:**17–37.
103. **Fuqua WC, Winans SC.** 1994. A LuxR-LuxI type regulatory system activates *Agrobacterium* Ti plasmid conjugal transfer in the presence of a plant tumor metabolite. *J Bacteriol* **176:**2796–2806.
104. **Hwang I, Li PL, Zhang L, Piper KR, Cook DM, Tate ME, Farrand SK.** 1994. TraI, a LuxI homologue, is responsible for production of conjugation factor, the Ti plasmid N-acylhomoserine lactone autoinducer. *Proc Natl Acad Sci USA* **91:**4639–4643.
105. **Luo ZQ, Qin Y, Farrand SK.** 2000. The antiactivator TraM interferes with the autoinducer-dependent binding of TraR to DNA by interacting with the C-terminal region of the quorum-sensing activator. *J Biol Chem* **275:**7713–7722.
106. **Haudecoeur E, Faure D.** 2010. A fine control of quorum-sensing communication in *Agrobacterium tumefaciens*. *Commun Integr Biol* **3:**84–88.
107. **Chai Y, Zhu J, Winans SC.** 2001. TrlR, a defective TraR-like protein of *Agrobacterium tumefaciens*, blocks TraR function *in vitro* by forming inactive TrlR:TraR dimers. *Mol Microbiol* **40:**414–421.
108. **Cho H, Winans SC.** 2005. VirA and VirG activate the Ti plasmid *repABC* operon, elevating plasmid copy number in response to wound-released chemical signals. *Proc Natl Acad Sci USA* **102:**14843–14848.
109. **Pappas KM, Winans SC.** 2003. A LuxR-type regulator from *Agrobacterium tumefaciens* elevates Ti plasmid copy number by activating transcription of plasmid replication genes. *Mol Microbiol* **48:**1059–1073.
110. **Platt TG, Bever JD, Fuqua C.** 2012. A cooperative virulence plasmid imposes a high fitness cost under conditions that induce pathogenesis. *Proc Biol Sci* **279:**1691–1699.
111. **Yamamoto S, Kiyokawa K, Tanaka K, Moriguchi K, Suzuki K.** 2009. Novel toxin-antitoxin system composed of serine protease and AAA-ATPase homologues determines the high level of stability and incompatibility of the tumor-inducing plasmid pTiC58. *J Bacteriol* **191:**4656–4666.
112. **Nautiyal CS, Dion P.** 1990. Characterization of the opine-utilizing microflora associated with samples of soil and plants. *Appl Environ Microbiol* **56:**2576–2579.
113. **Tzfira T, Citovsky V.** 2008. *Agrobacterium: From Biology to Biotechnology*. Springer Press, New York, NY.
114. **Vergunst AC, Jansen LE, Fransz PF, de Jong JH, Hooykaas PJ.** 2000. Cre/lox-mediated recombination in *Arabidopsis*: evidence for transmission of a translocation and a deletion event. *Chromosoma* **109:**287–297.
115. **Soltani J, van Heusden PH, Hooykaas PJJ.** 2008. *Agrobacterium*-mediated transformation of non-plant organisms. p 649–675. *In* Tzfira T, Citovsky V (ed), *Agrobacterium: From Biology to Biotechnology*. Springer Press, New York, NY.
116. **Kunik T, Tzfira T, Kapulnik Y, Gafni Y, Dingwall C, Citovsky V.** 2001. Genetic transformation of HeLa cells by *Agrobacterium*. *Proc Natl Acad Sci USA* **98:** 1871–1876.
117. **Christie PJ.** 2007. *Agrobacterium* and plant cell transformation. p 29–43. *In* Schaechter M (ed), *Desk Encyclopedia of Microbiology*, 2nd ed. Academic Press, San Diego, CA.
118. **Laverde-Gomez JA, Sarkar MK, Christie PJ.** 2012. Regulation of bacterial type IV secretion systems. p 335–362. *In* Vasil M, Darwin A (ed), *Regulation of Bacterial Virulence*. ASM Press, Washington, DC.
119. **Baxter JC, Funnell BE.** 2014. Plasmid partition mechanisms. *Microbiol Spectrum* **2**(5). doi:10.1128/microbiolspec.PLAS-0023-2014.

Plasmids—Biology and Impact in Biotechnology and Discovery
Edited by Marcelo E. Tolmasky and Juan C. Alonso

doi:10.1128/microbiolspec.PLAS-0012-2013

Laura C.C. Cook[1]
Gary M. Dunny[2]

18 The Influence of Biofilms in the Biology of Plasmids

INTRODUCTION

The natural state for many bacteria is not growth in liquid culture, but, rather, living as a community attached to a surface. These bacterial communities, termed biofilms, exist in the natural world as well as in the human host. The Centers for Disease Control and Prevention and the National Institutes of Health have estimated that approximately 65 to 80% of human infections are biofilm related. A recent burgeoning area of research has examined the role of plasmids in biofilms, including the effect of conjugative plasmid transfer on biofilm formation, as well as the role of biofilms in plasmid dissemination. In addition, heterogeneity in the biofilm population in terms of plasmid carriage has also been demonstrated. Most published studies of plasmid biology and conjugation in biofilms have focused on Gram-negative spp. such as *Pseudomonas aeruginosa* and *Escherichia coli*. In this article, we will review these studies in relation to recent work focusing on effects of biofilm growth on plasmid-related functions such as gene transfer and antimicrobial resistance in Gram-positive pathogens such as *Enterococcus faecalis* and *Staphylococcus aureus*.

The formation of bacterial biofilms involves three steps (Fig. 1). Initially, individual cells growing planktonically attach to a surface. Following surface adherence, additional cells may bind to previously attached cells. As the attached cells grow and divide, they produce an extracellular polymeric substance known as the biofilm matrix that stabilizes attachment of the cells to one another and to the surface. The biofilm matrix components may differ between species but frequently contain DNA (1), proteins (2), and polysaccharides (3), as well as other nutrients and cellular components (recently reviewed in reference 4). During and following formation of a fully structured biofilm, individual cells or even large pieces of the biofilm may break away. These cells may then revert back to a planktonic lifestyle or may attach to a surface elsewhere and seed a new biofilm (Fig. 1).

Many of the seminal studies of biofilm development that led to developmental models like the one shown in Fig. 1 utilized rod-shaped motile bacteria such as *E. coli, P. aeruginosa*, and *Bacillus subtilis* (5, 6, 7, 8, 9). In such bacteria, the sensing of surface attachment and transition from planktonic to biofilm growth may

[1]Department of Medicinal Chemistry, University of Illinois, Chicago, IL 60607; [2]Department of Microbiology, University of Minnesota, Minneapolis, MN 55455.

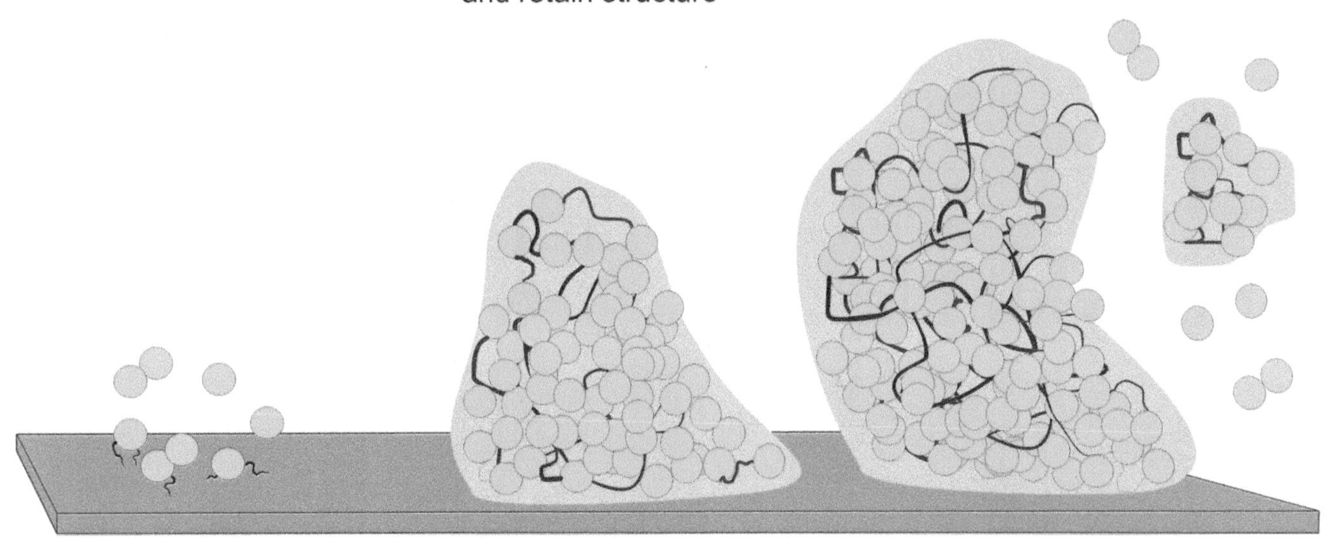

Figure 1 Formation of a bacterial biofilm. Bacterial biofilm development involves three stages. (**1**) Initial attachment of single group or small groups of bacteria to a surface, often aided by attachment structures such as pili. (**2**) Growth of these attached cells as well as attachment of additional cells increases the biomass of the biofilm. Concurrently, the bacteria produce an extracellular matrix made of up various components including DNA, protein, and polysaccharides that help the biofilm retain its structure and keep the biofilm cells attached to the surface and to each other. (**3**) During and after the formation of a large biofilm, individual cells or even large pieces of the biofilm may break off. These detached cells may go on to live a planktonic lifestyle or seed new biofilms. Dark lines indicate the components of the matrix used to attach cells to each other and the surface such as eDNA (1), while orange extracellular material indicates other matrix components used to retain biofilm structure and surface attachment. doi:10.1128/microbiolspec.PLAS-0012-2013.f1

involve components of the motility machinery and is accompanied by a loss of motility, whereas the dispersal phase can involve the reactivation of motility. It is interesting that nonmotile genera including staphylococci, streptococci, and enterococci show very similar patterns of biofilm development and dispersal, including characteristic cellular architecture and extracellular matrix in the biofilm structure. Because biofilms are believed to be highly heterogenous in terms of nutrient, pH, and oxygen gradients, it is likely that they are composed of heterogenous populations of cells that differ in terms of metabolic potential and phenotypic characteristics.

CONJUGATIVE PLASMIDS IN BIOFILMS

It is well established that the biofilm is an important niche for horizontal gene transfer (HGT) by transformation in naturally competent bacteria (10, 11), and that biofilm development and competence are mediated and regulated by many of the same gene products (10, 12, 13). HGT, the transferring genetic material between cells in which reproduction does not play a role, includes the processes of conjugation, transformation, and transduction. Increasingly, new studies have examined the interplay between conjugation and biofilm development (14, 15, 16, 17, 18, 19, 20). The seminal article by Ghigo describing the role of plasmids in biofilm formation described the effects of the well-studied conjugative F plasmid of *E. coli* biofilms (15). These experiments demonstrated that the addition of the F plasmid to *E. coli* cells greatly increased their ability to form biofilms in a conjugation-independent and plasmid-encoded, pilus-dependent fashion (15). This report documented an increase in biofilm formation by other Gram-negative bacteria when grown with pilus-encoding natural conjugative plasmids. In the case of F and other plasmids like it that express pili and other conjugation functions constitutively, the presence of the plasmid was associated with increased biofilm formation. Monocultures of donor strains carrying repressed plasmids such as R1 did not exhibit increased biofilm development. However, in biofilms formed from donor/recipient mixtures, or in recipient biofilms subsequently exposed to planktonic donor cells, Ghigo observed enhanced biofilm development. Ghigo hypothesized that a small number of spontaneously depressed pilus-producing bacteria in the planktonic donor cultures could adhere to the recipient biofilms and transfer their plasmids, followed by a period of "epidemic spread" through the entire biofilm. The increase in pilus-expressing bacteria could then aid in the formation of a large bacterial biofilm (15). The top portion of Fig. 2 illustrates the model proposed by Ghigo. The bottom depicts a variation on the theme of induction of conjugation and surface adhesins in a biofilm context for *E. faecalis*, where the expression of conjugative functions in donor cells can be activated by a peptide-mating pheromone produced by recipients, which is further discussed below. Both models illustrate how activation of conjugation in a biofilm context can lead to both plasmid transfer and increased biofilm biomass.

Transmission of a conjugative F plasmid also induces biofilm formation by a mixed population of laboratory and wild isolates of *E. coli* and plays a role in the overall structure of the biofilm (5, 20). Addition of an F-like conjugative plasmid, R1*drd*19, which, like F, constitutively synthesizes pili, was also shown to induce greater biofilm formation in *E. coli* cultures (19). Interestingly, the presence of the R1*drd*19 plasmid also increased the expression of numerous chromosomal genes including

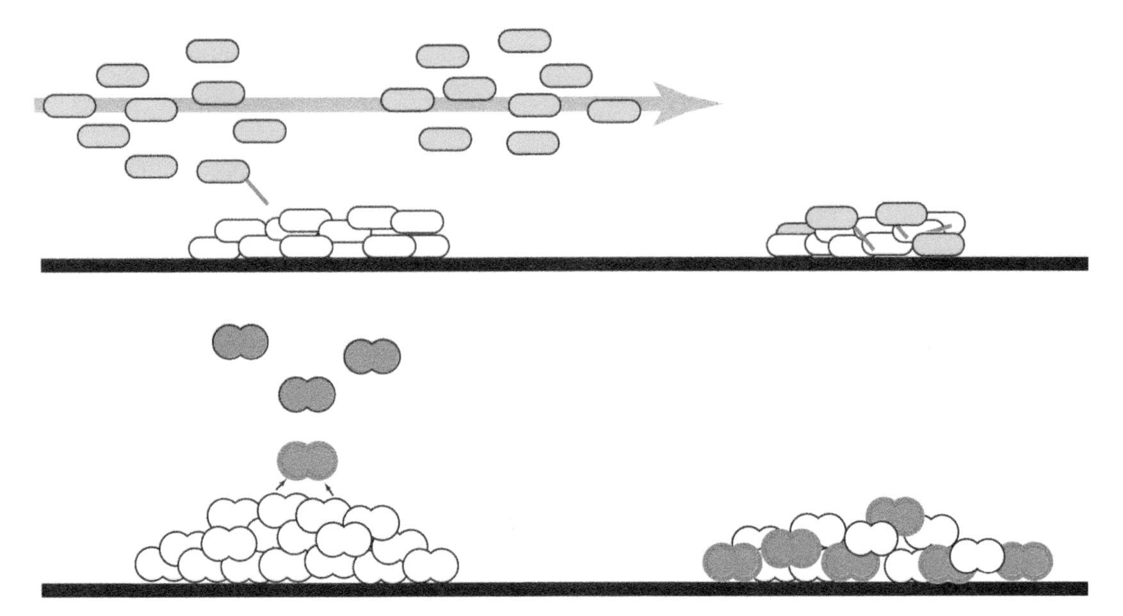

Figure 2 The interplay between conjugation and biofilm development. (Top) A model proposed by Ghigo (15), based on his analysis of conjugation between *E. coli* strains. Planktonic populations of donor cells (green), carrying plasmids such as R1, whose conjugation functions are normally repressed, contain a few spontaneously depressed individuals. When these depressed cells encounter a biofilm containing recipient cells (white), they can attach via their sex pili (red) and transfer the plasmid. In newly generated transconjugants, there is a transient period where repression of conjugation is not operative. This can be followed by "epidemic spread" of the plasmid through the biofilm population, and the associated production of sex pili also can increase the biofilm biomass directly. (Bottom) In *E. faecalis*, expression of conjugation is regulated by peptide-mating pheromones produced by recipient cells. In the scenario depicted on the left, the pheromone produced by recipient cells (white) in a biofilm turns on expression of conjugation in planktonic donor cells (blue) in close proximity, and the resulting synthesis of pheromone-induced surface adhesins (thick, gray layer) promotes both an increase in biofilm resulting from increased attachment of planktonic cells, and also leads to plasmid transfer within the biofilm. In the right, the development of a mixed biofilm as a result of attachment of both donors and recipients to the same surface may allow for signaling and conjugation between sessile donor and recipient cells in close proximity (34). doi:10.1128/microbiolspec.PLAS-0012-2013.f2

those related to envelope stress, motility, and other genes known to be involved in biofilm formation. It was also demonstrated that F pilus production caused increased colonic acid and curli (thin fimbriae) production (21). Curli have previously been shown to stimulate the attachment of *E. coli* cells to surfaces (22). Other conjugative plasmids of *E. coli*, including pOLA52 and pMAS2027, have been shown to enhance biofilm formation through type 3 fimbriae (14, 17). The pOLA52 plasmid can also be transferred to a variety of organisms and retains its ability to induce biofilm formation in *Salmonella enterica* serovar Typhimurium, *Kluyvera* spp., and *Enterobacter aerogenes* (14).

Studies on the TOL conjugative plasmid of *Pseudomonas putida* demonstrated that carriage of conjugative plasmids could also increase biofilm formation by increasing the amount of extracellular DNA (eDNA), thus aiding the formation of the biofilm matrix (16). Although not fully understood, the mechanism of increased eDNA does not appear to be caused by increased cell lysis and, alternatively, may be due to increased DNA secretion. These results demonstrate that the conjugative pili and fimbriae are not the only factors involved in plasmid-mediated enhancement of biofilm formation.

Coculture studies with *P. putida*, *E. coli*, and *Kluyvera* spp. also demonstrated the impact of conjugative plasmids on biofilm development. In these experiments, the conjugative plasmid pKJK5, an IncP-1 plasmid (23), altered biofilm development during growth in coculture. Cocultures of the three bacteria produced increased biofilms in comparison with any of the strains alone. Interestingly, when *P. putida* carried pKJK5, biofilm formation of a mixed coculture was decreased (24). In this case, the presence of a conjugative plasmid had a negative effect on biofilm formation rather than enhancing it. It is apparent from these studies that the role of conjugative plasmids in biofilm development is likely complex and varied and could depend strongly on host as well as plasmid-encoded factors.

The interplay between conjugative plasmids and biofilm formation is not a one-way street. Conjugative plasmids influence the development of biofilms, and, in turn, biofilms affect horizontal transfer of conjugative plasmids. Biofilm promotion of higher levels of HGT has been demonstrated for a variety of systems (18, 25, 26, 27). One of the first articles outlining HGT rates in biofilms used a green fluorescent protein (GFP)-tagged broad-host-range plasmid pRK415 (28) requiring a separate plasmid containing the cognate conjugation system, pRK2013. Using the GFP reporter and a TRITC counterstain, researchers were able to determine approximate conjugation rates in biofilms (25). Similarly, it was determined that transfer of the F-like plasmid R1*drd*19 in *E. coli* occurred at significantly higher rates in a biofilm, and transfer kinetics were similar between laboratory biofilms and growth in a mouse intestine (29). Another study looking at conjugative transfer of plasmid RP4 between *Pseudomonas* species demonstrated high transfer frequencies in biofilms and showed that shear force affected HGT, suggesting that altering various biofilm parameters affects rates of gene transfer (30).

A recent study attempted to determine the various factors involved in conjugative transfer of antibiotic resistance plasmids in *E. coli* biofilms in a comprehensive fashion (31). The transfer efficiency of 19 drug-resistant plasmids was measured by using biofilms of different ages and different times of exposure of biofilms to plasmid donor cells. Wide variation was observed in transfer efficiencies between different plasmids and efficiencies depended on conditions such as biofilm thickness and age, with older and thicker biofilms having a smaller proportion of transconjugants. Not surprisingly, allowing donor cells to incubate for longer periods of time with established biofilms also allowed for increased conjugation. Transfer was not only dependent on the particular plasmid and biofilm conditions, but transfer efficiency was also increased when the donor strain background had increased ability to attach to recipient cells, although this was also seen in liquid cultures (31).

HGT in Gram-positive bacterial biofilms has also been studied, although to a lesser extent. *S. aureus*, an important Gram-positive human pathogen, is known to form biofilms related to infections such as endocarditis, indwelling device-associated infections, and osteomyelitis. One report demonstrated that biofilm growth in *S. aureus* also facilitates higher levels of HGT with increased conjugation and mobilization frequencies leading to the spread of antibiotic resistance (27).

PLASMID COPY-NUMBER CONTROL AND MAINTENANCE IN BIOFILMS

Biofilm growth not only affects plasmid transfer, but it also appears to play a role in plasmid maintenance and copy-number control. In 1995, Davies and Geesey examined the regulation of alginate biosynthesis in *P. aeruginosa* biofilms using a reporter plasmid pNZ63 carrying a β-lactamase marker. During the course of their experiments they noted that, although the presence of the β-lactam antibiotic did not affect the average plasmid copy number in the population, biofilm growth coincided with an increase in plasmid number by ~1.5-fold (32). They also determined that the change in copy number was not due to plasmid loss because most of the cells retained antibiotic resistance. Additionally, it was shown that the copy number of pBR322, a plasmid carrying resistance genes against ampicillin and tetracycline, was increased approximately 2-fold in *E. coli* cells growing in a biofilm compared with planktonic copy numbers (33). Interestingly, an *E. coli* strain containing the pBR322 plasmid formed less biofilm than the plasmid-free cells, which the authors attributed to the presence of the *bla* ampicillin resistance gene. Addition of tetracycline or a combination of tetracycline and ampicillin at subinhibitory concentrations induced higher biofilm formation by *E. coli* cells harboring the plasmid but not plasmid-free cells (33). This study also found that the addition of antibiotics to planktonic populations caused an increase in plasmid copy number to levels seen in biofilm cells, supporting the hypothesis that increased plasmid copy number correlates with increased antibiotic resistance (33). Importantly, these studies examined the average copy number in biofilms which could significantly underestimate the copy-number heterogeneity that may exist in subpopulations of biofilm cells.

In 2011, a report by Cook et al. demonstrated that the copy number of the conjugative plasmid pCF10 of *E. faecalis* was increased during growth in a biofilm in comparison with planktonic growth (34). Not only was the average copy number of pCF10 increased in biofilm populations, but copy-number heterogeneity in the population was also highly increased, supporting the model of biofilms as complex communities in which not all cells are identical. While the average number of pCF10 copies/chromosome in biofilm cells was about twice that of planktonic cells, fluorescence-activated cell sorting experiments showed the existence of subpopulations of biofilm cells containing as many as 5 times as many plasmid copies as planktonic cells (34).

In the case of pCF10, an increase in plasmid copy number causes a subsequent increase in plasmid-borne negative regulators of conjugation including an inhibitory peptide, iCF10, and the negative regulator of conjugation, PrgX. This results in tighter control of conjugation induction; thus, biofilm cells carrying more copies of pCF10 require a higher concentration of inducer peptide, cCF10, to turn on conjugation. These same cells may show a significantly increased level of conjugation gene expression once the threshold-inducing concentration is exceeded, leading to a stronger response to peptide. It is hypothesized that this plasmid copy-number control allows restrictive regulation of conjugation, preventing the energetically costly process from occurring when potential plasmid recipient cells are not in the immediate vicinity (34). This is especially important in the case of *E. faecalis* because the nonmotile bacteria are fixed in the biofilm without the ability to migrate toward potential recipients.

Following this study, further analysis was completed on the copy number of various enterococcal plasmids to determine whether increased copy number was a pCF10-specific phenomenon. It was found that at least four other plasmids showed an increase in both plasmid copy number and copy-number heterogeneity when cells were grown in a biofilm (35). These four plasmids were not conjugative, were unrelated to pCF10, had different native copy numbers, and included both rolling circle and theta-replicating elements. In this study, increased plasmid copy number was also correlated with increased expression of plasmid-borne antibiotic resistance genes as well as increased plating efficiency on high concentrations of antibiotics (35). It was not possible to determine the heterogeneity of plasmid copy number in biofilm cells with all the plasmids analyzed, but the available data indicate that the heterogeneity observed with pCF10 could also exist for other, unrelated replicons.

The published studies of biofilm growth of *E. faecalis*, and its effect on plasmid copy number, plasmid transfer, and antibiotic resistance can be used to generate hypothetical mechanistic models that may inform the design of future experiments. Figure 3 illustrates three possible mechanisms, all of which assume that a subpopulation of plasmid-carrying cells in a biofilm display a significant increase in copy number. It has recently been shown that during early biofilm growth of *E. faecalis*, there is substantial secretion of eDNA into the extracellular matrix by a mechanism that does not require cell lysis and may be performed by a fraction of cells in the monospecies biofilm community (1). As shown in Fig. 3A, if chromosomal (rather than plasmid) DNA was a specific substrate for this novel secretion pathway, the relative intracellular concentration of plasmid would be increased. This could also indirectly increase the transcription of plasmid genes relative to chromosomal genes, by increasing the fraction of the cellular pool of RNA polymerase available to transcribe plasmid genes. Figure 3B depicts a scenario where the global physiological changes in a fraction of biofilm cells disrupts the initiation of chromosome replication, even when nutrients are not limiting and the overall growth rate of the community is high. In both Fig. 3A and B, the increased ratio of plasmids

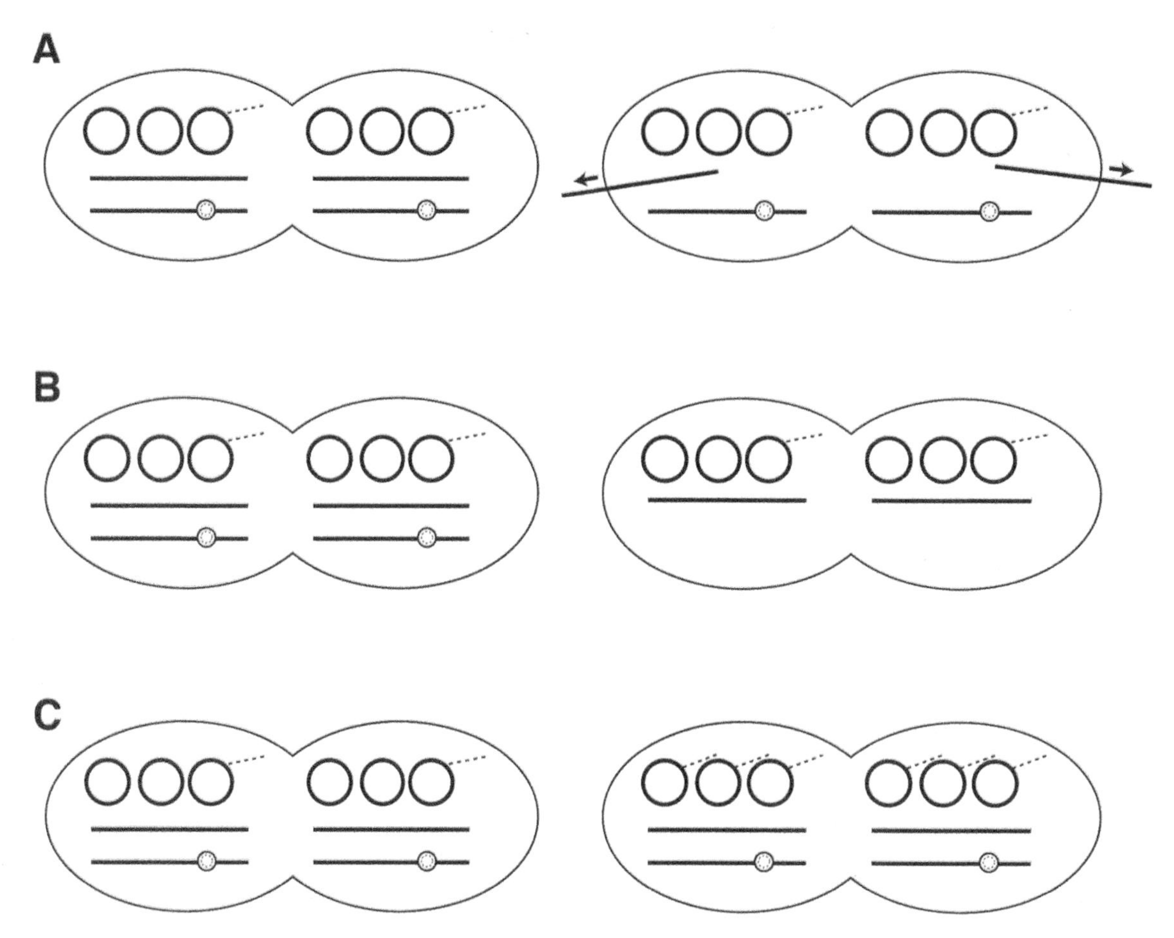

Figure 3 Three mechanistic models that could account for increased ratios of plasmids/chromosome in a fraction of *E. faecalis* biofilm cells. Figures depict actively growing "planktonic-like" diplococcal cells prior to cell division. Each cell half contains two copies of the chromosome (represented by thick lines), where a new round of replication has initiated on one chromosome (indicated by the "bubble"). Each cell half also contains three copies of a plasmid, with the dashed lines indicating rolling circle replication initiated by either a single-stranded replication initiator protein or a conjugative relaxase. (It should be noted that some plasmids showing increased copy number in biofilms actually use a "theta" mode of replication similar to the chromosome [35]). Each diagram contains one diplococcal cell (left) with a plasmid copy number similar to that of planktonic cells and a second cell with an altered copy number, representing a subpopulation of the biofilm community. The three diagrams illustrate three possible mechanisms by which the plasmid/chromosome ratio could be increased in these subpopulations. (A) Secretion of eDNA into the biofilm matrix by a fraction of cells reduces the ratio of chromosome/plasmid. (B) Biofilm growth reduces chromosomal copy number in subpopulations by inhibiting initiation of chromosome replication. (C) In a subpopulation of biofilm cells, changes in cellular physiology could disrupt normal mechanisms limiting conjugative nicking or vegetative plasmid replication initiation, leading to increases in copy number. Additionally, for some well-studied plasmids, such as ColE1, blockage of chromosome replication initiation (B) can lead to "runaway" plasmid replication (36). Based on published results (1, 34), all of these models are postulated to operate in the early stages of biofilm development, when the adherent cells are still growing, and there is no significant death or lysis. doi:10.1128/microbiolspec.PLAS-0012-2013.f3

actually results from a decrease in replication and reduced numbers of chromosomes/cell, and, in some cases, concomitant "runaway replication" (36) of plasmids residing in the same cell could occur. Runaway replication of plasmids has been demonstrated to occur in the presence of environmental factors such as antibiotics, nutrient levels, and temperature (36, 37, 38). In the previously published work (34, 35), significant changes in total DNA per cell were not found in the pooled biofilm cells, but the methods used may not have detected changes in subpopulations. In the model illustrated in Fig. 3C, the altered physiology of biofilm cells is postulated to increase the initiation of vegetative replication or conjugative nicking, both of which result in the generation of new copies of the plasmid. The models depicted in Fig. 3 all involve actively growing cells in early biofilm formation, when there is not a significant level of cell death or lysis. For cells in older biofilms experiencing nutrient limitation and other stresses, additional factors likely could have significant impacts on plasmid maintenance and copy control, as noted below.

Although plasmid copy number was increased in biofilms for numerous plasmids listed above, researchers have conversely demonstrated that, for some plasmids, plasmid loss is increased in biofilm populations, especially at specific foci within the biofilm. The TOL plasmid of *P. putida*, known for carrying genes allowing degradation of organic compounds, is used in bioremediation and the ability of this plasmid to be transferred stably in an environmental setting such as a biofilm has been examined. It was found that the probability of plasmid loss was increased in biofilms compared with planktonic cells. As was shown with plasmid copy number, plasmid loss was also heterogenous within the biofilm with the outside layers experiencing up to 80 times higher plasmid loss than the populations in the middle of the biofilm (39). Loss of plasmids in *P. putida* biofilm populations was attributed to segregational loss. In this study, the actual copy number of plasmids/cell was not examined, only the loss of the plasmid based on selective plating. It is important to remember that the *P. putida* biofilms that were examined ranged from 1 to 7 days of age, whereas plasmid copy-number analyses in *E. faecalis* were done on "young" biofilms at 4 or 24 hours after initial inoculation. The age of the biofilm could greatly impact the growth rate of the cells and thus affect plasmid copy-number control or plasmid loss.

The results of these studies demonstrate that the heterogeneous traits of biofilm cells extend to plasmid carriage. Upregulation of plasmid copy number presents a novel possible mechanism of the high level of antibiotic resistance observed in biofilms. From the available data, testable models (Fig. 3) can be generated to facilitate the design of new experiments focused on the mechanisms for alteration of copy number in biofilm growth. It has yet to be determined whether this phenomenon may be found in other human pathogens and whether it might impact antibiotic susceptibility of bacterial biofilm infections clinically.

To examine plasmid maintenance theoretically, Imran et al. derived a mathematical model to determine the interplay between the cost and benefits of plasmid carriage (40). This model assumed that the plasmid in question was beneficial for biofilm formation itself, which may not be the case for most plasmids but is applicable to some of the conjugative plasmids discussed above. Although this model does not provide definitive data on plasmid carriage in biofilms, it provides a good beginning framework for future work on this topic.

NONCONJUGATIVE PLASMIDS AND BIOFILM FORMATION

The majority of the studies examining plasmids and their relation to biofilm development have focused on conjugative plasmids, but recent research has demonstrated the influence of nonconjugative plasmids in biofilm formation as well. Teodosio et al. found that the addition of small nonconjugative plasmids pET28 and pUC8 to *E. coli* increased the concentration of cells growing in a biofilm compared with nontransformed cells (41). Additionally, the horizontal transfer of nonconjugative plasmids and genetic competence have also been demonstrated to occur in *E. coli* systems (42, 43, 44). Using DNase I to degrade DNA in F^- *E. coli* biofilms, it was demonstrated that some horizontal transfer (likely through transformation) occurred in biofilm cells at a higher level than in planktonically growing cells (45).

CONCLUSION

The role of plasmids and plasmid copy number in bacterial biofilm populations is likely much more complex than what is observed in homogenous batch cultures. Plasmid carriage in biofilms is probably highly heterogenous and may allow for the development of subgroups within the biofilm population that are specifically poised to react to different environments and stimuli. The development of improved methods for quantitative determination of plasmid copy number in individual cells, as well as improved visualization of individual conjugation events in real time, will be necessary for a full understanding of this topic. Recently, improved fluorescent reporter proteins and highly sensitive and specific reagents for fluorescent staining were used to demonstrate localization of the competence machinery and visualize individual genetic transformation events in *Streptococcus pneumoniae* (46). Similar approaches hold great promise for the analysis of plasmid copy number and conjugation in enterococci and staphylococci growing in biofilms.

The research summarized in this review details the importance of plasmids in the development of biofilms and that both single-species and multispecies biofilms are important ecological niches for HGT of plasmids by conjugation. Recent studies have also revealed that the physiological changes associated with the transition from planktonic to biofilm growth can impact the control of plasmid copy number, antibiotic resistance, and conjugative transfer (34, 35). A better understanding of these processes will be essential for the development of improved methods to prevent and control biofilm-related infections by resistant organisms, as well as for manipulation of biofilms to enhance the use of bacteria for biotechnological applications.

Acknowledgments. *Our biofilm and plasmid research was supported by NIH grants 1RO1AI58134 and 1RO1GM49530. L.C.C. was a predoctoral trainee under T32GM008347 from NIGMS (2007–2009), and received a fellowship from the American Academy of University Women (2010–2011). We thank Tim Leonard for making Figures 2 and 3. Conflicts of interest: We disclose no conflicts.*

Citation. Cook LCC, Dunny GM. 2014. The influence of biofilms in the biology of plasmids. Microbiol Spectrum **2**(5): PLAS-0012-2013.

References

1. **Barnes AM, Ballering KS, Leibman RS, Wells CL, Dunny GM.** 2012. *Enterococcus faecalis* produces abundant extracellular structures containing DNA in the absence of cell lysis during early biofilm formation. *MBio* **3:** e00193-12. doi:10.1128/mBio.00193-12
2. **Branda SS, Chu F, Kearns DB, Losick R, Kolter R.** 2006. A major protein component of the *Bacillus subtilis* biofilm matrix. *Mol Microbiol* **59:**1229–1238.
3. **Costerton JW, Cheng KJ, Geesey GG, Ladd TI, Nickel JC, Dasgupta M, Marrie TJ.** 1987. Bacterial biofilms in nature and disease. *Annu Rev Microbiol* **41:**435–464.
4. **Flemming H-C, Wingender J.** 2010. The biofilm matrix. *Nat Rev Microbiol* **8:**623–633.
5. **Reisner A, Haagensen JA, Schembri MA, Zechner EL, Molin S.** 2003. Development and maturation of *Escherichia coli* K-12 biofilms. *Mol Microbiol* **48:**933–946.
6. **Pratt LA, Kolter R.** Genetic analysis of *Escherichia coli* biofilm formation: roles of flagella, motility, chemotaxis and type I pili. *Mol Microbiol* **30:**285–293.
7. **Sauer K, Camper AK, Ehrlich GD, Costerton JW, Davies DG.** 2002. *Pseudomonas aeruginosa* displays multiple phenotypes during development as a biofilm. *J Bacteriol* **184:**1140–1154.
8. **Singh PK, Schaefer AL, Parsek MR, Moninger TO, Welsh MJ, Greenberg EP.** 2000. Quorum-sensing signals indicate that cystic fibrosis lungs are infected with bacterial biofilms. *Nature* **407:**762–764.
9. **Chai Y, Chu F, Kolter R, Losick R.** 2008. Bistability and biofilm formation in *Bacillus subtilis*. *Mol Microbiol* **67:** 254–263.
10. **Aspiras MB, Ellen RP, Cvitkovitch DG.** 2004. ComX activity of *Streptococcus mutans* growing in biofilms. *FEMS Microbiol Lett* **238:**167–174.
11. **Li T, Lau P, Lee J, Ellen R, Cvitkovitch D.** 2001. Natural genetic transformation of *Streptococcus mutans* growing in biofilms. *J Bacteriol* **183:**897–908.
12. **Li YH, Tang N, Aspiras MB, Lau PC, Lee JH, Ellen RP, Cvitkovitch DG.** 2002. A quorum-sensing signaling system essential for genetic competence in *Streptococcus mutans* is involved in biofilm formation. *J Bacteriol* **184:** 2699–2708.
13. **Eaton RE, Jacques NA.** 2010. Deletion of competence-induced genes over-expressed in biofilms caused transformation deficiencies in *Streptococcus mutans*. *Mol Oral Microbiol* **25:**406–417.
14. **Burmolle M, Bahl MI, Jensen LB, Sorensen SJ, Hansen LH.** 2008. Type 3 fimbriae, encoded by the conjugative plasmid pOLA52, enhance biofilm formation and transfer frequencies in Enterobacteriaceae strains. *Microbiology* **154:**187–195.
15. **Ghigo J.** 2001. Natural conjugative plasmids induce bacterial biofilm development. *Nature* **412:**442–445.
16. **D'Alvise PW, Sjoholm OR, Yankelevich T, Jin Y, Weurtz S, Smets BF.** 2010. TOL plasmid carriage enhances biofilm formation and increases extracellular DNA content in *Pseudomonas putida* KT2440. *FEMS Microbiol Lett* **312:**84–92.
17. **Ong CL, Beatson SA, McEwan AG, Schembri MA.** 2009. Conjugative plasmid transfer and adhesion dynamics in an *Escherichia coli* biofilm. *Appl Environ Microbiol* **75:** 6783–6791.
18. **Molin S, Tolker-Nielsen T.** 2003. Gene transfer occurs with enhanced efficiency in biofilms and induced enhanced stabilisation of the biofilm structure. *Curr Opin Biotechnol* **14:**255–261.
19. **Yang X, Ma Q, Wood TK.** 2008. The R1 conjugative plasmid increases *Escherichia coli* biofilm formation through an envelope stress response. *Appl Environ Microbiol* **74:**2690–2699.
20. **Reisner A, Holler BM, Molin S, Zechner EL.** 2006. Synergistic effects in mixed *Escherichia coli* biofilms: conjugative plasmid transfer drives biofilm expansion. *J Bacteriol* **188:**3582–3588.
21. **May T, Okabe S.** 2008. Escherichia coli harboring a natural IncF conjugative F plasmid develops complex mature biofilms by stimulating synthesis of colanic acid and Curli. *J Bacteriol* **190:**7479–7490.
22. **Castonguay MH, van der Schaaf S, Koester W, Krooneman J, van der Meer W, Harmsen H, Landini P.** 2006. Biofilm formation by *Escherichia coli* is stimulated by synergistic interactions and co-adhesion mechanisms with adherence-proficient bacteria. *Res Microbiol* **157:** 471–478.
23. **Bahl MI, Hansen LH, Goesmann A, Sorensen SJ.** 2007. The multiple antibiotic resistance IncP-1 plasmid pKJK5 isolated from a soil environment is phylogenetically divergent from members of the previously established alpha, beta and delta sub-groups. *Plasmid* **58:** 31–43.

24. **Roder HL, Hansen LH, Sorensen SJ, Burmolle M.** 2013. The impact of the conjugative IncP-1 plasmid pKJK5 on multispecies biofilm formation is dependent on the plasmid host. *FEMS Microbiol Lett* **344:**186–192.
25. **Hausner M, Wuertz S.** 1999. High rates of conjugation in bacterial biofilms as determined by quantitative in situ analysis. *Appl Environ Microbiol* **65:**3710–3713.
26. **Sorensen SJ, Bailey M, Hansen LH, Kroer N, Wuertz S.** 2005. Studying plasmid horizontal transfer in situ: a critical review. *Nat Rev Microbiol* **3:**700–710.
27. **Savage VJ, Chopra I, O'Neill AJ.** 2013. *Staphylococcus aureus* biofilms promote horizontal transfer of antibiotic resistance. *Antimicrob Agents Chemother* **57:**1968–1970.
28. **Keen NT, Tamaki S, Kobayashi D, Trollinger D.** 1988. Improved broad-host-range plasmids for DNA cloning in gram-negative bacteria. *Gene* **70:**191–197.
29. **Licht TR, Christensen BB, Krogfelt KA, Molin S.** 1999. Plasmid transfer in the animal intestine and other dynamic bacterial populations: the role of community structure and environment. *Microbiology* **145**(Pt 9)**:**2615–2622.
30. **Ehlers LJ, Bouwer EJ.** 1999. Rp4 plasmid transfer among species of *Pseudomonas* in a biofilm reactor. *Water Sci Tech* **39:**163–171.
31. **Krol JE, Wojtowicz AJ, Rogers LM, Heuer H, Smalla K, Krone SM, Top EM.** 2013. Invasion of *E. coli* biofilms by antibiotic resistance plasmids. *Plasmid* **70:**110–119.
32. **Davies D, Geesey G.** 1995. Regulation of the alginate biosynthesis gene *algC* in *Pseudomonas aeruginosa* during biofilm development in continuous culture. *Appl Environ Microbiol* **61:**860–867.
33. **May T, Ito A, Okabe S.** 2009. Induction of multidrug resistance mechanism in *Escherichia coli* biofilms by interplay between tetracycline and ampicillin resistance genes. *Antimicrob Agents Chemother* **53:**4628–4639.
34. **Cook L, Chatterjee A, Barnes A, Yarwood J, Hu W-S, Dunny G.** 2011. Biofilm growth alters regulation of conjugation by a bacterial pheromone. *Mol Microbiol* **81:** 1499–1510.
35. **Cook LC, Dunny GM.** 2013. Effects of biofilm growth on plasmid copy number and expression of antibiotic resistance genes in *Enterococcus faecalis*. *Antimicrob Agents Chemother* **57:**1850–1856.
36. **Clewell DB.** 1972. Nature of Col E 1 plasmid replication in *Escherichia coli* in the presence of the chloramphenicol. *J Bacteriol* **110:**667–676.
37. **Togna PA, Shuler ML, Wilson DB.** 1993. Effects of plasmid copy number and runaway plasmid replication on overproduction and excretion of B-lactamase from *Escherichia coli*. *Biotechnol Prog* **9:**31–39.
38. **Uhlin BE, Nordström K.** 1978. A runaway-replication mutant of plasmid R1drd-19: temperature-dependent loss of copy number control. *Mol Gen Genet* **165:** 167–179.
39. **Ma H, Katzenmeyer KN, Bryers JD.** 2013. Non-invasive in situ monitoring and quantification of TOL plasmid segregational loss within *Pseudomonas putida* biofilms. *Biotechnol Bioeng* **110:**2949–2958.
40. **Imran M, Jones D, Smith H.** 2005. Biofilms and the plasmid maintenance question. *Math Biosci* **193:**183–204.
41. **Teodosio JS, Simoes M, Mergulhao FJ.** 2012. The influence of nonconjugative *Escherichia coli* plasmids on biofilm formation and resistance. *J Appl Microbiol* **113:** 373–382.
42. **Baur B, Hanselmann K, Schlimme W, Jenni B.** 1996. Genetic transformation in freshwater: Escherichia coli is able to develop natural competence. *Appl Environ Microbiol* **62:**3673–3678.
43. **Tsen SD, Fang SS, Chen MJ, Chien JY, Lee CC, Tsen DH.** 2002. Natural plasmid transformation in *Escherichia coli*. *J Biomed Sci* **9:**246–252.
44. **Maeda S, Sawamura A, Matsuda A.** 2004. Transformation of colonial *Escherichia coli* on solid media. *FEMS Microbiol Lett* **236:**61–64.
45. **Maeda S, Ito M, Ando T, Ishimoto Y, Fujisawa Y, Takahashi H, Matsuda A, Sawamura A, Kato S.** 2006. Horizontal transfer of nonconjugative plasmids in a colony biofilm of *Escherichia coli*. *FEMS Microbiol Lett* **255:** 115–120.
46. **Berge MJ, Kamgoue A, Martin B, Polard P, Campo N, Claverys JP.** 2013. Midcell recruitment of the DNA uptake and virulence nuclease, EndA, for pneumococcal transformation. *PLoS Pathog* **9:**e1003596. doi:10.1371/journal.ppat.1003596.

Plasmids—Biology and Impact in Biotechnology and Discovery
Edited by Marcelo E. Tolmasky and Juan C. Alonso

doi:10.1128/microbiolspec.PLAS-0003-2013

Yen-Ting Liu,[1] Saumitra Sau,[1] Chien-Hui Ma,[1]
Aashiq H. Kachroo,[1] Paul A. Rowley,[1] Keng-Ming Chang,[1]
Hsiu-Fang Fan,[2] and Makkuni Jayaram[2]

19

The Partitioning and Copy Number Control Systems of the Selfish Yeast Plasmid: An Optimized Molecular Design for Stable Persistence in Host Cells

INTRODUCTION

Selfish genetic elements (1–4), widespread in nature, are characterized by their ability to replicate efficiently and maintain themselves stably in host cell populations. A subset of these elements harbors the capacity to spread within a genome or, via horizontal transmission, between genomes. Selfish elements can also be frequently acquired by sexual transmission. The degree of selfishness can vary significantly among different elements. Some may increase the host's fitness at least under certain conditions and, in doing so, add to their own fitness in a self-serving fashion. Others may be more decidedly selfish in that they contribute little toward the host's fitness. Their long-term persistence is sustained solely by their capacity for replication and transmission during growth and division of host cells.

Selfish elements may be broadly divided into two groups: those that are integrated into the chromosome (s) of the host and those that remain extrachromosomal (reviewed in reference 4). The integrated class includes insertion sequences, lysogenic states of several bacteriophages, mobile DNA elements, and families of repeated DNA found in eukaryotes. The extrachromosomal class encompasses plasmids, lysogenic forms of certain bacterial viruses, RNA intermediates involved in the retrotransposition of mobile elements, and episomes of mammalian viruses belonging to the gamma herpes and papilloma families. Plasmids, found abundantly among

[1]Section of Molecular Genetics and Microbiology, Institute for Cellular and Molecular Biology, The University of Texas at Austin, Austin, Texas 78712; [2]Department of Life Sciences and Institute of Genome Sciences, National Yang-Ming University, Taipei 112, Taiwan.

prokaryotes, are almost nonexistent among eukaryotes, except for those encountered among members of the budding yeast (*Saccharomycetaceae*) lineage. The gamma herpes and papilloma viruses, whose extended latent periods of infection are characterized by their stable episomal existence, may be regarded as plasmid impostors of the eukaryotic world (5, 6).

We review here the properties of the 2-micron plasmid found nearly ubiquitously in *Saccharomyces* strains that justify its inclusion under the selfish DNA moniker. Furthermore, we describe the biochemical features and applications of a site-specific recombination system harbored by the plasmid. Based on genetic organization and functional attributes, it is logical to posit that the 2-micron plasmid is an authentic representative of the yeast plasmid family with respect to replication, segregation to daughter cells during cell division, and maintenance and regulation of copy number (7).

THE 2-MICRON PLASMID: AN OPTIMIZED AND MINIATURIZED SELFISH DNA ELEMENT

The 2-micron plasmid is a relatively small double-stranded circular DNA genome (~6.3 kbp) present in the yeast nucleus at an average copy number of 40 to 60 per haploid cell (7–9). The stability of the plasmid is remarkably similar to that of the host chromosomes, the loss rate being as low as 10^{-5} to 10^{-4} per cell division. Four protein coding regions together with the *cis*-acting sequences important for replication, partitioning, and copy number control leave little room to spare in the compactly designed plasmid genome (Fig. 1A). Even minor disruptions of the genetic organization of the plasmid lead to prominent deleterious effects on its physiology. Two-micron based artificial plasmids routinely used for genetic manipulations in yeast are two to three orders of magnitude less stable than the native plasmid. The 2-micron plasmid may be looked upon as a minimalist selfish DNA tailored for maximum functional efficiency.

The plasmid replication origin (*ORI*) is functionally equivalent to a chromosomal origin, so that each plasmid molecule replicates once (and only once) during S phase by utilizing the host replication machinery (10). The duplicated plasmid population is distributed equally (or almost equally) to mother and daughter cells by the plasmid partitioning system comprised of the Rep1 and Rep2 proteins and the *STB* locus (7, 11). Doubling of the copy number followed by equal (or almost equal) segregation marks the normal steady state lifestyle of the plasmid.

The copy number control system comes into play only when there is a reduction in plasmid population due to a rare missegregation event. The restoration of the copy number is mediated by a site-specific recombination system consisting of the Flp protein and its target sites (*FRT*s) embedded within a 599-bp inverted repeat region. The Raf1 protein positively regulates amplification. According to the currently accepted model (12, 13), plasmid amplification is triggered by a recombination event early during bidirectional replication when the *ORI*-proximal *FRT* site, but not the distal one, has been duplicated (Fig. 1B). The DNA inversion resulting from a crossover between the unreplicated *FRT* and a copy of the duplicated one will cause the replication forks to travel in the same direction around the circular template. This nonstandard mode of replication by the two unidirectional forks spins out multiple tandem copies of the plasmid without the need for *ORI* to fire more than once. Amplification can be terminated by a second recombination event that restores bidirectional fork movement. The amplified plasmid DNA can be resolved into single plasmid molecules by recombination mediated by Flp or by the host homologous recombination machinery.

The mechanism for dealing with a higher than steady state copy number caused by missegregation, if such a mechanism exists, is not clear. It is possible that the copy number may be lowered by under-replication, although there is no experimental evidence to support this notion. Very high plasmid copy numbers have a strong negative effect on cellular function, leading to premature lethality of such cells (14–16). Thus, cells bearing an undue plasmid burden would be eliminated from a growing population over time.

In sum, the 2-micron plasmid genome is comprised of three functional modules devoted to the sole purpose of self-perpetuation. They ensure (i) precise duplication of the plasmid population during a cell cycle, (ii) equal plasmid segregation at the time of nuclear division, and (iii) preservation of the plasmid copy number in individual cells.

COPY NUMBER CONTROL: PLASMID- AND HOST-MEDIATED REGULATORY MECHANISMS

There is an intricate communication between the partitioning and amplification systems by which an untoward increase in plasmid copy number is avoided (17–19) (Fig. 1C). The Rep1 and Rep2 proteins are thought to form a bipartite negative regulator that represses the expression of the *FLP*, *RAF1*, and *REP1* genes. *REP2* appears to be constitutively expressed. The level of Rep1 not only provides a measure of the plasmid copy number but also determines the effective concentration of the Rep1-Rep2 repressor. When the plasmid copy number falls below the steady state value, the corresponding drop in Rep1 reduces the repressor concentration below the threshold required for turning off *FLP* and *RAF1* expression. As a result, the amplification system is activated. The Raf1 protein is thought to antagonize the repressor, thereby augmenting plasmid amplification. When the steady state copy number is restored, the increase in Rep1 raises the Rep1-Rep2 concentration above the threshold, thus turning off the amplification system. The design of this genetic regulatory circuit quickly triggers the amplification response when called for without incurring the danger of a runaway increase in the plasmid population.

In addition to the plasmid-instituted control of the copy number at the level of gene expression, there is also host-mediated posttranslational regulation of Flp (20, 21). Flp is modified by sumoylation, which may regulate its activity and/or its stability. Sumoylation could provide the signal for the secondary modification of Flp by ubiquitination and its subsequent degradation by the proteasome pathway. Under conditions that disrupt sumoylation homeostasis, the 2-micron plasmid copy number rises to very high levels, causing cell lethality. A plausible cause for this aberrant plasmid amplification is the conversion of a Flp-induced stand nick at the *FRT* site into a double-strand DNA break as a result of replication (Fig. 2). The invasion by the broken end of a circular plasmid template, promoted by the host's homologous recombination/repair machinery, and repeated rounds of replication would result in an amplified plasmid concatemer. The mechanism is analogous to that proposed for the generation of copy number variations in eukaryotic genomes and the maintenance of telomere length by a telomerase-independent alternative pathway (22–25).

The transcriptional and posttranslational controls of plasmid amplification exemplify strategies by which a selfish genome and its host genome establish long-term coexistence with minimal conflicts between them.

CHROMOSOME-COUPLED SEGREGATION OF THE 2-MICRON PLASMID

The organization of the 2-micron plasmid partitioning system is deceptively similar to that of well-characterized bacterial plasmid partitioning systems in that it is comprised of two plasmid-coded proteins (Rep1 and Rep2) and a *cis*-acting locus (*STB*) that contains iterations of a consensus sequence element (7). However, there is no evidence to suggest functional similarities between the yeast and bacterial systems. For example, neither Rep1 nor Rep2 harbor sequence motifs that would typify an ATP binding/hydrolyzing activity. The fact that the 2-micron plasmid houses a tyrosine family site-specific recombinase (Flp) would be consistent with a prokaryotic origin of the plasmid, as tyrosine recombinases are more abundant in prokaryotes than in eukaryotes. In members of the budding yeast lineage, the recombinases are coded for by 2-micron related plasmids. Assuming that a prokaryotic DNA element acquired by an ancestral budding yeast via horizontal transmission served as the progenitor of the extant 2-micron plasmid, the Rep-*STB* system must represent the evolutionary transformation of the original partitioning system during adaptation to a new biological niche.

Plasmids capable of autonomous replication in yeast but lacking an active partitioning system, called *ARS* plasmids, are rapidly lost in the absence of continuous selection. This high instability is caused by the strong tendency of plasmid molecules to stay in the mother nucleus, causing their depletion from daughter cells (26–29). A variety of observations suggest that the Rep-*STB* system overcomes the mother bias by coupling plasmid segregation to chromosome segregation. For example, a number of mutations that conditionally disrupt normal chromosome segregation, *ipl1-1*(T^s) being one, also cause the 2-micron plasmid to missegregate (Fig. 3A) (30, 31). Furthermore, the plasmid tends to cosegregate with the missegregating bulk of chromosomes. Under the same conditions, an *ARS* plasmid does not show any coupling to chromosomes (Fig. 3B). Instead, it missegregates frequently with the characteristic strong mother bias. Attempts to uncouple plasmid segregation from chromosome segregation, except by employing conditions that directly or indirectly disable the partitioning system, have been unsuccessful. The most parsimonious hypothesis consistent with currently available data is that the 2-micron plasmid physically attaches to chromosomes and segregates by a hitchhiking mechanism (7, 30, 32). However, a chromosome-independent mode of plasmid segregation cannot be ruled out unequivocally. Direct spindle-mediated segregation of the plasmid is highly unlikely. Components of the kinetochore complex, which is responsible for linking centromeres to the mitotic spindle, have not been detected at *STB*. The presence of two or more *STB* loci within a single plasmid does not lead to instabilities normally observed when two centromere (*CEN*) sequences are harbored by a plasmid. Delaying spindle assembly until metaphase during a cell cycle

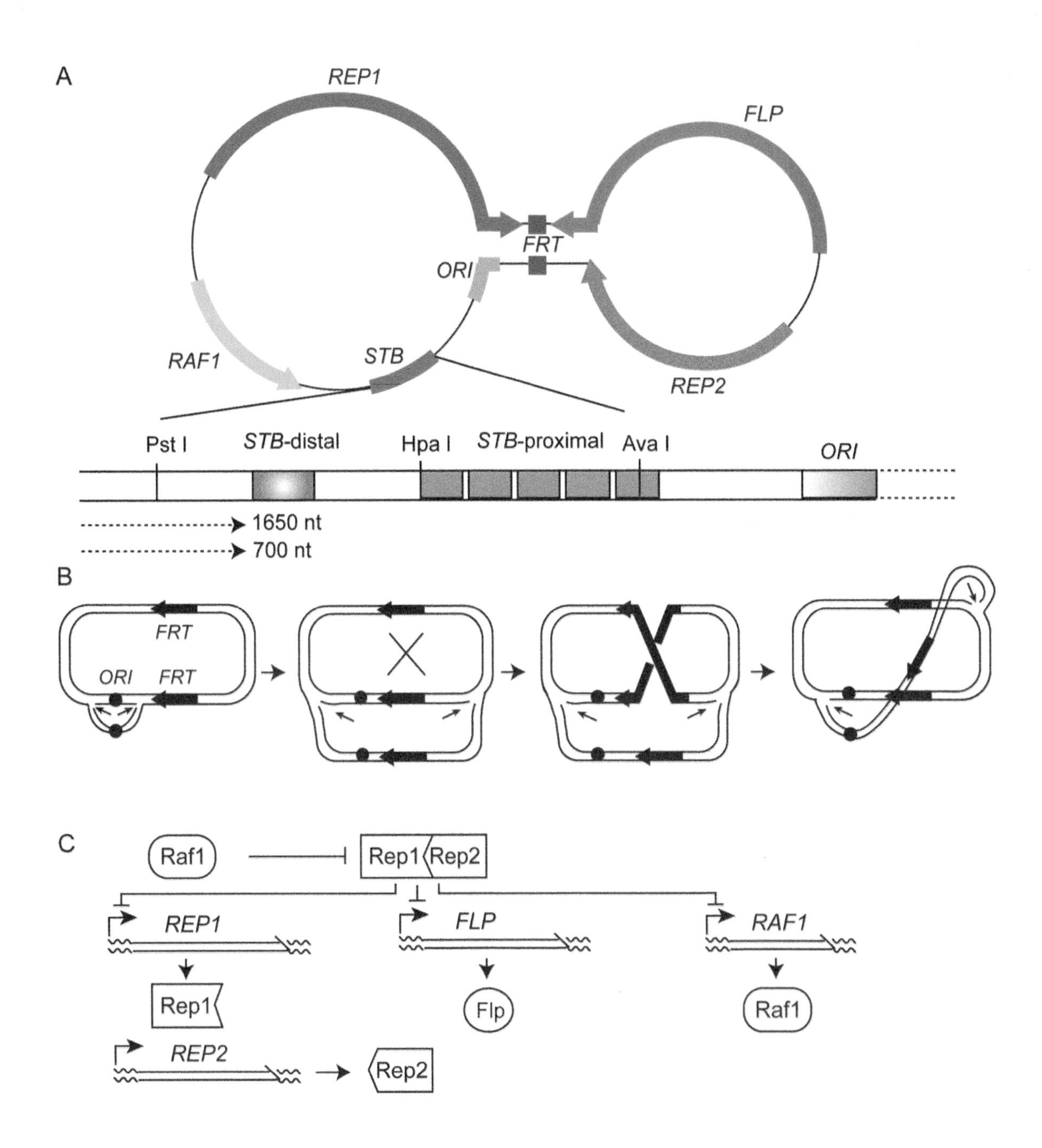
A
REP1
FLP
FRT
ORI
RAF1
STB
REP2
Pst I
STB-distal
Hpa I
STB-proximal
Ava I
ORI
1650 nt
700 nt
B
FRT
ORI
FRT
C
Raf1
Rep1
Rep2
REP1
FLP
RAF1
Rep1
Flp
Raf1
REP2
Rep2

does not affect plasmid or chromosome replication. However, upon spindle restoration, chromosomes form normal spindle attachments and segregate equally, whereas the plasmid missegregates (30). This result is more readily explained by the requirement of the pre-metaphase spindle for plasmid-chromosome coupling rather than by spindle-mediated plasmid segregation.

THE *STB* LOCUS: A SITE FOR HIGH-ORDER ASSEMBLY OF PROTEIN FACTORS

The Rep1 and Rep2 proteins interact with each other *in vivo* and *in vitro*, and they associate with the *STB* locus *in vivo* (33–36). The tripartite DNA-protein interactions are crucial for the plasmid partitioning function. Point mutations in Rep1 that disrupt its interaction with either Rep2 or *STB* lead to high plasmid instability (37). However, direct binding of either Rep1 or Rep2 to *STB in vitro* has been difficult to demonstrate by standard methods such as gel mobility shift assays. Evidence has been presented for the binding of Rep2 to *STB in vitro* (38). Perhaps, the interactions of these proteins with *STB* may be nucleated or stabilized by a host factor or factors (34). Alternatively, the chromatin organization of *STB* may be important for the recruitment of the Rep proteins.

The *STB* locus can be divided arbitrarily into two regions of approximately equal size with respect to their distances from the plasmid replication origin (39) (Fig. 1A). *STB*-proximal, consisting of five copies of a 60-bp consensus element (the iteron unit), appears to engender the partitioning function. *STB*-distal likely plays a secondary role in partitioning by helping to maintain *STB*-proximal in its active state. A transcription terminator sequence present within *STB*-distal serves to keep *STB*-proximal a transcription-free zone. *STB*-distal also harbors a silencer element that can turn down the activity of a promoter when placed adjacent to it. The potential contributions of the transcription terminator and the silencer toward plasmid partitioning have not been rigorously analyzed.

A number of host factors that associate with centromeres and contribute to normal chromosome segregation are also recruited at *STB* and promote equal plasmid segregation. These include the RSC2 chromatin remodeling complex, the Kip1 nuclear motor protein, the yeast cohesin complex, and the centromere-specific histone H3 variant Cse4 (40–44). At least a subset of these factors may affect plasmid stability indirectly through their effects on chromosome segregation. The lack of Rsc2 causes a significant increase in plasmid loss without obvious deleterious effects on chromosome segregation (44). Presumably, the RSC1 chromatin remodeling complex, in which Rsc2 is replaced by its functional homologue, Rsc1, can satisfy the requirements for chromosome segregation. The association of the cohesin

Figure 1 Genetic organization of the yeast plasmid and its copy number regulation. (**A**) The double-stranded DNA genome of the yeast 2-micron plasmid is generally represented as a dumbbell-shaped molecule to highlight the 599-bp inverted repeat (the handle of the dumbbell) that separates two unique regions. The four coding regions harbored by the plasmid are *REP1*, *REP2*, *FLP*, and *RAF1*. The directions in which these loci are transcribed are indicated by the arrowheads. The plasmid replication origin is indicated as *ORI*. Flp is a site-specific recombinase, whose target sites (*FRT*s) are embedded within the inverted repeat region. The plasmid partitioning locus *STB* can be divided into origin-proximal (*STB*-proximal) and origin-distal (*STB*-distal) segments. There are five repetitions of a 60-bp consensus sequence in *STB*-proximal. *STB*-distal, which harbors the termination signal for two origin-directed plasmid transcripts (1,650 and 700 nucleotides long) as well as a silencing element (shaded box), maintains *STB*-proximal as a transcription-free zone. The Rep1-Rep2-*STB* system ensures equal plasmid segregation. The (Flp-*FRT*)-Raf1 system is responsible for the maintenance of the steady state plasmid copy number. (**B**) The mechanism proposed for copy number correction of the plasmid by amplification (12) invokes a Flp-mediated recombination event that changes the direction of one of the replication forks (indicated by the thin short arrows) with respect to the other during bidirectional replication of the plasmid. The ensuing dual unidirectional mode of replication amplifies the plasmid as a concatemer of tandem plasmid units. There is a marked asymmetry in the location of the *FRT* sites (thick arrows) with respect to the replication origin (*ORI*). The consequent difference in their replication status, one duplicated and the other not, is responsible for the relative inversion of the replication forks as a result of recombination between them. (**C**) Efficient amplification without the danger of an unregulated increase in plasmid copy number is prevented by a transcriptional regulatory network. The putative [Rep1-Rep2] repressor negatively regulates *FLP*, *RAF1*, and *REP1* expression. Raf1 is thought to antagonize the action of the [Rep1-Rep2] repressor. doi:10.1128/microbiolspec.PLAS-0003-2013.f1

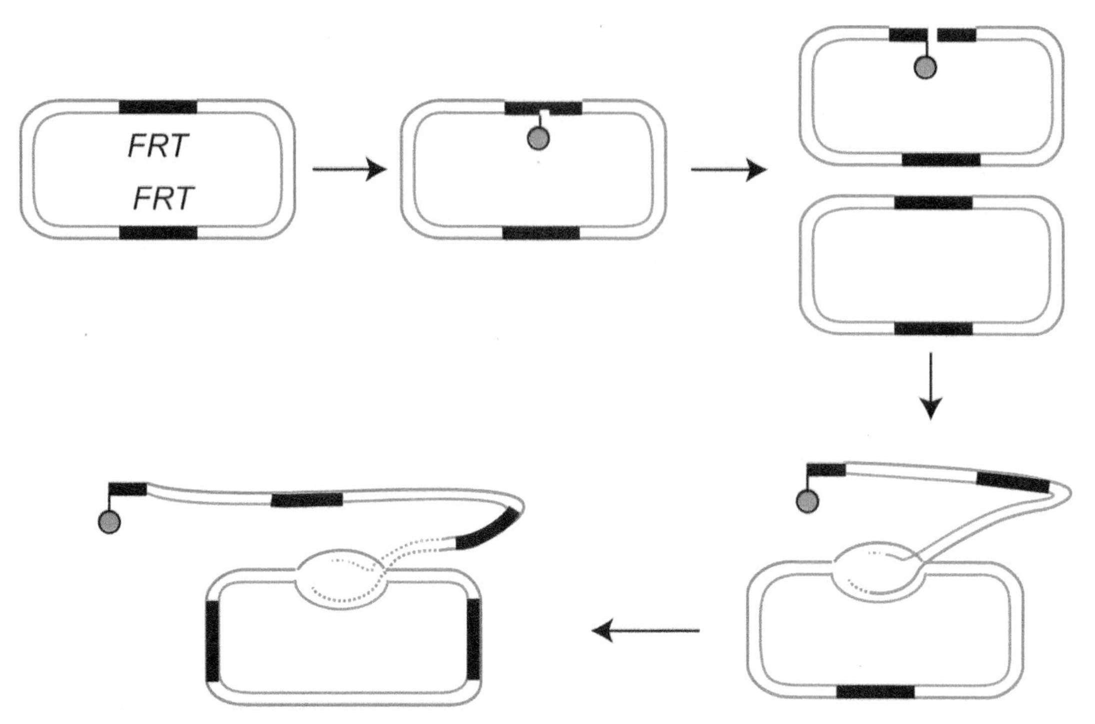

Figure 2 Aberrant plasmid amplification as a result of misregulation of Flp. Strand nicks formed at *FRT* by the action of Flp, if unrepaired, will give rise to double-strand breaks when they encounter replication forks. Such a broken end can invade an intact circular plasmid and trigger aberrant plasmid amplification by repair synthesis. Posttranslational modification of Flp by sumoylation is important in preventing this mode of plasmid amplification (20, 21). doi:10.1128/microbiolspec.PLAS-0003-2013.f2

complex as well as Cse4 at *STB* is highly substoichiometric (42, 45), raising concerns about their functional relevance to plasmid partitioning. However, the assembly of multiple plasmid molecules into one unit may help these factors to act in a catalytic manner by recycling among members of the unit. Such group behavior may be advantageous to a selfish genetic element by limiting its dependence on factors that are critical for the host's well-being but are in short supply.

The substoichiometric nature of Cse4 at *STB* also calls into question whether it is an authentic constituent of a specialized nucleosome assembled at *STB*. Cse4-*STB* association has been demonstrated by chromatin immunoprecipitation (ChIP) assays utilizing DNA-protein cross-linking or, alternatively, by plasmid pull-down assays in the absence of cross-linking (41, 42). Both types of analyses also reveal Cse4-centromere association, with the distinction that the efficiency of centromere (*CEN*) recovery in the plasmid pull-down assays is several-fold higher relative to *STB*. A Cse4 directed antibody can specifically bring down *CEN* DNA corresponding to a single nucleosome from micrococcal nuclease digested chromatin (46). In this assay, the immunoprecipitated *CEN* DNA is detected by Southern analysis, but *STB* DNA is not (Sue Biggins, personal communication). In analogous assays, we are able to use PCR to visualize both *CEN* and *STB* in the immunoprecipitated DNA but detect neither sequence by Southern analysis (42) (unpublished data). Whether Cse4-*STB* association is in the form of a nucleosome or whether this association is nonnucleosomal remains an open question. Nevertheless, the differences between *CEN* and *STB* in the requirements for Cse4 recruitment (41, 42), as well as the distinct life times of Cse4-*CEN* and Cse4-*STB* associations during the cell cycle (41), suggest that the interaction between Cse4 and *STB* is authentic.

The localization of Cse4 to noncentromeric regions of yeast chromosomes has been established by ChIP followed by hybridization to DNA microarrays and also by ChIP in conjunction with DNA sequencing (ChIp-Seq) (47, 48). In one of the assays (47), the regions enriched in Cse4 association include repeated loci such as Ty transposable elements, telomeres, and ribosomal DNA. In the other assay (48), the enriched loci,

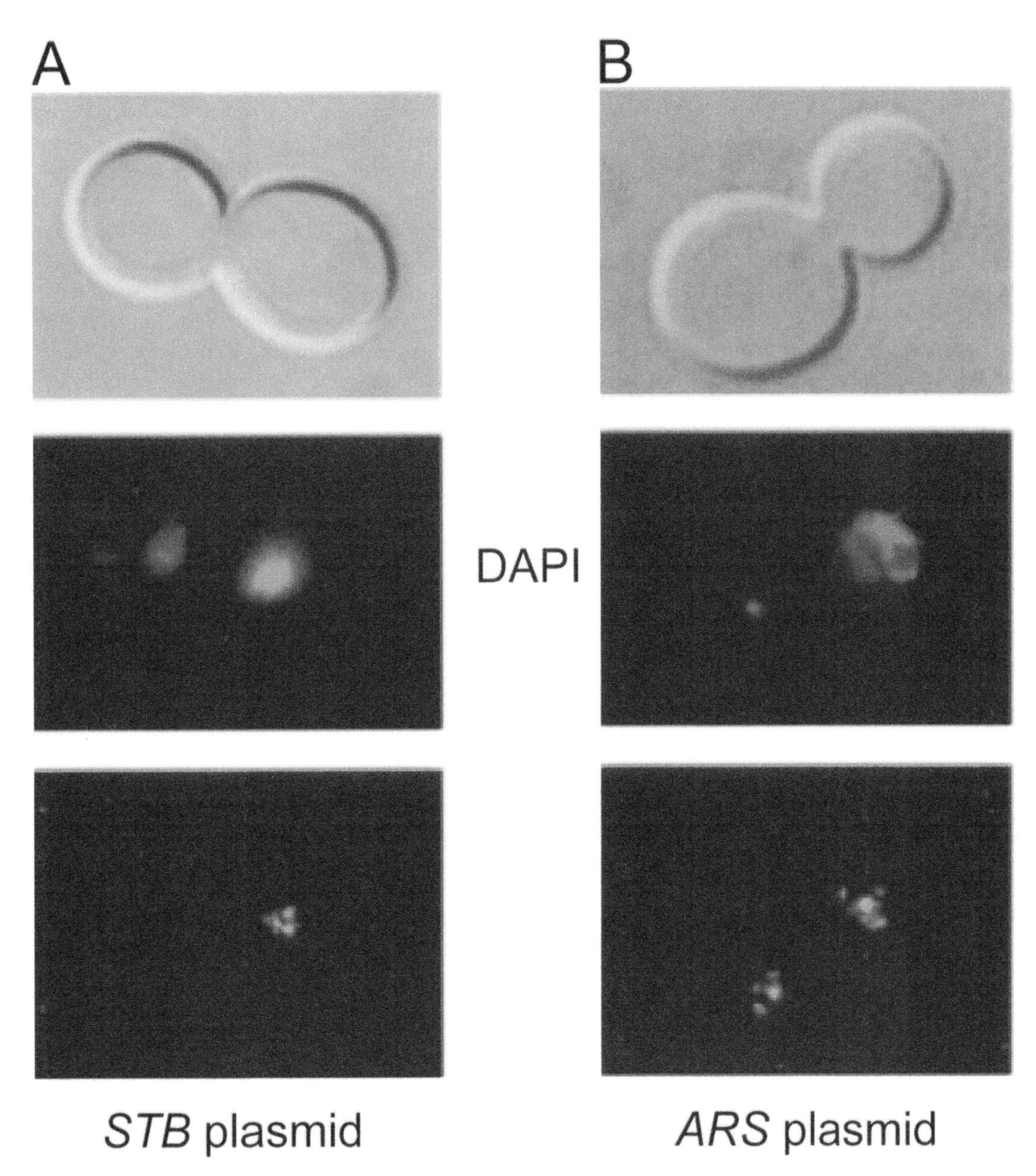

Figure 3 Segregation of the 2-micron plasmid in the *Ipl1-1* mutant. When the Ipl1 (aurora kinase) function is inactivated by a temperature-sensitive mutation, chromosomes frequently missegregate (shown here by the unequal DAPI staining in the mother and daughter compartments). (**A**) A multicopy 2-micron reporter plasmid, fluorescence tagged by (GFP-LacI)-LacO interaction, also shows high missegregation under this condition (31). Furthermore, the plasmid tends to be retained most often in the cell compartment that contains the bulk of the chromosomes. (**B**) The segregation of a plasmid lacking *STB* (an *ARS* plasmid) is not coupled to that of the chromosomes. However, such a plasmid missegregates frequently regardless of the Ipl1 status. doi:10.1128/microbiolspec.PLAS-0003-2013.f3

termed centromere-like regions, map near centromeres, and at least a subset of these sequences enhance the segregation potential of reporter plasmids and chromosomes. It is possible that centromere-like regions might denote ancient centromere sequences that became diverged or lost during the steps leading to the evolution of the present day budding yeast centromeres. As described in the succeeding section, the *STB* locus might have played a significant role in this evolutionary process.

THE BUDDING YEAST CENTROMERE (*CEN*) AND *STB*: A SHARED EVOLUTIONARY ANCESTRY?

The extremely short, and genetically defined, point centromere of the budding yeast stands out in sharp contrast to the much larger epigenetically specified regional centromeres common to nearly all eukaryotes (49). Interestingly, there is a strong correlation between the time of emergence of the point centromere and the partial or nearly complete loss of siRNA and heterochromatin machineries, which are central to the establishment and maintenance of regional centromeres, in the budding yeast lineage (50). Furthermore, this is the only lineage among eukaryotes whose members (though not all) harbor plasmids related to the 2-micron circle. Based on these circumstantial pieces of evidence, it has been speculated that an ancestral *STB* locus was domesticated by chromosomes as the point centromere to meet the evolutionary exigency that negated the assembly and function of regional centromeres (49). With the acquisition of a new type of centromere, the defunct centromeres would have diverged or become lost.

If the above model for the evolutionary transition to the point centromere is correct, it follows that the yeast chromosomes and the 2-micron plasmid utilized the same segregation mechanism at one point in their evolutionary history. The cost of bearing the plasmid burden, however small, would have provided the stimulus for the chromosome segregation machinery to evolve away from the plasmid segregation mechanism. The plasmid, in turn, might have evolved counterstrategies to indirectly exploit the logic of chromosome segregation to bolster its own propagation. The present day spindle-based chromosome segregation and chromosome-coupled plasmid segregation might thus represent two divergent solutions to the problem of achieving equal segregation, arrived at from a common start point.

The association of chromosome segregation factors at *STB* would be consistent with a shared evolutionary history between *STB* and *CEN*. While a subset of these associations may be relevant to the current plasmid segregation pathway, others may be evolutionary vestiges with little physiological significance. A recent analysis revealed that the functional states of *STB* and *Saccharomyces cerevisiae CEN* engender an unusual positive supercoil (51, 52). It has been argued that DNA is wrapped in a nonstandard right-handed fashion around the specialized Cse4-containing nucleosome core present at *CEN* (51). The net positive supercoiling contributed by *STB* also requires conditions that foster the association of Cse4 with *STB*. The substoichiometric nature of Cse4-*STB* interaction rules out the presence of a Cse4-containing nucleosome at every *STB*. Perhaps the short-lived assembly of a Cse4-containing nucleosome at *STB* or a transient nonnucleosomal interaction between Cse4 and *STB* may be sufficient to induce a positive supercoil. This DNA topology may then be stably entrapped by other proteins present at *STB*. We cannot strictly exclude the possibility that the positive supercoiling conferred by *STB* results from the loss of one or two standard nucleosomes. Nevertheless, the potential presence of a positive supercoil at *CEN* and *STB* chromatin adds another tantalizing piece of circumstantial evidence to a growing list suggesting a common ancestry for the partitioning loci of the 2-micron plasmid and the chromosomes of its host.

ORGANIZATION OF THE 2-MICRON PLASMID AND THE FUNCTIONAL PLASMID UNIT DURING SEGREGATION

Fluorescence-tagged *STB* reporter plasmids tend to form three to five foci in individual haploid nuclei. Given the mean copy number of 40 to 60 of the native plasmid, it has been assumed that each fluorescent focus comprises several coalesced plasmid molecules. The plasmid foci are often seen in close proximity to each other and appear to segregate in time lapse assays as a close-knit cluster (31). It is now clear that plasmid segregation as a single clustered unit is an incorrect impression conveyed by the small size of the haploid nucleus and the resolution limits of the microscopy assays (32) (unpublished data). Rather, each plasmid focus appears to function as an independent entity in segregation. This notion is further supported by the analysis of diploid cells with larger nuclei (and containing approximately double the plasmid copy number compared to haploid cells) going through meiotic cell divisions (unpublished data). In chromosome spreads prepared from cells at the pachytene stage of meiosis I, the 8 to 10 plasmid foci on average observed per cell are well resolved from each other.

SINGLE COPY FLUORESCENCE-TAGGED REPORTER PLASMIDS

One of the problems with assaying plasmid segregation precisely is the multicopy nature of *STB* plasmids and the attendant uncertainties in quantitating plasmids in mother and daughter nuclei by standard microscopy tools. This impediment has been solved by using fluorescence-tagged single copy reporter plasmids (32, 53). The general designs of such plasmids, placed in strains expressing GFP-LacI, are illustrated in Fig. 4. In one strategy, the plasmid copy number is reduced to one or close to one by incorporating a *CEN* into it (Fig. 4A). The *CEN* can be conditionally inactivated by driving copious transcription through it from a galactose-inducible promoter. In an alternative strategy, the reporter plasmid is excised from its integrated state in a haploid chromosome (Fig. 4B). In a single cell cycle, the segregation of plasmid sisters formed during S phase can be followed as 1:1 (equal), 2:0 (unequal with a mother bias), and 0:2 (unequal with a daughter bias) at the anaphase stage (Fig. 4C). Under this experimental regimen, equal plasmid segregation under *CEN* control is 85 to 90%, whereas that under *STB* control is ~70%. By contrast, the corresponding value for an *ARS* plasmid is ~25% with the missegregation biased strongly (90%) toward the mother (2:0).

Analyses using two single copy reporter plasmids, one tagged by green fluorescence [(GFP-LacI)-LacO interaction] and the other by red fluorescence [(TetR-RFP)-TetO interaction] in a strain expressing the two hybrid repressors, revealed that *STB* plasmids segregate in a sister-from-sister fashion (red from red and green from green) at a frequency significantly higher than

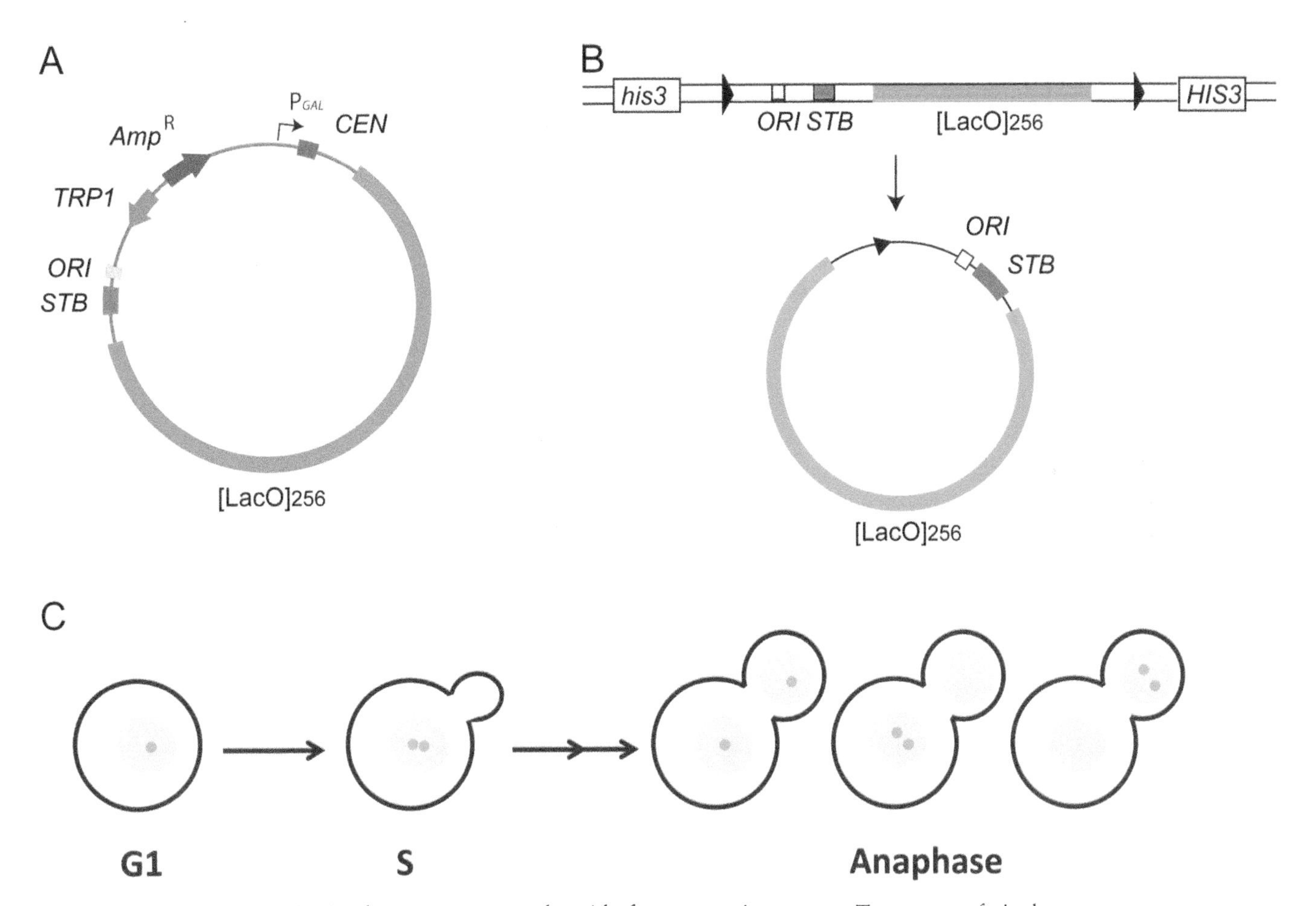

Figure 4 Single copy reporter plasmids for segregation assays. Two types of single copy reporter plasmids have been developed. (**A**) The copy number of a plasmid is maintained as one or nearly one by incorporating a *CEN* into it. The *CEN* can be conditionally inactivated by driving transcription through it from the inducible *GAL* promoter. (**B**) The reporter plasmid bordered by the target sites (shown by the arrowheads) for the R site-specific recombinase is integrated into a chromosome to keep the copy number strictly as one. The plasmid is excised by the inducible expression of the recombinase. (**C**) Following inactivation of the plasmid-borne *CEN*(**A**) or recombination-mediated plasmid excision (**B**) in G1 arrested cells, they are released into the cell cycle and plasmid segregation assayed at the anaphase stage. In either experimental scheme, the plasmid is visualized by operator-fluorescent repressor interaction. doi:10.1128/microbiolspec.PLAS-0003-2013.f4

that predicted for random association and segregation of the replicated plasmid molecules (Fig. 5) (53). This finding necessitates the following refinement of the hitchhiking model. If the model is valid, *STB* plasmid sisters must hitchhike on sister chromatids in order to segregate in a one-to-one fashion.

TESTS OF THE REFINED HITCHHIKING MODEL: DO SISTER PLASMIDS SEGREGATE BY ASSOCIATING WITH SISTER CHROMATIDS?

One way to test the hitchhiking model is to force the cosegregation of sister chromatids to the same cell pole during mitosis and scrutinize the behavior of plasmid sisters under this manipulation. The cosegregation of sister chromatids, which is antithetical to normal mitosis, and the separation of homologues is the norm during the first division of meiosis. This reductional mode of chromosome segregation is accomplished by the monopolin complex, which clamps down sister kinetochores with assistance from Sgo1 (shugoshin) and Ipl1 (aurora kinase) (54–59). By inappropriately expressing the monopolin complex early during a mitotic cell cycle, it is possible to direct sister chromatids to stay in the mother or migrate to the daughter (57) (Fig. 6A). The monopolin directed cosegregation of sister chromatids occurs at roughly 30% frequency without mother-daughter bias (Fig. 6B, right). *STB* plasmid sisters mimic sister chromatids in cosegregating under the influence of monopolin in a bias-free fashion (32) (Fig. 6B, left, and 6C). By contrast, *ARS* plasmids behave very differently from sister chromatids and *STB* plasmid sisters, and retain their strong mother bias. When monopolin expression is combined with the inactivation of Ipl1, a small but distinct daughter bias is imparted to sister chromatid cosegregation. A similar bias is also observed for *STB* plasmid sisters. Conditions that uncouple *STB* plasmid segregation from chromosome segregation, postponing spindle assembly until metaphase (60), for example, abolish the correlation between sister chromatids and *STB* plasmid sisters in their segregation patterns during monopolin-driven mitosis (Fig. 6D).

In more recent experiments, blocking the disassembly of the cohesin complex, which holds pairs of sister chromatids together until the onset of anaphase (61–64), has been utilized as the means for driving all chromosomes to the same cell compartment (unpublished data). Under this regimen, chromosome retention occurs in either the mother or the daughter. Even though spindle organization and separation of the duplicated spindle pole bodies are not normal under this condition, entry of the nucleus into the bud and nuclear division are not blocked. The nuclei lacking chromosomes are often malformed and have a collapsed appearance. Despite the perturbation of the mitotic cell cycle and nuclear segregation, this analysis clearly demonstrates the near perfect correlation between chromosomes and *STB* plasmid sisters in their cosegregation. No such correlation is observed for *ARS* plasmid sisters.

The segregation results obtained by enforcing limited or total missegregation of chromosomes strongly favor the hitchhiking model in which sister copies of the 2-micron plasmid adhere to sister chromatids in a one-to-one fashion. Whether this mechanism operates in the native context of a segregation unit comprising several plasmid molecules is not clear. One possibility is that a

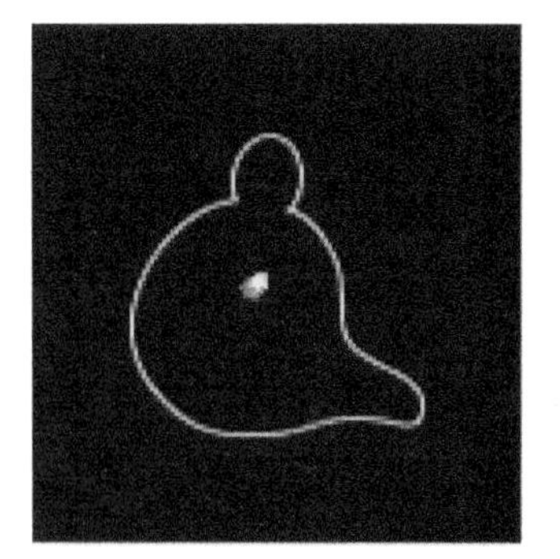

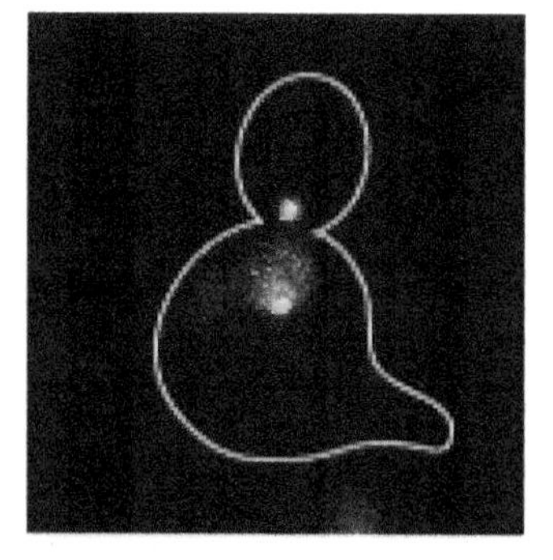
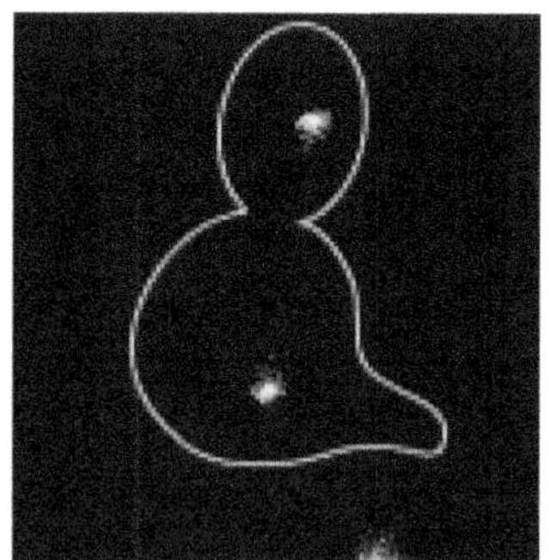

Figure 5 Segregation of two single copy *STB* reporter plasmids. Two single copy *STB* reporter plasmids cohabiting a nucleus are distinguished by tagging one with green fluorescence [(GFP-LacI)-LacO] and the other with red fluorescence [(TetR-RFP)-TetO]. Under a functional Rep-*STB* system, the red and green plasmid sisters segregate most of the time in a one-to-one fashion to yield individual cells containing one red plasmid and one green plasmid (53). doi:10.1128/microbiolspec.PLAS-0003-2013.f5

high-order organization within a plasmid group distributes each pair of plasmid sisters formed by replication into each of two sister plasmid groups. In other words, the plasmid group is duplicated in a precisely templated fashion. Although the segregation pattern of the single copy green and red fluorescence-tagged *STB* plasmids, referred to earlier, is consistent with this possibility, a more direct test of the model is beyond our present technical capability.

THE PLASMID AMPLIFICATION SYSTEM: *IN VIVO* ANALYSES

The Flp site-specific recombination system, responsible for the maintenance of the 2-micron plasmid copy number, has been studied only to a limited extent with respect to its *in vivo* physiological function. The absolute requirement of the recombination system for plasmid amplification has been demonstrated unequivocally (13). However, direct proof for the generally accepted amplification mechanism proposed by Futcher (12) is lacking. The replication intermediates predicted by the Futcher model have not been experimentally observed. An alternative pathway for amplification is the resolution of a plasmid dimer by Flp recombination while in the act of bidirectional replication (Fig. 7). Pince-nez structures, two-unit-length 2-micron circles interconnected by DNA chains of variable length, observed by electron microscopy under DNA elongation arrest by the *cdc8* mutation would be consistent with an amplifying plasmid dimer (65). Finally, the proposed overamplification of the 2-micron plasmid under misregualtion of Flp, initiated by a linear form of the plasmid (21) (Fig. 2), may represent a normal copy number control mechanism gone awry by perturbing posttranslational modification(s) of Flp. The available genetic and biochemical evidence, though rather limited, highlight the critical role of self-imposed and host-imposed safeguards against unrestricted increases in plasmid population, which would be harmful to the host and, in turn, to the plasmid.

Flp SITE-SPECIFIC RECOMBINATION SYSTEM: *IN VITRO* ANALYSES

In contrast to the paucity of *in vivo* analyses of Flp function, Flp-mediated recombination has been studied extensively *in vitro* as a model for phosphoryl transfer reactions in nucleic acids. Flp belongs to the tyrosine family of site-specific recombinases, whose members utilize an active-site tyrosine residue as the nucleophile to break the scissile phosphate ester bond during the strand cleavage step of recombination. The molecular-genetic, biochemical, and structural information derived from Flp and mechanistically related recombinases have revealed the conformational, mechanistic, and dynamic attributes that underlie tyrosine family site-specific recombination (66–72). While tyrosine recombinases follow the typical type IB topoisomerase chemical mechanism, they do so in the context of a recombination synapse containing two DNA partners, each bound by a pair of recombinase monomers (Fig. 8A). The reaction is completed in a carefully orchestrated sequence of two temporally separated steps of single-strand exchanges. The first step generates a Holliday junction as an obligatory intermediate; the second step resolves the junction into reciprocal recombinants.

Flp ACTIVE SITE: ORGANIZATION AND MECHANISM

The canonical active site of a tyrosine recombinase is characterized by an invariant tyrosine nucleophile utilized for strand cleavage as well as a highly conserved catalytic pentad (Arg-Lys-His-Arg-His/Trp) that helps stabilize the transition states involved in the strand scission and strand union reactions (72, 73). In addition, a conserved glutamic/aspartic acid residue appears to contribute structurally to the functional state of the active site (74). In Flp, the pentad residues of Arg-191, Lys-223, His-305, Arg-308, and Trp-330, assisted by Asp-194, serve to balance the negative charge on the scissile phosphate as well as to activate and orient the Tyr-343 nucleophile (75–78) (Fig. 8B). The strand cleavage step results in the covalent attachment of the tyrosine to the 3′-phopshate end and the exposure of the 5′-hydroxyl group adjacent to it. This 5-hydroxyl group is the nucleophile that attacks the phosphotyrosyl intermediate during the strand-joining reaction. When the joining reactions are directed across DNA partners, rather than within them, the result is the formation of reciprocal recombinant products.

While the active site mechanism described above is common to the tyrosine family, the organization of the Flp active site differs from that of other well-characterized members of this family such as phage λ integrase, phage P1 Cre, and *Escherichia coli* XerC-XerD recombinase. Whereas the norm is the assembly of a fully functional active site within a monomer (79–83), the Flp active site is assembled at the interface of two neighboring Flp monomers (75, 84). One Flp monomer provides the proactive site (shown in green in Fig. 8A, B) comprised of the catalytic pentad and Asp-194, while the second monomer (shown in magenta in

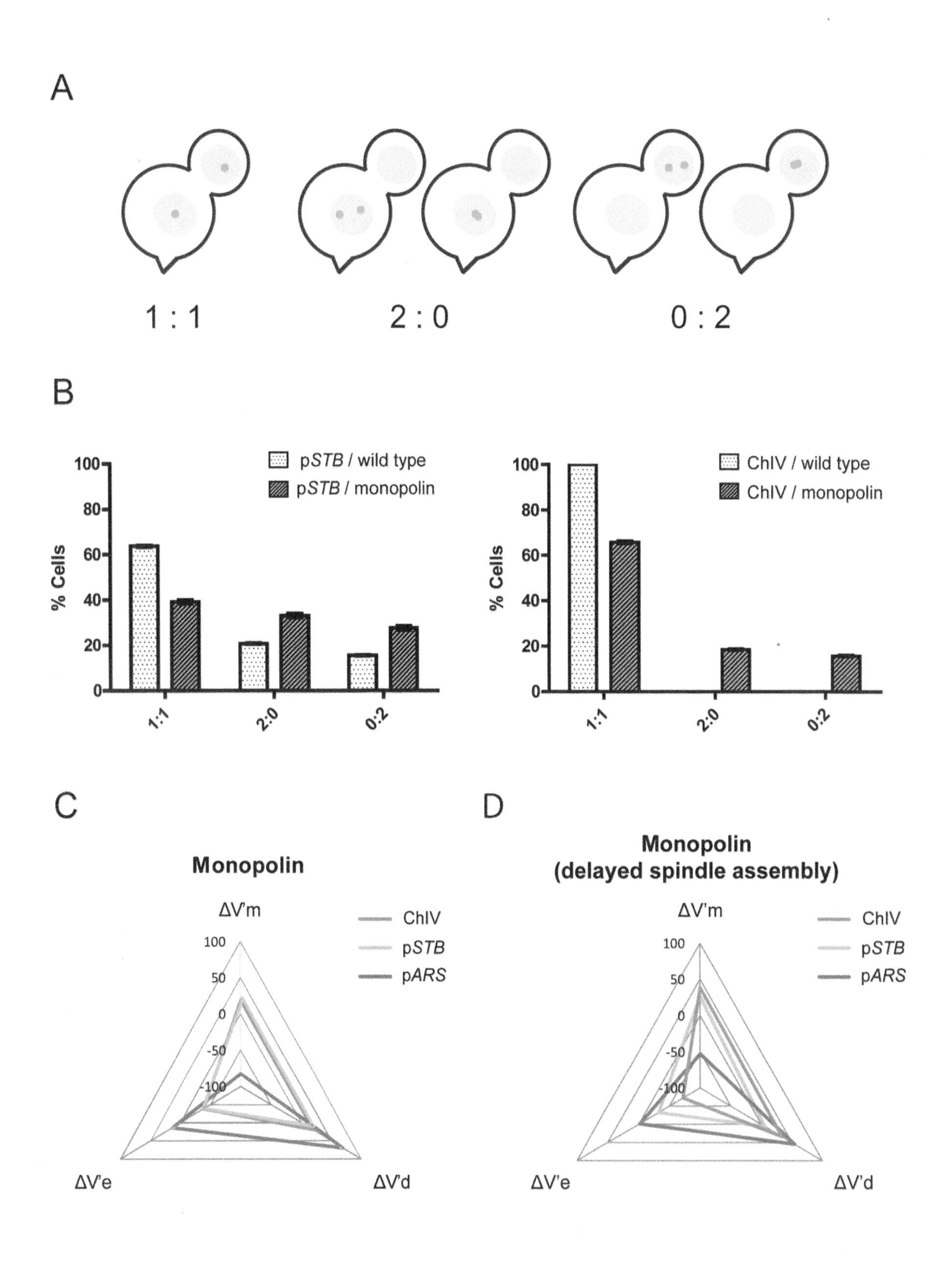
A
1 : 1
2 : 0
0 : 2
B
% Cells
pSTB / wild type
pSTB / monopolin
ChIV / wild type
ChIV / monopolin
1:1
2:0
0:2
C
Monopolin
D
Monopolin
(delayed spindle assembly)
ΔV'm
ΔV'e
ΔV'd
ChIV
pSTB
pARS

Fig. 8A, B) donates the tyrosine nucleophile in *trans*. The shared active site may be a common feature of the subfamily of tyrosine recombinases related to Flp housed by budding yeast plasmids (85). However, allosteric interaction with a neighboring monomer may be important for active site activation even in recombinases that assemble their active sites in *cis*. The unusual active site design of Flp has been particularly helpful in probing its active site mechanism using experimental strategies that are not applicable to recombinase active sites with a *cis* configuration. For example, exogenous nucleophiles that mimic the tyrosine nucleophile, in conjunction with the catalytically inactive Flp(Y343F), can recapitulate the mechanistic features of the normal strand cleavage reaction (86, 87).

Within the recombination synapse, the interactions among the four Flp monomers permit only two of the four Tyr-343 residues to be oriented in their reactive configuration (88). This half of the site's activity accounts for the cleavage/exchange of the DNA strands two at a time (88, 89). The formation of the Holliday junction intermediate is accompanied by the isomerization of its arm configuration, which results in the inactivation of the original pair of active sites and the activation of the new pair. The resulting cleavage/exchange of the second pair of strands resolves the junction. The majority of biochemical and topological evidence, as well as structural data, is consistent with the antiparallel arrangement of the target sites within the recombination synapse (Fig. 8A).

PROBING THE Flp ACTIVE SITE MECHANISM USING CHEMICAL MODIFICATIONS OF THE SCISSILE PHOSPHATE GROUP

Modifications of the nonbridging oxygen atoms of the phosphate group to alter its electronegativity and/or stereochemistry and of the bridging oxygen atoms to modulate leaving group potential (Fig. 9A, B) have provided a useful tool for analyzing mechanistic features of strand breakage and union reactions in nucleic acids. Shuman and colleagues have successfully exploited the replacement of one of the nonbridging oxygens on the scissile phosphate by sulfur (phosphorothioate) or the methyl group (methylphosphonate; MeP) as well as the 5′-bridging oxygen by sulfur (5′-thiolate) to unveil several active site attributes of vaccinia type IB topoisomerase (90–93). Analogous strategies have been employed for analyzing mechanistic features of the Flp active site (94, 95). Such studies have been performed predominantly using half-site substrates containing a single scissile phosphate or a modified scissile phosphate (Fig. 9C) together with Flp variants harboring specific active site mutations.

Reactions with a 5′-thiolate substituted DNA substrate have revealed a subtle role for Trp-330 of Flp in facilitating the departure of the 5′-hydroxyl group during the strand cleavage step (77). The primary function of Trp-330 is in promoting the proper alignment of the Tyr-343 nucleophile donated in *trans* through hydrophobic/van der Waals interactions (77, 96). Flp(W330H) is strongly compromised in its

Figure 6 Segregation of a single copy reporter plasmid during monopolin-directed mitosis. **(A, B)** The segregation of a fluorescence-tagged chromosome or a single copy reporter plasmid (p*STB*) is scored as 1:1 (equal segregation), 2:0 (mother-biased cosegregation), or 0:2 (daughter-biased cosegregation) during a normal or a monopolin-directed mitotic cell cycle. A subset of the missegregated plasmid sisters is often seen as coalesced or overlapping foci. The normal 1:1 segregation of sister chromatids is perturbed by monopolin toward 2:0 or 0:2 segregation. A similar effect is seen for p*STB* as well. **(C, D)** In these radar plots, the degree of correlation between a reporter plasmid and a chromosome in their segregation patterns under the influence of monopolin is represented in terms of three variables: deviation from equal segregation (V'_e), tendency toward mother segregation (V'_m), and tendency toward daughter segregation (V'_d). **(C)** The strong correlation between an *STB* plasmid and a chromosome in their bias-free cosegregation under the influence of monopolin is conveyed by the near congruence of the blue and green triangles. The absence of such correlation between an *ARS* plasmid (lacking *STB*) and a chromosome is evinced by the nonoverlapping disposition of the red triangle. **(D)** Under conditions that uncouple the segregation of the 2-micron plasmid from that of the chromosomes, the tight correlation between an *STB* reporter plasmid and a reporter chromosome breaks down during monopolin directed mitosis. In the example shown here, the G1 to G2/M phase of the cell cycle is contrived to proceed in the absence of a functional spindle (by treatment with nocodazole). The metaphase arrested cells are washed free of nocodazole to permit spindle assembly and continuation of the cell cycle. Further details can be found in reference 32.
doi:10.1128/microbiolspec.PLAS-0003-2013.f6

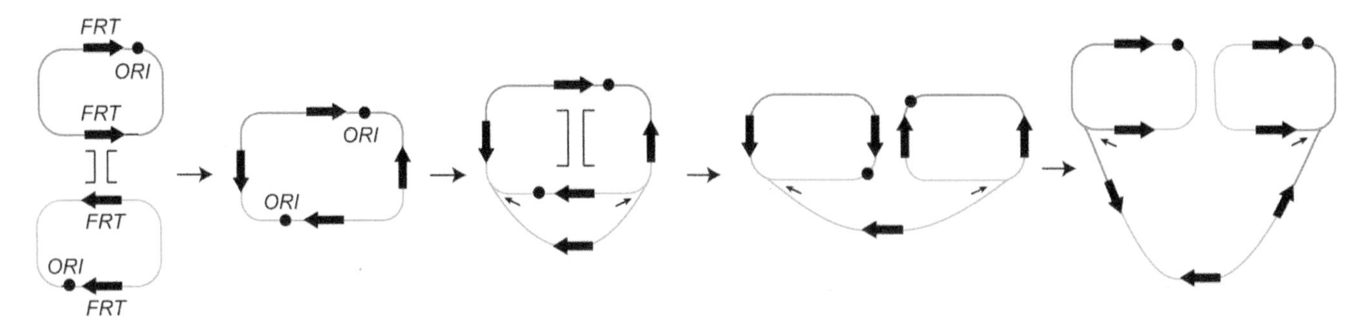

Figure 7 A possible alternative mechanism for 2-micron plasmid amplification. A plasmid dimer may be formed by Flp-mediated recombination between two monomers. Resolution of the dimer by Flp during the act of replication will give rise to two interconnected circles being replicated iteratively by unidirectional replication forks. Such structures, named pince-nez molecules, have been observed by electron microscopy (65).
doi:10.1128/microbiolspec.PLAS-0003-2013.f7

cleavage potential, while Flp(W330A) has barely detectable cleavage activity. However, both mutants show vastly improved activity on a substrate containing the 5′-thiolate modification. The thiol is a better leaving group than hydroxyl because of its lower pK_a. Thus, the catalytic deficiency caused by the absence of Trp-330 can be significantly ameliorated by a DNA modification that facilitates leaving-group departure.

The reactivity of Flp variants on MeP substrates demonstrates that neutralization of the phosphate negative charge in its ground state permits transition state stabilization in the absence of one of the two conserved arginines (Arg-191 or Arg-308) of the Flp active site (94, 95). Flp(R191A) and Flp(R308A) are reactive on half-site substrates containing MeP substitution at the scissile position (Fig. 10A, B), whereas both these variants are almost completely inactive on phosphate containing DNA substrates. The electrostatic suppression of the lack of a positive charge in the recombinase active site by a compensatory charge substitution in the scissile phosphate of the DNA substrate has also been demonstrated for the Cre recombinase (97, 98).

AN EXTRACATALYTIC ROLE FOR Arg-308 OF Flp REVEALED BY THE MeP REACTION

Flp(R308A) promotes direct hydrolysis of the MeP substrate without forming the tyrosyl intermediate (Fig. 10B) (94). In the absence of Arg-308, the abundant water nucleophile gains access to the reaction center, outcompeting Tyr-343 in the cleavage reaction to give a dead-end hydrolytic product. By contrast, Flp (R191A) does not mediate direct hydrolysis of the MeP bond; rather, it utilizes Tyr-343 as the cleavage nucleophile to yield the tyrosyl intermediate (95) (Fig. 10A). The absence of the tyrosyl intermediate during the breakage of the MeP bond by Flp(R308A) has been verified by the activity of the double mutant Flp(R308A, Y343F), which also yields the hydrolysis product with similar kinetics and V_{max} as Flp(R308A). Thus, in addition to its catalytic role in balancing the phosphate negative charge, Arg-308 appears to protect the normal reaction course from abortive hydrolysis, presumably by electrostatic misorientation of water. Unlike Arg-308 of Flp, the corresponding Arg-292 of Cre does not seem to perform a protective function against direct hydrolysis of the scissile phosphate. Cre(R292A) yields the MeP-tyrosyl intermediate by utilizing the active site tyrosine nucleophile (Tyr-324) (97, 98). This difference between Flp and Cre may be rationalized by the *trans* versus *cis* organization of their respective sites. When a Cre monomer binds to one of the two binding sites of a target site, the tyrosine nucleophile concomitantly engages the adjacent scissile phosphate (Fig. 10C). As a result, there is little danger from direct hydrolysis. By contrast, binding of a Flp monomer activates the adjacent scissile phosphate without immediate engagement by the tyrosine nucleophile. The donation of Tyr-343 in *trans* must await the binding of a second Flp monomer to the binding element across the strand exchange region. The potential time delay between the two binding events gives water the opportunity to attack the activated phosphate group (Fig. 10D). This untoward reaction is avoided by utilizing the positive charge on the Arg-308 side chain to misdirect water (which is electrically a dipole).

As the methyl substitution of one of the nonbridging oxygen atoms turns the normally symmetric phosphate

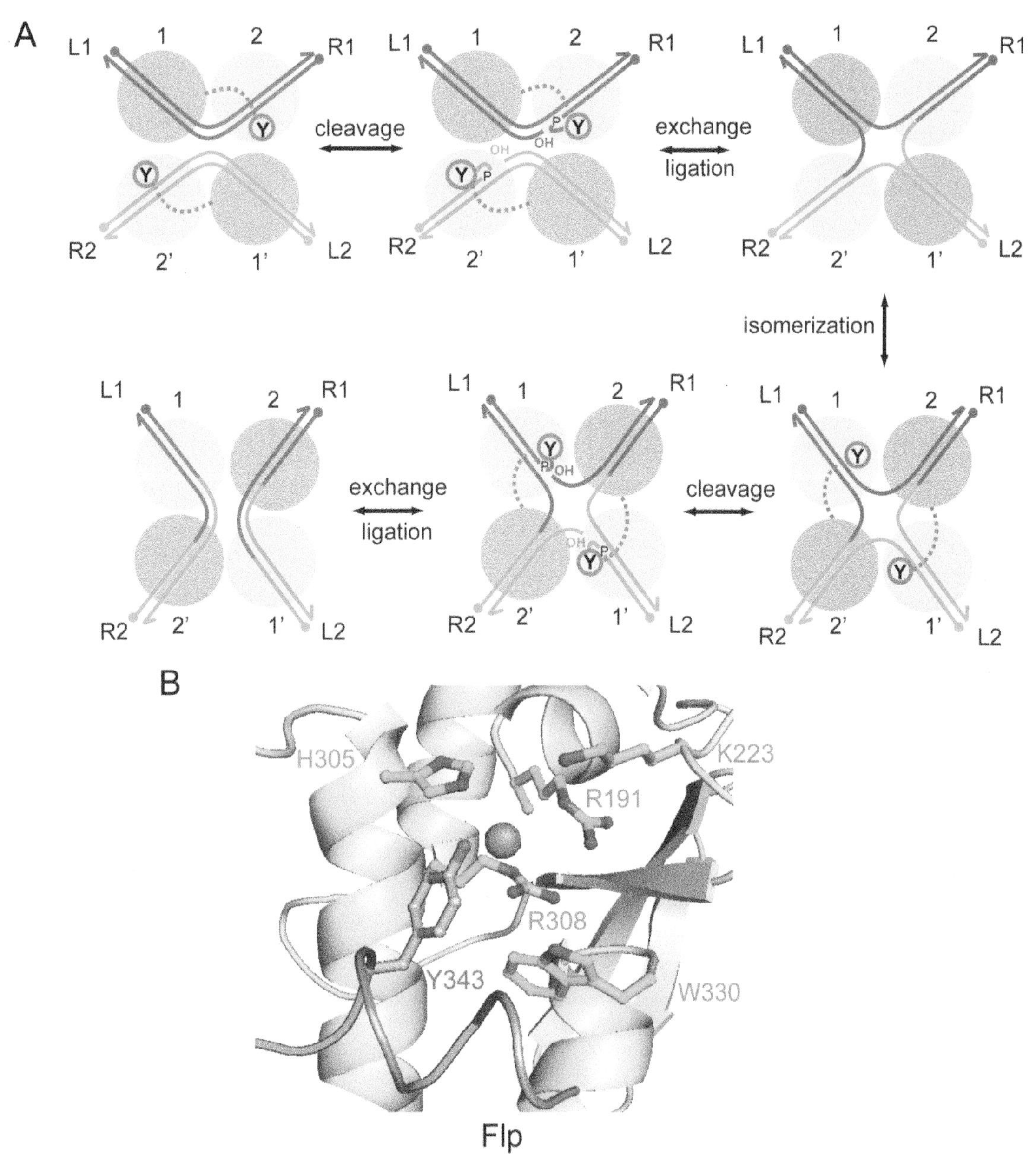

Figure 8 Flp-mediated site-specific recombination. (**A**) The recombination reaction is initiated by the synapsis of two *FRT* sites, L1-R1 and L2-R2 (L = left; R = right), each bound by two monomers of Flp (1, 2; 1′, 2′) across from the strand exchange region. The antiparallel arrangement of sites (left to right at the top and right to left at the bottom) within the recombination synapse is consistent with most (but not all) published data. The first pair of strand cleavage-exchange reactions gives rise to a Holliday junction intermediate; the second pair of analogous reactions resolves the junction into recombinant products, L1-R2 and L2-R1. The active Flp monomers, those adjacent to the scissile phosphates that are targeted by the active site tyrosine nucleophiles during a cleavage-exchange step, are shown in green. The switch between the active and inactive (magenta) pairs of Flp monomers accompanies the isomerization of the Holliday junction intermediate. (**B**) The organization of key active site residues within the Flp active site is shown (75, 89). The conserved catalytic pentad of the tyrosine family corresponds to Arg-191, Lys-223, His-305, Arg-308, and Trp-330 in Flp. The active site tyrosine (Tyr-343) is delivered by a second Flp monomer.
doi:10.1128/microbiolspec.PLAS-0003-2013.f8

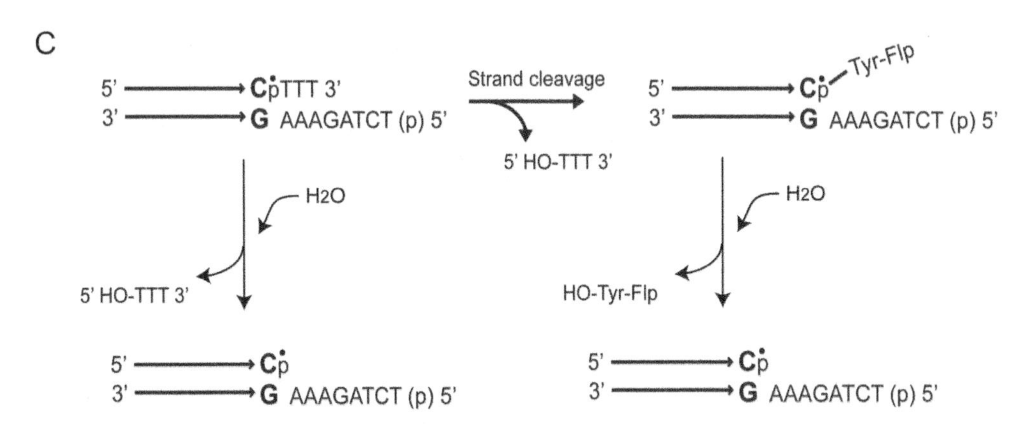

Figure 9 Chemical substitutions at the scissile phosphate position and half-site substrates for probing recombination mechanisms. (**A**) When the 5′ bridging oxygen atom of the phosphodiester bond is replaced by sulfur, the significantly lower pKa of the 5′-thiol (compared to the 5′-hydroxyl) makes it a stronger leaving group. The 5′ thiolate substitution, in conjunction with active site mutants, can shed light on the general acid and/or accessory residues involved in facilitating leaving group departure. (**B**) When one of the nonbridging oxygen atoms of the phosphate group is replaced by a methyl group, the resulting methylphosphonate (MeP) has no negative charge in the ground state. Furthermore, the methyl substitution introduces chirality at the phosphate center. The MeP substitution is useful for the analysis of electrostatic and stereochemical features of the recombination reaction. (**C**) Half-site substrates simplify the recombination reaction while retaining its intrinsic chemical features. A half-site contains a single Flp binding element and a single scissile phosphate (or MeP) on the cleavable strand followed by three (or two) nucleotides of the strand exchange region. The modified scissile phosphate in the MeP half-site is indicated by the dot placed over the "p." The other (noncleavable) strand contains all eight nucleotides of the strand exchange region. When the 5′-hydroxyl group of this strand is phosphorylated, it is blocked from partaking in a strand joining reaction. Tyr-343 mediated cleavage of the MeP bond will give rise to the Flp-linked DNA intermediate, which can be hydrolyzed slowly over time. Cleavage within a half-site is nearly irreversible, as the short tri- or dinucleotide product diffuses away from the reaction center. Since Flp-bound half-sites can associate to form dimers, trimers, and tetramers, Tyr-343 can be donated in *trans* as the cleavage nucleophile. In principle, the same hydrolysis product can also be formed by direct attack of water on the MeP bond (shown at the left). However, such a reaction is not observed during the action of Flp on native DNA substrates containing an unmodified phosphate at the scissile position. doi:10.1128/microbiolspec.PLAS-0003-2013.f9

group into an asymmetric center, an additional utility of the MeP substrates is in unveiling the stereochemical course of the recombination reaction. Stereochemically pure R_P and S_P forms of the MeP substrates are currently being used to dissect the individual stereochemical contributions of Arg-191 and Arg-308 and to probe how other members of the catalytic pentad might influence these contributions.

Flp REACTION PATHWAY: SINGLE MOLECULE ANALYSIS

Single molecule analysis of Flp recombination using real-time tethered particle motion (TPM) analysis has provided deeper insights into the prechemical and chemical steps of the reaction pathway (Fig. 11) (99). The results obtained reveal interesting similarities and contrasts of Flp with λ Int and phage P1 Cre, which

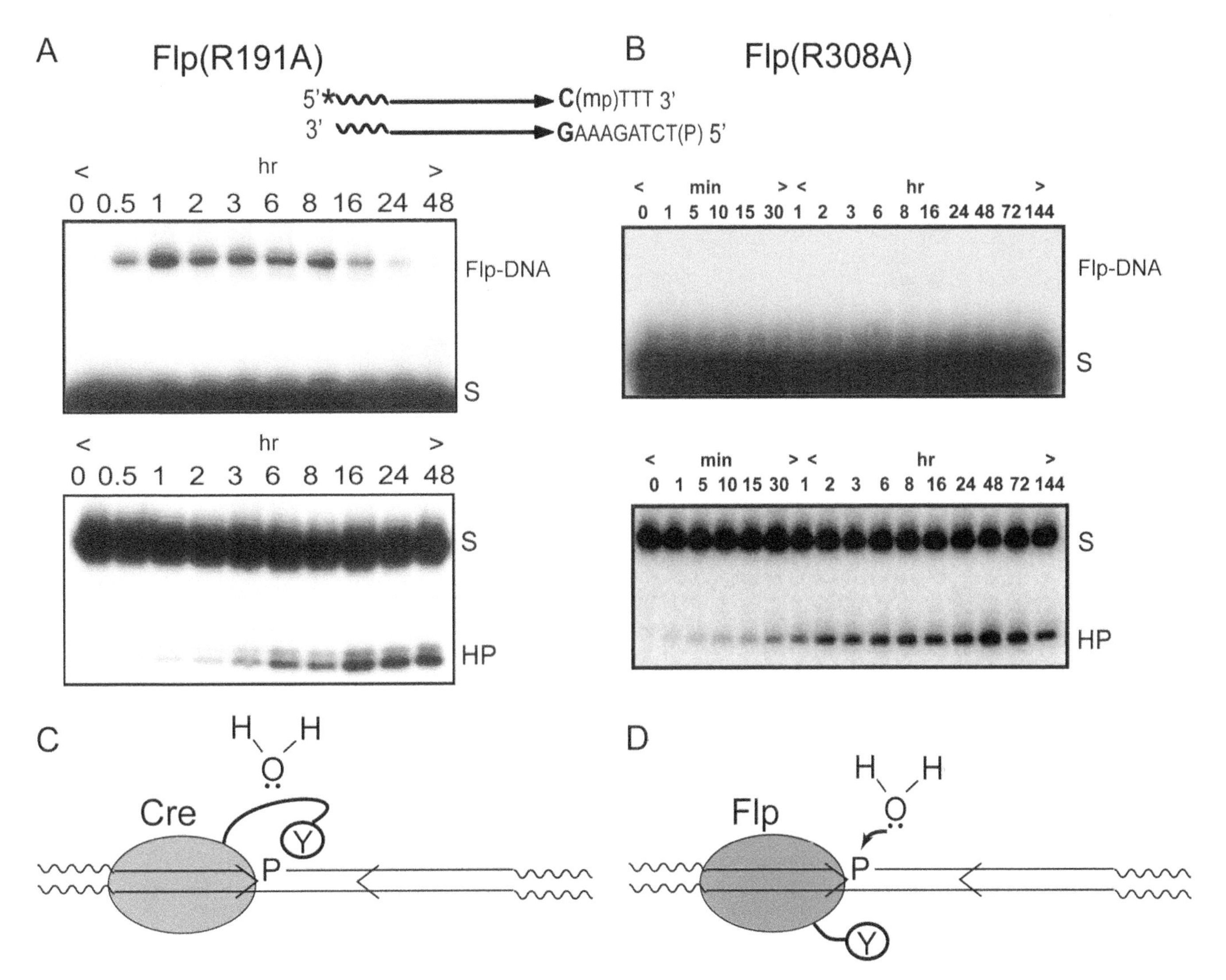

Figure 10 Activities of Flp(R191A) and Flp(R308A) on an MeP substituted half-site substrate; the difference between *cis*- and *trans*- active sites in protecting the scissile phosphate from abortive hydrolysis. The MeP half-site reactions are analyzed by electrophoresis in SDS-polyacrylamide gels (top panels) to detect the Flp linked tyrosyl-DNA intermediate or in urea-polyacrylamide gels (bottom panels) to visualize the hydrolysis product (HP). The unreacted substrate band is indicated by "S." The asterisk marks the ^{32}P label placed at the 5′ end of the cleavable strand of the half-site. (**A**) Flp(R191A) forms the tyrosyl intermediate, which then undergoes slow hydrolysis. (**B**) Flp(R308A) does not form the tyrosyl intermediate. Instead, it promotes direct hydrolysis of the MeP bond. (**C**) When a cis-acting recombinase (Cre, for example) monomer binds to its target site, the tyrosine nucleophile is oriented within the active site to engage the scissile phosphate. The binding of a second monomer allosterically activates this active site to trigger tyrosine-mediated strand cleavage. Thus, the scissile phosphate is not prone to attack by water acting as the nucleophile. (**D**) Binding of a Flp monomer activates the adjacent scissile phosphate even though the tyrosine nucleophile is not oriented within the active site. The engagement of the scissile phosphate by tyrosine must await the binding of a second Flp monomer. The time lag between the two binding events renders the scissile phosphate susceptible to direct hydrolysis. Electrostatic repulsion of water by Arg-308 appears to prevent this abortive reaction.
doi:10.1128/microbiolspec.PLAS-0003-2013.f10

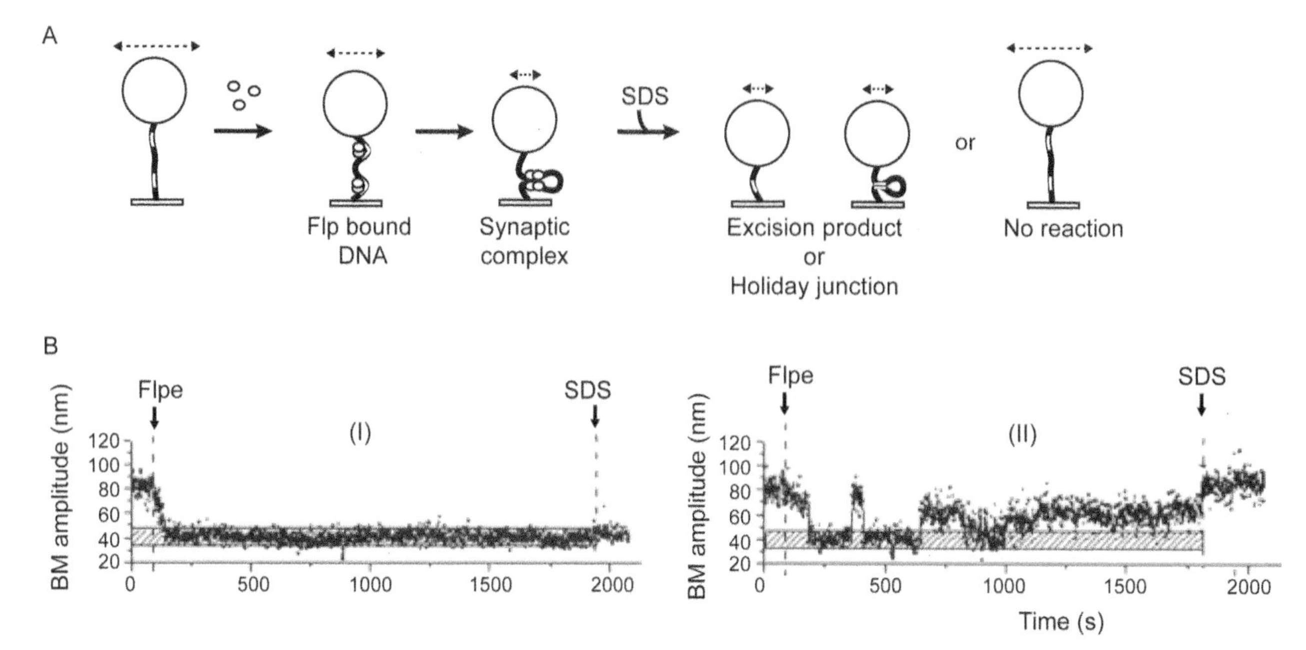

Figure 11 Single molecule TPM analysis of Flp site-specific recombination. The substrate DNA molecule containing a pair of *FRT* sites in direct (deletion substrate) or inverted (inversion substrate) orientation is tethered to a glass surface at one end and to a polystyrene bead at the other. The effective length of the DNA changes as a result of Flp binding to *FRT* sites or synapsis of the bound sites. Such changes can be assayed by changes in the Brownian motion (BM) amplitude of the attached bead (indicated schematically by the dashed line with arrowheads at either end). Within the recombination synapse, the chemical steps of recombination may or may not occur. The fate of the synapsed molecules can be revealed by the BM amplitudes they display after being stripped of noncovalently associated Flp. **(A)** The expected BM amplitudes following the addition of Flp to a deletion substrate. Binding of Flp to the *FRT* sites (unfilled rectangular boxes on the tethered DNA molecule) will cause a slight reduction in the BM amplitude because of the DNA bending induced by Flp. A more marked reduction in BM amplitude follows upon synapsis of the *FRT* sites. This low BM amplitude will be retained by the Holliday junction intermediate and the linear product of excision even after the addition of SDS. Flp-bound molecules that fail to synapse (nonproductive complexes) or synapsed molecules that fail to recombine (wayward complexes) will return to the high BM amplitude of the starting DNA substrate after SDS challenge. In the case of the inversion substrate, SDS challenge cannot distinguish a wayward complex from a completed recombination event. This is because the length of the inversion product is the same as that of the parental substrate. **(B)** Time traces of individual molecules illustrating two different states of the deletion substrate following Flp addition are shown. **(I)** The trace in the left panel indicates a molecule that formed a stable synapse of the Flp-bound *FRT* sites. The low BM amplitude of this molecule after SDS challenge indicates that it underwent Holliday junction formation or a complete recombination event. **(II)** The trace in the right panel indicates a molecule that underwent Flp binding and synapsis but failed to recombine or to form the Holliday junction, as indicated by its return to the starting high BM amplitude after SDS treatment (a wayward complex). The horizontal stippled bar represents the BM amplitude of the DNA in the synapsed state of *FRT* sites.
doi:10.1128/microbiolspec.PLAS-0003-2013.f11

have also been characterized by single molecule TPM or by single molecule TPM as well as TFM (tethered fluorophore motion) (100–102). The binding of recombinase to target sites and the pairing of bound sites is quite fast in all three cases, ruling out intrinsic barriers to synapsis during tyrosine family site-specific recombination, at least *in vitro*. There is a strong commitment to recombination following the association of Flp with the *FRT* sites. The formation of nonproductive complexes (those that do not synapse) and wayward complexes (those that do not form the Holliday junction intermediate or complete recombination after synapsis) constitute only minor detractions from the productive pathway. The stability of the synapse is enhanced by strand cleavage in the case of Flp and λ Int. However, Cre forms stable synapse even in the absence of strand cleavage. Despite the chemical reversibility of the individual reactions, a round of Flp recombination proceeds efficiently to completion, presumably aided by the conformational changes associated with each cleavage or joining step. Such irreversibility of an initiated recombination event would be a desirable attribute *in vivo* in bringing about the desired DNA rearrangement, namely, inversion of a replication fork in the case of Flp. However, the Holliday junction formed during Cre recombination is long lived, affording the opportunity for the reaction to reverse course, at least *in vitro*. It is possible that the *in vitro* Cre reaction fails to recapitulate the native regulatory features of recombination occurring within the phage genome organized into a nucleoprotein complex.

CONTRIBUTIONS OF THE 2-MICRON PLASMID TO BASIC BIOLOGY AND BIOENGINEERING APPLICATIONS

The 2-micron plasmid provides a model for how the collaboration between an efficient partitioning system and a copy number control system can confer chromosome-like stability on an extremely simple extrachromosomal genome (7). It exemplifies the evolutionary success of a selfish element through strategies that take advantage of the genetic endowments of its host, while at the same time moderating its selfishness to avoid harming the host (103). The plasmid partitioning system has provided the basis for propagating autonomously replicating plasmids with high stability in *S. cerevisiae*. Assuming that the proposed hitchhiking model for 2-micron plasmid segregation is correct, it may be possible to suitably engineer the Rep-*STB* system for the long-term maintenance of beneficial extrachromosomal elements in a variety of higher eukaryotes. The Flp site-specific recombination system harbored by the plasmid has been seminal to unveiling phosphoryl transfer mechanisms in nucleic acids and to understanding conformational dynamics associated with strand exchange between DNA partners (70, 76). The simple requirements of Flp and Cre reactions have been exploited to develop an analytical tool called "difference topology," which reveals the topological path of DNA within high-order DNA-protein complexes (7, 104–106). The rationale is to first assemble the complex of interest and then utilize Flp or Cre recombination to trap the DNA supercoils sequestered within it as crossings of DNA knots formed in an inversion reaction or DNA links (catenanes) formed in a deletion reaction. Care is taken to avoid the random entrapment of supercoils by placing the recombination target sites within close proximity of the complex. The knot or catenane crossings can be counted by suitable analytical methods. A simplified description of the principles and practice of difference topology can be found in reference 7. The method can potentially be applied to deduce the topological aesthetics of complex DNA-protein interactions to which standard biochemical and biophysical techniques are unfortunately blind. For example, it would be interesting to know whether the interactions among promoters, enhancers, and activator or repressor binding sites during transcriptional regulation engender specific DNA topologies. The targeted genetic manipulations that can be accomplished by Flp and related recombination systems (107–109) have been particularly helpful in addressing basic problems in gene regulation and developmental biology. They have also provided a powerful tool for genome engineering. The generation of altered target specificity recombinase variants (110–115), as well as the regulated and/or tissue-specific expression of recombinases (108, 116–118), has vastly expanded the utility of site-specific recombination in basic research and in biotechnological applications.

Acknowledgments. *Our studies on the partitioning and site-specific recombination systems have been supported by the National Institutes of Health, the National Science Foundation (MCB-1049925), and the Robert F Welch Foundation (F-1274). Conflicts of interest: We disclose no conflicts.*

Citation. Liu Y-T, Sau S, Ma C-H, Kachroo AH, Rowley PA, Chang K-M, Fan H-F, Jayaram M. 2014. The partitioning and copy number control systems of the selfish yeast plasmid: an optimized molecular design for stable persistence in host cells. Microbiol Spectrum **2**(5):PLAS-0003-2013.

References

1. **Dawkins R.** 1976. *The Selfish Gene*. Oxford University Press, Oxford, UK.
2. **Orgel LE, Crick FH.** 1980. Selfish DNA: the ultimate parasite. *Nature* **284:**604–607.
3. **Orgel LE, Crick FH, Sapienza C.** 1980. Selfish DNA. *Nature* **288:**645–646.
4. **Rowley PA, Kachroo AH, Jayaram M.** 2013. Selfish DNA, p 382–389. *In* Malloy S, Hughes K (ed), *Brenner's Encyclopedia of Genetics*, Vol **6**. Elsevier, Amsterdam.
5. **Frappier L.** 2004. Viral plasmids in mammalian cells, p 325–340. *In* Funnell BE, Phillips G (ed), *Plasmid Biology*. ASM Press, Washington, DC.
6. **McBride AA.** 2008. Replication and partitioning of papillomavirus genomes. *Adv Virus Res* **72:**155–205.
7. **Chang KM, Liu YT, Ma CH, Jayaram M, Sau S.** 2013. The 2 micron plasmid of *Saccharomyces cerevisiae*: a miniaturized selfish genome with optimized functional competence. *Plasmid* **70:**2–17.

8. **Broach JR, Volkert FC.** 1991. Circular DNA plasmids of yeasts, p 287–331. *In* Broach JR, Pringle JR, Jones EW (ed), *The Molecular Biology of the Yeast Saccharomyces. Genome Dynamics, Protein Synthesis and Energetics*. Cold Spring Harbor Laboratory Press, Cold Spring Harbor, New York.
9. **Jayaram M, Yang XM, Mehta S, Voziyanov Y, Velmurugan S.** 2004. The 2 micron plasmid of *Saccharomyces cerevisiae*, p 303–324. *In* Funnell BE, Phillips G (ed), *Plasmid Biology*. ASM Press, Washington, DC.
10. **Zakian VA, Brewer BJ, Fangman WL.** 1979. Replication of each copy of the yeast 2 micron DNA plasmid occurs during the S phase. *Cell* **4:**923–934.
11. **Jayaram M, Mehta S, Uzri D, Voziyanov Y, Velmurugan S.** 2004. Site-specific recombination and partitioning systems in the stable high copy propagation of the 2-micron yeast plasmid. *Prog Nucleic Acid Res Mol Biol* **77:**127–172.
12. **Futcher AB.** 1986. Copy number amplification of the 2 micron circle plasmid of *Saccharomyces cerevisiae*. *J Theor Biol* **119:**197–204.
13. **Volkert FC, Broach JR.** 1986. Site-specific recombination promotes plasmid amplification in yeast. *Cell* **46:** 541–550.
14. **Dobson MJ, Pickett AJ, Velmurugan S, Pinder JB, Barrett LA, Jayaram M, Chew JS.** 2005. The 2μm plasmid causes cell death in *Saccharomyces cerevisiae* with a mutation in Ulp1 protease. *Mol Cell Biol* **25:**4299–4310.
15. **Holm C.** 1982. Sensitivity to the yeast plasmid 2 μm DNA is conferred by the nuclear allele nib1. *Mol Cell Biol* **2:**985–992.
16. **Holm C.** 1982. Clonal lethality caused by the yeast plasmid 2 μm DNA. *Cell* **29:**85–94.
17. **Murray JA, Scarpa M, Rossi N, Cesareni G.** 1897. Antagonistic controls regulate copy number of the yeast 2 micron plasmid. *EMBO J* **6:**4205–4212.
18. **Reynolds AE, Murray AW, Szostak JW.** 1987. Roles of the 2 micron gene products in stable maintenance of the 2 micron plasmid of *Saccharomyces cerevisiae*. *Mol Cell Biol* **7:**3566–3573.
19. **Som T, Armstrong KA, Volkert FC, Broach JR.** 1988. Autoregulation of 2 micron circle gene expression provides a model for maintenance of stable plasmid copy levels. *Cell* **52:**27–37.
20. **Chen XL, Reindle A, Johnson ES.** 2005. Misregulation of 2 micron circle copy number in a SUMO pathway mutant. *Mol Cell Biol* **25:**4311–4320.
21. **Xiong L, Chen XL, Silver HR, Ahmed NT, Johnson ES.** 2009. Deficient SUMO attachment to Flp recombinase leads to homologous recombination-dependent hyperamplification of the yeast 2 micron circle plasmid. *Mol Biol Cell* **20:**1241–1251.
22. **Hastings PJ, Ira G, Lupski JR.** 2009. A microhomology-mediated break-induced replication model for the origin of human copy number variation. *PLoS Genet* **5:** e1000327.
23. **McEachern MJ, Haber JE.** 2006. Break-induced replication and recombinational telomere elongation in yeast. *Annu Rev Biochem* **75:**111–135.
24. **Tomaska L, Nosek J, Kramara J, Griffith JD.** 2009. Telomeric circles: universal players in telomere maintenance? *Nat Struct Mol Biol* **16:**1010–1015.
25. **Zhang F, Carvalho CM, Lupski JR.** 2009. Complex human chromosomal and genomic rearrangements. *Trends Genet* **25:**298–307.
26. **Gehlen LR, Nagai S, Shimada K, Meister P, Taddei A, Gasser SM.** 2011. Nuclear geometry and rapid mitosis ensure asymmetric episome segregation in yeast. *Curr Biol* **21:**25–33.
27. **Khmelinskii A, Meurer M, Knop M, Schiebel E.** 2011. Artificial tethering to nuclear pores promotes partitioning of extrachromosomal DNA during yeast asymmetric cell division. *Curr Biol* **21:**R17–R18.
28. **Murray AW, Szostak JW.** 1983. Pedigree analysis of plasmid segregation in yeast. *Cell* **34:**961–970.
29. **Shcheprova Z, Baldi S, Frei SB, Gonnet G, Barral Y.** 2008. A mechanism for asymmetric segregation of age during yeast budding. *Nature* **454:**728–734.
30. **Mehta S, Yang XM, Chan CS, Dobson MJ, Jayaram M, Velmurugan S.** 2002. The 2 micron plasmid purloins the yeast cohesin complex: a mechanism for coupling plasmid partitioning and chromosome segregation? *J Cell Biol* **158:**625–637.
31. **Velmurugan S, Yang XM, Chan CS, Dobson M, Jayaram M.** 2000. Partitioning of the 2-micron circle plasmid of *Saccharomyces cerevisiae*. Functional coordination with chromosome segregation and plasmid-encoded Rep protein distribution. *J Cell Biol* **149:** 553–566.
32. **Liu YT, Ma CH, Jayaram M.** 2013. Co-segregation of yeast plasmid sisters under monopolin-directed mitosis suggests association of plasmid sisters with sister chromatids. *Nucleic Acids Res* **41:**4144–4158.
33. **Ahn YT, Wu XL, Biswal S, Velmurugan S, Volkert FC, Jayaram M.** 1997. The 2 micron-plasmid-encoded Rep1 and Rep2 proteins interact with each other and colocalize to the *Saccharomyces cerevisiae* nucleus. *J Bacteriol* **179:**7497–7506.
34. **Hadfield C, Mount RC, Cashmore AM.** 1995. Protein binding interactions at the STB locus of the yeast 2 micron plasmid. *Nucleic Acids Res* **23:**995–1002.
35. **Scott-Drew S, Murray JA.** 1998. Localisation and interaction of the protein components of the yeast 2 micron circle plasmid partitioning system suggest a mechanism for plasmid inheritance. *J Cell Sci* **111:**1779–1789.
36. **Velmurugan S, Ahn YT, Yang XM, Wu XL, Jayaram M.** 1998. The 2 micron plasmid stability system: analyses of the interactions among plasmid- and host-encoded components. *Mol Cell Biol* **18:**7466–7477.
37. **Yang XM, Mehta S, Uzri D, Jayaram M, Velmurugan S.** 2004. Mutations in a partitioning protein and altered chromatin structure at the partitioning locus prevent cohesin recruitment by the *Saccharomyces cerevisiae* plasmid and cause plasmid missegregation. *Mol Cell Biol* **24:**5290–5303.
38. **Sengupta A, Blomqvist K, Pickett AJ, Zhang Y, Chew JS, Dobson MJ.** 2001. Functional domains of yeast plasmid-encoded Rep proteins. *J Bacteriol* **183:**2306–2315.

39. **Murray JA, Cesareni G.** 1986. Functional analysis of the yeast plasmid partition locus STB. *EMBO J* **5:** 3391–3399.
40. **Cui H, Ghosh SK, Jayaram M.** 2009. The selfish yeast plasmid uses the nuclear motor Kip1p but not Cin8p for its localization and equal segregation. *J Cell Biol* **185:** 251–264.
41. **Hajra S, Ghosh SK, Jayaram M.** 2006. The centromere-specific histone variant Cse4p (CENP-A) is essential for functional chromatin architecture at the yeast 2-micron circle partitioning locus and promotes equal plasmid segregation. *J Cell Biol* **174:**779–790.
42. **Huang CC, Hajra S, Ghosh SK, Jayaram M.** 2011. Cse4 (CenH3) association with the *Saccharomyces cerevisiae* plasmid partitioning locus in its native and chromosomally integrated states: implications for centromere evolution. *Mol Cell Biol* **31:**1030–1040.
43. **Ma CH, Cui H, Hajra S, Rowley PA, Fekete C, Sarkeshik A, Ghosh SK, Yates JR 3rd, Jayaram M.** 2013. Temporal sequence and cell cycle cues in the assembly of host factors at the yeast 2 micron plasmid partitioning locus. *Nucleic Acids Res* **41:**2340–2353.
44. **Wong MC, Scott-Drew SR, Hayes MJ, Howard PJ, Murray JA.** 2002. RSC2, encoding a component of the RSC nucleosome remodeling complex, is essential for 2 micron plasmid maintenance in *Saccharomyces cerevisiae*. *Mol Cell Biol* **22:**4218–4229.
45. **Ghosh SK, Huang CC, Hajra S, Jayaram M.** 2010. Yeast cohesin complex embraces 2 micron plasmid sisters in a tri-linked catenane complex. *Nucleic Acids Res* **38:**570–584.
46. **Furuyama S, Biggins S.** 2007. Centromere identity is specified by a single centromeric nucleosome in budding yeast. *Proc Natl Acad Sci USA* **104:**14706–14711.
47. **Camahort R, Shivaraju M, Mattingly M, Li B, Nakanishi S, Zhu D, Shilatifard A, Workman JL, Gerton JL.** 2009. Cse4 is part of an octameric nucleosome in budding yeast. *Mol Cell* **35:**794–805.
48. **Lefrancois P, Auerbach RK, Yellman CM, Roeder GS, Snyder M.** 2013. Centromere-like regions in the budding yeast genome. *PLoS Genet* **9:**e1003209.
49. **Malik HS, Henikoff S.** 2009. Major evolutionary transitions in centromere complexity. *Cell* **138:**1067–1082.
50. **Aravind L, Watanabe H, Lipman DJ, Koonin EV.** 2000. Lineage-specific loss and divergence of functionally linked genes in eukaryotes. *Proc Natl Acad Sci USA* **97:** 11319–11324.
51. **Furuyama T, Henikoff S.** 2009. Centromeric nucleosomes induce positive DNA supercoils. *Cell* **138:**104–113.
52. **Huang CC, Chang KM, Cui H, Jayaram M.** 2011. Histone H3-variant Cse4-induced positive DNA supercoiling in the yeast plasmid has implications for a plasmid origin of a chromosome centromere. *Proc Natl Acad Sci USA* **108:**13671–13676.
53. **Ghosh SK, Hajra S, Jayaram M.** 2007. Faithful segregation of the multicopy yeast plasmid through cohesin-mediated recognition of sisters. *Proc Natl Acad Sci USA* **104:**13034–13039.
54. **Kiburz BM, Amon A, Marston AL.** 2008. Shugoshin promotes sister kinetochore biorientation in *Saccharomyces cerevisiae*. *Mol Biol Cell* **19:**1199–1209.
55. **Kitajima TS, Kawashima SA, Watanabe Y.** 2004. The conserved kinetochore protein shugoshin protects centromeric cohesion during meiosis. *Nature* **427:**510–517.
56. **Meyer RE, Kim S, Obeso D, Straight PD, Winey M, Dawson DS.** 2013. Mps1 and Ipl1/Aurora B act sequentially to correctly orient chromosomes on the meiotic spindle of budding yeast. *Science* **339:**1071–1074.
57. **Monje-Casas F, Prabhu VR, Lee BH, Boselli M, Amon A.** 2007. Kinetochore orientation during meiosis is controlled by Aurora B and the monopolin complex. *Cell* **128:**477–490.
58. **Toth A, Rabitsch KP, Galova M, Schleiffer A, Buonomo SB, Nasmyth K.** 2000. Functional genomics identifies monopolin: a kinetochore protein required for segregation of homologs during meiosis I. *Cell* **103:**1155–1168.
59. **Yu HG, Koshland D.** 2007. The Aurora kinase Ipl1 maintains the centromeric localization of PP2A to protect cohesin during meiosis. *J Cell Biol* **176:**911–918.
60. **Mehta S, Yang XM, Jayaram M, Velmurugan S.** 2005. A novel role for the mitotic spindle during DNA segregation in yeast: promoting 2 micron plasmid-cohesin association. *Mol Cell Biol* **25:**4283–4298.
61. **Mehta GD, Rizvi SM, Ghosh SK.** 2012. Cohesin: a guardian of genome integrity. *Biochim Biophys Acta* **1823:**1324–1342.
62. **Nasmyth K.** 2011. Cohesin: a catenase with separate entry and exit gates? *Nat Cell Biol* **13:**1170–1177.
63. **Onn I, Heidinger-Pauli JM, Guacci V, Unal E, Koshland DE.** 2008. Sister chromatid cohesion: a simple concept with a complex reality. *Annu Rev Cell Dev Biol* **24:** 105–129.
64. **Remeseiro S, Losada A.** 2013. Cohesin, a chromatin engagement ring. *Curr Opin Cell Biol* **25:**63–71.
65. **Petes TD, Williamson DH.** 1994. A novel structural form of the 2 micron plasmid of the yeast *Saccharomyces cerevisiae*. *Yeast* **10:**1341–1345.
66. **Azaro MA, Landy A.** 2002. λ integrase and the λ int family, p 118–148. *In* Craig NL, Craigie R, Gellert M, Lambowitz AM (ed), *Mobile DNA II*. ASM Press, Washington, DC.
67. **Barre FX, Sherratt DJ.** 2002. Xer site-specific recombination: promoting chromosome segregation, p 149–161. *In* Craig NL, Craigie R, Gellert M, Lambowitz AM (ed), *Mobile DNA II*. ASM Press, Washington, DC.
68. **Biswas T, Aihara H, Radman-Livaja M, Filman D, Landy A, Ellenberger T.** 2005. A structural basis for allosteric control of DNA recombination by lambda integrase. *Nature* **435:**1059–1066.
69. **Grindley ND, Whiteson KL, Rice PA.** 2006. Mechanisms of site-specific recombination. *Annu Rev Biochem* **75:**567–605.
70. **Jayaram M, Grainge I, Tribble G.** 2002. Site-specific DNA recombination mediated by the Flp protein of *Saccharomyces cerevisiae*, p 192–218. *In* Craig NL, Craigie R, Gellert M, Lambowitz AM (ed), *Mobile DNA II*. ASM Press, Washington, DC.

71. **Rice PA.** 2002. Theme and variation in tyrosine recombinases: structure of a Flp-DNA complex, p 219–229. *In* Craig NL, Craigie R, Gellert M, Lambowitz AM (ed), *Mobile DNA II*. ASM Press, Washington, DC.
72. **Van Duyne GD.** 2002. A structural view of tyrosine recombinase site-specific recombination, p 93–117. *In* Craig NL, Craigie R, Gellert M, Lambowitz AM (ed), *Mobile DNA II*. ASM Press, Washington, DC.
73. **Grainge I, Jayaram M.** 1999. The integrase family of recombinase: organization and function of the active site. *Mol Microbiol* **33:**449–456.
74. **Gibb B, Gupta K, Ghosh K, Sharp R, Chen J, Van Duyne GD.** 2010. Requirements for catalysis in the Cre recombinase active site. *Nucleic Acids Res* **38:**5817–5832.
75. **Chen Y, Narendra U, Iype LE, Cox MM, Rice PA.** 2000. Crystal structure of a Flp recombinase-Holliday junction complex. Assembly of an active oligomer by helix swapping. *Mol Cell* **6:**885–897.
76. **Chen Y, Rice PA.** 2003. New insight into site-specific recombination from Flp recombinase-DNA structures. *Annu Rev Biophys Biomol Struct* **32:**135–159.
77. **Ma CH, Kwiatek A, Bolusani S, Voziyanov Y, Jayaram M.** 2007. Unveiling hidden catalytic contributions of the conserved His/Trp-III in tyrosine recombinases: assembly of a novel active site in Flp recombinase harboring alanine at this position. *J Mol Biol* **368:**183–196.
78. **Whiteson KL, Chen Y, Chopra N, Raymond AC, Rice PA.** 2007. Identification of a potential general acid/base in the reversible phosphoryl transfer reactions catalyzed by tyrosine recombinases: Flp H305. *Chem Biol* **14:**121–129.
79. **Arciszewska LK, Grainge I, Sherratt DJ.** 1997. Action of site-specific recombinases XerC and XerD on tethered Holliday junctions. *EMBO J* **16:**3731–3743.
80. **Blakely GW, Davidson AO, Sherratt DJ.** 1997. Binding and cleavage of nicked substrates by site-specific recombinases XerC and XerD. *J Mol Biol* **265:**30–39.
81. **Grainge I, Sherratt DJ.** Xer site-specific recombination. DNA strand rejoining by recombinase XerC. *J Biol Chem* **274:**6763–6769.
82. **Guo F, Gopaul DN, Van Duyne GD.** 1997. Structure of Cre recombinase complexed with DNA in a site-specific recombinase synapse. *Nature* **389:**40–46.
83. **Nunes-Duby SE, Tirumalai RS, Dorgai L, Yagil E, Weisberg RA, Landy A.** 1994. Lambda integrase cleaves DNA in cis. *EMBO J* **13:**4421–4430.
84. **Chen JW, Lee J, Jayaram M.** 1992. DNA cleavage in trans by the active site tyrosine during Flp recombination: switching protein partners before exchanging strands. *Cell* **69:**647–658.
85. **Yang SH, Jayaram M.** 1994. Generality of the shared active site among yeast family site-specific recombinases. The R site-specific recombinase follows the Flp paradigm. *J Biol Chem* **269:**12789–12796.
86. **Kimball AS, Lee J, Jayaram M, Tullius TD.** 1993. Sequence-specific cleavage of DNA via nucleophilic attack of hydrogen peroxide, assisted by Flp recombinase. *Biochem* **32:**4698–4701.
87. **Lee J, Jayaram M.** 1995. Functional roles of individual recombinase monomers in strand breakage and strand union during site-specific DNA recombination. *J Biol Chem* **270:**23203–23211.
88. **Lee J, Tonozuka T, Jayaram M.** 1997. Mechanism of active site exclusion in a site-specific recombinase: role of the DNA substrate in conferring half-of-the-sites activity. *Genes Dev* **11:**3061–3071.
89. **Conway AB, Chen Y, Rice PA.** 2003. Structural plasticity of the Flp-Holliday junction complex. *J Mol Biol* **326:**425–434.
90. **Krogh BO, Shuman S.** 2000. Catalytic mechanism of DNA topoisomerase IB. *Mol Cell* **5:**1035–1041.
91. **Stivers JT, Jagadeesh GJ, Nawrot B, Stec WJ, Shuman S.** 2000. Stereochemical outcome and kinetic effects of Rp- and Sp-phosphorothioate substitutions at the cleavage site of vaccinia type I DNA topoisomerase. *Biochem* **39:**5561–5572.
92. **Tian L, Claeboe CD, Hecht S, Shuman S.** 2003. Guarding the genome: electrostatic repulsion of water by DNA suppresses a potent nuclease activity of topoisomerase IB. *Mol Cell* **12:**199–208.
93. **Tian L, Claeboe CD, Hecht SM, Shuman S.** 2005. Mechanistic plasticity of DNA topoisomerase IB: phosphate electrostatics dictate the need for a catalytic arginine. *Structure* **13:**513–520.
94. **Ma CH, Rowley PA, Maciaszek A, Guga P, Jayaram M.** 2009. Active site electrostatics protect genome integrity by blocking abortive hydrolysis during DNA recombination. *EMBO J* **28:**1745–1756.
95. **Rowley PA, Kachroo AH, Ma CH, Maciaszek AD, Guga P, Jayaram M.** 2010. Electrostatic suppression allows tyrosine site-specific recombination in the absence of a conserved catalytic arginine. *J Biol Chem* **285:**22976–22985.
96. **Chen Y, Rice PA.** 2003. The role of the conserved Trp330 in Flp-mediated recombination. Functional and structural analysis. *J Biol Chem* **278:**24800–24807.
97. **Kachroo AH, Ma CH, Rowley PA, Maciaszek AD, Guga P, Jayaram M.** 2010. Restoration of catalytic functions in Cre recombinase mutants by electrostatic compensation between active site and DNA substrate. *Nucleic Acids Res* **38:**6589–6601.
98. **Ma CH, Kachroo AH, Macieszak A, Chen TY, Guga P, Jayaram M.** 2009. Reactions of Cre with methylphosphonate DNA: similarities and contrasts with Flp and vaccinia topoisomerase. *PLoS One* **4:**e7248.
99. **Fan HF, Ma CH, Jayaram M.** 2013. Real-time single-molecule tethered particle motion analysis reveals mechanistic similarities and contrasts of Flp site-specific recombinase with Cre and λ Int. *Nucleic Acids Res* **41:**7031–7047.
100. **Fan HF.** 2012. Real-time single-molecule tethered particle motion experiments reveal the kinetics and mechanisms of Cre-mediated site-specific recombination. *Nucleic Acids Res* **40:**6208–6222.
101. **Mumm JP, Landy A, Gelles J.** 2006. Viewing single lambda site-specific recombination events from start to finish. *EMBO J* **25:**4586–4595.

102. **Pinkney JN, Zawadzki P, Mazuryk J, Arciszewska LK, Sherratt DJ, Kapanidis AN.** 2012. Capturing reaction paths and intermediates in Cre-loxP recombination using single-molecule fluorescence. *Proc Natl Acad Sci USA* **109:**20871–20876.

103. **Velmurugan S, Mehta S, Jayaram M.** 2003. Selfishness in moderation: evolutionary success of the yeast plasmid. *Curr Top Dev Biol* **56:**1–24.

104. **Grainge I, Buck D, Jayaram M.** 2000. Geometry of site alignment during Int family recombination: antiparallel synapsis by the Flp recombinase. *J Mol Biol* **298:**749–764.

105. **Harshey RM, Jayaram M.** 2006. The Mu transpososome through a topological lens. *Crit Rev Biochem Mol Biol* **41:**387–405.

106. **Jayaram M, Harshey RM.** 2009. Difference topology: analysis of high-order DNA-protein assemblies, p 139–158. *In* Benham CJ, Harvey S, Olson WK, Sumners DW, Swigon D (ed), *Mathematics of DNA Structure, Function and Interactions. The IMA Volumes in Mathematics and Its Applications*, **Vol 150**. Springer, Dordrecht, The Netherlands.

107. **Garcia-Otin AL, Guillou F.** 2006. Mammalian genome targeting using site-specific recombinases. *Front Biosci* **11:**1108–1136.

108. **Turan S, Galla M, Ernst E, Qiao J, Voelkel C, Schiedlmeier B, Zehe C, Bode J.** 2011. Recombinase-mediated cassette exchange (RMCE): traditional concepts and current challenges. *J Mol Biol* **407:**193–221.

109. **Turan S, Zehe C, Kuehle J, Qiao J, Bode J.** 2013. Recombinase-mediated cassette exchange (RMCE): a rapidly-expanding toolbox for targeted genomic modifications. *Gene* **515:**1–27.

110. **Baldwin EP, Martin SS, Abel J, Gelato KA, Kim H, Schultz P, Santoro SW.** 2003. A specificity switch in selected Cre recombinase variants is mediated by macromolecular plasticity and water. *Chem Biol* **10:** 1085–1094.

111. **Bolusani S, Ma CH, Paek A, Konieczka JH, Jayaram M, Voziyanov Y.** 2006. Evolution of variants of yeast site-specific recombinase Flp that utilize native genomic sequences as recombination target sites. *Nucleic Acids Res* **34:**5259–5269.

112. **Buchholz F, Stewart AF.** 2001. Alteration of Cre recombinase site specificity by substrate-linked protein evolution. *Nat Biotechnol* **19:**1047–1052.

113. **Santoro SW, Schultz PG.** 2002. Directed evolution of the site specificity of Cre recombinase. *Proc Natl Acad Sci USA* **99:**4185–4190.

114. **Sarkar I, Hauber I, Hauber J, Buchholz F.** 2007. HIV-1 proviral DNA excision using an evolved recombinase. *Science* **316:**1912–1915.

115. **Voziyanov Y, Konieczka JH, Stewart AF, Jayaram M.** 2003. Stepwise manipulation of DNA specificity in Flp recombinase: progressively adapting Flp to individual and combinatorial mutations in its target site. *J Mol Biol* **326:**65–76.

116. **Hunter NL, Awatramani RB, Farley FW, Dymecki SM.** 2005. Ligand-activated Flpe for temporally regulated gene modifications. *Genesis* **41:**99–109.

117. **Kellendonk C, Tronche F, Monaghan AP, Angrand PO, Stewart F, Schutz G.** 1996. Regulation of Cre recombinase activity by the synthetic steroid RU 486. *Nucleic Acids Res* **24:**1404–1411.

118. **Zhang DJ, Wang Q, Wei J, Baimukanova G, Buchholz F, Stewart AF, Mao X, Killeen N.** 2005. Selective expression of the Cre recombinase in late-stage thymocytes using the distal promoter of the Lck gene. *J Immunol* **174:**6725–6731.

Plasmids—Biology and Impact in Biotechnology and Discovery
Edited by Marcelo E. Tolmasky and Juan C. Alonso

doi:10.1128/microbiolspec.PLAS-0027-2014

Patrick Forterre,[1,2] Mart Krupovic,[1]
Kasie Raymann,[1] and Nicolas Soler[3,4]

Plasmids from *Euryarchaeota* 20

INTRODUCTION

Archaea were confused with bacteria, under the term *prokaryotes*, until their originality was recognized in 1977 by Carl Woese and his collaborators of the "Urbana school" (1, 2). The classification of all cellular organisms into three domains based on rRNA was later confirmed by comparative genomic analyses that have shown that most universal proteins exist in three versions (*sensu* Woese), one in each domain: *Archaea*, *Bacteria*, and *Eukarya* (3). At the phenotypic level, archaea strikingly resemble bacteria in terms of size and shape, chromosome structure, and compact gene organization. However, when inspected at the molecular or biochemical level, archaea are either unique, for instance in terms of their lipids (4), or rather similar to eukaryotes (5, 6). Archaea resemble eukarya with respect to both their informational systems (DNA replication, transcription, translation) and their operational systems (ATP production, protein secretion, vesicle formation, cytoskeleton, protein modification machinery) (7). However, we will see in this chapter that archaeal plasmids (and mobilome in general) have a strong bacterial flavor, a paradox that remains to be explained (7).

Archaea encompass a collection of very diverse microorganisms that exhibit various phenotypes, use different types of metabolism, and can live in any kind of environment, from the coldest to the hottest places on our planet (8, 9). They were originally divided into two major phyla: *Crenarchaeota* and *Euryarchaeota*, based on phylogenetic analyses of rRNA sequences (10). This division has been validated by comparative genomic analyses that revealed major differences between these two phyla. For instance, DNA polymerase D, an archaea-specific DNA replicase, is present in *Euryarchaeota* but not in *Crenarchaeota*. All *Crenarchaeota* (from the Greek *Crenos*, for origin) are hyperthermophilic microbes thriving in volcanic hot springs. In contrast, *Euryarchaeota* (from the Greek *Euryos*, for diversity) exhibit most of the known phenotypes in *Archaea* (see below). Recently, a third major phylum of *Archaea*, the *Thaumarchaeota*, has been identified through phylogenetic analyses based on ribosomal proteins (11, 12). These archaea, which also contain a DNA polymerase of the D family, exhibit additional eukaryote-like features compared to *Crenarchaeota* and *Euryarchaeota*, such as the presence of a type IB DNA topoisomerase (13). Only a few genomes of *Thaumarchaeota* are presently available, and so far no plasmids have been detected in this phylum. However, a provirus corresponding to head and tailed viruses (*Caudovirales*) has been detected in the genome of the thaumarchaeon *Nitrososphaera viennensis*, an ammonia-oxidizing mesophilic microorganism (14).

[1]Institut Pasteur, 75015 Paris, France; [2]Institut de Génétique et Microbiologie, Université Paris-Sud 11, UMR 8621 CNRS, 91405 Orsay, France; [3]Université de Lorraine, DynAMic, UMR1128, Vandoeuvre-lès-Nancy, France; [4]INRA, DynAMic, UMR1128, Vandoeuvre-lès-Nancy, France.

Many viruses and plasmids have been detected and characterized in *Crenarchaeota* and *Euryarchaeota*. *Crenarchaeota* are infected by a great diversity of viruses producing virions with unique morphologies that have not yet been found in *Euryarchaeota* (15, 16). In contrast, *Euryarchaeota* are infected by head and tailed viruses (*Caudovirales*) that have never been observed among viruses infecting *Crenarchaeota* (17) but are widely found among *Bacteria*. In this article, we will focus on plasmids from *Euryarchaeota*.

DIVERSITY OF THE *EURYARCHAEOTA*

Euryarchaeota include microorganisms with very diverse phenotypes: halophiles, which are mesophilic aerobes thriving in very high salt biotopes (up to 4 M NaCl) and accumulate high concentrations of KCl (up to 4-5 M) and sometimes $MgCl_2$ (up to 1 M) in their cytoplasm; thermoacidophiles, which are heterotrophic aerobes living at moderately high temperatures (45 to 65°C) in very acidic environments (pH 1 to 3), with some of them even growing at pH 0; anaerobic hyperthermophiles living in terrestrial and submarine volcanic hot springs; and methanogens, which are strict anaerobes characterized by a unique type of metabolism, the production of methane from hydrogen and CO_2 (methanogenesis). Methanogens are especially versatile, with some species living in cold and dry Antarctic desert valleys, whereas others (e.g., *Methanopyrus kandleri*) thrive in hydrothermal vents at temperatures up to 110°C. Methanogens also inhabit animal guts and form a sizable component of the human microbiome (18). *Euryarchaeota* also include nanosized archaea that live as parasites of hyperthermophilic organisms, such as *Nanoarchaea* (19, 20) or as free-living organisms in acidic or salty environments (21–23). Finally, several lineages of uncharacterized and uncultured *Euryarchaea* have been detected by environmental PCR in both terrestrial and marine biotopes (24).

Figure 1 illustrates a tree of *Euryarchaeota* based on concatenated DNA replication proteins and rooted using homologous proteins from *Crenarchaea* and *Thaumarchaea* (25). In this analysis, which is roughly congruent with those based on ribosomal proteins (6), *Euryarchaeota* are clearly divided into two major subphyla, hereafter called groups I and II *Euryarchaeota*. Group I *Euryarchaeota* includes hyperthermophiles of the order *Thermococcales* and methanogens of the orders *Methanopyrales*, *Methanococcales*, and *Methanobacteriales* (corresponding to group I methanogens; see reference 26), which all use carbon dioxide reduction for methane formation. *Euryarchaeota* of group II include *Haloarchaea*; hyperthermophiles of the order *Archaeoglobales*; thermoacidophiles of the order *Thermoplasmatales*; methanogens of the orders *Methanomicrobiales*, *Methanosarcinales*, and *Methanocellales* (corresponding to group II methanogens); and the recently described *Methanomassilicoccales* (27); most of these methanogens use acetate or methylated compounds for methane formation.

Notably, all group II *Euryarchaeota* contain DNA gyrase, whereas this enzyme is missing in group I *Euryarchaeota*, as well as in all other *Archaea* (*Crenarchaeota*, *Thaumarchaeota*, and *Koryarchaeota*) (25). The archaeal DNA gyrases branch within *Bacteria* in a phylogenetic tree of type IIA DNA topoisomerases (25). DNA gyrase was thus most likely recruited from *Bacteria* in an ancestor of group II *Euryarchaeota* (Fig. 1). After this transfer, all group II euryarchaeotes retained DNA gyrase, and this enzyme became essential, as shown by their sensitivity to the DNA gyrase inhibitor, novobiocin (28–30). As a consequence of DNA gyrase acquisition, plasmids from group II *Euryarchaeota* are negatively supercoiled, as in *Bacteria* (31, 32). In contrast, plasmids from hyperthermophilic group I *Euryarchaeota* lacking DNA gyrase are relaxed (33). This is also the case for the plasmid pME2011 from the moderate thermophile *M. thermoautotrophicum*, which lacks both gyrase and reverse gyrase (an enzyme present in all hyperthermophiles and some thermophiles and that produces positive supercoiling; see the section "Plasmids and Genetic Tools for *Euryarchaeota*") (33). The unusual DNA topological state of relaxed plasmids isolated from hyperthermophilic species lacking DNA gyrase is shown in Fig. 1, in which preparations of the same recombinant shuttle plasmid vector extracted either from the bacterium *Escherichia coli* or from the euryarchaeon *Thermococcus nautili* were run in parallel on agarose gel (34). This example illustrates the critical role of DNA gyrase in DNA supercoiling *in vivo*.

Many plasmids have now been described in different groups of *Euryarchaeota* (Fig. 2) (a complete list of archaeal plasmids with identification numbers can be found at the NCBI site: http://mirrors.vbi.vt.edu/mirrors/ftp.ncbi.nih.gov/genomes/IDS/Archaea.ids). They resemble bacterial plasmids in terms of size (from small plasmids encoding only one gene up to large megaplasmids) and replication mechanisms (rolling circle or theta). Besides these free plasmids, many plasmids or related viruses are integrated in archaeal genomes (14, 35–37). In all cases studied, these plasmids and viruses encode typical tyrosine recombinases belonging to either type I

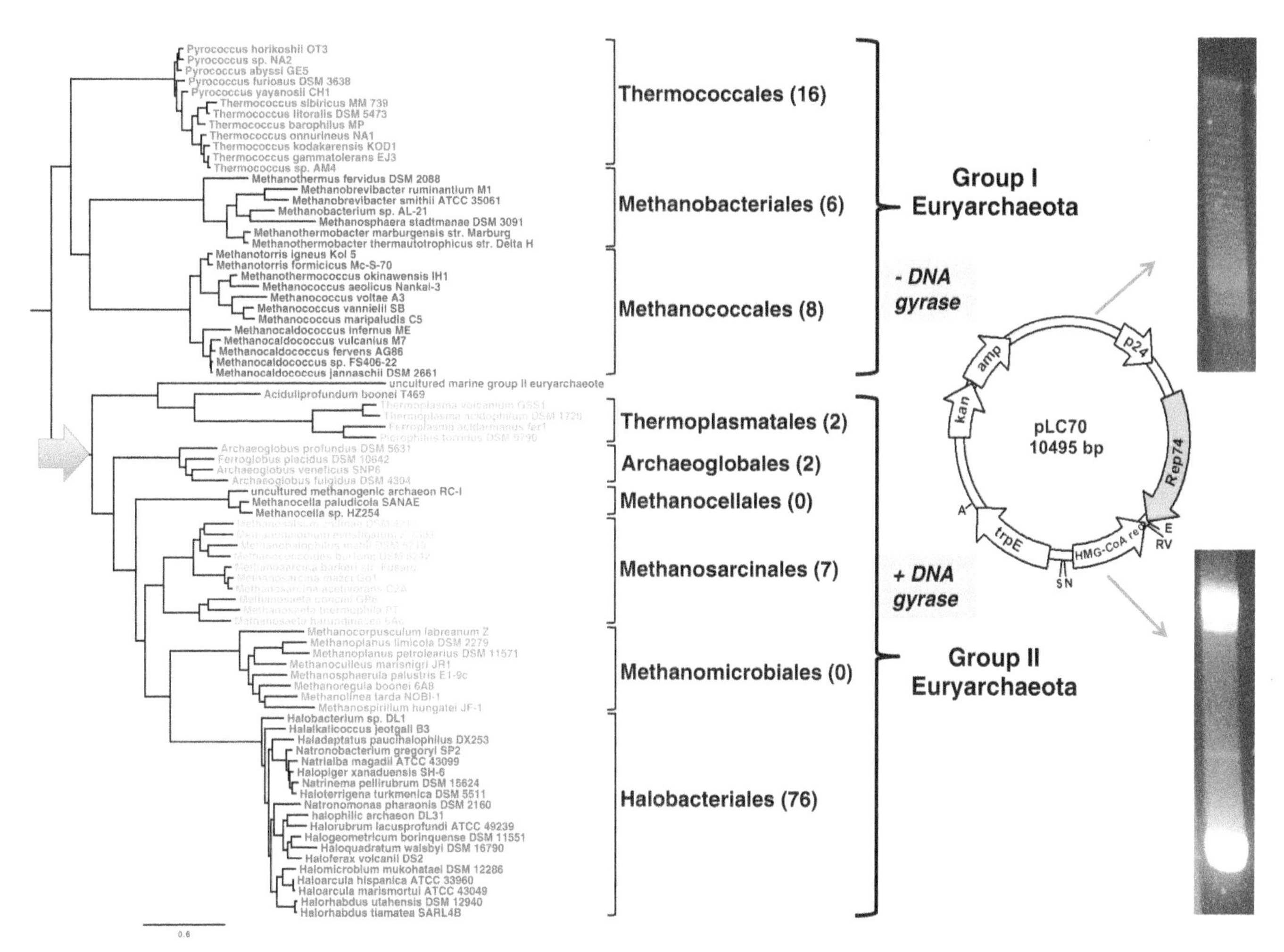

Figure 1 Phylogeny of *Euryarchaeota.* The tree is based on the concatenation of core DNA replication proteins present in the last archaeal common ancestor (adapted from reference 25); for each order, the number of identified plasmids is indicated in brackets. All group I *Euryarchaeota* lack DNA gyrase (similarly to *Crenarchaeota* and *Thaumarchaeota*), whereas all group II *Euryarchaeota* contain a DNA gyrase gene of bacterial origin. The blue arrow indicates the acquisition of this gyrase gene. Right panel: topology of plasmids present in organisms with and without DNA gyrase. The difference in topology is illustrated by comparing the electrophoretic mobility of the same plasmid, pLC70, a derivative of pTN1 from *T. nautili*, purified from *T. kodakaraensis* (lacking DNA gyrase, upper picture) or from *E. coli* (a bacterium containing DNA gyrase); adapted from reference 34.
doi:10.1128/microbiolspec.PLAS-0027-2014.f1

or II archaeal integrases. The prototype of type I is the integrase of fusellovirus SSV1 (38, 39). The integrated elements encoding this type of integrase are usually inserted into tRNA genes and are framed by the N- and C-terminal moieties of the SSV1-like integrase gene. The disruption of the integrase gene upon element integration is due to the location of the *attP* recombination site within the integrase gene itself. It is unclear how these integrated elements can be reactivated in the absence of detectable excisionase. In contrast, the mode of integration of plasmids and viruses encoding the type II integrases is indistinguishable from that characterized for various bacterial mobile elements, where the *attP* site is located outside of the integrase gene.

PLASMIDS FROM *THERMOCOCCALES* AND OTHER GROUP I *EURYARCHAEOTA*

Most well-characterized plasmids from group I *Euryarchaeota* have been isolated from *Thermococcales* (Tables 1 and 2). A few other plasmids have been isolated from different methanogens. Most other known

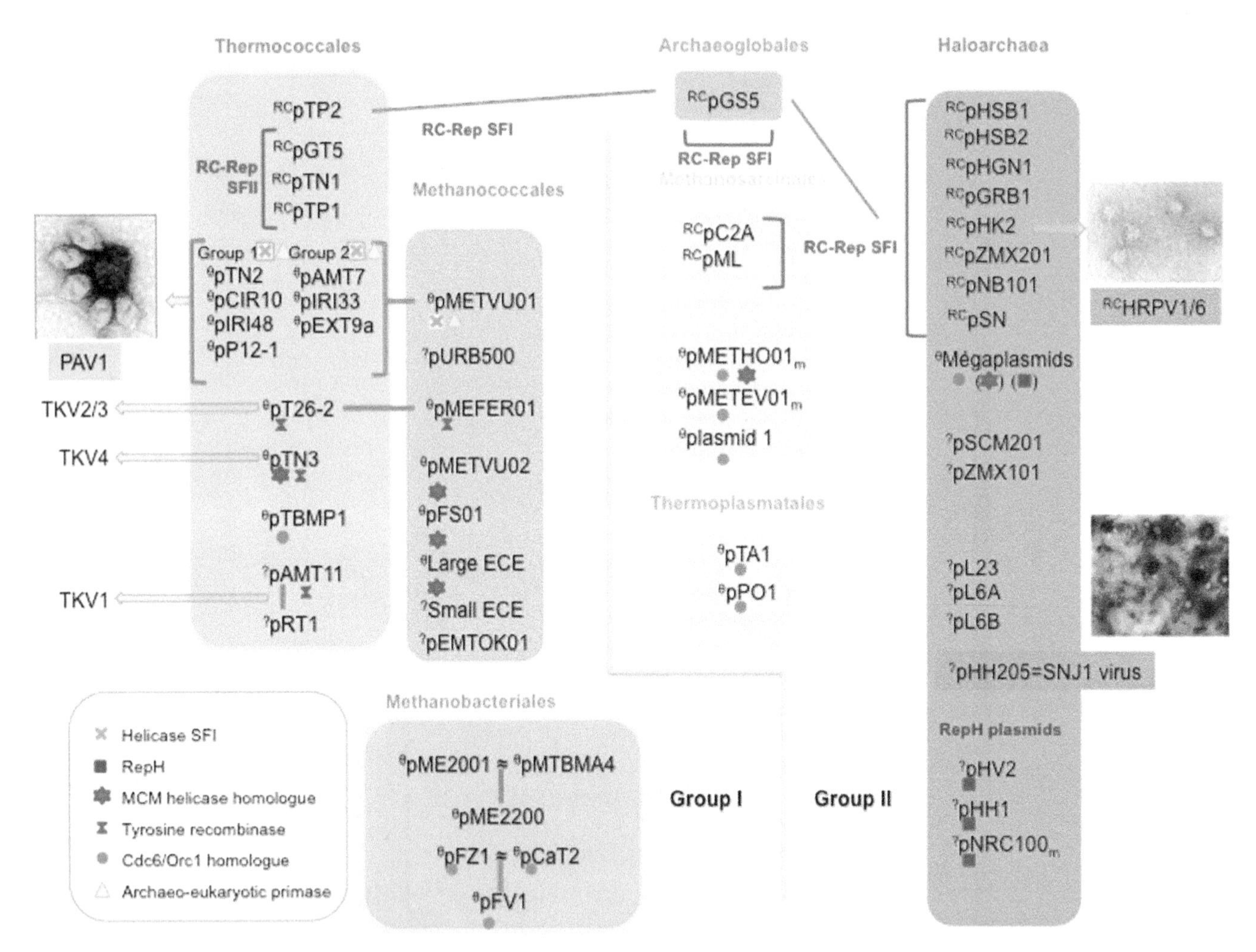

Figure 2 The wonderful world of euryarchaeal plasmids. General scheme representing most of the plasmids cited in the text. Relationships with viruses or other mobile elements are marked with green arrows. Dark grey lines link plasmids that are evolutionarily related. Keynote replication proteins are indicated by different colored symbols (Cf legend in the bottom-left box). The mode of replication of each plasmid is indicated in front of the plasmid names: RC, rolling-circle replication; θ, theta mode of replication; ?, undetermined or litigious mode of replication. Refer to tables for the size and host of the plasmids. doi:10.1128/microbiolspec.PLAS-0027-2014.f2

plasmids in this group, as well as several integrated mobile elements, have been detected in genome sequencing projects but have not been further characterized. We will first focus here on plasmids and viruses from *Thermococcales* that have been studied in detail and are still under active investigation in several laboratories.

Thermococcales (genera *Thermococcus, Pyrococcus,* and *Paleococcus*) are heterotrophs, anaerobes, growing in a temperature range between 70 and 105°C in either terrestrial or marine hot springs. They are relatively easy to cultivate and are typically the first to pop up in enrichment cultures derived from hot spring samples (40). Furthermore, efficient and versatile genetic tools have been developed for *Thermococcus kodakaraensis* (41, 42) and are on the way for a few others (43). As a consequence, many laboratories use them as model organisms to decipher molecular mechanisms that allow cells to live at extremely high temperatures or to exploit their biotechnological potential (44, 45). *Thermococcales* are indeed the source of robust "extremozymes," and some authors hope to use them as cell factories, for instance for hydrogen production (46). About one quarter to one half of *Thermococcales* strains isolated from deep-sea hydrothermal vents were found to harbor at least one extrachromosomal element (plasmid or extrachromosomal viral genome)

Table 1 Rolling circle plasmids from *Euryarchaeota*

Name	Size (kb)	Rep name	Rep size (aa)	Rep (GI)	Related viruses	Host strains	Reference
Rep-RC Superfamily II (Y)							
pGT5	3.9	Rep 75	654	2826813		*Pyrococcus abyssi* GE-5	54
pTN1	3.4	Rep 74	641	153012336		*T. nautili* 30-1	55
pTP1	3.1	Orf3	606	500998695		*Thermococcus prieurii*	56
Rep-RC Superfamily I (Y...Y)							
pTP2	2.0	RepTP2	252	499110067	HRPV	*T. prieurii*	56
pGS5(≈pArcpr01)	2.8	Rep	463	251781464	HRPV	*Archaeoglobus profundus*	31
pML	2.2	Orf1	328	42761455	HRPV	*Methanohalophilus mahii*	
pC2A	5.5		437	20091430	HRPV	*Methanosarcina acetivorans* C2A	126
pHK2	10.8	Orf1	544	410688694	HPPV	*Haloferax lucentense* Aa2	103
pZMX201	1.7	Rep	386	64204362	HRPV	*Natrinema* sp. CX2021	108
pNB101	2.5	Rep		38640510	HRPV	*Natronobacterium* sp. AS-7091	111
pGRB	1.7	Rep	316	140038	HRPV	*Halobacterium salinarium* sp. GRB	106
pHSB1	1.7			43672		*H. salinarium* sp. SB3	104
pHSB2	1.7			10954592			107
pHGN1	1,7	Rep	328	10954594	HRPV	*Halobacterium* sp. GN101	105
pSN	1,7	Rep	330	270208375	HRPV	*Haloterrigena thermotolerans* H13	

(40). (*Thermococcales* from terrestrial hot springs have not yet been screened for viruses or plasmids.)

Up until now, only two viruses infecting *Thermococcales* have been isolated and characterized: the virus PAV1 infecting *Pyrococcus abyssi* and the virus TPV1, infecting *Thermococcus prieurii* (47–49). The circular double-stranded DNA genomes of PAV1 (18 kb) and TPV1 (21.5 kb) share only three genes (49, 50). These viruses produce spindle-shaped virions similar to those of fuselloviruses infecting *Sulfolobus* species (crenarchaeota) and that of virus His1 infecting *Haloarcula hispanica* (euryarchaeota) (51). All these viruses can be grouped into an archaea-specific viral family, *Fuselloviridae*, based on the presence of related major capsid proteins and overlapping gene content (50).

The viruses PAV1 and TPV1 encode several proteins with homologues in plasmids and/or mobile elements integrated in the genomes of *Thermococcales* and *Methanococcales* (35, 49) (see below). Notably, TPV1 coexist in *T. prieurii* with two small rolling-circle plasmids, pTP1 and pTP2 (see below). However, none of these plasmids were observed in DNA preparations extracted directly from TPV1 viral particles. This is different from the situation observed with crenarchaeal fuselloviruses, which can package resident plasmids in their virions. Interestingly, TPV1 can infect different strains of *Thermococcus*, including the genetically tractable strain *T. kodakaraensis* that can be transformed with exogeneous plasmids (15). TPV1 and *T. prieurii* could thus become an interesting host-virus experimental system to study the tripartite plasmid-virus-host interactions in *Euryarchaeota*.

Several integrated mobile elements (TKV1, TKV2, TKV3, and TKV4) have been described as proviruses in *T. kodakaraensis* (37). TKV stands for T. kodakaraensis virus-related region, although it is unknown if the elements are indeed proviruses or integrated plasmids. We will see below that all of them are homologous to free plasmids or viral genomes that have been detected in other *Thermococcus* species (34, 35, 52, 53).

Rolling Circle Plasmids from *Thermococcales*

Four small rolling circle (RC) plasmids have been described in *Thermococcales* (Table 1): the plasmid pGT5 (3.4 kb) from *P. abyssi* GE5 (54), pTN1 (3.4 kb) from *T. nautili* sp. 30-1 (55), and the plasmids pTP1 (3.1 kb) and pTP2 (2 kb) from *T. prieurii* (56). The plasmids pTP1 and pTP2 coexist in the same cell without any apparent antagonistic effect on each other.

The plasmids pGT5, pTN1, and pTP1 share homologous Rep proteins (Rep75, Rep74, and RepTP1, respectively) containing the three amino acid motifs (motifs 1, 2, and 3) characteristic of the RC-Rep initiator proteins (57). PSI-BLAST analysis showed that these proteins are distantly related to transposases encoded by bacterial insertion sequence (IS) elements using the RC mechanism for transposition (families IS91, IS801, and IS1294) (55). The RC mode of replication was confirmed by the detection of circular single-stranded intermediates in *P. abyssi* cells for pGT5 (54) and by

Table 2 Plasmids from group I *Euryarchaeota* replicating (putatively) via the theta mechanism

Plasmid name	Size (kb)	Replication proteins, gi (aa size)	Viral homologue	Host strain	Reference
pTN2	13	P: PriSL-PolPrim, 295698814 (923) H: SFI helicase, 9112455 (569), 295698803 (569) h: HTH DBP, 295698798 (169)	PAV1	*T. nautili* 30-1	35
pCIR10	13.3	(P-H-h) PriSL type	PAV1	*Thermococcus* sp. CIR10	63
pP12-1	12.2	(P-H-h) PriSL type	PAV1	*Pyrococcus* sp. P12-1	35
pIR148	12.9	(P-H-h) PriS type	PAV1	*Thermococcus* sp. IR148	63
pEXT9	10.5	(P-H-h) PriL type	PAV1	*Thermococcus* sp. EXT9	63
pMETVU01	10.7	(P-H-h) PriL type	PAV1	*Methanocaldococcus vulcanius* M7	63
pAMT7	8.5	(P-H-h) PriL type	PAV1	*Thermococcus* sp. AMT7	63
pIRI33	11	(P-H-h) PriL type	PAV1	*Thermococcus* sp. IRI33	63
pRT1	3.3	Rep 63, 15004884 (543)	« TKV1 »[a]	*Pyrococcus* sp. JT1	72
pAMT11	20.5	Rep72, 257070385 (620)	« TKV1 »[a]	*Thermococcus* sp. AMT11	52
pT26-2	21.5	t26-22p AAA+ ATPase, 295698837 (701)	« TKV2/V3 »[a]	*Thermococcus* sp. 26-2	35
pMEFER01	27.2	MCM	« TKV2/V3 »[a]	*Methanocaldococcus fervens* AG86	64
pTN3	18.3	MCM	« TKV4 »[a] PRD1/Adenoviruses	*T. nautili* 30-1	53
pTBMP1	55.5	Cdc6,HTH DBP 331746804 (163)		*Thermococcus barophilus*	65
pURB500 ≈pMMC501	8.3	Orf1 AAA+ ATPase (MCM) 10954554 (870) Orf3 HTH DBP 10954556 (252)		*Methanococcus maripaludis*	78
pMETVU02	4.7	MCM		*M. vulcanius* M7	
pFS01	12.1	MCM		*Methanocaldococcus sp.* FS406-22	
Large ECE	58.4	MCM, ParA		*Methanocaldococcus jannashii*	77
Small ECE	18.6			*M. jannashii*	77
pEMTOK01	14.9	Helicase domain, 336120995 (1015)		*Methanothermococcus okinawensis*	
psiM1	26.1		psiM2	*Methanothermobacter marburgensis*	80
		Orf3, Rep p2 ATPase Walker		*M. marburgensis* Marburg	82
pMTBMA4	4.4	304315533 (804)			
≈pME2001	4.4	10954597 (531)			
pME2200	4.4	10954596 (273)		Uncultured archaeon	
pMTB1	4.4	10954575 (801) 556862601 (801)			
pFZ1 ≈pCaT2	11	Cdc6		*Methanothermobacter thermoautotrophicum*	84, 85
pFV1	13.5	Cdc6		*M. thermoautotrophicum*	84

[a]« », TKV are only putative viruses.

demonstration of the nicking activity of both Rep75 of pGT5 and Rep74 of pTN1 (55, 58). RC-Rep proteins are site-specific endonuclease ligases that initiate DNA replication by introducing a nick at the double-stranded replication origin (*dso*), producing a free 3′ OH that serves as the primer for the host DNA polymerase (59). This reaction is catalyzed by a conserved tyrosine residue present in motif 3. The RC-Rep proteins of pGT5, pTN1, and pTP1 contain a single tyrosine in motif 3, and can therefore be classified as members of the RC-Rep superfamily II (57). The Rep75 protein has been extensively characterized *in vitro* (60). The protein cleaves specifically an oligonucleotide containing an 11-bp sequence of the *dso* found within the *rep75* gene. Rep 75 also exhibits DNA ligase, type I DNA topoisomerase activity *in vitro*, and an unusual deoxynucleotidyl transferase activity (58, 60). Interestingly, the Rep proteins of pGT5 and pTN1 are much larger than typical RC proteins (74 to 75 kDa instead of about 30 to 40 kDa). Sequence analysis suggests that they are formed by the fusion of a central RC initiator module with two additional modules of uncharacterized function in the N- and C-termini, whereas pTP1 only contains the N-terminal module (55, 56).

In addition to their Rep proteins, pGT5 and pTN1 encode unrelated proteins of 46 and 24 kDa (p46 and p24), respectively (54, 55). The p24 protein harbors a zinc binding domain, a highly charged region, and a hydrophobic segment. A p24 truncated soluble protein (lacking the hydrophobic segment) was purified and shown to bind double-stranded and single-stranded DNA without sequence specificity (55). Deletion of the p24 gene has no effect on the stability of a shuttle vector derived from pTN1 (61), indicating that this protein is not essential for plasmid replication and/or segregation. Notably, homologues of the p24 protein are encoded by two elements from Thermococcales replicating via the theta mode, the plasmid pTN3, and the integrated mobile element TKV4. It has been suggested, but not demonstrated, that p24 could facilitate association of the pTN1 plasmid with membrane vesicles produced by *Thermococcales* (55). In agreement with this hypothesis, membrane vesicles produced by *T. nautili* are enriched in plasmids pTN1 and pTN3, compared to the third *T. nautili* plasmid, pTN2, which does not encode a p24 homologue (53).

Surprisingly, in addition to its complete RC-Rep gene, pTP1 harbors on the opposite strand two open reading frames (ORFs) encoding "mini-Reps" that match the central and 3′-distal regions of the main pTP1 RC-Rep gene, respectively (56). Sequence analysis suggests that these two mini-Rep fragments originated from an ancestral full Rep gene that was split in the course of a recombination event between two ancestral pTP1 plasmids, leading to integration of a new RC-Rep gene copy into the preexisting one. It was suggested that this integration was possibly catalyzed by RepTP1 itself, considering that this protein exhibits sequence similarities with the RC transposases (56). One of the two mini-Reps harbors the RC motives, but with mutations in conserved amino acid positions, indicating that it cannot be involved in the initiation of RC replication. However, sequence analysis has shown that this mini-Rep evolves under strong purifying selection, indicating that it is somehow involved in pTP1 propagation (56).

The Rep protein of pTP2 exhibits no similarity with those of pTP1 but with putative RC-Rep proteins of variable sizes (430 to 700 amino acids [aa]) encoded by several plasmids, viruses, and integrated mobile elements of diverse euryarchaeota (both groups I and II) (Table 1, Fig. 3). Unlike RC-Rep proteins of pGT5-related plasmids, all these Rep proteins contain two tyrosines in motif 3 and can therefore be classified as members of the RC-Rep superfamily I (57). Thus, these RC-Reps are more closely related to bacterial homologues, such as the ΦX174 protein A, than to RC-Rep of superfamily II from *Euryarchaea*. However, the Rep protein of pTP2 is much shorter (252 aa) than other members of this superfamily, lacking a variable N-terminal domain of its euryarchaeal homologues.

All archaeal Rep proteins of the RC-Rep superfamily I are evolutionarily related but very divergent. They can only be aligned around the three conserved RC motifs and in a short region (motif 4) located after motif 3. Notably, motif 4 of the pTP2 RC-Rep proteins share specific conserved amino acids with RC-Rep proteins from the small plasmids pGS5 from *Archaeoglobus profundus* and pC2A from *Methanosarcina acetivorans* (Fig. 3) and with Rep proteins of the viruses HRPV1 and HRPV6 infecting halophilic archaea that belong to the recently proposed family *Pleolipoviridae* (56, 62). A cladogram based on a concatenation of amino acid sequences of regions surrounding motifs 1 to 4 (Fig. 3) shows that these proteins form a subgroup of RC-Rep superfamily I (hereafter called pGS5-like, pGS5 being the first plasmid to have been characterized in that subgroup). They are clearly separated from another subgroup of RC-Rep superfamily I from small haloarchaeal plasmids, hereafter called the pGRB-like subgroup, and to RC-Rep proteins of the plasmid pML from *Methanohalophilus mahii* and the pleolipovirus HRPV2 (Fig. 3). Iterative Psi-BLAST analyses indicate that all these proteins exhibit weak similarities between each other but also with cellular proteins widespread in *Archaea*, including *Sulfolobales*. These data suggest the existence of a large family of RC-Rep initiator proteins of the superfamily I, with members widespread in *Euryarchaeota* and including all presently known archaeal RC plasmids, except those of class II, until now only present in *Thermococcales*.

Thermococcales Plasmids That Replicate by the Theta Mode

Several plasmids that probably replicate by the theta mode have been isolated from *Thermococcales* (Table 2). A family of plasmids, whose prototype is the pTN2 plasmid (13 kb) from *T. nautili*, shares an extensive gene content with the PAV1 virus (35, 63). Another family of mobile elements is exemplified by the plasmid pT26-2 (21.5 kb) from *Thermococcus* sp. 26-2 (35, 64). Several mobile elements of this family corresponding to integrated forms have been discovered in genomes of *Thermococcales* and *Methanococcales* (35, 64). They correspond to the TKV2 and TKV3 integrated elements in the genome of *T. kodakaraensis* (37). Two other plasmids, pAMT11 (20.5 kb) and pTN3 (18.3 kb) have been associated with two different

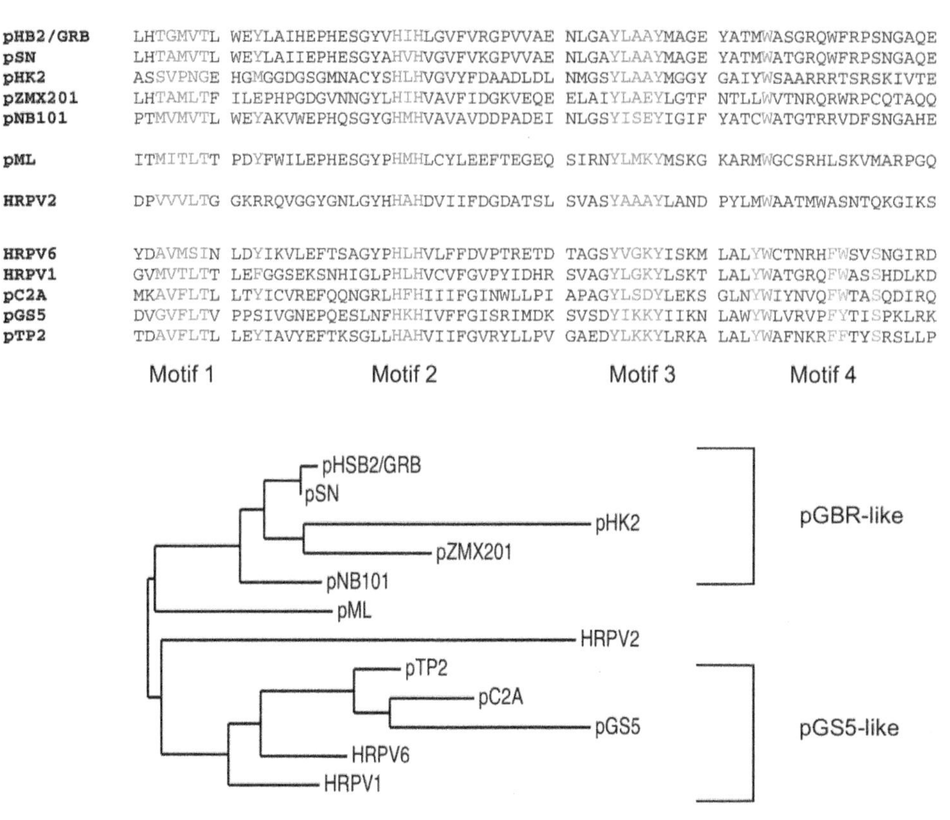

Figure 3 Analysis of archaeal RC-Rep proteins of superfamily I. Hand-made alignment of amino acid regions located around the four conserved previously known motifs (numbered 1 to 3) and the fourth motif detected in this analysis (motif 4). The cladogram was produced from this alignment (after concatenation) using the program Phylogeny.fr (166). doi:10.1128/microbiolspec.PLAS-0027-2014.f3

integrated elements in *T. kodakaraensis*, TKV1 and TKV4, respectively (52, 53). All these plasmids are prototypes of new plasmid families that are described below. Finally, a larger plasmid of 55.5 kb, pTBMP1, has been sequenced in the framework of the *Thermococcus barophilus* genome project (65).

Putative replication origins have been detected by the cumulative GC skew method in plasmids of the pTN2 and pT26-2 families (35, 63). These putative origins are AT rich, corresponding to regions of low stability, and contain multiple direct and inverted repeated sequences, similar to iterons present in the replication origins of bacterial theta plasmids. In all these plasmids, most genes are transcribed in the same direction. Notably, one or two genes transcribed in the opposite direction are often located close to the putative origin, suggesting that they could encode proteins involved in the control of plasmid replication (64). Interestingly, the pTBMP1 plasmid encodes a homologue of archaeal (eukaryote-like) DNA replication initiator protein Cdc6/Orc1, in agreement with a theta mode of replication and a homologue of the eukaryote-like transcription factor TFB, a rare occurrence of plasmid-encoded archaeal transcription factor (P. Forterre, unpublished observations)

The pTN2 Plasmid Family

Seven plasmids from *Thermococcales* strains and one plasmid from a *Methanococcales* strain can be grouped into a single family based on their gene content (35, 63). These plasmids, with sizes in the range of 8.5 to 13.3 kb, are characterized by the presence of a helicase of the superfamily I and a small protein with a winged helix-turn-helix (HTH) motif, both conserved in all members of the family (63). It is tempting to speculate that the small putative DNA binding protein is involved in the initiation of DNA replication in this plasmid family. All of them also encode different forms of a DNA polymerase/primase distantly related to Rep proteins encoded by some *Sulfolobus* plasmids. Secondary structure alignments have shown that these proteins form a new branch in the archaeo-eukaryotic primase (AEP) superfamily (66, 67).

The polymerase and primase activities of the pTN2 enzyme (PolpTN2) have been experimentally demonstrated (35, 67). This enzyme also exhibits nucleotidyl transferase activity, much like some other proteins of the AEP superfamily. PolpTN2 exhibits no sequence similarities with other proteins of the AEP superfamily, except for three amino acid motifs located in the N-terminal domain, which corresponds to the catalytic domain of cellular PriS-like primases. Secondary structure alignment revealed that PolpTN2 corresponds to the fusion of two domains homologous to PriS (the catalytic subunit) and PriL (the regulatory subunit) of archaeal and eukaryotic primases (67). Homologues of the complete PolpTN2 primase are present in the plasmids pP12-1 (12.2 kb) from *Pyrococcus* sp. P12-1 and pCIR10 from *Thermococcus* strain CIR10 (13.3 kb). Another plasmid, pIRI48 (12.9 kb) from *Thermococcus* strain IR148, harbors a primase domain (PriS-like) fused to a different C-terminal domain, whereas the three plasmids pEXT9a (10.5 kb), pAMT7 (8.5 kb), and pIRI33 (11 kb) from *Thermococcus* strains EXT9, AMT7, and IR133, respectively, only encoded the regulatory domain (63). The latter proteins could be used to recruit the host PriS for plasmid replication *in vivo*. In all cases, the genes encoding the small putative DNA binding protein, the helicase, and the primase are clustered together close to the putative replication origin, suggesting functional linkage. Indeed, the pTN2 helicase stimulates the priming activity of the pTN2 primase *in vitro* (Pierre Beguin, personal communication).

Interestingly, another member of the pTN2 plasmid family, pMETVU01 (10.7 kb), has been detected in *Methanocaldococcus vulcanius* M7 (63). This plasmid harbors a gene encoding a homologous protein of the C-terminal domain of the PolpTN2 primase. The four plasmids carrying this domain—pMETVU01, pEXT9a, pAMT7, and pIRI33—have been classified in a subfamily distinct from the subfamily including plasmid pTN2 based on gene content and phylogenetic analysis of their superfamily I helicases. This analysis groups together plasmids with the same type of primase and suggests that pMETVU01 was acquired by a member of *Methanoccales* from a *Thermococcus* species (63). Notably, seven genes of the virus PAV1 have homologues in plasmids of the pTN2 subfamily, and all of them are clustered in one half of the viral genome (63).

Until now, integrated forms of pTN2-like plasmids or PAV1 have not been detected; however, several close homologues of the pTN2 superfamily I helicase and primases that could be remnants of integrated mobile elements are present in the genomes of *Thermococcales* and *Methanococcales* (35, 67). Indeed, the superfamily I helicase has a homologue in the integrated element TGV1 of *Thermococcus gammatolerans*, whereas the primase is present on the element TGV2 (64), indicating that these replication proteins have been frequently exchanged between different mobile genetic elements of *Thermococcales*.

The pT26-2 (TKV2,3) Plasmid Family

A mid-sized plasmid, pT26-2 (21.5 kb), that probably replicates by the theta mode has been isolated from *Thermococcus* strain 26-2 (35). Notably, the plasmid pT26-2 encodes an ATPase that is replaced by a Mini Chromosome Maintenance (MCM) replicative helicase of cellular origin in the integrated element TKV2. Therefore, the ATPase encoded by pT26-2-like plasmids is likely a novel type of plasmid-encoded replicative helicase (35). Seven genes present in pT26-2 have homologues in the plasmid pMEFER01 (22.2 kb) from *Methanocaldococcus fervens* and are clustered in strings of genes that are syntenic in both plasmids (64). Several complete integrated elements related to these two plasmids have been detected in the genomes of some *Thermococcales* and other more reduced forms in the genomes of several *Methanococcales* (35). The two closely related elements correspond to TKV2 and TKV3. The plasmid pT26-2 shares 19 genes with TKV2 and TKV3 (including 7 common to pMEFER01) and 7 with another more distant element, TKV1 (itself related to pAMT11; see below) (35). Four genes are common to both pT26-2, TKV1, TKV2, and TKV3, indicating that gene exchange occurs between plasmid families.

The nine genes shared between pT26-2 and present both in TKV2/3 and in all homologous integrated elements of *Methanococcales* can be considered as core genes of the pT26-2 family (35). The two larger genes encoding proteins of unknown function are also present alone in a few other loci of *Methanococcales* genomes. The structure of one of these proteins (from pT26-2) has been solved, revealing three new folds in a single polypeptide (68). Phylogenetic analyses indicate that the conserved core genes of the pT26-2 family have coevolved with the plasmid hosts (35, 68).

A few genes present in plasmids and/or elements of the pT26-2 family have homologues in other plasmids or integrated elements, all within *Euryarchaeota* (except for their integrases that also have homologues among *Crenarchaeota*). The network of evolutionary relationships between pT26-2 and mobile elements sharing from two to nine genes with it has shown that all these elements coevolved with their hosts (35). Indeed, the number of genes common to different mobile elements in the network decreases with increasing phylogenetic distances between their hosts.

The pTN3 (TKV4) Plasmid Family

A third plasmid, pTN3 (18.3 kb), has been described in *T. nautili* (53). This plasmid encodes an integrase of the tyrosine recombinase superfamily and is present in the *T. nautili* genome in both episomal and integrated forms (69). The plasmid pTN3 is evolutionarily related to the integrated element TKV4 of *T. kodakaraensis* (35). Interestingly, pTN3 and TKV4 encode the two hallmark proteins of viruses belonging to the PRD1/adenovirus viral lineage: a capsid protein containing the double jelly roll fold and a putative genome packaging ATPase (70). Notably, this family of ATPases is specific to membrane-containing viruses and has never been previously found encoded by plasmids or cellular organisms. Viruses of the PRD1/adenovirus viral lineage are known to infect cells from the three domains of life and typically produce icosahedral capsids (70). Viral particles cannot be observed in cultures of *T. nautili* or *T. kodakaraensis*, suggesting that pTN3 and TKV4 correspond to defective viruses. Like many other archaeal mobile elements (71), pTN3 encodes an MCM-like helicase. Notably, pTN3 is strikingly enriched in preparations of membrane vesicles produced by *T. nautili*, suggesting that this element uses membrane vesicles, instead of virions, to propagate from cell to cell, leading to the concept of "viral vesicles" (53).

The pRT1/pAMT11 (TKV1) Plasmid Family

The plasmid pRT1 (3.3 kb) from *Pyrococcus* strain JT1 (72) (see above) resembles small RC plasmids of *Thermococcales* in size and genome organization. It was thus originally described as an RC plasmid. In particular, the Rep63 protein of pRT1 was first described as homologous to the RC-Rep75 from pGT5 (72). However, a homologue of the pRT1 Rep63 protein was later found in a larger plasmid, pAMT11 (20.5 kb), isolated from *Thermococcus* sp. AMT11 (52). Alignment of this pAMT11 protein, called Rep72, with the Rep63 protein of pRT1 clearly shows that these proteins are not homologous to the RC-Rep protein of pGT5 and pTN1 (52, 64). In particular, they lack the motifs 1, 2, and 3, which are typical of RC-Rep proteins in the proper orientation.

The plasmid pAMT11 encodes an integrase related to tyrosine recombinases and exhibits extensive similarities with the integrated mobile element TKV1 of *T. kodoakaraensis* (37). Interestingly, the gene encoding the Rep72 protein of pAMT11 is replaced in TKV1 by a gene encoding an MCM-like helicase. This suggests that Rep72 might be a novel type of helicase that could also be involved in the initiation of plasmid replication. Since MCM has until now always been associated with the theta type mode of replication, the functional replacement between MCM and Rep72 strongly suggests that both pAMT11 and pRT1 replicate via a theta mode too.

Plasmids from Group I Methanogens

Methanogens from group I are hydrogenotrophs, using carbon dioxide (CO_2) as a source of carbon and hydrogen as a reducing agent. They include three orders: *Methanopyrales*, *Methanococcales*, and *Methanobacteriales* (73). *Methanopyrales* are hyperthermophilic organisms growing at temperatures up to 110°C. There is presently only one strain (and species) that has been described, *M. kandleri*, and this strain does not harbor any plasmids, nor is it known to be infected by viruses. In contrast, *Methanococcales* and *Methanobacteriales* have been extensively studied, and one virus as well as several plasmids have been described for these two orders (Fig. 2, Tables 1 and 2).

Methanococcales are round-shaped organisms that can thrive in a wide range of temperatures (from 15 to 85°C). In some phylogenetic analyses based on DNA replication proteins, they form a monophyletic group with *Thermococcales* and *Methanobacteriales* (25) (Fig. 1). Viral particles looking like *Caudovirales* were detected early on by Bertani in the mesophile *Methanococcus voltae* (strains PS and A3) (74). These particles were called voltae transfer agents (VTAs) since they were shown to carry small fragments of cellular DNA rather than viral genomes (4.4 kb) and were able to promote gene transfer (74). These VTA are strongly reminiscent of GTAs (gene transfer agents) that are widespread in alpha proteobacteria (75). Independently, Konisky and coworkers isolated from the same strain (A3) of *M. voltae* a plasmid of 23 kb associated with virus-like particles resembling extracellular membrane vesicles, which were named A3 VLP (76). Unfortunately, these preliminary studies have not been pursued, and the VTA and plasmid from *M. voltae* strain PS have not been sequenced. However, a region in the *M. voltae* A3 genome, which could encode the observed VTA, has been identified and shown to be derived from head and tailed viruses of the order *Caudovirales* (17). Furthermore, the region corresponding to the A3 VLP has also been identified in the *M. voltae* A3 genome based on comparison of the restriction patterns (70). Interestingly, based on the presence of the gene for the major capsid protein and several other genes shared with spindle-shaped archaeal viruses, A3 VLP has been suggested to be classified in the family *Fuselloviridae* (50).

Plasmids from *Methanococcales*

More recently, seven plasmids have been sequenced and described in *Methanococcales* (Table 2): two of them, the plasmid pMEFER01 (27.2 kb, direct GenBank submission) from the hyperthermophile *M. fervens* AG86, and the plasmid pMETVU01 (10.7 kb, direct submission) from the thermophile *M. vulcanius* M7 are closely related to plasmids pT26-2 and pTN2/pAMT7 from *Thermococcales*, respectively (63, 64). A second plasmid, pMETVU02 (4.7 kb, direct GenBank submission) is present in *M. vulcanius* M7. The plasmid pMETVU02 encodes a putative MCM helicase and no RC-initiator protein, suggesting that it could replicate by the theta mode.

Two plasmids, called large extrachromosomal element (58.4 kb) and small extrachromosomal element (16.6 kb), have been briefly described in the genome paper of the hyperthermophile *Methanocaldococcus jannaschii* (formerly *Methanococcus jannaschii*), the first archaeal genome to be sequenced (77). These two plasmids share a restriction/modification system. In addition, the large extrachromosomal element encodes homologues of MCM and ParA that could be involved in its replication and segregation (77).

A plasmid, pURB500 (8.3 kb), has been described from the mesophile *Methanococcus maripaludis* and has been used to construct an *E. coli* shuttle vector (78) (see below). A plasmid quasi identical to pURB500, pMMC501, was recently detected in another strain of *M. maripaludis*. Finally, a plasmid was detected via sequencing of a *Methanothermococcus okinawensis* strain (Table 2); this plasmid, pEMTK01 (14.9 kb), encodes an atypical large protein with a nuclease domain fused to two helicase domains.

Plasmids from *Methanobacteriales*

Methanobacteriales are rod-shaped Gram-positive archaea that contain pseudomurein in their cell wall (79) and can grow in a wide range of temperatures. A single virus has been isolated and characterized in *Methanobacteriales*, the virus ΨM1 (and its deletion derivative ΨM2; 26.1 kb), infecting the moderate thermophile *Methanothermobacter marburgensis* (80). This virus encodes an endolysin that can degrade the pseudomurein cell wall typical of *Methanobacteriales*. Besides ΨM1/2, two integrated proviruses corresponding to the genome of *Caudovirales* have been detected and described in *Methanobacteriales* (17, 81): one in the genome of *Methanothermobacter wolfeii* (82) and the other in *Methanobrevibacter smithii*, the dominant archaeal symbiont found in the human gut (17).

Three related plasmids, pME2001 (nearly identical to pMTBMA4; 4.4 kb) and pME2200 (6.2 kb), have been described in *M. marburgensis* strains Marburg and ZH3, respectively (82, 83) (Table 2). They share a common backbone of 4.3 kb, supposedly corresponding to the minimal replication cassette. In addition, pME2200 encodes a protein (ORF6) homologous to the ORF6 protein of the virus ΨM2. This protein may confer the apparent immunity of its host strain to infection by the virus ΨM2 (82). One of the proteins shared between pME2001 and pME2200 harbors an AAA+ ATPase domain followed by the helix-turn-helix motif, which is also found in the Cdc6 and MCM proteins, making it a good candidate for being the replication initiator of these plasmids. The putative Rep protein of pME2001 was expressed in *E. coli*. Notably, both plasmids harbor four inserted sequences that are flanked by direct repeats 25 to 52 bp in length (82).

Three plasmids—pFV1 (13.5 kb), pFZ1 (11.0 kb), and pCaT2 (11.0 kb)—are present in three strains of another moderate thermophile, *Methanothermobacter thermoautotrophicum* (formerly *Methanobacterium thermoautotrophicum)* (84, 85) (Table 2). The plasmids pFZ1 and pCaT2 are nearly identical at the nucleotide level. All three plasmids share a common backbone with different modules encoding restriction/modification systems and a Cdc6-like replication initiation protein. A region very similar in sequence and organization to the chromosomal replication origin of *M. thermoautotrophicum* was detected by cumulative GC skew analysis in the pFZ1 plasmid, suggesting that pFV1 and pFZ1/pCaT2 could have recruited a *bona fide* chromosomal replication origin for their replication (86).

PLASMIDS FROM *HALOARCHAEA* AND OTHER GROUP II *EURYARCHAEOTA*

Plasmids and Megaplasmids from *Haloarchaea*

Haloarchaea (formerly halobacteria) have relatively large genomes (from 2.7 to 5.5 Mb) with high GC content (>60%) and a highly acidic proteome (pI 5). They are also especially rich in bacteria-like genes that were probably obtained by lateral gene transfer from salt-loving bacteria, such as *Salinobacter* species. Recent phylogenomic analyses have suggested that an ancestor common to all modern haloarchaea has recruited from bacteria around 1,000 genes, possibly in a single event (87).

Haloarchaea have been extensively investigated for plasmids and viruses, because they are both aerobic and mesophilic, which makes them easy to grow (with reasonable generation times) under laboratory conditions. Several viruses infecting haloarchaea have been isolated that belong to the order *Caudovirales* or to the families *Fuselloviridae*, *Pleolipoviridae*, and *Sphaerolipoviridae* (51, 88, 89). The first archaeovirus ever discovered has been the caudovirus ΦH from *Halobacterium salinarum* (90). This virus has been extensively studied in the Zillig laboratory (91 and references therein). Since then, a number of other caudovirales (including siphoviruses, myoviruses, and one podovirus) infecting haloarchaea, with genomes between 30 and 77 kb, have been described (15, 92), as well as two integrated proviruses (17). Recent analysis of the capsid structure of the haloarchaeal virus HTSV-1 (93) has confirmed the previous conclusion, based on comparative genomics and structural modeling of the capsid proteins, that archaeal *Caudovirales* are evolutionarily related to those infecting *Bacteria* (17). Indeed, the HTSV-1 capsid protein was shown to adopt the HK97 fold found in the capsid proteins of bacterial *Caudovirales* (93).

The virus ΦH from *H. salinarum* turned out to be evolutionarily related to two large plasmids from other *H. salinarum* strains (see below). Four viruses infecting haloarchaea and producing spherical virions with internal lipid membranes (SH1, PH1, HHIV-2, and SNJ1) are related to some viruses infecting thermophilic bacteria and have been grouped together into different genera of a novel viral family, the *Sphaerolipoviridae* (88, 94). They are also evolutionarily related to some archaeal plasmids (see below).

The most striking recent discovery was that of a new viral family, *Pleolipoviridae*, that groups very closely related viruses (prototype HRPV-1) with either single- or double-stranded DNA genomes (62, 95, 96). These viruses produce capsids resembling membrane vesicles decorated by glycoprotein spikes. Once again, the genome of these viruses can be found as free plasmids in strains closely related to viral host strains as well as integrated into archaeal genomes. This illustrates an evolutionary continuity between ssDNA viruses, dsDNA viruses, plasmids, and proviruses and challenges viral classification based on genome structure (97).

Finally, two other haloviruses, His1 (14.4 kb) and His2 (16 kb), infecting haloarchaea, have been described and initially classified into the genus *Salterprovirus*. However, in an unexpected turn, His1 was shown to be distantly related to spindle-shaped viruses infecting *Thermococcales* and *Sulfolobales* and was consequently reclassified into the family *Fuselloviridae* (50), whereas His2 was shown to be related to *Pleolipoviridae* (62).

Haloarchaea are especially rich in plasmids of all sizes (from small RC plasmids of a few kilobases up to large megaplasmids) (Tables 3 and 4). All haloarchaeal species contain at least one plasmid, but usually from 2 to 7, including both small and large ones (98). A few haloarchaeal plasmids have been characterized to various extents (Table 3) and will be discussed in more detail below; a selection of large haloarchaeal plasmids is listed in Table 4 (a complete list of archaeal plasmids with identification numbers can be found at the NCBI site: http://mirrors.vbi.vt.edu/mirrors/ftp.ncbi.nih.gov/genomes/IDS/Archaea.ids). Haloarchaea often harbor one or more large replicons that could be regarded as megaplasmids as well as minichromosomes (Chromid, *sensu* Harrison et al. [99]), as their essentiality has been poorly challenged experimentally (Table 4). Interestingly, haloarchaea are the only archaea harboring such minichromosomes and/or megaplasmids, whereas these large replicons seem to be more widespread in bacteria (99). Megaplasmids and chromosomes of haloarchaea are characterized by the presence of many IS elements that promote extensive recombination between plasmids and between plasmids and chromosomes (100). Megaplasmids have much lower GC content than the chromosome, suggesting a different origin. An appealing (but speculative) hypothesis could be that haloarchaea initially recruited their 1,000 bacterial genes from a bacterial megaplasmid, possibly following a fusion event, and that present-day haloarchaeal megaplasmids are relics of this initial event. Indeed, haloarchaea have been shown to extensively exchange genes and recombine chromosomal DNA between species by an unusual fusion mechanism that is triggered by the formation of tubular structures (nanotubes) between cells (99, 101). One could imagine that such fusion occurred with a wall-less form of a salt-loving bacterium, such as *Salinibacter*, containing a megaplasmid.

Most plasmids of haloarchaea, except for the smallest ones, encode at least one protein homologous to the archaeal DNA replication initiator protein Cdc6/Orc1 (and often several copies, up to five in the megaplasmid pHLAC01 of *Halorubrum lacusprofondis*) (Tables 3 and 4) (98, 25). Some of these proteins are likely used as initiators for plasmid replication. Several Cdc6/Orc1-associated plasmid replication origins have been predicted *in silico* and a few of them experimentally tested in haloarchaeal plasmids (98). Phylogenetic analyses indicate that Cdc6/Orc1 proteins encoded by plasmids from haloarchaea are very divergent from their cellular homologues, suggesting that the recruitment of these *cdc6/orc1*-like genes has not occurred recently (25) (Fig. 4). The chromosomes of haloarchaea also contain multiple *cdc6/orc1* genes and putative associated replication origins. In a few cases studied, it has

Table 3 Plasmids from group II *Euryarchaeota* replicating (putatively) via the theta mechanism

Name	Size (kb)	Putative replication and/or segregation proteins, gi (size aa)	Host strains	Reference
pHV2	6.3	RepH (orf3), 292494317 (881)	*Haloferax volcanii*	119
pSCM201	3.5	Rep201, 83743430 (399) Related to IS91 family transposon	*Haloarcula* sp. AS7094	113
pZMX101	3.9	RepA (homologous to Rep201), 27753937(383) Related to IS91 family transposon	*Halorubrum saccharovorum*	114
pHH205 (=SNJ1)	16		*H. salinarium* J7	122
pL23	23.5	XerCD	*Natrosomonas pharaonis*	124
pNRC100	191	2 RepH, 305350 (1009), 10803558 (1073)	*H. salinarium* NRC1	117
pHH1	143	RepH, 487075 (301)	*H. salinarium*	116
pL6A	66		*Haloquadratum walsbyi* C23	125
pL6B			*H. walsbyi* C23	125
pMETHO01	285.1	6 cdc6, 2 ParA, 2 RPA, 2 Top1A MCM, CRISPR locus Transposases, SFII helicases, Retron type reverse transcriptase, RNase HII	*Methanomethylovorans hollandica*	
pMETEV01	163.9	Cdc6, ParA, transposases	*Methanohalobium evestigatum*	
Plasmid 1	36.4	Cdc6, ParA	*Methanosarcina barkeri*	127
pH6Ac	12	PriS PolPrim fused to AAA+ATPase, 386003076 (1124)	*Methanosaeta harundinaceae*	
pGP6	18	AAA+ ATPase, 328929963 (657)	*Methanosaeta concilii* GP-6	130
pTA1	15.7	Cdc6	*Thermoplasma acidophilum*	131
pPO1	7.6	Cdc6	*Picrophilus oshimae*	132

been found that only two to four of these genes are associated with active chromosomal replication origins (98). Indeed, phylogenomic analyses have shown that two Cdc6 proteins, named Cdc6/Orc1-1 and Cdc6/Orc1-2, were probably present in the last archaeal common ancestor and that most haloarchaea still contain descendents of these two forms that likely correspond to the *bona fide* cellular Cdc6 (25). Other haloarchaeal "cellular" Cdc6/Orc1 proteins group with plasmidic and viral Cdc6/Orc1 proteins and are very divergent from both Cdc6/Orc1-1 and Cdc6/Orc1-2, suggesting that their genes are located on integrated plasmids or proviruses (25) (Fig. 4). The chromosome-encoded extra Cdc6/Orc1 proteins do not form a single monophyletic group but are distributed in several divergent clusters of sequences, some of them with no plasmid-encoded representative, suggesting that they are encoded by integrated mobile elements for which no free forms are presently known.

Haloarchaeal Rolling-Circle Plasmids of the pGRB Family

Several haloarchaeal RC plasmids have been characterized, and some of them have been used to build shuttle vectors that allowed the development of genetics tools for haloarchaea (see the section "Plasmids and Genetic Tools for Euryarchaeota") (102, 103) (Table 1).

Four small RC plasmids were found early on in *Halobacterium* strains (pHSB1, pHSB2, pHGN1, and pGRB1) and were sequenced in the late 1980s (104–107). They turned out to be very similar in sequence and size (around 1.7 kb), and single-stranded DNA intermediates were detected, confirming their RC mode of replication (32). They all encode RC-Rep proteins of superfamily I and are prototypes of the pGRB-like subgroup (57, 108) (Fig. 3). The plasmid pGRB1 was also extensively used in the 1980s to probe DNA topology in *Archaea*, as well as to test the effect of DNA topoisomerase inhibitors, either antibiotics or antitumoral drugs (28, 32, 109). Interestingly, inhibition of DNA gyrase by novobiocin induced a strong overproduction of the single-stranded replication intermediate characteristic of RC replication, a phenomenon that has not been reported (or tested) for other RC plasmids in *Archaea* or *Bacteria* (110).

Three small RC plasmids closely related to previous ones (pZMX201 [1.7 kb] from a *Natrinema* strain, pNB101 [2.5 kb] from a *Natronobacterium* strain, and pSN [1.7 kb] from *Haloterrigena thermotolerans*) have been described more recently (Table 3, Fig. 3) (108, 111; P. Forterre, unpublished observations). The single-stranded replication intermediate of pNB101 was detected (111), and the *dso* of plasmid pZMX201 was precisely identified by electron microscopy and *dso*

Table 4 Large plasmids from haloarchaea encoding RepH and/or replication proteins homologous from cellular ones[a,b]

Species	Plasmids	Cdc6/Orc	Other replication proteins	Plasmid size (bp)
Haloarcula jeotgali B3	Plasmid 1	3	PolB	406,285
H. jeotgali B3	Plasmid 2	3		363,534
H. salinarium R1	pHS3	3	PolB PriS 2 RPA	284,332
H. salinarium R1	pHS2	4		194,963
H. salinarium R1	pHS1	2		147,625
Halobacterium spNRC1	pNRC200	6	RepH PolB PriS RPA	365,425
Halobacterium spNRC1	pNRC100			191,346
H. volcanii DS4	pHV4	4	PolB PriS RPA	635,786
H. volcanii DS4	pHV3	1		437,906
H. volcanii DS4	pHV1	2		85,092
Haloarcula marismortui	Chromosome II	2		
H. marismortui	pNG700	1		410,554
H. marismortui	pNG600	2	PolB RPA	155,300
H. marismortui	pNG500	2	2 RepH	132,678
H. marismortui	pNG300	1	MCM Gins51	39,521
H. marismortui	pNG200	0	RPA	33,452
H. marismortui	pNG100	1	RPA	33,303
H. hispanica	Chromosome II	3	2 PCNA	488,918
H. hispanica	pHH400	1		405,816
Halogeometricum borinquense	pHBOR01	1		362,194
H. borinquense	pHBOR02	2		339,010
H. borinquense	pHBOR03	3	PCNA	210,350
H. borinquense	pHBOR04	1		194,834
Halomicrobium mukohataei	pHmuk01	1		221,862
Halopiger xanaduensis	pHALXA01	1		436,718
H. xanaduensis	pHALXA02	1	PriS RPA	181,778
H. xanaduensis	pHALXA03	1	MCM	68,763
H. walsbyi	pL47	0		46,867
H. lacusprofondis	Chromosome II	5	PolB 2 RPA	525,943
H. lacusprofondis	pHLAC01	5		431,338
Haloterrigena turkmenica	pHTUR01	1		698,495
H. turkmenica	pHTUR02	1		413,648
H. turkmenica	pHTUR03	1	PriS 2 RPA RepH	180,781
H. turkmenica	pHTUR04	3		171,934
H. turkmenica	pHTUR05	1	RepH	71,062
Natrialba magadii	pNMAG01	1		378,348
N. magadii	pNMAG02	2	RPA	254,950
N. magadii	pNMAG03	0	RepH	58,487
N. pharaonis	pL131	1		130,998
Natrinema J7-2	pJ7-1	?		95,000
H. xanaduensis	pHALXA01	1		436,718

(Continued)

Table 4 *(Continued)*

Species	Plasmids	Cdc6/Orc	Other replication proteins	Plasmid size (bp)
H. xanaduensis	pHALXA02	1		181,778
H. xanaduensis	pHALXA03	1		68,763
Natrinema pellirubrum	pNATPE01	3		278,800
N. pellirubrum	pNATPE02	3	PolB 2 RPA	275,821
Halophilic archaeon DL31	Plasmid phalar01	1	PolB PriS 3 RPA	705,810

[a]Data are from references 25 and 98.
[b]A few other large haloarchaeal plasmids recently detected have not yet been analyzed and are not listed in this table. A complete list of archaeal plasmids with identification numbers can be found at the NCBI site: http://mirrors.vbi.vt.edu/mirrors/ftp.ncbi.nih.gov/genomes/IDS/Archaea.ids

mapping (108). DNA nicking occurred within a heptanucleotide sequence (TCTC/GGC) located in the stem region of an imperfect hairpin structure. This nick site sequence was conserved among all haloarchaeal RC plasmids, suggesting that the *dso* nick site might be the same for all members of this plasmid family.

A larger RC plasmid member of the pGRB-like subgroup is pHK2 (10.8 kb) (Fig. 3). This plasmid was isolated in the early 1990s from a *Haloferax* strain to construct a shuttle vector based on a novobiocin-resistant gene (103). The rep gene of this plasmid was identified thanks to the sequencing of the minimal replicon fragment of pHK2 (112). Notably, recent sequencing of the entire pHK2 plasmid showed that it is closely related to the haloarchaeal viruses of the *Pleolipoviridae* family in terms of both gene content and synteny. Consequently, pHK2 likely represents an episomal provirus, possibly defective, rather than a *bona fide* plasmid (95).

The pSCM201 Family

The pSCM201 plasmid (3.5 kb) from *Haloarcula* sp. AS7094 and the pZMX101 plasmid (3.9 kb) from *Halorubrum saccharovorum* are closely related small plasmids that share three homologous proteins and constitute a new family of haloarchaeal plasmids (113, 114) (Table 3). Southern blot, electron microscopy, and initiation point mapping analyses all were consistent with a unidirectional theta mode of replication for pSCM201 (113). The minimal replicon of pSCM201 was shown to contain a functional replication origin (AT-rich region and repeats) and a gene encoding a single essential replication protein (Rep201) presumed to be a replication initiator protein. The authors reported that the Rep201 protein harbors a putative leucine zipper motif, an HTH motif, and an ATPase domain. We could not reproduce these observations. Instead, we found that the Rep proteins of pSCM201 and pZMX101 are distantly related to a protein called RepA encoded by the RC-plasmid pC2A from *M. acetivorans* (see below) and to transposases of the IS91 family that replicate via RC (with one conserved tyrosine). These observations, suggesting that Rep201 belongs to the RC-Rep superfamily II, are difficult to reconcile with the experimental evidence supporting a theta mode of replication for these plasmids (113). The mode of replication of plasmids pSCM201 and pZMX101 and the mode of action of their Rep proteins are worthy of future investigations.

The RepH Plasmids Family

Several large plasmids were isolated in the early 1980s from *Halobacterium* strains, but only a few of them were subsequently analyzed and sequenced (115). It was found early on that the plasmid pHH1 (143 kb) from *H. salinarum* (formerly *Halobacterium halobium*) encodes proteins involved in the production of gas vesicle production (116). The minimal replication fragment of pHH1 and of another megaplasmid, pNRC100 (191 kb), from *H. halobium* have been determined (117, 118). It was around 3 to 4 kb long and in both megaplasmids contained an AT-rich region of about 350 to 550 bp and a gene encoding a putative initiator protein called RepH. Mutagenesis experiments were performed, which demonstrated that the *repH* gene is essential for pNRC100 replication (117). A small plasmid from *Halobacterium volcanii*, pHV2 (6.3 kb), also encodes a *repH* gene (119). Homologues of RepH are encoded by megaplasmids of *Haloarcula marismortui* and *Halobacterium* (118, 120) (Tables 3 and 4) and by haloviruses of the order *Caudovirales*, specifically in ΦCh1 (*Natrilba magadii*) and ΦH (*H. salinarum*) (121), and in a large variety of haloarchaeal genomes. These proteins are very heterogeneous in size (from 300 to 1,100 amino acids), but they do not have any sequence similarities with other proteins in databases that could make it possible to infer a function.

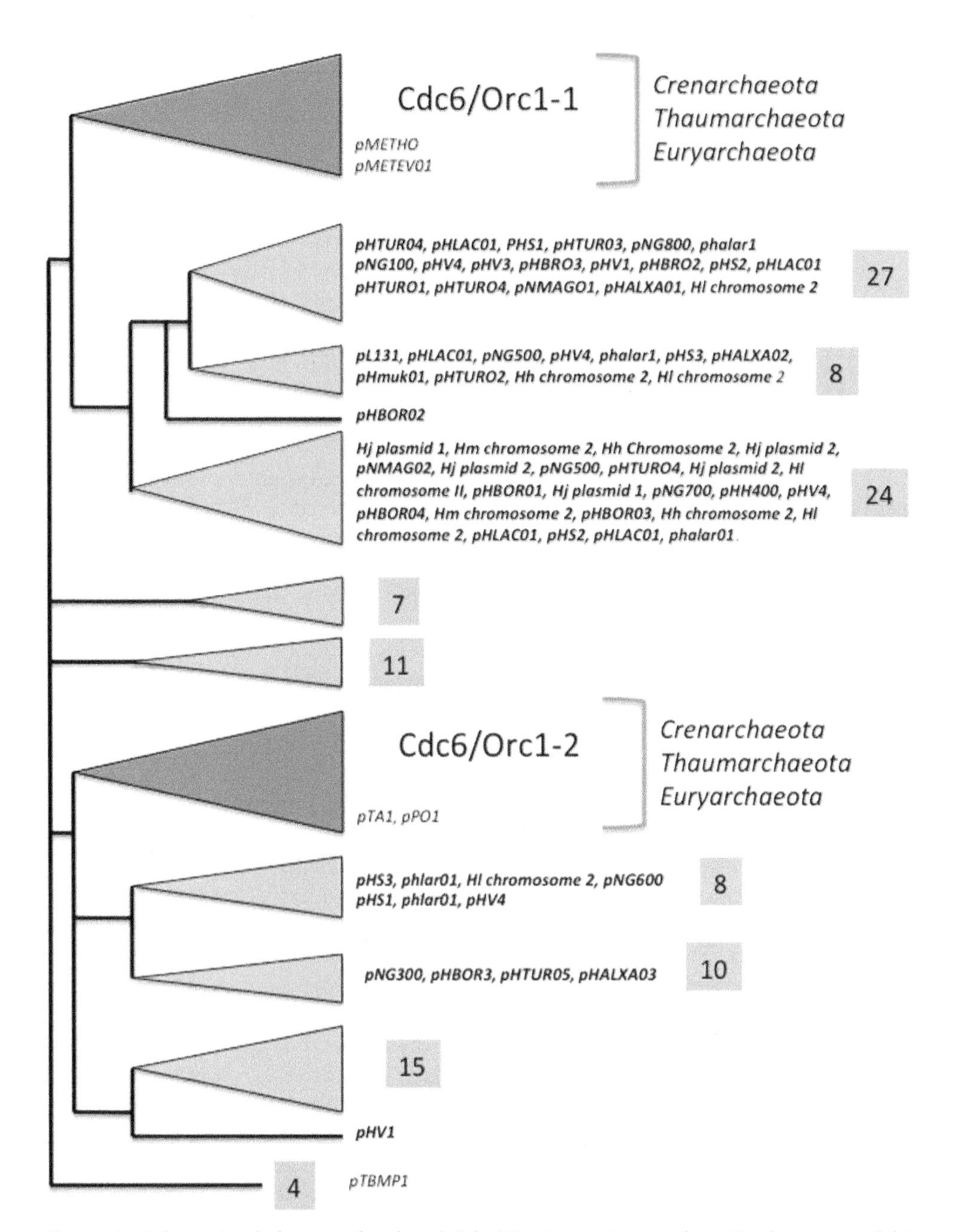

Figure 4 Schematic phylogeny of archaeal Cdc6/Orc1 proteins; see http://archaea.u-psud.fr/cdc6/cdc6.html for the original and complete phylogenies and reference 25 for material and methods. The Cdc6/Orc1-1 and Cdc6/Orc1-2 groups (large dark triangles) contain representatives from most archaeal orders, and internal phylogenies are roughly congruent with archaeal phylogenies based on ribosomal proteins (25). Other groups contain both plasmid-encoded members whose names are indicated beside the triangles (with haloarchaeal plasmids in bold) and chromosome-encoded members whose numbers are indicated in gray squares. doi:10.1128/microbiolspec.PLAS-0027-2014.f4

Other Putative Theta Plasmids from Haloarchaea (Table 3)

In 2003, a 16.3-kb extrachromosomal element was described as a plasmid named pHH205 from *H. salinarum* J7 (122). Thirty-eight putative ORFs were identified, with almost no functional prediction inferred from sequence comparisons in databases. Interestingly, a GC skew analysis allowed detection of a putative origin of replication, and a minimal replication fragment containing this putative origin is functional for replication (122). Recently, it was established that this element is the genome of the sphaerolipovirus SNJ1 (88, 123).

The complete sequencing of the *Natronomonas pharaonis* genome allowed the identification of two plasmids that, unfortunately, were not further characterized: a megaplasmid pL131 (131 kb) typical of *Halobacteriales* and a multicopy plasmid pL23 (23.5 kb) (124). The latter encodes a putative integrase of the XerCD family, and a copy has been found integrated into the chromosome next to a tRNA gene. Among the 35 other putative proteins of pL23, no significant similarities to the proteins in sequence databases could be found.

Recently, sequencing of the *Haloquadratum walsbyi* C23 genome revealed the presence of three plasmids: one large, PL100 (100 kb), and two small, closely related plasmids, PL6A and PL6B (6 kb) (125). The two smaller plasmids do not encode any identifiable homologues of the known genome replication proteins. However, several potential DNA binding proteins, containing predicted HTH or ribbon-helix-helix (RHH) motifs, could be identified. Intriguingly, proteins Hqrw_6007 (from PL6A) and Hqrw_7007 (from PL6B) contain N-terminal CopG-like RHH motifs and C-terminal membrane-spanning domains and were found by mass spectrometry to be associated *H. walsbyi* C23 membranes (125). This organization is equivalent to that found in the p24 protein of *Thermococcus* plasmid pTN1 (see above), although the latter has a different N-terminal DNA binding domain (zinc finger). Nevertheless, two, likely independent, cases of membrane-anchored DNA binding proteins encoded by euryarchaeal plasmids suggest that there might be certain biological significance for such subcellular localization of these proteins and possibly also the corresponding plasmids. One of the HTH motif-containing proteins also possesses a P-loop ATPase domain and is homologous to a protein encoded by the spindle-shaped haloarchaeal virus His1. Finally, numerous homologues of an uncharacterized gene Hqrw_6002 are found in the genomes of pleolipoviruses and related proviruses (125), once again showing that there is an active gene exchange between different types of mobile genetic elements occupying the same ecological niche.

Plasmids from Other Group II *Euryarchaea*

Only a few plasmids have been sequenced from other group II *Euryarchaea* (Tables 1 and 3). Below, we briefly present the essential data on the main plasmids from the orders *Methanosarcinales*, *Thermoplasmatales*, and *Archaeoglobales*. Unfortunately, there is presently no isolated and/or characterized virus for these important groups of archaea. However, one integrated *Caudovirales* has been described in a *Methanosarcinales* (17) and another in an *Archaeoglobales* (17).

Plasmids from *Methanosarcinales*

Methanosarcinales produce methane from methyl group-containing compounds and have the widest substrate range among methanogens. The cells can form aggregates (sarcines) with an outer layer composed of heteropolysaccharides. Seven different plasmids have been sequenced from *Methanosarcinales*, in some cases in the framework of whole-genome sequencing projects.

Two RC plasmids encoding the RC-Rep protein of superfamily I have been described in *Methanosarcina*: the plasmids pC2A from *M. acetivorans* (5.5 kb) and pML from *M. mahii* (2.2 kb) (Table 1). The plasmid pC2A was used to develop the first shuttle vector for *Methanosarcinales* (126). Notably, this plasmid (four putative ORFs) encodes a putative integrase of the XerDC family, which is unusual for RC plasmids. The RC-Rep protein of pC2A belongs to the pGS5-like subgroup of RCI-Rep protein (Fig. 3). In contrast, the RC-Rep protein of the pML plasmid cannot be clearly affiliated with either the pGRB- or pGS5-like subgroups (see Fig. 3).

Five plasmids that probably replicate via a theta mechanism have been detected in *Methanosarcinales*. Two of them, pMETHO01 from *Methanomethylovorans hollandica* DSM 15978 (285.1 kb) and pMETEV01 from *Methanohalobium evestigatum* Z-7303 (163.9 kb), are very large and, in this respect, resemble the megaplasmids of haloarchaea. The two plasmids have been deposited in GenBank and thus far have not been characterized in any detail. However, annotation of pMETEO01 (NC_019972) indicates that this plasmid is exceptionally rich in homologues of archaeal cellular DNA replication proteins, with six Cdc6 homologues, two homologues of the single-stranded DNA binding protein RPA, and one homologue of the MCM helicase. Notably, pMETEO01 is the only archaeal plasmid harboring genes (two) encoding a type IA DNA topoisomerase and one gene encoding a retron-type reverse transcriptase.

Plasmid pMETEV01 from *M. evestigatum* and plasmid 1 from *Methanosarcina barkeri* (36.4 kb) (127, 128) also encode one Cdc6 homologue and one homologue of the chromosomal partitioning ATPase ParA (but not ParB). The sequence of plasmid 1 includes a 5.6-kb noncoding region with long direct repeats that could be involved in plasmid replication (127).

Phylogenetic analysis indicates that, unlike Cdc6 from haloarchaeal plasmids, the Cdc6 protein encoded by these plasmids from *Methanosarcinales* belongs to the Cdc6/Orc-2 family of cellular Cdc6. It has therefore likely been recruited recently by these plasmids for the initiation and/or regulation of its replication.

Plasmids from *Methanosarcinales* also encode many transposases. Given that *Methanosarcinales* genomes themselves harbor many transposases, and transposons are supposed to play an important role in genome dynamics and evolution (128), these plasmid-encoded transposases could serve as vehicles for horizontal spread of the transposons in the population due to the inherent promiscuity of plasmids.

Methanosaeta harundinacea 6Ac carries a plasmid, pH6Ac (12 kb), that contains 18 predicted genes (128). Unfortunately, the gene content of this plasmid has not been thoroughly investigated. Preliminary analysis indicates that pH6Ac encodes a large protein (1,124 aa) that contains an N-terminal PriS-like primase domain followed by the AAA+ ATPase domain, which is likely to function as a helicase. This protein is likely to be responsible for the initiation of the pH6Ac replication, similar to the pTN2-like plasmids of *Thermococcales* (67) and the pRN1-like plasmids of *Sulfolobales* (129). Notably, the closest homologues of this primase (based on BLASTP analysis) are encoded in diverse bacteria. In addition, the plasmid carries a gene for N6 adenine-specific DNA methyltransferase, a homologue of which is also present on the larger plasmid (PL100) of *H. walsbyi* C23 (see above). With other methanosarcinal plasmids (specifically pMETHO01), pH6Ac appears to share only one gene, which encodes a protein with a C-terminal HTH motif. The presence of a primase-helicase protein suggests that pH6Ac is likely to replicate via the theta mode.

Another peculiar moderate-sized plasmid, pGP6 (18 kb), is present in *Methanosaeta concilii* GP-6 (130). The plasmid is not clearly related to other known plasmids of *Methanosarcinales*. It encodes a putative protein of 657 aa, which contains the AAA+ ATPase domain. However, unlike in the case of pH6Ac, the primase domain could not be located. Interestingly, fragments of the ATPase domain-containing gene have been partially duplicated and are found in three copies scattered along pGP6. Strikingly, the plasmid also carries a fragment of the methionyl-tRNA synthetase gene, which is most similar (93% identical at the protein level) to the corresponding gene of *M. concilii* GP-6. The gene fragment encompasses 155 codons (of the total 668 codons) and has likely been captured from the host genome. The gene is unlikely to produce a fully functional tRNA synthetase. Nevertheless, the very presence of such a gene on a plasmid is unexpected.

Plasmids from *Thermoplasmatales*

Thermoplasmales are thermoacidophiles growing around pH 0 to 2 and with optimal growth temperature between 50 and 70°C. Two plasmids have been described in *Thermoplasmatales*: pTA1 from *Thermoplasma acidophilum* (15.7 kb) and pPO1 from *Picrophilus oshimae* (7.6 kb) (131, 132) (Fig. 2, Table 3). They both supposedly replicate by the theta mode, as they encode homologues of the Cdc6 replication protein, and neither a RC Rep protein-coding gene nor an ssDNA replication intermediate were detected (experimentally tested for pPO1 [132]). However, pTA1 and pPO1 do not seem to be closely related, as the only shared gene is *cdc6*. Phylogenetic analysis indicates that these cdc6 genes were recently acquired from cellular genes encoding Cdc6/Orc1-2 proteins (25) (Fig. 4). Interestingly, the plasmid pPO1 encodes the first described archaeal restriction/modification system, with the closest homologues being detected within *Actinobacteria* (132).

Plasmid from *Archaeoglobales*

Archaeoglobales are the only hyperthermophilic organisms in the *Euryarchaea* group II. Accordingly, they have both gyrase and revere gyrase (see below), and a small plasmid, pGS5 (2.8 kb, direct GenBank submission), which was isolated from *A. profundus*, turned out to be negatively supercoiled (see the following section, "Archaeal Plasmids as Probes to Study DNA Topology," and reference 31). The plasmid pGS5 should replicate via the RC mechanism, as it encodes a putative RC-Rep initiator protein of superfamily I related to those of pTP2 and pC2A and to RC-Rep of pleolipoviruses HRPV1 and HRPV6 (Fig. 3). Recently, a nearly identical plasmid, pArcpr01, was detected in another strain of *A. profondus* (Table 1).

ARCHAEAL PLASMIDS AS PROBES TO STUDY DNA TOPOLOGY

Plasmids from *Euryarchaeota*, but also from *Crenarchaeota* of the genus *Sulfolobus*, have been used to probe intracellular DNA topology of various archaeal species. Early studies have shown that inhibitors of DNA gyrase modify the topology of the RC plasmid pGRB from *H. salinarum* (formerly *Halobium*) sp. GRB, suggesting for the first time the presence of a bacteria-like DNA gyrase in archaea (28, 32, 109). Notably, antitumoral drugs, such as epipodophylotoxins, that target human type II DNA topoisomerases promote intracellular linearization of pGRB (109). These drugs are known to induce DNA cleavage by type II DNA topoisomerases in stabilizing the reaction intermediate in which these enzymes are covalently linked in 5′ to the double-stranded breaks made during the first step of the reaction. This reveals that the archaeal DNA gyrase is also sensitive to these drugs and that plasmids from haloarchaea could be used *a priori* as probes in drug screening (109). As already mentioned in the introduction, plasmids from archaea containing DNA gyrase (*Euryarchaeota* group II) are all negatively supercoiled, whereas plasmids from archaea lacking this enzyme are usually relaxed. However, an intriguing exception is the plasmid pURB500 from the group I euryarchaeon *M. maripaludis*, which is negatively supercoiled, despite the absence of DNA gyrase in that strain (32). In that case, negative supercoiling could be due to the formation of toroidal negative superturns constrained by the Hmf histones that are present in all euryarchaea, but possibly with different levels of supramolecular organization.

Beside DNA gyrase, archaea contain a unique set of DNA topoisomerases compared to bacteria and eukarya (133). All archaea, with the exception of *Thermoplasmatales*, contain an archaeal specific type II DNA topoisomerase, DNA topoisomerase VI, also present in plants, some protists, and a few bacteria. This enzyme, which lacks DNA gyrase activity, is the prototype of a new family of type II enzyme (DNA topoisomerase IIB), unrelated to DNA gyrase and eukaryotic DNA topoisomerases IIA. In addition, all hyperthermophilic archaea contain reverse gyrase, an atipycal type I DNA topoisomerase that introduces positive supercoiling *in vitro* (134). Reverse gyrase is apparently essential for adaptation to high temperatures, since this enzyme is ubiquitous in hyperthermophiles, widespread in thermophiles, but never present in mesophiles (134, 135). Moreover, a *T. kodakaraensis* reverse gyrase deletion mutant is thermosensitive (136). Archaeal plasmids were thus also used as probes to test the intracellular topological state of strains containing or not containing reverse gyrase. Surprisingly, these topological analyses revealed that reverse gyrase does not determine intracellular DNA supercoiling, since plasmids isolated from group I *Euryarchaeota* (lacking DNA gyrase) are relaxed (not positively supercoiled) in species with and without reverse gyrase (33). Interestingly, the plasmid pGS5 isolated from the hyperthermophilic group II euryarchaeote *Archaeoglobus fulgidus*, which contains both gyrase and reverse gyrase, is negatively supercoiled (31). This indicates that gyrase overrides the reverse gyrase when both enzymes coexist in the same cell (31). This was also previously observed in the case of a plasmid from the hyperthermophilic bacterium *Thermotoga maritima* (137). The existence of hyperthermophiles with negatively supercoiled plasmids, such as *Archaeoglobus* and *Thermotoga*, is not so surprising, since a topologically closed circular plasmid, either positively or negatively supercoiled, is highly resistant to thermal denaturation (138). However, this indicates that positive supercoiling is not a prerequisite for life at very high temperatures. In fact, despite the importance of reverse gyrase in the history of life (7), the actual role of this enzyme in hyperthermophiles remains unknown.

Archaeal plasmids have also been used as probes to study the variation of the intracellular topological states of plasmids depending on salt, growth phase, and temperature (139–141). These results suggest that homeostatic control of DNA supercoiling occurs in archaea as well as in bacteria. They revealed that relaxed plasmids from *Euryarchaeota* or *Crenarchaeota* become more negatively supercoiled after a cold shock and slightly positively supercoiled after heat shock (140, 141). Interestingly, an increase in the linking number (relaxation or positive supercoiling) with increasing temperature turned out to be a general property of bacterial or archaeal plasmids, despite the very different DNA topoisomerases present in archaea and bacteria (for review see reference 142). These transient changes in DNA supercoiling might be essential for adaptation to environmental challenges by modulating the activities of promoters involved in stress responses.

PLASMIDS AND GENETIC TOOLS FOR *EURYARCHAEOTA*

Many plasmids from *Euryarchaeota* have been used to build shuttle vectors in order to develop the genetics of model organisms in archaea. The methanogens and halophiles are the two first groups of archaea for which genetic tools have become available (42). This is in part due to their earlier isolation during the last century in regard to the more recent isolation of thermoacidophilic and hyperthermophilic euryarchaea. Indeed, for some archaeal orders (i.e., *Methanosarcinales*, *Thermococcales*) the availability of a selective marker and the establishment of clonal colonies on plates have been difficult to set up.

Among *Methanococcales*, shuttle vectors and expression vectors have been constructed on the basis of plasmid pURB500 (78, 143), whereas plasmid pC2A was used for the same purpose in *Methanosarcinales* (126). In both cases strong promoters allow overexpression of genes and purification of tagged proteins from the natural host (144, 145). Among various processes in the understanding of molecular mechanisms in archaea, these genetic tools helped decipher the methanogenesis pathway and regulatory mechanisms such as nitrogen assimilation (145).

In 1987, Charlebois et al. succeeded, for the first time, in transforming an archaeon using a plasmid, the pHV2 from *Haloferax volcanii* (119). Other replication origins from *H. volcanii* plasmids have been used since then: those from pHK2 and pHV1/4 (103, 146). Even if two main genetic models exist for haloarchaea, *H. salinarum* and *H. volcanii*, a wider diversity of genetic tools have been developed for *H. volcanii*, which grows faster and harbors a more stable genome than *H. salinarum* (42). Many shuttle vectors were, however, also developed for the latter. These are based on plasmids pGRB1, pHH1, and pNRC100 and are especially employed for the studies of gas vesicle formation, a particular plasmid-encoded feature of some haloarchaea, allowing the cells to float on the water surface (147–149).

In recent years, great advances have been made for *Thermococcales* genetics, based on the discovery of naturally transformable strains and the isolation of *Thermococcales* plasmids (41, 42). Three main models have emerged: *T. kodakaraensis*, *P. abyssi*, and *Pyrococcus furiosus*. The small RC plasmids pGT5 and pTN1 from *P. abyssi* and *T. nautili*, respectively, have been used to design shuttle vectors for *Thermococcales*. The plasmid pGT5 was first engineered to produce pYS2, a plasmid that uses the *pyrE* gene of *Sulfolobus acidocaldarius* as a selection marker (150). This vector was later improved and engineered to transform *P. furiosus* (151). The plasmid pTN1 was engineered to produce pLC70, an *E. coli-Thermococcus* shuttle vector that could replicate in *T. kodakaraensis*, using the *trpE* gene of *T. kodakaraensis* and an HMG-CoA gene as selection markers (61). The pTN1-based vectors have also been successfully used as expression vectors (61) and to study various aspects of translation and transcription in archaea (for review see reference 41). The compatibility of the three genetic elements and the high copy number of pTP1 and pTP2 plasmids (50 copies/cell) might be useful for developing new genetic tools for studying hyperthermophilic euryarchaea and their viruses (56). In the future, these plasmids could also be useful to study different molecular systems (e.g., DNA replication) at high temperatures. More generally, the variety of plasmids found in *Euryarchaeota* could help to design more versatile vectors and other genetic tools for more organisms of this major archaeal phylum.

EVOLUTION

The studies of archaeal plasmids and viruses have revealed strong links between these mobile elements (15). Examples in *Euryarchaeota* include the relationships between the fusellovirus PAV1 and plasmids of the pTN2 family, between RC plasmids of superfamily I and pleolipoviruses, or else between plasmids and *Caudovirales* encoding RepH proteins. Several replicons first characterized as plasmids, such as the haloarchaeal plasmids pHH205 and pHK2, turned out later to be viral genomes. Similar observations have been made in the other two domains, such as the connection between RC plasmids of superfamily I and ssDNA viruses in *Bacteria* (they share homologous Rep proteins) or between retroviruses and retroposons in *Eukarya* (152). These observations have suggested the existence of a common sequence space partly shared by viruses and plasmids (the greater viral world, *sensu* Koonin and Dolja [152]). The only difference between viruses and plasmids is the capacity of the former to encode at least one protein required for the formation of a viral particle (capsid protein) (97). All types of plasmids previously described in bacteria have been reported in archaea, but with some specificity; i.e., rolling-circle plasmids have only been observed in *Euryarchaeota* and conjugative plasmids only in *Crenarchaeota*. It is unclear if this is due to a sampling bias or if it reflects the specific loss of some types of viruses and plasmids in these archaeal phyla.

Many viruses and plasmids are found integrated into archaeal genomes (36). Some of them have been well described when they are closely related to well-known free viruses and plasmids s (17, 35). However, many more probably exist. Indeed, many putative integrated viruses and plasmids have been detected in archaeal genomes by *in silico* analyses as clusters of atypical genes, i.e., genes with atypical dinucleotide composition compared to that of core genes common to a particular group of species (36). On average, between 4 and 18 clusters of atypical genes not previously described as viruses or plasmids were detected in various genomes of *Euryarchaeota* available at the time of the analysis. This suggests the existence of several families of viruses and plasmids that have not yet been characterized in this phylum.

Integrated viruses and plasmids that have been characterized in some detail are always integrated in the genomes of archaeal species closely related to species harboring episomal forms of these mobile elements. This strongly suggests that viruses and plasmids coevolve with their hosts, explaining, for instance, the close similarity observed between plasmids of

Thermococcales and *Methanococcales*, since these two orders are closely related in archaeal phylogeny (6, 25). Similarly, some similarities can be detected between some plasmids of haloarchaea and some plasmids of *Methanosarcinales*, two groups that, together with *Methanomicrobiales*, form a monophyletic clade in group II *Euryarchaeota* (Fig. 1).

Coevolution within the group formed by *Thermococcales* and *Methanococcales* has been confirmed in the case of a pT26-2-like plasmid by phylogenetic analysis of their core genes (35). Furthermore, network analysis of archaeal proteins homologous to pT26-2 proteins has shown that coevolution takes place at the whole-domain level. Indeed, the number of proteins homologous to pT26-2 proteins decreases with increasing phylogenetic distance from *Thermococcales* (35). All these homologues are only found in *Euryarchaeota*, with the exception of the integrases that also have homologues encoded by fusselloviruses infecting *Crenarchaeota* (*Sulfolobales*). The combined phylogenetic analysis of archaeal MCM helicases encoded by cellular genomes and by viruses and plasmids also argues in favor of coevolution, since MCM proteins encoded by viruses and plasmids always branch together with species close to those harboring these viruses and plasmids (71). One can infer from such analysis that MCM helicases were recruited recently by viruses and plasmids from their hosts and that viruses and plasmids are rarely involved in gene transfer between distantly related archaeal lineages. The latter observation contrasts with the frequent assumption that viruses and plasmids are extensively used as vehicles for gene transfer between distant cellular lineages or else that the gene flux from cells to viruses and plasmids is overwhelming compared to the gene flux from viruses and plasmids to cells (153). On the contrary, it seems that the gene flux mainly takes place from viruses and plasmids to cells (154). This is striking in the case of archaea, since a rather large fraction of archaeal genomes is made of recently integrated viruses and plasmids (36), whereas only a very limited portion of archaeal virus and plasmid proteins have cellular homologues. Most of them have no homologues in databases or only homologues encoded by other archaeal viruses and plasmids. When these proteins have cellular homologues, genome context analyses usually reveal that these homologues are encoded by integrated viruses and plasmids.

The few virus and plasmid proteins frequently recruited from cells are those involved in DNA replication (Cdc6 initiator protein, MCM helicase) or in virus- and plasmid-host interactions (restriction-modification or toxin-antitoxin systems). The origin of virus- and plasmid-associated Cdc6 proteins is unclear since they branch outside of cellular Cdc6 in phylogenetic analyses (25). It has been recently shown that a haloarchaeon can live without any Cdc6 by initiating DNA replication via homologous recombination (155). A mutant lacking Cdc6 grows even better than the wild type. The authors suggested that replication origins are selfish elements that invade cellular genomes, in agreement with replicon takeover hypotheses, which suggests a virus and plasmid origin for DNA replication proteins (156). One scenario could be that two ancestral *Cdc6* genes linked to replication origins invaded two ancestors of the last archaeal common ancestor, becoming the ancestor of the two paralogous cellular *Cdc6* genes that were present in the last archaeal common ancestor, whereas one or several other plasmid-borne *Cdc6* genes continue to coevolve with archaea, sporadically integrating into their genomes, explaining the presence of multiple *Cdc6* genes in several archaeal species, especially in haloarchaea.

Beside proteins recruited from cells, viruses and plasmids encode several specific replication proteins that have no cellular counterpart, such as the Rep proteins of RC plasmids or the superfamily III helicases often found in plasmid replication proteins. Several new families of virus and plasmid replication proteins probably remain to be discovered. For instance, it has been noticed that an MCM helicase has replaced in TKV1 a virus- and plasmid-specific ATPase present in pT26-2, TKV2 and TKV3, and a virus- and plasmid-specific Rep protein present in pAMT11, suggesting that these ATPases and Rep proteins could correspond to a new helicase family (35). The discovery of novel types of DNA replication proteins in viruses and plasmids illustrates the fact that mobile elements can be the cradle of new proteins involved in DNA replication, in agreement with the idea that DNA and DNA-manipulating enzymes originated first in an ancient viral world (156).

Biological Roles of Plasmids from *Euryarchaeota*

In bacteria, plasmids have been widely studied because of their role as carriers of antibiotic resistance genes, virulence genes, and other adaptation genes essential for their maintenance and dissemination. Plasmids can also carry genes encoding secondary metabolite pathways that can be useful for the host in order to benefit from additional sources of nutrients or energy. Finally, plasmids can also encode toxins (e.g., colicins) or pathogenicity islands important for pathogenicity or in the competition between bacterial species. As a consequence, plasmids have been considered as extrachromosomal elements useful for the cell and as vehicles for gene transfer, with the exception of a few "selfish" cryptic plasmids. The known repertoire for archaeal plasmids is much more limited, and most of them appear to be selfish indeed. A major exception is the presence of gene clusters encoding proteins for synthesis of buoyant gas-filled vesicles in several large plasmids of *Haloarchaea*, such as the plasmid pHH1 (150 kb) from *H. salinarum* or pNRC100 (200 kb) from *H. halobium* (157–159). Notably, archaeal gas vesicle proteins are homologous to bacterial gas vesicle proteins found in cyanobacteria or in the firmicute *Bacillus megaterium* (160). It is tempting to speculate that haloarchaea obtained their gas vesicle genes from a large bacterial megaplasmid, which could also bring more than 1,000 bacterial genes to the base of the haloarchaeal tree (87).

Most archaeal plasmids thus appear to emphasize the selfishness of these mobile elements, another trait that groups them with viruses as molecular organisms that can colonize the cellular ones, themselves becoming cellular in the process. Conjugation has been often viewed as a kind of prokaryotic sex, favoring cellular gene exchange for the benefit of the cell population. However, conjugation is first of all an efficient mechanism of selfish virus-like propagation for plasmids.

Several euryarchaea exhibit natural transformation, suggesting that transformation could be an efficient method of plasmid transfer in these archaea in the absence of conjugation. However, transformation may not be so efficient in natural settings, especially in hot environments, considering the instability of DNA at high temperatures (138). Interestingly, it has been shown that *Thermococcus* species produce extracellular membrane vesicles that can harbor plasmids (34, 53, 64). These vesicles may be an alternative to conjugation for these plasmids. Vesicle transport could be especially important for hyperthermophilic organisms since DNA associated with membrane vesicles is much more resistant to denaturation than free DNA (161). The existence of "viral vesicles," i.e., extracellular membrane vesicles carrying viral genomes, such as pTN3 in *T. nautili*, could also be another avenue for the dissemination of viral genomes and viral "resurrection" by facilitating recombination between viral, plasmid, and/or cellular chromosomes in the absence of viral infection (53).

PERSPECTIVES

The role played by plasmids in *Archaea* is essential for the study and comprehension of the biology of these microorganisms, as it has been for *Bacteria*. However, the number of plasmids described in *Euryarchaeota*—a large and very diverse phylum—is still rather small and limited to a few orders or genera. Many plasmids have been detected as fortunate byproducts of genome sequencing projects but have not been characterized (yet). One should expect that with the increasing number of complete genome sequences, the number of detected plasmids will increase too, providing new opportunities for plasmid hunters. To gain better insight into plasmid diversity, an attractive possibility would be to design genome sequencing programs focusing on archaeal strains in which plasmids were previously detected by systematic screening of large strain collections. Plasmids from a given group (e.g., *Thermococcales*) can be divided into a small number of families based on their putative replication proteins. One can wonder if further systematic screening in this group would reveal additional novel plasmid families or if the number of known plasmid families has already reached saturation. It is also important to get more plasmids from poorly sampled groups such as the *Thermoplasmatales* or methanogen group II (Fig. 1). There is still no plasmid isolated from *Methanomicrobiales*, *Methanocellales*, or the recently described "seventh order of methanogens," *Methanomassilicoccales*. Finally, the *Euryarchaeota* include many nanosized archaea, such as *Nanoarchaeota*, ARMAN, and nanohalobacteriales. At the moment, none of them seem to harbor plasmids. Therefore, it remains unclear whether the small size of these cells limits their ability to maintain plasmids. More generally, exploration of plasmid diversity and distribution might have an important impact on the general comprehension of the role of archaea in nature. For example, further understanding of the biology of the *Thaumarchaeota*—an important recently identified phylum with huge importance in the nitrogen cycle (162)—necessitates the development of genetic tools, and discovery of thaumarchaeal plasmids would be an essential step in that direction. For this purpose, the assembly of large, publicly available strain collections and their subsequent screening for extrachromosomal elements should be prioritized.

Plasmids from *Euryarchaeota* have already been extensively used for designing genetic tools in various archaeal groups (especially halophiles), and one can expect that more of them will be used in the future to increase the number of archaea amenable to genetic analysis. These plasmids also are a potentially rich source of new materials for molecular biologists interested in gene regulation, copy control, transposition, integration, and/or DNA replication. In many cases, these mechanisms should operate in extremophilic conditions—high salt or high temperature. Plasmids from *Euryarchaeota* are thus a rich source of new extremozymes or new models to study protein-nucleic acid interactions at extremely high salt concentrations.

Interestingly, most replication proteins that have been detected in euryarchaeal plasmids have homologues in cellular genomes. These cellular counterparts could be used as baits to identify integrated plasmids and/or related viruses in euryarchaeal genomes. Systematic identification of such mobile elements would be very useful to distinguish between true cellular and plasmid/viral genes in comparative genomic and phylogenomic analyses (163). Some of these replication proteins also have homologues in bacterial genomes. Besides replication proteins, euryarchaeal plasmids are also rich in modification/restriction systems and toxin/antitoxin components that are very similar to bacterial ones. They also often encode integrases that are closely related to their bacterial counterparts. Similarly, archaeal plasmids often carry bacteria-like IS elements. These observations raise major evolutionary questions: Were these proteins already present in plasmids present at the time of divergence between *Archaea* and *Bacteria*, or do they testify to the involvement of plasmids in gene transfer between domains? Indeed, interdomain conjugation and plasmid transfer from *E. coli* to the anaerobic archaeon *M. maripaludis* has been experimentally demonstrated (164), indicating that mobile elements, even if at low frequencies, can be transferred between the two prokaryotic domains. Nevertheless, the extent of such transfer and the diversity of underlying mechanisms remain obscure. The question is especially intriguing in the case of hyperthermophiles, considering the instability of "naked" DNA at high temperatures that complicates natural transformation with free DNA. It has been speculated that plasmids could use membrane vesicles as a vehicle to shuttle between hyperthermophiles (53), but this remains to be demonstrated. Conjugation is another important mechanism of horizontal gene transfer that could allow gene exchange in hot environments. However, this mechanism does not seem to be widespread in *Archaea*, as it was found only in one crenarchaeal order and, to date, not in *Euryarchaea*.

Another striking finding revealed by studies of archaeal plasmid diversity is the apparent relationship between archaeal plasmids and viruses (167). There seems to be an extensive gene pool shared between these two distinct types of mobile genetic elements. Further isolation of novel archaeal viruses and plasmids should deepen our understanding of the interconnectedness of the archaeal mobilome and provide an answer to the question of whether limits can be defined for certain groups (families) of mobile genetic elements or if all of them eventually coalesce into a giant genetic network, traversing different archaeal lineages.

Finally, considering the debate about the topology of the universal tree of life, it remains to be understood why the archaeal and bacterial mobilomes are so similar, compared to the eukaryotic one. This is especially striking considering that bacteria and archaea have completely different, nonhomologous, replication machineries, whereas these machineries (as well as most other basic molecular mechanisms) are built around closely related orthologous proteins in archaea and eukarya (7, 165).

Acknowledgments. *P.F.'s research is funded by the European Research Council under the European Union's Seventh Framework Program (FP/2007–2013)/Project EVOMOBIL —ERC Grant Agreement no. 340440 and member of the Institut Universitaire de France. K.R. is a scholar from the Pasteur-Paris University (PPU) International PhD program and received a stipend from the Paul W. Zuccaire Foundation. Conflicts of interest: We disclose no conflicts.*

Citation. Forterre P, Krupovic M, Raymann K, Soler N. 2014. Plasmids from Euryarchaeota. Microbiol Spectrum **2**(6): PLAS-0027-2014.

References

1. **Sapp J, Fox GE.** 2013. The singular quest for a universal tree of life. *Microbiol Mol Biol Rev* **77:**541–550.
2. **Woese CR, Fox EF.** 1977. Phylogenetic structure of prokaryotic domain: the primary kingdoms. *Proc Natl Acad Sci USA* **74:**5088–5090.
3. **Ciccarelli FD, Doerks T, von Mering C, Creevey CJ, Snel B, Bork P.** 2006. Toward automatic reconstruction of a highly resolved tree of life. *Science* **311:**1283–1287.
4. **Lombard J, López-García P, Moreira D.** 2012. Phylogenomic investigation of phospholipid synthesis in *Archaea*. *Archaea* **2012:**1–13.
5. **Garrett R, Klenk H-P.** 2007. *Archaea*. Blackwell Publishing, Oxford, UK.
6. **Brochier-Armanet C, Forterre P, Gribaldo S.** 2011. Phylogeny and evolution of the *Archaea*: one hundred genomes later. *Curr Opin Microbiol* **14:**274–281.
7. **Forterre P.** 2013. The common ancestor of *Archaea* and *Eukarya* was not an archaeon. *Archaea* **2013:**1–18.
8. **Reed CJ, Lewis H, Trejo E, Winston V, Evilia C.** 2013. Protein adaptations in archaeal extremophiles. *Archaea* **2013:**1–14.
9. **Mardanov AV, Ravin NV.** 2012. The impact of genomics on research in diversity and evolution of archaea. *Biochemistry (Mosc)* **77:**799–812.
10. **Woese CR, Kandler O, Wheelis ML.** Towards a natural system of organisms: proposal for the domains *Archaea*, *Bacteria*, and *Eucarya*. *Proc Natl Acad Sci USA* **87:** 4576–4579.

11. **Brochier-Armanet C, Boussau B, Gribaldo S, Forterre P.** 2008. Mesophilic crenarchaeota: proposal for a third archaeal phylum, the *Thaumarchaeota*. *Nat Rev Microbiol* **6:**245–252.
12. **Brochier-Armanet C, Gribaldo S, Forterre P.** 2012. Spotlight on the *Thaumarchaeota*. *ISME J* **6:**227–230.
13. **Brochier-Armanet C, Gribaldo S, Forterre P.** 2008. A DNA topoisomerase IB in *Thaumarchaeota* testifies for the presence of this enzyme in the last common ancestor of *Archaea* and *Eucarya*. *Biol Direct* **3:**54. doi: 10.1186/1745-6150-3-54.
14. **Krupovic M, Spang A, Gribaldo S, Forterre P, Schleper C.** 2011. A thaumarchaeal provirus testifies for an ancient association of tailed viruses with archaea. *Biochem Soc Trans* **39:**82–88.
15. **Pina M, Bize A, Forterre P, Prangishvili D.** 2011. The archeoviruses. *FEMS Microbiol Rev* **35:**1035–1054.
16. **Pietilä MK, Demina TA, Atanasova NS, Oksanen HM, Bamford DH.** 2014. Archaeal viruses and bacteriophages: comparisons and contrasts. *Trends Microbiol* **22:**334–344.
17. **Krupovič M, Forterre P, Bamford DH.** 2010. Comparative analysis of the mosaic genomes of tailed archaeal viruses and proviruses suggests common themes for virion architecture and assembly with tailed viruses of bacteria. *J Mol Biol* **397:**144–160.
18. **Dridi B, Henry M, El Khéchine A, Raoult D, Drancourt M.** 2009. High prevalence of *Methanobrevibacter smithii* and *Methanosphaera stadtmanae* detected in the human gut using an improved DNA detection protocol. *PLoS ONE* **4:**e7063. doi:10.1371/journal.pone.0007063.
19. **Huber H, Hohn MJ, Stetter KO, Rachel R.** 2003. The phylum *Nanoarchaeota*: present knowledge and future perspectives of a unique form of life. *Res Microbiol* **154:**165–171.
20. **Forterre P, Gribaldo S, Brochier-Armanet C.** 2009. Happy together: genomic insights into the unique *Nanoarchaeum/Ignicoccus* association. *J Biol* **8:**7.
21. **Comolli LR, Baker BJ, Downing KH, Siegerist CE, Banfield JF.** 2009. Three-dimensional analysis of the structure and ecology of a novel, ultra-small archaeon. *ISME J* **3:**159–167.
22. **Baker BJ, Comolli LR, Dick GJ, Hauser LJ, Hyatt D, Dill BD, Land ML, VerBerkmoes NC, Hettich RL, Banfield JF.** 2010. Enigmatic, ultrasmall, uncultivated *Archaea*. *Proc Natl Acad Sci USA* **107:**8806–8811.
23. **Narasingarao P, Podell S, Ugalde JA, Brochier-Armanet C, Emerson JB, Brocks JJ, Heidelberg KB, Banfield JF, Allen EE.** 2012. De novo metagenomic assembly reveals abundant novel major lineage of *Archaea* in hypersaline microbial communities. *ISME J* **6:**81–93.
24. **Schleper C, Jurgens G, Jonuscheit M.** 2005. Genomic studies of uncultivated archaea. *Nat Rev Microbiol* **3:** 479–488.
25. **Raymann K, Forterre P, Brochier-Armanet C, Gribaldo S.** 2014. Global phylogenomic analysis disentangles the complex evolutionary history of DNA replication in *Archaea*. *Genome Biol Evol* **6:**192–212.
26. **Bapteste É, Brochier C, Boucher Y.** 2005. Higher-level classification of the *Archaea*: evolution of methanogenesis and methanogens. *Archaea* **1:**353–363.
27. **Borrel G, O'Toole PW, Harris HMB, Peyret P, Brugere J-F, Gribaldo S.** 2013. Phylogenomic data support a seventh order of methylotrophic methanogens and provide insights into the evolution of methanogenesis. *Genome Biol Evol* **5:**1769–1780.
28. **Sioud M, Possot O, Elie C, Sibold L, Forterre P.** 1988. Coumarin and quinolone action in archaebacteria: evidence for the presence of a DNA gyrase-like enzyme. *J Bacteriol* **170:**946–953.
29. **Holmes ML, Dyall-Smith ML.** 1991. Mutations in DNA gyrase result in novobiocin resistance in halophilic archaebacteria. *J Bacteriol* **173:**642–648.
30. **Lopez-Garcia P, Anton J, Abad J, Amils R.** 1994. Halobacterial megaplasmids are negatively supercoiled. *Mol Microbiol* **11:**421–427.
31. **Lopez-Garcia P, Forterre P, van der Oost J, Erauso G.** 2000. Plasmid pGS5 from the hyperthermophilic archaeon *Archaeoglobus profundus* Is negatively supercoiled. *J Bacteriol* **182:**4998–5000.
32. **Sioud M, Baldacci G, de Recondo A-M, Forterre P.** 1988. Novobiocin induces positive supercoiling of small plasmids from halophilic archaebacteria *in vivo*. *Nucleic Acids Res* **16:**1379–1391.
33. **Charbonnier F, Forterre P.** 1994. Comparison of plasmid DNA topology among mesophilic and thermophilic eubacteria and archaebacteria. *J Bacteriol* **176:** 1251–1259.
34. **Gaudin M, Gauliard E, Schouten S, Houel-Renault L, Lenormand P, Marguet E, Forterre P.** 2013. Hyperthermophilic archaea produce membrane vesicles that can transfer DNA: membrane vesicles from *Thermococcales*. *Environ Microbiol Rep* **5:**109–116.
35. **Soler N, Marguet E, Cortez D, Desnoues N, Keller J, van Tilbeurgh H, Sezonov G, Forterre P.** 2010. Two novel families of plasmids from hyperthermophilic archaea encoding new families of replication proteins. *Nucleic Acids Res* **38:**5088–5104.
36. **Cortez D, Forterre P, Gribaldo S.** 2009. A hidden reservoir of integrative elements is the major source of recently acquired foreign genes and ORFans in archaeal and bacterial genomes. *Genome Biol* **10:**R65.
37. **Fukui T.** 2005. Complete genome sequence of the hyperthermophilic archaeon *Thermococcus kodakaraensis* KOD1 and comparison with *Pyrococcus* genomes. *Genome Res* **15:**352–363.
38. **She Q, Peng X, Zillig W, Garrett RA.** 2001. Genome evolution: gene capture in archaeal chromosomes. *Nature* **409:**478.
39. **Serre M-C.** 2002. Cleavage properties of an archaeal site-specific recombinase, the SSV1 integrase. *J Biol Chem* **277:**16758–16767.
40. **Lepage E, Marguet E, Geslin C, Matte-Tailliez O, Zillig W, Forterre P, Tailliez P.** 2004. Molecular diversity of new *Thermococcales* isolates from a single area of hydrothermal deep-sea vents as revealed by randomly amplified polymorphic DNA fingerprinting and 16S

rRNA gene sequence analysis. *Appl Environ Microbiol* **70:**1277–1286.

41. **Farkas JA, Picking JW, Santangelo TJ.** 2013. Genetic techniques for the *Archaea. Annu Rev Genet* **47:**539–561.
42. **Leigh JA, Albers S-V, Atomi H, Allers T.** 2011. Model organisms for genetics in the domain *Archaea*: methanogens, halophiles, *Thermococcales* and *Sulfolobales*: archaeal model organisms. *FEMS Microbiol Rev* **35:** 577–608.
43. **Thiel A, Michoud G, Moalic Y, Flament D, Jebbar M.** 2014. Genetic manipulations of the hyperthermophilic piezophilic archaeon *Thermococcus barophilus. Appl Environ Microbiol* **80:**2299–2306.
44. **Hileman TH, Santangelo TJ.** 2012. Genetics techniques for *Thermococcus kodakaraensis. Front Microbiol* **3:** 195. doi:10.3389/fmicb.2012.00195.
45. **Imanaka T.** 2011. Molecular bases of thermophily in hyperthermophiles. *Proc Jpn Acad Ser B* **87:**587–602.
46. **Kim M-S, Bae SS, Kim YJ, Kim TW, Lim JK, Lee SH, Choi AR, Jeon JH, Lee J-H, Lee HS, Kang SG.** 2013. CO-dependent H2 production by genetically engineered *Thermococcus onnurineus* NA1. *Appl Environ Microbiol* **79:**2048–2053.
47. **Geslin C, Le Romancer M, Erauso G, Gaillard M, Perrot G, Prieur D.** 2003. PAV1, the first virus-like particle isolated from a hyperthermophilic euryarchaeote, "*Pyrococcus abyssi.*" *J Bacteriol* **185:**3888–3894.
48. **Geslin C, Gaillard M, Flament D, Rouault K, Le Romancer M, Prieur D, Erauso G.** 2007. Analysis of the first genome of a hyperthermophilic marine virus-like particle, PAV1, isolated from *Pyrococcus abyssi. J Bacteriol* **189:**4510–4519.
49. **Gorlas A, Koonin EV, Bienvenu N, Prieur D, Geslin C.** 2012. TPV1, the first virus isolated from the hyperthermophilic genus *Thermococcus*: characterization of a new hyperthermophilic virus TPV1. *Environ Microbiol* **14:**503–516.
50. **Krupovic M, Quemin ERJ, Bamford DH, Forterre P, Prangishvili D.** 2014. Unification of the globally distributed spindle-shaped viruses of the *Archaea. J Virol* **88:** 2354–2358.
51. **Bath C, Cukalac T, Porter K, Dyall-Smith ML.** 2006. His1 and His2 are distantly related, spindle-shaped haloviruses belonging to the novel virus group, *Salterprovirus. Virology* **350:**228–239.
52. **Gonnet M, Erauso G, Prieur D, Le Romancer M.** 2011. pAMT11, a novel plasmid isolated from a *Thermococcus* sp. strain closely related to the virus-like integrated element TKV1 of the *Thermococcus kodakaraensis* genome. *Res Microbiol* **162:**132–143.
53. **Gaudin M, Krupovic M, Marguet E, Gauliard E, Cvirkaite-Krupovic V, Le Cam E, Oberto J, Forterre P.** 2013. Extracellular membrane vesicles harbouring viral genomes: viral vesicles. *Environ Microbiol.* doi: 10.1111/1462–2920.12235.
54. **Erauso G, Marsin S, Benbouzid-Rollet N, Baucher M-F, Barbeyron T, Zivanovic Y, Prieur D, Forterre P.** 1996. Sequence of plasmid pGT5 from the archaeon *Pyrococcus abyssi*: evidence for rolling-circle replication in a hyperthermophile. *J Bacteriol* **178:**3232–3237.
55. **Soler N, Justome A, Quevillon-Cheruel S, Lorieux F, Le Cam E, Marguet E, Forterre P.** 2007. The rolling-circle plasmid pTN1 from the hyperthermophilic archaeon *Thermococcus nautilus. Mol Microbiol* **66:** 357–370.
56. **Gorlas A, Krupovic M, Forterre P, Geslin C.** 2013. Living side by side with a virus: characterization of two novel plasmids from *Thermococcus prieurii*, a host for the spindle-shaped virus TPV1. *Appl Environ Microbiol* **79:**3822–3828.
57. **Ilyina TV, Koonin EV.** 1992. Conserved sequence motifs in the initiator proteins for rolling circle DNA replication encoded by diverse replicons from eubacteria, eucaryotes and archaebacteria. *Nucleic Acids Res* **20:** 3279–3285.
58. **Marsin S, Forterre P.** 1998. A rolling circle replication initiator protein with a nucleotidyl-transferase activity encoded by the plasmid pGT5 from the hyperthermophilic archaeon *Pyrococcus abyssi. Mol Microbiol* **27:** 1183–1192.
59. **Khan SA.** 1997. Rolling-circle replication of bacterial plasmids. *Microbiol Mol Biol Rev* **61:**442–455.
60. **Marsin S, Forterre P.** 2001. pGT5 replication initiator protein Rep75 from *Pyrococcus abyssi. Methods Enzymol* **334:**193–204.
61. **Santangelo TJ, Cubonova L, Reeve JN.** 2008. Shuttle vector expression in *Thermococcus kodakaraensis*: contributions of cis elements to protein synthesis in a hyperthermophilic archaeon. *Appl Environ Microbiol* **74:** 3099–3104.
62. **Pietila MK, Atanasova NS, Manole V, Liljeroos L, Butcher SJ, Oksanen HM, Bamford DH.** 2012. Virion architecture unifies globally distributed pleolipoviruses infecting halophilic archaea. *J Virol* **86:**5067–5079.
63. **Krupovic M, Gonnet M, Hania WB, Forterre P, Erauso G.** 2013. Insights into dynamics of mobile genetic elements in hyperthermophilic environments from five new thermococcus plasmids. *PLoS ONE* **8:**e49044. doi: 10.1371/journal.pone.0049044.
64. **Soler N, Gaudin M, Marguet E, Forterre P.** 2011. Plasmids, viruses and virus-like membrane vesicles from *Thermococcales. Biochem Soc Trans* **39:**36–44.
65. **Vannier P, Marteinsson VT, Fridjonsson OH, Oger P, Jebbar M.** 2011. Complete genome sequence of the hyperthermophilic, piezophilic, heterotrophic, and carboxydotrophic archaeon *Thermococcus barophilus* MP. *J Bacteriol* **193:**1481–1482.
66. **Iyer LM.** 2005. Origin and evolution of the archaeo-eukaryotic primase superfamily and related palm-domain proteins: structural insights and new members. *Nucleic Acids Res* **33:**3875–3896.
67. **Gill S, Krupovic M, Desnoues N, Beguin P, Sezonov G, Forterre P.** 2014. A highly divergent archaeo-eukaryotic primase from the *Thermococcus nautilus* plasmid, pTN2. *Nucleic Acids Res* **42:**3707–3719.
68. **Keller J, Leulliot N, Soler N, Collinet B, Vincentelli R, Forterre P, van Tilbeurgh H.** 2009. A protein encoded

by a new family of mobile elements from euryarchaea exhibits three domains with novel folds. *Protein Sci* **18:** 850–855.

69. **Oberto J, Gaudin M, Cossu M, Gorlas A, Slesarev A, Marguet E, Forterre P.** 2014. Genome sequence of a hyperthermophilic archaeon, *Thermococcus nautili* 30-1, that produces viral vesicles. *Genome Announc* **2:** e00243–14. doi:10.1128/genomeA.00243-14.
70. **Krupovič M, Bamford DH.** 2008. Archaeal proviruses TKV4 and MVV extend the PRD1-adenovirus lineage to the phylum *Euryarchaeota*. *Virology* **375:**292–300.
71. **Krupovic M, Gribaldo S, Bamford DH, Forterre P.** 2010. The evolutionary history of archaeal MCM helicases: a case study of vertical evolution combined with hitchhiking of mobile genetic elements. *Mol Biol Evol* **27:**2716–2732.
72. **Ward DE, Revet IM, Nandakumar R, Tuttle JH, de Vos WM, van der Oost J, DiRuggiero J.** 2002. Characterization of plasmid pRT1 from *Pyrococcus* sp. strain JT1. *J Bacteriol* **184:**2561–2566.
73. **Liu Y, Whitman WB.** 2008. Metabolic, phylogenetic, and ecological diversity of the methanogenic archaea. *Ann NY Acad Sci* **1125:**171–189.
74. **Eiserling F, Pushkin A, Gingery M, Bertani G.** 1999. Bacteriophage-like particles associated with the gene transfer agent of methanococcus voltae PS. *J Gen Virol* **80:**3305–3308.
75. **Lang AS, Zhaxybayeva O, Beatty JT.** 2012. Gene transfer agents: phage-like elements of genetic exchange. *Nat Rev Microbiol* **10:**472–482.
76. **Wood AG, Whitman WB, Konisky J.** 1989. Isolation and characterization of an archaebacterial viruslike particle from *Methanococcus voltae* A3. *J Bacteriol* **171:** 93–98.
77. **Bult CJ, White O, Olsen GJ, Zhou L, Fleischmann RD, Sutton GG, Blake JA, FitzGerald LM, Clayton RA, Gocayne JD.** 1996. Complete genome sequence of the methanogenic archaeon, *Methanococcus jannaschii*. *Science* **273:**1058–1073.
78. **Tumbula DL, Bowen TL, Whitman WB.** 1997. Characterization of pURB500 from the archaeon *Methanococcus maripaludis* and construction of a shuttle vector. *J Bacteriol* **179:**2976–2986.
79. **Wolfe RS.** 1992. Biochemistry of methanogenesis. *Biochem Soc Symp* **58:**41–49.
80. **Pfister P, Wasserfallen A, Stettler R, Leisinger T.** 1998. Molecular analysis of *Methanobacterium* phage ΨM2. *Mol Microbiol* **30:**233–244.
81. **Luo Y, Pfister P, Leisinger T, Wasserfallen A.** 2001. The genome of archaeal prophage M100 encodes the lytic enzyme responsible for autolysis of *Methanothermobacter wolfeii*. *J Bacteriol* **183:**5788–5792.
82. **Luo Y, Leisinger T, Wasserfallen A.** 2001. Comparative sequence analysis of plasmids pME2001 and pME2200 of *Methanothermobacter marburgensis* strains Marburg and ZH3. *Plasmid* **45:**18–30.
83. **Meile L, Kiener A, Leisinger T.** 1983. A plasmid in the archaebacterium *Methanobacterium thermoautotrophicum*. *Mol Gen Genet* **191:**480–484.
84. **Nölling J, van Eeden FJ, Eggen RI, de Vos WM.** 1992. Modular organization of related archaeal plasmids encoding different restriction-modification systems in *Methanobacterium thermoformicicum*. *Nucleic Acids Res* **20:**6501–6507.
85. **Kosaka T, Toh H, Toyoda A.** 2013. Complete genome sequence of a thermophilic hydrogenotrophic methanogen, *Methanothermobacter* sp. strain CaT2. *Genome Announc* **1:**e00672–13. doi:10.1128/genomeA.00672-13.
86. **Lopez P, Philippe H, Myllykallio H, Forterre P.** 1999. Identification of putative chromosomal origins of replication in archaea. *Mol Microbiol* **32:**883–886.
87. **Nelson-Sathi S, Dagan T, Landan G, Janssen A, Steel M, McInerney JO, Deppenmeier U, Martin WF.** 2012. Acquisition of 1,000 eubacterial genes physiologically transformed a methanogen at the origin of haloarchaea. *Proc Natl Acad Sci USA* **109:**20537–20542.
88. **Pawlowski A, Rissanen I, Bamford JKH, Krupovic M, Jalasvuori M.** 2014. Gammasphaerolipovirus, a newly proposed bacteriophage genus, unifies viruses of halophilic archaea and thermophilic bacteria within the novel family *Sphaerolipoviridae*. *Arch Virol* **159:**1541–1554. doi:10.1007/s00705-013-1970-6.Epub2014.
89. **Dyall-Smith M, Tang S-L, Bath C.** 2003. Haloarchaeal viruses: how diverse are they? *Res Microbiol* **154:** 309–313.
90. **Torsvik T, Dundas I.** 1974. Bacteriophage of *Halobacterium salinarum*. *Nature* **248:**680–681.
91. **Stolt P, Zillig W.** 1994. Transcription of the halophage H repressor gene is abolished by transcription from an inversely oriented lytic promoter. *FEBS Lett* **344:**125–128.
92. **Senčilo A, Jacobs-Sera D, Russell DA, Ko C-C, Bowman CA, Atanasova NS, Österlund E, Oksanen HM, Bamford DH, Hatfull GF, Roine E, Hendrix RW.** 2013. Snapshot of haloarchaeal tailed virus genomes. *RNA Biol* **10:**803–816.
93. **Pietila MK, Laurinmaki P, Russell DA, Ko C-C, Jacobs-Sera D, Hendrix RW, Bamford DH, Butcher SJ.** 2013. Structure of the archaeal head-tailed virus HSTV-1 completes the HK97 fold story. *Proc Natl Acad Sci USA* **110:**10604–10609.
94. **Porter K, Tang S-L, Chen C-P, Chiang P-W, Hong M-J, Dyall-Smith M.** 2013. PH1: An archaeovirus of *Haloarcula hispanica* related to SH1 and HHIV-2. *Archaea* **2013:**1–17.
95. **Roine E, Kukkaro P, Paulin L, Laurinavicius S, Domanska A, Somerharju P, Bamford DH.** 2010. New, closely related haloarchaeal viral elements with different nucleic acid types. *J Virol* **84:**3682–3689.
96. **Pietila MK, Laurinavicius S, Sund J, Roine E, Bamford DH.** 2010. The single-stranded DNA genome of novel archaeal virus halorubrum pleomorphic virus 1 is enclosed in the envelope decorated with glycoprotein spikes. *J Virol* **84:**788–798.
97. **Krupovic M, Bamford DH.** 2010. Order to the viral universe. *J Virol* **84:**12476–12479.
98. **Wu Z, Liu H, Liu J, Liu X, Xiang H.** 2012. Diversity and evolution of multiple orc/cdc6-adjacent replication origins in haloarchaea. *BMC Genomics* **13:**478.

99. **Harrison PW, Lower RPJ, Kim NKD, Young JPW.** 2010. Introducing the bacterial "chromid": not a chromosome, not a plasmid. *Trends Microbiol* **18:**141–148.
100. **Filee J, Siguier P, Chandler M.** 2007. Insertion sequence diversity in archaea. *Microbiol Mol Biol Rev* **71:** 121–157.
101. **Rosenshine I, Tchelet R, Mevarech M.** 1989. The mechanism of DNA transfer in the mating system of an archaebacterium. *Science* **245:**1387–1389.
102. **Zhou M, Xiang H, Sun C, Tan H.** 2004. Construction of a novel shuttle vector based on an RCR-plasmid from a haloalkaliphilic archaeon and transformation into other haloarchaea. *Biotechnol Lett* **26:**1107–1113.
103. **Holmes ML, Dyall-Smith ML.** 1990. A plasmid vector with a selectable marker for halophilic archaebacteria. *J Bacteriol* **172:**756–761.
104. **Kagramanova VK, Derckacheva NI, Mankin AS.** 1988. The complete nucleotide sequence of the archaebacterial plasmid pHSB from *Halobacterium*, strain SB3. *Nucleic Acids Res* **16:**4158.
105. **Hall MJ, Hackett NR.** 1989. DNA sequence of a small plasmid from *Halobacterium* strain GN101. *Nucleic Acids Res* **17:**10501.
106. **Hackett NR, Krebs MP, DasSarma S, Goebel W, RajBhandary UL, Khorana HG.** 1990. Nucleotide sequence of a high copy number plasmid from *Halobacterium* strain GRB. *Nucleic Acids Res* **18:**3408.
107. **Akhmanova AS, Kagramanova VK, Mankin AS.** 1993. Heterogeneity of small plasmids from halophilic archaea. *J Bacteriol* **175:**1081–1086.
108. **Zhou L, Zhou M, Sun C, Han J, Lu Q, Zhou J, Xiang H.** 2008. Precise determination, cross-recognition, and functional analysis of the double-strand origins of the rolling-circle replication plasmids in haloarchaea. *J Bacteriol* **190:**5710–5719.
109. **Sioud M, Forterre P, de Recondo A-M.** 1987. Effects of the antitumor drug VP16 (etoposide) on the archaebacterial *Halobacterium* GRB 1.7 kb plasmid *in vivo*. *Nucleic Acids Res* **15:**8217–8234.
110. **Sioud M, Baldacci G, Foeterre P, de Recondo A-M.** 1988. Novobiocin induces accumulation of a single strand of plasmid pGRB-1 in the archaebacterium *Halobacterium* GRB. *Nucleic Acids Res* **16:**7833–7842.
111. **Zhou M, Xiang H, Sun C, Li Y, Liu J, Tan H.** 2004. Complete sequence and molecular characterization of pNB101, a rolling-circle replicating plasmid from the haloalkaliphilic archaeon *Natronobacterium* sp. strain AS7091. *Extremophiles* **8:**91–98.
112. **Holmes M, Pfeifer F, Dyall-Smith ML.** 1995. Analysis of the halobacterial plasmid pHK2 minimal replicon. *Gene* **153:**117–121.
113. **Sun C, Zhou M, Li Y, Xiang H.** 2006. Molecular characterization of the minimal replicon and the unidirectional theta replication of pSCM201 in extremely halophilic archaea. *J Bacteriol* **188:**8136–8144.
114. **Zhou L, Zhou M, Sun C, Xiang H, Tan H.** 2007. Genetic analysis of a novel plasmid pZMX101 from *Halorubrum saccharovorum*: determination of the minimal replicon and comparison with the related haloarchaeal plasmid pSCM201. *FEMS Microbiol Lett* **270:**104–108.
115. **Pfeifer F, Weidinger G, Goebel W.** 1981. Characterization of plasmids in halobacteria. *J Bacteriol* **145:** 369–374.
116. **Weidinger G, Klotz G, Goebel W.** 1979. A large plasmid from *Halobacterium halobium* carrying genetic information for gas vacuole formation. *Plasmid* **2:**377–386.
117. **Ng W-L, DasSarma S.** 1993. Minimal replication origin of the 200-kilobase Halobacterium plasmid pNRC100. *J Bacteriol* **175:**4584–4596.
118. **Pfeifer F, Ghahraman P.** 1993. Plasmid pHH1 of *Halobacterium salinarium*: characterization of the replicon region, the gas vesicle gene cluster and insertion elements. *Mol Gen Genet* **238:**193–200.
119. **Charlebois RL, Lam WL, Cline SW, Doolittle WF.** 1987. Characterization of pHV2 from *Halobacterium volcanii* and its use in demonstrating transformation of an archaebacterium. *Proc Natl Acad Sci USA* **84:** 8530–8534.
120. **Ng WV, Kennedy SP, Mahairas GG, Berquist B, Pan M, Shukla HD, Lasky SR, Baliga NS, Thorsson V, Sbrogna J.** 2000. Genome sequence of *Halobacterium* species NRC-1. *Proc Natl Acad Sci USA* **97:**12176–12181.
121. **Klein R, Beranyl U, Rossler N, Greineder B, Scholz H, Witte A.** *Natrialba magadii* virus phiCh1: first complete nucleotide sequence and functional organization of a virus infecting a haloalkaliphilic archaeon. *Mol Microbiol* **45:**851–863.
122. **Ye X, Ou J, Ni L, Shi W, Shen P.** 2003. Characterization of a novel plasmid from extremely halophilic *Archaea*: nucleotide sequence and function analysis. *FEMS Microbiol Lett* **221:**53–57.
123. **Zhang Z, Liu Y, Wang S, Yang D, Cheng Y, Hu J, Chen J, Mei Y, Shen P, Bamford DH, Chen X.** 2012. Temperate membrane-containing halophilic archaeal virus SNJ1 has a circular dsDNA genome identical to that of plasmid pHH205. *Virology* **434:**233–241.
124. **Falb M, Pfeiffer F, Palm P, Rodewald K, Hickmann V, Tittor J, Oesterhelt D.** 2005. Living with two extremes: conclusions from the genome sequence of *Natronomonas pharaonis*. *Genome Res* **15:**1336–1343.
125. **Dyall-Smith ML, Pfeiffer F, Klee K, Palm P, Gross K, Schuster SC, Rampp M, Oesterhelt D.** 2011. *Haloquadratum walsbyi*: limited diversity in a global pond. *PLoS ONE* **6:**e20968. doi:10.1371/journal.pone.0020968.
126. **Metcalf WW, Zhang JK, Apolinario E, Sowers KR, Wolfe RS.** 1997. A genetic system for *Archaea* of the genus *Methanosarcina*: liposome-mediated transformation and construction of shuttle vectors. *Proc Natl Acad Sci USA* **94:**2626–2631.
127. **Maeder DL, Anderson I, Brettin TS, Bruce DC, Gilna P, Han CS, Lapidus A, Metcalf WW, Saunders E, Tapia R, Sowers KR.** 2006. The *Methanosarcina barkeri* genome: comparative analysis with *Methanosarcina acetivorans* and *Methanosarcina mazei* reveals extensive rearrangement within methanosarcinal genomes. *J Bacteriol* **188:** 7922–7931.

128. **Zhu J, Zheng H, Ai G, Zhang G, Liu D, Liu X, Dong X.** 2012. The genome characteristics and predicted function of methyl-group oxidation pathway in the obligate aceticlastic methanogens, *Methanosaeta* spp. *PLoS ONE* **7:**e36756. doi:10.1371/journal.pone.0036756.
129. **Lipps G.** 2011. Structure and function of the primase domain of the replication protein from the archaeal plasmid pRN1. *Biochem Soc Trans* **39:**104–106.
130. **Barber RD, Zhang L, Harnack M, Olson MV, Kaul R, Ingram-Smith C, Smith KS.** 2011. Complete genome sequence of *Methanosaeta concilii*, a specialist in aceticlastic methanogenesis. *J Bacteriol* **193:**3668–3669.
131. **Yamashiro K, Yokobori S, Oshima T, Yamagishi A.** 2006. Structural analysis of the plasmid pTA1 isolated from the thermoacidophilic archaeon *Thermoplasma acidophilum*. *Extremophiles* **10:**327–335.
132. **Angelov A, Voss J, Liebl W.** 2011. Characterization of plasmid pPO1 from the hyperacidophile *Picrophilus oshimae*. *Archaea* **2011:**1–4.
133. **Forterre P, Gadelle D.** 2009. Phylogenomics of DNA topoisomerases: their origin and putative roles in the emergence of modern organisms. *Nucleic Acids Res* **37:** 679–692.
134. **Forterre P.** 2002. A hot story from comparative genomics: reverse gyrase is the only hyperthermophile-specific protein. *Trends Genet* **18:**236–237.
135. **Brochier-Armanet C, Forterre P.** 2006. Widespread distribution of archaeal reverse gyrase in thermophilic bacteria suggests a complex history of vertical inheritance and lateral gene transfers. *Archaea* **2:**83–93.
136. **Atomi H, Matsumi R, Imanaka T.** 2004. Reverse gyrase is not a prerequisite for hyperthermophilic life. *J Bacteriol* **186:**4829–4833.
137. **Guipaud O, Marguet E, Noll KM, De La Tour CB, Forterre P.** 1997. Both DNA gyrase and reverse gyrase are present in the hyperthermophilic bacterium *Thermotoga maritima*. *Proc Natl Acad Sci USA* **94:**10606–10611.
138. **Marguet E, Forterre P.** 1994. DNA stability at temperatures typical for hyperthermophiles. *Nucleic Acids Res* **22:**1681–1686.
139. **Mojica FJ, Charbonnier F, Juez G, Rodriguez-Valera F, Forterre P.** 1994. Effects of salt and temperature on plasmid topology in the halophilic archaeon *Haloferax volcanii*. *J Bacteriol* **176:**4966–4973.
140. **Marguet E, Zivanovic Y, Forterre P.** 1996. DNA topological change in the hyperthermophilic archaeon *Pyrococcus abyssi* exposed to low temperature. *FEMS Microbiol Lett* **142:**31–36.
141. **Lopez-Garcia P, Forterre P.** 1997. DNA topology in hyperthermophilic archaea: reference states and their variation with growth phase, growth temperature, and temperature stresses. *Mol Microbiol* **23:**1267–1279.
142. **Lopez-Garcia P, Forterre P.** 2000. DNA topology and the thermal stress response, a tale from mesophiles and hyperthermophiles. *BioEssays* **22:**738–746.
143. **Gardner WL, Whitman WB.** 1999. Expression vectors for *Methanococcus maripaludis*: overexpression of acetohydroxyacid synthase and β-galactosidase. *Genetics* **152:**1439–1447.
144. **Guss AM, Rother M, Zhang JK, Kulkkarni G, Metcalf WW.** 2008. New methods for tightly regulated gene expression and highly efficient chromosomal integration of cloned genes for *Methanosarcina* species. *Archaea* **2:** 193–203.
145. **Dodsworth JA, Leigh JA.** 2006. Regulation of nitrogenase by 2-oxoglutarate-reversible, direct binding of a PII-like nitrogen sensor protein to dinitrogenase. *Proc Natl Acad Sci USA* **103:**9779–9784.
146. **Norais C, Hawkins M, Hartman AL, Eisen JA, Myllykallio H, Allers T.** 2007. Genetic and physical mapping of DNA replication origins in *Haloferax volcanii*. *PLoS Genet* **3:**e77. doi:10.1371/journal.pgen.0030077.
147. **DasSarma S, Arora P, Lin F, Molinari E, Yin LR.** 1994. Wild-type gas vesicle formation requires at least ten genes in the gvp gene cluster of *Halobacterium halobium* plasmid pNRC100. *J Bacteriol* **176:**7646–7652.
148. **Blaseio U, Pfeifer F.** 1990. Transformation of *Halobacterium halobium*: development of vectors and investigation of gas vesicle synthesis. *Proc Natl Acad Sci USA* **87:**6772–6776.
149. **Krebs MP, Hauss T, Heyn MP, RajBhandary UL, Khorana HG.** 1991. Expression of the bacterioopsin gene in *Halobacterium halobium* using a multicopy plasmid. *Proc Natl Acad Sci USA* **88:**859–863.
150. **Lucas S, Toffin L, Zivanovic Y, Charlier D, Moussard H, Forterre P, Prieur D, Erauso G.** 2002. Construction of a shuttle vector for, and spheroplast transformation of, the hyperthermophilic archaeon *Pyrococcus abyssi*. *Appl Environ Microbiol* **68:**5528–5536.
151. **Waege I, Schmid G, Thumann S, Thomm M, Hausner W.** 2010. Shuttle vector-based transformation system for *Pyrococcus furiosus*. *Appl Environ Microbiol* **76:** 3308–3313.
152. **Koonin EV, Dolja VV.** 2013. A virocentric perspective on the evolution of life. *Curr Opin Virol* **3:**546–557.
153. **López-García P, Moreira D.** 2008. Tracking microbial biodiversity through molecular and genomic ecology. *Res Microbiol* **159:**67–73.
154. **Forterre P, Prangishvili D.** 2013. The major role of viruses in cellular evolution: facts and hypotheses. *Curr Opin Virol* **3:**558–565.
155. **Hawkins M, Malla S, Blythe MJ, Nieduszynski CA, Allers T.** 2013. Accelerated growth in the absence of DNA replication origins. *Nature* **503:**544–547.
156. **Forterre P.** 2002. The origin of DNA genomes and DNA replication proteins. *Curr Opin Microbiol* **5:** 525–532.
157. **DasSarma S, Damerval T, Jones J, Tandeau de Marsac N.** 1987. A plasmid-encoded gas vesicle protein gene in a halophilic archaebacterium. *Mol Microbiol* **1:**365–370.
158. **Pfeifer F, Blaseio U, Horne M.** 1989. Genome structure of *Halobacterium halobium*: plasmid dynamics in gas vacuole deficient mutants. *Can J Microbiol* **35:** 96–100.
159. **Ng W-L, Kothakota S, DasSarma S.** 1991. Structure of the gas vesicle plasmid in *Halobacterium halobium*:

inversion isomers, inverted repeats, and insertion sequences. *J Bacteriol* **173:**1958–1964.

160. **Li N, Cannon MC.** 1998. Gas vesicle genes identified in *Bacillus megaterium* and functional expression in *Escherichia coli. J Bacteriol* **180:**2450–2458.

161. **Soler N, Marguet E, Verbavatz J-M, Forterre P.** 2008. Virus-like vesicles and extracellular DNA produced by hyperthermophilic archaea of the order *Thermococcales. Res Microbiol* **159:**390–399.

162. **Stahl DA, de la Torre JR.** 2012. Physiology and diversity of ammonia-oxidizing archaea. *Annu Rev Microbiol* **66:**83–101.

163. **Forterre P.** 2012. Darwin's goldmine is still open: variation and selection run the world. *Front Cell Infect Microbiol* **2:**106. doi:10.3389/fcimb.2012.00106.

164. **Dodsworth JA, Li L, Wei S, Hedlund BP, Leigh JA, de Figueiredo P.** 2010. Interdomain conjugal transfer of DNA from bacteria to archaea. *Appl Environ Microbiol* **76:**5644–5647.

165. **Forterre P.** 2013. Why are there so many diverse replication machineries? *J Mol Biol* **425:**4714–4726.

166. **Dereeper A, Guignon V, Blanc G, Audic S, Buffet S, Chevenet F, Dufayard J-F, Guindon S, Lefort V, Lescot M, Claverie J-M, Gascuel O.** 2008. Phylogeny.fr: robust phylogenetic analysis for the non-specialist. *Nucleic Acids Res* **36:**W465–W469.

167. **Greve B, Jensen S, Brügger K, Zillig W, Garrett RA.** 2004. Genomic comparison of archaeal conjugative plasmids from Sulfolobus. *Archaea* **1**(4):231–239.

Plasmid Ecology and Evolution

Plasmids—Biology and Impact in Biotechnology and Discovery
Edited by Marcelo E. Tolmasky and Juan C. Alonso

doi:10.1128/microbiolspec.PLAS-0039-2014

Val Fernández Lanza,[1,2,3] Ana P. Tedim,[1,2,3] José Luís Martínez,[2,4]
Fernando Baquero,[1,2,3] and Teresa M. Coque[1,2,3]

21

The Plasmidome of *Firmicutes*: Impact on the Emergence and the Spread of Resistance to Antimicrobials

INTRODUCTION

Firmicutes constitutes one of the dominant bacteria phyla of human and animal gut microbiota. It comprises a number of genera of outstanding relevance in health care and industry such as *Staphylococcus*, *Listeria*, and lactic acid bacteria (LAB), a group of microorganisms that ferment carbohydrates into lactic acid and that includes the genera *Enterococcus*, *Lactobacillus*, *Lactococcus*, *Leuconostoc*, *Pediococcus*, *Streptococcus*, and *Weisella*. Furthermore, species of *Negativicutes* (*Selenomonas*, *Veillonella*) and *Clostridium* have clinical interest for humans and animals (Table 1).

Antibiotic resistance (AbR) in this heterogeneous group of organisms constitutes a significant part of the public health problem. The most recent report by the Centers for Disease Control and Prevention in the United States provides a ranking list of AbR human pathogens according to their threat level to society and the attention that such a problem requires. Gram-positive organisms were grouped in the categories of "urgent" (*Clostridium difficile*), "serious" (methicillin-resistant *Staphylococcus aureus* [MRSA], antibiotic-resistant *Streptococcus pneumoniae*, vancomycin-resistant *Enterococcus* [VRE]), and "concerning" (erythromycin-resistant *Streptococcus pyogenes* and clindamycin-resistant *Streptococcus agalactiae*) on the basis of the limited therapeutic options to treat infections caused by these bacteria-resistant variants (1). LAB, which are used as probiotics and in the preparation of various products (dairy, fermented meat and seafood, fermented cereals and vegetables, wine), are defined as "generally regarded as safe" (GRAS) microorganisms by the U.S. Food and Drug Administration. However, the potential risk to transfer acquired AbR genes recently found in LAB species to animal and human pathogens is a cause for concern. AbR LAB may also contaminate industrial processes, leading to

[1]Servicio de Microbiología, Hospital Universitario Ramón y Cajal, Instituto Ramón y Cajal de Investigación Sanitaria (IRYCIS), Madrid, Spain; [2]Unidad de Resistencia a Antibióticos y Virulencia Bacteriana (HRYC-CSIC), Madrid, Spain; [3]Centro de Investigación en Red en Epidemiología y Salud Pública (CIBER-ESP), Spain; [4]Centro Nacional de Biotecnología (CNB-CSIC), Madrid, Spain.

TABLE 1 Fully characterized plasmids from low G+C bacteria available in GenBank database (updated September 2014)

Phylum/class	Order	Family	Plasmid			
			Total	AbR	Met^R	AbR+Met^R
Actinobacteria			**252**	**16**	**40**	7
	Actynomicetales	*Streptomycetaceae*	56	2	4	2
		Corynebacteriaceae	38	10	7	3
		Nocardiaceae[a]	30	0	9	0
		Mycobacteriaceae[b]	29	2	7	2
		Micrococcaceae[c]	27	1	7	0
		Pseudonocardiaceae[d]	9	0	1	0
		Propionibacteriaceae	7	0	0	0
		Gordonaciaee	6	0	3	0
		Microbacteriaceae[e]	4	0	0	0
		Streptosporangiaceae[f]	2	0	0	0
		Nocardiopsaceae	1	0	0	0
		Nocardioidaceae	1	0	0	0
		Promicromonosporaceae[g]	1	0	0	0
		Micromonosporaceae	1	0	0	0
		Kineosporiineae	2	0	1	0
		Brevibacteriaceae	2	0	0	0
		Frankiaceae	3	0	0	0
		Tsukamurellaceae	1	0	0	0
	Bifidobacteriales	*Bifidobacteriaceae*	29	0	0	0
Firmicutes			**1,073**	**244**	**178**	**85**
Negativicutes	*Selenomonadales*		17	0	1	0
Clostridia			86	7	4	0
Erysipelotrix			1	0	0	0
Bacilli			969	237	121	85
	Lactobacillales					
		Lactobacillaceae	172	16	19	3
		Streptococcaceae	133	15	4	0
		Enterococcaceae	74	24	3	2
		Leuconostoc	29	0	9	0
		Carnobacteriaceae	6	0	1	0
		Oenococcus	6	0	0	0
		Weisella	4	0	0	0
	Bacillales	*Staphylococcaceae*	275	175	118	80
		Bacillaceae	223	5	9	0
		Listeriaceae	14	1	8	0
		Paenobacillaceae	9	0	0	0
		Macrococcus	8	1	0	0
		Planococcaceae	6	0	1	0
		Bacillales group XII	4	0	0	0
		Alicyclobacillaceae	3	0	0	0
		Bhargaceae	2	0	0	0

[a]*Nocardiaceae* (4 *Nocardia*, 25 *Rhodococcus*)
[b]*Mycobacteriaceae* (*Mycobacterium* plus *Amycolicicoccus*)
[c]*Micrococcaceae* (4 *Micrococcus*, 23 *Arthrobacter*)
[d]*Pseudonocardiaceae* (2 *Amycolatopsis*, 6 *Pseudonocardia*, 1 *Saccharomonospora*)
[e]*Microbacteriaceae* (4 *Clavibacter*)
[f]*Streptosporangiaceae* (1 *Planobispora*, 1 *Streptosporangium*)
[g]*Promicromonosporaceae* (*Xylanimonas*)

economic losses (2). In addition, the possibility that opportunistic or commensal bacteria and nonpathogen organisms could serve as reservoirs of AbR genes is increasingly recognized (3). Consequently, several European and American regulatory agencies have recently recommended the mandatory screening of some species such as *Enterococcus faecalis* and *Enterococcus faecium* as indicators of the presence of AbR in foods and food animals and as a mirror of the patterns of antibiotic use in veterinary medicine and agriculture (4, 5). Finally, it is worth mentioning that AbR in a context of the wide use of antibiotics favors the selection of clonal lineages of multihost species with zoonotic potential (e.g., *S. aureus*, *E. faecium*, *Clostridium perfringens*) as well as emblematic zoonotic species such as *Listeria monocytogenes* (see below).

The presence of the same AbR genes in ecologically connected (but also in unconnected) bacterial genera, mentioned above, indicates a complex history of genetic interactions in which AbR genes have parasitized the natural circuits of adaptive gene flow. Plasmids have largely contributed to the spread of AbR and other adaptive genes among members of *Staphylococcus*, *Enterococcus*, and to a lesser extent, species of the *Streptococcus* pyogenic group (6–8), thus influencing the selection of particular subspecies populations due to the acquisition of AbR (8–10). However, the global adaptive role of plasmids of other genera remains largely unexplored outside single pathogens colonizing or infecting single "relevant" hosts. The "single centric" perspective, focusing on "gene tracking" or "vehicle centric" (plasmid, transposon, or other mobile genetic elements [MGEs]) in "single host-single pathogen" systems hampers a comprehensive view of gene and plasmid dynamics and their role in the evolvability of bacterial communities. An integrative view of plasmid ecology is needed to understand community evolvability.

In this work, we analyze the development of AbR in *Firmicutes* within an ecological framework using gene exchange networks. We also discuss the role of plasmids in the emergence, spread, and maintenance of genes encoding resistance to antimicrobials (antibiotics, heavy metals, and biocides) and their influence on the genomic diversity of the main Gram-positive opportunistic pathogens in the light of evolutionary ecology. Finally, a critical revision of plasmid classifications in this group of microorganisms is also provided under this eco-evo perspective by analyzing the 1,326 fully sequenced plasmids of Gram-positive bacteria (*Firmicutes* and *Actinobacteria*) available in the GenBank database at the time this article was written.

AN ECO-EVO PERSPECTIVE TO ANALYZE HGT IN *FIRMICUTES*

Recent phylogenomic analyses using networks revealed a history of horizontal gene transfer (HGT) events even among highly structured and ecologically disconnected groups of bacteria (11–13). These events are more likely to occur in the case of donors and recipients with a similar G+C content (differing in <5% for 86% of connected pairs) (14) and involving plasmids able to mediate exchange of information between close or distant chromosomal backgrounds (12, 15). Although limited by the current number of available genome sequences, such studies evidenced sound differences in "betweeness" among different bacterial groups and plasmids of *Firmicutes*. LAB frequently undergo HGT events among similar species (11), with streptococci acting as a hub for interactions with more distant ecological groups (12), and some plasmids of the Inc18 family possibly contributing to the spread of AbR genes among different bacterial species (15). To analyze this situation in more detail, we constructed a gene exchange network that comprises all genes conferring resistance to antibiotics and heavy metals described in *Firmicutes* so far (Fig. 1 and 2). This network clearly shows that many resistance genes in different bacterial genera can present plasmid and/or chromosomal locations, illustrating the diversity of interactions, often plasmid mediated, within bacterial communities (Fig. 1 and 2). Available (and often fragmented) knowledge from different fields enabled us to state that the dynamics of bacterial populations are influenced by the interplay of selection processes at different levels of organization (genes, MGEs, clones, species) and their associated environments (16–20). Because of that, the complexity resulting from such interplay cannot be understood using either single centric studies or the above-mentioned phylogenomic analysis of HGT networks.

The presence of the same genes in different genetic contexts implies contacts and exchanges between bacteria belonging to different genera, probably facilitated in complex biofilms and environments allowing high local bacterial densities. HGT via transduction or conjugative mechanisms has been extensively documented in *Lactobacillales* and is a prominent process for niche-specific adaptation in different genera (12, 21–23), with plasmids and conjugative transposons being the most relevant providers of communal adaptive gene pools in microbial ensembles sharing complex niches.

An important question is if resistance genes contributed to the recombination between different replicons and, consequently, to their evolvability. The frequent

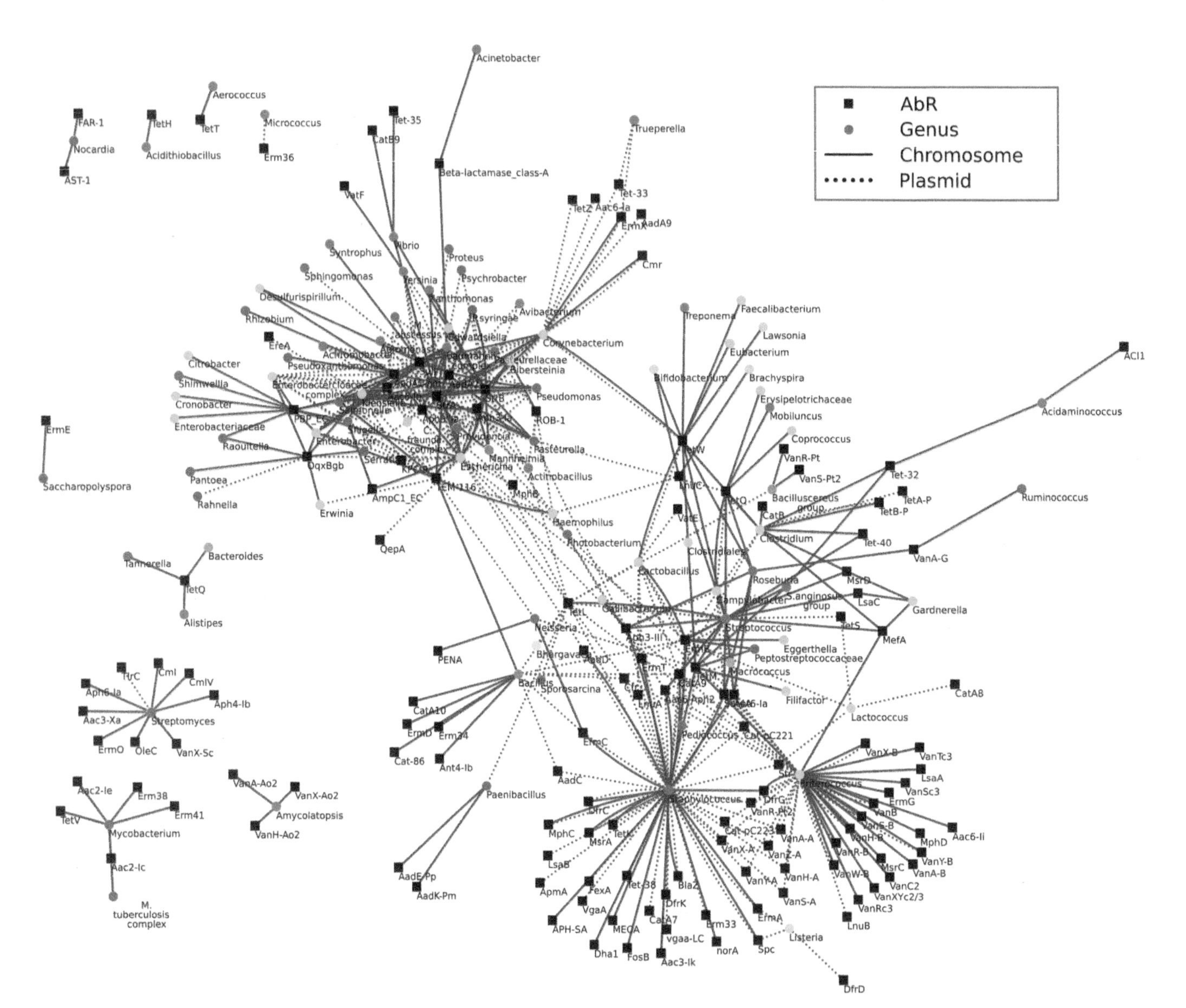

Figure 1 Protein content network (PCN) of AbR proteins found in plasmids and chromosomes of *Firmicutes* and *Actinobacteria*. To determine the AbR protein catalog of Gram-positive strains (chromosomes and plasmids), a Blastp search was performed of all their proteomes against the ARG-ANNOT database (http://en.mediterranee-infection.com/article.php?laref=283&titre=arg-annot) using a cut-off of 1e-30 and 85% of identity. The presence of the Gram-positive AbR proteins identified above in all bacterial species (only complete sequences, not partial) was determined using a similar Blast search (blastp, 1e-30 E-value and 85% identity) against the NCBI GenBank database. The nodes correspond to bacterial species (circular nodes; each color indicates one genus) and AbR proteins (square nodes). Nodes were connected by an edge when a positive hit between AbR proteins and one or more strains of a given species were identified. Edges further indicate the location of the AbR genes associated with each AbR protein of the Gram-positive catalog. Solid lines represent chromosomal location, and dotted lines represent plasmid location. When an AbR gene was located in both chromosomes and plasmids, both lines were plotted.
doi:10.1128/microbiolspec.PLAS-0039-2014.f1

association of resistance genes with site-specific recombination systems and insertion sequences located either in plasmids or in chromosomes favors homologous recombination and therefore different events of integration or excision, as well as the interplay among different elements (19, 24–28). Restriction-modification (RM) systems and clustered regularly interspaced short palindromic repeats (CRISPR) are the main posttransfer barriers protecting a given host cell from invasion by foreign DNA either by conjugation transformation or transduction (9, 29, 30). Some RM systems specifically limit the acquisition of plasmids to some pathogens which may influence their clonal structure (e.g., RM types I, III, and IV in *S. aureus*) (31, 32). This may also explain the lack of plasmids in certain species such as *S. pneumoniae* or the narrow host range of plasmids from some *Clostridium* species (33, 34). Anti-RM systems such as analogues of ArdA (alleviation of restriction of DNA) proteins that act against type I restriction systems (detected in Tn*916* and CTn*6000*) or other genes predicted to be involved in methylation (e.g., in CTn*6000* and Tn*1721*) are involved in the restricted spread of certain MGEs, as well as in certain clonal expansions. There is evidence that the presence of complete CRISPR loci is inversely proportional to the presence of MGEs in *Clostridium* and *Staphylococcus* (34, 35), a situation that has also been suggested to occur for *Streptococcus* and *Enterococcus* (34, 36, 37). In agreement with this statement, Fig. 1 and 2 reflect a heterogeneous distribution of AbR and heavy metal and biocide resistance (Met^R and Bc^R) genes in different genera. Particularly interesting is the confinement of vancomycin resistance within enterococci and of some AbR, Met^R, and Bc^R genes within staphylococci and clostridia, a situation that is in part due to the barriers shaping different populations (see next sections).

Fluctuating environments, concentration gradients, and high population sizes, all frequent in different "source-sink" ecologies such as bacterial populations under antimicrobial selective pressure, favor DNA transfer and the selection of some clonal and plasmid variants (38–42). Such selective processes favor the emergence of novel variants, resulting from genetic drift, or migration to heterogeneous environments. Recent studies demonstrated that plasmid transfer occurs *in vivo* more frequently than in experimental evolution assays (43), and that gradients in populations under selection pressure by diverse antimicrobials favor selection of multidrug resistance (MDR) plasmids (44, 45).

The influence of plasmids has been extensively discussed in the literature from different evolutionary perspectives (20, 46–48), but only limited information about plasmid ecology and the specific roles that plasmids actively play within microbial systems *in situ* is available. Mosaic MGEs have often been documented in *Firmicutes*, reflecting a different epidemiological history of contacts with strains of the same or different species. Some mosaic MGEs have been fixed, making the interpretation of results by using traditional classification schemes extremely difficult (49).

PLASMID DIVERSITY AND CLASSIFICATION SYSTEMS

Efforts in plasmid characterization and classification are justified for the understanding of plasmid biology. Nowadays, plasmid categorization is relevant from the public and environmental health perspective to follow the movement of genes coding for resistance to antimicrobials (antibiotics, heavy metals, biocides), colonization and virulence factors for humans and animals, and/or other adaptive traits that drive ecological success (bacteriocins, metabolic traits) and consequently increase the population size of bacteria harboring MGEs. In fact, only a "representative diversity" of bacterial plasmids has been systematically analyzed in a few genera of multihost opportunistic pathogens of interest in biomedicine, with a particular emphasis on species of the *Enterobacteriaceae*, *Pseudomonadaceae*, *Staphylococcaceae*, and *Enterococcaceae* families (7, 50–53). The diversity of plasmids from *Lactococcus* (54), *Lactobacillus* (55), *C. perfringens* (56), *Micrococcus* (57), and *Bifidobacterium* (58) has also been analyzed from different perspectives.

Plasmid diversity within a particular bacterial species in the *Firmicutes* phylum started to be comprehensively analyzed in the 1960s just after the discovery of staphylococcal plasmids. These elements were initially categorized into three main classes designated by roman numerals on the basis of size, replication machinery, ability to be transferred, phenotypic and functional characteristics, and host range (7, 51, 53, 59, 60). Class I comprised high copy number plasmids (10 to 60 copies per cell) of less than 5 kb with a rolling circle replication (RCR) mechanism that often harbored one or two AbR genes (usually conferring resistance to tetracycline, chloramphenicol, macrolides, and trimethoprim). Class II comprised low copy number plasmids (4 to 6 copies per cell) of 15 to 40 kb, with a theta replication mechanism, which typically carried resistance to antibiotics (β-lactams, aminoglycosides, and macrolides), heavy metals (arsenic, cadmium, and mercury), and/or antiseptics (quaternary ammonium compounds). Class III comprised plasmids similar to those found in

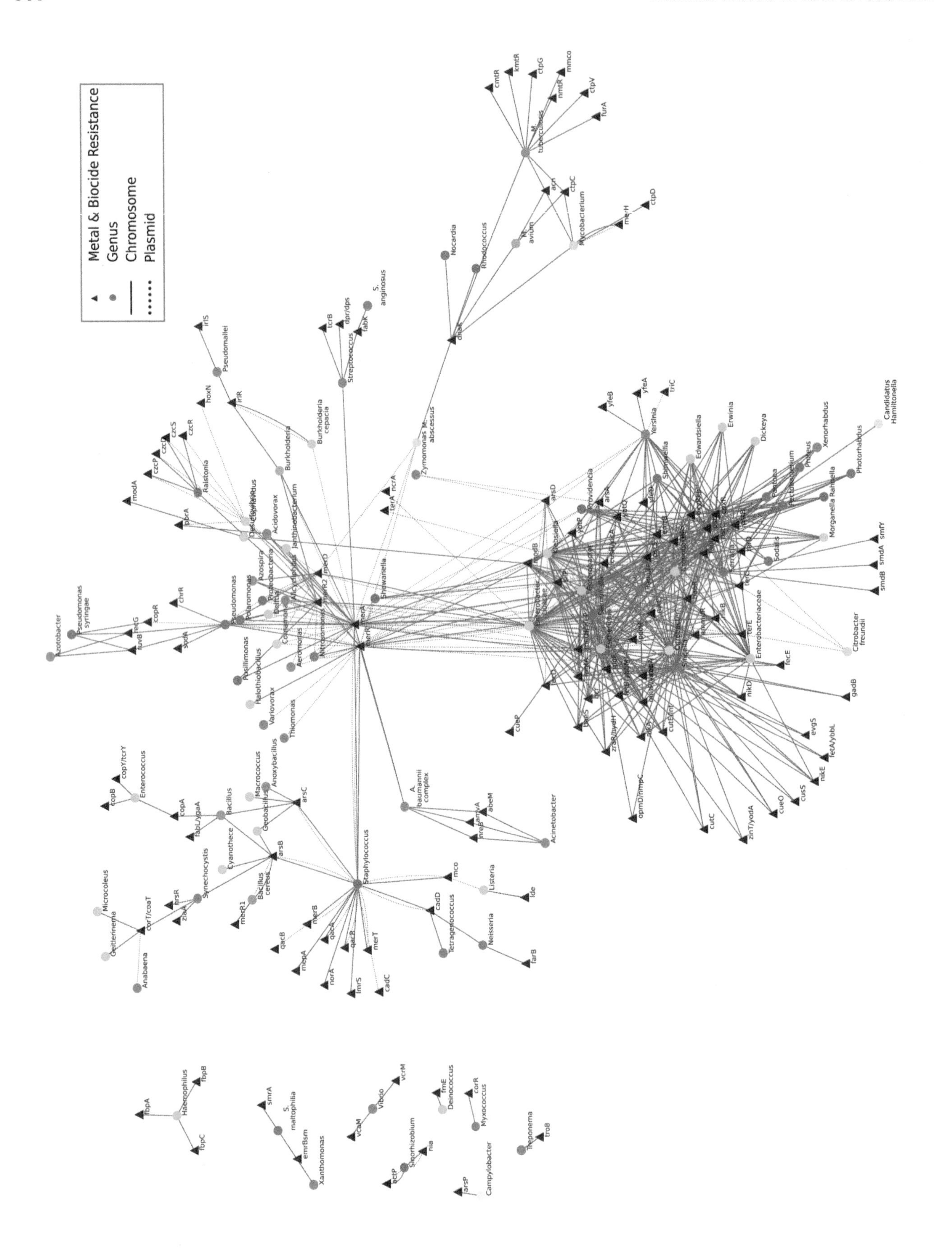
Metal & Biocide Resistance
Genus
Chromosome
Plasmid
M. tuberculosis
Mycobacterium
M. avium
Nocardia
Rhodococcus
Streptococcus
S. anginosus
Pseudomallei
Burkholderia
Burkholderia cepacia
Zymomonas
M. abscessus
Ralstonia
Pseudomonas
Pseudomonas syringae
Azotobacter
Acidovorax
Janthinobacterium
Azospira
Shewanella
Polaromonas
Comamonas
Aeromonas
Pusillimonas
Variovorax
Thiomonas
Yersinia
Edwardsiella
Erwinia
Dickeya
Xenorhabdus
Photorhabdus
Candidatus Hamiltonella
Providencia
Morganella
Sodalis
Enterobacteriaceae
Citrobacter freundii
Enterococcus
Macrococcus
Anoxybacillus
Bacillus
Geobacillus
Cyanothece
Synechocystis
Bacillus cereus
Microcoleus
Geitlerinema
Anabaena
Staphylococcus
A. baumannii complex
Acinetobacter
Listeria
Tetragenococcus
Neisseria
Haemophilus
S. maltophilia
Xanthomonas
Vibrio
Sinorhizobium
Deinococcus
Myxococcus
Campylobacter
Treponema
cmtR
kmtR
ctpG
nmtR
mmco
ctpV
furA
acn
ctpC
merH
ctpD
tcrB
dpr/dps
fabK
irlS
irlR
hoxN
czcP
czcD
czcS
czcR
modA
pbrA
chrR
copR
fabV
tcrA
terA
arsD
yfeB
yfeA
triC
smfY
smdA
smdB
terE
fecE
gadB
nikD
evgS
fetA/ybbL
nikE
cusS
cueO
zinT/yodA
cutC
opmD/nmpC
zraR/hydH
cueP
copY/tcrY
copB
copA
fabL/ygaA
arsC
arsB
arsR
corT/coaT
ziaA
merR1
merB
qacB
qacA
qacR
merT
mepA
norA
lmrS
cadC
cadD
mco
lde
farB
abeM
fbpA
fbpB
fbpC
smrA
emrBsm
vcrM
vcaM
fmE
corR
nia
actP
arsP
troB

class II which were transferred by conjugation (61). Afterward, Richard Novick and others classified staphylococcal plasmids in 15 incompatibility (Inc) groups based on the finding that two plasmids with the same replication (rep) proteins cannot be stably maintained in the same cell (50, 62, 63). Plasmids of most Inc groups correspond to class I (10 Inc groups of apparently closely related plasmids) and class II (diverse plasmids that belong to the same Inc group) (53). Following the same Inc numeral designation criteria, Brantl et al. categorized a few streptococcal plasmids that replicated via a theta mechanism and that were regulated by an antisense RNA that mediated transcriptional attenuation, such as the Inc18 family (64) (see below). Pheromone-responsive plasmids of enterococci were also subgrouped into different incompatibility groups on the basis of distinct responses to small peptides or pheromones which are secreted by plasmid-free donors (65).

A multiplex-PCR typing system based on the diversity of replication initiator proteins (RIPs) developed by Jensen et al. (59) has recently been applied for the characterization of *Firmicutes* plasmids, mainly staphylococci (66) and enterococci (67–71) of human, animal, and environmental origin. According to this typing system, RIP variants are designated as "rep" followed by a subindex number and are arbitrarily called Rep families. Although this system is very useful to enlarge the knowledge of scarcely explored plasmid diversity in contemporary isolates of enterococci and staphylococci, its application is limited to known plasmids, mainly AbR plasmids of these genera, as illustrated in various surveys and this study (59, 66, 68, 71).

The diversity of mobilization (MOB) systems has also recently been used to classify plasmids and other conjugative elements in different bacterial groups including *Firmicutes* (72, 73). The approach relies on the variability of relaxases (RELs), which form part of the plasmid MOB region, are involved in the initiation of DNA transfer, and that, aside from the origin of transfer (*oriT*), are present in both conjugative and mobilizable plasmids as well as in other conjugative elements (74). To date, only five (MOB_P, MOB_Q, MOB_V, MOB_C, and MOB_T) out of seven known REL families have been identified in *Firmicutes* (7, 72, 74). MOB_Q, MOB_C, and MOB_T are present in conjugative elements, and MOB_V is present in mobilizable plasmids. MOB_P has been identified in both conjugative and mobilizable elements (7, 72, 73). The application of this PCR-based classification scheme is obviously limited to the typing of known RELs. Frequent plasmid mosaicism, redundancy, and coexistence of different "core" genes, and the interplay of plasmids with other conjugative elements that contain homologs of RIPs and RELs, complicates the establishment of a robust plasmid core ontology and precludes the use of typing approaches similar to those used in Gram-negative organisms such as plasmid multilocus sequence type (http://pubmlst.org/plasmid/).

Whole-genome (plasmid/chromosome) sequencing provides accurate and nonbiased information on plasmid backbones. Although the number of fully sequenced plasmids in databases is still limited, we used a plasmid homology network analysis of the 1,326 fully sequenced plasmids of *Firmicutes* and *Actinobacteria* to study the diversity of plasmids carrying genes coding for AbR and Met^R/Bc^R and the impact of plasmids in the evolvability of contemporary AbR bacterial populations of *Firmicutes*.

Figures 3 and 4 illustrate the existence of group-specific plasmid populations, with a number of plasmids being shared between *Lactobacillales* (mainly *Enterococcus* and *Streptococcus*) and *Bacillales* (*Staphylococcus*), which are greatly implicated in the spread of AbR and Met^R/Bc^R. These shared plasmids include RCR and theta-replicating plasmids of different families, which

Figure 2 PCN of metal-biocide (Met^R/Bc^R) proteins found in plasmids and chromosomes of *Firmicutes* and *Actinobacteria*. To determine the Met^R/Bc^R protein catalog of Gram-positive strains (chromosomes and plasmids), a Blastp search was performed of all their proteomes against the BacMet database (http://bacmet.biomedicine.gu.se/) using a cut-off of 1e-30 and 85% of identity. The presence of the Gram-positive Met^R/Bc^R proteins identified above in all bacterial species (only complete sequences, not partial) was determined using a similar Blastp search (blastp, 1e-30 evalue and 85% identity) against the NCBI GenBank database. The nodes correspond to bacterial species (circular nodes) and Met^R/Bc^R proteins (triangular nodes). Nodes were connected by an edge when a positive hit between Met^R/Bc^R proteins on one or more strains of a given species was identified. Edges further indicate the location of the Met^R/Bc^R genes associated with each Met^R/Bc^R protein of the Gram-positive catalog. Solid lines represent chromosomal location, and dotted lines represent plasmid location. When a Met^R/Bc^R gene was located in both chromosomes and plasmids, both lines were plotted. doi:10.1128/microbiolspec.PLAS-0039-2014.f2

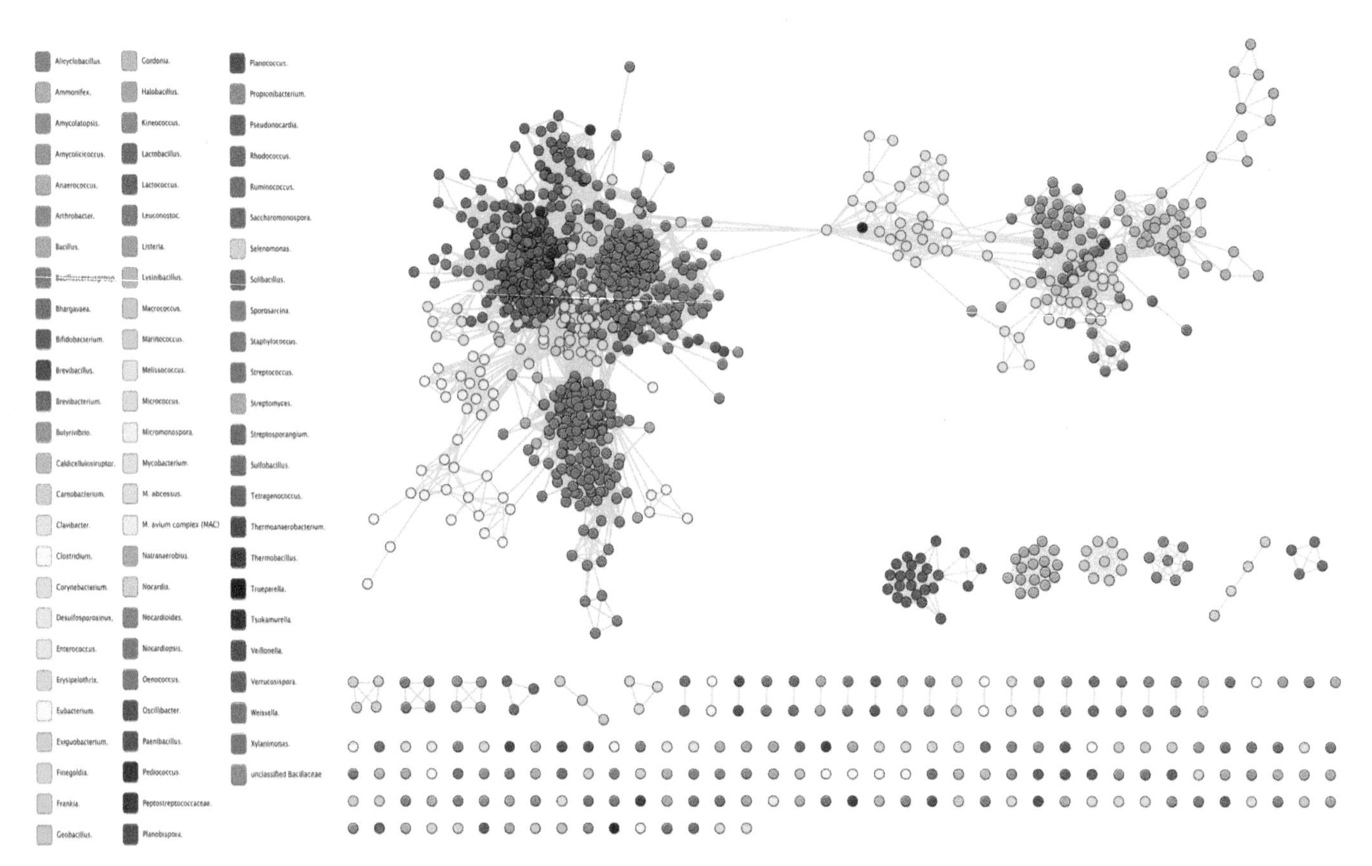

Figure 3 Plasmid homology network. The genomic homology network was performed using "All-versus-All" genomic Megablast (238) of 1,326 fully sequenced plasmids from low G+C bacterial species (*Firmicutes* and *Actinobacteria* phyla) available at public gene databases. The nodes correspond to bacterial plasmids (circular nodes; different colors representing different genera). Two nodes are connected by an edge if they share homologous DNA. doi:10.1128/microbiolspec.PLAS-0039-2014.f3

have been recently analyzed at the molecular level (7, 75, 76). Next in this section, we will analyze the diversity of these groups and highlight the usefulness of current typing systems for each group. However, it is of note that the main genera of *Firmicutes* carry a variable number of plasmids containing several replication and transfer systems, some of them being able to be transferred. The interplay between genes, plasmids, and populations will be analyzed under an ecological perspective in the section "Gene and Plasmid Flow Shapes the Evolutionary Ecology of Firmicutes."

Rolling Circle Replication Plasmids

RCR plasmids are classified in a few families according to the RIP and the double origin of replication (*dso*) (see comprehensive reviews in references 75, 77–79). Most of the RCR plasmids known to date have been found in species of *Firmicutes*, *Proteobacteria*, *Cyanobacteria*, and *Spirochaetes*, and some of them have been identified in genetically distant hosts. The production of single-stranded DNA and the mechanism of replication of these plasmids enhance their ability to recombine, by either homologous recombination or illegitimate recombination with other RCRs and theta replicating plasmids (Fig. 5 to 10 and text below). RCR plasmids are also frequently integrated into chromosomes (e.g., pUB110 within SCCmec cassettes in methicillin-resistant *S. aureus* or pC194/pUB110 [*catA*] in *S. pneumoniae* genomes) (80).

In *Firmicutes*, four groups of RCR plasmids have been defined according to RIP similarity, namely Rep_trans (PF025486), Rep_1 (PF14046), Rep_2 (PF01719), and Rep_L, which are historically represented by plasmids pT181, pUB110, pMV158, and pSN2, respectively (53, 75, 77, 81). Within these families, some members have been fixed by selection and might be maintained by the vertical expansion of certain clones, aside from HGT, with the emergence of

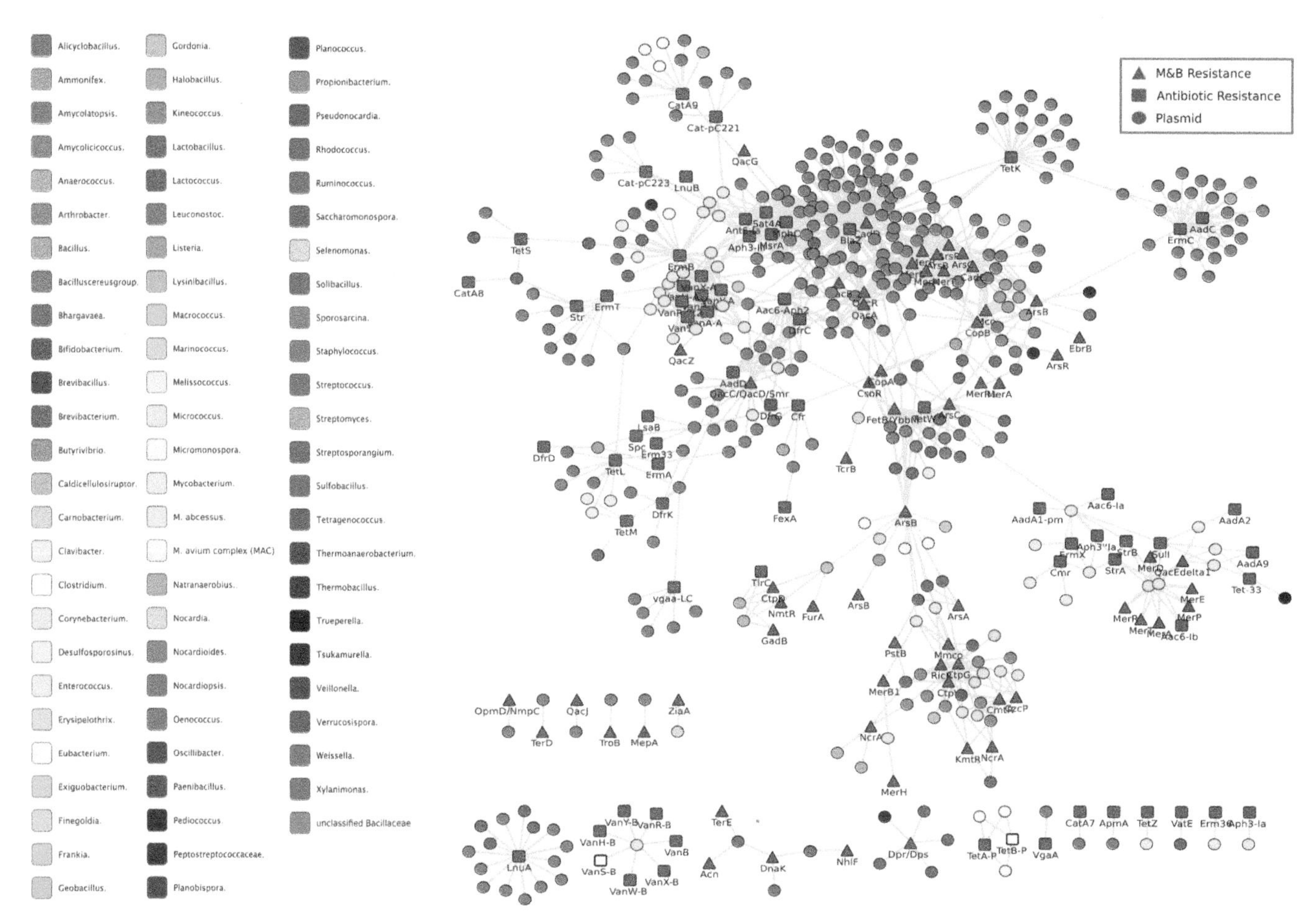

Figure 4 PCN of AbR and Met^R/Bc^R proteins located on plasmids of *Firmicutes* and *Actinobacteria*. PCN of AbR and Met^R proteins found in Gram-positive plasmids. We formed the PCN by representing plasmids as circular nodes, AbR as square nodes, and Met^R/Bc^R as triangular nodes, connecting two nodes (plasmid and AbR or Met^R/Bc^R) if one plasmid has this AbR or Met^R/Bc^R. The presence of the Met^R/Bc^R gene was determined by blasp of all plasmid proteins against the BacMet database. The presence of the AbR gene was determined by blasp of all plasmid proteins against the ARG-ANNOT database. The colors for the genus are the same as those in Fig. 3.
doi:10.1128/microbiolspec.PLAS-0039-2014.f4

variants from time to time. Figures 11 to 14 and Supplementary Table S1 show the similarity of genes encoding RIPs of all available fully sequenced plasmids and the correspondence to the Rep families described by Jensen et al. (59). These plasmids may contain different adaptive genes (AbR, heat shock proteins, or bacteriocins), although most of them are classified as "cryptic," without any clear adaptive function.

The Rep_1 family

The Rep_1 family comprises plasmids with RIPs of the families rep_{13} (associated with *catA7*, which encodes resistance to chloramphenicol), rep_{21} (cryptic or eventually carrying *lnuA*, coding for resistance to lincosamides), rep_{22} (carrying a variety of AbR genes), and other underrepresented members categorized as $rep_{Unique7}$. However, the available typing systems are unable to classify relevant Rep_1 plasmid members including plasmids containing heat shock proteins in *Streptococcus thermophilus*, plasmid-borne bacteriocins in *S. pyogenes* (82), or *S. pneumoniae* plasmids (80, 83), among others (Fig. 11). Remarkably, RIPs of this Rep_1 group are often detected in mosaic plasmids of staphylococci and enterococci (Fig. 5 and 7), some plasmid chimeras being fixed and persistently recovered for years. For example, emblematic mosaics theta/RCR plasmids of staphylococci (e.g., cointegrates of RepA_N/pSK41 and Rep_1/pUB110, which encode

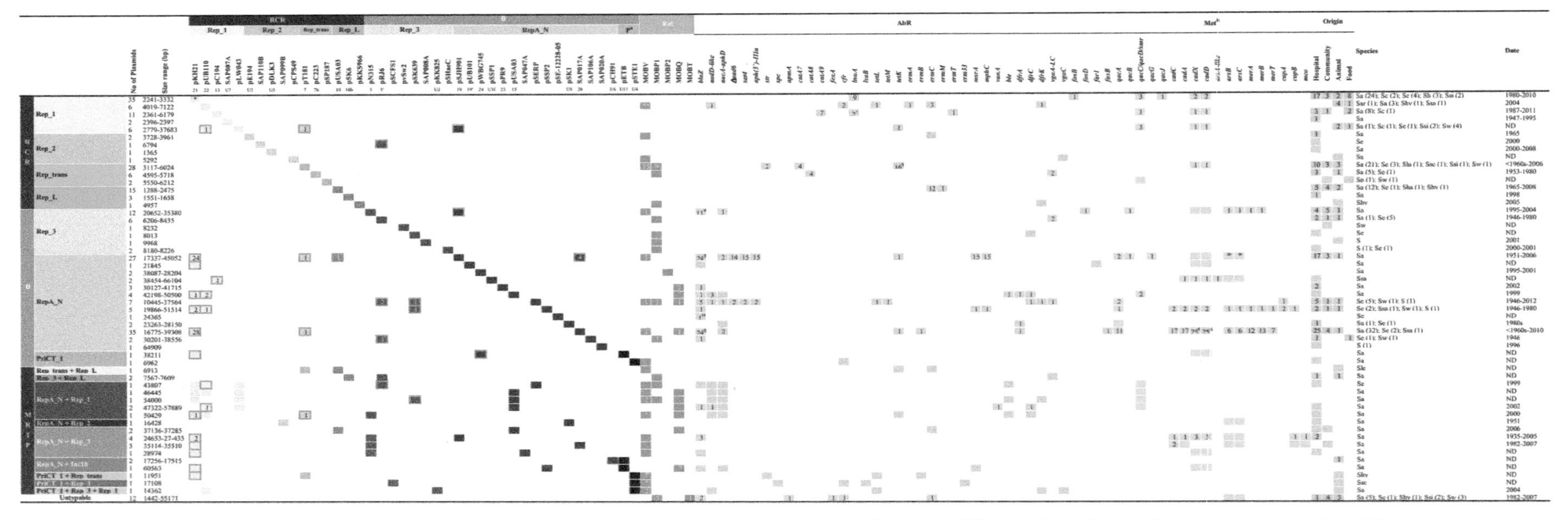

Figure 5 Plasmids from *Staphylococcus* spp. The presence of an orange border in the RIP family indicates that the corresponding RIP is truncated. [a]PriCT_1; *One of these plasmids (GenBank accession number NC_013381) has a truncated rep_pKH21 (rep_1) gene, and no other known RIPs were identified. **Two of these plasmids (GenBank accession number NC_016054 and NC_019144) appear to have two copies of the MOB_V gene. +One of these plasmids (GenBank accession number NC_008354) has two copies of the *lnuA* gene. §One plasmid (GenBank accession number NC_001393) has a truncated copy of the *tetK* gene. ¥One plasmid (GenBank accession number NC_010419) has a truncated copy of the *blaZ* gene. £The plasmid (GenBank accession number NC_005076) appears to have two copies of the MOB_V gene. $One plasmid (GenBank accession number NC_018959) has a truncated copy of the *blaZ* gene. &Two plasmids (GenBank accession numbers NC_007931 and NC_016942) have two copies of the *arsB* and *arsC* genes. ¢This plasmid (GenBank accession number NC_013320) appears to have two copies of the MOB_V gene. ΠThis plasmid (GenBank accession number NC_005004) has a truncated copy of the *blaZ* gene. ♀Three plasmids (GenBank accession numbers NC_013321, NC_019007, and NC_018976) have a truncated copy of the *blaZ* gene. Eleven of these plasmids have a truncated copy of the *cadD* gene (GenBank accession numbers NC_020531, NC_020538, NC_013337, NC_020534, NC_020565, NC_020567, NC_020530, NC_020539, NC_017352, NC_013323, and NC_022610). One plasmid (GenBank accession number NC_013334) has a truncated copy of the *cadX* gene. This plasmid (GenBank accession number NC_020237) appears to have two copies of the MOB_V gene. This plasmid (GenBank accession number NC_022598) appears to have two copies of the MOB_V gene. Abbreviations: MRIP, Multi-RIP; S, *Staphylococcus* spp; Sar, *Staphylococcus arlettae*; Sa, *S. aureus*; Sc, *Staphylococcus chromogenes*; Se, *Staphylococcus epidermidis*; Sha, *Shaemolyticus haemolyticus*; Shy, *Staphylococcus hyicus*; Sle, *Staphylococcus lentus*; Slu, *Staphylococcus lugdunensis*; Sp, *Staphylococcus pasteuri*; Ssa, *Staphylococcus saprophyticus*; Ssc, *Staphylococcus sciuri*; Ssi, *Staphylococcus simulans*; Sw, *Staphylococcus warneri*.
doi:10.1128/microbiolspec.PLAS-0039-2014.f5

resistance to gentamicin) and *E. faecalis* (e.g., pAMα1) have both been selected in those lineages since the early 1970s (7, 71, 84).

The Rep_trans group

Plasmids of the Rep_trans group are clustered in two large branches (Fig. 12). One branch comprises plasmids of *Staphylococcus* that harbor *tetK* (rep_7) and *catA8/catA7* (rep_{7b}) with different MOB genes. Such plasmids have been reported in *S. aureus* since their first detection in the early 1950s (84) and were eventually described in contemporary *E. faecalis* isolates (68, 85). A second branch contains pRI1-like plasmids (rep_{14}), which correspond to plasmids of different enterococcal species (*E. faecium, Enterococcus hirae, Enterococcus mundtii)* isolated from foodborne animals and hospital patients (7, 59, 71, 86). These plasmids can be mobilized by other AbR conjugative theta replicating plasmids present in the same cell (71, 87), and it seems they are widely spread among enterococcal populations.

The Rep_2 group

The Rep_2 group (Fig. 13) comprises numerous promiscuous elements able to replicate in distant hosts which have been extensively analyzed at the molecular level by Espinosa et al. using pMV158 as a model (75). Plasmids carrying *ermT* (an inducible methylase conferring resistance to first-line macrolide-lincosamide antibiotics such as erythromycin and clindamycin), from group A Streptococci (GAS) and group B Streptococci (GBS), are the sole representatives of AbR in this group. They appear to be responsible for the rise of macrolide resistance among GAS and GBS in hospitals since the mid-1990s (8).

The Rep_L group

In contrast to the above-mentioned RCR plasmid groups, proteins within the Rep_L family (Fig. 14) are represented in public gene databases by a very few RIPs of *Staphylococcus, Selenomonas* (class *Negativicutes*), and *Butyrivibrio* (Clostridia) species, all these genera being frequent components of the oral flora of humans and the rumen of some animal species. These plasmids are responsible for the widespread *ermC* in staphylococci (rep_{10}). Interestingly, the emergence of both *ermT*-Rep_2 and *ermC*-Rep_L plasmids seems to be associated with the abusive use of tylosin in cattle, amplified by the location of these AbR genes in RCR plasmids, and further transferred to other populations of *Firmicutes* (8, 88, 89).

RCR plasmids were associated with REL of the group MOB_{V1} (72, 73, 75), although representatives of all the RCR groups mentioned above that lack REL were detected in databases. Interestingly, RELs of MOB_{P1} and MOB_T families were also found, and their presence is probably due to the co-integration of RCR with theta-replicating plasmids (see below).

Theta-Replicating Plasmids

Four families of plasmids that replicate by a theta mechanism, three that comprise conjugative plasmids (RepA_N, Inc18, and pMG1) and one in groups small nonconjugative elements (Rep_3), are involved in the capture, spread, and maintenance of AbR among different genera of *Firmicutes*. Members of the RepA_N and Inc18 families are often enriched in insertion sequences, mainly IS*257*, IS*256*, IS*1216*, IS*L3*, and IS*431*, that facilitate co-integration, rearrangements, and deletions among elements of *Staphylococcus*, *Enterococcus*, LAB, and *Clostridium* of different origins (6, 7, 28, 90–95). Such recombination events seem to have facilitated the origin of the great mosaicism of MDR plasmids that often carry more than one RIP, lack transfer and maintenance modules, and eventually carry more than one REL (Figs. 5 and 7). The transfer mechanisms of RepA_N pSK41-like plasmids and the Inc18-like plasmids are similar and are categorized as type IV secretion systems (96).

The Rep_3 family

Plasmids containing RIPs with the Rep_3 domain (Fig. 15) are common among disparate bacterial genera including *Staphylococcus*, *Enterococcus*, *Streptococcus*, *Lactobacillus*, and *Enterobacteriaceae* (7). Figure 15 shows the diversity of RIPs among fully sequenced plasmids of *Firmicutes*, and Fig. 5 to 10 reflect the features of known members of this family within each genus of biomedical interest. In enterococci, Rep_3 plasmids (<15 kb) have been found in isolates recovered from hospitalized patients, animals (pigs, cows), cheese, milk, and dry-fermented sausage, frequently associated with the production of bacteriocins that are active against a variety of Gram-positive genera (7). In *Lactobacillus* and *Lactococcus*, they harbor bacteriocins and, eventually, AbR genes. Rep_3 plasmids play a relevant role as vehicles of AbR among staphylococci. Plasmids from *S. aureus* are overrepresented by closely related variants containing Rep_5, which are associated with genes coding for penicilinase and resistance to heavy metals (cadmium and arsenic) (51, 53, 66, 84). Staphylococcal plasmids within this group include AbR plasmids from coagulase-negative strains of animal origin, some of them with RIPs that would not be detected by current typing systems (97, 98).

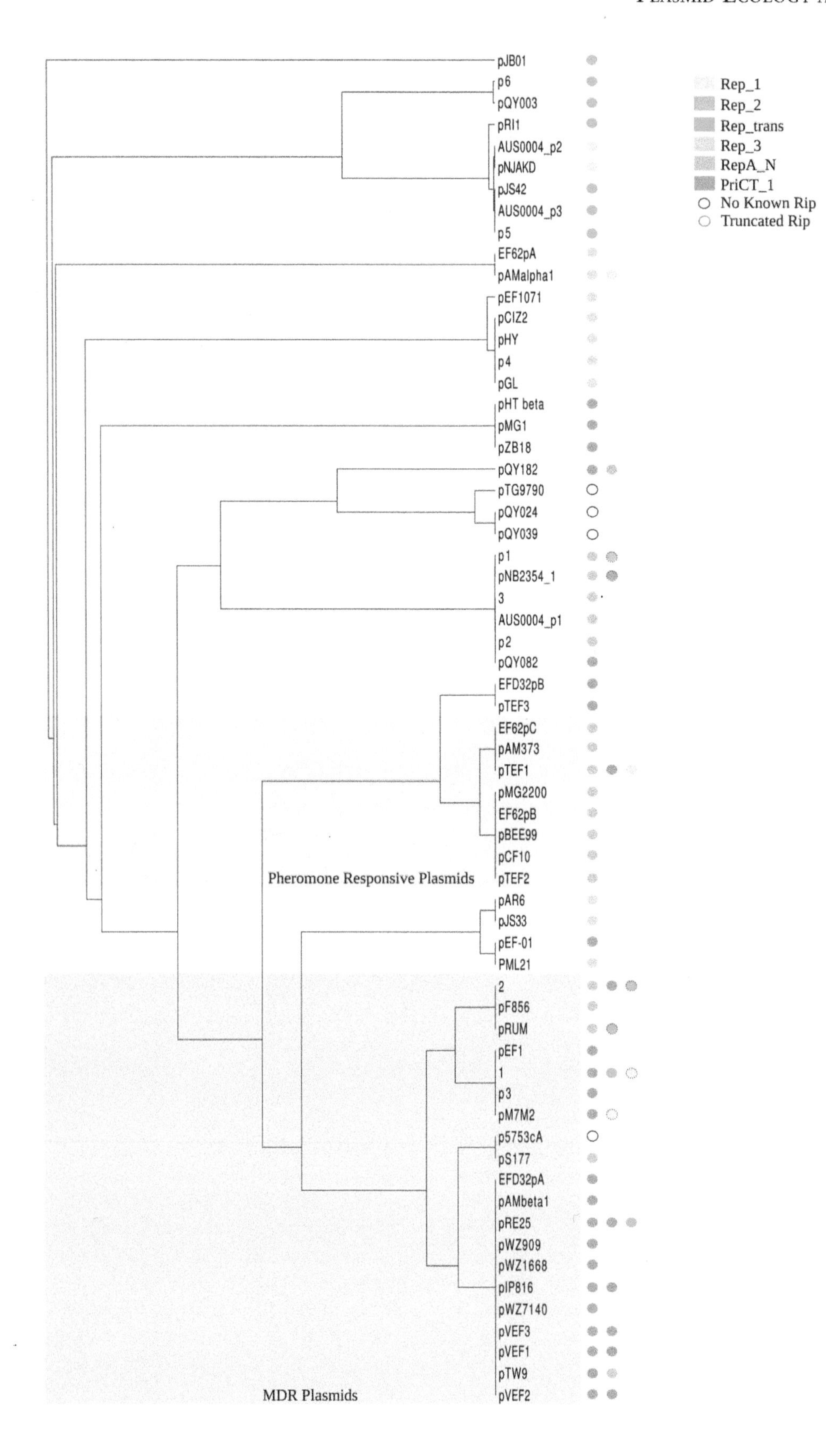

pJB01
p6
pQY003
pRI1
AUS0004_p2
pNJAKD
pJS42
AUS0004_p3
p5
EF62pA
pAMalpha1
pEF1071
pCIZ2
pHY
p4
pGL
pHT beta
pMG1
pZB18
pQY182
pTG9790
pQY024
pQY039
p1
pNB2354_1
3
AUS0004_p1
p2
pQY082
EFD32pB
pTEF3
EF62pC
pAM373
pTEF1
pMG2200
EF62pB
pBEE99
pCF10
pTEF2
Pheromone Responsive Plasmids
pAR6
pJS33
pEF-01
PML21
2
pF856
pRUM
pEF1
1
p3
pM7M2
p5753cA
pS177
EFD32pA
pAMbeta1
pRE25
pWZ909
pWZ1668
pIP816
pWZ7140
pVEF3
pVEF1
pTW9
pVEF2
MDR Plasmids
Rep_1
Rep_2
Rep_trans
Rep_3
RepA_N
PriCT_1
No Known Rip
Truncated Rip

The Inc18 family

First described in the 1990s, the Inc18 family comprises a highly heterogeneous group of broad host range, low copy number plasmids (<10 per cell) that replicate by a theta mechanism, regulated by an antisense RNA that mediates transcriptional attenuation and that are able to conjugate on solid media at high frequencies (64, 99). The transfer system of pIP501 has been extensively studied and constitutes a paradigm of conjugation systems, showing significant similarity with the *tra* regions of RepA_N plasmids pGO1 and pSK41 from *S. aureus* and pMRC01 from *Lactococcus lactis* (96, 100).

The Inc18 group is traditionally represented by three emblematic plasmids: pSM19035 (101) and pIP501 from *S. agalactiae* and pAMβ1 from *E. faecalis* (64, 101–104). It gets its name from the apparent incompatibility of these plasmids with each other described in seminal studies in the field and following the nomenclature of Inc groups started by Richard Novick for staphylococcal plasmids (50, 60, 64, 105). Inc18 plasmids frequently carry the postsegregational killing systems, ες, and type I partition cassette *prgPprgO*, which are associated with a variety of RIPs and seem to contribute to their persistence in different populations in the absence of antibiotic selection pressure (7, 106, 107). Detailed molecular characterization of such plasmids is described elsewhere (64, 99, 108) and shows a remarkably high modular interplay among different Inc18 plasmids, leading to the high modularity observed in plasmid sequences (see Fig. 5 to 10 and text below).

Inc18 plasmids have contributed remarkably to the spread of AbR (macrolides, chloramphenicol, aminoglycosides, and glycopeptides) and Met^{R} (copper and mercury) among streptococci and other phylogenetically distant genera of Gram-positive (*S. aureus*, *Listeria*, *Bacillus subtilis*, *Lactobacillus*, *Leuconostoc*, various *Clostridium* species) and Gram-negative bacteria (108–111). Plasmid relatives of pAMβ1 (harboring *ermAM*, later on recognized as *ermB*, and conferring resistance to macrolides, lincosamides, and streptogramines) and pIP501 (carrying *ermB* and $catA7_{pC221}$, which confers resistance to chloramphenicol) were rapidly spread during the 1970s and have frequently been detected among streptococci of groups A, B, and D (enterococci) since then (110, 112–114) (see also Supplementary Table S1 for contemporary representatives of this plasmid group). Initially, the successful spread of intact AbR plasmids among clones of various streptococcal genera, including *S. pneumoniae*, and *S. aureus* was reported, despite the lack of stability in these last two clonal backgrounds (110, 113). Inc18 plasmids conferring resistance to aminoglycosides (kanamycin, streptomycin, and neomycin) and to macrolides were also detected in 1972, in the emblematic *Streptococcus* (*Enterococcus*) *faecalis* strain JH1 that carried pJH1 (an MDR plasmid, presumably Inc18) and pJH2 (a RepA_N pheromone-responsive plasmid carrying hemolysin and bacteriocins). pJH1 represented the first description of conjugative transfer of AbR plasmids in enterococci (114). Aminoglycoside resistance in pJH1 relatives was due to the presence of Tn*5405*, a transposon comprising three genes in tandem (an aminoglycoside 6-adenyltransferase [*aad*], a streptothricin acetyltransferase [*sat*], and an aminoglycoside-phosphotransferase [*aph3*]). These genes were identified later on in *S. pyogenes, S. agalactiae, S. aureus, Campylobacter coli, C. perfringens, and C. difficile* (now *Peptoclostridium difficile)*.

More recently, diverse Inc18 plasmids carrying Tn*1546* in enterococci and staphylococci have emerged in different locations. In Europe, Inc18::Tn*1546* plasmids (such as pVEF1, pEVF2, pVEF3, and pVEF4) seem to have evolved from pIP816 (the first Inc18::Tn*1546* was isolated in France in 1987). They lack a transfer system and appear to be confined to *E. faecium* (70, 115, 116). Inc18::Tn*1546* plasmids from the United States are linked to *E. faecalis* isolates (pWZ909, pWZ1668, pWZ7140) and contain a complete transfer system (117, 118). A plethora of multiresistant mosaic Inc18 plasmids containing up to three RIPs, including RepR of pIP501 (CAA35647.1) and RepS of pRE25 (YP_783890.1), have been described in

Figure 6 Hierarchical clustering dendrogram of plasmids from enterococci. The matrix distance used for building the UPGMA dendrogram is based on the Raup-Crick distance of the orthologous protein profile of each plasmid. For each plasmid, a presence/absence protein profile was made using cut-off values of 80% identity and 80% coverage. Protein clustering was made by using CD-HIT (239). Different background colors are used to emphasize branches of related plasmids and are the same as those defined in Fig. 5. Names to the left of the dendrogram indicate the RIP family. Background colors were used to point out plasmid groups frequently involved in mobility of AbR genes and *E. faecalis* pheromone-responsive plasmids. Circles indicate RIPs identified in each plasmid according to data shown in Fig. 7.
doi:10.1128/microbiolspec.PLAS-0039-2014.f6

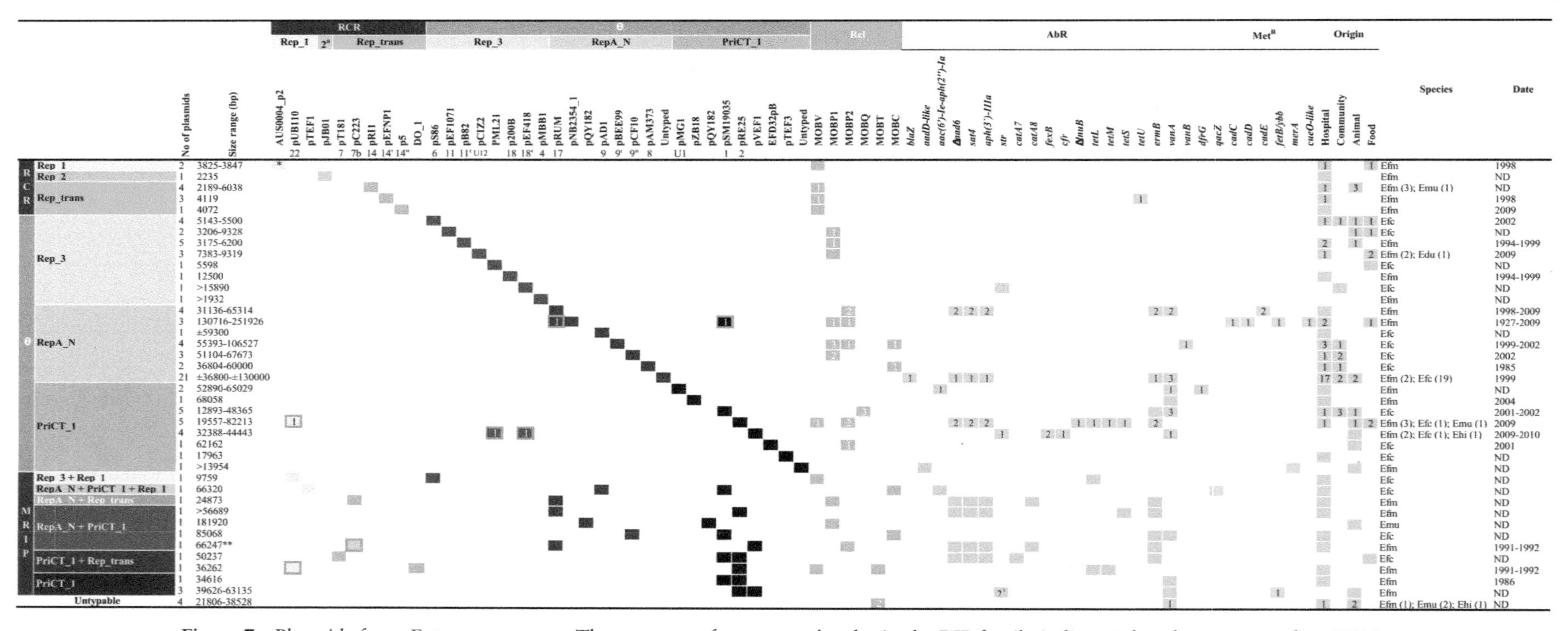

Figure 7 Plasmids from *Enterococcus* spp. The presence of an orange border in the RIP family indicates that the corresponding RIP is truncated. [a]Rep_2; *One of these plasmids (GenBank accession number NC_015849) has a truncated rep_AUS0004_p2 (Rep_1) gene, and no other known replication initiator proteins were found. **This plasmid (GenBank accession number NC_017962) has two copies of Tn*401*; in one of them the *add(6)* gene is not truncated; this plasmid also appears to have two copies of the MOB_{P1} gene. [+]These two plasmids (GenBank accession numbers NC_008768 and NC_008821) have a truncated copy of the *str* gene. Abbreviations: MRIP, multi-RIP; Efm, *E. faecium*; Efc, *E. faecalis*; Emu, *E. mundtii*; Edu, *E. durans*; Ehi, *E. hirae*.
doi:10.1128/microbiolspec.PLAS-0039-2014.f7

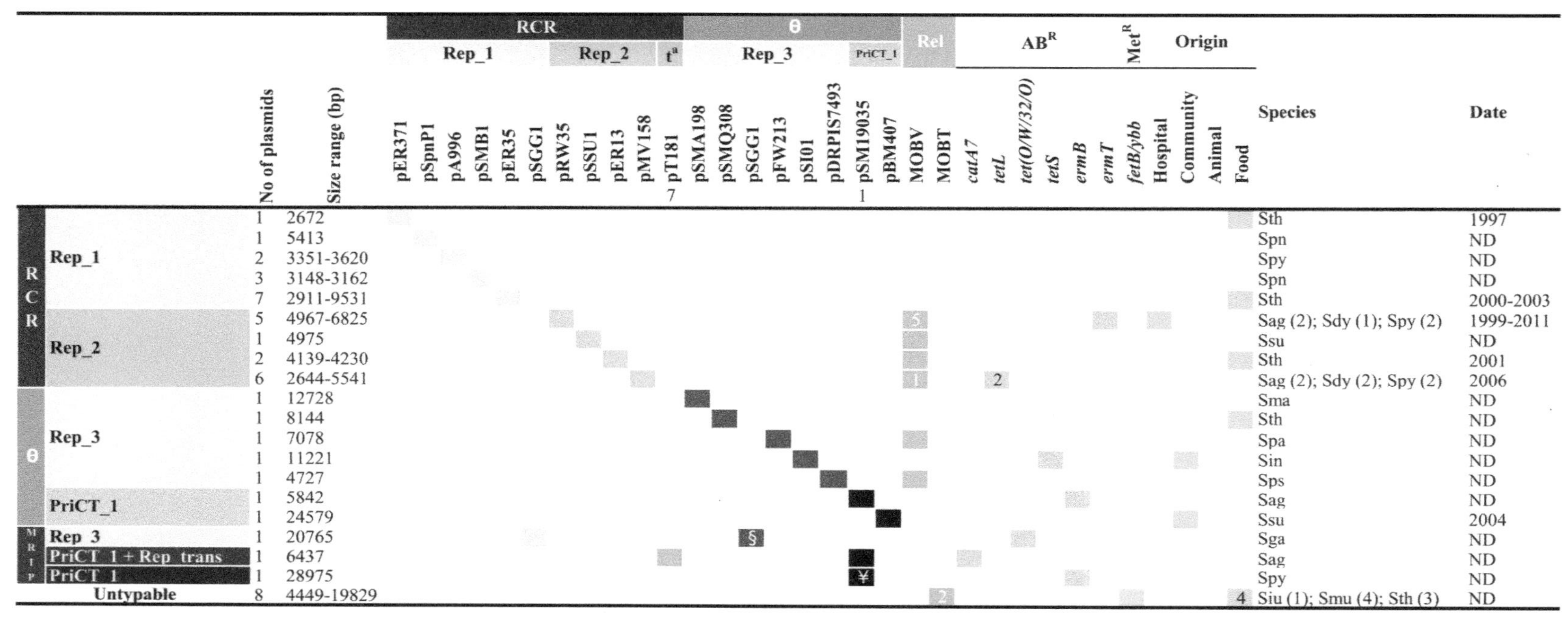

Figure 8 Plasmids from *Streptococcus* spp. [a]Rep_trans; [§]This plasmid (GenBank accession number NC_015219) has two similar replication genes belonging to the Rep_3 family. [¥]This plasmid (GenBank accession number NC_006979) has two similar replication genes belonging to the PriCT_1 family. Abbreviations: MRIP, Multi-RIP; Sag, *S. agalactiae*; Sdy, *Streptococcus dysgalactiae*; Sga, *Streptococcus gallolyticus*; Siu, *Streptococcus infantarius*; Sin, *Streptococcus infantis*; Sma, *Streptococcus macedonicus*; Smu, *Streptococcus mutans*; Spa, *Streptococcus parasanguinis*; Spn, *S. pneumoniae*; Sps, *Streptococcus pseudopneumoniae*; Spy, *S. pyogenes*; Ssu, *Streptococcus suis*; Sth, *Streptococcus thermophilus*. doi:10.1128/microbiolspec.PLAS-0039-2014.f8

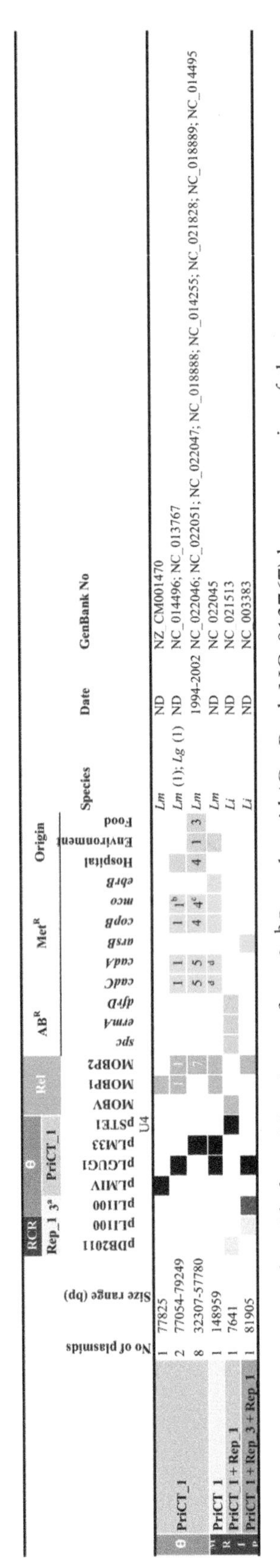

Figure 9 Plasmids from *Listeria* spp. [a]rep_3. [b]One plasmid (GenBank NC_013767) has two copies of the *mco* gene, one of which is truncated. [c]One plasmid (GenBank NC_018888) has a truncated copy of the *mco* gene. [d]This plasmid (GenBank NC_022045) has two copies of the *cadC-cadA* operon. Abbreviations: MRIP, multi-RIP; Lm, *L. monocytogenes*; Lg, *Listeria grayi*; Li, *Listeria innocua*.doi:10.1128/microbiolspec.PLAS-0039-2014.f9

different *Firmicutes* (70, 71, 116, 119). These plasmids have an arsenal of insertion sequences, mainly IS*1216* and IS*L3*, which facilitate genetic exchange with different genetic elements of different origins and the acquisition of different AbR (*tetS)* and Met^R (*tcrB*, *mer* operon). These ISs also facilitate the co-integration with other RCR (e.g., pC221, which is cointegrated in pRE25) or theta replication plasmids as pheromone responsive (116, 119–121) or some pSK41-like elements. Figure 16 shows the diversity of Inc18 RIPs that can be identified by typing systems. All these RIPs have a primase domain PriCT_1 that allowed their identification as belonging to the Inc18 family. Fig. 6, 7, and 9 illustrate the mosaicism of Inc18 plasmids in enterococci and *Listeria*.

The pMG1/pHT plasmids

The pMG1/pHT plasmids are related to those of the Inc18 family (RIP homolog approximately 30% identical to Inc18 initiators, including the PriCT_1 domain [Fig. 16]) (122), although they also show high homology with the pXO2 virulence plasmid from *Bacillus anthracis*. Because many open reading frames of pHT and pMG1 plasmids do not show significant homology with any reported proteins, they used to be categorized as a new type of highly efficient conjugative plasmids with a MOB$_P$ family REL. This plasmid group is represented by relatives of pHT (pHTα, pHTβ, and pHTγ) and pMG1, which have greatly facilitated the dissemination of resistance to glycopeptides (Tn*1546*_*vanA*) and high-level resistance to aminoglycosides (Tn*4001*-like elements) among human *E. faecium* and *E. avium* isolates from the United States and Japan (123, 124) and, to a lesser extent, European countries (7, 70, 71, 122).

The RepA_N family

This is a large family of plasmids (also including a few phages) that are widespread among the low G+C Gram-positive bacteria and which possess RIP homologs to the RepA protein of pAD1 (76). The five groups of RepA homologs identified are phylogenetically congruent with their host background (Fig. 17), suggesting that the replicons have evolved along with their current hosts and that intergenus movement of RepA_N plasmids does not often occur. Such RepA_N clusters correspond to plasmids from *Staphylococcus* (Met^R/*bla* pSK1 and pSK41 MDR plasmids), plasmids from *Enterococcus* (*E. faecalis* pheromone-responsive plasmids and *E. faecium* non-pheromone-responsive plasmids related to pRUM, pLG1, or untypeable megaplasmids), plasmids from *Lactobacillus* and *Lactococcus*, phage

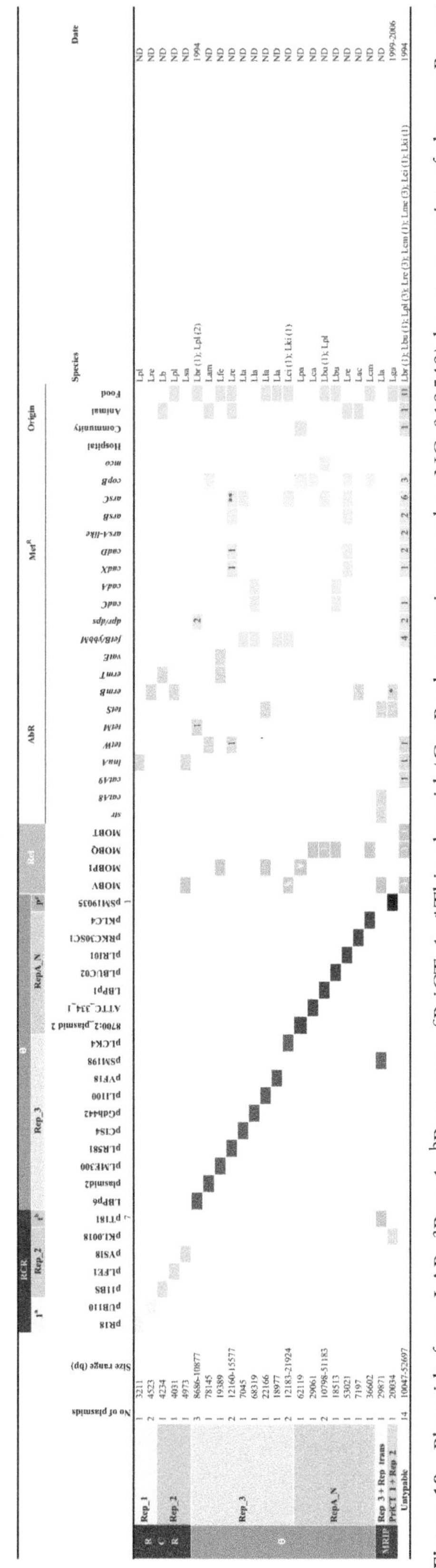

Figure 10 Plasmids from LAB. [a]Rep_1. [b]Rep_trans. [c]PriCT_1. *This plasmid (GenBank accession number NC_010540) has two copies of the *ermB* gene. **One of these plasmids (GenBank accession number NC_010603) has a truncated copy of the *arsC* gene. [§]One of these plasmids (GenBank accession number NC_014133) appears to have three copies of the MOBV gene. [¥]One of these plasmids (GenBank accession number NC_022123) appears to have two copies of the MOB_{P1} gene. Abbreviations: MRip, multi-RIP; Lb, *Lactobacillus* spp; Lca, *Lactobacillus acidophilus*; Lam, *Lactobacillus amylovorus*; Lbr, *Lactobacillus brevis*; Lbu, *Lactobacillus buchneri*; Lca, *Lactobacillus casei*; Lfe *Lactobacillus fermentum*; Lpa, *Lactobacillus paracasei*; Lpl, *Lactobacillus plantarum*; Lre, *Lactobacillus reuteri*; Lsa, *Lactobacillus sakei*; Lga, *Lactococcus garvieae*; Lla, *Lactococcus lactis*; Lcm, *Leuconostoc carnosum*; Lci, *Leuconostoc citreum*; Lki, *Leuconostoc kimchii*; Lme, *Leuconostoc mesenteroides*.doi:10.1128/microbiolspec.PLAS-0039-2014.f10

homologs from *Streptococcus* (*S. pneumoniae, S. thermophilus*), and plasmids from *B. subtilis* (e.g., pLS32). Staphylococcal and enterococcal RepA_N plasmids have greatly contributed to the spread of AbR genes among humans and, eventually, animals and will be further described below. They also facilitate the movement of other nonconjugative plasmids and large genomic regions (36, 125, 126).

RepA_N staphylococcal plasmids (Fig. 5)

Large staphylococcal MDR plasmids use evolutionarily related theta-mode replication, although they can be further divided into three types: the Met^R/beta-lactamase-producing plasmids, the pSK1 family, and pSK41-like conjugative elements. All these are compatible and can be identified as the rep_{19}, rep_{20}, and rep_{15} families, respectively, according to Jensen's plasmid typing system (59, 127, 128). The pSK41 family (rep_{15}) is the largest group of conjugative plasmids in staphylococci, traditionally represented by pSK41, pG01, and pJE1, which emerged in the early 1980s associated with resistance to gentamicin due to the presence of Tn*4001* (84, 129). They often confer resistance to other antibiotics such as neomycin, tobramycin and kanamycin (due to the integration of pUB110 plasmids that harbor the *aadD* gene), antiseptics (due to the presence of *qac* genes) (130), and eventually trimethoprim (mediated by Tn*4001*), penicillins (due to the presence of Tn*552*::*blaZ*), and others. Plasmids of this group may also confer resistance to mupirocin (131–133) and vancomycin (134, 135), represented by pUSA03 (which harbors *ileS* and *tetK*) and pWL1043 (which contains Tn*1546*, Tn*4001*, Tn*4002*, Tn*552*, and *qacC*). The pSK41-like plasmids are able to mobilize other plasmids present in the same bacterial cell (133, 136, 137). The pSK1 and Met^R/beta-lactamase plasmids belong to the same incompatibility groups and are also compatible with pSK41 plasmids. Despite their inability to self-transfer, these groups of plasmids have been detected in many staphylococcal species.

RepA_N enterococcal plasmids

This cluster groups pheromone-responsive plasmids of *E. faecalis* and pRUM- and pLG1-like plasmids of *E. faecium* (7) (Fig. 6 and 7).

Pheromone-responsive plasmids. Pheromone-responsive plasmids represent a paradigm of elements in the biology of MGEs and are, together with Inc18 plasmids, the best-known plasmids described to date. For details about the mechanism of replication, conjugation, and evolvability of this plasmid group

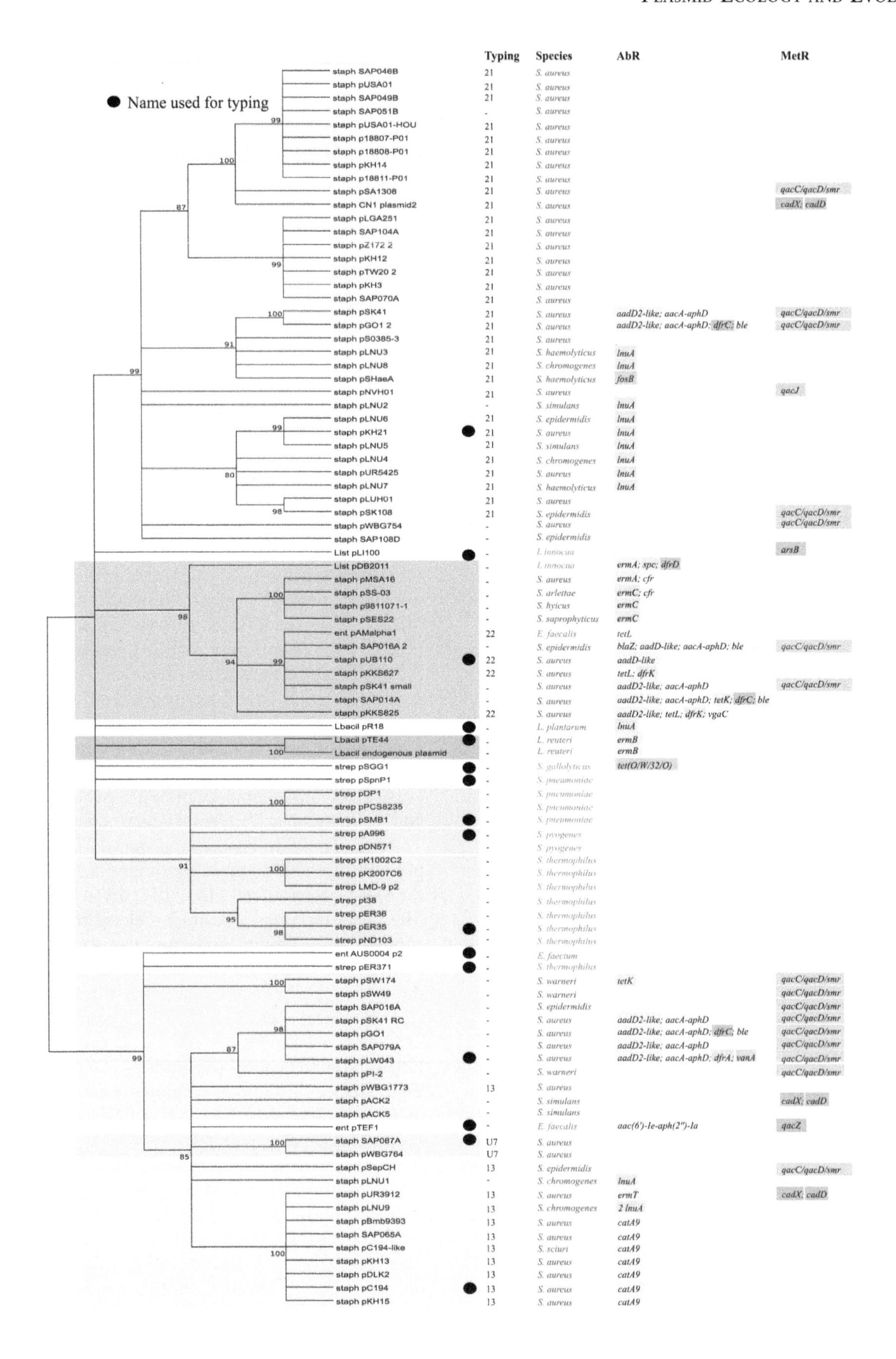
● Name used for typing
Typing Species AbR MetR
staph SAP046B 21 S. aureus
staph pUSA01 21 S. aureus
staph SAP049B 21 S. aureus
staph SAP051B - S. aureus
staph pUSA01-HOU 21 S. aureus
staph p18807-P01 21 S. aureus
staph p18808-P01 21 S. aureus
staph pKH14 21 S. aureus
staph p18811-P01 21 S. aureus
staph pSA1308 21 S. aureus qacC/qacD/smr
staph CN1 plasmid2 21 S. aureus cadX; cadD
staph pLGA251 21 S. aureus
staph SAP104A 21 S. aureus
staph pZ172 2 21 S. aureus
staph pKH12 21 S. aureus
staph pTW20 2 21 S. aureus
staph pKH3 21 S. aureus
staph SAP070A 21 S. aureus
staph pSK41 21 S. aureus aadD2-like; aacA-aphD qacC/qacD/smr
staph pGO1 2 21 S. aureus aadD2-like; aacA-aphD; dfrC; ble qacC/qacD/smr
staph pS0385-3 21 S. aureus
staph pLNU3 21 S. haemolyticus lnuA
staph pLNU8 21 S. chromogenes lnuA
staph pSHaeA 21 S. haemolyticus fosB
staph pNVH01 21 S. aureus qacJ
staph pLNU2 - S. simulans lnuA
staph pLNU6 21 S. epidermidis lnuA
staph pKH21 21 S. aureus lnuA
staph pLNU5 21 S. simulans lnuA
staph pLNU4 21 S. chromogenes lnuA
staph pUR5425 21 S. aureus lnuA
staph pLNU7 21 S. haemolyticus lnuA
staph pLUH01 21 S. aureus
staph pSK108 21 S. epidermidis qacC/qacD/smr
staph pWBG754 - S. aureus qacC/qacD/smr
staph SAP108D - S. epidermidis
List pLI100 - L. innocua arsB
List pDB2011 - L. innocua ermA; spc; dfrD
staph pMSA16 - S. aureus ermA; cfr
staph pSS-03 - S. arlettae ermC; cfr
staph p9811071-1 - S. hyicus ermC
staph pSES22 - S. saprophyticus ermC
ent pAMalpha1 22 E. faecalis tetL
staph SAP016A 2 - S. epidermidis blaZ; aadD-like; aacA-aphD; ble qacC/qacD/smr
staph pUB110 22 S. aureus aadD-like
staph pKKS627 22 S. aureus tetL; dfrK
staph pSK41 small - S. aureus aadD2-like; aacA-aphD qacC/qacD/smr
staph SAP014A - S. aureus aadD2-like; aacA-aphD; tetK; dfrC; ble
staph pKKS825 22 S. aureus aadD2-like; tetL; dfrK; vgaC
Lbacil pR18 - L. plantarum lnuA
Lbacil pTE44 - L. reuteri ermB
Lbacil endogenous plasmid - L. reuteri ermB
strep pSGG1 - S. gallolyticus tet(O/W/32/O)
strep pSpnP1 - S. pneumoniae
strep pDP1 - S. pneumoniae
strep pPCS8235 - S. pneumoniae
strep pSMB1 - S. pneumoniae
strep pA996 - S. pyogenes
strep pDN571 - S. pyogenes
strep pK1002C2 - S. thermophilus
strep pK2007C6 - S. thermophilus
strep LMD-9 p2 - S. thermophilus
strep pt38 - S. thermophilus
strep pER36 - S. thermophilus
strep pER35 - S. thermophilus
strep pND103 - S. thermophilus
ent AUS0004 p2 - E. faecium
strep pER371 - S. thermophilus
staph pSW174 - S. warneri tetK qacC/qacD/smr
staph pSW49 - S. warneri qacC/qacD/smr
staph SAP016A - S. epidermidis qacC/qacD/smr
staph pSK41 RC - S. aureus aadD2-like; aacA-aphD qacC/qacD/smr
staph pGO1 - S. aureus aadD2-like; aacA-aphD; dfrC; ble qacC/qacD/smr
staph SAP079A - S. aureus aadD2-like; aacA-aphD qacC/qacD/smr
staph pLW043 - S. aureus aadD2-like; aacA-aphD; dfrA; vanA qacC/qacD/smr
staph pPI-2 - S. warneri qacC/qacD/smr
staph pWBG1773 13 S. aureus
staph pACK2 - S. simulans cadX; cadD
staph pACK5 - S. simulans
ent pTEF1 - E. faecalis aac(6')-Ie-aph(2")-Ia qacZ
staph SAP087A U7 S. aureus
staph pWBG764 U7 S. aureus
staph pSepCH 13 S. epidermidis qacC/qacD/smr
staph pLNU1 - S. chromogenes lnuA
staph pUR3912 13 S. aureus ermT cadX; cadD
staph pLNU9 13 S. chromogenes 2 lnuA
staph pBmb9393 13 S. aureus catA9
staph SAP065A 13 S. aureus catA9
staph pC194-like 13 S. sciuri catA9
staph pKH13 13 S. aureus catA9
staph pDLK2 13 S. aureus catA9
staph pC194 13 S. aureus catA9
staph pKH15 13 S. aureus catA9

see references 7, 49, 65, 92, and 138. Plasmids that respond to pheromones are present in most contemporary *E. faecalis* isolates from humans and birds but are only occasionally found among *E. faecium*. Synthesis of pheromones is confined to *E. faecalis*, although *Enterococcus hirae*, *S. aureus*, and *Streptococcus gordonii* may secrete cAM373-like peptides that facilitate the conjugation of pAM373 from *E. faecalis* (139). The description of cAM373-responsive plasmids coding for resistance to glycopeptides (Tn*1546-vanA*) highlights the potential risk of the spread of glycopeptide resistance in staphylococci in institutions where VRE are endemic (134, 140). Although pheromone plasmids are unable to replicate in *S. aureus*, their transference and establishment in this host might occur by co-integration with other plasmids able to replicate in this species. In addition, some pAD1 relatives enhance the rate of mobilization of plasmids, conjugative transposons, and PAIs (125).

Plasmids of this family can be classified on the basis of responses to pheromones in different incompatibility groups (139) or according to RIP diversity (59, 68) within rep_8 (pAM373) and rep_9 (further split into subgroups $rep_{9a(pAD1)}$ and $rep_{9b(pTEF2)}$) families (59, 68). Transfer systems of MOBC or MOBP families have been detected in plasmids of this family.

Pheromone-responsive plasmids may encode putative virulence traits (aggregation substance, hemolysin/bacteriocin) and a diversity of AbR elements located on transposable elements such as Tn*916*-like (*tetM*), Tn*4001* (*aac-aph*), Tn*1546* (*vanA*), Tn*1549* (*vanB*), and a composite transposon containing a β-lactamase gene flanked by two IS4 copies (7). The *par* locus encodes a unique antisense-regulated toxin-antitoxin system present in the plasmid pAD1, but *par* homologs have been detected on other plasmids and chromosomes of *E. faecalis* and *Staphylococcus, Clostridium, Listeria*, and *Lactobacillus* species (141). Toxin-antitoxin systems associated with other plasmid families such as εζ and *relBE* have been detected on members of this plasmid group, reflecting rearrangements with representatives of other plasmid families (7). Even though to date, only a few members of pheromone-responsive plasmids have been fully sequenced, typing surveys reveal a wide diversity of plasmids among populations, often containing RIPs, RELs, or regions from plasmids of different origins (68, 71).

pRUM-like plasmids. pRUM-like plasmids (represented by pRUM, p5373c, pS177, and pDO2) are mosaic plasmids of variable size (>30 kb) that comprise diverse genetic elements of different origins (transposons, insertion sequences, small theta-replicating plasmids, bacteriocin clusters). They can be identified as the rep_{17} family according to PCR-based typing systems (59) but differ in the RIP sequence, the MOB system, and the presence of the toxin-antitoxin Axe-Txe locus (71, 142, 143). Both Inc18 and pRUM plasmids are driving the spread of glycopeptide resistance among contemporary isolates of *E. faecium* by carrying Tn*1546* (*vanA*) or Tn*1549* (*vanB*). Two types of pRUM plasmids are currently widespread among VRE and vancomycin-susceptible *E. faecium* isolates from different hosts. One contains RepA and Axe-Txe from pRUM and, eventually, the mobilization system of pC223 from *S. aureus* (70, 71, 142–144). The other type is characterized by a RepA protein that is 95% identical to RepA-pRUM, lacking postsegregational killing Axe-Txe and the presence of a MOB_{P7} relaxase originally detected in pEF1, a plasmid with an environmental origin. Tn*1546* is frequently located on both types of pRUM plasmids, frequently containing replicons of other plasmid families (author's unpublished results).

Large plasmids. Plasmids larger than 150 kb are widely distributed among *E. faecium*, *Enterococcus durans*, and *E. hirae* from different origins, but they have not been detected among *E. faecalis* (71, 144–150). To date, only a handful of *E. faecium* megaplasmids have been fully sequenced (AUS0085_p1 [NC_021987], pNB2354_1 [NC_020208], DO_3 [NC_017963], and pLG1, although this last one has

Figure 11 Similarity of *rep*-like sequences encoding RIPs of the Rep_1 family. A neighbor-joining tree of gene sequences coding for RIPs of the Rep_1 family was built using MEGA 6.06. A cut-off equal to or higher than 80% and a bootstrap analysis based on 1,000 permutations were applied to the analysis. Alignment of nucleotide sequences was performed using ClustalW2 (http://www.ebi.ac.uk/Tools/phylogeny/clustalw2_phylogeny/), and sequences showing an identity equal to or higher than 80% were clustered in groups that were highlighted by different backgrounds colors. Black dots indicate the RIP of the plasmid used for further comparison in Figs. 5 to 10.
doi:10.1128/microbiolspec.PLAS-0039-2014.f11

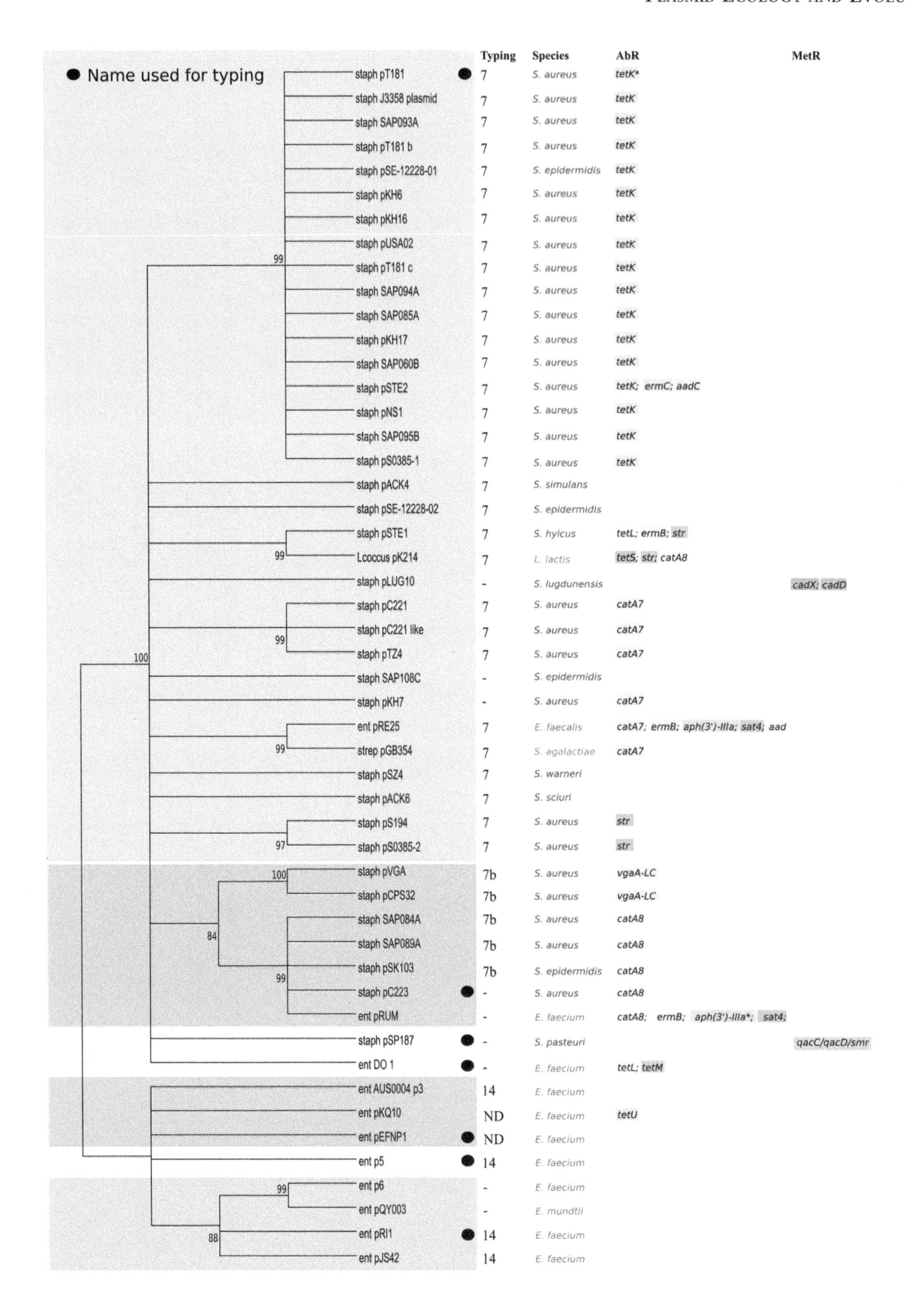

● Name used for typing
Typing
Species
AbR
MetR
staph pT181 7 S. aureus tetK*
staph J3358 plasmid 7 S. aureus tetK
staph SAP093A 7 S. aureus tetK
staph pT181 b 7 S. aureus tetK
staph pSE-12228-01 7 S. epidermidis tetK
staph pKH6 7 S. aureus tetK
staph pKH16 7 S. aureus tetK
staph pUSA02 7 S. aureus tetK
staph pT181 c 7 S. aureus tetK
staph SAP094A 7 S. aureus tetK
staph SAP085A 7 S. aureus tetK
staph pKH17 7 S. aureus tetK
staph SAP060B 7 S. aureus tetK
staph pSTE2 7 S. aureus tetK; ermC; aadC
staph pNS1 7 S. aureus tetK
staph SAP095B 7 S. aureus tetK
staph pS0385-1 7 S. aureus tetK
staph pACK4 7 S. simulans
staph pSE-12228-02 7 S. epidermidis
staph pSTE1 7 S. hyicus tetL; ermB; str
Lcoccus pK214 7 L. lactis tetS; str; catA8
staph pLUG10 - S. lugdunensis cadX; cadD
staph pC221 7 S. aureus catA7
staph pC221 like 7 S. aureus catA7
staph pTZ4 7 S. aureus catA7
staph SAP108C - S. epidermidis
staph pKH7 - S. aureus catA7
ent pRE25 7 E. faecalis catA7; ermB; aph(3')-IIIa; sat4; aad
strep pGB354 7 S. agalactiae catA7
staph pSZ4 7 S. warneri
staph pACK6 7 S. sciuri
staph pS194 7 S. aureus str
staph pS0385-2 7 S. aureus str
staph pVGA 7b S. aureus vgaA-LC
staph pCPS32 7b S. aureus vgaA-LC
staph SAP084A 7b S. aureus catA8
staph SAP089A 7b S. aureus catA8
staph pSK103 7b S. epidermidis catA8
staph pC223 - S. aureus catA8
ent pRUM - E. faecium catA8; ermB; aph(3')-IIIa*; sat4;
staph pSP187 - S. pasteuri qacC/qacD/smr
ent DO 1 - E. faecium tetL; tetM
ent AUS0004 p3 14 E. faecium
ent pKQ10 ND E. faecium tetU
ent pEFNP1 ND E. faecium
ent p5 14 E. faecium
ent p6 - E. faecium
ent pQY003 - E. mundtii
ent pRI1 14 E. faecium
ent pJS42 14 E. faecium
99
99
99
100
99
97
100
84
99
99
88

not been closed [148]). All of them contain a RIP similar to $RepA_{pAD1}$, making them part of the RepA_N family (Fig. 7 and 17, Supplementary Table S1) (59, 71, 148). A similar RIP has also been found in a 130-kb plasmid (NC_021987) from a VRE ST203 *E. faecium* strain isolated in 2009 in Australia (151). Although RIP sequences of pLG1 plasmids are often identified among enterococcal megaplasmids, most of them do not hybridize with known RIP genes included in published schemes (71, 148, 152). Enterococcal megaplasmids carry genes involved in sugar metabolism (mannitol, glycerol, sorbitol, raffinose, complex carbohydrates), AbR (macrolides, glycopeptides, aminoglycosides), Met^R (copper-*tcrYAZB*), and enhanced adhesion (71, 126, 144, 147–149, 152–154). They are frequently involved in the acquisition or persistence of AbR among *E. faecium* isolates from food animals (144, 150).

GENE AND PLASMID FLOW SHAPES THE EVOLUTIONARY ECOLOGY OF *FIRMICUTES*

As described in previous sections, the acquisition of novel traits encoding adaptive resistance to antimicrobials in *Firmicutes* is mainly due to genes located on plasmids and transposable elements. This acquisition is, certainly, regulated by interactions at genetic and ecological (social) levels. Interplay between genes, mobile genetic elements, and microbial populations and their relation with the host population and local or global environments shapes the plasmid flow. Such flow can be modified by "external" (supra-cellular) changes, including variations in the host population structure and size (e.g., mass rearing, crowding) and their associated chemical or behavioral landscape (e.g., use of different antimicrobials, immunization, global food supply, international travel). These changes ultimately determine the density and diversity of particular bacterial populations in particular habitats, leading to ecological specialization, clonalization, and gradual emergence of gene flow barriers (23, 155, 156), a process that mimics the general dynamics of speciation, as bacterial clones and species constitute ecological units of microbial biodiversity.

The challenge to define "units of biodiversity" in microbial community ecology has approached the concept of genes as "defining elements of networks and metacommunities" (155). In such a context, extra-chromosomal elements greatly influence the HGT interactions between microbial organisms and are the building forces for the establishment of "gene exchange communities" (155, 157). The selective power of antimicrobials (antibiotics [Ab], heavy metal, biocides) may then shape this multilevel bacterial population biology (158, 159), involving genes, plasmids (MGEs), bacterial clones and species, and gene exchange communities. The evolutionary tradeoff between early and late stages of adaptation to such selective pressures may determine the local evolvability of clonal and plasmid populations by increasing the number of genotypes resulting from chromosomal and plasmid recombination processes that facilitate further ecological differentiation (18). To establish effective public health interventions to fight the AbR problem in its eco-biological dimension, we then need to define the gene exchange communities relevant for the acquisition, evolution, and spread of resistance (160, 161). Below, we will specifically discuss the role of AbR genes and plasmids in the ecological differentiation of bacterial populations of the main *Firmicutes* genera.

Antimicrobial Resistance Genes and Bacterial Population Ecology

The environmental origin of AbR genes has been extensively discussed, but very few AbR genes identified in the environment are found in human or animal pathogens, which indicates severe bottlenecks for their acquisition and transmission (162, 163). However, the gut microbiota is increasingly considered a significant reservoir of AbR genes (3), which is supported by studies

Figure 12 Similarity of *rep*-like sequences encoding RIPs of the Rep_trans family. A neighbor-joining tree of gene sequences coding for RIPs of the Rep_trans family was built using MEGA 6.06. A cut-off equal to or higher than 80% and a bootstrap analysis based on 1,000 permutations were applied to the analysis. Alignment of nucleotide sequences was performed using ClustalW2 (http://www.ebi.ac.uk/Tools/phylogeny/clustalw2_phylogeny/), and sequences showing an identity equal to or higher than 80% were clustered in groups that were highlighted by different backgrounds colors. Black dots indicate the RIP of the plasmid used for further comparison in Figs. 5 to 10. *Truncated gene. **Similar to *E. faecalis ant6-Ia* and *aadE*. Abbreviations: ND, not determined.
doi:10.1128/microbiolspec.PLAS-0039-2014.f12

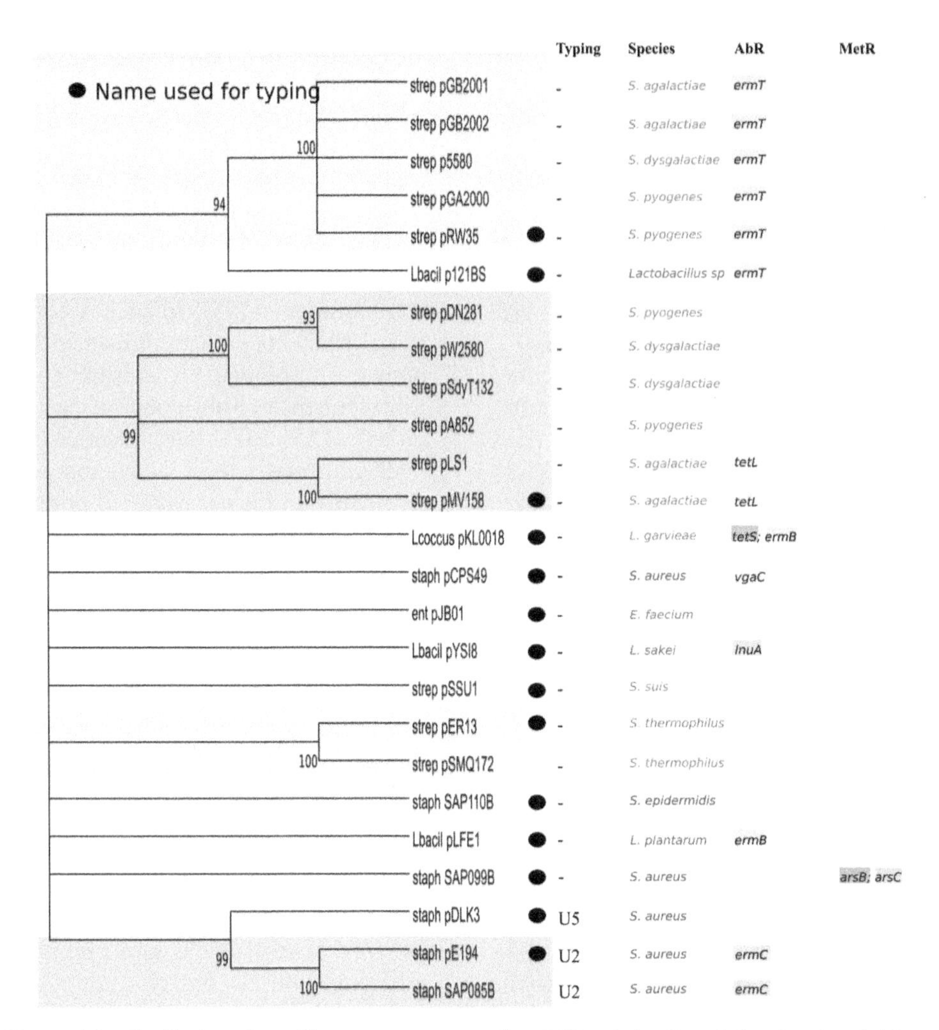

Figure 13 Similarity of *rep*-like sequences encoding RIPs of the Rep_2 family. A neighbor-joining tree of gene sequences coding for RIPs of the Rep_2 family was built using MEGA 6.06. A cut-off equal to or higher than 80% and a bootstrap analysis based on 1,000 permutations were applied to the analysis. Alignment of nucleotide sequences was performed using ClustalW2 (http://www.ebi.ac.uk/Tools/phylogeny/clustalw2_phylogeny/), and sequences showing an identity equal to or higher than 80% were clustered in groups that were highlighted by different background colors. Black dots indicate the RIP of the plasmid used for further comparison in Figs. 5 to 10.
doi:10.1128/microbiolspec.PLAS-0039-2014.f13

that associate widely spread AbR genes of relevance in clinical therapy, such as *ermB*, *ermT*, *ermC* (encoding resistance to macrolides), *vanB* (coding for resistance to glycopeptides), and *cfr* (coding for resistance to different antimicrobials), with members of the normal microbiota such as species of the *Clostridium* group XIVa now reclassified as family *Lachnospiraceae* (*Clostridium bolteae*, *Clostridium innocuum*–like, *Clostridium lavalense*, *Clostridium symbiosum*) and some lactic bacteria (3, 88, 164–168).

Recent work demonstrates that a given AbR gene (or genetic element such as Tn*1549_vanB*) may be independently acquired by different clonal populations in the intestine of a particular host (165). Once an AbR gene is present in gut commensals (independent of the origin of the gene), members of the normal intestinal flora of humans and animals can exchange such genes among themselves or with bacterial pathogens, which might be present in low numbers or just be passing through the intestine after being transferred from other body sites or with food intake, using different intermediates in the case of distant bacteria (3, 165, 169).

The rapid emergence in *Firmicutes* of genes coding for AbR, Met^R, and Bc^R immediately after their intro-

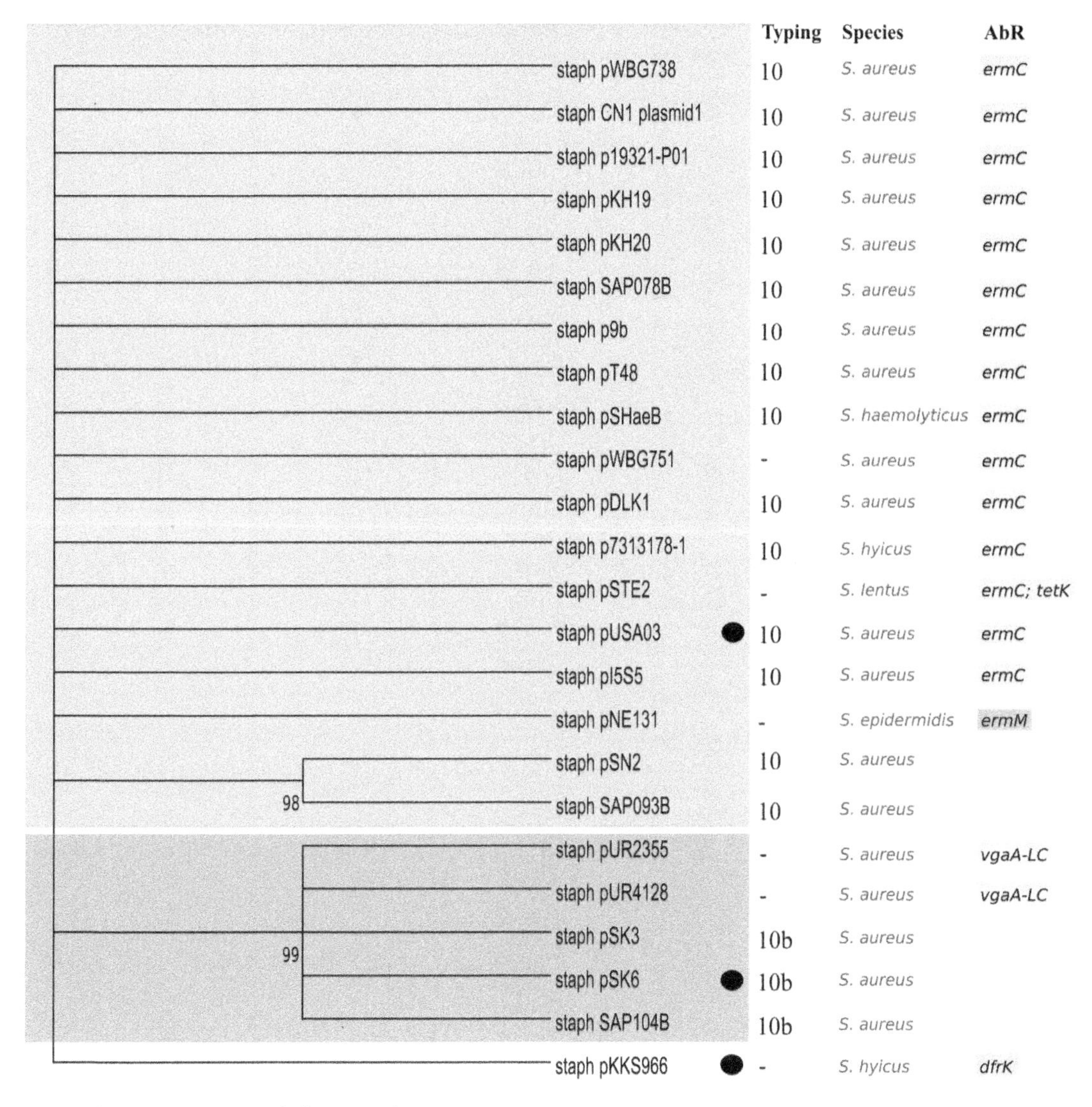

Figure 14 Similarity of *rep*-like sequences encoding RIPs of the Rep_L family. A neighbor-joining tree of gene sequences coding for RIPs of the Rep_L family was built using MEGA 6.06. A cut-off equal to or higher than 80% and a bootstrap analysis based on 1,000 permutations were applied to the analysis. Alignment of nucleotide sequences was performed using ClustalW2 (http://www.ebi.ac.uk/Tools/phylogeny/clustalw2_phylogeny/), and sequences showing an identity equal to or higher than 80% were clustered in groups that were highlighted by different background colors. Black dots indicate the RIP of the plasmid used for further comparison in Figs. 5 to 10.
doi:10.1128/microbiolspec.PLAS-0039-2014.f14

duction and significant (often massive) use in different settings has been demonstrated for chloramphenicol (*catA*), tetracyclines (*tetL*), macrolides (*ermB*), neomycin (*aad*), gentamicin (*aac6aph2*), trimethoprim (*dfr*), beta-lactams (*blaZ*), and antiseptics (*qac*) in hospitals during the 1950s to 1970s, and for tylosin (*ermC*, *ermT*), phenicols (*fex*), pleuromutilins (*cfr*), and zinc-bacitracin in animals during the 1990s, thus supporting the hypothesis of the existence of a previous gastrointestinal reservoir of genes that were selected for the first time as AbR genes (gene exaptation) (84, 88, 91).

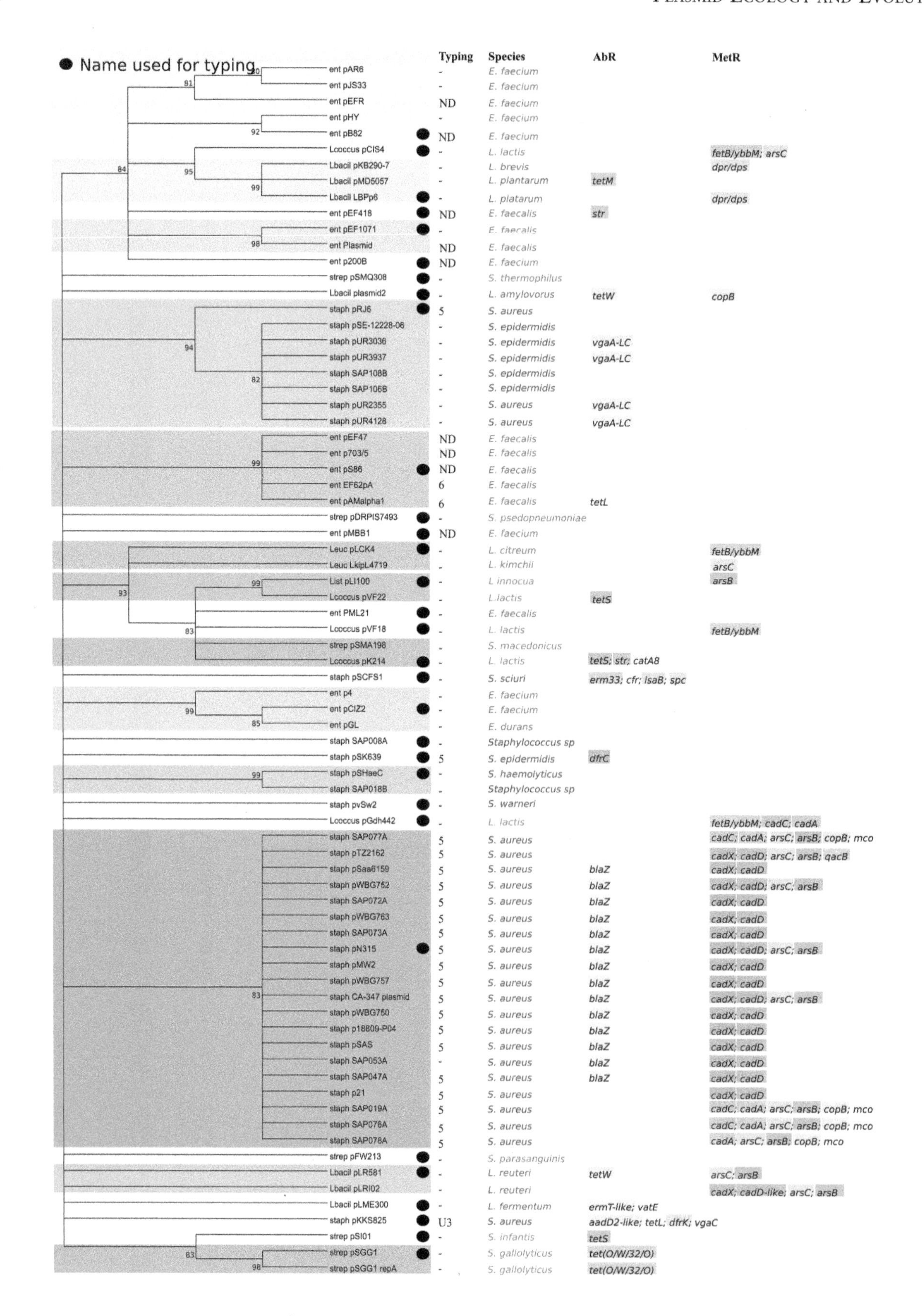

Name used for typing
Typing
Species
AbR
MetR
ent pAR6 - E. faecium
ent pJS33 - E. faecium
ent pEFR ND E. faecium
ent pHY - E. faecium
ent pB82 ND E. faecium
Lcoccus pCIS4 - L. lactis fetB/ybbM; arsC
Lbacil pKB290-7 - L. brevis dpr/dps
Lbacil pMD5057 - L. plantarum tetM
Lbacil LBPp6 - L. platarum dpr/dps
ent pEF418 ND E. faecalis str
ent pEF1071 - E. faecalis
ent Plasmid ND E. faecalis
ent p200B ND E. faecium
strep pSMQ308 - S. thermophilus
Lbacil plasmid2 - L. amylovorus tetW copB
staph pRJ6 5 S. aureus
staph pSE-12228-06 - S. epidermidis
staph pUR3036 - S. epidermidis vgaA-LC
staph pUR3937 - S. epidermidis vgaA-LC
staph SAP108B - S. epidermidis
staph SAP106B - S. epidermidis
staph pUR2355 - S. aureus vgaA-LC
staph pUR4128 - S. aureus vgaA-LC
ent pEF47 ND E. faecalis
ent p703/5 ND E. faecalis
ent pS86 ND E. faecalis
ent EF62pA 6 E. faecalis
ent pAMalpha1 6 E. faecalis tetL
strep pDRPIS7493 - S. psedopneumoniae
ent pMBB1 ND E. faecium
Leuc pLCK4 - L. citreum fetB/ybbM
Leuc LkipL4719 - L. kimchii arsC
List pLI100 - L innocua arsB
Lcoccus pVF22 - L.lactis tetS
ent PML21 - E. faecalis
Lcoccus pVF18 - L. lactis fetB/ybbM
strep pSMA198 - S. macedonicus
Lcoccus pK214 - L. lactis tetS; str; catA8
staph pSCFS1 - S. sciuri erm33; cfr; lsaB; spc
ent p4 - E. faecium
ent pCIZ2 - E. faecium
ent pGL - E. durans
staph SAP008A - Staphylococcus sp
staph pSK639 5 S. epidermidis dfrC
staph pSHaeC - S. haemolyticus
staph SAP018B - Staphylococcus sp
staph pvSw2 - S. warneri
Lcoccus pGdh442 - L. lactis fetB/ybbM; cadC; cadA
staph SAP077A 5 S. aureus cadC; cadA; arsC; arsB; copB; mco
staph pTZ2162 5 S. aureus cadX; cadD; arsC; arsB; qacB
staph pSaa6159 5 S. aureus blaZ cadX; cadD
staph pWBG752 5 S. aureus blaZ cadX; cadD; arsC; arsB
staph SAP072A 5 S. aureus blaZ cadX; cadD
staph pWBG763 5 S. aureus blaZ cadX; cadD
staph SAP073A 5 S. aureus blaZ cadX; cadD
staph pN315 5 S. aureus blaZ cadX; cadD; arsC; arsB
staph pMW2 5 S. aureus blaZ cadX; cadD
staph pWBG757 5 S. aureus blaZ cadX; cadD
staph CA-347 plasmid 5 S. aureus blaZ cadX; cadD; arsC; arsB
staph pWBG750 5 S. aureus blaZ cadX; cadD
staph p18809-P04 5 S. aureus blaZ cadX; cadD
staph pSAS 5 S. aureus blaZ cadX; cadD
staph SAP053A - S. aureus blaZ cadX; cadD
staph SAP047A 5 S. aureus blaZ cadX; cadD
staph p21 5 S. aureus cadX; cadD
staph SAP019A 5 S. aureus cadC; cadA; arsC; arsB; copB; mco
staph SAP076A 5 S. aureus cadC; cadA; arsC; arsB; copB; mco
staph SAP078A 5 S. aureus cadA; arsC; arsB; copB; mco
strep pFW213 - S. parasanguinis
Lbacil pLR581 - L. reuteri tetW arsC; arsB
Lbacil pLRI02 - L. reuteri cadX; cadD-like; arsC; arsB
Lbacil pLME300 - L. fermentum ermT-like; vatE
staph pKKS825 U3 S. aureus aadD2-like; tetL; dfrK; vgaC
strep pSI01 - S. infantis tetS
strep pSGG1 - S. gallolyticus tet(O/W/32/O)
strep pSGG1 repA - S. gallolyticus tet(O/W/32/O)
81
84
95
92
99
98
94
82
99
93
99
83
99
85
99
83
83
98

Plasmids and Bacterial Population Ecology

The number and types of *Firmicutes* plasmids and integrative-conjugative elements (currently considered as plasmids under the perspective of evolutionary biology [22]) greatly vary with the different bacterial species, certainly as a result of both ecological specialization and selective events resulting from exposure to different anthropogenic activities. Most (if not all) of the contemporary isolates belonging to different species of staphylococci, enterococci, lactobacilli, and others contain plasmids of different families in a consistent pattern (for instance, RCR, small theta, or megaplasmids in *E. faecium*; pheromone plasmids in *E. faecalis*) (7, 68, 71). Such frequent plasmid-bacteria host correspondence indicates a basic coadaptive evolutionary relation between two different types of organisms.

For a long time, plasmids were considered as "organisms," units of a continuous lineage with an individual evolutionary history, and hence producing evolved populations, in line with the Luria and other seminal works in the field (46, 240). However, plasmids are not necessarily discrete units or individuals as classically considered in evolutionary theory (20, 170, 240). Organisms are units of selection, evolutionary units in a sense "evolutionary individuals," defined as any entity that, independently from the number of elements that enters into its composition or from its hierarchical level of complexity, is selected and evolves as a unit (170, 171). The frequent out-of-equilibrium events that characterize the interplay between bacterial hosts, plasmids, and gene populations is explained because selective events might act independently on these different evolutionary individuals, as predicted in the "levels of selection" conceptual frame (20, 172–174). However, it is of note that we should recognize "levels of individuality"; for instance, a substantial number of *Firmicutes* plasmids have a lower-level self-identity than their bacterial hosts (18, 155) because of the more complex genetic interplay with other mobile genetic elements which in turn are also "leaky individuals," frequently mosaics of individuals with a partial or contingent self-identity, produced under the effect of adaptive challenges when confronting variable environments (155, 175). Even if this problem of "individual constancy" (176) makes it difficult to study the network of plasmids and hosts in *Firmicutes*, and such a network were biased by sampling, we should accept the existence of a certain interactive frame.

Valeria Souza, still following Maynard Smith's ideas about the population structure of bacteria, proposed in 1997 to classify plasmid-bacteria interactions in four patterns, namely, (i) the plasmid-host clonal pattern, where the plasmid phylogeny is mirrored by host phylogeny; (ii) the limited transfer pattern, in which the plasmid flow is limited to closely related (genetically and/or ecologically) strains; (iii) panmictic plasmid spread, in the case of plasmids that circulate among a variety of hosts (the stability of the association being dependent on the benefits and costs of plasmid carriage); and (iv) epidemic plasmid dispersal, in which "successful" plasmids spread in bacterial populations because they provide a clear advantage in high-potency selective landscapes (49, 170). Although illustrative and useful for epidemiological purposes, this single centric view should not replace the complex interplay between different elements that may result in the emergence of different chimeric configurations (49, 177). Therefore, these "patterns" should be currently understood as possible interactive states, even though some of them could be more ephemeral than others, depending on the coevolutionary history, the adaptive demands of the plasmids, and the bacterial populations and communities.

Plasmids and Population Biology of *Firmicutes*

This section will focus on the genera of *Firmicutes* that are relevant to the problem of AbR (1).

Streptococcus

The genus *Streptococcus*, a main hub in gene networks in this and other studies (11, 12), is one of the most heterogeneous groups within the phyla *Firmicutes*. Remarkably, the 138 known species of streptococci found as opportunistic pathogens or commensals (many of

Figure 15 Similarity of *rep*-like sequences encoding RIPs of the Rep_3 family. A neighbor-joining tree of gene sequences coding for RIPs of the Rep_3 family was built using MEGA 6.06. A cut-off equal to or higher than 80% and a bootstrap analysis based on 1,000 permutations were applied to the analysis. Alignment of nucleotide sequences was performed using ClustalW2 (http://www.ebi.ac.uk/Tools/phylogeny/clustalw2_phylogeny/), and sequences showing an identity equal to or higher than 80% were clustered in groups that were highlighted by different background colors. Black dots indicate the RIP of the plasmid used for further comparison in Figs. 5 to 10. Abbreviations: ND, not determined.
doi:10.1128/microbiolspec.PLAS-0039-2014.f15

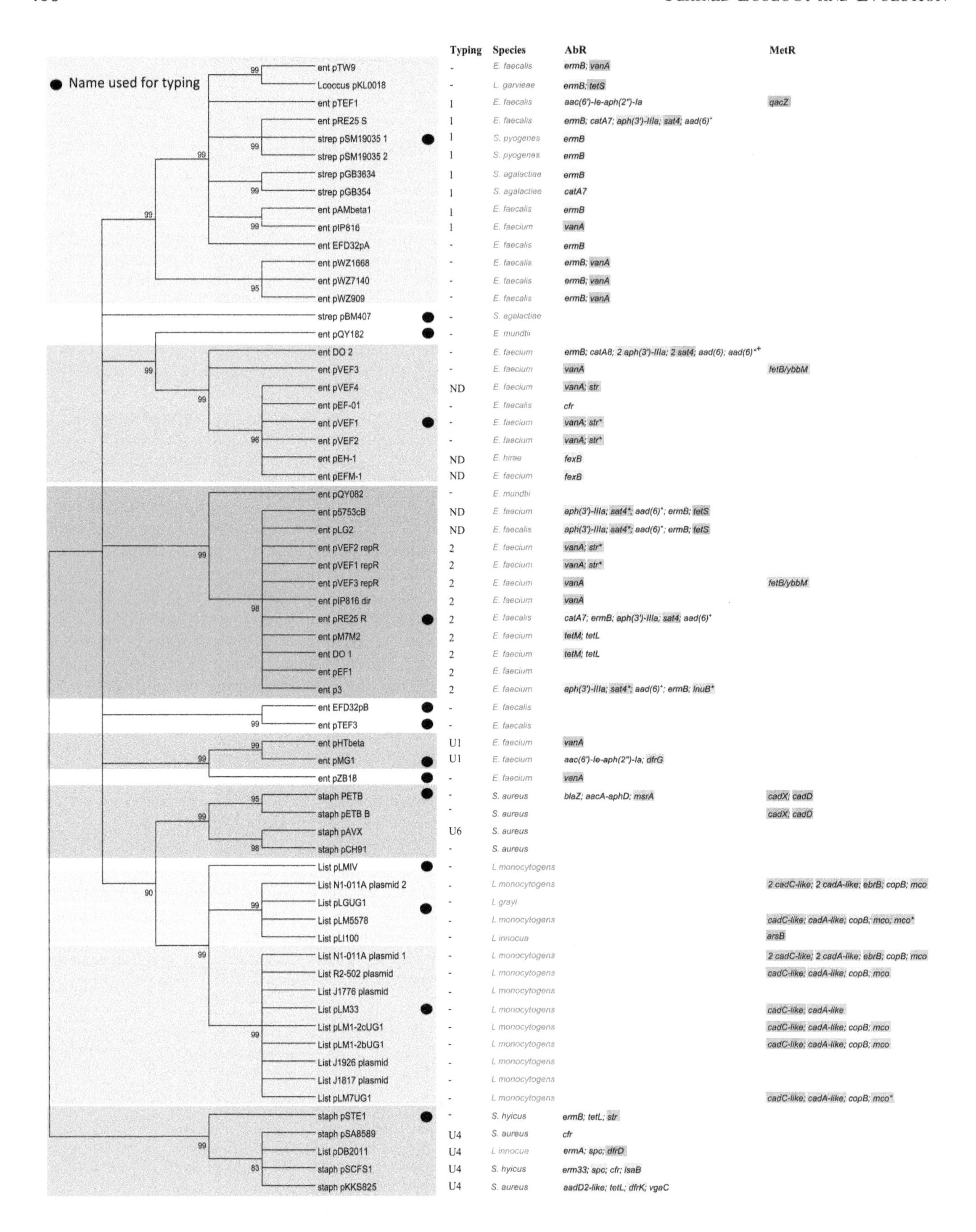
Name used for typing
Typing
Species
AbR
MetR
ent pTW9
Lcoccus pKL0018
ent pTEF1
ent pRE25 S
strep pSM19035 1
strep pSM19035 2
strep pGB3634
strep pGB354
ent pAMbeta1
ent pIP816
ent EFD32pA
ent pWZ1668
ent pWZ7140
ent pWZ909
strep pBM407
ent pQY182
ent DO 2
ent pVEF3
ent pVEF4
ent pEF-01
ent pVEF1
ent pVEF2
ent pEH-1
ent pEFM-1
ent pQY082
ent p5753cB
ent pLG2
ent pVEF2 repR
ent pVEF1 repR
ent pVEF3 repR
ent pIP816 dir
ent pRE25 R
ent pM7M2
ent DO 1
ent pEF1
ent p3
ent EFD32pB
ent pTEF3
ent pHTbeta
ent pMG1
ent pZB18
staph PETB
staph pETB B
staph pAVX
staph pCH91
List pLMIV
List N1-011A plasmid 2
List pLGUG1
List pLM5578
List pLI100
List N1-011A plasmid 1
List R2-502 plasmid
List J1776 plasmid
List pLM33
List pLM1-2cUG1
List pLM1-2bUG1
List J1926 plasmid
List J1817 plasmid
List pLM7UG1
staph pSTE1
staph pSA8589
List pDB2011
staph pSCFS1
staph pKKS825
E. faecalis
L. garvieae
S. pyogenes
S. agalactiae
E. faecium
E. mundtii
E. hirae
S. aureus
L. monocytogens
L. grayi
L. innocua
S. hyicus
ermB; vanA
ermB; tetS
aac(6')-Ie-aph(2")-Ia
ermB; catA7; aph(3')-IIIa; sat4; aad(6)+
catA7
vanA
ermB; catA8; 2 aph(3')-IIIa; 2 sat4; aad(6); aad(6)*+
vanA; str
cfr
vanA; str*
fexB
aph(3')-IIIa; sat4*; aad(6)*; ermB; tetS
catA7; ermB; aph(3')-IIIa; sat4; aad(6)+
tetM; tetL
aph(3')-IIIa; sat4*; aad(6)*; ermB; lnuB*
aac(6')-Ie-aph(2")-Ia; dfrG
blaZ; aacA-aphD; msrA
ermB; tetL; str
ermA; spc; dfrD
erm33; spc; cfr; lsaB
aadD2-like; tetL; dfrK; vgaC
qacZ
fetB/ybbM
cadX; cadD
2 cadC-like; 2 cadA-like; ebrB; copB; mco
cadC-like; cadA-like; copB; mco; mco*
arsB
cadC-like; cadA-like; copB; mco
cadC-like; cadA-like
cadC-like; cadA-like; copB; mco*

them zoonotic pathogens) in humans, horses, pigs, cows, and fish have recently been divided into seven species groups on the basis of 16S rDNA gene sequencing, chemotaxonomic approaches, and DNA hybridization, namely the bovis, pyogenic, mitis, mutans, salivarius, anginosus, and unknown groups (178–180). HGT seems to play a relevant role in the adaptation and cohesiveness of the groups (179). Available information about streptococcal plasmids is scarce, with only a few plasmids being fully sequenced, representing an unbalanced sample of species and ecological groups (Supplementary Table S1). Figure 10 illustrates the 20 AbR plasmids currently found in streptococci.

The streptococcal groups bovis and mutans rarely harbor plasmids, although they can be relevant in the adaptation of particular species. *S. thermophilus*, a nonpathogenic species in the bovis group that is used in the dairy industry (181), contains a set of plasmids harboring heat shock proteins; *Streptococcus mutans*, a member of the human indigenous flora that is transmitted mostly from mother to child, often carries 5- to 6-kb cryptic plasmids that parallel the evolution of lineages associated with racial cohorts and geographical locations (182). Megaplasmids in the group salivarius coding for different lantibiotics favor their persistence in the oral cavity (183). Conversely, the pyogenic group, which is represented by species of clinical interest such as *S. agalactiae* and *S. pyogenes* (also called GAS and GBS, respectively), frequently carry plasmids that code for AbR genes aside from bacteriocins. Inc18 plasmids are widely spread among streptococci and seem to have determined the selection of certain populations resistant to chloramphenicol, aminoglycosides, and macrolides since the late 1970s in different groups of streptococci and enterococci (105, 110, 184). Rep_2 plasmids carrying *erm(T)* seem to have recently spread among GAS and GBS clinical isolates of different countries, having contributed to the increase of macrolide resistance rates in these species since the mid-1990s, either by clonal expansion, in the case of GAS, or by plasmid transference among unrelated clonal backgrounds, in the case of GBS (8, 185).

These *erm(T)*-containing plasmids are also spread among other non-streptococcal species, such as *Enterococcus*, *Staphylococcus,* and *Lactobacillus* (89, 186, 187). Often, streptococcal plasmids are mobilized by coresident integrative-conjugative elements belonging to the ICESa2603 family (188). Resistance to macro lides (*ermB*, *mefA)*, tetracyclines (*tetM*, *tetS*, and other mosaic *tet* genes), aminoglycosides (*aph3*, *aadA6*, Tn*4001*), or vancomycin (*vanA*, *vanB*) is commonly detected among isolates of this group, but the location of determinants seems to be linked to transposable elements often involving insertion sequences (reviewed in reference 181). *Streptococcus suis*, a particularly virulent emerging zoonotic pathogen that remains an outlier to the mitis, sanguinis, and anginosus groups is known to carry plasmids, although they have been scarcely characterized (189, 190). Relevant AbR genes coding for chloramphenicol (*cfr* and *fexA*) and lincosamides (*lnu*) embedded in composite regions similar to those present in plasmids of *E. faecalis* have been located in streptococcal plasmids of approximately 100 kb (191). Smaller plasmids carrying *tetB* associated with Gram-negative species have been described (192).

Enterococcus

The genus *Enterococcus* comprises different species, members of the intestinal flora of animals and humans able to cause disease in their hosts (193). Although seminal works in the field of plasmid biology focus on particular enterococcal plasmids and transposons, such as pheromone-responsive plasmids or Tn*916*, which became paradigms of different mechanisms of conjugation, the plasmidome of enterococcal species has scarcely been analyzed (7). Recent studies revealed that most strains of *E. faecium* and *E. faecalis*, the two main species detected in humans and animals, carry a number of plasmids of different families that include species-specific plasmids (e.g., narrow host range RCRs and RepA_N plasmids such as megaplasmids in *E. faecium* and pheromone-responsive plasmids in *E. faecalis*) and broad host range plasmids (e.g., Inc18), plasmid chimeras being abundant and still difficult to

Figure 16 Similarity of *rep*-like sequences encoding RIPs with the PriCT_1 domain. A neighbor-joining tree of gene sequences coding for RIPs with PriCT_1 domains was built using MEGA 6.06. A cut-off equal to or higher than 80% and a bootstrap analysis based on 1,000 permutations were applied to the analysis. Alignment of nucleotide sequences was performed using ClustalW2 (http://www.ebi.ac.uk/Tools/phylogeny/clustalw2_phylogeny/), and sequences showing an identity equal to or higher than 80% were clustered in groups that were highlighted by different background colors. Black dots indicate the RIP of the plasmid used for further comparison in Figs. 5 to 10. Abbreviations: ND, not determined. doi:10.1128/microbiolspec.PLAS-0039-2014.f16

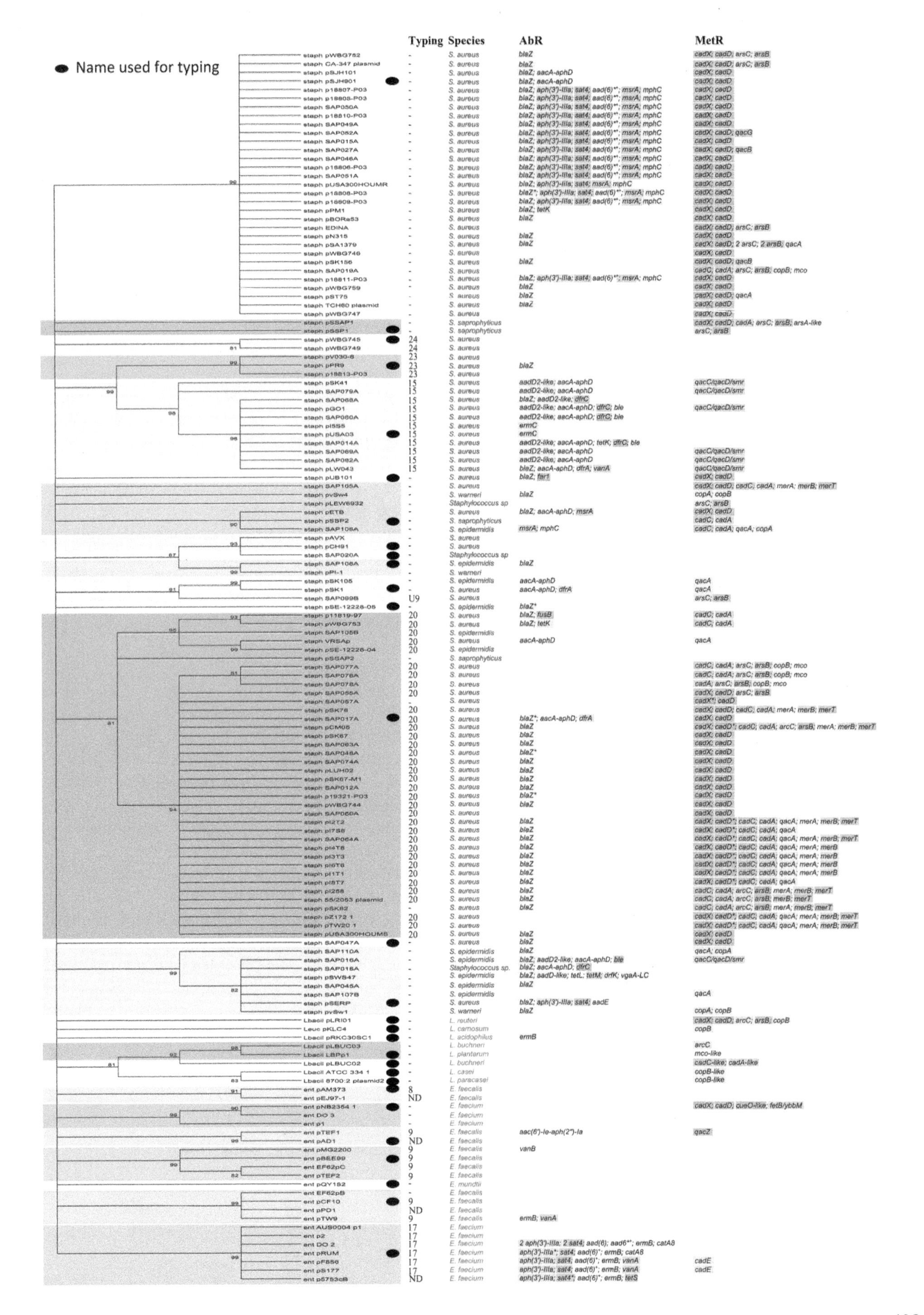

classify (Figs. 6 and 7; see previous section and comprehensive reviews in references 7, 141). Megaplasmids of *E. faecium* or pheromone-responsive *E. faecalis* plasmids enhance the ability to colonize, invade, and form biofilms (65, 126, 154). Conjugative plasmids may influence the mobilization of nonconjugative elements and chromosomal regions and facilitate the acquisition of different adaptive traits and genome evolvability (71, 125, 126). Most enterococcal plasmids are able to acquire and disseminate AbR genes by different mechanisms of genetic exchange. However, the role of plasmids in the population structure and evolvability of these enterococcal species has been poorly addressed (194–197) due to the overrepresentation of recent clinical and animal isolates of specific lineages commonly associated with AbR included in most studies (7, 141) and due to the lack of available plasmid sequences. Similar plasmids have been found in *E. faecium* and other enterococcal species that may play equivalent functional roles in the gastrointestinal tract such as *Enterococcus avium, Enterococcus raffinosus, E. durans*, and *E. hirae* (195, 198).

AbR genes are located on plasmids that often contain different replicons associated with different narrow (RCR, RepA_N) and broad host range (Inc18) plasmids. Inc18 streptococcal plasmids greatly influenced the worldwide increase of aminoglycoside-macrolide resistance among *E. faecalis* isolates from humans and animals during the 1970s (199). They also contributed to the spread of vancomycin resistance among *E. faecium* of animal origin in Europe and *E. faecalis* from hospitalized patients in the United States (7, 70, 71). Diverse narrow host range plasmids have been involved in local expansions of enterococci conferring resistance to first-line antibiotics such as gentamicin (Tn*4001*) or beta-lactams (Δ*Tn552 blaZ*) (152) and beta-lactamase-producing *E. faecalis* and *E. faecium* (152) (200–202), which highlights the role of endogenous plasmids and recombination in the adaptation of particular lineages (*E. faecium* ST17, ST18, ST78 and *E. faecalis* ST6 and ST16) (7, 67, 144, 203).

Analysis of the same AbR genes in different species (*cfr*, *bac*, lincosamide resistance genes) reflects the impact of recombination events between genes, MGEs, and different populations of *Firmicutes* (*Staphylococcus, Clostridium, Lactobacillus, Lactococcus*, and *Enterococcus*) and other Gram-negative organisms (200, 204) in the gastrointestinal tract of animals and humans (120, 144, 205, 206).

Staphylococci

These organisms are opportunistic pathogens and members of the commensal flora of skin and mucous membranes of humans and animals (207–209) and, thus, are part of a microbial community with limited contact with members of other main genera of *Firmicutes* that inhabit distinct body sites (210). Figures 1 and 2 show the limited plasmid connectivity of staphylococci with other genera. However, HGT and the acquisition of AbR and Met^R is relevant in the evolvability of this genus, mainly due to genetic exchange events between closely related species (Fig. 5) (9, 211–213). Comprehensive reviews address the essentially clonal population structure of *S. aureus* (214–216) and other staphylococcal species (207) and also the impact of HGT in the evolutionary history of staphylococcal populations (9, 217–219), with emphasis on the description of the plasmids associated with AbR genetic elements (9, 51, 84, 220) and their influence on the variability of lineages (217, 219, 221–224).

Plasmids, transposons, and staphylococcal chromosomal cassettes (SCC*mec*) are infrequently transferred among isolates of a different origin. A close association of MGE and particular staphylococcal lineages has been suggested (31, 225), with country-specific variations (208, 226). This highlights the relevance of local conditions and the emergence of gene flow barriers in the ecological differentiation of staphylococcal lineages such as in the case of *S. aureus* CC30 (219, 227). The origin, rapid spread, and evolution of staphylococcal populations resistant to beta-lactams was mainly influenced by the interplay of genetic elements including plasmids (84, 177).

Figure 17 Similarity of *rep*-like sequences encoding RIPs of the RepA_N family. A neighbor-joining tree of gene sequences coding for replication initiator proteins of the RepA_N family was built using MEGA 6.06. A cut-off equal to or higher than 80% and a bootstrap analysis based on 1,000 permutations were applied to the analysis. Alignment of nucleotide sequences was performed using ClustalW2 (http://www.ebi.ac.uk/Tools/phylogeny/clustalw2_phylogeny/), and sequences showing an identity equal to or higher than 80% were clustered in groups that were highlighted by different background colors. Black dots indicate the RIP of the plasmid used for further comparison in Figs. 5 to 10. *Truncated gene. +Similar to *E. faecalis ant6-Ia* and *aadE*. Abbreviations: ND, not determined.
doi:10.1128/microbiolspec.PLAS-0039-2014.f17

Clostridium

Clostridium is a large and extremely heterogeneous genus that has traditionally grouped more than 100 species widely distributed in the gut microbiota of mammals, amphibians, and insects and in soils. An extensive update of clostridial classification is included in the latest edition of *Bergey's Manual*, although many unrelated species still retain the *Clostridium* name, causing major confusion in the clostridial taxonomy (228). To date, only 60 plasmids have been fully sequenced, mainly corresponding to *C. perfringens*, *Clostridium botulinum*, and other group I clostridia species (1 *Clostridium butyricum*, 2 *Clostridium kluyveri*, 3 *Clostridium acetobutylicum*). Some species in which plasmids were analyzed have been moved to other genera such as *Clostridium aciduricidi* (now *Anaerococcus prevotii* type XII) and *Clostridium thermocellum* (now belonging to the family *Ruminococcaceae*). Several sequenced plasmids correspond to the same strain and are mostly from contemporary isolates, thus limiting the possible knowledge about the role of plasmids in the evolution of these species (Supplementary Table S1). Only narrow host range conjugative plasmids of *C. perfringens* (CpCP) or linear megaplasmids from *C. butyricum* have been associated with AbR.

CpCP plasmids belong to the pCW3 family and are widely spread among isolates of *C. perfringens*, carrying genes encoding AbR (tetracycline [*tetAB(P)*], chloramphenicol [*catP*-Tn*4451*], lincomycin [*lnuP*-tISCpe8]) and/or enterotoxins, ε-toxin, or iota-toxin production that determine different toxinotypes (56, 229–231). All pCW3-like plasmids have a conjugative transfer locus of 11 open reading frames (*orfs*) (*tcp* [transfer *C. perfringens*]) that includes an integrase and a T4CP protein but lacks relaxase (73, 231). A transposable origin similar to that of Tn*916* has been suggested for the *tcp* module of pCW3-like plasmids, which would have acquired a replication machinery specific to this species. Often, *C. perfringens* isolates harbor more than one pCW3 plasmid, which carry different adaptative traits and partition machineries. The presence of different partition systems explains the coexistence of different plasmids with the same type of RIP in the same cell (56, 231–233).

These plasmids can be transferred (and eventually serve as donors of AbR genes) but cannot replicate in other species such as *P. difficile*, *Clostridium sordelli*, or *Clostridium septicum*, which could explain the confinement of some AbR genes in these populations (234). An evolutionary scenario for CpCP has been reported, with pCW3 (*tetAB-P*) and pIP401 (*tetAB-P* and Tn*4451*) being suggested as the precursors of this family, which would have acquired different toxins by homologous recombination involving composite transposons flanked by insertion sequences (56). Large linear plasmids containing AbR have recently been described in neurotoxigenic *C. butyricum*, one of the six phylogroups able to produce the botulinum toxin (34, 235). These plasmids contain four beta-lactamase genes, transcriptional regulators and two-component regulatory systems, involved in the regulation of expression of the *bont/A* gene and a region with a functional CRISPR-cas locus that provides a defense against invading genetic elements present in the intestinal environment.

Acquired resistance to tetracyclines (*tetM*, *tetL*, *tetK*, *tetO*, *tetW*), chloramphenicol, macrolides (*ermB*, *lnu*), and bacitracin (a bacitracin efflux pump and an overproduced undercaprenol kinase gene located on a genetic island flanked by copies of IS*1216*) has been reported in human and animal clostridium species including *C. perfringens*, often associated with conjugative transposons and plasmids widespread in other species (234, 236, 237). A detailed analysis of AbR networks suggests further ecological connections with mobile genetic elements of other prokaryotic groups (Figs. 1 and 2).

CONCLUSION

This work offers for the first time an integrated and comprehensive analysis of the dynamics of AbR genes in Gram-positive bacteria and highlights the need for a population view to analyze the problem of antibiotic resistance. The article analyzed the relevance of the plasmidome in the emergence, spread, and maintenance of genes encoding resistance to antimicrobials (antibiotics, heavy metals, and biocides) and their influence on the structure of bacterial populations in the light of evolutionary ecology. A critical revision of plasmid typing systems highlights the limitation of available knowledge about plasmid diversity in this group of bacteria.

Acknowledgments. *Research in the authors' laboratories is funded by the European Commission (EvoTAR-282004 for F.B., J.L.M., and T.M.C.); the Plan Nacional de I+D+i 2008-2011: (i) the Ministry of Economy and Competitiveness (BIO2011-25255 for J.L.M.), (ii) the Instituto de Salud Carlos III (PI12-01581 for T.M.C., PI10-02588 for F.B.), (iii) CIBER (CB06/02/0053) cofinanced by the European Development Regional Fund "A way to achieve Europe" (ERDF), and (iv) the Spanish Network for Research on Infectious Diseases (REIPI RD12/0015 for J.L.M.); the Regional Government of Madrid in Spain (PROMPT-S2010/ BMD2414 for J.L.M. and F.B.); the Spanish Network for the Study of*

Plasmids and Extrachromosomal Elements (REDEEX, BFU2011-14145-E for T.M.C.).T.M.C. wishes to express a deep gratitude to Ana Freitas, Carla Novais, and Luísa Peixe for fruitful discussions about epidemiology of enterococcal mobile genetic elements throughout the years.Val F. Lanza and Ana P. Tedim contributed equally to the elaboration of this article. Conflicts of interest: We declare no conflicts.

Citation. Fernández Lanza V, Tedim AP, Martínez JL, Baquero F, Coque TM. 2015. The plasmidome of *firmicutes*: impact on the emergence and the spread of resistance to antimicrobials. Microbiol Spectrum 3(2):PLAS-0039-2014.

References

1. **Centers for Disease Control and Prevention.** 2013. *Threat Report 2013. Antimicrobial Resistance.* http://www.cdc.gov/drugresistance/threat-report-2013/
2. **Murphree CA, Heist EP, Moe LA.** 2014. Antibiotic resistance among cultured bacterial isolates from bioethanol fermentation facilities across the United States. *Curr Microbiol* **69:**277–285.
3. **Salyers AA, Gupta A, Wang Y.** 2004. Human intestinal bacteria as reservoirs for antibiotic resistance genes. *Trends Microbiol* **12:**412–416.
4. **European Food Safety Authority.** 2012. Technical specifications on the harmonised monitoring and reporting of antimicrobial resistance in *Salmonella*, *Campylobacter* and indicator *Escherichia coli* and *Enterococcus* spp. bacteria transmitted through food. *EFSA J* **10:**2742.
5. **European Food Safety Authority.** 2012. Technical specifications for the analysis and reporting of data on antimicrobial resistance (AMR) in the European Union Summary Report. *EFSA J* **10:**2587.
6. **Schwarz S, Shen J, Wendlandt S, Feßler AT, Wang Y, Kadlec K, Wu C.** 2015. Plasmid-mediated antimicrobial resistance in staphylococci and other *Firmicutes*. *In* Tolmasky ME, Alonso JC (ed), *Plasmids: Biology and Impact in Biotechnolgy and Discovery.* ASM Press, Washington, DC, in press.
7. **Clewell DB, Weaver KE, Dunny GM, Coque TM, Francia MV, Hayes F.** 2014. Extrachromosomal and mobile elements in enterococci: transmission, maintenance, and epidemiology. *In* Gilmore MS, Clewell DB, Ike Y, Shankar N (ed), *Enterococci: From Commensals to Leading Causes of Drug Resistant Infection.* Massachusetts Eye and Ear Infirmary, Boston, MA.
8. **DiPersio LP, DiPersio JR, Beach JA, Loudon AM, Fuchs AM.** 2011. Identification and characterization of plasmid-borne erm(T) macrolide resistance in group B and group A *Streptococcus*. *Diagn Microbiol Infect Dis* **71:**217–223.
9. **Lindsay JA.** 2014. *Staphylococcus aureus* genomics and the impact of horizontal gene transfer. *Int J Med Microbiol* **304:**103–109.
10. **Lebreton F, van Schaik W, McGuire AM, Godfrey P, Griggs A, Mazumdar V, Corander J, Cheng L, Saif S, Young S, Zeng Q, Wortman J, Birren B, Willems RJL, Earl AM, Gilmore MS.** 2013. Emergence of epidemic multidrug-resistant *Enterococcus faecium* from animal and commensal strains. *MBio* **4**(4):e00534-13. doi:10.1128/mBio.00534-13.
11. **Beiko RG, Harlow TJ, Ragan MA.** 2005. Highways of gene sharing in prokaryotes. *Proc Natl Acad Sci USA* **102:**14332–14337.
12. **Halary S, Leigh JW, Cheaib B, Lopez P, Bapteste E.** 2010. Network analyses structure genetic diversity in independent genetic worlds. *Proc Natl Acad Sci USA* **107:**127–132.
13. **Popa O, Dagan T.** 2011. Trends and barriers to lateral gene transfer in prokaryotes. *Curr Opin Microbiol* **14:**615–623.
14. **Popa O, Hazkani-Covo E, Landan G, Martin W, Dagan T.** 2011. Directed networks reveal genomic barriers and DNA repair bypasses to lateral gene transfer among prokaryotes. *Genome Res* **21:**599–609.
15. **Tamminen M, Virta M, Fani R, Fondi M.** 2012. Large-scale analysis of plasmid relationships through gene-sharing networks. *Mol Biol Evol* **29:**1225–1240.
16. **Nogueira T, Rankin DJ, Touchon M, Taddei F, Brown SP, Rocha EPC.** 2009. Horizontal gene transfer of the secretome drives the evolution of bacterial cooperation and virulence. *Curr Biol* **19:**1683–1691.
17. **McGinty SE, Rankin DJ, Brown SP.** 2011. Horizontal gene transfer and the evolution of bacterial cooperation. *Evolution* **65:**21–32.
18. **Friedman J, Alm EJ, Shapiro BJ.** 2013. Sympatric speciation: when is it possible in bacteria? *PLoS One* **8:**e53539. doi:10.1371/journal.pone.0053539.
19. **Hobman JL.** 2015. *Molecular Life Sciences.* Springer, New York, NY.
20. **Eberhard WG.** 1990. Evolution in bacterial plasmids and levels of selection. *Q Rev Biol* **65:**3–22.
21. **Kloesges T, Popa O, Martin W, Dagan T.** 2011. Networks of gene sharing among 329 proteobacterial genomes reveal differences in lateral gene transfer frequency at different phylogenetic depths. *Mol Biol Evol* **28:**1057–1074.
22. **Guglielmini J, Quintais L, Garcillán-Barcia MP, de la Cruz F, Rocha EPC.** 2011. The repertoire of ICE in prokaryotes underscores the unity, diversity, and ubiquity of conjugation. *PLoS Genet* **7:**e1002222. doi:10.1371/journal.pgen.1002222.
23. **Shapiro BJ, Friedman J, Cordero OX, Preheim SP, Timberlake SC, Szabó G, Polz MF, Alm EJ.** 2012. Population genomics of early events in the ecological differentiation of bacteria. *Science* **336:**48–51.
24. **DeMaere MZ, Williams TJ, Allen MA, Brown MV, Gibson JAE, Rich J, Lauro FM, Dyall-Smith M, Davenport KW, Woyke T, Kyrpides NC, Tringe SG, Cavicchioli R.** 2013. High level of intergenera gene exchange shapes the evolution of haloarchaea in an isolated Antarctic lake. *Proc Natl Acad Sci USA* **110:**16939–16944.
25. **Toussaint A, Chandler M.** 2012. Prokaryote genome fluidity: toward a system approach of the mobilome. *Methods Mol Biol* **804:**57–80.
26. **Shen J, Wang Y, Schwarz S.** 2013. Presence and dissemination of the multiresistance gene *cfr* in Gram-positive and Gram-negative bacteria. *J Antimicrob Chemother* **68:**1697–1706.

27. **Bonafede ME, Carias LL, Rice LB.** 1997. Enterococcal transposon Tn5384: evolution of a composite transposon through cointegration of enterococcal and staphylococcal plasmids. *Antimicrob Agents Chemother* **41:** 1854–1858.
28. **Hung W-C, Takano T, Higuchi W, Iwao Y, Khokhlova O, Teng L-J, Yamamoto T.** 2012. Comparative genomics of community-acquired ST59 methicillin-resistant *Staphylococcus aureus* in Taiwan: novel mobile resistance structures with IS1216V. *PLoS One* **7:** e46987. doi:10.1371/journal.pone.0046987.
29. **Marraffini LA, Sontheimer EJ.** 2010. CRISPR interference: RNA-directed adaptive immunity in bacteria and archaea. *Nat Rev Genet* **11:**181–190.
30. **Johnston C, Martin B, Polard P, Claverys J-P.** 2013. Postreplication targeting of transformants by bacterial immune systems? *Trends Microbiol* **21:**516–521.
31. **McCarthy AJ, Lindsay JA.** 2012. The distribution of plasmids that carry virulence and resistance genes in *Staphylococcus aureus* is lineage associated. *BMC Microbiol* **12:**104.
32. **Roberts GA, Houston PJ, White JH, Chen K, Stephanou AS, Cooper LP, Dryden DTF, Lindsay JA.** 2013. Impact of target site distribution for type I restriction enzymes on the evolution of methicillin-resistant *Staphylococcus aureus* (MRSA) populations. *Nucleic Acids Res* **41:** 7472–7484.
33. **Attaiech L, Olivier A, Mortier-Barrière I, Soulet A-L, Granadel C, Martin B, Polard P, Claverys J-P.** 2011. Role of the single-stranded DNA-binding protein SsbB in pneumococcal transformation: maintenance of a reservoir for genetic plasticity. *PLoS Genet* **7:**e1002156. doi:10.1371/journal.pgen.1002156.
34. **Iacobino A, Scalfaro C, Franciosa G.** 2013. Structure and genetic content of the megaplasmids of neurotoxigenic clostridium butyricum type E strains from Italy. *PLoS One* **8:**e71324. doi:10.1371/journal.pone.0071324.
35. **Hatoum-Aslan A, Maniv I, Samai P, Marraffini LA.** 2014. Genetic characterization of antiplasmid immunity through a type III-A CRISPR-Cas system. *J Bacteriol* **196:**310–317.
36. **Palmer KL, Gilmore MS.** 2010. Multidrug-resistant enterococci lack CRISPR-cas. *MBio* **1**(4):e00227-10. doi:10.1128/mBio.00227-10.
37. **Serbanescu MA, Cordova M, Krastel K, Flick R, Beloglazova N, Latos A, Yakunin AF, Senadheera DB, Cvitkovitch DG.** 2014. Role of the *Streptococcus mutans* CRISPR/Cas systems in immunity and cell physiology. *J Bacteriol* [Epub ahead of print.] doi:10.1128/JB.02333-14.
38. **Hermsen R, Deris JB, Hwa T.** 2012. On the rapidity of antibiotic resistance evolution facilitated by a concentration gradient. *Proc Natl Acad Sci USA* **109:**10775–10780.
39. **Shu C-C, Chatterjee A, Hu W-S, Ramkrishna D.** 2013. Role of intracellular stochasticity in biofilm growth. Insights from population balance modeling. *PLoS One* **8:**e79196. doi:10.1371/journal.pone.0079196.
40. **Król JE, Wojtowicz AJ, Rogers LM, Heuer H, Smalla K, Krone SM, Top EM.** 2013. Invasion of *E. coli* biofilms by antibiotic resistance plasmids. *Plasmid* **70:**110–119.
41. **Hermsen R, Hwa T.** 2010. Sources and sinks: a stochastic model of evolution in heterogeneous environments. *Phys Rev Lett* **105:**248104.
42. **Nielsen KM, Bøhn T, Townsend JP.** 2014. Detecting rare gene transfer events in bacterial populations. *Front Microbiol* **4:**415.
43. **McCarthy AJ, Loeffler A, Witney AA, Gould KA, Lloyd DH, Lindsay JA.** 2014. Extensive horizontal gene transfer during *Staphylococcus aureus* co-colonization *in vivo*. *Genome Biol Evol* **6:**2697–2708.
44. **Gullberg E, Albrecht LM, Karlsson C, Sandegren L, Andersson DI.** 2014. Selection of a multidrug resistance plasmid by sublethal levels of antibiotics and heavy metals. *MBio* **5:**e01918-14. doi:10.1128/mBio.01918-14.
45. **Baquero F, Coque TM.** 2014. Widening the spaces of selection: evolution along sublethal antimicrobial gradients. *MBio* **5:**e02270-14. doi:10.1128/mBio.02270-14
46. **Datta N.** 1985. Plasmids in bacteria, p 3–16. *In* Helinski DR, Cohen SN, Clewell DB, Jackson DA, Hollaender A (ed), *Plasmids in Bacteria*. Basic Life Sciences, **Vol. 30**. Springer US, New York, New York.
47. **Eberhard WG.** 1989. Why do bacterial plasmids carry some genes and not others? *Plasmid* **21:**167–174.
48. **Souza V, Eguiarte LE.** 1997. Bacteria gone native vs. bacteria gone awry?: plasmidic transfer and bacterial evolution. *Proc Natl Acad Sci USA* **94:**5501–5503.
49. **Kozlowicz BK, Dworkin M, Dunny GM.** 2009. Pheromone-inducible conjugation in *Enterococcus faecalis*: a model for the evolution of biological complexity? *Int J Med Microbiol* **296:**141–147.
50. **Taylor DE, Gibreel A, Lawley TD, Tracz DM.** 2004. Antibiotic resistance plasmids, p 473–491. *In* Funnell B, Phillips G (ed), *Plasmid Biology*. ASM Press, Washington, DC.
51. **Shearer JES, Wireman J, Hostetler J, Forberger H, Borman J, Gill J, Sanchez S, Mankin A, Lamarre J, Lindsay JA, Bayles K, Nicholson A, O'Brien F, Jensen SO, Firth N, Skurray RA, Summers AO.** 2011. Major families of multiresistant plasmids from geographically and epidemiologically diverse staphylococci. *G3* **1:** 581–591.
52. **Lacey RW.** 1975. Antibiotic resistance plasmids of *Staphylococcus aureus* and their clinical importance. *Bacteriol Rev* **39:**1–32.
53. **Novick RP.** 1989. Staphylococcal plasmids and their replication. *Annu Rev Microbiol* **43:**537–565.
54. **Mills S, McAuliffe OE, Coffey A, Fitzgerald GF, Ross RP.** 2006. Plasmids of lactococci: genetic accessories or genetic necessities? *FEMS Microbiol Rev* **30:**243–273.
55. **Wang TT, Lee BH.** 1997. Plasmids in *Lactobacillus*. *Crit Rev Biotechnol* **17:**227–272.
56. **Li J, Adams V, Bannam TL, Miyamoto K, Garcia JP, Uzal FA, Rood JI, McClane BA.** 2013. Toxin plasmids of *Clostridium perfringens*. *Microbiol Mol Biol Rev* **77:** 208–233.

57. **Dib JR, Liebl W, Wagenknecht M, Farías ME, Meinhardt F.** 2013. Extrachromosomal genetic elements in *Micrococcus*. *Appl Microbiol Biotechnol* **97:**63–75.
58. **Guglielmetti S, Mayo B, Álvarez-Martín P.** 2013. Mobilome and genetic modification of bifidobacteria. *Benef Microbes* **4:**143–166.
59. **Jensen LB, Garcia-Migura L, Valenzuela AJ, Løhr M, Hasman H, Aarestrup FM.** 2010. A classification system for plasmids from enterococci and other Gram-positive bacteria. *J Microbiol Methods* **80:**25–43.
60. **Iordanescu S, Surdeanu M, Della Latta P, Novick R.** 1978. Incompatibility and molecular relationships between small staphylococcal plasmids carrying the same resistance marker. *Plasmid* **1:**468–479.
61. **Macrina FL, Archer GL.** 1993. *Bacterial Conjugation.* Plenum, New York, NY.
62. **Novick RP.** 1987. Plasmid incompatibility. *Microbiol Rev* **51:**381–395.
63. **Projan SJ, Novick R.** 1988. Comparative analysis of five related staphylococcal plasmids. *Plasmid* **19:**203–221.
64. **Brantl S, Behnke D, Alonso JC.** 1990. Molecular analysis of the replication region of the conjugative *Streptococcus agalactiae* plasmid pIP501 in *Bacillus subtilis.* Comparison with plasmids pAM beta 1 and pSM19035. *Nucleic Acids Res* **18:**4783–4790.
65. **Dunny GM.** 2013. Enterococcal sex pheromones: signaling, social behavior, and evolution. *Annu Rev Genet* **47:**457–482.
66. **Lozano C, García-Migura L, Aspiroz C, Zarazaga M, Torres C, Aarestrup FM.** 2012. Expansion of a plasmid classification system for Gram-positive bacteria and determination of the diversity of plasmids in *Staphylococcus aureus* strains of human, animal, and food origins. *Appl Environ Microbiol* **78:**5948–5955.
67. **Sadowy E, Luczkiewicz A.** 2014. Drug-resistant and hospital-associated *Enterococcus faecium* from wastewater, riverine estuary and anthropogenically impacted marine catchment basin. *BMC Microbiol* **14:**66.
68. **Wardal E, Gawryszewska I, Hryniewicz W, Sadowy E.** 2013. Abundance and diversity of plasmid-associated genes among clinical isolates of *Enterococcus faecalis*. *Plasmid* **70:**329–342.
69. **Wardal E, Markowska K, Zabicka D, Wróblewska M, Giemza M, Mik E, Połowniak-Pracka H, Woźniak A, Hryniewicz W, Sadowy E.** 2014. Molecular analysis of vanA outbreak of *Enterococcus faecium* in two Warsaw hospitals: the importance of mobile genetic elements. *Biomed Res Int* **2014:**575367.
70. **Rosvoll TCS, Pedersen T, Sletvold H, Johnsen PJ, Sollid JE, Simonsen GS, Jensen LB, Nielsen KM, Sundsfjord A.** 2010. PCR-based plasmid typing in *Enterococcus faecium* strains reveals widely distributed pRE25-, pRUM-, pIP501- and pHTbeta-related replicons associated with glycopeptide resistance and stabilizing toxin-antitoxin systems. *FEMS Immunol Med Microbiol* **58:** 254–268.
71. **Freitas AR, Novais C, Tedim AP, Francia MV, Baquero F, Peixe L, Coque TM.** 2013. Microevolutionary events involving narrow host plasmids influences local fixation of vancomycin-resistance in *Enterococcus* populations. *PLoS One* **8:**e60589. doi:10.1371/journal.pone.0060589.
72. **Francia MV, Varsaki A, Garcillán-Barcia MP, Latorre A, Drainas C, de la Cruz F.** 2004. A classification scheme for mobilization regions of bacterial plasmids. *FEMS Microbiol Rev* **28:**79–100.
73. **Garcillan-Barcia MP, Francia MV, de la Cruz F.** 2009. The diversity of conjugative relaxases and its application in plasmid classification. *FEMS Microbiol Rev* **33:** 657–687.
74. **Smillie C, Garcillán-Barcia MP, Francia MV, Rocha EPC, de la Cruz F.** 2010. Mobility of plasmids. *Microbiol Mol Biol Rev* **74:**434–452.
75. **Lorenzo-Díaz F, Fernández-López C, Garcillán-Barcia MP, Espinosa M.** 2014. Bringing them together: plasmid pMV158 rolling circle replication and conjugation under an evolutionary perspective. *Plasmid* **74:**15–31.
76. **Weaver KE, Kwong SM, Firth N, Francia MV.** 2009. The RepA_N replicons of Gram-positive bacteria: a family of broadly distributed but narrow host range plasmids. *Plasmid* **61:**94–109.
77. **Gruss A, Ehrlich SD.** 1989. The Family of highly interrelated single-stranded deoxyribonucleic acid plasmids. *Microbiol Rev* **53:**231–241.
78. **del Solar G, Moscoso M, Espinosa M.** 1993. Rolling circle-replicating plasmids from Gram-positive and Gram-negative bacteria: a wall falls. *Mol Microbiol* **8:** 789–796.
79. **Khan SA.** 1997. Rolling-circle replication of bacterial plasmids. *Microbiol Mol Biol Rev* **61:**442–455.
80. **Widdowson CA, Adrian PV, Klugman KP.** 2000. Acquisition of Chloramphenicol resistance by the linearization and integration of the entire staphylococcal plasmid pC194 into the chromosome of *Streptococcus pneumoniae*. *Antimicrob Agents Chemother* **44:**393–395.
81. **Del Solar G, Moscoso M, Espinosa M.** 1993. Rolling circle-replicating plasmids from Gram-positive and Gram-negative bacteria: a wall falls. *Mol Microbiol* **8:** 789–796.
82. **Heng NCK, Burtenshaw GA, Jack RW, Tagg JR.** 2004. Sequence analysis of pDN571, a plasmid encoding novel bacteriocin production in M-type 57 *Streptococcus pyogenes*. *Plasmid* **52:**225–229.
83. **Romero P, Llull D, García E, Mitchell TJ, López R, Moscoso M.** 2007. Isolation and characterization of a new plasmid pSpnP1 from a multidrug-resistant clone of *Streptococcus pneumoniae*. *Plasmid* **58:**51–60.
84. **Gillespie MT, May JW, Skurray RA.** 1985. Antibiotic resistance in *Staphylococcus aureus* isolated at an Australian hospital between 1946 and 1981. *J Med Microbiol* **19:**137–147.
85. **Schiwon K, Arends K, Rogowski KM, Fürch S, Prescha K, Sakinc T, Van Houdt R, Werner G, Grohmann E.** 2013. Comparison of antibiotic resistance, biofilm formation and conjugative transfer of *Staphylococcus* and *Enterococcus* isolates from International Space Station and Antarctic Research Station Concordia. *Microb Ecol* **65:**638–651.

86. **Shiwa Y, Yanase H, Hirose Y, Satomi S, Araya-Kojima T, Watanabe S, Zendo T, Chibazakura T, Shimizu-Kadota M, Yoshikawa H, Sonomoto K.** 2014. Complete genome sequence of *Enterococcus mundtii* QU 25, an efficient L-(+)-lactic acid-producing bacterium. *DNA Res* **21:**369–377.

87. **Garcia-Migura L, Hasman H, Jensen LB.** 2009. Presence of pRI1: a small cryptic mobilizable plasmid isolated from *Enterococcus faecium* of human and animal origin. *Curr Microbiol* **58:**95–100.

88. **De Vries LE, Christensen H, Agersø Y.** 2012. The diversity of inducible and constitutively expressed erm(C) genes and association to different replicon types in staphylococci plasmids. *Mob Genet Elements* **2:**72–80.

89. **Gómez-Sanz E, Kadlec K, Feßler AT, Zarazaga M, Torres C, Schwarz S.** 2013. Novel erm(T)-carrying multiresistance plasmids from porcine and human isolates of methicillin-resistant *Staphylococcus aureus* ST398 that also harbor cadmium and copper resistance determinants. *Antimicrob Agents Chemother* **57:** 3275–3282.

90. **Firth N, Skurray RA.** 1998. Mobile elements in the evolution and spread of multiple-drug resistance in staphylococci. *Drug Resist Updat* **1:**49–58.

91. **Charlebois A, Jalbert L-A, Harel J, Masson L, Archambault M.** 2012. Characterization of genes encoding for acquired bacitracin resistance in *Clostridium perfringens. PLoS One* **7:**e44449. doi:10.1371/journal.pone.0044449.

92. **Palmer KL, Kos VN, Gilmore MS.** 2010. Horizontal gene transfer and the genomics of enterococcal antibiotic resistance. *Curr Opin Microbiol* **13:**632–639.

93. **Montilla A, Zavala A, Cáceres Cáceres R, Cittadini R, Vay C, Gutkind G, Famiglietti A, Bonofiglio L, Mollerach M.** 2014. Genetic environment of the lnu(B) gene in a *Streptococcus agalactiae* clinical isolate. *Antimicrob Agents Chemother* **58:**5636–5637.

94. **Zong Z.** 2013. Characterization of a complex context containing mecA but lacking genes encoding cassette chromosome recombinases in *Staphylococcus haemolyticus. BMC Microbiol* **13:**64.

95. **Locke JB, Zuill DE, Scharn CR, Deane J, Sahm DF, Denys GA, Goering RV, Shaw KJ.** 2014. Linezolid-resistant *Staphylococcus aureus* strain 1128105, the first known clinical isolate possessing the cfr multidrug resistance gene. *Antimicrob Agents Chemother* **58:** 6592–6598.

96. **Goessweiner-Mohr N, Arends K, Keller W, Grohmann E.** 2013. Conjugative type IV secretion systems in Gram-positive bacteria. *Plasmid* **70:**289–302.

97. **Li J, Li B, Wendlandt S, Schwarz S, Wang Y, Wu C, Ma Z, Shen J.** 2014. Identification of a novel vga(E) gene variant that confers resistance to pleuromutilins, lincosamides and streptogramin A antibiotics in staphylococci of porcine origin. *J Antimicrob Chemother* **69:** 919–923.

98. **Lozano C, Aspiroz C, Rezusta A, Gómez-sanz E, Simon C, Gómez P, Ortega C, José M, Zarazaga M, Torres C.** 2012. Identification of novel vga(A)-carrying plasmids and a Tn 5406-like transposon in meticillin-resistant *Staphylococcus aureus* and *Staphylococcus epidermidis* of human and animal origin. *Int J Antimicrob Agents* **40:**306–312.

99. **Brantl S.** 2014. Antisense-RNA mediated control of plasmid replication: pIP501 revisited. *Plasmid.* [Epub ahead of print.] doi:10.1016/j.plasmid.2014.07.004.

100. **Grohmann E, Muth G, Espinosa M.** 2003. Conjugative plasmid transfer in Gram-positive bacteria. *Microbiol Mol Biol Rev* **67:**277–301.

101. **Behnke D, Malke H, Hartmann M, Walter F.** 1979. Post-transformational rearrangement of an *in vitro* reconstructed group-A streptococcal erythromycin resistance plasmid. *Plasmid* **2:**605–616.

102. **Clewell DB, Yagi Y, Dunny GM, Schultz SK.** 1974. Characterization of three plasmid deoxyribonucleic acid molecules in a strain of *Streptococcus faecalis*: identification of a plasmid determining erythromycin resistance. *J Bacteriol* **117:**283–289.

103. **Horodniceanu T, Bouanchaud DH, Bieth G, Chabbert YA.** 1976. R plasmids in *Streptococcus agalactiae* (group B). *Antimicrob Agents Chemother* **10:**795–801.

104. **Thompson JK, Collins MA.** 2003. Completed sequence of plasmid pIP501 and origin of spontaneous deletion derivatives. *Plasmid* **50:**28–35.

105. **Horaud T, Le Bouguenec C, Pepper K.** 1985. Molecular genetics of resistance to macrolides, lincosamides and streptogramin B (MLS) in streptococci. *J Antimicrob Chemother* **16:**111–135.

106. **Derome A, Hoischen C, Bussiek M, Grady R, Adamczyk M, Kedzierska B, Diekmann S, Barillà D, Hayes F.** 2008. Centromere anatomy in the multidrug-resistant pathogen *Enterococcus faecium. Proc Natl Acad Sci USA* **105:**2151–2156.

107. **Li X, Alvarez V, Harper WJ, Wang HH.** 2011. Persistent, toxin-antitoxin system-independent, tetracycline resistance-encoding plasmid from a dairy *Enterococcus faecium* isolate. *Appl Environ Microbiol* **77:**7096–7103.

108. **Abajy MY, Kopeć J, Schiwon K, Burzynski M, Döring M, Bohn C, Grohmann E.** 2007. A type IV-secretion-like system is required for conjugative DNA transport of broad-host-range plasmid pIP501 in Gram-positive bacteria. *J Bacteriol* **189:**2487–2496.

109. **Krah ER, Macrina FL.** 1991. Identification of a region that influences host range of the streptococcal conjugative plasmid pIP501. *Plasmid* **25:**64–69.

110. **Schaberg DR, Zervos MJ.** 1986. Intergeneric and interspecies gene exchange in Gram-positive cocci. *Antimicrob Agents Chemother* **30:**817–822.

111. **Teuber M, Meile L, Schwarz F.** 1999. Acquired antibiotic resistance in lactic acid bacteria from food. *Antonie Van Leeuwenhoek* **76:**115–137.

112. **El-Solh N, Bouanchaud DH, Horodniceanu T, Roussel A, Chabbert YA.** 1978. Molecular studies and possible relatedness between R plasmids from groups B and D streptococci. *Antimicrob Agents Chemother* **14:** 19–23.

113. **Engel HW, Soedirman N, Rost JA, van Leeuwen WJ, van Embden JD.** 1980. Transferability of macrolide, lincomycin, and streptogramin resistances between group A, B, and D streptococci, *Streptococcus pneumoniae*, and *Staphylococcus aureus*. *J Bacteriol* **142:**407–413.

114. **Clewell DB.** 1981. Plasmids, drug resistance, and gene transfer in the genus *Streptococcus*. *Microbiol Rev* **45:** 409–436.

115. **Sletvold H, Johnsen PJ, Simonsen GS, Aasnaes B, Sundsfjord A, Nielsen KM.** 2007. Comparative DNA analysis of two vanA plasmids from *Enterococcus faecium* strains isolated from poultry and a poultry farmer in Norway. *Antimicrob Agents Chemother* **51:** 736–739.

116. **Sletvold H, Johnsen PJ, Wikmark O-G, Simonsen GS, Sundsfjord A, Nielsen KM.** 2010. Tn1546 is part of a larger plasmid-encoded genetic unit horizontally disseminated among clonal *Enterococcus faecium* lineages. *J Antimicrob Chemother* **65:**1894–1906.

117. **Zhu W, Clark NC, McDougal LK, Hageman J, McDonald LC, Patel JB.** 2008. Vancomycin-resistant *Staphylococcus aureus* isolates associated with Inc18-like vanA plasmids in Michigan. *Antimicrob Agents Chemother* **52:**452–457.

118. **Zhu W, Murray PR, Huskins WC, Jernigan JA, McDonald LC, Clark NC, Anderson KF, McDougal LK, Hageman JC, Olsen-Rasmussen M, Frace M, Alangaden GJ, Chenoweth C, Zervos MJ, Robinson-Dunn B, Schreckenberger PC, Reller LB, Rudrik JT, Patel JB.** 2010. Dissemination of an *Enterococcus* Inc18-Like vanA plasmid associated with vancomycin-resistant *Staphylococcus aureus*. *Antimicrob Agents Chemother* **54:**4314–4320.

119. **Teuber M, Schwarz F, Perreten V.** 2003. Molecular structure and evolution of the conjugative multiresistance plasmid pRE25 of *Enterococcus faecalis* isolated from a raw-fermented sausage. *Int J Food Microbiol* **88:**325–329.

120. **Maki T, Santos MD, Kondo H, Hirono I, Aoki T.** 2009. A transferable 20-kilobase multiple drug resistance-conferring R plasmid (pKL0018) from a fish pathogen (*Lactococcus garvieae*) is highly homologous to a conjugative multiple drug resistance-conferring enterococcal plasmid. *Appl Environ Microbiol* **75:** 3370–3372.

121. **Lim S-K, Tanimoto K, Tomita H, Ike Y.** 2006. Pheromone-responsive conjugative vancomycin resistance plasmids in *Enterococcus faecalis* isolates from humans and chicken feces. *Appl Environ Microbiol* **72:** 6544–6553.

122. **Tanimoto K, Ike Y.** 2008. Complete nucleotide sequencing and analysis of the 65-kb highly conjugative *Enterococcus faecium* plasmid pMG1: identification of the transfer-related region and the minimum region required for replication. *FEMS Microbiol Lett* **288:** 186–195.

123. **Tomita H, Pierson C, Lim SK, Clewell DB, Ike Y.** 2002. Possible connection between a widely disseminated conjugative gentamicin resistance (pMG1-like) plasmid and the emergence of vancomycin resistance in *Enterococcus faecium*. *J Clin Microbiol* **40:**3326–3333.

124. **Tomita H, Tanimoto K, Hayakawa S, Morinaga K, Ezaki K, Oshima H, Ike Y.** 2003. Highly conjugative pMG1-like plasmids carrying Tn1546-like transposons that encode vancomycin resistance in *Enterococcus faecium*. *J Bacteriol* **185:**7024–7028.

125. **Manson JM, Hancock LE, Gilmore MS.** 2010. Mechanism of chromosomal transfer of *Enterococcus faecalis* pathogenicity island, capsule, antimicrobial resistance, and other traits. *Proc Natl Acad Sci USA* **107:** 12269–12274.

126. **Arias CA, Panesso D, Singh KV, Rice LB, Murray BE.** 2009. Cotransfer of antibiotic resistance genes and a hylEfm-containing virulence plasmid in *Enterococcus faecium*. *Antimicrob Agents Chemother* **53:**4240–4246.

127. **Firth N, Apisiridej S, Berg T, Rourke BAO, Curnock S, Dyke KGH, Skurray RA.** 2006. Replication of staphylococcal multiresistance plasmids. *J Bacteriol* **182:** 2170–2178.

128. **Kwong SM, Lim R, Lebard RJ, Skurray RA, Firth N.** 2008. Analysis of the pSK1 replicon, a prototype from the staphylococcal multiresistance plasmid family. *Microbiology* **154:**3084–3094.

129. **Berg T, Firth N, Apisiridej S, Leelaporn A, Skurray RA, Hettiaratchi A.** 2006. Complete nucleotide sequence of pSK41: evolution of staphylococcal conjugative multiresistance plasmids. *J Bacteriol* **180:**4350–4359.

130. **Littlejohn TG, DiBerardino D, Messerotti LJ, Spiers SJ, Skurray RA.** 1991. Structure and evolution of a family of genes encoding antiseptic and disinfectant resistance in *Staphylococcus aureus*. *Gene* **101:**59–66.

131. **Pérez-Roth E, Kwong SM, Alcoba-Florez J, Firth N, Méndez-Alvarez S.** 2010. Complete nucleotide sequence and comparative analysis of pPR9, a 41.7-kilobase conjugative staphylococcal multiresistance plasmid conferring high-level mupirocin resistance. *Antimicrob Agents Chemother* **54:**2252–2257.

132. **Pérez-Roth E, Potel-Alvarellos C, Espartero X, Constela-Caramés L, Méndez-Álvarez S, Alvarez-Fernández M.** 2013. Molecular epidemiology of plasmid-mediated high-level mupirocin resistance in methicillin-resistant *Staphylococcus aureus* in four Spanish health care settings. *Int J Med Microbiol* **303:**201–204.

133. **Udo EE, Jacob LE.** 1998. Conjugative transfer of high-level mupirocin resistance and the mobilization of non-conjugative plasmids in *Staphylococcus*. *Microb Drug Resist* **4:**185–193.

134. **Weigel LM, Clewell DB, Gill SR, Clark NC, McDougal LK, Flannagan SE, Kolonay JF, Shetty J, Killgore GE, Tenover FC.** 2003. Genetic analysis of a high-level vancomycin-resistant isolate of *Staphylococcus aureus*. *Science* **302:**1569–1571.

135. **Diep BA, Gill SR, Chang RF, Phan TH, Chen JH, Davidson MG, Lin F, Lin J, Carleton HA, Mongodin EF, Sensabaugh GF, Perdreau-Remington F.** 2006. Complete genome sequence of USA300, an epidemic clone of community-acquired meticillin-resistant *Staphylococcus aureus*. *Lancet* **367:**731–739.

136. **Thomas WD, Archer GL.** 1992. Mobilization of recombinant plasmids from *Staphylococcus aureus* into coagulase negative *Staphylococcus* species. *Plasmid* **27:** 164–168.
137. **Projan SJ, Archer GL.** 1989. Mobilization of the relaxable *Staphylococcus aureus* plasmid pC221 by the conjugative plasmid pGO1 involves three pC221 loci. *J Bacteriol* **171:**1841–1845.
138. **Wardal E, Sadowy E, Hryniewicz W.** 2010. Complex nature of enterococcal pheromone-responsive plasmids. *Pol J Microbiol* **59:**79–87.
139. **Clewell DB, Francia MV.** 2004. Conjugation in Gram-positive bacteria. *In* Funnell B, Phillips JG (ed), *Plasmid Biology*. ASM Press, Washington, DC.
140. **Clewell DB, Francia MV, Flannagan SE, An FY.** 2002. Enterococcal plasmid transfer: sex pheromones, transfer origins, relaxases, and the *Staphylococcus aureus* issue. *Plasmid* **48:**193–201.
141. **Weaver KE.** 2014. The type I toxin-antitoxin par locus from *Enterococcus faecalis* plasmid pAD1: RNA regulation by both cis- and trans-acting elements. *Plasmid.* [Epub ahead of print.] doi:10.1016/j.plasmid.2014.10.001.
142. **Grady R, Hayes F.** 2003. Axe-Txe, a broad-spectrum proteic toxin-antitoxin system specified by a multidrug-resistant, clinical isolate of *Enterococcus faecium. Mol Microbiol* **47:**1419–1432.
143. **Halvorsen EM, Williams JJ, Bhimani AJ, Billings EA, Hergenrother PJ.** 2011. Txe, an endoribonuclease of the enterococcal Axe-Txe toxin-antitoxin system, cleaves mRNA and inhibits protein synthesis. *Microbiology* **157:**387–397.
144. **Freitas AR, Coque TM, Novais C, Hammerum AM, Lester CH, Zervos MJ, Donabedian S, Jensen LB, Francia MV, Baquero F, Peixe L.** 2011. Human and swine hosts share vancomycin-resistant *Enterococcus faecium* CC17 and CC5 and *Enterococcus faecalis* CC2 clonal clusters harboring Tn1546 on indistinguishable plasmids. *J Clin Microbiol* **49:**925–931.
145. **Werner G, Klare I, Witte W.** 1999. Large conjugative vanA plasmids in vancomycin-resistant *Enterococcus faecium. J Clin Microbiol* **37:**2383–2384.
146. **Hasman H, Villadsen AG, Aarestrup FM.** 2005. Diversity and stability of plasmids from glycopeptide-resistant *Enterococcus faecium* (GRE) isolated from pigs in Denmark. *Microb Drug Resist* **11:**178–184.
147. **Biavasco F, Foglia G, Paoletti C, Zandri G, Magi G, Guaglianone E, Sundsfjord A, Pruzzo C, Donelli G, Facinelli B.** 2007. VanA-type enterococci from humans, animals, and food: species distribution, population structure, Tn1546 typing and location, and virulence determinants. *Appl Environ Microbiol* **73:**3307–3319.
148. **Laverde Gomez JA, van Schaik W, Freitas AR, Coque TM, Weaver KE, Francia MV, Witte W, Werner G.** 2011. A multiresistance megaplasmid pLG1 bearing a hylEfm genomic island in hospital *Enterococcus faecium* isolates. *Int J Med Microbiol* **301:**165–175.
149. **Zhang X, Vrijenhoek JEP, Bonten MJM, Willems RJL, van Schaik W.** 2011. A genetic element present on megaplasmids allows *Enterococcus faecium* to use raffinose as carbon source. *Environ Microbiol* **13:** 518–528.
150. **Gordoncillo MJN, Donabedian S, Bartlett PC, Perri M, Zervos M, Kirkwood R, Febvay C.** 2013. Isolation and molecular characterization of vancomycin-resistant *Enterococcus faecium* from swine in Michigan, USA. *Zoonoses Public Health* **60:**319–326.
151. **Johnson PDR, Ballard SA, Grabsch EA, Stinear TP, Seemann T, Young HL, Grayson ML, Howden BP.** 2010. A sustained hospital outbreak of vancomycin-resistant *Enterococcus faecium* bacteremia due to emergence of vanB *E. faecium* sequence type 203. *J Infect Dis* **202:**1278–1286.
152. **Rosvoll TCS, Lindstad BL, Lunde TM, Hegstad K, Aasnaes B, Hammerum AM, Lester CH, Simonsen GS, Sundsfjord A, Pedersen T.** 2012. Increased high-level gentamicin resistance in invasive *Enterococcus faecium* is associated with aac(6′)Ie-aph(2′′)Ia-encoding transferable megaplasmids hosted by major hospital-adapted lineages. *FEMS Immunol Med Microbiol* **66:** 166–176.
153. **Panesso D, Montealegre MC, Rincón S, Mojica MF, Rice LB, Singh KV, Murray BE, Arias CA.** 2011. The hylEfm gene in pHylEfm of *Enterococcus faecium* is not required in pathogenesis of murine peritonitis. *BMC Microbiol* **11:**20.
154. **Kim DS, Singh KV, Nallapareddy SR, Qin X, Panesso D, Arias CA, Murray BE.** 2010. The fms21 (pilA)-fms20 locus encoding one of four distinct pili of *Enterococcus faecium* is harboured on a large transferable plasmid associated with gut colonization and virulence. *J Med Microbiol* **59:**505–507.
155. **Boon E, Meehan CJ, Whidden C, Wong DH-J, Langille MGI, Beiko RG.** 2014. Interactions in the microbiome: communities of organisms and communities of genes. *FEMS Microbiol Rev* **38:**90–118.
156. **Smillie CS, Smith MB, Friedman J, Cordero OX, David LA, Alm EJ.** 2011. Ecology drives a global network of gene exchange connecting the human microbiome. *Nature* **480:**241–244.
157. **Skippington E, Ragan MA.** 2011. Within-species lateral genetic transfer and the evolution of transcriptional regulation in *Escherichia coli* and *Shigella. BMC Genomics* **12:**532.
158. **Baquero F, Coque TM, Canto R.** 2003. Antibiotics, complexity, and evolution. *ASM News* **69:**547–552.
159. **Baquero F, Tedim AP, Coque TM.** 2013. Antibiotic resistance shaping multilevel population biology of bacteria. *Front Microbiol* **4:**1–15.
160. **Baquero F, Lanza VF, Canton R, Coque TM.** 2014. Public health evolutionary biology of antimicrobial resistance: priorities for intervention. *Evol Appl.* [Epub ahead of print.] doi:10.1111/eva.12235.
161. **Baquero F, Coque TM, Cantón R.** 2014. Counteracting antibiotic resistance: breaking barriers among antibacterial strategies. *Expert Opin Ther Targets* **18:**1–11.
162. **Martínez JL.** 2008. Antibiotics and antibiotic resistance genes in natural environments. *Science* **321:**365–367.

163. **Martínez JL.** 2011. Bottlenecks in the transferability of antibiotic resistance from natural ecosystems to human bacterial pathogens. *Front Microbiol* **2:**265.

164. **Berryman DI, Rood JI.** 1995. The closely related ermB-ermAM genes from *Clostridium perfringens*, *Enterococcus faecalis* (pAM beta 1), and *Streptococcus agalactiae* (pIP501) are flanked by variants of a directly repeated sequence. *Antimicrob Agents Chemother* **39:**1830–1834.

165. **Howden BP, Holt KE, Lam MMC, Seemann T, Ballard S, Coombs GW, Tong SYC, Grayson ML, Johnson PDR, Stinear TP.** 2013. Genomic insights to control the emergence of vancomycin-resistant enterococci. *MBio* **4.** doi:10.1128/mBio.00412-13.

166. **Marvaud J-C, Mory F, Lambert T.** 2011. Clostridium clostridioforme and *Atopobium minutum* clinical isolates with vanB-type resistance in France. *J Clin Microbiol* **49:**3436–3438.

167. **Ballard SA, Grabsch EA, Johnson PDR, Grayson ML.** 2005. Comparison of three PCR primer sets for identification of vanB gene carriage in feces and correlation with carriage of vancomycin-resistant enterococci: interference by vanB-containing anaerobic bacilli. *Antimicrob Agents Chemother* **49:**77–81.

168. **Domingo M-C, Huletsky A, Boissinot M, Hélie M-C, Bernal A, Bernard KA, Grayson ML, Picard FJ, Bergeron MG.** 2009. *Clostridium lavalense* sp. nov., a glycopeptide-resistant species isolated from human faeces. *Int J Syst Evol Microbiol* **59:**498–503.

169. **Marshall BM, Ochieng DJ, Levy SB.** 2009. Commensals: underappreciated reservoir of antibiotic resistance. *Microbe* **4:**231–238.

170. **Maynard-Smith J.** 1982. The century since Darwin. *Nature* **296:**599–601.

171. **Baquero F.** 2014. Genetic hyper-codes and multidimensional Darwinism: replication modes and codes in evolutionary individuals of the bacterial world. *In* Trueba G (ed), *Why does Evolution Matter?* Cambridge Scholars Publishing, Newcastle Upon Tyne, UK.

172. **Baquero F.** 2004. From pieces to patterns: evolutionary engineering in bacterial pathogens. *Nat Rev Microbiol* **2:**510–518.

173. **Okasha S.** 2006. *Evolution and the Levels of Selection.* Clarendon Press, Oxford, UK.

174. **Paulsson J.** 2002. Multileveled selection on plasmid replication. *Genetics* **161:**1373–1384.

175. **Cordero OX, Polz MF.** 2014. Explaining microbial genomic diversity in light of evolutionary ecology. *Nat Rev Microbiol* **12:**263–273.

176. **Norton B.** 1996. Change, constancy, and creativity: the new ecology and some old problems. *Duke Environ Law Policy Forum* **7:**49–70.

177. **Hiramatsu K, Ito T, Tsubakishita S, Sasaki T, Takeuchi F, Morimoto Y, Katayama Y, Matsuo M, Kuwahara-Arai K, Hishinuma T, Baba T.** 2013. Genomic basis for methicillin resistance in *Staphylococcus aureus. Infect Chemother* **45:**117–136.

178. **Haines AN, Gauthier DT, Nebergall EE, Cole SD, Nguyen KM, Rhodes MW, Vogelbein WK.** 2013. First report of *Streptococcus parauberis* in wild finfish from North America. *Vet Microbiol* **166:**270–275.

179. **Richards VP, Palmer SR, Pavinski Bitar PD, Qin X, Weinstock GM, Highlander SK, Town CD, Burne RA, Stanhope MJ.** 2014. Phylogenomics and the dynamic genome evolution of the genus *Streptococcus. Genome Biol Evol* **6:**741–753.

180. **Gao X-Y, Zhi X-Y, Li H-W, Klenk H-P, Li W-J.** 2014. Comparative genomics of the bacterial genus *Streptococcus* illuminates evolutionary implications of species groups. *PLoS One* **9:**e101229. doi:10.1371/journal.pone.0101229.

181. **Jans C, Meile L, Lacroix C, Stevens MJA.** 2014. Genomics, evolution, and molecular epidemiology of the *Streptococcus bovis/Streptococcus equinus* complex (SBSEC). *Infect Genet Evol.* [Epub ahead of print.] doi:10.1016/j.meegid.2014.09.017.

182. **Caufield PW, Saxena D, Fitch D, Li Y.** 2007. Population structure of plasmid-containing strains of *Streptococcus mutans*, a member of the human indigenous biota. *J Bacteriol* **189:**1238–1243.

183. **Wescombe PA, Burton JP, Cadieux PA, Klesse NA, Hyink O, Heng NCK, Chilcott CN, Reid G, Tagg JR.** 2006. Megaplasmids encode differing combinations of lantibiotics in *Streptococcus salivarius. Antonie Van Leeuwenhoek* **90:**269–280.

184. **Pepper K, Le Bouguénec C, de Cespédès G, Horaud T.** 1986. Dispersal of a plasmid-borne chloramphenicol resistance gene in streptococcal and enterococcal plasmids. *Plasmid* **16:**195–203.

185. **Woodbury RL, Klammer KA, Xiong Y, Bailiff T, Glennen A, Bartkus JM, Lynfield R, Van Beneden C, Beall BW.** 2008. Plasmid-borne erm(T) from invasive, macrolide-resistant *Streptococcus pyogenes* strains. *Antimicrob Agents Chemother* **52:**1140–1143.

186. **Whitehead TR, Cotta MA.** 2001. Sequence analyses of a broad host-range plasmid containing ermT from a tylosin-resistant *Lactobacillus* sp. isolated from swine feces. *Curr Microbiol* **43:**17–20.

187. **DiPersio LP, DiPersio JR.** 2006. High rates of erythromycin and clindamycin resistance among OBGYN isolates of group B *Streptococcus. Diagn Microbiol Infect Dis* **54:**79–82.

188. **Palmieri C, Magi G, Creti R, Baldassarri L, Imperi M, Gherardi G, Facinelli B.** 2013. Interspecies mobilization of an ermT-carrying plasmid of *Streptococcus dysgalactiae* subsp. equisimilis by a coresident ICE of the ICESa2603 family. *J Antimicrob Chemother* **68:**23–26.

189. **Holden MTG, Hauser H, Sanders M, Ngo TH, Cherevach I, Cronin A, Goodhead I, Mungall K, Quail MA, Price C, Rabbinowitsch E, Sharp S, Croucher NJ, Chieu TB, Mai NTH, Diep TS, Chinh NT, Kehoe M, Leigh JA, Ward PN, Dowson CG, Whatmore AM, Chanter N, Iversen P, Gottschalk M, Slater JD, Smith HE, Spratt BG, Xu J, Ye C, Bentley S, Barrell BG, Schultsz C, Maskell DJ, Parkhill J.** 2009. Rapid evolution of virulence and drug resistance in the emerging zoonotic pathogen *Streptococcus suis. PLoS One* **4:** e6072. doi:10.1371/journal.pone.0006072.

190. **Cantin M, Harel J, Higgins R, Gottschalk M.** 1992. Antimicrobial resistance patterns and plasmid profiles of *Streptococcus suis* isolates. *J Vet Diagn Invest* **4:** 170–174.
191. **Wang Y, Li D, Song L, Liu Y, He T, Liu H, Wu C, Schwarz S, Shen J.** 2013. First report of the multiresistance gene *cfr* in *Streptococcus suis*. *Antimicrob Agents Chemother* **57:**4061–4063.
192. **Chander Y, Oliveira SR, Goyal SM.** 2011. Identification of the tet(B) resistance gene in *Streptococcus suis*. *Vet J* **189:**359–360.
193. **Lebreton F, Willems RJL, Gilmore MS.** 2014. *Enterococcus* diversity, origins in nature, and gut colonization. *In* Gilmore MS, Clewell DB, Ike Y, Shankar N (ed), *Enterococci: From Commensals to Leading Causes of Drug Resistant Infection*. Massachusetts Eye and Ear Infirmary, Boston, MA.
194. **Willems RJL, Top J, van Schaik W, Leavis H, Bonten M, Sirén J, Hanage WP, Corander J.** 2012. Restricted gene flow among hospital subpopulations of *Enterococcus faecium*. *MBio* **3:**e00151-12. doi:10.1128/mBio.00151-12.
195. **Tedim AP, Ruiz-Garbajosa P, Corander J, Rodríguez CM, Cantón R, Willems R, Baquero F, Coque TM.** 2014. Population biology of *Enterococcus* from intestinal colonization in hospitalized and non-hospitalized individuals in different age groups. *Appl Environ Microbiol* **81.** [Epub ahead of print.] doi:10.1128/AEM.03661-14.
196. **McBride SM, Fischetti VA, Leblanc DJ, Moellering RC, Gilmore MS.** 2007. Genetic diversity among *Enterococcus faecalis*. *PLoS One* **2:**e582. doi:10.1371/journal.pone.0000582.
197. **Kuch A, Willems RJL, Werner G, Coque TM, Hammerum AM, Sundsfjord A, Klare I, Ruiz-Garbajosa P, Simonsen GS, van Luit-Asbroek M, Hryniewicz W, Sadowy E.** 2012. Insight into antimicrobial susceptibility and population structure of contemporary human *Enterococcus faecalis* isolates from Europe. *J Antimicrob Chemother* **67:**551–558.
198. **López M, Tedim AP, Baquero F, Torres C, Coque TM.** 2011. Plasmid diversity among *Enterococcus* spp from humans and animals (1991-2010) Poster presented at 4th Congress of European Microbiologist, FEMS. Geneva, Switzerland.
199. **LeBlanc DJ, Inamine JM, Lee LN.** 1986. Broad geographical distribution of homologous erythromycin, kanamycin, and streptomycin resistance determinants among group D streptococci of human and animal origin. *Antimicrob Agents Chemother* **29:**549–555.
200. **Murray BE, An FY, Clewell DB.** 1988. Plasmids and pheromone response of the beta-lactamase producer *Streptococcus* (*Enterococcus*) *faecalis* HH22. *Antimicrob Agents Chemother* **32:**547–551.
201. **Murray BE, Singh KV, Markowitz SM, Lopardo HA, Patterson JE, Zervos MJ, Rubeglio E, Eliopoulos GM, Rice LB, Goldstein FW.** 1991. Evidence for clonal spread of a single strain of beta-lactamase-producing *Enterococcus* (*Streptococcus*) *faecalis* to six hospitals in five states. *J Infect Dis* **163:**780–785.
202. **Patterson JE, Singh KV, Murray BE.** 1991. Epidemiology of an endemic strain of beta-lactamase-producing *Enterococcus faecalis*. *J Clin Microbiol* **29:**2513–2516.
203. **Werner GDA, Strassmann JE, Ivens ABF, Engelmoer DJP, Verbruggen E, Queller DC, Noë R, Johnson NC, Hammerstein P, Kiers ET.** 2014. Evolution of microbial markets. *Proc Natl Acad Sci USA* **111:**1237–1244.
204. **Datta N, Hughes VM.** Plasmids of the same Inc groups in *Enterobacteria* before and after the medical use of antibiotics. *Nature* **306:**616–617.
205. **Wang X-M, Li X-S, Wang Y-B, Wei F-S, Zhang S-M, Shang Y-H, Du X-D.** 2015. Characterization of a multidrug resistance plasmid from *Enterococcus faecium* that harbours a mobilized bcrABDR locus. *J Antimicrob Chemother* **70:**609–611.
206. **Devirgiliis C, Zinno P, Perozzi G.** 2013. Update on antibiotic resistance in foodborne *Lactobacillus* and *Lactococcus* species. *Front Microbiol* **4:**301.
207. **Becker K, Heilmann C, Peters G.** 2014. Coagulase-negative staphylococci. *Clin Microbiol Rev* **27:**870–926.
208. **Lindsay JA, Holden MTG.** 2004. *Staphylococcus aureus*: superbug, super genome? *Trends Microbiol* **12:** 378–385.
209. **Lindsay JA.** 2010. Genomic variation and evolution of *Staphylococcus aureus*. *Int J Med Microbiol* **300:** 98–103.
210. **Dethlefsen L, McFall-Ngai M, Relman DA.** 2007. An ecological and evolutionary perspective on human-microbe mutualism and disease. *Nature* **449:**811–818.
211. **Lindsay JA, Knight GM, Budd EL, McCarthy AJ.** 2012. Shuffling of mobile genetic elements (MGEs) in successful healthcare-associated MRSA (HA-MRSA). *Mob Genet Elements* **2:**239–243.
212. **Livermore DM.** 2000. Antibiotic resistance in staphylococci. *Int J Antimicrob Agents* **16**(Suppl 1):S3–S10.
213. **Sidhu MS, Heir E, Leegaard T, Wiger K, Holck A.** 2002. Frequency of disinfectant resistance genes and genetic linkage with beta-lactamase transposon Tn552 among clinical staphylococci. *Antimicrob Agents Chemother* **46:**2797–2803.
214. **Willems RJL, Hanage WP, Bessen DE, Feil EJ.** 2011. Population biology of Gram-positive pathogens: high-risk clones for dissemination of antibiotic resistance. *FEMS Microbiol Rev* **35:**872–900.
215. **Lindsay JA, Holden MTG.** 2006. Understanding the rise of the superbug: investigation of the evolution and genomic variation of *Staphylococcus aureus*. *Funct Integr Genomics* **6:**186–201.
216. **Lutz C, Erken M, Noorian P, Sun S, McDougald D.** 2013. Environmental reservoirs and mechanisms of persistence of *Vibrio cholerae*. *Front Microbiol* **4:**375.
217. **Lindsay JA.** 2014. Evolution of *Staphylococcus aureus* and MRSA during outbreaks. *Infect Genet Evol* **21:** 548–553.
218. **McDougal LK, Fosheim GE, Nicholson A, Bulens SN, Limbago BM, Shearer JES, Summers AO, Patel JB.** 2010. Emergence of resistance among USA300 methicillin-resistant *Staphylococcus aureus* isolates causing invasive

disease in the United States. *Antimicrob Agents Chemother* **54**:3804–3811.

219. **McCarthy AJ, Witney AA, Gould KA, Moodley A, Guardabassi L, Voss A, Denis O, Broens EM, Hinds J, Lindsay JA.** 2011. The distribution of mobile genetic elements (MGEs) in MRSA CC398 is associated with both host and country. *Genome Biol Evol* **3**:1164–1174.
220. **Alibayov B, Baba-Moussa L, Sina H, Zdeňková K, Demnerová K.** 2014. *Staphylococcus aureus* mobile genetic elements. *Mol Biol Rep* **41**:5005–5018.
221. **Uhlemann A-C, Dordel J, Knox JR, Raven KE, Parkhill J, Holden MTG, Peacock SJ, Lowy FD.** 2014. Molecular tracing of the emergence, diversification, and transmission of *S. aureus* sequence type 8 in a New York community. *Proc Natl Acad Sci USA* **111**:6738–6743.
222. **Rossi F, Diaz L, Wollam A, Panesso D, Zhou Y, Rincon S, Narechania A, Xing G, Di Gioia TSR, Doi A, Tran TT, Reyes J, Munita JM, Carvajal LP, Hernandez-Roldan A, Brandão D, van der Heijden IM, Murray BE, Planet PJ, Weinstock GM, Arias CA.** 2014. Transferable vancomycin resistance in a community-associated MRSA lineage. *N Engl J Med* **370**:1524–1531.
223. **Strommenger B, Bartels MD, Kurt K, Layer F, Rohde SM, Boye K, Westh H, Witte W, De Lencastre H, Nübel U.** 2014. Evolution of methicillin-resistant *Staphylococcus aureus* towards increasing resistance. *J Antimicrob Chemother* **69**:616–622.
224. **Locke JB, Zuill DE, Scharn CR, Deane J, Sahm DF, Goering RV, Jenkins SG, Shaw KJ.** 2014. Identification and characterization of linezolid-resistant cfr-positive *Staphylococcus aureus* USA300 isolates from a New York City Medical Center. *Antimicrob Agents Chemother* **58**:6949–6952.
225. **Sung JM-L, Lloyd DH, Lindsay JA.** 2008. *Staphylococcus aureus* host specificity: comparative genomics of human versus animal isolates by multi-strain microarray. *Microbiology* **154**:1949–1959.
226. **Holden MTG, Feil EJ, Lindsay JA, Peacock SJ, Day NPJ, Enright MC, Foster TJ, Moore CE, Hurst L, Atkin R, Barron A, Bason N, Bentley SD, Chillingworth C, Chillingworth T, Churcher C, Clark L, Corton C, Cronin A, Doggett J, Dowd L, Feltwell T, Hance Z, Harris B, Hauser H, Holroyd S, Jagels K, James KD, Lennard N, Line A, Mayes R, Moule S, Mungall K, Ormond D, Quail MA, Rabbinowitsch E, Rutherford K, Sanders M, Sharp S, Simmonds M, Stevens K, Whitehead S, Barrell BG, Spratt BG, Parkhill J.** 2004. Complete genomes of two clinical *Staphylococcus aureus* strains: evidence for the rapid evolution of virulence and drug resistance. *Proc Natl Acad Sci USA* **101**: 9786–9791.
227. **Robinson DA, Kearns AM, Holmes A, Morrison D, Grundmann H, Edwards G, O'Brien FG, Tenover FC, McDougal LK, Monk AB, Enright MC.** 2005. Re-emergence of early pandemic *Staphylococcus aureus* as a community-acquired meticillin-resistant clone. *Lancet* **365**:1256–1258.
228. **Yutin N, Galperinr MY.** 2014. A genomic update on clostridial phylogeny: Gram-negative spore-formers and other misplaced clostridia. *Environ Microbiol* **15**: 2631–2641.
229. **Magot M.** 1984. Physical characterization of the *Clostridium perfringens* tetracycline-chloramphenicol resistance plasmid pIP401. *Ann Microbiol* **135B**:269–282.
230. **Lyras D, Adams V, Ballard SA, Teng WL, Howarth PM, Crellin PK, Bannam TL, Songer JG, Rood JI.** 2009. tISCpe8, an IS1595-family lincomycin resistance element located on a conjugative plasmid in *Clostridium perfringens*. *J Bacteriol* **191**:6345–6351.
231. **Bannam TL, Teng WL, Bulach D, Lyras D, Rood JI.** 2006. Functional identification of conjugation and replication regions of the tetracycline resistance plasmid pCW3 from *Clostridium perfringens*. *J Bacteriol* **188**: 4942–4951.
232. **Gurjar A, Li J, McClane BS.** 2010. Characterization of toxin plasmids in *Clostridium perfringens* type C isolates. *Infect Immun* **78**:4860–4869.
233. **Nowell VJ, Kropinski AM, Songer JG, MacInnes JI, Parreira VR, Prescott JF.** 2012. Genome sequencing and analysis of a type A *Clostridium perfringens* isolate from a case of bovine clostridial abomasitis. *PLoS One* **7**:e32271. doi:10.1371/journal.pone.0032271.
234. **Sasaki Y, Yamamoto K, Tamura Y, Takahashi T.** 2001. Tetracycline-resistance genes of *Clostridium perfringens*, *Clostridium septicum* and *Clostridium sordellii* isolated from cattle affected with malignant edema. *Vet Microbiol* **83**:61–69.
235. **Franciosa G, Scalfaro C, Di Bonito P, Vitale M, Aureli P.** 2011. Identification of novel linear megaplasmids carrying a ß-lactamase gene in neurotoxigenic *Clostridium butyricum* type E strains. *PLoS One* **6**:e21706. doi: 10.1371/journal.pone.0021706.
236. **Slavić D, Boerlin P, Fabri M, Klotins KC, Zoethout JK, Weir PE, Bateman D.** 2011. Antimicrobial susceptibility of *Clostridium perfringens* isolates of bovine, chicken, porcine, and turkey origin from Ontario. *Can J Vet Res* **75**:89–97.
237. **Farrow KA, Lyras D, Rood JI.** 2000. The macrolide-lincosamide-streptogramin B resistance determinant from *Clostridium difficile* 630 contains two erm(B) genes. *Antimicrob Agents Chemother* **44**:411–413.
238. **Altschul SF, Madden TL, Schäffer AA, Zhang J, Zhang Z, Miller W, Lipman DJ.** 1997. Gapped BLAST and PSI-BLAST: a new generation of protein database search programs. *Nucleic Acids Res* **25**:3389–3402.
239. **Li W, Godzik A.** 2006. Cd-hit: a fast program for clustering and comparing large sets of protein or nucleotide sequences. *Bioinformatics* **22**:1658–1659.
240. **Lederberg J.** 1998. Plasmid (1952–1997). *Plasmid* **39**: 1–9.

Plasmids—Biology and Impact in Biotechnology and Discovery
Edited by Marcelo E. Tolmasky and Juan C. Alonso

doi:10.1128/microbiolspec.PLAS-0020-2014

Stefan Schwarz,[1] Jianzhong Shen,[2] Sarah Wendlandt,[1]
Andrea T. Feßler,[1] Yang Wang,[2] Kristina Kadlec,[1] and Cong-Ming Wu[1]

22 Plasmid-Mediated Antimicrobial Resistance in Staphylococci and Other *Firmicutes*

INTRODUCTION

Antimicrobial resistance plasmids of staphylococci are circular, double-stranded DNA molecules that range in size from about 2 kilobase pairs (kbp) to larger than 100 kbp. Antimicrobial resistance plasmids share a number of basic properties with other plasmids: (i) they are present in variable copy numbers per bacterial cell, (ii) they replicate independently from the chromosomal DNA by using specific replication (*rep*) genes, (iii) they are commonly distributed to daughter cells during bacterial cell division, and (iv) they can be exchanged between bacteria by horizontal gene transfer mechanisms, such as conjugation, mobilization, or transduction (1–8). Staphylococcal antimicrobial resistance plasmids may carry *mob* genes for mobilization or a *tra* gene complex for conjugative transfer (1, 2, 4). Furthermore, antimicrobial resistance plasmids can undergo interplasmid recombination that may result in structurally diverse plasmids consisting of parts of both parental plasmids (9). Such recombined plasmids may be better adapted to the conditions in the host bacterium and harbor additional genetic information. In contrast to genes that play essential roles in replication, recombination, and horizontal gene transfer, antimicrobial resistance genes are not essential for the survival of the bacteria under physiological conditions. However, they may be helpful for the bacteria to survive under specific conditions, e.g., in the presence of antimicrobial agents (1, 7–9).

To date, numerous plasmids, which differ considerably in size and structure, have been described to mediate antimicrobial resistance in staphylococci of human and animal origin. The range of staphylococcal resistance plasmids varies from small plasmids that carry only a single antimicrobial resistance gene to large plasmids that carry a number of different antimicrobial resistance genes occasionally coupled with genes that confer resistance to heavy metal ions, disinfectants, and/or biocides. Analysis of plasmid-borne antimicrobial resistance genes in staphylococci also identified

[1]Institute of Farm Animal Genetics, Friedrich-Loeffler-Institut (FLI), Neustadt-Mariensee, Germany; [2]Beijing Key Laboratory of Detection Technology for Animal-Derived Food Safety, College of Veterinary Medicine, China Agricultural University, Beijing 100193, P. R. China.

resistance genes that were first detected in other bacteria of the orders *Bacillales* and *Lactobacillales*, such as members of the genera *Bacillus*, *Streptococcus*, and *Enterococcus* (10–15). In addition, several resistance plasmids (which carry novel resistance genes and plasmid replication and mobilization genes), which are only distantly related to those known from staphylococci, have recently been detected in staphylococci of animal origin (16, 17), and their original host remains to be identified. However, certain antimicrobial resistance genes, identified on plasmids in *Staphylococcus aureus* and other staphylococcal species, have in the meantime also been detected and shown to be functionally active in a wide range of other *Firmicutes* and *Proteobacteria*, respectively (10, 11, 18). These observations suggest that staphylococci can not only acquire and express plasmid-borne resistance genes from various known or even unknown sources, but can also pass such plasmid-borne resistance genes to other bacteria. This situation underlines the role of antimicrobial resistance plasmids in the gene pool to which staphylococci, but also other bacteria, have access.

This article reviews—without the aim of being exhaustive—the knowledge of plasmid-borne antimicrobial resistance among staphylococci of human and animal origin. If appropriate, examples are shown that illustrate the structural similarities between staphylococcal plasmids or parts thereof with corresponding resistance plasmids of other *Firmicutes*. Moreover, examples are provided of how staphylococcal plasmids can recombine, form cointegrates, or integrate into other plasmids to give rise to new types of multiresistance plasmids. In addition, examples are also given of how staphylococcal plasmids can function as vectors for transposon-borne antimicrobial resistance genes and how plasmids carrying antimicrobial resistance genes can integrate into the chromosomal DNA.

PLASMID-BORNE ANTIMICROBIAL RESISTANCE GENES IN STAPHYLOCOCCI

A wide variety of plasmid-borne genes that mediate resistance to antimicrobial agents have been identified in staphylococci of human and animal origin. Table 1 provides an overview of the most relevant resistance genes, the resistance mechanisms specified by them, and the associated resistance phenotypes. It should be noted that not the same amount of information is available for every resistance gene. Some resistance genes and the mobile elements that carry them have been studied extensively, whereas only limited information is currently available for other resistance genes and their vectors. In staphylococci, numerous small antimicrobial resistance plasmids have been identified during the past 30 years (1–3, 6–9). These plasmids are usually less than 15 kbp in size and often carry only a single resistance gene. Most of these plasmids confer resistance to antimicrobial agents that interfere with the bacterial protein biosynthesis (9). In addition, a considerable number of larger plasmids are known to occur in staphylococci. These plasmids often carry several resistance genes. Closer inspection often revealed a mosaic structure, which was very likely due to recombination and integration processes.

Tetracycline Resistance Plasmids

In staphylococci, three plasmid-borne tetracycline resistance genes have been identified. The genes *tet*(K) and *tet*(L) code for tetracycline-specific exporters of the major facilitator superfamily, which can export tetracyclines except minocycline. In contrast, the gene *tet*(M) codes for a ribosome protective protein that also confers minocycline resistance (11, 19) (Table 1).

Small tetracycline resistance plasmids carrying the gene *tet*(K) are most widespread among staphylococci (9, 11, 19, 20). These plasmids are usually about 4.5 kbp in size, are structurally closely related, and carry the gene *tet*(K) as the only resistance gene. The prototype plasmid is pT181, first identified in *S. aureus* of human origin (21–23) (Fig. 1). Plasmid pT181 carries a *repC* gene for plasmid replication and a plasmid recombination/mobilization gene *pre/mob*, as well as the staphylococcal recombination site A (RS_A) at which the Pre/Mob protein acts (23, 24). Such pT181-like plasmids have been identified in the coagulase-positive *S. aureus* and *Staphylococcus (pseud)intermedius*, in the coagulase-variable *Staphylococcus hyicus*, and in numerous coagulase-negative staphylococci (CoNS), including among others *Staphylococcus epidermidis*, *Staphylococcus haemolyticus*, *Staphylococcus saprophyticus*, *Staphylococcus sciuri*, and *Staphylococcus xylosus* (1, 3, 7–9, 20, 25–28). The 6,913-base-pair (bp) plasmid pSTE2 from *Staphylococcus lentus* was shown to represent an RS_A-mediated fusion product of a small *tet*(K)-carrying plasmid and a small plasmid carrying the macrolide-lincosamide-streptogramin B (MLS_B) resistance gene *erm*(C) (29).

Tetracycline resistance plasmids that carry the gene *tet*(L) were described to be less widely disseminated in staphylococci (20). They have been initially found in *S. hyicus* and CoNS species such as *S. epidermidis*, *S. lentus*, *S. sciuri*, and *S. xylosus* of animal origin (30, 31). Plasmids that harbor only the *tet*(L) gene are structurally more diverse than the *tet*(K)-carrying plasmids and vary in size from 4.3 to 7.5 kbp (31). The ca. 5.5-kbp plasmid pSTS7 that carried the kanamycin/neomycin resistance gene *aadD* in addition to the *tet*(L) gene was identified in porcine *S. epidermidis* (32). In connection with the increased interest in livestock-associated methicillin-resistant *S. aureus* (LA-MRSA) since 2005, numerous *tet*(L)-carrying plasmids have been identified in LA-MRSA, mainly in isolates of the clonal complex 398 (33). These novel plasmids are larger and often carry one or more additional resistance genes (9, 34, 35). In LA-MRSA, the tetracycline resistance gene *tet*(L) was often found to be physically linked to the trimethoprim resistance gene *dfrK* (33–36).

In staphylococci, *tet*(M) genes have been detected frequently and represent the second most prevalent *tet* genes in animal staphylococci (20). However, if present, they are commonly found in the chromosomal DNA, whereas *tet*(M)-carrying transposons are often located on plasmids in other *Firmicutes*, such as enterococci or streptococci. Most recently, a functionally active *tet*(M) gene was identified on the 28,743-bp multiresistance plasmid pSWS47 from a feline methicillin-resistant *S. epidermidis* (MRSE) isolate (37). The *tet*(M) gene of pSWS47 is located within a largely truncated Tn*916*-like transposon. It should be noted that pSWS47 also carried a functionally active *tet*(L) gene along with three other resistance genes (37). The presence of two functionally active tetracycline resistance genes on the same plasmid is a rare observation.

Phenicol Resistance Plasmids

In staphylococci, plasmid-borne phenicol resistance can be due to either chloramphenicol acetyltransferase (*cat*) genes that confer resistance to nonfluorinated phenicols (e.g., chloramphenicol), the *fexA* gene that encodes a phenicol-specific efflux pump of the major facilitator superfamily, or the gene *cfr* that encodes an rRNA methylase that also confers resistance to lincosamides, oxazolidinones, pleuromutilins, and streptogramin A antibiotics (9, 38) (Table 1). The gene products of *fexA* and *cfr* confer resistance to nonfluorinated and fluorinated phenicols (e.g., florfenicol). Information on plasmids carrying the gene *cfr* is given in the subsection on oxazolidinone resistance plasmids.

Three prototype *cat* genes have been identified in staphylococci and named according to the plasmids on which they were first identified: cat_{pC221}, cat_{pC223}, and cat_{pC194} (39–42) (Fig. 1). Each *cat* gene codes for a chloramphenicol acetyltransferase monomer of 215, 215, and 216 amino acids, respectively, whereas the functional Cat enzyme is a trimer (38). These *cat* genes are inducibly expressed via translational attenuation, with chloramphenicol acting as an inducer (38). The 4,555-bp plasmid pC221 has four genes, two of which code for RepD and Cat proteins, while the remaining two in part overlapping genes code for MobA and MobB proteins that play a role in mobilization (39, 40). Various pC221-related chloramphenicol resistance plasmids of 2.9 to 4.6 kbp have been identified in staphylococci of animal origin (38, 43–48). The 4,608-bp plasmid pC223 carries five genes for RepJ, a Cat protein, a relaxase accessory protein Rlx (MobC), and the mobilization proteins MobA and MobB (41). Other pC223-like plasmids were—except for the 3.75-kbp plasmid pSCS5 from canine *S. haemolyticus* (49)—virtually indistinguishable from pC223 (46, 48, 50). The 2,910-bp plasmid pC194 differs distinctly in its structure from the other types of chloramphenicol resistance plasmids. It harbors genes for a Rep protein and a Cat protein (42). Database comparisons also identified a 120-amino acid protein, which was closely related to Cop proteins of *S. aureus* and *S. sciuri* that are assumed to play a role in maintaining the plasmid copy number. In contrast to pC221-like and pC223-like plasmids, which are widely disseminated among animal staphylococci, only a single pC194-like plasmid, pSCS34, has been identified in an *S. sciuri* isolate of animal origin (51).

The gene *fexA* is part of the small nonconjugative transposon Tn*558* (52, 53). Tn*558* is a member of the Tn*554* family of transposons and integrates site-specifically into the chromosomal *radC* gene, but also into plasmids. Surveys of staphylococci with elevated MICs of florfenicol (54, 55) showed that the *fexA* gene is widely disseminated in staphylococci. Several plasmids that harbor intact or truncated copies of Tn*558* have been sequenced (18, 56); in many of the truncated copies, Tn*558* was interrupted by the integration of a segment that carried the multiresistance gene *cfr* (18, 54–58). Most recently, a *fexA* variant gene has been identified that conferred chloramphenicol resistance but not florfenicol resistance (59). This gene had two mutations in the domain obviously involved in substrate recognition. Reversion of these two positions to the original *fexA* sequence by site-directed mutagenesis restored the florfenicol resistance phenotype (59).

Oxazolidinone Resistance Plasmids

The gene *cfr* is currently the only transferable oxazolidinone resistance gene (18). It codes for an rRNA methylase that targets the adenine residue at position 2503 in the 23S rRNA (60), which is located in the overlapping binding domain of phenicols, lincosamides, oxazolidinones, pleuromutilins, and streptogramin A

Table 1 Overview of plasmid-borne antimicrobial resistance genes in staphylococci[a]

Resistance to. . .	Mechanism	Gene	Location[b]	Mobile genetic element	GenBank accession no.
β-Lactams (except isoxazolyl-penicillins)	Enzymatic inactivation (β-lactamase)	*blaZ*	Tn, P::Tn, C::Tn	Tn*552*	X52734
Tetracyclines (except minocycline and glycylcyclines)	Active efflux (major facilitator superfamily)	*tet*(K)	P, C::P	pT181	J01764.1
		tet(L)	P	pKKS2187	FM207105.1
Tetracyclines (including minocycline but excluding glycylcyclines)	Target site protection (ribosome protective protein)	*tet*(M)	Tn, C::Tn, P::Tn	Tn*916*	U09422.1
Nonfluorinated phenicols	Enzymatic inactivation (acetylation)	cat_{pC221}	P	pC221	NC_006977
		cat_{pC223}	P	pC223	NC_005243
		cat_{pC194}	P	pC194	NC_002013
All phenicols	Active efflux (major facilitator superfamily)	*fexA*	Tn, P::Tn, C::Tn	Tn*558*	AM086211.1
Aminoglycosides (gentamicin, kanamycin, tobramycin, amikacin)	Enzymatic inactivation (acetylation and phosphorylation)	*aacA-aphD*	Tn, P::Tn, C::Tn	Tn*4001*	AJ536196.1
Aminoglycosides (kanamycin, neomycin, tobramycin)	Enzymatic inactivation (adenylation)	*aadD*	P, C::P	pUB110	NC_001384.1
Aminoglycosides (kanamycin, neomycin, amikacin)	Enzymatic inactivation (phosphorylation)	*aphA3*	Tn, P::Tn, C::Tn	Tn*5405*	AF299292
Aminoglycosides (streptomycin)	Enzymatic inactivation (adenylation)	*aadE*	Tn, P::Tn, C::Tn	Tn*5405*	AF299292
		str	P	pS194	NC_005564
Aminocyclitols (spectinomycin)	Enzymatic inactivation (adenylation)	*spc*	Tn, P::Tn, C::Tn	Tn*554*	X03216.1
		spw	P, C::P	pV7037	JX560992.1
		spd	P	pDJ91S	KC895984.1
Aminocyclitols/aminoglycosides (apramycin, decreased susceptibility to gentamicin)	Enzymatic inactivation (acetylation)	*apmA*	P	pAFS11	FN806789.1
Streptothricins	Enzymatic inactivation (acetylation)	*sat4*	Tn, P::Tn	Tn*5405*	AF299292.1
Macrolides, lincosamides, streptogramin B	Target site modification (rRNA methylation)	*erm*(A)	Tn, P::Tn, C::Tn	Tn*554*	X03216.1
		erm(B)	Tn, P::Tn, C::Tn	Tn*917*	U35228.1
		erm(C)	P	pE194, pNE131	NC_005908.1, NC_001390.1
		erm(T)	P, C::P	pKKS25	FN390947.1
		erm(Y)	P	pMS97	AB179623.1
		erm(33)	P	pSCFS1	NC_005076.1

Macrolides, streptogramin B	Active efflux (ABC transporter)	*msr*(A)	C, P	p18805-P03	NC_019150.1
Macrolides	Enzymatic inactivation (phosporylation)	*mph*(C)	C, P	p18805-P03	NC_019150.1
Lincosamides	Enzymatic inactivation (nucleotidylation)	*lnu*(A)	P	pLNU1	NC_007768.1
		lnu(B)	P, C::P	pV7037	JX560992.1
	Active efflux (ABC transporter)	*lsa*(B)	P	pSCFS1	NC_005076.1
Streptogramin A	Enzymatic inactivation (acetylation)	*vat*(A)	P	pIP680	L07778.1
		vat(B)	P	pIP1156	U19459.1
		vat(C)	P	pIP1714	AF015628
Streptogramin B	Enzymatic inactivation (hydrolyzation)	*vgb*(A)	P	pIP524	M20129.1
		vgb(B)	P	pIP1714	AF015628.1
Lincosamides, pleuromutilins, streptogramin A	Active efflux (ABC transporter)	*vga*(A)	Tn, P::Tn	Tn*5406*	M90056.1
		vga(A) variant	C, P	unknown	AF186237
		vga(C)	P	pKKS825	FN377602.2
		vga(E) variant	P	pSA-7	KF540226.1
		lsa(E)	P, C::P	pV7037	JX560992.1
All phenicols, lincosamides, oxazolidinones, pleuromutilins, streptogramin A	Target site modification (rRNA methylation)	*cfr*	P, C	pSCFS1	NC_005076.1
Trimethoprim	Target replacement (trimethoprim-resistant dihydrofolate reductase)	*dfrA* (*dfrS1*)	Tn, P::Tn, C::Tn	Tn*4003*	GU565967.1
		dfrD	P	pABU17	Z50141.1
		dfrK	Tn, P, C::Tn	Tn*559* pKKS2187	FN677369.2, FM207105.1
Fusidic acid	Target site protection (ribosome protective protein)	*fusB*	P	pUB101	NC_005127.1
		fusC	P, C	SCC*mec*	HE980450.1
Mupirocin	Target replacement (mupirocin-insensitive isoleucyl-tRNA synthase)	*mupA* (*ileS*, *ileS2*)	P	pPR9	NC_013653.1
Vancomycin	Target modification (synthesis of D-Ala-D-Lac depsipeptide low-affinity peptidoglycan precursors)	*vanA* gene cluster	Tn, P::Tn	pLW043	AE017171.1
Bleomycin	Bleomycin binding protein	*ble*	P, C::P	pUB110	NC_001384.1

[a]Modified from Wendlandt S, Feßler AT, Monecke S, Ehricht R, Schwarz S, Kadlec K, *Int J Med Microbiol* **303**:338–349, 2013 [8], copyright 2013, with permission from Elsevier.
[b]Abbreviations: P, plasmid; T, transposon; C::P, plasmid integrated in part or completely into the chromosomal DNA; P::Tn, transposon integrated in part or completely into a plasmid.

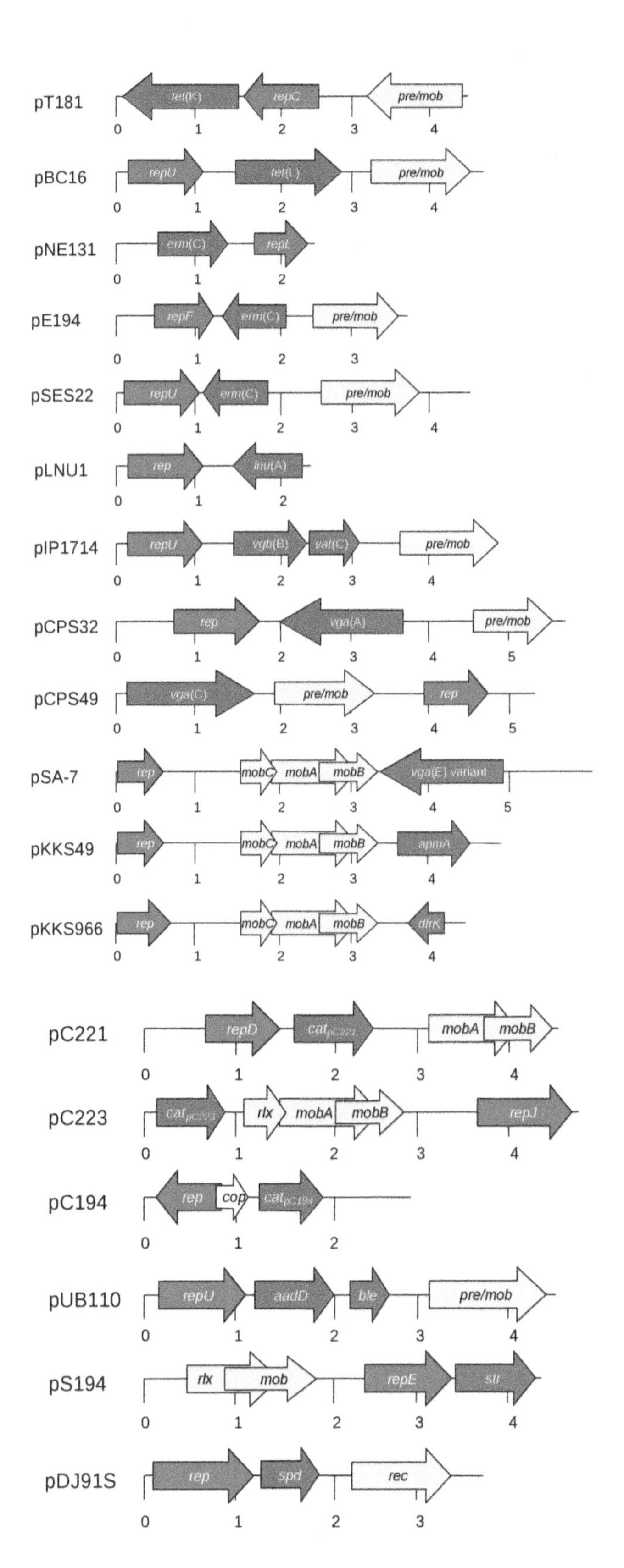

antibiotics, and thereby confers resistance to the aforementioned classes of antimicrobial agents (61). Initially identified on the 17,108-bp multiresistance plasmid pSCFS1 (62, 63) (Table 2) from bovine *S. sciuri*, this gene has been detected on a number of plasmids in various staphylococcal species. A recent review described in detail the structural variability of the so far sequenced *cfr*-carrying plasmids or the *cfr*-carrying regions thereof (18, 56). This analysis revealed that the smallest *cfr*-carrying plasmids known to date are 7,054 bp (pMSA16 from bovine MRSA) or 7,057 bp (pSS-03 from porcine *S. sciuri*, *Staphylococcus cohnii*, *Staphylococcus arlettae*, and *S. saprophyticus*) (55, 64). Usually, the *cfr* gene is located on larger plasmids that range in size from 30 to 55 kbp and often carry one or more additional resistance genes (18, 56). Maps of these plasmids or their *cfr*-carrying sequenced segments are shown in Fig. 2.

Plasmids carrying the gene *cfr* occasionally also harbor complete or truncated transposons such as the *fexA*-carrying transposon Tn*558*, the *aacA-aphD*-carrying transposon Tn*4001*, the *erm*(B)-carrying transposon Tn*917*, or the *erm*(A)-carrying transposon Tn*554* (18) (Fig. 2). Various insertion sequences, such as IS*26-558*, IS*431*, IS*257*, and IS*Enfa4* (formerly known as IS*256*-like), have been detected in the proximity of the *cfr* gene and are believed to play a role in the mobility of *cfr* but also in the mosaic structure of the *cfr*-carrying plasmids (18, 56) (Fig. 2). Moreover, various additional resistance genes have been detected to be colocated with *cfr* on the same staphylococcal plasmids and may play a role in the dissemination and persistence of the *cfr*-carrying plasmids via coselection (18).

Figure 1 Schematic presentation of the maps of small plasmids below 6 kbp in size carrying one or two antimicrobial resistance gene(s): pT181 (database accession no. J01764), pBC16 (U32369), pNE131 (M12730), pE194 (NC_005908), pSES22 (AM159501), pLNU1 (AM184099), pIP1714 (AF015628), pCPS32 (FN806791), pCPS49 (FN806792), pSA-7 (KF540226), pKKS49 (HE611647), pKKS966 (FN677368), pC221 (X02166), pC223 (AY355285), pC194 (V01277), pUB110 (M37273), pS194 (X06627), and pDJ91S (KC895984). The arrows indicate the positions of the genes and their directions of transcription. Resistance genes are marked in red; genes involved in plasmid recombination, mobilization, and relaxation (*pre/mob, rlx, mob*) in yellow; and plasmid replication genes (*rep*) in blue. Genes that exhibit other functions are shown in white. A distance scale in kbp is shown below each map. (This figure has been modified and expanded from Fig. 1 in reference 9 [Schwarz S, Feßler AT, Hauschild T, Kehrenberg C, Kadlec K, *Ann NY Acad Sci* **1241:**82–103, 2011] with permission.)
doi:10.1128/microbiolspec.PLAS-0020-2014.f1

Table 2 Examples of larger staphylococcal plasmids that carry several antimicrobial resistance genes occasionally together with metal or biocide/disinfectant resistance genes[a]

Plasmid	Origin	Size (bp)	Antimicrobial resistance gene(s)	Accession no.
pLW043	*S. aureus*	57,889	*aacA-aphD* (Tn*4001*), *vanA* gene cluster (Tn*1546*), *dfrA* (Tn*4003*), *blaZ* (Tn*552*), *qacC*	AE017171.1
pGO1	*S. aureus*	54,000	*aacA-aphD* (Tn*4001* with IS*256* truncated by IS*257*), *aadD*, *ble* (pUB110 integrated via IS*257*), *dfrA* (Tn*4003*), *qac*	NC_012547.1
pAFS11	*S. aureus*	49,192	*aadD*, *erm*(B), *tet*(L), *dfrK*, *apmA*	FN806789.1
pSK41	*S. aureus*	46,445	*aacA-aphD* (Tn*4001* with IS*256* truncated by IS*257*), *aadD*, *ble* (pUB110 integrated via IS*257*)	NC_005024.1
pSCFS6	*S. warneri*	ca. 43	*lsa*(B) + *cfr* (bracketed by IS*21-558* and integrated into Tn*558*), *fexA* (Tn*558*)	AM408573.1
pPR9	*S. aureus*	41,715	*mupA* (bracketed by IS*257*), *blaZ* (Tn*552*)	NC_013653.1
pV7037	*S. aureus*	40,971	*aacA-aphD* (ΔTn*4001*), *erm*(B) (ΔTn*917*), *tet*(L), *cadDX* operon, *aadE*, *spw*, *lsa*(E), *lnu*(B), Δ*blaZ* (ΔTn*552*)	JX560992.1, HF586889.1
pUSA03	*S. aureus*	37,136	*mupA* (bracketed by IS*257*), *erm*(C) (pNE131-like plasmid integrated via IS*257*)	NC_007792.1
pTZ2162	*S. aureus*	35,380	*aacA-aphD* (Tn*4001*), *blaZ*, *ars* operon, *cadDX* operon, *fosB*, *qacC*	AB304512.1
pMS97	*S. aureus*	33,347	*msr*(A), *mph*(C), *erm*(Y), *blaZ* (Tn*552*), *mer* operon	AB179623.1
pI258	*S. aureus*	29,254	*blaZ* (ΔTn*552*), *erm*(B) (ΔTn*917*), *mer* operon, *cadDX* operon, *ars* operon,	GQ900378.1
pSWS47	*S. epidermidis*	28,743	*vga*(A), *aadD*, *tet*(L), *dfrK*, *tet*(M), *blaZ* (*Tn552*)	HG_380319.1
pSK1	*S. aureus*	28,150	*aacA-aphD* (Tn*4001*), *dfrA* (Tn*4003*), *qacA*	NC_014369.1
pSERP	*S. epidermidis*	27,310	*aphA3* + *sat* + *aadE* (Tn*5405*), *blaZ*, *cadDX* operon	NC_006663.1
pLAC-P03	*S. aureus*	27,068	*msr*(A), *mph*(C), *aphA-3* + *sat* + *aadE* (Tn*5405*), *blaZ*, *cadDX* operon, *bcrB-bcrA*	CP002149.1
pUB101	*S. aureus*	21,845	*fusB*, *blaZ* (truncated Tn*552*), *cadDX* operon	AY373761.1
pSCFS1	*S. sciuri*	17,108	*cfr*, *lsa*(B), *erm*(33), *spc*	NC_005076.1
pKKS825	*S. aureus*	14,365	*aadD*, *tet(L)*, *dfrK*, *vga*(C)	FN377602.1s

[a]This table has been modified from Schwarz S, Feßler AT, Hauschild T, Kehrenberg C, Kadlec K, *Ann NY Acad Sci* **1241**:82–103, 2011 [9], with permission.

Macrolide-Lincosamide-Streptogramin Resistance Plasmids

Over the years, various genes have been detected that confer resistance to macrolides, lincosamides, streptogramins, or combinations thereof (8, 9) (Table 1). Many of these genes have been identified primarily on plasmids or on transposons that have integrated into plasmids.

Among staphylococci of human and animal origin, small plasmids of 2.3 to 4.0 kbp that carry the rRNA methylase gene *erm*(C) have been detected most frequently (1, 3, 8, 9). Based on their structure, at least three different types of *erm*(C)-carrying plasmids can be differentiated: plasmids that resemble the 2,355-bp plasmid pNE131 or the 2,475-bp plasmid pT48 (65, 66), plasmids that resemble the 3,728-bp plasmid pE194 and its relative pSES6 (67, 68), and plasmids that resemble the 4,040-bp plasmid pSES22 (69) (Fig. 1). The small pNE131-like plasmids have only two genes, which code for the replication protein RepL and the rRNA methylase Erm(C) (65, 66). Plasmid pE194 has a *repF* gene for plasmid replication and a *pre*/*mob* gene for plasmid recombination/mobilization in addition to *erm*(C) (67), whereas plasmid pSES22 consists of a *repU* gene, a different *pre*/*mob* gene, and the *erm*(C) gene (69). Recent analysis identified an MRSA CC398 isolate that harbored two *erm*(C)-carrying plasmids, a pNE131-like plasmid, and a pE194-like plasmid, stably side-by-side (70).

In another study of LA-MRSA, a small *erm*(C)-carrying plasmid was found to be integrated into a larger plasmid that already had an *erm*(T) gene (71). The gene *erm*(T) was found among MRSA or methicillin-susceptible *S. aureus* (MSSA) CC398 from human, animal, and food sources (35, 71–75). One such plasmid, pUR3912, which also carried a *cadDX* operon for cadmium resistance, proved to be able to integrate in and excise from the staphylococcal chromosome (73). Other *erm* genes, such as the Tn*554*-associated gene *erm*(A) and the Tn*917*-associated gene *erm*(B), are more rarely seen on plasmids in staphylococci. Occasionally, intact transposons were located on plasmids (55, 76). More often only parts of these transposons including the respective *erm* gene were found on staphylococcal plasmids (64, 77, 78). The gene *erm*(33), which represents an *in vivo* recombination product of the genes *erm*(A) and *erm*(C) (79) was found on the multiresistance plasmid pSCFS1 from *S. sciuri* (63). The 33,347-bp plasmid pMS97 has been reported to carry the three macrolide resistance genes

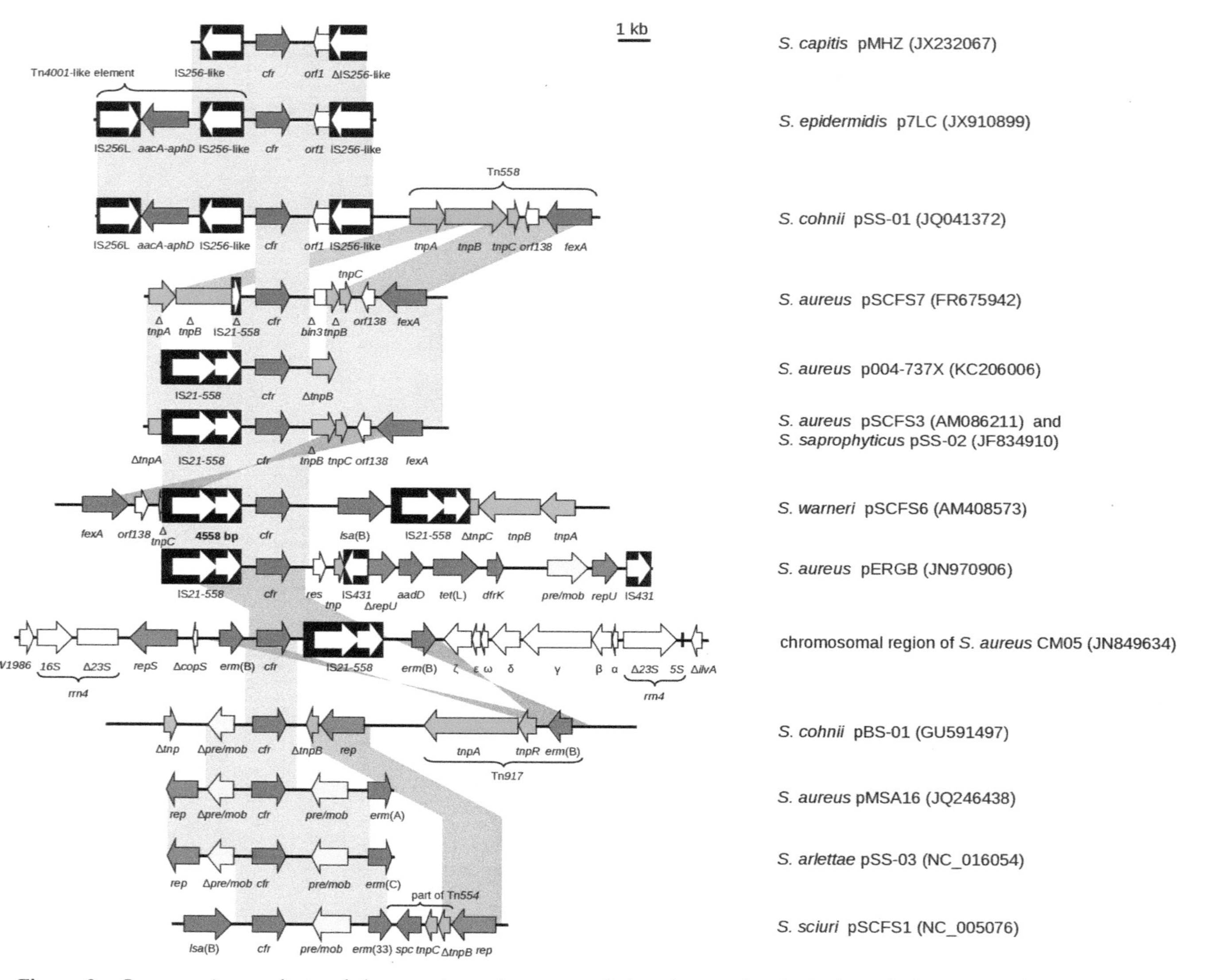

Figure 2 Comparative analysis of the genetic environment of the *cfr* gene in plasmids and chromosomal DNA from staphylococci. The arrows indicate the positions of the genes and their directions of transcription. Resistance genes (including the *cfr* gene) are marked in red, *pre/mob* genes in yellow, *rep* genes in blue, and transposase genes in green. Genes that exhibit other functions are shown in white. Insertion sequences are displayed as black boxes, with their transposase genes as white arrows. The regions of >95% homology are shaded in gray. Different gray shadings were used to better illustrate the homologous regions. Δ indicates a truncated gene. A 1-kbp distance scale is displayed in the upper center. (This figure has been modified from Fig. 1a in reference 18 [Shen J, Wang Y, Schwarz S, *J Antimicrob Chemother* **68**:1697–1706, 2013] with permission.) doi:10.1128/microbiolspec.PLAS-0020-2014.f2

erm(Y), *msr*(A), and *mph*(C) (80) (Table 2). The genes *msr*(A) and *mph*(C) are often found together (81), although little is known about the plasmids on which they are located. Besides their presence in bovine staphylococci from cases of mastitis, these two genes have also been detected among CoNS and MRSA CC398 from humans in Germany and Spain, respectively (82, 83), and in plasmids of the *S. aureus* clone USA300 (84).

Plasmids that carry the lincosamide nucleotidyltransferase gene *lnu*(A) have mainly been identified in *S. aureus* and CoNS (e.g., *Staphylococcus chromogenes*, *S. epidermidis*, *S. haemolyticus*, *Staphylococcus simulans*) mainly from cases of bovine mastitis (85, 86). These plasmids range in size from 2.3 to 3.8 kbp and usually carry only a *rep* gene and the *lnu*(A) gene (Fig. 1). Solely in plasmid pLNU9, two copies of the *lnu*(A) gene were present due to a duplication of a 1,050-bp segment (86). More recently, small *lnu*(A)-carrying plasmids similar to pLNU1 and pLNU4 (86) were also detected in LA-MRSA (13). In the same study, the authors also identified the gene *lnu*(B) either on a large plasmid or in the chromosomal DNA (13) of MRSA isolates. This observation was extended by the finding of the *lnu*(B) gene on plasmid pV7037 from a porcine MRSA ST9 or in the chromosomal DNA of MRSA ST9 isolates from China (87, 88). The gene *lsa*(B)—also located on the multiresistance plasmid pSCFS1 (63)—was shown to confer only slightly elevated MICs for lincosamides.

The 4,978-bp plasmid pIP1714 from *S. cohnii* was found to carry the streptogramin resistance genes *vgb*(B) and *vat*(C), together with a *repU* gene and a *pre/mob* gene (89) (Fig. 1). The Vgb(B) and Vat(C) proteins act synergistically in mediating resistance to streptogramin B and A compounds, respectively. In *S. aureus* of human origin, plasmids ranging from 26 to 45 kbp in size were found that carried the three resistance genes *vat*(A), *vgb*(A), and *vga*(A) (90). The gene *vat*(B) was found on the ca. 60-kbp plasmid pIP1156 from *S. aureus* (91). Information on plasmid-borne resistance genes that confer streptogramin A resistance along with resistance to lincosamides and pleuromutilins is provided in the subsection on pleuromutilin resistance plasmids.

Pleuromutilin Resistance Plasmids

Several *vga* and *lsa* ABC transporter genes have been identified in staphylococci of human and animal origin that confer combined resistance to pleuromutilins, lincosamides, and streptogramin A antibiotics (8, 9) (Table 1).

The first *vga* gene, *vga*(A), was detected on a plasmid in 1992 (92). At that time, it was believed that this gene exclusively confers resistance to streptogramin A antibiotics. Further studies showed that the *vga*(A) gene is part of Tn*5406*, a small nonconjugative transposon of the Tn*554* family (93). Tn*5406* preferably integrated into the chromosomal *radC* gene (93), although the *vga*(A) gene was also found on plasmids of different sizes. In *S. epidermidis* of human origin, it was detected on plasmids of 7.5 kbp and 14.4 kbp, in the latter together with the trimethoprim resistance gene *dfrA* (94). More recently, the *vga*(A) gene was detected as part of the multiresistance plasmid pSWS47 from feline *S. epidermidis* (37). The 5,718-bp plasmid pCPS32 carried only the *vga*(A) gene and was identified in a Portuguese MRSA ST398 from a dust sample taken in a hog house (95) (Fig. 1). This plasmid showed 99.9% nucleotide sequence identity to the 5,713-bp plasmid pVGA from a human clinical *S. aureus*, also identified in Portugal (96). Similar plasmids ranging in size from 5.7 to 7.6 kbp were found in MRSA ST398, MRSE, and methicillin-susceptible *S. epidermidis* of human and animal origin in Spain (97). The *vga*(A) gene was also detected on a ca. 24-kbp plasmid from swine and a swine farmer in the United States (98). A variant gene that showed 83.2% nucleotide sequence identity to the *vga*(A) gene has been detected either in the chromosomal DNA or on plasmids in staphylococci (99). Chesneau and colleagues confirmed that the *vga*(A) gene and its variant conferred resistance not only to streptogramin A, but also to lincosamides (100). Another variant gene, *vga*$(A)_{LC}$, with shifted substrate specificity toward lincosamides was detected in *S. haemolyticus* (101). During a study of clinical *S. aureus* isolates for pleuromutilin susceptibility, Gentry and colleagues showed that the gene *vga*(A) and its variants also conferred high MICs to pleuromutilins (96).

A different *vga* gene, *vga*(B), was found in human staphylococci to be present on plasmids of 50 to 90 kbp that also carried the gene *vat*(B) for a streptogramin A acetyltransferase (102). The gene *vga*(C) was first identified on the 14,362-bp multiresistance plasmid pKKS825, which also carried the resistance genes *tet*(L), *dfrK*, and *aadD*, from a porcine MRSA CC398 isolate in Germany (34) (Table 2). Shortly thereafter, it was also detected as the only resistance gene on the 5,292-bp plasmid pCPS49 from porcine MRSA CC398 in Portugal (95). The transposon Tn*6133*-associated gene *vga*(E) has so far been found exclusively in the chromosomal DNA of MRSA CC398 from various sources in Switzerland and Germany (103, 104). However, most recently, a novel *vga*(E) variant gene has been identified on the 5,584-bp plasmid pSA-7 in porcine *S. cohnii* and *S. simulans* isolates from China

(17) (Fig. 1). This gene, which shared 85.8% nucleotide sequence identity with the Tn*6133*-associated *vga*(E) gene, also confers the pleuromutilin-lincosamide-streptogramin A resistance phenotype. Besides the *vga*(E) variant gene, this plasmid harbored a *rep* gene and three in part overlapping *mob* genes (17).

In addition to the aforementioned *vga* genes, a different gene, designated *lsa*(E), also mediated combined resistance to pleuromutilins, lincosamides, and streptogramin A antibiotics (14). This gene has been detected upstream of the lincosamide resistance gene *lnu*(B) in MRSA ST398, MSSA ST9, and MRSA ST9 of human and porcine origin, as well as on the 40,971-bp plasmid pV7037 from porcine MRSA ST9 (13, 14, 87, 88) (Fig. 3 and Table 2).

Aminoglycoside-Aminocyclitol-Streptothricin Resistance Plasmids

Several plasmid-borne genes that confer resistance to various aminoglycosides, aminocyclitols, or streptothricins have been detected in staphylococci of human and

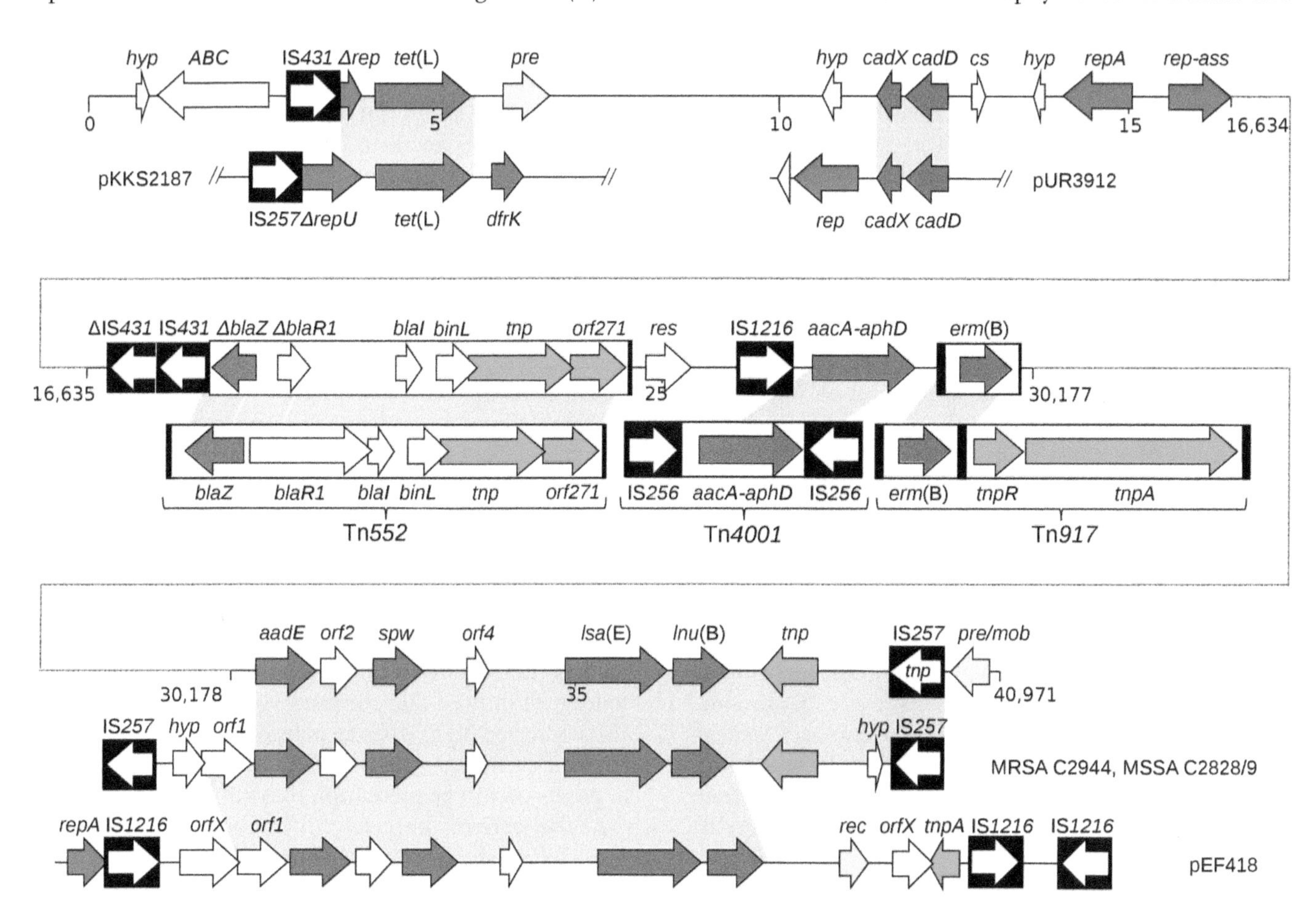

Figure 3 Schematic presentation of the organization of the 40,971-bp plasmid pV7037 from porcine MRSA ST9 as an example of a multiresistance plasmid composed of different segments previously identified on other plasmids or transposons from Gram-positive bacteria. The arrows indicate the positions of the genes and their directions of transcription. Resistance genes are marked in red, *pre/mob* genes in yellow, *rep* genes in blue, and transposase genes in green. Genes that exhibit other functions or no known functions are shown in white. Insertion sequences are displayed as black boxes, with their transposase genes as white arrows. Plasmids, transposons, and chromosomal fragments that share similarity with plasmid pV7037 are displayed below the map of pV7037 with the homologous region(s) indicated by gray shading. A distance scale in bp is given below the maps. A dotted line connects the different parts of pV7037. The Δ symbol indicates a truncated gene. (This figure has been modified from Fig. 1 in reference 88 [Wendlandt S, Li B, Ma Z, Schwarz S, *Vet Microbiol* **166:**650–654, 2013], copyright 2013, with permission from Elsevier.) doi:10.1128/microbiolspec.PLAS-0020-2014.f3

animal origin (Table 1) (1, 7–9, 105–107). Some of these genes represent parts of transposons that can integrate into plasmids as well as into the chromosomal DNA (9).

The 4,548-bp plasmid pUB110 from *S. aureus* carries the aminoglycoside adenyltransferase gene *aadD* and the bleomycin resistance gene *ble* in addition to the plasmid replication gene *rep* and the plasmid recombination/mobilization gene *pre/mob* (108, 109) (Fig. 1). The gene *aadD* [also known as *aad(4′)-Ia*] confers resistance to kanamycin, neomycin, and tobramycin (106, 107). The 4,397-bp plasmid pS194, also from *S. aureus*, carries a *repE* gene, the streptomycin adenyltransferase gene *str*, and two, in part overlapping genes for relaxation/mobilization proteins (110) (Fig. 1). Streptomycin resistance plasmids that are structurally closely related to pS194 have also been identified in staphylococci of animal origin (111). Small plasmids that carry the streptomycin resistance gene *str* downstream of a cat_{pC221} gene have been detected in *S. hyicus* and *S. sciuri* (112, 113). The 28,150-bp multiresistance plasmid pSK1 (1, 7, 114) harbors a complete copy of the composite aminoglycoside resistance transposon Tn*4001*. Tn*4001* is composed of a central region, consisting of the *orf132* of unknown function and the aminoglycoside resistance gene *aacA-aphD* (also known as *aac[6′]-aph[2″]*), which is bracketed by two complete copies of the insertion sequence IS*256* that are situated in opposite orientations (115, 116). The *aacA-aphD* gene codes for a bifunctional enzyme with acetyltransferase and phosphotransferase functions and confers resistance to gentamicin, kanamycin, tobramycin, and, when overexpressed, to amikacin (9, 106, 117). Truncated Tn*4001* elements, in which the terminal IS*256* sequences were partially deleted, have been found on multiresistance plasmids in human *S. aureus* and in avian CoNS isolates, including *Staphylococcus warneri* and *S. sciuri*. In the latter plasmids, the terminal IS*256* elements of Tn*4001* were truncated by the insertion of complete or incomplete copies of IS*257* (118–120). Transposon Tn*5405* from *S. aureus* is, similar to Tn*4001*, also a composite transposon. Tn*5405* has a central resistance gene region encompassing the streptomycin resistance gene *aadE* [also known as *ant(6)-Ia*], the streptothricin acetyltransferase gene *sat4*, and the kanamycin/neomycin/amikacin resistance gene *aphA3* (also known as *aph[3′]-IIIa*). The central region is bracketed by the insertion sequences IS*1182L* and ΔIS*1182R* (121). While the complete Tn*5405* was initially discovered on the ca. 39.2-kbp plasmid pIP1085B (121), the Tn*5405*-associated resistance gene cluster has also been detected on other staphylococcal plasmids, e.g., pSERP from *S. epidermidis* (122) or p18806-P03 and related plasmids from the *S. aureus* clone USA300 (84) (Table 2). The *aadE* gene—apart from *sat4* and *aphA3*—was shown to be part of the aforementioned chromosomal or plasmidic multiresistance gene cluster in MRSA ST398, MSSA ST9, and MRSA ST9 of human and porcine origin, respectively (13, 14, 87, 88) (Fig. 3).

Staphylococcal resistance to the aminocyclitol antibiotic spectinomycin is frequently mediated by the Tn*554*-associated gene *spc* (also known as *ant[9]*-Ia). If complete copies of Tn*554* insert into plasmids, the *spc* gene is usually distributed as part of these plasmids (76). The *spc* gene has also been found as part of a largely truncated Tn*554* copy on the *cfr*-carrying multiresistance plasmid pSCFS1 (63). The analysis of the aforementioned multiresistance gene cluster (13, 14, 87, 88) also identified the presence of a novel spectinomycin resistance gene, designated *spw* (15). Most recently, a third spectinomycin resistance gene, designated *spd*, on a 3,928-bp plasmid pDJ91S from MRSA ST398 of chicken origin has been described (123) (Fig. 1). A slightly smaller plasmid, designated pSWS2889, with a divergent *rep* gene but the same *spd* and *pre* (*rec*) gene, has also been detected among MRSA CC398 from diverse sources (124). The first and so far only gene that confers resistance to the aminocyclitol antibiotic apramycin in staphylococci, *apmA*, was identified on the 49,192-bp multiresistance plasmid pAFS11 from a bovine MRSA CC398 isolate (78) (Table 2). This plasmid also carried the resistance genes *erm*(B), *aadD*, *tet*(L), and *dfrK*. Plasmids closely related to pAFS11 were also detected in MRSA CC398 isolates from swine and food of poultry origin. In addition, the 4,809-bp plasmid pKKS49 obtained from a porcine MRSA isolate in Portugal was found to carry the *apmA* gene as the only resistance gene (16) (Fig. 1). Apart from the *apmA* gene, plasmids pAFS11 and pKKS49 did not reveal any significant sequence similarities (16).

Mupirocin Resistance Plasmids

Plasmid-borne resistance to mupirocin is based on the presence of a mupirocin-insensitive isoleucyl-tRNA synthase encoded by the gene *mupA* (also known as *ileS* or *ileS2*) (125–128) (Table 1). One such completely sequenced plasmid is the 41,715-bp plasmid pPR9 from *S. aureus*, which in addition to the *mupA* gene also carried a copy of transposon Tn*552* with the β-lactamase gene *blaZ* (128) (Table 2). The structural relationship of plasmid pPR9 with other conjugative plasmids carrying or not carrying the *mupA* gene has been described in detail by Pérez-Roth and coworkers (128). The conjugative plasmid pKM01 from a canine

Staphylococcus pseudintermedius has recently been shown to carry the *mupA* gene along with the *aacA-aphD* gene (129). Large plasmids carrying the *mupA* gene have also been described to occur in *S. haemolyticus* of human origin (130). Several plasmids ranging in size between 37.1 and 47.3 kbp and obtained from MRSA USA300 isolates have been described to carry the *mupA* gene alone or in combination with other resistance genes, most frequently *aacA-aphD* and *aadD* (131).

Fusidic Acid Resistance Plasmids

Plasmid-borne resistance to fusidic acid is based on the expression of the genes *fusB* and *fusC* (132) (Table 1). The FusB protein has been shown to bind to staphylococcal elongation factor G (EF-G) and thereby to protect the staphylococcal translation system from the inhibitory effects of fusidic acid. Inducible expression of *fusB* is regulated by translational attenuation (133). The 21,845-bp plasmid pUB101 (database accession no. NC_005127) represents the prototype *fusB*-carrying plasmid in staphylococci (134) (Table 2). This plasmid also carried a *cadDX* operon for cadmium resistance and a copy of transposon Tn*552* with the β-lactamase gene *blaZ* (134). The analysis of bovine CoNS for fusidic acid resistance identified the *fusB* gene on a ca. 40-kbp plasmid in a single *S. haemolyticus* isolate (135).

Trimethoprim Resistance Plasmids

At least four different genes (*dfrA*, *dfrD*, *dfrG*, and *dfrK*) play a role in trimethoprim resistance of staphylococci (1, 7, 8, 36). Among them, *dfrA*, *dfrD*, and *dfrK* are transferable and often located on plasmids (Table 1). The first trimethoprim resistance gene identified in staphylococci was *dfrA* (also referred to as *dfrS1*) (136). This gene is part of the composite transposon Tn*4003*, in which the genes for a dihydrofolate reductase (*dfrA*) and a thymidylate synthetase (*thyE*) are flanked by three copies of IS*257* (137). Tn*4003* integrates into the chromosomal DNA but has also been found on large conjugative plasmids in *S. aureus* and CoNS (114, 120, 138–140). The 8,013-bp plasmid pSK639 from *S. epidermidis* also carried an IS*257*-flanked segment that contained the genes *dfrA* and *thyE*, which closely resembled a large part of Tn*4003* (141). A different *dfr* gene, *dfrD*, was found on the 3.8-kbp plasmid pABU17 from *S. haemolyticus* (142). Little is known about plasmids that carry the gene *dfrD* in staphylococci.

The most recent *dfr* gene detected in staphylococci is *dfrK*. This gene was initially identified on the ca. 40-kbp plasmid pKKS2187 from porcine MRSA CC398, where it was physically linked to the tetracycline resistance gene *tet*(L) (36). Clusters comprising *tet*(L)-*dfrK* but also other resistance genes have subsequently been detected on various large multiresistance plasmids, mainly from MRSA CC398 (34, 35, 71) but also from MRSA ST125 (143) and *S. epidermidis* ST5 (37). In addition, the *dfrK* gene was identified as the only resistance gene on the 4,957-bp plasmid pKKS966 from a porcine *Staphylococcus hyicus* isolate (16) (Fig. 1). In addition to its location on structurally diverse plasmids, the *dfrK* gene was also shown to be part of the small nonconjugative transposon Tn*559* (144), another member of the Tn*554* family.

Vancomycin Resistance Plasmids

Vancomycin resistance is still very rare in staphylococci (145). The first clinical highly vancomycin-resistant *S. aureus* (VRSA) isolate detected in a human patient in 2002 harbored the 57,889-bp conjugative multiresistance plasmid pLW043, which consisted of a pSK41-like *S. aureus* resistance plasmid (*dfrA*, *aacA-aphD*, *blaZ*, and *qacC*) with an insertion of a Tn*1546*-like element carrying the *vanA* gene cluster (139) (Table 2). In other VRSA isolates, the Tn*1546*-associated *vanA* genes were located on larger plasmids of ca. 120 kbp (146) or ca. 100 kbp (147). Further VRSA isolates showed the presence of an Inc18-like *vanA*-carrying plasmid similar to those of coresident vancomycin-resistant enterococci (12, 148). Recent studies showed that a pSK41-like plasmid is necessary for Inc18-like *vanA* plasmid transfer from *Enterococcus faecalis* to *S. aureus* under *in vitro* conditions (149). Detailed analysis of the Tn*1546* elements in VRSA isolates revealed structural alterations due to the insertion of IS*1216V* and IS*1251*, which, however, did not affect the structural genes required for vancomycin resistance (145).

Penicillin Resistance Plasmids

In staphylococci, the *blaZ*-encoded β-lactamase confers resistance to penicillins except isoxazolyl-penicillins (8). The *blaZ-blaI-blaR1* operon has been identified on transposon Tn*552* (150, 151), which has been detected on plasmids as well as in the chromosomal DNA (1, 7, 152). The complete Tn*552* is 6,545 bp in size and characterized by two transposase genes, *p271* and *tnp* (*p480*), a recombinase gene, *bin*, and three β-lactamase-associated genes, the β-lactamase repressor gene *blaI*, the β-lactamase regulator gene *blaR1*, and the β-lactamase gene *blaZ*. A recent analysis of staphylococcal plasmids showed that Tn*552* can be the subject of deletions. Two variants, ΔTn*552* and Tn*552*Δ, were detected (152). ΔTn*552* lacks the transposase genes, while the three *bla* genes have been lost in Tn*552*Δ (152). It should be noted that ΔTn*552* was first described on plasmid pI258 (151) (Table 2). As reviewed by Lyon and Skurray (1) as well as by Jensen and Lyon (7), most of the older *blaZ*-carrying plasmids also harbored genes for resistance to inorganic ions such as antimony, arsenate, arsenite, bismuth, cadmium, lead, mercury, and zinc (153); organomercurial compounds such as phenylmercuric acetate (154); antiseptics, disinfectants, and dyes such as acriflavine (155); ethidium bromide (156); and quaternary ammonium compounds (155, 157). Additional antimicrobial resistance genes, such as *erm*(B) (158), *fusB* (134), and *aacA-aphD* (159) were also detected on these β-lactamase plasmids. More recently, *blaZ*-carrying multiresistance plasmids that harbored the enterotoxin genes *sed*, *sej*, and *ser* (152) or the gene for the exfoliative toxin B (160) have been described.

It should be noted that the *blaZ* gene has been identified in numerous staphylococci of animal origin as reviewed by Wendlandt and coworkers (8); however, in most cases information about a plasmid location and/or the size and structure of the respective plasmids is not available. The analysis of 137 penicillin-resistant *S. aureus* and CoNS strains from cases of bovine mastitis identified *blaZ* genes in all isolates tested. However, in only 14 isolates, which represented the species *S. aureus*, *S. epidermidis*, *S. chromogenes*, *S. haemolyticus*, and *S. xylosus*, the *blaZ* gene was located on plasmids that ranged in size between 24 and 58 kbp (161). Further recent studies identified copies of Tn*552* on multiresistance plasmids of porcine MRSA ST9 (88) and feline MRSE ST5 (37), in which the *blaZ* gene was functionally deleted by frame-shift mutations that caused premature stop codons.

PROCESSES INVOLVED IN THE FORMATION OF OLIGO- OR MULTIRESISTANCE PLASMIDS

Different mechanisms by which antimicrobial resistance plasmids in staphylococci can recombine, form cointegrates, and integrate—in part or *in toto*—into other plasmids have already been reviewed in detail (2, 7, 9). The following sections will provide a few examples of these processes.

Recombination of Resistance Plasmids via Site-Specific Recombination

Staphylococcal plasmid cointegrates can be formed by host- and phage-mediated general *rec* systems that act on short regions of homology (162). Two such regions of homology, named recombination sites A and B (RS_A, RS_B), have been identified on several small staphylococcal plasmids (2, 24, 162). It has been shown that the plasmid recombination protein Pre mediates a site-specific recombination that involves RS_A but not RS_B. Heterologous cointegrates between the *tet*(K)-carrying tetracycline resistance plasmid pT181 and the *erm*(C)-carrying plasmid pE194, both of which carry RS_A sites and *pre* genes, have already been derived *in vitro* (24, 162). At least one example of *in vivo* RS_A-mediated recombination between a small pT181-like *tet*(K)-carrying plasmid and a small pPV141-like *erm*(C)-carrying plasmid has been described (29). The plasmid pSTS7 from *S. epidermidis* is assumed to have developed from an *aadD-ble*-encoding pUB110-like resistance plasmid and a pNS1981-like *tet*(L)-encoding tetracycline resistance plasmid by RS_A-mediated recombination and subsequent intraplasmid recombination (32). Naturally occurring plasmids such as pSCS12 and pSCGp3EB, which carry a pC221-analogous *repD* gene, the cat_{pC221} gene, and the pS194-associated *str* and *rlxA/mob* genes, are believed to have developed by RS_B-mediated recombination and been followed by intraplasmid recombination (112).

Exchange of Antimicrobial Resistance Genes via Illegitimate Recombination

Several small plasmids, that replicate via a rolling circle mechanism, carry closely related *repU*-like plasmid replication and *pre/mob* genes, but differ in their resistance genes, have been identified in staphylococci and other *Firmicutes*. These include the *aadD-ble*-carrying plasmid pUB110 from *S. aureus* (108, 109), the *vgb*(B)-*vat*(C)-carrying plasmid pIP1714 from *S. cohnii* (89), the *aadD-tet*(L)-carrying plasmid pSTS7 from *S. epidermidis* (32), the *tet*(L)-*dfrK*-carrying plasmid pKKS627 from MRSA ST398 (GenBank accession no. NC_014156.1), the *erm*(C)-carrying plasmid pSES22 from *S. saprophyticus* (69), the *tet*(L)-carrying plasmids pNS1981 and pBC16 from *Bacillus subtilis* and *Bacillus cereus* (163–165), and the *dfrD*-carrying plasmid pIP823 from *Listeria monocytogenes* (166). As previously reported (9, 69, 89), 4-bp direct repeat 5′-GGGC-3′ and adjacent more or less perfect inverted repeat sequences were found at the junctions of the *repU*-*pre/mob*-homologous and nonhomologous segments (Fig. 4). These sequences are believed to play a role in the acquisition and exchange of resistance genes by illegitimate recombination. Illegitimate recombination processes do not require long stretches of identical

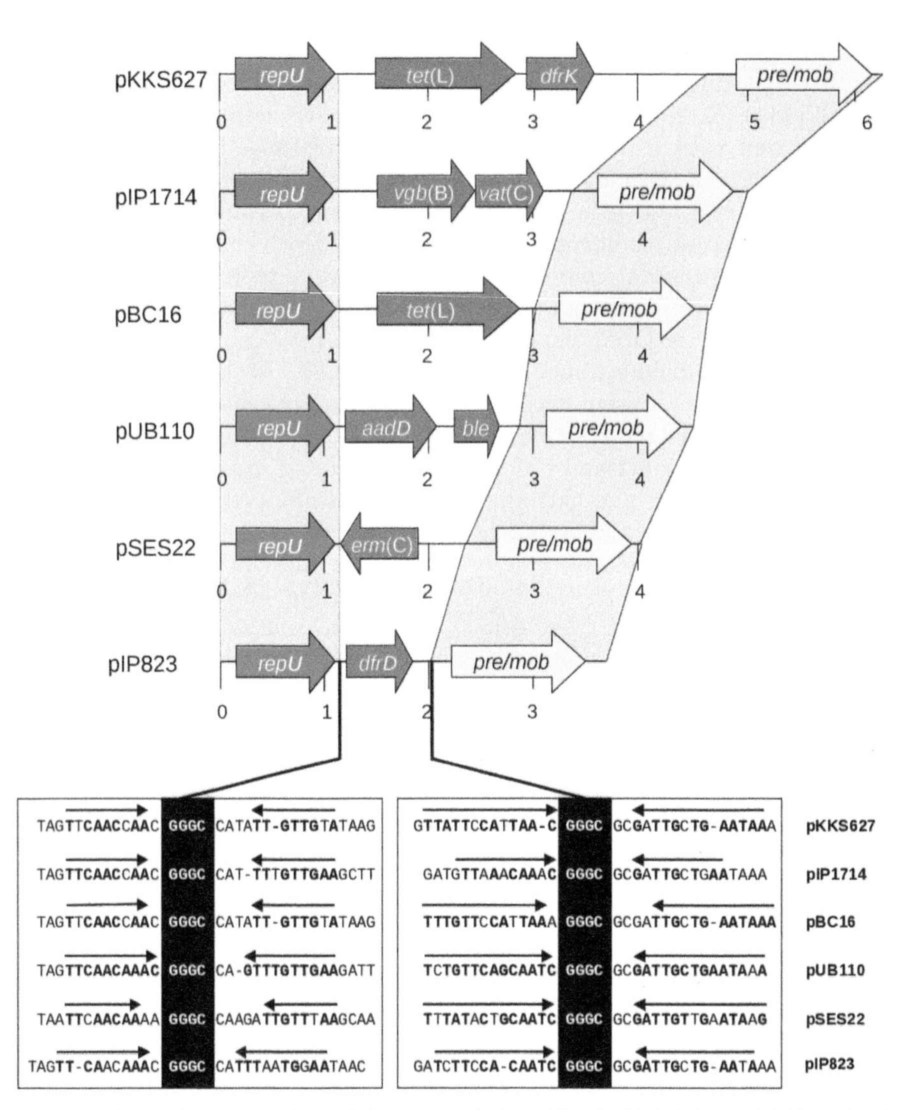

Figure 4 Schematic presentation of the maps of plasmids pKKS627 (FN390948), pIP1714 (AF015628), pBC16 (U32369), pUB110 (M37273), pSES22 (CAJ43792), and pIP823 (U40997) and comparison of the sequences at the boundaries of *repU-pre/mob*-homologous and -nonhomologous regions. A distance scale in kilobases is shown below each map. Resistance genes [*tet*(L), *dfrK*, *vgb*(B), *vat*(C), *aadD*, *ble*, *erm*(C), and *dfrD*] are marked in red, *pre/mob* genes in yellow, and *repU* genes in blue. The areas of extended sequence similarity between the six plasmids are gray-shaded. The 4-bp direct repeats GGGC are shown as black boxes; the flanking regions are aligned, and the small arrows indicate the imperfect inverted repeats. In the imperfect inverted repeats, matching bases are displayed in bold type. (This figure was reproduced from reference 9 [Schwarz S, Feßler AT, Hauschild T, Kehrenberg C, Kadlec K, *Ann NY Acad Sci* **1241**:82–103, 2011] with permission.) doi:10.1128/microbiolspec.PLAS-0020-2014.f4

sequences, and regions characterized by inverted repeat sequences are considered to be preferential areas for illegitimate recombination events (167). Moreover, it may also be possible that the largely common plasmid backbones serve as targets for homologous recombination events that enable the exchange of resistance gene regions. Another example is the MLS_B resistance gene *erm*(33), which represents an *in vivo*-derived hybrid between a plasmid-borne *erm*(C) gene and a transposon-borne *erm*(A) gene (79). At the junction of *erm*(C)-homologous and *erm*(A)-homologous sequences, an area of 45 bp was detected, which displayed 95.6%

identity to *erm*(A) and 93.3% identity to *erm*(C) and is believed to have served for recombination between the genes *erm*(C) and *erm*(A) (79).

Plasmid and Resistance Gene Integration via Insertion Sequences

A number of insertion sequences have been identified in staphylococci of human and animal origin, some of which, like IS*257* or the closely related IS*431* and IS*Sau10*, have proved to play an important role in the integration of small plasmids into larger plasmids or into the chromosomal DNA (9) (Fig. 5). IS*257* elements might have been involved in the integration of a small pKKS627-like *tet*(L)-*dfrK*-carrying plasmid into the ca. 40-kbp plasmid pKKS2187 from a porcine MRSA ST398 isolate (36). IS*257*-mediated integration of pT181-like plasmids into larger plasmids was first described in the 34.2-kbp plasmid pJ3358 from *S. aureus* (126). Other examples are the plasmids pSTS23 (ca. 49 kbp) from *S. epidermidis*, pSTS20 (ca. 67 kbp) from *S. haemolyticus*, pSTS21 (ca. 82 kbp) from *S. aureus*, and pSTS22 (ca. 31 kbp) from *S. warneri* (168). Moreover, IS*257* seems to be involved in the mobility of mupirocin resistance genes. For example, the mupirocin resistance gene *mupA* was initially described to be bracketed by IS*257* copies in the 23.6- to 34.2-kbp plasmids pJ3355-pJ3358 from *S. aureus* (126). Comparative sequence analysis of various staphylococcal mupirocin resistance plasmids of 34.0 to 41.7 kbp confirmed this observation (128). IS*257*-mediated integration of a small *erm*(C)-carrying plasmid was described in plasmid pUSA03 (169), whereas IS*Sau10*-mediated integration of a similar plasmid was seen in plasmid pUR2940 (71). IS*Sau10* was initially identified on the ca. 40-kbp plasmid pKKS25 from porcine MRSA ST398. Two copies of IS*Sau10* located in the same orientation flanked a 4,675-bp resistance gene region that contains the genes *tet*(L), *dfrK*, and *erm*(T) (35). A very similar situation was observed in plasmid pUR2940, while regions carrying *tet*(L) and *erm*(T) and flanked by IS*Sau10* were detected on plasmids pUR1902 and pUR2941 (71). The gene *aadD* was also detected on plasmids pUR1902 and pUR2941 to be flanked by IS*Sau10* copies in the same orientation (71). In contrast, a pUB110-like plasmid with the *aadD* gene was inserted via IS*257* in plasmids pGO1 and pSK41 (118, 170). Various complete or truncated IS*257*, IS*431*, and IS*1216* elements have been identified on plasmids pSWS47 and pV7037 and are believed to account for the mosaic structures of these multiresistance plasmids (37, 88). The insertion sequence IS*431* plays an

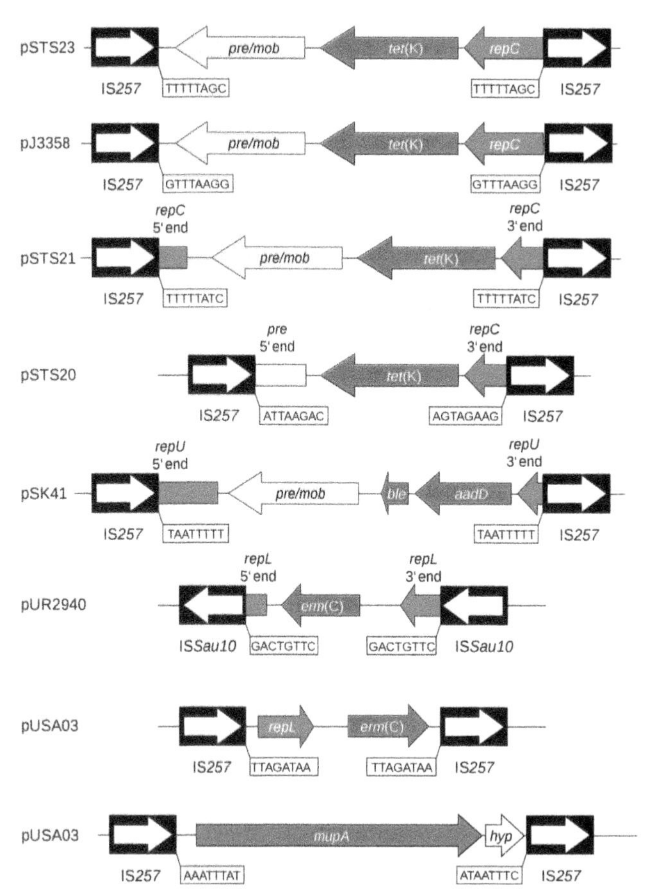

Figure 5 Examples of IS*257*- and IS*Sau10*-mediated recombination processes identified on plasmids pJ3358 (database accession no. U36910), pSK41 (NC_005024), pUR2940 (HF583292), pUSA03 (NC_007792), and pSTS20, pSTS21, and pSTS23 (168). The arrows indicate the positions of the genes and their directions of transcription. Resistance genes are marked in red, *pre*/*mob* genes in yellow, *rep* genes in blue, and genes that exhibit other functions or no known functions in white. Insertion sequences are displayed as black boxes, with their transposase genes as white arrows. The 8-bp sequences at the boundaries of the IS-flanked segments are shown in boxes. (This figure has been modified and expanded from Fig. 5b in reference 9 [Schwarz S, Feßler AT, Hauschild T, Kehrenberg C, Kadlec K, *Ann NY Acad Sci* **1241:**82–103, 2011] with permission.)
doi:10.1128/microbiolspec.PLAS-0020-2014.f5

important role in the insertion/excision of plasmid pUR3912 in/from the chromosomal DNA of MSSA ST398 (72, 73). The insertion sequence IS*21-558* was first identified on the ca. 43-kbp plasmid pSCFS6 in *S. warneri* and *S. simulans* isolates, where it bracketed a region containing the resistance genes *cfr* and *lsa*(B) (54) (Table 2). IS*21-558* has exclusively been found in connection with the gene *cfr* and appears to have an integration preference for the transposase genes of the *fexA*-carrying transposon Tn*558* (18, 57).

DISSEMINATION OF MOBILE RESISTANCE GENES WITHIN THE *FIRMICUTES* GENE POOL

Staphylococci of human and animal origin do not live in isolation. Instead, they exchange genetic material, including resistance genes, with a wide variety of other bacteria. In this regard, transfer across species boundaries is mainly mediated by plasmids or plasmid-located transposons. For some of the resistance genes currently known to occur in staphylococci, the origin and the direction of transfer seem to be clear. Examples for the transfer of resistance genes from *Enterococcus* to *Staphylococcus* are the VRSA isolates detected so far. These isolates harbor a plasmid-borne Tn*1546* that is believed to have been acquired by conjugation from a vancomycin-resistant *Enterococcus* isolate. In cases where the enterococcal plasmid was replication-deficient in the *S. aureus* host, Tn*1546* transposed from the enterococcal plasmid to a plasmid already present in the new staphylococcal host cell (145, 148, 149). Another more recent example of an *Enterococcus* to *Staphylococcus* transfer is the presence of a multiresistance gene cluster that comprised the genes *aadE*, *spw*, *lsa*(E), and *lnu*(B) in staphylococci. This multiresistance gene cluster was initially detected on plasmid pEF418 from *E. faecalis* (GenBank accession no. AF408195.1) and has recently been found on plasmids and in the chromosomal DNA of MRSA ST398 and MRSA/MSSA ST9 (13–15, 87, 88) (Fig. 3). The multiresistance gene *cfr* has recently been assumed to originate from members of the order *Bacillales* (171).

For other resistance genes, which are widespread within the *Firmicutes* gene pool, it is more difficult to state from which donor the staphylococci have acquired them. As such, the *erm*(B) gene located on the nonconjugative transposons Tn*917* and Tn*551* (10) has been identified in at least 34 different genera, including *Staphylococcus*, *Streptococcus*, *Enterococcus*, *Lactobacillus*, and *Bacillus,* while the *erm*(T) gene has been found in the four genera *Staphylococcus*, *Streptococcus*, *Enterococcus*, and *Lactobacillus* (http://faculty.washington.edu/marilynr/). Most likely, *erm*(B) and *erm*(T) are not indigenous staphylococcal genes, but it is not possible to state from which bacterium staphylococci have acquired these genes first or from which bacterium the genes originate. The same is true for the tetracycline resistance gene *tet*(L). This gene is believed to originate from *Bacillus*, but has also been identified in various genera of the *Firmicutes*, including *Staphylococcus*, *Streptococcus*, *Enterococcus*, *Lactobacillus*, *Clostridium*, and *Listeria*. One of the most widespread resistance genes, the tetracycline resistance gene *tet*(M) (http://faculty.washington.edu/marilynr/), is located on conjugative transposons, with Tn*916* representing the prototype *tet*(M)-harboring transposon. Complete or truncated copies of *tet*(M) are commonly found on plasmids of streptococci and enterococci, but the first functionally active plasmid-borne *tet*(M) gene was detected on the *S. epidermidis* plasmid pSWS47 most recently (37). This *tet*(M) gene and its flanking regions showed highest homology to plasmid pMD5057 from *Lactobacillus plantarum* (37, 172).

For some of the novel resistance genes described over the last few years, very limited—if any—information is available about their origin (33). This refers in particular to the genes *vga*(C) (34, 95), *apmA* (16, 78), *dfrK* (16, 36), and the novel *vga*(E) variant (17). Analysis of the plasmid backbone of the small plasmids carrying, *apmA*, *dfrK*, or the *vga*(E) variant gene revealed that they have a similar overall structure that is composed of three in part overlapping *mob* genes and a *rep* gene (16, 17). However, comparisons at the amino acid level showed that the respective Rep proteins showed negligible similarity to staphylococcal Rep proteins. Its best match to sequences deposited in the databases was 48% amino acid identity to Rep proteins of rumen bacteria of the genus *Selenomonas* (16). Similarly, the Mob proteins showed only 29 to 56% identity to Mob from the genera *Pediococcus*, *Lactobacillus*, and *Staphylococcus* (16). Analysis of the 5,292-bp *vga*(C)-carrying plasmid pCPS49 revealed the presence of a *pre/mob* gene and a *rep* gene. The respective Pre/Mob protein was only distantly related to staphylococcal Pre/Mob proteins and showed a highest amino acid identity of 50.2% and 49.2% to mobilization proteins of *Lactococcus lactis* and *Lactobacillus sakei*, respectively (95). The Rep protein of pCPS49 also displayed a highest amino acid identity of only 63.9% to a plasmid replication protein of *L. sakei* (95). These examples show that staphylococci can acquire resistance plasmids from bacteria of unknown origin. Surprisingly, some of these plasmids can replicate without problems in the staphylococcal host and also express their resistance genes.

In addition to accepting foreign resistance genes and their associated mobile genetic elements, staphylococci may also pass plasmid- or transposon-borne resistance genes to other bacteria. Thus, the *dfrK* gene and the *dfrK*-carrying transposon Tn*559*, both first identified in MRSA (36, 144) and soon thereafter in other staphylococci (16), have recently also been detected in enterococci (173). An even more impressive situation is the spread of the multiresistance gene *cfr*, also first identified in staphylococci, to other bacteria (18). In recent years, *cfr* was detected in *Bacillus* spp. (174–176), in various *Enterococcus* spp. (177–179), in *Macrococcus caseolyticus* and *Jeotgalicoccus pinnipedialis* (180), as well as in *Streptococcus suis* (181). The *cfr* gene was even identified in the chromosomal DNA of *Proteus vulgaris* (182) and on plasmids of *Escherichia coli* (183, 184).

Finally, it should be noted that some resistance plasmids found in staphylococci, such as pUB110, proved to be able to replicate in other *Firmicutes*, for example *Bacillus*, and have also been isolated from *Bacillus* hosts (185). This seems to be particularly true for *tet* (L)-carrying plasmids that harbor a *repU* gene (163–165). In addition to pUB110 and tetracycline resistance plasmids, the chloramphenicol resistance plasmids pC221, pC223, and pC194 have been shown to replicate in *Bacillus* hosts (186). Moreover, the 2,246-bp *erm*(C)-carrying plasmid pIM13 from *B. subtilis* closely resembled the similar-sized staphylococcal *erm*(C)-carrying plasmids (187). The 16,492-bp plasmid pBS-01 initially detected in *Bacillus* spp. (174) has subsequently been found in porcine *S. cohnii*, *S. saprophyticus*, and *S. sciuri* isolates (55). The *dfrD*-carrying plasmid pIP823 has a broad host range that includes *L. monocytogenes*, *E. faecalis*, *S. aureus*, *B. subtilis*, and even *E. coli* (166). Mobilization of pIP823 by using either self-transferable plasmids or the conjugative transposon Tn*1545* was shown to occur between *L. monocytogenes* and *E. faecalis* and between *L. monocytogenes* and *E. coli* (166). In contrast, other plasmids may be restricted to replication in their specific host bacteria or their very close relatives. When transferred to a distantly related new host bacterium, cointegrate formation and recombination with plasmids indigenous to the new host may ensure the maintenance of the corresponding resistance genes in the new host.

Acknowledgments. *We apologize in advance to all the investigators whose research could not be appropriately cited owing to space limitations. The work of Andrea T. Feler, Kristina Kadlec, Sarah Wendlandt, and Stefan Schwarz on antimicrobial resistance genes in staphylococci is financially supported by the German Federal Ministry of Education and Research (BMBF) through the German Aerospace Center (DLR), grant number 01KI1301D (MedVet-Staph 2). Sarah Wendlandt is supported by scholarships of the Akademie für Tiergesundheit (AfT) e.V. and the University of Veterinary Medicine Hannover Foundation. The work of Jianzhong Shen, Yang Wang, and Cong-Ming Wu is supported by the Major State Basic Research Development Program of China (973 Program, No. 2013CB127200). Conflicts of interest: We disclose no conflicts of interest.*

Citation. Schwarz S, Shen J, Wendlandt S, Feßler AT, Wang Y, Kadlec K, Wu C. 2014. Plasmid-mediated antimicrobial resistance in staphylococci and other *firmicutes*. Microbiol Spectrum 2(6):PLAS-0020-2014.

References

1. **Lyon BR, Skurray R.** 1987. Antimicrobial resistance of *Staphylococcus aureus*: Genetic basis. *Microbiol Rev* **51:**88–134.
2. **Novick RP.** 1989. Staphylococcal plasmids and their replication. *Annu Rev Microbiol* **43:**537–565.
3. **Werckenthin C, Cardoso M, Martel JL, Schwarz S.** 2001. Antimicrobial resistance in staphylococci from animals with particular reference to bovine *Staphylococcus aureus*, porcine *Staphylococcus hyicus*, and canine *Staphylococcus intermedius*. *Vet Res* **32:**341–362.
4. **Taylor DE, Gibreel A, Lawley TD, Tracz DM.** 2004. Antibiotic resistance plasmids, p 473–491. *In* Phillips G, Funnell B (ed), *Plasmid Biology*. ASM Press, Washington DC.
5. **Clewell DB, Francia MV.** 2004. Conjugation in Gram-positive bacteria, p 227–256. *In* Phillips G, Funnell G (ed), *Plasmid Biology*. ASM Press, Washington DC.
6. **Aarestrup FM, Schwarz S.** 2006. Antimicrobial resistance in staphylococci and streptococci of animal origin, p 187–212. *In* Aarestrup FK (ed), *Antimicrobial Resistance in Bacteria of Animal Origin*. ASM Press, Washington, DC.
7. **Jensen SO, Lyon BR.** 2009. Genetics of antimicrobial resistance in *Staphylococcus aureus*. *Future Microbiol* **4:** 565–582.
8. **Wendlandt S, Feler AT, Monecke S, Ehricht R, Schwarz S, Kadlec K.** 2013. The diversity of antimicrobial resistance genes among staphylococci of animal origin. *Int J Med Microbiol* **303:**338–349.
9. **Schwarz S, Feler AT, Hauschild T, Kehrenberg C, Kadlec K.** 2011. Plasmid-mediated resistance to protein biosynthesis inhibitors in staphylococci. *Ann NY Acad Sci* **1241:**82–103.
10. **Roberts MC, Sutcliffe J, Courvalin P, Jensen LB, Rood J, Seppala H.** 1999. Nomenclature for macrolide and macrolide-lincosamide-streptogramin B resistance determinants. *Antimicrob Agents Chemother* **43:**2823–2830.
11. **Roberts MC.** 1996. Tetracycline resistance determinants: mechanisms of action, regulation of expression, genetic mobility, and distribution. *FEMS Microbiol Rev* **19:**1–24.
12. **Zhu W, Murray PR, Huskins WC, Jernigan JA, McDonald LC, Clark NC, Anderson KF, McDougal LK, Hageman JC, Olsen-Rasmussen M, Frace M, Alangaden GJ, Chenoweth C, Zervos MJ, Robinson-Dunn B, Schreckenberger PC, Reller LB, Rudrik JT, Patel JB.** 2010. Dissemination of an *Enterococcus* Inc18-Like *vanA* plasmid associated with vancomycin-resistant *Staphylococcus aureus*. *Antimicrob Agents Chemother* **54:**4314–4320.
13. **Lozano C, Aspiroz C, Sáenz Y, Ruiz-García M, Royo-García G, Gómez-Sanz E, Ruiz-Larrea F, Zarazaga M, Torres C.** 2012. Genetic environment and location of the *lnu*(A) and *lnu*(B) genes in methicillin-resistant *Staphylococcus aureus* and other staphylococci of animal and human origin. *J Antimicrob Chemother* **67:** 2804–2808.

14. **Wendlandt S, Lozano C, Kadlec K, Gómez-Sanz E, Zarazaga M, Torres C, Schwarz S.** 2013. The enterococcal ABC transporter gene *lsa*(E) confers combined resistance to lincosamides, pleuromutilins and streptogramin A antibiotics in methicillin-susceptible and methicillin-resistant *Staphylococcus aureus. J Antimicrob Chemother* **68:**473–475.
15. **Wendlandt S, Li B, Lozano C, Ma Z, Torres C, Schwarz S.** 2013. Identification of the novel spectinomycin resistance gene *spw* in methicillin-resistant and methicillin-susceptible *Staphylococcus aureus* of human and animal origin. *J Antimicrob Chemother* **68:**1679–1680.
16. **Kadlec K, Feler AT, Couto N, Pomba CF, Schwarz S.** 2012. Unusual small plasmids carrying the novel resistance genes *dfrK* or *apmA* isolated from methicillin-resistant or -susceptible staphylococci. *J Antimicrob Chemother* **67:**2342–2345.
17. **Li J, Li B, Wendlandt S, Schwarz S, Wang Y, Wu C, Ma Z, Shen J.** 2014. Identification of a novel *vga*(E) gene variant that confers resistance to pleuromutilins, lincosamides and streptogramin A antibiotics in staphylococci of porcine origin. *J Antimicrob Chemother* **69:** 919–923.
18. **Shen J, Wang Y, Schwarz S.** 2013. Presence and dissemination of the multiresistance gene *cfr* in Gram-positive and Gram-negative bacteria. *J Antimicrob Chemother* **68:**1697–1706.
19. **Chopra I, Roberts M.** 2001. Tetracycline antibiotics: mode of action, applications, molecular biology, and epidemiology of bacterial resistance. *Microbiol Mol Biol Rev* **65:**232–260.
20. **Schwarz S, Roberts MC, Werckenthin C, Pang Y, Lange C.** 1998. Tetracycline resistance in *Staphylococcus* spp. from domestic animals. *Vet Microbiol* **63:**217–227.
21. **Mojumdar M, Khan SA.** 1988. Characterization of the tetracycline resistance gene of plasmid pT181 of *Staphylococcus aureus. J Bacteriol* **170:**5522–5528.
22. **Guay GG, Khan SA, Rothstein DM.** 1993. The *tet*(K) gene of plasmid pT181 of *Staphylococcus aureus* encodes an efflux protein that contains 14 transmembrane helices. *Plasmid* **30:**163–166.
23. **Khan SA, Novick RP.** 1983. Complete nucleotide sequence of pT181, a tetracycline-resistance plasmid from *Staphylococcus aureus. Plasmid* **10:**251–259.
24. **Gennaro ML, Kornblum J, Novick RP.** 1987. A site-specific recombination function in *Staphylococcus aureus* plasmids. *J Bacteriol* **169:**2601–2610.
25. **Greene RT, Schwarz S.** 1992. Small antibiotic resistance plasmids in *Staphylococcus intermedius. Zentralbl Bakteriol* **276:**380–389.
26. **Schwarz S, Blobel H.** 1990. Isolation and restriction endonuclease analysis of a tetracycline resistance plasmid from *Staphylococcus hyicus. Vet Microbiol* **24:**113–122.
27. **Schwarz S, Cardoso M, Grölz-Krug S, Blobel H.** 1990. Common antibiotic resistance plasmids in *Staphylococcus aureus* and *Staphylococcus epidermidis* from human and canine infections. *Zentralbl Bakteriol* **273:**369–377.
28. **Hauschild T, Kehrenberg C, Schwarz S.** 2003. Tetracycline resistance in staphylococci from free-living rodents and insectivores. *J Vet Med B Infect Dis Vet Public Health* **50:**443–446.
29. **Hauschild T, Lüthje P, Schwarz S.** 2005. Staphylococcal tetracycline-MLS_B resistance plasmid pSTE2 is the product of an RS_A-mediated *in vivo* recombination. *J Antimicrob Chemother* **56:**399–402.
30. **Schwarz S, Cardoso M, Wegener HC.** 1992. Nucleotide sequence and phylogeny of the *tet*(L) tetracycline resistance determinant encoded by plasmid pSTE1 from *Staphylococcus hyicus. Antimicrob Agents Chemother* **36:**580–588.
31. **Schwarz S, Noble WC.** 1994. Tetracycline resistance genes in staphylococci from the skin of pigs. *J Appl Bacteriol* **76:**320–326.
32. **Schwarz S, Gregory PD, Werckenthin C, Curnock S, Dyke KG.** 1996. A novel plasmid from *Staphylococcus epidermidis* specifying resistance to kanamycin, neomycin and tetracycline. *J Med Microbiol* **45:**57–63.
33. **Kadlec K, Feler AT, Hauschild T, Schwarz S.** 2012. Novel and uncommon antimicrobial resistance genes in livestock-associated methicillin-resistant *Staphylococcus aureus. Clin Microbiol Infect* **18:**745–755.
34. **Kadlec K, Schwarz S.** 2009. Identification of a novel ABC transporter gene, *vga*(C), located on a multiresistance plasmid from a porcine methicillin-resistant *Staphylococcus aureus* ST398 strain. *Antimicrob Agents Chemother* **53:**3589–3591.
35. **Kadlec K, Schwarz S.** 2010. Identification of a plasmid-borne resistance gene cluster comprising the resistance genes *erm*(T), *dfrK*, and *tet*(L) in a porcine methicillin-resistant *Staphylococcus aureus* ST398 strain. *Antimicrob Agents Chemother* **54:**915–918.
36. **Kadlec K, Schwarz S.** 2009. Identification of a novel trimethoprim resistance gene, *dfrK*, in a methicillin-resistant *Staphylococcus aureus* ST398 strain and its physical linkage to the tetracycline resistance gene *tet* (L). *Antimicrob Agents Chemother* **53:**776–778.
37. **Weiss S, Kadlec K, Feler AT, Schwarz S.** 2014. Complete sequence of a multiresistance plasmid from a methicillin-resistant *Staphylococcus epidermidis* ST5 isolated in a small animal clinic. *J Antimicrob Chemother* **69:**847–849.
38. **Schwarz S, Kehrenberg C, Doublet B, Cloeckaert A.** 2004. Molecular basis of bacterial resistance to chloramphenicol and florfenicol. *FEMS Microbiol Rev* **28:** 519–542.
39. **Brenner DG, Shaw WV.** 1985. The use of synthetic oligonucleotides with universal templates for rapid DNA sequencing: results with staphylococcal replicon pC221. *EMBO J* **4:**561–568.
40. **Projan SJ, Kornblum J, Moghazeh SL, Edelman I, Gennaro ML, Novick RP.** 1985. Comparative sequence and functional analysis of pT181 and pC221, cognate plasmid replicons from *Staphylococcus aureus. Mol Gen Genet* **3:**452–464.
41. **Smith MC, Thomas CD.** 2004. An accessory protein is required for relaxosome formation by small staphylococcal plasmids. *J Bacteriol* **186:**3363–3373.

42. **Horinouchi S, Weisblum B.** 1982. Nucleotide sequence and functional map of pC194, a plasmid that specifies inducible chloramphenicol resistance. *J Bacteriol* **150:** 815–825.

43. **Schwarz S, Cardoso M, Blobel M.** 1989. Plasmid-mediated chloramphenicol resistance in *Staphylococcus hyicus*. *J Gen Microbiol* **135:**3329–3336.

44. **Schwarz S, Cardoso M, Blobel H.** 1990. Detection of a novel chloramphenicol resistance plasmid from "equine" *Staphylococcus sciuri*. *J Vet Med B Infect Dis Vet Public Health* **37:**674–679.

45. **Schwarz S, Spies U, Cardoso M.** 1991. Cloning and sequence analysis of a plasmid-encoded chloramphenicol acetyltransferase gene from *Staphylococcus intermedius*. *J Gen Microbiol* **137:**977–981.

46. **Cardoso M, Schwarz S.** 1992. Chloramphenicol resistance plasmids in *Staphylococcus aureus* isolated from bovine subclinical mastitis. *Vet Microbiol* **30:**223–232.

47. **Schwarz S, Werckenthin C, Pinter L, Kent LE, Noble WC.** 1995. Chloramphenicol resistance in *Staphylococcus intermedius* from a single veterinary centre: evidence for plasmid and chromosomal location of the resistance genes. *Vet Microbiol* **43:**151–159.

48. **Schwarz S.** 1994. Emerging chloramphenicol resistance in *Staphylococcus lentus* from mink following chloramphenicol treatment: characterisation of the resistance genes. *Vet Microbiol* **41:**51–61.

49. **Schwarz S, Cardoso M.** 1991. Molecular cloning, purification, and properties of a plasmid-encoded chloramphenicol acetyltransferase from *Staphylococcus haemolyticus*. *Antimicrob Agents Chemother* **35:**1277–1283.

50. **Schwarz S, Cardoso M.** 1991. Nucleotide sequence and phylogeny of a chloramphenicol acetyltransferase encoded by the plasmid pSCS7 from *Staphylococcus aureus*. *Antimicrob Agents Chemother* **35:**1551–1556.

51. **Hauschild T, Stepanović S, Vuković D, Dakić I, Schwarz S.** 2009. Occurrence of chloramphenicol resistance and corresponding resistance genes in members of the *Staphylococcus sciuri* group. *Int J Antimicrob Agents* **33:**383–384.

52. **Kehrenberg C, Schwarz S.** 2004. *fexA*, a novel *Staphylococcus lentus* gene encoding resistance to florfenicol and chloramphenicol. *Antimicrob Agents Chemother* **48:**615–618.

53. **Kehrenberg C, Schwarz S.** 2005. Florfenicol-chloramphenicol exporter gene *fexA* is part of the novel transposon Tn*558*. *Antimicrob Agents Chemother* **49:** 813–815.

54. **Kehrenberg C, Schwarz S.** 2006. Distribution of florfenicol resistance genes *fexA* and *cfr* among chloramphenicol-resistant *Staphylococcus* isolates. *Antimicrob Agents Chemother* **50:**1156–1163.

55. **Wang Y, Zhang W, Wang J, Wu C, Shen Z, Fu X, Yan Y, Zhang Q, Schwarz S, Shen J.** 2012. Distribution of the multidrug resistance gene *cfr* in *Staphylococcus* species isolates from swine farms in China. *Antimicrob Agents Chemother* **56:**1485–1490.

56. **He T, Wang Y, Schwarz S, Zhao Q, Shen J, Wu C.** 2014. Genetic environment of the multi-resistance gene *cfr* in methicillin-resistant coagulase-negative staphylococci from chickens, ducks, and pigs in China. *Int J Med Microbiol* **304:**257–261.

57. **Kehrenberg C, Aarestrup FM, Schwarz S.** 2007. IS*21-558* insertion sequences are involved in the mobility of the multiresistance gene *cfr*. *Antimicrob Agents Chemother* **51:**483–487.

58. **Shore AC, Brennan OM, Ehricht R, Monecke S, Schwarz S, Slickers P, Coleman DC.** 2010. Identification and characterization of the multidrug resistance gene *cfr* in a Panton-Valentine leukocidin-positive sequence type 8 methicillin-resistant *Staphylococcus aureus* IVa (USA300) isolate. *Antimicrob Agents Chemother* **54:**4978–4984.

59. **Gómez-Sanz E, Kadlec K, Feler AT, Zarazaga M, Torres C, Schwarz S.** 2013. A novel FexA variant from a canine *Staphylococcus pseudintermedius* isolate that does not confer florfenicol resistance. *Antimicrob Agents Chemother* **57:**5763–5766.

60. **Kehrenberg C, Schwarz S, Jacobsen L, Hansen LH, Vester B.** 2005. A new mechanism for chloramphenicol, florfenicol and clindamycin resistance: methylation of 23S ribosomal RNA at A2503. *Mol Microbiol* **57:** 1064–1073.

61. **Long KS, Poehlsgaard J, Kehrenberg C, Schwarz S, Vester B.** 2006. The Cfr rRNA methyltransferase confers resistance to phenicols, lincosamides, oxazolidinones, pleuromutilins, and streptogramin A antibiotics. *Antimicrob Agents Chemother* **50:**2500–2505.

62. **Schwarz S, Werckenthin C, Kehrenberg C.** 2000. Identification of a plasmid-borne chloramphenicol-florfenicol resistance gene in *Staphylococcus sciuri*. *Antimicrob Agents Chemother* **44:**2530–2533.

63. **Kehrenberg C, Ojo KK, Schwarz S.** 2004. Nucleotide sequence and organization of the multiresistance plasmid pSCFS1 from *Staphylococcus sciuri*. *J Antimicrob Chemother* **54:**936–939.

64. **Wang XM, Zhang WJ, Schwarz S, Yu SY, Liu H, Si W, Zhang RM, Liu S.** 2012. Methicillin-resistant *Staphylococcus aureus* ST9 from a case of bovine mastitis carries the genes *cfr* and *erm*(A) on a small plasmid. *J Antimicrob Chemother* **67:**1287–1289.

65. **Lampson BC, Parisi JT.** 1986. Nucleotide sequence of the constitutive macrolide-lincosamide-streptogramin B resistance plasmid pNE131 from *Staphylococcus epidermidis* and homologies with *Staphylococcus aureus* plasmids pE194 and pSN2. *J Bacteriol* **167:**888–892.

66. **Catchpole I, Thomas C, Davies A, Dyke KG.** 1988. The nucleotide sequence of *Staphylococcus aureus* plasmid pT48 conferring inducible macrolide-lincosamide-streptogramin B resistance and comparison with similar plasmids expressing constitutive resistance. *J Gen Microbiol* **134:**697–709.

67. **Horinouchi S, Weisblum B.** 1982. Nucleotide sequence and functional map of pE194, a plasmid that specifies inducible resistance to macrolide, lincosamide, and streptogramin type B antibodies. *J Bacteriol* **150:**804–814.

68. **Lodder G, Schwarz S, Gregory P, Dyke KG.** 1996. Tandem duplication in *ermC* translational attenuator of the

macrolide-lincosamide-streptogramin B resistance plasmid pSES6 from *Staphylococcus equorum*. *Antimicrob Agents Chemother* **40**:215–217.

69. **Hauschild T, Lüthje P, Schwarz S.** 2006. Characterization of a novel type of MLS_B resistance plasmid from *Staphylococcus saprophyticus* carrying a constitutively expressed *erm*(C) gene. *Vet Microbiol* **115**:258–263.

70. **Wendlandt S, Kadlec K, Feler AT, van Duijkeren E, Schwarz S.** 2014. Two different *erm*(C)-carrying plasmids in the same methicillin-resistant *Staphylococcus aureus* CC398 isolate from a broiler farm. *Vet Microbiol* **171**:382–387.

71. **Gómez-Sanz E, Kadlec K, Feler AT, Zarazaga M, Torres C, Schwarz S.** 2013. Novel *erm*(T)-carrying multiresistance plasmids from porcine and human isolates of methicillin-resistant *Staphylococcus aureus* ST398 that also harbor cadmium and copper resistance determinants. *Antimicrob Agents Chemother* **57**:3275–3282.

72. **Gómez-Sanz E, Kadlec K, Feler AT, Billerbeck C, Zarazaga M, Schwarz S, Torres C.** 2013. Analysis of a novel *erm*(T)- and *cadDX*-carrying plasmid from methicillin-susceptible *Staphylococcus aureus* ST398-t571 of human origin. *J Antimicrob Chemother* **68**: 471–473.

73. **Gómez-Sanz E, Zarazaga M, Kadlec K, Schwarz S, Torres C.** 2013. Chromosomal integration of the novel plasmid pUR3912 from methicillin-susceptible *Staphylococcus aureus* ST398 of human origin. *Clin Microbiol Infect* **19**:E519–E522.

74. **Feler A, Scott C, Kadlec K, Ehricht R, Monecke S, Schwarz S.** 2010. Characterization of methicillin-resistant *Staphylococcus aureus* ST398 from cases of bovine mastitis. *J Antimicrob Chemother* **65**:619–625.

75. **Feler AT, Kadlec K, Hassel M, Hauschild T, Eidam C, Ehricht R, Monecke S, Schwarz S.** 2011. Characterization of methicillin-resistant *Staphylococcus aureus* isolates from food and food products of poultry origin in Germany. *Appl Environ Microbiol* **77**:7151–7157.

76. **Townsend DE, Bolton S, Ashdown N, Annear DI, Grubb WB.** 1986. Conjugative staphylococcal plasmids carrying hitch-hiking transposons similar to Tn*554*: intra- and interspecies dissemination of erythromycin resistance. *Aust J Exp Biol Med Sci* **64**:367–379.

77. **Werckenthin C, Schwarz S, Dyke KG.** 1996. Macrolide-lincosamide-streptogramin B resistance in *Staphylococcus lentus* results from the integration of part of a transposon into a small plasmid. *Antimicrob Agents Chemother* **40**:2224–2225.

78. **Feler AT, Kadlec K, Schwarz S.** 2011. Novel apramycin resistance gene *apmA* in bovine and porcine methicillin-resistant *Staphylococcus aureus* ST398 isolates. *Antimicrob Agents Chemother* **55**:373–375.

79. **Schwarz S, Kehrenberg C, Ojo KK.** 2002. *Staphylococcus sciuri* gene *erm*(33), encoding inducible resistance to macrolides, lincosamides, and streptogramin B antibiotics, is a product of recombination between *erm*(C) and *erm*(A). *Antimicrob Agents Chemother* **46**:3621–3623.

80. **Matsuoka M, Inoue M, Endo Y, Nakajima Y.** 2003. Characteristic expression of three genes, *msr*(A), *mph* (C) and *erm*(Y), that confer resistance to macrolide antibiotics on *Staphylococcus aureus*. *FEMS Microbiol Lett* **220**:287–293.

81. **Lüthje P, Schwarz S.** 2006. Antimicrobial resistance of coagulase-negative staphylococci from bovine subclinical mastitis with particular reference to macrolide-lincosamide resistance phenotypes and genotypes. *J Antimicrob Chemother* **57**:966–969.

82. **Gatermann SG, Koschinski T, Friedrich S.** 2007. Distribution and expression of macrolide resistance genes in coagulase-negative staphylococci. *Clin Microbiol Infect* **13**:777–781.

83. **Lozano C, Rezusta A, Gómez P, Gómez-Sanz E, Báez N, Martin-Saco G, Zarazaga M, Torres C.** 2012. High prevalence of *spa* types associated with the clonal lineage CC398 among tetracycline-resistant methicillin-resistant *Staphylococcus aureus* strains in a Spanish hospital. *J Antimicrob Chemother* **67**:330–334.

84. **Kennedy AD, Porcella SF, Martens C, Whitney AR, Braughton KR, Chen L, Craig CT, Tenover FC, Kreiswirth BN, Musser JM, DeLeo FR.** 2010. Complete nucleotide sequence analysis of plasmids in strains of *Staphylococcus aureus* clone USA300 reveals a high level of identity among isolates with closely related core genome sequences. *J Clin Microbiol* **48**:4504–4511.

85. **Loeza-Lara PD, Soto-Huipe M, Baizabal-Aguirre VM, Ochoa-Zarzosa A, Valdez-Alarcón JJ, Cano-Camacho H, López-Meza JE.** 2004. pBMSa1, a plasmid from a dairy cow isolate of *Staphylococcus aureus*, encodes a lincomycin resistance determinant and replicates by the rolling-circle mechanism. *Plasmid* **52**:48–56.

86. **Lüthje P, von Köckritz-Blickwede M, Schwarz S.** 2007. Identification and characterization of nine novel types of small staphylococcal plasmids carrying the lincosamide nucleotidyltransferase gene *lnu*(A). *J Antimicrob Chemother* **59**:600–606.

87. **Li B, Wendlandt S, Yao J, Liu Y, Zhang Q, Shi Z, Wei J, Shao D, Schwarz S, Wang S, Ma Z.** 2013. Detection and new genetic environment of the pleuromutilin-lincosamide-streptogramin A resistance gene *lsa*(E) in methicillin-resistant *Staphylococcus aureus* of swine origin. *J Antimicrob Chemother* **68**:1251–1255.

88. **Wendlandt S, Li B, Ma Z, Schwarz S.** 2013. Complete sequence of the multi-resistance plasmid pV7037 from a porcine methicillin-resistant *Staphylococcus aureus*. *Vet Microbiol* **166**:650–654.

89. **Allignet J, Liassine N, El Solh N.** 1998. Characterization of a staphylococcal plasmid related to pUB110 and carrying two novel genes, *vatC* and *vgbB*, encoding resistance to streptogramins A and B and similar antibiotics. *Antimicrob Agents Chemother* **42**:1794–1798.

90. **Allignet J, El Solh N.** 1999. Comparative analysis of staphylococcal plasmids carrying three streptogramin-resistance genes: *vat-vgb-vga*. *Plasmid* **42**:134–138.

91. **Allignet J, El Solh N.** 1995. Diversity among the Gram-positive acetyltransferases inactivating streptogramin A and structurally related compounds and characterization

of a new staphylococcal determinant, *vatB*. *Antimicrob Agents Chemother* **39**:2027–2036.

92. **Allignet J, Loncle V, El Sohl N.** 1992. Sequence of a staphylococcal plasmid gene, *vga*, encoding a putative ATP-binding protein involved in resistance to virginiamycin A-like antibiotics. *Gene* **117**:45–51.
93. **Haroche J, Allignet J, El Solh N.** 2002. Tn*5406*, a new staphylococcal transposon conferring resistance to streptogramin A and related compounds including dalfopristin. *Antimicrob Agents Chemother* **46**:2337–2343.
94. **Aubert S, Dyke KG, El Solh N.** 1998. Analysis of two *Staphylococcus epidermidis* plasmids coding for resistance to streptogramin A. *Plasmid* **40**:238–242.
95. **Kadlec K, Pomba CF, Couto N, Schwarz S.** 2010. Small plasmids carrying *vga*(A) or *vga*(C) genes mediate resistance to lincosamides, pleuromutilins and streptogramin A antibiotics in methicillin-resistant *Staphylococcus aureus* ST398 from swine. *J Antimicrob Chemother* **65**: 2692–2693.
96. **Gentry DR, McCloskey L, Gwynn MN, Rittenhouse SF, Scangarella N, Shawar R, Holmes DJ.** 2008. Genetic characterization of Vga ABC proteins conferring reduced susceptibility of *Staphylococcus aureus* to pleuromutilins. *Antimicrob Agents Chemother* **52**:4507–4509.
97. **Lozano C, Aspiroz C, Rezusta A, Gómez-Sanz E, Simon C, Gómez P, Ortega C, Revillo MJ, Zarazaga M, Torres C.** 2012. Identification of novel *vga*(A)-carrying plasmids and a Tn*5406*-like transposon in methicillin-resistant *Staphylococcus aureus* and *Staphylococcus epidermidis* of human and animal origin. *Int J Antimicrob Agents* **40**:306–312.
98. **Mendes RE, Smith TC, Deshpande L, Diekema DJ, Sader HS, Jones RN.** 2011. Plasmid-borne *vga*(A)-encoding gene in methicillin-resistant *Staphylococcus aureus* ST398 recovered from swine and a swine farmer in the United States. *Diagn Microbiol Infect Dis* **71**: 177–180.
99. **Haroche J, Allignet J, Buchrieser C, El Solh N.** 2000. Characterization of a variant of *vga*(A) conferring resistance to streptogramin A and related compounds. *Antimicrob Agents Chemother* **44**:2271–2275.
100. **Chesneau O, Ligeret H, Hosan-Aghaie N, Morvan A, Dassa E.** 2005. Molecular analysis of resistance to streptogramin A compounds conferred by the Vga proteins of staphylococci. *Antimicrob Agents Chemother* **49**:973–980.
101. **Novotna G, Janata J.** 2006. A new evolutionary variant of the streptogramin A resistance protein, Vga(A)$_{LC}$, from *Staphylococcus haemolyticus* with shifted substrate specificity towards lincosamides. *Antimicrob Agents Chemother* **50**:4070–4076.
102. **Allignet J, El Solh N.** 1997. Characterization of a new staphylococcal gene, *vgaB*, encoding a putative ABC transporter conferring resistance to streptogramin A and related compounds. *Gene* **202**:133–138.
103. **Schwendener S, Perreten V.** 2011. New transposon Tn*6133* in methicillin-resistant *Staphylococcus aureus* ST398 contains *vga*(E), a novel streptogramin A, pleuromutilin, and lincosamide resistance gene. *Antimicrob Agents Chemother* **55**:4900–4904.
104. **Hauschild T, Feler AT, Kadlec K, Billerbeck C, Schwarz S.** 2012. Detection of the novel *vga*(E) gene in methicillin-resistant *Staphylococcus aureus* CC398 isolates from cattle and poultry. *J Antimicrob Chemother* **67**:503–504.
105. **Shaw KJ, Rather PN, Hare SR, Miller GH.** 1993. Molecular genetics of aminoglycoside resistance genes and familial relationships of the aminoglycoside-modifying enzymes. *Microbiol Rev* **57**:138–163.
106. **Mingeot-Leclercq MP, Glupczynski Y, Tulkens PM.** 1999. Aminoglycosides: activity and resistance. *Antimicrob Agents Chemother* **43**:727–737.
107. **Ramirez MS, Tolmasky ME.** 2010. Aminoglycoside modifying enzymes. *Drug Resist Update* **13**:151–171.
108. **McKenzie T, Hoshino T, Tanaka T, Sueoka N.** 1987. Correction. A revision of the nucleotide sequence and functional map of pUB110. *Plasmid* **17**:83–85.
109. **McKenzie T, Hoshino T, Tanaka T, Sueoka N.** 1986. The nucleotide sequence of pUB110: some salient features in relation to replication and its regulation. *Plasmid* **15**:93–103.
110. **Projan SJ, Moghazeh S, Novick RP.** 1988. Nucleotide sequence of pS194, a streptomycin-resistance plasmid from *Staphylococcus aureus*. *Nucleic Acids Res* **16**: 2179–2187.
111. **Schwarz S, Blobel H.** 1990. A new streptomycin-resistance plasmid from *Staphylococcus hyicus* and its structural relationship to other staphylococcal resistance plasmids. *J Med Microbiol* **32**:201–205.
112. **Schwarz S, Noble WC.** 1994. Structure and putative origin of a plasmid from *Staphylococcus hyicus* that mediates chloramphenicol and streptomycin resistance. *Lett Appl Microbiol* **18**:281–284.
113. **Schwarz S, Grölz-Krug S.** 1991. A chloramphenicol-streptomycin-resistance plasmid from a clinical strain of *Staphylococcus sciuri* and its structural relationships to other staphylococcal resistance plasmids. *FEMS Microbiol Lett* **66**:319–322.
114. **Jensen SO, Apisiridej S, Kwong SM, Yang YH, Skurray RA, Firth N.** 2010. Analysis of the prototypical *Staphylococcus aureus* multiresistance plasmid pSK1. *Plasmid* **64**:135–142.
115. **Rouch DA, Byrne ME, Kong YC, Skurray RA.** 1987. The *aacA-aphD* gentamicin and kanamycin resistance determinant of Tn*4001* from *Staphylococcus aureus*: expression and nucleotide sequence analysis. *J Gen Microbiol* **133**:3039–3052.
116. **Byrne ME, Rouch DA, Skurray RA.** 1989. Nucleotide sequence analysis of IS*256* from the *Staphylococcus aureus* gentamicin-tobramycin-kanamycin-resistance transposon Tn*4001*. *Gene* **81**:361–367.
117. **Ferretti JJ, Gilmore K, Courvalin P.** 1986. Nucleotide sequence analysis of the gene specifying the bifunctional 6′-aminoglycoside acetyltransferase 2″-aminoglycoside phosphotransferase enzyme in *Streptococcus faecalis* and identification and cloning of gene regions specifying the two activities. *J Bacteriol* **167**:631–638.
118. **Caryl JA, O'Neill AJ.** 2009. Complete nucleotide sequence of pGO1, the prototype conjugative plasmid from the staphylococci. *Plasmid* **62**:35–38.

119. **Lange CC, Werckenthin C, Schwarz S.** 2003. Molecular analysis of the plasmid-borne *aacA/aphD* resistance gene region of coagulase-negative staphylococci from chickens. *J Antimicrob Chemother* **51:**1397–1401.
120. **Skurray RA, Rouch DA, Lyon BR, Gillespie MT, Tennent JM, Byrne ME, Messerotti LJ, May JW.** 1988. Multiresistant *Staphylococcus aureus*: genetics and evolution of epidemic Australian strains. *J Antimicrob Chemother* **21**(Suppl C)**:**19–39.
121. **Derbise A, Dyke KG, El Solh N.** 1996. Characterization of a *Staphylococcus aureus* transposon, Tn*5405*, located within Tn*5404* and carrying the aminoglycoside resistance genes, *aphA-3* and *aadE*. *Plasmid* **35:**174–188.
122. **Gill SR, Fouts DE, Archer GL, Mongodin EF, Deboy RT, Ravel J, Paulsen IT, Kolonay JF, Brinkac L, Beanan M, Dodson RJ, Daugherty SC, Madupu R, Angiuoli SV, Durkin AS, Haft DH, Vamathevan J, Khouri H, Utterback T, Lee C, Dimitrov G, Jiang L, Qin H, Weidman J, Tran K, Kang K, Hance IR, Nelson KE, Fraser CM.** 2005. Insights on evolution of virulence and resistance from the complete genome analysis of an early methicillin-resistant *Staphylococcus aureus* strain and a biofilm-producing methicillin-resistant *Staphylococcus epidermidis* strain. *J Bacteriol* **187:**2426–2438.
123. **Jamrozy D, Coldham N, Butaye P, Fielder M.** 2014. Identification of a novel plasmid-associated spectinomycin adenyltransferase gene *spd* in methicillin-resistant *Staphylococcus aureus* ST398 isolated from animal and human sources. *J Antimicrob Chemother* **69:**1193–1196.
124. **Wendlandt S, Feler AT, Kadlec K, van Duijkeren E, Schwarz S.** 2014. Identification of the novel spectinomycin resistance gene *spd* in a different plasmid background among methicillin-resistant *Staphylococcus aureus* CC398 and methicillin-susceptible *S. aureus* ST433. *J Antimicrob Chemother* **69:**2000–2008.
125. **Hodgson JE, Curnock SP, Dyke KG, Morris R, Sylvester DR, Gross MS.** 1994. Molecular characterization of the gene encoding high-level mupirocin resistance in *Staphylococcus aureus* J2870. *Antimicrob Agents Chemother* **38:**1205–1208.
126. **Needham C, Rahman M, Dyke KG, Noble WC.** 1994. An investigation of plasmids from *Staphylococcus aureus* that mediate resistance to mupirocin and tetracycline. *Microbiology* **140:**2577–2583.
127. **Yoo JI, Shin ES, Chung GT, Lee KM, Yoo JS, Lee YS.** 2010. Restriction fragment length polymorphism (RFLP) patterns and sequence analysis of high-level mupirocin-resistant methicillin-resistant staphylococci. *Int J Antimicrob Agents* **35:**50–55.
128. **Pérez-Roth E, Kwong SM, Alcoba-Florez J, Firth N, Mendez-Alvarez S.** 2010. Complete nucleotide sequence and comparative analysis of pPR9, a 41.7-kilobase conjugative staphylococcal multiresistance plasmid conferring high-level mupirocin resistance. *Antimicrob Agents Chemother* **54:**2252–2257.
129. **Matanovic K, Pérez-Roth E, Pintarić S, Šeol Martinec B.** 2013. Molecular characterization of high-level mupirocin resistance in *Staphylococcus pseudintermedius*. *J Clin Microbiol* **51:**1005–1007.
130. **do Carmo Ferreira N, Schuenck RP, dos Santos KR, de Freire Bastos Mdo C, Giambiagi-deMarval M.** 2011. Diversity of plasmids and transmission of high-level mupirocin *mupA* resistance gene in *Staphylococcus haemolyticus*. *FEMS Immunol Med Microbiol* **61:**147–152.
131. **McDougal LK, Fosheim GE, Nicholson A, Bulens SN, Limbago BM, Shearer JE, Summers AO, Patel JB.** 2010. Emergence of resistance among USA300 methicillin-resistant *Staphylococcus aureus* isolates causing invasive disease in the United States. *Antimicrob Agents Chemother* **54:**3804–3811.
132. **O'Neill AJ, McLaws F, Kahlmeter G, Henriksen AS, Chopra I.** 2007. Genetic basis of resistance to fusidic acid in staphylococci. *Antimicrob Agents Chemother* **51:**1737–1740.
133. **O'Neill AJ, Chopra I.** 2006. Molecular basis of *fusB*-mediated resistance to fusidic acid in *Staphylococcus aureus*. *Mol Microbiol* **59:**664–676.
134. **O'Brien FG, Price C, Grubb WB, Gustafson JE.** 2002. Genetic characterization of the fusidic acid and cadmium resistance determinants of *Staphylococcus aureus* plasmid pUB101. *J Antimicrob Chemother* **50:**313–321.
135. **Yazdankhah SP, Asli AW, Sørum H, Oppegaard H, Sunde M.** 2006. Fusidic acid resistance, mediated by *fusB*, in bovine coagulase-negative staphylococci. *J Antimicrob Chemother* **58:**1254–1256.
136. **Young HK, Skurray RA, Amyes SG.** 1987. Plasmid-mediated trimethoprim-resistance in *Staphylococcus aureus*. Characterization of the first Gram-positive plasmid dihydrofolate reductase (type S1). *Biochem J* **243:** 309–312.
137. **Rouch DA, Messerotti LJ, Loo LS, Jackson CA, Skurray RA.** 1989. Trimethoprim resistance transposon Tn*4003* from *Staphylococcus aureus* encodes genes for a dihydrofolate reductase and thymidylate synthetase flanked by three copies of IS*257*. *Mol Microbiol* **3:**161–175.
138. **Tennent JM, Young HK, Lyon BR, Amyes SG, Skurray RA.** 1988. Trimethoprim resistance determinants encoding a dihydrofolate reductase in clinical isolates of *Staphylococcus aureus* and coagulase-negative staphylococci. *J Med Microbiol* **26:**67–73.
139. **Weigel LM, Clewell DB, Gill SR, Clark NC, McDougal LK, Flannagan SE, Kolonay JF, Shetty J, Killgore GE, Tenover FC.** 2003. Genetic analysis of a high-level vancomycin-resistant isolate of *Staphylococcus aureus*. *Science* **302:**1569–1571.
140. **Galetto DW, Johnston JL, Archer GL.** 1987. Molecular epidemiology of trimethoprim resistance among coagulase-negative staphylococci. *Antimicrob Agents Chemother* **31:**1683–1688.
141. **Apisiridej S, Leelaporn A, Scaramuzzi CD, Skurray RA, Firth N.** 1997. Molecular analysis of a mobilizable theta-mode trimethoprim resistance plasmid from coagulase-negative staphylococci. *Plasmid* **38:**13–24.
142. **Dale GE, Langen H, Page MG, Then RL, Stüber D.** 1995. Cloning and characterization of a novel, plasmid-encoded trimethoprim-resistant dihydrofolate reductase from *Staphylococcus haemolyticus* MUR313. *Antimicrob Agents Chemother* **39:**1920–1924.

143. **Gopegui ER, Juan C, Zamorano L, Pérez JL, Oliver A.** 2012. Transferable multidrug resistance plasmid carrying *cfr* associated with *tet*(L), *ant(4′)-Ia*, and *dfrK* genes from a clinical methicillin-resistant *Staphylococcus aureus* ST125 strain. *Antimicrob Agents Chemother* **56:** 2139–2142.

144. **Kadlec K, Schwarz S.** 2010. Identification of the novel *dfrK*-carrying transposon Tn*559* in a porcine methicillin-susceptible *Staphylococcus aureus* ST398 strain. *Antimicrob Agents Chemother* **54:**3475–3477.

145. **Périchon B, Courvalin P.** 2009. VanA-type vancomycin-resistant *Staphylococcus aureus*. *Antimicrob Agents Chemother* **53:**4580–4587.

146. **Tenover FC, Weigel LM, Appelbaum PC, McDougal LK, Chaitram J, McAllister S, Clark N, Killgore G, O'Hara CM, Jevitt L, Patel JB, Bozdogan B.** 2004. Vancomycin-resistant *Staphylococcus aureus* isolate from a patient in Pennsylvania. *Antimicrob Agents Chemother* **48:**275–280.

147. **Weigel LM, Donlan RM, Shin DH, Jensen B, Clark NC, McDougal LK, Zhu W, Musser KA, Thompson J, Kohlerschmidt D, Dumas N, Limberger RJ, Patel JB.** 2007. High-level vancomycin-resistant *Staphylococcus aureus* isolates associated with a polymicrobial biofilm. *Antimicrob Agents Chemother* **51:**231–238.

148. **Zhu W, Clark NC, McDougal LK, Hageman J, McDonald LC, Patel JB.** 2008. Vancomycin-resistant *Staphylococcus aureus* isolates associated with Inc18-like *vanA* plasmids in Michigan. *Antimicrob Agents Chemother* **52:**452–457.

149. **Zhu W, Clark N, Patel JB.** 2013. pSK41-like plasmid is necessary for Inc18-like *vanA* plasmid transfer from *Enterococcus faecalis* to *Staphylococcus aureus in vitro*. *Antimicrob Agents Chemother* **57:**212–219.

150. **Rowland SJ, Dyke KG.** 1989. Characterization of the staphylococcal beta-lactamase transposon Tn*552*. *EMBO J* **8:**2761–2773.

151. **Rowland SJ, Dyke KG.** 1990. Tn*552*, a novel transposable element from *Staphylococcus aureus*. *Mol Microbiol* **4:**961–975.

152. **Shearer JE, Wireman J, Hostetler J, Forberger H, Borman J, Gill J, Sanchez S, Mankin A, Lamarre J, Lindsay JA, Bayles K, Nicholson A, O'Brien F, Jensen SO, Firth N, Skurray RA, Summers AO.** 2011. Major families of multiresistant plasmids from geographically and epidemiologically diverse staphylococci. *G3 (Bethesda)* **1:**581–591.

153. **Novick RP, Roth C.** 1968. Plasmid-linked resistance to inorganic salts in *Staphylococcus aureus*. *J Bacteriol* **95:** 1335–1342.

154. **Weiss AA, Murphy SD, Silver S.** 1977. Mercury and organomercurial resistances determined by plasmids in *Staphylococcus aureus*. *J Bacteriol* **132:**197–208.

155. **Gillespie MT, May JW, Skurray RA.** 1986. Plasmid-encoded resistance to acriflavine and quaternary ammonium compounds in methicillin-resistant *Staphylococcus aureus*. *FEMS Microbiol Lett* **34:**47–51.

156. **Johnston LH, Dyke KG.** 1969. Ethidium bromide resistance, a new marker on the staphylococcal penicillinase plasmid. *J Bacteriol* **100:**1413–1414.

157. **Emslie KR, Townsend DE, Bolton S, Grubb WB.** 1985. Two distinct resistance determinants to nucleic acid binding compounds in *Staphylococcus aureus*? *FEMS Microbiol Lett* **27:**61–64.

158. **Novick RP, Edelman I, Schwesinger MD, Gruss AD, Swanson EC, Pattee PA.** 1979. Genetic translocation in *Staphylococcus aureus*. *Proc Natl Acad Sci USA* **76:** 400–404.

159. **Gillespie MT, Skurray RA.** 1986. Plasmids in multiresistant *Staphylococcus aureus*. *Microbiol Sci* **3:**53–58.

160. **Hisatsune J, Hirakawa H, Yamaguchi T, Fudaba Y, Oshima K, Hattori M, Kato F, Kayama S, Sugai M.** 2013. Emergence of *Staphylococcus aureus* carrying multiple drug resistance genes on a plasmid encoding exfoliative toxin *B*. *Antimicrob Agents Chemother* **57:** 6131–6140.

161. **Olsen JE, Christensen H, Aarestrup FM.** 2006. Diversity and evolution of *blaZ* from *Staphylococcus aureus* and coagulase-negative staphylococci. *J Antimicrob Chemother* **57:**450–460.

162. **Novick RP, Projan SJ, Rosenblum W, Edelman I.** 1984. Staphylococcal plasmid cointegrates are formed by host- and phage-mediated general *rec* systems that act on short regions of homology. *Mol Gen Genet* **195:** 374–377.

163. **Palva A, Vigren G, Simonen M, Rintala H, Laamanen P.** 1990. Nucleotide sequence of the tetracycline resistance gene of pBC16 from *Bacillus cereus*. *Nucleic Acids Res* **18:**1635.

164. **Shishido K, Tanaka Y.** 1984. A restriction map of *Bacillus subtilis* tetracycline-resistance plasmid pNS1981. *Plasmid* **12:**65–66.

165. **Sakaguchi R, Shishido K, Hoshino T, Furukawa K.** 1986. The nucleotide sequence of the tetracycline resistance gene of plasmid pNS1981 from *Bacillus subtilis* differs from pTHT15 from a thermophilic *Bacillus* by two base pairs. *Plasmid* **16:**72–73.

166. **Charpentier E, Gerbaud G, Courvalin P.** 1999. Conjugative mobilization of the rolling-circle plasmid pIP823 from *Listeria monocytogenes* BM4293 among Gram-positive and Gram-negative bacteria. *J Bacteriol* **18:** 3368–3374.

167. **Leach DRF.** 1996. *Genetic Recombination*. Blackwell Science Ltd., Oxford, UK.

168. **Werckenthin C, Schwarz S, Roberts MC.** 1996. Integration of pT181-like tetracycline resistance plasmids into large staphylococcal plasmids involves IS*257*. *Antimicrob Agents Chemother* **40:**2542–2544.

169. **Diep BA, Gill SR, Chang RF, Phan TH, Chen JH, Davidson MG, Lin F, Lin J, Carleton HA, Mongodin EF, Sensabaugh GF, Perdreau-Remington F.** 2006. Complete genome sequence of USA300, an epidemic clone of community-acquired meticillin-resistant *Staphylococcus aureus*. *Lancet* **367:**731–739.

170. **Berg T, Firth N, Apisiridej S, Hettiaratchi A, Leelaporn A, Skurray RA.** 1998. Complete nucleotide sequence of pSK41: evolution of staphylococcal conjugative multiresistance plasmids. *J Bacteriol* **180:**4350–4359.

171. **Hansen LH, Planellas MH, Long KS, Vester B.** 2012. The order *Bacillales* hosts functional homologs of the worrisome *cfr* antibiotic resistance gene. *Antimicrob Agents Chemother* **56:**3563–3567.

172. **Danielsen M.** 2002. Characterization of the tetracycline resistance plasmid pMD5057 from *Lactobacillus plantarum* 5057 reveals a composite structure. *Plasmid* **48:** 98–103.

173. **López M, Kadlec K, Schwarz S, Torres C.** 2012. First detection of the staphylococcal trimethoprim resistance gene *dfrK* and the *dfrK*-carrying transposon Tn*559* in enterococci. *Microb Drug Resist* **18:**13–18.

174. **Dai L, Wu C, Wang M, Wang Y, Wang Y, Huang S, Xia LN, Li B, Shen J.** 2010. First report of the multidrug resistance gene *cfr* and the phenicol resistance gene *fexA* in a *Bacillus* strain from swine feces. *Antimicrob Agents Chemother* **54:**3953–3955.

175. **Zhang WJ, Wu C, Wang Y, Shen Z, Dai L, Han J, Foley S, Shen J, Zhang Q.** 2011. The new genetic environment of *cfr* on plasmid pBS-02 in a *Bacillus* strain. *J Antimicrob Chemother* **66:**1174–1175.

176. **Wang Y, Schwarz S, Shen Z, Zhang W, Qi J, Liu Y, He T, Shen J, Wu C.** 2012. Co-location of the multiresistance gene *cfr* and the novel streptomycin resistance gene *aadY* on a small plasmid in a porcine *Bacillus* strain. *J Antimicrob Chemother* **67:**1547–1549.

177. **Liu Y, Wang Y, Wu C, Shen Z, Schwarz S, Du XD, Dai L, Zhang W, Zhang Q, Shen J.** 2012. First report of the multidrug resistance gene *cfr* in *Enterococcus faecalis* of animal origin. *Antimicrob Agents Chemother* **56:**1650–1654.

178. **Liu Y, Wang Y, Schwarz S, Li Y, Shen Z, Zhang Q, Wu C, Shen J.** 2013. Transferable multiresistance plasmids carrying *cfr* in *Enterococcus* spp. from swine and farm environment. *Antimicrob Agents Chemother* **57:**42–48.

179. **Diaz L, Kiratisin P, Mendes RE, Panesso D, Singh KV, Arias CA.** 2012. Transferable plasmid-mediated resistance to linezolid due to *cfr* in a human clinical isolate of *Enterococcus faecalis*. *Antimicrob Agents Chemother* **56:**3917–3922.

180. **Wang Y, Wang Y, Schwarz S, Shen Z, Zhou N, Lin J, Wu C, Shen J.** 2012. Detection of the staphylococcal multiresistance gene *cfr* in *Macrococcus caseolyticus* and *Jeotgalicoccus pinnipedialis*. *J Antimicrob Chemother* **67:**1824–1827.

181. **Wang Y, Li D, Song L, Liu Y, He T, Liu H, Wu C, Schwarz S, Shen J.** 2013. First report of the multiresistance gene *cfr* in *Streptococcus suis*. *Antimicrob Agents Chemother* **57:**4061–4063.

182. **Wang Y, Wang Y, Wu C, Schwarz S, Shen Z, Zhang W, Zhang Q, Shen J.** 2011. Detection of the staphylococcal multiresistance gene *cfr* in *Proteus vulgaris* of food animal origin. *J Antimicrob Chemother* **66:**2521–2526.

183. **Wang Y, He T, Schwarz S, Zhou D, Shen Z, Wu C, Wang Y, Ma L, Zhang Q, Shen J.** 2012. Detection of the staphylococcal multiresistance gene *cfr* in *Escherichia coli* of domestic-animal origin. *J Antimicrob Chemother* **67:**1094–1098.

184. **Zhang WJ, Xu XR, Schwarz S, Wang XM, Dai L, Zheng HJ, Liu S.** 2014. Characterization of the IncA/C plasmid pSCEC2 from *Escherichia coli* of swine origin that harbours the multiresistance gene *cfr*. *J Antimicrob Chemother* **69:**385–389.

185. **Polak J, Novick RP.** 1982. Closely related plasmids from *Staphylococcus aureus* and soil bacilli. *Plasmid* **7:** 152–162.

186. **Ehrlich SD.** 1977. Replication and expression of plasmids from *Staphylococcus aureus* in *Bacillus subtilis*. *Proc Natl Acad Sci USA* **74:**1680–1682.

187. **Monod M, Denoya C, Dubnau D.** 1986. Sequence and properties of pIM13, a macrolide-lincosamide-streptogramin B resistance plasmid from *Bacillus subtilis*. *J Bacteriol* **167:**138–147.

Plasmids—Biology and Impact in Biotechnology and Discovery
Edited by Marcelo E. Tolmasky and Juan C. Alonso

doi:10.1128/microbiolspec.PLAS-0038-2014

Kornelia Smalla[1]
Sven Jechalke[1]
Eva M. Top[2]

23 Plasmid Detection, Characterization, and Ecology

PLASMIDS: ANCIENT MEANS OF BACTERIAL ADAPTATION AND DIVERSIFICATION

Plasmid-mediated horizontal gene transfer is recognized as a major driving force for bacterial adaptation and diversification. Different environmental settings have distinct bacterial community compositions, which determine—possibly with the exception of broad host range plasmids—the type of dominant plasmids that can be found. It is assumed that only a fraction of a population carries plasmids, which ensures a rapid adaptation of the population to changing environmental conditions (1). Without a doubt, plasmid-mediated spread of antibiotic resistance genes among bacteria of different taxa is one of the most impressive examples of bacterial plasticity in response to various selective pressures (2, 3). While the molecular biology of the plasmid-encoded replication, maintenance, and transfer processes of some plasmids has been studied for decades, little attention has been paid to their dissemination in the environment, their ecology, and the factors that drive their spread and diversification. In an overwhelming number of studies, the investigated plasmid-carrying strains originate from clinical specimens or diseased plant material, mostly human or plant pathogens. The reason for the lack of studies of the ecology of plasmids in natural settings was mainly the lack of tools to detect and quantify plasmids and to successfully culture their hosts.

Thanks to rapidly advancing sequencing technologies, the number of completely sequenced plasmids increased in the last decades. Comparative plasmid sequence analysis has provided insights into the evolution of plasmids and their relatedness, their modular structure, and the existence of hot spots for the insertion of accessory genes (4–6). The growing plasmid sequence database is also the prerequisite for the development of primers and probes to detect or classify plasmids. With the development of cultivation-independent DNA-based methods, it became possible to study the occurrence of plasmids in various environmental samples and to quantify their abundance. Here we emphasize that studying plasmid ecology is, without a doubt, a prerequisite for a better understanding of the role of plasmids and their contribution to bacterial diversification and adaptation. In particular, the disentanglement

[1]Julius Kühn-Institut, Federal Research Centre for Cultivated Plants (JKI), Institute for Epidemiology and Pathogen Diagnostics, Messeweg 11-12, 38104 Braunschweig, Germany; [2]University of Idaho, Department of Biological Sciences, 875 Perimeter MS 3051, Moscow, Idaho 83844-3051.

of factors that foster the proliferation of plasmid-carrying strains, horizontal gene exchange, and determine the costs and benefits plasmids bring to their hosts needs further consideration. This article aims to provide an overview of state-of-the-art methods being used to detect, isolate, and characterize plasmids and to study various aspects of their ecology. Examples from recent studies will be given to illustrate the potential and limitations of the methods employed to study plasmid occurrence and diversity, as well as some insights obtained. Although the methods described are applicable to Gram-negative and Gram-positive bacteria from any kind of environment, the major focus of the article is on plasmids in Gram-negative bacteria.

DETECTION AND QUANTIFICATION OF PLASMID-SPECIFIC SEQUENCES IN TOTAL COMMUNITY DNA

In the 1980s, Staley and Konopka (7) described the great plate count anomaly as the discrepancy between microscopic and colony forming counts, the latter being often much lower than the microscopic counts observed for the same sample. In general, the more oligotrophic an environment, the smaller is the proportion of bacteria able to form colonies on plates (8). In addition, many bacteria that are well known to form colonies on plates can, under environmental conditions, enter a state called viable but nonculturable (*vbnc*) (9). The *vbnc* status was reported for many human and plant pathogens.

In the last two decades, the development and application of cultivation-independent methods provided a better and more comprehensive picture of the occurrence and distribution of plasmids in different environmental settings. In particular, the use of total community DNA (TC-DNA), extracted directly from environmental samples, became a more and more widely used approach for the detection and quantification of plasmid occurrence and abundance. For most types of environmental samples, PCR-amplification with primers targeting replication or transfer-related segments of the backbone of particular plasmid groups is required because of low plasmid abundance. When TC-DNA is analyzed for the presence of particular plasmid-specific sequences, it is important that the absence of PCR-inhibiting substances is confirmed, e.g., by the amplification of the 16S rRNA gene fragments and sample dilutions. Furthermore, PCR-amplicons from TC-DNA should be analyzed either by cloning and sequencing, by amplicon sequencing, or by Southern blot hybridization with labeled probes generated from PCR-amplicons obtained from reference strains to exclude false-positive detection. This strategy has recently been used to screen TC-DNA from various types of environments originating from different geographic origins for the presence of IncP-1, IncP-7, and IncP-9 plasmids (10). These plasmid groups are known to carry degradative genes or complete operons encoding degradative pathways (11, 12).

The study by Dealtry et al. (10) showed a remarkably wide distribution of these plasmids and enabled the authors to identify so-called hot spots of plasmid occurrence. For example, samples from various pesticide bio-purification systems (BPSs) and river sediments were identified as environments with a high abundance of bacterial populations carrying IncP-1, IncP-7, and IncP-9 plasmids. Most remarkable was the presence of all IncP-1 subgroups in samples from BPSs, except for the ζ subgroup that was recently described by Norberg et al. (5). The strength of the hybridization signals obtained with probes targeting particular IncP-1 subgroups clearly differed in intensity, indicating differences in their abundances, although different amplification efficiencies of the different primer systems used could not be excluded. Cloning and sequencing of amplicons obtained with a newly designed IncP-9 primer system from TC-DNA of BPS samples revealed not only the presence of known plasmid groups but also the presence of unknown, yet to be isolated IncP-9 plasmids in these samples (10). In many studies, cloning of PCR-amplicons provided insights into the sequence diversity of the amplicons. Bahl et al. (13) were the first to show the presence of the different IncP-1 subgroups in the inflow of a wastewater treatment plant. These authors had designed primers targeting the plasmid replication initiation gene *trfA* of the different IncP-1 groups, also including the γ, ε, and δ subgroups that were amplified by the primers used in the study by Götz et al. (14). Due to the sequence divergence of the *trfA* gene, three different primer systems had to be mixed to guarantee the parallel detection of all IncP-1 plasmid subgroups discovered at the time.

In another approach, 454 pyrosequencing of amplicons was employed to study IncP-1 plasmid diversity. In contrast to the traditional cloning and sequencing approaches, the number of sequences achieved by amplicon pyrosequencing at relatively low costs is impressive, but limitations, such as limitations of available sequences in databases, remain (15, 16). The 454 amplicon sequencing of *trfA* genes amplified from the TC-DNA of samples taken from an on-farm BPS showed changes in the relative abundance of different IncP-1 subgroups over the agricultural season. While the relative abundance of IncP-1ε plasmids decreased over the season, the relative abundance of IncP-1β plasmids increased (15). These results strongly point to an enrichment of populations carrying IncP-1β plasmids that code for enzymes likely involved in the degradation of pesticides under field conditions. Other studies showed a high abundance of plasmids in different types of environments such as manure, manure-treated soils, river water sediments, and sea sediments by PCR Southern blot hybridization (17–19) or quantitative real-time PCR (qPCR) (20, 21).

qPCR became an important tool in plasmid ecology such as for the elucidation of environmental factors influencing the relative abundance of plasmids in microbial communities. Although the enormous diversity of plasmid groups is certainly not covered by the primer systems available, recent examples have demonstrated the applicability of this tool in ecological studies (22–24). Different biotic and abiotic factors can influence the relative abundance of plasmids and their hosts, and these factors can now be investigated in experiments comprising sufficient replicates and proper controls. This type of experiment is needed to test hypotheses and confirm correlations. For example, a study by Smalla et al. (18) investigated the effects of mercury pollution on the abundance of IncP-1 plasmids in river sediments. Replicated river sediment samples were taken along a gradient of mercury pollution and, although different hybridization signal intensities were only semi-quantitative, PCR Southern blot analysis indicated that the abundance of the mercury resistance gene *merRTP* and the IncP-1-specific *trfA* gene correlated with the concentration of mercury pollution in the sediment samples. However, the authors could not exclude the possibility that additional factors could have influenced the bacterial community composition and the relative abundance of IncP-1-carrying bacterial cells.

Another example is the study by Jechalke et al. (23) investigating plasmids in on-farm BPSs over the agricultural season. This study showed that the relative abundance of IncP-1 plasmids, which was quantified by qPCR with primers targeting IncP-1 *korB*, increased and that this effect was correlated with an increased concentration of a wide diversity of pesticides. The *korB* gene encodes a protein involved in the partitioning system and the regulatory network of the plasmid (25, 26). As no controls were available, no causal relationship could be demonstrated. However, in microcosm experiments in which linuron was added to BPS material the relative abundance of IncP-1 plasmids was significantly increased compared to the controls, providing evidence for the previously assumed correlation (S. Dealtry et al., in preparation). It was also shown by qPCR that the application of manure containing antibiotics to arable soils increased the relative abundance of plasmids and class 1 integrons in the rhizosphere and in bulk soil (27, 28).

Furthermore, it is important to note that the abundance of plasmid-carrying populations can increase in response to various triggers, e.g., root exudates, as recently shown by Jechalke et al. (22). In the rhizosphere of lettuce grown in three different soil types a significantly increased relative abundance of the IncP-1 *korB* gene was found, while in the rhizosphere of potato plants grown in the same type of soil no enrichment of IncP-1 plasmids was observed by PCR Southern blot. The 454 pyrosequencing of 16S rRNA gene fragments amplified from TC-DNA of lettuce rhizosphere samples revealed that several genera were enriched that are known to include strains with the potential to degrade aromatic compounds (29), and the detection of aromatic compounds in the root exudates of lettuce supported the assumption that degraders of these molecules were enriched (30). Most likely, the bacterial populations that were enriched in response to the lettuce root exudates carried IncP-1 plasmids with degradative genes. To test this hypothesis it will be necessary to isolate the IncP-1 plasmids either by a traditional cultivation approach or by exogenous isolation.

In summary, the detection and quantification of plasmid-related sequences in TC-DNA allows for the first time surveys of the dissemination of plasmid-specific sequences and, moreover, makes it possible to link plasmid abundance with environmental factors and pollutant concentrations. However, experiments with treatments and controls performed in a sufficient number of independent replicates are needed to unequivocally test interdependencies.

PLASMID GENOME SEQUENCING

Comparative analysis of whole-plasmid genome sequences is rapidly becoming a standard approach to increase our understanding of the genetic diversity and evolutionary history of plasmids. In November 2014, a total of 4,638 complete plasmid genome sequences are available in Genbank. Just over 6 years earlier, their number was only 1,490. Many of them have been sequenced as part of entire bacterial genomes, but an almost equal number were submitted as plasmid genomes. The data not only help to define the molecular events that took place during the evolution of these plasmids, but also give us a more complete overview of the enormous collection of accessory genes encoded on plasmids. Future work will have to focus on annotating many of these genes that still have unknown or hypothetical functions.

Methods to determine the complete genome sequence of plasmids have evolved very rapidly over the last decade. To illustrate with an example, in a joint project to determine the genome sequence of 100 broad host range plasmids with the U.S. Department of Energy Joint Genome Institute (JGI, Walnut Creek, CA), the institute used three different methods during the lifetime of the project. While in 2008 the plasmids were still being sequenced using Sanger sequencing of ~3-kb clone libraries, the following year a switch was made to the Roche/454 platform with GS FLX Titanium Sequencing chemistry and, soon thereafter, to Illumina sequencing technology (6, 31). As a consequence of the novel sequence technology, the cost of DNA sequencing per nucleotide has drastically decreased in recent years. Particularly for plasmids, which have smaller genomes than bacterial chromosomes but require the same library preparation procedure, library preparation is relatively expensive compared to the actual sequencing run. Moreover, improved bioinformatics pipelines are now available to separate chromosomal from plasmid DNA sequences. This has allowed us and others to simply determine the entire bacterial genome sequence and bioinformatically extract the plasmid genome sequence rather than invest a lot of labor and cost in purifying the plasmid DNA. Previously, for many bacteria other than *Escherichia coli*, labor-intensive large-scale plasmid extraction methods were required to generate sufficient plasmid DNA low in contaminating chromosomal DNA. Today we can perform rapid total-genomic DNA extractions instead. This approach works especially well in resequencing projects where the wild-type bacterial strain with its plasmid(s) has been completely sequenced previously, and the researcher wants to learn about the genetic changes after experimental evolution or other genetic manipulations. The method also works well for *de novo* plasmid sequence analysis when some information on the plasmids of interest is already available, and especially when the plasmid has first been transferred out of its native host into a previously sequenced laboratory strain.

For *de novo* sequencing projects of truly novel plasmids in their native host where the research question requires a correct assembly, we still recommend at least to enrich if not to purify the plasmid DNA. Very often, manually closing the sequence gaps between contigs by PCR amplification and subsequent Sanger sequencing of the PCR product is required. It is also highly advisable to compare experimental restriction fragment length patterns of the plasmid DNA with *in silico* digests, to confirm correct assembly. In our experience, with plasmids that have a well-known backbone structure, we have encountered a few cases where automated assembly was incorrect. This was even the case for a single IncP-1 plasmid, where plasmid DNA had been purified but a large duplication impeded the correct assembly of the sequence without additional experimental work (32). In general, large duplications make it nearly impossible to correctly determine a plasmid genome sequence by short read sequencing such as Illumina HiSeq and MiSeq. Methods such as SMRT (single-molecule real-time sequencing) applied by Pacific Biosciences (http://www.pacificbiosciences.com) are very promising for plasmid genome sequencing, as fragments as long as 40 kb can be sequenced in one read. For many plasmids, one to three reads could thus be sufficient, and assembly is no longer an issue. A recent example is the complete genome sequencing in a single run of a *Klebsiella pneumoniae* strain with four plasmids that encode NDM-1 and oxa-232 carbapenemases and other types of drug resistance (12).

When plasmids do not have a marker gene that can be used in the laboratory to select for the presence of the plasmid, we and others have marked the plasmid with a mini-transposon. The first examples were described by Tauch et al. (33) and Gstalder et al. (34) for the plasmids pIPO2 and pMOL98. The drawback of the marking is that the obtained genome sequence does not represent the native, unmarked plasmid. In more recent work, the DNA sequences of the mini-transposons were removed from each plasmid sequence to reconstruct the genome sequences of the originally captured plasmids. By confirming the presence of repeats at each side of the transposon, we knew that the transposition had not caused deletions of flanking regions (31).

Annotation of plasmids is still a great challenge due to the confusing nomenclature of backbone genes, which is different for different plasmid incompatibility groups, and due to the large numbers of accessory genes with unknown or hypothetical functions. Automatic annotation is a first step but always requires a follow-up by careful manual annotation. In our previous plasmid sequencing projects, automatic annotation was carried out by the J. Craig Venter Institute Annotation Service (www.jcvi.org) or the Institute for Genome Sciences (IGS) (http://www.igs.umaryland.edu). Problems with the annotation of plasmid-encoded genes have been recently discussed by plasmid biology experts (35), but the debate continues.

DETECTION OF PLASMID-SPECIFIC SEQUENCES IN METAGENOMES AND METAMOBILOMES

With the rapidly advancing sequencing technologies, deep sequencing of metagenomes or plasmidomes has opened a new path to discover mobile genetic elements by means of bioinformatics tools. Recently, the first studies have been published that attempted to describe plasmid genome diversity within a natural microbial community based on metagenomic approaches. In short, genomic or plasmid DNA is extracted directly from the sample lysates, for example, by CsCl density gradient centrifugation, or bacterial populations are first isolated from the samples, which can be followed by removal of linear chromosomal DNA and a nonspecific amplification step to increase the amount of circular plasmid DNA (36–38).

Plasmid extraction and purification directly from the microbial cell fraction of a sample by alkaline lysis, followed by size-dependent DNA separation methods such as ultracentrifugation or column-based binding assays, was proposed by Li et al. (39) as a straightforward approach to studying plasmids in environmental samples. To minimize contamination with chromosomal fragments, sheared genomic and linear plasmid DNA is degraded by a plasmid-safe ATP-dependent DNase, residual plasma DNA is amplified by 29 DNA polymerase to ensure optimal amounts (micrograms), and subsequently sequenced. Although traditional cultivation biases were avoided and a large richness of previously unknown replicons was revealed, the method was found to favor small circular plasmids, most likely due to the employed multiple displacement amplification. The finding was confirmed using pKJK10 and pBR322 as model plasmids (38). The inclusion of an additional electroelution step provided improved coverage of larger plasmids and thus could make the method more suitable to study the plasmid content of environmental bacteria. However, this method will obviously exclude the detection of linear plasmids.

Metagenomics will definitely shed more light on the diversity of plasmids and the accessory genes they carry, but due to the mosaic nature of plasmids with similar or identical sequences of considerable length, obtaining correctly closed genomes of large plasmids remains a challenge for complex microbial communities (see also previous section).

Another drawback of metagenomics methods is the lack of information on plasmid-host associations, just as for the plasmid capture methods described in the following section. Although an enormous diversity of novel and known resistance genes was revealed, based on metagenomic information, it is currently impossible to determine which plasmid belongs to which host. Interestingly, promising techniques are being tested to physically link plasmid DNA in one cell to parts of chromosomal DNA, based on Hi-C. This is a method that relies on cross-linking molecules in close physical proximity and consequently identifies both, thus reflecting the spatial arrangement of DNA at the time of cross-linking within cells (40, 41) (Fig. 1). Proof of principle was recently shown by Beitel et al. (42).

Another culture-independent method to isolate novel plasmids from microbial communities is the transposon-aided capture method (TRACA) (43). In brief, this method uses purified plasmid DNA extracted from bacterial cells, cell cultures, or environmental samples, which is then amended with an EZ-*Tn5 OriV* Kan2 transposon and a transposase for the *in vitro* transposition reaction. Following the transposition reaction, the transposition reaction mix is diluted, purified, and electroporated into *E. coli* EPI300 cells. Transformants are plated on Luria Bertani broth containing 50 g/ml kanamycin to select for EZ-*Tn5*, and captured plasmids can be further investigated or sequenced. Using this method, plasmids were successfully captured and characterized from different environments such as the human gut, activated sludge from a sewage treatment plant, and human dental plaque associated with periodontal disease (43–45). The advantage of the TRACA method is that plasmids from Gram-positive and Gram-negative species can be captured without the requirement for selectable markers, mobilization, or conjugative functions. However, the results suggest that small plasmids (<10 kb) are preferentially isolated, which might be due to a lower copy number and lower transformation frequency of larger plasmids (43–45).

In the future a combination of different plasmid isolation, sequencing, and bioinformatics methods will be needed to improve our view of the diversity and accessory gene content of these important mobile genetic elements in the horizontal gene pool.

EXOGENOUS CAPTURING OF PLASMIDS BY MEANS OF BIPARENTAL AND TRIPARENTAL MATINGS

These methods allow the capture of conjugative as well as mobilizable plasmids from environmental bacteria without the need to culture their original host. The bacterial communities associated with an environmental sample are mixed with recipient cells, and after a filter mating on nonselective agar the cells are resuspended and plated on media containing rifampicin, kanamycin

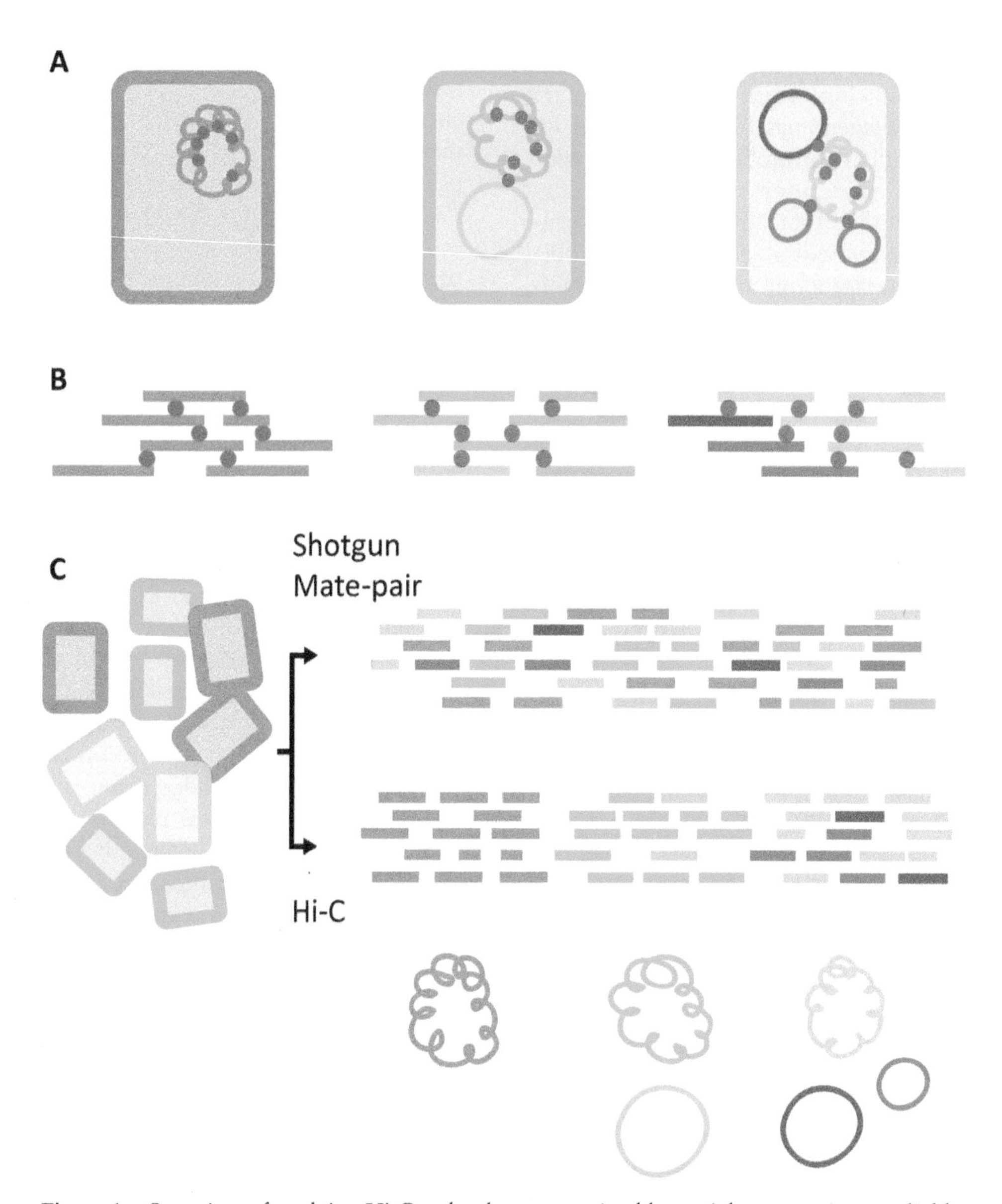

Figure 1 Overview of applying Hi-C technology to a mixed bacterial community to reliably associate plasmids with the chromosomes of their hosts (modified from Burton et al. [41]). (A) Rectangles indicate different cells carrying plasmids or not. Plasmids are cross-linked with bacterial chromosomes in close proximity (red circles). (B) The DNA in the cross-linked protein complexes is digested with *Hind*III endonuclease following cell lysis, and free DNA ends are tagged with biotin. After ligation of blunt-ended DNA fragments under highly dilute conditions, which preferentially ligates fragments that are within the same cross-linked DNA/protein complex, cross-links are removed, DNA is purified, biotin is eliminated from unligated ends, DNA is size-selected, and ligation products are selected for through a biotin pull-down. The resulting Hi-C library is further analyzed by sequencing. (C) Workflow to create individual species/plasmid assemblies from a metagenome sample by combining shotgun, Hi-C, and (optionally) mate-pair libraries.
doi:10.1128/microbiolspec.PLAS-0038-2014.f1

(to select for the recipient), and antibiotics or heavy metals to which the recipient is sensitive (Fig. 2A). However, the few cells which received a mobile genetic element from the indigenous donor bacteria which confer the types of resistance required form colonies on plates, and the transfer frequencies are in general given as the quotient of transconjugant and recipient numbers. The successful detection of transconjugants depends on the transfer and replication range as well as the presence and expression of the respective resistance or degradative functions. In contrast to the biparental matings, the so-called triparental matings involve a second donor carrying a small mobilizable IncQ plasmid, and the plasmid capturing is exclusively based on plasmid mobilizing capacity (Fig. 2B). Thus, both methods can be seen as cultivation-independent methods, although a successful transfer event requires cells of sufficient metabolic activity, as conjugative type IV secretion (T4S) to translocate from a donor to a recipient bacterium is an energetically costly process (46). A disadvantage of both capturing methods is that the original host remains unknown. But both methods allow the capture of novel types of plasmids as shown for plasmids belonging to the PromA group (33, 47, 48) or the so-called low G+C plasmid family (49, 50).

A range of Gram-negative recipients have been successfully used in exogenous plasmid-capturing experiments. Very often, independent from the environmental sample analyzed, exogenously captured plasmids belonged to different types of IncP-1 plasmids. The frequent isolation of IncP-1 plasmids is remarkable, as their relative abundance is typically low, with the exception of some hot spots (e.g., sewage or BPS). However, many other broad and narrow host range plasmids have also been isolated with this method and subsequently characterized. Biparental matings with *E. coli gfp* as the recipient strain were used to monitor the effect of the presence of antibiotics in manure added to soil on the frequency of capturing transferable antibiotic resistance (51). Transfer frequencies from soils that did not receive manure were very low or below detection. In the presence of manure the transfer frequencies were several orders of magnitude higher compared to mating with control soil or the piggery manure. Interestingly, two months after the treatments from the sandy soil, transconjugants were only captured when the manure was spiked with sulfadiazine (51). Thus, from this type of experiment, we could once again show the importance of nutrient availability and selective pressure. Obviously, the latter depends on the physico-chemical properties of the antibiotic, as these determine the interaction with the different anorganic and organic components of a soil or sediment sample (24). It remains to be determined whether the higher plasmid-capturing frequencies were observed because of higher transfer rates in the presence of antibiotics or because of the higher abundance of potential plasmid donors. In soil microcosms and in mesocosms planted with maize, or in field experiments with maize and grass, the majority of the plasmids, which were captured based on the sulfadiazine resistance conferred, belonged to the low G+C plasmids and the IncP-1ε group (28, 49, 50, 52).

When different recipient strains were used to exogenously capture plasmids, typically distinct sets of plasmids were obtained (53, 54). An exception was the capturing of similar IncP-1 plasmids in recipient strains that belonged to three different classes of the *Proteobacteria* (*Agrobacterium tumefaciens*, *Cupriavidus necator*, *Pseudomonas putida*, *E. coli*) from sewage sludge in Belgium (17). Thus, the type of plasmids captured in recipients obviously depends on the type and abundance of plasmid donor strains, the plasmid transfer, and the replication range but might be also influenced by the mating conditions (liquid or surface mating, duration, temperature, O_2 availability, pH, chemotaxis). Variations of these factors might favor the transfer/capture of distinct sets of plasmids, as they influence the metabolic activity of the plasmid donor and host and the plasmid encoded transfer apparatus. Thus, exogenous plasmid isolation done in surface mating will obviously miss IncF and IncI types of plasmids that transfer better in broth matings. These plasmids seem to be very important for antibiotic resistance and pathogenicity in *E. coli*.

ANALYSES OF THE PLASMID CONTENT IN BACTERIAL ISOLATES

Several chapters of this book discuss the in-depth characterization of plasmids from bacterial isolates such as enteropathogenic *E. coli* strains. That is why this article focuses on some aspects of the epidemiology and ecology of plasmids in cultivable bacteria. There is no doubt that plasmid ecology is tightly linked to the ecology of the host, which remains unknown when plasmid occurrence and abundance is studied by TC-DNA extraction. We picked some very recent studies to discuss the tools used for plasmid typing and diversity studies. Isolates were obtained after selective plating on nutrient media supplemented with a range of antibiotics. In other studies the plasmids originated from human, animal, or plant pathogenic strains or from isolates with degradative functions. Many of the studies, for obvious

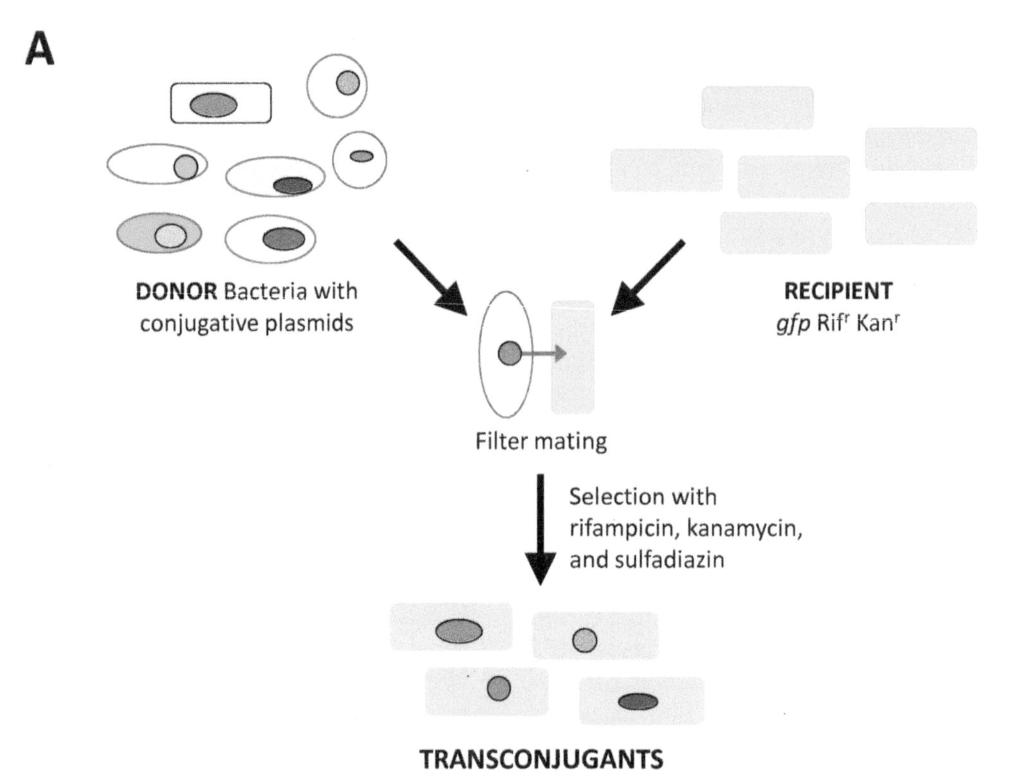
A
DONOR Bacteria with conjugative plasmids
RECIPIENT
gfp Rif^r Kan^r
Filter mating
Selection with rifampicin, kanamycin, and sulfadiazin
TRANSCONJUGANTS
with plasmids carrying sulfonamide resistance

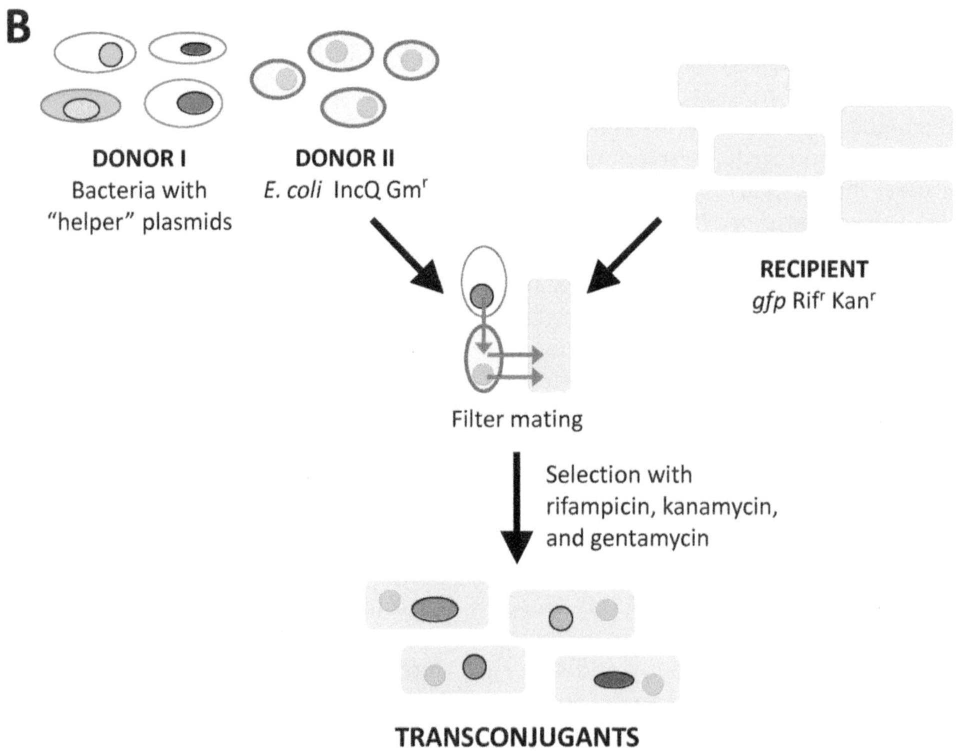
B
DONOR I
Bacteria with "helper" plasmids
DONOR II
E. coli IncQ Gm^r
RECIPIENT
gfp Rif^r Kan^r
Filter mating
Selection with rifampicin, kanamycin, and gentamycin
TRANSCONJUGANTS
carrying IncQ and mobilizing plasmid

reasons, focus on the detection of antibiotic resistance genes and their genetic context either by using a functional genomics approach or by investigating single isolates. While the fraction of bacteria forming colonies on plates likely does not represent the complete plasmid content of a given sample, the great advantage of the cultivation-based approach is that the localization of antibiotic resistance genes on the chromosome (integrons, integrative and conjugative elements [ICE]) or on plasmids and the genetic context of the acquired gene load can be analyzed. The presence and diversity of plasmids is traditionally assessed by plasmid DNA extraction and restriction analysis. To date, the initial plasmid typing is done by PCR with primers targeting either plasmid replication (10, 13, 55) or transfer regions (56). More details on relaxase-based typing are given in reference 57. Plasmid multilocus sequence typing is used for a few plasmid families within the *Enterobacteriaceae*, e.g., IncI1 plasmids (http://pubmlst.org/plasmid/).

With prices for sequencing decreasing further, it is likely that future plasmid typing will be done exclusively by sequencing. Moreover, the reconstruction of plasmid content from whole-genome sequencing data sets recently became feasible thanks to tools such as PLACNET (plasmid constellation networks). This tool was recently employed to analyze the plasmidome of 10 *E. coli* lineage ST131 strains from whole-genome sequencing data sets (75). The study showed that *E. coli* ST131 strains, which are practically identical in their core genomes, contain a striking variety of different plasmids including IncF and IncI.

Amos et al. (58) isolated *Enterobacteriaceae* resistant to extended-spectrum beta-lactams from river sediments, taken at two time points upstream and downstream of a sewage treatment plant. Third-generation cephalosporin-resistant *Enterobacteria* were enumerated and screened by PCR for bla_{TEM}, bla_{SHV}, and $bla_{CTX\text{-}M}$. Isolates which were positive for CTX-M-15 were subjected to conjugation to a new *E. coli* background. These authors reported for the first time the occurrence of CTX-M-15 localized on IncF and IncI type plasmids in enterobacterial isolates from sediments. Several novel genetic contexts flanking the CTX-M-15 gene were discovered with many mobile genetic elements such as IS*Ecp1* and IS26 (58). Although the number of samples investigated was limited, the data suggested that the sewage treatment process increased the abundance of $bla_{CTX\text{-}M\text{-}15}$-carrying *Enterobacteriaceae* in river sediments.

Many Gram-negative plant pathogenic bacteria carry genes involved in the interaction with plants, localized on pathogenicity islands or plasmids. We recently characterized plant pathogenic *Pseudomonas savastanoi* strains isolated from different host plants and noticed that these strains could not be differentiated based on their 16S rRNA gene sequence and their BOX fingerprints. However, the isolates clearly differed in their plasmid content. Based on their *repA* sequence, the plasmids were assigned to the pAT family (59).

As described by de Toro et al. (57), more and more plasmids are being discovered and assembled from whole-genome sequences of isolates. The assembly of plasmid replicons is particularly challenging due to the short read length of Illumina sequences and the high abundance of IS, IS*CR*, transposons, and truncated sequences due to multiple insertion processes. Bioinformatics tools have been developed which will assist not only in the assembly of novel plasmids and comparative analyses, but also in their visualization. Publicly available web-based tools are urgently needed to manage the enormous plasmid-related sequence data sets. A few recently mentioned tools, in addition to PLACNET, introduced in reference 57 are Seqfindr and plasmid barcodes, which were recently used for rapid plasmid profiling and comparison of plasmids encoding CTX-M-15 beta-lactamase genes from globally disseminated *E. coli* ST131 strains.

PLASMID HOST RANGE DETERMINATION AND PLASMID COSTS AND BENEFITS

The host range of a plasmid is thought to be a key parameter that determines plasmid ecology (60). While traditionally plasmid host range studies were done by mating with a few selected recipient strains under optimal conditions, a comparative plasmid host range determination in the rhizosphere of grass was undertaken

Figure 2 Exogenous capturing of plasmids by means of (A) biparental and (B) triparental mating. For biparental mating, environmental bacteria are mixed with recipient cells, and after a filter mating the cells are resuspended and plated on media containing rifampicin (Rif), kanamycin (Kan) (to select for the recipient), and antibiotics or heavy metals to which the recipient is sensitive. Triparental matings involve a second donor carrying a small mobilizable IncQ plasmid, and the plasmid capturing is exclusively based on the plasmid-mobilizing capacity. To facilitate the identification of transconjugants the recipient is labeled with the green fluorescent protein (*gfp*). doi:10.1128/microbiolspec.PLAS-0038-2014.f2

by Pukall et al. (61). In this study *E. coli* donor strains carrying IncP-1 (pTH10, pTH16, RP4), IncQ (pIE639), IncN (pIE1037), IncW (pIE1056), IncI1 (pIE1040), or IncFII (pIE1055) plasmids, all carrying the Tn7-like transposon Tn*1826*, were applied together with an *E. coli* recipient strain into soil. The highest transfer frequencies were observed for IncP-1, followed by IncN and IncW, whereas under the same conditions no transfer was detected for IncI1 and IncFII. Fifty days after the *E. coli* donor pTH16 and recipients were introduced into nonsterile soil, rhizosphere bacteria that captured the nourseothricin resistance plasmid were isolated and identified by BIOLOG as *Agrobacterium*, *Pseudomonas*, and *Flavobacterium*. A similar approach was taken to identify the host range of the IncP-1ε pHH3414 in soil microcosms planted with *Acacia caven* (62). In this experiment soil bacteria that received pHH3414 were identified by 16S rRNA gene sequencing as *Gammaproteobacteria* (*Enterobacter amnigenus*, *Xanthomonas codiaei*) and *Betaproteobacteria* (*Cupriavidus campinensis*, *Alcaligenes* spp.). Soil microcosm experiments were also used by Goris et al. (36) to determine the host range of the catabolic plasmids pJP4 and pEMT1, conferring the ability to degrade the herbicide 2,4 dichlorophenoxyacetic acid. Recipients were mainly identified as *Burkholderia* species when no additional nutrients were added, while the amendment of the soils with nutrients resulted in additional transconjugants identified as *Stenotrophomonas* species. Thus, again the hosts of both plasmids comprised *Beta-* and *Gammaproteobacteria*.

Fluorescent marker–tagged plasmids have been used in different studies to elucidate the host range of plasmids. De Gelder et al. (63) showed for another *rfp*-marked IncP-1 plasmid (pB10) that the recipients among sewage sludge obtaining the plasmid depended on the donor (*Ensifer meliloti*, *C. necator*, *Pseudomonas putida*) and the mating conditions (liquid versus plate matings). Recently, a molecular approach based on qPCR was presented to detect rare plasmid transfer events following a low initial pB10-donor inoculation to environmental samples, indicating that eukaryotic predation can affect plasmid transfer events (64). Surprisingly, in the study by De Gelder et al. (63) the recipients of pB10 were affiliated with *Alpha-* and *Gammaproteobacteria*. Also, in studies by Musovic et al. (65) *P. putida* KT2440::*lacI*q1, harboring the IncP-ε conjugative plasmid pKJK10, was used to determine its host range in soil-barley microcosms. Plasmid pKJK10 had the *gfp*(*mut3b*) gene inserted downstream of the *lac*-repressible promoter *PA1-04/03*. Because this promoter is *LacI* repressed, the inserted *gfp* gene is silent in the donor strain. After seven days the bacterial fraction, collected by a Nycodenz centrifugation step, was subjected to cell sorting by flow cytometry. Transconjugants were obtained after cell sorting. 16S rRNA gene fragments amplified from *gfp*-positive cells revealed a broad diversity of transconjugants affiliated with *Alpha-*, *Beta-*, and *Gammaproteobacteria* and, most remarkably, with *Actinobacteria* (*Arthrobacter*). In a follow-up study Musovic et al. (66) developed a minimal-cultivation approach in combination with zygotic fluorescence expression and microscopy to quantify the recipient fraction of a soil bacterial community permissive to the *gfp*-tagged IncP-1α plasmid RP4. The authors showed that the host range, which again comprised *Alpha-*, *Beta-*, and *Gammaproteobacteria*, varied strongly depending on the nutrient media used.

More recently, cultivation-independent methods have been used to assess the *in situ* host range of plasmids. Klümper et al. (60) combined high-throughput cell sorting of donor and transconjugant and used 454 sequencing of 16S rRNA gene fragments amplified from the transconjugant pools to determine the diversity of plasmid recipients in soil bacterial communities under conditions optimized for cell-to-cell contact. The plasmids RP4 (IncP-1α), pKJK5 (IncP-1ε), and pIPO2 (pPromA) were shown to have an unexpectedly broad transfer range, and more than 300 transconjugant operational taxonomic units (OTUs) were detected. Transconjugants not only comprised all proteobacterial classes but also included diverse members of 10 additional phyla including *Verrucomicrobia*, *Bacteriodetes*, and *Actinobacteria*. Thus, several studies often showed a transfer range beyond the Gram-negative bacteria. It remains to be seen whether all these transconjugants also replicate the plasmid as a separate mobile element or if integration in the chromosome allowed some or all of the genes to persist in these novel diverse hosts.

This finding was confirmed by subsequent studies using the same technology, but Shintani et al. (67) investigated the conjugative transfer ranges of three different *gfp*-tagged plasmids of the incompatibility groups IncP-1 (pBP136), IncP-7 (pCAR1), and IncP-9 (NAH7) in soil bacterial communities using both cultivation-dependent and cultivation-independent methods. GFP-expressing transconjugants sorted by flow cytometry were characterized by sequencing of 16S rRNA genes following either whole-genome amplification or cultivation. In accordance with other studies, the recipients of pBP136 belonged to diverse species within the phylum *Proteobacteria* as revealed by both culture-dependent and culture-independent methods. Transconjugants belonging to the phyla *Actinobacteria*,

Bacteroidetes, and *Firmicutes* were detected only by the culture-independent method. Furthermore, the transconjugants of pCAR1 and NAH7 were identified by both methods as *Pseudomonas*, indicating a rather narrow host range of the plasmids. The cultivation-independent methods indicated that *Delftia* species (class *Betaproteobacteria*) were "transient" hosts of pCAR1. Thus, several studies showed that the transfer range of different plasmids studied might be far broader than previously expected, which has implications for the spread of plasmid-encoded accessory genes. This was already known from matings in the laboratory (68) and is now confirmed for *in situ* plasmid spread. Additionally, it was shown that the permissiveness of a bacterial community, and even of isolates, to receive and maintain plasmids can change by several orders of magnitude in response to irregular environmental change such as manure fertilization (62, 69).

It is well known that most plasmids impose a cost on their host when their accessory genes are not providing a benefit to that host, such as antibiotic resistance plasmids in the absence of antibiotics. In addition to the more obvious cases of fitness due to expression of costly proteins, recent studies are also trying to understand the more subtle interactions between plasmids and their hosts at the molecular level. For example, the impact of plasmid pCAR1 carriage on the expression of chromosomal genes of three different *Pseudomonas* hosts was investigated by transcriptome analyses of plasmid-free and pCAR1-carrying strains (11, 70). The transcriptomic studies revealed that carriage of the IncP-7 plasmid pCAR1 altered the gene expression of iron acquisition in all strains, likely due to the expression of the carbazole degradative genes carried on pCAR1. In *P. putida* KT2440 the expression of the pyoverdine gene was higher in the plasmid-carrying strain, and genes involved in SOS response (*lexA*, *recA*) and genes on a putative prophage were upregulated due to plasmid carriage. Furthermore, an increased chloramphenicol resistance phenotype was observed. *In vitro* evolution analyses showed that the fitness cost of pCAR1 might be partially due to the pCAR1-induced constitutive expression of chromosomal genes and iron shortage induced due to *car* gene expression (H. Nojiri et al., in preparation).

In addition to the many competition studies performed in liquid cultures, recent work has shown the effect of antibiotic presence on the competitiveness of plasmid-bearing strains in soil. In two independent competition experiments with plasmid-free and plasmid-carrying *Acinetobacter baylyi* BD413 performed in soil microcosms, Jechalke et al. (71) demonstrated that the low G+C-type plasmid pHHV216 conferred a fitness advantage to *A. baylyi* BD413 in soils that had received manure spiked with the sulfonamide antibiotic sulfadiazine (SDZ). Plasmid carriage had a disadvantage when the selective pressure by SDZ was absent. A very interesting study recently showed that even sublethal levels of antibiotics, far below the MIC, selected for plasmid-bearing strains in laboratory competitions with plasmid-free isogenic strains (72). These findings suggest that even when growth inhibition is not detected by the traditional MIC assays, the antibiotic impedes growth enough to allow the drug resistance plasmid–bearing strain to outcompete its plasmid-free counterpart. To determine the true costs and benefits of plasmids to their hosts, more studies need to be performed in the hosts' natural environments and at various concentrations of selecting agents.

At the rate at which new plasmid sequence information is being released, it is no longer possible to empirically test the host range of all newly described plasmids. Therefore, genomics-based methods may be used to predict the likely host and host range of a specific plasmid based on its DNA sequence. Bacteria clearly differ in the relative abundance of di-, tri-, or tetranucleotides (also referred to as their genomic sequence signature), and it appears that plasmids that have a long-term association with hosts of a similar signature tend to acquire that signature (73). In contrast, broad host range plasmids thought to move between distantly related bacteria show distinct genomic signatures (74). Therefore, we can now easily examine the genomic signature of uncharacterized plasmids and infer their likely host or host range. While in the case of a well-characterized plasmid, the method often accurately predicts its likely hosts, more experimental validation is needed for some novel key plasmids that are being sequenced as part of genome and metagenome projects.

CONCLUSIONS

During the past two decades, significant progress has been made in our ability to detect and quantify specific groups of plasmids in environmental and clinical samples and to determine and compare their entire genome sequences. This expansion in studies and methods has greatly increased our understanding of the diversity of plasmids that exist, even within already well-studied groups such as the IncP-1 plasmids. While until 2004 only two IncP-1 subgroups had been described, there are now, just a decade later, likely seven phylogenetically distinct clades (5, 31). We expect that the fast pace of development of new genome sequencing methods and creative approaches such as applying Hi-C to metagenomic studies will further enhance our insight into the important role of plasmids in the rapid adaptation of microbial communities to man-made changes in their environment. Obvious examples of such environmental changes are the increased presence of antibiotics, heavy metals, and xenobiotics, but also increased worldwide travel, high densities of food animals and people, exploration of new territories, and climate change, which likely affect the spread of bacterial pathogens and their often mobile virulence genes. There are plenty of challenges ahead of us, but given the progress made in the past decades, these are exciting times for plasmid ecology.

Acknowledgments. *S.J. was funded by the Federal Environment Agency (Umweltbundesamt) (FKZ 3713 63 402). We are also grateful for funding from NIH grant no. R01 AI084918 from the National Institute of Allergy and Infectious Diseases (NIAID) and DOD grant DM110149 to E.M.T.We thank Wesley Loftie-Eaton for drawing Fig. 1 and Thibault Stadler for useful comments. Conflicts of interest: We declare no conflicts.*

Citation. Smalla K, Jechalke S, Top EM. 2015. Plasmid detection, characterization, and ecology. Microbiol Spectrum 3(1): PLAS-0038-2014.

References

1. **Heuer H, Abdo Z, Smalla K.** 2008. Patchy distribution of flexible genetic elements in bacterial populations mediates robustness to environmental uncertainty. *FEMS Microbiol Ecol* **65:**361–371.
2. **Heuer H, Smalla K.** 2012. Plasmids foster diversification and adaptation of bacterial populations in soil. *FEMS Microbiol Rev* **36:**1083–1104.
3. **Djordjevic SP, Stokes HW, Roy Chowdhury P.** 2013. Mobile elements, zoonotic pathogens and commensal bacteria: conduits for the delivery of resistance genes into humans, production animals and soil microbiotia. *Front Microbiol* **4:**86. doi:10.3389/fmicb.2013.00086.
4. **Schlüter A, Szczepanowski R, Pühler A, Top EM.** 2007. Genomics of IncP-1 antibiotic resistance plasmids isolated from wastewater treatment plants provides evidence for a widely accessible drug resistance gene pool. *FEMS Microbiol Rev* **31:**449–477.
5. **Norberg P, Bergström M, Jethava V, Dubhashi D, Hermansson M.** 2011. The IncP-1 plasmid backbone adapts to different host bacterial species and evolves through homologous recombination. *Nat Commun* **2.** doi:10.1038/ncomms1267.
6. **Sen D, Brown CJ, Top EM, Sullivan J.** 2012. Inferring the evolutionary history of IncP-1 plasmids despite incongruence among backbone gene trees. *Mol Biol Evol* [Epub ahead of print.] doi:10.1093/molbev/mss210.
7. **Staley JT, Konopka A.** 1985. Measurement of *in-situ* activities of nonphotosynthetic microorganisms in aquatic and terrestrial habitats. *Annu Rev Microbiol* **39:**321–346.
8. **Amann RI, Ludwig W, Schleifer KH.** 1995. Phylogenetic identification and *in situ* detection of individual microbial cells without cultivation. *Microbiol Rev* **59:**143–169.
9. **Oliver JD.** 2010. Recent findings on the viable but nonculturable state in pathogenic bacteria. *FEMS Microbiol Rev* **34:**415–425.
10. **Dealtry S, Ding GC, Weichelt V, Dunon V, Schlüter A, Martini MC, Del Papa MF, Lagares A, Amos GCA, Wellington EMH, Gaze WH, Sipkema D, Sjöling S, Springael D, Heuer H, van Elsas JD, Thomas C, Smalla K.** 2014. Cultivation-independent screening revealed hot spots of IncP-1, IncP-7 and IncP-9 plasmid occurrence in different environmental habitats. *PLoS One* **9.** doi: 10.1371/journal.pone.0089922.
11. **Shintani M, Takahashi Y, Yamane H, Nojiri H.** 2010. The behavior and significance of degradative plasmids belonging to Inc groups in *Pseudomonas* within natural environments and microcosms. *Microbes Environ* **25:**253–265.
12. **Dennis JJ.** 2005. The evolution of IncP catabolic plasmids. *Curr Opin Biotechnol* **16:**291–298.
13. **Bahl MI, Burmølle M, Meisner A, Hansen LH, Sørensen SJ.** 2009. All IncP-1 plasmid subgroups, including the novel ε subgroup, are prevalent in the influent of a Danish wastewater treatment plant. *Plasmid* **62:**134–139.
14. **Götz A, Pukall R, Smit E, Tietze E, Prager R, Tschäpe H, Van Elsas JD, Smalla K.** 1996. Detection and characterization of broad-host-range plasmids in environmental bacteria by PCR. *Appl Environ Microbiol* **62:**2621–2628.
15. **Dealtry S, Holmsgaard PN, Dunon V, Jechalke S, Ding GC, Krögerrecklenfort E, Heuer H, Hansen LH, Springael D, Zühlke S, Sørensen SJ, Smalla K.** 2014. Shifts in abundance and diversity of mobile genetic elements after the introduction of diverse pesticides into an on-farm biopurification system over the course of a year. *Appl Environ Microbiol* **80:**4012–4020.
16. **Holmsgaard PN, Sørensen SJ, Hansen LH.** 2013. Simultaneous pyrosequencing of the 16S rRNA, IncP-1 *trfA*, and *merA* genes. *J Microbiol Methods* **95:**280–284.
17. **Heuer H, Krögerrecklenfort E, Wellington EMH, Egan S, van Elsas JD, van Overbeek L, Collard JM, Guillaume G, Karagouni AD, Nikolakopoulou TL, Smalla K.** 2002. Gentamicin resistance genes in environmental bacteria: prevalence and transfer. *FEMS Microbiol Ecol* **42:**289–302.
18. **Smalla K, Haines AS, Jones K, Krögerrecklenfort E, Heuer H, Schloter M, Thomas CM.** 2006. Increased abundance of IncP-1 beta plasmids and mercury resistance genes in mercury-polluted river sediments: first discovery of IncP-1 beta plasmids with a complex mer transposon as the sole accessory element. *Appl Environ Microbiol* **72:**7253–7259.
19. **Binh CTT, Heuer H, Kaupenjohann M, Smalla K.** 2008. Piggery manure used for soil fertilization is a reservoir for transferable antibiotic resistance plasmids. *FEMS Microbiol Ecol* **66:**25–37.
20. **Jutkina J, Heinaru E, Vedler E, Juhanson J, Heinaru A.** 2011. Occurrence of plasmids in the aromatic degrading bacterioplankton of the Baltic Sea. *Genes* **2:**853–868.
21. **Zhu Y-G, Johnson TA, Su J-Q, Qiao M, Guo G-X, Stedtfeld RD, Hashsham SA, Tiedje JM.** 2013. Diverse and abundant antibiotic resistance genes in Chinese swine farms. *Proc Natl Acad Sci USA* **110:**3435–3440. doi:10.1073/pnas.1222743110.
22. **Jechalke S, Schreiter S, Wolters B, Dealtry S, Heuer H, Smalla K.** 2014. Widespread dissemination of class 1

integron components in soils and related ecosystems as revealed by cultivation-independent analysis. *Front Microbiol* 4. doi:10.3389/fmicb.2013.00420.

23. **Jechalke S, Dealtry S, Smalla K, Heuer H.** 2013. Quantification of IncP-1 plasmid prevalence in environmental samples. *Appl Environ Microbiol* **79:**1410–1413.
24. **Jechalke S, Heuer H, Siemens J, Amelung W, Smalla K.** 2014. Fate and effects of veterinary antibiotics in soil. *Trends Microbiol* **22:**536–545.
25. **Rosche TM, Siddique A, Larsen MH, Figurski DH.** 2000. Incompatibility protein IncC and global regulator KorB interact in active partition of promiscuous plasmid RK2. *J Bacteriol* **182:**6014–6026.
26. **Herman D, Thomas CM, Stekel DJ.** 2011. Global transcription regulation of RK2 plasmids: a case study in the combined use of dynamical mathematical models and statistical inference for integration of experimental data and hypothesis exploration. *BMC Syst Biol* **5:**119. doi:10.1186/1752-0509-5-119.
27. **Jechalke S, Focks A, Rosendahl I, Groeneweg J, Siemens J, Heuer H, Smalla K.** 2013. Structural and functional response of the soil bacterial community to application of manure from difloxacin-treated pigs. *FEMS Microbiol Ecol* **87:**78–88.
28. **Kopmann C, Jechalke S, Rosendahl I, Groeneweg J, Krögerrecklenfort E, Zimmerling U, Weichelt V, Siemens J, Amelung W, Heuer H, Smalla K.** 2013. Abundance and transferability of antibiotic resistance as related to the fate of sulfadiazine in maize rhizosphere and bulk soil. *FEMS Microbiol Ecol* **83:**125–134.
29. **Schreiter S, Ding G-C, Heuer H, Neumann G, Sandmann M, Grosch R, Kropf S, Smalla K.** 2014. Effect of the soil type on the microbiome in the rhizosphere of field-grown lettuce. *Front Microbiol* **5:**144.
30. **Neumann G, Bott S, Ohler MA, Mock HP, Lippmann R, Grosch R, Smalla K.** 2014. Root exudation and root development of lettuce (*Lactuca sativa* L. cv. Tizian) as affected by different soils. *Front Microbiol* 5. doi:10.3389/fmicb.2014.00002.
31. **Brown CJ, Sen DY, Yano H, Bauer ML, Rogers LM, Van der Auwera GA, Top EM.** 2013. Diverse broad-host-range plasmids from freshwater carry few accessory genes. *Appl Environ Microbiol* **79:**7684–7695.
32. **Król JE, Penrod JT, McCaslin H, Rogers LM, Yano H, Stancik AD, Dejonghe W, Brown CJ, Parales RE, Wuertz S, Top EM.** 2012. Role of IncP-1 beta plasmids pWDL7::rfp and pNB8c in chloroaniline catabolism as determined by genomic and functional analyses. *Appl Environ Microbiol* **78:**828–838.
33. **Tauch A, Schneiker S, Selbitschka W, Pühler A, van Overbeek LS, Smalla K, Thomas CM, Bailey MJ, Forney LJ, Weightman A, Ceglowski P, Pembroke T, Tietze E, Schröder G, Lanka E, van Elsas JD.** 2002. The complete nucleotide sequence and environmental distribution of the cryptic, conjugative, broad-host-range plasmid pIPO2 isolated from bacteria of the wheat rhizosphere. *Microbiology* **148:**1637–1653.
34. **Gstalder ME, Faelen M, Mine N, Top EM, Mergeay M, Couturier M.** 2003. Replication functions of new broad host range plasmids isolated from polluted soils. *Res Microbiol* **154:**499–509.
35. **Frost LS, Thomas CM.** 2014. Naming and annotation of plasmids. *Mol Life Sci.* doi:10.1007/978-1-4614-6436-5_568-2.
36. **Goris J, Dejonghe W, Falsen E, De Clerck E, Geeraerts B, Willems A, Top EM, Vandamme P, De Vos P.** 2002. Diversity of transconjugants that acquired plasmid pJP4 or pEMT1 after inoculation of a donor strain in the A- and B-horizon of an agricultural soil and description of *Burkholderia hospita* sp nov and *Burkholderia terricola* sp nov. *Syst Appl Microbiol* **25:**340–352.
37. **Sentchilo V, Mayer AP, Guy L, Miyazaki R, Tringe SG, Barry K, Malfatti S, Goessmann A, Robinson-Rechavi M, Van der Meer JR.** 2013. Community-wide plasmid gene mobilization and selection. *ISME J* **7:**1173–1186.
38. **Norman A, Riber L, Luo WT, Li LL, Hansen LH, Sørensen SJ.** 2014. An improved method for including upper size range plasmids in metamobilomes. *PLoS One* 9. doi:10.1371/journal.pone.0104405.
39. **Li LL, Norman A, Hansen LH, Sørensen SJ.** 2012. Metamobilomics: expanding our knowledge on the pool of plasmid encoded traits in natural environments using high-throughput sequencing. *Clin Microbiol Infect* **18:**5–7.
40. **Umbarger MA, Toro E, Wright MA, Porreca GJ, Bau D, Hong SH, Fero MJ, Zhu LJ, Marti-Renom MA, McAdams HH, Shapiro L, Dekker J, Church GM.** 2011. The three-dimensional architecture of a bacterial genome and its alteration by genetic perturbation. *Mol Cell* **44:**252–264.
41. **Burton JN, Liachko I, Dunham MJ, Shendure J.** 2014. Species-level deconvolution of metagenome assemblies with Hi-C-based contact probability maps. *G3:Genes, Genomes, Genetics* **4:**1339–1346.
42. **Beitel CW, Froenicke L, Lang JM, Korf IF, Michelmore RW, Eisen JA, Darling AE.** 2014. Strain- and plasmid-level deconvolution of a synthetic metagenome by sequencing proximity ligation products. *PeerJ* **2:**e415. doi:10.7717/peerj.415.
43. **Jones BV, Marchesi JR.** 2007. Transposon-aided capture (TRACA) of plasmids resident in the human gut mobile metagenome. *Nat Methods* **4:**55–61.
44. **Warburton PJ, Allan E, Hunter S, Ward J, Booth V, Wade WG, Mullany P.** 2011. Isolation of bacterial extra-chromosomal DNA from human dental plaque associated with periodontal disease, using transposon aided capture (TRACA). *FEMS Microbiol Ecol* **523:**349–354.
45. **Zhang T, Zhang X-X, Ye L.** 2011. Plasmid metagenome reveals high levels of antibiotic resistance genes and mobile genetic elements in activated sludge. *PLoS One* **6:** e26041. doi:10.1371/journal.pone.0026041.
46. **Low HH, Gubellini F, Rivera-Calzada A, Braun N, Connery S, Dujeancourt A, Lu F, Redzej A, Fronzes R, Orlova EV, Waksman G.** 2014. Structure of a type IV secretion system. *Nature* **508:**550–553.
47. **van Elsas JD, Gardener BBM, Wolters AC, Smit E.** 1998. Isolation, characterization, and transfer of cryptic gene- mobilizing plasmids in the wheat rhizosphere. *Appl Environ Microbiol* **64:**880–889.
48. **Schneiker S, Keller M, Dröge M, Lanka E, Pühler A, Selbitschka W.** 2001. The genetic organization and evolution of the broad host range mercury resistance plasmid pSB102 isolated from a microbial population residing in the rhizosphere of alfalfa. *Nucleic Acids Res* **29:**5169–5181.

49. **Heuer H, Kopmann C, Binh CTT, Top EM, Smalla K.** 2009. Spreading antibiotic resistance through spread manure: characteristics of a novel plasmid type with low %G plus C content. *Environ Microbiol* **11:**937–949.

50. **Jechalke S, Kopmann C, Rosendahl I, Groeneweg J, Weichelt V, Krögerrecklenfort E, Brandes N, Nordwig M, Ding G-C, Siemens J, Heuer H, Smalla K.** 2013. Increased abundance and transferability of resistance genes after field application of manure from sulfadiazine-treated pigs. *Appl Environ Microbiol* **79:**1704–1711.

51. **Heuer H, Smalla K.** 2007. Manure and sulfadiazine synergistically increased bacterial antibiotic resistance in soil over at least two months. *Environ Microbiol* **9:**657–666.

52. **Heuer H, Binh CTT, Jechalke S, Kopmann C, Zimmerling U, Krögerrecklenfort E, Ledger T, González B, Top EM, Smalla K.** 2012. IncP-1ε plasmids are important vectors of antibiotic resistance genes in agricultural systems: diversification driven by class 1 integron gene cassettes. *Front Microbiol* **3.** doi:10.3389/fmicb.2012.00002.

53. **Smalla K, Heuer H, Götz A, Niemeyer D, Krögerrecklenfort E, Tietze E.** 2000. Exogenous isolation of antibiotic resistance plasmids from piggery manure slurries reveals a high prevalence and diversity of IncQ-like plasmids. *Appl Environ Microbiol* **66:**4854–4862.

54. **Drønen AK, Torsvik V, Top EM.** 1999. Comparison of the plasmid types obtained by two distantly related recipients in biparental exogenous plasmid isolations from soil. *FEMS Microbiol Lett* **176:**105–110.

55. **Carattoli A.** 2009. Resistance plasmid families in *Enterobacteriaceae*. *Antimicrob Agents Chemother* **53:** 2227–2238.

56. **Alvarado A, Garcillán-Barcia MP, de la Cruz F.** 2012. A degenerate primer MOB typing (DPMT) method to classify gamma-Proteobacterial plasmids in clinical and environmental settings. *Plos One* **7.** doi:10.1371/journal.pone.0040438.

57. **de Toro M, Garcillán-Barcia MP, de la Cruz F.** 2015. Plasmid diversity and adaptation analyzed by massive sequencing of *Escherichia coli* plasmids. *In* Tolmasky ME, Alonso JC (ed), *Plasmids: Biology and Impact in Biotechnology and Discovery*. ASM Press, Washington, DC, in press.

58. **Amos GCA, Hawkey PM, Gaze WH, Wellington EM.** 2014. Waste water effluent contributes to the dissemination of CTX-M-15 in the natural environment. *J Antimicrob Chemother* **69:**1785–1791.

59. **Eltlbany N, Prokscha ZZ, Castañeda-Ojeda MP, Krogerrecklenfort E, Heuer H, Wohanka W, Ramos C, Smalla K.** 2012. A new bacterial disease on *Mandevilla sanderi*, caused by *Pseudomonas savastanoi*: lessons learned for bacterial diversity studies. *Appl Environ Microbiol* **78:**8492–8497.

60. **Klümper U, Riber L, Dechesne A, Sannazzarro A, Hansen LH, Sørensen SJ, Smets BF.** 2014. Broad host range plasmids can invade an unexpectedly diverse fraction of a soil bacterial community. *ISME J.* doi: 10.1038/ismej.2014.191.

61. **Pukall R, Tschäpe H, Smalla K.** 1996. Monitoring the spread of broad host and narrow host range plasmids in soil microcosms. *FEMS Microbiol Ecol* **20:**53–66.

62. **Heuer H, Ebers J, Weinert N, Smalla K.** 2010. Variation in permissiveness for broad-host-range plasmids among genetically indistinguishable isolates of *Dickeya* sp. from a small field plot. *FEMS Microbiol Ecol* **73:**190–196.

63. **De Gelder L, Vandecasteele FPJ, Brown CJ, Forney LJ, Top EM.** 2005. Plasmid donor affects host range of promiscuous IncP-1 beta plasmid pB10 in an activated-sludge microbial community. *Appl Environ Microbiol* **71:**5309–5317.

64. **Bellanger X, Guilloteau H, Bonot S, Merlin C.** 2014. Demonstrating plasmid-based horizontal gene transfer in complex environmental matrices: a practical approach for a critical review. *Sci Total Environ* **493:**872–882.

65. **Musovic S, Oregaard G, Kroer N, Sørensen SJ.** 2006. Cultivation-independent examination of horizontal transfer and host range of an IncP-1 plasmid among Gram-positive and Gram-negative bacteria indigenous to the barley rhizosphere. *Appl Environ Microbiol* **72:**6687–6692.

66. **Musovic S, Dechesne A, Sørensen J, Smets BF.** 2010. Novel assay to assess permissiveness of a soil microbial community toward receipt of mobile genetic elements. *Appl Environ Microbiol* **76:**4813–4818.

67. **Shintani M, Matsui K, Inoue J, Hosoyama A, Ohji S, Yamazoe A, Nojiri H, Kimbara K, Ohkuma M.** 2014. Single-cell analyses revealed transfer ranges of IncP-1, IncP-7, and IncP-9 plasmids in a soil bacterial community. *Appl Environ Microbiol* **80:**138–145.

68. **Thomas CM, Smith CA.** 1987. Incompatibility group-P plasmids: genetics, evolution, and use in genetic manipulation. *Annu Rev Microbiol* **41:**77–101.

69. **Musovic S, Klumper U, Dechesne A, Magid J, Smets BF.** 2014. Long-term manure exposure increases soil bacterial community potential for plasmid uptake. *Environ Microbiol Rep* **6:**125–130.

70. **Takahashi Y, Shintani M, Takase N, Kazo Y, Kawamura F, Hara H, Nishida H, Okada K, Yamane H, Nojiri H.** 2014. Modulation of primary cell function of host *Pseudomonas* bacteria by the conjugative plasmid pCAR1. *Environ Microbiol* [Epub ahead of print.] doi:10.1111/1462-2920.12515.

71. **Jechalke S, Kopmann C, Richter M, Moenickes S, Heuer H, Smalla K.** 2013. Plasmid-mediated fitness advantage of *Acinetobacter baylyi* in sulfadiazine-polluted soil. *FEMS Microbiol Lett* **348:**127–132.

72. **Gullberg E, Albrecht LM, Karlsson C, Sandegren L, Andersson DI.** 2014. Selection of a multidrug resistance plasmid by sublethal levels of antibiotics and heavy metals. *mBio* **5.** doi:10.1128/mBio.01918-14.

73. **Suzuki H, Sota M, Brown CJ, Top EM.** 2008. Using Mahalanobis distance to compare genomic signatures between bacterial plasmids and chromosomes. *Nucleic Acids Res* **36:**e147.

74. **Suzuki H, Yano H, Brown CJ, Top EM.** 2010. Predicting plasmid promiscuity based on genomic signature. *J Bacteriol* **192:**6045–6055.

75. **Lanza VF, de Toro M, Garcillán-Barcia MP, Mora A, Blanco J, Coque TM, de la Cruz F.** 2014. Plasmid flux in *Escherichia coli* ST131 sublineages, analyzed by Plasmid Constellation Network (PLACNET), a new method for plasmid reconstruction from whole genome sequences. *PLoS Genet* **10:**e1004766.

Plasmids—Biology and Impact in Biotechnology and Discovery
Edited by Marcelo E. Tolmasky and Juan C. Alonso

doi:10.1128/microbiolspec.PLAS-0016-2013

Maria S. Ramirez,[1,2] German M. Traglia,[2] David L. Lin,[1]
Tung Tran,[1] and Marcelo E. Tolmasky[1]

24 Plasmid-Mediated Antibiotic Resistance and Virulence in Gram-Negatives: The *Klebsiella pneumoniae* Paradigm

The study of plasmids and their biology has had a decisive impact on the advance of molecular genetics, contributing numerous fundamental discoveries beyond the field of plasmid biology (1). Interestingly, the study of plasmids was already well under way before the structure of DNA was known, with the experiments that led to the discovery of conjugation and recombination in bacteria using as a system the plasmid F, known at that time as the "F factor" (2, 3). The continuation of these studies showed that bacterial plasmids are responsible for many of the particular properties of bacteria that are of medical, industrial, and agricultural interest. Their fundamental role in shaping the characteristics of the host bacteria and their ability to propagate led some authors to propose the somewhat controversial idea that they should be considered independent organisms (4). The role of plasmids in antibiotic resistance was first recognized in Japan when strains that were susceptible or multiresistant were isolated from the same patient during a single epidemic of dysentery. This fact suggested that susceptible strains were becoming multiresistant, not through successive mutational steps, but rather by acquisition of the necessary genetic determinants in a single step. Watanabe and Fukasawa reported that this process was due to transfer of a plasmid (at that time called the resistance transfer factor, RTF, or R-factor) that harbored the resistance genes (5, 6). Later it became clear that plasmids were carriers of not only antibiotic resistance genes but also genes or groups of genes that specify properties that are essential or contribute to the virulence of the host bacteria (7–20). Studies during the following few decades revealed in some detail numerous biological characteristics of plasmids, as well as the high diversity of existing plasmids and their association with other genetic mobile elements.

There is currently an epidemic of antibiotic-resistant bacterial infections, which the World Health Organization

[1]Center for Applied Biotechnology Studies, Department of Biological Science, College of Natural Sciences and Mathematics, California State University Fullerton, Fullerton, CA; [2]Institute of Microbiology and Medical Parasitology, National Scientific and Technical Research Council (CONICET), University of Buenos Aires, Buenos Aires, Argentina.

has identified as one of the greatest threats to human health (http://www.who.int/drugresistance/activities/wha66_side_event/en/index.html) (21–24). Within this epidemic, a group of pathogens has been individualized and collectively named ESKAPE (*Enterococcus faecium, Staphylococcus aureus, Klebsiella pneumoniae, Acinetobacter baumannii, Pseudomonas aeruginosa*, and *Enterobacter*). These pathogens are the causative agents of the majority of hospital infections because they "escape" the antibiotic treatment by becoming resistant or persistent to antibiotic treatment (21, 25–27). Plasmids play a central role in the dissemination and acquisition of the resistant determinants in these bacteria. In this article we describe aspects of plasmids of *K. pneumoniae*, a representative of problematic Gram-negative opportunistic pathogens belonging to the ESKAPE group (28–30), with emphasis on their role in virulence and resistance to antibiotics.

K. pneumoniae is the causative agent of serious community- and hospital-acquired infections including but not limited to urinary tract infections, pneumonia, septicemias, meningitis, and soft tissue infections (31–38). *K. pneumoniae* has also been identified as a causative agent of other less common, yet serious, infections such as liver abscess and invasive syndrome (39, 40), septic arthritis (41), and generalized pustulosis (42) and as the triggering factor in the initiation and development of ankylosing spondylitis and Crohn's disease (43–45). *K. pneumoniae* strains have accumulated plasmids that carry virulence and resistance genes that keep increasing its ability to resist the main antibiotics used for treatment such as cephalosporins, carbapenems, penicillins, aminoglycosides, and fluoroquinolones (26, 37, 38, 46, 47).

K. pneumoniae strains usually harbor more than one plasmid, including small high-copy-number plasmids and low-copy-number plasmids that are usually large. Of all completed genomes so far, the multiresistant strain HS11286, isolated from human sputum, harbors the most plasmids, with sizes 1.31, 3.35, 3.75, 105.97, 111.19, and 122.80 kbp (48). More than 70 *K. pneumoniae* plasmids have been completely sequenced, and some have been further analyzed with particular attention to the presence of virulence and resistance genes as well as mobile elements.

SMALL PLASMIDS

A large group of the plasmids found in *K. pneumoniae* isolates are small, with sizes spanning between less than 2 kbp and 25 kbp (Table 1). Most of these plasmids, which share homology at the replication regions (Fig. 1a and b), replicate through the ColE1-type mechanism (49–53), are non-self-transmissible, and do not always encode resistance genes as do plasmids pKPN2 (54) and pKlebB-k17/80 (55), which encode a restriction-modification system and a bacteriocin, respectively (Table 1). The backbones of the plasmids within this group have a common general organization consisting of the ColE1-type replication region that includes the genes coding for RNA I and RNA II and sometimes the negative regulator *rom* (or *rop*), a transfer region consisting of *oriT* or this locus accompanied by the genes coding for the remaining relaxosome components (56), and a Xer site-specific recombination site (Fig. 1a). Probably, these plasmids are not confined to *K. pneumoniae*, but they are able to replicate and be stably maintained in other *Enterobacteriaceae* or Gram-negatives.

One of the most thoroughly studied ColE1-type plasmids from *K. pneumoniae* is pJHCMW1 (Table 1), isolated from a neonate with meningitis (38, 57). This plasmid is 11,354 bp long, of which 7,992 bp are Tn*1331*, a multiresistance transposon that includes *aac(6′)-Ib*, *ant(3′′)-Ia*, *bla*$_{OXA-9}$, and *bla*$_{TEM-1}$ (58–61). The backbone of pJHCMW1 is composed of a ColE1-type replication region lacking *rom*, a functional *oriT* that mediates mobilization when the remaining components of the relaxosome and the transferosome are supplied *in trans* by a helper plasmid (62), and a Xer site-specific recombination site named *mwr*, which has been studied in some detail (see below) (63–66). The *aac(6′)-Ib*, *ant(3′′)-Ia*, and *bla*$_{OXA-9}$ genes are organized in a region resembling the variable region of integrons (59, 60, 67–69). Tn*1331*, first found in pJHCMW1 (58), or its variations have been found in plasmids hosted by other Gram-negatives such as *Enterobacter*, *Salmonella*, *Serratia*, and *Pseudomonas* (37, 58, 68–78). pJHCMW1 was recently used as model of ColE1-type plasmids in microscopy studies to determine the mobility of the molecules inside the cells as well as their location and the implications in partition at the moment of cell division (79). Plasmid molecules were highly mobile but were mainly found located at the poles of the cell because they tend to be excluded from the nucleoid occupied space. In fact, in experiments where the nucleoid-free space was increased by using a *dnaN159*(ts) mutant at the nonpermissive temperature or by treatment with cephalexin, the plasmid molecules occupied all the nucleoid-free space (79). These results confirmed that the pJHCMW1 plasmid molecules freely move and are not specifically targeted to the pole, but rather they tend to occupy the nucleoid-free space. The molecules did not form clusters and occasionally were able to move between poles (79).

TABLE 1 Completely sequenced ColE1-type plasmids of *K. pneumoniae*

Plasmid[a]	Size (bp)	Relevant genotype[b]	Accession number	Reference
pKPN2	4,196	*kpn2klR, kpn2klM*	AF300473	54
pIP843	7,086	$bla_{CTX\text{-}M\text{-}17}$	AY033516	98
pH205	8,197	$bla_{CMY\text{-}36}$	EU331426	155
pKlebB-k17/80	5,258	*kba, kbi*	AF156893	55
pJHCMW1	11,354	*aac(6′)-Ib, ant(3″)-Ia,* $bla_{OXA\text{-}9}$, bla_{TEM}	AF479774	61
p15S	24,296	*aac(6′)-Ib,* $bla_{KPC\text{-}2}$, cloacin, immunity protein	FJ223606	74
pColEST258	13,636	*aac(6′)-Ib,* cloacin, immunity protein	JN247853	82

[a]Plasmids that have been subjected to characterization studies beyond just sequencing.
[b]*kpn2klR* and *kpn2klM,* type II restriction-modification system; *kba and kbi,* klebicin B (a bacteriocin) and immunity.

This tendency to preferentially locate at the nucleoid-free poles ensures that at cell division both daughter cells host the plasmid.

A BLAST analysis of the backbone region of pJHCMW1 shows that the *Salmonella enterica* subsp. *enterica* serovar Typhimurium plasmid pFPTB1, which includes the $bla_{TEM\text{-}135}$ and *tetR/tetA* genes in Tn*3*- and Tn*1721*-related transposons, is the most closely related, with identity coverage including the replication region (99% identity) and the Xer site-specific recombination site *fpr* but leaving a gap between these two regions (nucleotides 1770 and 3081) where the *oriT* and a deficient Xer site-specific recombination site are located in pJHCMW1 (Fig. 2) (80, 81).

Fig. 1b shows the alignment of the replication regions of the seven plasmids listed in Table 1. Plasmid p15S completely contains pColEST258, and therefore they share a common block of DNA including the replication region (82). These two plasmids include the region of Tn*1331* that encompasses the genes *tnpA, tnpR,* and *aac(6′)-Ib* (82). All other replication regions are related but not identical. Since small variations in the nucleotide sequence of the replication regions are sufficient to result in compatible plasmids, only experimental assays will determine whether these plasmids are incompatible. All seven plasmids include a Xer site-specific recombination site consisting of the core recombination site and the accessory sequences where the architectural proteins, usually ArgR and PepA, bind to help formation of the synaptic complex (Fig. 1c). The function of this site, at least in some plasmids, is to maximize stabilization by mediating resolution of multimers, a process that leads to a reduction in the effective number of molecules and results in segregational loss of the plasmid (83–87). Other plasmid-related functions of these sites may include exchange of DNA regions among plasmids (54, 88) as well as mediating the insertion of integrative mobile elements into chromosomes and plasmids (83, 89, 90).

The XerD binding sites, usually the most conserved fraction of Xer recombination target sites (91, 92), are identical in all seven plasmids listed in Table 1 (Fig. 1c). XerC binding sites are more divergent, and the central regions are considerably different in sequence and size. While six sites include all eight nucleotides that are highly conserved in all ARG boxes (Fig. 1c), pJHCMW1 and pH205 include a C instead of a T in one of those positions (arrowhead in Fig. 1c). This substitution has been shown to impair the efficiency of a site to act as a target for Xer site-specific recombination (64). In particular, the pJHCMW1 Xer site-specific recombination site, called *mwr*, has been studied in more detail and shows some interesting characteristics. The resolution of dimers harboring this site is inefficient when the *Escherichia coli* host cells are cultured in standard L broth (osmolality 209 mmol/kg) (63, 64). However, the efficiency of resolution is inversely proportional to the osmolarity of the medium, and all molecules appear as monomers when cells are cultured in L broth without NaCl added (osmolality 87 mmol/kg) (66). Less than ideal interaction of ArgR with the *mwr* ARG-box at higher osmolarity may, at least in part, be responsible for deficient formation of the synaptic complex—a problem that may be compensated for when the cells grow in medium with lower osmolarity (Fig. 3). This compensation seems to occur through an increase in negative supercoiling density (Fig. 3) (66).

In vitro recombination experiments suggest that the increase in efficiency of resolution occurs at the level of formation of the Holliday junction, rather than at the level of Holliday junction resolution (Fig. 3) (66). It is of interest that numerous Xer site-specific recombination sites with identical deficiency in the ARG box have been detected, and in some cases experiments demonstrated that they mediate dimer resolution at low efficiency (81). This fact, taken together with other findings such as the presence of Xer site-specific recombination sites flanking bla_{OXA} genes in *Acinetobacter*

a

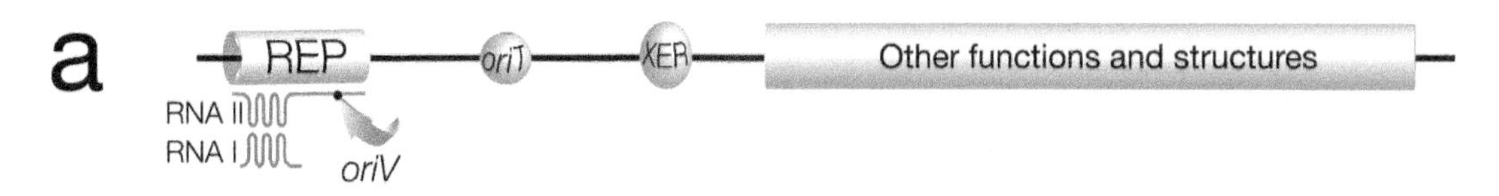

b

```
pJHCMW1    TTGAGATCCTGTCCGTTGCGTCGTAATCTCTTTCTCTGGAAACGAAAAAACCGCCCTGCAAGGCGGTTTTT-CGAAGGTTCTCAGAGCT-ACCAACTCTTTGAACCGCGGTAA-CTGGCT
pKLeb      TTGAGATCCCATTTGGATCGTCGTAATCTCTTGCTCTGTAAACGAAAAAACCGCCTTGGCGGGCGGTTTTTTCGAAGGTTCGAGGAGTT-GGCG-CTCTTTGAACCGAGGTAA-CTGGCT
pIP843     TTGAGATCTGACGTTGGACGTCGTAGTCTCTTGCCATGTAAACGAAAAAACCACCTTGCAGGGTGGTTTTT-CGAAGGTTCAGTAAACC-GGCAAGTTTCTGAACCGTGGTAA-CAGTCT
pKPN2      -TGAGATCCTTTTTTTCTGCGCGTAATCTTGGACC-TGTAAACGAAAAAACCACCCTGGCAGGTGGTTTTTTCGAAGGTTAGCTAATCCTGGCAGATTATCTAACCGAGGTAATCTGGCT
p15S       TTGAGATCCTTTTTTTCTGCGCGTAATCTTTTGCCCTGTAAACGAAAAAACCACCTGGGGAGGTGGTTTGATCGAAGGTTAAGTCAGTT-GGGGAACTGCTTAACCGTGGTAA-CTGGCT
pColEST258 TTGAGATCCTTTTTTTCTGCGCGTAATCTTTTGCCCTGTAAACGAAAAAACCACCTGGGGAGGTGGTTTGATCGAAGGTTAAGTCAGTT-GGGGAACTGCTTAACCGTGGTAA-CTGGCT
pH205      TTGAGATCCTTTTATTCTGCGCGTAATCTCTTGTCCTGAAAACGAAAAAACCACCTGGGGAGGTGGTTTTT-TCGAAGGTTCAGTAAGTT-GGGGAACTTCTGAACCGTGGTAA-CAGGGT
            *******          **** ***      ** ************* **  *   ** *****   ********     *                 ***** ***** * *  *

pJHCMW1    TG-GAGGAACGCAGTAACCAAAT-CTGTCCTTTCAGTTTAGCCTTAACCGGCGCATAACTTCAAGAC-----------------------TAACTCCTCTAAATCA-GTTACCAGTGGC
pKLeb      TG-GAGGAGCGCAGTAACCAAAT-TCGTTCTTTCAGTTTAGCCTTAACTGGCACATAACTTCAAGAC-----------------------TAACTCCTCTAAATCA-GTTACCAGTGGC
pIP843     TGTGCGAGAC---GTCACCAAAT-CTGTCCTTTTAGTGTAGCCTCAGTCTGGCCACCACTTCAAGAACTCTCGATACATCT---------CTCGCAC----ATCCTG-TTTACCAGTGGC
pKPN2      TCAGCAGAGCACAGATACCAAATACTGTCCTTCCAGTGTAGCCGTAGTTAGGCCATCACTTCAAGAACTCTGTAAGCATCTGGATAAATCCTCGCTCTGCTAATCCG-GTTACCAGTGGC
p15S       TTCGCAGAGCACAGCAACCAAAT-CTGTCCTTCCAGTGTAGCCGGACTTTGGCGCACACTTCAAGAGCAACCGCGTGTTTAGCTAAACAAATCCTCTGCGAACTCCCAGTTACCAATGGC
pColEST258 TTCGCAGAGCACAGCAACCAAAT-CTGTCCTTCCAGTGTAGCCGGACTTTGGCGCACACTTCAAGAGCAACCGCGTGTTTAGCTAAACAAATCCTCTGCGAACTCCCAGTTACCAATGGC
pH205      GTACAAGACCGCTGCCACCAAAT-ACGTCCTTTCAGTTTAGCCGTAGTTAGGCTTCAACTTCAAGAAC----------TCTGCACCAGTAATCTCTTGTACACCCCT--TTACCAGTGGC
                   *  *  *******   ** ***  *** *****  *    *     *********                         *        *         ****** ****

pJHCMW1    TGCTGCCAGTGGCGCTTTTGCATGTCTTTCCGGGTTGGACTCAAGATGATAGTTACCGGACAAGGCGCAGCGGTCGGACTGAACGGGGGGTTCGTGCATACAGTCCAGCTTGGAGCGAAC
pKLeb      TGCTGCCAGTGGCGCTTTTGCATGCCTTTCCGGGTTGGACTCAAGATGACAGTTACCGGATAAGGCGCAGCAGTCGGACTGAACGGGGGGTTCGTGCATACAGTCCAGCTTGGAGCGAAC
pIP843     CGCTGCCAGTGGCGTTAAGTCGTGTCTTTCCGGGTTGGACTCAAGACGATAGTTACCGGAAAAGGCGCAGCGGTCGGGCTGAACGGGGGGTTCTTGCATACAGCCCAGCTTGGAGCGAAC
pKPN2      TGCTGCCAGTGGCGTTAAGGCGTGTCTTACTGGGTTGGACTCAAGACGATAGTTACCGGATAAGGCGCAGCGGTCGGGCTGAACGGGGGGTTCGTGCACACAGCCCAGCTTGGAGCGAAC
p15S       TGCTGCCAGTGGCGTTTTG-CGTGCTTTTCCGGGTTGGACTCAAGTGAACAGTTACCGGATAAGGCGCAGCAGTCGGGCTGAACGGGGGGTTCTTGCTTACAGCCCAGCTTGGAGCGAAC
pColEST258 TGCTGCCAGTGGCGTTTTG-CGTGCTTTTCCGGGTTGGACTCAAGTGAACAGTTACCGGATAAGGCGCAGCAGTCGGGCTGAACGGGGGGTTCTTGCTTACAGCCCAGCTTGGAGCGAAC
pH205      TACCGCCAATGGCGATTTGACGTGTCTGTAAGGGTTGGACTCAAGACGATAGTTACCTTACATGGCGCGGTAGTCGGACTGAACGGGGGGTTCGTGCATACAGTCCAGCTTGGAGCGAAC
            * **** ***** *    * **  *    **************   * *******  * * ***** *  ***** *************** ***  **** ***************

pJHCMW1    TGCCTACCCGGAACTGAG-TGTCA--GGCGTGGAATGAGATAAA------CGCGGCC--------ATAACAGCGGAAT-GACAC-CGGTAAACCGAAAGGCAGGAACAGGAGAGCGCACG
pKLeb      TGCCTACCCGGAACTGAG-TGTCA--GGCGTGGAATGAGACAAA------CGCGGCC--------ATAACAGCGGAAT-GACAC-CGGTAAACCGAATGGCAGGAACAGGAGAGCGCACG
pIP843     TGTCTAAACGGAACGGGG-------CGTGGTGATTTAGGTAAA--------GCTTCC-----------------ACC-ACTACACGG-ATGCCGCAGGACGGGAACAGGAGAGCGCAAG
pKPN2      GACCTACACCGGGCCGAGATACCAACAGCGTGAGCTATGAGAAAGCGCCACGCTTCCCG-AAGGGAGAAAGGCGGACA-GGTATCCGGTAAGCGGCAGGGTCGGAACAGGAGAGCGCACG
p15S       GACCTACACCGAGCCGAGATACCA-GAGTGTGAGCTATGAGAAAGCGCCACACTTCCCGCAAGGGAGAAAGGCGGAACAGGTATCCGGGAAACGGCAGGGTTGGAACAGGAGAGCGCAAG
pColEST258 GACCTACACCGAGCCGAGATACCA-GAGTGTGAGCTATGAGAAAGCGCCACACTTCCCGCAAGGGAGAAAGGCGGAACAGGTATCCGGGAAACGGCAGGGTTGGAACAGGAGAGCGCAAG
pH205      TGCCTACCCGGAACTGAG-TGTCA-G-GCGTGGA--ATGAGACAA----ACGCGGCC--------ATAACAGCGGAAT-GACA-CCGGTAAACCGAATGGCAGGAACAGGAGAGCGCACG
              ***  * *  * * *           ***      *  *          *  **              *      *  *** *  * * * *   *************** *

pJHCMW1    AGGGAGCCACCAGGGGGAAACGCCTGGTATCTTTA-AGTCCTGTCGGGTTTCGCCACCACTGAT----------TTGAGC-----------------------------GTCCGATTCT
pKLeb      AGGGAGCCATCAGGGGGAAACGCCTGGTATCTTTATAGTCCTGTCGGGGTTCGCCACCACTGAT----------TTGAGC-----------------------------GTCAAATTCT
pIP843     AGGGAGCCACCAGGGGGAAACGCCTGGTATCTTTATAGTCCTGTCGGGTTTCGCCACCACTGAT----------TTGAGC-----------------------------GTCAGATTTC
pKPN2      AGGGAGCTTCCAGGGGGAAACGCCTGGTATCTTTATAGTCCTGTCGGGTTTCGCCACCCCTGAC----------TTGAGC-----------------------------GTCGATTTTT
p15S       AGGGAGCGATTCATCGGAAACGGTGATGATCTTTA-AGTCCTGTCGGGTTTCGCCACTCCTGACTGATTCATGGTTGAGCCACCGGCTCCCACAGATGCACCGAAAAAGCGTCTGTTTAT
pColEST258 AGGGAGCGATTCATCGGAAACGGTGATGATCTTTA-AGTCCTGTCGGGTTTCGCCACTCCTGACTGATTCATGGTTGAGCCACCGGCTCCCACAGATGCACCGAAAAAGCGTCTGTTTAT
pH205      AGGGAGCGACCCATCGGAAACGGTGGGGATCTTTA-AGTCCTGTCGGGTTTCGTCACCGCTGTCGGATTCATGGTTGAGCT-CAGGCTCCCACAGATGCACCGAAAAAGTGTCTGTTTAT
           *******       *******     ******* ************ **** ***  ***           ******                              ***   **

pJHCMW1    GTGATGCTTGTCAGGGGGGCGGAGCCTATGGAAAA
pKLeb      GTGATGCTTGTCAGGGGGGCGGAGCCTATGGAAAA
pIP843     GTGATGCTTGTCAGGGGGGCGGAGCCTATGGAAAA
pKPN2      GTGATGCTCGTCAGGGGGGCGGAGCCTATGGAAAA
p15S       GTGAACTCAGTCAGGAGGGCGGAGCCTATGGAAAA
pColEST258 GTGAACTCAGTCAGGAGGGCGGAGCCTATGGAAAA
pH205      --GAATTCCGGCAGGGGGGCGGAGCCTATGAAAAA
             **     * **** ************** ****
```

c

```
                ARG box
            consensus sequence
            a ▼     aatt      t
            tN  A   ttaa   T  Na
p15S        CGTGCATAGGCATGCATTAGGATAAAATTTATCGGG-CGCGTTTCCGGCAGTTTTCCGGGGGGGTTGTTGCCTG
pColEST258  CGTGCATAGGCATGCATTAGGATAAAATTTATCGGG-CGCGTTTCCGGCAGTTTTCCGGGGGGGTTGTTGCCTG
pJHCMW1     CGCGCATACTCATGCATGCCGTAAAAACAGAGCCTG-CGCGTTTCTGGCGGGTTTTCGGGTGGTTTGTTGCCTG
pH205       TGCGCATACTCATGCATGCCGTAAAAACAGAGCCTG-CGCGTTTCTGGCGGGTTTTCGGGTGGTTTGTTGCCTG
pKLeb       CGTGCATACTCATGCATGCCGTAAAAACAGAGCCTG-CGCGTTTCTGGCGGGTTTTCGGGTGGTTTGTTGCCTG
pIP843      CGTGCATACTCATGCATGCCGTAAAAACAGAGCCAG-CGCGTTTCTGGCGGATTTTCGGGTGGTTTGTTGCCTG
pKPN2       CATGCATAGCTATGCAGTGAGCTGAAAGCGATCCTGACGCATTTTTC-CGGTTTACCCCGGGGAGAACATCTCT
               *****   *****    *    ***    * *  * *** ***    * * **  *  * **       *

                                                 Core Recombination Site
                                              XerC          central    XerD
                                              binding site  region     binding site
p15S        TTTTATCCCGTAGCCGCCGGAAACGCCCTGAGCCCGTCTGAG-CGGTGCGCGTAAT-GACGCGTTATGGTAAAT
pColEST258  TTTTATCCCGTAGCCGCCGGAAACGCCCTGAGCCCGTCTGAG-CGGTGCGCGTAAT-GACGCGTTATGGTAAAT
pJHCMW1     TTTTACCGGTTTCCCGTCAGAAACGCCCTGAGGGCCTCTCAGGCGGTGCACGCAA--CAGATGTTATGGTAAAT
pH205       TTTTACCGGTTTCCCGTCAGAAACGCCCTCAGAGGCTCTGAGGCGGTGCCGATAA--GGGATGTTATGGTAAAT
pKLeb       TTTTACCGGTTTCCCGTCGGAAACGCCCTGAGGCCGTTTTAG-CGGTGCGTACAATTAAGGGATTATGGTAAAT
pIP843      TTTTACCGGTTTCCCGTCAGAAACGCCCTGAGGCCGTTTTCG-CGGTGCGCGTAAT-GAGACGTTATGGTAAAT
pKPN2       TTTTGCGGTGTTCGCGGCAGAA-TGCTTTCAGCGCGTTTTAG-CGGTGCGCGCAAT-GCGACGTTATGGTAAAT
            ****      *   ** * ***   **  * **      * *  * ******    **       ***********
```

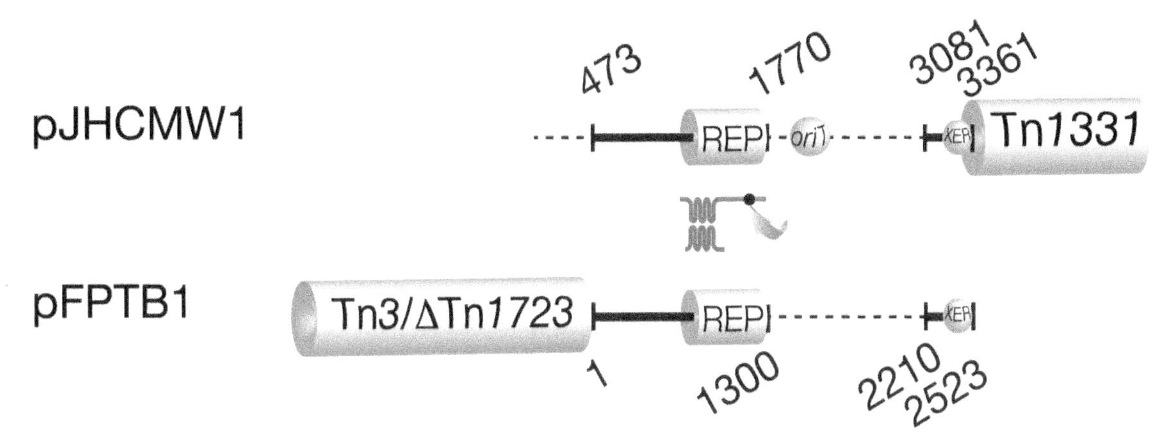

Figure 2 Comparison of pFPTB1 and pJHCMW1. The black lines, which represent regions of homology (coordinates 473 to 3361 in pJHCMW1), are drawn to scale. The Tn*3*-like transposons, Tn*1331* and Tn*3*/DeltaTn*1723*, as well as the dots indicating *oriT* and the Xer target sites, are shown at the correct locations but are not drawn to scale. The replication regions (REP) share 97% homology. The numbers indicate the coordinates in the GenBank database (pJHCMW1, accession number AF479774; pFPTB1, accession number AJ634602). The location of the similar but not identical Xer site-specific recombination sites (81) is indicated. doi:10.1128/microbiolspec.PLAS-0016-2013.f2

plasmids (90, 93–96) or the presence of different DNA fragments flanked by a Xer recombination site and an *oriT* in otherwise identical plasmids (54, 88), leads to the idea that not all Xer recombination sites stabilize plasmids by multimer resolution but may play other, or additional, roles related to plasmid evolution.

Small plasmids may be less than ideal vehicles for dissemination of resistance genes because (i) some of the genes are located outside gene cassettes or mobile elements, reducing the versatility shown by these elements to promote dissemination at the molecular level, and (ii) the plasmids may lack an *oriT* or possess one but lack all other conjugation functions including the proteins that form the specific relaxosome, making dissemination at the cellular level by conjugation less efficient (97). Taken together, these factors must reduce the ability of some genes to be mobilized between molecules and cells. A recent report describes a way to reduce this constraint by cointegration of the small plasmid with another plasmid that can provide the machinery for conjugation. The *K. pneumoniae* plasmid pIP843, which harbors the extended spectrum β-lactamase gene $bla_{\text{CTX-M-17}}$ (98), in spite of including an *oriT* locus, was not transferred in mating experiments between the original *K. pneumoniae* isolate and a recipient *E. coli* isolate (98). This result suggested that the presence in the same cells of helper plasmids that provide all necessary components, specific and nonspecific, to mobilize plasmids such as pIP843 might not be the most usual situation. However, dissemination of $bla_{\text{CTX-M-17}}$ is dramatically enhanced by its presence in a large conjugative plasmid, pE66An, isolated from an *E. coli* clinical strain (99). Interestingly, pE66An has the structure of a cointegrate formed between an original ~73-kbp plasmid and pIP843 (Fig. 4). The point of cointegration is the pIP843 RNA II gene. Therefore, this event resulted in inactivation of the ColE1-type replicon (Fig. 4), which must have been important for generating a stable large plasmid.

Figure 1 Small plasmids. (**a**) General genetic organization of small ColE1-type plasmids from *K. pneumoniae* and other *Enterobacteriaceae*. (**b**) Alignment of the nucleotide sequences of the replication regions of *K. pneumoniae* ColE1-type plasmids using CLUSTAL W (146). (**c**) Alignment of the nucleotide sequences of Xer site-specific recombination sites of *K. pneumoniae* ColE1-type plasmids using CLUSTAL W. The ARG box, XerC, and XerD binding sites are shown in color, and the central regions are boxed. Blue capital letters indicate the most important conserved nucleotides in the ARG box. The downward pointing arrowhead shows the conserved T nucleotide that is substituted by a C in several Xer site-specific recombination sites (64, 81, 88). doi:10.1128/microbiolspec.PLAS-0016-2013.f1

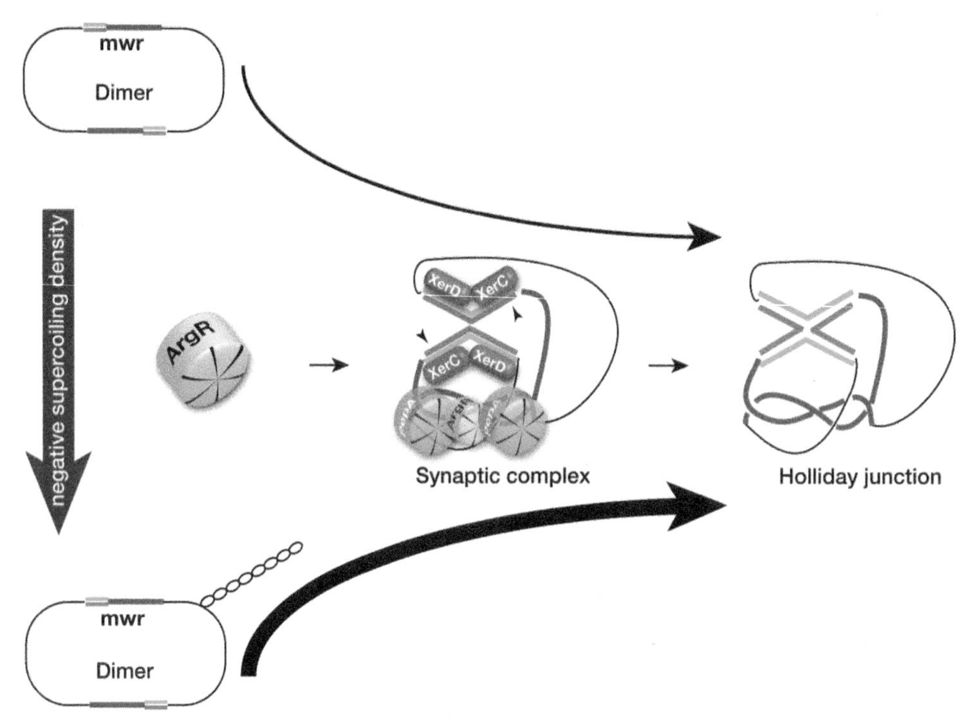

Figure 3 Effect of changes in osmolarity of the culture medium on Xer site-specific recombination at *mwr*. Schematic representation of the possible chain of events that lead to a higher efficiency of Xer site-specific recombination at the *K. pneumoniae* plasmid pJHCMW1 site *mwr*. A decrease in the NaCl concentration in the growth medium (L broth containing 0.5% NaCl added to no NaCl added) is correlated with an increase in supercoiling density, which facilitates interaction of ArgR with the substandard *mwr* ARG box leading to a more efficient formation of a productive synaptic complex and Holliday junction (66). Molecular models of the interwrapped synaptic complex are available in references 147–149. The two strands are shown only in the core recombination site (red and green lines); blue lines represent the accessory sequences.
doi:10.1128/microbiolspec.PLAS-0016-2013.f3

LARGE PLASMIDS

Other plasmids belonging to diverse incompatibility groups, usually larger than those discussed in the previous section, are found in *K. pneumoniae*. While it was well known that antibiotic resistance as well as virulence genes are housed in several of these plasmids, the interest in studying them has recently increased with the realization that they host genes responsible for resistance to last resort antibiotics such as bla_{KPC}, bla_{NDM-1}, and bla_{OXA} (34, 46, 100). The virulence phenotype of a *K. pneumoniae* strain was first associated with the presence of a plasmid when it was determined that the 180-kbp plasmid, pKP100, harbors the genes coding for the aerobactin iron uptake system and the mucoid phenotype (101–104). Iron uptake systems are well-known virulence factors of numerous bacterial pathogens (12, 105–108). Loss of pKP100 resulted in concomitant loss of virulence, and the transfer of a mobilizable derivative of pKP100 resulted in reacquisition of the virulent phenotype (103). In another instance, a 185-kbp plasmid was isolated by conjugation using as donor a *K. pneumoniae* isolate that contains three plasmids. This plasmid includes several genes conferring resistance to β-lactams, kanamycin, neomycin, streptomycin, sulfonamides, and tetracyclines, the genes coding for the aerobactin system, and a gene encoding a 29-kDa protein responsible for the ability of this strain to adhere to intestinal cells (109). The association of the mucoid phenotype and production/utilization of aerobactin and their relation to virulence has been observed in several but not all studies (101, 110, 111). A 219-kbp virulence plasmid, pLVPK, that carries the genes coding for the aerobactin system and is most probably related to those mentioned above was isolated from a highly virulent clinical isolate of K2 serotype and completely sequenced (112). In fact, it has been suggested that numerous *K. pneumoniae* blood isolates harbor a large virulence plasmid of about

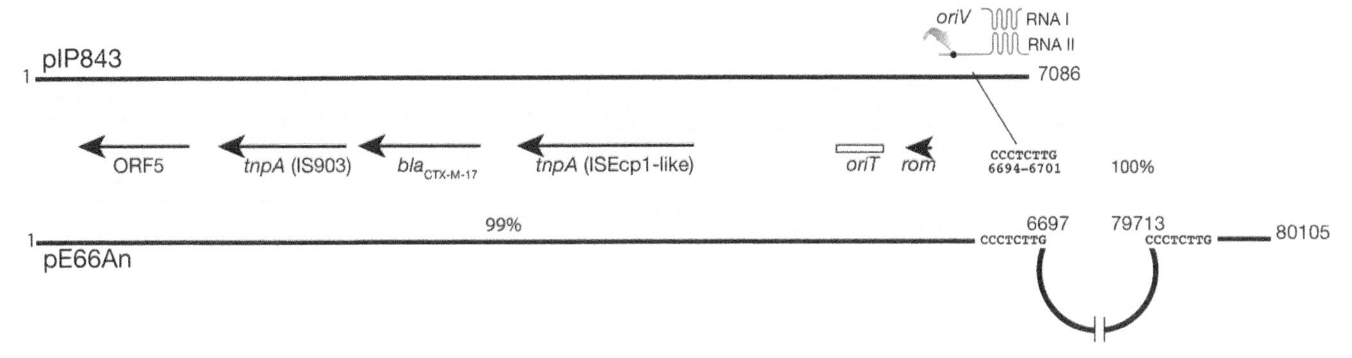

Figure 4 Genetic maps comparing the *K. pneumoniae* pIP843 and the *E. coli* pE66An. The shadowed areas show regions of homology (6681/6701 identities and six gaps in the region with 99% homology). The ColE1-type replication region is schematically shown on top of the pIP843 map. The semicircle in pE66An represents the region encompassing nucleotides 6697 to 79713. doi:10.1128/microbiolspec.PLAS-0016-2013.f4

200 kbp that includes the aerobactin system, the ability to express the mucoid phenotype, and resistance to antimicrobials (112).

In addition to aerobactin, other iron uptake systems may be included in these plasmids. In the case of pLVPK, the plasmid also includes two more iron-transport systems: the *iroBCDN* cluster that mediates iron uptake through a catecholate siderophore (113), and a homolog of *fecIRA*, which encodes a Fur-dependent regulatory system for iron uptake (114). However, the FecR encoded by this plasmid is truncated at the C-proximal end, and the FecA is active in transport, but it is induction inactive (115). Interestingly, incomplete FecIRA systems are commonly found in plasmids in *K. pneumoniae* and are chromosomally mediated in *Enterobacter* (115). BLAST analyses against *K. pneumoniae* sequences using the pLVPK regions encompassing the *iutA* and *iucDCBA*, the *iroBCDN*, and the *fecIRA* genes showed 100% homology to the plasmids pK2044 (116) and pKCTC2242 (117) in all three regions and 99% homology to the plasmids pKN-LS6 (accession number JX442974.1), pKPN-IT (82), and pKPN_CZ (118) in the *fecIRA* region. A comparison of pLVPK, pK2044, and pKCTC2242 using the MAUVE aligner version in which different colors represent local collinear blocks (LCB) with the location of key genes or clusters is shown in Fig. 5.

While pLVPK and pK2044 are the most related including all LCBs in the same location, in pKCTC2242 the LCB that in the other two plasmids includes a *ter* cluster similar to that found in *E. coli* O157:H7 (119) is truncated, losing the cluster. Furthermore, the LCBs in pKCTC2242 are rearranged with respect to pVLKP and pK2044 (Fig. 5). All three plasmids also include an *rmpA2* homolog and *rmpA*, the genes involved in the mucoid phenotype (41, 102, 104, 112). In addition, three physically linked gene clusters coding for resistance to lead, copper, and silver homologs to other known clusters found in *Ralstonia metallidurans* (120), *E. coli* (121), and *Salmonella enterica* serovar Typhimurium (122), respectively, were found in pLVPK as well as pK2044 and pKCTC2242 (Fig. 5). Several other *K. pneumoniae* plasmids (pKPX-1 [123], pUUH239.2 [124], pKN-LS6 [accession number JX442974], pBK32179 [71], pKPN-IT [82], and pKPN-CZ [118]) include highly related but not identical fragments.

Replication of pLVPK probably occurs through the iteron mechanism, as a *repA* homolog located between two sets of iterons was found. The region shows higher than 90% homology to the *K. pneumoniae* plasmids, pKCTC2242 (117), pK2044 (116), and pNDM-MAR (77), and the RepA amino acid sequence shows similarity to numerous proteins from plasmids isolated from enterobacteria (112). Another putative replication protein was found in pLVPK with homology to several replicator proteins in the database. In addition, homologs to *sopA* and *sopB* strongly suggested the presence of an F plasmid-like partitioning system.

An interesting case is that of pClpk, a 150-kbp self-transferable plasmid isolated from *K. pneumoniae* C132-98, a strain that caused critical infections over a two-year period in a hospital and that is characterized for an unusual thermotolerance (125). *K. pneumoniae* C132-98 harbors at least eight plasmids, four of which are 100 kbp or larger. The plasmid pClpk encodes ClpK, an ATPase responsible for the increased heat resistance that characterizes *K. pneumoniae* C132-98 (125). The *clpK* gene was then found in other *K. pneumoniae* strains and in at least another plasmid, the 220-kbp IncFIIk pUUH239.2, which also harbors several

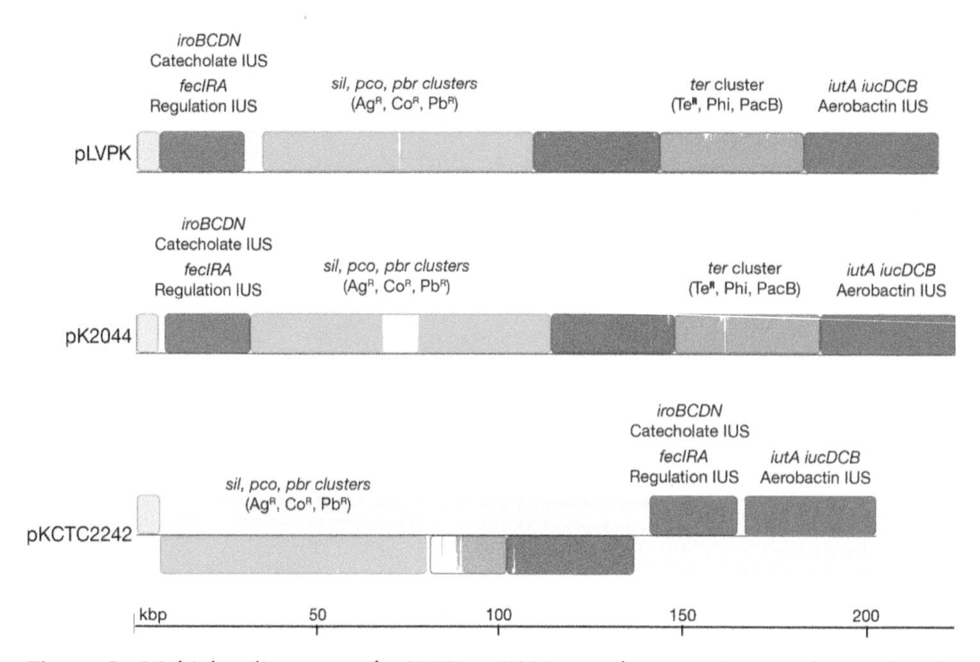

Figure 5 Multiple alignment of pLVPK, pK2044, and pKCTC2242. The nucleotide sequences of pLVPK (accession number AY378100.1) (112), pK2044 (accession number AP006726.1) (116), and pKCTC2242 (accession number CP002911.1) (117) were compared using the MAUVE aligner version 2.3.1 (150). Different colors represent local LCBs. Inside each block there is a similarity profile of the sequence; the height corresponds to the average level of conservation. Completely white areas are not aligned and probably contain sequences specific to the particular molecule. In pKCTC2242 the LCBs drawn below the black line are inverted with respect to their homologs in pLVPK and pK2044. Some genes or clusters present in these blocks are identified by name. The *terZ* gene has been reported as "truncated" (112). The truncation is a consequence of an extra T in the sequence that could also be a sequencing error. The *terBCDE* genes are sufficient for the tellurite resistance phenotype (Te^R). The *ter* cluster is also responsible for the phage inhibition (Phi) and colicin resistance (PacB) phenotypes (151). Copper (*pco*), silver (*sil*), lead (*pbr*), and tellurite (*ter*) resistance related genes; IUS, iron uptake system.
doi:10.1128/microbiolspec.PLAS-0016-2013.f5

resistance genes including $bla_{\text{CTX-M-15}}$ and *aac(6′)-Ib-cr* (124, 126).

Although cases such as that of the small plasmid pJHCMW1 (38, 57, 58, 61) are well known, numerous plasmids including multiple resistance genes are large and self-transmissible. This conjugation machinery requires genes coding for the relaxosome, the type IV coupling protein, and the type IV secretion system (127). As illustrated in previous paragraphs the resistance genes may be included in various genetic elements and coexist with other genetic elements and genes coding for resistance to metals or virulence factors.

A well-studied multiresistance *K. pneumoniae* plasmid is pMET1, a plasmid isolated from a clinical strain responsible for a high-mortality hospital outbreak (37, 128). Its replication was extensively studied by cloning and deletion assays that permitted the identification of a 1,655-bp replication region that includes *repA*, which codes for a protein belonging to the RepA IncFII superfamily, and an AT-rich sequence likely to include *ori* (128). BLAST analysis of this region showed extensive homology only with the *Yersinia pestis* plasmid pCRY (129), the *Cronobacter turicensis* plasmid pCTU2 (130), and the *Cronobacter sakazakii* plasmid pESA2 (accession number CP000784). In addition, a pMET1 segment including the type IV secretion system, which is related to those found in the ICE_{Kp1} and the HPI_{ECOR31} (131), and the *mobB* and *mobC* genes shared high similarity with the cryptic *Y. pestis* plasmid pCRY. The homology shared between the cryptic pCRY and the multidrug-resistant pMET1 suggests that the latter plasmid can be replicated and stably maintained in *Yersinia* species, representing a high public health and biodefense threat due to transfer of multiple resistance genes to pathogenic *Yersinia* strains. Absent in pCRY are the pMET1 partition genes *parF* and *parG*.

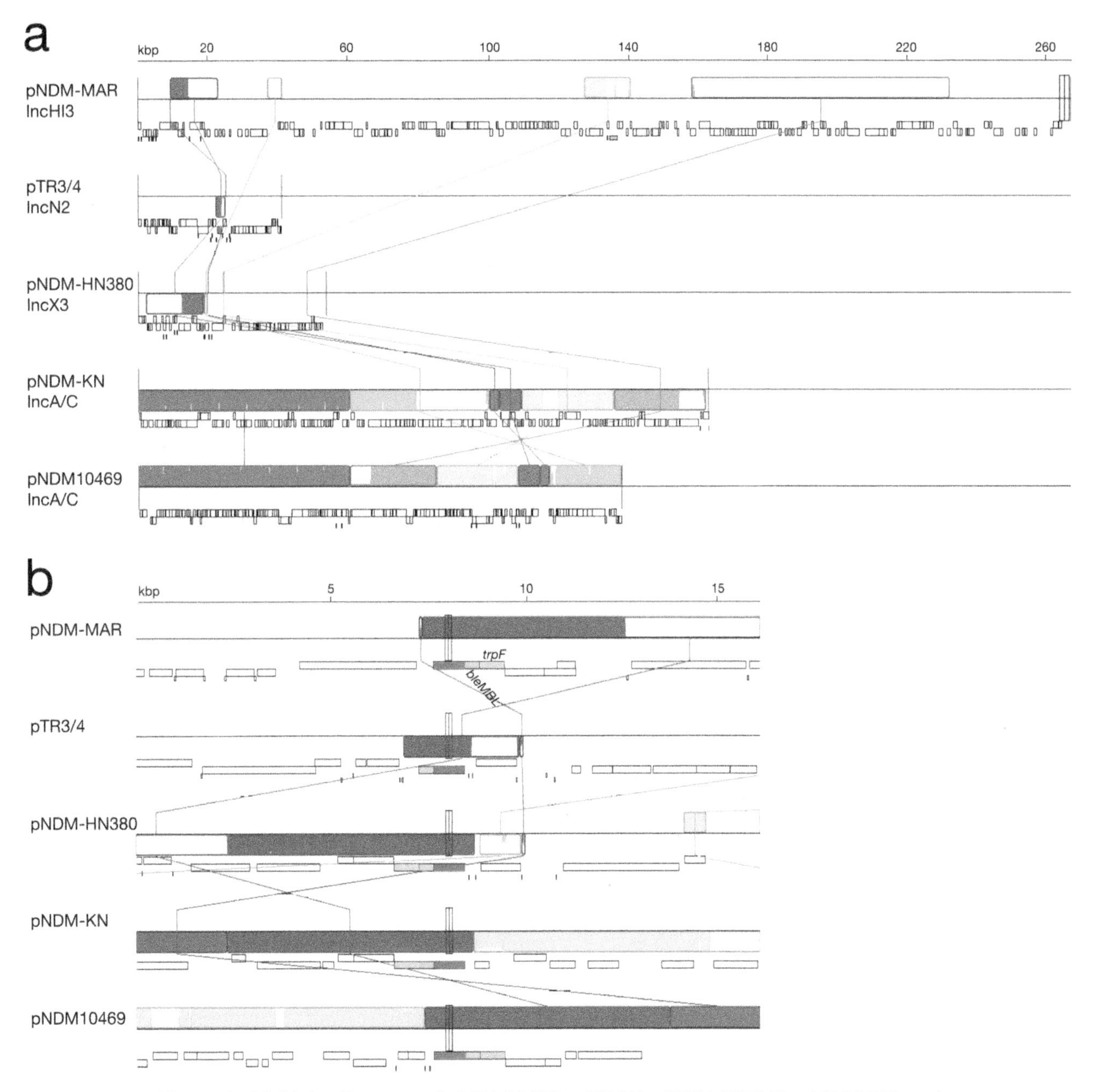

Figure 6 Multiple alignment of pNDM-MAR, pTR3/4, pNDM-HN380, pNDM-KN, and pNDM10469. The nucleotide sequences of pNDM-MAR (accession number JN420336) (77), pTR3/4 (accession number JQ349086) (152), pNDM-HN380 (accession number JX104760) (153), pNDM-KN (accession number JN157804) (154), and pNDM10469 (accession number JN861072) were compared using the MAUVE aligner version 2.3.1 (150). The bla_{NDM-1} gene is represented in red; genes *ble*MBL and *trpF* are represented in light blue and light brown, respectively. Plasmids pTR3 and pTR4, originally thought to be similar but not identical were later proved to be identical and were renamed pTR3/4 (152). (**a**) The comparison of the complete nucleotide sequence is shown with LCBs represented in blocks of different colors. (**b**) Zoom-in of the region including the bla_{NDM-1} gene. doi:10.1128/microbiolspec.PLAS-0016-2013.f6

The resistance genes in pMET1 reside in Tn*1331.2*, an 11,042-bp transposon highly related to Tn*1331*, the transposon found in pJHCMW1 (see above). Tn*1331.2* has a perfect duplication of the 3,047-bp DNA region that includes *aac(6′)-Ib, aadA1*, and bla_{OXA-9} (37, 128).

The resistance to antibiotics encoded by some of the large *K. pneumoniae* plasmids plays a central role in enhancing the morbidity and mortality of *K. pneumoniae* due to the advent of the carbapenemases, in particular KPC (*K. pneumoniae*carbapenemase) and NDM-1 (New Delhi metallo-β-lactamase-1), creating a significant public health threat (71, 132, 133). Both bla_{KPC} and bla_{NDM-1} were first detected in *K. pneumoniae*, but they are present in other Gram-negatives. The emergence of carbapenemases has virtually eliminated the possibility of using β-lactams to treat multidrug-resistant Gram-negative bacteria infections (134). Furthermore, the presence of genes coding for carbapenemases is usually accompanied by others that code for resistance to other β-lactams, aminoglycosides, and fluoroquinolones, which leaves few treatment options (34, 100, 135). In particular, the common localization of carbapenemase-coding genes in plasmids has renewed the interest in their study, and numerous large plasmids' coding for carbapenemases (KPC, VIM, NDM, IMP, and OXA types) and CTX-M-type extended spectrum β-lactamases have been fully sequenced (71, 132, 136–138). Since there is consensus that currently we are witnessing a convergence of two plasmid-driven epidemics, one of bla_{KPC} and the other of bla_{NDM} carrying *K. pneumoniae* (100, 134, 139–141), in the following paragraphs we describe some representative characteristics about plasmids including bla_{NDM} and bla_{KPC} genes.

The carbapenemase NDM-1 is a broad-spectrum metallo-β-lactamase that mediates inactivation of nearly all β-lactams, with the exception of aztreonam. It was first identified in a *K. pneumoniae* isolate from a urinary tract infection (142) and has rapidly disseminated among *Enterobacteriaceae* and *Acinetobacter* spp. (143). Plasmids carrying bla_{NDM-1} belonging to different incompatibility groups have already been isolated. Figure 6 shows a comparative diagram of sequenced *K. pneumoniae* plasmids that harbor this gene. Although plasmids carrying the gene can be largely unrelated, belonging to different incompatibility groups (Fig. 6a), the immediate environment of bla_{NDM-1} is highly related in all plasmids (Fig. 6b). The bla_{KPC} gene is found as part of the 10-kbp transposon Tn*4401*, which is present in numerous multiresistance plasmids belonging to different incompatibility groups. It may be worth mentioning that Tn*4401* has not yet been found in any chromosome. The large number of different conjugative plasmids allowed the rapid and successful spreading of bla_{KPC} to other Gram-negatives. Of special interest is the recent report of the IncI plasmid pBK15692 in which Tn*4401* has been inserted within the *tnpA* gene of Tn*1331* (Fig. 7), creating a transposon with the capability to confer resistance to virtually all β-lactams and aminoglycosides. The dissemination of pBK15692 will increase the already complicated landscape of Gram-negative bacterial infections.

CONCLUDING REMARKS

Most interactions between humans and bacteria do not result in disease, and many of them are beneficial for one or both interacting partners. When the interaction results in disease, the relation between humans and pathogenic bacteria has been one resembling an arms race through evolution. As the human body was developing and evolving numerous strategies to defend against bacterial infections, bacteria were doing the same to counter those defenses that limit or prevent the establishment of invading bacteria. In addition, an artificial defense that humans have counted on for a few decades is the utilization of antibiotics. The evading strategies that permit bacteria to colonize or cause

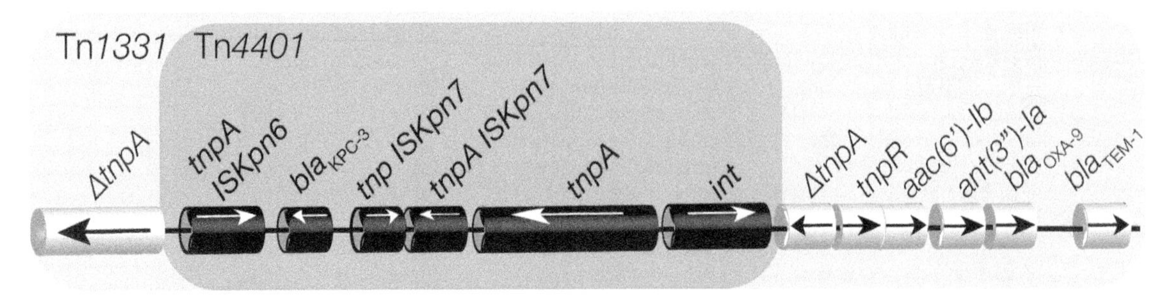

Figure 7 Genetic map of the Tn*1331*: :Tn*4401* region in the *K. pneumoniae* plasmid pBK15692. doi:10.1128/microbiolspec.PLAS-0016-2013.f7

damage to the host are known as virulence factors. The presence or absence of a virulence factor can determine whether a bacterium behaves as a pathogen in normal conditions. Numerous virulence factors as well as antibiotic resistance genes are usually part of plasmids or genetic elements located in plasmids that have the capability to disseminate at the molecular level such as integrons or transposons. This intense activity permits the generation of new variations of plasmid molecules that can accumulate genetic determinants for several virulence factors and resistance. This dissemination in combination with the tremendous tendency that plasmids have to disseminate at the cellular level results in the virtual elimination of barriers among different kinds of bacteria, permitting these genes to reach nearly all bacteria (144, 145). Plasmids play an essential role in this fluid situation in which genes behave as a pool being able to reach any bacteria, transforming them from friend into foe or from susceptible into resistant. This chapter illustrates the rich variety of plasmids that can harbor numerous virulence factors and resistance genes in *K. pneumoniae*. However, careful examination of the existing knowledge shows that we are barely scratching the surface. There is a wealth of information that still needs to be acquired about these plasmids. Their study using the most modern technologies will enhance the possibilities to design new strategies to deal with emerging infectious diseases that represent a serious threat to human health.

Acknowledgments. *The authors' work cited in this review article was funded by Public Health Service grant 2R15AI047115 (to MET) from the National Institutes of Health and PIP2011/2013 number 11420100100152 (to MSR). Conflicts of interest: We disclose no conflicts.*

Citation. Ramirez MS, Traglia GM, Lin DL, Tran T, Tolmasky ME. 2014. Plasmid-mediated antibiotic resistance and virulence in Gram-negatives: the *Klebsiella pneumoniae* paradigm. Microbiol Spectrum 2(5):PLAS-0016-2013.

References

1. **Cohen SN.** 1993. Bacterial plasmids: their extraordinary contribution to molecular genetics. *Gene* **135:**67–76.
2. **Lederberg J, Tatum EL.** 1946. Gene recombination in *Escherichia coli*. *Nature* **158:**558.
3. **Tatum EL, Lederberg J.** 1947. Gene recombination in the bacterium *Escherichia coli*. *J Bacteriol* **53:**673–684.
4. **Datta N.** 1985. Plasmids as organisms, p 3–16. *In* Helinski D, Cohen S, Clewell D, Jackson D, Hollaender A (ed), *Plasmids in Bacteria*, vol. **30.** Plenum Press, New York, NY.
5. **Watanabe T, Fukasawa T.** 1961. Episome-mediated transfer of drug resistance in *Enterobacteriaceae*. I. Transfer of resistance factors by conjugation. *J Bacteriol* **81:**669–678.
6. **Watanabe T.** 1963. Infective heredity of multiple drug resistance in bacteria. *Bacteriol Rev* **27:**87–115.
7. **Hammerl JA, Freytag B, Lanka E, Appel B, Hertwig S.** 2012. The pYV virulence plasmids of *Yersinia pseudotuberculosis* and *Y. pestis* contain a conserved DNA region responsible for the mobilization by the self-transmissible plasmid pYE854. *Environ Microbiol Rep* **4:**433–438.
8. **Stephens C, Murray W.** 2001. Pathogen evolution: how good bacteria go bad. *Curr Biol* **11:**R53–R56.
9. **Elwell LP, Shipley PL.** 1980. Plasmid-mediated factors associated with virulence of bacteria to animals. *Annu Rev Microbiol* **34:**465–496.
10. **Guiney DG, Fang FC, Krause M, Libby S.** 1994. Plasmid-mediated virulence genes in non-typhoid *Salmonella* serovars. *FEMS Microbiol Lett* **124:**1–9.
11. **Johnson TJ, Nolan LK.** 2009. Pathogenomics of the virulence plasmids of *Escherichia coli*. *Microbiol Mol Biol Rev* **73:**750–774.
12. **Tolmasky ME, Crosa JH.** 1991. Regulation of plasmid-mediated iron transport and virulence in *Vibrio anguillarum*. *Biol Met* **4:**33–35.
13. **Actis LA, Tolmasky ME, Crosa JH.** 2011. Vibriosis, p 570–605. *In* Woo PT, Bruno DW (ed), Fish Diseases and Disorders, vol. **3**, *Viral, Bacterial and Fungal Infections.* Cab International Publishing, Wallingford, UK.
14. **Shannon JG, Hasenkrug AM, Dorward DW, Nair V, Carmody AB, Hinnebusch BJ.** 2013. *Yersinia pestis* subverts the dermal neutrophil response in a mouse model of bubonic plague. *MBio* **4:**e00170–13.
15. **Matsui H, Bacot CM, Garlington WA, Doyle TJ, Roberts S, Gulig PA.** 2001. Virulence plasmid-borne *spvB* and *spvC* genes can replace the 90-kilobase plasmid in conferring virulence to *Salmonella enterica* serovar Typhimurium in subcutaneously inoculated mice. *J Bacteriol* **183:**4652–4658.
16. **Fabrega A, Vila J.** 2013. *Salmonella enterica* serovar Typhimurium skills to succeed in the host: virulence and regulation. *Clin Microbiol Rev* **26:**308–341.
17. **Waters VL, Crosa JH.** 1991. Colicin V virulence plasmids. *Microbiol Rev* **55:**437–450.
18. **Drancourt M.** 2012. Plague in the genomic area. *Clin Microbiol Infect* **18:**224–230.
19. **Wajima T, Sabui S, Kano S, Ramamurthy T, Chatterjee NS, Hamabata T.** 2013. Entire sequence of the colonization factor coli surface antigen 6-encoding plasmid pCss165 from an enterotoxigenic *Escherichia coli* clinical isolate. *Plasmid* **70:**343–352.
20. **Crosa JH, Actis LA, Mitoma Y, Perez-Casal J, Tolmasky ME, Valvano M.** 1985. Plasmid-mediated iron sequestering systems in pathogenic strains of *Vibrio anguillarum* and *Escherichia coli*, p 759–774. *In* Helinski D, Cohen S, Clewell D, Jackson D, Hollaender A (ed), *Plasmids in Bacteria.* Plenum Press, New York, NY.
21. **Infectious Diseases Society of America.** 2010. The 10 x '20 Initiative: pursuing a global commitment to develop 10 new antibacterial drugs by 2020. *Clin Infect Dis* **50:** 1081–1083.
22. **Spellberg B, Guidos R, Gilbert D, Bradley J, Boucher HW, Scheld WM, Bartlett JG, Edwards J Jr, Infectious**

Diseases Society of America. 2008. The epidemic of antibiotic-resistant infections: a call to action for the medical community from the Infectious Diseases Society of America. *Clin Infect Dis* **46:**155–164.

23. **Shlaes DM, Sahm D, Opiela C, Spellberg B.** 2013. The FDA reboot of antibiotic development. *Antimicrob Agents Chemother* **57:**4605–4607.

24. **Spellberg B, Bartlett JG, Gilbert DN.** 2013. The future of antibiotics and resistance. *N Engl J Med* **368:** 299–302.

25. **Boucher HW, Talbot GH, Benjamin DK Jr, Bradley J, Guidos RJ, Jones RN, Murray BE, Bonomo RA, Gilbert D, Infectious Diseases Society of America.** 2013. 10 x '20 progress: development of new drugs active against gram-negative bacilli: an update from the Infectious Diseases Society of America. *Clin Infect Dis* **56:** 1685–1694.

26. **Boucher HW, Talbot GH, Bradley JS, Edwards JE, Gilbert D, Rice LB, Scheld M, Spellberg B, Bartlett J.** 2009. Bad bugs, no drugs: no ESKAPE! An update from the Infectious Diseases Society of America. *Clin Infect Dis* **48:**1–12.

27. **Rice LB.** 2010. Progress and challenges in implementing the research on ESKAPE pathogens. *Infect Control Hosp Epidemiol* **31**(Suppl 1)**:**S7–S10.

28. **Kuehn BM.** 2013. "Nightmare" bacteria on the rise in US hospitals, long-term care facilities. *JAMA* **309:** 1573–1574.

29. **Rice LB.** 2009. The clinical consequences of antimicrobial resistance. *Curr Opin Microbiol* **12:**476–481.

30. **Perez F, Endimiani A, Hujer KM, Bonomo RA.** 2007. The continuing challenge of ESBLs. *Curr Opin Pharmacol* **7:**459–469.

31. **Coque TM, Oliver A, Perez-Diaz JC, Baquero F, Canton R.** 2002. Genes encoding TEM-4, SHV-2, and CTX-M-10 extended-spectrum beta-lactamases are carried by multiple *Klebsiella pneumoniae* clones in a single hospital (Madrid, 1989 to 2000). *Antimicrob Agents Chemother* **46:**500–510.

32. **Daza R, Gutierrez J, Piedrola G.** 2001. Antibiotic susceptibility of bacterial strains isolated from patients with community-acquired urinary tract infections. *Int J Antimicrob Agents* **18:**211–215.

33. **Liam CK, Lim KH, Wong CM.** 2001. Community-acquired pneumonia in patients requiring hospitalization. *Respirology* **6:**259–264.

34. **Nordmann P, Cuzon G, Naas T.** 2009. The real threat of *Klebsiella pneumoniae* carbapenemase-producing bacteria. *Lancet Infect Dis* **9:**228–236.

35. **Poulou A, Voulgari E, Vrioni G, Koumaki V, Xidopoulos G, Chatzipantazi V, Markou F, Tsakris A.** 2013. Outbreak caused by an ertapenem-resistant, CTX-M-15-producing *Klebsiella pneumoniae* ST101 clone carrying an OmpK36 porin variant. *J Clin Microbiol* **51:**3176–3182.

36. **Ramirez MS, Xie G, Marshall SH, Hujer KM, Chain PS, Bonomo RA, Tolmasky ME.** 2012. Multidrug-resistant (MDR) *Klebsiella pneumoniae* clinical isolates: a zone of high heterogeneity (HHZ) as a tool for epidemiological studies. *Clin Microbiol Infect* **18:** E254–E258.

37. **Tolmasky ME, Chamorro RM, Crosa JH, Marini PM.** 1988. Transposon-mediated amikacin resistance in *Klebsiella pneumoniae*. *Antimicrob Agents Chemother* **32:**1416–1420.

38. **Woloj M, Tolmasky ME, Roberts MC, Crosa JH.** 1986. Plasmid-encoded amikacin resistance in multiresistant strains of *Klebsiella pneumoniae* isolated from neonates with meningitis. *Antimicrob Agents Chemother* **29:** 315–319.

39. **Hsueh PR, Wu JJ, Teng LJ, Chen YC, Yang PC, Ho SW, Luh KT.** 2002. Primary liver abscess caused by one clone of *Klebsiella pneumoniae* with two colonial morphotypes and resistotypes. *Emerg Infect Dis* **8:** 100–102.

40. **Siu LK, Yeh KM, Lin JC, Fung CP, Chang FY.** 2012. *Klebsiella pneumoniae* liver abscess: a new invasive syndrome. *Lancet Infect Dis* **12:**881–887.

41. **Suzuki K, Nakamura A, Enokiya T, Iwashita Y, Tomatsu E, Muraki Y, Kaneko T, Okuda M, Katayama N, Imai H.** 2013. Septic arthritis subsequent to urosepsis caused by hypermucoviscous *Klebsiella pneumoniae*. *Intern Med* **52:**1641–1645.

42. **Huang HY, Wu YH, Kuo CF.** 2013. *Klebsiella pneumoniae* sepsis with unusual cutaneous presentation of generalized pustulosis. *Clin Exp Dermatol* **38:**626–629.

43. **Rashid T, Wilson C, Ebringer A.** 2013. The link between ankylosing spondylitis, Crohn's disease, *Klebsiella*, and starch consumption. *Clin Dev Immunol* **2013:**872632.

44. **Ebringer A, Rashid T, Tiwana H, Wilson C.** 2007. A possible link between Crohn's disease and ankylosing spondylitis via *Klebsiella* infections. *Clin Rheumatol* **26:** 289–297.

45. **Rashid T, Ebringer A.** 2007. Ankylosing spondylitis is linked to *Klebsiella*: the evidence. *Clin Rheumatol* **26:** 858–864.

46. **Patel G, Bonomo RA.** 2013. "Stormy waters ahead": global emergence of carbapenemases. *Front Microbiol* **4:**48.

47. **Pendleton JN, Gorman SP, Gilmore BF.** 2013. Clinical relevance of the ESKAPE pathogens. *Expert Rev Anti Infect Ther* **11:**297–308.

48. **Liu P, Li P, Jiang X, Bi D, Xie Y, Tai C, Deng Z, Rajakumar K, Ou HY.** 2012. Complete genome sequence of *Klebsiella pneumoniae* subsp. *pneumoniae* HS11286, a multidrug-resistant strain isolated from human sputum. *J Bacteriol* **194:**1841–1842.

49. **Tolmasky ME, Actis LA, Crosa JH.** 2010. Plasmid DNA replication, p 3931–3953. *In* Flickinger M (ed), *Encyclopedia of Industrial Biotechnology: Bioprocess, Bioseparation, and Cell Technology*, vol. 6. John Wiley and Sons, New York, NY.

50. **Actis LA, Tolmasky ME, Crosa JH.** 1999. Bacterial plasmids: replication of extrachromosomal genetic elements encoding resistance to antimicrobial compounds. *Front Biosci* **4:**D43–D62.

51. **Polisky B.** 1988. ColE1 replication control circuitry: sense from antisense. *Cell* **55:**929–932.

52. **Allen JM, Simcha DM, Ericson NG, Alexander DL, Marquette JT, Van Biber BP, Troll CJ, Karchin R, Bielas JH, Loeb LA, Camps M.** 2011. Roles of DNA polymerase I in leading and lagging-strand replication defined by a high-resolution mutation footprint of ColE1 plasmid replication. *Nucleic Acids Res* **39:**7020–7033.
53. **Eguchi Y, Tomizawa J.** 1991. Complexes formed by complementary RNA stem-loops. Their formations, structures and interaction with ColE1 Rom protein. *J Mol Biol* **220:**831–842.
54. **Zakharova MV, Beletskaya IV, Denjmukhametov MM, Yurkova TV, Semenova LM, Shlyapnikov MG, Solonin AS.** 2002. Characterization of pECL18 and pKPN2: a proposed pathway for the evolution of two plasmids that carry identical genes for a type II restriction-modification system. *Mol Genet Genomics* **267:**171–178.
55. **Riley MA, Pinou T, Wertz JE, Tan Y, Valletta CM.** 2001. Molecular characterization of the klebicin B plasmid of *Klebsiella pneumoniae. Plasmid* **45:**209–221.
56. **Smillie C, Garcillan-Barcia MP, Francia MV, Rocha EP, de la Cruz F.** 2010. Mobility of plasmids. *Microbiol Mol Biol Rev* **74:**434–452.
57. **Tolmasky ME, Roberts M, Woloj M, Crosa JH.** 1986. Molecular cloning of amikacin resistance determinants from a *Klebsiella pneumoniae* plasmid. *Antimicrob Agents Chemother* **30:**315–320.
58. **Tolmasky ME, Crosa JH.** 1987. Tn*1331*, a novel multiresistance transposon encoding resistance to amikacin and ampicillin in *Klebsiella pneumoniae. Antimicrob Agents Chemother* **31:**1955–1960.
59. **Tolmasky ME.** 1990. Sequencing and expression of *aadA*, *bla*, and *tnpR* from the multiresistance transposon Tn*1331*. *Plasmid* **24:**218–226.
60. **Tolmasky ME, Crosa JH.** 1993. Genetic organization of antibiotic resistance genes (*aac(6′)-Ib*, *aadA*, and *oxa9*) in the multiresistance transposon Tn*1331*. *Plasmid* **29:** 31–40.
61. **Sarno R, McGillivary G, Sherratt DJ, Actis LA, Tolmasky ME.** 2002. Complete nucleotide sequence of *Klebsiella pneumoniae* multiresistance plasmid pJHCMW1. *Antimicrob Agents Chemother* **46:**3422–3427.
62. **Dery KJ, Chavideh R, Waters V, Chamorro R, Tolmasky LS, Tolmasky ME.** 1997. Characterization of the replication and mobilization regions of the multiresistance *Klebsiella pneumoniae* plasmid pJHCMW1. *Plasmid* **38:**97–105.
63. **Tolmasky ME, Colloms S, Blakely G, Sherratt DJ.** 2000. Stability by multimer resolution of pJHCMW1 is due to the Tn*1331* resolvase and not to the *Escherichia coli* Xer system. *Microbiology* **146:**581–589.
64. **Pham H, Dery KJ, Sherratt DJ, Tolmasky ME.** 2002. Osmoregulation of dimer resolution at the plasmid pJHCMW1 *mwr* locus by *Escherichia coli* XerCD recombination. *J Bacteriol* **184:**1607–1616.
65. **Bui D, Ramiscal J, Trigueros S, Newmark JS, Do A, Sherratt DJ, Tolmasky ME.** 2006. Differences in resolution of *mwr*-containing plasmid dimers mediated by the *Klebsiella pneumoniae* and *Escherichia coli* XerC recombinases: potential implications in dissemination of antibiotic resistance genes. *J Bacteriol* **188:**2812–2820.
66. **Trigueros S, Tran T, Sorto N, Newmark J, Colloms SD, Sherratt DJ, Tolmasky ME.** 2009. *mwr* Xer site-specific recombination is hypersensitive to DNA supercoiling. *Nucleic Acids Res* **37:**3580–3587.
67. **Ramirez MS, Parenteau TR, Centron D, Tolmasky ME.** 2008. Functional characterization of Tn*1331* gene cassettes. *J Antimicrob Chemother* **62:**669–673.
68. **Ramirez MS, Tolmasky ME.** 2010. Aminoglycoside modifying enzymes. *Drug Resist Updat* **13:**151–171.
69. **Ramirez MS, Nikolaidis N, Tolmasky ME.** 2013. Rise and dissemination of aminoglycoside resistance: the *aac(6′)-Ib* paradigm. *Front Microbiol* **4:**121.
70. **Alavi MR, Antonic V, Ravizee A, Weina PJ, Izadjoo M, Stojadinovic A.** 2011. An *Enterobacter* plasmid as a new genetic background for the transposon Tn*1331*. *Infect Drug Resist* **4:**209–213.
71. **Chen L, Chavda KD, Al Laham N, Melano RG, Jacobs MR, Bonomo RA, Kreiswirth BN.** 2013. Complete nucleotide sequence of a *bla*KPC-harboring IncI2 plasmid and its dissemination in New Jersey and New York hospitals. *Antimicrob Agents Chemother* **57:**5019–5025.
72. **Garcia DC, Woloj M, Kaufman S, Sordelli DO, Pineiro S.** 1995. Sequences related to Tn*1331* associated with multiple antimicrobial resistance in different *Salmonella* serovars. *Int J Antimicrob Agents* **5:**199–202.
73. **Garcia DC, Catalano M, Pineiro S, Woloj M, Kaufman S, Sordelli DO.** 1996. The emergence of resistance to amikacin in *Serratia marcescens* isolates from patients with nosocomial infection. *Int J Antimicrob Agents* **7:** 203–210.
74. **Gootz TD, Lescoe MK, Dib-Hajj F, Dougherty BA, He W, Della-Latta P, Huard RC.** 2009. Genetic organization of transposase regions surrounding *bla*KPC carbapenemase genes on plasmids from *Klebsiella* strains isolated in a New York City hospital. *Antimicrob Agents Chemother* **53:**1998–2004.
75. **Poirel L, Cabanne L, Collet L, Nordmann P.** 2006. Class II transposon-borne structure harboring metallo-beta-lactamase gene *bla*VIM-2 in *Pseudomonas putida. Antimicrob Agents Chemother* **50:**2889–2891.
76. **Rice LB, Carias LL, Hutton RA, Rudin SD, Endimiani A, Bonomo RA.** 2008. The KQ element, a complex genetic region conferring transferable resistance to carbapenems, aminoglycosides, and fluoroquinolones in *Klebsiella pneumoniae. Antimicrob Agents Chemother* **52:**3427–3429.
77. **Villa L, Poirel L, Nordmann P, Carta C, Carattoli A.** 2012. Complete sequencing of an IncH plasmid carrying the *bla*NDM-1, *bla*CTX-M-15 and *qnrB1* genes. *J Antimicrob Chemother* **67:**1645–1650.
78. **Warburg G, Hidalgo-Grass C, Partridge SR, Tolmasky ME, Temper V, Moses AE, Block C, Strahilevitz J.** 2012. A carbapenem-resistant *Klebsiella pneumoniae* epidemic clone in Jerusalem: sequence type 512 carrying a plasmid encoding *aac(6′)-Ib. J Antimicrob Chemother* **67:**898–901.
79. **Reyes-Lamothe R, Tran T, Meas D, Lee L, Li AM, Sherratt DJ, Tolmasky ME.** 2014. High-copy bacterial

plasmids diffuse in the nucleoid-free space, replicate stochastically and are randomly partitioned at cell division. *Nucleic Acids Res* **42:**1042–1051.

80. **Pasquali F, Kehrenberg C, Manfreda G, Schwarz S.** 2005. Physical linkage of Tn*3* and part of Tn*1721* in a tetracycline and ampicillin resistance plasmid from *Salmonella* Typhimurium. *J Antimicrob Chemother* **55:** 562–565.
81. **Tran T, Sherratt DJ, Tolmasky ME.** 2010. *fpr*, a deficient Xer recombination site from a *Salmonella* plasmid, fails to confer stability by dimer resolution: comparative studies with the pJHCMW1 *mwr* site. *J Bacteriol* **192:**883–887.
82. **Garcia-Fernandez A, Villa L, Carta C, Venditti C, Giordano A, Venditti M, Mancini C, Carattoli A.** 2012. *Klebsiella pneumoniae* ST258 producing KPC-3 identified in Italy carries novel plasmids and OmpK36/OmpK35 porin variants. *Antimicrob Agents Chemother* **56:**2143–2145.
83. **Das B, Martinez E, Midonet C, Barre FX.** 2013. Integrative mobile elements exploiting Xer recombination. *Trends Microbiol* **21:**23–30.
84. **Summers DK, Sherratt DJ.** 1984. Multimerization of high copy number plasmids causes instability: ColE1 encodes a determinant essential for plasmid monomerization and stability. *Cell* **36:**1097–1103.
85. **Colloms SD, Sykora P, Szatmari G, Sherratt DJ.** 1990. Recombination at ColE1 *cer* requires the *Escherichia coli xerC* gene product, a member of the lambda integrase family of site-specific recombinases. *J Bacteriol* **172:** 6973–6980.
86. **Cornet F, Mortier I, Patte J, Louarn JM.** 1994. Plasmid pSC101 harbors a recombination site, *psi*, which is able to resolve plasmid multimers and to substitute for the analogous chromosomal *Escherichia coli* site *dif*. *J Bacteriol* **176:**3188–3195.
87. **Summers D.** 1998. Timing, self-control and a sense of direction are the secrets of multicopy plasmid stability. *Mol Microbiol* **29:**1137–1145.
88. **Tran T, Andres P, Petroni A, Soler-Bistue A, Albornoz E, Zorreguieta A, Reyes-Lamothe R, Sherratt DJ, Corso A, Tolmasky ME.** 2012. Small plasmids harboring *qnrB19*: a model for plasmid evolution mediated by site-specific recombination at *oriT* and Xer sites. *Antimicrob Agents Chemother* **56:**1821–1827.
89. **Val ME, Bouvier M, Campos J, Sherratt D, Cornet F, Mazel D, Barre FX.** 2005. The single-stranded genome of phage CTX is the form used for integration into the genome of *Vibrio cholerae*. *Mol Cell* **19:**559–566.
90. **Grosso F, Quinteira S, Poirel L, Novais A, Peixe L.** 2012. Role of common *bla*OXA-24/OXA-40-carrying platforms and plasmids in the spread of OXA-24/OXA-40 among *Acinetobacter* species clinical isolates. *Antimicrob Agents Chemother* **56:**3969–3972.
91. **Blakely GW, Sherratt DJ.** 1994. Interactions of the site-specific recombinases XerC and XerD with the recombination site *dif*. *Nucleic Acids Res* **22:**5613–5620.
92. **Hayes F, Sherratt DJ.** 1997. Recombinase binding specificity at the chromosome dimer resolution site *dif* of *Escherichia coli*. *J Mol Biol* **266:**525–537.
93. **D'Andrea MM, Giani T, D'Arezzo S, Capone A, Petrosillo N, Visca P, Luzzaro F, Rossolini GM.** 2009. Characterization of pABVA01, a plasmid encoding the OXA-24 carbapenemase from Italian isolates of *Acinetobacter baumannii*. *Antimicrob Agents Chemother* **53:**3528–3533.
94. **Merino M, Acosta J, Poza M, Sanz F, Beceiro A, Chaves F, Bou G.** 2010. OXA-24 carbapenemase gene flanked by XerC/XerD-like recombination sites in different plasmids from different *Acinetobacter* species isolated during a nosocomial outbreak. *Antimicrob Agents Chemother* **54:**2724–2727.
95. **Povilonis J, Seputiene V, Krasauskas R, Juskaite R, Miskinyte M, Suziedelis K, Suziedeliene E.** 2013. Spread of carbapenem-resistant *Acinetobacter baumannii* carrying a plasmid with two genes encoding OXA-72 carbapenemase in Lithuanian hospitals. *J Antimicrob Chemother* **68:**1000–1006.
96. **Montealegre MC, Maya JJ, Correa A, Espinal P, Mojica MF, Ruiz SJ, Rosso F, Vila J, Quinn JP, Villegas MV.** 2012. First identification of OXA-72 carbapenemase from *Acinetobacter pittii* in Colombia. *Antimicrob Agents Chemother* **56:**3996–3998.
97. **Francia MV, Varsaki A, Garcillan-Barcia MP, Latorre A, Drainas C, de la Cruz F.** 2004. A classification scheme for mobilization regions of bacterial plasmids. *FEMS Microbiol Rev* **28:**79–100.
98. **Cao V, Lambert T, Courvalin P.** 2002. ColE1-like plasmid pIP843 of *Klebsiella pneumoniae* encoding extended-spectrum beta-lactamase CTX-M-17. *Antimicrob Agents Chemother* **46:**1212–1217.
99. **Le TM, Baker S, Le TP, Le TP, Cao TT, Tran TT, Nguyen VM, Campbell JI, Lam MY, Nguyen TH, Nguyen VV, Farrar J, Schultsz C.** 2009. High prevalence of plasmid-mediated quinolone resistance determinants in commensal members of the *Enterobacteriaceae* in Ho Chi Minh City, Vietnam. *J Med Microbiol* **58:**1585–1592.
100. **Nordmann P, Naas T, Poirel L.** 2011. Global spread of carbapenemase-producing *Enterobacteriaceae*. *Emerg Infect Dis* **17:**1791–1798.
101. **Nassif X, Sansonetti PJ.** 1986. Correlation of the virulence of *Klebsiella pneumoniae* K1 and K2 with the presence of a plasmid encoding aerobactin. *Infect Immun* **54:**603–608.
102. **Nassif X, Honore N, Vasselon T, Cole ST, Sansonetti PJ.** 1989. Positive control of colanic acid synthesis in *Escherichia coli* by *rmpA* and *rmpB*, two virulence-plasmid genes of *Klebsiella pneumoniae*. *Mol Microbiol* **3:**1349–1359.
103. **Nassif X, Fournier JM, Arondel J, Sansonetti PJ.** 1989. Mucoid phenotype of *Klebsiella pneumoniae* is a plasmid-encoded virulence factor. *Infect Immun* **57:** 546–552.
104. **Wacharotayankun R, Arakawa Y, Ohta M, Tanaka K, Akashi T, Mori M, Kato N.** 1993. Enhancement of extracapsular polysaccharide synthesis in *Klebsiella pneumoniae* by RmpA2, which shows homology to NtrC and FixJ. *Infect Immun* **61:**3164–3174.
105. **Perry RD, Fetherston JD.** 2011. Yersiniabactin iron uptake: mechanisms and role in *Yersinia pestis* pathogenesis. *Microbes Infect* **13:**808–817.

106. **Zimbler DL, Penwell WF, Gaddy JA, Menke SM, Tomaras AP, Connerly PL, Actis LA.** 2009. Iron acquisition functions expressed by the human pathogen *Acinetobacter baumannii*. *Biometals* **22:**23–32.

107. **Crosa JH.** 1997. Signal transduction and transcriptional and posttranscriptional control of iron-regulated genes in bacteria. *Microbiol Mol Biol Rev* **61:**319–336.

108. **Chen Q, Wertheimer AM, Tolmasky ME, Crosa JH.** 1996. The AngR protein and the siderophore anguibactin positively regulate the expression of iron-transport genes in *Vibrio anguillarum*. *Mol Microbiol* **22:**127–134.

109. **Darfeuille-Michaud A, Jallat C, Aubel D, Sirot D, Rich C, Sirot J, Joly B.** 1992. R-plasmid-encoded adhesive factor in *Klebsiella pneumoniae* strains responsible for human nosocomial infections. *Infect Immun* **60:**44–55.

110. **Yu VL, Hansen DS, Ko WC, Sagnimeni A, Klugman KP, von Gottberg A, Goossens H, Wagener MM, Benedi VJ, International Klebseilla Study Group.** 2007. Virulence characteristics of *Klebsiella* and clinical manifestations of *K. pneumoniae* bloodstream infections. *Emerg Infect Dis* **13:**986–993.

111. **Vernet V, Madoulet C, Chippaux C, Philippon A.** 1992. Incidence of two virulence factors (aerobactin and mucoid phenotype) among 190 clinical isolates of *Klebsiella pneumoniae* producing extended-spectrum betalactamase. *FEMS Microbiol Lett* **75:**1–5.

112. **Chen YT, Chang HY, Lai YC, Pan CC, Tsai SF, Peng HL.** 2004. Sequencing and analysis of the large virulence plasmid pLVPK of *Klebsiella pneumoniae* CG43. *Gene* **337:**189–198.

113. **Sorsa LJ, Dufke S, Heesemann J, Schubert S.** 2003. Characterization of an iroBCDEN gene cluster on a transmissible plasmid of uropathogenic *Escherichia coli*: evidence for horizontal transfer of a chromosomal virulence factor. *Infect Immun* **71:**3285–3293.

114. **Braun V, Mahren S, Ogierman M.** 2003. Regulation of the FecI-type ECF sigma factor by transmembrane signalling. *Curr Opin Microbiol* **6:**173–180.

115. **Mahren S, Schnell H, Braun V.** 2005. Occurrence and regulation of the ferric citrate transport system in *Escherichia coli* B, *Klebsiella pneumoniae*, *Enterobacter aerogenes*, and *Photorhabdus luminescens*. *Arch Microbiol* **184:**175–186.

116. **Wu KM, Li LH, Yan JJ, Tsao N, Liao TL, Tsai HC, Fung CP, Chen HJ, Liu YM, Wang JT, Fang CT, Chang SC, Shu HY, Liu TT, Chen YT, Shiau YR, Lauderdale TL, Su IJ, Kirby R, Tsai SF.** 2009. Genome sequencing and comparative analysis of *Klebsiella pneumoniae* NTUH-K2044, a strain causing liver abscess and meningitis. *J Bacteriol* **191:**4492–4501.

117. **Shin SH, Kim S, Kim JY, Lee S, Um Y, Oh MK, Kim YR, Lee J, Yang KS.** 2012. Complete genome sequence of the 2,3-butanediol-producing *Klebsiella pneumoniae* strain KCTC 2242. *J Bacteriol* **194:**2736–2737.

118. **Dolejska M, Villa L, Dobiasova H, Fortini D, Feudi C, Carattoli A.** 2013. Plasmid content of a clinically relevant *Klebsiella pneumoniae* clone from the Czech Republic producing CTX-M-15 and QnrB1. *Antimicrob Agents Chemother* **57:**1073–1076.

119. **Taylor DE, Rooker M, Keelan M, Ng LK, Martin I, Perna NT, Burland NT, Blattner FR.** 2002. Genomic variability of O islands encoding tellurite resistance in enterohemorrhagic *Escherichia coli* O157:H7 isolates. *J Bacteriol* **184:**4690–4698.

120. **Borremans B, Hobman JL, Provoost A, Brown NL, van Der Lelie D.** 2001. Cloning and functional analysis of the pbr lead resistance determinant of *Ralstonia metallidurans* CH34. *J Bacteriol* **183:**5651–5658.

121. **Brown NL, Barrett SR, Camakaris J, Lee BT, Rouch DA.** 1995. Molecular genetics and transport analysis of the copper-resistance determinant (*pco*) from *Escherichia coli* plasmid pRJ1004. *Mol Microbiol* **17:** 1153–1166.

122. **Gupta A, Matsui K, Lo JF, Silver S.** 1999. Molecular basis for resistance to silver cations in *Salmonella*. *Nat Med* **5:**183–188.

123. **Huang TW, Chen TL, Chen YT, Lauderdale TL, Liao TL, Lee YT, Chen CP, Liu YM, Lin AC, Chang YH, Wu KM, Kirby R, Lai JF, Tan MC, Siu LK, Chang CM, Fung CP, Tsai SF.** 2013. Copy number change of the NDM-1 sequence in a multidrug-resistant *Klebsiella pneumoniae* clinical isolate. *PLoS One* **8:**e62774.

124. **S andegren L, Linkevicius M, Lytsy B, Melhus A, Andersson DI.** 2012. Transfer of an *Escherichia coli* ST131 multiresistance cassette has created a *Klebsiella pneumoniae*-specific plasmid associated with a major nosocomial outbreak. *J Antimicrob Chemother* **67:** 74–83.

125. **Bojer MS, Struve C, Ingmer H, Hansen DS, Krogfelt KA.** 2010. Heat resistance mediated by a new plasmid encoded Clp ATPase, ClpK, as a possible novel mechanism for nosocomial persistence of *Klebsiella pneumoniae*. *PLoS One* **5:**e15467.

126. **Bojer MS, Hammerum AM, Jorgensen SL, Hansen F, Olsen SS, Krogfelt KA, Struve C.** 2012. Concurrent emergence of multidrug resistance and heat resistance by CTX-M-15-encoding conjugative plasmids in *Klebsiella pneumoniae*. *APMIS* **120:**699–705.

127. **Guglielmini J, de la Cruz F, Rocha EP.** 2013. Evolution of conjugation and type IV secretion systems. *Mol Biol Evol* **30:**315–331.

128. **Soler Bistue AJ, Birshan D, Tomaras AP, Dandekar M, Tran T, Newmark J, Bui D, Gupta N, Hernandez K, Sarno R, Zorreguieta A, Actis LA, Tolmasky ME.** 2008. *Klebsiella pneumoniae* multiresistance plasmid pMET1: similarity with the *Yersinia pestis* plasmid pCRY and integrative conjugative elements. *PLoS One* **3:**e1800.

129. **Song Y, Tong Z, Wang J, Wang L, Guo Z, Han Y, Zhang J, Pei D, Zhou D, Qin H, Pang X, Han Y, Zhai J, Li M, Cui B, Qi Z, Jin L, Dai R, Chen F, Li S, Ye C, Du Z, Lin W, Wang J, Yu J, Yang H, Wang J, Huang P, Yang R.** 2004. Complete genome sequence of *Yersinia pestis* strain 91001, an isolate avirulent to humans. *DNA Res* **11:**179–197.

130. **Stephan R, Lehner A, Tischler P, Rattei T.** 2011. Complete genome sequence of *Cronobacter turicensis* LMG 23827, a food-borne pathogen causing deaths in neonates. *J Bacteriol* **193:**309–310.

131. **Schubert S, Dufke S, Sorsa J, Heesemann J.** 2004. A novel integrative and conjugative element (ICE) of *Escherichia coli*: the putative progenitor of the *Yersinia* high-pathogenicity island. *Mol Microbiol* **51:**837–848.

132. **Carattoli A.** 2013. Plasmids and the spread of resistance. *Int J Med Microbiol* **303:**298–304.

133. **Schultsz C, Geerlings S.** 2012. Plasmid-mediated resistance in *Enterobacteriaceae*: changing landscape and implications for therapy. *Drugs* **72:**1–16.

134. **Munoz-Price LS, Poirel L, Bonomo RA, Schwaber MJ, Daikos GL, Cormican M, Cornaglia G, Garau J, Gniadkowski M, Hayden MK, Kumarasamy K, Livermore DM, Maya JJ, Nordmann P, Patel JB, Paterson DL, Pitout J, Villegas MV, Wang H, Woodford N, Quinn JP.** 2013. Clinical epidemiology of the global expansion of *Klebsiella pneumoniae* carbapenemases. *Lancet Infect Dis* **13:**785–796.

135. **Nordmann P, Poirel L, Walsh TR, Livermore DM.** 2011. The emerging NDM carbapenemases. *Trends Microbiol* **19:**588–595.

136. **Chen L, Chavda KD, Melano RG, Hong T, Rojtman AD, Jacobs MR, Bonomo RA, Kreiswirth BN.** 2014. Molecular survey of the dissemination of two *bla*KPC-harboring IncFIA plasmids in New Jersey and New York hospitals. *Antimicrob Agents Chemother* **58:**2289–2294.

137. **Chen L, Chavda KD, Melano RG, Jacobs MR, Koll B, Hong T, Rojtman AD, Levi MH, Bonomo RA, Kreiswirth BN.** 2014. Comparative genomic analysis of KPC-encoding pKpQIL-like plasmids and their distribution in New Jersey and New York Hospitals. *Antimicrob Agents Chemother* **58:**2871–2877.

138. **Deleo FR, Chen L, Porcella SF, Martens CA, Kobayashi SD, Porter AR, Chavda KD, Jacobs MR, Mathema B, Olsen RJ, Bonomo RA, Musser JM, Kreiswirth BN.** 2014. Molecular dissection of the evolution of carbapenem-resistant multilocus sequence type 258 *Klebsiella pneumoniae*. *Proc Natl Acad Sci USA* **111:**4988–4993.

139. **Giakkoupi P, Papagiannitsis CC, Miriagou V, Pappa O, Polemis M, Tryfinopoulou K, Tzouvelekis LS, Vatopoulos AC.** 2011. An update of the evolving epidemic of *bla*KPC-2-carrying *Klebsiella pneumoniae* in Greece (2009-10). *J Antimicrob Chemother* **66:**1510–1513.

140. **Johnson AP, Woodford N.** 2013. Global spread of antibiotic resistance: the example of New Delhi metallo-beta-lactamase (NDM)-mediated carbapenem resistance. *J Med Microbiol* **62:**499–513.

141. **Snitkin ES, Zelazny AM, Thomas PJ, Stock F, NISC Comparative Sequencing Program Group, Henderson DK, Palmore TN, Segre JA.** 2012. Tracking a hospital outbreak of carbapenem-resistant *Klebsiella pneumoniae* with whole-genome sequencing. *Sci Transl Med* **4:**148ra116.

142. **Yong D, Toleman MA, Giske CG, Cho HS, Sundman K, Lee K, Walsh TR.** 2009. Characterization of a new metallo-beta-lactamase gene, *bla*(NDM-1), and a novel erythromycin esterase gene carried on a unique genetic structure in *Klebsiella pneumoniae* sequence type 14 from India. *Antimicrob Agents Chemother* **53:**5046–5054.

143. **Nordmann P, Poirel L, Toleman MA, Walsh TR.** 2011. Does broad-spectrum beta-lactam resistance due to NDM-1 herald the end of the antibiotic era for treatment of infections caused by Gram-negative bacteria? *J Antimicrob Chemother* **66:**689–692.

144. **Tolmasky ME.** 2007. Overview of dissemination mechanisms of genes coding for resistance to antibiotics, p 267–270. *In* Bonomo RA, Tolmasky ME (ed), *Enzyme-Mediated Resistance to Antibiotics: Mechanisms, Dissemination, and Prospects for Inhibition.* ASM Press, Washington, DC.

145. **Levy S.** 2002. *The Antibiotic Paradox. How the Misuse of Antibiotics Destroys Their Curative Powers*, 2nd ed. Perseus Publishing, Cambridge, MA.

146. **Thompson JD, Higgins DG, Gibson TJ.** 1994. CLUSTAL W: improving the sensitivity of progressive multiple sequence alignment through sequence weighting, position-specific gap penalties and weight matrix choice. *Nucleic Acids Res* **22:**4673–4680.

147. **Colloms SD.** 2013. The topology of plasmid-monomerizing Xer site-specific recombination. *Biochem Soc Trans* **41:**589–594.

148. **Minh PN, Devroede N, Massant J, Maes D, Charlier D.** 2009. Insights into the architecture and stoichiometry of *Escherichia coli* PepA*DNA complexes involved in transcriptional control and site-specific DNA recombination by atomic force microscopy. *Nucleic Acids Res* **37:**1463–1476.

149. **Reijns M, Lu Y, Leach S, Colloms SD.** 2005. Mutagenesis of PepA suggests a new model for the Xer/cer synaptic complex. *Mol Microbiol* **57:**927–941.

150. **Darling AE, Mau B, Perna NT.** 2010. progressiveMauve: multiple genome alignment with gene gain, loss and rearrangement. *PLoS One* **5:**e11147.

151. **Whelan KF, Colleran E, Taylor DE.** 1995. Phage inhibition, colicin resistance, and tellurite resistance are encoded by a single cluster of genes on the IncHI2 plasmid R478. *J Bacteriol* **177:**5016–5027.

152. **Chen YT, Lin AC, Siu LK, Koh TH.** 2012. Sequence of closely related plasmids encoding *bla*(NDM-1) in two unrelated *Klebsiella pneumoniae* isolates in Singapore. *PLoS One* **7:**e48737.

153. **Ho P, Li Z, Lo W, Cheung Y, Lin C, Sham P, Cheng V, Ng T, Que T, Chow K.** 2012. Identification and characterization of a novel incompatibility group X3 plasmid carrying *bla*NDM-1 in *Enterobacteriaceae* isolates with epidemiological links to multiple geographical areas in China. *Emerg Microbes Infect* **1:**e39.

154. **Carattoli A, Villa L, Poirel L, Bonnin RA, Nordmann P.** 2012. Evolution of IncA/C *bla*CMY-(2)-carrying plasmids by acquisition of the *bla*NDM-(1) carbapenemase gene. *Antimicrob Agents Chemother* **56:**783–786.

155. **Zioga A, Whichard JM, Kotsakis SD, Tzouvelekis LS, Tzelepi E, Miriagou V.** 2009. CMY-31 and CMY-36 cephalosporinases encoded by ColE1-like plasmids. *Antimicrob Agents Chemother* **53:**1256–1259.

Plasmids—Biology and Impact in Biotechnology and Discovery
Edited by Marcelo E. Tolmasky and Juan C. Alonso

doi:10.1128/microbiolspec.PLAS-0006-2013

George A. Jacoby[1]
Jacob Strahilevitz[2]
David C. Hooper[3]

25 Plasmid-Mediated Quinolone Resistance

INTRODUCTION

Plasmid-mediated quinolone resistance (PMQR) was late in being discovered. Nalidixic acid, the first quinolone to be used clinically, was introduced in 1967 for urinary tract infections. Resistance was soon observed and could also be readily selected in the laboratory. It was produced by amino acid substitutions in the cellular targets of quinolone action: DNA gyrase and topoisomerase IV (1–3). Later, decreased quinolone accumulation due to pump activation and porin loss was added as an additional resistance mechanism. The search for transferable nalidixic acid resistance in over 500 Gram-negative strains in the 1970s was unrevealing (4). In the 1980s fluoroquinolones became available that were more potent and broader in spectrum. Quinolone usage increased, with subsequent parallel increases in quinolone resistance (5, 6). In 1987 PMQR was reported to be present in a nalidixic acid-resistant isolate of *Shigella dysenteriae* from Bangladesh (7), but this claim was later withdrawn (8). True PMQR was reported in 1998 in a multiresistant urinary *Klebsiella pneumoniae* isolate at the University of Alabama that could transfer low-level resistance to nalidixic acid, ciprofloxacin, and other quinolones to a variety of Gram-negative recipients (9). The responsible gene was termed *qnr*, later amended to *qnrA* as additional *qnr* alleles were discovered. Investigation of a *qnrA* plasmid from Shanghai that provided more than the expected level of ciprofloxacin resistance led to the discovery in 2006 of a second mechanism for PMQR: modification of certain quinolones by a particular aminoglycoside acetyltransferase, AAC(6′)-Ib-cr (10). A third mechanism for PMQR was added in 2007 with the discovery of plasmid-mediated quinolone efflux pumps QepA (11, 12) and OqxAB (13). A multiplex PCR assay for eight PMQR genes (lacking only *qnrVC*) has recently been perfected (14). In the past decade these genes have been found in bacterial isolates from around the world. They reduce the susceptibility of bacteria to quinolones, usually not to the level of nonsusceptibility but facilitating the selection of more quinolone-resistant mutants and treatment failure. PMQR has been frequently reviewed (15–20).

Qnr STRUCTURE AND FUNCTION

Cloning and sequencing *qnrA* revealed that it encoded a 218-residue protein with a tandemly repeating unit of

[1]Lahey Hospital and Medical Center, Burlington, MA 01805; [2]Hadassah-Hebrew University, Jerusalem 91120 Israel; [3]Massachusetts General Hospital, Boston, MA 02114.

five amino acids that indicated membership in the large (more than 1,000 members) pentapeptide repeat family of proteins (21). Knowledge of the sequence allowed the search for *qnrA* by PCR, and it was soon discovered in a growing number of organisms, including other *K. pneumoniae* strains in the United States (22, 23), *Escherichia coli* isolates in Shanghai (24), and *Salmonella enterica* strains in Hong Kong (25). *qnrA* was subsequently followed by discovery of plasmid-mediated *qnrS* (26), *qnrB* (27), *qnrC* (28), and *qnrD* (29). The *qnrVC* gene from *Vibrio cholerae* can also be located in a plasmid (30–33) or in transmissible form as part of an integrating conjugative element (34). These *qnr* genes generally differ in sequence by 35% or more from *qnrA* and each other. Allelic variants have also been described in each family differing by 10% or less: 5 alleles for *qnrVC*, 7 alleles for *qnrA*, 9 for *qnrS*, and 71 for *qnrB* (35) (http://www.lahey.org/qnrstudies/, accessed 12/09/13). *qnr* genes are also found on the chromosome of both Gram-negative and Gram-positive bacteria from both clinical and environmental sources (36–38).

The sequence of pentapeptide repeat proteins can be represented as [S,T,A,V][D,N][L,F][S,T,R][G] (39). The first such protein to have its structure determined by X-ray crystallography was MfpA, encoded on the chromosome of mycobacterial species including *Mycobacterium smegmatis*, where its deletion increased fluoroquinolone susceptibility (40). MfpA is a dimer linked C-terminus to C-terminus and folded into a right-handed quadrilateral β helix with size, shape, and charge mimicking the β form of DNA (41). The middle, usually hydrophobic, amino acid (i) of the pentapeptide repeat and the first polar or hydrophobic residue (i-2) point inward, while the remaining amino acids (i-1, i+1, i+2) are oriented outward, presenting a generally anionic surface. Extensive hydrogen bonding between backbone atoms of neighboring coils stabilizes the helix. The structures of three Qnr proteins are known: EfsQnr from *Enterococcus faecalis* (42), AhQnr from *Aeromonas hydrophila* (43), and plasmid-mediated QnrB1 (44). All are rod-like dimers (Fig. 1). The monomers of QnrB1 and AhQnr have projecting loops of 8 and 12 amino acids that are important for their activity (Fig. 1). Deletion of the smaller A loop reduces quinolone protection, while deletion of the larger B loop or both loops destroys protective activity (43, 44). Deletion of even a single amino acid in the larger loop compromises protective activity (45). MfpA and EfsQnr lack loops, but EfsQnr differs from MfpA in having an additional β-helical rung, a capping peptide, and a 25-amino acid flexible extension required for full protective activity.

Topoisomerases twist and untwist the DNA helix by binding to it and introducing a pair of staggered, single-strand breaks in one segment, through which a second DNA segment is passed (46). Quinolones bind to the complex of enzyme and DNA, stabilizing the cleavage or cleaved complex, blocking religation, and leading ultimately to lethal double-stranded breaks (47). In cell-free systems QnrA (21, 48), QnrB (27, 45, 49), QnrS (50), AhQnr (43), and EfsQnr (42) have been shown to protect *E. coli* DNA gyrase from quinolone inhibition (Fig. 2). Protection of topoisomerase IV by QnrA has been demonstrated as well (51). Protection occurs at low concentrations of Qnr relative to quinolone. Figure 2 shows that for DNA gyrase inhibited by 6 M (2 g/ml) ciprofloxacin, half protection required only 0.5 nM QnrB1, and some protective effect was seen with as little as 5 pM (27). At high QnrB concentrations (25–30 μM in the same system) gyrase inhibition occurs (27, 49). In contrast, MfpA inhibits *M. tuberculosis* or *E. coli* gyrase with an IC_{50} of 1.75 to 3 M and lacks any protective effect against ciprofloxacin. EfsQnr is intermediate. It partially protects *E. coli* gyrase against ciprofloxacin inhibition but also inhibits ATP-dependent supercoiling activity of gyrase with an IC_{50} of 1.2 M (42). In a gel displacement assay QnrA binds to DNA gyrase and its GyrA and GyrB subunits and also to topoisomerase IV and its ParC and ParE subunits (48, 51). Competition between MfpA and substrate DNA for binding to gyrase has been proposed as the mechanism for its inhibitory effect (41). Qnr proteins with their additional structural features (loops, N-terminal extension) are proposed to bind to gyrase and topoisomerase IV targets in such a way as to destabilize the cleavage complex between enzyme, DNA, and quinolone, causing quinolone release, religation of DNA, and regeneration of active topoisomerase (43, 44).

Qnr ORIGIN

Qnr homologs can be found on the chromosome of many *γ-Proteobacteria*, *Firmicutes*, and *Actinomycetales*, including species of *Bacillus*, *Enterococcus*, *Listeria*, and *Mycobacteria*, as well as anaerobes such as *Clostridium difficile* and *Clostridium perfringens* (36, 38, 52, 53). Nearly 50 allelic variants have been found on the chromosome of *Stenotrophomonas maltophilia* (36, 54–57). Aquatic bacteria are especially well represented including species of *Aeromonas*, *Photobacterium*, *Shewanella*, and Vibrio (17, 58, 59). QnrA1 is 98% identical to the chromosomally determined Qnr of *Shewanella algae* (58), QnrS1 is 83% identical to

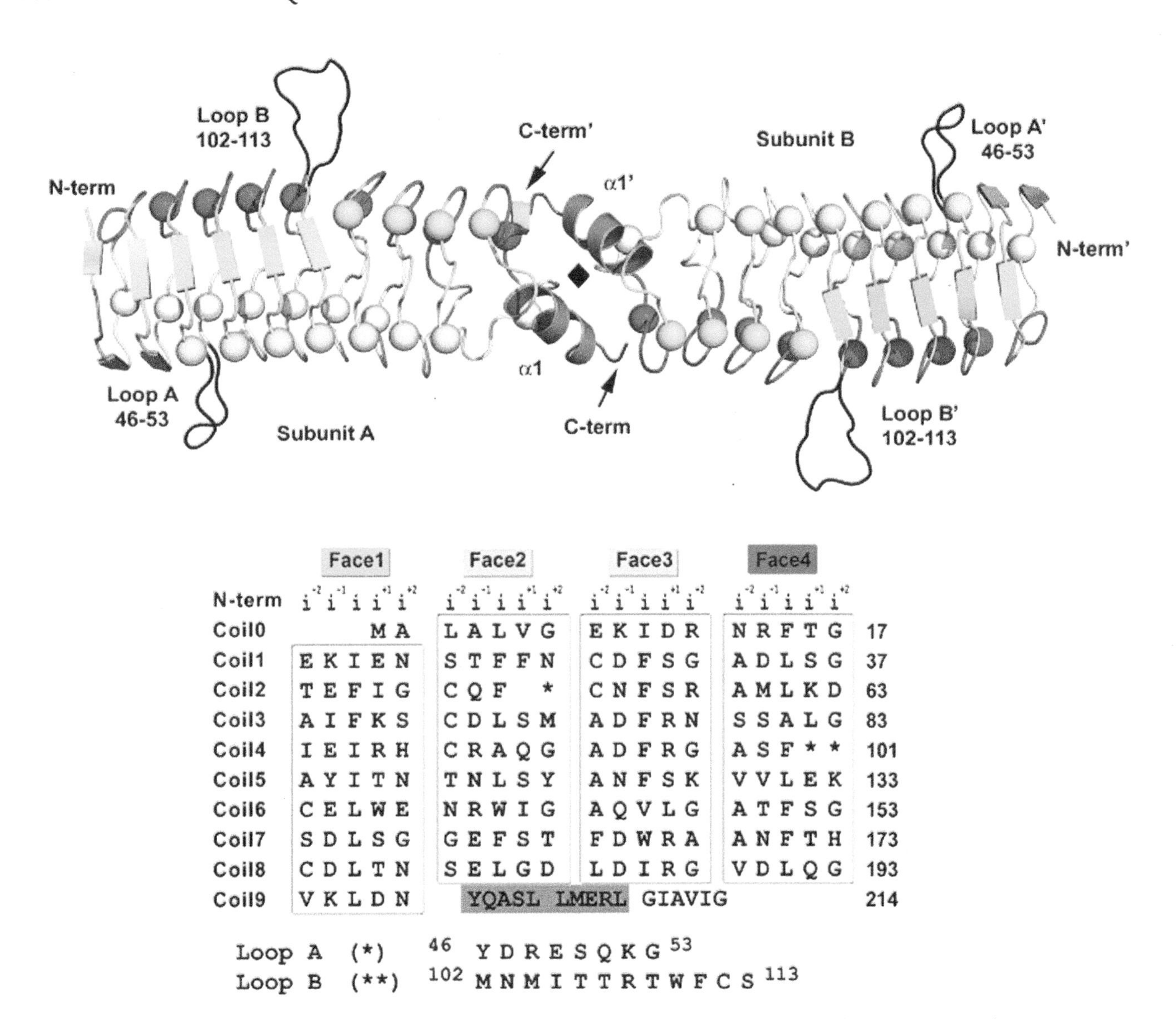

Figure 1 The rod-like structure of the QnrB1 dimer is shown above, with the sequence of the monomer below. The sequence is divided into four columns representing the four faces of the right-handed quadrilateral β-helix. Face names and color are shown at the top along with the naming convention for the five residues of the pentapeptide repeats. Loops A and B are indicated by one and two asterisks, respectively, with their sequences indicated below and the loops shown as black traces on the diagram. The N-terminal α-helix is colored pink. The molecular 2-fold symmetry is indicated with a black diamond. Type II turn-containing faces are shown as spheres, and type IV-containing faces as strands (235). Adapted from the *Journal of Biological Chemistry* (44), copyright 2011, the American Society for Biochemistry and Molecular Biology. doi:10.1128/microbiolspec.PLAS-0006-2013.f1

Qnr from *Vibrio splendidus* (60), and QnrC is 72% identical to chromosomal Qnr in *Vibrio orientalis* or *V. cholerae*. (28). QnrB homologs, on the other hand, are found on the chromosome of members of the *Citrobacter freundii* complex, including *Citrobacter braakii*, *Citrobacter werkmanii*, and *Citrobacter youngae* of both clinical (61) and environmental origin. The small, nonconjugative plasmids that carry *qnrD* can be found in other *Enterobacteriaceae* but are especially likely to be found in *Proteeae*, such as *Proteus mirabilis*, *Proteus vulgaris*, and *Providencia rettgeri* (62), and may have originated there (63, 64).

The wide distribution of *qnr* suggests an origin well before quinolones were discovered. Indeed, *qnrB* genes

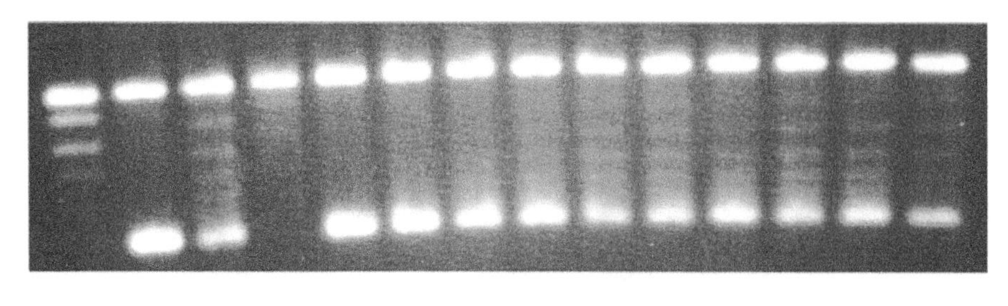

Figure 2 QnrB1 protection of DNA gyrase from ciprofloxacin inhibition of supercoiling. Reaction mixtures of 30 l were analyzed by agarose gel electrophoresis. Reaction mixtures contained 0.2 g relaxed pBR322 DNA (lanes 1 to 14), 6.7 nM gyrase (lanes 2 to 14), 2 g/ml ciprofloxacin (lanes 3 to 14), and QnrB-His6 fusion protein at 25 M (lane 4), 5 M (lane 5), 2.5 M (lane 6), 0.5 M (lane 7), 50 nM (lane 8), 5 nM (lane 9), 0.5 nM (lane 10), 50 pM (lane 11), 5 pM (lane 12), or 0.5 pM (lane 13). Reprinted from reference 27.
doi:10.1128/microbiolspec.PLAS-0006-2013.f2

and pseudogenes have been discovered on the chromosome of *Citrobacter freundii* strains collected in the 1930s (65).

Qnr PLASMIDS

Genes for PMQR have been found on plasmids varying in size and incompatibility specificity (Table 1), indicating that the spread of multiple plasmids has been responsible for the dissemination of this resistance around the world. Such plasmid heterogeneity may also have contributed to the variety of bacterial hosts for PMQR and indicates that plasmid acquisition of *qnr* and other quinolone resistance determinants occurred independently multiple times.

A mobile or transposable element is almost invariably associated with *qnr* genes (Table 2 and Fig. 3). *qnrA1* has usually been associated with IS*CR1* (66), although 63% of *qnrA1*-positive *K. pneumoniae* strains in a study from South Korea were negative for IS*CR1* by PCR (67). The IS*CR1* element is not only involved in gene mobilization. It also provides an active promoter for resistance gene expression (68). Often a single copy of IS*CR1* is found upstream from *qnrA1*, but in pMG252 and related plasmids, the *qnrA1* gene is bracketed by two copies of IS*CR1* (69, 70). The *qnrA1* IS*CR1* complex is inserted in turn into a *sul1*-type integron containing several other resistance gene cassettes (24). In pSZ50 from Mexico the integron containing IS*CR1* and *qnrA1* is duplicated in tandem (71). The *qnrA3* and *qnrA6* alleles are also linked to IS*CR1*. The gene for *qnrB1*, however, is often associated not with IS*CR1* but with *orf1005*, encoding a putative transposase (27). *qnrB1* has also been found linked to an upstream truncated *orf1005* and a downstream IS26 (72, 73), while the *qnrB20* allele is sandwiched between an upstream IS26 and a downstream *orf1005* (72). Alleles *qnrB2, qnrB4, qnrB6,* and *qnrB10* are associated with IS*CR1*, usually as a single copy (73–75), but in some plasmids two copies of IS*CR1* surround *qnrB2* (76, 77). *qnrB19* has been found in three genetic environments: within large plasmids associated with IS*Ecp1C*-based transposons, in large plasmids bracketed by IS26, and in small ColE1-type plasmids (~3 kb) lacking insertion sequences in which a flanking *oriT* locus and Xer recombination site have been proposed to be involved in site-specific recombination (73, 78–82).

In all three settings *qnrB19* is linked to a fragment of *pspF*, implying that the putative mobilization pathways may be related. In plasmid pLRM24 *qnrB19* linked to IS*Ecp1* has inserted into a Tn*3*-like element also containing a mobile element encoding KPC-3 carbapenemase (83). The *qnrS1* gene is not linked to IS*CR1* but is associated with an upstream Tn*3*-like transposon in several plasmids containing an active TEM-1 gene (26, 73, 84–86). In other plasmids *qnrS1* is associated with IS26 (87), IS2 (88), or IS*Ecl2*, a novel insertion element belonging to the IS3 family (89). On the other hand, *qnrS2* has been found as part of a mobile insertion cassette, an element with bracketing inverted repeats but lacking a transposase (90). The *qnrC* gene is found downstream from IS*Pmi1*, an insertion sequence also belonging to the IS3 family (28). *qnrD* has typically been found on small, nonconjugative plasmids and is also located inside a mobile insertion cassette (29, 62–64, 91). *qnrVC* is so far the only *qnr* gene located in a cassette with a linked *attC* site (92). *qnrVC* genes have been found on plasmids in *A. punctata* (30) and *Vibrio fluvialis* (32), within integrons in *Acinetobacter baumannii* (93) and *Pseudomonas aeruginosa*

Table 1 Representative plasmids and transmissible PMQR genes

Plasmid	PMQR gene	Host	Year of isolation	Size (kb)	Inc group	Country	Linked *bla* genes[a]	Reference
pMG252	*qnrA1*	*K. pneumoniae*	1994	~180		USA	FOX-5	9
pHSH2	*qnrA1*	*E. coli*	2000-2001	85		China		24
pQR1	*qnrA1*	*E. coli*	2003	180	A/C_2	France	VEB-1	126, 128
—[b]	*qnrA1*	*Enterobacter cloacae*	2002-2005	75	H12	France	SHV-12	119, 126
—[b]	*qnrA1*	*Enterobacter aerogenes*	2002-2005	150	FII	France	SHV-12	119, 126
pSZ50	*qnrA1*	*E. coli*	2004	~50	N	Mexico		71
pHE96	*qnrA3*	*K. pneumoniae*	2004	70	N	France		236
—[b]	*qnrA6*	*Providencia stuartii*	2011	100	A/C	Tunisia	OXA-48, PER-1, CMY-4	237
pMG298	*qnrB1*	*K. pneumoniae*	2002-2003	340		India	CTX-M-15	27
pJIBE401	*qnrB2*	*K. pneumoniae*	2003	>150	L/M	Australia	IMP-4	74
—[b]	*qnrB2*	*S. enterica*	2000			Senegal	SHV-12	77
pMG319	*qnrB4*	*E. cloacae*	1999-2004	200		USA	DHA-1	238
pPMDHA	*qnrB4*	*Klebsiella oxytoca*	2002	(Tra⁻)		France	DHA-1	134
—[b]	*qnrB4*	*E. cloacae*	—[b]	119		China	DHA-1	108
pHND2	*qnrB6*	*K. pneumoniae*	2006			China	CTX-M-9G[c]	75
pARCF702	*qnrB10*	*Citrobacter freundii*	2005			Argentina		239
pR4525	*qnrB19*	*E. coli*	2002	40		Columbia	SHV-12 CTX-M-12	78
pLRM24	*qnrB19*	*K. pneumoniae*	2007	80		USA	KPC-3	83, 117
pPAB19-1	*qnrB19*	*S. enterica*	2006	2.7	Col-E1	Argentina		82
pQAR2078	*qnrB19*	*E. coli*	2005	42.4	N	Germany		81
pAH0376	*qnrS1*	*Shigella flexneri*	2003	~50		Japan		26
pINF5	*qnrS1*	*S. enterica*	2004	58		Europe		84
TP*qnrS*-2a	*qnrS1*	*S. enterica*	2004	44	N	UK		240
pK245	*qnrS1*	*K. pneumoniae*	2002	98		Taiwan	SHV-2	87
—[b]	*qnrS1*	*S. enterica*	1999-2006	>250	HI2	Netherlands	LAP-2	120
—[b]	*qnrS2*	—[d]	2004	8.5	Q	Germany		241
p37	*qnrS2*	*A. punctata*	2006	55	U	France		242
pHS10	*qnrC*	*P. mirabilis*	2006	~120		China		28
p2007057	*qnrD1*	*S. enterica*	2006-2007	4.3		China		29
—[e]	*qnrVC1*	*V. cholerae*	2002-2008			Bangladesh		34
—[b]	*qnrVC4*	*A. punctata*	2008			China	PER-1	30
pBD146	*qnrVC5*	*Vibrio fluvialis*	1998-2002	7.5		India		243
pHSH10-2	*aac(6′)-Ib-cr*	*E. coli*	2000-2001			China		10
pC15-1a	*aac(6′)-Ib-cr*	*E. coli*	2000-2002	92	FII	Canada	CTX-M-15	244
pHPA	*qepA1*	*E. coli*	2002		FII	Japan	CTX-M-12	11
pIP1206	*qepA1*	*E. coli*	2000-2005	168	FI	Belgium		197
pQep	*qepA2*	*E. coli*	2007	90	FI	France		199
pOLA52	*oqxAB*	*E. coli*	—[b]	52	X1	Denmark		200, 202
pHXY0908	*oqxAB*	*S. typhimurium*	2009		HI2	China		207

[a]Only unusual *bla* genes are shown.
[b]Not specified.
[c]CTX-M-9 group. See reference for details.
[d]Unidentified bacteria in activated sludge.
[e]Transmitted as part of an integrating conjugative element.

Table 2 Distribution of PMQR genes

PMQR gene	Source country	Organism	Mobilizing element	Reference
qnrA1	Algeria, Australia, Belgium, Brazil, Central African Republic, China, Denmark, Egypt, France, Germany, Hungary, India, Israel, Ivory Coast, Japan, Kenya, Mexico, Morocco, Netherlands, Nigeria, Portugal, Romania, Saudi Arabia, Singapore, South Korea, Spain, Sweden, Taiwan, Thailand, Turkey, UK, Uruguay, USA, Vietnam	*Acinetobacter baumannii, C. freundii, E. aerogenes, Enterobacter cloacae, Enterobacter sakazakii, E. coli, Haemophilus parasuis, K. oxytoca, K. pneumoniae, P. mirabilis, P. aeruginosa, Pseudomonas oryzihabitans, Pseudomonas putida, S. enterica, Serratia marcescens, Shigella sonnei, S. maltophilia, V. fluvialis*	ISC*R1*	9, 22–24, 67, 98, 104–108, 113, 115, 119, 122–125, 127–130, 138–141, 143, 144, 147, 155, 226, 227, 238, 245–290
qnrA3	China, France, Hong Kong	*K. pneumoniae, Kluyvera ascorbata, Kluyvera* spp., *S. enterica, S. algae*	ISC*R1*, IS*26*	25, 153, 156, 236, 291
qnrA6	France, Tunisia	*C. freundii, K. pneumoniae, P. mirabilis, P. stuartii*	ISC*R1*	237, 292, 293
qnrB1	Algeria, Argentina, Brazil, China, Czech Republic, Denmark, France, Egypt, India, Italy, Ivory Coast, Malaysia, Mexico, Morocco, Netherlands, Nigeria, Norway, Saudi Arabia, Scotland, Singapore, South Korea, Spain, Sweden, Thailand, Tunisia, Turkey, UK	*C. freundii, Citrobacter koseri, E. cloacae, Enterobacter gergoviae, E. coli, Klebsiella ornithinolytica, K. pneumoniae, S. enterica, S. marcescens*	Orf1005, IS*26*	27, 67, 72, 73, 98, 102, 105, 109, 110, 113, 162, 186, 251, 259, 263, 268, 269, 271, 279, 281, 284, 285, 287, 288, 292, 294–302
qnrB2	Argentina, Australia, Bolivia, Brazil, Czech Republic, China, France, Germany, Hungary, Ireland, Israel, Kuwait, Mexico, Morocco, Netherlands, Portugal, Peru, Scotland, Senegal, South Korea, Spain, Sweden, Switzerland, Taiwan, Tunisia, UK, USA	*C. freundii, E. cloacae, E. coli, K. oxytoca, K. pneumoniae, S. enterica, S. typhi*	ISC*R1*	27, 61, 67, 73, 77, 96, 98, 108, 115, 116, 120, 125, 132, 139, 145, 155, 157, 161, 227, 238, 251, 257, 259, 265, 268, 271, 273, 274, 276, 279, 281, 290, 292, 299, 301, 303–316
qnrB3	USA	*E. coli*		238
qnrB4	Algeria, Australia, China, France, Germany, Ivory Coast, Japan, Morocco, Netherlands, Saudi Arabia, Singapore, South Korea, Spain, Sweden, Switzerland, Taiwan, Thailand, UK, USA	*C. freundii, C. koseri, E. aerogenes, E. cloacae, E. gergoviae, E. coli, K. ornithinolytica, K. oxytoca, K. pneumoniae, S. enterica, S. marcescens, Shigella* spp.	ISC*R1*	61, 67, 72, 75, 96, 98, 104, 105, 108, 109, 113, 121, 127, 132, 133, 145, 155, 186, 192, 196, 220, 238, 251, 256, 257, 266, 268, 271, 276, 280, 281, 285, 287, 290, 299, 309, 310, 316–319
qnrB5	Denmark, France, Mexico, South Korea, UK, USA	*E. coli, K. pneumoniae, S. enterica*		96, 97, 145, 259, 265, 279, 288, 295, 299, 303, 305, 320
qnrB6	China, France, Germany, Japan, Malaysia, Mexico, Netherlands, Poland, Singapore, South Korea, Spain, Sweden, Thailand, USA	*C. freundii, E. aerogenes, E. cloacae, E. coli, H. parasuis, K. oxytoca, K. pneumoniae, S. enterica, S. fonticola, S. marcescens, S. flexneri*	ISC*R1*	67, 72, 75, 98, 100, 104, 108, 127, 139, 141, 148, 153, 155, 167, 169, 192, 196, 204, 227, 251, 257, 266–268, 273, 279, 285, 287, 288, 290, 302, 309, 321, 322

Table 2 *(Continued)*

PMQR gene	Source country	Organism	Mobilizing element	Reference
qnrB7	Kuwait, Netherlands, Norway, South Korea	*C. freundii, E. cloacae, K. pneumoniae, S. enterica*		98, 155, 296, 313
qnrB8	Brazil, China, France, Kuwait, South Korea, UK	*C. freundii, E. aerogenes*		98, 104, 269, 299, 306, 313
qnrB9	China, South Korea	*C. freundii*		98, 290, 323
qnrB10	Argentina, Bolivia, China, Malaysia, Nigeria, Peru, South Korea	*C. braakii, C. freundii, Enterobacter amnigenus, E. cloacae, E. coli, K. pneumoniae, Salmonella choleraesuis, S. marcescens*	IS*CR1*	73, 127, 145, 153, 281, 302, 314, 324
qnrB12	Netherlands, South Korea	*C. freundii, S. enterica*		98, 155
qnrB16	South Korea	*K. pneumoniae*		281
qnrB19	Argentina, Bolivia, Brazil, Columbia, Czech Republic, Finland, Denmark, Germany, Italy, Mexico, Netherlands, Nigeria, Peru, Poland, South Korea, UK, USA, Venezuela	*E. aerogenes, Escherichia fergusonii, E. coli, K. oxytoca, K. pneumoniae, K. ascorbata, S. enterica, S. sonnei*	IS*Ecp1*, IS*26*	73, 78–83, 86, 120, 148, 155, 158, 166, 169, 288, 289, 301, 314, 315, 322, 324–329
qnrB20	Mexico, Singapore	*K. pneumoniae*	Orf1005, IS*26*	72, 288
qnrB22	South Korea	*C. werkmanii*[a]		330
qnrB23	South Korea	*C. freundii*[a]		330
qnrB26	China	*P. vulgaris*		290
qnrB31	China	*K. pneumoniae*		331
qnrB32	China	*K. pneumoniae*		331, 332
qnrS1	Algeria, Argentina, Belgium, Bolivia, Brazil, Canada, China, Czech Republic, Denmark, Egypt, France, Germany, Greece, Hungary, Israel, Italy, Ivory Coast, Japan, Malaysia, Mexico, Morocco, Netherlands, Nigeria, Norway, Poland, Peru, Romania, Serbia, South Africa, South Korea, Slovakia, Spain, Sweden, Switzerland, Taiwan, Thailand, Tunisia, Turkey, UK, USA, Vietnam	*C. freundii, C. koseri, E. aerogenes, E. cloacae, E. coli, K. oxytoca, K. pneumoniae, Morganella morganii, P. mirabilis, S. enterica, S. typhi, Shigellaa boydii, S. flexneri, S. dysenteriae*		26, 67, 73, 75, 84, 86, 98, 104, 105, 107–109, 115, 119–121, 125, 127, 131, 132, 143, 145, 148, 150, 151, 153–156, 158, 166, 167, 169, 172, 184, 186, 192, 196, 204, 220, 227, 240, 251, 255–257, 259, 261, 262, 264, 265, 268–273, 276, 279, 281, 284, 285, 287, 290, 292, 294–296, 299, 301–303, 305, 307, 308, 314–316, 318–322, 324, 333–341
qnrS2	China, Czech Republic, Denmark, France, Germany, India, South Korea, Spain, Switzerland, USA	*Aeromonas allosaccharophila, Aeromonas caviae, A. hydrophila, Aeromonas media, A. puctata, Aeromonas veronii, E. cloacae, Pseudoalteromonas* spp., *Pseudomonas* spp., *S. enterica, S. marcescens, Shigella* spp.		90, 100, 127, 148, 156, 167, 241, 242, 266, 272, 295, 299, 303, 319, 323, 342–345
qnrS3	China	*E. coli*		162
qnrS4	Denmark	*S. enterica*		295
qnrS5	South Korea	*Aeromonas* spp.		345
qnrC	China	*P. mirabilis*	IS*Pmi1*	28
qnrD1				

Table 2 Distribution of PMQR genes *(Continued)*

PMQR gene	Source country	Organism	Mobilizing element	Reference
	China, Czech Republic, France, India, Italy, Netherlands, Nigeria, Poland, Spain	*E. coli, C. freundii, K. pneumoniae, M. morganii, P. mirabilis, P. vulgaris, P. rettgeri, P. aeruginosa, S. enterica*		29, 62, 63, 91, 138, 148, 155, 204, 323, 340, 346, 347
qnrVC1	Bangladesh, Brazil, India, Tunisia	*P. aeruginosa, V. cholerae*		33, 34, 92, 94, 95
qmrVC3	India	*V. cholerae*		95
qnrVC4	China, Haiti, Portugal	*A. hydrophila, A. punctata, Aeromonas* spp., *Pseudomonas* spp., *V. cholera*	ISC*R1*	30, 33
qnrVC5	China, Haiti, India	*Vibrio parahaemolyticus, V. cholera, V. fluvialis*		31–33
qnrVC6	China	*A. baumannii*		93
aac(6′)-Ib-cr	Algeria, Argentina, Australia, Bolivia, Brazil, Bulgaria, Canada, China, Croatia, Czech Republic, Denmark, Egypt, France, Germany, Hungary, India, Israel, Italy, Japan, Kenya, Mexico, Netherlands, Nigeria, Norway, Portugal, Peru, Saudi Arabia, Serbia, Singapore, Slovenia, South Korea, Spain, Switzerland, Taiwan, Thailand, UK, Uruguay, USA, Vietnam	*C. braakii, C. freundii, C. koseri, E. aerogenes, E. cloacae, E. coli, H. parasuis, K. pneumoniae, Kluyvera* spp., *Laribacter hongkongensis, M. morganii, P. aeruginosa, Pseudomonas luteola, P. oryzihabitans, P. mirabilis, Raoultella* spp., *S. enterica, S. marcescens, Serratia odorifera, S. flexneri, S. sonnei, Shigella* spp., *V. fluvialis*	IS26	10, 14, 67, 73, 75, 100–102, 104, 110, 113, 116, 125, 127, 133, 138, 141, 145, 148, 150, 153, 155–157, 161, 167, 169, 172, 182, 184–187, 189, 190, 192, 194–196, 220, 226, 256, 261, 263, 266, 267, 270, 272, 275, 278, 284, 287, 288, 290, 296, 297, 301, 315, 316, 318, 319, 322, 336, 348–362
oqxAB	Argentina, Czech Republic, China, Denmark, Hong Kong, Italy, Japan, Poland, Serbia, South Korea, Spain	*E. coli, K. oxytoca, K. pneumoniae, S. enterica*	IS26	73, 101, 148, 156, 168, 202–206, 208, 339, 360, 363, 364
qepA1	Belgium, Bolivia, Canada, China, Egypt, India, Japan, Mexico, Nigeria, South Korea, Spain, UK, USA, Vietnam	*E. cloacae, E. coli, K. pneumoniae, M. morgnaii, P. oryzihabitans, S. enterica, S. flexneri*	IS26, ISC*R3C*	11, 12, 75, 114, 138, 145, 150, 153, 156, 167, 187, 192, 196–198, 220, 261, 270, 272, 284, 285, 290, 321, 324, 361, 365–367
qepA2	France	*E. coli*	ISC*R3C*	199

[a]*Citrobacter* spp. contain both plasmid-mediated and chromosomal *qnr* genes. Genes for *qnrB22* and *qnrB23* were transferred by conjugation to *E. coli* and hence proved to be plasmid determined.

(94), and within the transmissible SXT integrating element of *V. cholerae* (34, 95).

qnr genes are usually found in multiresistance plasmids linked to other resistance determinants. β-lactamase genes have been conspicuously common, including genes for AmpC β-lactamases (22, 96–101), CTX-M enzymes (96, 97, 99, 100, 102–114), IMP enzymes (74, 115), KPC enzymes (116–118), LAP-1 or LAP-2 (88, 89, 119–121), SHV-12 (96, 99, 104, 105, 107, 109, 115, 119, 121–127), VEB-1 (126, 128–130), and VIM-1 (111, 131). *qnrB4* and bla_{DHA-1} have been found near each other on similar plasmids from around the world (96, 108, 127, 132–135). *qnrB* alleles are also frequently found in plasmids linked to variable portions of the operons for *psp* (phage shock protein) and *sap* (peptide ABC transporter, ATP-binding protein) genes. These genes flank *qnrB* on the chromosome of several *Citrobacter* spp., and their coacquisition with *qnrB* is one of the arguments for *Citrobacter* as the source of *qnrB* alleles (61). Molecular studies with I-CeuI and S1

pMG252 USA 1994

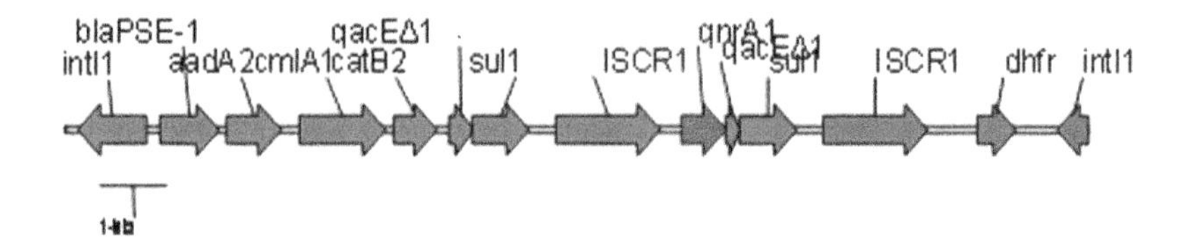

Unnamed plasmid Senegal 2000

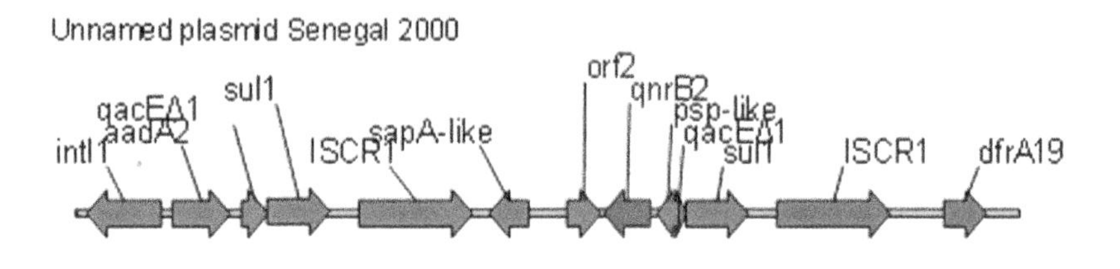

pPMDHA France 2002

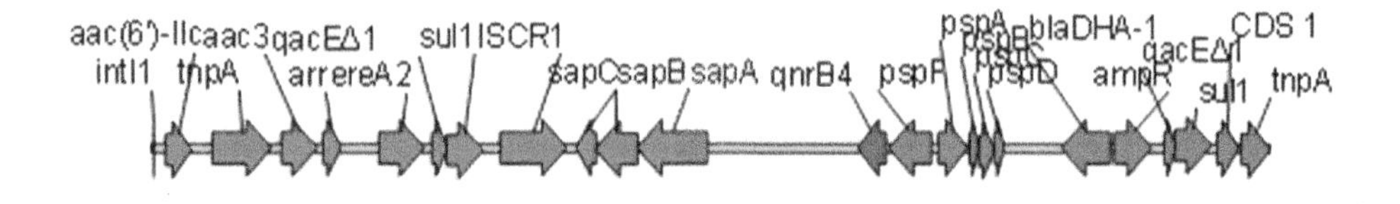

Tn2012 in pR4525 Columbia 2002

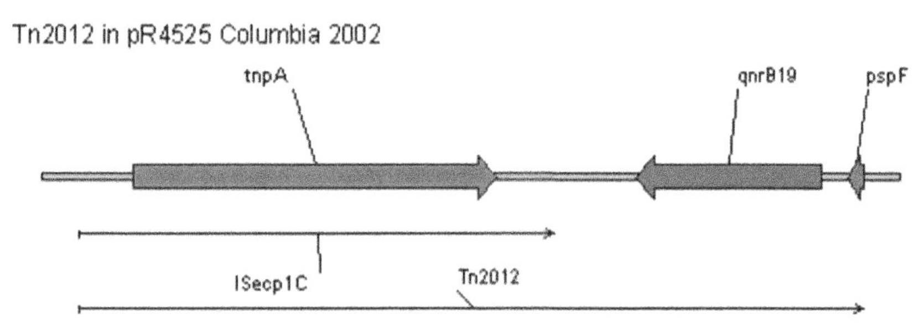

Figure 3 Genetic environment of *qnr* alleles. doi:10.1128/microbiolspec.PLAS-0006-2013.f3

nuclease followed by double hybridization for *qnr* and 23s rRNA genes have identified alleles for *qnrB6* (136), *qnrB12* (137), and *qnrB16* (136) on the chromosome of several *Citrobacter* spp. Note that these same alleles have also been reported on plasmids in species other than *Citrobacter* (Table 2).

SPREAD OF *qnr* PLASMIDS

PMQR genes have been found in a variety of *Enterobacteriaceae*, especially *E. coli* and species of *Enterobacter, Klebsiella*, and *Salmonella* (Table 2). They have been conspicuously rare in nonfermenters but have occasionally been reported in *P. aeruginosa*, other *Pseudomonas* spp. (138, 139), *A. baumannii* (139, 140), and *S. maltophilia* (139). Genes for *qnrA1* and *qnrB6* have also been found in *Haemophilus parasuis* from pigs in South China (141). *qnr* genes are found in a variety of Gram-positive organisms but are chromosomal and not plasmid-mediated (38, 53). Of the various *qnr* varieties, *qnrB* seems somewhat more common than *qnrA* and *qnrS*, which are more common than *qnrD*. Only a single isolate of *qnrC* is known (28). The relative frequency of various alleles can be judged by the

pQNR2078 Germany 2005

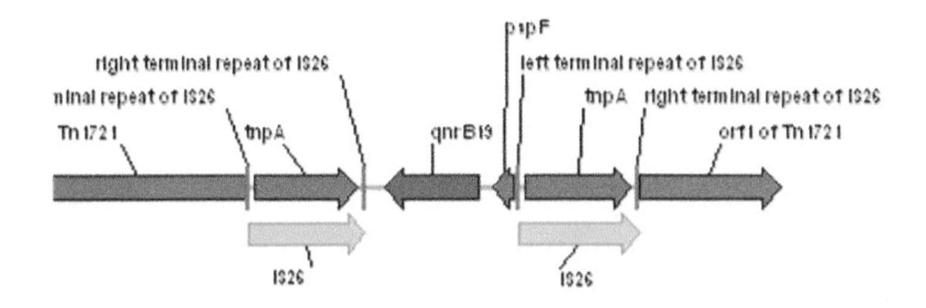

p2007057 China 2006-2007

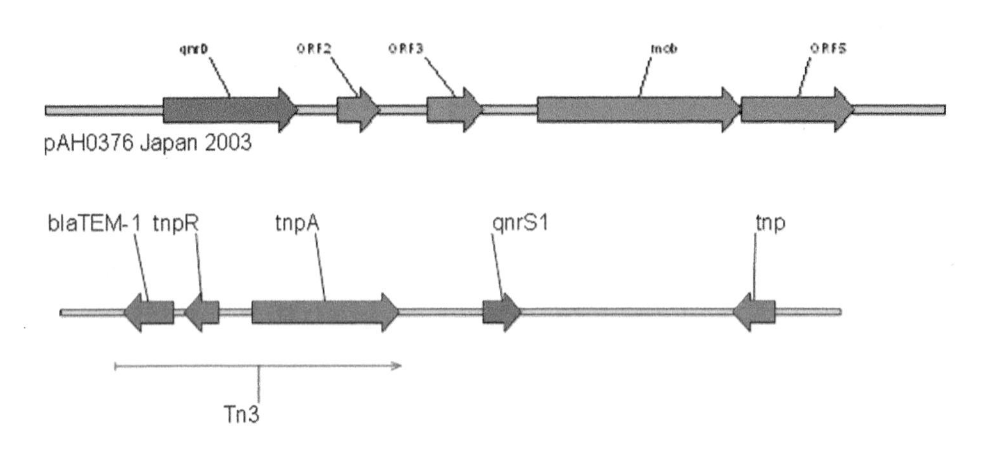

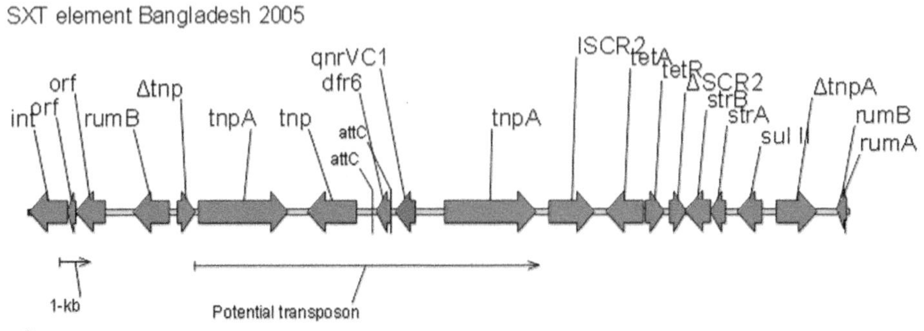

Figure 3 *(Continued)*

number of references in Table 2: i.e., for *qnrB* the most frequently detected alleles are *qnrB1, qnrB2, qnrB4, qnrB6, qnrB10,* and *qnrB19*. The earliest known *qnr* outside of *Citrobacter* spp. dates from 1988 (142). Studies in the last decade suggest that *qnr* detection is increasing but is still less than 10% in unselected clinical isolates, with usually the greatest prevalence in *Enterobacter cloacae*, less in *K. pneumoniae*, and the least in *E. coli* (127, 143–146). Higher frequencies result if samples are preselected for extended-spectrum β-lactamase or other resistance phenotypes (127, 147), but the prevalence of *qnr* genes has reached as high as 39% in an unselected sample of *E. cloacae* isolates at one hospital in China (127).

Although most prevalence studies have surveyed hospital isolates, animals have not been neglected. PMQR genes have been found in samples from domestic or wild birds (75, 86, 148, 149), cats (75, 81, 112, 150), cattle (151, 152), chickens (75, 86, 153–160), dogs (75, 81, 112, 150, 156, 161), ducks (101, 156, 160, 162), fish (163–165), geese (101, 156, 160, 162), horses (81, 86, 166), pigs (75, 101, 155, 156, 160, 162, 167), rabbits (168), reptiles (155, 169), sheep (155), turkeys (155), and zoo animals (170–172).

REGULATION OF *qnr*

Not surprisingly, *qnrA* expression is influenced by the strength of its promoter. A *qnrA1* plasmid from Shanghai was found to give 8-fold higher ciprofloxacin MICs than other *qnrA1* plasmids and had a 12-fold stronger promoter attributed to a 7-bp deletion between the +1 transcription initiation site and the start of the qnrA1 gene (144). Environmental conditions have also been found to affect expression of *qnr* genes and may offer clues concerning the native function of these genes. Expression of the *qnrA* gene of *S. algae*, an organism adapted to growth at low temperature, is stimulated up to 8-fold by cold shock but not by other conditions such as DNA damage, oxidative or osmotic stress, starvation, or heat shock (173). Expression of *qnrB* alleles, on the other hand, is augmented up to 9-fold by exposure to DNA-damaging agents such as ciprofloxacin or mitomycin C via an upstream LexA binding site and the classical SOS system (174, 175). *qnrD* and the chromosomal *qnr* of *Serratia marcescens* are similarly regulated (176). Expression of plasmid-mediated *qnrS1* or the related chromosomal *qnrVS1* of *V. splendidus* is also stimulated by ciprofloxacin up to 30-fold, but by a mechanism independent of the SOS system. No LexA binding site is found upstream from these *qnr* genes, but an upstream sequence is required for quinolone stimulation to occur (177). Some naturally occurring quinolone-like compounds such as quinine, 2-hydroxyquinoline, 4-hydroxyquinoline, or the *Pseudomonas* quinolone signal for quorum sensing also induce *qnrS1*, but not *qnrVS1* (178).

AAC(6´)-Ib-cr

AAC(6´)-Ib-cr is a bifunctional variant of a common acetyltransferase active on such aminoglycosides as amikacin, kanamycin, and tobramycin but is also able to acetylate those fluoroquinolones with an amino nitrogen on the piperazinyl ring, such as ciprofloxacin and norfloxacin (10). Compared to other AAC(6´)-Ib enzymes, the -cr variant has two unique amino acid substitutions: Trp102Arg and Asp179Tyr, both of which are required for quinolone acetylating activity. Models of enzyme action suggest that the Asp179Tyr replacement is particularly important in permitting π-stacking interactions with the quinolone ring to facilitate quinolone binding. The role of Trp102Arg is to position the Tyr face for optimal interaction (179) or to hydrogen-bond to keto or carboxyl groups of the quinolone to fix it in place (180). Both AGG and CGG have been found as the Arg codon at 102, allowing variants of the *aac(6´)-Ib-cr* gene to be distinguished (73). A lower level of *aac(6´)-Ib-cr* expression has been found in a strain with an upstream 12-bp deletion displacing the promoter -10 box (181). A 26-amino acid larger AAC(6´)-Ib-cr4 enzyme with consequent Trp128Arg and Asp205Tyr substitutions (182) and nonfunctional truncated *aac(6´)-Ib-cr* genes (183) have also been reported.

The *aac(6´)-Ib-cr* gene is usually found in a cassette as part of an integron in a multiresistance plasmid, which may contain other PMQR genes. Association with extended-spectrum β-lactamase CTX-M-15 is particularly common (110, 184–190). A mobile genetic element, especially IS26, is often associated (191). *aac(6´)-Ib-cr* may also be chromosomal (192, 193). The gene has been found world-wide (Table 2) in a variety of *Enterobacteriaceae* and even in *P. aeruginosa* (138). It is more prevalent in *E. coli* than in other *Enterobacteriaceae* (145, 184, 194, 195) and has often been more common than *qnr* alleles (14, 75, 184, 192, 194, 196).

QepA and OqxAB

QepA is a plasmid-mediated efflux pump in the major facilitator family that decreases susceptibility to hydrophilic fluoroquinolones, especially ciprofloxacin and norfloxacin (11, 197). *qepA* has often been found on plasmids also encoding aminoglycoside ribosomal methylase *rmtB* (12, 167, 198). Substantial differences in quinolone resistance produced by different *qepA* transconjugants suggest variability in the level of *qepA* expression, by mechanisms as yet to be defined (167). IS26 elements and IS*CR3C* have been implicated in mobilizing the *qepA* gene to plasmids (199). A variant differing in two amino acids (QepA2) has also been described (199).

OqxAB is an efflux pump in the resistance-nodulation-division family that was initially recognized on transmissible plasmids responsible for resistance to olaquindox used for growth enhancement in pigs (200, 201). It has a wide substrate specificity including chloramphenicol, trimethoprim, and quinolones such as ciprofloxacin, flumequin, norfloxacin, and nalidixic acid (13). *oqxAB* has been found on plasmids in clinical isolates of *E. coli* and *K. pneumoniae* and in the chromosome and on plasmids of *S. enteritis* flanked in both locations by IS26-like elements (202–207). In *E. coli* isolates from farms in China where olaquindox was in use, *oqxAB* was found on transmissible plasmids in 39% of isolates from animals and 30% of isolates from farm workers (204). Linkage of *oqxAB* with genes for CTX-M-14 and other plasmid-mediated CTX-M alleles has been noted (160). It is common (usually 75% or more) on the chromosome of *K. pneumoniae* isolates, where up to 20-fold variation in expression implies the presence of regulatory control (73, 203, 206, 208, 209). In *K. pneumoniae*, overexpression of the nearby *rarA* gene is associated with increased *oqxAB* expression, while increased expression of the adjacent *oqxR* gene downregulates OqxAB production (210, 211). Sequence variants *oqxA2, oqxB2*, and *oqxB3* have been described (208).

Other plasmid-mediated efflux pumps active on quinolones have been reported but as yet little studied. Plasmid pRSB101 isolated from an uncultivated organism in activated sludge at a wastewater treatment plant contained a multidrug resistance (MDR) transport system with a resistance-nodulation-division-type membrane fusion protein conferring resistance to nalidixic acid and norfloxacin (212). It differs in sequence from QepA and OqxAB. Plasmids in *Staphylococcus aureus* (especially multiresistant *S. aureus*) encoding the QacBIII variant belonging to the major facilitator family confer decreased susceptibility to norfloxacin and ciprofloxacin (213).

RESISTANCE PRODUCED BY PMQR DETERMINANTS

Table 3 shows the MIC produced in a common *E. coli* strain by PMQR genes. *qnr* genes produce about the same resistance to ciprofloxacin and levofloxacin as single mutations in *gyrA* but have less affect on susceptibility to nalidixic acid. *aac(6′)-Ib-cr* and *qepA* give lower levels, which is confined to ciprofloxacin in the case of *aac(6′)-Ib-cr* because of its substrate specificity. All provide a decrease in susceptibility that does not reach the CLSI breakpoint for even intermediate resistance. How then can PMQR genes be clinically important?

Table 3 Effect of different quinolone resistance mechanisms on quinolone susceptibility of *E. coli*

	MIC g/ml		
E. coli strain	Ciprofloxacin	Levofloxacin	Nalidixic acid
J53	0.008	0.015	4
J53 *gyrA* (S83L)	0.25	0.5	≥256
J53 pMG252 (*qnrA1*)	0.25	0.5	16
J53 pMG298 (*qnrB1*)	0.25	0.5	16
J53 pMG306 (*qnrS1*)	0.25	0.38	16
J53 pMG320 (*aac(6′)-Ib-cr*)	0.06	0.015	4
J53 pAT851 (*qepA*)	0.064	0.032	4
CLSI susceptibility breakpoint	≤1.0	≤2.0	≤16

The answer is that PMQR genes facilitate the selection of higher levels of quinolone resistance. Figure 4 shows the effect of plasmid pMG252 encoding QnrA on survival of *E. coli* J53 at increasing concentrations of ciprofloxacin. Survivors occur until a concentration of more than 1 g/ml ciprofloxacin is reached. This limiting concentration has been termed the mutant prevention concentration (MPC), and the concentration between the MIC and MPC at which mutants are selected is the mutant selection window (214). PMQR genes exert their influence by widening the mutant selection window and elevating the MPC, as shown for both *qnr* (215, 216) and *aac(6′)-Ib-cr* (10). The same augmentation of resistance selection has been found in clinical isolates of *E. cloacae* (147). Surprisingly, in *qnr*-harboring *E. coli gyrA* resistance mutants are rarely selected (217); although resistance produced by *qnr* and *gyrA* is additive (218–220), clinical isolates with *qnr* alleles or *aac(6′)-Ib-cr* and *gyrA, parC*, and other resistance mutations are common (221), and *gyrA* mutants have developed in *qnr*-positive isolates from patients treated with quinolone for *E. coli* (222) or *S. enterica* (223) infections.

It should be noted that higher levels of quinolone resistance are seen if a plasmid or strain carries two or more genes for quinolone resistance, such as both *qnr* and *aac(6′)-Ib-cr*, and that ciprofloxacin MICs of 2 g/ml can be reached with *qnrA* in *E. coli* overexpressing the AcrAB multidrug efflux pump (224). Fully resistant *E. coli* with a ciprofloxacin MIC of 4 g/ml has been reported with plasmid-mediated *qnrS1* and *oqxAB* as well as overexpression of AcrAB and other efflux pumps (225).

While the frequency of quinolone resistance in clinical isolates has paralleled quinolone usage, the appearance of PMQR has also played a role. At Hadassah hospitals in Israel ciprofloxacin resistance was uncommon in *E. coli, K. pneumoniae*, and *Enterobacter* spp. until the mid-1990s, just when *qnr* and *aac(6′)-Ib-cr* genes became prevalent in these strains (143, 226). Similarly, in Korea the increasing frequency of ciprofloxacin resistance in *Enterobacteriaceae* since 2000 has been associated with an increasing prevalence of PMQR genes (145). In Spain, also, the prevalence of PMQR in clinical isolates of *E. coli* and *K. pneumoniae* increased between 2000 and 2006 (227). In Canada, as

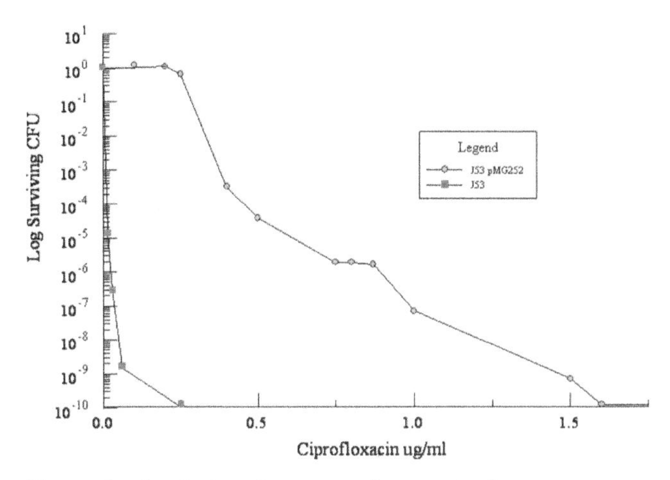

Figure 4 Survival at increasing fluoroquinolone concentrations for *E. coli* J53 and J53 pMG252. A large inoculum (10^{10} colony forming units) and appropriate dilutions were applied to Mueller-Hinton agar plates containing the indicated concentration of ciprofloxacin, and surviving colonies were counted after incubation for 72 h at 37°C. doi:10.1128/microbiolspec.PLAS-0006-2013.f4

well, the prevalence of *aac(6′)-Ib-cr* increased in the Calgary Health Region between 2004 and 2007 (184).

CLINICAL IMPORTANCE

In animal model infections the presence of a *qnr* gene makes an infecting agent harder to treat with quinolones. This detrimental effect has been shown in mice with pneumonia produced by *K. pneumoniae* or *E. coli* (228, 229) and in *E. coli* UTI models (230, 231).

Patients treated with levofloxacin for bloodstream infections caused by Gram-negative organisms with elevated quinolone MICs that were still within the susceptible category had worse outcomes than similar patients infected with more susceptible organisms (232), but a specific effect of PMQR carriage on outcome has been harder to document. Chong et al. evaluated 351 blood culture isolates of *Enterobacter* or *Klebsiella* at a health center in Korea and found 26 positive for *qnrA*, *qnrB*, or *qnrS* genes. The *qnr*-positive patients were hospitalized longer, but there was no difference in in-hospital or 30-day mortality between the *qnr*-positive or -negative patient groups (233). Liao et al. studied 227 blood culture isolates of *Klebsiella* from a hospital in Taiwan and found 9 positive for *qnrB* or *qnrS*. The 14-day mortality was similar in patients infected with or without *qnr*-containing isolates, but there was a trend for increased in-hospital mortality (234). Further studies are needed to distinguish a specific effect of *qnr* from the effect of other resistance genes so often linked to it.

CODA

The varieties of PMQR that have emerged exemplify the three general mechanisms of bacterial resistance to any antimicrobial agents: target alteration (Qnr), drug modification [AAC(6′)-Ib-cr], and efflux pump activation (QepA and OqxAB). AAC(6′)-Ib-cr arose by mutations altering two amino acids encoded by a common aminoglycoside resistance gene. Acquisition of the other PMQR genes illustrates the variety of genetic elements now available on plasmids for resistance gene mobilization and the frequent obscurity of the ultimate gene donors. Who would have guessed that aquatic bacteria harbored *qnr* genes without evident selective advantage, thus encoding a protein that blocked the action of a synthetic antimicrobial agent that they would never be expected to encounter? Unfortunately, knowledge of resistance mechanisms has not led to new therapeutic strategies. Acetylation of ciprofloxacin is inhibited competitively by aminoglycoside substrates of AAC(6′)-Ib-cr (179), but other inhibitors of the enzyme are not yet known, and no efflux pump inhibitors are commercially available. Overcoming Qnr blockage of quinolone inhibition will require deeper knowledge of how this DNA mimic interacts with topoisomerase as well as the development of other bacterial topoisomerase poisons that escape the action of Qnr.

Acknowledgments. *This work was supported by grant R01 AI057576 from the U.S. Public Health Service, National Institutes of Health, to D.C.H. and G.A.J. Conflicts of interest: We disclose no conflicts.*

Citation. Jacoby GA, Strahilevitz J, Hooper DC. 2014. Plasmid-mediated quinolone resistance. Microbiol Spectrum 2(5):PLAS-0006-2013.

References

1. **Gellert M, Mizuuchi K, O'Dea MH, Itoh T, Tomizawa JI.** 1977. Nalidixic acid resistance: a second genetic character involved in DNA gyrase activity. *Proc Natl Acad Sci USA* **74:**4772–4776.
2. **Yoshida H, Kojima T, Yamagishi J, Nakamura S.** 1988. Quinolone-resistant mutations of the *gyrA* gene of *Escherichia coli*. *Mol Gen Genet* **211:**1–7.
3. **Hooper DC.** 2003. Mechanisms of quinolone resistance, p 41–67. *In* Hooper DC, Rubinstein E (ed), *Quinolone Antimicrobial Agents*, 3rd ed, ASM Press, Washington, DC.
4. **Burman LG.** 1977. Apparent absence of transferable resistance to nalidixic acid in pathogenic Gram-negative bacteria. *J Antimicrob Chemother* **3:**509–516.
5. **Neuhauser MM, Weinstein RA, Rydman R, Danziger LH, Karam G, Quinn JP.** 2003. Antibiotic resistance among Gram-negative bacilli in US intensive care units:

implications for fluoroquinolone use. *JAMA* **289:** 885–888.

6. **Linder JA, Huang ES, Steinman MA, Gonzales R, Stafford RS.** 2005. Fluoroquinolone prescribing in the United States: 1995 to 2002. *Am J Med* **118:**259–268.
7. **Munshi MH, Sack DA, Haider K, Ahmed ZU, Rahaman MM, Morshed MG.** 1987. Plasmid-mediated resistance to nalidixic acid in *Shigella dysenteriae* type 1. *Lancet* **2:**419–421.
8. **Ashraf MM, Ahmed ZU, Sack DA.** 1991. Unusual association of a plasmid with nalidixic acid resistance in an epidemic strain of *Shigella dysenteriae* type 1 from Asia. *Can J Microbiol* **37:**59–63.
9. **Martínez-Martínez L, Pascual A, Jacoby GA.** 1998. Quinolone resistance from a transferable plasmid. *Lancet* **351:**797–799.
10. **Robicsek A, Strahilevitz J, Jacoby GA, Macielag M, Abbanat D, Park CH, Bush K, Hooper DC.** 2006. Fluoroquinolone-modifying enzyme: a new adaptation of a common aminoglycoside acetyltransferase. *Nat Med* **12:**83–88.
11. **Yamane K, Wachino J, Suzuki S, Kimura K, Shibata N, Kato H, Shibayama K, Konda T, Arakawa Y.** 2007. New plasmid-mediated fluoroquinolone efflux pump, QepA, found in an *Escherichia coli* clinical isolate. *Antimicrob Agents Chemother* **51:**3354–3360.
12. **Périchon B, Courvalin P, Galimand M.** 2007. Transferable resistance to aminoglycosides by methylation of G1405 in 16S rRNA and to hydrophilic fluoroquinolones by QepA-mediated efflux in *Escherichia coli*. *Antimicrob Agents Chemother* **51:**2464–2469.
13. **Hansen LH, Jensen LB, Sorensen HI, Sorensen SJ.** 2007. Substrate specificity of the OqxAB multidrug resistance pump in *Escherichia coli* and selected enteric bacteria. *J Antimicrob Chemother* **60:**145–147.
14. **Ciesielczuk H, Hornsey M, Choi V, Woodford N, Wareham DW.** 2013. Development and evaluation of a multiplex PCR for eight plasmid-mediated quinolone-resistance determinants. *J Med Microbiol* **62:**1823–1827.
15. **Guan X, Xue X, Liu Y, Wang J, Wang Y, Wang J, Wang K, Jiang H, Zhang L, Yang B, Wang N, Pan L.** 2013. Plasmid-mediated quinolone resistance: current knowledge and future perspectives. *J Int Med Res* **41:**20–30.
16. **Ruiz J, Pons MJ, Gomes C.** 2012. Transferable mechanisms of quinolone resistance. *Int J Antimicrob Agents* **40:**196–203.
17. **Poirel L, Cattoir V, Nordmann P.** 2012. Plasmid-mediated quinolone resistance; interactions between human, animal, and environmental ecologies. *Front Microbiol* **3:**24.
18. **Rodriguez-Martinez JM, Cano ME, Velasco C, Martinez-Martinez L, Pascual A.** 2011. Plasmid-mediated quinolone resistance: an update. *J Infect Chemother* **17:** 149–182.
19. **Hernandez A, Sanchez MB, Martinez JL.** 2011. Quinolone resistance: much more than predicted. *Front Microbiol* **2:**22.
20. **Strahilevitz J, Jacoby GA, Hooper DC, Robicsek A.** 2009. Plasmid-mediated quinolone resistance: a multi-faceted threat. *Clin Microbiol Rev* **22:**664–689.
21. **Tran JH, Jacoby GA.** 2002. Mechanism of plasmid-mediated quinolone resistance. *Proc Natl Acad Sci USA* **99:**5638–5642.
22. **Rodríguez-Martínez JM, Pascual A, García I, Martínez-Martínez L.** 2003. Detection of the plasmid-mediated quinolone resistance determinant *qnr* among clinical isolates of *Klebsiella pneumoniae* producing AmpC-type β-lactamase. *J Antimicrob Chemother* **52:**703–706.
23. **Wang M, Sahm DF, Jacoby GA, Hooper DC.** 2004. Emerging plasmid-mediated quinolone resistance associated with the *qnr* gene in *Klebsiella pneumoniae* clinical isolates in the United States. *Antimicrob Agents Chemother* **48:**1295–1299.
24. **Wang M, Tran JH, Jacoby GA, Zhang Y, Wang F, Hooper DC.** 2003. Plasmid-mediated quinolone resistance in clinical isolates of *Escherichia coli* from Shanghai, China. *Antimicrob Agents Chemother* **47:** 2242–2248.
25. **Cheung TK, Chu YW, Chu MY, Ma CH, Yung RW, Kam KM.** 2005. Plasmid-mediated resistance to ciprofloxacin and cefotaxime in clinical isolates of *Salmonella enterica* serotype Enteritidis in Hong Kong. *J Antimicrob Chemother* **56:**586–589.
26. **Hata M, Suzuki M, Matsumoto M, Takahashi M, Sato K, Ibe S, Sakae K.** 2005. Cloning of a novel gene for quinolone resistance from a transferable plasmid in *Shigella flexneri* 2b. *Antimicrob Agents Chemother* **49:** 801–803.
27. **Jacoby GA, Walsh KE, Mills DM, Walker VJ, Oh H, Robicsek A, Hooper DC.** 2006. *qnrB*, another plasmid-mediated gene for quinolone resistance. *Antimicrob Agents Chemother* **50:**1178–1182.
28. **Wang M, Guo Q, Xu X, Wang X, Ye X, Wu S, Hooper DC, Wang M.** 2009. New plasmid-mediated quinolone resistance gene, *qnrC*, found in a clinical isolate of *Proteus mirabilis*. *Antimicrob Agents Chemother* **53:** 1892–1897.
29. **Cavaco LM, Hasman H, Xia S, Aarestrup FM.** 2009. *qnrD*, a novel gene conferring transferable quinolone resistance in *Salmonella enterica* serovar Kentucky and Bovismorbificans strains of human origin. *Antimicrob Agents Chemother* **53:**603–608.
30. **Xia R, Guo X, Zhang Y, Xu H.** 2010. *qnrVC*-like gene located in a novel complex class 1 integron harboring the IS*CR1* element in an *Aeromonas punctata* strain from an aquatic environment in Shandong Province, China. *Antimicrob Agents Chemother* **54:**3471–3474.
31. **Ming L.** 2013. *Mechanisms of fluoroquinolone resistance in* Vibrio parahaemolyticus. GenBank KC540630. http://www.ncbi.nlm.nih.gov/.
32. **Singh R, Rajpara N, Tak J, Patel A, Mohanty P, Vinothkumar K, Chowdhury G, Ramamurthy T, Ghosh A, Bhardwaj AK.** 2012. Clinical isolates of *Vibrio fluvialis* from Kolkata, India, obtained during 2006: plasmids, the *qnr* gene and a mutation in gyrase A as mechanisms of multidrug resistance. *J Med Microbiol* **61:**369–374.
33. **Fonseca EL, Vicente AC.** 2013. Epidemiology of qnrVC alleles and emergence out of the *Vibrionaceae* family. *J Med Microbiol* **62:**1628–1630.

34. **Kim HB, Wang M, Ahmed S, Park CH, LaRocque RC, Faruque AS, Salam MA, Khan WA, Qadri F, Calderwood SB, Jacoby GA, Hooper DC.** 2010. Transferable quinolone resistance in *Vibrio cholerae*. *Antimicrob Agents Chemother* **54:**799–803.

35. **Jacoby G, Cattoir V, Hooper D, Martínez-Martínez L, Nordmann P, Pascual A, Poirel L, Wang M.** 2008. *qnr* gene nomenclature. *Antimicrob Agents Chemother* **52:** 2297–2299.

36. **Sánchez MB, Hernández A, Rodríguez-Martínez JM, Martínez-Martínez L, Martínez JL.** 2008. Predictive analysis of transmissible quinolone resistance indicates *Stenotrophomonas maltophilia* as a potential source of a novel family of Qnr determinants. *BMC Microbiol* **8:** 148–161.

37. **Boulund F, Johnning A, Pereira MB, Larsson DG, Kristiansson E.** 2012. A novel method to discover fluoroquinolone antibiotic resistance (qnr) genes in fragmented nucleotide sequences. *BMC Genomics* **13:**695.

38. **Jacoby GA, Hooper DC.** 2013. Phylogenetic analysis of chromosomally determined Qnr and related proteins. *Antimicrob Agents Chemother* **57:**1930–1934.

39. **Vetting MW, Hegde SS, Fajardo JE, Fiser A, Roderick SL, Takiff HE, Blanchard JS.** 2006. Pentapeptide repeat proteins. *Biochemistry* **45:**1–10.

40. **Montero C, Mateu G, Rodriguez R, Takiff H.** 2001. Intrinsic resistance of *Mycobacterium smegmatis* to fluoroquinolones may be influenced by new pentapeptide protein MfpA. *Antimicrob Agents Chemother* **45:** 3387–3392.

41. **Hegde SS, Vetting MW, Roderick SL, Mitchenall LA, Maxwell A, Takiff HE, Blanchard JS.** 2005. A fluoroquinolone resistance protein from *Mycobacterium tuberculosis* that mimics DNA. *Science* **308:**1480–1483.

42. **Hegde SS, Vetting MW, Mitchenall LA, Maxwell A, Blanchard JS.** 2011. Structural and biochemical analysis of the pentapeptide repeat protein *Efs*Qnr, a potent DNA gyrase inhibitor. *Antimicrob Agents Chemother* **55:**110–117.

43. **Xiong X, Bromley EH, Oelschlaeger P, Woolfson DN, Spencer J.** 2011. Structural insights into quinolone antibiotic resistance mediated by pentapeptide repeat proteins: conserved surface loops direct the activity of a Qnr protein from a gram-negative bacterium. *Nucleic Acids Res* **39:**3917–3927.

44. **Vetting MW, Hegde SS, Wang M, Jacoby GA, Hooper DC, Blanchard JS.** 2011. Structure of QnrB1, a plasmid-mediated fluoroquinolone resistance factor. *J Biol Chem* **286:**25265–25273.

45. **Jacoby GA, Corcoran MA, Mills DM, Griffin CM, Hooper DC.** 2013. Mutational analysis of quinolone resistance protein QnrB1. *Antimicrob Agents Chemother* **57:**5733–5736.

46. **Kampranis SC, Bates AD, Maxwell A.** 1999. A model for the mechanism of strand passage by DNA gyrase. *Proc Natl Acad Sci USA* **96:**8414–8419.

47. **Drlica K, Malik M, Kerns RJ, Zhao X.** 2008. Quinolone-mediated bacterial death. *Antimicrob Agents Chemother* **52:**385–392.

48. **Tran JH, Jacoby GA, Hooper DC.** 2005. Interaction of the plasmid-encoded quinolone resistance protein Qnr with *Escherichia coli* DNA gyrase. *Antimicrob Agents Chemother* **49:**118–125.

49. **Mérens A, Matrat S, Aubry A, Lascols C, Jarlier V, Soussy CJ, Cavallo JD, Cambau E.** 2009. The pentapeptide repeat proteins $MfpA_{Mt}$ and QnrB4 exhibit opposite effects on DNA gyrase catalytic reactions and on the ternary gyrase-DNA-quinolone complex. *J Bacteriol* **191:**1587–1594.

50. **Tavio MM, Jacoby GA, Hooper DC.** 2014. QnrS1 structure-activity relationships. *J Antimicrob Chemother* **69:**2102–2109.

51. **Tran JH, Jacoby GA, Hooper DC.** 2005. Interaction of the plasmid-encoded quinolone resistance protein QnrA with *Escherichia coli* topoisomerase IV. *Antimicrob Agents Chemother* **49:**3050–3052.

52. **Arsène S, Leclercq R.** 2007. Role of a *qnr*-like gene in the intrinsic resistance of *Enterococcus faecalis* to fluoroquinolones. *Antimicrob Agents Chemother* **51:** 3254–3258.

53. **Rodríguez-Martínez JM, Velasco C, Briales A, García I, Conejo MC, Pascual A.** 2008. Qnr-like pentapeptide repeat proteins in Gram-positive bacteria. *J Antimicrob Chemother* **61:**1240–1243.

54. **Shimizu K, Kikuchi K, Sasaki T, Takahashi N, Ohtsuka M, Ono Y, Hiramatsu K.** 2008. Sm*qnr*, a new chromosome-carried quinolone resistance gene in *Stenotrophomonas maltophilia*. *Antimicrob Agents Chemother* **52:**3823–3825.

55. **Sánchez MB, Martínez JL.** 2010. SmQnr contributes to intrinsic resistance to quinolones in *Stenotrophomonas maltophilia*. *Antimicrob Agents Chemother* **54:** 580–581.

56. **Gordon NC, Wareham DW.** 2010. Novel variants of the Sm*qnr* family of quinolone resistance genes in clinical isolates of *Stenotrophomonas maltophilia*. *J Antimicrob Chemother* **65:**483–489.

57. **Zhang R, Sun Q, Hu YJ, Yu H, Li Y, Shen Q, Li GX, Cao JM, Yang W, Wang Q, Zhou HW, Hu YY, Chen GX.** 2012. Detection of the *Smqnr* quinolone protection gene and its prevalence in clinical isolates of *Stenotrophomonas maltophilia* in China. *J Med Microbiol* **61:**535–539.

58. **Poirel L, Rodriguez-Martinez JM, Mammeri H, Liard A, Nordmann P.** 2005. Origin of plasmid-mediated quinolone resistance determinant QnrA. *Antimicrob Agents Chemother* **49:**3523–3525.

59. **Poirel L, Liard A, Rodriguez-Martinez JM, Nordmann P.** 2005. *Vibrionaceae* as a possible source of Qnr-like quinolone resistance determinants. *J Antimicrob Chemother* **56:**1118–1121.

60. **Cattoir V, Poirel L, Mazel D, Soussy CJ, Nordmann P.** 2007. *Vibrio splendidus* as the source of plasmid-mediated QnrS-like quinolone resistance determinants. *Antimicrob Agents Chemother* **51:**2650–2651.

61. **Jacoby GA, Griffin CM, Hooper DC.** 2011. *Citrobacter* spp. as a source of *qnrB* alleles. *Antimicrob Agents Chemother* **55:**4979–4984.

62. **Zhang S, Sun J, Liao XP, Hu QJ, Liu BT, Fang LX, Deng H, Ma J, Xiao X, Zhu HQ, Liu YH.** 2013. Prevalence and plasmid characterization of the *qnrD* determinant in *Enterobacteriaceae* isolated from animals, retail meat products, and humans. *Microb Drug Resist* **19:** 331–335.
63. **Guillard T, Cambau E, Neuwirth C, Nenninger T, Mbadi A, Brasme L, Vernet-Garnier V, Bajolet O, de Champs C.** 2012. Description of a 2,683-base-pair plasmid containing *qnrD* in two *Providencia rettgeri* isolates. *Antimicrob Agents Chemother* **56:**565–568.
64. **Guillard T, Grillon A, de Champs C, Cartier C, Madoux J, Lozniewski A, Berçot B, Riahi J, Vernet-Garnier V, Cambau E.** 2014. Mobile insertion cassetts as a source of *qnrD* mobilization onto small non-transmissible plasmids in *Proteeae*. *PLoS One* **9**(2): e87801. doi:10.1371/journal.pone.0087801.
65. **Saga T, Sabtcheva S, Mitsutake K, Ishii Y, Tateda K, Yamaguchi K, Kaku M.** 2013. Characterization of *qnrB*-like genes in *Citrobacter* species of the American Type Culture Collection. *Antimicrob Agents Chemother* **57:**2863–2866.
66. **Toleman MA, Bennett PM, Walsh TR.** 2006. IS*CR* elements: novel gene-capturing systems of the 21st century? *Microbiol Mol Biol Rev* **70:**296–316.
67. **Park YJ, Yu JK, Kim SY, Lee S, Jeong SH.** 2010. Prevalence and characteristics of *qnr* determinants and *aac(6')-Ib-cr* among ciprofloxacin-susceptible isolates of *Klebsiella pneumoniae* in Korea. *J Antimicrob Chemother* **65:**2041–2043.
68. **Rodriguez-Martinez JM, Poirel L, Canton R, Nordmann P.** 2006. Common region CR1 for expression of antibiotic resistance genes. *Antimicrob Agents Chemother* **50:**2544–2546.
69. **Robicsek A, Jacoby GA, Hooper DC.** 2006. The worldwide emergence of plasmid-mediated quinolone resistance. *Lancet Infect Dis* **6:**629–640.
70. **Rodriguez-Martinez JM, Velasco C, Garcia I, Cano ME, Martinez-Martinez L, Pascual A.** 2007. Characterisation of integrons containing the plasmid-mediated quinolone resistance gene *qnrA1* in *Klebsiella pneumoniae*. *Int J Antimicrob Agents* **29:**705–709.
71. **Garza-Ramos U, Barrios H, Hernandez-Vargas MJ, Rojas-Moreno T, Reyna-Flores F, Tinoco P, Othon V, Poirel L, Nordmann P, Cattoir V, Ruiz-Palacios G, Fernandez JL, Santamaria RI, Bustos P, Castro N, Silva-Sanchez J.** 2012. Transfer of quinolone resistance gene *qnrA1* to *Escherichia coli* through a 50 kb conjugative plasmid resulting from the splitting of a 300 kb plasmid. *J Antimicrob Chemother* **67:**1627–1634.
72. **Teo JW, Ng KY, Lin RT.** 2009. Detection and genetic characterisation of *qnrB* in hospital isolates of *Klebsiella pneumoniae* in Singapore. *Int J Antimicrob Agents* **33:**177–180.
73. **Andres P, Lucero C, Soler-Bistue A, Guerriero L, Albornoz E, Tran T, Zorreguieta A, PMQR Group, Galas M, Corso A, Tolmasky ME, Petroni A.** 2013. Differential distribution of plasmid-mediated quinolone resistance genes in clinical enterobacteria with unusual phenotypes of quinolone susceptibility from Argentina. *Antimicrob Agents Chemother* **57:**2467–2675.
74. **Espedido BA, Partridge SR, Iredell JR.** 2008. bla_{IMP-4} in different genetic contexts in *Enterobacteriaceae* isolates from Australia. *Antimicrob Agents Chemother* **52:** 2984–2987.
75. **Ma J, Zeng Z, Chen Z, Xu X, Wang X, Deng Y, Lu D, Huang L, Zhang Y, Liu J, Wang M.** 2009. High prevalence of plasmid-mediated quinolone resistance determinants *qnr*, *aac(6')-Ib-cr* and *qepA* among ceftiofur-resistant *Enterobacteriaceae* isolates from companion and food-producing animals. *Antimicrob Agents Chemother* **53:**519–524.
76. **Chen YT, Liao TL, Liu YM, Lauderdale TL, Yan JJ, Tsai SF.** 2009. Mobilization of *qnrB2* and IS*CR1* in plasmids. *Antimicrob Agents Chemother* **53:**1235–1237.
77. **Garnier F, Raked N, Gassama A, Denis F, Ploy MC.** 2006. Genetic environment of quinolone resistance gene *qnrB2* in a complex *sul1*-type integron in the newly described *Salmonella enterica* serovar Keurmassar. *Antimicrob Agents Chemother* **50:**3200–3202.
78. **Cattoir V, Nordmann P, Silva-Sanchez J, Espinal P, Poirel L.** 2008. IS*Ecp1*-mediated transposition of *qnrB*-like gene in *Escherichia coli*. *Antimicrob Agents Chemother* **52:**2929–2932.
79. **Dionisi AM, Lucarelli C, Owczarek S, Luzzi I, Villa L.** 2009. Characterization of the plasmid-borne quinolone resistance gene *qnrB19* in *Salmonella enterica* serovar Typhimurium. *Antimicrob Agents Chemother* **53:**4019–4021.
80. **Hordijk J, Bosman AB, van Essen-Zandbergen A, Veldman K, Dierikx C, Wagenaar JA, Mevius D.** 2011. *qnrB19* gene bracketed by IS26 on a 40-kilobase IncR plasmid from an *Escherichia coli* isolate from a veal calf. *Antimicrob Agents Chemother* **55:**453–454.
81. **Schink AK, Kadlec K, Schwarz S.** 2012. Detection of *qnr* genes among *Escherichia coli* isolates of animal origin and complete sequence of the conjugative *qnrB19*-carrying plasmid pQNR2078. *J Antimicrob Chemother* **67:**1099–1102.
82. **Tran T, Andres P, Petroni A, Soler-Bistue A, Albornoz E, Zorreguieta A, Reyes-Lamothe R, Sherratt DJ, Corso A, Tolmasky ME.** 2012. Small plasmids harboring *qnrB19*: a model for plasmid evolution mediated by site-specific recombination at *oriT* and Xer sites. *Antimicrob Agents Chemother* **56:**1821–1827.
83. **Rice LB, Carias LL, Hutton RA, Rudin SD, Endimiani A, Bonomo RA.** 2008. The KQ element, a complex genetic region conferring transferable resistance to carbapenems, aminoglycosides, and fluoroquinolones in *Klebsiella pneumoniae*. *Antimicrob Agents Chemother* **52:**3427–3429.
84. **Kehrenberg C, Friederichs S, de Jong A, Michael GB, Schwarz S.** 2006. Identification of the plasmid-borne quinolone resistance gene *qnrS* in *Salmonella enterica* serovar Infantis. *J Antimicrob Chemother* **58:**18–22.
85. **Kehrenberg C, Hopkins KL, Threlfall EJ, Schwarz S.** 2007. Complete nucleotide sequence of a small *qnrS1*-carrying plasmid from *Salmonella enterica* subsp.

enterica Typhimurium DT193. *J Antimicrob Chemother* **60**:903–905.

86. **Dolejska M, Villa L, Hasman H, Hansen L, Carattoli A.** 2013. Characterization of IncN plasmids carrying $bla_{CTX-M-1}$ and *qnr* genes in *Escherichia coli* and *Salmonella* from animals, the environment and humans. *J Antimicrob Chemother* **68**:333–339.
87. **Chen YT, Shu HY, Li LH, Liao TL, Wu KM, Shiau YR, Yan JJ, Su IJ, Tsai SF, Lauderdale TL.** 2006. Complete nucleotide sequence of pK245, a 98-kilobase plasmid conferring quinolone resistance and extended-spectrum-beta-lactamase activity in a clinical *Klebsiella pneumoniae* isolate. *Antimicrob Agents Chemother* **50**: 3861–3866.
88. **Hu FP, Xu XG, Zhu DM, Wang MG.** 2008. Coexistence of *qnrB4* and *qnrS1* in a clinical strain of *Klebsiella pneumoniae. Acta Pharmacol Sin* **29**:320–324.
89. **Poirel L, Cattoir V, Soares A, Soussy CJ, Nordmann P.** 2007. Novel Ambler class A β-lactamase LAP-1 and its association with the plasmid-mediated quinolone resistance determinant QnrS1. *Antimicrob Agents Chemother* **51**:631–637.
90. **Picão RC, Poirel L, Demarta A, Silva CS, Corvaglia AR, Petrini O, Nordmann P.** 2008. Plasmid-mediated quinolone resistance in *Aeromonas allosaccharophila* recovered from a Swiss lake. *J Antimicrob Chemother* **62**: 948–950.
91. **Mazzariol A, Kocsis B, Koncan R, Kocsis E, Lanzafame P, Cornaglia G.** 2012. Description and plasmid characterization of *qnrD* determinants in *Proteus mirabilis* and *Morganella morganii. Clin Microbiol Infect* **18**: E46–E48.
92. **Fonseca EL, Dos Santos Freitas F, Vieira VV, Vicente AC.** 2008. New *qnr* gene cassettes associated with superintegron repeats in *Vibrio cholerae* O1. *Emerg Infect Dis* **14**:1129–1131.
93. **Wu K, Wang F, Sun J, Wang Q, Chen Q, Yu S, Rui Y.** 2012. Class 1 integron gene cassettes in multidrug-resistant Gram-negative bacteria in southern China. *Int J Antimicrob Agents* **40**:264–267.
94. **Dubois V, Thabet L, Laffargue A, Andre C, Coulange-Mayonnave L, Quentin C.** 2013. *Overlapping outbreaks of imipenem resistant* Pseudomonas aeruginosa *in a Tunisian burn unit: emergence of new integrons encompassing* bla_{VIM-2}. GenBank accession number JX861889. http://www.ncbi.nlm.nih.gov/.
95. **Kumar P, Thomas S.** 2011. Presence of *qnrVC3* gene cassette in SXT and class 1 integrons of *Vibrio cholerae. Int J Antimicrob Agents* **37**:280–281.
96. **Pai H, Seo MR, Choi TY.** 2007. Association of QnrB determinants and production of extended-spectrum β-lactamases or plasmid-mediated AmpC β-lactamases in clinical isolates of *Klebsiella pneumoniae. Antimicrob Agents Chemother* **51**:366–368.
97. **Castanheira M, Mendes RE, Rhomberg PR, Jones RN.** 2008. Rapid emergence of bla_{CTX-M} among *Enterobacteriaceae* in U.S. medical centers: molecular evaluation from the MYSTIC Program (2007). *Microb Drug Resist* **14**:211–216.
98. **Jeong HS, Bae IK, Shin JH, Jung HJ, Kim SH, Lee JY, Oh SH, Kim HR, Chang CL, Kho WG, Lee JN.** 2011. Prevalence of plasmid-mediated quinolone resistance and its association with extended-spectrum beta-lactamase and AmpC beta-lactamase in *Enterobacteriaceae. Korean J Lab Med* **31**:257–264.
99. **Luo Y, Yang J, Zhang Y, Ye L, Wang L, Guo L.** 2011. Prevalence of β-lactamases and 16S rRNA methylase genes amongst clinical *Klebsiella pneumoniae* isolates carrying plasmid-mediated quinolone resistance determinants. *Int J Antimicrob Agents* **37**:352–355.
100. **Yang HF, Cheng J, Hu LF, Ye Y, Li JB.** 2012. Plasmid-mediated quinolone resistance in extended-spectrum-β-lactamase- and AmpC β-lactamase-producing *Serratia marcescens* in China. *Antimicrob Agents Chemother* **56**:4529–4531.
101. **Liu BT, Liao XP, Yue L, Chen XY, Li L, Yang SS, Sun J, Zhang S, Liao SD, Liu YH.** 2013. Prevalence of β-lactamase and 16S rRNA methylase genes among clinical *Escherichia coli* isolates carrying plasmid-mediated quinolone resistance genes from animals. *Microb Drug Resist* **19**:237–245.
102. **Soge OO, Adeniyi BA, Roberts MC.** 2006. New antibiotic resistance genes associated with CTX-M plasmids from uropathogenic Nigerian *Klebsiella pneumoniae. J Antimicrob Chemother* **58**:1048–1053.
103. **Cattoir V, Weill FX, Poirel L, Fabre L, Soussy CJ, Nordmann P.** 2007. Prevalence of *qnr* genes in *Salmonella* in France. *J Antimicrob Chemother* **59**:751–754.
104. **Jiang Y, Zhou Z, Qian Y, Wei Z, Yu Y, Hu S, Li L.** 2008. Plasmid-mediated quinolone resistance determinants *qnr* and *aac(6′)-Ib-cr* in extended-spectrum β-lactamase-producing *Escherichia coli* and *Klebsiella pneumoniae* in China. *J Antimicrob Chemother* **61**: 1003–1006.
105. **Guessennd N, Bremont S, Gbonon V, Kacou-Ndouba A, Ekaza E, Lambert T, Dosso M, Courvalin P.** 2008. Résistance aux quinolones de type qnr chez les entérobactéries productrices de bêta-lactamases à spectre élargi à Abidjan en Côte d'Ivoire. *Pathol Biol (Paris)* **56**: 439–446.
106. **Lavigne JP, Marchandin H, Delmas J, Bouziges N, Lecaillon E, Cavalie L, Jean-Pierre H, Bonnet R, Sotto A.** 2006. *qnrA* in CTX-M-producing *Escherichia coli* isolates from France. *Antimicrob Agents Chemother* **50**: 4224–4228.
107. **Lavilla S, Gonzalez-Lopez JJ, Sabate M, Garcia-Fernandez A, Larrosa MN, Bartolome RM, Carattoli A, Prats G.** 2008. Prevalence of *qnr* genes among extended-spectrum β-lactamase-producing enterobacterial isolates in Barcelona, Spain. *J Antimicrob Chemother* **61**: 291–295.
108. **Yang H, Chen H, Yang Q, Chen M, Wang H.** 2008. High prevalence of plasmid-mediated quinolone resistance genes *qnr* and *aac(6′)-Ib-cr* in clinical isolates of *Enterobacteriaceae* from nine teaching hospitals in China. *Antimicrob Agents Chemother* **52**:4268–4273.
109. **Iabadene H, Messai Y, Ammari H, Ramdani-Bouguessa N, Lounes S, Bakour R, Arlet G.** 2008. Dissemination

of ESBL and Qnr determinants in *Enterobacter cloacae* in Algeria. *J Antimicrob Chemother* **62:**133–136.

110. **Perilli M, Forcella C, Celenza G, Frascaria P, Segatore B, Pellegrini C, Amicosante G.** 2009. Evidence for *qnrB1* and *aac(6′)-Ib-cr* in CTX-M-15-producing uropathogenic *Enterobacteriaceae* in an Italian teaching hospital. *Diagn Microbiol Infect Dis* **64:**90–93.

111. **Miró E, Segura C, Navarro F, Sorlí L, Coll P, Horcajada JP, Álvarez-Lerma F, Salvadó M.** 2010. Spread of plasmids containing the bla_{VIM-1} and bla_{CTX-M} genes and the *qnr* determinant in *Enterobacter cloacae, Klebsiella pneumoniae* and *Klebsiella oxytoca* isolates. *J Antimicrob Chemother* **65:**661–665.

112. **Albrechtova K, Dolejska M, Cizek A, Tausova D, Klimes J, Bebora L, Literak I.** 2012. Dogs of nomadic pastoralists in northern Kenya are reservoirs of plasmid-mediated cephalosporin- and quinolone-resistant *Escherichia coli*, including pandemic clone B2-O25-ST131. *Antimicrob Agents Chemother* **56:**4013–4017.

113. **Shibl AM, Al-Agamy MH, Khubnani H, Senok AC, Tawfik AF, Livermore DM.** 2012. High prevalence of acquired quinolone-resistance genes among *Enterobacteriaceae* from Saudi Arabia with CTX-M-15 β-lactamase. *Diagn Microbiol Infect Dis* **73:**350–353.

114. **Li DX, Zhang SM, Hu GZ, Wang Y, Liu HB, Wu CM, Shang YH, Chen YX, Du XD.** 2012. Tn3-associated *rmtB* together with *qnrS1, aac(6′)-Ib-cr* and $bla_{CTX-M-15}$ are co-located on an F49:A-:B- plasmid in an *Escherichia coli* ST10 strain in China. *J Antimicrob Chemother* **67:** 236–238.

115. **Wu JJ, Ko WC, Tsai SH, Yan JJ.** 2007. Prevalence of plasmid-mediated quinolone resistance determinants QnrA, QnrB, and QnrS among clinical isolates of *Enterobacter cloacae* in a Taiwanese hospital. *Antimicrob Agents Chemother* **51:**1223–1227.

116. **Chmelnitsky I, Navon-Venezia S, Strahilevitz J, Carmeli Y.** 2008. Plasmid-mediated *qnrB2* and carbapenemase gene bla_{KPC-2} carried on the same plasmid in carbapenem-resistant ciprofloxacin-susceptible *Enterobacter cloacae* isolates. *Antimicrob Agents Chemother* **52:**2962–2965.

117. **Endimiani A, Carias LL, Hujer AM, Bethel CR, Hujer KM, Perez F, Hutton RA, Fox WR, Hall GS, Jacobs MR, Paterson DL, Rice LB, Jenkins SG, Tenover FC, Bonomo RA.** 2008. Presence of plasmid-mediated quinolone resistance in *Klebsiella pneumoniae* isolates possessing bla_{KPC} in the United States. *Antimicrob Agents Chemother* **52:**2680–2682.

118. **Mendes RE, Bell JM, Turnidge JD, Yang Q, Yu Y, Sun Z, Jones RN.** 2008. Carbapenem-resistant isolates of *Klebsiella pneumoniae* in China and detection of a conjugative plasmid (bla_{KPC-2} plus *qnr*B4) and a bla_{IMP-4} gene. *Antimicrob Agents Chemother* **52:**798–799.

119. **Poirel L, Leviandier C, Nordmann P.** 2006. Prevalence and genetic analysis of plasmid-mediated quinolone resistance determinants QnrA and QnrS in *Enterobacteriaceae* isolates from a French university hospital. *Antimicrob Agents Chemother* **50:**3992–3997.

120. **García-Fernández A, Fortini D, Veldman K, Mevius D, Carattoli A.** 2009. Characterization of plasmids harbouring *qnrS1, qnrB2* and *qnrB19* genes in *Salmonella. J Antimicrob Chemother* **63:**274–281.

121. **Potron A, Poirel L, Bernabeu S, Monnet X, Richard C, Nordmann P.** 2009. Nosocomial spread of ESBL-positive *Enterobacter cloacae* co-expressing plasmid-mediated quinolone resistance Qnr determinants in one hospital in France. *J Antimicrob Chemother* **64:** 653–654.

122. **Cambau E, Lascols C, Sougakoff W, Bebear C, Bonnet R, Cavallo JD, Gutmann L, Ploy MC, Jarlier V, Soussy CJ, Robert J.** 2006. Occurrence of *qnrA*-positive clinical isolates in French teaching hospitals during 2002–2005. *Clin Microbiol Infect* **12:**1013–1020.

123. **Corkill JE, Anson JJ, Hart CA.** 2005. High prevalence of the plasmid-mediated quinolone resistance determinant *qnrA* in multidrug-resistant *Enterobacteriaceae* from blood cultures in Liverpool, UK. *J Antimicrob Chemother* **56:**1115–1117.

124. **Rodriguez-Martinez JM, Poirel L, Pascual A, Nordmann P.** 2006. Plasmid-mediated quinolone resistance in Australia. *Microb Drug Resist* **12:**99–102.

125. **Szabó D, Kocsis B, Rókusz L, Szentandrássy J, Katona K, Kristóf K, Nagy K.** 2008. First detection of plasmid-mediated, quinolone resistance determinants *qnrA, qnrB, qnrS* and *aac(6′)-Ib-cr* in extended-spectrum β-lactamase (ESBL)-producing *Enterobacteriaceae* in Budapest, Hungary. *J Antimicrob Chemother* **62:** 630–632.

126. **Poirel L, Villa L, Bertini A, Pitout JD, Nordmann P, Carattoli A.** 2007. Expanded-spectrum β-lactamase and plasmid-mediated quinolone resistance. *Emerg Infect Dis* **13:**803–805.

127. **Zhao X, Xu X, Zhu D, Ye X, Wang M.** 2010. Decreased quinolone susceptibility in high percentage of *Enterobacter cloacae* clinical isolates caused only by Qnr determinants. *Diagn Microbiol Infect Dis* **67:** 110–113.

128. **Mammeri H, Van De Loo M, Poirel L, Martinez-Martinez L, Nordmann P.** 2005. Emergence of plasmid-mediated quinolone resistance in *Escherichia coli* in Europe. *Antimicrob Agents Chemother* **49:**71–76.

129. **Nazic H, Poirel L, Nordmann P.** 2005. Further identification of plasmid-mediated quinolone resistance determinant in *Enterobacteriaceae* in Turkey. *Antimicrob Agents Chemother* **49:**2146–2147.

130. **Poirel L, Van De Loo M, Mammeri H, Nordmann P.** 2005. Association of plasmid-mediated quinolone resistance with extended-spectrum β-lactamase VEB-1. *Antimicrob Agents Chemother* **49:**3091–3094.

131. **Aschbacher R, Doumith M, Livermore DM, Larcher C, Woodford N.** 2008. Linkage of acquired quinolone resistance (*qnrS1*) and metallo-beta-lactamase (bla_{VIM-1}) genes in multiple species of *Enterobacteriaceae* from Bolzano, Italy. *J Antimicrob Chemother* **61:**515–523.

132. **Wu JJ, Ko WC, Wu HM, Yan JJ.** 2008. Prevalence of Qnr determinants among bloodstream isolates of *Escherichia coli* and *Klebsiella pneumoniae* in a Taiwanese hospital, 1999–2005. *J Antimicrob Chemother* **61:** 1234–1239.

133. **Lee CH, Liu JW, Li CC, Chien CC, Tang YF, Su LH.** 2011. Spread of IS*CR1* elements containing bla_{DHA-1} and multiple antimicrobial resistance genes leading to increase of flomoxef resistance in extended-spectrum-β-lactamase-producing *Klebsiella pneumoniae*. *Antimicrob Agents Chemother* **55:**4058–4063.

134. **Verdet C, Benzerara Y, Gautier V, Adam O, Ould-Hocine Z, Arlet G.** 2006. Emergence of DHA-1-producing *Klebsiella* spp. in the Parisian region: genetic organization of the *ampC* and *ampR* genes originating from *Morganella morganii*. *Antimicrob Agents Chemother* **50:**607–617.

135. **Hidalgo L, Gutierrez B, Ovejero CM, Carrilero L, Matrat S, Saba CK, Santos-Lopez A, Thomas-Lopez D, Hoefer A, Suarez M, Santurde G, Martin-Espada C, Gonzalez-Zorn B.** 2013. *Klebsiella pneumoniae* sequence type 11 from companion animals bearing ArmA methyltransferase, DHA-1 β-lactamase, and QnrB4. *Antimicrob Agents Chemother* **57:**4532–4534.

136. **Sanchez-Cespedes J, Marti S, Soto SM, Alba V, Melción C, Almela M, Marco F, Vila J.** 2009. Two chromosomally located *qnrB* variants, *qnrB6* and the new *qnrB16*, in *Citrobacter* spp. isolates causing bacteraemia. *Clin Microbiol Infect* **15:**1132–1138.

137. **Kehrenberg C, Friederichs S, de Jong A, Schwarz S.** 2008. Novel variant of the *qnrB* gene, *qnrB12*, in *Citrobacter werkmanii*. *Antimicrob Agents Chemother* **52:**1206–1207.

138. **Ogbolu DO, Daini OA, Ogunledun A, Alli AO, Webber MA.** 2011. High levels of multidrug resistance in clinical isolates of Gram-negative pathogens from Nigeria. *Int J Antimicrob Agents* **37:**62–66.

139. **Wang F, Wu K, Sun J, Wang Q, Chen Q, Yu S, Rui Y.** 2012. Novel IS*CR1*-linked resistance genes found in multidrug-resistant Gram-negative bacteria in southern China. *Int J Antimicrob Agents* **40:**404–408.

140. **Touati A, Brasme L, Benallaoua S, Gharout A, Madoux J, De Champs C.** 2008. First report of *qnrB*-producing *Enterobacter cloacae* and *qnrA*-producing *Acinetobacter baumannii* recovered from Algerian hospitals. *Diagn Microbiol Infect Dis* **60:**287–290.

141. **Guo L, Zhang J, Xu C, Zhao Y, Ren T, Zhang B, Fan H, Liao M.** 2011. Molecular characterization of fluoroquinolone resistance in *Haemophilus parasuis* isolated from pigs in South China. *J Antimicrob Chemother* **66:**539–542.

142. **Jacoby GA, Gacharna N, Black TA, Miller GH, Hooper DC.** 2009. Temporal appearance of plasmid-mediated quinolone resistance genes. *Antimicrob Agents Chemother* **53:**1665–1666.

143. **Strahilevitz J, Engelstein D, Adler A, Temper V, Moses AE, Block C, Robicsek A.** 2007. Changes in *qnr* prevalence and fluoroquinolone resistance in clinical isolates of *Klebsiella pneumoniae* and *Enterobacter* spp. collected from 1990 to 2005. *Antimicrob Agents Chemother* **51:**3001–3003.

144. **Xu X, Wu S, Ye X, Liu Y, Shi W, Zhang Y, Wang M.** 2007. Prevalence and expression of the plasmid-mediated quinolone resistance determinant *qnrA1*. *Antimicrob Agents Chemother* **51:**4105–4110.

145. **Kim HB, Park CH, Kim CJ, Kim EC, Jacoby GA, Hooper DC.** 2009. Prevalence of plasmid-mediated quinolone resistance determinants over a 9-year period. *Antimicrob Agents Chemother* **53:**639–645.

146. **Kim NH, Choi EH, Sung JY, Oh CE, Kim HB, Kim EC, Lee HJ.** 2013. Prevalence of plasmid-mediated quinolone resistance genes and ciprofloxacin resistance in pediatric bloodstream isolates of *Enterobacteriaceae* over a 9-year period. *Jpn J Infect Dis* **66:**151–154.

147. **Robicsek A, Sahm DF, Strahilevitz J, Jacoby GA, Hooper DC.** 2005. Broader distribution of plasmid-mediated quinolone resistance in the United States. *Antimicrob Agents Chemother* **49:**3001–3003.

148. **Literak I, Micudova M, Tausova D, Cizek A, Dolejska M, Papousek I, Prochazka J, Vojtech J, Borleis F, Guardone L, Guenther S, Hordowski J, Lejas C, Meissner W, Marcos BF, Tucakov M.** 2012. Plasmid-mediated quinolone resistance genes in fecal bacteria from rooks commonly wintering throughout Europe. *Microb Drug Resist* **18:**567–573.

149. **Tausova D, Dolejska M, Cizek A, Hanusova L, Hrusakova J, Svoboda O, Camlik G, Literak I.** 2012. *Escherichia coli* with extended-spectrum β-lactamase and plasmid-mediated quinolone resistance genes in great cormorants and mallards in Central Europe. *J Antimicrob Chemother* **67:**1103–1107.

150. **Shaheen BW, Nayak R, Foley SL, Boothe DM.** 2013. Chromosomal and plasmid-mediated fluoroquinolone resistance mechanisms among broad-spectrum-cephalosporin-resistant *Escherichia coli* isolates recovered from companion animals in the USA. *J Antimicrob Chemother* **68:**1019–1024.

151. **Kirchner M, Wearing H, Teale C.** 2011. Plasmid-mediated quinolone resistance gene detected in *Escherichia coli* from cattle. *Vet Microbiol* **148:**434–435.

152. **Madec JY, Poirel L, Saras E, Gourguechon A, Girlich D, Nordmann P, Haenni M.** 2012. Non-ST131 *Escherichia coli* from cattle harbouring human-like $bla_{CTX-M-15}$-carrying plasmids. *J Antimicrob Chemother* **67:**578–581.

153. **Wu CM, Wang Y, Cao XY, Lin JC, Qin SS, Mi TJ, Huang SY, Shen JZ.** 2009. Emergence of plasmid-mediated quinolone resistance genes in *Enterobacteriaceae* isolated from chickens in China. *J Antimicrob Chemother* **63:**408–411.

154. **Cerquetti M, Garcia-Fernandez A, Giufre M, Fortini D, Accogli M, Graziani C, Luzzi I, Caprioli A, Carattoli A.** 2009. First report of plasmid-mediated quinolone resistance determinant *qnrS1* in an *Escherichia coli* strain of animal origin in Italy. *Antimicrob Agents Chemother* **53:**3112–3114.

155. **Veldman K, Cavaco LM, Mevius D, Battisti A, Franco A, Botteldoorn N, Bruneau M, Perrin-Guyomard A, Cerny T, De Frutos Escobar C, Guerra B, Schroeter A, Gutierrez M, Hopkins K, Myllyniemi AL, Sunde M, Wasyl D, Aarestrup FM.** 2011. International collaborative study on the occurrence of plasmid-mediated quinolone resistance in *Salmonella enterica* and *Escherichia coli* isolated from animals, humans, food and the

environment in 13 European countries. *J Antimicrob Chemother* **66:**1278–1286.

156. **Chen X, Zhang W, Pan W, Yin J, Pan Z, Gao S, Jiao X.** 2012. Prevalence of *qnr, aac(6')-Ib-cr, qepA,* and *oqxAB* in *Escherichia coli* isolates from humans, animals, and the environment. *Antimicrob Agents Chemother* **56:**3423–3427.

157. **Du XD, Li DX, Hu GZ, Wang Y, Shang YH, Wu CM, Liu HB, Li XS.** 2012. Tn*1548*-associated *armA* is co-located with *qnrB2*, *aac(6')-Ib-cr* and $bla_{CTX-M-3}$ on an IncFII plasmid in a *Salmonella enterica* subsp. *enterica* serovar Paratyphi B strain isolated from chickens in China. *J Antimicrob Chemother* **67:** 246–248.

158. **Literak I, Reitschmied T, Bujnakova D, Dolejska M, Cizek A, Bardon J, Pokludova L, Alexa P, Halova D, Jamborova I.** 2013. Broilers as a source of quinolone-resistant and extraintestinal pathogenic *Escherichia coli* in the Czech Republic. *Microb Drug Resist* **19:**57–63.

159. **Kim JH, Cho JK, Kim KS.** 2013. Prevalence and characterization of plasmid-mediated quinolone resistance genes in *Salmonella* isolated from poultry in Korea. *Avian Pathol* **42:**221–229.

160. **Liu BT, Yang QE, Li L, Sun J, Liao XP, Fang LX, Yang SS, Deng H, Liu YH.** 2013. Dissemination and characterization of plasmids carrying *oqxAB*-bla_{CTX-M} genes in *Escherichia coli* isolates from food-producing animals. *PLoS One* **8:**e73947. doi:10.1371/journal.pone.0073947.

161. **Pomba C, da Fonseca JD, Baptista BC, Correia JD, Martinez-Martinez L.** 2009. Detection of the pandemic O25-ST131 human virulent *Escherichia coli* CTX-M-15-producing clone harboring the *qnrB2* and *aac (6')-Ib-cr* genes in a dog. *Antimicrob Agents Chemother* **53:**327–328.

162. **Yue L, Jiang HX, Liao XP, Liu JH, Li SJ, Chen XY, Chen CX, Lu DH, Liu YH.** 2008. Prevalence of plasmid-mediated quinolone resistance *qnr* genes in poultry and swine clinical isolates of *Escherichia coli. Vet Microbiol* **132:**414–420.

163. **Verner-Jeffreys DW, Welch TJ, Schwarz T, Pond MJ, Woodward MJ, Haig SJ, Rimmer GS, Roberts E, Morrison V, Baker-Austin C.** 2009. High prevalence of multidrug-tolerant bacteria and associated antimicrobial resistance genes isolated from ornamental fish and their carriage water. *PLoS One* **4:**e8388. doi: 10.1371/journal.pone.0008388.

164. **Jiang HX, Tang D, Liu YH, Zhang XH, Zeng ZL, Xu L, Hawkey PM.** 2012. Prevalence and characteristics of β-lactamase and plasmid-mediated quinolone resistance genes in *Escherichia coli* isolated from farmed fish in China. *J Antimicrob Chemother* **67:**2350–2353.

165. **Han JE, Kim JH, Choresca CH Jr, Shin SP, Jun JW, Chai JY, Park SC.** 2012. First description of ColE-type plasmid in *Aeromonas* spp. carrying quinolone resistance (*qnrS2*) gene. *Lett Appl Microbiol* **55:**290–294.

166. **Dolejska M, Duskova E, Rybarikova J, Janoszowska D, Roubalova E, Dibdakova K, Maceckova G, Kohoutova L, Literak I, Smola J, Cizek A.** 2011. Plasmids carrying $bla_{CTX-M-1}$ and *qnr* genes in *Escherichia coli* isolates from an equine clinic and a horseback riding centre. *J Antimicrob Chemother* **66:**757–764.

167. **Liu JH, Deng YT, Zeng ZL, Gao JH, Chen L, Arakawa Y, Chen ZL.** 2008. Coprevalence of plasmid-mediated quinolone resistance determinants QepA, Qnr, and AAC(6')-Ib-cr among 16S rRNA methylase RmtB-producing *Escherichia coli* isolates from pigs. *Antimicrob Agents Chemother* **52:**2992–2993.

168. **Dotto G, Giacomelli M, Grilli G, Ferrazzi V, Carattoli A, Fortini D, Piccirillo A.** 2013. High prevalence of *oqxAB* in *Escherichia coli* isolates from domestic and wild lagomorphs in Italy. *Microb Drug Resist* **20:** 118–123.

169. **Guerra B, Helmuth R, Thomas K, Beutlich J, Jahn S, Schroeter A.** 2010. Plasmid-mediated quinolone resistance determinants in *Salmonella* spp. isolates from reptiles in Germany. *J Antimicrob Chemother* **65:** 2043–2045.

170. **Ahmed AM, Motoi Y, Sato M, Maruyama A, Watanabe H, Fukumoto Y, Shimamoto T.** 2007. Zoo animals as reservoirs of Gram-negative bacteria harboring integrons and antimicrobial resistance genes. *Appl Environ Microbiol* **73:**6686–6690.

171. **Wang Y, He T, Han J, Wang J, Foley SL, Yang G, Wan S, Shen J, Wu C.** 2012. Prevalence of ESBLs and PMQR genes in fecal *Escherichia coli* isolated from the non-human primates in six zoos in China. *Vet Microbiol* **159:**53–59.

172. **Dobiasova H, Dolejska M, Jamborova I, Brhelova E, Blazkova L, Papousek I, Kozlova M, Klimes J, Cizek A, Literak I.** 2013. Extended spectrum beta-lactamase and fluoroquinolone resistance genes and plasmids among *Escherichia coli* isolates from zoo animals, Czech Republic. *FEMS Microbiol Ecol* **85:**604–611.

173. **Kim HB, Park CH, Gavin M, Jacoby GA, Hooper DC.** 2011. Cold shock induces *qnrA* expression in *Shewanella algae. Antimicrob Agents Chemother* **55:**414–416.

174. **Wang M, Jacoby GA, Mills DM, Hooper DC.** 2009. SOS regulation of *qnrB* expression. *Antimicrob Agents Chemother* **53:**821–823.

175. **Da Re S, Garnier F, Guerin E, Campoy S, Denis F, Ploy MC.** 2009. The SOS response promotes *qnrB* quinolone-resistance determinant expression. *EMBO Rep* **10:**929–933.

176. **Briales A, Rodriguez-Martinez JM, Velasco C, Machuca J, Diaz de Alba P, Blazquez J, Pascual A.** 2012. Exposure to diverse antimicrobials induces the expression of *qnrB1*, *qnrD* and *smaqnr* genes by SOS-dependent regulation. *J Antimicrob Chemother* **67:**2854–2859.

177. **Okumura R, Liao CH, Gavin M, Jacoby GA, Hooper DC.** 2011. Quinolone induction of *qnrVS1* in *Vibrio splendidus* and plasmid-carried *qnrS1* in *Escherichia coli*, a mechanism independent of the SOS system. *Antimicrob Agents Chemother* **55:**5942–5945.

178. **Kwak YG, Jacoby GA, Hooper DC.** 2013. Induction of plasmid-carried *qnrS1* in *Escherichia coli* by naturally occurring quinolones and quorum-sensing signal molecules. *Antimicrob Agents Chemother* **57:**4031–4034.

179. **Vetting MW, Park CH, Hegde SS, Jacoby GA, Hooper DC, Blanchard JS.** 2008. Mechanistic and structural analysis of aminoglycoside *N*-acetyltransferase AAC (6′)-Ib and its bifunctional fluoroquinolone-active AAC (6′)-Ib-cr variant. *Biochemistry* **47:**9825–9835.

180. **Maurice F, Broutin I, Podglajen I, Benas P, Collatz E, Dardel F.** 2008. Enzyme structural plasticity and the emergence of broad-spectrum antibiotic resistance. *EMBO Rep* **9:**344–349.

181. **Ruiz E, Ocampo-Sosa AA, Alcoba-Florez J, Roman E, Arlet G, Torres C, Martinez-Martinez L.** 2012. Changes in ciprofloxacin resistance levels in *Enterobacter aerogenes* isolates associated with variable expression of the *aac(6′)-Ib-cr* gene. *Antimicrob Agents Chemother* **56:**1097–1100.

182. **de Toro M, Rodriguez I, Rojo-Bezares B, Helmuth R, Torres C, Guerra B, Saenz Y.** 2013. pMdT1, a small ColE1-like plasmid mobilizing a new variant of the *aac(6′)-Ib-cr* gene in *Salmonella enterica* serovar Typhimurium. *J Antimicrob Chemother* **68:**1277–1280.

183. **Guillard T, Duval V, Moret H, Brasme L, Vernet-Garnier V, de Champs C.** 2010. Rapid detection of *aac(6′)-Ib-cr* quinolone resistance gene by pyrosequencing. *J Clin Microbiol* **48:**286–289.

184. **Pitout JD, Wei Y, Church DL, Gregson DB.** 2008. Surveillance for plasmid-mediated quinolone resistance determinants in *Enterobacteriaceae* within the Calgary Health Region, Canada: the emergence of *aac(6′)-Ib-cr*. *J Antimicrob Chemother* **61:**999–1002.

185. **Jones GL, Warren RE, Skidmore SJ, Davies VA, Gibreel T, Upton M.** 2008. Prevalence and distribution of plasmid-mediated quinolone resistance genes in clinical isolates of *Escherichia coli* lacking extended-spectrum β-lactamases. *J Antimicrob Chemother* **62:**1245–1251.

186. **Oteo J, Cuevas O, López-Rodríguez I, Banderas-Florido A, Vindel A, Pérez-Vázquez M, Bautista V, Arroyo M, García-Caballero J, Marin-Casanova P, González-Sanz R, Fuentes-Gómez V, Oña-Compán S, García-Cobos S, Campos J.** 2009. Emergence of CTX-M-15-producing *Klebsiella pneumoniae* of multilocus sequence types 1, 11, 14, 17, 20, 35 and 36 as pathogens and colonizers in newborns and adults. *J Antimicrob Chemother* **64:** 524–528.

187. **Baudry PJ, Nichol K, DeCorby M, Lagace-Wiens P, Olivier E, Boyd D, Mulvey MR, Hoban DJ, Zhanel GG.** 2009. Mechanisms of resistance and mobility among multidrug-resistant CTX-M-producing *Escherichia coli* from Canadian intensive care units: the 1st report of QepA in North America. *Diagn Microbiol Infect Dis* **63:**319–326.

188. **Coelho A, Mirelis B, Alonso-Tarrés C, Nieves Larrosa M, Miró E, Clivillé Abad R, Bartolomé RM, Castañer M, Prats G, Johnson JR, Navarro F, González-López JJ.** 2009. Detection of three stable genetic clones of CTX-M-15-producing *Klebsiella pneumoniae* in the Barcelona metropolitan area, Spain. *J Antimicrob Chemother* **64:**862–864.

189. **Literacka E, Bedenic B, Baraniak A, Fiett J, Tonkic M, Jajic-Bencic I, Gniadkowski M.** 2009. bla_{CTX-M} genes in *Escherichia coli* strains from Croatian hospitals are located in new ($bla_{CTX-M-3a}$) and widely spread ($bla_{CTX-M-3a}$ and $bla_{CTX-M-15}$) genetic structures. *Antimicrob Agents Chemother* **53:**1630–1635.

190. **Sabtcheva S, Kaku M, Saga T, Ishii Y, Kantardjiev T.** 2009. High prevalence of the *aac(6′)-Ib-cr* gene and its dissemination among *Enterobacteriaceae* isolates by CTX-M-15 plasmids in Bulgaria. *Antimicrob Agents Chemother* **53:**335–336.

191. **Ruiz E, Rezusta A, Saenz Y, Rocha-Gracia R, Vinue L, Vindel A, Villuendas C, Azanedo ML, Monforte ML, Revillo MJ, Torres C.** 2011. New genetic environments of *aac(6′)-Ib-cr* gene in a multiresistant *Klebsiella oxytoca* strain causing an outbreak in a pediatric intensive care unit. *Diagn Microbiol Infect Dis* **69:**236–238.

192. **Ruiz E, Saenz Y, Zarazaga M, Rocha-Gracia R, Martinez-Martinez L, Arlet G, Torres C.** 2012. *qnr, aac(6′)-Ib-cr* and *qepA* genes in *Escherichia coli* and *Klebsiella* spp.: genetic environments and plasmid and chromosomal location. *J Antimicrob Chemother* **67:** 886–897.

193. **Musumeci R, Rausa M, Giovannoni R, Cialdella A, Bramati S, Sibra B, Giltri G, Vigano F, Cocuzza CE.** 2012. Prevalence of plasmid-mediated quinolone resistance genes in uropathogenic *Escherichia coli* isolated in a teaching hospital of northern Italy. *Microb Drug Resist* **18:**33–41.

194. **Park CH, Robicsek A, Jacoby GA, Sahm D, Hooper DC.** 2006. Prevalence in the United States of *aac(6′) Ib-cr* encoding a ciprofloxacin-modifying enzyme. *Antimicrob Agents Chemother* **50:**3953–3955.

195. **Kim ES, Jeong JY, Jun JB, Choi SH, Lee SO, Kim MN, Woo JH, Kim YS.** 2009. Prevalence of *aac(6′)-Ib-cr* encoding a ciprofloxacin-modifying enzyme among *Enterobacteriaceae* blood isolates in Korea. *Antimicrob Agents Chemother* **53:**2643–2645.

196. **Yang J, Luo Y, Li J, Ma Y, Hu C, Jin S, Ye L, Cui S.** 2010. Characterization of clinical *Escherichia coli* isolates from China containing transferable quinolone resistance determinants. *J Antimicrob Chemother* **65:** 453–459.

197. **Périchon B, Bogaerts P, Lambert T, Frangeul L, Courvalin P, Galimand M.** 2008. Sequence of conjugative plasmid pIP1206 mediating resistance to aminoglycosides by 16S rRNA methylation and to hydrophilic fluoroquinolones by efflux. *Antimicrob Agents Chemother* **52:**2581–2592.

198. **Yamane K, Wachino J, Suzuki S, Arakawa Y.** 2008. Plasmid-mediated *qepA* gene among *Escherichia coli* clinical isolates from Japan. *Antimicrob Agents Chemother* **52:**1564–1566.

199. **Cattoir V, Poirel L, Nordmann P.** 2008. Plasmid-mediated quinolone resistance pump QepA2 in an *Escherichia coli* isolate from France. *Antimicrob Agents Chemother* **52:**3801–3804.

200. **Sorensen AH, Hansen LH, Johannesen E, Sorensen SJ.** 2003. Conjugative plasmid conferring resistance to olaquindox. *Antimicrob Agents Chemother* **47:**798–799.

201. **Hansen LH, Johannesen E, Burmolle M, Sorensen AH, Sorensen SJ.** 2004. Plasmid-encoded multidrug efflux

pump conferring resistance to olaquindox in *Escherichia coli*. *Antimicrob Agents Chemother* **48**:3332–3337.

202. **Norman A, Hansen LH, She Q, Sørensen SJ.** 2008. Nucleotide sequence of pOLA52: a conjugative IncX1 plasmid from *Escherichia coli* which enables biofilm formation and multidrug efflux. *Plasmid* **60**:59–74.
203. **Kim HB, Wang M, Park CH, Kim EC, Jacoby GA, Hooper DC.** 2009. *oqxAB* encoding a multidrug efflux pump in human clinical isolates of *Enterobacteriaceae*. *Antimicrob Agents Chemother* **53**:3582–3584.
204. **Zhao J, Chen Z, Chen S, Deng Y, Liu Y, Tian W, Huang X, Wu C, Sun Y, Zeng Z, Liu JH.** 2010. Prevalence and dissemination of *oqxAB* in *Escherichia coli* isolates from animals, farmworkers, and the environment. *Antimicrob Agents Chemother* **54**:4219–4224.
205. **Wong MH, Chen S.** 2013. First detection of *oqxAB* in *Salmonella* spp. isolated from food. *Antimicrob Agents Chemother* **57**:658–660.
206. **Rodríguez-Martínez JM, Díaz de Alba P, Briales A, Machuca J, Lossa M, Fernández-Cuenca F, Rodríguez Baño J, Martínez-Martínez L, Pascual A.** 2013. Contribution of OqxAB efflux pumps to quinolone resistance in extended-spectrum-β-lactamase-producing *Klebsiella pneumoniae*. *J Antimicrob Chemother* **68**:68–73.
207. **Li L, Liao X, Yang Y, Sun J, Li L, Liu B, Yang S, Ma J, Li X, Zhang Q, Liu Y.** 2013. Spread of oqxAB in *Salmonella enterica* serotype Typhimurium predominantly by IncHI2 plasmids. *J Antimicrob Chemother* **68**: 2263–2268.
208. **Yuan J, Xu X, Guo Q, Zhao X, Ye X, Guo Y, Wang M.** 2012. Prevalence of the *oqxAB* gene complex in *Klebsiella pneumoniae* and *Escherichia coli* clinical isolates. *J Antimicrob Chemother* **67**:1655–1659.
209. **Perez F, Rudin SD, Marshall SH, Coakley P, Chen L, Kreiswirth BN, Rather PN, Hujer AM, Toltzis P, van Duin D, Paterson DL, Bonomo RA.** 2013. OqxAB, a quinolone and olaquindox efflux pump, is widely distributed among multidrug-resistant *Klebsiella pneumoniae* isolates of human origin. *Antimicrob Agents Chemother* **57**:4602–4603.
210. **Veleba M, Higgins PG, Gonzalez G, Seifert H, Schneiders T.** 2012. Characterization of RarA, a novel AraC family multidrug resistance regulator in *Klebsiella pneumoniae*. *Antimicrob Agents Chemother* **56**: 4450–4458.
211. **De Majumdar S, Veleba M, Finn S, Fanning S, Schneiders T.** 2013. Elucidating the regulon of multidrug resistance regulator RarA in *Klebsiella pneumoniae*. *Antimicrob Agents Chemother* **57**:1603–1609.
212. **Szczepanowski R, Krahn I, Linke B, Goesmann A, Puhler A, Schluter A.** 2004. Antibiotic multiresistance plasmid pRSB101 isolated from a wastewater treatment plant is related to plasmids residing in phytopathogenic bacteria and carries eight different resistance determinants including a multidrug transport system. *Microbiology* **150**:3613–3630.
213. **Nakaminami H, Noguchi N, Sasatsu M.** 2010. Fluoroquinolone efflux by the plasmid-mediated multidrug efflux pump QacB variant QacBIII in *Staphylococcus aureus*. *Antimicrob Agents Chemother* **54**: 4107–4111.
214. **Drlica K.** 2003. The mutant selection window and antimicrobial resistance. *J Antimicrob Chemother* **52**: 11–17.
215. **Jacoby GA.** 2005. Mechanisms of resistance to quinolones. *Clin Infect Dis* **41**(Suppl 2):S120–S126.
216. **Rodríguez-Martínez JM, Velasco C, García I, Cano ME, Martínez-Martínez L, Pascual A.** 2007. Mutant prevention concentrations of fluoroquinolones for *Enterobacteriaceae* expressing the plasmid-carried quinolone resistance determinant *qnrA1*. *Antimicrob Agents Chemother* **51**:2236–2239.
217. **Cesaro A, Bettoni RR, Lascols C, Merens A, Soussy CJ, Cambau E.** 2008. Low selection of topoisomerase mutants from strains of *Escherichia coli* harbouring plasmid-borne *qnr* genes. *J Antimicrob Chemother* **61**: 1007–1015.
218. **Martínez-Martínez L, Pascual A, García I, Tran J, Jacoby GA.** 2003. Interaction of plasmid and host quinolone resistance. *J Antimicrob Chemother* **51**:1037–1039.
219. **Briales A, Rodriguez-Martinez JM, Velasco C, Diaz de Alba P, Dominguez-Herrera J, Pachon J, Pascual A.** 2011. *In vitro* effect of *qnrA1*, *qnrB1*, and *qnrS1* genes on fluoroquinolone activity against isogenic *Escherichia coli* isolates with mutations in *gyrA* and *parC*. *Antimicrob Agents Chemother* **55**:1266–1269.
220. **Luo Y, Li J, Meng Y, Ma Y, Hu C, Jin S, Zhang Q, Ding H, Cui S.** 2011. Joint effects of topoisomerase alterations and plasmid-mediated quinolone-resistant determinants in *Salmonella enterica* Typhimurium. *Microb Drug Resist* **17**:1–5.
221. **Lascols C, Robert J, Cattoir V, Bébéar C, Cavallo JD, Podglajen I, Ploy MC, Bonnet R, Soussy CJ, Cambau E.** 2007. Type II topoisomerase mutations in clinical isolates of *Enterobacter cloacae* and other enterobacterial species harbouring the *qnrA* gene. *Int J Antimicrob Agents* **29**:402–409.
222. **Poirel L, Pitout JD, Calvo L, Rodriguez-Martinez JM, Church D, Nordmann P.** 2006. *In vivo* selection of fluoroquinolone-resistant *Escherichia coli* isolates expressing plasmid-mediated quinolone resistance and expanded-spectrum β-lactamase. *Antimicrob Agents Chemother* **50**:1525–1527.
223. **de Toro M, Rojo-Bezares B, Vinué L, Undabeitia E, Torres C, Sáenz Y.** 2010. *In vivo* selection of *aac(6′)-Ib-cr* and mutations in the *gyrA* gene in a clinical *qnrS1*-positive *Salmonella enterica* serovar Typhimurium DT104B strain recovered after fluoroquinolone treatment. *J Antimicrob Chemother* **65**:1945–1949.
224. **Jeong JY, Kim ES, Choi SH, Kwon HH, Lee SR, Lee SO, Kim MN, Woo JH, Kim YS.** 2008. Effects of a plasmid-encoded *qnrA1* determinant in *Escherichia coli* strains carrying chromosomal mutations in the *acrAB* efflux pump genes. *Diagn Microbiol Infect Dis* **60**: 105–107.
225. **Sato T, Yokota S, Uchida I, Okubo T, Usui M, Kusumoto M, Akiba M, Fujii N, Tamura Y.** 2013. Fluoroquinolone resistance mechanisms in an *Escherichia*

coli isolate, HUE1, without quinolone resistance-determining region mutations. *Front Microbiol* **4**:125. doi:10.3389/fmicb.2013.00125.

226. **Warburg G, Korem M, Robicsek A, Engelstein D, Moses AE, Block C, Strahilevitz J.** 2009. Changes in *aac(6′)-Ib-cr* prevalence and fluoroquinolone resistance in nosocomial isolates of *Escherichia coli* collected from 1991 through 2005. *Antimicrob Agents Chemother* **53**: 1268–1270.

227. **Briales A, Rodriguez-Martinez JM, Velasco C, de Alba PD, Rodriguez-Bano J, Martinez-Martinez L, Pascual A.** 2012. Prevalence of plasmid-mediated quinolone resistance determinants *qnr* and *aac(6′)-Ib-cr* in *Escherichia coli* and *Klebsiella pneumoniae* producing extended- spectrum β-lactamases in Spain. *Int J Antimicrob Agents* **39**:431–434.

228. **Rodríguez-Martínez JM, Pichardo C, García I, Pachón-Ibañez ME, Docobo-Pérez F, Pascual A, Pachón J, Martínez-Martínez L.** 2008. Activity of ciprofloxacin and levofloxacin in experimental pneumonia caused by *Klebsiella pneumoniae* deficient in porins, expressing active efflux and producing QnrA1. *Clin Microbiol Infect* **14**:691–697.

229. **Domínguez-Herrera J, Velasco C, Docobo-Pérez F, Rodríguez-Martínez JM, López-Rojas R, Briales A, Pichardo C, Díaz-de-Alba P, Rodríguez-Baño J, Pascual A, Pachón J.** 2013. Impact of *qnrA1*, *qnrB1* and *qnrS1* on the efficacy of ciprofloxacin and levofloxacin in an experimental pneumonia model caused by *Escherichia coli* with or without the GyrA mutation Ser83Leu. *J Antimicrob Chemother* **68**:1609–1615.

230. **Allou N, Cambau E, Massias L, Chau F, Fantin B.** 2009. Impact of low-level resistance to fluoroquinolones due to *qnrA1* and *qnrS1* genes or a *gyrA* mutation on ciprofloxacin bactericidal activity in a murine model of *Escherichia coli* urinary tract infection. *Antimicrob Agents Chemother* **53**:4292–4297.

231. **Jakobsen L, Cattoir V, Jensen KS, Hammerum AM, Nordmann P, Frimodt-Moller N.** 2012. Impact of low-level fluoroquinolone resistance genes *qnrA1*, *qnrB19* and *qnrS1* on ciprofloxacin treatment of isogenic *Escherichia coli* strains in a murine urinary tract infection model. *J Antimicrob Chemother* **67**:2438–2444.

232. **Defife R, Scheetz MH, Feinglass JM, Postelnick MJ, Scarsi KK.** 2009. Effect of differences in MIC values on clinical outcomes in patients with bloodstream infections caused by Gram-negative organisms treated with levofloxacin. *Antimicrob Agents Chemother* **53**: 1074–1079.

233. **Chong YP, Choi SH, Kim ES, Song EH, Lee EJ, Park KH, Cho OH, Kim SH, Lee SO, Kim MN, Jeong JY, Woo JH, Kim YS.** 2010. Bloodstream infections caused by *qnr*-positive *Enterobacteriaceae:* clinical and microbiologic characteristics and outcomes. *Diagn Microbiol Infect Dis* **67**:70–77.

234. **Liao CH, Hsueh PR, Jacoby GA, Hooper DC.** 2013. Risk factors and clinical characteristics of patients with *qnr*-positive *Klebsiella pneumoniae* bacteraemia. *J Antimicrob Chemother* **68**:2907–2914.

235. **Buchko GW, Ni S, Robinson H, Welsh EA, Pakrasi HB, Kennedy MA.** 2006. Characterization of two potentially universal turn motifs that shape the repeated five-residues fold: crystal structure of a lumenal pentapeptide repeat protein from *Cyanothece* 51142. *Protein Sci* **15**:2579–2595.

236. **Lascols C, Podglajen I, Verdet C, Gautier V, Gutmann L, Soussy CJ, Collatz E, Cambau E.** 2008. A plasmid-borne *Shewanella algae* gene, *qnrA3*, and its possible transfer *in vivo* between *Kluyvera ascorbata* and *Klebsiella pneumoniae*. *J Bacteriol* **190**:5217–5223.

237. **Mnif B, Ktari S, Chaari A, Medhioub F, Rhimi F, Bouaziz M, Hammami A.** 2013. Nosocomial dissemination of *Providencia stuartii* isolates carrying bla_{OXA-48}, bla_{PER-1}, bla_{CMY-4} and *qnrA6* in a Tunisian hospital. *J Antimicrob Chemother* **68**:329–332.

238. **Robicsek A, Strahilevitz J, Sahm DF, Jacoby GA, Hooper DC.** 2006. *qnr* prevalence in ceftazidime-resistant *Enterobacteriaceae* isolates from the United States. *Antimicrob Agents Chemother* **50**:2872–2874.

239. **Quiroga MP, Andres P, Petroni A, Soler Bistué AJ, Guerriero L, Vargas LJ, Zorreguieta A, Tokumoto M, Quiroga C, Tolmasky ME, Galas M, Centrón D.** 2007. Complex class 1 integrons with diverse variable regions, including *aac(6′)-Ib-cr*, and a novel allele, *qnrB10*, associated with IS*CR1* in clinical enterobacterial isolates from Argentina. *Antimicrob Agents Chemother* **51**: 4466–4470.

240. **Hopkins KL, Wootton L, Day MR, Threlfall EJ.** 2007. Plasmid-mediated quinolone resistance determinant *qnrS1* found in *Salmonella enterica* strains isolated in the UK. *J Antimicrob Chemother* **59**:1071–1075.

241. **Bönemann G, Stiens M, Pühler A, Schlüter A.** 2006. Mobilizable IncQ-related plasmid carrying a new quinolone resistance gene, *qnrS2*, isolated from the bacterial community of a wastewater treatment plant. *Antimicrob Agents Chemother* **50**:3075–3080.

242. **Cattoir V, Poirel L, Aubert C, Soussy CJ, Nordmann P.** 2008. Unexpected occurrence of plasmid-mediated quinolone resistance determinants in environmental *Aeromonas* spp. *Emerg Infect Dis* **14**:231–237.

243. **Rajpara N, Patel A, Tiwari N, Bahuguna J, Antony A, Choudhury I, Ghosh A, Jain R, Bhardwaj AK.** 2009. Mechanism of drug resistance in a clinical isolate of *Vibrio fluvialis*: involvement of multiple plasmids and integrons. *Int J Antimicrob Agents* **34**:220–225.

244. **Boyd DA, Tyler S, Christianson S, McGeer A, Muller MP, Willey BM, Bryce E, Gardam M, Nordmann P, Mulvey MR.** 2004. Complete nucleotide sequence of a 92-kilobase plasmid harboring the CTX-M-15 extended-spectrum β-lactamase involved in an outbreak in long-term-care facilities in Toronto, Canada. *Antimicrob Agents Chemother* **48**:3758–3764.

245. **Jacoby G, Chow N, Waites K.** 2003. Prevalence of plasmid-mediated quinolone resistance. *Antimicrob Agents Chemother* **47**:559–562.

246. **Jonas D, Biehler K, Hartung D, Spitzmüller B, Daschner FD.** 2005. Plasmid-mediated quinolone

resistance in isolates obtained in German intensive care units. *Antimicrob Agents Chemother* **49:**773–775.

247. **Jeong JY, Yoon HJ, Kim ES, Lee Y, Choi SH, Kim NJ, Woo JH, Kim YS.** 2005. Detection of *qnr* in clinical isolates of *Escherichia coli* from Korea. *Antimicrob Agents Chemother* **49:**2522–2524.
248. **Honoré S, Lascols C, Malin D, Targaouchi R, Cattoir V, Legrand P, Soussy CJ, Cambau E.** 2006. Émergence et diffusion chez les entérobactéries du nouveau mécanisme de résistance plasmidique aux quinolones Qnr (résultats hôpital Henri-Mondor 2002–2005). *Pathol Biol (Paris)* **54:**270–279.
249. **Saito R, Kumita W, Sato K, Chida T, Okamura N, Moriya K, Koike K.** 2007. Detection of plasmid-mediated quinolone resistance associated with *qnrA* in an *Escherichia coli* clinical isolate producing CTX-M-9 β-lactamase in Japan. *Int J Antimicrob Agents* **29:** 600–602.
250. **Poirel L, Nordmann P, De Champs C, Eloy C.** 2007. Nosocomial spread of QnrA-mediated quinolone resistance in *Enterobacter sakazakii. Int J Antimicrob Agents* **29:**223–224.
251. **Park YJ, Yu JK, Lee S, Oh EJ, Woo GJ.** 2007. Prevalence and diversity of *qnr* alleles in AmpC-producing *Enterobacter cloacae, Enterobacter aerogenes, Citrobacter freundii* and *Serratia marcescens*: a multicentre study from Korea. *J Antimicrob Chemother* **60:** 868–871.
252. **Paauw A, Verhoef J, Fluit AC, Blok HE, Hopmans TE, Troelstra A, Leverstein-van Hall MA.** 2007. Failure to control an outbreak of *qnrA1*-positive multidrug-resistant *Enterobacter cloacae* infection despite adequate implementation of recommended infection control measures. *J Clin Microbiol* **45:**1420–1425.
253. **Ellington MJ, Hope R, Turton JF, Warner M, Woodford N, Livermore DM.** 2007. Detection of *qnrA* among *Enterobacteriaceae* from South-East England with extended-spectrum and high-level AmpC β-lactamases. *J Antimicrob Chemother* **60:**1176–1178.
254. **Castanheira M, Pereira AS, Nicoletti AG, Pignatari AC, Barth AL, Gales AC.** 2007. First report of plasmid-mediated *qnrA1* in a ciprofloxacin-resistant *Escherichia coli* strain in Latin America. *Antimicrob Agents Chemother* **51:**1527–1529.
255. **Cavaco LM, Hansen DS, Friis-Møller A, Aarestrup FM, Hasman H, Frimodt-Møller N.** 2007. First detection of plasmid-mediated quinolone resistance (*qnrA* and *qnrS*) in *Escherichia coli* strains isolated from humans in Scandinavia. *J Antimicrob Chemother* **59:**804–805.
256. **Xiong Z, Wang P, Wei Y, Wang H, Cao H, Huang H, Li J.** 2008. Investigation of *qnr* and *aac(6′)-Ib-cr* in *Enterobacter cloacae* isolates from Anhui Province, China. *Diagn Microbiol Infect Dis* **62:**457–459.
257. **Tamang MD, Seol SY, Oh JY, Kang HY, Lee JC, Lee YC, Cho DT, Kim J.** 2008. Plasmid-mediated quinolone resistance determinants *qnrA, qnrB*, and *qnrS* among clinical isolates of *Enterobacteriaceae* in a Korean hospital. *Antimicrob Agents Chemother* **52:**4159–4162.
258. **Öktem IM, Gülay Z, Biçmen M, Gür D.** 2008. *qnrA* prevalence in extended-spectrum β-lactamase-positive *Enterobacteriaceae* isolates from Turkey. *Jpn J Infect Dis* **61:**13–17.
259. **Murray A, Mather H, Coia JE, Brown DJ.** 2008. Plasmid-mediated quinolone resistance in nalidixic-acid-susceptible strains of *Salmonella enterica* isolated in Scotland. *J Antimicrob Chemother* **62:**1153–1155.
260. **Cavaco LM, Frimodt-Møller N, Hasman H, Guardabassi L, Nielsen L, Aarestrup FM.** 2008. Prevalence of quinolone resistance mechanisms and associations to minimum inhibitory concentrations in quinolone-resistant *Escherichia coli* isolated from humans and swine in Denmark. *Microb Drug Resist* **14:** 163–169.
261. **Amin AK, Wareham DW.** 2009. Plasmid-mediated quinolone resistance genes in *Enterobacteriaceae* isolates associated with community and nosocomial urinary tract infection in East London, UK. *Int J Antimicrob Agents* **34:**490–491.
262. **Usein CR, Palade AM, Tatu-Chitoiu D, Ciontea S, Ceciu S, Nica M, Damian M.** 2009. Identification of plasmid-mediated quinolone resistance *qnr*-like genes in Romanian clinical isolates of *Escherichia coli* and *Klebsiella pneumoniae. Roum Arch Microbiol Immunol* **68:** 55–57.
263. **Kim SY, Park YJ, Yu JK, Kim YS, Han K.** 2009. Prevalence and characteristics of *aac(6′)-Ib-cr* in AmpC-producing *Enterobacter cloacae, Citrobacter freundii,* and *Serratia marcescens*: a multicenter study from Korea. *Diagn Microbiol Infect Dis* **63:**314–318.
264. **Gunell M, Webber MA, Kotilainen P, Lilly AJ, Caddick JM, Jalava J, Huovinen P, Siitonen A, Hakanen AJ, Piddock LJ.** 2009. Mechanisms of resistance in non-typhoidal *Salmonella enterica* strains exhibiting a non-classical quinolone resistance phenotype. *Antimicrob Agents Chemother* **53:**3832–3836.
265. **Sjölund-Karlsson M, Folster JP, Pecic G, Joyce K, Medalla F, Rickert R, Whichard JM.** 2009. Emergence of plasmid-mediated quinolone resistance among non-Typhi *Salmonella enterica* isolates from humans in the United States. *Antimicrob Agents Chemother* **53:** 2142–2144.
266. **Cui S, Li J, Sun Z, Hu C, Jin S, Li F, Guo Y, Ran L, Ma Y.** 2009. Characterization of *Salmonella enterica* isolates from infants and toddlers in Wuhan, China. *J Antimicrob Chemother* **63:**87–94.
267. **Ode T, Saito R, Kumita W, Sato K, Okugawa S, Moriya K, Koike K, Okamura N.** 2009. Analysis of plasmid-mediated multidrug resistance in *Escherichia coli* and *Klebsiella oxytoca* isolates from clinical specimens in Japan. *Int J Antimicrob Agents* **34:** 347–350.
268. **Fang H, Huang H, Shi Y, Hedin G, Nord CE, Ullberg M.** 2009. Prevalence of *qnr* determinants among extended- spectrum β-lactamase-positive *Enterobacteriaceae* clinical isolates in southern Stockholm, Sweden. *Int J Antimicrob Agents* **34:**268–270.
269. **Naqvi SM, Jenkins C, McHugh TD, Balakrishnan I.** 2009. Identification of the *qnr* family in *Enterobacteriaceae* in clinical practice. *J Antimicrob Chemother* **63:**830–832.

270. **Le TM, Baker S, Le TP, Cao TT, Tran TT, Nguyen VM, Campbell JI, Lam MY, Nguyen TH, Nguyen VV, Farrar J, Schultsz C.** 2009. High prevalence of plasmid-mediated quinolone resistance determinants in commensal members of the *Enterobacteriaceae* in Ho Chi Minh City, Vietnam. *J Med Microbiol* **58:** 1585–1592.

271. **Bouchakour M, Zerouali K, Gros Claude JD, Amarouch H, El Mdaghri N, Courvalin P, Timinouni M.** 2010. Plasmid-mediated quinolone resistance in expanded spectrum beta lactamase producing *enterobacteriaceae* in Morocco. *J Infect Dev Ctries* **4:** 779–803.

272. **Xiong Z, Li J, Li T, Shen J, Hu F, Wang M.** 2010. Prevalence of plasmid-mediated quinolone-resistance determinants in *Shigella flexneri* isolates from Anhui Province, China. *J Antibiot (Tokyo)* **63:**187–189.

273. **Kim YS, Kim ES, Jeong JY.** 2010. Genetic organization of plasmid-mediated Qnr determinants in cefotaxime-resistant *Enterobacter cloacae* isolates in Korea. *Diagn Microbiol Infect Dis* **68:**318–321.

274. **Ferreira S, Paradela A, Velez J, Ramalheira E, Walsh TR, Mendo S.** 2010. Carriage of *qnrA1* and *qnrB2*, $bla_{CTX-M15}$, and complex class 1 integron in a clinical multiresistant *Citrobacter freundii* isolate. *Diagn Microbiol Infect Dis* **67:**188–190.

275. **Kiiru J, Kariuki S, Goddeeris BM, Revathi G, Maina TW, Ndegwa DW, Muyodi J, Butaye P.** 2011. *Escherichia coli* strains from Kenyan patients carrying conjugatively transferable broad-spectrum beta-lactamase, *qnr, aac(6′)-Ib-cr* and 16S rRNA methyltransferase genes. *J Antimicrob Chemother* **66:**1639–1642.

276. **Herrera-Leon S, Gonzalez-Sanz R, Herrera-Leon L, Echeita MA.** 2011. Characterization of multidrug-resistant *Enterobacteriaceae* carrying plasmid-mediated quinolone resistance mechanisms in Spain. *J Antimicrob Chemother* **66:**287–290.

277. **Garcia-Fulgueiras V, Bado I, Mota MI, Robino L, Cordeiro NF, Varela A, Algorta G, Gutkind G, Ayala JA, Vignoli R.** 2011. Extended-spectrum β-lactamases and plasmid-mediated quinolone resistance in enterobacterial clinical isolates in the paediatric hospital of Uruguay. *J Antimicrob Chemother* **66:**1725–1729.

278. **Chowdhury G, Pazhani GP, Nair GB, Ghosh A, Ramamurthy T.** 2011. Transferable plasmid-mediated quinolone resistance in association with extended-spectrum β-lactamases and fluoroquinolone-acetylating aminoglycoside-6′-N-acetyltransferase in clinical isolates of *Vibrio fluvialis*. *Int J Antimicrob Agents* **38:** 169–173.

279. **Silva-Sanchez J, Barrios H, Reyna-Flores F, Bello-Diaz M, Sanchez-Perez A, Rojas T, Consortium BR, Garza-Ramos U.** 2011. Prevalence and characterization of plasmid-mediated quinolone resistance genes in extended-spectrum β-lactamase-producing *Enterobacteriaceae* isolates in Mexico. *Microb Drug Resist* **17:** 497–505.

280. **Muller S, Oesterlein A, Frosch M, Abele-Horn M, Valenza G.** 2011. Characterization of extended-spectrum beta-lactamases and *qnr* plasmid-mediated quinolone resistance in German isolates of *Enterobacter* species. *Microb Drug Resist* **17:**99–103.

281. **Jeong HS, Bae IK, Shin JH, Kim SH, Chang CL, Jeong J, Kim S, Lee CH, Ryoo NH, Lee JN.** 2011. Fecal colonization of *Enterobacteriaceae* carrying plasmid-mediated quinolone resistance determinants in Korea. *Microb Drug Resist* **17:**507–512.

282. **Frank T, Mbecko JR, Misatou P, Monchy D.** 2011. Emergence of quinolone resistance among extended-spectrum β-lactamase-producing *Enterobacteriaceae* in the Central African Republic: genetic characterization. *BMC Res Notes* **4:**309.

283. **Tran QT, Nawaz MS, Deck J, Nguyen KT, Cerniglia CE.** 2011. Plasmid-mediated quinolone resistance in *Pseudomonas putida* isolates from imported shrimp. *Appl Environ Microbiol* **77:**1885–1887.

284. **Hassan WM, Hashim A, Domany R.** 2012. Plasmid mediated quinolone resistance determinants *qnr, aac (6′)*-Ib-cr, and *qep* in ESBL-producing *Escherichia coli* clinical isolates from Egypt. *Indian J Med Microbiol* **30:** 442–447.

285. **Paltansing S, Kraakman ME, Ras JM, Wessels E, Bernards AT.** 2013. Characterization of fluoroquinolone and cephalosporin resistance mechanisms in *Enterobacteriaceae* isolated in a Dutch teaching hospital reveals the presence of an *Escherichia coli* ST131 clone with a specific mutation in *parE*. *J Antimicrob Chemother* **68:**40–45.

286. **Wang Y, Ying C, Wu W, Ye Y, Zhang H, Yu J.** 2009. Investigation of quinolone resistance in *Shigella sonnei*. *Chin J Infect Chemother* **9:**27–31.

287. **Pasom W, Chanawong A, Lulitanond A, Wilailuckana C, Kenprom S, Puang-Ngern P.** 2013. Plasmid-mediated quinolone resistance genes, *aac(6′)-Ib-cr, qnrS, qnrB*, and *qnrA*, in urinary isolates of *Escherichia coli* and *Klebsiella pneumoniae* at a teaching hospital, Thailand. *Jpn J Infect Dis* **66:**428–432.

288. **Silva-Sánchez J, Cruz-Trujillo E, Barrios H, Reyna-Flores F, Sánchez-Pérez A, Consortium BR, Garza-Ramos U.** 2013. Characterization of plasmid-mediated quinolone resistance (PMQR) genes in extended-spectrum β-lactamase-producing *Enterobacteriaceae* pediatric clinical isolates in Mexico. *PLoS One* **8:**e77968. doi: 10.1371/journal.pone.0077968.

289. **Ferrari R, Galiana A, Cremades R, Rodriguez JC, Magnani M, Tognim MC, Oliveira TC, Royo G.** 2013. Plasmid-mediated quinolone resistance (PMQR) and mutations in the topoisomerase genes of *Salmonella enterica* strains from Brazil. *Braz J Microbiol* **44:** 651–656.

290. **Xia R, Ren Y, Xu H.** 2013. Identification of plasmid-mediated quinolone resistance *qnr* genes in multidrug-resistant Gram-negative bacteria from hospital wastewaters and receiving waters in the Jinan area, China. *Microb Drug Resist* **19:**446–456.

291. **Chu YW, Cheung TK, Ng TK, Tsang D, To WK, Kam KM, Lo JY.** 2006. Quinolone resistance determinant *qnrA3* in clinical isolates of *Salmonella* in 2000–2005 in Hong Kong. *J Antimicrob Chemother* **58:**904–905.

292. **Dahmen S, Poirel L, Mansour W, Bouallegue O, Nordmann P.** 2010. Prevalence of plasmid-mediated quinolone resistance determinants in *Enterobacteriaceae* from Tunisia. *Clin Microbiol Infect* **16:**1019–1023.

293. **Arpin C, Thabet L, Yassine H, Messadi AA, Boukadida J, Dubois V, Coulange-Mayonnove L, Andre C, Quentin C.** 2012. Evolution of an incompatibility group IncA/C plasmid harboring bla_{CMY-16} and *qnrA6* genes and its transfer through three clones of *Providencia stuartii* during a two-year outbreak in a Tunisian burn unit. *Antimicrob Agents Chemother* **56:**1342–1349.

294. **Nazik H, Ongen B, Kuvat N.** 2008. Investigation of plasmid-mediated quinolone resistance among isolates obtained in a Turkish intensive care unit. *Jpn J Infect Dis* **61:**310–312.

295. **Torpdahl M, Hammerum AM, Zachariasen C, Nielsen EM.** 2009. Detection of *qnr* genes in *Salmonella* isolated from humans in Denmark. *J Antimicrob Chemother* **63:**406–408.

296. **Karah N, Poirel L, Bengtsson S, Sundqvist M, Kahlmeter G, Nordmann P, Sundsfjord A, Samuelsen O.** 2010. Plasmid-mediated quinolone resistance determinants *qnr* and *aac(6′)-Ib-cr* in *Escherichia coli* and *Klebsiella* spp. from Norway and Sweden. *Diagn Microbiol Infect Dis* **66:**425–431.

297. **Meradi L, Djahoudi A, Abdi A, Bouchakour M, Perrier Gros Claude JD, Timinouni M.** 2011. Résistance aux quinolones de types qnr, aac (6′)-Ib-cr chez les entérobactéries isolées à Annaba en Algérie. *Pathol Biol (Paris)* **59:**e73–e78.

298. **Villa L, Poirel L, Nordmann P, Carta C, Carattoli A.** 2012. Complete sequencing of an IncH plasmid carrying the bla_{NDM-1}, $bla_{CTX-M-15}$ and *qnrB1* genes. *J Antimicrob Chemother* **67:**1645–1650.

299. **Guillard T, de Champs C, Moret H, Bertrand X, Scheftel JM, Cambau E.** 2012. High-resolution melting analysis for rapid characterization of *qnr* alleles in clinical isolates and detection of two novel alleles, *qnrB25* and *qnrB42*. *J Antimicrob Chemother* **67:**2635–2639.

300. **Dolejska M, Villa L, Dobiasova H, Fortini D, Feudi C, Carattoli A.** 2013. Plasmid content of a clinically relevant *Klebsiella pneumoniae* clone from the Czech Republic producing CTX-M-15 and QnrB1. *Antimicrob Agents Chemother* **57:**1073–1076.

301. **Viana AL, Cayo R, Avelino CC, Gales AC, Franco MC, Minarini LA.** 2013. Extended-spectrum β-lactamases in *Enterobacteriaceae* isolated in Brazil carry distinct types of plasmid-mediated quinolone resistance genes. *J Med Microbiol* **62:**1326–1331.

302. **Saiful Anuar AS, Mohd Yusof MY, Tay ST.** 2013. Prevalence of plasmid-mediated *qnr* determinants and gyrase alteration in *Klebsiella pneumoniae* isolated from a university teaching hospital in Malaysia. *Eur Rev Med Pharmacol Sci* **17:**1744–1747.

303. **Gay K, Robicsek A, Strahilevitz J, Park CH, Jacoby G, Barrett TJ, Medalla F, Chiller TM, Hooper DC.** 2006. Plasmid-mediated quinolone resistance in non-Typhi serotypes of *Salmonella enterica*. *Clin Infect Dis* **43:** 297–304.

304. **Whichard J, Gay K, Stevenson JE, Joyce K, Cooper K, Omondi M, Medalla F, Jacoby GA, Barrett TJ.** 2007. Human *Salmonella* and concurrent decreased susceptibility to quinolones and extended-spectrum cephalosporins. *Emerg Infect Dis* **13:**1681–1688.

305. **Hopkins KL, Day M, Threlfall EJ.** 2008. Plasmid-mediated quinolone resistance in *Salmonella enterica*, United Kingdom. *Emerg Infect Dis* **14:**340–342.

306. **Minarini LA, Poirel L, Cattoir V, Darini AL, Nordmann P.** 2008. Plasmid-mediated quinolone resistance determinants among enterobacterial isolates from outpatients in Brazil. *J Antimicrob Chemother* **62:** 474–478.

307. **Wu JJ, Ko WC, Chiou CS, Chen HM, Wang LR, Yan JJ.** 2008. Emergence of Qnr determinants in human *Salmonella* isolates in Taiwan. *J Antimicrob Chemother* **62:** 1269–1272.

308. **Veldman K, van Pelt W, Mevius D.** 2008. First report of *qnr* genes in *Salmonella* in The Netherlands. *J Antimicrob Chemother* **61:**452–453.

309. **Shin JH, Jung HJ, Lee JY, Kim HR, Lee JN, Chang CL.** 2008. High rates of plasmid-mediated quinolone resistance *QnrB* variants among ciprofloxacin-resistant *Escherichia coli* and *Klebsiella pneumoniae* from urinary tract infections in Korea. *Microb Drug Resist* **14:** 221–226.

310. **Liassine N, Zulueta-Rodriguez P, Corbel C, Lascols C, Soussy CJ, Cambau E.** 2008. First detection of plasmid-mediated quinolone resistance in the community setting and in hospitalized patients in Switzerland. *J Antimicrob Chemother* **62:**1151–1152.

311. **Gutierrez B, Herrera-Leon S, Escudero JA, Hidalgo L, Gonzalez-Sanz R, Arroyo M, San Millan A, Echeita MA, Gonzalez-Zorn B.** 2009. Novel genetic environment of *qnrB2* associated with TEM-1 and SHV-12 on pB1004, an IncHI2 plasmid, in *Salmonella* Bredeney BB1047 from Spain. *J Antimicrob Chemother* **64:** 1334–1336.

312. **Pfeifer Y, Matten J, Rabsch W.** 2009. *Salmonella enterica* serovar Typhi with CTX-M β-lactamase, Germany. *Emerg Infect Dis* **15:**1533–1535.

313. **Cattoir V, Poirel L, Rotimi V, Soussy CJ, Nordmann P.** 2007. Multiplex PCR for detection of plasmid-mediated quinolone resistance *qnr* genes in ESBL-producing enterobacterial isolates. *J Antimicrob Chemother* **60:** 394–397.

314. **Pallecchi L, Riccobono E, Mantella A, Bartalesi F, Sennati S, Gamboa H, Gotuzzo E, Bartoloni A, Rossolini GM.** 2009. High prevalence of *qnr* genes in commensal enterobacteria from healthy children in Peru and Bolivia. *Antimicrob Agents Chemother* **53:**2632–2635.

315. **Sjölund-Karlsson M, Howie R, Rickert R, Krueger A, Tran TT, Zhao S, Ball T, Haro J, Pecic G, Joyce K, Fedorka-Cray PJ, Whichard JM, McDermott PF.** 2010. Plasmid-mediated quinolone resistance among non-Typhi *Salmonella enterica* isolates, USA. *Emerg Infect Dis* **16:**1789–1791.

316. **Lin CJ, Siu LK, Ma L, Chang YT, Lu PL.** 2012. Molecular epidemiology of ciprofloxacin-resistant extended-

spectrum β-lactamase-producing *Klebsiella pneumoniae* in Taiwan. *Microb Drug Resist* **18:**52–58.

317. **Nordmann P, Poirel L, Mak JK, White PA, McIver CJ, Taylor P.** 2008. Multidrug-resistant *Salmonella* strains expressing emerging antibiotic resistance determinants. *Clin Infect Dis* **46:**324–325.

318. **Kanamori H, Yano H, Hirakata Y, Endo S, Arai K, Ogawa M, Shimojima M, Aoyagi T, Hatta M, Yamada M, Nishimaki K, Kitagawa M, Kunishima H, Kaku M.** 2011. High prevalence of extended-spectrum β-lactamases and *qnr* determinants in *Citrobacter* species from Japan: dissemination of CTX-M-2. *J Antimicrob Chemother* **66:**2255–2262.

319. **Zhu YL, Yang HF, Liu YY, Hu LF, Cheng J, Ye Y, Li JB.** 2013. Detection of plasmid-mediated quinolone resistance determinants and the emergence of multidrug resistance in clinical isolates of *Shigella* in SiXian area, China. *Diagn Microbiol Infect Dis* **75:**327–329.

320. **Cavaco LM, Korsgaard H, Sorensen G, Aarestrup FM.** 2008. Plasmid-mediated quinolone resistance due to *qnrB5* and *qnrS1* genes in *Salmonella enterica* serovars Newport, Hadar and Saintpaul isolated from turkey meat in Denmark. *J Antimicrob Chemother* **62:**632–634.

321. **Lunn AD, Fabrega A, Sanchez-Cespedes J, Vila J.** 2010. Prevalence of mechanisms decreasing quinolone-susceptibility among Salmonella spp. clinical isolates. *Int Microbiol* **13:**15–20.

322. **Folster JP, Pecic G, Bowen A, Rickert R, Carattoli A, Whichard JM.** 2011. Decreased susceptibility to ciprofloxacin among *Shigella* isolates in the United States, 2006 to 2009. *Antimicrob Agents Chemother* **55:**1758–1760.

323. **Zhao JY, Dang H.** 2012. Coastal seawater bacteria harbor a large reservoir of plasmid-mediated quinolone resistance determinants in Jiaozhou Bay, China. *Microb Ecol* **64:**187–199.

324. **Fortini D, Fashae K, Garcia-Fernandez A, Villa L, Carattoli A.** 2011. Plasmid-mediated quinolone resistance and β-lactamases in *Escherichia coli* from healthy animals from Nigeria. *J Antimicrob Chemother* **66:**1269–1272.

325. **Hammerl JA, Beutlich J, Hertwig S, Mevius D, Threlfall EJ, Helmuth R, Guerra B.** 2010. pSGI15, a small ColE-like *qnrB19* plasmid of a *Salmonella enterica* serovar Typhimurium strain carrying *Salmonella* genomic island 1 (SGI1). *J Antimicrob Chemother* **65:**173–175.

326. **Pallecchi L, Riccobono E, Mantella A, Fernandez C, Bartalesi F, Rodriguez H, Gotuzzo E, Bartoloni A, Rossolini GM.** 2011. Small *qnrB*-harbouring ColE-like plasmids widespread in commensal enterobacteria from a remote Amazonas population not exposed to antibiotics. *J Antimicrob Chemother* **66:**1176–1178.

327. **Jeong HS, Kim JA, Shin JH, Chang CL, Jeong J, Cho JH, Kim MN, Kim S, Kim YR, Lee CH, Lee K, Lee MA, Lee WG, Lee JN.** 2011. Prevalence of plasmid-mediated quinolone resistance and mutations in the gyrase and topoisomerase IV genes in Salmonella isolated from 12 tertiary-care hospitals in Korea. *Microb Drug Resist* **17:**551–557.

328. **Riveros M, Riccobono E, Durand D, Mosquito S, Ruiz J, Rossolini GM, Ochoa TJ, Pallecchi L.** 2012. Plasmid-mediated quinolone resistance genes in enteroaggregative *Escherichia coli* from infants in Lima, Peru. *Int J Antimicrob Agents* **39:**540–542.

329. **Gonzalez F, Pallecchi L, Rossolini GM, Araque M.** 2012. Plasmid-mediated quinolone resistance determinant *qnrB19* in non-typhoidal *Salmonella enterica* strains isolated in Venezuela. *J Infect Dev Ctries* **6:**462–464.

330. **Bae IK, Park I, Lee JJ, Sun HI, Park KS, Lee JE, Ahn JH, Lee SH, Woo GJ.** 2010. Novel variants of the *qnrB* gene, *qnrB22* and *qnrB23*, in *Citrobacter werkmanii* and *Citrobacter freundii. Antimicrob Agents Chemother* **54:**3068–3069.

331. **Wang D, Wang H, Qi Y, Liang Y, Zhang J, Yu L.** 2011. Novel variants of the *qnrB* gene, *qnrB31* and *qnrB32*, in *Klebsiella pneumoniae. J Med Microbiol* **60:**1849–1852.

332. **Wang D, Wang H, Qi Y, Liang Y, Zhang J, Yu L.** 2012. Characteristics of *Klebsiella pneumoniae* harboring QnrB32, Aac(6′)-Ib-cr, GyrA and CTX-M-22 genes. *Folia Histochem Cytobiol* **50:**68–74.

333. **Poirel L, Nguyen TV, Weintraub A, Leviandier C, Nordmann P.** 2006. Plasmid-mediated quinolone resistance determinant *qnrS* in *Enterobacter cloacae. Clin Microbiol Infect* **12:**1021–1023.

334. **Avsaroglu MD, Helmuth R, Junker E, Hertwig S, Schroeter A, Akcelik M, Bozoglu F, Guerra B.** 2007. Plasmid-mediated quinolone resistance conferred by *qnrS1* in *Salmonella enterica* serovar Virchow isolated from Turkish food of avian origin. *J Antimicrob Chemother* **60:**1146–1150.

335. **Vasilaki O, Ntokou E, Ikonomidis A, Sofianou D, Frantzidou F, Alexiou-Daniel S, Maniatis AN, Pournaras S.** 2008. Emergence of the plasmid-mediated quinolone resistance gene *qnrS1* in *Escherichia coli* isolates in Greece. *Antimicrob Agents Chemother* **52:**2996–2997.

336. **Pu XY, Pan JC, Wang HQ, Zhang W, Huang ZC, Gu YM.** 2009. Characterization of fluoroquinolone-resistant *Shigella flexneri* in Hangzhou area of China. *J Antimicrob Chemother* **63:**917–920.

337. **Keddy KH, Smith AM, Sooka A, Ismail H, Oliver S.** 2010. Fluoroquinolone-resistant typhoid, South Africa. *Emerg Infect Dis* **16:**879–880.

338. **Le TM, AbuOun M, Morrison V, Thomson N, Campbell JI, Woodward MJ, Van Vinh Chau N, Farrar J, Schultsz C, Baker S.** 2011. Differential phenotypic and genotypic characteristics of *qnrS1*-harboring plasmids carried by hospital and community commensal enterobacteria. *Antimicrob Agents Chemother* **55:**1798–1802.

339. **Sato T, Yokota S, Uchida I, Okubo T, Ishihara K, Fujii N, Tamura Y.** 2011. A fluoroquinolone-resistant *Escherichia coli* clinical isolate without quinolone resistance-determining region mutations found in Japan. *Antimicrob Agents Chemother* **55:**3964–3965.

340. **Mokracka J, Gruszczynska B, Kaznowski A.** 2012. Integrons, β-lactamase and *qnr* genes in multidrug resistant clinical isolates of *Proteus mirabilis* and *P. vulgaris*. *APMIS* **120:**950–958.

341. **Kanamori H, Yano H, Hirakata Y, Hirotani A, Arai K, Endo S, Ichimura S, Ogawa M, Shimojima M, Aoyagi T, Hatta M, Yamada M, Gu Y, Tokuda K, Kunishima H, Kitagawa M, Kaku M.** 2012. Molecular characteristics of extended-spectrum beta-lactamases and *qnr* determinants in *Enterobacter* species from Japan. *PLoS One* **7:**e37967. doi:10.1371/journal.pone.0037967.

342. **Sánchez-Céspedes J, Blasco MD, Marti S, Alba V, Alcalde E, Esteve C, Vila J.** 2008. Plasmid-mediated QnrS2 determinant from a clinical *Aeromonas veronii* isolate. *Antimicrob Agents Chemother* **52:**2990–2991.

343. **Arias A, Seral C, Navarro F, Miro E, Coll P, Castillo FJ.** 2010. Plasmid-mediated QnrS2 determinant in an *Aeromonas caviae* isolate recovered from a patient with diarrhoea. *Clin Microbiol Infect* **16:**1005–1007.

344. **Majumdar T, Das B, Bhadra RK, Dam B, Mazumder S.** 2011. Complete nucleotide sequence of a quinolone resistance gene (*qnrS2*) carrying plasmid of *Aeromonas hydrophila* isolated from fish. *Plasmid* **66:**79–84.

345. **Han JE, Kim JH, Cheresca CH Jr, Shin SP, Jun JW, Chai JY, Han SY, Park SC.** 2012. First description of the *qnrS*-like (*qnrS5*) gene and analysis of quinolone resistance-determining regions in motile *Aeromonas* spp. from diseased fish and water. *Res Microbiol* **163:** 73–79.

346. **Kristiansson E, Fick J, Janzon A, Grabic R, Rutgersson C, Weijdegard B, Soderstrom H, Larsson DG.** 2011. Pyrosequencing of antibiotic-contaminated river sediments reveals high levels of resistance and gene transfer elements. *PLoS One* **6:**e17038. doi:10.1371/ journal.pone.0017038.

347. **Li J, Wang T, Shao B, Shen J, Wang S, Wu Y.** 2012. Plasmid-mediated quinolone resistance genes and antibiotic residues in wastewater and soil adjacent to swine feedlots: potential transfer to agricultural lands. *Environ Health Perspect* **120:**1144–1149.

348. **Karisik E, Ellington MJ, Pike R, Warren RE, Livermore DM, Woodford N.** 2006. Molecular characterization of plasmids encoding CTX-M-15 β-lactamases from *Escherichia coli* strains in the United Kingdom. *J Antimicrob Chemother* **58:**665–668.

349. **Machado E, Coque TM, Cantón R, Baquero F, Sousa JC, Peixe L.** 2006. Dissemination in Portugal of CTX-M-15-, OXA-1-, and TEM-1-producing *Enterobacteriaceae* strains containing the *aac(6′)-Ib-cr* gene, which encodes an aminoglycoside- and fluoroquinolone-modifying enzyme. *Antimicrob Agents Chemother* **50:** 3220–3221.

350. **Pallecchi L, Bartoloni A, Fiorelli C, Mantella A, Di Maggio T, Gamboa H, Gotuzzo E, Kronvall G, Paradisi F, Rossolini GM.** 2007. Rapid dissemination and diversity of CTX-M extended-spectrum β-lactamase genes in commensal *Escherichia coli* isolates from healthy children from low-resource settings in Latin America. *Antimicrob Agents Chemother* **51:**2720–2725.

351. **Ambrožič Avguštin J, Keber R, Žerjavič K, Oražem T, Grabnar M.** 2007. Emergence of the quinolone resistance-mediating gene *aac(6′)-Ib-cr* in extended-spectrum-β-lactamase-producing *Klebsiella* isolates collected in Slovenia between 2000 and 2005. *Antimicrob Agents Chemother* **51:**4171–4173.

352. **Cordeiro NF, Robino L, Medina J, Seija V, Bado I, Garcia V, Berro M, Pontet J, Lopez L, Bazet C, Rieppi G, Gutkind G, Ayala JA, Vignoli R.** 2008. Ciprofloxacin-resistant enterobacteria harboring the *aac (6′)-Ib-cr* variant isolated from feces of inpatients in an intensive care unit in Uruguay. *Antimicrob Agents Chemother* **52:**806–807.

353. **Chmelnitsky I, Hermesh O, Navon-Venezia S, Strahilevitz J, Carmeli Y.** 2009. Detection of *aac (6′)-Ib-cr* in KPC-producing *Klebsiella pneumoniae* isolates from Tel Aviv, Israel. *J Antimicrob Chemother* **64:**718–722.

354. **Deepak RN, Koh TH, Chan KS.** 2009. Plasmid-mediated quinolone resistance determinants in urinary isolates of *Escherichia coli* and *Klebsiella pneumoniae* in a large Singapore hospital. *Ann Acad Med Singapore* **38:**1070–1073.

355. **Morgan-Linnell SK, Becnel Boyd L, Steffen D, Zechiedrich L.** 2009. Mechanisms accounting for fluoroquinolone resistance in *Escherichia coli* clinical isolates. *Antimicrob Agents Chemother* **53:**235–241.

356. **Bell JM, Turnidge JD, Andersson P.** 2010. *aac(6′)-Ib-cr* genotyping by simultaneous high-resolution melting analyses of an unlabeled probe and full-length amplicon. *Antimicrob Agents Chemother* **54:**1378–1380.

357. **Hidalgo-Grass C, Strahilevitz J.** 2010. High-resolution melt curve analysis for identification of single nucleotide mutations in the quinolone resistance gene *aac(6′)-Ib-cr*. *Antimicrob Agents Chemother* **54:**3509–3511.

358. **Christiansen N, Nielsen L, Jakobsen L, Stegger M, Hansen LH, Frimodt-Moller N.** 2011. Fluoroquinolone resistance mechanisms in urinary tract pathogenic *Escherichia coli* isolated during rapidly increasing fluoroquinolone consumption in a low-use country. *Microb Drug Resist* **17:**395–406.

359. **Frasson I, Cavallaro A, Bergo C, Richter SN, Palu G.** 2011. Prevalence of *aac(6′)-Ib-cr* plasmid-mediated and chromosome-encoded fluoroquinolone resistance in *Enterobacteriaceae* in Italy. *Gut Pathog* **3:**12.

360. **Park KS, Kim MH, Park TS, Nam YS, Lee HJ, Suh JT.** 2012. Prevalence of the plasmid-mediated quinolone resistance genes, *aac(6′)-Ib-cr*, *qepA*, and *oqxAB* in clinical isolates of extended-spectrum β-lactamase (ESBL)-producing *Escherichia coli* and *Klebsiella pneumoniae* in Korea. *Ann Clin Lab Sci* **42:**191–197.

361. **Bartoloni A, Pallecchi L, Riccobono E, Mantella A, Magnelli D, Di Maggio T, Villagran AL, Lara Y, Saavedra C, Strohmeyer M, Bartalesi F, Trigoso C, Rossolini GM.** 2013. Relentless increase of resistance to fluoroquinolones and expanded-spectrum cephalosporins in *Escherichia coli*: 20 years of surveillance in resource-limited settings from Latin America. *Clin Microbiol Infect* **19:**356–361.

362. **Chen DQ, Yang L, Luo YT, Mao MJ, Lin YP, Wu AW.** 2013. Prevalence and characterization of quinolone resistance in *Laribacter hongkongensis* from grass carp and Chinese tiger frog. *J Med Microbiol* **62:**1559–1564.

363. **Liu BT, Wang XM, Liao XP, Sun J, Zhu HQ, Chen XY, Liu YH.** 2011. Plasmid-mediated quinolone resistance determinants *oqxAB* and *aac(6′)-Ib-cr* and extended-spectrum β-lactamase gene $bla_{CTX-M-24}$ co-located on the same plasmid in one *Escherichia coli* strain from China. *J Antimicrob Chemother* **66:** 1638–1639.

364. **Yang H, Duan G, Zhu J, Zhang W, Xi Y, Fan Q.** 2013. Prevalence and characterisation of plasmid-mediated quinolone resistance and mutations in the gyrase and topoisomerase IV genes among *Shigella* isolates from Henan, China, between 2001 and 2008. *Int J Antimicrob Agents* **42:**173–177.

365. **Rocha-Gracia R, Ruiz E, Romero-Romero S, Lozano-Zarain P, Somalo S, Palacios-Hernández JM, Caballero-Torres P, Torres C.** 2010. Detection of the plasmid-borne quinolone resistance determinant *qepA1* in a CTX-M-15-producing *Escherichia coli* strain from Mexico. *J Antimicrob Chemother* **65:**169–171.

366. **Deng Y, He L, Chen S, Zheng H, Zeng Z, Liu Y, Sun Y, Ma J, Chen Z, Liu JH.** 2011. F33:A-:B- and F2:A-:B- plasmids mediate dissemination of *rmtB*-$bla_{CTX-M-9}$ group genes and *rmtB-qepA* in *Enterobacteriaceae* isolates from pets in China. *Antimicrob Agents Chemother* **55:**4926–4929.

367. **Tian GB, Rivera JI, Park YS, Johnson LE, Hingwe A, Adams-Haduch JM, Doi Y.** 2011. Sequence type ST405 *Escherichia coli* isolate producing QepA1, CTX-M-15, and RmtB from Detroit, Michigan. *Antimicrob Agents Chemother* **55:**3966–3967.

Specialized Functions Mediated by Plasmids

VI

Plasmids—Biology and Impact in Biotechnology and Discovery
Edited by Marcelo E. Tolmasky and Juan C. Alonso

doi:10.1128/microbiolspec.PLAS-0013-2013

Ana Segura[1]
Lázaro Molina[2]
Juan Luis Ramos[1]

26

Plasmid-Mediated Tolerance Toward Environmental Pollutants

INTRODUCTION

Since the Industrial Revolution, there has been an increasing pace in the production of environmentally hazardous compounds that deliberately or accidentally have reached waters and soils, polluting them. The survival capacity of microorganisms in a contaminated environment is limited by the concentration and/or toxicity of the pollutant. Some contaminants are able to disrupt the normal development of the cell, others induce mutations, and some of these can kill cells at very low concentrations. Through evolutionary processes, some bacteria have developed or acquired mechanisms to cope with the deleterious effects of toxic compounds, permitting normal cellular subsistence in polluted environments—a phenomenon known as tolerance. Common mechanisms of tolerance include the extrusion of contaminants to the outer media and, when concentrations of pollutants are low, the degradation of the toxic compound. For both of these approaches, plasmids play an important role in the evolution and dissemination of the catabolic pathways and efflux pumps.

In this article, we will briefly describe catabolic and tolerance plasmids and advances in the knowledge and biotechnological applications of tolerance plasmids.

ENVIRONMENTAL CATABOLIC PLASMIDS

The mineralization of pollutants decreases their concentration in the environment and therefore allows better survival of the organism; at the same time, pollutants are used to obtain energy for growth. This mechanism of resistance is only useful when the contaminant concentration is moderate or low enough to allow normal bacterial metabolism; for example, while *Pseudomonas putida* mt-2 is able to mineralize toluene through the TOL pathway, the strain fails to thrive in high concentrations of toluene (1).

Environmental plasmids are often conjugative and can transfer their genes between different strains; furthermore, many catabolic genes are often associated with transposable elements, making them one of the major players in bacterial evolution. Genes for the

[1]Estación Experimental del Zaidin (CSIC), Environmental Protection Department, Profesor Albareda, 1, 18008 Granada, Spain; [2]CIDERTA, Laboratorio de Investigación y Control Agroalimentario (LICAH), Parque Huelva Empresarial, 21007 Huelva, Spain.

degradation of contaminants such as toluene, xylenes, alkanes, naphthalene, phenol, nitrobenzene, triazine, and others have been reported to be encoded in plasmids (2, 3, 4). Catabolic plasmids are generally large in size (60 to 200 kb), making them difficult to isolate and complicating the ascription of genes to these extrachromosomal elements. Recent advances in sequencing technologies have resulted in a growing number of completely sequenced catabolic plasmids. Metagenomic techniques have significantly advanced our knowledge of genes that are present in noncultivatable bacteria; however, it has been suggested that the presence of plasmid-encoded genes in metagenomic libraries is often underrepresented (5), leaving a knowledge gap regarding the role and importance of plasmids harbored by noncultivatable bacteria.

It should be noted that the presence of catabolic genes has not only been identified in plasmids, but also in the chromosome. Here, we describe plasmid-based systems that have been studied in detail.

Plasmids That Encode Pathways for the Degradation of Monoaromatic Compounds

Toluene is a natural product and, although aerobic bacteria have evolved several different pathways to degrade it, they all share a common strategy. In many cases, an *upper* pathway is able to activate the aromatic ring throughout mono- or dioxygenases, or by oxidation of the alkylic substituent; subsequently, the *lower* pathway breaks down the activated aromatic ring to produce intermediates that can enter the tricarboxylic acid cycle (6, 7). The most studied plasmid for the degradation of monoaromatic compounds is the TOL plasmid (pWW0), which encodes the genes necessary for the degradation of toluene, *m*- and *p*-xylene. This plasmid was originally identified in *P. putida* mt-2 and has since been the subject of intensive research (8). pWW0, a 117-kbp self-transmissible plasmid belonging to the IncP-9 incompatibility group, was completely sequenced in 2002 (9). Together with toluene catabolic genes, the plasmid encodes all the functions necessary for its replication, stable inheritance, and conjugation. The catabolic genes within the pWW0 plasmid are organized in two operons, the "upper pathway" (*xylUWCMABN*), which converts toluene and xylenes into benzoate and methyl benzoates, and the "meta pathway" (*xylXYZLTEGFJQKIH*), which transforms these intermediates into Krebs cycle intermediates (Fig. 1). Two regulatory proteins, XylR and XylS, control the expression of these two operons through a fine-tuned regulatory circuit that has also been studied in great detail (10, 11). In fact, this regulatory system represents a paradigm for signal integration in gene regulatory networks (12). These two operons, together with the 14 open reading frames between them, are located between two identical repeats of 1275 bp. This complete region has been named IS*1246* because of its insertion sequence characteristics. Spontaneous deletion of a 39-kbp region comprising the catabolic genes has been observed, probably as a consequence of recombination between the two repeated sequences (13, 14). Two transposable elements, with characteristics of class II transposons, have been identified in the pWW0 plasmid (Tn*4651* and Tn*4653*) (15, 16); and another insertion sequence (IS*Ppu*12) is also present and active in this plasmid (17). The presence of multiple insertion sequences, transposases, and recombinases is a common element in most catabolic plasmids, and it is thought to be related to the acquisition of new genes and microbial evolution.

Plasmids that carry *xyl* genes homologous to those of the pWW0 plasmids are normally referred to as TOL plasmids and have been isolated from different locations (18, 19, 20, 21). They differ in the organization of the catabolic pathways, in size, and even in replicon type; e.g., plasmids pWW53 and pDK1 belong to the IncP-7 plasmid family (22, 23). Comparison between multiple TOL plasmids indicate that transposition-related genes and sites have contributed to the diversification of plasmid structures and to the dissemination of common gene clusters to various plasmids (22).

Plasmid pTOM (latter designated pBV1E04) of *Burkholderia vietnamensis* G4 (24; CP_000620) encodes the toluene 2-monooxygenase (T2MO) that catalyzes the two initial oxidation steps in toluene degradation, transforming toluene into 3-methylcatechol via *o*-cresol (25). Ring cleavage proceeds via a *meta* cleavage pathway that is also encoded in the pBV1E04 plasmid. Although it is known that T2MO activity is inducible, the regulation of this pathway has not been studied; however, a gene with homology to the XylR/NtrC family of transcriptional regulators that is located upstream of the T2MO gene cluster has been hypothesized to be the regulatory protein of the operon (6). *B. vietnamensis* G4 was initially isolated because it oxidized trichloroethylene (TCE), but it was later demonstrated that it can grow in the presence of toluene, phenol, *o*- , *m*-, and *p*-cresol, and benzoate (26). This strain and the pTOM plasmid have been used extensively for different biotechnological approaches, including TCE and toluene rhizoremediation (see below). Interestingly, the backbone of the pBV1E04 plasmid is quite similar to that of pGRT1 from the solvent tolerant *P. putida* DOT-T1E strain (27). Several transposases, insertion sequences, and integrases are encoded in the pBVIE04 plasmid, many of them surrounding the area where the catabolic genes are located, suggesting transfer of catabolic genes to/from other strains.

Plasmids That Encode Pathways for the Degradation of Polycyclic Aromatic Hydrocarbons and Heteroaromatic Compounds

Since the first report of a plasmid that encoded the catabolism of camphor, the CAM plasmid (28, 29), various plasmids that confer the ability to grow in polycyclic aromatic compounds have been isolated. Naphthalene has served as a model compound for the study of polycyclic aromatic hydrocarbons (PAH) degradation and the associated plasmids have been named as NAH plasmids. Naphthalene degradation enzymes are encoded within two operons; the *upper pathway* encodes functions for the conversion of naphthalene to salicylate (Fig. 1), and the *lower pathway* encodes genes for the *meta*-ring fission pathway and a predicted methyl-accepting chemotaxis protein, NahY (30, 31, 32, 33, 34, 35). The *nahR* gene encodes the regulatory protein of the system, which belongs to the LysR family (36) and induces the two operons in the presence of salicylate.

Plasmids pDTG1 from *P. putida* strain NCIB 9816-4, pND6-1 from *Pseudomonas* sp. strain ND6, and NAH7 from *P. putida* G7 have been completely sequenced (37, 38, 39). These three plasmids encode the *nahR* gene, as well as two catabolic operons: the *nah* operon, which encodes the *upper* pathway, and the *lower* operon for salicylate degradation (Fig. 1). Proteins for naphthalene degradation are almost identical in the three strains (99% to 100% identity in amino acid sequences), with the exception of two duplicated genes in pND6. Furthermore, nucleotide identity in the catabolic region of plasmids pND6 and pDTG1 is >99%, including a 15-kbp region between the two operons that contain a number of functionally unrelated genes (38). Despite this homology, *P. putida* strain NCIB 9816-4 degrades naphthalene through a chromosomally encoded *ortho*-pathway due to the presence of an IS*Pre1* between the *nahG* (encoding the salicylate hydroxylase) and *nahT* genes of the pDTG1 plasmid (37). In pND6 and pDTG1, the two operons are transcribed in opposite directions, while, in NAH7, both operons are transcribed in the same direction. *nahR* in NAH7 is not located upstream of the *lower* operon, but in between the two operons (38). In this plasmid, the naphthalene degradation genes are located within a class II transposon (Tn*4655*), which lacks a *tnpA* gene, but that can be mobilized by the action of the TnpA from pWW0 Tn*4653* supplied in *trans* (40). Within these three plasmids, numerous transposases, resolvases, and integrases are located in the vicinity of the catabolic genes, and, in many cases, they are similar to elements found in catabolic plasmids pWW0 and pCAR1. The "backbone" of the pDTG1 and NAH7 plasmids is homologous to that of the pWW0 plasmid, suggesting that the exchange of catabolic genes by means of horizontal gene transfer may have occurred between these plasmids.

Plasmids involved in the degradation of polycyclic aromatic hydrocarbons with three aromatic rings have also been described (e.g., the NAH-plasmid-encoded pathway is able to mineralize phenanthrene and anthracene [41]). Not only pseudomonads, but many other bacteria are able to degrade naphthalene and phenanthrene through similar biochemical steps. The enzymes of the naphthalene pathway are also able to degrade phenanthrene and anthracene via the 1-hydroxy-2-naphthoic acid intermediate, which is subsequently oxidized through salicylate and catechol (in other cases, this intermediate can be channeled to *o*-phthalate and protocatechuate [42]). The *Burkholderia* sp. strain RP007 contains a plasmid that encodes the upper pathway for the degradation of naphthalene and phenanthrene (43). The *Sphingomonas aromaticivorans* F199 plasmid known as pNL1 encodes genes associated with the degradation of biphenyl, naphthalene, *m*-xylene, and *p*-cresol. Interestingly, although the related genes are organized in operons under a commonly occurring pseudomonad regulatory scheme, in sphingomonads the genes are scattered across the plasmids (44, 45). In the pNL1 plasmid, genes associated with aromatic degradation are distributed among at least 11 transcriptional units (Fig. 1). The unusual coclustering of genes associated with different catabolic pathways (biphenyl, toluene, xylenes, and naphthalene) observed in this plasmid is likely due to evolutionary modifications applied to similar biochemical mechanisms for the degradation of intermediates in the different pathways (46). As in many other catabolic plasmids, the presence of a recombinase, an excisionase, a phage-type integrase-recombinase, and two transposons have been identified.

The genus *Sphingomonas* is able to degrade a wide variety of xenobiotics (biphenyl, PAHs, and substituted PAHs, carbazole, diphenyl ethers, furans, dibenzo-*p*-dioxins, and others) and many of the catabolic genes for these compounds are located on plasmids that can

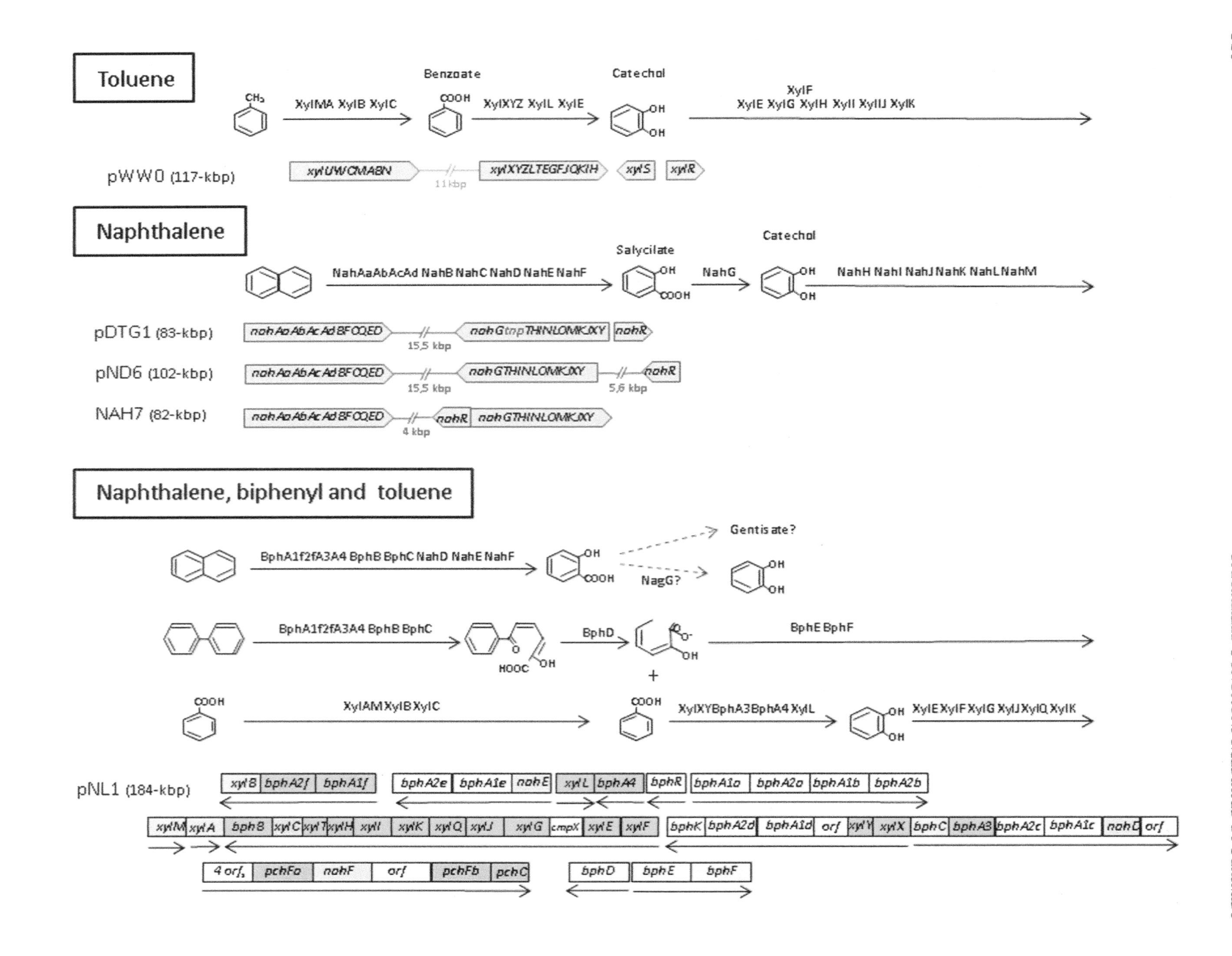

Toluene
Benzoate
Catechol
XylMA XylB XylC
XylXYZ XylL XylE
XylF
XylE XylG XylH XylI XylJ XylK
pWW0 (117-kbp)
11 kbp
Naphthalene
Salycilate
Catechol
NahAaAbAcAd NahB NahC NahD NahE NahF
NahG
NahH NahI NahJ NahK NahL NahM
pDTG1 (83-kbp)
15,5 kbp
pND6 (102-kbp)
15,5 kbp
5,6 kbp
NAH7 (82-kbp)
4 kbp
Naphthalene, biphenyl and toluene
BphA1f2fA3A4 BphB BphC NahD NahE NahF
Gentisate?
NagG?
BphA1f2fA3A4 BphB BphC
BphD
BphE BphF
XylAM XylB XylC
XylXYBphA3BphA4 XylL
XylE XylF XylG XylJ XylQ XylK
pNL1 (184-kbp)

be mobilized to other *Sphingomonas* strains (45, 47, 48). Plasmid pBN6 of *S. xenophaga* BN6 encodes the pathway involved the degradation of naphthalene sulfonate, and the related genes show high similarity to those encoded in the pNL1 plasmid of *S. aromaticivorans* F199 and also share the structural organization of the naphthalene, biphenyl, and phenanthrene genes within three transcriptional units. Not surprisingly, these genes are flanked with genes that encode putative mobile genetic elements (49).

Another group of plasmids that have been extensively studied are the pCAR plasmids that encode genes for the degradation of carbazole/dioxin. pCAR plasmids that have been fully sequenced include pCAR1 from *Pseudomonas resinovorans* CA10 (50), pCAR2 from *P. putida* HS01 (51), and pCAR3 from *Sphingomonas* sp. KA1 (52). Enzymes encoded by the *car* operon transform carbazole to anthranilate, and the proteins encoded by the *ant* operon are responsible for the conversion of anthranilate to catechol (Fig. 2). Carbazole degradation starts with the angular deoxygenation of the compound by the CARDO complex, a three-component dioxygenase that has been extensively characterized (53, 54, 55). In pCAR1, the *car* and *ant* operons are found within a 72.8-kbp transposon named Tn*4676* whose transposon-related genes show homology with Tn*4651* of the TOL plasmid pWW0. After the introduction of plasmid pCAR1 plasmid into *P. fluorescens* Pf0-1 many rearrangements were observed; one of these is the insertion of Tn*4676* into the chromosome (56). Numerous mobile genetic elements are found within the Tn*4676* transposon and its flanking region (50). Insertion sequences identical to IS*1162* (57), with homology to IS*Ec8* (58), IS*1491* (59), and IS*Pre* (60) are also encoded in pCAR1. The pCAR3 plasmid is bigger than pCAR1 (254,797 versus 199,035 bp); encoded in pCAR3 are the *car* and *and* operons, as well as a second cluster of *car* genes (*car*-II) and putative genes for the degradation of catechol, protocatechuate, and phthalate. Five different types of insertion sequences and transposons were identified in pCAR3. The backbone of this plasmid is similar to that of pNL1 of *Novosphingobium aromaticivorans* F1999 and, although it encoded all the functions necessary for conjugation, attempts to transfer it to a pCAR3-cured strain have been unsuccessful (52).

Plasmid pARUE113 (pAL1) from the *Actinobacteria Arthrobacter* sp. Rue61a encodes genes for the degradation of quinaldine (61). Quinaldine is transformed to anthranilate through the enzymes encoded by the *meqABC* (*qoxLMS*) operon that encodes quinaldine 4-oxidase and the divergently transcribed *meqDEF* (formerly named *moq*, *hod*, and *amp*) (62, 63) (Fig. 2). The plasmid also contains a *lower pathway* for the degradation of anthranilate through CoA intermediates to produce 2-amino-5-oxo-cyclohex-1-ene-carbonyl-CoA. This product is then thought to be degraded via a β-oxidation-like pathway encoded in the chromosome (63, 64). The enzymes of the pathway are induced in the presence of the substrate, and two putative transcriptional regulators, belonging to the GntR superfamily, are located near the catabolic cluster involved in quinaldine degradation. The function of MeqR1 is currently unknown, while MeqR2 is able to bind the promoter region of *mepC*, *meqD*, *orf1*, and to its own promoter and shows high binding specificity for anthraniloyl-CoA. These results suggest that MeqR2 may be involved in the regulation of the pathway, although other regulatory systems exist that are known to govern the expression of the *meq* genes (65). Although conserved gene clustering has been observed in pAL1, suggesting a modular structure, only one insertion sequence has been detected (63). Interestingly, pAL1 is a conjugative linear plasmid (61). It has been suggested that the replication of linear plasmids proceeds bidirectionally from an internal origin toward the telomeres generating replicative intermediates that contain 3′-strand overhangs (66). The left and right ends of pAL1 contain palindromic sequences that could be important for telomere patching; furthermore, pAL1 encodes putative proteins that could also be associated with the telomere patching.

Plasmids That Encode Pathways for the Degradation of Chlorinated Compounds

Catabolic pathways for the degradation of 2,4-dichlorophenoxyacetic acid (2,4-D), a xenobiotic herbi-

Figure 1 Degradation pathways of mono- and biaromatic compounds. Major intermediates of the pathways are depicted. Genes or operons in different plasmids are colored to indicate their role: blue for toluene degradation genes, pink for naphthalene degradation genes, and yellow for biphenyl degradation genes. In green are the genes that can function in different degradation pathways. Genes and operons are not drawn to scale. Operon organization in some cases has not been experimentally demonstrated.
doi:10.1128/microbiolspec.PLAS-0013-2013.f1

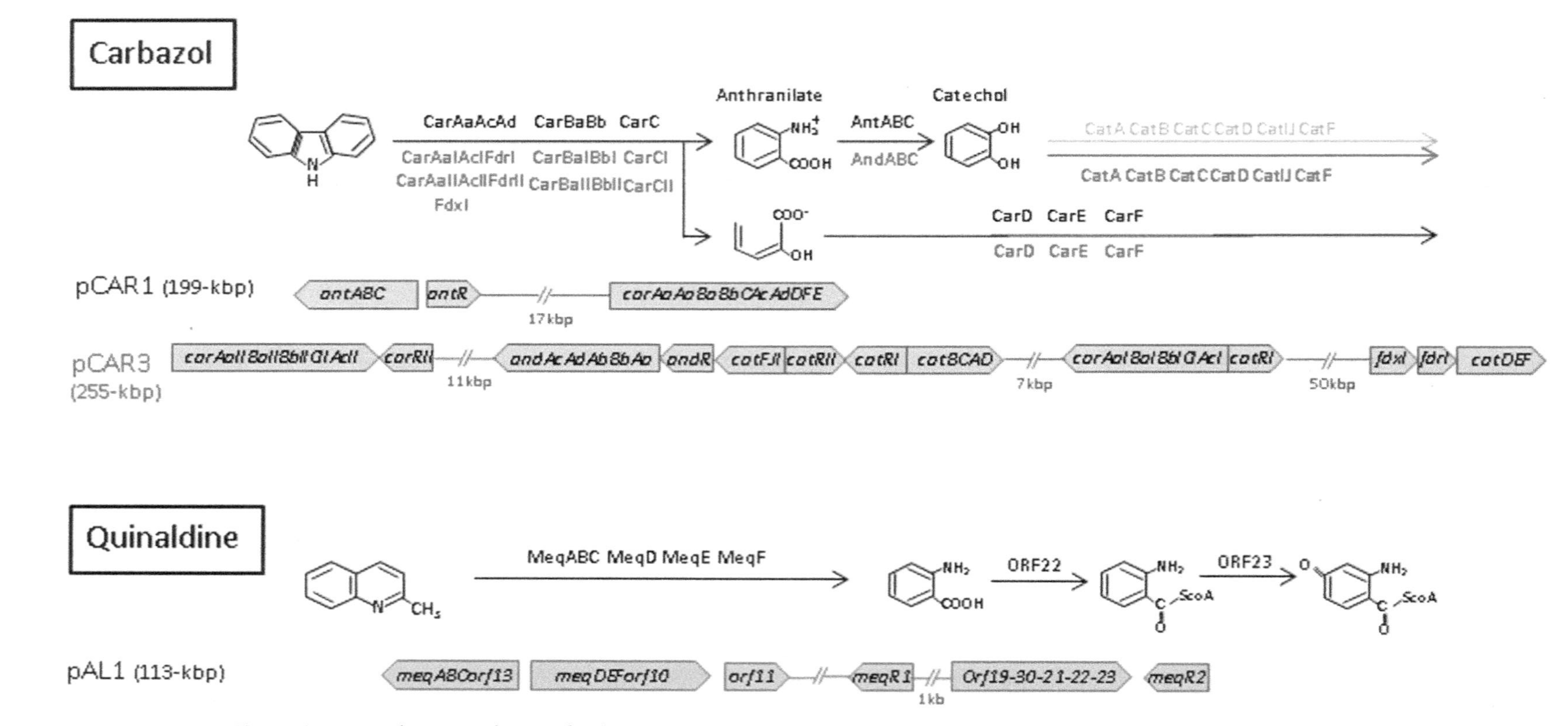

Figure 2 Degradation pathways for heteroaromatic compounds. Major intermediates of the pathways are depicted. Genes or operons in different plasmids are shown in different colors. Genes and operons are not drawn to scale. doi:10.1128/microbiolspec.PLAS-0013-2013.f2

cide, have been studied for almost 30 years. One of the best characterized pathways for the degradation of this herbicide is the pathway encoded by the pJP4 plasmid from *Wautersia eutropha* JMP134 (formerly *Ralstonia eutropha*), a strain that is able to mineralize 2,4-D and 3-chlorobenzoic acid (67) (Fig. 2). Two chlorocatechol-degrading gene clusters (*tfd*I and *tfd*II) are required for efficient degradation of the compounds (68, 69). The IncP1-β conjugative plasmid pJP4 was sequenced in 2004 (70) and was reported to contain inactive transposons and remnants of lateral gene transfer events. A putative IS*JP4*-based transposon encompassing the *tfd*II gene cluster and a putative IS*1071*-based transposon that flanks all the catabolic genes are found in the pJP4 sequence. However, it has been suggested that these putative transposons are not active (70). The pEST4011 plasmid from *Achromobacter xylosoxidans* subsp. *denitrificans* also carries genes for the degradation of 2,4-D (71). Based on differences among backbone proteins of the IncP1 group, it has been hypothesized that this plasmid belongs to a new category, known as the δ subgroup. Within this group, degradative genes are inserted into the backbone at a 48-kbp catabolic transposon that is very similar to Tn*5530*, which was identified in plasmid pIJB1 from *Burkholderia cepacia* 2a (72). The *tfd* operon is quite similar to *tfdII* of pJP4, suggesting a common origin, although *tfdD* and *tfdF* have been deleted in the pEST4011 operon (Fig. 2). These genes could have been recruited during evolution to replace the genes lost from the original operon.

Chlorobenzoates are intermediates within the bacterial degradation of polychlorinated biphenyls. Plasmid pA81, from *A. xylosoxidans* A8 is an IncP1-β plasmid that harbors genes for the degradation of *ortho*-substituted chlorobenzoates (*ohbRAB*, *mocpRABCD*) in a class I transposon named Tn*AxI* (73). This transposon also carries an operon predicted to function in salicylate degradation, known as *hybRABCD* (74) (Fig. 3). The pA81 plasmid carries another transposon, known as Tn*AxII*, that is involved in heavy metal resistance.

Atrazine is another widespread herbicide that many bacteria are able to degrade (75). Genes for the degradation of atrazine in *Pseudomonas* sp. ADP are located in the 109-kbp pADP-1 plasmid, which was sequenced in 2001 (76, 77). *atzA*, *atzB*, and *atzC* encode the enzymes for the transformation of atrazine to cyanuric acid. Each of these genes is flanked by elements with homology to IS*1071*, suggesting that they may have been acquired at different times by independent transposition events. The *atzDEF* operon encodes enzymes that convert cyanuric acid to ammonia and carbon dioxide (Fig. 3). The plasmid belongs to the IncPβ plasmid and has full capabilities for replication, stable maintenance, and conjugation (76, 77).

Chloroaniline degradation is linked to the presence of plasmids pWDL7::*rfp* (a derivative of plasmid pWDL7 of *Comamonas testosteroni* strain WDL7) and pNB8c (from *Delftia acidovorans* strain B8c). Contained within these plasmids are genes for the *upper pathway* (*dcaQTA1A2BR*) of chloroaniline degradation, which mediates its transformation to chlorocatechol. This product is then converted to tricarboxylic acid (TCA) intermediates by modified *ortho*- or *meta*-pathways encoded by the chromosomes. Both plasmids belong to the IncP-1β incompatibility group and their sequences are very similar, although pWDL7::*rfp* contains two Tn*6063* transposons that carry *dcaQTA1A2BR* genes (78). These plasmids can be transferred between *Betaproteobacteria* and *Gammaproteobacteria* indicating a broad host range. Transfer of plasmid pNB8c into *Cupriavidus pinatubonensis* JMP228 confers upon the strain the ability to degrade anilines but not chloroanilines. This deficiency is due to the lack of induction of the 3-CA pathway in this strain. Sequencing of the two plasmids allowed the identification of a difference in the *dca* promoter region responsible for this lack of induction (78).

Chlorinated nitroaromatic compounds are among the most difficult compounds to degrade because of the electron-withdrawing properties of the nitro and chloro groups. However, some genes associated with chloronitrobenzene degradation are encoded in the pCNB-1 plasmid of *Comamonas* sp. CNB-1 (79, 80, 81). The degradation pathway converts chloronitrobenzene into 2-amino-5-chloromuconate, which is then degraded to TCA intermediates (Fig. 3). The *cnb* genes, together with an operon that encodes for resistance to chromate and arsenate are contained within a large transposon belonging to the class I transposon named Tn*CNB1* (82). pCNB-1 contains 45 open reading frames (ORFs) that provide capabilities for replication, transfer and stability, and similar to other catabolic plasmids, belongs to the IncP1-β incompatibility group.

Other Catabolic Genes Located in Plasmids

A number of genes associated with degradation of lindane (γ-hexachlorocyclohexane) have been found on plasmids. The *lin* genes of *Sphingobium japonicum* UT26, which transform lindane to β-ketoadipate, are located in three replicons, with *linDE* (for the conversion of 2,5-dichlorohydroquinone to maleylacetate) and the regulatory gene *linR* found within the 185-kbp

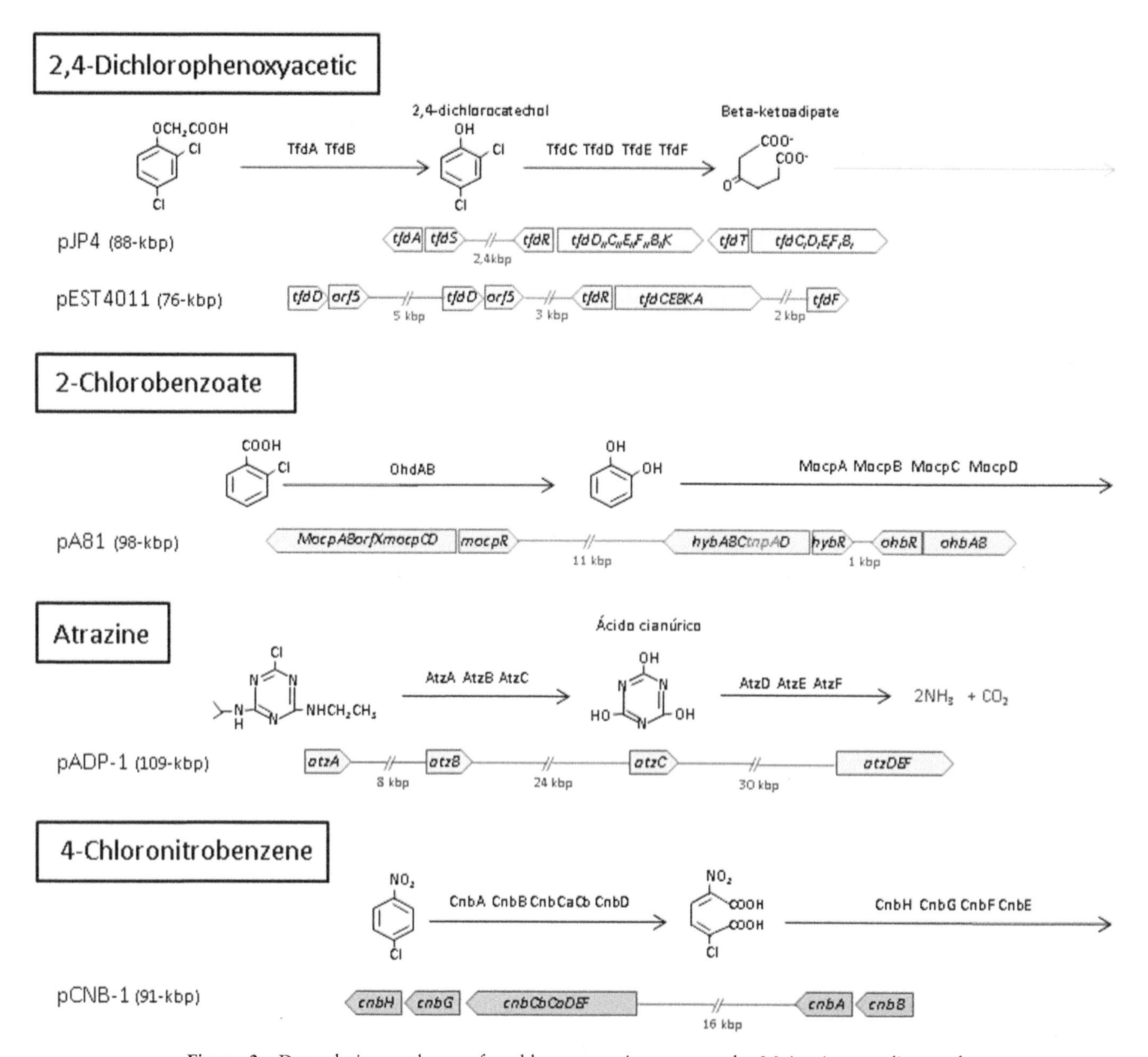

Figure 3 Degradation pathways for chloroaromatic compounds. Major intermediates of the pathways are depicted. Genes or operons in different plasmids are shown in different colors. Genes and operons are not drawn to scale.
doi:10.1128/microbiolspec.PLAS-0013-2013.f3

conjugative pCHQ1 plasmid (83). *linB* (which encodes a halidohydrolase) has been found within the 66-kbp pLB1 plasmid (84). Interestingly, two copies of the gene are located within an IS*6100* composite transposon (84). Based on its homology with the RepA proteins of other plasmids, pLB1 may represent a new plasmid incompatibility group. It is noteworthy that pLB1 was isolated from an unknown bacteria via an "exogenous plasmid isolation" technique, which is used to capture plasmids directly from the environmental microbial community (85).

The *tsaMVCD*$_1$, *tsaQ*$_1$, and *tsaT* operons, together with the *tsaMBCD* operon encoding proteins involved in the degradation of *p*-toluenesulfonate in *C. testosteroni* T-2, are linked to the IncP1-β conjugative plasmid known as pTSA (86, 87). The *tsaMBCD* operon contains no promoter-like sequence and is not expressed. The *tsa* region forms part of a composite transposon that is flanked by two IS*1071* elements. This transposon has been found in several bacterial strains that were isolated from samples enriched in toluenesulfonate on three different continents and was found to be inserted within plasmids or chromosomes (87).

Degradation of phenol has been linked with the presence of plasmids in *Pseudomonas* sp. EST1001 and *Pseudomonas* sp. CF600. The Phe^{+} plasmid of strain EST1001 allows it to grow on phenol (88, 89) and, although it has not been completely sequenced, it is known that the *pheAB* operon is flanked by two IS elements that are involved in the activation of phenol genes (90) and in horizontal gene transfer (91). *Pseudomonas* sp. CF600 harbors the pVI150 megaplasmid that belongs to the IncP-2 incompatibility group and encodes all the genetic information for the degradation of phenol, cresols, and 3,4-dimethylphenol (92, 93).

The alkane degradation pathway encoded in the OCT plasmid of *P. putida* GPo1 is organized into two operons, *alkBFGHJKL* and *alkST*. The first operon encodes all, except for one, of the enzymes required for the terminal oxidation of alkanes to their corresponding fatty acids. The *alkST* cluster codes for the transcriptional regulator AlkS and for a rubredoxin reductase (AlkT) that is part of the alkane hydroxylase complex (94, 95). The OCT plasmid was first described in the 1970s, and its operon expression has served as a model for the study of global regulation responses (96). The two *alk* operons are flanked by IS*Ppu4*, forming a class I transposon.

TOLERANCE PLASMIDS IN BIODEGRADATION

Although catabolic genes can contribute to the survival of the bacteria in the presence of contaminants, when these contaminants are present at high concentrations, additional mechanisms—in addition to the degradation of the compounds—are required to avoid toxicity. This is the case with organic solvents, which are highly hydrophobic and accumulate in and disrupt cell membranes causing cellular death (97). If the toxic compounds are not degradable (i.e., heavy metals), survival is provided by the action of efflux pumps that eliminate the compounds from the cytoplasm or the membranes of the microorganisms. These mechanisms are normally known as tolerance or resistance mechanisms and many of them are also encoded in plasmids, although, as is sometimes the case with catabolic genes, they can also be located on the chromosome.

Plasmids Involved in Solvent/Aromatic Compound Tolerance

P. putida DOT-T1E, a strain isolated from a seawater treatment plant in Granada, provides a well-studied example of survival in extreme conditions. This strain is able to degrade different aromatic compounds including toluene (98), and to survive at solvent concentrations up to 90% (vol/vol) (99). The genes that encode the toluene degradation pathway are located on the chromosome (100, 101), while its ability to survive in the presence of high concentrations of toluene is enabled by a 133-kbp plasmid named pGRT1 (102). Solvent tolerance is a multifactorial process that involves a wide range of physiological changes to overcome solvent damage (103); however, mutational experiments have demonstrated that the TtgGHI efflux pump is the most important determinant of solvent tolerance (102, 104). The *ttgGHI* operon of the pGRT1 plasmid encodes an efflux pump of the RND family (resistance, nodulation, cell division) that extrudes a wide variety of compounds, including antibiotics, although not all of them with the same efficiency (104). This family of efflux pumps has been extensively studied because they confer multidrug resistance to some relevant clinical strains (105, 106). The efflux system is formed by three components: an efflux pump transporter, located in the cytoplasmic membrane that recognizes substrates in the periplasm or in the cytoplasmic membrane (107, 108); an outer membrane protein that forms a trimeric channel capable of penetrating into the periplasm and contacting directly with the efflux pump transporter (109); and a lipoprotein anchored to the inner membrane that expands into the periplasmic space and may serve as a bracket for the other two components (110, 111). In the pGRT1 system, the *ttgH* gene encodes for the efflux pump, *ttgI* for the outer membrane protein, and *ttgG* for the periplasmic adaptor protein. Two other RND efflux pumps, chromosomally encoded, have been shown to participate in solvent tolerance in *P. putida* strain DOT-T1E, TtgABC (112), and TtgDEF (113). Despite the contribution of TtgABC and TtgDEF to the solvent-tolerant phenotype, the loss of the TtgGHI efflux pump renders the cells unable to survive in the presence of high concentrations of toluene (102, 104).

The pGRT1 plasmid sequence confirms that it is not a catabolic plasmid (27); however, it shares many features typical of catabolic plasmids. First, the toluene tolerance genes are located within a Tn*4653*-like transposon (Fig. 4A), which is similar to those found in TOL and pCAR plasmids. This transposon also contains *ttgV*, which encodes the regulatory protein that controls the expression of the efflux pumps (114), two genes that encode methyl-accepting chemotaxis proteins (*mcpT1* and *mcpT2*), and a gene that is homologous to *uvrD*. The two MCP proteins are almost identical and allow this bacterium to move toward a broad variety of aromatic compounds and crude oil (115). Monocopies of McpT homologues (99% sequence identity) have been found on other catabolic plasmids of hydrocarbon-degrading strains like pCAR1 of *P. resinovorans* (116), the TOL plasmid pWW53 of *P. putida* (22), and the pMAQU02 plasmid of *Marinobacter aquaeolei* VT8 (NC_008739). All of these strains showed a chemo-attractive response to toluene. Other examples of MCPs located on plasmids are the NahY protein located on the NAH7 catabolic plasmid of *P. putida* that responds to naphthalene (35) and NbaY from *P. fluorescens* KU-7 that responds to 2-nitrobenzoate (117). UvrD is a DNA helicase that regulates the activity of the *ruvA* and *ruvB* genes, which are present on the pGRT1 plasmid (Fig. 4), and which is involved in the repair of DNA cross-links produced by exposure to UV light and by exposure to toluene (118). UV-resistance genes are located within a broad set of catabolic plasmids. Examples of catabolic plasmids bearing close homologues to the *uvrD*, *ruvA*, and *ruvB* genes from pGRT1 are the naphthalene-degradative pND6 plasmid and the toluene-degradative plasmids pWW0 and pWW53 from *P. putida* (9, 22, 38).

As well as these traits, other characteristics of pGRT1 that are shared with catabolic plasmids include the fact that it encodes proteins related to DNA replication and modification, plasmid maintenance, mobilization, and transfer (27) with high similarity to that of the pBVIE04 plasmid *of B. vietnamensis* G4. Inserted within the pGRT1 backbone, in addition to the Tn*4653*-like transposon, is a second DNA segment that contains invertases, transposases, and recombinase (Fig. 4B). This segment contains *ruvA*, *ruvB*, and other genes related to stress responses. As well, the protein encoded by *orf32* is a new toluene tolerance modulator in DOT-T1E. This protein contains a domain identified in several SdiA-regulated proteins, and it is known that chromosomally encoded SdiA proteins in *Escherichia coli* and *Salmonella enterica* were able to modulate AcrAB activity (119, 120). Knockout mutants of ORF32 showed a reduction of the expression of the *ttgV* repressor gene that in turn enhances the expression of the *ttgGHI* efflux pump and concomitantly increases solvent tolerance (27). ORF32 is highly homologous to a protein of unknown function (p032) from the pBS228 antibiotic resistance plasmid of *Pseudomonas aeruginosa* (121), and proteins encoded by naphthalene-degradative plasmids pND071, pDTG1, and pND6-1 (37, 38). One homologue to the universal stress protein (UspA) of *E. coli* (122, 123) is also encoded in this pGRT1 fragment. It was demonstrated that a pGRT1 mutant in UspA was more sensitive to UV light than wild type, indicating that UspA has a role in the cells' response to UV stress, possibly in cooperation with UvrD, RuvA, and RuvB homologues (27). The presence of the pGRT1 plasmid was also required for the release of siderophores into the media in response to iron deficiency. This phenotype was linked to the presence of ORF35, which is also encoded by pGRT1. Surprisingly, a BLAST search using ORF35 returns a putative and chromosomally encoded sulfate permease (*sulP*) found within *Pseudomonas* and *Burkholderia* strains (27), although one homologue of

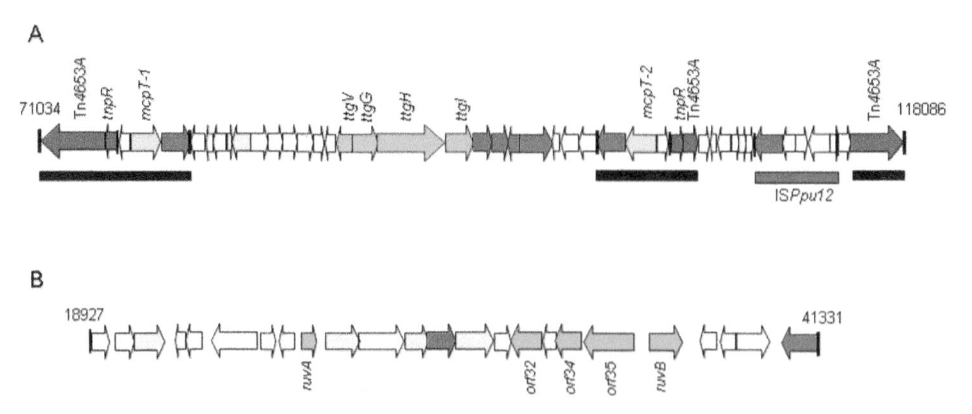

Figure 4 Schematic representation of Tn*4653*-like region (A) and second region encoding stress resistance genes in pGRT1 (B). In red are indicated transposition-related genes, in green are stress-related functions, and in blue are putative recombinases or integrases. doi:10.1128/microbiolspec.PLAS-0013-2013.f4

this protein has been identified in the antibiotic resistance plasmid pOZ176 of *P. aeruginosa* 96 (124).

pGRT1 was the first plasmid that was identified to confer cells with solvent tolerant traits. Genes with high sequence homology to those of the *ttgGHI* operon have been found in other *Pseudomonas* strains but they are located either within the chromosome (L. Molina, A. Segura, Z. Udaondo, C. Molina-Santiago, J.L. Ramos, unpublished results) or are at undefined locations (125). Because the spectrum of substrates transported by the efflux pump is so wide and extrusion is so effective, plasmid pGRT1 has been considered a paradigm and a model for solvent tolerance. Furthermore, pGRT1 encodes determinants for solvent tolerance, strong chemotaxis toward pollutants, survival under different stress conditions, and genetic transfer to other *Pseudomonas* strains, thus providing degrader bacteria the ability to survive in heterogeneously polluted environments and to achieve more efficient *in situ* bioremediation (126). The localization of the efflux pump genes in a transposon and the multiple residues of transposases, recombinases, and integrases found in pGRT1 suggest that genes associated with stress endurance have been recruited from different origins during evolution.

Although TtgGHI is an extreme example of the contributions of an efflux pump toward the resistance to high concentrations of aromatic compounds, many other bacteria have developed systems to cope with toxic compounds present in their niche. An example is the nitrogen-fixing rhizobia bacteria, which are well adapted to live in polyphenol-rich environments. Polyphenols are produced by plants under biotic and abiotic stresses such as water stress, bacterial or fungal infections, UV radiation, and others (127). Some of the polyphenols exuded by legume seeds and roots are necessary to establish a symbiotic association between rhizobia with their host plant (reviewed in reference 128). However, rhizobia have also developed several mechanisms to thrive in the presence of otherwise toxic concentrations of these compounds, including degradation (129) and tolerance (130). One of the tolerance mechanisms in *Rhizobium etli* CFN42 is mediated by the RmrAB efflux pump that is encoded by one of the strain's seven plasmids (131). The rhizobium multiresistance genes (*rmrAB*) of *R. etli* are located within the 184-kbp plasmid known as p42b. This operon is induced by bean exudates and is able to extrude naringenin (a flavonoid inductor of nodulation), coumaric, and salicylic acid, a well-known inductor of plant defense mechanisms that is important for the survival of *R. etli* in rhizospheric environments (130).

While extrusion of the toxic compounds forms part of the defensive mechanisms against pollutants, stress response mechanisms also play a role. Heat shock proteins (HSPs), ubiquitous proteins that are the key players in the general stress response system, act by mediating the folding and transport of proteins. Organic solvents cause protein damage leading to misfolding. It has been demonstrated that overexpression of heat shock proteins, such as the GroESL system in *Clostridium acetobutylicum* (a butanol-producing bacteria), results in improved solvent tolerance and solvent production (132). HSP genes can be found in a broad range of antibiotic-resistant plasmids from the *Enterobacteriaceae* family and in plasmids involved in rhizobia-plant symbiosis. The ORFs known as *groES* and *groEL* encode small heat shock proteins that in *Rhizobium* sp. NGR234 are located within the pNGR234b plasmid (133). Interestingly, genes involved in the degradation of protocatechuate, opine, and naphthalene are also encoded in this plasmid as well as 26 integrases and recombinases. Fragments with different guanine and cytosine (G+C) content have been identified, suggesting that lateral transfer of genetic material may have occurred. The byi_1p (BY123_D) plasmid of *Burkholderia* sp. YI23 also contains genes for heat shock proteins. This strain was isolated because it was able to degrade the organophosphorus insecticide fenitrothion. With three chromosomes and three plasmids, the strain has degradative genes located in plasmid byi_2p (BY123_E) and byi_3p (BY123_F) (134). Whether or not the heat shock proteins identified within this strain are involved in tolerance has not yet been investigated.

Plasmids Involved in Heavy Metal Tolerance

Metal-containing minerals are abundant on Earth. Natural events such as volcanic emissions, forest fires, deep-sea vents, and geysers, together with anthropogenic activities (mining, smelting, the creation of industrial chemical waste, etc.) have contributed to the distribution of these metals across the world. Some metals are necessary to sustain life (calcium and sodium are essential micronutrients; cobalt, copper, nickel, and zinc are vital cofactors for enzymes and metalloproteins); however, at high concentrations they can have toxic effects on the organisms. Other nonessential heavy metals, such as mercury, lead, and cadmium are considered toxic at any concentration (135). Most organisms have developed homeostasis systems in order to maintain optimal intracellular concentrations of metals. This is achieved through the control of the processes of transport (import and export) and intracellular trafficking to prevent cellular damage and ensure cellular bioavailability. Metal transporters move metal ions or chelates through membranes, metallochaperones transfer the metal to appropriate cellular compartments or acceptor proteins, and efflux pumps can eliminate excess and unwanted metals. Regulatory proteins serve to control the expression of all these proteins in response to metal deprivation or overload (136). While normal heavy metal concentrations in soils are generally nontoxic, there are specific locations where metal concentrations are high enough to prevent "normal" organisms from existing (137) and where only organisms with special mechanisms of tolerance can thrive. Efflux pumps are one of the main mechanisms for metal resistance in several bacteria, but metal complexation and metal reduction can also contribute to tolerance (135).

Cupriavidus metallidurans CH34 has served as a model microorganism for heavy metal resistance. The bacterium was isolated from the sludge from a zinc decantation tank in Belgium (138) and tolerates high concentrations of heavy metal ions, including but not necessarily limited to Cu^{+}, Cu^{2+}, Ni^{2+}, Zn^{2+}, Co^{2+}, Cd^{2+}, CrO_4^{2-}, Pb^{2+}, Ag^{+}, Au^{+}, Au^{3+}, $HAsO_4^{2-}$, AsO_2^{-}, Hg^{2+}, Cs^{+}, Bi^{3+}, Tl^{+}, SeO_3^{2-}, SeO_4^{2-}, and Sr^{2+}. This strain is a facultative chemolithoautotroph that is able to grow using toluene, benzene, or xylene as the sole carbon source. This ability is conferred by a number of chromosomally encoded genes as well as uncharacterized dioxygenases located on plasmids. The genome of *C. metallidurans* CH34 is composed of four replicons: two main chromosomes and two megaplasmids named pMOL28 and pMOL30 (135, 139). This strain accomplishes metal detoxification via the action of a wide variety of efflux systems of the RND, P-type ATPase, and Cation Diffusion Facilitation (CDF) families. It also contains transporters that are specific to certain metal ions, whereby efflux may be followed by metal sequestration or complexation. Many of these systems are localized in two large plasmids known as pMOL28 and pMOL30 (135).

The pMOL28 plasmid (139) has been associated with tolerance to Ni(II), Co(II), $CrO^{2-}{}_4$, and Hg(II). The nickel/cobalt tolerance determinant present in pMOL28 is the *cnr* operon, which is composed of the RND efflux pump CnrABC, and a regulatory complex formed by the regulatory proteins CnrXY and the sigma factor, CnrH. Additionally, CnrT, a cation diffusion facilitator, is encoded by this region. Cation diffusion facilitators are transport systems driven by a chemiosmotic gradient (140) and represent another line of defense against excess metal cations. These proteins form a family of membrane-bound secondary transport systems for divalent transition metal cations (141). The *Cupriavidus* CH34 strain contains two additional CDF proteins, known as DmeF and FieF (140), that are chromosomally encoded (142). Chromate tolerance is mediated by the efflux pump ChrA, which belongs to the chromate-efflux-related protein family. These systems are driven by proton-motive force and remove chromate by efflux in cooperation with other proteins (143). Proteins related to this efflux pump are the regulatory proteins ChrF, ChrI, and ChrB; ChrE, which is involved in the processing of chromium-glutathione complexes; and ChrC, an iron superoxide dismutase (Fe-SOD). Homologues of the *chr* operon have also been found in a plasmid of the potential PAH degrader *Arthrobacter* sp. FB24 (144); in pCNB1, which provides *Comamonas* CNB-1 the ability to degrade chloronitrobenzenes (82); and in the pRA4000 plasmid, which allows *P. putida* NCIMB 9866 to degrade 2,4-xylenol and *p*-cresol (145). Mercury resistance is widespread in plasmids. Although mercury concentration in soils and waters is normally very low, since the start of the industrial revolution, the amount of mercury mobilized and released into the biosphere has increased, and, in some places, local mercury levels have increased by several thousandfold above ground. The heightened levels of mercury in the environment may be responsible for the widespread occurrence of the *mer* operon in nature (146). In pMOL28, mercury resistance is associated with the presence of the *merRTPADE* operon (147). MerR acts as the key regulator of this operon (148). The highly toxic Hg^{2+} cation is bound in the periplasm by MerP, imported into the cytoplasm by MerT, and reduced to metallic mercury by the MerA NADPH-dependent flavoprotein. While these enzymes are essential for mercury detoxification, in some organisms the operon contains additional genes including *merB*, which encodes an organomercurial lyase; *merC*, a mercury transporter; and MerD, which is involved in regulation of the *mer* operon (reviewed in reference 149). The presence of the *mer* operon in other plasmids has been associated with other degradative pathways. Examples include the pW2 plasmid from *P. putida* W2 (NC_013176), involved in bisphenol A degradation; the byi_1p plasmid from *Burkholderia* sp. YI23, that degrades fenitrothion (134); the pI2 plasmid from *Comamonadaceae* for the degradation of aniline (150); and in the pJP4 plasmid from *R. eutropha* JMP134, which allows degradation of chloroaromatics (70).

The backbone genes of the pMOL28 plasmid are highly homologous with the pHG1 plasmid core genes, which contain essential genes required for the facultative lithoautotrophic and facultative anaerobic lifestyles of *R. eutropha* H16 (151) and pSym of *Rhizobium taiwanensis* LMG19424 (152). The presence of one genomic island in pMOL28 that contains all the determinants for heavy metal resistance has been described (139). This region is flanked by IS*1071* and a partial IS from the Tn*3* family.

The pMOL30 plasmid (234-kbp) encodes determinants for tolerance to Ag(I), Cd(II), Co(II), Cu(II), Hg(II), Pb(II), and Zn(II) (139). In addition to containing a *mer* operon that is similar to that found in pMOL28, pMOL30 contains the *czcABC* operon (one of the best studied zinc-tolerance systems), which encodes an RND efflux pump for cobalt, zinc, and cadmium resistance in *C. metallidurans* (140). The *czcABC* operon includes three components of the proton antiporter efflux system: *czcD*, which encodes a cation diffusion facilitator; and *czcSR*, which encodes a two-component regulatory system. These regulators are homologues to those involved in copper resistance, and it is likely that this operon mediates copper extrusion (139). Three additional proteins, CzcN, CzcI, and CzCE, are putative components of the resistance mechanisms, although their function is currently unknown. Extrusion systems that are homologous to *czc* have been found in catabolic plasmids such pCAR3 of *Sphingomonas* sp. KA1, which degrades chlorinated dibenzo-*p*-dioxins (52), byi_1p of *Burkholderia* sp. YI23 (134), and pCMU01 from the chloromethane degrader *Methylobacterium extorquens* (153). Belonging to the same RND subgroup, the *nccCBA* efflux system (154) is also encoded by pMOL30. The *cnr* (of pMOL28) and *ncc* systems are related to the NccA protein of the nickel-cobalt-cadmium determinant of the pTOM9 plasmid from *Achromobacter* sp. 31A (155). The *sil/cus* system encoded by pMOL30 may be involved in the efflux of silver and copper ions. The *silABC* operon encodes an efflux pump for the transport of these ions (140). Other factors that are not found on the pMOL30 plasmid include: SilE, a protein that acts as an extracellular metal-binding protein; and *silRS*, which encodes a two-component sensor and transcriptional responder. The *silRS* operon is widespread in plasmids originating from clinical strains probably because silver compounds are used as antimicrobial agents against bacterial infections (156).

Two of the 13 predicted P-type ATPases of *C. metallidurans* CH34 are present in pMOL30. P-type ATPases constitute a ubiquitous superfamily of transport proteins that are driven by ATP hydrolysis. Among their substrates are inorganic cations such as H^+, Na^+, K^+, $Mg2^+$, $Ca2^+$, Cu^+, Ag^+, Zn^{2+}, and Cd^{2+}. In contrast to RND pumps, these ATPases are also able to detoxify heavy metal cations bound to thiols. One of the P-type ATPases of pMOL30 is encoded by the large cluster *copVTMKNS1R1A1B1C1D1IJGFLQHE* in which CopF is the P-type ATPase involved in Cu(II) efflux from the cytoplasm. CopT is believed to be involved in metal transport from the periplasm to the cytoplasm; CopR and CopS are members of the two-component family of sensor regulators; CopA is a multicopper oxidase; CopB is thought to be involved in Cu(I) fixation; CopC contains binding sites for Cu(I) and Cu(II) and may be involved in detoxification of both ions; *copD* encodes a putative channel involved in loading CopA with Cu(II); CopI is a putative oxidoreductase; CopJ showed similarity with cytochrome *c* proteins; CopL may have a role in regulation; and CopH has homology with *czcE*. No function has been assigned to CopV, CopK, CopM, CopN, CopG, CopQ, and CopE (157). The *cop* system is responsible for copper and silver resistance of *E. coli* (158). Homologous *cop* genes have been identified in the metal tolerance plasmid pSPHCH01 of *Sphingobium chlorophenolicum* L-1 (159), and in the plasmid pISP0 from *Sphingomonas* sp. MM-1, which also harbors genes involved in the degradation of γ-hexachlorocyclohexane (160). Homologues of *copA*, *mco* (multicopper oxidase gene), *cadD*, and *cadX* (cadmium resistance) have also been recently identified in multiresistance plasmids from *Staphylococcus aureus* ST398 (161). The colocalization of antimicrobial resistance genes and genes that confer tolerance to heavy metals can facilitate the persistence and dissemination of these traits and could pose a serious problem for human health. Because of the widespread use of copper bactericides to control bacterial infections in crop plants, the *copABCD* system (present in the chromosome of *C. metallidurans* CH34 and sharing low sequence homology with the *cop* operon of plasmid pMOL30) and the *cusCBA* operon are found in plasmids of *P. syringae* pv. *tomato* (162) and pv. *syringae* UMAF0081 (163). The second P-type ATPase system present in pMOL30 is the lead resistance operon, *pbr*, which contains the following structural genes: *pbrT*, which encodes a Pb(II) uptake protein; *pbrA*, which encodes a P-type Pb(II) efflux ATPase; *pbrB*, which encodes a predicted integral membrane protein of unknown function; and *pbrC*, which encodes a predicted prolipoprotein signal peptidase. Downstream of *pbrC*, the *pbrD* gene encodes a Pb(II)-binding protein. The P_{pbrA} promoter is regulated by *PbrR*, which belongs to the MerR family of metal ion-sensing regulatory proteins. This operon is also found in antibiotic/

heavy metal tolerance plasmids of *Enterobacter cloacae* and *Klebsiella pneumoniae* strain KCTC 2242 (164).

pMOL30 *parAB* genes are very similar to those in *B. vietnamensis* G4, and belong to a different incompatibility group than pMOL28, as they are stably maintained. pMOL30 was found to be able to transfer at very low frequency. Many recombinases, IS, and truncated IS have been identified in pMOL30, most of them close to tolerance determinants. The *czc* and *pbr* clusters are flanked by the mercury transposon Tn*4380* on one side and by three *mer* genes that might be remnants of former rearrangements. The region containing the *cop*, *sil*, and *mre-ncc* clusters is flanked by a complete IS*Rme10* element and a remnant of another named IS*Rme10* (139).

Resistance to the toxic divalent heavy metal cations of cobalt, nickel, cadmium, and zinc is widespread among bacteria, although in many of them the genetic determinants of this characteristic are associated with chromosomal genes (165). As mentioned above, *C. metallidurans* CH34 also has many tolerance determinants located on the chromosome.

Resistance to arsenite {As[III], $As(OH)_3$}, arsenate (As[V], AsO_4^{3-}), and antimony [Sb(III)] is widely found among Gram-negative and Gram-positive bacteria. Usually this resistance is determined by the presence of an *ars* operon with a minimum of three cotranscribed genes that include *arsR* (a regulatory repressor), *arsB* (a membrane transport pump), and *arsC* (a small intracellular arsenate reductase). Additionally, in this operon other proteins are encoded, including: ArsA, which is an ATPase coupled to ArsB (together, these increase arsenite resistance); ArsD, which acts as an arsenite chaperon; and ArsP, which is a putative membrane permease (166). Although the *ars* operon is mostly located chromosomally, components are also distributed in plasmids, especially those encoding antibiotic resistance (167). One example of *ars* operon within a plasmid is found in *Arthrobacter* sp. Rue61, which degrades quinaldine through a pathway encoded by the pARUE113 (pAL1) linear plasmid. This strain also has a circular plasmid, named pARUE232, which harbors *ars* genes (64). Another example of a plasmid that has the *ars* operon is the metal tolerance plasmid pOC167 from *Oligotropha carboxidovorans* OM5 (168).

GENOMIC BIOAUGMENTATION AS A BIOTECHNOLOGICAL APPLICATION

The existence of the described solvent, heavy metal tolerance, and catabolic plasmids provides a useful battery of biotechnological tools for remediation of polluted environments. Contaminants are normally present in complex mixtures that make degradation difficult. Some of these compounds, as described earlier, are toxic and are able to avoid the action of the catabolic pathways of pollutant degrader microorganisms. Tolerance plasmids and genes, once introduced into degrader strains, will allow the survival and the metabolic activity of bacteria living in polluted environments. One of the strategies to bioremediate contaminated environments is bioaugmentation with strains that have the capacity to degrade pollutants. However, it has been observed that laboratory strains, on many occasions, were not able to thrive in the new environment, leading to unsuccessful bioremediation (169). Given the possibilities of plasmid exchange among strains, bioaugmentation in catabolic genes, not in bacteria, has taken relevance in the past decade (170). This approach is called "genetic bioaugmentation" and involves the introduction of bacteria harboring a relevant catabolic self-transmissible plasmid that stimulates the horizontal gene transfer of the plasmids into indigenous microorganisms with better fitness for survival in the corresponding niche (171). To design an optimal genetic bioaugmentation is important to choose the appropriate type of plasmid, to study the transfer capacity of the microorganism and the stability of the plasmid in the new bacteria, as well as to study the expression of the catabolic genes (171, 172). Plasmid stability in the recipient strains may impact the effectiveness of bioremediation; highly stable plasmids could be necessary to clean up sites that continuously receive contaminant input (171). Horizontal gene transfer from donor to indigenous bacteria following the deliberate release of phenol-degrading laboratory bacteria was shown to be important in the degradation of phenol in river waters continuously polluted by phenolic compounds (91).

It is well documented that the TOL plasmid can be transferred from *P. putida* to other *Pseudomonas* and *Erwinia* strains (173), and to *Enterobacteriaceae* (174). Ikuma et al. (175) showed that the soil organic carbon present in sterilized soil slurries was sufficient for transfer of TOL plasmid using different mixtures of recipient bacteria (including *E. coli*, *E. cloacae*, *Serratia marcescens*, *P. fluorescens*, and *P. putida* BBC443). However, they also observed that addition of glucose sometimes improved the specific toluene degradation rates of *Enterobacteriaceae* transconjugants. The G+C genomic content of the recipient strains have a clear influence on the expression of the toluene degradation genes in TOL transconjugants and the presence of alternative carbon sources (such as glucose) was shown to alleviate the limitations of the expression of the acquired genes in some transconjugants (176). Observations made under laboratory conditions (177) indicated the importance of the initial recipient-to-donor cell density; however, in the experiments conducted by Ikuma et al. (175) this factor had only a minor impact on plasmid conjugation. The spatial separation between donor and recipient strains can pose a problem for the plasmid transfer and efficiency of biodegradation; this problem has been tackled with the use of different approaches. The use of earthworms has been reported to increase the dispersal of donor (*A. eutrophus* [pJP4]) and recipient bacteria (*P. fluorescens*) and to increase the frequency of transconjugants in soil microcosms (178). Similarly, biofilm structures, with open channels and pores, allow the efficient transport of donor cells facilitating collision between bacteria and enhancing the introduction of mobile elements into an existing microbial community (179). Interestingly, transfer of the TOL plasmid in sequencing batch biofilm reactors (SBBR) used for treating synthetic wastewater containing benzyl alcohol was observed on a laboratory scale but not in the pilot scale bioreactor (180). This failure can be attributed to different operational conditions in the two bioreactors and to the lack of selective pressure in the pilot scale bioreactor, because complete benzyl alcohol removal was achieved during the first 60 minutes of operation. While the presence of contaminant has been reported to exert a positive effect on the efficiency of the pJP4 plasmid transfer under nonsterile conditions (181), the presence of toluene did not influence the TOL transfer frequency when tested in filter matings (177) or in slurry soils at environmentally relevant concentrations (176).

Other approaches to bring together donor and recipient bacteria take advantage of the high number of bacteria living near the plant roots; roots additionally provide a solid surface for conjugation (172). Although the rhizosphere can improve the survival of microorganisms in soil (182) and thus promote the elimination of contaminants (183), it is also a complex environment where plant and microorganisms establish different kinds of relationships. Because the rhizosphere is highly populated, microbes have to compete for niches and nutrients (184); furthermore, plants secrete harmful compounds such as phytoalexins, phenolic derivatives, and others. Transcriptional experiments performed with the rhizospheric strain *P. putida* KT2440 showed that they sense the rhizosphere as a stressful environment (185). Some genetic traits involved in coping with this high-stress environment have been shown to be encoded on plasmids; the best characterized examples are the symbiotic plasmids from *Rhizobium* strains. These plasmids encode chemotaxis systems that recognize rhizosphere nutrients, genes involved in catabolism of these nutrients, nutrient uptake, attachment to roots, and detoxification of harmful molecules produced by plants (reviewed in reference 186). To avoid competition with other rhizospheric microorganisms, Barac et al. (187) used a new approach; it involved the introduction of the pTOM plasmid into *B. cepacia* L.S.2.4, a natural endophyte of the legume, *Lupinus luteus* (yellow lupine). This approach improved the biodegradation of toluene. Inoculation of poplar trees with another yellow lupine endophyte, *B. cepacia* VM1468 containing the pTOM plasmid, also resulted in decreased toluene toxicity toward the plant. Although *B. cepacia* VM1468 was not able to successfully establish in the plant at high levels, it was able to transfer the pTOM plasmid to other indigenous endophytes, demonstrating the horizontal transfer of the plasmid (188).

In recent years some "rhizospheric" plasmids have been identified. One of them is plasmid pQBR103 from *P. fluorescens* SBW25. It belongs to a large group of plasmids known to persist in the sugar beet phytosphere (rhizosphere and phyllosphere) and to be confined within the *Pseudomonas* group. The pQBR103 is a 425-kbp plasmid that encodes one mercury resistance operon located in Tn*5042* type II transposon, and also encodes the RulAB proteins that confer UV light resistance. Field release trials have shown that pQBR103 confers a significant advantage to the SBW25 strain, 3 to 5 months after planting (189). Preliminary studies demonstrated that the advantages provided by this plasmid are imposed by the modulation of plant responses, although the mechanism of the adaptive advantage remains unclear (190). It has been shown that pQBR103 regulates up to 48 proteins encoded on the chromosome of its host strain.

PERSPECTIVES

Even though the decreasing costs of sequencing have allowed more information to be obtained about catabolic or tolerance plasmids in the environment, there is still a considerable lack of knowledge regarding plasmids from noncultivatable bacteria. Several approaches have been followed to fill this gap. The most commonly used technique has probably been "exogenous plasmid isolation." In this technique, a cultivatable recipient strain is used to acquire plasmids from different environments through conjugation (191, 192, 193, 194, 195). However, transfer is limited by the incompatibility of the indigenous bacteria for establishing interactions with the recipient bacteria and the capacity of the plasmid to be transferred and replicate in the new strain. To avoid the dependence on the plasmid-encoded traits two different techniques have been described. One of them is called transposon-aided capture (TRACA) of plasmids. Genomic extracts from environmental samples are treated with plasmid-safe DNase that digest linear but not circular DNA, then circular plasmids are subjected to transposition with insertions that contain an *E. coli* origin of replication and a selectable marker later transformed into *E. coli* strains (5). This technique has been successfully used for the identification of plasmids resident in the human gut meta-genome but, to date, we have not found any report describing the utilization of this technique in the isolation of catabolic or tolerance plasmids. We can envisage certain drawbacks for TRACA utilization in the identification of catabolic or tolerance plasmids; because of the large size of the catabolic plasmids, the DNA can be physically broken or damaged during the extraction process and thus later on digested by the DNase; also because of this large size transformation into *E. coli* strains would be difficult. The approach followed by Kav and colleagues (196) is also based on the treatment of genomic environmental DNA with a plasmid-safe DNase, but the resultant circular plasmid DNA is subjected to amplification with DNA polymerase from phage ϕ29 and sequenced. With the use of this technique, the plasmidome of the bovine rumen has been identified (196). Implementation and improvements in both protocols could circumvent the plasmid size issue of catabolic and tolerance plasmids and will allow the analysis of the plasmidome of environmentally relevant niches.

The impact of catabolic plasmids on host cell physiology is another area of research that has to be further developed (review in reference 197), because this information is important for improving the biotechnological utilization of bacteria carrying catabolic genes. It is known that plasmid carriage generally leads to loss of host fitness (198), but catabolic and tolerance plasmids are quite stable in their corresponding natural host even in the absence of selective pressure. The pCAR1 plasmid is stably maintained not only in its natural host, *P. resinovorans*, but also in *P. putida* KT2440. The transcriptome of *P. putida* KT2440 harboring pCAR1 was compared with that of the plasmid-free strain when growing on succinate (116). Interestingly, plasmid pCAR1 did not significantly interfere with the host transcriptional patterns, and only *parI*, a homologue of the ParA family of plasmid partitioning proteins from a cryptic genomic island, showed a significant induction in expression. It was latter demonstrated that ParI interferes with the IncP-7 plasmid partitioning system (199), so the reason why pCAR1 is stable in KT2440 remains unclear. Transcriptional analysis of pCAR1-carrying and pCAR1-free strains of *P. putida* KT2440, *P. aeruginosa* PAO1, and *P. fluorescens* Pf0-1 was performed by using high-density tiling arrays in order to identify changes in transcriptional patterns. Although 70 to 100 genes (depending on the strain) were found to be altered in their expression levels in the pCAR1-carrying strains compared with the pCAR1-free strains, only four genes were found to be upregulated in the three strains. Three of the genes form an operon that contained a Fur-associated gene (Fur is a global regulator in response to iron limitation); the fourth gene (*phuR*) is known to be upregulated under iron-limited conditions in *P. aeruginosa*. Among the commonly regulated genes in two of the hosts, most of them were related to iron acquisition and transport systems. Notably, pCAR1-carrying *P. putida* KT2440 and *P. aeruginosa* PAO1 have higher levels of pyoverdine than the plasmid-free cells, which suggests that iron concentration may be key factor in the maintenance of pCAR1 in host cells (200). Backbone-related functions are also important in plasmid survival in different strains; more than 90% of ORFs on the pCAR1 backbone were transcribed in six different host strains (201).The influence of the NAH7 plasmid on the *P. putida* KT2440 transcriptome has also been analyzed. Despite the fact that the presence of the plasmid relieved the stress caused by the presence of naphthalene in the host strain, few genes were differentially expressed between the containing and non-containing strains when naphthalene was not provided (202). The effect that nucleoid-associated proteins (NAPs) have on the expression of plasmids and host chromosomal genes has to be further explored. Disruption of Pmr, a histone-like protein H1(H-NS) encoded on pCAR1, significantly alters the expression levels of several genes in *P. putida* KT2440 (203). The knockout inactivation of *TurA*, a chromosomally encoded protein with structural similarity to H-NS proteins, resulted in enhanced transcription initiation from the *Pu* promoter, suggesting a negative regulatory role of TurA on *Pu* expression (204).

Many different biotechnological applications that rely on plasmid activities in a community are currently being developed for the elimination of contaminants (205). Although plasmid stability is usually high, segregants can become a majority if plasmid-free cells have faster growing rates (206). To monitor the abundance of plasmids, new rapid, noninvasive, *in situ* monitoring techniques will have to be improved to detect the maintenance of plasmids within the reactors. Traditional monitoring techniques, such as replica plating, selective markers, and PCR detection (207) have, in general, low sensitivity and, in some cases, they can disrupt the function of the system. Although techniques based on the introduction of *gfp*-labeled plasmids have been proposed as a noninvasive, *in situ* monitoring solution (208), the use of recombinant plasmids may carry some legal restrictions in real-life applications.

CONCLUSIONS

Although many catabolic pathways and resistance operons are located on plasmids, these genes can also on many occasions be found on the chromosome. Nevertheless, catabolic and tolerance plasmids that are generally self-transmissible, are valuable tools for bioremediation applications. Although in the 1980s there was avid interest in the discovery of contaminant degradation pathways and their regulatory circuits, it was at the beginning of the 21st century that the simplification of sequencing techniques allowed a better understanding of the plasmids that encode these pathways. It is now known that most of the catabolic and tolerance genes encoded on plasmids are associated with mobile elements, and that transfer to or from the host bacteria has occurred not only as a consequence of plasmid transfer, but also because of transposition and recombination. The backbones of many of these plasmids are related to the IncP incompatibility group, and, in many cases, plasmids share a similar backbone but are loaded with different catabolic or tolerance genes (i.e., the backbone of pBVIE04 that carries the toluene monoxygenase genes and the backbone of pGRT1 that carries the toluene tolerance operon).

There is a lack of information about plasmids in the environment that are harbored by noncultivatable strains; this knowledge gap needs to be filled. Awareness of the abundance and significance of these plasmids in the environment will allow us to better understand the function of ecosystems. Understanding the mechanisms behind plasmid stability will improve the outcomes of the biotechnological application of these plasmids.

Acknowledgments. *Conflicts of interest: We declare no conflicts.*

Citation. Segura A, Molina L, Ramos JL. 2014. Plasmid-mediated tolerance toward environmental pollutants. Microbiol Spectrum **2**(6):PLAS-0013-2013.

References

1. **Segura A, Rojas A, Hurtado A, Huertas MJ, Ramos JL.** 2003. Comparative genomic analysis of solvent extrusion pumps in *Pseudomonas* strains exhibiting different degrees of solvent tolerance. *Extremophiles* **7:** 371–376.
2. **Williams PA, Jones RM, Zysltra GJ.** 2004. Genomics of catabolic plasmids, p 165–195. *In* Ramos JL (ed), *Pseudomonas*, **vol 1.** Kluwer Academic, Plenum Publishers, New York.
3. **Ogawa N, Chackrabarty AM, Zaborina O.** 2004. Degradative plasmids, p 341–392. *In* Funnell BE, Phillips GJ (ed), *Plasmid Biology*. ASM Press, Washington, DC.
4. **Springael D, Top EM.** 2004. Horizontal gene transfer and microbial adaptation to xenobiotics: new types of mobile genetic elements and lessons from ecological studies. *Trends Microbiol* **12:**53–58.
5. **Jones BV, Marchesi JR.** 2007. Transposon-aided capture (TRACA) of plasmids resident in the human gut mobile metagenome. *Nat Methods* **4:**55–61.
6. **Parales RE, Parales JV, Pelletier DA, Ditty JL.** 2008. Diversity of microbial toluene degradation pathways. *Adv Appl Microbiol* **64:**1–73.
7. **Bertini L, Calafaro V, Proietti S, Caporale C, Capasso P, Caruso C, Di Donato A.** 2013. Deepening TOL and TAU catabolic pathways of *Pseudomonas* sp. OX1: Cloning, sequencing and characterization of the *lower* pathways. *Biochimie* **95:**241–250.
8. **Williams PA, Murray K.** 1974. Metabolism of benzoate and the methylbenzoates by *Pseudomonas putida (arvilla)* mt-2: evidence for the existence of a TOL plasmid. *J Bacteriol* **120:**416–423.
9. **Greated A, Lambertsen L, Williams PA, Thomas CM.** 2002. Complete sequence of the IncP-9 TOL plasmid pWW0 from *Pseudomonas putida*. *Environ Microbiol* **4:**856–871.
10. **Gallegos MT, Williams PA, Ramos JL.** 1997. Transcriptional control of the multiple catabolic pathways encoded on the TOL plasmid pWW53 of *Pseudomonas putida* MT53. *J Bacteriol* **179:**5024–5029.
11. **Ramos JL, Marques S, Timmis KN.** 1997. Transcriptional control of the *Pseudomonas* TOL plasmid catabolic operons is achieved through an interplay of host factors and plasmid-encoded regulators. *Annu Rev Microbiol* **51:**341–371.
12. **Silva-Roche R, de Lorenzo V.** 2013. The TOL network of *Pseudomonas putida* mt-2 processes multiple environmental inputs into a narrow response space. *Env Microbiol* **15:**271–286.
13. **Bayley SA, Duggleby CJ, Worsey MJ, Williams PA, Hardy KB, Broda P.** 1977. Two modes of loss of the Tol function from *Pseudomonas putida* mt-2. *Mol Gen Genet* **154:**203–204.
14. **Muñoz R, Hernández M, Segura A, Gouveia J, Rojas A, Ramos JL, Villaverde S.** 2009. Continuous cultures of *Pseudomonas putida* mt-2 overcome catabolic function loss under real case operating conditions. *Appl Microbiol Biotech* **83:**189–198.

15. **Tsuda M, Iino T.** 1987. Genetic analysis of a transposon carrying toluene degrading genes on a TOL plasmid pWW0. *Mol Gen Genet* **210:**270–276.
16. **Tsuda M, Iino T.** 1988. Identification and characterization of Tn4653, a transposon covering the toluene transposon Tn4651 on TOL plasmid pWW0. *Mol Gen Genet* **213:**72–77.
17. **Williams PA, Jones RM, Shaw LE.** 2002. A third transposable element, IS*Ppu12*, from the toluene-xylene catabolic plasmid pWW0 of *Pseudomonas putida* mt-2. *J Bacteriol* **184:**6572–6580.
18. **Williams PA, Worsey MJ.** 1976. Ubiquity of plasmids in coding for toluene and xylene metabolism in soil bacteria: evidence for the existence of new TOL plasmids. *J Bacteriol* **125:**818–828.
19. **Keil H, Keil S, Pickup W, Williams PA.** 1985. Evolutionary conservation of genes coding for *meta* pathway enzymes within TOL plasmids pWW0 and pWW53. *J Bacteriol* **164:**887–895.
20. **Chatfield LK, Williams PA.** 1986. Naturally occurring TOL plasmids in *Pseudomonas* strains carry either two homologous or two nonhomologous catechol 2,3-oxygenase genes. *J Bacteriol* **168:**878–885.
21. **Sentchilo VS, Perebituk AN, Zehnder AJ, van der Meer JR.** 2000. Molecular diversity of plasmids bearing genes that encode toluene and xylene metabolism in *Pseudomonas* strains isolated from different contaminated sites in Belarus. *Appl Environ Microbiol* **66:**2842–2852.
22. **Yano H, Garruto CE, Sota M, Ohtsubo Y, Nagata Y, Zylstra GJ, Williams PA, Tsuda M.** 2007. Complete sequence determination combined with analysis of transposition/site-specific recombination events to explain genetic organization of IncP-7 TOL plasmid pWW53 and related mobile genetic elements. *J Mol Biol* **369:**11–26.
23. **Yano H, Miyakoshi M, Ohshima K, Tabata M, Nagata Y, Hattori M, Tsuda M.** 2010. Complete nucleotide sequence of TOL plasmid pDK1 provides evidence for evolutionary history of IncP-7 catabolic plasmids. *J Bacteriol* **192:**4337–4347.
24. **Shields MS, Reagin MJ, Gerger RR, Campbell R, Somerville C.** 1995. TOM, a new aromatic degradative plasmid from *Burkholderia (Pseudomonas) cepacia* G4. *Appl Environ Microbiol* **61:**1352–1356.
25. **Shields MS, Montgomery SO, Cuskey SM, Chapman PJ, Pritchard PH.** 1991. Mutants of *Pseudomonas cepacia* G4 defective in catabolism of aromatic compounds and trichloroethylene. *Appl Environ Microbiol* **57:**1935–1941.
26. **Nelson MJK, Montgomery SO, Mahaffey WR, Pritchard PH.** 1987. Biodegradation of trichloroethylene and involvement of an aromatic biodegradative pathway. *Appl Environ Microbiol* **53:**949–954.
27. **Molina L, Duque E, Gómez MJ, Krell T, Lacal J, García-Puente A, García V, Matilla MA, Ramos JL, Segura A.** 2011. The pGRT1 plasmid of *Pseudomonas putida* DOT-T1E encodes functions relevant for survival under harsh conditions in the environment. *Environ Microbiol* **13:**2315–2327.
28. **Rheinwald JG, Chakrabarty AM, Gunsalus IC.** 1973. A transmissible plasmid controlling camphor oxidation in *Pseudomonas putida. Proc Natl Acad Sci USA* **70:** 885–889.
29. **Dunn NW, Gunsalus IC.** 1973. Transmissible plasmid coding early enzymes of naphthalene oxidation in *Pseudomonas putida. J Bacteriol* **114:**974–979.
30. **Yen KM, Gunsalus IC.** 1982. Plasmid gene organization: naphthalene/salicylate oxidation. *Proc Natl Acad Sci USA* **79:**874–879.
31. **Yen KM, Serdar CM.** 1988. Genetics of naphthalene catabolism in pseudomonads. *Crit Rev Microbiol* **15:** 247–268.
32. **Denome SA, Stanley DC, Olson ES, Young KD.** 1993. Metabolism of dibenzothiophene and naphthalene in *Pseudomonas* strains: complete DNA sequence of upper naphthalene catabolic pathway. *J Bacteriol* **175:**6890–6901.
33. **Simon MJ, Osslund TD, Saunders R, Ensley BD, Suggs S, Harcourt A, Suen WC, Cruden DL, Gibson DT, Zylstra GJ.** 1993. Sequences of genes encoding naphthalene dioxygenase in *Pseudomonas putida* strains G7 and NCIB 9816-4. *Gene* **127:**31–37.
34. **Eaton RW.** 1994. Organization and evolution of naphthalene catabolic pathways: sequence of the DNA encoding 2-hydroxy-chromene-2-carboxylate isomerase and trans-o-hydroxybenzylidenepyruvate hydratase-aldolase from the NAH7 plasmid. *J Bacteriol* **176:** 7757–7762.
35. **Grimm AC, Harwood CS.** 1999. NahY, a catabolic plasmid-encoded receptor required for chemotaxis of *Pseudomonas putida* to the aromatic hydrocarbon naphthalene. *J Bacteriol* **181:**3310–3316.
36. **Schell MA, Sukordhaman M.** 1989. Evidence that the transcription activator encoded by the *Pseudomonas putida nahR* gene is evolutionarily related to the transcription activators encoded by the *Rhizobium nodD* genes. *J Bacteriol* **171:**1952–1959.
37. **Dennis JJ, Zylstra GJ.** 2004. Complete sequence and genetic organization of pDTG1, the 83 kilobase naphthalene degradation plasmid from *Pseudomonas putida* strain NCIB 9816-4. *J Mol Biol* **341:**753–768.
38. **Li W, Shi J, Wang X, Han Y, Tong W, Ma L, Liu B, Cai B.** 2004. Complete nucleotide sequence and organization of the naphthalene catabolic plasmid pND6-1 from *Pseudomonas* sp. strain ND6. *Gene* **336:**231–240.
39. **Sota M, Yano H, Ono A, Miyazaki R, Ishii H, Genka H, Top EM, Tsuda M.** 2006. Genomic and functional analysis of the IncP-9 naphthalene-catabolic plasmid NAH7 and its transposon Tn*4655* suggests catabolic gene spread by a tyrosine recombinase. *J Bacteriol* **188:** 4057–4067.
40. **Tsuda M, Iino T.** 1990. Naphthalene degrading genes on plasmid NAH7 are on a defective transposon. *Mol Gen Genet* **223:**33–39.
41. **Sanseverino J, Applegate BM, King JMH, Sayler GS.** 1993. Plasmid-mediated mineralization of naphthalene, phenanthrene and anthracene. *Appl Environ Microbiol* **59:**1931–1937.

42. **Kiyohara H, Nagao K.** 1978. The catabolism of phenanthrene and naphthalene in bacteria. *J Gen Microbiol* **105:**69–75.

43. **Laurie AD, Lloyd-Jones G.** 1999. The *phn* genes of *Burkholderia* sp. strain RP007 constitute a divergent gene cluster for polycyclic aromatic hydrocarbon catabolism. *J Bacteriol* **181:**531–540.

44. **Pinyakong O, Habe H, Omori T.** 2003. The unique aromatic catabolic genes in sphingomonads degrading polycyclic aromatic hydrocarbons (PAHs). *J Gen Appl Microbiol* **49:**1–19.

45. **Basta T, Keck A, Klein J, Stolz A.** 2004. Detection and characterization of conjugative degradative plasmids in xenobiotic-degrading *Sphingomonas* strains. *J Bacteriol* **186:**3862–3872.

46. **Romine MF, Stillwell LC, Wong K-K, Thurston SJ, Sisk EC, Sensen C, Gaasterland T, Fredrickson JK, Saffer JD.** 1999. Complete sequence of a 184-kilobase catabolic plasmid from *Sphingomonas aromaticivorans* F199. *J Bacteriol* **181:**1585–1602.

47. **Feng X, Ou X-L, Ogram A.** 1997. Plasmid-mediated mineralization of carbofuran by *Sphingomonas* sp. CF-06. *Appl Environ Microbiol* **63:**1332–1337.

48. **Basta I, Buerger S, Stolz A.** 2005. Structural and replicative diversity of large plasmids from sphingomonads that degrade polycyclic aromatic compounds and xenobiotics. *Microbiology* **151:**2025–2037.

49. **Keck A, Conradt D, Mahler A, Stolz A, Mattes R, Klein J.** 2006. Identification and functional analysis of the genes for naphthalenesulfonate catabolism by *Sphingomonas xenophaga* BN6. *Microbiology* **152:**1929–1940.

50. **Maeda K, Nojiri H, Shintani M, Yoshida T, Habe H, Omori T.** 2003. Complete nucleotide sequence of carbazole/dioxin-degrading plasmid pCAR1 in *Pseudomonas resinovorans* strain CA10 indicates its mosaicity and the presence of large catabolic transposon Tn4676. *J Mol Biol* **326:**21–33.

51. **Takahasi Y, Shintani M, Yamane H, Nojiri H.** 2009. The complete nucleotide sequence of pCAR2: pCAR2 and pCAR1 were structurally identical IncP-7 carbazole degradative plasmids. *Biosci Biotechnol Biochem* **73:**744–746.

52. **Shintani M, Urata M, Inoue K, Eto K, Habe H, Omori T, Yamane H, Nojiri H.** 2007. The *Sphingomonas* plasmid pCAR3 is involved in complete mineralization of carbazole. *J Bacteriol* **189:**2007–2020.

53. **Nam J-W, Nojiri H, Noguchi H, Uchimura H, Yoshida T, Habe H, Yamane H, Omori T.** 2002. Purification and characterization of carbazole 1,9a-dioxigenase, a three-component dioxygenase system of *Pseudomonas resinovorans* strain CA10. *Appl Environ Microbiol* **68:**5882–5890.

54. **Nojiri H.** 2012. Structural and molecular genetic analyses of the bacterial carbazole degradation system. *Biosci Biotechnol Biochem* **76:**1–18.

55. **Ashikawa Y, Fujimoto Z, Usami Y, Inoue K, Noguchi H, Yamane H, Nojiri H.** 2012. Structural insight into the substrate- and dioxygen-binding manner in the catalytic cycle of Rieske nonheme iron oxygenase system, carbazole 1,9a-dioxygenase. *BMC Struct Biol* **12:**15. doi:10.1186/1472-6807-12-15.

56. **Shintani M, Matsumoto T, Yoshikawa H, Yamane H, Ohkuma M, Nojiri H.** 2011. DNA rearrangement has occurred in the carbazole-degradative plasmid pCAR1 and the chromosome of its unsuitable host, *Pseudomonas fluorescens* PF0-1. *Microbiology* **157:**3405–3416.

57. **Solinas F, Marconi AM, Ruzzi M, Zennaro E.** 1995. Characterization and sequence of a novel insertion sequence, IS*1162*, from *Pseudomonas fluorescens*. *Gene* **155:**77–82.

58. **Schneiker S, Kosier B, Puhler A, Selbitschka W.** 1999. The *Sinorhizobium meliloti* insertion sequence (IS) element ISRm14 is related to a previously unrecognized IS element located adjacent to the *Escherichia coli* locus of enterocyte effacement (LEE) pathogenicity island. *Curr Microbiol* **39:**274–281.

59. **Yao CC, Wong DTS, Poh CL.** 1998. IS*1491* from *Pseudomonas alcaligenes* NCIB 9867: characterization and distribution among *Pseudomonas* species. *Plasmid* **39:**187–195.

60. **Nojiri H, Sekiguchi H, Maeda K, Urata M, Nakai S, Yoshida T, Habe H, Omori T.** 2001. Genetic characterization and evolutionary implications of a car gene cluster in carbazole degrader *Pseudomonas* sp. strain CA10. *J Bacteriol* **183:**3663–3679.

61. **Overhage J, Sielker S, Hombrug S, Parschat K, Fetzner S.** 2005. Identification of large linear plasmids in *Arthrobacter* spp. encoding the degradation of quinaldine to anthranilate. *Microbiology* **151:**491–500.

62. **Parschat K, Hauer B, Kappl R, Kraft R, Hüttermann J, Fetzner S.** 2003. Gene cluster of *Arthrobacter ilicis* Rü61a involved in the degradation of quinaldine to anthranilate. Characterization and functional expression of the quinaldine 4-oxidase *qoxLMS* genes. *J Biol Chem* **278:**27483–27494.

63. **Parschat K, Overhage J, Strittmatter AW, Henne A, Gottschalk G, Fetzner S.** 2007. Complete nucleotide sequence of the 113-kilobase linear catabolic plasmid pAL1 of *Arthrobacter nitroguajacolicus* Rü61a and transcriptional analysis of genes involved in quinaldine degradation. *J Bacteriol* **189:**3855–3867.

64. **Niewerth H, Schuldes J, Parschat K, Kiefer P, Vorholt JA, Daniel R, Fetzner S.** 2012. Complete genome sequence and metabolic potential of the quinaldine-degrading bacterium *Arthrobacter* sp. Rue61a. *BMC Genomics* **13:**534. doi:10.1186/1471-2164-13-534.

65. **Niewerth H, Parschat K, Rauschenberg M, Ravoo BJ, Fetzner S.** 2013. The PaaX-type repressor MeqR2 of *Arthrobacter* sp. strain Rue61a, involved in the regulation of quinaldine catabolism, binds to its own promoter and to catabolic promoters and specifically responds to anthraniloyl coenzyme A. *J Bacteriol* **195:**1068–1080.

66. **Wu W, Leblanc SKD, Piktel J, Jensen SE, Roy KL.** 2006. Prediction and functional analysis of the replication origin of the linear plasmid pSCL2 in *Streptomyces clavuligerus*. *Can J Microbiol* **52:**293–300.

67. **Don RH, Pemberton JM.** 1981. Properties of six pesticide degradation plasmids isolated from *Alcaligenes paradoxus* and *Alcaligenes eutrophus*. *J Bacteriol* **145:** 681–686.

68. **Laemmli CM, Leveau JHJ, Zehnder AJB, van der Meer JR.** 2000. Characterization of a second *tfd* gene cluster for chlorophenol and chlorocatechol metabolism on plasmid pJP4 in *Ralstonia eutropha* JMP134 (pJP4). *J Bacteriol* **182:**4165–4172.

69. **Pérez-Pantoja D, Guzmán L, Manzano M, Pieper DH, González B.** 2000. Role of $tfdC_ID_IE_IF_I$ and $tfdD_{II}C_{II}E_{II}F_{II}$ gene modules in catabolism of 3-chlorobenzoate by *Ralstonia eutropha* JMP134 (pJP4). *Appl Environ Microbiol* **66:**1602–1608.

70. **Trefault N, De la Iglesia R, Molina AM, Manzano M, Ledger T, Perez-Pantoja D, Sánchez MA, Stuardo M, Gonzalez B.** 2004. Genetic organization of the catabolic plasmid pJP4 from *Ralstonia eutropha* JMP134 (pJP4) reveals mechanisms of adaptation to chloroaromatic pollutants and evolution of specialized chloroaromatic degradation pathways. *Environ Microbiol* **6:**655–668.

71. **Vedler E, Vahter M, Heinaru A.** 2004. The completely sequenced plasmid pEST4011 contains a novel IncP1 backbone and a catabolic transposon harboring *tfd* genes for 2,4-dichlorophenoxyacetic acid degradation. *J Bacteriol* **186:**7161–7174.

72. **Poh RP-C, Smith ARW, Bruce IJ.** 2002. Complete characterization of Tn*5530* from *Burkholderia cepacia* strain 2a (pIJB1) and studies of 2,4-dichlorophenoxyacetate uptake by the organism. *Plasmid* **48:**1–12.

73. **Tsoi TV, Plotnikova EG, Cole JR, Guerin WF, Bagdasarian M, Tiedje JM.** 1999. Cloning, expression and nucleotide sequence of the *Pseudomonas aeruginosa* 142 *ohb* genes coding for oxygenolytic ortho dehalogenation of halobenzoates. *Appl Environ Microbiol* **65:**2151–2162.

74. **Hickey WJ, Sabat G, Yuroff AS, Arment AR, Perez-Lesher J.** 2001. Cloning, nucleotide sequencing and functional analysis of a novel, mobile cluster of biodegradation genes from *Pseudomonas aeruginosa* strain JB2. *Appl Environ Microbiol* **67:**4603–4609.

75. **de Souza ML, Sadowsky MJ, Seffernick J, Martinez B, Wackett LP.** 1998. The atrazine catabolism genes are widespread and highly conserved. *J Bacteriol* **180:** 1951–1954.

76. **de Souza ML, Wackett LP, Sadowsky MJ.** 1998. The *atzABC* genes encoding atrazine catabolism are located on a self-transmissible plasmid in *Pseudomonas* sp. strain ADP. *Appl Environ Microbiol* **64:**2323–2326.

77. **Martinez B, Tomkins J, Wackett LP, Wing R, Sadowsky MJ.** 2001. Complete nucleotide sequence and organization of the atrazine catabolic plasmid pADP-1 from *Pseudomonas* sp. strain ADP. *J Bacteriol* **183:**5684–5697.

78. **Król JE, Penrod JT, McCaslin H, Rogers LM, Yano H, Stancik AD, Dejonghe W, Brown CJ, Parales RE, Wuertz S, Top EM.** 2011. Role of IncP-1ß plasmids pWDL7*::rfp* and pNB8c in chloroaniline catabolism as determined by genomic and functional analyses. *Appl Environ Microbiol* **78:**828–838.

79. **Wu JF, Sun CW, Jiang CY, Liu ZP, Liu SJ.** 2005. A novel 2-aminophenol 1,6-dioxygenase involved in the degradation of *p*-chloronitrobenzene by *Comamonas* strain CNB-1 purification, properties, genetic cloning and expression in *Escherichia coli*. *Arch Microbiol* **183:** 1–8.

80. **Wu JF, Jiang CY, Wang BJ, Ma YF, Liu ZP, Liu SJ.** 2006. Novel partial reductive pathway for 4-chloronitrobenzene and nitrobenzene degradation in *Comamonas* sp. strain CNB-1. *Appl Environ Microbiol* **72:**1759–1765.

81. **Liu L, Wu JF, Ma YF, Wang SY, Zhao GP, Liu SJ.** 2007. A novel deaminase involved in chloronitrobenzene and nitrobenzene degradation with *Comamonas* sp. strain CNB-1. *J Bacteriol* **189:**2677–2682.

82. **Ma YF, Wu JF, Wang SY, Jiang CY, Zhang Y, Qi SW, Liu L, Zhao GP, Liu SJ.** 2007. Nucleotide sequence of plasmid pCNB1 from *Comamonas* strain CNB-1 reveals novel genetic organization and evolution for 4-chloronitrobenzene degradation. *Appl Environ Microbiol* **73:**4477–4483.

83. **Nagata Y, Kamakura M, Endo R, Miyazaki R, Ohtsubo Y, Tsuda M.** 2006. Distribution of γ-hexachlorocyclohexane-degrading genes on three replicons in *Sphingobium japonicum* UT26. *FEMS Microbiol Lett* **256:**112–118.

84. **Miyazaki R, Sato Y, Ito M, Ohtsubo Y, Nagata Y, Tsuda M.** 2006. Complete nucleotide sequence of an exogenously isolated plasmid pLB1, involved in γ-hexachlorocyclohexane degradation. *Appl Environ Microbiol* **72:**6923–6933.

85. **Smalla K, Osborn AM, Wellington EMH.** 2000. Isolation and characterization of plasmids from bacteria, p 207–248. *In* Thomas CM (ed), *The Horizontal Gene Pool–Bacterial Plasmids and Gene Spread.* Harwood Academic Publishers, Amsterdam, The Netherlands.

86. **Junker F, Cook AM.** 1997. Conjugative plasmids and the degradation of arylsulfonates in *Comamonas testosteroni*. *Appl Environ Microbiol* **63:**2403–2410.

87. **Tralau T, Cook AM, Ruff J.** 2001. Map of the IncP1ß plasmid pTSA encoding the widespread genes (*tsa*) for *p*-toluenesulfonate degradation in *Comamonas testosteroni* T-2. *Appl Environ Microbiol* **67:**1508–1516.

88. **Kivisaar MA, Habicht JK, Heinaru AL.** 1989. Degradation of phenol and *m*-toluate in *Pseudomonas* sp. strain EST1001 and its *Pseudomonas putida* transconjugants is determined by a multiplasmid system. *J Bacteriol* **171:** 5111–5116.

89. **Kivisaar M, Hõrak R, Kasak L, Heinaru A, Habicht J.** 1990. Selection of independent plasmids determining phenol degradation in *Pseudomonas putida* and the cloning and expression of genes encoding phenol monooxygenase and catechol 1,2-dioxygenase. *Plasmid* **24:**25–36.

90. **Kallastu A, Hõrak R, Kivisaar M.** 1998. Identification and characterization of IS*1411*, a new insertion sequence which causes transcriptional activation of the phenol degradation genes in *Pseudomonas putida*. *J Bacteriol* **180:**5306–5312.

91. **Peters M, Heinaru E, Talpsep E, Wand H, Stottmeister U, Heinaru A, Nurk A.** 1997. Acquisition of a deliberately introduced phenol degradation operon, *pheBA*, by different indigenous *Pseudomonas* species. *Appl Environ Microbiol* **63:**4899–4906.
92. **Shingler V, Franklin FC, Tsuda M, Holroyd D, Bagdasarian M.** 1989. Molecular analysis of a plasmid-encoded phenol hydroxylase from *Pseudomonas* CF600. *J Gen Microbiol* **135:**1083–1092.
93. **Powlowski J, Shingler V.** 1994. Genetics and biochemistry of phenol degradation by *Pseudomonas* sp. CF600. *Biodegradation* **5:**219–236.
94. **van Beilen JB, Wubbolts MG, Witholt B.** 1994. Genetics of alkane oxidation by *Pseudomonas oleovorans. Biodegradation* **5:**161–174.
95. **van Beilen JB, Panke S, Lucchini S, Franchini AG, Röthlisberger M, Witholt B.** 2001. Analysis of *Pseudomonas putida* alkane degradation gene clusters and flanking insertion sequences: evolution and regulation of the *alk* genes. *Microbiology* **147:**1621–1630.
96. **Dinamarca MA, Aranda-Olmedo I, Puyet A, Rojo F.** 2003. Expression of the *Pseudomonas putida* OCT plasmid alkane degradation pathway is modulated by two different global control signals: evidence from continuous cultures. *J Bacteriol* **185:**4772–4778.
97. **Sikkema J, de Bont JA, Poolman B.** 1995. Mechanisms of membrane toxicity of hydrocarbons. *Microbiol Rev* **59:**201–222.
98. **Udaondo Z, Molina L, Daniels C, Gómez MJ, Molina-Henares MA, Matilla MA, Roca A, Fernández M, Duque E, Segura A, Ramos JL.** 2013. Metabolic potential of the organic-solvent tolerant *Pseudomonas putida* DOT-T1E deduced from its annotated genome. *Microb Biotechnol* **6:**598–611.
99. **Ramos JL, Duque E, Huertas MJ, Haïdour A.** 1995. Isolation and expansion of the catabolic potential of a *Pseudomonas putida* strain able to grow in the presence of high concentrations of aromatic hydrocarbons. *J Bacteriol* **177:**3911–3916.
100. **Mosqueda G, Ramos-Gonzalez MI, Ramos JL.** 1999. Toluene metabolism by the solvent tolerant *Pseudomonas putida* DOT-T1 strain, and its role in solvent impermeabilization. *Gene* **232:**69–76.
101. **Udaondo Z, Duque E, Fernández M,Molina L, de la Torre J, Bernal P, Niqui JL, Pini C, Roca A, Matilla MA, Molina-Henares MA, Silva-Jiménez H, Navarro-Avilés G, Busch A, Lacal J, Krell T, Segura A, Ramos JL.** 2012. Analysis of solvent tolerance in *Pseudomonas putida* DOT-T1E based on its genome sequence and a collection of mutants. *FEBS Lett* **586:**2932–2938.
102. **Rodríguez-Herva JJ, García V, Hurtado A, Segura A, Ramos JL.** 2007. The ttgGHI solvent efflux pump operon of *Pseudomonas putida* DOT-T1E is located on a large self-transmissible plasmid. *Environ Microbiol* **9:** 1550–1561.
103. **Segura A, Molina L, Fillet S, Krell T, Bernal P, Muñoz-Rojas J, Ramos JL.** 2012. Solvent tolerance in Gram-negative bacteria. *Curr Opin Biotechnol* **23:** 415–421.
104. **Rojas A, Duque E, Mosqueda G, Golden G, Hurtado A, Ramos JL, Segura A.** 2001. Three efflux pumps are required to provide efficient tolerance to toluene to toluene in *Pseudomonas putida* DOT-TIE. *J Bacteriol* **183:** 3967–3973.
105. **Blair JM, Piddock LJ.** 2009. Structure, function and inhibition of RND efflux pumps in Gram-negative bacteria: an update. *Curr Opin Microbiol* **5:**512–519.
106. **Hinchliffe P, Symmons MF, Hughes C, Koronakis V.** 2013. Structure and operation of bacterial tripartite pumps. *Annu Rev Microbiol* **67:**221–242.
107. **Murakami S, Nakashima R, Yamashita E, Yamaguchi A.** 2002. Crystal structure of bacterial multidrug efflux transporter AcrB. *Nature* **419:**587–593.
108. **Sennhauser G, Bukowska MA, Briand C, Grütter MG.** 2009. Crystal structure of the multidrug exporter MexB from *Pseudomonas aeruginosa. J Mol Biol* **389:** 134–145.
109. **Koronakis V, Sharff A, Koronakis E, Luisi B, Hughes C.** 2000. Crystal structure of the bacterial membrane protein TolC central to multidrug efflux and protein export. *Nature* **405:**914–919.
110. **Nikaido H, Zgurskaya HI.** 2001. AcrAB and related multidrug efflux pumps of *Escherichia coli. J Mol Microbiol Biotechnol* **3:**215–218.
111. **Symmons MF, Bokma E, Koronakis E, Hughes C, Koronakis V.** 2009. The assembled structure of a complete tripartite bacterial multidrug efflux pump. *Proc Natl Acad SciUSA* **106:**7173–7178.
112. **Ramos JL, Duque E, Godoy P, Segura A.** 1998. Efflux pumps involved in toluene tolerance in *Pseudomonas putida* DOT-T1E. *J Bacteriol* **180:**3323–3329.
113. **Mosqueda G, Ramos JL.** 2000. A set of genes encoding a second toluene efflux system in *Pseudomonas putida* DOT-T1E is linked to the *tod* genes for toluene metabolism. *J Bacteriol* **182:**937–943.
114. **Rojas A, Segura A, Guazzaroni ME, Terán W, Hurtado A, Gallegos MT, Ramos JL.** 2003. In vivo and in vitro evidence that TtgV is the specific regulator of the TtgGHI multidrug and solvent efflux pump of *Pseudomonas putida. J Bacteriol* **185:**4755–4763.
115. **Lacal J, Muñoz-Martínez F, Reyes-Darías JA, Duque E, Matilla M, Segura A, Calvo JJ, Jímenez-Sánchez C, Krell T, Ramos JL.** 2011. Bacterial chemotaxis towards aromatic hydrocarbons in *Pseudomonas. Environ Microbiol* **13:**1733–1744.
116. **Miyakoshi M, Shintani M, Terabayashi T, Kai S, Yamane H, Nojiri H.** 2007. Transcriptome analysis of *Pseudomonas putida* KT2440 harboring the completely sequenced IncP-7 plasmid pCAR1. *J Bacteriol* **189:** 6849–6860.
117. **Iwaki H, Muraki T, Ishihara S, Hasegawa Y, Rankin KN, Sulea T, Boyd J, Lau PC.** 2007. Characterization of a pseudomonad 2-nitrobenzoate nitroreductase and its catabolic pathway-associated 2-hydroxylaminobenzoate mutase and a chemoreceptor involved in 2-nitrobenzoate chemotaxis. *J Bacteriol* **189:**3502–3514.
118. **Yoakum GH, Cole RS.** 1977. Role of ATP in removal of psoralen cross-links from DNA of *Escherichia coli*

permeabilized by treatment with toluene. *J Biol Chem* 252:7023–7030.

119. **Rahmati S, Yang S, Davidson AL, Zechiedrich EL.** 2002. Control of the AcrAB multidrug efflux pump by quorum-sensing regulator SdiA. *Mol Microbiol* **43:** 677–685.

120. **Nikaido E, Yamaguchi A, Nishino K.** 2008. AcrAB multidrug efflux pump regulation in *Salmonella enterica* serovar Typhimurium by RamA in response to environmental signals. *J Biol Chem* **283:**24245–24253.

121. **Haines AS, Jones K, Batt SM, Kosheleva IA, Thomas CM.** 2007. Sequence of plasmid pBS228 and reconstruction of the IncP-1a phylogeny. *Plasmid* **58:** 76–83.

122. **Kvint K, Nachin L, Diez A, Nyström T.** 2003. The bacterial universal stress protein: function and regulation. *Curr Opin Microbiol* **6:**140–145.

123. **Nachin L, Nannmark U, Nyström T.** 2005. Differential roles of the universal stress proteins of *Escherichia coli* in oxidative stress resistance, adhesion, and motility. *J Bacteriol* **187:**6265–6272.

124. **Xiong J, Alexander DC, Ma JH, Déraspe M, Low DE, Jamieson FB, Roy PH.** 2013. Complete sequence of pOZ176, a 500-kilobase IncP-2 plasmid encoding IMP-9-mediated carbapenem resistance, from outbreak isolate *Pseudomonas aeruginosa* 96. *Antimicrob Agents Chemother* **57:**3775–3782.

125. **Kieboom J, Dennis JJ, de Bont JA, Zylstra GJ.** 1998. Identification and molecular characterization of an efflux pump involved in *Pseudomonas putida* S12 solvent tolerance. *J Biol Chem* **273:**85–91.

126. **Lacal J, Reyes-Darias JA, García-Fontana C, Ramos JL, Krell T.** 2013. Tactic responses to pollutants and their potential to increase biodegradation efficiency. *J Appl Microbiol* **114:**923–933.

127. **Kuc J.** 1995. Phytoalexins, stress metabolism, and disease resistance in plants. *Annu Rev Phytopathol* **33:** 275–297.

128. **Spaink HP.** 1995. The molecular basis of infection and nodulation by rhizobia: the ins and outs of sympathogenesis. *Annu Rev Phytopathol* **33:**345–368.

129. **Rao JR, Cooper JE.** 1994. Rhizobia catabolize nod gene-inducing flavonoids via C-ring fission mechanisms. *J Bacteriol* **176:**5409–5413.

130. **González-Pasayo R, Martínez-Romero E.** 2000. Multiresistance genes of *Rhizobium etli* CFN42. *Mol Plant Microbe Interact* **13:**572–577.

131. **Gonzalez V, Santamaria RI, Bustos P, Hernandez-Gonzalez I, Medrano-Soto A, Moreno-Hagelsieb G, Janga SCRamirez MA, Jimenez-Jacinto V, Collado-Vides J, Davila G.** 2006. The partitioned *Rhizobium etli* genome: genetic and metabolic redundancy in seven interacting replicons. *Proc Natl Acad Sci USA* **103:**3834–3839.

132. **Mann MS, Dragovic Z, Schirrmacher G, Lütke-Eversloh T.** 2012. Over-expression of stress protein-encoding genes helps *Clostridium acetobutylicum* to rapidly adapt to butanol stress. *Biotechnol Lett* **34:** 1643–1649.

133. **Streit WR, Schmitz RA, Perret X, Staehelin C, Deakin WJ, Raasch C, Liesegang H, Broughton WJ.** 2004. An evolutionary hot spot: the pNGR234b replicon of *Rhizobium* sp. strain NGR234. *J Bacteriol* **186:**535–542.

134. **Lim JS, Choi BS, Choi AY, Kim KD, Kim DI, Choi IY, Ka JO.** 2012. Complete genome sequence of the fenitrothion-degrading *Burkholderia* sp. strain YI23. *J Bacteriol* **194:**896. doi:10.1128/JB.06479-11.

135. **Janssen PJ, Van Houdt R, Moors H, Monsieurs P, Morin N, Michaux A, Benotmane MA, Leys N, Vallaeys T, Lapidus A, Monchy S, Médigue C, Taghavi S, McCorkle S, Dunn J, van der Lelie D, Mergeay M.** 2010. The complete genome sequence of *Cupriavidus metallidurans* strain CH34, a master survivalist in harsh and anthropogenic environments. *PLoS One* **5:**e10433. doi:10.1371/journal.pone.0010433.

136. **Ma Z, Jacobsen FE, Giedroc DP.** 2009. Metal transporters and metal sensors: How coordination chemistry controls bacterial metal homeostasis. *Chem Rev* **13:** 4644–4681.

137. **Adriano DC.** 2001. *Trace Elements in Terrestrial Environments: Biogeochemistry, Bioavailability, and Risks of Metals.* Springer-Verlag, New York, NY.

138. **Mergeay M, Nies D, Schlegel HG, Gerits J, Charles P, Van Gijsegem F.** 1985. *Alcaligenes eutrophus* CH34 is a facultative chemolithotroph with plasmid-bound resistance to heavy metals. *J Bacteriol* **162:**328–334.

139. **Monchy S, Benotmane MA, Janssen P, Vallaeys T, Taghavi S, van der Lelie D, Mergeay M.** 2007. Plasmids pMOL28 and pMOL30 of *Cupriavidus metallidurans* are specialized in the maximal viable response to heavy metals. *J Bacteriol* **189:**7417–7425.

140. **Nies DH.** 2003. Efflux-mediated heavy metal resistance in prokaryotes. *FEMS Microbiol Rev* **27:**313–339.

141. **Paulsen IT, Saier MH Jr.** 1997. A novel family of ubiquitous heavy metal ion transport proteins. *J Membr Biol* **156:**99–103.

142. **Munkelt D, Grass G, Nies DH.** 2004. The chromosomally encoded cation diffusion facilitator proteins DmeF and FieF from *Wautersia metallidurans* CH34 are transporters of broad metal specificity. *J Bacteriol* **186:** 8036–8043.

143. **Ramírez-Díaz MI, Díaz-Pérez C, Vargas E, Riveros-Rosas H, Campos-García J, Cervantes C.** 2008. Mechanisms of bacterial resistance to chromium compounds. *Biometals* **21:**321–332.

144. **Henne KL, Nakatsu CH, Thompson DK, Konopka AE.** 2009. High-level chromate resistance in *Arthrobacter* sp. strain FB24 requires previously uncharacterized accessory genes. *BMC Microbiol* **16:**199. doi:10.1186/1471-2180-9-199.

145. **Chen YF, Chao H, Zhou NY.** 2014. The catabolism of 2,4-xylenol and *p*-cresol share the enzymes for the oxidation of para-methyl group in *Pseudomonas putida* NCIMB 9866. *Appl Microbiol Biotechnol* **98:**1349–1356.

146. **Misra TK.** 1992. Bacterial resistances to inorganic mercury salts and organomercurials. *Plasmid* **27:**4–16.

147. **Diels L, Faelen M, Mergeay M, Nies D.** 1985. Mercury transposons from plasmids governing multiple resistance to heavy metals in *Alcaligenes eutrophus* CH34. *Arch Intern Physiol Biochem* **93:**27–28.
148. **Silver S, Phung LT.** 1996. Bacterial heavy metal resistance: new surprises. *Annu Rev Microbiol* **50:**753–789.
149. **Nascimento AM, Chartone-Souza E.** 2003. Operon *mer*: bacterial resistance to mercury and potential for bioremediation of contaminated environments. *Genet Mol Res* **2:**92–101.
150. **Boon N, Goris J, De Vos P, Verstraete W, Top EM.** 2001. Genetic diversity among 3-chloroaniline- and aniline-degrading strains of the *Comamonadaceae*. *Appl Environ Microbiol* **67:**1107–1115.
151. **Schwartz E, Henne A, Cramm R, Eitinger T, Friedrich B, Gottschalk G.** 2003. Complete nucleotide sequence of pHG1: a *Rastonia eutropha* H16 megaplasmid encoding key enzymes of H(2)-based lithoautotrophy and anaerobiosis. *J Mol Biol* **332:**369–383.
152. **Chen WM, Moulin L, Bontemps C, Vandamme P, Bena G, Boivin-Masson C.** 2003. Legume symbiotic nitrogen fixation by beta-proteobacteria is widespread in nature. *J Bacteriol* **185:**7266–7272.
153. **Roselli S, Nadalig T, Vuilleumier S, Bringel F.** 2013. The 380 Kbp pCMU01 plasmid encodes chloromethane utilization genes and redundant genes for vitamin B12- and tetrahydrofolate-dependent chloromethane metabolism in *Methylobacterium extorquens* CM4: A proteomic and bioinformatics study. *PLoS One* **8:** e56598. doi:10.1371/journal.pone.0056598.
154. **Liesegang H, Lemke K, Siddiqui RA, Schlegel HG.** 1993. Characterization of the inducible nickel and cobalt resistance determinant *cnr* from pMOL28 of *Alcaligenes eutrophus* CH34. *J Bacteriol* **175:**767–778.
155. **Schmidt T, Schlegel HG.** 1994. Combined nickel-cobalt-cadmium resistance encoded by the ncc locus of *Alcaligenes xylosoxidans* 31A. *J Bacteriol* **176:**7045–7054.
156. **Silver S, Gupta A, Matsui K, Lo JF.** 1999. Resistance to Ag(I) Cations in bacteria: environments, genes and proteins. *Met Based Drugs* **6:**315–320.
157. **Monchy S, Benotmane MA, Wattiez R, van Aelst S, Auquier V, Borremans B, Mergeay M, Taghavi S, van der Lelie D, Vallaeys T.** 2006. Transcriptomic and proteomic analyses of the pMOL30-encoded copper resistance in *Cupriavidus metallidurans* strain CH34. *Microbiology* **152:**1765–1776.
158. **Mealman TD, Blackburn NJ, McEvoy MM.** 2012. Metal export by CusCFBA, the periplasmic Cu(I)/Ag(I) transport system of *Escherichia coli*. *Curr Top Membr* **69:**163–196.
159. **Copley SD, Rokicki J, Turner P, Daligault H, Nolan M, Land M.** 2012. The whole genome sequence of *Sphingobium chlorophenolicum* L-1: insights into the evolution of the pentachlorophenol degradation pathway. *Genome Biol Evol* **4:**184–198.
160. **Tabata M, Ohtsubo Y, Ohhata S, Tsuda M, Nagata Y.** 2013. Complete genome sequence of the γ-hexachlorocyclohexane-degrading bacterium *Sphingomonas* sp. strain MM-1. *Genome Announc* **1:**pii e00247-13. doi:10.1128/genomeA.00247-13.
161. **Gómez-Sanz E, Kadlec K, Feßler AT, Zaragoza M, Torre C, Schwarz S.** 2013. Novel *erm*(T)-carrying multiresistance plasmids from porcine and human isolates of methicillin-resistant *Staphylococcus aureus* ST398 that also harbor cadmium and copper resistance determinants. *Antimicrob Agents Chemother* **57:**3275–3282.
162. **Bender CL, Cooksey DA.** 1987. Molecular cloning of copper resistance genes from *Pseudomonas syringae* pv. tomato. *J Bacteriol* **169:**470–474.
163. **Gutiérrez-Barranquero JA, de Vicente A, Carrión VJ, Sundin GW, Cazorla FM.** 2013. Recruitment and rearrangement of three different genetic determinants into a conjugative plasmid increase copper resistance in *Pseudomonas syringae*. *Appl Environ Microbiol* **79:** 1028–1033.
164. **Shin SH, Kim S, Kim JY, Lee S, Um Y, Oh MK, Kim YR, Lee J, Yang KS.** 2012. Complete genome sequence of the 2,3-butanediol-producing *Klebsiella pneumoniae* strain KCTC 2242. *J Bacteriol* **194:**2736–2737.
165. **Diels L, Mergeay M.** 1990. DNA probe-mediated detection of resistance bacteria from soils highly polluted by heavy metals. *Appl Environ Microbiol* **56:**1485–1491.
166. **Slyemi D, Bonnefoy V.** 2012. How prokaryotes deal with arsenic. *Environ Microbiol Reports* **4:**571–586.
167. **Dhuldhaj UP, Yadav IC, Singh S, Sharma NK.** 2013. Microbial interactions in the arsenic cycle: adoptive strategies and applications in environmental management. *Rev Environ Cont Toxicol* **224:**1–38.
168. **Volland S, Rachinger M, Strittmatter A, Daniel R, Gottschalk G, Meyer O.** 2011. Complete genome sequences of the chemolithoautotrophic *Oligotropha carboxidovorans* strains OM4 and OM5. *J Bacteriol* **193:**5043. doi:10.1128/JB.05619-11.
169. **Vogel TM.** 1996. Bioaugmentation as a soil bioremediation approach. *Curr Opin Biotechnol* **7:**311–316.
170. **Top EM, Springael D, Boon N.** 2002. Catabolic mobile genetic elements and their potential use in bioaugmentation of polluted soils and waters. *FEMS Microbiol Ecol* **42:**199–208.
171. **Ikuma K, Gunsch CK.** 2012. Genetic bioaugmentation as an effective method for in situ bioremediation. Functionality of catabolic plasmids following conjugal transfers. *Bioengineered* **3:**236–241.
172. **Jussila MM, Zhao J, Souminen L, Lindström K.** 2007. TOL plasmid transfer during bacterial conjugation *in vitro* and rhizoremediation of oil compounds *in vivo*. *Environ Pollut* **146:**510–524.
173. **Ramos-Gonzalez MI, Duque E, Ramos JL.** 1991. Conjugational transfer of recombinant DNA in cultures and in soils: host range of *Pseudomonas putida* TOL plasmids. *Appl Environ Microbiol* **57:**3020–3027.
174. **Molbak L, Licht TR, Kvist T, Kroer N, Andersen SR.** 2003. Plasmid transfer from *Pseudomonas putida* to the indigenous bacteria on alfalfa sprouts: characterization, direct quantification and in situ location of transconjugant cells. *Appl Environ Microbiol* **69:**5536–5542.

175. **Ikuma K, Holzem RM, Gunsch CK.** 2012. Impacts of organic carbon availability and recipient bacteria characteristics on the potential for TOL plasmid genetic bioaugmentation on soil slurries. *Chemosphere* **89:**158–163.

176. **Ikuma K, Gunsch CK.** 2013. Functionality of the TOL plasmid under varying environmental conditions following conjugal transfer. *Appl Microbiol Biotechnol* **97:** 395–408.

177. **Pinedo CA, Smets BF.** 2005. Conjugal TOL transfer from *Pseudomonas putida* to *Pseudomonas aeruginosa*: effects of restriction proficiency, toxicant exposure, cell density ratios, and conjugation detection method on observed transfer efficiencies. *Appl Environ Microbiol* **71:** 51–57.

178. **Daane LL, Molina J, Sadowsky MJ.** 1997. Plasmid transfer between spatially separated donor and recipient bacteria en earthworm-containing soil microcosms. *Appl Environ Microbiol* **63:**679–686.

179. **Wuertz S, Okabe S, Hausner M.** 2004. Microbial communities and their interactions in biofilm systems: an overview. *Water Sci Technol* **49:**327–336.

180. **Mohan SV, Falkentoft C, Nancharaiah YV, McSwain Sturm BS, Wattiau P, Wilderer PA, Wuertz S, Hausner M.** 2009. Bioaugmentation of microbial communities in laboratory and pilot scale sequencing batch biofilm reactors using the TOL plasmid. *Bioresour Technol* **100:**1746–1753.

181. **Neilson JW, Josephson KL, Pepper IL, Arnold RB, Di Giovanni GD, Sinclair NA.** 1994. Frequency of horizontal gene transfer of a large catabolic plasmid (pJP4) in soil. *Appl Environ Microbiol* **60:**4053–4058.

182. **Molina L, Ramos C, Duque E, Ronchel MC, Garcöla JM, Wyke L, Ramos JL.** 2000. Survival of Pseudomonas putida KT2440 in soil and in the rhizosphere of plants under greenhouse and environmental conditions. *Soil Biol Biochem* **32:**315–321.

183. **Mackova M, Dowling D, Macek T (ed).** 2006. *Phytoremediation and Rhizoremediation. Theoretical Background.* Series: Focus on Biotechnology. Springer, New York, NY.

184. **Walker TS, Bais HP, Grotewold E, Vivanco JM.** 2003. Root exudation and rhizosphere biology. *Plant Physiol* **132:**44–51.

185. **Matilla MA, Espinosa-Urgel M, Rodriguez-Hervá JJ, Ramos JL, Ramos-González MI.** 2007. Genomic analysis reveals the major driving forces of bacterial life in the rhizosphere. *Genome Biol* **8:**R179. doi:10.1186/gb-2007-8-9-r179.

186. **López-Guerrero MG, Ormeño-Orrillo E, Acosta JL, Mendoza-Vargas A, Rogel MA, Ramírez MA, Rosenblueth M, Martínez-Romero J, Martínez-Romero E.** 2012. Rhizobial extrachromosomal replicon variability, stability and expression in natural niches. *Plasmid* **68:**149–158.

187. **Barac T, Taghavi S, Borremans B, Provoost A, Oeyen L, Colpaert JV, Vangronsveld J, van der Lelie D.** 2004. Engineered endophytic bacteria improve phytoremediation of water-soluble, volatile, organic pollutants. *Nat Biotech* **22:**583–588.

188. **Taghavi S, Barac T, Greenberg B, Borremans B, Vangronsveld J, van der Lelie D.** 2005. Horizontal gene transfer to endogenous endophytic bacteria from poplar improves phytoremediation of toluene. *Appl Environ Microbiol* **71:**8500–8505.

189. **Lilley AK, Bailey MJ.** 1997. Impact of plasmid pQBR103 acquisition and carriage on the phytosphere fitness of *Pseudomonas fluorescens* SBW25: burden and benefit. *Appl Environ Microbiol* **63:**1584–1587.

190. **Tett A, Spiers AJ, Crossman LC, Ager D, Ciric L, Dow JM, Fry JC, Harris D, Lilley A, Oliver A, Parkhill J, Quail MA, Rainey PB, Saunders NJ, Seeger K, Snyder LA, Squares R, Thomas CM, Turner SL, Zhang XX, Field D, Bailey MJ.** 2007. Sequence-based analysis of pQBR103; a representative of a unique, transfer-proficient mega plasmid resident in the microbial community of sugar beet. *ISME J* **1:**331–340.

191. **Bale MJ, Fry JC, Day MJ.** 1987. Plasmid transfer between strains of *Pseudomonas aeruginosa* on membrane filters attached to river stones. *J Gen Microbiol* **133:** 3099–3107.

192. **Top EM, Holben WE, Forney LJ.** 1995. Characterization of diverse 2,4-dichlorophenoxyacetic acid-degradative plasmids isolated from soil by complementation. *Appl Environ Microbiol* **61:**1691–1698.

193. **Lilley AK, Bailey MJ.** 1997. The acquisition of indigenous plasmids by a generically marked pseudomonad population colonizing the sugar beet phytosphere is related to local environmental conditions. *Appl Environ Microbiol* **63:**1577–1583.

194. **Dahlberg C, Linberg C, Torsvik VL, Hermansson M.** 1997. Conjugative plasmids isolated from bacteria in marine environments show various degrees of homology to each other and are not closely related to well characterized plasmids. *Appl Environ Microbiol* **63:**4692–4697.

195. **Ono A, Miyazaki R, Sota M, Ohtsubo Y, Nagata Y, Tsuda M.** 2007. Isolation and characterization of naphthalene-catabolic genes and plasmids from oil-contaminated soil by using two cultivation-independent approaches. *Appl Microbiol Biotechnol* **74:**501–510.

196. **Kav AB, Sasson G, Jami E, Doron-Faigenboim A, Benhar I, Mizrahi I.** 2012. Insights into the bovine rumen plasmidome. *Proc Natl Acad Sci USA* **109:** 5452–5457.

197. **Nojiri H.** 2013. Impact of catabolic plasmids on host cell physiology. *Curr Opin Biotech* **24:**423–430.

198. **Diaz-Ricci JC, Hernández ME.** 2000. Plasmid effects on *Escherichia coli* metabolism. *Crit Rev Biotechnol* **20:** 79–108.

199. **Miyakoshi M, Shitani M, Inoue K, Terabayashi T, Sai F, Ohkuma M, Nojiri H, Nagata Y, Tsuda M.** 2012. ParI, an orphan ParA family protein from *Pseudomonas putida* KT2440-specific genomic island, interferes with the partition system of IncP-7 plasmids. *Environ Microbiol* **14:**2946–2959.

200. **Shintani M, Takahashi Y, Tokumaru H, Kadota K, Hara H, Miyakoshi M, Naito K, Yamane H, Nishida H, Nojiri H.** 2010. Response of the *Pseudomonas* host

chromosomal transcriptome to carriage of the IncP-7 plasmid pCAR1. *Environ Microbiol* **12:**1413–1426.

201. **Shintani M, Tokumaru H, Takahasi Y, Miyakoshi M, Yamane H, Nishida H, Nojiri H.** 2011. Alterations of RNA maps of IncP-7 plasmid pCAR1 in various *Pseudomonas* bacteria. *Plasmid* **66:**85–92.
202. **Fernández M, Niqui-Arroyo JL, Conde S, Ramos JL, Duque E.** 2012. Enhanced tolerance to naphthalene and enhanced rhizoremediation performance for *Pseudomonas putida* KT2440 via the NAH7 catabolic plasmid. *Appl Environ Microbiol* **78:**5104–5110.
203. **Yun CS, Suzuki C, Naito K, Takeda T, Takahashi Y, Sai F, Terabayashi T, Miyakoshi M, Shintani M, Nishida H, Yamane H, Nojiri H.** 2010. Pmr, a histone-like protein H1 (H-NS) family protein encoded by the IncP-7 plasmid pCAR1, is a key global regulator that alters host function. *J Bacteriol* **192:**4720–4731.
204. **Rescalli E, Saini S, Bartocci C, Rychlewski L, de Lorenzo V, Bertoni G.** 2004. Novel physiological modulation of the Pu promoter of TOL plasmid: negative regulatory role of the TurA protein of *Pseudomonas putida* in the response to suboptimal growth temperatures. *J Biol Chem* **279:**7777–7784.
205. **Bathe S, Mohan TV, Wuertz S, Hausner M.** 2004. Bioaugmentation of a sequencing batch biofilm reactor by horizontal gene transfer. *Water Sci Technol* **49:**337–344.
206. **De Gelder L, Vandecasteele FP, Brown CJ, Forney LJ, Top EM.** 2005. Plasmid donor affects host range of promiscuous IncP-1 beta plasmid pB10 in an activated-sludge microbial community. *Appl Environ Microbiol* **71:**5309–5317.
207. **Smalla K, Sobecky PA.** 2002. The prevalence and diversity of mobile genetic elements in bacterial communities of different environmental habitats: insights gained from different methodological approaches. *FEMS Microbiol Ecol* **42:**165–175.
208. **Ma H, Katzenmeyer KN, Bryers JD.** 2013. Non-invasive *in situ* monitoring and quantification of TOL plasmid segregational loss within *Pseudomonas putida* biofilms. *Biotech Bioeng* **110:**2949–2958.

Plasmids—Biology and Impact in Biotechnology and Discovery
Edited by Marcelo E. Tolmasky and Juan C. Alonso

doi:10.1128/microbiolspec.PLAS-0024-2014

Vicki Adams,[1] Jihong Li,[2] Jessica A. Wisniewski,[1] Francisco A. Uzal,[3]
Robert J. Moore,[1,4] Bruce A. McClane,[1,2] and Julian I. Rood[1]

27 Virulence Plasmids of Spore-Forming Bacteria

INTRODUCTION

Spore-forming bacteria cause some of the most significant diseases of both humans and animals, including tetanus, botulism, gas gangrene, anthrax, and many different enteric or gastroenteritis syndromes. The pathogenesis of most of these diseases involves the production of potent protein toxins, including tetanus and botulinum toxins, anthrax toxin, and alpha-toxin, epsilon-toxin (ETX), and enterotoxin (CPE) from *Clostridium perfringens*. The genes for many of these toxins, as well as other virulence factors such as the capsule biosynthesis genes of *Bacillus anthracis*, are located on plasmids, with examples including the tetanus toxin plasmid, the conjugative toxin plasmids of *C. perfringens*, and the pXO1 and pXO2 virulence plasmids from *B. anthracis* (1). In this chapter we will review our knowledge of these plasmids, the virulence factors that they encode, and their role in disease.

TOXIN PLASMIDS OF *C. PERFRINGENS*

C. perfringens causes many histotoxic and gastrointestinal diseases in both humans and domestic livestock, and strains are divided into five toxinotypes (A to E) based on their ability to produce one or more of four typing toxins: alpha-toxin, beta-toxin (CPB), ETX, and iota-toxin (ITX) (2). This classification system is based functionally on the presence or absence of toxin plasmids, specifically plasmids encoding CPB (*cpb* gene), ETX (*etx*), or ITX (*iap* and *ibp*) (3). Most of these plasmids are very closely related, sharing core plasmid replication, maintenance, and conjugation regions (Fig. 1) (4). These regions are also shared by a group of almost identical conjugative tetracycline resistance plasmids typified by pCW3 (5). Accordingly, the nature of these common regions will be described before the various toxin plasmids are discussed.

Replication and Maintenance of Toxin Plasmids of *C. perfringens*

Determination of the first complete sequences of *C. perfringens* toxin plasmids, those of the CPE (*cpe* gene) encoding plasmids pCPF5603 and pCPF4969, did not reveal any obvious plasmid replication genes (6). However, functional studies of the sequenced conjugative tetracycline resistance plasmid, pCW3, led to the

[1]Australian Research Council Centre of Excellence in Structural and Functional Microbial Genomics, Department of Microbiology, Monash University, Clayton, Vic 3800, Australia; [2]Department of Microbiology and Molecular Genetics, University of Pittsburgh School of Medicine, Pittsburgh, PA; [3]California Animal Health and Food Safety Laboratory, San Bernardino Branch, School of Veterinary Medicine, University of California—Davis, San Bernardino, CA; [4]CSIRO Biosecurity Flagship, Australian Animal Health Laboratory, Geelong 3220, Australia.

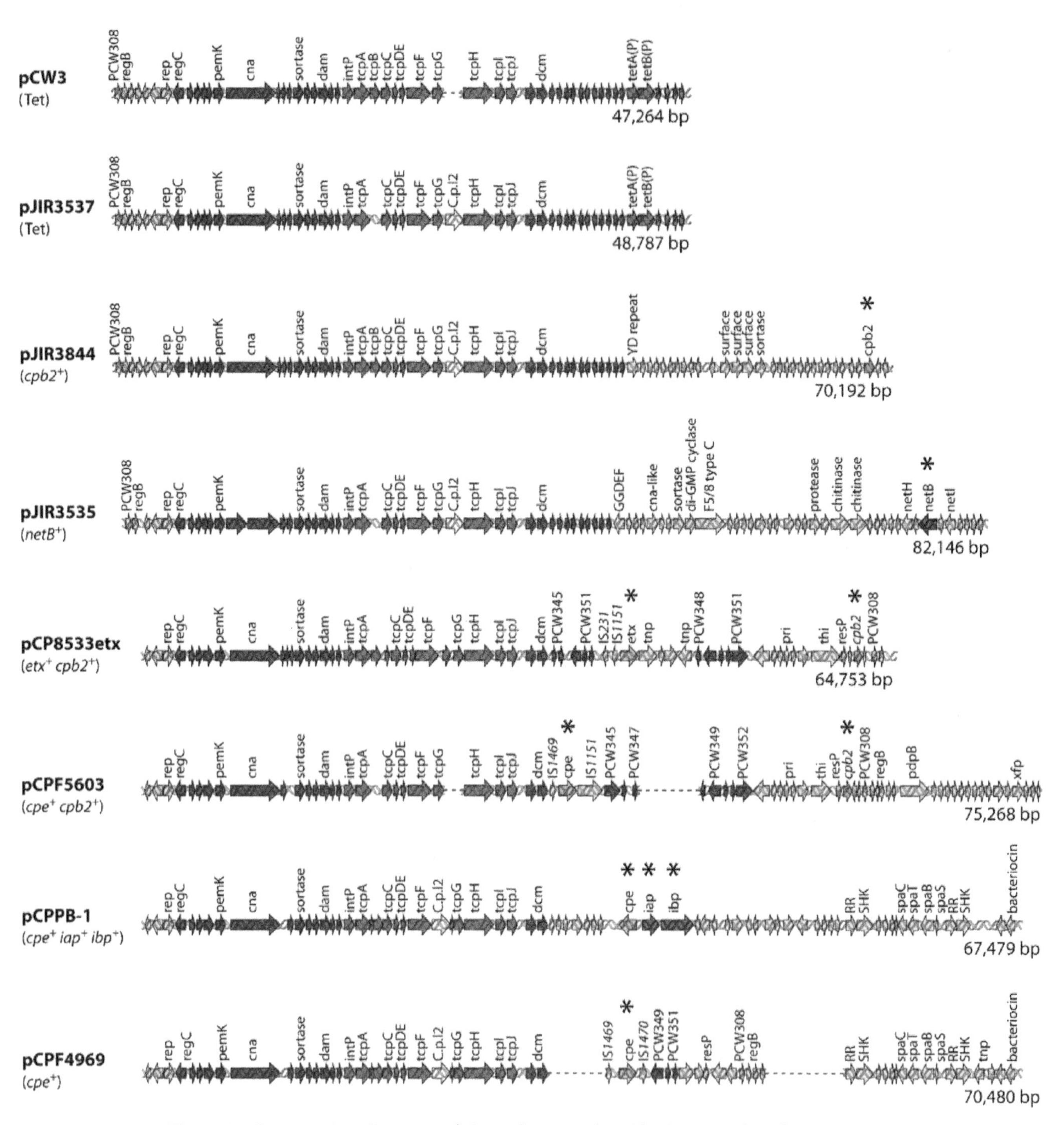

Figure 1 Comparative alignment of *C. perfringens* plasmids. Open reading frames (ORFs) are indicated by arrows as follows: red, the *tcp* locus; dark blue, other shared ORFs; light purple, tetracycline resistance genes; green, the *cpb2* toxin gene; purple, the *netB* toxin gene; pink, the *etx* gene; gray, the *cpe* gene; dark gray, the iota-toxin gene; yellow, plasmid replication region; light blue, regions unique to each plasmid. Asterisks denote a toxin gene. Reproduced with permission from reference 4.
doi:10.1128/microbiolspec.PLAS-0024-2014.f1

identification of the plasmid replication protein, Rep (5). The Rep protein is not related to any other plasmid replication proteins and, although disruption of the *rep* gene abolishes plasmid replication, the precise mode of action of Rep remains unclear (5). The *rep* gene subsequently has been shown to be conserved (at least 97% Rep amino acid sequence identity) in all plasmids related to pCW3, including pCPF5603 and pCPF4969, that have been sequenced to date (Fig. 1). Therefore, it is postulated that all members of this plasmid family replicate by the same Rep-dependent mechanism.

In all of these plasmids the gene immediately upstream of the *rep* gene has similarity to *parM* genes. The *parM* gene encodes an actin-like protein that, in conjunction with a downstream *parR* gene and an upstream *parC* centromere-like site, constitutes the plasmid partitioning system of several large, low-copy-number plasmids from Gram-negative bacteria (7). The ParMR system relies on the ability of the ParM protein to form long ATP-dependent filaments, the ends of which bind to ParR adaptor proteins that are in turn bound to the centromere-like *parC* region located upstream of the *parM* gene (8). Filament extension subsequently results in the separation of similar plasmid molecules with the same *parCMR* locus to opposite ends of a ParM filament, and therefore to opposite ends of the cell, prior to cell division, resulting in stable plasmid inheritance (8). Potential *parR* genes have been annotated on several of the sequenced toxin plasmids, and on the other plasmids an unannotated *parR* gene can be identified downstream of *parM*. These data suggest that a ParMRC-like partitioning system is utilized by pCW3, the toxin plasmids, and other members of this plasmid family (5, 9, 10).

The other factor relevant to plasmid maintenance is the ability of many spore-forming bacteria to carry multiple plasmids within the same cell. In *C. perfringens*, many of these multiplasmid strains contain several plasmids that are closely related and encode the same replication and conjugation functions (4). For these very similar plasmids to be stably inherited, distinct plasmid maintenance systems need to be employed. The detailed study of one such multiplasmid strain, EHE-NE18, has demonstrated that three very closely related plasmids are able to coexist in this strain and that they are capable of independent transfer (9). It was postulated that allele differences that can be identified between *parC* and the ParM and ParR proteins encoded by these different toxin and resistance plasmids determine plasmid incompatibility. Therefore, plasmids with different *parCMR* alleles can be stably maintained in the same cell; plasmids with the same alleles are predicted to be incompatible (9, 10).

Subsequent analysis of published sequence data has led to the assignment of ParM sequences into one of four distinct groups (10), which have now been designated as the ParMRC clades A to D (J. Rood, V. Adams, and J. Prescott, unpublished). A dendrogram of the ParM proteins from all available *C. perfringens* plasmid sequences clearly indicates the distinct $ParM_A$, $ParM_B$, $ParM_C$, and $ParM_D$ clades and leads to the identification of a fifth *C. perfringens*-derived group, designated here as the $ParM_E$ clade (Fig. 2). Recently, another clostridial toxin plasmid has been shown to utilize a ParMR-like system for plasmid maintenance, the pE88 plasmid from *Clostridium tetani*, which encodes the tetanus toxin structural gene (11). This ParM protein is only distantly related to the *C. perfringens* ParM proteins and serves to anchor the phylogenetic tree (Fig. 2). There is also a ParM homologue present in *Clostridium botulinum* Eklund strain 17B, which contains a plasmid-encoded neurotoxin locus (Fig. 3). This putative ParM protein has a much closer relationship to the $ParM_B$ family of *C. perfringens* ParM proteins (Fig. 2). Overall, the data suggest that ParMR systems represent a highly conserved mechanism of plasmid maintenance, not just in the clostridia, but in many other bacterial genera (12).

Mechanism of Conjugative Transfer of Toxin Plasmids of *C. perfringens*

Conjugative transfer in *C. perfringens* was recognized first in experiments that involved the transfer of closely related tetracycline resistance plasmids (13, 14). Subsequently, the first toxin plasmid shown to transfer independently was a marked clinically derived CPE plasmid, pMRS4969 (15). Transfer of pMRS4969 resulted in transconjugants that were able to act as plasmid donors, highly suggestive that transfer was mediated by a conjugation-like mechanism encoded on the plasmid. Conjugative transfer has now been demonstrated for several toxin plasmids, including plasmids encoding ETX, NetB toxin, and CPB2, as well as a lincomycin resistance plasmid (9, 16, 17).

Early analysis of the conjugative plasmids from *C. perfringens* suggested that they shared large regions of similarity (15, 18, 19). Ten toxin and resistance plasmids from *C. perfringens* have been completely sequenced, including known conjugative plasmids; the data confirmed that these plasmids share a highly conserved 35-kb region that includes the transfer of the clostridial plasmid (*tcp*) conjugation locus (Fig. 1) (4–6, 9, 10, 20, 21).

The *tcp* locus has been shown to mediate conjugative transfer of the 47-kb tetracycline resistance plas-

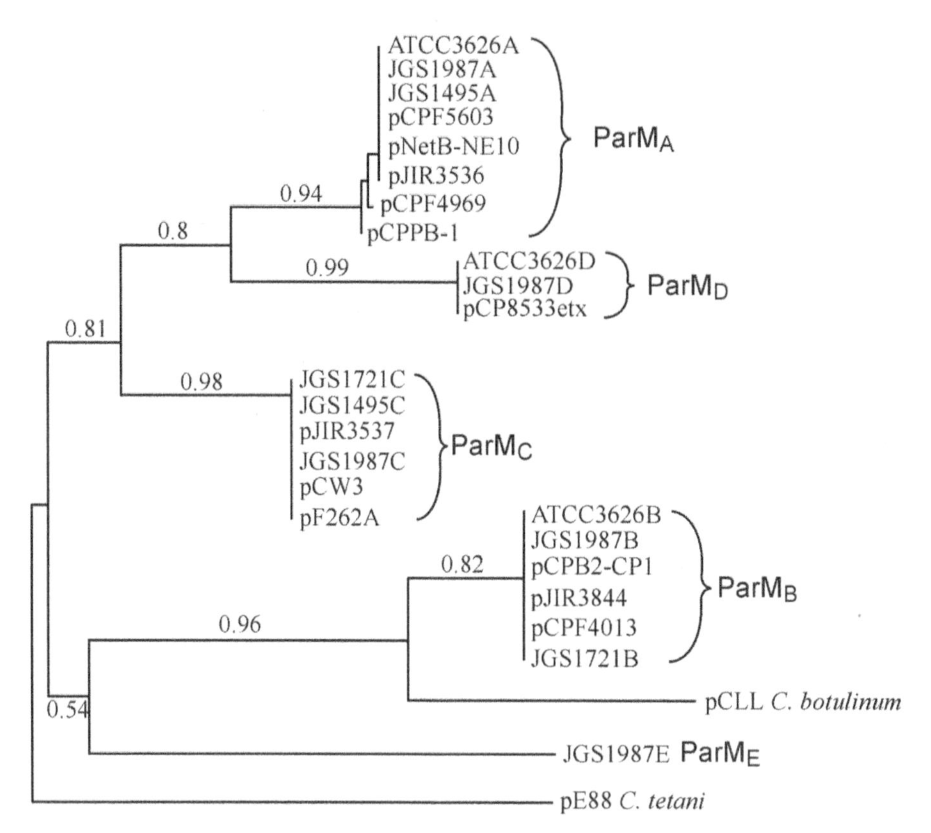

Figure 2 Phylogenetic analysis of ParM variants. The phylogenetic tree was constructed using the amino acid sequences of ParM proteins identified using BlastP searches of the nonredundant NCBI protein database. The phylogenetic tree was constructed using the phylogeny analysis software: http://www.phylogeny.fr/version2_cgi/index.cgi (171, 172). The JGS1495, JGS1987, and ATCC3626 sequences are from genome sequencing projects and yielded multiple ParM homologues from putative plasmid sequences; each ParM homologue was named according to its ParM group. doi:10.1128/microbiolspec.PLAS-0024-2014.f2

mid pCW3 (5). The locus originally was identified based on low levels of amino acid identity between several of the encoded Tcp proteins and conjugation proteins from the conjugative transposon Tn*916*. Significant genetic rearrangements between the conjugation regions from Tn*916* and pCW3, as well as limited primary amino acid sequence similarity to Gram-negative systems, suggested that the mechanism of conjugation in *C. perfringens* was unique. Since this locus is conserved in most of the toxin plasmids, is present in all known conjugative *C. perfringens* plasmids, and appears to be responsible for the conjugative transfer of toxin genes, it is important to review what is currently known about the function of this conjugation system.

The *tcp* locus comprises an 11-gene operon (*intP*, *tcpA-tcpJ*), and mutagenesis studies have shown that several of the Tcp proteins encoded by this locus are involved in pCW3 transfer (5, 22–24). The genes in the *tcp* locus are conserved in the toxin plasmids, except that in some plasmids the nonessential *tcpB* gene is missing and/or there may be insertions of small hypothetical open reading frames (ORFs) and group II introns (5, 6, 9, 16, 17). A model for the *C. perfringens* conjugation apparatus has been proposed based on functional studies of the Tcp proteins involved in the conjugative transfer of pCW3 (4, 25). Analysis of these Tcp proteins has identified some functional similarity to protein families that form the Gram-negative inner membrane secretion channel (26).

Coupling proteins are an essential component of conjugation systems and are involved in substrate recognition and translocation (27). Two proteins encoded on the *tcp* locus, TcpA and TcpB, were identified as possessing FtsK-like domains belonging to the FtsK superfamily of DNA translocases (23). TcpA was shown to be essential for transfer, while TcpB was not required, a result consistent with the absence of the *tcpB* gene in some of the toxin plasmids. Bioinformatic analysis of the TcpA protein identified two N-terminal transmembrane domains (TMDs) and three ATP-binding and hydrolysis motifs in the C-terminal cytoplasmic domain. Supporting the hypothesis that TcpA is an ATPase-dependent DNA translocase, the Walker A, Walker B, and RAAG ATP-binding motifs were shown to be essential for TcpA function (23). Similar to other coupling proteins, the TMDs of TcpA were shown to be required for wild-type TcpA function and to be involved in TcpA homo-oligomerization as well as protein-protein interactions between TcpA and TcpC, and TcpH and TcpG (25). Deletion of 61 amino acids from the C-terminus of TcpA reduced TcpA-TcpC interactions and transfer efficiency, providing evidence that this interaction was important for the transfer process (25). Although TcpA was not identified as a classical coupling protein, these studies provide convincing evidence that TcpA is the DNA translocase of the *C. perfringens* conjugation system.

One of the first proteins shown to be essential for conjugation was TcpH, with transfer not detected in *tcpH* mutants (5). Predicted to be an integral membrane protein, this 832-amino acid protein contains 8 TMDs and a large cytoplasmic domain. TcpH has a VirB6 domain between TMD5-TMD8 and was therefore proposed to have a similar role as the VirB6 protein, which forms part of the central channel in the inner membrane of the Gram-negative conjugation apparatus (28). In agreement with a proposed role for TcpH as the major component of the conjugation apparatus in the donor cell wall, TcpH was shown to localize to the cell membrane fraction at the poles of *C. perfringens* cells (29). The first 581 amino acids of TcpH, which include the TMDs and VirB6 domain, comprised the minimal TcpH derivative required to complement a *tcpH* mutant. This region was also essential for TcpH homo-oligomerization and interactions with TcpA and TcpC (25, 29). TcpH interactions were not dependent on the VirB6 domain or the conserved $_{242}$VQQPW$_{246}$ motif, also identified as being essential for TcpH function (29). The unique C-terminal cytoplasmic domain was required for wild-type TcpH function but was not involved in TcpH protein-protein interactions. This cytoplasmic domain is not present in other VirB6-like conjugation proteins, demonstrating the functional complexity of TcpH and the unique nature of the *C. perfringens* conjugation system.

TcpC was originally identified as a bitopic membrane protein with 24% amino acid identity to ORF13 from Tn*916* (5). A very significant reduction in transfer efficiency was observed in a *tcpC* mutant, indicating that TcpC plays a key role in conjugative transfer (24). Determination of the crystal structure of the soluble TcpC$_{99-359}$ derivative to 1.8 Å identified two domains that each unexpectedly shared a similar fold to members of the nuclear transport factor-2 superfamily, which includes VirB8 proteins that form part of the transfer apparatus at the inner membrane in Gram-negative conjugation systems. VirB8-like proteins act as scaffolding proteins promoting transfer complex assembly and stabilization; TcpC is therefore proposed to have a similar role in *C. perfringens*. Consistent with this role, TcpC was shown to localize independently to the cell wall of *C. perfringens* cells as well as interacting with itself, TcpA, TcpH, and TcpG, with all of these interactions requiring the presence of the essential N-terminal TMD. TcpC forms a trimeric structure with the internal central domain shown to be involved in TcpC self-interaction (24). The essential C-terminal domain forms the major exterior surface of the trimer and is required for TcpC interactions with TcpA, TcpH, and TcpG. It is through these protein-protein interactions that TcpC is postulated to direct the formation of the transfer apparatus.

TcpF was identified as a putative ATPase with distant similarity to the VirB4 ATPase of Gram-negative conjugation proteins (29). Like ATPases from other conjugation systems, TcpF was shown to be essential for conjugative transfer (5). Although TcpF was demonstrated to colocalize with TcpH at the poles of *C. perfringens* donor cells, it remains unclear how TcpF interacts with the conjugation apparatus, with no protein-protein interactions identified between TcpF and the other Tcp proteins (29).

Peptidoglycan hydrolases are another family of proteins commonly present in conjugation systems, playing a role in the assembly of the conjugation apparatus in the cell wall (22). Two genes encoding potential peptidoglycan hydrolases were identified in the *tcp* locus, *tcpG* and *tcpI* (5). Genetic studies showed that the *tcpI* gene product was not required for conjugative transfer (22). Mutagenesis of the *tcpG* gene resulted in a reduced transfer frequency, consistent with results for peptidoglycan hydrolases from other conjugation systems. The role of TcpG as a peptidoglycan hydrolase was confirmed by the demonstration that purified TcpG had hydrolase-like activity on cognate *C. perfringens* peptidoglycan (22).

In the model of the *C. perfringens* conjugation apparatus (4, 25) the integral membrane protein TcpH, with the atypical VirB8-like protein TcpC acting as a scaffold, forms the core channel of the conjugation apparatus. As well as stabilizing the complex, TcpC is proposed to be involved in recruiting TcpA, TcpH, and

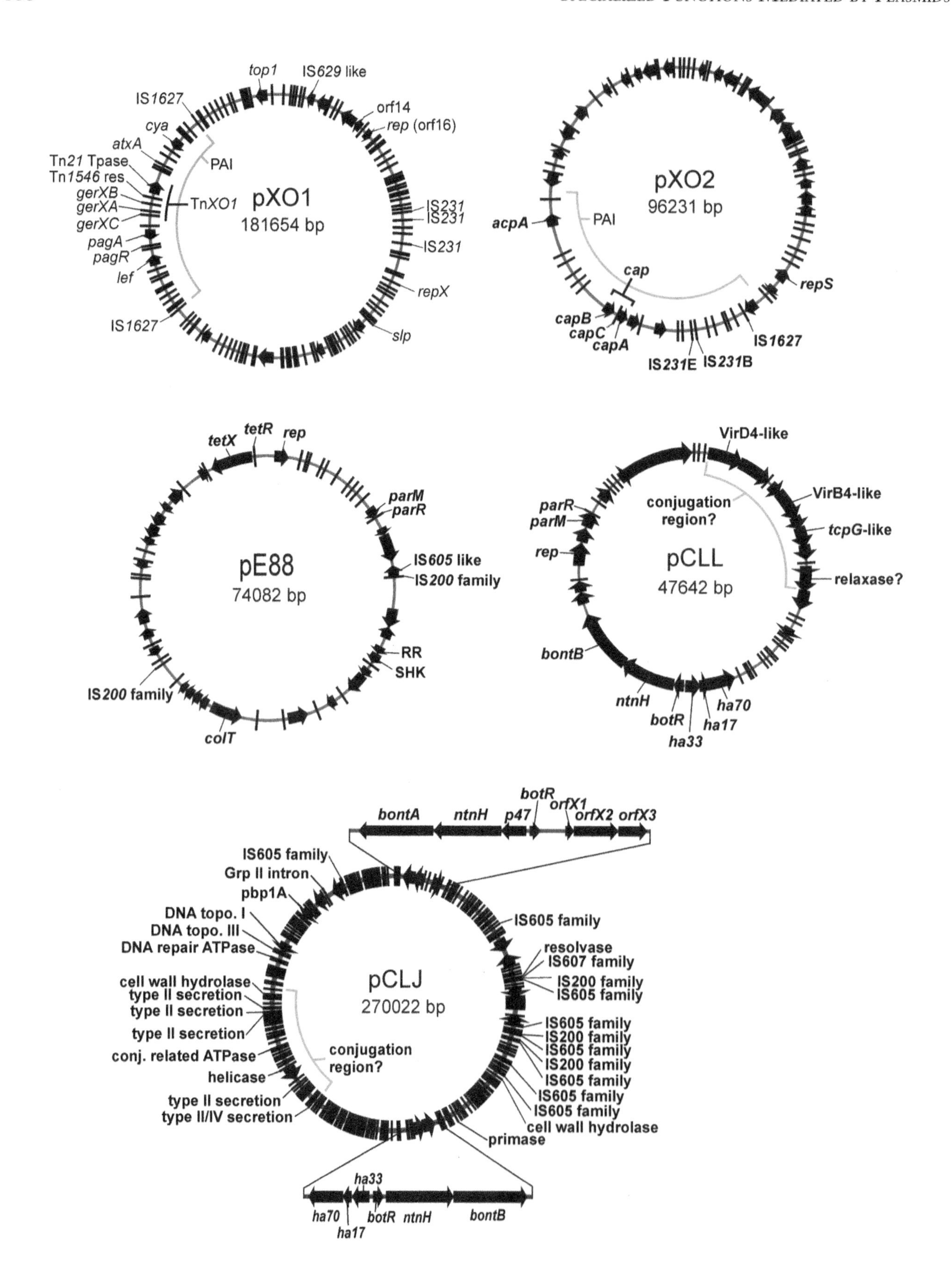
pXO1
181654 bp
top1
IS629 like
IS1627
orf14
rep (orf16)
cya
atxA
Tn21 Tpase
Tn1546 res
gerXB
gerXA
gerXC
pagA
pagR
lef
PAI
TnXO1
IS231
IS231
IS231
repX
slp
IS1627
pXO2
96231 bp
acpA
PAI
cap
repS
capB
capC
capA
IS231E
IS231B
IS1627
pE88
74082 bp
tetX
tetR
rep
parM
parR
IS605 like
IS200 family
RR
SHK
IS200 family
colT
pCLL
47642 bp
VirD4-like
VirB4-like
tcpG-like
relaxase?
conjugation
region?
parR
parM
rep
bontB
ntnH
botR
ha33
ha17
ha70
pCLJ
270022 bp
bontA
ntnH
p47
botR
orfX1
orfX2
orfX3
IS605 family
Grp II intron
pbp1A
DNA topo. I
DNA topo. III
DNA repair ATPase
cell wall hydrolase
type II secretion
type II secretion
type II secretion
conj. related ATPase
helicase
type II secretion
type II/IV secretion
conjugation
region?
IS605 family
resolvase
IS607 family
IS200 family
IS605 family
IS605 family
IS200 family
IS605 family
IS200 family
IS605 family
IS605 family
IS605 family
cell wall hydrolase
primase
ha70
ha17
ha33
botR
ntnH
bontB

TcpG (24). The putative motor ATPases TcpA and TcpF are predicted to energize the system, although it is currently unknown how TcpF interacts with the other components (25, 29). Similarly, the early steps in plasmid transfer are not clear since neither a classical relaxase protein gene nor an origin of transfer (*oriT*) have been identified on these plasmids (5). The relaxase protein and *oriT* site are usually essential components that are involved in plasmid recognition and nicking prior to transfer (30). The first gene in the *tcp* locus, *intP*, which is conserved in the toxin plasmids, encodes a putative tyrosine recombinase that is postulated to be a potentially novel relaxase (5). Genetic and functional studies are required to determine whether IntP, as well as the remaining highly conserved proteins TcpD, TcpE, and TcpJ, are involved in *C. perfringens* conjugative transfer. These studies will further define our understanding of the unique conjugation process in *C. perfringens*, a process that is central to the transfer of both antibiotic resistance and toxin genes.

Enterotoxin (CPE) Plasmids of *C. perfringens*

C. perfringens type A strains that produce *C. perfringens* CPE are causative agents of human food poisoning and non-food-borne diarrhea (31). CPE is encoded by the *cpe* gene and is only produced during sporulation, where its synthesis is controlled by Spo0A and the alternative sigma factors SigF, SigE, and SigK (32–34). The toxin then accumulates intracellularly until it is released when the mother cell lyses to free the mature endospore at the completion of sporulation. CPE then binds to claudin receptors present in the gastrointestinal tract, which induces toxin oligomerization and pore formation in enterocytes. The resultant Ca^{2+} influx causes cytotoxicity that leads to villus blunting, necrosis, and epithelial desquamation in all sections of the small intestines, along with luminal fluid accumulation (31).

CPE, a 35-kDa polypeptide, has a highly conserved amino acid sequence in all CPE-producing *C. perfringens* strains, with the exception of type E strains. Structurally, this toxin belongs to the aerolysin family of pore-forming toxins, which also includes ETX (35, 36). The CPE protein consists of a C-terminal domain that confers binding to claudin receptors and an N-terminal domain that mediates both the oligomerization of toxin monomers and membrane insertion during pore formation (37–39).

While some *C. perfringens* type A, C, D, and E isolates produce CPE, no CPE-producing type B strains have been identified (20, 21, 40–44). In the 1 to 5% of global type A strains that are *cpe*-positive (45), the *cpe* gene is located either on the chromosome or on large plasmids that range in size from ~50 to ~75 kb (6). Virulence studies testing molecular Koch's postulates have confirmed the importance of CPE expression for the gastrointestinal virulence of both chromosomal and plasmid *cpe* type A strains (46).

Type A strains with a chromosomal *cpe* gene are estimated to cause ~75 to 80% of all *C. perfringens* type A food poisoning cases (31). This illness is now considered the second most common bacterial food-borne disease in the United States, where it involves one million cases per year (31, 47). The remaining cases of *C. perfringens* type A food poisoning are caused by strains carrying a *cpe* plasmid. Type A strains with a *cpe* plasmid are also responsible for 5 to 10% of all cases of human non-food-borne gastrointestinal diseases, which include both antibiotic-associated and sporadic diarrheas (31, 48). These diseases occur more frequently in the elderly and are more severe and long-lasting than typical type A food poisoning (49). Type A strains carrying a *cpe* plasmid also reportedly cause enteric disease in domestic animals, particularly dogs (50).

Based upon results from sequencing, pulse-field Southern blot analyses, and overlapping PCR assays, it has been shown that there are two major families of *cpe* plasmids in type A strains (4). The pCPF5603 *cpe* plasmid family includes plasmids that are usually ~75 kb in size and carry genes encoding both CPE and CPB2 (encoded

Figure 3 Plasmid maps of toxin plasmids from spore-forming bacteria. The circular maps of the *B. anthracis* toxin plasmids, pXO1 (AF065404) and pXO2 (AF188935), the *C. tetani* neurotoxin plasmid pE88 (AF528077), and the conjugative group I (pCLJ, CP001081) and group II (pCLL, CP001057) *C. botulinum* neurotoxin plasmids are shown. Predicted ORFs are depicted as black arrows or bars along the circular maps. Regions of interest are indicated inside the plasmid circles, such as the pathogenicity islands present on pXO1 and pXO2. Genes of interest are indicated on the outside of the plasmid maps. Gene names are italicized, while ORFs with similarity to known proteins (such as IS elements) are not italicized. The two neurotoxin loci encoded on the pCLJ plasmid are enlarged showing an example of an *orfX* neurotoxin locus (*bontA* region, top) and a *ha* neurotoxin locus (*bontB* region, bottom); see text. doi:10.1128/microbiolspec.PLAS-0024-2014.f3

by the *cpb2* gene) (6) (Fig. 1). In contrast, the ~70-kb pCPF4969 *cpe* plasmid family has the *cpe* gene, but not the *cpb2* gene. Both *cpe* plasmid families share an ~35-kb conserved region (6) that contains the *tcp* region, which mediates the conjugative transfer of several *C. perfringens* plasmids (5). The presence of this *tcp* region explains the conjugative transfer of a tagged F4969 plasmid derivative (15). The variable region of pCPF5603-like plasmids contains a cluster of metabolic genes in addition to a functional *cpb2* gene. In contrast, the pCPF4969 variable region includes two putative bacteriocin genes and genes encoding a putative two-component regulator with similarity to VirS and VirR from *C. perfringens* (6).

Whether chromosomal or plasmid-borne, the *cpe* gene is closely associated with insertion sequences (IS). However, there is considerable variation among type A strains regarding the specific IS elements present in the *cpe* locus, as well as their arrangement relative to the *cpe* gene (Fig. 4). In the pCPF5603-like plasmids, the *cpe* gene is bounded by an upstream IS*1469* sequence and a downstream IS*1151* sequence, while in pCPF4969-like plasmids; the *cpe* gene is upstream of an IS*1470*-like sequence rather than IS*1151* (6). By comparison, the chromosomal *cpe* locus has an IS*1469* immediately upstream of the *cpe* gene (51). This chromosomal IS*1469-cpe* region is then flanked by IS*1470* sequences, suggesting it could be part of an integrated transposon (51).

C. perfringens type C Darmbrand strains produce CPE, as well as CPB, and are genetically related to type A food poisoning strains carrying a chromosomal *cpe* gene (52). However, these type C strains carry a plasmid-borne *cpe* gene, as well as a plasmid-borne *cpb* gene located on ~75- or ~85-kb plasmids. In other Darmbrand strains the *cpe* gene is present on an ~110-kb plasmid that does not carry the *cpb* gene. The *cpe* plasmids in all studied CPE-positive type C strains have the *tcp* region, suggesting that they are conjugative. Interestingly, the plasmid *cpe* locus in type C strains is similar to the chromosomal *cpe* locus of type A isolates, except that an IS*1469* sequence is located upstream of IS*1470*, while an IS*1151*-like sequence is located downstream of the *cpe* gene (52) (Fig. 4). Experiments have indicated that IS mobile genetic elements can excise the *cpe* gene as circular DNA molecules that could be transposition intermediates (52). Those results suggest, but do not yet prove, that the chromosomal *cpe* locus in type A strains may be derived from insertion of a *cpe*-carrying transposon originating from a *cpe* plasmid

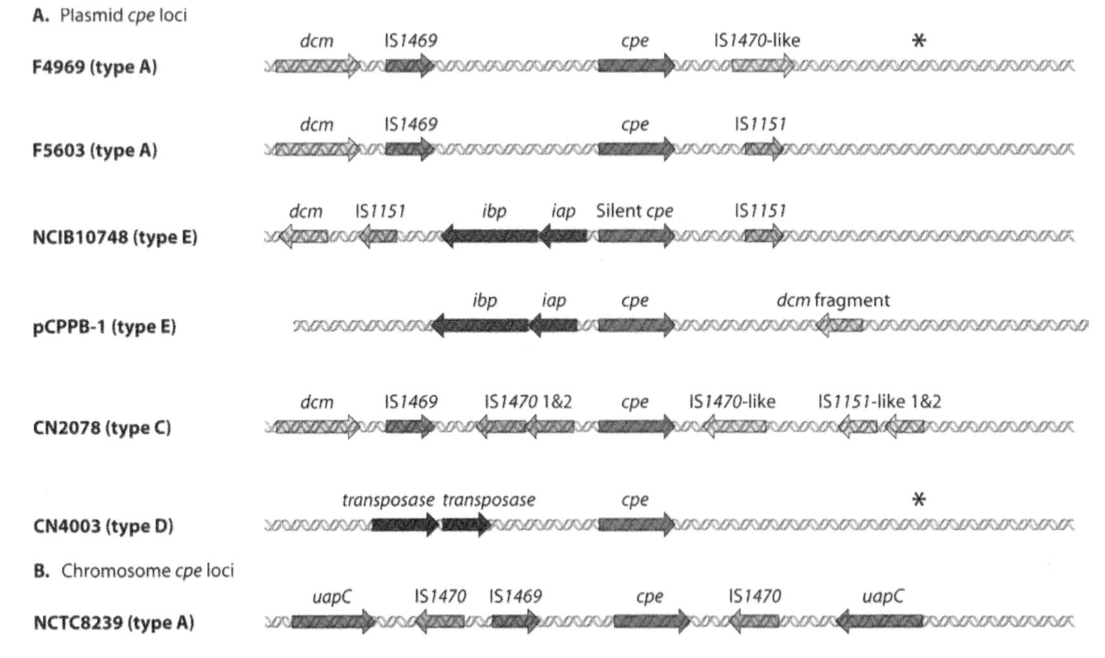

Figure 4 Genetic organization of the *cpe* gene regions. Organization of plasmid-borne (**A**) and (**B**) chromosomal *cpe* loci from *C. perfringens*. *cpe* genes are indicated by red arrows, *iap* and *ibp* genes by purple arrows, and *dcm* genes by yellow arrows. Related IS elements are indicated by identical colors. Reproduced with permission from reference 4. doi:10.1128/microbiolspec.PLAS-0024-2014.f4

similarto the *cpe* plasmid found in type C Darmbrand strains.

Many type D isolates also carry a *cpe* gene on large plasmids that range in size from 75 to 110 kb (40). In some of these strains, the *cpe* gene is located on the same plasmid as the *etx* gene, but other CPE-positive type D strains carry separate *cpe* and *etx* plasmids (40). Like *cpe* plasmids in type A and C strains, most *cpe* plasmids of type D isolates have a *tcp* region, suggesting they are conjugative (53). However, in *cpe*-positive type D strains, the *cpe* locus has a unique genetic organization differing from the *cpe* locus of *cpe*-positive type A or type C strains (53). While the region downstream of the *cpe* gene in type D strains is similar to the sequences downstream of the pCPF4969 *cpe* gene, except for the absence of an IS*1470*-like sequence, two copies of an ORF sharing 67% identity with a Tn*1456*-like transposase gene are located upstream of the *cpe* gene in the type D strains that have been studied (Fig. 4). All type E isolates studied to date were shown to carry *cpe* sequences on large plasmids that also encode ITX (21, 54), which will be discussed in the section describing the ITX plasmids.

Beta Toxin (CPB) Plasmids of *C. perfringens*

CPB is encoded on large plasmids in both toxinotype B and C isolates (41, 52, 55). The CPB-encoding toxinotypes are responsible for a range of highly significant and often fatal animal diseases. In addition, type C isolates are the only non-type A strains known to cause disease in humans.

Type C-mediated human disease affects undernourished individuals existing on a protein-deficient diet and is often accentuated by the consumption of foods rich in trypsin inhibitors (56). It is known as Darmbrand in Germany (52) and pig bel in Papua New Guinea (56). Symptoms include severe upper abdominal pain, bloody diarrhea, and vomiting (57). As the disease progresses, diarrhea ceases and the bowel becomes obstructed due to damage in the small intestine (56, 57). Enterotoxemia and necrotizing enteritis of livestock are also caused by CPB-producing strains of *C. perfringens* (58). Type C disease occurs mostly in neonates of several animal species. This age predisposition is believed to be related to the presence of colostrum, which inhibits trypsin activity and in turn allows CPB to remain active in the intestinal lumen. Animal diseases attributed to type B isolates resemble both type C- and/or type D-mediated enteric and/or neurologic syndromes (58). Like type C diseases of livestock, type B infections can involve necrotic enteritis or sudden death, but animals with type B infections may display neurological signs similar to type D disease (59).

Mutational studies combined with the use of a rabbit small intestinal loop model have shown that CPB is essential for the intestinal pathology observed in type C-mediated disease (60). In addition, these type C mutants were used to identify a key role for CPB in lethal enterotoxemia in a mouse model (61), which was confirmed in a large animal model (goat) more recently (62). When cultures of a wild-type strain, a *cpb* null mutant, and a genetically reversed mutant, were introduced into the goat small intestine, disease progression for the wild type and the complemented mutant occurred in a manner indistinguishable from the natural disease. By contrast, goats inoculated with the *cpb* mutant showed no sign of disease (62). These studies (60, 62) provide compelling evidence that plasmid-encoded CPB is the most important toxin in type C-mediated disease.

The pathology of type B infection is complicated, as type B isolates typically express the largest number of different lethal toxins: alpha-toxin, CPB, and ETX (41, 58, 59). The only virulence study of type B isolates to date (59) correlated toxin levels in culture supernatants with mouse intravenous lethality. The results indicated that (i) CPB is a major contributor to lethality, (ii) ETX is also implicated, and (iii) other toxins may also be involved. The complexity of disease progression when caused by type B isolates is likely dependent on a number of factors including the protease status of the host and the toxin expression profile of the infecting isolate (59). Such studies are complicated by the fact that while CPB is highly sensitive to trypsin, ETX requires trypsin for activation.

CPB is a 35-kDa pore-forming toxin with about 22 to 28% amino acid sequence similarity to several pore-forming toxins of *Staphylococcus aureus* (63), 38% amino acid sequence identity to *C. perfringens* NetB toxin (64), and 43% amino acid sequence identity with delta-toxin of *C. perfringens* (65). CPB forms an oligomer after binding to an unidentified receptor on sensitive cells, which results in pore formation that produces an ~12-Å channel that is selective for monovalent cations (66). This toxin is very sensitive to degradation by trypsin or chymotrypsin (67), and intestinal trypsin is therefore the main natural defense against type C infections.

All type B and C isolates produce CPB toxin during late log-phase growth. Beta toxin production by type C strains is controlled by the chromosomal two-component VirS/VirR regulatory system (68). CPB production is also regulated by a chromosomal Agr-like quorum sensing (QS) system in both the type C strain CN3685 and the type B strains CN1793 and CN1795 (69, 70). CPB production increases when the bacteria closely contact enterocyte-like Caco-2 cells, and this effect requires both the VirS/VirR and Agr-like QS systems (68, 69).

Furthermore,both the VirS/VirR and Agr-like QS systems are essential for CN3685 to cause either necrotic enteritis or lethal enterotoxemia in animal models because these regulatory systems are required for *in vivo* CPB production (68, 69).

In all type B and C strains studied to date, the *cpb* genes localized on large plasmids (41, 55). Southern hybridization of pulsed-field gels showed that the *cpb* gene in most type B strains was located on ~90-kb plasmids, although a few type B strains carried an ~65-kb *cpb* plasmid (41). These studies also showed that the *cpb* plasmid in these type B strains is distinct from their *etx* plasmid, although a few type B strains possess *cpb* and *etx* plasmids of the same (65 kb) size, just to make the picture more confusing (41). While type B strains carry only 65-kb or 90-kb *cpb* plasmids, the *cpb* plasmids in type C strains display more diversity, ranging in size from ~65 to ~110 kb. The ~65-kb and 90-kb *cpb* plasmids present in these type C strains are very similar to the plasmids of matching size found in type B isolates (55). Sequencing and overlapping PCR analysis revealed that in these 65-kb and 90-kb *cpb* plasmids, a *tpeL* gene is located ~3 kb downstream from their *cpb* gene. This gene encodes a large clostridial glycosylating toxin, TpeL, that is related to the large glucosylating toxins from *Clostridium difficile* (71). This *cpb*- and *tpeL*-containing locus is associated with IS*1151* sequences that can excise to form circular molecules that may be transposition intermediates (41, 55).

Type C strains can carry other potential virulence plasmids (55). Type C strains carry either plasmid-borne *cpe* or *tpeL* genes, but never both genes. Some type C strains have ~65- to ~90-kb plasmids that carry *cpb2* genes (72), but these *cpb2* plasmids never carry the *cpb* gene (55).

Almost without exception, all CPB-encoding plasmids appear to encode the *tcp* conjugation locus (41, 55). Since there have been no reports of a *tcp* locus-containing plasmid that is nonconjugative, it is highly likely that all *cpb* plasmids are conjugative. It is postulated that type B strains most likely are derived from a conjugative transfer event from a type C to a type D strain (or vice versa) (55), but more detailed genomic studies are required to validate this hypothesis.

Epsilon-Toxin (ETX) Plasmids of *C. perfringens*

ETX is produced by *C. perfringens* type B and type D strains and is expressed as an inactive prototoxin that is activated by cleavage with proteases such as trypsin (73). Livestock diseases caused by *C. perfringens* type B isolates include both enterotoxemias with neurological clinical signs and lesions and necrotic enteritis (59). *C. perfringens* type D strains are the most common cause of enterotoxemia in sheep and goats (74), and genetic studies recently have shown that this syndrome is mediated by ETX (75). The disease has been reported, albeit rarely, in cattle. Type D disease in sheep, goats, and cattle is characterized by neurological and respiratory clinical signs and lesions. Enterocolitis with severe diarrhea also occurs in goats.

ETX is a neurotoxin that is able to be absorbed from the gastrointestinal tract, where it is produced, to spread systemically through the circulatory system, where the toxin mostly affects the heart, lungs, kidneys, and brain (74). It is able to bind to vascular endothelial cells and cause cell lysis, leading to increased vascular permeability that causes brain and pulmonary edema. The latter eventually leads to degeneration and necrosis of surrounding tissues. In addition, ETX is also able to cross the blood-brain barrier and target neuronal cells directly. The neurological signs of type B and D disease are believed to be a consequence of the brain damage produced by ETX acting on the vasculature and on neurons (74, 75).

Regulation of ETX production in a type D isolate is controlled by an *agr*-like QS system (76) and the global virulence regulator CodY (77). How these two regulatory systems interact is not known, but CodY is also responsible for the upregulation of adherence to Caco2 cells *in vitro* as well as regulating spore production under nutrient-rich conditions. By contrast, regulation studies in two type B isolates have indicated that ETX is not regulated by the *agr*-like QS system in these strains, unlike other toxins (70). The role of CodY in type B strains is not known.

In all strains examined to date the *etx* gene is located on a large plasmid (4, 40, 41). In type B strains the *cpb* and *etx* plasmids show limited plasmid variation, and no strains have been identified to date that encode both toxins on the same plasmid (20, 41). The *cpb* gene, as discussed earlier, is encoded on either a 65-kb plasmid or, more commonly, on a 90-kb plasmid in these strains (41). There is no discernable variation in the 65-kb *etx* plasmids of type B isolates; these plasmids also carry the *cpb2* gene (20).

The only complete sequence available for an *etx* plasmid is from the type B isolate NCTC8533. This plasmid, pCP8533etx (64,753 bp), is closely related at the sequence level to the pCPF5603 family of CPE-encoding plasmids, although it lacks the *cpe* locus and another large region that encodes putative metabolic genes that are more commonly found on the chromosome (20). Instead of the *cpe* locus, pCP8533etx carries an *etx* locus that includes a Mu-like element downstream of the *etx* gene and IS*1151*- and IS*231*-like elements directly upstream of the *etx* gene (Fig. 5). This *etx* locus is largely conserved in all type B *etx* plasmids and some type D *etx* plasmids (20). The *etx* gene is located approximately 9 kb downstream of the *dcm* gene in these isolates, and there appears to have been an IS element-mediated duplication of several genes located downstream of the *cpe* locus in pCPF5603. These genes now flank the *etx* locus in pCP8533etx (Fig. 5). Another variant of the *etx* locus, represented by a plasmid in strain CN1675, has been identified in type D isolates that do not carry the *cpe* or *cpb2* genes. These isolates share the same downstream sequences as the pCP8533etx-like loci, but the upstream sequence is significantly different, encoding an IS*1151* sequence and a Tn*3* family transposase upstream of the *etx* gene. In addition, the *etx* locus is located closer to the *dcm* gene in these strains (20) (Fig 5).

Type D *etx* plasmids show a marked degree of plasmid diversity, in direct contrast to the situation in type B isolates (20, 40). It is clear that a minority of surveyed type D strains contain a type B-like 65-kb *etx* plasmid that also carries the *cpb2* gene (20). Type D isolates that lack the *cpe* and *cpb2* genes mostly contain ~48-kb plasmids (although a few contain 75-kb plasmids) that are likely to be conjugative since they appear to carry a *tcp* locus (40). Most of these strains contain a CN1675-like *etx* locus (20). Type D isolates that encode the *etx*, *cpe*, and/or *cpb2* gene(s) are larger, ranging in size from ~65 to 110 kb. In one type D isolate, carrying all three of these toxin genes, each toxin gene is located on a separate plasmid, while other isolates encode all three genes on the one plasmid (40).

Conjugative transfer of *etx* plasmids has been demonstrated in two type D isolates, CN1020 and CN3718 (16); these strains do not encode *cpe* or *cpb2* genes and contain CN1675-like *etx* loci (20, 40). Transfer was initially demonstrated using marked plasmids, where the *etx* gene was deleted and replaced with a chloramphenicol resistance gene (16). Transfer frequencies from both primary matings (a mating between two unrelated strains) and secondary matings (a mating between two genetically distinct but isogenic strains) were very high, ranging from 1-5 x 10^{-1} transconjugants/donor cell. Due to the very high frequency of transfer, a wild-type *etx* plasmid was subsequently transferred to a *C. perfringens* type A laboratory strain without any genetic selection (16). Plasmid transfer was detected at a frequency of 0.8% and effectively converted a toxinotype

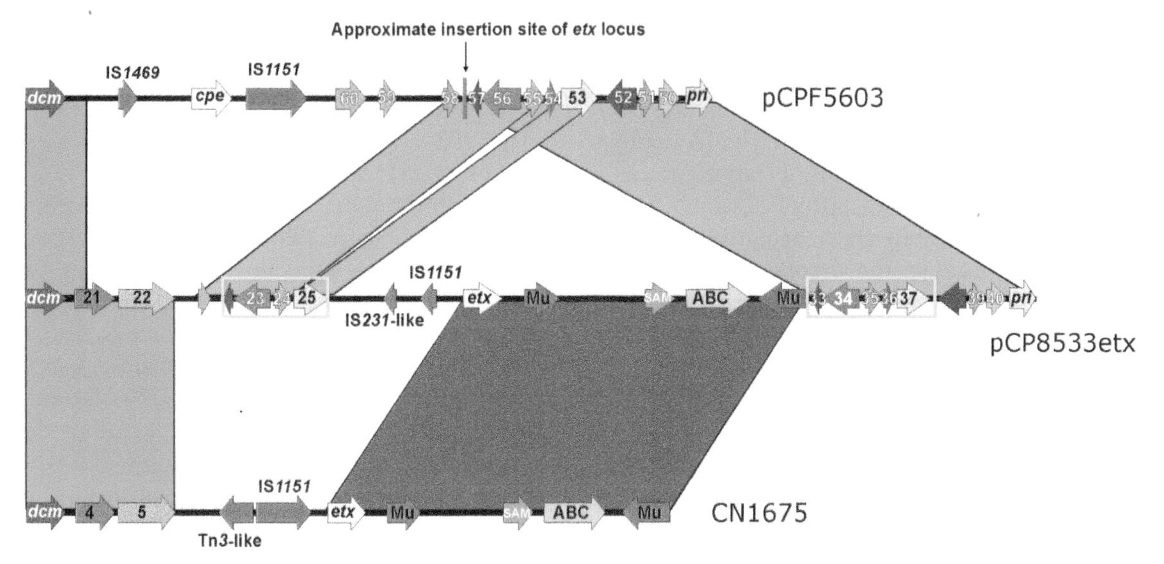

Figure 5 Comparative alignment of *etx* loci. The two *etx* (yellow arrows) loci from plasmid pCP8533etx and strain CN1675 are compared to the location of the *cpe* (yellow arrow) gene and surrounding sequences in plasmid pCPF5603. The aligned region begins with the *dcm* gene (left side) found downstream of the *tcp* locus in all sequenced plasmids. Genes (or DNA sequences) with greater than 90% nucleotide identity are colored alike (except for the *cpe* and *etx* genes; both are in yellow but are not related). The genes inside the green boxes appear to have been duplicated and flank the *etx* locus in plasmid pCP8533etx. Numbers are the CDS designations from the respective sequences: pCPF5603 (accession number: NC_007773), pCP8533etx (accession number: NC_011412), and CN1675 (accession number: EU852100). Arrows without numbers represent pseudogenes.
doi:10.1128/microbiolspec.PLAS-0024-2014.f5

A strain to a toxinotype D strain. These results demonstrated for the first time that one of the toxins that is used to determine the toxinotype of an isolate was mobile and therefore that strains that do not contain the *etx* gene may be able to acquire it laterally (and at high frequency) from a coresident type D isolate perhaps encountered within the milieu of the gastrointestinal tract.

Iota-Toxin (ITX) Plasmids of *C. perfringens*

C. perfringens type E isolates produce ITX and have been associated with enteric disease in rabbits, sheep, and cattle, although conclusive evidence of the role of type E strains and ITX in the pathogenesis of animal disease is lacking. Type E strains are the only *C. perfringens* isolates that produce ITX, a member of the clostridial binary toxin family (4, 78–80). This toxin family also includes *C. botulinum* C2 toxin, as well as the *Clostridium spiroforme* and *C. difficile* binary toxins. Consistent with its designation as a binary toxin, ITX is comprised of two polypeptides, IA and IB, that are encoded by the *iap* and *ibp* genes, respectively. IA and IB are secreted from *C. perfringens* as proproteins and must be proteolytically activated by removal of their N-terminal regions by either host proteases or *C. perfringens* lambda-toxin (78–80).

IB, the receptor-binding component of ITX (80), interacts with the host lipolysis-stimulated lipoprotein (81). Lipolysis-stimulated lipoprotein is also a receptor for *C. difficile* and *C. spiroforme* binary toxins, although it is not a C2 toxin receptor (82). The IA component interacts with cell-bound IB, promoting the endocytosis of ITX. IA is transported through the endocytic vacuole, presumably via a heptameric pore formed by IB, and once in the cytoplasm it ADP-ribosylates actin at Arg177, which depolymerizes actin and collapses the cytoskeleton of host cells (78–80). The multifunctional mammalian surface protein CD44 also increases ITX binding, perhaps by acting as a second receptor or a coreceptor (83).

The *iap* and *ibp* genes comprise an operon that in "classical" type E strains is located on the largest (~100 to 135 kb) known toxin plasmids found in *C. perfringens* (54, 84). These ITX plasmids generally have a pCPF5603-like backbone, plus additional genes that encode several other potential virulence factors, including lambda-toxin and urease. In these classical type E strains, IS*1151* sequences are closely associated with the *iap/ibp* genes, where they may be involved in their movement by a transposition process. These toxin genes can be detected on small circular DNA molecules that could represent transposition intermediates (84). On this basis, it has been suggested that the ITX plasmids in classical type E strains originated from the integration of an IS*1151*-based, *iap/ibp*-carrying mobile genetic element into an F5603-like *cpe* plasmid in a *C. perfringens* type A strain (54). This putative integration event apparently occurred near the promoter region of a *cpe* gene (Fig. 4) (84) and thereby inactivated *cpe* expression, such that these strains produce ITX but not CPE. Subsequent to this silencing of the *cpe* gene by promoter inactivation, the silent *cpe* genes have accumulated additional frame-shift and nonsense mutations, converting them to pseudogenes (54).

C. perfringens type E strains have traditionally been viewed as the least important of the *C. perfringens* types with respect to causing human or animal disease. However, the disease potential of type E strains may require some reevaluation given a recent study reporting that the standard *C. perfringens* multiplex PCR toxinotyping assay incorrectly classifies some "nonclassical" type E strains as type A due to variations in their *iap/ibp* gene sequences (21). Since those variant ITX genes are expressed, some strains classified by PCR as type A may actually be type E strains.

The ITX-encoding pCPPB-1 plasmid from one nonclassical type E strain has been sequenced (21). The analysis of this sequence revealed four fundamental differences between pCPPB-1 and the classical ITX plasmids: (i) pCPPB-1 is smaller (~65 kb), (ii) the backbone of pCPPB-1 is similar to that of pCPF4969, in contrast to the pCPF5603-like backbone of classical ITX plasmids, (iii) pCPPB-1 does not encode lambda-toxin or urease, and (iv) an IS*1151* element carrying the *iap/ibp* genes has apparently inserted near the most upstream of three *cpe* promoters in pCPPB-1; however, this *cpe* gene is still expressed, presumably from the other two more downstream *cpe* promoters. Interestingly, the CPE encoded by pCPPB-1 is a variant compared to the CPE produced by other CPE-positive strains. A *tcp* locus is present on both the pCPPB-1-like ITX plasmid and the classical ITX plasmids, strongly suggesting that all of these plasmids are conjugative (21, 84). However, conjugative transfer has not yet been experimentally demonstrated.

NetB Plasmids of *C. perfringens*

NetB, a plasmid-encoded 33-kDa β-barrel pore-forming toxin, is a major virulence factor in *C. perfringens*-mediated necrotic enteritis of poultry (85). This disease is manifested in two forms: as a frank clinical disease characterized by a sudden increase in flock mortality and as a milder subclinical form characterized by chronic damage to the intestinal mucosa, leading to poor digestion and nutrient absorption and resulting in a loss of bird performance. The subclinical form is of greatest consequence to the industry because of its wider prevalence and the difficulty of detection and amelioration before economic damage occurs. Treatment and control measures are estimated to cost the global poultry industry $2 billion per annum (86).

Structural studies of the secreted soluble monomeric form (87) and the heptameric pore form (88) of NetB have revealed that the conformation of the membrane-binding domain of NetB is significantly different from other beta-barrel pore-forming toxins such as the alpha-hemolysin of *S. aureus*, indicating that there are likely to be differences in the way these toxins interact with cell membranes. There are clear differences in the toxic activity of NetB toward different cell types; among a number of cultured cell lines, only chicken LMH cells were susceptible (85), and avian red blood cells were much more susceptible than mammalian red blood cells (87). NetB interacts directly with cholesterol (88), and the pores demonstrate a high single-channel conductance and have a preference for cations (87). Mutagenesis studies have shown that key residues in structural elements, such as those required for oligomerization and membrane binding, reduce toxic activity (87, 88). Several recent studies have shown that NetB can be effectively used as an antigen to induce a protective immune response in vaccinated birds (89–92).

Strain surveys have shown that *netB* is restricted to poultry-derived isolates of *C. perfringens* (93), apart from a single example of a *netB*-carrying strain isolated from cattle (94). All *C. perfringens* isolates that have been rigorously proven to cause necrotic enteritis in experimental infections carry the *netB* gene (85, 93, 95). Some *C. perfringens* strains isolated from clinical cases of necrotic enteritis lack the *netB* gene; however, none of these strains can reproducibly induce disease in experimental infection models (93, 95). We postulate that these disease-derived strains lacking the *netB* gene may have lost the plasmid carrying the gene during the initial strain isolation from clinical samples. In all isolates for which information is available the *netB* gene is carried on a large plasmid.

In the extensively studied strain, EHE-NE18, the *netB* plasmid pJIR3535 is one of three large conjugative plasmids (9). Each of these plasmids has been sequenced, as have two plasmids from a Canadian necrotic enteritis isolate (10), and all have very similar conjugation loci and plasmid replication and maintenance regions, as already discussed (Figs. 1 and 2). The plasmids are differentiated by the other genes they encode: pJIR3535 carries *netB* and a series of other genes that could potentially have a role in virulence, pJIR3844 carries a *cpb2* gene encoding an atypical form of CPB2 plus genes that encode putative surface proteins of unknown function, and pJIR3537 carries the *tet*(P) tetracycline resistance operon (Fig. 1). All of these plasmids can transfer at high frequency. The extensive region of sequence similarity (~35 kb) made it difficult to assemble whole-genome sequence data and to complete the sequence of each of these plasmids. This problem is common to the analysis of many *C. perfringens* strains, of all toxin types, where the presence of multiple, closely related large plasmids is very common. Sophisticated bioinformatic approaches, such as the construction of de Bruijn graphs (9), were used to assist in determining whole-plasmid sequences, but the clearest and most readily assembled results can be obtained by transferring each of the plasmids to *C. perfringens* strains lacking endogenous plasmids and then sequencing the whole genomes of transconjugant clones. Multiple large plasmid carriage has been found in all necrotic enteritis-derived strains investigated, with some reported to have up to five large plasmids (96).

Two *netB* plasmids (pJIR3535 and pNetB-NE10) have been sequenced and shown to be very similar 82-kb plasmids despite being derived from geographically distinct *C. perfringens* strains (9, 10). Analysis of the genomes of other strains by microarray hybridization suggests that the *netB* plasmid gene content is largely conserved across strains carrying the plasmid (96). One smaller *netB* plasmid has been reported (in strain CP2 [88]), but the alterations in the plasmid have not been characterized. The *netB* plasmids contain other genes that may have a role in virulence, for example, genes homologous to internalins, chitinases, leukocidins, proteases, and carbohydrate binding proteins. Expression of *netB* is under the control of the VirS/VirR two-component regulatory system (97), and the adjacent *netI* gene also has a predicted VirR binding box in the upstream promoter region and thus may also be under VirS/VirR control (96).

The mobility of virulence factors such as NetB is potentially important in the origin and spread of pathogenic strains. It appears that NetB is absolutely required for the pathogenesis of avian necrotic enteritis, but it is not yet clear what other plasmid- or chromosomally encoded genes may also be required to arm *C. perfringens* with the ability to cause necrotic enteritis. Is it simply enough for a *C. perfringens* strain capable of colonizing chickens to carry a *netB* plasmid, or are there wider requirements? Other genes that are associated to varying degrees with strains isolated from diseased birds have been identified (98), and it will be important to determine the contribution that these genes make toward the pathogenic phenotype. Multilocus sequence typing has shown that necrotic enteritis-derived *C. perfringens* isolates cluster into two major clonal groups that are associated with carriage of *netB* (99). It will be of interest to determine whether the clonal grouping is related to other properties, besides *netB*, that pathogenic strains may require (e.g., the ability to colonize the appropriate gastrointestinal niche) or whether it indicates some selective ability of these clonal groups to stably maintain the plasmids carrying *netB*.

Beta2-Toxin (CPB2) Plasmids of *C. perfringens*

CPB2 is a putative pore-forming cytolysin, but it is generally regarded as an accessory toxin that is not yet associated with any specific disease. Although there has been much speculation about the role of CPB2 in animal disease, most studies were based on detection of *cpb2*-positive isolates in the intestinal tract. However, the prevalence of such isolates seems to be similar in diseased and healthy individuals of most animal species (with the possible exception of the pig), which renders this diagnostic criterion unreliable. Most experimental studies have failed find any role for CPB2 in the pathogenesis of animal disease (100–102).

The *cpb2* gene has been identified in some strains of all five *C. perfringens* toxinotypes and is always located on a large plasmid, although there is considerable variation in plasmid size (45 to 90 kb). In *cpe*/*cpb2*-positive type A strains the two toxin genes can be carried on separate plasmids that are characterized by *cpe* genes with downstream IS*1470*-like sequences. In some of these isolates they are on the same plasmid, which is characterized by IS*1151* sequences downstream of the *cpe* gene (103, 104).

Several *cpb2*-encoding plasmids have been fully sequenced (6, 9, 10, 105). The *cpb2* gene locus is conserved (Fig. 6) between sequenced plasmids from diverse origins such as the human gas gangrene isolate type A strain 13 (105), the lamb dysentery-derived type B strain NCTC8533 (20), the pig necrotizing enterocolitis type C strain CWC245 (72), two type A human gastrointestinal disease isolates (F5603 and F4013 [6]), a type A strain (F262) from a bovine clostridial abomasitis case (106), and two type A chicken necrotic enteritis strains isolated from two different countries (9, 10).

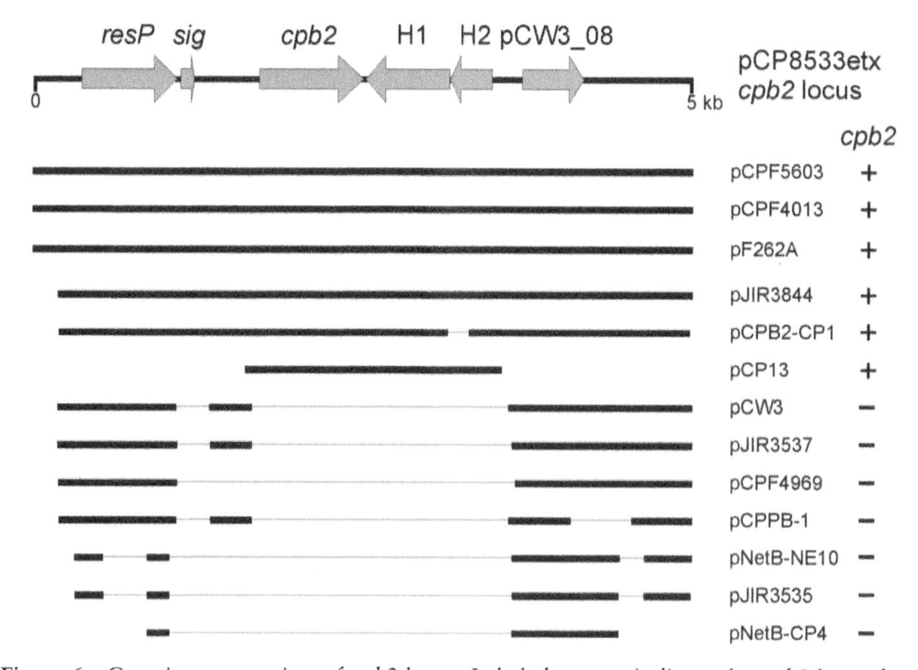

Figure 6 Genetic conservation of *cpb2* locus. Labeled arrows indicate the *cpb2* locus from pCP8533etx: *resP* encoding a putative resolvase, *sig* encoding a putative signal peptidase I, the *cpb2* gene, two hypothetical genes (H1 and H2), and a conserved hypothetical (pCW3_08). Below the locus are heavy lines indicating nucleotide sequence homology of 68% (pCP13) or >95% (all others). The gray lines indicate gaps in the sequence alignment where the sequence is absent from that particular plasmid sequence. Whether the *cpb2* gene is present in each plasmid is indicated on the right.
doi:10.1128/microbiolspec.PLAS-0024-2014.f6

The location of the *cpb2* gene is also conserved in these plasmids. It is found downstream of a putative resolvase gene, *resP*, and is flanked by the conserved hypothetical gene, *pcw308* (pCW3 nomenclature [5]), that is located downstream of the *resP* gene in pCW3-like *cpb2*-negative plasmids (Fig. 6). pCP13 from strain 13 (107) has a silent *cpb2* gene and is different from the conjugative toxin and antibiotic resistance plasmids described in detail in this chapter; for example, it does not carry the *tcp* conjugation locus and has a very different plasmid replication and maintenance region, but it still has some sequence conservation with this plasmid family. Despite the more distant relationship between pCP13 and other *cpb2* plasmids, the genes located downstream of the *cpb2* gene are still conserved (Fig. 6), suggesting that a *cpb2* cassette, consisting of the *cpb2* gene and the two downstream hypothetical genes, may have been mobile at some stage. The plasmid encoding the *cpb2* gene from the chicken necrotic enteritis strain EHE-NE18 encodes an intact *tcp* locus, and conjugative transfer at a high frequency has been demonstrated both from the original host as well as from secondary hosts, in the absence of the other conjugative plasmids (9). Finally, expression of the *cpb2* gene has been shown to be under the regulatory control of both the VirS/VirR two-component regulatory system (108) and the Agr-like QS system (109).

THE NEUROTOXIN PLASMIDS OF *C. BOTULINUM* AND *C. TETANI*

The most potent bacterial toxins are the neurotoxins encoded by *C. botulinum* and *C. tetani*. These bacteria are the causative agents of botulism and tetanus, respectively, and the pathogenesis of these often fatal conditions is mediated by similar neurotoxins: botulinum toxin (BoNT) and tetanus toxin (TeNT). Both BoNT and TeNT are related zinc metalloproteases that cleave either *v*SNARE or *t*SNARE proteins that are associated with neuronal synaptic vesicles. Cleavage of SNARE proteins stops neural signal transmission; for BoNT, the neuromuscular junctions are affected, leading to flaccid paralysis, while TeNT affects the relaxation pathway in the spinal cord, leading to the characteristic spastic paralysis of tetanus (110).

TeNT is encoded by the *tetX* gene, which is highly conserved between strains and is located on a 74-kb plasmid in the sequenced strain E88 (111, 112) (Fig. 3). This plasmid, pE88, encodes 61 ORFs that include a collagenase gene, *colT*, genes encoding a two-component signal transduction system that may be involved in the regulation of TeNT production (113), and several putative alternative sigma factors. Regulation of toxin production is mediated, at least in part, by the *tetR* gene, which is located immediately upstream of the *tetX* gene (Fig. 3). TetR is an alternative sigma factor that is closely related to the BotR protein, which in *C. botulinum* regulates the expression of BoNT and associated proteins. TetR and BotR also are similar to the TcdR and UviA proteins, which positively regulate the expression of toxins A and B in *C. difficile* and bacteriocin production in *C. perfringens,* respectively. These alternative sigma factors interact with RNA polymerase to facilitate toxin gene expression (114).

In addition, it was recently shown that several two-component signal transduction systems are involved in both positive and negative regulation of the toxin locus in *C. botulinum* (115, 116). *C. botulinum* group I organisms have two *agr*-like quorum sensing systems, one regulating sporulation and the other BoNT production (117). It is likely that the regulation of TeNT production in *C. tetani* is also complex (118).

C. botulinum isolates were historically designated to belong to the species *C. botulinum* based on the single phenotype of BoNT production. However, it now is clear that strains able to produce BoNT are, in some instances, only distantly related at the phylogenetic level (119). As a result, BoNT-producing bacteria are classified into four groups: group I or proteolytic *C. botulinum*, group II or nonproteolytic *C. botulinum*, group III encoding the BoNT/C and BoNT/D serotypes, and group IV (often referred to as *Clostridium argentinense). In addition, *Clostridium baratii* strains producing BoNT/F and strains of *Clostridium butyricum* producing BoNT/E have been identified (120). Until very recently there were seven serologically distinct BoNT types (BoNT/A to BoNT/G), but an eighth serotype, BoNT/H, recently was identified from a case of infant botulism (121, 122). Most of the serotypes have been further divided into subtypes (for example A1-A5). The BoNT serotype and the mechanism of toxin carriage often are common among members of the same group.

The BoNT locus is also variable and consists of two polycistronic operons: the *orfX* and HA loci. The *bont* and *ntnh* genes are common to all strains; the *botR* gene is also well conserved, but is not present in serotype E strains, and is often located between the two operons (123). Some bivalent strains have more than one BoNT locus, encoding BoNT proteins of different serotypes. In this situation, one BoNT serotype is generally found to be more abundant than the other, and the toxin designation used is, for example, Bf, where serotype B will demonstrate higher expression levels than toxin serotype F (123). Some chimeric Cd and Dc strains have been identified that have undergone recombination within the *bont* gene (113, 123). The variability of BoNT distribution is most likely due to mobile genetic elements such as plasmids and bacteriophages as well as transposons (120, 124).

BoNT loci can be chromosomally encoded, located on large plasmids, or encoded by lysogenic bacteriophages (123). In group III *C. botulinum* type C and D strains, the BoNT locus is located on large 186- to 203-kb pseudolysogenic bacteriophages (125, 126). These bacteriophage genomes are not integrated into the chromosome but exist as plasmids, although functional virus particles can be purified and used to infect susceptible strains (126). Group III isolates may also encode the C2 toxin, a binary toxin related to the ITX toxin of *C. perfringens* (127). The C2 toxin in these strains is also plasmid encoded, on large plasmids in excess of 100 kb. Several type C and D strains have been shown to contain other plasmids that encode putative toxin genes including a homologue of *C. septicum* alpha-toxin and two orthologues of ETX from *C. perfringens* (125).

For many years it was thought that, apart from the association of type C and D BoNT loci with bacteriophages, most other serotypes encoded chromosomal neurotoxin loci, an assumption borne out by the release of the first type A genome sequence—strain Hall (128). However, it is now clear that the toxin genes in all of the toxin serotypes studied to date may be either chromosomally encoded or plasmid-determined. This variability was first demonstrated in *C. botulinum* type A strains such as Loch Maree and 657Ba (pCLJ; Fig. 3), which encode BoNT/A loci on 267- and 270-kb plasmids, respectively (115, 129, 130). Plasmid-borne toxin loci in type B strains have also been demonstrated. In group I BoNT/B isolates, the plasmids range in size from 140 to 245 kb, while group II BoNT/B strains encode the toxin locus on 48- to 55-kb plasmids (Fig. 3) (130, 131). Some BoNT/E strains have been shown to harbor BoNT/E plasmids, although most appear to have chromosomally located neurotoxin genes (130, 132).

Notwithstanding the variation in the BoNT loci and their very different locations (chromosome, plasmid, or bacteriophage) the available sequence data indicate that the BoNT gene regions integrate into only one of three chromosomal locations or one of two plasmid insertion points, indicating that this locus, which is flanked by IS elements in many strains, potentially constitutes a mobile genetic element. Transfer of these loci could easily be facilitated by their carriage on elements such as bacteriophages, and recently it has been demonstrated that three type A and B toxin plasmids, two from group I organisms (pCLJ from strain 657Ba and pBotCDC-A3 from strain CDC-A3) and one from group II (pCLL from strain Eklund 17B), are capable of conjugative transfer (133). Transfer rates varied between the three plasmids, but all transferred at a relatively low level; however, plasmid transfer frequencies increased for the group I strains when a different nontoxigenic *C. botulinum* recipient was used. The plasmid genes responsible for the conjugative transfer remain to be identified, although pCLL showed some limited homology to the nonconjugative *C. perfringens* plasmid pCP13 as well as to regions of the conjugative *C. perfringens* plasmids (133). The similarity of several predicted ORFs to components of type II and IV bacterial secretion systems has suggested potential regions of pCLL and pCLJ that may be involved in conjugative transfer (Fig. 3). Conjugation adds yet another mode of potential transfer for these highly potent neurotoxins, which are clearly not restricted to a single clostridial strain.

TOXIN AND CAPSULE PLASMIDS OF *B. ANTHRACIS*

B. anthracis belongs to a family of six closely related pathogenic bacilli that includes *Bacillus cereus* and *Bacillus thuringiensis* and is known as the *B. cereus* group; members of this group are very closely related and are probably derived from a common ancestor (134). *B. anthracis* is the causative agent of anthrax, an important disease of both humans and domestic livestock. In recent years it has had increased attention as a result of its use as a bioterrorism agent. There are two major forms of anthrax: pulmonary anthrax, which is often fatal, and cutaneous anthrax, which is a more localized infection that is often restricted to workers in the wool, hide, or meat industries. The epidemiology of anthrax involves the acquisition of spores or vegetative cells by inhalation or by contact with contaminated animal material or soil (135). Disease pathogenesis is somewhat unusual, as it involves two major toxin subunits that enter the target host cell by forming heterooligomeric complexes with the third, binding and pore-forming component, protective antigen (PA, encoded by the *pagA* gene). The toxic components are edema factor (EF, *cya* gene), a calmodulin-dependent adenylyl cyclase, and lethal factor (LF, *lef* gene), a zinc metalloprotease that cleaves the N-terminal region of MAPKK. Another important *B. anthracis* virulence factor is its ability under appropriate conditions to produce a poly-d-glutamic acid capsule that protects the invading bacterial cell from phagocytosis (136).

Toxic activity is initiated by the binding of the 83-kDa PA protein to one of two receptors, TEM8 or CMG2, on the host cell surface (137, 138). The PA protein then is proteolytically cleaved by host furins, and the resultant activated PA_{63} derivative then oligomerizes to form a heptameric prepore. This PA_{63} complex can bind up to three EF or LF monomers, in any combination. After subsequent endocytosis, acidification of the endosome leads to the formation of a heptameric PA_{63} pore in the membrane; conformational changes then lead to the active translocation of EF and LF into the cytoplasm, where they have their lethal biological effects (139, 140).

The genes for the three components of anthrax toxin and an operon that encodes capsule production (*capBCADE*), together with genes involved in the regulation of their production, are all carried on two virulence plasmids, pXO1 (181.6 kb) and pXO2 (96.2 kb) (141, 142) (Fig. 3). Both plasmids are required for virulence; strains without pXO1 are avirulent because they do not produce the anthrax toxins, whereas strains lacking pXO2, such as the Sterne vaccine strain, are greatly attenuated because they do not have the capsule.

pXO1 has 203 ORFs including the structural genes for all three components of anthrax toxin (142). These toxin genes are located on a 44.8-kb pathogenicity island that is flanked by inverted copies of IS*1627* and includes an 8.7-kb Tn*3*-like transposon, Tn*XO1*, that carries the *gerX* operon, which is involved in spore germination (143). Toxin and capsule production, and therefore virulence, is regulated by AtxA, which is encoded by the pathogenicity island located on pXO1, and AcpA, which is encoded by pXO2. There is regulatory crosstalk between genes that are encoded on the plasmids and the chromosome; for example, AtxA, which has been designated as a global regulator, also regulates chromosomal genes, including S-layer genes (144, 145). The pXO1 pathogenicity island also encodes an S-layer adhesin, BslA, which is essential for the adherence of vegetative *B. anthracis* cells to human cells (146).

pXO2 has 110 ORFS, including the *cap* capsule operon and the *acpAB* regulatory genes (141, 147, 148). Capsule production is tightly regulated and is only produced in the presence of bicarbonate ions.

Virulence Plasmids of *B. cereus* and *B. thuringiensis*

The carriage of different virulence plasmids traditionally was considered a major contributor to the phenotypic properties that were critical for the speciation of *B. anthracis*, *B. cereus*, and *B. thuringiensis*. For example, pXO1 and pXO2 were regarded as major defining characteristics of *B. anthracis*. However, recent findings suggest a need to reconsider traditional species assignments that were based largely upon plasmid-mediated pathogenic phenotypes. *B. thuringiensis* strains have been identified that produce the pXO2-encoded polyglutamate capsule that was historically associated with *B. anthracis* (149). Additionally, several *B. cereus* strains have been isolated from patients in Louisiana and Texas who were suffering from severe pneumonias that clinically presented like inhalational anthrax (150). Analysis of G9241, one of those *B. cereus* strains, reveals that it carries a pXO1 plasmid that is 99.6% identical to the *B. anthracis* pXO1 plasmid and expresses all three toxin components (151). Interestingly, G9241 is encapsulated but lacks pXO2; instead, it carries another large plasmid, pBC210, that is predicted to encode a polysaccharide (not polypeptide) capsule. Animal studies indicate that both pBC210 and pXO1 are required for the full virulence of G9241 (151). Finally, *B. cereus* strains carrying both pXO1 and pXO2 have been isolated in Africa from great apes suffering from anthrax (152, 153). Collectively, these studies indicate that possession of pXO1 and pXO2 no longer defines *B. anthracis*.

Although pXO1 and pXO2 do not appear to be conjugative, they carry genes encoding components of a type IV secretion system similar to those of a classical conjugation apparatus (154) and are capable of being mobilized by conjugative plasmids (155, 156). While pXO2 does not encode its own transfer, a plasmid from *B. thuringiensis*, pAW63, which is closely related to pXO2, is conjugative (148). Comparative analysis has shown that 50 of the 76 ORFs found on pXO2 have significant sequence similarity to genes on pAW63; the genetic organization of these genes is also very similar and is closely related to that of pBT9727 from *B. thuringiensis* (Fig. 7). By contrast, pXO2 carries a 37-kb pathogenicity island that is absent from the *B. thuringiensis* plasmids (148). pAW63 carries a 42-kb conjugation region, and most of these genes are also present on pXO2, but the precise reason why pXO2 is not conjugative is not known. These data provide clear evidence that although pXO1 and pXO2 are not conjugative, there are conjugative plasmids that are closely related to these plasmids and that conjugation and plasmid mobilization have played a major role in the wide distribution of these plasmids throughout the *B. cereus* group.

B. cereus is commonly responsible for two different types of human food poisoning, an emetic form and a diarrheal form (157). The diarrheal form is caused by several enterotoxins that are encoded by genes with a chromosomal location in *B. cereus*, although genes encoding one of these enterotoxins (Nhe) have been iden-

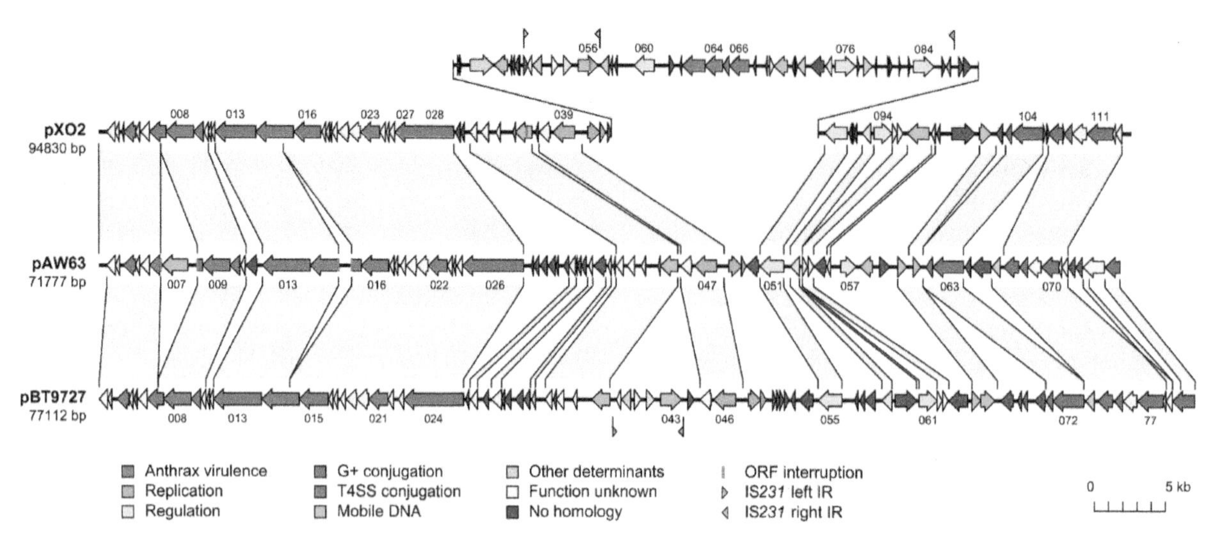

Figure 7 Comparative genetic organization of pXO2, pAW63, and pBT9727. Shared regions are indicated by shaded segments. The pathogenicity island on pXO2 is raised above the map. Reproduced from reference 148 with permission of the authors. doi:10.1128/microbiolspec.PLAS-0024-2014.f7

tified on a large plasmid in *Bacillus weihenstephanensis* (157). The emetic form of *B. cereus* food poisoning is caused by cereulide, a cyclic dodecapeptide with a molecular mass of 1.2 kDa (157). The *ces* genes responsible for cereulide synthesis are present on pCER270 (originally named pBCE4810). This large (~270 kb), low-copy-number plasmid is related to pXO1 but lacks the pathogenicity island encoding EF and LF (pCER270 is apparently restricted to only a few clonal clusters of *B. cereus*). *B. cereus* also causes opportunistic human infections, but the involvement of plasmids in those diseases remains to be determined (157).

B. thuringiensis is an important insect and nematode pathogen but only very rarely a mammalian pathogen. The *B. thuringiensis* Cry (crystal) or Cyt (cytolytic) toxins are produced during stationary phase and then localize in paracrystalline bodies (158–160). Sequestration of these toxins in crystalline inclusions offers protection from bacterial cytoplasmic proteases; these inclusions are eventually solubilized by the alkaline pH of the insect midgut, and the solubilized toxins are then proteolytically activated by insect intestinal proteases (159). Different members of the Cry and Cyt toxin family exhibit toxicity for specific insects or nematodes, and some of these toxins act synergistically. Receptors for these toxins vary and can include insect plasma membrane proteins such as cadherin-like proteins, aminopeptidase N, and alkaline phosphatase (161). The Cry and Cyt toxins are considered beta pore-forming toxins that create channels in epithelial cells to damage the insect midgut (162). A single B. *thuringiensis* strain can carry up to seven different genes encoding crystal-associated toxins (163). Expression of these toxin genes occurs during stationary phase, and these toxins can represent up to 20 to 30% of the protein in sporulating cells (159), due to a combination of transcriptional, posttranscriptional, and posttranslational regulatory effects (159, 160, 164).

B. thuringiensis strains carry from 1 to 17 plasmids, with sizes ranging from 3 to ~120 kb (164, 165). Most *cry* and *cyt* genes are encoded on large (>70 kb) plasmids that are often transmissible, due to either their own conjugative ability or to mobilization by helper plasmids (164, 166–168). A single plasmid can encode up to six different Cry and Cyt toxins (169). Recent studies showed that pBMB0228, which is an unusual 18-kb plasmid encoding two nematicidal Cry toxins, is a co-integrate of two plasmids, each with a *cry* gene (168). pBMB0228 is mobilizable and can resolve into its two separate plasmids after conjugative transfer; those plasmids can later fuse back together. In addition to often being located on conjugative plasmids, the dispersion of *cry* genes is probably assisted by their association with various IS elements, including IS*231*, IS*232*, IS*240*, IS*Bt1*, and IS*Bt2*, and transposons, including Tn*4430* and Tn*5401* (159, 160, 170).

CONCLUSIONS

We now know that toxin plasmids play an important role in the virulence of the spore-forming bacteria, but there is still a lot to learn about the biology of these plasmids and their molecular epidemiology. These plasmids represent fertile ground for future research. We need to elucidate the precise mechanism by which the conjugation process occurs in *C. perfringens* and how *C. perfringens* cells can maintain so many closely related but distinct plasmids. The role of several *C. perfringens* toxins in disease still remains to be determined, as do any synergistic effects on disease. There is little functional data available about the mechanism of replication and conjugative transfer of the toxin plasmids of the neurotoxic clostridia or the bacilli. Finally, there is a need for a systematic plasmid sequencing approach to determine the molecular epidemiology and evolution of the virulence plasmids of the *B. cereus* group.

Note Added in Proof.. Recently other workers have identified a novel binary toxin, BEC, that is produced by *C. perfringens* isolates from two outbreaks of human food-borne gastroenteritis in Japan. The BECa and BECb components are related to the equivalent ITX proteins and the becAB genes are encoded on 54.5-kb plasmids that are related to pCP13 (173).

Acknowledgments. *The authors gratefully acknowledge research support provided by U.S. Public Health Service grants R37AI19844-30 (B.A.M.) and R01AI056177-09 (B.A.M., J.I.R., and F.A.U.), by the Australian Research Council (ARC) through funding of the ARC Centre of Excellence in Structural and Functional Microbial Genomics (J.I.R.,), by the Australian National Health and Medical Research Council (J.I.R.), and by the Australian Poultry CRC (R.J.M. and J.I.R.). J.A.W. was supported by the provision of an Australian Postgraduate Award. Conflicts of interest: We disclose no conflicts.*

Citation. Adams V, Li J, Wisniewski JA, Uzal FA, Moore RJ, McClane BA, Rood JI. 2014. Virulence plasmids of spore-forming bacteria. Microbiol Spectrum **2(6)**:PLAS-0024-2014.

References

1. **Rood JI.** 2004. Virulence plasmids of spore-forming bacteria, p 413–422. *In* Funnell BE, Phillips GJ (ed), *The Biology of Plasmids*. ASM Press, Washington, DC.
2. **Rood JI.** 1998. Virulence genes of *Clostridium perfringens*. *Annu Rev Microbiol* **52:**333–360.
3. **Petit L, Gibert M, Popoff MR.** 1999. *Clostridium perfringens*: toxinotype and genotype. *Trends Microbiol* **7:** 104–110.
4. **Li J, Adams V, Bannam TL, Miyamoto K, Garcia JP, Uzal FA, Rood JI, McClane BA.** 2013. Toxin plasmids of *Clostridium perfringens*. *Microbiol Mol Biol Rev* **77:** 208–233.
5. **Bannam TL, Teng WL, Bulach D, Lyras D, Rood JI.** 2006. Functional identification of conjugation and replication regions of the tetracycline resistance plasmid pCW3 from *Clostridium perfringens*. *J Bacteriol* **188:** 4942–4951.
6. **Miyamoto K, Fisher DJ, Li J, Sayeed S, Akimoto S, McClane BA.** 2006. Complete sequencing and diversity analysis of the enterotoxin-encoding plasmids in *Clostridium perfringens* type A non-food-borne human gastrointestinal disease isolates. *J Bacteriol* **188:**1585–1598.
7. **Gerdes K, Howard M, Szardenings F.** 2010. Pushing and pulling in prokaryotic DNA segregation. *Cell* **141:** 927–942.
8. **Salje J, Gayathri P, Lowe J.** 2010. The ParMRC system: molecular mechanisms of plasmid segregation by actin-like filaments. *Nat Rev Microbiol* **8:**683–692.
9. **Bannam TL, Yan XX, Harrison PF, Seemann T, Keyburn AL, Stubenrauch C, Weeramantri LH, Cheung JK, McClane BA, Boyce JD, Moore RJ, Rood JI.** 2011. Necrotic enteritis-derived *Clostridium perfringens* strain with three closely related independently conjugative toxin and antibiotic resistance plasmids. *mBio* **2:** e00190-00111.
10. **Parreira VR, Costa M, Eikmeyer F, Blom J, Prescott JF.** 2012. Sequence of two plasmids from *Clostridium perfringens* chicken necrotic enteritis isolates and comparison with *C. perfringens* conjugative plasmids. *PLoS One* **7:**e49753.
11. **Popp D, Narita A, Lee LJ, Ghoshdastider U, Xue B, Srinivasan R, Balasubramanian MK, Tanaka T, Robinson RC.** 2012. Novel actin-like filament structure from *Clostridium tetani*. *J Biol Chem* **287:**21121–21129.
12. **Derman AI, Becker EC, Truong BD, Fujioka A, Tucey TM, Erb ML, Patterson PC, Pogliano J.** 2009. Phylogenetic analysis identifies many uncharacterized actin-like proteins (Alps) in bacteria: regulated polymerization, dynamic instability and treadmilling in Alp7A. *Mol Microbiol* **73:**534–552.
13. **Brefort G, Magot M, Ionesco H, Sebald M.** 1977. Characterization and transferability of *Clostridium perfringens* plasmids. *Plasmid* **1:**52–66.
14. **Rood JI, Maher EA, Somers EB, Campos E, Duncan CL.** 1978. Isolation and characterization of multiple antibiotic-resistant *Clostridium perfringens* strains from porcine feces. *Antimicrob Agents Chemother* **13:**871–880.
15. **Brynestad S, Sarker MR, McClane BA, Granum PE, Rood JI.** 2001. Enterotoxin plasmid from *Clostridium perfringens* is conjugative. *Infect Immun* **69:**3483–3487.
16. **Hughes ML, Poon R, Adams V, Sayeed S, Saputo S, Uzal FA, McClane BA, Rood JI.** 2007. Epsilon toxin plasmids of *Clostridium perfringens* type D are conjugative. *J Bacteriol* **189:**7531–7538.
17. **Lyras D, Adams V, Ballard SA, Teng WL, Howarth PM, Crellin PK, Bannam TL, Songer JG, Rood JI.** 2009. t*ISCpe8*, an *IS1595*-family lincomycin resistance element located on a conjugative plasmid in *Clostridium perfringens*. *J Bacteriol* **191:**6345–6351.
18. **Abraham LJ, Rood JI.** 1985. Molecular analysis of transferable tetracycline resistance plasmids from *Clostridium perfringens*. *J Bacteriol* **161:**636–640.
19. **Abraham LJ, Wales AJ, Rood JI.** 1985. Worldwide distribution of the conjugative *Clostridium perfringens* tetracycline resistance plasmid, pCW3. *Plasmid* **14:**37–46.
20. **Miyamoto K, Li J, Sayeed S, Akimoto S, McClane BA.** 2008. Sequencing and diversity analyses reveal extensive

similarities between some epsilon-toxin-encoding plasmids and the pCPF5603 *Clostridium perfringens* enterotoxin plasmid. *J Bacteriol* **190:**7178–7188.

21. **Miyamoto K, Yumine N, Mimura K, Nagahama M, Li J, McClane BA, Akimoto S.** 2011. Identification of novel *Clostridium perfringens* type E strains that carry an iota toxin plasmid with a functional enterotoxin gene. *PLoS One* **6:**e20376.
22. **Bantwal R, Bannam TL, Porter CJ, Quinsey NS, Lyras D, Adams V, Rood JI.** 2012. The peptidoglycan hydrolase TcpG is required for efficient conjugative transfer of pCW3 in *Clostridium perfringens. Plasmid* **67:** 139–147.
23. **Parsons JA, Bannam TL, Devenish RJ, Rood JI.** 2007. TcpA, an FtsK/SpoIIIE homolog, is essential for transfer of the conjugative plasmid pCW3 from *Clostridium perfringens. J Bacteriol* **189:**7782–7790.
24. **Porter CJ, Bantwal R, Bannam TL, Rosado CJ, Pearce MC, Adams V, Lyras D, Whisstock JC, Rood JI.** 2012. The conjugation protein TcpC from *Clostridium perfringens* is structurally related to the type IV secretion system protein VirB8 from Gram-negative bacteria. *Mol Microbiol* **83:**275–288.
25. **Steen JA, Bannam TL, Teng WL, Devenish RJ, Rood JI.** 2009. The putative coupling protein TcpA interacts with other pCW3-encoded proteins to form an essential part of the conjugation complex. *J Bacteriol* **191:**2926–2933.
26. **Goessweiner-Mohr N, Arends K, Keller W, Grohmann E.** 2013. Conjugative type IV secretion systems in Gram-positive bacteria. *Plasmid* **70:**289–302.
27. **Gomis-Rüth FX, Coll M.** 2001. Structure of TrwB, a gatekeeper in bacterial conjugation. *Int J Biochem Cell Biol* **33:**839–843.
28. **Fronzes R, Christie PJ, Waksman G.** 2009. The structural biology of type IV secretion systems. *Nat Rev Microbiol* **7:**703–714.
29. **Teng WL, Bannam TL, Parsons JA, Rood JI.** 2008. Functional characterization and localization of the TcpH conjugation protein from *Clostridium perfringens. J Bacteriol* **190:**5075–5086.
30. **de la Cruz F, Frost LS, Meyer RJ, Zechner EL.** 2010. Conjugative DNA metabolism in Gram-negative bacteria. *FEMS Microbiol Rev* **34:**18–40.
31. **McClane BA, Robertson SL, Li J.** 2013. *Clostridium perfringens*, p 465–490. *In* Doyle MP, Buchanan RA (ed), *Food Microbiology: Fundamentals and Frontiers.* ASM Press, Washington, DC.
32. **Huang IH, Waters M, Grau RR, Sarker MR.** 2004. Disruption of the gene (*spo0A*) encoding sporulation transcription factor blocks endospore formation and enterotoxin production in enterotoxigenic *Clostridium perfringens* type A. *FEMS Microbiol Lett* **233:**233–240.
33. **Harry KH, Zhou R, Kroos L, Melville SB.** 2009. Sporulation and enterotoxin (CPE) synthesis are controlled by the sporulation-specific sigma factors SigE and SigK in *Clostridium perfringens. J Bacteriol* **191:**2728–2742.
34. **Li J, McClane BA.** 2010. Evaluating the involvement of alternative sigma factors SigF and SigG in *Clostridium perfringens* sporulation and enterotoxin synthesis. *Infect Immun* **78:**4286–4293.
35. **Briggs DC, Naylor CE, Smedley JG 3rd, Lukoyanova N, Robertson S, Moss DS, McClane BA, Basak AK.** 2011. Structure of the food-poisoning *Clostridium perfringens* enterotoxin reveals similarity to the aerolysin-like pore-forming toxins. *J Mol Biol* **413:**138–149.
36. **Kitadokoro K, Nishimura K, Kamitani S, Fukui-Miyazaki A, Toshima H, Abe H, Kamata Y, Sugita-Konishi Y, Yamamoto S, Karatani H, Horiguchi Y.** 2011. Crystal structure of *Clostridium perfringens* enterotoxin displays features of beta-pore-forming toxins. *J Biol Chem* **286:**19549–19555.
37. **Katahira J, Sugiyama H, Inoue N, Horiguchi Y, Matsuda M, Sugimoto N.** 1997. *Clostridium perfringens* enterotoxin utilizes two structurally related membrane proteins as functional receptors *in vivo. J Biol Chem* **272:** 26652–26658.
38. **Hanna PC, Mietzner TA, Schoolnick GK, McClane BA.** 1991. Localization of the receptor-binding region of *Clostridium perfringens* enterotoxin utilizing cloned toxin fragments and synthetic peptides. The 30 C-terminal amino acids define a functional binding region. *J Biol Chem* **266:**11037–11043.
39. **Kokai-Kun JF, Benton K, Wieckowski EU, McClane BA.** 1999. Identification of a *Clostridium perfringens* enterotoxin region required for large complex formation and cytotoxicity by random mutagenesis. *Infect Immun* **67:**5634–5641.
40. **Sayeed S, Li J, McClane BA.** 2007. Virulence plasmid diversity in *Clostridium perfringens* type D isolates. *Infect Immun* **75:**2391–2398.
41. **Sayeed S, Li J, McClane BA.** 2010. Characterization of virulence plasmid diversity among *Clostridium perfringens* type B isolates. *Infect Immun* **78:** 495–504.
42. **Fisher DJ, Fernandez-Miyakawa ME, Sayeed S, Poon R, Adams V, Rood JI, Uzal FA, McClane BA.** 2006. Dissecting the contributions of *Clostridium perfringens* type C toxins to lethality in the mouse intravenous injection model. *Infect Immun* **74:**5200–5210.
43. **Collie R, McClane B.** 1998. Phenotypic characterization of enterotoxigenic *Clostridium perfringens* isolates associated with nonfoodborne human gastrointestinal diseases. *Anaerobe* **4:**69–79.
44. **Fernandez-Miyakawa ME, Sayeed S, Fisher DJ, Poon R, Adams V, Rood JI, McClane BA, Saputo J, Uzal FA.** 2007. Development and application of an oral challenge mouse model for studying *Clostridium perfringens* type D infection. *Infect Immun* **75:**4282–4288.
45. **Kokai-Kun JF, Songer JG, Czeczulin JR, Chen F, McClane BA.** 1994. Comparison of Western immunoblots and gene detection assays for identification of potentially enterotoxigenic isolates of *Clostridium perfringens. J Clin Microbiol* **32:**2533–2539.
46. **Sarker MR, Carman RJ, McClane BA.** 1999. Inactivation of the gene (*cpe*) encoding *Clostridium perfringens* enterotoxin eliminates the ability of two *cpe*-positive *C. perfringens* type A human gastrointestinal disease

isolates to affect rabbit ileal loops. *Mol Microbiol* **33:** 946–958.

47. **Scallan E, Griffin PM, Angulo FJ, Tauxe RV, Hoekstra RM.** 2011. Foodborne illness acquired in the United States: unspecified agents. *Emerg Infect Dis* **17:**16–22.
48. **Carman RJ.** 1997. *Clostridium perfringens* in spontaneous and antibiotic-associated diarrhoea of man and other animals. *Rev Med Microbiol* 8(Suppl. 1):S43–S45.
49. **Bos J, Smithee L, McClane B, Distefano R, Uzal F, Songer JG, Mallonee S, Crutcher JM.** 2005. Fatal necrotizing enteritis following a foodborne outbreak of enterotoxigenic *Clostridium perfringens* type A infection. *Clin Infect Dis* **15:**e78–e83.
50. **Marks SL, Kather EJ, Kass PH, Melli AC.** 2002. Genotypic and phenotypic characterization of *Clostridium perfringens* and *Clostridium difficile* in diarrheic and healthy dogs. *J Vet Intern Med* **16:**533–540.
51. **Brynestad S, Synstad B, Granum PE.** 1997. The *Clostridium perfringens* enterotoxin gene is on a transposable genetic element in type A human food poisoning strains. *Microbiology* **143:**2109–2115.
52. **Ma M, Li J, McClane BA.** 2012. Genotypic and phenotypic characterization of *Clostridium perfringens* isolates from darmbrand cases in post-World War II Germany. *Infect Immun* **80:**4354–4363.
53. **Li J, Miyamoto K, Sayeed S, McClane BA.** 2010. Organization of the *cpe* locus in CPE-positive *Clostridium perfringens* type C and D isolates. *PLoS One* **5:**e10932.
54. **Billington SJ, Wieckowski EU, Sarkar MR, Bueschel D, Songer JG, McClane BA.** 1998. *Clostridium perfringens* type E animal enteritis isolates with highly conserved, silent enterotoxin gene sequences. *Infect Immun* **66:** 4531–4536.
55. **Gurjar A, Li J, McClane BA.** 2010. Characterization of toxin plasmids in *Clostridium perfringens* type C isolates. *Infect Immun* **78:**4860–4869.
56. **Lawrence G.** 2005. The pathogenesis of pig-bel in Papua New Guinea. *P N G Med J* **48:**39–49.
57. **Murrell TG, Walker PD.** 1991. The pigbel story of Papua New Guinea. *Trans R Soc Trop Med Hyg* **85:** 119–122.
58. **McClane BA, Uzal FA, Fernandez Miyakawa ME, Lyerly D, Wilkins TD.** 2006. The enterotoxic clostridia, p 698–752. *In* Dworkin M, Falkow S, Rosenberg E, Schleifer K, Stackebrandt E (ed), *The Prokaryotes*. Springer, New York, NY.
59. **Fernandez-Miyakawa ME, Fisher DJ, Poon R, Sayeed S, Adams V, Rood JI, McClane BA, Uzal FA.** 2007. Both epsilon-toxin and beta-toxin are important for the lethal properties of *Clostridium perfringens* type B isolates in the mouse intravenous injection model. *Infect Immun* **75:**1443–1452.
60. **Sayeed S, Uzal FA, Fisher DJ, Saputo J, Vidal JE, Chen Y, Gupta P, Rood JI, McClane BA.** 2008. Beta toxin is essential for the intestinal virulence of *Clostridium perfringens* type C disease isolate CN3685 in a rabbit ileal loop model. *Mol Microbiol* **67:**15–30.
61. **Vidal JE, Ohtani K, Shimizu T, McClane BA.** 2009. Contact with enterocyte-like Caco-2 cells induces rapid upregulation of toxin production by *Clostridium perfringens* type C isolates. *Cell Microbiol* **11:**1306–1328.
62. **Garcia JP, Beingesser J, Fisher DJ, Sayeed S, McClane BA, Posthaus H, Uzal FA.** 2012. The effect of *Clostridium perfringens* type C strain CN3685 and its isogenic beta toxin null mutant in goats. *Vet Microbiol* **157:** 412–419.
63. **Hunter SEC, Brown JE, Oyston PCF, Sakurai J, Titball RW.** 1993. Molecular genetic analysis of beta-toxin of *Clostridium perfringens* reveals sequence homology with alpha-toxin, gamma-toxin, and leukocidin of *Staphylococcus aureus*. *Infect Immun* **61:**3958–3965.
64. **Keyburn AL, Bannam TL, Moore RJ, Rood JI.** 2010. NetB, a pore-forming toxin from necrotic enteritis strains of *Clostridium perfringens*. *Toxins* **2:**1913–1927.
65. **Manich M, Knapp O, Gibert M, Maier E, Jolivet-Reynaud C, Geny B, Benz R, Popoff MR.** 2008. *Clostridium perfringens* delta toxin is sequence related to beta toxin, NetB, and *Staphylococcus* pore-forming toxins, but shows functional differences. *PLoS One* **3:**e3764.
66. **Shatursky O, Bayles R, Rogers M, Jost BH, Songer JG, Tweten RK.** 2000. *Clostridium perfringens* beta-toxin forms potential-dependent, cation-selective channels in lipid bilayers. *Infect Immun* **68:**5546–5551.
67. **Vidal JE, McClane BA, Saputo J, Parker J, Uzal FA.** 2008. Effects of *Clostridium perfringens* beta-toxin on the rabbit small intestine and colon. *Infect Immun* **76:** 4396–4404.
68. **Ma M, Vidal J, Saputo J, McClane BA, Uzal F.** 2011. The VirS/VirR two-component system regulates the anaerobic cytotoxicity, intestinal pathogenicity, and enterotoxemic lethality of *Clostridium perfringens* type C isolate CN3685. *MBio* **2:**e00338-00310.
69. **Vidal JE, Ma M, Saputo J, Garcia J, Uzal FA, McClane BA.** 2012. Evidence that the Agr-like quorum sensing system regulates the toxin production, cytotoxicity and pathogenicity of *Clostridium perfringens* type C isolate CN3685. *Mol Microbiol* **83:**179–194.
70. **Chen J, McClane BA.** 2012. Role of the Agr-like quorum-sensing system in regulating toxin production by *Clostridium perfringens* type B strains CN1793 and CN1795. *Infect Immun* **80:**3008–3017.
71. **Amimoto K, Noro T, Oishi E, Shimizu M.** 2007. A novel toxin homologous to large clostridial cytotoxins found in culture supernatant of *Clostridium perfringens* type C. *Microbiology* **153:**1198–1206.
72. **Gibert M, Jolivet-Renaud C, Popoff MR.** 1997. Beta2 toxin, a novel toxin produced by *Clostridium perfringens*. *Gene* **203:**65–73.
73. **Popoff MR, Bouvet P.** 2009. Clostridial toxins. *Future Microbiol* **4:**1021–1064.
74. **Uzal FA, Songer JG.** 2008. Diagnosis of *Clostridium perfringens* intestinal infections in sheep and goats. *J Vet Diagn Invest* **20:**253–265.
75. **Garcia JP, Adams V, Beingesser J, Hughes ML, Poon R, Lyras D, Hill A, McClane BA, Rood JI, Uzal FA.** 2013. Epsilon toxin is essential for the virulence of

Clostridium perfringens type D infection in sheep, goats, and mice. *Infect Immun* **81:**2405–2414.

76. **Chen J, Rood JI, McClane BA.** 2011. Epsilon toxin production by *Clostridium perfringens* type D strain CN3718 is dependent on the *agr* operon but not the VirS/VirR two component regulatory system. *MBio* **2:** e00275-00211.
77. **Li J, Ma M, Sarker MR, McClane BA.** 2013. CodY is a global regulator of virulence-associated properties for *Clostridium perfringens* type D strain CN3718. *MBio* **4:** e00770-00713.
78. **Stiles BG, Wigelsworth DJ, Popoff MR, Barth H.** 2011. Clostridial binary toxins: iota and C2 family portraits. *Front Cell Infect Microbiol* **1:**11.
79. **Aktories K, Schwan C, Papatheodorou P, Lang AE.** 2012. Bidirectional attack on the actin cytoskeleton. Bacterial protein toxins causing polymerization or depolymerization of actin. *Toxicon* **60:**572–581.
80. **Sakurai J, Nagahama M, Oda M, Tsuge H, Kobayashi K.** 2009. *Clostridium perfringens* iota-toxin: structure and function. *Toxins* **1:**208–228.
81. **Papatheodorou P, Carette JE, Bell GW, Schwan C, Guttenberg G, Brummelkamp TR, Aktories K.** 2011. Lipolysis-stimulated lipoprotein receptor (LSR) is the host receptor for the binary toxin *Clostridium difficile* transferase (CDT). *Proc Natl Acad Sci USA* **108:**16422–16427.
82. **Papatheodorou P, Wilczek C, Nolke T, Guttenberg G, Hornuss D, Schwan C, Aktories K.** 2012. Identification of the cellular receptor of *Clostridium spiroforme* toxin. *Infect Immun* **80:**1418–1423.
83. **Wigelsworth DJ, Ruthel G, Schnell L, Herrlich P, Blonder J, Veenstra TD, Carman RJ, Wilkins TD, Van Nhieu GT, Pauillac S, Gibert M, Sauvonnet N, Stiles BG, Popoff MR, Barth H.** 2012. CD44 promotes intoxication by the clostridial iota-family toxins. *PLoS One* **7:**e51356.
84. **Li J, Miyamoto K, McClane BA.** 2007. Comparison of virulence plasmids among *Clostridium perfringens* type E isolates. *Infect Immun* **75:**1811–1819.
85. **Keyburn AL, Boyce JD, Vaz P, Bannam TL, Ford ME, Parker D, Di Rubbo A, Rood JI, Moore RJ.** 2008. NetB, a new toxin that is associated with avian necrotic enteritis caused by *Clostridium perfringens*. *PLoS Pathog* **4:**e26.
86. **van der Sluis W.** 2000. Clostridial enteritis is an often underestimated problem. *World Poultry* **16:**42–43.
87. **Yan X, Porter CJ, Hardy SP, Steer D, Smith AI, Quinsey NS, Hughes V, Cheung JK, Keyburn AL, Kaldhusdal M, Moore RJ, Bannam TL, Whisstock JC, Rood JI.** 2013. Structural and functional analysis of the pore-forming toxin NetB from *Clostridium perfringens*. *mBio* **4:** e00019-00013.
88. **Savva CG, Fernandes da Costa SP, Bokori-Brown M, Naylor CE, Cole AR, Moss DS, Titball RW, Basak AK.** 2013. Molecular architecture and functional analysis of NetB, a pore-forming toxin from *Clostridium perfringens*. *J Biol Chem* **288:**3512–3522.
89. **Fernandes da Costa SP, Savva CG, Bokori-Brown M, Naylor CE, Moss DS, Basak AK, Titball RW.** 2014. Identification of a key residue for oligomerisation and pore-formation of *Clostridium perfringens* NetB. *Toxins* **6:**1049–1061.
90. **Keyburn AL, Portela RW, Ford ME, Bannam TL, Yan XX, Rood JI, Moore RJ.** 2013. Maternal immunization with vaccines containing recombinant NetB toxin partially protects progeny chickens from necrotic enteritis. *Vet Res* **44:**108.
91. **Keyburn AL, Portela RW, Sproat K, Ford ME, Bannam TL, Yan X, Rood JI, Moore RJ.** 2013. Vaccination with recombinant NetB toxin partially protects broiler chickens from necrotic enteritis. *Vet Res* **44:**54.
92. **Jang SI, Lillehoj HS, Lee SH, Lee KW, Lillehoj EP, Hong YH, An DJ, Jeong W, Chun JE, Bertrand F, Dupuis L, Deville S, Arous JB.** 2012. Vaccination with *Clostridium perfringens* recombinant proteins in combination with Montanide ISA 71 VG adjuvant increases protection against experimental necrotic enteritis in commercial broiler chickens. *Vaccine* **30:**5401–5406.
93. **Keyburn AL, Yan XX, Bannam TL, Van Immerseel F, Rood JI, Moore RJ.** 2010. Association between avian necrotic enteritis and *Clostridium perfringens* strains expressing NetB toxin. *Vet Res* **41:**21.
94. **Martin TG, Smyth JA.** 2009. Prevalence of *netB* among some clinical isolates of *Clostridium perfringens* from animals in the United States. *Vet Microbiol* **136:**202–205.
95. **Smyth JA, Martin TG.** 2010. Disease producing capability of *netB* positive isolates of *C. perfringens* recovered from normal chickens and a cow, and *netB* positive and negative isolates from chickens with necrotic enteritis. *Vet Microbiol* **146:**76–84.
96. **Lepp D, Roxas B, Parreira VR, Marri PR, Rosey EL, Gong J, Songer JG, Vedantam G, Prescott JF.** 2010. Identification of novel pathogenicity loci in *Clostridium perfringens* strains that cause avian necrotic enteritis. *PLoS One* **5:**e10795.
97. **Cheung JK, Low L-Y, Hiscox TJ, Rood JI.** 2013. Regulation of extracellular toxin production in *Clostridium perfringens*, p 281–294. *In* Vasil ML, Darwin AJ (ed), *Regulation of Bacterial Virulence*. ASM Press, Washington, DC.
98. **Lepp D, Gong J, Songer JG, Boerlin P, Parreira VR, Prescott JF.** 2013. Identification of accessory genome regions in poultry *Clostridium perfringens* isolates carrying the NetB plasmid. *J Bacteriol* **195:**1152–1166.
99. **Hibberd MC, Neumann AP, Rehberger TG, Siragusa GR.** 2011. Multilocus sequence typing subtypes of poultry *Clostridium perfringens* isolates demonstrate disease niche partitioning. *J Clin Microbiol* **49:**1556–1567.
100. **Manteca C, Daube G, Jauniaux T, Linden A, Pirson V, Detilleux J, Ginter A, Coppe P, Kaeckenbeeck A, Mainil JG.** 2002. A role for the *Clostridium perfringens* beta2 toxin in bovine enterotoxaemia? *Vet Microbiol* **86:**191–202.
101. **Dray T.** 2004. *Clostridium perfringens* type A and beta2 toxin associated with enterotoxemia in a 5-week-old goat. *Can Vet J* **45:**251–253.

102. **Uzal FA, Vidal JE, McClane BA, Gurjar AA.** 2010. *Clostridium perfringens* toxins involved in mammalian veterinary diseases. *Open Toxinology J* **2:**24–42.

103. **Fisher DJ, Miyamoto K, Harrison B, Akimoto S, Sarker MR, McClane BA.** 2005. Association of beta2 toxin production with *Clostridium perfringens* type A human gastrointestinal disease isolates carrying a plasmid enterotoxin gene. *Mol Microbiol* **56:**747–762.

104. **Harrison B, Raju D, Garmory HS, Brett MM, Titball RW, Sarker MR.** 2005. Molecular characterization of *Clostridium perfringens* isolates from humans with sporadic diarrhea: evidence for transcriptional regulation of the beta2-toxin-encoding gene. *Appl Environ Microbiol* **71:**8362–8370.

105. **Shimizu T, Ohtani K, Hirakawa H, Ohshima K, Yamashita A, Shiba T, Ogasawara N, Hattori M, Kuhara S, Hayashi H.** 2002. Complete genome sequence of *Clostridium perfringens*, an anaerobic flesh-eater. *Proc Natl Acad Sci USA* **99:**996–1001.

106. **Nowell VJ, Kropinski AM, Songer JG, MacInnes JI, Parreira VR, Prescott JF.** 2012. Genome sequencing and analysis of a type A *Clostridium perfringens* isolate from a case of bovine clostridial abomasitis. *PLoS One* **7:**e32271.

107. **Shimizu T, Shima K, Yoshino K, Yonezawa K, Hayashi H.** 2002. Proteome and transcriptome analysis of the virulence genes regulated by the VirR/VirS system in *Clostridium perfringens*. *J Bacteriol* **184:**2587–2594.

108. **Ohtani K, Kawsar HI, Okumura K, Hayashi H, Shimizu T.** 2003. The VirR/VirS regulatory cascade affects transcription of plasmid-encoded putative virulence genes in *Clostridium perfringens* strain 13. *FEMS Microbiol Lett* **222:**137–141.

109. **Li J, Chen J, Vidal JE, McClane BA.** 2011. The Agr-like quorum-sensing system regulates sporulation and production of enterotoxin and beta2 toxin by *Clostridium perfringens* type A non-food-borne human gastrointestinal disease strain F5603. *Infect Immun* **79:**2451–2459.

110. **Caleo M, Schiavo G.** 2009. Central effects of tetanus and botulinum neurotoxins. *Toxicon* **54:**593–599.

111. **Finn C Jr, Silver R, Habig W, Hardegree M, Zon G, Garon C.** 1984. The structural gene for tetanus neurotoxin is on a plasmid. *Science* **224:**881–884.

112. **Brüggemann H, Bäumer S, Fricke WF, Wiezer A, Liesegang H, Decker I, Herzberg C, Martinez-Arias R, Merkl R, Henne A, Gottschalk G.** 2003. The genome sequence of *Clostridium tetani*, the causative agent of tetanus disease. *Proc Natl Acad Sci USA* **100:**1316–1321.

113. **Bruggemann H, Bauer R, Raffestin S, Gottschalk G.** 2004. Characterization of a heme oxygenase of *Clostridium tetani* and its possible role in oxygen tolerance. *Arch Microbiol* **182:**259–263.

114. **Raffestin S, Dupuy B, Marvaud JC, Popoff MR.** 2005. BotR/A and TetR are alternative RNA polymerase sigma factors controlling the expression of the neurotoxin and associated protein genes in *Clostridium botulinum* type A and *Clostridium tetani*. *Mol Microbiol* **55:**235–249.

115. **Zhang Z, Korkeala H, Dahlsten E, Sahala E, Heap JT, Minton NP, Lindstrom M.** 2013. Two-component signal transduction system CBO0787/CBO0786 represses transcription from botulinum neurotoxin promoters in *Clostridium botulinum* ATCC 3502. *PLoS Pathog* **9:**e1003252.

116. **Connan C, Brueggemann H, Mazuet C, Raffestin S, Cayet N, Popoff MR.** 2012. Two-component systems are involved in the regulation of botulinum neurotoxin synthesis in *Clostridium botulinum* type A strain Hall. *PLoS One* **7:**e41848.

117. **Cooksley CM, Davis IJ, Winzer K, Chan WC, Peck MW, Minton NP.** 2010. Regulation of neurotoxin production and sporulation by a putative agrBD signaling system in proteolytic *Clostridium botulinum*. *Appl Environ Microbiol* **76:**4448–4460.

118. **Connan C, Deneve C, Mazuet C, Popoff MR.** 2013. Regulation of toxin synthesis in *Clostridium botulinum* and *Clostridium tetani*. *Toxicon* **75:**90–100.

119. **Hill KK, Smith TJ, Helma CH, Ticknor LO, Foley BT, Svensson RT, Brown JL, Johnson EA, Smith LA, Okinaka RT, Jackson PJ, Marks JD.** 2007. Genetic diversity among botulinum neurotoxin-producing clostridial strains. *J Bacteriol* **189:**818–832.

120. **Hill KK, Xie G, Foley BT, Smith TJ, Munk AC, Bruce D, Smith LA, Brettin TS, Detter JC.** 2009. Recombination and insertion events involving the botulinum neurotoxin complex genes in *Clostridium botulinum* types A, B, E and F and *Clostridium butyricum* type E strains. *BMC Biol* **7:**66.

121. **Barash JR, Arnon SS.** 2014. A novel strain of *Clostridium botulinum* that produces type B and type H botulinum toxins. *J Infect Dis* **209:**183–191.

122. **Dover N, Barash JR, Hill KK, Xie G, Arnon SS.** 2014. Molecular characterization of a novel botulinum neurotoxin type H gene. *J Infect Dis* **209:**192–202.

123. **Hill KK, Smith TJ.** 2013. Genetic diversity within *Clostridium botulinum* serotypes, botulinum neurotoxin gene clusters and toxin subtypes. *Curr Top Microbiol Immunol* **364:**1–20.

124. **Popoff MR, Bouvet P.** 2013. Genetic characteristics of toxigenic clostridia and toxin gene evolution. *Toxicon* **75:**63–89.

125. **Skarin H, Segerman B.** 2011. Horizontal gene transfer of toxin genes in *Clostridium botulinum*: involvement of mobile elements and plasmids. *Mobile Genetic Elements* **1:**213–215.

126. **Sakaguchi Y, Hayashi T, Kurokawa K, Nakayama K, Oshima K, Fujinaga Y, Ohnishi M, Ohtsubo E, Hattori M, Oguma K.** 2005. The genome sequence of *Clostridium botulinum* type C neurotoxin-converting phage and the molecular mechanisms of unstable lysogeny. *Proc Natl Acad Sci USA* **102:**17472–17477.

127. **Sakaguchi Y, Hayashi T, Yamamoto Y, Nakayama K, Zhang K, Ma S, Arimitsu H, Oguma K.** 2009. Molecular analysis of an extrachromosomal element containing the C2 toxin gene discovered in *Clostridium botulinum* type C. *J Bacteriol* **191:**3282–3291.

128. **Sebaihia M, Peck MW, Minton NP, Thomson NR, Holden MT, Mitchell WJ, Carter AT, Bentley SD,**

Mason DR, Crossman L, Paul CJ, Ivens A, Wells-Bennik MH, Davis IJ, Cerdeno-Tarraga AM, Churcher C, Quail MA, Chillingworth T, Feltwell T, Fraser A, Goodhead I, Hance Z, Jagels K, Larke N, Maddison M, Moule S, Mungall K, Norbertczak H, Rabbinowitsch E, Sanders M, Simmonds M, White B, Whithead S, Parkhill J. 2007. Genome sequence of a proteolytic (group I) *Clostridium botulinum* strain Hall A and comparative analysis of the clostridial genomes. *Genome Res* **17:**1082–1092.

129. **Marshall KM, Bradshaw M, Pellett S, Johnson EA.** 2007. Plasmid encoded neurotoxin genes in *Clostridium botulinum* serotype A subtypes. *Biochem Biophys Res Commun* **361:**49–54.

130. **Zhang Z, Hintsa H, Chen Y, Korkeala H, Lindstrom M.** 2013. Plasmid-borne type E neurotoxin gene clusters in *Clostridium botulinum* strains. *Appl Environ Microbiol* **79:**3856–3859.

131. **Franciosa G, Maugliani A, Scalfaro C, Aureli P.** 2009. Evidence that plasmid-borne botulinum neurotoxin type B genes are widespread among *Clostridium botulinum* serotype B strains. *PLoS One* **4:**e4829.

132. **Wang X, Maegawa T, Karasawa T, Kozaki S, Tsukamoto K, Gyobu Y, Yamakawa K, Oguma K, Sakaguchi Y, Nakamura S.** 2000. Genetic analysis of type E botulinum toxin-producing *Clostridium butyricum* strains. *Appl Environ Microbiol* **66:**4992–4997.

133. **Marshall KM, Bradshaw M, Johnson EA.** 2010. Conjugative botulinum neurotoxin-encoding plasmids in *Clostridium botulinum*. *PLoS One* **5:**e11087.

134. **Pilo P, Frey J.** 2011. *Bacillus anthracis*: molecular taxonomy, population genetics, phylogeny and pathoevolution. *Infect Genet Evol* **11:**1218–1224.

135. **Koehler TM.** 2009. *Bacillus anthracis* physiology and genetics. *Mol Aspects Med* **30:**386–396.

136. **Fouet A.** 2009. The surface of *Bacillus anthracis*. *Mol Aspects Med* **30:**374–385.

137. **Bradley KA, Mogridge J, Mourez M, Collier RJ, Young JA.** 2001. Identification of the cellular receptor for anthrax toxin. *Nature* **414:**225–229.

138. **Scobie HM, Rainey GJ, Bradley KA, Young JA.** 2003. Human capillary morphogenesis protein 2 functions as an anthrax toxin receptor. *Proc Natl Acad Sci USA* **100:** 5170–5174.

139. **Young JA, Collier RJ.** 2007. Anthrax toxin: receptor binding, internalization, pore formation, and translocation. *Annu Rev Biochem* **76:**243–265.

140. **Collier RJ.** 2009. Membrane translocation by anthrax toxin. *Mol Aspects Med* **30:**413–422.

141. **Okinaka R, Cloud K, Hampton O, Hoffmaster A, Hill K, Keim P, Koehler T, Lamke G, Kumano S, Manter D, Martinez Y, Ricke D, Svensson R, Jackson P.** 1999. Sequence, assembly and analysis of pX01 and pX02. *J Appl Bacteriol* **87:**261–262.

142. **Okinaka RT, Cloud K, Hampton O, Hoffmaster AR, Hill KK, Keim P, Koehler TM, Lamke G, Kumano S, Mahillon J, Manter D, Martinez Y, Ricke D, Svensson R, Jackson PJ.** 1999. Sequence and organization of pXO1, the large *Bacillus anthracis* plasmid harboring the anthrax toxin genes. *J Bacteriol* **181:**6509–6515.

143. **Van der Auwera G, Mahillon J.** 2005. Tn*XO1*, a germination-associated class II transposon from *Bacillus anthracis*. *Plasmid* **53:**251–257.

144. **Bourgogne A, Drysdale M, Hilsenbeck SG, Peterson SN, Koehler TM.** 2003. Global effects of virulence gene regulators in a *Bacillus anthracis* strain with both virulence plasmids. *Infect Immun* **71:**2736–2743.

145. **Mignot T, Mock M, Fouet A.** 2003. A plasmid-encoded regulator couples the synthesis of toxins and surface structures in *Bacillus anthracis*. *Mol Microbiol* **47:** 917–927.

146. **Kern JW, Schneewind O.** 2008. BslA, a pXO1-encoded adhesin of *Bacillus anthracis*. *Mol Microbiol* **68:** 504–515.

147. **Candela T, Mock M, Fouet A.** 2005. CapE, a 47-amino-acid peptide, is necessary for *Bacillus anthracis* polyglutamate capsule synthesis. *J Bacteriol* **187:**7765–7772.

148. **Van der Auwera GA, Andrup L, Mahillon J.** 2005. Conjugative plasmid pAW63 brings new insights into the genesis of the *Bacillus anthracis* virulence plasmid pXO2 and of the *Bacillus thuringiensis* plasmid pBT9727. *BMC Genomics* **6:**103.

149. **Kolstø A, Tourasse NJ, Økstad OA.** 2009. What sets *Bacillus anthracis* apart from other *Bacillus* species? *Annu Rev Microbiol* **63:**451–476.

150. **Hoffmaster AR, Ravel J, Rasko DA, Chapman GD, Chute MD, Marston CK, De BK, Sacchi CT, Fitzgerald C, Mayer LW, Maiden MC, Priest FG, Barker M, Jiang L, Cer RZ, Rilstone J, Peterson SN, Weyant RS, Galloway DR, Read TD, Popovic T, Fraser CM.** 2004. Identification of anthrax toxin genes in a *Bacillus cereus* associated with an illness resembling inhalation anthrax. *Proc Natl Acad Sci USA* **101:**8449–8454.

151. **Wilson MK, Vergis JM, Alem F, Palmer JR, Keane-Myers AM, Brahmbhatt TN, Ventura CL, O'Brien AD.** 2011. *Bacillus cereus* G9241 makes anthrax toxin and capsule like highly virulent *B. anthracis* Ames but behaves like attenuated toxigenic nonencapsulated *B. anthracis* Sterne in rabbits and mice. *Infect Immun* **79:**3012–3019.

152. **Klee SR, Brzuszkiewicz EB, Nattermann H, Bruggemann H, Dupke S, Wollherr A, Franz T, Pauli G, Appel B, Liebl W, Couacy-Hymann E, Boesch C, Meyer FD, Leendertz FH, Ellerbrok H, Gottschalk G, Grunow R, Liesegang H.** 2010. The genome of a *Bacillus* isolate causing anthrax in chimpanzees combines chromosomal properties of *B. cereus* with *B. anthracis* virulence plasmids. *PLoS One* **5:**e10986.

153. **Leendertz FH, Ellerbrok H, Boesch C, Couacy-Hymann E, Matz-Rensing K, Hakenbeck R, Bergmann C, Abaza P, Junglen S, Moebius Y, Vigilant L, Formenty P, Pauli G.** 2004. Anthrax kills wild chimpanzees in a tropical rainforest. *Nature* **430:**451–452.

154. **Grynberg M, Li Z, Szczurek E, Godzik A.** 2007. Putative type IV secretion genes in *Bacillus anthracis*. *Trends Microbiol* **15:**191–195.

155. **Green BD, Battisti L, Koehler TM, Thorne CB, Ivins BE.** 1985. Demonstration of a capsule plasmid in *Bacillus anthracis*. *Infect Immun* **49:**291–297.
156. **Reddy A, Battisti L, Thorne CB.** 1987. Identification of self-transmissible plasmids in four *Bacillus thuringiensis* subspecies. *J Bacteriol* **169:**5263–5270.
157. **Stenfors Arnesen LP, Fagerlund A, Granum PE.** 2008. From soil to gut: *Bacillus cereus* and its food poisoning toxin. *FEMS Microbiol Rev* **32:**579–606.
158. **Bravo A, Gill SS, Soberon M.** 2007. Mode of action of *Bacillus thuringiensis* Cry and Cyt toxins and their potential for insect control. *Toxicon* **49:**423–435.
159. **Schnepf E, Crickmore N, Van Rie J, Lereclus D, Baum J, Feitelson J, Zeigler DR, Dean DH.** 1998. *Bacillus thuringiensis* and its pesticidal crystal proteins. *Microbiol Mol Biol Rev* **62:**775–806.
160. **Lereclus D, Agaisse H, Grandvalet C, Salamitou S, Gominet M.** 2000. Regulation of toxin and virulence gene transcription in *Bacillus thuringiensis*. *Int J Med Microbiol* **290:**295–299.
161. **Pardo-Lopez L, Soberon M, Bravo A.** 2013. *Bacillus thuringiensis* insecticidal three-domain Cry toxins: mode of action, insect resistance and consequences for crop protection. *FEMS Microbiol Rev* **37:**3–22.
162. **Soberon M, Lopez-Diaz JA, Bravo A.** 2013. Cyt toxins produced by *Bacillus thuringiensis*: a protein fold conserved in several pathogenic microorganisms. *Peptides* **41:**87–93.
163. **Doggett NA, Stubben CJ, Chertkov O, Bruce DC, Detter JC, Johnson SL, Han CS.** 2013. Complete genome sequence of *Bacillus thuringiensis* serovar *israelensis* strain HD-789. *Genome Announc* **1:**e01023-13.
164. **Baum JA, Malvar T.** 1995. Regulation of insecticidal crystal protein production in *Bacillus thuringiensis*. *Mol Microbiol* **18:**1–12.
165. **Reyes-Ramirez A, Ibarra JE.** 2008. Plasmid patterns of *Bacillus thuringiensis* type strains. *Appl Environ Microbiol* **74:**125–129.
166. **Gonzalez JM Jr, Brown BJ, Carlton BC.** 1982. Transfer of *Bacillus thuringiensis* plasmids coding for delta-endotoxin among strains of *B. thuringiensis* and *B. cereus*. *Proc Natl Acad Sci USA* **79:**6951–6955.
167. **Gammon K, Jones GW, Hope SJ, de Oliveira CM, Regis L, Silva Filha MH, Dancer BN, Berry C.** 2006. Conjugal transfer of a toxin-coding megaplasmid from *Bacillus thuringiensis* subsp. *israelensis* to mosquitocidal strains of *Bacillus sphaericus*. *Appl Environ Microbiol* **72:**1766–1770.
168. **Wang P, Zhang C, Zhu Y, Deng Y, Guo S, Peng D, Ruan L, Sun M.** 2013. The resolution and regeneration of a cointegrate plasmid reveals a model for plasmid evolution mediated by conjugation and *oriT* site-specific recombination. *Environ Microbiol* **15:**3305–3318.
169. **Hu X, Hansen BM, Yuan Z, Johansen JE, Eilenberg J, Hendriksen NB, Smidt L, Jensen GB.** 2005. Transfer and expression of the mosquitocidal plasmid pBtoxis in *Bacillus cereus* group strains. *FEMS Microbiol Lett* **245:**239–247.
170. **Huang T, Liu J, Song F, Shu C, Qiu J, Guan X, Huang D, Zhang J.** 2004. Identification, distribution pattern of IS*231* elements in *Bacillus thuringiensis* and their phylogenetic analysis. *FEMS Microbiol Lett* **241:**27–32.
171. **Dereeper A, Guignon V, Blanc G, Audic S, Buffet S, Chevenet F, Dufayard JF, Guindon S, Lefort V, Lescot M, Claverie JM, Gascuel O.** 2008. Phylogeny.fr: robust phylogenetic analysis for the non-specialist. *Nucleic Acids Res* **36:**W465–W469.
172. **Dereeper A, Audic S, Claverie JM, Blanc G.** 2010. BLAST-EXPLORER helps you building datasets for phylogenetic analysis. *BMC Evol Biol* **10:**8.
173. **Yonogi S, Matsuda S, Kawai T, Yoda T, Harada T, Kumeda Y, Gotoh K, Hiyoshi H, Nakamura S, Kodama T, Ilda T.** 2014. BEC, a novel enterotoxin of *Clostridium perfringens* found in human clinical isolates from acute gastroenteritis outbreaks. *Infect Immun* **82:**2390–2399.

Plasmids—Biology and Impact in Biotechnology and Discovery
Edited by Marcelo E. Tolmasky and Juan C. Alonso

doi:10.1128/microbiolspec.PLAS-0002-2013

Daria Van Tyne[1]
Michael S. Gilmore[1]

28

Virulence Plasmids of Nonsporulating Gram-Positive Pathogens

VIRULENCE PLASMIDS IN *STAPHYLOCOCCUS AUREUS*

S. aureus: Virulence and Pathogenesis

Infection with *S. aureus* can result in a wide variety of diseases, including wound infections, toxic shock, food poisoning, endocarditis, pneumonia, and septicemia (1). Virulence and drug resistance often occur together, as recent outbreak strains of methicillin-resistant *S. aureus* also produce a number of different virulence factors (2). It is perhaps not surprising that a bacterium capable of causing such a wide array of diseases possesses a diverse repertoire of virulence factors. A consequence of this versatility is that the pathogenesis of *S. aureus* is usually multifactorial (3).

S. aureus is capable of producing a number of extracellular toxins, including cytolytic toxins (α-toxin, β-toxin, γ-toxin), enterotoxins, toxic shock syndrome toxin (TSST-1), and exfoliative toxins. Although staphylococcal virulence is seldom attributable to one factor alone, different toxins have been linked to different types of staphylococcal infection. For example, staphylococcal enterotoxins are associated with food poisoning (4, 5), and exfoliative toxins have been linked to staphylococcal scalded-skin syndrome (SSSS) (6–9). Many of these toxins function as superantigens, causing widespread T-lymphocyte activation and resulting in systemic shock (10–15). Cytolytic toxins, enterotoxins, and TSST-1 have all been shown to function as superantigens, while the association of superantigenic activity with exfoliative toxins is less clear (9, 16). Several staphylococcal antigens have been found to occur on mobile genetic elements, including the plasmid-encoded staphylococcal enterotoxin D (SED), staphylococcal enterotoxin J (SEJ), and exfoliative toxin B (ETB), which are discussed in greater detail below.

Enterotoxin

Staphylococcal enterotoxins (SEs) comprise a large family of related proteins similar to streptococcal pyrogenic exotoxins (15). These heat-stabile, pepsin-resistant proteins include SEA-SEE, SEG-SEI, and SER-SET (4, 5). Staphylococcal food poisoning results primarily from the ingestion of contaminated meat or poultry, and to a lesser

[1]Department of Ophthalmology, Massachusetts Eye and Ear Infirmary, Boston, MA 02114, and Department of Microbiology and Immunobiology, Harvard Medical School, Boston, MA 02114.

degree from contaminated fish, shellfish, and milk (4, 5, 17). Enterotoxin production within different *S. aureus* strains may vary, but food poisoning often results from the ingestion of any one or a combination of preformed enterotoxins (18). Historically, the most common enterotoxin associated with staphylococcal food poisoning is staphylococcal enterotoxin A (SEA), followed by SED and SEB (19, 20).

The gene encoding SED, *sed*, was determined to be present on a 27.6-kilobase penicillinase plasmid designated pIB485, which was isolated from strain KSI1410 and also contains genes for resistance to penicillin and cadmium sulfate (Fig. 1) (21). Upon curing the strain of this plasmid, the ability to produce SED was lost. Cloning and expression of SED in *Escherichia coli* from a restriction fragment of pIB485 proved that *sed* was encoded by pIB485. Transcription of *sed* was found to be regulated by the accessory gene regulator system (*agr*), a two-component system that regulates the transcription of virulence factors in a cell density-dependent manner (22–24). Regulation of *sed* by *agr* results in its induction during postexponential growth (23). A study by Zhang et al. (23) identified a second enterotoxin-like gene within pIB485. This gene was termed *selj* and encodes staphylococcal enterotoxin-like protein J (SEJ) (Fig. 1).

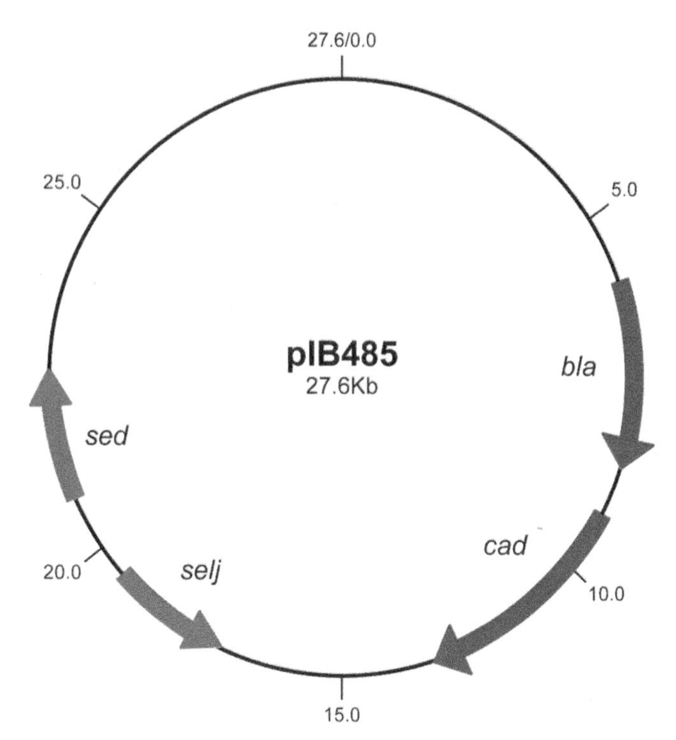

Figure 1 Plasmid map of pIB485, encoding staphylococcal enterotoxins SED (*sed*) and SEJ (*selj*). Toxin genes are colored red. Genes encoding resistance to cadmium sulfate (*cad*) as well as resistance to beta-lactams (*bla*) are colored dark blue. doi:10.1128/microbiolspec.PLAS-0002-2013.f1

More recently, additional plasmid-encoded enterotoxins have been found in *S. aureus*. These include the enterotoxin-like protein SER, which shows sequence similarity to SEG and is encoded by a plasmid similar to pIB485 named pF5, for the Fukuoka 5 strain from which it was isolated (25). pF5 was also found to encode staphylococcal enterotoxins S and T (SES and SET), as well as several accessory genes. The contributions of SEJ, SER, SES, and SET to the pathogenesis of staphylococcal food poisoning, either singularly or in combination with SED, have not yet been addressed.

Exfoliative Toxin

S. aureus is capable of producing four exfoliative toxins: exfoliative toxin A (ETA), ETB, ETC, and ETD. Exfoliative toxins are most common among phage group II isolates of *S. aureus* (26, 27), and cause peeling and blistering lesions of the skin (28). ETA, ETB, and ETD are glutamate-specific serine proteases that cleave desmoglein I, a desmosomal intracellular adhesion molecule expressed in the superficial layers of the epidermis (29). ETC was first isolated from a horse lesion and appears to affect horses, chickens, and mice (30), but its role in human disease, if any, has not been characterized. Exfoliative toxins have been associated with SSSS, a group of diseases including Ritter's disease, toxic epidermal necrosis, bullous impetigo, and erythema (7, 31). It has been suggested that high nasal carriage rates of *S. aureus* in adults, as well as the apparent protective nature of exfoliative toxin antibodies in SSSS (32–35), account for why SSSS is seen primarily in young children and the immunocompromised (36, 37). Finally, fewer than 5% of *S. aureus* clinical isolates harbor genes for exfoliative toxins (38, 39), suggesting that these genes are acquired by horizontal gene transfer and can be passed between bacterial strains on mobile genetic elements.

While ETA, ETB, and ETD share similar gene sequences and modes of action, they are found in distinct genetic contexts. The gene encoding ETA is normally localized to the *S. aureus* chromosome (40), but it has also been found within the genome of an integrated temperate phage (41). ETD was identified on a novel *S. aureus* chromosomal pathogenicity island along with a gene for the epidermal cell differentiation inhibitor-B (EDIN-B) (42). ETD appears to act similarly to ETA, but it has also been found in patients with infections other than bullous impetigo or SSSS. Additionally, ETD was found to be present in approximately 10% of *S. aureus* clinical isolates in France (43), suggesting that it may play a role in a wider array of staphylococcal infections than ETA and ETB.

In contrast to ETA and ETD, the locus for ETB is typically plasmid-encoded (Fig. 2) (44, 45). Formal evidence that the structural gene for ETB, designated *etb*, occurred on a plasmid was provided by Jackson and Iandolo (46). It was also discovered that the plasmid carries a cadmium sulfate resistance gene, which is part of a putative transposon (47). ETB is found on plasmids that vary in size from approximately 35 to 60 kilobases, yet the genetic organization of these plasmids appears to be conserved (Fig. 2B) (48). Interestingly, the exfoliative toxins of *Staphylococcus hycius*, SHETA and SHETB, which are associated with exudative epidermitis in pigs and chickens, are similar to ETA and ETB; SHETA is produced by plasmid-free strains, while SHETB production is dependent on the presence of a large plasmid (49, 50).

A staphylococcal bacteriocin two-component lantibiotic system, named BacR1/C55, is also located on various ETB-producing plasmids of *S. aureus* (51). BacR1 was initially characterized by Rogolsky and Wiley (52) and was more extensively examined by Crupper et al. (53). Staphylococcin activity was attributed to the synergistic activity of two peptides, called C55α and C55β. The respective structural genes *sacA* and *sacβA*, as well as potential processing genes *sacM1* and *sacT*, are all organized within the lantibiotic operon shown in Fig. 2 (51). The *sacA* and *sacA* genes were detected by PCR only in strains that were also positive for ETB. Although found on different plasmids, C55 and BacR1 appear to constitute the same lantibiotic system, which also resembles the two-component bacteriocin lacticin 3147 from *Lactococcus lactis* (51).

In addition to bacteriocin activity attributable to either BacR1 or C55, a novel virulence factor has also been linked to ETB-encoding plasmids. Through sequencing, Yamaguchi et al. (48) identified a protein potentially capable of ADP-ribosylating Rho GTPases, which are members of the Ras superfamily of proteins involved in cytoskeletal network regulation within eukaryotic cells. Inactivation of Rho GTPases has been shown to inhibit the chemotactic and phagocytic activities of immune cells during infection (54), inhibit the differentiation of structural cells (55), and contribute to the dissemination of bacteria through the vasculature and tissues (56). Inactivation of Rho GTPases by *S. aureus* is accomplished by exotoxins of the EDIN family (48, 55, 57, 58). The EDIN determinant on ETB-containing plasmids was designated EDIN-C (*ednC*) (Fig. 2). EDIN-C has been found in the vast majority of EDIN-carrying *S. aureus* isolates and was enriched among isolates recovered from deep-seated infections (59). This further suggests that EDIN-C is an important contributor to virulence of *S. aureus* infections in humans.

More recently, next-generation DNA sequencing has allowed for a systematic analysis of virulence factor-encoding plasmids in *S. aureus* (60). This effort has more than doubled the number of known plasmid groups, giving a fuller picture of the dissemination of plasmids among different strains of *S. aureus*, as well as the drug resistance and virulence traits that are associated with particular plasmid groups. The diversity among *S. aureus* plasmids is so great that plasmids can be used to track the spread of drug-resistant isolates within a hospital setting (61). Importantly, this effort has identified many new *S. aureus* plasmids, which likely contain additional virulence factors that have yet to be described.

VIRULENCE PLASMIDS IN *ENTEROCOCCUS FAECALIS*

E. faecalis: Virulence and Pathogenesis

E. faecalis and *Enterococcus faecium* are members of the normal commensal flora of the gastrointestinal tracts of humans and other animals (62). However, *E. faecalis* and *E. faecium* are also leading causes of hospital-acquired infection, including urinary tract infections, bacteremia, endocarditis, and intra-abdominal infections (63–68). The treatment of enterococcal infections is particularly challenging because antibiotic resistance enables the bacteria to survive standard therapies (69). Furthermore, the ability of enterococci to survive harsh environments, such as low pH, detergents, and bile salts, enables them to persist in the hospital environment (70–72). Vancomycin has been considered a drug of last resort for many enterococcal infections, but vancomycin-resistant enterococci, particularly *E. faecium*, are now commonly found in hospital settings (73). Although the majority of vancomycin-resistant enterococci infections are due to *E. faecium*, nearly 75% of all enterococcal infections are caused by *E. faecalis* (66, 74). This may be attributable to the additional virulence determinants that *E. faecalis* possesses, some of which are described below.

E. faecalis produces a number of factors that are presumably involved in adhesion to host tissues and thereby contribute to host-pathogen interactions, as well as to the virulence and pathogenesis of enterococcal infections. These factors include enterococcal surface protein (Esp), aggregation substance, and the cytolysin. *E. faecalis* also expresses a capsular polysaccharide, which helps form the physical interface between bacteria and

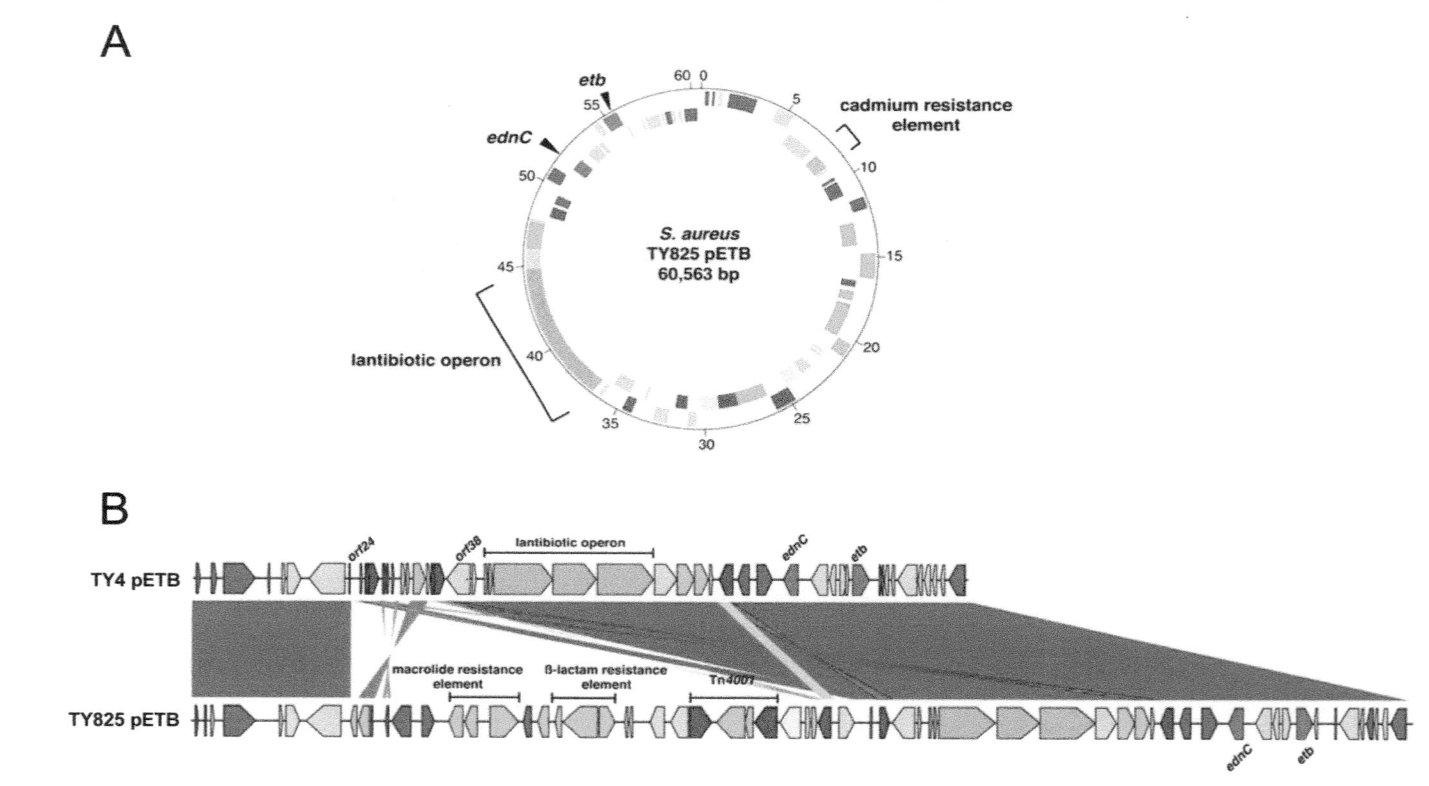

Figure 2 (**A**) Plasmid map of a representative staphylococcal enterotoxin B (ETB)-expressing plasmid, TY825 pETB. The outer circle shows genes that are transcribed clockwise; genes in the inner circle are transcribed counterclockwise. Genes in red are pathogenic factors; genes in green encode antibiotic resistances; genes in blue are involved in DNA replication, recombination, and repair; genes in light blue are transcriptional regulators; genes in purple are transposases; genes in yellow are involved in conjugal transfer; genes in orange encode the BacR1/C55 lantibiotic operon; and genes in gray are conserved ORFs. (**B**) Structural comparison of TY4 pETB and TY825 pETB plasmids of *S. aureus*. Shading indicates homologous regions. Figure is adapted from reference 209. doi:10.1128/microbiolspec.PLAS-0002-2013.f2

host. The capsular polysaccharide varies in structure, possibly because of immune pressure (75, 76). Because it is easily detectable on the surface of the bacterial cell and is the target of complement-mediated opsonophagocytosis, the capsular polysaccharide has been used in typing schemes to classify different serotypes of *E. faecalis* (77). While the structure of the capsular polysaccharide is variable among different *E. faecalis* strains, the genes encoding the capsule have thus far been localized to the bacterial chromosome (78, 79).

The enterococcal surface protein Esp is an adhesin that has been localized to a 150-kilobase pathogenicity island in strains of *E. faecalis* (80, 81). Esp has been associated with outbreaks of vancomycin-resistant enterococci in hospitals (82), antibiotic resistance and biofilm formation (83, 84), and adherence of *E. faecalis* in the pathogenesis of urinary tract infections (80, 85). Aggregation substance is a surface protein that has similarly been shown to contribute to disease pathogenesis, also via enhanced adhesion and biofilm formation (86–92). Finally, some *E. faecalis* strains produce a cytolysin with both bacteriocidal and toxin activities (93, 94). The cytolysin operon is found, along with aggregation substance, on both pheromone-responsive plasmids (95), as well as within the chromosomal pathogenicity island on which Esp is also found (81). Pheromone-responsive plasmids in enterococci are briefly reviewed below, followed by a discussion of the *E. faecalis* pheromone-responsive virulence plasmids that encode for aggregation substance and the cytolysin.

Pheromone-Responsive Plasmids in Enterococci

Plasmid-free enterococci secrete over a dozen distinct peptide pheromones of seven to eight amino acids in length, which induce a mating response from donors carrying plasmids that specifically respond to each of these peptides (96–101) (Fig. 3). Upon pheromone peptide secretion by recipient cells, donor cells containing the corresponding pheromone-responsive plasmids will express factors that promote physical interaction between donor and recipient cells, and thereby allow for conjugal DNA transfer. Aggregation substance on the surface of donor cells mediates clumping between donors and recipients via binding substance, a constituent of the cell wall of both plasmid-containing and plasmid-free enterococci (102). It is believed that binding substance consists in part of lipoteichoic acid (103–105). Aggregation provides the initial cell-cell contact required for conjugal transfer of plasmids from donor to recipient. In broth, the mating potential of donor cells upon exposure to peptide pheromones is at least 100,000 times greater than that exhibited by noninduced donor cells (106). Transfer dynamics likely differ between bacteria growing in biofilms

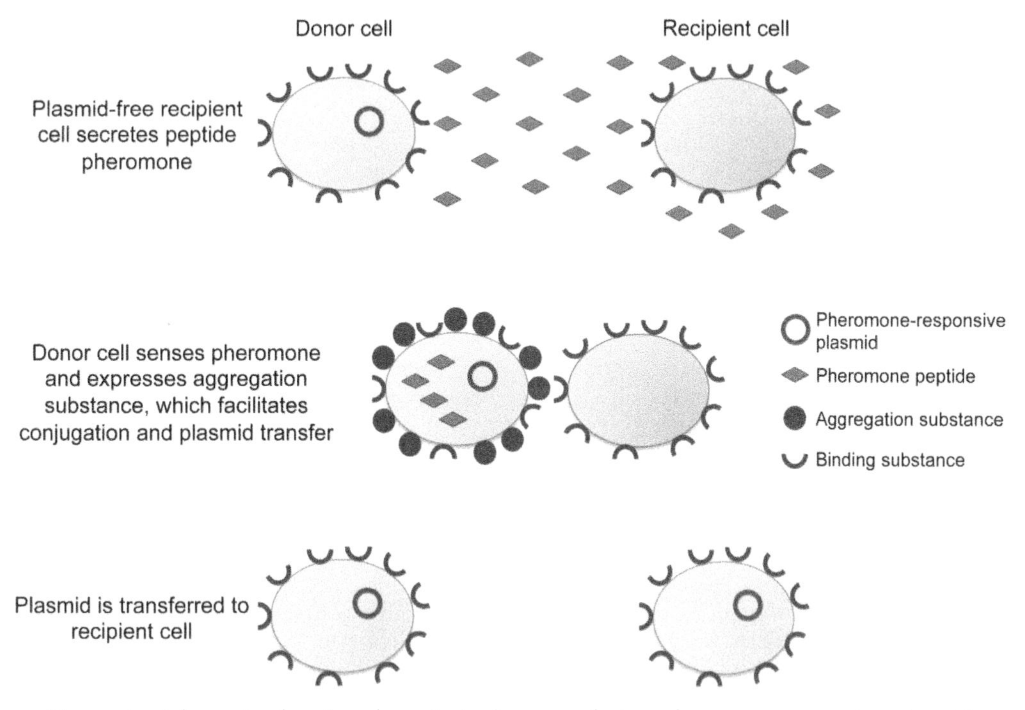

Figure 3 Schematic showing the principal events during pheromone-responsive plasmid transfer between *E. faecalis* cells. doi:10.1128/microbiolspec.PLAS-0002-2013.f3

compared to broth (107), but in both cases plasmid transfer depends upon cells sensing that they are in close proximity to one another. More than a dozen distinct pheromone-responsive plasmids have been identified within *E. faecalis* (108); the most well characterized of these include pAD1, pCF10, and pPD1. The mechanistic details of pheromone-responsive plasmid transfer in enterococci is covered in considerable detail elsewhere. However, the mechanism underlying transfer of pAD1 is also covered below.

Aggregation Substance

The gene encoding aggregation substance (*asa1*) is often found on pAD1, a 60-kilobase, pheromone-responsive, transmissible plasmid (98, 109). Aggregation substances of other pheromone-responsive plasmids are homologous to the aggregation substance of pAD1 (108), although these other plasmids respond to different pheromones. Aggregation substance found on pAD1 is a 137-kDa surface protein and has often been implicated in the adherence and virulence of enterococci (86, 87, 89, 110, 111). Aggregation substance also appears to contribute to the spread of cytolysin and antibiotic resistance determinants, via its role in the conjugal transfer of pheromone-responsive plasmids (112). The pheromone eliciting a response from pAD1-containing donor cells is designated cAD1 (113).

Genes encoding aggregation substance, functions involved in pheromone sensing and response, maintenance of plasmid pAD1, and the cytolysin operon, are shown in Fig. 4. The *asa1* gene is located adjacent to *sea1*, which encodes a surface exclusion protein that inhibits transfer between cells containing similar plasmids and thereby limits redundant plasmid transfer (Fig. 4A) (108). The *asa1* and *sea1* genes are adjacent to the Tra regulon, which encodes a number of open reading frames (ORFs) involved in pheromone sensing and regulation. Products of the *traA*, *traB*, *traD*, and *traE1*

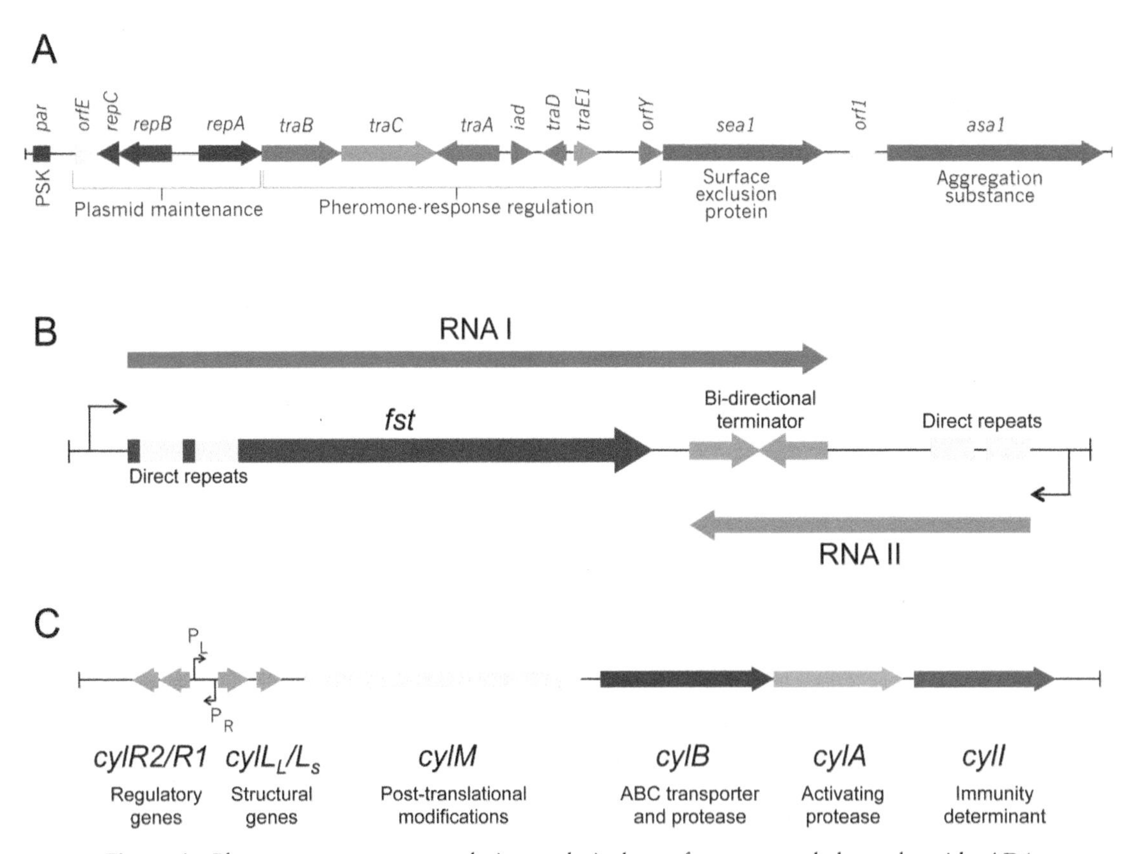

Figure 4 Pheromone-response regulation and virulence factors encoded on plasmid pAD1 **(A)** Genes encoded by pAD1 that are involved in plasmid transfer and pheromone sensing. Blue genes are involved in replication and maintenance, red genes are negative regulators of pheromone response, and green genes are positive regulators of pheromone response. **(B)** Detail of the postsegregation killing (PSK) *par* locus of pAD1, which encodes the Fst toxin. **(C)** Detailed schematic of individual genes within the cytolysin operon.
doi:10.1128/microbiolspec.PLAS-0002-2013.f4

genes are involved in the regulation of the pheromone response (106, 114–118), while TraC is a surface protein involved in pheromone sensing and functions as a specific binding factor (119). The *iad* gene encodes the pheromone inhibitor iAD1, which prevents self-induction of the pheromone response by inhibiting it at lower pheromone concentrations (120). Low concentrations of pheromone may arise from plasmid-containing or plasmid-free cells, at such a distance that an interaction allowing for conjugation is unlikely to occur.

Adjacent to genes associated with the transfer of pAD1 lies the *par* locus, a postsegregational killing mechanism (PSK) (Fig. 4B) (121). A region exhibiting a high degree of homology to the *par* determinant of pAD1 has also been identified in pCF10, another pheromone-responsive plasmid of *E. faecalis* (121). The primary function of PSKs is to maintain low-copy plasmids within a population of bacteria by killing plasmid-free segregants. The *par* locus within pAD1 encodes a toxin, Fst, which is believed to target chromosomal separation and/or cell division. Transcript RNA I contains the ORF believed to encode Fst toxin, a 33-amino acid peptide (122). Transcript RNA II is capable of occluding the 5′ end of RNA I via antisense transcription of direct repeats within RNA I, as well as the 3′ end of RNA I via the interaction of complementary transcriptional terminator stem-loops (122). RNA II is less stable than RNA I, so cells that do not maintain plasmids will not be able to inhibit toxin activity through the synthesis of antidote RNA II. PSKs are discussed in more detail elsewhere. The gene products of *repA* and *repB* are also involved in plasmid replication and maintenance (123, 124). RepA is responsible for initiation of replication, while RepBC, along with flanking repeat sequences, probably represents a ParAB/SopAB partitioning system, which is covered in detail elsewhere.

Although the exact occurrence of events, as well as the interplay of the various Tra regulon factors in the coordination of the pheromone response, continue to be somewhat controversial topics, a generalized scheme is presented in Fig. 5. In the uninduced state, TraA and a small RNA molecule resulting from the transcription of *traD* both negatively regulate levels of TraE1 (125). TraA accomplishes this negative regulation through DNA binding upstream of *iad* and *traE1*. When levels of the pheromone cAD1 surpass inhibition by iAD1, cAD1 binds directly to TraA, which loses its affinity for DNA and allows for transcription of *traE1* (126). TraE1 then goes on to activate production of other factors involved in conjugation (127).

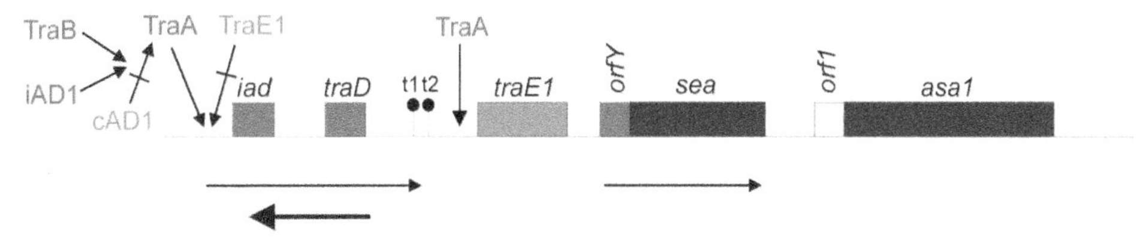

INDUCED:

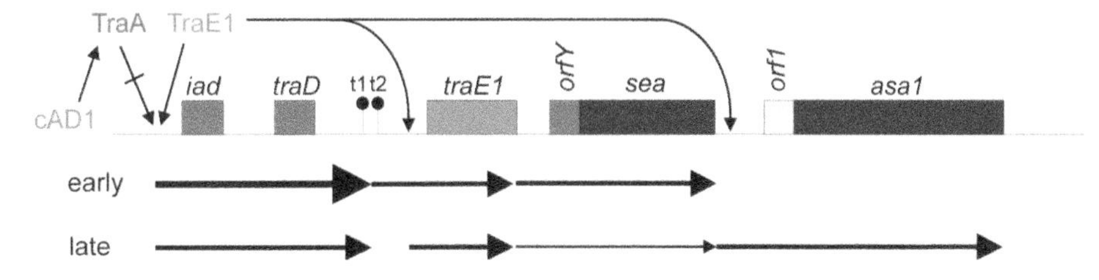

Figure 5 Regulation of pheromone sensing and plasmid transfer by the enterococcal Tra regulon. Genes encoding surface exclusion protein (*sea*) and aggregation substance (*asa1*) are shown in dark blue. Positive regulators are shown in green, and negative regulators are shown in red. Straight arrows below the genes indicate transcripts detected in the uninduced and induced states, and arrow thickness indicates relative transcript abundance. (Adapted in part from *Cell*, 73:9–12, 1993, with permission from Elsevier [87], and from the *Journal of Bacteriology*, **182**:3816–3825, 2000, with permission from ASM [117].)
doi:10.1128/microbiolspec.PLAS-0002-2013.f5

Apart from its role in conjugal plasmid transfer, aggregation substance has also been linked to the virulence of *E. faecalis*. Aggregation substance was postulated to have a role in the adherence of enterococci to host cells, based on the presence of two Arg-Gly-Asp (RGD) motifs within its protein sequence. These motifs were hypothesized to mediate interactions between fibronectin and integrins on mammalian cell surfaces (128). *E. faecalis* expressing aggregation substance was shown to more readily bind renal epithelial cells in culture, and this enhanced adherence was blocked by the addition of a synthetic RGD peptide, but not by a peptide of divergent sequence (86). Mutating the RGD motifs in the pCF10-encoded aggregation substance gene (*prgB*) affected the virulence of *E. faecalis* in a rabbit endocarditis model (129), but not the invasion of human intestinal epithelial cells *in vitro* (130). The latter study focused instead on the aggregation domain being critical for internalization. Finally, expression of aggregation substance has been reported to be induced *in vivo*, via an interaction between plasma components and the inhibitor peptide pheromone, which renders the latter unable to inhibit endogenous pheromone (91).

Few studies of the role of aggregation substance in the pathogenesis of infection have distinguished between the ability of aggregation substance to promote bacterial clumping versus its potential contribution to the direct interaction between bacteria and specific types of human or animal cells (131). The aggregation domain of aggregation substance has been shown to mediate internalization of bacteria by enterocytes (130, 132). Expression of aggregation substance also correlates with enhanced uptake of enterococci by intestinal epithelial cells, but this increase in uptake did not result in an increase in bacterial translocation across the intestinal epithelium *in vitro* (89). Expression of aggregation substance was also found to affect the function of polymorphonuclear neutrophils and macrophages, by promoting the phagocytosis of enterococci by these cells (133, 134). Interestingly, aggregation substance was observed to promote the intracellular survival of enterococci within polymorphonuclear neutrophils by inhibiting acidification of the phagolysosome (110) and within macrophages by inhibiting respiratory burst (134). In addition, the peptide pheromones involved in the induction of aggregation substance expression are also potentially able to alter the host immune response, in that the pheromones appear to be chemotactic for polymorphonuclear neutrophils (135).

Aggregation substance does not appear to contribute to the virulence of *E. faecalis* in either a rabbit model of endophthalmitis (136) or in *Caenorhabditis elegans* infection (137). However, in a rabbit model of endocarditis, a disease in which vegetations of platelets and fibrin resulting from bacterial infection are associated with inflammation of the heart valves and lining, aggregation substance from both pAD1 and pCF10 caused an increase in the size of cardiac vegetations (88, 90, 129). No effect of aggregation substance on vegetation size was observed in a rat endocarditis model (138), suggesting that the effects of aggregation substance on pathogenesis may be species- or model-specific. The role of aggregation substance in an infection may be to promote the formation of a quorum of cells in a localized microenvironment through clumping, which could affect the expression of factors now known to be quorum regulated, such as the cytolysin (139). Aggregation substance in the absence of cytolysin expression was shown to increase the size of vegetations in endocarditis (88). When expressed from a plasmid that also encodes the cytolysin, such as pAD1, vegetations were observed to be smaller in mass but much more toxic. Thus, aggregation substance, by increasing the number of bacteria at a site of infection, could enhance production of cytolysin, though this has not yet been proven directly.

Apart from its potential to aid in achieving a quorum, the ability of aggregation substance to spread pheromone-responsive plasmids throughout a bacterial population could also enhance the virulence of *E. faecalis*, via the concomitant spread of cytolysin and antibiotic resistance determinants. In a study by Huycke et al. (140), an *in vivo* model was used to examine pheromone-responsive plasmid transfer between *E. faecalis* strains in the intestine. This study found that pAD1 and its derivatives were transferred between bacteria at high frequency, independent of antibiotic selection. While investigating the contribution of aggregation substance to endocarditis in a rabbit model, Hirt et al. (91) found that pCF10 was also transferred at high frequency *in vivo*. Although the conjugative spread of pheromone-responsive plasmids *in vivo* has been well documented, the precise consequence of the spread of these plasmids and their determinants during infection is not well understood.

Cytolysin

The cytolysin produced by *E. faecalis* is structurally and functionally unique (141, 142). It is capable of lysing erythrocytes, eukaryotic cells, and bacteria and may also function as a signaling molecule to communicate with other bacteria and probe the environment (143, 144). Since its identification by Todd in 1934 from group D streptococci (*Streptococcus faecalis* subspecies *zymogenes* has since been reclassified as *E. faecalis*)

(93), it has been characterized genetically and shown to contribute to virulence in virtually all models of *E. faecalis* infection tested (64, 88, 137, 145, 146). The cytolytic phenotype has also been reported to be more prevalent among infection-derived isolates of *E. faecalis* compared to noninfection isolates (147–149).

The determinant for cytolysin was originally suspected to be plasmid-encoded, because the hemolytic phenotype was observed to be a variable trait of *E. faecalis*, and this trait could be transferred to plasmid-free recipients (150, 151). Clewell noted that a transposon insertion into pAD1 correlated with the loss of hemolysin/bacteriocin activity (152). Since then, a highly conserved cytolysin determinant has been found within plasmids of various *E. faecalis* isolates, including the pheromone-responsive plasmids pAD1 (152–154), pJH2 (151, 155), pOB1 (155, 156), pAMδ1 (157), pX98 (158), and pIP964 (159). Plasmids encoding cytolysin have been determined to fall within the same plasmid incompatibility group (160).

The organization of the cytolysin operon was elucidated through transposon mutagenesis of pAD1 and extracellular complementation, showing that eight genes are involved in the production of the toxin (Fig. 4C) (95, 139, 161–163). A schematic of cytolysin expression, including posttranslational modifications, is depicted in Fig. 6. Cytolysin activity is attributed to two structural components designated $CylL_L$ and $CylL_S$. Both peptides are posttranslationally modified within *E. faecalis* by CylM (164), resulting in $CylL_L^*$ and $CylL_S^*$. After modification, the subunits are secreted by CylB, an ATP-binding transporter (161, 164). While both $CylL_L^*$ and $CylL_S^*$ require CylB for secretion, only $CylL_L^*$ secretion is dependent upon ATP hydrolysis (161, 164). During secretion, the serine protease domain of CylB removes the leader sequences associated with the modified $CylL_L^*$ and $CylL_S^*$ peptides (165), resulting in extracellular CyL_L' and $CylL_S'$. Finally, CyL_L' and $CylL_S'$ are further cleaved by CylA, an extracellular serine protease, to generate the active toxin subunits CyL_L'' and $CylL_S''$ (162, 165). Expression of cytolysin is tightly regulated within *E. faecalis*. Its production was determined to be density dependent, with $CylL_S''$ serving as a signaling molecule for the induction of cytolysin expression (139, 166). Two genes involved in autoinduction by $CylL_S''$, *cylR1* and *cylR2*, have also been identified (Fig. 4C) (139). CylR2 appears to repress expression of the operon in the absence of autoinducer via direct binding to the cytolysin promoter region (167, 168). The precise role of CylR1 is still undetermined.

Although the exact mechanism of action of the cytolysin bacteriocin has yet to be determined, cytolysin struc-

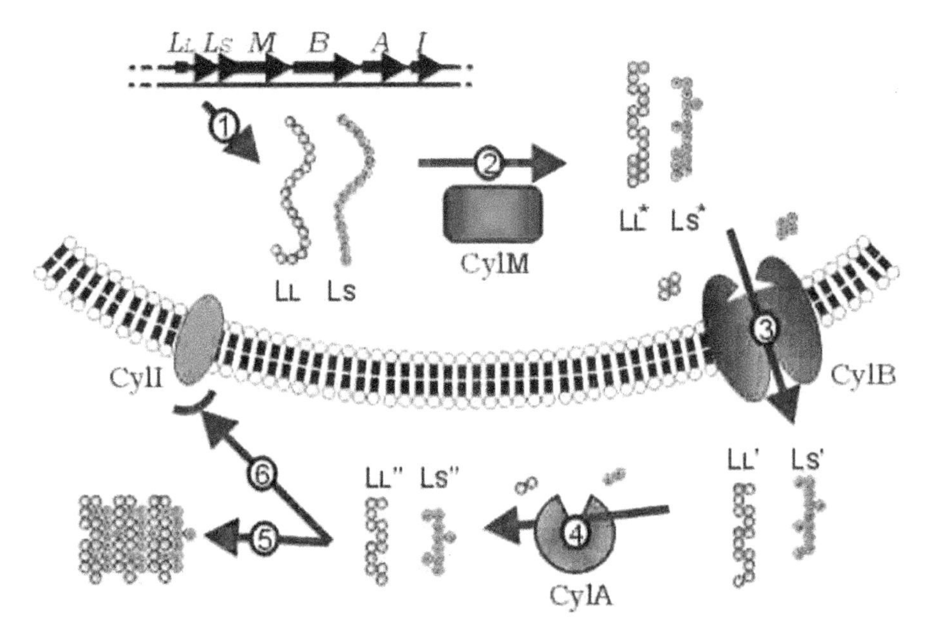

Figure 6 Schematic of *E. faecalis* cytolysin expression, posttranslational modification, processing, and export. (**1**) $CylL_L$ and $CylL_s$ precursor peptides are synthesized (**2**) and are intracellularly modified by CylM to create $CylL_L^*$ and $CylL_s^*$. (**3**) $CylL_L^*$ and $CylL_s^*$ are secreted and further modified by CylB, resulting in $CylL_L'$ and $CylL_s'$, (**4**) which are cleaved extracellularly by CylA to form the active cytolysin components $CylL_L''$ and $CylL_s''$. (**5**) $CylL_L''$ and $CylL_s''$ are capable of forming aggregates and are prevented from lysing cytolysin-expressing cells via CylI (**6**). doi:10.1128/microbiolspec.PLAS-0002-2013.f6

tural components resemble lantibiotics, a class of bactericidal peptides produced by Gram-positive bacteria (164, 165, 169). The composition, size, and presence of lanthionine linkages (posttranslational condensation of cysteine and serine or threonine side chains) are characteristics shared with other lantibiotics (170–173). Cytolysin was found to kill a wide variety of Gram-positive species including staphylococci, streptococci, clostridia, and enterococci (174, 175). It has been shown that CylI, a putative membrane metalloprotease encoded within the cytolysin operon, confers immunity to the cytolysin-producing cell (163), although the precise mechanism of immunity conferred by CylI has not yet been determined.

The contribution of the cytolysin to the virulence of *E. faecalis* has been well demonstrated in a number of infection models. Cytolysin-expressing bacterial strains exhibit a significantly lower LD_{50} in a murine intraperitoneal challenge model than strains deficient in cytolysin (145, 146). In a rabbit model of endocarditis, lethality attributable to *E. faecalis* expressing aggregation substance in the absence of cytolysin was only 7% (88). However, 55% lethality was observed in *E. faecalis* infections where the bacteria expressed both aggregation substance and cytolysin (88). Cytolysin was also determined to contribute to the virulence of *E. faecalis* in a rabbit model of endophthalmitis, in which significant retinal damage and loss of vision was attributable to cytolysin expression (176, 177). Furthermore, cytolysin was shown to increase the lethality of *E. faecalis* in *C. elegans* infection (137), a model currently used as an initial rapid screen for potential virulence factors.

The ability of cytolysin to contribute to enhanced colonization of the intestine by pathogenic *E. faecalis*, as well as the spread of enterococci into the bloodstream, has been addressed in several studies. The potential role of cytolysin in competition with other enterococci within the intestinal niche was examined by Huycke et al. (178). In the same study, cytolytic strains were able to outcompete noncytolytic strains *in vitro*. However, in mice that were fed a mixture of cytolytic and noncytolytic *E. faecalis*, no significant difference in the proportion of cytolytic and noncytolytic isolates was observed in stool samples. Although a cytolytic strain has been shown to be capable of translocation across the intestinal epithelium (179), this activity has not been conclusively determined to be dependent upon expression of cytolysin. The exact contribution of cytolysin to *E. faecalis* bacteremia has yet to be determined, but cytolytic *E. faecalis* was found to spread from the peritoneum into the bloodstream of mice more readily than noncytolytic isogenic mutants (66). It has also been suggested that cytolysin-expressing bacteria may be able to acquire exogenous heme, resulting in enhanced growth within the bloodstream, although this hypothesis has yet to be tested (180).

It is clear that plasmids encoding cytolysin contribute to the virulence of *E. faecalis*. However, the precise contributions of cytolysin to the pathogenesis of infection remain to be described.

VIRULENCE PLASMIDS IN OTHER NONSPORULATING, GRAM-POSITIVE PATHOGENS

Rhodococcus equi

R. equi, a facultative intracellular pathogen that can persist and multiply within macrophages, is associated with severe equine pneumonia in foals and is also capable of causing pneumonia in humans (181). It has been shown that equine and human clinical isolates of *R. equi* contain large plasmids (182, 183), and these plasmids have been determined to be essential for the ability of *R. equi* to survive within macrophages (184). In the absence of these plasmids, the virulence of *R. equi* is significantly attenuated (184–186).

Virulence-associated proteins (VapA-M) encoded by large plasmids (>50 kilobases) from *R. equi* clinical isolates are localized to the bacterial cell surface and appear to be critical for virulence (182, 185, 187–190). VapA in particular appears to play a major role in *R. equi* pathogenesis (184, 191). Expression of VapA was found to be dependent on both temperature and pH (192, 193), a feature shared with proteins encoded by virulence plasmids of *Yersinia pestis* and *Shigella flexneri* (194–196). VapB is encoded by a separate plasmid from the one encoding VapA (197, 198) and is associated with less virulent *R. equi* infections in humans (199). VapA-containing plasmids also encode the additional Vap proteins VapC-H (200), while VapB-containing plasmids encode Vap proteins VapI-M (198). A map of the VapA-containing plasmid p33701 is shown in Fig. 7 (200). The plasmid also contains genes that are associated with conjugation and maintenance functions (Fig. 7; red ORFs). The region containing the genes encoding VapC-H (Fig. 7; blue ORFs) is bounded by two transposon resolvases. p33701 also contains a putative two-component regulator (*tcr*) and a putative lysyl-tRNA synthetase (Fig. 7; yellow ORFs) that both appear from G+C content to be foreign to *R. equi* (200). All Vaps resemble each other in sequence, and Vap homologs have not yet been identified in other organisms. The specific roles of the different Vap proteins in *R. equi* infection remain unknown.

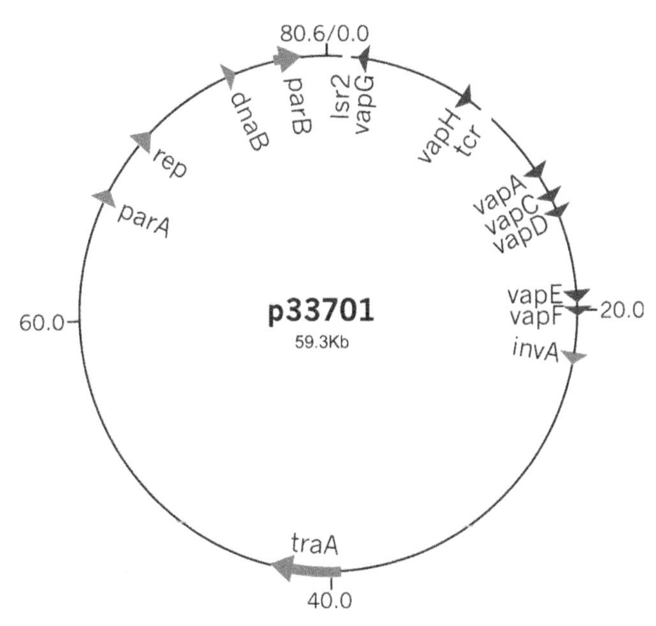

Figure 7 Plasmid map of the virulence plasmid p33701 of *R. equi*. Genes believed to be involved in plasmid maintenance and conjugation are shown in red, genes encoding Vaps are shown in blue, and putative genes within the proposed pathogenicity island are depicted in yellow. Figure is adapted from reference 200.
doi:10.1128/microbiolspec.PLAS-0002-2013.f7

Lactococcus garvieae

L. garvieae is a widely distributed lactic acid bacterium that is a well-known pathogen of fish but is also occasionally found as an opportunistic pathogen of humans (201, 202). *L. garvieae* is considered an emerging zoonotic pathogen, and the number of human cases of infection has increased in recent years (203). A recently isolated human strain of *L. garvieae* was found to contain five plasmids, one of which appears to encode putative virulence factors such as toxins and possible surface proteins (204). These proteins may aid the bacteria in adhering to host tissue; future studies should aim to determine if these genes do in fact play a role in *L. garvieae* virulence.

CONCLUDING REMARKS

In general, virulence plasmids are uncommon among nonsporulating Gram-positive pathogens, with notable exceptions in staphylococci and enterococci, as well as the opportunistic pathogens *R. equi* and *L. garvieae*. Staphylococci and enterococci are leading causes of antibiotic-resistant, hospital-acquired infections (205). Although these organisms are major causes of nosocomial infection, numerically only a miniscule population of cells is involved in infection, with the vast majority of bacteria occurring in peaceful coexistence with the host as part of the commensal flora. Virulence plasmids may therefore represent "selfish DNA" (206), taking advantage of the otherwise stable association between commensal organism and host to propagate throughout a population. Alternatively, virulence plasmids may represent a recent acquisition by these species, which at this point has failed to reach a deep penetrance in the population, potentially because of niche subspecialization and isolation. The occurrence of clonal lineages possessing virulence traits in these species argues for the latter (207, 208), but these two prospects are not mutually exclusive.

Acknowledgments. *The authors thank Chris Pillar, Philip Coburn, and Wolfgang Haas for their contributions to the prior edition of this chapter.Portions of this work were funded by NIH grant AI072360 and the Harvard-wide Program on Antibiotic Resistance (NIH grant AI083214). D.V.T. is supported by Grant EY007145 from the National Institutes of Health. Conflicts of interest: We disclose no conflicts.*

Citation. Van Tyne D, Gilmore MS. 2014. Virulence plasmids of nonsporulating Gram-positive pathogens. Microbiol Spectrum 2(5):PLAS-0002-2013.

References

1. **Novick RP.** 2000. Pathogenicity factors and their regulation, p 392–407. *In* Fischetti V (ed), *Gram-Positive Pathogens*. American Society for Microbiology, Washington, DC.
2. **Otto M.** 2012. MRSA virulence and spread. *Cell Microbiol* **14:**1513–1521.
3. **Projan SJ, Novick RP.** 1997. The molecular basis of virulence, p 55–81. *In* Archer G, Crossley K (ed), *Staphylococci in Human Disease*. Churchill Livingstone, New York, NY.
4. **Balaban N, Rasooly A.** 2000. Staphylococcal enterotoxins. *Int J Food Microbiol* **61:**1–10.
5. **Argudin MA, Mendoza MC, Rodicio MR.** 2010. Food poisoning and *Staphylococcus aureus* enterotoxins. *Toxins (Basel)* **2:**1751–1773.
6. **Lowney ED, Baublis JV, Kreye GM, Harrell ER, McKenzie AR.** 1967. The scalded skin syndrome in small children. *Arch Dermatol* **95:**359–369.
7. **Melish ME, Glasgow LA.** 1970. The staphylococcal scalded-skin syndrome. *N Engl J Med* **282:**1114–1119.
8. **Melish ME, Glasgow LA, Turner MD, Lillibridge CB.** 1974. The staphylococcal epidermolytic toxin: its isolation, characterization, and site of action. *Ann NY Acad Sci* **236:**317–342.
9. **Bukowski M, Wladyka B, Dubin G.** 2010. Exfoliative toxins of *Staphylococcus aureus*. *Toxins (Basel)* **2:** 1148–1165.
10. **Choi YW, Herman A, DiGiusto D, Wade T, Marrack P, Kappler J.** 1990. Residues of the variable region of the T-cell-receptor beta-chain that interact with *S. aureus* toxin superantigens. *Nature* **346:**471–473.

11. **Chintagumpala MM, Mollick JA, Rich RR.** 1991. Staphylococcal toxins bind to different sites on HLA-DR. *J Immunol* **147:**3876–3881.
12. **Mollick JA, Chintagumpala M, Cook RG, Rich RR.** 1991. Staphylococcal exotoxin activation of T cells. Role of exotoxin-MHC class II binding affinity and class II isotype. *J Immunol* **146:**463–468.
13. **Monday SR, Bohach GA.** 1999. Properties of *Staphylococcus aureus* enterotoxins and toxic shock syndrome toxin-1, p 589–610. *In* Alouf J, Freer J (ed), *The Comprehensive Sourcebook of Bacterial Protein Toxins.* Academic Press, London, England.
14. **Rajagopalan G, Sen MM, Singh M, Murali NS, Nath KA, Iijima K, Kita H, Leontovich AA, Gopinathan U, Patel R, David CS.** 2006. Intranasal exposure to staphylococcal enterotoxin B elicits an acute systemic inflammatory response. *Shock* **25:**647–656.
15. **Spaulding AR, Salgado-Pabon W, Kohler PL, Horswill AR, Leung DY, Schlievert PM.** 2013. Staphylococcal and streptococcal superantigen exotoxins. *Clin Microbiol Rev* **26:**422–447.
16. **Ladhani S, Joannou CL, Lochrie DP, Evans RW, Poston SM.** 1999. Clinical, microbial, and biochemical aspects of the exfoliative toxins causing staphylococcal scalded-skin syndrome. *Clin Microbiol Rev* **12:**224–242.
17. **Wieneke AA, Roberts D, Gilbert RJ.** 1993. Staphylococcal food poisoning in the United Kingdom, 1969–90. *Epidemiol Infect* **110:**519–531.
18. **Holmberg SD, Blake PA.** 1984. Staphylococcal food poisoning in the United States. New facts and old misconceptions. *JAMA* **251:**487–489.
19. **Casman EP.** 1965. Staphylococcal enterotoxin. *Ann NY Acad Sci* **128:**124–131.
20. **Bergdoll MS, Czop JK, Gould SS.** 1974. Enterotoxin synthesis by the staphylococci. *Ann NY Acad Sci* **236:** 307–316.
21. **Bayles KW, Iandolo JJ.** 1989. Genetic and molecular analyses of the gene encoding staphylococcal enterotoxin D. *J Bacteriol* **171:**4799–4806.
22. **Kornblum J, Kreiswirth BN, Projan SJ, Ross H, Novick RP.** 1991. Agr: a polycistronic locus regulating exoprotein synthesis in *Staphylococcus aureus*, p 373–402. *In* Novick RP (ed), *Molecular Biology of the Staphylococci.* VCH Publishers, New York, NY.
23. **Zhang S, Iandolo JJ, Stewart GC.** 1998. The enterotoxin D plasmid of *Staphylococcus aureus* encodes a second enterotoxin determinant (sej). *FEMS Microbiol Lett* **168:**227–233.
24. **Tseng CW, Zhang S, Stewart GC.** 2004. Accessory gene regulator control of staphyloccoccal enterotoxin d gene expression. *J Bacteriol* **186:**1793–1801.
25. **Omoe K, Hu DL, Takahashi-Omoe H, Nakane A, Shinagawa K.** 2003. Identification and characterization of a new staphylococcal enterotoxin-related putative toxin encoded by two kinds of plasmids. *Infect Immun* **71:**6088–6094.
26. **Dajani AS.** 1972. The scalded-skin syndrome: relation to phage-group II staphylococci. *J Infect Dis* **125:**548–551.
27. **Kondo I, Sakurai S, Sarai Y.** 1973. Purification of exfoliatin produced by *Staphylococcus aureus* of bacteriophage group 2 and its physicochemical properties. *Infect Immun* **8:**156–164.
28. **Iwatsuki K, Yamasaki O, Morizane S, Oono T.** 2006. Staphylococcal cutaneous infections: invasion, evasion and aggression. *J Dermatol Sci* **42:**203–214.
29. **Nishifuji K, Sugai M, Amagai M.** 2008. Staphylococcal exfoliative toxins: "molecular scissors" of bacteria that attack the cutaneous defense barrier in mammals. *J Dermatol Sci* **49:**21–31.
30. **Sato H, Matsumori Y, Tanabe T, Saito H, Shimizu A, Kawano J.** 1994. A new type of staphylococcal exfoliative toxin from a *Staphylococcus aureus* strain isolated from a horse with phlegmon. *Infect Immun* **62:** 3780–3785.
31. **Melish ME, Glasgow LA.** 1971. Staphylococcal scalded skin syndrome: the expanded clinical syndrome. *J Pediatr* **78:**958–967.
32. **Miller MM, Kapral FA.** 1972. Neutralization of *Staphylococcus aureus* exfoliatin by antibody. *Infect Immun* **6:** 561–563.
33. **Elias PM, Fritsch P, Tappeiner G, Mittermayer H, Wolff K.** 1974. Experimental staphylococcal toxic epidermal necrolysis (TEN) in adult humans and mice. *J Lab Clin Med* **84:**414–424.
34. **McLay AL, Arbuthnott JP, Lyell A.** 1975. Action of staphylococcal epidermolytic toxin on mouse skin: an electron microscopic study. *J Invest Dermatol* **65:**423–428.
35. **Wiley BB, Glasgow LA, Rogolsky M.** 1976. Staphylococcal scalded-skin syndrome: development of a primary binding assay for human antibody to the exfoliative toxin. *Infect Immun* **13:**513–520.
36. **Kapral FA.** 1974. *Staphylococcus aureus*: some host-parasite interactions. *Ann NY Acad Sci* **236:**267–276.
37. **Todd JK.** 1985. Staphylococcal toxin syndromes. *Annu Rev Med* **36:**337–347.
38. **Dancer SJ, Noble WC.** 1991. Nasal, axillary, and perineal carriage of *Staphylococcus aureus* among women: identification of strains producing epidermolytic toxin. *J Clin Pathol* **44:**681–684.
39. **Becker K, Friedrich AW, Lubritz G, Weilert M, Peters G, Von Eiff C.** 2003. Prevalence of genes encoding pyrogenic toxin superantigens and exfoliative toxins among strains of *Staphylococcus aureus* isolated from blood and nasal specimens. *J Clin Microbiol* **41:**1434–1439.
40. **Lee CY, Schmidt JJ, Johnson-Winegar AD, Spero L, Iandolo JJ.** 1987. Sequence determination and comparison of the exfoliative toxin A and toxin B genes from *Staphylococcus aureus*. *J Bacteriol* **169:**3904–3909.
41. **Yamaguchi T, Hayashi T, Takami H, Nakasone K, Ohnishi M, Nakayama K, Yamada S, Komatsuzawa H, Sugai M.** 2000. Phage conversion of exfoliative toxin A production in *Staphylococcus aureus*. *Mol Microbiol* **38:** 694–705.
42. **Yamaguchi T, Nishifuji K, Sasaki M, Fudaba Y, Aepfelbacher M, Takata T, Ohara M, Komatsuzawa H, Amagai M, Sugai M.** 2002. Identification of the

Staphylococcusaureus etd pathogenicity island which encodes a novel exfoliative toxin, ETD, and EDIN-B. *Infect Immun* 70:5835–5845.

43. **Yamasaki O, Tristan A, Yamaguchi T, Sugai M, Lina G, Bes M, Vandenesch F, Etienne J.** 2006. Distribution of the exfoliative toxin D gene in clinical *Staphylococcus aureus* isolates in France. *Clin Microbiol Infect* **12:** 585–588.
44. **Warren R, Rogolsky M, Wiley BB, Glasgow LA.** 1974. Effect of ethidium bromide on elimination of exfoliative toxin and bacteriocin production in *Staphylococcus aureus*. *J Bacteriol* **118:**980–985.
45. **Rogolsky M, Warren R, Wiley BB, Nakamura HT, Glasgow LA.** 1974. Nature of the genetic determinant controlling exfoliative toxin production in *Staphylococcus aureus*. *J Bacteriol* **117:**157–165.
46. **Jackson MP, Iandolo JJ.** 1986. Cloning and expression of the exfoliative toxin B gene from *Staphylococcus aureus*. *J Bacteriol* **166:**574–580.
47. **Jackson MP, Iandolo JJ.** 1986. Sequence of the exfoliative toxin B gene of *Staphylococcus aureus*. *J Bacteriol* **167:**726–728.
48. **Yamaguchi T, Hayashi T, Takami H, Ohnishi M, Murata T, Nakayama K, Asakawa K, Ohara M, Komatsuzawa H, Sugai M.** 2001. Complete nucleotide sequence of a *Staphylococcus aureus* exfoliative toxin B plasmid and identification of a novel ADP-ribosyltransferase, EDIN-C. *Infect Immun* **69:**7760–7771.
49. **Sato H, Tanabe T, Kuramoto M, Tanaka K, Hashimoto T, Saito H.** 1991. Isolation of exfoliative toxin from *Staphylococcus hyicus* subsp. *hyicus* and its exfoliative activity in the piglet. *Vet Microbiol* **27:**263–275.
50. **Sato H, Watanabe T, Murata Y, Ohtake A, Nakamura M, Aizawa C, Saito H, Maehara N.** 1999. New exfoliative toxin produced by a plasmid-carrying strain of *Staphylococcus hyicus*. *Infect Immun* **67:**4014–4018.
51. **Navaratna MA, Sahl HG, Tagg JR.** 1999. Identification of genes encoding two-component lantibiotic production in *Staphylococcus aureus* C55 and other phage group II *S. aureus* strains and demonstration of an association with the exfoliative toxin B gene. *Infect Immun* **67:**4268–4271.
52. **Rogolsky M, Wiley BB.** 1977. Production and properties of a staphylococcin genetically controlled by the staphylococcal plasmid for exfoliative toxin synthesis. *Infect Immun* **15:**726–732.
53. **Crupper SS, Gies AJ, Iandolo JJ.** 1997. Purification and characterization of staphylococcin BacR1, a broad-spectrum bacteriocin. *Appl Environ Microbiol* **63:**4185–4190.
54. **Aktories K.** 1997. Rho proteins: targets for bacterial toxins. *Trends Microbiol* **5:**282–288.
55. **Sugai M, Enomoto T, Hashimoto K, Matsumoto K, Matsuo Y, Ohgai H, Hong YM, Inoue S, Yoshikawa K, Suginaka H.** 1990. A novel epidermal cell differentiation inhibitor (EDIN): purification and characterization from *Staphylococcus aureus*. *Biochem Biophys Res Commun* **173:**92–98.
56. **Munro P, Benchetrit M, Nahori MA, Stefani C, Clement R, Michiels JF, Landraud L, Dussurget O, Lemichez E.** 2010. The *Staphylococcus aureus* epidermal cell differentiation inhibitor toxin promotes formation of infection foci in a mouse model of bacteremia. *Infect Immun* **78:**3404–3411.
57. **Inoue S, Sugai M, Murooka Y, Paik SY, Hong YM, Ohgai H, Suginaka H.** 1991. Molecular cloning and sequencing of the epidermal cell differentiation inhibitor gene from *Staphylococcus aureus*. *Biochem Biophys Res Commun* **174:**459–464.
58. **Aktories K.** 2011. Bacterial protein toxins that modify host regulatory GTPases. *Nat Rev Microbiol* **9:**487–498.
59. **Munro P, Clement R, Lavigne JP, Pulcini C, Lemichez E, Landraud L.** 2011. High prevalence of edin-C encoding RhoA-targeting toxin in clinical isolates of *Staphylococcus aureus*. *Eur J Clin Microbiol Infect Dis* **30:**965–972.
60. **McCarthy AJ, Lindsay JA.** 2012. The distribution of plasmids that carry virulence and resistance genes in *Staphylococcus aureus* is lineage associated. *BMC Microbiol* **12:**104.
61. **Lindsay JA, Knight GM, Budd EL, McCarthy AJ.** 2012. Shuffling of mobile genetic elements (MGEs) in successful healthcare-associated MRSA (HA-MRSA). *Mob Genet Elements* **2:**239–243.
62. **Mundt JO.** 1963. Occurrence of enterococci in animals in a wild environment. *Appl Microbiol* **11:**136–140.
63. **Murray BE.** 1990. The life and times of the *Enterococcus*. *Clin Microbiol Rev* **3:**46–65.
64. **Jett BD, Huycke MM, Gilmore MS.** 1994. Virulence of enterococci. *Clin Microbiol Rev* **7:**462–478.
65. **Moellering R.** 1995. *Enterococcus* species, *Streptococcus bovis*, and *Leuconostac* species, p 1826–1835. *In* Mandell G, Bennett J, Dolin R (ed), *Principles and Practices of Infectious Diseases*, 4th ed. Churchill Livingston, New York, NY.
66. **Huycke MM, Sahm DF, Gilmore MS.** 1998. Multiple-drug resistant enterococci: the nature of the problem and an agenda for the future. *Emerg Infect Dis* **4:**239–249.
67. **Willems RJ, van Schaik W.** 2009. Transition of *Enterococcus faecium* from commensal organism to nosocomial pathogen. *Future Microbiol* **4:**1125–1135.
68. **Gilmore MS, Lebreton F, van Schaik W.** 2013. Genomic transition of enterococci from gut commensals to leading causes of multidrug-resistant hospital infection in the antibiotic era. *Curr Opin Microbiol* **16:**10–16.
69. **Hollenbeck BL, Rice LB.** 2012. Intrinsic and acquired resistance mechanisms in enterococcus. *Virulence* **3:**421–433.
70. **Flahaut S, Frere J, Boutibonnes P, Auffray Y.** 1996. Comparison of the bile salts and sodium dodecyl sulfate stress responses in *Enterococcus faecalis*. *Appl Environ Microbiol* **62:**2416–2420.
71. **Flahaut S, Hartke A, Giard JC, Benachour A, Boutibonnes P, Auffray Y.** 1996. Relationship between stress response toward bile salts, acid and heat treatment in *Enterococcus faecalis*. *FEMS Microbiol Lett* **138:**49–54.
72. **Flahaut S, Hartke A, Giard JC, Auffray Y.** 1997. Alkaline stress response in *Enterococcus faecalis*: adaptation,

cross-protection, and changes in protein synthesis. *Appl Environ Microbiol* **63**:812–814.

73. **Arias CA, Murray BE.** 2012. The rise of the *Enterococcus*: beyond vancomycin resistance. *Nat Rev Microbiol* **10**:266–278.

74. **Mundy LM, Sahm DF, Gilmore M.** 2000. Relationships between enterococcal virulence and antimicrobial resistance. *Clin Microbiol Rev* **13**:513–522.

75. **Maekawa S, Yoshioka M, Kumamoto Y.** 1992. Proposal of a new scheme for the serological typing of *Enterococcus faecalis* strains. *Microbiol Immunol* **36**:671–681.

76. **Thurlow LR, Thomas VC, Fleming SD, Hancock LE.** 2009. *Enterococcus faecalis* capsular polysaccharide serotypes C and D and their contributions to host innate immune evasion. *Infect Immun* **77**:5551–5557.

77. **Hufnagel M, Hancock LE, Koch S, Theilacker C, Gilmore MS, Huebner J.** 2004. Serological and genetic diversity of capsular polysaccharides in *Enterococcus faecalis*. *J Clin Microbiol* **42**:2548–2557.

78. **Hancock LE, Gilmore MS.** 2002. The capsular polysaccharide of *Enterococcus faecalis* and its relationship to other polysaccharides in the cell wall. *Proc Natl Acad Sci USA* **99**:1574–1579.

79. **Teng F, Singh KV, Bourgogne A, Zeng J, Murray BE.** 2009. Further characterization of the epa gene cluster and Epa polysaccharides of *Enterococcus faecalis*. *Infect Immun* **77**:3759–3767.

80. **Shankar N, Lockatell CV, Baghdayan AS, Drachenberg C, Gilmore MS, Johnson DE.** 2001. Role of *Enterococcus faecalis* surface protein Esp in the pathogenesis of ascending urinary tract infection. *Infect Immun* **69**: 4366–4372.

81. **Shankar N, Baghdayan AS, Gilmore MS.** 2002. Modulation of virulence within a pathogenicity island in vancomycin-resistant *Enterococcus faecalis*. *Nature* **417**: 746–750.

82. **Willems RJ, Homan W, Top J, van Santen-Verheuvel M, Tribe D, Manzioros X, Gaillard C, Vandenbroucke-Grauls CM, Mascini EM, van Kregten E, van Embden JD, Bonten MJ.** 2001. Variant esp gene as a marker of a distinct genetic lineage of vancomycin-resistant *Enterococcus faecium* spreading in hospitals. *Lancet* **357**: 853–855.

83. **Toledo-Arana A, Valle J, Solano C, Arrizubieta MJ, Cucarella C, Lamata M, Amorena B, Leiva J, Penades JR, Lasa I.** 2001. The enterococcal surface protein, Esp, is involved in *Enterococcus faecalis* biofilm formation. *Appl Environ Microbiol* **67**:4538–4545.

84. **Foulquie Moreno MR, Sarantinopoulos P, Tsakalidou E, De Vuyst L.** 2006. The role and application of enterococci in food and health. *Int J Food Microbiol* **106**: 1–24.

85. **Borgmann S, Niklas DM, Klare I, Zabel LT, Buchenau P, Autenrieth IB, Heeg P.** 2004. Two episodes of vancomycin-resistant *Enterococcus faecium* outbreaks caused by two genetically different clones in a newborn intensive care unit. *Int J Hyg Environ Health* **207**:386–389.

86. **Kreft B, Marre R, Schramm U, Wirth R.** 1992. Aggregation substance of *Enterococcus faecalis* mediates adhesion to cultured renal tubular cells. *Infect Immun* **60**: 25–30.

87. **Clewell DB.** 1993. Bacterial sex pheromone-induced plasmid transfer. *Cell* **73**:9–12.

88. **Chow JW, Thal LA, Perri MB, Vazquez JA, Donabedian SM, Clewell DB, Zervos MJ.** 1993. Plasmid-associated hemolysin and aggregation substance production contribute to virulence in experimental enterococcal endocarditis. *Antimicrob Agents Chemother* **37**:2474–2477.

89. **Olmsted SB, Dunny GM, Erlandsen SL, Wells CL.** 1994. A plasmid-encoded surface protein on *Enterococcus faecalis* augments its internalization by cultured intestinal epithelial cells. *J Infect Dis* **170**:1549–1556.

90. **Schlievert PM, Gahr PJ, Assimacopoulos AP, Dinges MM, Stoehr JA, Harmala JW, Hirt H, Dunny GM.** 1998. Aggregation and binding substances enhance pathogenicity in rabbit models of *Enterococcus faecalis* endocarditis. *Infect Immun* **66**:218–223.

91. **Hirt H, Schlievert PM, Dunny GM.** 2002. *In vivo* induction of virulence and antibiotic resistance transfer in *Enterococcus faecalis* mediated by the sex pheromone-sensing system of pCF10. *Infect Immun* **70**: 716–723.

92. **Chuang-Smith ON, Wells CL, Henry-Stanley MJ, Dunny GM.** 2010. Acceleration of *Enterococcus faecalis* biofilm formation by aggregation substance expression in an *ex vivo* model of cardiac valve colonization. *PLoS One* **5**:e15798.

93. **Todd E.** 1934. A comparative serological study of streptolysins derived from human and from animal infections, with notes on pneumococcal haemolysin, tetanolysin and staphylococcus toxin. *J Pathol Bacteriol* **39**:299–321.

94. **Sherwood N, Russell B, Jay A, Bowman K.** 1949. Studies on streptococci. III. New antibiotic substances produced by beta hemolytic streptococci. *J Infect Dis* **84**:88–91.

95. **Ike Y, Clewell DB, Segarra RA, Gilmore MS.** 1990. Genetic analysis of the pAD1 hemolysin/bacteriocin determinant in *Enterococcus faecalis*: Tn917 insertional mutagenesis and cloning. *J Bacteriol* **172**:155–163.

96. **Wirth R.** 1994. The sex pheromone system of *Enterococcus faecalis*. More than just a plasmid-collection mechanism? *Eur J Biochem* **222**:235–246.

97. **Dunny GM, Leonard BA, Hedberg PJ.** 1995. Pheromone-inducible conjugation in *Enterococcus faecalis*: interbacterial and host-parasite chemical communication. *J Bacteriol* **177**:871–876.

98. **Clewell D.** 1999. Sex pheromone systems in enterococci, p 47–65. *In* Dunny G, Winans S (ed), *Cell-Cell Signaling in Bacteria*. American Society for Microbiology, Washington, DC.

99. **Wirth R.** 2000. Sex pheromones and gene transfer in *Enterococcus faecalis*. *Res Microbiol* **151**:493–496.

100. **Palmer KL, Kos VN, Gilmore MS.** 2010. Horizontal gene transfer and the genomics of enterococcal antibiotic resistance. *Curr Opin Microbiol* **13**:632–639.

101. **Wardal E, Sadowy E, Hryniewicz W.** 2010. Complex nature of enterococcal pheromone-responsive plasmids. *Pol J Microbiol* **59**:79–87.

102. **Galli D, Wirth R, Wanner G.** 1989. Identification of aggregation substances of *Enterococcus faecalis* cells after induction by sex pheromones. An immunological and ultrastructural investigation. *Arch Microbiol* **151:** 486–490.

103. **Ehrenfeld EE, Kessler RE, Clewell DB.** 1986. Identification of pheromone-induced surface proteins in *Streptococcus faecalis* and evidence of a role for lipoteichoic acid in formation of mating aggregates. *J Bacteriol* **168:** 6–12.

104. **Bensing BA, Dunny GM.** 1993. Cloning and molecular analysis of genes affecting expression of binding substance, the recipient-encoded receptor(s) mediating mating aggregate formation in *Enterococcus faecalis. J Bacteriol* **175:**7421–7429.

105. **Waters CM, Hirt H, McCormick JK, Schlievert PM, Wells CL, Dunny GM.** 2004. An amino-terminal domain of *Enterococcus faecalis* aggregation substance is required for aggregation, bacterial internalization by epithelial cells and binding to lipoteichoic acid. *Mol Microbiol* **52:** 1159–1171.

106. **Ike Y, Clewell DB.** 1984. Genetic analysis of the pAD1 pheromone response in *Streptococcus faecalis*, using transposon Tn917 as an insertional mutagen. *J Bacteriol* **158:**777–783.

107. **Cook L, Chatterjee A, Barnes A, Yarwood J, Hu WS, Dunny G.** 2011. Biofilm growth alters regulation of conjugation by a bacterial pheromone. *Mol Microbiol* **81:**1499–1510.

108. **Hirt H, Wirth R, Muscholl A.** 1996. Comparative analysis of 18 sex pheromone plasmids from *Enterococcus faecalis*: detection of a new insertion element on pPD1 and implications for the evolution of this plasmid family. *Mol Gen Genet* **252:**640–647.

109. **Francia MV, Haas W, Wirth R, Samberger E, Muscholl-Silberhorn A, Gilmore MS, Ike Y, Weaver KE, An FY, Clewell DB.** 2001. Completion of the nucleotide sequence of the *Enterococcus faecalis* conjugative virulence plasmid pAD1 and identification of a second transfer origin. *Plasmid* **46:**117–127.

110. **Rakita RM, Vanek NN, Jacques-Palaz K, Mee M, Mariscalco MM, Dunny GM, Snuggs M, Van Winkle WB, Simon SI.** 1999. *Enterococcus faecalis* bearing aggregation substance is resistant to killing by human neutrophils despite phagocytosis and neutrophil activation. *Infect Immun* **67:**6067–6075.

111. **Fisher K, Phillips C.** 2009. The ecology, epidemiology and virulence of *Enterococcus. Microbiology* **155:**1749–1757.

112. **Clewell DB.** 2007. Properties of *Enterococcus faecalis* plasmid pAD1, a member of a widely disseminated family of pheromone-responding, conjugative, virulence elements encoding cytolysin. *Plasmid* **58:**205–227.

113. **Mori M, Sakagami Y, Narita M, Isogai A, Fujino M, Kitada C, Craig RA, Clewell DB, Suzuki A.** 1984. Isolation and structure of the bacterial sex pheromone, cAD1, that induces plasmid transfer in *Streptococcus faecalis. FEBS Lett* **178:**97–100.

114. **de Freire Bastos MC, Tanimoto K, Clewell DB.** 1997. Regulation of transfer of the *Enterococcus faecalis* pheromone-responding plasmid pAD1: temperature-sensitive transfer mutants and identification of a new regulatory determinant, traD. *J Bacteriol* **179:**3250–3259.

115. **Tanimoto K, Clewell DB.** 1993. Regulation of the pAD1-encoded sex pheromone response in *Enterococcus faecalis*: expression of the positive regulator TraE1. *J Bacteriol* **175:**1008–1018.

116. **An FY, Clewell DB.** 1994. Characterization of the determinant (traB) encoding sex pheromone shutdown by the hemolysin/bacteriocin plasmid pAD1 in *Enterococcus faecalis. Plasmid* **31:**215–221.

117. **Muscholl-Silberhorn AB.** 2000. Pheromone-regulated expression of sex pheromone plasmid pAD1-encoded aggregation substance depends on at least six upstream genes and a cis-acting, orientation-dependent factor. *J Bacteriol* **182:**3816–3825.

118. **Weaver KE, Clewell DB.** 1988. Regulation of the pAD1 sex pheromone response in *Enterococcus faecalis*: construction and characterization of lacZ transcriptional fusions in a key control region of the plasmid. *J Bacteriol* **170:**4343–4352.

119. **Tanimoto K, An FY, Clewell DB.** 1993. Characterization of the traC determinant of the *Enterococcus faecalis* hemolysin-bacteriocin plasmid pAD1: binding of sex pheromone. *J Bacteriol* **175:**5260–5264.

120. **Mori M, Tanaka H, Sakagami Y, Isogai A, Fujino M, Kitada C, White BA, An FY, Clewell DB, Suzuki A.** 1986. Isolation and structure of the *Streptococcus faecalis* sex pheromone, cAM373. *FEBS Lett* **206:**69–72.

121. **Weaver KE, Jensen KD, Colwell A, Sriram SI.** 1996. Functional analysis of the *Enterococcus faecalis* plasmid pAD1-encoded stability determinant par. *Mol Microbiol* **20:**53–63.

122. **Weaver K.** 2000. Enterococcal genetics, p 259–271. *In* Fischetti V, Novick R, Ferretti J, Portnoy D, Rood J (ed), *Gram-Positive Pathogens*. American Society for Microbiology, Washington, DC.

123. **Weaver KE, Clewell DB.** 1989. Construction of *Enterococcus faecalis* pAD1 miniplasmids: identification of a minimal pheromone response regulatory region and evaluation of a novel pheromone-dependent growth inhibition. *Plasmid* **22:**106–119.

124. **Weaver KE, Clewell DB, An F.** 1993. Identification, characterization, and nucleotide sequence of a region of *Enterococcus faecalis* pheromone-responsive plasmid pAD1 capable of autonomous replication. *J Bacteriol* **175:**1900–1909.

125. **Tomita H, Clewell DB.** 2000. A pAD1-encoded small RNA molecule, mD, negatively regulates *Enterococcus faecalis* pheromone response by enhancing transcription termination. *J Bacteriol* **182:**1062–1073.

126. **Fujimoto S, Clewell DB.** 1998. Regulation of the pAD1 sex pheromone response of *Enterococcus faecalis* by direct interaction between the cAD1 peptide mating signal and the negatively regulating, DNA-binding TraA protein. *Proc Natl Acad Sci USA* **95:**6430–6435.

127. **Pontius LT, Clewell DB.** 1992. Conjugative transfer of *Enterococcus faecalis* plasmid pAD1: nucleotide sequence and transcriptional fusion analysis of a region involved in positive regulation. *J Bacteriol* **174:**3152–3160.

128. **Galli D, Lottspeich F, Wirth R.** 1990. Sequence analysis of *Enterococcus faecalis* aggregation substance encoded by the sex pheromone plasmid pAD1. *Mol Microbiol* **4:** 895–904.

129. **Chuang ON, Schlievert PM, Wells CL, Manias DA, Tripp TJ, Dunny GM.** 2009. Multiple functional domains of *Enterococcus faecalis* aggregation substance Asc10 contribute to endocarditis virulence. *Infect Immun* **77:** 539–548.

130. **Waters CM, Wells CL, Dunny GM.** 2003. The aggregation domain of aggregation substance, not the RGD motifs, is critical for efficient internalization by HT-29 enterocytes. *Infect Immun* **71:**5682–5689.

131. **Muscholl-Silberhorn A.** 1998. Analysis of the clumping-mediating domain(s) of sex pheromone plasmid pAD1-encoded aggregation substance. *Eur J Biochem* **258:**515–520.

132. **Waters CM, Dunny GM.** 2001. Analysis of functional domains of the *Enterococcus faecalis* pheromone-induced surface protein aggregation substance. *J Bacteriol* **183:**5659–5667.

133. **Vanek NN, Simon SI, Jacques-Palaz K, Mariscalco MM, Dunny GM, Rakita RM.** 1999. *Enterococcus faecalis* aggregation substance promotes opsonin-independent binding to human neutrophils via a complement receptor type 3-mediated mechanism. *FEMS Immunol Med Microbiol* **26:**49–60.

134. **Sussmuth SD, Muscholl-Silberhorn A, Wirth R, Susa M, Marre R, Rozdzinski E.** 2000. Aggregation substance promotes adherence, phagocytosis, and intracellular survival of *Enterococcus faecalis* within human macrophages and suppresses respiratory burst. *Infect Immun* **68:**4900–4906.

135. **An FY, Clewell DB.** 2002. Identification of the cAD1 sex pheromone precursor in *Enterococcus faecalis*. *J Bacteriol* **184:**1880–1887.

136. **Jett BD, Atkuri RV, Gilmore MS.** 1998. *Enterococcus faecalis* localization in experimental endophthalmitis: role of plasmid-encoded aggregation substance. *Infect Immun* **66:**843–848.

137. **Garsin DA, Sifri CD, Mylonakis E, Qin X, Singh KV, Murray BE, Calderwood SB, Ausubel FM.** 2001. A simple model host for identifying Gram-positive virulence factors. *Proc Natl Acad Sci USA* **98:**10892–10897.

138. **Berti M, Candiani G, Kaufhold A, Muscholl A, Wirth R.** 1998. Does aggregation substance of *Enterococcus faecalis* contribute to development of endocarditis? *Infection* **26:**48–53.

139. **Haas W, Shepard BD, Gilmore MS.** 2002. Two-component regulator of *Enterococcus faecalis* cytolysin responds to quorum-sensing autoinduction. *Nature* **415:** 84–87.

140. **Huycke MM, Gilmore MS, Jett BD, Booth JL.** 1992. Transfer of pheromone-inducible plasmids between *Enterococcus faecalis* in the Syrian hamster gastrointestinal tract. *J Infect Dis* **166:**1188–1191.

141. **Van Tyne D, Martin MJ, Gilmore MS.** 2013. Structure, function, and biology of the *Enterococcus faecalis* cytolysin. *Toxins (Basel)* **5:**895–911.

142. **Tang W, van der Donk WA.** 2013. The sequence of the enterococcal cytolysin imparts unusual lanthionine stereochemistry. *Nat Chem Biol* **9:**157–159.

143. **Cox CR, Coburn PS, Gilmore MS.** 2005. Enterococcal cytolysin: a novel two component peptide system that serves as a bacterial defense against eukaryotic and prokaryotic cells. *Curr Protein Pept Sci* **6:**77–84.

144. **Roux A, Payne SM, Gilmore MS.** 2009. Microbial telesensing: probing the environment for friends, foes, and food. *Cell Host Microbe* **6:**115–124.

145. **Ike Y, Hashimoto H, Clewell DB.** 1984. Hemolysin of *Streptococcus faecalis* subspecies zymogenes contributes to virulence in mice. *Infect Immun* **45:**528–530.

146. **Dupont H, Montravers P, Mohler J, Carbon C.** 1998. Disparate findings on the role of virulence factors of *Enterococcus faecalis* in mouse and rat models of peritonitis. *Infect Immun* **66:**2570–2575.

147. **Ike Y, Hashimoto H, Clewell DB.** 1987. High incidence of hemolysin production by *Enterococcus (Streptococcus) faecalis* strains associated with human parenteral infections. *J Clin Microbiol* **25:**1524–1528.

148. **Huycke MM, Gilmore MS.** 1995. Frequency of aggregation substance and cytolysin genes among enterococcal endocarditis isolates. *Plasmid* **34:**152–156.

149. **Booth MC, Hatter KL, Miller D, Davis J, Kowalski R, Parke DW, Chodosh J, Jett BD, Callegan MC, Penland R, Gilmore MS.** 1998. Molecular epidemiology of *Staphylococcus aureus* and *Enterococcus faecalis* in endophthalmitis. *Infect Immun* **66:**356–360.

150. **Dunny GM, Clewell DB.** 1975. Transmissible toxin (hemolysin) plasmid in *Streptococcus faecalis* and its mobilization of a noninfectious drug resistance plasmid. *J Bacteriol* **124:**784–790.

151. **Jacob AE, Douglas GJ, Hobbs SJ.** 1975. Self-transferable plasmids determining the hemolysin and bacteriocin of *Streptococcus faecalis* var. zymogenes. *J Bacteriol* **121:** 863–872.

152. **Clewell DB, Tomich PK, Gawron-Burke MC, Franke AE, Yagi Y, An FY.** 1982. Mapping of *Streptococcus faecalis* plasmids pAD1 and pAD2 and studies relating to transposition of Tn917. *J Bacteriol* **152:**1220–1230.

153. **Dunny GM, Craig RA, Carron RL, Clewell DB.** 1979. Plasmid transfer in *Streptococcus faecalis*: production of multiple sex pheromones by recipients. *Plasmid* **2:** 454–465.

154. **Tomich PK, An FY, Damle SP, Clewell DB.** 1979. Plasmid-related transmissibility and multiple drug resistance in *Streptococcus faecalis* subsp. zymogenes strain DS16. *Antimicrob Agents Chemother* **15:**828–830.

155. **Clewell DB.** 1981. Plasmids, drug resistance, and gene transfer in the genus *Streptococcus*. *Microbiol Rev* **45:** 409–436.

156. **Oliver DR, Brown BL, Clewell DB.** 1977. Characterization of plasmids determining hemolysin and bacteriocin

production in *Streptococcus faecalis* 5952. *J Bacteriol* **130:**948–950.

157. **Clewell D, Yagi Y, Ike Y, Craig R, Brown B, An F.** 1982. Sex pheromones in *Streptococcus faecalis*: multiple pheromone systems in strain DS5, similarities of pAD1 and pAMd1, and mutants of pAD1 altered in conjugative properties, p 97–100. *In* Schlessinger D (ed), *Microbiology*. American Society for Microbiology, Washington, DC.
158. **Jett BD, Gilmore MS.** 1990. The growth-inhibitory effect of the *Enterococcus faecalis* bacteriocin encoded by pAD1 extends to the oral streptococci. *J Dent Res* **69:** 1640–1645.
159. **Le Bouguenec C, de Cespedes G, Horaud T.** 1988. Molecular analysis of a composite chromosomal conjugative element (Tn3701) of *Streptococcus pyogenes*. *J Bacteriol* **170:**3930–3936.
160. **Colmar I, Horaud T.** 1987. *Enterococcus faecalis* hemolysin-bacteriocin plasmids belong to the same incompatibility group. *Appl Environ Microbiol* **53:**567–570.
161. **Gilmore MS, Segarra RA, Booth MC.** 1990. An HlyB-type function is required for expression of the *Enterococcus faecalis* hemolysin/bacteriocin. *Infect Immun* **58:**3914–3923.
162. **Segarra RA, Booth MC, Morales DA, Huycke MM, Gilmore MS.** 1991. Molecular characterization of the *Enterococcus faecalis* cytolysin activator. *Infect Immun* **59:**1239–1246.
163. **Coburn PS, Hancock LE, Booth MC, Gilmore MS.** 1999. A novel means of self-protection, unrelated to toxin activation, confers immunity to the bactericidal effects of the *Enterococcus faecalis* cytolysin. *Infect Immun* **67:** 3339–3347.
164. **Gilmore MS, Segarra RA, Booth MC, Bogie CP, Hall LR, Clewell DB.** 1994. Genetic structure of the *Enterococcus faecalis* plasmid pAD1-encoded cytolytic toxin system and its relationship to lantibiotic determinants. *J Bacteriol* **176:**7335–7344.
165. **Booth MC, Bogie CP, Sahl HG, Siezen RJ, Hatter KL, Gilmore MS.** 1996. Structural analysis and proteolytic activation of *Enterococcus faecalis* cytolysin, a novel lantibiotic. *Mol Microbiol* **21:**1175–1184.
166. **Coburn PS, Pillar CM, Jett BD, Haas W, Gilmore MS.** 2004. *Enterococcus faecalis* senses target cells and in response expresses cytolysin. *Science* **306:**2270–2272.
167. **Rumpel S, Razeto A, Pillar CM, Vijayan V, Taylor A, Giller K, Gilmore MS, Becker S, Zweckstetter M.** 2004. Structure and DNA-binding properties of the cytolysin regulator CylR2 from *Enterococcus faecalis*. *EMBO J* **23:**3632–3642.
168. **Rumpel S, Becker S, Zweckstetter M.** 2008. High-resolution structure determination of the CylR2 homodimer using paramagnetic relaxation enhancement and structure-based prediction of molecular alignment. *J Biomol NMR* **40:**1–13.
169. **Patton GC, van der Donk WA.** 2005. New developments in lantibiotic biosynthesis and mode of action. *Curr Opin Microbiol* **8:**543–551.
170. **Banerjee S, Hansen JN.** 1988. Structure and expression of a gene encoding the precursor of subtilin, a small protein antibiotic. *J Biol Chem* **263:**9508–9514.
171. **Schnell N, Entian KD, Schneider U, Gotz F, Zahner H, Kellner R, Jung G.** 1988. Prepeptide sequence of epidermin, a ribosomally synthesized antibiotic with four sulphide-rings. *Nature* **333:**276–278.
172. **Kaletta C, Entian KD.** 1989. Nisin, a peptide antibiotic: cloning and sequencing of the nisA gene and posttranslational processing of its peptide product. *J Bacteriol* **171:**1597–1601.
173. **Willey JM, van der Donk WA.** 2007. Lantibiotics: peptides of diverse structure and function. *Annu Rev Microbiol* **61:**477–501.
174. **Stark J.** 1960. Antibiotic activity of haemolytic enterococci. *Lancet* **i:**733–734.
175. **Brock T, Peacher B, Pierson D.** 1963. Survey of the bacteriocines of enterococci. *J Bacteriol* **86:**702–707.
176. **Stevens SX, Jensen HG, Jett BD, Gilmore MS.** 1992. A hemolysin-encoding plasmid contributes to bacterial virulence in experimental *Enterococcus faecalis* endophthalmitis. *Invest Ophthalmol Vis Sci* **33:**1650–1656.
177. **Jett BD, Jensen HG, Nordquist RE, Gilmore MS.** 1992. Contribution of the pAD1-encoded cytolysin to the severity of experimental *Enterococcus faecalis* endophthalmitis. *Infect Immun* **60:**2445–2452.
178. **Huycke MM, Joyce WA, Gilmore MS.** 1995. *Enterococcus faecalis* cytolysin without effect on the intestinal growth of susceptible enterococci in mice. *J Infect Dis* **172:**273–276.
179. **Wells CL, Jechorek RP, Erlandsen SL.** 1990. Evidence for the translocation of *Enterococcus faecalis* across the mouse intestinal tract. *J Infect Dis* **162:**82–90.
180. **Gilmore MS, Coburn PS, Nallapareddy SR, Murray BE.** 2002. Enterococcal virulence. *In* Gilmore MS (ed), *The Enterococci. Pathogenesis, Molecular Biology, and Antimicrobial Resistance*. American Society for Microbiology, Washington, D.C.
181. **Yamshchikov AV, Schuetz A, Lyon GM.** 2010. *Rhodococcus equi* infection. *Lancet Infect Dis* **10:**350–359.
182. **Tkachuk-Saad O, Prescott J.** 1991. *Rhodococcus equi* plasmids: isolation and partial characterization. *J Clin Microbiol* **29:**2696–2700.
183. **Takai S, Imai Y, Fukunaga N, Uchida Y, Kamisawa K, Sasaki Y, Tsubaki S, Sekizaki T.** 1995. Identification of virulence-associated antigens and plasmids in *Rhodococcus equi* from patients with AIDS. *J Infect Dis* **172:** 1306–1311.
184. **Giguere S, Hondalus MK, Yager JA, Darrah P, Mosser DM, Prescott JF.** 1999. Role of the 85-kilobase plasmid and plasmid-encoded virulence-associated protein A in intracellular survival and virulence of *Rhodococcus equi*. *Infect Immun* **67:**3548–3557.
185. **Takai S, Sekizaki T, Ozawa T, Sugawara T, Watanabe Y, Tsubaki S.** 1991. Association between a large plasmid and 15- to 17-kilodalton antigens in virulent *Rhodococcus equi*. *Infect Immun* **59:**4056–4060.
186. **Yager JA, Prescott CA, Kramar DP, Hannah H, Balson GA, Croy BA.** 1991. The effect of experimental infec-

tion with *Rhodococcus equi* on immunodeficient mice. *Vet Microbiol* 28:363–376.

187. **Takai S, Koike K, Ohbushi S, Izumi C, Tsubaki S.** 1991. Identification of 15- to 17-kilodalton antigens associated with virulent *Rhodococcus equi*. *J Clin Microbiol* **29:** 439–443.

188. **Takai S, Watanabe Y, Ikeda T, Ozawa T, Matsukura S, Tamada Y, Tsubaki S, Sekizaki T.** 1993. Virulence-associated plasmids in *Rhodococcus equi*. *J Clin Microbiol* **31:**1726–1729.

189. **Tan C, Prescott JF, Patterson MC, Nicholson VM.** 1995. Molecular characterization of a lipid-modified virulence-associated protein of *Rhodococcus equi* and its potential in protective immunity. *Can J Vet Res* **59:** 51–59.

190. **Takai S, Anzai T, Fujita Y, Akita O, Shoda M, Tsubaki S, Wada R.** 2000. Pathogenicity of *Rhodococcus equi* expressing a virulence-associated 20 kDa protein (VapB) in foals. *Vet Microbiol* **76:**71–80.

191. **Jain S, Bloom BR, Hondalus MK.** 2003. Deletion of vapA encoding Virulence Associated Protein A attenuates the intracellular actinomycete *Rhodococcus equi*. *Mol Microbiol* **50:**115–128.

192. **Takai S, Iie M, Watanabe Y, Tsubaki S, Sekizaki T.** 1992. Virulence-associated 15- to 17-kilodalton antigens in *Rhodococcus equi*: temperature-dependent expression and location of the antigens. *Infect Immun* **60:** 2995–2997.

193. **Takai S, Fukunaga N, Kamisawa K, Imai Y, Sasaki Y, Tsubaki S.** 1996. Expression of virulence-associated antigens of *Rhodococcus equi* is regulated by temperature and pH. *Microbiol Immunol* **40:**591–594.

194. **Dorman CJ, Porter ME.** 1998. The *Shigella* virulence gene regulatory cascade: a paradigm of bacterial gene control mechanisms. *Mol Microbiol* **29:**677–684.

195. **Lindler LE, Plano GV, Burland V, Mayhew GF, Blattner FR.** 1998. Complete DNA sequence and detailed analysis of the *Yersinia pestis* KIM5 plasmid encoding murine toxin and capsular antigen. *Infect Immun* **66:**5731–5742.

196. **Perry RD, Straley SC, Fetherston JD, Rose DJ, Gregor J, Blattner FR.** 1998. DNA sequencing and analysis of the low-Ca2+-response plasmid pCD1 of *Yersinia pestis* KIM5. *Infect Immun* **66:**4611–4623.

197. **Byrne BA, Prescott JF, Palmer GH, Takai S, Nicholson VM, Alperin DC, Hines SA.** 2001. Virulence plasmid of *Rhodococcus equi* contains inducible gene family encoding secreted proteins. *Infect Immun* **69:**650–656.

198. **Letek M, Ocampo-Sosa AA, Sanders M, Fogarty U, Buckley T, Leadon DP, Gonzalez P, Scortti M, Meijer WG, Parkhill J, Bentley S, Vazquez-Boland JA.** 2008. Evolution of the *Rhodococcus equi* vap pathogenicity island seen through comparison of host-associated vapA and vapB virulence plasmids. *J Bacteriol* **190:**5797–5805.

199. **Ribeiro MG, Takai S, de Vargas AC, Mattos-Guaraldi AL, Ferreira Camello TC, Ohno R, Okano H, Silva AV.** 2011. Short report: identification of virulence-associated plasmids in *Rhodococcus equi* in humans with and without acquired immunodeficiency syndrome in Brazil. *Am J Trop Med Hyg* **85:**510–513.

200. **Takai S, Hines SA, Sekizaki T, Nicholson VM, Alperin DA, Osaki M, Takamatsu D, Nakamura M, Suzuki K, Ogino N, Kakuda T, Dan H, Prescott JF.** 2000. DNA sequence and comparison of virulence plasmids from *Rhodococcus equi* ATCC 33701 and 103. *Infect Immun* **68:**6840–6847.

201. **Vendrell D, Balcazar JL, Ruiz-Zarzuela I, de Blas I, Girones O, Muzquiz JL.** 2006. *Lactococcus garvieae* in fish: a review. *Comp Immunol Microbiol Infect Dis* **29:** 177–198.

202. **Chan JF, Woo PC, Teng JL, Lau SK, Leung SS, Tam FC, Yuen KY.** 2011. Primary infective spondylodiscitis caused by *Lactococcus garvieae* and a review of human *L. garvieae* infections. *Infection* **39:**259–264.

203. **Wang CY, Shie HS, Chen SC, Huang JP, Hsieh IC, Wen MS, Lin FC, Wu D.** 2007. *Lactococcus garvieae* infections in humans: possible association with aquaculture outbreaks. *Int J Clin Pract* **61:**68–73.

204. **Aguado-Urda M, Gibello A, Blanco MM, Lopez-Campos GH, Cutuli MT, Fernandez-Garayzabal JF.** 2012. Characterization of plasmids in a human clinical strain of *Lactococcus garvieae*. *PLoS One* **7:**e40119.

205. **National Nosocomial Infections Surveillance (NNIS) System..** NNIS Report, Data Summary from January 1992-June 2001, Issues August 2001. *Am J Infect Control* **29:**404–421.

206. **Doolittle WF, Sapienza C.** 1980. Selfish genes, the phenotype paradigm and genome evolution. *Nature* **284:** 601–603.

207. **Huycke MM, Spiegel CA, Gilmore MS.** 1991. Bacteremia caused by hemolytic, high-level gentamicin-resistant *Enterococcus faecalis*. *Antimicrob Agents Chemother* **35:**1626–1634.

208. **Booth MC, Pence LM, Mahasreshti P, Callegan MC, Gilmore MS.** 2001. Clonal associations among *Staphylococcus aureus* isolates from various sites of infection. *Infect Immun* **69:**345–352.

209. **Hisatsune J, Hirakawa H, Yamaguchi T, Fudaba Y, Oshima K, Hattori M, Kato F, Kayama S, Sugai M.** 2013. Emergence of *Staphylococcus aureus* carrying multiple drug resistance genes on a plasmid encoding exfoliative toxin B. *Antimicrob Agents Chemother* **57:** 6131–6140.

Plasmids—Biology and Impact in Biotechnology and Discovery
Edited by Marcelo E. Tolmasky and Juan C. Alonso

doi:10.1128/microbiolspec.PLAS-0030-2014

Manuela Di Lorenzo[1]
Michiel Stork[2]

29 Plasmid-Encoded Iron Uptake Systems

INTRODUCTION

Iron is one of the most important metals for life, as it is necessary for the proper functioning of proteins that mediate essential cellular processes such as DNA precursor synthesis, respiration, photosynthesis, and nitrogen fixation. Iron is one of the most abundant elements on earth; however, its bioavailability is very low. In the presence of oxygen, ferrous iron oxidizes to ferric iron that is poorly soluble at neutral pH. Additionally, free iron is toxic due to the formation of free oxygen radicals that can cause cell damage (1). Therefore, in biological systems the iron is complexed to keep it soluble and to reduce the toxicity of free iron.

To overcome the low availability of iron in their environment, bacteria have evolved specialized mechanisms to take up either the scarce soluble ions or to compete with iron-chelating complexes (2). In particular, pathogenic bacteria face extremely low iron concentrations due to the presence of high-affinity iron-binding proteins in the host that perform a dual function: they protect cells from the toxic effect of free iron while inhibiting bacterial growth (3). Competition for iron within the host is thus a critical factor in host-pathogen interaction (4).

Mechanisms for iron uptake range from direct binding of transferrin, lactoferrin, and heme by outer membrane receptors, to compounds with a high affinity for iron that strip the metal ions chelated to protein and complexes (5). To take up iron from the host iron-binding proteins transferrin and lactoferrin, bacteria express specific outer membrane receptors. Upon binding to the receptor, iron is released and transported through the receptor to the periplasmic space and subsequently to the cytoplasm (6, 7). Bacteria can also obtain iron from red blood cells via similar though more complex mechanisms. Hemoglobin and heme are released from hematocytes by the action of cytolysins and hemolysins. Receptors at the bacterial surface recognize the free heme or hemoproteins and transport heme to the periplasm. Once inside the cell, iron is released from heme (8). More elaborate heme acquisition systems secrete heme-scavenging molecules, the hemophores, which extract heme from host hemoproteins and deliver it to an outer membrane receptor for internalization (8).

Siderophore-mediated systems are the most studied iron uptake mechanism of bacteria. Siderophores are low-molecular-weight compounds that have a very high association constant for iron, up to 10^{50} (9). Most of these

[1]Netherlands Institute of Ecology (NIOO-KNAW), Department of Microbial Ecology, 6708 PB Wageningen, The Netherlands; [2]Institute for Translational Vaccinology, Process Development, 3720 AL Bilthoven, The Netherlands.

secondary metabolites are synthesized via a nonribosomal peptide synthetase mechanism (10). Siderophores are produced in the cytoplasm and then secreted to the extracellular space. Outside the cell, the siderophore competes for iron with iron complexes, such as the iron-binding proteins lactoferrin and transferrin in the host (11, 12). Iron-loaded siderophores are bound by cognate receptors, and iron is internalized via a specific uptake system in the membrane of the bacterium (13). The uptake of ferric-siderophores by the bacterium is initiated by a chance encounter of the bacterial outer membrane receptor with the iron-loaded siderophore.

In all cases, transport of iron across the membranes needs energy. In Gram-negative bacteria there is no energy generation in the outer membrane, so the energy required to cross it is provided by the inner membrane. The proton motive force of the inner membrane is transduced to the outer membrane receptors by the TonB complex. This complex consists of several proteins (TonB, ExbB, ExbD, and in some cases TtpC) that are always chromosomally encoded (14–17). Once the iron or iron complex is in the periplasm, periplasmic binding proteins, either soluble or membrane-bound through a lipid anchor, mediate delivery to a complex in the inner membrane that usually belongs to the ATP-binding cassette (ABC) transporter family. The ABC transporters consist of two inner membrane permeases and an ATPase that hydrolyzes ATP to energize transport of the ligand across the cytoplasmic membrane (18).

While outer membrane receptors are highly specific for the system they belong to, the rest of the proteins involved in transport into the cytosol are often interchangeable between different iron uptake systems (18). Once it is in the cytoplasm, iron can be used for immediate cellular needs or stored in bacterioferritins (19).

The gene products involved in iron uptake are tightly regulated by iron availability. In iron-replete environments the genes are repressed by the ferric uptake regulator (Fur) (20–22). Fur in complex with iron binds to specific DNA sequences at the promoter regions of iron-regulated genes and blocks transcription. When iron is limiting, Fur no longer contains iron and therefore cannot bind the DNA. As a consequence, Fur repression is released and transcription proceeds (22). In addition to Fur, expression of iron uptake genes is also controlled by numerous small RNAs as well as regulatory proteins and two component systems (23, 24).

Most bacteria have at least one iron uptake mechanism, although the presence of multiple systems is quite common. Although the majority of these systems are chromosomally encoded, some of them are plasmid-mediated (25). The most studied plasmids harboring iron uptake systems are the ColV plasmids, found mainly in *Enterobacteriaceae*, and the pJM1-type plasmids, found in *Vibrio anguillarum* strains. Interestingly, there is no homology between these two plasmids and the systems they carry (26). In this article, we describe these two plasmid types and their iron uptake systems.

ColV PLASMIDS

ColV plasmids are large and belong to the IncFI incompatibility group. Plasmids in this group are heterogeneous in size (typically from 80 to 180 kb), genetic composition, ability to conjugate, and virulence properties (27).

These plasmids often possess more than one IncF replicon, but the replicons and their genetic organization are not always conserved across the different plasmids. The *repFIB* replicon is the most widespread among ColV plasmids, and it is located between the *sitABCD* and the *etsABC* regions (27–31). The *repFIB* replicon of pColV-K30 encodes two essential factors for replication: an initiator protein, RepI, and five 18-bp direct repeats (32–34). It has been proposed that upon binding to the direct repeats at the origin of replication, RepI facilitates the opening of the DNA helix, initiating replication. Furthermore, when bound to the origin, RepI also represses its own expression (35, 36). Another replicon is commonly found in proximity to the transfer region; while in most cases this is a *repFIIA* replicon, in plasmids pAPEC-1 and pAPEC-O1-ColBM a composite *repFIIA-repFIC* origin is present (28, 31). Some ColV plasmids also carry a replicon homologous to *repFIA* of plasmid F (37). In these instances, a genetic organization with the two replicons *repFIA* and *repFIB* flanking the aerobactin system seems to be conserved even in non-ColV plasmids, such as *Escherichia coli* R plasmids (38).

ColV plasmids have long been associated with the virulence of extraintestinal pathogenic *E. coli*, ExPEC (29, 39). This group of pathogenic *E. coli* has acquired numerous virulence traits, which confer the ability to cause an infection outside the host intestinal tract. The ExPEC group includes human and animal pathogens that cause urinary tract infections, neonatal meningitis, and septicemia. Avian pathogenic *E. coli* (APEC) strains are also ExPEC, and ColV plasmids are a defining feature of the APEC pathotype (40). ColV plasmids have also been identified at high frequencies in strains of *Salmonella enterica* serovar Kentucky isolated from chickens (41, 42).

ColV plasmids owe their name to colicin V, a toxin that is active against *E. coli*, *Shigella*, and *Salmonella* strains. It was first described by Andre Gratia in 1925

(43) and later shown to be transferable and linked to F-type plasmids (44, 45). Colicin V, the first identified colicin, is a microcin, a low-molecular-weight peptide antibiotic secreted through a dedicated ABC transporter (46, 47). Colicin V exerts bactericidal activity only at the periplasmic face of the inner membrane, where it interferes with energy production by disrupting the membrane potential (48). Four plasmid-encoded genes (*cvaA*, *cvaB*, *cvaC*, and *cvi*) are involved in colicin synthesis, export, and immunity (Fig. 1). In addition, two chromosome-encoded genes (*cvpA* and *tolC*) are required for production and secretion (46, 49). Colicin V is often found in association with colicin Ia (50). Plasmid pS88 is an example of a plasmid harboring colicin V and colicin Ia operons (51). Colicin Ia has the same mode of action as colicin V, but while the production of colicin Ia is induced under stress conditions, the expression of colicin V is regulated by iron (49, 52).

Other colicin-encoding plasmids have been identified among ExPEC and APEC, the ColBM plasmids. These plasmids have received less attention than the more widespread ColV plasmids, but recent studies have shown that the ColBM plasmids harbor virulence clusters similar to those present on ColV plasmids. Analyses of the available sequences of ColBM plasmids suggest that they have evolved from an ancestral ColV plasmid by insertion of the ColBM region within the ColV operon. In fact, high DNA homology between the plasmids pAPEC-O2-ColV and pAPEC-O1-ColBM is observed in the virulence clusters, and 67% of the predicted proteins of pAPEC-O1-ColBM are also found on pAPEC-O2-ColV (28). These two plasmids also have nearly identical transfer regions and possess *repFIB* and *repFIIA* replicons (28). Alternatively, transfer of the colicin V genes and associated virulence factors onto ColBM plasmids could have occurred (53).

The ColBM region harbors the ColB and ColM operons. These two operons encode colicin B and colicin M, respectively. Each operon consists of a structural gene for the colicin (*cba* in ColB and *cma* in ColM) and an immunity gene (*cbi* in ColB and *cmi* in ColM) encoding a protein that binds and inactivates the colicin. Colicin B is a pore-forming colicin and has a mode of action similar to colicin V (49), while colicin M, the smallest known colicin, inhibits peptidoglycan biosynthesis (54). Interestingly, both colicins, as well as colicin V and Ia, gain access to sensitive cells via TonB-dependent receptors: the enterobactin receptor FepA for colicin B, the ferrichrome receptor FhuA for colicin M, and the catechol receptor Cir for colicin V and colicin Ia (47, 55). The use of iron-regulated receptors for internalization and their iron-regulated expression show the intimate link between colicins and iron. Colicin-producing cells might have an advantage in iron-limited environments, since development of resistance in sensitive cells might occur at the cost of iron uptake efficiencies since it could mean the loss of TonB-dependent receptors (47).

Despite their name, it was shown early on that the virulence phenotype conferred by the ColV and ColBM plasmids is not colicin-mediated (56, 57). Several other determinants that are present in different combinations in the various plasmids of the ColV group have been shown to contribute to virulence. ColV plasmids harbor genes that encode serum resistance, adherence, resistance to chlorine and disinfectants, bacteriophage resistance, hemagglutination properties, and several iron acquisition systems (29). The presence of several clusters involved in iron metabolism is not surprising when considering the extraintestinal lifestyle of ExPEC strains, where iron is one of the main limiting factors for colonizing bacteria.

Aerobactin Iron Uptake System

The first ColV iron uptake system identified was the one that codes for the hydroxamate siderophore aerobactin and its outer membrane receptor (58). The aerobactin system is, together with colicin V, the hallmark of ColV plasmids. Aerobactin was first discovered in the supernatant of *Aerobacter aerogenes* 62-I and was later shown to be encoded by plasmid pSMN1 (59, 60). Even though they are both plasmid-encoded, the aerobactin cluster of pSMN1 shares only a weak homology with the genes encoding the aerobactin system found on the ColV plasmids (61). These two genetically distinct systems nonetheless produce siderophores with identical structures. Interestingly, the aerobactin cluster is also found on non-ColV R plasmids of *Salmonella* and *E. coli* (38, 60, 62, 63). In contrast to the case of pSMN1, the aerobactin sequences found on ColV and *E. coli* and *Salmonella* R plasmids are conserved and linked to the *repFIB* replication region (37, 62). It is worth noting that the origin of replication of plasmid pSMN1 is not of the IncFI incompatibility group (61), which suggests a linkage between the *repFIB* origin of replication and the aerobactin cluster found on ColV and R plasmids. They could constitute a virulence factor-replication unit that ensures perpetuation of the aerobactin genes even in the event of a deletion (37). The aerobactin system was also found in close proximity to the *repFIB* replicon in *Chronobacter* virulence plasmid and in a plasmid isolated from a sewage treatment plant (64–66).

The aerobactin genes in ColV and *Salmonella* R plasmids are flanked by the inverted IS*1* sequences that could contribute to the spread of this system either by homologous recombination with other IS*1* sequences or

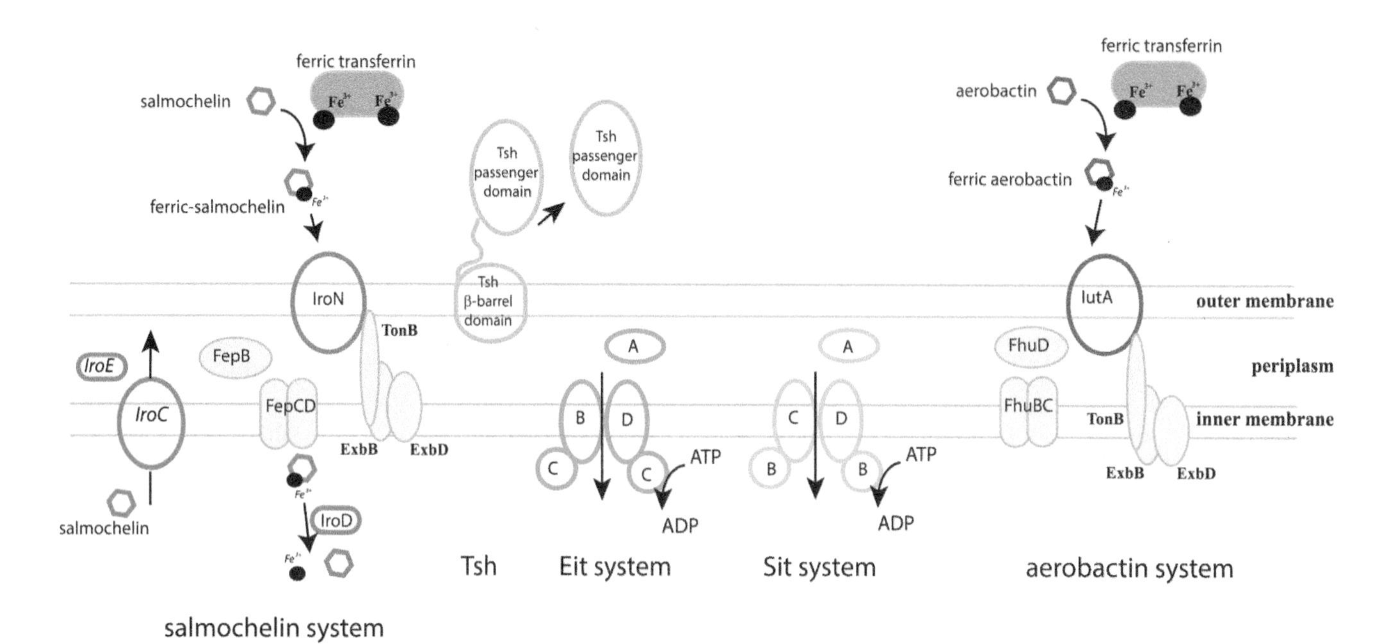

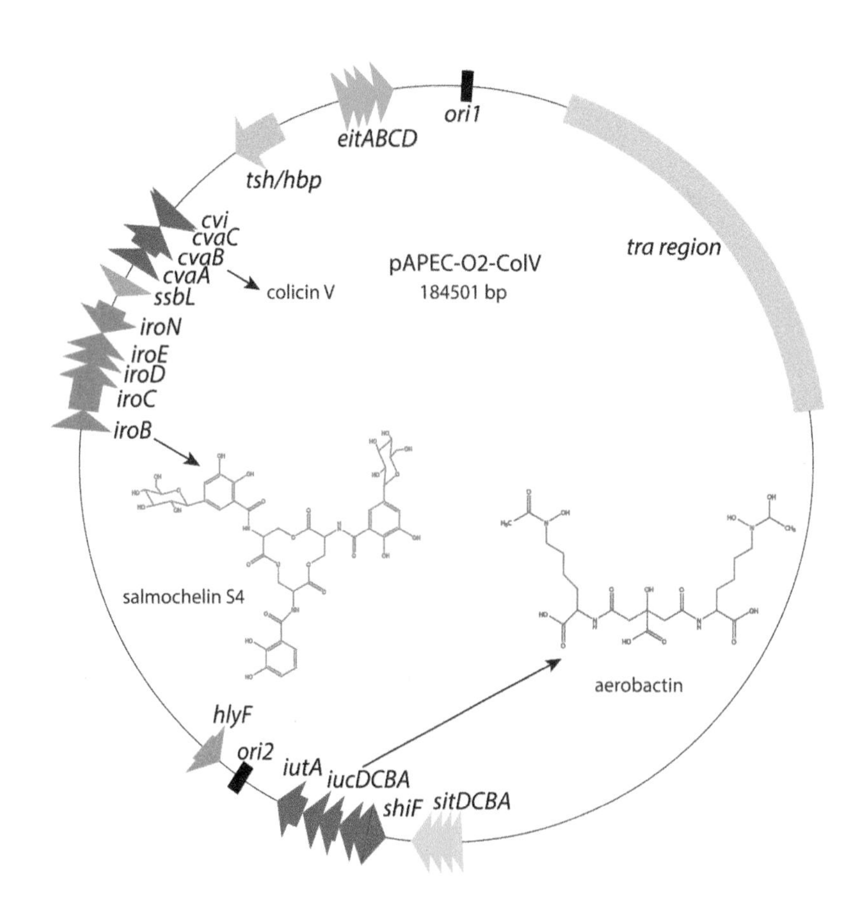

Figure 1 Schematic representation of a ColV plasmid (30) showing all open reading frames related to iron uptake, their function, and their membrane localization when relevant. Each system is color-coded. The *tra* region is shown as a gray box, and the origins of replication are shown as black boxes. Structures of the two siderophores aerobactin and salmochelin S4 are shown within the plasmid. doi:10.1128/microbiolspec.PLAS-0030-2014.f1

by transposition (37, 67, 68). The increased mobility conferred by the flanking IS*1* sequences could explain the occurrence of the aerobactin system also on chromosomes of *E. coli*, *Salmonella*, and *Shigella* strains (60, 69–73). The presence of a colicin system within the pathogenicity island (PAI) of *Shigella flexneri* upstream of the aerobactin genes further supports the hypothesis of a ColV origin for the chromosomal systems (73). On the other hand, the absence of flanking IS*1* sequences both on plasmid- and chromosome-encoded aerobactin systems as well as some sequence divergence, might argue against an IS*1*-mediated mechanism of dissemination and a ColV origin (74, 75).

The gene cluster of the aerobactin system consists of five open reading frames, *iucABCDiutA* (Fig. 1) (76). The products of the *iuc* genes are involved in the biosynthesis of aerobactin from lysine and citrate. The citrate molecule serves as a central linker for two molecules of N^6-acetyl-N^6-hydroxy lysine. These molecules are produced from lysine in two steps. In the first step, the IucD monooxygenase utilizes molecular oxygen to generate N^6-hydroxy lysine. In the subsequent step, the acetylase IucB *N*-acetylates the hydroxy lysine using acetyl-CoA (77). The final coupling of two molecules of N^6-acetyl-N^6-hydroxy lysine to citrate is catalyzed by the two synthetases IucA and IucC. Each synthetase assembles one modified lysine molecule onto the citrate, forming an amide bond between the α-amino group of lysine and the carbon of one of the carboxyl groups of citrate (78). The last gene of the cluster, *iutA*, encodes the outer membrane receptor for ferric aerobactin that functions in a TonB-dependent manner (79). Once in the periplasm, aerobactin is bound and transported across the inner membrane by the generic chromosomally encoded hydroxamate transport system, which consists of a periplasmic binding protein, FhuD, and inner membrane permeases, FhuBC (80–82). The clusters coding for the TonB complex and the hydroxamate transport system are both located on the chromosome (83).

Like other iron uptake systems, expression of the aerobactin cluster is negatively regulated by Fur and the iron state of the cell (84). Fur binds in an iron-dependent manner to multiple sites within a large region of the aerobactin operon promoter (85).

Several pieces of evidence point to aerobactin as a virulence factor. A large percentage of *E. coli* strains isolated from human extraintestinal infections tested positive for aerobactin, while only one-third of normal fecal isolates carried genes of the aerobactin system (86, 87). Among avian *E. coli* isolates, only the most virulent harbor the aerobactin genes, while these same genes are absent from nonvirulent and nonlethal isolates (88, 89). Furthermore, aerobactin genes were expressed during infection, and defined deletion mutants in biosynthetic and transport genes resulted in decreased virulence in a chicken infection model (90–92). In whole-plasmid transcriptional studies, an additional gene, *shiF*, was upregulated during growth in human serum and urine and in a neonatal rat sepsis model (93, 94). The *shiF* gene has been found to be conserved upstream of the aerobactin cluster, in some cases in association with another gene, *viuB* (51, 64–66). Based on their homology, the products of these two genes have been proposed to be part of the aerobactin iron uptake system (64–66, 93).

Salmochelin Iron Uptake System

The salmochelin iron uptake system, encoded on the chromosome of *Salmonella* species and of several uropathogenic *E. coli* (UPEC) (95–98), has been recently associated with the ColV plasmids of APEC strains (90, 92, 99, 100) as well as with non-ColV transmissible plasmids of UPEC and *Klebsiella* strains (101, 102).

Salmochelins are catechol siderophores that consist of enterobactin glucosylated at one or two of the 2,3-dihydroxybenzoyl (DHB) rings (103). Salmochelin S4 (Fig. 1), the most common form, has the macrolactone of enterobactin intact, while this is hydrolyzed in salmochelin S2, where the unglycosylated DHB-serine moiety is located at the C-terminal end. Salmochelins S1 and SX are, respectively, the dimer DHB(gucosyl)-seryl-DHB-serine and the monomeric DHB(glucosyl)-serine molecule. All these salmochelin forms have been isolated from *S. enterica* growth medium, but as in the case of enterobactin, these linear forms are probably the degradation products of salmochelin S4, the key metabolite (103). Since the substrate for salmochelin S4 synthesis is enterobactin itself, genes of the enterobactin cluster are also essential for salmochelin biosynthesis (103, 104). Glycosylation of enterobactin results in a siderophore with reduced hydrophobicity that is no longer recognized and bound by the host protein siderocalin and serum albumin (103, 105, 106). Perhaps immunologic pressure in serum resulted in a selective advantage for strains harboring salmochelin over those producing enterobactin.

Genes coding for salmochelins and the outer membrane receptor were first identified in the chromosome of *Salmonella* serovar Typhimurium in the Fur-regulated locus, *iroA* (96). The *iroA* locus consists of two divergently transcribed sets of genes, *iroBCDE* and *iroN* (95).

IroN is an outer-membrane protein with homology to TonB-dependent receptors. IroN is the receptor for salmochelins, as shown by uptake experiments with radiolabeled ^{55}F-salmochelin and growth promotion assays (107). Salmochelin uptake can also be facilitated by the enterobactin receptor FepA and the Cir receptor in *Salmonella* (107). On the other hand, IroN can transport, besides salmochelins, other catechol siderophores, such as enterobactin and corynebactin (102, 108). An additional role as an internalization factor has been proposed for IroN during invasion of urothelial cells by ExPEC strains (109).

The *iroB* gene is the only one in the *iroA* cluster involved in salmochelin biosynthesis (Fig. 1). IroB encodes a glucosyltransferase that glucosylates the DHB rings of enterobactin using UDP-Glc as the glycosyl donor. A remarkable characteristic of IroB is the ability to catalyze formation of a C-C bond during enterobactin glucosylation. In this regard, IroB differs from other glycosyltransferases, such as those identified in the antibiotics vancomycin and ramoplanin, which glycosylate oxygen atoms (104).

The *iroC* gene encodes a four-domain ABC exporter similar to eukaryotic multidrug resistance transporters (Fig. 1). IroC is involved in salmochelin S4 secretion rather than in the uptake of ferric-salmochelin, as shown by deletion of the *iroC* gene (110, 111).

The last two genes in the *iroBCDE* operon encode two proteins, IroD and IroE, which are predicted to have hydrolytic activity. IroD is a cytoplasmic protein similar to Fes, the esterase of the enterobactin system. IroE is instead predicted to be a periplasmic esterase. The function of these two esterases in the salmochelin system is still not well established as well as their substrate specificity. Most likely, IroD is the esterase responsible for release of iron from ferric-salmochelin into the cytoplasm. IroE acts in the periplasm and might be more important for export or processing of cyclic salmochelins prior to release (99, 110, 112).

Besides the *iroBCDE* and *iroN* plasmid-encoded genes, the entire chromosomally encoded enterobactin cluster is required for salmochelin biosynthesis and transport (113). In addition, the chromosomally encoded TonB complex is needed to transport salmochelin across the outer membrane.

It has been shown that salmochelin is required for full virulence in a chicken infection model (90). Mutations in the salmochelin system slightly affect virulence, while mutations in multiple iron uptake systems result in a dramatic decrease of chicken tissue colonization (91, 114). This is probably due to the high redundancy of iron uptake systems in pathogenic strains. Levels of salmochelin produced during infection of host tissues are similar to those produced in iron-poor culture medium (110). Although salmochelin levels are lower than the aerobactin levels, both *in vitro* and *in vivo*, production of salmochelin *in vivo* further supports a role in virulence for this siderophore. Transcriptional studies in human serum and urine point to a similar role of the salmochelin system found on plasmid pS88 of an *E. coli* strain that causes neonatal meningitis (93). Interestingly, in the same study a gene, *ssbL*, was identified that displayed the highest upregulation *in vivo*. The *ssbL* gene shows a strong association with the *iro* cluster (115). The *ssbL* gene is involved in the shikimate pathway, and its product has been proposed to enhance production of salmochelin by boosting the catechol metabolic pathway (115).

SitABCD Iron Transport System

Operons encoding the SitABCD system have been found on the reference colicin V plasmid pColV-K30 and on ColV-like plasmids of APEC strains (28, 30, 31, 116, 117).

The SitABCD system is an ATP-binding cassette (ABC) transporter of divalent metal cations, originally identified on the chromosome of serovar Typhimurium in *Salmonella* PAI1, SPI1 (118). As in typical ABC transport systems, the *sit* genes encode a periplasmic protein, SitA, that binds the ligand and transfers it to the inner membrane permeases, SitC and SitD (Fig. 1). SitB, the ATPase of the system, catalyzes hydrolysis of ATP to energize transport across the cytoplasmic membrane of the ligand (119).

The *sitABCD* operon is expressed *in vivo* during the systemic stages of *Salmonella* serovar Typhimurium infection in mice. A mutation in the *sit* locus resulted in a defect in virulence of the mutant strain, although it was subtle, probably due to the presence of redundant systems for iron and manganese uptake (120). After its initial identification as a siderophore-independent iron ABC transporter, the Sit system was shown to be involved in manganese and ferrous iron transport from the periplasm to the cytoplasm (116, 121, 122). The ability of the SitABCD transport system to transport manganese might enhance resistance to oxidative stress and survival in extraintestinal tissues (123).

In APEC strains, *sit* genes are highly associated with virulence in 1-day-old chicks, and there is a prevalence of this system in virulent isolates when compared with fecal commensals (40, 123). Interestingly, although *sit* genes are also found on the chromosome of some *E. coli* strains, the *sitABCD* operon in APEC strains is mainly plasmid-encoded (124).

EitABCD System

An additional ABC transport system, EitABCD, has been found on pAPEC-O2-ColV and on pAPEC-O1-ColBM plasmids (28, 30). EitABCD is similar at the protein level, with an iron transport system of the plant pathogen *Pseudomonas syringae* (30). The functionality of this ABC system in iron uptake has yet to be determined. The homology between the Eit and Sit ABC transporters present on the pAPEC-O2-ColV plasmid is very low, and the functional genes also differ in the order within the clusters (Fig. 1). While the SitABCD system has been shown to be a metal ion transporter, EitABCD shares similarity with other ABC transporters involved in siderophore transport.

Other ColV plasmids sequenced so far do not harbor the *eitABCD* cluster, while this system is found on non-ColV *repFIB* and *repFII* plasmids (29). Interestingly, these two replicons are those harbored by several ColV plasmids. The *repFIB* plasmids, found in 97% of 229 *Chronobacter* spp., emerging neonatal pathogens, harbor the Eit system. The *eit* gene arrangement in the *Chronobacter* plasmids, pESA3 and pCTU1, differs from the typical single-operon *eitABCD* found in several enteric pathogens. In these *Chronobacter* plasmids the genes are organized as two divergent operons, *eitABC* and *eitD* (64). Expression of these *eit* operons does not seem to be regulated by iron despite the presence of putative Fur boxes upstream of the *eitA* and *eitD* genes (65). IncFII plasmids harboring the *eitABCD* cluster are instead found in APEC and *Klebsiella* strains (125–127).

The prevalence of *eit* genes among ExPEC of human and avian origin is low, which could indicate a recent acquisition of these genes by ExPEC strains (126).

Temperature-Sensitive Hemagglutinin Tsh

The Tsh protein of APEC strain χ7122 was characterized as a hemagglutinin with a proteolytic domain expressed at lower temperatures (128). This hemagglutinin was the first identified member of an expanding subclass of the IgA protease family of autotransporters present in *Shigella* and numerous pathotypes of *E. coli* (129). Autotransporters consist of three functional domains: a sec-dependent amino-terminal leader sequence, a passenger domain, and an outer membrane-associated carboxy-terminal β-barrel domain (see Fig. 1) (129). The latter domain mediates secretion of the passenger domain, the extracellular or surface-secreted mature protein. The 106-kDa passenger domain of Tsh contains a serine protease motif whose function could not be demonstrated (130). Tsh is almost identical to Hbp (only two amino acid differences), an autotransporter from an *E. coli* clinical isolate from a wound infection. Hbp was shown to specifically degrade human hemoglobin and bind heme and it was proposed to be part of a hemophore-dependent heme acquisition system (131). A similar role was speculated for the other members of the Tsh family (131).

The *tsh* genes as well as the *hbp* genes are often located on plasmids (131, 132). In a screening study of 300 avian *E. coli* isolates, the *tsh* gene was more prevalent in high-lethality isolates, where it was always plasmid-encoded (132). In the same study, it was shown that the majority of the plasmids harboring *tsh* were of the ColV-type, while a restricted minority was not (132). The association of the *tsh* gene with large plasmids containing the colicin V gene cluster is confirmed by the growing number of complete sequences of ColV plasmids, although not all of the ColV plasmids harbor the *tsh* gene (117). The *tsh* gene has also been found within PAI III of UPEC strain 536 together with the *iro* cluster and a putative heme receptor (98). The different distribution of *tsh* compared to the widespread aerobactin system supports their association with virulence of APEC strains; it is worth noting that although the *tsh* gene is associated with high lethality, the highest level of virulence seems to require the aerobactin siderophore and not Tsh (89). In agreement with this is the minor role of iron acquisition through heme during infection, which is confirmed by studies with mutant strains in a chicken infection model (91). On the other hand, Tsh might have a role in the air sacs, as supported by infection studies and expression of the gene in the air sacs (90, 132). Other iron uptake systems, such as aerobactin and salmochelin siderophores, might be more important in deeper tissues (90).

Hemolysin

Another gene involved in iron uptake and associated with ColV plasmids is the *hlyF* gene (Fig. 1). This gene has been found in the majority of the APEC ColV plasmids and in several ExPEC strains (29, 31, 133). The *hlyF* gene, its product HlyF, and its mode of action still need to be characterized. HlyF has hemolytic activity, but its amino acid sequence shows no significant homology to HlyA, the RTX hemolysin of several *E. coli* pathotypes, or to HlyE, a more recently identified hemolysin distinct from HlyA (134–137). Besides HlyF, the HlyA hemolysin is associated with plasmids. In fact, while the *hlyCABD* operon is found on the chromosome in UPEC strains, large plasmids harbor hemolysin clusters in human enterohemorrhagic *E. coli* (EHEC) strains and enterotoxigenic, shigatoxigenic, and enteropathogenic *E. coli* strains isolated from animals (138–141). The HlyA hemolysin found in EHEC, the enterohemolysin, is distinct from the HlyA hemolysin found in the other *E. coli* pathogroups, the α-hemolysin (134, 137). Plasmids of EHEC strains are, like ColV plasmids, F-like plasmids, and they share a similar backbone for what concerns origins of replication and transfer region, although the majority of EHEC plasmids are transfer-defective (29, 142, 143). On the other hand, plasmids harboring the genetic determinants for production of α-hemolysin were found to be heterogeneous also in their incompatibility group (144, 145). DNA hybridization experiments indicate that plasmid-encoded α-hemolysin and enterohemolysin have evolved separately in EHEC and enteropathogenic strains, although they might exert the same function (141, 146). On the other hand, the high similarity at the sequence level of α-hemolysin operons present on plasmids in enterotoxigenic, shigatoxigenic, and enteropathogenic strains indicates a common origin for these genes (138, 146).

The *hlyA* gene encodes the pro-hemolysin that is converted to its active form by the product of the *hlyC* gene by transfer of two fatty-acyl residues onto pro-HlyA. A type-I secretory complex comprising the ATP-binding cassette HlyB, the membrane-fusion protein HlyD, and the outer-membrane protein TolC secrete the active HlyA as well as the pro-HlyA. HlyA interaction with the target cell membrane occurs in two steps: the α-hemolysin absorbs onto the membrane in a reversible manner that is followed by irreversible insertion within the membrane. Once HlyA is inserted in the membrane, oligomerization occurs, leading to pore formation and cell lysis (147).

The role of hemolysins in iron uptake is still under debate. It has been proposed that secretion of hemolysins and cytolysins can quickly increase the local concentration of free heme and hemoglobin. This proposal is supported by indirect evidence such as hemolysin expression often repressed by iron and the presence of heme transport systems in strains with hemolytic activity (134, 148, 149). The presence of *hlyF* together with the *tsh* gene on the same plasmid further supports the possibility of their products acting as a hemophore-dependent heme acquisition system in combination with a chromosomally encoded heme receptor, such as ChuA. Besides colocalization in some cases, *hly* genes and *chuA* share a positive regulator, as can be expected for genes that function in the same system (147, 150).

Heterogeneity of ColV Determinants

As proposed by Waters and Crosa (27), a "constant" and a "variable" region, together spanning a 94-kb cluster of putative virulence genes, can be distinguished within the ColV PAI. The constant region contains the *repFIB* replicon, the aerobactin operon, the Sit iron and manganese transport system, a putative outer membrane protease gene, *ompT*, the hemolysin gene, *hlyF*, a novel ABC transport system known as Ets, the salmochelin siderophore system, and *iss* that encodes serum resistance (30). The temperature-sensitive hemagglutinin gene *tsh* and another novel transport system known as Eit are part of the variable portion, as they are less frequently associated with ColV plasmids than are the genes in the constant region (29). The divide between the two regions appears to be within the *colV* operon, with a clear higher cooccurrence of genes within the constant region (about 60%) than the 26% for genes of the variable region (30).

From the increasing number of large plasmid complete sequences, a remarkable high nucleotide similarity (>95%) in the common regions of ColV plasmids is becoming evident. This suggests that ColV plasmids have evolved by adding and deleting large blocks of DNA, and these rearrangements are most likely the product of IS-mediated recombination events that affect the core as well as the noncore regions. As a consequence, a great heterogeneity in gene content is found within the ColV plasmid group (29).

As a result of frequent recombination events, widespread diversity in the noncore components is found in the sequenced plasmids, none of which are identical to each other with regard to gene content. In extreme cases, plasmids can lack the entire variable region. There is also an example of a strain harboring the constant region on one plasmid and part of the variable region (the *eit* operon) on another (31, 51, 126).

IS-mediated recombination events may also account for the occurrence of some of the ColV plasmid components on non-ColV plasmids or on the chromosome of *E. coli*, *Shigella*, *Salmonella*, and possibly other bacterial strains. It is not yet clear if this transfer of genetic information proceeds from plasmid to chromosome or from chromosome to plasmid. The fact that in *S. flexneri* PAI SHI-2 harboring the aerobactin system also encodes immunity to colicins I and V suggests that the aerobactin genes found in SHI-2 originated from a ColV plasmid (73). In contrast, it has been proposed that ColV plasmids function as transposon traps for virulence-related determinants, pointing to a flow of genetic information to the plasmid rather than from it (131).

Aerobactin is also found on *Klebsiella* high-molecular-weight plasmids, and the presence of this system has been associated with virulent isolates (75). Although the system found in *Klebsiella* plasmids shows homology with the ColV genes, a certain degree of divergence and the absence of IS*1* insertion sequences downstream of the genes in *Klebsiella* might indicate different origins of the two systems (75).

Genes coding for aerobactin, salmochelin, and the Sit systems have been found on non-ColV plasmids of *E. coli*, *Salmonella*, and *Klebsiella* that are often associated with antibiotic resistance genes (38, 51, 62, 75, 102, 151). Occurrence of resistance determinants on ColV plasmids from different bacterial hosts has also been reported (29, 41, 42). The high degree of conservation of these resistance elements and their presence on unrelated plasmids suggest a shared horizontal resistance gene pool available to bacteria from different environments (152). The association of drug resistance determinants with virulence plasmids is quite worrisome since it may enhance pathogenicity in environments where antibiotics are used.

pJM1-TYPE PLASMIDS

Plasmids of the pJM1 type are only found in the bacterium *V. anguillarum*, and more specifically, only in serotype O1 strains of this bacterium. *V. anguillarum* is the causative agent of vibriosis in fish, a highly fatal hemorrhagic septicemic disease (153). Several studies showed that *V. anguillarum* strains harboring pJM1 have a reduced infectivity in the absence of the plasmid (153–158). Plasmids of the pJM1 family range between 65 and 70 kb, and their difference in size is mainly due to the number of insertion sequences present. Plasmids in this group are highly similar, as shown in independent studies using restriction length polymorphism, where only a few patterns were identified among numerous different isolates from the United States, Japan, Spain, and Denmark (159–162).

To date, three complete annotated sequences of pJM1-type plasmids are available: the pJM1 plasmid from *V. anguillarum* 775 (65,009 nucleotides) and pEIB1 from *V. anguillarum* MVM425 and plasmid M3 from *V. anguillarum* M3, which are both 66,164 nucleotides (161, 163, 164). These plasmids are highly conserved and differ only in few nucleotides over the 65 kb they have in common.

The vast majority of the genes encoded on the pJM1-type plasmids are involved in iron uptake, either directly or indirectly (Fig. 2). The remaining genes code for proteins involved in plasmid replication, insertion sequences, and those for which no function can be assigned based on their homology (161, 163, 164).

Replication and Transfer

The origin of replication was first identified on the pJM1 plasmid and later was further characterized on pJM1 and the pJM1-type plasmid pEIB1 (164–166). A 1,400-bp Sau3A fragment from pEIB1 was identified that can replicate in *E. coli* and *V. anguillarum* as well as in other *Vibrio* species such as *Vibrio alginolyticus*. The region contains replication features such as DnaA binding sites and inverted repeats (164). Recombinant plasmids consisting of different pJM1 fragments containing this origin of replication (ori1) and a resistance marker could replicate in a plasmid-less derivative of *V. anguillarum* 775 but resulted in different copy numbers (166). However, in *E. coli* all fragments tested had similar copy numbers, in all instances higher than in *V. anguillarum*. Hence, the pJM1 plasmid ori1 has some kind of copy number control that only functions in *V. anguillarum* (166). In the same study, it was shown that DnaA is not required for replication of pJM1, in contrast with the results obtained with the pEIB1 plasmid (164, 166). Interestingly, deletion of ori1 resulted in a pJM1 derivative still able to replicate, indicating the presence of a second origin of replication (ori2). Ori2 is located at ORF25 (Fig. 2), but unlike ori1, it cannot replicate in *E. coli* (164, 166).

Genes such as those encoding DnaB, DnaC, and DnaG, which are needed for DNA replication, are provided by the chromosome as well as a polymerase since no polymerase is encoded by the plasmid (161, 166). The incompatibility group for both replicons is unknown, and no homology with known origins of replication has been found. Ori1 is compatible with plasmids such as pUC18 (pMB1), pKA3 (pSC101), and p15A in *E. coli* (164). On the other hand, conjugation into *V. anguillarum* of some plasmids was prevented by the presence of plasmids of the pJM1 group (167). Singer and colleagues hypothesized that the lack of transconjugants in their study resulted from the presence of a restriction-modification system on pJM1-type plasmids (167). From the several complete sequences of pJM1-type plasmids available, it is evident that no restriction-modification system is present. An alternative explanation for the lack of transconjugants could be incompatibility between the replicon of the tested plasmids and one of the origins of replication of pJM1. The fact that ultraviolet treatment restored plasmid introduction by conjugation could then be explained by a mutation in the incompatible origin.

No transfer genes have been identified in the available complete sequences of pJM1-type plasmids (161, 163, 164). The lack of *tra* genes might explain the

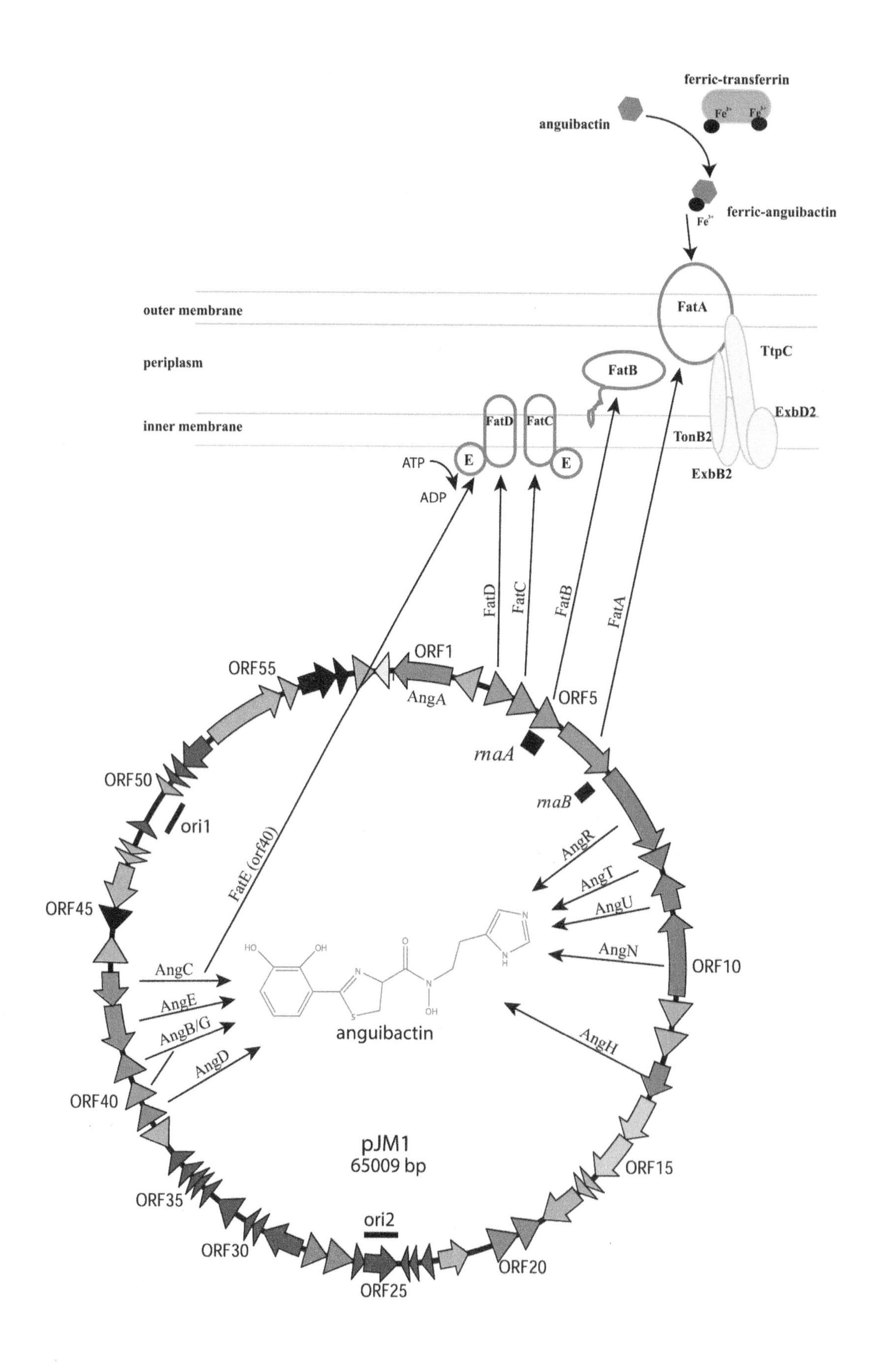
ferric-transferrin
anguibactin
ferric-anguibactin
outer membrane
periplasm
inner membrane
FatA
TtpC
FatB
FatD
FatC
ExbD2
TonB2
ExbB2
ATP
ADP
E
E
FatD
FatC
FatB
FatA
FatE (orf40)
ORF1
ORF55
AngA
ORF5
rnaA
rnaB
ORF50
ori1
AngR
AngT
AngU
AngN
ORF10
ORF45
AngC
AngE
AngB/G
AngD
AngH
anguibactin
ORF40
pJM1
65009 bp
ORF15
ORF35
ori2
ORF30
ORF20
ORF25

restrictionof these plasmids to serotype O1 strains of *V. anguillarum* (168).

Iron Uptake System

V. anguillarum 775, the prototype serogroup O1 strain, secretes a siderophore, a low-molecular-weight high-affinity iron-binding compound (169). This siderophore, called anguibactin, is linked to the plasmid, since siderophore production is abolished in a strain lacking the plasmid (157, 170, 171). In addition, the proteins needed for the transport of the ferric-siderophore across the membranes are also located on the plasmid (170–172). Fish infectivity studies showed that both the siderophore and the transport system are needed for infection. In strains harboring only the transport genes, establishment of an infection and replication in the bloodstream can occur only in the presence of strains producing the siderophore (158).

Anguibactin Biosynthesis

The siderophore anguibactin is produced via a nonribosomal peptide synthetase mechanism. Nonribosomal peptide synthetases (NRPSs) catalyze peptide bond formation in the absence of an RNA template (10). Via this mechanism non-amino acid substrates, such as carboxylic acid, can also be incorporated. NRPSs are multimodular, and the order of the modules determines the order of the amino/carboxilic acids in the final product. A minimal NRPS consists of an adenylation domain (A) to activate the substrate, a peptidyl or aryl carrier domain (PCP/ArCP) to tether the activated substrate, and a condensation domain (C) to form a peptide bond between two activated substrates (173). In addition, there are domains that can modify the substrates, bringing even more variability to the peptide structure. In the case of antibiotic synthesis these modules are often located on one or two large polypeptides, while in the case of siderophore biosynthesis these modules are scattered over multiple proteins.

The building blocks of anguibactin are *N*-hydroxyhistamine, L-cysteine, and the carboxylic acid dihydroxy benzoic acid (DHBA) (169). All enzymes needed for the production of anguibactin are encoded on the pJM1- and pJM1-type plasmids (Fig. 2) (161, 163, 164). Genes for the biosynthesis of the anguibactin precursor DHBA, *angB* and *angC*, are also carried by the plasmid, with the exception of *angA*, for which only a truncated nonfunctional gene is present on the plasmid. As a consequence, for DHBA production, a functional AngA protein must be encoded on the chromosome. In some *V. anguillarum* strains, the chromosomal *angA* gene is present in a cluster with the other DHBA biosynthetic genes since these strains still produce DHBA when cured from the plasmid (174, 175). In the case of *V. anguillarum* 531A, curing of the plasmid abolishes DHBA production, pointing to an incomplete chromosomally encoded DHBA cluster (176). The plasmid also carries the genes for the synthesis of *N*-hydroxyhistamine from histidine. The *angH* and the *angU* genes encode, respectively, a histine decarboxylase and a histamine monooxygenase. Mutations in *angH* or curing of the plasmid abolished histamine production, indicating that there are no chromosomal copies of this gene (177, 178).

The synthesis of anguibactin begins with activation of DHBA by the adenylation domain of AngE. Activated DHBA is then tethered to the ArCP domain of AngG to further continue its assembly in the anguibactin molecule (100, 175, 176, 179). The *angG* gene is transcribed from the same promoter as *angB* and can be found in one polypeptide with AngB but also as a separate polypeptide (176). Tethering of activated substrates to ArCP and PCP domains of NRPSs occurs at a phosphopantetheinyl arm. This arm is posttranslationally added by AngD to a conserved serine residue within the ArCP and PCP domains (180).

L-cysteine is activated by the A domain of AngR and tethered to the PCP domain of AngM (181). The first amide bond forms between the DHBA tethered onto the ArCP domain of AngB and the amino group of the cysteine loaded on the PCP domain of AngM. This step is followed by cyclodehydration of cysteine to a thiazoline ring. Both steps are mediated by the Cy domains of AngN (182). The final product, anguibactin, is then released by transfer of the dihydroxyphenyl thiazolyl intermediate from the PCP domain of AngM to the secondary amine of hydroxyhistamine in a reaction that requires the activity of the C domain of AngM (181). DHBA and histamine are dedicated metabolites for the synthesis of anguibactin, while L-cysteine is the only proteinogenic

Figure 2 Schematic representation of the pJM1 plasmid showing all open reading frames, the structure of anguibactin, and the transport proteins in the membranes. Genes that are involved in siderophore synthesis are shown in red; those involved in transport are blue. Black boxes indicate the location of the antisense RNAs. The shaded proteins in transport are chromosomally encoded. Location of the origins of replication is indicated by a black line.
doi:10.1128/microbiolspec.PLAS-0030-2014.f2

amino acid of the siderophore. Anguibactin has a quite compact structure, with three different chemical groups, all possessing iron-chelating properties (183). Furthermore, anguibactin belongs to the hydroxamate as well as the catecholate siderophores due to the presence of both functional groups.

Transport

The uptake of anguibactin (Fig. 2) initiates with the binding of the ferric-siderophore to the outer membrane receptor FatA (184–187). FatA is a typical outer membrane receptor containing 22 antiparallel beta strands that form the barrel. It has a plug domain that blocks the open channel and extracellular loops that are needed for the initial binding to the siderophore (187). FatA then facilitates the transport over the outer membrane via an energy-dependent mechanism that requires the TonB system (16, 188). The genes of the TonB system are encoded on the chromosome (16, 188). *V. anguillarum* has two TonB systems, and both can transduce the energy to FatA. However, one of the two systems is much more efficient in the transport of anguibactin and requires an additional protein, TtpC, in the complex (16, 188). Once in the periplasm, ferric-anguibactin is bound by FatB, a lipoprotein anchored to the outer leaflet of the inner membrane (189–191). Anchoring of the periplasmic binding protein is not essential since transport still occurs when the lipid moiety is removed (191). However, transport is abolished in the absence of the FatB protein. From the periplasmic binding protein FatB, the ferric-siderophore is transferred to FatC and FatD. FatC and FatD form a typical ABC transporter and are both essential for internalization of anguibactin (191). The energy for this internalization is provided by the ATPase FatE, which is encoded by a gene within the cluster for DHBA at a different locus than the other transport genes. A FatE mutant was still proficient in the uptake of the siderophore, while a double FvtE and FatE strain was not capable of internalizing the siderophore (192). The FvtE protein is involved in the uptake of vanchrobactin, a chromosomally encoded siderophore found in many *V. anguillarum* strains (193, 194). In strains harboring pJM1-type plasmids, genes of the vanchrobactin system, including the *fvtE* gene, are present on the chromosome, but this system is inactivated by an insertion in one of the vanchrobactin biosynthesis genes (195).

Regulation

The best-studied region of pJM1-type plasmids is the so-called iron transport and biosynthesis (ITB) operon. This operon consists of six genes transcribed from a single promoter (161, 170). The first four genes, *fatD*, *C*, *B*, and *A*, are involved in iron transport, and the last two, *angR* and *angT*, in siderophore biosynthesis—hence the name.

Three plasmid-encoded factors act as positive regulators enhancing the expression of the ITB operon. One of these factors is the AngR protein, which is needed for the synthesis of the siderophore and encoded by a gene that is part of the ITB operon (196, 197). It was shown that the AngR protein enhances the expression of the ITB operon by 2.5-fold (198). The AngR polypeptide contains predicted helix-turn-helix and leucine zipper motifs common to DNA binding proteins and transcriptional regulators (199). In addition to AngR, another factor, called TAF, enhances gene expression of the ITB operon by about 4-fold (198). TAF stands for trans acting factor because it is located outside the ITB operon. If AngR- and TAF-positive regulation are combined, a 23-fold increase in expression is seen, indicating some synergistic/cooperative action between the two regulators (198). Besides the ITB operon promoter, AngR and TAF also regulate the expression of the *angN* gene (182).

The third factor is the anguibactin siderophore itself, which acts as a positive regulator of the ITB operon promoter (23).

Like most iron uptake systems, the iron uptake genes on the pJM1-type plasmid are regulated by the chromosomally encoded protein Fur (200–203). Fur binds to two loci upstream of the ITB operon promoter and bends the DNA by dimerization of the proteins at the two loci (202). Under iron-limiting conditions, the Fur protein dimer loses its iron and the dimer falls apart, abolishing DNA binding and thereby relieving repression (202). The regulation by Fur extends to other genes involved in anguibactin biosynthesis such as *angM* and *angN* (181, 182).

Besides this general regulator, the plasmid itself encodes two antisense RNAs that act as negative regulators of gene expression.

The first antisense RNA, RNAα, was discovered on the opposite strand of the *fatB* gene (203, 204). Transcription of this antisense RNA requires the Fur protein, and under iron-replete conditions RNAα negatively regulates the *fatA* and *fatB* transcripts (203).

The second antisense RNA is located within the ITB operon, complementary to the *fatA* and *angR* genes (205). This RNA, called RNAβ, controls the differential gene expression between the transport and the biosynthesis genes within the ITB operon (205). In fact, although they are transcribed from the same promoter, the levels of *fatA* and *angR* differ by 17-fold (206). The difference is due to transcription termination between the *fatA* and *angR* genes that is mediated by RNAβ (206, 207). It has been proposed that RNAβ binds to the newly transcribed message, destabilizing the transcription complex, which leads to transcription termination (206). It is unknown if the two antisense RNAs control gene expression at loci other than their sense transcripts.

Heterogeneity of the pJM1 Determinants

As already mentioned, the pJM1-type plasmids are highly similar. The six restriction length polymorphism patterns identified differ mainly in the number of insertion sequences (the ISV-A2 elements) present (161, 162). However, there are some minor changes at the nucleotide level that have major impacts. For instance, the DNA sequence of the *angR* gene differs by a single nucleotide between pJM1 plasmid from *V. anguillarum* 775 and the pJM1-type plasmid pJHC1 from *V. anguillarum* 531A (208, 209). Where a histidine is present at position 267 in the AngR protein encoded by the pJM1 plasmid, an asparagine is present in the AngR protein of pJHC1 (208). This nucleotide change results in higher anguibactin production in the 531A strain. The LD_{50} values are similar for both bacteria, making it unclear what is the benefit for the bacterium to produce increased levels of siderophore (199).

When the three sequenced pJM1-type plasmids are aligned, only 16 nucleotide changes are revealed. Besides these single-nucleotide polymorphisms, the pJM1 plasmid has an extra 359-nucleotide HindIII fragment in the *angN* gene. The absence of this fragment in the M3 and pEIB1 plasmids results in the deletion of one of the two cyclization domains in AngN. Site-directed mutagenesis in the *angN* gene of pJM1 showed that at least one of the two Cy domains of AngN must be functional for anguibactin biosynthesis (182). However, one missing Cy domain reduced the production of anguibactin. The anguibactin-producing phenotypes of M3 and pEIB1 are not known. It would be of interest to see the effect of the deletion in AngN on anguibactin production, especially since the M3 plasmid and pEIB1 carry the point mutation in the *angR* gene that causes increased production of anguibactin in strain 531A. Another major change is the presence of an additional ISV-A2 element in the pEIB1 and M3 plasmids (163, 164).

While pJM1-type plasmids are found only in serogroup O1 strains, some genes are found outside this specific reservoir. The *fatA* and *fatD* genes have been identified on the chromosome of serotype O2a and NT4 *V. anguillarum* strains by PCR and DNA sequencing. These strains did not contain the *angR* gene (210).

A more striking discovery was the identification of an *angR* gene homolog in a whole-genome sequence of *Vibrio harveyi* (211). Further detailed analysis revealed that the chromosome of this bacterium encodes the complete ITB operon including the upstream *angM* gene and downstream genes *angH, angU*, and *angN*. In the case of *V. harveyi* the *angBG* gene is located in proximity to the cluster, while on the pJM1 plasmids the two clusters are located in noncontiguous regions (211).

CONCLUDING REMARKS

The first identified plasmid harboring an iron uptake system was a ColV plasmid, which was followed soon after by the discovery of pJM1 (58, 157). Plasmid types encoding iron uptake systems are rare, and despite their concomitant discovery, ColV- and pJM1-type plasmids are quite different. The most striking difference is the presence of numerous iron and non-iron-related determinants on ColV against the single iron uptake system encoded on pJM1. The anguibactin system is the only plasmid trait identified so far on pJM1-type plasmids. Furthermore, the latter are restricted to serotype O1 *V. anguillarum* strains, while the ColV plasmids are more widespread even across genera (29, 41, 42, 51, 152). The restriction of pJM1-type plasmids could be the result of their inability to be transferred, whereas the ColV plasmids are conjugative. As a consequence, it is not surprising that, over 50 years after the first description of a plasmid-linked colicin V, ColV plasmids are still identified in isolates from human and animal infections (45, 117, 212, 213). Another important difference between the ColV- and pJM1-type plasmids is the recent association of the former with antibiotic-resistance determinants. Non-ColV plasmids harboring multidrug resistance and ColV virulence factors have been described in the past, but recently, a ColV plasmid encoding antibiotic resistance has been identified (212). The lack of such determinants on pJM1-type plasmids could be a consequence of lower levels of antibiotics in environments where *V. anguillarum* is present. Alternatively, the limited interaction with other plasmids harboring resistance genes could have prevented the spread of multidrug resistance on pJM1-type plasmids.

The evolution of both plasmid types has most likely proceeded by the addition and deletion of large blocks of DNA. Entire systems are either absent or highly similar at the nucleotide level when present (214). Within this mosaic structure the iron uptake systems present are highly conserved across the members of the ColV and pJM1 types.

Acknowledgments. *We thank Jorge H. Crosa, who passed away too soon. He not only formed the basis of both of our careers, but also of the plasmid-mediated iron uptake research. Publication 5716 of the Netherlands Institute of Ecology (NIOO-KNAW). Conflicts of interest: We disclose no conflicts.*

Citation. Di Lorenzo M, Stork M. 2014. Plasmid-encoded iron uptake systems. Microbiol Spectrum 2(6):PLAS-0030-2014.

REFERENCES

1. **Galaris D, Pantopoulos K.** 2008. Oxidative stress and iron homeostasis: mechanistic and health aspects. *Crit Rev Clin Lab Sci* **45:**1–23.
2. **Crosa JH, Mey AR, Payne SM.** 2004. *Iron Transport in Bacteria*. ASM Press, Washington, DC.
3. **Weinberg ED.** 1993. The development of awareness of iron-withholding defense. *Perspect Biol Med* **36:**215–221.
4. **Skaar EP.** 2010. The battle for iron between bacterial pathogens and their vertebrate hosts. *PLoS Pathog* **6:** e1000949. doi:10.1371/journal.ppat.1000949.
5. **Porcheron G, Garenaux A, Proulx J, Sabri M, Dozois CM.** 2013. Iron, copper, zinc, and manganese transport and regulation in pathogenic *Enterobacteria*: correlations between strains, site of infection and the relative importance of the different metal transport systems for virulence. *Front Cell Infect Microbiol* **3:**90.
6. **Noinaj N, Buchanan SK, Cornelissen CN.** 2012. The transferrin-iron import system from pathogenic *Neisseria* species. *Mol Microbiol* **86:**246–257.
7. **Noinaj N, Cornelissen CN, Buchanan SK.** 2013. Structural insight into the lactoferrin receptors from pathogenic *Neisseria*. *J Struct Biol* **184:**83–92.
8. **Runyen-Janecky LJ.** 2013. Role and regulation of heme iron acquisition in Gram-negative pathogens. *Front Cell Infect Microbiol* **3:**55.
9. **Raymond KN, Dertz EA.** 2004. Biochemical and physical properties of siderophores, p 3–17. *In* Crosa JH, Mey AR, Payne SM (ed), *Iron Transport in Bacteria*. ASM Press, Washington, DC.
10. **Crosa JH, Walsh CT.** 2002. Genetics and assembly line enzymology of siderophore biosynthesis in bacteria. *Microbiol Mol Biol Rev* **66:**223–249.
11. **Crosa JH.** 1989. Genetics and molecular biology of siderophore-mediated iron transport in bacteria. *Microbiol Rev* **53:**517–530.
12. **Ratledge C, Dover LG.** 2000. Iron metabolism in pathogenic bacteria. *Annu Rev Microbiol* **54:**881–941.
13. **Chakraborty R, Storey E, van der Helm D.** 2007. Molecular mechanism of ferricsiderophore passage through the outer membrane receptor proteins of *Escherichia coli*. *Biometals* **20:**263–274.
14. **Kuehl CJ, Crosa JH.** 2010. The TonB energy transduction systems in *Vibrio* species. *Future Microbiol* **5:** 1403–1412.
15. **Kustusch RJ, Kuehl CJ, Crosa JH.** 2012. The *ttpC* gene is contained in two of three TonB systems in the human pathogen *Vibrio vulnificus*, but only one is active in iron transport and virulence. *J Bacteriol* **194:**3250–3259.
16. **Stork M, Otto BR, Crosa JH.** 2007. A novel protein, TtpC, is a required component of the TonB2 complex for specific iron transport in the pathogens *Vibrio anguillarum* and *Vibrio cholerae*. *J Bacteriol* **189:**1803–1815.
17. **Postle K, Larsen RA.** 2007. TonB-dependent energy transduction between outer and cytoplasmic membranes. *Biometals* **20:**453–465.
18. **Krewulak KD, Peacock RS, Vogel HJ.** 2004. Perisplasmic binding proteins involved in bacterial iron uptake, p 113–129. *In* Crosa JH, Mey AR, Payne SM (ed), *Iron Transport in Bacteria*. ASM Press, Washington, DC.
19. **Andrews SC, Robinson AK, Rodriguez-Quinones F.** 2003. Bacterial iron homeostasis. *FEMS Microbiol Rev* **27:**215–237.
20. **Bagg A, Neilands JB.** 1987. Ferric uptake regulation protein acts as a repressor, employing iron (II) as a cofactor to bind the operator of an iron transport operon in *Escherichia coli*. *Biochemistry* **26:**5471–5477.
21. **Hantke K.** 1981. Regulation of ferric iron transport in *Escherichia coli* K12: isolation of a constitutive mutant. *Mol Gen Genet* **182:**288–292.
22. **Troxell B, Hassan HM.** 2013. Transcriptional regulation by ferric uptake regulator (Fur) in pathogenic bacteria. *Front Cell Infect Microbiol* **3:**59.
23. **Chen Q, Wertheimer AM, Tolmasky ME, Crosa JH.** 1996. The AngR protein and the siderophore anguibactin positively regulate the expression of iron-transport genes in *Vibrio anguillarum*. *Mol Microbiol* **22:**127–134.
24. **Mahren S, Braun V.** 2003. The FecI extracytoplasmic-function sigma factor of *Escherichia coli* interacts with the beta′ subunit of RNA polymerase. *J Bacteriol* **185:** 1796–1802.
25. **Crosa JH, Actis LA, Mitoma Y, Perez-Casal J, Tolmasky ME, Valvano MA.** 1985. Plasmid-mediated iron sequestering systems in pathogenic strains of *Vibrio anguillarum* and *Escherichia coli*. *Basic Life Sci* **30:** 759–774.
26. **Walter MA, Bindereif A, Neilands JB, Crosa JH.** 1984. Lack of homology between the iron transport regions of two virulence-linked bacterial plasmids. *Infect Immun* **43:**765–767.
27. **Waters VL, Crosa JH.** 1991. Colicin V virulence plasmids. *Microbiol Rev* **55:**437–450.
28. **Johnson TJ, Johnson SJ, Nolan LK.** 2006. Complete DNA sequence of a ColBM plasmid from avian pathogenic *Escherichia coli* suggests that it evolved from closely related ColV virulence plasmids. *J Bacteriol* **188:** 5975–5983.
29. **Johnson TJ, Nolan LK.** 2009. Pathogenomics of the virulence plasmids of *Escherichia coli*. *Microbiol Mol Biol Rev* **73:**750–774.
30. **Johnson TJ, Siek KE, Johnson SJ, Nolan LK.** 2006. DNA sequence of a ColV plasmid and prevalence of selected plasmid-encoded virulence genes among avian *Escherichia coli* strains. *J Bacteriol* **188:**745–758.

31. **Mellata M, Touchman JW, Curtiss R.** 2009. Full sequence and comparative analysis of the plasmid pAPEC-1 of avian pathogenic *E. coli* χ7122 (O78:K80:H9). *PLoS One* **4:**e4232. doi:10.1371/journal.pone.0004232.
32. **Perez-Casal JF, Crosa JH.** 1987. Novel incompatibility and partition loci for the REPI replication region of plasmid ColV-K30. *J Bacteriol* **169:**5078–5086.
33. **Perez-Casal JF, Gammie AE, Crosa JH.** 1989. Nucleotide sequence analysis and expression of the minimum REPI replication region and incompatibility determinants of pColV-K30. *J Bacteriol* **171:**2195–2201.
34. **Gammie AE, Crosa JH.** 1991. Roles of DNA adenine methylation in controlling replication of the REPI replicon of plasmid pColV-K30. *Mol Microbiol* **5:**495–503.
35. **Gammie AE, Crosa JH.** 1991. Co-operative autoregulation of a replication protein gene. *Mol Microbiol* **5:** 3015–3023.
36. **Gammie AE, Tolmasky ME, Crosa JH.** 1993. Functional characterization of a replication initiator protein. *J Bacteriol* **175:**3563–3569.
37. **Waters VL, Crosa JH.** 1986. DNA environment of the aerobactin iron uptake system genes in prototypic ColV plasmids. *J Bacteriol* **167:**647–654.
38. **Colonna B, Nicoletti M, Visca P, Casalino M, Valenti P, Maimone F.** 1985. Composite IS*1* elements encoding hydroxamate-mediated iron uptake in FI*me* plasmids from epidemic *Salmonella* spp. *J Bacteriol* **162:**307–316.
39. **Fernandez-Beros ME, Kissel V, Lior H, Cabello FC.** 1990. Virulence-related genes in ColV plasmids of *Escherichia coli* isolated from human blood and intestines. *J Clin Microbiol* **28:**742–746.
40. **Rodriguez-Siek KE, Giddings CW, Doetkott C, Johnson TJ, Nolan LK.** 2005. Characterizing the APEC pathotype. *Vet Res* **36:**241–256.
41. **Fricke WF, McDermott PF, Mammel MK, Zhao S, Johnson TJ, Rasko DA, Fedorka-Cray PJ, Pedroso A, Whichard JM, Leclerc JE, White DG, Cebula TA, Ravel J.** 2009. Antimicrobial resistance-conferring plasmids with similarity to virulence plasmids from avian pathogenic *Escherichia coli* strains in *Salmonella enterica* serovar Kentucky isolates from poultry. *Appl Environ Microbiol* **75:**5963–5971.
42. **Johnson TJ, Thorsness JL, Anderson CP, Lynne AM, Foley SL, Han J, Fricke WF, McDermott PF, White DG, Khatri M, Stell AL, Flores C, Singer RS.** 2010. Horizontal gene transfer of a ColV plasmid has resulted in a dominant avian clonal type of *Salmonella enterica* serovar Kentucky. *PLoS One* **5:**e15524. doi:10.1371/journal.pone.0015524.
43. **Gratia A.** 1925. Sur un remarquable exemple d'antagisme entre deux souches de collibacille. *Crit Rev Soc Biol* **93:**1041–1042.
44. **MacFarren AC, Clowes RC.** 1967. A comparative study of two F-like colicin factors, ColV2 and ColV3, in *Escherichia coli* K-12. *J Bacteriol* **94:**365–377.
45. **antagismeNagel De Zwaig R.** 1966. Association between colicinogenic and fertility factors. *Genetics* **55:** 381–390.
46. **Gilson L, Mahanty HK, Kolter R.** 1987. Four plasmid genes are required for colicin V synthesis, export, and immunity. *J Bacteriol* **169:**2466–2470.
47. **Gordon DM, O'Brien CL.** 2006. Bacteriocin diversity and the frequency of multiple bacteriocin production in *Escherichia coli*. *Microbiology* **152:**3239–3244.
48. **Yang CC, Konisky J.** 1984. Colicin V-treated *Escherichia coli* does not generate membrane potential. *J Bacteriol* **158:**757–759.
49. **Cascales E, Buchanan SK, Duche D, Kleanthous C, Lloubes R, Postle K, Riley M, Slatin S, Cavard D.** 2007. Colicin biology. *Microbiol Mol Biol Rev* **71:**158–229.
50. **Jeziorowski A, Gordon DM.** 2007. Evolution of microcin V and colicin Ia plasmids in *Escherichia coli*. *J Bacteriol* **189:**7045–7052.
51. **Peigne C, Bidet P, Mahjoub-Messai F, Plainvert C, Barbe V, Medigue C, Frapy E, Nassif X, Denamur E, Bingen E, Bonacorsi S.** 2009. The plasmid of *Escherichia coli* strain S88 (O45:K1:H7) that causes neonatal meningitis is closely related to avian pathogenic *E. coli* plasmids and is associated with high-level bacteremia in a neonatal rat meningitis model. *Infect Immun* **77:** 2272–2284.
52. **Chehade H, Braun V.** 1988. Iron-regulated synthesis and uptake of colicin V. *FEMS Microbiol Lett* **52:** 177–182.
53. **Christenson JK, Gordon DM.** 2009. Evolution of colicin BM plasmids: the loss of the colicin B activity gene. *Microbiology* **155:**1645–1655.
54. **El Ghachi M, Bouhss A, Barreteau H, Touze T, Auger G, Blanot D, Mengin-Lecreulx D.** 2006. Colicin M exerts its bacteriolytic effect via enzymatic degradation of undecaprenyl phosphate-linked peptidoglycan precursors. *J Biol Chem* **281:**22761–22772.
55. **Braun V, Patzer SI, Hantke K.** 2002. Ton-dependent colicins and microcins: modular design and evolution. *Biochimie* **84:**365–380.
56. **Quackenbush RL, Falkow S.** 1979. Relationship between colicin V activity and virulence in *Escherichia coli*. *Infect Immun* **24:**562–564.
57. **Williams PH, Warner PJ.** 1980. ColV plasmid-mediated, colicin V-independent iron uptake system of invasive strains of *Escherichia coli*. *Infect Immun* **29:**411–416.
58. **Williams PH.** 1979. Novel iron uptake system specified by ColV plasmids: an important component in the virulence of invasive strains of *Escherichia coli*. *Infect Immun* **26:**925–932.
59. **Gibson F, Magrath DI.** 1969. The isolation and characterization of a hydroxamic acid (aerobactin) formed by *Aerobacter aerogenes* 62-I. *Biochim Biophys Acta* **192:** 175–184.
60. **McDougall S, Neilands JB.** 1984. Plasmid- and chromosome-coded aerobactin synthesis in enteric bacteria: insertion sequences flank operon in plasmid-mediated systems. *J Bacteriol* **159:**300–305.
61. **Waters VL, Crosa JH.** 1988. Divergence of the aerobactin iron uptake systems encoded by plasmids pColV-K30 in *Escherichia coli* K-12 and pSMN1 in *Aerobacter aerogenes* 62-1. *J Bacteriol* **170:**5153–5160.

62. **Colonna B, Ranucci L, Fradiani PA, Casalino M, Calconi A, Nicoletti M.** 1992. Organization of aerobactin, hemolysin, and antibacterial resistance genes in lactose-negative *Escherichia coli* strains of serotype O4 isolated from children with diarrhea. *Infect Immun* **60:**5224–5231.
63. **Riley PA, Threlfall EJ, Cheasty T, Wooldridge KG, Williams PH, Phillips I.** 1993. Occurrence of FI*me* plasmids in multiple antimicrobial-resistant *Escherichia coli* isolated from urinary tract infection. *Epidemiol Infect* **110:**459–468.
64. **Franco AA, Hu L, Grim CJ, Gopinath G, Sathyamoorthy V, Jarvis KG, Lee C, Sadowski J, Kim J, Kothary MH, McCardell BA, Tall BD.** 2011. Characterization of putative virulence genes on the related RepFIB plasmids harbored by *Cronobacter* spp. *Appl Environ Microbiol* **77:**3255–3267.
65. **Grim CJ, Kothary MH, Gopinath G, Jarvis KG, Beaubrun JJ, McClelland M, Tall BD, Franco AA.** 2012. Identification and characterization of *Cronobacter* iron acquisition systems. *Appl Environ Microbiol* **78:**6035–6050.
66. **Szczepanowski R, Braun S, Riedel V, Schneiker S, Krahn I, Puhler A, Schluter A.** 2005. The 120 592 bp IncF plasmid pRSB107 isolated from a sewage-treatment plant encodes nine different antibiotic-resistance determinants, two iron-acquisition systems and other putative virulence-associated functions. *Microbiology* **151:** 1095–1111.
67. **de Lorenzo V, Herrero M, Neilands JB.** 1988. IS*1*-mediated mobility of the aerobactin system of pColV-K30 in *Escherichia coli. Mol Gen Genet* **213:**487–490.
68. **Perez-Casal JF, Crosa JH.** 1984. Aerobactin iron uptake sequences in plasmid ColV-K30 are flanked by inverted IS*1*-like elements and replication regions. *J Bacteriol* **160:**256–265.
69. **Valvano MA, Crosa JH.** 1988. Molecular cloning, expression, and regulation in *Escherichia coli* K-12 of a chromosome-mediated aerobactin iron transport system from a human invasive isolate of *E. coli* K1. *J Bacteriol* **170:**5529–5538.
70. **Lawlor KM, Payne SM.** 1984. Aerobactin genes in *Shigella* spp. *J Bacteriol* **160:**266–272.
71. **Purdy GE, Payne SM.** 2001. The SHI-3 iron transport island of *Shigella boydii* 0-1392 carries the genes for aerobactin synthesis and transport. *J Bacteriol* **183:** 4176–4182.
72. **Valvano MA, Crosa JH.** 1984. Aerobactin iron transport genes commonly encoded by certain ColV plasmids occur in the chromosome of a human invasive strain of *Escherichia coli* K1. *Infect Immun* **46:**159–167.
73. **Vokes SA, Reeves SA, Torres AG, Payne SM.** 1999. The aerobactin iron transport system genes in *Shigella flexneri* are present within a pathogenicity island. *Mol Microbiol* **33:**63–73.
74. **Marolda CL, Valvano MA, Lawlor KM, Payne SM, Crosa JH.** 1987. Flanking and internal regions of chromosomal genes mediating aerobactin iron uptake systems in enteroinvasive *Escherichia coli* and *Shigella flexneri. J Gen Microbiol* **133:**2269–2278.
75. **Nassif X, Sansonetti PJ.** 1986. Correlation of the virulence of *Klebsiella pneumoniae* K1 and K2 with the presence of a plasmid encoding aerobactin. *Infect Immun* **54:**603–608.
76. **Carbonetti NH, Williams PH.** 1984. A cluster of five genes specifying the aerobactin iron uptake system of plasmid ColV-K30. *Infect Immun* **46:**7–12.
77. **de Lorenzo V, Bindereif A, Paw BH, Neilands JB.** 1986. Aerobactin biosynthesis and transport genes of plasmid ColV-K30 in *Escherichia coli* K-12. *J Bacteriol* **165:** 570–578.
78. **de Lorenzo V, Neilands JB.** 1986. Characterization of *iucA* and *iucC* genes of the aerobactin system of plasmid ColV-K30 in *Escherichia coli. J Bacteriol* **167:** 350–355.
79. **Wooldridge KG, Morrissey JA, Williams PH.** 1992. Transport of ferric-aerobactin into the periplasm and cytoplasm of *Escherichia coli* K12: role of envelope-associated proteins and effect of endogenous siderophores. *J Gen Microbiol* **138:**597–603.
80. **Coulton JW, Mason P, Allatt DD.** 1987. *fhuC* and *fhuD* genes for iron (III)-ferrichrome transport into *Escherichia coli* K-12. *J Bacteriol* **169:**3844–3849.
81. **Fecker L, Braun V.** 1983. Cloning and expression of the *fhu* genes involved in iron(III)-hydroxamate uptake by *Escherichia coli. J Bacteriol* **156:**1301–1314.
82. **Koster W, Braun V.** 1990. Iron (III) hydroxamate transport into *Escherichia coli.* Substrate binding to the periplasmic FhuD protein. *J Biol Chem* **265:**21407–21410.
83. **Braun V, Burkhardt R, Schneider R, Zimmermann L.** 1982. Chromosomal genes for ColV plasmid-determined iron(III)-aerobactin transport in *Escherichia coli. J Bacteriol* **151:**553–559.
84. **Braun V, Burkhardt R.** 1982. Regulation of the ColV plasmid-determined iron (III)-aerobactin transport system in *Escherichia coli. J Bacteriol* **152:**223–231.
85. **Escolar L, Perez-Martin J, de Lorenzo V.** 2000. Evidence of an unusually long operator for the fur repressor in the aerobactin promoter of *Escherichia coli. J Biol Chem* **275:**24709–24714.
86. **Carbonetti NH, Boonchai S, Parry SH, Vaisanen-Rhen V, Korhonen TK, Williams PH.** 1986. Aerobactin-mediated iron uptake by *Escherichia coli* isolates from human extraintestinal infections. *Infect Immun* **51:**966–968.
87. **Johnson JR, Moseley SL, Roberts PL, Stamm WE.** 1988. Aerobactin and other virulence factor genes among strains of *Escherichia coli* causing urosepsis: association with patient characteristics. *Infect Immun* **56:** 405–412.
88. **Lafont JP, Dho M, D'Hauteville HM, Bree A, Sansonetti PJ.** 1987. Presence and expression of aerobactin genes in virulent avian strains of *Escherichia coli. Infect Immun* **55:**193–197.
89. **Tivendale KA, Allen JL, Ginns CA, Crabb BS, Browning GF.** 2004. Association of *iss* and *iucA*, but not *tsh*, with plasmid-mediated virulence of avian pathogenic *Escherichia coli. Infect Immun* **72:**6554–6560.
90. **Dozois CM, Daigle F, Curtiss R 3rd.** 2003. Identification of pathogen-specific and conserved genes expressed

in vivo by an avian pathogenic *Escherichia coli* strain. *Proc Natl Acad Sci USA* **100:**247–252.

91. **Gao Q, Wang X, Xu H, Xu Y, Ling J, Zhang D, Gao S, Liu X.** 2012. Roles of iron acquisition systems in virulence of extraintestinal pathogenic *Escherichia coli*: salmochelin and aerobactin contribute more to virulence than heme in a chicken infection model. *BMC Microbiol* **12:**143.
92. **Ling J, Pan H, Gao Q, Xiong L, Zhou Y, Zhang D, Gao S, Liu X.** 2013. Aerobactin synthesis genes *iucA* and *iucC* contribute to the pathogenicity of avian pathogenic *Escherichia coli* O2 strain E058. *PLoS One* **8:** e57794. doi:10.1371/journal.pone.0057794.
93. **Lemaitre C, Bidet P, Bingen E, Bonacorsi S.** 2012. Transcriptional analysis of the *Escherichia coli* ColV-Ia plasmid pS88 during growth in human serum and urine. *BMC Microbiol* **12:**115.
94. **Lemaitre C, Mahjoub-Messai F, Dupont D, Caro V, Diancourt L, Bingen E, Bidet P, Bonacorsi S.** 2013. A conserved virulence plasmidic region contributes to the virulence of the multiresistant *Escherichia coli* meningitis strain S286 belonging to phylogenetic group C. *PLoS One* **8:**e74423. doi:10.1371/journal.pone.0074423.
95. **Baumler AJ, Norris TL, Lasco T, Voight W, Reissbrodt R, Rabsch W, Heffron F.** 1998. IroN, a novel outer membrane siderophore receptor characteristic of *Salmonella enterica*. *J Bacteriol* **180:**1446–1453.
96. **Baumler AJ, Tsolis RM, van der Velden AW, Stojiljkovic I, Anic S, Heffron F.** 1996. Identification of a new iron regulated locus of *Salmonella typhi*. *Gene* **183:**207–213.
97. **Brzuszkiewicz E, Bruggemann H, Liesegang H, Emmerth M, Olschlager T, Nagy G, Albermann K, Wagner C, Buchrieser C, Emody L, Gottschalk G, Hacker J, Dobrindt U.** 2006. How to become a uropathogen: comparative genomic analysis of extraintestinal pathogenic *Escherichia coli* strains. *Proc Natl Acad Sci USA* **103:** 12879–12884.
98. **Dobrindt U, Blum-Oehler G, Nagy G, Schneider G, Johann A, Gottschalk G, Hacker J.** 2002. Genetic structure and distribution of four pathogenicity islands (PAI I (536) to PAI IV(536)) of uropathogenic *Escherichia coli* strain 536. *Infect Immun* **70:**6365–6372.
99. **Lin H, Fischbach MA, Liu DR, Walsh CT.** 2005. *In vitro* characterization of salmochelin and enterobactin trilactone hydrolases IroD, IroE, and Fes. *J Am Chem Soc* **127:**11075–11084.
100. **Liu Q, Ma Y, Wu H, Shao M, Liu H, Zhang Y.** 2004. Cloning, identification and expression of an *entE* homologue *angE* from *Vibrio anguillarum* serotype O1. *Arch Microbiol* **181:**287–293.
101. **Chen YT, Chang HY, Lai YC, Pan CC, Tsai SF, Peng HL.** 2004. Sequencing and analysis of the large virulence plasmid pLVPK of *Klebsiella pneumoniae* CG43. *Gene* **337:**189–198.
102. **Sorsa LJ, Dufke S, Heesemann J, Schubert S.** 2003. Characterization of an iroBCDEN gene cluster on a transmissible plasmid of uropathogenic *Escherichia coli*: evidence for horizontal transfer of a chromosomal virulence factor. *Infect Immun* **71:**3285–3293.
103. **Bister B, Bischoff D, Nicholson GJ, Valdebenito M, Schneider K, Winkelmann G, Hantke K, Sussmuth RD.** 2004. The structure of salmochelins: C-glucosylated enterobactins of *Salmonella enterica*. *Biometals* **17:**471–481.
104. **Fischbach MA, Lin H, Liu DR, Walsh CT.** 2005. *In vitro* characterization of IroB, a pathogen-associated C-glycosyltransferase. *Proc Natl Acad Sci USA* **102:** 571–576.
105. **Fischbach MA, Lin H, Zhou L, Yu Y, Abergel RJ, Liu DR, Raymond KN, Wanner BL, Strong RK, Walsh CT, Aderem A, Smith KD.** 2006. The pathogen-associated *iroA* gene cluster mediates bacterial evasion of lipocalin 2. *Proc Natl Acad Sci USA* **103:**16502–16507.
106. **Goetz DH, Holmes MA, Borregaard N, Bluhm ME, Raymond KN, Strong RK.** 2002. The neutrophil lipocalin NGAL is a bacteriostatic agent that interferes with siderophore-mediated iron acquisition. *Mol Cell* **10:**1033–1043.
107. **Hantke K, Nicholson G, Rabsch W, Winkelmann G.** 2003. Salmochelins, siderophores of *Salmonella enterica* and uropathogenic *Escherichia coli* strains, are recognized by the outer membrane receptor IroN. *Proc Natl Acad Sci USA* **100:**3677–3682.
108. **Rabsch W, Voigt W, Reissbrodt R, Tsolis RM, Baumler AJ.** 1999. *Salmonella typhimurium* IroN and FepA proteins mediate uptake of enterobactin but differ in their specificity for other siderophores. *J Bacteriol* **181:** 3610–3612.
109. **Feldmann F, Sorsa LJ, Hildinger K, Schubert S.** 2007. The salmochelin siderophore receptor IroN contributes to invasion of urothelial cells by extraintestinal pathogenic *Escherichia coli in vitro*. *Infect Immun* **75:**3183–3187.
110. **Caza M, Lepine F, Milot S, Dozois CM.** 2008. Specific roles of the *iroBCDEN* genes in virulence of an avian pathogenic *Escherichia coli* O78 strain and in production of salmochelins. *Infect Immun* **76:**3539–3549.
111. **Crouch ML, Castor M, Karlinsey JE, Kalhorn T, Fang FC.** 2008. Biosynthesis and IroC-dependent export of the siderophore salmochelin are essential for virulence of *Salmonella enterica* serovar Typhimurium. *Mol Microbiol* **67:**971–983.
112. **Zhu M, Valdebenito M, Winkelmann G, Hantke K.** 2005. Functions of the siderophore esterases IroD and IroE in iron-salmochelin utilization. *Microbiology* **151:** 2363–2372.
113. **Muller SI, Valdebenito M, Hantke K.** 2009. Salmochelin, the long-overlooked catecholate siderophore of *Salmonella*. *Biometals* **22:**691–695.
114. **Skyberg JA, Johnson TJ, Nolan LK.** 2008. Mutational and transcriptional analyses of an avian pathogenic *Escherichia coli* ColV plasmid. *BMC Microbiol* **8:**24.
115. **Lemaitre C, Bidet P, Benoist JF, Schlemmer D, Sobral E, d'Humieres C, Bonacorsi S.** 2014. The *ssbL* gene harbored by the ColV plasmid of an *Escherichia coli* neonatal meningitis strain is an auxiliary virulence factor boosting the production of siderophores through the shikimate pathway. *J Bacteriol* **196:**1343–1349.

116. **Sabri M, Leveille S, Dozois CM.** 2006. A SitABCD homologue from an avian pathogenic *Escherichia coli* strain mediates transport of iron and manganese and resistance to hydrogen peroxide. *Microbiology* **152:**745–758.

117. **Tivendale KA, Allen JL, Browning GF.** 2009. Plasmid-borne virulence-associated genes have a conserved organization in virulent strains of avian pathogenic *Escherichia coli. J Clin Microbiol* **47:**2513–2519.

118. **Zhou D, Hardt WD, Galan JE.** 1999. *Salmonella typhimurium* encodes a putative iron transport system within the centisome 63 pathogenicity island. *Infect Immun* **67:**1974–1981.

119. **Runyen-Janecky LJ, Reeves SA, Gonzales EG, Payne SM.** 2003. Contribution of the *Shigella flexneri* Sit, Iuc, and Feo iron acquisition systems to iron acquisition *in vitro* and in cultured cells. *Infect Immun* **71:**1919–1928.

120. **Janakiraman A, Slauch JM.** 2000. The putative iron transport system SitABCD encoded on SPI1 is required for full virulence of *Salmonella typhimurium. Mol Microbiol* **35:**1146–1155.

121. **Boyer E, Bergevin I, Malo D, Gros P, Cellier MF.** 2002. Acquisition of Mn(II) in addition to Fe(II) is required for full virulence of *Salmonella enterica* serovar Typhimurium. *Infect Immun* **70:**6032–6042.

122. **Runyen-Janecky L, Dazenski E, Hawkins S, Warner L.** 2006. Role and regulation of the *Shigella flexneri* Sit and MntH systems. *Infect Immun* **74:**4666–4672.

123. **Sabri M, Caza M, Proulx J, Lymberopoulos MH, Bree A, Moulin-Schouleur M, Curtiss R 3rd, Dozois CM.** 2008. Contribution of the SitABCD, MntH, and FeoB metal transporters to the virulence of avian pathogenic *Escherichia coli* O78 strain chi7122. *Infect Immun* **76:** 601–611.

124. **Ewers C, Li G, Wilking H, Kiessling S, Alt K, Antao EM, Laturnus C, Diehl I, Glodde S, Homeier T, Bohnke U, Steinruck H, Philipp HC, Wieler LH.** 2007. Avian pathogenic, uropathogenic, and newborn meningitis-causing *Escherichia coli*: how closely related are they? *Int J Med Microbiol* **297:**163–176.

125. **Mellata M, Ameiss K, Mo H, Curtiss R 3rd.** 2010. Characterization of the contribution to virulence of three large plasmids of avian pathogenic *Escherichia coli* chi7122 (O78:K80:H9). *Infect Immun* **78:**1528–1541.

126. **Mellata M, Maddux JT, Nam T, Thomson N, Hauser H, Stevens MP, Mukhopadhyay S, Sarker S, Crabbe A, Nickerson CA, Santander J, Curtiss R 3rd.** 2012. New insights into the bacterial fitness-associated mechanisms revealed by the characterization of large plasmids of an avian pathogenic *E. coli. PLoS One* **7:**e29481. doi: 10.1371/journal.pone.0029481.

127. **Yi H, Xi Y, Liu J, Wang J, Wu J, Xu T, Chen W, Chen B, Lin M, Wang H, Zhou M, Li J, Xu Z, Jin S, Bao Q.** 2010. Sequence analysis of pKF3-70 in *Klebsiella pneumoniae*: probable origin from R100-like plasmid of *Escherichia coli. PLoS One* **5:**e8601. doi:10.1371/journal.pone.0008601.

128. **Provence DL, Curtiss R 3rd.** 1994. Isolation and characterization of a gene involved in hemagglutination by an avian pathogenic *Escherichia coli* strain. *Infect Immun* **62:**1369–1380.

129. **Leo JC, Grin I, Linke D.** 2012. Type V secretion: mechanism(s) of autotransport through the bacterial outer membrane. *Philos Trans R Soc Lond B Biol Sci* **367:** 1088–1101.

130. **Stathopoulos C, Provence DL, Curtiss R 3rd.** 1999. Characterization of the avian pathogenic *Escherichia coli* hemagglutinin Tsh, a member of the immunoglobulin A protease-type family of autotransporters. *Infect Immun* **67:**772–781.

131. **Otto BR, van Dooren SJ, Nuijens JH, Luirink J, Oudega B.** 1998. Characterization of a hemoglobin protease secreted by the pathogenic *Escherichia coli* strain EB1. *J Exp Med* **188:**1091–1103.

132. **Dozois CM, Dho-Moulin M, Bree A, Fairbrother JM, Desautels C, Curtiss R 3rd.** 2000. Relationship between the Tsh autotransporter and pathogenicity of avian *Escherichia coli* and localization and analysis of the Tsh genetic region. *Infect Immun* **68:**4145–4154.

133. **Kaczmarek A, Budzynska A, Gospodarek E.** 2012. Prevalence of genes encoding virulence factors among *Escherichia coli* with K1 antigen and non-K1 *E. coli* strains. *J Med Microbiol* **61:**1360–1365.

134. **Beutin L.** 1991. The different hemolysins of *Escherichia coli. Med Microbiol Immunol* **180:**167–182.

135. **Morales C, Lee MD, Hofacre C, Maurer JJ.** 2004. Detection of a novel virulence gene and a *Salmonella virulence* homologue among *Escherichia coli* isolated from broiler chickens. *Foodborne Pathog Dis* **1:**160–165.

136. **Reingold J, Starr N, Maurer J, Lee MD.** 1999. Identification of a new *Escherichia coli* She haemolysin homolog in avian *E. coli. Vet Microbiol* **66:**125–134.

137. **Welch RA.** 1991. Pore-forming cytolysins of gram-negative bacteria. *Mol Microbiol* **5:**521–528.

138. **Burgos Y, Beutin L.** 2010. Common origin of plasmid encoded alpha-hemolysin genes in *Escherichia coli. BMC Microbiol* **10:**193.

139. **Muller D, Hughes C, Goebel W.** 1983. Relationship between plasmid and chromosomal hemolysin determinants of *Escherichia coli. J Bacteriol* **153:**846–851.

140. **Schmidt H, Beutin L, Karch H.** 1995. Molecular analysis of the plasmid-encoded hemolysin of *Escherichia coli* O157:H7 strain EDL 933. *Infect Immun* **63:**1055–1061.

141. **Schmidt H, Kernbach C, Karch H.** 1996. Analysis of the EHEC *hly* operon and its location in the physical map of the large plasmid of enterohaemorrhagic *Escherichia coli* O157:H7. *Microbiology* **142:**907–914.

142. **Brunder W, Karch H, Schmidt H.** 2006. Complete sequence of the large virulence plasmid pSFO157 of the sorbitol-fermenting enterohemorrhagic *Escherichia coli* O157:H- strain 3072/96. *Int J Med Microbiol* **296:**467–474.

143. **Burland V, Shao Y, Perna NT, Plunkett G, Sofia HJ, Blattner FR.** 1998. The complete DNA sequence and analysis of the large virulence plasmid of *Escherichia coli* O157:H7. *Nucleic Acids Res* **26:**4196–4204.

144. **Cavalieri SJ, Bohach GA, Snyder IS.** 1984. *Escherichia coli* alpha-hemolysin: characteristics and probable role in pathogenicity. *Microbiol Rev* **48:**326–343.

145. **de la Cruz F, Muller D, Ortiz JM, Goebel W.** 1980. Hemolysis determinant common to *Escherichia coli* hemolytic plasmids of different incompatibility groups. *J Bacteriol* **143:**825–833.

146. **Burgos YK, Pries K, Pestana de Castro AF, Beutin L.** 2009. Characterization of the alpha-haemolysin determinant from the human enteropathogenic *Escherichia coli* O26 plasmid pEO5. *FEMS Microbiol Lett* **292:** 194–202.

147. **Bakás L, Maté S, Vazquez R, Herlax V.** 2012. *E. coli* alpha hemolysin and properties. *In* Ekinci PD (ed), *Biochemistry*. InTech ship, Rijeka, Croatia.

148. **Lebek G, Gruenig HM.** 1985. Relation between the hemolytic property and iron metabolism in *Escherichia coli*. *Infect Immun* **50:**682–686.

149. **Nataro JP, Kaper JB.** 1998. Diarrheagenic *Escherichia coli*. *Clin Microbiol Rev* **11:**142–201.

150. **Nagy G, Dobrindt U, Kupfer M, Emody L, Karch H, Hacker J.** 2001. Expression of hemin receptor molecule ChuA is influenced by RfaH in uropathogenic *Escherichia coli* strain 536. *Infect Immun* **69:**1924–1928.

151. **Han J, Lynne AM, David DE, Tang H, Xu J, Nayak R, Kaldhone P, Logue CM, Foley SL.** 2012. DNA sequence analysis of plasmids from multidrug resistant *Salmonella enterica* serotype Heidelberg isolates. *PLoS One* **7:** e51160. doi:10.1371/journal.pone.0051160.

152. **Fricke WF, Wright MS, Lindell AH, Harkins DM, Baker-Austin C, Ravel J, Stepanauskas R.** 2008. Insights into the environmental resistance gene pool from the genome sequence of the multidrug-resistant environmental isolate *Escherichia coli* SMS-3-5. *J Bacteriol* **190:**6779–6794.

153. **Naka H, Crosa JH.** 2011. Genetic determinants of virulence in the marine fish pathogen *Vibrio anguillarum*. *Fish Pathol* **46:**1–10.

154. **Crosa JH, Schiewe MH, Falkow S.** 1977. Evidence for plasmid contribution to the virulence of fish pathogen *Vibrio anguillarum*. *Infect Immun* **18:**509–513.

155. **Crosa JH, Hodges LL, Schiewe MH.** 1980. Curing of a plasmid is correlated with an attenuation of virulence in the marine fish pathogen *Vibrio anguillarum*. *Infect Immun* **27:**897–902.

156. **Crosa JH.** 1984. The relationship of plasmid-mediated iron transport and bacterial virulence. *Annu Rev Microbiol* **38:**69–89.

157. **Crosa JH.** 1980. A plasmid associated with virulence in the marine fish pathogen *Vibrio anguillarum* specifies an iron-sequestering system. *Nature* **284:**566–568.

158. **Wolf MK, Crosa JH.** 1986. Evidence for the role of a siderophore in promoting *Vibrio anguillarum* infections. *J Gen Microbiol* **132:**2949–2952.

159. **Mitoma Y, Aoki T, Crosa JH.** 1984. Phylogenetic relationships among *Vibrio anguillarum* plasmids. *Plasmid* **12:**143–148.

160. **Tolmasky ME, Actis LA, Toranzo AE, Barja JL, Crosa JH.** 1985. Plasmids mediating iron uptake in *Vibrio anguillarum* strains isolated from turbot in Spain. *J Gen Microbiol* **131:**1989–1997.

161. **Di Lorenzo M, Stork M, Tolmasky ME, Actis LA, Farrell D, Welch TJ, Crosa LM, Wertheimer AM, Chen Q, Salinas P, Waldbeser L, Crosa JH.** 2003. Complete sequence of virulence plasmid pJM1 from the marine fish pathogen *Vibrio anguillarum* strain 775. *J Bacteriol* **185:**5822–5830.

162. **Olsen JE, Larsen JL.** 1990. Restriction fragment length polymorphism of the *Vibrio anguillarum* serovar O1 virulence plasmid. *Appl Environ Microbiol* **56:**3130–3132.

163. **Li G, Mo Z, Li J, Xiao P, Hao B.** 2013. Complete genome sequence of *Vibrio anguillarum* M3, a serotype O1 strain isolated from Japanese flounder in China. *Genome Announc* **1:**e00769-13. doi:10.1128/genomeA.00769-13.

164. **Wu H, Ma Y, Zhang Y, Zhang H.** 2004. Complete sequence of virulence plasmid pEIB1 from the marine fish pathogen *Vibrio anguillarum* strain MVM425 and location of its replication region. *J Appl Microbiol* **97:** 1021–1028.

165. **Chen Q.** 1995. *Molecular Microbiology and Immunology, Ph.D. thesis*. Oregon Health and Science University, Portland, OR.

166. **Naka H, Chen Q, Mitoma Y, Nakamura Y, McIntosh-Tolle D, Gammie AE, Tolmasky ME, Crosa JH.** 2012. Two replication regions in the pJM1 virulence plasmid of the marine pathogen *Vibrio anguillarum*. *Plasmid* **67:** 95–101.

167. **Singer JT, Choe W, Schmidt KA, Makula RA.** 1992. Virulence plasmid pJM1 prevents the conjugal entry of plasmid DNA into the marine fish pathogen *Vibrio anguillarum* 775. *J Gen Microbiol* **138:**2485–2490.

168. **Larsen JL, Olsen JE.** 1991. Occurrence of plasmids in Danish isolates of *Vibrio anguillarum* serovars O1 and O2 and association of plasmids with phenotypic characteristics. *Appl Environ Microbiol* **57:**2158–2163.

169. **Actis LA, Fish W, Crosa JH, Kellerman K, Ellenberger SR, Hauser FM, Sanders-Loehr J.** 1986. Characterization of anguibactin, a novel siderophore from *Vibrio anguillarum* 775(pJM1). *J Bacteriol* **167:**57–65.

170. **Tolmasky ME, Actis LA, Crosa JH.** 1988. Genetic analysis of the iron uptake region of the *Vibrio anguillarum* plasmid pJM1: molecular cloning of genetic determinants encoding a novel trans activator of siderophore biosynthesis. *J Bacteriol* **170:**1913–1919.

171. **Tolmasky ME, Crosa JH.** 1984. Molecular cloning and expression of genetic determinants for the iron uptake system mediated by the *Vibrio anguillarum* plasmid pJM1. *J Bacteriol* **160:**860–866.

172. **Walter MA, Potter SA, Crosa JH.** 1983. Iron uptake system medicated by *Vibrio anguillarum* plasmid pJM1. *J Bacteriol* **156:**880–887.

173. **Walsh CT, Marshall CG.** 2004. Siderophore biosynthesis in bacteria, p 18–37. *In* Crosa JH, Mey AR, Payne SM (ed), *Iron Transport in Bacteria*. ASM Press, Washington, DC.

174. **Chen Q, Actis LA, Tolmasky ME, Crosa JH.** 1994. Chromosome-mediated 2,3-dihydroxybenzoic acid is a precursor in the biosynthesis of the plasmid-mediated siderophore anguibactin in *Vibrio anguillarum*. *J Bacteriol* **176:**4226–4234.
175. **Alice AF, Lopez CS, Crosa JH.** 2005. Plasmid- and chromosome-encoded redundant and specific functions are involved in biosynthesis of the siderophore anguibactin in *Vibrio anguillarum* 775: a case of chance and necessity? *J Bacteriol* **187:**2209–2214.
176. **Welch TJ, Chai S, Crosa JH.** 2000. The overlapping *angB* and *angG* genes are encoded within the trans-acting factor region of the virulence plasmid in *Vibrio anguillarum*: essential role in siderophore biosynthesis. *J Bacteriol* **182:**6762–6773.
177. **Barancin CE, Smoot JC, Findlay RH, Actis LA.** 1998. Plasmid-mediated histamine biosynthesis in the bacterial fish pathogen *Vibrio anguillarum*. *Plasmid* **39:**235–244.
178. **Tolmasky ME, Actis LA, Crosa JH.** 1995. A histidine decarboxylase gene encoded by the *Vibrio anguillarum* plasmid pJM1 is essential for virulence: histamine is a precursor in the biosynthesis of anguibactin. *Mol Microbiol* **15:**87–95.
179. **Di Lorenzo M, Stork M, Crosa JH.** 2011. Genetic and biochemical analyses of chromosome and plasmid gene homologues encoding ICL and ArCP domains in *Vibrio anguillarum* strain 775. *Biometals* **24:**629–643.
180. **Liu Q, Ma Y, Zhou L, Zhang Y.** 2005. Gene cloning, expression and functional characterization of a phosphopantetheinyl transferase from *Vibrio anguillarum* serotype O1. *Arch Microbiol* **183:**37–44.
181. **Di Lorenzo M, Poppelaars S, Stork M, Nagasawa M, Tolmasky ME, Crosa JH.** 2004. A nonribosomal peptide synthetase with a novel domain organization is essential for siderophore biosynthesis in *Vibrio anguillarum*. *J Bacteriol* **186:**7327–7336.
182. **Di Lorenzo M, Stork M, Naka H, Tolmasky ME, Crosa JH.** 2008. Tandem heterocyclization domains in a nonribosomal peptide synthetase essential for siderophore biosynthesis in *Vibrio anguillarum*. *Biometals* **21:**635–648.
183. **Jalal M, Hossain D, van der Helm D, Sanders-Loehr J, Actis LA, Crosa JH.** 1989. Structure of anguibactin, a unique plasmid-related bacterial siderophore from the fish pathogen *Vibrio anguillarum*. *J Am Chem Soc* **111:**292–296.
184. **Actis LA, Tolmasky ME, Farrell DH, Crosa JH.** 1988. Genetic and molecular characterization of essential components of the *Vibrio anguillarum* plasmid-mediated iron-transport system. *J Biol Chem* **263:**2853–2860.
185. **Actis LA, Potter SA, Crosa JH.** 1985. Iron-regulated outer membrane protein OM2 of *Vibrio anguillarum* is encoded by virulence plasmid pJM1. *J Bacteriol* **161:**736–742.
186. **Crosa JH, Hodges LL.** 1981. Outer membrane proteins induced under conditions of iron limitation in the marine fish pathogen *Vibrio anguillarum* 775. *Infect Immun* **31:**223–227.
187. **Lopez CS, Alice AF, Chakraborty R, Crosa JH.** 2007. Identification of amino acid residues required for ferric-anguibactin transport in the outer-membrane receptor FatA of *Vibrio anguillarum*. *Microbiology* **153:**570–584.
188. **Stork M, Di Lorenzo M, Mourino S, Osorio CR, Lemos ML, Crosa JH.** 2004. Two *tonB* systems function in iron transport in *Vibrio anguillarum*, but only one is essential for virulence. *Infect Immun* **72:**7326–7329.
189. **Actis LA, Tolmasky ME, Crosa LM, Crosa JH.** 1995. Characterization and regulation of the expression of FatB, an iron transport protein encoded by the pJM1 virulence plasmid. *Mol Microbiol* **17:**197–204.
190. **Koster WL, Actis LA, Waldbeser LS, Tolmasky ME, Crosa JH.** 1991. Molecular characterization of the iron transport system mediated by the pJM1 plasmid in *Vibrio anguillarum* 775. *J Biol Chem* **266:**23829–23833.
191. **Naka H, Lopez CS, Crosa JH.** 2010. Role of the pJM1 plasmid-encoded transport proteins FatB, C and D in ferric anguibactin uptake in the fish pathogen *Vibrio anguillarum*. *Environ Microbiol Rep* **2:**104–111.
192. **Naka H, Liu M, Crosa JH.** 2013. Two ABC transporter systems participate in siderophore transport in the marine pathogen *Vibrio anguillarum* 775 (pJM1). *FEMS Microbiol Lett* **341:**79–86.
193. **Balado M, Osorio CR, Lemos ML.** 2006. A gene cluster involved in the biosynthesis of vanchrobactin, a chromosome-encoded siderophore produced by *Vibrio anguillarum*. *Microbiology* **152:**3517–3528.
194. **Iglesias E, Brandariz I, Jimenez C, Soengas RG.** 2011. Iron(III) complexation by vanchrobactin, a siderophore of the bacterial fish pathogen *Vibrio anguillarum*. *Metallomics* **3:**521–528.
195. **Naka H, Lopez CS, Crosa JH.** 2008. Reactivation of the vanchrobactin siderophore system of *Vibrio anguillarum* by removal of a chromosomal insertion sequence originated in plasmid pJM1 encoding the anguibactin siderophore system. *Environ Microbiol* **10:**265–277.
196. **Farrell DH, Mikesell P, Actis LA, Crosa JH.** 1990. A regulatory gene, *angR*, of the iron uptake system of *Vibrio anguillarum*: similarity with phage P22 *cro* and regulation by iron. *Gene* **86:**45–51.
197. **Salinas PC, Crosa JH.** 1995. Regulation of angR, a gene with regulatory and biosynthetic functions in the pJM1 plasmid-mediated iron uptake system of *Vibrio anguillarum*. *Gene* **160:**17–23.
198. **Salinas PC, Tolmasky ME, Crosa JH.** 1989. Regulation of the iron uptake system in *Vibrio anguillarum*: evidence for a cooperative effect between two transcriptional activators. *Proc Natl Acad Sci USA* **86:**3529–3533.
199. **Wertheimer AM, Verweij W, Chen Q, Crosa LM, Nagasawa M, Tolmasky ME, Actis LA, Crosa JH.** 1999. Characterization of the *angR* gene of *Vibrio anguillarum*: essential role in virulence. *Infect Immun* **67:**6496–6509.

200. **Tolmasky ME, Wertheimer AM, Actis LA, Crosa JH.** 1994. Characterization of the *Vibrio anguillarum* fur gene: role in regulation of expression of the FatA outer membrane protein and catechols. *J Bacteriol* **176:** 213–220.
201. **Wertheimer AM, Tolmasky ME, Actis LA, Crosa JH.** 1994. Structural and functional analyses of mutant Fur proteins with impaired regulatory function. *J Bacteriol* **176:**5116–5122.
202. **Chai S, Welch TJ, Crosa JH.** 1998. Characterization of the interaction between Fur and the iron transport promoter of the virulence plasmid in *Vibrio anguillarum. J Biol Chem* **273:**33841–33847.
203. **Chen Q, Crosa JH.** 1996. Antisense RNA, Fur, iron, and the regulation of iron transport genes in *Vibrio anguillarum. J Biol Chem* **271:**18885–18891.
204. **Waldbeser LS, Chen Q, Crosa JH.** 1995. Antisense RNA regulation of the *fatB* iron transport protein gene in *Vibrio anguillarum. Mol Microbiol* **17:**747–756.
205. **Salinas PC, Waldbeser LS, Crosa JH.** 1993. Regulation of the expression of bacterial iron transport genes: possible role of an antisense RNA as a repressor. *Gene* **123:** 33–38.
206. **Stork M, Di Lorenzo M, Welch TJ, Crosa JH.** 2007. Transcription termination within the iron transport-biosynthesis operon of *Vibrio anguillarum* requires an antisense RNA. *J Bacteriol* **189:**3479–3488.
207. **McIntosh-Tolle D, Stork M, Alice A, Crosa JH.** 2012. Secondary structure of antisense RNAβ, an internal transcriptional terminator of the plasmid-encoded iron transport-biosynthesis operon of *Vibrio anguillarum. Biometals* **25:**577–586.
208. **Tolmasky ME, Actis LA, Crosa JH.** 1993. A single amino acid change in AngR, a protein encoded by pJM1-like virulence plasmids, results in hyperproduction of anguibactin. *Infect Immun* **61:**3228–3233.
209. **Tolmasky ME, Salinas PC, Actis LA, Crosa JH.** 1988. Increased production of the siderophore anguibactin mediated by pJM1-like plasmids in *Vibrio anguillarum. Infect Immun* **56:**1608–1614.
210. **Bay L, Larsen JL, Leisner JJ.** 2007. Distribution of three genes involved in the pJM1 iron-sequestering system in various *Vibrio anguillarum* serogroups. *Syst Appl Microbiol* **30:**85–92.
211. **Naka H, Actis LA, Crosa JH.** 2013. The anguibactin biosynthesis and transport genes are encoded in the chromosome of *Vibrio harveyi*: a possible evolutionary origin for the pJM1 plasmid-encoded system of *Vibrio anguillarum*? *MicrobiolOgyopen* **2:**182–194.
212. **Johnson TJ, Jordan D, Kariyawasam S, Stell AL, Bell NP, Wannemuehler YM, Alarcon CF, Li G, Tivendale KA, Logue CM, Nolan LK.** 2010. Sequence analysis and characterization of a transferable hybrid plasmid encoding multidrug resistance and enabling zoonotic potential for extraintestinal *Escherichia coli. Infect Immun* **78:**1931–1942.
213. **Johnson TJ, Wannemuehler Y, Johnson SJ, Stell AL, Doetkott C, Johnson JR, Kim KS, Spanjaard L, Nolan LK.** 2008. Comparison of extraintestinal pathogenic *Escherichia coli* strains from human and avian sources reveals a mixed subset representing potential zoonotic pathogens. *Appl Environ Microbiol* **74:**7043–7050.
214. **Tivendale KA, Noormohammadi AH, Allen JL, Browning GF.** 2009. The conserved portion of the putative virulence region contributes to virulence of avian pathogenic *Escherichia coli. Microbiology* **155:**450–460.

Plasmids as Genetic Tools VIII

Plasmids—Biology and Impact in Biotechnology and Discovery
Edited by Marcelo E. Tolmasky and Juan C. Alonso

doi:10.1128/microbiolspec.PLAS-0014-2013

Chang-Ho Baek,[1] Michael Liss,[1] Kevin Clancy,[1] Jonathan Chesnut,[1] and Federico Katzen[1]

30

DNA Assembly Tools and Strategies for the Generation of Plasmids

INTRODUCTION

Recombinant plasmids are possibly the biological reagents most frequently used in molecular biology. Myriad lasmid assembly strategies have been devised since the first recombinant DNA molecule was generated over 40 years ago (1). Cloning protocols are now so profuse that it is not always trivial to choose one that is the most suitable for a particular purpose. The aspects to consider at the time of choosing an assembly strategy include, among others, the nature of the sequences, fragment size and number, template availability, plasmid capacity and stability, and selection against a background of other unwanted assemblies.

Before the emergence of PCR in the late 1980s, recombinant DNA approaches were based exclusively on restriction enzyme-based protocols. These sequence-dependent procedures rely on unique and specific sites in the target molecules and do not allow the seamless cloning of various DNA fragments at the same time. During the past two decades, numerous sequence-dependent and -independent strategies have been developed that permit the simultaneous assembly of multiple sequences of various sizes with high efficiencies, flexible throughputs, and reduced hands-on and overall times.

In this article, we attempt to provide a summary of useful tools for constructing plasmids, emphasizing on those that are most recent and innovating, stating their advantages and disadvantages. We refer the reader to complementary articles for those technologies not mentioned or reviewed here in sufficient detail (2, 3, 4, 5, 6, 7).

CHOICE OF REPLICON

Usually, three major factors are the determinants of what type of plasmid to use: (i) the downstream application, (ii) the size of the expected construct, and (iii) the desired copy number. If the cloning approach (see below) is predetermined, then it would also play a significant role in the decision. The plasmid type is largely imposed by the nature of its origin of replication (replicon). In this section, we intend to provide examples of available replicon alternatives geared toward the categories below. For deeper information on origins of replication, we suggest consulting earlier literature (8, 9, 10).

[1]Life Technologies, Carlsbad, CA 92008.

By far, the most common replicon in *Escherichia coli* is the ColE1 type derived from the plasmid pMB1, which evolved via plasmids pBR322 and pUC18/19 (11, 12). Thus, vectors containing ColE1/pMB1-derived origins of replication fall into the same incompatibility group (13). The major advantages of these replicons are their multicopy nature, which ensures high plasmid DNA yield and makes them desirable for many downstream processes. On the other hand, derivatives of these plasmids have a limited host range and capacity, usually between 15 and 20 kb, which may hinder their value in applications that require large genetic elements as is the case of the construction of circuits for synthetic biology. Other shortcomings are the potential side effects of their multicopy nature, such as (i) the toxicity of certain plasmid elements, or (ii) the difficulty on the downregulation of gene expression, necessary to balance the expression of different components in complex gene networks.

These shortcomings could be avoided by using compatible replicons with a different copy number, host range, and capacity (Table 1). Comprehensive reviews on functional replicons for *E. coli* and other Gram-negative bacteria have been published previously (8, 14). When copy number must be strictly limited to 1, the solutions are either (i) to recombine the DNA fragment into the bacterial chromosome, or (ii) to use the *E. coli* F factor (Table 1) (for a review see reference 10). The latter option has the advantage of being a large-size-capacity conjugative episome. However, the host range is limited to *E. coli* and some related bacteria. For broad-range Gram-negative host options, plasmids from the IncP and IncQ incompatibility groups are recommended, yet their size limits are in the order of a few ten thousand base pairs (Table 1).

The emerging synthetic biology and bioengineering fields require the cloning of large constructs, such as genetic pathways or even chromosomes that, because of their sheer size and number of input segments, are difficult (often impossible) to assemble in *E. coli*. In these cases, the only viable cloning organism has been shown to be *Saccharomyces cerevisiae* (Table 1). Large sequences can be cloned circularly in yeast as circular episomes or yeast artificial chromosomes (YACs). The two major functional requirements of these elements are (i) a yeast centromere (CEN) and (ii) an autonomously replicating sequence (ARS) (for a review, see reference 15). By combining different CEN and ARS elements, stable yeast episomes can be used to clone sequences in the mega base pair scale (Table 1) (16).

SEQUENCE-DEPENDENT CLONING *IN VITRO*

A common feature of sequence-dependent cloning approaches is the requirement of specific sites in the inserts and/or vector. Many of these leave behind operational sequences that could interfere with the function of the genetic assembly.

The classical example of a sequence-dependent cloning strategy is that of one built upon restriction enzymes and ligases. Although this method is allegedly the most widely used tool in any molecular biology laboratory, it is relatively inefficient in the sense that (i) it is based on particular target sequences, (ii) it requires multiple stages and verification steps, (iii) it is not universal

Table 1 Replicons most commonly used in plasmids

Replicons	Related vectors	Copy number	Capacity	Host	Reference
pMB1/ColE1	pBR322 and derivatives	15–20	Up to, at least, 15 kb	*E. coli*	11
Mutated pMB1	pUC vectors	500–700	Up to, at least, 15 kb	*E. coli*	12
p15A	pACYC184	18–22	Up to, at least, 10 kb	*E. coli*	80
pSC101	pSC101	Approximately 5	Up to, at least, 16 kb	*E. coli*	81
IncP	RK2	4–40 (adjustable by mutations)	Up to, at least, 60 kbp	Broad host range	82, 83
IncQ	RSF1010	12	Up to, at least, 14 kbp	Broad host range	84
R6K	R6K and its derivatives	15–30	Up to, at least, 38 kbp	*E. coli*	85
F factor	F plasmid and its derivatives (i.e., bacterial artificial chromosomes [BACs])	1	Often up to 350 kb	*E. coli*	8
2-micron	2-micron and derivatives	Approximately 100	At least 3.9 kb	*S. cerevisiae*	86
Cen6/ARSH4	pRS series	Possibly 1	1–2 Mbp	*S. cerevisiae*	51, 87
Cen4/ARS1	pYAC2, pYAC4	>12	>1 Mbp	*S. cerevisiae*	88, 89

because each particular experiment uses a specific enzyme and conditions, (iv) it leaves behind scar sequences, and (v) it can assemble a very limited number of fragments at once. Given the vast array of information accumulated on this topic, we recommend consulting earlier publications (5, 7, 17).

With the emergence of PCR, a number of sequence-dependent methods were proposed, such as TA cloning (18), topoisomerase-based cloning (TOPO cloning) (19), and site-specific recombination approaches such as Gateway (20, 21) and Cre/LoxP recombination (22). Main providers for commercial products using the methodologies above are New England Biolabs (Ipswich, MA), Agilent Technologies (Santa Clara, CA), and Life Technologies (Carlsbad, CA). For further details on these technologies, refer to earlier reviews (23, 24).

More recently, restriction ligation cloning has found an application that standardizes the DNA assembly process, the BioBrick assembly (25). Each BioBrick part consist of a DNA sequence flanked by two different and unique restriction sites, which allows the generation of larger BioBrick parts by chaining together smaller ones. This process generates scar junctions that lack the original sites, still allowing for idempotent cloning cycles. A bundled BioBrick Assembly Kit is being commercialized by New England Biolabs (Ipswich, MA). In some specific cases, the creation of seamless connections between arbitrary DNA sequences in cloning vectors can be established by using a BioBrick part (called a BioScaffold) that can be excised by a type IIB restriction enzyme to leave a gap into which other DNA elements can be cloned (26).

Another way to resolve the presence of undesired scars in the final construct is by the use of type IIS restriction enzymes. These endonucleases cleave the DNA at distal locations from the recognition sites. The recognition sites can be strategically placed into regions of PCR-amplified fragments that will be excluded from the final construct, and, therefore, the restriction and ligation steps can be consolidated into a single reaction (27, 28). The strategy requires that these particular target sites must be absent in the individual parts. The method has been termed "Golden Gate cloning" (Fig. 1A) and has been used for (i) assembling gene variants or repetitive sequences (29, 30), (ii) generating iterative cloning systems (31), and (iii) obtaining standardized parts for multigene assemblies (32). Kits using this technology accompanied by a design web tool are being commercialized under the name of GeneArt type IIs by Life Technologies (Carlsbad, CA). Variants of this method have been proposed where the type IIS endonuclease is replaced either by a nicking enzyme (33) or by a type IIM restriction enzyme (34). In this latter case, the enzyme recognizes a methylated site, virtually transforming this approach into a sequence-independent strategy.

SEQUENCE-INDEPENDENT CLONING *IN VITRO*

During the past few years, a number of methods that do not rely on specific target sites and do not leave unwanted sequences in the final construct have been developed. Some of them are based on the ability of exonucleases to "chew back" one of the strands of double-stranded (ds) DNA, thereby exposing complementary single-stranded (ss) DNA sequences that can anneal to each other (Fig. 1B). The remaining gaps or nicks can be repaired *in vitro* or within the host organism. Examples of these include ligation-independent cloning (LIC) (35), sequence-and-ligation-independent cloning (SLIC) (36), fusion assembly (37), isothermal/Gibson assembly (38), and GeneArt Seamless and Cloning Assembly (39). Several of these technologies have been turned into products, commercialized by a variety of companies such as New England Biolabs (Ipswich, MA), EMD (San Diego, CA), Clontech (Mountain View, CA), and Life Technologies (Carlsbad, CA).

A similar recombination principle is used by systems that rely on different modes of exposing single-strand complementary sequences. For example, the uracil-specific excision reagent (USER), composed of uracil DNA glycosylase and endonuclease VIII, catalyzes the excision of a deoxyuracil residue, incorporated into a PCR fragment by a corresponding oligonucleotide (40) (Fig. 1C). In a different approach, DNA amplification is performed by using oligonucleotides containing phosphorothioate nucleotides in which a phosphodiester bond is replaced by a phosphothioester bond. These bonds can be chemically cleaved in an iodine/ethanol solution at elevated temperatures (41, 42). By strategically positioning the modified nucleotides in both methods above, fragments with compatible recessed ends can be generated and self-organized in a predefined order. Finally, in a related approach, compatible ssDNA overhangs are generated by PCR amplification with the use of oligonucleotides that contain regions of RNA sequence that cannot be copied by certain thermostable DNA polymerases (43). From the three approaches above, only the USER method is commercially available (marketed by New England Biolabs, Ipswich, MA).

A third approach relies entirely on polymerase extension of one or multiple overlapping fragments without calling for ssDNA complementary regions for

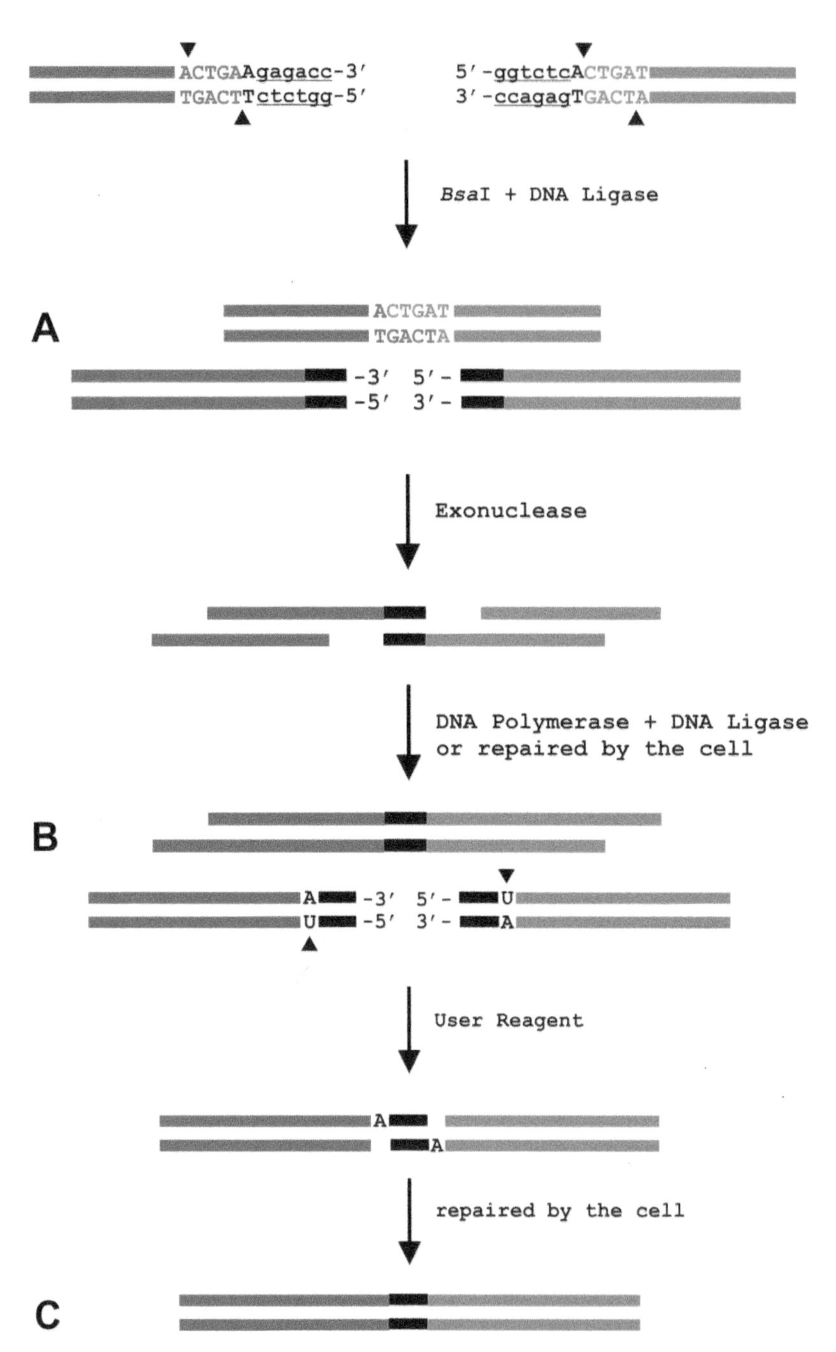

Figure 1 Schematic representation of different DNA assembly methodologies. (**A**) Golden Gate Cloning. Fragments to be assembled (red and green) have strategically placed terminal type IIS endonuclease recognition sites (in this case *Bsa*I sites shown in lowercase and underlined). Black arrowheads point toward the *Bsa*I cleavage sites. Simultaneous incubation with BsaI and DNA ligase results in covalently linked fragments. (**B**) Chew-back and repair-based assembly. Adjacent DNA fragments (red and green) sharing terminal sequence overlaps are incubated with DNA exonuclease, thereby exposing complementary DNA strands. The strands are annealed and the gaps can be sealed either *in vitro* by DNA polymerase and DNA ligase, or by the cell upon transformation. (**C**) USER assembly. Adjacent fragments (red and green), amplified with compatible uracil-containing primers, are incubated with the USER enzyme mix, which removes the uracils. The small terminal complementary DNA strands (in black) anneal to each other, outcompeting the small terminal loose strand. Gaps are repaired and sealed by the cell upon transformation.
doi:10.1128/microbiolspec.PLAS-0014-2013.f1

recombination. In its simplest conception, a DNA fragment can be seamlessly inserted into a recipient vector by two consecutive PCR amplification assays (44). During the first one, tails complementary to the insertion region are added to the fragment to be inserted. The resulting dsDNA fragment is used as a giant primer to PCR amplify the recipient plasmid, thereby generating the desired molecule. This method can be used for site-directed mutagenesis as well (see below). In a more recent related idea, it was shown that a similar polymerase extension strategy can be applied to join overlapping DNA fragments into a double-stranded circular form without a template (45).

IN VIVO CLONING

The "chew-back and repair" mechanism has also been applied to strategies where the assembly process occurs entirely within a living cell. Several hosts are particularly proficient in homologous recombination, a trait that could be used for cloning purposes. *E. coli* cells exhibit poor homologous recombination activity. However, the expression of the phage lambda *redET* genes (lambda red system) strongly promotes homologous recombination between a linear and circular DNA molecule (46) or between two linear DNA molecules (47). Gene replacement and simple cloning reactions can be performed by using *E. coli* strains harboring the *redET* genes, however, multiple fragments cannot easily be assembled into a single construct (unpublished data). The system is commercially available from Gene Bridges (Heidelberg, Germany).

Another organism used for cloning purposes is *Bacillus subtilis*, where DNA fragments can be naturally transformed and "stitched" together in a stepwise manner, generating large episomes or DNA structures integrated into a plasmid or the cell's chromosome (48). The mechanism of DNA uptake by *B. subtillis* has not completely worked out, but some reports indicate that, during the process, the incoming DNA is cleaved and generates smaller ssDNA fragments (for a review, see reference 49). These features might reduce the overall efficiency of the system compared with other *in vivo* cloning methods.

Possibly the most widely used *in vivo* cloning approach is the one involving the budding yeast *S. cerevisiae*. This organism has the ability to assemble and maintain constructs larger than 1 Mbp starting from dozens of overlapping DNA fragments of varying sizes with as few as 30 bp in common at their ends (50; reference 51 and references therein). The technology was first described in the early 1990s (52), but it really took off during the 2000s when further applications were reported. For example, it was shown that the system can join DNA fragments with non-homologous ends by means of double stranded oligonucleotides (53). This property allows the reuse of existing fragments without the need to reamplify them to incorporate end-homology. More recently, it has been demonstrated that yeast can also assemble constructs starting from overlapping oligonucleotides, with overlaps as short as 20 bp (54). Depending on the final size, assembled constructs can be transferred back to *E. coli* by "electroporating" lysed yeast colonies into electrocompetent *E. coli* cells (39). Finally, the combination of homing endonuclease recognition sites with compatible yeast markers has been shown to allow the sequential insertion of an indefinite number of DNA fragments in a predetermined locus (55). The *S. cerevisiae* recombination system is commercialized by Life Technologies (Carlsbad, CA) under the name of GeneArt High-Order Genetic Assembly System.

SITE-DIRECTED MUTAGENESIS

In some plasmid construction workflows, rather than reassembling the episome from a variety of constituent parts, it is much simpler to apply minor modifications to an existing molecule. During the past three decades, site-directed mutagenesis has become one of the most powerful tools in genetics. Its power lies in its ability, by chemical and/or enzymatic manipulation, to change a specific DNA target in a definable and often predetermined way. With the advent of synthetic biology and rational design, the manipulation of genes to produce enzymes with subtle differences from wild type has increased significantly. An array of methods has been described in the literature (for extensive information on this topic, see reference 56). Several of these technologies are commercialized by a variety of companies (for some examples, see Table 2). These products use at least one of the following approaches: (i) the isolation of a single- or double-stranded circular DNA template to create the mutation by using one or more primers that anneal to the same strand (57, 58, 59, 60); (ii) the design of two sets of PCR primers that overlap the mutation site followed by the amplification of the template by two PCR reactions and subsequent cloning (61); (iii) the PCR amplification of a plasmid by using complementary oligonucleotides and the subsequent elimination of the template molecule (44, 62, 63); (iv) a ligation during amplification strategy by inward PCR amplification of circular templates using phosphorylated primers (64); (v) the divergent PCR amplification

Table 2 Examples of commercial site-directed mutagenesis kits

Name of the product	Source	Reference
Change-IT Multiple Mutation Site Directed Mutagenesis Kit	Affymetrix (Santa Clara, CA)	64
QuikChange Lightning Site Directed Mutagenesis Kits	Agilent Technologies (Santa Clara, CA)	59, 60, 62, 63
Phusion Site-Directed Mutagenesis Kit	Thermo Fisher (Waltham, MA) and New England Biolabs (Ipswich, MA)	
GeneArt Site-Directed Mutgenesis Kits	Life Technologies (Carlsbad, CA)	65
Altered Sites II in vitro Mutagenesis System	Promega (Madison, WI)	57, 58
GeneEditor in vitro Site-Directed Mutagenesis System	Promega (Madison, WI)	57, 58

of circular templates by using phosphorylated primers followed by ligation; and (vi) the *in vitro* recombination of the ends of one or more PCR fragments (65). Preferred systems are those that use unmodified desalted oligonucleotides, dsDNA as a substrate, and a minimum number of steps. Some of the products listed in Table 2 work for single-site and multi-site mutagenesis. For further details, we suggest consulting the particular manufacturer's protocols.

DE NOVO GENE SYNTHESIS

What options, other than those mentioned above, are available for scenarios where (i) no DNA templates are available or (ii) the most straightforward cloning strategy to reach the desired goal is too tedious or convoluted? Perhaps the simple answer is artificial gene synthesis.

The first example of *de novo* synthesis of a DNA sequence was demonstrated by Khorana and coworkers in 1970 (66). In an effort that took several years, they assembled a 77-bp gene encoding a yeast alanine transfer RNA by using short oligonucleotides obtained through organic chemistry methods. In 1990, Mandecki and colleagues crossed the 1000-bp size barrier with the synthesis of a 2.1-kbp fully synthetic plasmid (67). Yet, another 20 years later, Venter's group designed, synthesized, and assembled the 1.08-Mbp *Mycoplasma mycoides* JCVI-syn1.0 genome starting from digitized genome sequence information, resulting in the first self-replicating organism derived from a fully synthetic genome (68).

Today, *de novo* gene synthesis enhances the biotechnology field concerning: safety, availability, reliability, throughput, flexibility, and, last but not least, total cost. Its application not only completely changed the way in which scientists think when designing cloning strategies, but also facilitated the outsourcing of related experimental steps in order to concentrate on less trivial scientific operations. Researchers can electronically access DNA sequences through comprehensive databases, redesign constructs *in silico* to specifically fit given requirements, and then order online the designed genes being synthesized and have them shipped within a matter of days.

Given the flexibility to synthesize any conceivable string of nucleotides, it is reasonable to alter a natural gene sequence before synthesis to ensure its best performance in the required application or experiment—a process known as gene optimization. The most commonly used modification of protein-coding genes is codon usage adaptation. However, further variables to consider are (i) GC content, (ii) mRNA half-life and RNA secondary structures, avoiding direct and reverse repeats, (iii) restriction sites, (iv) ribosomal entry sites, (v) cryptic splice motifs, and (vi) polyadenylation signals, among others. Taken together, this results in a multiparameter optimization, requiring sophisticated algorithms and significant computational speed (for an example, see reference 69).

Most protocols for gene synthesis are based on the assembly of chemically synthesized oligonucleotides to yield longer contiguous sequences. Their template-less production is still a significant cost factor, and different strategies have been conceived to reduce these expenses. It is, however, particularly important not to compromise on the accuracy of the applied chemical synthesis process, because the large number of sequential chemical reactions on the elongated oligonucleotide, together with the inherent imperfection of each step, lead to an increasing probability of incorporating a mutation within the molecule. These are usually single-nucleotide deletions, insertions, or depurinations that occur with a frequency of 0.1 to 0.5% (0.1% = every 1,000th nucleotide has an error, or one of fifty 20-mers carries a mutation). When assembling many oligonucleotides into a longer contiguous molecule, the statistical clustering of mutations within a synthetic gene increases exponentially. The length of the gene and the error rate of the oligonucleotides both have a major effect on the final sequence accuracy. For example, a synthetic gene of 1,000 bp made from oligonucleotides with an error rate of 0.1% will have a total accuracy of $(100\% - 0.1\%)^{1000} = 37\%$, while an error rate of 0.3%

decreases the final accuracy to 5%. In the first case, one of three sequenced clones will contain the accurate sequence, whereas, in the second case, 20 clones will have to be sequenced, on average, to find a correct one.

Taking these factors into account, the maximal final length of the initial assembly constructs must be considered carefully, in order to find the best economical ratio between the likelihood of errors in the product and the number of transformants to screen. Currently, the most cost-effective size of these synthetic building blocks is between 1 and 3 kbp.

The assembly process itself is basically a multiplex primer extension reaction under controlled temperature cycling conditions (Fig. 2). In the first cycling round, overlapping primers anneal to each other and are filled in by polymerase to form short double strands. These can again anneal to each other in the subsequent cycle and are extended to fragments bridging four oligonucleotides. This progression continues until fragments arise containing the complete length of the intended product. Once achieved, the excess of terminal primers amplifies the full-length product exponentially (70).

Different techniques can then be applied to reduce the remaining errors in the assembly product before ligation and transformation. Denaturation and reannealing

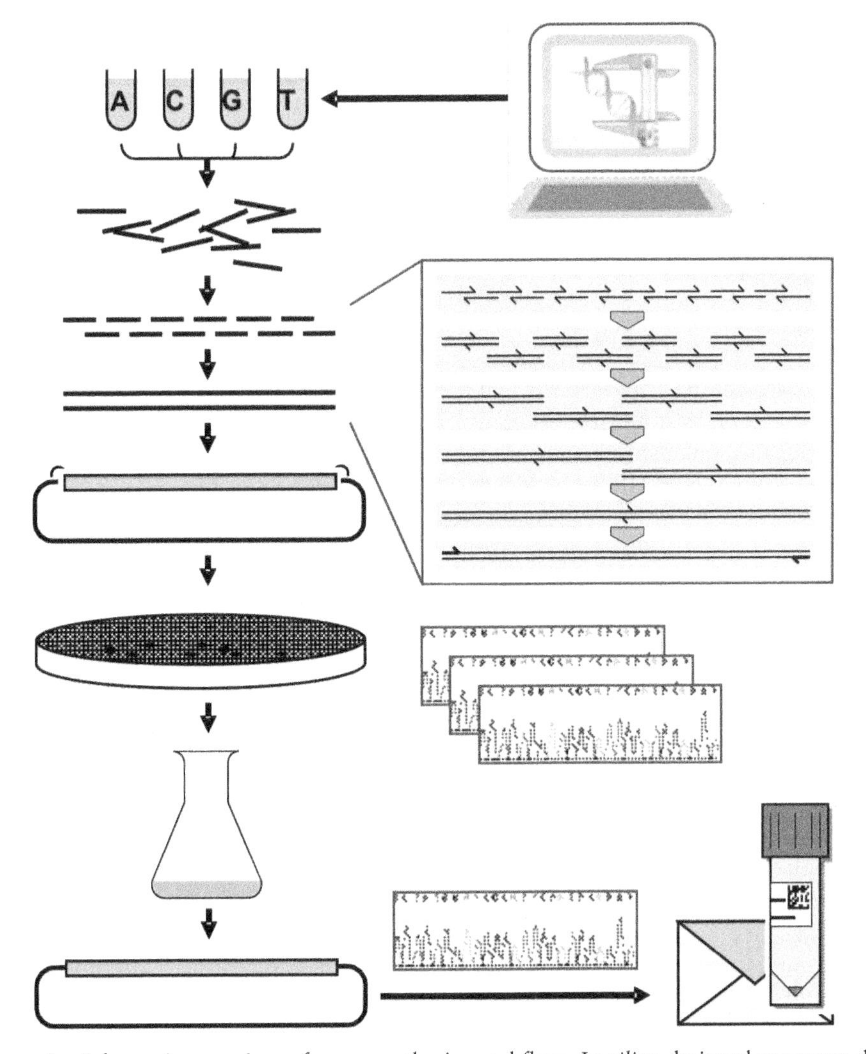

Figure 2 Schematic overview of gene synthesis workflow. *In silico* designed sequence data are converted into a set of oligonucleotides by automated organic chemistry. These are stepwise assembled, elongated, and amplified into a full-length fragment (see box), which is then ligated into a cloning vector. After transformation, *E. coli* colonies are screened for error-free insert sequences and a correct colony is cultivated for plasmid isolation. After a final sequence verification of the plasmid preparation, the construct is ready to be used or to be further assembled into larger constructs. doi:10.1128/microbiolspec.PLAS-0014-2013.f2

separates random mutations in complementary positions and results in dsDNA carrying mismatches, while strands with the correct sequence can form perfect duplex DNA. Mismatches in imperfect heteroduplexes then serve as entry points for digestion by different enzyme systems or affinity depletion by mismatch binding proteins (71, 72, 73).

After assembly, amplification, and error reduction, the linear gene synthesis product is ligated into a minimal cloning vector by using classical restriction endonuclease techniques or other assembly techniques (see preceding sections). After transformation of *E. coli*, some colonies are selected for plasmid preparation, and the accuracy of the synthesized DNA construct is verified by sequencing. Altogether, conditions for mass production are chosen to have a >95% chance of picking at least one correct fragment with a single screen, limiting the size of the initial fragment to 1 to 3 kbp. Compiling synthetic constructs exceeding 3 kbp requires the further sequence-independent and scarless assembly of cloned and sequence-verified building blocks (see preceding sections).

Type II class S restriction sites can produce sticky ends outside their recognition sequence, while the nucleotides of the adjacent cohesive stretch can be chosen freely, representing a common part of 4 bp for ligation (74). Designing this common part to have a length of ~20 bp allows flexible and specific attachment of two or more DNA fragments by fusion PCR, but is limited to moderate overall size and inherits an additional source of sequence errors through PCR (75). The DISEC-TRISEC and LIC-PCR methods use the exonuclease activity of Klenow or T4 DNA polymerase to generate compatible single-stranded overhangs, which are then combined with or without ligase, respectively (35, 76) (see preceding sections for additional methodologies). *In vitro* recombination extends this technology by annealing the overhangs under more stringent conditions at elevated temperatures and then filling and closing gaps with a heat-stable polymerase and ligase. This already allows for the efficient assembly of molecules in the range of 100 to 200 kbp (38). To access even larger fragment sizes, recent protocols have taken advantage of the recombination efficiency of yeast, enabling the assembly of complete operons and genomes of sizes exceeding 1 Mbp (see above sections).

IN SILICO DESIGN TOOLS

Bioinformatics software is an important component of designing and planning assembly of plasmids. Good bioinformatics tools assist users with devising, planning, and testing their ideas *in silico* as well as verifying and validating their experimental results. Bioinformatics software fulfills seven main needs by providing:

1. Curation capabilities—means of storing and managing data on plasmid sequences;
2. Discovery capabilities—provision of series of tools to allow users to find, compare, and analyze their sequences for biological insights;
3. Design capabilities—provision of series of tools to design or modify existing sequences, identify needed materials such as amplimers and oligonucleotides for experimental purposes, and highlight and resolve likely experimental issues *in silico*;
4. Inventory capabilities—provision of tools to manage their experimental materials, especially when reusing or repurposing existing plasmids, oligonucleotides, amplimers, etc.;
5. Confirmation capabilities—provision of tools to model, validate, and verify the plasmids they have created;
6. Visualization capabilities—provision of easy-to-use interfaces that allow users to easily understand what they are working on, find the tools to analyze these sequences, and transition their projects through discovery, design, and confirmation of experimental materials;
7. Sharing capabilities—coordination of groups of users working together, most simply through the use of common file formats, more extensively through user-to-user sharing of electronic records or databases.

With the recent advent of synthetic biology, scientists are increasingly using design-based approaches for developing and managing their plasmid collections (77). To achieve this, the traditional sequence and feature-based view of plasmids will be updated to regard features as discrete parts that can be reused in plasmid design projects (78). Parts, in turn, are assembled into devices that help the plasmid interact with the cell, such as antibiotic resistance devices, origin of replication devices, or expression devices (79). Devices themselves are controlled through circuits, such as the use of one or more inducer-based elements to control the expression of genes or combinations of different origins of replication to stably transform host cell lines. Bioinformatics software will need to enable users to switch back and forth between feature- and functional-based views of the same molecules.

Table 3 contains a list of pieces of software that have been used by the scientific community for cloning-based

Table 3 Most commonly used bioinformatics software

Product name	Platform	Website	Plasmid construction support
ApE	Mac/Windows/Unix	http://biologylabs.utah.edu/jorgensen/wayned/ape/	Restriction, Golden Gate
UGENE	Mac/Windows/Unix	http://ugene.unipro.ru/	Restriction, Golden Gate
Gene inspector[a]	Mac/Windows/Unix	http://www.textco.com	Restriction, Golden Gate
Gene Construction Kit[a]	Mac/Windows/Unix	http://www.textco.com	Restriction, Golden Gate
DNA Dynamo[a]	Mac/Windows/Unix	http://www.bluetractorsoftware.co.uk/	Restriction, Golden Gate
GENtle[a]	Mac/Windows/Unix	http://gentle.magnusmanske.de/	Restriction, Golden Gate
QuickGene[a]	Mac/Windows/Unix	http://www.crimsonbase.com	Restriction, Golden Gate
pDRAW32	Mac/Windows/Unix	http://www.acaclone.com/	Restriction, Golden Gate
Benchling[a]	Mac/Windows/Unix/Amazon Cloud	https://benchling.com/	Restriction, Golden Gate
DNA Strider/Lab Strider[a]	Mac/Windows/Unix	http://sourceforge.net/projects/dnastrider/	Restriction, Golden Gate
Plasma DNA	Mac/Windows/Unix	http://research.med.helsinki.fi/plasmadna/	Restriction, Golden Gate
Serial Cloner	Mac/Windows/Unix	http://serialbasics.free.fr	Restriction, Golden Gate, Gateway
CLC Bio Main Workbench[a]	Mac/Windows/Unix	http://www.clcbio.com/products/clc-main-workbench/	Restriction, Golden Gate, Gateway
Clone Manager[a]	Windows	http://www.scied.com	Restriction, Golden Gate, Gateway, TOPO
Lasergene[a]	Mac/Windows/Unix/Amazon Cloud	http://www.dnastar.com	Restriction, Golden Gate, Gateway, TOPO, TA
Sim Vector[a]	Mac/Windows/Unix	http://www.premierbiosoft.com/plasmid_maps/	Restriction, Golden Gate, Gateway, TOPO, TA
Mac Vector[a]	Mac	http://macvector.com/	Restriction, Golden Gate, Gateway, TOPO, TA
Geneious[a]	Mac/Windows/Unix	http://www.geneious.com/	Restriction, Golden Gate, Gateway, TOPO, TA, homologous recombination
TinkerCell	Mac/Windows/Unix	http://www.tinkercell.com	Restriction, synthetic biology support
iBioSim	Mac/Windows/Unix	http://www.async.ece.utah.edu/iBioSim/	Restriction, synthetic biology support
Lab Genius[a]	Mac/Windows/Unix/Amazon Cloud	http://beta.labgeni.us/welcome/	Restriction, synthetic biology support
Clotho	Mac/Windows/Unix	http://www.clothocad.org	Restriction, Golden Gate, homologous recombination, synthetic biology support
j5[a]	Mac/Windows/Unix	http://www.teselagen.com/	Restriction, Golden Gate, homologous recombination, synthetic biology support
Genome Compiler[a]	Mac/Windows/Unix/Amazon Cloud	http://www.genomecompiler.com/	Restriction, Golden Gate, homologous recombination, synthetic biology support
Vector Friends[a]	Mac/Windows/Unix	http://www.vectorfriends.com	Restriction, Golden Gate, Gateway, TOPO, TA, homologous recombination
SnapGene[a]	Mac/Windows/Unix	http://www.snapgene.com/	Restriction, Golden Gate, Gateway, TOPO, TA, homologous recombination
Vector NTI[a]	Mac/Windows/Unix	http://www.lifetechnologies.com/vectornti	Restriction, Golden Gate, Gateway, TOPO, TA, homologous recombination, synthetic biology support

[a]Commercial product.

tasks. All of these pieces of software have user interfaces that assist the end user with managing their data, and they vary most in terms of the types of tools offered for modeling different sequence assembly strategies. Due to the prevalence of restriction enzyme-based cloning, all support these types of assemblies. The main differentiators in the collection consist of the use of more recent sequence-independent cloning methodologies, highlighted by the availability of Gateway cloning, homologous recombination, and synthetic biology tools. Simpler tools such as ApE or pDRAW32 provide free software for users with basic capabilities. Pieces of software like Clone Manager and Gene Construction kit provide more functionality but with basic data management capabilities. Software like LaserGene, Geneious, CLC Main Workbench, Vector NTI Advance and Vector NTI Express provide sophisticated analysis platforms fulfilling all the main bioinformatics software needs. Finally, software such as TinkerCell, j5, iBioSim, Clotho, and Vector NTI Express Designer provide access to synthetic biology design tools.

FINAL REMARKS

Since the generation of the first recombinant DNA molecule took place (1) remarkable progress has been accomplished in the way plasmids and large DNA molecules are made. However, there is still a technological gap between the ability to assemble and make DNA molecules and to assemble meaningful operating DNA. Although we have acquired reasonably good knowledge on how to make functional plasmids, we still struggle working with genetic pathways on a mere 1-kb scale. Many technologies must materialize before we can fully exploit the remarkable advances of molecular cloning.

Acknowledgments. *Conflicts of interest: The authors hold financial interest in Life Technologies.*

Citation. Baek C-H, Liss M, Clancy K, Chesnut J, Katzen F. 2014. DNA assembly tools and strategies for the generation of plasmids. Microbiol Spectrum **2**(5):PLAS-0014-2013.

References

1. **Jackson DA, Symons RH, Berg P.** 1972. Biochemical method for inserting new genetic information into DNA of Simian Virus 40: circular SV40 DNA molecules containing lambda phage genes and the galactose operon of Escherichia coli. *Proc Natl Acad Sci USA* **69:** 2904–2909.
2. **Lu Q.** 2005. Seamless cloning and gene fusion. *Trends Biotechnol* **23:**199–207.
3. **Ellis T, Adie T, Baldwin GS.** 2011. DNA assembly for synthetic biology: from parts to pathways and beyond. *Integr Biol (Camb)* **3:**109–118.
4. **Durani V, Sullivan BJ, Magliery TJ.** 2012. Simplifying protein expression with ligation-free, traceless and tag-switching plasmids. *Protein Expr Purif* **85:**9–17.
5. **Tolmachov O.** 2009. Designing plasmid vectors. *Methods Mol Biol* **542:**117–129.
6. **Wang X, Sa N, Tian PF, Tan TW.** 2011. Classifying DNA assembly protocols for devising cellular architectures. *Biotechnol Adv* **29:**156–163.
7. **Green MR, Sambrook J.** 2012. *Molecular Cloning: A Laboratory Manual*, 4th ed. Cold Spring Harbor Laboratory Press, Cold Spring Harbor, NY.
8. **Preston A.** 2003. Choosing a cloning vector. *Methods Mol Biol* **235:**19–26.
9. **del Solar G, Giraldo R, Ruiz-Echevarria MJ, Espinosa M, Diaz-Orejas R.** 1998. Replication and control of circular bacterial plasmids. *Microbiol Mol Biol Rev* **62:** 434–464.
10. **Scott JR.** 1984. Regulation of plasmid replication. *Microbiol Rev* **48:**1–23.
11. **Bolivar F, Rodriguez RL, Greene PJ, Betlach MC, Heyneker HL, Boyer HW, Crosa JH, Falkow S.** 1977. Construction and characterization of new cloning vehicles. II. A multipurpose cloning system. *Gene* **2:**95–113.
12. **Vieira J, Messing J.** 1982. The pUC plasmids, an M13mp7-derived system for insertion mutagenesis and sequencing with synthetic universal primers. *Gene* **19:** 259–268.
13. **Hashimoto-Gotoh T, Timmis KN.** 1981. Incompatibility properties of Col E1 and pMB1 derivative plasmids: random replication of multicopy replicons. *Cell* **23:** 229–238.
14. **Kues U, Stahl U.** 1989. Replication of plasmids in gram-negative bacteria. *Microbiol Rev* **53:**491–516.
15. **Ramsay M.** 1994. Yeast artificial chromosome cloning. *Mol Biotechnol* **1:**181–201.
16. **Den Dunnen JT, Grootscholten PM, Dauwerse JG, Walker AP, Monaco AP, Butler R, Anand R, Coffey AJ, Bentley DR, Steensma HY, et al.** 1992. Reconstruction of the 2.4 Mb human DMD-gene by homologous YAC recombination. *Hum Mol Genet* **1:**19–28.
17. **Loenen WA, Raleigh EA.** 2014. The other face of restriction: modification-dependent enzymes. *Nucleic Acids Res* **42:**56–69.
18. **Zhou MY, Clark SE, Gomez-Sanchez CE.** 1995. Universal cloning method by TA strategy. *Biotechniques* **19:** 34–35.
19. **Shuman S.** 1994. Novel approach to molecular cloning and polynucleotide synthesis using vaccinia DNA topoisomerase. *J Biol Chem* **269:**32678–32684.
20. **Hartley JL, Temple GF, Brasch MA.** 2000. DNA cloning using in vitro site-specific recombination. *Genome Res* **10:**1788–1795.
21. **Cheo DL, Titus SA, Byrd DR, Hartley JL, Temple GF, Brasch MA.** 2004. Concerted assembly and cloning of multiple DNA segments using in vitro site-specific

recombination: functional analysis of multi-segment expression clones. *Genome Res* **14**:2111–2120.

22. **Buchholz F, Bishop M.** 2001. LoxP-directed cloning: use of Cre recombinase as a universal restriction enzyme. *Biotechniques* **31**:906–908, 910, 912, 914, 916, 918.
23. **Zhou MY, Gomez-Sanchez CE.** 2000. Universal TA cloning. *Curr Issues Mol Biol* **2**:1–7.
24. **Katzen F.** 2007. Gateway recombinational cloning: a biological operating system. *Expert Opin Drug Discov* **2**: 571–589.
25. **Knight TF.** 2003. *Idempotent Vector Design for Standard Assembly of BioBricks*. MIT Synthetic Biology Working Group Technical Reports. http://hdl.handle.net/1721.1/21168.
26. **Norville JE, Derda R, Gupta S, Drinkwater KA, Belcher AM, Leschziner AE, Knight TF Jr.** 2010. Introduction of customized inserts for streamlined assembly and optimization of BioBrick synthetic genetic circuits. *J Biol Eng* **4**: 17. doi:10.1186/1754-1611-4-17.
27. **Engler C, Kandzia R, Marillonnet S.** 2008. A one pot, one step, precision cloning method with high throughput capability. *PLoS One* **3**:e3647. doi:10.1371/journal.pone.0003647.
28. **Engler C, Gruetzner R, Kandzia R, Marillonnet S.** 2009. Golden gate shuffling: a one-pot DNA shuffling method based on type IIs restriction enzymes. *PLoS One* **4**: e5553. doi:10.1371/journal.pone.0005553.
29. **Engler C, Marillonnet S.** 2011. Generation of families of construct variants using golden gate shuffling. *Methods Mol Biol* **729**:167–181.
30. **Weber E, Gruetzner R, Werner S, Engler C, Marillonnet S.** 2011. Assembly of designer TAL effectors by Golden Gate cloning. *PLoS One* **6**:e19722. doi:10.1371/journal.pone.0019722.
31. **Sarrion-Perdigones A, Falconi EE, Zandalinas SI, Juarez P, Fernandez-del-Carmen A, Granell A, Orzaez D.** 2011. GoldenBraid: an iterative cloning system for standardized assembly of reusable genetic modules. *PLoS One* **6**: e21622. doi:10.1371/journal.pone.0021622.
32. **Weber E, Engler C, Gruetzner R, Werner S, Marillonnet S.** 2011. A modular cloning system for standardized assembly of multigene constructs. *PLoS One* **6**:e16765. doi:10.1371/journal.pone.0016765.
33. **Wang RY, Shi ZY, Guo YY, Chen JC, Chen GQ.** 2013. DNA fragments assembly based on nicking enzyme system. *PLoS One* **8**:e57943. doi:10.1371/journal.pone.0057943.
34. **Chen WH, Qin ZJ, Wang J, Zhao GP.** 2013. The MASTER (methylation-assisted tailorable ends rational) ligation method for seamless DNA assembly. *Nucleic Acids Res* **41**:e93. doi:10.1093/nar/gkt122.
35. **Aslanidis C, de Jong PJ.** 1990. Ligation-independent cloning of PCR products (LIC-PCR). *Nucleic Acids Res* **18**:6069–6074.
36. **Li MZ, Elledge SJ.** 2007. Harnessing homologous recombination in vitro to generate recombinant DNA via SLIC. *Nat Methods* **4**:251–256.
37. **Zhu B, Cai G, Hall EO, Freeman GJ.** 2007. In-fusion assembly: seamless engineering of multidomain fusion proteins, modular vectors, and mutations. *Biotechniques* **43**: 354–359.
38. **Gibson DG, Young L, Chuang RY, Venter JC, Hutchison CA 3rd, Smith HO.** 2009. Enzymatic assembly of DNA molecules up to several hundred kilobases. *Nat Methods* **6**:343–345.
39. **Tsvetanova B, Peng L, Liang X, Li K, Yang JP, Ho T, Shirley J, Xu L, Potter J, Kudlicki W, Peterson T, Katzen F.** 2011. Genetic assembly tools for synthetic biology. *Methods Enzymol* **498**:327–348.
40. **Bitinaite J, Rubino M, Varma KH, Schildkraut I, Vaisvila R, Vaiskunaite R.** 2007. USER friendly DNA engineering and cloning method by uracil excision. *Nucleic Acids Res* **35**:1992–2002.
41. **Blanusa M, Schenk A, Sadeghi H, Marienhagen J, Schwaneberg U.** 2010. Phosphorothioate-based ligase-independent gene cloning (PLICing): an enzyme-free and sequence-independent cloning method. *Anal Biochem* **406**:141–146.
42. **Marienhagen J, Dennig A, Schwaneberg U.** 2012. Phosphorothioate-based DNA recombination: an enzyme-free method for the combinatorial assembly of multiple DNA fragments. *Biotechniques* **0**:1–6.
43. **Coljee VW, Murray HL, Donahue WF, Jarrell KA.** 2000. Seamless gene engineering using RNA- and DNA-overhang cloning. *Nat Biotechnol* **18**:789–791.
44. **Miyazaki K, Takenouchi M.** 2002. Creating random mutagenesis libraries using megaprimer PCR of whole plasmid. *Biotechniques* **33**:1033–1034, 1036–1038.
45. **Quan J, Tian J.** 2011. Circular polymerase extension cloning for high-throughput cloning of complex and combinatorial DNA libraries. *Nat Protoc* **6**:242–251.
46. **Zhang Y, Buchholz F, Muyrers JP, Stewart AF.** 1998. A new logic for DNA engineering using recombination in *Escherichia coli*. *Nat Genet* **20**:123–128.
47. **Fu J, Bian X, Hu S, Wang H, Huang F, Seibert PM, Plaza A, Xia L, Muller R, Stewart AF, Zhang Y.** 2012. Full-length RecE enhances linear-linear homologous recombination and facilitates direct cloning for bioprospecting. *Nat Biotechnol* **30**:440–446.
48. **Itaya M, Fujita K, Kuroki A, Tsuge K.** 2008. Bottom-up genome assembly using the *Bacillus subtilis* genome vector. *Nat Methods* **5**:41–43.
49. **Kruger NJ, Stingl K.** 2011. Two steps away from novelty–principles of bacterial DNA uptake. *Mol Microbiol* **80**: 860–867.
50. **Gibson DG, Benders GA, Axelrod KC, Zaveri J, Algire MA, Moodie M, Montague MG, Venter JC, Smith HO, Hutchison CA 3rd.** 2008. One-step assembly in yeast of 25 overlapping DNA fragments to form a complete synthetic *Mycoplasma genitalium* genome. *Proc Natl Acad Sci USA* **105**:20404–20409.
51. **Tagwerker C, Dupont CL, Karas BJ, Ma L, Chuang RY, Benders GA, Ramon A, Novotny M, Montague MG, Venepally P, Brami D, Schwartz A, Andrews-Pfannkoch C, Gibson DG, Glass JI, Smith HO, Venter JC, Hutchison CA 3rd.** 2012. Sequence analysis of a complete 1.66 Mb Prochlorococcus marinus MED4 genome cloned in yeast. *Nucleic Acids Res* **40**:10375–10383.

52. **Larionov V, Kouprina N, Eldarov M, Perkins E, Porter G, Resnick MA.** 1994. Transformation-associated recombination between diverged and homologous DNA repeats is induced by strand breaks. *Yeast* **10:**93–104.
53. **Raymond CK, Sims EH, Olson MV.** 2002. Linker-mediated recombinational subcloning of large DNA fragments using yeast. *Genome Res* **12:**190–197.
54. **Gibson DG.** 2009. Synthesis of DNA fragments in yeast by one-step assembly of overlapping oligonucleotides. *Nucleic Acids Res* **37:**6984–6990.
55. **Wingler LM, Cornish VW.** 2011. Reiterative Recombination for the in vivo assembly of libraries of multigene pathways. *Proc Natl Acad Sci USA* **108:** 15135–15140.
56. **Braman J.** 2002. *In Vitro Mutagenesis Protocols.* Humana Press, New York City, NY.
57. **Hutchison CA 3rd, Phillips S, Edgell MH, Gillam S, Jahnke P, Smith M.** 1978. Mutagenesis at a specific position in a DNA sequence. *J Biol Chem* **253:** 6551–6560.
58. **Zoller MJ, Smith M.** 1987. Oligonucleotide-directed mutagenesis: a simple method using two oligonucleotide primers and a single-stranded DNA template. *Methods Enzymol* **154:**329–350.
59. **Hogrefe HH, Cline J, Youngblood GL, Allen RM.** 2002. Creating randomized amino acid libraries with the QuikChange Multi Site-Directed Mutagenesis Kit. *Biotechniques* **33:**1158–1160, 1162, 1164–1155.
60. **Bauer JC, Wright DA, Braman JC, Geha RS.** 1998. *Circular site-directed mutagenesis.* US patent 5789166, August.
61. **Stemmer WP, Crameri A, Ha KD, Brennan TM, Heyneker HL.** 1995. Single-step assembly of a gene and entire plasmid from large numbers of oligodeoxyribonucleotides. *Gene* **164:**49–53.
62. **Kunkel TA.** 1985. Rapid and efficient site-specific mutagenesis without phenotypic selection. *Proc Natl Acad Sci USA* **82:**488–492.
63. **Hemsley A, Arnheim N, Toney MD, Cortopassi G, Galas DJ.** 1989. A simple method for site-directed mutagenesis using the polymerase chain reaction. *Nucleic Acids Res* **17:**6545–6551.
64. **Chen Z, Ruffner DE.** 1998. Amplification of closed circular DNA in vitro. *Nucleic Acids Res* **26:**1126–1127.
65. **Liang X, Peng L, Li K, Peterson T, Katzen F.** 2012. A method for multi-site-directed mutagenesis based on homologous recombination. *Anal Biochem* **427:**99–101.
66. **Agarwal KL, Buchi H, Caruthers MH, Gupta N, Khorana HG, Kleppe K, Kumar A, Ohtsuka E, Rajbhandary UL, Van de Sande JH, Sgaramella V, Weber H, Yamada T.** 1970. Total synthesis of the gene for an alanine transfer ribonucleic acid from yeast. *Nature* **227:** 27–34.
67. **Mandecki W, Hayden MA, Shallcross MA, Stotland E.** 1990. A totally synthetic plasmid for general cloning, gene expression and mutagenesis in *Escherichia coli. Gene* **94:**103–107.
68. **Gibson DG, Glass JI, Lartigue C, Noskov VN, Chuang RY, Algire MA, Benders GA, Montague MG, Ma L, Moodie MM, Merryman C, Vashee S, Krishnakumar R, Assad-Garcia N, Andrews-Pfannkoch C, Denisova EA, Young L, Qi ZQ, Segall-Shapiro TH, Calvey CH, Parmar PP, Hutchison CA 3rd, Smith HO, Venter JC.** 2010. Creation of a bacterial cell controlled by a chemically synthesized genome. *Science* **329:**52–56.
69. **Raab D, Graf M, Notka F, Schodl T, Wagner R.** 2010. The GeneOptimizer Algorithm: using a sliding window approach to cope with the vast sequence space in multiparameter DNA sequence optimization. *Syst Synth Biol* **4:**215–225.
70. **Graf M, Schoedl T, Wagner R.** 2009. Rationales of gene design and de novo gene construction, p 411–438. *In* Fu P, Panke S (ed), *Systems Biology and Synthetic Biology,* **vol 1.** John Wiley & Sons Inc, Hoboken, NJ.
71. **Greger B, Kemper B.** 1998. An apyrimidinic site kinks DNA and triggers incision by endonuclease VII of phage T4. *Nucleic Acids Res* **26:**4432–4438.
72. **Smith J, Modrich P.** 1997. Removal of polymerase-produced mutant sequences from PCR products. *Proc Natl Acad Sci USA* **94:**6847–6850.
73. **Young L, Dong Q.** 2004. Two-step total gene synthesis method. *Nucleic Acids Res* **32:**e59. doi:10.1093/nar/gnh058.
74. **Padgett KA, Sorge JA.** 1996. Creating seamless junctions independent of restriction sites in PCR cloning. *Gene* **168:**31–35.
75. **Mullinax RL, Gross EA, Hay BN, Amberg JR, Kubitz MM, Sorge JA.** 1992. Expression of a heterodimeric Fab antibody protein in one cloning step. *Biotechniques* **12:** 864–869.
76. **Dietmaier W, Fabry S, Schmitt R.** 1993. DISEC-TRISEC: di- and trinucleotide-sticky-end cloning of PCR-amplified DNA. *Nucleic Acids Res* **21:**3603–3604.
77. **Clancy K, Voigt CA.** 2010. Programming cells: towards an automated 'Genetic Compiler.' *Curr Opin Biotechnol* **21:**572–581.
78. **Chen YJ, Clancy K, Voigt CA.** 2012. Modeling genetic parts for synthetic biology, p 197–231. *In* Wall ME (ed), *Quantitative Biology: From Molecular to Cellular Systems,* **vol 1.** CRC Press, Boca Raton, FL.
79. **D C, Bergmann FT, Sauro HM, Densmore D.** 2011. Computer-aided design for synthetic biology, p 203–224. *In* Koeppl H, Densmore D, Setti G, di Bernardo M (ed), *Design and Analysis of Biomolecular Circuits. Engineering Approaches to Systems and Synthetic Biology,* **vol 1.** Springer, New York, NY.
80. **Chang AC, Cohen SN.** 1978. Construction and characterization of amplifiable multicopy DNA cloning vehicles derived from the P15A cryptic miniplasmid. *J Bacteriol* **134:**1141–1156.
81. **Cohen SN, Chang AC, Boyer HW, Helling RB.** 1973. Construction of biologically functional bacterial plasmids in vitro. *Proc Natl Acad Sci USA* **70:**3240–3244.
82. **Thomas CM, Smith CA.** 1987. Incompatibility group P plasmids: genetics, evolution, and use in genetic manipulation. *Annu Rev Microbiol* **41:**77–101.
83. **Keen NT, Tamaki S, Kobayashi D, Trollinger D.** 1988. Improved broad-host-range plasmids for DNA cloning in gram-negative bacteria. *Gene* **70:**191–197.

84. **Rawlings DE, Tietze E.** 2001. Comparative biology of IncQ and IncQ-like plasmids. *Microbiol Mol Biol Rev* **65:**481–496.

85. **Kontomichalou P, Mitani M, Clowes RC.** 1970. Circular R-factor molecules controlling penicillinase synthesis, replicating in *Escherichia coli* under either relaxed or stringent control. *J Bacteriol* **104:**34–44.

86. **Murray JA.** 1987. Bending the rules: the 2-mu plasmid of yeast. *Mol Microbiol* **1:**1–4.

87. **Sikorski RS, Hieter P.** 1989. A system of shuttle vectors and yeast host strains designed for efficient manipulation of DNA in Saccharomyces cerevisiae. *Genetics* **122:** 19–27.

88. **Bitoun R, Zamir A.** 1986. Spontaneous amplification of yeast CEN ARS plasmids. *Mol Gen Genet* **204:**98–102.

89. **Burke DT, Carle GF, Olson MV.** 1987. Cloning of large segments of exogenous DNA into yeast by means of artificial chromosome vectors. *Science* **236:**806–812.

Plasmids—Biology and Impact in Biotechnology and Discovery
Edited by Marcelo E. Tolmasky and Juan C. Alonso

doi:10.1128/microbiolspec.PLAS-0011-2013

José L. García[1]
Eduardo Díaz[1]

Plasmids as Tools for Containment

31

Biotechnology provides a large number of environmental applications (e.g., bioremediation, biofilters, bioleaching, biopesticides, biofuels, biotransformations [green chemistry], live vaccines, etc.) that support the development of the bioeconomy (1–3). Nevertheless, the biotechnological processes planned to work in the open field are not easy to implement since they have to cope with a wide range of chemical, physical, and biological variations that could lead to a low level of predictability, making such processes difficult to control (4). In addition, the biotechnological processes that introduce large quantities of microorganisms into the ecosystem have raised concerns about their potential impact on the environment. Such concerns have prompted the creation of risk assessment research programs oriented to developing new strategies in order to increase the safety and predictability of the microorganisms released into the environment and especially those that have been genetically modified (5–10).

To better control the behavior of a genetically modified organism (GMO) introduced in a target habitat, two different and complementary containment approaches have to be considered: (i) the survival of the GMO has to be limited by engineering a controlled life cycle to reduce its dissemination (biological containment), and (ii) the ability of the GMO to spread the new genetic information to potential recipients has to be eliminated (gene containment) (Fig. 1).

Plasmids harboring biological containment circuits to control the life cycle of the GMO and contained plasmids endowed with a gene containment system that reduces their horizontal transfer frequencies have been successfully engineered. Biological and gene containment strategies are of great interest not only for their biotechnological applications but also for being useful tools for fundamental research. In this chapter we will review and discuss different aspects of the containment systems, including their fundamentals, major shortcomings and the challenges to reaching absolute reliability ("certainty of containment"), and their different applications in biology.

FUNDAMENTALS OF CONTAINMENT AND GENETIC SAFEGUARD CIRCUITS

There are two main strategies to diminish the potential risks associated with the deliberate or unintentional release of GMOs into the environment. These two strategies have been called passive and active containment systems.

[1]Department of Environmental Biology, Centro de Investigaciones Biológicas (CSIC), 28040 Madrid, Spain.

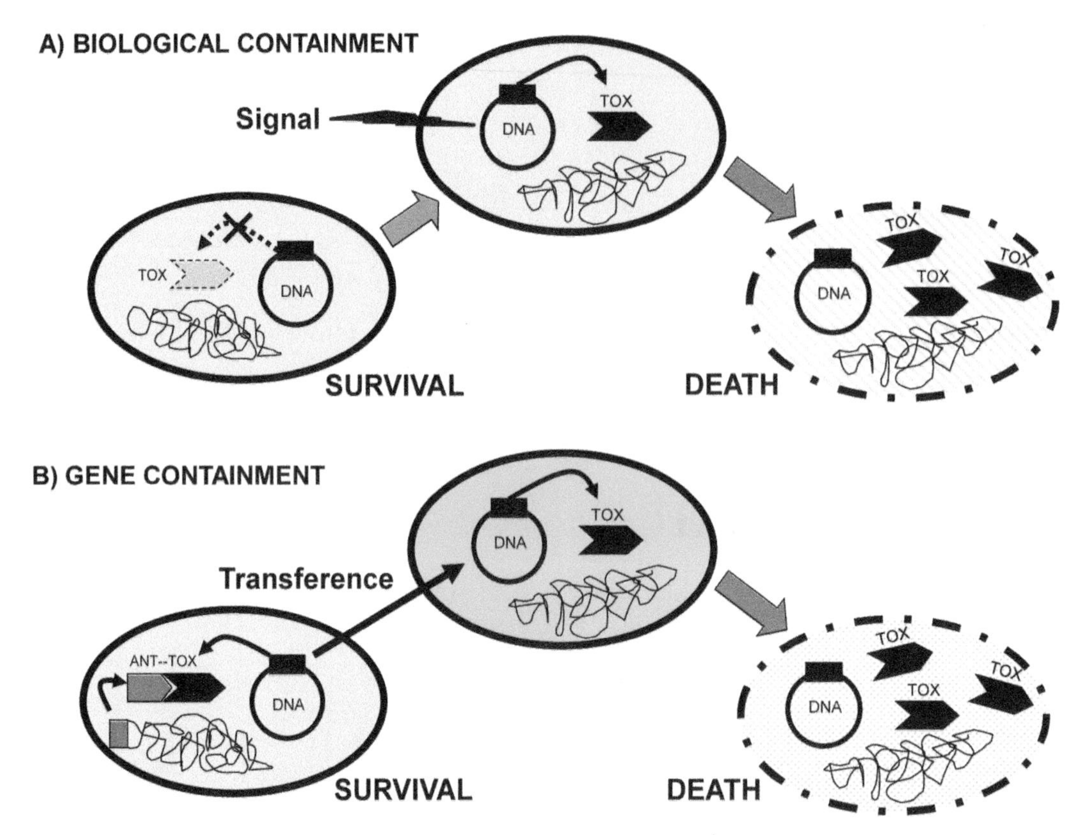

Figure 1 Concept of biological and gene containment. (**A**) In biological containment the GMO is restricted to the target habitat for a limited period of time. To accomplish this, the organism is engineered with a suicide circuit encoding a toxin (TOX) that usually is located in a plasmid and switched off (SURVIVAL), but it becomes activated in response to a specific environmental signal leading to cell death (DEATH). (**B**) In gene containment, it is the recombinant DNA rather than the organism itself that is the subject of containment. To create a barrier that restricts dispersal of such novel DNA from the GMO to the indigenous microbiota, a genetic circuit based on a toxic function (TOX) closely linked to the recombinant DNA needs to be engineered in such a way that the lethal function is inactive in the host GMO (e.g., formation of an antitoxin [ANT]-toxin [TOX] complex) but becomes activated (TOX) in the potential recipients of the contained DNA that lack the antitoxin function, leading to the death of such cells.
doi:10.1128/microbiolspec.PLAS-0011-2013.f1

Passive Containment

Passive containment consists of creating disabled strains through the induction of debilitating mutations. One method widely used for passive containment is to engineer auxotrophic organisms unable to synthesize an essential compound required for their survival. These auxotrophic organisms die or cannot divide when they escape the bioreactor where the essential compound is provided (11). Most passive containment systems have the drawback that they are bacteriostatic rather than bactericidal. Nevertheless, two auxotrophies have been described that have a high potential for human applications since they are bactericidal (12), i.e., alanine racemase mutants that have a requirement for D-Ala (13) and thymidilate synthase mutants that have a requirement for thymidine or thymine (14).

However, the disabled organisms engineered for passive containment may not compete successfully with the wild-type microbiota, and therefore they cannot be ideal candidates for some environmental applications (15). Moreover, engineered auxotrophy can remain effective as long as gene transfer does not compensate for the mutation and/or the nutrient required for survival is not available outside the target area.

Active Containment

An alternative to passive containment is to use competitive wild-type organisms and to equip them with controllable lethal functions that will not interfere with their life cycle unless some environmental signals of interest are present (5, 6, 15). The two basic elements of an active containment system are (i) lethal functions and (ii) control elements (Fig. 2).

Although the lethal functions usually have a broad host range, cell death efficiency depends upon the host cell and its metabolism. Usually, the lethal effect is higher during the exponential phase and declines progressively during the stationary phase (16). The level of resistance to the toxic protein depends on the strain under study; e.g., whereas *Pseudomonas putida* is sensitive to the toxic effect of the Gef protein, *Pseudomonas aeruginosa* seems to be particularly resistant to such lethal function (17).

On the other hand, when GMOs are destined to carry out a function in the open environment, the best strategy for containment is to engineer a control element that takes this into account. The introduced organisms should persist in the environment as long as they are needed, and cell death should be induced upon completion of the application. It becomes necessary, therefore, to analyze the task of the organism to identify the changes that take place when this task is completed and, then, to design a control element that responds to the most appropriate environmental changes (5, 15).

The Lethal Functions

The properties that a lethal function should have to be used in the design of active containment systems are (i) high efficiency of killing, (ii) broad host range targeting central cellular functions, (iii) low mutation rate, and (iv) no interference with the biotechnological application. The lethal functions used in active containment systems can be classified according to the cellular compartment that they target.

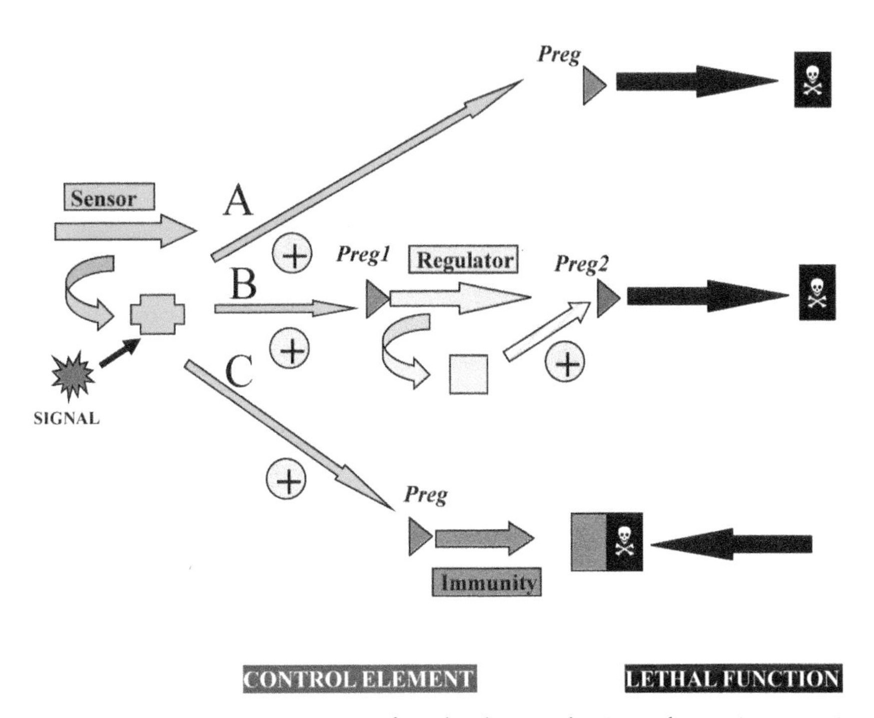

Figure 2 Schematic representations of molecular mechanisms for active containment systems. (**A**) A control element responds to the appropriate environmental signal through a sensor protein and a cognate regulated promoter (*Preg*), and regulates at the transcriptional level the expression of a lethal function. (**B**) The control element can be engineered as a double transcriptional regulatory circuit. The sensor protein recognizes an environmental signal and interacts with the cognate regulated promoter (*Preg1*), which in turn drives transcription of a regulator that controls the expression of a second regulated promoter (*Preg2*) running transcription of the lethal gene. (**C**) The control element may involve posttranslational regulation through an immunity protein that specifically neutralizes the killing effect of the constitutively expressed lethal function. To this end, the expression of the immunity gene is driven by the *Preg* promoter under control of the sensor protein that responds to the environmental signal. doi:10.1128/microbiolspec.PLAS-0011-2013.f2

Cell envelope

Toxin-antitoxin loci, which are ubiquitous in *Archaea* and *Bacteria*, play important roles in several cellular processes and can be used for developing containment systems. The toxin-antitoxin module consists of labile antitoxin and a stable toxin (18). The lethal functions most extensively used for developing active containment circuits are those that disrupt the membrane potential, especially the type I two-component toxin-antitoxin systems involved in postsegregational killing of plasmid-free cells (*hok* genes from plasmids R1 and F) and their chromosomal counterparts (*relF* [*hokD*] and *gef* [*hokC*] genes from *Escherichia coli*) (15, 19) (Table 1). The cytoplasmic membrane is also the cellular target of other lethal functions such as holins, e.g., the S and Ejh holins (16, 20) (Table 1). Holins are small proteins produced by bacteriophages that permeabilize the cytoplasmic membrane at a programmed time leading to cell death and to the release of a lytic protein (endolysin) that degrades the bacterial peptidoglycan, provoking lysis of the host cell and the release of the phage progeny (21). Endolysins such as protein E and protein L from the phages ΦX174 and MS2, respectively (21), have been used as lethal functions (22, 23) (Table 1). Chimeric *E-L* lytic genes showed a strong killing effect (22). The amidase from T7 phage, as well as phage lytic cassettes containing both holin and endolytic genes, e.g., *S/R/Rz* cassette from λ phage, are additional lethal elements that target the cell envelope (21) used in active containment systems (16, 23, 24) (Table 1). The PezT/ξ toxins of the epsilon/zeta antitoxin-toxin systems from *Firmicutes* are an example of lethal functions that cause programmed cell death by inhibiting peptidoglycan biosynthesis (25, 26), and therefore they could also be used to engineer active containment systems.

Periplasm

A lethal function that targets the periplasmic space of Gram-negative and some Gram-positive bacteria (corynebacteria and mycobacteria) is the levansucrase coded by the *sacB* gene from *Bacillus subtilis*. The *sacB* gene is induced in the presence of sucrose, and the SacB protein results in the synthesis of lethal amounts of levan that accumulate in the periplasm causing cell lysis. A conditional suicide system based on this toxic function has been developed (27) (Table 1).

Cytoplasm

The *relBE* genes from *E. coli* encode a type II toxin-antitoxin couple that affects protein synthesis (18, 19). While expression of the *relE* gene is highly toxic to *Saccharomyces cerevisiae*, this toxicity can be partially counteracted by expressing the *relB* gene. Therefore, the *relE-relB* genes, as well as similar toxin-antitoxin pairs, can be used to design active containment systems in genetically modified yeasts (28). The *ccdB*

Table 1 Lethal functions used in active containment systems

Protein	Origin	Function	Cellular target	Reference
Hok	R1 and F plasmids	Postsegregational cell killing	Membrane	15, 19
RelF	*E. coli*	Toxin-antitoxin module	Membrane	15, 19
Gef	*E. coli*	Toxin-antitoxin module	Membrane	15, 19
S	8 phage	Holin	Membrane	16
Ejh	EJ-1 phage	Holin	Membrane	20
E	ΦX174 phage	Lysis protein	Membrane	22, 23
L	MS2 phage	Lysis protein	Membrane	22, 23
E-L	Chimeric	Lysis protein	Membrane	22, 23
T7 amidase	T7 phage	Endolysin	Peptidoglycan	24
PezT	*Streptococcus pneumoniae*	Toxin-antitoxin module	Peptidoglycan	25
S/R/Rz cassette	8 phage	Lysis module	Cellular envelope	24
SacB	*Bacillus subtilis*	Levansucrase	Periplasm	27
RelE	*E. coli*	Translation inhibition	Ribosomes	28
RelA	*E. coli*	ppGpp synthetase I	Cytoplasm	23
CcdB	*E. coli* plasmid	DNA gyrase inhibitor	DNA	29
Streptavidin	*Streptomyces avidinii*	D-biotin inactivation	Cytoplasm	30
NucA	*Serratia marcescens*	Endonuclease	RNA, DNA	31, 33
*Eco*RI	*E. coli*	Restriction endonuclease	DNA	35
*Lla*IR	*Lactococcus lactis*	Restriction cassette	DNA	34
I-*Ceu*I	*Chlamydomonas moewusii*	Homing endonuclease	23S rDNA	36
ColE3	*E. coli*	RNase	16S rRNA	38, 39

gene, which is the toxin of another type II toxin-antitoxin module, encodes a protein that inhibits the DNA gyrase complex, and it has also been used to construct circuits for programmed death in response to changes in the environment (29) (Table 1).

Overexpression of the *relA* gene product (ribosome-associated ppGpp synthetase I) produces growth arrest of *E. coli* cultures, and the participation of the *relA* gene has been proposed in a concept for a biological containment system (23) (Table 1).

A highly efficient biological containment system is based on the tightly regulated derepression of the streptavidin gene (*stv*) from *Streptomyces avidinii* (Table 1). Streptavidin binds to D-biotin (vitamin H) and depletes this essential prosthetic group involved in one-carbon unit metabolism, causing cell death (30).

Nucleases have two main advantages over other lethal functions used in active containment systems: (i) their cellular target (DNA and/or RNA) is universal, and (ii) the genetic material itself is degraded, which eliminates the main source of concern. A modified *nucA* gene encoding the mature form of the extracellular nuclease from *Serratia marcescens*, which endonucleolytically cleaves RNA and DNA, has been used to construct several effective conditional suicide systems for bacteria and yeasts (31–33) (Table 1). The three-gene restriction cassette *Lla*IR^{+} from *Lactococcus lactis* (34), as well as the *ecoRIR* lethal gene encoding the type II *Eco*RI restriction endonuclease (35), and the I-*Ceu*I homing endonuclease from the green alga *Chlamydomonas moewusii* that breaks chromosomal 23S rDNA (36), were also used to engineer containment systems (Table 1).

RNases of bacterial origin have been used as lethal functions in both eukaryotes and prokaryotes. Barnase from *Bacillus amyloliquefaciens* was used for the inhibition of fungal diseases and for induction of male sterility in plants (37). Colicin E3 is a 58-kDa protein encoded by plasmid pColE3-CA38 that specifically cleaves the 16S rRNA of *E. coli* at a position 49 nucleotides from its 3′ end, in a region that is conserved among the 16S and 16S-like rRNAs in all three primary kingdoms (38). Colicin E3-mediated cleavage prevents protein synthesis and causes cell death. The *colE3* gene has been engineered to develop efficient gene and biological containment systems (38, 39) (Table 1).

The Control Element

When GMOs are used under controlled conditions (e. g., in the laboratory or in a fermentation plant) it is possible to couple the expression of the lethal gene to well-known regulatory circuits. Tight control of the lethal gene is crucial to design an efficient containment system, and this control can be performed at different levels.

Chemical control

The first conditional suicide system was engineered to contain microorganisms that are killed if they accidentally escape the bioreactor (40). To accomplish this, the *hok* gene was cloned under the control of the *E. coli* tryptophan promoter (*Ptrp*) in a pBR322 derivative such that the presence of tryptophan repressed *hok* expression and allowed the survival of recombinant *E. coli* cells in industrial fermentations. A deliberate or unintentional release of the recombinant bacteria to the environment, where the concentration of tryptophan is negligible, will cause derepression of the *hok* gene, leading to cell death (40) (Table 2). An analogous containment system was developed with the promoter of the *E. coli* alkaline phosphatase gene (*PphoA*), which is active under phosphate limitation, fused to the complete *parB* locus (*hok/sok*) of plasmid R1. The *parB* locus in combination with *PphoA* ensures both plasmid stabilization by postsegregational killing of plasmid free cells during fermentation and killing of the cells if they are released from the bioreactor to a phosphate-free or limited environment (phosphate is frequently a growth-limiting factor in nature) (41) (Table 2). The combination of *E. coli* K-12 *relA* mutants, which die very fast under nutrient-poor environmental conditions because they lack the stringent response (passive containment), with a plasmid-encoded suicide system based on the T7 endolysin-encoding gene expressed from the *PphoA* promoter resulted in effective containment of GMOs outside the laboratory (24). Biological containment systems have also been engineered on plasmids using the *Plac* promoter and the LacI repressor from *E. coli*, and they are triggered by addition of isopropyl-β-D-thiogalactopyranoside (Table 2). These suicide plasmids restricted survival of the *E. coli* host cells in different growth media as well as in soil (42, 43). Whereas the *Plac* promoter is effective in *E. coli*, to develop containment systems in nonenteric bacteria of environmental relevance (e.g., pseudomonads) it was necessary to replace *Plac* with some broad host range derivatives that retain the LacI-mediated repression (e.g., *Ptac* and *Ptrc* promoters [16, 30]) or the strong $P_{A1/04/03}$ synthetic promoter (39, 44, 45) (Table 2).

The *PsacB* promoter/SacR regulator and the *sacB* lethal gene have also been used to induce suicide of *E. coli* cells in a sucrose-containing medium. This containment system was tested under *in vitro* conditions and

Table 2 Control elements used in active containment systems

Signal	Type of control	Sensor	Regulated promoter	Immunity protein	Reference
Tryptophan limitation	Transcriptional	TrpR	*Ptrp*		40
Phosphate limitation	Transcriptional	PhoRB	*PphoA*		24, 41
IPTG	Transcriptional	LacI	*Plac*		42, 43
IPTG	Transcriptional	LacI	*Ptac, Ptrc*		16, 30
IPTG	Transcriptional	LacI	$P_{A1/04/03}$		39, 44, 45
Sucrose	Transcriptional	SacR	*PsacB*		27
Arabinose	Transcriptional	AraC	P_{BAD}		16, 33
Glucose limitation	Transcriptional		$P_{ADH2GAPDH}$		32
Iron limitation	Transcriptional	Fur	P_{viuB}		46
Rhamnose	Transcriptional	RhaRS	P_{RHA}		36
Benzoate and analogues	Transcriptional	XylS	*Pm*		16, 20
Temperature	Transcriptional	CI857	P_R, P_L		22, 36
Carbon starvation	Transcriptional	ppGpp/RNA Pol[a]	*rrnB* P1		23
Root exudates (proline)	Transcriptional	PutA	*PputA*		48, 49
Phage infection	Transcriptional	N.D.[b]	*31P*		34
Cell density	Transcriptional	LuxR	P_{luxI}		29
Gene transfer	Posttranscriptional			ImmE3 protein	38
Gene transfer	Posttranscriptional			EcoRIM protein	35
Time	Transcriptional	FimBE	*PfimA*		47
Limitation of benzoate and analogues	Transcriptional	XylS2/LacI	$Pm/P_{A1/04/03}$		51
Limitation of benzoate and analogues	Transcriptional	XylS2/LacI/ T7 RNA Pol[a]	*Pm/Ptac/10P*		69

[a]RNA Pol, RNA polymerase
[b]N.D., not determined

in soil microcosms (27) (Table 2). The *Pm* promoter from *P. putida*, which is activated through the XylS regulator in the presence of benzoate or benzoate analogues, was used in combination with the lethal *ejh* gene or the gene *E* for developing suicide plasmids that upon induction caused death in a broad range of Gram-negative bacteria (16, 20) (Table 2). The AraC/P_{BAD} regulatory system that responds to arabinose (Table 2) has been used for the control of the lethal *nucA* gene to engineer a suicide cassette activated by the presence of arabinose in recombinant *E. coli* cells (33), as well as to engineer passive containment circuits (16). A hybrid ADH2/GAPDH promoter that is repressed by glucose (Table 2) in *S. cerevisiae* has been fused to *nucA* to engineer a conditional suicide system that kills the yeast cells and destroys the genetic material whenever the host cells escape into the environment where glucose is scarce (32).

The strict iron-regulated promoter *PviuB* from *Vibrio cholerae* was fused to lysis gene *E* from phage ΦX174 to establish an *in vivo* inducible lysis circuit (Table 2). An isopropyl-β-D-thiogalactopyranoside-inducible PT7-controlled antigen expression circuit was integrated into the lysis system to build a novel *E. coli*-based antigen delivery system. The strain lyses in response to an iron-limiting signal *in vivo*, implementing antigen release and biological containment (46).

Different conditional suicide systems that respond to the absence of certain aromatic compounds in the environment, most of which are formed during the degradation of toxic pollutants, have been developed to engineer contained biodegraders (Table 2), and they will be discussed in the section "Biotechnological Applications of Containment Circuits."

Environmental stress control

Although most of the containment systems are triggered by a chemical signal that controls the expression of the lethal function, physical control based on temperature has also been reported (Table 2). Most of the regulatory circuits that respond to temperature changes are based on the P_L and P_R promoters from the lytic module of the λ phage and the cognate thermosensitive CI857 mutant repressor that becomes inactivated at temperatures higher than 30°C. Suicide plasmids with different resistance markers, origins of replication, and P_L::gene *E* or P_R::gene *E* fusions have been constructed (22). Although bacterial lysis due to expression of gene *E* is usually induced by a temperature shift of the growing culture from 28 to 42°C, the construction of a mutated P_R promoter/operator region resulted in a new expression system that stably repressed gene *E* expression at temperatures up to 37°C but still allowed induction of cell lysis at a temperature range of 39 to 42°C (22). By also using the heat-sensitive CI857 repressor, an efficient bacterial suicide system based on an I-*Ceu*I homing endonuclease from the green alga *C. moewusii* was developed. Thermo-induction of the expression of I-*Ceu*I endonuclease caused irreparable chromosomal DNA damage and cell death in recombinant *E. coli* cells (36). L-rhamnose induction of I-*Ceu*I was also engineered as an alternative suicide system that could be useful for manufacturing heat-sensitive products (36) (Table 2).

On the other hand, cold-sensitive suicide plasmids have also been developed to contain bacteria that escape into the environment from bioreactors during fermentation processes or from humans and animals treated with bacterial live vaccines. To construct such cold-sensitive lethal circuits, the λ *cI857*/P_R expression system was combined with either the *lacI*/*Plac* or the phage 434 *cI*/P_R regulatory elements that control the expression of the lysis gene *E*. *E. coli* cells harboring these suicide plasmids are able to grow at 37°C, but cell lysis takes place at temperatures below 30°C (22).

Environmental stresses other than temperature, e.g., starvation conditions, pH gradients, etc., are also signaling factors that can be used to develop conditional suicide systems for active containment of microorganisms in the open field. In this sense, the P1 promoter of the *E. coli* ribosomal *rrnB* operon (*rrnB* P1) is an efficient biosensor for poor growth conditions expected to occur after the accidental release of GMOs from a bioreactor to the open environment, where the microorganisms usually face starvation conditions. A concept for a conditional suicide system employing the *rrnB* P1 promoter, the *relA* gene, and phage-derived cell wall lytic genes has been proposed (23) (Table 2).

Stochastic control

For some purposes it might be useful to develop control systems for containment dependent on time (stochastic induction) rather than on the composition of the environment. If killing is induced just as a function of time, many unpredictable factors that can influence killing induction in real-life scenarios can be overcome. Such a stochastic induction system has been based on DNA recombination events (switches) that randomly result in activation of the suicide function (15). The invertible switch promoter *PfimA* directing the synthesis of type 1 fimbriae of *E. coli* has been used as a model system for the stochastic induction of suicide (Table 2). The orientation of the phase switch is controlled *in trans* by the products of two genes, *fimB* and *fimE*, located upstream of the *PfimA::gef* fusion. When the resulting system was expressed in a plasmid, it was observed as a reduction of the bacterial population with increasing significance as the cell growth rate decreased. When an *E. coli* culture harboring the suicide plasmid was left as stationary cells in suspension without nutrients, viability dropped exponentially over a period of several days. Moreover, in competition with noncontained cells, the contained cells are always outcompeted during growth (47). Although this system is only valid for *E. coli*, similar strategies can be applied to other organisms by using species-specific stochastic expression circuits or recombination systems that are functional in a broad spectrum of bacteria. To this end, a recombinational system based on the broad host range RP4 resolvase and two *res* sites embracing a transcription terminator placed between a strong promoter and a lethal gene was designed for suicide purposes (15).

Biological interaction-dependent control

For microorganisms interacting with plants (biocontrol strains, plant growth-promoting inoculants, rhizoremediation biocatalysts, etc.) it would be interesting to develop containment systems that respond to plant-derived signals, thus avoiding dispersion of such microorganisms outside the plant environment. In this sense, the *putA* and *putP* promoters of the proline catabolic pathway from *P. putida* are induced upon exposure of cells to proline or root exudates (48). By combining the $P_{A1/04/03}$::gene *E* fusion with the $Pput_A$::*lacI* regulatory element, a containment system was developed that restricts dispersal of bacteria to the presence of proline from the rhizosphere of plants of agronomic importance (49) (Table 2). Thus, the rhizobacterium *P. putida* strain CS-4 endowed with the active biological containment system colonized the rhizosphere at a level similar to that of the wild-type strain. However, when the plants were removed, the CS-4 population decreased in bulk soil at a higher rate than the wild-type strain (49).

Sometimes the signal that triggers the suicide system is a biological agent. This is the case of a regulatory circuit based on a phage-inducible promoter isolated from the lytic phage Φ31 from *L. lactis* (Table 2). This promoter is used to activate the expression of a lethal function after phage infection, thus protecting the *L. lactis* bacterial population as a whole against the infecting bacteriophage (34).

Cell density-dependent control

A population control circuit that autonomously regulates the density of an *E. coli* population has been engineered by coupling gene expression driven by cell-cell communication to cell survival and death. The cell density is broadcasted and detected by elements from a bacterial quorum-sensing system (LuxRI) (Table 2), which in turn regulate the death rate caused by the *ccdB* lethal gene (Table 1). This type of regulatory circuit makes it possible to program the dynamics of a population despite variability in the behavior of individual cells (29).

Gene transfer control

Plasmids are frequently transferred among microbes, and cell-free DNA released after the death of engineered organisms can remain functional and transferable even after long periods of time. Thus, gene containment systems have to be developed to prevent uptake and inheritance of engineered DNA (10). Although usually the expression of the lethal gene is controlled at the transcriptional level, posttranslational control of the lethal function has also been engineered in gene containment circuits (Fig. 2). In this sense, the immunity E3 protein is a plasmid-encoded acidic 9.3-kDa protein that binds stoichiometrically to the C-terminal RNase domain of colicin E3, preventing its RNase activity. The *colE3* and *immE3* genes have been engineered to develop a gene containment system in which gene (plasmid) transfer is the signal that activates the lethal function (38) (Table 2). A posttranslational control in which the cognate inhibitor acts at the target site of the lethal function has been developed with the *Eco*RI methytransferase, a monomeric *S*-adenosyl-L-methionine-dependent enzyme that delivers a methyl group to the internal adenine of the *Eco*RI recognition site and thus protects DNA of cleavage by the *Eco*RI restriction endonuclease. As indicated above for the *colE3/immE3* genes, the *ecoRIR/ecoRIM* genes have been used as a toxin-antitoxin genetic circuit that significantly reduces lateral gene spread (35) (Table 2).

STRATEGIES TO ENHANCE CONTAINMENT

A general feature of all of the containment systems described so far is that a surviving subpopulation of cells, in the range of 10^{-7} to 10^{-3} per cell per generation depending on the particular system under study, ceases to respond to the toxic effect of the lethal function. Although a rate of escape of killing of 10^{-7} is not too high, the number of survivors escaping suicide can be significant in environmental applications that use large quantities of cells (15, 30). The recommended limit of survival or engineered DNA transmission is less than 1 cell per 10^{8} cells (50) or less than 1,000 cells per 2 liters, according to the National Institutes of Health.

The analysis of the survivors revealed that mutations are the main drawback in containment. Mutations that inactivate the containment system have been located either in the lethal function (39, 51) or in the control element (42, 44). A detailed molecular analysis revealed that spontaneous transposition of insertion elements within the lethal gene encoding the *Eco*RI endonuclease was the most frequent cause of survival to the acquisition of a contained plasmid in *E. coli* cells (35). In contrast, deletions involving the lethal gene seem to be the most frequent cause of acquisition of a contained plasmid in other Gram-negative bacteria such as *P. putida*, *Agrobacterium tumefaciens*, and *Ralstonia eutropha* (35).

A way to reduce the problem of mutations and increase the efficiency of containment is to engineer genetic circuits with more than one lethal function. When the *gef* gene was used in a system responding to the absence of benzoate effectors, the rate of escape from killing decreased from 10^{-6} (one copy of *gef*) to 10^{-8} (two copies of *gef*) (51). Duplication of the *relF* gene also increased plasmid containment by about three orders of magnitude (43). The two copies of the *relF* gene were arranged in the contained plasmid in such a way that no single mutational event (deletion in particular) can lead to inactivation of both lethal genes. Protection against plasmid transfer was observed in test tubes and in rat intestine. Moreover, in soil and seawater a more than 7 orders of magnitude reduction in suicide bacteria (biological containment) was achieved (44). Duplication of the *nucA* gene encoding the *S. marcescens* nuclease in a conditional suicide plasmid was also reported to improve the reliability of containment (33). Nevertheless, the reduction in the number of survivors by duplicating a lethal function is always smaller than that expected by theoretical calculations, which predict that the efficiency of a dual containment system should be the product of the containment efficiencies due to each individual lethal function (44). This reduction in the efficiency of containment can be due to several factors such as (i) homologous recombination and gene conversion between the two copies of the suicide function, (ii) the existence of mutations that inactivate the regulatory element controlling the expression of both lethal genes, and (iii) the existence of mutations in the cellular target of the lethal function. To circumvent these limitations, the use of nonidentical suicide functions with different cellular target sites whose expression is under the control of different regulatory circuits may be a suitable strategy (44). In this sense, the *colE3* gene, encoding an RNase, and the *ecoRIR* gene, encoding a DNase, have been placed under different transcriptional signals and combined in the same plasmid. This dual containment circuit significantly reduced the spread of a gene cloned between the two lethal genes, and the system achieved the anticipated level of containment (52).

Biological containment systems may induce slow death of bacterial cells in soil versus the fast killing rate in laboratory assays, as observed with the *xylS/Pm::lacI/*$P_{A1/04/03}$*::gef* system (see above) in *P. putida* (45). A way to improve the performance of the containment system is by linking the expression of a gene that gives rise to essential metabolites (e.g., the *asd* gene product for biosynthesis of diaminopimelic acid, methionine, lysine, and threonine) to that of the repressor that prevents the expression of the lethal gene. This would guarantee at the same time both the synthesis of essential metabolites and the repression of the killing element (53). Upon triggering by the appropriate environmental signal, expression of the lethal gene and debilitation of the strain should lead to a faster disappearance rate. A *P. putida* Δ*asd* strain harboring a *Pm::asd* fusion and the suicide system (*xylS/Pm::lacI/*$P_{A1/04/03}$*::gef*) in the chromosome survived and colonized rhizosphere soil containing the effector 3-methylbenzoate at a level similar to that of the wild-type strain. However, whereas the *asd*$^{+}$-contained strain was still detectable in soils without 3-methylbenzoate after 100 days, the contained Δ*asd* mutant disappeared in less than 25 days, and the rate of mutations leading to escape from the lethal effect of the *gef* gene product was reduced by at least one order of magnitude ($<10^{-9}$ mutants per cell per generation) (53). These results indicate that an active containment system can be reinforced by using a host strain with a genetically engineered background that, under the desired conditions, debilitates the cell and increases its rate of suicide (53).

An important issue in the containment of GMOs is the maintenance of the viability of the contained cells when the specific environmental signal reaches a level that cannot be detected by the regulatory element that controls the lethal gene. For instance, in the case of GMOs destined to remove toxic compounds, the target pollutant may not be bioavailable or may be present at a concentration that cannot activate the regulatory circuit that controls the lethal function. The combination of the containment system with a transport protein that increases the uptake of the pollutant could decrease the threshold to which suicide is induced and therefore increase the survival of the contained cells in slightly polluted environments or in environments where the pollutant has been partially removed by the GMO (J. L. García, and E. Díaz, unpublished data).

THE NEW GENERATION OF CONTAINMENT SYSTEMS: XENOBIOLOGY

A major conclusion of the large amount of work done in the past decades on systems for containing GMOs is that barriers based on conditional suicide circuits are under Darwinian evolution, and therefore they are sooner or later defeated. The probability of escape from containment and unintended interaction with the environment is now far from what has been called "certainty of containment" or absolute reliability (54). Scientists have been recently developing creative new strategies to address this problem, and the shortcomings of the current genetic safeguards were presented above (3). Thus, the construction of minimal genomes that contain only the genes that are necessary to sustain life may make many random mutations lethal, and therefore the likelihood of unpredictable behavior after the microbe is released into the environment would be reduced (55). A different strategy is based on developing orthogonal life forms that use artificial genetic languages that, by definition, cannot communicate with the extant biochemistry of the existing live world, what could be called a "genetic firewall" (3). The orthogonal life form approach uses biochemical building blocks, e.g., nucleic acids and amino acids, that are incompatible with the natural biological systems and, thus, bring genetic interactions to practically zero. Xenonucleic acids (XNAs) are DNA-like molecules in which the nucleotide bases, double helix geometries, or nucleic acid backbones are nonnatural and cannot be read or replicated by natural polymerases, and therefore they are useless to DNA-based organisms. This scenario is optimal to maintain the distance between the engineered and the existing biological world. Some examples of XNAs that can be replicated in cell-free systems, e.g., PCR, or used to express functional proteins, e.g., green fluorescent protein, have been reported (56, 57). Recently, living bacteria that look and behave like normal have been evolved to use the synthetic 5-chlorouracil instead of the natural base thymine as a component of their DNA. Since these cells are auxotrophic for a nonnatural chemical and contain a form of DNA that cannot be deciphered by other organisms, they are endowed with two concomitant firewalls for any interaction with other bacteria (58).

Orthogonal life can be also engineered at the protein level. Thus, synthetic mRNA codons (new genetic alphabet [e.g., quadruplet codons], 6-letter genetic code) (59), artificial tRNA/aminoacyl tRNA synthetase pairs, and the evolution of new ribosomes able to decode new genetic codes (60), can be used to design synthetic organisms made of unnatural polymers that do not interfere with the existing biological world.

BIOTECHNOLOGICAL APPLICATIONS OF CONTAINMENT CIRCUITS

Containment Strategies for Bioremediation

The most advanced containment systems are those developed for bacteria that degrade pollutants. To develop a suitable containment system for biodegraders, the expression of the lethal gene should be coupled with the regulatory system of a pathway for biodegradation of the pollutant in such a way that cells survive in the presence of the pollutant but die after completion of the degradation of the toxic compound or after dispersing to locations outside the polluted area. Since the first biological containment system for bacteria that degrade pollutants was reported (61), different improvements have been carried out, from decreasing the rate of escape to the toxic effect of the lethal function. The control element of the system is a double regulatory circuit (Fig. 2, Table 2) based on the *xylS* regulatory gene and its cognate *Pm* promoter from the *meta*-cleavage pathway for degradation of toluene and xylenes of the *P. putida* pWW0 (TOL) plasmid. Gene *E* or *gef* was used as a lethal gene under the control of the LacI-controlled $P_{A1/04/03}$ promoter (62). The system predicts survival of the biologically contained strain in the presence of XylS effectors, e.g., a wide variety of alkyl- and halo-substituted benzoates (intermediates formed during the biodegradation of toluene, xylenes, and their halo-derivatives), because expression of the *Pm::lacI* fusion gives rise to the production of the LacI repressor, which in turn prevents the expression of the lethal gene. Once the pollutant is exhausted or the bacteria move to locations outside the polluted area, the LacI repressor is no longer produced, leading to the expression of the lethal gene and subsequent cell death (62) (Fig. 3). Although the system was initially associated with plasmids, the lower rate of killing escape was obtained by integrating the suicide circuit into the chromosome of *P. putida* cells. This rate of escape (in the range of 10^{-8} per cell per generation) was considered satisfactory, and field-release assays were carried out with both the contained and the control uncontained

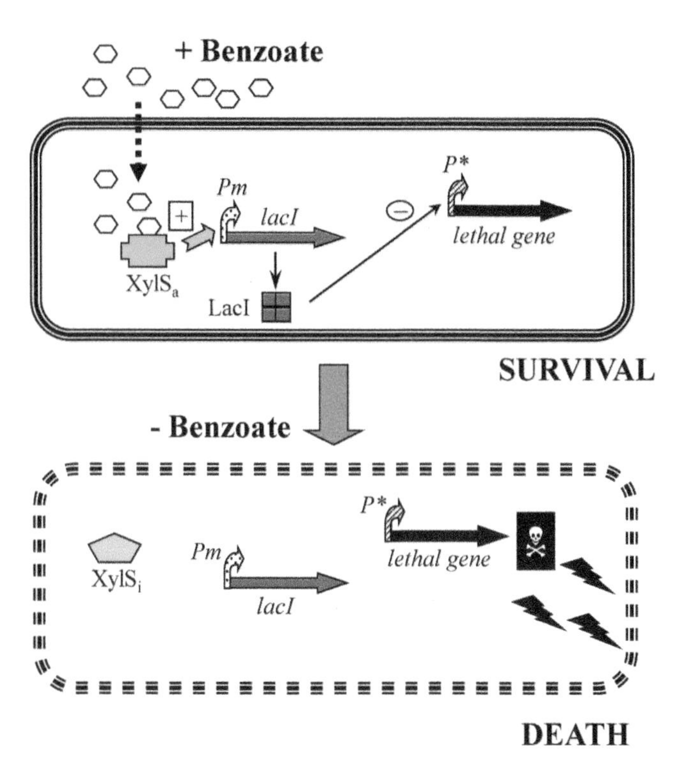

Figure 3 Rationale of a model biological containment system for biodegraders. The control element consists of a double regulatory circuit based on (i) the XylS sensor protein that recognizes benzoate or benzoate analogues (alkyl- and halo-benzoates) and stimulates gene expression from the *Pm* promoter and (ii) the *lacI* gene, whose expression is driven by the *Pm* promoter, coding for the LacI repressor protein that inhibits the $P_{A1/04/03}$ promoter (P^*). The expression of the lethal gene is under control of the $P_{A1/04/03}$ (P^*) promoter. In the presence of benzoate or benzoate analogues (+ Benzoate) the control element is switched on and the lethal function is not produced (SURVIVAL). Once bacteria complete the degradation of the aromatic compound or spread to a nonpolluted site (– Benzoate), the control system is switched off and, as a consequence, the lethal function is produced (DEATH). XylS$_a$, active conformation of XylS protein. XylS$_i$, inactive conformation of XylS protein.
doi:10.1128/microbiolspec.PLAS-0011-2013.f3

strains. While the biologically contained bacteria were able to colonize the rhizosphere of plants in soil with 3-methylbenzoate but not in nonpolluted soils, the parental *P. putida* strain colonized the rhizosphere of plants grown in both polluted and nonpolluted soils. Moreover, no evidence of dispersal of the test strains outside the experimental plots was observed (45).

The XylS/*Pm::lacI* regulatory loop has also been combined with the *Ptac*::T7 RNA polymerase gene and the *Φ10P* (T7 RNA polymerase-dependent promoter of gene 10 of T7 phage)::*stv* (streptavidin encoding lethal gene) fusions to engineer a triple regulatory circuit that tightly regulates efficient biological containment for bacteria that degrade aromatic compounds (30) (Table 2). An additional level of regulation to decrease uninduced expression of the lethal *stv* gene was achieved by generating an antisense RNA complementary to the *stv* transcript (30).

Although the biological containment systems increase the predictability of GMOs, one of the main concerns about the release of such GMOs to the environment is how recombinant DNA (often located in plasmids) can spread among indigenous bacterial populations. In particular, the transfer of genes to pathogenic bacteria that could increase their virulence and the impact of such transfer on natural populations of organisms are topics of major concern. Moreover, horizontal DNA transfer from transgenic organisms may also be undesirable for process protection and process optimization reasons (5, 15, 39, 63). Therefore, there is a need to reduce gene spread to ecologically insignificant levels. A first development toward this goal was the construction of mini-transposon cloning vectors derived from Tn*5* and Tn*10* transposons for the stable integration of heterologous DNA segments into the chromosome of Gram-negative bacteria. Genes introduced into the chromosome on mini-transposons exhibited transfer frequencies ($<10^{-9}$ per cell per generation) several orders of magnitude lower than those on plasmid vectors (64). To reduce further gene spread, gene containment systems have been engineered (15).

The first systems were based on the transcriptional control of the lethal gene by the *Ptrp* or the $P_{A1/04/03}$ promoters (40, 44). An alternative system was based on a posttranslational inhibition of a lethal function by a cognate immunity function (38). Since the containment systems should be effective in the open field, the lethal function should be active in a wide range of potential recipients, but the immunity function should exist naturally in a very restricted number of organisms. In the GMO the lethal gene is closely linked to the novel trait that needs to be contained and constitutively expressed from a promoter that is functional in a broad spectrum of bacteria. In contrast, the immunity gene is located at a distant location in the GMO, such that cotransfer of the lethal and immunity functions will be an extremely low-frequency event. The transfer of the novel trait to a nonimmune organism will be accompanied by transfer of the closely linked lethal gene that, in the absence of immunity, leads to the rapid killing of the recipient cells and prevents the spread of the novel trait (Fig. 4). A plasmid containment system based on a toxin-antitoxin couple was engineered with the colicin E3/immunity E3 proteins (38). Whereas the *immE3* gene was stably inserted into the chromosome of different *E. coli* and *P. putida* strains, the *colE3* gene was cloned in a promiscuous plasmid under the control of the *Ptac* promoter. The effectiveness of containment was tested using different DNA transfer mechanisms (transformation and conjugation) and a variety of recipient microorganisms of environmental importance. The *colE3*-based gene containment system decreased the transfer frequencies of the contained plasmid with respect to those of a control plasmid by 4 to 5 orders of magnitude in all recipients checked (38). A similar containment strategy was developed using the *Eco*RI endonuclease/*Eco*RI methylase functions, and this system was shown to be effective also in a broad range of microorganisms (35).

The *colE3*-based genetic circuit has been used to develop *P. putida* strains able to degrade pollutants, whose relevant genotype is subject to a powerful containment system (65). Thus, the *colE3* gene was closely linked to the *bph* genes for the catabolism of biphenyl and toxic polychlorinated biphenyls, and the resulting *bph-colE3* cassette was stably inserted into the chromosome of an immune *P. putida* strain in a location far from that of the *immE3* gene. The resulting strain was able to grow on biphenyl as the sole carbon and energy source, and it did not show any growth disadvantage with respect to the uncontained strain. Moreover, transfer of the *bph* genes was below detection levels using optimal laboratory conditions for transfer (65). The *colE3*-based containment circuit was also combined with the *sty* genes responsible for styrene degradation in *Pseudomonas* sp. Y2, and the *colE3-sty* cassette was engineered in an immune *P. putida* strain to develop a contained biocatalyst for styrene removal (66).

Containment Strategies for Vaccines

Although active containment systems are usually designed to prevent the potential risks associated with the accidental or intentional release of GMOs to the open environment, they also have attractive elements for developing vaccines (Table 3). Suicide plasmids harboring tightly regulated holin-like genes, e.g., gene *E* or chimeric *E-L* lysis genes, have been used to induce formation of empty cells with intact envelopes (bacterial ghosts) that are of interest as nonliving candidate vaccines. The bacterial ghosts share all functional and antigenic determinants of the envelope with their living counterparts and thus are able to induce strong local immunity (22). In the recombinant ghost system, foreign proteins can be targeted to the inner membrane, to the periplasmic space, or to S-layer proteins prior to *E*-mediated lysis. As ghosts have inherent adjuvant properties, they can be used as adjuvants in combination with subunit vaccines. The capacity of all species of

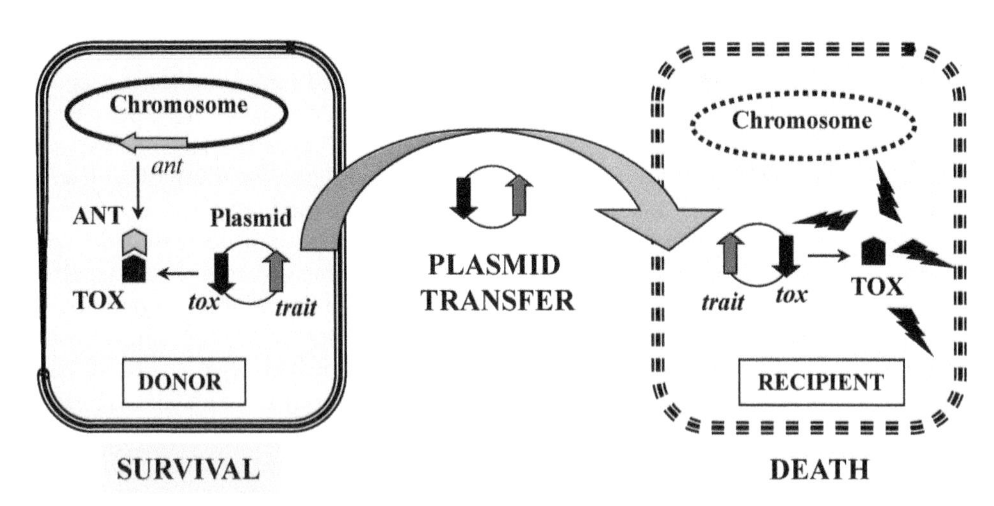

Figure 4 Rationale of a model plasmid containment system. The scheme has been modified from Diaz et al. (38). Gene containment is achieved by a lethal donation of a toxic function. In the GMO (donor cell), the lethal gene (*tox*) is closely linked to the novel trait (*trait*) in a plasmid, and the toxic effect of the lethal function (TOX) is neutralized by the product of an antitoxin gene (*ant*) located at the chromosome, such that cotransfer of the *tox* and *ant* genes will be an extremely low-frequency event. Plasmid transfer to a nonimmune organism (recipient cell) will lead to the activation of the lethal function and to cell death, thus preventing the spread of the plasmid and the novel associated trait.
doi:10.1128/microbiolspec.PLAS-0011-2013.f4

ghosts (membranes, periplasm, and internal lumen) to carry foreign antigens seems to be surprisingly large and can be exploited to design new combination vaccines. By using derivatives of the λP_r promoter and CI repressor it was possible to engineer regulatory circuits that allowed expression of the lethal gene, i.e., ghost formation, at different temperatures (22).

A regulated programmed lysis of recombinant attenuated *Salmonella* live vaccine for antigen delivery after colonization of host lymphoid tissues has been developed (11). The system is composed of two parts. The first component is an engineered strain with deletion of the *asdA* gene, involved in diaminopimelic acid synthesis, and arabinose-regulated expression of *murA* gene, involved in muramic acid synthesis, two genes required for peptidoglycan synthesis, and additional mutations to enhance complete lysis and antigen delivery. The second component is a plasmid that encodes arabinose-regulated *murA* and *asdA* expression and C2 repressor-regulated synthesis of antisense *asdA* and *murA* mRNA transcribed from the P22 phage P_R promoter. An arabinose-regulated *c2* gene is present in the chromosome. The recombinant strain exhibits arabinose-dependent growth. Upon invasion of host tissues (i.e., an arabinose-free environment) transcription of *asdA*, *murA*, and *c2* ceases, and the drop in C2 results in activation of P_R, driving synthesis of antisense mRNA to block translation of any residual *asdA* and *murA* mRNA. This passive containment system has potential applications with other bacteria in which biological containment would be desirable (11). Passive containment strategies for lactic acid bacteria used as vaccine vectors have also been reported (12, 14).

An active containment system that responds to iron limitation was expressed in an *E. coli*-based protein delivery system to develop a recombinant vector vaccine that was shown to lyse in response to iron-limiting signal *in vivo*, implementing antigen release and biological containment (46).

Table 3 Applications of active containment systems

Applied research
Open environments
Biological containment
Gene containment
Non-living vaccines
Contained environments
Protection of industrial processess
Optimization of industrial processes
Facilitation of downstream processing
Fundamental research
Genetic tools for:
Cloning and manipulation of genes
Studying natural transformation and the environmental fate of DNA
Assessing the role of plasmid transfer inmicrobial communities

Containment Strategies for Fermentation Processes

Active containment systems are also of great utility for biological processes carried out in contained environments (Table 3). Biotechnology companies have tools to satisfy the biosafety requirements of the products they want to commercialize. Furthermore, the use of genetic circuits that limit DNA transfer are of interest for process protection and process optimization reasons. For instance, phage infections threaten most fermentation bioprocesses. A phage protection strategy relies on designing mechanisms that will capture an emerging phage and prevent its proliferation to disruptive population levels. Sacrificing infected cells through phage-induced suicide strategies establishes an altruistic goal of protecting the bacterial population as a whole (34). To this end, a suicide plasmid was constructed that harbors the phage-inducible promoter (*31P*) from the lytic phage Φ31 of *L. lactis*, driving the expression of a lethal restriction cassette (*Lla*IR$^+$) for the simultaneous killing of the propagation host cell and the infecting phage. Infection assays revealed that only 15% of infected cells released the phage progeny. The *31P*/*Lla*IR$^+$ suicide plasmid also inhibited four Φ31-derived recombinant phages at levels at least 10-fold greater than that of Φ31 (34). This type of phage-triggered suicide strategy can be applied to any bacterium-phage combination in fermentation environments, given a regulatory element that is triggered by the infecting phage and a lethal gene that causes effective killing of the propagation host (34). On the other hand, the design of biocatalysts whose lysis is artificially triggered by a cheap inducer, e.g., benzoate and its derivatives, might facilitate downstream processing operations for industrial applications. Thus, in bioprocesses that involve lysis of the cells, the release of chromosomal DNA increases the sample viscosity and reduces the efficiency of the subsequent centrifugation, filtration, and other downstream processing steps. The use of suicide circuits based on nucleases may be of interest for the reduction of such cell lysate viscosity (67).

A programmed self-disruptive *P. putida* strain was constructed to control and facilitate the release of polyhydroxyalkanoate granules to the extracellular medium. The autolytic cell disruption system was based on two simultaneous strategies: the coordinated action of two proteins from the pneumococcal bacteriophage EJ-1, an endolysin (Ejl) and a holin (Ejh), and the mutation of the *tolB* gene, which exhibits alterations in outer membrane integrity that induce lysis hypersensitivity. The *ejl* and *ejh* genes were expressed under an XylS/*Pm* monocopy expression system inserted into the chromosome of the *tolB* mutant strain in the presence of 3-methylbenzoate as the inducer molecule (68).

Containment Strategies for Fundamental Research

Conditional suicide systems are useful genetic tools in the laboratory for safe cloning and manipulation of genes encoding highly toxic proteins as well as for developing positive selection cloning vectors and various forms of gene replacement (69) (Table 3).

Suicide plasmids constitute interesting tools for fundamental research in microbial ecology (Table 3). Genetic flux in microbial communities is a critical component of their metabolic potential and, therefore, it is important to acquire a fundamental understanding of the parameters that influence it in nature (70). Suicide plasmids allow the release of biochemically unmanipulated DNA into a selected environment at a specific time point. The lytic cassette *S/R/Rz* from λ phage was placed under the control of the *Ptac* promoter in different Gram-negative bacteria, and the resulting plasmid caused, after addition of the inducer, controlled cell death and release of large amounts of nucleic acids that persisted into the surrounding medium for several days. This system has been used to study natural transformation in bacteria and the environmental fate of DNA released by cell death (16).

To study horizontal gene transfer in environmental samples from an introduced donor bacterium to a complex microbial community, selection of the recipient cells harboring the gene of interest requires counter-selection of the donor cells. For this purpose, the introduction of an inducible suicide system in such donor cells would be very useful (15, 71). A *P. putida* strain with two chromosomal copies of the *gef* gene was used as the donor of the conjugative RP4 plasmid. Although the donor strain was successfully eliminated in filter-matings, elimination of the donor in microcosms by induction of the suicide genes did not succeed (72), and therefore this strategy requires further development.

Gene transfer plays an important role in adaptive responses of microbial communities to environmental changes, in genetic stability, and in species identity. For instance, the ability of some bacteria to degrade toxic pollutants is usually plasmid encoded. Even if the introduced strains are unable to compete with the preadapted microbiota and disappear, transfer of the degradative plasmids to the indigenous population of microorganisms contributes significantly to the improved rate of *in situ* degradation (70). To assess the importance of plasmid transfer in adaptation of microbial communities to different stress conditions, it would be very helpful to use contained plasmids that possess the appropriate traits to adapt to such conditions but that cannot spread from an introduced donor organism to the indigenous microbiota. Lethal donation circuits, such as those described above for gene containment, constitute interesting tools to explore the ecological and evolutionary consequences of shifting the natural equilibrium between genetic change and genetic constancy toward the latter (35, 38).

To conclude this section it is worth mentioning that since 1995 several patents have been granted for the use of the containment systems discussed above (US 5,670,370; US 5,672,345; US 5,679,533; US 5,681,745; US 5,702,916; US 6,287,844; US 7,183,097; US 7,780,961; US 2003/026883; US 2005/0276788; US 2013/0023035).

CONCLUSIONS AND OUTLOOK

Active containment systems are a major tool to reduce the uncertainty associated with the introduction of monocultures, genetically engineered or not, into target habitats. While biological containment reduces the survival of the introduced organism outside the target habitat and/or upon completion of the projected task, gene containment strategies reduce the lateral spread of the key genetic determinants to indigenous microorganisms. Whereas trials of such contained microorganisms have been carried out in experimental microcosms, validation of the containment systems in real field releases constitutes the next step.

Many lethal functions and regulatory circuits have been used and combined to design efficient containment systems. As many new genomes are being sequenced, novel lethal genes and regulatory elements are available, e.g., new toxin-antitoxin modules, and they could be used to increase further the current containment efficiencies and to expand containment to other organisms. In this sense, since most of the existing suicide systems have been designed for aerobic proteobacteria, it becomes necessary to develop similar systems for other microorganisms, e.g., Gram-positive bacteria, anaerobes, etc., that also play an important role in the environment and are the subject of genetic manipulations for biotechnological purposes. Recently, synthetic gene circuits have been constructed to develop counters that can be used in different settings for a variety of purposes across a range of timescales. For example, if inputs to the synthetic counter were coupled to the cell cycle, one might program cell death to occur after a user-defined number of cell divisions to limit the life as a safety mechanism in engineered strains (73). Engineering cell death controlled by bacterial quorum-sensing systems that respond to cell density, and that do not require the addition of external inducers, would be another strategy for containment of GMOs intended for large-scale use in the environment.

In fundamental research, suicide circuits become relevant tools to address the role of gene transfer, mainly plasmid transfer, in evolution and how this transfer contributes to genome plasticity and to the rapid adaptation of microbial communities to environmental changes. Nevertheless, the use of containment strategies to address key questions in molecular microbial ecology requires further development.

We should be aware that although the current containment systems can increase the predictability of GMOs in the environment, containment will never be absolute, due to the existence of mutations that lead to the appearance of surviving subpopulations. In this sense, orthogonal systems appear to be the solution for setting a functional genetic firewall that will allow absolute containment of recombinant organisms. However, we are still a long way from having such orthogonal systems. First, we have to engineer cells in which DNA has been replaced by XNAs and the appropriate replication machinery. Then, XNAs should be combined with cognate RNA polymerases that generate transcripts with expanded or alternative genetic codes that incorporate nonnatural amino acids to proteins. From the point of view of containment, fully orthogonal organisms will only survive with a constant supply of orthogonal building blocks from a controlled environment. Nevertheless, the efficiency of such synthetic organisms for a particular biotechnological process is still unknown. Moreover, recent findings suggest that XNAs might interfere with replication and gene expression of natural DNA and RNA, thus creating the possibility that escaped orthogonal organisms might impact natural ecosystems (10).

Acknowledgments. *We thank K.N. Timmis for encouraging our work in active containment. This work was supported by EU FP7 Grant 311815 and by grants BIO2012-39501, CSD2007-00005, BFU2009-11545-C03-03, and BIO2012-39695-C02-01 from the Ministry of Economy and Competitiveness of Spain. Conflicts of interest: We disclose no conflicts.*

Citation. García JL, Díaz E. 2014. Plasmids as tools for containment. Microbiol Spectrum **2**(5):PLAS-0011-2013.

References

1. **Gavrilescu M.** 2010. Environmental biotechnology: achievements, opportunities and challenges. *Dynamic Biochem Process Biotechnol Mol Biol* **4:**1–36.
2. **Singh JS, Abhilash PC, Singh HB, Singh RP, Singh DP.** 2011. Genetically engineered bacteria: an emerging tool

for environmental remediation and future research perspectives. *Gene* **480:**1–9.

3. **Schmidt M, de Lorenzo V.** 2012. Synthetic constructs in/for the environment: managing the interplay between natural and engineered biology. *FEBS Lett* **586:**2199–2206.
4. **de Lorenzo V.** 2009. Recombinant bacteria for environmental release: what went wrong and what we have learnt from it. *Clin Microbiol Infect* **15:**63–65.
5. **Ramos JL, Andersson P, Jensen LB, Ramos C, Ronchel MC, Díaz E, Timmis KN, Molin S.** 1995. Suicide microbes on the loose. *Biotechnology* **13:**35–37.
6. **Davison J.** 2002. Towards safer vectors for the environmental release of recombinant bacteria. *Environ Biosafety Res* **1:**9–18.
7. **Davison J.** 2005. Risk mitigation of genetically modified bacteria and plants designed for bioremediation. *J Ind Microbiol Biotechnol* **32:**639–650.
8. **Paul D, Pandey G, Jain RK.** 2005. Suicidal genetically engineered microorganisms for bioremediation: need and perspectives. *Bioessays* **27:**563–573.
9. **de Lorenzo V.** 2010. Environmental biosafety in the age of synthetic biology: do we really need a radical new approach? *Bioessays* **32:**926–931.
10. **Moe-Behrens GH, Davis R, Haynes KA.** 2013. Preparing synthetic biology for the world. *Front Microbiol* **4:**5.
11. **Kong W, Wanda SY, Zhang X, Bollen W, Tinge SA, Roland KL, Curtiss R 3rd.** 2008. Regulated programmed lysis of recombinant *Salmonella* in host tissues to release protective antigens and confer biological containment. *Proc Natl Acad Sci USA* **105:**9361–9366.
12. **Lee P.** 2010. Biocontainment strategies for live lactic acid bacteria vaccine vectors. *Bioeng Bugs* **1:**75–77.
13. **Bron PA, Benchimol MG, Lambert J, Palumbo E, Deghorain M, Delcour J, De Vos WM, Kleerebezem M, Hols P.** 2002. Use of the *alr* gene as a food-grade selection marker in lactic acid bacteria. *Appl Environ Microbiol* **68:**5663–5670.
14. **Steidler L, Neirynck S, Huyghebaert N, Snoeck V, Vermeire A, Goddeeris B, Cox E, Remon JP, Remaut E.** 2003. Biological containment of genetically modified *Lactococcus lactis* for intestinal delivery of human interleukin 10. *Nat Biotechnol* **21:**785–789.
15. **Molin S, Boe L, Jensen LB, Kristensen CS, Givskov M, Ramos JL, Bej AK.** 1993. Suicidal genetic elements and their use in biological containment of bacteria. *Annu Rev Microbiol* **47:**139–166.
16. **Kloos D-U, Strätz M, Güttler A, Steffan RJ, Timmis KN.** 1994. Inducible cell lysis system for the study of natural transformation and environmental fate of DNA released by cell death. *J Bacteriol* **176:**7352–7361.
17. **Soberón-Chávez G.** 1996. Evaluation of the biological containment system based on the *Escherichia coli gef* gene in *Pseudomonas aeruginosa* W51D. *Appl Microbiol Biotechnol* **46:**549–553.
18. **Yamaguchi Y, Park JH, Inouye M.** 2011. Toxin-antitoxin systems in bacteria and archaea. *Annu Rev Genet* **45:**61–79.
19. **Gerdes K.** 2000. Toxin-antitoxin modules may regulate synthesis of macromolecules during nutritional stress. *J Bacteriol* **182:**561–572.
20. **Díaz E, Munthali M, Lünsdorf H, Höltje J-V, Timmis KN.** 1996. The two-step lysis system of pneumococcal bacteriophage EJ-1 is functional in Gram-negative bacteria: triggering of the major pneumococcal autolysin in *Escherichia coli*. *Mol Microbiol* **19:**667–681.
21. **Young R, Wang I-N, Roof WD.** 2000. Phages will out: strategies of host cell lysis. *Trends Microbiol* **8:**120–128.
22. **Lubitz W, Witte A, Eko FO, Kamal M, Jechlinger W, Brand E, Marchart J, Haidinger W, Huter V, Felnerova D, Stralis-Alves N, Lechleitner S, Melzer H, Szostak MP, Resch S, Mader H, Kuen B, Mayr B, Mayrhofer P, Geretsschläger R, Haslberger A, Hensel A.** 1999. Extended recombinant bacterial ghost system. *J Biotechnol* **73:** 261–273.
23. **Tedin K, Witte A, Reisinger G, Lubitz W, Bläsi U.** 1995. Evaluation of the *E. coli* ribosomal *rrnB* P1 promoter and phage-derived lysis genes for the use in a biological containment system: a concept study. *J Biotechnol* **39:** 137–148.
24. **Schweder T, Hofmann K, Hecker M.** 1995. *Escherichia coli* K12 *relA* strains as safe hosts for expression of recombinant DNA. *Appl Microbiol Biotechnol* **42:** 718–723.
25. **Mutschler H, Gebhardt M, Shoeman RL, Meinhart A.** 2011. A novel mechanism of programmed cell death in bacteria by toxin-antitoxin systems corrupts peptidoglycan synthesis. *PLoS Biol* **9:**e1001033. doi:101371/journal.pbio.1001033.
26. **Lioy VS, Machon C, Tabone M, Gonzalez-Pastor JE, Daugelavicius R, Ayora S, Alonso JC.** 2012. The ζ toxin induces a set of protective responses and dormancy. *PLoS One* **7:**e30282. doi:101371/journal.pone.30282.
27. **Recorbet G, Robert C, Givaudan A, Kudla B, Normand P, Faurie G.** 1993. Conditional suicide system of *Escherichia coli* released into soil that uses the *Bacillus subtilis sacB* gene. *Appl Environ Microbiol* **59:**1361–1366.
28. **Kristoffersen P, Jensen GB, Gerdes K, Piskur J.** 2000. Bacterial toxin-antitoxin gene system as containment control in yeast cells. *Appl Environ Microbiol* **66:**5524–5526.
29. **You L, Cox RS 3rd, Weiss R, Arnold FH.** 2004. Programmed population control by cell-cell communication and regulated killing. *Nature* **428:**868–871.
30. **Szafranski P, Mello CM, Sano T, Smith CL, Kaplan DL, Cantor CR.** 1997. A new approach for containment of microorganisms: dual control of streptavidin expression by antisense RNA and the T7 transcription system. *Proc Natl Acad Sci USA* **94:**1059–1063.
31. **Ahrenholtz I, Lorenz MG, Wackernagel W.** 1994. A conditional suicide system in *Escherichia coli* based on the intracellular degradation of DNA. *Appl Environ Microbiol* **60:**3746–3751.
32. **Balan A, Schenberg ACG.** 2005. A conditional suicide system for *Saccharomyces cerevisiae* relying on the intracellular production of the *Serratia marcescens* nuclease. *Yeast* **22:**203–212.
33. **Li Q, Wu YJ.** 2009. A fluorescent, genetically engineered microorganism that degrades organophosphates and commits suicide when required. *Appl Microbiol Biotechnol* **82:**749–756.

34. **Djordjevic GM, O'Sullivan DJ, Walker SA, Conkling MA, Klaenhammer TR.** 1997. A triggered-suicide system designed as a defense against bacteriophages. *J Bacteriol* **179:**6741–6748.

35. **Torres B, Jaenecke S, Timmis KN, García JL, Díaz E.** 2000. A gene containment strategy based on a restriction-modification system. *Environ Microbiol* **2:**555–563.

36. **Tsuji S, Naili I, Authement NR, Segall AM, Hernandez V, Hancock BM, Giacalone MJ, Maloy SR.** 2010. An efficient thermoinducible bacterial suicide system. Elimination of viable parental bacteria from minicells. *BioProcess Int* **April:**28–40.

37. **Ulyanova V, Vershinina V, Ilinskaya O.** 2011. Barnase and binase: twins with distinct fates. *FEBS J* **278:**3633–3643.

38. **Díaz E, Munthali M, de Lorenzo V, Timmis KN.** 1994. Universal barrier to lateral spread of specific genes among microorganisms. *Mol Microbiol* **13:**855–861.

39. **Munthali MT, Timmis KN, Díaz E.** 1996. Use of colicin E3 for biological containment of microorganisms. *Appl Environ Microbiol* **62:**1805–1807.

40. **Molin S, Klemm P, Poulsen LK, Biehl H, Gerdes K, Andersson P.** 1987. Conditional suicide system for containment of bacteria and plasmids. *Biotechnology* **5:** 1315–1318.

41. **Schweder T, Schmidt I, Herrmann H, Neubauer P, Hecker M, Hofmann K.** 1992. An expression vector system providing plasmid stability and conditional suicide of plasmid-containing cells. *Appl Microbiol Biotechnol* **38:**91–93.

42. **Bej AK, Perlin MH, Atlas RM.** 1988. Model suicide vector for containment of genetically engineered microorganisms. *Appl Environ Microbiol* **54:**2472–2477.

43. **Knudsen SM, Karlström OH.** 1991. Development of efficient suicide mechanisms for biological containment of bacteria. *Appl Environ Microbiol* **57:**85–92.

44. **Knudsen S, Saadbye P, Hansen LH, Collier A, Jacobsen BL, Schlundt J, Karlström OH.** 1995. Development and testing of improved suicide functions for biological containment of bacteria. *Appl Environ Microbiol* **61:**985–991.

45. **Molina L, Ramos C, Ronchel M-C, Molin S, Ramos JL.** 1998. Construction of an efficient biologically contained *Pseudomonas putida* strain and its survival in outdoor assays. *Appl Environ Microbiol* **64:**2072–2078.

46. **Guan L, Mu W, Champeimont J, Wang Q, Wu H, Xiao J, Lubitz W, Zhang Y, Liu Q.** 2011. Iron-regulated lysis of recombinant *Escherichia coli* in host releases protective antigen and confers biological containment. *Infect Immun* **79:**2608–2618.

47. **Klemm P, Jensen LB, Molin S.** 1995. A stochastic killing system for biological containment of *Escherichia coli*. *Appl Environ Microbiol* **61:**481–486.

48. **Vílchez S, Molina L, Ramos C, Ramos JL.** 2000. Proline catabolism by *Pseudomonas putida*: cloning, characterization, and expression of the *put* genes in the presence of root exudates. *J Bacteriol* **182:**91–99.

49. **van Dillewijn P, Vílchez S, Paz JA, Ramos JL.** 2004. Plant-dependent active biological containment system for recombinant rhizobacteria. *Environ Microbiol* **6:**88–92.

50. **Wilson DJ.** 1993. NIH guidelines for research involving recombinant DNA molecules. *Account Res* **3:**177–185.

51. **Jensen LB, Ramos JL, Kaneva Z, Molin S.** 1993. A substrate-dependent biological containment system for *Pseudomonas putida* based on the *Escherichia coli gef* gene. *Appl Environ Microbiol* **59:**3713–3717.

52. **Torres B.** 2002. *Ph.D. thesis*. Universidad Autónoma de Madrid, Madrid, Spain.

53. **Ronchel MC, Ramos JL.** 2001. Dual system to reinforce biological containment of recombinant bacteria designed for rhizoremediation. *Appl Environ Microbiol* **67:**2649–2656.

54. **Marlière P.** 2009. The farther, the safer: a manifesto for securely navigating synthetic species away from the old living world. *Syst Synth Biol* **3:**77–84.

55. **Dewall MT, Cheng DW.** 2011. The minimal genome: a metabolic and environmental comparison. *Brief Funct Genomics* **10:**312–315.

56. **Krueger AT, Peterson LW, Chelliserry J, Kleinbaum DJ, Kool ET.** 2011. Encoding phenotype in bacteria with an alternative genetic set. *J Am Chem Soc* **133:** 18447–18451.

57. **Malyshev DA, Dhami K, Quach HT, Lavergne T, Ordoukhanian P, Torkamani A, Romesberg FE.** 2012. Efficient and sequence-independent replication of DNA containing a third base pair establishes a functional six-letter genetic alphabet. *Proc Natl Acad Sci USA* **109:** 12005–12010.

58. **Marlière P, Patrouix J, Döring V, Herdewijn P, Tricot S, Cruveiller S, Bouzon M, Mutzel R.** 2011. Chemical evolution of a bacterium's genome. *Angew Chem Int Ed Engl* **50:**7109–7114.

59. **Yang Z, Chen F, Alvarado JB, Benner SA.** 2011. Amplification, mutation, and sequencing of a six-letter synthetic genetic system. *J Am Chem Soc* **133:**15105–15112.

60. **Neumann H, Wang K, Davis L, Garcia-Alai M, Chin JW.** 2010. Encoding multiple unnatural amino acids via evolution of a quadruplet-decoding ribosome. *Nature* **464:** 441–444.

61. **Contreras A, Molin S, Ramos JL.** 1991. Conditional-suicide containment system for bacteria which mineralize aromatics. *Appl Environ Microbiol* **57:**1504–1508.

62. **Ronchel MC, Ramos C, Jensen LB, Molin S, Ramos JL.** 1995. Construction and behavior of biologically contained bacteria for environmental applications in bioremediation. *Appl Environ Microbiol* **61:**2990–2994.

63. **Dröge M, Pühler A, Selbitschka W.** 1998. Horizontal gene transfer as a biosafety issue: a natural phenomenon of public concern. *J Biotechnol* **64:**75–90.

64. **de Lorenzo V, Herrero M, Sánchez JM, Timmis KN.** 1998. Mini-transposons in microbial ecology and environmental biotechnology. *FEMS Microbiol Ecol* **27:**211–224.

65. **Munthali MT, Timmis KN, Díaz E.** 1996. Restricting the dispersal of recombinant DNA: design of a contained biological catalyst. *Biotechnology* **14:**189–191.

66. **Lorenzo P, Alonso S, Velasco A, Díaz E, García JL, Perera J.** 2003. Design of catabolic cassettes for styrene biodegradation. *Antonie van Leeuwenhoek* **84:**17–24.

67. **Boynton ZL, Koon JJ, Brennan EM, Clouart JD, Horowitz DM, Gerngross TU, Huisman GW.** 1999. Reduction of cell lysate viscosity during processing of

poly(3-hydroxyalkanoates) by chromosomal integration of the staphylococcal nuclease gene in *Pseudomonas putida*. *Appl Environ Microbiol* **65**:1524–1529.

68. **Martínez V, García P, García JL, Prieto MA.** 2011. Controlled autolysis facilitates the polyhydroxyalkanoate recovery in *Pseudomonas putida* KT2440. *Microb Biotechnol* **4**:533–547.

69. **Davison J.** 2002. Genetic tools for Pseudomonads, Rhizobia, and other Gram-negative bacteria. *Biotechniques* **32**: 386–394.

70. **Davison J.** 1999. Genetic exchange between bacteria in the environment. *Plasmid* **42**:73–91.

71. **Clerc S, Simonet P.** 1998. A review of available systems to investigate transfer of DNA to indigenous soil bacteria. *Antonie van Leeuwenhoek* **73**:15–23.

72. **Sengelov G, Sorensen SJ.** 1998. Methods for detection of conjugative plasmid transfer in aquatic environments. *Curr Microbiol* **37**:274–280.

73. **Lu TK, Khalil AS, Collins JJ.** 2009. Next-generation synthetic gene networks. *Nat Biotechnol* **27**:1139–1150.

Plasmids—Biology and Impact in Biotechnology and Discovery
Edited by Marcelo E. Tolmasky and Juan C. Alonso

doi:10.1128/microbiolspec.PLAS-0033-2014

Esteban Martínez-García,[1] Ilaria Benedetti,[1]
Angeles Hueso,[1] and Víctor de Lorenzo[1]

32 Mining Environmental Plasmids for Synthetic Biology Parts and Devices

INTRODUCTION

In its most widespread meaning, synthetic biology has been defined as the engineering of biology (1, 2). There are two sides to this definition. The first deals with fundamental science, as synthesis is the counterpart of analysis, the second leg of any rigorous research endeavor. Synthesizing something by rational assembly of its individual components is, in fact, the ultimate proof of understanding, as the celebrated statement of R. Feynman posthumously declared ("What I cannot create, I do not understand."). On the other hand, the definition above implicitly announces that biological objects can indeed be engineered for a practical purpose. While in the field of molecular biology the term "genetic engineering" is normally used as a metaphor or an analogy, synthetic biology adopts "engineering" as an authentic conceptual and technical frame for repurposing existing biological entities and for creating new-to-nature properties. The underlying idea is that any biological system can be regarded as a combination of individual functional elements that are comparable to those found in man-made devices. These can be described as wholes of a limited number of components that can possibly be combined in novel configurations to modify existing properties or to create new ones—and so can their biological counterparts. Paramount in this concept is the identification and rigorous description of biological parts (the shortest DNA sequences encoding unique, stand-alone, unambiguous biological functions) and devices (an assembly of parts that runs a specified action with a definite input and output governed by a fixed transfer function). Although parts and devices are the basis of any extant biological system, the angle of synthetic biology is that at least some of them can be excised from their natural context, reformatted to meet some compositional standards, and rewired with a different genetic connectivity to create synthetic systems that bring about novel phenotypic traits in a host organism (often called "the chassis" in the synthetic biology jargon) (3). For this to happen, it is essential that parts and devices maintain the functions and the parameters that they possess in their natural context once they are placed somewhere else.

Alas, any biologist knows this is not likely to be the case (4). Unlike components used in industrial manufacturing, the behavior of biological constituents is

[1]Systems and Synthetic Biology Programs, Centro Nacional de Biotecnología, CNB-CSIC, 28049 Madrid, Spain.

extremely context-dependent—let alone that they can mutate and evolve erratically. Furthermore, their combinations may create emergent properties that are often unpredictable, thereby hindering any rigorous engineering based on them. Not surprisingly, many contemporary synthetic biology efforts try to tackle this challenge. One of the avenues to this end is the pursuit of "orthogonality" of the engineered constructs in respect to their host. This concept, imported from mathematics and computer science, indicates zero or a minimal influence of the engineered devices and systems on their biological carrier and *vice versa*. One approach to orthogonality involves eliminating as many as possible of the undesired connections that a part or device may have in its natural context or to set new information codes that only the designed construct can decipher (5). But one can also ask whether nature has in some cases favored the evolution of functional modules with basic functions that are to an extent autonomous in respect to the biological container where they reside.

This is where plasmids and other mobile genetic elements come into play in the synthetic biology scenario, as they are able to travel through different types of microorganisms and express their genetic complement in diverse genomic and metabolic backgrounds. Phages, in particular, are programmed to take over the entire expression machinery of the host for its own sake, often by means of alternative RNA polymerases (RNAPs) that recognize promoter sequences altogether different from those of the host. Not surprisingly, the RNAP of bacteriophage T7 (T7 pol) is one of the most used biological parts in synthetic biology: the pair T7pol/P_{T7} (T7 RNAP and the cognate promoter of gene *10* of phage T7) suffices to trigger transcription of downstream DNA sequences whether *in vivo* or *in vitro*, with the only condition being that nucleoside triphosphates are available. This happens regardless of the biological carrier, from bacteria to animal cells, thus providing a reasonably orthogonal tool for manipulation gene expression at the user's will. Also, repressors found in transposons are excellent assets for biological engineering. Repression is a less demanding mechanism of transcriptional regulation in prokaryotes (6) and is thus far easier to enter in a heterologous context. The tetracycline-dependent TetR repressor of the *tetA* gene of Tn*10* and its target DNA sequence has been employed and refactored in many ways to this end (7), and it is also one of the most popular parts for construction of artificial genetic circuits. Nevertheless, synthetic biology is not only about heterologous, orthogonal transcription initiation; it requires many other engineer-able functions. These are not found in either phages or transposons but, as discussed below, on broad host range (BHR) conjugative or mobilizable plasmids.

PROMISCUOUS PLASMIDS BEAR KEY HOST-INDEPENDENT BIOLOGICAL FUNCTIONS

BHR conjugative (or at least mobilizable) plasmids are objectively the most favorable source of natural orthogonal parts and devices, as long as they deploy their full or partial genetic complement in diverse biological carriers. The key component to this end is the vegetative origin of replication. It is remarkable that many environmental plasmids seem to have just such an *oriV* that either uses the host replication machinery or is accompanied by a cognate, plasmid-encoded replication protein(s) gene(s), with no other recognizable attribute encoded, in what would look like a perfect example of "selfish" DNA. One can consider such minimalist plasmids as the evolutionary solution to the engineering challenge of having an optimal DNA-carrying vector. Note that the bottlenecks that even the simplest plasmids must overcome for their propagation involve not only their replication proper (for which they have a suite of mechanisms; see, e.g., references 8, 9) but also signaling replication termination and resolution of replication intermediates. By default, this can happen through homologous recombination between DNA sequences, but in other instances, specific molecular mechanisms take care of unknotting the otherwise intractable replicating molecules.

But just having autonomous replication is no guarantee of doing well unless it is accompanied by other complementary functions. Environmental plasmids often ensure their maintenance by having active countersegregation systems that force bacteria to retain them or otherwise be killed. But in other cases, stable plasmid-host parterships occur without any conspicuous selective advantage. In many instances, however, plasmids end up recruiting to their frame some traits that are clearly beneficial to the host. These last include a large suite of generally dispensable genes which, under certain circumstances, can endow the carrier with some recompense in return for the physicological load of being carried as an extrachromosomal element. These genes range from resistance to antibiotics (the best known plasmid-borne features) to novel metabolic capabilities (e.g., routes for biodegradation of environmental pollutants). As discussed below, such pathways are themselves a source of both blocks for metabolic engineering as well as the source of a wealth of conditional expression systems that can be co-opted in synthetic constructs. To assemble whatever circuit of choice, such functional DNA fragments are typically recruited to the plasmid frame through a suite of site-specific recombination (SSR) devices which either remain active in the extant plasmid structures or become vestigial or even erased after a period of stable coexistence with a given host.

In a further turn of the screw, some plasmids deliver to the host novel functions that change their lifestyle (e.g., the thermal window of optimal viability) and even switch the order of preference of carbon source consumption (10–12). Spreading these properties in a population depends on plasmids' ability to move between different hosts. By default, this can occur through natural transformation, but it usually happens through conjugation. To this end, some plasmids carry a complete set of genes for conjugal self-transmission, while others just have an origin of transfer (*oriT*) which can be recognized by other plasmids' machinery and thus bring about mobilization. All these components can be found in various configurations in extant plasmids accompanied by what we could call "genetic garbage," i.e., remnants of insertion sequences and nonfunctional pseudogenes for which no clear role may be assigned upfront (13).

It is not surprising that some synthetic biology efforts have been recently directed to re-create a sort of quintessential BHR plasmid frame, devoid of parasitic-looking elements. Along this line, Hansen (14) constructed an all-synthetic IncX1 plasmid based on the components just mentioned, although some of them are somewhat compressed in very close or overlapping DNA sequences: *tra* (conjugal transfer), *mob1* (mobilization 1), *rep/stb* (replication initiation/toxin-antitoxin plasmid addiction), *mob2* (mobilization 2), *par* (plasmid partitioning), and *res* (resistance marker). The synthetic plasmid kept all the properties shared by the natural members of the IncX1 group but, as long as it was artificially simplified, created a construct much easier to further engineer (15). Table 1 summarizes the key functions that environmental plasmids are typically endowed with and which, following some formatting and repurposing, become useful assets for biosystem engineering along the line of contemporary synthetic biology. Note that the list is by no means exhaustive, and the reader is directed to the ACLAME database

Table 1 Principal biological parts found in BHR environmental plasmids

Functional segment	Description	Synthetic biology application	References
Origin of replication	Narrow	Physical assembly, gene cloning in *E. coli*	16–24
	Broad	Analysis and engineering of diverse Gram-negative bacteria	25–47
Selection marker	Antibiotic resistance	Plasmid retention, biomarker	49–53
	Auxotrophy compensation	Self-selecting plasmid maintenance	54–56
Toxin/antitoxin systems	Plasmid countersegregation	Marker-free vectors, biosafety circuits, conditional viability	62–73
Conjugative transfer	Origin of transfer	Plasmid transmission between *E. coli* and diverse recipients	89, 90
	Transfer and mobilization	Suicide delivery of transposon and transposon-vectors	91–93
		Biological computing and execution of logic (Boolean) programs	94
Regulated promoters	Substrate-induced transcription of metabolic operons	Regulatory devices turned on by chemical effectors	98–113
Cis-acting antisense RNAs	Formation of asRNA/target RNA duplexes that are degraded by RNase III	Gene expression silencing	116–123
Metabolic genes	Catabolism of unusual carbon sources	Models for metabolic engineering and implant-chassis retroactivity	130, 131
Site-specific recombination systems	Resolution of replication intermediates	Excision of undesired DNA segments in genomes of interest	74–80
		Engineering of operative systems, memory, and logic gates	82, 84
	Recruitment of functional sequences	(Combinatorial) assembly of functional DNA in a gene cluster	83, 96

(http://aclame.ulb.ac.be) for a thorough compilation of what genes and functions one can find in a large collection of plasmids and other mobile elements. In the following sections we discuss in more detail some of these parts and how they have been incorporated into the synthetic biology toolbox.

NARROW HOST RANGE ORIGINS OF REPLICATION

Plasmid vectors are the basis of recombinant DNA technology and the workhorses of any genetic engineering or synthetic biology project. The one key function that enables their existence as extrachromosomal elements is their origin of replication. Most of those employed for gene cloning are intended to work in *Escherichia coli* and thus have a narrow host range. The most used to this end are those based on the vegetative origin of replication (*oriV*) of plasmid ColE1/pMB1, which is the type of default origin of replication found in a large number of plasmid vectors (16). The main advantage is its small size and its mechanism of replication, which is entirely enabled by host functions (i.e., *E. coli*). This makes the DNA sequences needed for autonomous propagation of plasmids with a ColE1 origin compressed in a very small fragment that can be easily moved according to the user's needs. Most ColE1-derived vectors bear variants of this *oriV* with mutations that cause a very high copy number. While this eases DNA extraction and preparation from host cells, it may become detrimental if the plasmid carries cloned genes—and more so if they are toxic. A second type of narrow host range vectors are those based on the p15A replicon (16). These have a lower copy number than their ColE1 counterparts (but still high) and are compatible with them, thereby enabling expression of cloned genes from two coexisting vectors. Another *oriV* quite popular in plasmids for *E. coli* is SC101, whose advantage is its low copy number. The downside in this case is its lower yield when extracted from host bacteria and the need for an extra plasmid-encoded replication protein, which increases the size of the DNA required for its autonomous propagation.

Finally, one narrow host range replication origin that has been heavily refactored for different purposes since its description (17) is that of plasmid R6K. The property that makes this system so attractive is that the plasmid-borne replication protein Π (encoded by the so-called *pir* gene) can also actuate in *trans* on the origin of replication (*oriV*R6K). This allows the physical separation as well as the mutual dependence of the two components (*pir* gene and *oriV*) for plasmid functioning (Fig. 1). By artificially placing the *pir* gene in the chromosome and the *oriV* in a plasmid, one not only ensures a narrow-range replication system, but fulfills the complete requirement of a strain-specific host for vector propagation (18). This feature has been exploited in a large number of synthetic biology circuits aimed at suicide delivery of transposon vectors (19–21) (see below) and engineering of biosafety circuits by designing mutual host-plasmid addiction (22). Not surprisingly, the small collection of narrow host range origins of replication just mentioned have been incorporated into the various repositories of biological parts available to synthetic biologists (23, 24) (http://parts.igem.org).

BHR ORIGINS OF REPLICATION

As long as synthetic biology projects go beyond model *E. coli* strains and toy circuits and enter deployment of deeply engineered traits for industrial, medical, and environmental settings, new hosts (i.e., chassis) different from *E. coli* are required, and BHR plasmid vectors

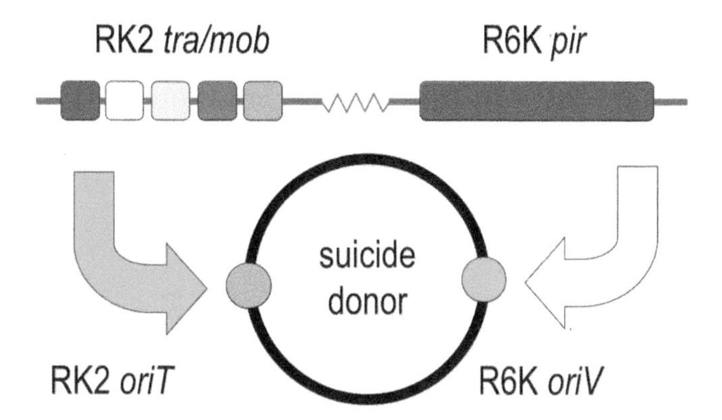

Figure 1 Conditional replication and suicide delivery device. Replication of some plasmids requires (apart from host factors) the action of a specific replication protein on the vegetative *oriV* sequence. For the R6K plasmid, it is possible to reshape the natural arrangement in *cis* of the *pir* gene that encodes the replication protein Π and the *oriV* so that they are in *trans*. In this instance, replication of the resulting plasmid depends entirely on expression of *pir*, e.g., from an engineered chromosomal location. On the other hand, plasmids with an *oriT* can be mobilized into conjugal recipients through the action of the *tra/mob* genes of a self-conjugative plasmid (e.g., RK2). By combining the two traits in the same covalently closed circular DNA, one creates a plasmid that is both entirely dependent on a specialized host for replication and can be delivered to a recipient where it can transiently stay but not replicate. This is the basis of the many conditional suicide systems for insertion of transposons that are found in the genetic engineering and synthetic biology literature.
doi:10.1128/microbiolspec.PLAS-0033-2014.f1

become mandatory. BHR origins of replication have been known since the onset of the recombinant DNA era (25, 26), but their generally more complex replication mechanism (and therefore their larger size) have limited their use in vector development. Such origins afford autonomous propagation of the replicon on a variety of bacterial carriers. Most BHRs of this kind have been developed for Gram-negative microorganisms, although some of them could also be useful in Gram-positive bacteria (see below). One widely used BHR origin is that of plasmid RSF1010. Although its replication mechanism is somewhat intricate and the corresponding functional segment contains intermingled replication and transfer functions (27, 28), RSF1010-based plasmids are among the most host-promiscuous vectors. The segment of the natural plasmid that suffices for autonomous replication has been recently streamlined (29) and formatted as a separate module for vector assembly (23). A second origin very frequently found in BHR vectors is that of the RK2/RP4 plasmid (30). Despite the (comparatively) small size of the cognate functional segment, the RK2 origin also counts among the most promiscuous. A large number of vectors—before and after the onset of synthetic biology—based on RK2 have been instrumental in analyzing and constructing complex phenotypes in bacteria such as *Pseudomonas*, *Rhizobium*, *Agrobacterium*, and a suite of Gram-negative species (31–33).

But other origins of replication are increasingly used for BHR vector development as well. For instance, pBBR1, a plasmid first isolated from *Bordetella bronchiseptica* (34), was adopted as the basis of a considerable number of genetic tools (24, 35–37). One of the advantages is that BBR1-derived plasmids generally have a higher copy number than their RK2-based and RSF1010-based counterparts (with all the pluses and minuses discussed above). By the same token, the origin of environmental IncW plasmid pSa (38), which was adopted in the late 1980s for constructing BHR vectors for *Agrobacterium* (39) and *Xanthomonas* (40), has been recently reshaped and formatted as a building block for generating standardized vectors for *Ralstonia* (41). In contrast, genus-specific (e.g., *Pseudomonas*) origins of replication from, e.g., *Pseudomonas savastanoi* (42) which were proposed at some point for generating a suite of new BHR vectors (43) never made it to mainstream use. One way of broadening the range of such otherwise narrow-range plasmids is to join origins that function alternatively in two or more hosts in the same DNA segment, as has been done with the originally *P. aeruginosa*-only plasmid pRO1614 (44, 45) and the *E. coli/Pseudomonas putida/Bacillus* shuttle named pEBP (46). However, despite some incursions into other functional origins of replication, vectors based on RSF1010, RK2, BBR1, and pSa have been the pillars of the genetic engineering of Gram-negative organisms other than *E. coli* for a long time, and now they remain the basis of standardized modules for assembly of synthetic devices for the same type of organisms (23, 24) (http://parts.igem.org). Alas, the same standardization effort has been more limited thus far in the Gram-positive realm, although some exciting developments are under way at the time of writing this article (47, 48).

SELECTION MARKERS

Antibiotic resistance genes have been systematically the favorite resource for enabling selection of recombinant plasmids. By adding the cognate agent to the selection medium, one can not only pick out transformants after DNA ligation or assembly but also ensure retention of a given plasmid if no endogenous countersegregation system is in operation. A variety of resistance genes for ampicillin (*bla*, Ap) tetracycline (*tet*, Tc), chloramphenicol (*cat*, Cm), kanamycin (*kan*, Km), streptomycin/spectinomycin (Sm/Sp), and gentamycin (Gm) have been retrieved repeatedly from their original natural context (often a clinical isolate) and reused in a large variety of vectors (49, 50). More recently, their corresponding sequences have been formatted as modules for assembling all types of synthetic biology constructs (23, 24) (http://parts.igem.org). But the ease of use of these markers has two major drawbacks for applications involving industrial or environmental settings. First, the products of these genes act on antibiotics that are in clinical use (or are similar to them), and engineering bacterial biomass which carries resistance genes may thus raise safety concerns (51). Second, antibiotics themselves are expensive, and using them as a bulk additive to any large-scale growth medium might be economically prohibitive.

This has prompted the quest for alternative markers or strategies to ensure stable plasmid maintenance without having to use typical antibiotics. One such stratagem is the adoption of resistance to herbicides (e.g., bialaphos) (52) or heavy metals (e.g., tellurite) (53) as markers. Another useful approach is placing in the plasmid the sequence of an essential gene that complements a deletion of the same gene excised from the chromosome of the host strain. In this way, any bacterium that loses the plasmid is eliminated. The genes *infA* (encoding translation initiation factor 1) (54), *asd* (aspartate-beta-semialdehyde dehydrogenase)

(55), *dapA* (dihydrodipicolinate synthase) (22), and *thyA* (thymidylate synthase) (56) have been used to this end, and many others are surely eligible for the same purpose. This type of marker eliminates the need for drug resistance markers in the vector and is thus appealing whenever stable maintenance of engineered traits is required in the absence of external selection. An alternative approach to ensure recombinant plasmid retention without antibiotics relies on the refactoring of toxin-antitoxin systems (see below). In addition, users can consider the definite insertion in the chromosome of the DNA segment of choice by using one of the many mini-transposon vectors based on either Tn5 (20, 21) or Tn7 (57, 58) that have been designed to that end in recent years. Some of these transposon vectors incorporate in their design the possibility of removing the antibiotic markers employed for selection of insertion events once transposition has occurred (59, 60).

TOXIN-ANTITOXIN SYSTEMS

Although many naturally occurring plasmids, in particular the smallest, seem to be maintained on the sheer basis of their high numbers, others with lower copy doses have developed sophisticated molecular mechanisms to coordinate chromosomal and plasmid replication and/or to penalize cells that may accidentally lose the extra replicon. These so-called plasmid addiction systems (61) seem to be a special case of a more general method for protecting the integrity of the genomic complement under environmental and nutritional stress conditions (62–64). These devices systematically involve a DNA segment that encodes a toxin and its antidote. Toxin-antitoxin modules usually are composed of two cognate proteins with different expression levels and/or half lifetimes (e.g., the *parD* locus encoding Kis/Kid proteins in plasmid R1) (65). Under normal conditions, the antidote inhibits the toxin, whereas plasmid loss makes proteolytic degradation of the antidote faster, titrating the remedy out and killing the host cells. Alternatively, the antidote of the toxin-antitoxin pair is an RNA that prevents translation of a toxic gene in such a way that cells that lose the plasmid are actively eliminated by the system. The archetypal example of such a scenario is that of the *parB* system of R1, which encodes two small genes, *hok* and *sok* (66). The expression of the first (a toxin) is regulated by the complementary *hok* antisense RNA to the *hok* mRNA. Postsegregation killing in single cells is brought about by the differential decay of the *hok* and *sok* RNA. The modularity of the system stabilizes any plasmid that is inserted with a *parB*-encoding DNA segment (67). This feature has made the *parB* system a remarkable tool for ensuring plasmid maintenance in individual bacteria without any external selection (68).

A different type of toxin-antitoxin system is that of colicins (69, 70), large protein poisons that are released into the medium to kill related bacteria which may have lost the colicin locus (i.e., the DNA segment encoding the structural gene for the toxin and a colicin immunity gene) or similar strains that have a receptor for the colicin in question. Although the biological role of colicins is still debated (71), they provide a powerful tool to engineer population-wide circuits for genetic and biological containment of recombinant DNA (72, 73). In reality, one of the most sophisticated containment host-plasmid mutual systems designed thus far combines conditional plasmid replication (i.e., initiators for the R6K or ColE2-P9 origins are provided in *trans* by a specified host), rich-media-compatible auxotrophies (*dapA* and *thyA*; see above), and toxin-antitoxin pairs (22). Such constructs ensure a potentially perfect genetic firewall to engineered genetic devices without significantly affecting (at least in the short run) growth parameters.

SITE-SPECIFIC RECOMBINATION

While homologous recombination typically requires a considerable degree of DNA identity through a good length of the sequences in question, SSR occurs between shorter segments sharing limited similarity. This phenomenon, which is widespread through all biological kingdoms, is brought about by the action of the so-called site-specific recombinases on short DNA sites which they manage to recognize, cleave, exchange between DNA strands, and rejoin. This basic device, which by default just needs a recombinase and a suite of target sites, is a key component of a large collection of biological phenomena in both eukaryotic and prokaryotic organisms. One of these functions is the resolution of replication intermediates in plasmids with an active partitioning mechanism. An SSR which has been repeatedly reshaped and repurposed for genetic engineering is that of the multimer resolution system (*mrs*) of BHR plasmid RP4. This system is one of the functions encoded within the *par* region, which determines the major RP4 stabilization system. The product of the *parA* gene is a site-specific resolvase that catalyzes *recA*-independent intramolecular recombination between two directly oriented resolution (*res*) sites. If these sites flank a supercoiled DNA segment, the result of the process is the excision of the intervening sequence. Transient expression of *parA* has been exploited to delete specific DNA segments (e.g., antibiotic resistance genes) delivered with transposons to the chromosome of target bacteria (e.g., *E. coli* or *Pseudomonas*) (74). This permits the retention of heterologous DNA segments devoid of any phenotypic marker to the genome of the target bacterium, an important consideration for microorganisms destined for environmental release (75). Further plasmid-based site-specific recombination systems have been extensively employed for the same or similar purposes, e.g., the Cre-*lox* resolvase of the phage/plasmid P1 (76, 77). But others are available and well suited as tools for deep refactoring of both prokaryotic and eukaryotic systems. Among them is the β-recombination system of pSM19035 (78), a virtually orthogonal SSR that originates in the complex replication mechanism of this plasmid of Gram-positive bacteria (79, 80).

Integrons are a second type of useful SSR system often found in environmental plasmids and other mobile elements (81). They are DNA elements with an ability to capture genes with a mechanism that typically involves an integrase gene (*int*) and a nearby recombination site (*att*). The recruitable DNA is located on gene cassettes in which the functional sequences are adjacent to a matching *att* recombination site. The essence of the integron is that such DNA sequences must occur in covalently closed circular DNA but cannot be replicated as such; they can only be replicated when recombination occurs between the two *att* sites, thereby entering the cassette into the cognate target. As the *att* sites are maintained after the integration, the resulting molecular product can again act as a landing pad for further integration rounds. Furthermore, this is a reversible process, so that DNA sequences can be further excised and integrated in a very dynamic fashion. Although the activity of integrons is one of the most dreaded mechanisms of acquisition of antibiotic resistance in clinical settings (81), they provide an extraordinary platform for creating genetic diversity in synthetic biology constructs (82). In one case, a tryptophan production operon was reconstructed by separately delivering individual genes as recombination cassettes to a synthetic integron platform in *E. coli.* Integrase-mediated recombination generated a large number of genetic combinations *in vivo* with a large number of combinations of thereby arrayed genes, some with increasing production of the amino acid by 11-fold compared to the natural gene order of the same pathway (83).

In a further twist of their natural properties, SSR systems can not only be used as secondary tools for construction of synthetic biology devices but also can be engineered as actuators able to record information and implement a gene expression program. For instance, by flanking transcriptional terminators or promoters with SSR targets one can engineer complex circuits based on the control of the flow of RNAP along DNA. Integrase-mediated inversion or deletion of such transcription checkpoints allows construction of AND, NAND, OR, XOR, NOR, and XNOR logic gates that use common signals and enable design of sequential logic operations (84). Similarly, it is possible to assemble genetic circuits that use recombinases for running Boolean logic functions by means of DNA-encoded memory of events (85). One appealing possibility in this context is the engineering of circuits in which stable gene expression (output) is the result of transient environmental or chemical inputs (86). If one combines SSR-based gates with suitable reporter systems (e.g., the green fluorescent protein gene [GFP]) and PCR, it becomes possible to interrogate the state of the synthetic devices for past events.

CONJUGATION AND TRANSFER FUNCTIONS

Plasmids are the main agents of horizontal gene transfer in nature. This can occur through natural transformation (sometimes electro-transformation due to atmospheric lighting) (87, 88), but usually plasmid transit from one bacterium to another occurs through conjugation. This phenomenon includes different mechanisms of DNA delivery that require a physical contact between a donor and a recipient, often facilitated by specific pili which bridge and bring together mating partners. Complete conjugation systems share a basic layout composed of four constituents (89). The first is the so-called relaxase, the key protein in conjugation, which recognizes the second component: a short DNA sequence that acts as the *oriT*. Having a relaxase and an *oriT* in *cis* or in *trans* generally suffices for a plasmid to be conjugally transmissible (Fig. 1). The relaxase catalyzes (i) a site-specific cleavage of the *oriT* sequence in the donor, thereby producing a DNA strand that will be transferred, and (ii) the eventual religation of DNA once it reaches the recipient bacteria and transfer is completed, thereby reconstituting the whole plasmid in the new host. Depending on the plasmid in question, the relaxase may work in concert with other proteins that form a complex called relaxosome. To complete the conjugation process, one needs two additional functions. One is the formation of a mating channel through which DNA can pass from one cell to the other. This structure consists of a type IV secretion system (T4SS), which transports the relaxase attached to the DNA to be transferred. The second is the so-called type IV coupling protein (T4CP), which makes the essential connection of the relaxosome/DNA complex and the transport channel. In addition, some conjugative plasmids encode structural and regulatory genes for formation of pili on the cell surface, which eases and tightens the physical connection between the donor and the recipient.

Typical self-transmissible environmental plasmids (e.g., RK2/RP4 or R388) encode the entire complement (*oriT/tra/mob* genes) for all functions necessary to bring about the complete transfer process. Other plasmids may carry only the relaxosomal components (i.e., *oriT*, the relaxase gene, and auxiliary proteins). In this case (e.g., RSF1010) the DNA can initiate its own transfer but has to be complemented by functions of other conjugative plasmids for finishing the task. Finally, other plasmids may just bear DNA sequences that can be recognized by relaxases encoded by other replicons and used as *oriT* for lateral transfer.

Some components of the conjugation and mobilization machineries just mentioned have been repurposed for diverse genetic engineering and synthetic biology applications. The most salient involves the inclusion of the RK2/RP4 *oriT* in a variety of plasmid vectors for making them mobilizable to new hosts by means of the transfer functions of the same plasmid provided in *trans* (Fig. 1). Note that this conjugative system is the most promiscuous known, as it is able not only to move plasmids among bacteria, but also to deliver them to yeasts and mammalian cells (90). Such a broad capability has been formatted in various ways. The first methods to mobilize *oriT*-containing plasmids from a donor host (typically *E. coli*) to other Gram-negative recipient bacteria required the action of a third mating partner which transiently provided such *tra* functions to the donor but could not replicate in the final recipient (91). The products encoded by the helper plasmid could thereby escort the transfer of the *oriT* construct without itself being inherited in the destination strain (Fig. 2). In a subsequent advance, specialized strains

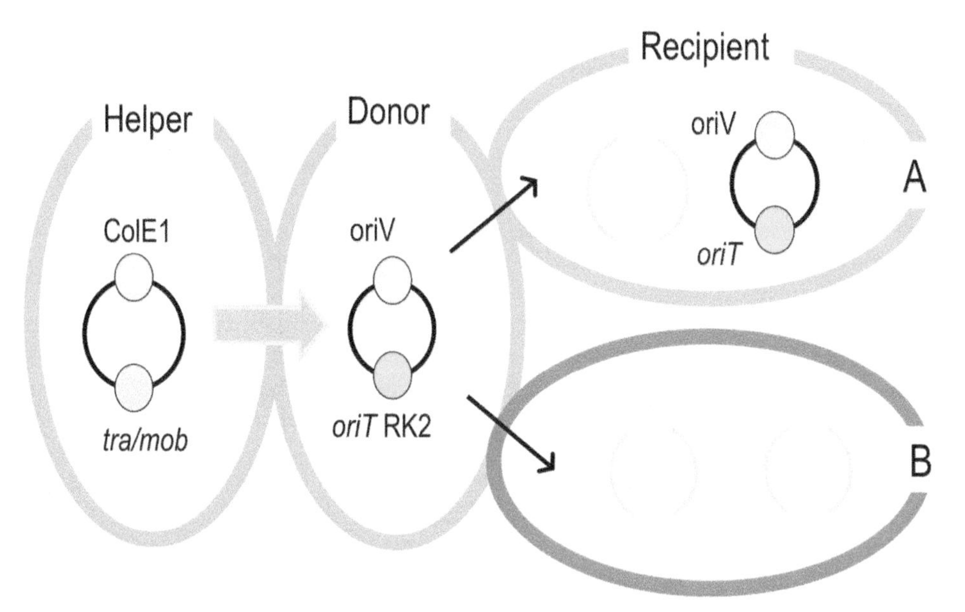

Figure 2 Tripartite matings for plasmid mobilization. The transfer of an *oriT*-containing plasmid from a donor to a target recipient can be brought about by setting a mating between the two partners (donor and recipient) plus a helper strain that transiently delivers the functions that are necessary for the passage of the plasmid from one to the other. If the *oriV* replication origin of the thereby transferred plasmid is BHR, its DNA can further proliferate in the non-*E. coli* recipient cells of the triparental mating (case A). The helper plasmid is not inherited in any case, as its replication origin (e.g., ColE1 in this example) is not functional in nonenteric recipients. Alternatively, neither the plasmid of interest nor the helper construct can replicate in the destination strain (case B), thereby providing an excellent scenario for suicide delivery, e.g., for selecting insertions of a transposon borne by the mobilized plasmid. doi:10.1128/microbiolspec.PLAS-0033-2014.f2

were designed such that the *tra* functions of RP4 were integrated in the chromosome of a donor *E. coli* strain where the *oriT*-plasmids could be placed and poised for conjugation with available target bacteria (92). This allowed users to get the desired plasmid transfer by setting a simple two-partner mating (rather than the triparental mating above). A side benefit of such a scheme was the possibility to combine plasmid mobilization with conditional replication of the same plasmid (possible in the donor bacterium, not possible in the recipient) as a way to deliver transposons engineered in the same circular DNA frame (18). Alas, the genetic technology available at the time obliged the donor bacteria to carry the Mu phage also (as a carrier of the RP4 *tra* functions), thus creating a chance of artifactual insertions and misleading phenotypes. Only recently has this problem been completely solved by engineering an *E. coli* Mu-free donor strain that stably harbors all of the RP4 conjugative functions and sustains replication of Π-dependent suicide vectors (93) (see above).

However, conjugation has been exploited in synthetic biology not only for the sake of using its functional parts, but also as a phenomenon that can be utilized for biological computing and digital cell-to-cell communication (94). One can see DNA transfer as a scenario where individual bacterial types perform specific subtasks, the results of which are then communicated to other cell types for further processing. Specific cell-cell conjugation allows direct transfer of genetic information in a digital fashion; i.e., cells in a population can only have or not have a given plasmid. Mixing different strains in a single population allows a multicellular pool to execute the Boolean XOR function. Models predict this approach to be more powerful than other biological computation approaches using, e.g., quorum sensing. A conjugation-wired computing system could thus become a realistic methodology to implement new-to-nature properties in biological systems (94).

To finish this section we need to refer to the so-called integrative and conjugative elements (ICEs), which combine interesting properties of transfer systems and site-specific recombination in the same DNA segment. ICEs are mobile genetic elements that by default stay in the host chromosome but which can occasionally be activated to adopt a plasmid-like form that can be transferred between cells by conjugation (95). ICEs can encode genes for antibiotic resistance, metabolic functions, virulence factors, and many other advantageous traits. The organization of ICEs varies enormously, but *inter alia* they encode an integrase that plays a role in different steps of the element's life cycle. One interesting ICE specimen is the so-called ICE*clc*, an element found in *Pseudomonas knackmussii* B13 that provides the host with the capacity to metabolize chlorocatechols (95). The integrase activity of this ICE has been recently used for developing a genetic tool able to stably implant large DNA fragments into one or more specific sites of a target chromosome (96). The genetic kit to this end involves only a cloning vector and a mini-transposable element with which large DNA segments engineered in *E. coli* can be tagged with the integrase gene. The system has been instrumental in introducing cosmids and bacterial artificial chromosome DNAs from *E. coli* into *P. putida* in a site-specific manner. These developments illustrate how fundamental knowledge of the different steps and components of horizontal gene transfer (HGT) become a treasure trove of building blocks for biosystems engineering.

EXPRESSION SYSTEMS

Environmental plasmids found in bacteria that inhabit niches with a history of pollution by industrial chemicals frequently carry in their DNA one or more gene clusters that expand the range of compounds that can be used as carbon sources (97). It is often the case that transcription of such genes responds to either the substrates of the corresponding pathways or to some of their metabolic intermediates. The bottom line is that in the very competitive environments where these bacteria thrive, expression of extra genes must be tightly controlled for being triggered only when and where they are needed. This scenario is thus an excellent ground in which to mine pairs of transcriptional factors (TFs)/cognate chemical inducers which can be reformatted for creating new conditional expression systems (98). This approach is helping to overcome the dearth of alternative inducible devices for synthetic biology circuits (Fig. 3), which is thus far limited to those typically used in *E. coli*: LacI/*Plac* (and its many variants), the thermo-inducible cI-repressed λ phage promoters, AraC/P_{BAD}, TetR/*Ptet* (24), and more recently, the RhaR/*Prha* system (99), either by themselves or in combination with the T7 RNAP.

A number of alternative effector-triggered systems from various origins have been assembled for utilization in a variety of hosts including (but not limited to) *E. coli*. The basic expression node includes a low-activity promoter that transcribes the effector-responsive TF gene and the target cognate promoter that is activated by the TF/inducer complex. Promoters regulated by TFs of this kind endowed with various degrees of standardization include AlkS/*PalkB* (activated by *n*-octane) (100, 101), NahR/*Psal* (activated by salicylate) (102), CprK/*Pcpr*

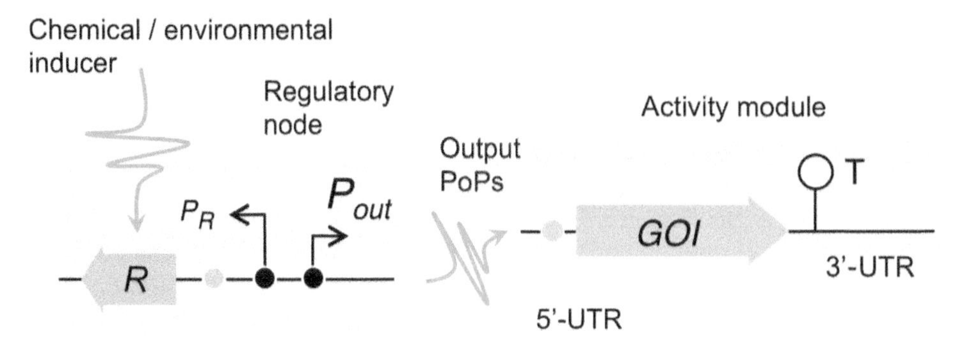

Figure 3 The basic organization of synthetic biology constructs. Genetic devices are usually composed of a regulatory node that includes a transcriptional factor (an activator or a repressor) responsive to a physicochemical signal (e.g., a chemical inducer) that triggers its ability to stimulate transcription. *R*, gene encoding the regulatory protein; P_R, promoter of the regulatory *R* gene; T, transcriptional terminator; UTR, untranslated regions of mRNA. The output of the regulatory node is a given level of PoPS (polymerase per second), which represents the count of RNAP molecules that pass through the promoter DNA each second. PoPS is then wired to a gene of interest (*GOI*), which is punctuated by 5′-UTR and 3′-UTR regions. doi:10.1128/microbiolspec.PLAS-0033-2014.f3

(activated by *o*-chlorophenylacetic acid) (103), ChnR/*PchnB* (activated by cyclohexanone) (104), and MekR/*PmekA* (activated by methylethyl ketone) (105). The benzoate-inducible pair XylS/*Pm*, which stems from the toluene/*m*-xylene degradation plasmid pWW0 of *P. putida*, has been the basis of a large number of recombinant expression systems both in plasmids (106–108) and in transposon vectors (60, 102). It is noteworthy that the first rationally engineered feed-forward loop described in the literature was built on the basis of genetically rewiring two salicylate-responsive regulators for forming a transcriptional amplification cascade (109, 110). Another regulator of the pWW0 plasmid, the *m*-xylene-responsive TF called XylR, has also been adopted alone or in combination with the T7 RNAP for engineering whole-cell biosensors of aromatic compounds (111–113).

The list of chemically inducible TFs that can be exploited to the same or similar ends is very large (114, 115) and likely to be expanded in the near future. An added bonus of these expression systems is that owing to being encoded by promiscuous plasmids, they are likely to be operative in a large variety of hosts. Furthermore, when the devices are placed in a host without the corresponding metabolic genes, then the biochemical network of the recipient is usually *blind* to the effector, and the induction method has minimal or no impact (i.e., is orthogonal) on the host physiology.

Apart from transcriptional regulators, plasmids often also encode antisense RNAs (asRNAs) involved in various functions that can be recruited for engineering post-transcriptional circuits and expression systems based on RNA (116–118). To this end, a large number of *cis*-encoded asRNAs can be found in plasmids, phages, and transposons (119). Some of these asRNAs have been repurposed for engineering gene silencing (120, 121) and artificially halting bacteriophage infections (122, 123).

REPOSITORIES OF PLASMID-RELATED PARTS AND DEVICES: THE SEVA COLLECTION

One of the key synthetic biology tenets is the adoption of rules for the physical (and, wherever possible, functional) composition of biological parts to create devices with clear boundaries and defined input-output transfer functions (124). The advantages, downsides, and challenges of this view have been extensively discussed elsewhere (4) and will not be addressed here. Different synthetic biology communities have adopted different strategies for building increasingly complex genetic constructs on the basis of their own repositories of biological components. Table 2 shows a nonexhaustive list of web resources where some of the most popular materials available for synthetic biology purposes are compiled. Note that each collection has a different degree of curation and reliability and also a very diverse retrieve policy—from complete open access to pay-per-request to intricate material transfer agreement (MTA) procedures. One of the most accessed is the Registry of Standard Biological Parts (http://parts.igem.org). This collection, founded in 2003 at MIT, has accumulated a

Table 2 Databases, repositories, and suppliers of synthetic biology parts and devices

Database/repository webpage	Description/application
Public/open source/nonprofit	
http://parts.igem.org	The largest collection of standardized synthetic biology parts and devices. Not a curated repository
http://biobricks.org	Large number of resources for the open source synthetic biology community
http://seva.cnb.csic.es	Curated repository of standardized plasmid vectors for engineering non-*E. coli* bacteria
http://www.addgene.org	The largest distributor of vectors, plasmids, and cloned genes to the academic community
http://www.shigen.nig.ac.jp	Cloning vectors for *E. coli*, preserved as purified DNA samples
http://bccm.belspo.be	Strains, plasmids, libraries, and other biological materials contributed by Belgian researchers
http://www.bioss.uni-freiburg.de/toolbox	Curated collection of vectors (mostly plasmids) for biosystems engineering
http://www.science.co.il	Compilation of databases of mobile genetic elements, plasmids, and vectors
http://dnasu.org	Repository for plasmid clones and collections with links to collections of individual researchers
https://public-registry.jbei.org	Software and platform for managing information about biological parts and host strains
http://www.beiresources.org	Biological materials and tools with emphasis on infectious diseases
Commercial/private	
https://www.snapgene.com	Software for simulation and guiding of DNA cloning in different types of vectors
http://cms.plasmidfactory.com/en/	Plasmid vectors made *á la carte* for specific bioengineering projects
http://www.lgcstandards-atcc.org	ATCC-linked repository of biological materials including vectors and strains
http://www.lifesciences.sourcebioscience.com	Tailored DNA clones, vectors, and bioinformatic services
http://www.promega.es/products/vectors	Large collection of vectors and other genetic tools for gene expression and reporter systems
http://www.bioclone.us	Commercial provider of molecular tools for optimization of gene expression
http://plasmid.med.harvard.edu/PLASMID/	Harvard-based provider of sequence-verified plasmid constructs
https://www.neb.com	Materials and general support for iGEM projects following standardized assembly methods
http://www.lifetechnologies.com	Chemical synthesis, cloning, and sequence verification of genetic sequences *á la carte*
http://www.genscript.com	Gene synthesis, DNA libraries, and building blocks for synthetic biology
http://www.labguru.com/features/molecular/plasmids	Plasmid vector database along with tools for cloning projects and plasmid retrieval

very large number of parts and devices that have been employed in educational synthetic biology projects (iGEM: http://igem.org). Although the collection is mostly used and produced by undergraduate students, and quality control is not optimal, the large number of items and their inclusion as standardized parts has made the registry a very valuable resource for many synthetic biology endeavors (125–128).

The Standard European Vector Architecture database (SEVA-DB) deserves a separate comment (Fig. 4). This platform is both a web resource and a compendium of standardized and modular plasmid vectors. This collection originated in our efforts to expand the realm of synthetic biology toward Gram-negative bacteria different from *E. coli* but endowed with a biotechnological or environmental value. The idea and design behind this plasmid vector compilation is simple but has strict rules. This repository is composed of vectors that have up to four different modules decorating a fixed plasmid backbone. The rules of the SEVA compositional standard are as follows: (i) the DNA elements that comprise the plasmids have to be devoid of certain restriction sites (for specific details see reference 23; http://seva.cnb.csic.es), and (ii) the different modules have to be flanked by other sets of rare restriction sequences. In brief, the plasmid backbone is common to all constructs and includes an origin of transfer (*oriT*) and two transcriptional terminators (T_1 and T_0) to keep the adjacent DNA segments transcriptionally insulated. A typical plasmid from the collection is composed of an antibiotic resistance marker module, an origin of replication segment, and a cargo part. Optimally, vectors may also carry a genetic gadget to endow plasmids with additional properties. The antibiotic module comprises the six most common antibiotic resistance markers, and the origin of replication can be one out of nine choices with different functionalities in terms of host range (broad and narrow) and copy number (low, medium, and high copy), leaving their optimal combination to the user's choice. The cargo

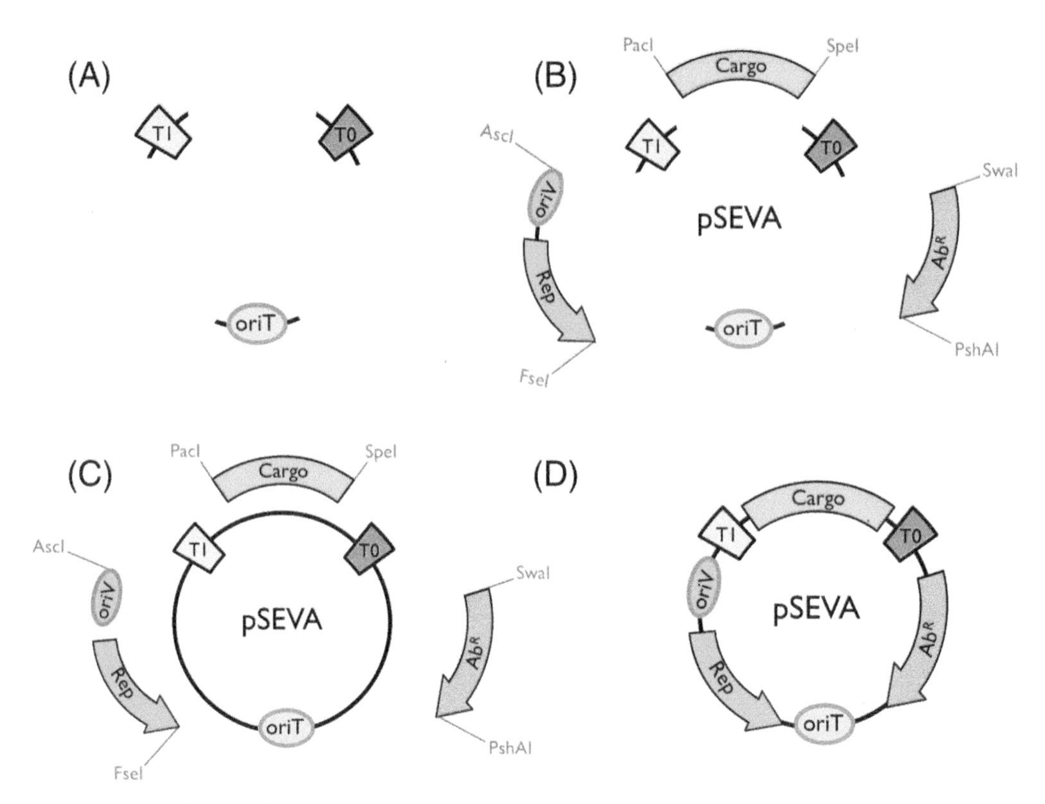

Figure 4 The Standard European Vector Architecture (SEVA) pipeline. (A) The organization of all SEVA constructs is framed within three basic DNA parts of reference, i.e., two transcriptional terminators and one *oriT*. (B) The positions of these parts leave three openings flanked by specific and unusual restriction sites which are then used to insert synthetic DNA fragments encoding an antibiotic resistance marker, an origin of replication, and a cargo segment. The orientation and specifications that such segments must follow to acquire the SEVA standard are described at http://seva.cnb.csic.es. (C) Segments are assembled on the formatted frame, and (D) a pSEVA vector is added to the database and to the material repository. doi:10.1128/microbiolspec.PLAS-0033-2014.f4

endows functionalities to the plasmid, e.g., by means of a polylinker, different promoter-probe elements, or different expression systems. Since modules are punctuated by uncommon restriction sites, their exchange for creation of different plasmid combinations is straightforward. The collection is updated in real time and is increasingly adopting the synthetic biology open language (SBOL) for describing all vectors of the SEVA repository in a language that can be interfaced with other platforms and understood by robotic DNA synthesis platforms (129). While the manifest destiny of genetic constructs is having them made through chemical synthesis (for which assembly vectors and material DNA repositories may not be necessary any longer), these collections of synthetic biology building blocks will still be required for some time, in particular for addressing basic biological questions and engineering simple phenotypes in non-*E. coli* bacteria.

OUTLOOK

As discussed above, environmental plasmids offer synthetic biology a diverse palette of gene-encoded activities that once excised from its natural context, minimized, and standardized, become an extraordinary asset for a wide range of applications. But there is surely more that plasmid biology can deliver beyond providing building blocks for genetic constructs. First, as mentioned earlier, plasmids, especially those that are promiscuous, move through the microbial population, and their encoded genes are to be expressed in different types of hosts. What determines such ability? Both the plasmid-borne promoters involved, their regulation, and the proteins they encode must operate in a fashion that is relatively independent of the host, thereby creating a natural case of implant-chassis orthogonality that must be ruled by principles that deserve further research. Second, the acquisition of environmental plasmids encoding new metabolic capabilities is merely a natural case of biochemical engineering where the preexisting molecular network of the receiving cells has to cope with implantation of a new genetic and enzymatic system (97). Studies of such scenarios (e.g., the interplay of the pWW0-encoded system for toluene and *m*-xylene degradation and the *P. putida* host) reveal both the type of biochemical conflicts that such situations create and how bacteria have found a solution to them (130, 131). More instances of this sort could be worth examining for the sake of handling the problem of implant-chassis retroactivity, one of the central challenges in biological engineering (132, 133).

Finally, once plasmids make it to a new host, their encoded proteins have to be placed somewhere in the tridimensional architecture of the receiving cells, where molecular crowding imposes considerable constraints. Are plasmid-encoded functions, cognate protein structures, and processes associated with such three-dimensional addresses? It will be important to classify plasmid-encoded proteins into those that distribute evenly versus those that go to a specific cell site and to see what the difference is, thereby understanding the principles that may rule such a positioning. In this sense, promiscuous environmental plasmids will continue enriching the material toolbox for engineering biological systems as well as providing new conceptual insights on how to program microorganisms to do (new) things that they normally do not do.

Acknowledgments. *The work in the authors' laboratory was supported by the BIO program of the Spanish Ministry of Economy and Competitiveness, the ST-FLOW, ARISYS, and EVOPROG contracts of the EU, the ERANET-IB Program, and funding from the Autonomous Community of Madrid (PROMPT). Conflicts of interest: We declare no conflicts.*

Citation. Martínez-García E, Benedetti I, Hueso A, de Lorenzo V. 2015. Mining environmental plasmids for synthetic biology parts and devices. Microbiol Spectrum 3(1): PLAS-0033-2014.

References

1. **Bio FAB Group, Baker D, Church G, Collins J, Endy D, Jacobson J, Keasling J, Modrich P, Smolke C, Weiss R.** 2006. Engineering life: building a fab for biology. *Sci Am* **294:**44–51.
2. **Endy D.** 2005. Foundations for engineering biology. *Nature* **438:**449–453.
3. **de Lorenzo V.** 2011. Beware of metaphors: chasses and orthogonality in synthetic biology. *Bioeng Bugs* **2:** 3–7.
4. **Porcar M, Danchin A, de Lorenzo V.** 2015. Confidence, tolerance and allowance in biological engineering: the nuts and bolts of living things. *BioEssays*. doi:10.1002/bies.201400091.
5. **Purnick PEM, Weiss R.** 2009. The second wave of synthetic biology: from modules to systems. *Nat Rev Mol Cell Biol* **10:**410–422.
6. **Alon U.** 2006. *An Introduction to Systems Biology: Design Principles of Biological Circuits*. CRC Press, London.
7. **Ramos JL, Martinez-Bueno M, Molina-Henares AJ, Teran W, Watanabe K, Zhang X, Gallegos MT, Brennan R, Tobes R.** 2005. The TetR family of transcriptional repressors. *Microbiol Mol Biol Rev* **69:**326–356.
8. **Konieczny I, Bury K, Wawrzycka A, Wegrzyn K.** 2015. Iteron plasmids. *In* Tolmasky ME, Alonso JC (ed), *Plasmids: Biology and Impact in Biotechnology and Discovery*. ASM Press, Washington, DC, in press.
9. **Lilly J, Camps M.** 2015. Mechanisms of theta plasmid replication. *In* Tolmasky ME, Alonso JC (ed), *Plasmids: Biology and Impact in Biotechnology and Discovery*. ASM Press, Washington, DC, in press.
10. **Basu A, Phale PS.** 2008. Conjugative transfer of preferential utilization of aromatic compounds from *Pseudomonas putida* CSV86. *Biodegradation* **19:**83–92.
11. **Basu A, Shrivastava R, Basu B, Apte SK, Phale PS.** 2007. Modulation of glucose transport causes preferential utilization of aromatic compounds in *Pseudomonas putida* CSV86. *J Bacteriol* **189:**7556–7562.
12. **Shrivastava R, Basu B, Godbole A, Mathew MK, Apte SK, Phale PS.** 2011. Repression of the glucose-inducible outer-membrane protein OprB during utilization of aromatic compounds and organic acids in *Pseudomonas putida* CSV86. *Microbiology* **157:**1531–1540.
13. **Trefault N, De la Iglesia R, Molina AM, Manzano M, Ledger T, Perez-Pantoja D, Sanchez MA, Stuardo M, Gonzalez B.** 2004. Genetic organization of the catabolic plasmid pJP4 from *Ralstonia eutropha* JMP134 (pJP4) reveals mechanisms of adaptation to chloroaromatic pollutants and evolution of specialized chloroaromatic degradation pathways. *Environ Microbiol* **6:**655–668.
14. **Hansen LH, Bentzon-Tilia M, Bentzon-Tilia S, Norman A, Rafty L, Sorensen SJ.** 2011. Design and synthesis of a quintessential self-transmissible IncX1 plasmid, pX1.0. *PloS One* **6:**e19912. doi:10.1371/journal.pone.0019912.
15. **Bentzon-Tilia M, Sørensen S, Hansen L.** 2014. Synthetic plasmid biology, p 1–5. *In* Bell E (ed), *Molecular Life Sciences*. Springer, New York.
16. **Sambrook J, Fritsch EF, Maniatis T.** 1989. *Molecular Cloning*, vol. 2. Cold Spring Harbor Laboratory Press, New York.
17. **Kolter R, Inuzuka M, Helinski DR.** 1978. Trans-complementation-dependent replication of a low molecular weight origin fragment from plasmid R6K. *Cell* **15:** 1199–1208.
18. **Miller VL, Mekalanos JJ.** 1988. A novel suicide vector and its use in construction of insertion mutations: osmoregulation of outer membrane proteins and virulence determinants in *Vibrio cholerae* requires *toxR*. *J Bacteriol* **170:**2575–2583.
19. **de Lorenzo V, Herrero M, Jakubzik U, Timmis KN.** 1990. Mini-Tn5 transposon derivatives for insertion mutagenesis, promoter probing, and chromosomal

insertion of cloned DNA in Gram-negative eubacteria. *J Bacteriol* **172:**6568–6572.

20. **Martinez-Garcia E, Calles B, Arevalo-Rodriguez M, de Lorenzo V.** 2011. pBAM1: an all-synthetic genetic tool for analysis and construction of complex bacterial phenotypes. *BMC Microbiol* **11:**38.
21. **Martinez-Garcia E, de Lorenzo V.** 2012. Transposon-based and plasmid-based genetic tools for editing genomes of Gram-negative bacteria. *Methods Mol Biol* **813:**267–283.
22. **Wright O, Delmans M, Stan GB, Ellis T.** 2014. GeneGuard: a modular plasmid system designed for biosafety. *ACS Synth Biol.* doi:10.1021/sb500234s.
23. **Silva-Rocha R, Martinez-Garcia E, Calles B, Chavarria M, Arce-Rodriguez A, de Las Heras A, Paez-Espino AD, Durante-Rodriguez G, Kim J, Nikel PI, Platero R, de Lorenzo V.** 2013. The Standard European Vector Architecture (SEVA): a coherent platform for the analysis and deployment of complex prokaryotic phenotypes. *Nucleic Acids Res* **41:**D666–D675.
24. **Lee TS, Krupa RA, Zhang F, Hajimorad M, Holtz WJ, Prasad N, Lee SK, Keasling JD.** 2011. BglBrick vectors and datasheets: a synthetic biology platform for gene expression. *J Biol Eng* **5:**12.
25. **Bagdasarian M, Lurz R, Rückert B, Franklin F, Bagdasarian M, Frey J, Timmis K.** 1981. Specific-purpose plasmid cloning vectors. II. Broad host range, high copy number, RSF 1010-derived vectors, and a host-vector system for gene cloning in *Pseudomonas*. *Gene* **16:**237–247.
26. **Labes M, Pühler A, Simon R.** 1990. A new family of RSF1010-derived expression and *lac* fusion broad-host-range vectors for Gram-negative bacteria. *Gene* **89:** 37–46.
27. **Scholz P, Haring V, Wittmann-Liebold B, Ashman K, Bagdasarian M, Scherzinger E.** 1989. Complete nucleotide sequence and gene organization of the broad-host-range plasmid RSF1010. *Gene* **75:**271–288.
28. **Meyer R.** 2009. Replication and conjugative mobilization of broad host-range IncQ plasmids. *Plasmid* **62:** 57–70.
29. **Katashkina JI, Kuvaeva TM, Andreeva IG, Skorokhodova AY, Biryukova IV, Tokmakova IL, Golubeva LI, Mashko SV.** 2007. Construction of stably maintained non-mobilizable derivatives of RSF1010 lacking all known elements essential for mobilization. *BMC Biotechnol* **7:**80.
30. **Kolatka K, Kubik S, Rajewska M, Konieczny I.** 2010. Replication and partitioning of the broad-host-range plasmid RK2. *Plasmid* **64:**119–134.
31. **Bierman M, Logan R, O'Brien K, Seno ET, Rao RN, Schoner BE.** 1992. Plasmid cloning vectors for the conjugal transfer of DNA from *Escherichia coli* to *Streptomyces* spp. *Gene* **116:**43–49.
32. **Ditta G, Schmidhauser T, Yakobson E, Lu P, Liang XW, Finlay DR, Guiney D, Helinski DR.** 1985. Plasmids related to the broad host range vector, pRK290, useful for gene cloning and for monitoring gene expression. *Plasmid* **13:**149–153.
33. **Santos PM, Di Bartolo I, Blatny JM, Zennaro E, Valla S.** 2001. New broad-host-range promoter probe vectors based on the plasmid RK2 replicon. *FEMS Microbiol Lett* **195:**91–96.
34. **Antoine R, Locht C.** 1992. Isolation and molecular characterization of a novel broad-host-range plasmid from *Bordetella bronchiseptica* with sequence similarities to plasmids from Gram-positive organisms. *Mol Microbiol* **6:**1785–1799.
35. **Kovach ME, Elzer PH, Hill DS, Robertson GT, Farris MA, Roop RM 2nd, Peterson KM.** 1995. Four new derivatives of the broad-host-range cloning vector pBBR1MCS, carrying different antibiotic-resistance cassettes. *Gene* **166:**175–176.
36. **Lale R, Brautaset T, Valla S.** 2011. Broad-host-range plasmid vectors for gene expression in bacteria. *Methods Mol Biol* **765:**327–343.
37. **Obranic S, Babic F, Maravic-Vlahovicek G.** 2013. Improvement of pBBR1MCS plasmids, a very useful series of broad-host-range cloning vectors. *Plasmid* **70:** 263–267.
38. **Tait RC, Close TJ, Rodriguez RL, Kado CI.** 1982. Isolation of the origin of replication of the IncW-group plasmid pSa. *Gene* **20:**39–49.
39. **Close TJ, Zaitlin D, Kado CI.** 1984. Design and development of amplifiable broad-host-range cloning vectors: analysis of the *vir* region of *Agrobacterium tumefaciens* plasmid pTiC58. *Plasmid* **12:**111–118.
40. **DeFeyter R, Kado CI, Gabriel DW.** 1990. Small, stable shuttle vectors for use in *Xanthomonas*. *Gene* **88:** 65–72.
41. **Gruber S, Hagen J, Schwab H, Koefinger P.** 2014. Versatile and stable vectors for efficient gene expression in *Ralstonia eutropha* H16. *J Biotechnol* **186C:**74–82.
42. **Nieto C, Giraldo R, Fernandez-Tresguerres E, Diaz R.** 1992. Genetic and functional analysis of the basic replicon of pPS10, a plasmid specific for *Pseudomonas* isolated from *Pseudomonas syringae* patovar *savastanoi*. *J Mol Biol* **223:**415–426.
43. **Nieto C, Fernandez-Tresguerres E, Sanchez N, Vicente M, Diaz R.** 1990. Cloning vectors, derived from a naturally occurring plasmid of *Pseudomonas savastanoi*, specifically tailored for genetic manipulations in *Pseudomonas*. *Gene* **87:**145–149.
44. **Lee BU, Hong JH, Kahng HY, Oh KH.** 2006. Construction of an *Escherichia-Pseudomonas* shuttle vector containing an aminoglycoside phosphotransferase gene and a lacZ′′ Gene for alpha-complementation. *J Microbiol* **44:**671–673.
45. **West SE, Schweizer HP, Dall C, Sample AK, Runyen-Janecky LJ.** 1994. Construction of improved *Escherichia-Pseudomonas* shuttle vectors derived from pUC18/19 and sequence of the region required for their replication in *Pseudomonas aeruginosa*. *Gene* **148:**81–86.
46. **Troeschel SC, Thies S, Link O, Real CI, Knops K, Wilhelm S, Rosenau F, Jaeger KE.** 2012. Novel broad host range shuttle vectors for expression in *Escherichia coli*, *Bacillus subtilis* and *Pseudomonas putida*. *J Biotechnol* **161:**71–79.

47. **Sorg RA, Kuipers OP, Veening JW.** 2014. Gene expression platform for synthetic biology in the human pathogen *Streptococcus pneumoniae. ACS Synth Biol.* doi: 10.1021/sb500229s.
48. **Overkamp W, Beilharz K, Detert Oude Weme R, Solopova A, Karsens H, Kovacs A, Kok J, Kuipers OP, Veening JW.** 2013. Benchmarking various green fluorescent protein variants in *Bacillus subtilis, Streptococcus pneumoniae*, and *Lactococcus lactis* for live cell imaging. *Appl Environ Microbiol* **79:**6481–6490.
49. **Foster T.** 1983. Plasmid-determined resistance to antimicrobial drugs and toxic metal ions in bacteria. *Microbiolol Rev* **47:**361.
50. **Reece KS, Phillips GJ.** 1995. New plasmids carrying antibiotic-resistance cassettes. *Gene* **165:**141–142.
51. **Carattoli A.** 2013. Plasmids and the spread of resistance. *Int J Med Microbiol* **303:**298–304.
52. **Herrero M, de Lorenzo V, Timmis KN.** 1990. Transposon vectors containing non-antibiotic resistance selection markers for cloning and stable chromosomal insertion of foreign genes in Gram-negative bacteria. *J Bacteriol* **172:**6557–6567.
53. **Sanchez-Romero JM, Diaz-Orejas R, De Lorenzo V.** 1998. Resistance to tellurite as a selection marker for genetic manipulations of *Pseudomonas* strains. *Appl Environ Microbiol* **64:**4040–4046.
54. **Hägg P, De Pohl JW, Abdulkarim F, Isaksson LA.** 2004. A host/plasmid system that is not dependent on antibiotics and antibiotic resistance genes for stable plasmid maintenance in *Escherichia coli. J Biotechnol* **111:**17–30.
55. **Nakayama K, Kelly SM, Curtiss R.** 1988. Construction of an ASD$^+$ expression-cloning vector: stable maintenance and high level expression of cloned genes in a *Salmonella vaccine strain. Nat Biotechol* **6:**693–697.
56. **Ross P, O'Gara F, Condon S.** 1990. Thymidylate synthase gene from *Lactococcus lactis* as a genetic marker: an alternative to antibiotic resistance genes. *Appl Environ Microbiol* **56:**2164–2169.
57. **Choi KH, Gaynor JB, White KG, Lopez C, Bosio CM, Karkhoff-Schweizer RR, Schweizer HP.** 2005. A Tn7-based broad-range bacterial cloning and expression system. *Nat Methods* **2:**443–448.
58. **Koch B, Jensen LE, Nybroe O.** 2001. A panel of Tn7-based vectors for insertion of the gfp marker gene or for delivery of cloned DNA into Gram-negative bacteria at a neutral chromosomal site. *J Microbiol Methods* **45:**187–195.
59. **Jittawuttipoka T, Buranajitpakorn S, Fuangthong M, Schweizer HP, Vattanaviboon P, Mongkolsuk S.** 2009. Mini-Tn7 vectors as genetic tools for gene cloning at a single copy number in an industrially important and phytopathogenic bacteria, *Xanthomonas* spp. *FEMS Microbiol Lett* **298:**111–117.
60. **Nikel PI, de Lorenzo V.** 2013. Implantation of unmarked regulatory and metabolic modules in Gram-negative bacteria with specialised mini-transposon delivery vectors. *J Biotechnol* **163:**143–154.
61. **Rawlings DE.** 1999. Proteic toxin-antitoxin, bacterial plasmid addiction systems and their evolution with special reference to the pas system of pTF-FC2. *FEMS Microbiol Lett* **176:**269–277.
62. **Buts L, Lah J, Dao-Thi M-H, Wyns L, Loris R.** 2005. Toxin–antitoxin modules as bacterial metabolic stress managers. *Trends Biochem Sci* **30:**672–679.
63. **Gerdes K, Christensen SK, Lobner-Olesen A.** 2005. Prokaryotic toxin-antitoxin stress response loci. *Nat Rev Microbiol* **3:**371–382.
64. **Hernández-Arriaga AM, Chan WT, Espinosa M, Díaz-Orejas R.** 2015. Conditional activation of toxin-antitoxin systems: postsegregational killing and beyond. *In* Tolmasky ME, Alonso JC (ed), *Plasmids: Biology and Impact in Biotechnology and Discovery.* ASM Press, Washington, DC, in press.
65. **de la Cueva-Méndez G, Pimentel B.** 2007. Gene and cell survival: lessons from prokaryotic plasmid R1. *EMBO Rep* **8:**458–464.
66. **Thisted T, Gerdes K.** 1992. Mechanism of post-segregational killing by the *hok/sok* system of plasmid R1: Sok antisense RNA regulates *hok* gene expression indirectly through the overlapping *mok* gene. *J Mol Biol* **223:**41–54.
67. **Gerdes K.** 1988. The parB (hok/sok) locus of plasmid R1: a general purpose plasmid stabilization system. *Nat Biotechnol* **6:**1402–1405.
68. **Galen JE, Nair J, Wang JY, Wasserman SS, Tanner MK, Sztein MB, Levine MM.** 1999. Optimization of plasmid maintenance in the attenuated live vector vaccine strain *Salmonella typhi* CVD 908-*htrA. Infect Immun* **67:** 6424–6433.
69. **Jakes KS, Cramer WA.** 2012. Border crossings: colicins and transporters. *Annu Rev Genet* **46:**209–231.
70. **Papadakos G, Wojdyla JA, Kleanthous C.** 2012. Nuclease colicins and their immunity proteins. *Q Rev Biophys* **45:**57–103.
71. **Cascales E, Buchanan SK, Duche D, Kleanthous C, Lloubes R, Postle K, Riley M, Slatin S, Cavard D.** 2007. Colicin biology. *Microbiol Mol Biol Rev* **71:**158–229.
72. **Munthali MT, Timmis KN, Diaz E.** 1996. Use of colicin E3 for biological containment of microorganisms. *Appl Environ Microbiol* **62:**1805–1807.
73. **Torres B, Jaenecke S, Timmis KN, Garcia JL, Diaz E.** 2003. A dual lethal system to enhance containment of recombinant micro-organisms. *Microbiology* **149:**3595–3601.
74. **Kristensen CS, Eberl L, Sanchez-Romero JM, Givskov M, Molin S, de Lorenzo V.** 1995. Site-specific deletions of chromosomally located DNA segments with the multimer resolution system of broad-host-range plasmid RP4. *J Bacteriol* **177:**52–58.
75. **Panke S, Sanchez-Romero JM, de Lorenzo V.** 1998. Engineering of quasi-natural *Pseudomonas putida* strains for toluene metabolism through an ortho-cleavage degradation pathway. *Appl Environ Microbiol* **64:**748–751.
76. **Hoess RH, Abremski K.** 1990. The Cre-lox recombination system, p 99–109. *In* Eckstein F, Lilley DJ (ed), *Nucleic Acids and Molecular Biology*, **vol. 4**. Springer, Berlin, Germany.

77. **Enyeart PJ, Chirieleison SM, Dao MN, Perutka J, Quandt EM, Yao J, Whitt JT, Keatinge-Clay AT, Lambowitz AM, Ellington AD.** 2013. Generalized bacterial genome editing using mobile group II introns and Cre-*lox*. *Mol Syst Biol* **9:**685.

78. **Rojo F, Alonso JC.** 1994. A novel site-specific recombinase encoded by the *Streptococcus pyogenes* plasmid pSM19035. *J Mol Biol* **238:**159–172.

79. **Diaz V, Rojo F, Martinez AC, Alonso JC, Bernad A.** 1999. The prokaryotic beta-recombinase catalyzes site-specific recombination in mammalian cells. *J Biol Chem* **274:**6634–6640.

80. **Diaz V, Servert P, Prieto I, Gonzalez MA, Martinez AC, Alonso JC, Bernad A.** 2001. New insights into host factor requirements for prokaryotic beta-recombinase-mediated reactions in mammalian cells. *J Biol Chem* **276:**16257–16264.

81. **Partridge SR, Tsafnat G, Coiera E, Iredell JR.** 2009. Gene cassettes and cassette arrays in mobile resistance integrons. *FEMS Microbiol Rev* **33:**757–784.

82. **Bikard D, Mazel D.** 2013. Shuffling of DNA cassettes in a synthetic integron. *Methods Mol Biol* **1073:**169–174.

83. **Bikard D, Julie-Galau S, Cambray G, Mazel D.** 2010. The synthetic integron: an *in vivo* genetic shuffling device. *Nucleic Acids Res* **38:**e153.

84. **Bonnet J, Yin P, Ortiz ME, Subsoontorn P, Endy D.** 2013. Amplifying genetic logic gates. *Science* **340:** 599–603.

85. **Ham TS, Lee SK, Keasling JD, Arkin AP.** 2008. Design and construction of a double inversion recombination switch for heritable sequential genetic memory. *PloS One* **3:**e2815. doi:10.1371/journal.pone.0002815.

86. **Siuti P, Yazbek J, Lu TK.** 2013. Synthetic circuits integrating logic and memory in living cells. *Nat Biotechnol* **31:**448–452.

87. **Ceremonie H, Buret F, Simonet P, Vogel TM.** 2004. Isolation of lightning-competent soil bacteria. *Appl Environ Microbiol* **70:**6342–6346.

88. **Demaneche S, Bertolla F, Buret F, Nalin R, Sailland A, Auriol P, Vogel TM, Simonet P.** 2001. Laboratory-scale evidence for lightning-mediated gene transfer in soil. *Appl Environ Microbiol* **67:**3440–3444.

89. **Smillie C, Garcillán-Barcia MP, Francia MV, Rocha EPC, de la Cruz F.** 2010. Mobility of plasmids. *Microbiol Mol Biol Rev* **74:**434–452.

90. **Waters VL.** 2001. Conjugation between bacterial and mammalian cells. *Nat Genet* **29:**375–376.

91. **de Lorenzo V, Timmis KN.** 1994. Analysis and construction of stable phenotypes in Gram-negative bacteria with Tn*5* and Tn*10*-derived minitransposons. *Methods Enzymol* **235:**386–405.

92. **Simon R, Priefer U, Puhler A.** 1983. A broad host range mobilization system for *in vivo* genetic engineering: transposon mutagenesis in Gram-negative bacteria. *Nat Biotechnol* **1:**784–791.

93. **Ferrieres L, Hemery G, Nham T, Guerout AM, Mazel D, Beloin C, Ghigo JM.** 2010. Silent mischief: bacteriophage Mu insertions contaminate products of *Escherichia coli* random mutagenesis performed using suicidal transposon delivery plasmids mobilized by broad-host-range RP4 conjugative machinery. *J Bacteriol* **192:** 6418–6427.

94. **Goñi-Moreno A, Amos M, de la Cruz F.** 2013. Multicellular computing using conjugation for wiring. *PloS One* **8:**e65986. doi:10.1371/journal.pone.0065986.

95. **Minoia M, Gaillard M, Reinhard F, Stojanov M, Sentchilo V, van der Meer JR.** 2008. Stochasticity and bistability in horizontal transfer control of a genomic island in *Pseudomonas*. *Proc Natl Acad Sci USA* **105:** 20792–20797.

96. **Miyazaki R, van der Meer JR.** 2013. A new large-DNA-fragment delivery system based on integrase activity from an integrative and conjugative element. *Appl Environ Microbiol* **79:**4440–4447.

97. **Nojiri H.** 2013. Impact of catabolic plasmids on host cell physiology. *Curr Opin Biotechnol* **24:**423–430.

98. **Brautaset T, Lale R, Valla S.** 2009. Positively regulated bacterial expression systems. *Microb Biotechnol* **2:** 15–30.

99. **Terpe K.** 2006. Overview of bacterial expression systems for heterologous protein production: from molecular and biochemical fundamentals to commercial systems. *Appl Microbiol Biotechnol* **72:**211–222.

100. **Kumari R, Tecon R, Beggah S, Rutler R, Arey JS, van der Meer JR.** 2011. Development of bioreporter assays for the detection of bioavailability of long-chain alkanes based on the marine bacterium *Alcanivorax borkumensis* strain SK2. *Environ Microbiol* **13:**2808–2819.

101. **Smits TH, Seeger MA, Witholt B, van Beilen JB.** 2001. New alkane-responsive expression vectors for *Escherichia coli* and *Pseudomonas*. *Plasmid* **46:**16–24.

102. **de Lorenzo V, Fernandez S, Herrero M, Jakubzik U, Timmis KN.** 1993. Engineering of alkyl- and haloaromatic-responsive gene expression with mini-transposons containing regulated promoters of biodegradative pathways of *Pseudomonas*. *Gene* **130:**41–46.

103. **Levy C, Pike K, Heyes DJ, Joyce MG, Gabor K, Smidt H, van der Oost J, Leys D.** 2008. Molecular basis of halorespiration control by CprK, a CRP-FNR type transcriptional regulator. *Mol Microbiol* **70:**151–167.

104. **Steigedal M, Valla S.** 2008. The *Acinetobacter* sp. *chnB* promoter together with its cognate positive regulator ChnR is an attractive new candidate for metabolic engineering applications in bacteria. *Metab Eng* **10:**121–129.

105. **Graf N, Altenbuchner J.** 2013. Functional characterization and application of a tightly regulated MekR/*PmekA* expression system in *Escherichia coli* and *Pseudomonas putida*. *Appl Microbiol Biotechnol* **97:**8239–8251.

106. **Sletta H, Nedal A, Aune TE, Hellebust H, Hakvag S, Aune R, Ellingsen TE, Valla S, Brautaset T.** 2004. Broad-host-range plasmid pJB658 can be used for industrial-level production of a secreted host-toxic single-chain antibody fragment in *Escherichia coli*. *Appl Environ Microbiol* **70:**7033–7039.

107. **Zwick F, Lale R, Valla S.** 2012. Strong stimulation of recombinant protein production in *Escherichia coli* by

combining stimulatory control elements in an expression cassette. *Microb Cell Fact* **11:**133.

108. **Zwick F, Lale R, Valla S.** 2013. Combinatorial engineering for heterologous gene expression. *Bioengineered* **4:**431–434.

109. **Cebolla A, Royo JL, de Lorenzo V, Santero E.** 2002. Improvement of recombinant protein yield by a combination of transcriptional amplification and stabilization of gene expression. *Appl Environ Microbiol* **68:** 5034–5041.

110. **Cebolla A, Sousa C, de Lorenzo V.** 2001. Rational design of a bacterial transcriptional cascade for amplifying gene expression capacity. *Nucleic Acids Res* **29:** 759–766.

111. **de las Heras A, Carreno CA, de Lorenzo V.** 2008. Stable implantation of orthogonal sensor circuits in Gram-negative bacteria for environmental release. *Environ Microbiol* **10:**3305–3316.

112. **de Las Heras A, de Lorenzo V.** 2012. Engineering whole-cell biosensors with no antibiotic markers for monitoring aromatic compounds in the environment. *Methods Mol Biol* **834:**261–281.

113. **de Las Heras A, Fraile S, de Lorenzo V.** 2012. Increasing signal specificity of the TOL network of *Pseudomonas putida* mt-2 by rewiring the connectivity of the master regulator XylR. *PLoS Genet* **8:**e1002963. doi: 10.1371/journal.pgen.1002963.

114. **Tropel D, van der Meer JR.** 2004. Bacterial transcriptional regulators for degradation pathways of aromatic compounds. *Microbiol Mol Biol Rev* **68:**474–500.

115. **Carbajosa G, Trigo A, Valencia A, Cases I.** 2009. Bionemo: molecular information on biodegradation metabolism. *Nucleic Acids Res* **37:**D598–D602.

116. **Win MN, Liang JC, Smolke CD.** 2009. Frameworks for programming biological function through RNA parts and devices. *Chem Biol* **16:**298–310.

117. **Isaacs FJ, Dwyer DJ, Collins JJ.** 2006. RNA synthetic biology. *Nat Biotechnol* **24:**545–554.

118. **Lioliou E, Romilly C, Romby P, Fechter P.** 2010. RNA-mediated regulation in bacteria: from natural to artificial systems. *N Biotechnol* **27:**222–235.

119. **Brantl S.** 2007. Regulatory mechanisms employed by cis-encoded antisense RNAs. *Curr Opin Biotechnol* **10:** 102–109.

120. **Engdahl HM, Hjalt TA, Wagner EG.** 1997. A two unit antisense RNA cassette test system for silencing of target genes. *Nucleic Acids Res* **25:**3218–3227.

121. **Nakashima N, Tamura T.** 2009. Conditional gene silencing of multiple genes with antisense RNAs and generation of a mutator strain of *Escherichia coli*. *Nucleic Acids Res* **37:**e103.

122. **Sturino JM, Klaenhammer TR.** 2002. Expression of antisense RNA targeted against *Streptococcus thermophilus* bacteriophages. *Appl Environ Microbiol* **68:**588–596.

123. **Sturino JM, Klaenhammer TR.** 2004. Antisense RNA targeting of primase interferes with bacteriophage replication in *Streptococcus thermophilus*. *Appl Environ Microbiol* **70:**1735–1743.

124. **Canton B, Labno A, Endy D.** 2008. Refinement and standardization of synthetic biological parts and devices. *Nat Biotechnol* **26:**787–793.

125. **Peccoud J, Blauvelt MF, Cai Y, Cooper KL, Crasta O, DeLalla EC, Evans C, Folkerts O, Lyons BM, Mane SP, Shelton R, Sweede MA, Waldon SA.** 2008. Targeted development of registries of biological parts. *PloS One* **3:**e2671. doi:10.1371/journal.pone.0002671.

126. **Galdzicki M, Rodriguez C, Chandran D, Sauro HM, Gennari JH.** 2011. Standard biological parts knowledgebase. *PloS One* **6:**e17005. doi:10.1371/journal.pone.0017005.

127. **Peccoud J, Anderson JC, Chandran D, Densmore D, Galdzicki M, Lux MW, Rodriguez CA, Stan GB, Sauro HM.** 2011. Essential information for synthetic DNA sequences. *Nat Biotechnol* **29:**22.

128. **Hesselman MC, Koehorst JJ, Slijkhuis T, Odoni DI, Hugenholtz F, van Passel MW.** 2012. The Constructor: a web application optimizing cloning strategies based on modules from the registry of standard biological parts. *J Biol Eng* **6:**14.

129. **Galdzicki M, Clancy KP, Oberortner E, Pocock M, Quinn JY, Rodriguez CA, Roehner N, Wilson ML, Adam L, Anderson JC, Bartley BA, Beal J, Chandran D, Chen J, Densmore D, Endy D, Grunberg R, Hallinan J, Hillson NJ, Johnson JD, Kuchinsky A, Lux M, Misirli G, Peccoud J, Plahar HA, Sirin E, Stan GB, Villalobos A, Wipat A, Gennari JH, Myers CJ, Sauro HM.** 2014. The Synthetic Biology Open Language (SBOL) provides a community standard for communicating designs in synthetic biology. *Nat Biotechnol* **32:** 545–550.

130. **Jimenez JI, Perez-Pantoja D, Chavarria M, Diaz E, de Lorenzo V.** 2014. A second chromosomal copy of the *catA* gene endows *Pseudomonas putida* mt-2 with an enzymatic safety valve for excess of catechol. *Environ Microbiol* **16:**1767–1778.

131. **Perez-Pantoja D, Kim J, Silva-Rocha R, de Lorenzo V.** 2014. The differential response of the *Pben* promoter of *Pseudomonas putida* mt-2 to BenR and XylS prevents metabolic conflicts in *m*-xylene biodegradation. *Environ Microbiol* [Epub ahead of print.] doi:10.1111/1462-2920.12443.

132. **Jayanthi S, Nilgiriwala KS, Del Vecchio D.** 2013. Retroactivity controls the temporal dynamics of gene transcription. *ACS Synth Biol* **2:**431–441.

133. **Nilgiriwala KS, Jimenez J, Rivera PM, Del Vecchio D.** 2014. A synthetic tunable amplifying buffer circuit in *E. coli*. *ACS Synth Biol* [Epub ahead of print.] doi: 10.1021/sb5002533.

Plasmids—Biology and Impact in Biotechnology and Discovery
Edited by Marcelo E. Tolmasky and Juan C. Alonso

doi:10.1128/microbiolspec.PLAS-0028-2014

John S. Tregoning[1]
Ekaterina Kinnear[1]

Using Plasmids as DNA Vaccines for Infectious Diseases

33

INTRODUCTION: HISTORY, ADVANTAGES, REGULATION, AND LICENSED PRODUCTS

The observation that antigens encoded by DNA could induce an immune response when injected *in vivo* was first made in the 1960s using DNA from mouse tumors (1). Humoral (antibody) responses were subsequently observed following injection of DNA from polyoma (2) and hepatitis B viruses (3). The first study to specifically look at DNA as an immunogen was performed in the early 1990s using human growth hormone as a model antigen, delivered by a biolistic transfer device (gene gun) (4). This was quickly followed by studies showing that not only antibody, but also cytotoxic T cells could be induced by DNA delivery (5–7).

Since then, DNA vaccines have been proposed for a diverse range of diseases, from cancer (8) to allergy (9) and autoimmunity (10), but this review will focus on prophylactic DNA vaccines for infections. Recently, a database (DNAVAXDB) was developed that encompasses the wide range of approaches tested (http://www.violinet.org/dnavaxdb/index.php) (11). This broad range of targets and diseases is reflective of the fact that DNA vaccines are significantly easier and cheaper to develop for any desired antigen than protein vaccines. The speed of development is a key advantage of DNA vaccines, especially for pandemic outbreaks. The switch of many labs from cloning target genes directly from the pathogen to custom synthesis (12) has further hastened the process. In addition, it has made it safer and easier. DNA vaccines are widely described as being more thermostable than other vaccine approaches. This assumption is based on characteristics of the DNA molecule, but no studies have been published that formally demonstrate this. DNA vaccines should have a better safety profile than live attenuated vaccines, with no risk of reversion to virulence, as they only encode individual genes from the pathogen. There is also little reported inflammation following DNA vaccination in clinical trials (Table 1). Expression from eukaryotic cells means DNA vaccines may mimic the posttranslational modifications that viral proteins undergo, potentially inducing a more appropriate immune response. This is advantageous in the induction of antibodies that recognize the tertiary structure of viral proteins but may be a disadvantage for antigens from bacterial pathogens. With the correct targeting sequences, DNA vaccines may also be effective for the expression of membrane proteins, which are traditionally hard to express by conventional recombinant protein approaches.

[1]Mucosal Infection and Immunity Group, Section of Virology, Imperial College London, St Mary's Campus, London W2 1PG, United Kingdom.

Table 1 Published DNA vaccine clinical trials.[a,b]

Trial identifier	Date posted	Phase	Pathogen	Antigen	Additional protocols	Ref	Safe?	Immunogenic?
NCT00047931	Oct 2002	I	HIV	Clade B Gag-Pol-Nef fusion protein and modified envelope (Env) (VRC-HIVDNA009-00-VP)		128	Safe	Some antibody and T cell responses
NCT00054860	Feb 2003	I	HIV	Poly epitope		129	Safe	Not immunogenic
NCT00069030	Sep 2003	I	HIV	Gag-Pol-Nef–multiclade Env (VRC-HIVDNA009-00-VP)	IL-2 DNA	116	Safe	Timing of IL-2 boost mattered, 2 day delay increased response
NCT00109629	Apr 2005	I	HIV (clades A–C)	Env (A–C), Gag, Pol, Nef (VRC-HIVDNA016-00-VP and VRC-HIVDNA014-00-VP)	Needle-free (Bioject) Ad boost	130	Well tolerated	Needle free delivery improved immune response
NCT00111605	May 2005	I	HIV	Gag	IL-12 DNA	123	Well tolerated	No detectable effect of cytokines
NCT00115960	Jun 2005	I	HIV	Gag	IL-12 and/or IL-15 DNA	123	Safe	No detectable effect of cytokines
NCT00125099	Jul 2005	I/II	HIV	Gag, Pol, Nef, Env (VRC-HVDNA 009-00-VP)		131	Safe	Poorly Immunogenic
NCT00141024	Aug 2005	I	HIV	Poly epitope		132		
NCT00249106	Nov 2005	I	HIV clade C/B′	Env, Gag, Pol, Nef, and Tat (ADVAX)		133	Safe	Modest but transient
NCT00270465	Dec 2005	I	HIV	Env (A–C), Gag, Pol, Nef (VRC-HIVDNA016-00-VP and VRC-HIVDNA014-00-VP)	Needle-free (Bioject)	134	Safe and well tolerated	Increased T cell breadth and magnitude. No effect on viral control
NCT00290147	Feb 2006	I	Dengue virus	Env	Needle-free (Bioject)	135	Some local pain/swelling	Modest: 5 of 12 subjects in high dose group had neutralizing antibody. 10/12 had Interferon-γ (IFN) positive responses
NCT00301184	Mar 2006	I	HIV	VLP (pGA2/JS7)	Modified vaccinia Ankara (MVA) boost	136	Mild or moderate local responses	Modest responses, different patterns of response according to prime boost regime
NCT00321061	May 2006	I	HIV (clades A–C)	Env (A–C), Gag, Pol, Nef (VRC-HIVDNA016-00-VP)	Ad boost	60	Local responses after i.d. or s.c.	No difference by route of administration
NCT00384787	Oct 2006	I	HIV	Env (A–C), Gag, Pol, Nef (VRC-HIVDNA009-00-VP)	Ad boost	137	Safe	Modest
NCT00428337	Jan 2007	I	HIV	CTL epitopes	MVA boost	138	Safe	Ineffective
NCT00536627	Sep 2007	I	HBV	Env		139, 140	Well tolerated	Induction of antigen specific T cells, no protection from HBV reactivation after cessation of anti-viral therapy

NCT00545987	Oct 2007	I	HIV clade C/B′	Env, Gag, Pol, Nef, and Tat (ADVAX)	Electroporation (Ichor)	95	Safe	Electroporation increased T cell responses
NCT00694213	Jun 2008	I	Influenza H5	HA	Lipid-based adjuvant (Vaxfectin) and needle-free injection (Bioject)	76	Safe	No difference to needle delivery above
NCT00709800	Feb 2009	I	Influenza H5	HA	Lipid-based adjuvant (Vaxfectin)	76	Safe	Antibody responses induced
NCT00870987	Mar 2009	I	Malaria	Circumsporozoite protein (*Pf*CSP) and apical membrane antigen-1 (AMA1)	Ad boost	141	Safe	4/15 volunteers protected against malaria challenge
NCT00865566	Mar 2009	II	HIV	Clade B Gag, Pol, and Nef and Env from clades A, B, and C (VRC-HIVDNA016-00-VP)	Ad boost	142	Tolerated	No effect on HIV acquisition or viral set point
NCT00988767	Oct 2009	I	HBV	Env		143		Changes in NK population
NCT01086657 NCT00776711	Mar 2010	I	Influenza H5 strain	HA	Inactivated virus boost	144, 145	Safe	Higher titer antibody compared to inactivated only groups (144). Increased antibody repertoire in DNA primed groups (145). Interval in boost important in effect (144).
		I	Malaria	*Pf*CSP		146		CTL induced
		I	Malaria	*Pf*CSP		147	Mild	CTL induced, no antibody
		I	Malaria, strain T9/96	Multiple epitopes-TRAP	MVA boost	148		Prime-boost increased efficacy, partial protection
		I	Malaria, strain T9/96	Multiple epitopes-TRAP	MVA boost	149		Some T cell induced
BB-IND 8687		I	Malaria	*Pf*CSP, *Pf*SSP2/TRAP, *Pf*EXP1, *Pf*LSA1, and *Pf*LSA3	GM-CSF DNA	118	Safe	No Protection, some T cell responses
		I	Malaria	CSP, TRAP	MVA boost	150		T cell responses. 1 of 8 protected. TRAP better than CSP
		I	Malaria	*Pf*CSP, *Pf*SSP2/TRAP, *Pf*EXP1, *Pf*LSA1, and *Pf*LSA3	GM-CSF DNA	151		No Protection, some T cell responses
		I/II	Malaria	Multiple epitopes-TRAP	MVA boost	152	Safe	Some T cell responses 10% efficacy
		I	Malaria	Multiple epitopes-TRAP	MVA boost Fowlpox boost	153	Safe	T cell responses induced
		I	Malaria	*Pf*CSP	RTS,S/AS02A (protein) boost	154	Safe	Heterologous prime boost increased responses

(Continued)

Table 1 Published DNA vaccine clinical trials.[a,b] *(Continued)*

Trial identifier	Date posted	Phase	Pathogen	Antigen	Additional protocols	Ref	Safe?	Immunogenic?
CTRI/2009/ 091/000051		I	HIV-1 subtype C	HIV-1 subtype C	MVA boost	155	Safe	Less antibody in DNA alone than MVA alone Slightly more T cells
		I	HIV clade C/B′	Env, Gag, Pol, Nef, and Tat (ADVAX)	Needle-free injection (Bioject) MVA boost	155		Less antibody in DNA alone than MVA alone Slightly more T cells
		I	Influenza H5	HA	Needle-free injection (Bioject)	156	Safe	Some antibody and T cells i.d. not better than i.m.
NCT00408109 NCT00489931		I	HIV clade C/B′	Env, Gag, Pol, Nef, and Tat (ADVAX)	Electroporation (Ichor)	157	Safe	Electroporation better than none

[a]Articles were retrieved either using the term "DNA vaccine" on clinicaltrials.org to access NCT reference numbers that were used in PubMed or using the term "DNA vaccine" and the clinical trial filter on PubMed. This is an incomplete picture due to the limited publication of the clinical trials, the exclusion of EU phase I clinical trials from the public record, and the failure to clearly link publications to the clinical trial records.

[b]Abbreviations: i.m., intramuscular; s.c., subcutaneous; i.d., intradermal; Env, envelope; HA, hemagglutinin; IL, interleukin; IFN, interferon; HBV, hepatitis B virus; CSP, circumsporozoite protein; TRAP, thrombospondin-related adhesive protein; Ad, adenovirus; MVA, modified vaccinia Ankara; GM-CSF, granulocyte macrophage colony-stimulating factor.

Three DNA vaccines have been successfully licensed in veterinary practice. The first licensed DNA vaccine was against West Nile virus (Fort Dodge Animal Health, part of Wyeth/Pfizer), approved for use in horses by the USDA (13), though it is not commercially available. Another DNA vaccine, for infectious hematopoietic necrosis virus (Apex-IHN, Novartis Animal Health), has been approved for use in farmed salmon and trout (14) in Canada, but not in the United States or Europe. Apex-IHN is delivered by a single intramuscular injection of a small volume (50 l). A therapeutic cancer vaccine expressing human tyrosinase (Oncept, Merial), used for the treatment of oral canine melanoma, was licensed by the USDA in 2010 (15), but a randomized, placebo-controlled clinical trial is required to confirm efficacy in improving postsurgery survival times (16). Oncept is delivered using a transdermal/needle-free system and requires four shots at 2-week intervals.

Initially, additional safety concerns with DNA vaccines led to specific regulatory requirements compared to other vaccines. These included concerns over the generation of an anti-DNA autoimmune response, the induction of tolerance to the vaccine antigen, and the transfer of antibiotic resistance. One other consideration about DNA vaccines is integration into the host genome. This has been a potential safety concern, as integration may induce tumors by activating oncogenes, but this appears to be a rare event (17). The FDA guidelines only require preclinical integration studies when more than 30,000 copies of plasmid are detected per microgram of host DNA. While some studies suggest that integrative DNA vaccine constructs can improve the T cell response (18), constitutive expression of vaccine antigen may actually be counterproductive for the optimum immune response, especially memory responses that require a contraction of the effector pool (19). The success of phase I clinical trials in demonstrating DNA vaccines to be safe and well tolerated has led to the relaxation of the initial safety concerns (Table 1). Specific guidelines for DNA vaccines have been published by the FDA (20) and the WHO (21). These guidelines cover issues of good manufacturing practice (GMP) specific to DNA vaccines, including sequencing of the plasmid, the required degree of supercoiling, and removal of other macromolecules from the bacterial preparation used to generate the plasmid, particularly lipopolysaccharide. The preclinical checks for DNA vaccines are very similar to other vaccine types including toxicity and immunogenicity and, specifically, examining the effect of any genetic adjuvant included. We would argue for the further relaxation of regulation of DNA vaccines that come from the same manufacturer targeting variants of the same antigen, e.g., influenza hemagglutinin or the HIV envelope protein, being treated as a single product for preclinical toxicity. This would allow for the rapid development to market of a vaccine in response to a pandemic (influenza) or multiple iterations of similar antigens for experimental human vaccine studies (HIV).

In spite of the relaxation of FDA guidelines around DNA vaccines, there is still a drive to improve DNA vaccine manufacture. There are some safety concerns about the presence of antibiotic-resistance genes needed for selection and growth in bacteria, particularly related horizontal transfer to commensals (22). Additionally, antibiotic selection can reduce the plasmid yield (23). While large-scale, cost-effective production of DNA vaccines through bacterial fermentation is relatively simple, a limiting step in production lies with the purification of the DNA, particularly supercoiled DNA, which is similar in size to many contaminants (24). An additional concern is that bacterial fermentation steps currently used in plasmid preparation introduce endotoxins, and stringent certified quality control checks are required to produce a GMP-quality product (25). A number of strategies have been proposed to overcome these issues (26) and are being incorporated into the current generation of DNA vaccines. New methods based either on acellular production of expression cassettes or using alternative selection methods have been developed to avoid antibiotic resistance (Table 2).

In spite of the perceived advantages of DNA vaccines, the enormous amount of preclinical research (>55,000 articles [27]), and the limited successes in veterinary vaccines, DNA vaccines have not translated into human vaccines. As of July 2014, there were 121 registered phase I clinical trials, 38 registered phase II clinical trials, and no phase III clinical trials listed on www.clinicaltrials.gov and 3 phase I and 9 phase II clinical trials listed on www.clinicaltrialsregister.eu (using the search term "DNA vaccine"). Some of this data is captured by a database of gene therapy clinical trials (http://www.abedia.com/wiley/). Of the registered trials, 66 use HIV antigens and 15 use influenza antigen, the others comprising a mixture of other pathogens, including dengue, malaria, and hepatitis B and C viruses. The trials using infectious disease antigens that have published data are described in Table 1. In spite of good safety and tolerability, the clinical trial results have been disappointing, showing limited immunogenicity. Where immunogenicity has been reported, it has mainly been the development of T cell responses and very limited antibody responses.

DISADVANTAGES: WHY DON'T THEY WORK

A number of issues have limited the translation of DNA vaccines from successful preclinical studies into the clinic, and a better understanding of these issues will greatly help the development of a clinically effective DNA vaccine. One of the simpler issues is scale: the original volumes and doses used in murine studies (often 50 g in a 50-l volume) were large relative to the size of the muscle. The delivery of such a large bolus of DNA would have a biophysical effect on the delivery of the DNA, creating shear forces on the cells, inducing inflammation, and generating hydrodynamic pressure increasing DNA uptake (28). Scaling this up into the human muscle (based on values of muscle size from reference 29), an equivalent dose of 19.4 mg in a volume of 19.4 ml would be required, equating to approximately 40 times the normal vaccine volume. Second, the immune response required to protect a small rodent against infection may not be the same as that required to protect a human. Furthermore, the induction of immune responses, particularly the innate immune recognition of cytosolic and extracellular DNA, may differ among species (30). Inadequate animal models are an ongoing issue with all preclinical vaccine development and not limited to DNA vaccines. Third, the way in which DNA is taken up, processed, and expressed appears to be different in different species; for example, the DNA scavenger serum amyloid P is more active in humans than mice (31). Interestingly, inhibition of serum amyloid P is being targeted in an ongoing clinical trial to improve DNA vaccine responses (project ref: MR/J008605/1 on http://gtr.rcuk.ac.uk/).

Steps have been made in recent years to overcome these issues for clinical use. Before exploring the approaches used to optimize DNA vaccines for clinical use, it is necessary to understand how they work, both as gene delivery systems and as vaccine immunogens.

HOW DNA VACCINES WORK: EXPRESSION

To induce an immune response, the gene encoded on the DNA plasmid needs to be expressed in the host cell, and there are a number of barriers to this (32). While a large amount of DNA that is injected into the animal enters the blood circulation and is degraded (33, 34), the remainder enters cells that are local to the injection site, for example, myocytes after intramuscular injections or keratinocytes after subcutaneous injections. Some of the DNA directly enters and is expressed by local antigen-presenting cells (APCs) (35). It is not entirely clear how naked DNA enters the cell, but caveolae-dependent endocytosis (36), macropinocytosis (37), and clathrin-mediated endocytosis (38) have all

TABLE 2 Novel DNA vaccine technologies to remove antibiotic resistance

Name of approach	Description	Reference
Alternate vectors		
Mini-circle DNA vaccines	Produced by a site-specific recombination event of a parental plasmid producing a mini-circle containing an expression cassette and a mini-plasmid comprised of backbone regions.	158
Minimalistic immunogenically defined gene expression (MIDGE) vectors	Use a restriction digestion system to excise the backbone, resulting in a linear expression cassette that is capped by hairpin loop.	159
Doggybones	Capped linear DNA constructs made by a novel one-step enzymatic process where antibiotic resistance genes are first included in parental plasmids and then eliminated once capped linear constructs are produced. Can be generated independently of a bacterial fermentation step.	160
Alternative selection systems (22)		
Operator-repressor-titration systems	Essential gene under repressor control is incorporated into bacterial host chromosome. Plasmid encodes several operator cassettes, e.g., Lac or RNAII; these titrate out the repressor, allowing gene expression.	161, 162
Auxotrophic complementation	Rely on deletion of an essential or conditionally essential gene where bacterial growth is only restored upon introduction of a plasmid containing the altered gene.	163
Toxin-antitoxin-based systems	Toxins remain in bacterial cells and lead to cell death after loss of plasmid.	164
RNA-out	Use antisense RNA on the plasmid to silence toxin genes expressed by the bacteria.	165
Overexpression of growth genes	Certain bacterial growth genes circumvent toxic effects of antibacterials when overexpressed, e.g., the enoyl ACP reductase encoded by the *fabI* gene.	166

been proposed as mechanisms. There are differences depending on the cell type that is taking up the material and whether the DNA is packaged chemically or naked. Having entered the cell, the DNA travels to the nucleus (39); movement to the nucleus has been suggested to occur either as free DNA (40) or in vesicles associated with the cytoskeleton (41), and progression through the cytoplasm can be associated with some degradation of the DNA by nucleases (42).

The nuclear membrane is a significant barrier to the expression of DNA: less than 0.1% of plasmid DNA that enters the cytosol is transcribed (43). Nuclear entry of plasmid DNA occurs either during nuclear envelope breakdown and reformation in mitosis (44) or via nuclear pores (45). Due to its small size, supercoiled plasmid DNA can transit through nuclear pores, but it is more likely that the plasmid DNA piggybacks onto proteins with nuclear localization signals (46), for example, transcription factors. This process can be optimized by encoding sequences in the DNA vaccine that increase binding to these proteins, for example, the enhancer region from simian virus 40 (SV40) (47). Finally, the gene(s) encoded on the plasmid need to be transcribed and translated. Expression level is a function of a number of features including the promoter used, the presence of a Kozak consensus sequence, and codon usage.

Transit into and across the cell also has an impact on the activation of the innate immune response to DNA vaccines and therefore the immunogenicity of the vaccine.

HOW DNA VACCINES WORK: IMMUNE RESPONSE

Innate Response to DNA Vaccines

DNA sensing by pattern recognition receptors has been shown to be essential for vaccine responses (30). Many of the pattern recognition receptors used by cells to detect infection are based on the recognition of chemicals that are outside their usual location, so the presence of naked DNA outside of the nucleus is inherently inflammatory. Transfected immune and somatic cells are able to sense the presence of "foreign DNA" in the cytosol, using a range of intracellular pattern recognition receptors including RIG-I, AIM2, ZBP1/DAI, and HB2 histones (48). However, their role in the initiation of an innate signaling cascade to DNA vaccination is unclear, as knockout studies showed minimal contributions, suggesting redundancy (30). The detection of unmethylated CpG motifs inherently present in DNA vaccines by toll-like receptor (TLR) 9 is assumed to have a role in the detection of DNA vaccines, but *TLR9*$^{-/-}$ mice had immune responses to DNA vaccines similar to control mice (49), again suggesting some redundancy. Studies have shown that the downstream signaling molecules STING and TBK1 are essential for instigation of an innate immune response to DNA vaccines, as their deletion abrogated type I interferon production (50, 51). The induction of an innate immune response to the DNA vaccine is critical in activating the APCs that present the expressed antigen to the T cells.

Adaptive Responses to DNA Vaccines

The adaptive immune response to a DNA vaccine is heavily influenced by the cell that is transfected by the DNA. As described above, antigen is either expressed by non-antigen-presenting cells such as myocytes or keratinocytes or by APCs that have taken up the DNA. Which cell expresses the DNA is influenced by a number of factors including the route of delivery, the device used to deliver the vaccine, the formulation of the vaccine, and the use of adjuvants. Speculatively, if APCs are directly transfected, they are most likely to present antigen on major histocompatibility complex class I (MHC-I) molecules, thereby initiating a CD8+ T cell response. If somatic cells are transfected, the antigen will be either displayed on MHC-I, secreted as processed antigen, or released upon cell death. Antigen that is displayed upon MHC-I acts as a trigger for activated CD8+ T cells to kill the transfected cell, reducing the expression of the vaccine and potentially dampening the immunogenicity (52). Secreted antigen is either taken up by APCs and presented on MHC-II molecules to CD4+ T cells or interacts with B cells, inducing an antibody response. Finally, antigen released after cell death most likely enters APCs and is loaded onto MHC-I by the cross-presentation pathway. Cross-presentation of DNA vaccine derived antigen has been demonstrated *in vitro* (53), and *in vivo* studies suggest that antigen expressed from non-antigen-presenting cells is more important in the induction of the immune response to DNA vaccines than antigen from APCs (54).

The clinical trial experience (Table 1) suggests that DNA vaccines are better at inducing T cell responses than B cell responses. There are a number of reasons why this might be the case, including incorrectly folded antigen, low expression levels, poor activation of the innate response via pattern recognition receptors, and differences in MHC-I and MHC-II loading. Our understanding of the immune response to expressed antigen can be used to target either the DNA vaccine or the expressed antigen to different cells to alter the immune response (55). Fusing the antigen to an anti-MHC-II single chain fragment variable molecule that targets it to APCs has been shown to increase the response (56). Optimizing the expression, presentation, and secretion of the antigen to engage different arms of the adaptive response is critical for improving immunogenicity.

OPTIMIZING FOR THE CLINIC

Route of Delivery

Altering the route of delivery can have a marked effect on the downstream immune response to DNA vaccines. Various routes have been proposed for DNA vaccine delivery including intramuscular, subcutaneous, intranasal, and oral (57). Ultimately, route selection is a balance between optimum expression, immunogenicity, cost, and acceptability for patients. Considerations of route and delivery are dependent upon the target of the vaccine, cancer vaccines because of the individual nature of the treatment may be able to exploit more niche approaches than infectious disease vaccines that need to be cheap and easy to deliver. Conventionally, DNA vaccines are delivered intramuscularly, which has the advantage of matching currently available vaccines and therefore requiring less training of staff administering the vaccine. Expression of DNA following intramuscular delivery is mostly by myocytes, which because they are energy rich, low turnover cells, may lead to increased expression levels. The subcutaneous route is also relatively easy for vaccine delivery and may be advantageous because of the high numbers of APCs in the skin (58). Intradermal immunization is slightly more challenging, requiring specific needles and training of healthcare staff, but may be advantageous because of the specific dendritic cell subsets present in the dermis (59). While small animal studies suggest a switching in the immune response when different routes are compared (57), a recent clinical trial comparing DNA vaccines delivered by these routes found no difference in immunogenicity (60) but found mildly greater reactogenicity after intradermal or subcutaneous delivery. In addition to delivering vaccines via these routes by injection, a number of innovative approaches have been proposed including tape stripping of the hair follicles and "painting" the vaccine on (61).

Other more exotic routes have been explored to improve DNA vaccine immunogenicity. One method proposed is hydrodynamic delivery (62). This method is based on the delivery of high volumes of DNA solution intravenously at high pressure, which effectively forces the DNA into the cells and mainly targets the liver (63). Beyond the complexity of the procedure, there is an additional scale-up complication: a dose in humans equivalent to that used in mice would be 7 liters, based on 100 ml/kg in mice (63). While this approach has been demonstrated to be effective in nonhuman primates (64), to date no clinical trials have been performed using this route.

Another, more attractive, prospect is the targeting of mucosal surfaces because of the high incidence of infections at mucosal sites and the potential for generating local immunity. DNA vaccines have been tested in small animal models using the nasal (65), vaginal (66), rectal (67), sublingual (68), and oral routes. Of these only an orally delivered cancer vaccine has made it to a phase I clinical trial (69). Mucosally delivered DNA vaccines are particularly sensitive to degradation because the tissues have a number of chemical and enzymatic barriers including nucleases and mucus. Therefore, one of the considerations for DNA vaccine delivery by the mucosal route is how the DNA is packaged, but formulation of DNA vaccines is also important for other routes.

Packaging for Delivery

A lot of research has focused on the packaging of DNA to optimize delivery and expression. Formulation of the DNA vaccine can alter the routes of cellular entry and intracellular trafficking, increasing uptake and expression. Two broad approaches have been used: compaction with cationic polymers and liposomes (70). The phosphate backbone of DNA imparts a significant negative charge to the macromolecule, which enables the formation of complexes with cationic polymers. A number of different molecules have been proposed including polyethylenimine, chitosan, and cationic poly(lactic-co-glycolic acid) nanoparticles. Interestingly, both polyethylenimine (71) and chitosan (72) have an additional adjuvant effect, causing local inflammation. Of these agents, poly(lactic-co-glycolic acid) has been tested in a clinical trial (73), with a good safety profile but limited immunogenicity. Manipulating the chemistry of the polymer can alter the immune response to these carriers, and the ratio of polymer to DNA can be altered, affecting particle size, which may have an effect on the immune response to the vaccine, though the effect of particle size on immune response is mixed (74). In addition to stabilizing the plasmid in the extracellular space, formulating appears to alter the route of uptake of the plasmid and transit from the endosome into the cell (32). Alternatively, DNA vaccines can be packaged in phospholipid bilayers (liposomes), with the aim being to increase fusion with the cell membrane and enable cellular entry of the DNA (75). A number of different synthetic lipids have been suggested for this approach, but most commonly they include DOPE/DOTAP. DNA can either be bound to the surface through electrostatic forces or can be encapsulated within the liposome. A lipid formulation, Vaxfectin (Vical), has been tested in a number of clinical trials of DNA vaccines (76). There are issues with the stability and quality assurance of liposomes, limiting their translation to the clinic.

Another approach is to use virosomes, which have viral proteins studded in the lipid membrane, enabling the targeting of particles to specific cell types (77). However, the virosome approach would negate a key advantage of DNA vaccines—their low manufacturing cost—because the addition of proteins would increase the price significantly. Alternatively, the use of attenuated bacteria as carriers of DNA vaccines (78) has been proposed, as some bacteria are able to transfer the plasmid DNA across phylogenetic borders. There are issues with the development of a GMP-quality, live attenuated, genetically modified vaccine, which again may limit its use as a prophylactic vaccine. While these approaches are often described as DNA or genetic vaccines, they should be considered vectored vaccines.

Much of the thinking behind formulation approaches comes from transfection of mammalian cells *in vitro*, which may not translate *in vivo*. The inclusion of extra chemical components may delay the licensure of the vaccine due to concerns over the inflammatory profile of the agents used. Rather than chemically modifying each DNA vaccine to enhance uptake, a more universal approach is to use alternative delivery devices to enhance the uptake and immunogenicity.

Device-Mediated Delivery

DNA vaccines have to enter cells and be expressed prior to inducing an immune response. This is fundamentally different from conventional vaccines, which exploit the normal surveillance and uptake processes of the immune response. This necessity for antigen expression prior to immunogenicity means that alternative delivery approaches may be required to achieve optimal responses. To this end, a number of different devices have been developed and tested to improve DNA vaccine delivery. The acceptability profile of these devices may be very different from conventional needle-delivered vaccines, as they can be perceived to be more invasive.

One of the earliest techniques proposed is biolistic delivery (also referred to as a gene gun) (79), where DNA is coated onto inert microscopic particles (often made of gold) that are then "fired" at the skin by the biolistic device. This was literally the case in the initial devices, as gunpowder was used to power the transit of the particles, though helium is now used. These approaches have been licensed by PowderJect (Chiron/Novartis) and PowderMed (Pfizer). DNA vaccines delivered by these devices have gone into the clinic with moderate efficacy but a key advantage of dose sparing (80–83). Alternatively, liquid jet systems that use compressed gas or springs to force liquid through a small orifice (0.1 to 0.5 mm) are also able to force DNA into cells (84), with systems made by a number of companies (including Bioject and PharmaJet). A clinical trial comparing forced DNA vaccine delivery with and without microparticles (using the PowderMed and Bioject devices, respectively) showed no difference in immunogenicity between the two approaches (85). Alternatively, the use of tattooing has been suggested as an approach (86), with some efficacy in preclinical studies (87), though again, scale-up and acceptability may be an issue; for example, in a recent study using rhesus macaques a 30 cm^2 area was used (88). Finally, some of the formulation approaches described above can be combined with delivery approaches, for example, microneedles, which are soluble patches of multiple needles approximately 100 μm long coated in DNA (89). Microneedles have a number of advantages, including controlling the depth of vaccine delivery, combining adjuvant and antigen in the same device, stability, and reduced risk to the physician or patient. These are all conceptually interesting ideas and have been demonstrated in a number of preclinical studies, but whether they can be translated to efficacious clinical approaches remains to be seen.

The biggest advance in DNA vaccine delivery has been the use of electroporation (also described as electropermeabilization) to improve uptake and transfection efficiency (90, 91). Electroporation has long been used for *in vitro* transformation of cells (both bacterial and mammalian) and was first proposed for use *in vivo* in 1996 (92) and for the delivery of DNA vaccines in 2000 (93). Though not entirely understood, electroporation is believed to work mainly by inducing and stabilizing pores in the cell membrane and then moving the DNA along the gradient of the applied current (toward the positive electrode). It probably also induces local inflammation, boosting the immune response (94). A range of variations on the theme of electroporation have been explored, including changing the parameters of the electric pulse (length, voltage, and current) delivered and the type of electrode used. It has been used in a number of clinical trials (Table 1) and is well tolerated, with a short muscle contraction that can cause some discomfort (95). These approaches all require the development of practical and affordable devices that can be used in the community, particularly in resource-poor settings.

OPTIMIZING EXPRESSION

Trials have shown that the primary problem with the use of DNA vaccines is the lack of immunogenicity in humans and primates. There is some evidence that immune responses to DNA vaccines increase with dose (96, 97), but there is an upper dose limit that is feasible for administration to humans. Therefore, optimizing the expression of antigen from current DNA vaccines is key to improving performance (98). All aspects of the transcription and translation of the encoded gene can be optimized to improve expression and can be combined for additive effect. Promoter elements placed upstream of the open reading frame in the plasmid backbone have been optimized, with the majority containing the cytomegalovirus promoter, which allows for constitutive expression in many mammalian cell lines, and being more effective than tissue-specific promoters (99, 100). Protein expression is increased several-fold after codon optimization, when the codon use of the gene of interest is matched to the species vaccinated (101). The manipulation of the termination sequence (e.g., adding a polyA tail) has been shown to increase the amount of mRNA exported out of the nucleus by allowing efficient termination of transcription (102). Relaxing of the supercoiled form of the plasmid DNA upon delivery makes it more susceptible to endonucleases that degrade the DNA, and any DNA secondary structures exacerbate this degradation, so they are now avoided (103). Other molecular approaches to increasing expression include the addition of signal sequences that target the antigen to intracellular compartments involved in MHC-I or MHC-II processing (104). Nuclear localization signals have also been tested, aiming to direct DNA to the nucleus for transcription (105). In addition to improving the transfection efficiency of DNA vaccines, various strategies have been evaluated to improve their immunogenicity.

OPTIMIZING IMMUNOGENICITY

Electroporation and biolistic and liquid jet delivery all have the additional effect of causing cell damage, which induces a local inflammatory immune response (94), which in turn may potentiate the immunogenicity of the delivered DNA. Boosting the immunogenicity to delivered antigen is critical in maximizing the efficacy of vaccines. Adjuvants are chemicals used to increase the local inflammatory response, thereby increasing the uptake and presentation of the antigen by APCs and also increasing the recruitment of cells to the site. Two broad adjuvant approaches are used in DNA vaccines: the addition of inflammatory agents (chemical adjuvants) and the inclusion of genes that trigger the immune response (genetic adjuvants).

Chemical Adjuvants

DNA vaccines by their nature are inflammatory because they introduce extra-nuclear DNA to the cell. Unmethylated CpG oligodinucleotides are used as an adjuvant in their own right (106), and the effect of the addition of these motifs has been explored in DNA vaccines. When the CpG motifs were methylated there was a reduction in the immunogenicity of DNA vaccines (107), and the introduction of CpG motifs increased antibody responses in vaccinated fish (108), but other studies have suggested that TLR9-/- mice are still able to respond to DNA vaccines (49). The understanding of other sensors of cytosolic DNA is not as comprehensive as the understanding of the TLR pathway. As yet, apart from the STING pathway (109), little has been done to exploit cytosolic DNA sensing for the development of adjuvants, but conceivably, DNA vaccine plasmids could be engineered to target these receptors. In addition to increasing the inherent immunogenicity of the plasmid vaccine, a number of agents have been coformulated with DNA vaccines, including conventional chemical adjuvants such as alum (110) or MF59 (111) and more immunologically derived adjuvants including TLR7/8/9 ligands (112), resiquimod (113), or extracted material from BCG (114). However, these approaches appear to have had a limited impact on DNA vaccine immunogenicity. This may be due to timing issues since adjuvant induced inflammation is very acute and may not match the peak of antigen expression. Alternatively, the inflammation induced by the adjuvant may lead to the clearance of the transfected cells before sufficient antigen is expressed.

Genetic Adjuvants

One elegant strategy to overcome the poor immunogenicity of DNA vaccines is to include genes that boost or modulate the immune response. This may be advantageous because the timing of expression will match that of the antigen and therefore lead to a coordinated response. The effect may also be more focused than TLR-type adjuvants, which can be reactogenic. A wide range of genes have been tested in this context (described in reference 115). They fit into the following broad categories: (i) cytokines, which are molecules that attract or modulate the responses of cells in the immune response, e.g., IL-2 (116) and IL-12 (117); (ii) factors that enhance APC function, e.g., GM-CSF (118) and Flt3 (112); (iii) costimulatory molecules, e.g., CD86 but not CD80 (119); (iv) adhesion molecules, e.g., ICAM1 (120); (v) bacterial ligands, e.g., flagellin (121); and (vi) molecules in pattern recognition pathways, e.g., RIG-I (122) or DAI (122). A number of these agents have entered clinical trials, including IL-12 (117), IL-15 (123), GM-CSF (118), and IL-2 (116), with acceptable safety profiles but modest immunogenicity.

PRIME BOOST

The apparent failure of DNA vaccines as individual agents in clinical trials has led to a large body of work using them as the priming immunization in various heterologous prime boost regimes (where the initial vaccination is different from subsequent ones) (86). These have usually been associated with a range of recombinant attenuated viruses expressing the vaccine antigen, including adenovirus, modified vaccinia Ankara, and fowlpox (Table 1). The rationale for this approach is that priming with the DNA vaccine may reduce the negative effect of antivector immunity. It has also been suggested that DNA priming can increase the breadth of the immune response (124). An array of combinations has been tried in various orders and sequences, and greater responses have been reported for prime boost combinations than individual immunizations (Table 1). One issue is that the regimes required to achieve the moderate immune response observed with these approaches are often long and complex, again limiting the efficacy in a mass vaccination campaign.

NONCONVENTIONAL VACCINES

Given the limited efficacy of DNA vaccines in humans, there may be alternative uses for DNA vaccines beyond the classic prophylactic vaccine. The first is for experimental vaccines, to screen and identify the best candidate antigens prior to developing costly GMP-quality proteins (125). In this context the use of complex delivery devices or large doses may not be an issue. This may be particularly effective for membrane proteins (for example, HIV envelope), which are extremely difficult to express as proteins in high quantities in the correct confirmation. They may also be more appropriate for epitope screening, particularly in chimeric or mosaic vaccines. Alternatively, the expression of monoclonal neutralizing antibodies by DNA may be used as a form of passive immunization (126), which could be used for at-risk patients, especially when short-term protection is required. A third use may be the generation of monoclonal antibodies, either in mice (127) or in humans. These could be developed postimmunization with a hybridoma approach or with novel sorting and expression techniques. This approach would also benefit from the ease of production of the constructs and the bypassing of the requirement to make a protein antigen. Finally, due to their individual nature and high costs, DNA vaccines may ultimately be better suited to cancer treatment than prophylactic vaccines to infectious diseases.

CONCLUSIONS

DNA vaccines in humans have yet to live up to the excitement generated by the preclinical studies. This is due to issues with scaling up the dose and differences in both the expression of foreign nucleic acids and the initiation of an immune response to DNA between mice and humans. Making the preclinical models more relevant to humans is a key priority, particularly dose reduction, but novel approaches to improve expression levels and immunogenicity in clinical studies are also required. Of the current platforms, electroporation appears to have the largest impact, but current-generation devices do not lend themselves to mass vaccination campaigns. If the immunogenicity and expression issues can be resolved, then DNA vaccines will revolutionize vaccines.

Acknowledgments. *We thank Dr. Adam Walters (Oxford University) for proofreading.E.K. is funded by an MRC CASE studentship MR/J006548/1. This work was supported by the European Community's European 7th Framework Program ADITEC (HEALTH-F4-2011-18 280873). Conflicts of interest: We disclose no conflicts.*

Citation. Tregoning JS, Kinnear E. 2014. Using plasmids as DNA vaccines for infectious diseases. Microbiol Spectrum 2(6):PLAS-0028-2014.

References

1. **Orth G, Atanasiu P, Boiron M, Rebiere JP, Paoletti C.** 1964. Infectious and oncogenic effect of DNA extracted from cells infected with polyoma virus. *Proc Soc Exp Biol Med* **115:**1090–1095.
2. **Israel MA, Chan HW, Hourihan SL, Rowe WP, Martin MA.** 1979. Biological activity of polyoma viral DNA in mice and hamsters. *J Virol* **29:**990–996.
3. **Will H, Cattaneo R, Darai G, Deinhardt F, Schellekens H, Schaller H.** 1985. Infectious hepatitis B virus from cloned DNA of known nucleotide sequence. *Proc Natl Acad Sci USA* **82:**891–895.

4. **Tang DC, DeVit M, Johnston SA.** 1992. Genetic immunization is a simple method for eliciting an immune response. *Nature* **356:**152–154.
5. **Wang B, Ugen KE, Srikantan V, Agadjanyan MG, Dang K, Refaeli Y, Sato AI, Boyer J, Williams WV, Weiner DB.** 1993. Gene inoculation generates immune responses against human immunodeficiency virus type 1. *Proc Natl Acad Sci USA* **90:**4156–4160.
6. **Fynan EF, Webster RG, Fuller DH, Haynes JR, Santoro JC, Robinson HL.** 1993. DNA vaccines: protective immunizations by parenteral, mucosal, and gene-gun inoculations. *Proc Natl Acad Sci USA* **90:** 11478–11482.
7. **Ulmer JB, Donnelly JJ, Parker SE, Rhodes GH, Felgner PL, Dwarki VJ, Gromkowski SH, Deck RR, DeWitt CM, Friedman A, et al.** 1993. Heterologous protection against influenza by injection of DNA encoding a viral protein. *Science* **259:**1745–1749.
8. **Rice J, Ottensmeier CH, Stevenson FK.** 2008. DNA vaccines: precision tools for activating effective immunity against cancer. *Nat Rev Cancer* **8:**108–120.
9. **Chua KY, Kuo IC, Huang CH.** 2009. DNA vaccines for the prevention and treatment of allergy. *Curr Opin Allergy Clin Immunol* **9:**50–54.
10. **Ramshaw IA, Fordham SA, Bernard CC, Maguire D, Cowden WB, Willenborg DO.** 1997. DNA vaccines for the treatment of autoimmune disease. *Immunol Cell Biol* **75:**409–413.
11. **He Y, Racz R, Sayers S, Lin Y, Todd T, Hur J, Li X, Patel M, Zhao B, Chung M, Ostrow J, Sylora A, Dungarani P, Ulysse G, Kochhar K, Vidri B, Strait K, Jourdian GW, Xiang Z.** 2014. Updates on the web-based VIOLIN vaccine database and analysis system. *Nucleic Acids Res* **42:**D1124–D1132.
12. **Czar MJ, Anderson JC, Bader JS, Peccoud J.** 2009. Gene synthesis demystified. *Trends Biotechnol* **27:**63–72.
13. **Davis BS, Chang GJ, Cropp B, Roehrig JT, Martin DA, Mitchell CJ, Bowen R, Bunning ML.** 2001. West Nile virus recombinant DNA vaccine protects mouse and horse from virus challenge and expresses *in vitro* a noninfectious recombinant antigen that can be used in enzyme-linked immunosorbent assays. *J Virol* **75:**4040–4047.
14. **Garver KA, LaPatra SE, Kurath G.** 2005. Efficacy of an infectious hematopoietic necrosis (IHN) virus DNA vaccine in Chinook *Oncorhynchus tshawytscha* and sockeye *O. nerka* salmon. *Dis Aquat Organ* **64:** 13–22.
15. **Grosenbaugh DA, Leard AT, Bergman PJ, Klein MK, Meleo K, Susaneck S, Hess PR, Jankowski MK, Jones PD, Leibman NF, Johnson MH, Kurzman ID, Wolchok JD.** 2011. Safety and efficacy of a xenogeneic DNA vaccine encoding for human tyrosinase as adjunctive treatment for oral malignant melanoma in dogs following surgical excision of the primary tumor. *Am J Vet Res* **72:**1631–1638.
16. **Denies S, Sanders NN.** 2012. Recent progress in canine tumor vaccination: potential applications for human tumor vaccines. *Exp Rev Vaccines* **11:**1375–1386.
17. **Nichols WW, Ledwith BJ, Manam SV, Troilo PJ.** 1995. Potential DNA vaccine integration into host cell genome. *Ann NY Acad Sci* **772:**30–39.
18. **Bertino P, Urschitz J, Hoffmann FW, You BR, Rose AH, Park WH, Moisyadi S, Hoffmann PR.** 2014. Vaccination with a piggyBac plasmid with transgene integration potential leads to sustained antigen expression and CD8(+) T cell responses. *Vaccine* **32:** 1670–1677.
19. **Hovav AH, Panas MW, Rahman S, Sircar P, Gillard G, Cayabyab MJ, Letvin NL.** 2007. Duration of antigen expression *in vivo* following DNA immunization modifies the magnitude, contraction, and secondary responses of CD8+ T lymphocytes. *J Immunol* **179:**6725–6733.
20. **USA FDA.** 2007. Guidance for industry: considerations for plasmid DNA vaccines for infectious disease indications. *Biotechnol Law Rep* **26:**641–647.
21. **Roberston J, Ackland J, Holm A.** 2007. Guidelines for assuring the quality and nonclinical safety evaluation of DNA vaccines. *WHO Tech Rep Ser* **941:**57–81.
22. **Vandermeulen G, Marie C, Scherman D, Preat V.** 2011. New generation of plasmid backbones devoid of antibiotic resistance marker for gene therapy trials. *Mol Ther* **19:**1942–1949.
23. **Cunningham DS, Koepsel RR, Ataai MM, Domach MM.** 2009. Factors affecting plasmid production in *Escherichia coli* from a resource allocation standpoint. *Microb Cell Fact* **8:**27.
24. **Josefsberg JO, Buckland B.** 2012. Vaccine process technology. *Biotechnol Bioeng* **109:**1443–1460.
25. **Prather KJ, Sagar S, Murphy J, Chartrain M.** 2003. Industrial scale production of plasmid DNA for vaccine and gene therapy: plasmid design, production, and purification. *Enzyme Microb Tech* **33:**865–883.
26. **Glenting J, Wessels S.** 2005. Ensuring safety of DNA vaccines. *Microb Cell Fact* **4:**26.
27. **Racz R, Li X, Patel M, Xiang Z, He Y.** 2014. DNAVaxDB: the first web-based DNA vaccine database and its data analysis. *BMC Bioinformatics* **15:**S2.
28. **Dupuis M, Denis-Mize K, Woo C, Goldbeck C, Selby MJ, Chen M, Otten GR, Ulmer JB, Donnelly JJ, Ott G, McDonald DM.** 2000. Distribution of DNA vaccines determines their immunogenicity after intramuscular injection in mice. *J Immunol* **165:**2850–2858.
29. **Leamy VL, Martin T, Mahajan R, Vilalta A, Rusalov D, Hartikka J, Bozoukova V, Hall KD, Morrow J, Rolland AP, Kaslow DC, Lalor PA.** 2006. Comparison of rabbit and mouse models for persistence analysis of plasmid-based vaccines. *Hum Vaccin* **2:**113–118.
30. **Coban C, Kobiyama K, Jounai N, Tozuka M, Ishii KJ.** 2013. DNA vaccines: a simple DNA sensing matter? *Hum Vaccin Immunother* **9:**2216–2221.
31. **Wang Y, Guo Y, Wang X, Huang J, Shang J, Sun S.** 2011. Human serum amyloid P functions as a negative regulator of the innate and adaptive immune responses to DNA vaccines. *J Immunol* **186:**2860–2870.
32. **Wiethoff CM, Middaugh CR.** 2003. Barriers to nonviral gene delivery. *J Pharm Sci* **92:**203–217.

33. **Kim BM, Lee DS, Choi JH, Kim CY, Son M, Suh YS, Baek KH, Park KS, Sung YC, Kim WB.** 2003. *In vivo* kinetics and biodistribution of a HIV-1 DNA vaccine after administration in mice. *Arch Pharm Res* **26:** 493–498.
34. **Zhang HY, Sun SH, Guo YJ, Chen ZH, Huang L, Gao YJ, Wan B, Zhu WJ, Xu GX, Wang JJ.** 2005. Tissue distribution of a plasmid DNA containing epitopes of foot-and-mouth disease virus in mice. *Vaccine* **23:** 5632–5640.
35. **Condon C, Watkins SC, Celluzzi CM, Thompson K, Falo LD Jr.** 1996. DNA-based immunization by *in vivo* transfection of dendritic cells. *Nat Med* **2:**1122–1128.
36. **Wolff JA, Dowty ME, Jiao S,, Repetto G, Berg RK, Ludtke JJ, Williams P, Slautterback DB.** 1992. Expression of naked plasmids by cultured myotubes and entry of plasmids into T tubules and caveolae of mammalian skeletal muscle. *J Sci* **103**(Pt 4):1249–1259.
37. **Budker V, Budker T, Zhang G, Subbotin V, Loomis A, Wolff JA.** 2000. Hypothesis: naked plasmid DNA is taken up by cells *in vivo* by a receptor-mediated process. *J Gene Med* **2:**76–88.
38. **Latz E, Schoenemeyer A, Visintin A, Fitzgerald KA, Monks BG, Knetter CF, Lien E, Nilsen NJ, Espevik T, Golenbock DT.** 2004. TLR9 signals after translocating from the ER to CpG DNA in the lysosome. *Nat Immunol* **5:**190–198.
39. **Lechardeur D, Verkman AS, Lukacs GL.** 2005. Intracellular routing of plasmid DNA during non-viral gene transfer. *Adv Drug Del Rev* **57:**755–767.
40. **Mui B, Ahkong QF, Chow L, Hope MJ.** 2000. Membrane perturbation and the mechanism of lipid-mediated transfer of DNA into cells. *Biochim Biophys Acta* **1467:** 281–292.
41. **Trombone AP, Silva CL, Lima KM, Oliver C, Jamur MC, Prescott AR, Coelho-Castelo AA.** 2007. Endocytosis of DNA-Hsp65 alters the pH of the late endosome/lysosome and interferes with antigen presentation. *PloS One* **2:**e923. doi:10.1371/journal.pone.0000923.
42. **Lechardeur D, Sohn KJ, Haardt M, Joshi PB, Monck M, Graham RW, Beatty B, Squire J, O'Brodovich H, Lukacs GL.** 1999. Metabolic instability of plasmid DNA in the cytosol: a potential barrier to gene transfer. *Gene Ther* **6:**482–497.
43. **Capecchi MR.** 1980. High efficiency transformation by direct microinjection of DNA into cultured mammalian cells. *Cell* **22:**479–488.
44. **Wilke M, Fortunati E, van den Broek M, Hoogeveen AT, Scholte BJ.** 1996. Efficacy of a peptide-based gene delivery system depends on mitotic activity. *Gene Ther* **3:**1133–1142.
45. **van der Aa MA, Mastrobattista E, Oosting RS, Hennink WE, Koning GA, Crommelin DJ.** 2006. The nuclear pore complex: the gateway to successful nonviral gene delivery. *Pharm Res* **23:**447–459.
46. **Talcott B, Moore MS.** 1999. Getting across the nuclear pore complex. *Trends Cell Biol* **9:**312–318.
47. **Dean DA.** 1997. Import of plasmid DNA into the nucleus is sequence specific. *Exp Cell Res* **230:**293–302.
48. **Gurtler C, Bowie AG.** 2013. Innate immune detection of microbial nucleic acids. *Trends Microbiol* **21:**413–420.
49. **Babiuk S, Mookherjee N, Pontarollo R, Griebel P, van Drunen Littel-van den Hurk S, Hecker R, Babiuk L.** 2004. TLR9-/- and TLR9+/+ mice display similar immune responses to a DNA vaccine. *Immunology* **113:** 114–120.
50. **Ishii KJ, Kawagoe T, Koyama S, Matsui K, Kumar H, Kawai T, Uematsu S, Takeuchi O, Takeshita F, Coban C, Akira S.** 2008. TANK-binding kinase-1 delineates innate and adaptive immune responses to DNA vaccines. *Nature* **451:**725–729.
51. **Ishikawa H, Ma Z, Barber GN.** 2009. STING regulates intracellular DNA-mediated, type I interferon-dependent innate immunity. *Nature* **461:**788–792.
52. **Davis HL, Millan CL, Watkins SC.** 1997. Immune-mediated destruction of transfected muscle fibers after direct gene transfer with antigen-expressing plasmid DNA. *Gene Ther* **4:**181–188.
53. **Palumbo RN, Zhong X, Wang C.** 2012. Polymer-mediated DNA vaccine delivery via bystander cells requires a proper balance between transfection efficiency and cytotoxicity. *J Control Release* **157:**86–93.
54. **Cho JH, Youn JW, Sung YC.** 2001. Cross-priming as a predominant mechanism for inducing CD8(+) T cell responses in gene gun DNA immunization. *J Immunol* **167:**5549–5557.
55. **Fredriksen AB, Sandlie I, Bogen B.** 2012. Targeted DNA vaccines for enhanced induction of idiotype-specific B and T cells. *Front Oncol* **2:**154.
56. **Grodeland G, Mjaaland S, Roux KH, Fredriksen AB, Bogen B.** 2013. DNA vaccine that targets hemagglutinin to MHC class II molecules rapidly induces antibody-mediated protection against influenza. *J Immunol* **191:** 3221–3231.
57. **McCluskie MJ, Brazolot Millan CL, Gramzinski RA, Robinson HL, Santoro JC, Fuller JT, Widera G, Haynes JR, Purcell RH, Davis HL.** 1999. Route and method of delivery of DNA vaccine influence immune responses in mice and non-human primates. *Mol Med* **5:**287–300.
58. **Combadiere B, Liard C.** 2011. Transcutaneous and intradermal vaccination. *Hum Vaccin* **7:**811–827.
59. **Romani N, Flacher V, Tripp CH, Sparber F, Ebner S, Stoitzner P.** 2012. Targeting skin dendritic cells to improve intradermal vaccination. *Curr Top Microbiol Immunol* **351:**113–138.
60. **Enama ME, Ledgerwood JE, Novik L, Nason MC, Gordon IJ, Holman L, Bailer RT, Roederer M, Koup RA, Mascola JR, Nabel GJ, Graham BS, VRC 011 Study Team.** 2014. Phase I randomized clinical trial of VRC DNA and rAd5 HIV-1 vaccine delivery by intramuscular (i.m.), subcutaneous (s.c.) and intradermal (i.d.) administration (VRC 011). *PloS One* **9:**e91366. doi:10.1371/journal.pone.0091366.
61. **Mahe B, Vogt A, Liard C, Duffy D, Abadie V, Bonduelle O, Boissonnas A, Sterry W, Verrier B, Blume-Peytavi U, Combadiere B.** 2008. Nanoparticle-based targeting of vaccine compounds to skin antigen-presenting cells by

hair follicles and their transport in mice. *J Invest Dermatol* **129:**1156–1164.

62. **Budker V, Zhang G, Danko I, Williams P, Wolff J.** 1998. The efficient expression of intravascularly delivered DNA in rat muscle. *Gene Ther* **5:**272–276.
63. **Hodges BL, Scheule RK.** 2003. Hydrodynamic delivery of DNA. *Exp Opin Biol Ther* **3:**911–918.
64. **Hegge JO, Wooddell CI, Zhang G, Hagstrom JE, Braun S, Huss T, Sebestyen MG, Emborg ME, Wolff JA.** 2010. Evaluation of hydrodynamic limb vein injections in nonhuman primates. *Hum Gene Ther* **21:**829–842.
65. **Bivas-Benita M, Ottenhoff TH, Junginger HE, Borchard G.** 2005. Pulmonary DNA vaccination: concepts, possibilities and perspectives. *J Control Release* **107:**1–29.
66. **Schautteet K, Stuyven E, Beeckman DS, Van Acker S, Carlon M, Chiers K, Cox E, Vanrompay D.** 2011. Protection of pigs against *Chlamydia trachomatis* challenge by administration of a MOMP-based DNA vaccine in the vaginal mucosa. *Vaccine* **29:**1399–1407.
67. **Hamajima K, Hoshino Y, Xin KQ, Hayashi F, Tadokoro K, Okuda K.** 2002. Systemic and mucosal immune responses in mice after rectal and vaginal immunization with HIV-DNA vaccine. *Clin Immunol* **102:**12–18.
68. **Mann JF, McKay PF, Arokiasamy S, Patel RK, Tregoning JS, Shattock RJ.** 2013. Mucosal application of gp140 encoding DNA polyplexes to different tissues results in altered immunological outcomes in mice. *PloS One* **8:**e67412. doi:10.1371/journal.pone.0067412.
69. **Niethammer AG, Lubenau H, Mikus G, Knebel P, Hohmann N, Leowardi C, Beckhove P, Akhisaroglu M, Ge Y, Springer M, Grenacher L, Buchler MW, Koch M, Weitz J, Haefeli WE, Schmitz-Winnenthal FH.** 2012. Double-blind, placebo-controlled first in human study to investigate an oral vaccine aimed to elicit an immune reaction against the VEGF-receptor 2 in patients with stage IV and locally advanced pancreatic cancer. *BMC Cancer* **12:**361.
70. **Greenland JR, Letvin NL.** 2007. Chemical adjuvants for plasmid DNA vaccines. *Vaccine* **25:**3731–3741.
71. **Wegmann F, Gartlan KH, Harandi AM, Brinckmann SA, Coccia M, Hillson WR, Kok WL, Cole S, Ho LP, Lambe T, Puthia M, Svanborg C, Scherer EM, Krashias G, Williams A, Blattman JN, Greenberg PD, Flavell RA, Moghaddam AE, Sheppard NC, Sattentau QJ.** 2012. Polyethyleneimine is a potent mucosal adjuvant for viral glycoprotein antigens. *Nat Biotechnol* **30:**883–888.
72. **Klein K, Mann JF, Rogers P, Shattock RJ.** 2014. Polymeric penetration enhancers promote humoral immune responses to mucosal vaccines. *J Control Release* **183:**43–50.
73. **Klencke B, Matijevic M, Urban RG, Lathey JL, Hedley ML, Berry M, Thatcher J, Weinberg V, Wilson J, Darragh T, Jay N, Da Costa M, Palefsky JM.** 2002. Encapsulated plasmid DNA treatment for human papillomavirus 16-associated anal dysplasia: a phase I study of ZYC101. *Clin Cancer Res* **8:**1028–1037.
74. **Oyewumi MO, Kumar A, Cui Z.** 2010. Nano-microparticles as immune adjuvants: correlating particle sizes and the resultant immune responses. *Exp Rev Vaccin* **9:**1095–1107.
75. **Fenske DB, Cullis PR.** 2008. Liposomal nanomedicines. *Exp Opin Drug Deliv* **5:**25–44.
76. **Smith LR, Wloch MK, Ye M, Reyes LR, Boutsaboualoy S, Dunne CE, Chaplin JA, Rusalov D, Rolland AP, Fisher CL, Al-Ibrahim MS, Kabongo ML, Steigbigel R, Belshe RB, Kitt ER, Chu AH, Moss RB.** 2010. Phase 1 clinical trials of the safety and immunogenicity of adjuvanted plasmid DNA vaccines encoding influenza A virus H5 hemagglutinin. *Vaccine* **28:**2565–2572.
77. **Kheiri MT, Jamali A, Shenagari M, Hashemi H, Sabahi F, Atyabi F, Saghiri R.** 2012. Influenza virosome/DNA vaccine complex as a new formulation to induce intra-subtypic protection against influenza virus challenge. *Antiviral Res* **95:**229–236.
78. **Schoen C, Stritzker J, Goebel W, Pilgrim S.** 2004. Bacteria as DNA vaccine carriers for genetic immunization. *Int J Med Microbiol* **294:**319–335.
79. **Yang NS, Burkholder J, Roberts B, Martinell B, McCabe D.** 1990. *In vivo* and *in vitro* gene transfer to mammalian somatic cells by particle bombardment. *Proc Natl Acad Sci USA* **87:**9568–9572.
80. **Roy MJ, Wu MS, Barr LJ, Fuller JT, Tussey LG, Speller S, Culp J, Burkholder JK, Swain WF, Dixon RM, Widera G, Vessey R, King A, Ogg G, Gallimore A, Haynes JR, Heydenburg Fuller D.** 2000. Induction of antigen-specific CD8+ T cells, T helper cells, and protective levels of antibody in humans by particle-mediated administration of a hepatitis B virus DNA vaccine. *Vaccine* **19:**764–778.
81. **Boudreau EF, Josleyn M, Ullman D, Fisher D, Dalrymple L, Sellers-Myers K, Loudon P, Rusnak J, Rivard R, Schmaljohn C, Hooper JW.** 2012. A phase 1 clinical trial of Hantaan virus and Puumala virus M-segment DNA vaccines for hemorrhagic fever with renal syndrome. *Vaccine* **30:**1951–1958.
82. **Rottinghaus ST, Poland GA, Jacobson RM, Barr LJ, Roy MJ.** 2003. Hepatitis B DNA vaccine induces protective antibody responses in human non-responders to conventional vaccination. *Vaccine* **21:**4604–4608.
83. **Tacket CO, Roy MJ, Widera G, Swain WF, Broome S, Edelman R.** 1999. Phase 1 safety and immune response studies of a DNA vaccine encoding hepatitis B surface antigen delivered by a gene delivery device. *Vaccine* **17:**2826–2829.
84. **Shergold OA, Fleck NA, King TS.** 2006. The penetration of a soft solid by a liquid jet, with application to the administration of a needle-free injection. *J Biomech* **39:**2593–2602.
85. **Ginsberg BA, Gallardo HF, Rasalan TS, Adamow M, Mu Z, Tandon S, Bewkes BB, Roman RA, Chapman PB, Schwartz GK, Carvajal RD, Panageas KS, Terzulli SL, Houghton AN, Yuan JD, Wolchok JD.** 2010. Immunologic response to xenogeneic gp100 DNA in melanoma patients: comparison of particle-mediated epidermal delivery with intramuscular injection. *Clin Cancer Res* **16:**4057–4065.
86. **Kutzler MA, Weiner DB.** 2008. DNA vaccines: ready for prime time? *Nat Rev Genet* **9:**776–788.

87. **Grunwald T, Tenbusch M, Schulte R, Raue K, Wolf H, Hannaman D, de Swart RL, Uberla K, Stahl-Hennig C.** 2014. Novel vaccine regimen elicits strong airway immune responses and control of respiratory syncytial virus in nonhuman primates. *J Virol* **88:**3997–4007.

88. **Verstrepen BE, Bins AD, Rollier CS, Mooij P, Koopman G, Sheppard NC, Sattentau Q, Wagner R, Wolf H, Schumacher TNM, Heeney JL, Haanen JBAG.** 2008. Improved HIV-1 specific T-cell responses by short-interval DNA tattooing as compared to intramuscular immunization in non-human primates. *Vaccine* **26:**3346–3351.

89. **DeMuth PC, Min Y, Huang B, Kramer JA, Miller AD, Barouch DH, Hammond PT, Irvine DJ.** 2013. Polymer multilayer tattooing for enhanced DNA vaccination. *Nat Mat* **12:**367–376.

90. **van Drunen Littel-van den Hurk S, Hannaman D.** 2010. Electroporation for DNA immunization: clinical application. *Exp Rev Vaccines* **9:**503–517.

91. **Gothelf A, Gehl J.** 2012. What you always needed to know about electroporation based DNA vaccines. *Hum Vaccines Immunother* **8:**1694–1702.

92. **Heller R, Jaroszeski M, Atkin A, Moradpour D, Gilbert R, Wands J, Nicolau C.** 1996. *In vivo* gene electroinjection and expression in rat liver. *FEBS Lett* **389:** 225–228.

93. **Widera G, Austin M, Rabussay D, Goldbeck C, Barnett SW, Chen M, Leung L, Otten GR, Thudium K, Selby MJ, Ulmer JB.** 2000. Increased DNA vaccine delivery and immunogenicity by electroporation *in vivo*. *J Immunol* **164:**4635–4640.

94. **Ahlen G, Soderholm J, Tjelle T, Kjeken R, Frelin L, Hoglund U, Blomberg P, Fons M, Mathiesen I, Sallberg M.** 2007. *In vivo* electroporation enhances the immunogenicity of hepatitis C virus nonstructural 3/4A DNA by increased local DNA uptake, protein expression, inflammation, and infiltration of CD3+ T cells. *J Immunol* **179:**4741–4753.

95. **Vasan S, Hurley A, Schlesinger SJ, Hannaman D, Gardiner DF, Dugin DP, Boente-Carrera M, Vittorino R, Caskey M, Andersen J, Huang Y, Cox JH, Tarragona-Fiol T, Gill DK, Cheeseman H, Clark L, Dally L, Smith C, Schmidt C, Park HH, Kopycinski JT, Gilmour J, Fast P, Bernard R, Ho DD.** 2011. *In vivo* electroporation enhances the immunogenicity of an HIV-1 DNA vaccine candidate in healthy volunteers. *PloS One* **6:**e19252. doi:10.1371/journal.pone.0019252.

96. **Pavlenko M, Roos AK, Lundqvist A, Palmborg A, Miller AM, Ozenci V, Bergman B, Egevad L, Hellstrom M, Kiessling R, Masucci G, Wersall P, Nilsson S, Pisa P.** 2004. A phase I trial of DNA vaccination with a plasmid expressing prostate-specific antigen in patients with hormone-refractory prostate cancer. *Br J Cancer* **91:** 688–694.

97. **Trimble CL, Peng S, Kos F, Gravitt P, Viscidi R, Sugar E, Pardoll D, Wu TC.** 2009. A phase I trial of a human papillomavirus DNA vaccine for HPV16+ cervical intraepithelial neoplasia 2/3. *Clin Cancer Res* **15:** 361–367.

98. **Garmory HS, Brown KA, Titball RW.** 2003. DNA vaccines: improving expression of antigens. *Genet Vaccines Ther* **1:**2.

99. **Manoj S, Babiuk LA, van Drunen Littel-van den Hurk S.** 2004. Approaches to enhance the efficacy of DNA vaccines. *Crit Rev Clin Lab Sci* **41:**1–39.

100. **Yew NS, Wysokenski DM, Wang KX, Ziegler RJ, Marshall J, McNeilly D, Cherry M, Osburn W, Cheng SH.** 1997. Optimization of plasmid vectors for high-level expression in lung epithelial cells. *Hum Gene Ther* **8:** 575–584.

101. **Nagata T, Uchijima M, Yoshida A, Kawashima M, Koide Y.** 1999. Codon optimization effect on translational efficiency of DNA vaccine in mammalian cells: analysis of plasmid DNA encoding a CTL epitope derived from microorganisms. *Biochem Biophys Res Commun* **261:**445–451.

102. **Goodwin EC, Rottman FM.** 1992. The 3′-flanking sequence of the bovine growth hormone gene contains novel elements required for efficient and accurate polyadenylation. *J Biol Chem* **267:**16330–16334.

103. **Williams JA, Carnes AE, Hodgson CP.** 2009. Plasmid DNA vaccine vector design: impact on efficacy, safety and upstream production. *Biotechnol Adv* **27:**353–370.

104. **Carvalho JA, Azzoni AR, Prazeres DMF, Monteiro GA.** 2010. Comparative analysis of antigen-targeting sequences used in DNA vaccines. *Mol Biotechnol* **44:** 204–212.

105. **Luo D, Saltzman WM.** 2000. Synthetic DNA delivery systems. *Nat Biotechnol* **18:**33–37.

106. **Scheiermann J, Klinman DM.** 2014. Clinical evaluation of CpG oligonucleotides as adjuvants for vaccines targeting infectious diseases and cancer. *Vaccine* **32:**6377–6389. doi:10.1016/j.vaccine.2014.06.065.

107. **Klinman DM, Yamshchikov G, Ishigatsubo Y.** 1997. Contribution of CpG motifs to the immunogenicity of DNA vaccines. *J Immunol* **158:**3635–3639.

108. **Martinez-Alonso S, Martinez-Lopez A, Estepa A, Cuesta A, Tafalla C.** 2011. The introduction of multi--copy CpG motifs into an antiviral DNA vaccine strongly up-regulates its immunogenicity in fish. *Vaccine* **29:** 1289–1296.

109. **Miyabe H, Hyodo M, Nakamura T, Sato Y, Hayakawa Y, Harashima H.** 2014. A new adjuvant delivery system 'cyclic di-GMP/YSK05 liposome' for cancer immunotherapy. *J Controll Release* **184:**20–27.

110. **Khosroshahi KH, Ghaffarifar F, Sharifi Z, D'Souza S, Dalimi A, Hassan ZM, Khoshzaban F.** 2012. Comparing the effect of IL-12 genetic adjuvant and alum non-genetic adjuvant on the efficiency of the cocktail DNA vaccine containing plasmids encoding SAG-1 and ROP-2 of *Toxoplasma gondii*. *Parasitol Res* **111:**403–411.

111. **Ott G, Singh M, Kazzaz J, Briones M, Soenawan E, Ugozzoli M, O'Hagan DT.** 2002. A cationic submicron emulsion (MF59/DOTAP) is an effective delivery system for DNA vaccines. *J Controll Release* **79:**1–5.

112. **Kwissa M, Amara RR, Robinson HL, Moss B, Alkan S, Jabbar A, Villinger F, Pulendran B.** 2007. Adjuvanting a

DNA vaccine with a TLR9 ligand plus Flt3 ligand results in enhanced cellular immunity against the simian immunodeficiency virus. *J Exp Med* **204:**2733–2746.

113. **Otero M, Calarota SA, Felber B, Laddy D, Pavlakis G, Boyer JD, Weiner DB.** 2004. Resiquimod is a modest adjuvant for HIV-1 gag-based genetic immunization in a mouse model. *Vaccine* **22:**1782–1790.

114. **Sun J, Hou J, Li D, Liu Y, Hu N, Hao Y, Fu J, Hu Y, Shao Y.** 2013. Enhancement of HIV-1 DNA vaccine immunogenicity by BCG-PSN, a novel adjuvant. *Vaccine* **31:**472–479.

115. **Abdulhaqq SA, Weiner DB.** 2008. DNA vaccines: developing new strategies to enhance immune responses. *Immunol Res* **42:**219–232.

116. **Baden LR, Blattner WA, Morgan C, Huang Y, Defawe OD, Sobieszczyk ME, Kochar N, Tomaras GD, McElrath MJ, Russell N, Brandariz K, Cardinali M, Graham BS, Barouch DH, Dolin R, NIAID HIV Vaccine Trials Network 044 Study Team.** 2011. Timing of plasmid cytokine (IL-2/Ig) administration affects HIV-1 vaccine immunogenicity in HIV-seronegative subjects. *J Infect Dis* **204:**1541–1549.

117. **Kalams SA, Parker SD, Elizaga M, Metch B, Edupuganti S, Hural J, De Rosa S, Carter DK, Rybczyk K, Frank I, Fuchs J, Koblin B, Kim DH, Joseph P, Keefer MC, Baden LR, Eldridge J, Boyer J, Sherwat A, Cardinali M, Allen M, Pensiero M, Butler C, Khan AS, Yan J, Sardesai NY, Kublin JG, Weiner DB, NIAID HIV Vaccine Trials Network.** 2013. Safety and comparative immunogenicity of an HIV-1 DNA vaccine in combination with plasmid interleukin 12 and impact of intramuscular electroporation for delivery. *J Infect Dis* **208:** 818–829.

118. **Richie TL, Charoenvit Y, Wang R, Epstein JE, Hedstrom RC, Kumar S, Luke TC, Freilich DA, Aguiar JC, Sacci JB Jr, Sedegah M, Nosek RA Jr, De La Vega P, Berzins MP, Majam VF, Abot EN, Ganeshan H, Richie NO, Banania JG, Baraceros MF, Geter TG, Mere R, Bebris L, Limbach K, Hickey BW, Lanar DE, Ng J, Shi M, Hobart PM, Norman JA, Soisson LA, Hollingdale MR, Rogers WO, Doolan DL, Hoffman SL.** 2012. Clinical trial in healthy malaria-naive adults to evaluate the safety, tolerability, immunogenicity and efficacy of MuStDO5, a five-gene, sporozoite/hepatic stage *Plasmodium falciparum* DNA vaccine combined with escalating dose human GM-CSF DNA. *Hum Vaccin Immunother* **8:**1564–1584.

119. **Kim JJ, Bagarazzi ML, Trivedi N, Hu Y, Kazahaya K, Wilson DM, Ciccarelli R, Chattergoon MA, Dang K, Mahalingam S, Chalian AA, Agadjanyan MG, Boyer JD, Wang B, Weiner DB.** 1997. Engineering of *in vivo* immune responses to DNA immunization via codelivery of costimulatory molecule genes. *Nat Biotechnol* **15:** 641–646.

120. **Zhai YZ, Zhou Y, Ma L, Feng GH.** 2013. The dominant roles of ICAM-1-encoding gene in DNA vaccination against Japanese encephalitis virus are the activation of dendritic cells and enhancement of cellular immunity. *Cell Immunol* **281:**1–10.

121. **Saha S, Takeshita F, Matsuda T, Jounai N, Kobiyama K, Matsumoto T, Sasaki S, Yoshida A, Xin KQ, Klinman DM, Uematsu S, Ishii KJ, Akira S, Okuda K.** 2007. Blocking of the TLR5 activation domain hampers protective potential of flagellin DNA vaccine. *J Immunol* **179:**1147–1154.

122. **Luke JM, Simon GG, Soderholm J, Errett JS, August JT, Gale M Jr, Hodgson CP, Williams JA.** 2011. Co-expressed RIG-I agonist enhances humoral immune response to influenza virus DNA vaccine. *J Virol* **85:** 1370–1383.

123. **Kalams SA, Parker S, Jin X, Elizaga M, Metch B, Wang M, Hural J, Lubeck M, Eldridge J, Cardinali M, Blattner WA, Sobieszczyk M, Suriyanon V, Kalichman A, Weiner DB, Baden LR, NIAID HIV Vaccine Trials Network.** 2012. Safety and immunogenicity of an HIV-1 gag DNA vaccine with or without IL-12 and/or IL-15 plasmid cytokine adjuvant in healthy, HIV-1 uninfected adults. *PloS One* **7:**e29231. doi:10.1371/journal.pone.0029231.

124. **Ledgerwood JE, Wei CJ, Hu Z, Gordon IJ, Enama ME, Hendel CS, McTamney PM, Pearce MB, Yassine HM, Boyington JC, Bailer R, Tumpey TM, Koup RA, Mascola JR, Nabel GJ, Graham BS, VRC 306 Study Team.** 2011. DNA priming and influenza vaccine immunogenicity: two phase 1 open label randomised clinical trials. *Lancet Infect Dis* **11:**916–924.

125. **Liu MA.** 2011. DNA vaccines: an historical perspective and view to the future. *Immunol Rev* **239:**62–84.

126. **Muthumani K, Flingai S, Wise M, Tingey C, Ugen KE, Weiner DB.** 2013. Optimized and enhanced DNA plasmid vector based *in vivo* construction of a neutralizing anti-HIV-1 envelope glycoprotein Fab. *Hum Vaccin Immunother* **9:**2253–2262.

127. **Tan GS, Krammer F, Eggink D, Kongchanagul A, Moran TM, Palese P.** 2012. A pan-H1 anti-hemagglutinin monoclonal antibody with potent broad-spectrum efficacy *in vivo*. *J Virol* **86:**6179–6188.

128. **Graham BS, Koup RA, Roederer M, Bailer RT, Enama ME, Moodie Z, Martin JE, McCluskey MM, Chakrabarti BK, Lamoreaux L, Andrews CA, Gomez PL, Mascola JR, Nabel GJ, Vaccine Research Center 004 Study Team.** 2006. Phase 1 safety and immunogenicity evaluation of a multiclade HIV-1 DNA candidate vaccine. *J Infect Dis* **194:**1650–1660.

129. **Gorse GJ, Baden LR, Wecker M, Newman MJ, Ferrari G, Weinhold KJ, Livingston BD, Villafana TL, Li H, Noonan E, Russell ND, HIV Vaccine Trials Network.** 2008. Safety and immunogenicity of cytotoxic T-lymphocyte poly-epitope, DNA plasmid (EP HIV-1090) vaccine in healthy, human immunodeficiency virus type 1 (HIV-1)-uninfected adults. *Vaccine* **26:**215–223.

130. **Graham BS, Enama ME, Nason MC, Gordon IJ, Peel SA, Ledgerwood JE, Plummer SA, Mascola JR, Bailer RT, Roederer M, Koup RA, Nabel GJ, VRC 008 Study Team.** 2013. DNA vaccine delivered by a needle-free injection device improves potency of priming for antibody and CD8+ T-cell responses after rAd5 boost in a randomized clinical trial. *PloS One* **8:**e59340. doi: 10.1371/journal.pone.0059340.

131. Rosenberg ES, Graham BS, Chan ES, Bosch RJ, Stocker V, Maenza J, Markowitz M, Little S, Sax PE, Collier AC, Nabel G, Saindon S, Flynn T, Kuritzkes D, Barouch DH, AIDS Clinical Trials Group A5187 Team. 2010. Safety and immunogenicity of therapeutic DNA vaccination in individuals treated with antiretroviral therapy during acute/early HIV-1 infection. *PloS One* **5**: e10555. doi:10.1371/journal.pone.0010555.

132. Jin X, Newman MJ, De-Rosa S, Cooper C, Thomas E, Keefer M, Fuchs J, Blattner W, Livingston BD, McKinney DM, Noonan E, Decamp A, Defawe OD, Wecker M, NIAID HIV Vaccine Trials Network. 2009. A novel HIV T helper epitope-based vaccine elicits cytokine-secreting HIV-specific CD4+ T cells in a phase I clinical trial in HIV-uninfected adults. *Vaccine* **27**: 7080–7086.

133. Vasan S, Schlesinger SJ, Huang Y, Hurley A, Lombardo A, Chen Z, Than S, Adesanya P, Bunce C, Boaz M, Boyle R, Sayeed E, Clark L, Dugin D, Schmidt C, Song Y, Seamons L, Dally L, Ho M, Smith C, Markowitz M, Cox J, Gill DK, Gilmour J, Keefer MC, Fast P, Ho DD. 2010. Phase 1 safety and immunogenicity evaluation of ADVAX, a multigenic, DNA-based clade C/B′ HIV-1 candidate vaccine. *PloS One* **5**:e8617. doi:10.1371/journal.pone.0008617.

134. Casazza JP, Bowman KA, Adzaku S, Smith EC, Enama ME, Bailer RT, Price DA, Gostick E, Gordon IJ, Ambrozak DR, Nason MC, Roederer M, Andrews CA, Maldarelli FM, Wiegand A, Kearney MF, Persaud D, Ziemniak C, Gottardo R, Ledgerwood JE, Graham BS, Koup RA, VRC 101 Study Team. 2013. Therapeutic vaccination expands and improves the function of the HIV-specific memory T-cell repertoire. *J Infect Dis* **207**: 1829–1840.

135. Beckett CG, Tjaden J, Burgess T, Danko JR, Tamminga C, Simmons M, Wu SJ, Sun P, Kochel T, Raviprakash K, Hayes CG, Porter KR. 2011. Evaluation of a prototype dengue-1 DNA vaccine in a phase 1 clinical trial. *Vaccine* **29**:960–968.

136. Goepfert PA, Elizaga ML, Sato A, Qin L, Cardinali M, Hay CM, Hural J, DeRosa SC, DeFawe OD, Tomaras GD, Montefiori DC, Xu Y, Lai L, Kalams SA, Baden LR, Frey SE, Blattner WA, Wyatt LS, Moss B, Robinson HL, National Institute of Allergy and Infectious Diseases HIV Vaccine Trials Network. 2011. Phase 1 safety and immunogenicity testing of DNA and recombinant modified vaccinia Ankara vaccines expressing HIV-1 virus-like particles. *J Infect Dis* **203**:610–619.

137. Koblin BA, Casapia M, Morgan C, Qin L, Wang ZM, Defawe OD, Baden L, Goepfert P, Tomaras GD, Montefiori DC, McElrath MJ, Saavedra L, Lau CY, Graham BS, NIAID HIV Vaccine Trials Network. 2011. Safety and immunogenicity of an HIV adenoviral vector boost after DNA plasmid vaccine prime by route of administration: a randomized clinical trial. *PloS One* **6**: e24517. doi:10.1371/journal.pone.0024517.

138. Gorse GJ, Newman MJ, deCamp A, Hay CM, De Rosa SC, Noonan E, Livingston BD, Fuchs JD, Kalams SA, Cassis-Ghavami FL, NIAID HIV Vaccine Trials Network. 2012. DNA and modified vaccinia virus Ankara vaccines encoding multiple cytotoxic and helper T-lymphocyte epitopes of human immunodeficiency virus type 1 (HIV-1) are safe but weakly immunogenic in HIV-1-uninfected, vaccinia virus-naive adults. *Clin Vaccine Immunol* **19**:649–658.

139. Godon O, Fontaine H, Kahi S, Meritet J, Scott-Algara D, Pol S, Michel M, Bourgine M. 2014. Immunological and antiviral responses after therapeutic DNA immunization in chronic hepatitis B patients efficiently treated by analogues. *Mol Ther* **22**:675–684.

140. Mancini-Bourgine M, Fontaine H, Scott-Algara D, Pol S, Brechot C, Michel ML. 2004. Induction or expansion of T-cell responses by a hepatitis B DNA vaccine administered to chronic HBV carriers. *Hepatology* **40**: 874–882.

141. Chuang I, Sedegah M, Cicatelli S, Spring M, Polhemus M, Tamminga C, Patterson N, Guerrero M, Bennett JW, McGrath S, Ganeshan H, Belmonte M, Farooq F, Abot E, Banania JG, Huang J, Newcomer R, Rein L, Litilit D, Richie NO, Wood C, Murphy J, Sauerwein R, Hermsen CC, McCoy AJ, Kamau E, Cummings J, Komisar J, Sutamihardja A, Shi M, Epstein JE, Maiolatesi S, Tosh D, Limbach K, Angov E, Bergmann-Leitner E, Bruder JT, Doolan DL, King CR, Carucci D, Dutta S, Soisson L, Diggs C, Hollingdale MR, Ockenhouse CF, Richie TL. 2013. DNA prime/adenovirus boost malaria vaccine encoding *P. falciparum* CSP and AMA1 induces sterile protection associated with cell-mediated immunity. *PloS One* **8**:e55571. doi: 10.1371/journal.pone.0055571.

142. Hammer SM, Sobieszczyk ME, Janes H, Karuna ST, Mulligan MJ, Grove D, Koblin BA, Buchbinder SP, Keefer MC, Tomaras GD, Frahm N, Hural J, Anude C, Graham BS, Enama ME, Adams E, DeJesus E, Novak RM, Frank I, Bentley C, Ramirez S, Fu R, Koup RA, Mascola JR, Nabel GJ, Montefiori DC, Kublin J, McElrath MJ, Corey L, Gilbert PB, HVTN 505 Study Team. 2013. Efficacy trial of a DNA/rAd5 HIV-1 preventive vaccine. *N Engl J Med* **369**:2083–2092.

143. Scott-Algara D, Mancini-Bourgine M, Fontaine H, Pol S, Michel ML. 2010. Changes to the natural killer cell repertoire after therapeutic hepatitis B DNA vaccination. *PloS One* **5**:e8761. doi:10.1371/journal.pone.0008761.

144. Ledgerwood JE, Zephir K, Hu Z, Wei CJ, Chang L, Enama ME, Hendel CS, Sitar S, Bailer RT, Koup RA, Mascola JR, Nabel GJ, Graham BS, VRC 310 Study Team. 2013. Prime-boost interval matters: a randomized phase 1 study to identify the minimum interval necessary to observe the H5 DNA influenza vaccine priming effect. *J Infect Dis* **208**:418–422.

145. Khurana S, Wu J, Dimitrova M, King LR, Manischewitz J, Graham BS, Ledgerwood JE, Golding H. 2013. DNA priming prior to inactivated influenza A(H5N1) vaccination expands the antibody epitope repertoire and increases affinity maturation in a boost-interval-dependent manner in adults. *J Infect Dis* **208**:413–417.

146. Wang R, Doolan DL, Le TP, Hedstrom RC, Coonan KM, Charoenvit Y, Jones TR, Hobart P, Margalith M, Ng J, Weiss WR, Sedegah M, De Taisne C, Norman JA,

Hoffman SL. 1998. Induction of antigen-specific cytotoxic T lymphocytes in humans by a malaria DNA vaccine. *Science* **282:**476–480.

147. **Le TP, Coonan KM, Hedstrom RC, Charoenvit Y, Sedegah M, Epstein JE, Kumar S, Wang R, Doolan DL, Maguire JD, Parker SE, Hobart P, Norman J, Hoffman SL.** 2000. Safety, tolerability and humoral immune responses after intramuscular administration of a malaria DNA vaccine to healthy adult volunteers. *Vaccine* **18:**1893–1901.
148. **McConkey SJ, Reece WH, Moorthy VS, Webster D, Dunachie S, Butcher G, Vuola JM, Blanchard TJ, Gothard P, Watkins K, Hannan CM, Everaere S, Brown K, Kester KE, Cummings J, Williams J, Heppner DG, Pathan A, Flanagan K, Arulanantham N, Roberts MT, Roy M, Smith GL, Schneider J, Peto T, Sinden RE, Gilbert SC, Hill AV.** 2003. Enhanced T-cell immunogenicity of plasmid DNA vaccines boosted by recombinant modified vaccinia virus Ankara in humans. *Nat Med* **9:**729–735.
149. **Moorthy VS, Pinder M, Reece WH, Watkins K, Atabani S, Hannan C, Bojang K, McAdam KP, Schneider J, Gilbert S, Hill AV.** 2003. Safety and immunogenicity of DNA/modified vaccinia virus Ankara malaria vaccination in African adults. *J Infect Dis* **188:**1239–1244.
150. **Dunachie SJ, Walther M, Epstein JE, Keating S, Berthoud T, Andrews L, Andersen RF, Bejon P, Goonetilleke N, Poulton I, Webster DP, Butcher G, Watkins K, Sinden RE, Levine GL, Richie TL, Schneider J, Kaslow D, Gilbert SC, Carucci DJ, Hill AV.** 2006. A DNA prime-modified vaccinia virus Ankara boost vaccine encoding thrombospondin-related adhesion protein but not circumsporozoite protein partially protects healthy malaria-naive adults against *Plasmodium falciparum* sporozoite challenge. *Infect Immun* **74:**5933–5942.
151. **Wang R, Richie TL, Baraceros MF, Rahardjo N, Gay T, Banania JG, Charoenvit Y, Epstein JE, Luke T, Freilich DA, Norman J, Hoffman SL.** 2005. Boosting of DNA vaccine-elicited gamma interferon responses in humans by exposure to malaria parasites. *Infect Immun* **73:**2863–2872.
152. **Moorthy VS, Imoukhuede EB, Milligan P, Bojang K, Keating S, Kaye P, Pinder M, Gilbert SC, Walraven G, Greenwood BM, Hill AS.** 2004. A randomised, double-blind, controlled vaccine efficacy trial of DNA/MVA ME-TRAP against malaria infection in Gambian adults. *PLoS Med* **1:**e33. doi:10.1371/journal.pmed.0010033.
153. **Moorthy VS, Imoukhuede EB, Keating S, Pinder M, Webster D, Skinner MA, Gilbert SC, Walraven G, Hill AV.** 2004. Phase 1 evaluation of 3 highly immunogenic prime-boost regimens, including a 12-month reboosting vaccination, for malaria vaccination in Gambian men. *J Infect Dis* **189:**2213–2219.
154. **Epstein JE, Charoenvit Y, Kester KE, Wang R, Newcomer R, Fitzpatrick S, Richie TL, Tornieporth N, Heppner DG, Ockenhouse C, Majam V, Holland C, Abot E, Ganeshan H, Berzins M, Jones T, Freydberg CN, Ng J, Norman J, Carucci DJ, Cohen J, Hoffman SL.** 2004. Safety, tolerability, and antibody responses in humans after sequential immunization with a PfCSP DNA vaccine followed by the recombinant protein vaccine RTS,S/AS02A. *Vaccine* **22:**1592–1603.
155. **Mehendale S, Thakar M, Sahay S, Kumar M, Shete A, Sathyamurthi P, Verma A, Kurle S, Shrotri A, Gilmour J, Goyal R, Dally L, Sayeed E, Zachariah D, Ackland J, Kochhar S, Cox JH, Excler JL, Kumaraswami V, Paranjape R, Ramanathan VD.** 2013. Safety and immunogenicity of DNA and MVA HIV-1 subtype C vaccine prime-boost regimens: a phase I randomised trial in HIV-uninfected Indian volunteers. *PloS One* **8:**e55831. doi:10.1371/journal.pone.0055831.
156. **Ledgerwood JE, Hu Z, Gordon IJ, Yamshchikov G, Enama ME, Plummer S, Bailer R, Pearce MB, Tumpey TM, Koup RA, Mascola JR, Nabel GJ, Graham BS, VRC 304 and VRC 305 Study Teams.** 2012. Influenza virus h5 DNA vaccination is immunogenic by intramuscular and intradermal routes in humans. *Clin Vaccine Immunol* **19:**1792–1797.
157. **Kopycinski J, Cheeseman H, Ashraf A, Gill D, Hayes P, Hannaman D, Gilmour J, Cox JH, Vasan S.** 2012. A DNA-based candidate HIV vaccine delivered via *in vivo* electroporation induces CD4 responses toward the alpha4beta7-binding V2 loop of HIV gp120 in healthy volunteers. *Clin Vaccine Immunol* **19:**1557–1559.
158. **Kay MA.** 2011. State-of-the-art gene-based therapies: the road ahead. *Nat Rev Genetics* **12:**316–328.
159. **Moreno S, Lopez-Fuertes L, Vila-Coro AJ, Sack F, Smith CA, Konig SA, Wittig B, Schroff M, Juhls C, Junghans C, Timon M.** 2004. DNA immunisation with minimalistic expression constructs. *Vaccine* **22:**1709–1716.
160. **Walters AA, Kinnear E, Shattock RJ, McDonald JU, Caproni LJ, Porter N, Tregoning JS.** 2014. Comparative analysis of enzymatically produced novel linear DNA constructs with plasmids for use as DNA vaccines. *Gene Ther* **21:**645–652.
161. **Cranenburgh RM, Hanak JA, Williams SG, Sherratt DJ.** 2001. *Escherichia coli* strains that allow antibiotic-free plasmid selection and maintenance by repressor titration. *Nucleic Acids Res* **29:**E26.
162. **Mairhofer J, Cserjan-Puschmann M, Striedner G, Nobauer K, Razzazi-Fazeli E, Grabherr R.** 2010. Marker-free plasmids for gene therapeutic applications: lack of antibiotic resistance gene substantially improves the manufacturing process. *J Biotechnol* **146:**130–137.
163. **Vidal L, Pinsach J, Striedner G, Caminal G, Ferrer P.** 2008. Development of an antibiotic-free plasmid selection system based on glycine auxotrophy for recombinant protein overproduction in *Escherichia coli. J Biotechnol* **134:**127–136.
164. **Szpirer CY, Milinkovitch MC.** 2005. Separate-component-stabilization system for protein and DNA production without the use of antibiotics. *Biotechniques* **38:**775–781.
165. **Luke J, Carnes AE, Hodgson CP, Williams JA.** 2009. Improved antibiotic-free DNA vaccine vectors utilizing a novel RNA based plasmid selection system. *Vaccine* **27:**6454–6459.
166. **Goh S, Good L.** 2008. Plasmid selection in *Escherichia coli* using an endogenous essential gene marker. *BMC Biotechnol* **8:**61.

Plasmids—Biology and Impact in Biotechnology and Discovery
Edited by Marcelo E. Tolmasky and Juan C. Alonso

doi:10.1128/microbiolspec.PLAS-0022-2014

Duarte Miguel F. Prazeres[1]
Gabriel A. Monteiro[1]

Plasmid Biopharmaceuticals

34

INTRODUCTION

The contributions of plasmids to biology and their impact in biotechnology and discovery have been immense. Together with restriction enzymes, plasmids were one of the key molecular tools at the heart of the invention and development of DNA cloning and recombinant DNA by Hebert Boyer and Stanley Cohen (1, 2). These fundamental technologies shaped molecular biology and paved the way to the development of the modern, multibillion dollar biotechnology industry (2, 3). The ability to produce unlimited amounts of proteins via the cloning of the corresponding gene into a plasmid and subsequent transformation of a microbial host made it possible to develop a range of medically and industrially relevant products and applications. The development of molecular diagnostics and protein biopharmaceuticals, for example, would have been impossible without plasmids. However, few would have suspected in the earlier years of recombinant DNA that plasmids could one day assume the role of biopharmaceuticals themselves (4).

The breakthrough that sparked the development of plasmid biopharmaceuticals came in 1990, when Wolff and colleagues injected saline solutions of plasmids containing genes for chloramphenicol acetyltransferase, luciferase, and β-galactosidase into the skeletal muscle of live mice (Fig. 1) (5). The authors found that the reporter transgenes encoded in such a "naked" plasmid DNA molecule were expressed within the muscle cells and concomitantly envisaged the use of plasmid-mediated gene transfer into human muscle as a means of improving the effects of genetic diseases of muscle. Transfection by direct injection of naked DNA was subsequently found in tissues other than skeletal muscle, like liver (6), heart (7), and brain (8), and in species as varied as fish (9), chicken (10), and cattle (11). The proximate discovery that mice could elicit antibodies (12) and generate cytotoxic T lymphocytes (13) in response to the direct administration of naked plasmid DNA molecules encoding an antigen showed that, in principle, plasmids could also be used to immunize animals against pathogens. This innovative approach, later termed DNA vaccination, constituted a radical departure from conventional immunization methodologies, which relied on the industrial production of the vaccinating antigens before their administration.

The seminal discoveries of the early 1990s that opened up the possibility of using plasmids as biopharmaceuticals for therapy (5) or prophylaxis (12, 13) were followed by 10 years of major innovations (Fig. 1) (14). The following are noteworthy examples of those milestones: the delivery of plasmid DNA by particle bombardment (15), the application of electroporation for *in vivo* delivery (16), the coexpression of cytokines alongside with the target genes (17), the addition of

[1]IBB, Institute for Biotechnology and Bioengineering, Centre for Biological and Chemical Engineering, Department of Bioengineering, Instituto Superior Técnico, Universidade de Lisboa, 1049-001 Lisboa, Portugal.

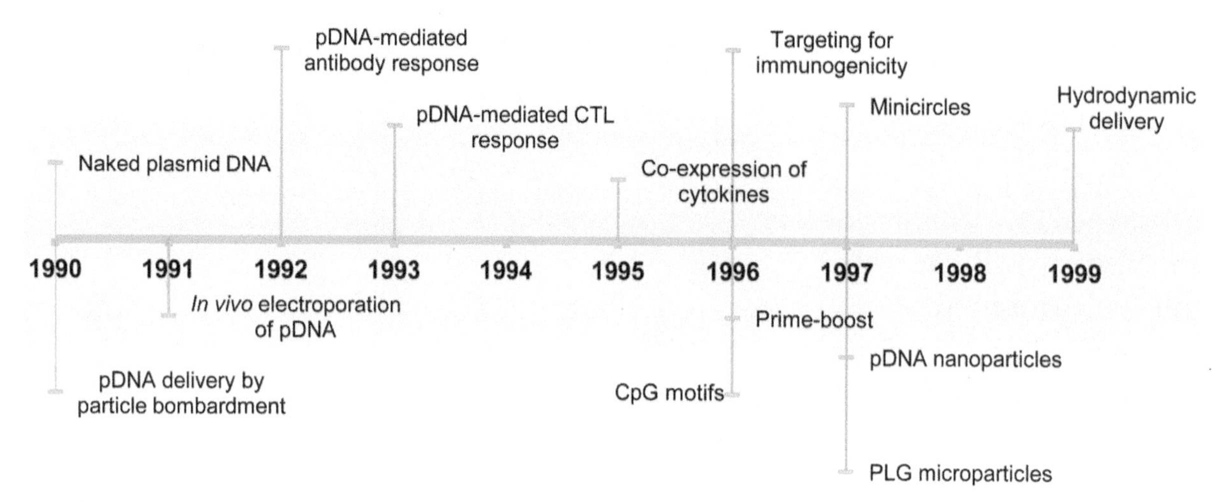

Figure 1 Plasmid biopharmaceuticals: the formative years. doi:10.1128/microbiolspec.PLAS-0022-2014.f1

immunostimulatory CpG motifs to plasmid backbones (18), prime (with DNA vaccine)-boost (with non-DNA vaccine) vaccination (19), the use of targeting sequences to enhance the immunogenicity of DNA vaccines (20), the encapsulation of plasmids in microparticles (21), the design of minimal plasmids (so-called minicircles) containing only the functional elements required for expression of the transgene (22), the compaction of single molecules of DNA into minimally sized nanoparticles (23), and the systemic *in vivo* administration of plasmid DNA by rapid injection of large volumes of solution (24). These formative years (Fig. 1) were followed by intensive research efforts directed toward the adaptation of the major concepts developed earlier to new applications and to the expansion and accumulation of the scientific know-how related to the mechanisms of action of plasmid biopharmaceuticals *in vivo*, via laboratorial, preclinical, and clinical experimentation. Major innovations also took place on the "process" side as the industry sensed the increase in the maturity of the product prototypes and concepts and prepared for clinical development and manufacturing (14).

PLASMIDS IN DISEASE MANAGEMENT

Plasmids versus Viral Vectors

The use of plasmids as carriers of medically relevant genes is usually considered on par with viral vectors. Viral vectors are very effective at transferring genes because of their natural ability to deliver and express genes, while avoiding the different defense barriers of the host organism and cells. For this reason, recombinant viral vectors are the gene carriers of choice in more than 65% of the clinical trials of gene therapy registered as of January 2014. (Data were extracted from *The Journal of Gene Medicine* Gene Therapy Clinical Trials Worldwide website, http://www.wiley.co.uk/genmed/clinical [last accessed on 4 February 2014].) However, even though recombinant viruses used in gene transfer are designed to minimize the toxicity and immunogenicity of their natural counterparts, safety concerns associated with the use of viral vectors remain high as a result of a number of incidents and serious adverse events recorded during a number of gene therapy clinical trials (25). Plasmids, on the other hand, are characterized by an excellent safety profile (see below). For this reason, close to 20% of the gene therapy clinical trials recorded up to 2014 had used plasmid DNA as a carrier of the therapeutic/prophylactic transgenes (*The Journal of Gene Medicine* Gene Therapy Clinical Trials Worldwide website mentioned above).

Role of Transgene Products

Generically speaking, plasmid biopharmaceuticals are used to transfer genes with the goal of managing disease in humans and animals. The rationale behind this approach is that, once expressed in the target cells/tissues, the products coded in the plasmid-borne transgenes will act in such a way as to tackle and resolve the specific disease or clinical condition under study. The different functions exerted by plasmid-borne transgene products can be broadly divided into five categories, as briefly described next (26).

Boosting

Plasmids can be used to increase the expression of a specific endogenous protein, whose level is otherwise normal, by adding more copies of the coding genes. This could contribute to accelerating the endogenous response generated by our bodies in the context of a specific disease (26). For example, an increase in the expression of vascular endothelial growth factor (VEGF) can accelerate the vascularization of ischemic tissue in arterial diseases (27), and an increase in the expression of the hepatocyte growth factor (HGF) gene may enhance the function of dopaminergic neurons in Parkinson's disease (28).

Replacement

When a hereditary defect in a single gene prevents the body from functioning normally, e.g., as in cystic fibrosis (29) or Duchenne muscular dystrophy (30), regular levels of the normal protein can be supplemented by transferring the correct gene via plasmids (26). Replacement can also be explored to compensate for the deterioration of normal levels of a protein as a consequence of disease (e.g., insulin in type 1 diabetes mellitus [31], erythropoietin in anemia [32]).

Immune Stimulation

DNA vaccines can be designed on the basis of plasmids that transfer genes whose products are able to recruit the immune system (26, 33). These vaccines can be administered to prevent future episodes of the target disease (prophylactic vaccines) or to motivate the immune system to fight cancer (therapeutic vaccines). In the first case, the DNA vaccine carries the gene that codes for a specific antigen of the causative infectious agent (e.g., AIDS [34], malaria [35], tuberculosis [36], influenza [37]), whereas, in the second case, genes that code for products that increase tumor immunogenicity and mobilize immune cells to fight cancer are used (38). Unlike in the case of traditional vaccines, antigens delivered by DNA vaccines are synthesized endogenously, and, thus, the process of antigen presentation that ensues may mimic natural infection more closely.

Cytotoxicity

The plasmid-mediated transfer of genes can be used to kill malignant cells. The therapeutic strategy is usually designed so that the gene product plays an intermediate role (e.g., by replacing a missing key protein, stimulating the immune system into recognizing harmful cells, or introducing a new functionality that contributes to kill cells) in a more complex network of events that ultimately result in the death of the target cells (26).

Blocking

The genetic information in a plasmid can also code for short hairpin RNAs (shRNAs), which once expressed will knock down the expression of the disease-related target genes via RNA interference pathway (39, 40, 41).

Plasmids for Therapy

Gene transfer via plasmid molecules has been studied as a possibility to treat both hereditary disorders that are characterized by deficiencies at the single-gene level and diseases that are caused by a combination of environmental factors and genetic predisposition.

Hereditary Disorders

In this case, the expectation is that the plasmid-mediated delivery of the correct genes results in the restoration of normal levels of the faulty protein and hence in the halting of the course of the disease. The management of a genetic disorder by using plasmids will inevitably rely on chronic administration, since plasmids are typically cleared by the cell machinery after a certain amount of time has elapsed. The possibility of the development of autoimmune responses or immune tolerance is thus a cause for concern. Examples of genetic disorders that have been addressed by plasmid-based gene transfer include (i) hemophilia, a coagulation disorder associated with defects in factor VIII (hemophilia A) and factor IX (hemophilia B) (42, 43); (ii) cystic fibrosis, a multiorgan disease caused by an abnormal cystic fibrosis transmembrane regulator gene (29, 44); and (iii) Duchenne muscular dystrophy, a neuromuscular disorder associated with defects in the dystrophin gene (30, 45). A key challenge in the management of these diseases is to ensure that the corrective genes are delivered to the proper cell. One of the strategies studied to achieve this targeting relies on the use of plasmid delivery vehicles modified with ligands for specific receptors of the target cells. In other situations, the relevant tissue can be targeted by direct administration of the plasmid to the relevant tissue (e.g., aerosol delivery of plasmids to the lungs in the case of cystic fibrosis [29], intramuscular injection in the case of muscular dystrophy [30]).

Multifactorial Diseases

Clinical research on plasmid biopharmaceuticals has focused strongly on multifactorial diseases that result from a combination of environmental factors and genetic predisposition. For example, in the context of coronary and peripheral arterial diseases, plasmids have been used to deliver specific genes such as fibroblast growth factor (FGF), VEGF, HGF, and AGGF1 that promote the formation of new blood vessels and thus increase blood flow to the affected ischemic tissues (myocardium, limbs) (46, 47, 48, 49).

Additionally, many clinical trials and a large body of scientific research have been directed at treating cancer with plasmids. Several therapeutic strategies can be devised to kill cancer cells on the basis of genes including tumor suppression, suicide gene therapy, antiangiogenesis, and immune stimulation. For example, cancer cells can be driven to commit suicide by delivering a missing or defective tumor suppressor gene like p53 (50, 51, 52, 53). The suicide gene therapy approach calls for plasmid-encoded proteins to convert a coadministered prodrug into an active, cytotoxic agent that kills the tumor cells. The coupling of the cytosine deaminase gene with the prodrug 5-fluorocytosine (54) and of the herpes simplex virus thymidine kinase gene with the antiviral drug ganciclovir (55) are two of the most researched enzyme-prodrug systems. In this case, the cytotoxicity is not limited to the transfected cells, since the active drug can diffuse and act on neighboring cells (bystander effect). Cancer can hypothetically be treated by inhibiting the formation of the network of blood vessels (i.e., angiogenesis) that supply nutrients and oxygen to tumoral cells. Plasmids have been used in this context to deliver genes that code for antiangiogenic proteins, such as the VEGF receptor sFLT147 (56), endostatin (41), or interleukin-12 to the affected tissues (57). Immune cells can be stimulated to fight cancer cells by using DNA vaccines that promote the expression of cell surface markers or cytokines. In the first case, plasmids are used to deliver the genes that code for tumor-associated antigens, like the prostate-specific antigen (58) in prostate cancer and gp75/tyrosinase-related proteins in melanoma (59). Once expressed, these antigenic markers are adequately processed and displayed, recruiting cytotoxic T lymphocytes (CTLs) that eventually kill the neoplastic cells (38, 60). In the second case, the regulatory role of interleukins, interferons, tumor necrosis factors, and colony-stimulating factors in the pathogenesis of cancer is explored as a means of generating potent antitumor responses (42, 61, 62).

Further examples of multifactorial diseases in which plasmid vectors have been used to deliver therapeutic genes include Alzheimer's (63), anemia (32), arthritis (64), burn wounds (65), dental caries (66), diabetes mellitus (31), glaucoma (67), lupus (68), sepsis (69), spinal cord injury (70), and wound healing (71).

Plasmids for Prophylaxis

One of the most appealing applications of plasmid biopharmaceuticals is in the prevention of infectious diseases. In principle, DNA vaccines can be designed to immunize humans and animals against an enormous range of diseases caused by viruses, bacteria, protozoans, and fungi (33). Studies performed with animal models have demonstrated that the protection conferred by DNA vaccines against infectious agents occurs via the activation of the innate immune system and the induction of CTLs, T-helper (Th) cells, and neutralizing antibodies that are antigen specific (72). DNA vaccines from the earlier generation, in general, were poorly immunogenic (33). Over the past years, several strategies have been pursued to improve the immunogenicity of DNA vaccines that include the optimization of codons in the antigen gene (73), the coadministration of genes coding for immunostimulatory functions (e.g., cytokines [74]), the use of CpG motifs in plasmid backbones (75), the fusion of antigens with sequences that target specific major histocompatibility complex (MHC) pathways and Th-cell responses (76), the design of heterologous prime/boost immunization modalities (77), and the use of delivery methodologies such as the gene gun (78). DNA vaccine prototypes have been constructed by cloning genes coding for antigens associated with a range of diseases including AIDS (77), dengue (79), human papillomavirus (80), influenza (81), sleeping sickness (82), and tuberculosis (83).

MOLECULAR ASPECTS

Basic Components

Plasmid-based gene therapy and DNA vaccination rely on an effective delivery and expression (in terms of level and duration) of the transgene to the target cells. The typical plasmid vector (Fig. 2) is a covalently closed, double-stranded DNA molecule derived from natural plasmids, which is mostly found as a tightly twisted, supercoiled topoisomer (84). It contains a set of prokaryotic sequences necessary for plasmid amplification in a bacterial host (replication origin, antibiotic resistance gene), and an eukaryotic expression cassette, which includes the therapeutic gene and the regulatory elements required for expression in eukaryotic cells (e.g., promoter, polyadenylation sequence [Fig. 3]). The design of such a plasmid vector should take into consideration aspects such as the stability of plasmids *in vivo*, the profile of transgene expression, the impact of prokaryotic sequences, and the response of the host immune system to the vector.

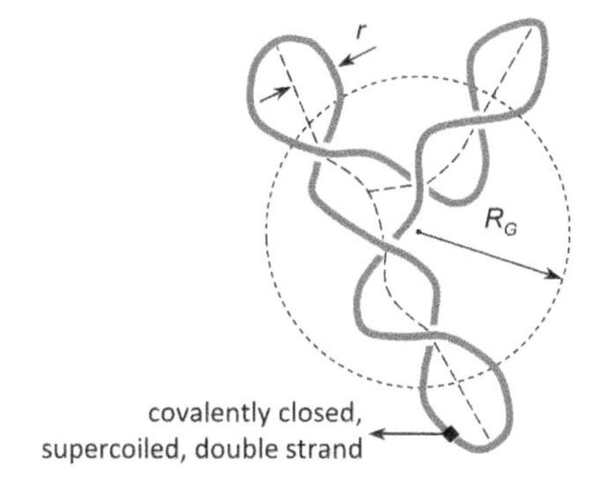

Characteristic	Typical value
N. base pairs	2000-10000
Molecular weight	1320-6600
Charge, z	+2000 - +10000
Superhelix axis, L (nm)	280-1400
Superhelix radius, r (Å)	53
Radius of gyration, R_G (nm)	500-1500
Diffusion coefficient, D (cm^2/s)	6.2-2.1 × 10^{-8}

Figure 2 Basic physical characteristics of plasmid vectors. Data presented is for 2,000- to 10,000-bp plasmids with a typical degree of supercoiling (Prazeres, 2011). Image is reprinted with permission from reference 84 with permission from Wiley. Copyright 2011, John Wiley and Sons, Inc. doi:10.1128/microbiolspec.PLAS-0022-2014.f2

In Vivo Plasmid Stability

The plasmid journey to the cell nucleus is hindered by several physical barriers, including the cytoplasmic membrane, the network of cytoskeleton proteins and organelles that overcrowd the cytoplasm, and the nucleus envelope (85). Moreover, the degradation of plasmid DNA by intra- and extracellular exo/endonucleases constitutes a major barrier to gene expression (86, 87). Only very small amounts (0.1%) of the plasmid molecules that enter cells reach the nucleus (86). The removal of secondary forming sequences (e.g., homopurine-rich and cruciforms) from the plasmid backbone (87, 88) and the coadministration of nuclease inhibitors like aurintricarboxylic acid (89) and DMI-2 (90) have both been advocated as a means to improve the resistance of plasmids to nucleases. Moreover, plasmids containing direct and inverted repeats, insertion sequences, and regions similar to genomic DNA can suffer genetic rearrangements, such as deletions, duplications, inversions, translocations, and insertions, albeit at very low frequencies (91, 92, 93, 94). These rearrangements can affect both plasmid production in *Escherichia coli* and the efficiency of transgene expression. These and other unstable regions should be removed from the vector or at least changed whenever possible. Removal of nonessential sequences from plasmids also reduces the size

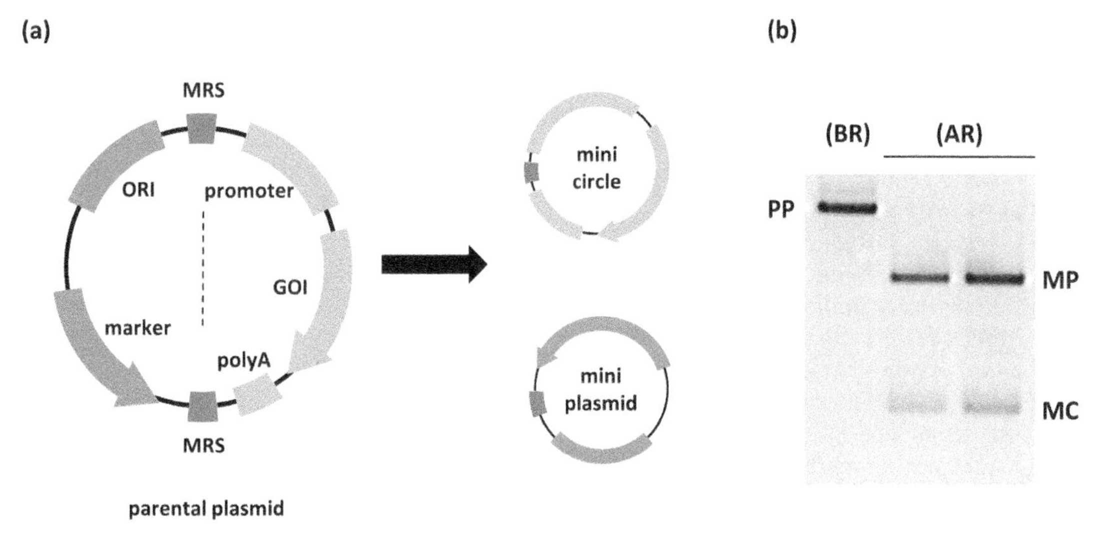

Figure 3 (A) Schematic representation of the recombination of a parental plasmid (PP) into a minicircle (MC) and a miniplasmid (MP) via the excision of the eukaryotic expression cassette that is flanked by two multimer resolution sites (MRS). (B) Agarose gel electrophoresis showing a parental plasmid before the induction of recombination (BR) and minicircle and miniplasmid species after recombination (AR). Abbreviations: ORI, origin of replication; GOI, gene of interest. doi:10.1128/microbiolspec.PLAS-0022-2014.f3

which *per se* decreases the number of intrinsic nuclease-susceptible or inhibitory regions and might increase the molecular stability.

Transgene Expression

Upon successful arrival of the transgene to the nucleus of the target cells, a reasonable amount of gene expression during a more or less extended time period is required to elicit the wanted therapeutic effect. The transgene expression profile can be modulated by judiciously selecting the best promoter and transcriptional regulators. A few promoters are currently used to drive transgene expression in the context of gene therapy and DNA vaccination. The cytomegalovirus (CMV) immediate early promoter is the most widely used in many vectors because of its high strength in many different tissues. The activity of the CMV promoter can be modulated by the presence of specific cytokines (95) or by other proteins like p53 and Mekk1 (96). The downregulation of the promoter activity might prevent transgene expression. Therefore, the use of alternative promoters is often considered, e.g., viral promoters (e.g., Rous sarcoma virus, simian virus 40 [SV40]) and some cellular promoters (e.g., human ubiquitin B [UbB], ubiquitin C, human elongation factor 1α), or chimeric promoters (e.g., CMV-chicken-β-actin, CMV-UbB) (97, 98, 99).

The transcriptional inactivation of the promoters that regulate transgenes is expected during the normal homeostasis of the cells. A sustained and long-term expression depends on the promoter type that includes regulatory elements such as enhancers, boundary elements, and silencers (98, 99) and also on the cell type and physiological state at the time of transcription (96). For example, the Kozak sequence (gccRccAUGG), which includes the start codon, helps the transcript to bind to ribosome to start translation. Another important regulatory sequence is the polyadenylation site (AAUAAA) located at the 3′ end of the mRNA that allows termination of transcription and is important for the nuclear export, stability of mRNA, and, consequently, translation. Many eukaryotic expression vectors use the bovine growth hormone, SV40, or rabbit β-globin terminator sequences (88, 98, 99), or endogenous terminators that are downstream from the open reading frame of the gene of interest to ensure proper transcriptional termination.

Immune Response to Plasmid Vectors

The administration of plasmid vectors that contain bacterial sequences is likely to generate immune responses. The innate immune system is able to discriminate microbial components and self-components by identifying the pathogen-associated molecular patterns (PAMPs). Depending on their composition (lipopolysaccharides, nucleic acids, proteins, etc.), PAMPs are recognized by different pattern-recognition receptors (PRRs), triggering single, multiple, cooperative, or redundant specific signaling pathways that set up an immune response (100, 101). Toll-like receptors (TLRs), like others PRRs, are expressed in the cell surface or intracellularly (e.g., TLR9) by various immune cell types (e.g., macrophages, dendritic cells, B cells) and also by nonprofessional immune cells (e.g., fibroblasts, epithelial cells) (100, 101). Usually, sensing of PAMPs by PRRs upregulates the transcription of type I interferons and proinflammatory cytokines.

DNAs of bacterial origin show a high frequency of unmethlylated cytosine-phosphate-guanine (CpG) dinucleotides. In contrast, CpG motifs are infrequent in mammalian DNAs, and when present, are highly methylated (5mCpG). Overall, DNA of bacterial origin is a PAMP that is recognized by intracellular TLR9 and by stimulator of interferon genes (STING) proteins, activating innate immunity response (101, 102, 103). The TLR9 signaling cascade induces expression of type I interferons and inflammatory genes, mainly through activation of the transcription factors interferon regulatory factor 7 (IRF7) and NF-kB (101, 102). STINGs are endoplasmic reticulum translocon-associated transmembrane dimer proteins that are critical for regulating the production of interferon in response to cytoplasmic DNA (102, 104). STINGs bind to cytosolic double-stranded DNA without a requirement for accessory molecules (103), activating the TANK-binding kinase 1 (TBK1). STING appears essential for escorting TBK1 to endosomal compartments for activation of the transcription factors IRF3/IRF7 and NF-kB, leading to the expression of type I interferon and inflammatory cytokines (103).

Although this immunostimulatory property of CpG motifs is undesirable when plasmids are used for gene therapy purposes, it can be used favorably as an adjuvant in DNA vaccination (105). The idea is that the type and the number of CpG motifs in a plasmid backbone can be designed to modulate the humoral immune response triggered by DNA vaccines (106, 107, 108, 109). Several human clinical trials have used such CpG adjuvants in the context of preventive and therapeutic vaccination (105, 110).

Optimized Vectors

Some of the functional elements (origin of replication, prokaryotic resistance marker) found in a plasmid are required only during the replication process that takes place during the growth of the prokaryotic production host. Once the cell culture is halted, those prokaryotic sequences are no longer needed and may actually decrease stability, uptake, and efficacy. Furthermore, even though the presence of some prokaryotic sequences in plasmids is approved by the FDA and European Medicines Agency, their use can be detrimental both from a clinical and environmental point of view. For example, concerns have been raised that a widely used selection marker like the kanamycin resistance gene may be horizontally transmitted to the recipient's enteric bacteria (111, 112). Thus, and in line with the recommendations of regulatory agencies to eliminate antibiotic resistance markers from plasmid vectors, a number of antibiotic-free selection systems have been developed to produce safer and smaller (and eventually more efficient) plasmids (99, 112, 113). Noteworthy examples include plasmids with conditional origin of replication and plasmids free of antibiotic resistance, vectors expressing small RNA-OUT antisense RNA (114), RNAI/RNAII-based plasmids containing ColE1-type origin (115), and operator-repressor titration systems (116). Additionally, the absence of antibiotic resistance genes from bacterial origin in these minimized plasmids may lead to reductions in innate immune responses and minimize the risk of silencing of transgene expression.

The backbone of a plasmid-based DNA vaccine can be designed with the amount of correct unmethylated CpG sequences that maximizes immune response. On the contrary, an effective plasmid vector for gene therapy can be designed by avoiding unmethylated CpGs or by adding CpG antagonists of TLR9 (e.g., containing a (5-methyl-dC)p(7-deaza-dG) or (5-methyl-dC)p(arabino-G) motif) (117). Another strategy involves the methylation of the plasmid before its administration. While several methylases (SssI, HpaII, or HaeIII) have been used *in vitro*, the costs associated with this strategy have prompted researchers to explore the possibility of performing methylation *in vivo*, for example, using SssI methylase. This approach has shown prolonged transgene expression by circumventing immune recognition (118).

Avoiding the presence of CpG sequences within regulatory sequences can also increase the level and the duration of expression (97, 119). Long-term gene expression (at least 19 months) has been reported by Wolff et al. (120), suggesting that plasmid DNA could stably persist and be expressed in nondividing muscle cells. However, strategies like chromosomal integration or episomal replication are usually required to obtain long expression periods. Interestingly, the Kay's group (121) has shown that the administration of plasmid DNA containing CpG motifs (methylated or not) to the mouse liver will only lead to transgene silencing and innate immune responses if those sequences are covalently linked to the transgene. These authors hypothesized that the absence of DNA sequences devoid of transcriptional enhancers that maintain an active transcription state are prone to form repressive heterochromatin on the plasmid DNA backbone, which then spreads and inactivates the transgene in *cis*, but not in *trans* (121). Kay's group also suggests that it is the length (>1 kb) and not the sequence of the extragenic DNA flanking the transgene expression cassette that leads to transgene silencing (122). This is a very complex subject from which much more insights are needed to understand the complex nature of regulation of gene expression.

The presence of bacterial sequences triggers an association with inactive forms of chromatin. Episomal DNA constructs with persistent expression have a slight greater abundance of histone H3 lysine 4 dimethylated, while unexpressed constructs showed enrichment in histone H3 lysine 9 trimethylated (123, 124, 125). Interestingly, AT-rich scaffold or matrix attachment regions, which facilitate opening and maintenance of euchromatin, can be incorporated near promoters to allow enhanced and persistent transgene expression (126).

Other plasmid derivatives that are devoid of bacterial sequences have been designed and tested successfully. Minicircles, for example, are double-stranded and supercoiled expression eukaryotic vectors devoid of bacterial sequences such as the origin of replication and the antibiotic resistant marker (22). Minicircles are produced in *E. coli* by excising the desired expression cassette from a parental plasmid. This excision takes place by promoting the *in vivo* recombination between two recombinase target sites strategically located in the parental plasmid backbone (Fig. 3). Several recombinases acting under the regulation of inducible promoters have been used to catalyze this excision, including λ-integrase, Cre recombinase, C31 integrase, and Par resolvase (112, 113). The recombination event generates two products, a replication-deficient minicircle, which contains the mammalian expression cassette, and a miniplasmid, which contains the undesired antibiotic resistance gene and the bacterial origin of replication (Fig. 3). The selection of the recombination system should be aimed at providing the best balance between recombination efficiency and the yield of supercoiled minicircle species. Since both products will coexist inside *E. coli* cells once the process is terminated, adequate methodologies must be developed to isolate the therapeutically useful minicircle from its counterpart (as well as from the usual bacterial impurities). This separation constitutes a challenge on its own owing to the similarity of the physical-chemical characteristics of the two DNA rings. One of the strategies devised to purify minicircles relies on the placement of lactose operator sites in the minicircle moiety and on the use of an affinity chromatography matrix with bound lac repressor (LacI). When a mixture of minicircles and miniplasmids is contacted with this affinity column, minicircles will bind to LacI, whereas miniplasmids are washed away in the flowthrough. Minicircles can subsequently be recovered by eluting the column with isopropyl-β-D-1-thiogalactopyranoside (127). Experimental evidence has shown that adequately purified minicircles are able to generate a persistent and high-level transgene expression *in vivo* (22).

The concept of transforming plasmids into minimal-size gene transfer units also fostered the development of short eukaryotic expression cassettes called Minimalistic Immunogenically Defined Gene Expression (MIDGE) vectors. Unlike the case of minicircles, however, the final vector is a small, linear molecule that is covalently closed at the extremities by two short hairpin oligonucleotide sequences. The linear DNA fragments are generated either by restriction digestion of conventional therapeutic vectors or by PCR-mediated amplification (128, 129).

ADMINISTRATION AND DELIVERY

Barriers

The transportation of plasmids from the outside of the body of a patient into the cell nuclei and the subsequent expression of the gene cargo are critical for the success of a plasmid-mediated gene transfer intervention (130). However, this process is difficult to achieve because of the existence of a series of barriers that DNA molecules encounter during their journey across the entry route, capillaries, interstitial spaces, tissues, body fluids, membranes, and cells cytoplasm and that contribute to reduce the number of molecules arriving at the cell nucleus (131). Examples of such barriers include mononuclear phagocytes, blood components, low pH, plasma and cellular endonucleases, cellular membranes, endosomes, lysosomes, and narrow nuclear pore complexes (131, 132). The efficiency of plasmid vectors is critically dependent of the use of delivery systems adequate to overcome these barriers and guarantee that a significant fraction of the administered pool of plasmid molecules arrives safely and ready for transcription into the cell nucleus (Fig. 4).

Plasmid biopharmaceuticals can be administered through different routes depending on the disease and therapeutic intervention planned. For example, intratumoral administration is used in the treatment of solid tumors (133), the airways are preferred when tackling lung diseases (134), and muscular or skin tissues are favored when administering prophylactic DNA vaccines (72). Virtually all organs and tissues in the human body have been used as entry points for plasmids (130). The administration route, to a large extent, will constrain the choice of the delivery system used to carry the plasmid from the vial on the shelf into the cell nucleus. Likewise, certain delivery systems and devices are specifically designed to serve defined entry routes.

Plasmid DNA

Plasmids can be delivered to cells of a living recipient via the needle injection of a saline solution of the plasmid into muscle (Fig. 5), as originally experimented by Wolff and coworkers (5). While this naked DNA approach is simple, easy to execute, and safe, in most cases, the efficiency of expression of transgenes in the skeletal muscle of nonhuman primates is inferior in comparison with alternative delivery methodologies (135). The key problem relies on the fact that plasmids are rapidly cleared from the injection site because of the action of endogenous nucleases. This means that only a fraction of the injected molecules will transfect cells and that this is essentially restricted to the injection site. The exact mechanism by which naked plasmids cross cell membranes is unclear, but suggestions have been made regarding a receptor-mediated formation of plasmid vesicles at the cell surface and subsequent formation of endosomes and then fusion to lysosomes (136, 137). Plasmids must then escape endosomes/lysosomes and travel through the cytoplasm toward the cell nucleus, most likely via a mechanism of active transport involving microtubule networks (138). The final crossing of the nuclear envelope is facilitated if cells are engaged in division, but in nondividing cells the nuclear pore complex (NPC) becomes the only gate to access the nucleus (138, 139). In the latter case, the presence of nuclear localization signals has been shown to increase the translocation of plasmids through NPCs to a certain extent (140).

Naked DNA can be delivered more effectively via the rapid intravenous injection of a large volume (8% to 10% of the body weight) of a plasmid-containing saline solution, a procedure that favors the transfection of hepatocytes (24, 141). However, this so-called hydrodynamic injection is an inherently invasive and complex procedure difficult to transfer to the clinic

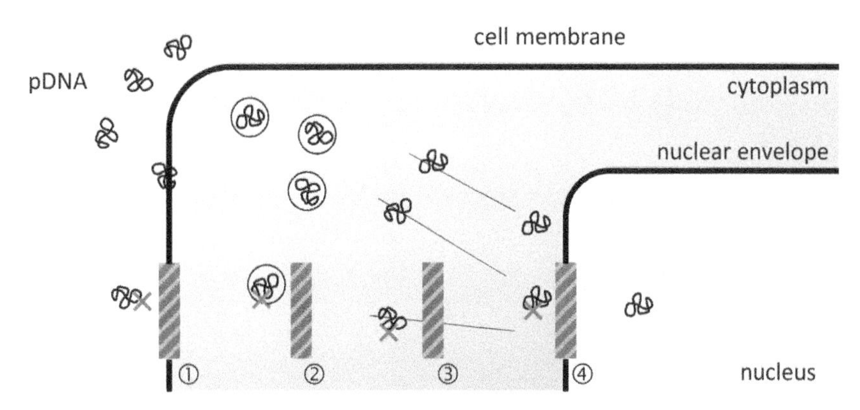

Figure 4 The intracellular barriers to plasmid-based gene transfer. In their journey to the nucleus, plasmids have to cross the phospholipidic cell membrane through endocytosis (1), escape entrapment and degradation in endosomes and lysosomes (2), survive degradation by cytosolic nucleases, traffic the overcrowded cytoplasm (3), and translocate across the nuclear envelope (4). doi:10.1128/microbiolspec.PLAS-0022-2014.f4

(142). Naked plasmid DNA can also be delivered via liquid jet injectors, devices that generate fine (~76 to 360 μm) high-pressure jets that puncture through the skin at high velocities (100 m s^{-1}) and deposit solutions in the tissue beneath (141, 143).

Gene Gun

Particle bombardment (*aka* biolistics) is one of the most effective ways to deliver plasmids to living cells and tissues (Fig. 5) (72, 144, 145). The method relies on a hand-held device (the gene gun) that uses pressurized gases such as helium to propel plasmid-coated nonporous metallic microparticles (0.1 to 5 m). Cartridges must first be prepared with a dry powder of the plasmid-coated particles and then inserted into the gene gun. Gold has been the metal of choice for medical applications. When the device is triggered, a gas jet crosses the cartridge, releasing and accelerating the particles with a speed that allows penetration of target tissues or organs. Gene guns have been often used to obtain strong immune responses on delivery of DNA vaccines to the antigen-presenting cells residing in the top layers of skin. The microparticles ejected by the gene gun cross past the outermost layer of the epidermis and puncture through the membranes and across the cytoplasm and nuclei of cells (72, 144, 145). Following expression, the antigens encoded in the plasmid are then processed, eliciting primary cellular responses and fostering the production of antibodies (72, 145). A number of preclinical and human clinical trials have been conducted to study the outcome of gene-gun-delivered DNA vaccines in the context of immunization against infectious diseases. These studies have indicated that substantially smaller doses of DNA vaccine (~1 to 10 μg) are required to obtain immune responses (antibody titers and CD8+ T cells) in mice and primates in comparison with intramuscular or intradermal injections (72, 145, 146).

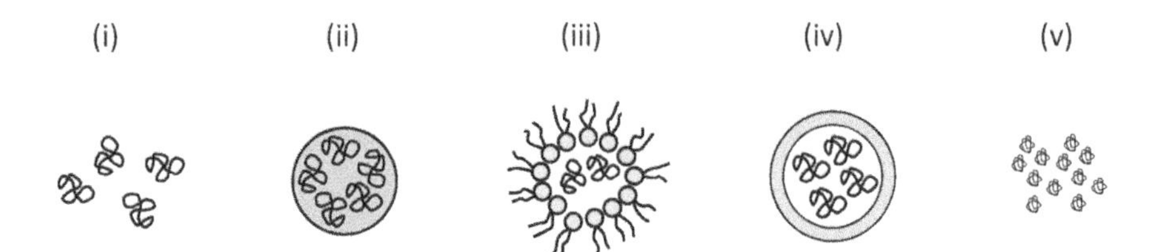

Figure 5 *In vivo* plasmid delivery. Plasmid DNA can be combined and formulated with buffers, stabilizers, and inorganic or organic matrices and molecules to produce: (i) a saline solution of plasmid, (ii) gold particles coated with plasmid, (iii) plasmids complexed with cationic lipids or polymers, (iv) polymeric microparticles with encapsulated or surface-adsorbed plasmid, or (v) nanoparticles of compacted plasmid. doi:10.1128/microbiolspec.PLAS-0022-2014.f5

Electroporation

The low efficiency of gene expression that is associated with the injection of naked plasmid DNA can be improved with electric fields generated by high-energy pulses, i.e., via electroporation. Allegedly, such electric fields transiently increase the transmembrane potential, leading to the opening up of ephemeral (microseconds to seconds) transbilayer electropores (<10 nm), or to the creation of structural defects in the membranes (147, 148). Suggestions have also been made that electrophoretic effects are created that actively drive the negatively charged plasmids and foster their passage across pores and into the cell cytoplasm (147, 148). Other authors, however, advocate that transit across the destabilized membranes occurs by passive diffusion (149) or that charged plasmid vesicles or stable DNA/membrane complexes are formed and subsequently endocytized (150). Critical electric parameters that can be manipulated to control plasmid delivery by electroporation include the number, length (few microseconds to milliseconds), voltage (50 to 1500 V), and waveform (exponential decay or square wave) of the pulses (148). A few companies have designed and developed electroporation devices to meet the requirements for a safe and consistent clinical delivery of plasmid DNA. Such devices typically combine a system that delivers the required electric pulses with an injection needle for intramuscular administration (151, 152). In general, preclinical and clinical data have shown that *in vivo* uptake of plasmids by tissues like the skin and muscle and transgene expression can be increased by electroporation (153, 154). Negative aspects that have been associated with electroporation include muscle stimulation, patient discomfort, and tissue damage (148).

Cationic Lipids and Polymers

The transfection ability of plasmid biopharmaceuticals can be improved by formulating the plasmids with specific molecules such as cationic lipids and soluble polymers (Fig. 5). The methodology relies on the electrostatic interaction between the polyanionic plasmids with a cationic lipid (e.g., DOTAP-1,2-dioleoyl-3-trimethylammonium propane [155]) or polymer (e.g., polyethyleneimine [PEI], polylysine [156]). As a result of this interaction, plasmids collapse and condense, acquiring dimensions which are substantially smaller than the size of the individual plasmids. This coating of the negatively charged plasmids with cationic "envelopes" facilitates the fusion of the complexes with the negatively charged cell membranes, thus favoring internalization by endocytosis, and also protects plasmids against the attack of lysosomal and cytosolic nucleases.

Micro- and Nanoparticles

Gene delivery can also be accomplished by plasmid-loaded polymeric microparticles of a defined size (0.5 to 10 µm; Fig. 5). A key advantage of these microparticles is that they allow a more prolonged release of plasmids instead of the bolus type of delivery that is characteristic of the submicron plasmid/polymer complexes described above (21, 157, 158). Two of the most popular polymers used in this context are poly(DL-lactide-co-glycolide) and poly(DL-lactic acid) owing to their biocompatible and biodegradable nature. Plasmid molecules can be either encapsulated (21, 159, 160) or adsorbed to the surface of the microparticles (158). Plasmid-loaded microparticles can be administered via subcutaneous or intramuscular needle injection. Once *in vivo*, the particles are phagocytosed by professional antigen-presenting cells (macrophages, dendritic cells) and then transported to the lymph nodes where plasmids are gradually released (161). The usefulness of microparticles as *in vivo* plasmid delivery agents has been described in the context of several diseases. including cancer (162), hepatitis B (163), and tuberculosis (164). In general, plasmid/microparticle formulations are safe and able to increase gene expression and the immunogenicity of DNA vaccines (162, 163).

The lipid/polymer complexes and microparticles described above have sizes in the 200-nm to 5-µm range and typically contain several plasmid molecules. This means that some kind of disaggregation or dismantling process must take place in the cytoplasm for plasmid molecules to be able to pass through the 25-nm-wide nuclear pore complexes of the nuclear envelope of cells. One way to overcome this need for dismantling before nuclear entry is to produce plasmid nanoparticles with sizes smaller than 100 nm (Fig. 5). These nanoparticles can be prepared, for example, by using chitosan (165), peptide-polyethylene glycol conjugates (134), or protamine sulfate-calcium carbonate (166).

SAFETY ISSUES

Like other biopharmaceuticals, plasmids hold in them the potential to injure recipients. The specific safety issues that have been raised in association with the clinical use of plasmids include (i) the potential of plasmids and derived fragments to integrate into the host genomic DNA (167) and (ii) the stimulation of anti-DNA antibodies and autoimmune reactions (168). These questions have been addressed during preclinical development by performing pharmacological and toxicological studies with adequate animal models in line with the recommendations of regulatory bodies (169, 170).

The goals of such studies include the definition of safe starting doses and escalation regimens and the identification of organs at risk and parameters to monitor toxicity.

The potential for integration in the genome is minimal since sequences that might drive homologous recombination and direct integration (e.g., insertion sequences, retroviral-like long terminal repeats, sequences homologous to the packaging sequences of retroviruses) are removed during the design of the plasmid molecule (171). So far, results show that the risk for integration of plasmid sequences is much lower than natural, random mutations (172, 173, 174, 175). Furthermore, biodistribution and persistence studies have indicated that most plasmids that are administered intramuscularly (e.g., by needle injection, needleless jet, or particle-mediated delivery) remain close to the injection site and are rapidly degraded by endogenous nuclease within the first minutes (174, 176, 177), reducing even further the likelihood of integration.

A number of animal experiments have been conducted to investigate whether the administration of plasmids and the concomitant expression of the encoded transgenes *in vivo* could generate and promote the development of autoimmunity and other deleterious immunological responses (168, 178, 179, 180). One of the specific safety concerns is whether plasmids can induce the production of anti-DNA antibodies. Such anti-DNA antibodies could form immune complexes with circulating DNA, damaging various tissues and blood vessels in critical areas of the body, as is characteristic of systemic lupus erythematosus (171). However, no link between plasmid administration and changes in clinical markers of autoimmunity has been found yet (33).

So far, none of the concerns highlighted above have materialized, with scientific and clinical studies indicating that plasmid biopharmaceuticals are in general well tolerated and safe (174, 176, 181, 182, 183). Another reason for the favorable appreciation surrounding plasmid biopharmaceuticals is related to the fact that they are, in the vast majority of cases, designed to promote transient expression of the encoded protein in the target human tissues.

PLASMID MANUFACTURING

Overview

The development of plasmid-manufacturing processes is an undertaking that must occur in parallel with product development, not only because it is required to generate material for preclinical and clinical trials, but also because the methodology that will ultimately produce the plasmid biopharmaceuticals for sale must be established before market approval is received (184). The manufacturing of a plasmid biopharmaceutical product will consist of a string of activities (Fig. 6) that are set up and carried out with the aim of consistently producing a defined amount (e.g., measured as biological activity or mass) of a product that is safe and efficacious (184). The preparation of cell banks containing the plasmid of interest and the selection and testing of raw materials are at the forefront of the activities. Cell culture and downstream processing unit operations are then selected, arranged, designed, and operated to manufacture unformulated (i.e., bulk) plasmid DNA (Fig. 6). This purified plasmid product must then be adequately formulated by considering aspects such as the method of delivery, the final product form, ingredients (excipients, adjuvants, stabilizers), dosage details, packaging, etc. After the "filling and finishing" stage, the product is ready for clinical testing or marketing (Fig. 6).

Cell Culture

Plasmids are produced by promoting replication in *E. coli*. Before routine culture, a strain has to be chosen (e.g., DH5a, JM109) or developed (e.g., GALG20 [185]). While the genetic background of these producer strains may vary, mutations in the *recA* and *endA* genes to minimize recombination events and plasmid DNA degradation, respectively, are close to universal. Other important genetic traits are related to genetic modifications (e.g., *pkyA*, *pkyF*, *pgi*) that increase the metabolic flux toward the formation of nucleotide precursors (185). Once a strain has been selected and transformed with the target plasmid, high-production clones must be carefully selected and isolated and used to establish master and working cell banks with stocks of vials of the plasmid-bearing cells. Growth medium composition, bioreactor-operating variables, and cultivation strategy must then be selected that maximize plasmid production. By adequately combining these parameters with a high-copy-number plasmid, volumetric plasmid yields up to 2.0 g/l can be obtained (186).

Downstream Processing

The train of unit operations in the downstream processing section is designed to recover plasmid DNA and remove host impurities (genomic DNA, RNA, proteins, etc.) until a level of purity compatible with human use is met (184). The unit operations can be grouped into three stages: primary isolation, intermediate purification, and

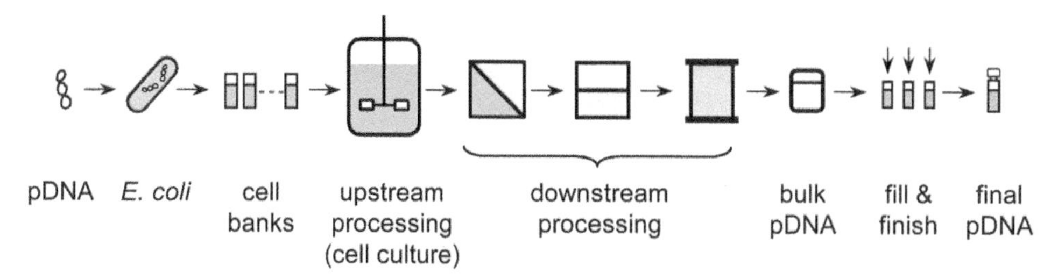

Figure 6 An overview of the different activities and steps involved in the manufacturing of plasmid biopharmaceuticals. doi:10.1128/microbiolspec.PLAS-0022-2014.f6

final purification. The starting point is typically a broth with high cell densities (>55 g of dry cell weight/l) and plasmid concentrations (1.6 to 2.0 g/l [186, 187]]). In the first stage, cells are harvested (e.g., by microfiltration) and lysed (e.g., alkaline lysis, thermal lysis) to release plasmid DNA. In the subsequent intermediate purification stage, clarified lysates are processed by using operations such as tangential flow filtration, precipitation (188), adsorption (189), and aqueous two-phase systems (190) to reduce the impurity load and concentrate the plasmid. Final purification aims to remove the more recalcitrant impurities such as gDNA and pDNA variants. Traditional chromatographic modalities such as gel filtration, anion exchange, and hydrophobic interaction (191) have all been used, mostly in the fixed-bed mode, to purify plasmid. Efforts have also been made to develop chromatographic operations based on amino acid (192), thiophilic (193), and phenyl boronate (189) ligands. Although dominating, chromatography is faced with limitations (poor selectivity, coelution, low capacity, and slow internal diffusion) that are related to the structural nature of the stationary phases and molecules (molecular mass $>10^6$ g mol^{-1}, $D = 10^8$ cm^2/s) involved (194, 195). Larger capacity and faster internal mass transfer can be achieved if chromatographic membranes and monoliths are used instead of beads (195). Once impurities have been reduced to levels below the specifications, corrections to the plasmid concentration and buffer exchange can be accomplished by operations like ultrafiltration and alcohol precipitation (196). The final step in the downstream processing train is usually sterile filtration with 0.22-μm filters (197).

THE ROAD TO THE MARKET

A handful of plasmid biopharmaceuticals have already found their way into the market (198). In 2005, a veterinary DNA vaccine designed and developed jointly by the CDC and Fort Dodge Animal Health (Fort Dodge, Iowa) to protect horses against West Nile virus was licensed by the Center for Veterinary Biologics of the U.S. Department of Agriculture, thus becoming the first DNA vaccine to be registered with a governmental regulatory body (199). The vaccine was subsequently launched in the market in December 2008, under the trade name West Nile Innovator DNA (200). In the same year, a DNA vaccine developed by Novartis Animal Health (Victoria, Canada) to protect farm-raised salmon against Infectious Hematopoietic Necrosis virus (Apex-IHN) also obtained regulatory approval and license (201). In early 2007, the U.S. Department of Agriculture conditionally approved a therapeutic DNA vaccine that delivers the MHC gene to dog tumors to treat melanoma in dogs (202). The vaccine hit the market under the trade name Oncept (198). Finally, in 2008, an injectable plasmid DNA encoding for porcine Growth Hormone Releasing Hormone (GHRH) developed by VGX Animal Health, Inc. (The Woodlands, Texas) to decrease perinatal mortality and morbidity in pigs obtained market entrance approval from the Australian Pesticides and Veterinary Medicines Authority (203).

CONCLUDING REMARKS

Plasmid-mediated gene transfer has slowly materialized as a possible solution for the management of an entire constellation of veterinary and human diseases. The investment made during the past 20 years in the development of this new class of biopharmaceuticals has generated a substantial amount of scientific and technological knowledge. Furthermore, plasmid biopharmaceutical prototypes are currently at the clinical stage of development to tackle multifactorial diseases like cancer and cardiovascular disorders and to prevent the onset of infections like AIDS or influenza. So far, the data accumulated have shown that plasmids, in general, are well tolerated and safe. However, progress must be made to increase the potency and efficacy of plasmid molecules *in vivo*. Advances are clearly needed in the delivery methodologies used to increase the number of administered plasmids that reach the cell nucleus. The manipulation of plasmids and plasmid-related molecules (e.g., minicircles, MIDGEs) is also likely to originate molecules better adapted to bypass cell barriers and to mediate the expression of therapeutic genes.

Acknowledgments. Conflicts of interest: We declare no conflicts.

Citation. Prazeres DMF, Monteiro GA. 2014. Plasmid biopharmaceuticals. Microbiol Spectrum 2(6):PLAS-0022-2014.

References

1. **Cohen SN, Chang AC, Boyer HW, Helling RB.** 1973. Construction of biologically functional bacterial plasmids *in vitro*. *Proc Natl Acad Sci USA* **70:**3240–3244.
2. **Russo E.** 2003. Special report: the birth of biotechnology. *Nature* **421:**456–457.
3. **Hughes SS.** 2001. Making dollars out of DNA: the first major patent in biotechnology and the commercialization of molecular biology, 1974–1980. *Isis* **92:** 541–575.
4. **Prazeres DMF.** 2011. Historical perspective, p 3–34. *In Plasmid Biopharmaceuticals: Basics Applications and Manufacturing*. John Wiley & Sons, Inc., Hoboken, NJ.
5. **Wolff JA, Malone RW, Williams P, Chong W, Acsadi G, Jani A, Felgner PL.** 1990. Direct gene transfer into mouse muscle *in vivo*. *Science* **247:**1465–1468.
6. **Hickman MA, Malone RW, Lehmann-Bruinsma K, Sih TR, Knoell D, Szoka FC, Walzem R, Carlson DM, Powell JS.** 1994. Gene expression following direct injection of DNA into liver. *Hum Gene Ther* **5:**1477–1483.
7. **Ardehali A, Fyfe A, Laks H, Drinkwater DC, Qiao J-H, Lusis AJ.** 1995. Direct gene transfer into donor hearts at the time of harvest. *J Thor Cardiovasc Surg* **109:** 716–720.
8. **Schwartz B, Benoist C, Abdallah B, Rangara R, Hassan A, Scherman D, Demeneix BA.** 1996. Gene transfer by naked DNA into adult mouse brain. *Gene Ther* **3:** 405–411.
9. **Hansen E, Fernandes K, Goldspink G, Butterworth P, Umeda PK, Chang KC.** 1991. Strong expression of foreign genes following direct injection into fish muscle. *FEBS Lett* **290:**73–76.
10. **Fynan EF, Webster RG, Fuller DH, Haynes JR, Santoro JC, Robinson HL.** 1993. DNA vaccines: protective immunizations by parenteral, mucosal, and gene-gun inoculations. *Proc Natl Acad Sci USA* **90:**11478–11482.
11. **Cox GJ, Zamb TJ, Babiuk LA.** 1993. Bovine herpesvirus 1: immune responses in mice and cattle injected with plasmid DNA. *J Virol* **67:**5664–5667.
12. **Tang DC, DeVit M, Johnston SA.** 1992. Genetic immunization is a simple method for eliciting an immune response. *Nature* **356:**152–154.
13. **Ulmer JB, Donnelly JJ, Parker SE, Rhodes GH, Felgner PL, Dwarki VJ, Gromkowski SH, Deck RR, Dewitt CM, Friedman A, Hawe LA, Leander KR, Martinez D, Perry HC, Shiver JW, Montgomery DL, Liu MA.** 1993. Heterologous protection against influenza by injection of DNA encoding a viral protein. *Science* **259:**1745–1749.
14. **Prazeres DMF.** 2011. Concluding remarks and outlook, p 565–578. *In Plasmid Biopharmaceuticals: Basics Applications and Manufacturing*. John Wiley & Sons, Inc., Hoboken, NJ.
15. **Yang NS, Burkholder J, Roberts B, Martinell B, McCabe D.** 1990. *In vivo* and *in vitro* gene transfer to mammalian somatic cells by particle bombardment. *Proc Natl Acad Sci USA* **87:**9568–9572.
16. **Titomirov AV, Sukharev S, Kistanova E.** 1991. *In vivo* electroporation and stable transformation of skin cells of newborn mice by plasmid DNA. *Biochim Biophys Acta* **1088:**131–134.
17. **Xiang Z, Ertl HC.** 1995. Manipulation of the immune response to a plasmid-encoded viral antigen by coinoculation with plasmids expressing cytokines. *Immunity* **2:**129–135.
18. **Sato Y, Roman M, Tighe H, Lee D, Corr M, Nguyen MD, Silverman GJ, Lotz M, Carson DA, Raz E.** 1996. Immunostimulatory DNA sequences necessary for effective intradermal gene immunization. *Science* **273:** 352–354.
19. **Davies HL, Mancini M, Michel M-L, Whalen RG.** 1996. DNA-mediated immunization to hepatitis B surface antigen: longevity of primary response and effect of boost. *Vaccine* **14:**910–915.
20. **Ciernik IF, Berzofsky JA, Carbone DP.** 1996. Induction of cytotoxic T lymphocytes and antitumor immunity with DNA vaccines expressing single T cell epitopes. *J Immunol* **156:**2369–2375.
21. **Jones DH, Corris S, McDonald S, Clegg JCS, Farrar GH.** 1997. Poly(DL-lactide-co-glycolide)-encapsulated plasmid DNA elicits systemic and mucosal antibody responses to encoded protein after oral administration. *Vaccine* **15:**814–817.
22. **Darquet AM, Cameron B, Wils P, Scherman D, Crouzet J.** 1997. A new DNA vehicle for nonviral gene delivery: supercoiled minicircle. *Gene Ther* **4:**1341–1349.
23. **Perales JC, Grossmann GA, Molas M, Liu G, Ferkol T, Harpst J, Oda H, Hanson RW.** 1997. Biochemical and functional characterization of DNA complexes capable of targeting genes to hepatocytes via the asialoglycoprotein receptor. *J Biol Chem* **272:**7398–7407.
24. **Liu F, Song Y, Liu D.** 1999. Hydrodynamics-based transfection in animals by systemic administration of plasmid DNA. *Gene Ther* **6:**1258–1266.
25. **Raper SE, Chirmule N, Lee FS, Wivel NA, Bagg A, Gao GP, Wilson JM, Batshaw ML.** 2003. Fatal systemic inflammatory response syndrome in a ornithine transcarbamylase deficient patient following adenoviral gene transfer. *Mol Genet Metab* **80:**148–158.
26. **Prazeres DMF.** 2011. Gene transfer with plasmid biopharmaceuticals, p 35–68. *In Plasmid Biopharmaceuticals: Basics Applications and Manufacturing*. John Wiley & Sons, Inc., Hoboken, NJ.
27. **Zhong J, Eliceiri B, Stupack D, Penta K, Sakamoto G, Quertermous T, Coleman M, Boudreau N, Varner JA.** 2003. Neovascularization of ischemic tissues by gene delivery of the extracellular matrix protein Del-1. *J Clin Invest* **112:**30–41.
28. **Koike H, Ishida A, Shimamura M, Mizuno S, Nakamura T, Ogihara T, Kaneda Y, Morishita R.** 2006. Prevention

of onset of Parkinson's disease by *in vivo* gene transfer of human hepatocyte growth factor in rodent model: a model of gene therapy for Parkinson's disease. *Gene Ther* 13:1639–1644.

29. **Alton EW, Stern M, Farley R, Jaffe A, Chadwick SL, Phillips J, Davies J, Smith SN, Browning J, Davies MG, Hodson ME, Durham SR, Li D, Jeffery PK, Scallan M, Balfour R, Eastman SJ, Cheng SH, Smith AE, Meeker D, Geddes DM.** 1999. Cationic lipid-mediated CFTR gene transfer to the lungs and nose of patients with cystic fibrosis: a double-blind placebo-controlled trial. *Lancet* 353:947–954.
30. **Romero NB, Braun S, Benveniste O, Braun S, Benveniste O, Leturcq F, Hogrel JY, Morris GE, Barois A, Eymard B, Payan C, Ortega V, Boch AL, Lejean L, Thioudellet C, Mourot B, Escot C, Choquel A, Recan D, Kaplan JC, Dickson G, Klatzmann D, Molinier-Frenckel V, Guillet JG, Squiban P, Herson S, Fardeau M.** 2004. Phase I study of dystrophin plasmid-based gene therapy in Duchenne/Becker muscular dystrophy. *Hum Gene Ther* 15:1065–1076.
31. **Croze F, Prud'homme GJ.** 2003. Gene therapy of streptozotocin-induced diabetes by intramuscular delivery of modified preproinsulin genes. *J Gene Med* 5: 425–437.
32. **Sebestyén MG, Hegge JO, Noble MA, Lewis DL, Herweijer H, Wolff JA.** 2007. Progress toward a nonviral gene therapy protocol for the treatment of anemia. *Hum Gene Ther* 18:269–285.
33. **Ferraro B, Morrow MP, Hutnick NA, Shin TH, Lucke CE, Weiner DB.** 2011. Clinical applications of DNA vaccines: current progress. *Clin Infect Dis* 53:296–302.
34. **Giri M, Ugen KE, Weiner DB.** 2004. DNA vaccines against human immunodeficiency virus type 1 in the past decade. *Clin Microbiol Rev* 17:370–389.
35. **Rainczuk A, Scorza T, Spithill TW, Smooker PM.** 2004. A bicistronic DNA vaccine containing apical membrane antigen 1 and merozoite surface protein 4/5 can prime humoral and cellular immune responses and partially protect mice against virulent *Plasmodium chabaudi adami* DS malaria. *Infect Immun* 72:5565–5573.
36. **Haile M, Kallenius G.** 2005. Recent developments in tuberculosis vaccines. *Curr Opin Infect Dis* 18:211–215.
37. **Drape RJ, Macklin MD, Barr LJ, Jones S, Haynes JR, Dean HJ.** 2006. Epidermal DNA vaccine for influenza is immunogenic in humans. *Vaccine* 24:4475–4481.
38. **Lowe DB, Shearer MH, Kennedy RC.** 2006. DNA vaccines: successes and limitations in cancer and infectious disease. *J Cell Biochem* 98:235–242.
39. **Kaykas A, Moon RT.** 2004. A plasmid-based system for expressing small interfering RNA libraries in mammalian cells. *BMC Cell Biol* 5:16. doi:10.1186/1471-2121-5-16.
40. **Zeitelhofer M, Karra D, Vessey JP, Jaskic E, Macchi P, Thomas S, Riefler J, Kiebler M, Dahm R.** 2009. High-efficiency transfection of short hairpin RNAs-encoding plasmids into primary hippocampal neurons. *J Neurosci Res* 87:289–300.
41. **Li Z, Yang S, Chang T, Cao X, Shi L, Fang G.** 2012. Anti-angiogenesis and anticancer effects of a plasmid expressing both ENDO-VEGI151 and small interfering RNA against surviving. *Int J Mol Med* 29: 485–490.
42. **Fewell JG, MacLaughlin F, Mehta V, Gondo M, Nicol F, Wilson E, Smith LC.** 2001. Gene therapy for the treatment of hemophilia B using PINC-formulated plasmid delivered to muscle with electroporation. *Mol Ther* 3:574–583.
43. **Lee JS, Lee M, Kim SW.** 2004. A new potent hFIX plasmid for hemophilia B gene therapy. *Pharm Res* 21: 1229–1232.
44. **Pringle IA, Hyde SC, Connolly MM, Lawton AE, Xu B, Nunez-Alonso G, Davies LA, Sumner-Jones SG, Gill DR.** 2012. CpG-free plasmid expression cassettes for cystic fibrosis gene therapy. *Biomaterials* 33: 6833–6842.
45. **Bertoni C, Jarrahian S, Wheeler TM, Li Y, Olivares EC, Calos MP, Rando TA.** 2006. Enhancement of plasmid-mediated gene therapy for muscular dystrophy by directed plasmid integration. *Proc Natl Acad Sci USA* 103:419–424.
46. **Makinen K, Manninen H, Hedman M, Matsi P, Mussalo H, Alhava E, Ylä-Herttuala S.** 2002. Increased vascularity detected by digital subtraction angiography after VEGF gene transfer to human lower limb artery: a randomized, placebo-controlled, double-blinded phase II study. *Mol Ther* 6:127–133.
47. **Nikol S, Baumgartner I, Van Belle E, Diehm C, Visoná A, Capogrossi MC, Ferreira-Maldent N, Gallino A, Wyatt MG, Wijesinghe LD, Fusari M, Stephan D, Emmerich J, Pompilio G, Vermassen F, Pham E, Grek V, Coleman M, Meyer F.** 2008. Therapeutic angiogenesis with intramuscular NV1FGF improves amputation-free survival in patients with critical limb ischemia. *Mol Ther* 16:972–978.
48. **Vera Janavel G, Crottogini A, Cabeza Meckert P, Cuniberti L, Mele A, Papouchado M, Fernández N, Bercovich A, Criscuolo M, Melo C, Laguens R.** 2006. Plasmid-mediated VEGF gene transfer induces cardiomyogenesis and reduces myocardial infarct size in sheep. *Gene Ther* 13:1133–1142.
49. **Lu Q, Yao Y, Yao Y, Liu S, Huang Y, Lu S, Bai Y, Zhou B, Xu Y, Li L, Wang N, Wang L, Zhang J, Cheng X, Qin G, Ma W, Xu C, Tu X, Wang Q.** 2012. Angiogenic factor AGGF1 promotes therapeutic angiogenesis in a mouse limb ischemia model. *PLoS One* 7:e46998. doi:10.1371/journal.pone.0046998.
50. **Xu L, Pirollo KF, Chang EH.** 2001. Tumor-targeted p53-gene therapy enhances the efficacy of conventional chemo/radiotherapy. *J Control Release* 74:115–128.
51. **Xu L, Huang CC, Huang W, Tang WH, Rait A, Yin YZ, Cruz I, Xiang LM, Pirollo KF, Chang EH.** 2002. Systemic tumor-targeted gene delivery by anti-transferrin receptor scFv-immunoliposomes. *Mol Cancer Ther* 1: 337–346.
52. **Kim CK, Choi EJ, Choi SH, Park JS, Haider KH, Ahn WS.** 2003. Enhanced p53 gene transfer to human ovarian cancer cells using the cationic nonviral vector, DDC. *Gynecol Oncol* 90:265–272.

53. **Nakase M, Inui M, Okumura K, Kamei T, Nakamura S, Tagawa T.** 2005. *p53* gene therapy of human osteosarcoma using a transferrin-modified cationic liposome. *Mol Cancer Ther* **4:**625–631.

54. **Bil J, Wlodarski P, Winiarska M, Kurzaj Z, Issat T, Jozkowicz A, Wegiel B, Dulak J, Golab J.** 2010. Photodynamic therapy-driven induction of suicide cytosine deaminase gene. *Cancer Lett* **290:**216–222.

55. **Maruyama-Tabata H, Harada Y, Matsumura T, Satoh E, Cui F, Iwai M, Kita M, Hibi S, Imanishi J, Sawada T, Mazda O.** 2000. Effective suicide gene therapy *in vivo* by EBV-based plasmid vector coupled with polyamidoamine dendrimer. *Gene Ther* **7:**53–60.

56. **Kendall RL, Thomas KA.** 1993. Inhibition of vascular endothelial cell growth factor activity by an endogenously encoded soluble receptor. *Proc Natl Acad Sci USA* **90:**10705–10709.

57. **Fewell JG, Matar MM, Rice JS, Brunhoeber E, Slobodkin G, Pence C, Worker M, Lewis DH, Anwer K.** 2009. Treatment of disseminated ovarian cancer using nonviral interleukin-12 gene therapy delivered intraperitoneally. *J Gene Med* **11:**718–728.

58. **Roos AK, King A, Pisa P.** 2008. DNA vaccination for prostate cancer. *Methods Mol Biol* **423:**463–472.

59. **Wolchok JD, Yuan J, Houghton AN, Gallardo HF, Rasalan TS, Wang J, Zhang Y, Ranganathan R, Chapman PB, Krown SE, Livingston PO, Heywood M, Riviere I, Panageas KS, Terzulli SL, Perales MA.** 2007. Safety and immunogenicity of tyrosinase DNA vaccines in patients with melanoma. *Mol Ther* **5:**2044–2050.

60. **Condon C, Watkins SC, Celluzzi CM, Thompson K, Falo LD.** 1996. DNA-based immunization by *in vivo* transfection of dendritic cells. *Nature Med* **2:** 1122–1128.

61. **Daud AI, DeConti RC, Andrews S, Urbas P, Riker AI, Sondak VK, Munster PN, Sullivan DM, Ugen KE, Messina JL, Heller R.** 2008. Phase I trial of interleukin-12 plasmid electroporation in patients with metastatic melanoma. *J Clinic Oncol* **26:**5896–5903.

62. **Heinzerling L, Burg G, Dummer R, Maier T, Oberholzer PA, Schultz J, Elzaouk L, Pavlovic J, Moelling K.** 2005. Intratumoral injection of DNA encoding human interleukin 12 into patients with metastatic melanoma: clinical efficacy. *Human Gene Ther* **16:**35–48.

63. **Tuszynski MH, Thal L, Pay M, Salmon DP, U HS, Bakay R, Patel P, Blesch A, Vahlsing HL, Ho G, Tong G, Potkin SG, Fallon J, Hansen L, Mufson EJ, Kordower JH, Gall C, Conner J.** 2005. A phase 1 clinical trial of nerve growth factor gene therapy for Alzheimer disease. *Nature Med* **11:**551–555.

64. **Fernandes JC, Wang H, Jreyssaty C, Benderdour M, Lavigne P, Qiu X, Winnik FM, Zhang X, Dai K, Shi Q.** 2008. Bone-protective effects of nonviral gene therapy with folate-chitosan DNA nanoparticle containing interleukin-1 receptor antagonist gene in rats with adjuvant-induced arthritis. *Mol Ther* **16:**1243–1251.

65. **Steinstraesser L, Hirsch T, Beller J, Mittler D, Sorkin M, Pazdierny G, Jacobsen F, Eriksson E, Steinau HU.** 2007. Transient non-viral cutaneous gene delivery in burn wounds. *J Gene Med* **9:**949–955.

66. **Liu C, Fan M, Bian Z, Chen Z, Li Y.** 2008. Effects of targeted fusion anti-caries DNA vaccine pGJA-P/VAX in rats with caries. *Vaccine* **26:**6685–6689.

67. **Ishikawa H, Takano M, Matsumoto N, Sawada H, Ide C, Mimura O.** 2005. Effect of GDNF gene transfer into axotomized retinal ganglion cells using *in vivo* electroporation with a contact lens-type electrode. *Gene Ther* **12:**289–298.

68. **Hayashi T, Hasegawa K, Sasaki Y, Mori T, Adachi C, Maeda K.** 2007. Systemic administration of interleukin-4 expressing plasmid DNA delays the development of glomerulonephritis and prolongs survival in lupus-prone female NZB x NZW F1 mice. *Nephrol Dial Transplant* **22:**3131–3138.

69. **Tohyama S, Onodera S, Tohyama H, Yasuda K, Nishihira J, Mizue Y, Hamasaka A, Abe R, Koyama Y.** 2008. A novel DNA vaccine-targeting macrophage migration inhibitory factor improves the survival of mice with sepsis. *Gene Ther* **15:**1513–1522.

70. **De Laporte L, Yang Y, Zelivyanskaya ML, Cummings BJ, Anderson AJ, Shea LD.** 2009. Plasmid releasing multiple channel bridges for transgene expression after spinal cord injury. *Mol Ther* **17:**318–326.

71. **Ferraro B, Cruz YL, Coppola D, Heller R.** 2009. Intradermal delivery of plasmid VEGF(165) by electroporation promotes wound healing. *Mol Ther* **17:**651–657.

72. **Dean HJ, Fuller D, Osorio JE.** 2003. Powder and particle-mediated approaches for delivery of DNA and protein vaccines into the epidermis. *Comp Immunol Microbiol Infect Dis* **26:**373–388.

73. **Uchijima M, Yoshida A, Nagata T, Koide Y.** 1998. Optimization of codon usage of plasmid DNA vaccine is required for the effective MHC class I-restricted T cell responses against intracellular bacterium. *J Immunol* **161:**5594–5599.

74. **Babiuk S, Babiuk LA, van Drunen Littel-van den Hurk S.** 2006. DNA vaccination: A simple concept with challenges regarding implementation. *Int Rev Immunol* **25:**51–81.

75. **Eo SK, Lee S, Chun S, Rouse BT.** 2001. Modulation of immunity against herpes simplex virus infection via mucosal genetic transfer of plasmid DNA encoding chemokines. *J Virol* **75:**569–578.

76. **Leifert JA, Rodriguez-Carreno MP, Rodriguez F, Whitton JL.** 2004. Targeting plasmid-encoded proteins to the antigen presentation pathways. *Immunol Rev* **199:**40–53.

77. **Vaine M, Wang S, Hackett A, Arthos J, Lu S.** 2010. Antibody responses elicited through homologous or heterologous prime-boost DNA and protein vaccinations differ in functional activity and avidity. *Vaccine* **28:** 2999–3007.

78. **Yager EJ, Dean HJ, Fuller DH.** 2009. Prospects for developing an effective particle-mediated DNA vaccine against influenza. *Expert Rev Vaccines* **8:**1205–1220.

79. **Imoto J, Konishi E.** 2007. Dengue tetravalent DNA vaccine increases its immunogenicity in mice when mixed with a dengue type 2 subunit vaccine or an inactivated Japanese encephalitis vaccine. *Vaccine* **25:**1076–1084.

80. **Yan J, Harris K, Khan AS, Draghia-Akli R, Sewell D, Weiner DB.** 2008. Cellular immunity induced by a novel HPV18 DNA vaccine encoding an E6/E7 fusion consensus protein in mice and rhesus macaques. *Vaccine* **26:**5210–5215.
81. **Jones S, Evans K, McElwaine-Johnn H, Sharpe M, Oxford J, Lambkin-Williams R, Mant T, Nolan A, Zambon M, Ellis J, Beadle J, Loudon PT.** 2009. DNA vaccination protects against an influenza challenge in a double-blind randomized placebo-controlled phase 1b clinical trial. *Vaccine* **27:**2506–2512.
82. **Silva MS, Prazeres DMF, Lança A, Atouguia J, Monteiro GA.** 2009. Trans-sialidase from *Trypanosoma brucei* as a potential target for DNA vaccine development against African trypanosomiasis. *Parasitol Res* **105:**1223–1229.
83. **Wang QM, Kang L, Wang XH.** 2009. Improved cellular immune response elicited by a ubiquitin-fused ESAT-6 DNA vaccine against *Mycobacterium tuberculosis*. *Microbiol Immunol* **53:**384–390.
84. **Prazeres DMF.** 2011. Structure, p 85–128. *In Plasmid Biopharmaceuticals: Basics Applications and Manufacturing*. John Wiley & Sons, Inc., Hoboken, NJ.
85. **Faurez F, Dory D, Le Moigne V, Gravier R, Jestin A.** 2010. Biosafety of DNA vaccines: New generation of DNA vectors and current knowledge on the fate of plasmids after injection. *Vaccine* **28:**3888–3895.
86. **Lechardeur D, Sohn KJ, Haardt M, Joshi PB, Monck M, Graham RW, Beatty B, Squire J, O'Brodovich H, Lukacs GL.** 1999. Metabolic instability of plasmid DNA in the cytosol: a potential barrier to gene transfer. *Gene Ther* **6:**482–497.
87. **Ribeiro SC, Monteiro GA, Prazeres DMF.** 2004. The role of polyadenylation signal secondary structures on the resistance of plasmid vectors to nucleases. *J Gene Med* **6:**565–573.
88. **Azzoni AR, Ribeiro SC, Monteiro GA, Prazeres DMF.** 2007. The impact of polyadenylation signals on plasmid nuclease-resistance and transgene expression. *J Gene Med* **9:**392–402.
89. **Walther W, Stein U, Siegel R, Fichtner I, Schlag PM.** 2005. Use of the nuclease inhibitor aurintricarboxylic acid (ATA) for improved non-viral intratumoral *in vivo* gene transfer by jet-injection. *J Gene Med* **7:**477–485.
90. **Ross GF, Bruno MD, Uyeda M, Suzuki K, Nagao K, Whitsett JA, Korfhagen TR.** 1998. Enhanced reporter gene expression in cells transfected in the presence of DMI-2, an acid nuclease inhibitor. *Gene Ther* **5:** 1244–1250.
91. **Prather KL, Edmonds MC, Herod JW.** 2006. Identification and characterization of IS1 transposition in plasmid amplification mutants of E. coli clones producing DNA vaccines. *Appl Microbiol Biotechnol* **73:**815–826.
92. **Ribeiro SC, Oliveira PH, Prazeres DMF, Monteiro GA.** 2008. High frequency plasmid recombination mediated by 28 bp direct repeats. *Mol Biotechnol* **40:**252–260.
93. **Oliveira PH, Prather KJ, Prazeres DMF, Monteiro GA.** 2009. Structural instability of plasmid biopharmaceuticals: challenges and implications. *Trends Biotechnol* **27:**503–511.
94. **Oliveira PH, Prazeres DMF, Monteiro GA.** 2009. Deletion formation mutations in plasmid expression vectors are unfavored by runaway amplification conditions and differentially selected under kanamycin stress. *J Biotechnol* **143:**231–238.
95. **Ritter T, Brandt C, Prösch S, Vergopoulos A, Vogt K, Kolls J, Volk HD.** 2000. Stimulatory and inhibitory action of cytokines on the regulation of hCMV-IE promoter activity in human endothelial cells. *Cytokine* **12:** 1163–1170.
96. **Rodova M, Jayini R, Singasani R, Chipps E, Islam MR.** 2013. CMV promoter is repressed by p53 and activated by JNK pathway. *Plasmid* **69:**223–230.
97. **Yew NS, Przybylska M, Ziegler RJ, Liu D, Cheng SH.** 2001. High and sustained transgene expression *in vivo* from plasmid vectors containing a hybrid ubiquitin promoter. *Mol Ther* **4:**75–82.
98. **Kutzler MA, Weiner DB.** 2008. DNA vaccines: ready for prime time? *Nat Rev Genet* **9:**776–788.
99. **Williams JA, Carnes AE, Hodgson CP.** 2009. Plasmid DNA vaccine vector design: Impact on efficacy, safety and upstream production. *Biotechnol Adv* **27:**353–370.
100. **Takeuchi O, Akira S.** 2010. Pattern recognition receptors and inflammation. *Cell* **140:**805–820.
101. **Kawai T, Akira S.** 2010. The role of pattern-recognition receptors in innate immunity: update on Toll-like receptors. *Nat Immunol* **11:**373–384.
102. **Barber GN.** 2011. Innate immune DNA sensing pathways: STING, AIMII and the regulation of interferon production and inflammatory responses. *Curr Opin Immunol* **23:**10–20.
103. **Abe T, Harashima A, Xia T, Konno H, Konno K, Morales A, Ahn J, Gutman D, Barber GN.** 2013. STING recognition of cytoplasmic DNA instigates cellular defense. *Mol Cell* **50:**5–15.
104. **Ishikawa H, Barber GN.** 2011. The STING pathway and regulation of innate immune signaling in response to DNA pathogens. *Cell Mol Life Sci* **68:**1157–1165.
105. **Bode C, Zhao G, Steinhagen F, Kinjo T, Klinman DM.** 2011. CpG DNA as a vaccine adjuvant. *Expert Rev Vaccines* **10:**499–511.
106. **Krieg AM.** 1999. Direct immunologic activities of CpG DNA and implications for gene therapy. *J Gene Med* **1:** 56–63.
107. **Gürsel M, Verthelyi D, Gürsel I, Ishii KJ, Klinman DM.** 2002. Differential and competitive activation of human immune cells by distinct classes of CpG oligodeoxynucleotide. *J Leukoc Biol* **71:**813–820.
108. **Rutz M, Metzger J, Gellert T, Luppa P, Lipford GB, Wagner H, Bauer S.** 2004. Toll-like receptor 9 binds single-stranded CpG-DNA in a sequence- and pH-dependent manner. *Eur J Immunol* **34:**2541–2550.
109. **Klinman DM.** 2006. Adjuvant activity of CpG oligodeoxynucleotides. *Int Rev Immunol* **25:**135–154.
110. **Krieg AM.** 2012. CpG still rocks! Update on an accidental drug. *Nucleic Acid Ther* **22:**77–89.
111. **Sørensen SJ, Bailey M, Hansen LH, Kroer N, Wuertz S.** 2005. Studying plasmid horizontal transfer in situ: a critical review. *Nat Rev Microbiol* **3:**700–710.

112. **Oliveira PH, Mairhofer J.** 2013. Marker-free plasmids for biotechnological applications – implications and perspectives. *Trends Biotechnol* **31:**539–547.

113. **Li L, Saade F, Petrovsky N.** 2012. The future of human DNA vaccines. *J Biotechnol* **162:**171–182.

114. **Luke J, Carnes AE, Hodgson CP, Williams JA.** 2009. Improved antibiotic-free DNA vaccine vectors utilizing a novel RNA based plasmid selection system. *Vaccine* **27:**6454–6459.

115. **Mairhofer J, Cserjan-Puschmann M, Striedner G, Nöbauer K, Razzazi-Fazeli E, Grabherr R.** 2010. Marker-free plasmids for gene therapeutic applications–lack of antibiotic resistance gene substantially improves the manufacturing process. *J Biotechnol* **146:**130–137.

116. **Williams SG, Cranenburgh RM, Weiss AM, Wrighton CJ, Sherratt DJ, Hanak JA.** 1998. Repressor titration: a novel system for selection and stable maintenance of recombinant plasmids. *Nucleic Acids Res* **26:**2120–2124.

117. **Kandimalla ER, Bhagat L, Wang D, Yu D, Sullivan T, La Monica N, Agrawal S.** 2013. Design, synthesis and biological evaluation of novel antagonist compounds of Toll-like receptors 7, 8 and 9. *Nucleic Acids Res* **41:** 3947–3961.

118. **Reyes-Sandoval A, Ertl HCJ.** 2004. CpG methylation of a plasmid vector results in extended transgene product expression by circumventing induction of immune responses. *Mol Ther* **9:**249–261.

119. **Hyde SC, Pringle IA, Abdullah S, Lawton AE, Davies L, Varathalingam A, Nunez-Alonso G, Green AM, Bazzani RP, Sumner-Jones SG, Chan M, Li H, Yew NS, Cheng SH, Boyd AC, Davies JC, Griesenbach U, Porteous DJ, Sheppard DN, Munkonge FM, Alton EW, Gill DR.** 2008. CpG-free plasmids confer reduced inflammation and sustained pulmonary gene expression. *Nat Biotechnol* **26:**549–551.

120. **Wolff JA, Ludtke JJ, Acsadi G, Williams P, Jani A.** 1992. Long-term persistence of plasmid DNA and foreign gene expression in mouse muscle. *Hum Mol Genet* **1:**363–369.

121. **Chen ZY, Riu E, He CY, Xu H, Kay MA.** 2008. Silencing of episomal transgene expression in liver by plasmid bacterial backbone DNA is independent of CpG methylation. *Mol Ther* **16:**548–556.

122. **Lu J, Zhang F, Xu S, Fire AZ, Kay MA.** 2012. The extragenic spacer length between the 5′ and 3′ ends of the transgene expression cassette affects transgene silencing from plasmid-based vectors. *Mol Ther* **20:**2111–2119.

123. **Suzuki M, Kasai K, Saeki Y.** 2006. Plasmid DNA sequences present in conventional herpes simplex virus amplicon vectors cause rapid transgene silencing by forming inactive chromatin. *J Virol* **80:**3293–3300.

124. **Riu E, Chen ZY, Xu H, He CY, Kay MA.** 2007. Histone modifications are associated with the persistence or silencing of vector-mediated transgene expression *in vivo*. *Mol Ther* **15:**1348–1355.

125. **Gracey Maniar LE, Maniar JM, Chen ZY, Lu J, Fire AZ, Kay MA.** 2013. Minicircle DNA vectors achieve sustained expression reflected by active chromatin and transcriptional level. *Mol Ther* **21:**131–138.

126. **Argyros O, Wong SP, Fedonidis C, Tolmachov O, Waddington SN, Howe SJ, Niceta M, Coutelle C, Harbottle RP.** 2011. Development of S/MAR minicircles for enhanced and persistent transgene expression in the mouse liver. *J Mol Med* **89:**515–529.

127. **Mayrhofer P, Blaesen M, Schleef M, Jechlinger W.** 2008. Minicircle-DNA production by site specific recombination and protein-DNA interaction chromatography. *J Gene Med* **10:**1253–1269.

128. **Schakowski F, Gorschlüter M, Junghans C, Schroff M, Buttgereit P, Ziske C, Schöttker B, König-Merediz SA, Sauerbruch T, Wittig B, Schmidt-Wolf IG.** 2001. A novel minimal-size vector (MIDGE) improves transgene expression in colon carcinoma cells and avoids transfection of undesired DNA. *Mol Ther* **3:**793–800.

129. **Schakowski F, Gorschlüter M, Buttgereit P, Märten A, Lilienfeld-Toal MV, Junghans C, Schroff M, König-Merediz SA, Ziske C, Strehl J, Sauerbruch T, Wittig B, Schmidt-Wolf IG.** 2007. Minimal size MIDGE vectors improve transgene expression *in vivo*. *In Vivo* **21:** 17–23.

130. **Prazeres DMF.** 2011. Delivery, p 167–210. *In Plasmid Biopharmaceuticals: Basics Applications and Manufacturing*. John Wiley & Sons, Inc., Hoboken, NJ.

131. **Nishikawa M, Huang L.** 2001. Nonviral vectors in the new millennium: delivery barriers in gene transfer. *Hum Gene Ther* **12:**861–870.

132. **Grigsby CL, Leong KW.** 2010. Balancing protection and release of DNA: tools to address a bottleneck of non-viral gene delivery. *J R Soc Interface* **7:**S67–S82.

133. **Mahvi DM, Henry MB, Albertini MR, Weber S, Meredith K, Schalch H, Rakhmilevich A, Hank J, Sondel P.** 2007. Intratumoral injection of IL-12 plasmid DNA–results of a phase I/IB clinical trial. *Cancer Gene Ther* **14:**717–723.

134. **Konstan MW, Davis PB, Wagener JS, Hilliard KA, Stern RC, Milgram LJ, Kowalczyk TH, Hyatt SL, Fink TL, Gedeon CR, Oette SM, Payne JM, Muhammad O, Ziady AG, Moen RC, Cooper MJ.** 2004. Compacted DNA nanoparticles administered to the nasal mucosa of cystic fibrosis subjects are safe and demonstrate partial to complete cystic fibrosis transmembrane regulator reconstitution. *Hum Gene Ther* **15:**1255–1269.

135. **Wolff JA, Budker V.** 2005. The mechanism of naked DNA uptake and expression. *Adv Genet* **54:**3–20.

136. **Satkauskas S, Bureau MF, Mahfoudi A, Mir LM.** 2001. Slow accumulation of plasmid in muscle cells: supporting evidence for a mechanism of DNA uptake by receptor-mediated endocytosis. *Mol Ther* **4:**317–323.

137. **Budker V, Budker T, Zhang G, Subbotin VM, Loomis A, Wolff JA.** 2000. Hypothesis: naked plasmid DNA is taken up by cells *in vivo* by a receptor-mediated process. *J Gene Med* **2:**76–88.

138. **Vaughan EE, DeGiulio JV, Dean DA.** 2006. Intracellular trafficking of plasmids for gene therapy: mechanisms of cytoplasmic movement and nuclear import. *Curr Gene Ther* **6:**671–681.

139. **Dean DA, Strong DD, Zimmer WE.** 2005. Nuclear entry of nonviral vectors. *Gene Ther* **12:**881–890.

140. **van der Aa MA, Mastrobattista E, Oosting RS, Hennink WE, Koning GA, Crommelin DJ.** 2006. The nuclear pore complex: the gateway to successful nonviral gene delivery. *Pharm Res* **23:**447–459.

141. **Villemejane J, Mir LM.** 2009. Physical methods of nucleic acid transfer: general concepts and applications. *Br J Pharmacol* **157:**207–219.

142. **Suda T, Suda K, Liu D.** 2008. Computer-assisted hydrodynamic gene delivery. *Mol Ther* **16:**1098–1104.

143. **Mitragotri S.** 2006. Current status and future prospects of needle-free liquid jet injectors. *Nat Rev Drug Discov* **5:**543–548.

144. **Wang S, Joshi S, Lu S.** 2004. Delivery of DNA to skin by particle bombardment. *Meth Mol Biol* **245:**185–193.

145. **Dean HJ, Haynes J, Schmaljohn C.** 2005. The role of particle-mediated DNA vaccines in biodefense preparedness. *Adv Drug Deliv Rev* **57:**1315–1342.

146. **Fuller DH, Loudon P, Schmaljohn C.** 2006. Preclinical and clinical progress of particle-mediated DNA vaccines for infectious diseases. *Methods* **40:**86–97.

147. **Escoffre JM, Portet T, Wasungu L, Teissie J, Dean D, Rols MP.** 2009. What is (still not) known of the mechanism by which electroporation mediates gene transfer and expression in cells and tissues. *Mol Biotechnol* **41:** 286–295.

148. **Denet AR, Vanbever R, Preat V.** 2004. Skin electroporation for transdermal and topical delivery. *Adv Drug Deliv Rev* **56:**659–674.

149. **Chiarella P, Massi E, De Robertis M, Sibilio A, Parrella P, Fazio VM, Signori E.** 2008. Electroporation of skeletal muscle induces danger signal release and antigen-presenting cell recruitment independently of DNA vaccine administration. *Expert Opin Biol Ther* **8:** 1645–1657.

150. **Faurie C, Rebersek M, Golzio M, Kanduser M, Escoffre JM, Pavlin M, Teissie J, Miklavcic D, Rols MP.** 2010. Electro-mediated gene transfer and expression are controlled by the life-time of DNA/membrane complex formation. *J Gene Med* **12:**117–125.

151. **Luxembourg A, Hannaman D, Ellefsen B, Nakamura G, Bernard R.** 2006. Enhancement of immune responses to an HBV DNA vaccine by electroporation. *Vaccine* **24:**4490–4493.

152. **Luxembourg A, Evans CF, Hannaman D.** 2007. Electroporation-based DNA immunisation: translation to the clinic. *Expert Opin Biol Ther* **7:**1647–1664.

153. **Heller LC, Heller R.** 2006. *In vivo* electroporation for gene therapy. *Hum Gene Ther* **17:**890–897.

154. **Bodles-Brakhop AM, Heller R, Draghia-Akli R.** 2009. Electroporation for the delivery of DNA-based vaccines and immunotherapeutics: current clinical developments. *Mol Ther* **17:**585–592.

155. **Templeton NS, Lasic DD, Frederik PM, Strey HH, Roberts DD, Pavlakis GN.** 1997. Improved DNA: liposome complexes for increased systemic delivery and gene expression. *Nat Biotechnol* **15:**647–652.

156. **Farrell LL, Pepin J, Kucharski C, Lin X, Xu Z, Uludag H.** 2007. A comparison of the effectiveness of cationic polymers poly-L-lysine (PLL) and polyethylenimine (PEI) for non-viral delivery of plasmid DNA to bone marrow stromal cells (BMSC). *Eur J Pharm Biopharm* **65:**388–397.

157. **Aral C, Akbuga J.** 2003. Preparation and *in vitro* transfection efficiency of chitosan microspheres containing plasmid DNA:poly(L-lysine) complexes. *J Pharm Pharm Sci* **6:**321–326.

158. **Basarkar A, Devineni D, Palaniappan R, Singh J.** 2007. Preparation, characterization, cytotoxicity and transfection efficiency of poly(DL-lactide-co-glycolide) and poly (DL-lactic acid) cationic nanoparticles for controlled delivery of plasmid DNA. *Int J Pharm* **343:**247–254.

159. **Mok H, Park TG.** 2008. Direct plasmid DNA encapsulation within PLGA nanospheres by single oil-in-water emulsion method. *Eur J Pharm Biopharm* **68:**105–111.

160. **Hao T, McKeever U, Hedley ML.** 2000. Biological potency of microsphere encapsulated plasmid DNA. *J Control Release* **69:**249–259.

161. **Singh M, Briones M, Ott G, O'Hagan D.** 2000. Cationic microparticles: a potent delivery system for DNA vaccines. *Proc Natl Acad Sci USA* **97:**811–816.

162. **Luo Y, O'Hagan D, Zhou H, Singh M, Ulmer J, Reisfeld RA, James Primus F, Xiang R.** 2003. Plasmid DNA encoding human carcinoembryonic antigen (CEA) adsorbed onto cationic microparticles induces protective immunity against colon cancer in CEA-transgenic mice. *Vaccine* **21:**1938–1947.

163. **He X, Jiang L, Wang F, Xiao Z, Li J, Liu LS, Li D, Ren D, Jin X, Li K, He Y, Shi K, Guo Y, Zhang Y, Sun S.** 2005. Augmented humoral and cellular immune responses to hepatitis B DNA vaccine adsorbed onto cationic microparticles. *J Control Release* **107:**357–372.

164. **Mollenkopf HJ, Dietrich G, Fensterle J, Grode L, Diehl KD, Knapp B, Singh M, O'Hagan DT, Ulmer JB, Kaufmann SH.** 2004. Enhanced protective efficacy of a tuberculosis DNA vaccine by adsorption onto cationic PLG microparticles. *Vaccine* **22:**2690–2695.

165. **Bozkir A, Saka OM.** 2004. Chitosan nanoparticles for plasmid DNA delivery: effect of chitosan molecular structure on formulation and release characteristics. *Drug Deliv* **11:**2690–2695.

166. **Wang C-Q, Wu J-L, Zhuo R-X, Cheng S-X.** 2014. Protamine sulfate–calcium carbonate–plasmid DNA ternary nanoparticles for efficient gene delivery. *Mol BioSyst* **10:**672–678.

167. **Nichols WW, Ledwith BJ, Manam SV, Troilo PJ.** 1995. Potential DNA vaccine integration into host cell genome. *Ann N Y Acad Sci* **772:**30–39.

168. **Mor G, Singla M, Steinberg AD, Hoffman SL, Okuda K, Klinman DM.** 1997. Do DNA vaccines induce autoimmune disease? *Hum Gene Ther* **8:**293–300.

169. **European Agency for the Evaluation Medicinal Products.** 24 April 2001. *Note for guidance on quality, preclinical and clinical aspects of gene transfer medicinal products (CPMP/BWP/3088/99).* European Medicines Agency, London, UK.

170. **US Food and Drug Administration.** 2007. *Guidance for industry: considerations for plasmid DNA vaccines for*

preventive infectious disease indications. US Food and Drug Administration, Rockville, MD.

171. **Cichutek K.** 2000. DNA vaccines: development, standardization and regulation. *Intervirology* **43:**331–338.

172. **Ledwith BJ, Manam S, Troilo PJ, Barnum AB, Pauley CJ, Griffiths TG, Harper LB, Beare CM, Bagdon WJ, Nichols WW.** 2000. Plasmid DNA vaccines: investigation of integration into host cellular DNA following intramuscular injection in mice. *Intervirology* **43:** 258–272.

173. **Wang Z, Troilo PJ, Wang X, Griffiths TG, Pacchione SJ, Barnum AB, Harper LB, Pauley CJ, Niu Z, Denisova L, Follmer TT, Rizzuto G, Ciliberto G, Fattori E, Monica NL, Manam S, Ledwith BJ.** 2004. Detection of integration of plasmid DNA into host genomic DNA following intramuscular injection and electroporation. *Gene Ther* **11:**711–721.

174. **Sheets RL, Stein J, Manetz TS, Duffy C, Nason M, Andrews C, Kong WP, Nabel GJ, Gomez PL.** 2006. Biodistribution of DNA plasmid vaccines against HIV-1, Ebola, Severe Acute Respiratory Syndrome, or West Nile virus is similar, without integration, despite differing plasmid backbones or gene inserts. *Toxicol Sci* **91:** 610–619.

175. **Iiizumi S, Kurosawa A, So S, Ishii Y, Chikaraishi Y, Ishii A, Koyama H, Adachi N.** 2008. Impact of non-homologous end-joining deficiency on random and targeted DNA integration: implications for gene targeting. *Nucleic Acids Res* **36:**6333–6342.

176. **Pilling AM, Harman RM, Jones SA, McCormack NA, Lavender D, Haworth R.** 2002. The assessment of local tolerance, acute toxicity, and DNA biodistribution following particle-mediated delivery of a DNA vaccine to minipigs. *Toxicol Pathol* **30:**298–305.

177. **Manam S, Ledwith BJ, Barnum AB, Troilo PJ, Pauley CJ, Harper LB, Griffiths TG, Niu Z, Denisova L, Follmer TT, Pacchione SJ, Wang Z, Beare CM, Bagdon WJ, Nichols WW.** 2000. Plasmid DNA vaccines: tissue distribution and effects of DNA sequence, adjuvants and delivery method on integration into host DNA. *Intervirology* **43:**273–281.

178. **Nabel EG, Gordon D, Yang ZY, Xu L, San H, Plautz GE, Wu BY, Gao X, Huang L, Nabel GJ.** 1992. Gene transfer *in vivo* with DNA-liposome complexes: lack of autoimmunity and gonadal localization. *Hum Gene Ther* **3:**649–656.

179. **Mor G, Eliza M.** 2001. Plasmid DNA vaccines. Immunology, tolerance, and autoimmunity. *Mol Biotechnol* **19:**245–250.

180. **Choi SM, Lee DS, Son MK, Sohn YS, Kang KK, Kim CY, Kim BM, Kim WB.** 2003. Safety evaluation of GX-12. A new DNA vaccine for HIV infection in rodents. *Drug Chem Toxicol* **26:**271–284.

181. **Schalk JAC, Mooi FR, Berbers GAM, van Aerts LAGJM, Ovelgönne H, Kimman TG.** 2006. Preclinical and clinical safety studies on DNA vaccines. *Hum Vaccines* **2:**45–53.

182. **Sheets RL, Stein J, Manetz TS, Andrews C, Bailer R, Rathmann J, Gomez PL.** 2006. Toxicological safety evaluation of DNA plasmid vaccines against HIV-1, Ebola, Severe Acute Respiratory Syndrome, or West Nile virus is similar despite differing plasmid backbones or gene-inserts. *Toxicol Sci* **91:**620–630.

183. **Tavel JA, Martin JE, Kelly GG, Enama ME, Shen JM, Gomez PL, Andrews CA, Koup RA, Bailer RT, Stein JA, Roederer M, Nabel GJ, Graham BS.** 2007. Safety and immunogenicity of a Gag-Pol candidate HIV-1 DNA vaccine administered by a needle-free device in HIV-1-seronegative subjects. *J Acquir Immune Defic Syndr* **44:**601–605.

184. **Prazeres DMF.** 2011. Product and process development, p 69–84. *In Plasmid Biopharmaceuticals: Basics Applications and Manufacturing*. John Wiley & Sons, Inc., Hoboken, NJ.

185. **Gonçalves GAL, Prazeres DMF, Monteiro GA, Prather KLJ.** 2013. *De Novo* Creation of MG1655-derived *Escherichia coli* strains specifically designed for plasmid DNA production. *App Microbiol Biotechnol* **97:** 611–620.

186. **Carnes AE, Williams JA.** 2007. Plasmid DNA manufacturing technology. *Recent Pat Biotechnol* **1:**151–166.

187. **Listner K, Bentley L, Okonkowski J, Kistler C, Wnek R, Caparoni A, Junker B, Robinson D, Salmon P, Chartrain M.** 2006. Development of a highly productive and scalable plasmid DNA production platform. *Biotechnol Prog* **22:**1335–1345.

188. **Freitas SS, Santos JAL, Prazeres DMF.** 2006. Optimisation of isopropanol and ammonium sulphate precipitation steps in the purification of plasmid DNA. *Biotechnol Prog* **22:**1179–1186.

189. **Gomes GA, Azevedo AM, Aires-Barros MR, Prazeres DMF.** 2010. Clearance of host-cell impurities from plasmid-containing lysates by boronate adsorption. *J Chromatogr A* **1217:**2262–2266.

190. **Gomes GA, Azevedo AM, Aires-Barros MR, Prazeres DMF.** 2009. Purification of plasmid DNA using aqueous two-phase systems with PEG 600 and sodium citrate/ammonium sulphate. *Sep Pur Technol* **65:**22–30.

191. **Freitas S, Santos JAL, Prazeres DMF.** 2009. Plasmid purification by hydrophobic interaction chromatography using sodium citrate in the mobile phase. *Sep Pur Technol* **65:**95–104.

192. **Sousa F, Freitas SS, Azzonni A, Prazeres DMF, Queiroz JA.** 2006. Selective purification of supercoiled plasmid DNA from cell lysates with a single histidine-agarose chromatography step. *Biotechnol App Biochem* **45:** 131–140.

193. **Lemmens R, Olsson U, Nyhammar T, Stadler J.** 2003. Supercoiled plasmid DNA: selective purification by thiophilic/aromatic adsorption. *J Chromatog B* **784:** 291–300.

194. **Diogo MM, Queiroz JA, Prazeres DMF.** 2005. Chromatography of plasmid DNA. *J Chrom A* **1069:**3–22.

195. **Prazeres DMF.** 2009. Chromatographic separation of plasmid DNA using macroporous beads, p 335–361. *In* Mattiasson B, Kumar A, Galaev IY (ed), *Macroporous Polymers: Production, Properties and Biotechnological/Biomedical Applications*. CRC Press, New York, NY.

196. **Urthaler J, Buchinger W, Necina R.** 2005. Industrial scale cGMP purification of pharmaceutical-grade plasmid DNA. *Chem Eng Technol* **28:**1408–1420.

197. **Watson MP, Winters MA, Sagar SL, Konz JO.** 2006. Sterilizing filtration of plasmid DNA: effects of plasmid concentration, molecular weight, and conformation. *Biotechnol Prog* **22:**465–470.

198. **Prazeres DMF.** 2011. Veterinary case studies: West Nile, infectious hematopoietic necrosis and melanoma, p 69–84. *In Plasmid Biopharmaceuticals: Basics Applications and Manufacturing*. John Wiley & Sons, Inc., Hoboken, NJ.

199. **Powell K.** 2004. DNA vaccines—back in the saddle again? *Nat Biotechnol* **22:**799–801.

200. **Anonymous.** 2008. *West Nile-Innovator DNA—The first USDA approved DNA vaccine*. Fort Dodge Animal Health, Fort Dodge, IA.

201. **Novartis.** 19 July 2005. *Novel Novartis vaccine to protect Canadian salmon farms from devastating viral disease*. Novartis Animal Health Inc., Basel, Switzerland.

202. **Merial.** 26 March 2007. *USDA grants conditional approval for first therapeutic vaccine to treat cancer*. Merial Limited, Duluth, GA.

203. **Person R, Bodles-Brakhop AM, Pope MA, Brown PA, Khan AS, Draghia-Akli R.** 2008. Growth hormone-releasing hormone plasmid treatment by electroporation decreases offspring mortality over three pregnancies. *Mol Ther* **16:**1891–1897.

Index

Q

U

V

W

X

Y